HANDBOOK OF ENERGY SYSTEMS ENGINEERING

HANDBOOK OF ENERGY SYSTEMS ENGINEERING

PRODUCTION AND UTILIZATION

Edited by

LESLIE C. WILBUR, M.S., P.E., Fellow, ASME
Department of Mechanical Engineering
Worcester Polytechnic Institute
Worcester, Massachusetts

WILEY SERIES IN MECHANICAL ENGINEERING PRACTICE

CONSULTING EDITOR

Marvin D. Martin
President, Marvin D. Martin, Inc., Consulting Engineers, Tucson, Arizona

A Wiley-Interscience Publication
JOHN WILEY & SONS
New York • Chichester • Brisbane • Toronto • Singapore

Copyright © 1985 by John Wiley & Sons, Inc.

Library of Congress Cataloging in Publication Data:

Main entry under title:

Handbook of energy systems engineering
    (Wiley series in mechanical engineering practice)
    "A Wiley-Interscience publication."
    Bibliography: p.
    Includes index.
    1. Power (Mechanics)—Handbooks, manuals, etc.

I. Wilbur, Leslie C.   II. Series.
TJ163.9.H35   1985   621.042   84-20830
ISBN 0-471-86633-4

Printed in the United States of America

10 9 8 7 6 5 4 3 2 1

CONTRIBUTORS

Hemendra K. Acharya, *Consultant, Stone and Webster Engineering Corporation, Boston, Massachusetts*

Roy P. Allen, *Manager of Application Engineering, Gas Turbine Department, General Electric Company, Schenectady, New York*

Amiya Basu, *Arizona Public Service Company, Phoenix, Arizona*

William W. Bathie, *Department of Mechanical Engineering, Iowa State University, Ames, Iowa*

Robert D. Bessette, *Manager, Sales Engineering, Old Ben Coal Company, Lexington, Kentucky*

Kenneth J. Bell, *Department of Chemical Engineering, Oklahoma State University, Stillwater, Oklahoma*

Joseph K. Benner, *Department of Nuclear Engineering, The Pennsylvania State University, University Park, Pennsylvania*

Frank P. Bleier, *Fan Design Consultant, 536 West Cornelia Avenue, Chicago, Illinois*

Harold B. Boyd, *M. W. Kellogg Company, Houston, Texas*

Ernie B. Branch, *Design Director, Sargent and Lundy, Chicago, Illinois*

John Briedis, *Assistant Chief Geotechnical Engineer, Stone and Webster Engineering Corporation, Boston, Massachusetts*

Dominique Brocard, *Assistant Director, Alden Research Laboratories, Holden, Massachusetts*

Warren H. Brown, *Consultant, Byron Jackson Pump Division, Borg Warner Corporation, Long Beach, California*

John M. Burns, *Consultant, Power Division, Stone and Webster Engineering Corporation, Boston, Massachusetts*

C. Cheng, *General Motors Corporation, LaGrange, Illinois*

Melvin H. Chiogioji, *Deputy Assistant Secretary, State and Local Assistance Programs, U.S. Department of Energy, Washington, D.C.*

Chong Chiu, *Combustion Engineering, Inc., Windsor, Connecticut*

Cyrus Clark, *CBI Industries, Inc., Oak Brook, Illinois*

Richard L. P. Custer, *Associate Director, Center for Fire Safety Studies, Worcester Polytechnic Institute, Worcester, Massachusetts*

Joseph Duff, *Manager of Chemical Services Engineering, Burns and Roe, Inc., Oradell, New Jersey*

William W. Durgin, *Department of Mechanical Engineering, Worcester Polytechnic Institute, Worcester, Massachusetts*

Robert A. Evans, *Helium Breeders Association, La Jolla, California*

Rocco Fazzolare, *Department of Nuclear and Energy Engineering, University of Arizona, Tucson, Arizona*

Rudolf Fehlau, *Production Manager, Horizontal Pumps, Byron Jackson Pump Division, Borg Warner Corporation, Long Beach, California*

Richard S. Fein, *Richard S. Fein Associates, Poughkeepsie, New York*

Robert W. Fitzmaurice, *Allen Sherman Hoff Company, Malvery, Pennsylvania*

James J. Garrity, *Chief Hydraulic Engineer, Stone and Webster Engineering Corporation, Boston, Massachusetts*

Marc W. Goldsmith, *President, Energy Research Group, Inc., Waltham, Massachusetts*

V. E. Grimshaw, *Manager, Logistics Planning, Union Oil Company of California, Los Angeles, California*

Okan Gurel, *IBM Cambridge Scientific Center, Cambridge, Massachusetts*

Daniel F. Hang, *Department of Nuclear Engineering, University of Illinois, Urbana, Illinois*

Frederick R. Harty, Jr., *Senior Hydraulic Engineer, Stone and Webster Engineering Corporation, Boston, Massachusetts*

Harlan T. Holmes, *Vice President, Energy Engineering, Garver and Garver, Inc., Little Rock, Arkansas*

Arthur Karlin, *Booz-Allen and Hamilton, Washington, D.C.*

E. H. Kennedy, *Manager, Project and Generic Licensing, Combustion Engineering, Inc., Windsor, Connecticut*

Donald L. Klass, *Assistant Research Director, Institute of Gas Technology, Chicago, Illinois*

Ivan V. LaFave, *CBI Industries, Inc., Oak Brook, Illinois*

Thomas J. Lamb, *Senior Soils Engineer, Stone and Webster Engineering Corporation, Boston, Massachusetts*

David Lanning, *Department of Nuclear Engineering, Massachusetts Institute of Technology, Cambridge, Massachusetts*

Thomas Lepley, *Arizona Public Service Company, Phoenix, Arizona*

Jack B. Levy, *General Electric Company, Research and Development Center, Schenectady, New York*

Edward S. Lipinsky, *Battelle-Columbus Laboratories, Columbus, Ohio*

C. Hardy Long, *Department of Mechanical Engineering, Virginia Polytechnic Institute and State University, Blacksburg, Virginia*

A. W. Loomis, *Private Consultant, Stewartsville, New Jersey*

Michael McDonald, *Booz-Allen and Hamilton, Washington, D.C.*

Robert A. McElroy, Jr., *Pipeline Engineer, Union Oil Company of California, Los Angeles, California*

John A. Mayer, Jr., *Department of Mechanical Engineering, Worcester Polytechnic Institute, Worcester, Massachusetts*

L. D. Mears, *General Manager, Gas Cooled Reactor Associates, La Jolla, California*

E. H. Miller, *Manager of Thermal Development Engineering, General Electric Company, Schenectady, New York*

Chester Nachtigal, *Weyerhaeuser Company, Tacoma, Washington*

Ravindra M. Nadkarni, *Vice President, Technology, Leach & Garner, Attleboro, Massachusetts*

Gordon Oates, *Department of Aeronautics and Astronautics, University of Washington, Seattle, Washington*

Francisco Palacios, *Riley Stoker Corporation, Worcester, Massachusetts*

C. A. Podczerwinski, *Chief Computer Application Engineer, Sargent and Lundy, Chicago, Illinois*

Ryszard J. Pryputniewicz, *Department of Mechanical Engineering, Worcester Polytechnic Institute, Worcester, Massachusetts*

Frank J. Rahn, *Electric Power Research Institute, Palo Alto, California*

A. G. Randol III, *Exxon Corporation USA, New Orleans, Louisiana*

Norman Rasmussen, *Department of Nuclear Engineering, Massachusetts Institute of Technology, Cambridge, Massachusetts*

Albert H. Rawdon, *Riley Stoker Corporation, Worcester, Massachusetts*

Robert R. Reeves, *Department of Chemistry, Rensselaer Polytechnic Institute, Troy, New York*

Archie Riviera, *Booz-Allen and Hamilton, Washington, D.C.*

Walton A. Rodger, *Chairman, OTHA, Inc., Glen Echo, Maryland*

Charles W. Ruoff, *Supervisor, Energy and Utilities Engineering, ARCO Chemical Company, Newtown Square, Pennsylvania*

Michel A. Saad, *Department of Mechanical Engineering, University of Santa Clara, Santa Clara, California*

Chakra J. Santhanam, *Manager, Chemical & Metallurgical Engineering, A. D. Little Company, Cambridge, Massachusetts*

Robert L. Scheffler, *Program Director, Wind Energy Conversion, Southern California Edison, Rosemead, California*

A. E. Scherer, *Director, Nuclear Licensing, Combustion Engineering, Inc., Windsor, Connecticut*

CONTRIBUTORS vii

Paul F. Shiers, *Senior Hydraulic Engineer, Stone and Webster Engineering Corporation, Boston, Massachusetts*

James K. Shilenn, *Department of Nuclear Engineering, The Pennsylvania State University, University Park, Pennsylvania*

Richard D. Shultz, *Department of Electrical Engineering, University of Illinois, Urbana, Illinois*

Robert H. Simon, *GA Technologies, Inc., San Diego, California*

John Simonson, *Vice President, Engineering, Valtek, Inc., Springville, Utah*

Eric Smith, *Consultant, Holden, Massachusetts*

Richard A. Smith, *Department of Electrical Engineering, University of Illinois, Urbana, Illinois*

Stephen E. Smith, *Consultant, Cogeneration and Energy Systems, Energy Technologies, Inc., Tucson, Arizona*

Joong Min Soh, *ARCO Chemical Company, Newtown Square, Pennsylvania*

John Soma, *Private Consultant, Brooklyn, New York*

S. Srinivasan, *Institute for Hydrogen Systems, Mississauga, Ontario, Canada*

Herbert R. Stewart, *Private Consultant, Waban, Massachusetts*

R. Peter Stickles, *A. D. Little Company, Cambridge, Massachusetts*

Thomas E. Stott, *Thomas Stott & Associates, Cummaquid, Massachusetts*

James Swisher, *Director, Division of Energy Storage Technology, Office of Energy Systems Research, Department of Energy, Washington, D.C.*

J. B. Thomas, *Director, Technical Marketing, Ingersol-Rand Company, Allentown, Pennsylvania*

Owen D. Thomas, *President, Owen Thomas & Associates, Salt Lake City, Utah*

B. V. Tilak, *Occidental Chemical Corporation, Hooker Research Center, Grand Island, New York*

David A. Tillman, *Environsphere Company, Bellevue, Washington*

David Turner, *Booz-Allen and Hamilton, Washington, D.C.*

Thomas C. Varljen, *Westinghouse Electric Fusion Power Systems, Advanced Reactor Division, Madison, Pennsylvania*

Thomas F. Weaver, *Department of Resource Economics, University of Rhode Island, Kingston, Rhode Island*

Eric Weber, *Arizona Public Service Company, Phoenix, Arizona*

Michel C. Wehrey, *WTG Test Program Manager, Southern California Edison, Rosemead, California*

Leslie C. Wilbur, *Department of Mechanical Engineering, Worcester Polytechnic Institute, Worcester, Massachusetts*

John Winner, *Booz-Allen and Hamilton, Bethesda, Maryland*

Warren Witzig, *Department of Nuclear Engineering, The Pennsylvania State University, University Park, Pennsylvania*

Bernard D. Wood, *Department of Mechanical and Aerospace Engineering, Syracuse University, Syracuse, New York*

Douglas Woods, *Head, Department of Social Science and Policy Studies, Worcester Polytechnic Institute, Worcester, Massachusetts*

Paul C. Yuen, *Dean, College of Engineering, University of Hawaii, Honolulu, Hawaii*

Russell Zub, *Transportation Systems Center, Cambridge, Massachusetts*

SERIES PREFACE

The Wiley Series in Mechanical Engineering Practice is written for the practicing engineer. Students and academicians may find it useful, but its primary thrust is for the working engineer who needs a convenient and comprehensive reference on hand.

Two kinds of information are contained in the several volumes:

1. Numerical information such as strengths of materials, thermodynamic properties of fluids, standard pipe sizes, thread systems, and so on.
2. Descriptive and mathematical information typical of the "state of the art" of the many facets and specialties encompassed by the broad term "mechanical engineering."

The profession has expanded to cover such a broad range of engineering activities that no one can be knowledgeable in more than a fraction of the whole field. Yet, in day-to-day work, practicing engineers frequently have to use, or at least interface with, specialty areas outside their normal sphere of competence. This book is written to provide readers with the state of the art information and standard practices in these other areas.

The task of covering such a vast amount of material has dictated the decision to split the series into five separate volumes:

Design and Manufacturing
Fluids and Fluid Machinery
Mechanics, Materials, and Structures
Power and Energy Systems
Instrumentation and Control

Each volume is designed to stand alone but the five complement each other in providing the broad coverage mentioned above. Within each volume chapter and section headings are designed to help the user in finding the material being sought.

A serious attempt was made to provide state of the art material at the time of writing. Since many of the areas are in a state of rapid change, there will be some obsolescence by the time printing is complete. It is planned to revise and update at reasonable intervals so that users may purchase newer editions and keep their references up to date.

The many editors and contributors who have made this series possible join me in the hope that the several volumes will turn out to be really useful tools for the practicing engineer.

MARVIN D. MARTIN

Tucson, Arizona
January 1985

PREFACE

The intent of this volume is to provide professionally trained persons with a concentrated store of user-oriented information on a broad spectrum of energy applications. The contributing authors were instructed to select the optimum material in their respective areas, taking into consideration severe space constraints. Each section is written as a miniprimer adequate to enable the reader to grasp vital concepts at a decision-making level and to give the nonexpert in a given discipline a reasonable degree of literacy.

Carefully selected bibliographies have been provided to expedite a follow-up where more detailed information is needed. Chapters of mathematical relationships and fundamental data are included. Contributors were permitted to use the system of units they considered most convenient for current practitioners in their own fields. Extensive conversion factors are provided in Chapter 18.

As Editor-in-Chief I wish to state my gratitude to the contributors for their professional, dedicated efforts, and to the numerous workers who contributed to the preparation of this material. In particular, Nancy Stanhope and Catherine Marinelli provided outstanding and tireless support. Special recognition is due Professor Donald N. Zwiep, past president of ASME, and Chairman, Department of Mechanical Engineering, Worcester Polytechnic Institute, for his vigorous assistance and encouragement. Most of all I wish to express my appreciation to my wife Gertrude for her unfailing inspiration and support.

LESLIE C. WILBUR

Berlin, Massachusetts
May 1985

CONTENTS

HANDBOOK OF ENERGY SYSTEMS ENGINEERING

CHAPTER 1
DEMOGRAPHICS

DOUGLAS W. WOODS

Worcester Polytechnic Institute
Worcester, Massachusetts

HEMENDRA K. ACHARYA

Stone & Webster Engineering Corporation
Boston, Massachusetts

The energy demographics chapter of the handbook presents statistics on energy demand, production, and resources. Section 1.1 examines energy consumption trends in the United States and the rest of the world. Sections 1.2–1.5 present data on reserves, resources, and production of four major conventional fuels: coal, oil, natural gas, and uranium. In general, these sections present reserve estimates for the United States and the world as a whole covering a number of past years, current reserve figures distributed by country, and current estimates of total U.S. and world resources of each fuel. The quality of the deposits and accuracy of the resource estimates are examined. Total reserves and potential resources are compared with current production levels and trends to evaluate the adequacy of U.S. and world supplies of conventional fossil fuels. Section 1.6 discusses the potential of renewable energy resources (such as solar, wind, wave, and hydroelectric) to contribute to meeting the world's energy needs. Section 1.7 describes the world's resources of alternative nonconventional fossil fuels: shale oil, tar sands, and peat.

1.1 ENERGY DEMAND

This section presents data on world and U.S. energy consumption trends over the last several decades. The growing world demand for energy is illustrated graphically in Fig. 1.1-1. The levels of total world primary energy consumption and the consumption of oil, coal, natural gas, hydroelectric energy, and nuclear energy in millions of tons of oil equivalent are graphed by decade from 1928 to 1965 and annually thereafter to 1981. Primary energy consumption and the consumption of the major fuels in 1981 are listed in Table 1.1-1 for the major countries of the world.

Data on U.S. energy demand is presented in two tables. Tables 1.1-2 and 1.1-3 give energy consumption by type of energy and energy end use, respectively, for each year from 1951 to 1981. Table 1.1-3 lists the U.S. domestic price of each type of energy for the years 1951–1981.

H. K. Acharya is the author of Section 1.6-9.

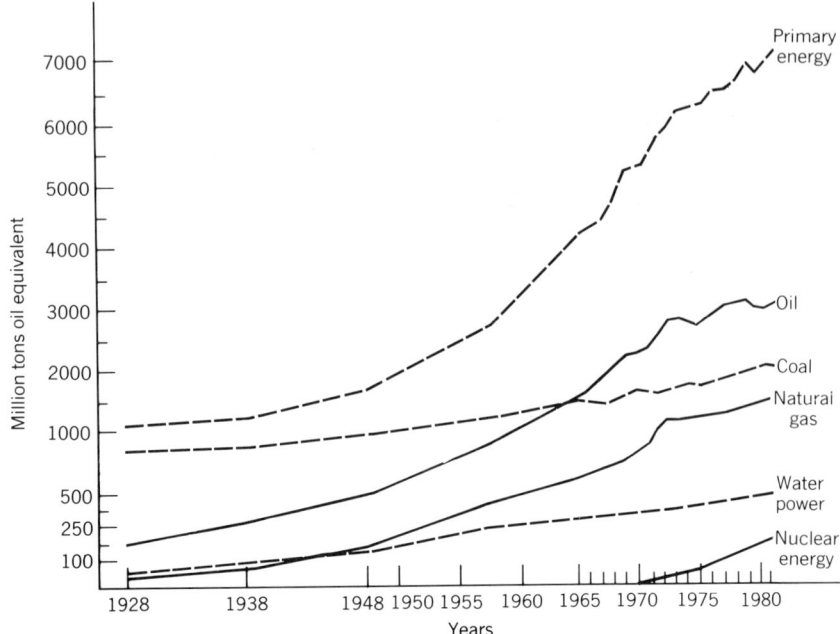

Fig. 1.1-1 World energy consumption. (Source: The British Petroleum Co. Ltd., *BP Statistical Review of the World Oil Industry*, Various Issues. London: The British Petroleum Co. Ltd., 1981.)

Since 1928 total world primary energy consumption has grown from 1.166 billion[†] tons of oil equivalent to nearly 6 times that amount, 6.849 billion in 1981. The annual compound rate of growth has been 3.4%, about the same as the rate of growth in industrial production. Growth in energy consumption was particularly rapid during the first three decades of the postwar period. From 1948 to 1973 energy consumption grew from 1.739 billion tons of oil equivalent to 5.92 billion tons—an annual compound rate of growth of 5%. Since 1973 the growth rate in world primary energy consumption has slowed considerably, dropping to 1.8% per annum. In the non-Communist world the growth rate since 1973 has been even lower, averaging only 1% per annum according to figures presented in the British Petroleum Company *BP Statistical Review of the World Oil Industry* (1982).

The reason for the sharp drop in the rate of growth in energy consumption since 1973 has, of course, been the energy crisis—the large increase in the cost of petroleum and most other sources of energy that has occurred since 1973. Table 1.1-4 shows that U.S. crude oil prices, for example, increased by a factor of 8 from 1973 to 1981 and that the combined price index for all fuels increased from $0.394 per million Btu to $2.70 over this period.

Although all forms of energy have become more costly since 1973, the price of oil has increased at a much faster rate than that of other types of energy. For example, the price of oil was approximately double that of coal in 1973 and is currently about 5 times the price of coal.

The increase in the relative costliness of oil compared with other forms of energy has resulted in a substantial shift in energy shares away from oil to coal and natural gas. From 1928 to 1973 the share of world primary energy consumption accounted for by oil increased steadily at the expense of coal. Since 1973 this pattern has changed dramatically, with oil consumption actually declining in the non-Communist world and growing at a rate of only a 0.4% per annum in the world as a whole. In contrast, world coal consumption has grown at 2.3% per annum since 1973—greater than the *average* growth rate in demand for primary energy.

In the United States the pattern has been very similar. Since 1973 coal consumption has resumed growing after a 50-year hiatus. Petroleum consumption, on the other hand, declined from 6.3 billion barrels in 1973 to 5.8 billion in 1981. Hydro and nuclear power have grown substantially since 1973, particularly the latter. Natural gas consumption, on the other hand, has declined slightly, primarily due to price regulation, which has kept gas limited in supply. In the United States total energy consumption declined slightly, from 74.6 quadrillion[‡] Btu in 1973 to 73.9 quatrillion Btu in 1981.

[†] Billion is here defined as 10^9.
[‡] Quadrillion is here defined as 10^{15}.

TABLE 1.1-1   ENERGY CONSUMPTION, 1981[a]

Country/Area	Primary Energy	Coal	Oil	Natural Gas	Water Power	Nuclear Energy
U.S.A.	1 806.2	406.3	743.2	509.4	74.0	73.3
Canada	221.5	22.9	81.6	48.1	59.2	9.7
Total North America	2 027.7	429.2	824.8	557.5	133.2	83.0
Latin America	348.5	16.9	227.8	54.4	48.6	0.8
Total Western Hemisphere	2 376.2	446.1	1 052.6	611.9	181.8	83.8
Western Europe						
Austria	24.8	3.1	10.9	4.0	6.8	—
Belgium & Luxembourg	48.3	10.3	25.6	9.5	0.1	2.8
Denmark	17.1	5.4	11.7	—	—	—
Finland	21.3	1.8	12.0	0.6	3.4	3.5
France	188.1	26.2	99.3	24.7	15.8	22.1
Greece	16.7	3.9	11.9	—	0.9	—
Iceland	1.8	—	0.5	—	1.3	—
Republic of Ireland	8.2	1.7	5.4	0.9	0.2	—
Italy	143.8	13.5	95.5	22.8	10.7	1.3
Netherlands	69.3	4.3	35.2	28.8	—	1.0
Norway	28.6	0.5	7.5	—	20.6	—
Portugal	11.6	0.4	8.9	—	2.3	—
Spain	74.3	16.4	48.0	1.9	6.2	1.8
Sweden	39.4	1.4	21.8	—	9.9	6.3
Switzerland	26.4	0.7	11.9	0.8	9.3	3.7
Turkey	25.4	7.8	15.4	—	2.2	—
United Kingdom	195.7	69.7	74.6	42.0	1.3	8.1
West Germany	259.2	82.7	117.6	41.2	5.8	11.9
Yugoslavia	40.0	14.6	14.4	3.7	7.3	—
Cyprus/Gibraltar/Malta	1.4	b	1.4	—	—	—
Total Western Europe	1 241.4	264.4	629.5	180.9	104.1	62.5
Middle East	121.0	b	84.7	35.3	1.0	—
Africa	175.1	67.1	75.9	18.4	13.7	—
South Asia	154.7	88.7	43.4	8.9	12.8	0.9
South East Asia	200.3	57.2	123.7	7.3	8.6	3.5
Japan	353.6	63.2	224.3	24.2	20.7	21.2
Australasia	89.0	31.5	36.3	11.6	9.6	—
USSR	1 198.1	336.8	444.1	353.7	48.5	15.0
Eastern Europe	439.8	258.0	102.4	69.4	5.8	4.2
China	499.6	394.2	84.8	10.4	10.2	—
Total Eastern Hemisphere	4 472.6	1 561.1	1 849.1	720.1	235.0	107.3
World (excl. USSR, E. Europe & China)	4 711.3	1 018.2	2 270.4	898.5	352.3	171.9
World	6 848.8	2 007.2	2 901.7	1 332.0	416.8	191.1

Source. The British Petroleum Co. Ltd., *BP Statistical Review of the World Oil Industry 1982*, The British Petroleum Co. Ltd., London, 1983.

[a] In million tonnes oil equivalent.
[b] Less than 0.05 million tonnes oil equivalent.

TABLE 1.1-2   CONSUMPTION OF ENERGY BY TYPE, 1951–1981

Year	Coal[a] Quadrillion Btu	Coal[a] Million Short Tons	Natural Gas Quadrillion Btu	Natural Gas Trillion Cubic Feet	Petroleum[b] Quadrillion Btu	Petroleum[b] Million Barrels	Hydropower[c] Quadrillion Btu	Hydropower[c] Billion kW · h[f]	Nuclear Power Quadrillion Btu	Nuclear Power Billion kW · h[f]
1951	13.20	505.9	7.05	6.81	14.43	2,561	1.45	106.6	0	0
1952	11.84	454.1	7.55	7.29	14.96	2,661	1.50	112.0	0	0
1953	11.87	454.8	7.91	7.64	15.56	2,774	1.44	111.6	0	0
1954	10.17	389.9	8.33	8.05	15.84	2,831	1.39	114.0	0	0
1955	11.52	447.0	9.00	8.69	17.25	3,086	1.41	120.3	0	0
1956	11.72	456.9	9.61	9.29	17.94	3,212	1.49	129.8	0	0
1957	11.14	434.5	10.19	9.85	17.93	3,215	1.56	137.0	0	0
1958	9.83	385.7	10.66	10.30	18.53	3,328	1.63	146.9	(^{g})	0.2
1959	9.79	385.1	11.72	11.32	19.32	3,477	1.59	144.7	(^{g})	0.2
1960	10.12	398.0	12.39	11.97	19.92	3,586	1.65	153.7	0.01	0.5
1961	9.89	390.3	12.93	12.49	20.22	3,641	1.68	157.5	0.02	1.7
1962	10.17	402.2	13.73	13.27	21.05	3,796	1.82	172.2	0.03	2.3
1963	10.69	423.5	14.40	13.97	21.70	3,921	1.77	169.1	0.04	3.2
1964	11.25	445.7	15.29	14.81	22.30	4,034	1.91	182.3	0.04	3.3
1965	11.89	472.0	15.77	15.28	23.25	4,202	2.06	196.8	0.04	3.7
1966	12.48	497.7	17.00	16.45	24.40	4,411	2.07	199.0	0.06	5.5
1967	12.24	491.4	17.94	17.39	25.28	4,585	2.34	224.6	0.09	7.7
1968	12.66	509.8	19.21	18.63	26.98	4,902	2.34	225.2	0.14	12.5
1969	12.72	516.4	20.68	20.06	28.34	5,160	2.66	254.5	0.15	13.9
1970	12.66	523.2	21.79	21.14	29.52	5,364	2.65	252.9	0.24	21.8
1971	12.01	501.6	22.47	21.79	30.56	5,553	2.86	273.1	0.41	38.1
1972	12.45	524.3	22.70	22.10	32.95	5,990	2.94	283.6	0.58	54.1
1973	13.30	562.6	22.51	22.05	34.84	6,317	3.01	289.7	0.91	83.5
1974	12.88	558.4	21.73	21.22	33.45	6,078	3.31	316.9	1.27	114.0
1975	12.82	562.6	19.95	19.54	32.73	5,958	3.22	309.3	1.90	172.5
1976	13.73	603.8	20.35	19.95	35.17	6,391	3.07	295.5	2.11	191.1
1977	13.96	625.3	19.93	19.52	37.12	6,727	2.51	241.0	2.70	250.9
1978	13.85	625.2	20.00	19.63	37.97	6,879	3.14	303.2	3.02	276.4
1979	15.11	680.5	20.67	20.24	37.12	6,757	3.14	303.4	2.71	255.2
1980	15.46	702.7	20.39	19.88	34.20	6,242	3.11	300.1	2.67	251.1
1981[i]	16.01	727.7	19.93	19.42	32.00	5,840	2.97	287.0	2.90	272.3

Source. U.S. Dept. of Energy, *1981 Annual Report to Congress Vol. 2: Energy Statistics*, U.S. Dept. of Energy, Washington, D.C., 1982 (Table 3, p. 7).

[a] Bituminous coal, lignite, and anthracite.
[b] Refined petroleum products supplied including natural gas plant liquids and crude oil burned as fuel.
[c] Electric utility and industrial generation of hydropower and net electricity imports.
[d] Consumed by electric utilities.
[e] Wood, refuse, and other vegetal fuels consumed by electric utilities. Converted to Btu by applying national average heat rates for fossil fuel steam electric plants. Data do not include the consumption of wood-derived fuel (other than that consumed by the

TABLE 1.1-2 (*Continued*)

Year	Geothermal[d] Quadrillion Btu	Geothermal[d] Billion kW·h[e]	Wood and Waste[e] Quadrillion Btu	Wood and Waste[e] Billion kW·h[f]	Net Imports of Coal Coke Quadrillion Btu	Net Imports of Coal Coke Thousand Short Tons	Total Energy Consumption Quadrillion Btu	Change from Previous Year Percent[f]
1951	0	0	0	0	−0.02	−865	36.11	7.4
1952	0	0	0	0	−0.01	−479	35.83	−0.8
1953	0	0	0	0	−0.01	−363	36.76	2.6
1954	0	0	0	0	−0.01	−272	35.73	−2.8
1955	0	0	0	0	−0.01	−405	39.17	9.6
1956	0	0	0	0	−0.01	−525	40.75	4.0
1957	0	0	0	0	−0.02	−704	40.80	0.1
1958	0	0	0	0	−0.01	−271	40.65	−0.4
1959	0	0	0	0	−0.01	−337	42.41	4.3
1960	(g)	0	(g)	0.1	−0.01	−227	44.08	3.9
1961	(g)	0.1	(g)	0.1	−0.01	−318	44.72	1.5
1962	(g)	0.1	(g)	0.1	−0.01	−222	46.80	4.6
1963	(g)	0.2	(g)	0.1	−0.01	−298	48.61	3.9
1964	(g)	0.2	(g)	0.1	−0.01	−421	50.78	4.5
1965	(g)	0.2	(g)	0.3	−0.02	−744	52.99	4.4
1966	(g)	0.2	(g)	0.3	−0.03	−1,006	55.99	5.7
1967	0.01	0.3	(g)	0.3	−0.02	−618	57.89	3.4
1968	0.01	0.4	(g)	0.4	−0.02	−698	61.32	5.9
1969	0.01	0.6	(g)	0.3	−0.04	−1,456	64.53	5.2
1970	0.01	0.5	(g)	0.4	−0.06	−2,325	66.83	3.6
1971	0.01	0.5	(g)	0.3	−0.03	−1,335	68.30	2.2
1972	0.03	1.5	(g)	0.3	−0.03	−1,047	71.63	4.9
1973	0.04	2.0	(g)	0.3	−0.01	−317	74.61	4.2
1974	0.05	2.5	(g)	0.3	0.06	2,262	72.76	−2.5
1975	0.07	3.2	(g)	0.2	0.01	546	70.71	−2.8
1976	0.08	3.6	(g)	0.3	(g)	−4	74.51	5.4
1977	0.08	3.6	0.01	0.5	0.02	588	76.33	2.4
1978	0.06	3.0	(g)	0.3	0.13	5,029	78.18	2.4
1979	0.08	3.9	0.01	0.5	0.07	2,534	78.91	0.9
1980	0.11	5.1	(g)	0.4	−0.04	−1,412	75.91	−3.8
1981[h]	0.12	5.7	(g)	0.4	−0.02	−643	73.91	−2.6

electric utility industry) which amounted to an estimated 2.2 quadrillion Btu (1981). This table excludes small quantities of energy forms for which consistent historical data are not available, such as solar energy obtained by the use of thermal and photovoltaic collectors; wind energy; and geothermal, biomass, and waste energy other than that consumed at electric utilities.

[f] Percent change calculated from data prior to rounding.
[g] Less than 0.005 quadrillion Btu.
[h] Preliminary.

Note: Sum of components may not equal total due to independent rounding.

TABLE 1.1-3 CONSUMPTION OF ENERGY BY END-USE SECTOR, 1951–1981[a]

	Residential and Commercial		Industrial		Transportation			
Year	Without Electricity Distributed	With Electricity Distributed[b]	Without Electricity Distributed	With Electricity Distributed[b]	Without Electricity Distributed	With Electricity Distributed[b]	Electric Utilities	Total Energy Consumption
1951	7.00	9.60	14.66	17.41	8.99	9.11	5.45	36.11
1952	7.04	9.80	14.18	16.99	8.94	9.04	5.67	35.83
1953	6.83	9.75	14.83	17.86	9.05	9.15	6.06	36.76
1954	7.02	10.04	13.76	16.77	8.83	8.91	6.12	35.73
1955	7.47	10.62	15.44	18.99	9.47	9.55	6.79	39.17
1956	7.78	11.19	15.88	19.70	9.79	9.86	7.30	40.75
1957	7.54	11.17	15.61	19.46	9.84	9.90	7.55	40.80
1958	8.04	11.83	15.16	18.82	9.97	10.02	7.51	40.65
1959	8.23	12.33	15.80	19.74	10.30	10.35	8.08	42.41
1960	8.91	13.22	16.46	20.34	10.48	10.52	8.23	44.08
1961	9.13	13.63	16.47	20.44	10.62	10.66	8.51	44.72
1962	9.60	14.43	17.04	21.23	11.10	11.14	9.06	46.80
1963	9.62	14.86	17.79	22.17	11.54	11.58	9.66	48.61
1964	9.72	15.35	18.82	23.50	11.89	11.92	10.34	50.78
1965	10.13	16.19	19.50	24.47	12.30	12.33	11.07	52.99
1966	10.53	17.13	20.39	25.78	13.04	13.08	12.03	55.99
1967	11.09	18.17	20.38	26.00	13.69	13.72	12.73	57.89
1968	11.44	19.30	21.16	27.20	14.80	14.83	13.92	61.32
1969	11.94	20.66	21.91	28.40	15.43	15.46	15.25	64.53
1970	12.18	21.76	22.32	29.00	16.03	16.06	16.29	66.83
1971	12.38	22.67	22.05	28.96	16.65	16.68	17.22	68.30
1972	12.64	23.73	22.77	30.24	17.63	17.66	18.58	71.63
1973	12.24	24.20	23.86	31.88	18.49	18.52	20.01	74.61
1974	11.74	23.77	22.85	30.94	18.00	18.03	20.16	72.76
1975	11.58	23.92	20.57	28.60	18.14	18.18	20.42	70.71
1976	12.25	25.01	21.68	30.44	19.03	19.07	21.55	74.51
1977	11.83	25.41	21.97	31.18	19.70	19.74	22.82	76.33
1978	11.93	26.00	22.12	31.56	20.58	20.61	23.55	78.18
1979	11.79	26.08	22.58	32.39	20.40	20.43	24.14	78.91
1980	10.98	25.87	20.85	30.36	19.65	19.68	24.44	75.91
1981[c]	10.69	25.64	19.39	29.02	19.18	19.22	24.63	73.91

Source. U.S. Dept. of Energy, *1981 Annual Report to Congress, Vol. 2: Energy Statistics*, U.S. Dept. of Energy, Washington, D.C., 1982 (Table 4, p. 9).

[a] Data do not include consumption of wood-derived fuel (other than that consumed by the electric utility industry), which amounted to an estimated 2.2 quadrillion Btu in 1981. Also, small quantities of other energy forms for which consistent historical data are not available, such as solar energy obtained by the use of thermal and photovoltaic collectors; wind energy; and geothermal, biomass, and waste energy other than that consumed at electric utilities, are not included. In quadrillion Btu.
[b] Energy consumption by electric utilities is allocated to the three major end-use sectors in proportion to electricity sales.
[c] Preliminary.
Note: Sum of components may not equal total due to independent rounding.

Table 1.1-3 shows that the fastest growing energy end-use sector in the United States was residential and commercial, which grew from 24.2 quadrillion Btu in 1973 to 25.64 in 1981. Energy consumption also grew in transportation, from 18.5 quatrillion Btu to 19.2. In the industrial sector energy consumption declined slightly over the 1973–1981 period. From 1951 to 1981 total energy consumption in the United States grew from 36.1 to 73.9 quadrillion Btu, a growth rate of nearly 3%. The most rapidly growing sector was residential and commercial, which increased practically threefold, from 9.6 to 25.64 quadrillion Btu.

1.2 COAL RESOURCES AND PRODUCTION

1.2-1 Introduction

Coal is a dark brown or black combustible substance consisting of carbonized vegetable matter. It was created during past geologic ages when plant material in swamps was compacted under successive layers of vegetation and transformed first into peat and later, as marine or continental deposits covered the coal swamps, into coal. This transformation was marked by a progressive decrease in the amount of volatile matter and moisture resulting from the increased compression and temperature associated with greater depth of burial.[1] The geologic origin of coal is discussed in greater detail in Chapter 6.

The chief constituents of coal are fixed carbon, moisture, ash, volatile material, and sulfur. The ash and volatile material contain hydrocarbons, nitrogen, and polyacyclic organic matter as well as inorganic trace elements and radio nuclites.[2] The composition of coal is of great economic importance. The higher the carbon and heat content and the lower the percentage of ash, volatile material, and sulfur, the more valuable coal is to consumers.

1.2-2 U.S. Coal Resources

Coal is a very abundant fuel both in the United States and in the world as a whole. The total resources of coal in the United States have been estimated at approximately 4 trillion tons (see Table 1.2-1). In 1980 U.S. coal production was 835 million short tons or less than $\frac{1}{4500}$ of total resources.[3] A recent U.S. Geological Survey estimate[4] put the world total of identified plus hypothetical coal resources at 17 trillion short tons. In 1980 world production of coal was approximately 4 billion short tons or less than $\frac{1}{4000}$ of total resources.

Tables 1.2-1 and 1.2-2 present the U.S. Geological Survey estimates of the total remaining coal resources and the coal reserve base of the United States as of January 1, 1974.[5] The resource figures in Table 1.2-1 are divided into three categories—identified, hypothetical, and additional hypothetical—according to the methods of estimation and expected accuracy.

The estimates of identified resources shown in Table 1.2-2 are based on detailed information accumulated by mapping outcrops of coal beds and drilling holes to test coal bed thickness. They are subject to increase in the future as mapping, prospecting, and development are continued.[6] The resources included in the identified category satisfy certain constraints with respect to the minimum thickness of coal seams (14 in. for anthracite and bituminous coal and $2\frac{1}{2}$ ft for subbituminous coal and lignite) and the maximum depth of overburden (3000 ft in most states). In addition, the maximum ash and sulfur content is limited to 32.6 and 7.7%, respectively. It should be noted that the bulk of identified coal resources fall well within these limits.[7]

The percentage distribution of total identified resources by depth of overburden, thickness of beds, and rank are presented in Fig. 1.2-1. Overburden and seam thickness are major determinants of the cost of mining coal. Fig. 1.2-1 shows that the bulk, over 91% of resources, lies within 1000 ft of the surface and 58% is contained in seams of 5 ft or thicker for subbituminous coal and lignite and 28 in. or thicker for anthracite and bituminous coal. Conservative procedures were employed in estimating average seam thickness between points of measurement and in estimating the areal extent of coal beds around isolated points of information.

The hypothetical resources shown in Table 1.2-1 are estimates of coal in areas of known coal fields that are unmapped and unexplored. Hypothetical resources contain coal located in areas of coal-bearing rock that were excluded from consideration in the identified category because of lack of specific information about the occurrence and thickness of coal. Most exploration and mining of coal in the United States is concentrated along outcrops. As a result, only general information is available about coal in the centers of large coal basins. Moreover, many coal-bearing areas are so remote that they have been examined only by reconnaissance. The estimated hypothetical resources include an allowance for coal, which should be discovered when detailed geological mapping is extended into such areas.[8]

The estimates of coal in the hypothetical category are subject to the same constraints with respect to seam thickness and overburden applied to identified resources. The bulk of hypothetical resources lies in the range of 1000–2000 ft deep. Estimated hypothetical resources located below the 3000 ft

TABLE 1.1-4   PRICES OF DOMESTICALLY PRODUCED FOSSIL FUELS, 1951–1981 (CENTS PER MILLION BTU)

Year	Crude Oil[a] Current	Constant[d]	Natural Gas[b] Current	Constant[d]	Bituminous Coal and Lignite Current	Constant[d]	Anthracite Current	Constant	Composite[c] Current	Constant[d]
1951	43.6	76.4	6.6	11.6	18.8	32.9	40.2	70.4	25.5	44.7
1952	43.6	75.3	7.0	12.1	18.7	32.3	38.9	67.2	25.7	44.4
1953	46.2	77.5	8.2	13.9	18.8	32.0	40.2	68.3	26.9	45.7
1954	47.9	80.4	9.1	15.3	17.3	29.1	35.6	59.8	27.4	46.0
1955	47.8	78.6	9.3	15.3	17.3	28.4	32.6	53.6	27.0	44.4
1956	48.1	76.6	9.7	15.4	18.6	29.6	34.2	54.5	27.5	43.8
1957	53.3	82.1	10.2	15.7	19.6	30.2	37.6	57.9	29.7	45.7
1958	51.9	78.6	10.7	16.2	18.7	28.3	37.3	56.5	28.9	43.8
1959	50.0	74.0	11.6	17.2	18.6	27.5	35.1	51.9	28.3	41.9
1960	49.7	72.3	12.6	18.3	18.3	26.6	33.0	48.0	28.1	40.9
1961	49.8	71.8	13.6	19.6	17.9	25.8	33.8	48.8	28.5	41.1
1962	50.0	70.8	14.0	19.8	17.5	24.8	32.8	46.5	28.4	40.2
1963	49.8	69.5	14.3	20.0	17.2	24.0	35.7	49.8	28.0	39.1
1964	49.7	68.3	14.0	19.2	17.5	24.0	37.0	50.8	27.7	38.1
1965	49.3	66.3	14.2	19.1	17.5	23.5	35.3	47.5	27.5	37.0
1966	49.7	64.7	14.2	18.5	17.9	23.3	33.7	43.9	27.7	36.1
1967	50.3	63.6	14.5	18.3	18.4	23.3	34.7	43.9	28.3	35.8
1968	50.7	61.4	14.7	17.8	18.6	22.5	37.6	45.6	28.5	34.5
1969	53.3	61.4	15.1	17.4	20.0	23.0	42.3	48.7	29.6	34.1

Year										
1970	54.8	59.9	15.5	16.9	25.5	27.9	47.1	51.5	31.6	34.6
1971	58.4	60.8	16.5	17.2	29.2	30.4	51.4	53.5	33.9	35.3
1972	58.4	58.4	16.9	16.9	31.9	31.9	52.9	52.9	34.6	34.6
1973	67.1	63.5	19.8	18.7	35.5	33.6	58.9	55.7	39.4	37.3
1974	118.4	103.0	27.7	24.1	66.4	57.8	98.4	85.6	67.3	58.6
1975	132.2	105.3	40.6	32.3	82.9	66.0	137.9	109.8	82.0	65.3
1976	141.2	106.9	53.1	40.2	83.9	63.5	149.0	112.8	89.8	68.0
1977	147.8	105.7	72.3	51.7	87.3	62.4	150.4	107.6	100.6	71.9
1978	155.2	103.4	83.2	55.4	97.1	64.7	149.9	99.9	111.1	74.0
1979	217.9	133.9	107.9	66.3	104.7	64.3	174.1	107.0	141.3	86.8
1980	365.3	206.0	145.9	82.3	106.1	59.8	188.4	106.2	200.2	112.9
1981[e]	535.0	277.6	187.7	97.4	112.3	58.3	203.4	105.6	270.4	140.3

Source. U.S. Dept. of Energy, *1981 Annual Report to Congress, Vol. 2: Energy Statistics* (Washington, D.C.: U.S. Dept. of Energy, 1982) Table 10, p. 21.

[a]Includes lease condensate.
[b]Wet natural gas, prior to extraction of natural gas plant liquids.
[c]Derived by multiplying the price per Btu of each fossil fuel by the total Btu content of the production of each fossil fuel and dividing the accumulated price of total fossil fuel production by the accumulated Btu content of total fossil fuel production.
[d]Constant 1972 prices calculated using GNP implicit price deflators, 1972 = 100.
[e]Estimated.

Note: All fuel prices taken as close as possible to the point of production.

TABLE 1.2-1 TOTAL ESTIMATED REMAINING COAL RESOURCES OF THE UNITED STATES, JANUARY 1, 1974[a]

State	Remaining Identified Resources, Jan. 1, 1974				Overburden 0–3000 ft			Overburden 3000–6000 ft	Overburden 0–6000 ft
	Bituminous Coal	Subbituminous Coal	Lignite	Anthracite and Semianthracite	Total	Estimated Hypothetical Resources in Unmapped and Unexplored Areas[b]	Estimated Total Identified and Hypothetical Resources Remaining in the Ground	Estimated Additional Hypothetical Resources In Deeper Structural Basins[b]	Estimated Total Identified and Hypothetical Resources Remaining in the Ground
Alabama	13,262	0	2,000	0	15,262	20,000	35,262	6,000	41,262
Alaska	19,413	110,666	(c)	(d)	130,079	130,000	260,079	5,000	265,079
Arizona	21,234[e]	(e)	0	0	21,234	0	21,234	0	21,234
Arkansas	1,638	0	350	428	2,416	4,000[f]	6,416	0	6,416
Colorado	109,117	19,733	20	78	128,948	161,272	290,220	143,991	434,211
Georgia	24	0	0	0	24	60	84	0	84
Illinois	146,001	0	0	0	146,001	100,000	246,001	0	246,001
Indiana	32,868	0	0	0	32,868	22,000	54,868	0	54,868
Iowa	6,505	0	0	0	6,505	14,000	20,505	0	20,505
Kansas	18,668	0	(g)	0	18,668	4,000	22,668	0	22,668
Kentucky:									
Eastern	28,226	0	0	0	28,226	24,000	52,226	0	52,226
Western	36,120	0	0	0	36,120	28,000	64,120	0	64,120
Maryland	1,152	0	0	0	1,152	400	1,552	0	1,552
Michigan	205	0	0	0	205	500	705	0	705
Missouri	31,184	0	0	0	31,184	17,489	48,673	0	48,673
Montana	2,299	176,819	112,521	0	291,639	180,000	471,639	0	471,639
New Mexico	10,748	50,639	0	4	61,391	65,556[h]	126,947	74,000	200,947
North Carolina	110	0	0	0	110	20	130	5	135
North Dakota	0	0	350,602	0	350,602	180,000	530,602	0	530,602
Ohio	41,166	0	0	0	41,166	6,152	47,318	0	47,318
Oklahoma	7,117	0	(g)	0	7,117	15,000	22,117	5,000[i]	27,117
Oregon	50	284	0	0	334	100	434	0	434
Pennsylvania	63,940	0	0	18,812	82,752	4,000[j]	86,752	3,600[k]	90,352

State							
South Dakota	0	0	2,185	0	2,185	1,000	3,185
Tennessee	2,530	0	0	0	2,530	2,000	4,530
Texas	6,048	0	10,293	0	16,341	112,100[l]	128,441
Utah	23,186[m]	173	0	0	23,359	22,000[n]	80,359
Virginia	9,216	0	0	335	9,551	5,000	14,651
Washington	1,867	4,180	117	5	6,169	30,000	51,169
West Virginia	100,150	0	0	0	100,150	0	100,150
Wyoming	12,703	123,240	([c])	0	135,943	700,000	935,943
Other States[o]	610	32[p]	46[q]	0	688	1,000	1,688
Total	747,357	485,766	478,134	19,662	1,730,919	1,849,649	3,968,264

Source. Paul Averitt, *Coal Resources of the United States, January 1, 1974: Geological Survey Bulletin 1412,* U.S. Government Printing Office, Washington, D.C. (Table 3, pp. 14–15).

[a] In millions (10^6) of short tons. Estimates include beds of bituminous coal and anthracite generally 14 in. or more thick, and beds of subbituminous coal and lignite generally $2\frac{1}{2}$ ft or more thick, to overburden depths of 3000 and 6000 ft. Figures are for resources in the ground.

[b] Source of estimates: Alabama, W. C. Culbertson; Arkansas, B. R. Haley; Colorado, Holt (1975); Illinois, M. E. Hopkins and J. A. Simon; Indiana, C. W. Wier; Iowa, E. R. Landis; Kentucky, K. J. Englund; Missouri, Robertson (1971, 1973); Montana, R. E. Matson; New Mexico, Fassett and Hinds (1971); North Dakota, R.A. Brant; Ohio, H. R. Collins and D. O. Johnson from data in Struble and others (1971); Oklahoma, S. A. Friedman; Oregon, R. S. Mason; Pennsylvania anthracite, Arndt and others (1968); Pennsylvania bituminous coal, W. E. Edmunds; Tennessee, E. T. Luther; Texas lignite, Kaiser (1974); Virginia, K. J. Englund; Utah, H. H. Doelling, Washington, H. M. Beikman; Wyoming, N. M. Denson, G. B. Glass, W. R. Keefer, and E. M. Schell; remaining States, by the author.

[c] Small resources of lignite included under subbituminous coal.

[d] Small resources of anthracite in the Bering River field believed to be too badly crushed and faulted to be economically recoverable (Barnes, 1951).

[e] All tonnage is in the Black Mesa field. Some coal in the Dakota Formation is near the rank boundary between bituminous and subbituminous coal. Does not include small resources of thin and impure coal in the Deer Creek and Pinedale fields.

[f] Lignite.

[g] Small resources of lignite in western Kansas and western Oklahoma in beds generally less than 30 in. thick.

[h] After Fassett and Hinds (1971), who reported 85,222 million tons "inferred by zone" to an overburden depth of 3000 ft in the Fruitland Formation of the San Juan basin. Their figure has been reduced by 19,666 million tons as reported by Read and others (1950) for coal in all categories also to an overburden depth of 3000 ft in the Fruitland Formation of the San Juan basin. The figure of Read and others was based on measured surface sections.

[i] Includes 100 million tons inferred below 3000 ft.

[j] Bituminous coal.

[k] Anthracite.

[l] Lignite, overburden 200–5000 ft; identified and hypothetical resources undifferentiated. All beds assumed to be 2 ft thick, although many are thicker.

[m] Excludes coal in beds less than 4 ft thick.

[n] Includes coal in beds 14 in. or more thick, of which 15,000 million tons is in beds 4 ft or more thick.

[o] California, Idaho, Nebraska, and Nevada.

[p] California and Idaho.

[q] California, Idaho, Louisiana, and Mississippi.

TABLE 1.2-2   RESERVE BASE OF THE UNITED STATES BY SULFUR CONTENT, JANUARY 1, 1974

State	≤ 1.0%	1.1–3.0%	> 3.0%	Unknown	Total[a]
			Sulfur Content		
		Surface Mining			
Alabama	35.4	83.2	1.6	1,063.2	1,183.7
Alaska	7,377.6	21.0	0.0	0.0	7,399.0
Arizona	173.3	176.7	0.0	0.0	350.0
Arkansas	37.9	152.8	17.1	55.2	263.3
Colorado	724.2	146.2	0.0	0.0	870.0
Illinois	60.4	1,493.0	9,321.3	1,347.8	12,222.9
Indiana	105.3	559.2	907.3	101.6	1,674.1
Kansas	0.0	309.2	695.6	383.2	1,388.1
Kentucky, East	1,515.7	929.9	86.8	915.3	3,450.2
Kentucky, West	0.2	177.8	2,017.5	1,708.8	3,904.0
Maryland	28.6	66.6	16.2	34.6	146.3
Michigan	0.1	0.5	0.1	0.0	0.6
Missouri	0.1	47.8	1,635.8	1,730.0	3,413.7
Montana	38,182.4	2,175.2	46.4	2,166.7	42,561.9
New Mexico	1,681.0	579.3	0.0	0.0	2,258.3
North Carolina	0.0	0.0	0.0	0.4	0.4
North Dakota	5,389.0	10,325.4	268.7	15.0	16,003.0
Ohio	18.9	991.0	2,524.9	117.9	3,653.9
Oklahoma	120.5	88.1	38.8	186.2	434.1
Oregon	0.5	0.3	0	0.0	0.8
Pennsylvania	138.6	718.4	231.5	89.5	1,181.4
South Dakota	103.1	287.9	35.9	1.0	428.0
Tennessee	65.5	163.2	55.2	34.1	319.6
Texas	659.8	1,884.6	284.1	444.0	3,271.9
Utah	52.3	149.1	42.6	18.0	262.0
Virginia	411.6	218.1	2.1	46.7	679.2
Washington	172.5	307.7	25.8	2.2	508.1
West Virginia	3,005.5	1,422.8	270.4	509.6	5,212.0
Wyoming	13,192.8	10,122.3	425.5	105.3	23,845.3
Total[a]	73,252.3	33,597.4	18,950.9	11,076.1	136,885.7

maximum depth are included in Table 1.2-1 in the column headed "Estimated Additional Hypothetical Resources in Deeper Structural Basins."

Table 1.2-2 presents estimates of the U.S. coal reserve base distributed by state and sulfur content and divided into two categories according to potential mining method (i.e., underground or surface). The reserve base is that portion of the identified resources considered suitable for mining by current methods. It includes only coal in beds 28 in. or more thick for bituminous coal and anthracite and 60 in. or more for subbituminous coal and lignite and at depths of 0–120 ft for lignite and 0–1000 ft for all higher ranks of coal.[9] The reserve base consists of coal deposits that have been measured and surveyed with sufficient accuracy to serve as an inventory of coal available for mining.

U.S. coal reserves consist of that part of the reserve base that is recoverable. Based on past mining experiences, at least half of the coal in the ground is recoverable. The rate of recovery is determined by the method of mining. The room and pillar method, typical of most underground mining in the United States today, achieves an average recovery rate of approximately 50%. The rate of recovery is much higher for longwall mining, which is growing in popularity. The average recoverability in strip-mining operations is on the order of 80%.[10] Applying the 50% rate of recovery to the reserve base for underground mining and 80% to the reserve base for surface mining yields a total estimate of U.S. reserves based on the reserve base figures shown in Table 1.2-3 of 259,000 million short tons.

TABLE 1.2-2 (*Continued*)

State	\u2264 1.0%	1.1–3.0%	> 3.0%	Unknown	Total[a]
		Sulfur Content			
		Underground Mining			
Alabama	589.3	1,016.7	14.8	176.2	1,798.1
Alaska	4,080.8	163.2	0.0	0.0	4,246.4
Arkansas	43.3	310.3	29.2	19.1	402.4
Colorado	6,751.3	640.0	47.3	6,547.3	13,999.2
Georgia	0.3	0.0	0.0	0.2	0.5
Illinois	1,034.7	5,848.4	33,647.6	12,908.4	53,441.9
Indiana	443.5	2,746.6	4,355.1	1,402.5	8,948.5
Iowa	1.5	226.7	2,105.9	549.2	2,884.9
Kentucky, East	5,042.7	2,391.9	212.7	1,814.0	9,466.5
Kentucky, West	0.0	386.6	7,226.4	1,107.1	8,719.9
Maryland	106.5	623.9	171.2	0.0	901.9
Michigan	4.6	84.9	20.8	7.0	117.6
Missouri	0.0	134.2	3,590.2	2,350.5	6,073.6
Montana	63,464.2	1,939.8	456.2	0.0	65,834.3
New Mexico	1,894.3	214.1	0.9	27.5	2,136.5
North Carolina	0.0	0.0	0.8	31.3	31.3
Ohio	115.5	5,449.9	10,109.4	1,754.1	17,423.3
Oklahoma	154.5	238.5	202.6	264.3	860.1
Oregon	1.0	0.0	0.0	0.0	1.0
Pennsylvania	7,179.7	16,195.2	3,568.1	2,864.7	29,819.2
Tennessee	139.3	370.0	101.4	53.9	667.1
Utah	1,916.2	1,397.6	6.8	460.3	3,780.5
Virginia	1,728.5	945.4	12.0	283.3	2,970.7
Washington	431.0	957.8	13.2	42.9	1,445.9
West Virginia	11,086.6	12,583.4	6,552.9	4,142.9	34,377.8
Wyoming	20,719.5	4,535.1	1,275.6	2,955.0	29,490.8
Total[a]	126,928.8	59,400.2	73,720.2	39,761.6	299,839.7

Source. The President's Commission on Coal, *Coal Data Book*, The President's Commission on Coal, Washington, D.C., February 1980.

[a] Data may not add to totals shown due to rounding.

1.2-3 World Coal Resources

World geologic resources and recoverable reserves of coal are presented in Table 1.2-3. Separate resource and reserve estimates are presented for both hard coal (anthracite and bituminous) and brown coal (subbituminous and lignite). U.S. Geological Survey estimates of the total of identified and hypothetical resources for the major regions of the world are presented in Table 1.2-4. The figures in Table 1.2-3 were compiled from a country by country survey of coal resources conducted for the World Energy Conference (WEC) in 1979 by the Federal Institute for Geosciences and Natural Resources of the Federal Republic of Germany and were published in a 1980 report. The 1980 WEC Survey of Energy Resources report also contains information on the portion of reserves that are surface mineable and the portion of coking quality.[11] The term *additional resources* as used in Table 1.2-3 corresponds roughly to the definition of estimated total identified and hypothetical resources (as used in Table 1.2-1) less reserves. Recoverable reserves are comparable to the recoverable portion of the reserve base in Table 1.2-3. The maximum depth limits specified by most countries for additional resources of hard coal and brown coal were 2000 and 1500 m, respectively. The corresponding limits for recoverable reserves were 1500 and 600 m, respectively. The minimum seam thicknesses for recoverable reserves were generally around 0.6 m for hard coal and 1.5 m for brown coal. The precise

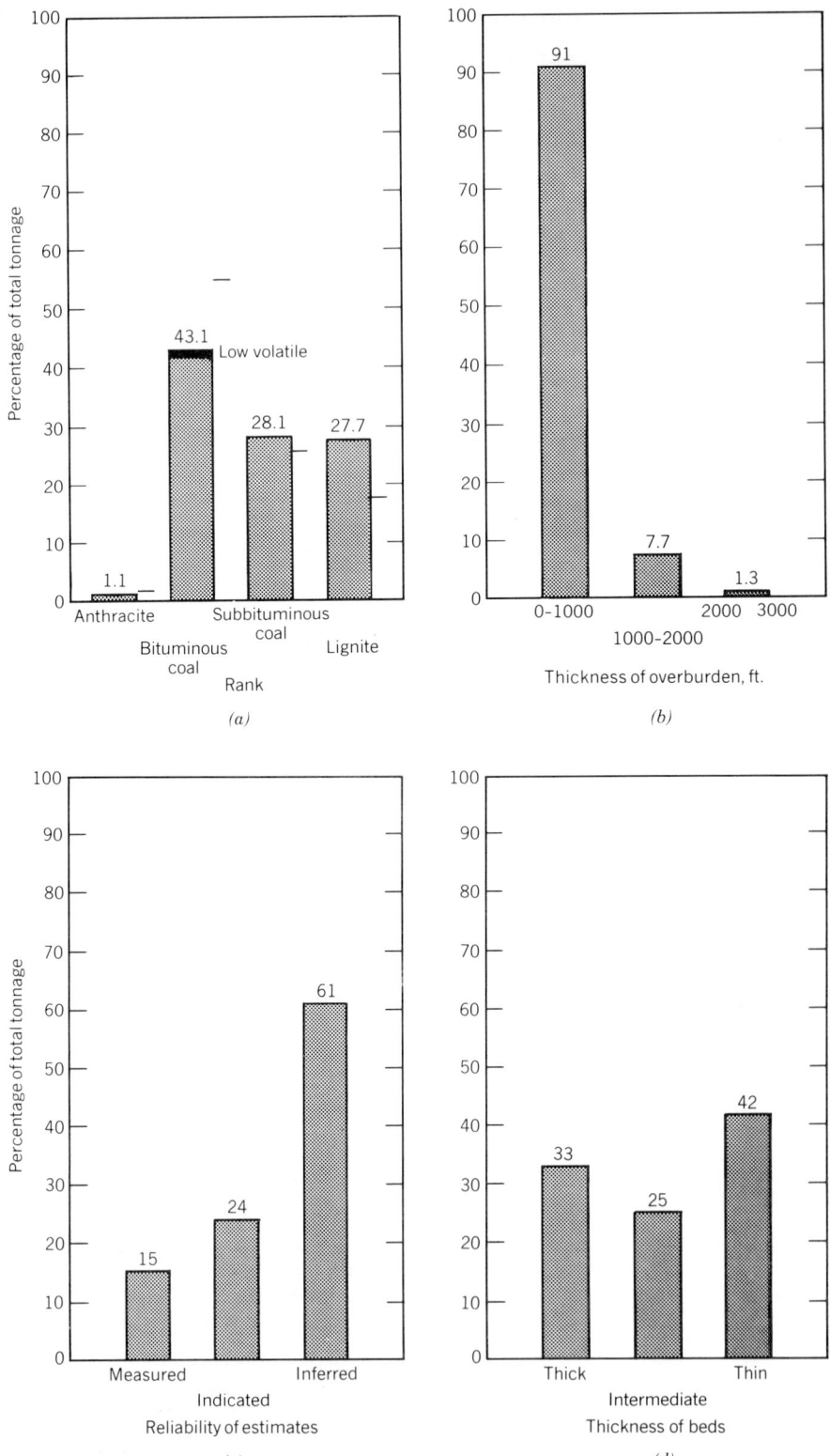

Fig. 1.2-1 Identified coal resources of the United States: Percentage distribution. (From ref. 1, p. 37.)

TABLE 1.2-3   WORLD COAL RESOURCES AND RESERVES (10^6 t.c.e.)

Country and Continent	Proved Recoverable Reserves[a]	Additional Resources[b]	Total
Bituminous Coal and Anthracite			
America			
Argentina	—	—	—
Brazil	189.0	1,447.0	1,636.0
Canada	1,607.0	93,413.0	95,020.0
Chile	26.5	291.3	317.8
Colombia	1,010.0	7,200.0	8,210.0
Mexico	1,200.0	1,300.0	2,500.0
Peru	—	835.0	835.0
United States	107,183.0	1,072,000.0	1,179,183.0
Venezuela	134.0	4,700.0	4,834.0
Other	NA	NA	NA
Total America	111,349.5	1,181,186.3	1,292,535.8
Europe			
Belgium	440.0	2,617.0	3,057.0
Bulgaria	30.0	1,200.0	1,230.0
Czechoslovakia	2,700.0	5,500.0	8,200.0
France	550.0	200.0	750.0
Germany (DR)	—	—	—
Germany (FR)	23,991.0	186,300.0	210,291.0
Greece	—	—	—
Hungary	225.0	350.0	575.0
Netherlands	—	1,194.0	1,194.0
Poland	27,000.0	84,000.0	111,000.0
Romania	50.0	520.0	570.0
Spain	398.0	2,375.0	2,773.0
United Kingdom	45,000.0	145,000.0	190,000.0
Yugoslavia	70.0	22.0	92.0
Other Countries	78.3	199.0	277.3
Total Europe	100,532.3	429,477.0	530.009.3
Africa			
Mozambique	240.0	155.0	395.0
Nigeria	—	21.0	21.0
Republic of Botswana	3,500.0	100,000.0	103,500.0
Republic of South Africa	25,290.0	33,762.0	59,052.0
Rhodesia	734.0	5,820.0	6,554.0
Swaziland	1,820.0	3,000.0	4,820.0
Zambia	24.0	98.0	122.0
Other	917.9	1,500.0	2,417.9
Total Africa	32,525.9	144,356.0	176,881.9
Australia and Pacific South Sea			
Australia	25,400.0	503,000.0	528,400.0
New Zealand	35.0	125.0	160.0

TABLE 1.2-3   (*Continued*)

Country and Continent	Proved Recoverable Reserves[a]	Additional Resources[b]	Total
Other	2.0	8.0	10.0
Total Australia and Pacific South Sea	25,437.0	503,133.0	528,570.0
Asia			
Bangladesh	242.0	—	242.0
China (PR)	99,000.0	1,326,000.0	1,425,000.0
India	12,610.0	91,139.0	103.749.0
Indonesia	10.9	4.0	14.9
Iran	193.0	—	193.0
Japan	1,050.0	0	1,050.0
North Korea	300.0	2,700.0	3,000.0
South Korea	116.3	1,049.0	1,165.3
Turkey	186.2	924.0	1,110.2
USSR	104,000.0	2,480,000.0	2,584,000.0
Other	218.3	1,397.0	1,630.2
Total Asia	217,926.7	3,903,213.0	4,121,139.7
Total World	487,771.4	6,161,365.3	6,649,136.7

Subbituminous Coal and Lignite

Country and Continent	Proved Recoverable Reserves	Additional Resources	Total
America			
Argentina	117.0	3,397.5	3,514.5
Brazil	720.7	9,960.6	10,681.3
Canada	2,760.5	272,802.0	275,562.5
Chile	897.0	3,225.3	4,122.3
Colombia	19.1	526.2	545.3
Mexico	299.5	390.0	689.5
Peru	—	33.0	33.0
United States	83,707.0	1,447,200.0	1,530,907.0
Venezuela	4.9	3,354.0	3,358.9
Other	—	37.0	37.0
Total America	88,525.7	1,740,925.6	1,829,451.3
Europe			
Belgium	—	—	—
Bulgaria	1,850.0	350.0	2,200.0
Czechoslovakia	1,716.0	972.0	2,688.0
France	24.3	27.3	51.6
Germany (DR)	7,500.0	—	7,500.0
Germany (FR)	10,545.0	—	10,545.0
Greece	511.5	379.5	891.0
Hungary	1,320.0	1,386.0	2,706.0
Netherlands	—	—	—
Poland	3,600.0	7,200.0	10,800.0
Romania	363.0	—	363.0
Spain	237.8	884.0	1,121.8

TABLE 1.2-3 (*Continued*)

Country and Continent	Proved Recoverable Reserves[a]	Additional Resources[b]	Total
United Kingdom	—	—	—
Yugoslavia	8,670.0	1,964.5	10,634.5
Other	64.7	60.7	125.4
Total Europe	36,402.3	13,224.0	49,626.3
Africa			
Mozambique	—	—	—
Nigeria	131.8	780.0	911.8
Republic of Botswana	—	—	—
Republic of South Africa	—	—	—
Rhodesia	—	—	—
Swaziland	—	—	—
Zambia	—	—	—
Other	1.3	14.4	15.7
Total Africa	133.1	794.4	927.5
Australia and Pacific South Sea			
Australia	10,902.0	108,600.0	119,502.0
New Zealand	126.9	2,053.9	2,180.8
Other	—	—	—
Total Australia and Pacific South Sea	11,028.9	110,653.9	121,682.8
Asia			
Bangladesh	—	1.0	1.0
China (PR)	—	13,365.0	13,365.0
India	524.0	92.7	616.7
Indonesia	223.2	6,288.5	6,511.7
Iran	—	—	—
Japan	5.9	—	5.9
North Korea	234.0	1,716.0	1,950.0
South Korea	—	—	—
Turkey	570.3	92.3	662.6
USSR	61,470.0	1,952,400.0	2,013,870.0
Other	601.0	108.9	709.9
Total Asia	63,628.4	1,974,064.4	2,037,692.8
Total World	199,718.4	3,839,662.3	4,039,380.7

Source. *International Coal*, National Coal Association, 1982.

[a] The part of proved reserves that can be recovered under present and expected local economic conditions with existing available technology.
[b] All resources that are of foreseeable economic interest (reflect a reasonable level of confidence).

TABLE 1.2-4 ESTIMATED TOTAL ORIGINAL COAL RESOURCES OF THE WORLD, BY CONTINENTS[a]

Continent	Identified Resources	Estimated Hypothetical Resources	Estimated Total Resources
Asia[b]	4,000[c]	7,000	11,000[d]
North America	1,900	2,500	4,400
Europe[e]	300	500	800
Africa	90	160	250
Oceania[f]	70	60	130
South and Central America	30	10	40
Total	6,390	10,230	16,620

Source. Paul Averitt, *Coal Resources of the United States, January 1, 1974: Geological Survey Bulletin 1412*, U.S. Government Printing Office, Washington, D.C. (Table 9, p. 91).

[a] In billion of short tons. Original resources in the ground in beds 12 in. or more thick and generally less than 4000 ft below surface but includes small amounts between 4000 and 6000 ft.
[b] Includes European USSR.
[c] Includes about 2,300 billion short tons in the USSR (Mel'nikov, 1972, p. 78).
[d] Includes about 9,500 billion short tons in the USSR (Mel'nikov, 1972, p. 79).
[e] Includes Turkey.
[f] Australia, New Zealand, New Caledonia.

criteria and methods of estimation, of course, varied from country to country. The criteria for most countries were included in the 1980 WEC report.

It is worth noting that the figures given for both additional resources and recoverable reserves for the United States in Table 1.2-3 appear conservative compared with figures presented in Table 1.2-1. The estimates of resources and reserves of coal compiled by the WEC periodically over the last decade have trended steadily upward as a result of the ongoing mapping and surveying of coal fields, particularly in areas that have been relatively unexplored.

The figures in Table 1.2-3 show that the countries with the largest resources and reserves of coal are the USSR, the United States, and China. These three countries account for more than two-thirds of total world resources. Other countries with substantial resources and/or recoverable reserves include Canada, Germany, Poland, the United Kingdom, Botswana, The Republic of South Africa, India, and Australia.

1.2-4 Coal Production

Table 1.2-5 gives the production of coal by country through the years 1973–1980 as reported by the U.S. Energy Information Administration. Table 1.2-6 presents statistics on U.S. coal production from 1951 to 1981. In Table 1.2-6 production figures are given separately for each type of coal, method of mining, and location.

Coal has been mined and used as a source of energy for centuries. It has been a major fuel for industry since the beginnings of the industrial revolution in the early seventeenth century. Figures 1.2-2 and 1.2-3 illustrate plots of U.S. and world production of coal and lignite from the early 1800s to the present. World coal production has grown exponentially from 1860 to the present. The average annual rate of growth from 1860 to 1960 has been around 3%. In the last two decades, from 1960 to 1980, the average percentage rate of growth has slowed to around 2%. However, since the rapid increase in petroleum prices beginning in 1973, the rate of growth in world coal production has increased slightly to a little over $2\frac{1}{2}$% per annum.

In the United States the pattern of change in coal production over time has been somewhat different. From 1850 to 1907 the curve of U.S. coal production followed a constant annual compound growth rate of 6% per year. From about 1910 to the early 1960s U.S. coal production fluctuated around a level of 500 million tons per year. Since the mid-1960s, however, it has increased steadily, reaching 830 million tons in 1980. The rate of increase since 1973 has been particularly rapid at 4.8% per year.

The trends in U.S. coal consumption by end use sector since 1951 are presented in Table 1.2-7. The difference between the total coal consumption in Table 1.2-7 and U.S. production in Table 1.2-6 is

TABLE 1.2-5 WORLD COAL PRODUCTION, 1973–1980[a]

Region and Country	1973	1974	1975	1976	1977	1978	1979	1980[b]
North America								
Canada	23	23	28	28	32	34	33	37
Mexico	5	6	6	6	7	8	8	8
United States	599	610	655	685	697	670	781	835
Total	626	639	688	719	736	711	822	800
Central and South America								
Brazil	2	3	3	4	4	4	5	5
Colombia	4	3	4	4	4	4	6	6
Other	2	2	2	2	2	2	2	1
Total	8	9	9	10	10	10	13	12
Western Europe								
Belgium	10	9	8	8	8	7	7	7
France	31	28	28	28	27	26	23	23
Germany, West	238	244	238	247	229	228	239	239
Greece	14	16	20	25	26	25	26	26
Spain	14	15	15	16	19	22	24	26
Turkey	12	11	12	11	13	15	22	23
United Kingdom	143	121	142	137	135	136	135	141
Yugoslavia	36	37	39	41	43	44	46	50
Other	8	8	7	7	6	6	6	6
Total	508	488	509	518	506	509	529	541
Eastern Europe and U.S.S.R.								
Bulgaria	30	27	31	28	28	26	31	28
Czechoslovakia	120	122	127	130	134	136	137	133
Germany, East	272	269	272	273	280	279	282	282
Hungary	29	28	27	28	28	28	28	28
Poland	216	222	233	241	250	258	264	254
Romania	27	30	30	28	30	32	36	35
USSR	736	755	773	784	796	798	792	789
Other	1	1	1	1	1	1	1	1
Total	1431	1454	1494	1513	1546	1560	1571	1550
Africa								
South Africa	69	73	77	85	94	100	114	126
Zimbabwe	3	3	3	3	3	3	4	3
Other	3	3	3	3	3	3	3	3
Total	75	79	82	91	100	106	120	132
Middle East, Far East and Oceania								
Australia	94	100	105	117	119	124	139	140
China	520	548	570	586	606	681	699	682
India	89	96	109	116	115	116	118	109
Japan	25	22	26	20	20	21	19	20
Korea, North	41	43	44	45	45	45	48	48
Korea, South	15	17	19	18	19	20	20	20
Taiwan	4	3	3	4	3	3	3	3
Other	12	13	13	15	16	16	18	19
Total	800	842	890	921	942	1026	1063	1040
World Total	3447	3511	3673	3772	3841	3922	4119	4155

Source. U.S. Dept. of Energy, *International Energy Annual 1979*, NTIS September 1981, p. 11.

[a] In million short tons.
[b] Preliminary. Includes anthracite, subanthracite, bituminous, subbituminous, lignite, and brown coal. Sum of components may not equal total due to independent rounding.

TABLE 1.2-6   U.S. COAL PRODUCTION, 1951–1981[a]

	Type		Method of Mining		Location		
Year	Bituminous Coal and Lignite	Anthracite	Underground	Surface	West of the Mississippi	East of the Mississippi	Total
1951	534	43	442	134	35	542	576
1952	467	41	381	126	33	475	507
1953	457	31	367	121	31	458	488
1954	392	29	306	115	25	395	421
1955	465	26	358	133	27	464	491
1956	501	29	381	149	26	504	530
1957	493	25	374	144	25	493	518
1958	410	21	298	134	20	411	432
1959	412	21	293	140	20	412	433
1960	416	19	293	142	21	413	434
1961	403	17	280	141	22	399	420
1962	422	17	288	151	21	418	439
1963	459	18	309	168	24	453	477
1964	487	17	328	176	26	478	504
1965	512	15	338	189	27	500	527
1966	534	13	343	204	28	519	547
1967	553	12	352	212	29	536	565
1968	545	11	347	210	30	527	557
1969	561	10	349	222	33	538	571
1970	603	10	341	272	45	568	613
1971	552	9	277	284	51	510	561
1972	595	7	305	297	64	538	602
1973	592	7	300	298	76	522	599
1974	603	7	278	332	92	518	610
1975	648	6	293	361	111	544	655
1976	679	6	295	389	136	549	685
1977	691	6	267	431	164	533	697
1978	665	5	243	427	183	487	670
1979	776	5	321	460	221	560	781
1980	824	6	338	492	251	579	830
1981[b]	802	6	319	489	270	538	808

Source. U.S. Dept. of Energy, *1981 Annual Report to Congress, Vol. 2: Energy Statistics*, Washington, D.C.: U.S. Dept. of Energy, 1982 (Table 54, p. 125).

[a] In million short tons. Sum of components may not equal total due to rounding.
[b] Preliminary.

accounted for by exports. The figures in Table 1.2-7 reveal that coal consumption by industry and in the transportation, commercial, and residential sectors has been trending downward—very rapidly in the case of transportation, residential, and commercial consumption. Since 1973 coal consumption in the industrial sector has essentially leveled off at about half of its 1951 value. The end use that has accounted for the substantial growth in consumption since the mid-1960s is electricity generation. The electric utilities now account for 82% of total U.S. coal consumption.

The statistics on world and U.S. coal production show that coal use has been growing steadily worldwide since the mid-nineteenth century and that this growth has continued into the modern era but at a slower rate than other energy sources. In recent years the production growth rate has been

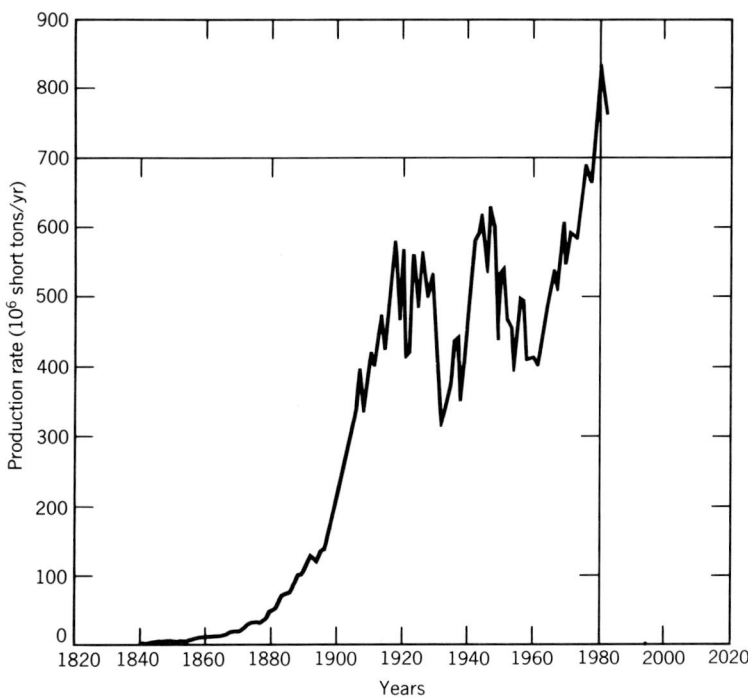

Fig. 1.2-2 U.S. production of coal and lignite. (Source: M. King Hubbert, "Energy Resources," *Resources and Man*, National Academy of Sciences–National Research Council, 1969, Fig. 8.4, p. 164.)

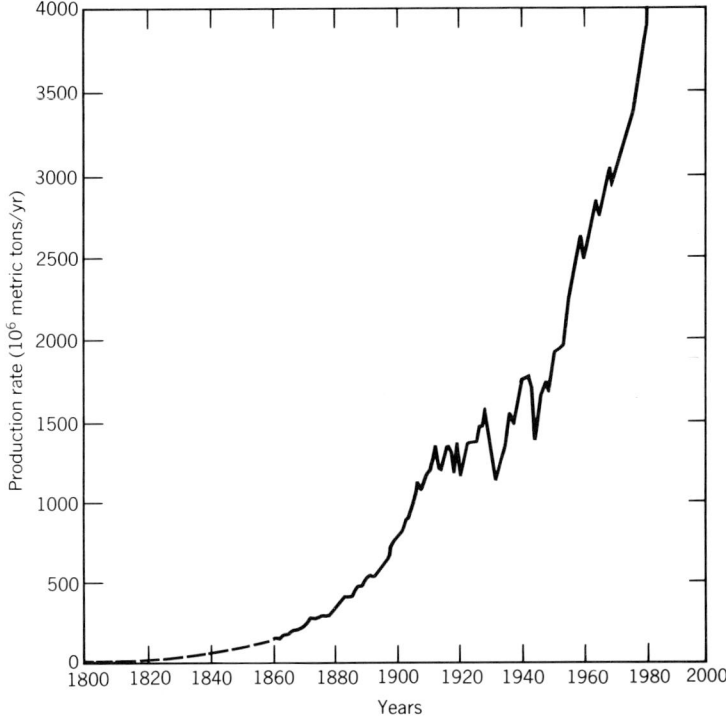

Fig. 1.2-3 World production of coal and lignite. (Source: M. King Hubbert, "Energy Resources," *Resources and Man*, National Academy of Sciences–National Research Council, 1969, Fig. 8.1, p. 161.)

TABLE 1.2-7   U.S. COAL CONSUMPTION BY END USE SECTOR,[a] **1959–1981**

| | | | Industry and Miscellaneous | | | | |
Year	Electric Utilities	Coke Plants	Other Industry and Miscellaneous	Total	Transportation	Residential and Commercial	Total
1951	105.8	113.7	128.7	242.4	56.2	101.5	505.9
1952	107.1	97.8	117.1	214.9	39.8	92.3	454.1
1953	115.9	113.1	117.0	230.1	29.6	79.2	454.8
1954	118.4	85.6	98.2	183.9	18.6	69.1	389.9
1955	143.8	107.7	110.1	217.8	17.0	68.4	447.0
1956	158.3	106.3	114.3	220.6	13.8	64.2	456.9
1957	160.8	108.4	106.5	214.9	9.8	49.0	434.5
1958	155.7	76.8	100.5	177.4	4.7	47.9	385.7
1959	168.4	79.6	92.7	172.3	3.6	40.8	385.1
1960	176.6	81.4	96.0	177.4	3.0	40.9	398.0
1961	182.1	74.2	95.9	170.1	0.8	37.3	390.3
1962	193.2	74.7	97.1	171.7	0.7	36.5	402.2
1963	211.3	78.1	101.9	180.0	0.7	31.5	423.5
1964	225.4	89.2	103.1	192.4	0.7	27.2	445.7
1965	244.8	95.3	105.6	200.8	0.7	25.7	472.0
1966	266.5	96.4	108.7	205.1	0.6	25.6	497.7
1967	274.2	92.8	101.8	194.6	0.5	22.1	491.4
1968	297.8	91.3	100.4	191.6	0.4	20.0	509.8
1969	310.6	93.4	93.1	186.6	0.3	18.9	516.4
1970	320.2	96.5	90.2	186.6	0.3	16.1	523.2
1971	327.3	83.2	75.6	158.9	0.2	15.2	501.6
1972	351.8	87.7	72.9	160.6	0.2	11.7	524.3
1973	389.2	94.1	68.0	162.1	0.1	11.1	562.6
1974	391.8	90.2	64.9	155.1	0.1	11.4	558.4
1975	406.0	83.6	63.6	147.2	[b]	9.4	562.6
1976	448.4	84.7	61.8	146.5	[b]	8.9	603.8
1977	477.1	77.7	61.5	139.2	[b]	9.0	625.3
1978	481.2	71.4	63.1	134.5	[b]	9.5	625.2
1979	527.1	77.4	67.7	145.1	[b]	8.4	680.5
1980	569.3	66.7	60.3	127.0	[b]	6.5	702.7
1981[c]	596.2	60.8	64.4	125.1	[b]	6.4	727.7

Source. U.S. Dept. of Energy, *1981 Annual Report to Congress, Vol. 2: Energy Statistics*, Washington, D.C.: U.S. Dept. of Energy, 1981 (Table 55, p. 127).

[a] Data in million short tons. Sum of components may not equal total due to independent rounding.
[b] Less than 0.05 million short tons. Quantities are included in the Other Industry and Miscellaneous category.
[c] Preliminary.

moving up slowly in response to the rise in the cost of competing sources of energy. This trend is most pronounced in the United States where production growth has resumed after a 50-year pause.

A comparison of current production levels with the figures on reserves and resources presented above both in the United States and worldwide makes it very clear that coal availability will not set a constraint on the rate of growth in coal production in the immediate future. Moreover, at an average price in 1981 for bituminous coal, F.O.B. the mine head of approximately $1 per million Btu, coal is inexpensive relative to other forms of energy.

Unfortunately, environmental factors may retard the growth of coal production and use. Emissions of ash and sulfur from coal combustion can be largely eliminated by current abatement technology at a reasonable cost. But other forms of pollution from coal are less amenable to control. The removal of many trace elements is difficult. Nitrogen dioxide emissions from coal burning can be sharply reduced but not eliminated. However, unquestionably the most serious, long-range concern about the effect of coal burning on the environment is its contribution to rising atmospheric levels of carbon dioxide. Some scientists fear that high levels of CO_2 in the earth's atmosphere produced by burning fossil fuels will trap heat radiating from the earth's surface into space and produce a "greenhouse effect," which will alter the earth's climate significantly by the middle of the next century. Unfortunately, coal releases 11% more carbon dioxide when burned than oil and 67% more than natural gas.[12]

REFERENCES

1.2-1 Paul Averitt, *Coal Resources of the United States, January 1, 1974*, Geological Survey Bulletin 1412, U.S. Government Printing Office, Washington, D.C., 1975, p. 16.

1.2-2 Office of Technology Assessment, *The Direct Use of Coal: Prospects and Problems of Production and Combustion*, Office of Technology Assessment. Washington, D.C., 1979, p. 60.

1.2-3 Averitt, p. 16.

1.2-4 Averitt, p. 91.

1.2-5 Averitt, pp. 14–15.

1.2-6 Averitt, p. 10.

1.2-7 Averitt, pp. 10–39.

1.2-8 Averitt, pp. 43–44.

1.2-9 Averitt, p. 32.

1.2-10 Averitt, pp. 30–31.

1.2-11 J. Koch, "Solid Fossil Fuels," in *Survey of Energy Resources, 1980 Part B Appendix*, 79th World Energy Conference at Munich, September 8–12, 1980, Federal Institute For Geosciences and Natural Resources. Hanover, Federal Republic of Germany, 1980, Summary Tables.

1.2-12 Office of Technology Assessment, p. 226.

1.3   PETROLEUM RESOURCES AND PRODUCTION

1.3-1   Introduction

Petroleum is a mixture or solution of raw chemical compounds of hydrogen and carbon. About half of the weight of average crude oil consists of hydrocarbon compounds containing from 3 to 14 carbon atoms per molecule.[1] These chemicals are of biological origin. Petroleum was formed from the remains of dead plants and animals that were deposited in the bottom sediments of seas and lakes in past ages. Commercial deposits of petroleum are found in close association with sedimentary rocks of marine and lacustrine origin.[2] The great bulk of hydrocarbons contained in sedimentary rocks is disseminated in low concentrations through large rock volumes. Petroleum reservoirs occur where hydrocarbons migrating through permeable sedimentary rock toward regions of lower pressure are trapped by a layer of impermeable rock. Petroleum geology is discussed in detail in Chapter 8.

The quality of crude oil is affected by the length and depth of burial. Older crude oils that have been buried at greater depths and, therefore, exposed to higher temperatures are chemically simple, light in weight, and contain a high percentage of low-molecular-weight fractions. They have low contents of oxygen- and nitrogen-containing materials.[3] For these reasons, they are less costly to refine and produce fewer emissions when burned. The specific gravity of crude oil ranges from 0.8 to 1 and is positively correlated with sulfur content.[4] The energy content of crude oil averages about 19,000 Btu/lb or 5.8 million Btu per 42-gal barrel.[5]

From small beginnings in the late nineteenth century, the use of petroleum has expanded enormously, coming in the last few decades to account for almost half (44%) of the world's total energy consumption. Petroleum is the most widely used and versatile of our fuels, meeting virtually all requirements for energy in transportation as well as substantial shares of the requirements for space heating, industrial process heat, and electricity generation.

1.3-2   U.S. Resources and Reserves

U.S. resources and reserves of petroleum distributed by petroleum region are presented in Table 1.3-1. The resource estimates in Table 1.3-1 were prepared by the U.S. Geological Survey in 1980 and the reserve figures were derived from American Petroleum Institute (API) data and pertain to 1979. Total U.S. reserves are presented annually for the years 1948–1982 in Table 1.3-2. The terms *measured reserves* as used in Table 1.3-1 and *proved reserves* as used in Table 1.3-2 are identical and refer to that portion of economic or recoverable resources that is estimated from geologic evidence supported directly by engineering measurements.[6] Economic or recoverable resources are those resources that are expected to be economically extractable given existing prices and technology.

TABLE 1.3-1 CRUDE OIL PRODUCTION, RESERVES, AND ESTIMATES OF UNDISCOVERED RECOVERABLE RESOURCES OF THE UNITED STATES[a]

Petroleum Region	Cumulative Production[b]	Identified Resources[b] Measured Reserves	Indicated Reserves	Inferred Reserves	Undiscovered Recoverable Resources[c] Low F_{95}	High F_5	Mean	Standard Deviation
		Onshore						
1 Alaska	1.2	8.7	0	5.0	2.5	14.6	6.9	4.3
2 Pacific Coast	16.5	3.2	1.6	1.2	2.1	7.9	4.4	2.0
3 Colorado Plateau and Basin and Range	1.7	0.3	Negl.	1.0	6.9	25.9	14.2	8.0
4 Rocky Mountains and northern Great Plains	6.9	1.1	0.2	2.9	6.0	14.0	9.4	2.6
5 West Texas and eastern New Mexico	25.2	5.4	1.3	4.0	2.7	9.4	5.4	2.2
6 Gulf Coast	34.9	3.8	0.2	5.3	3.6	12.6	7.1	2.8
7 Mid-continent	18.2	1.5	0.2	1.4	2.3	7.7	4.4	1.8
8 Michigan Basin	0.8	0.2	Negl.	0.8	0.3	2.7	1.1	0.8
9 Eastern Interior	4.3	0.2	Negl.	0.1	0.3	1.9	0.9	0.6
10 Appalachians	2.8	0.2	Negl.	0.1	0.1	1.5	0.6	0.5
11 Atlantic Coast	0.1	Negl.	0	0.1	0.1	0.8	0.3	0.3
Entire onshore	112.6	24.7	3.6	21.8	41.7	71.0	54.6	10.5
		Offshore—Shelf						
1A Alaska[d]	0.7	0.2	0	0.1	3.8	22.0	10.8	6.4
2A Pacific Coast	1.9	1.2	0	0.5	0.6	3.0	1.5	0.8
6A Gulf of Mexico	5.6	1.7[e]	Negl.[e]	1.0[e]	1.3	7.9	4.0	2.1
11A Atlantic Coast	0	0	0	0	0	3.9	1.3	1.4
Entire shelf	8.2	3.1	Negl.	1.5	9.2	30.2	17.6	6.9
		Offshore–Slope						
1A Alaska[d]	0	0	0	0	0	5.2	1.4	2.3
2A Pacific Coast	0	0	0	0	0.6	6.0	2.4	2.0
6A Gulf of Mexico	Negl.	No data[e]	0[e]	No data[e]	0.9	5.2	2.5	1.4
11A Atlantic Coast	0	0	0	0	0	10.7	4.1	3.6
Entire slope	0	0	0	0	4.2	19.2	10.4	4.9
		Offshore—Combined Shelf and Slope						
1A Alaska[d]	0.7	0.2	0	0.1	4.6	24.2	12.2	6.8
2A Pacific Coast	1.9	1.2	0	0.5	1.7	7.9	3.8	2.2
6A Gulf of Mexico	5.6	1.7	Negl.	1.0	3.1	11.1	6.5	2.5
11A Atlantic Coast	0	0	0	0	1.1	12.9	5.4	3.9
Entire offshore	8.2	3.1	Negl.	1.5	16.9	43.5	28.0	8.5
		Combined Onshore and Offshore						
Entire United States	120.7	27.8	3.6	23.4	64.3	105.1	82.6	13.4

Source. G. L. Dalton et al., *Estimates of Undiscovered Recoverable Conventional Resources of Oil and Gas in the United States*, Geological Survey Circular 860, U.S. Government Printing Office, Washington, D.C., 1981 (Table 4, p. 22).

[a]All tabulated values are rounded numbers; therefore, values for production and reserves and for means of undiscovered resources, all of which are additive, may not be precisely so. Values shown are in billions of barrels. Negl., negligible, less than or equal to 0.05 billion barrels of oil.

[b]Cumulative production and reserves are as of December 31, 1979. Production and reserve figures were derived from API and AGA data (American Petroleum Institute, American Gas Association, and Canadian Petroleum Association, 1980) except for California for which production and reserve data were taken from California Division of Oil and Gas (1980) and the U.S. Geological Survey (Kalil, 1980).

[c]F_{95} denotes the 95th fractile; the probability of *more than* the amount F_{95} is 95%. F_5 is defined similarly. Fractile values are not additive.

[d]Includes quantities considered recoverable only if technology permits their exploitation beneath Arctic pack ice—a condition not yet met.

[e]API and AGA reserve data for the Gulf of Mexico are not available within separate shelf and slope classifications. However, the declared reserves probably represent only the shelf and are so treated.

TABLE 1.3-2 ESTIMATED PROVED WORLD RESERVES OF CRUDE OIL ANNUALLY AS OF JANUARY 1[a]

Year	United States	Canada	Middle East	Total Free World	Total World
1948	21,487,685	125,000	28,550,000	61,697,685	68,197,685
1949	23,280,444	500,000	32,621,000	68,954,444	73,599,444
1950	24,649,489	1,200,000	32,413,000	71,711,739	76,452,739
1951	25,268,398	1,202,607	41,567,000	81,961,405	89,926,405
1952	27,468,031	1,376,600	51,320,000	95,517,481	103,444,481
1953	27,960,554	1,679,509	64,825,000	109,046,263	118,454,263
1954	28,944,828	1,845,422	78,160,000	125,514,750	134,969,750
1955	29,560,746	2,207,614	97,459,000	147,452,560	157,499,560
1956	30,012,170	2,509,534	126,271,000	178,737,704	189,570,704
1957	30,434,649	2,849,370	144,470,000	202,958,019	227,958,019
1958	30,300,405	2,874,454	169,566,000	234,440,359	260,640,359
1959	30,535,917	3,165,904	173,951,000	244,697,821	272,402,821
1960	31,719,347	3,497,124	181,436,000	260,033,071	290,035,071
1961	31,613,211	3,678,542	183,160,000	264,241,553	297,743,553
1962	31,758,505	4,173,569	188,204,000	271,155,174	305,407,174
1963	31,389,223	4,480,702	193,975,000	279,538,025	309,514,025
1964	30,969,990	4,881,492	207,368,000	297,446,782	326,946,782
1965	30,990,510	6,177,546	212,180,000	307,917,956	338,667,956
1966	31,352,391	6,711,237	215,360,000	314,736,878	348,118,878
1967	31,452,127	7,791,751	235,614,600	348,047,250	381,885,250
1968	31,376,670	8,168,924	249,209,000	371,780,039	407,553,039
1969	30,707,117	8,381,613	270,760,000	398,857,505	454,734,505
1970	29,631,862	8,619,805	333,506,000	470,534,067	530,534,067
1971	39,001,335	8,558,980	344,574,900	511,195,133	611,195,133
1972	38,062,957	8,333,087	367,386,000	534,053,494	632,553,494
1973	36,339,408	8,020,141	355,852,000	566,219,559	664,219,559
1974	35,299,839	7,674,150	350,162,500	523,706,479	626,706,479
1975	34,249,956	7,171,229	403,858,200	601,018,535	712,418,535
1976	32,682,127	6,653,002	368,410,570	554,920,849	657,920,849
1977	30,942,166	6,257,082	367,681,220	538,989,568	640,089,568
1978	29,486,402	5,970,872	366,166,000	547,805,194	645,805,194
1979	27,803,760	6,860,000	369,996,000	547,771,585	641,771,585
1980	27,051,289	6,800,000	361,947,300	522,174,789	642,174,789
1981	29,805,000	6,400,000	362,071,000	565,629,712	651,929,712
1982	29,785,000	7,300,000	362,839,950	584,864,150	670,709,150

Source. American Petroleum Institute, *Basic Petroleum Data Book: Petroleum Industry Statistics*, Vol. III, Number 1, American Petroleum Institute, January 1983.

[a] In thousands of barrels.

The category of identified resources as used in Table 1.3-1 contains, in addition to measured reserves, indicated and inferred reserves. *Indicated reserves* are economic reserves in existing fields that are expected to be recoverable with the use of advanced recovery techniques such as fluid injection. *Inferred reserves* are comprised of oil that is expected to be discovered through the continued exploration of existing fields. The U.S. Geological Survey made estimates of the ultimate production from known U.S. oil fields and subtracted the API estimates of cumulative production, proved reserves, and indicated reserves to obtain the figures shown in Table 1.3-1 for inferred reserves.[7]

When an exploratory well succeeds in finding oil or gas, an estimate can be made of the minimum amount (the proved reserve) that will be producible by the well and the ultimate potential of the entire field. As additional wells are drilled, both estimates are likely to be revised up or down. But the estimate of ultimate potential will vary over a much wider range than the figure of proved reserves. "For a typical field, it takes approximately five or six years before the proved reserve estimate of remaining oil plus past production begins to approach a true estimate of ultimate production."[8] Proved reserve data is considered to be accurate within $\pm 20\%$. Inferred reserves, the additional oil that will be produced from known fields above the level currently reported as proved, is typically on the order of 80% of proved reserves.[9]

The undiscovered recoverable resources listed in Table 1.3-1 are resources "outside known fields estimated from broad geologic knowledge and theory. Also included are resources from undiscovered pools that occur as unrelated accumulations controlled by distinctly separate structural features and (or) stratigraphic conditions within known fields."[10]

The potential quantity of undiscovered resources is typically estimated in one of three ways:

1. Through extrapolation of past finding rates (volumes discovered in relation to footage of wells drilled).

2. Volumetric yield methods in which the amount of discovered oil per unit volume of rock in well-explored districts is applied to the volume of rock in less explored districts.

3. Play analysis methods in which the geological variables that determine the probable quantity of undiscovered petroleum in an area are inputed into a "reservoir engineering equation" in the form of probability distributions. A Monte Carlo simulation is frequently used to generate the probability distribution of the amount of oil.[11]

In calculating the undiscovered petroleum resource estimates that appear in Table 1.3-1, the U.S. Geological Survey indirectly employed all three approaches (with emphasis on the first two). The subjective assessments of teams of experts familiar with the geology of each region were surveyed using a Delphi approach to determine the following for each petroleum province in the United States:

1. the marginal probability that recoverable oil was present at all;

2. a low estimate of the quantity of the resource conditional on the recoverable resource being present (the low-resource estimate corresponds to a 95% probability of more than that amount being present);

3. a high-resource estimate corresponding to a 5% probability of more than that amount being present;

4. a modal, most likely estimate of the quantity, of the resource conditional upon recoverable resources being present.

The high, low, and modal conditional estimates of undiscovered resources were used to determine a conditional probability distribution of the quantity of undiscovered recoverable resources for each province. The product of the marginal probability of resources being present and the corresponding conditional probability distribution for the province determined the judgmental probability distribution of the quantity of undiscovered recoverable resources. These distributions were then aggregated over the individual provinces within each oil region to obtain the high, low, and mean estimates of undiscovered/recoverable resources for each petroleum region presented in Table 1.3-1.[12]

Great differences exist between the estimates of undiscovered recoverable resources produced by different experts in the past. These differences are traceable in the main to different assumptions concerning the economic and technological conditions that determine if oil can be recovered profitably.[13] The U.S. Geological Survey estimates are based on standardized assumptions concerning recoverability. The U.S. Geological Survey geologists "assume that undiscovered resources of oil and gas will be recoverable under conditions represented by a continuation of price/cost relationships and technological trends that prevail at the time of the assessment (1980)."[14]

The sum of cumulative production to date, the identified resources, and the U.S. Geological Survey's mean estimate of undiscovered/recoverable resources is 258.1 billion barrels (Table 1.3-1). This figure represents the total amount of oil that will ultimately be recovered from U.S. oil fields

TABLE 1.3-3 WORLD CRUDE OIL PRODUCTION BY AREAS[a]

Year	United States[b]	Canada	Middle East	Total Free World	Total World
1947	1,856,987	7,692	306,320	2,791,259	3,022,139
1948	2,020,185	12,287	416,780	3,165,311	3,433,234
1949	1,841,940	21,305	511,507	3,117,616	3,404,142
1950	1,973,574	29,044	640,862	3,489,030	3,803,027
1951	2,247,711	47,615	700,643	3,951,984	4,282,730
1952	2,289,836	61,237	760,599	4,122,269	4,531,114
1953	2,357,082	80,899	884,736	4,338,708	4,798,055
1954	2,314,988	96,080	998,932	4,501,366	5,016,591
1955	2,484,428	129,440	1,185,519	5,016,962	5,625,659
1956	2,617,283	171,981	1,260,610	5,409,403	6,124,676
1957	2,616,901	181,848	1,293,099	5,617,319	6,438,444
1958	2,449,016	165,496	1,558,351	5,676,265	6,607,750
1959	2,574,590	184,778	1,684,636	6,074,425	7,133,238
1960	2,574,933	189,534	1,923,285	6,466,460	7,674,460
1961	2,621,758	220,861	2,055,933	6,819,352	8,186,213
1962	2,676,189	244,136	2,257,951	7,362,284	8,881,858
1963	2,752,723	257,662	2,486,348	7,865,825	9,538,346
1964	2,786,822	275,364	2,786,199	8,487,719	10,309,644
1965	2,848,514	296,997	3,053,941	9,085,034	11,062,515
1966	3,027,763	320,467	3,408,361	9,856,722	12,021,786
1967	3,215,742	351,287	3,667,393	10,586,778	12,914,340
1968	3,329,042	435,906	4,135,044	11,675,468	14,146,318
1969	3,371,751	410,814	4,542,994	12,583,611	15,222,511
1970	3,517,450	461,177	5,094,380	13,843,870	16,718,708
1971	3,453,914	491,846	5,979,115	14,564,001	17,662,793
1972	3,455,368	560,693	6,630,500	15,349,856	18,600,745
1973	3,360,903	648,348	7,744,933	16,764,730	20,367,981
1974	3,202,585	616,532	7,986,831	16,542,571	20,537,727
1975	3,056,779	520,666	7,160,993	15,174,305	19,502,335
1976	2,976,180	488,680	8,115,708	16,575,174	21,191,540
1977	3,009,265	482,021	8,186,851	17,108,078	21,900,695
1978	3,175,927	483,260	7,884,927	17,195,001	22,158,251
1979	3,114,545	546,040	7,868,670	17,752,140	22,870,170
1980	3,146,502	521,184	6,735,498	16,559,304	21,765,654
1981	3,128,780	469,025	5,749,480	15,181,445	20,373,570

Source. American Petroleum Institute, *Basic Petroleum Data Book: Petroleum Industry Statistics*, Vol. III, Number 1, American Petroleum Institute, January 1983.

[a] In thousands of barrels.
[b] Includes lease condensate.

(ultimate production). As 120 billion barrels have already been produced, the remaining resources total 137.4 billion barrels.

Many other estimates of undiscovered resources and ultimate production have been published. In the late 1960s M. King Hubbert published two estimates of the amount of oil that will ultimately prove recoverable in the United States.[15] The first estimate of 173 billion barrels for the United States (excluding Alaska) was based on complete-cycle analysis. Complete-cycle analysis assumes that the peak rates in the discovery and production of oil will occur in time at about the halfway point between initial discovery and production and final exhaustion of the resource. According to Hubbert, the peak in the rate of discovery of oil in the United States occurred in 1957 and the peak in production would occur in 1970. The figures on crude oil production presented in Table 1.3-3 indicate that U.S. production did indeed peak in 1970 and has declined steadily since that date. The peak in proved reserves can be expected to occur midway between the peaks in the rate of discovery and the rate of production. Reserves will peak when the rising rate of production becomes equal to a rate of discovery that has already begun to decrease. API-proved reserve figures (Table 1.3-2) confirm that U.S. crude oil reserves did indeed peak in the early 1960s and have declined steadily ever since.

Hubbert's second estimate of ultimate U.S. cumulative production and discovery was based on extrapolation of finding rate data. The quantity of crude oil discovered per foot of exploratory drilling was plotted against cumulative exploratory drilling footage in the United States. The result showed that discoveries per foot had decreased steadily with cumulative drilling at a negative exponential rate. By fitting a curve through the plotted observations of discoveries per foot versus cumulative drilling footage and extrapolating this curve until it intersected the axis, an estimate was obtained of the amount of additional oil that would be discovered by additional exploratory drilling. Using this method, Hubbert's estimate of ultimate cumulative discovery and production for the "coterminous" United States was 165 billion barrels. To that he added a maximum of 25 billion barrels for Alaska to obtain a total figure of 190 billion barrels for the United States as a whole.[16]

Regardless of which of the estimates cited above is accepted, it is clear that, in contrast to the situation with coal, U.S. domestic crude oil resources are not large compared with domestic requirements. Using the U.S. Geological Survey's estimate of 137.4 billion barrels of recoverable oil remaining to be extracted and an *Oil and Gas Journal* estimate of current 1982 production of 3.16 billion barrels (see Table 1.3-3) yields a figure of 43 years of domestic supply at the current production rate. Based on an assumed exponential decline in production rates and his 165-billion-barrel estimate of cumulative production for the "coterminous United States," Hubbert calculated that cumulative production will reach 90% of its ultimate value by about the year 2000.

1.3-3   World Reserves and Resources

API and *Oil and Gas Journal* estimates of proved petroleum reserves worldwide are presented in Tables 1.3-2 and 1.3-4, respectively. API reserve figures are given for the United States, Canada, the Mideast, the Free World, and the entire world for the years 1948–1982. *Oil and Gas Journal* estimates by country are given for January 1, 1983.

Table 1.3-2 reveals that the level of proved world reserves has been declining since 1975; a downturn that is due to the increasing difficulty of finding oil. The rapid increase in the international price of oil since 1973 has stimulated exploration and tended to increase the rate of discovery while simultaneously reducing consumption and production relative to what they otherwise would have been.

Table 1.3-5 contains three sets of estimates of conventional oil resources of the world. The first was prepared for the World Energy Conference (WEC) in 1980 and the second by Richard Nehring for the CIA in 1978. The third set of estimates was obtained from the report of the 1978 WEC. Conventional petroleum can be defined as crude oil for which exploration and exploitation is currently being carried on with technology that is now considered classical at a cost that is acceptable today. It excludes petroleum from tar sands and shale.

The 1980 WEC estimates shown in Table 1.3-5 are based on a 1976 study conducted by the West German Federal Institute for Geosciences and National Resources updated from the results of a survey of the national committees of the member countries of the WEC. The 1980 WEC and Nehring estimates of ultimate conventional resources (Table 1.3-5) include cumulative production and proved reserves as well as resources remaining to be discovered.

The 1978 WEC estimates of recoverable oil resources remaining in the ground presented in Table 1.3-5 are the mean estimates obtained from a survey of world petroleum experts conducted in 1977. They are "an evaluation of technical maxima in respect of what nature and available technology will make it possible to produce within the limits of cost and price increasing up to $20 (1976) per barrel in the year 2000."[17] The estimates assume that the present world recovery rate of 25% is raised to 40% by the end of the century.

TABLE 1.3-4 ESTIMATED PROVED OIL RESERVES BY COUNTRY, JANUARY 1, 1983[a]

Asia-Pacific		*Middle East (Continued)*	
Australia	1,622,077	Iran	55,308,000
Bangladesh	—	Iraq	41,000,000
Brunei	1,240,000	Israel	893
Burma	32,000	Jordan	—
China, Taiwan	6,700	Kuwait	64,230,000
Guam	—	Lebanon	—
India	3,416,400	Oman	2,730,000
Indonesia	9,550,000	Qatar	3,425,000
Japan	60,000	Saudi Arabia	162,400,000
Korea, South	—	Sharjah	404,000
Malaysia	3,325,000	South Yemen	
New Zealand	169,000	(Aden)	—
Okinawa	—	Syria	1,521,000
Pakistan	196,300	Turkey	280,000
Philippines	35,600	Total Middle East	369,285,893
Singapore	—		
Sri Lanka	—	*Africa*	
Thailand	103,000	Algeria	9,440,000
Total Asia-Pacific	19,756,077	Angola-Cabinda	1,635,000
		Cameroon	530,000
Western Europe		Congo Republic	1,550,000
Austria	128,200	Egypt	3,325,000
Belgium	—	Ethiopia	—
Cyprus	—	Gabon	460,000
Denmark	473,000	Ghana	5,400
Finland	—	Ivory Coast	110,500
France	124,300	Kenya	—
Germany, West	310,000	Liberia	—
Greece	59,780	Libya	21,500,000
Ireland	—	Madagascar	—
Italy-Sicily	703,000	Morocco	290
Netherlands	294,000	Mozambique	—
Norway	6,800,000	Nigeria	16,750,000
Portugal	—	Senegal	—
Spain	131,400	Sierra Leone	—
Sweden	—	South Africa	116,500
Switzerland	—	Sudan	400,000
United Kingdom	13,900,000	Tanzania	—
Total Western		Togo	—
Europe	22,923,680	Tunisia	1,860,000
		Zaire	139,000
Middle East		Zambia	—
Abu Dhabi	30,510,000	Total Africa	57,621,690
Bahrain	197,000		
Divided (Neutral)		*Western Hemisphere*	
Zone	5,840,000	Antigua	—
Dubai	1,440,000	Argentina	2,590,000

TABLE 1.3-4   (*Continued*)

Western Hemisphere (*Continued*)		Western Hemisphere (*Continued*)	
Bahamas	—	Peru	835,336
Barbados	730	Puerto Rico	—
Bolivia	180,000	Trinidad & Tobago	580,000
Brazil	1,750,000	Uruguay	—
Chile	760,000	Venezuela	21,500,000
Colombia	536,000	Virgin Islands	—
Costa Rica	—	United States	29,785,000
Dominican Republic	—	Canada	7,020,000
Ecuador	1,400,000	Total Western Hemisphere	115,287,066
El Salvador	—	Total Non-Communist	585,074,406
Guatemala	50,000		
Honduras	—	*Communist Areas*	
Jamaica	—	China	19,485,000
Martinique	—	USSR	63,000,000
Mexico	48,300,000	Other	2,630,000
Netherlands Antilles	—	Total Communist	85,115,000
Nicaragua	—		
Panama	—	Total World 1982	670,189,406
Paraguay	—		

Source. "Worldwide Oil and Gas at a Glance," *Oil and Gas Journal*, December 27, 1982, pp. 78–79.
[a] In thousands of barrels.

There is considerable consistency among the estimates of ultimate recovery and recoverable resources presented in Table 1.3-5. Converting the 1980 WEC estimate of ultimate recoverable resources from metric tons to barrels using the 7.3-barrels to 1-ton conversion factor cited in the 1980 WEC report yields a figure of 2584 billion barrels, which is approximately 10% above the high end of the range provided by Nehring.

As noted, the 1980 WEC and Nehring figures for ultimate recovery include oil in the ground plus past production. Subtracting cumulative production of $52,800 \times 10^6$ metric tons from the 1980 WEC estimate of total world recovery of $353,940 \times 10^6$ metric tons gives $301,140 \times 10^6$ metric tons of remaining recoverable oil, essentially identical to the 1978 WEC figure. Using the 7.3 barrels to 1 ton conversion factor, the WEC estimates of remaining oil both turn out to be about 2200 billion barrels.

1.3-4   Crude Oil Production

Crude oil production in the United States, Canada, the Mideast, the Free World, and the entire world each year from 1947 to 1982 is presented in Table 1.3-3. United States crude oil production from 1875 to the present and total world production of crude oil from 1880 to the present are also graphed in Figs. 1.3-1 and 1.3-2, respectively.

The graph of U.S. crude oil production clearly shows that output peaked in the early 1970s and has since entered a period of decline. Given the undiscovered resources remaining in the ground, the rate at which U.S. or world crude oil production will decline in the future depends on economic pressure. However, the decline in U.S. production over the last few years has occurred despite the intense demand for domestic oil as imported oil increased in price several-fold. The high price of imported oil has also greatly stimulated exploration activity in the United States as well as other regions of the world (see Section 1.3-5).

World oil production has continued to increase since 1973, except for the last three years in which the combined effect of the original 1973 quadrupling of oil prices and the 1979 doubling plus the current recession has finally caused production to decline.

Comparing the resource estimates in Table 1.3-5 with the world production figures presented in Table 1.3-3 reveals that petroleum is a relatively scarce energy resource worldwide as well as in the United States. Using the WEC estimate of 2200 billion barrels of recoverable oil remaining in the ground gives 108 years world supply at the 1981 rate of production of 20.4 billion barrels.

TABLE 1.3-5 THE CUMULATIVE PRODUCTION, RESERVES, RESOURCES, AND ULTIMATE RECOVERY FOR OIL ACCORDING TO REGION: THREE ESTIMATES[a]

Region	Cumulative Production[b] (10^6 t)	Proved Recoverable Resources[b] (10^6 t)	Estimated Additional Recoverable Resources[b] (10^6 t)	Ultimate Recovery[b] (10^6 t)	Estimated Ultimate Conventional Resources[c] (bbls)	Total Remaining Recoverable Resources[d] (10^6 t)
Africa	3,750	8,040	34,000	45,790	120–170	—
Africa South of the Sahara	—	—	—	—	—	11,300
North America	17,520	4,480	24,000	46,000	280–380	28,500
Latin America	7,040	7,770	12,000	26,810	120–160	22,900
Far East/Pacific	1,720	2,390	12,000	16,110	150–155	15,100
Middle East	14,680	51,050	52,000	117,730	860–1140	—
Middle East and North Africa	—	—	—	—	—	109,100
Western Europe	560	2,710	10,000	13,270	50–70	11,200
USSR, China, and Eastern Europe	7,530	12,700	64,000	84,230	165–225	59,400
Antarctic	—	—	4,000	4,000	—	—
Deep Offshore and Polar areas	—	—	—	—	—	38,700
Total	52,800	89,140	212,000	353,940	1,700–2,300	302,400

[a] The last three columns of this table present three alternative estimates of total world oil resources. The first two of these (columns 5 and 6) include cumulative production to date.
[b] Dr. E. Schubert, *Survey of Energy Resources, 1980, Part A Text Section A2*, 11th World Energy Conference at Munich 8–12 September 1980, Federal Institute for Geosciences and Natural Resources, Hanover, Federal Republic of Germany, 1980, Table 2.4, p. 93.
[c] Richard Nehring, *Giant Oil Fields and World Oil Resources*, R-2284 CIA, The Rand Corporation, Santa Monica, CA, June 1978, Table 5.2, p. 88.
[d] Pierre Despraires, "Worldwide Petroleum Supply Limits," *World Energy Resources 1985–2020*, The World Energy Conference, London, 1978, p. 24.

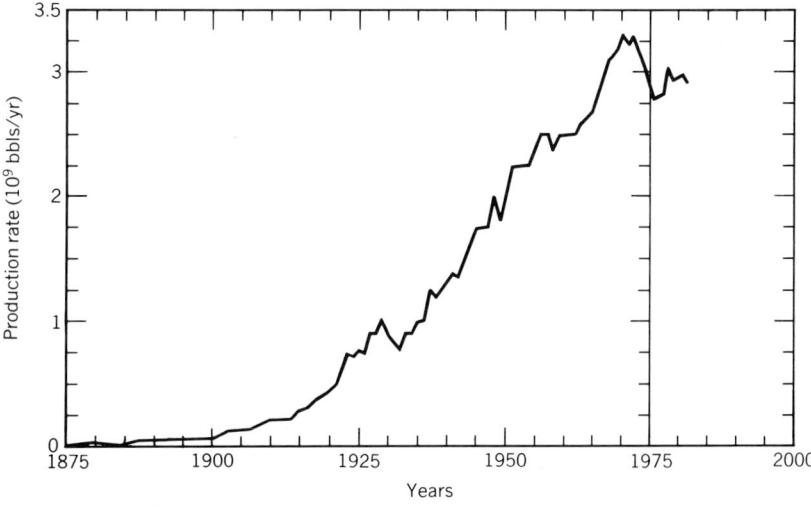

Fig. 1.3-1 Production of crude oil in the United States, exclusive of Alaska. (Source: M. King Hubbert, "Energy Resources," *Resources and Man*, National Academy of Sciences–National Research Council, 1969, Fig. 8.5.)

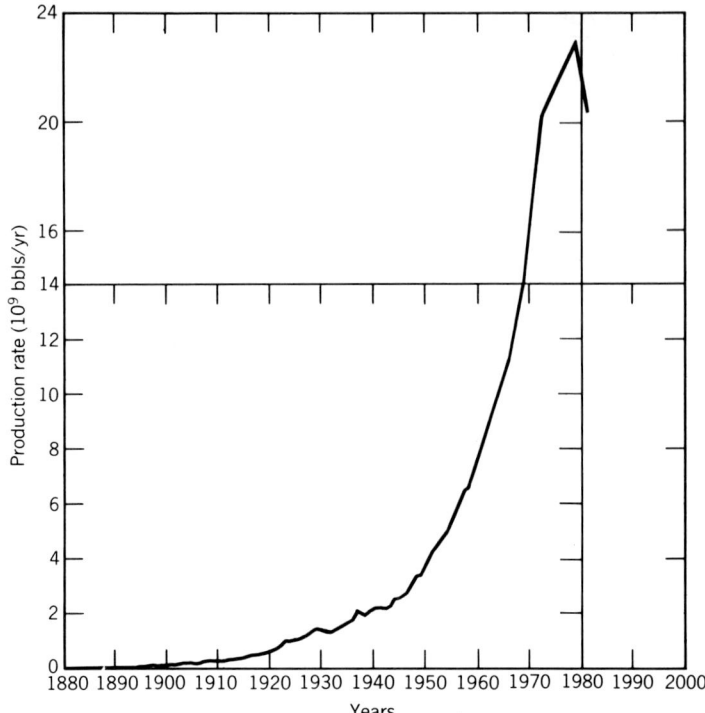

Fig. 1.3-2 World production of crude oil. (Source: M. King Hubbert, "Energy Resources," *Resources and Man*, National Academy of Sciences–National Research Council, 1969, Fig. 8.2, p. 162.)

It is clear that the rapid rate of increase in the price of petroleum worldwide since 1973 has been the major factor in causing the growth of production to slow down and finally decline. Prior to the "energy crisis," from 1947 to 1972, oil production grew at an annual compound rate of 7.5% per year. From 1973 to 1979 oil production grew at less than 2% per annum, and since 1979 it has actually declined to less than the 1973 level.

Although the recent great increases in the price of oil and the current recession have caused world production to drop off in the last 3 years, it is generally believed that the world oil production rate will continue to increase at least slowly in the next few decades. The U.S. Energy Information Administration currently forecasts that world oil production rate will increase at 0.4% per year to 1995.[18] Given this rate of growth, total recoverable oil resources of 2200 billion barrels are sufficient to last for 89 years.

Based on complete-cycle analysis and an estimate of ultimate recoverable resources of oil remaining in the ground of 2000 billion barrels, Hubbert and Root estimated that world crude oil production would peak around 1997 and that 80% of the oil available for extraction will have been produced by about the year 2027.[19]

1.3-5 Exploration

This section examines past and current exploration for both oil and natural gas. Oil and natural gas are often found in close association. Discussions of petroleum exploration in the literature typically deal with both, and the statistics of exploration pertaining to oil and gas are frequently combined.

Statistics on oil and gas exploration activity are presented for the United States in Table 1.3-6 and for the world in Table 1.3-7. Table 1.3-6 contains figures on the number and footage of dry holes and exploratory wells yielding oil and gas drilled each year in the United States from 1951 to 1981. Table 1.3-7 contains figures for the total number of exploratory well completions in the free world from 1970 to 1978.

Figures in both tables indicate that exploration activity has increased markedly since the fourfold increase in petroleum prices in 1973–1974. The total number of oil and gas wells drilled in the United States (including dry holes) declined from a maximum of 16,210 in 1956 to a low of 6920 in 1971. From 1973 to 1981 the total number of exploratory wells drilled grew from 7470 to 15,200. The

TABLE 1.3-6   EXPLORATORY WELLS DRILLED FOR OIL, 1951–1982[a]

Year	Wells Drilled (thousands)				Footage Drilled (million feet)				Average Depth (feet)				Successful Wells (%)
	Oil	Gas	Dry Holes	Total	Oil	Gas	Dry Holes	Total	Oil	Gas	Dry Holes	Total	
1951	1.76	0.45	9.54	11.76	8.1	2.5	38.7	49.3	4,609	5,497	4,059	4,197	18.9
1952	1.78	0.56	10.09	12.42	8.5	3.4	43.7	55.6	4,781	6,071	4,334	4,476	18.8
1953	1.98	0.70	10.63	13.31	9.4	4.0	47.3	60.7	4,761	5,654	4,447	4,557	20.1
1954	1.98	0.73	10.39	13.10	9.4	4.4	45.8	59.6	4,740	6,059	4,408	4,550	20.7
1955	2.24	0.87	11.83	14.94	10.8	5.2	53.2	69.2	4,819	5,964	4,498	4,632	20.8
1956	2.27	0.82	13.12	16.21	11.1	5.2	58.0	74.3	4,901	6,301	4,425	4,587	19.1
1957	1.94	0.86	11.90	14.71	9.8	6.0	53.4	69.2	5,036	6,898	4,488	4,702	19.1
1958	1.74	0.82	10.63	13.20	8.7	5.5	47.3	61.5	4,993	6,657	4,449	4,658	19.4
1959	1.70	0.91	10.58	13.19	8.5	6.0	48.7	63.3	5,021	6,613	4,602	4,795	19.8
1960	1.32	0.87	9.52	11.70	6.8	5.5	43.5	55.8	5,170	6,298	4,575	4,770	18.7
1961	1.16	0.81	9.02	10.99	5.9	5.2	43.3	54.4	5,099	6,457	4,799	4,953	17.9
1962	1.21	0.77	8.82	10.80	6.2	5.2	42.2	53.6	5,124	6,728	4,790	4,966	18.4
1963	1.31	0.66	8.69	10.66	6.4	4.2	42.8	53.5	4,878	6,370	4,933	5,016	18.5
1964	1.22	0.56	8.95	10.73	6.7	4.2	44.6	55.5	5,509	7,547	4,980	5,174	16.6
1965	0.95	0.52	8.00	9.47	5.4	3.8	40.1	49.2	5,672	7,295	5,007	5,198	15.4
1966	1.20	0.70	8.42	10.31	6.8	5.8	43.1	55.7	5,700	8,321	5,117	5,402	18.4
1967	0.99	0.53	7.36	8.88	5.7	4.0	38.2	47.8	5,758	7,478	5,188	5,388	17.1
1968	0.95	0.49	7.44	8.88	5.6	3.7	41.6	51.0	5,914	7,697	5,589	5,739	16.2
1969	1.08	0.62	8.00	9.70	6.6	5.0	45.9	57.5	6,054	8,092	5,739	5,924	17.5
1970	0.79	0.48	6.42	7.69	5.1	3.7	36.5	45.3	6,399	7,639	5,687	5,882	16.5
1971	0.65	0.44	5.83	6.92	3.7	3.3	33.3	40.4	5,702	7,616	5,716	5,835	15.7
1972	0.68	0.60	6.25	7.54	4.0	4.6	36.4	45.0	5,850	7,641	5,828	5,975	17.0
1973	0.62	0.90	5.95	7.47	3.9	6.2	34.8	44.8	6,226	6,856	5,844	5,997	20.3
1974	0.81	1.20	6.61	8.62	4.9	7.7	37.7	50.3	5,961	6,421	5,709	5,832	23.3
1975	0.97	1.17	7.07	9.21	5.7	8.0	40.1	53.8	5,863	6,831	5,678	5,844	23.3
1976	1.05	1.40	6.78	9.23	6.1	9.2	37.5	52.8	5,864	6,550	5,525	5,719	26.5
1977	1.21	1.48	7.28	9.96	7.1	9.7	40.9	57.7	5,834	6,550	5,626	5,788	27.0
1978	1.12	1.60	7.95	10.68	6.8	10.8	45.6	63.2	6,039	6,747	5,740	5,923	25.6
1979	1.24	1.78	7.46	10.48	7.5	11.8	43.2	62.5	6,023	6,599	5,794	5,958	28.8
1980	1.60	1.97	8.34	11.92	9.3	13.7	46.9	69.9	5,787	6,974	5,625	5,870	30.0
1981[b]	2.22	2.38	10.62	15.23	13.3	16.0	60.5	89.8	5,989	6,694	5,697	5,896	30.3

Source. U.S. Dept. of Energy, *1981 Annual Report to Congress*, Vol. 2: *Energy Statistics*, Washington, D.C.: U.S. Dept. of Energy, 1982 (Table 17, p. 37).

[a]Sum of components may not equal total due to independent rounding. Average depth may not equal average of components due to independent rounding. For the period 1960 forward, data are for well completion reports received by the American Petroleum Institute during the reporting year.
[b]Preliminary.

number of wells that succeeded in finding oil and gas followed a similar pattern. The number of successful oil wells peaked at 2270 in 1956 and reached a minimum of 650 in 1973. The 1981 figure is 2219. The number of successful gas wells rose to 910 in 1959, declined to a minimum of 440 in 1971, and has subsequently increased to 2380.

Abroad the pattern is similar, in part, however, because Canada and the United States account for a large amount of the total activity: 9685 out of 10,975 wells drilled in 1973. The total number of exploratory wells drilled in the free world outside the United States and Canada remained essentially

TABLE 1.3-7 EXPLORATORY DRILLING FOR OIL AND GAS[a]

Area/Country	1971	1972	1973	1974	1975	1976	1977	1978
United States	6,922	7,539	7,466	8,619	9214	9,234	9,961	10,677
Canada	1,534	1,633	2,219	1,735	1649	2,481	2,801	2,144
Mexico	129	143	104	98	87	79	79	83
Total North America	8,585	9,315	9,789	10,452	10,950	11,794	12,841	13,904
Argentina	157	110	139	117	78	83	143	81
Brazil	87	80	78	86	87	105	155	89
Ecuador	15	20	8	5	3	3	5	3
Peru	21	25	31	43	43	36	37	28
Trinidad-Tobago	35	23	15	13	14	23	17	8
Venezuela	44	64	63	76	35	45	48	60
Other	60	77	50	65	61	67	56	60
Total Latin America	419	399	384	405	321	362	406	329
Total Western Hemisphere	9,004	9,714	10,173	10,857	11,271	12,156	13,247	14,233
Total North Sea	53	51	70	92	124	109	21	38
Other European	146	159	147	151	172	163	231	259
Total Europe	199	210	217	243	296	272	252	297
Algeria	25	10	13	9	17	11	11	3
Gabon	11	17	15	13	23	19	19	20
Libya	41	34	25	32	36	22	33	41
Nigeria	55	61	45	51	33	21	24	35
Other	125	101	80	86	91	89	117	102
Total Africa	257	223	178	191	200	162	204	201
Abu Dhabi	8	21	7	3	0	9	5	19
Dubai	0	0	1	0	0	5	9	5
Sharjah	2	1	0	0	0	1	0	0
Iran	6	2	18	27	26	26	25	10
Iraq	0	0	3	2	0	2	NA	NA
Kuwait	2	0	0	0	0	2	0	0
Saudi Arabia	2	6	3	4	9	24	4	25
Neutral Zone	0	0	0	0	0	0	0	0
Qatar	1	6	1	0	0	0	0	0
Oman	0	16	0	0	0	0	10	24
Other	27	29	32	34	65	51	46	39
Total Middle East	48	81	65	70	100	120	99	122
Indonesia	135	137	169	176	186	130	115	146
Other	126	175	173	162	127	119	132	249
Total Far East	261	312	342	338	313	249	247	395
Total Eastern Hemisphere	765	826	802	842	909	803	802	1,015
Total Free World	9,769	10,540	10,975	11,699	12,180	12,959	14,049	15,248

Source. American Petroleum Institute: *Basic Petroleum Data Book: Petroleum Industry Statistics*, Vol. III, Number 1, American Petroleum Institute, January 1983.

[a] The figures in this table are the total number of wells drilled by country.

unchanged through the 1970–1973 period. Since 1973 the number has increased slightly, from 1290 to 1427 in 1978. The relatively slow growth in drilling activity in countries other than the United States and Canada reflects the fact that the largest free world producers abroad belong to OPEC and are already limiting production to levels well below capacity in order to support high prices.

Although exploratory efforts to locate new oil and gas reservoirs in the United States have increased since 1973, the rate of discovery has not kept pace; consequently, the number of barrels discovered per foot of wells drilled has continued to decrease.

The large decline in discoveries per foot since the beginning of U.S. petroleum exploration has occurred despite the effect of improved exploration technology. It results from the manner in which oil is distributed in the earth's crust. The bulk of U.S. and world oil is located in a relatively small number of giant oil fields. These fields tend to be discovered fairly early in the exploration process because their areal projection on the earth's surface is proportionate to their large size and because exploration is directed initially toward areas with the highest probability of containing petroleum based on geologic factors. The bulk of the new oil currently being discovered in the United States is contained in small fields. As there are a very large number of small fields, the rate of discovery per foot is likely to decline fairly slowly over the next few years.[20]

Despite the recent discovery of some major fields, such as the Mexican and North Sea oil and natural gas fields, the rate of discovery per foot of well drilled appears to be decreasing abroad as well as in the United States. World reserves have declined despite the stepped-up level of exploration activity and the recent decline in production.

REFERENCES

1.3-1 T. H. McCulloh, "Oil and Gas" in *United States Mineral Resources: Geological Survey Professional Paper 820*, Donald A. Brobst, and Walden P. Pratt (eds.), U.S. Government Printing Office, Washington, D.C., 1974, p. 478.

1.3-2 McCulloh, p. 478.

1.3-3 McCulloh, p. 478.

1.3-4 James G. Speight, *The Chemistry and Technology of Petroleum*, Vol. 3, Heinemann and Heinz (eds.), Marcel Deckker, New York, 1980.

1.3-5 Speight, pp. 106–108.

1.3-6 G. L. Dalton, et al., *Estimates of Undiscovered Recoverable Conventional Resources of Oil and Gas in the United States: Geological Survey Circular 860*, U.S. Government Printing Office, Washington, D.C., 1981, p. 7.

1.3-7 Dalton et al., pp. 7 and 22.

1.3-8 John J. Schranz, Jr., "Oil and Gas Resources Welcome to Uncertainty," *Resources: No. 58*, Resources for the Future, March 1978, p. 79.

1.3-9 Schranz, p. 80.

1.3-10 Dalton et al., p. 7.

1.3-11 Dalton et al., p. 14.

1.3-12 Dalton et al., pp. 14–20.

1.3-13 Schranz, p. 86.

1.3-14 Dalton et al., p. 7.

1.3-15 M. King Hubbert, "Energy Resources," in *Resources and Man*, National Academy of Sciences–National Research Council, 1969, p. 176.

1.3-16 Hubbert, pp. 182–184.

1.3-17 Pierre Desprairies, "Worldwide Petroleum Supply Limits," in *World Energy Resources 1985–2020*, The World Energy Conference, London, 1978, p. 8.

1.3-18 Department of Energy, Energy Information Agency, *1981 Annual Report to Congress, Vol. 3*, Supplement 2, February 1982, Table 1, p. 20.

1.3-19 M. King Hubbert and David H. Root, "Outlook For Fuel Reserves," in *Resources and Man*, National Academy of Sciences–National Research Council, 1969.

1.3-20 David H. Root and Lawrence J. Drew, "The Pattern of Petroleum Discovery Rates," *American Scientist*, **67**, November–December, 1979.

1.4 NATURAL GAS RESOURCES AND PRODUCTION

1.4-1 Introduction

Natural gas is a mixture of gaseous hydrocarbons of relatively low molecular weight trapped in reservoir rocks in the earth's crust. Natural gas was created (along with oil) by the burial of organic wastes beneath sediments of ancient seas and lakes. The heat and pressure resulting from the subsequent burial of these sediments at great depth in the earth's crust converted the contained biological material into oil and gas. Natural gas is frequently found in close association with oil as "associated gas," which is free natural gas occurring as a gas cap above an oil accumulation, and as "dissolved gas," which is natural gas dissolved in crude oil.[1] Natural gas also occurs as nonassociated gas in reservoirs not containing crude oil, typically at relatively great depths. There is also a category of hydrocarbon midway between oil and gas classified as natural gas liquids. These hydrocarbons are gaseous at atmospheric pressure and liquid at higher pressures. They include, but are not limited to, ethane, propane, butane, pentane, natural gasoline, and condensate.[2]

In the early decades of the petroleum industry most natural gas was flared off at the well head. But with the development of gas pipelines and distributing networks linking the fields to potential users and more recently pressurized LNG tankers, a large market has opened up for natural gas. Gas is used extensively for residential and commercial space heating and in industry to meet requirements for clean process heat. Natural gas currently supplies 19% of world energy requirements.

1.4-2   U.S. Reserves and Resources

The U.S. Geological Survey's estimates of proved reserves and undiscovered resources of natural gas in the United States are presented in Table 1.4-1. The definition and calculation of measured and inferred reserves and the procedures used to develop the low, high, and mean estimates of undiscovered recoverable resources are all identical to those used for crude petroleum in Table 1.3-1 and discussed in Section 1.3-2.

TABLE 1.4-1   TOTAL GAS PRODUCTION, RESERVES, AND ESTIMATES OF UNDISCOVERED RECOVERABLE RESOURCES OF THE UNITED STATES[a]

	Petroleum Region	Cumulative Production[b]	Identified Resources[b,c] Measured Reserves	Identified Resources[b,c] Inferred Reserves	Undiscovered Recoverable Resources[d] Low F_{95}	Undiscovered Recoverable Resources[d] High F_5	Undiscovered Recoverable Resources[d] Mean	Undiscovered Recoverable Resources[d] Standard Deviation
	Onshore							
1	Alaska	1.2	30.0	4.4	19.8	62.3	36.6	14.0
2	Pacific Coast	27.0	4.2	3.8	8.2	24.9	14.7	5.5
3	Colorado Plateau and Basin and Range	15.0	11.4	4.5	53.5	142.4	90.1	29.7
4	Rocky Mountains and northern Great Plains	13.2	7.1	5.5	29.6	69.0	45.7	12.6
5	West Texas and eastern New Mexico	70.7	15.9	18.1	22.4	75.2	42.8	17.8
6	Gulf Coast	228.5	45.3	74.3	56.5	249.1	124.4	63.6
7	Mid-continent	127.8	32.0	18.8	22.9	80.8	44.5	18.4
8	Michigan Basin	1.3	1.1	1.1	1.8	10.9	5.1	3.1
9	Eastern Interior	1.8	Negl.	0.1	1.2	5.0	2.7	1.3
10	Appalachians	34.2	6.2	5.9	6.4	45.8	20.1	13.2
11	Atlantic Coast	Negl.	Negl.	Negl.	Negl.	0.4	0.1	0.3
	Entire onshore	520.6	153.3	136.5	322.5	567.9	426.8	78.5
	Offshore—Shelf							
1A	Alaska[e]	0.6	2.0	1.2	28.5	99.0	57.4	23.6
2A	Pacific Coast	1.5	1.9	0.4	0.9	5.2	2.5	1.5
6A	Gulf of Mexico	55.3	34.4[f]	39.5[f]	22.0	79.2	45.3	18.5
11A	Atlantic Coast	0	0	0	2.2	17.9	8.2	5.0
	Entire shelf	57.5	38.2	41.0	72.0	166.8	113.4	30.4
	Offshore—Slope							
1A	Alaska[e]	0	0	0	0	20.2	7.2	12.3
2A	Pacific Coast	0	0	0	1.9	10.2	4.4	3.0
6A	Gulf of Mexico	Negl.	No data[f]	No data[f]	11.1	51.8	26.5	13.8
11A	Atlantic Coast	0	0	0	5.1	34.5	15.4	9.6
	Entire slope	0	0	0	28.6	87.1	53.6	21.0

TABLE 1.4-1 (*Continued*)

Petroleum Region	Cumulative Production[b]	Identified Resources[b,c]		Undiscovered Recoverable Resources[d]			
		Measured Reserves	Inferred Reserves	Low F_{95}	High F_5	Mean	Standard Deviation
Offshore—Combined Shelf and Slope							
1A Alaska[e]	0.6	2.0	1.2	33.3	109.6	64.6	26.6
2A Pacific Coast	1.5	1.9	0.4	3.7	13.6	6.9	3.3
6A Gulf of Mexico	55.3	34.4	39.5	41.7	114.2	71.8	23.1
11A Atlantic Coast	0	0	0	9.2	42.8	23.7	10.8
Entire offshore	57.5	38.2	41.0	117.4	230.6	167.0	37.0
Combined Onshore and Offshore							
Entire United States	578.0	191.5	177.5	474.6	739.3	593.8	86.8

Source. G. L. Dalton et al., *Estimates of Undiscovered Recoverable Conventional Reserves of Oil and Gas in the U.S.*, U.S. Government Printing Office, Washington, D.C., 1981 (Table 7, p. 25).

[a]All tabulated values are rounded numbers; therefore values for production and reserves and for means of undiscovered resources, all of which are additive, may not be precisely so. Values shown are in trillion cubic feet. Negl., negligible, less than or equal to 0.05 trillion cubic feet of gas.
[b]Cumulative production and reserves are as of December 31, 1979. Production and reserve figures were derived from API and AGA data (American Petroleum Institute, American Gas Association, and Canadian Petroleum Association, 1980) except for California for which production and reserve data were taken from California Division of Oil and Gas (1980) and the U.S. Geological Survey (Kalil, 1980).
[c]Does not include gas in storage.
[d]F_{95} denotes the 95th fractile; the probability of *more than* the amount F_{95} is 95%. F_5 is defined similarly. Fractile values are not additive.
[e]Includes quantities considered recoverable only if technology permits their exploitation beneath Arctic pack ice—a condition not yet met.
[f]API and AGA reserve data for the Gulf of Mexico are not available within separate shelf and slope classifications. However, the declared reserves probably represent only the shelf and are so treated.

The U.S. Geological Survey estimate of identified resources (measured and inferred reserves) plus undiscovered recoverable resources is 962 trillion cubic feet. Adding the cumulative production to date of 578 trillion cubic feet to the total of recoverable reserves and resources gives a figure of 1540 trillion cubic feet for ultimate cumulative production.

Several other estimates of ultimate U.S. gas resources have appeared in the literature. For example, Hubbert (1969) estimated ultimate cumulative natural gas production in the coterminous United States (excluding Alaska and Hawaii) to be 1044 trillion cubic feet. This figure was based on his estimate of ultimate cumulative production of crude oil (discussed above) and an extrapolation of the historical relationship between natural gas and crude oil production. The most recent report of the Potential Gas Agency places ultimate gas supply in the coterminous United States at 1396–1446 trillion cubic feet. This range is consistent with the U.S. Geological Survey estimate of 1400 trillion cubic feet.[3]

API, American Gas Association, and Department of Energy estimates of approved U.S. reserves (equivalent to the measured reserves in Table 1.4-1) are presented for the years 1947–1982 in Table 1.4-2. Department of Energy estimates in Table 1.4-2 that cover only the years 1977–1981 are based on an analysis of data filed by the operators of oil and gas wells in the Energy Information Administration's (EIA's) annual survey of domestic oil and gas reserves. According to the EIA, "the crude oil and natural gas proved reserves estimates are associated with sampling errors less than 1.4 percent at a 95 percent confidence level."[4]

1.4-3 World Gas Reserves and Resources

Table 1.4-3 contains estimates of world cumulative production, recoverable reserves, estimated additional resources, and the projected ultimate world recovery of gas and natural gas liquids as of 1979. The estimates in Table 1.4-3 were prepared for the World Energy Conference in 1980 by the West German Federal Institute for Geosciences and National Resources based in part on responses to a questionnaire sent to all WEC member countries in 1979. The resources listed in Table 1.4-3 are conventional natural gas (and gas liquids) excluding gas from coal, biomass, or pressured resources.

TABLE 1.4-2   NATURAL GAS AND NATURAL GAS LIQUID RESERVES[a]

Year	United States	Year	United States	Canada	Middle East	Total Free World	Total World
1947	159,703,813	1967	289,333	43,450	215,070	891,590	1,041,590
1948	165,025,765	1968	292,908	45,682	220,670	967,599	1,183,099
1949	172,925,056	1969	287,350	47,666	223,775	983,936	1,326,936
1950	179,401,693	1970	275,109	51,951	235,275	1,140,928	1,490,928
1951	184,584,745	1971	290,746[b]	53,376	354,262	1,167,061	1,607,061
1952	192,758,910	1972	278,806[b]	55,462	343,930	1,176,966	1,734,966
1953	198,631,566	1973	266,085[b]	52,936	344,150	1,211,040	1,875,440
1954	210,298,763	1974	249,950[b]	52,457	413,325	1,302,770	2,038,170
1955	210,560,931	1975	237,132[b]	56,708	672,670	1,700,404	2,546,404
1956	222,482,544	1976	228,200[b]	56,975	538,648	1,413,897	2,248,897
1957	236,483,215	1977	216,026[b]	58,282	536,460	1,372,085	2,325,085
1958	245,230,137	1978	208,878[b]	59,472	719,660	1,564,909	2,519,909
1959	252,761,792	1979	200,302[b]	59,000	730,660	1,552,312	2,497,312
1960	261,170,431	1980	194,917[b]	85,500	740,330	1,639,158	2,574,158
1961	262,326,326	1981	199,021[b]	87,300	752,415	1,692,622	2,646,522
1962	266,273,642	1982	198,000[b]	89,900	762,490	1,716,646	2,911,346
1963	272,278,858						
1964	276,151,233						
1965	281,251,454						
1966	286,468,923						

Source. American Petroleum Institute, *Basic Petroleum Data Book: Petroleum Industry Statistics*, Vol. III, Number 1, American Petroleum Institute, January 1983.

[a] In millions of cubic feet, 14.72 psia at 60°F.
[b] Figures include 26 trillion cubic feet in Prudhoe Bay, Alaska (discovered in 1968) for which transportation facilities are not yet available.

TABLE 1.4-3   WORLD RESERVES AND RESOURCES OF NATURAL GAS

$(1.000 \times 10^9 \text{ m}^3)$	Cumulative Production Up to 1-1-79	Proved Rec. Reserves on 1-1-79	Estimated Additional Resources	"Ultimate Recovery"
Africa	0.1	7.3	26	33.4
North America	16.9	7.5	42	66.4
Latin America	1.8	4.7	10	16.5
Far East/Pacific	0.2	3.3	10	13.5
Middle East	1.1	20.5	30	51.6
Western Europe	1.5	3.9	6	11.4
USSR, China, Eastern Europe	5.2	26.9	64	96.1
Antarctic			4	4.0
Total	26.8	74.1	192	292.9

Source. Dr. E. Schubert, *Survey of Energy Resources, 1980, Part A Text Section A2*, 11th World Energy Conference at Munich 8–12 September 1980, Federal Institute for Geosciences and Natural Resources, Hanover, Federal Republic of Germany, 1980 (Table 2.11, p. 128).

TABLE 1.4-4   ESTIMATED PROVED GAS RESERVES BY COUNTRY, JANUARY 1, 1983[a]

Asia-Pacific		Middle East (Continued)	
Australia	17,768	Jordan	—
Bangladesh	7,000	Kuwait	29,900
Brunei	6,800	Lebanon	—
Burma	190	Oman	2,690
China, Taiwan	560	Qatar	62,000
Guam	—	Saudi Arabia	117,000
India	14,508	Sharjah	5,000
Indonesia	29,600	South Yemen	
Japan	720	(Aden)	—
Korea, South	—	Syria	1,260
Malaysia	34,000	Turkey	545
New Zealand	5,545	Total Middle East	769,730
Okinawa	—	*Africa*	
Pakistan	18,540		
Philippines	16	Algeria	111,250
Singapore	—	Angola-Cabinda	1,470
Sri Lanka	—	Cameroon	4,450
Thailand	11,000	Congo Republic	2,700
Total Asia-Pacific	146,247	Egypt	7,180
		Ethiopia	—
Western Europe		Gabon	485
Austria	337	Ghana	—
Belgium	—	Ivory Coast	3,040
Cyprus	—	Kenya	—
Denmark	2,300	Liberia	—
Finland	—	Libya	21,500
France	2,700	Madagascar	—
Germany, West	6,319	Morocco	—
Greece	3,500	Mozambique	—
Ireland	—	Nigeria	32,400
Italy-Sicily	4,320	Senegal	—
Netherlands	51,920	Sierra Leone	—
Norway	58,000	South Africa	400
Portugal	—	Sudan	—
Spain	1,940	Tanzania	200
Sweden	—	Togo	—
Switzerland	—	Tunisia	4,300
United Kingdom	25,400	Zaire	48
Total Western		Zambia	—
Europe	156,736	Total Africa	189,423
Middle East		*Western Hemisphere*	
Abu Dhabi	19,260	Antigua	—
Bahrain	7,890	Argentina	25,200
Divided (Neutral)		Bahamas	—
Zone	8,380	Barbados	1
Dubai	4,320	Bolivia	5,700
Iran	482,600	Brazil	2,330
Iraq	28,800	Chile	2,515
Israel	85	Colombia	4,580

TABLE 1.4-4   (*Continued*)

Western Hemisphere (Continued)		*Western Hemisphere (Continued)*	
Costa Rica	—	Venezuela	54,079
Dominican Republic	—	Virgin Islands	—
Ecuador	4,100	United States	204,000
El Salvador	—	Canada	97,000
Guatemala	35	Total Western	
Honduras	—	Hemisphere	487,591
Jamaica	—	Total Non-Communist	1,749,727
Martinique	—		
Mexico	75,850	*Communist Areas*	
Netherlands Antilles	—	China	29,800
Nicaragua	—	USSR	1,240,000
Panama	—	Other	14,000
Paraguay	—	Total Communist	1,283,800
Peru	1,201		
Puerto Rico	—	Total World	
Trinidad & Tobago	11,000	1982	3,023,527
Uruguay	—		

Source.   "Worldwide Oil and Gas at a Glance," *Oil and Gas Journal*, December 27, 1982, pp. 78–79.

a In millions of cubic feet.

They include only resources recoverable with present technology and assume a continuation of the present average recovery factor of about 80%.[5] The WEC estimate of ultimate world recovery of natural gas is 10,344 trillion cubic feet. Subtracting cumulative production to date of 946 trillion cubic feet gives 9398 trillion cubic feet as the WEC estimate of recoverable resources remaining in the ground.

Hubbert's 1969 estimate of ultimate world recovery of natural gas is 8000–12,000 trillion cubic feet. This range was based on two estimates of the ultimately recoverable resources of crude oil of 1350 million barrels and 2000 million barrels and the current domestic U.S. ratio of natural gas production to crude oil production.[6]

Table 1.4-2 presents the API estimates of proved world reserves of natural gas for the United States, Canada, the Mideast, the free world, and the entire world for the years 1967–1982. API's *Basic Petroleum Data Book* contains reserve figures for other regions as well.[7] Table 1.4-4 gives the reserve figures by country for 1983 as estimated by *Oil and Gas Journal*. The reserve figures in both tables are for proved reserves recoverable with present technology and prices. World proved reserves are currently about 3000 trillion cubic feet. Total world proved reserves have been growing steadily since 1967 and reserve growth has occurred in most regions of the world. The two major exceptions are the United States and Western Europe. U.S. reserves have declined steadily since the late 1960s. West European reserves peaked in 1975.

1.4-4   Natural Gas Production

The annual production of natural gas in the United States and other regions of the world is given in Table 1.4-5 for the years 1950–1982. Table 1.4-5 reveals that U.S. natural gas production peaked at 22.6 trillion cubic feet in 1973. The rate of growth from 1950 to 1973 was 5.7% per year. Since 1973 U.S. natural gas production has declined slightly to 20.2 trillion cubic feet in 1981.

In the United States the relative scarcity of natural gas parallels that of crude oil. Current proved reserves are adequate for 10 years at the current rate of production and the total remaining identified and undiscovered resources is sufficient for 50 years at the current rate of production. The shortage of natural gas reserves is, however, only one factor underlying the constant rate of production of natural gas that has occurred since 1973 in the United States. Another important factor is the regulation of natural gas prices, which has held the price of natural gas well below the energy-equivalent price of oil.

Worldwide the situation is somewhat different than in the United States. From 1950 to 1972 world production of natural gas grew at an annual rate of 8.9%. The rate of increase has declined to 3% per

TABLE 1.4-5 WORLD MARKETED PRODUCTION OF NATURAL GAS BY AREA[a]

Year	United States	Canada	Middle East	Total Free World	Total World
1950	6,282,060	71,692	0	6,702,357	6,702,357
1951	7,457,359	83,970	0	7,867,431	8,122,669
1952	8,013,457	93,710	0	8,477,779	8,783,169
1953	8,396,916	106,735	0	8,958,841	9,462,400
1954	8,742,546	127,597	0	9,381,090	9,919,430
1955	9,405,351	150,772	4	10,184,886	10,709,769
1956	10,081,923	169,153	15,552	10,964,587	11,621,504
1957	10,680,258	220,007	25,578	11,750,864	12,677,019
1958	11,030,298	337,804	26,288	12,402,058	13,781,726
1959	12,046,115	417,335	44,520	13,716,582	15,404,605
1960	12,771,038	522,972	57,265	14,724,677	16,916,609
1961	13,254,025	655,738	172,065	15,639,201	18,387,603
1962	13,876,622	946,703	173,312	16,726,384	20,092,409
1963	14,666,559	1,111,478	149,155	17,340,755	21,018,948
1964	15,462,143	1,327,664	162,593	18,471,359	22,868,453
1965	16,039,753	1,442,448	171,219	19,340,595	24,480,160
1966	17,206,628	1,341,833	193,147	20,640,719	26,394,231
1967	18,171,325	1,471,725	229,334	22,074,776	28,397,772
1968	19,322,400	1,692,301	388,208	24,330,287	31,403,142
1969	20,698,240	1,977,838	441,702	26,815,668	34,385,070
1970	21,920,642	2,277,109	454,125	29,365,043	38,093,961
1971[b]	22,493,012	2,499,024	697,997	31,715,796	41,188,396
1972[b]	22,531,698	2,913,537	905,622	33,386,384	43,513,884
1973[b]	22,647,549	3,119,461	1,276,853	35,169,412	46,127,788
1974[b]	21,600,522	3,045,506	1,423,416	35,054,188	47,171,491
1975	20,108,661	3,075,693	1,436,210	33,931,050	47,207,325
1976	19,952,438	3,067,353	1,492,772	34,657,932	49,459,213
1977	20,025,463	3,230,672	1,506,982	35,379,833	50,100,090
1978	19,974,033	3,128,056	1,579,478	35,930,904	51,749,302
1979	20,471,260	3,646,500	1,624,700	38,075,060	57,666,460
1980[b]	20,378,787	2,668,300	1,221,300	37,677,687	58,747,587
1981[b]	20,177,701	2,623,000	1,542,000	39,187,701	58,397,701

Source. American Petroleum Institute: *Basic Petroleum Data Book: Petroleum Industry Statistics*, Vol. III, Number 1, American Petroleum Institute, January 1983.

[a] In millions of cubic feet.
[b] Revised.

annum since 1973. The production of natural gas has continued to grow abroad despite the increase in price brought about by the rise in the price of oil. Natural gas prices have not increased as rapidly as oil prices because the OPEC countries are not the predominant world producers of natural gas. Many countries have sought to reduce oil imports by substituting less expensive and indigenous supplies of natural gas, and this process is continuing. In the world as a whole, natural gas is considerably more abundant relative to consumption than is the case in the United States. Proved reserves of 3023 trillion cubic feet are equivalent to a 52-year supply at the current world production rate of 58.4 trillion cubic feet per annum. And total world reserves and resources of 9398 trillion cubic feet are equivalent to a 161-year supply. Obviously, if world production continues to increase at recent rates, natural gas resources will not last that long. At the 3% rate of growth in world production that has occurred since 1973, natural gas resources would be exhausted in 59 years. For additional material on natural gas see Chapter 9.

REFERENCES

1.4-1 G. L. Dalton, et al. *Estimates of Undiscovered Recoverable Conventional Resources of Oil and Gas in the United States: Geological Survey Circular 860*, U.S. Government Printing Office, Washington, D.C., 1981, p. 6.

1.4-2 Dalton et al., p. 6.

1.4-3 Potential Gas Committee, *Potential Supply of Natural Gas in the United States* (*as of December 31, 1976*), Potential Gas Agency, Colorado School of Mines, Golden, CO, 1976.

1.4-4 Department of Energy, Energy Information Agency, *United States Crude Oil and Natural Gas Reserves 1977*, NTIS, Springfield, VA, August 1981, p. 83.

1.4-5 E. Schubert, "Hydrocarbons," *Survey of Energy Resources, 1980, Part A Text, Section A2*, 11th World Energy Conference at Munich, 8–12 September 1980, Federal Institute for Geosciences and Natural Resources, Hanover, Federal Republic of Germany, 1980, p. 124.

1.4-6 M. King Hubbert, "Energy Resources," in *Resources and Man*, National Academy of Sciences–National Research Council, 1969, p. 197, Table 8.3.

1.4-7 American Petroleum Institute, *Basic Petroleum Data Book: Petroleum Industry Statistics*, Vol. III, No. 1, American Petroleum Institute, January 1983.

1.5   NUCLEAR FUELS: RESOURCES AND PRODUCTION

1.5-1   Introduction

The principal fuels currently consumed in the nuclear reactors of the electric utility industry are uranium and thorium; chiefly the former. This section is mainly devoted to the major fuel, uranium. (Reasonably assured resources and estimated additional resources of thorium are, however, listed by country in Table 1.5-5).

Uranium is a silvery white metal that consists of three semistable radioactive isotopes: U 238, constituting 99% of the total, U 235, and U 234. Fission of the readily fissionable isotope U 235, accounting for only 0.7% of natural uranium, releases large amounts of energy. Uranium was discovered in 1789 in pitchblende from a mine in Germany and the element was first isolated in 1842. Prior to 1942 uranium was used chiefly for coloring glass and ceramic glazes.

The principal uranium ore is uranium oxide U_3O_8, or yellow cake, and is found in many different types of deposits throughout the world. These deposits vary widely in the percentage of uranium oxide contained. Although all deposits contain concentrations of uranium many times crustal abundance, which is about 2 ppm, the ore grades are generally less than half of 1%. A few small deposits contain ore grades as high as several percent. The great bulk of U.S. uranium reserves are contained in sandstone beds having been precipitated from groundwater during the late Paleozoic to the Tertiary geologic ages.[1] Elsewhere in the world many uranium deposits are in pre-Cambrian quartz pebble conglomerates. Uranium deposits also occur as veins, as fissure fillings, and as false joints in fracture zones.[2] Uranium exists in low concentrations in uraniferous igneous rocks such as pegmatites, in phosphatic rocks, and in marine black shales.[3]

1.5-2   U.S. Reserves and Resources of Uranium

Tables 1.5-1 and 1.5-2 present estimates of U.S. reserves and resources of uranium. Both reserve and resource estimates are shown separately for five different "forward" costs of mining: $8–10, $15, $30, $50, and $100 per pound of U_3O_8. These estimates were prepared by the Department of Energy and were published in the DOE's *Statistical Data of the Uranium Industry* (1982). The "forward" cost figures are the marginal costs of producing uranium oxide exclusive of capital costs. By way of perspective, uranium oxide prices are currently (1983) around $23 per ton per pound.

As is the normal practice in reserve and resource appraisal, the reserve estimates shown in Table 1.5-1 are based on drilling information. On the other hand, the resource estimates are derived from geologic assessments of the components of the uranium endowment. Typically, the endowment in unexplored or underexplored regions is estimated by analogy with regions that have been well explored.[4] The "probable" category of potential resource contains resources expected to occur in known productive uranium areas in (1) extensions of known deposits or (2) undiscovered deposits within known geologic trends or areas of mineralization.[5] The total potential resource in 1982 of 3.2

TABLE 1.5-1   HISTORICAL AND CURRENT ESTIMATES OF URANIUM RESERVES[a]

Year	Thousand Tons U_3O_8				
	\$8/lb U_3O_8	\$15/lb U_3O_8	\$30/lb U_3O_8	\$50/lb U_3O_8	\$100/lb U_3O_8
1/1/65	151	—	—	—	—
1/1/66	145	—	—	—	—
1/1/67	141	—	—	—	—
1/1/68	148	248	—	—	—
1/1/69	161	265	—	—	—
1/1/70	204	317	—	—	—
1/1/71	246	391	—	—	—
1/1/72	273	520	—	—	—
1/1/73	273	520	—	—	—
1/1/74	277	520	634	—	—
1/1/75	200	420	600	—	—
1/1/76	—	430	640	—	—
1/1/77	—	410	680	840	—
1/1/78	—	370	690	890	—
1/1/79	—	290	690	920	—
1/1/80	—	225	645	936	1122
1/1/81	—	112	470	787	1034
1/1/82	—	—	205	594	894

Source. U.S. Dept. of Energy, "Historical and Current Estimates of Uranium Reserves," *Statistical Data of the Uranium Industry*, U.S. Dept. of Energy, Washington, D.C., 1982 (Table 1-1, p. 10).

The resource estimates listed in this table for each forward cost of mining are inclusive of the estimates given for all lower forward costs.

[a] The January 1, 1982, \$30 reserves estimate is 205,000 tons U_3O_8, a 56% reduction from the 470,000 tons estimated in 1981. The \$50 reserves estimate was reduced from 787,000 tons in 1981 to the current 594,000 tons U_3O_8, a 25% decrease, and the \$100 reserves estimate declined 14% from the 1,034,000 tons U_3O_8 reported in 1981 to the current 894,000 tons.

Except for depletion of reserves by mining, these reductions are due to rising production costs and do not indicate a decrease in the amount of uranium present in the ground. Depletion consists of two reductions to reserves: (1) a prorated quantity of actual mine production based on production grade compared to the reserve grades at various cost categories and (2) erosion, that is, the amount of uranium-bearing material not recoverable in the future at the given cost level as a result of the mining of lower cost reserves.

million tons at a forward cost of \$100/lb is the mean of a probability distribution ranging from 1.8 million tons at the 95th percentile to 5.2 million tons at the 5th percentile.

The reserve and resource estimates in Tables 1.5-1 and 1.5-2 are given by year. The estimates for each forward cost show a tendency to initially increase over time and then decline. The decrease in recent years is "due to rising production costs and do not indicate any decrease in the amount of uranium present in the ground."[6] Depletion by mining has been a relatively small factor.

Because of the nature of uranium deposits, the richest are not necessarily discovered first. The richest deposits containing, in many instances, very large amounts of uranium often cover relatively small areas. Deposits covering large areas such as Conway, New Hampshire, granite and Chattanooga, Tennessee, shales have ore grades that are too low to make them exploitable even at the \$100/lb figure.

As the U.S. uranium drilling statistics presented in Table 1.5-4 illustrate, exploration languished following the collapse of the uranium market in the mid-1960s. With the resumption of active exploration in recent years, discoveries and reserves have increased steadily.

1.5-3   World Uranium Reserves and Resources

In Table 1.5-3 estimates of world uranium resources are given in three categories: reasonably assured resources (RAR), estimated additional resources (EAR), and total uranium potential. Resources in the first two categories are listed by country and are divided between those available at a uranium oxide

TABLE 1.5-2 POTENTIAL URANIUM RESOURCES, 1975–1982

| Forward-Cost Category | Thousand Tons U_3O_8 | | | | | | |
	1/1/75	1/1/76	1/1/77	1/1/78	1/1/79 and 1/1/80[a]	1/1/81	1/1/82
$10/lb U_3O_8							
Probable	460	440	275	—	—	—	—
Possible	390	420	115	—	—	—	—
Speculative	110	145	100	—	—	—	—
$15/lb U_3O_8							
Probable	680	655	585	540	415	295	—
Possible	640	675	490	490	210	87	—
Speculative	210	290	190	165	75	74	—
$30/lb U_3O_8							
Probable	1140	1060	1090	1015	1005	885	596
Possible	1340	1270	1120	1135	675	346	227
Speculative	410	590	480	415	300	311	236
$50/lb U_3O_8							
Probable	—	—	1370	1395	1505	1426	1080
Possible	—	—	1420	1515	1170	641	473
Speculative	—	—	540	565	550	482	421
$100/lb U_3O_8							
Probable	—	—	—	—	—	2080	1740
Possible	—	—	—	—	—	1005	784
Speculative	—	—	—	—	—	696	685

Source. U.S. Department of Energy, "Historical and Current Estimates of Uranium Reserves," *Statistical Data of the Uranium Industry*, U.S. Department of Energy, Washington, D.C., 1982 (Table 11.2, p. 26).

The resource estimates listed in this table for each forward cost of mining are inclusive of the estimates given for all lower forward costs.

[a] No new estimates were released for January 1, 1980, since the NURE program was to publish comprehensive potential resource estimates by October 1980.

cost of less than $80/kg and those available at a cost of between $80 and $130 per kilogram. The source for each estimate is also indicated in the table. The bulk of the estimates were taken from a recent report jointly authored by the Nuclear Energy Agency of OECD and the International Atomic Energy Agency.[7] The remainder were obtained from the WEC 1980 report and are based on the results of a WEC survey in 1979 of the national committees of member states, nonmember states, and international organizations. The reasonably assured resources category corresponds to the U.S. Department of Energy's proved reserves while the EAR are expected to occur on the basis of geological evidence and "are considered as having a potential for later conversion to RAR through further exploration."[8] Total world reasonably assured resources recoverable at costs of less than $130/kg are 2.3 million tons of uranium, while estimated additional resources amount to 2.7 million tons.

Table 1.5-3 also contains several estimates of the total world uranium potential listed by sources. The first estimate in the table is a range of 9.9–22.1 million metric tons[9] prepared by the International Atomic Energy Agency. The estimate represents the views of a panel of experts who gave general consideration to the geological favorability of each of the regions. The other three estimates in the table were prepared by DeVerle Harris (1979).[10] The first estimate of 15.2 million tons of uranium oxide was calculated by multiplying the ratio of U.S. potential supply to reserves at a forward cost of $50/lb by a 1975 OECD/IAEA estimate of reasonably assured resources at $130/kg ($50/lb). This approach assumes that the ratio of potential resources to reserves is the same for all regions as it is in the United States. Harris's second estimate of 27.7 million tons of uranium oxide at $50/lb was calculated by assuming that the United States is representative of the rest of the world in terms of the density of uranium deposits per unit area. Harris obtained his third estimate of 20 million tons by reducing the second estimate based on relative areas to account for differences between the United States and other parts of the world in exploration intensity.

TABLE 1.5-3 WORLD RESOURCES OF URANIUM (WOCA COUNTRIES)

Cost Range	Reasonably Assured Resources[a]		Estimated Additional Resources[a]	
	Up to $80/kg U	$80–130/kg U	Up to $80/kg U	$80–130/kg U
Algeria	26	0	0	0
Argentina	25	5.3	3.8	9.6
Australia	294	23	264	21
Austria	—	0.3	0.7	1
Brazil	119.1	0	81.2	0
Canada	230	28	358	402
Central African Republic	18	0	0	0
Chile	0	0.02	0	6.7
Denmark	0	27	0	16
Egypt	0	0	0	5
Finland	0	3.4	0	0
France	59.3	15.6	28.4	18.1
Gabon	19.4	2.2	0	9.9
Germany (FR)	1	41	1.5	7
Greece	1.4	4	2	5.3
India	32	0	0.9	24.2
Italy	0	2.4	0	2
Japan	7.7	0	0	0
Korea, Republic of	0.04	11	NA	NA
Mexico	2.9	0	3.5	2.6
Namibia	119	16	30	23
Niger	160	0	53	0
Philippines	0.3	0	0	3.4
Portugal	6.7	1.5	2.5	0
Somalia	0	6.6	0	3.4
South Africa	247	109	84	91
Spain	12.5	3.9	8.5	0
Sweden	0	38	0	44
Turkey	2.5	2.1	0	0
United Kingdom	0	0	0	7.4
United States	362	243	681	416
Yugoslavia	4.5	2.0	0	7.4
Zaire	1.8	0	1.7	0
Total (rounded)	1751	548	1605	1125

Total World Uranium Potential[b, c]	
NEA/IAEA	9.9–22.1
DeVerle Harris (1978)	
1. Uniform Potential Reserve Ratio	15.2
2. Uniform Endowment/Per Unit Area	27.7
3. Item 2 Adjusted by Exploration Intensity	20.0

Source. For the Philippines and Yugoslavia: J. Meyer, "Nuclear Resources" *Survey of Energy Resources, 1980*, Federal Institute for Geosciences and Natural Resources, Federal Republic of Germany, 1980 (Table 3.2, p. 192 and Table 3.3, p. 194). All other data: OECD Nuclear Energy Agency and the International Atomic Energy Agency, *Uranium: Resources, Production and Demand*, OECD, February 1982, (Table 1, p. 18 and Table 2, p. 19).

[a] In thousands of tons of uranium.
[b] Deverle P. Harris, "World Uranium Resources," *Annual Review of Energy 1979*, Vol. 4, Palo Alto, CA, Annual Reviews, Inc., 1979.
[c] In millions of tons of uranium.

1.5-4   Production and Exploration Activity

Annual and cumulative world production of uranium is graphed in Fig. 1.5-1. Annual production figures are shown from 1957 to 1978 and projected to 1980. Cumulative production is graphed from 1971 to 1978 and projected to 1980. U.S. annual production and exploratory drilling footage annually is presented in Table 1.5-4.

The world's dominant uranium producers are the United States, Canada, and South Africa. These three countries account for three-quarters of annual world production and along with Australia also have the largest reserves.

The pattern of change in production and in exploration activity both in the United States and in the rest of the world since the 1950s is similar. Production in the world and the United States peaked around 1960 in response to the demand for uranium for defense purposes, declined steadily until the late 1960s, and began to increase slowly as commercial nuclear power developed. In the last few years production has been increasing very rapidly.

Current annual world production of 47,000 lb of uranium is obviously a tiny fraction of both reserves and resources. It is less than 2.0% of world reserves and 0.3% of the minimum estimate of the total world uranium potential. U.S. production of approximately 20,000 tons is 3.3% of $50 U.S. reserves and 0.6% of total potential resources.

Whether these reserve–production ratios are adequate to assure the continued availability of uranium supplies over the next few decades depends on two factors that are quite uncertain. The first is the rate of growth in installed world capacity of nuclear power plants and the second is the rate at which breeder reactors are substituted for conventional light-water reactors.

Due to the controversy concerning the safety of nuclear power plants, particularly in the United States and to a lesser extent in other countries, the rate of increase in nuclear power plant capacity slowed dramatically in the late seventies and early eighties. A sharp drop in the rate of growth in demand for electricity, and the perception that the cost advantage of nuclear power over the cheapest alternative—typically coal—has narrowed substantially, has also contributed to a cessation of orders for new nuclear power plants and the abandonment of construction plans by a number of utilities in the United States. How long these factors will continue to impede the growth of the nuclear power industry is very difficult to predict.

By converting more of the uranium to fissionable materials than they consume, breeder reactors have the theoretical capability of converting virtually 100% of the uranium fuel supply to fissionable materials, from which energy can be obtained, compared with the current 1% conversion obtained by conventional light-water reactors. Substitution of breeder for conventional reactors would obviously

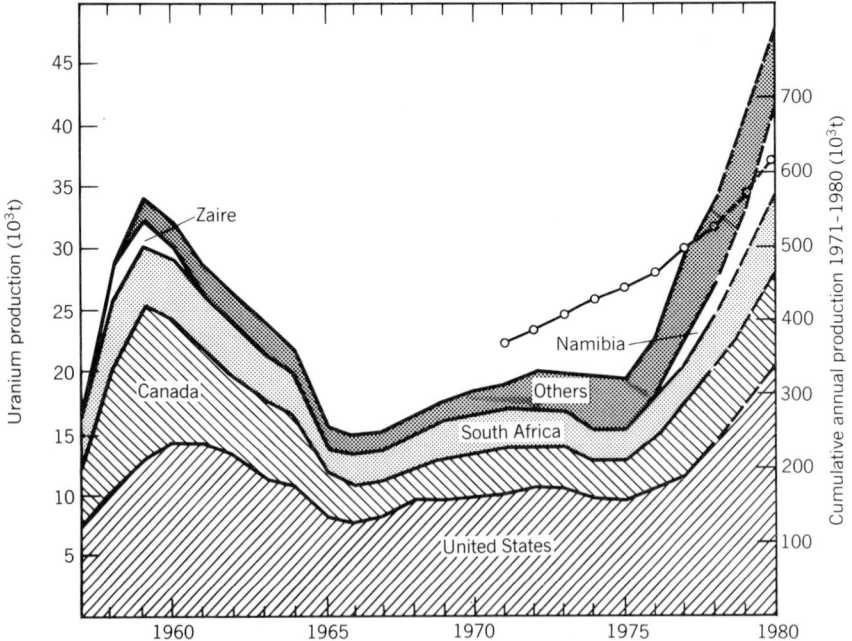

Fig. 1.5-1   Uranium production of non-Communist countries: annual 1957–1978 (estimated to 1980), cumulative 1971–1978 (estimated to 1980). (Source: J. Meyer, "Nuclear Resources," *Survey of Energy Resources, 1980*, Federal Institute for Geosciences and Natural Resources, Federal Republic of Germany, 1980, p. 218, Fig. 5.3.)

TABLE 1.5-4 URANIUM PRODUCTION AND SURFACE DRILLING

Year	Domestic Production[a]	Exploration[b]	Development[b]	Total Surface Drilling[b]
1951	0.77	1.080	0.348	1.428
1952	0.87	1.362	0.300	1.662
1953	1.16	3.648	0.367	4.015
1954	1.70	4.057	0.553	4.610
1955	2.78	5.267	0.762	6.029
1956	5.96	7.287	1.503	8.790
1957	8.48	7.352	1.848	9.200
1958	12.44	3.759	3.494	7.253
1959	16.24	2.368	3.282	5.650
1960	17.64	1.399	4.211	5.610
1961	17.35	1.319	3.190	4.509
1962	17.01	1.483	2.431	3.914
1963	14.22	0.880	1.977	2.857
1964	11.85	0.967	1.245	2.212
1965	10.44	1.164	0.949	2.113
1966	10.59	1.800	2.400	4.200
1967	11.25	5.435	5.329	10.764
1968	12.37	16.277	7.527	23.754
1969	11.61	20.470	9.385	29.855
1970	12.90	17.981	5.547	23.528
1971	12.27	11.400	4.052	15.452
1972	12.90	11.815	3.609	15.424
1973	13.24	10.831	5.590	16.421
1974	11.53	16.000	6.000	22.000
1975	11.60	16.538	9.004	25.542
1976	12.75	19.527	14.704	34.231
1977	14.94	25.924	14.626	40.550
1978	18.49	32.203	14.800	47.003
1979	18.73	26.842	13.926	40.768
1980	21.85	19.950	7.905	27.855
1981	19.00[c]	9.752	4.366	14.118

Source. U.S. Dept. of Energy, *1981 Annual Report to Congress*, Vol. 2, Energy Statistics U.S. Dept. of Energy, Washington, D.C., 1982 (Table 77, p. 175).

[a] Thousand short tons of U_3O_8.
[b] Surface drilling in million feet.
[c] Import quantities through 1970 are reported for fiscal years. Until 1971 the Atomic Energy Commission was the sole purchaser of all imported U_3O_8.

greatly increase the potential energy obtainable from the world's uranium resources. The chief objection to the development of breeder reactors on a commercial scale is that they produce plutonium, which in the wrong hands is regarded as a potentially much more dangerous material than the uranium fuel itself.

Based on their assessment of the rate at which world nuclear power plant capacity will increase and breeder reactors be deployed, the authors of the 1980 International Nuclear Fuel Cycle Evaluation (INFCE) study concluded that total annual demand for uranium by non-Communist countries would lie between 80,000 and 161,000 long tons of uranium by the year 2000.[11] This estimate implies a rapid rate of growth for the uranium mining industry (more rapid than some analysts believe can be achieved) and the need for the industry to utilize higher cost ores. Nevertheless, uranium resources appear adequate to support the projected growth.

1.5-5   Conclusion

The potential energy contained in the world's resources of uranium is enormous. The thermal energy per gram that can be obtained from natural uranium or thorium by means of conversion or breeding is about 8.2×10^{10} J/g.[12] Therefore, the world's potential uranium resource of something on the order of 15 million metric tons of uranium contains $15 \times 10^{12} \times 8.2 \times 10^{10} = 123 \times 10^{22}$ J energy, given the use of breeder reactors to obtain virtually 100% of the available energy. Given the use of breeders, the energy potential of the world's uranium resources is about 4 times that of total coal reserves and resources and about 20 times that of the total reserves and resources of hydrocarbons. Moreover, the 15-million-ton estimate of the world's total uranium endowment does not include vast quantities of resources presently regarded as subeconomic because of low ore grade or thick overburden. In

TABLE 1.5-5   SUMMARY TABLES OF THORIUM RESOURCES (WOCA COUNTRIES)[a]

	Lower-Cost Resources (Recovery costs up to $75/kg)		Higher-Cost Resources (Recovery costs $75/kg)		
World Region	RAR	EAR	RAR	EAR	Sources
1. North America					
Canada	0	293	0	0	1
USA[b]	109	260	n.a.(14)	n.a.(17)	1 (2)
2. Western Europe					
Denmark (Greenland)	0	0	n.a.(54)	n.a.(32)	1 (2)
Finland	0	0(60)	0	0	1 (2)
Norway	132	0	132	0	1 (2)
3. Australia[c]	—	—	17.6	0.2	
7. Latin America					
Argentina[c]	0	0	1.1	0	1
Brazil	−(68)	−(1200)	70.3	1200	1 (2)
Uruguay	0	0	0.8	1.5	3
8. Middle East & N. Africa					
Egypt	15.0	280.0	0	0	2
Turkey	—	—	380.0(330)	0(440)	1 (2)
9. Africa South of Sahara					
South Africa	n.a.(11)	n.a.(0)	n.a.(0)	n.a.(0)	1 (2)
Malagasy	0	0	2.0	n.a.	3
Malawi	0	0	0	8.8	3
Other	0	0	17.2	30.0	3
10. East Asia					
Malaysia	18.3	n.a.	n.a.	n.a.	3
11. South Asia					
India	319	0	0	0	1
Iran	0	30	0	0	2
Total (rounded)	725	860	490	1240	

Source. J. Meyer, "Nuclear Resources," *Survey of Energy Resources, 1980,* Federal Institute for Geosciences and Natural Resources, Federal Republic of Germany, 1980 (Table 3.4, p. 199).

[a] In thousand tonnes. n.a.: Information is not available; information insufficient to allocate resources to lower cost category.
[b] Thorium resources are recoverable up to $80/kg Th.
[c] No cost category made available; resources are allocated to higher-cost category.

addition, the world has substantial resources of thorium capable of yielding as much energy per pound as uranium. The world reserves and resources of thorium are listed in Table 1.5-5.

Resource scarcity will not play a significant role in limiting the world's ability to derive energy from fission given the eventual use of breeder reactors. Like coal, the problem with nuclear fuel is environmental concerns, not inadequate resources. At every stage of the nuclear fuel cycle the fuel itself, the reactors, and the waste products must be carefully isolated from the environment. The technology for accomplishing this isolation economically exists. The factor that now appears to block the growth of nuclear power is the fear that man, with his fallible nature, will not succeed in implementing this technology with the high degree of reliability required.

REFERENCES

1.5-1 Warren I. Finch, et al., "Nuclear Fuels," *United States Mineral Resources: Geological Survey Professional Paper 820*, U.S. Government Printing Office, Washington, D.C., 1974, p. 460.
1.5-2 Finch, pp. 460–461.
1.5-3 Finch, pp. 460–462.
1.5-4 United States Department of Energy, *Statistical Data of the Uranium Industry*, U.S. Department of Energy, Washington, D.C., p. 82.
1.5-5 United States Department of Energy, Statistical Data.
1.5-6 United States Department of Energy, Statistical Data.
1.5-7 OCED Nuclear Energy Agency and International Atomic Energy Agency, *Uranium Resources, Production and Demand*, Paris, February 1982.
1.5-8 J. Meyer, "Nuclear Resources," in *Survey of Energy Resources, 1980, Part A Text: Section A3*, 11th World Energy Conference at Munich 8–12 September 1980, Federal Institute for Geosciences and National Resources, Hanover, Federal Republic of Germany, 1980, p. 195.
1.5-9 Meyer, p. 214.
1.5-10 DeVerle P. Harris, "World Uranium Resources," in *Annual Review of Energy, 1979*, Vol. 4, Annual Review Inc., Palo Alto, CA, 1979, pp. 420–426.
1.5-11 Meyer, pp. 235–238.
1.5-12 M. King Hubbert, "Energy Resources" in *Resources and Man*, National Academy of Sciences–National Research Council, 1969, p. 221.

1.6 RENEWABLE ENERGY RESOURCES

1.6-1 Introduction

Renewable energy resources are sources of energy whose output will remain constant over a span of millions of years. They include direct sunlight, wind, wave, tidal current, biomass, geothermal, and hydraulic energy. While the internal heat of the earth, gravitational force of the moon and sun, and the rotation of the earth play a role in creating many of these energy resources, renewable energy sources are derived principally from the radiation of the sun. (See Chapter 11 for additional information.)

The aggregate amount of solar radiant energy reaching the earth is enormous: 1.75×10^{17} W.[1] Over a year this amounts to approximately 20,000 times the total annual world energy consumption.[1] About two-thirds of this energy reaches the earth's surface, the remainder being absorbed in the outer atmosphere.

The preceding sections have discussed energy sources that are both inexpensive and relatively environmentally benign but limited in supply (the hydrocarbons) and those sources that are inexpensive and abundant but perceived to have serious environmental drawbacks (coal and nuclear energy). The renewable energy resources provide energy that is abundant and free from serious environmental problems but is, at the present time, very expensive to harness. The potential supply of energy from those renewable resources that can be harnessed at costs close to conventional—wind and tidal energy at favorable sites and geothermal energy—is very limited, only a tiny fraction of the total and small relative to our needs.

The difficulty with the greater part of the renewable energy available is that it is very diffuse. The maximum average power density of the sunlight reaching the earth is about 250 W/m^2.[1] Very large and, therefore, costly mechanical devices are required to capture this energy relative to the energy obtained.

In this section the energy potential of the world's renewable energy resources is discussed and analyzed. The capacity of each renewable source to contribute to the solution of the world's energy problem is assessed.

Figure 1.6-1 shows how sunlight, planetary motion, and geothermal energy contribute to the world's supply of renewable energy. The figure also shows the proportion of solar radiation that is reflected back into space; that generates wind, waves, ocean currents, evaporation, and precipitation on earth; and that is absorbed by photosynthesis.

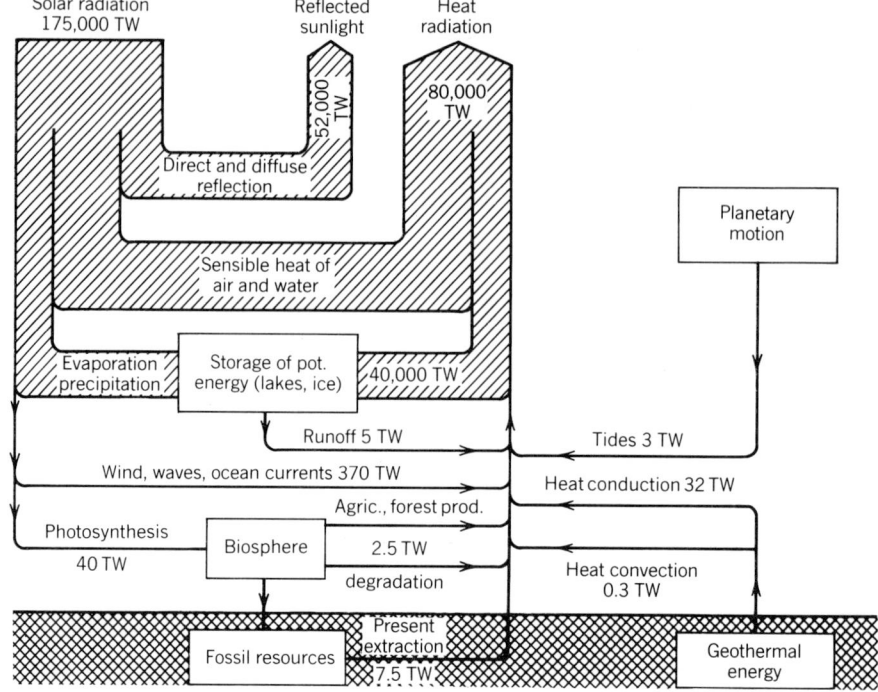

Fig. 1.6-1   Energy flux diagram of solar radiation at the earth. (From ref. 6, p. 299.)

1.6-2   Direct Energy from Sunlight

Only a portion of the vast amount of solar energy striking the outer atmosphere reaches the surface of the earth and the atmosphere acts as both an absorbing and reflecting medium. The maximum peak solar energy density of the most favorable locations on earth is about 1 kW/m^2 at sea level while, as noted above, the average maximum energy density is about one-quarter of the peak density.[1] Figure 1.6-2 and Table 1.6-1 show how insolation varies around the world. Each line in the figure is a line of equal insolation measured in kilowatt hours per square meter annually. Insolation in much of the United States ranges from 1400 to 2000 kW · h/m^2 annually. In Europe the range is from about 900–1800 kW · h/m^2. Insolation is highest in the world's deserts including southwestern United States, the Sahara, and Western Australia. The range is from 2000 to 2500 kW · h/m^2 annually.

Solar energy not only varies from region to region but from hour to hour, seasonally, and with weather conditions. Consequently, employment of solar energy on a massive scale will entail extensive use of energy storage, long-distance transmission, or conversion from electricity to an alternate form such as hydrogen. Nevertheless, sunlight does have attractive qualities as an energy source. It is globally abundant, renewable, nondepletable, and environmentally benign.

Solar energy has the potential capability of supplying the energy requirements of a large world population with a higher average per capita consumption of energy than exists today. If solar energy can be collected with an average efficiency of 10% (the efficiency of today's photovoltaic energy conversion systems), the world's present energy consumption of about 85 trillion[†] kilowatt hours annually could be supplied with the energy collected from sunlight on 43,500 km^2 of desert area. Present U.S. energy consumption of about 23 trillion kilowatt hours would require about 12,000 km^2, or 4700 mi^2, an area measuring 70 × 67 mi^2.

Energy from sunlight can be collected in the form of heat or electricity in a wide variety of ways. One of the most mature and currently cost-effective technologies is the use of active solar systems with flat plate collectors for space and water heating. Active systems require the use of pumps or fans to circulate water or air through the collector and heat storage unit. Passive solar heating systems in which structural components of the building—windows, floors, and walls—are used as solar collectors and heat storage reservoirs are quite economical and are gaining widespread use. Unfortunately,

---

[†] Trillion is here defined as 10^{12}.

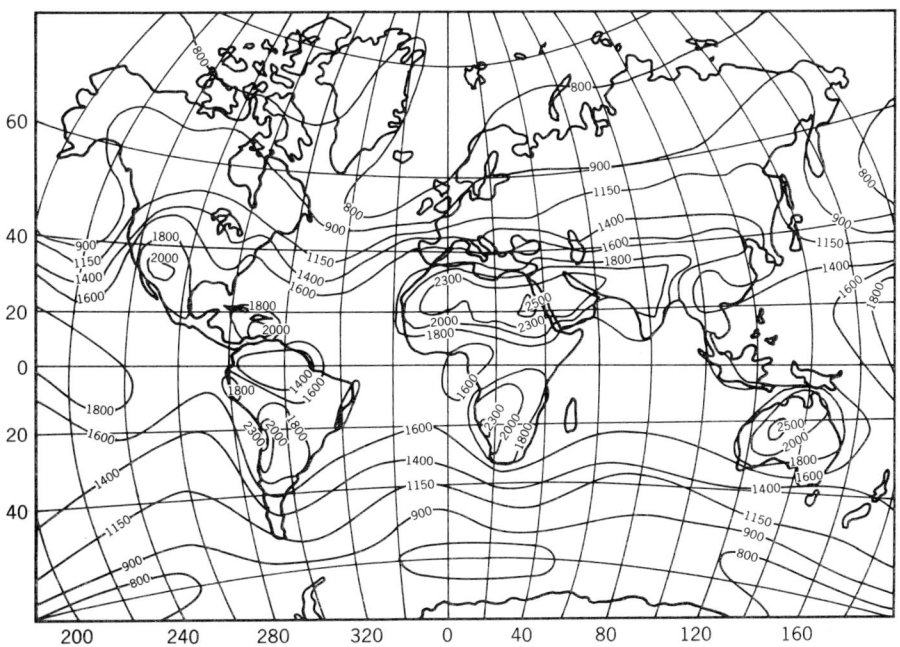

Fig. 1.6-2 World map of mean annual solar radiation (kW · h/m² year). (From ref. 6, p. 302.)

TABLE 1.6-1 SOLAR INSOLATION OF DIFFERENT REGIONS

Regions	Insolation kW · h/m² day	W/m² Average
Annual average (horizontal surface)		
Tropic regions, deserts	5–6	210–250
Temperate zones	3–5	130–210
Less sunny regions (e.g., northern Europe)	2–3	80–130
Annual average (inclined surface)		
Sunny regions	7–8	290–330

Source. W. H. Bloss and O. Kappelmeyer, "Renewable Energy Resources," *Survey of Energy Resources, 1980* (Table 4.3, p. 300).

passive solar systems are not well suited for retrofitting on existing dwellings. Moreover, only a fraction of existing and new houses are or will be located where they can take advantage of either form of solar space heating. A survey of the location and orientation of houses in Denver, Colorado (an area of high insolation), showed that only 50% were suitable for active-energy space heating systems.[2]

The two other principal technologies for capturing the energy of the sun and the only two suitable for the use of solar energy on a massive scale are direct solar thermal electrical conversion and photovoltaic energy conversion. In a solar thermal electric conversion plant solar energy is converted into heat, which provides steam or hot gas to drive an electric turbine. The plant consists of a central receiving tower surrounded by a large area covered by mirrors that reflect the sun's rays directly on top of the tower. Overall efficiencies of these plants are expected to be in the range of 15–20%.[3]

Sunlight may be converted directly into energy by means of photovoltaic cells constructed of thin layers of silicon or other semiconductor materials. Photovoltaics have numerous advantages: no moving parts, long lifetimes, minimum maintenance, and an ability to use both direct and diffuse solar radiation. Moreover, they are inherently modular and lend themselves to the design of virtually any system size, large or small. Current efficiencies of individual cells are on the order of 12–15%.

At present the costs of both solar thermal electric and photovoltaic conversion substantially exceed the cost of conventional electricity generation; in the case of photovoltaics by approximately an order of magnitude.

1.6-3 Ocean Thermal Energy Resources

Scientists have been aware for more than a century that the temperature difference between the surface and the deep waters of the ocean could be used as a source of energy. In the tropics about 90% of the earth's surface is water, with an average surface temperature of about 82°F. Water reaches its maximum density at about 39°F; consequently, water at the bottom of the ocean below depths of 2000 ft in the tropics has a temperature of about 39°F. To utilize the temperature differential between the surface and deep waters to produce energy, the warm sea water of the surface is pumped through an evaporator to vaporize a working fluid. The vapor drives a low-pressure turbine to create electricity. The vapor is condensed back to a fluid by cooling it with cold water pumped up from the depths.

The potential supply of energy from ocean thermal energy conversion is theoretically very large. If ocean thermal plants were placed approximately 15 km apart throughout the oceans in the tropics, the theoretical maximum limit for electricity production would be 150 trillion watts (thermal).[4] However, if consideration is limited to those sites where deep water is located close enough to shore to make transmission of the power feasible with present technology, the potential contribution of ocean thermal energy is greatly reduced. Suitable locations may be found on the coast of Africa, the Pacific coast of the Americas, in the Carribbean, and in the Gulf Stream off the coasts of Florida and South Carolina. In the latter location the available power has been estimated at 0.32 trillion watts or 2.8 trillion kilowatt hours annually, assuming continuous operation.[5]

Ocean thermal energy conversion plants face numerous problems. The energy contained in ocean thermal gradients is very low in density. Ocean energy conversion plants would have to be of enormous size to generate relatively small amounts of power. Their output would be about 1 W/m².[6] In addition, the fouling of equipment by marine organisms that occurs near the surface would pose a very severe maintenance problem. At the present time ocean thermal conversion energy is not comparable in technical and economic feasibility to solar thermal conversion or photovoltaics.

1.6-4 Wave Energy

Proposals have been made for the construction of mechanical devices that would convert the mechanical energy in ocean waves into electricity. The waves are created by wind and the friction of air on the surface of the water. Thus, wave energy is derived from wind energy, which in turn is

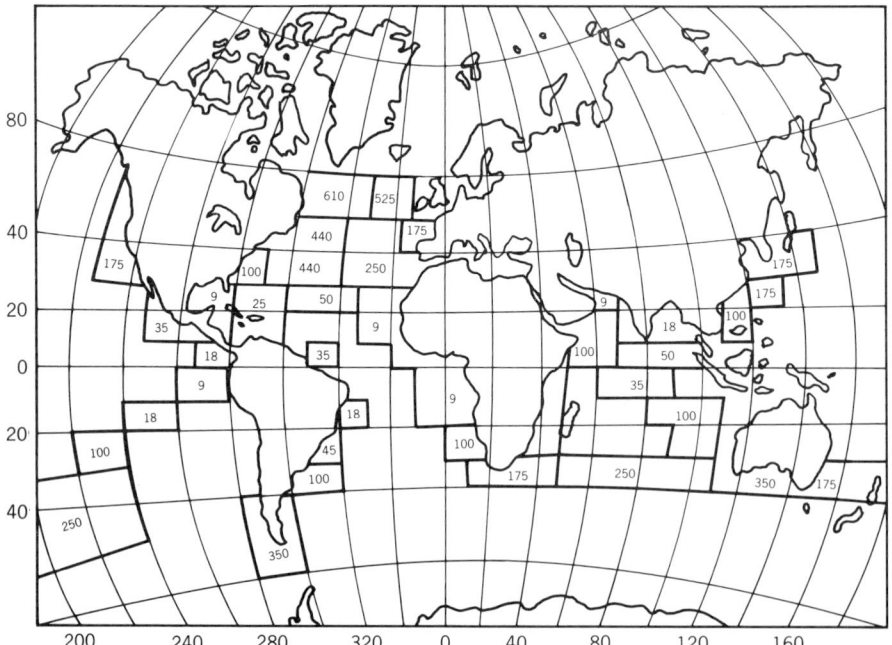

Fig. 1.6-3 Global annual availability of waves with height more than 5 m (n/y). (From ref. 6, p. 323.)

created by the rotation of the earth and heat from the sun. Global wave power is estimated to amount to between 2 and 2.7 trillion watts.[7] Figure 1.6-3 shows the global annual frequency of waves with heights exceeding 5 m. The figure reveals that the magnitude of waves varies with location and that the North Atlantic is clearly one of the best areas for locating wave energy plants. Measurements made in the northeast Atlantic off Scotland indicate that the annual power in the waves there is 91 kW/m.[7] If this power could be captured within an efficiency of 20%, 2×10^4 kW of electricity could be produced by converters covering 1 km of coastline. Annual energy output would be 17.5×10^7 kW · h of electricity. A 1000-MW plant would require 500 km of coastline. Obviously, the potential contribution from wave energy is quite small compared with either ocean thermal or direct solar.

TABLE 1.6-2 TIDAL POWER SITES AND MAXIMUM POTENTIAL POWER

Location	Average Range R (m)	R^2 (m^2)	Basin Area S (km^2)	R^2S (m^2)(km^2)	Average Potential Power P (10^3 kW)	Potential Annual Energy E (10^6 kW · h)
North America						
Bay of Fundy						
Passamaquoddy	5.52	30.5	262	7,990	1,800	15,800
Cobscook	5.5	30.3	106	3,210	722	6,330
Annapolis	6.4	41.0	83	3,440	765	6,710
Minas-Cobequid	10.7	114	777	88,600	19,900	175,000
Amherst Point	10.7	114	10	1,140	256	2,250
Shepody	9.8	96	117	11,200	2,520	22,100
Cumberland	10.1	102	73	7,450	1,680	14,700
Petitcodiac	10.7	114	31	3,530	794	6,960
Memramcook	10.7	114	23	2,620	590	5,170
Subtotal					29,027	255,020
South America						
Argentina						
San José	5.9	34.8	750	26,100	5,870	51,500
Europe						
England						
Severn	9.8	96.0	70	7,460	1,680	14,700
France						
Aber-Benoit	5.2	27.0	2.9	78	18	158
Aber-Wrac'h	5.0	25.0	1.1	28	6	53
Arguenon & Lancieux	8.4	70.6	28.0	1,980	446	3,910
Frênaye	7.4	54.8	12.0	658	148	1,300
La Rance	8.4	70.6	22.0	1,550	349	3,060
Rothéneuf	8.0	64.0	1.1	70	16	140
Mont Saint-Michel	8.4	70.6	610	43,100	9,700	85,100
Somme	6.5	42.3	49	2,070	466	4,090
Subtotal					11,149	97,811
USSR						
Kislaya Inlet	2.37	5.62	2.0	11	2	22
Lumbovskii Bay	4.20	17.6	70	1,230	277	2,430
White Sea	5.65	31.9	2,000	63,800	14,400	126,000
Mezen Estuary	6.60	43.6	140	6,100	1,370	12,000
Subtotal					16,049	140,452
Grand Total					63,775	559,483

Source. M. K. Hubbert, "Energy Resources," *Resources and Man*, National Academy of Sciences, National Research Council, 1969.

1.6-5 Tidal Energy

The ocean tides, which are produced by the rotation of the earth and the gravitational pull of the moon and sun, have been used as a source of power since the Middle Ages. In 1580 water wheels were installed under the London Bridge to pump water for the city. With the development of lower-cost sources of power, the use of tide mills gradually disappeared. Total global tidal energy has been estimated at about 3 trillion watts.[8] However, tidal power can be utilized only at certain favorable coastal sites having sufficient tidal amplitude and appropriate littoral topography. It has been estimated that only about 25 sites worldwide meet these conditions (see Table 1.6-2). The total average potential hydraulic power available at these sites amounts to approximately 0.06 trillion watts. Of this potential, it is unlikely that more than 10–25% could be converted to electricity.[9]

At the present time there are only two tidal power plants in operation: a 400-kW plant in Kislaya Bay in the Soviet Union and a 240-MW plant on the Rance River on the north coast of France. The Rance plant is capable of generating power from both the incoming and outgoing tides; moreover, the turbines are capable of pumping water as well as generating power. During periods of low power demand, the pumping capability is used to boost the level of water in the storage basin behind the dam. As a result of these features, the Rance station is capable of achieving an efficiency of about 25%.[10]

Extensive studies have been conducted of the engineering problems and economic feasibility of tidal plants in a number of sites around the world. Although the Canadian government recently decided against proceeding with a large-scale plant on the Bay of Fundy, the studies undertaken for that site show that a plant located there could produce power almost as cheaply as conventional power plants.

1.6-6 Wind Power

Windmills have been used for centuries to grind grain and pump water. With the coming of the industrial revolution and the invention of the steam engine, use of windmills went into decline. Recently the energy crisis has revived interest in windmills. The development of large-scale windmills to generate electricity is underway in a number of countries. Windmills with a generating capacity of between 0.2 and 2.5 MW have been constructed under a cooperative DOE–NASA program and are in operation. These windmills all have horizontal axes and two blades. The largest have blades

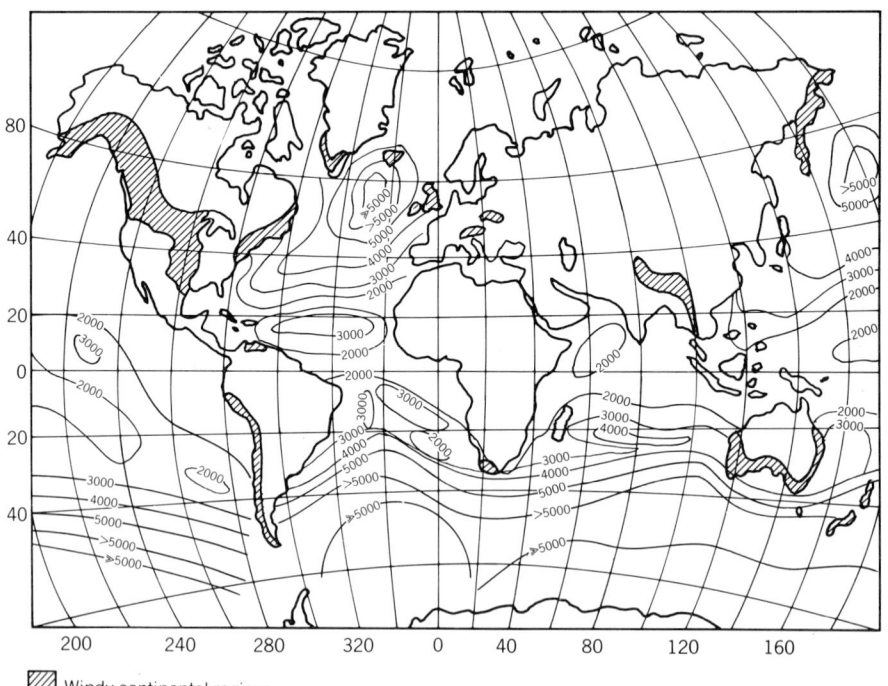

Windy continental regions

Fig. 1.6-4   Annual energy yields from the wind. (From ref. 6, p. 315.)

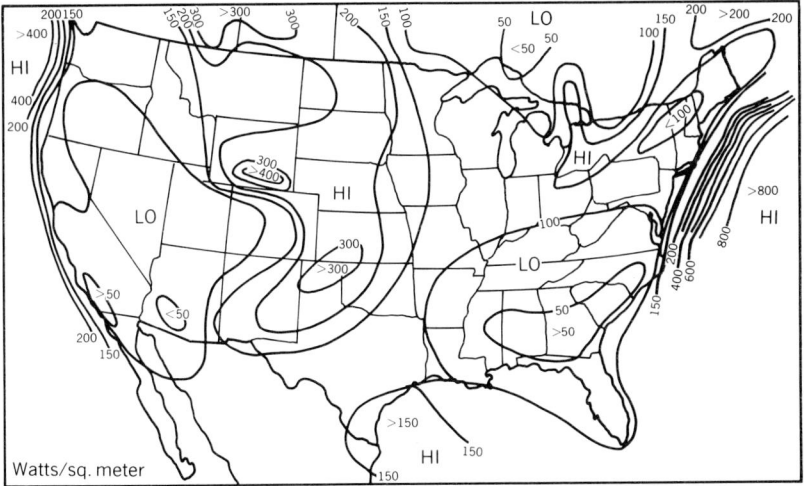

Fig. 1.6-5   Available wind power, annual average. (Source: ref. 32.)

constructed of a wood-epoxy laminate that sweep a circle with a radius of 150 ft and are mounted on a tower 300 ft high.[11]

Winds are caused by the rotation of the earth and the heating of the atmosphere by the sun. The total annual kinetic energy of air movement in the atmosphere is estimated to be about 3×10^{15} kW · h or about 0.2% of the solar energy reaching the earth.[12] The maximum technically usable potential is estimated to be theoretically 30 trillion kilowatt-hours per year,[13] or about 35% of current world total energy consumption. Because of limitations on the spacing of windmills, wind generators will require between 4 and 5 times the amount of land area to generate a given amount of electricity than that required by solar electric converters. However, it might be possible to use the land for other purposes such as agriculture.

The power in the wind blowing at a speed of 16 mi/h is about 200 W/m^2 of the area swept by the windmill. Approximately 35% of this power can be captured by the windmill and converted into electricity. However, it is important to note that the power output from the windmill varies with the cube of the wind speed. Consequently, only windy locations on mountains and coasts are suitable for the economical generation of electricity by wind power. Windmills are typically designed to begin functioning at a wind speed of about 12–14 mi/h and to reach peak output at a wind speed of around 20–30 mi/h. The suitability of a site is a function of the proportion of time that the wind blows at speeds of 14–20 mi or more per hour. Figure 1.6-4 shows the annual output that would be achieved by a 1250-kW turbine 55 m in diameter at various locations around the world. Figure 1.6-5 shows the annual average wind power available throughout the United States measured in watts per square meter.

1.6-7 Hydroelectric Power

The use of water power to grind grain dates from Roman times and was used extensively in the Middle Ages. In the early years of the industrial revolution water power was the chief source of energy for grist mills, saw mills, textile mills, and other factories. However, with the invention of the steam engine in the 1800s, the use of water power went into decline until, at about the beginning of this century, the development of electric power transmission made possible the large-scale generation and transmission of hydraulic energy. Since then, world hydroelectric power production has grown to 1744 billion net kilowatt-hours annually. This output is equivalent in energy content to about 414 million barrels of oil or 6% of total world annual energy consumption.[14]

The annual potential output at a 50% capacity factor of all installed or installable conventional hydroelectric plants in the world is presented in Table 1.6-3. The second column of the table contains the figures for the maximum potential that is technically and economically feasible. The third column lists the annual hydroelectric outputs of plants presently in operation; columns 4 and 5 list the expected annual outputs of plants under construction and planned, respectively.

The total annual output of plants operating, under construction, or planned for the entire world is 6.4 trillion kilowatt-hours compared with a technically and economically feasible maximum of 19 trillion kilowatt-hours. Thus, even when all plants that are under construction or planned are completed and in operation, world output of hydroelectric power will be approximately one-third the

TABLE 1.6-3   ANNUAL HYDRAULIC POTENTIALS

Region	Theoretical Potential (10^{12} kW · h)	Technical Usable Potential (10^{12} kW · h)	Operating Potential (10^{12} kW · h)	Potential Under Construction (10^{12} kW · h)	Planned Potential (10^{12} kW · h)
Africa	10.118	3.14	0.151	0.047	0.201
America (North)	6.15	3.12	1.129	0.303	0.342
America (Latin)	5.67	3.78	0.299	0.355	0.809
Asia (excluding USSR)	16.486	5.34	0.465	0.080	0.368
Oceania	1.5	0.39	0.059	0.020	0.032
Europe	4.36	1.43	0.842	0.094	0.197
USSR	3.94	2.19	0.265	0.191	0.17[a]
Total	44.28	19.39	3.207	1.090	2.12

Source. W. H. Bloss and O. Kappelmeyer, "Renewable Energy Resources," *Survey of Energy Resources, 1980* (Table 4.7, p. 347).

[a]Estimated.

output that is technically feasible. If the technically feasible potential were fully realized, the contribution from hydroelectric power would amount to the equivalent of 65% of *current* world annual energy requirements.

The principal question is how rapidly can world hydroelectric power output be increased toward the upper limit set by the technically feasible potential of 19 trillion kilowatt-hours per year. A recent study has indicated that 80% of the total installable capacity will be in operation by the year 2020 and that operating capacity will be brought up to 35% of the technically feasible maximum by the turn of the century.[15]

The energy that can be obtained from flowing water depends on the volume of the water and the height through which it falls before reaching the electric turbines. Construction of hydroelectric dams requires a steep drop in the elevation of the river at the chosen site, suitable topography for the storage dam, and river banks that will provide an adequate support for the dam's foundation. The estimate of maximum world hydroelectric potential given in Table 1.6-3 was determined by calculating the capacities at all potential sites on the world's rivers of the plants installed or those which could be installed to utilize the available energy.[16]

The question as to whether dams will in fact be built at all sites that meet the necessary criteria depends on the attractiveness of hydroelectric power compared with other alternatives.

There is no doubt that hydroelectric power has many attractive aspects. It is continually renewable by the cycle of evaporation and precipitation generated by the sun. It is nonpolluting: hydroelectric generation releases no emissions into the atmosphere or water. Hydroelectric energy can be an important part of the multipurpose utilization of water resources. Hydroelectric dams provide water for irrigation and are useful for flood control.

The characteristics of the energy supplied by hydroelectric plants is an important element in their favor. Hydroelectric turbines are very reliable and flexible in operation. They have long lives and low operating costs. The technology is well developed and fully proven. Hydroelectric turbines with efficiencies as high as 95% are now available.[17] Hydroelectric installations can provide convenient and economical peaking power and power storage capabilities that are unavailable with other types of electricity generation. At favorable sites the cost of hydroelectric energy is low compared with oil or even coal and nuclear power.

Hydroelectric power does, however, have some important drawbacks. There is a tendency for the impoundment areas of hydroelectric dams to silt up over time. Although their operation creates no pollution, the construction of hydroelectric dams does have significant environmental impacts on plant and animal ecosystems and on the aesthetic and recreational attributes of the affected land and waterways. These impacts have proved to be of sufficient concern to environmentalists to result in the postponement of projects in the United States. The fate of the proposed Dickey-Lincoln Dam on the St. John's River in Maine is a good example.

There is no doubt that hydroelectric power will be developed over the next few decades at sites around the world where the environmental impacts do not generate great opposition and that are close enough to industrial and urban centers for the economical transmission of power. Nevertheless, it is exceedingly difficult to predict what proportion of the sites that are technically suitable will in fact be utilized. Table 1.6-3 reveals that Africa, Asia, the Soviet Union, and to a lesser extent South America

TABLE 1.6-4   PRELIMINARY INVENTORY OF HYDROELECTRIC RESOURCES REGIONAL SUMMARIES

Existing,[a] Potential Incremental,[b] and Undeveloped[c] Capacity Ranges

Section	Small-Scale (0.05–15 MW)				Intermediate (15–25 MW)				Large-Scale (Greater Than 25 MW)				Total (All Sizes)			
	Exist.	Incre.	Undev.	Total	Exist.	Incre.	Undev.	Total	Exist.	Incre.	Undev.	Total	Exist.	Incre.	Undev.	Total
Vol. 1																
Pacific N. West																
No. of Sites	93	282	745	1,120	13	36	208	257	73	83	896	1,052	179	401	1,849	2,429
Cap. (MW)	430	642	3,702	4,774	234	700	4,069	5,003	26,141	31,919	259,709	317,769	26,804	33,262	267,480	327,546
Ener. (GWH)	2,441	2,234	16,390	21,065	1,216	1,943	14,738	17,897	130,365	33,999	673,918	838,282	134,022	38,175	705,045	877,242
Vol. 2																
Pacific S. West																
No. of Sites	111	354	272	737	9	17	26	52	69	43	110	222	189	414	408	1,011
Cap. (MW)	410	574	632	1,616	171	345	509	1,025	9,347	5,109	16,043	30,499	9,928	6,028	17,184	33,140
Ener. (GWH)	2,176	1,569	1,640	5,385	837	550	1,059	2,446	37,311	8,729	31,877	77,917	40,325	10,849	34,577	85,751
Vol. 3																
Mid-Continent																
No. of Sites	54	779	666	1,499	11	15	63	89	44	59	234	337	109	853	963	1,925
Cap. (MW)	184	850	1,182	2,216	218	317	1,311	1,846	6,087	6,589	27,376	40,052	6,488	7,758	29,868	44,114
Ener. (GWH)	1,372	2,138	3,074	6,584	1,006	524	3,142	4,672	22,403	12,481	64,274	99,158	24,781	15,144	70,491	110,416
Vol. 4																
Lake Central																
No. of Sites	204	601	551	1,356	10	43	16	69	17	88	59	164	231	732	626	1,389
Cap. (MW)	734	914	926	2,574	180	875	319	1,374	1,689	14,038	6,552	22,279	2,602	15,830	7,799	26,231
Ener. (GWH)	3,439	3,128	2,859	9,426	940	2,124	763	3,827	5,475	39,514	17,380	62,369	9,854	44,766	21,004	75,624

TABLE 1.6-4 (Continued)

| | Existing,[a] Potential Incremental,[b] and Undeveloped[c] Capacity Ranges | | | | | | | | | | | | Total | | | |
| | Small-Scale (0.05–15 MW) | | | | Intermediate (15–25 MW) | | | | Large-Scale (Greater Than 25 MW) | | | | (All Sizes) | | | |
Section	Exist.	Incre.	Undev.	Total	Exist.	Incre.	Undev.	Total	Exist.	Incre.	Undev.	Total	Exist.	Incre.	Undev.	Total
Vol. 5																
Southeast																
No. of Sites	110	566	265	941	19	29	54	102	98	87	146	331	227	682	465	1,374
Cap. (MW)	285	704	1,077	2,066	360	559	1,114	2,033	11,182	11,758	20,969	43,909	11,827	13,021	23,160	48,008
Ener (GWH)	1,000	2,189	3,349	6,538	1,105	1,185	2,863	5,153	36,409	21,466	67,460	125,335	38,514	24,840	73,672	137,026
Vol. 6[d]																
Northeast																
No. of Sites	270	2,231	143	2,644	19	26	20	65	27	85	58	170	316	2,342	221	2,879
Cap. (MW)	914	1,771	491	3,176	354	524	400	1,278	4,784	16,446	7,568	28,798	6,053	18,737	8,457	33,247
Ener (GWH)	4,620	6,009	1,531	12,160	1,613	1,533	938	4,084	26,276	81,898	28,610	136,784	32,508	89,440	31,078	153,026
National Total																
No. of Sites	842	4,813	2,642	8,297	81	166	387	634	328	445	1,503	2,276	1,251	5,424	4,532	11,207
Cap. (MW)	2,957	5,455	8,010	16,442	1,517	3,320	7,722	12,559	59,230	85,859	338,217	483,306	63,702	94,636	353,948	512,286
Ener (GWH)	15,048	17,267	28,843	61,158	6,717	7,859	23,503	38,079	258,239	198,087	883,519	1,339,845	280,004	223,214	935,867	1,439,085

Source. U.S. Army Corps of Engineers, *National Hydroelectric Power Resources Study*, Vol. 4, 1979 (Table 1, pp. 8–9).

[a] Existing hydroelectric power facilities currently generating power.
[b] Existing dams and/or other water resource projects with the potential for new and/or additional hydroelectric capacity.
[c] Undeveloped sites where no dam or other engineering structure presently exists.
[d] Data on undeveloped sites in the New England states are not available (NA).

contain the world's greatest unexploited hydroelectric potential. Unfortunately, many of the sites in South America, Asia, and Africa that are suitable for hydroelectric power generation are located a very long distance from the urban and industrial centers capable of utilizing the power output. Development of this potential is likely to have to wait for the growth of industry in the developing countries of these areas.

In the industrialized world the available hydroelectric potential has been much more heavily utilized, although even in North America as much as 40% of the total shown in Table 1.6-3 is unexploited. Much of this is located in areas of Canada sufficiently close to U.S. markets to make possible economical long-distance transmission of the power, yet remote enough so that concern over the environmental impacts will not prevent development.

Substantial unexploited potential remains even in the United States, as indicated by the figures on current U.S. operating capacity, annual production, and potential capacity by region given in Table 1.6-4.

1.6-8   Energy from Biomass

Plants convert sunlight to energy through photosynthesis at an efficiency of only about 5–6%.[18] However, because the amount of energy in the sunlight reaching the earth's surface is so enormous and the area of the earth covered by plant life of one kind or another is so great, the total chemical energy fixed in the biomass produced annually far exceeds the energy needs of mankind. Table 1.6-5 shows the total area of the earth's surface covered by oceans, fresh water, and the principal types of terrestrial fauna together with the annual production of biomass from these regions. Figure 1.6-6 shows how the annual production of fixed carbon in biomass is distributed around the world. As only 3% of the exploitable production of biomass is required to supply the nutritional requirements of people and animals today, it is clear that energy from biomass has a significant potential for contributing to mankind's energy requirements.[19]

Energy can be obtained from biomass in a variety of ways: pyrolysis, direct combustion, and fermentation and distillation into alcohol. The vegetable matter converted into energy by these processes can be obtained from a great variety of sources.

It has been estimated that municipal and agricultural organic wastes could be utilized to supply a substantial fraction of U.S. energy needs. Most of this waste material would be suitable for combustion to produce electricity. Many Third World countries, of which India is a prime example, meet a significant proportion of their energy needs by burning animal dung.[20]

A portion of U.S. agricultural wastes could be fermented and distilled into alcohol for use as motor fuel. Crops of sugar beets, Jerusalem artichokes, corn, and sugar cane can be and in some parts of the world are grown to produce alcohol for use as fuel.

TABLE 1-6.5   GLOBAL ANNUAL PRODUCTION RATE OF BIOMASS AND EQUIVALENT OF ENERGY

Biome	Area (10^6 km^2)	Annual Production of Biomass (10^9 t)	Energy Equivalent (10^9 MW · h)
Forests and woodland	57	68	345
Brushwoods	26	2.4	12
Meadowlands	24	15	70
Agricultural regions	14	9	44
Deserts	24	—	—
Freshwater regions	4	5	25
Total land regions	149	100	495
Oceanic regions	361	55	303
Total	510	155	799

Source. W. H. Bloss and O. Kappelmeyer, "Renewable Energy Resources," *Survey of Energy Resources, 1980* (Table 4.6, p. 332).

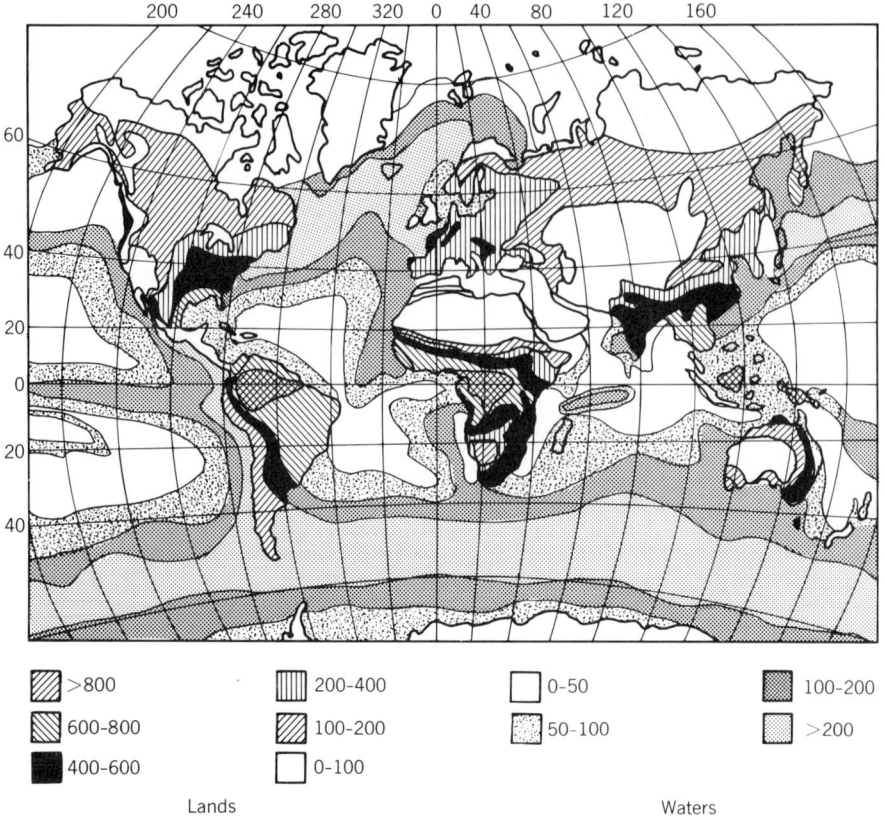

Fig. 1.6-6 Annual production rate of fixed carbon (g/m^2 year). (From ref. 6, p. 336.)

Firewood was the chief fuel of industry at the dawn of the industrial revolution. It is still being used extensively in many parts of the world today. Its contribution to the world's energy supply could be substantially increased. Table 1.6-6 presents the United Nations' statistics on the annual production of firewood, charcoal, and bagasse (the refuse from sugar beets and sugar cane) for the years 1976–1980 in various regions of the world. The statistics show that the Third World in particular, and the Communist countries to a lesser extent, make significant use of wood and charcoal to generate energy, while the United States and the rest of the industrialized West burn very little wood at the present time. This is clearly subject to change. New technologies have been developed to automate wood burning and to reduce the labor involved in harvesting wood. A recent study concluded that in New England energy from wood could contribute up to an equivalent of 20% of that area's current energy requirements by the year 2000.[21]

1.6-9 Geothermal Energy

Hemendra K. Acharya

History of Utilization of Geothermal Energy

Geothermal waters have been used since ancient times for recreation and therapeutic purposes. Geothermal energy has been used for municipal heating, in Iceland since the 1930s. Electric power production from geothermal energy began in 1904 at Larderello, a dry-steam field in Italy. A 250-kW generator was installed at the Geysers in California in the 1920s but was later abandoned. It was not until 1960 that a large-scale commercial geothermal power plant was built and operated at the Geysers.

The geothermal fields at Larderello and the Geysers are dry-steam reservoirs and, as such, are relatively easy to exploit as a source of electric power. Most geothermal reservoirs are, however, not as amenable to use since they are liquid dominated, that is, they yield a mixture of hot water and steam

TABLE 1.6-6 SELECTED SERIES OF STATISTICS ON FUELWOOD, CHARCOAL, AND BAGASSE

Country or area	Year	Fuelwood[a]			Charcoal[b]		Bagasse Production
		Production	Imports	Exports	Imports	Exports	
World	1976	1,507,675	1,711	542	230	229	166,790
	1977	1,533,157	1,788	801	264	240	178,072
	1978	1,565,503	1,074	379	268	219	178,806
	1979	1,598,997	1,102	423	343	239	177,862
	1980	1,627,880	1,083	491	410	281	174,339
Developed Marked Economy	1976	61,299	1,657	486	170	92	51,286
	1977	56,910	1,723	725	201	100	56,970
	1978	56,668	1,021	267	215	89	51,661
	1979	56,598	1,027	380	244	89	50,481
	1980	57,071	1,005	443	308	102	53,794
Developing Market Economy	1976	1,143,124	54	26	55	127	111,103
	1977	1,170,596	65	24	58	129	115,951
	1978	1,201,641	53	29	48	120	121,114
	1979	1,233,728	75	17	95	135	119,785
	1980	1,260,914	78	27	98	168	113,210
Centrally Planned Economy	1976	303,252	0	30	5	10	4,401
	1977	305,651	0	52	5	11	5,151
	1978	307,194	0	83	5	10	6,031
	1979	308,671	0	26	4	15	7,596
	1980	309,895	0	21	4	11	7,335
Canada	1976	3,900	—	—	21	—	—
	1977	3,682			36		
	1978	3,682			23		
	1979	3,682			29		
	1980	3,682			29		
United States	1976	14,868	—	—	9	27	8,192
	1977	14,150	—	—	14	33	7,806
	1978	14,150	—	—	36	28	7,002
	1979	14,150	—	—	34	16	8,029
	1980	14,150	—	—	37	11	8,010

Source. United Nations Dept. of International Economic and Social Affairs (Statistical Office), *1980 Yearbook of World Energy Statistics*, United Nations, New York, 1981.

[a] In thousand cubic meters.
[b] In thousand metric tons.

at the well head. Exploitation of liquid-dominated hydrothermal reservoirs on a large scale did not take place until the Wairakei plant was built in New Zealand in 1958. This plant site now has an installed capacity of 192,600 kW. Although this technological breakthrough increased the utilization of geothermal energy, it was at a slow pace and not widespread. In 1971 the total power production from geothermal energy was about 900 MW, and four nations—Italy, New Zealand, the United States, and Japan—had 99% of the world's geothermal generating capacity.[22]

The onset of the oil crisis in 1973 accelerated the pace of exploration and utilization, and many other countries began to utilize their geothermal fields for power production. At present 11 countries generate some of their electricity from geothermal energy. The contribution ranges from a negligible amount to as much as 30% depending on the circumstances in each case.[23] The present annual growth rate worldwide is about 15%, a rate that is expected to hold in the near future. Table 1.6-7 shows the generation of electricity from geothermal energy in various countries as of 1980. Figure 1.6-7 shows the past and future growth in the utilization of geothermal energy for power production. This projected growth, based on the U.N. report,[24] reflects the conversion of increased exploration effort in the 1970s into increased exploitation potential in the 1980s. Historical growth for nonelectric use is discussed in Section 12.1.

TABLE 1.6-7   INSTALLED GEOTHERMAL GENERATING CAPACITY

Country	No. Units	Generating Capacity (MW) Present, July 1981	Projected, 1983
United States	18	932.2	1297.2
Philippines	11	446.0	781.0
Italy	39	439.6	463.6
New Zealand	14	202.6	202.6
Mexico	5	180.0	425.0
Japan	7	168.0	223.0
El Salvador	3	95.0	95.0
Iceland	5	41.0	50.0
Kenya	1	15.0	30.0
Soviet Union	1	11.0	21.0
Azores	1	3.0	3.0
Indonesia	2	2.25	32.25
China	7	1.936	7.936
Turkey	1	0.5	5.5
Totals	115	2538.086	3636.986

Worldwide Geothermal Utilization

The primary utilization of geothermal energy worldwide is for power production. The Philippines is rapidly emerging as a major geothermal-energy-producing nation, and Indonesia appears to be on the threshold of significant growth in the geothermal power industry. Except for small projects in Kenya, Vietnam, and China, there is no direct utilization of geothermal energy in the developing countries. Countries in Eastern Europe and Iceland continue to pursue nonelectric uses. The development and utilization of geothermal energy in different countries for major power production are discussed below based on excellent surveys of Muffler[25] and Di Pippo.[26]

United States.   Hundreds of localities in the United States have been examined to one degree or another in an attempt to assess their geothermal potential.[27] Convective hydrothermal systems have been exploited at a number of places to produce electric power. Figure 1.6-8 identifies the locations and type of environment where geothermal energy is exploited at present or exploitation is planned in the near future. All of these systems are located in areas of high heat flow.

The largest geothermal electric power complex in the world is located at the Geysers in northern California. Power production at present is about 1005 MW[28] and the present proved capacity of the Geysers exceeds 2000 MW. The geothermal reservoir is of the vapor-dominated (dry) type. The Imperial Valley of southern California holds a huge reserve of liquid-dominated geothermal energy. A recent conservative estimate of the potential of this area suggests that 8700 MW may be possible, assuming 20–30 year plant lifetimes.[29] A number of power plants of various designs are either under construction or in the advanced stages of planning. Outside California, power plants have been proposed in the rift zone near Puna, Hawaii, Raft River, Idaho, and Rossevelt Hot Springs, Utah.

Research and development is proceeding at a haphazard pace to exploit energy in hot dry rock, geopressured zones, and in low-porosity conductive environments modified by circulation of meteoric water. Large-scale experiments have been conducted in New Mexico to determine the feasibility of exploitation of hot dry rock,[30] and the results are encouraging. Research and development for the exploitation of energy in geopressured zones in the Gulf Coast area (Fig. 1.6-8) is proceeding at a slow pace. No wells have yet been drilled and produced from potential geopressured geothermal energy prospects. A vast reservoir of liquid-dominated low-temperature systems may exist in the eastern United States and may be used to supply housing and industrial complexes with economical heat.[31]

El Salvador.   El Salvador is the first country in Central America to construct and operate geothermal power plants. At Ahuachapan, in western El Salvador, about 95 MW is generated, constituting about 15% of the total power production in El Salvador. Exploration for geothermal energy is going on at Chinameca, Chipilapa, and San Vincente, while a proven field exists at Berlin in eastern El Salvador.

Japan.   Commercial power generation by geothermal energy commenced in 1966 at Matsukawa, making Japan the fourth country behind Italy, New Zealand, and the United States to convert this

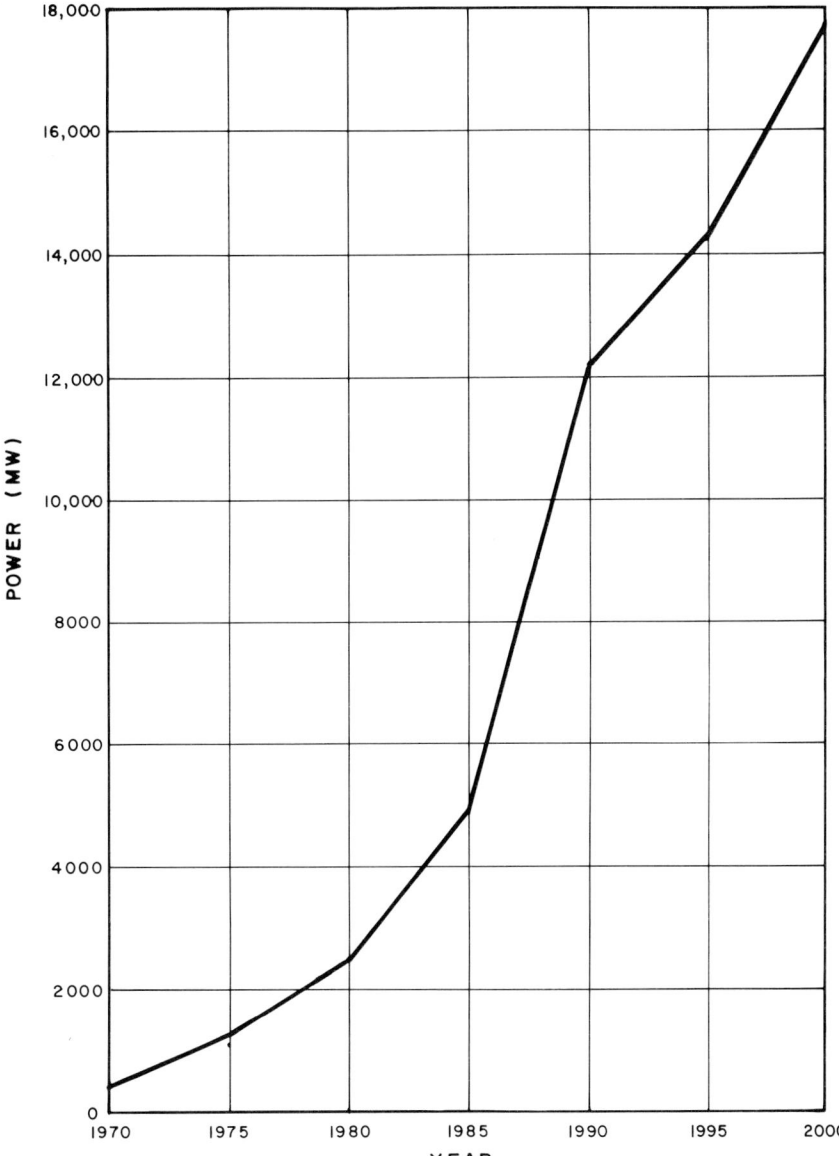

Fig. 1.6-7 Past and future growth in the utilization of geothermal energy for power production in the world. (From ref. 23.)

natural resource into electricity. At present six major geothermal plants are in operation and another is under construction. Exploitation of geothermal energy has nevertheless been slow because nearly all of the outstanding prospects are located in national parks, which are enthusiastically protected for their natural beauty. The construction and operation of geothermal power plants is thus subject to rigid and stringent controls.

Philippines. As noted earlier, the Philippines is the second largest producer of electricity from geothermal energy in the world. Based on present construction plans, geothermal energy would constitute nearly one-quarter of the total electric-generating capacity in the Philippines by 1985. The principal producing geothermal fields are located at Tiwi and Los Banos in Luzon and Tongonan in Leyte. Numerous other fields are being explored with power generation plans dependent on financing and availability of skilled personnel.

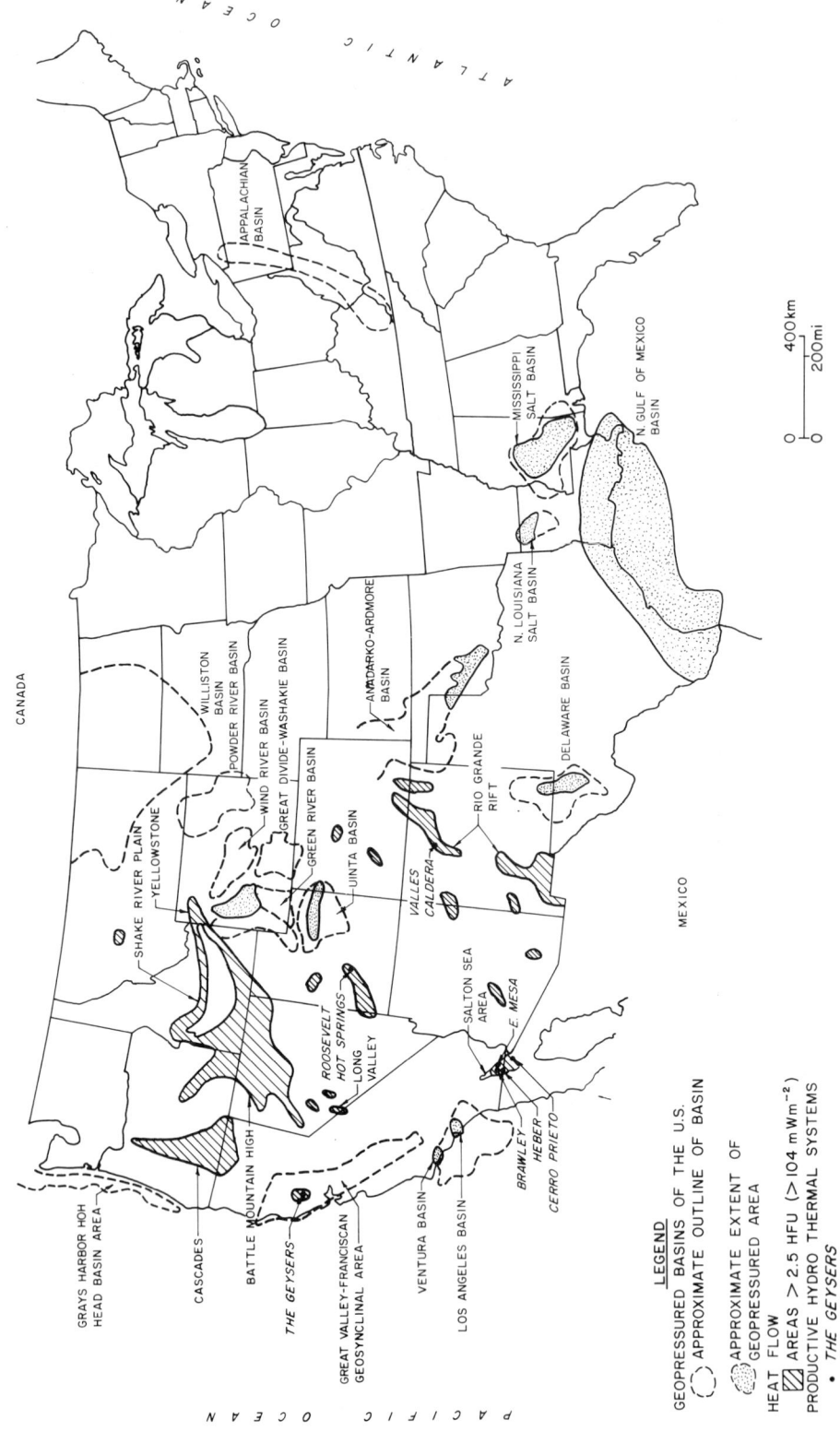

Fig. 1.6-8 Geothermal resources potential in the United States. (From ref. 25.)

LEGEND

GEOPRESSURED BASINS OF THE U.S.
APPROXIMATE OUTLINE OF BASIN

APPROXIMATE EXTENT OF
GEOPRESSURED AREA

HEAT FLOW

AREAS > 2.5 HFU (>104 mWm^{-2})

PRODUCTIVE HYDRO THERMAL SYSTEMS

• *THE GEYSERS*

New Zealand.   New Zealand pioneered the use of liquid-dominated geothermal resources for the production of electricity on a commercial scale at Wairakei. Two other areas are also being used or developed for electric power generation: Kawerau and Ohaki (Broadlands). At Kawerau multiple use is being made of geothermal energy for process heating, clean steam generation, and electricity production. Concern about environmental impact has slowed the development of this natural resource in New Zealand.

USSR.   Although the Soviet Union has a large potential of moderate-temperature waters, and some of these sites are being exploited for space or process heating, the only known sites at which geothermal energy is used for power production are located in Kamchatka Peninsula, well removed from the main population centers.

Italy.   The power of natural steam was first harnessed in Larderello, Italy, in 1904 to produce electricity. Steam has been tapped at two other sites, Monte Amiata and Travale, and the total installed geothermal electric-generating capacity has grown to 440 MW.

Iceland.   Iceland is perhaps better known for its direct use of geothermal energy in space-heating applications than for power generation. Roughly 65% of the population of this island heats its homes with geothermal hot water, and plans call for its utilization for almost 80% of space-heating needs. There are only a few locations in Iceland where electric power is generated from geothermal energy: at Svartsengi in the southwestern part of Iceland and at Namafjall and Krafla in northern Iceland. These sites are located in the mid-Atlantic Ridge spreading system which passes through Iceland.

Mexico.   Since 1973 the geothermal power plants at Cerro Prieto in northern Mexico have been reliable producers. The full potential of the field is known to be at least 400 MW, although power

TABLE 1.6-8   ESTIMATES OF U.S. AVAILABLE THERMAL ENERGY IN HYDROTHERMAL SYSTEMS (90°C)[a]

	System	Location	Thermal Energy $(10^{18}$ J)
1.	Surprise Valley	California	79
2.	Lassen	California	42
3.	The Geysers	California	100
4.	Long Valley Caldera	California	75
5.	Coso	California	25
6.	Salton Sea	California	97
7.	Brawley	California	22
8.	Westmoreland	California	67
9.	East Mesa	California	16.3
10.	Heber	California	31
11.	Crane Creek; Cove Creek	Idaho	16.4
12.	Bruneau	Idaho	450
13.	Hewdale area	Idaho	20
14.	Raft River	Idaho	7.4
15.	Steamboat Springs	Nevada	14.4
16.	Stillwater area	Nevada	23
17.	Desert Peak	Nevada	29
18.	Valles Caldera	New Mexico	87
19.	Vale Hot Springs	Oregon	45
20.	Klamath Falls	Oregon	30
21.	Cove Port-Sulphurdale	Utah	16
22.	Roosevelt	Utah	32
23.	Yellowstone Caldera	Wyoming	1240

[a]Estimates by C. A. Brook and others (1979) in "Assessment of Geothermal Resources of the United States—1978," L. J. P. Muffler (ed.), U.S. Geological Survey Circular 790, 1979.

production at present is only 150 MW. Although there are over 130 geothermal regions in Mexico, for foreseeable future geothermal power development will center on Cerro Prieto.

Exploration for the utilization of geothermal energy is proceeding in Indonesia, India, China, Kenya, Ethiopia, Turkey, Guadalupe, Dominica, Taiwan, Chile, Bolivia, Peru, and many other countries.

Geothermal Resource Assessment

The geothermal resource potential of the United States has been assessed by the U.S. Geological Survey[27] based on available geological, geophysical, geochemical, and hydrological data and refined, if possible, by drilling. The resource potential (q_R) for hydrothermal systems with temperatures of greater than 90°C has been calculated using the reservoir temperature, (t) reservoir area (a), and thickness (d) as follows:

$$q_R = \rho_c \times a \times d \times (t - t_{ref})$$

where ρ_c = specific heat,

t_{ref} = reference temperature (15°C).

Statistical methods were incorporated into the calculation of thermal energy for each system, reflecting uncertainties in values estimated for various parameters. Table 1.6-8 shows the computed thermal energy for different hydrothermal systems with temperatures of greater than 90°C in the United States. The thermal energy so computed is then converted into utilizable electric energy on the basis of several assumptions not fully verified at all producing fields.

Table 1.6-8 shows that the Yellowstone Caldera is the biggest reservoir of geothermal energy in the United States. The location of this reservoir in the National Park System, however, precludes its exploitation.

The assessment by the U.S. Geological Survey also shows an immense quantity of thermal energy in conduction-dominated and igneous-related environments, but the technology to utilize most of this energy has not yet been demonstrated.

REFERENCES

1.6-1 P. L. Auer, et al., "Unconventional Energy Resources," in *World Energy Resources, 1985–2020*, The World Energy Conference, London, 1978, p. 140.

1.6-2 Craig H. Peterson, "The Market Capture Potential of Single vs. Multiphase Solar Energy Space Systems: 1975–2010."

1.6-3 Auer et al., p. 144.

1.6-4 Auer et al., p. 152.

1.6-5 W. P. Goss, et al., "Summary of University of Massachusetts Research on Gulf Stream Based Ocean Thermal Power Plant," Proceedings of the Third Workshop in Ocean Thermal Energy Conversion, Houston, Texas, May 1975.

1.6-6 W. H. Bloss and O. Kappelmeyer, "Renewable Energy Resources," in *Survey of Energy Resources, 1980, Part A Text, Section A4*, 11th World Energy Conference at Munich 8–12 September 1980, F. Bender et al., (eds.) Federal Institute for Geosciences and Natural Resources, Hanover, Federal Republic of Germany, 1980, p. 148.

1.6-7 Auer et al., p. 148.

1.6-8 Bloss and Kappelmeyer, p. 324.

1.6-9 Auer et al., p. 150.

1.6-10 Hubbert, pp. 214–215.

1.6-11 W. H. Robbins and R. L. Thomas, *Large Horizontal Axis Wind Turbine Development*, NASA, Cleveland, Ohio, 1979.

1.6-12 Bloss and Kappelmeyer, p. 312.

1.6-13 Auer et al., p. 147.

1.6-14 The British Petroleum Company Limited, *BP Statistical Review of the World Oil Industry, 1980*, The British Petroleum Company Limited, London, 1980.

1.6-15 Ellis L. Armstrong, "Hydraulic Resources," in *World Energy Resources: 1985–2020*, The World Energy Conference, London, 1978, p. 102, Table 1.

1.6-16 Armstrong, p. 93.

1.6-17 Armstrong, p. 91.

1.6-18 Bloss and Kappelmeyer, p. 331.

1.6-19 Bloss and Kappelmeyer, p. 338.

1.6-20 Auer et al., p. 155.

1.6-21 Final Report of the New England Energy Congress, May 1979, p. xxxiv.

1.6-22 J. R. Koenig, J. R. McNitt, and M. C. Gardner, "Geothermal Power Developments in the Third World," *Transactions, Geothermal Resources Council* **5**, 23–27 (1981).

1.6-23 R. Di Pippo, "Geothermal Power Plants: Worldwide Survey as of July 1981," *Transactions, Geothermal Resources Council* **5**, 5–8 (1981).

1.6-24 U.N. Report on Geothermal Energy produced by the Secretariat for the U.N. Conference on New and Renewable Sources of Energy 1981.

1.6-25 L. J. P. Muffler, "Summary of Section I: Present Status of Resources Development," Proceedings Second U.N. Symposium on the Development and Use of Geothermal Resources, San Francisco, Vol. 1, pp. xxxiii–xLiv, 1976.

1.6-26 R. Di Pippo, "Geothermal energy as a source of electricity—A Worldwide Survey of the Design and Operation of Geothermal Power Plants," U.S. Dept. of Energy, DOE/RA/28320-1, January 1980.

1.6-27 L. J. P. Muffler, "Assessment of geothermal resources of the United States—1978," U.S. Geological Survey Circular 790, 1979.

1.6-28 *Electrical World*, *Electrical World Directory of Electric Utilities*, 90th ed., McGraw-Hill, New York, 1982, p. 78.

1.6-29 L. Younker and P. Kasameyer, "1978, A revised estimate of recoverable thermal energy in the Salton Sea Geothermal Resource Area," Lawrance Livermore Laboratory Rept. No. UCRL-52450, Livermore, CA, 1978.

1.6-30 J. H. Hill, "The LASL hot dry rock geothermal energy development project," *Geothermal Resources Council Transactions* **2**, 275–277 (1978).

1.6-31 D. N. Anderson and J. W. Lund (eds.), "Direct utilization of geothermal energy," A Technical Handbook, Special Report No. 7, Davis, California, Geothermal Resources Council, 1979.

1.6-32 Jack W. Reed, Wind Power Climatology of the United States, SAND74-0348, Sandia Laboratories, Albuquerque, NM, June 1975.

TABLE 1.7-1   THE AVAILABILITY OF OIL FROM OIL AND BITUMINOUS SANDS[a]

Region	Oil Shales Proved Recoverable Reserves (1-1-79)	Additional Resources (1-1-79)	Total	Bituminous Sands Proved Recoverable Reserves (1-1-79)	Additional Resources (1-1-79)	Total
Canada				19,300	16,300	35,600
United States	28,000	236,000	264,000	1		1
Australia		490	490			
New Zealand	1		1			
Thailand	2,015		2,015			
Brasil	84		84			
Venezuela				20,000	50,000	70,000
Jordan	800	7,000	7,800	700	10,000	10,700
Morocco	7,400		7,400			
Federal Republic of Germany	250		250	50		50
Austria						
Sweden	880		880			
Spain	12		12			
Totals	46,262	292,670	338,932	40,051	76,300	116,350
Desprairies[b]			400,000			300,000

Source. E. Schubert, "Hydro Carbons," *Survey of Energy Resources, 1980 Part A Text*, Section A2, 11th World Energy Conference at Munich, 8–12 September 1980, Federal Institute for Geosciences and Natural Resources, Hanover, Federal Republic of Germany, 1980 (Table 2.17, p. 157, and Table 2.16, p. 155).

[a] In million tons of recoverable oil.
[b] *Source.* Desprairies, Pierre, "Worldwide Petroleum Supply Limts," in *World Energy Resources 1985–2020*, The World Energy Conference, London, England, 1978.

1.7   ALTERNATIVE FOSSIL FUELS

In addition to coal, conventional petroleum, and natural gas the world has large quantities of lower-grade fossilized fuels to which it can turn when the conventional fuels approach exhaustion. These alternative fossil fuels are oil shale, tar sands, and peat.

1.7-1   World Resources of Oil Shale and Tar Sands

Oil shale is a sedimentary rock containing organic matter that will yield a substantial amount of oil when heated.[1] Shale oil contains kerogen, a chemical akin to oil. In the methods developed to date, oil shale is mined as a solid, crushed, and heated in a retort to separate the shale oil from the solid. Large volumes of by-products are created, consisting mostly of spent shale. These wastes constitute a significant environmental problem. Large quantities of water are required in the production process, and water availability in the relatively arid west, where the shale is located in the United States, poses a problem for the development of a large-scale shale-mining industry.

TABLE 1.7-2   SHALE OIL RESOURCES OF THE WORLD AND THE UNITED STATES[a]

	Identified[b]		Hypothetical[c]		Speculative[d]	
	25–100 gal/ton	10–25 gal/ton	25–100 gal/ton	10–25 gal/ton	25–100 gal/ton	10–25 gal/ton
Deposit						
Green River Formation, Colorado, Utah, and Wyoming	418[e]	1,400	50	600	—	—
Chattanooga Shale and equivalent formations, Central and Eastern United States	—	200	—	800	—	—
Marine shale, Alaska	Small	Small	250	200	—	—
Other shale deposits	—	Small	Ne	Ne	600	23,000
Total	418	1,600	300	1,600	600	23,000
Continents						
Africa	100	Small	Ne	Ne	4,000	80,000
Asia	90	14	2	3,700	5,400	110,000
Australia and New Zealand	Small	1	Ne	Ne	1,000	20,000
Europe	70	6	100	200	1,200	26,000
North America (excluding United States)	Small	Small	50	100	1,000	23,000
South America	Small	800	Ne	3,200	2,000	36,000
Total	260	820	150	7,200	14,600	295,000

Source.   William C. Culbertson and Janet K. Pitman, "Oil Shale," *United States Mineral Resources: Geological Survey Professional Paper 820*, U.S. Government Printing Office, (Washington, D.C., 1974 (Tables 95 and 96, pp. 500–501).

[a] In billions of barrels, by grade (oil yield) of oil shale.
[b] Identified resources: Specific, identified mineral deposits that may or may not be evaluated as to extent and grade, and whose contained minerals may or may not be profitably recoverable with existing technology and economic conditions.
[c] Hypothetical resources: Undiscovered mineral deposits, whether of recoverable or subeconomic grade, that are geologically predictable as existing in known districts.
[d] Speculative resources: Undiscovered mineral deposits, whether of recoverable or subeconomic grade, that may exist in unknown districts or in unrecognized or unconventional form.
[e] The 25–100 gal/ton category is considered virtually equivalent to the category "average of 30 or more gallons per ton."

TABLE 1.7-3 DEPOSITS OF BITUMEN-BEARING ROCKS IN THE UNITED STATES WITH RESOURCES OF AT LEAST 1 MILLION BARRELS

Age	Name of Deposit	Type of Deposit	Bitumen Resources (Millions of barrels, rounded)
California			
Late Miocene and early Pliocene	Edna	Sandstone	141–166
	South Casmalia	Diatomaceous mudstone and vein deposits	46
	North Casmalia	Diatomaceous mudstone and vein deposits	40
Pliocene	Sisquoc	Sandstone	26–50
Miocene	Santa Cruz	Sandstone	10
Pliocene and Pleistocene	McKittrick	Sandstone	5–9
Miocene	Point Arena	Sandstone	1
Kentucky			
Pennsylvanian	Kyrock area	Sandstone	18
	Davis-Dismal area	Sandstone	7–11
	Bee Spring area	Sandstone	8
New Mexico			
Triassic	Santa Rosa	Sandstone	57
Texas			
Cretaceous	Uvalde	Limestone	124–141
Utah			
Permian and Triassic	Tar Sand Triangle	Sandstone	10,000–18,100
Eocene	P.R. Springs	Sandstone	3,700–4,000
	Sunnyside	Sandstone	2,000–3,000
Triassic	Circle Cliffs	Sandstone	1,000–1,300
Cretaceous and Oligocene	Asphalt Ridge	Sandstone	1,000–1,200
Jurassic	Whiterocks	Sandstone	65–125
Eocene	Hill Creek	Sandstone	300–400
	Lake Fork	Sandstone	15–20
	Raven Ridge	Sandstone	100–125
	Rim Rock	Sandstone	30–35
Total			18,693–28,862

Source. W. B. Cashion, "Bitumen-Bearing Rocks," *United States Mineral Resources: Geological Survey Professional Paper 820*, U.S. Government Printing Office, Washington, D.C., 1974 (Table 19, p. 101).

**TABLE 1.7-4   TOTAL RESOURCES AND
RECOVERABLE RESERVES OF PEAT**[a]

Proved reserves	56.7
Additional resources	262.6
Total	318.3
Proved recoverable resources	15.8–16
Calculated static lifetime, years	326

Source. J. Koch, "Solid Fossil Fuels," *Survey
of Energy Resources Part A Text*, 11th World
Energy Conference at Munich 8–12 Sept. 1980,
Federal Republic of Germany, Federal Institute for Geosciences and Natural Resources,
1980 (Table 1.3, p. 52).

[a] In billion tons.

Tar sands or bituminous sands are sands or rock containing nonfluid bitumen that occurs in disseminated deposits and small pore basins spread throughout the rock material. The oil in tar sands is of high viscosity and specific gravity. The methods of extraction currently in use involve conventional mining of sands followed by mixing hot water with the sands to dissolve out the bitumen.[2]

The 1980 WEC estimates of world resources of oil shale and bituminous sand are presented in Table 1.7-1. The U.S. Geological Survey estimates of U.S. and world oil shale resources are presented in Table 1.7-2. The 1980 WEC figures are based on a survey of member countries. The exploration of oil shales and tar sands outside of North America has been very limited to date. Consequently, it is believed that the figures obtained from the questionnaire are incomplete. Table 1.7-1 also contains earlier estimates of the total oil contained in oil shales and tar sands as reported by Despraires in the 1978 WEC report. The latter estimates are considerably higher than those obtained from the 1980 survey, particularly in the case of tar sands.

The U.S. Geological Survey estimates reported in Table 1.7-2 are given in billions of barrels as opposed to the millions of metric tons employed in Table 1.7-1. Using a conversion factor of 7 barrels per metric ton yields an estimate of proved reserves of U.S. shale oil as reported by WEC in Table 1.7-1 which is approximately equal to the U.S. Geological Survey's figure for identified resources. However, the WEC estimate of 328 trillion barrels of proved recoverable reserves plus additional resources is far less than the figure given by the U.S. Geological Survey for total U.S. identified, hypothetical, and speculative resources in the 25–100 gal/ton category.

Among the richest and most thoroughly explored deposits of oil shale and tar sands are those of the United States and Canada. A large proportion of the identified and hypothetical U.S. resources of shale oil are located in the Green River formation of Colorado, Utah, and Wyoming. In places the oil-bearing shales are 300 m thick. Nearly one-quarter of the resource is located in zones more than 100 ft thick.[3] The oil content can be over 100 gal/ton of rock but on average is considerably lower (Table 1.7-3).

The Athabaska tar sands in Alberta, Canada, cover an area of approximately 50,000 km^2. The maximum oil content of the oil sands is around 18% by volume.[4] The deposits are estimated to contain something on the order of 270 trillion barrels of oil.

To date, production of oil from oil shales or tar sands has been very limited. Shale oil production has reached the pilot plant stage in the United States, and according to Table 1.7-1, production has reached 260 million barrels a year in the Soviet Union. Two companies are producing oil from the Canadian tar sands and their combined production is expected to reach 63 million barrels per year.[4]

1.7-2   Peat

Peat consists of partly decayed vegetable matter. It is formed in swamps or bogs as a result of the accumulation of plant debris without complete decomposition due to the presence of water. Dried peat can be burned as fuel and is currently used to generate electricity on a small scale in Ireland and the Soviet Union.[5] The peat resources of the world are listed in Table 1.7-4. Total reserves of peat are 15.8 trillion tons and resources are 318.3 trillion tons. The bulk of this is located in the Soviet Union followed by Canada and the United States.[6]

REFERENCES

1.7-1   William C. Culbertson and Janet K. Pitman, "Oil Shale," in *United States Mineral Resources: Geological Survey Professional Paper 820*, Donald A. Brobst and Walden P. Pratt, (eds.), U.S. Government Printing Office, Washington, D.C., 1974, p. 497.

1.7-2 W. B. Cashion, "Bitumen-Bearing Rocks," in *United States Mineral Resources: Geological Survey Professional Paper 820*, Donald A. Brobst and Walden P. Pratt (eds.), U.S. Government Printing Office, Washington, D.C., 1974, p. 99.

1.7-3 Culberston and Pitman, p. 500.

1.7-4 E. Schubert, "Hydrocarbons," in *Survey of Energy Resources, 1980 Part A Text, Section A2*, 11th World Energy Conference at Munich 8–12 September, 1980, F. Bender, et al., (eds.), Federal Institute for Geosciences and National Resources, Hanover Federal Republic of Germany, 1980, p. 158.

1.7-5 Cornelia C. Cameron, "Peat," in *United States Mineral Resources: Geological Survey Professional Paper 820*, U.S. Government Printing Office, Washington, D.C., 1974, p. 506.

1.7-6 Cameron, p. 507.

BIBLIOGRAPHY

Bender, F. (Ed.) *Survey of Energy Resources, 1980, Part B Appendix*, 11th World Energy Conference at Munich 8–12 September 1980, Federal Institute for Geosciences and Natural Resources, Hanover, Federal Republic of Germany, 1980.

Nehring, Richard, *Giant Oil Fields and World Oil Resources*, R-2284 CIA. The Rand Corporation, Santa Monica, CA, June 1978.

Peters, W., et al. "An Appraisal of Coal Resources and Their Future Availability," in *World Energy Resources* 1985–2020. The World Energy Conference, London, 1978.

The President's Commission on Coal, *Coal Data Book*, The President's Commission on Coal, Washington, D.C., February 1980.

United Nations Department of International Economic and Social Affairs (Statistical Office), *1980 Yearbook of World Energy Statistics*, United Nations, New York, 1981.

CHAPTER 2

OPTIMIZATION OF ENERGY USE

MELVIN H. CHIOGIOJI

U.S. Department of Energy
Washington, D.C.

JOHN SOMA

619 Union Avenue
Brooklyn, New York

ROCCO FAZZOLARE

University of Arizona
Tucson, Arizona

STEPHEN E. SMITH

Energy Technologies, Inc.
Tucson, Arizona

JAMES H. SWISHER

U.S. Department of Energy
Washington, D.C.

ROBERT R. REEVES

Rensselaer Polytechnic Institute
Troy, New York

JOHN WINNER, ARTHUR KARLIN, ARCHIE RIVIERA, MICHAEL McDONALD, DAVID TURNER

Booz-Allen and Hamilton
Washington, D.C.

RUSSELL W. ZUB

Transportation Systems Center
Cambridge, Massachusetts

J. Soma contributed to the "Energy Audits" section.

2.1   ENERGY CONSERVATION TECHNIQUES

Melvin H. Chiogioji and John Soma

2.1-1   Introduction

Most of the energy-consuming facilities and processes now in use in the United States were designed and built in an era of such cheap, abundant energy sources as fossil fuels. However, with rapid increases in fuel cost, with changes in fuel availability, and with the present ecological situation, old "rules of thumb" and "accepted practices" are being critically reexamined. Many of the current energy technologies must be modernized and perfected. For example, plant designs that optimize feedstock consumption will indirectly reduce overall energy requirements for processes in which feedstock is also utilized as fuel.

Over the past two decades most of the major industrial firms have been able to decrease the amount of energy they use per unit of output by substituting capital-embodying technology for energy. These firms have also seen a shift in manufacturing trends—away from the energy-intensive industries and toward the less energy-intensive industries. Since the former group of industries is primarily composed of basic materials producers, this shift is part of a long-term historical pattern of development toward higher degrees of fabrication and, as such, has contributed to the decline in the energy output ratio for all of manufacturing.

In an era of cheap easily available energy supplies and abundant sources of cooling water, energy economics and energy conservation have not necessarily been synonymous or even compatible. Although some of the more intensive industrial users of energy, such as the chemical, paper, and petroleum-refining industries, have long found it competitively advantageous to design their plants for energy conservation, until recently, the savings realizable from the use of energy recovery equipment have not always offset the cost of its installation. In this era of uncertain energy supply and steeply rising fuel costs, energy conservation must become not only a corporate virtue but an object of national concern.

Two obvious economic incentives for the development of an energy conservation program on a plant-by-plant basis can be identified: (1) the savings realized by reducing energy use and (2) the economic losses prevented by avoiding loss of production when fuel supply is curtailed. Demand reduction constitutes conservation; however, it is economically attractive only if the energy cost savings over an acceptable period of time are greater than the cost of implementing the conservation measures. It remains difficult to convince a plant manager that he should make an investment in energy conservation if such a program holds no promise of economic return.

The second incentive—energy security—is only important if the loss associated with process downtime or facility closure that is caused by energy curtailment is great. Many factors—production schedule, materials and product shelf lives, unit costs, plant shutdown and start-up costs, union agreements concerning temporary plant closure, backup fuel availability and anticipated length of the shutdown—must be examined to evaluate the impact of such an event.

In the past, in many cases, the cost of energy was ignored because it was low in relation to such other manufacturing inputs as manpower and materials. With today's spiraling energy prices, however, more attention must be given to energy input. By improving energy conversion processes, by recycling waste energy, or by reusing waste materials, we can make more efficient use of our energy supply. Our existing technology also offers us opportunities to accrue large energy savings, but in order to identify these areas for savings, we must first determine the answers to two primary questions:

1. What are the areas of activity in which there may be significant potential for the better use of energy?
2. Within these areas, what are the specific measures or alternative options that could lead to better and more efficient use of energy?

2.1-2   Energy Management Principles

Establishing the Program

In any organization the purpose of management is to achieve, through optimal use of such inputs as raw materials, energy, money, equipment, technology, and people, specific end results or goals as effectively and as efficiently as possible. Since a significant cost element in many operating systems is the energy used in obtaining the raw materials, in processing the materials, in selling the product, and in distributing them to consumers, the manager must, in order to reduce costs and to increase overall efficiency and profitability, cut down on energy usage.

The achievement of meaningful energy savings in existing processes hinges upon management's ability to provide effective personnel motivation, planning, and administration. To this end the establishment of a formalized energy management program will provide both the focus and direction required. Business managers must learn what their energy use and costs are, what the future energy supply is likely to be as well as what it is going to cost, what problems or opportunities the energy crunch is going to present, and what alternative solutions should be pursued.

The following discussion is an attempt to identify and describe the steps needed to implement an effective energy management program. It combines items suggested by many analysts of energy technology and of energy economics.

Top Management Commitment. The successful implementation of an effective energy management program rests upon a firm commitment to that program. Management's commitment must be clearly communicated to all organizational levels and must be made visible, and its commitment must be indicated by words and *actions*. Management's participation must be *active*, not passive, and should incorporate the following:

1. Informing line supervisors of the economic reasons for the need to conserve energy and of their responsibility for implementing energy-saving actions in the areas of their accountability.
2. Establishing a committee that has the responsibility for formulating and conducting an energy conservation program and that is comprised of representatives from each department in the plant and a coordinator appointed by and reporting to management.
3. Providing the committee with guidelines as to what is expected of them.
4. Setting goals in energy saving that are specific, consistent with production goals, and attainable.
5. Employing external assistance in surveying the plant and making recommendations for changes in operation, if necessary.
6. Communicating periodically to employees management's emphasis on energy conservation action and reporting to them the progress that has been made.

Organizing the Program. An organizational plan for the implementation and ongoing surveillance of energy conservation programs should consist of a definition of the responsibilities of the energy conservation coordinator or committee; a proposal for an effective communication system between coordinator and major divisions, departments, and employees; an elaboration of an energy accounting and monitoring system, and guidance for the education and motivation of employees.

Individuals responsible for the energy management at each major facility should serve on the corporate-level committee. This energy management committee should also have representation from the corporate management and planning staff and from the engineering department. It should enlist the aid of those employees whose abilities could contribute to a more effective energy management program. It should be vested with enough authority to facilitate its investigation of prevailing energy supply and demand situations and its implementation of policy recommendations. The specific tasks of the energy management committee should be assigned a priority that is consistent with the current or potential importance of energy problems in the company operation.

Energy Audits

John Soma

The first step to meaningful energy conservation is the measurement of all the energy that enters and leaves a plant during a given period. This energy audit is a management tool used to determine the energy status of a facility. Basically, it identifies and quantifies how energy is being utilized. By thus characterizing the energy intensity of a particular activity and by comparing the results of the audit with baseline data or norms for that activity, management can determine the potential for greater energy savings. Further engineering and economic analyses can then ascertain what changes—the implementation of procedural changes, the addition of automatic controls, or the installation of retrofit or new equipment—are within the facility's return on investment criteria.

Energy audits are conducted on many technological levels, from the simple to the complex. Eclectic model audits have been expediently developed to treat classes of facilities such as residential[1] and commercial[2] buildings and commerce and light industry,[3-8] from general guidelines,[9] specialized literature and industrial searches, and from proprietary information. Each facility is unique; hence, each facility should have an energy audit planned for it by a registered professional engineer or equivalent.

The audit itself is very labor intensive. A well-engineered one can easily run in the 5–10% of annual energy costs range, depending on complexity. An audit usually consists of three parts: a historic energy audit, an energy usage audit, and an energy efficiency audit. The historic audit examines past energy purchases and usage, associated costs, local weather conditions, levels of activity (production, hours of production), energy contracts, and the security of energy supply, past and future. The energy usage audit serves to identify where and quantify how much purchased energy is ultimately used within a facility. Correlation analyses, both economic and energetic, are performed in these two audit parts to determine the energy usage that is based on level of activity and environmental parameters.[10]

Two of the first-order linear models that can be used in energy audits are

$$\text{Energy} = \alpha + \beta \text{ production}$$

and

$$\text{Energy} = \alpha + \beta \text{ production} + \gamma \text{ heating DD} + \delta \text{ cooling DD}$$

where Energy is total or partial energy, α is a measure of the no-production energy (the energy overhead), β, γ, and δ are sensitivity coefficients, and DD represents degree days. These models serve to describe how energy is being used, indicating sensitivity to production and environmental parameters. They form the basis of monitoring energy management activity and, if periodically updated, form the basis for forecasting energy use.

The first two audit parts, the historic energy audit and the energy usage audit, serve to rank order the energy-intensive activities that are to be addressed by the energy efficiency audit. This energy efficiency audit constitutes a series of process studies wherein a normalized measure of energy efficiency (energy/unit) is sought. Ideal energy/unit and losses/unit information is obtained especially to identify the sources and the magnitudes of energy losses. A fully instrumented facility survey is usually undertaken in order to determine the values of the thermodynamic variables needed to solve the energy balances.

The energy efficiency (Eff.) is then

$$\text{Eff.} = \frac{(\text{ideal energy})}{(\text{actual energy})}$$

After an energy conservation activity is implemented, the savings is given by

$$\text{Savings} = \frac{(\text{new eff.}) - (\text{old eff.})}{(\text{new eff.})} \times 100\%$$

If the changes result in increased production at the same total energy usage, then the

$$\text{Productivity Gain} = \frac{(\text{new eff.}) - (\text{old eff.})}{(\text{new eff.})} \times 100\%$$

The thermoeconomic evaluation of energy activities requires consideration of the second law of thermodynamics.[11] The energy audit, which is based on the first law of thermodynamics, treats all energy as interconvertible and regards all energy units as equal, even though their thermodynamic potential to effect change (exergy or available energy) is different. First-law analysis is a necessary, but insufficient, energy diagnostic system.

The emerging field of exergy management[12] treats exergy as the energy commodity to be managed, vis-à-vis the second law of thermodynamics. However, although second-law inferences may be abstracted from definitive energy audits,[13] the preferred assessment tool for the management of exergy is the exergy assay.[14] The exergy assay uses the first and second laws and certain economic principles, in concert with productivity measures, to arrive at a full energy assessment. It does not treat energy in isolation, nor does it regard energy conservation as the end goal. Instead, the exergy assay, which seeks the synergism of all energy elements, yields optimal productivity within a facility's constraints.[16]

There is a direct correlation between the extent of data collection and consolidation and the evaluation of energy conservation opportunities. While an insufficient data base may prevent the identification of several energy-saving opportunities, too extensive a baseline survey may prove unnecessary and wasteful and may divert funds and time from more rewarding conservation opportunities. There is also a direct relationship between the extent of data collection and consolidation and the cost and time required for implementing an energy conservation program. While the best results can be obtained by conducting a thorough, comprehensive survey and analysis of all site energy and utility systems, time and budgetary constraints may impose limitations on the extent of the survey and appraisal for various site energy systems.

A comprehensive energy survey should be conducted whenever time and funds permit, since the primary advantage of such a survey is that it provides a complete baseline energy picture of the site, a picture which can be used as a basis for the definition of the interrelationships among energy systems, for the identification of energy conservation opportunities, and as a reference for the improvement or modification of future systems. Typically, the comprehensive energy survey involves the following

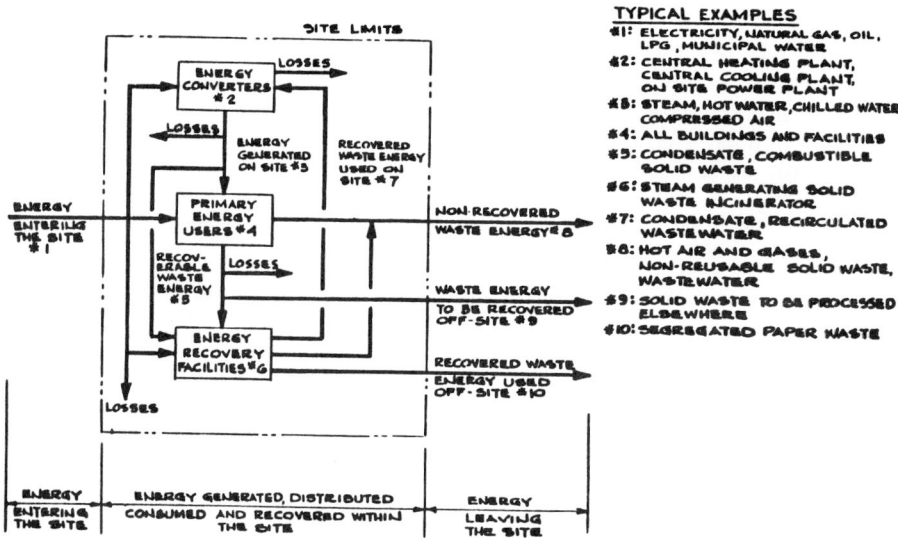

Fig. 2.1-1 Typical site energy and utility diagram.

steps:

1. Identify all site energy systems (see Fig. 2.1-1).
2. Survey each energy system entering the site, converted on site, generated on site, distributed throughout the site, and consumed on site, as well as each recoverable waste energy system.
3. Evaluate all site energy and utility systems.
4. Identify applicable energy conservation opportunities.

When a comprehensive energy survey cannot be implemented, a limited energy survey, which covers only significant site energy systems, should be performed. The steps in a limited energy survey are as follows:

1. Identify all forms of energy that enter the site, that are converted, generated, distributed and consumed on the site, and identify all recoverable waste energy.
2. Select significant (more than 5% of total) energy systems within each of the above categories. The selection should be based on total quantities, expressed in equivalent Btu.
3. Perform a detailed baseline survey only for the significant energy systems.
4. Perform an energy appraisal of the significant energy systems.
5. Identify energy conservation opportunities within the significant energy system.

Energy Balance

The next step in evaluating opportunities for energy conservation is to develop and analyze major energy balances. A basic energy balance is shown in Fig. 2.1-2. An energy balance is based on measurement and calculation. Energy is fed to an operational system that produces something in the form of output, so that the difference between the energy input and the energy of the output is waste.

Energy balances should be developed for each process in order to define in detail the energy input, the amount of raw materials and utilities required, the amount of energy consumed in waste disposal, the amount of energy credit for by-products, the amount of net energy charged to the product, and the amount of energy dissipated or wasted. Furthermore, all process energy balances should be analyzed

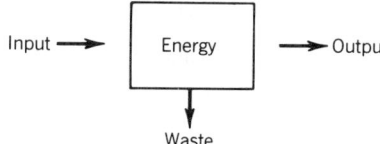

Fig. 2.1-2 Energy balance.

in-depth, with consideration being given to the following key points:

1. Can waste heat be recovered to generate steam or to heat water or a raw material?
2. Can a process step be eliminated or modified in some way to reduce energy use?
3. Can an alternate raw material with lower energy content be used?
4. Is there a way to improve yield?
5. Is there justification for:
    (a) Replacing old equipment with new equipment that requires less energy?
    (b) Replacing an obsolete, inefficient process plant with a whole new and different process that uses less energy?

The energy audit and energy balances identify energy-wasting situations and differentiate between those energy wastes that can be corrected by maintenance and operations measures and those wastes that require capital expenditures. The former can be corrected in a short time, with almost immediate results—dollars saved with little effort or delay. The latter require some investment dollars and delivery time for materials and equipment.

Each individual plant should make its own decision as to who should conduct the survey. In some cases the survey may be undertaken by one person; in others a team may be required; and in others still each plant department may be made responsible for its own audit. Nevertheless, the person or persons conducting the survey must be aware of the importance of finding energy wastes, of determining the amount of dollars involved, and of reducing the energy bill. The initial survey must be periodically followed up with additional surveys in order to ensure that wastes are kept to a minimum and new problems do not occur.

Energy Monitoring and Accounting

An effective energy conservation program, like any activity, requires feedback on performance. Hence, an energy accounting system is essential for the control and evaluation of an energy conservation program. However, the need for making measurements may differ from one plant to another or even within a single facility, depending upon the number of applications for using energy and the potential for variation in each application. In some cases, for example, periodic or irregular fuel deliveries or meter readings will not provide information adequate to determine the variation of energy use with daily, weekly, or monthly production cycles or with seasonal climate changes. In other cases energy in the form of fuels, electricity, and process steam will be used in two or more major applications throughout the plant. Then, in some energy applications, such as in combustion processes, measurement and/or control at each point of application may be necessary to ensure efficient use of energy.

Since companywide energy conservation is relatively new in most companies, the best techniques for doing so are not yet apparent. However, effective measuring methods are necessary to control, to evaluate, and to manage efforts to conserve energy. One effort, described by R. E. Doerr, incorporates the methods that Monsanto uses to maintain energy conservation performance at its plants and offices worldwide.[15] In the engineering of new facilities, potential conservation is monitored by using:

Percent reduction energy rate method.
Design energy savings report.

For existing facilities energy conservation is monitored by using:

Activity method.
Energy rate method.
Variable energy rate method.
Tracking charts.

Percent Reduction Energy Rate Method. The percent reduction energy rate is determined by comparing the project's product energy rate, expressed in Btu/lb, with the plant's product energy rate. An example is shown in Table 2.1-1.

The product energy requirements consist of all energy supplied, minus credit for all usable energy exported. The requirements include purchased energy, such as gas, oil, and electricity, plus the energy generated by in-plant utilities, such as steam, refrigeration, compressed air, and cooling water. All energy, including those generated by utilities, has to be expressed in equivalent purchased energy—in Btu. The manufacturing department is given credit for usable energy that is exported, for example, condensate and steam.

Design Energy-Savings Report. The design energy-savings report refers to an energy-savings idea that is incorporated in the design of a project. It is prepared by a project participant whenever he

TABLE 2.1-1   PRODUCT ENERGY RATE

Current plant product energy rate	1950 Btu/lb
Project product energy rate	1480 Btu/lb

Percent reduction in product energy rate

$$\frac{1950 - 1480}{1950} \times 100 = 24\%$$

Source. R. E. Doerr, "Six Ways to Keep Score on Energy Savings," *Oil and Gas Journal*, May 17, 1976, pp. 130–145. Reprinted with permission.

incorporates an idea or innovation that will reduce energy requirements, an idea that deviates from present design or plant practices for the product line. For a new product the reference point is the pilot plant or project definition report.

The report serves three purposes:

1. It serves as a means of information exchange on energy-saving ideas.
2. It provides an opportunity to monitor energy conservation capital requirements versus energy savings on projects.
3. It assists in an energy-awareness program in the engineering department.

A quarterly report on the number of energy-saving ideas submitted by each design group encourages competition between design groups.

Activity Method.   The activity method measures the anticipated Btu of energy saved in relation to the amount of energy purchased. The percentage of energy saved is based on the energy savings per year, expressed in Btu.

Since the activity method provides an immediate indication of conservation results and since savings are not affected by changes in product energy efficiency as the production rate varies, the activity method is an excellent method for monitoring performance. It provides a good visual account of activities that have been completed and planned to save energy. However, because the savings are only realized on an annual basis, the activity method does not reflect the true reduction in energy rate. (Btu/lb of product). A typical quarterly report for a company using the activity method could look like that shown in Table 2.1-2.

A corresponding report, expressing the savings in Btu and dollars statistics, both of which are most useful for internal communications and external publicity, can also be prepared.

Energy Rate Method.   The chemical industry energy rate method for reporting energy conservation results was developed by the Manufacturing Chemists Association (MCA). This method compensates for product mix and for the addition and deletion of products and allows a plant to compensate for OSHA and environmental energy requirements.

An example of the calculations entailed in the energy rate method is shown in Table 2.1-3. The base period rate is calculated by using the total pounds of product manufactured in year 0 and the

TABLE 2.1-2   ACTIVITY METHOD REPORT

XYZ Manufacturing Co.
Energy-conservtion results
Activity method
(% energy savings)
1975 Completed activities

Plant	First 3 quarters	Fourth quarter	Total	Planned activities 1976	1977
Fulton......................	2.0	0.2	2.2	2.1	1.0
Grace......................	1.8	0.1	1.9	0.8	0.4
Nixon......................	1.5	0.2	1.7	2.3	0.3
St. James.................	0.8	0.4	1.2	0.1	0
Company total.......	1.7	0.3	2.0	1.8	0.4

Source. R. E. Doerr, "Six Ways to Keep Score on Energy Savings," *Oil and Gas Journal*, May 17, 1976, pp. 130–145. Reprinted with permission.

TABLE 2.1-3 ENERGY RATE METHOD

Fulton plant Products manufactured	Year 0 (base period) Total production (x 10⁹ lb)	Total energy (x 10⁹ BTU)	Base Energy rate (x 10⁹ BTU/lb)	Year 3 Total production (x 10⁹ lb)	Total energy (1 x 10⁹ STU)	Comparison base period energy (x 10⁹ BTU)	% Reduction energy consumption rate
	1	2	3	4	5	3 x 4=6	7
A	200	10,000	50	300		15,000	
B	10,000	30,000	3	12,000		36,000	
C	2,000	20,000	10	3,000		30,000	
D	3,000	50,000	20	5,000		120,000	
	15,200	120,000		21,300		201,000	
Adjustments to base new products (+) or discontinued products after 12/31/72							
E (1973)	1,000	10,000	10	2,000		20,000	
F (1974)	1,000	5,000	5	1,000		5,000	
	2,000	15,000		3,000		25,000	
Total products	17,200	135,000		24,300	209,000	226,000	7.5*
Adjustments for environment % OSHA	Base				(1,000)		
Adjusted grand total	17,200	135,000		24,300	208,000	226,000	8.0*

*% Reduction energy consumption rate $\dfrac{6-5}{6} \times 100 = 7$

Source. R. E. Doerr, "Six Ways to Keep Score on Energy Savings," *Oil and Gas Journal*, May 17, 1976, pp. 130–145. Reprinted with permission.

[a] % Reduction energy consumption rate $= 6 - 5/6 \times 100 = 7$

equivalent amount of purchased energy consumed by the department manufacturing the product. Included in the equivalent amount of purchased energy is the product's share of energy that is consumed by plant services and by unaccounted-for losses. For year 3, or any reporting year, the percent reduction in energy consumption rate is calculated by comparing the amount of energy used during the base period with the total amount of energy, excluding feedstock energy, purchased by the plant. The comparison base period energy for each product is calculated by multiplying the base year product energy rate by the weight of the product manufactured in the current year. This is duplicated for each product manufactured in the current year. Changes in production rate affect conservation results because the product energy rate requirements are made up of fixed and variable energies.

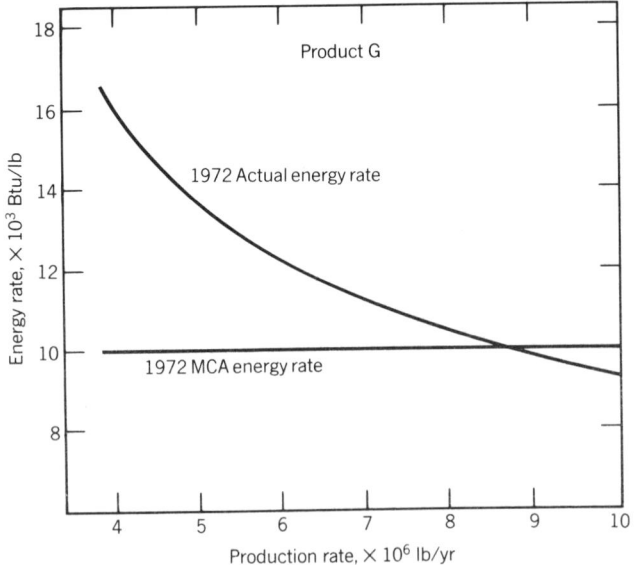

Fig. 2.1-3 Change in production. (Source: R. E. Doerr, "Six Ways to Keep Score on Energy Savings," *Oil and Gas Journal*, May 17, 1976, pp. 130–145. Reprinted with permission.)

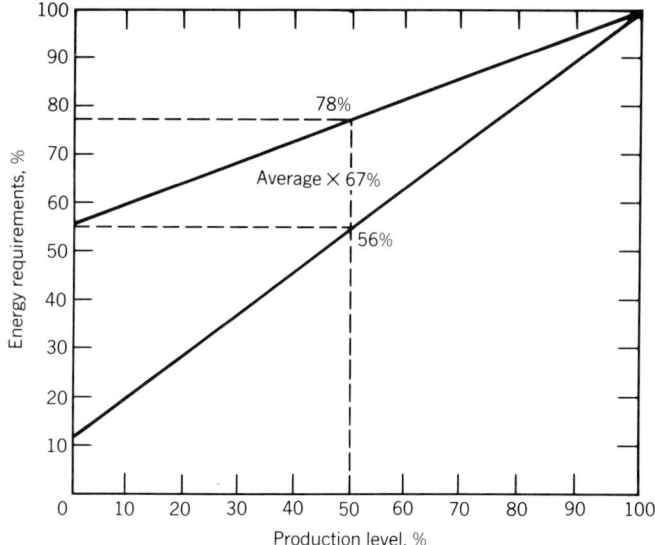

Fig. 2.1-4   Energy production. (Source: R. E. Doerr, "Six Ways to Keep Score on Energy Savings," *Oil and Gas Journal*, May 17, 1976, pp. 130–145, Reprinted with permission.)

Variable Energy Rate Method.   Among the many variables that must be considered in a conservation monitoring method are:

1. Changes in energy rate as production rate varies (see Figs. 2.1-3 and 2.1-4).
2. Product energy requirement changes that result from ambient temperature changes during the year.
3. Changes in raw material quality.
4. Changes in the plant heat balance, which may require the process to operate turbine drives instead of motor drives, and vice versa.
5. Minor changes in product quality or specifications for different customers.
6. A change in either equipment or process, for the purpose of increasing output.

To illustrate, the only variable considered at a couple of Monsanto plants is production rate and its effect on energy rate. From this variable a base year product energy rate versus production rate curve can be developed for each product, as is shown in Fig. 2.1-3.

To calculate the conservation performance for any year, the product production rate for that year is used to establish the base period energy rate, using the energy rate curve in Fig. 2.1-5. For example, using the MCA energy rate method for year 0, the product energy rate is 10,000 Btu/lb, and the actual base year energy rate is 11,500 Btu/lb. The amount 11,500 Btu/lb, instead of 10,000 Btu/lb is used in calculating the comparison base period energy rate for product G. This same procedure is then duplicated for each product.

This method, although not 100% accurate in compensating for all variables, does compensate for the major variable, production rate, and does prevent wide swings in conservation results as the production rate varies.

Tracking Charts.   Tracking charts can be used to establish trends in the energy usage associated with conservation, with maintenance of the plant, or the lack thereof, and with changes in operational techniques (see Fig. 2.1-6). This method constitutes a speedy, useful technique to track department and total plant production, purchased energies, generated utilities, and product energy rates. Also, tracking charts can be a method for monitoring trends in conservation and energy usage and for identifying potential problem areas.

Two types of charts are generally used. The first is a tracking chart that monitors the shift, daily, or weekly energy usage spanning a month to a year's time period. Usually maintained in the utilities or manufacturing department, this chart can be reviewed by shift operating personnel for quick feedback on operational performance.

The second type of tracking chart monitors weekly, biweekly, or monthly energy usage over a longer period of time, generally for one or more years. This type of chart tends to average out

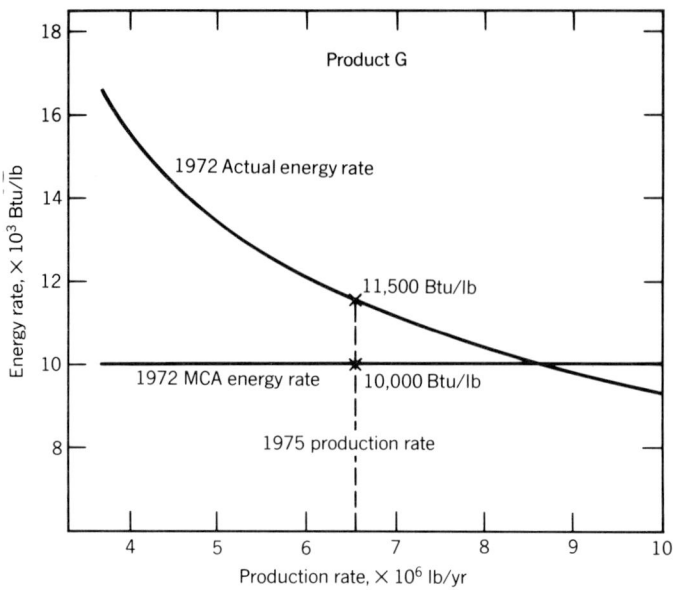

Fig. 2.1-5 Actual and MCA energy rate versus production rate. (Source: R. E. Doerr, "Six Ways to Keep Score on Energy Savings," *Oil and Gas Journal*, May 17, 1976, pp. 130–145. Reprinted with permission.)

Fuel rate (oil)

Electric rate

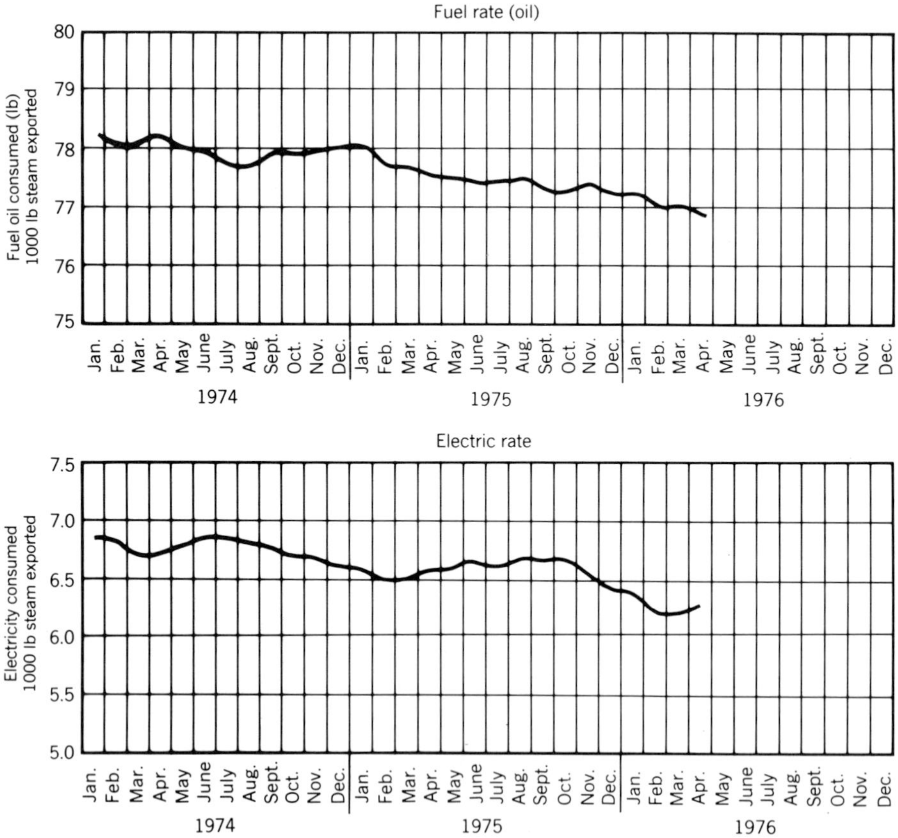

Fig. 2.1-6 Tracking chart. (Source: R. E. Doerr, "Six Ways to Keep Score on Energy Savings," *Oil and Gas Journal*, May 17, 1976, pp. 130–145. Reprinted with permission.)

fluctuations in energy usage since consideration is given to trends over a more lengthy period of time. Any analysis by tracking charts, however, must take into consideration production rate and the effect it has on energy usage.

Employee Participation

An energy conservation program can only be successful if it arouses and maintains the participative interest of the employees. Employees who feel themselves partners in the planning and implementation of the program will share pride in the results. To this end, communicating to the potential participants the importance of energy conservation can be done in several ways: in face-to-face discussions, in seminars and workshops, through the distribution of informative and descriptive literature, through slide presentations and moving pictures, and, most important of all, through example—the constant observation of conservation measures by management.

The use of company newsletters, bulletin boards, or posters for pictorializing energy conservation objectives and accomplishments will help impress employees with the importance of such matters. Employee participation can be encouraged by citing examples of energy conservation ideas being implemented, by posting photographs of persons who submitted the idea, and by publishing information on the savings realized. Competition between departments, sections, or groups within the company—programs that foster competition toward the attainment of conservation goals—can also generate enthusiasm among employees and should be encouraged. The acknowledgement of good ideas is a key to the success of this approach.

Employees can be made more aware of the value of conservation measures through workshops and training courses for supervisory personnel, through articles published in the company newsletter, and through energy conservation checklists given to each person on the work force. This last, a clear, concise list of energy "dos and don'ts," will guide employees in the performance of their work and can be helpful in establishing energy conservation practices. Such lists, distributed to all employees whose jobs involve the use or control of energy, should be updated periodically. Supervisors should have the responsibility of seeking adherence to all items on the lists.

2.1-3 Energy Conservation Opportunities

A key element of the energy management process is the identification and analysis of energy conservation opportunities. These opportunities can be categorized as follows:

1. Improving energy efficiency by improving maintenance of energy-consuming devices, and by changing such operational procedures as turning off lights and equipment, changing thermostat settings, and analyzing means for reducing demand charges for electricity.
2. Implementing such off-the-shelf technology as waste heat recovery, improved combustion controls, and computerized energy management systems.
3. Utilizing longer-range technologies that require R & D.

In any manufacturing process the total amount of energy can be broken down according to the various items of plant equipment utilized in the manufacture of the final products and to the transmission systems that convey energy to those items of equipment. Any change in the gross or unit energy consumption for a given product line can be associated with changes in the energy consumption of the various items of equipment used to produce that line. Thus, equipment plays a major role in energy conservation. Understanding the relative energy consumption by equipment type and the opportunities that exist for decreasing this consumption must constitute the primary effort in an effective conservation program.

All potential energy conservation projects should be followed by an evaluation of each individual project before it is implemented. The following methodology may be used:

1. Calculate annual energy savings for each project.
2. Project future energy costs and calculate annual dollar savings.
3. Estimate project capital or expense cost.
4. Evaluate investment merit of projects using such measures as return on investment, and so forth.
5. Assign priorities to projects based on investment merit.
6. Select conservation projects for implementation, and request capital authorization.
7. Implement authorized projects.

The implementation of an energy conservation opportunity requires careful evaluation, for the possibility exists that under certain circumstances, that opportunity could be counterproductive. In some cases the fact that existing equipment will have operating limits must be considered.

ECO Evaluation Procedures

Energy conservation opportunities (ECOs) that generate benefits greater than costs, without sacrificing product quality, are generally profitable and are therefore attractive. However, many ECOs require initial capital outlays, which must be amortized over the expected lifetimes of the energy savings generated. This section reviews some of the basic tools of financial analysis, which may be useful in the economic evaluation of such ECOs.

Sound, consistent economic criteria for evaluating energy conservation opportunities are quite important. Before any investment is made, some quantitative measure of profitability is needed to enable the investor to compare his expected return with the return from alternative investment opportunities. Because true economic cost includes opportunity costs of foregone investments, ECOs should be considered to be profitable only when their expected rate of return is greater than that which could be realized from alternative investment opportunities, whether in energy conservation or elsewhere.

The following are a few examples of measures of performance for energy conservation investments.

Payback Period.   Payback period is defined as the first cost divided by the net annual savings, or

$$\text{Payback period} = \frac{\text{first cost}}{(\text{annual fuel savings}) \times (\text{projected fuel price}) - \text{annual operating cost}}$$

Comparing the payback period to the expected lifetime of the investment allows the investor to make some rough judgment as to the investment's potential for recoupment. A payback period of less than one-half the lifetime of an investment, where the lifetime is 10 years or less, would generally be considered profitable.

The payback period as a measure of performance gives rise to problems, however, because dollars saved in future years are considered to be equivalent in value to dollars saved in current years and because comparisons between alternative investment opportunities of different lifetimes cannot be made.

Return on Investment.   Return on investment (ROI) considers the depletion of the economic life of an investment by providing for renewal of the investment through a depreciation charge. By using a straight-line depreciation charge (DC), in which

$$\text{DC} = \frac{\text{first cost}}{\text{estimated lifetime}}$$

the percent return on investment can be calculated with the equation

$$\text{ROI, } (\%/\text{yr}) = \frac{(\text{Net annual savings} - \text{depreciation charge})}{\text{first cost}} \times 100\%$$

ROI evaluates investments with different life expectancies on a comparable basis. It is frequently used in the financial analysis of potential investments because of its simplicity of calculation.

Benefit–Cost Analysis.   Benefit–cost analysis entails a direct comparison between the present value benefits (savings) generated by a given investment and its costs. Generally, this is formulated in terms of a benefit–cost ratio (B–C). A ratio greater than unity implies that the expected net benefits (properly discounted and summed over the lifetime of the investment) will exceed the initial costs and that, therefore, such an investment is profitable. Conversely, a benefit–cost ratio less than unity implies that such an investment is not profitable.

Breakeven Analysis.   The period of time in which capital investments can be recouped, or the "breakeven" period, is similar in concept to the payback period, except that the breakeven period takes discount rates into consideration. Again, the chief drawback of such a measurement is that investments of unequal lifetimes cannot be compared.

Internal Rate of Return.   The internal rate of return (IRR) is defined as the discount rate that reduces the stream of net returns associated with the investment to a present value of zero. While the IRR is not always a good measurement of economic performance, it will give good results when used to evaluate a project whose fixed first cost is followed by a stream of positive net benefits.

Marginal Analysis.   Typical of many investments are those energy conservation opportunities whose rates of return decrease as the levels of investment increase. The installation of insulation is a case in point; each additional increment generates less savings than the last. In considering such investments,

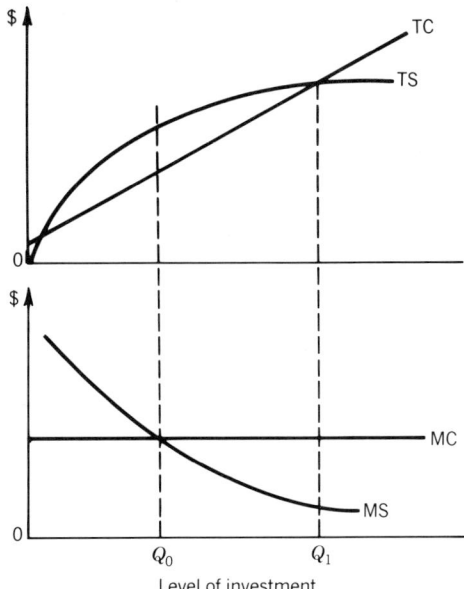

Fig. 2.1-7 Determination of optimal investment size based on the marginal savings–marginal cost relationship.

one may wish to estimate the optimal level of installation since no other level will generate greater net savings (savings minus costs).

The following equation will be useful in estimating the optimal investment amount for any given ECO that has more than one installation level:

$$MS = MC$$

where MS = marginal savings, the present value savings generated by the last increment of the project
MC = marginal cost, the present value cost of this last increment

The diagrams in Figure 2.1-7 bear out the equation.

The upper portion of Figure 2.1-7 shows total cost (TC) and total savings (TS) as functions of investment size. While any level of investment between Q and Q_1 is profitable (since TS > TC), the most profitable level is at that point where the distance between TS and TC is maximized, at Q_0. This occurs when the slope (or rate of change) of the TC and TS functions are equal. Directly below this point, on the lower diagram, we see that MS = MC. At any investment point less than Q_0, increasing investment will earn greater profit since the additional savings generated are greater than their cost. At any point beyond Q_0, however, increased savings are not covered by their costs and, thus, profits will be diminished.

Marginal analysis will not only aid in determining the point at which increasing investment ceases to be profitable, it may also indicate that by reducing the level of investment to the point at which MS = MC, a project that appears to be unprofitable may indeed be made profitable.

Industrial Processes

The amount of energy consumed for the manufacture of any given set of products depends upon such basic variables as raw materials selected, the manufacturing process adopted, and the level of production decided upon. Beyond this threshold value, an economic balance between the incremental cost of more energy-efficient equipment or technologies and the value of the energy that can be saved by these means must be maintained.

Energy savings in industrial processes can be achieved in the following ways:

1. Housekeeping measures. Significant reductions in energy usage can be attained by better maintenance and housekeeping (e.g., shutting off standby furnaces, practicing better space condition-ing, eliminating steam and heat leaks) and by greater emphasis on the optimization of energy usage.

Other savings can be achieved through improved operating practices (e.g., operating at a lower acceptable temperature that does not affect productivity).

2. Equipment and process modification. Equipment and product modifications can be applied to existing equipment (retrofitting), incorporated in the design of new equipment, or both. The modifications could be the result of better quality control, of the use of more durable or more efficient components, or of the implementation of a novel or formerly neglected, more efficient design concept. Changes for realizing greater energy efficiency can also be made in the industrial process or in the replacement of that process, producing the desired amount and quality of goods while using less energy.

3. Integrated operations. A careful examination of production processes, schedules, and operating practices can result in more efficient utilization of equipment. Other improvements in plant efficiency can be achieved through:

1. Proper sequencing of process operations (e.g., using high-pressure steam or gas for either power generation or motivating purposes, before extracting the heat content of the steam or burning the gas, will make maximum use of the available energy content).

2. Rearranging schedules to utilize process equipment for continuous periods of operation in order to avoid numerous short runs and minimize heat-up losses.

3. Scheduling process operations during off-peak periods in order to level electrical demands, and conserving the use of energy during peak demand periods.

4. Utilizing waste heat. Methods for utilizing waste heat can be categorized as follows:

   (a) Direct utilization (e.g., for drying or preheating process materials when no external heat exchanged is employed).

   (b) Recuperation, in which the waste gases and air or other gas used for preheating are separated by a metallic or, in cases of very high temperatures, a refractory heat exchange surface. Transfer of energy from one fluid to another occurs continuously.

   (c) Regeneration, in which heat from waste gas is conducted to and stored in a heat exchange medium, in a refractory, or in metallic materials and are subsequently used to heat air for preheating.

   (d) Waste heat boiler, a form of recuperation, in which hot waste gases generate process steam or hot water. Both water tube or fire tube designs can be used.

   (e) Cogeneration, in which electricity and process steam are generated together (e.g., a steam turbine driving an electrical generator). After the energy necessary for doing work is removed, the steam turbine exhausts partially spent steam at a lower pressure than that of the inlet pressure. The energy in the turbine exhaust steam can then be used, in the usual ways, for process heat.

   (f) Energy cascading, in which the energy is used at its highest quality first and then at lower qualities in other associated processes, until the energy is of such low quality that it is no longer useful. In the chemical industry's use of energy cascading, heat condensation from one distillation column operates a second unit.

Motors and Drives

Motors and drives consume a large percentage of the electrical energy used in U.S. industry today. A survey conducted by Arthur D. Little showed that motors comprised 75% of the electrical usage in industrial applications (see Table 2.1-4). It is imperative, then, that the efficiency of motors be improved.

Table 2.1-5 reveals that there are high-efficiency motors currently available on the marketplace and that future motors will be even more efficient. These high-efficiency motors are now competitive with conventional types of motors, considering the cost of motor losses.

The operation of a motor requires energy for (1) the mechanical load of the motor; (2) the mechanical losses in the motor; and (3) the electrical losses in the motor and the electrical supply system. Energy efficiency improvements can be achieved in any of the above areas.

The following are suggestions for energy conservation opportunities for motors and drives:

1. Select the most efficient motor available for the application.
2. Match the load with motor used.
3. Turn off and restart the motor, rather than allowing it to idle when it is not in use.
4. Recycle or recover motor heat for other uses.
5. Select the most efficient type of drive for the job.
6. Clean and maintain the motor in order to lower heat buildup.
7. Use two (or more) motors in tandem if doing so will more efficiently operate a variable load.
8. Keep the motor in good repair.

TABLE 2.1-4   TOTAL INDUSTRIAL ELECTRICAL CONSUMPTION (1972)[a]

Industrial Motor Drive (except HVAC)		458
Pumps	143	
Compressors	83	
Blowers and fans	73	
Machine tools	40	
Other integral hp applications	52	
dc drives	47	
Fractional hp applications	20	
Other Industrial Electrical Usage		142
Electrolytic		
Direct heat		
HVAC		
Transportation		
Lighting		
Total Industrial		600

Source. Federal Energy Administration, *Energy Efficiency and Electric Motors*, Conservation Paper 58, August 1976, p. 26.
[a] Billions kW · h

TABLE 2.1-5   CURRENT AND FUTURE MOTOR EFFICIENCIES IN INTEGRAL hp ac-POLYPHASE MOTORS

	Current			Future
	Worst	Best	Average	Improved Efficiency Models
1 hp	68	78	73	85.5
5 hp	78	81.5	80	89
10 hp	81	88	85	90
50 hp	88.5	.0	90	92.5
100 hp	90.5	92.5	91.5	93
200 hp	94	95	94.5	95

Source. Federal Energy Administration, *Energy Efficiency and Electrical Motors*, Conservation Paper 58, August 1976, p. 43.

Electrical Service and Distribution System

Utility rate schedules are generally structured in such a way that the initial blocks of monthly kilowatt demand and monthly kilowatt-hour consumption are charged at the highest rate. With separately metered services, the customer must pay the high costs of initial blocks of demand and energy for each meter. By physically rewiring the system, so that separately metered services are combined into one, or by arranging with the utility for conjunctive billing of two or more metered services, significant dollar savings can usually be effected.

Obtain Service at Highest Voltage at the Site.   Most utility rate schedules offer a cost incentive to customers who accept services at the highest voltage that is available at the site. To take advantage of such a rate incentive, the customer is usually required to purchase and maintain all switchgear, transformers, buses, and cables on his side of the utility's metering transformers.

Reduce Monthly Demand Charges.   Utility rate schedules generally fall into two types, as follows:

1. Separate charges, developed for demand and energy consumption, respectively, are combined to become the total that appears on the monthly invoice.
2. Monthly kilowatt-hour consumption is divided by the monthly kilowatt demand, to arrive at a monthly value of kilowatt-hour per kilowatt demand. The unit price per kilowatt-hour is then applied to this value to calculate monthly charges.

Assuming a constant monthly kilowatt-hour energy consumption, significant dollar savings can be realized by reducing the peak kilowatt billing demand. In effect, this means that the monthly kilowatt-hour load must be spread more evenly over the hours of operation. This, in turn, may entail rescheduling the operation of high-demand loads to evenings or weekends, at a time at which the kilowatt demand of other loads on site may be substantially reduced.

Eliminate Monthly Power Factor Penalty Charges.   Many utility rate schedules contain a power factor penalty clause that increases demand charges for the months that the customer's load operates at a lower power factor. Improvement of a system's power factor can eliminate monthly utility penalty charges, can increase the power-carrying capacity, and can reduce line losses (energy losses).

A low system power factor usually results from a preponderance of lightly loaded induction motors. In this case power factor correction measures may include the following:

1. Install switched or unswitched line capacitors on the customer's primary distribution system.
2. Install switched or unswitched line capacitors at various points on the secondary distribution system.
3. Install switched capacitors on motors or on other loads that operate at low power factor, and connect them in such a way that the capacitors are energized simultaneously with the load.
4. Substitute or specify synchronous motor drives on large motor loads.
5. Select motor-driven equipment and induction motors that are rated close to the requirements of the load served, rather than oversize the motors. Motor power factor is higher when operating at or near full load.
6. Insure that fluorescent ballasts are of a high power factor type.

Primary system line capacitors often afford an economical means of improving system power factor. Capacitors placed on the primary system constitute a constant reactive load, regardless of the kilowatt load served. Hence, if the kilowatt load varies significantly, it may be necessary to regulate segments of line capacitors in order to keep the system power factor within an acceptable range.

Locate Utility Meter Close to Site.   In some instances the location of a utility meter can be too distant from the distribution center on the site. Relocating the meters (arrangement to do so should be made with the utility company) to a point closer to the distribution center will reduce the line losses billed to the customer and improve the voltage regulation.

Reduction of Transformer Losses.   Transformer losses can be either no-load or load losses. No-load losses are constant for a given transformer, regardless of whether that transformer is carrying a load. Load losses constitute the I^2R losses that are caused by secondary current flow under load conditions. For any existing distribution system, opportunities for reducing transformer losses without capital expenditure are limited to deenergizing the transformers, whenever they carry no load. However, when system changes or additions are designed, the rating of any proposed transformer replacement or addition should be carefully selected, so that the system can operate most efficiently, with the lowest overall energy loss at the load levels.

Building Shell

Most of the building stock in existence today was built in the pre-embargo era when energy was cheaper than investment capital. Then, the architect's primary aim was to minimize cost per square foot, so that a minimum was spent by the contractor or owner on the structure of the building. This practice, combined with a preference for the "international style" of structure, which placed a premium on large windows and clean exteriors, resulted in buildings that, typically, have extremely poor thermal performance and are extremely sensitive to weather, sun, and wind.

The walls, roofs, and windows of these buildings can be made substantially more energy efficient, often at low capital cost to the owner. In addition, careful siting and orientation of the building can result in large savings at zero or very low cost. The following are recommendations for improving energy efficiency in the building shell.

Siting.   The physical orientation of a building can influence energy consumption significantly. If the building is rectangular, it should be sited with the long axis east–west, since east and west walls receive more direct sunlight than the south wall, and for longer periods of time. If the building requires cooling, however, wall and glass exposures should be minimized on the west and southwest. The impact of solar heat gain on the building can be reduced substantially by locating unconditioned spaces on this side. Plantings near the walls and below the window can "anchor" a layer of air near the wall and help insulate the building. The same effect can be obtained with clinging vines.

Improved Resistance to Heat Transfer.   Much can be done to limit the heating and cooling loads transmitted to the interior of a building. The thermal quality of the roof should be evaluated, for example, and additional insulation considered. Exterior insulation may be cost-effective if the roof needs repair. In some cases insulation can be applied to the underside of the roof, but the practicality

TABLE 2.1-6   TYPES OF THERMAL BUILDING INSULATION

Loose Fill Insulations	**Fibrous Wool**	Rock, Glass, Slag, Wool, Wood Fiber
	Granular	Perlite, Vermiculite, Granulate Cork
Blanket Insulations	1. Plain (No Covering) 2. Open on One Side: Vapor Barrier Paper on Other 3. Enclosed with Paper on One Side and Vapor Barrier Paper on Other 4. Reflective Vapor Barrier on One Side: Other Side Open or Enclosed with Paper	Rock, Glass or Slag Mineral Wool or Wood Fiber or Cotton
Batt Insulations	Same as Blanket Insulations	Same as Blanket Insulations
Insulation Board	Interior Boards Tile, Plank, Sheathing, Roof Insulation, Insulating Roof Deck Shingle Backer, Sound Insulation Board Acoustical Tile	Vegetable Fibers, Mineral Fibers, Plastic Foams
Slab or Block	Small Rigid Units Usually 1 in. or More Thick	1. Corkboard 2. Wood Fiber and Cement 3. Mineral Wool 4. Insulating Board (Fiberboard) 5. Perlite and Binder 6. Cellular Glass

Reflective Insulations	1. Sheets and Blankets
	  (a) Plain (No Paper Backing)
	  (b) Paper-Backed Foil in Single or Multiple Layers or Accordian Types
		Aluminum Foil Plus Other Materials

	2. Aluminum Foil Surfaced Gypsum Board or Other Materials	Aluminum Foil Plus Other Materials
	3. Foil-Surfaced Blanket and Batt Insulations	Aluminum Foil Plus Blanket and Batt Insulations
	4. Reflective Coatings Applied to Paper, Etc.	Coatings Applied to Paper in Single or Multiple Sheets
Plastic Foam Insulation	Available in Slab or Block Forms and Other Types Including Sandwich Panels	Polystyrene, Urethane and Other Types of Plastic

Source.   Enviro-Management and Research, *Promotional Information for Energy Conservation*, U.S. Department of Energy, March 1978, pp. 7–9.

of doing so depends on the accessibility of that area and on the presence of other structures, such as ductwork.

In slab-on-grade buildings significant heat loss (or gain) occurs at the perimeters of the floor. To counteract this situation, insulation can be added to the outside walls, extended down the sides of the foundation and into the ground. In cold climates particular attention must be paid to north walls.

There are many different types of insulation, both in terms of form (rigid boards, loose fill, etc.) and in terms of the materials that comprise them. Table 2.1-6 gives a comprehensive breakdown of the variety available. The reflective insulation referenced in the table refer to insulation to which reflective surfaces have been added, usually as vapor barriers. All insulation applied to floors, roofs, and walls must have vapor barriers to prevent moisture from condensing in the insulation and ruining its insulating qualities.

The optimal amount of insulation to be used depends on the weather and on the other efficiency improvements one may be considering for the particular building, and the calculation of this amount usually requires a detailed simulation of the building in its microclimate. Moreover, exterior insulation of walls, a growing practice in both the United States and Europe, is complicated by aesthetic issues and by window and door openings.

Insulation should be measured in terms of R (for resistance) value. The higher a given material's R value (shown on the package label), the more effectively it resists heat flow. Accordingly, the amount of insulation needed depends on the amount required to achieve the R value desired. The important point, however, is not a particular type of insulation's R value; rather, it is the value needed to reduce, in a cost-effective manner, the amount of energy consumed by heating and cooling.

Table 2.1-7 lists the insulating value of most of the common materials found in construction. The R value shown in the right-hand column indicates the effectiveness, or resistance value, of the material. When building sections are made of several materials, the resistance value of each of the individual materials can be added together to obtain the overall resistance value. This overall R value can be used to determine the amount of heat loss.

TABLE 2.1-7 INSULATION VALUE OF COMMON MATERIALS

Material	Thickness (in.)	R Value
Air film and spaces		
Air space, bounded by ordinary materials	$\frac{3}{4}$ or more	1.01
Air space, bounded by aluminum foil	$\frac{3}{4}$ or more	2.77
Exterior surface resistance	—	0.17
Interior surface resistance	—	0.68
Masonry		
Sand and gravel concrete block	8	1.11
	12	1.28
Lightweight concrete block	8	2.00
	12	2.27
Face brick	4	0.44
Concrete cast in place	8	0.64
Building Materials—General		
Wood sheathing or subfloor	$\frac{3}{4}$	0.94
Fiber board insulating sheathing	$\frac{3}{4}$	2.06
Plywood	$\frac{5}{8}$	0.77
	$\frac{1}{2}$	0.62
	$\frac{3}{8}$	0.47
Bevel-lapped siding	$\frac{1}{2} \times 8$	0.81
	$\frac{3}{4} \times 10$	1.05
Vertical tonge and groove board	$\frac{3}{4}$	1.00
Drop siding	$\frac{3}{4}$	0.79

TABLE 2.1-7 *(Continued)*

Material	Thickness (in.)	R Value
Asbestos board	$\frac{1}{4}$	0.06
$\frac{3}{8}$ in. gypsum lath and $\frac{3}{8}$ in. plaster	$\frac{3}{4}$	0.42
Gypsum board (sheet rock)	$\frac{3}{8}$	0.32
Interior plywood panel	$\frac{1}{4}$	0.31
Building paper	—	0.06
Vapor barrier	—	0.00
Wood shingles	—	0.87
Asphalt shingles	—	0.44
Linoleum	—	0.08
Carpet with fiber pad	—	2.08
Hardwood floor	—	0.68
Insulation materials (mineral wool, glass wool, wood wool)		
Blanket or batts	1	3.20
	$3\frac{1}{2}$	11.00
	6	19.00
Loose fill	1	3.35
Rigid insulation board (sheathing)	$\frac{3}{4}$	2.10
Windows and doors		
Single window	—	Approx. 1.00
Double window	—	Approx. 2.00
Exterior window	—	Approx. 2.00

Source. ASHRAE 1981 Fundamentals Handbook: Thermal Properties of Typical Building and Insulation Materials. Table 3-A, pp. 23.14–23.17. Printed with Permission of the American Society of Heating, Refrigerating and Air Conditioning Engineers, Atlanta, GA.

The amount of additional R value needed depends on many different factors. Generally speaking, the primary determinant is cost. But while more insulation will reduce heat flow, this principle is subject to the "law of diminishing return." In other words, beyond a certain critical point, the cost of additional insulation will not justify the increasingly smaller savings it will provide. When the correct amount of insulation is used, the savings can be substantial. Table 2.1-8 provides information on the amount of insulation needed for each R value shown.

TABLE 2.1-8 *R*-VALUE CHART

	Batts or Blankets		Loose Fill (Poured In)		
	Glass Fiber	Rock Wool	Glass Fiber	Rock Wool	Cellulosic Fiber
R-11	3½"x4"	3"	5"	4"	3"
R-13	4"	4½"	6"	4½"	3½"
R-19	6"-6½"	5¼"	8"-9"	6"-7"	5"
R-22	6½"	6"	10"	7"-8"	6"
R-26	8"	8½"	12"	9"	7"-7½"
R-30	9½"-10½"	9"	13"-14"	10"-11"	9"
R-33	11"	10"	15"	11"-12"	9"
R-38	12"-13"	10½"	17"-18"	13"-14"	10"-11"

Source. Tips for Energy Savers, U.S. Department of Energy, 1980.

Windows. From the standpoint of energy-efficient building operation, the ideal window would allow, with very low thermal loss, both the visible sunlight and the invisible infrared rays that comprise over half of the sunlight to enter the building whenever heating is needed. During the summer, when no heating is needed while providing a view of the outside, the window would admit only visible light, and only in the quantities needed for lighting a room.

A number of basic design and operational factors are critical to a window's performance. These include size of the window, its orientation and exterior shading, its ability for thermal resistance, the amount of air leakage, whether internal and external window coverings are used, and the amount of daylight to be used.

The following items should be considered for the construction and operation of more energy-efficient buildings.

1. Orient the window away from the north and away from prevailing winds in order to cut energy costs substantially.

2. Encourage the use of double-glazed and/or storm windows in moderate-to-cold climates, particularly on north walls and for large windows.

3. Encourage the use of designs that provide good natural lighting, particularly for rooms that will predominantly be used in the daytime.

4. Encourage the use of window coverings, such as draperies, blinds, shades, or shutters, that have good thermal resistance and shading potential and that complement daylight utilization.

TABLE 2.1-9 SUMMARY OF NEW FENESTRATION TECHNOLOGIES

Name of technology	Winter Radiative gain	Winter Heat loss (all types)	Summer Radiative gain	Summer Other gains	Effect on lighting/ outlook when operating	Opera- bility	Estimated cost- effectiveness	Applica- bility to retrofit	Current status
1. Double-sided blind	≅90% of un-shaded; u.v. is controlled	moderate reduction	moderate reduction	moderate reduction	no outlook; very low light	manual; seasonal reversal of blind required	very high	very high	ready for commercialization
2. Triple blind	≅90% of un-shaded; u.v. is controlled	moderate-high reduction	moderate reduction	moderate reduction	no outlook; very low light	manual, somewhat complex	high	high	on market
3. Shades between glazing with heat recovery	some loss but distribution/storage capability; u.v. is controlled	moderate reduction	moderate-high reduction	moderate reduction	some outlook; lighting pos-sible when in operation	manual	high	low	ready commercialization
4. Between glazing convection and radiation control	≅90% of un-shaded; slats can direct Sun away from furnishings	high reduction	moderate-high reduction	high reduction	no outlook or light; some degradation of outlook at all times	none	moderate-high	low	early R & D
5. Insulating shades and shutters	N/A	very high reduction	high reduction	very high reduction if closed	no outlook; no light	manual; or automatic	low-moderate	low-moderate	products marketed
6. Skylid®	N/A	high reduction	high reduction if closed	high reduction	no outlook; no light	automatic	low-moderate	low	on market
7. Beadwall®	N/A	very high reduction	high reduction if filled	very high reduction if filled	no outlook; no light	automatic	low-moderate	low	on market
8. Beam daylighting	beneficial distribution effects and u.v. control	N/A	no effect	reduced lighting load	increased daylighting; affects up-per part of window only	manual; only seasonal adjustment is essential	low-moderate	low	advanced R & D
9. Selective solar control reflective film	significantly reduced	N/A	moderate-high reduction	N/A	none	none	high in pre-dom, cooling climates	very high	R & D
10. Heat mirror	low reduction	low-moderate reduction	low reduction	some reduction	none	none	moderate-high	very high	on market; retrofit package near marketing
11. Optical shutter	low reduction	N/A	high reduction	N/A	no outlook when translucent	automatic	probably low	probably low	R & D
12. Weather panel®	low reduction unless overheating occurs	very high reduction	high reduction	very high reduction	no outlook	passive/ automatic			R & D
13. Optimization of fenestration design (size, ori-entation)	significant optimization over current practice	N/A	high reduction through proper shad-ing, etc.	reduction	N/A	N/A	very high	very low	early R & D

Source. Office of Technology Assessment, *Residential Energy Conservation*, U.S. Congress, July 1979, p. 239.

In the last several years increasing attention has been devoted to new window products that will add to the flexibility of window use and substantially improve its overall energy efficiency. Some of the salient features of these new technologies are pointed out in Table 2.1-9, which also presents the performance parameters, operating requirements, estimated cost-effectiveness, applicability to retrofit, and current status of each new technology. The estimates of cost-effectiveness are qualitative and approximate.

The exterior shading devices that have, for a long time, been used on residential buildings are beginning to appear on commercial buildings. An "eyelid," which can be placed over a south window to shade it from the strong, high sun of summer, is one such device. In winter, when the sun's angle is lower and warmth is wanted, solar energy can reach the window under the lid. Vertical fins or louvers can also be fitted to the outside of the building to admit sunlight at some angles but not at others. Louvers are particularly effective on east and west windows that must cope with low morning and evening sun angles.

Lighting

Lighting represents a major percentage of commercial energy use. Commercial buildings, however, have been notorious for their inefficient and oversized lighting systems, because the designs for these wasteful lighting systems were originally prompted by the desire to reduce building cost and produce good working conditions rather than by the need to conserve energy.

To reduce energy waste in lighting, lighting levels should first be reduced overall. A recent study by the Environmental Protection Agency predicted that in many buildings, lighting levels could be reduced to 50%, without impairing comfort. Lighting levels should then be tailored to the individual requirements of the task to be performed ("task lighting").

The General Services Administration (GSA) has recommended lighting levels for task lighting that are shown in Table 2.1-10. The GSA also recommends general lighting levels for different types of rooms (see Table 2.1-11). Task lighting may actually increase working efficiency since it produces a working environment with less glare and more contrast. As much as 15–20% of the energy currently used for lighting in commercial buildings can be saved. At constant illumination (footcandle) levels, considerable energy can be saved by installing more efficient lamps and lighting reflectors. For example, fluorescent and mercury lamps can be twice as energy efficient as incandescent lamps.

In addition, lights should be wired so that they can only be turned on over part of a floor. This reduces waste after hours, when commercial buildings are being cleaned, or on weekends, when only a few people are present. Timed switches can be installed to turn lights out automatically in infrequently used areas.

Finally, maintenance is very important for avoiding waste in lighting. Lighting reflectors and fixtures must be cleaned regularly. Lamp efficiency degenerates over the life of a lamp, so lamps must be replaced before their efficiencies fall too low.

Site Lighting. Lighting is available from a variety of sources—from incandescent to the most efficient source, high-pressure sodium vapor. Between these extremes exist other alternatives, such as

TABLE 2.1-10   GSA RECOMMENDED LIGHTING LEVELS

Task or Area	Visual Difficulty (VDF)	Design Level (FC)	Average Level Range (FC)
Service or Public Areas	-	15	12-18
Circulation Areas within Office Space, but not at Work Stations	-	30	24-36
Normal Office Work, Reading, Writing, etc.	1-39	50	40-60
Office Work, Prolonged, Visually Difficult or Critical in Nature	40-59	75	60-90
Office Work, Prolonged, Visually Difficult and Critical in Nature	60 & Up	100	80-120

Source. General Services Administration, *Energy Conservation Guidelines for Existing Office Buildings*, Superintendent of Documents, U.S. Government Printing Office, Washington, D.C., February 1977, pp. 4–14.

TABLE 2.1-11   GENERAL LIGHTING LEVELS

Areas	Design Level (FC)	Range (FC)
Auditoriums	30	20-40
Cafeteria	30	20-40
Conference Rooms	30	25-35
Corridors, Lobbies and Means of Egress	15	10-18
Kitchen (Average)	50	30-70
Mechanical Rooms (General Areas)	10	5-15
Storage Areas (General Storage)	10	5-15
Storage Areas (Fine Details Required)	30	25-35
Toilets	20	15-30

Source. General Services Administration, *Energy Conservation Guidelines for Existing Office Buildings*, U.S. Government Printing Office, Washington, D.C., February 1977, pp. 4–14.

TABLE 2.1-12   LAMP EFFICIENCY COMPARISON—STREET LIGHTING

Lamp Type	Lamp	Wattage Ballast	Total	Initial Lumens	Lumens per Watt	Approx. Hours Life
Incandescent	189	None	189	2,500	13	3,000
Quartz	250	None	250	4,850	19	2,000
Mercury	175	33	208	7,950	38	24,000
Fluorescent (very high output)	215	13	228	16,000	70	12,000
Metal Halide	175	40	215	15,000	70	7,500
High-pressure sodium vapor	150	33	183	13,000	71	12,000
Incandescent	1,000	None	1,000	19,800	20	2,500
Quartz	1,000	None	1,000	19,800	20	4,000
Mercury	1,000	100	1,100	63,000	57	24,000
Metal halide	1,000	90	1,090	100,000	92	10,000
High-pressure sodium vapor	1,000	130	1,130	140,000	124	15,000

quartz lamps, mercury vapor lamps, fluorescent lamps, and multivapor (metal halide) lamps. The relative efficiency of the available light sources may be determined from Table 2.1-12.

Mercury vapor lamps can sometimes be replaced, without replacing the ballast, by either metal halide or high-pressure sodium vapor lamps, which are more efficient light sources. This approach to energy conservation may make it possible for some fixtures to be left unlamped, without sacrificing acceptable light levels.

Heating, Ventilation, and Air Conditioning

Space conditioning requires a substantial amount of energy in the industrial and commercial sectors. As is the case with lighting, ventilation standards in the United States have had little to do with the real needs of the inhabitants. Standard practice has specified ventilation levels much higher than are really needed for most of the activities in the building. Again as with lighting, efficiency can be improved by lowering the overall level of ventilation and by adapting ventilation levels to specific tasks.

Current HVAC systems in many commercial and industrial buildings are inefficient in design and operation and are also poorly maintained. They abound with leaks, dirt, and equipment that should be replaced. There is, thus, a need for more thorough maintenance of the HVAC system and of the entire building than is now usually the case.

Energy conservation practices in maintenance and operating procedures and in capital improvements are described below:

Improved maintenance and operating procedures:

1. Optimize ventilation air only to what is really needed.
2. Reduce building exhausts, which will reduce the requirement for makeup air.
3. Shut down the ventilation during unoccupied hours (it is usually possible to conduct business in the building during the first hour in the morning and the last hour before quitting time without using any forced ventilation).
4. Reduce the heating level or raise the cooling level whenever the building is not in use.
5. Air condition only the space being used.
6. Recycle, to the maximum extent, the air needed for heating, ventilating, and air conditioning.
7. Replace the air filters at periodic intervals.
8. Periodically clean the air-conditioning refrigerant condensers in order to reduce the amount of horsepower used by the compressor.
9. Test and balance the air distribution system.

Capital improvement to conserve energy:

1. Install such heat recovery devices as a heat wheel, heat pipes, and run-around coils to transfer the heat that is wasted in the exhaust to the makeup air.
2. Install centralized controls to improve system regulations and to permit shutting down of unused sections in remote locations.

2.1-4 Management Summary

Energy management is a broader concept than energy conservation per se. In energy management one attempts to achieve equivalent levels of production with a lower expenditure of energy, an adequate energy supply, and at the lowest possible cost for completing the job.

The achievement of meaningful energy savings in existing processes is a function of management committing itself to do the job through effective personnel motivation, planning, and administration. To this end, the establishment of a formalized energy management responsibility will give the effort both focus and direction. Business managers must learn about energy use and costs, about the future availability of our energy supply, about the problems and opportunities the energy crunch is likely to present, and about alternative solutions to those problems. Local plant operators in key positions need to participate in the program, since these are the persons who know where the energy "leaks" or wastage are and who are generally in positions to act quickly if given the authority and resources they need.

In summary, we can make the following observations:

1. Business managers are aware of the energy situation and are receptive to energy conservation methods.
2. The implementation of well-established basic principles and off-the-shelf technology can show appreciable energy reductions, as much as 20–30%, in many cases.
3. Plant engineers, management, and consultants must recognize and develop, on a wider basis, the specific possibilities engendered by energy conservation efforts.
4. The present guidelines for capital expenditures on energy conservation systems must be examined since they may restrict the implementation of capital investments.
5. A positive, well-planned energy conservation program with specific manpower responsibilities can show definite results.

Our nation's industrial success shows that corporate individualism, operating under the profit motive in a free marketplace, has been a major factor contributing to that success, but industrial progress has also been a result of the manager's dedication to the substituting of inexpensive electric and thermal energy for human labor in the production of goods and services. Today, however, we can no longer assume that energy will be cheap or readily available.

Energy conservation is not the cure-all to our energy problems, but it does provide us with the opportunity to put good engineering in practice.

REFERENCES

2.1-1 U.S. Department of Housing and Urban Development, *In the Bank ... or up the Chimney?* U.S. Government Printing Office, Washington, D.C., Stock No. 023-000-00411-9 August 1977.

2.1-2 *Total Energy Management—A Practical Handbook on Energy Conservation and Management*, National Electrical Manufacturers Association, New York, n.d.

2.1-3 U.S. Department of Energy, *Energy Audit Workbook for Bakeries*, DOE/CS-0041/10, U.S. Government Printing Office, Washington, D.C., Stock No. 061-00-00165-4, September 1978.

2.1-4 U.S. Department of Energy, *Energy Audit Workbook for Die Casting Plants*, DOE/CS-0041/5, U.S. Government Printing Office, Washington, D.C., Stock No. 061-000-00167-1, September 1978.

2.1-5 U.S. Department of Energy, *Energy Audit Workbook for Office Buildings*, DOE/CS-0041/6, U.S. Government Printing Office, Washington, D.C., Stock No. 061-000-00162-0, September 1978.

2.1-6 U.S. Department of Energy, *Energy Audit Workbook for Restaurants*, DOE/CS-0041/7, U.S. Government Printing Office, Washington, D.C., Stock No. 061-000-00163-8, September 1978.

2.1-7 U.S. Department of Energy, *Energy Audit Workbook for Retail Stores*, DOE/CS-0041/11, U.S. Government Printing Office, Washington, D.C., Stock No. 061-000-00164-6, September 1978.

2.1-8 U.S. Department of Energy, *Energy Audit Workbook for Warehouses*, DOE/CS-0041/9, U.S. Government Printing Office, Washington, D.C., Stock No. 061-000-00161-1, September 1978.

2.1-9 Albert Thumann, *Handbook of Energy Audits*, Fairmount, Atlanta, 1979.

2.1-10 Oliver Pyle, *Efficient Use of Steam*, Her Majesty's Stationery Office, London, 1968.

2.1-11 J. H. Keenan, *Transactions A.S.M.E.* (195), 54, 1932.

2.1-12 John Soma, "Enter Exergy Management," *Plant Energy Management*, **6**(2), March 1982.

2.1-13 John Soma, "Rapid Estimation of Second Law Inferences After a First Law Energy Audit," *Energy Engineering*, **79**(1), Dec./Jan. 1982.

2.1-14 John Soma and Harvey N. Morris, "Exergy Management: The Seminal Synergism of Thermodynamics and Economics," *Energy Economics, Policy and Management*, **1**(4) 1982.

2.1-15 R. E. Doerr, "Six Ways to Keep Score on Energy Savings," *The Oil and Gas Journal*, May 17, 1976, pp. 130–145.

BIBLIOGRAPHY

John Soma, "Energy and Productivity," *Energy Engineering*, Vol. 80, No. 2, Feb./Mar., 1983.

Numerous energy conservation opportunities are described in the following publications: Robert Gatts, Robert Massey, and John Robertson, *Energy Conservation Program Guide for Industry and Commerce*, NBS Handbook 115, U.S. Department of Commerce, Washington, D.C., September 1974; *Site Energy Handbook*, Energy Research and Development Administration, Washington, D.C., October 1976.

2.2 EFFECTIVE HEAT UTILIZATION

Rocco Fazzolare and Stephen E. Smith

The improvement of energy utilization in industrial applications can be achieved through sound management concepts and thermodynamic principles. Approximately 26% of the energy resources utilized in the United States are for industrial process heat applications. Most of the industrial energy is used to produce process steam (16%), and the remainder is for direct process heating (10%). Another 10% of the U.S. energy requirement goes for producing electricity for industrial operations; most of this energy being used for motive operations.

The effective utilization of heat in industry can lead to significant cost reductions as well as conserving vital resources. The management of heat in industrial operations not only improves the productivity of labor and equipment but also improves the price competitiveness of the products produced.

2.2-1 The Management of Energy Flow

The management of energy flow is quite similar in approach to financial cash flow management. By means of audits, the examination of historical records, and monitoring, one can determine the energy flow profile of the industrial facility or plant. Once a profile of energy flow is established along with the details of fuel consumption by the various unit operations in the process, one then searches for technical ways and means to improve energy use and reduce waste.

Effective heat utilization is based both on minimizing the quantitative waste of energy in terms of joules or Btus, and also in maximizing the extraction of work from the fuel (thermodynamic availability). Both the first and second laws of thermodynamics should be considered to effectively utilize heat sources. An improvement in first-law efficiency can be achieved by simply reducing the rejection of waste heat as much as possible. This goal is accomplished by making improvements to the process itself and by the recovery of heat from the waste streams. Waste heat reduction and recovery is generally cost-effective at today's fuel prices.

Improvements in second-law efficiency can be accomplished by the conversion of thermal energy to work prior to or after utilizing the heat in the process operations. The work can be used to generate motive power and electricity. The addition of power conversion equipment requires an added investment of capital and operational costs. This simultaneous generation of thermal energy and power is commonly known as cogeneration. The trade-offs between costs and benefits of cogeneration are often site specific and depend greatly on fuel costs. Cogeneration will be discussed in a following section.

2.1-Waste Heat Recovery

In order to assess the possibilities for waste heat recovery, it is essential to first identify the sources of waste heat at a given site. This can be accomplished by developing an energy profile for the plant

TABLE 2.2-1   PLANT ENERGY PROFILE

Tri State Brewery 15 Park Pl. Milwaukee, Wisconsin	Fossil Fuel Use (Btu/yr) 9.05×10^{11}		Electricity Use (kW · h/yr) 36.2×10^6	
Unit Operations	Temperature (°F)	Steam (lbm/h)	Furnace (Btu/h)	Electric (kW)
Process Energy Profile				
Grinding	—	—	—	420
Cooker	250	14,000 (40% HW)	—	—
Filter	180	4,700 (HW)	—	570
Feedstuff dryer	700	—	38×10^6	160
Brewing	250	9,100	—	—
Cooler	Refrig.	—	—	2,930
Clean	160	12,000 (HW)	—	—
Pasteurize	180	34,000 (HW)	—	—
Package	—	—	—	480
Other	—	8,700	—	1,480
		82,500 l/h	38×10^6	6,040

Source	Temperature (°F)	Flow (lbm/h)
Waste Energy Profile		
Boiler stack	450	116×10^3
Dryer stack	650	200×10^3
Nonreturned condensate	180	26×10^3
Contaminated water	150	270×10^3

Source. H. Brown and B. Hamel, "The Pre-Audit-A Tool for Industrial Management," in R. A. Fazzolare and C. B. Smith, *Changing Energy Use Futures*, Pergamon, New York, 1979.

through performance of a physical audit of energy use in the facility and by examining historical utility and fuel records. The process is essentially one of deriving a plant energy balance: examining energy inputs, the distribution and flow of energy within the process, and the characterization of waste heat streams. All of the energy sources entering the plant must ultimately flow out. In general, very little of the heat leaves with the end products.

An energy profile of a hypothetical and typical brewery operation is shown in Table 2.2-1. The table displays a plant energy flow profile for the process unit operations and a waste energy profile. The temperature of each stream is indicated, as well as the estimated flow of energy. This is a composite profile based on actual data from existing brewery operations in the United States developed by Hamel & Brown from a national data base of industrial energy use. The profile illustrates the type of desired summary that should be developed for an existing operation. Once the process energy and waste energy profiles have been developed, it is possible to examine various scenarios for either recycling the waste streams into existing unit operations or to suggest alternatives to reduce energy waste and energy input to the plant. The utility of the waste stream depends both on the amount of energy flow and its temperature. In general, the higher the temperature, the more valuable the waste stream will be in a recycling application. An example of a number of fuel-saving retrofit scenarios applied to the waste streams of the brewery described in Table 2.2-1 is shown in Table 2.2-2. Anticipated cash paybacks are indicated for the various options.

Table 2.2-3 is a list of industrial waste heat sources and associated characteristic temperatures. The possible uses and applications for the waste stream are a function of the temperatures. When the exhaust temperatures are above 920°C ($\approx$ 1200°F), they are more suitable for power generation through the use of waste heat boilers or high-temperature heat exchangers. Medium temperature waste streams between 530 and 920 K ($\approx$ 500–1200°F) can be used for power generation as well as

TABLE 2.2-2   WASTE STREAM PAYBACK SUMMARY[a]

Waste Stream	End Use	Technology	Payback (yr)
Boiler stack	comb. preheat	HX	2.1
Boiler stack	hot water	HX	.9
Boiler stack	steam	HX	.8
Dryer stack	comb. preheat	HX	1.9
Dryer stack	hot water	HX	.7
Dryer stack	steam	HX	.7
NR cond.	preheat	HX	5.6
NR cond.	hot water	HX	.8
Waste water	preheat	HX	6.9
Waste water	hot water	HX	1.0
Boiler stack	elec	ORC	13.6
Dryer stack	elec	ORC	7.5
NR cond.	elec	ORC	18.8
Cond.	HW	HP	6.0
Waste	HW	HP	4.6
	HW	5% solar	15.6
	HW	10% solar	15.1
	HW	20% solar	15.1
Boiler conv.			11.1
Boiler stack	refrig	ABS	1.8
Dryer stack	refrig	ABS	1.6
NR Cond. + WW	refrig	ABS	2.5

HX = heat exchanger
ORC = organic cycle
HW = hot water
HP = heat pump
ABS = absorption cooling
NR Cond. = nonreturned condensate

[a] See Hamel and Brown in the bibliography.

TABLE 2.2-3 INDUSTRIAL WASTE HEAT: SOURCES, TEMPERATURES, COMMON USES, AND RECOVERY EQUIPMENT

	Source of Waste Heat	Temperature (°F)	Common Uses	Equipment for Recovery
High temperature	Nickel-refining furnace	2500–3000	Preheat combustion air for boilers, furnaces, ovens, gas turbines	Radiation recuperators
	Glass-melting furnace	1800–2800		Ceramic heat wheels
	Fume incinerators	1200–2600	Generate steam	Waste heat boilers
	Copper reverberatory furnace	1650–2000	Generate mechanical power	Convection recuperators
	Zinc-refining furnace	1400–2000	Generate electric power	
	Steel-heating furnace	1700–1900		
	Hydrogen plants	1200–1800		
	Solid waste incinerators	1200–1800		
	Copper-refining furnace	1400–1500		
	Aluminum-refining furnace	1200–1400		
	Cement kiln (dry process)	1150–1350		
	Open hearth furnace	1200–1300		
Medium temperature	Annealing furnace-cooling systems	800–1200	Preheat combustion air for boilers, furnaces, ovens, gas turbines	Convection recuperators
				Metallic heat wheels
	Catalytic crackers	800–1200		Ceramic heat wheels
	Heat-treating furnaces	800–1200	Generate steam	Passive regenerators
	Drying and baking ovens	450–1100	Generate mechanical power	Finned tube heat exchangers
	Reciprocating engine exhausts	600–1100	Generate electric power	Tube and shell heat exchangers
	Gas turbine exhausts	700–1000	Preheat boiler feedwater or makeup water	Waste heat boilers
	Steam boiler exhausts	450–800		Heat pipes
	Reciprocating engine exhausts (turbocharged)	450–700		

TABLE 2.2-3 *(Continued)*

	Source of Waste Heat	Temperature (°F)	Common Uses	Equipment for Recovery
Low temperature	Drying, baking, and curing ovens	200–450	Supply domestic hot water	Metallic heat wheels
	Hot processed solids	200–450	Provide hot water space heating	Hygroscopic heat wheels
	Hot processed liquids	90–450		Passive regenerators
	Process steam condensate	130–190	Preheat boiler feedwater or makeup water	Finned tube heat exchangers
	Liquid still condensers	90–190		Shell and tube heat exchangers
	Air-conditioning and refrigeration condensers	90–110		Waste heat boilers
	Cooling water from:			heat pipes
	Annealing furnaces	150–450		
	Internal combustion engines	150–250		
	Bearings	90–190		
	Welding machines	90–190		
	Injection molding machines	90–190		
	Forming dies	80–190		
	Pumps	80–190		
	Furnace doors	90–130		
	Air compressors	80–120		

Source. Derived from National Bureau of Standards, *Waste Heat Management Guidebook*, U.S. Government Printing Office, Washington, D.C., 1977.

preheating combustion air and feedwater. At the lower temperatures, from 310 to 530 K ($\approx$ 100–500°F), power generation is not generally appropriate, and the transfer of heat from one stream to another is more suitable.

Some typical uses for waste heat are:

1. Preheating combustion air for furnaces, turbines, boilers, and ovens.
2. Preheating feedwater to steam boilers.
3. Generating steam for further process heat applications or for the generation of electrical and mechanical power.
4. Preheating raw materials to be used in the process.
5. Environmental cooling and refrigeration through absorption units.
6. Space heating and domestic water heating and some special applications in agriculture and aquaculture.

2.2-3 Heat Recovery Equipment

A number of factors must be considered in the proper selection of equipment for recycling heat. One of the most important is the nature of the waste stream. Corrosive wastes will require suitable technology and materials.

A wide variety of heat recovery equipment is available on the market. The type of equipment to be selected depends on the specific application in terms of the nature of the waste stream and the characteristics of the fluid that will recover the process heat. Heat exchange equipment is very often designed for a particular recovery operation. Flue gases can present an excellent opportunity for heat recovery and often may be used to preheat air to combusters. These are known as air preheaters, or recuperators. Air preheaters depend on direct heat exchange between the exhaust gases and the combustion air. An example of an air preheater or recuperator for combustion gas heat recovery is shown in Figs. 2.2-1 and 2.2-2.

A device containing a thermal storage material that receives the heat and later transfers it to another fluid, most often air, is a regenerator. One type of regenerator is of a rotary type (also called a heat wheel) and is illustrated in Figs. 2.2-3 and 2.2-4.

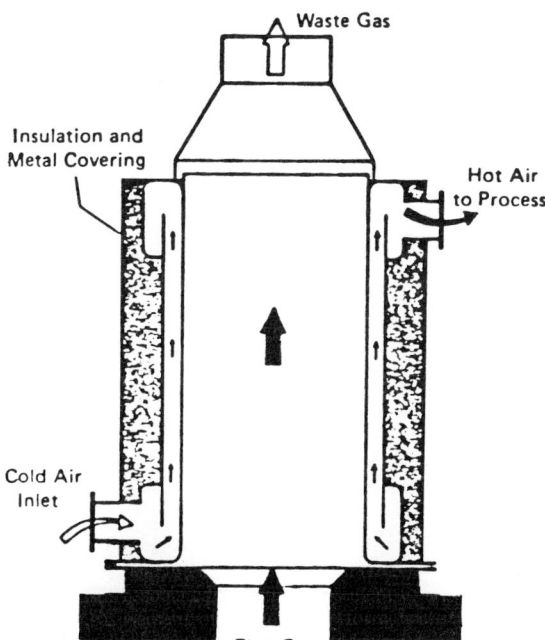

Fig. 2.2-1 Metallic radiation recuperator. (Source: National Bureau of Standards, *Waste Heat Management Guidebook*, Handbook 121, 1977.)

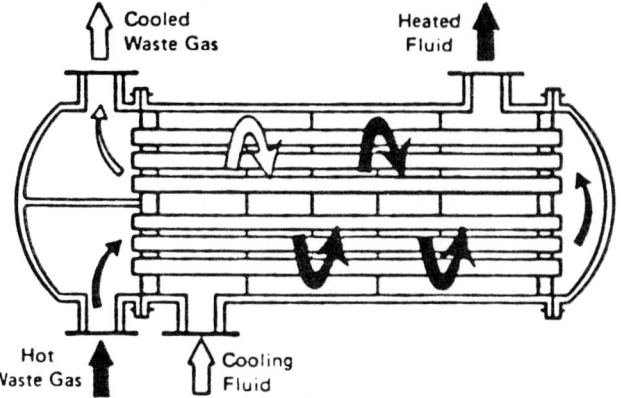

Fig. 2.2-2 Convective-type recuperator. (Source: National Bureau of Standards, *Waste Heat Management Guidebook*, Handbook 121, 1977.)

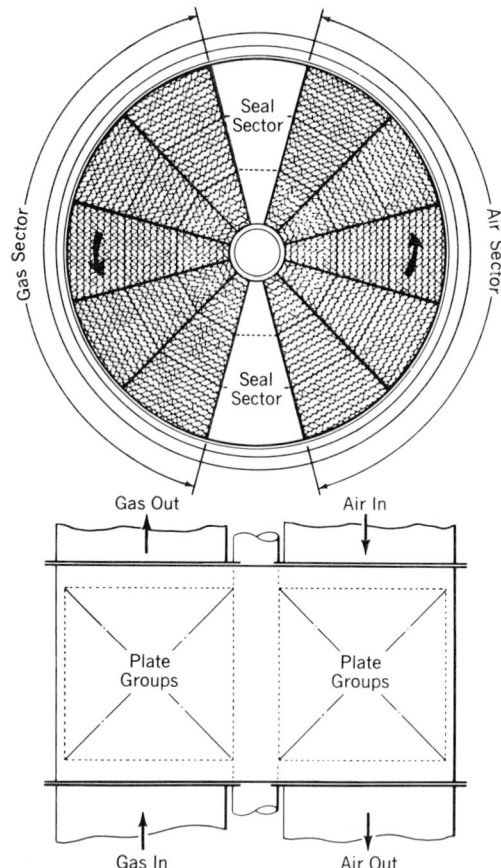

Fig. 2.2-3 Elements of a rotary regenerative air heater source. (Source: *Steam*, Babcock & Wilcox, 39th ed., 1978, Courtesy of Babcock & Wilcox.)

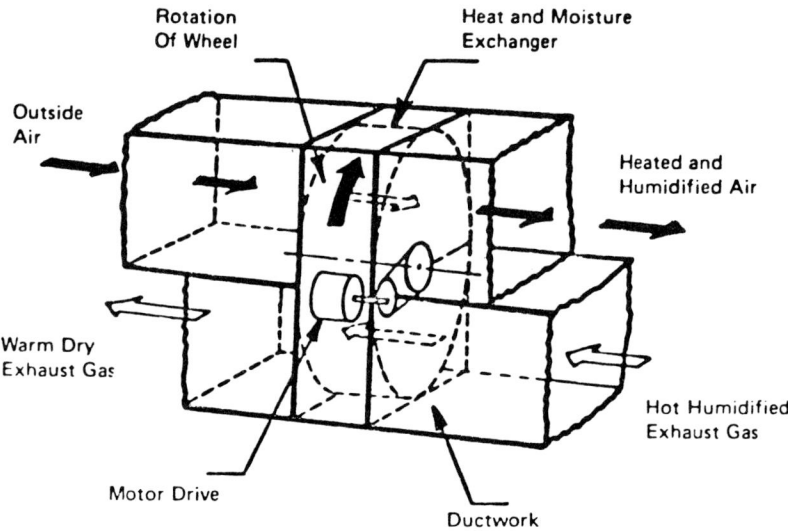

Fig. 2.2-4   Heat wheel. (Source: National Bureau of Standards, *Waste Heat Management Guidebook*, Handbook 121, 1977.)

Heat pipes represent a concept in heat exchange equipment that can be useful for polluted or corrosive streams of gases because the source and receiving streams flow in separate duct systems. The exchange of heat between the two ducts is accomplished by vaporizing a refrigerant, which then cycles back as a liquid by capillary action after the vapor has been cooled. The heat pipe system has no moving parts and requires minimal maintenance. An example of a heat pipe system is shown in Fig. 2.2-5.

Waste heat boilers are used in many installations to recover the heat of hot flue gases resulting from high-temperature processes such as roasting. The flue gas is diverted to a special boiler where steam is produced for further use as a heat source or for power generation. A typical waste heat boiler arrangement is shown in Fig. 2.2-6.

In low-temperature applications, below 400 K ($\approx$ 250°F), the recovery of heat may not be very practical in a straight heat exchange device. In such a situation it is possible to use a mechanical heat pump based on a compression refrigeration system to boost the temperature to a level suitable for water or environmental heating. A heat pump schematic diagram is shown in Fig. 2.2-7.

The operation and application characteristics of various industrial heat exchangers are shown in Table 2.2-4, which summarizes the significant attributes of the most common types of equipment available.

One method of conservation that has been receiving increased attention is a practice known as cogeneration. Cogeneration in its most general form may be defined as the simultaneous or sequential production of mechanical and useful thermal energy from a single energy source. In most cases the mechanical energy is directly utilized for the generation of electricity, and it is is to this form—the coproduction of electricity and heat—that the term *cogeneration* is most commonly applied. Cogeneration systems derive technical justification from the fact that it is more efficient to simultaneously produce electric power and thermal energy than to produce each independently. Exploitation of this fact requires the presence of a concurrent demand for both forms of energy.

Legislative and regulatory reforms have sparked a resurgence of interest in cogeneration. Potential users heretofore intimidated by the possibility of utilitylike regulation have begun to reexamine the advantages of on-site power production. Currently, there are numerous successfully operating installations located throughout the United States.

2.2-4   Cogeneration Cycles

Potential applications for cogeneration can be classified into bottoming and topping cycles. In a bottoming cycle energy leaving a process heat operation is fed to a conversion device such as a turbine to produce electricity. In a topping cycle mechanical power or electricity is produced first, and then the by-product waste heat is supplied to the process.

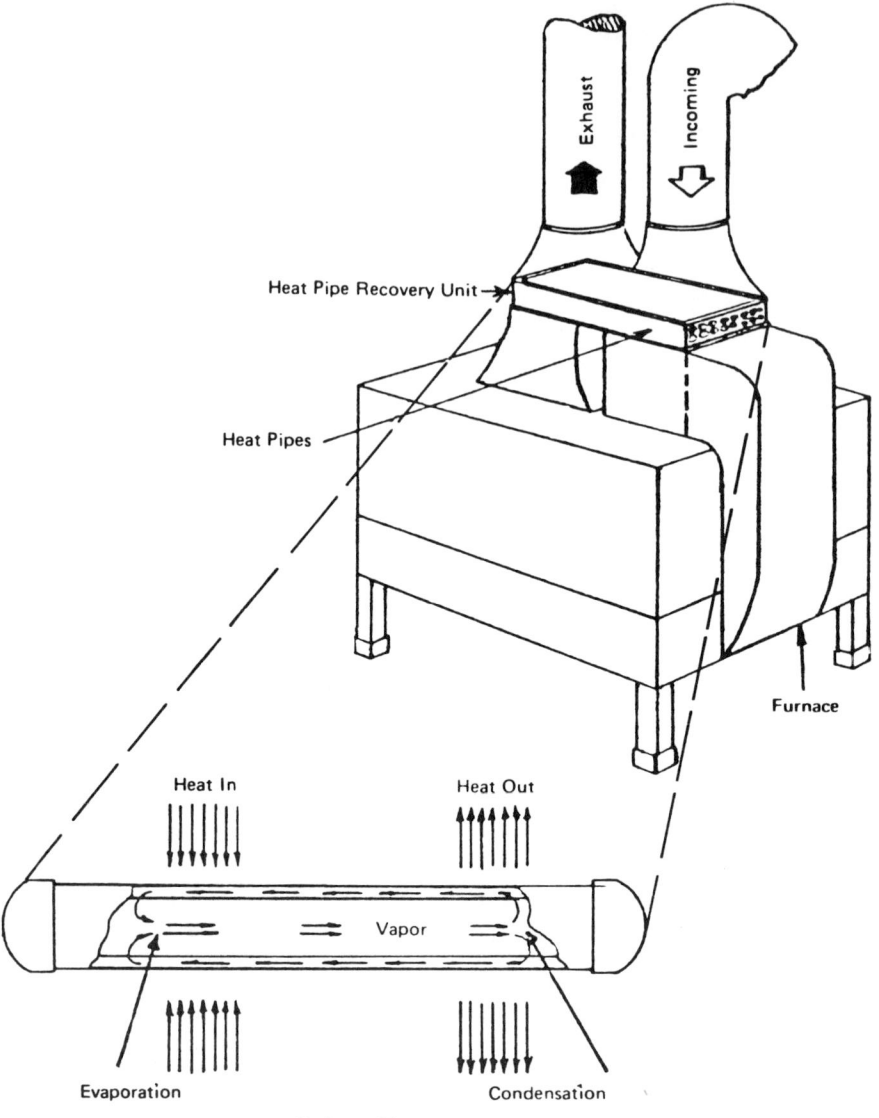

Fig. 2.2-5 An industrial furnace heat pipe installation. (Source: C. B. Smith, *Efficient Electricity Use*, p. 64, Pergamon Press. Used by permission.)

The cogeneration cycle applicable depends upon the temperature requirements of the process and the availability and temperature of a waste stream. Bottoming cycles are normally associated with high-temperature industrial process operations as found in the production of many chemicals (e.g., ethylene or ammonia) or in roasting and smelting. Such processes typically require the rejection of high-temperature, approximately 1075 K (1500°F), process streams that may be used to produce steam and power. In contrast, topping cycles may be applied in many situations requiring low-to-moderate temperature process heat. Consequently, topping cycles have a much wider range of applicability and allow greater flexibility in equipment selection. In some cases waste streams employed in bottoming cycles may be in the form of corrosive effluents that require the use of exotic and expensive heat exchangers. Alternatively, the waste stream may be derived from a prime mover such as a high-temperature gas turbine. The turbine exhaust is used to produce steam, which is then used to drive a steam turbogenerator; this is known as a combined cycle.

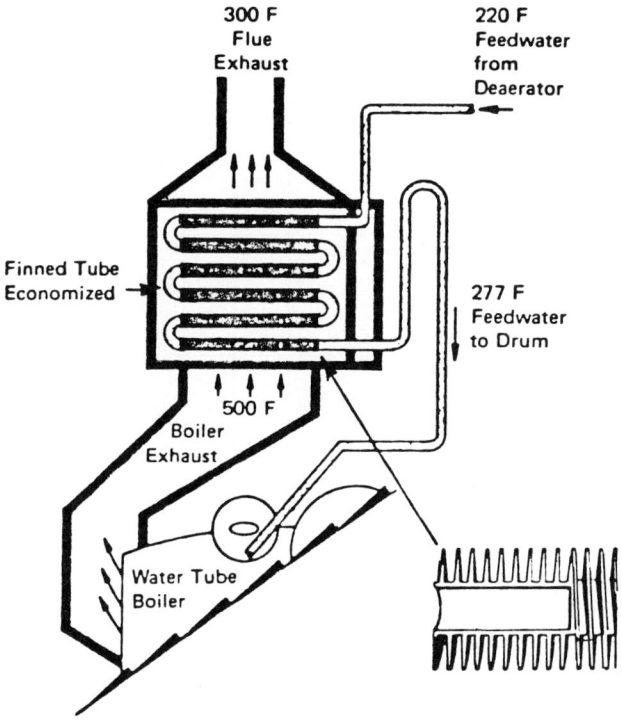

Fig. 2.2-6 Finned tube regenerator. (Source: National Bureau of Standards, *Waste Heat Management Guidebook*, Handbook 121, 1977.)

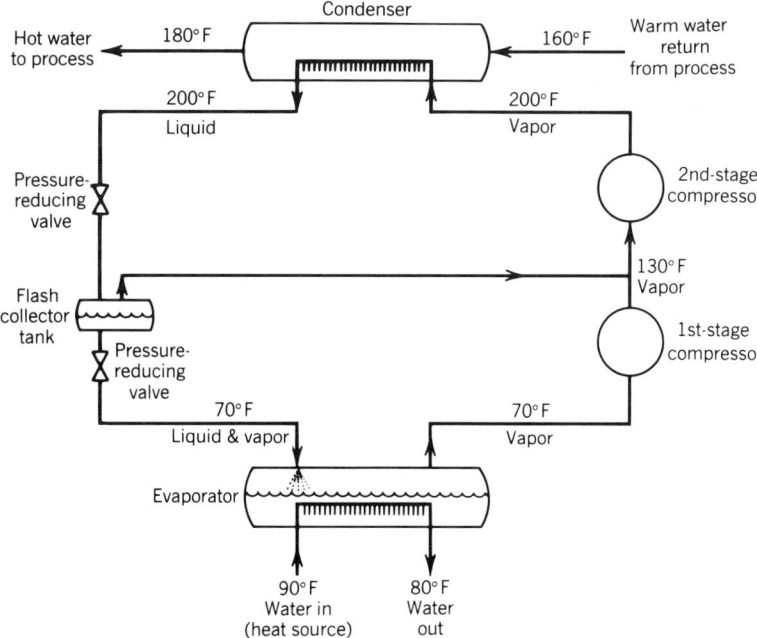

Fig. 2.2-7 Heat pump schematic diagram. (Source: National Bureau of Standards, *Waste Heat Management Guidebook*, Handbook 121, 1977.)

TABLE 2.2-4 OPERATION AND APPLICATION CHARACTERISTICS OF INDUSTRIAL HEAT EXCHANGERS

Commercial Heat Transfer Equipment	Low Temperature Subzero–250°F	Intermediate Temperature 250–1200°F	High Temperature 1200–2000°F	Recovers Moisture	Large Temperature Differentials Permitted	Packaged Units Available	Can Be Retrofit	No-Cross Contamination	Compact Size	Gas-to-Gas Heat Exchange	Gas-to-Liquid Heat Exchanger	Liquid-to-Liquid Heat Exchanger	Corrosive Gases Permitted With Special Construction
Radiation Recuperator			•		•	a	•	•		•			•
Convection Recuperator		•	•		•	•	•	•		•			•
Metallic Heat Wheel	•	•		b		•	•	c	•				•
Hygroscopic Heat Wheel	•			•		•	•	c	•				
Ceramic Heat Wheel		•	•		•	•	•	•	•				•
Passive Regenerator	•	•			•	•	•	•	•				•
Finned-Tube Heat Exchanger	•	•			•	•	•	•	•		•		d
Tube Shell-and-Tube Exchanger	•	•			•	•	•	•	•		•	•	
Waste Heat Boilers		•				•	•	•		•	•		d
Heat Pipes	•	•			e	•	•	•	•				•

a Off-the-shelf items available in small capacities only.
b Controversial subject. Some authorities claim moisture recovery. Do not advise depending on it.
c With a purge section added, cross-contamination can be limited to less than 1% by mass.
d Can be constructed of corrosion-resistant materials, but consider possible extensive damage to equipment caused by leaks or tube ruptures.
e Allowable temperatures and temperature differential limited by the phase equilibrium properties of the internal fluid.

TABLE 2.2-5   CHARACTERISTICS OF SELECTED COGENERATION TECHNOLOGIES

System	Electric Generation Efficiency[a]	Capital Cost $/kW[a]	Fuels Used
Steam turbines	.11–.20	1150	Any suitable boiler fuel
Gas turbine	.20–.33	1050	
Open cycle			Natural gas or distillates
Closed cycle			Any
Diesel	.32–.40	1400	
High speed			Natural gas or distillates
Low speed			Low-grade boiler fuel
Fuel Cell	.38	3000	Hydrogen, distillates
Stirling	.24	1150	Any
Thermionics	.17–.23	6000	Distillates and lower grades

[a] Based upon projections given in Gerluah et al., and Baumeister (see bibliography).

A wide variety of equipment configurations may be considered for a specific cogeneration application. Table 2.2-5 presents the technologies that are considered prime candidates for cogeneration today and in the near future. The table also presents comparative data on three of the more important parameters associated with each system: capital costs, fuels used, and typical electric conversion efficiency.

Selection of a specific system for a candidate site should proceed from a general review of feasibility constraints to a detailed examination of the technical, economic, regulatory, and institutional issues. The feasibility study begins with a detailed energy audit and development of an energy profile of the facility. The profile includes a review of historical process heat and electric demand, projections of future load schedules, specific characteristics of the process demands, and opportunities for energy savings through effective energy management.

The dual-purpose nature of cogeneration systems emphasizes the need for a high degree of reliability since both electricity and heat production are lost during downtime. Failure of a prime mover in a cogeneration installation may seriously impair the proper functioning of the facility as a whole. Facilities with standby utility power may face considerably higher power costs. Fortunately, experience has demonstrated that the reliability of cogeneration plants is high, exceeding 99%.[1,2] Capital costs, however, are strongly correlated to the degree of reliability required. Since cogeneration plants are normally located at facilities not primarily engaged in electric generation, selected technologies should be proven, readily available, and relatively unsophisticated in operational requirements. Three basic approaches emerge as the candidates of choice in the majority of situations: the back-pressure and extraction steam turbine, the gas turbine, and the diesel engine.

2.2-5   Back-Pressure Turbine Topping Cycle

Figure 2.2-8 presents a schematic diagram of a cogeneration topping cycle based upon a back-pressure (noncondensing) steam turbine. In this system steam enters the turbine at boiler exit conditions and exhausts at process demand conditions. The latent heat of the steam, which would normally be lost in the condenser of a condensing power cycle, is available to the process, resulting in an increase in overall energy utilization. Back-pressure turbines may be obtained in a number of configurations. These include straight noncondensing, single and double automatic extraction.

The maximum amount of power that can be generated by a back-pressure turbine is primarily dependent upon four factors: (1) the steam entry and exit enthalpy, (2) the mass flow rate, (3) the internal thermodynamic efficiency of the turbine, and (4) the mechanical efficiency. Factors 1 and 2 are process dependent, while factors 3 and 4 are a function of, among other things, the size and speed of the turbine. One drawback of this type of turbine is the relative inefficiency of small size units. In the cogeneration situation this factor is not necessarily detrimental since low efficiency simply translates to a higher ratio of thermal-to-electric power production.

The use of a back-pressure turbine in a cogeneration system implies that steam will be extracted from the turbine at the required process temperature(s) and pressure(s). Since power generation is limited by the difference between inlet and exit enthalpies, it is evident that, for a given steam mass flow rate, power generation can be increased by increasing the entering steam conditions. The results of this may be seen in Fig. 2.2-9, which displays the effect of turbine inlet and outlet conditions upon by-product power generation.

A certain amount of flexibility is available in adjusting the ratio of thermal-to-electric power production of the steam turbine cogeneration system to allow closer approach to process requirements.

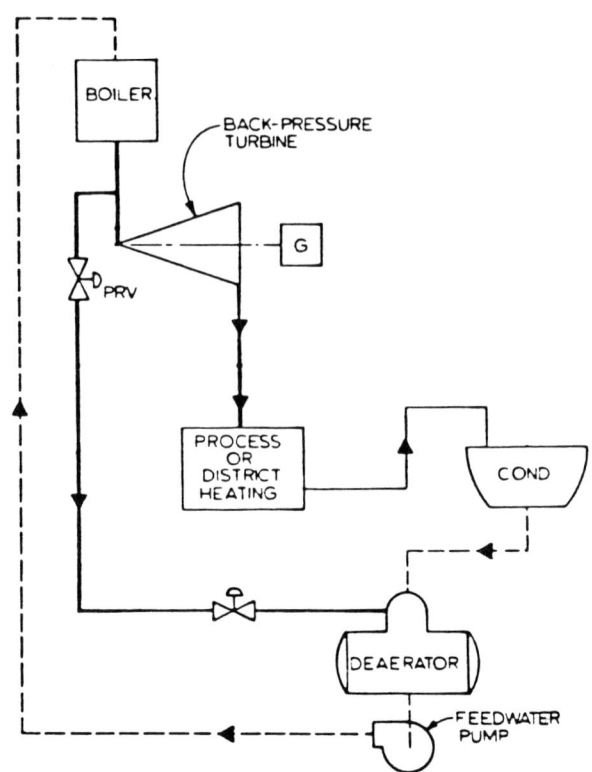

Fig. 2.2-8 Back-pressure steam turbine topping cycle for cogeneration. (Source: G. Polimeros, *Energy Cogeneration Handbook—Criteria for Central Plant Design*, Industrial Press, 1981.)

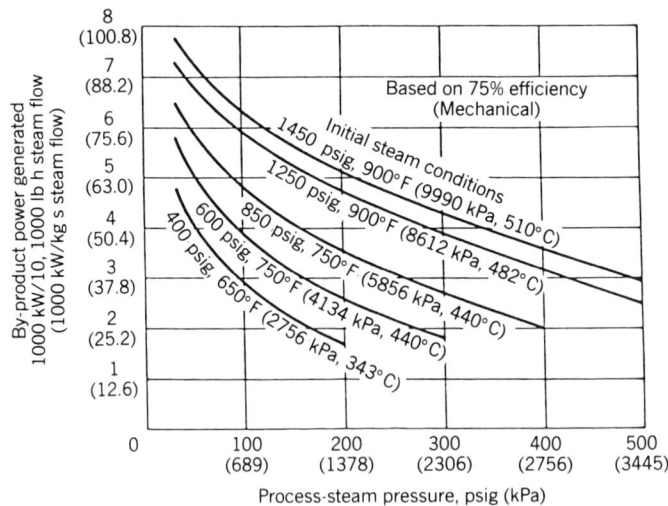

Fig. 2.2-9 By-product power generation versus initial steam conditions. (Source: G. Polimeros, *Energy Cogeneration Handbook—Criteria for Central Plant Design*, Industrial Press, 1981.)

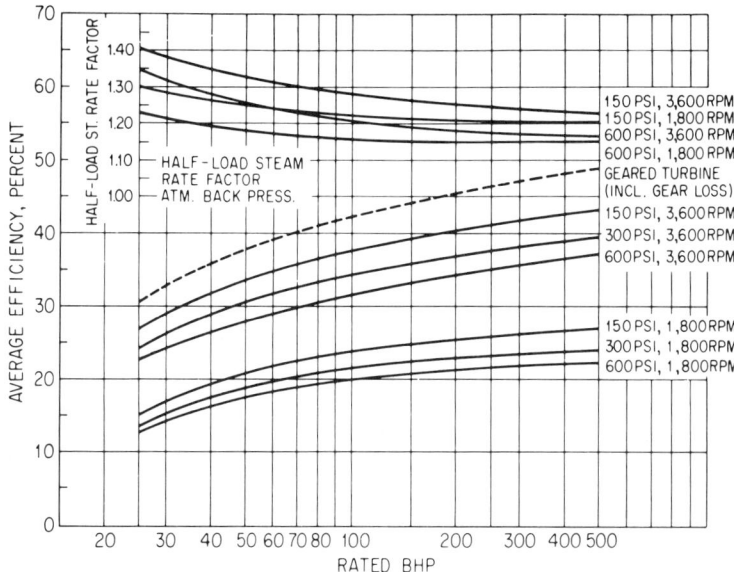

Fig. 2.2-10 Representative performance data for small single-stage steam turbines. (Source: *DeLaval Engineering Handbook*, 3rd ed. Used by permission of McGraw-Hill Book Co.)

The raising of the inlet steam conditions will be accomplished at the expense of increased boiler costs and operational expenses. Additionally, a requirement for higher flow rates will imply a larger turbine, increasing costs, and generally lead to a requirement for higher initial steam conditions in order to reach a satisfactory economic return.

The design of a cogeneration system involving the use of either back-pressure or condensing turbines will require an examination of performance characteristics of each candidate turbine unit. In general, these data should be obtained directly from the manufacturer and are supplied in light of the application. It is, however, possible to generalize to some extent the performance characteristics of basic turbine designs.

Figure 2.2-10 displays representative efficiency data for single-stage noncondensing turbines operating at various inlet steam conditions. The graphs show the trend of average internal efficiency η_i as a function of the turbine size. Figure 2.2-11 displays similar data for the internal efficiency of multistage condensing and noncondensing turbines. The figure includes superheat and half-load steam rate correction factor curves.

In order to predict the electric production of a cogeneration system, it is necessary to predict the part-load performance of the back-pressure turbine. Part-load steam flow rates for both noncondensing and condensing extraction turbines are normally determined by use of a plot of steam flow rate in pound-mass/hour against horsepower or kilowatt shaft output. This plot, known as a Willans line, is supplied by the turbine manufacturer and depends upon the specific design. Conventionally, the steam flow rates for a turbine operating at constant rpm under full- and half-load or no-load conditions are used to plot the Willans line, which is approximately linear and is normally displayed as such. Figure 2.2-12 is a typical Willans line plot for a condensing automatic extraction turbine. The figure indicates the relationship between the throttle flow and the output of the turbine, both in percent. While the plot shown is for a condensing turbine, the general relationship also holds for a back-pressure unit.

2.2-6 Extraction Turbine Topping Cycle

Figure 2.2-13 presents a schematic diagram of a cogeneration topping cycle based upon a single extraction condensing turbine. This is the most frequently used type of extraction turbine and for design purposes may be treated as a back-pressure and a condensing turbine operating in series.

Referring to the figure, steam exits the boiler and enters the high-pressure stage of the turbine. An amount of steam equal to the instantaneous process load is extracted and flows to the process demand. If the boiler flow is greater than the process demand flow, the balance of the steam flow proceeds through the second stage of the turbine where it exits at condenser inlet conditions.

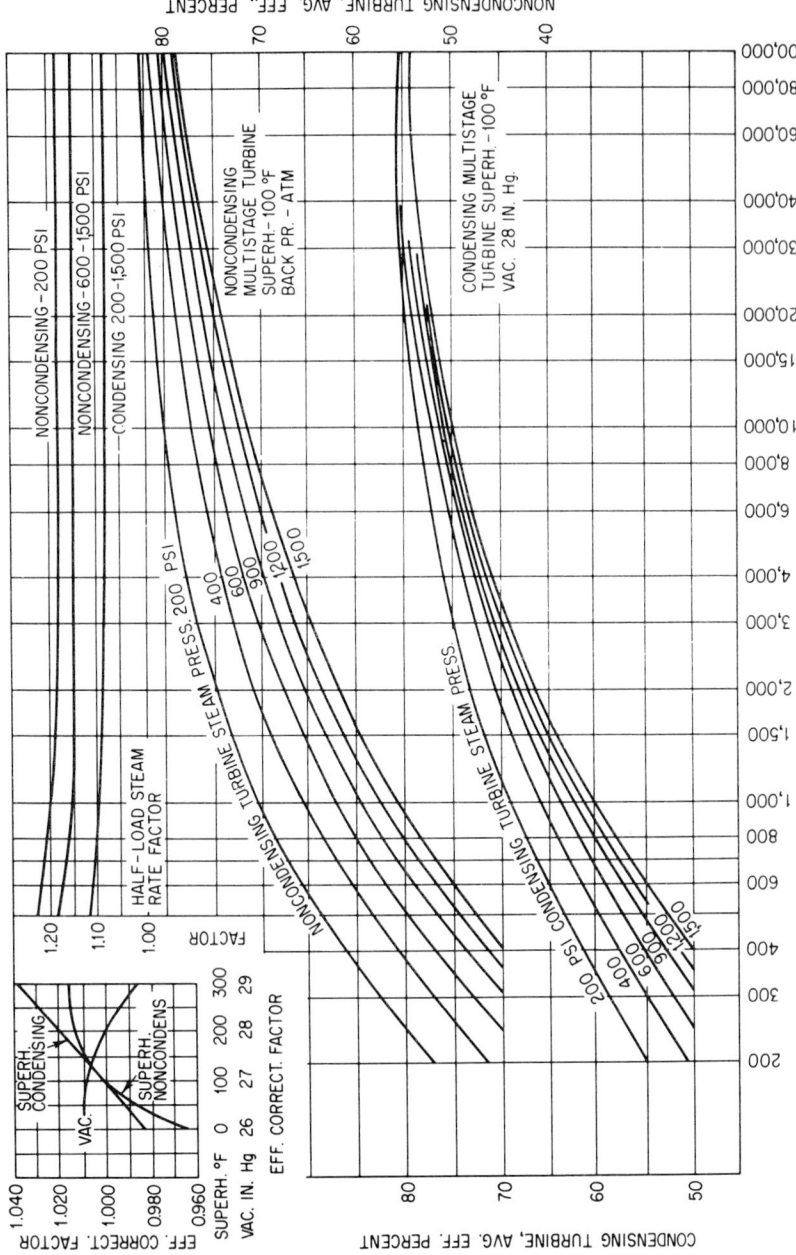

Fig. 2.2-11 Representative performance data for condensing and noncondensing multistage steam turbines. (Source: *DeLaval Engineering Handbook*, 3rd ed. Used by permission of McGraw Hill Book Co.)

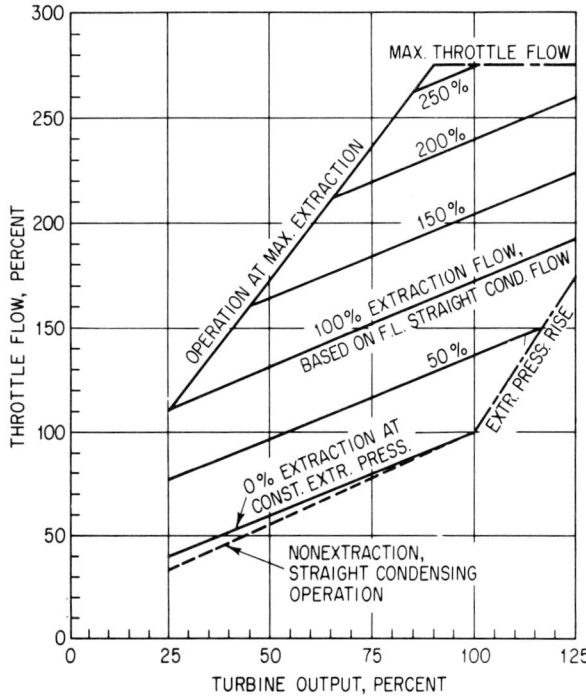

Fig. 2.2-12 Throttle flow versus output for condensing automatic extraction turbine. (Source: *DeLaval Engineering Handbook*, Gartman, 1970. Used by permission of McGraw-Hill Book Co.)

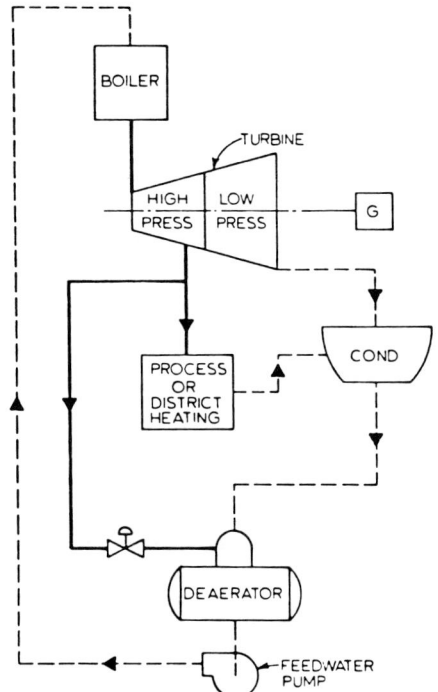

Fig. 2.2-13 Single-extraction steam turbine topping cycle for cogeneration. (Source: G. Polimeros, *Energy Cogeneration Handbook—Criteria for Central Plant Design*, Industrial Press, 1981.)

The utility of an extraction condensing turbine lies with its ability to meet a relatively wide range of power-to-steam demand ratios. When the demand for process steam is high, the unit operates at 100% extraction and is essentially a back-pressure turbine. When process demand is low, the excess steam may be diverted to the condensing section where additional electric power is generated. Without the condensing section, excess steam would be vented as waste or the boiler would be turned down with an associated efficiency loss.

Since this turbine configuration is essentially two turbines in tandem, each section will have individual performance characteristics dependent upon the steam flow. The back-pressure section will have the highest efficiency at 100% extraction and maximum load. The condensing section will be most efficient at zero extraction and high load. The overall turbine performance will result from a composite efficiency. In some cases efficiency can be maximized by the installation of two turbines—one back pressure and one condensing. When the condensing turbine is not needed, it may be shut down (along with its auxiliaries) resulting in energy savings. In contrast, the two-stage turbine will always incur performance losses from the condensing section, even when operating at 100% extraction. Balancing the improved efficiency of the two-turbine system is the necessity for warm up when bringing the condensing turbine on-line and the additional capital expenses over the single-turbine configuration.

The electric production of an extraction turbine depends upon the relative flow of steam through the two turbine stages, the steam conditions at the inlet and extraction points, and the efficiency of the two stages.

It should be noted that, for the purposes of cogeneration, only that power production associated with steam flowing through the back-pressure section of the turbine is cogenerated. Power production by the condensing section is considered self-generation. If, for example, the turbine was operating at zero extraction flow, no steam would be flowing to the process and by definition cogeneration would not exist. Proper account of this fact must be taken in the computation of by-product power costs. Since a significant portion of the energy passing through the condensing section of the turbine will ultimately be lost in the condenser, operation below 100% extraction will significantly degrade the overall energy effectiveness of the system.

2.2-7   Combustion Turbine Cogeneration

Turbine Cycles

In a simple gas turbine fuel enters a combustion chamber where it is burned in the presence of compressed inlet air. The hot combustion products are allowed to expand through a turbine section, producing shaft work. The potential for cogeneration resides in the exhaust gases from the turbine, which possess a considerable amount of thermal energy. The relatively low thermal efficiency of simple open-cycle gas turbines, combined with the large amount of excess air required to ensure acceptable turbine inlet temperature, leads to a large potential for heat recovery from the exhaust gases.

Gas turbines are available in open- and closed-cycle arrangements. Both are suitable for cogeneration. The primary advantage of a closed cycle is that the turbine components are essentially isolated from the combustion products, minimizing corrosion and wear. A much wider variety of fuels can thus be burned in the closed cycle. Closed-cycle combustion turbines have been operated on fuels as diverse as coal, blast furnace gas, and wood chips. While the fuel flexibility of the closed cycle offers a number of advantages, the increased complexity of the system along with the necessity of environmental controls for lower quality fuels results in higher capital costs.

Characteristics of the Open Cycle

The open-cycle gas turbine is a direct outgrowth of extensive research and development performed mainly for the military. It is compact and embodies a number of characteristics that make it ideally suited for use in cogeneration. A number of modifications on the open cycle can be added to improve the overall efficiency of the machine by up to 10%. These include regeneration, intercooling, and reheat. In spite of this, the majority of gas turbine systems manufactured in the United States do not employ these options.

Figure 2.2-14 displays three gas turbine configurations commonly found in cogeneration applications. All three employ a turbogenerator to produce electricity, differing only in the utilization of the exhaust heat. In the first case exhaust gases are used directly as process heat for a drying application. In the second and third cases the exhaust is passed through a waste heat recovery boiler, generating steam for process or use in an absorption chiller. The latter is most applicable to commercial buildings where both electricity and comfort space conditioning is required. The turbine is not restricted to electric generation. The mechanical output may be used instead to drive pumps, centrifugal chillers, or any other drive; however, these applications are less common.

A major problem with natural gas use can arise when the gas supply available is at low delivery pressures. Since the gas must be supplied to the turbine at a pressure above that of the combustor

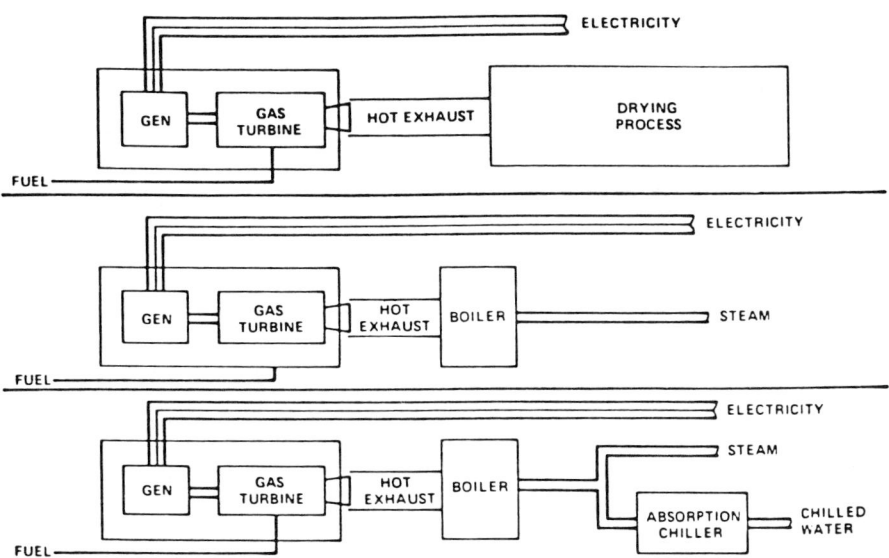

Fig. 2.2-14 Example application configurations for gas turbine cogeneration. (Source: R. Mackay, *Gas Turbines and Cogeneration*, 1979. Courtesy of the Garrett Corporation.)

($\approx$ 1.2 MPa), a considerable amount of compression power may be required for fuel delivery. Additionally, if by utility regulation the gas supplies are interruptible, the user may have to install equipment capable of burning backup fuels such as propane or diesel.

Estimating Turbine Performance

Estimation of the performance of a gas turbine in any particular cogeneration application should always be made on the basis of data supplied by the turbine manufacturer. Manufacturer's specifications generally describe turbine performance at 101.4 kPa and 15°C with zero inlet and exhaust duct losses. Since performance is degraded at higher altitudes and ambient temperatures, it is necessary to employ altitude and temperature performance correction factors applicable to the proposed site environment. Figure 2.2-15 is a typical manufacturer's performance specification plot for a gas turbogenerator set. This type of plot displays rated output, exhaust mass flow, exhaust temperature, and fuel flow as a function of turbine inlet air temperature. The plot can also be used to determine part-load performance. In the absence of manufacturer's specifications the potential user can estimate the performance of a turbine using the approximation procedure outlined in Table 2.2-6 and the average characteristics presented in Table 2.2-7. Steam production may be projected using Q_a as heat input and selected process steam conditions. Note that no correction has been made for changes in inlet air temperature. Power output decreases as the ambient air temperature increases and vice versa. Increased power output can be obtained by precooling turbine inlet air. An example of the effect of inlet air temperature on power output can be seen in Fig. 2.2-16.

Partial-load operation of the gas turbine impacts the cogeneration plant in two ways. First, as the turbine loading decreases, the thermal efficiency drops, resulting in an increase in the fuel-to-electric ratio. Each kilowatt-hour becomes progressively more fuel expensive. Second, as load decreases, exhaust temperature and mass flow decrease, lowering the heat available to process. This latter effect is partially offset by the increase in heat rejection due to the dropping thermal efficiency. The overall effect of operation at partial load can be a significant increase in generated electricity costs.

How well the performance of the turbine matches the characteristics of the process thermal and electric demand will strongly influence the economic viability of the cogeneration installation. In general, applications take one of two approaches: load following or block loading.

For the load-following situation the turbine may track either the steam or electric loads. In the first case, when electric demand exceeds production, the difference is purchased from the utility or load shaving is performed. When electric production exceeds demand, the excess may be sold to the utility or in special cases (very high thermal/electric demand ratio) the excess may be dumped to electric boilers for additional process heat production. In the second case the turbine can be made to follow electric demand. In this instance excess process heat may be dumped to the atmosphere and any deficit supplied by an auxiliary boiler.

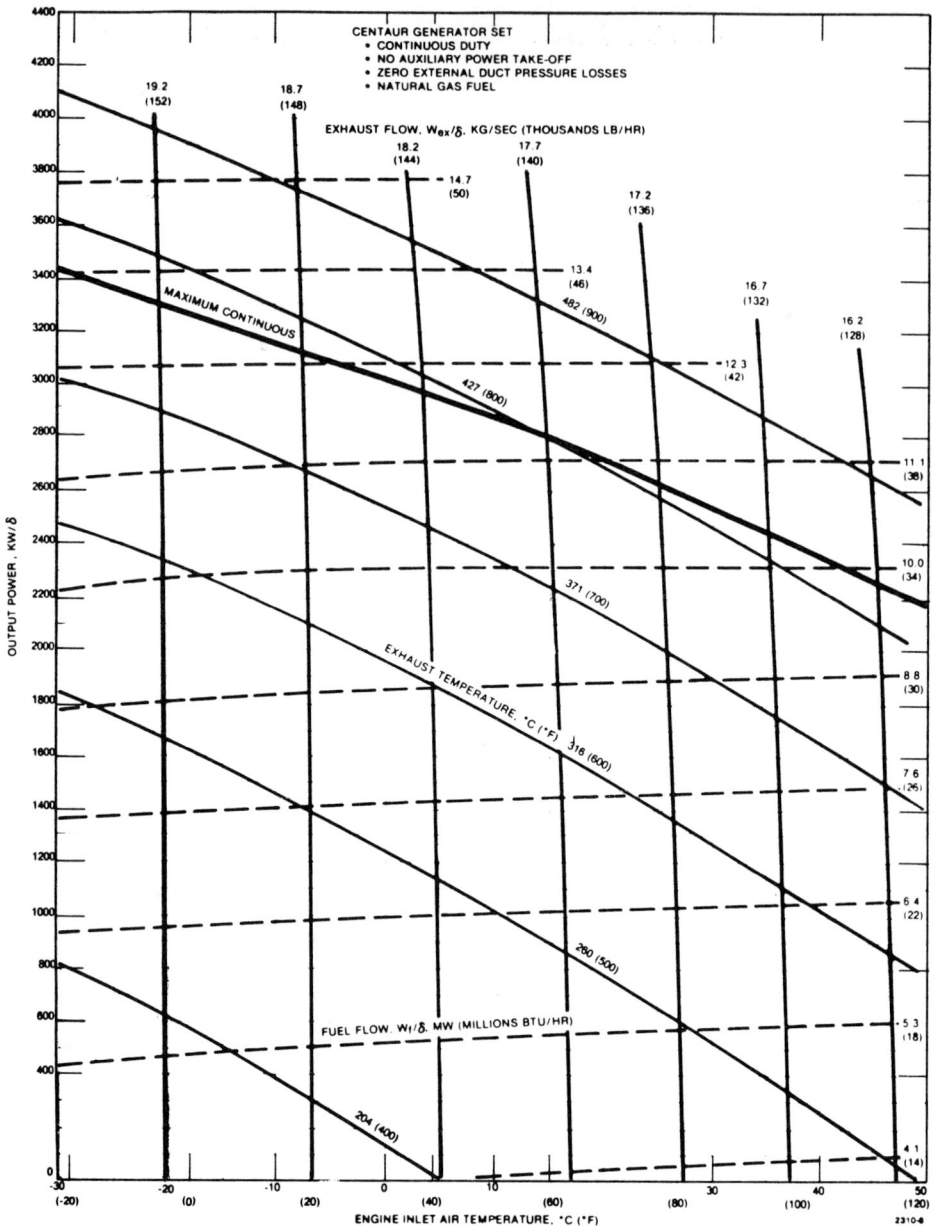

Fig. 2.2-15 Typical output power available, gas fuel. (Source: *Solar Centaur Gas Turbine Generator Set GS 4000 Centaur Continuous Duty Generator Set Performance*, International Harvester Group, San Diego, 1977.)

The load-following mode suffers from two major drawbacks. First, tracking either electric or steam demand requires additional equipment, which increases the system capital cost. Additionally, the overall system performance becomes more difficult to predict as the ratio of thermal-to-electric load varies. The second drawback occurs as a result of the effect of load modulation on equipment operation. Load following may increase operation and maintenance and decrease the turbine lifetime significantly.

The block-loading scheme is an attempt to mitigate some of the problems associated with load modulation. In block loading the turbine is run continually at full output. When the exhaust heat

TABLE 2.2-6 GAS TURBINE PERFORMANCE APPROXIMATION PROCEDURE FOR USE WHEN MANUFACTURER'S DATA IS NOT AVAILABLE

1. Select continuous shaft power (P_{max}) rating of turbine at 101.4 kPa and 15°C (14.7 psia, 60°F) from Table 2.2-7.

2. Compute altitude derating factor d_a from

$$d_a = \frac{\text{ambient pressure (kPa)(psia)}}{101.4 \text{ kPa } (14.7 \text{ psia})}$$

3. Compute altitude losses from

$$L_a = (1.0 - d_a) P_{max}$$

4. Estimate inlet and exhaust duct losses (L_d) from

$$L_d = (0.02)(P_{max})$$

5. Compute minimum available shaft power (P_s) froM

$$P_s = P_{max} - L_a - L_d$$

6. Compute electric power generated (P_e)

$$P = (\eta_{gen})(P_s); \text{ where } \eta_{gen} = \text{generator efficiency} \cong 0.95$$

7. Estimate fuel consumption of turbine (F) from

$$F(\text{J/h}) = \frac{(d_a)(P_s)(3.6 \times 10^6 \text{ J/kW} \cdot \text{h})}{\eta_{Sh}}$$

where η_{Sh} = turbine shaft efficiency; estimate from Table 2.2-7
[Fuel consumption is specified on the basis of lower heating value (LHV).]

8. Estimate exhaust heat (E) produced by turbine from

$$E(\text{J/h}) = F(\text{J/h})\left[1.0 - \eta_{Sh} - l_g - l_o - l_r - l_m\right]$$

where l_g = generator losses $\cong (0.05)(\eta_{Sh})$
l_o = oil cooler losses $\cong 0.01$
l_r = radiation losses $\cong 0.02$
l_m = miscellaneous losses $\cong 0.01$

9. Estimate heat recoverable for process application Q_a from

$$Q_a = \left[\frac{T_e - T_s}{T_e - T_a}\right](\eta_b)(E)$$

where Q_a = heat to process (J/h)
T_e = turbine exhaust temperature, K (see Table 2.2-7)
T_s = minimum exhaust stack exit temperature $\cong 425$ K
T_a = ambient temperature, K
η_b = heat recovery boiler efficiency $\cong 0.98$

available exceeds the process heat requirement, a bypass damper routes the excess exhaust up the stack. When a shortfall occurs, a makeup boiler may be used to supply the deficit. For an installation with utility backup, excess electricity can be exported. For isolated installations block loading implies that the turbine is sized to meet a minimum facility electric load.

The United States Public Utility Regulatory Policies Act of 1978 (P.U.R.P.A.) establishes efficiency standards for cogeneration systems that effectively inhibit both thermal dumping and turbine modulation (load following). The act establishes that for a system using oil or natural gas (for

TABLE 2.2-7   AVERAGED CHARACTERISTICS FOR GAS TURBINE COGENERATION ANALYSIS

Shaft Output (kW)	Shaft Efficiency (η_{Sh})	Exhaust Temperature (K)	Exhaust Temperature (°F)
530	0.212	772	930
2,950	0.259	702	805
2,960	0.212	766	920
3,290	0.281	789	960
4,835	0.302	858	1085
5,720	0.320	827	1030
7,790	0.321	689	780
20,510	0.367	766	920

systems with installation begun after March 13, 1980), the useful electric energy output of the facility plus one-half of the useful thermal energy output, when averaged over any calendar year, must be greater than or equal to 42.5% of the fuel energy input. If a facility fails to meet this criteria, the user's utility is not required to meet Federal Energy Regulatory Commission guidelines concerning interconnection and power buyback.

In order to minimize the waste of exhaust heat at times of low thermal demand, two turbines may be installed. In this situation one turbine normally operates continuously at full output, and the second is brought on- and off-line according to a set of load criteria. The interturbine load, that is, heat or steam loads greater than the supply capability of the first turbine and less than the minimum load at which the second turbine comes on-line, may be met by a backup boiler or afterburner.

2.2-8   Internal Combustion Engine Cogeneration

The internal combustion engine is one of the most versatile of all the prime movers. It is readily available in a wide range of sizes from a large number of manufacturers. It is capable of utilizing a number of different types of fuels of varying quality. It is compact in size and can be started and brought into service quickly with a minimum of operator intervention. For these and other reasons the internal combustion engine, and in particular the compression ignition diesel engine, has become widely accepted as a supplier of both primary and standby mechanical power.

Diesel engines are normally categorized by fuel use, combustion cycle, and rotational speed. High-speed engines (over 720 rpm) are generally more compact, weigh less, and are lower in cost per kilowatt than low-speed sets. High-speed engines also generally require higher quality and thus more costly fuels. Low- and medium-speed engines are larger in physical size and higher in capital and installation costs. However, low-speed engines normally attain somewhat higher efficiencies, are more dependable, and operate for up to at least twice as long between overhauls. Additionally, low-speed engines are able to use fuels of considerably lower quality and thus lower cost.

During the operation of a diesel engine, the energy input in the fuel is distributed among four end points. A portion of the fuel energy is converted into shaft power. The remainder is either absorbed by the water used to cool the cylinders, by the engine lubricating oil, or exits with the hot exhaust gases. A small portion is lost as radiation. For cogeneration the energy and work capability of the exhaust gases can be recovered to produce steam. The jacket cooling water heat content can be used to produce hot water or, in some cases, low-temperature steam.

Figure 2.2-17 presents a heat balance for a typical turbocharged diesel engine. This figure shows that the net thermal efficiency of the engine remains fairly constant down to approximately 50% load, where it begins to drop rapidly. Additionally, it is seen that as loading drops the majority of additional relative heat production is absorbed in the cooling water. The exhaust heat content, as a percent of fuel input, remains fairly constant over this load range within about 5%. Also it might be noted that, over the range 50–110% of rated output, the exhaust temperature may be expected to remain constant within ±5%.[3]

The full-load efficiency of a diesel engine is dependent upon the size of the unit, the engine cycle, the type of fuel used, and whether the engine is turbocharged. In general, full-load efficiency will increase as the size of the engine increases, within the lower output power ranges. However, size effects become less significant at higher ratings ($\approx$ 1 MW). High-speed ($>$ 900 rpm) four-stroke diesel engines may be expected to attain shaft efficiencies on the order of 30–34%, while medium-speed (400–900 rpm) two- and four-stroke engines will normally be slightly more efficient (34–38%). Very large ($\approx$ 36 MW) slow-speed (100–150 rpm) diesel engines are capable of attaining thermal efficien-

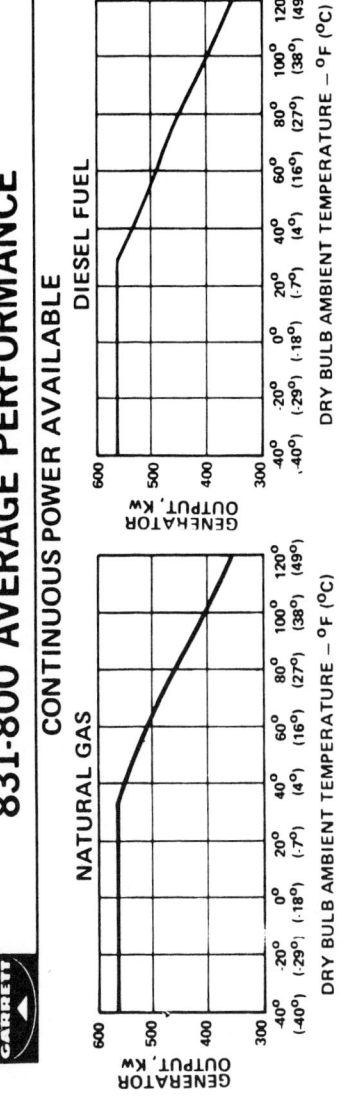

831-800 AVERAGE PERFORMANCE

CONTINUOUS POWER AVAILABLE

NATURAL GAS

DIESEL FUEL

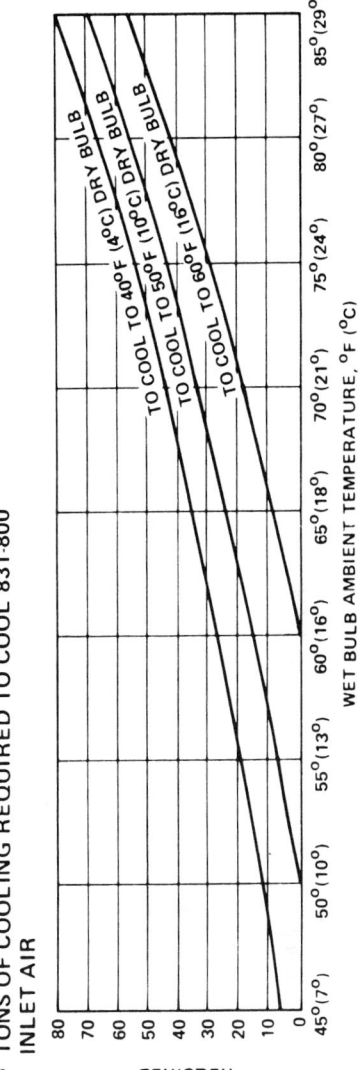

- TONS OF COOLING REQUIRED TO COOL 831-800 INLET AIR

Fig. 2.2-16 Effect of ambient temperature on representative gas turbine performance. (Source: R. Mackay, *Gas Turbines and Cogeneration*, 1979. Courtesy of the Garrett Corporation.)

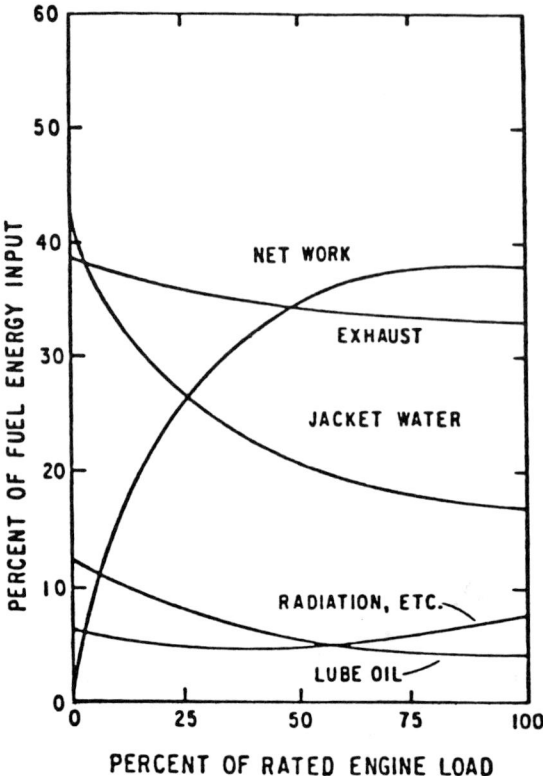

Fig. 2.2-17   Representative heat balance curve for a turbocharged compression ignition diesel engine. (Source: Charles L. Seagaser, *Internal Combustion Piston Engines*, I.C.E.S. Technology Evaluations, Argonne, IL, Argonne National Laboratory, 1977.)

cies on the order of 40%. Although these engines were originally developed for marine applications, they are also available for stationary use. Such engines are designed to use low-quality heavy-residual fuels such as Bunker-C and are employed throughout the world for power production.

The use of a diesel engine in a cogeneration application is somewhat restricted by the capability to recover waste heat in a useful form. Since the major opportunity for heat recovery is associated with the jacket cooling water and the exhaust, process applications are limited by the temperature and heat rate characteristics of these sources.

Process heat can be recovered in three forms: hot air, hot water, or steam. When hot air is required, it may be captured indirectly though the jacket water radiator and/or from a heat exchanger through which the engine exhaust flows. Where contamination is not of concern, the exhaust may be used directly. When hot water is required, heat can be recovered from both the exhaust gases and the engine jacket. Heat exchangers must be used since the primary engine coolant circuit must remain closed to minimize the possibility of coolant contamination. When steam is required at a pressure above approximately 205 kPa, it must be recovered through use of an exhaust recovery boiler. When the pressure of the demand process steam is less than 205 kPa, the heat energy rejected to the jacket cooling water can be recovered as an additional steam supply. This may be accomplished either through the use of a flash boiler or by ebullient cooling. A flash boiler is normally mounted above the engine. As the jacket water approaches the boiler, the static pressure drops and the water flashes to steam. In the ebullient system vapor formation is allowed in the engine cooling jacket, with natural circulation being used to continually remove the steam bubbles from the cooling surfaces (see Fig. 2.2-18). A steam separator is required at some point above the engine. In either case jacket water steam production is limited to approximately 205 kPa due to the high-speed engine jacket operating temperature limit of approximately 90 K. Current low-speed diesel engine generator sets operate with cooling water jacket temperatures of about 340 K, with some advanced designs capable of temperatures up to 400 K.[4]

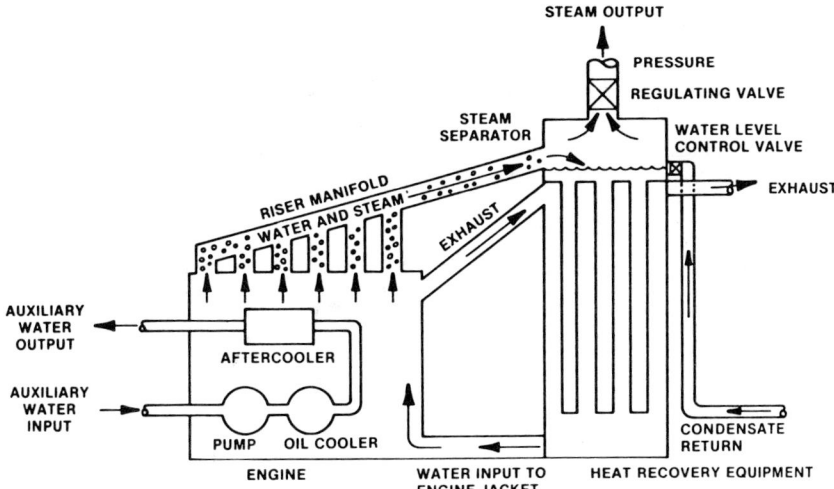

Fig. 2.2-18 Ebulient system. (Source: *On Site Power Generation Handbook*, Catepillar Engine Division.)

Examination of Fig. 2.2-17 reveals that, in situations demanding steam above 205 kPa, a significant amount of engine heat production potential will be unusable. However, it should be noted that conventional absorption air-conditioning chillers are designed to use low-pressure steam at approximately 184 kPa and represent an extensive application potential.

The steam production potential of exhaust heat is dependent upon the exhaust flow rate and its temperature. The exhaust temperature for high- and medium-speed engines using natural gas will generally be in the range of 700–1000 K, while low-speed engines will operate with exhaust temperatures between 620 and 730 K, depending upon load and fuel used. An approximation of the performance of a diesel engine for cogeneration may be made in a manner similar to that previously outlined for gas turbines.

The performance of diesel cycle engines, like gas turbines, is affected by altitude and ambient temperatures. Also, the recovery of exhaust heat is limited by a minimum allowable stack exit temperature of approximately 440 K.

REFERENCES

2.2-1 Jerome H. Rothenburg and N. Richard Friedman. "The H.U.D. Total Energy Demonstration in Jersey City, New Jersey: A Report on Operation," in *Cogeneration and Central Station Generation*. Electric Power Research Institute, Palo Alto, 1981.

2.2-2 Robin Mackay, "Gas Turbines and Cogeneration." Cogeneration and Central Station Generation, Electric Power Research Institute, Palo Alto, 1981. EPRI 2084.

2.2-3 Walter Murgatroyd, "The Diesel Engine as a Cogeneration Option." Cogeneration and Central Station Generation, Electric Power Research Institute, Palo Alto, 1981. EPRI 2084.

2.2-4 United Technologies Corporation. Power Systems Division. *Cogeneration Technology Alternatives Study (CTAS) United Technologies Corporation Final Report*. Vol. 1—Summary Report. U.S. Department of Energy, Fossil Fuel Utilization Division, Washington, D.C., 1980.

BIBLIOGRAPHY

Babcock and Wilcox Co. *Steam, Its Generation and Use*, 39th ed. New York, 1978.

Baumeister, T. (Ed.) *Mark's Standard Handbook for Mechanical Engineers*, 8th ed. McGraw-Hill, New York, 1978.

Caterpillar Engine Division. "On Site Power Generation Handbook." Pub. LEBX 9212.

Fazzolare, Rocco "Industry," in Craig B. Smith (Ed.), *Efficient Electricity Use: A Reference Book on Energy Management for Engineers, Architects, Planners and Managers*, 2nd ed. Pergamon, New York, 1978.

Gartmann, Hans (Ed.) *DeLaval Engineering Handbook*, 3rd ed. McGraw-Hill, New York, 1970.

Gerlaugh, H. E. et al. *Cogeneration Technology Alternatives Study (CTAS) General Electric Company Final Report*. Volume 1—Summary Report. U.S. Department of Energy, Washington, D.C., Fossil Fuel Utilization Division, 1980.

Hamel, B. and H. Brown "The Pre-audit—A Tool for Industrial Management," in *Changing Energy Use Futures*, R. A. Fazzolare and C. B. Smith (Eds.), Pergamon, New York, 1979.

Kreider, Kenneth and Michael McNeil (Eds.) *Waste Heat Management Guidebook*, NBS Handbook 121. U.S. Government Printing Office, Washington, D.C., January 1977.

National Bureau of Standards. *Waste Heat Management Guidebook*, U.S. Government Printing Office, Washington, D.C., 1977.

Polimeros, George *Energy Cogeneration Handbook—Criteria for Central Plant Design*. Industrial Press, New York, 1981.

Seagaser, Charles L. *Internal Combustion Piston Engines*. I.C.E.S Technology Evaluations. Argonne National Laboratory, Argonne, IL, 1977.

Solar Turbines Inc./International Harvester Group. "Solar Centaur Gas Turbine Generator Set GS-4000." San Diego, CA, 1977.

United States Congress. *Public Utility Regulatory Policies Act of 1978*. Nov. 9, 1978. P.L. 95-617, 92 Stat. 3117.

Wilson, W. B. "How to Use Steam Turbines in Refining, Petrochemical and Chemical Industries," in *Power Magazine*, McGraw-Hill, New York, February 1960.

2.3 ENERGY STORAGE TECHNOLOGY

James H. Swisher and Robert R. Reeves

2.3-1 Introduction

The need for energy storage exists if there is a mismatch between the times when energy is readily available and when it is required for use. In the past the abundance of inexpensive petroleum fuels and the ease with which they can be stored made the development of new technologies for storing energy relatively unimportant. Now, however, the need for reliable and cost-effective methods for storing nonfossil energy has become of critical importance. Energy may be stored using physical, chemical, and electrochemical methods.

Energy storage can assist in reducing the use of petroleum fuels in a number of ways. Enhanced use of solar and wind energy becomes possible because storage permits continuous output from these intermittent sources. Waste heat from industrial processes can be recovered, stored, and reused to improve overall process efficiency. Electric utilities can make greater use of nuclear and coal-fired base-load plants if storage is used instead of inefficient peaking turbines, which burn oil or natural gas. Another important possibility for using efficient base-load power is to charge batteries in electric vehicles. All of these storage options can assist in reducing oil imports and increasing energy self-sufficiency in the United States.

In principle, storage batteries could provide the needed storage capability for any energy system. However, while batteries are efficient devices and have a rapid response time, they do not benefit as much from economy of scale as some of the physical methods as, for example, compressed air energy storage. This point is illustrated in Fig. 2.3-1, where capital cost is plotted versus discharge time (a measure of storage capacity) for batteries, flywheels, and compressed air energy storage. For small systems (low discharge time), batteries are the lowest cost option, while for very large systems compressed air is the most economical approach. The data used in Fig. 2.3-1 were derived from "An Assessment of Energy Storage Systems Suitable for Use by Electric Utilities."[1] While somewhat out of date, this report is still considered a reliable source of information on energy storage options for electric utility systems. An escalation factor of 1.8 was used to convert the cost figures given at the time of the study to 1982 dollars, but no improvements in technology were considered in preparing the figure.

Since many sources of energy and uses of energy exist, the choice of the storage technology to be used to couple these together in the most efficient and cost-effective manner will be dependent on the overall system. In the selection of storage option for a thermal system, it is generally undesirable to convert from thermal energy to electricity for storage, and then subsequently convert back to thermal energy for heating applications. The same principle applies in maintaining high efficiency in other systems as well. It is generally desirable to minimize the number of conversion processes from one energy form to another when storage is used to correct the timing mismatch between energy supply and demand. Thus, to meet the variety of requirements, several storage technologies are needed because of the multiplicity of energy system requirements and the different characteristics of each technology.

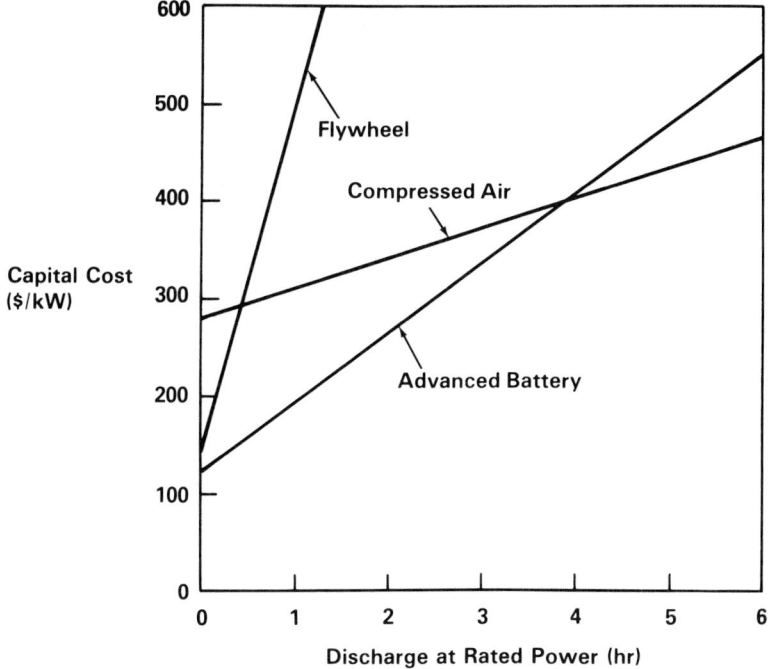

Fig. 2.3-1   Capital cost as a function of system size.

The importance of the cost of providing storage capability cannot be overemphasized. A considerable amount of energy is wasted today because it is actually more costly to provide a storage capability than to discard the energy. The wasting of flare gas in oil fields is an example. So-called low-grade waste heat, which is available only at low temperatures, is particularly difficult to store and reuse in a cost-effective manner.

Cost goals have been set for virtually all storage development projects. Cost and efficiency goals for several storage technologies and time frames are listed in Table 2.3-1. Corresponding values for state-of-the-art devices are included for comparison. Data from Table 2.3-1 will be mentioned in the following sections without specific references to the table. Some of the data are explained or qualified in the footnotes to the table.

2.3-2   Pumped Hydrostorage

The only type of nonfossil energy storage that is in widespread use today, other than batteries for emergency power, automobile starting, and so forth, is pumped hydrostorage. There are approximately 25 pumped hydroplants in commercial operation in the United States at present, and several others are under construction. In operation, energy is stored by pumping water from a river, lake, or reservoir to a second body of water at a higher elevation during the night and on weekends when the demand for electricity is relatively low. This process is reversed on weekdays when the demand for electricity is high and so-called peaking power is needed. A reversible pump turbine is the key component in the system. It takes electricity for pumping from the utility grid during the off-peak periods and then generates electricity during high-demand periods. The electricity from the grid that provides input to the storage plant is generated from efficient base-load plants, which are operated continuously at full power.[2]

These systems have a distinct advantage over the use of modified gas turbines for the generation of peaking power because peaking turbines have a thermal efficiency of less than 30%, and they use either oil or natural gas as fuel, which means they have a high operating cost. The pumped hydrosystems are more than twice as efficient. Even with a higher capital cost of $400/kW for pumped hydro versus $250/kW for the peaking turbine, the overall cost of producing electricity is lower with pumped hydrosystems.

It is unlikely, however, that there will be many more pumped hydroplants built in the future for two reasons. The most important one is the scarcity of additional sites near large bodies of water. To compound this problem, there is frequently public resistance to building such plants in areas of

TABLE 2.3-1 CAPITAL COSTS AND EFFICIENCIES FOR ENERGY STORAGE DEVICES

Technology	State-of-the-Art Technology		Technology to Be Available by 1990		Long-Range Technology Goal	
	Capital Cost ($/kW)	Efficiency (%)	Capital Cost ($/kW)	Efficiency (%)	Capital Cost ($/kW)	Efficiency (%)
Pumped hydro						
Aboveground	300–500	71–74	—	—	—	—
Underground	—	—	600[a]	70	500[b]	75
Compressed air	770[c]	50–52	600[d, e]	58	600[d, f]	65–70
Superconducting magnetic	—	—	—	—	2600[g]	80–90
Batteries	600[h]	75	500[h]	75	35–55[i]	50–60
Flywheels	—	—	1200	70	—	—
Thermal						
Residential Lo temp	200	100	150	100	—	—
Industrial[d] Hi temp	400	70	250	80–90	200	85–95
Commercial-cooling	400	85–100	400	100	—	—
Seasonal storage	—	—	200	60–70	—	—
Hydrogen						
Daily storage	40–50	68–84	45	68	—	—
Electrolytic production[j]	400–500	60–65	300–350	70–75	200–300	80–85

[a] Two stages of pump-turbine equipment.
[b] Single high-head pump turbine.
[c] For first-of-a-kind system of less than optimal size.
[d] Cost basis is system with 8-h storage capacity.
[e] Hybrid system with thermal storage of heat of compression.
[f] Adiabatic system with no fuel combustion.
[g] Cost basis is system with 11-h storage capacity.
[h] Cost basis is stationary battery with 5-h storage capacity.
[i] Cost basis is vehicle battery with 2-h storage capacity.
[j] Efficiencies for electrolysis plant only.

natural beauty, which is often the case in the vicinity of large lakes and rivers. The second problem is that demand projections for peaking power have softened to the point where an individual utility has difficulty in justifying the need for a plant with a capacity to store several thousands of megawatt-hours of electricity. This problem, however, can be solved if two or more utilities share the cost and the use of the plant. There are now examples of such joint ventures on both conventional pumped hydrosystems and planned construction of plants that use newer technology.[3]

An option for taking advantage of the desirable features of conventional pumped hydrostorage, but with a much smaller land-use requirement, is underground pumped hydrostorage.[4] Here one of the reservoirs is several hundred feet underground. Not only is there one rather than two aboveground reservoirs, but the one aboveground can be smaller because the vertical drop between reservoirs can be much larger for the same storage capacity. Thus, while there are siting limitations with underground pumped hydrosystems, they are less serious than with conventional surface systems. No underground pumped hydroplants have been built and none are presently under construction. The technology is available for such plants, if pump turbines are placed at two or more elevations beneath the surface. Development work is in progress on a high-head pump turbine, which will provide an improvement in both cost, efficiency, and availability of sites in the long term.

2.3-3 Compressed Air Storage

Compressed air energy storage (CAES) is another option for large-scale storage of off-peak electrical energy.[5,6] In this method of storing energy, a compressor is used to inject air into an underground cavern. There are three cavern types: excavated rock formations, solution-mined salt deposits, and aquifers. In the latter water is displaced in a porous underground rock formation by injection of compressed air. When the energy stored in the compressed air is converted back into electricity, it is expanded through a modified gas turbine. The expanding air in state-of-the-art systems must be heated by burning fuel to reach the normal turbine air inlet temperature.

The only CAES plant in commercial use today is located in Huntdorf, West Germany. Its power rating is 290 MW, and the storage cavern is a salt cavity. Natural gas is used as a fuel to heat the air during the power generation step. No information has been made public on the capital cost. The heat rate, which is an inverse measure of system efficiency, is 5500 Btu/kW · h or 5800 J/W · h.

Specifying a value for efficiency of a compressed air system is not a simple task because the overall cycle combines storage with fuel combustion; and the energy output of the turbine is greater than the amount of energy originally injected into the storage reservoir. For a state-of-the-art system, 1 kW · h of electricity is generated by the turbine from 0.75 kW · h of base-load electricity used for air compression. If a weighted average is taken for the storage efficiency and the fuel combustion efficiency, the overall efficiency is 50%.

In the United States designs for plants using all three types of storage caverns are available, and plans have been implemented for the construction of a plant with storage in an excavated rock cavern. The technology is basically the same as used in the Huntdorf plant. An efficiency improvement from 50 to 52% is made possible through the use of a heat recuperator, which returns some of the waste heat from the turbine exhaust to the air inlet section of the turbine. The capital cost is expected to be $770/kW for a 220 MW plant. This size is less than optimum. For a plant in the 500–1000 MW range, the capital cost could be as low as $500/kW.

The incentive for developing new technology for compressed air systems is not primarily to obtain improvements in capital cost and efficiency but to reduce or eliminate the need to use premium fuels.[7] By 1990, technology should be available to use underground thermal storage beds to make use of the heat of compression from the two low-pressure compressors. The stored heat would then be used to substitute for some of the fuel combustion that would otherwise be needed in the gas expansion step. In these so-called hybrid systems, there are possibilities for using fluidized bed coal combustion and/or solar energy during air expansion to reduce further the need for combustion of oil and natural gas. For hybrid systems the expected improvement in system efficiency is 8% (from 50 to 58%) and there is a decrease in oil or gas use of one-third.

It is possible to eliminate entirely the need for fuel combustion by recovering heat from all compressor stages and then storing this energy efficiently in thermal storage beds. The turbine in this proposed system must be redesigned so that it will operate with a lower gas inlet temperature. The capital cost of this "adiabatic" system is projected to be $600/kW, which is the same as for a hybrid system, but the system efficiency is calculated to be 65–70%, compared to 58% for a nonadiabatic hybrid system. Even if the technology for an adiabatic system were available today, it is questionable whether or not it would be the system of choice for a plant to be built within the next 10 years. The reasons for selecting the adiabatic option only become compelling with the escalation of premium fuel prices expected in the future.

2.3-4 Superconducting Magnetic Energy Storage

The potential for highly efficient electrical energy storage is particularly attractive for utilities as the cost of energy continues to increase. Superconducting magnetic energy storage (SMES) may be as high as 90% efficient, according to some estimates. The storage coil would be helical and located belowground. In order to obtain high current densities in the coil, and thereby reduce the amount of costly superconductor needed, the proposed storage system is recommended for operation at 1.85 °K.

The storage unit could be charged during off-peak hours, with the energy discharged back to the grid later to meet peaking needs. The unit would generally operate on one charging and discharging cycle per day and would be connected to a three-phase utility transmission line. The storage system requires conversion of alternating current to direct current for storage in the superconducting coil. A 1-GW · h (3.6×10^{-2} J) capacity is typical of the size storage system that has been considered.[8] The costs are projected to be high, however, and no large-scale SMES device is actually expected to be built in the foreseeable future.

A small SMES prototype system is now being built for use on transmission lines to dampen rapid voltage variations occurring with a periodicity on the order of seconds.[9] The system, 10 kW · h (30 MJ) in size, is designed to respond in fractions of a second.

If line variations are reliably damped, the effective total capacity of the transmission system increases. The value of this increase in capacity is expected to more than compensate for the cost of the storage system. The stabilization unit itself is similar to the large-scale system for daily energy storage, but operates at the normal boiling point of helium. The cost is on the order of $400/kW, but with only a few seconds of storage at the designed power rating of 10 MW. The SMES stabilization unit now under construction is expected to be installed for testing by the Bonneville Power Administration in 1983.

2.3-5 Batteries

Certainly the most familiar energy storage device is the rechargeable battery. The engineering principle, that of converting electrical into chemical energy and back again, can be demonstrated rather simply with a power source, two electrodes, and an electrolyte. Conformance to Faraday's law, which quantifies the relationship between the amounts of electric and chemical energy converted, is easily obtained with simple cells operating at low current. However, high-capacity batteries operating under high current conditions must be designed with many complexities to achieve high efficiency under normal operating conditions. Otherwise polarization at the electrodes, unwanted secondary reactions such as the generation of hydrogen, and internal cell resistance have large deleterious effects on the efficiency of the batteries.

Batteries are commercially available in a multitude of types and sizes. Two common types are the lead–acid battery (used for starting vehicles) and rechargeable nickel–cadmium batteries (used in small appliances). Present-day batteries are not suitable for most energy applications because of their weight, cost, and/or performance characteristics.

There are three categories of energy applications for which batteries are potentially attractive: electric utility load management, electric vehicles, and storage for renewable energy systems (e.g., photovoltaic and wind systems). In the first three of these,[10] batteries are preferable to pumped hydro and compressed air systems if there is a need for a smaller capacity than is optimum for a mechanical system. Other advantages are that savings can be realized in transmission system costs and in lead time for construction. Also a battery facility would be modular in construction, making it easier to match storage capacity with utility system requirements. Batteries have a higher efficiency than mechanical systems (70–80% versus 50–70%), but they do not benefit as much from economy of scale. For a large utility system it is conceivable that two or more types of storage could be used. The compressed air option would fit the needs for electricity generation over an 8–12-h period, while a battery facility would provide electricity over a 3–5-h period. For batteries used in utility systems life-cycle cost and service life are important characteristics, while weight, volume, and power density are of secondary importance.

Most of the assessments and design studies on battery storage for utility systems have focused on facilities for storage of approximately 1000 MW · h of energy at a so-called substation, which is located closer to the customers of the electricity rather than to large generating plants. Another option is to install batteries on the customer's premises. In many parts of the United States, utility companies offer lower rates for electricity used at night and on weekends. Even with state-of-the-art lead–acid batteries, the savings in electricity cost by charging batteries at night and discharging them during the day can result in a payback time on the batteries of just a few years. Where lower nighttime rates are not available to industrial and commercial customers, substantial savings can still be realized because billing for electricity is based to a large extent on peak rather than average load. Batteries and other storage methods can be used to levelize the load that is measured on the demand meter. A subway is a prime example of a system that could benefit greatly by using batteries on the customer-side of the meter. Batteries would levelize the heavy electrical loads resulting from the large number of trains operating during rush hours.

For electric utility and other stationary applications, one of the newer electrochemical technologies looks particularly promising. So-called flow batteries are being developed that have separate compartments for the anode and cathode. A pump is used to circulate the electrolyte solution through the battery. Although the volume and weight of these batteries are high, the projected cost is low compared to competing options. Zinc–chlorine and zinc–bromine[11] are examples of flow batteries.

For electric vehicle applications low initial cost, high specific energy (maximum energy storage capacity per unit weight), and high power density (maximum power per unit weight) are the most important battery characteristics. The limited number of electric cars that have been marketed to date are powered by lead–acid batteries. They are basically scaled-up golf cart batteries. They are charged with off-peak electrical power, which, combined with the high efficiency of the power train, make electric cars less expensive to operate than cars with gasoline engines. However, the initial cost is much higher, range is limited to approximately 50 mi between charges, and poor acceleration is a problem.

It is generally accepted that improvements in battery technology are required before a large market for electric vehicles can develop. In addition to the efficiency and cost characteristics listed in Table 2.3-1, improvements in specific energy, power density, and cycle life are needed. A matrix of goals for these performance characteristics in shown in Table 2.3-2. The battery types that are expected to be commercially available by 1990 are an improved lead–acid battery, a nickel–iron battery, and perhaps a nickel–zinc battery. Another battery that shows promise for the 1990–1995 time frame is a high-temperature sodium–sulfur battery,[12] which is also a leading candidate for electric utility applications (see Fig. 2.3-2).

In the long term the technology goal in the vehicle area is to provide an electric drive train that will have comparable range and performance to today's general-purpose automobile. It is not at all clear that *any* of the electrically rechargeable batteries under development will be able to reach this goal. It

TABLE 2.3-2 PERFORMANCE CHARACTERISTICS OF ELECTRIC VEHICLE BATTERIES

	State-of-the-Art Technology	Technology Expected by 1990	Long-Range Technology Goal
Specific energy[a] (W · h/kg)	30	120	260
Power density[b] (W/kg)	100	150	150
Cycle life[c]	400	800	800

[a] Measure of vehicle range; 260 W · h/kg equivalent to 250 mi between charges.
[b] Measure of vehicle acceleration.
[c] Measure of service life of battery.

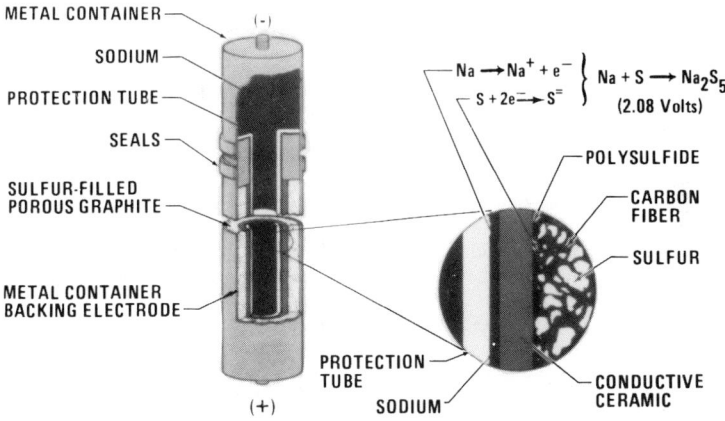

$$Na \longrightarrow Na^+ + e^-$$
$$S + 2e^- \longrightarrow S^=$$
$$Na + S \longrightarrow Na_2S_5$$
$$(2.08 \text{ Volts})$$

Fig. 2.3-2 High-temperature sodium–sulfur cell.

is more likely that a mechanically rechargeable battery will be able to provide the required range of 200–300 mi between charges. The aluminum–air battery, for example, is recharged by the periodic addition of an aluminum plate to the anode section of the cell (see Fig. 2.3-3). While this feature is attractive, the product of the cell reaction is an aluminum hydroxide precipitate, and it may be necessary to reclaim the aluminum, if the technology is to be economic.

The third application of batteries is in renewable energy systems. As with utility applications, weight, volume, and power density of the battery are of secondary importance. By far the most important need is for the development of a low-cost system. In the absence of lower-cost batteries, most photovoltaic and wind systems will have to be connected to a large electric utility system. The

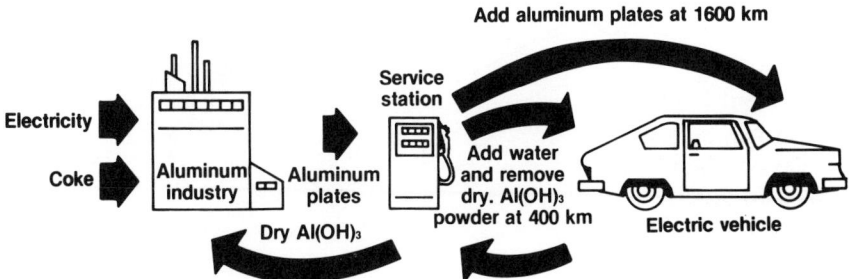

Fig. 2.3-3 Aluminum–air battery-powered vehicle.

larger system could then provide the needed storage function at another location where, in most instances, pumped hydro or compressed air storage would be used. If this storage capability is not provided at all, the photovoltaic and wind devices cannot replace more conventional equipment, such as peaking turbines, since the solar and wind energy is not always available when needed. Therefore improved batteries are needed to make "stand-alone" systems with photovoltaic and wind generating equipment cost-effective. The lack of satisfactory storage technology will seriously limit the market potential of all solar–electric and wind systems.

There are a few unique requirements for solar–electric batteries.[13] One is the ability to operate at widely varying charge-to-discharge rates. A second is the ability to stand at a partial state-of-charge for long periods of time. Improvements are being made to lead–acid batteries to meet these requirements. These needs are also being factored into the development projects on flow batteries.

2.3-6   Flywheels

In its simplest form the flywheel principle has been used for many years in such devices as potter's wheels and toy friction cars. One of the most attractive characteristics of a flywheel is the rate at which it can absorb and deliver energy. Because of this feature, it can be used effectively for providing intermittent bursts of energy to the drive train of a vehicle. One of the limitations of electric vehicles now on the market is that they lack sufficient power for rapid acceleration and hill climbing. When used in combination with batteries in vehicles, flywheels allow braking energy to be stored for subsequent use (regenerative braking). Flywheels may also be used to extend battery life by allowing the battery to operate with nearly constant power output.[14]

Flywheels may also be used in conjunction with an internal combustion engine (ICE) in a vehicle power train (see Fig. 2.3-4). The flywheel could be regeneratively charged during braking and could then smooth out acceleration demands so that the ICE is allowed to operate at a constant speed with maximum efficiency. This would be particularly attractive in fleet vehicles, which are used for urban driving where the fuel saving that can be realized through regenerative braking is significant. For example, it is estimated that at least a 50% improvement in fuel economy is possible in normal urban driving, and a fleet of taxicabs could achieve as much as a factor of 3 improvement.[15] The use of flywheels in vehicles is limited by the present state-of-the-art in transmissions and other drive-train components.

The idea of using a flywheel as the only propulsion device has merit for special-purpose vehicles. Trolley buses powered by flywheels were used in Switzerland in the 1950s. After traveling approximately 1 km, the flywheels were recharged using electricity from overhead trolley wires. This mode of operation is now being pursued further in the United States and abroad. With improved technology, there is a possibility that a flywheel costing $20,000 could be used to power a bus that normally uses a 235-hp diesel engine.[16] This cost is equivalent to a capital cost (of the flywheel component) of $120/kW.

For small photovoltaic systems flywheels are comparable to batteries in estimated cost.[17] One potential advantage of flywheels for residential use is in environmental impact. The flywheel could be housed without difficulty in an underground tank. Many feel this type of installation would be

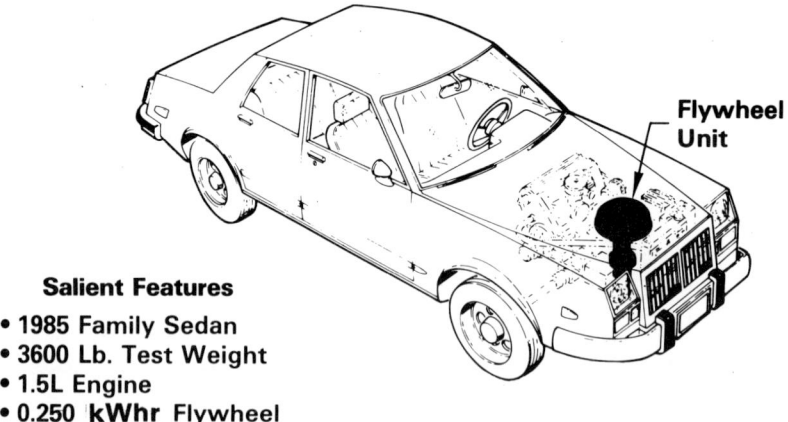

Salient Features
- 1985 Family Sedan
- 3600 Lb. Test Weight
- 1.5L Engine
- 0.250 kWhr Flywheel
- 40 MPG
- 0-60 MPH 13 Seconds

Fig. 2.3-4   Internal combustion engine vehicle with flywheel energy storage unit.

preferable to installing batteries in a separate room or small building to avoid problems with acid fumes. Neither storage system will find extensive use in the near future because of the present high cost of photovoltaic cells.

Unless major advances are made in flywheel technology, it is unlikely that flywheels can compete favorably with other storage options for electric utility load management. The needed storage capacity, even at dispersed substations, is sufficiently large, (e.g., 1000 MW · h) that batteries and compressed air systems are more cost-effective. Flywheels, however could be competitive in smaller systems on a customer's premises. This approach would be applicable if the customers were on a demand charge billing system for electricity use.

State-of-the-art flywheel systems would use rotors made from ordinary or specialty steels. The energy density of steel rotors is 15–20 W · h/kg, which is acceptable for stationary applications but is lower than desired for vehicle use. Energy densities of 80 W · h/kg appear possible with rotors made from organic-matrix composite materials. The fibers that show promise for these composites are glass, Kevlar, and graphite.

Another advantage of composite over steel rotors is in operational safety. The rotational speed of a flywheel is approximately 30,000 rpm. In the event of a failure, a thick steel housing would be needed to contain fragments of a steel rotor, while a lighter housing could be used with a composite rotor.

2.3-7 Thermal Energy Storage

Thermal energy represents the largest part of the energy supply in the United States. Until recently the cost of thermal energy has been low, and the economic driving force for conservation has been minimal. Within recent years the escalating price of fuels has stimulated conservation mainly through reduced usage. For example, major savings were readily obtained in heating and cooling of homes by adjusting thermostats and by installing insulation.

Residential Thermal Storage

Thermal energy storage can provide another means of reducing the cost of energy in a variety of applications. The most familiar form of thermal energy storage is the hot water tank, which provides a reservoir to supply hot water on demand where a relatively small heat source, such as an electric heater or oil or gas burner, is used to heat the water.

In Europe electric heating is common in homes and frequently thermal storage is included.[18] By using electrically heated thermal storage units, the utility loads can be managed more effectively since

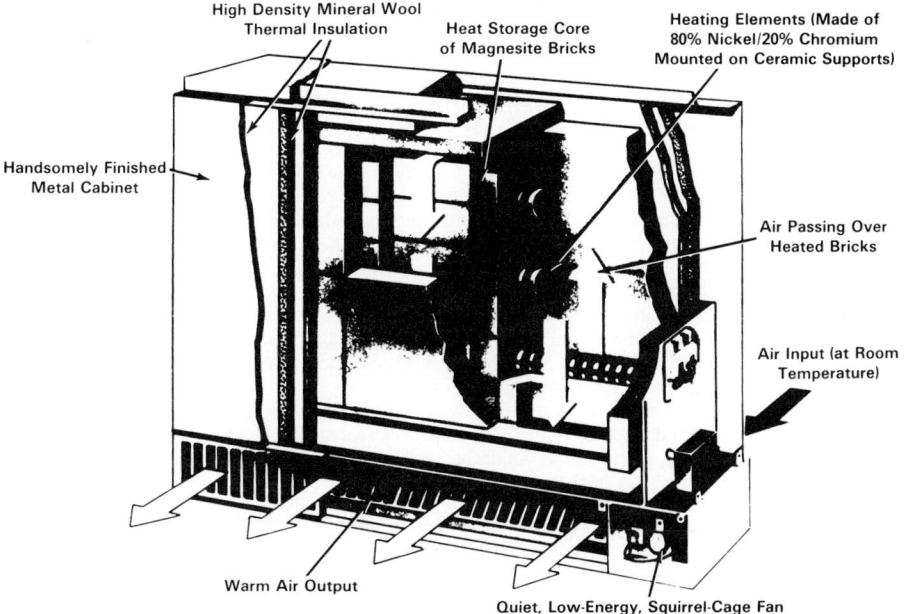

Fig. 2.3-5   Electric storage heater.

the storage units can be heated during off-peak periods (see Fig. 2.3-5). European residential rate structures provide for lower charges for use of off-peak electricity, thereby making the thermal storage units cost-effective and encouraging their widespread use. The number of units installed in some areas is sufficient to result in an almost flat electrical load during the heating season. Such units are not practical at present in most areas of the United States since the required rate structure is not available.[19,20] However, some units have been installed in Maine and Vermont to determine the feasibility of more widespread use. Cost data for residential units used daily are included in Table 2.3-1

Thermal storage can be coupled to solar energy sources for home heating and hot water supply. A variety of hot water storage systems are in use, and eventually alternative systems may be available to reduce costs by providing more compact units.[21,22]

Thermal Energy Storage for Commercial Applications

Nearly 300 commercial thermal storage installations have been identified in the United States and Canada.[23,24] Cold water and ice are used for cold storage, while hot water, sand, brick, and concrete are used for hot storage.

Cool storage is frequently a practical method for improving the cost-effectiveness of air-conditioning systems in commercial facilities. During the night the cool storage can be charged and then used during the day to meet the cooling load. About half the load can be supplied by chillers during the day, with the storage system supplying the balance of the cooling. This reduces the size of chillers required by about half.

Current cost of storage using cold water is approximately $400/kW covering a 10-h load period. However, if the size of the chillers needed is reduced by about 50%, there is a substantial saving in chiller costs, and the net cost for storage would be below $300/kW. This cost can be defrayed through savings on electrical demand charges.[25]

Ice storage requires less volume and may have an advantage in retrofit applications. Ice storage costs are similar to those for cold water, but the energy efficiency is about 85% because the chiller must now operate below the freezing point of water. This loss of efficiency can be overcome by using materials other than ice which have a solid–liquid transition at higher temperatures. Using cold water storage, the chiller system, operating during the cooler parts of the night compared to the heat of the day, may have a higher efficiency than under conventional operation and, even with some losses from storage, the energy used to meet the load remains constant and the efficiency approaches 100%.

Alternate storage methods are being developed that should reduce the capital cost of thermal storage for commercial applications. Many of these incorporate the use of salts as phase-change material (which liquify or dehydrate on heating) with higher energy densities than water and therefore require less storage volume.

Industrial–Solar Thermal Power

For industrial, utility, and solar thermal power applications several mid-to-high temperature thermal storage methods are under development.[26] Oil and rock heat storage systems are applicable for temperatures up to 330°C. This storage technology is being incorporated in the solar central receiver power plant at Barstow, California. Molten salts as heat storage materials are potentially more stable than oils. As an example, a 7-MW · h thermal storage tank has recently been successfully tested using a mixture of potassium and sodium nitrates. The system is designed for operation up to 566°C and uses an Inconel liner and insulating firebrick on the inside of the tank to reduce costs.[27]

Phase-change materials may provide high-energy density storage and thereby reduce container volume and associated costs. An approach that holds promise is the incorporation of alkali carbonate materials in a ceramic matrix.[28] Such materials may be useful over a wide temperature range (> 800°C). Since the carbonate in the ceramic matrix is structurally stable, hot gases can be passed directly over the pelletized material, eliminating the need for a heat exchanger and thereby potentially reducing the costs of the storage systems significantly.

Estimated cost and efficiency of near-term low- and high-temperature storage systems such as an oil and rock system are indicated in Table 2.3-1. Improved storage technology capable of operating to higher temperatures, using molten salts, should be available by 1990, and ultimately the use of phase change materials should permit further cost reductions.[29,30]

Seasonal Thermal Energy Storage

Storing heat during the summer and using it during the winter appears to be an ideal way to make use of waste heat energy. Accomplishing this requires very inexpensive storage as any container becomes too expensive to build when used only seasonally. Using available containers that have become obsolete or otherwise can be considered "free," is one way to consider long-term storage. Under-

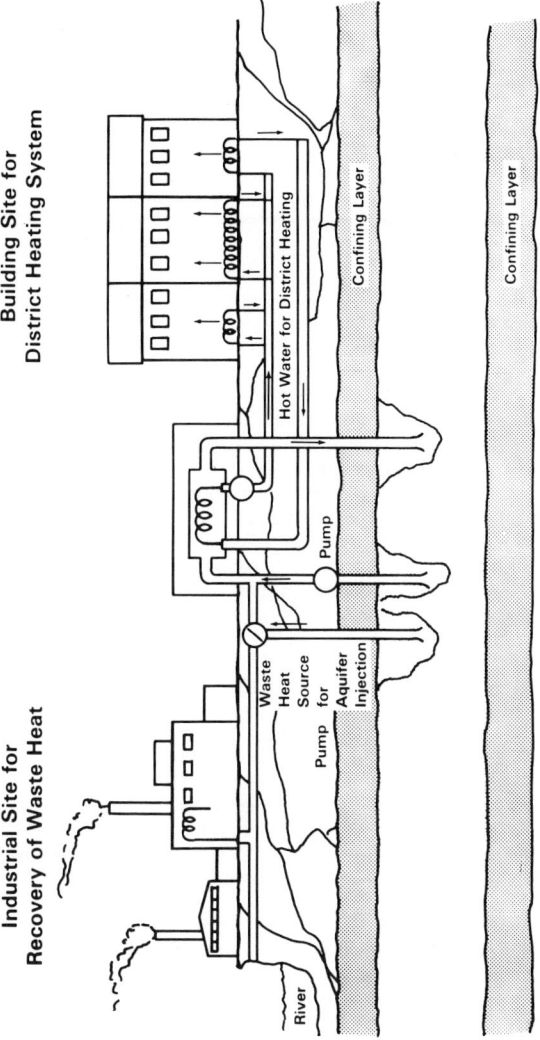

Fig. 2.3-6 Seasonal thermal energy storage.

129

ground natural formations such as aquifers may provide a good alternative, and many such formations exist across the country which may be suitable. Storage in aquifers has been under study for some years and appears to be one of the more promising candidates for both hot and cold seasonal storage (see Fig. 2.3-6). However, economic and technical viability has yet to be proven.

The delivered cost of thermal energy stored in an aquifer includes the cost of the energy placed into storage, plus capital investment, operating and maintenance, and electric energy. A base case has been considered with an energy source approximately 3 km from a user with an aquifer located at an intermediate point. Water is assumed to be drawn from the aquifer, heated at the energy source (typically from waste heat of an industrial facility), and the hot water is then injected back into the aquifer at a second site for storage. The hot water can then be withdrawn and piped as needed to the user site for heating, and the returning water reinjected into the original aquifer site to await reheating at a later time in the year. Assuming a depth of 400 ft for the injection wells, the levelized cost of supplied heat is estimated at $4–20 per million Btu, dependent on system size and the source temperature (ranging from 175 to 325°F).[31] The costs decrease with increasing system size and source temperature and increase if there is a value assigned to the waste heat. Capital costs for the 325°F source temperature have been estimated at $200/kW as noted in Table 2.3-5, with efficiency of heat storage in the aquifer of approximately 60%. At lower storage temperatures the efficiency increases.

Increasing the cost of the source energy directly affects the cost of the energy supplied to the user. For large systems on the order of 50 MW, varying costs of well drilling and source-to-user distances are not major factors in the final cost of the delivered energy. In the base case the user is assumed to be located at a single site. The cost of any distribution system will be additional and several examples were calculated.[31] If the energy is essentially waste heat and obtained at only nominal cost, the economics appear favorable. The aquifer also must be evaluated with regard to environmental aspects. These aspects have been considered in some detail.[32]

2.3-8   Hydrogen Energy Storage

Hydrogen is a fuel that can be produced using various energy sources; thus hydrogen can be considered as a chemical form of energy storage.[33] Currently, hydrogen is used mainly to produce ammonia and methanol and is consumed essentially as fast as it is produced. However, some hydrogen is produced and marketed for a variety of other applications in limited quantities. State-of-the-art storage technology is available for both compressed gas and liquid. Hydrogen stored in compressed gas cylinders at 2200 psi is in common use at a calculated cost of $50/kW with 84% efficiency. The efficiency loss is determined by the energy required to compress the gas from a reference pressure (100 psi supply was assumed) and includes the losses in thermal energy conversion to mechanical energy. Liquid hydrogen is somewhat less costly to store in large quantities, but the energy efficiency is lower because of losses in converting the gas to a liquid at 20 K. Hydrogen stored in metal hydrides may be a convenient, safe storage method, if hydrogen is used as a vehicle fuel sometime in the future (see Fig. 2.3-7). Cost projections appear high at present ($120/kW). Energies required for compressing hydrogen to charge the hydride and to dissociate the hydride for release of the hydrogen result in a relatively low overall estimated efficiency (60%). If heat from a vehicle exhaust were used to dissociate the hydride, the effective efficiency could be substantially increased.[34]

A relatively new method of storing hydrogen employing hollow glass microspheres is being investigated and may represent 1990 state-of-the-art technology for vehicle applications. Hydrogen at high pressure and elevated temperatures diffuses into the spheres. Hydrogen is retained in the spheres by cooling. (At low temperatures the hydrogen no longer can diffuse through the glass walls.) Reheating the spheres allows hydrogen to be released. The efficiency losses are mainly in the high-pressure compression process.[35]

Using hydrogen for energy storage and as an energy carrier may have application in the decades to come as fossil fuel supplies become depleted. Hydrogen can be readily produced by electrolyzing water, and the hydrogen may be used in gas pipelines to supplement or replace natural gas. Coal can be used to produce hydrogen or other gaseous and liquid fuels, but coal contains impurities such as sulfur which can cause air pollution problems. The cost of removing impurities from coal or coal-derived fuels is sufficiently high that it is a major factor in economic competitiveness. More extensive use of coal in the future may cause a harmful increase in global temperatures due to a high carbon dioxide level in the atmosphere.[36] Therefore, hydrogen produced from nonpolluting sources such as solar energy has advantages. Solar and wind sources could be useful for powering electrolyzers to produce hydrogen. The hydrogen could be stored and used as needed, permitting solar energy to meet energy requirements even at night.

Storing energy by electrolyzing water can be accomplished using present-day electrolyzer technology. Since energy has become so costly, significant R&D has been undertaken recently to increase the efficiency of electrolyzers. Capital costs and efficiencies are indicated in Table 2.3-1 for current units and those to be available in the future as a result of R&D.

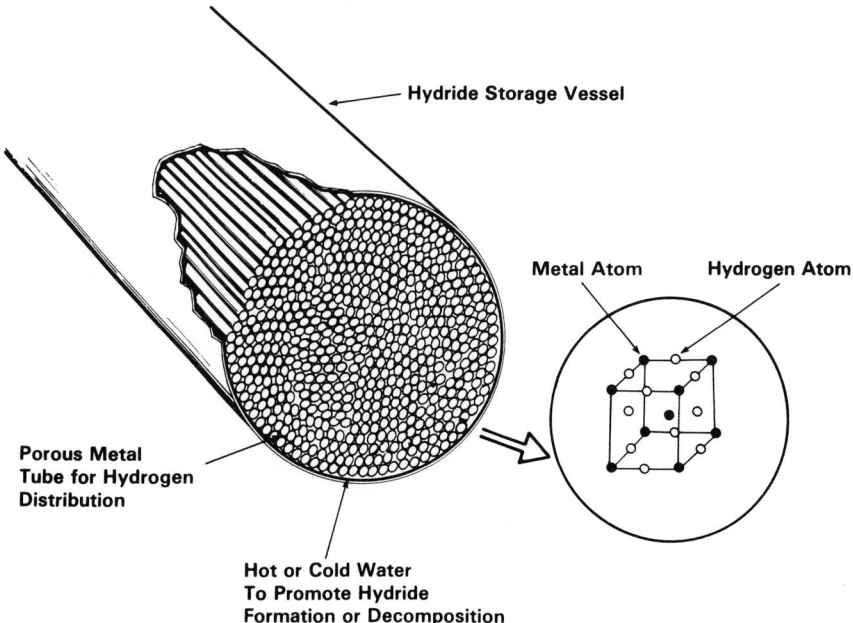

Fig. 2.3-7 Storage of hydrogen with metal hydrides.

Currently, in the United States, hydrogen is made from natural gas, which is a form of chemical energy storage found in nature. Hydrogen could also be produced by dissociating water using chemical methods that require high temperatures such as can be obtained with solar concentrators or gas-cooled nuclear reactors.

See Chapter 15.1 for a more comprehensive treatment of hydrogen production and utilization.

REFERENCES

2.3-1 "An Assessment of Energy Storage Systems Suitable for Use by Electric Utilities," Electric Power Research Institute, Palo Alto, CA, Report Number EM-264, July 1976.

2.3-2 "A Vast New Warehouse for Electricity," *Fortune*, December 1971, pp. 88–93.

2.3-3 "Survey and Analysis of Selected Jointly Owned Large-Scale Electric Utility Storage Projects," Pacific Northwest Laboratories, Richland, WA, Report Number PNL-4124, May 1982.

2.3-4 "Report on Technical Feasibility of Underground Pumped Hydroelectric Storage in a Marble Quarry Site in the Northeast United States," Pacific Northwest Laboratory, Richland, WA, Report Number PNL-4248, March 1982.

2.3-5 "CAES—It's More Than Hot Air," *Mechanical Engineering*, January 1982, pp. 20–23.

2.3-6 F. Kalhammer, "Compressed Air Energy Storage," *EPRI Journal*, Electric Power Research Institute, Palo Alto, CA, May 1981, pp. 48–50.

2.3-7 F. R. Zaloudek, "An Assessment of Second Generation Compressed Air Energy Storage Concepts," Pacific Northwest Laboratory, Richland, WA, Report Number PNL-3978, July 1982.

2.3-8 J. D. Rogers, '1-GWh Diurnal Load-leveling Superconducting Magnetic Energy Storage System Reference Design," in *Proceedings of the 1979 Mechanical and Magnetic Energy Storage Contractors' Review Meeting*, Department of Energy, Washington, D.C., Report Number Conf-790854, December 1979, pp. 28–46.

2.3-9 R. I. Schermer, "30 MJ Superconducting Magnetic Energy Storage for BPA Transmission Line Stabilizer," in *Proceedings of the Mechanical, Magnetic and Underground Energy Storage 1981 Contractors' Review*, Department of Energy, Washington, D.C., Report Number Conf-8-10833, February 1982, pp. 82–86.

2.3-10 A. R. Landgrebe and I. B. Weinstock, "Application of Electrochemical Technologies in the Utility Industry," *Modern Power Systems*, Vol. 1, No. 5, May 1981, pp. 42–47.

2.3-11 R. A. Putt, "Zinc/Bromine Batteries for Stationary Energy Storage," in *Proceedings 16th*

Intersociety Energy Conversion Engineering Conference, Atlanta, GA, August 8–14, 1981, pp. 793–797.

2.3-12 D. W. Bridges and R. W. Minck, "Evaluation of Small Sodium-Sulfur Batteries for Load Leveling," in *Proceedings 16th Intersociety Energy Conversion Engineering Conference*, Atlanta, GA, August 8–14, 1981, pp. 830–835.

2.3-13 S. Ruby, R. P. Clark, and D. L. Caskey, "Battery Technology for Solar and Wind Systems," in *Solar Storage Workshop Proceedings*, Midwest Research Institute, Kansas City, MO, Report Number CP 270-1521, 1982, pp. 46–60.

2.3-14 M. W. Schwartz, "Assessment of the Applicability of Mechanical Energy Storage Devices to Electric and Hybrid Vehicles," Lawrence Livermore Laboratory, Livermore, CA, Report Number UCRL-52773, Vol. 2, May 1979.

2.3-15 L. Forrest and L. H. Kubo, "Assessment of Flywheel System Benefits in Selected Vehicle Applications," Aerospace Corporation, El Segundo, CA, Report Number ATR-892(7842)-1, October 1981.

2.3-16 B. Noton, Private Communication, Battelle-Columbus Laboratories, June, 1982.

2.3-17 R. D. Hay, A. R. Millner, and P. O. Jarvinen, "Performance Testing and Economic Analysis of a Photovoltaic Flywheel Energy Storage and Conversion System," in *Proceedings of the 1980 Flywheel Technology Symposium*, U.S. Department of Energy, Washington, D.C., Report Number Conf-801022, 1980, pp. 259–267.

2.3-18 J. G. Asbury and A. Konvalis, "Electric Storage Heating: The Experience in England and Wales in the Federal Republic of Germany," Argonne National Laboratory, Argonne, IL, Report Number ANL/ES-54, May 1976.

2.3-19 H. N. Hersh, "Optimal Sizing of Heating Systems that Store and Use Thermal Energy," Argonne National Laboratory, Argonne, IL, Report Number ANL/SPG-18, June 1981.

2.3-20 H. N. Hersh "Field Evaluation and Assessment of Thermal Energy Storage for Residential Space Heating," in *Proceedings of the Sixth Annual Thermal and Chemical Storage Contractors' Review Meeting*, U.S. Department of Energy, Washington, D.C., Report Number Conf-810940, February 1982, pp. 161–165.

2.3-21 Roger Cole, K. J. Neild, R. R. Rohde, and R. M. Wolosewicz, "Design and Instruction Manual for Thermal Energy Storage," Argonne National Laboratory, Argonne, IL, Report Number ANL 79-15 (2nd ed.), January 1980.

2.3-22 *Proceedings of Solar Energy Storage Options*, San Antonio, Texas, U.S. Department of Energy Report Number Conf-790328, March 1979.

2.3-23 H. G. Lorsch and M. A. Baker, "Survey of Commercial Thermal Storage Installations in the United States and Canada," in *Proceedings of the Sixth Annual Thermal and Chemical Storage Contractors' Review Meeting*, U.S. Department of Energy, Washington, D.C., Report Number Conf-810940, February 1982, pp. 205–210.

2.3-24 "Thermal Energy Storage: Cooling Commercial Buildings Using Off-Peak Energy," in *Seminar Proceedings*, The Electric Power Research Institute, Palo Alto, CA, Report Number EPRI EM-2244, February 1982.

2.3-25 Some current research and development efforts are described in a series of papers in the proceedings cited in ref. 23.

2.3-26 Howard Edde, "Study of Thermal Energy Storage: System Information for Pulp and Paper Industry," Oak Ridge National Laboratory, Oak Ridge, TN, Report Number ORNL-SUB/86X-95003/81, 1981.

2.3-27 P. B. Wells and A. P. Nassopoulos, "Molten Salt Thermal Energy Storage Subsystem for Solar Thermal Central Receiver Plants," in *Proceedings of the Sixth Annual Thermal and Chemical Storage Contractors' Review Meeting*, U.S. Department of Energy, Washington, D.C., Report Number Conf-810840, February 1982, pp. 279–285.

2.3-28 T. D. Claar and R. T. Waibel, "Advanced High-Temperature Thermal Energy Storage Media for Industrial Applications," in *Proceedings of the Sixth Annual Thermal and Chemical Storage Contractors' Review Meeting*, U.S. Department of Energy, Washington, D.C., Report Number Conf-810840, February 1982, pp. 263–268.

2.3-29 R. J. Copeland, "Analyses of Thermal Energy Storage Systems," Solar Energy Research Institute, Golden, CO, Report Number SERI/TP-252-1685, August 1982.

2.3-30 R. J. Copeland, "Advanced, High Temperature Molten Salt Storage," Solar Energy Research Institute, Golden, CO, Report Number SERI/TP-252-1684, August 1982.

2.3-31 R. W. Reilly, D. R. Brown, and H. D. Huber, "Economic Assessment, Aquifer Thermal Energy Storage (ATES) Systems," in *Proceedings of the Mechanical, Magnetic and Underground Energy Storage 1981 Annual Contractors' Review*, U.S. Department of Energy, Washington, D.C., Report Number Conf-810833, February 1982, pp. 132–138.

2.3-32 "Environmental Assessment, Aquifer Thermal Energy Storage Program," U.S. Department of Energy, Washington, D.C., Report Number DOE/EA-0131, January 1981.

2.3-33 J. O'M Bockris, *Energy Options, Real Economics and the Solar Hydrogen System*, Halstead Press, Wiley, New York, 1980.

2.3-34 S. L. Robinson and J. Iannucci, "Technologies and Economics of Small-Scale Hydrogen Storage," Sandia Laboratories, Livermore, California Report Number SAND 79-8646, December 1979.

2.3-35 R. J. Teitel, "Microcavity Storage Update—1980," in *Proceedings of the DOE Thermal and Chemical Storage annual Contractors' Meeting*, U.S. Department of Energy, Washington, D.C. Report Number Conf.-801055, March 1981, pp. 197–200.

2.3-36 Roger Revelle, "Carbon Dioxide and World Climate," *Scientific American*, August 1982, p. 35.

2.4 TRANSPORTATION SYSTEMS[†]

John Winner, Arthur Karlin, Archie Riviera, Michael McDonald, David Turner, and Russell W. Zub

2.4-1 Energy Use in Transportation Systems: Overview

Over the past 30 years energy used in transportation has remained at approximately 25% of the country's total energy consumption (Fig. 2.4-1). While the transportation share of U.S. energy consumption has remained relatively constant, annual consumption has increased from 7.2 quads (1 quad is 10^{15} Btu) in 1950 to 20.1 quads in 1980. Petroleum derivatives are the primary source of energy for transportation purposes, supplying nearly 97% of transportation energy requirements and making up 52% of total U.S. petroleum consumption.

Automobiles consumed approximately 49% and trucks about 26% of all transportation energy in 1980 (Table 2.4-1). Gasoline was the primary fuel used in transportation, representing 68% of all transportation energy.

Table 2.4-2 shows a range of values of primary fuel energy use for different modes. Fuel use per vehicle-mile is influenced by vehicle technology, system operating characteristics, and terrain. Fuel use per passenger-mile also depends on the typical vehicle load. Rail transit and commuter rail have the lowest fuel cost per passenger-mile, followed by buses. Autos have the highest costs, although new high-mileage vehicles can approach the costs of the mass-transit modes in normal operating conditions.

2.4-2 Automotive Systems

Russell W. Zub

Energy efficiency, as related to the automobile and its associated subsystems, has been expressed in various ways. Because the main purpose of the automobile is to transport people over a given distance with the subsequent expense of petroleum consumption, its overall efficiency can be expressed as passenger-miles per gallon. This implies that the efficiency of the automobile can be increased not only by increasing the passenger load but also by reducing the fuel consumption. Since this section is limited to automobiles, the efficiency discussed will be fuel economy, which is the accepted standard value to compare the fuel use of automobiles.

Because of various political and economic events over the previous 10 years, the automobile has been modified to produce higher fuel economy. The most widely known plan to increase the fuel economy of domestic automobiles was the Energy Policy and Conservation Act of 1975 passed by Congress. The law established a corporate average fuel economy (CAFE) value for automobile manufacturers for each year, as shown in Table 2.4-3. The law is not scheduled to be extended beyond 1985.

The miles that a vehicle can cover on a gallon of fuel are based on results of the U.S. Environmental Protection Agency (EPA) test procedures. Each year manufacturers submit new vehicles to the EPA for testing. The vehicles are tested under controlled laboratory conditions by using a dynamometer to simulate driving conditions. The results of these emission tests are used to calculate fuel economy for an urban and highway drive cycle. The urban drive (Fig. 2.4-2) schedule, which represents city driving, is 7.5 mi long, averages 20 mph and has 18 stops. The highway cycle (Fig. 2.4-2) is 10.3 mi long, averages 48 mph, and has only 2 stops. The composite fuel economy, which is based on a weighted average of the urban and highway fuel economy, is calculated by

$$\text{Composite (mpg)} = \frac{1}{\dfrac{0.55}{\text{urban (mpg)}} + \dfrac{0.45}{\text{highway (mpg)}}}$$

[†] With the exception of Section 2.4-2, which was contributed by Russell W. Zub, this material was gathered for the U.S. Department of Transportation and will be assembled in Report Number MA-06-0153-1 entitled "Summary of Advanced Developments in Mass Transportation Energy Efficiency."

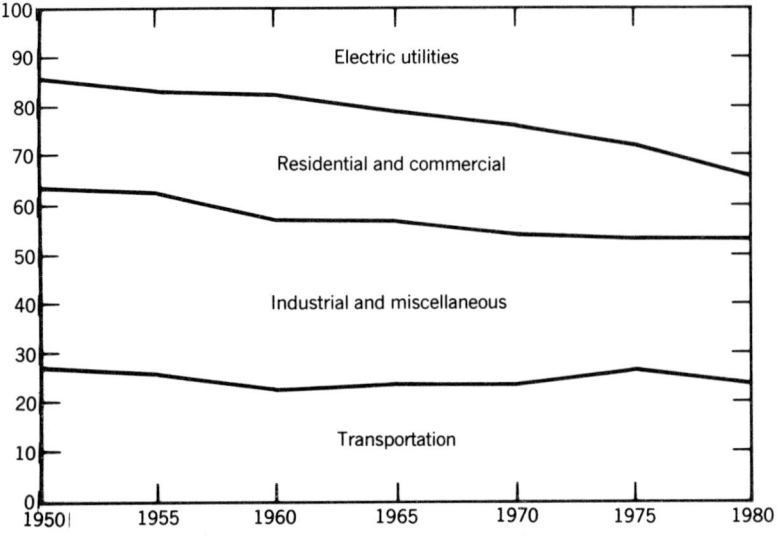

Fig. 2.4-1 U.S. energy consumption and use shares.

TABLE 2.4-1 ENERGY CONSUMPTION BY MODE AND FUEL TYPE (1980, QUADS)

Mode	Gasoline	Distillate Fuel	Liquefied Gases	Jet Fuel	Residual Fuel	Natural Gas	Electricity	Total	Percent of Total
Motorcycle	0.1	—	—	—	—	—	—	0.1	0.5
Automobile	9.9	—	—	—	—	—	—	9.9	49.3
Recreational vehicle	0.1	—	—	—	—	—	—	0.1	0.5
Bus	—	0.1	—	—	—	—	—	0.1	0.5
Truck	3.3	1.8	0.1	—	—	—	—	5.2	25.9
Air	0.1	—	—	1.6	—	—	—	1.7	8.5
Marine	0.2	0.2	—	—	1.4	—	—	1.8	9.0
Pipeline	—	—	—	—	—	0.5	0.1	0.6	3.0
Rail	—	0.6	—	—	—	—	—	0.6	3.0
Total	13.7	2.7	0.1	1.6	1.4	0.5	0.1	20.1	1000
Percent of total	68.2	13.4	0.5	8.0	7.0	2.5	0.5	100%	

The specific areas for improving automotive fuel economy, which will be discussed individually in the next sections, are based in part on the relative energy usage of these areas over the urban and highway drive schedules. For example, the energy distribution of selected components for a 1981 2400-lb vehicle is shown in Table 2.4-4. The influence of these components on fuel economy can be comprehended by utilizing Fig. 2.4-3, which is a representation of a spark ignition engine. This steady-state engine map is depicted by engine load versus speed for constant lines of brake-specific fuel consumption (BSFC), which indicate how efficiently an engine is converting fuel into work. As a vehicle traverses a particular drive cycle, the engine operating points can be recorded as incremental steady-state points, as shown in Fig. 2.4-4. These engine operating points used in conjunction with the engine map enable the vehicle fuel economy to be analytically determined. One instantaneous engine operating point of the urban drive schedule in Fig. 2.4-4 is 10 bhp at 2400 rpm, which corresponds to a point on Fig. 2.4-3 of 0.59 lb/bhp · h. If the corresponding vehicle speed is 25 mph, the instantaneous fuel economy is calculated to be 25.4 mpg. The fuel economy for a complete drive schedule is based on the sum of the instantaneous fuel consumed over a specified distance. The weight,

TABLE 2.4-2 ENERGY USE AND ENERGY COST PER PASSENGER-MILE BY MODE[a]

Mode	Equivalent Vehicle-Miles per Gasoline Gallon[b] (range of typical values)	Btu per Mile	Load: Passengers per Vehicle		Btu per Passenger-Mile[b]		Fuel Cost (cents) per Passenger-Mile[c]	
			Typical (range)	Maximum	At Typical Load	At Maximum Load	At Typical Load	At Maximum Load
Automobile	11.4	10,970	1.1–1.6	5	6,856–9,972	2,194	6.9–10.0	2.2
	15	8,330			5,206–7,572	1,666	5.2–7.6	1.7
	22.5	5,260			3,288–4,782	1,052	3.3–4.8	1.1
Bus	3.8	32,660	9.8–19.1	50	1,709–3,333	653	1.7–3.3	0.7
	4.8	26,310			1,377–2,685	526	1.4–2.7	0.5
Light-rail transit	1.45	86,400	17–27	63	3,200–5,082	1,371	1.9–3.0	0.8
	2.17	57,750			2,139–3,397	917	1.3–2.0	0.6
Heavy-rail transit	1.71	73,180	18–28	72	2,613–4,065	1,016	1.6–2.4	0.6
	2.23	56,175			2,006–3,120	780	1.2–1.9	0.5
Commuter rail	1.8	69,540	24–60	100	1,159–2,898	695	1.2–2.9	0.7
	2.90	43,150			719–1,798	432	0.7–1.8	0.4
Air (certified carrier)	0.26	490,740	86	150	5,708	3,272	5.7	3.3

Source. Committee on Environment and Public Works, "Urban Transportation and Energy: The Potential Savings of Different Modes," U.S. Senate, September 1977. "National Transportation Statistics," September 1980.

[a]Note: 1 mpg = 0.425 km/L; 1 Btu/mi = 0.000181 kW · h/km.

[b]Values are expressed in terms of energy content of primary fuels. Thus, for electrical modes (transit), energy requirements are three times the operating system electrical energy use to reflect 33% efficiency of electrical generation and distribution. One gallon of gasoline = 125,071 Btu.

[c]Fuel cost to operating authority. Electrical energy requirements for transit are divided by 3 to convert to actual electricity use. Costs are $1.30/gal of liquid fuel (about 1 cent/1000 Btu) and 6 cents/kW · h of electricity (about 1.8 cents/1,000 Btu).

TABLE 2.4-3   CORPORATE AVERAGE FUEL ECONOMY

Year	Automobile Fuel Economy (mpg)	Year	Automobile Fuel Economy (mpg)
1978	18.0	1982	24.0
1979	190	1983	26.0
1980	20.0	1984	27.0
1981	22.0	1985	27.5

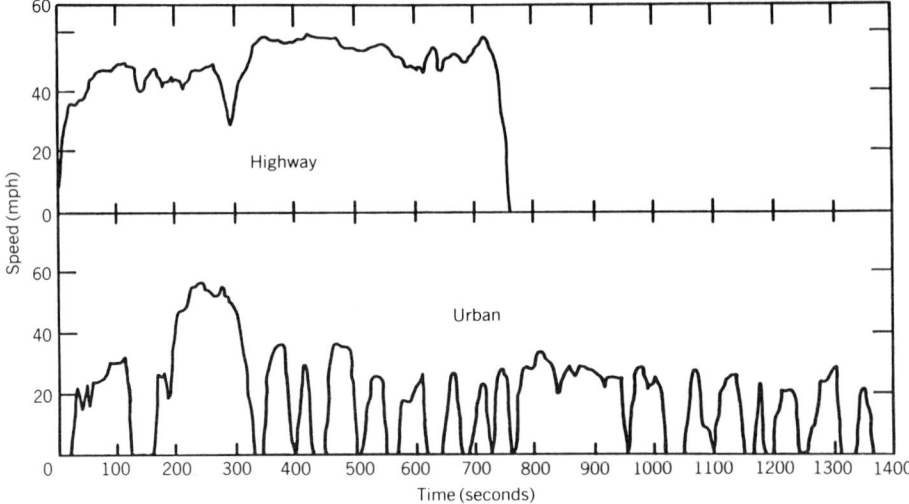

Fig. 2.4-2   Speed versus time trace of urban and highway drive cycles.

TABLE 2.4-4   PERCENT OF TOTAL ENGINE ENERGY OUTPUT

Component	Urban Cycle	Highway Cycle
Accessories	8	4
Transmission (manual)	3	2
Differential	3	3
Aerodynamics	25	56
Tire rolling resistance	27	28
Braking energy	34	7

aerodynamic drag, tire rolling resistance, and power train influence the results because a change in vehicle components will cause the engine to operate at different speed and load points, thereby changing its fuel rate and subsequently its fuel economy.

Weight

The braking energy value shown in Table 2.4-4 is a function of drive cycle and vehicle weight. For the urban cycle weight is the most important parameter affecting fuel economy.[1] The vehicle propulsion system must overcome not only rolling resistance but also vehicle inertia, which are both functions of vehicle weight. A 10% reduction in vehicle weight can result in a fuel economy improvement of 2–9%, depending on the weight reduction technique.[2] The overall fleet weight reduction and fuel economy improvement trend is shown in Fig. 2.4-5.

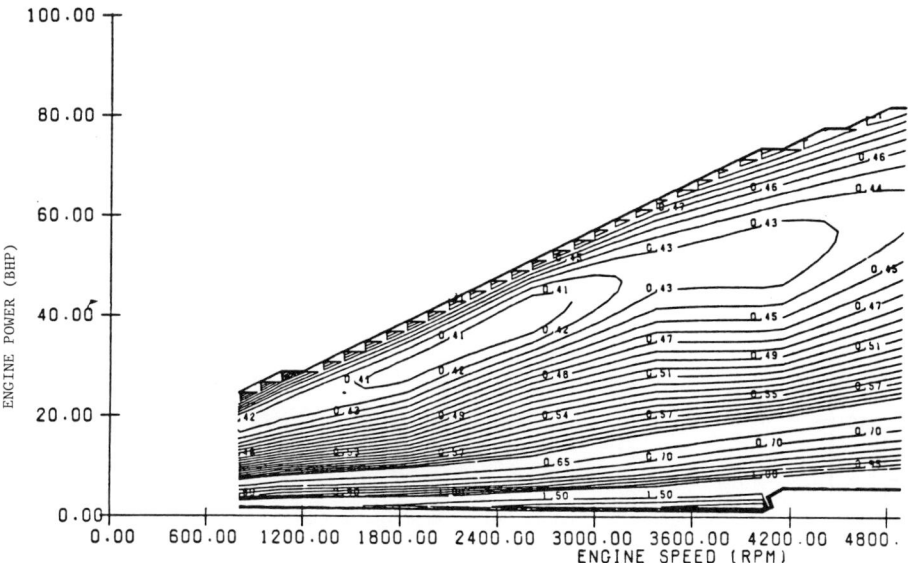

Fig. 2.4-3 1981 1.95-L engine; BSFC (lb/bhp-h).

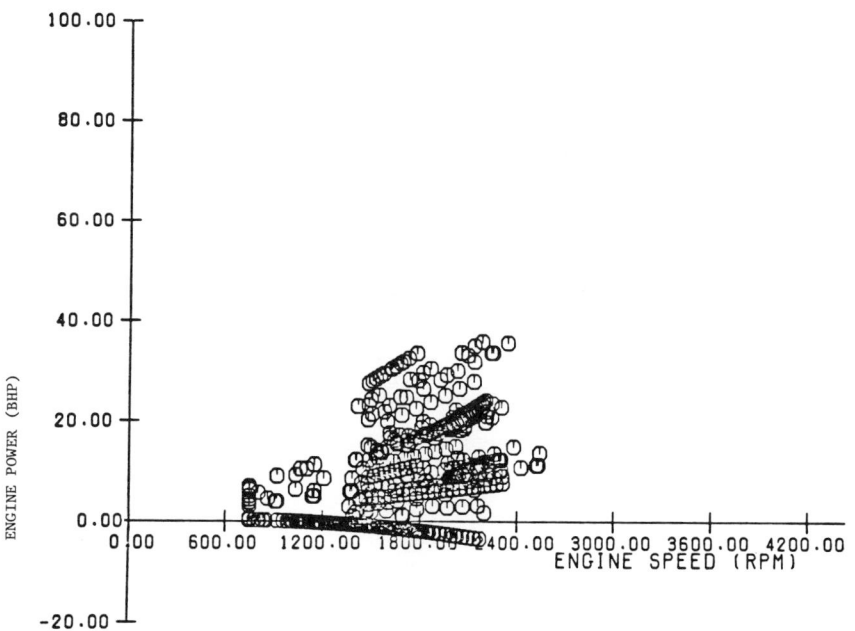

Fig. 2.4-4 Engine operating points for urban drive schedule.

The weight of vehicles can be reduced by three techniques: vehicle redesign, material substitution, and new vehicle and component design. Vehicle redesign, which includes redimensioning or downsizing a vehicle, was introduced by a domestic manufacturer in 1977. The newly introduced vehicle design was approximately 13% lighter and about 6% shorter, yet maintained about the same interior space. Because domestic automobiles were comparatively large in the late 1970s, downsizing was a viable and attractive technique. As the vehicle is reduced in size, material substitution, which involves the replacement of standard components with lighter-weight components, can be introduced. Candi-

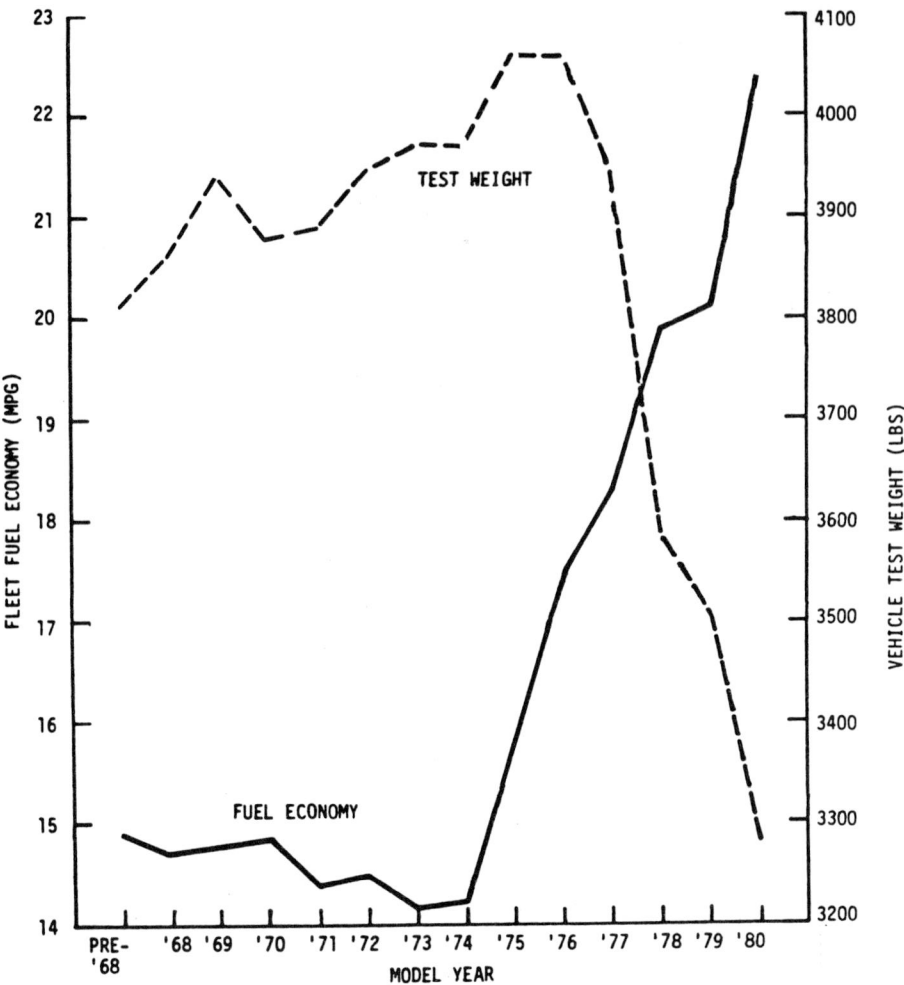

Fig. 2.4-5 Trends in sales—weighted fleet fuel economy and vehicle test weight for passenger cars. (Source: H. Hsia and J. A. Kidd, "Weight Reduction Potential of Automobiles and Light Trucks," Report No. DOT-TSC-NHTSA-79-54, Transportation Systems Center, Cambridge, MA, October, 1979.)

date materials include high-strength steels,[3] aluminum alloys, fiberglass-reinforced plastics,[4] and hybrid-reinforced plastics. The complete design of a vehicle for minimum weight at the expense of other parameters is a complicated process that is beyond the scope of this chapter. However, when considering weight reduction, factors such as safety, performance, consumer attributes, and costs must be taken into account.

Engine

The engine influences fuel economy through the shape and absolute value of the BSFC lines as shown in Fig. 2.4-3. Engine features (e.g., compression ratio, fuel delivery, and timing) determine engine map characteristics and consequently fuel economy. As vehicles become smaller, the engines have also been reduced in size. The sales-weighted engine displacement was 210 CID in 1980 but decreased to 182 CID in 1981.[5] These smaller engines coupled with more efficient fuel delivery systems have contributed substantially to an overall vehicle fuel economy increase. The increase in fuel economy attributed to individual engine features is difficult to estimate because features are often implemented concurrently and comparisons claims do not always use identical base vehicles. Features such as electronic engine controls, optimized combustion chamber design, and higher compression ratios have increased the efficiency of the smaller engines. Simultaneously, to maintain performance, the power density of engines has been increased by means of fuel injection, turbocharging, and supercharging.

Electronic engine controls, which can improve fuel economy and reduce emissions, involve computer systems that control spark timing, idle speed, fuel delivery, and torque converter clutch engagement, where applicable. One new engine configuration that improves combustion efficiency utilizes fast-burn combustion, increased swirl, and two plugs per cylinder to obtain a 30% increase in fuel economy.[6] The use of higher compression ratio engines is accomplished by employing knock sensors. General Motors has developed a throttle body injection (TBI) system that uses precise fuel controls. By 1985, this comparatively simple system will be used on 40% of all General Motors cars.[7] Turbocharging and supercharging, which increase the power density of the engine, can also yield fuel economy increases,[8] although some structural modifications may be required. Further improvements in engine efficiency are accomplished by reducing engine friction through the use of roller tappets and lighter reciprocating parts.

The diesel engine, which is inherently more efficient than the spark ignition engine due to its higher compression ratio, is offered as an option on some passenger cars. The diesel-equipped automobile will achieve approximately a 25% greater fuel economy than a comparably equipped spark ignition automobile. This fuel economy advantage has increased the market share of diesel passenger cars. In 1980 U.S. diesel car sales represented 4.4% of total new car sales. In 1981 this figure increased to 6.1%.[9] Future improvements in the diesel engine involve direct injection diesels. For light-duty applications the fuel economy of a direct injection diesel can be 15–20% greater than their indirect injected counterparts.[10]

Alternative power plants such as the Stirling[11] and rotary[12] engines are being developed extensively. Like other alternative power plants, these engines have benefits and disadvantages and must be evaluated with respect to their applications.

Transmission and Differential

The typical automobile transmission is either a manual transmission or automatic transmission with a torque converter, although some transmissions use a combination of both. The fuel economy of an automobile can be improved by increasing the efficiency of the transmission components, extending the gear ratio range (first-gear ratio divided by final-gear ratio), and/or modifying the shift logic.

The efficiency of the automatic transmission can be improved by reducing energy losses in the torque converter. The torque converter, which multiplies engine torque, allows the engine to idle with the gear selector in the drive position and improves shift smoothness at the expense of converter slip and subsequent energy loss. More efficient torque converters can be designed at the expense of start-up drivability, shift smoothness, and torque multiplication. A more effective method of reducing torque converter losses is by bypassing the converter. Usually, the upper gears are locked up (converter-bypassed) completely or locked up at some predetermined speed and the first gear power path remains through the torque converter. This allows the torque converter losses at cruise to be reduced while low-speed drivability and torque multiplication are maintained. Depending on the type of lockup scheme used, the composite fuel economy improvement for a three-speed transmission can be 1.5–3.5% over a baseline three-speed automatic transmission. The fuel economy benefit of locking up a torque converter is dependent upon a number of factors, which include the torque converter efficiency profile, vehicle parameters, and the effect of lockup on engine BSFC. The torque converter efficiency profile determines the losses and thus limits the potential of eliminating those losses.

The gear ratio range can be extended to increase fuel economy. The increase, which is limited by the number of gears, gear spacing, and engine–transmission matching, allows the engine to operate at a lower BSFC in the top gears. As shown in Fig. 2.4-6, this improvement can approach 20%. In determining the effect of transmission modifications on fuel economy, the main parameter is the span of the gearbox.[13] The improvement of a four-speed versus a three-speed has been measured at 3% in the urban cycle and 21% in the highway cycle.[14] More recently[15] the introduction of four-speed automatic transmissions has improved fuel economy 2.5–3 mpg.

The shift logic for an automatic transmission is set by the manufacturer. Essentially, the vehicle shifts "automatically" based on predetermined inputs of speed and load. By making the transmission shift earlier, fuel economy can be improved. The shift logic for a manual transmission is determined by the driver. However, for EPA testing, the shift logic or schedule is based on a fixed vehicle speed, unless a manufacturer can demonstrate that the driver will shift differently. For example, a vehicle with a five-speed transmission must upshift from first to second at 15 mph, from second to third at 25 mph, from third to fourth at 40 mph, and from fourth to fifth at 45 mph. Forcing the vehicle to shift in every gear can, in some instances, force the engine to operate at a higher speed than is needed, thus reducing the fuel economy versus the optimum shift schedule.[16] Nevertheless, a driver with a manual transmission may shift any time, and generally the quicker the upshift the better the fuel economy. One manufacturer has installed a light on the dash that signals the driver to shift for best fuel economy. The improvement with this upshift indicator is 7% in the urban cycle.[17]

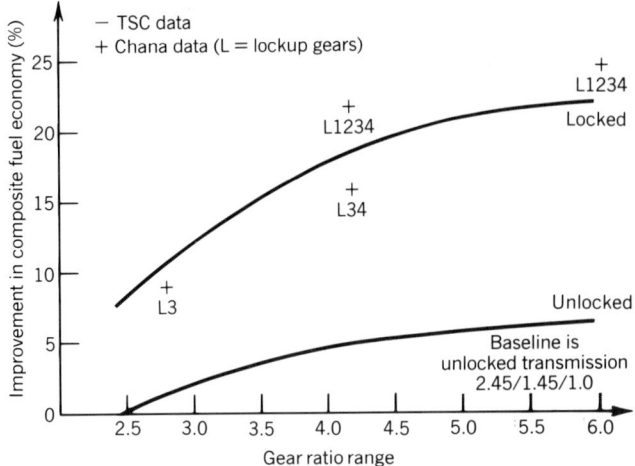

Fig. 2.4-6 Effect of gear ratio range and lockup on fuel economy. (Source: R. W. Zub and R. P. Meisner, "Effect of Vehicle Design Variables on Top Speed, Performance and Fuel Economy—1981." Transportation Systems Center, Report No. DOT-TSC-NHTSA-81-23, April, 1981.)

A recently introduced transmission concept is the split-torque transmission.[18] This transmission utilizes a fuel-efficient torque converter, a 60% mechanical and 40% hydrodynamic split in third gear, and a fully mechanical overdrive fourth gear. Composite fuel economy savings over a three-speed automatic average 7–8% for the composite cycle.

Another transmission concept is the continuously variable transmission (CVT). This transmission allows the engine to operate at minimum BSFC by continuously varying the transmission gear ratio. Fuel economy improvement utilizing a CVT is on the order of 10–20% over an automatic.[19] This concept has been combined with the elimination of engine idle, recovery of energy from braking, and minimizing transmission losses to form a flywheel energy management vehicle that obtains 50 mpg in the urban cycle.[20]

Aerodynamics

In terms of obtaining the most miles from a gallon of gasoline, aerodynamics must be considered, after accounting for downsizing and weight reduction.[21] The average European car has a drag coefficient of 0.46,[22] but research vehicles, such as the Volkswagen Auto 2000 and the Ford Probe III, have drag coefficients of 0.25 and 0.22, respectively. The techniques used to achieve these low coefficients include a squared-off trail, rear spoiler, underbody panel, and flush side windows. These methods of reducing aerodynamic drag will be implemented on future production vehicles because of the fuel savings they offer.

The aerodynamic drag of a typical vehicle consumes approximately 25% of the energy output of the engine over the EPA urban drive schedule and 56% over the highway drive schedule. The effect of aerodynamic drag on fuel economy over the composite drive schedule is characterized by a sensitivity of approximately 0.2 (%Δ fuel economy/%Δ drag reduction) for automobiles and 0.25 for light-duty trucks.[23] This sensitivity can be increased by modifying the drive train to maintain constant performance.

There is no question that reducing the aerodynamic drag will improve fuel economy, however, the possibility of all future low-drag automobiles looking similar could present a problem. Volkswagenwerk AG[24] concluded that even at low values of drag there is enough latitude for stylists to give the vehicles a diversified appearance. Previously, constraints of styling, production costs, and safety requirements produced a procedure[25] in which the aerodynamicist was limited to small drag reductions. However, now the importance of aerodynamics has to be enhanced and new development procedures reflecting this increased importance need to be implemented.

Tire Rolling Resistance

The majority of tire rolling resistance loss is hysteretic, which is influenced by inflation pressure, load, tread composition, carcass design, and inflation temperature.[23] By properly modifying these variables, the hysteretic losses can be reduced and fuel economy increased. In general, a 10% change in the rolling resistance results in a 2% change in composite vehicle fuel economy.[26]

In an effort to reduce fuel consumption, tire pressures have been increased. The sensitivity of rolling resistance to inflation pressure is a function of tire material and design. Tire pressure and rolling resistance information can be obtained from a "carpet plot" of rolling resistance versus reciprocal inflation pressure for constant-load lines. Road tests involving tire pressure changes have been conducted.[27]

A reduction in rolling resistance will improve fuel economy, but the relationship among rolling resistance tire parameters and cornering force properties[28] is being investigated from a safety viewpoint.

The next generation of passenger car tires will involve compromises. Since the tire is an integral part of the vehicle, future tire specifications will depend on the buyers' needs and car performance.

Accessories

The accessories of a vehicle can contribute significantly to fuel consumption, particularly during low-speed operation. The most energy-intensive accessory is the air-conditioning compressor. There is no consensus on the magnitude of the effect of air conditioning on fuel economy because of the wide variation of air-conditioning operation. For example, the duty cycle (percent time on/total time) will be greater in hot climates than in relatively mild climates. The dynamometer test procedure simulates air conditioning by increasing the road load horsepower absorbed by the dynamometer by 10%.[29] Since the sensitivity of fuel economy to aerodynamic drag is approximately 0.2, the 10% increase in road load horsepower would translate into a 2% fuel economy penalty. Actual road tests to determine the effect of air conditioners on fuel economy usually yield a 10% reduction in fuel economy.

Because there is usually no need for a radiator cooling fan above engine speeds of about 2500 rpm, belt-driven fans waste energy. Two solutions to this energy waste are thermostatically activated clutch fans and electric fans; both reduce the duty cycle of the fan. Efforts to improve fans further include more efficient blade design and lighter blade materials.

The use of power steering will depend upon the size of the vehicle, drive train configuration, and consumer demand. The energy used in the power steering pump is influenced by its design. For example, one manufacturer,[30] utilizing a radial piston design, has developed a pump that consumes only about 61% of the power of those commercially used in the United States in 1979. Further development of this system is projected to reduce pump fuel consumption to 0.04 gal per 100 mi.

The energy used by the alternator is dependent upon the electrical demand of the vehicle. Computer-simulated alternator loads representing night driving lighting conditions yielded a 1–3% decrease in fuel economy over the composite cycle.[31]

The use of the secondary air pump for emission control depends upon the emission standards and the type of emission control. The oil and water pumps, which are directly connected to the engine, will probably not be modified substantially for energy savings because they are necessary for engine operation.

One technique for lowering belt-driven accessories is to limit the accessory speed, thereby lowering parasitic horsepower. This technique is accomplished by utilizing a controlled-speed accessory drive (CSAD), which serves as a power takeoff for all accessories normally driven from the engine crankshaft.[32] Results of this system are not available, although it is known that this method would not be effective in an EPA dynamometer test because of the test procedure simulating air conditioning previously mentioned.

Investigation of Vehicle Parameters

Although the previous sections were concerned primarily with energy efficiency and fuel economy, there are other factors that must be considered when evaluating automobiles for fuel economy improvement. Compromises among performance, emissions, safety, and size have to be achieved. The effect of vehicle design variables on the above criteria should be discerned particularly if a vehicle is being designed primarily for fuel economy.[23,33] Therefore, when assessing the effect of design variables on fuel economy, the resulting changes in other evaluation criteria should conform to acceptable or mandated levels before any fuel economy improvements are claimed. Inevitably, there will be trade-offs.

2.4-3 Bus Systems

The basic vehicle power train relationships in buses are nearly the same as those in automobiles, but with fundamental differences in the physical characteristics of size, shape, and payload. To gain a better understanding of the energy intensity of the transit bus, the previous section on automotive systems should be reviewed before this section is examined.

Duty Cycles

The standard duty cycle used in most testing procedures to estimate power utilization and fuel consumption is the composite advanced design bus (ADB) cycle. This cycle is a combination of three operating phases: central business district (CBD), arterial (ART), and commuter (COM). The cycle is performance based; acceleration is limited only by the capabilities of the vehicle.

The CBD phase consists of seven consecutive stops per mile (4.4/km) over a 2-mi (3.2-km) course with accelerations from 0 to 20 mph (0 to 32 kph). This phase simulates the boarding and exiting of passengers in the business district, where frequent stops must be made and heavy traffic is encountered. The ART phase consists of accelerations from 0 to 40 mph (0 to 64 kph) and two stops per mile (0.8/km) over a 2-mi (3.2-km) course. Passenger activity in less congested areas where traffic is lighter and higher vehicle speeds are attained is represented in this phase. Finally, the COM phase has one 0–55-mph (88.5-kph) acceleration and deceleration activity and close to 4 mi (6.4 km) of highway-speed operation over a 4-mi (6.4-km) course. This phase models boarding of passengers in suburban areas and transportation to metropolitan areas. Table 2.4-5 and Fig. 2.4-7 present the composite ADB cycle and its components in greater detail.

The Society of Automotive Engineers (SAE) has developed a series of procedures for measuring fuel economy. Currently, the most common of these procedures for measuring basic highway vehicle engine performance and fuel consumption for both spark ignition and diesel vehicles are the June 1980 SAE J1312 and the joint TMC/SAE fuel consumption test procedure, type II, SAE J1321 of October 1981. The most recent procedure, SAE J1321, has been gaining wide acceptance within the industry as an accurate method of monitoring fuel use in trucks and buses, addressing consumption for both the entire vehicle and its components.

These SAE standards are testing procedures and depend on a realistic duty cycle selection. The specific duty cycle used to measure the fuel economy of a bus is very important since fuel economy measurements can vary by more than 100%, depending on the bus duty cycle.

TABLE 2.4-5   COMPOSITE ADB DUTY CYCLE[a]

	Phase							
	CBD	Idle	Arterial	CBD	Arterial	CBD	Commuter	Total
Stops/mile	7	—	2	7	2	7	1 stop for phase	
Top speed (mph)	20	—	40	20	40	20	Maximum or 55	
Miles	2	—	2	2	2	2	4	14
Approximate acceleration distance (ft)	155	—	1035	155	1035	155	5500	
Approximate acceleration time (s)	10	—	29	10	35	10	90	
Approximate cruise distance (ft)	540	—	1350	540	1350	540	2 mi + 4580 ft	
Approximate cruise time (s)	18.5	—	22.5	18.5	22.5	18.5	188	
Approximate deceleration rate (fpsps)	6.78	—	6.78	6.78	6.78	6.78	6.78	
Approximate deceleration distance (ft)	60	—	255	60	255	60	480	
Approximate deceleration time (s)	4.5	—	9	4.5	9	4.5	12	
Approximate dwell time (s)	7	—	7	7	7	7	20	
Approximate cycle time (min-s)	9–20	5–0	4–30	9–20	4–30	9–20	5–10	47–10
Total stops	14	—	4	14	4	14	1	51

[a] Note: 1 mi = 1.609344 km; 1 mph = 1.62 kph; 1 ft = 0.31 m.

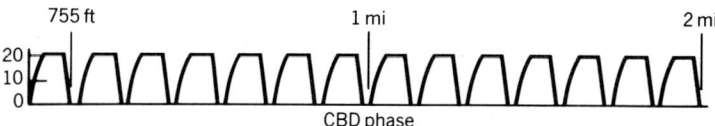

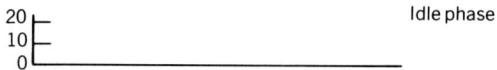

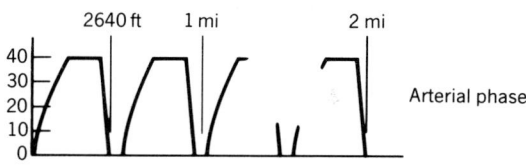

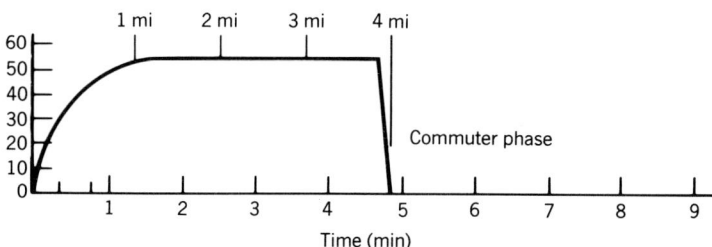

Fig. 2.4-7 ADB duty cycle components.

Bus Energy Management and Performance Measures

The ADB cycle averages 3.6 stops per mile (2.3/km) compared with a little over 1 stop per mile (1.6 km) for the EPA automotive combined cycle. Thus, acceleration energy use is more critical in buses than in cars. In addition, since vehicle speeds are kept lower, aerodynamics are relatively less important with buses.

A simplified overall energy balance equation for automobiles or buses can be expressed as

$$P_{eng} = \left[\underset{(1)}{\frac{C_D}{2}\rho A V^2} + \underset{(2)}{C_R Mg} + \underset{(3)}{Mg\sin\theta} + \underset{(4)}{Ma} \right] V + \underset{(5)}{P_a} + \underset{(6)}{P_L} + \underset{(7)}{P_I}$$

where P_{eng} = power from engine,
  V = vehicle velocity,
  A = vehicle frontal area,
  ρ = density of air,
  M = vehicle mass,
  g = acceleration of gravity,
  a = vehicle acceleration,
  P_a = accessory power,
  P_L = drive train friction losses,
  P_I = power required to accelerate drive train,
  θ = grade angle,
  C_D = drag coefficient,
  C_R = rolling resistance coefficient.

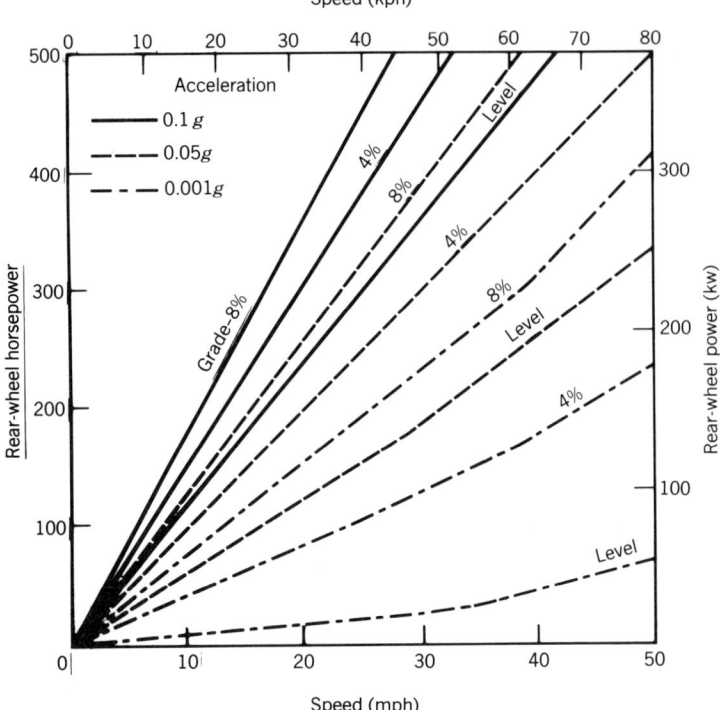

Fig. 2.4-8 Effect of grade on required propulsive power for constant acceleration. (Vehicle weight = 32,000 lb; Mass = 14,500 kg.)

Term (1) represents the aerodynamic resistance and is a function of the velocity squared and the vehicle frontal area. The drag coefficient C_D is determined by the vehicle's shape, load distribution, and external items. Term (2) represents the rolling resistance. This term is predominantly proportional to vehicle weight, although there is some effect from vehicle speed. The rolling resistance coefficient C_R will also be affected by road surface conditions and tire pressure. Term (3) is the force required to move a vehicle on a grade, and Term (4) is the force needed to accelerate a vehicle. Both of these terms are directly proportional to vehicle weight. Term (5) is the power of accessories such as fans, pumps, air conditioning, and alternators and depends on the specific accessories and their use. This is not a simple function of weight or vehicle speed, although it can be affected by driving cycle. For instance, certain cooling fans are now designed to shut off at high vehicle speeds. Air-conditioning power requirements vary with the duration of compressor operation. Term (6) represents friction losses in the torque converter, transmission, and other parts to the drive train. These losses are also not simple functions of weight or speed but depend on gear, vehicle speed, and load. Term (7) represents power required to accelerate the engine, drive train, and wheels.

Figure 2.4-8 shows the bus power required for various grade and acceleration levels. At 30 mph (48 kph), a bus accelerating at 0.05g will be using less than 40 hp to overcome rolling resistance and aerodynamic resistance (from the 0.001g curve) and about 150 hp for acceleration.

One measure of performance for the transit bus is its ability to accelerate rapidly but smoothly. Gradeability, an additional performance measure, is the percentage of grade (feet of vehicle rise per 100 ft of horizontal distance) that a vehicle can negotiate at a sustained road speed from a running start.

The maximum engine power on a bus is generally 190–270 hp (142–201 kW). The power-to-weight ratio is in the range of 0.006–0.008 hp/lb (0.009–0.013 kW/kg), about one-third of typical automotive values. Thus, buses have much less available power than cars, and performance measures such as acceleration and gradeability become critical in bus specification. Table 2.4-6 shows the acceleration rate for a V6 bus compared with an automobile. The bus is considerably slower past 10 mph (16 kph). The top speed of a bus on a 4.5% grade is typically about 25 mph (40 kph).

TABLE 2.4-6   ACCELERATION
FROM STOP FOR A CAR AND
A V6 BUS ON LEVEL GROUND[a]

Vehicle Speed (mph)	Passenger Car (g)	V6 Bus (g)
0	—	0.138
10	0.079	0.082
20	0.072	0.059
30	0.066	0.023
40	0.046	0.017

Source. John S. Ludwick, Jr., and
George F. Swetnam, Jr., "A Preliminary Review of Propulsion Requirements for an Urban Transit Bus,"
MTR 6688, June 1974.

[a]Note: 1 mph = 1.62 kph; $1g = 32.2$
ft/s^2.

Chassis and Body Characteristics

Buses have both larger frontal areas and larger drag coefficients than automobiles and thus, for
equivalent speeds, bus aerodynamic drag can be substantially larger than that for passenger cars.
Table 2.4-7 shows aerodynamic drag coefficients for a number of vehicles.

Recent work on bus aerodynamics has focused on the use of add-on devices for bus bodies to
reduce drag. One study has shown that drag can be reduced as much as 27% by modifying body shape
and improving overall aerodynamic characteristics. This can reduce fuel consumption on the ADB
cycle by 2%.

As with cars, vehicle weight affects rolling resistance, acceleration, and grade power requirements.
Figure 2.4-9 shows the effect of bus weight on fuel economy over the ADB cycle.

Bus tire rolling resistance coefficients at various speeds are shown in Fig. 2.4-10. Although this
coefficient behaves the same as that for automotive tires, its magnitude may be 30% less because of the
higher bus tire pressure. Most buses also use bias rather than radial tires for improved durability.

TABLE 2.4-7 AIR RESISTANCE
DRAG   COEFFICIENT   OF
SEVERAL   REPRESENTATIVE
TYPES OF VEHICLES

Vehicle Type	C_D (Dimensionless)
Racing car	0.25–0.3
Passenger car	0.40–0.55
Intercity bus	0.65–0.75
Urban transit bus	0.55–0.80
Truck	0.80–1.60
Tractor-trailer truck	1.30–2.00
Geometrical bodies:	
Streamlined body	0.13
Sphere	0.47
Square (flat) plate	1.2

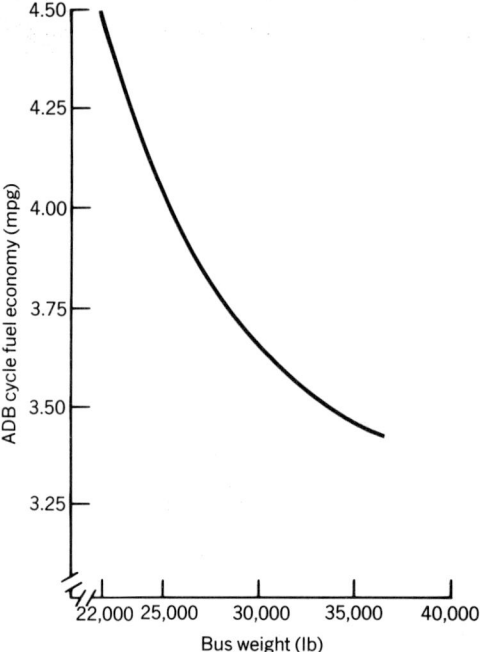

Fig. 2.4-9 Fuel economy versus bus weight.

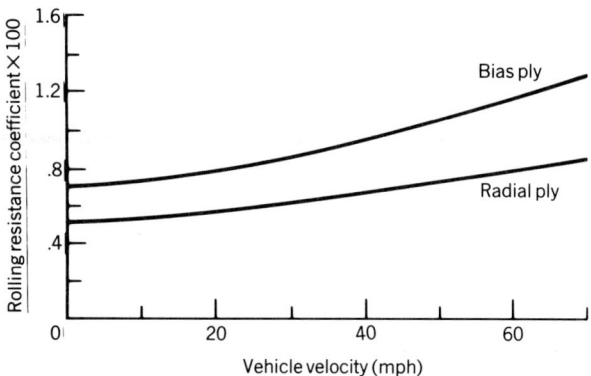

Fig. 2.4-10 Rolling resistance coefficients. (Source: Kevin A. Copeland, "Fuel Economy on Automatic Equipped Transit Busses", General Motors Institute, April 1980.)

Power Train Characteristics

The predominant engine used in transit buses is the 8V-71 Detroit Diesel Allison eight-cylinder, two-cycle diesel engine. The diesel engine has been the main power source for transit buses since the 1950s. The two-cycle diesels are characterized by high low-speed torque and high power-to-weight ratio. The diesels also substantially improve bus fuel economy compared to gasoline engines. However, the two-stroke diesel cycle is generally not as efficient as the four-stroke diesel cycle primarily because of the volumetric inefficiency in the intake system and mechanical losses in the blower system. The engine does, however, provide high power for its size and weight.

Other engines in bus use include the DDA 6V-92TA, which is a turbocharged and aftercooled six-cylinder, two-cycle engine. The "aftercooling" refers to a cooling of the intake air after turbocharge compression. This further increases the air density and thus the power potential of the engine. The

TABLE 2.4-8   FUEL EFFICIENCY OF DIFFERENT ENGINES

Configuration[a]	Engine	Fuel Economy (mpg)				0–15 mph Performance (s)
		CBD	ART	COM	ADB	
A	8V-71	3.23	3.37	4.97	3.70	5.9
B	6V-92TA	3.59	3.66	5.05	3.98	6.5
A	VTB-903	3.45	3.64	5.30	3.95	6.8

[a] Configuration = Bus type/transmission/torque converter/rear axle ratio/weight
     A = RTS-II/V-730/TSC-490/5.29/31,498 lb
     B = RTS-II/V-730/TC-470/5.857/31,498 lb
Note: 1 mpg = 0.425 km/L; 1 mph = 1.62 kph.

Cummins VTB-903 is a turbocharged, eight-cylinder, four-cycle diesel engine. All three of these engines have maximum power around 270 hp (201 kW). The DDA 6V-71 is a naturally aspirated six-cylinder diesel with maximum power around 190 hp.

Table 2.4-8 compares the fuel efficiency of several of these engines with other standard features on a bus, such as a three-speed transmission with a torque converter, a rear-wheel axle ratio of approximately 5.13, and a weight of 34,000 lb (15,400 kg). The turbocharged 6V-92TA shows improved fuel economy because of its improved efficiency at high loads. The VTB-903's improved fuel economy is due to the inherently greater efficiency of a four-cycle engine.

Most buses use lockup torque converters for fuel efficiency. The effect on fuel economy of transmission gearing for buses depends heavily on the driving cycle. A transmission with advantageous gearing in low-speed driving will yield better fuel economy on an urban-type cycle than will a transmission optimized for higher speeds.

One of the major users of energy on a bus is the air-conditioning system. Such systems can degrade fuel economy by 8–20%. They also extract a performance penalty, increasing 0–30-mph (0–48-kph) acceleration times by 18%.

Other Recent Developments

Alternate fuels can be used to reduce petroleum fuel dependency. Candidates for diesel engines in buses include methanol, ethanol, methane, propane, ammonia, and hydrogen. None of these fuels can be used directly in existing engines since their octane numbers in all cases are too low to cause self-ignition on the diesel cycle. Table 2.4-9 shows five types of internal combustion engines that can be developed by modifying existing engines. The gasoline engine (Otto cycle) appears to be the most suitable power plant across existing alternative fuel types with potential use for all but diesel fuel.

TABLE 2.4-9   TYPES OF INTERNAL COMBUSTION ENGINES DEVELOPED BY MODIFYING EXISTING ENGINES

Engine Type	Characteristic		
	Method of Fuel Delivery	Method of Fuel Ignition	Compression Ratio
Diesel	High-pressure injection	Compression	17–20
Gasoline	Carburetion, low-pressure injection	Spark	8–11 (depending on fuel)
Stratified charge	High-pressure injection	Spark	10–16 (depending on fuel)
Fumigated diesel	Carburetation, high-pressure injection	Compression	17–20
Dual injection	High-pressure injection	Compression	17–20

Methanol, ethanol, and gasoline could potentially be used in three of the engine types—gasoline, spark-assisted stratified charge, and dual fuel injection.

Energy storage devices offer another possible area of bus fuel conservation. Research is being conducted on hydraulic retarders that are capable of storing braking energy. These retarders use a single or dual hydraulic motor and pump arrangement to compress air in hydraulic accumulator storage vessels. The stored energy is then used to propel the vehicle forward during acceleration. When used in stop-and-go driving; this type of retarder can greatly improve the fuel efficiency of a bus. A 30% fuel consumption reduction (50% fuel economy improvement) in transit buses has been achieved using these systems.

2.4-4 Urban Rail Transit Systems

The two principal submodes of urban rail transit are heavy (commuter) and light rail; automated guideway transit (AGT) is a third submode. Heavy-rail transit is provided by trains of electrically self-propelled railcars, 50–80 ft (15–24 m) long, operating on tracks using an exclusive, separated right-of-way. The New York, Chicago, and San Francisco systems are examples. Light-rail urban transit is provided by lighter electrically self-propelled railcars operating individually or in trains on tracks using city streets or on semiexclusive or exclusive rights-of-way. Examples of light-rail systems can be found in Pittsburgh, San Francisco, and San Diego. AGT systems are small, electrically self-propelled, automatic, driverless vehicles or trains that operate on exclusive, separated guideways. Examples of AGT systems are those at the Atlanta and Houston airports and the urban system in Morgantown, West Virginia.

Energy Use and Load Factors

Urban rail transit accounts for only about 0.04% of U.S. transportation energy consumption, virtually all of which is delivered as electric power. While efficiency of rail transit equipment is intrinsically high, it is strongly dependent on commuter traffic patterns. A crush-loaded transit car has a passenger transit efficiency of more than 30 times that of an average car. When passenger loadings drop to one-tenth of crush capacity or below, the energy efficiency of urban rail transit is heavily eroded.

Table 2.4-10 gives key indicators of performance and energy use for selected U.S. fixed guideway systems. There is a remarkable variation in equipment efficiency, as shown by the range of kilowatt-hour per car-mile—between 4.98 and 131.5. This variation is further amplified by the distribution of passenger use of trains, yielding transport efficiencies in the range of 0.15–24.8 kW · h per passenger-mile. Some systems shut down during off-peak periods, while others continue operating. Systems like the New York subway are characterized by wide ranges in load factor—crush loads during travel peaks and very light loads during most off-peak hours. Patterns of passenger use and operating service characteristics act on absolute vehicle efficiency to generate the wide range of efficiencies observed.

Rail Transit Energy Budget

Figure 2.4-11 is an energy reconciliation for electric power distributed to the San Francisco BART system. BART's use of electric power (kW · h/car-mile) is typical of new high-performance heavy-rail transit systems. In the figure the energy use percentages are related to the original value of the energy used at the source. Thus, the input is the input to the power plant, where a 62% loss is associated with the conversion of petroleum or coal to electricity.

In a typical transit cycle a train dwells at a station using energy only for auxiliaries, accelerates out of the station to line speed, cruises at that speed while overcoming friction and windage, and decelerates into the next station. The energy uses associated with this cycle are shown in Fig. 2.4-12 for a typical case.

As the train accelerates, electric power is stored in the train as kinetic energy. When the train decelerates at the next station, that energy must either be dissipated as heat by friction or dynamic brakes, be regenerated back into a receptive load on the power distribution system by chopper, or be stored aboard the vehicle. In the example in Fig. 2.4-12, the stored energy of motion at the moment deceleration begins is 83% of the energy used on the 1-mi (1.6-km) run. Techniques for energy management in rail transit address both reduction of this energy requirement and provision for reuse of the energy of motion.

Energy Reduction

The energy needed to move rail transit cars may be reduced absolutely by reducing the mass or maximum velocity of the train. Energy may also be reduced relative to the amount of work done by lengthening the distance between stations. Figure 2.4-13 shows the relationship between transit system

TABLE 2.4-10 KEY PERFORMANCE INDICATORS AND ENERGY USE OF U.S. FIXED GUIDEWAY SYSTEMS[a]

City	Route Miles	Station Spacing	Passengers Capacity Total	Passengers Capacity Seated	Passengers Average Use	Vehicle Miles (million/yr)	Energy Consumed kW·h/car-mile	Energy Consumed kW·h/pass.-mile
Heavy rail								
Boston	34.2	0.68	239	64	NA	11.10	5.95	—
Chicago	89.0	0.62	150	49	15.9	49.65	4.98	0.31
Cleveland	19.0	1.11	140	80	NA	5.50	8.80	—
New York (NYCTA)	230.6	0.50	350	76	38.3	248.50	5.84	0.15
Philadelphia (SEPTA)	24.4	0.50	250	67	25.7	13.10	9.31	0.36
Light rail								
Shaker Heights	13.1	NA	NA	82	29.0	1.25	4.30	—
Philadelphia	20.8	NA	NA	68	18.2	2.85	6.06	—
Newark	4.3	NA	NA	73	10.2	0.55	5.70	—
AGT								
Tampa	0.7	0.09	100	—	6.0	0.41	57.6	9.6
Dallas/Ft. Worth	12.8	0.09	40	—	—	0.00	81.5	—
Morgantown	2.1	0.7	21	—	5.3	0.58	131.5	24.8
Houston	1.5	0.16	36	18	12.2	0.22	1.0	0.09

Sources. *APTA Transit Fact Book*, APTA, 1981; "The World in Transit," *Railway Age*, September 28, 1981; *Light Rail Transit*, U.S. Department of Transportation, Spring 1976; Pushkarev, *Urban Rail in America*, Indiana University Press, 1982; Turner and Wolf, *Houston Airport Peoplemover*, IEEE Vehicular Technology Conference, May 1982.

[a]Note: NA = Not Available; 1 mi = 1.609344 km.

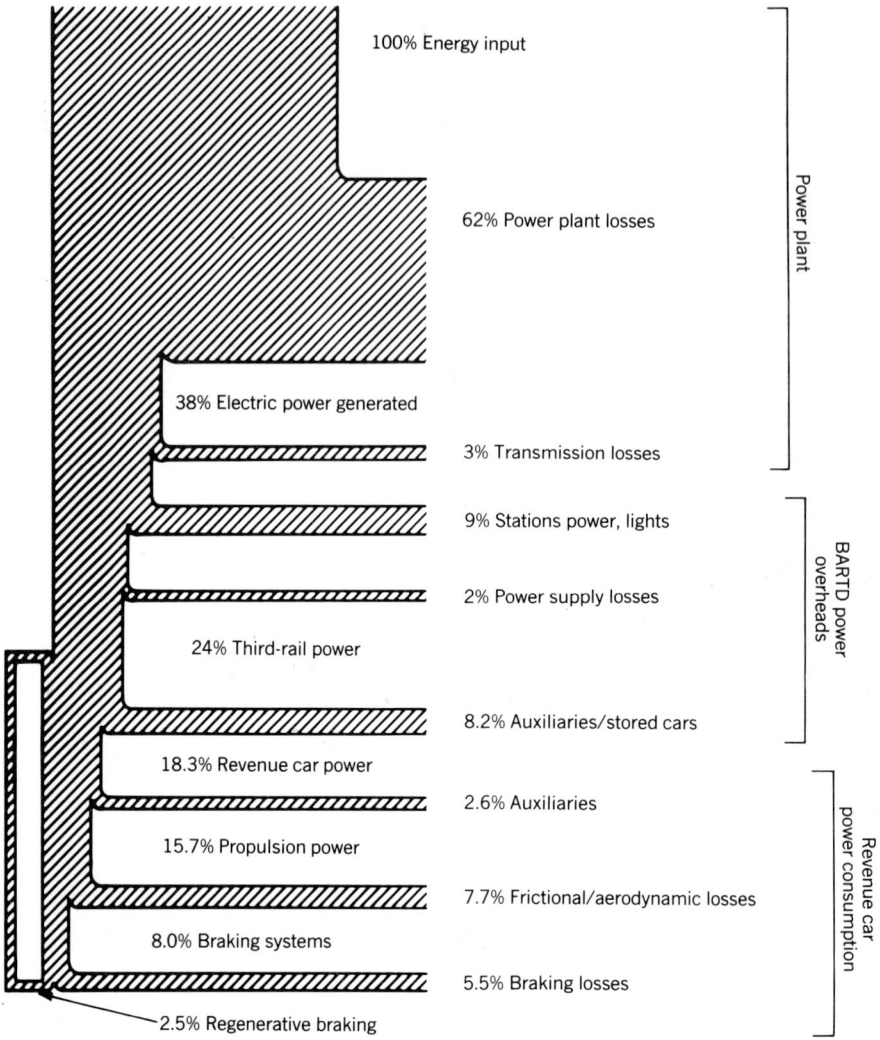

Fig. 2.4-11 BARTD energy utilization. (Source: Bolger, et al. "Application of Energy Storage Power Systems to NonHighway Transportation," Lawrence Livermore Laboratory, May 1977.)

energy consumption and station spacing for several weights of vehicles. The heaviest vehicle's variation in transit efficiency is 2.15 : 1 for the typical range of heavy-rail station spacing.

An operating policy that matches train length or number of cars (and therefore total weight) and schedule to passenger demand is necessary to control energy costs. A consequence of such a policy is the need for trains that can be lengthed or shortened quickly, without multiple train crews.

Another technique for reducing energy consumption is in reducing top operating speed—that is, degrading the performance of the system by increasing schedule times. Further energy savings can be achieved from coasting, if schedule times can be increased. Under a coasting strategy, a train is accelerated to its top speed as rapidly as possible, within the constraints of the performance limits and the electric utility demand charge. From that point the vehicle is allowed to coast to some lower speed. Figure 2.4-14 shows the potential improvements in transit car efficiency that can be attained by decreasing top speeds and by adopting a suitable top speed and coast operating strategy. In both cases schedule times must increase.

Control of dc traction motors for rail transit has historically been accomplished by switching resistors in series with the motor, by changing field and armature connections, and by field weakening.

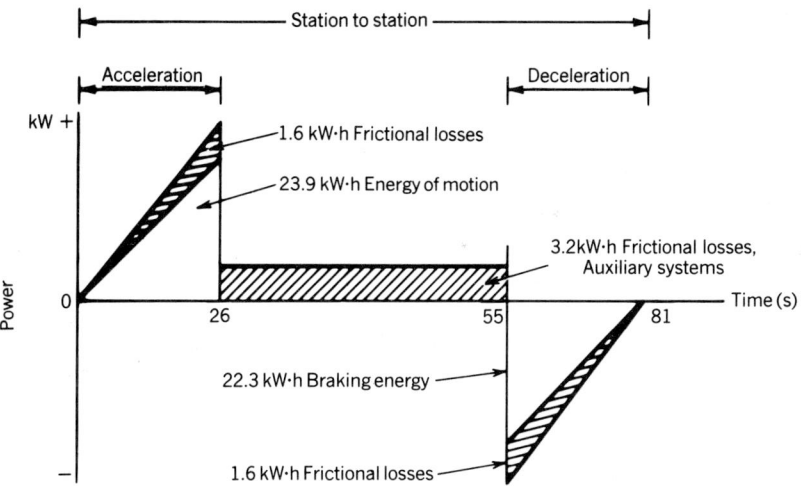

Fig. 2.4-12 Typical energy requirement for six-car 225-ton train over 1-mi run. (Source: R. A. Uher, "Energy Management for Electric Rail System," Carnegie-Mellon University.)

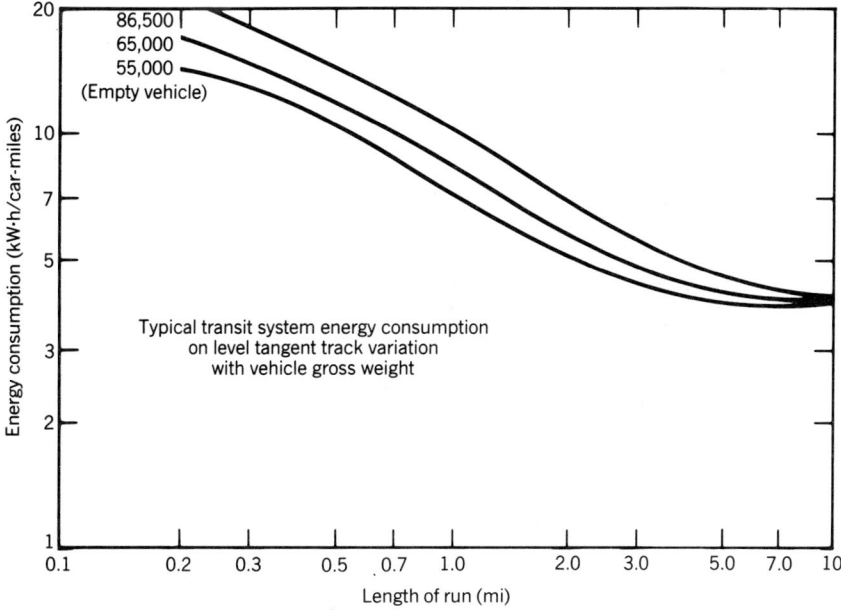

Fig. 2.4-13 Relationship between transit system efficiency station spacing and vehicle weight. (Source: R. A. Uher, "Energy Management for Electric Rail System," Carnegie-Mellon University.)

This technique, known as cam control, is inherently dissipative of energy during acceleration. The advent of thyristor chopper motor controls provided the capability to control the dc motor starting torque without dissipative current limiting. For a station spacing of 0.6 mi (1 km), the transit power requirement (kW · h/car-mile) decreases by 15–40% with chopper controls. However, for station spacing on the order of 2 mi (3.2 km) or greater, the influence of this loss is negligible.

The thyristor chopper brings another substantial advantage to rail transit—the ability to regenerate the energy of motion into dc power, which can then be recoupled to the dc power distribution system (third rail). If a suitable load is attached to the line, some of the kinetic energy can be reused.

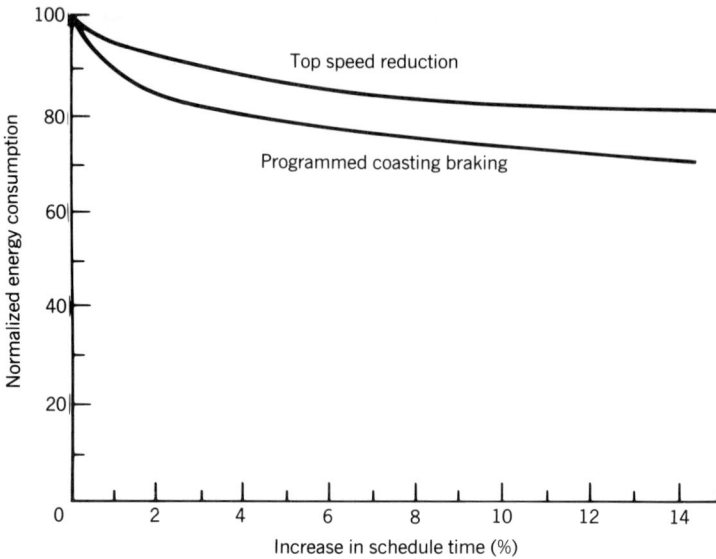

Fig. 2.4-14 Comparison of top speed reduction. (Source: R. A. Uher, "Energy Management for Electric Rail System," Carnegie-Mellon University.)

Energy Storage

A perfect system could recover about 50% of the third-rail energy. As shown in Fig. 2.4-11, the contribution of regeneration in the BART system is 10% of the third-rail energy use. Energy recovery is strongly influenced by system design, which must provide loads on the third rail to make it receptive to regenerated energy of motion. The receptivity of the line can be increased by running trains close together, which increases the probability that a nearby train will be accelerating when another is decelerating. Energy transfer between trains is increased by raising the maximum permissible

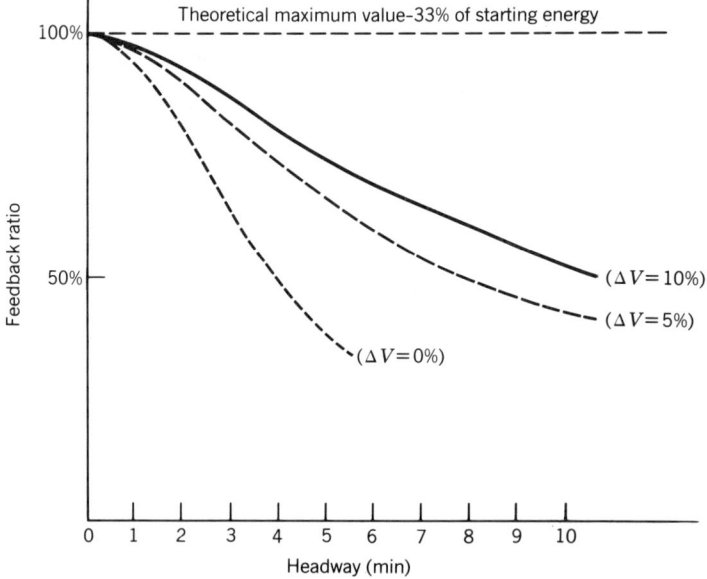

Fig. 2.4-15 Energy feedback ratio relative to headways. (Source: J. Amber, "The Effects of DC-Chopper Technology in Rapid Transit Passenger Transport," Federal German Ministry for Research and Technology, January 1978.)

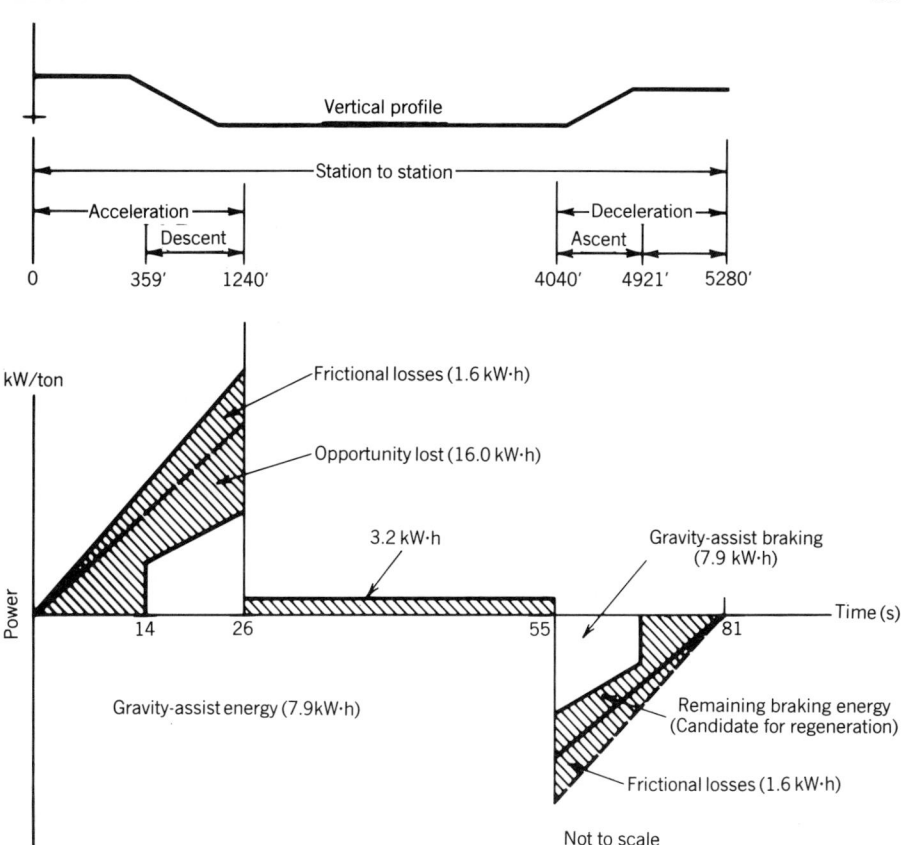

Fig. 2.4-16 Energy map for vertical-profile system (energy for six-car train with car weight of 75,000 lb). (Source: Booz, Allen & Hamilton, "Study of Energy Management Alternatives for SCRTD," April 1982.)

regeneration voltage with respect to the open-circuit substation voltage, by decreasing the power rail impedances, and by placing substations far from decelerating trains. Figure 2.4-15 shows the calculated values of recovered energy for a range of headways for three voltage differences between substation and regeneration voltage.

The receptivity of the system to regeneration can also be assured by providing storage for the energy on the vehicle or along the wayside, most likely by using flywheels. Studies and tests suggest that energy savings in the range of 25–35% are feasible. The receptivity of the system can also be increased by making the substations convert power bidirectionally. Regenerated power would then be transferred back into the utility supply. However, this approach does not appear to be cost-effective except on sustained downhill runs.

Energy can also be reused by providing a vertical profile track, where acceleration out of the station is aided by a downgrade and deceleration into a station is aided by an upgrade. The kinetic energy is partially recovered as gravitational potential energy. Figure 2.4-16 shows a gravity profile energy map of the 1-mi (1.6-km) station run shown in Fig. 2.4-12. In this example 31% of the accelerating energy is provided by the gravity profile. Further, even without regeneration, 35% of the energy of motion is recovered; additional gains can be achieved with a coasting strategy. However, special civil work and propulsion and train controls must be provided that recognize and integrate the gravity energy contribution; the cost of such civil work and additional control systems often outweigh the energy savings potential.

REFERENCES

2.4-1 T. C. Austin and K. H. Hellman, "Passenger Car Fuel Economy—Trends and Influencing Factors," SAE Paper 730790, February 1973.

2.4-2 H. Hsia and J. A. Kidd, "Weight Reduction Potential of Automobiles and Light Trucks," Report No. DOT-TSC-NHTSA-79-54, Transportation Systems Center, Cambridge, MA, October 1979.

2.4-3 S. Dinda, et al., "High Strength Steels in Production Automobiles," SAE Paper 770139.

2.4-4 E. P. Trueman, "Weight Saving Approaches Through the Use of Fiber Glass-Reinforced Plastic," SAE Paper 750155.

2.4-5 J. A. Foster, et al., "Light Duty Automotive Fuel Economy . . . Trends Through 1981," SAE Paper 810386, 1981.

2.4-6 Masanoi Harada, et al., "Nissan Naps 'Z' Engine Realizes Better Fuel Economy and Low Nox Emission," SAE Paper 810010, February 1981.

2.4-7 *Ward's Engine Update* **7**(13), July 1981.

2.4-8 J. Buike, et al., "Supercharging for Fuel Economy," SAE Paper 810006, February 1981.

2.4-9 *Ward's Automotive Reports*, January 18, 1982, p. 19.

2.4-10 R. Cichocki and W. Cartillieri, "The Passenger Car Direct Injection Diesel—A Performance and Emissions Update," SAE Paper 810480, February 1981.

2.4-11 M. W. Dowdy and N. P. Nightingale, "Mod. I Automotive Stirling Engine System Performance," SAE Paper 820353, February 1982.

2.4-12 Charles Jones, "An Update of Applicable Automotive Engine Rotary Stratified Charge Developments," SAE Paper 820347, February 1982.

2.4-13 R. H. Thring, "Engine Transmission Matching," SAE Paper 810446, February 1981.

2.4-14 A. Numazawa, et al., "Toyota Four Speed Automatic Transmission With Overdrive" SAE Paper 780097, March 1978.

2.4-15 "GM 4 Speed Automatics," *Ward's Engine Update* **7**(16), August 1981.

2.4-16 D. J. Bickerstaff, "Light Truck Fuel Economy by Design Efficiency," SAE Paper 781063, February 1978.

2.4-17 "Volkswagen Formula E," *Ward's Engine Update* **7**(12), June 1981.

2.4-18 Sam Dabich, "Ford Motor Company Automatic Overdrive Transmission," SAE Paper 800004, February 1980.

2.4-19 "Automatics vs. CVT," *Ward's Engine Update* **7**(20), October 1981.

2.4-20 N. H. Beachly and A. A. Frank, "Design Consideration for Flywheel-Transmission Automobiles," SAE Paper 800886, August 1980.

2.4-21 "VW ARVW," *Ward's Automotive Update* **7**(1), January 1981.

2.4-22 L. T. Janssen and H. J. Emmelmann, "Aerodynamic Improvement—A Great Potential for Better Fuel Economy," SAE Paper 780265, March 1978.

2.4-23 R. W. Zub and R. G. Colello, "Effect of Vehicle Design Variables on Top Speed, Performance and Fuel Economy," SAE Paper 800215, February 1980.

2.4-24 W. H. Hocho, et al., "The Optimization of Body Details—A Method for Reducing the Aerodynamic Drag of Road Vehicles," SAE Paper 760185, February 1976.

2.4-25 R. Buchheim, et al., "Necessity and Premises for Reducing the Aerodynamic Drag of Future Passenger Cars," SAE Paper 810185, February 1981.

2.4-26 G. D. Thompson and M. E. Reineman, "Tire Rolling Resistance and Vehicle Fuel Consumption," SAE Paper 810168, February 1981.

2.4-27 B. C. Grugett, et al., "The Effect of Tire Inflation Pressure on Passenger Car Fuel Consumption," SAE Paper 810069, February 1981.

2.4-28 M. A. Yurjevich, "The Effect of Stabilizer Ply Geometry on Rolling Resistance and Cornering Force Properties," SAE Paper 810065, February 1981.

2.4-29 U.S. Environmental Protection Agency, "OMSAPC Advisory Circular," A/C No. 55B, December 1978.

2.4-30 H. Oetting, "Passenger Car Spark Ignition Data Base," Report No. DOT-TSC-NHTSA-79-16.II, Transportation Systems Center, 1979.

2.4-31 R. W. Zub and R. P. Meisner, "Effect of Vehicle Design Variables on Top Speed, Performance and Fuel Economy—1981," Transportation Systems Center, Report No. DOT-TSC-NHTSA-81-23, April 1981.

2.4-32 G. A. Woollard, "Morse Controlled Speed Accessory Drive," Morse Chain Division, Borg Warner Corporation, Ithaca, NY.

2.4-33 F. C. Porter, "Design for Fuel Economy—The New GM Front Drive Cars," SAE Paper 790721, June 1979.

BIBLIOGRAPHY

Amler, J. "The Effects of the DC-Chopper Technology in Public Rapid Transit Passenger Transport," Federal German Ministry for Research and Technology, January 1978.

APTA Transit Fact Book, 1981.

Bolger, et al. "Application of Energy Storage Power Systems to Non-highway Applications." Lawrence Livermore Laboratory, May 1977.

Changes in the Future Use and Characteristics of the Automobile Transportation System, Vol. II, p. 328. Office of Technology Assessment, Washington, D.C.

"Future Passenger Car Diesels May Be Direct Injection," *Automotive Engineering*, June 1981, p. 51.

Hafter and Hanocq. *Energy in Metropolitan Railways*. International Union of Public Transport, Brussels, 1979.

Henderson, et al. *Energy Study of Automated Guideway Transit Systems*. U.S. Department of Energy, August 1979.

Hob, Daniel J. "Aerodynamics: The New Automotive Frontier." *Automotive Engineering*, January 1981, p. 50.

McNutt, B. D. "Comparison of Gasoline and Diesel Automotive Fuel Economy as Seen by the Consumer," SAE Paper 810387, 1981.

"Microprocessor Control Improves Economy, Reduces Emissions," *Automotive Engineering*, February 1980, p. 51. Test was on the Japanese 10-mode driving cycle.

"Passenger Car Fuel Economy: EPA and Road," EPA 46013-80-010. September 1980, pp. 135 and 160.

Pushkarev. *Urban Rail in America*. Indiana University Press, 1982.

Uher. *Energy Management for Electric Rail Systems*, Carnegie-Mellon University, June 1980.

U.S. Department of Transportation, *Light Rail Transit*, Spring 1976.

CHAPTER 3

ENERGY UTILIZATION LAWS AND PRINCIPLES

BERNARD D. WOOD

Syracuse University
Syracuse, New York

MICHEL A. SAAD

University of Santa Clara
Santa Clara, California

WILLIAM W. DURGIN

Worcester Polytechnic Institute
Worcester, Massachusetts

3.1 THERMODYNAMICS

Bernard D. Wood

3.1-1 Explanation of Terms

A system to be considered in a thermodynamic analysis is either determined by obvious physical boundaries or may be established by imaginary boundaries chosen arbitrarily for convenience. All things with which this system interacts by exchange of mass or energy make up the surroundings. The system plus surroundings therefore make an isolated system, which does not interact with anything else, and is sometimes called the universe of this analysis. A closed system does not exchange mass with its surroundings.

Property values establish the thermodynamic state point of the system and are therefore point functions. Because property values depend only on the state point and not on the path taken to reach the point, infinitesimal changes in property values are treated as exact differentials, written dX. Properties such as pressure and temperature, which are not affected by the amount of mass being considered, are intensive properties; properties such as volume and internal energy are extensive properties, the values of which do depend on the amount considered. Specific values of extensive properties, such as volume per unit mass, are intensive properties.

In a quasistatic process the property values change progressively through a series of equilibrium conditions from one state point to another. Thermodynamic equilibrium implies mechanical, chemical, and thermal equilibrium, the last implying an absence of temperature difference. A completely reversible process (impossible in nature) can be returned along the same path (succession of state points) to the starting point, leaving no change in the system and surroundings. In an internally reversible process the system can be made to retrace its path exactly. Processes may be adiabatic (no heat transfer), isothermal (constant temperature), isentropic (constant enthalpy), isobaric (constant pressure), isochoric (constant volume), and polytropic (several property values changing simultaneously).

3.1-2 Units of Energy and Power

The basic unit of energy in the SI system is the newton-meter (N · m); in the English Engineering system, it is the foot-pound, force (ft · lbf). Conversion to other systems of units will be found in Chapter 18. A newton-meter is also called a joule (J); 1 kilojoule is 1000 N · m.

Power is the rate of energy transfer. The basic SI unit is the watt (W), which is one joule per second. One kilowatt is equal to 1 kilojoule per second. Also by definition, one horsepower (hp) is energy transfer at a rate of 550 foot-pound (force) per second: 1 hp = 0.7457 kW.

The concept of force (F) derives from Newton's second law of motion: $F = ma/g_c$. By definition, one newton will accelerate one kilogram (mass) at the rate of one (meter per second) per second, so that the value of the dimensional constant (g_c) in SI units is 1.0 kg · m/N · s^2. Similarly, in the English Engineering system of units, one pound (force) will accelerate a pound (mass) at the rate of standard gravitational acceleration, which is 9.80665 m/s^2 or approximately 32.174 ft/s^2. Thus, the dimensional constant in this system is 32.174 lb · ft/lbf · s^2.

3.1-3 Energy Terms

Work ($\mathcal{W}$) is defined as the product of a force (F) acting through a collinear distance (x) so that

$$\delta\mathcal{W} = F\,dx + PA\,dx = P\,dV$$

In this section, work done by the system will be considered positive so that when the change of volume is positive, the work will be positive. (This is an arbitrary sign convention. Some authorities prefer to consider work done on the system to be positive, which would change the sign of $\mathcal{W}$ throughout this section.)

Displacement work ($\mathcal{W}_D$) occurs at the boundary of a system when the volume changes: $\mathcal{W}_D = \int P\,dV$. This can be integrated when a functional relationship is known between pressure and volume.

1. When P is constant,

$$\mathcal{W}_D = P(V_2 - V_1) \tag{3.1-1}$$

2. When PV is constant,

$$\mathcal{W}_D = PV \ln\frac{V_2}{V_1} = PV \ln\frac{P_1}{P_2} \tag{3.1-2}$$

3. When PV^n is constant,

$$\mathcal{W}_D = \frac{P_1V_1 - P_2V_2}{n - 1} \tag{3.1-3}$$

Indicated power is the rate at which work is done by or on a working fluid inside a machine. The term derives from the fact that this was commonly determined by means of a mechanical device called an *indicator*, which automatically produced a plot of pressure on volume for a cycle in a reciprocating engine or compressor. From such a diagram the indicated mean effective pressure (P_I) is the enclosed area (a) divided by the diagram length (l) multiplied by the vertical pressure scale (s), so that $P_I = (a/l) \cdot s$. Then the indicated power in SI units is $P_I ALN$ kilowatts

where P_I = indicated mean effective pressure, kPa
  A = piston area, m^2
  L = stroke length, m
  N = working cycles per second

Alternatively, indicated power is $(P_I ALN)/33,000$ hp

where P_I = indicated mean effective pressure, lbf/in.^{2}
  A = piston area, in.^{2}
  L = stroke length, ft
  N = working cycles per minute

Shaft work ($\mathcal{W}_{Sh}$) is work done by a rotating shaft or work that could be imagined to be easily convertible to work at a rotating shaft, such as net displacement work converted through a mechanism or electrical work converted through an ideal motor. In one revolution, for a force F at a radius R,

$$\mathcal{W}_{Sh} = F(2\pi R) = 2\pi(\text{torque}) \tag{3.1-4}$$

The units of shaft work will be the same as the units of torque, either newton-meters or foot-pound (force).

Shaft power is the rate at which shaft work is obtained from or delivered to a working system and is

$$2\pi(\text{torque}) \cdot (\text{rpm})/60 = (\text{torque})\,\omega \tag{3.1-5}$$

where ω is the rate of rotation in radians per second and torque is in newton-meters.

Brake power is synonymous with shaft power and may be either input or output shaft power. Thus, brake horsepower is $2\pi(\text{torque}) \cdot (\text{rpm})/33,000$ where torque is in foot-pound (force).

The brake mean effective pressure (P_B) is a hypothetical pressure related to brake power or shaft power in the same way that indicated mean effective pressure is related to indicated horsepower. Thus,

$$P_B = \frac{\text{shaft power}}{ALN} \qquad P_B = \frac{\text{shaft power, kw}}{ALN} \quad \text{kPa} \tag{3.1-6}$$

or

$$P_B = \frac{(\text{shaft horsepower})(33,000)}{ALN} \quad \text{lbf/in.}^2 \tag{3.1-7}$$

where A, L, and N are in appropriate units.

Flow work ($\mathcal{W}_F$) is the work necessary to force a fluid across a boundary against a pressure P. For a volume V, $\mathcal{W}_F = PV$. Flow work per unit mass ($\mathcal{W}_F$) is $P(V/m) = Pv$.

Electrical work ($\mathcal{W}_E$) is generally expressed as power multiplied by time. Since 1 watt is 1 joule per second, $\mathcal{W}_E = (\text{power}) \cdot (\text{time})$, for example, watt-second or kilowatt-hour.

Heat transfer (Q) is energy transfer from one part of a system to another part or from the surroundings to the system, solely because of temperature differences. By convention, heat transfer to a system is considered positive. Heat transfer, like work, is energy in transit. Both are path functions, their values depending on the path taken between state points, in contrast with properties that are point functions. For this reason infinitesimal amounts of heat transfer or work are inexact differentials, written δQ or $\delta\mathcal{W}$.

3.1-4 Efficiency and Coefficient of Performance

In general, efficiency is simply the ratio of the desired output actually obtained to the input required. For a thermodynamic cycle the purpose of which is to produce useful work or power, the thermal efficiency is

$$\eta_{th} = \frac{\text{net useful work}}{\text{energy supplied}} = \frac{\sum \mathcal{W}}{Q_H}$$

In any mechanical system there will be friction loss between the work done at some location and the useful work resulting from this at another location. For instance, the power produced by the working fluid in an engine or turbine (commonly called indicated power) is reduced by friction before the energy is available as shaft power (sometimes called brake power). The ratio is the mechanical efficiency, which is

$$\eta_M = \frac{\text{shaft power}}{\text{indicated power}}$$

On the other hand, for a compressor or pump, the friction reduces the input shaft power before useful work is actually done on the working fluid, still conventionally called indicated power. In this type of machine

$$\eta_M = \frac{\text{indicated power}}{\text{shaft power}}$$

For a process requiring work input, for example, compression, the efficiency is the ratio of the ideal (minimum) work required for the intended process to the actual work supplied:

$$\eta_C = \frac{\text{ideal work}}{\text{actual work}}$$

For a process delivering useful work, for example, expansion, the efficiency is the ratio of actual work delivered to the ideal (maximum) work obtainable from the intended process:

$$\eta_T = \frac{\text{actual work}}{\text{ideal work}}$$

Where the process is intended to be isothermal or isentropic, the efficiency should be designated with a subscript t or s, respectively, for example, $\eta_{C(t)}$.

When the term *efficiency* is applied to a heat exchanger where two substances flow through the device in thermal contact, the efficiency is the ratio of the enthalpy change achieved in one substance to the maximum possible enthalpy change in that substance. The maximum change would be achieved if its final temperature were to be brought to the inlet temperature of the other substance:

$$\eta_{HT} = \frac{\Delta h\,(\text{actual})}{\Delta h\,(\text{max})}$$

When specific heat values are practically constant, this may be written as

$$\eta_{HT} = \frac{\Delta T(\text{actual})}{\Delta T(\text{ideal})}$$

where ΔT is the temperature change of the one substance.

For a refrigeration or heat pump system or cycle the term *efficiency* is inappropriate because the desired effect may be greater than the energy supplied to achieve it. The appropriate term is *coefficient of performance* (COP). For a refrigerator the desired effect is heat transfer from a low-temperature source (Q_L). For a vapor compression refrigeration cycle the energy input is work ($\mathcal{W}$), and therefore, $\text{COP}_R = Q_L/\mathcal{W}$. For an absorption refrigeration system the energy input is generally heat transfer from a high-temperature source (Q_H), and then $\text{COP}_R = Q_L/Q_H$. For a heat pump the desired effect is generally heat transfer to a higher temperature (Q_H), and the energy input is generally work ($\mathcal{W}$). Thus, $\text{COP}_{HP} = Q_H/\mathcal{W}$.

3.1-5 Basic Laws of Thermodynamics

The laws of thermodynamics, like all laws of nature, can have no proof; they are the formal statements of the eventual conclusions derived from repeated observation and verification. Natural laws are confirmed by the inability of repeated scientific research to find exceptions. Therefore, it is to be expected that there should be various statements or formulations of these laws. The statements given here are the most useful for the material in this section.

Zeroth Law

The zeroth law precedes and is basic to the concept of a thermodynamic temperature scale, which will be dealt with later. It states:

If two bodies are each in thermal equilibrium with (at the same temperature as) a third body, they are in thermal equilibrium with (at the same temperature as) each other.

First Law

The first law of thermodynamics is essentially a law of conservation of energy and assumes a conservation of mass. It states:

For a closed system executing a thermodynamic cycle, the cyclic integral of work done by the system is exactly equal to the cyclic integral of heat transfer to the system.

$$\oint Q = \oint \mathcal{W}$$

The implication of this statement for a closed, stationary system following any process from state point 1 to state point 2 is the existence of a property called internal energy (U) such that

$$U_2 - U_1 = \int_1^2 \delta Q - \int_1^2 \delta \mathcal{W} \tag{3.1-8}$$

or

$$dU = \delta Q - \delta \mathcal{W} \tag{3.1-9}$$

The law of conservation of energy and mass, which is an essential part of the first law of thermodynamics, when applied to an open system between time 1 and time 2 dictates that:

From the beginning to the end of a particular time period, the difference between the sums of all the energy values associated with the masses within the control volume of a system must be accounted for by the net sum of all the energies associated with all the masses crossing the control surface, plus the net heat transfer to the system, minus the net sum of all the forms of work done by the system.

Mathematically, this may be stated as

$$\sum m_2 \left(u_2 + \frac{\overline{V}_2^2}{2g_c} + \frac{g}{g_c} z_2 \right) - \sum m_1 \left(u_1 + \frac{\overline{V}_1^2}{2g_c} + \frac{g}{g_c} z_1 \right)$$

$$= \left[\sum m_i \left(u_i + \frac{\overline{V}_i^2}{2g_c} + \frac{g}{g_c} z_i \right) - \sum m_e \left(u_e + \frac{\overline{V}_e^2}{2g_c} + \frac{g}{g_c} z_e \right) \right]$$

$$+ \sum Q - \left[\sum \mathcal{W}_{Sh} + \int_1^2 P\, dV + \sum m_e (P_e v_e) - \sum m_i (P_i v_i) \right] \tag{3.1-10}$$

where internal energy (u), kinetic energy ($\overline{V}^2/2g_c$), potential energy [$(g/g_c)z$], heat transfer (Q), and shaft work, displacement work, and flow work are the energies of interest. For each identified mass entering the system (m_i) or leaving the system (m_e), the flow work term (Pv) may conveniently be grouped with the internal energy term (u) as enthalpy (h), which is defined as $h = u + Pv$.

For an open system in which there is a uniform state throughout at time 1 and again at time 2, and for which the boundaries are fixed ($dV = 0$), the first law may be simplified to

$$m_2 \left(u_2 + \frac{\overline{V}_2^2}{2g_c} + \frac{g}{g_c} z_2 \right) - m_1 \left(u_1 + \frac{\overline{V}_1^2}{2g_c} z_1 \right)$$

$$= \sum m_i \left(h_i + \frac{\overline{V}_i^2}{2g_c} + \frac{g}{g_c} z_i \right) - \sum m_e \left(h_e + \frac{\overline{V}_e^2}{2g_c} + \frac{g}{g_c} z_e \right) + \sum Q - \sum \mathcal{W}_{Sh} \tag{3.1-11}$$

When there is steady state within the control volume (no change in energy values) and steady flow ($m_i = m_e$) as well as fixed boundaries ($dV = 0$), this reduces to

$$\sum m_i \left(h_i + \frac{\overline{V}_i^2}{2g_c} + \frac{g}{g_c} z_i \right) + \sum Q = \sum m_e \left(h_e + \frac{\overline{V}_e^2}{2g_c} + \frac{g}{g_c} z_e \right) + \sum \mathcal{W}_{Sh} \tag{3.1-12}$$

Finally, when there is steady state, steady flow, fixed boundaries, and only one inlet and one exit so that $m_i = m_e = m$, then mass can be eliminated, and the first law becomes

$$\left(h_i + \frac{\overline{V}_i^2}{2g_c} + \frac{g}{g_c} z_i \right) + q = \left(h_e + \frac{\overline{V}_e^2}{2g_c} + \frac{g}{g_c} z_e \right) + \sum \mathcal{W}_{Sh} \tag{3.1-13}$$

Where changes in kinetic and potential energies are negligible, this reduces to

$$(h_i - h_e) + q = \mathcal{W}_{Sh} \tag{3.1-14}$$

When the definition of enthalpy is combined with the first law and applied to the property values of a unit mass, which is a closed system, this last equation can be expressed in terms of pressure and specific volume:

$$h = u + Pv$$

$$dh = du + P\,dv + v\,dP = \delta q + v\,dP$$

Between the inlet (i) and the exit point (e)

$$(h_e - h_i) - \sum q = \int_i^e v\,dP = -\sum W_{\text{Sh}} \tag{3.1-15}$$

Of course, changes in kinetic and potential energies, when significant, must be added.

Second Law

There are three statements of the second law of thermodynamics that will be useful in this section. It can be shown that they are in fact equivalent to one another because a violation of any one statement will result in a violation of the other two.

The Kelvin–Plank Statement.   No system operating continuously in a cycle can receive heat transfer from a single source at a uniform temperature and convert all of that energy to useful work.

The Clausius Statement.   No system can cause net heat transfer to flow from a source at some temperature to a sink at a higher temperature spontaneously, that is, without some other effect being required.

The Caratheodory Statement.   In the vicinity of any equilibrium state point of a system, there are other state points that cannot be reached by a reversible adiabatic path.

Consequences or Corollaries of the Second Law.   As a consequence of the second law of thermodynamics, a number of secondary principles must follow. In fact, some of these, to be stated below, could serve as satisfactory statements of the second law, from which the statements above would follow.

   1.   The existence of the property internal energy followed from the first law. The second law dictates that there is a property, entropy (S), defined as $dS = (\delta Q/T)_R$. The subscript R indicates that this definition is restricted to a reversible process. Entropy is the only thermodynamic property defined with such a restriction.
   2.   The thermal efficiency of any working cycle is less than unity, $\eta_{\text{th}} < 1.0$.
   3.   The Carnot cycle is an ideal reversible cycle composed of two isothermal processes and two adiabatic processes. All the positive heat transfer (Q_H) is at the higher temperature, and all the negative heat transfer (Q_L) is at the lower temperature. From the first law $W = Q_H - Q_L$. Thus

$$\eta_{\text{th}} = \frac{Q_H - Q_L}{Q_H} = 1 - \frac{Q_L}{Q_H}$$

The thermodynamic temperature scale is defined by the equality

$$\frac{T_L}{T_H} = \frac{Q_L}{Q_H}$$

where Q_H and Q_L are the positive and negative heat transfers in a Carnot cycle.
   Only one point need be established on this scale, the triple point of pure water, which is 273.16 K, defined as 0.01°C (see Fig. 3.1-1).
   The thermodynamic temperature scale is compatible with the familiar ideal-gas temperature scale, which can be defined as

$$\theta = 273.16 \lim_{P_3 \to 0} \frac{P}{P_3} \quad \text{(at constant volume)}$$

where P is the gas pressure at temperature θ and P_3 is the pressure of the same gas in the same volume but at the triple point of water, as P_3 is extrapolated to zero. For an ideal gas (Section 3.1-6) the pressure would approach zero as the temperature approaches zero, at constant volume.
   4.   The maximum possible thermal efficiency of a working cycle between a source at T_H and a sink at T_L is the efficiency of a reversible cycle receiving heat only at T_H and rejecting heat only at T_L. The Carnot cycle is an example of such a reversible cycle. Its thermal efficiency is $\eta_{\text{th}} = 1 - (T_L/T_H)$.

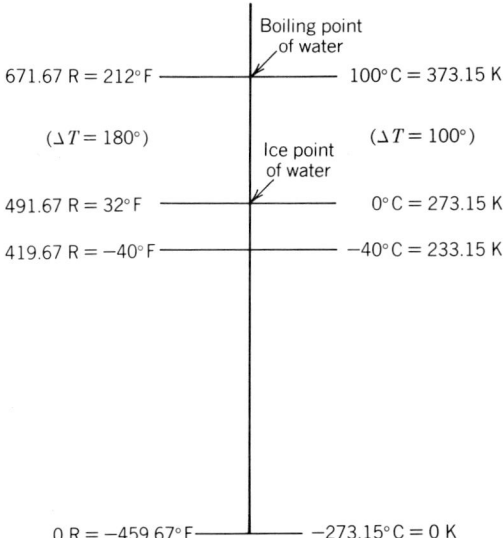

Fig. 3.1-1   Exact equivalents on the Kelvin, Celsius, Rankine, and Fahrenheit temperature scales.

5.   All reversible cycles receiving heat at T_H and rejecting heat at T_L will have the same thermal efficiency. Other such cycles are the ideal Stirling cycle with perfect regeneration and the ideal Ericsson cycle with perfect regeneration.

6.   For a refrigeration system or heat pump, the maximum possible coefficient of performance is achieved by a reversible cycle such as the Carnot cycle, receiving heat only at T_L and rejecting heat only at T_H.

$$\text{Maximum COP}_R = \frac{T_L}{T_H - T_L} \qquad \text{Maximum COP}_{HP} = \frac{T_H}{T_H - T_L}$$

7.   The inequality of Clausius states that for any cycle, $\oint(\delta Q/T) \leq 0$, and consequently, for any process, $dS \geq \delta Q/T$. The equalities hold only for a completely reversible cycle or process.

8.   The entropy principle or the principle of increase of entropy states that, for a system plus its surroundings, $\Sigma\,dS \geq 0$. Consequently, for an isolated system or any adiabatic process, $\Sigma\,dS \geq 0$. Again, the equalities hold only for reversible situations.

Third Law

As with the second law of thermodynamics, there are several statements of the third law. All of these are essentially equivalent.

1. The entropy change associated with any isothermal reversible process of a condensed system approaches zero as the temperature approaches zero.
2. It is impossible to reduce the entropy of a system to its zero-point value by any finite number of operations.
3. It is impossible to reduce the temperature of a system to absolute zero in any finite number of operations.

This last statement can be called the "principle of unattainability of absolute-zero temperature."

3.1-6 Thermodynamic Properties of a Pure Substance

A pure substance is one of nonvarying composition. For a constant mass of such a substance, the values of two independent properties establish the thermodynamic state point and therefore the values of all dependent properties. That is, there is a functional relationship among the values of three properties, and these may be treated by the normal rules of mathematics. For instance, if some function of x, y, and z is equal to zero,

$$dz = \left(\frac{\partial z}{\partial x}\right)_y dx + \left(\frac{\partial z}{\partial y}\right)_x dy = M\,dx + N\,dy$$

and

$$\left(\frac{\partial M}{\partial y}\right)_x = \left(\frac{\partial N}{\partial x}\right)_y$$

Also,

$$\left(\frac{\partial x}{\partial y}\right)_z \left(\frac{\partial y}{\partial z}\right)_x \left(\frac{\partial z}{\partial x}\right)_y = -1$$

Properties are point functions, so that changes in property values can be evaluated along any path. If a reversible path is chosen, $\delta Q = T\,dS$. For a pure substance of constant mass, the work is simply displacement work ($P\,dV$). Consequently, the first law may be written: $T\,dS + dU + P\,dV$. When this is combined with the mathematical principles above, the various energy terms can be converted to other property values.

Internal Energy. From the first law $dU = T\,dS - P\,dV$. Thus,

$$T = \left(\frac{\partial U}{\partial S}\right)_V \qquad P = -\left(\frac{\partial U}{\partial V}\right)_S$$

and

$$\left(\frac{\partial T}{\partial V}\right)_S = -\left(\frac{\partial P}{\partial S}\right)_V \quad \text{(Maxwell's equation I)}$$

Enthalpy. By definition, $H = U + PV$

$$dH = dU + P\,dV + V\,dP = T\,dS + V\,dP$$

Thus,

$$T = \left(\frac{\partial H}{\partial S}\right)_P \qquad V = \left(\frac{\partial H}{\partial P}\right)_S$$

and,

$$\left(\frac{\partial T}{\partial P}\right)_S = \left(\frac{\partial V}{\partial S}\right)_P \quad \text{(Maxwell's equation II)}$$

Helmholtz Function or Helmholtz "Free Energy." By definition,

$$F = U - TS$$

$$dF = dU - T\,dS - S\,dT = -P\,dV - S\,dT$$

Thus,

$$P = -\left(\frac{\partial F}{\partial V}\right)_T \qquad S = -\left(\frac{\partial F}{\partial T}\right)_V$$

and

$$\left(\frac{\partial S}{\partial V}\right)_T = \left(\frac{\partial P}{\partial T}\right)_V \quad \text{(Maxwell's equation III)}$$

Gibbs Function or Gibbs "Free Energy." By definition,

$$G = H - TS$$

$$dG = dH - T\,dS - S\,dT = T\,dS + V\,dP - T\,dS - S\,dT = V\,dP - S\,dT$$

Thus,

$$V = \left(\frac{\partial G}{\partial P}\right)_T \qquad S = -\left(\frac{\partial G}{\partial T}\right)_P$$

and

$$\left(\frac{\partial V}{\partial T}\right)_P = -\left(\frac{\partial S}{\partial P}\right)_T \quad \text{(Maxwell's equation IV)}$$

Specific Heats

The specific heat of a substance is a property of that substance, and its value is a function of other property values. In addition, a process is implied, so that the value of specific heat must depend on the implied process. Two particular specific heats are of interest.

Specific Heat at Constant Volume. By definition

$$C_V = \left(\frac{\partial u}{\partial T}\right)_V$$

From the First Law applied to a unit mass of a given substance

$$du = \delta q - P\,dv = T\,ds - P\,dv$$

Thus at constant volume ($P\,dv = 0$), and in the absence of any other work term

$$du = T\,ds, \quad \text{and} \quad C_V = \left(\frac{\delta q}{dT}\right)_V$$

This is the historic concept of specific heat as related to heat transfer.

Specific Heat at Constant Pressure. By definition

$$C_P = \left(\frac{\partial h}{\partial T}\right)_P$$

From the definition of enthalpy

$$dh = du + P\,dv + v\,dP = \delta q + v\,dP = T\,ds + v\,dP$$

Again it is seen that the historic concept of specific heat, as related only to heat transfer, is compatible with the definition of C_P in the absence of work other than $P\,dv$. When $dP = 0$, then $dh = T\,ds$, and $C_P = (\delta q/T)_P$

Volume Expansivity (β)

Volume expansivity is the rate of change of volume with temperature at constant pressure:

$$\beta = \frac{1}{V}\left(\frac{\partial V}{\partial T}\right)_P$$

Compressibility (κ)

Compressibility is the rate of change of volume with pressure in a compression or expansion process.

Isothermal compressibility: $\qquad \kappa_T = -\frac{1}{V}\left(\frac{\partial V}{\partial P}\right)_T$

Isentropic compressibility: $\qquad \kappa_S = -\frac{1}{V}\left(\frac{\partial V}{\partial P}\right)_S$

T ds Equations

Entropy (S), being a property, can be considered as a function of any two others, for example, temperature (T), volume (V), or pressure (P). When this is stated mathematically and the appropriate

Maxwell equation and definition of specific heat are substituted for the partial derivatives, three $T\,ds$ equations result:

$$T\,ds = C_V\,dT + T\left(\frac{\partial P}{\partial T}\right)_V dV$$

$$T\,ds = C_P\,dT - T\left(\frac{\partial V}{\partial T}\right)_P dP$$

$$T\,dS = C_P\left(\frac{\partial T}{\partial V}\right)_P dV + C_V\left(\frac{\partial T}{\partial P}\right)_V dP$$

Energy Equations

Similarly, from the first-law property relationship with appropriate substitutions from the Maxwell equations, two energy equations are derived:

$$\left(\frac{\partial U}{\partial V}\right)_T = T\left(\frac{\partial P}{\partial V}\right)_V - P$$

$$\left(\frac{\partial U}{\partial P}\right)_T = -T\left(\frac{\partial V}{\partial T}\right)_P - P\left(\frac{\partial V}{\partial P}\right)_T$$

Specific Heat Relationships

When the first and second energy equations are combined mathematically, and the definitions of volume expansivity and compressibility are inserted, two important specific heat relationships result:

$$C_P - C_V = -T\left(\frac{\partial V}{\partial T}\right)_P^2 \left(\frac{\partial P}{\partial V}\right)_T = \frac{TV\beta^2}{\kappa}$$

$$\gamma = \frac{C_P}{C_V} = \frac{\kappa_T}{\kappa_S}$$

The Thermodynamic Surface

Because of the functional relationships among property values, the thermodynamic conditions at which a unit mass (1 kg or 1 lb) of a particular simple substance can exist are strictly limited. On pressure–volume–temperature coordinates, the possible state points form a thermodynamic surface, as shown in Fig. 3.1-2. The shape is for a substance that expands on melting. (Only water, because of its unusual crystal structure as a solid, expands when it solidifies.) The coordinates are logarithmic because of the very wide range of values between significant points for most substances.

The thermodynamic conditions at which two or more phases (solid, liquid, and vapor) can exist in equilibrium are saturation conditions. Because the pressure at which a phase change occurs establishes the temperature (or vice versa), the saturation conditions are lines rather than areas on pressure–temperature coordinates. A phase change generally implies a volume change, so that saturation conditions form a surface on pressure–volume or temperature–volume coordinates.

A pressure and temperature at which three phases can exist in equilibrium is a triple point, which is a line on $P\text{-}v$ or $T\text{-}v$ coordinates. The substance can exist as a liquid and a vapor in equilibrium only in the region below the critical-point values of pressure and temperature. Above the critical point, the two phases cannot be distinguished.

When thermodynamic property values at saturation conditions are tabulated, it is conventional to use the subscript f to indicate saturated liquid conditions, the subscript g to indicate saturated vapor conditions, and the subscript fg to indicate the change from liquid to vapor. For instance, $v_f + v_{fg} = v_g$. When liquid and vapor are in equilibrium, the mass fraction in the vapor phase is the dryness fraction, sometimes called *quality*, conventionally designated as x. Thus, the specific volume (for instance) of the mixture would be

$$v_{\text{mix}} = (1 - x) \cdot v_f + x \cdot v_g = v_f + x \cdot v_{fg}$$

Similarly,

$$h_{\text{mix}} = h_f + x \cdot h_{fg}, \quad \text{and so on.}$$

log P

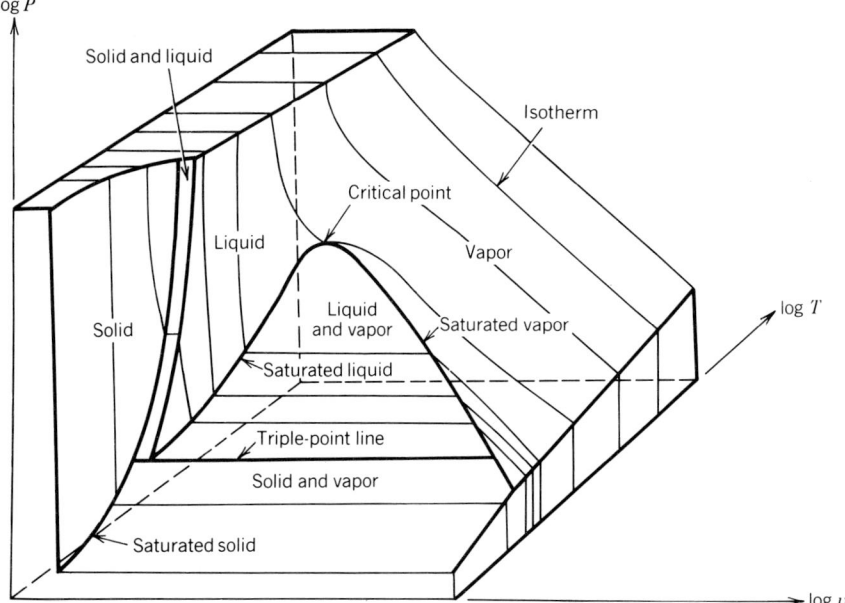

Fig. 3.1-2   The thermodynamic surface for an ordinary substance.

Saturated vapor conditions exist only on the line so identified on Fig. 3.1-2. Above and to the right of the saturated liquid–vapor region, the vapor is superheated. That is, the vapor is at a temperature higher than the saturation temperature corresponding to its pressure.

The Ideal Gas

An ideal gas is defined simply by its equation of state, which may be written as

$$Pv = RT = \frac{\overline{R}}{\mathscr{M}}T, \quad \text{or} \quad PV = mRT = n\overline{R}T$$

where P = absolute pressure, N/m^2 or lbf/ft^2
  v = specific volume, m^3/kg or ft^3/lb
  T = absolute temperature, Kelvin (K) or Rankine (R) degrees
  $\mathscr{M}$ = molecular weight of gas
  m = mass, kg or lb
  n = number of moles = $m/\mathscr{M}$
  R = individual gas constant, $J/kg \cdot K$ or $ft \cdot lbf/lb \cdot R$
  $\overline{R}$ = universal gas constant = $\mathscr{M} \cdot R$
   $= 8.31434 \times 10^3$ $J/kg\text{-mol} \cdot K$
   $= 1545$ $ft \cdot lbf/lb\text{-mol} \cdot R$
   $= 1.9859$ Btu (IST)$/lb\text{-mol} \cdot R$

Note that the numerical value of the molecular weight ($\mathscr{M}$) is the same in all systems of units. For instance, the molecular "weight" of oxygen (O_2) is 31.999 whether expressed as kilograms per kilogram-mole (kg-mol), grams per gram-mole (g-mol), or pounds per pound-mole (lb-mol). Thus, the specific volume of oxygen at a pressure of 1 atm (1.01325×10^5 Pa or 2116.22 lb/ft^2) and 15°C (59°F) is

$$v = \frac{(8.3143 \times 10^3) \cdot (273.15 + 15)}{(31.999) \cdot (1.01325 \times 10^5)} = 0.7389 \text{ m}^3/\text{kg}$$

$$= \frac{(1545) \cdot (459.67 + 59)}{(31.999) \cdot (2116.22)} = 11.834 \text{ ft}^3 \text{lb}$$

By a combination of the ideal-gas equation of state ($Pv = RT$) and the appropriate energy and specific heat relationships developed above, it can be shown that the following statements can be made for an ideal gas:

1. Internal energy (u), enthalpy (h), specific heat at constant pressure (C_P), and specific heat at constant volume (C_V) are all functions of temperature only. That is, these terms are not constant, but they are not affected by pressure or specific volume. The values of C_P and C_V are constant for monatomic gases only (see Chapter 18).

2. The difference between the two specific heats is the gas constant R. Thus,

$$C_P - C_V = R$$

$$\gamma = \frac{C_P}{C_V} = 1 + \frac{R}{C_V}$$

$$C_V = \frac{R}{\gamma - 1} \quad \text{and} \quad C_P = \frac{\gamma}{\gamma - 1} \cdot R$$

3. For a polytropic process in which $PV^n = $ a constant,

$$\frac{T_1}{T_2} = \left(\frac{v_2}{v_1}\right)^{n-1} = \left(\frac{P_1}{P_2}\right)^{(n-1)/n}$$

Here, n has any value appropriate to the process. For a reversible adiabatic process, which is an isentropic process, γ is substituted for the exponent n.

4. The change of entropy for an ideal gas is given by either of the following two equations:

$$s_2 - s_1 = \int_1^2 C_V \frac{dT}{T} + R \int_1^2 \frac{dV}{V}$$

$$s_2 - s_1 = \int_1^2 C_P \frac{dT}{T} - R \int_1^2 \frac{dP}{P}$$

In these two equations the first term on the right-hand side is a function of temperature only, and the last term is easily calculated as $\ln(V_2/V_1)$ or $\ln(P_2/P_1)$, respectively. Clearly, then, at constant volume or constant pressure the change of entropy depends only on the temperature limits, being either $\int_1^2 C_V(dT/T)$ or $\int_1^2 C_P(dT/T)$, respectively.

5. Insofar as real gases follow the ideal-gas relationship with reasonable accuracy, the temperature-dependent terms can be tabulated against temperature only, which is done in the *Gas Tables* of Keenan, Chao, and Kaye (see Chapter 18). The terms conventionally tabulated are

$T = $ absolute temperature
$h = \int_{T_0}^T C_P \, dT$
$P_r = $ relative pressure $= P/P_0$ in an isentropic process
$u = \int_{T_0}^T C_V \, dT$
$v_r = $ relative volume $= v/v_0$ in an isentropic process
$\phi = $ constant pressure entropy function $= \int_{T_0}^T C_P(dT/T)$

The use of these tables will be illustrated in Section 3.1-7.

The deviation of the properties of a real gas from those predicted by the ideal-gas equation of state is indicated by the value of the "compressibility factor" (Z), defined as

$$Z = \frac{Pv}{RT}$$

The value of Z would be unity for the ideal gas, and that is approached for each substance at pressures well below its critical-point pressure and temperatures well above its critical-point temperature. When good empirical data are not available for some substance, but the molecular weight and the critical-point values of pressure and temperature are known, an appropriate value of the compressibility factor is useful to determine, for instance, specific volume values from values of pressure and temperature.

With an accuracy generally better than $\pm 10\%$, it is found that the compressibility factor is the same function of the reduced pressure (P_R) and the reduced temperature (T_R) for a large number of substances in their vapor phase. Here, $P_R = P/P_{cr}$, and $T_R = T/T_{cr}$, when P_{cr} and T_{cr} are the critical-point pressure and temperature values of the individual substances. A generalized compressibility chart is given in Chapter 18. The use of this chart is best illustrated with an example using a known substance.

Example 3.1-1   Use of the Generalized Compressibility Chart.   Consider water vapor at a pressure of 10 MPa and a temperature of 600°C (873 K). The critical-point values for water are approximately 22.1 MPa and 374°C (647 K). Thus, at the point chosen, $P_R = 0.453$ and $T_R = 1.35$. From the chart, at these values of P_R and T_R, $Z = 0.95$. Therefore,

$$v \approx 0.95 \left(\frac{8.314 \times 10^3}{18} \right) \cdot 873 \cdot \frac{1}{10 \times 10^6}$$

$$\approx 3.83 \times 10^{-2} \ \text{m}^3/\text{kg}$$

The *Steam Tables* give the correct value as 3.837×10^{-2}. Thus, the chart value is very nearly correct for water vapor at this point.

3.1-7   Process Analysis

In this section the laws of thermodynamics (Section 3.1-5) and property values of pure substances (Section 3.1-6) will be utilized for the calculation of heat transfer quantities, the amount of work done, and changes in state-point values during various processes. Only pure, nonreacting substances will be considered here; reactive mixtures and the combustion process will be dealt with in Sections 3.1-10 and 3.1-11. Most processes are defined by the identification of one or more property values that are held constant or energy transfer terms that are found to be zero throughout.

Changes in state-point values in a process may be calculated from (1) an equation of state if it is known for the substance and is reasonably simple, (2) from tabulated property values, or (3) from chart values when those are available. Examples will be given of each.

The Adiabatic Process

Adiabatic simply means "no heat transfer," that is, $Q = 0$. For a closed, stationary system, the first law [Eqs. (3.1-8) and (3.1-9)] dictates that the adiabatic work, which is entirely displacement work, is equal to the decrease in the internal energy of the system.

$$\mathscr{W} = \int_{V_1}^{V_2} P \, dV = -(U_2 - U_1)$$

If the system is open, Eqs. (3.1-10) to (3.1-15) apply. Where a working machine has only one inlet and one exit, as in a turbine or compressor, Eq. (3.1-13) applies. For an adiabatic process this reduces to

$$\mathscr{W}_{Sh} = (h_i - h_e) + \frac{\bar{V}_i^2 - \bar{V}_e^2}{2g_c} + \frac{g}{g_c}(z_i - z_e)$$

Frequently changes in kinetic and potential energies also can be ignored. Then Eqs. (3.1-10) to (3.1-15) reduce to

$$\mathscr{W}_{Sh} = h_i - h_e$$

Example 3.1-2   Use of Air Tables for an Adiabatic Compression Process.   Consider the adiabatic compression of air from a pressure of 1 atm and a temperature of 300 K to 5 atm and 500 K. If the process were reversible as well as adiabatic, the process would be isentropic and the work required would be a minimum. For isentropic compression the end point would be determined by the relative pressure (P_r) where $P_{r2}/P_{r1} = P_2/P_1$ (see Fig. 3.1-3).
From the *Air Tables* of Keenan, Chao and Kaye:

Point	T (K)	h (kJ/kg)	P_r	u (kJ/kg)	ϕ (kJ/kg-K)
1	300	300.19	1.3860	214.09	2.5153
			$\times \ \ 5$		
2(s)	473.6	475.91	6.930	339.98	2.9770
2	500	503.02		359.53	3.0328

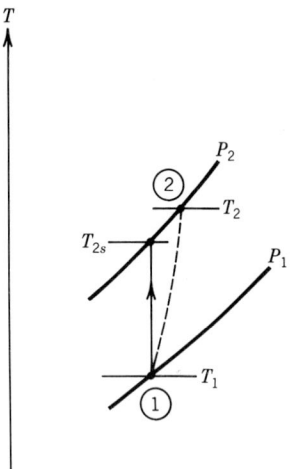

Fig. 3.1-3   Adiabatic compression of a gas (Example 3.1-2).

1.   **Closed System**

$$\mathscr{W}_C = u_2 - u_1 = 359.53 - 214.09$$

$$= 145.44 \text{ kJ/kg}$$

$$\mathscr{W}_{(s)} = u_{2(s)} - u_1 = 339.98 - 214.09$$

$$= 125.98 \text{ kJ/kg}$$

$$\eta_{C(s)} = \frac{\mathscr{W}_{\text{Sh}(s)}}{\mathscr{W}_{\text{Sh}}} = 86.6\%$$

2.   **Open System**

$$\mathscr{W}_{\text{Sh}} = h_2 - h_1 = 503.02 - 300.19$$

$$= 202.83 \text{ kJ/kg}$$

$$\mathscr{W}_{\text{Sh}(s)} = h_{2(s)} - h_1 = 475.91 - 300.19$$

$$= 175.75 \text{ kJ/kg}$$

$$\eta_{C(s)} = \frac{\mathscr{W}_{\text{Sh}(s)}}{\mathscr{W}_{\text{Sh}}} = 86.6\%$$

The change of entropy in this compression process is most easily calculated as the difference at P_2 between ϕ_2 and $\phi_{2(s)}$, which is 0.0558 kJ/kg · K. Alternatively, it may be calculated from point 1 to point 2 using the ideal-gas equation:

$$s_2 - s_1 = (\phi_2 - \phi_1) - R \ln \frac{P_2}{P_1}$$

$$= (3.0328 - 2.5153) - \frac{8.314}{28.96} \ln 5$$

$$= (0.5175) - (0.4620) = 0.0555 \text{ kJ/kg} \cdot \text{K}$$

which is close. The difference results only from the limitations in significant figures.

Example 3.1-3   Use of Steam Tables for an Adiabatic Expansion Process.   Consider the adiabatic expansion of steam in a turbine from 1.0 MPa (which is 1000 kPa or 10 bars) and 400°C, to 0.05 MPa

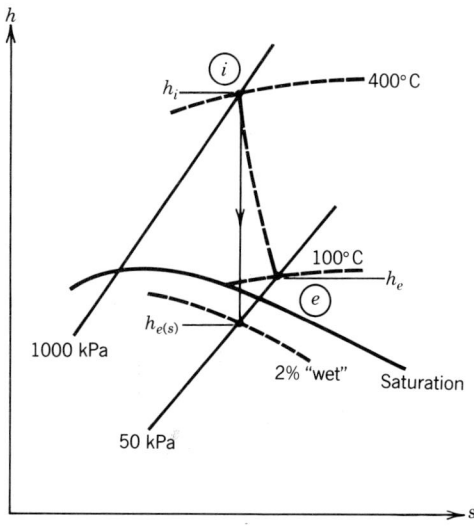

Fig. 3.1-4 Adiabatic expansion of steam (Example 3.1-3).

(50 kPa or 0.5 bar) and 100°C. The end points as well as the ideal process can be found most easily on a "Mollier diagram" (h-s) for steam (see Fig. 18.1-3). For this example, values will be taken from the *Steam Tables* of Keenan, Keyes, Hill, and Moore. For a steam turbine changes in kinetic and potential energies can usually be ignored so that the shaft work is simply the change of enthalpy. For a reversible adiabatic process the exit entropy would be equal to the inlet entropy (see Fig. 3.1-4).

At 1.0 MPa and 400°C, h_i = 3263.9 kJ/kg and s_i = 7.4651 kJ/kg · K. At 0.05 MPa and 100°C, h_e = 2682.5 kJ/kg, s_e = 7.6947 kJ/kg · K.

$$\mathscr{W}_{Sh} = 3263.9 - 2682.5 = 581.4 \text{ kJ/kg}$$

For the ideal expansion, s_e = 7.4651 kJ/kg · K. At 0.05 MPa, h_f = 340.49 and h_{fg} = 2305.4 kJ/kg; s_f = 1.0910 and s_{fg} = 6.5029 kJ/kg · K.

$$x = \frac{7.4651 - 1.0910}{6.5029} = 0.98$$

$$h_{e(s)} = 340.49 + 0.98(2305.4)$$

$$= 2600.23 \text{ kJ/kg}$$

$$\text{Ideal } \mathscr{W} = 3263.9 - 2600.2 = 663.7 \text{ kJ/kg}$$

$$\eta_{T(s)} = \frac{581.4}{663.7} = 87.6\%$$

The Isothermal Compression Process

When an ideal gas is compressed isothermally, there is no change in internal energy or enthalpy, since both are functions of temperature only. For a closed system the first law [Eqs. (3.1-8) to (3.1-9)] dictates that the cooling by heat transfer (q) must be equal to the displacement work ($\mathscr{W}$). The work is given by Eq. (3.1-2):

$$\mathscr{W}_D = Pv \ln\frac{V_2}{V_1} = RT \ln\frac{V_2}{V_1}$$

$$= Pv \ln\frac{P_1}{P_2} = RT \ln\frac{P_1}{P_2}.$$

For an open system the cooling required is again equal to the work done as shown by Eq. (3.1-14). In addition, from Eq. (3.1-15),

$$\mathscr{W}_{\text{Sh}} = -\int_i^e v\,dP$$

For an ideal gas in an isothermal process, Pv = a constant; then

$$\int P\,dv = -\int v\,dP$$

Consequently, for an ideal gas undergoing an isothermal process, the work and the heat transfer are equal in sign and magnitude and are the same for a closed and open system.

For the isothermal compression of substances that do not follow the ideal gas equation of state, the first-law equations apply, but with changes in internal energy (closed system) or enthalpy (open system) taken into account. Either tabulated or plotted values of thermodynamic properties may be used. Also, for an internally reversible process, the heat transfer can be calculated as $\int T\,ds$ which, for an isothermal process, is simply $T(\Delta s)$. The ratio of ideal work to actual work is the compression efficiency for the given pressure ratio.

Example 3.1-4 Isothermal Compression of a Gas, Moderate Pressure Ratio. Over a moderate pressure range and near normal ambient temperatures, air can be dealt with as an ideal gas. Consider the compression of air from a pressure of 1.0 atm to a pressure of 5.0 atm, isothermally at 300 K. Both the work done and the heat transfer should be given by

$$q = \mathscr{W}_C = RT\ln\frac{P_1}{P_2}$$

$$= \left(\frac{8.314}{28.96}\right)\cdot(300)\cdot\ln\left(\frac{1}{5}\right)$$

$$= -138.6 \text{ kJ/kg air}$$

The negative sign indicates work done *on* the gas and heat transfer *from* the gas.

If it is found that it requires a power input of 25 kW to deliver 8.0 kg air per minute, then the isothermal efficiency of compression is

$$\eta_{C(t)} = \frac{8.0 \times 138.6}{60}\cdot\frac{1}{25} = 73.9\%$$

Example 3.1-5 Isothermal compression, Large Pressure Ratio. For gas compression through a very large pressure ratio, multiple-stage compression with intercooling will generally be used in an attempt to approach isothermal compression and to avoid the high end temperature and large work input of adiabatic compression. For an open system (suction, compression, and delivery) with negligible changes in kinetic and potential energies, Eq. (3.1-14) applies. For an internally reversible isothermal process, this can be written as

$$\mathscr{W}_{C(t)} = (h_i - h_e) + T\cdot(s_e - s_i)$$

Property values are most conveniently obtained from the temperature–entropy chart for air. Figure 3.1-5 shows the points of interest for this example.

Consider the isothermal compression of air from 1.0 to 200 atm at 300 K.

$$\mathscr{W}_{C(t)} = (12,400 - 11,400) + 300\cdot(65 - 113)$$

$$= -13,400 \text{ kJ/kg-mol}$$

$$= -\frac{13,400}{28.96} = -463 \text{ kJ/kg air}$$

Again, the negative sign indicates work done on the fluid.

If it is found that it requires a power input of 85 kW to deliver 8 kg of air per minute, the isothermal efficiency of compression is

$$\eta_{C(t)} = \frac{8 \times 463}{60}\cdot\frac{1}{85} = 72.6\%$$

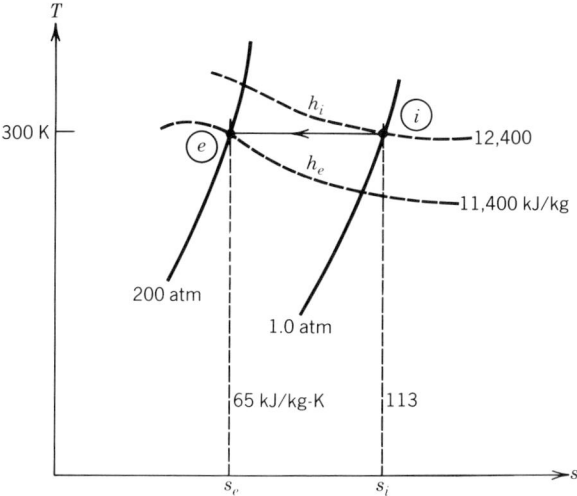

Fig. 3.1-5   Isothermal compression of a gas through a large pressure ratio (Example 3.1-5).

The Throttling Process

When a flowing fluid is allowed to expand from a higher to a lower pressure without doing useful work, the process is called *throttling*. Equation (3.1-13) applies with $W_{Sh} = 0$. When this process occurs through a short distance as in a valve or an orifice, there is not enough time for significant heat transfer, and change of elevation is negligible. If flow areas are appropriately adjusted, changes in kinetic energy are also generally negligible. Thus it is usually accurate to say that the final enthalpy h_e is equal to the initial enthalpy h_i.

For an ideal gas (Section 3.1-6) for which enthalpy is a function of temperature only, there would be no temperature change in a throttling process without heat transfer. For vapors deviating significantly from ideal-gas properties, there generally is a change in temperature with pressure at constant enthalpy. This change is indicated by the Joule–Thomson or Joule–Kelvin coefficient defined as

$$\mu = \left(\frac{\partial T}{\partial P} \right)_h$$

It can be shown that there are two components to this partial derivative:

$$\mu = \frac{1}{C_P} \left[-\left(\frac{\partial u}{\partial P} \right)_T - \left(\frac{\partial Pv}{\partial P} \right)_T \right]$$

For an ideal gas both components are zero. For vapors below the *maximum inversion temperature* either component may predominate and μ may be positive or negative or, at the inversion temperature, zero. Above the maximum inversion temperature (which is different for every substance), μ is always negative so that the temperature rises when the vapor is throttled from a higher to a lower pressure (see Fig. 3.1-6).

The drop in temperature in a throttling process is important in many cryogenic devices, particularly in the liquefaction of "gases." Only hydrogen, helium, and neon have maximum inversion temperatures below normal ambient temperature. For these gases precooling is necessary if the Joule–Thomson effect is to be utilized.

The throttling process is important in the conventional vapor compression refrigeration cycle (see Chapter 13.4-2). When liquid at or above its saturation pressure is throttled to a sufficiently low pressure, the final state point is in the liquid–vapor region and at the saturation temperature corresponding to that lower pressure. Evaporation at the lower temperature produces the desired cooling effect. Conventionally, properties of refrigerants are plotted on pressure–enthalpy coordinates. The end point of the throttling process is easily found from the inlet conditions and the final pressure, as shown on Fig. 3.1-7.

The equality of enthalpy values at the beginning and end of a throttling process is also useful in the determination of inlet point values in the limited part of the liquid–vapor region from which throttling

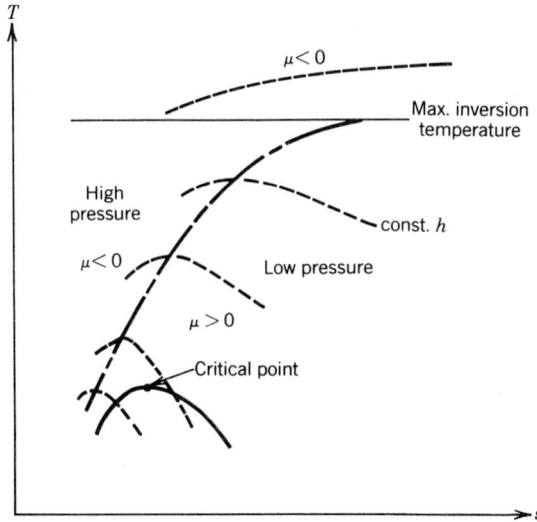

Fig. 3.1-6   The Joule–Thomson (Joule–Kelvin) coefficient on *T-s* coordinates.

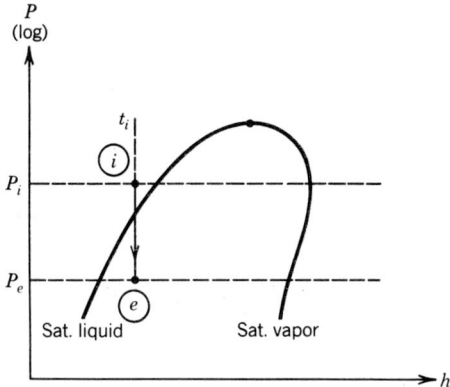

Fig. 3.1-7   The throttling process in vapor compression refrigeration.

will put the end point in the superheat region where temperature and pressure, being independent properties, will establish the enthalpy value. This is the principle of the so-called throttling calorimeter used to determine the wetness or dryness of steam, as in the following example.

Example 3.1-6   The Throttling Steam Calorimeter.   A sample of steam from a pipeline at a pressure of 2.0 MPa is extracted though a properly insulated throttling valve. The exhaust steam at 100 kPa is found to be at a temperature of 150°C, at which the Steam Table value of enthalpy is 2776.4 kJ/kg. At 2.0 MPa, $h_f = 908.79$ and $h_{fg} = 1890.7$ kJ/kg. Equating enthalpy values at inlet and exit gives

$$h_{fi} + x_i h_{fgi} = h_e$$

$$x_i = \frac{2776.4 - 908.79}{1890.7} = 0.988$$

Therefore, the steam in the pipe is 98.8% dry vapor and 1.2% liquid. This is illustrated in Fig. 3.1-8.

3.1-8   Thermodynamic Cycles

In order to produce a thermodynamic effect continuously, such as the production of power or refrigeration, it is usually necessary for the system to operate in a reproducible cycle. The only

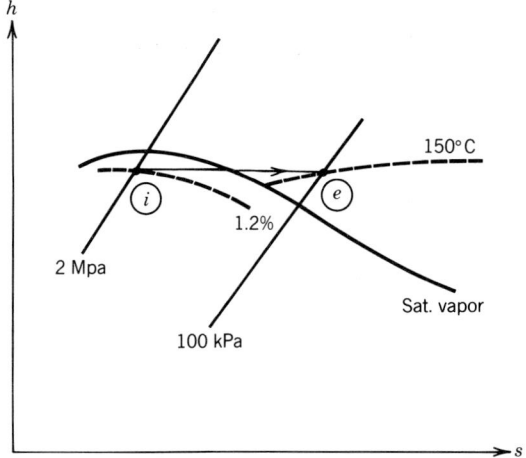

Fig. 3.1-8 The throttling process in a steam "calorimeter."

alternative is some type of direct energy conversion system, which is much less common. A cycle is any sequence of processes that returns the system to its original thermodynamic state.

Gas Power Cycles

The term *gas cycle* applies when the working fluid does not change phase, as it does for instance in the Rankine cycle, but remains in the vapor phase. Actually, the properties of most working fluids deviate significantly from the ideal-gas equation of state, and the specific heat values for most increase as temperatures rise above normal ambient. These deviations generally act to reduce the attainable values of thermal efficiencies. Therefore, in the development of expressions for maximum ideal thermal efficiencies of the various gas cycles, an ideal frictionless gas with constant specific heat values is assumed. With these assumptions, relatively simple expressions for thermal efficiency can be developed. Recall that

$$\eta_{th} = \frac{\sum \mathcal{W}}{q_H} = 1 - \frac{|q_L|}{|q_H|}$$

Heat addition at constant pressure: $q_P = C_P \cdot (T_2 - T_1)$

Heat addition at constant volume: $q_V = C_V \cdot (T_2 - T_1)$

The Carnot, Stirling, and Ericsson Cycles

Two corollaries of the second law of thermodynamics (Section 3.1-5) state that the maximum possible thermal efficiency of a working cycle between a source at T_H and a sink at T_L is the efficiency of a reversible cycle receiving heat only at T_H and rejecting heat only at T_L, and that all reversible cycles so constrained will have the same thermal efficiency. The ideal Carnot cycle and the ideal Stirling and Ericsson cycles with perfect regeneration are three such cycles. Each has a thermal efficiency (η_{th}) equal to $(1 - T_L/T_H)$. That is the maximum possible value between T_L and T_H. The three cycles are shown on *P-v* and on *T-s* coordinates in Fig. 3.1-9, with the same volume and temperature ranges.

The Carnot cycle is composed of two reversible adiabatic (isentropic) processes, *c-e* and *g-a*, and two reversible isothermal processes, *e-g* at T_H and *a-c* at T_L. The thermodynamic temperature scale is defined by the equality $q_L/q_H = T_L/T_H$ in the Carnot cycle. Therefore,

$$\eta_{th} = 1 - \frac{|q_L|}{|q_H|} = 1 - \frac{T_L}{T_H}$$

The Stirling cycle with ideal regeneration has its heat supply on the isothermal line *e-h* at T_H and its heat rejection on the isothermal line *a-d* at T_L. With ideal regeneration all the heat rejected from the working fluid along the constant-volume line *h-a* is stored in a regenerator and returned to the working fluid along the constant-volume line *d-e*. The regenerator energy is equal to $C_V \cdot (T_H - T_L)$.

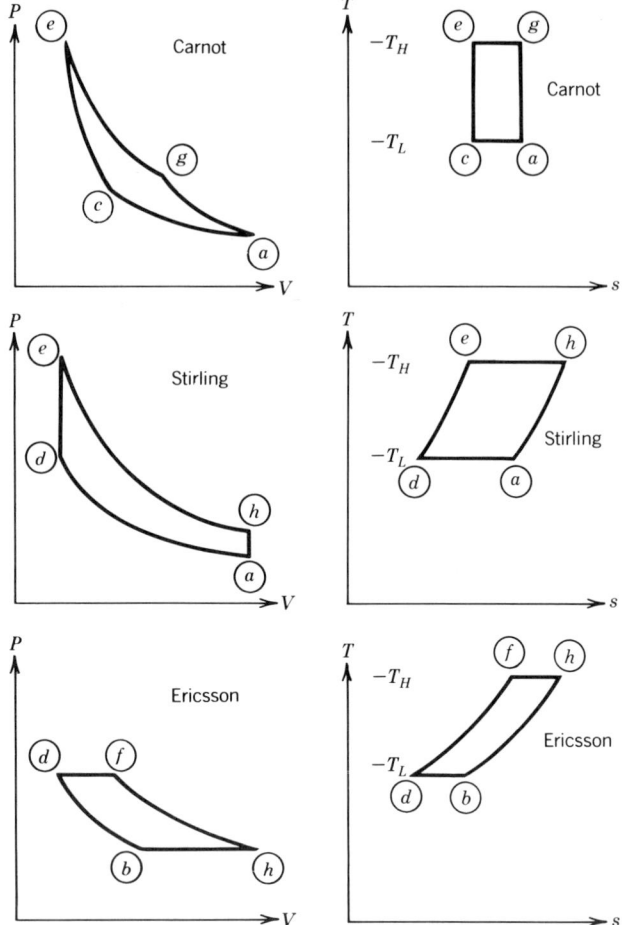

Fig. 3.1-9 Gas power cycles—Carnot, Stirling, and Ericsson—between T_H and T_L and between V_a and V_d.

The Ericsson cycle with ideal regeneration also receives heat transfer isothermally at T_H along f-h and rejects heat transfer isothermally at T_L along b-d. The energy stored in the regenerator during the constant-pressure process h-b is $C_P \cdot (T_H - T_L)$. The same quantity is returned to the cycle during the constant-pressure process d-f.

Other Gas Power cycles

The P-v and T-s diagrams of Figs. 3.1-10 and 3.1-11 serve to define, by the four reversible processes in each cycle, the Otto, Diesel, mixed, Brayton (sometimes called Joule), and Atkinson cycles. In each of these the process a-b is reversible adiabatic (isentropic) compression of the gas. Conventionally, the ratio $V_a : V_b$ is called the compression ratio (CR). This process is followed in each cycle by addition of heat along a constant-volume process, a constant-pressure process, or a combination of the two as in the mixed cycle. In each cycle there is then adiabatic expansion. Finally there is either constant-volume or constant-pressure heat rejection. Expressions for thermal efficiency of each cycle are noted on the diagrams in terms of pressure ratio or volume ratio, as appropriate.

Most spark ignition internal combustion engines, whether two-stroke or four-stroke versions, operate on the Otto cycle. Modern high-speed diesel engines, more properly called compression ignition engines, operate on mixed cycles, more closely approaching the Otto than the Diesel. Gas turbine power plants operate on the Brayton (Joule) cycle, often with internal heat transfer (regeneration). An approach to the Atkinson cycle is found in a combination of a free-piston reciprocating engine and a gas turbine.

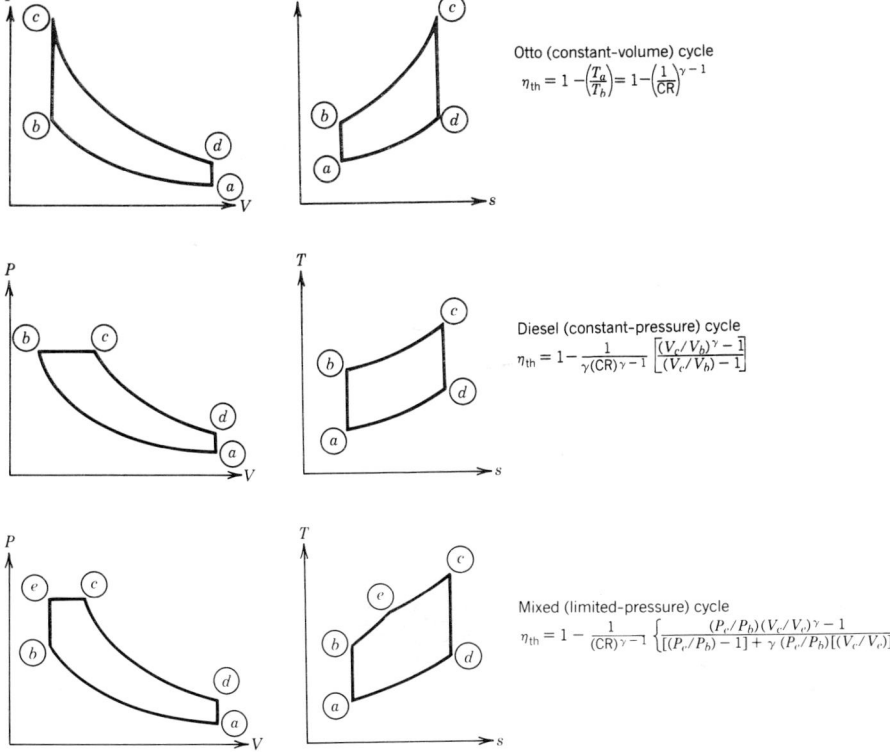

Otto (constant-volume) cycle

$$\eta_{th} = 1 - \left(\frac{T_a}{T_b}\right) = 1 - \left(\frac{1}{CR}\right)^{\gamma - 1}$$

Diesel (constant-pressure) cycle

$$\eta_{th} = 1 - \frac{1}{\gamma (CR)^{\gamma - 1}} \left[\frac{(V_c/V_b)^\gamma - 1}{(V_c/V_b) - 1}\right]$$

Mixed (limited-pressure) cycle

$$\eta_{th} = 1 - \frac{1}{(CR)^{\gamma - 1}} \left\{\frac{(P_c/P_b)(V_c/V_c)^\gamma - 1}{[(P_c/P_b) - 1] + \gamma (P_c/P_b)[(V_c/V_c)]}\right\}$$

Fig. 3.1-10 Gas power cycles: Otto, Diesel, and mixed.

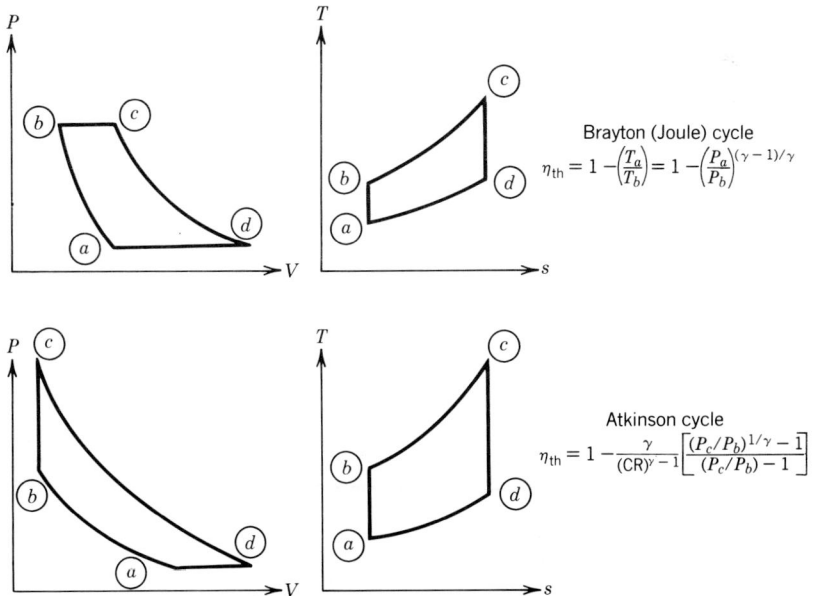

Brayton (Joule) cycle

$$\eta_{th} = 1 - \left(\frac{T_a}{T_b}\right) = 1 - \left(\frac{P_a}{P_b}\right)^{(\gamma - 1)/\gamma}$$

Atkinson cycle

$$\eta_{th} = 1 - \frac{\gamma}{(CR)^{\gamma - 1}} \left[\frac{(P_c/P_b)^{1/\gamma} - 1}{(P_c/P_b) - 1}\right]$$

Fig. 3.1-11 Gas power cycles: Brayton (Joule) and Atkinson.

177

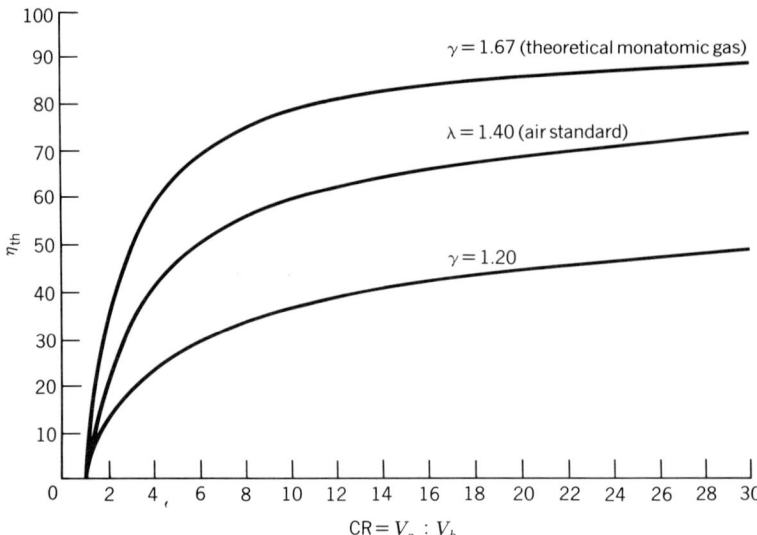

Fig. 3.1-12 Ideal thermal efficiencies of the Carnot, Otto, and Brayton cycles as a function of compression ratio (CR) and specific heat ratio (γ).

Air-Standard Cycle. Because so many gas cycles use air as the working fluid, maximum values of thermal efficiency are often calculated for the various ideal cycles with the assumption that the working fluid is a hypothetical frictionless substance with the thermodynamic properties of real air at standard conditions: 15°C and 1 atm pressure. It is assumed that the specific heat values remain constant and therefore γ is constant at 1.40.

For the Carnot, Otto, and Brayton cycles the thermal efficiency is a function of only γ and the compression ratio, $V_a : V_b$. Values are plotted in Fig. 3.1-12 for three values of γ including the air-standard value.

Gas Cycles for Refrigeration or Heat Pumping. The ideal-gas power cycles described above are composed of reversible processes. Therefore, theoretically, each could be reversed to receive energy by heat transfer at a low temperature and reject energy by heat transfer at a higher temperature. They would then be operating as refrigerators or heat pumps. Again, as corollaries of the second law, the Carnot, Stirling and Ericsson cycles would all have the same coefficient of performance (COP), and that would be the maximum value between T_L and T_H:

$$\text{COP}_{R(\text{max})} = \frac{T_L}{T_H - T_L} \qquad \text{COP}_{\text{HP(max)}} = \frac{T_H}{T_H - T_L}$$

No practical power or refrigeration system actually attempts to operate on the Carnot cycle at this time. Both power and refrigeration machines do attempt the Stirling cycle. Certain power cycles, such as a gas turbine system with several stages of intercooling during compression and reheating during expansion and with regeneration, approach the Ericsson cycle, as do some refrigeration systems.

The "reversed" Brayton (Joule) cycle is used in practice for refrigeration as well as for power. In fact, a Brayton power cycle driving a Brayton refrigeration cycle is a practical combined system.

Vapor Power Cycles

A "vapor" cycle is one in which the working fluid changes phase from liquid to vapor and, in a later process, back to liquid again. The simple Rankine cycle is the basis of all vapor power cycles. For large, central power stations the working fluid is almost always water, but many different substances are possible for smaller power output.

Many additions to the simple Rankine cycle, such as reheat and regeneration (see Chapter 14.2), are employed either to improve the thermal efficiency of the cycle or to reduce the required mass rate of circulation or both. Also, the Rankine cycle can be combined with other power cycles, either for "topping" or for "bottoming," in which rejected heat from one power cycle becomes the input heat to another. In a combined power–heating system, called *cogeneration*, extracted working fluid from any

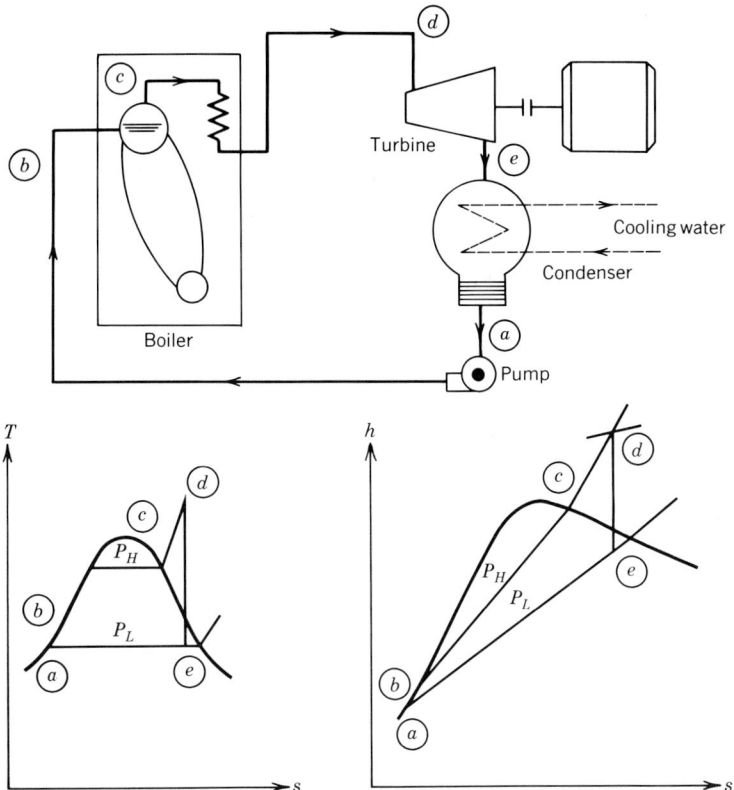

Fig. 3.1-13 The simple, ideal Rankine power cycle.

part of the cycle or the heat rejected in the condenser of the basic cycle can be utilized for building or process heat (see Chapter 2.2).

The basic components of a simple Rankine cycle are shown schematically in Fig. 3.1-13. The accompanying property diagrams illustrating the four ideal processes utilize the shape of the liquid–vapor dome for water because that is most commonly used. Other substances display liquid–vapor regions with inherent advantages.

Each component of the system can be considered separately as an open system for which Eq. (3.1-13) applies. In vapor power systems changes in kinetic and potential energies are generally negligible compared with the large heat transfer and work terms. Therefore, Eq. (3.1-14) will be used in this simplified analysis of the ideal Rankine cycle. Furthermore, in each component—boiler, turbine, condenser, and feedwater pump—there is either heat transfer or work done, but not both. Thus, the heat transfer or the work done is equal to the change in enthalpy of the working fluid through each component. In the following analysis energy terms will be expressed as kJ/kg or Btu/lb.

Boiler. The energy supply to the boiler is most frequently from the combustion of some fuel in air. It may also be from a nuclear source, a solar energy collector, or the energy rejection from some process or some other power-producing system. It is heat transfer to the working fluid at constant pressure, P_H.

$$q_H = h_d - h_b$$

Turbine. The expansion of steam through a turbine was used as an example of adiabatic expansion in Example 3.1-3. In the ideal cycle this is reversible and adiabatic, and therefore isentropic expansion from P_H to P_L. Whether the process is reversible or not, the heat transfer is generally negligible so that

$$\mathcal{W}_T = h_d - h_e$$

Condenser. All the heat rejection from the ideal cycle occurs during the condensation of the working fluid to saturated liquid at the lower pressure, P_L. This energy transfer may be directly or indirectly to ambient water or air, or it may be to some heating system in a cogeneration plant or to another power-producing cycle in a "bottoming" arrangement.

$$q_L = -(h_e - h_a)$$

Pump. The power input to the cycle required to raise the liquid pressure from P_L to P_H is generally very small compared to the power output from the turbine. It can sometimes be ignored in a first-approach analysis.

$$\mathcal{W}_P = -(h_b - h_a)$$

Because the specific volume of the liquid changes very little during pressurization, Eq. (3.1-15) can be used with a mean value of specific volume of the liquid at saturation (v_f).

$$\mathcal{W}_P = -\int_a^b v\,dP \simeq -v_{\text{mean}}(P_H - P_L)$$

Thermal Efficiency of the Ideal Rankine Cycle.

$$n_{\text{th}} = \frac{\sum \mathcal{W}}{q_H} = \frac{(h_d - h_e) - (h_b - h_a)}{h_d - h_b} = 1 - \frac{h_e - h_a}{h_d - h_b} \simeq \frac{h_d - h_e}{h_d - h_a}$$

Vapor Cycles for Refrigeration or Heat Pumping. The most common cycle for conventional refrigeration or heat pump systems is the vapor cycle shown schematically in Fig. 3.1-14. The working fluid is selected for its thermodynamic properties to achieve evaporation at a reasonable low pressure (P_L) for the desired heat-absorbing temperature, and to achieve condensation at a reasonable high pressure (P_H) at the desired heat-rejection temperature.

As with the ideal Rankine power cycle, there is either heat transfer or work done in each component, so that Eq. (3.1-14) will be used in a simplified analysis of the simple, ideal cycle. One obvious difference is that the vapor refrigeration cycle includes a throttling process for which the enthalpy value at exit is equal to the value at inlet.

Compressor. The work of compression (supplied by an electric motor in this illustration) per unit mass of the working fluid is

$$\mathcal{W}_C = -(h_b - h_a)$$

Condenser. For most (but not all) working fluids point b is in the superheat region. Thus, the heat transfer to the surroundings (q_H) includes removal of superheat ($h_b - h_c$) as well as latent heat of condensation ($h_c - h_d$), and perhaps some subcooling ($h_d - h_e$).

$$q_H = -(h_b - h_e)$$

Throttling Valve (TV). The work of expansion in the liquid region or on the liquid side of the liquid–vapor region is not large, and no attempt is made to utilize it in the conventional vapor refrigeration system. The Joule–Thomson (Joule–Kelvin) effect, however, is large, so that the fluid exits at a lower temperature when throttled from P_H to P_L:

$$h_f = h_e$$

Evaporator. Heat transfer to the working fluid (q_L) takes place in the evaporator as the major part of the latent heat of evaporation and perhaps some superheat is added from the surroundings:

$$q_L = h_a - h_f$$

Coefficient of Performance. Whether a system is considered a refrigerator or a heat pump depends only on the purpose for which it is being used. Where the purpose is to remove q_L from some object or space, it is a refrigerator, and its coefficient of performance is

$$\text{COP}_{\text{HP}} = \frac{h_a - h_f}{h_b - h_a}$$

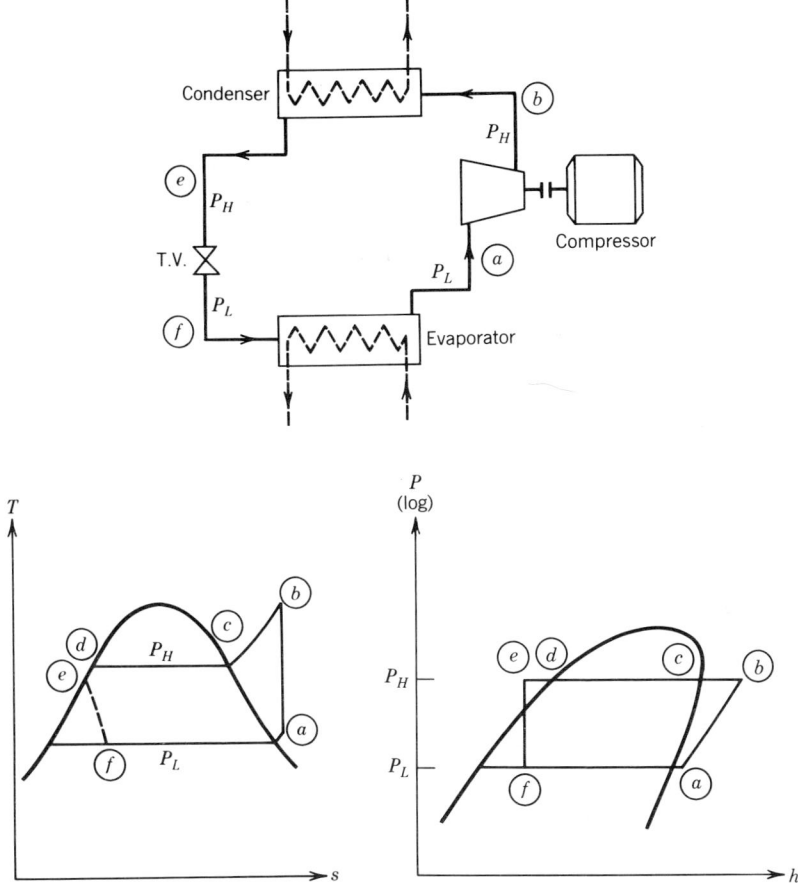

Fig. 3.1-14   A simple, ideal vapor compression refrigeration cycle.

When its purpose is to deliver the heating effect q_H, it is called a heat pump, and its coefficient of performance is

$$\text{COP}_{HP} = \frac{h_b - h_e}{h_b - h_a}$$

In the simple system, where the only external heat transfers are q_L and q_H,

$$\text{COP}_{HP} = \text{COP}_R + 1.0$$

Maximum Possible Overall COP.   A vapor compression refrigeration cycle requires the input of shaft work, which can be obtained from a conventional power-producing cycle. For the cooling effect (q_L) in the refrigeration cycle, the original input of energy has been q_H in the power cycle. The overall coefficient of performance is $\text{COP}_{oa} = q_L/q_H$ (see Fig. 3.1-15). The entropy principle, which is a corollary of the second law of thermodynamics, can be used to show that the maximum possible overall COP is equal to the product of the thermal efficiency of a Carnot cycle operating as an engine between T_H and the ambient temperature T_A and the COP of a Carnot cycle refrigerator operating between T_L and the same T_A.

$$\text{COP}_{oa(max)} = \frac{T_H - T_A}{T_H} \cdot \frac{T_L}{T_A - T_L}$$

This limiting COP_{oa} applies to a compound system where the output from some obvious power system is the input to some obvious refrigeration system. It applies as well where the two separate

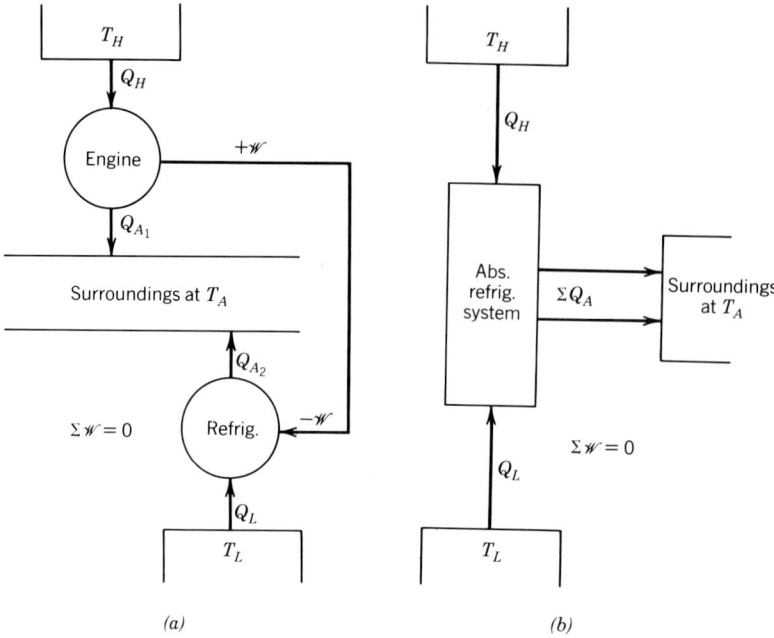

Fig. 3.1-15   The maximum possible overall coefficient of performance for a self-driven refrigeration system between T_H and T_L with ambient temperature T_A.

$$\text{COP}_{OA} = \frac{Q_L}{Q_H} \le \left(\frac{T_H - T_A}{T_H} \right) \cdot \left(\frac{T_L}{T_A - T_L} \right)$$

(*a*)   Refrigerator driven by an engine. (*b*) Absorption refrigeration system.

systems are not identifiable as is the case for an absorption refrigeration plant (see Chapter 13.4-3). In such a system there is heat transfer from some high-temperature source, cooling at a temperature below ambient, and heat rejection to the surroundings at T_A.

Minimum Work Required to Cool a Finite System.   A refrigeration device can be required to cool some finite body or system from temperature $T_1 = T_A$ to some lower temperature T_2 (see Fig. 3.1-16). The change of entropy of the finite system is $S_2 - S_1$. The change of entropy of the surroundings at T_A, to which the sum of heat transfer from the system plus the work done to drive the refrigerator is delivered, must be $(_1Q_2 + W)/T_A$. The entropy principle dictates that $\Sigma\,\Delta S$ be equal to or greater than zero.

$$\Sigma\,\Delta S = \frac{_1Q_2 + W}{T_A} + (S_2 - S_1)$$

Therefore,

$$W \ge T_A(S_1 - S_2) - _1Q_2$$

These energy quantities are shown as areas on the *T-S* diagram.

Relative Efficiency and Cycle Efficiency.   Ideal cycles, other than the Carnot, Stirling, and Ericsson cycles, will have efficiencies less than the maximum between the top and bottom temperatures (T_H and T_L) if there is any heat transfer to the system at a temperature below T_H or any heat transfer from the system above T_L. The relative efficiency (η_r) compares an ideal cycle, such as the Otto or the Rankine, with the best possible:

$$\eta_r = \frac{\eta_{th(ideal)}}{\eta_{th(Carnot)}}$$

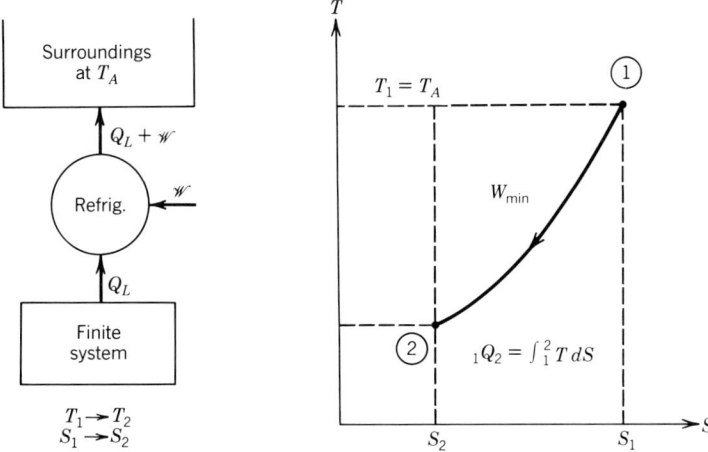

Fig. 3.1-16 The minimum work required to cool a finite body.

The cycle efficiency (η_{cy}) compares the efficiency of an actual cycle with that of the ideal cycle presumably being attempted:

$$\eta_{cy} = \frac{\eta_{th(actual)}}{\eta_{th(ideal)}}$$

When these two efficiency ratios are combined,

$$\eta_{th(actual)} = \eta_{th(Carnot)} \cdot \eta_r \cdot \eta_{cy}$$

3.1-9 Availability, Effectiveness, and Second-Law Efficiency

Because a working cycle receiving energy from some source above ambient temperature must eventually reject heat at ambient temperature, there is a limit to the amount of useful work that can possibly be obtained from the heat transfer received. Also, energy in any system can be converted to useful work only until that system comes into mechanical and thermal equilibrium with the surroundings at ambient temperature. The maximum possible useful work obtainable in any situation is called the available energy or the availability of that system in that environment. In recent years, some authorities have used the term "exergy" in place of available energy.

Maximum output of work (or minimum work required) depends on all cycles and processes being reversible. Any irreversibility results in lost work. Either term is the difference between the available energy and the work actually done (or the difference between the work actually required and the minimum possible work required). Lost work is not a loss in the usual sense; it is a lost opportunity to harness some part of the available energy (or the expenditure of more than the minimum work required).

Kinetic and potential energies can be converted directly to useful work until the velocity and elevation, respectively, of the mass of the system become zero relative to the environment of the system. Those terms, then, are part of the available energy of the system.

Nonuseful work must be subtracted from the maximum work to give the maximum useful work. A distinction is made here between the terms *surroundings* and *environment*. The surroundings include everything with which the system interacts, including other systems on which useful work can be done. The environment is the atmosphere or other medium around the system with which the system exchanges energy by heat transfer. If the system expands, pushing against the environment, the work done is not generally useful. The pressure of the environment will be designated P_0. For a finite change in system volume ΔV, the nonuseful work is $P_0 \Delta V$ and may be positive or negative depending on the sign of ΔV.

Available Energy (A) and Lost Work (LW) in Various Situations

For each situation the environment is at T_0 and P_0. Note that, for the system plus the environment, $\Sigma \Delta S \geq 0$ (see Fig. 3.1-17).

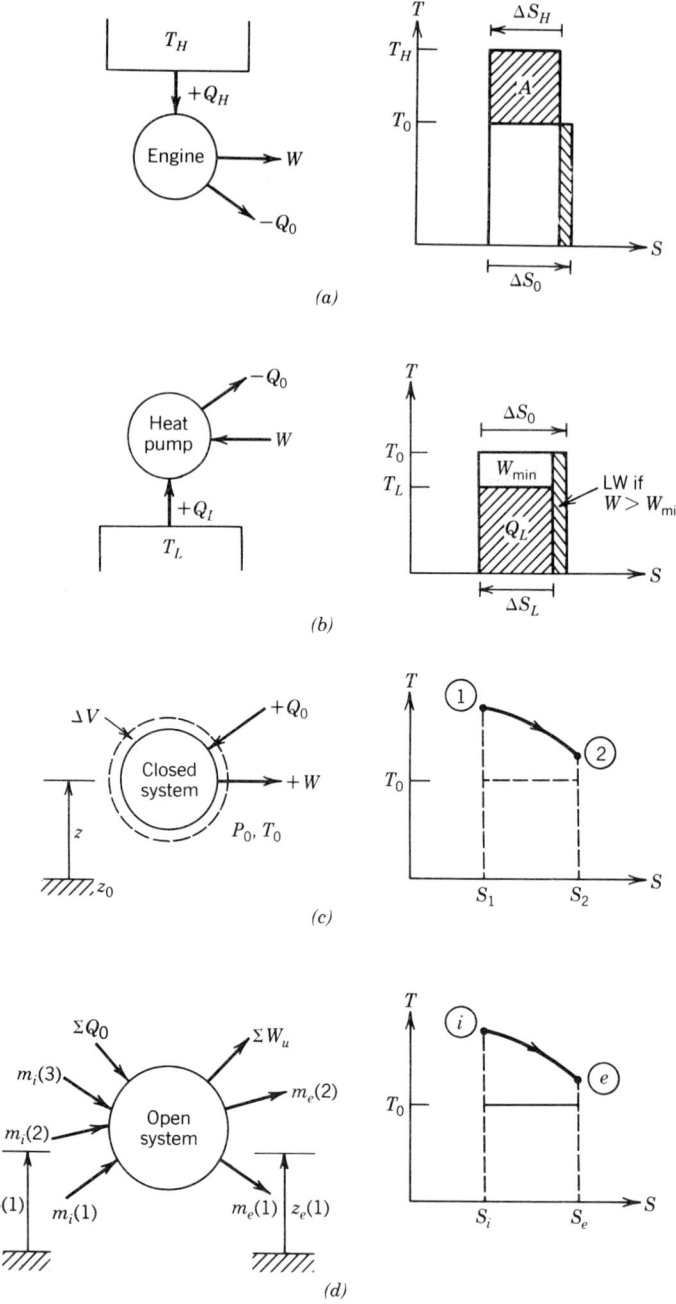

Fig. 3.1-17 Available energy and lost work in various situations. (*a*) Engine with source at T_H. (*b*) Minimum work to maintain T_L. (*c*) Closed finite system. (*d*) Open system, steady flow, steady state.

1. Available fraction of heat transfer (Q_H) from an infinite source at T_H.

$$A = Q_H \left(1 - \frac{T_0}{T_H}\right) \qquad \sum \Delta S = -\frac{Q_H}{T_H} + \frac{Q_0}{T_0}$$

$$W = Q_H - Q_0 \qquad LW = A - W = T_0 \sum \Delta S$$

2. Minimum work to maintain a temperature $T_L < T_0$.

$$A = W_{\min} = Q_L \left(\frac{T_0}{T_L} - 1\right) \qquad \sum \Delta S = -\frac{Q_L}{T_L} + \frac{Q_0}{T_0}$$

$$W = Q_0 - Q_L \qquad LW = W - A = T_0 \sum \Delta S$$

3. Available energy from a closed, finite system changing from state-point 1 to state-point 2 while receiving heat transfer (Q_0) from the environment. According to the first law, the work of the system must be

$$_1W_2 = Q_0 - (E_2 - E_1)$$

where

$$E = U + \frac{m}{g_c}\left(\frac{\bar{V}^2}{2} + gz\right)$$

Of this work done, $P_0(V_2 - V_1)$ is not useful. Also, $S_2 - S_1 \geq Q_0/T_0$, so that $Q_0 \leq T_0 \cdot (S_2 - S_1)$. Consequently, the maximum possible useful work between conditions 1 and 2, which is the availability change between 1 and 2, is

$$_1A_2 = T_0 \cdot (S_2 - S_1) - (E_2 - E_1) - P_0 \cdot (V_2 - V_1)$$

$$= (E_1 + P_0 V_1 + T_0 S_1) - (E_2 + P_0 V_2 + T_0 S_2)$$

$$= \phi_1 - \phi_2$$

where ϕ is the "availability function."

When the closed, finite system undergoes no change in volume or position,

$$_1A_2 = (U_1 - T_0 S_1) - (U_2 - T_0 S_2)$$

In certain systems, such as a closed calorimeter for the determination of "heat of reaction," there is heat transfer to the environment and the system is in thermal equilibrium with the environment before and after the process. Then $T_1 = T_2 = T_0$, and

$$_1A_2 = (U_1 - T_1 S_1) - (U_2 - T_2 S_2) = (F_1 - F_2)_{T_0 V}$$

where F is the Helmholtz function, sometimes called the "nonflow" or "batch" energy function because of its significance in the situation described. When the Helmholtz function was defined in Section 3.1-6, it was noted that

$$dF = -P\,dV - S\,dT$$

For a simple, nonreacting substance with no change in volume or temperature, $dF = 0$, and change of availability has no significance.

4. Available energy from steady flow through an open system at steady state. The first law for this system [Eq. (3.1-13)] gives the useful work (which is the shaft work) as

$$\sum W_u = \sum m_i \left(h_i + \frac{\bar{V}_i^2}{2g_c} + \frac{g}{g_c} z_i\right) + \sum Q_0 - \sum m_e \left(h_e + \frac{\bar{V}_e^2}{2g_c} + \frac{g}{g_c} z_e\right)$$

$$= \left(\sum H_i - \sum H_e\right) + \sum Q_0 - \sum \Delta KE - \sum \Delta PE$$

Again, from the entropy principle, $\sum Q_0 \leq T_0(S_0 - S_i)$. Consequently, the maximum possible useful work, which is the availability, would be

$$_iA_e = \left(\sum H_i - \sum H_e\right) - T_0 \cdot (S_i - S_e) - \sum \Delta KE - \sum \Delta PE$$

Frequently there is only one inlet and one outlet and ΔKE and ΔPE are negligible. Then,

$$_iA_e = (H_i - T_0 S_i) - (H_e - T_0 S_e) = B_i - B_e$$

where B is the Darrieus function.

To generalize to more than one inlet, more than one exhaust, and significant changes in kinetic and potential energies,

$$\sum {}_iA_e = \left(\sum B_i - \sum B_e\right) - \sum KE - \sum PE$$

In certain systems, such as a fuel cell or an open "calorimeter" for determining the energy of reaction of a mixture of fluids, both the inlet and exhaust streams are in equilibrium with the environment at T_0 and P_0. When ΔKE and ΔPE can be ignored, the availability of the process is

$$_iA_e = (H_i - T_iS_i) - (H_e - T_eS_e) = (G_i - G_e)_{T_0 P_0}$$

where G is the Gibbs function, sometimes called the steady-state, steady-flow energy function. In Section 3.1-6 it was noted that

$$dG = V\,dP - S\,dT$$

Thus, the availability for a nonreacting open system with $T_i = T_e$ and $P_i = P_e$ would be zero.

Effectiveness (ε')

Since only a part of the energy associated with or delivered to a system is available for conversion to useful work, the value of thermal efficiency (η_{th}) does not truly indicate the effectiveness of a device used to harness the energy. The concept of availability does provide a meaningful comparison. The effectiveness (ε') of a device or a system is defined as

$$\varepsilon' = \frac{\text{useful work actually performed}}{\text{energy available for useful work}}$$

Note that this term evaluates the performance of a device without consideration of whether an alternative system might have been more effective.

Consider the adiabatic expansion of a fluid through a turbine, as shown in Fig. 3.1-18. The adiabatic expansion efficiency of the turbine in the process from i to e is

$$\eta_{T(s)} = \frac{h_i - h_e}{h_i - h_{e(s)}} = \frac{h_i - h_e}{(h_i - h_e) + (h_e - h_{e(s)})}$$

The turbine "loss" would be $h_e - h_{e(s)}$, which is the difference between the actual turbine work and the isentropic work.

The available energy in this process is the change in the Darrieus function (b kJ/kg), so that the effectiveness is

$$\varepsilon' = \frac{h_i - h_e}{\phi_i - \phi_e} = \frac{h_i - h_e}{b_i - b_e}$$

$$= \frac{h_i - h_e}{(h_i - T_0 S_i) - (h_e - T_0 S_e)}$$

$$= \frac{h_i - h_e}{(h_i - h_e) + T_0 \cdot (s_e - s_i)}$$

The "lost work" in this process it $T_0 \cdot (s_e - s_i)$. Because $T_0 \cdot (s_e - s_i)$ is less than $h_e - h_{e(s)}$, the effectiveness is greater than the efficiency. This recognizes the fact that, although the work done was less than the work of isentropic expansion, this loss was retained by the fluid; a part of this loss (above T_0) is still theoretically available for work production, as in a subsequent stage of a multistage turbine.

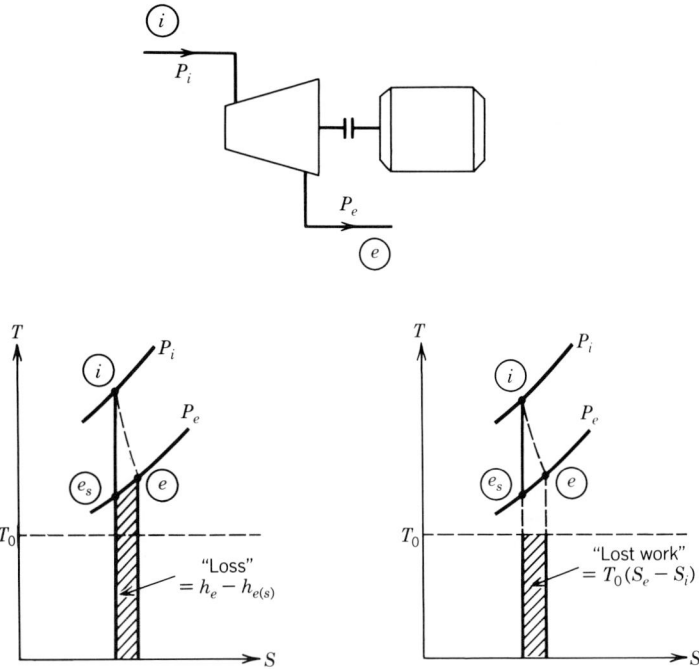

Fig. 3.1-18 Comparison between loss and lost work in turbine expansion.

Example 3.1-7 Effectiveness (ε′) and Lost Work (LW) in an Expansion Process. The adiabatic efficiency of expansion in a steam turbine was illustrated in Example 3.1-3, for which significant points were shown on Fig. 3.1-4. All the state-point values at inlet *i* and exhaust *e* were given in that example. The efficiency of the process ($\eta_{T(s)}$) was found to be 87.6%.

To the conditions of Example 3.1-3, add that the ambient temperature (T_0) is 288 K. Then

$$b_i = h_i - T_0 s_i = 1113.9 \text{ kJ/kg}$$

$$b_e = h_e - T_0 s_e = 466.4 \text{ kJ/kg}$$

$$\varepsilon' = \frac{h_i - h_e}{b_i - b_e} = \frac{581.4}{647.5} = 89.8\%$$

Note that the turbine loss in this example, which is $h_e - h_{e(s)}$, is 82.3 kJ/kg, while the lost work, which is $T_0(s_e - s_i)$, is 66.1 kJ/kg. The difference between loss and lost work and between η_T and ε' is not great in this situation because the exhaust temperature was close to ambient. The difference would be much more significant for a typical gas turbine.

Example 3.1-8 Effectiveness (ε′) and Lost Work (LW) in a Compression Process. The adiabatic efficiency of compression of a gas was calculated in Example 3.1-2 and illustrated in Fig. 3.1-3. Consider the open system. The additional work required to compress from P_1 to P_2 was $h_2 - h_{2(s)}$, which was $503.0 - 475.9 = 27.1$ kJ/kg. This additional energy was, however, retained by the fluid. Part of that could ideally be converted to useful work. The lost work in the compression process was $T_0(s_2 - s_1)$. If $T_0 = 288$ K,

$$\text{LW} = 288(0.0558) = 16.1 \text{ kJ/kg}$$

It would be logical then to define the effectiveness of this process as

$$\varepsilon' = \frac{\text{minimum work to compress}}{\text{minimum work} + \text{lost work}} = \frac{175.75}{175.75 + 16.1} = 91.6\%$$

This is significantly greater than the efficiency of adiabatic compression (86.6%).

Second-Law Efficiency (ε'')

The term *second-law efficiency* (ε'') goes beyond the term *effectiveness* (ε') in the application of the concept of availability. While effectiveness evaluates how well a system or device has utilized the available energy, the second-law efficiency considers whether some other system might accomplish the same task with a smaller expenditure of available energy. It therefore evaluates the task.

$$\varepsilon'' = \frac{\text{heat or work usefully transferred}}{\substack{\text{maximum possible heat or work} \\ \text{transfer for the same energy input}}}$$

$$= \frac{\text{minimum availability required}}{\text{actual availability expended}}$$

When the sole purpose of a device is the production of useful work, then the amount of work actually produced is the minimum availability required. In that case the second-law efficiency is the same as the effectiveness. Many systems provide both power and heat transfer. Generally there are alternative means of accomplishing this task. A simple illustration of the calculation of second-law efficiency is provided by an examination of alternative means used to heat a building.

Example 3.1-9 Heating a Building with a Supply of Hot Gas. When a simple furnace is used to heat a building, the gaseous products of combustion transfer heat to the building by cooling close to building temperature. Those hot gases, however, could have been the source of energy to produce shaft power (high-grade energy) to drive a heat pump.

Suppose products of combustion with a specific heat of 1.0 kJ/kg · K cool from 1200 K (927°C) to 400 K (127°C) transferring heat to a heating system at 320 K (47°C) when the ambient temperature is 270 K (-3°C).

Since the gases might have cooled to 320 K with perfect heat transfer, the first-law efficiency of this furnace, ignoring imperfect combustion, would be

$$\eta = \frac{1.0(1200 - 400)}{1.0(1200 - 320)} = 90.9\%$$

Alternatively, the available energy in this gas might ideally have driven an engine, which in turn would drive a heat pump delivering energy from the atmosphere at 270 K to the building's heating system at 320 K. The available energy per kilogram of gas is

$$a = (h_{1200} - h_{270}) - T_0(s_{1200} - s_{270})$$

$$= 1.0(1200 - 270) - 270\left(1.0\ln\frac{1200}{270}\right)$$

$$= 527 \text{ kJ/kg gas}$$

The maximum possible coefficient of performance of a heat pump between 270 and 320 K would be

$$\text{COP}_{HP} = \frac{320}{320 - 270} = 6.40$$

Thus, the maximum possible energy delivery to the building under these circumstances would be

$$Q = 6.4(527) = 3373 \text{ kJ/kg gas}$$

With the furnace, only $1200 - 400 = 800$ kJ were delivered per kilogram of hot gas by the expenditure of the same available energy.

Therefore, the second-law efficiency of the furnace is

$$\varepsilon'' = \frac{\text{actual energy delivered}}{\text{maximum possible energy delivered}}$$

$$= \frac{\text{minimum available energy expended}}{\text{actual available energy expended}}$$

$$= \frac{800}{3373} = 23.7\% = \frac{800/6.40}{527}$$

Obviously, in this example only energy quantities have been dealt with. The complexity and cost of the apparatus have not been considered.

3.1-10 Nonreactive Mixtures and Solutions

In Sections 3.1-6 to 3.1-8 the thermodynamic properties of pure substances and processes and cycles involving those substances were dealt with. In this section mixtures of pure substances are considered in which there is no chemical reaction but in which the proportions might change. In Section 3.1-11 the thermodynamics of a mixture in which there is a chemical reaction that results in changes in molecular structure will be introduced.

Mixtures of Ideal Gases

When two or more nonreactive (inert) ideal gases are mixed, it is assumed, in the idealization, that the mixture itself continues to be an ideal gas. That is,

$$PV = n\bar{R}T$$

where n is the total number of moles of the component gases and $\bar{R}$ is the universal gas constant as defined in Section 3.1-6. Also, the various individual component gases continue to conform to ideal-gas relationships as if the other components of the mixture were not present. Two models are used to represent these assumptions.

The Dalton model treats each component gas (i) as if it occupies the total volume (V) at its own partial pressure (P_i). All components are, of course, in equilibrium at the same temperature (T). Thus,

$$P_i V = n_i \bar{R} T, \quad P = \sum P_i, \quad n = \sum n_i, \quad \text{and} \quad \frac{P_i}{P} = \frac{n_i}{\sum n},$$

and

$$\frac{PV}{\bar{R}T} = \sum \frac{P_i V}{\bar{R}T} = \frac{V}{\bar{R}T} \sum P_i$$

The Amagat model treats the component gases as if each occupies its own partial volume (V_i) but at the total pressure (P) and at the same temperature (T). Thus,

$$PV_i = n_i \bar{R} T, \quad V = \sum V_i, \quad n = \sum n_i, \quad \text{and} \quad \frac{V_i}{V} = \frac{n_i}{\sum n},$$

and

$$\frac{PV}{\bar{R}T} = \sum \frac{PV_i}{\bar{R}T} = \frac{P}{\bar{R}T} \sum V_i$$

In both models Avogadro's constant is assumed to apply to the individual components and to the mixture. Avogadro's rule states that there are 6.022169×10^{26} molecules of any gas in a kilogram-mole, and therefore the mass of a given volume at the same temperature and pressure is directly proportional to the molecular weight of that gas. For the mixture, as for the individual gases,

$$PV = mRT = m\frac{\bar{R}}{\mathcal{M}}T$$

in which R is the individual gas constant for the mixture and $\mathcal{M}$ is the weighted average of the molecular "weights."

Example 3.1-10 Molecular Fraction and Mass Fraction in a Mixture of Gases. The reliability of the assumptions in the Dalton and Amagat models for gases is illustrated by the application of these models to the accepted definition of "dry air" by, among others, the American Society of Heating, Refrigerating, and Air Conditioning Engineers. The conversion from molecular fraction, which is identical to volume fraction, to mass fraction is shown in Table 3.1-1. Note that values are given to greater than five significant figures for these common gases.

From the calculated value of molecular weight for dry air, the individual gas constant is found to be $8.31434/28.9645 = 0.28705$ kJ/kg-mol $\cdot$ K.

TABLE 3.1-1

Substance	Molecular Weight Based on C = 12 (Definition)	Molecular Fraction in Dry Air (Definition)	Partial Molecular Weight in Dry Air (Calculated)	Mass Fraction in Dry Air (Calculated)
Nitrogen (N_2)	28.013 4	0.780 84	21.873 98	0.755 20
Oxygen (O_2)	31.998 8	0.209 476	6.702 98	0.231 42
Argon (A)	39.948	0.009 34	0.373 11	0.012 88
Carbon dioxide (CO_2)[a]	44.009 95	0.000 314	0.013 82	0.000 48
Other[a]	—	0.000 03	0.000 61	0.000 02
Dry air		1.0	28.9645	1.0

[a] Variable.

Where the nonreactive ideal-gas assumptions apply, energy values for the mixture at a particular temperature can be summed as shown.

$$\bar{u} = \frac{\sum n_i \bar{u}_i}{\sum n_i} \qquad \bar{h} = \frac{\sum n_i \bar{h}_i}{\sum n_i}$$

This summation is possible for the Dalton or for the Amagat model because internal energy and enthalpy do not depend on pressure and volume for ideal gases.

For a value of entropy of the mixture of ideal gases, the gases must be assumed to exist at their partial pressures (the Dalton model) because entropy is a function of both temperature and pressure. With each component gas at the given temperature and at its partial pressure,

$$\bar{s} = \frac{\sum n_i \bar{s}_i}{\sum n_i}$$

The summations indicated are called the Gibbs–Dalton formulations. The results are in error insofar as the molecules of one component do interact with those of the others. That error is small where the compressibility factor Z for each component at its temperature and partial pressure is close to 1.0 (see Section 3.1-6). The summation of partial values of entropy is useful only where mixture ratios remain constant because the significant entropy change associated with mixing is ignored.

The entropy of mixing for ideal gases is readily found as the sum of the entropy changes as each component expands from its partial volume (V_i) and the total pressure (P) to the total volume (V) and its partial pressure (P_i), at constant temperature (T). The entropy change of an ideal gas is

$$\bar{s}_2 - \bar{s}_1 = \int_1^2 \bar{C}_p \frac{dT}{T} - \bar{R} \int_1^2 \frac{dP}{P}$$

Then

$$\Delta S_i = +\bar{R} \ln \frac{P}{P_i}$$

$$= +R \ln \frac{\sum n}{n_i}$$

and

$$\sum \Delta S = \bar{R} \sum n_i \ln \frac{\sum n}{n_i}$$

which must always be positive.

Mixtures of a Gas and a Vapor

When one of the components in a mixture is close to or at the saturation temperature for its partial pressure, it must be considered as a vapor, and the terms *humidity ratio, degree of saturation, relative humidity*, and *dew point* have relevance. When one of the components condenses, the mass ratio of the gas–vapor mixture changes. The most common example of a mixture of a gas and a vapor is "moist air" where the components of the dry air can be assumed constant (see Example 3.1-10) and acting as a single gas, and the mass of water vapor present per unit volume can be anywhere between zero and the density for saturation conditions at the existing temperature. The term *saturation* has exactly the same significance here as in the discussion of thermodynamic properties of a pure substance (Section 3.1-6). The thermodynamic properties of moist air are called psychrometric properties.

Humidity ratio (W) of a mixture of a gas and a vapor is simply the mass ratio.

$$W = \frac{\text{mass of vapor}}{\text{mass of gas}}$$

In psychrometrics

$$W = \frac{\text{mass of water vapor}}{\text{mass of dry air}}$$

Degree of saturation (μ) is the ratio of the existing humidity ratio (W) to the humidity ratio that would exist at saturation (W_s) for the same temperature (T) and total pressure (P). Thus

$$\mu = \frac{W}{W_s}(T, P)$$

Relative humidity (RH) is a term used most frequently in connection with the properties of moist air. It is applicable, however, to any mixture of gas and vapor, or in fact to any vapor alone. Relative humidity is the existing mole fraction of vapor in the mixture divided by the mole fraction of vapor for saturation conditions of the vapor at the same temperature and total pressure. This is the same as the ratio of the moles or mass of vapor to the moles or mass of vapor for saturation, per unit volume, at the same temperature.

$$RH = \frac{n_v / \sum n}{n_{vs} / \sum n}(T, P)$$

$$= \frac{n_v}{n_{vs}}(T, V) = \frac{m_v}{m_{vs}}(T, V)$$

When the vapor is at a low value of partial pressure so that it acts as an ideal gas ($PV = n\bar{R}T$),

$$RH \simeq \frac{P_v}{P_{vs}}$$

The relationship between relative humidity and degree of saturation, when ideal-gas conditions exist, is

$$RH \simeq \frac{\mu}{1 - (1 - \mu)(P_{vs}/P)}$$

The dew-point temperature of a mixture of gas and vapor is the saturation temperature of the vapor component corresponding to its partial pressure in the mixture.

Adiabatic Combination or Separation of Two-Component Mixtures

Figure 3.1-19 shows two streams (1 and 2) being brought together with no heat transfer, and the combination of the two leaving as a single stream (3). Each stream is a mixture of two substances, A and B, with different mixture ratios (x) and enthalpy values (h).

If x is defined as the mass fraction $(m_A / \sum m)$ for each stream and h is defined as the enthalpy per unit mass of each stream $(H / \sum m)$, it is easily shown that the following linear relationships hold true:

$$\frac{x_2 - x_3}{x_3 - x_1} = \frac{h_2 - h_3}{h_3 - h_1} = \frac{m_1}{m_2}$$

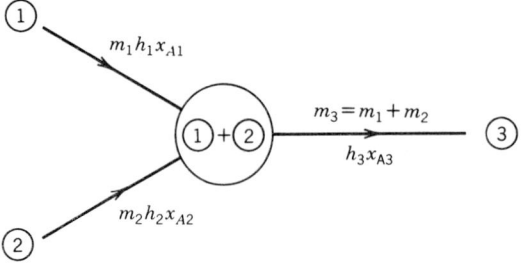

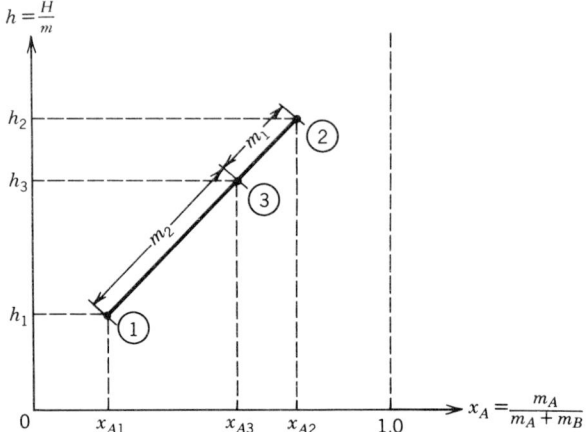

Fig. 3.1-19 Adiabatic combination of two different two-component streams.

This is illustrated on the h-x diagram of Fig. 3.1-19, where point 3 is found on a straight line between points 1 and 2, and the line is divided on the horizontal scale and on the vertical scale in the ratio m_1/m_2. The same would be true if x were defined as the mole fraction $n_A/\Sigma\, n$ and h as the enthalpy per mole $H/\Sigma\, n$.

The same relationships apply for a separation of one stream (3) into two streams (1 and 2). While internal heat transfer may be required to achieve separation in some situations, the whole system is adiabatic if there is no heat exchange with the surroundings. Adiabatic separation occurs where a mixture of liquid and vapor is separated into the two phases simply by gravity, and in the two-component system the liquid phase has a different value of component strength (x) from that of the vapor phase.

The straight-line relationship holds as well on a psychrometric chart, used in air conditioning, where enthalpy per unit mass of the dry air (h_a) typically is plotted against humidity ratio (W), which is a mass ratio (m_v/m_a) rather than a mass fraction.

Partial Molal Properties

The value of any extensive property varies directly with the mass or the number of moles of the substance being considered. In a mixture of different substances the value of the extensive property is the sum of the products of the partial molal value of each times the number of moles of each. For instance, for a mixture of several substances the total volume at a particular temperature and pressure is the sum of the volumes of the components in the mixture:

$$V = \sum n_i \left(\frac{\partial V}{\partial n_i} \right)_{P,T,n_j}$$

where $\partial V/\partial n_i$ is the partial molal volume of component i at the prescribed temperature (T) and pressure (P) when all other components (j) are at prescribed values.

For ideal gases, in accordance with the Amagat model, the partial molal volume is the same as the specific volume (mole basis) for each component. For nonideal substances the volume of the mixture is typically less than the sum of the volumes of the components separately at the same temperature and pressure.

A solution may be in the solid, liquid, or vapor phase. An ideal solution is one in which the total volume is the same as the sum of the volumes of the separate parts at the same temperature and pressure. An ideal mixture of ideal gases is a special case of the ideal solution.

Other extensive thermodynamic properties such as internal energy (U), enthalpy (H), Gibbs function (G), and Helmholtz function (F), when related to a mixture or solution, must be expressible in partial molal form, as was illustrated above for volume (V).

Mixtures with Variable Composition

When a small amount of one or more components in a mixture is added or removed, each extensive property value is changed in proportion to the number of moles added or removed, and in proportion to the partial molal values of that property for those components. Consider the internal energy (U) of a system of fixed mass:

$$dU = T\,dS - P\,dV$$

When a relatively small amount of any one component is added,

$$dU = T\,dS - P\,dV + \left[\left(\frac{\partial U}{\partial n_i} \right)_{S,V,n_j} \right] \cdot dn_i$$

If there is a change in the number of moles of several components,

$$dU = T\,dS - P\,dV + \sum \left[\left(\frac{\partial U}{\partial n_i} \right)_{S,V,n_j} \right] \cdot dn_i$$

Similarly, the changes in the other energy terms are

$$dH = T\,dS + V\,dP + \sum \left(\frac{\partial H}{\partial n_i} \right)_{S,P,n_j} \cdot dn_i$$

$$dG = -S\,dT + V\,dP + \sum \left(\frac{\partial G}{\partial n_i} \right)_{T,P,n_j} \cdot dn_i$$

$$dF = -S\,dT - P\,dV + \sum \left(\frac{\partial F}{\partial n_i} \right)_{T,V,n_j} \cdot dn_i$$

It can be shown that each of the partial-energy properties in the above equations has the same value. Each is given the name chemical potential (u).

$$u_i = \left(\frac{\partial U}{\partial n_i} \right)_{S,V,n_j} = \left(\frac{\partial H}{\partial n_i} \right)_{S,P,n_j}$$

$$= \left(\frac{\partial G}{\partial n_i} \right)_{T,P,n_j} = \left(\frac{\partial F}{\partial n_i} \right)_{T,V,n_j}$$

The chemical potential (u), defined here for a nonreactive mixture, becomes very significant as a "driving force" in a reactive system. It is an intensive property.

Of the four different expressions for u_i only the one concerning Gibbs function (G) is a partial molal property, because for that one it is specified that temperature (T) and pressure (P) do not change.

Evaporation and Condensation of a Two-Component Mixture

When the temperature of a two-component liquid solution is increased, evaporation will begin when the saturation temperature is reached for that particular pressure and mixture strength. In general, the two components will have different vapor pressures at that temperature and total pressure, so that one

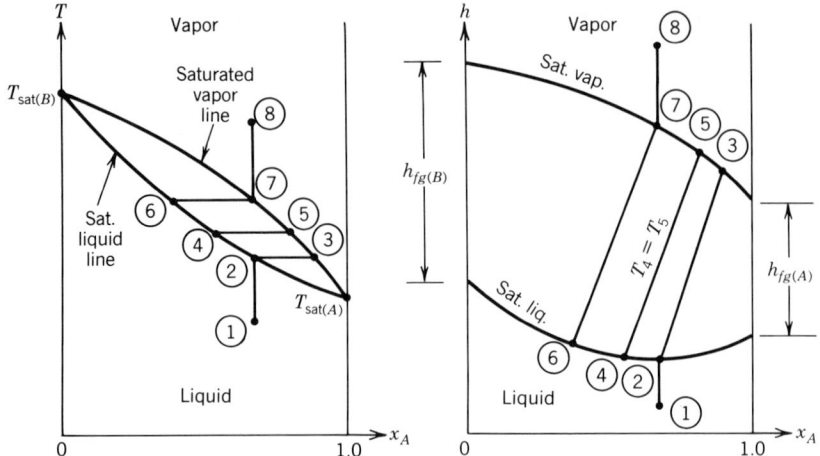

Fig. 3.1-20 Evaporation of a two-component solutions at constant pressure.

component will evaporate "preferentially." Consequently, the vapor driven off initially will have a concentration that is different from the liquid with which it is in equilibrium. Equilibrium implies the same temperature and total pressure.

Figure 3.1-20 shows the evaporation process for a liquid solution starting at point 1 at which the temperature is below saturation temperature for the existing total pressure. Both components, A and B, are assumed to have significant but different vapor pressures. If no component leaves the system, the eventual superheated vapor will have the same mass fraction, x_A, at point 8 that the solution had at point 1. As shown, the saturated liquid line and the saturated vapor line will not be the same either on T-x or on h-x coordinates.

The process from 1 to 8 on both sets of coordinates implies that component A has the higher vapor pressure—evaporates preferentially—when the saturated liquid line is reached at 2. Consequently, the saturated vapor at 3 that is in equilibrium with saturated liquid at 2 has a higher concentration of component A. The concentration of the remaining liquid moves toward point 6, and the concentration of the equilibrium vapor moves toward 7. At any temperature before evaporation is completed, liquid at some point (4) will be in equilibrium with saturated vapor at 5. The isotherm $(T_4 = T_5)$ between these two points is a straight line on the h-x coordinates since this is an example, at equilibrium, of adiabatic mixing or separation (see Fig. 3.1-19). Above point 7 the mixture can be superheated at a constant concentration value.

Starting with superheated vapor at point 8, the process can be reversed to show cooling, condensation, and eventual subcooling to point 1.

Various substances in solution exhibit azeotropic concentration values, over certain temperature and pressure ranges, at which the solution will change phase as a single substance, maintaining the same concentration value through the evaporation or condensation process. The azeotropic temperature may be higher or lower (and the corresponding pressure lower or higher) than the saturation value for either component separately. The azeotropic mixture strength (x value) is a function of temperature and pressure. At sufficiently low pressures the azeotropic point disappears for all mixtures.

Equilibrium Between Phases

In Section 3.1-6 the Gibbs function was defined as

$$G = H - TS$$

and it was shown that

$$dG = V\,dP - S\,dT$$

Since a pure substance or an ideal nonreactive mixture that changes from phase 1 to phase 2 at constant pressure does so also at constant temperature, $dG = 0$.

It is noted above that the chemical potential of one component in a mixture (μ_i) is the partial molal value of the Gibbs function, $(\partial G/\partial n_i)_{T,P,n_j}$. During a phase change from phase 1 to phase 2, the decrease in one $(-dn_1)$ is the increase in the other $(+dn_2)$. At constant T and P

$$dG = \mu_1\, dn_1 + \mu_2\, dn_2 = 0 \quad \text{and} \quad \mu_1 = \mu_2$$

Thus, the specific value of the Gibbs function is the same for the two phases:

$$\bar{g}_1 = \bar{g}_2$$

This remains true as the pressure and corresponding temperature change:

$$d\bar{g}_1 = d\bar{g}_2$$

$$v_1\, dP - s_1\, dT = v_2\, dP - s_2\, dT$$

$$\frac{dP}{dT} = \frac{s_2 - s_1}{v_2 - v_1} = \frac{h_2 - h_1}{T(v_2 - v_1)}$$

This is the Clausius–Clapeyron equation, applicable to changes between any two phases. If the change is from liquid to vapor, and the conventional subscripts are employed,

$$\frac{dP}{dT} = \frac{h_{fg}}{T_{sat}v_{fg}}$$

which is the slope of the saturation line on the $P - T$ coordinates. At a triple point, where three phases are in equilibrium,

$$g_1 = g_2 = g_3$$

3.1-11   Reactive Systems

Reactive systems are those in which there is a change in the chemical composition of the components. An example is the combustion of fuel with oxygen alone or with air.

Stoichiometry

The stoichiometric mixture of fuel and oxidizer is the mixing proportion that is just sufficient to produce complete oxidation of the components of the fuel with no excess of either fuel or oxidizer. This may also be called the chemically correct mixture strength, the theoretically correct mixture strength, or the combining proportions.

For example, a hydrocarbon fuel with the chemical formula C_nH_m will require sufficient oxygen (O_2) to produce only carbon dioxide (CO_2) and water (H_2O) molecules. The reaction equation would be

$$C_nH_m + \left(n + \tfrac{1}{4}m\right)O_2 \rightarrow nCO_2 + \tfrac{1}{2}mH_2O$$

Example 3.1-11.   The stoichiometric mixture strength of the hydrocarbon fuel octane (C_8H_{18}) with oxygen would be

$$C_8H_{18} + 12.5O_2 \rightarrow 8CO_2 + 9H_2O$$

On a mole basis,

$$1.0 + 12.5 \rightarrow 8 + 9$$

On a mass basis (approximately),

$$1.0(8 \times 12 + 18 \times 1) + 12.5(32) \rightarrow 8(12 + 32) + 9(2 + 16)$$

$$114 + 400 \rightarrow 352 + 162$$

Note that there must be a mass balance, but that the number of moles will not necessarily balance. Also, if both fuel and oxidizer could be considered as ideal gases, which is not so in this example at normal temperature, a mole basis would also be a volume basis.

In this example the fuel-to-oxidizer ratio, mass basis, is $114/400 = 0.285$.

Frequently, of course, the oxygen is supplied in air. It is conventional in this type of calculation to consider that air is 21.0% oxygen and 79.0% nitrogen on a mole or volume basis, the other inert constituents of air being grouped with the nitrogen. (Compare the constituents of standard dry air as given in Example 3.1-10.) The ratio 79/21 is approximately 3.76.

The stoichiometric or chemically correct mixture proportions for octane and air would be

$$C_8H_{18} + 12.5O_2 + (3.76)(12.5)N_2 \rightarrow 8CO_2 + 9H_2O + (3.76)(12.5)N_2$$

On a mole basis,

$$1 + 12.5 + 47 \rightarrow 8 + 9 + 47$$

On a mass basis,

$$114 + 400 + 47(28) \rightarrow 1830$$

The stoichiometric fuel-to-air ratio then is $114/1716 = 0.0664$, and the air-to-fuel ratio would be $15.05/1.0$, on a mass basis.

A mixture different from the stoichiometric proportions is either rich (insufficient oxygen for complete combustion) or lean (an excess of oxygen). For the example of octane in air, a 10% lean mixture or a mixture with 10% excess air would have an air-to-fuel ratio of $1.10 \times 15.05 = 16.56/1.0$. A mixture 10% rich would have a fuel-to-air ratio of $1.10 \times 0.0664 = 0.0730/1.0$. Unless otherwise stated, fuel-to-air and air-to-fuel ratios should be given on a mass basis.

Enthalpy of Formation

For most tables and charts giving thermodynamic properties of pure substances, the enthalpy is given the value zero at some arbitrary temperature chosen for convenience, for example, the ice point of water for the Steam Tables and $-40°C$ for most refrigerants. For reacting systems, where substances previously dealt with as having fixed chemical formulas will combine with others or perhaps dissociate, the energy transfers must involve the energy supplied or released in the original formation of the substance. It is conventional to take the value of enthalpy as equal to zero at the standard temperature and pressure, 25°C (298 K) and 1.0 bar (10^5 Pa or 0.1 MPa) for an element or for a particular molecular structure. Thus, carbon (C), diatomic oxygen (O_2), monatomic oxygen (O), and so on, all have the same value (zero) at that standard temperature and pressure.

The enthalpy of formation is the energy released when two elements or molecules at 25°C and 1.0 bar, at which their enthalpy values are zero, combine to produce another molecular structure delivered at the same temperature and pressure in a steady-state, steady-flow process. Consider the formation of water from hydrogen and oxygen.

$$\left(\sum n_i h_i\right)_{25°C,1\ bar} + Q = \left(\sum n_e h_e\right)_{25°C,1\ bar}$$

$$H_2 + \tfrac{1}{2}O_2 + Q = H_2O$$

When the water is in the vapor phase, the value of Q is approximately -2.418×10^5 kJ/kmol of water. That value, then, is the enthalpy of formation per mole ($\bar{h}_f^\circ$) for water vapor, $H_2O(g)$. The superscript (°) indicates standard conditions, 25°C and 1.0 bar.

If the product were to be $H_2O(l)$, which is water in the liquid phase, then $\bar{h}_f^\circ$ would be -2.858×10^5 kJ/kmol. The difference is the latent heat of vaporization for water, per kg-mol ($\bar{h}_{fg}$) at 25°C. This example was taken, because of our familiarity with the properties of water, to illustrate the importance of phase and to note that, in fact, water would not be vapor at these conditions. This conventional adjustment of the $\bar{h}_f^\circ$ value simplifies reaction calculations for various substances when the products might be found in one phase or in another.

Enthalpy values at 1.0 bar and at temperatures above or below 298 K are tabulated as

$$\bar{h}_T^\circ - \bar{h}_f^\circ = \int_{298}^{T} \bar{C}_p\, dT$$

Conventionally, entropy values are tabulated at 1.0 bar pressure as absolute values. That is, entropy is zero at 0 K in accordance with the third law of thermodynamics.

Heat of Reaction

When several substances (reactants) combine chemically to produce several different substances (products), the first law of thermodynamics can be applied to the process. For a closed system,

$$\sum U_R + Q_V = \sum U_P + \mathscr{W}$$

For an open system

$$\sum H_R + Q_P = \sum H_P + \mathcal{W}$$

In the summation of internal energies or enthalpies for both products and reactants, the energy of formation at standard temperature and pressure as well as changes above or below those conditions must be taken into account. The heat of reaction (T, P) would be the value of Q when the work $(\mathcal{W})$ is zero.

This is illustrated in Fig. 3.1-21. For the open system, $\sum H_R = \sum n_i \bar{h}_i^{\circ})_R$ and $\sum H_p = \sum n_i \bar{h}_i^{\circ})_p$ are both plotted as a function of temperature (T). All the components supplied and all the products including unburned fuel and excess oxidizer as well as inert components must be included in the summations. For simplicity, it is assumed that all components are ideal gases so that pressure changes are ignored. For the conditions shown (arbitrarily selected) the temperature (T_R) at which the reactants are supplied is greater than the standard temperature, 298 K, and the temperature of the products (T_p) is higher than T_R. The value of Q for the conditions shown is negative, which indicates heat transfer from the system; this is the most commonly expected situation.

For what is generally referred to as the heating value of a fuel, as determined by a flow calorimeter at constant pressure, the products are assumed to leave at $T_P = T_R$, and the Q_P value is the vertical distance between the two $\sum H$ lines at that temperature, which may or may not be standard temperature.

For a constant-volume reaction, as for the heating value determined in a closed calorimeter, the diagram is similar, as shown, but $Q_V = \sum U_P - \sum U_R$ is the significant term.

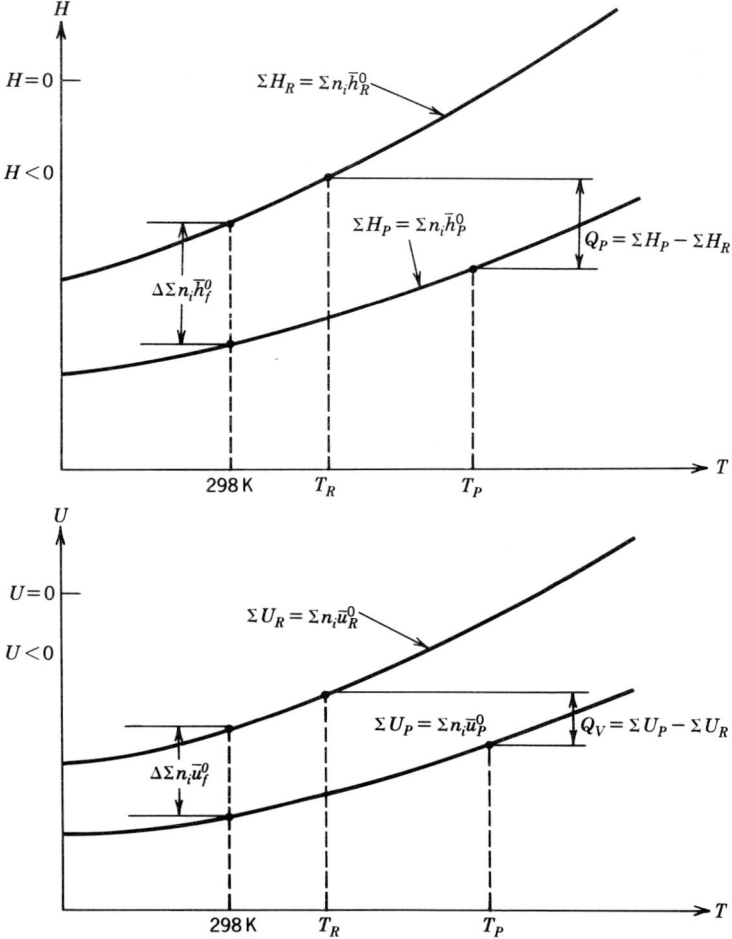

Fig. 3.1-21 Heat of formation and heat of reaction at constant pressure and at constant volume.

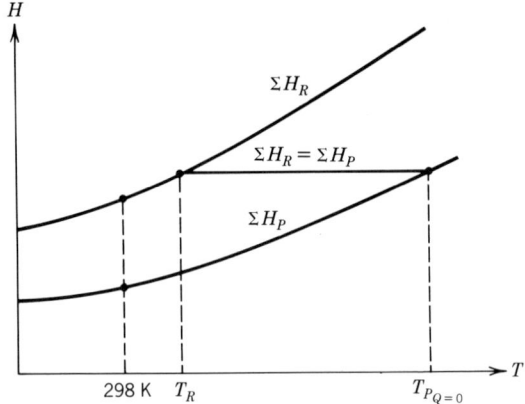

Fig. 3.1-22 Adiabatic flame temperature at constant pressure.

Adiabatic Flame Temperature

When there is zero heat transfer, the first-law equations still apply. The temperature of the products that satisfies the energy balance equation with $Q = 0$ is the adiabatic flame temperature as shown in Fig. 3.1-22. Unless otherwise specified, this is given for an open system where $\Sigma\ H_R = \Sigma\ H_p$.

The maximum adiabatic flame temperature is achieved for the stoichiometric mixture proportions. It is possible, therefore, to control the combustion temperature by an adjustment of the mixture ratio of the reactants. Either a lean mixture (excess air) or a rich mixture (excess fuel) will produce an adiabatic flame temperature less than the maximum because part of the energy released must raise the temperature of these inert components.

Note that dissociation of products at the higher temperatures has not been considered in these definitions. That will be discussed below.

Chemical Equilibrium

In the foregoing discussions of stoichiometric mixtures and adiabatic flame temperatures, it has been assumed that reactions have gone to completion. In fact, reactions will proceed in one direction only until an equilibrium situation is reached, at which the driving forces (chemical potentials) are equal in both directions.

Consider the reactions involving four constituents described by the equation

$$v_A A + v_B B \rightleftharpoons v_C C + v_D D$$

The values of v_i are the individual stoichiometric coefficients for each, in other words the number of moles of each that would produce a balance for all elements in the equation. On the other hand, the number of moles of each (n_i) need not be the correct combining proportions. As the reaction proceeds to the right or to the left, the changes in the number of moles of each (dn_i) will be of importance. Also, the total number of moles present will determine partial pressures, which are significant.

The degree of reaction will be ϵ, and the degree of dissociation will be $1 - \epsilon$, where reaction means movement toward the right and dissociation is movement toward the left, as defined by the following equations:

$$dn_A = -v_A\, d\epsilon \qquad dn_B = -v_B\, d\epsilon$$

$$dn_C = +v_C\, d\epsilon \qquad dn_D = +v_D\, d\epsilon$$

For simplicity, in the following discussion it will be assumed that each constituent acts as an ideal gas so that its mole fraction ($y_i = n_i / \Sigma\ n$) is the ratio of its partial pressure to the total pressure (P_i / P). Where this is not so, the fugacity of the components must be substituted for pressure.

Equilibrium is achieved in any reaction when the Gibbs function values are the same on both sides of the reaction equation. The value of the Gibbs function per mole for an ideal gas is

$$\bar{g} = \bar{h} - T\bar{s},$$

and

$$\bar{s} = \int_0^T \bar{C}p \frac{dT}{T} - \bar{R} \ln \frac{P}{P^\circ}$$

The change in the Gibbs function for the mixture, at a given temperature and total pressure, depends on the change in the degree of reaction, so that

$$dG_{T,P} = d\epsilon \sum v_i \left[\bar{g}_i + \bar{R}T \ln y_i \left(\frac{P}{P^\circ} \right) \right]$$

$$= d\epsilon \left[\Delta G + \sum v_i \bar{R}T \ln y_i \left(\frac{P}{P^\circ} \right) \right]$$

where

$$\Delta G = -v_A \mu_A - v_B \mu_B + v_C \mu_C + v_D \mu_D$$

and

$$\sum v_i \bar{R}T \ln y_i = \bar{R}T \ln \left[\frac{y_C^{v_C} \cdot y_D^{v_D}}{y_A^{v_A} \cdot y_B^{v_B}} \right] \left(\frac{P}{P^\circ} \right)^{v_C + v_D - v_A - v_B}$$

$$= \bar{R}T \ln K$$

As defined by these equations, K is the equilibrium constant. At equilibrium, $dG_{T,P} = 0$, so that for equilibrium,

$$\ln K = -\frac{\Delta G}{\bar{R}T}$$

The equilibrium constant (K) is a function of temperature as is the heat of reaction (ΔH) defined as

$$\Delta H = v_C \bar{h}_C + v_D \bar{h}_D - v_A \bar{h}_A - v_B \bar{h}_B$$

The differential of $\ln K$ with respect to temperature at constant pressure produces the van't Hoff equation, or van't Hoff isobar:

$$\frac{d}{dT} \ln K = \frac{\Delta H}{\bar{R}T^2} \bigg|_P$$

In the defining equation for K, the ratio P/P° can be written simply as P if its units are bars and $P^\circ = 1$ bar, as it conventionally is.

As noted above, the development was presented with the assumption that all constituents in the reaction equation are ideal gases. Where that is not so, fugacity (f) must be substituted for pressure. Fugacity has been referred to as a "pseudo pressure" defined as follows:
For an ideal gas

$$P v = RT \quad \text{and} \quad dg_T = v \, dP_T = \left(\frac{RT}{P} \right) dP_T = RT \, d(\ln P)_T$$

For a real gas

$$P v = ZRT \quad \text{and} \quad dg_T = ZRT \left(\frac{dP_T}{P} \right) = ZRT \, d(\ln P)_T$$

$$= RT \, d(\ln f)_T,$$

in which the fugacity (f) is such that the ratio f/P approaches unity as P and f approach zero.

When it is necessary to use fugacity rather than pressure in the determination of the equilibrium constant (K), the development of the defining equation is more complicated and results in the following definition in which P° is unity and $P/P^\circ = P$:

$$K = \left[\frac{y_C^{v_C} \cdot y_D^{v_D}}{y_A^{v_A} \cdot y_B^{v_B}} \right] \cdot P^{v_C + v_D - v_A - v_B} \left[\frac{(f/P)_C^{v_C} \cdot (f/P)_D^{v_D}}{(f/P)_A^{v_A} \cdot (f/P)_B^{v_B}} \right]$$

The change in the equilibrium constant (K_f) with temperature is again given by the van't Hoff equation. Reasonable accuracy is obtained when the heat of reaction in taken at standard pressure, as indicated by the symbol $\Delta H°$. Thus,

$$\frac{d \ell n K_f}{dT} \simeq \frac{\Delta H°}{\overline{R}T^2}$$

For a more detailed discussion of fugacity, see a reference text such as those by Zemansky or Van Wylen and Sonntag, listed in the Bibliography.

Note that dissociation, movement toward the left in the reaction equation, includes dissociation of polyatomic molecules into monatomic molecules; it also includes ionization of molecules, which is the separation of negatively charged electrons from molecules to leave positively charged ions. These two phenomena occur at relatively high temperatures.

LIST OF SYMBOLS

Symbol[†]	Meaning	Units
A	Area	m^2
	Availability	kJ
a	Acceleration	m/s^2
B	Darrieus function	kJ
Btu	British thermal unit (IST value)	
bhp	Brake (shaft) horsepower	
bmep	Brake mean effective pressure	kPa
°C	Degrees Celsius	
C_P	Specific heat at constant pressure	kJ/kg · K
C_V	Specific heat at constant volume	kJ/kg · K
d	Differential (exact)	
E	Energy	kJ
e	High-grade energy, electrical or shaft work (subscript)	
°F	Degrees Fahrenheit	
F	Helmholtz function	kJ
	Force	N
ft	Feet	ft
f	Saturated liquid (subscript)	
f	Fugacity	kPa
G	Gibbs function	kJ
g	Acceleration of gravity Saturated vapor (subscript)	m/s^2
g_c	Dimensional constant in $F = ma/g_c$	kg · m/N · s^2
H	Enthalpy	kJ
	Higher temperature (subscript)	
h	Hour	
hp	Horsepower (1.0 hp = 550 ft · lbf/s)	hp
ihp	Indicated horsepower	
imep	Indicated mean effective pressure	kPa
J	Joule	N · m
K	Degrees Kelvin	degrees
k	Kilo, 1000 (prefix)	
kg	Kilogram (unit of mass)	kg
L	Length Lower temperature (subscript)	m
LW	Lost work	kJ
lb	Pound mass	
lbf	Pound force	

LIST OF SYMBOLS   (*Continued*)

Symbol[†]	Meaning	Units
$\mathcal{M}$	Molecular "weight"	
m	Meter (unit of length)	m
	Mass	kg
N	Newton (unit of force)	N
n	Number of moles	
	A general exponent	
o	Ambient or base condition (subscript)	
	Standard Conditions (superscript)	
P	Power	W
	Pressure	kPa
Pa	Pascal (unit of pressure)	N/m^2
Q	Heat transfer	kJ
R	Degrees Rankine	degrees
	Radius	m
	Individual gas constant	$kJ/kg \cdot K$
$\overline{R}$	Universal gas constant	$kJ/kg\text{-}mol \cdot K$
S	Entropy	kJ/K
s	Seconds (unit of time)	s
T	Temperature	K
V	Volume	m^3
$\overline{V}$	Velocity	m/s
U	Internal energy	kJ
W	Watt	J/s
$\mathcal{W}$	Work	kJ
x	Dryness fraction	
	Displacement	m
Z	Compressibility factor	
γ	Specific heat ratio, C_P/C_V	
Δ	Finite difference	
δ	Inexact differential	
	Infinitesimal difference	
ϵ	Degree of reaction	
ε'	Effectiveness	
ε''	Second-law efficiency	
η	First-law efficiency	
κ	Compressibility	
μ	Joule–Thomson (Joule–Kelvin) coefficient	
	Chemical potential	$kJ/kg\text{-}mol$
ν	Stoichiometric coefficient	
ρ	Density	kg/m^3
Σ	Sum, algebraic	
ϕ	Availability function	kJ
	Entropy function $\int C_p \, dT/T$	$kJ/kg \cdot K$
ω	Rotational speed, rad/s	s^{-1}
∂	Partial derivative	
ln	Natural logarithm	
$\oint$	Cyclic integral (integration around a complete cycle)	

[†]*Note:* Lowercase letters frequently mean specific values; for example, $v = V/m$ m^3/kg; $h = H/m$ kJ/kg

BIBLIOGRAPHY

Benson, Rowland S. *Advanced Engineering Thermodynamics*, 2nd ed. Pergamon, Oxford, 1977.

Denbigh, Kenneth. *The Principles of Chemical Equilibrium*, 3rd ed. Cambridge University Press, London, 1971.

Hatsopoulos, G. N., and J. H. Keenan, *Principles of General Thermodynamics*. Wiley, New York, 1965.

Keenan, Joseph H. *Thermodynamics*. Wiley, New York, 1941.

Kestin, Joseph, *A Course in Thermodynamics*. Blaisdell, Waltham, MA, Vol. 1, 1966; Vol. II, 1968.

Lee, J. F., and F. W. Sears. *Thermodynamics*, 2nd ed. Addison-Wesley, Reading, MA, 1963.

Reynolds, W. C., and H. C. Perkins. *Engineering Thermodynamics*, 2nd ed. McGraw-Hill, New York, 1977.

Roberts, J. K., and A. R. Miller, *Heat and Thermodynamics*, 4th ed. Blackie and Son, London, 1951.

Van Wylen, G. J., and R. E. Sonntag. *Fundamentals of Classical Thermodynamics*, 2nd ed. Wiley, New York, 1973, S.I. version, 1976.

Wood, Bernard D. *Applications of Thermodynamics*, 2nd ed. Addison-Wesley, Reading, MA, 1982.

Zemansky, Mark W. *Heat and Thermodynamics*, 5th ed. McGraw-Hill, New York, 1957.

Zemansky, M. W., M. M. Abbott, and H. C. VanNess. *Basic Engineering Thermodynamics*, 2nd ed. McGraw-Hill, New York, 1975.

3.2   HEAT TRANSFER

Michel A. Saad

3.2-1   Introduction

Heat transfer takes place between material systems as a result of a temperature difference. The transmission process involves energy conversions governed by the first and second laws of thermodynamics. The heat transfer proceeds from a high-temperature region to a low-temperature region, and because of the finite thermal potential, there is an increase in entropy. Thermodynamics, however, is concerned with equilibrium states, which includes thermal equilibrium, irrespective of the time necessary to attain these equilibrium states. But heat transfer is a result of thermal nonequilibrium conditions, therefore, the laws of thermodynamics alone cannot describe completely the heat transfer process. In practice, most engineering problems are concerned with the rate of heat transfer rather than the quantity of heat being transferred. Resort then is directed to the particular laws governing the transfer of heat.

There are three distinct modes of heat transfer: conduction, convection, and radiation. Although these modes are herein discussed separately, all three types may occur simultaneously.

3.2-2   Conduction

Heat transfer by conduction in a solid body is due to the existence of a temperature gradient in the body. This mode of heat transfer is attributed to the collisions of the molecules as a result of their vibration. The fundamental equation of heat transfer by conduction was formulated by Joseph Fourier in 1822. For steady one-dimensional heat transfer, Fourier's law takes the form

$$q = -kA\frac{\partial T}{\partial x} \tag{3.2-1}$$

where q is the heat transfer per unit time (W), k is the thermal conductivity of the substance (W/m · K), A is the area of heat flow measured perpendicular to the direction of heat flow (m^2), $\partial T/\partial x$ is the temperature gradient in the x direction (K/m). The negative sign in Eq. (3.2-1) indicates that heat flows in the direction of a negative temperature gradient.

The Fourier equation states that the rate of heat transfer across an area A of a solid slab of thickness dx is proportional to the temperature gradient. The thermal conductivity k is an experimentally determined proportionality factor and varies widely from one material to another. It depends on the state of the substance and also varies with temperature. Figures 3.2-1 to 3.2-3 show the variation of thermal conductivity with temperature for various substances, and Tables 3.2-1 to 3.2-6 give thermophysical properties for various solids, liquids, and gases.

The general Fourier equation, which predicts the temperature distribution in a homogeneous solid during unsteady heat transfer, is

$$\frac{\partial}{\partial x}\left(k\frac{\partial T}{\partial x}\right) + \frac{\partial}{\partial y}\left(k\frac{\partial T}{\partial y}\right) + \frac{\partial}{\partial z}\left(k\frac{\partial T}{\partial z}\right) + \dot{q} = c\rho\frac{\partial T}{\partial t} \tag{3.2-2}$$

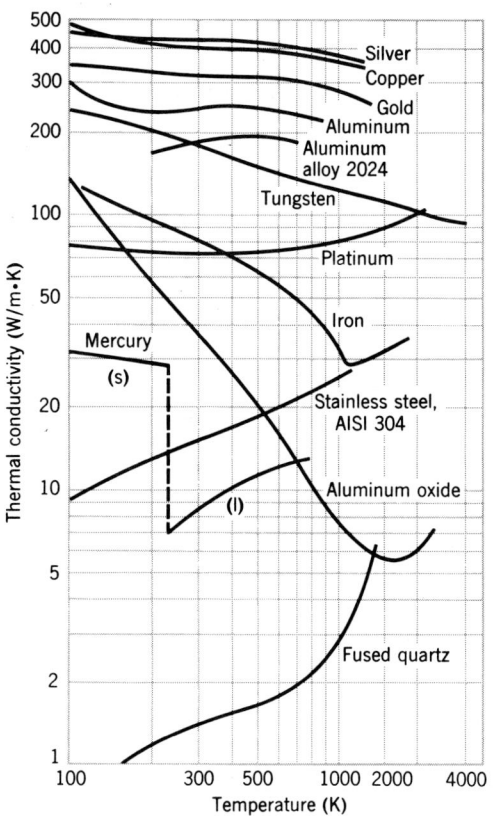

Fig. 3.2-1 The temperature dependence of the thermal conductivity of selected solids. (From ref. 1; used by permission.).

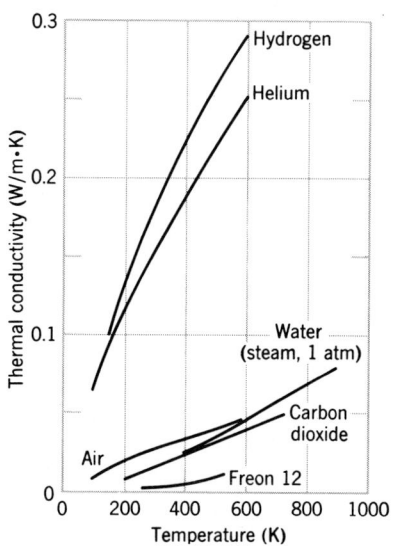

Fig. 3.2-2 The temperature dependence of the thermal conductivity of selected gases at normal pressures. (From ref. 1; used by permission.)

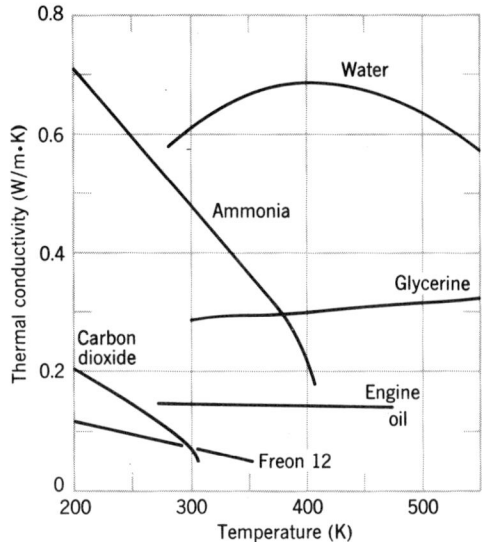

Fig. 3.2-3 The temperature dependence of the thermal conductivity of selected nonmetallic liquids under saturated conditions. (From ref. 1; used by permission.)

where $\dot{q}$ is the rate at which energy is generated per unit volume (W/m³), c is the specific heat (J/kg · K), and ρ is the density (kg/m³).

If the thermal conductivity k is assumed constant, then Eq. (3.2-2) becomes

$$\frac{\partial^2 T}{\partial x^2} + \frac{\partial^2 T}{\partial y^2} + \frac{\partial^2 T}{\partial z^2} + \frac{\dot{q}}{k} = \frac{c\rho}{k}\frac{\partial T}{\partial t}$$

$$= \frac{1}{\alpha}\frac{\partial T}{\partial t} \qquad (3.2\text{-}3)$$

where $\alpha = k/c\rho$ is the thermal diffusivity of the substance (m²/h). Three cases may be considered:

1. If there is no energy generation within the body, then

$$\frac{\partial^2 T}{\partial x^2} + \frac{\partial^2 T}{\partial y^2} + \frac{\partial^2 T}{\partial z^2} = \frac{1}{\alpha}\frac{\partial T}{\partial t} \quad \text{(Fourier's equation)} \qquad (3.2\text{-}4)$$

2. If steady state prevails, then

$$\frac{\partial^2 T}{\partial x^2} + \frac{\partial^2 T}{\partial y^2} + \frac{\partial^2 T}{\partial z^2} + \frac{\dot{q}}{k} = 0 \quad \text{(Poisson's equation)} \qquad (3.2\text{-}5)$$

3. If in addition to steady state, no energy is absorbed or generated within the body, then

$$\frac{\partial^2 T}{\partial x^2} + \frac{\partial^2 T}{\partial y^2} + \frac{\partial^2 T}{\partial z^2} = 0 \quad \text{(Laplace's equation)} \qquad (3.2\text{-}6)$$

The heat transfer equations, when expressed in cylindrical coordinates (r, ϕ, z), take the form

$$\frac{1}{r}\frac{\partial}{\partial r}\left(kr\frac{\partial T}{\partial r}\right) + \frac{1}{r^2}\frac{\partial}{\partial \phi}\left(k\frac{\partial T}{\partial \phi}\right) + \frac{\partial}{\partial z}\left(k\frac{\partial T}{\partial z}\right) + \dot{q} = \rho c\frac{\partial T}{\partial t} \qquad (3.2\text{-}7)$$

TABLE 3.2-1 THERMOPHYSICAL PROPERTIES OF SELECTED METALLIC SOLIDS[a,b]

Each cell under "Properties at various temperatures" lists the values as k, W/m·K (top) / c_p, J/kg·K (bottom).

COMPOSITION	MELTING POINT K	ρ kg/m³	c_p J/kg·K	k W/m·K	$\alpha \cdot 10^6$ m²/s	100	200	400	600	800	1000	1200	1500	2000	2500
Aluminum Pure	933	2702	903	237	97.1	302 / 482	237 / 798	240 / 949	231 / 1033	218 / 1146					
Alloy 2024-T6 (4.5% Cu, 1.5% Mg, 0.6% Mn)	775	2770	875	177	73.0	65 / 473	163 / 787	186 / 925	186 / 1042						
Alloy 195, Cast (4.5% Cu)		2790	883	168	68.2			174 / —	185 / —						
Beryllium	1550	1850	1825	200	59.2	990 / 203	301 / 1114	161 / 2191	126 / 2604	106 / 2823	90.8 / 3018	78.7 / 3227	/ 3519		
Bismuth	545	9780	122	7.86	6.59	16.5 / 112	9.69 / 120	7.04 / 127							
Boron	2573	2500	1107	27.0	9.76	190 / 128	55.5 / 600	16.8 / 1463	10.6 / 1892	9.60 / 2160	9.85 / 2338				
Cadmium	594	8650	231	96.8	48.4	203 / 198	99.3 / 222	94.7 / 242							
Chromium	2118	7160	449	93.7	29.1	159 / 192	111 / 384	90.9 / 484	80.7 / 542	71.3 / 581	65.4 / 616	61.9 / 682	57.2 / 779	49.4 / 937	
Cobalt	1769	8862	421	99.2	26.6	167 / 236	122 / 379	85.4 / 450	67.4 / 503	58.2 / 550	52.1 / 628	49.3 / 733	42.5 / 674		
Copper Pure	1358	8933	385	401	117	482 / 252	413 / 356	393 / 397	379 / 417	366 / 433	352 / 451	339 / 480			
Commercial bronze (90% Cu, 10% Al)	1293	8800	420	52	14		42 / 785	52 / 460	59 / 545						
Phosphor gear bronze (89% Cu, 11% Sn)	1104	8780	355	54	17		41	65	74						
Cartridge brass (70% Cu, 30% Zn)	1188	8530	380	110	33.9	75	95 / 360	137 / 395	149 / 425						
Constantan (55% Cu, 45% Ni)	1493	8920	384	23	6.71	17 / 237	19 / 362								

205

TABLE 3.2-1 (*Continued*)

COMPOSITION	MELTING POINT K	PROPERTIES AT 300 K				PROPERTIES AT VARIOUS TEMPERATURES (K) k, W/m·K / c_p, J/kg·K									
		ρ kg/m³	c_p J/kg·K	k W/m·K	$\alpha \cdot 10^6$ m²/s	100	200	400	600	800	1000	1200	1500	2000	2500
Germanium	1211	5360	322	59.9	34.7	232 190	96.8 290	43.2 337	27.3 348	19.8 357	17.4 375	17.4 395			
Gold	1336	19300	129	317	127	327 109	323 124	311 131	298 135	284 140	270 145	255 155			
Iridium	2720	22500	130	147	50.3	172 90	153 122	144 133	138 138	132 144	126 153	120 161	111 172		
Iron Pure	1810	7870	447	80.2	23.1	134 216	94.0 384	69.5 490	54.7 574	43.3 680	32.8 975	28.3 609	32.1 654		
Armco (99.75% pure)		7870	447	72.7	20.7	95.6 215	80.6 384	65.7 490	53.1 574	42.2 680	32.3 975	28.7 609	31.4 654		
Carbon steels Plain carbon (Mn ≤ 1%, Si ≤ 0.1%)		7854	434	60.5	17.7			56.7 487	48.0 559	39.2 685	30.0 1169				
AISI 1010		7832	434	63.9	18.8			58.7 487	48.8 559	39.2 685	31.3 1168				
Carbon-silicon (Mn ≤ 1%, 0.1% < Si ≤ 0.6%)		7817	446	51.9	14.9			49.8 501	44.0 582	37.4 699	29.3 971				
Carbon-manganese-silicon (1% < Mn ≤ 1.65%, 0.1% < Si ≤ 0.6%)		8131	434	41.0	11.6			42.2 487	39.7 559	35.0 685	27.6 1090				
Chromium (low) steels ½Cr-¼Mo-Si (0.18% C, 0.65% Cr, 0.23% Mo, 0.6% Si)		7822	444	37.7	10.9			38.2 492	36.7 575	33.3 688	26.9 969				
1Cr-½Mo (0.16% C, 1% Cr, 0.54% Mo, 0.39% Si)		7858	442	42.3	12.2			42.0 492	39.1 575	34.5 688	27.4 969				
1 Cr-V (0.2% C, 1.02% Cr, 0.15% V)		7836	443	48.9	14.1			46.8 492	42.1 575	36.3 688	28.2 969				

Material	Melting Point (K)	ρ (kg/m³)	cₚ (J/kg·K)	k (W/m·K)	α·10⁶ (m²/s)	k 100K	cₚ 100K	k 200K	cₚ 200K	k 400K	cₚ 400K	k 600K	cₚ 600K	k 800K	cₚ 800K	k 1000K	cₚ 1000K	k 1200K	cₚ 1200K	k 1500K	cₚ 1500K	k 2000K	cₚ 2000K	k 2500K	cₚ 2500K
Stainless steels																									
AISI 302		8055	480	15.1	3.91					17.3	512	20.0	559	22.8	585	25.4	606								
AISI 304	1670	7900	477	14.9	3.95	9.2	272	12.6	402	16.6	515	19.8	557	22.6	582	25.4	611	28.0	640	31.7	682				
AISI 316		8238	468	13.4	3.48					15.2	504	18.3	550	21.3	576	24.2	602								
AISI 347		7978	480	14.2	3.71					15.8	513	18.9	559	21.9	585	24.7	606								
Lead	601	11340	129	35.3	24.1	39.7	118	36.7	125	34.0	132	31.4	142												
Magnesium	923	1740	1024	156	87.6	169	649	159	934	153	1074	149	1170	146	1267										
Molybdenum	2894	10240	251	138	53.7	179	141	143	224	134	261	126	275	118	285	112	295	105	308	98	330	90	380	86	459
Nickel Pure	1728	8900	444	90.7	23.0	164	232	107	383	80.2	485	65.6	592	67.6	530	71.8	562	76.2	594	82.6	616				
Nichrome (80% Ni, 20% Cr)	1672	8400	420	12	3.4					14	480	16	525	21	545										
Inconel X-750 (73% Ni, 15% Cr, 6.7% Fe)	1665	8510	439	11.7	3.1	8.7	—	10.3	372	13.5	473	17.0	510	20.5	546	24.0	626	27.6	—	33.0	—				
Niobium	2741	8570	265	53.7	23.6	55.2	188	52.6	249	55.2	274	58.2	283	61.3	292	64.4	301	67.5	310	72.1	324	79.1	347		
Palladium	1827	12020	244	71.8	24.5	76.5	168	71.6	227	73.6	251	79.7	261	86.9	271	94.2	281	102	291	110	307				
Platinum Pure	2045	21450	133	71.6	25.1	77.5	100	72.6	125	71.8	136	73.2	141	75.6	146	78.7	152	82.6	157	89.5	165	99.4	179		
Alloy 60Pt-40Rh (60% Pt, 40% Rh)	1800	16630	162	47	17.4					52	—	59	—	65	—	69	—	73	—	76	—				
Rhenium	3453	21100	136	47.9	16.7	58.9	97	51.0	127	46.1	139	44.2	145	44.1	151	44.6	156	45.7	162	47.8	171	51.9	186		
Rhodium	2236	12450	243	150	49.6	186	147	154	220	146	253	136	274	127	293	121	311	116	327	110	349	112	376		
Silicon	1685	2330	712	148	89.2	884	259	264	556	98.9	790	61.9	867	42.2	913	31.2	946	25.7	967	22.7	992				

207

TABLE 3.2-1 (*Continued*)

COMPOSITION	MELTING POINT K	PROPERTIES AT 300 K				PROPERTIES AT VARIOUS TEMPERATURES (K)									
		ρ kg/m³	c_p J/kg·K	k W/m·K	$\alpha \cdot 10^6$ m²/s	100	200	400	600	800	1000	1200	1500	2000	2500
Silver	1235	10500	235	429	174	444 / 187	430 / 225	425 / 239	412 / 250	396 / 262	379 / 277	361 / 292			
Tantalum	3269	16600	140	57.5	24.7	59.2 / 110	57.5 / 133	57.8 / 144	58.6 / 146	59.4 / 149	60.2 / 152	61.0 / 155	62.2 / 160	64.1 / 172	65.6 / 189
Thorium	2023	11700	118	54.0	39.1	59.8 / 99	54.6 / 112	54.5 / 124	55.8 / 134	56.9 / 145	56.9 / 156	58.7 / 167			
Tin	505	7310	227	66.6	40.1	85.2 / 188	73.3 / 215	62.2 / 243							
Titanium	1953	4500	522	21.9	9.32	30.5 / 300	24.5 / 465	20.4 / 551	19.4 / 591	19.7 / 633	20.7 / 675	22.0 / 620	24.5 / 686		
Tungsten	3660	19300	132	174	68.3	208 / 87	186 / 122	159 / 137	137 / 142	125 / 145	118 / 148	113 / 152	107 / 157	100 / 167	95 / 176
Uranium	1406	19070	116	27.6	12.5	21.7 / 94	25.1 / 108	29.6 / 125	34.0 / 146	38.8 / 176	43.9 / 180	49.0 / 161			
Vanadium	2192	6100	489	30.7	10.3	35.8 / 258	31.3 / 430	31.3 / 515	33.3 / 540	35.7 / 563	38.2 / 597	40.8 / 645	44.6 / 714	50.9 / 867	
Zinc	693	7140	389	116	41.8	117 / 297	118 / 367	111 / 402	103 / 436						
Zirconium	2125	6570	278	22.7	12.4	33.2 / 205	25.2 / 264	21.6 / 300	20.7 / 322	21.6 / 342	23.7 / 362	26.0 / 344	28.8 / 344	33.0 / 344	

$^a\rho$ = density; c_p = specific heat at constant pressure; k = thermal conductivity; and α = thermal diffusivity.
^{b}See refs. 1–3.

208

TABLE 3.2-2 THERMOPHYSICAL PROPERTIES OF SELECTED NONMETALLIC SOLIDS[a]

| | | PROPERTIES AT 300 K | | | | PROPERTIES AT VARIOUS TEMPERATURES (K) | | | | | | | | | |
| | | | | | | k, W/m·K / c_p, J/kg·K | | | | | | | | | |
COMPOSITION	MELTING POINT K	ρ kg/m³	c_p J/kg·K	k W/m·K	$\alpha \cdot 10^6$ m²/s	100	200	400	600	800	1000	1200	1500	2000	2500
Aluminum oxide, sapphire	2323	3970	765	46	15.1	450	82	32.4	18.9	13.0	10.5				
								940	1110	1180	1225				
Aluminum oxide, polycrystalline	2323	3970	765	36.0	11.9	133	55	26.4	15.8	10.4	7.85	6.55	5.66	6.00	
								940	1110	1180	1225				
Beryllium oxide	2725	3000	1030	272	88.0			196	111	70	47	33	21.5	15	
								1350	1690	1865	1975	2055	2145	2750	
Boron	2573	2500	1105	27.6	9.99	190	52.5	18.7	11.3	8.1	6.3	5.2			
								1490	1880	2135	2350	2555			
Boron fiber epoxy (30% vol) composite	590	2080													
k, || to fibers				2.29		2.10	2.23	2.28							
k, ⊥ to fibers				0.59		0.37	0.49	0.60							
c_p			1122			364	757	1431							
Carbon Amorphous	1500	1950	—	1.60	—	0.67	1.18	1.89	2.19	2.37	2.53	2.84	3.48		
Diamond, type IIa insulator	—	3500	509	2300		10000	4000	1540							
						21	194	853							
Graphite, pyrolytic	2273	2210		1950											
k, || to layers				1950		4970	3230	1390	892	667	534	448	357	262	
k, ⊥ to layers				5.70		16.8	9.23	4.09	2.68	2.01	1.60	1.34	1.08	0.81	
c_p			709			136	411	992	1406	1650	1793	1890	1974	2043	
Graphite fiber epoxy (25% vol) composite	450	1400													
k, heat flow || to fibers				11.1		5.7	8.7	13.0							
k, heat flow ⊥ to fibers				0.87		0.46	0.68	1.1							
c_p			935			337	642	1216							
Pyroceram, Corning 9606	1623	2600	808	3.98	1.89	5.25	4.78	3.64	3.28	3.08	2.96	2.87	2.79		
								908	1038	1122	1197	1264	1498		
Silicon carbide	3100	3160	675	490	230			880	1050	1135	1195	1243	1310		

TABLE 3.2-2 (Continued)

COMPOSITION	MELTING POINT K	PROPERTIES AT 300 K ρ kg/m³	c_p J/kg·K	k W/m·K	$\alpha \cdot 10^6$ m²/s	PROPERTIES AT VARIOUS TEMPERATURES (K) k, W/m·K / c_p, J/kg·K 100	200	400	600	800	1000	1200	1500	2000	2500
Silicon dioxide, crystalline (quartz)	1883	2650													
k, ∥ to c axis				10.4		39	16.4	7.6	5.0	4.2					
k, ⊥ to c axis				6.21		20.8	9.5	4.70	3.4	3.1					
c_p			745			—	—	885	1075	1250					
Silicon dioxide, polycrystalline (fused silica)	1883	2220	745	1.38	0.834	0.69	1.14	1.51	1.75	2.17	2.87	4.00			
						—	—	905	1040	1105	1155	1195			
Silicon nitride	2173	2400	691	16.0	9.65	—	—	13.9	11.3	9.88	8.76	8.00	7.16	6.20	
							578	778	937	1063	1155	1226	1306	1377	
Sulphur	392	2070	708	0.206	0.141	0.165	0.185								
						403	606								
Thorium dioxide	3573	9110	235	13	6.1			10.2	6.6	4.7	3.68	3.12	2.73	2.5	
								255	274	285	295	303	315	330	
Titanium dioxide, polycrystalline	2133	4157	710	8.4	2.8			7.01	5.02	3.94	3.46	3.28			
								805	880	910	930	945			

[a]See refs. 1–5.

In spherical coordinates (r, ϕ, θ)

$$\frac{1}{r^2}\frac{\partial}{\partial r}\left(kr^2\frac{\partial T}{\partial r}\right) + \frac{1}{r^2\sin^2\theta}\frac{\partial}{\partial \phi}\left(k\frac{\partial T}{\partial \phi}\right) + \frac{1}{r^2\sin\theta}\frac{\partial}{\partial \theta}\left(k\sin\theta\frac{\partial T}{\partial \theta}\right) + \dot q = \rho c\frac{\partial T}{\partial t} \qquad (3.2\text{-}8)$$

In one-dimensional steady state, heat transfer by conduction is given by

$$q = -kA\frac{\Delta T}{\Delta x} = -\frac{\Delta T}{\Delta x/kA} = -\frac{\Delta T}{R} \qquad (3.2\text{-}9)$$

where

$$R = \frac{\Delta x}{kA} = \text{thermal resistance} \qquad (3.2\text{-}10)$$

k being an average thermal conductivity for the temperature difference ΔT.

TABLE 3.2-3 THERMOPHYSICAL PROPERTIES OF COMMON MATERIALS[a]

Structural Building Materials

DESCRIPTION/COMPOSITION	TYPICAL PROPERTIES AT 300 K		
	DENSITY, ρ kg/m^3	THERMAL CONDUCTIVITY, k W/m·K	SPECIFIC HEAT, c_p J/kg·K
Building Boards			
Asbestos-cement board	1,920	0.58	—
Gypsum or plaster board	800	0.17	—
Plywood	545	0.12	1,215
Sheathing, regular density	290	0.055	1,300
Acoustic tile	290	0.058	1,340
Hardboard, siding	640	0.094	1,170
Hardboard, high density	1,010	0.15	1,380
Particle board, low density	590	0.078	1,300
Particle board, high density	1,000	0.170	1,300
Woods			
Hardwoods (oak, maple)	720	0.16	1,255
Softwoods (fir, pine)	510	0.12	1,380
Masonry Materials			
Cement mortar	1,860	0.72	780
Brick, common	1,920	0.72	835
Brick, face	2,083	1.3	—
Clay tile, hollow			
1 cell deep, 10 cm thick	—	0.52	—
3 cells deep, 30 cm thick	—	0.69	—
Concrete block, 3 oval cores			
sand/gravel, 20 cm thick	—	1.0	—
cinder aggregate, 20 cm thick	—	0.67	—
Concrete block, rectangular core			—
2 core, 20 cm thick, 16 kg	—	1.1	—
same with filled cores	—	0.60	—
Plastering Materials			
Cement plaster, sand aggregate	1,860	0.72	—
Gypsum plaster, sand aggregate	1,680	0.22	1,085
Gypsum plaster, vermiculite aggregate	720	0.25	—

(Continued)

TABLE 3.2-3 (*Continued*)

Insulating Materials and Systems

	TYPICAL PROPERTIES AT 300 K		
DESCRIPTION/COMPOSITION	DENSITY, ρ kg/m^3	THERMAL CONDUCTIVITY, k W/m·K	SPECIFIC HEAT, c_p J/kg·K
Blanket and Batt			
Glass fiber, paper faced	16	0.046	—
	28	0.038	—.
	40	0.035	—
Glass fiber, coated; duct liner	32	0.038	835
Board and Slab			
Cellular glass	145	0.058	1,000
Glass fiber, organic bonded	105	0.036	795
Polystyrene, expanded			
extruded (R-12)	55	0.027	1,210
molded beads	16	0.040	1,210
Mineral fiberboard; roofing material	265	0.049	—
Wood, shredded/cemented	350	0.087	1,590
Cork	120	0.039	1,800
Loose Fill			
Cork, granulated	160	0.045	—
Diatomaceous silica, coarse powder	350	0.069	—
	400	0.091	—
Diatomaceous silica, fine powder	200	0.052	—
	275	0.061	—
Glass fiber, poured or blown	16	0.043	835
Vermiculite, flakes	80	0.068	835
	160	0.063	1,000
Formed/Foamed-in-Place			
Mineral wool granules with asbestos/inorganic binders, sprayed	190	0.046	—
Polyvinyl acetate cork mastic; sprayed or troweled	—	0.100	—
Urethane, two-part mixture; rigid foam	70	0.026	1,045
Reflective			
Aluminum foil separating fluffy glass mats; 10-12 layers; evacuated; for cryogenic application (150 K)	40	0.00016	—
Aluminum foil and glass paper laminate; 75-150 layers; evacuated; for cryogenic application (150 K)	120	0.000017	—
Typical silica powder, evacuated	160	0.0017	—

(*Continued*)

TABLE 3.2-3 (*Continued*)

Industrial Insulation

DESCRIPTION/COMPOSITION	MAX SERVICE TEMP, K	TYPICAL DENSITY kg/m³	TYPICAL THERMAL CONDUCTIVITY, k(W/m·K), AT VARIOUS TEMPERATURES (K)													
			200	215	230	240	255	270	285	300	310	365	420	530	645	750
Blankets																
Blanket, mineral fiber, metal reinforced	920	96–192									0.038	0.046	0.056	0.078		
reinforced	815	40–96									0.035	0.045	0.058	0.088		
Blanket, mineral fiber, glass; fine fiber, organic bonded	450	10				0.036	0.038	0.040	0.043	0.048	0.052	0.076				
		12				0.035	0.036	0.039	0.042	0.046	0.049	0.069				
		16				0.033	0.035	0.036	0.039	0.042	0.046	0.062				
		24				0.030	0.032	0.033	0.036	0.039	0.040	0.053				
		32				0.029	0.030	0.032	0.033	0.036	0.038	0.048				
		48				0.027	0.029	0.030	0.032	0.033	0.035	0.045				
Blanket, alumina-silica fiber	1,530	48												0.071	0.105	0.150
		64												0.059	0.087	0.125
		96												0.052	0.076	0.100
		128												0.049	0.068	0.091
Felt, semi-rigid; organic bonded	480	50–125						0.035	0.036	0.038	0.039	0.051	0.063			
Felt, laminated; no binder	730	50		0.025	0.026			0.030	0.032	0.033	0.035	0.051	0.079			
	920	120											0.051	0.065	0.087	
Blocks, Boards, and Pipe Insulations																
Asbestos paper, laminated and corrugated																
4-ply	420	190								0.078	0.082	0.098				
6-ply	420	255								0.071	0.074	0.085				
8-ply	420	300								0.068	0.071	0.082				
Magnesia, 85%	590	185									0.051	0.055	0.061			
Calcium silicate	920	190									0.055	0.059	0.063			
Cellular glass	700	145			0.046	0.048	0.051	0.052	0.055	0.058	0.062	0.069	0.079			
Diatomaceous silica	1,145	345												0.075	0.089	0.104
														0.092	0.098	0.104
	1,310	385												0.101	0.100	0.115
Polystyrene rigid																
Extruded (R-12)	350	56	0.023	0.023	0.022	0.023	0.023	0.025	0.026	0.027	0.029					
Extruded (R-12)	350	35	0.023	0.023	0.025	0.023	0.025	0.026	0.027	0.029						
Molded beads	350	16	0.023		0.030	0.033	0.035	0.036	0.038	0.040						
Rubber, rigid foamed	340	70	0.026	0.029				0.029	0.030	0.032	0.033					

(*Continued*)

TABLE 3.2-3 (*Continued*)

Industrial Insulation (*Continued*)

DESCRIPTION/COMPOSITION	MAX SERVICE TEMP, K	TYPICAL DENSITY kg/m³	TYPICAL THERMAL CONDUCTIVITY, k(W/m·K), AT VARIOUS TEMPERATURES (K)													
			200	215	230	240	255	270	285	300	310	365	420	530	645	750
Insulating Cement																
Mineral fiber (rock, slag or glass)																
with clay binder	1,255	430									0.071	0.079	0.088	0.105	0.123	
with hydraulic setting binder	922	560									0.108	0.115	0.123	0.137		
Loose Fill																
Cellulose, wood or paper pulp	—	45							0.038	0.039	0.042					
Perlite, expanded	—	105	0.036	0.039	0.042	0.043	0.046	0.049	0.051	0.053	0.056					
Vermiculite, expanded	—	122			0.056	0.058	0.061	0.063	0.065	0.068	0.071					
		80			0.049	0.051	0.055	0.058	0.061	0.063	0.066					

(*Continued*)

TABLE 3.2-3 (*Continued*)

Other Materials

DESCRIPTION/COMPOSITION	TEMPERATURE K	DENSITY, ρ kg/m^3	THERMAL CONDUCTIVITY, k W/m·K	SPECIFIC HEAT, c_p J/kg·K
Asphalt	300	2,115	0.062	920
Bakelite	300	1,300	1.4	1,465
Brick, refractory				
Carborundum	872	—	18.5	—
	1,672	—	11.0	—
Chrome brick	473	3,010	2.3	835
	823		2.5	
	1,173		2.0	
Diatomaceous silica, fired	478	—	0.25	—
	1,145	—	0.30	
Fire clay, burnt 1600 K	773	2,050	1.0	960
	1,073	—	1.1	
	1,373	—	1.1	
Fire clay, burnt 1725 K	773	2,325	1.3	960
	1,073		1.4	
	1,373		1.4	
Fire clay brick	478	2,645	1.0	960
	922		1.5	
	1,478		1.8	
Magnesite	478	—	3.8	1,130
	922	—	2.8	
	1,478		1.9	
Clay	300	1,460	1.3	880
Coal, Anthracite	300	1,350	0.26	1,260
Concrete (stone mix)	300	2,300	1.4	880
Cotton	300	80	0.06	1,300
Foodstuffs				
Banana (75.7% water content)	300	980	0.481	3,350
Apple, red (75% water content)	300	840	0.513	3,600
Cake, batter	300	720	0.223	—
Cake, fully done	300	280	0.121	—
Chicken meat, white	198	—	1.60	—
(74.4% water content)	233	—	1.49	
	253		1.35	
	263		1.20	
	273		0.476	
	283		0.480	
	293		0.489	
Glass				
Plate (soda lime)	300	2,500	1.4	750
Pyrex	300	2,225	1.4	835
Ice	273	920	0.188	2,040
	253	—	0.203	1,945
Leather (sole)	300	998	0.159	—
Paper	300	930	0.180	1,340
Paraffin	300	900	0.240	2,890

(*Continued*)

TABLE 3.2-3 (*Continued*)

Other Materials (*Continued*)

DESCRIPTION/COMPOSITION	TEMPERATURE K	DENSITY, ρ kg/m^3	THERMAL CONDUCTIVITY, k W/m·K	SPECIFIC HEAT, c_p J/kg·K
Rock				
Granite, Barre	300	2,630	2.79	775
Limestone, Salem	300	2,320	2.15	810
Marble, Halston	300	2,680	2.80	830
Quartzite, Sioux	300	2,640	5.38	1,105
Sandstone, Berea	300	2,150	2.90	745
Rubber, vulcanized				
Soft	300	1,100	0.13	2,010
Hard	300	1,190	0.16	—
Sand	300	1,515	0.027	800
Soil	300	2,050	0.52	1,840
Snow	273	110	0.049	—
		500	0.190	—
Teflon	300	2,200	0.35	—
	400		0.45	—
Tissue, human				
Skin	300	—	0.37	—
Fat layer (adipose)	300	—	0.2	—
Muscle	300	—	0.41	—
Wood, cross grain				
Balsa	300	140	0.055	—
Cypress	300	465	0.097	—
Fir	300	415	0.11	2,720
Oak	300	545	0.17	2,385
Yellow pine	300	640	0.15	2,805
White pine	300	435	0.11	—
Wood, radial				
Oak	300	545	0.19	2,385
Fir	300	420	0.14	2,720

[a]See refs. 1, 2, and 6–11.

Equation (3.2-9) is analogous to Ohm's law in electricity where the heat flow q replaces the electric current, the temperature difference (thermal potential) replaces the electrical potential, and the thermal resistance R replaces the electric resistance. This analogy permits the use of simple electric circuits to describe heat flow problems.

In the following section Eq. (3.2-9) is applied to simple geometrical configurations.

Plane Wall

Consider a homogeneous wall of thickness Δx and of a uniform temperature T_1 at one surface and a uniform temperature T_2 at the opposite surface ($T_1 > T_2$). The temperature gradient for the wall is shown in Fig. 3.2-4. If the thermal conductivity k is constant, the heat flow by conduction is given by

$$q = \frac{kA}{\Delta x}(T_1 - T_2)$$

If the variation of k due to temperature change cannot be ignored, an average or a mean value k_m should be used:

$$k_m = \frac{1}{(T_1 - T_2)} \int_{T_1}^{T_2} k(T)\, dT \tag{3.2-11}$$

TABLE 3.2-4   THERMOPHYSICAL PROPERTIES OF GASES AT ATMOSPHERIC PRESSURE[a,b]

T K	ρ kg/m^3	c_p kJ/kg·K	$\mu \cdot 10^7$ N·s/m^2	$\nu \cdot 10^6$ m^2/s	$k \cdot 10^3$ W/m·K	$\alpha \cdot 10^6$ m^2/s	Pr
Air							
100	3.5562	1.032	71.1	2.00	9.34	2.54	0.786
150	2.3364	1.012	103.4	4.426	13.8	5.84	0.758
200	1.7458	1.007	132.5	7.590	18.1	10.3	0.737
250	1.3947	1.006	159.6	11.44	22.3	15.9	0.720
300	1.1614	1.007	184.6	15.89	26.3	22.5	0.707
350	0.9950	1.009	208.2	20.92	30.0	29.9	0.700
400	0.8711	1.014	230.1	26.41	33.8	38.3	0.690
450	0.7740	1.021	250.7	32.39	37.3	47.2	0.686
500	0.6964	1.030	270.1	38.79	40.7	56.7	0.684
550	0.6329	1.040	288.4	45.57	43.9	66.7	0.683
600	0.5804	1.051	305.8	52.69	46.9	76.9	0.685
650	0.5356	1.063	322.5	60.21	49.7	87.3	0.690
700	0.4975	1.075	338.8	68.10	52.4	98.0	0.695
750	0.4643	1.087	354.6	76.37	54.9	109	0.702
800	0.4354	1.099	369.8	84.93	57.3	120	0.709
850	0.4097	1.110	384.3	93.80	59.6	131	0.716
900	0.3868	1.121	398.1	102.9	62.0	143	0.720
950	0.3666	1.131	411.3	112.2	64.3	155	0.723
1000	0.3482	1.141	424.4	121.9	66.7	168	0.726
1100	0.3166	1.159	449.0	141.8	71.5	195	0.728
1200	0.2902	1.175	473.0	162.9	76.3	224	0.728
1300	0.2679	1.189	496.0	185.1	82	238	0.719
1400	0.2488	1.207	530	213	91	303	0.703
1500	0.2322	1.230	557	240	100	350	0.685
1600	0.2177	1.248	584	268	106	390	0.688
1700	0.2049	1.267	611	298	113	435	0.685
1800	0.1935	1.286	637	329	120	482	0.683
1900	0.1833	1.307	663	362	128	534	0.677
2000	0.1741	1.337	689	396	137	589	0.672
2100	0.1658	1.372	715	431	147	646	0.667
2200	0.1582	1.417	740	468	160	714	0.655
2300	0.1513	1.478	766	506	175	783	0.647
2400	0.1448	1.558	792	547	196	869	0.630
2500	0.1389	1.665	818	589	222	960	0.613
3000	0.1135	2.726	955	841	486	1570	0.536
Ammonia, NH$_3$							
300	0.6894	2.158	101.5	14.7	24.7	16.6	0.887
320	0.6448	2.170	109	16.9	27.2	19.4	0.870
340	0.6059	2.192	116.5	19.2	29.3	22.1	0.872
360	0.5716	2.221	124	21.7	31.6	24.9	0.872
380	0.5410	2.254	131	24.2	34.0	27.9	0.869
400	0.5136	2.287	138	26.9	37.0	31.5	0.853
420	0.4888	2.322	145	29.7	40.4	35.6	0.833
440	0.4664	2.357	152.5	32.7	43.5	39.6	0.826
460	0.4460	2.393	159	35.7	46.3	43.4	0.822
480	0.4273	2.430	166.5	39.0	49.2	47.4	0.822
500	0.4101	2.467	173	42.2	52.5	51.9	0.813
520	0.3942	2.504	180	45.7	54.5	55.2	0.827

TABLE 3.2-4 (*Continued*)

T K	ρ kg/m^3	c_p kJ/kg·K	$\mu \cdot 10^7$ N·s/m^2	$\nu \cdot 10^6$ m^2/s	$k \cdot 10^3$ W/m·K	$\alpha \cdot 10^6$ m^2/s	Pr
Ammonia, NH$_3$	**Continued**						
540	0.3795	2.540	186.5	49.1	57.5	59.7	0.824
560	0.3708	2.577	193	52.0	60.6	63.4	0.827
580	0.3533	2.613	199.5	56.5	63.8	69.1	0.817
Carbon Dioxide, CO$_2$							
280	1.9022	0.830	140	7.36	15.20	9.63	0.765
300	1.7730	0.851	149	8.40	16.55	11.0	0.766
320	1.6609	0.872	156	9.39	18.05	12.5	0.754
340	1.5618	0.891	165	10.6	19.70	14.2	0.746
360	1.4743	0.908	173	11.7	21.2	15.8	0.741
380	1.3961	0.926	181	13.0	22.75	17.6	0.737
400	1.3257	0.942	190	14.3	24.3	19.5	0.737
450	1.1782	0.981	210	17.8	28.3	24.5	0.728
500	1.0594	1.02	231	21.8	32.5	30.1	0.725
550	0.9625	1.05	251	26.1	36.6	36.2	0.721
600	0.8826	1.08	270	30.6	40.7	42.7	0.717
650	0.8143	1.10	288	35.4	44.5	49.7	0.712
700	0.7564	1.13	305	40.3	48.1	56.3	0.717
750	0.7057	1.15	321	45.5	51.7	63.7	0.714
800	0.6614	1.17	337	51.0	55.1	71.2	0.716
Carbon Monoxide, CO							
200	1.6888	1.045	127	7.52	17.0	9.63	0.781
220	1.5341	1.044	137	8.93	19.0	11.9	0.753
240	1.4055	1.043	147	10.5	20.6	14.1	0.744
260	1.2967	1.043	157	12.1	22.1	16.3	0.741
280	1.2038	1.042	166	13.8	23.6	18.8	0.733
300	1.1233	1.043	175	15.6	25.0	21.3	0.730
320	1.0529	1.043	184	17.5	26.3	23.9	0.730
340	0.9909	1.044	193	19.5	27.8	26.9	0.725
360	0.9357	1.045	202	21.6	29.1	29.8	0.725
380	0.8864	1.047	210	23.7	30.5	32.9	0.729
400	0.8421	1.049	218	25.9	31.8	36.0	0.719
450	0.7483	1.055	237	31.7	35.0	44.3	0.714
500	0.67352	1.065	254	37.7	38.1	53.1	0.710
550	0.61226	1.076	271	44.3	41.1	62.4	0.710
600	0.56126	1.088	286	51.0	44.0	72.1	0.707
650	0.51806	1.101	301	58.1	47.0	82.4	0.705
700	0.48102	1.114	315	65.5	50.0	93.3	0.702
750	0.44899	1.127	329	73.3	52.8	104	0.702
800	0.42095	1.140	343	81.5	55.5	116	0.705
Helium, He							
100	0.4871	5.193	96.3	19.8	73.0	28.9	0.686
120	0.4060	5.193	107	26.4	81.9	38.8	0.679
140	0.3481	5.193	118	33.9	90.7	50.2	0.676
160	—	5.193	129	—	99.2	—	—
180	0.2708	5.193	139	51.3	107.2	76.2	0.673
200	—	5.193	150	—	115.1	—	—
220	0.2216	5.193	160	72.2	123.1	107	0.675
240	—	5.193	170	—	130	—	—

TABLE 3.2-4 (*Continued*)

T K	ρ kg/m³	c_p kJ/kg·K	$\mu \cdot 10^7$ N·s/m²	$\nu \cdot 10^6$ m²/s	$k \cdot 10^3$ W/m·K	$\alpha \cdot 10^6$ m²/s	Pr
Carbon Monoxide, CO	Continued						
260	0.1875	5.193	180	96.0	137	141	0.682
280	—	5.193	190	—	145	—	—
300	0.1625	5.193	199	122	152	180	0.680
350	—	5.193	221	—	170	—	—
400	0.1219	5.193	243	199	187	295	0.675
450		5.193	263	—	204	—	—
500	0.09754	5.193	283	290	220	434	0.668
550	—	5.193	—	—	—	—	—
600	—	5.193	320	—	252	—	—
650	—	5.193	332	—	264	—	—
700	0.06969	5.193	350	502	278	768	0.654
750	—	5.193	364	—	291	—	—
800	—	5.193	382	—	304	—	—
900	—	5.193	414		330	—	—
1000	0.04879	5.193	446	914	354	1400	0.654
Hydrogen, H₂							
100	0.24255	11.23	42.1	17.4	67.0	24.6	0.707
150	0.16156	12.60	56.0	34.7	101	49.6	0.699
200	0.12115	13.54	68.1	56.2	131	79.9	0.704
250	0.09693	14.06	78.9	81.4	157	115	0.707
300	0.08078	14.31	89.6	111	183	158	0.701
350	0.06924	14.43	98.8	143	204	204	0.700
400	0.06059	14.48	108.2	179	226	258	0.695
450	0.05386	14.50	117.2	218	247	316	0.689
500	0.04848	14.52	126.4	261	266	378	0.691
550	0.04407	14.53	134.3	305	285	445	0.685
600	0.04040	14.55	142.4	352	305	519	0.678
700	0.03463	14.61	157.8	456	342	676	0.675
800	0.03030	14.70	172.4	569	378	849	0.670
900	0.02694	14.83	186.5	692	412	1030	0.671
1000	0.02424	14.99	201.3	830	448	1230	0.673
1100	0.02204	15.17	213.0	966	488	1460	0.662
1200	0.02020	15.37	226.2	1120	528	1700	0.659
1300	0.01865	15.59	238.5	1279	568	1955	0.655
1400	0.01732	15.81	250.7	1447	610	2230	0.650
1500	0.01616	16.02	262.7	1626	655	2530	0.643
1600	0.0152	16.28	273.7	1801	697	2815	0.639
1700	0.0143	16.58	284.9	1992	742	3130	0.637
1800	0.0135	16.96	296.1	2193	786	3435	0.639
1900	0.0128	17.49	307.2	2400	835	3730	0.643
2000	0.0121	18.25	318.2	2630	878	3975	0.661
Nitrogen, N₂							
100	3.4388	1.070	68.8	2.00	9.58	2.60	0.768
150	2.2594	1.050	100.6	4.45	13.9	5.86	0.759
200	1.6883	1.043	129.2	7.65	18.3	10.4	0.736
250	1.3488	1.042	154.9	11.48	22.2	15.8	0.727
300	1.1233	1.041	178.2	15.86	25.9	22.1	0.716
350	0.9625	1.042	200.0	20.78	29.3	29.2	0.711
400	0.8425	1.045	220.4	26.16	32.7	37.1	0.704

TABLE 3.2-4 (*Continued*)

T K	ρ kg/m^3	c_p kJ/kg·K	$\mu \cdot 10^7$ N·s/m^2	$\nu \cdot 10^6$ m^2/s	$k \cdot 10^3$ W/m·K	$\alpha \cdot 10^6$ m^2/s	Pr
Hydrogen, H$_2$	**Continued**						
450	0.7485	1.050	239.6	32.01	35.8	45.6	0.703
500	0.6739	1.056	257.7	38.24	38.9	54.7	0.700
550	0.6124	1.065	274.7	44.86	41.7	63.9	0.702
600	0.5615	1.075	290.8	51.79	44.6	73.9	0.701
700	0.4812	1.098	321.0	66.71	49.9	94.4	0.706
800	0.4211	1.122	349.1	82.90	54.8	116	0.715
900	0.3743	1.146	375.3	100.3	59.7	139	0.721
1000	0.3368	1.167	399.9	118.7	64.7	165	0.721
1100	0.3062	1.187	423.2	138.2	70.0	193	0.718
1200	0.2807	1.204	445.3	158.6	75.8	224	0.707
1300	0.2591	1.219	466.2	179.9	81.0	256	0.701
Oxygen, O$_2$							
100	3.945	0.962	76.4	1.94	9.25	2.44	0.796
150	2.585	0.921	114.8	4.44	13.8	5.80	0.766
200	1.930	0.915	147.5	7.64	18.3	10.4	0.737
250	1.542	0.915	178.6	11.58	22.6	16.0	0.723
300	1.284	0.920	207.2	16.14	26.8	22.7	0.711
350	1.100	0.929	233.5	21.23	29.6	29.0	0.733
400	0.9620	0.942	258.2	26.84	33.0	36.4	0.737
450	0.8554	0.956	281.4	32.90	36.3	44.4	0.741
500	0.7698	0.972	303.3	39.40	41.2	55.1	0.716
550	0.6998	0.988	324.0	46.30	44.1	63.8	0.726
600	0.6414	1.003	343.7	53.59	47.3	73.5	0.729
700	0.5498	1.031	380.8	69.26	52.8	93.1	0.744
800	0.4810	1.054	415.2	86.32	58.9	116	0.743
900	0.4275	1.074	447.2	104.6	64.9	141	0.740
1000	0.3848	1.090	477.0	124.0	71.0	169	0.733
1100	0.3498	1.103	505.5	144.5	75.8	196	0.736
1200	0.3206	1.115	532.5	166.1	81.9	229	0.725
1300	0.2960	1.125	588.4	188.6	87.1	262	0.721
Water Vapor (steam)							
380	0.5863	2.060	127.1	21.68	24.6	20.4	1.06
400	0.5542	2.014	134.4	24.25	26.1	23.4	1.04
450	0.4902	1.980	152.5	31.11	29.9	30.8	1.01
500	0.4405	1.985	170.4	38.68	33.9	38.8	0.998
550	0.4005	1.997	188.4	47.04	37.9	47.4	0.993
600	0.3652	2.026	206.7	56.60	42.2	57.0	0.993
650	0.3380	2.056	224.7	66.48	46.4	66.8	0.996
700	0.3140	2.085	242.6	77.26	50.5	77.1	1.00
750	0.2931	2.119	260.4	88.84	54.9	88.4	1.00
800	0.2739	2.152	278.6	101.7	59.2	100	1.01
850	0.2579	2.186	296.9	115.1	63.7	113	1.02

[a]ρ = density, c_p = specific heat at constant pressure, μ = viscosity, ν = kinematic viscosity, k = thermal conductivity, α = thermal diffusivity, Pr = Prandlt number.
[b]See refs. 6, 12, and 13.

TABLE 3.2-5 THERMOPHYSICAL PROPERTIES OF SATURATED WATER[a]

TEMP. (K) T	PRESSURE (BAR)[c] P	SPECIFIC VOLUME (m³/kg) $v_f \cdot 10^3$	v_g	HEAT OF VAPORIZATION (kJ/kg) h_{fg}	SPECIFIC HEAT (kJ/kg·K) $c_{p,f}$	$c_{p,g}$	VISCOSITY (N·s/m²) $\mu_f \cdot 10^6$	$\mu_g \cdot 10^6$	THERMAL CONDUCTIVITY (W/m·K) $k_f \cdot 10^3$	$k_g \cdot 10^3$	PRANDTL NO. Pr_f	Pr_g	SURFACE TENSION (N/m) $\sigma_f \cdot 10^3$	EXPANSION COEFFICIENT (K⁻¹) $\beta_f \cdot 10^6$	TEMP. (K) T
273.15	0.00611	1.000	206.3	2502	4.217	1.854	1750	8.02	659	18.2	12.99	0.815	75.5	-68.05	273.15
275	0.00697	1.000	181.7	2497	4.211	1.855	1652	8.09	574	18.3	12.22	0.817	75.3	-32.74	275
280	0.00990	1.000	130.4	2485	4.198	1.858	1422	8.29	582	18.6	10.26	0.825	74.8	46.04	280
285	0.01387	1.000	99.4	2473	4.189	1.861	1225	8.49	590	18.9	8.81	0.833	74.3	114.1	285
290	0.01917	1.001	69.7	2461	4.184	1.864	1080	8.69	598	19.3	7.56	0.841	73.7	174.0	290
295	0.02617	1.002	51.94	2449	4.181	1.868	959	8.89	606	19.5	6.62	0.849	72.7	227.5	295
300	0.03531	1.003	39.13	2438	4.179	1.872	855	9.09	613	19.6	5.83	0.857	71.7	276.1	300
305	0.04712	1.005	27.90	2426	4.178	1.877	769	9.29	620	20.1	5.20	0.865	70.9	320.6	305
310	0.06221	1.007	22.93	2414	4.178	1.882	695	9.49	628	20.4	4.62	0.873	70.0	361.9	310
315	0.08132	1.009	17.82	2402	4.179	1.888	631	9.69	634	20.7	4.16	0.883	69.2	400.4	315
320	0.1053	1.011	13.98	2390	4.180	1.895	577	9.89	640	21.0	3.77	0.894	68.3	436.7	320
325	0.1351	1.013	11.06	2378	4.182	1.903	528	10.09	645	21.3	3.42	0.901	67.5	471.2	325
330	0.1719	1.016	8.82	2366	4.184	1.911	489	10.29	650	21.7	3.15	0.908	66.6	504.0	330
335	0.2167	1.018	7.09	2354	4.186	1.920	453	10.49	656	22.0	2.88	0.916	65.8	535.5	335
340	0.2713	1.021	5.74	2342	4.188	1.930	420	10.69	660	22.3	2.66	0.925	64.9	566.0	340
345	0.3372	1.024	4.683	2329	4.191	1.941	389	10.89	668	22.6	2.45	0.933	64.1	595.4	345
350	0.4163	1.027	3.846	2317	4.195	1.954	365	11.09	668	23.0	2.29	0.942	63.2	624.2	350
355	0.5100	1.030	3.180	2304	4.199	1.968	343	11.29	671	23.3	2.14	0.951	62.3	652.3	355
360	0.6209	1.034	2.645	2291	4.203	1.983	324	11.49	674	23.7	2.02	0.960	61.4	697.9	360
365	0.7514	1.038	2.212	2278	4.209	1.999	306	11.69	677	24.1	1.91	0.969	60.5	707.1	365
370	0.9040	1.041	1.861	2265	4.214	2.017	289	11.89	679	24.5	1.80	0.978	59.5	728.7	370
373.15	1.0133	1.044	1.679	2257	4.217	2.029	279	12.02	680	24.8	1.76	0.984	58.9	750.1	373.15
375	1.0815	1.045	1.574	2252	4.220	2.036	274	12.09	681	24.9	1.70	0.987	58.6	761	375
380	1.2869	1.049	1.337	2239	4.226	2.057	260	12.29	683	25.4	1.61	0.999	57.6	788	380
385	1.5233	1.053	1.142	2225	4.232	2.080	248	12.49	685	25.8	1.53	1.004	56.6	814	385
390	1.794	1.058	0.980	2212	4.239	2.104	237	12.69	686	26.3	1.47	1.013	55.6	841	390
400	2.455	1.067	0.731	2183	4.256	2.158	217	13.05	688	27.2	1.34	1.033	53.6	896	400
410	3.302	1.077	0.553	2153	4.278	2.221	200	13.42	688	28.2	1.24	1.054	51.5	952	410
420	4.370	1.088	0.425	2123	4.302	2.291	185	13.79	688	29.8	1.16	1.075	49.4	1010	420
430	5.699	1.099	0.331	2091	4.331	2.369	173	14.14	685	30.4	1.09	1.10	47.2		430
440	7.333	1.110	0.261	2059	4.36	2.46	162	14.50	682	31.7	1.04	1.12	45.1		440
450	9.319	1.123	0.208	2024	4.40	2.56	152	14.85	678	33.1	0.99	1.14	42.9		450
460	11.71	1.137	0.167	1989	4.44	2.68	143	15.19	673	34.6	0.95	1.17	40.7		460
470	14.55	1.152	0.136	1951	4.48	2.79	136	15.54	667	36.3	0.92	1.20	38.5		470
480	17.90	1.167	0.111	1912	4.53	2.94	129	15.88	660	38.1	0.89	1.23	36.2		480

TABLE 3.2-5 (*Continued*)

TEMP. (K) T	PRESSURE (BAR)[c] P	SPECIFIC VOLUME (m³/kg) $v_f \cdot 10^3$	v_g	HEAT OF VAPORIZATION (kJ/kg) h_{fg}	SPECIFIC HEAT (kJ/kg·K) $c_{p,f}$	$c_{p,g}$	VISCOSITY (N·s/m²) $\mu_f \cdot 10^6$	$\mu_g \cdot 10^6$	THERMAL CONDUCTIVITY (W/m·K) $k_f \cdot 10^3$	$k_g \cdot 10^3$	PRANDTL NO. Pr_f	Pr_g	SURFACE TENSION (N/m) $\sigma_f \cdot 10^3$	EXPANSION COEFFICIENT (K⁻¹) $\beta_f \cdot 10^6$	TEMP. (K) T
490	21.83	1.184	0.0922	1870	4.59	3.10	124	16.23	651	40.1	0.87	1.25	33.9	—	490
500	26.40	1.203	0.0766	1825	4.66	3.27	118	16.59	642	42.3	0.86	1.28	31.6	—	500
510	31.66	1.222	0.0631	1779	4.74	3.47	113	16.95	631	44.7	0.85	1.31	29.3	—	510
520	37.70	1.244	0.0525	1730	4.84	3.70	108	17.33	621	47.5	0.84	1.35	26.9	—	520
530	44.58	1.268	0.0445	1679	4.95	3.96	104	17.72	608	50.6	0.85	1.39	24.5	—	530
540	52.38	1.294	0.0375	1622	5.08	4.27	101	18.1	594	54.0	0.86	1.43	22.1	—	540
550	61.19	1.323	0.0317	1564	5.24	4.64	97	18.6	580	58.3	0.87	1.47	19.7	—	550
560	71.08	1.355	0.0269	1499	5.43	5.09	94	19.1	563	63.7	0.90	1.52	17.3	—	560
570	82.16	1.392	0.0228	1429	5.68	5.67	91	19.7	548	76.7	0.94	1.59	15.0	—	570
580	94.51	1.433	0.0193	1353	6.00	6.40	88	20.4	528	76.7	0.99	1.68	12.8	—	580
590	108.3	1.482	0.0163	1274	6.41	7.35	84	21.5	513	84.1	1.05	1.84	10.5	—	590
600	123.5	1.541	0.0137	1176	7.00	8.75	81	22.7	497	92.9	1.14	2.15	8.4	—	600
610	137.3	1.612	0.0115	1068	7.85	11.1	77	24.1	467	103	1.30	2.60	6.3	—	610
620	159.1	1.705	0.0094	941	9.35	15.4	72	25.9	444	114	1.52	3.46	4.5	—	620
625	169.1	1.778	0.0085	858	10.6	18.3	70	27.0	430	121	1.65	4.20	3.5	—	625
630	179.7	1.856	0.0075	781	12.6	22.1	67	28.0	412	130	2.0	4.8	2.6	—	630
635	190.9	1.935	0.0066	683	16.4	27.6	64	30.0	392	141	2.7	6.0	1.5	—	635
640	202.7	2.075	0.0057	560	26	42	59	32.0	367	155	4.2	9.6	0.8	—	640
645	215.2	2.351	0.0045	361	90		54	37.0	331	178	12	26	0.1	—	645
647.3[b]	221.2	3.170	0.0032	0	∞	∞	45	45.0	238	238	∞	∞	0.0	—	647.3[b]

[a] See refs. 14 and 15.
[b] Critical temperature.
[c] 1 bar = 10^5 N/m².

TABLE 3.2-6   THERMOPHYSICAL PROPERTIES OF LIQUID METALS^a

COMPOSITION	MELTING POINT K	T K	ρ kg/m^3	c_p kJ/kg·K	$\nu \cdot 10^7$ m^2/s	k W/m·K	$\alpha \cdot 10^5$ m^2/s	Pr
Bismuth	544	589	10,011	0.1444	1.617	16.4	0.138	0.0142
		811	9,739	0.1545	1.133	15.6	1.035	0.0110
		1033	9,467	0.1645	0.8343	15.6	1.001	0.0083
Lead	600	644	10,540	0.159	2.276	16.1	1.084	0.024
		755	10,412	0.155	1.849	15.6	1.223	0.017
		977	10,140	—	1.347	14.9	—	—
Potassium	337	422	807.3	0.80	4.608	45.0	6.99	0.0066
		700	741.7	0.75	2.397	39.5	7.07	0.0034
		977	674.4	0.75	1.905	33.1	6.55	0.0029
Sodium	371	366	929.1	1.38	7.516	86.2	6.71	0.011
		644	860.2	1.30	3.270	72.3	6.48	0.0051
		977	778.5	1.26	2.285	59.7	6.12	0.0037
NaK, (45%/55%)	292	366	887.4	1.130	6.522	25.6	2.552	0.026
		644	821.7	1.055	2.871	27.5	3.17	0.0091
		977	740.1	1.043	2.174	28.9	3.74	0.0058
NaK, (22%/78%)	262	366	849.0	0.946	5.797	24.4	3.05	0.019
		672	775.3	0.879	2.666	26.7	3.92	0.0068
		1033	690.4	0.883	2.118	—	—	—
PbBi (44.5%/55.5%)	398	422	10,524	0.147	—	9.05	0.586	—
		644	10,236	0.147	1.496	11.86	0.790	0.189
		922	9,835	—	1.171	—	—	—
Mercury	234	See table A.5						

^{a}Adapted from *Liquid Materials Handbook*, 23rd. ed, the Atomic Energy Commission, Department of the Navy, Washington, D.C., 1952.

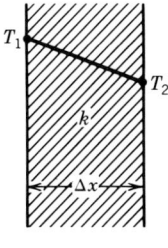

$$R = \frac{\Delta x}{kA}$$

T_1 —WW— T_2 **Fig. 3.2-4**   Heat flow by conduction in a slab.

Conduction through Composite Walls

Many problems in conduction involve composite walls or laminations of constant cross-sectional area but different thicknesses and conductivities, as shown in Fig. 3.2-5. If steady-state conditions prevail, then the rate of heat flow through a given area A is the same for all walls and is given by

$$q = \frac{T_1 - T_4}{\dfrac{\Delta x_1}{k_1 A} + \dfrac{\Delta x_2}{k_2 A} + \dfrac{\Delta x_3}{k_3 A}} = \frac{\Delta T}{\sum_i R_i} \qquad (3.2\text{-}12)$$

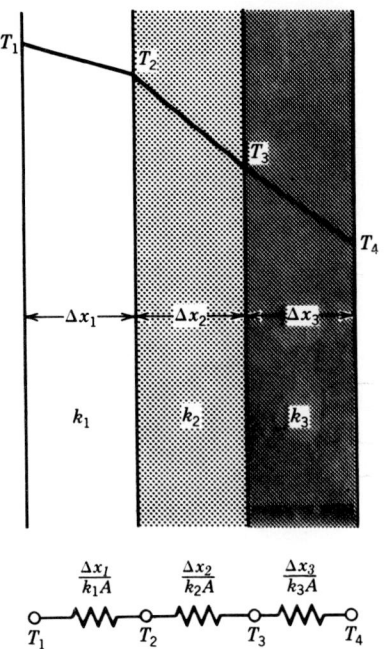

Fig. 3.2-5 Temperature distribution for a composite wall.

where ΔT is the overall temperature difference, R_i is the thermal resistance of the ith wall, and $\sum_i R_i$ is the total thermal resistance.

Equation (3.2-12) indicates that for the same q, the temperature drop is proportional to the thermal resistance. This equation can also be written in terms of conductances rather than resistances in the form:

$$q = UA(T_1 - T_4) \tag{3.2-13}$$

where U is the average overall heat transfer coefficient and UA is the overall conductance defined as

$$UA = \frac{1}{\dfrac{\Delta x_1}{Ak_1} + \dfrac{\Delta x_2}{Ak_2} + \dfrac{\Delta x_3}{Ak_3}} = \frac{1}{\displaystyle\sum_i R_i} \tag{3.2-14}$$

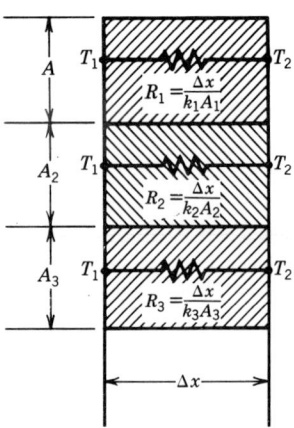

Fig. 3.2-6 Heat flow through bodies in parallel.

Conduction through Bodies in Parallel

As shown in Fig. 3.2-6, the rate of heat flow is equal to the sum of the rates of heat flow through each body:

$$q = \frac{\Delta T}{R_1} + \frac{\Delta T}{R_2} + \cdots + \frac{\Delta T}{R_n}$$

or

$$q = \Delta T \sum_i \frac{1}{R_i} \tag{3.2-15}$$

where n is the number of resistances in parallel.

Hollow Cylinder

As heat is conducted radially in a hollow cylinder, the area through which heat flows varies. If the outside area of the cylinder is not more than twice the inside area, the arithmetic average of the two areas may be used in Eq. (3.2-1) and the error introduced is only 4%.

The heat transfer in a hollow cylinder of inner and outer radii r_i and r_o and of length L (Fig. 3.2-7) is $q = -kA(dT/dr) = -k2\pi rL(dT/dr)$ where $2\pi rL$ is the area of heat flow. Separating variables and integration yields

$$q = \frac{2\pi Lk}{\ln(r_0/r_i)}(T_i - T_o) = \frac{T_i - T_o}{\dfrac{\ln(r_o/r_i)}{2\pi Lk}} = \frac{T_i - T_o}{R} \tag{3.2-16}$$

where R is the thermal resistance.

In the case of n coaxial cylinders as shown in Fig. 3.2-8, this equation can be applied to each of the coaxial cylinders so that

$$q = \frac{T_i - T_o}{R_{\text{total}}}$$

where

$$R_{\text{total}} = \frac{\ln(r_1/r_i)}{2\pi Lk_1} + \frac{\ln(r_2/r_1)}{2\pi Lk_2} + \cdots + \frac{\ln(r_o/r_n)}{2\pi Lk_n} \tag{3.2-17}$$

The overall heat transfer coefficient U can be expressed in terms of either the inner or the outer heat flow area:

$$q = U_i A_i \Delta T = U_o A_o \Delta T$$

where

$$U_i = \frac{1}{\dfrac{r_i \ln(r_1/r_i)}{k_1} + \dfrac{r_i \ln(r_2/r_1)}{k_2} + \cdots + \dfrac{r_i \ln(r_o/r_n)}{k_n}} \tag{3.2-18}$$

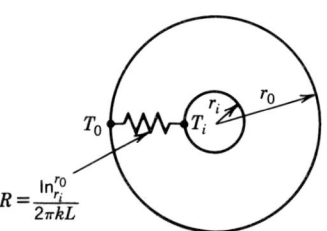

$$R = \frac{\ln \dfrac{r_o}{r_i}}{2\pi kL}$$

Fig. 3.2-7 Heat flow through the walls of a cylinder.

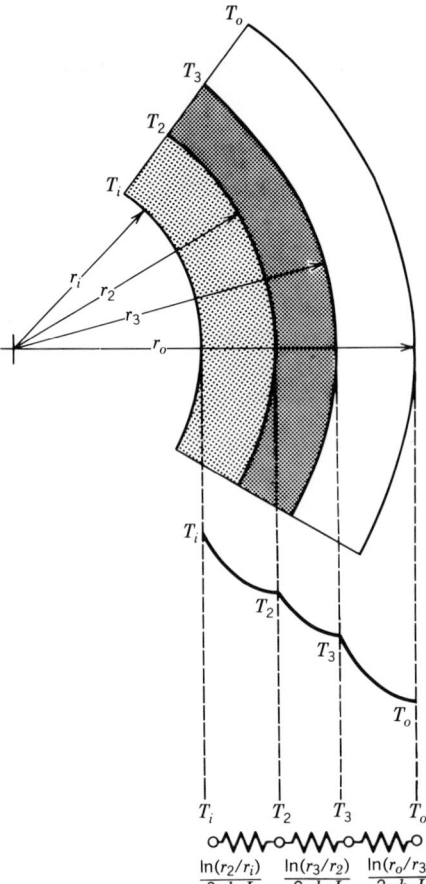

Fig. 3.2-8 Temperature distribution for a composite cylindrical wall.

and

$$U_0 = \cfrac{1}{\cfrac{r_o \ln(r_1/r_i)}{k_1} + \cfrac{r_o \ln(r_2/r_1)}{k_2} + \cdots + \cfrac{r_o \ln(r_o/r_n)}{k_n}} \qquad (3.2\text{-}19)$$

Note that

$$U_i A_i = U_2 A_2 = U_3 A_3 = U_o A_o = \frac{1}{\sum R} \qquad (3.2\text{-}20)$$

Hollow Sphere

The heat transfer by conduction along the radial direction in a sphere is

$$q = -kA \frac{dT}{dr} = -k \left(4\pi r^2\right) \frac{dT}{dr}$$

where $A = 4\pi r^2$. Integration gives

$$q = \frac{4\pi k r_0 r_i}{r_o - r_i} (T_i - T_o) = \frac{T_i - T_o}{R_{\text{sphere}}} \qquad (3.2\text{-}21)$$

To calculate the radial heat transfer in concentric spheres, the procedure is analogous to that used for a composite cylinder.

3.2-3   Heat Transfer with Thermal Energy Generation

Consider a plane wall of thickness $2L$ and a constant thermal conductivity k in which there is a uniform energy generation per unit volume $\dot{q}$ (Fig. 3.2-9). Let the surface temperatures T_{s1} and T_{s2} be maintained constant. For one-dimensional heat flow Poisson equation is

$$\frac{d^2 T}{dx^2} + \frac{\dot{q}}{k} = 0 \qquad (3.2\text{-}22)$$

Solution to this equation is subject to the boundary conditions and is given by

$$T(x) = -\frac{\dot{q}}{2k} x^2 + C_1 + C_2 \qquad (3.2\text{-}23)$$

where

$$C_1 = \frac{T_{s_2} - T_{s_1}}{2L} \quad \text{and} \quad C_2 = \frac{\dot{q}}{2k} L^2 + \frac{T_{s_1} + T_{s_2}}{2}$$

so that

$$T(x) = \frac{\dot{q} L^2}{2k} \left(1 - \frac{x^2}{L^2} \right) + \frac{T_{s_2} - T_{s_1}}{2} \frac{x}{L} + \frac{T_{s_1} + T_{s_2}}{2} \qquad (3.2\text{-}24)$$

For the symmetric case where $T_{s_1} = T_{s_2} = T_s$, the temperature gradient is 0 at $x = 0$ and the temperature distribution is

$$T(x) = \frac{\dot{q} L^2}{2k} \left(1 - \frac{x^2}{L^2} \right) + T_s \qquad (3.2\text{-}25)$$

The maximum (or minimum) temperature occurs at the midplane and is equal to

$$T_{(x=0)} = \frac{\dot{q} L^2}{2k} + T_s = T_0 \qquad (3.2\text{-}26)$$

Combining the two previous equations gives the temperature distribution as

$$\frac{T(x) - T_0}{T_s - T_0} = \left(\frac{x}{L} \right)^2 \qquad (3.2\text{-}27)$$

For a system such as the long solid cylinder shown in Fig. 3.2-10, the analogous equations in cylindrical coordinates are

$$\frac{1}{r} \frac{d}{dr} \left(r \frac{dT}{dr} \right) + \frac{\dot{q}}{k} = 0 \qquad (3.2\text{-}28)$$

$$T(r) = -\frac{\dot{q}}{4k} r^2 + C_1 \ln r + C_2 \qquad (3.2\text{-}29)$$

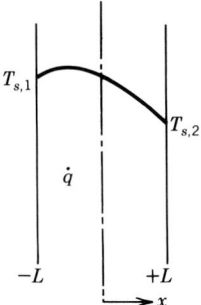

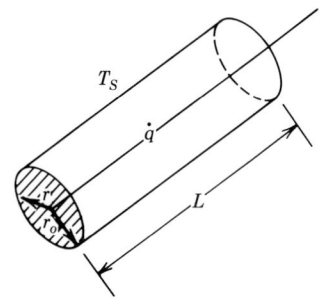

Fig. 3.2-9 Conduction in a solid cylinder with uniform heat generation.

Fig. 3.2-10 Conduction in a solid cylinder with uniform heat generation.

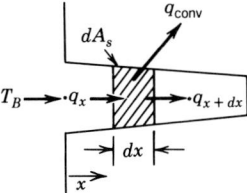

Fig. 3.2-11   Conduction and convection in a fin.

where $C_1 = 0$ and $C_2 = T_s + (\dot{q}/4k)r_o^2$ so that the temperature distribution is

$$T(r) = \frac{\dot{q}r^2}{4k}\left(1 - \frac{r^2}{r_o^2}\right) + T_s \tag{3.2-30}$$

Combining this equation with centerline temperature T_0 (at $r = 0$) gives

$$\frac{T(r) - T_s}{T_0 - T_s} = 1 - \left(\frac{r}{r_0}\right)^2 \tag{3.2-31}$$

3.2-4   Heat Transfer from Extended Surfaces

The heat transfer from a surface may be increased by increasing the effective surface area across which convection takes place. This situation often occurs when the convection coefficient h is small. Increasing the surface area is usually accomplished by using fins that extend from the wall. Referring to Fig. 3.2-11 and equating the net heat transfer across a differential element of the fin by conduction to the heat transfer by convection yields the following general equation:

$$\frac{d^2T}{dx^2} + \left(\frac{1}{A}\frac{dA}{dx}\right)\frac{dT}{dx} - \left(\frac{1}{A}\frac{h}{k}\frac{dA_s}{dx}\right)(T - T_\infty) = 0 \tag{3.2-32}$$

where A is the cross-sectional area of the fin which may vary with x and dA_s is the surface area of the differential element. In this equation steady-state conditions and one-dimensional conduction were assumed. It was further assumed that the thermal conductivity k is constant and the heat transfer coefficient by convection h is uniform over the surface of the fin.
  For a uniform cross-sectional area the above equation reduces to

$$\frac{d^2T}{dx^2} - \frac{hP}{kA}(T - T_\infty) = 0 \tag{3.2-33}$$

where P is the wetted perimeter, equal to A_s/x.

TABLE 3.2-7   TEMPERATURE DISTRIBUTION AND HEAT LOSS FOR FINS OF UNIFORM CROSS SECTION[a, b]

Case	Tip Condition ($x = L$)	Temperature Distribution, θ/θ_b	Fin Heat Transfer Rate, q_f
A	Convection heat transfer: $h\theta(L) = -k\,d\theta/dx\|_{x=L}$	$\dfrac{\cosh m(L - x) + (h/mk)\sinh m(L - x)}{\cosh mL + (h/mk)\sinh mL}$	$M\dfrac{\sinh mL + (h/mk)\cosh mL}{\cosh mL + (h/mk)\sinh mL}$
B	Adiabatic: $d\theta/dx\|_{x=L} = 0$	$\dfrac{\cosh m(L - x)}{\cosh mL}$	$M \tanh mL$
C	Prescribed temperature: $\theta(L) = \theta_L$	$\dfrac{(\theta_L/\theta_b)\sinh mx + \sinh m(L - x)}{\sinh mL}$	$M\dfrac{(\cosh mL - \theta_L/\theta_b)}{\sinh mL}$
D	Infinite fin ($L \to \infty$): $\theta(L) = 0$	e^{-mx}	M

[a] $\theta \equiv T - T_\infty$ $\qquad m^2 \equiv hP/kA$
$\theta_b = \theta(0) = T_b - T_\infty \qquad M \equiv \sqrt{hPkA}\,\theta_b$
[b] See ref. 1.

Solution of Eq. (3.2-33) gives the temperature distribution along the fin, and the gradient of this temperature, dT/dx, at the base of fin is then used to determine the heat transfer by the fin. The solution of Eq. (3.2-33) is subject to two boundary conditions. One condition is usually specified in terms of the temperature at the base of the fin T_b, and the second condition is specified at the fin tip ($x = L$). In Table 3.2-7 the temperature distribution and heat transfer from fins of uniform cross-sectional areas are indicated.

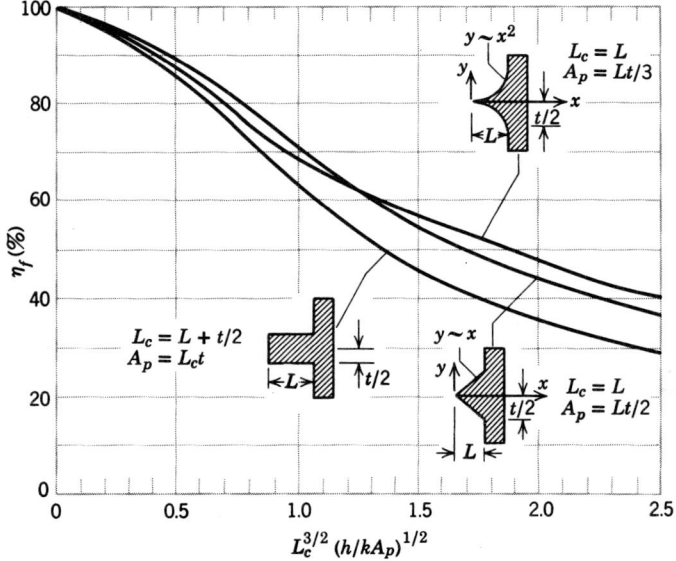

Fig. 3.2-12 Efficiency of straight fins (rectangular, triangular, and parabolic profiles). (From ref. 1; used by permission.)

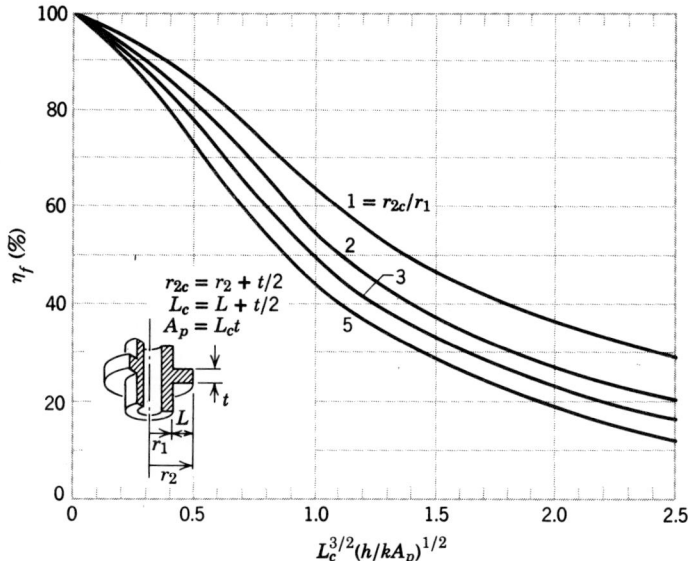

Fig. 3.2-13 Efficiency of annular fins of rectangular profile. (From ref. 1; used by permission.)

TABLE 3.2-8 CONDUCTION SHAPE FACTORS FOR SELECTED TWO-DIMENSIONAL SYSTEMS $[q = Sk(T_1 - T_2)]^a$

SYSTEM	SCHEMATIC	RESTRICTIONS	SHAPE FACTOR
Isothermal sphere buried in a semiinfinite medium		$z > D/2$	$\dfrac{2\pi D}{1 - D/4z}$
Horizontal isothermal cylinder of length L buried in a semiinfinite medium		$L \gg D$	$\dfrac{2\pi L}{\cosh^{-1}(2z/D)}$
		$L \gg D$ $z > 3D/2$	$\dfrac{2\pi L}{\ln(4z/D)}$
Vertical cylinder in a semiinfinite medium		$L \gg D$	$\dfrac{2\pi L}{\ln(4L/D)}$
Conduction between two cylinders of length L in infinite medium		$L \gg D_1, D_2$ $L \gg w$	$\dfrac{2\pi L}{\cosh^{-1}\left(\dfrac{4w^2 - D_1^2 - D_2^2}{2D_1 D_2}\right)}$
Horizontal circular cylinder of length L midway between parallel planes of equal length and infinite width		$z > D/2$	$\dfrac{2\pi L}{\ln(8z/\pi D)}$

Description	Schematic	Restrictions	Shape Factor
Circular cylinder of length L in a square solid of equal length	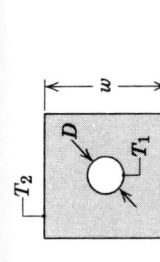	$w > D$	$\dfrac{2\pi L}{\ln(1.08\,w/D)}$
Plane wall	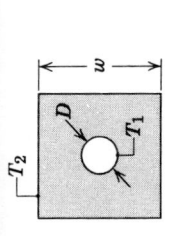	One-dimensional conduction	$\dfrac{A}{L}$
Conduction through the edge of adjoining walls	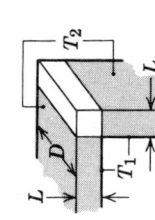	$D > L/5$	$0.54D$
Conduction through corner of three walls with a temperature difference ΔT_{1-2} across the walls		$L \ll$ length and width of wall	$0.15L$
Disc of diameter D and T_1 on a semiinfinite medium of thermal conductivity k and T_2	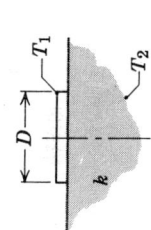	None	$2D$

[a]See ref. 1.

The effectiveness of the fin, defined as the ratio of the fin heat transfer rate to the heat transfer rate that would exist without the fin, is

$$e_f = \frac{q_f}{hA_b(T_b - T_\infty)} \tag{3.2-34}$$

where A_b is the fin cross-sectional area at the base. The efficiency of the fin η_f is another measure of its performance. The fin efficiency is defined as the ratio of the fin heat transfer rate divided by the maximum heat transfer rate, which occurs if the entire fin surface were at the base temperature, so that

$$\eta_f = \frac{q_f}{q_{max}} = \frac{q_f}{hA_f(T_b - T_\infty)} \tag{3.2-35}$$

where A_f is the total surface area of the fin. Figures 3.2-12 and 3.2-13 indicate η_f for straight fins and annular fins. For the first figure $q_{max} = hPL_c(T_b - T_\infty)$, whereas for the second figure $q_{max} = 2\pi h[(r_2)_c^2 - r_1^2](T_b - T_\infty)$.

3.2-5 Two-Dimensional Steady-State Conduction

The steady-state two-dimensional heat transfer equation with no heat generation and constant thermal conductivity is

$$\frac{\partial^2 T}{\partial x^2} + \frac{\partial^2 T}{\partial y^2} = 0 \tag{3.2-36}$$

This is Laplace's equation in two dimensions, for which solution can be obtained using analytical, graphical, or numerical techniques.

In the graphical solution a network of isotherms and heat flow lines is constructed. First, lines of symmetry (which are also adiabatic lines) and constant temperature lines (isotherms) are identified. Uniformly spaced isotherms and heat flow lines are sketched within the system, realizing that both lines intersect at right angles. Note also that adiabatic lines are perpendicular to isotherms. By trial and error a network of curvilinear squares is drawn so that the sums of the opposite sides of each square are equal. The heat transfer rate is

$$q = \sum_{i=1}^{m} q_i = mq_i \tag{3.2-37}$$

where m is the number of heat flow passages, each bounded by heat flow lines. The value of q_i is given by

$$q_i \approx kA_i \frac{\Delta T_j}{\Delta x}$$

But $A_i = \Delta y l$, where l is the depth of the heat flow passage. The total temperature difference

$$\Delta T = \sum_{j}^{m} \Delta T_j = n \Delta T_j$$

where n is the number of temperature increments between isotherms. If Δx is taken to be equal to Δy, Eq. (3.2-37) becomes

$$q \approx \frac{ml}{n} k \Delta T = Sk \Delta T \tag{3.2-38}$$

where S is called the shape factor. Table 3.2-8 gives values of S for several two-dimensional systems.

3.2-6 Finite-Difference Method

This technique utilizes numerical methods to determine temperature at discrete points or regions into which the material is divided. Indices m and n are used to identify the nodal points in the network as shown in Fig. 3.2-14. The temperature at each nodal point is expressed in terms of the temperatures of the neighboring nodal points. With the use of digital computers the accuracy of calculations can be greatly improved by increasing the number of nodal points.

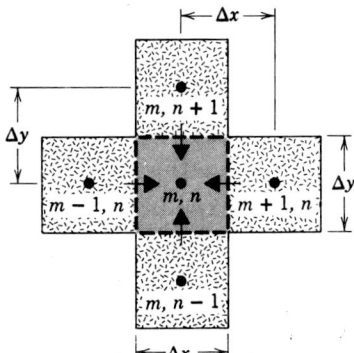

Fig. 3.2-14 Conduction to an interior node from its adjoining nodes.

Using a network for which $\Delta x = \Delta y$, an interior point temperature T_{mn} is the average of the surrounding temperatures.

$$T_{m,n+1} + T_{m,n-1} + T_{m+1,n} + T_{m-1,n} - 4T_{mn} = 0 \qquad (3.2\text{-}39)$$

The same technique can be used for points at the boundaries of the body, which may be insulated, exposed to convective heat transfer, or at a constant temperature. Energy balance equations at nodes for common geometrics are presented in Table 3.2-9.

The finite-difference equations written for each node can be solved by techniques such as the relaxation method, Gauss–Seidel iteration, or the matrix inversion method. In the relaxation method values are assigned to the different nodal temperature so that

$$T_{m,n+1} + T_{m,n-1} + T_{m+1,n} + T_{m-1,n} - 4T_{mn} = R_{m,n} \qquad (3.2\text{-}40)$$

where $R_{m,n}$ is called the residual at the node. Similarly, residuals appear on the right-hand side of each of the nodal equations. By iteration, attempts are made to reduce the residual to a minimum value close to zero.

3.2-7 Unsteady Heat Conduction

The temperature distribution within a solid as a function of time during a transient process can be obtained by solving the general Fourier equation. With no internal energy generation and the assumption of constant thermal conductivity, this equation in one dimension is

$$\frac{\partial^2 T}{\partial x^2} = \frac{1}{\alpha}\frac{\partial T}{\partial T} \qquad (3.2\text{-}41)$$

When the temperature gradient within the solid is small, a simple method of solution called the lumped capacitance method is used. By equating the heat transfer at the boundary to the energy change of the solid, the following equation is obtained:

$$\frac{T - T_\infty}{T_i - T_\infty} = e^{-(hA_s/\rho Vc)t} \qquad (3.2\text{-}42)$$

where T is the temperature of the solid at time t, T_∞ is the temperature of the surrounding medium, T_i is the temperature of the solid at time $t = 0$, h is the convective heat transfer coefficient, A_s is the area of the solid, ρ its density, V volume, and c thermal capacity. The factor $\rho Vc/hA_s$ is called the thermal time constant and ρVc is the lumped thermal capacitance of the solid. Equation (3.2-42) indicates that the temperature varies exponentially with time. The total heat transfer is given by

$$Q = hA_s \int_0^t (T - T_\infty)\, dt$$

$$= (\rho Vc)(T - T_i)[1 - e^{-(hA_s/\rho Vc)t}] \qquad (3.2\text{-}43)$$

Equations (3.2-42) and (3.2-43) are valid provided that the temperature of the solid is uniform at any instant of time during the transient process. This phenomenon is characterized by the dimension-

TABLE 3.2-9 SUMMARY OF NODAL FINITE-DIFFERENCE EQUATIONS[a,b]

CONFIGURATION	FINITE-DIFFERENCE EQUATION FOR $\Delta x = \Delta y$
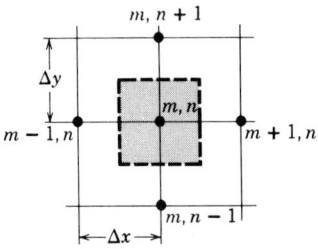	$T_{m,n+1} + T_{m,n-1} + T_{m+1,n} + T_{m-1,n}$ $-4T_{m,n} = 0$ 1. Interior node.
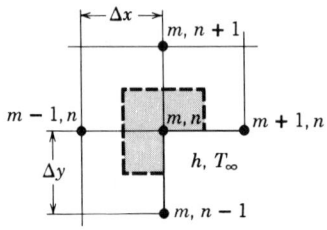	$2(T_{m-1,n} + T_{m,n+1}) + (T_{m+1,n} + T_{m,n-1})$ $+2\dfrac{h\Delta x}{k}T_\infty - 2\left(3 + \dfrac{h\Delta x}{k}\right)T_{m,n} = 0$ 2. Node at an internal corner with convection.
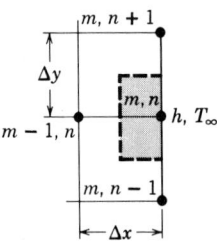	$(2T_{m-1,n} + T_{m,n+1} + T_{m,n-1}) + \dfrac{2h\Delta x}{k}T_\infty$ $-2\left(\dfrac{h\Delta x}{k} + 2\right)T_{m,n} = 0$ 3. Node at a plane surface with convection.
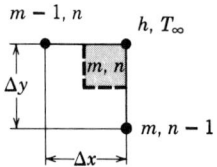	$(T_{m,n-1} + T_{m-1,n}) + 2\dfrac{h\Delta x}{k}T_\infty$ $-2\left(\dfrac{h\Delta x}{k} + 1\right)T_{m,n} = 0$ 4. Node at an external corner with convection.
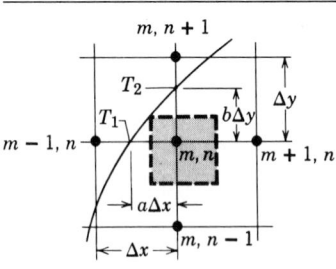	$\dfrac{2}{a+1}T_{m+1,n} + \dfrac{2}{b+1}T_{m,n-1}$ $+ \dfrac{2}{a(a+1)}T_1 + \dfrac{2}{b(b+1)}T_2$ $-\left(\dfrac{2}{a} + \dfrac{2}{b}\right)T_{m,n} = 0$ 5. Node near a curved surface maintained at a nonuniform temperature.

[a] To obtain the finite-difference equation for an adiabatic surface (or surface of symmetry), simply set h equal to zero.
[b] See ref. 1.

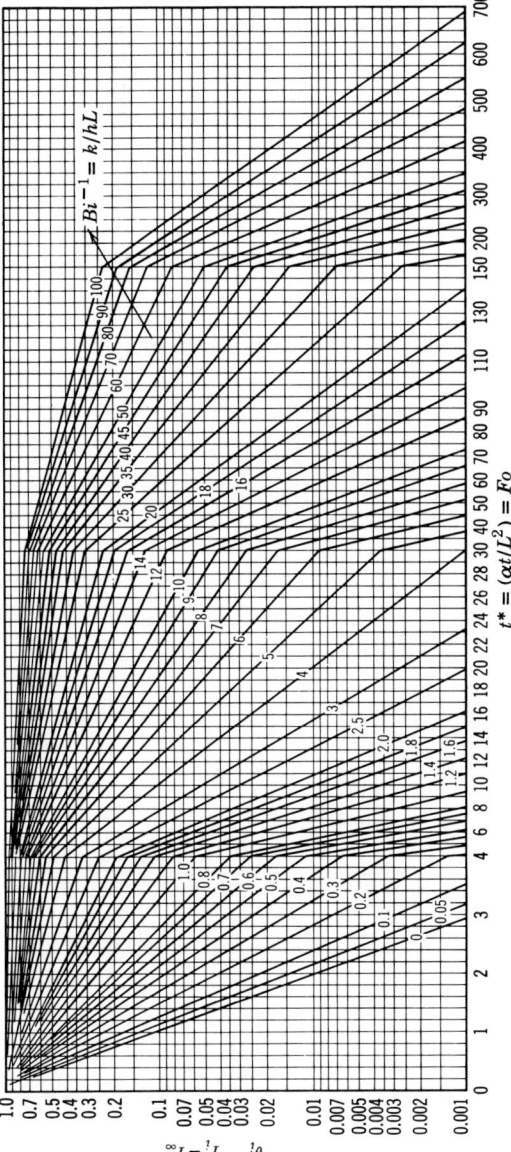

Fig. 3.2-15 Midplane temperature as a function of time for a plane wall of thickness $2L$. (From ref. 16; used by permission.).

235

less Biot number, which indicates the temperature drop in a solid relative to the temperature difference between the solid and the surrounding fluid. The Bi is defined as

$$\text{Bi} = \frac{hL}{k} = \frac{R_{\text{cond}}}{R_{\text{conv}}} \tag{3.2-44}$$

where L is the thickness of the solid. The above equation also indicates that the Bi is the ratio of the conduction resistance within the solid to the convection resistance to the surrounding fluid. The Bi is also defined in terms of a characteristic length $L_c = V/A_s$ so that if $\text{Bi} = hL_c/k$ is small (say ≤ 0.1), the error associated with the lumped capacitance method is small. Introducing the Bi into Eq. (3.2-42)

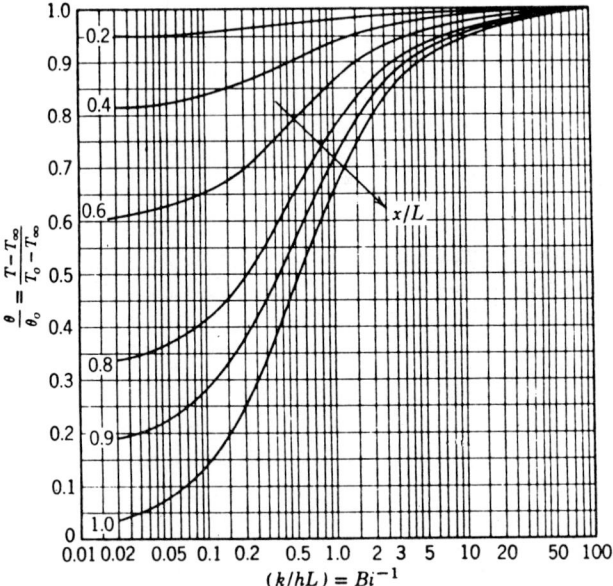

Fig. 3.2-16 Temperature distribution in a plane wall of thickness $2L$. (From ref. 16; used by permission.)

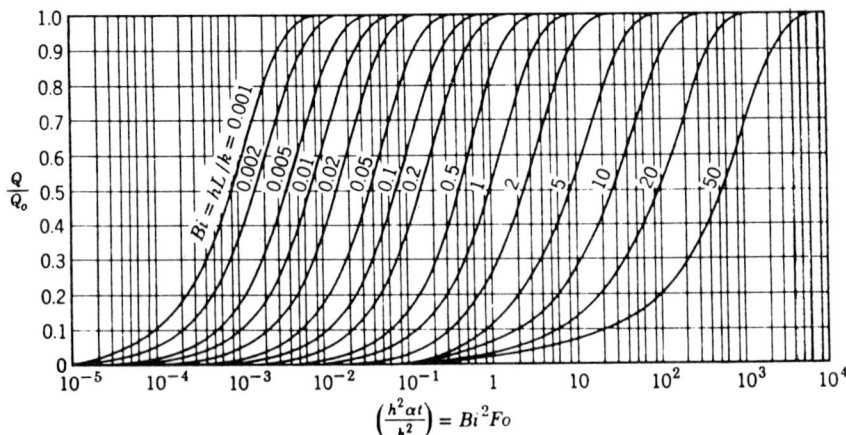

Fig. 3.2-17 Internal energy change as function of time for a plane wall of thickness $2L$. (From ref. 17; used by permission.)

gives

$$\frac{T - T_\infty}{T_i - T_\infty} = e^{-\mathrm{Bi} \cdot \mathrm{Fo}}$$ (3.2-45)

where Fo is called the Fourier number defined as

$$\mathrm{Fo} = \frac{k}{\rho c}\frac{t}{L_c^2} = \alpha\frac{t}{L_c^2}$$ (3.2-46)

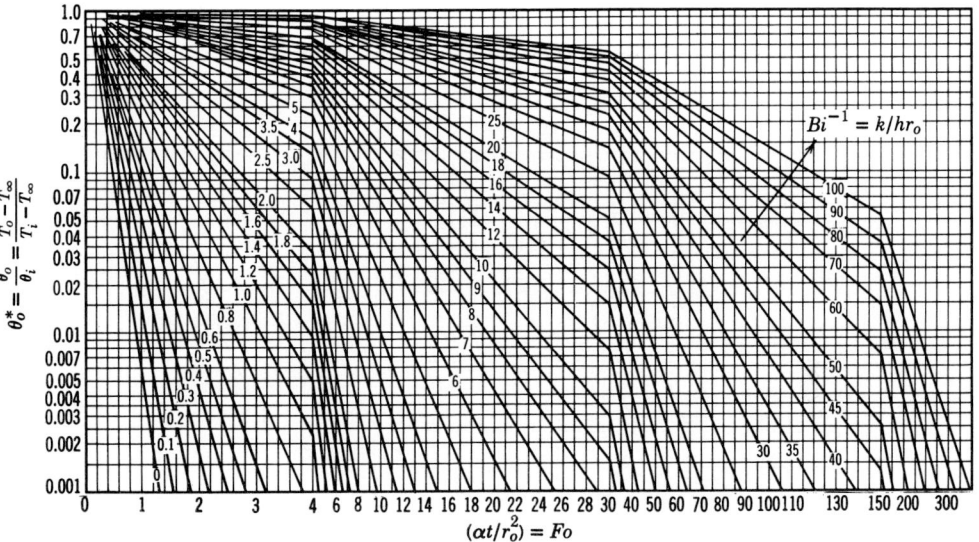

Fig. 3.2-18 Centerline temperature as a function of time for an infinite cylinder of radius r_0. (From ref. 16; used by permission.)

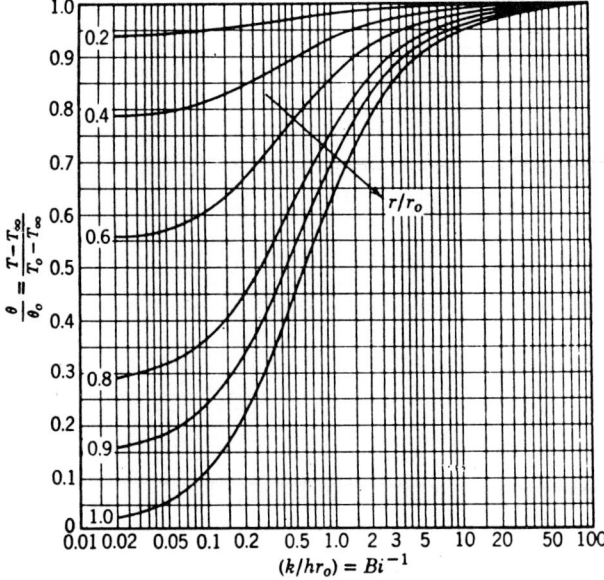

Fig. 3.2-19 Temperature distribution in an infinite cylinder of radius r_0. (From ref. 16; used by permission.)

When the temperature distribution in the solid cannot be ignored, recourse is made to the Fourier equation. Simplification of the solution is afforded by nondimensionalizing the equation and applying the initial and the boundary conditions. This results in the following relationship:

$$\theta^* = f(x^*, Fo, Bi) \tag{3.2-47}$$

where

$$\theta = \frac{T - T_\infty}{T_i - T_\infty} \tag{3.2-48}$$

and

$$x^* = \frac{x}{L} \tag{3.2-49}$$

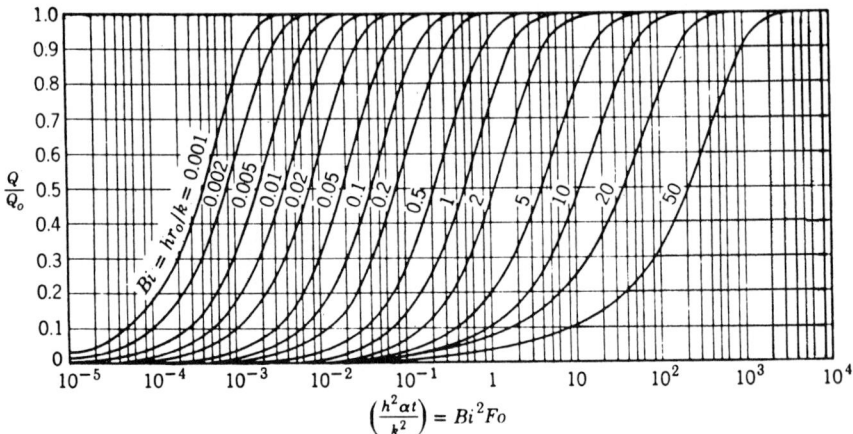

Fig. 3.2-20 Internal energy change as a function of time for an infinite cylinder of radius r_0. (From ref. 17; used by permission.)

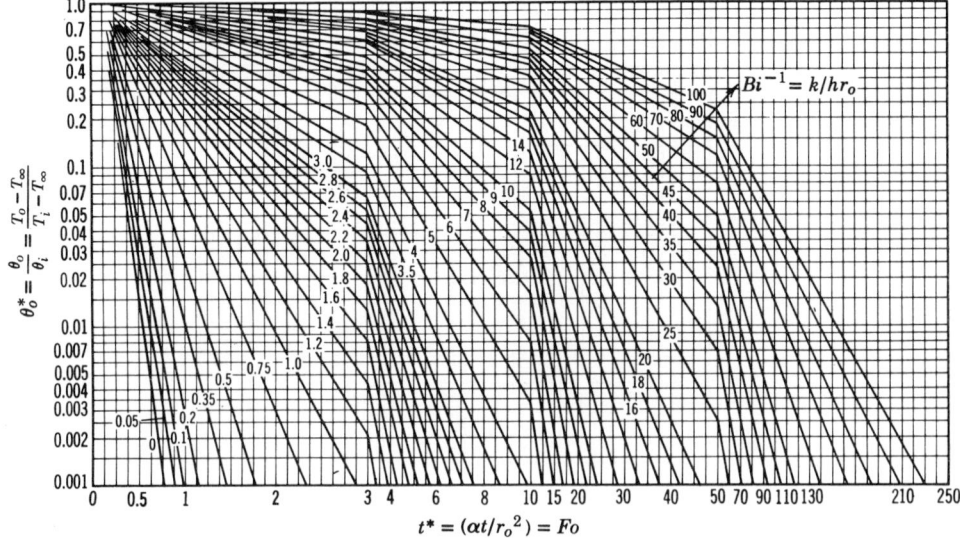

Fig. 3.2-21 Center temperature as a function of time in a sphere of radius r_0. (From ref. 16; used by permission.)

Figures 3.2-15 and 3.2-16 indicate graphically these functional relationships for a finite slab. The temperature T_0 is the temperature at the midplane of the wall.

The total amount of heat transfer Q during any time t is given by Fig. 3.2-17 where Q/Q_0 is plotted versus Bi^2Fo for various Bi numbers. The quantity $Q_0 = \rho c V(T_i - T_\infty)$. Similar results are presented in Figs. 3.2-18 to 3.2-23 for an infinite cylinder and a sphere.

Semi-Infinite Solid

Closed-form solutions of the Fourier equation are available for semi-infinite solids. The earth may be approximated as such a solid where interest lies in the transient heat transfer resulting from variation of temperature near its surface. Solutions for three common cases subject to initial and boundary

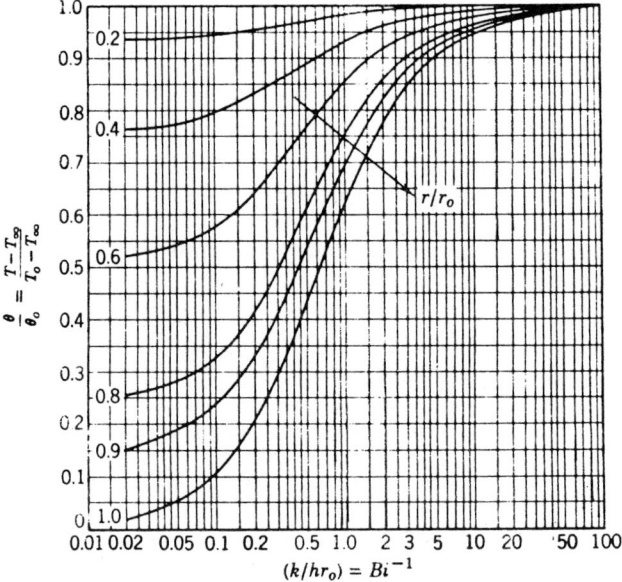

Fig. 3.2-22 Temperature distribution in a sphere of radius r_0. (From ref. 16; used by permission.)

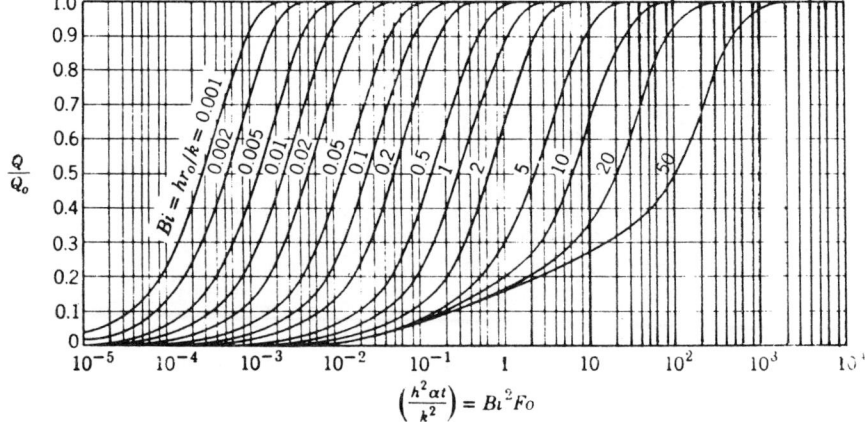

Fig. 3.2-23 Internal energy change as a function of time for a sphere of radius r_0. (From ref. 17; used by permission.)

conditions are given below:

1. Constant surface temperature:

$$T(x,0) = T_i, (0,t) = T_s$$

$$\frac{T(x,t) - T_s}{T_i - T_s} = \text{erf}\frac{x}{2\sqrt{\alpha t}} \qquad (3.2\text{-}50)$$

$$q_s(t) = -k\left.\frac{\partial T}{\partial x}\right|_{x=0} = \frac{k(T_s - T_i)}{\sqrt{\pi \alpha t}} \qquad (3.2\text{-}51)$$

2. Constant surface heat flux:

$$T(x,0) = T_i, q_0 = -k\left.\frac{\partial T}{\partial x}\right|_{x=0} = \text{const}$$

$$T(x,t) - T_i = \frac{2q_0(\alpha t/\pi)^{1/2}}{k}e^{-x^2/4\alpha t} - q_0\frac{x}{k}\text{erf}\left(\frac{x}{2\sqrt{\alpha t}}\right) \qquad (3.2\text{-}52)$$

3. Surface convection:

$$T(x,0) = T_i, \qquad -k\left.\frac{\partial T}{\partial x}\right|_{x=0} = h[T_\infty - T(0,t)]$$

$$\frac{T(x,t) - T_i}{T_\infty - T_i} = \text{erfc}\left(\frac{x}{2\sqrt{\alpha t}}\right) - \left[\exp\left(\frac{hx}{k} + \frac{h^2\alpha t}{k^2}\right)\right]$$

$$\times\left[\text{erfc}\left(\frac{x}{2\sqrt{\alpha t}} + \frac{h\sqrt{\alpha t}}{k}\right)\right] \qquad (3.2\text{-}53)$$

TABLE 3.2-10   GAUSSIAN ERROR FUNCTION[a]

w	erf w	w	erf w	w	erf w
0.00	0.00000	0.36	0.38933	1.04	0.85865
0.02	0.02256	0.38	0.40901	1.08	0.87333
0.04	0.04511	0.40	0.42839	1.12	0.88679
0.06	0.06762	0.44	0.46622	1.16	0.89910
0.08	0.09008	0.48	0.50275	1.20	0.91031
0.10	0.11246	0.52	0.53790	1.30	0.93401
0.12	0.13476	0.56	0.57162	1.40	0.95228
0.14	0.15695	0.60	0.60386	1.50	0.96611
0.16	0.17901	0.64	0.63459	1.60	0.97635
0.18	0.20094	0.68	0.66378	1.70	0.98379
0.20	0.22270	0.72	0.69143	1.80	0.98909
0.22	0.24430	0.76	0.71754	1.90	0.99279
0.24	0.26570	0.80	0.74210	2.00	0.99532
0.26	0.28690	0.84	0.76514	2.20	0.99814
0.28	0.30788	0.88	0.78669	2.40	0.99931
0.30	0.32863	0.92	0.80677	2.60	0.99976
0.32	0.34913	0.96	0.82542	2.80	0.99992
0.34	0.36936	1.00	0.84270	3.00	0.99998

[a] The Gaussian error function is defined as

$$\text{erf } w \equiv \frac{2}{\sqrt{\pi}}\int_0^w e^{-v^2}\,dv$$

The complementary error function is defined as

$$\text{erfc } w \equiv 1 - \text{erf } w$$

where erf is the Gaussian error function tabulated in Table 3.2-10 and erfc is the complimentary error function defined for a function w as

$$\text{erfc } w = 1 - \text{erf } w \qquad (3.2\text{-}54)$$

For multidimensional cases, solutions to the Fourier equation are expressed as a product of one-dimensional solutions. Results are given in Fig. 3.2-24 where S, P, and C are one-dimensional

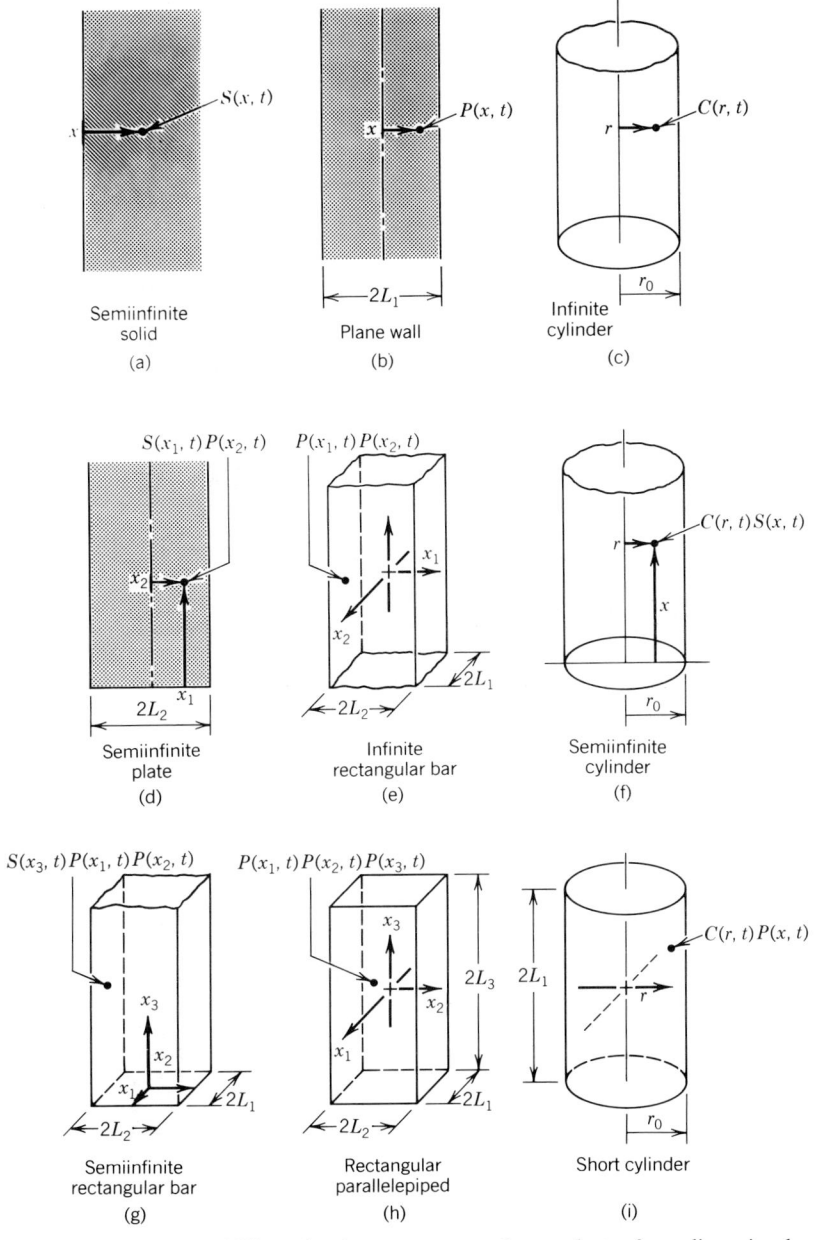

Fig. 3.2-24 Solutions for multidimensional systems expressed as products of one-dimensional results. (From ref. 1; used by permission.)

solutions defined by

$$S(x,t) = \frac{T(x,t) - T_\infty}{T_i - T_\infty} \quad \text{semi-infinite solid, } x \text{ measured from surface} \qquad (3.2\text{-}55)$$

$$P(x,t) = \frac{T(x,t) - T_\infty}{T_i - T_\infty} \quad \text{plane wall, } x \text{ measured from midplane} \qquad (3.2\text{-}56)$$

$$C(r,t) = \frac{T(r,t) - T_\infty}{T_i - T_\infty} \quad \text{infinite cylinder} \qquad (3.2\text{-}57)$$

For example, for a short cylinder the temperature distribution is

$$\frac{T(r,x,t) - T_\infty}{T_i - T_\infty} = P(x,t) \cdot C(r,t)$$

and for a parallelpiped

$$\frac{T(x_1, x_2, x_3, t) - T_\infty}{T_i - T_\infty} = P(x_1,t) \cdot P(x_2,t) \cdot P(x_3,t)$$

In the case of nonsimple geometrics recourse is made to finite-difference techniques for solution.

3.2-8 Convection

The heat transfer by convection predominates in the transfer of heat energy between a solid and a fluid. The amount of energy transfer depends on the temperature difference and the fluid motion.

Consider a hot body surrounded by a fluid at a lower temperature. As heat is transferred to the fluid by conduction, its temperature increases while its density decreases. The resulting buoyancy forces move the heated fluid upward against gravity forces. As the hot fluid is continuously replaced by the adjacent cold fluid, a free convection current is established in the vertical direction. This results in heat transfer by "natural convection." To increase the rate of heat transfer, the fluid motion is induced by a pump or a blower, and in this case, the heat transfer takes place by "forced convection." When the flow is laminar, no mixing occurs, and heat is transferred predominantly by conduction. In turbulent flow there is always a laminar sublayer adjacent to the boundary, and heat is transferred by both conduction and convection. The temperature change in the laminar sublayer is large in comparison with the turbulent flow outside this layer. Since the rate of heat transfer is directly proportional to temperature drop and inversely proportional to the resistance, the laminar sublayer provides most of the resistance to heat transfer.

Under steady-state conditions the rate of heat transfer by convection is given by

$$q = hA\,\Delta t \qquad (3.2\text{-}58)$$

where h is the coefficient of heat transfer by convection, A is the area of heat flow, and Δt is the temperature difference.

The problem of heat transfer by convection is extremely complicated and analytical solutions of only a few problems are available. The difficulty arises from the fact that the heat transfer is intimately related to a rather complex fluid flow phenomenon. Experiments indicate that h is a function of many factors such as the physical properties and phase of the contact materials, whether the flow is laminar or turbulent, the direction of heat flow, that is, from the wall to the fluid or vice versa, whether the flow is horizontal or vertical, the presence of scale and local changes of phase such as boiling. In order to reduce the number of variables, resort is made to dimensional analysis, which when supplemented by experimental data, yields approximate solutions for convective heat transfer problem. Results are usually expressed in the form of empirical formulas and graphical charts.

3.2-9 Natural Convection

In natural or free convection the heat transfer coefficient is generally given by an equation of the form

$$\text{Nu} = C(\text{Gr Pr})^n = C\,\text{Ra}^m \qquad (3.2\text{-}59)$$

where Nu is the Nusselt number $= hx/k$ and C is a constant, Gr is the Grashof number, Pr is the Prandtl number, and their product is the Rayleigh number Ra. In this equation properties are

TABLE 3.2-11 FREE CONVECTION EMPERICAL CORRELATIONS FOR EXTERNAL FLOW GEOMETRIES

Geometry	Ra_L	Recommended Correlation
1. Vertical plates or vertical cylinders if $(D/L) \geq (35/Gr_L^{1/4})$.	—	$Nu_L = \left\{ 0.825 + \dfrac{0.387\,Ra_L^{1/4}}{\left[1 + (0.492/Pr)^{9/16}\right]^{8/27}} \right\}^2$
	$0\text{--}10^9$	$Nu_L = 0.68 + \dfrac{0.670\,Ra_L^{1/4}}{\left[1 + (0.492/Pr)^{9/16}\right]^{4/9}}$
	$10^{-1}\text{--}10^4$	Fig. 3.2-25
	$10^4\text{--}10^9$	$Nu_L = 0.59\,Ra_L^{1/4}$
	$10^9\text{--}10^{13}$	$Nu_L = 0.021\,Ra_L^{2/5}$
	$10^9\text{--}10^{13}$	$Nu_L = 0.10\,Ra_L^{1/3}$
2. Horizontal plates[a]		
a. Upper surface of heated plates or lower surface of cooled plates.	$10^5\text{--}10^7$ $10^7\text{--}10^{10}$	$Nu_L = 0.54\,Ra_L^{1/4}$ $Nu_L = 0.15\,Ra_L^{1/3}$
b. Lower surface of heated plates or upper surface of cooled plates.	$10^5\text{--}10^{10}$	$Nu_L = 0.27\,Ra_L^{1/4}$
3. Horizontal Cylinders.	$10^{-5}\text{--}10^{12}$ $0\text{--}10^{-5}$	$Nu_D = \left\{ 0.60 + \dfrac{0.387\,Ra_D^{1/6}}{\left[1 + (0.559/Pr)^{9/16}\right]^{8/27}} \right\}^2$
	$10^{-5}\text{--}10^4$	$Nu_D = 0.4$
	$10^4 - 10^9$	Fig. 3.2-26
	$10^9\text{--}10^{12}$	$Nu_D = 0.53\,Ra_D^{1/4}$ $Nu_D = 0.13\,Ra_D^{1/3}$
4. Inclined plates (angle θ from the vertical)	$0\text{--}10^9,$ $\theta < 60°,$ $g \rightarrow g\cos\theta$ $> 10^9, \theta < 60°$	$Nu_L = 0.68 + \dfrac{0.670\,Ra_L^{1/4}}{\left[1 + (0.492/Pr)^{9/16}\right]^{4/9}}$ $Nu_L = \left\{ 0.825 + \dfrac{0.387\,Ra_L^{1/6}}{\left[1 + (0.492/Pr)^{9/16}\right]^{8/27}} \right\}^2$
5. Sphere	$1\text{--}10^5$, $Pr \sim 1$	$Nu_D = 2 + 0.43\,Ra_D^{1/4}$

[a] For horizontal plates the characteristic length L is given by A_s/P, where A_s and P are the plate surface area and perimeter respectively.

evaluated at the film temperature, which is an average between the wall temperature and the temperature of the bulk of the fluid. The Grashof number is

$$Gr = \frac{g\beta(T_\omega - T_\infty)x^3}{\nu^2} \tag{3.2-60}$$

where β is the volumetric thermal expansion defined by $\beta = -(1/\rho)(\partial\rho/\partial T)_p$, ν is the kinematic viscosity, and x is a characteristic length. The Grashof number indicates the ratio of the buoyancy force relative to the viscous force acting on the fluid.

The Rayleigh number Ra is used to correlate the transition from laminar to turbulent flow in free convection boundary layer. For vertical plates this transition takes place when $Ra \geq 10^9$. Table 3.2-11 gives empirical correlations for free convection for different surface configurations. Figures 3.2-25 and 3.2-26 are used for vertical plates and horizontal cylinders for low values of Ra. Table 3.2-12 gives simplified equation for free convection from various surfaces to air at atmospheric pressure and moderate temperatures.

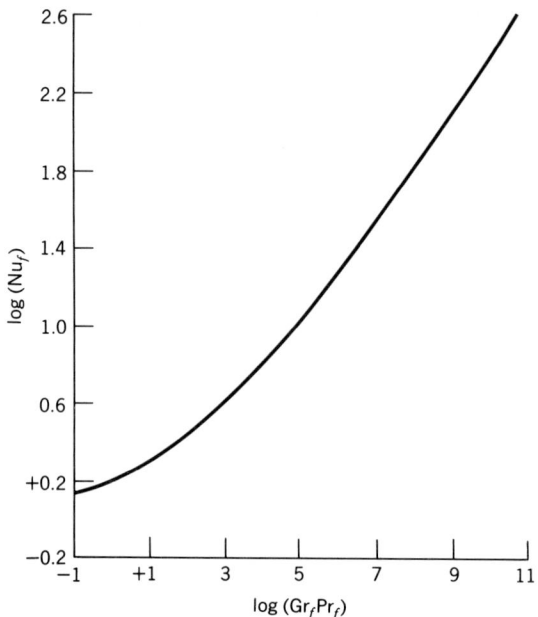

Fig. 3.2-25 Free-convection heat transfer correlation for heat transfer from heated vertical plates. (From ref. 18; used by permission.)

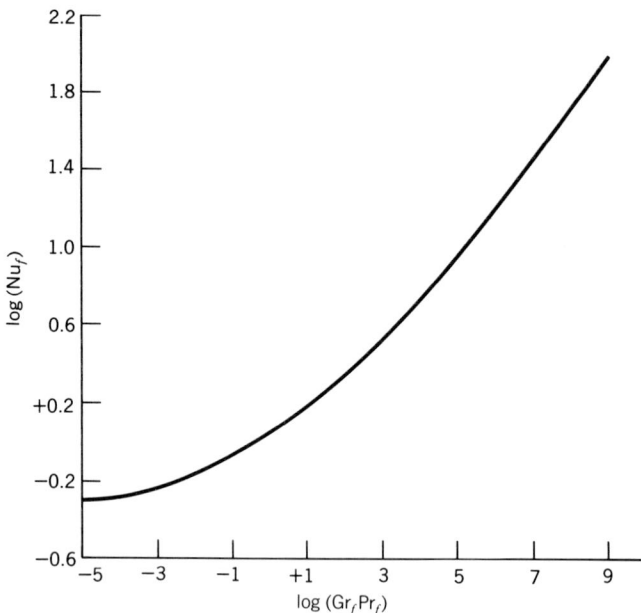

Fig. 3.2-26 Free-convection heat transfer correlations for heat transfer from heated horizontal cylinders. (From ref. 18; used by permission.)

TABLE 3.2-12 SIMPLIFIED EQUATIONS FOR FREE CONVECTION FROM VARIOUS SURFACES TO AIR AT ATMOSPHERIC PRESSURE

Surface	Laminar, $10^4 < Gr_f Pr_f < 10^9$	Turbulent, $Gr_f Pr_f > 10^9$
Vertical plane or cylinder	$h = 1.42 \left(\dfrac{\Delta T}{L}\right)^{1/4}$	$h = 0.95(\Delta T)^{1/3}$
Horizontal cylinder	$h = 1.32 \left(\dfrac{\Delta T}{d}\right)^{1/4}$	$h = 1.24(\Delta T)^{1/3}$
Horizontal plate:		
Heated plate facing upward or cooled plate facing downward	$h = 1.32 \left(\dfrac{\Delta T}{L}\right)^{1/4}$	$h = 1.43(\Delta T)^{1/3}$
Heated plate facing downward or cooled plate facing upward	$h = 0.61 \left(\dfrac{\Delta T}{L^2}\right)^{1/5}$	

where h = heat-transfer coefficient, W/m² °K
$\Delta T = T_w - T_x$, °K
L = vertical or horizontal dimension, m
d = diameter, m

Free Convection in Enclosures

The heat transfer by convection in enclosed spaces is characterized by the Prandtl and Rayleigh numbers. The Rayleigh number is the product of the Grashof and Prandtl numbers and indicates the ratio of the buoyancy to viscous forces. When $Ra_L < 1700$, heat transfer occurs predominantly by conduction and $Nu_L = 1$, and when $Ra_L > 1700$, heat transfer takes place by free convection. For narrow horizontal enclosures heated from below, the average value of the Nusselt number is given by

$$\overline{Nu}_L = \frac{\overline{h}L}{k} = 0.069 Ra_L^{1/3} Pr^{0.074} \qquad (3 \times 10^5 < Ra_L < 7 \times 10^9) \qquad (3.2\text{-}61)$$

For the vertical rectangular cavity shown in Fig. 3.2-27, the following relations apply:

$$\overline{Nu}_L = 0.22\left(\frac{Pr}{0.2 + Pr} Ra_L\right)^{0.28}\left(\frac{H}{L}\right)^{-1/4} \quad \begin{cases} 2 < H/L < 10 \\ Pr < 10^5 \\ Ra_L < 10^{10} \end{cases} \qquad (3.2\text{-}62)$$

$$\overline{Nu}_L = 0.18\left(\frac{Pr}{0.2 + Pr} Ra_L\right)^{0.29} \quad \begin{cases} 1 < H/L < 2 \\ 10^{-3} < Pr < 10^5 \\ 10^3 < (Ra_L Pr)/(0.2 + Pr) \end{cases} \qquad (3.2\text{-}63)$$

$$\overline{Nu}_L = 0.42\, Ra_L^{1/4} Pr^{0.012}\left(\frac{H}{L}\right)^{-0.3} \quad \begin{cases} 10 < H/L < 40 \\ 1 < Pr < 2 \times 10^4 \\ 10^4 < Ra_L < 10^7 \end{cases} \qquad (3.2\text{-}64)$$

$$\overline{Nu}_L = 0.046 Ra_L^{1/3} \quad \begin{cases} 1 < H/L < 40 \\ 1 < Pr < 20 \\ 10^6 < Ra_L < 10^9 \end{cases} \qquad (3.2\text{-}65)$$

Properties are evaluated at the average temperature $(T_1 + T_2)/2$

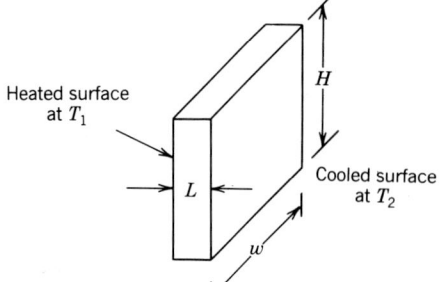

Fig. 3.2-27 Free-convection in a rectangular enclosure.

For inclined rectangular enclosures Nu_L depends on the tilt angle. For an angle θ (measured from the horizontal) less than a critical value θ^* indicated in Table 3.2-13, the following formulas are used:

$$\overline{\mathrm{Nu}}_L = 1 + 1.44\left[1 - \frac{1708}{\mathrm{Ra}_L\cos\theta}\right]\left[1 - \frac{1708(\sin 1.8\theta)^{1.6}}{\mathrm{Ra}_L\cos\theta}\right]$$

$$+\left[\left(\frac{\mathrm{Ra}_L\cos\theta}{5830}\right)^{1/3} - 1\right] \quad \begin{cases} H/L \geq 12 \\ 0 < \theta \leq \theta^* \end{cases} \tag{3.2-66}$$

In this equation, if the quantity in the first and third brackets is negative, it is set equal to zero resulting in $\overline{\mathrm{Nu}}_L = 1$. For small values of H/L,

$$\overline{\mathrm{Nu}}_L = \overline{\mathrm{Nu}}_L|_{\theta=0}\left[\frac{\overline{\mathrm{Nu}}_L|_{\theta=90}}{\overline{\mathrm{Nu}}_L|_{\theta=0}}\right]^{\theta/\theta^*}(\sin\theta^*)^{\theta/4\theta^*} \quad \begin{cases} H/L \leq 12 \\ 0 < \theta < \theta^* \end{cases} \tag{3.2-67}$$

When $\theta > \theta^*$, the following relations are recommended

$$\overline{\mathrm{Nu}}_L = \overline{\mathrm{Nu}}_L|_{\theta=90°}(\sin\theta)^{1/4} \quad (\theta^* < \theta < 90°) \tag{3.2-68}$$

$$\overline{\mathrm{Nu}}_L = 1 + [\overline{\mathrm{Nu}}_L|_{\theta=90°} - 1]\sin\theta \quad (90° < \theta < 180) \tag{3.2-69}$$

The heat transfer in the annular space between long, horizontal concentric cylinders is evaluated in terms of an effective thermal conductivity k_{eff}, defined as that of a stationary fluid capable of transferring the same amount of heat as the moving fluid.

The heat transfer is

$$q = \frac{2\pi k_{\mathrm{eff}}}{\ln(D_o/D_i)}(T_i - T_o) \tag{3.2-70}$$

where

$$\frac{k_{\mathrm{eff}}}{k} = 0.386\left(\frac{\mathrm{Pr}}{0.861 + \mathrm{Pr}}\right)^{1/4}(\mathrm{Ra}^*_c)^{1/4} \quad (10^2 \leq \mathrm{Ra}^*_c \leq 10^7) \tag{3.2-71}$$

$$\mathrm{Ra}^*_c = \frac{[\ln(D_o/D_i)]^4}{L^3(D_i^{-3/5} + D_o^{-3/5})^5}\mathrm{Ra}_L$$

$$L = \frac{D_o - D_i}{2} \tag{3.2-72}$$

and subscripts i and o refer to the inside and outside surfaces of the annular space.

TABLE 3.2-13 CRITICAL ANGLES FOR INCLINED RECTANGULAR ENCLOSURES[a]

H/L	1	3	6	12	< 12
θ^*	25°	53°	60°	67°	70°

[a]See ref. 1.

For concentric spheres the corresponding formulas are

$$q = k_{\text{eff}}\, \pi \left(\frac{D_i D_o}{L} \right)(T_i - T_o) \tag{3.2-73}$$

$$\frac{k_{\text{eff}}}{k} = 0.74 \left(\frac{\text{Pr}}{0.861 + \text{Pr}} \right)^{1/4} (\text{Ra}_s^*)^{1/4} \qquad (10^2 \leq \text{Ra}_s^* \leq 10^4) \tag{3.2-74}$$

$$\text{Ra}_s^* = \left[\frac{L}{(D_o D_i)^4} \frac{\text{Ra}_L}{\left(D_i^{-7/5} + D_o^{-7/5} \right)^5} \right]^{1/4} \tag{3.2-75}$$

3.2-10 Forced Convection

In forced convection the velocity of the fluid relative to the solid surface is induced by external means. Empirical correlations are used to determine the convection coefficient for different flow geometries. The resulting expressions are of the form

$$\overline{\text{Nu}}_L = C\, \text{Re}_L^m \text{Pr}^n \tag{3.2-76}$$

Values of the constant C and the exponents m and n depend on the surface geometry and whether the flow is laminar or turbulent. In the following sections expressions for the heat transfer coefficients are presented for external flow and internal flow.

3.2-11 External Flow

For laminar flow over a flat plate the local Nusselt number is

$$\text{Nu}_x \equiv \frac{h_x x}{k} = 0.332\, \text{Re}_x^{1/2} \text{Pr}^{1/3} \qquad (\text{Pr} \geq 0.6) \tag{3.2-77}$$

and the average value of Nusselt number over a flat plate of length L is obtained by integrating the above expression:

$$\overline{\text{Nu}}_L \equiv \frac{\overline{h}_L L}{k} = 0.664\, \text{Re}_L^{1/2} \text{Pr}^{1/3} \qquad (\text{Pr} \geq 0.6) \tag{3.2-78}$$

For liquid metals Pr is low and the Nusselt number is given by

$$\text{Nu}_x = 0.565\, \text{Re}_x^{1/2} \text{Pr}^{1/2} \qquad (\text{Pr} \leq 0.05) \tag{3.2-79}$$

For turbulent flow over a flat plate the local Nusselt number is given by

$$\text{Nu}_x = 0.0296\, \text{Re}_x^{4/5} \text{Pr}^{1/3} \qquad (0.6 < \text{Pr} < 60) \tag{3.2-80}$$

But the turbulent boundary layer on a flat plate is generally preceded by a laminar boundary layer, resulting in a mixed boundary layer. Under this condition Nusselt number is given by

$$\overline{\text{Nu}}_L = \left(0.037\, \text{Re}_L^{4/5} - A \right) \text{Pr}^{1/3} \tag{3.2-81}$$

where

$$A = 0.037\, \text{Re}_{x,c}^{4/5} - 0.664\, \text{Re}_{x,c}^{1/2} \tag{3.2-82}$$

The subscript c indicates the critical state at which the transition of the flow from laminar to turbulent occurs. When the typical value of $\text{Re}_{x,c} = 5 \times 10^5$ is substituted in Eq. (3.2-82), the following is obtained:

$$\text{Nu}_L = \left(0.037\, \text{Re}_L^{4/5} - 871 \right) \text{Pr}^{1/3} \qquad \begin{cases} 0.6 < \text{Pr} < 60 \\ 5 \times 10^{-5} < \text{Re}_L < 10^8 \\ \text{Re}_{x,c} = 5 \times 10^5 \end{cases} \tag{3.2-83}$$

If further $L \gg x_c$ then

$$\overline{\text{Nu}}_L = 0.037\, \text{Re}_L^{4/5} \text{Pr}^{1/3} \tag{3.2-84}$$

For wider range of Pr the following correlation can be used:

$$\mathrm{Nu}_L = 0.036\left(\mathrm{Re}_L^{4/5} - 9200\right)\mathrm{Pr}^{0.43}\left(\frac{\mu_\infty}{\mu_s}\right)^{1/4} \quad \begin{cases} 0.7 < \mathrm{Pr} < 380 \\ 10^5 < \mathrm{Re}_L < 5.5 \times 10^6 \\ 0.26 < (\mu_\infty/\mu_s) < 3.5 \end{cases} \quad (3.2\text{-}85)$$

For a cylinder in cross flow the heat transfer coefficient is calculated according to the relation

$$\mathrm{Nu}_D = \frac{hD}{k_f} = C\left(\frac{V_\infty D}{\nu_f}\right)^n \mathrm{Pr}^{1/3} \qquad\qquad (3.2\text{-}86)$$

where C and n are given in Table 3.2-14 and properties are evaluated at the film temperature. The subscript ∞ indicates free stream conditions. Figure 3.2-28 gives results for air flowing normal to a single cylinder.

For cross flow over cylinders of noncircular cross section, Eq. (3.2-86) is still applicable provided a characteristic dimension D (Table 3.2-15) is used for the indicated range of conditions. Another equation applicable for the entire range of Re_D and a wide range of Pr is

$$\overline{\mathrm{Nu}}_D = 0.3 + \frac{0.62\,\mathrm{Re}_D^{1/2}\mathrm{Pr}^{1/3}}{\left[1 + (0.4/\mathrm{Pr})^{2/3}\right]^{1/4}} + \left[1 + \left(\frac{\mathrm{Re}_D}{28200}\right)^{5/8}\right]^{4/5} \qquad (\mathrm{Re}_D\mathrm{Pr} > 0.2) \quad (3.2\text{-}87)$$

For flow over a sphere the following equation is recommended:

$$\overline{\mathrm{Nu}}_D = 2 + \left(0.4\,\mathrm{Re}_D^{1/2} + 0.06\,\mathrm{Re}_D^{2/3}\right)\mathrm{Pr}^{0.4}\left(\frac{\mu_\infty}{\mu_s}\right)^{1/4} \quad \begin{cases} 0.71 < \mathrm{Pr} < 380 \\ 3.5 < \mathrm{Re}_D < 7.6 \times 10^4 \\ 1.0 < (\mu_\infty/\mu_s) < 3.2 \end{cases} \quad (3.2\text{-}88)$$

TABLE 3.2-14   CONSTANTS OF EQ. (3.2-86)

Re_{D_f}	C		n		
0.4–4	0.989	0.330			
4–40	0.911	0.385			
40–4,000	0.683	0.466			
4,000–40,000	0.193	0.618	40,000–400,000	0.0266	0.805

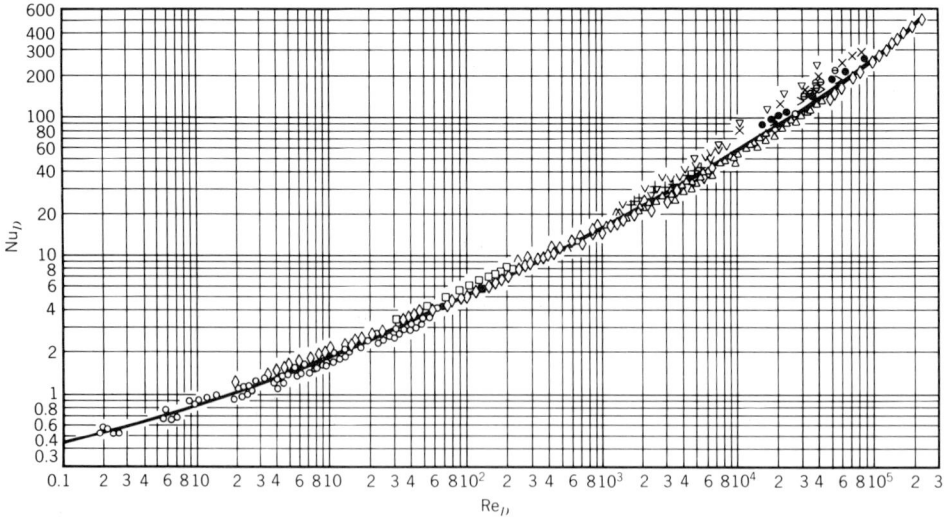

Fig. 3.2-28 Nu versus Re for flow normal to single cylinders. (From ref. 18; used by permission.)

TABLE 3.2-15 CONSTANTS FOR HEAT TRANSFER FROM NONCIRCULAR CYLINDERS[a]

Geometry	Re_{D_f}	C	n
$V_\infty \longrightarrow \diamondsuit \updownarrow D$	$5 \times 10^3\text{–}10^5$	0.246	0.588
$V_\infty \longrightarrow \square \updownarrow D$	$5 \times 10^3\text{–}10^5$	0.102	0.675
$V_\infty \longrightarrow \hexagon \updownarrow D$	$5 \times 10^3\text{–}1.95 \times 10^4$ $1.95 \times 10^4\text{–}10^5$	0.160 0.0385	0.638 0.782
$V_\infty \longrightarrow \hexagon \updownarrow D$	$5 \times 10^3\text{–}10^5$	0.153	0.638
$V_\infty \longrightarrow \mid \updownarrow D$	$4 \times 10^3\text{–}1.5 \times 10^4$	0.228	0.731

[a] See ref. 20.

Flow across Banks of Tubes

The heat transfer by convection as a result of cross flow over a bank of tubes has many industrial applications especially in heat exchangers. The tubes may be staggered or aligned in the direction flow. The geometric configuration is characterized by the tube diameter, the transverse pitch S_T, and longitudinal pitch S_L, measured between the tube centers. The average heat transfer coefficient for air flow across a bank of tubes of 10 or more rows ($N \geq 10$) is given by the relation

$$\overline{\mathrm{Nu}}_D = C_1 \mathrm{Re}_{D,\max}^m \tag{3.2-89}$$

where

$$\mathrm{Re}_{D,\max} = \frac{\rho V_{\max} D}{\mu} \tag{3.2-90}$$

and $V_{\max}$ is the maximum fluid velocity occurring in the tube bank. The above equation is valid for $2000 < \mathrm{Re}_{D,\max} < 40{,}000$ and $\mathrm{Pr} = 0.7$. Values of C_1 and m are given in Table 3.2-16.

For fluids other than air ($N \geq 10$, $2000 < \mathrm{Re}_{D,\max} < 40{,}000$, $\mathrm{Pr} \geq 0.7$):

$$\overline{\mathrm{Nu}}_D = 1.13 C_1 \mathrm{Re}_{D,\max}^m \mathrm{Pr}^{1/3} \tag{3.2-91}$$

When the number of rows of tubes is less than 10 the following relation is used:

$$\overline{\mathrm{Nu}}_D|_{N<10} = C_2 \overline{\mathrm{Nu}}_D|_{N \geq 10} \tag{3.2-92}$$

Values of C_2 are given in Table 3.2-17.

Another equation applicable for $N < 20$, $0.7 < \mathrm{Pr} < 500$ and $1000 < \mathrm{Re}_{D,\max} < 2 \times 10^6$ is

$$\overline{\mathrm{Nu}}_D = C \mathrm{Re}_{D,\max}^m \mathrm{Pr}^{0.36} \left(\frac{\mathrm{Pr}_\infty}{\mathrm{Pr}_s}\right)^{1/4} \tag{3.2-93}$$

Properties in this equation, except Pr_s, are evaluated at the bulk fluid temperature. Values of C and m are given in Table 3.2-18.

TABLE 3.2-16   CONSTANTS OF EQ. (3.2-89 AND (3.2-91) FOR A TUBE BANK OF 10 OR MORE ROWS[a]

	S_T/D							
	1.25		1.5		2.0		3.0	
S_L/D	C_1	m	C_1	m	C_1	m	C_1	m
Aligned								
1.25	0.348	0.592	0.275	0.608	0.100	0.704	0.0633	0.752
1.50	0.367	0.586	0.250	0.620	0.101	0.702	0.0678	0.744
2.00	0.418	0.570	0.299	0.602	0.229	0.632	0.198	0.648
3.00	0.290	0.601	0.357	0.584	0.374	0.581	0.286	0.608
Staggered								
0.600	—	—	—	—	—	—	0.213	0.636
0.900	—	—	—	—	0.446	0.571	0.401	0.581
1.000	—	—	0.497	0.558	—	—	—	—
1.125	—	—	—	—	0.478	0.565	0.518	0.560
1.250	0.518	0.556	0.505	0.554	0.519	0.556	0.522	0.562
1.500	0.451	0.568	0.460	0.562	0.452	0.568	0.488	0.568
2.000	0.404	0.572	0.416	0.568	0.482	0.556	0.449	0.570
3.000	0.310	0.592	0.356	0.580	0.440	0.562	0.428	0.574

[a]See ref. 17.

TABLE 3.2-17   CORRECTION FACTOR C_2 OF EQ. (3.2-92) FOR $N < 10$[a]

N	1	2	3	4	5	6	7	8	9
Aligned	0.64	0.80	0.87	0.90	0.92	0.94	0.96	0.98	0.99
Staggered	0.68	0.75	0.83	0.89	0.92	0.95	0.97	0.98	0.99

[a]See ref. 21.

TABLE 3.2-18   CONSTANTS OF EQ. (3.2-93) FOR THE TUBE BANK IN CROSS FLOW[a]

CONFIGURATION	$Re_{D,max}$	C	m
Aligned	$10^3 - 2 \times 10^5$	0.27	0.63
Staggerred $(S_T/S_L < 2)$	$10^3 - 2 \times 10^5$	$0.35(S_T/S_L)^{1/5}$	0.60
Staggered $(S_T/S_L > 2)$	$10^3 - 2 \times 10^5$	0.40	0.60
Aligned	$2 \times 10^5 - 2 \times 10^6$	0.021	0.84
Staggered	$2 \times 10^5 - 2 \times 10^6$	0.022	0.84

[a]See ref. 22.

The power required to move a fluid across a tube bank is proportional to the pressure drop given by

$$\Delta p = Nx \frac{\left(\rho V_{max}^2 \right)}{2} f \tag{3.2-94}$$

The correction factor x and the friction factor f for an in-line and an equilaterally staggered tube arrangements can be determined from Figs. 3.2-29 and 3.2-30.

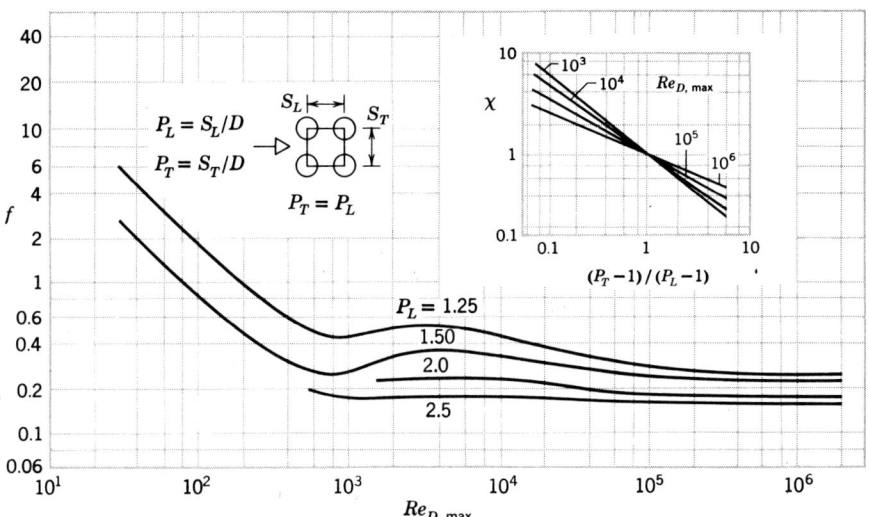

Fig. 3.2-29 Friction factor f and correction factor x for Eq. (3.2-94); in-line tube bundle arrangement. (From ref. 22; used by permission.)

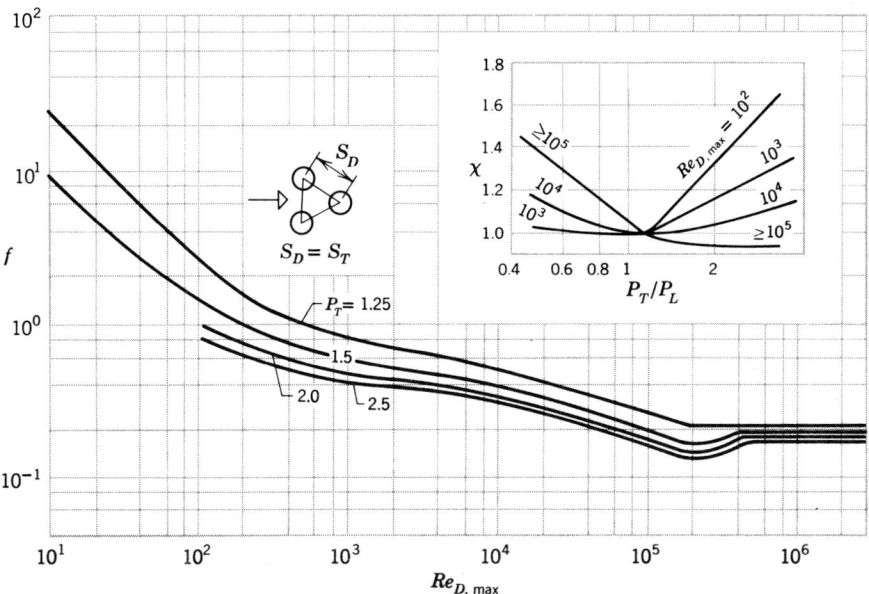

Fig. 3.2-30 Friction factor f and correction factor x for Eq. (3.2-94); staggered tube bundle arrangement. (From ref. 22; used by permission.)

3.2-12   Internal Flow

For fully developed laminar flow in a tube at constant wall temperature, the following relation is used to calculate the average heat transfer coefficient over the entire length of the tube.

$$\overline{\mathrm{Nu}}_D = 3.66 + \frac{0.0668(D/L)\mathrm{Re}_D\mathrm{Pr}}{1 + 0.04[(D/L)\mathrm{Re}_D\mathrm{Pr}]^{2/3}} \tag{3.2-95}$$

Another equation is

$$Nu_D = 1.86(Re_D Pr)^{1/3}\left(\frac{D}{L}\right)^{1/3}\left(\frac{\mu}{\mu_s}\right)^{0.14}$$

$$\begin{cases} T_s = \text{const.} \\ 0.48 < Pr < 16,700 \\ 0.0044 < (\mu/\mu_s) < 9.75 \end{cases} \tag{3.2-96}$$

This equation is valid for $Re_D Pr(D/L) > 10^8$.

For noncircular channels heat transfer correlations are based on the hydraulic diameter defined as four times the cross-sectional area divided by the wetted perimeter. Table 3.2-19 gives the value of Nu for fully developed laminar flow in ducts of various cross sections.

For fully developed turbulent flow in smooth circular tubes for $0.7 \leq Pr \leq 160$, $Re_D \geq 10,000$, and $L/D \geq 60$, the following equations are recommended for heating:

$$Nu_D = 0.023\,Re_D^{0.8}Pr^{0.4} \tag{3.2-97}$$

for cooling:

$$Nu_D = 0.023\,Re_D^{0.8}Pr^{0.3} \tag{3.2-98}$$

The properties in these equations are evaluated at the fluid bulk temperature. For a wide temperature difference ($0.7 < Pr < 16,700$) in the flow field, the following equation is recommended:

$$Nu_D = 0.027\,Re_D^{0.8}Pr^{1/3}\left(\frac{\mu}{\mu_s}\right)^{0.14} \tag{3.2-99}$$

TABLE 3.2-19 NUSSELT NUMBERS FOR FULLY DEVELOPED LAMINAR FLOW IN TUBES OF DIFFERENT CROSS SECTION[a]

CROSS SECTION	$\dfrac{b}{a}$	$Nu_D \equiv \dfrac{hD_h}{k}$	
		(Constant q_s'')	(Constant T_s)
○	—	4.36	3.66
$a\square b$	1.0	3.63	2.98
$a\square b$	1.4	3.78	—
$a\square b$	2.0	4.11	3.39
$a\square b$	3.0	4.77	—
$a\square b$	4.0	5.35	4.44
$a\square b$	8.0	6.60	5.95
	∞	8.23	7.54
△	—	3.00	2.35

[a] See ref. 23.

where μ_s is evaluated at the wall temperature. If the flow is not fully developed at the entrance of the tube, the following equation is used:

$$\mathrm{Nu}_D = 0.036\,\mathrm{Re}_D^{0.8}\,\mathrm{Pr}^{1/3}\left(\frac{D}{L}\right)^{0.055} \qquad \text{for } 10 < \frac{L}{D} = 400 \qquad (3.2\text{-}100)$$

where L is the tube length.

A more accurate equation including the friction factor f is

$$\mathrm{Nu}_D = \frac{(f/8)\mathrm{Re}_D\,\mathrm{Pr}}{1.07 + 12.7(f/8)^{1/2}(\mathrm{Pr}^{2/3} - 1)}(\mu_b/\mu_s)^n \qquad (3.2\text{-}101)$$

where $n = 0.11$ for $T_s > T_b$
    $= 0.25$ for $T_s < T_b$
    $= 0$ for constant heat flux or for gases.

Properties in this equation are evaluated at $(T_s + T_b)/2$ except for μ_b and μ_s.

For fully developed flow of a liquid metal $(3 \times 10^{-3} \le \mathrm{Pr} \le 5 \times 10^{-2})$ in a smooth circular duct with constant surface heat flux, the following relation is recommended:

$$\mathrm{Nu}_D = 4.82 + 0.0185(\mathrm{Pr}\,\mathrm{Re}_D)^{0.827} \qquad \begin{cases} 3.6 \times 10^3 < \mathrm{Re}_D < 9.05 \times 10^5 \\ 10^2 < \mathrm{Re}_D\,\mathrm{Pr} < 10^4 \end{cases} \qquad (3.2\text{-}102)$$

and for constant surface temperature

$$\mathrm{Nu}_D = 5 + 0.025(\mathrm{Pr}\,\mathrm{Re}_D)^{0.8} \qquad (3.2\text{-}103)$$

3.2-13 Boiling and Condensation

The heat interaction at a solid–liquid interface as the fluid undergoes changes in phase takes place without changing the fluid temperature. The coefficient of heat transfer as well as the rates of heat transfer are significantly larger than with normal convection without phase change. In boiling, the temperature of the solid surface is higher than the saturation temperature of the liquid, whereas the temperature of the surface is less than that of the liquid during condensation.

For a given liquid–surface combination, several boiling mechanisms exist depending on the excess temperature $\Delta T = T_{\text{surface}} - T_{\text{saturated}}$. Figure 3.2-31 shows the boiling curve for water where the surface heat flux is plotted as a function of the excess temperature for the different boiling regions. Other fluids show similar behavior, and the boiling regime varies from free convection, nucleate,

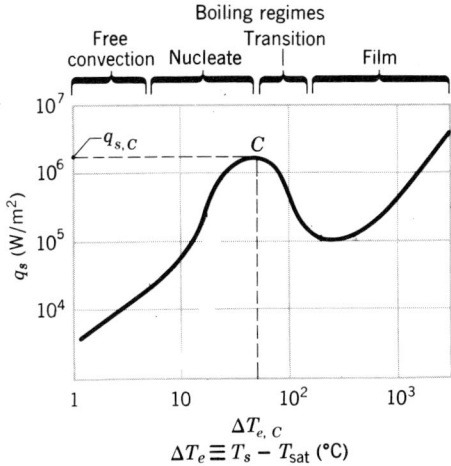

Fig. 3.2-31 The boiling curve for water: surface heat flux, q_s, as a function of excess temperature, $\Delta T \equiv T_s - T_{\text{sat}}$.

transition, and film boiling depending on the value of ΔT. Point c corresponds to a maximum heat flux called the critical heat flux, which is generally in excess of 1 MW/m². Operation in the nucleate boiling regime close to point c is desirable as it involves a large heat flux with small values of excess temperature. Characteristic values of the convection heat transfer coefficient in this regime are in excess of 10^4 W/m² · K. As ΔT is further increased, bubbles formation creates a vapor film resulting in a reduction in the heat transfer coefficient. Values of $\dot{q}_s$ in excess $\dot{q}_{sc}$ result in an abrupt increase in excess temperature to ΔT. This is associated with a large surface temperature, and for this reason point c is called the burnout point and operation should not exceed the critical heat flux $\dot{q}_{sc}$.

For the nucleate boiling regime the following relation is recommended:

$$\dot{q}_s = \mu_l h_{fg} \left[\frac{g(\rho_l - \rho_v)}{\sigma} \right]^{1/2} \left(\frac{c_{p_l} \Delta T}{C_{s,f} h_{fg} \mathrm{Pr}_l^{1.7}} \right)^3 \tag{3.2-104}$$

where σ is the surface tension and subscripts l and v denote saturated liquid and vapor states. Values of the coefficient $C_{s,f}$ depends on the surface–liquid conditions and are given in Table 3.2-20.

A simpler expression for nucleate boiling in water is given by

$$h = C(\Delta T)^n \left(\frac{p}{p_a} \right)^{0.4} \tag{3.2-105}$$

where p is the pressure of the system and p_a is atmospheric pressure. Values of C and n in the above equation are given in Table 3.2-21.

An expression for the critical heat flux in agreement with experiment is,

$$\dot{q}_{s,c} = 0.18 h_{fg} \rho_v \left[\frac{\sigma g(\rho_l - \rho_v)}{\rho_v^2} \right]^{1/4} \tag{3.2-106}$$

In the film-boiling regime the surface of the solid is completely covered by a vapor blanket and heat transfer takes place by both convection and radiation.

TABLE 3.2-20 VALUES OF $C_{s,f}$ FOR VARIOUS WATER SURFACE COMBINATIONS^a

Surface	$C_{s,f}$
Copper	
Scored	0.0068
Polished	0.0130
Stainless Steel	
Chemically etched or	
mechanically polished	0.0130
Ground and polished	0.0080
Brass	0.0060
Platinum	0.0130

^{a}See refs. 24–26.

TABLE 3.2-21 VALUES OF C AND N FOR USE WITH EQ. (3.2-8-105)^a

Surface	$\dot{q}_s$ (kW/m²)	C	n
Horizontal	$\dot{q}_s < 15.8$	1040	1/3
	$15.8 < \dot{q}_s < 236$	5.56	3
Vertical	$\dot{q}_s < 3.15$	539	1/7
	$3.15 < \dot{q}_s < 63.1$	7.95	3

^{a}See ref. 27.

For film boiling on the outer surface of a horizontal tube of diameter D, the heat transfer coefficient can be calculated from the equation

$$h^{4/3} = h_{conv}^{4/3} + h_{rad}^{1/3} \tag{3.2-107}$$

where

$$h_{conv} = 0.62\left[\frac{k_v^3 \rho_v (\rho_l - \rho_v) g (h_{fg} + 0.4 c_{p_v} \Delta T)}{\mu_v D \Delta T}\right] \tag{3.2-108}$$

and

$$h_{rad} = \frac{5.67 \times 10^{-8} \, \varepsilon \left(T_s^4 - T_{sat}^4\right)}{T_s - T_{sat}} \tag{3.2-109}$$

Properties are estimated at the average temperature $(T_s + T_{sat})/2$ and ε is the emissivity of the solid surface.

Laminar Film Condensation

When a vapor comes in contact with a cold surface its temperature is reduced below the saturation temperature. Condensation then takes place and heat is transferred to the surface. If the surface is clean, condensation takes place in the form of a film that coats the whole surface, otherwise dropwise condensation occurs. Higher heat transfer rates result with dropwise condensation than with film condensation (about an order of magnitude larger).

For film condensation the theoretically derived expressions for the heat transfer coefficient are usually modified as a result of experimental studies to give

$$\bar{h}_L = 1.13\left[\frac{g\rho_l(\rho_l - \rho_v)k_l^3 h_{fg}}{\mu_l(T_{sat} - T_s)L}\right]^{1/4} \tag{3.2-110}$$

In this equation L is the length of the surface, and the enthalpy of vaporization h_{fg} is evaluated at T_{sat}. This equation can also be used for condensation on the inner or outer surface of a vertical tube if the film thickness is small compared to the diameter of the tube.

Turbulent Film Condensation

For low values of Reynolds number ($Re_\delta < 1800$), Eq. (3.2-110) is recommended to estimate the heat transfer coefficient for turbulent film condensation. When Re_δ exceeds 1800 the following equation is used:

$$\bar{h}_L = 0.0077\left[\frac{g\rho_l(\rho_l - \rho_v)k_l^3}{\mu_l^2}\right]Re_\delta^{0.4} \tag{3.2-111}$$

The Reynolds number is given by

$$Re_\delta = \frac{\rho_l V_m D_h}{\mu_l} = \frac{4\delta\rho_l V_m}{\mu_l} \tag{3.2-112}$$

where V_m is the average velocity of the film, D_h is the hydraulic diameter, and δ is the film thickness.

Film Condensation on the Outside and Inside Horizontal Tubes

For condensation on a single horizontal tube

$$\bar{h}_D = 0.728\left[\frac{g\rho_l(\rho_1 - \rho_v)k_l^3 h_{fg}}{\mu_l(T_{sat} - T_s)D}\right]^{1/4} \tag{3.2-113}$$

and for a vertical tier of N horizontal tubes

$$\bar{h}_D = 0.728\left[\frac{g\rho_l(\rho_1 - \rho_v)k_l^3 h_{fg}}{N\mu_l(T_{sat} - T_s)D}\right]^{1/4} \tag{3.2-114}$$

The reduction in the heat transfer coefficient is attributed to the increase in the average thickness of the film over the tubes.

The heat transfer by condensation inside a horizontal tube depends on the velocity of the vapor inside the tube. For low vapor velocities corresponding to Re less than 35,000 the following equation is recommended.

$$\bar{h}_D = 0.555 \left[\frac{g\rho_l(\rho_l - \rho_n)k_l^3 h_{fg}'}{\mu_l(T_{sat} - T_s)D} \right]^{1/4} \tag{3.2-115}$$

where

$$h_{fg}' = h_{fg} + \tfrac{3}{8}\left[c_{p,l}(T_{sat} - T_s) \right] \tag{3.2-116}$$

3.2-14 Thermal Radiation

The transfer mechanism of radiant heat energy is electromagnetic waves emitted from a radiating body due to thermal excitation of its atoms or molecules. In contrast to conduction and convection, heat transfer by thermal radiation does not require the presence of any transmitting medium. Thermal radiation is one example of electromagnetic radiation. Radio waves, light, X-rays, gamma rays, and so forth are other examples transmitted by the same phenomenon, and they differ from each other only by the wavelength or the frequency at which they propagate. Visible light, for example, has wavelengths ranging from 0.35 to 0.75 μm, while thermal radiation has wavelengths ranging from 0.1 to 100 μm. Electromagnetic waves follow the laws of optics and travel in straight lines through a transparent medium or vacuum, but not through opaque bodies. Figure 3.2-32 shows the spectrum of some electromagnetic waves that comprises several orders of magnitudes of frequency or wavelength. The electromagnetic spectrum is composed of monochromatic (single wavelength) radiations, each corresponding to periodic phenomena of frequency ν and a wavelength λ such that $\lambda = c/\lambda$ where c is the velocity of light in vacuum (2.998 $\times$ 10^8 m/s).

Thermal radiation has also been considered as the propagation of discrete quanta of energy (photons). The energy of each photon is $h\nu$, where h is Planck's constant (6.625 $\times$ 10^{-34} J · s) and ν the frequency of propagation.

Radiant heat transfer is the net electromagnetic radiant energy exchange that takes place between bodies due to temperature difference. Bodies at any temperature above the absolute zero emit a spectrum of electromagnetic waves in all directions. Part of the internal energy of the body is converted into electromagnetic waves that travel at the speed of light until they impinge on another body. The impinging energy is partially absorbed by the intercepting body, thus increasing its internal energy and consequently its temperature. Solids and liquids emit continuous spectrum of electromagnetic waves of all wavelengths and different intensities. Gases generally emit a characteristic line and/or band spectrum.

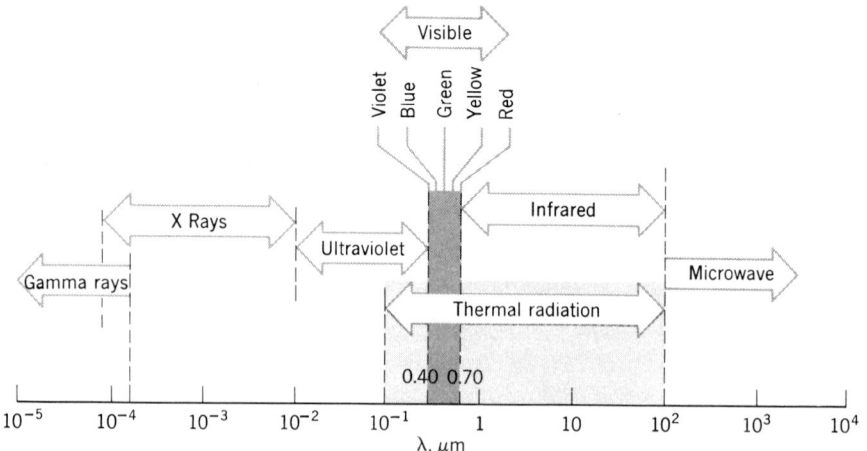

Fig. 3.2-32 Spectrum of electromagnetic radiation. (From ref. 1; used by permission.)

Blackbody Radiation

When a body intercepts radiant energy, part of the energy is absorbed, part is reflected, and part is transmitted. An energy balance thus yields $E = E_a + E_r + E_t$, where E is the total hemispherical emissive power defined as the rate at which radiation is emitted per unit area at all possible wavelengths and in all possible directions. Dividing by E gives

$$\frac{E_a}{E} + \frac{E_r}{E} + \frac{E_t}{E} = 1 \qquad (3.2\text{-}117)$$

or

$$\alpha + \rho + \tau = 1 \qquad (3.2\text{-}118)$$

where α is the absorptivity, ρ the reflectivity, and τ the transmissivity. Each of these terms represent the fractions of the incident energy that is absorbed, reflected, and transmitted by the body, and they are the integrated average of their spectral values over both direction and wavelengths.

If the body receiving radiant energy is opaque, no radiant energy can be transmitted through it, and all the incident energy has to be absorbed or reflected by the body. In this case

$$\alpha + \rho = 1 \qquad (3.2\text{-}119)$$

The concept of a blackbody was conceived by Kirchhoff as one that absorbs all radiation incident upon it at all wavelengths and directions ($\alpha = 1$) and reflects or transmits none ($\rho = 0$, $\tau = 0$). The blackbody can also be defined as one which at a certain temperature, emits the maximum possible amount of radiation at all wavelengths. Such a body can be conceived as a hollow cavity provided with a very small aperture. Incident radiation entering the cavity through the opening is continuously reflected and absorbed and never leaves the cavity. In this sense the radiation is totally absorbed by the cavity, and the area of the aperture can be considered as a surface of absorbtivity $\alpha = 1$. This is in accordance with the definition of a blackbody as a perfect absorber. The alternative definition of a blackbody as a perfect emitter can be visualized if a blackbody is placed within the cavity, its temperature being the same as that of the cavity. The blackbody, by definition, will absorb all the energy radiated from the cavity that impinges on it; but since its temperature has to remain the same, the rate at which it emits radiation must be equivalent to the rate at which it absorbs energy from the cavity. Thus the radiant energy leaving the cavity through the aperture is only due to emittance from the cavity. The area of the aperture can then be considered as a surface of emissivity $\varepsilon = 1$. It is also noted that the amount of energy emitted by the blackbody and its spectral distribution is independent of the nature of the blackbody but rather depends on its temperature and wavelength. Therefore it is concluded that a blackbody is a perfect emitter as well as a perfect absorber.

Emissivity and Absorptivity

The thermal radiation of nonblackbodies depends on the nature of the emitting source as well as on its absolute temperature. A rough surface, for example, emits more radiant energy than a smooth surface of the same material at the same temperature. A graybody is defined such that its emissivity (and absorbtivity) is less than 1 but is independent of wavelength.

The total emissive power E is defined as the time rate of the energy emitted from a body per unit area of its surface. The value of E depends on the wavelength λ, and at a particular λ, E_λ is called the monochromatic emissive power. Since a body at a certain temperature radiates energy at all wavelengths, the total emissive power is obtained by integrating E over all the wavelengths or

$$E = \int_{\lambda=0}^{\lambda=\infty} E_\lambda \, d\lambda \qquad (3.2\text{-}120)$$

For a blackbody the emissive power $E_{b\lambda}$ at a certain wavelength λ is a maximum, and the ratio $E_\lambda/E_{b\lambda}$ is called the monochromatic emissivity ε_λ or

$$\varepsilon_\lambda = \frac{E_\lambda}{E_{b\lambda}} \qquad (3.2\text{-}121)$$

The total emissivity is defined as

$$\varepsilon = \frac{E}{E_b} = \frac{1}{E_b} \int_{\lambda=0}^{\lambda=\infty} \varepsilon_\lambda E_{b\lambda} \, d\lambda \qquad (3.2\text{-}122)$$

An energy balance between two bodies exchanging radiation gives

$$A_1(\delta\dot{q}\alpha_{1\lambda}) = A_1 E_{1\lambda}$$

and

$$A_2(\delta\dot{q}\alpha_{2\lambda}) = A_2 E_{2\lambda}$$

where $\delta\dot{q}$ is the rate at which the radiant energy is exchanged between the two bodies per unit area, α_λ is the monochromatic coefficient of absorption, and subscripts 1 and 2 refer to the two bodies.

The monochromatic absorptivity α_λ is defined as the ratio of the absorbed energy to the incident energy at a wavelength λ. The total absorptivity α is the fraction of energy incident on a surface that is absorbed. It is a factor less than unity and its value depends on λ, the absolute temperature, the nature of the medium, and the direction along which radiant energy arrives at the body. From the above two equations

$$\frac{E_{1\lambda}}{\alpha_{1\lambda}} = \frac{E_{2\lambda}}{\alpha_{2\lambda}} = \frac{E_\lambda}{\alpha_\lambda} = \text{const} \qquad (3.2\text{-}123)$$

Equation (3.2-123) is called Kirchhoff's law for monochromatic radiation. It states that the ratio of the emissive power to the absorptivity is a constant for all thermal radiators at the same temperature and at any wavelength. It also follows that the radiation emitted by a graybody is equal to the radiation emitted by a blackbody at the same temperature and wavelength multiplied by α_λ. Since α is always less than unity, the radiant energy of a graybody is always less than that of a blackbody at the same temperature. By integrating Eq. (3.2-123) over the whole range of wavelengths from $\lambda = 0$ to $\lambda = \infty$, it is easily seen that Kirchhoff's law holds also for the total radiation or

$$\frac{E_1}{\alpha_1} = \frac{E_2}{\alpha_2} = \frac{E}{\alpha} \qquad (3.2\text{-}124)$$

The total emissivity is defined as the ratio of the emissive power of the body to that of a blackbody at the same temperature:

$$\varepsilon = \frac{E}{E_b} \qquad (3.2\text{-}125)$$

If one of the bodies were black ($\alpha = 1$), then

$$\frac{E}{E_b} = \alpha$$

From the above equation and Eq. (3.2-124) it follows that at thermal equilibrium

$$\varepsilon = \alpha \qquad (3.2\text{-}126)$$

That is, the total (integrated value over all wavelengths and directions) emissivity and absorptivity of gray bodies are equal at the same temperature, which is another statement of Kirchhoff's law. Table 3.2-22 gives average values of emissivities for different materials. Note that the emissivities of nonconductors for different materials are higher than those of metallic conductors.

Radiation of a Blackbody

The distribution of the emissive power of a blackbody with respect to wavelength is given by Planck's law:

$$E_{b\lambda} = \frac{C_1}{\lambda^5(e^{C_2/\lambda T} - 1)} \qquad (3.2\text{-}127)$$

where $E_{b\lambda}$ = monochromatic emissive power of a blackbody
λ = wavelength, μm
T = absolute temperature, K
$C_1 = 3.743 \times 10^8 \ \text{W} \cdot \mu\text{m}^4/\text{m}^2$
$C_2 = 1.4387 \times 10^4 \ \mu\text{m} \cdot \text{K}$

TABLE 3.2-22 TOTAL, NORMAL (n) OR HEMISPHERICAL (h) EMISSIVITY OF SELECTED MATERIALS[a]

Nonmetallic Solids

DESCRIPTION/COMPOSITION		TEMPERATURE (K)	EMISSIVITY ε
Aluminum oxide	(n)	600	0.69
		1000	0.55
		1500	0.41
Asphalt pavement	(h)	300	0.85 0.93
Building materials			
Asbestos sheet	(h)	300	0.93 0.96
Brick, red	(h)	300	0.93 0.96
Gypsum or plaster board	(h)	300	0.90 0.92
Wood	(h)	300	0.82 0.92
Cloth	(h)	300	0.75 0.90
Concrete	(h)	300	0.88 0.93
Glass, window	(h)	300	0.90 0.95
Ice	(h)	273	0.95 0.98
Paints			
Black (Parsons)	(h)	300	0.98
White, acrylic	(h)	300	0.90
White, zinc oxide	(h)	300	0.92
Paper, white	(h)	300	0.92 0.97
Pyrex	(n)	300	0.82
		600	0.80
		1000	0.71
		1200	0.62
Pyroceram	(n)	300	0.85
		600	0.78
		1000	0.69
		1500	0.57
Refractories (furnace liners)			
Alumina brick	(n)	800	0.40
		1000	0.33
		1400	0.28
		1600	0.33
Magnesia brick	(n)	800	0.45
		1000	0.36
		1400	0.31
		1600	0.40
Kaolin insulating brick	(n)	800	0.70
		1200	0.57
		1400	0.47
		1600	0.53
Sand	(h)	300	0.90
Silicon carbide	(n)	600	0.87
		1000	0.87
		1500	0.85
Skin	(h)	300	0.95
Snow	(h)	273	0.82 0.90
Soil	(h)	300	0.93 0.96
Rocks	(h)	300	0.88 0.95
Teflon	(h)	300	0.85
		400	0.87
		500	0.92
Vegetation	(h)	300	0.92 0.96
Water	(h)	300	0.96

(*Continued*)

TABLE 3.2-22 (*Continued*)

Metallic Solids and Their Oxides[a]

DESCRIPTION/COMPOSITION		ε_n or ε_h at various temperatures (K)										
		100	200	300	400	600	800	1000	1200	1500	2000	2500
Aluminum												
Highly polished, film	(h)	0.02	0.03	0.04	0.05	0.06						
Foil, bright	(h)	0.06	0.06	0.07								
Anodized	(h)			0.82	0.76							
Chromium												
Polished or plated	(n)	0.05	0.07	0.10	0.12	0.14						
Copper												
Highly polished	(h)			0.03	0.03	0.04	0.04	0.04				
Stably oxidized	(h)					0.50	0.58	0.80				
Gold												
Highly polished or film	(h)	0.01	0.02	0.03	0.03	0.04	0.05	0.06				
Foil, bright	(h)	0.06	0.07	0.07								
Molybdenum												
Polished	(h)					0.06	0.08	0.10	0.12	0.15	0.21	0.26
Shot-blasted, rough	(h)					0.25	0.28	0.31	0.35	0.42		
Stably oxidized	(h)					0.80	0.82					
Nickel												
Polished	(h)					0.09	0.11	0.14	0.17			
Stably oxidized	(h)					0.40	0.49	0.57				
Platinum												
Polished	(h)						0.10	0.13	0.15	0.18		
Silver												
Polished	(h)			0.02	0.02	0.03	0.05	0.08				
Stainless steels												
Typical, polished	(n)			0.17	0.17	0.19	0.23	0.30				
Typical, cleaned	(n)			0.22	0.22	0.24	0.28	0.35				
Typical, lightly oxidized	(n)						0.33	0.40				
Typical, highly oxidized	(n)						0.67	0.70	0.76			
AISI 347, stably oxidized	(n)					0.87	0.88	0.89	0.90			
Tantalum												
Polished	(h)								0.11	0.17	0.23	0.28
Tungsten												
Polished	(h)							0.10	0.13	0.18	0.25	0.29

[a]See refs. 2, 7, 28, and 29.

Figure 3.2-33 shows $E_{b\lambda}$ versus λ for various temperatures. The expression $E_\lambda\,d\lambda$ represents the quantity of heat radiated per unit area and unit time from a blackbody between the wavelengths λ and $\lambda + d\lambda$. The peak points of the isotherms of the curves of Fig. 3.2-33 shift toward shorter wavelengths as the temperature increases. Differentiating Eq. (3.2-127) with respect to λ and equating to zero gives the value of λ, which corresponds to the maximum emissive power:

$$\lambda_{max}T = 2897.6\ \mu m \cdot K \tag{3.2-128}$$

where λ_{max} is the wavelength corresponding to the maximum value of $E_{b\lambda}$ and T is the absolute temperature. Equation (3.2-128) is called Wien's displacement law. Integration of Eq. (3.2-127) over all wavelengths gives the Stefan–Boltzmann law, which states that the emissive power of a blackbody is proportional to the fourth power of its absolute temperature or

$$E_b = \int_{\lambda=0}^{\lambda=\infty} E_{b\lambda}\,d\lambda = \sigma T^4 \tag{3.2-129}$$

where $\sigma=$ Stefan–Boltzmann constant $= 5.699 \times 10^{-8}$ W/m²·K⁴
  $E_b=$ emissive power for blackbody, W/m²
  $T=$ absolute temperature, K

If a blackbody at temperature T is placed in an enclosure whose walls at a temperature T_0, the net interchange of radiation is given by

$$\frac{q}{A} = \sigma\left(T^4 - T_0^4\right) \tag{3.2-130}$$

If the body is gray and has an emissivity ε and absorptivity α, the above equation becomes

$$q_{net} = \sigma A \varepsilon T^4 - \sigma A \alpha T_0^4$$

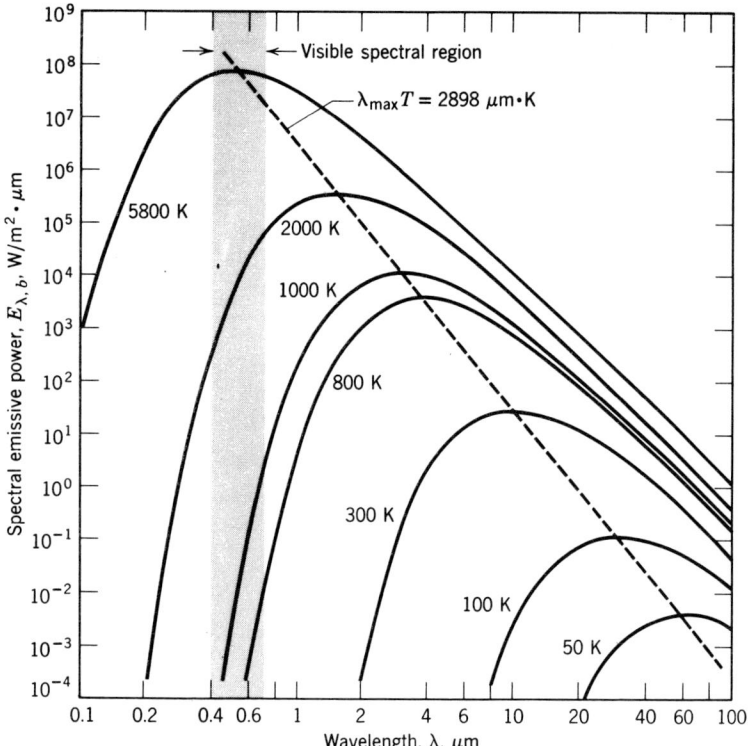

Fig. 3.2-33 Spectral blackbody emissive power. (From ref. 1; used by permission.)

TABLE 3.2-23 RADIATION FUNCTIONS[a]

λT		$E_{b\lambda}/T^5$		
$\mu m \cdot °R$	$\mu m \cdot K$	$\dfrac{Btu}{h \cdot ft^2 \cdot °R^5 \cdot \mu m}$ $\times 10^{15}$	$\dfrac{W}{m^2 \cdot K^5 \cdot \mu m}$ $\times 10^{11}$	$\dfrac{E_{b_{0-\lambda T}}}{\sigma T^4}$
1,000	555.6	0.000671	0.400×10^{-5}	0.170×10^{-7}
1,200	666.7	0.0202	0.120×10^{-3}	0.756×10^{-6}
1,400	777.8	0.204	0.00122	0.106×10^{-4}
1,600	888.9	1.057	0.00630	0.738×10^{-4}
1,800	1,000.0	3.544	0.02111	0.321×10^{-3}
2,000	1,111.1	8.822	0.05254	0.00101
2,200	1,222.2	17.776	0.10587	0.00252
2,400	1,333.3	30.686	0.18275	0.00531
2,600	1,444.4	47.167	0.28091	0.00983
2,800	1,555.6	66.334	0.39505	0.01643
3,000	1,666.7	87.047	0.51841	0.02537
3,200	1,777.8	108.14	0.64404	0.03677
3,400	1,888.9	128.58	0.76578	0.05059
3,600	2,000.0	147.56	0.87878	0.06672
3,800	2,111.1	164.49	0.97963	0.08496
4,000	2,222.2	179.04	1.0663	0.10503
4,200	2,333.3	191.05	1.1378	0.12665
4,400	2,444.4	200.51	1.1942	0.14953
4,600	2,555.6	207.55	1.2361	0.17337
4,800	2,666.7	212.32	1.2645	0.19789
5,000	2,777.8	215.06	1.2808	0.22285
5,200	2,888.9	216.00	1.2864	0.24803
5,400	3,000.0	215.39	1.2827	0.27322
5,600	3,111.1	213.46	1.2713	0.29825
5,800	3,222.2	210.43	1.2532	0.32300
6,000	3,333.3	206.51	1.2299	0.34734
6,200	3,444.4	201.88	1.2023	0.37118
6,400	3,555.6	196.69	1.1714	0.39445
6,600	3,666.7	191.09	1.1380	0.41708
6,800	3,777.8	185.18	1.1029	0.43905
7,000	3,888.9	179.08	1.0665	0.46031
7,200	4,000.0	172.86	1.0295	0.48085
7,400	4,111.1	166.60	0.99221	0.50066
7,600	4,222.2	160.35	0.95499	0.51974
7,800	4,333.3	154.16	0.91813	0.53809
8,000	4,444.4	148.07	0.88184	0.55573
8,200	4,555.6	142.10	0.84629	0.57267
8,400	4,666.7	136.28	0.81163	0.58891
8,600	4,777.8	130.63	0.77796	0.60449
8,800	4,888.9	125.15	0.74534	0.61941
9,000	5,000.0	119.86	0.71383	0.63371
9,200	5,111.1	114.76	0.68346	0.64740
9,400	5,222.2	109.85	0.65423	0.66051
9,600	5,333.3	105.14	0.62617	0.67305
9,800	5,444.4	100.62	0.59925	0.68506
10,000	5,555.6	96.289	0.57346	0.69655
10,200	5,666.7	92.145	0.54877	0.70754
10,400	5,777.8	88.181	0.52517	0.71806

TABLE 3.2-23 (*Continued*)

λT		$E_{b\lambda}/T^5$		$\dfrac{E_{b_{0-\lambda T}}}{\sigma T^4}$
$\mu m \cdot {}^\circ R$	$\mu m \cdot K$	$\dfrac{\text{Btu}}{h \cdot ft^2 \cdot {}^\circ R^5 \cdot \mu m}$ $\times 10^{15}$	$\dfrac{W}{m^2 \cdot K^5 \cdot \mu m}$ $\times 10^{11}$	
10,600	5,888.9	84.394	0.50261	0.72813
10,800	6,000.0	80.777	0.48107	0.73777
11,000	6,111.1	77.325	0.46051	0.74700
11,200	6,222.2	74.031	0.44089	0.75583
11,400	6,333.3	70.889	0.42218	0.76429
11,600	6,444.4	67.892	0.40434	0.77238
11,800	6,555.6	65.036	0.38732	0.78014
12,000	6,666.7	62.313	0.37111	0.78757
12,200	6,777.8	59.717	0.35565	0.79469
12,400	6,888.9	57.242	0.34091	0.80152
12,600	7,000.0	54.884	0.32687	0.80806
12,800	7,111.1	52.636	0.31348	0.81433
13,000	7,222.2	50.493	0.30071	0.82035
13,200	7,333.3	48.450	0.28855	0.82612
13,400	7,444.4	46.502	0.27695	0.83166
13,600	7,555.6	44.645	0.26589	0.83698
13,800	7,666.7	42.874	0.25534	0.84209
14,000	7,777.8	41.184	0.24527	0.84699
14,200	7,888.9	39.572	0.23567	0.85171
14,400	8,000.0	38.033	0.22651	0.85624
14,600	8,111.1	36.565	0.21777	0.86059
14,800	8,222.2	35.163	0.20942	0.86477
15,000	8,333.3	33.825	0.20145	0.86880
16,000	8,888.9	27.977	0.16662	0.88677
17,000	9,444.4	23.301	0.13877	0.90168
18,000	10,000.0	19.536	0.11635	0.91414
19,000	10,555.6	16.484	0.09817	0.92462
20,000	11,111.1	13.994	0.08334	0.93349
21,000	11,666.7	11.949	0.07116	0.94104
22,000	12,222.2	10.258	0.06109	0.94751
23,000	12,777.8	8.852	0.05272	0.95307
24,000	13,333.3	7.676	0.04572	0.95788
25,000	13,888.9	6.687	0.03982	0.96207
26,000	14,444.4	5.850	0.03484	0.96572
27,000	15,000.0	5.139	0.03061	0.96892
28,000	15.555.6*	4.532	0.02699	0.97174
29,000	16,111.1	4.012	0.02389	0.97423
30,000	16,666.7	3.563	0.02122	0.97644
40,000	22,222.2	1.273	0.00758	0.98915
50,000	27,777.8	0.560	0.00333	0.99414
60,000	33,333.3	0.283	0.00168	0.99649
70,000	38,888.9	0.158	0.940×10^{-3}	0.99773
80,000	44,444.4	0.0948	0.564×10^{-3}	0.99845
90,000	50,000.0	0.0603	0.359×10^{-3}	0.99889
100,000	55,555.6	0.0402	0.239×10^{-3}	0.99918

[a]See ref. 30.

or

$$E = \sigma\left(\varepsilon T^4 - \alpha T_0^4\right) \tag{3.2-131}$$

Note that ε is the emissivity of the particular surface at T, and the radiation incident on that surface depends on an emitting body at T_0. But Kirchhoff's law states that $\alpha = \varepsilon$ for a graybody; therefore, the above equation becomes

$$E = \varepsilon\sigma\left(T^4 - T_0^4\right) \tag{3.2-132}$$

For graybodies the values of α and ε are independent of temperature and wavelength.

Band Emission

The fraction of the total energy radiated between a wavelength interval (band) from 0 to λ is,

$$F_{0 \to \lambda} = \frac{E_{b,0 \to \lambda}}{E_{b,0 \to \infty}} = \frac{\int_0^\lambda E_{b\lambda}\, d\lambda}{\int_0^\infty E_{b\lambda}\, d\lambda} \tag{3.2-133}$$

Dividing Eq. (3.2-127) by T^5 gives

$$\frac{E_{b\lambda}}{T^5} = \frac{C_1}{(\lambda T)^5\left(e^{C_2/\lambda T} - 1\right)} \tag{3.2-134}$$

Hence for a given temperature the integrals of Eq. (3.2-133) are expressed in terms of one variable λT. The radiant energy emitted between two wavelengths is given by

$$E_{b,\lambda_1 \to \lambda_2} = E_{b,0 \to \infty}\left(\frac{E_{b,0 \to \lambda_2}}{E_{b,0 \to \infty}} - \frac{E_{b,0 \to \lambda_1}}{E_{b,0 \to \infty}}\right) \tag{3.2-135}$$

where $E_{b,0 \to \infty} = \sigma T^4$. The results are tabulated in Table 3.2-23.

3.2-15 Radiation View Factor

Referring to Fig. 3.2-34, the radiation view factor F_{12} (also called configuration or shape factor) from surface 1 to surface 2, is defined as the fraction of the total radiant flux leaving surface 1 that is intercepted by surface 2. It is a geometric function having the limiting values of zero and unity, and is given by

$$dA\, dF_{dA_1 - dA_2} = \frac{I(\theta_1)\cos\theta_1\, dA_1\, d\omega_1}{\pi I_m} \tag{3.2-136}$$

where $I(\theta)$ is the intensity at the angle θ, that is, the radiation emitted per unit area and per unit solid angle at the angle θ. The value of $I(\theta_1)$ depends on direction unless the surface is perfectly diffuse. Figure 3.2-35 gives the directional emittance for several materials.

I_m is the mean value of intensity defined by

$$I_m = \frac{E}{\pi} \tag{3.2-137}$$

A directional distribution function $D(\theta)$ can then be defined as

$$D(\theta) = \frac{I(\theta)}{I_m} \tag{3.2-138}$$

When D is independent of θ, Lambert's cosine principle holds. This principle states the function $D(\theta)$ is a constant equal to unity and invariable with θ.

In Eq. (3.2-136) $d\omega_1$ is the solid angle subtended by dA_2 at dA_1 and is given by

$$d\omega_1 = \frac{dA_2\cos\theta_2}{r^2} \tag{3.2-139}$$

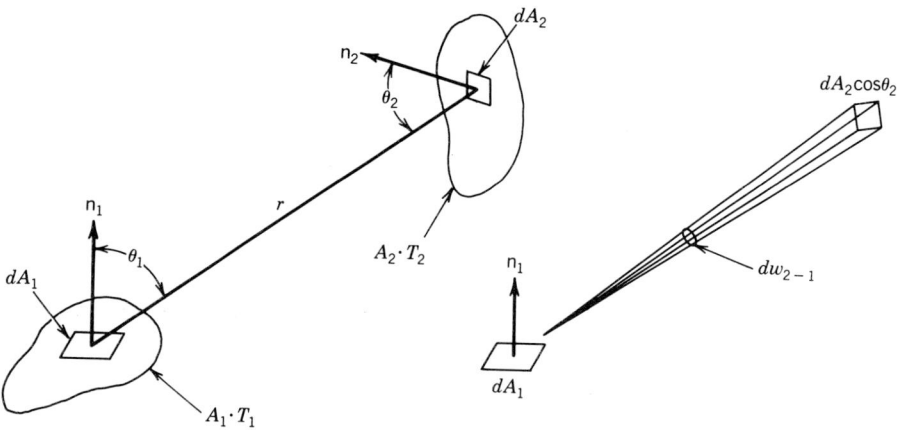

Fig. 3.2-34 View factor associated with radiation exchange between elemental surfaces of area dA and dA.

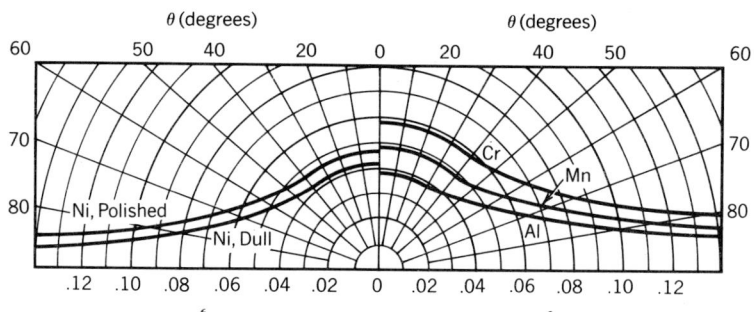

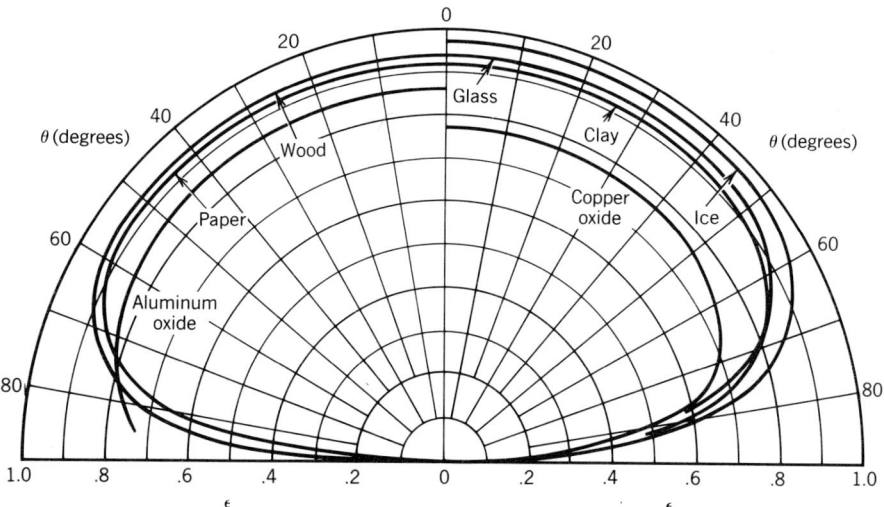

Fig. 3.2-35 Distribution of the total directional emittance for several materials.

Combining Eqs. (3.2-136) and (3.2-139) gives

$$dA_1 \, dF_{dA_1 \to dA_2} = D_1(\theta_1) \frac{\cos\theta_1 \cos\theta_2 \, dA_1 \, dA_2}{\pi r^2}$$

$$dA_2 \, dF_{dA_2 \to dA_1} = D_2(\theta_2) \frac{\cos\theta_1 \cos\theta_2 \, dA_1 \, dA_2}{\pi r^2}$$

If $D_1(\theta_1) = D_2(\theta_2)$, then

$$dA_1 \, dF_{dA_1 \to dA_2} = dA_2 \, dF_{dA_2 \to dA_1} \qquad (3.2\text{-}140)$$

Equation (3.2-140) is known as the reciprocity relation. If the Lambertian distribution is postulated ($D = 1$), then $F_{dA_1 \to A_2}$ is

$$dA_1 \, dF_{dA_1 \to A_2} = dA_1 \int_{A_2} dF_{dA_1 \to dA_2}$$

$$= dA_1 \int \frac{\cos\theta_1 \cos\theta_2 \, dA_2}{\pi r^2} \qquad (3.2\text{-}141)$$

Integration of this equation over A_1 gives $F_{A_1 \to A_2}$ (or F_{12}) so that

$$F_{12} = \frac{1}{A_1} \int_{A_1} \int_{A_2} \frac{\cos\theta_1 \cos\theta_2}{\pi r^2} \, dA_1 \, dA_2 \qquad (3.2\text{-}142)$$

Similarly F_{21} is given by

$$F_{21} = \frac{1}{A_2} \int_{A_1} \int_{A_2} \frac{\cos\theta_1 \cos\theta_2}{\pi r^2} \, dA_1 \, dA_2 \qquad (3.2\text{-}143)$$

The integrated form of the reciprocity relation is

$$A_1 F_{12} = A_2 F_{12} \qquad (3.2\text{-}144)$$

Also from the definition of the view factor,

$$\sum_{i=1}^{N} F_{ij} = 1 \qquad (3.2\text{-}145)$$

where F_{ij} is the fraction of the total radiation leaving surface i, which arrives at surface j, and N is the number of surfaces in an enclosure.

The net energy exchange between the two black surfaces is $q_{12} = A_1 F_{12} E_{b1} - A_2 f_{21} E_{b2}$. But $A_1 F_{12} = A_2 F_{21}$, so that

$$q_{12} = A_1 F_{12} \sigma \left(T_1^4 - T_2^4 \right)$$

$$= \sigma \left(T_1^4 - T_2^4 \right) \int_{A_1} \int_{A_2} \frac{\cos\theta_1 \cos\theta_2}{\pi r^2} \, dA_1 \, dA_2$$

In an enclosure of black surfaces, the net transfer of radiation from surface i to N surfaces maintained at different temperatures in the enclosure is

$$q_{i,\text{net}} = \sum_{i=1}^{N} A_i F_{ij} \sigma \left(T_i^4 - T_j^4 \right) \qquad (3.2\text{-}146)$$

For the spherical surfaces shown in Fig. 3.2-36 $F_{12} = 1$ and $F_{11} + F_{12} = 1$ or $F_{11} = 0$ (surface 1 does not see itself). Also, $F_{21} + F_{22} = 1$, but from the reciprocity relation

$$F_{21} = \frac{A_1}{A_2} F_{12} = \frac{A_1}{A_2}$$

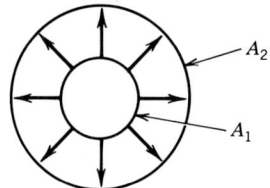

Fig. 3.2-36 View factor for concentric spherical surfaces.

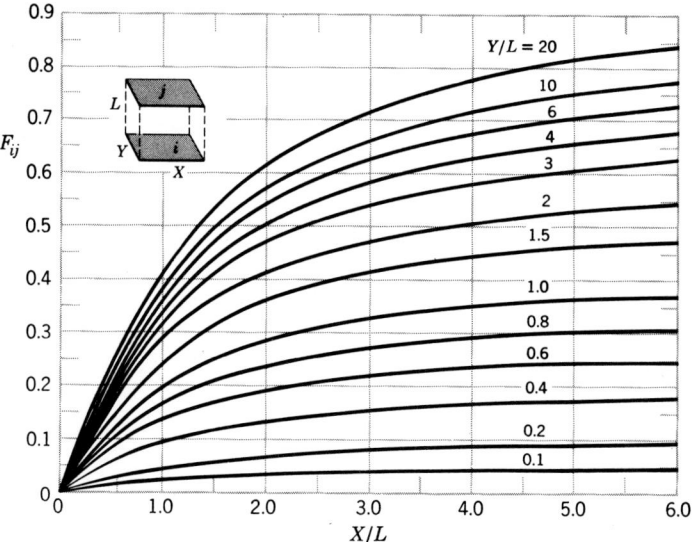

Fig. 3.2-37 View factor for aligned parallel rectangles. (From ref. 1; used by permission.)

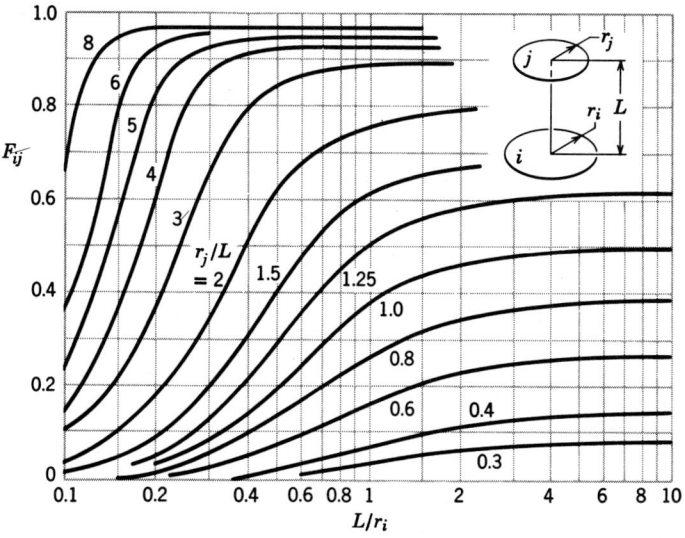

Fig. 3.2-38 View factor for coaxial parallel disks. (From ref. 1; used by permission.)

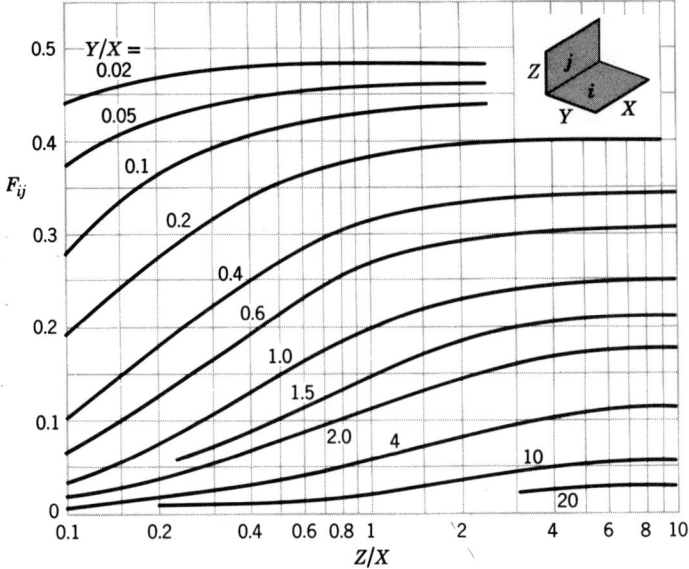

Fig. 3.2-39 View factor for perpendicular rectangles with a common edge. (From ref. 1; used by permission.)

so that

$$F_{22} = 1 - \frac{A_1}{A_2}$$

View factors for several geometries are given in Figs. 3.2-37 to 3.2-40. Note that by using the reciprocity relations other shape factors can be obtained. As an example the shape factor between areas A_1 and A_3 of Fig. 3.2-41 can be determined as follows:

$$F_{1-2,3} = F_{1-2} + F_{1-3}$$

$F_{1-1,3}$ and F_{1-2} can be determined from Fig. 3.2-39 so that F_{1-3} can be calculated from the dimensions given.

3.2-16 Heat Exchange between Nonblackbodies

Expressions for the heat exchange between nonblack surfaces can be obtained using electrical analogy. Two terms are first defined:

1. Irradiation G is the total radiation incident on a surface per unit time and unit area.
2. Radiosity J is the total radiation leaving a surface per unit time and unit area.

Assuming no energy transmitted through the surface ($\tau = 0$), the radiosity, equal to the sum of the radiation emitted and the radiation reflected, is

$$J = \varepsilon E_b + \rho G$$

where

$$\rho = 1 - \alpha = 1 - \varepsilon \qquad\qquad (3.2\text{-}147)$$

The net energy transfer from the surface is

$$\frac{q}{A} = J - G = \varepsilon E_b + (1 - \varepsilon)G$$

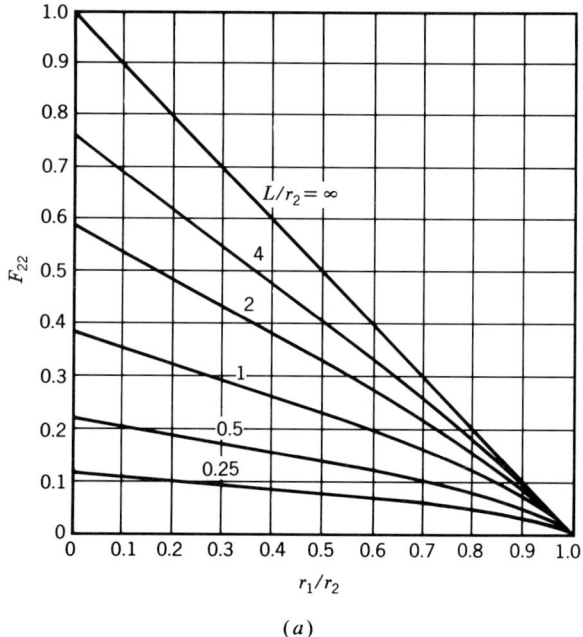

(a)

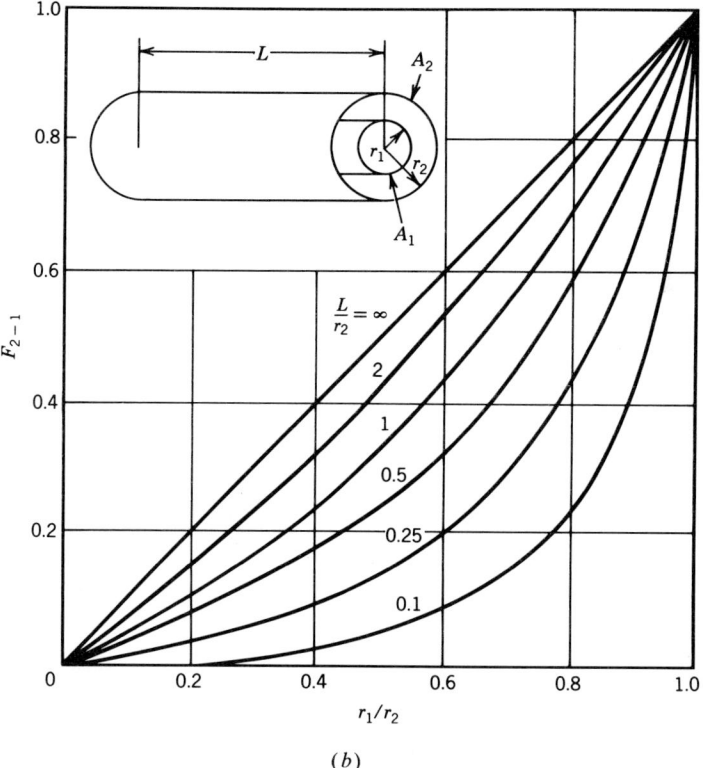

(b)

Fig. 3.2-40 Radiation shape factors for two concentric cylinders of finite length (a) outer cylinder to itself and (b) outer cylinder to inner cylinder. (From ref. 30; used by permission.)

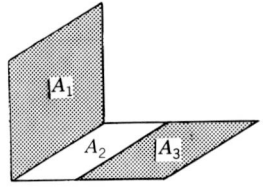

Fig. 3.2-41 Application of radiation view factor algebra to rectangular surfaces.

or

$$q = (E_b - J)/\frac{(1 - \varepsilon)}{\varepsilon A} \qquad (3.2\text{-}148)$$

The net heat exchange between two nonblack surfaces is

$$q_{1-2} = J_1 A_1 F_{12} - J_2 A_2 F_{21}$$

But $A_1 F_{12} = A_2 F_{21}$ (reciprocity relation), so that

$$q = \frac{J_1 - J_2}{1/A_1 F_{12}} \qquad (3.2\text{-}149)$$

Equations (3.2-148) and (3.2-149) can be represented, analogous to electrical systems, by the radiation network shown in Figure 3.2-42.

The net heat exchange, based on the overall potential, is

$$q_{\text{net}} = \frac{E_{b_1} - E_{b_2}}{\dfrac{1 - \varepsilon_1}{\varepsilon_1 A_1} + \dfrac{1}{A_1 F_{12}} + \dfrac{1 - \varepsilon_2}{\varepsilon_2 A_2}}$$

$$= \frac{\sigma\left(T_1^4 - T_2^4\right)}{\dfrac{1 - \varepsilon_1}{\varepsilon_1 A_1} + \dfrac{1}{A_1 F_{12}} + \dfrac{1 - \varepsilon_2}{\varepsilon_2 A_2}} \qquad (3.2\text{-}150)$$

For two infinite parallel plates ($A_1 = A_2$ and $F_{12} = 1$), this expression becomes

$$q = \frac{\sigma A\left(T_1^4 - T_2^4\right)}{\dfrac{1}{\varepsilon_1} + \dfrac{1}{\varepsilon_2} - 1} \qquad (3.2\text{-}151)$$

For two long concentric cylinders

$$q = \frac{\sigma A_1\left(T_1^4 - T_2^4\right)}{\dfrac{1}{\varepsilon_1} + \dfrac{A_1}{A_2}\left(\dfrac{1}{\varepsilon_2} - 1\right)} \qquad (3.2\text{-}152)$$

where subscripts 1 and 2 apply to the inner and outer cylinders, respectively.

In order to reduce radiation between surfaces, one or more radiation shields made of low-emissivity (high-reflectivity) material is used. For parallel and equal surfaces, the heat exchange is

$$\left(\frac{q}{A}\right)_{\text{with shields}} = \frac{1}{n + 1}\left(\frac{q}{A}\right)_{\text{without shields}} \qquad (3.2\text{-}153)$$

where n is the number of shields placed between the two surfaces.

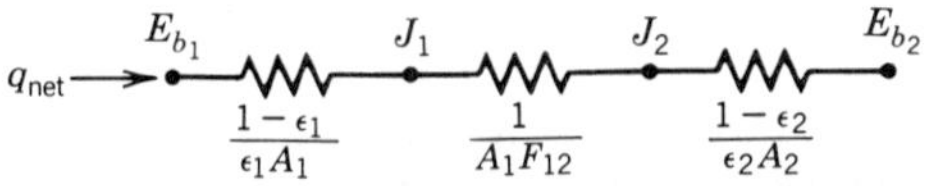

Fig. 3.2-42 Network representation of radiative exchange between nonblackbodies.

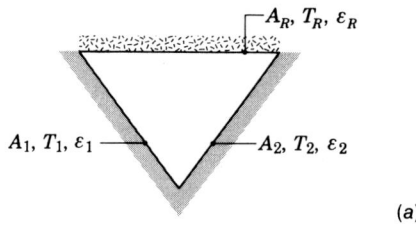

(a)

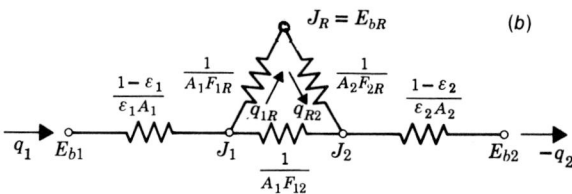

Fig. 3.2-43 A three-surface enclosure with one surface reradiating: (*a*) schematic and (*b*) network representation.

Reradiating Surface

The heat transfer between two surfaces in the presence of a reradiating surface, that is, one which does not exchange heat, is

$$q_{net} = \frac{\sigma A_1 \left(T_1^4 - T_2^4 \right)}{\left(\dfrac{1}{\varepsilon_1} - 1 \right) + \dfrac{A_1 + A_2 - 2 A_1 F_{12}}{A_2 - A_1 \left(F_{12} \right)^2} + \dfrac{A_1}{A_2} \left(\dfrac{1}{\varepsilon_2} - 1 \right)} \qquad (3.2\text{-}154)$$

In the derivation of this equation, use is made of the following relations:

$$F_{1R} = 1 - F_{12}, \qquad F_{2R} = 1 - F_{21}, \qquad A_1 F_{12} = A_2 F_{21}$$

where subscript R refers to the reradiating surface. The corresponding network is shown in Fig. 3.2-43. This equation applies only to surfaces which do not see themselves ($F_{11} = F_{22} = 0$).

The surface temperature of the reradiating surface at equilibrium is given by

$$T_R = \sqrt[4]{\frac{\left(A_1 - A_1 F_{12} \right) T_1^4 + \left(A_2 - A_1 F_{12} \right) T_2^4}{\left(A_1 - A_1 F_{12} \right) + \left(A_2 - A_1 F_{12} \right)}} \qquad (3.2\text{-}155)$$

3.2-17 Radiation from Nonluminous Gases

Several gases such as H_2O and CO_2 absorb and emit radiation, but unlike solids or liquids, they do so in certain narrow wavelength intervals or bands. Figures 3.2-44 and 3.2-45 give the gas emittances for CO_2 and H_2O for a total pressure of 1 atm as a function of temperature and the product pL, where p is the partial pressure of the gas and L is the mean beam length or the radius of a hemispherical gas mass. An approximation of the mean beam length is given by

$$L_e = 3.6 \frac{V}{A} \qquad (3.2\text{-}156)$$

where V is the total volume of the gas and A is the total surface area. Values of L_e for some geometries are given in Table 3.2-24. Figures 3.2-46 and 3.2-47 give correction factors for CO_2 and H_2O emissivities for total pressures other than 1 atm. Figure 3.2-48 gives an additional correction

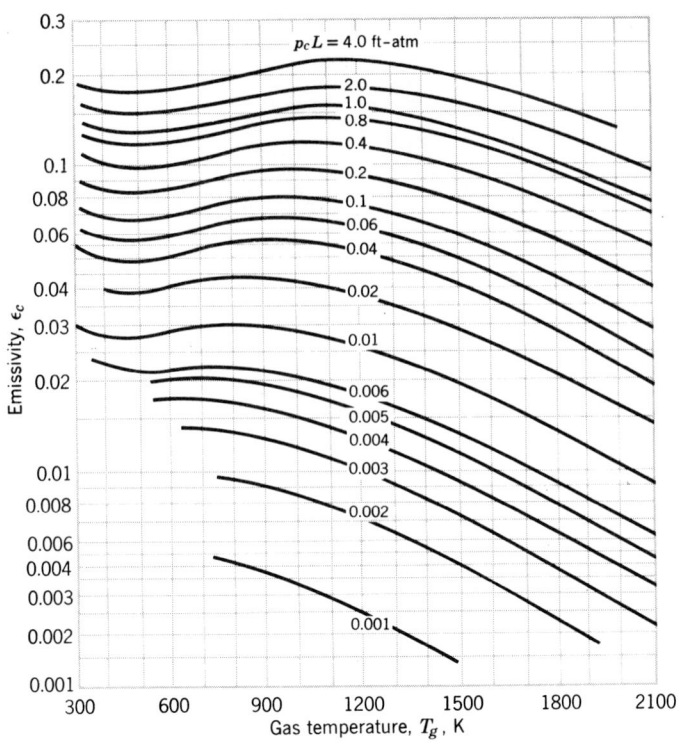

Fig. 3.2-44 Emissivity of carbon dioxide in a mixture with nonradiating gases at 2 atm total pressure and of hemispherical shape. (From ref. 31; used by permission.)

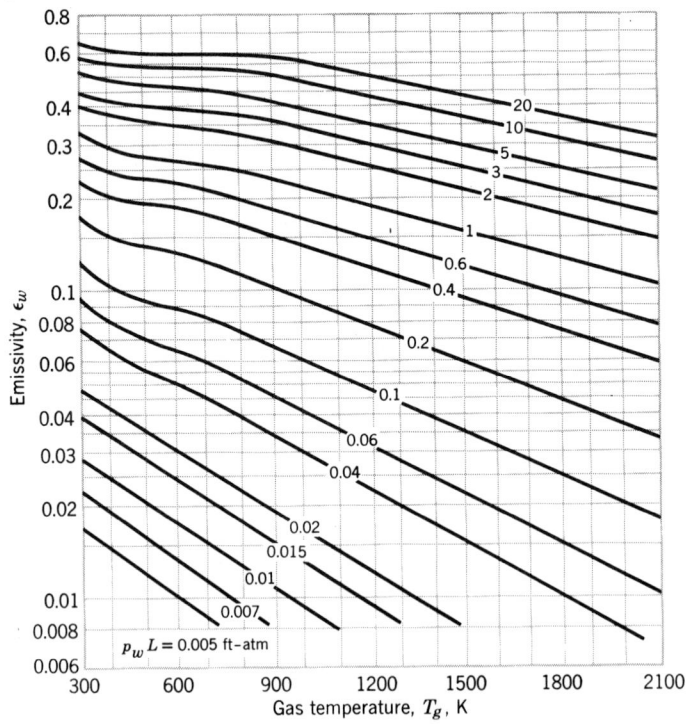

Fig. 3.2-45 Emissivity of water vapor in a mixture with nonradiating gases at 1 atm total pressure and of hemispherical shape. (From ref. 31; used by permission.)

TABLE 3.2-24 MEAN EQUIVALENT LENGTH L, FOR RADIATION FROM ENTIRE GAS VOLUME[a]

Gas volume	Characteristic dimension	L_e
Volume between two infinite planes	Separation distance L	$1.8L$
Circular cylinder with the height = diameter, radiation to center of base	Diameter D	$0.71D$
Hemisphere, radiation to element in center of base	Radius R	R
Sphere, radiation to entire surface	Diameter D	$0.65D$
Infinite circular cylinder, radiation to convex bounding surface	Diameter D	$0.95D$
Circular cylinder with height = diameter, radiation to entire surface	Diameter D	$0.60D$
Circular cylinder, semi-infinite height, radiation to entire base	Diameter D	$0.65D$
Cube, radiation to any face	Edge L	$0.60L$
Volume surrounding infinite tube bundle, radiation to a single tube	Tube diameter D, distance between tube centers S	
Equilateral-triangle arrangement:		
$\quad S = 2D$		$3.0(S - D)$
$\quad S = 3D$		$3.8(S - D)$
$\quad$ Square arrangement		$3.5(S - D)$

[a]See refs. 13 and 31.

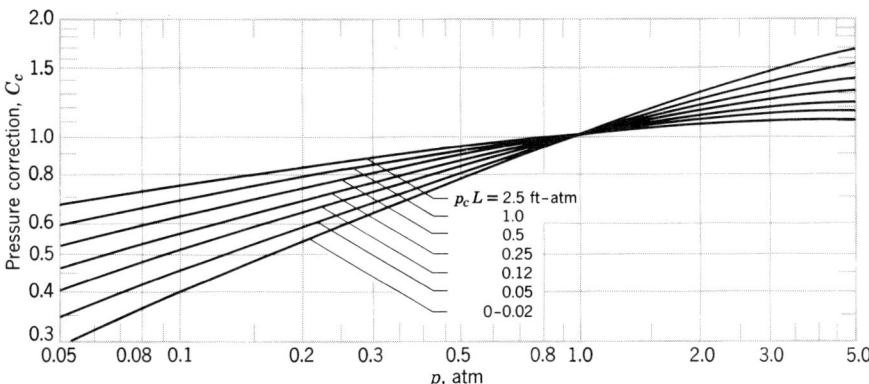

Fig. 3.2-46 Correction factor for obtaining carbon dioxide emissivities at pressures other than 1 atm ($\varepsilon_{c,\, p=1\text{ atm}} = C_c \varepsilon_{c,\, p=1\text{ atm}}$). (From ref. 31; used by permission.)

factor when CO_2 and H_2O are both present in an enclosure so that the total gas emissivity is expressed as

$$\varepsilon_g = C_c \varepsilon_c + C_w \varepsilon_w - \Delta \varepsilon \qquad (3.2\text{-}157)$$

The correction factors account for the reduction in emissivities due to the mutual absorption of radiation between the two gaseous species. The heat exchange between a gas mixture of CO_2 and H_2O at a temperature T_g and its enclosure at a temperature T_s is equal to the difference between the energy emitted by the gas and the energy absorbed by the gas due to the radiation from the enclosure:

$$\frac{q}{A} = \varepsilon_g \sigma T_g^4 - \alpha_g \sigma T_s^4 \qquad (3.2\text{-}158)$$

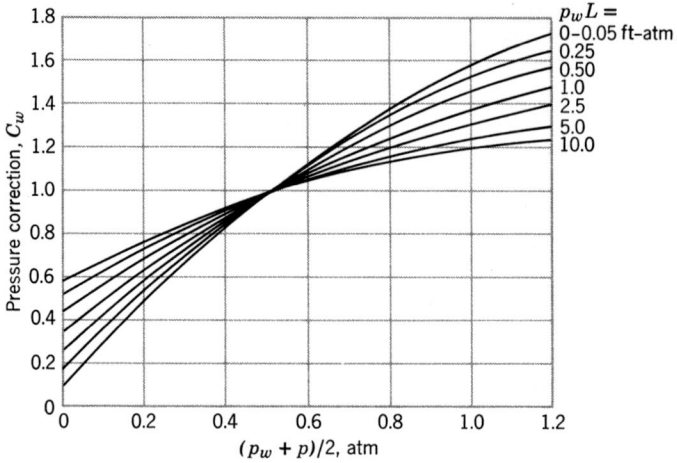

Fig. 3.2-47 Correction factor for obtaining water vapor emissivities at pressures other than 1 atm ($\varepsilon_{w, p-1 \text{ atm}} = C_w \varepsilon_{w, p=1 \text{ atm}}$). (From ref. 31; used by permission.)

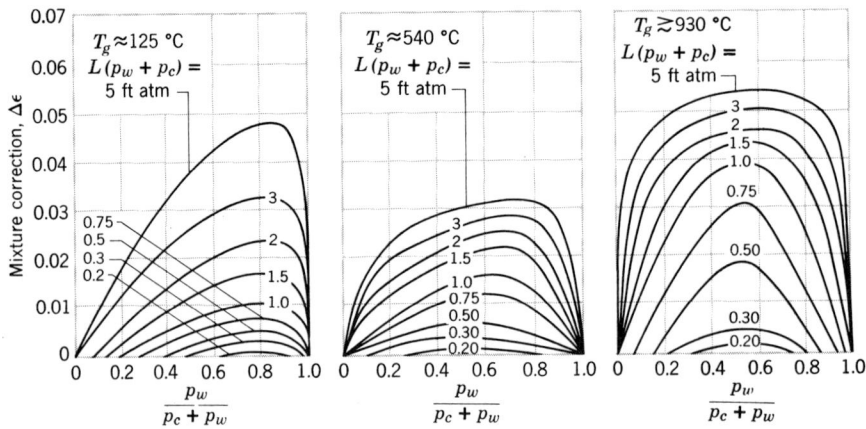

Fig. 3.2-48 Correction factor associated with mixtures of water vapor and carbon dioxide. (From ref. 31; used by permission.)

where ε_g is evaluated at T_g from Eq. (3.2-157). The absorbtivity of the gas α_g is determined by the expressions

$$\alpha_g(T_s) = \alpha_c + \alpha_w - \Delta\alpha$$

where

$$\alpha_c = C_c \varepsilon_c' \left(\frac{T_g}{T_s}\right)^{0.65}$$

and

$$\alpha_w = C_w \varepsilon_w' \left(\frac{T_g}{T_s}\right)^{0.45}$$

ϵ'_c and ϵ'_w are evaluated from Figs. 3.2-44 and 3.2-45 at T_s and pressure–beam length parameters of $p_c L(T_s/T_g)$ and $p_w L(T_s/T_g)$, respectively.

REFERENCES

3.2-1 F. F. Incropera and D. P. DeWitt, *Fundamentals of Heat Transfer*, Wiley, New York, 1981

3.2-2 Y. S. Touloukian and C. Y. Ho (eds.), *Thermophysical Properties of Matter*, Plenum Press, New York, Vol. 1, *Thermal Conductivity of Metallic Solids*; Vol. 2, *Thermal Conductivity of Non-metallic Solids*; Vol. 4, *Specific Heat of Metallic Solids*; Vol. 5, *Specific Heat of Nonmetallic Solids*; Vol. 7, *Thermal Radiative Properties of Metallic Solids*; Vol. 8, *Thermal Radiative Properties of Nonmetallic Solids*; Vol. 9, *Thermal Radiative Properties of Coatings*, 1972.

3.2-3 R. Hultgren, P. D. Desai, D. T. Hawkins, M. Gleiser, K. K. Kelley, and D. D. Wagman, *Selected Values of Thermodynamic Properties of Elements*, American Society of Metals, 1973.

3.2-4 Y. S. Touloukian and C. Y. Ho (eds.), *Thermophysical Properties of Selected Aerospace Materials*, Part I: Thermal Radiative Properties; Part II: Thermophysical Properties of Seven Materials; Thermophysical and Electronic Properties Information Analysis Center, CINDAS, Purdue University, West Lafayette, IN, 1976.

3.2-5 C. Y. Ho, R. W. Powell, and P. E. Liley, "Thermal Conductivity of the Elements: A Comprehensive Review," *J. Phys. and Chem. Reference Data* **3** (Supp. 1), (1974).

3.2-6 American Society of Heating, Refrigerating and Air Conditioning Engineers, *ASHRAE Handbook of Fundamentals*, 1972.

3.2-7 J. F. Mallory, *Thermal Insulation*, Van Nostrand Rheinhold, New York, 1969.

3.2-8 E. J. Hanley, D. P. DeWitt, and R. E. Taylor, "The Thermal Transport Properties at Normal and Elevated Temperature of Eight Representative Rocks," Proceedings of the Seventh Symposium on Thermophysical Properties, American Society of Mechanical Engineers, 1977.

3.2-9 V. E. Sweat, "A Miniature Thermal Conductivity Probe for Foods," American Society of Mechanical Engineers, Paper 76-HT-60, August 1976.

3.2-10 C. P. Kothandaraman and S. Subramanyan, *Heat and Mass Transfer Data Book*, Halsted Press, Wiley, New York, 1975.

3.2-11 A. J. Chapman, *Heat Transfer*, 3rd ed., Macmillan, New York, 1974.

3.2-12 N. B. Vargaftik, *Tables of Thermophysical Properties of Liquids and Gases*, 2nd ed., Hemisphere Publishing, Washington, D.C., 1975.

3.2-13 E. R. G. Eckert and R. M. Drake, *Analysis of Heat and Mass Transfer*, McGraw-Hill, New York, 1972.

3.2-14 M. P. Vukalovich, A. I. Ivanov, L. R. Fokin, and A. T. Yakovelev, *Thermophysical Properties of Mercury*, State Committee on Standards, State Service for Standards and Handbook Data, Monograph Series No. 9, Izd. Standartov, Moscow, 1971.

3.2-15 P. E. Liley, Steam Tables in SI Units, School of Mechanical Engineering, Purdue University, December 1978.

3.2-16 M. P. Heisler, "Temperature Charts for Conduction and Constant Temperature Heating," *Trans. ASME* **69**, 227–236.

3.2-17 E. D. Grimison, *Trans. ASME* **59**, 583 (1937).

3.2-18 W. H. McAdams, *Heat Transmission*, 3rd ed., McGraw-Hill, New York, 1954.

3.2-19 F. J. Bayley, "An Analysis of Turbulent Free Convection Heat Transfer," *Proc. Inst. Mech. Eng.*, **169**, 361, 1955.

3.2-20 M. Jakob, *Heat Transfer*, Vol. 1, Wiley, New York, 1949.

3.2-21 W. M. Kays and R. K. Lo, Stanford University Technical Report No. 15, 1952.

3.2-22 A. Zhukauskas, "Heat Transfer from Tubes in Cross Flow," in J. P. Hartnett and T. F. Irvine, Jr. (eds), *Advances in Heat Transfer*, Vol. 8, Academic Press, New York, 1972.

3.2-23 W. M. Kays, "Convection Heat and Mass Transfer," McGraw-Hill, New York, 1966.

3.2-24 E. L. Piret and H. S. Isbin, "Natural Circulation Evaporation Two-Phase Heat Transfer," *Chem. Eng. Prog.* **50**, 305, 1954.

3.2-25 R. I. Vachon, G. H. Nix, and G. E. Tanger, "Evaluation of Constants for the Rohsenow Pool-Boiling Correlation," *J. Heat Transfer* **90**, 239, 1968.

3.2-26 D. S. Cryder and A. C. Finalbargo, "Heat Transmission from Metal Surfaces to Boiling Liquids: Effect of Temperature on the Liquid on Film Coefficient," *Trans. AIChE* **33**, 346, 1937.

3.2-27 M. Jakob and G. A. Hawkins, *Elements of Heat Transfer*, 3rd ed. Wiley, New York, 1957.

3.2-28 G. G. Gubaref, J. E. Janssen, and R. H. Torborg, *Thermal Radiation Properties Survey*, Minneapolis-Honeywell Regulator Company, Minneapolis, MN, 1960.

3.2-29 F. Kreith and J. F. Kreider, *Principles of Solar Energy*, Hemisphere, Washington, D.C., 1978.

3.2-30 J. P. Holman, *Heat Transfer*, 5th ed., McGraw-Hill, New York, 1981.

3.2-31 H. C. Hottel, "Radiant Heat Transmission," in W. H. McAdams, *Heat Transmission*, 3rd ed., McGraw-Hill, New York, 1954.

3.3   FLUID DYNAMICS

William W. Durgin

3.3-1   Introduction

Although the equations governing the motion of a fluid are known, their solution is known in relatively few cases. The solution to fluid mechanic problems often necessitates experiments of some sort. The solution for laminar flow in a circular pipe is, for example, known but that for turbulent flow is not. Through clever use of theory to guide the interpretation of experimental results, a semiempirical method is available for general pipe flow problems. It may be necessary, for example, to determine the forces on a large structure for which it is impractical to perform direct measurement. In this case the testing of a scale model using similarity principles would be indicated.

In this section the general governing equations are given along with application in selected instances. There are sections on similitude, incompressible pipe flow, boundary layer flow, and compressible flow in ducts; all frequently encountered in energy applications.

3.3-2   Descriptive Method

A portion of fluid small enough to be considered mathematically infinitesimal but larger than the molecular scale is termed a *fluid particle*. Such a particle is small enough to be represented by single-valued macroscopic thermodynamic properties.

The geometric descriptive method adopted in fluid mechanics is to focus attention on spatial points rather than fluid particles. The velocity at the spatial point is the velocity of the particle instantaneously at that point. The velocity of all particles that constitute the flow is

$$\mathbf{v} = \mathbf{v}(\mathbf{r}, t) \tag{3.3-1}$$

where $\mathbf{r}$ is the position vector of a spatial point and t is time. If the flow is *steady*, then

$$\mathbf{v} = \mathbf{v}(r) \tag{3.3-2}$$

In order to subsequently find the motion of any fluid particle, the equation

$$\dot{\mathbf{r}} = \mathbf{v} \tag{3.3-3}$$

may be solved if the velocity is everywhere known.

The acceleration at the spatial point is given by

$$\mathbf{a} = \frac{d\mathbf{v}}{dt} = (\mathbf{v} \cdot \nabla)\mathbf{v} + \frac{\partial \mathbf{v}}{\partial t} \tag{3.3-4}$$

where d/dt is the substantial derivative and consists of both the convective term $(\mathbf{v} \cdot \nabla)\mathbf{v}$ and the local term $\partial \mathbf{v}/\partial t$. The former arises because the acceleration varies between spatial points at any given time. The latter results from the changing velocity of a particle instantaneously at a spatial point.

Streamlines are lines everywhere tangent to the velocity vectors, so that

$$d\mathbf{r} \times \mathbf{v} = 0 \tag{3.3-5}$$

gives the streamline equations. Pathlines depict the paths of particles and are given by solving Eq. (3.3-3). In the cartesian coordinates x, y, z, the velocity components are u, v, w. Thus

$$u = u(x, y, z, t)$$
$$v = v(x, y, z, t) \tag{3.3-6}$$
$$w = w(x, y, z, t)$$

and the acceleration is

$$a_x = u\frac{\partial u}{\partial x} + v\frac{\partial u}{\partial y} + w\frac{\partial u}{\partial z} + \frac{\partial u}{\partial t}$$

$$a_y = u\frac{\partial v}{\partial x} + v\frac{\partial v}{\partial y} + w\frac{\partial v}{\partial z} + \frac{\partial v}{\partial t} \tag{3.3-7}$$

$$a_z = u\frac{\partial w}{\partial x} + v\frac{\partial w}{\partial y} + w\frac{\partial w}{\partial z} + \frac{\partial w}{\partial t}$$

The streamline equations are

$$\frac{dx}{u} = \frac{dy}{z} = \frac{dz}{w}$$

(3.3-8)

with the particle pathlines given by

$$dx = u\,dt, \qquad dy = v\,dt, \qquad dz = w\,dt$$

(3.3-9)

Conservation Equations

The motion of a fluid is assumed to satisfy the physical laws of mass conservation (continuity), Newton's second law, and the first law of thermodynamics. Usually Newton's law is cast as a linear momentum equation, but an angular momentum equation might be more appropriate in some situations. The first law of thermodynamics is formulated as an equation of energy conservation.

Integral Forms

When the details of fluid motion within a region are not required and reasonable assumptions regarding the motion of some boundaries can be made, integral or control volume formulations are appropriate.

The *continuity equation* states that mass must be conserved. For a control volume V enclosed by a surface S, the equation is

$$\frac{\partial}{\partial t}\int \rho\,dV + \int_S \rho\mathbf{V}\cdot d\mathbf{s} = 0$$

(3.3-10)

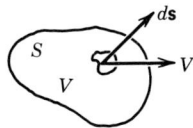

which states that the rate of mass increase inside the control volume plus the total mass outflow must be zero. The control volume may be moving as long as it is measured relative to the control volume.

If the control volume is fixed in time, then

$$\int_V \frac{\partial \rho}{\partial t}\,dV + \int_S \rho\mathbf{V}\cdot d\mathbf{s} = 0$$

(3.3-11)

and $\mathbf{V}$ is measured relative to the (fixed) coordinates.

For example, the steady flow of a fluid through a reducing elbow may be evaluated by assuming the velocity and density to be uniform over the entrance and exit so that

$$\rho_1 V_1 A_1 = \rho_2 V_2 A_2$$

(3.3-12)

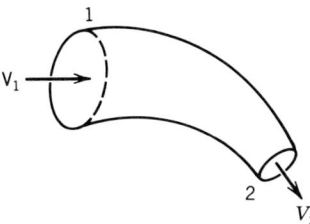

The *linear momentum* equation states that forces acting on or in a control volume produce changes in the momentum. If the surface forces are $\mathbf{F}_S$ and the body forces $\mathbf{F}_B$, then

$$\mathbf{F}_S + \mathbf{F}_B = \frac{\partial}{\partial t}\int_V \rho\mathbf{V}\,dV + \int_S \rho\mathbf{V}\mathbf{V}\cdot d\mathbf{s}$$

(3.3-13)

The surface forces include contributions from pressure and shear stresses, while the body forces are typically those of gravity. These forces act to increase the momentum inside the control volume via the first term on the right and to change the net momentum flux across the surface via the last term. The momentum equation is a vector equation and so can be decomposed into three scalar equations.

The component of the surface force due to shear stresses can often be neglected at inlets and outlets to the control volume so that

$$F_S = -\int_S p\,ds \tag{3.3-14}$$

The body force term often includes only the weight of the fluid so that

$$F_B = \int_V \rho g\,dV \tag{3.3-15}$$

In the steady flow through a 90° reducing elbow, there is no momentum change inside the control volume so that the surface force is equal to the integral of momentum flux.

$$F_S = -\rho_1 V_1 \hat{i}(V_1 \hat{i} \cdot A_1 \hat{i}) + \rho_2 V_2 \hat{j}(V_2 \hat{j} \cdot A_2 \hat{j}) \tag{3.3-16}$$

where the velocity and density have been assumed constant across the entrance and exit and the weight has been neglected. The surface force is composed of the force required to restrain the elbow and the pressure force on the inlet and outlet:

$$F_S = R + p_1 A_1 \hat{i} - p_2 A_2 \hat{j} \tag{3.3-17}$$

so that

$$R = \left(-p_1 - \rho_1 V_1^2 \right) A_1 \hat{i} + \left(p_2 + \rho_2 V_2^2 \right) A_2 \hat{j} \tag{3.3-18}$$

The *energy conservation* equation gives the rate of change of specific internal energy

$$e = u + \frac{V^2}{2} + gz \tag{3.3-19}$$

in terms of the heat transferred and the work done on the control volume:

$$\frac{\partial}{\partial t}\int_V e\rho\,dV + \int_S e\rho V \cdot ds + \int_S pV \cdot ds = \dot{Q} - \dot{W}_S - \dot{W}_V + \dot{S} \tag{3.3-20}$$

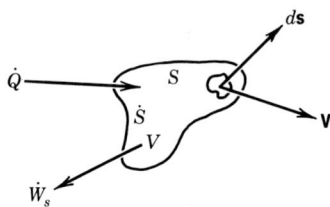

The first term represents the rate, of change within the control, volume, while the second represents, the flux across the control surface. The third term is the work done by pressure at the control surface. The heat transfer across the control surface, $\dot{Q}$, is taken as positive in and the shaft work done by the control volume, $\dot{W}_S$, as positive out. The internal heat sources $\dot{S}$ may represent chemical reactions, discrete heaters, or viscous dissipation.

Differential Forms

By considering the control volume fixed in time and applying Gauss's theorem, all of the conservation equations can be reformulated so that they apply to a fluid element of differential size. The *continuity equation* is

$$\frac{\partial \rho}{\partial t} + \nabla \cdot (\rho V) = 0 \tag{3.3-21}$$

If shear stresses are neglected (invicid flow), then the *linear momentum* equation is

$$\rho \frac{d\mathbf{v}}{dt} = -\nabla p + \rho \mathbf{g} \tag{3.3-22}$$

This is Euler's equation, which states that the rate of change of momentum is caused by the pressure gradient and gravitational acceleration. The acceleration $d\mathbf{v}/dt$ is, of course, composed of the local and convective terms, making this a nonlinear differential equation in three spatial directions.

If the viscous stress terms are included, the *linear momentum* equation becomes

$$\rho \frac{dv}{dt} = \rho \mathbf{g} + \frac{\partial}{\partial x_j} \tau_{ij} \tag{3.3-23}$$

where for a Newtonian fluid

$$\tau_{ij} = -p\delta_{ij} + \mu \left(\frac{\partial u_i}{\partial x_j} + \frac{\partial u_j}{\partial x_i} \right) + \delta_{ij} \lambda \nabla \cdot \mathbf{V}$$

with μ the coefficient of viscosity and λ the coefficient of bulk viscosity. It is often assumed that the mean pressure deforming the fluid is equal to the thermodynamic pressure, which leads to

$$\lambda + \tfrac{2}{3}\mu = 0 \tag{3.3-24}$$

This is the Stokes hypothesis, which is usually inconsequential because the third term in the stress formulation is negligible for many practical problems.

Using the Stokes hypothesis in conjunction with the assumption of incompressibility, the equations of motion are, in cartesian coordinates,

$$\varrho \left(\frac{\partial u}{\partial t} + u\frac{\partial u}{\partial x} + v\frac{\partial u}{\partial y} + w\frac{\partial u}{\partial z} \right) = X - \frac{\partial p}{\partial x} + \mu \left(\frac{\partial^2 u}{\partial x^2} + \frac{\partial^2 u}{\partial y^2} + \frac{\partial^2 u}{\partial z^2} \right)$$

$$\varrho \left(\frac{\partial v}{\partial t} + u\frac{\partial v}{\partial x} + v\frac{\partial v}{\partial y} + w\frac{\partial v}{\partial z} \right) = Y - \frac{\partial p}{\partial y} + \mu \left(\frac{\partial^2 v}{\partial x^2} + \frac{\partial^2 v}{\partial y^2} + \frac{\partial^2 v}{\partial z^2} \right) \tag{3.3-25}$$

$$\varrho \left(\frac{\partial w}{\partial t} + u\frac{\partial w}{\partial x} + v\frac{\partial w}{\partial y} + w\frac{\partial w}{\partial z} \right) = Z - \frac{\partial p}{\partial z} + \mu \left(\frac{\partial^2 w}{\partial x^2} + \frac{\partial^2 w}{\partial y^2} + \frac{\partial^2 w}{\partial z^2} \right)$$

where X, Y, and Z represent the components of the body force. These equations are exceedingly difficult with only a few hundred exact laminar flow solutions known. No exact turbulent flow solutions are known, and numerical solutions are generally limited to low Reynolds numbers. Useful solutions are usually approximate with some empiricism built into the solution.

3.3-3 Similitude

Geometrically similar flow situations may, under certain circumstances, have similar streamline patterns or kinematic similarity. Furthermore, such flows may be dynamically similar if the fluid forces act in the same directions and bear the same ratio at corresponding points in the flows. Under these conditions unique mathematical relationships between suitably nondimensionalized variables exist.

There are two methods of discovering the nondimensional groups that govern a problem. One is to list the variables of importance and find the groups using Buckingham's Π theorem. Another is to nondimensionalize the governing differential equations and boundary conditions. Both techniques depend on inclusion of important variables and terms at the outset. In addition, there will be no unique set of groups as output. It is possible to generate infinite variations of groups that are not independent. By convention, certain groups are used in fluid mechanics.

For example, the drag (D) on a smooth sphere in incompressible flow is a function of diameter d, free-stream velocity V, fluid density ρ, and fluid viscosity μ. The two independent nondimensional groups can be formed as

$$C_D = \frac{D}{\frac{1}{2}\rho U^2 \frac{1}{4}\pi d^2}$$

$$R = \frac{\rho u d}{\mu} \tag{3.3-26}$$

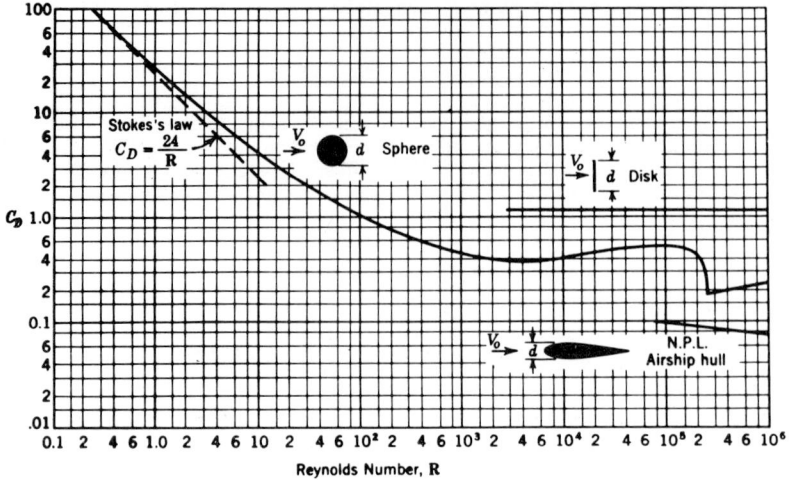

Fig. 3.3-1   Drag coefficient versus Reynolds number. (From ref. 16; used by permission.)

where the drag coefficient C_D includes numerical factors of $\frac{1}{2}$ and $\frac{1}{4}\pi$ so that the denominator is recognizable as the dynamic pressure times the frontal area. These factors have no effect on the nondimensionalization process and could have been omitted. By convention, however, they are included. The principles of similitude give the relationship between these groups as

$$C_D = f(R) \tag{3.3-27}$$

so that the drag coefficient is a function only of the Reynolds number. The relationship must be determined by other means, for example, experimentally (Fig. 3.3-1). Some important and commonly used nondimensional groups are as follows:

Reynolds number	$R = \dfrac{\rho Vl}{\mu}$	Ratio of inertia to friction
Mach number	$M = \dfrac{U}{c}$	Ratio of velocity to sound speed
Froude number	$F = \dfrac{U}{\sqrt{gL}}$	Ratio of inertia to gravity force
Weber number	$W = \dfrac{\rho U^2 L}{\sigma}$	Ratio of inertia to surface tension force
Euler number	$E = \dfrac{p}{\rho U^2}$	Ratio of pressure to inertia
Drag coefficient	$C_D = \dfrac{D}{\frac{1}{2}\rho U^2 A}$	Ratio of drag to dynamic pressure times area
Lift coefficient	$C_L = \dfrac{L}{\frac{1}{2}\rho U^2 A}$	Ratio of lift to dynamic pressure times area

where l = typical dimension
U = stream velocity
ρ = fluid density
μ = absolute velocity
c = speed of sound
g = gravitational acceleration
σ = surface tension
p = pressure
D = drag
L = lift
A = area

As an example, consider the gas flow in a complex duct system such as a boiler or exhaust system. In the simplest case there may be so many bends, expansions, contractions, shape changes, and so on, that a designer cannot rely on published values for loss coefficients because of interaction effects or because they simply are not available. In order to determine the expected pressure drop through such a system, a model of the prototype system could be constructed and flow-tested. It is desirable to use a geometric scale ratio that yields a model of convenient laboratory size, but other scaling requirements must also be met.

Supposing that compressibility and buoyancy effects are not predominant, then if we could make

$$R_m = R_p \tag{3.3-28}$$

we would ensure that

$$E_m = E_p \tag{3.3-29}$$

The first relation requires that

$$\left(\frac{\rho Ul}{\mu}\right)_{model} = \left(\frac{\rho Ul}{\mu}\right)_{prototype} \tag{3.3-30}$$

which becomes on rearrangement

$$\frac{(U/\nu)_{model}}{(U/\nu)_{prototype}} = \frac{lp}{lm} = r \tag{3.3-31}$$

where $\nu = \mu/\rho$ is the kinematic viscosity and r is the scale ratio and might be on the order of 10 for this type of model. If the model were to use air at temperature and pressure similar to the prototype, $\nu_m \approx \nu_p$, then the model would have to operate at velocities of rU_p to achieve dynamic similarity. It is often not possible to achieve such high velocities without encountering significant compressible flow effects in the model. Since $\nu_{air}/\nu_{water} \approx 16$ at STP, it might be appropriate to use water as the working fluid in the model, and the water velocities would be the same order of magnitude as the air velocities of the prototype.

Practitioners often take advantage of Euler number dependence on Reynolds number of the type shown for the sphere, disc, and airship. Above certain Reynolds number values the Euler number (C_D in this case) is quite constant. If the prototype operates at $R_p = 10^5$ and we use the same working fluid but at a geometric scale ratio of 10, then we should at least use $R_m = 10^4$. However, we could probably use Reynolds numbers as low as 10^3 and verify, perhaps, the independence of pressure coefficient with Reynolds number using the model itself.

The Reynolds number and pressure coefficient (C_p) can be calculated from measured data at each of several model-operating velocities. By plotting C_p versus R, the pressure coefficient for the prototype can be estimated and the pressure drop calculated as

$$p_2 - p_1 = \tfrac{1}{2}\rho_p U_p^2 C_p(R_p) \tag{3.3-32}$$

The model would probably be used to develop flow guidance devices, predict temperature or concentration distributions, or evaluate particulate sedimentation. Suitable similarity relationships would have to be developed in each case.

3.3-4   Flow in Pipes and Ducts

For steady, incompressible flow in a circular pipe between two stations, the continuity equation gives

$$Q_1 = Q_2 \tag{3.3-33}$$

where Q, the volumetric flow rate, is computed as the magnitude of the integral

$$Q = \left|\int_A \mathbf{V} \cdot d\mathbf{A}\right| \tag{3.3-34}$$

and the average velocity is

$$V = \frac{Q}{A} \tag{3.3-35}$$

The energy equation can be written as

$$\left(\frac{p_1}{\rho} + \frac{1}{2}\alpha_1 V_1^2 + gZ_1\right) - \left(\frac{p_2}{\rho} + \frac{1}{2}\alpha_2 V_2^2 + gZ_2\right) = u_2 - u_1 - \frac{\dot{Q}}{\dot{M}} \tag{3.3-36}$$

with α_1 and α_2 being energy correction factors. The specific mechanical energy at station 2 differs from that of station 1 by the amount

$$gh_l = u_2 - u_1 - \frac{\dot{Q}}{\dot{m}} \tag{3.3-37}$$

where h_l is called the head loss and represents the mechanical energy converted into internal energy or heat transfered by viscous stresses.

If the flow is fully developed, the energy equation becomes

$$h_l = \left(Z_1 + \frac{p_1}{\rho g}\right) - \left(Z_2 + \frac{p_2}{\rho g}\right) \tag{3.3-38}$$

Applying the momentum equation to the control volume between stations 1 and 2 gives

$$\left(Z_1 + \frac{p_1}{\rho g}\right) - \left(Z_2 + \frac{p_2}{\rho g}\right) = \frac{2\tau_w}{\rho g}\frac{L}{R} \tag{3.3-39}$$

where τ_w is the wall shear stress. The head loss is thus related to the shear stress as

$$h_l = \frac{2\tau_w}{\rho g}\frac{L}{R} \tag{3.3-40}$$

So far the equations apply either to laminar or turbulent flow and only assume that the profiles at 1 and 2 are the same. Dimensional considerations give

$$\tau_w = \tau_w(\rho, U, \mu, d, \varepsilon) \tag{3.3-41}$$

where ε is the wall roughness. Dimensional analysis gives

$$\frac{8\tau_w}{\rho U^2} = f\left(R_d, \frac{\varepsilon}{d}\right) \tag{3.3-42}$$

where f is the Darcy friction factor, which depends on the pipe Reynolds number and the relative roughness.

For laminar flow the differential equations of motion can be solved exactly to give

$$u = \frac{1}{4\mu}\left[-\frac{d}{dx}(p + \rho gz)\right](R^2 - r^2) \tag{3.3-43}$$

which is the Hagen–Poisuille parabolic velocity profile and

$$f = \frac{64}{R_d} \tag{3.3-44}$$

which is the laminar flow friction factor. The head loss becomes

$$h_l = \frac{128\mu LQ}{\pi \rho gd^4} \tag{3.3-45}$$

and is seen to be proportional to the pipe velocity.

For turbulent flow with smooth walls, the law of the wall can be used to determine the velocity profile[1]

$$\frac{u}{u_*} = \frac{1}{\kappa}\ln\frac{(R - r)u_*}{\mu} + B \tag{3.3-46}$$

where $u_* = \sqrt{\tau w/\rho}$ is the shear velocity, $\kappa \approx 0.41$, and $B \approx 5.0$. In this case only an implicit equation for friction factor can be obtained.

$$f^{-1/2} = 1.99 \log\left(R_d f^{1/2}\right) - 1.02 \tag{3.3-47}$$

The slightly modified form due to Prandtl is often used:

$$f^{-1/2} = 2.0 \log\left(R_d f^{1/2}\right) - 0.8 \tag{3.3-48}$$

For fully rough walls the factor B can be correlated with $\varepsilon u_*/\nu$, giving

$$f^{-1/2} = -2.0 \log\frac{\varepsilon/d}{3.7} \tag{3.3-49}$$

Colebrook combined the smooth and rough wall relations to cover the transitional range as

$$f^{-1/2} = -2.0 \log\left(\frac{\varepsilon/d}{3.7} + \frac{2.51}{R_d f^{1/2}}\right) \tag{3.3-50}$$

which was plotted by Moody in what is now known as the Moody chart (Fig. 3.3-2). Table 3.3-1 gives some roughness values by Moody.[2]

The pressure drop due to typical pipeline fittings can be conveniently calculated as

$$\Delta p = K\rho\frac{V^2}{2} \tag{3.3-51}$$

where K is the loss coefficient. Typically used values for common fittings are listed in Table 3.3-2. Loss coefficients depend on roughness, fastenings, and other details so that for critical calculations, more refined values should be used (e.g., see Benedict[3]).

For noncircular ducts the hydraulic diameter concept may be used wherein the hydraulic diameter

$$d_h = \frac{4A}{p} \tag{3.3-52}$$

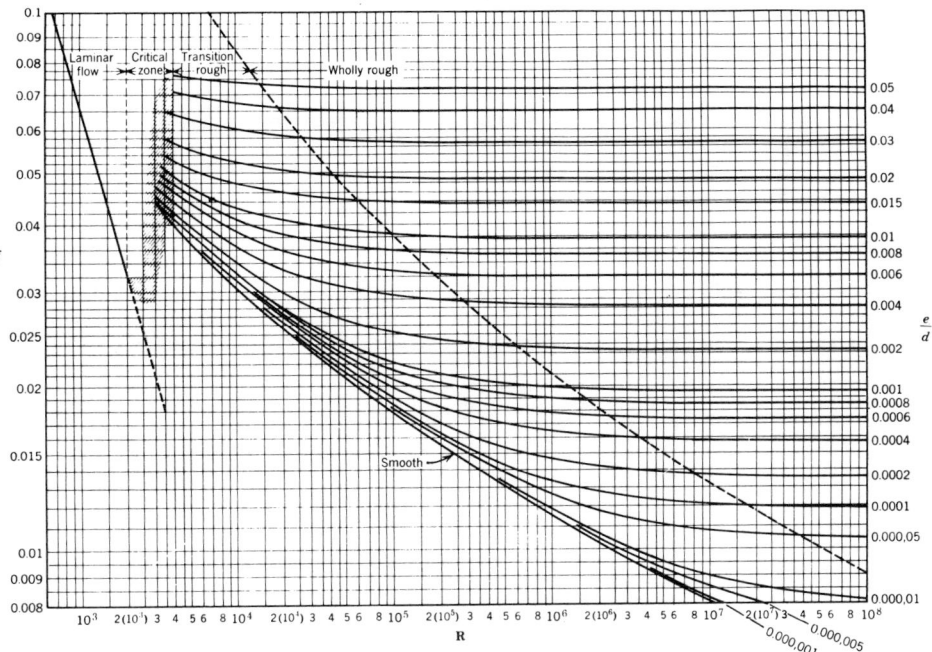

Fig. 3.3-2 Friction factor versus Reynolds number for various relative roughnesses. (As adapted from ref. 2 and ref. 16; used by permission of John Wiley & Sons and ASME.)

TABLE 3.3-1   ROUGHNESS OF NEW COMMERCIAL PIPE

Material (new)	ft	mm
Riveted steel	0.003–0.03	0.9–9.0
Concrete	0.001–0.01	0.3–3.0
Wood stave	0.0006–0.003	0.18–0.9
Cast iron	0.00085	0.26
Galvanized iron	0.0005	0.15
Asphalted cast iron	0.0004	0.12
Commercial steel or wrought iron	0.00015	0.046
Drawn tubing	0.000005	0.0015
Glass	"Smooth"	"Smooth"

Source.   Adapted from Moody, "Friction Factors for Pipe Flow,"
ASME Transactions **66**, 1944.

TABLE 3.3-2   LOSS COEFFICIENTS

Sharp entrance	0.50
Slightly rounded entrance	0.23
Well-rounded entrance	0.04
Inward-projecting entrance	0.78
Rounded, sharp, or projecting exit	1.00
Sudden enlargement $d_1/d_2 = 0.5$	0.56
Sudden contraction $d_2/d_1 = 0.5$	0.34
Globe valve, fully open	10.0
Angle valve, fully open	5.0
Butterfly valve, fully open	0.5
Standard sweep elbow	0.9
Medium sweep elbow	0.75

is substituted for the circular pipe diameter in the head loss equation. The duct area is A, and p is the wetted perimeter. Figure 3.3-3 gives values of d_h for several duct cross sections. For turbulent flow, use of the hydraulic diameter produces head losses accurate to about 2%. In laminar flow, errors can be as high as 23% (concentric annulus).

Boundary Layer

For two-dimensional boundary layer flow on a flat-plate (zero pressure gradient), the equations of motion can be solved under the boundary layer hypothesis, namely that the layer of fluid for which viscous effects are important is thin. The variation of pressure across the layer is then small:

$$\frac{\partial p}{\partial y} \approx 0 \tag{3.3-53}$$

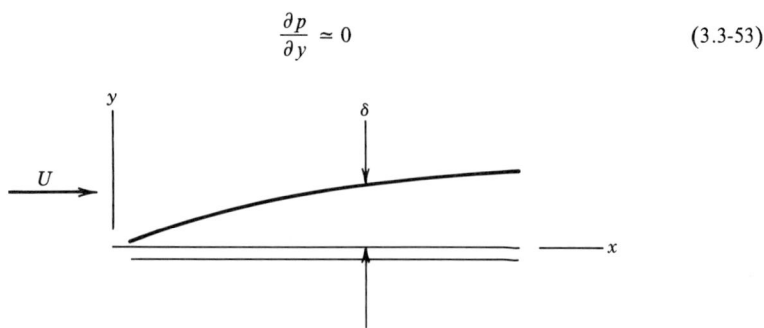

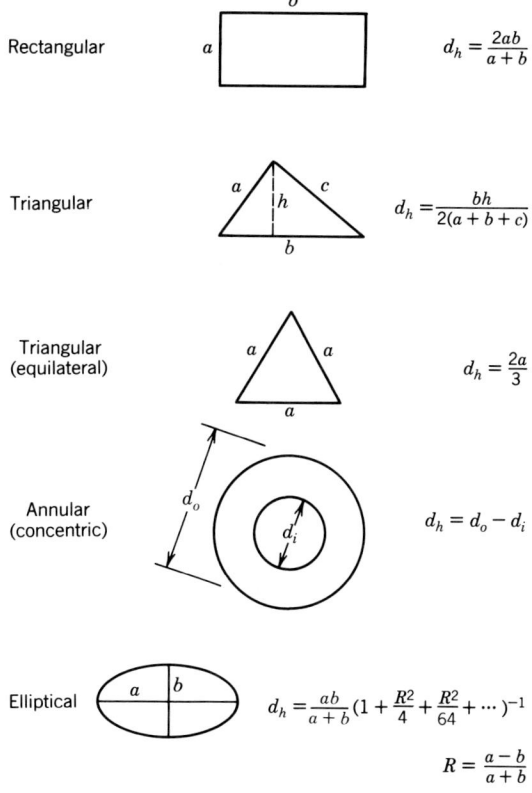

Fig. 3.3-3   Hydraulic diameters.

and the shear stress is primarily due to the variation of u with y so that

$$\frac{\partial^2 u}{\partial x^2} \ll \frac{\partial^2 u}{\partial y^2} \qquad (3.3\text{-}54)$$

The equations of motion then become

$$\frac{\partial u}{\partial x} + \frac{\partial v}{\partial y} = 0$$

$$u\frac{\partial u}{\partial x} + v\frac{\partial u}{\partial y} \simeq \frac{1}{\rho}\frac{\partial \tau}{\partial y} \qquad (3.3\text{-}55)$$

where

$$\tau = \mu\frac{\partial u}{\partial y} \qquad \text{laminar flow} \qquad (3.3\text{-}56)$$

$$\tau = \mu\frac{\partial u}{\partial y} - \rho\overline{u'v'} \quad \text{turbulent flow} \qquad (3.3\text{-}57)$$

For laminar flow the equations can be solved using the similarity transform of Blasius (cf. White[1]), so that the displacement thickness

$$\delta_* = \int_0^\infty \left(1 - \frac{u}{U}\right) dy \qquad (3.3\text{-}58)$$

can be found and normalized to give

$$\frac{\delta_*}{x} = \frac{1.721}{R_x^{1/2}}$$

(3.3-59)

and the skin friction is

$$C_f = \frac{\tau_w}{\frac{1}{2}\rho U^2} = \frac{0.664}{R_x^{1/2}}$$

(3.3-60)

For turbulent flow the law of the wall can be introduced and the equations integrated to give

$$R_x = \kappa^{-3}e^{-\kappa B}\left[e^z(z^2 - 4z + 6) - 2z\right]$$

(3.3-61)

where

$$z = \kappa\sqrt{2/C_f}$$

$$R = \frac{Ux}{\nu}$$

which cannot be solved explicitly for C_f but, nevertheless, can be tabulated, plotted, or calculated as necessary. In a similar manner the displacement-thickness-based Reynolds number becomes

$$R_\delta = \frac{1}{\kappa}e^{-\kappa B}(e^2 - 1)$$

(3.3-62)

As a laminar boundary layer grows in thickness, it becomes susceptible to growth of small disturbances. The transition of laminar to turbulent flow on a flat plate takes place at a nominal $R_x \approx 2.8 \times 10^6$. Free-stream turbulence, boundary roughness, and pressure gradient also affect the value. A free-stream turbulence intensity of 1% causes transition at $R_x \approx 0.5 \times 10^6$.[4] Wall roughness elements approximately one-half the displacement thickness cause transition at $R_x \approx 1.4 \times 10^6$, while roughness elements the same height as the displacement thickness can cause immediate transition.[5]

The free-stream pressure gradient not only affects transition but also separation of flow from a wall. Figure 3.3-4 illustrates boundary layer velocity profiles for flow, which might occur about an object or in a flow element. When the flow outside the boundary layer is increasing, the pressure is decreasing. There can be no flow separation in such regions of favorable pressure gradient, and a laminar boundary is very resistant to transition to turbulence. Farther along the surface, the rate of change of outer flow velocity is zero so that the pressure gradient is zero or neutral. The point of inflection in velocity profile is at the wall, and there can be no separation. Transition to turbulence will take place at a Reynolds number of no greater than 2.8×10^6, depending on free-stream turbulence and wall roughness.

When the outer flow velocity is decreasing, the pressure gradient is positive and is called an adverse gradient. The velocity profile inflection point moves away from the wall so that the boundary layer is very susceptible to transition. At moderate adverse pressure gradients the slope of the profile at the wall becomes zero so that the wall shear stress vanishes. This is the point of incipient separation, and farther along the surface reversed flow occurs.

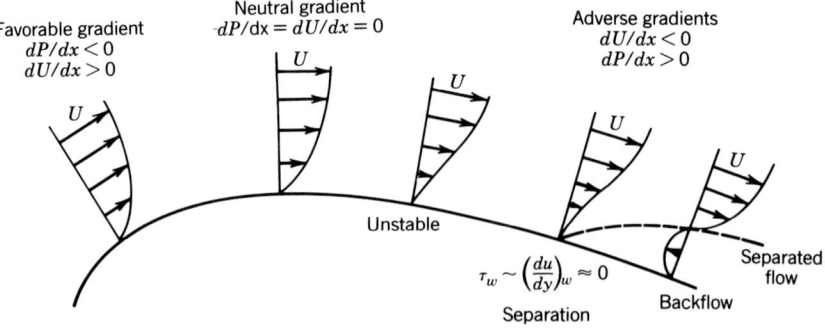

Fig. 3.3-4   Effect of pressure gradient.

Lift and Drag

Figure 3.3-5 illustrates the flow about a circular cylinder. If the boundary layer is laminar, separation takes place approximately 82° from the stagnation line. If the boundary layer is turbulent, because of roughness, free-stream turbulence, or a slightly higher Reynolds number, separation takes place approximately 120° from the stagnation point. In the wake region full pressure recovery is never achieved. For turbulent separation, however, the width of the wake is less. Typical pressure coefficient distributions are shown, with the invicid theory also shown for comparison. It is easy to see that in either laminar or turbulent separation a net force due to pressure is produced in the flow direction. This is often called pressure or form drag and acts in addition to that produced by shear stress at the surface.

Many objects have sharp corners that essentially fix the separation location so that the drag coefficient is not so dependent on the state of the boundary layer. Figure 3.3-6 shows a variety of shapes and typical drag coefficient values from various sources.

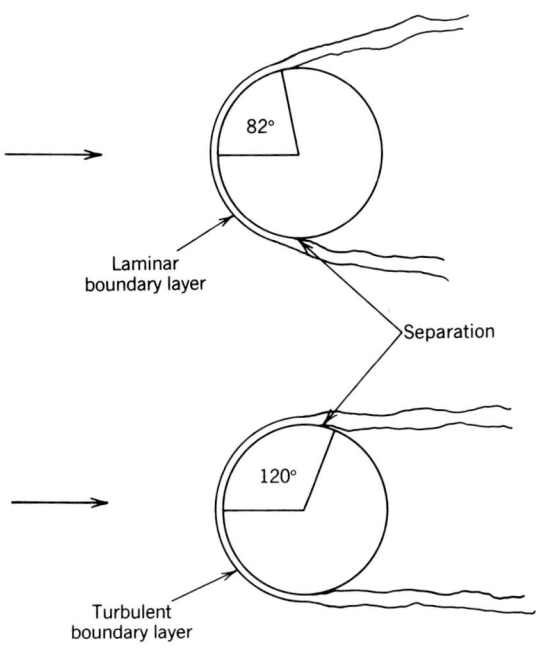

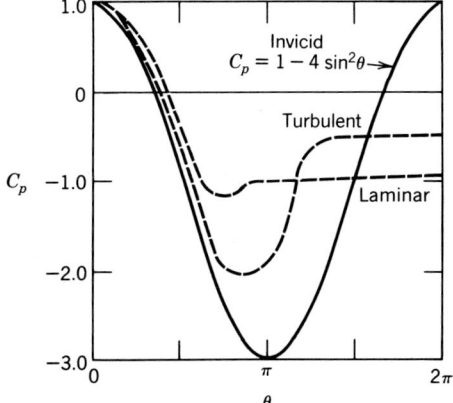

Fig. 3.3-5 Separation from cylinder.

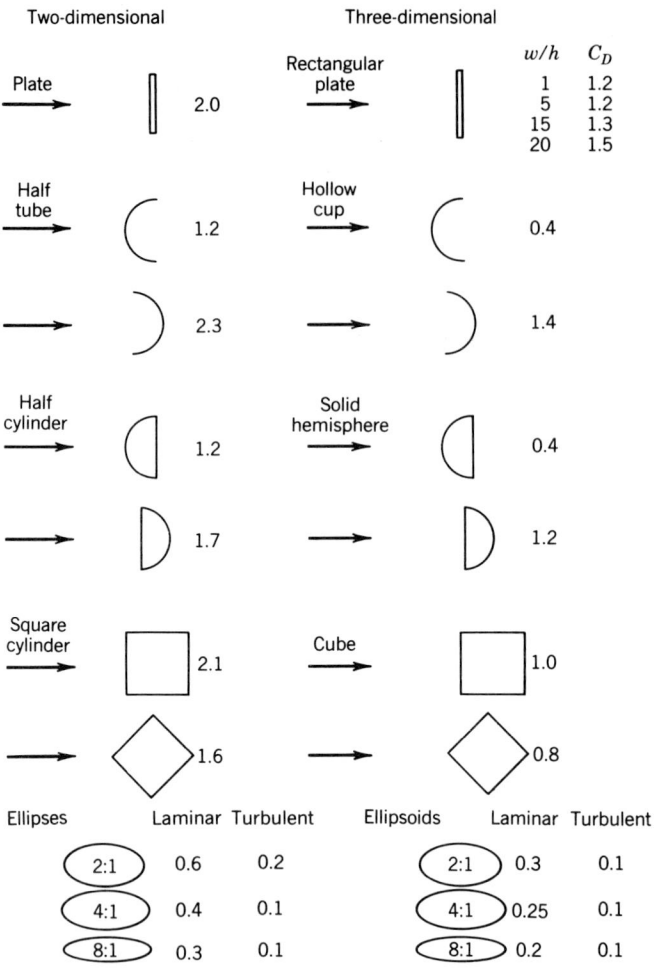

Fig. 3.3-6 Drag coefficients at $R = 10^5$.

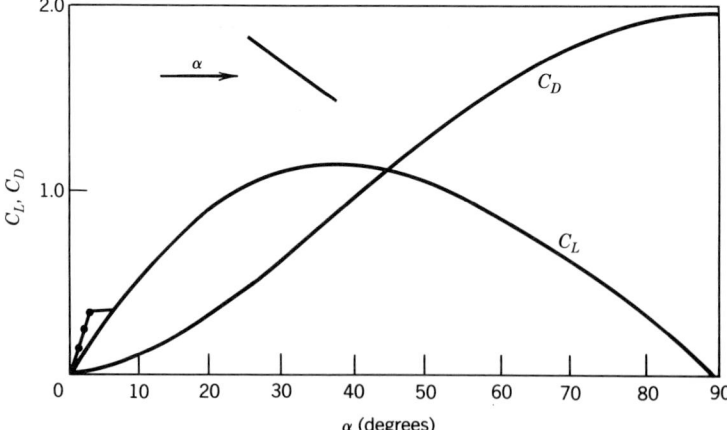

Fig. 3.3-7 Force coefficients for flat plate.

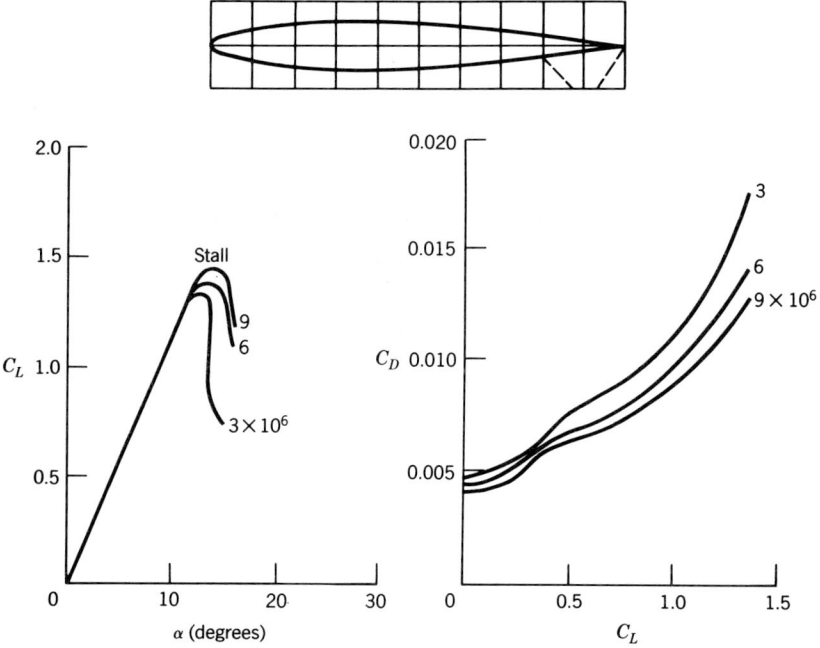

Fig. 3.3-8 Lift and drag data for airfoil.

Objects in a fluid stream can also experience forces perpendicular to the approach velocity, called lift. Hoerner[6] developed data for the flat plate, which has been plotted in Fig. 3.3-7 for attack angles from 0° to 90°. At very small angles of attack the flat plate develops good lift, but separation quickly forms at the sharp leading edge and the lift coefficient switches to the curve shown. When employing flat plates in fluid machinery and apparatus, it is important to consider the lift forces resulting from imperfect alignment with the velocity. The coefficients utilize the plate chord for nondimensionalization.

When production of lift is an objective, an airfoil section should be employed. Abbott and von Doenhoff,[7] give measured data for many airfoil shapes. Figure 3.3-8 shows lift coefficient versus angle of attack and a lift–drag polar plot for a NASA 63-012 airfoil. This is a symmetric airfoil of 12% thickness-to-chord ratio of the laminar flow type. The maximum thickness is quite far off, and the front portions are designed to maintain laminar flow, which is reflected by the "bucket" in the drag curve at small lift. The effect of a Reynolds number is illustrated and shows that the drag coefficient is reduced and stall is delayed as R increases. The lift and drag coefficient data utilize the airfoil chord for nondimensionalization.

Compressible Flow in Ducts

For the compressible flow of a fluid through a duct of varying area without heat transfer, friction, or other losses, the flow can be considered as one-dimensional. The continuity equation then states that the mass flow is the same at each station:

$$\dot{m} = \rho A u \tag{3.3-63}$$

The energy equation gives the stagnation enthalpy as constant

$$h_0 = h + \frac{u^2}{2} \tag{3.3-64}$$

which can be written

$$C_p T_0 = C_p T + \frac{u^2}{2} \tag{3.3-65}$$

for a perfect gas. The isentropic relation

$$\frac{p}{\rho^{\kappa}} = \text{const} \tag{3.3-66}$$

where κ, the ratio of specific heats, gives the speed of sound as

$$c = \sqrt{\kappa R T} \tag{3.3-67}$$

By convention, working relationships are expressed in terms of Mach number

$$M = \frac{u}{c} \tag{3.3-68}$$

the stagnation reference state 0, and the sonic reference state *. Defining $M^* = u/c^*$, the equations become

$$M^{*2} = \frac{[(\kappa + 1)/2]M^2}{1 + [(\kappa - 1)/2]M^2} \tag{3.3-69}$$

$$\frac{T_0}{T} = 1 + \frac{\kappa - 1}{2} M^2 \tag{3.3-70}$$

$$\frac{p_0}{p} = \left(1 + \frac{\kappa - 1}{2} M^2\right)^{\kappa/(\kappa - 1)} \tag{3.3-71}$$

$$\frac{\rho_0}{\rho} = \left(1 + \frac{\kappa - 1}{2} M^2\right)^{1/(\kappa - 1)} \tag{3.3-72}$$

$$\frac{A}{A^*} = \frac{1}{M}\left[\left(\frac{2}{\kappa + 1}\right)\left(1 + \frac{\kappa - 1}{2} M^2\right)\right]^{(\kappa + 1)/2(\kappa - 1)} \tag{3.3-73}$$

These equations are tabulated for air, $\kappa = 1.4$, in Table 18.1-16.
    At the sonic throat the relations become

$$\frac{T^*}{T_0} = \frac{c^{*2}}{c_0^2} = \frac{2}{\kappa + 1} \tag{3.3-74}$$

$$\frac{p^*}{p_0} = \left(\frac{1}{\kappa + 1}\right)^{\kappa/(\kappa - 1)} \tag{3.3-75}$$

$$\frac{\rho^*}{\rho_0} = \left(\frac{2}{\kappa + 1}\right)^{1/(\kappa - 1)} \tag{3.3-76}$$

If a sonic throat is present, then the maximum flow per unit area which can occur is given by

$$\frac{\dot{m}}{A^*} = \sqrt{\frac{\kappa}{R}\left(\frac{2}{\kappa + 1}\right)^{\kappa + 1/(\kappa - 1)}} \frac{p_0}{\sqrt{T_0}} \tag{3.3-77}$$

    For the case of adiabatic flow through ducts of constant area with friction, the continuity and energy equations for the isentropic case still apply. The equation of state is

$$P = \rho R T \tag{3.3-78}$$

and the momentum equation is

$$-A\, dp - \tau_w \pi D\, dx = \dot{m}\, du$$

Expressing the wall shear stress τ_w in terms of the Darcy friction factor f (which can be obtained from the Moody chart),

$$\tau_w = \frac{\frac{1}{4}f\rho U^2}{2}$$

(3.3-79)

gives

$$-dp - f\frac{\rho u^2}{2}\frac{dx}{D} = \rho\frac{u^2}{u}\frac{du}{u}$$

(3.3-80)

The equations can be solved using the Mach number as the independent variable. If $x = 0$ at some Mach number M, then the distance L_{max} to the station where M = 1 is given by

$$\bar{f}\frac{L_{max}}{D} = \frac{1 - M}{\kappa M^2} + \frac{\kappa + 1}{2\kappa}\ln\frac{(\kappa + 1)M^2}{2(1 + (\kappa - 1)/2M^2)}$$

(3.3-81)

where $\bar{f}$ is the mean friction coefficient over the length.

The length required to change the Mach number from M_1 to M_2 can be found as

$$\bar{f}\frac{L}{D} = \left(\bar{f}\frac{L_{max}}{D}\right)_{M_1} - \left(\bar{f}\frac{L_{max}}{D}\right)_{M_2}$$

(3.3-82)

Equations for other variables can be similarly found with respect to the sonic section M = 1 and are denoted by *. In addition, the flow of any station can be imagined to be isentropically brought to rest to the stagnation state 0. The relations are

$$\frac{p}{p^*} = \frac{1}{M}\left[\frac{\kappa + 1}{2\left(1 + \frac{\kappa - 1}{2}M^2\right)}\right]^{1/2}$$

(3.3-83)

$$\frac{\rho}{\rho^*} = \frac{V^*}{V} = \frac{1}{M}\left[\frac{2\left(1 + \frac{\kappa - 1}{2}M^2\right)}{\kappa + 1}\right]^{1/2}$$

(3.3-84)

$$\frac{T}{T^*} = \frac{c^2}{c^{*2}} = \frac{\kappa + 1}{2\left(1 + \frac{\kappa - 1}{2}M^2\right)}$$

(3.3-85)

$$\frac{p_0}{p_0^*} = \frac{1}{M}\left\{\left[\frac{2\left(1 + \frac{\kappa - 1}{2}M^2\right)}{\kappa + 1}\right]^{(\kappa+1)/(\kappa-1)}\right\}^{1/2}$$

(3.3-86)

These functions are tabulated for $k = 1.4$ in Table 18.1-17.

When a duct is fed with subsonic flow at M_1, there exists a maximum fL/D that will cause sonic conditions to exist. The flow is then said to be "choked." When a duct is fed with supersonic flow and is initially producing sonic flow of the exit, further increases in fL/D will cause a normal shock to form near the exit and move upstream as fL/D increases.

If flow occurs out of a large vessel through an isentropic nozzle, two flow regimes are possible (Fig. 3.3-9) depending on the pressure ratio. If the back pressure is lowered relative to the upstream pressure, p_0, flow begins (1) and increases as p_B decreases (Fig. 3.3-9). For pressure ratios of less than 1 but greater than p_*/p_0, the pressure drop through the nozzle is smooth (2). When $p_B = p_*$, conditions at the nozzle throat are sonic (3), and the mass flow parameter can no longer be increased by decreasing p_B. If p_B is reduced to less than p_* (4), the mass flow parameter remains constant and the flow is said to be choked. Flow through the nozzle is identical to 3, with the additional pressure adjustment through supersonic jet expansion in the receiver.

If instead the nozzle feeds a long adiabatic duct (Fig. 3.3-10), the flow again increases as p_B decreases when p_B is not much less than p_0 (conditions 1, 2, and 3). Because the Mach number increases with distance along the duct, sonic conditions will be reached at the exit while the nozzle is subsonic. The flow chokes when the exit flow Mach number becomes unity (state 4), which occurs at p_B/p_0 less than p_*/p_0 because of the pressure drop in the duct. Further reductions of back pressure

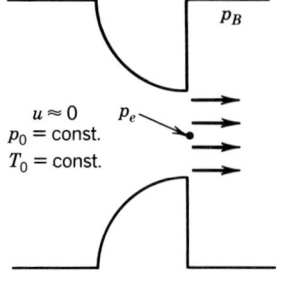

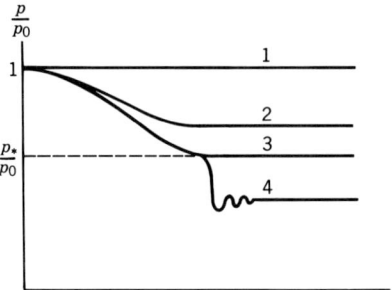

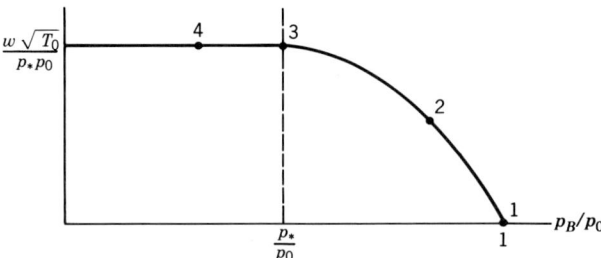

Fig. 3.3-9   Isentropic nozzle.

do not cause the flow to increase and the additional pressure drop takes place in the expansion waves at the outlet.

If the exit is not choked, calculation of flow rate can easily be achieved for a given pressure ratio by iterating from an initial guess using the isentropic and adiabatic tables (Tables 18.1-16 and 18.1-17). If the exit is choked, the Mach number in the entrance can immediately be found from Table 18.1-17. Table 18.1-16 would then be used for the nozzle calculation and hence mass flow rate.

Irrotational Flow

When a flow field or portions of a flow field are irrotational, $\omega = 0$, a potential function exists. This potential function satisfies a linear equation so that superposition of solutions is possible. For example, the flow external to a boundary layer is essentially irrotational so that the velocity potential can be utilized. In addition, potential flow solutions are often used as a first approximation to viscous flows, although it must be realized that boundary layers, wakes, and separated zones will not be properly represented. The velocity potential is defined so that by taking its gradient the velocity field results:

$$\mathbf{V} = \nabla \phi \qquad (3.3\text{-}87)$$

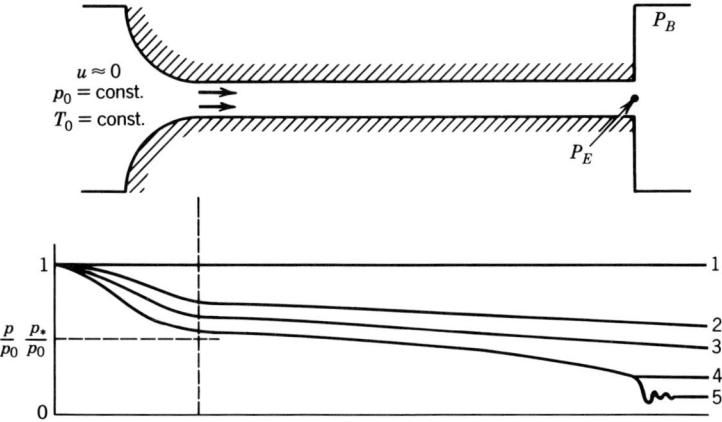

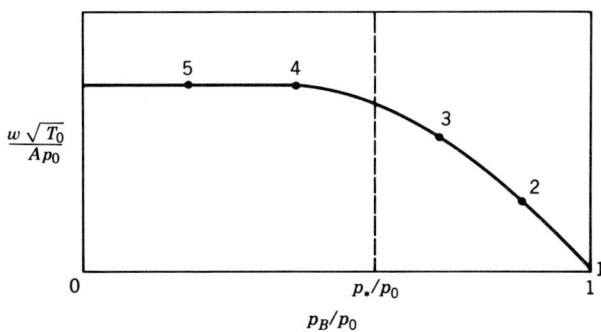

Fig. 3.3-10 Nozzle inlet to adiabatic duct.

In fact, to find the velocity component in a given direction, the derivative of ϕ in that direction is all that need be calculated. For example,

$$u = \frac{\partial \phi}{\partial x} \qquad v_r = \frac{\partial \phi}{\partial r}$$

$$v = \frac{\partial \phi}{\partial y} \qquad v_\theta = \frac{1}{r}\frac{\partial \phi}{\partial \theta}$$

$$w = \frac{\partial \phi}{\partial z} \qquad v_z = \frac{\partial \phi}{\partial z}$$

The velocity potential also has the property that it is constant on surfaces perpendicular to the streamlines.

Substituting for $\mathbf{V}$ in the incompressible form of the continuity equation yields

$$\nabla^2 \phi = 0 \qquad\qquad (3.3\text{-}88)$$

which is the Laplace equation. Solvers for this equation exist for many computer systems. The proper boundary conditions are that normal velocities at surfaces vanish:

$$\frac{\partial \phi}{\partial n}\bigg|_{surface} = 0 \qquad\qquad (3.3\text{-}89)$$

Uniform flow:

$$f(z) = Uz$$

$$u = U \qquad v = 0$$

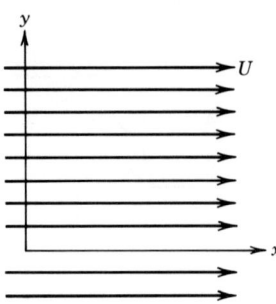

Source/sink:

$$f(z) = \frac{Q}{2\pi}\ln z$$

$$v_r = \frac{1}{r}\frac{\partial\psi}{\partial\theta} = \frac{Q}{2\pi r}$$

$$v_\theta = -\frac{\partial\psi}{\partial r} = 0$$

Q = volume flux through surface

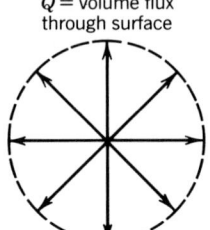

If the source is located at z_0 (instead of at the origin), then

$$t = \frac{Q}{2\pi}\ln(z - z_0)$$

For sink flow, the sign of Q is simply made negative.

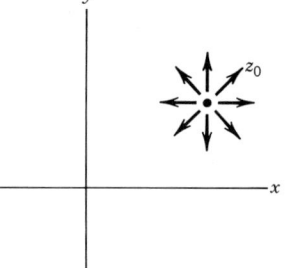

Doublet at origin:

$$f(z) = \frac{K}{z}$$

$$\phi = \frac{K}{r}\cos\theta$$

$$\psi = \frac{K}{r}\sin\theta$$

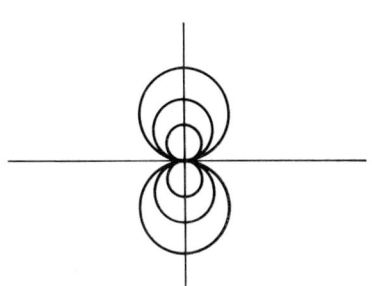

Fig. 3.3-11   Potential functions.

Stagnation point flow:

$$f = \tfrac{1}{2} K z^2$$

$$u = Kx$$

$$v = -Ky$$

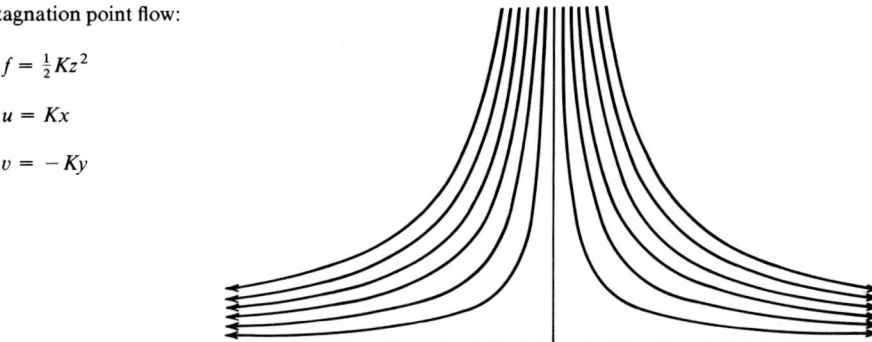

Two-Dimensional vortex:

$$f(z) = \frac{\Gamma}{2\pi i} \ln z \quad (\text{at origin})$$

$$u_r = 0$$

$$u_\theta = \frac{\Gamma}{2\pi r}$$

$$f(z) = \frac{\Gamma}{2\pi i} \ln(z - z_0) \quad (\text{at } z_0)$$

Fig. 3.3-11 *(Continued)*

By substituting for **V** in the linear momentum equation, the unsteady Bernoulli equation is obtained:

$$\frac{\partial \phi}{\partial t} + \frac{p}{\rho} + gz + \frac{V^2}{2} = C(t) \tag{3.3-90}$$

For example, to calculate the pressure difference along a length of duct L due to accelerating flow note

$$\phi = ux + f(t) \tag{3.3-91}$$

The Bernoulli equation becomes

$$x \frac{\partial u}{\partial t} + f'(t) + \frac{p}{\rho} + \frac{u^2}{2} = C(t) \tag{3.3-92}$$

If the velocity is spatially uniform, evaluation between $x = 0$ and $x = L$ gives

$$p_0 - p_l = \rho L \frac{\partial u}{\partial t} \tag{3.3-93}$$

The utility of the irrotational flow approach is that it allows separate consideration of continuity and momentum equations as well as superposition of potentials.

When the flow is two-dimensional, a simple stream function may also be introduced:

$$u = \frac{\partial \psi}{\partial y} \qquad v = -\frac{\partial \psi}{\partial x} \tag{3.3-94}$$

which identically satisfies the continuity equation. Applying the condition of irrotationality gives

$$\nabla^2 \psi = 0 \tag{3.3-95}$$

and it can be shown that lines of constant ψ are streamlines that form an orthogonal system with the equipotential lines.

Two-dimensional flows of this type are a natural application of complex variable theory. If a new potential, the complex potential $f(z)$, is formed as

$$f(z) = \phi(z) + i\psi(z) \tag{3.3-96}$$

where $z = x + iy$, any analytic $f(z)$ represents a potential flow. Since ϕ and ψ can be superimposed by virtue of their satisfying the (linear) Laplace equation, it follows that complex potentials can also be superimposed. The velocity components are found as

$$u - iv = f' \tag{3.3-97}$$

Figure 3.3-11 lists several complex functions and the flows they represent.

Although the flowfield was taken as irrotational, a finite number of singularities such as the two-dimensional vortex can be allowed. Care must be exercised as the velocity becomes infinite at such a singularity.

To illustrate the principle of superposition, the uniform stream and doublet may be added to give the flow about a circular cylinder:

$$f(z) = Uz + \frac{K}{z} \tag{3.3-98}$$

By setting $K = Ua^2$, the cylinder radius is a. The complex potential is

$$f(z) = U\left(z + \frac{a^2}{z}\right) \tag{3.3-99}$$

so that

$$\phi = U\cos\theta\left(r + \frac{a^2}{r}\right) \tag{3.3-100}$$

$$\psi = U\sin\theta\left(r - \frac{a^2}{r}\right) \tag{3.3-101}$$

and the complex velocity is

$$u - iv = U\left(1 - \frac{a^2}{z^2}\right) \tag{3.3-102}$$

Evaluating this on the cylinder surface $z = a\exp(i\theta)$ gives $|v| = 2U\sin\theta$. The Bernoulli equation can be used to calculate the pressure distribution, giving a pressure coefficient of

$$C_p = \frac{p - p_\infty}{\frac{1}{2}\rho U^2} = 1 - 4\sin^2\theta \tag{3.3-103}$$

Figure 3.3-12 illustrates the streamline pattern. As observed previously, in a viscous flow the flow separates near the cylinder extremes altering the downstream flow substantially.

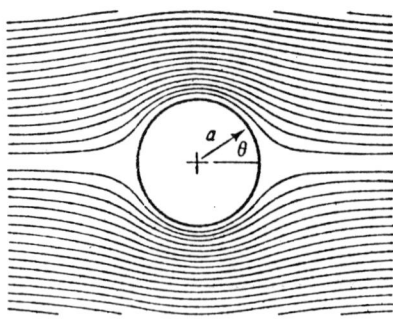

Fig. 3.3-12   Flow about circular cylinder.

It is easy to let a continuous distribution of sources, sinks, and doublets represent the shape of a symmetric body in a stream. In the same way a continuous distribution of vortices can represent a thin airfoil. By combining both superpositions, the flows about thick, curved foils can be generated. Milne-Thompson[8] is very useful for irrotational flows in general, while Kuethe and Schetzer[9] provides a good treatment of foil theory.

Fluid Transients

When liquid is flowing in a duct and the flow is suddenly changed, say by closing a valve, compression waves propagate in the liquid. Figure 3.3-13 shows a compression wave viewed from a fixed reference system in which the motion is unsteady and from a coordinate system that moves with the wave so that the motion is steady. For simplicity, assume that the downstream valve is suddenly fully closed. If c is the wave propagation speed in fluid at rest, then it propagates upstream at a speed $c - u$. In the moving coordinates flow approaches the wave at speed c and leaves at $c - u$. The continuity equation is

$$\rho cA = (\rho + d\rho)(c + dc)(A + dA) \tag{3.3-104}$$

If dp is the pressure difference across the wave, the momentum equation becomes

$$-A\,dp = \rho cA(c - u) - \rho cAc \tag{3.3-105}$$

By combining these two equations and retaining only first-order terms, the pressure rise and wave speed are

$$dp = \rho uc \tag{3.3-106}$$

$$c^2 = \frac{dp/\rho}{d\rho/\rho + dA/A} \tag{3.3-107}$$

The bulk modulus of the liquid is defined as

$$K = \rho\frac{dp}{d\rho} \tag{3.3-108}$$

so that

$$c^2 = \frac{dp/\rho}{dp/K + dA/A} \tag{3.3-109}$$

For perfectly rigid pipe, $dA/A = 0$, so that

$$c^2 = \frac{K}{\rho} \quad \text{(rigid)} \tag{3.3-110}$$

while for elastic pipe dA/A must be evaluated. This evaluation depends on the pipe material and its freedom of movement. If longitudinal movement is prohibited, then only the hoop stress must be considered. This gives

$$c^2 = \frac{K'}{\rho} \tag{3.3-111}$$

$$\frac{1}{K'} = \frac{1}{K} + \frac{d}{tE} \quad \text{(hoop stress)} \tag{3.3-112}$$

where d is the pipe diameter, t the wall thickness, and E the elastic modulus.

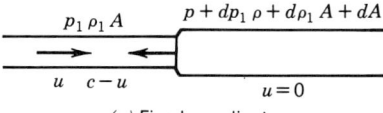

(a) Fixed coordinates

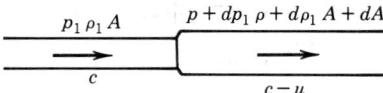

(b) Moving coordinates

Fig. 3.3-13 Compression wave.

If the flow is not shut off by valve motion but only adjusted from u_1 to u_2, the pressure rise is

$$dp = \rho c (u_1 - u_2) \qquad (3.3\text{-}113)$$

A slow valve closure may be analyzed by considering a series of such changes. Increases in flow may be similarly analyzed and give, of course, pressure decreases as expansion waves.

In the fixed coordinate system wave transmission and reflections must be considered. For example, assume that a long pipeline from a tank terminates at a valve that is quickly closed (Fig. 3.3-14). A

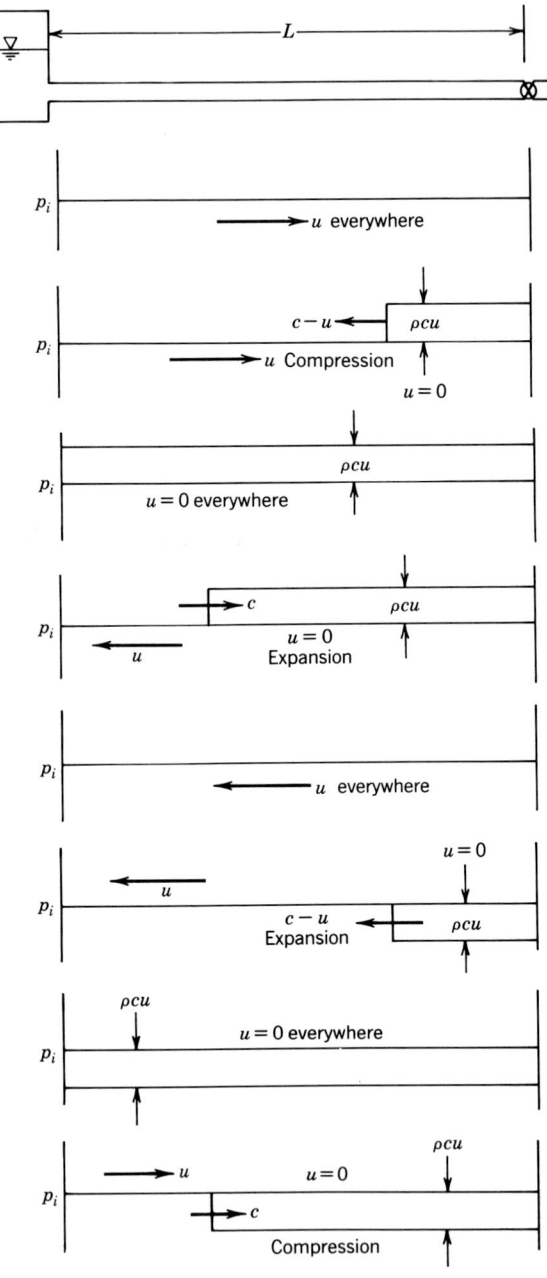

Fig. 3.3-14 Wave propagation.

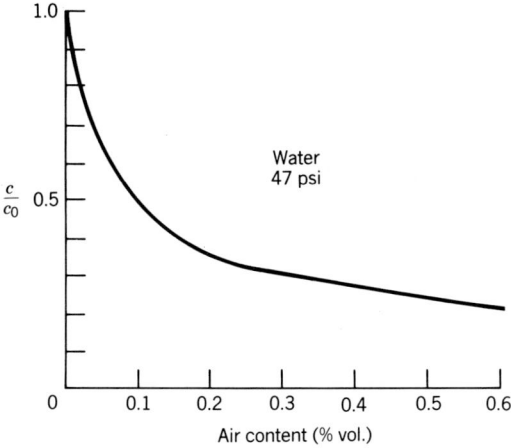

Fig. 3.3-15 Effect of gas bubbles on wave speed.

compression wave travels upstream with speed $c - u$ into the moving fluid. Behind the wave the pressure is increased ρUc above the gradeline and the fluid velocity is zero. There has not necessarily been energy dissipation in arresting the fluid velocity. Energy is instead stored in compressing the liquid and expanding the pipe walls. As the wave reaches the (large) tank, reflection of the wave takes place as an expansion wave equal in magnitude to the compression wave. As it propagates toward the valve, it cancels the high pressure created by the compression wave and starts fluid flowing toward the tank. When the expansion wave reaches the valve, the fluid is all moving away from the valve so that an expansion wave begins to propagate upstream. When this wave reaches the tank, it is reflected as a compression wave back toward the valve, which places the system in its initial state and the cycle repeats indefinitely.

It is important to observe that the total pressure excursion in the system is $2\rho uc$, which can be important for design considerations. In addition, the pressure can be reduced by ρUc below its steady-flow value. This often exceeds the depressurization required to vaporize the liquid leading to column separation. Condensation of such vapor pockets on subsequent compression waves can cause very large pressures.

Because the time required to make a round trip from the valve and back is $2L/c$, the pressure waveform at the valve is that of a square wave with period $2L/c$ and peak-to-peak amplitude of $2\rho uc$. At a distance D from the reservoir, however, there are a series of alternating positive and negative rectangular pressure pulses of width $2D/c$ at period $2L/c$.

Friction serves to reduce the amplitude of the pressure excursions with time. Since frictional loss vanishes during those parts of the motion without flow velocity, additional energy is available for the pressure excursions. This causes the peaks of the square waves to slope slightly above the value ρUc.

Gas bubbles in the liquid can drastically reduce the wave propagation speed. Figure 3.3-15, using data from Kobori et al.,[10] shows that for water at 47 psi with a 0.1% air bubble concentration by volume, the speed is reduced approximately 50%.

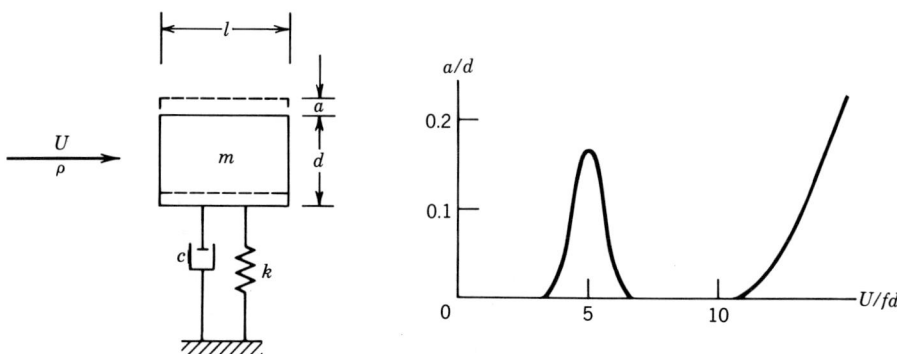

Fig. 3.3-16 Lateral vibration and response.

Flow-induced Vibration

Objects in a fluid stream can experience a variety of forced vibrations due to unsteady fluid forces that act on them. In addition, the vibratory motion of such an object can act to organize and control fluid motion such that large amplitudes are built up.

Figure 3.3-16 illustrates the lateral or across-stream vibration of a bluff object such as a structural beam, tall building, or strut. The nondimensional response amplitude as a function of reduced velocity U/fd is also shown. The first peak is resonance of the vortex-shedding frequency with the structural natural frequency. The second peak is due to an aeroelastic instability known as galloping.

The nondimensional variables used to describe flow-induced vibrations are as follows:

l/d	Fineness ratio
a/d	Amplitude ratio
U/fd	reduced velocity, U_r
$m/\rho d^2$	Mass ratio
Ud/ν	Reynolds number, R
$f_s d/U$	Strouhal number, S
ξ	damping factor

Here, f is the frequency of vibration $(\sqrt{K/m})$ and f_s is the vortex-shedding frequency for a rigid body, and

$$2\pi\xi = \ln\frac{a_i}{a_i+1} \tag{3.3-114}$$

where a_i and $a_i + 1$ are two successive amplitudes of the vibratory decay of object motion. At Reynolds numbers above 1000 most shapes have Strouhal numbers between 0.1 and 0.3 with the vast majority just near 0.2. For the example above

$$f_s = S\frac{U}{d} = 0.2\frac{U}{d} \tag{3.3-115}$$

so that

$$U_r = \frac{U}{d}\frac{d}{0.2U} = 5 \tag{3.3-116}$$

and hence the response peak at reduced velocity is equal to 5. The Strouhal number is a function of the Reynolds number, as shown in Fig. 3.3-17 for a circular cylinder and in Fig. 3.3-18 for other shapes.

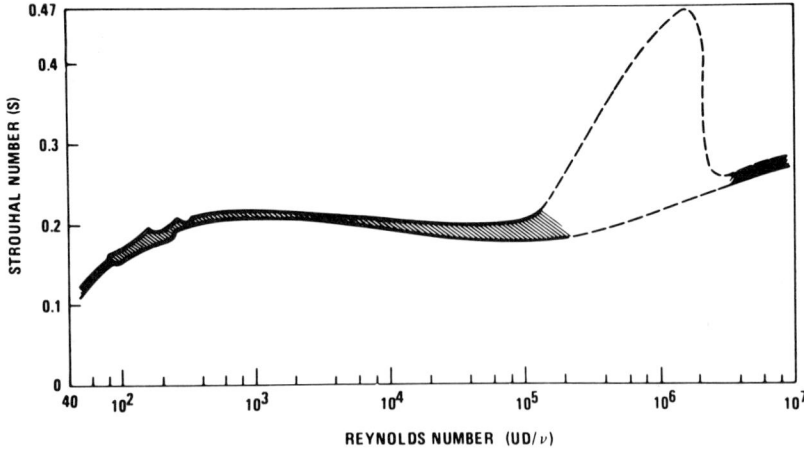

Fig. 3.3-17 The Strouhal–Reynolds number relationship for circular cylinders. (From ref. 11; reprinted by permission.)

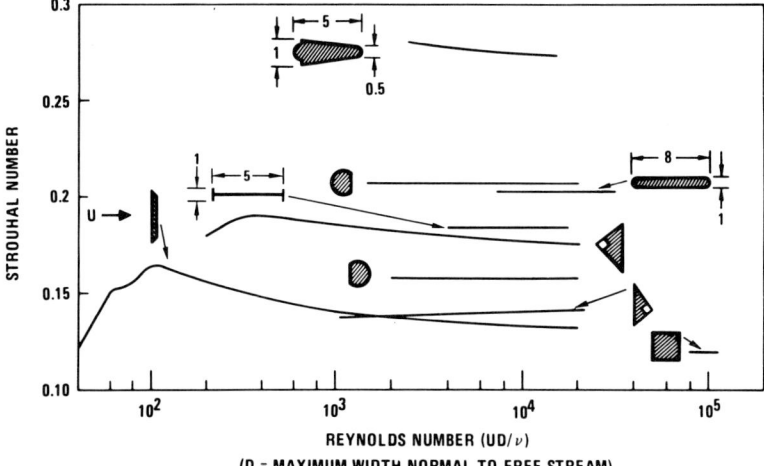

Fig. 3.3-18 Strouhal numbers for noncircular sections. (From ref. 11; reprinted by permission.)

The objective in conducting a flow-induced vibration analysis or experiment is to predict the response amplitude,

$$\frac{a}{d} = F\left(\frac{l}{d}, \frac{U_d}{\nu}, \frac{U}{f_d}, \frac{m}{\rho d^2}, \xi\right) \tag{3.3-117}$$

In any given situation some of these may have latitude for adjustment. For example, if flow-induced vibration is a problem, it may be feasible to increase damping, raise natural frequency, or change the object's cross-sectional shape.

In vortex-excited vibrations the regular shedding of vortices (Fig. 3.3-19) gives rise to alternating low-pressure regions near the object. Thus a periodic force is present at the shedding frequency. As velocity increases, so does shedding frequency. When the shedding frequency approaches the natural frequency, substantial oscillation amplitude can result. This may be augmented by the fact that the object motion can entrain the shedding frequency to the natural frequency over a range, causing frequency locking and greater amplitudes.

The galloping instability arises because, under certain circumstances, the work done by the aerodynamic forces can exceed the energy dissipated by the damping mechanisms. Some cross-sectional shapes are stable to such dynamic instabilities while others are unstable for free-stream velocities in excess of a critical value. For the lateral or plunge mode, the stability criterion is

$$\frac{U_c}{fd} = -\frac{4m(2\pi\xi)/\rho d^2}{(\partial C_L/\partial\alpha + C_D)} \tag{3.3-118}$$

where the damping factor and natural frequency refer to the lateral vibratory mode. The aerodynamic coefficients $\partial C_L/\partial\alpha$ and C_D are, respectively, the lift curve slope and drag coefficient at the appropriate attack angle.

Galloping can also occur for torsional oscillation modes (Fig. 3.3-20). The stability criterion is

$$\frac{U}{fd} = -\frac{4I(2\pi\xi)/\rho d^3 R}{\partial C_m/\partial\alpha} \tag{3.3-119}$$

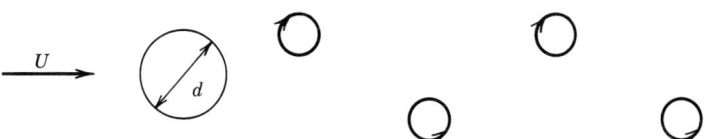

Fig. 3.3-19 Vortex shedding from circular cylinder.

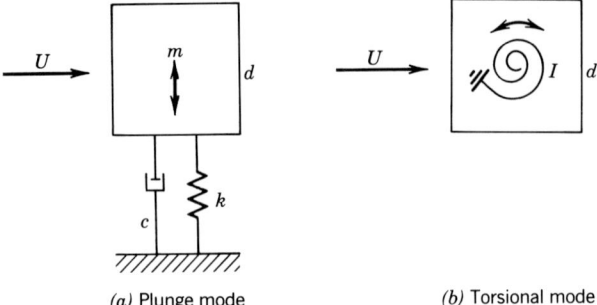

(a) Plunge mode   *(b)* Torsional mode

Fig. 3.3-20   Galloping instability.

**TABLE 3.3-3   FORCE COEFFICIENTS AND
MOMENT COEFFICIENTS FOR RECTANGLES**

a/b	$\partial C_y/\partial \alpha + C_D$	C_m
1	-2.7	-0.18
2	0	—
0.5	-3.0	-0.64
0.25	10.0	-18.00
0.20	—	-26.00

Source.   Robert D. Blevins, *Flow Induced Vibration*
Van Nostrand Reinhold, New York, 1977. Reprinted
by permission of the publisher.

where the damping and natural frequency are for the appropriate torsional mode. The characteristic radius R is usually taken as one-half the chord for structural shapes, but for aerodynamic shapes other conventions are in use. Table 3.3-3 gives typical data for plunge and torsional coefficients of rectangular shapes.

There are a wide variety of situations in which flow-induced vibrations can arise due to the above-mentioned and other phenomena. For example, tube banks in heat exchanges, tall buildings, offshore structures, guy cables, and so on. For a primer on flow-induced vibrations, the reader is referred to Blevins.[11]

3.3-5   Measuring Devices

Many devices for measuring fluid velocity exist, but many also have specialized applications. Three types that have broad areas of application are discussed below.

The Pitot static tube was, perhaps, the first instrument that yielded detailed measurements of fluid motion. Figure 3.3-21 illustrates such a device. The tip of the probe opens to the oncoming stream so that it senses the stagnation pressure of the stream, u,

$$p_t = p_\infty + \tfrac{1}{2}\rho u^2 \tag{3.3-120}$$

which is connected by an internal tube to a connection point. A series of holes is located about 3 diameters back from the tip to sense static pressure p_s, which is communicated within the probe body to the connection point. It is critical to locate the static pressure ports so that they most nearly sense the free-stream static pressure p_∞. Displacement of flow by the probe tip tends to lower the pressure around the tip below the free-stream value. The supporting shank, however, blocks the flow and tends to raise the pressure above that of the free stream. By properly positioning the ports, a good representation of the free-stream static pressure is achieved.

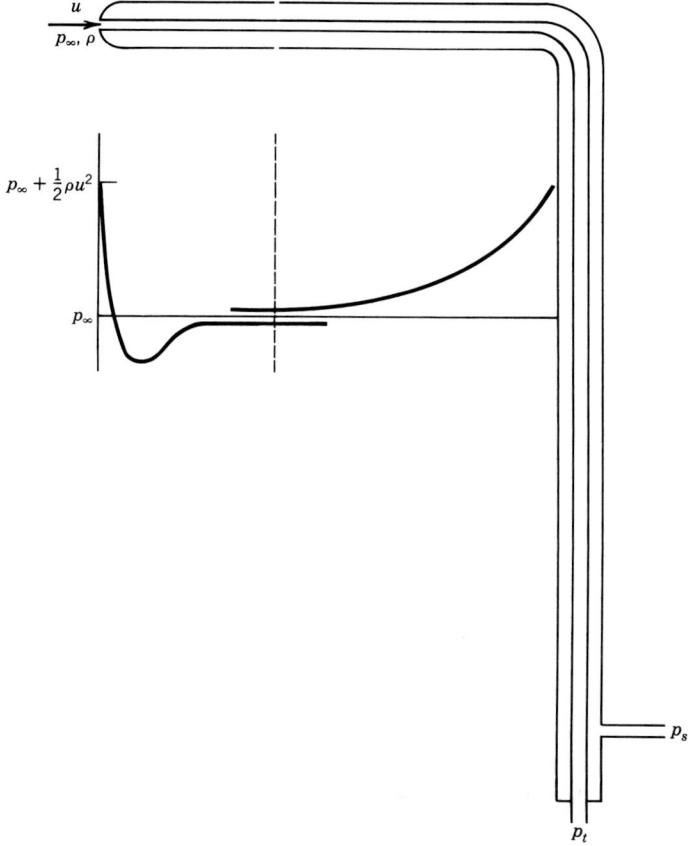

Fig. 3.3-21 Pitot static probe.

The pressure-sensing ports are usually connected to a device for measuring the pressure difference such as a manometer or pressure transducer. The velocity is then

$$u = \left[\frac{2}{\rho}(p_t - p_s)\right]^{1/2} \qquad (3.3\text{-}121)$$

Pitot static tubes and their associated pressure measurement systems usually suffer from an inability to respond to fluctuating velocities, especially turbulent velocities, but form some kind of average. As discussed in Hinze,[12] this value is not necessarily the mean of the velocity. The pressure coefficient for Pitot static tubes at small misalignments to the flow direction,

$$C_p = \frac{p_t - p_s}{\frac{1}{2}\rho u^2} \qquad (3.3\text{-}122)$$

is usually between 0.99 and 1.00. However, at Reynolds numbers (based on probe diameter) below 200 viscous effects become very important, causing the coefficient to increase substantially above unity. A wide variety of tip shapes have been investigated to reduce yaw angle sensitivity and many probe types are available commercially.

At high subsonic Mach numbers significant compressibility effects are present. The isentropic flow relations give

$$p_t - p_s = \frac{1}{2}\rho u^2 \frac{2}{\kappa M^2}\left[1 + \left(\frac{\kappa - 1}{2}\right)M^2\right]^{\frac{\kappa}{\kappa - 1}} \qquad (3.3\text{-}123)$$

TABLE 3.3-4

M	$\dfrac{p_0 - p}{\frac{1}{2}\rho u^2}$
0.1	1.003
0.2	1.010
0.3	1.023
0.4	1.041
0.5	1.064
0.6	1.093
0.7	1.129
0.8	1.170
0.9	1.219
1.0	1.276

which may be expanded as

$$p_t - p_s = \frac{1}{2}\rho u^2 \left[1 + \frac{M^2}{4} + \left(\frac{2 - \kappa}{24} \right)M^2 + \cdots \right] \tag{3.3-124}$$

Table 3.3-4 lists the bracketed quantity as a function of Mach number.

Pressure-producing probes are also designed to maximize sensitivity to yaw angle. These usually have three pressure-sensing ports: a central port for p_t and two side ports whose difference D_p varies with the angle. The average-side port pressure is also computed and used to determine p_s. Figure 3.3-22 shows the angular pressure coefficient versus yaw angle for two types of probe. The three-hole cylindrical probe is simpler to manufacture than the wedge-type probe, which has greater sensitivity. The sensitivity of wedge probes is of the order of 0.06 velocity heads per degree.

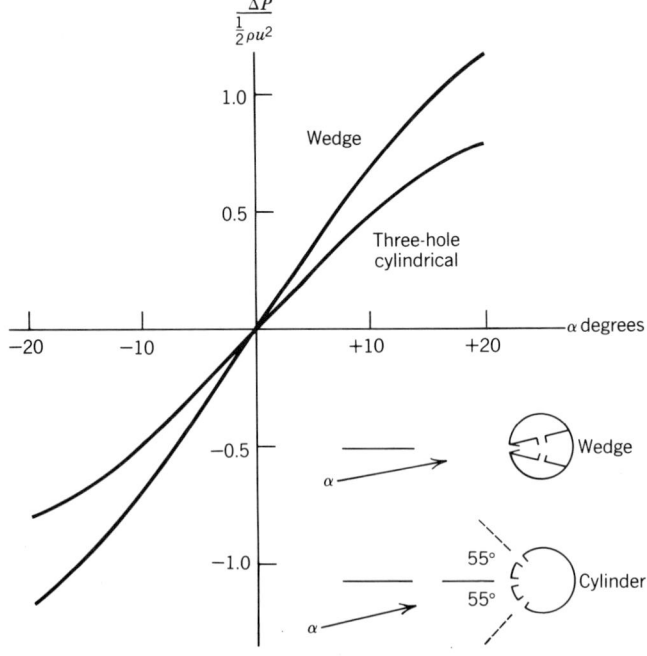

Fig. 3.3-22 Yaw probes.

Hot-wire and hot-film probes utilize the dependence of heat transfer on fluid velocity. Typically electric current is passed through a wire of small diameter or a thin metal film on a substrate so that there is a power input. The heat transfer to the fluid must equal the input power. A controller is used to adjust the current to maintain the sensor at constant temperature. The system is called a constant-temperature anemometer.

The sensor is installed in a bridge circuit (Fig. 3.3-23). A differential amplifier senses the voltage difference between a and b and provides current to the bridge in response. If the bridge is balanced at some velocity and the velocity increases, then R_w will tend to decrease. This causes e_b to exceed e_a so that the amplifier generates more current, raising the wire resistance through heating. Decreasing velocities cause opposite effects. The amplifier produces an output voltage e, which is a function of the velocity and is available for subsequent processing, say, on an oscilloscope, recorder, or computer (Fig. 3.3-24).

Equating the electric power to the heat transfer gives

$$I^2 R = \pi l K (\theta - \theta_a) N_u \tag{3.3-125}$$

where l is the wire length, K the thermal conductivity of the fluid, θ the wire temperature, and θ_a the fluid temperature. The temperature dependence of wire resistance can be written

$$R = R_0 [1 + b(\theta - \theta_0) + \cdots] \tag{3.3-126}$$

where $b = 3.5 \times 10^{-3} °C^{-1}$ for platinum and $b = 5.2 \times 10^{-3} °C^{-1}$ for tungsten. The heat transfer correlation should be chosen according to the mechanisms of importance. Compte-Bellot[13] reviews

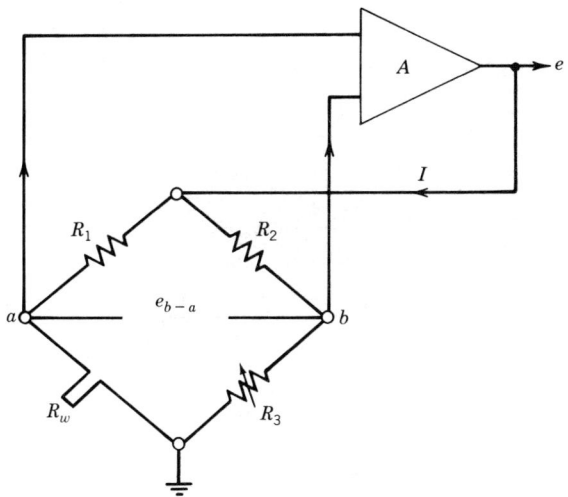

Fig. 3.3-23 Anemometer circuit.

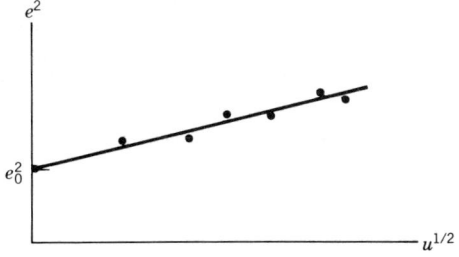

Fig. 3.3-24 Typical calibration curve.

many that are suitable for hot-wire anemometry. For forced convection and incompressible flow

$$N_u = (C + DR_e^n)\left(\frac{\theta + \theta_a}{2\theta_a}\right)^{0.17} \tag{3.3-127}$$

For Reynolds numbers, R_e, between 44 and 100 the constants are

$$n = 0.51 \qquad C = 0 \qquad D = 0.48$$

Defining the overheat ratio as the temperature rise divided by ambient temperature,

$$a = \frac{\theta - \theta_a}{\theta_a} \tag{3.3-128}$$

gives

$$\frac{I^2 R}{R - R_a} = A(1 + 0.49a) + B(1 + 0.12a)U^n \tag{3.3-129}$$

where constants A and B depend only on wire and fluid properties. For small overheat ratios

$$\frac{I^2 R}{R - R_a} = A + BU^n \tag{3.3-130}$$

Since $n \sim 0.5$ and the bridge output e is proportional to I, this law is often written

$$e^2 = A + BU^{1/2} \tag{3.3-131}$$

where the dependence of resistance on ambient fluid temperature has been absorbed. Alternatively, this equation can be inverted to give

$$u^{1/2} = c(e^2 - e_0^2) \tag{3.3-132}$$

where e_0 is the bridge output voltage when $u = 0$ and c is a calibration constant.

In practice, a small current is passed through the bridge and R_3 is adjusted for bridge nul, thus determining R_{wc}, the "cold" wire resistance. The overheat ratio is selected and R_3 adjusted to a value

$$R_3 = R_{wc}(1 + a) \tag{3.3-133}$$

The feedback amplifier is then switched on and immediately causes

$$R_w = R_3 \tag{3.3-134}$$

Commercial systems are able to track velocity fluctuations very well in air up to approximately 10 kHz. One drawback of thermal anemometers is the fourth-order transfer function. Another is difficulty with probe contamination in liquids. Figure 3.3-25 shows a hot-film sensor that is very rugged and a hot-wire that is more fragile but has better frequency response and causes less flow disturbance.

Laser doppler anemometry utilizes the monochromatic and coherent properties of laser light to measure the velocity of small particles in a flow field. Figure 3.3-26 illustrates one common optical arrangement, the dual-beam mode. A laser beam is divided into two beams of roughly equal intensity in a beam splitter. Each beam is then directed through a lens such that the beams cross at their focal point. This crossing forms the measuring volume, which is approximately that volume of space occupied coincidentally by the beams. The transmitting optics usually contains adjustments to steer the beams slightly to optimize the measuring volume.

A useful artifice for understanding the measuring principle is to imagine the interference pattern of the two beams in the measuring volume. If a concave lens were placed at the crossing point, it would project an image of the fringes on a screen such as that shown. The fringes run perpendicular to the plane of the two incident beams and are spaced a distance

$$d_f = \frac{\lambda}{2\sin\phi/2} \tag{3.3-135}$$

where λ is the wavelength of incident light and ϕ is the angle between the beams. The interference pattern consists of alternating light and dark regions. As a particle passes through, these variations in

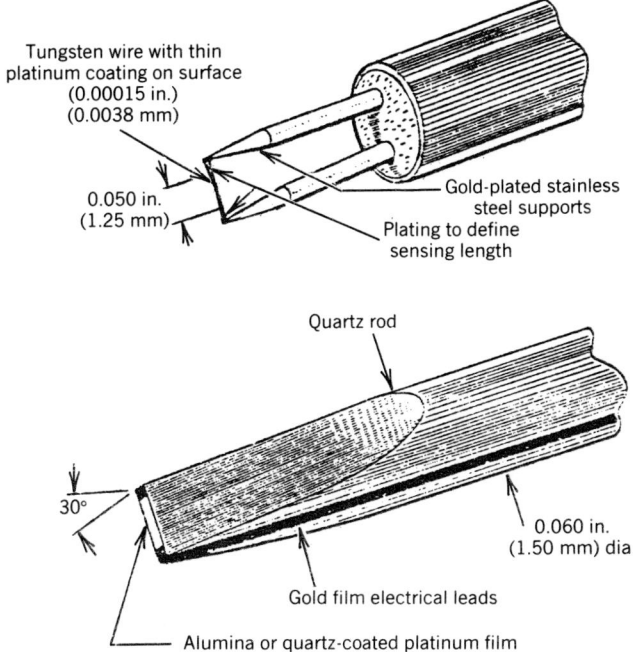

Fig. 3.3-25 Hot-wire and hot-film sensors. (Courtesy TSI, Inc., St. Paul, MN.)

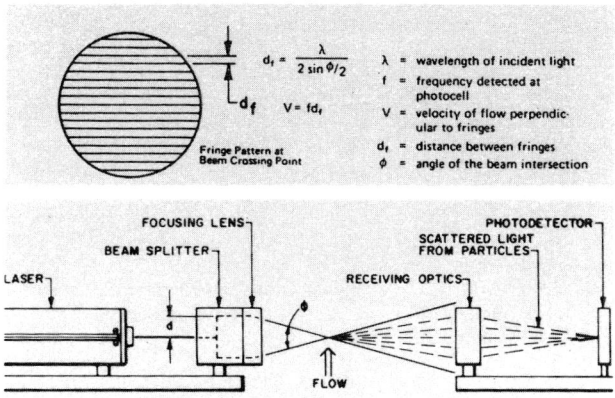

Fig. 3.3-26 Dual-beam optical configuration. (Courtesy TSI, Inc., St. Paul, MN.)

intensity cause the light scattered by the particle to vary in intensity. If a detector is provided to sense this variation, it will produce a signal whose frequency is proportional to the rate at which the particle is crossing fringes, that is, the component of particle velocity perpendicular to the fringes. Although the signal quality is dependent on the location of the receiving optics and photodetector, the measured frequency is not.

It is particularly convenient if the receiving optics can be located on the opposite side of the flow from the transmitting optics because particles tend to scatter light in a forward direction at higher intensity than other directions. An aperture is usually used in front of the photodetector so that only light from the measuring volume is incident on the detector. If good scattering intensity is available, photodiodes can be used. For low-intensity scattering, especially back scattering, then photomultipliers are used.

The dual-beam mode measures the component of a particle's velocity in the plane of the incident beams and perpendicular to their bisector as

$$V = f d_f \qquad (3.3\text{-}136)$$

where f is the frequency of the photodetector output. The beam angle is set by the beam separation produced by the beam splitter and the focal length of the transmitting lens. Angles from 4° to 30° are commonly used. In selecting the angle, consideration must be given to the desired penetration into the fluid, the refractive index of the fluid, and the tolerable size of the measuring volume. Almost any continuous wave laser can be used, but those in the visible spectrum minimize alignment problems. The power required is dictated by the speed of the flow, the particle density, and the direction of scattering detection. The helium–neon and argon–ion lasers have received wide usage. The output wavelengths are

Helium–neon $\lambda = 632.8$ nm

Argon–ion $\lambda = 488.0$ and 514.5 nm

These are in the red, blue, and green portions of the visible spectrum, respectively.

The wavelength and beam angle are the only selectable parameters. Once these are set, the proportionality constant between signal frequency and velocity is set. It is important to observe that the relationship is linear as contrasted to the nonlinear transfer functions of Pitot tube and hot-wire anemometer. There are many other possible optical arrangements, some of which can measure all three velocity components simultaneously. Other types are more appropriate for certain physical situations such as boundary layers, jets, near-wall flows, high-speed flows, and so forth. For unsteady flows that at times might have very small or reversing velocities, it is not difficult to shift the frequency of the incident beams relative to each other. This can be done with acoustic modulation or moving diffraction gratings and serves to "roll" the fringe pattern such that the photodetector output frequency is the shift frequency of zero velocity. Positive velocities cause the frequency to increase while negative velocities cause a decrease.

Detection of the doppler frequency requires specialized equipment designed to process the sequence of "bursts" from the detector. As a particle travels through the measuring volume, the detector output typically looks like that shown in Fig. 3.3-27. Because the fringe intensity is lower near the outer portions of the volume and because the particle probably overlaps more than one fringe, the signal grows in amplitude with a mean offset from the baseline. The function of the processor is to provide an analog or digital output representative of the fringe-crossing frequency.

The frequency tracker (Fig. 3.3-28) uses an error detector to constantly tune an internal oscillator to the doppler frequency of the incoming signal. It provides filtering to remove the lower-frequency component of the signal (the pedestal) and sample-hold circuits to lock the output when there is no incoming signal such as between particle arrivals. The output is typically an analog voltage proportional to doppler frequency. These instruments are designed for very wide bandwidth and require high particle concentrations in the fluid.

Counter-type processors (Fig. 3.3-29), provide digital measurement of the doppler frequency. Again, the input is filtered to remove the pedestal, and the resulting signal is amplified and clipped to form a pulse train. A selected number of pulses is counted over a selected time interval and the result converted to a binary representation of frequency.

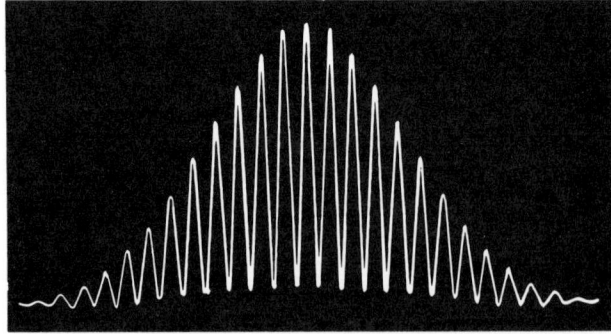

Fig. 3.3-27   Doppler signal. (Courtesy TSI, Inc., St. Paul, MN.)

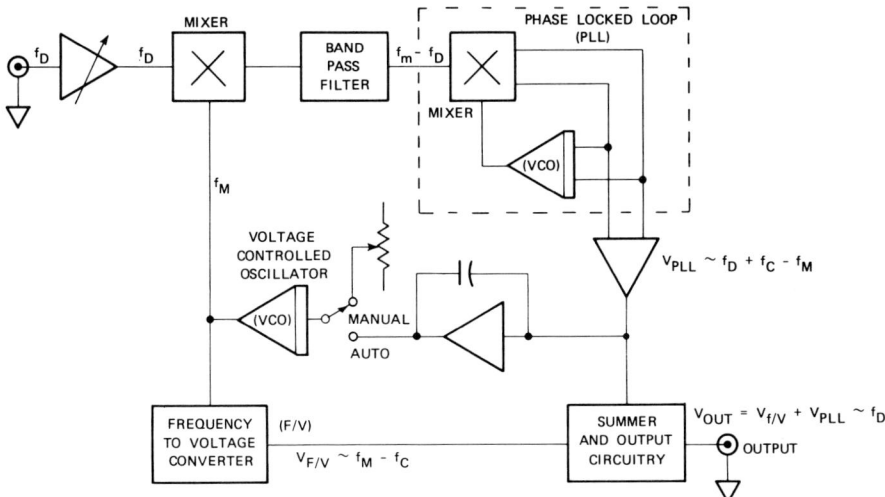

Fig. 3.3-28 Frequency tracker. (Courtesy TSI, Inc., St. Paul, MN.)

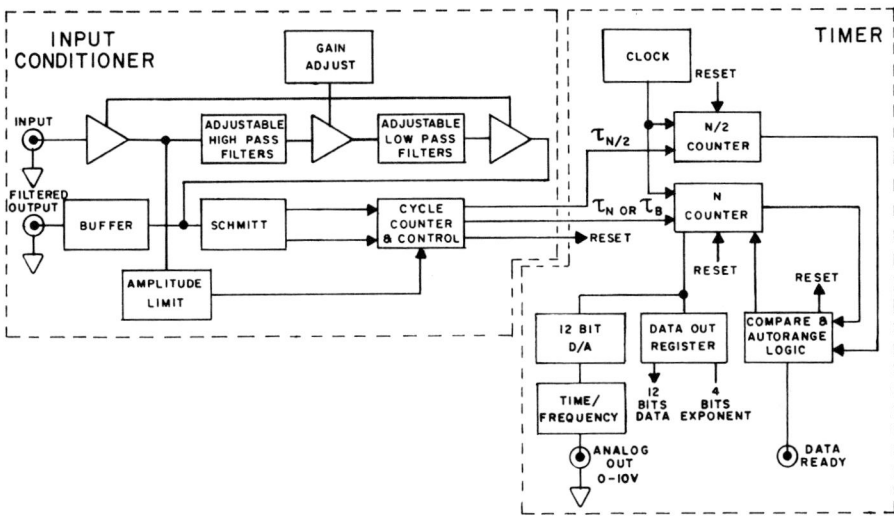

Fig. 3.3-29 Counter processor. (Courtesy TSI, Inc., St. Paul, MN.)

In order to avoid error due to extraneous or missed counts, timing of different numbers of pulses over different time intervals is usually provided concurrently. The results must agree in order to validate the signal. The digital result can be directly fed to a computing system or converted to a voltage for more conventional processing.

The scattering particles used should have a high refractive index, be uniform in size, and be as large as possible (up to 5μm). Care must be exercised, however, in selecting particles, that their velocity is essentially that of the fluid, that is, that they "follow the flow." If the concentration is too high, then substantial numbers might be in the measuring volume at a given instant causing frequency measurement problems. If there are too few particles, then the processor cannot update the output often enough to give a true representation of velocity. Tap water usually contains enough natural particles for continuous signal output. If not, addition of a small amount of milk, silicon carbide ($1.5\ \mu$m), or titanium dioxide ($0.2\ \mu$m) will usually suffice. For gas flows oil aerosols work well and can be generated using various atomizers. Alternatively, solutions can be atomized wherein the solute evaporates and leaves solid particles. Salt and sugar are often used. At higher temperatures aluminum oxide is satisfactory.

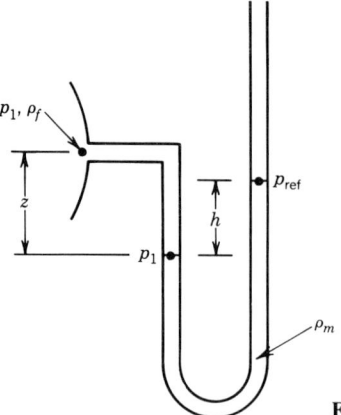

Fig. 3.3-30 Manometer.

The laser doppler anemometer is inherently linear and noninvasive, making it a unique measuring instrument for fluid measurement. Its principal disadvantage is that is requires high-quality windows to the fluid being measured. This is usually not a serious problem in the laboratory but can be extremely troublesome in industrial plant environments.

The two principal methods of measuring pressure are the raising of a fluid column in the gravitational field and the deformation of a metal material. The first, the manometer, is shown in Fig. 3.3-30 for comparison of pressure p at a point in a fluid, ρ_f, of elevation, z, with respect to the first manometer interface. The manometer contains liquid of density, ρ_m, and the pressure is to be determined relative to a reference, p_{ref}, which might be a vacuum or atmospheric pressure. Writing the Bernoulli equation gives

$$p_1 = p + \rho_f gz \tag{3.3-137}$$

or

$$p_1 - p_{ref} = p + \rho_f gz - p_{ref} \tag{3.3-138}$$

For the manometer fluid

$$p_1 = p_{ref} = \rho_m gh \tag{3.3-139}$$

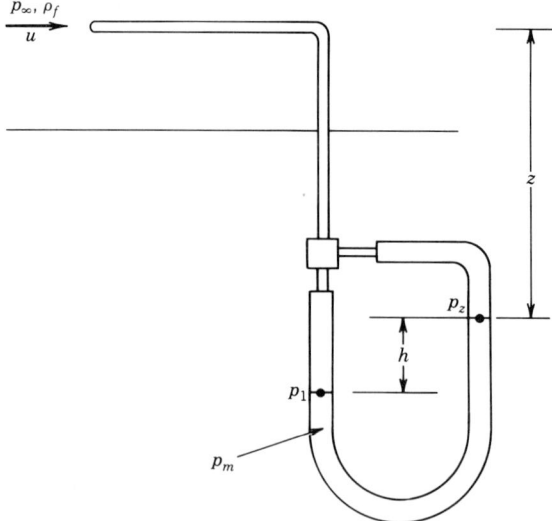

Fig. 3.3-31 Differential manometer.

so that

$$p_1 - p_{ref} = \rho_m g \Delta h - \rho_f g z \tag{3.3-140}$$

It is often necessary to measure pressure differences within the flowing fluid such as when the Pitot static or similar probe is used. Figure 3.3-31 illustrates the connection of a differential manometer.

The pressures at p_1 and p_2 can be calculated as

$$p_1 = p_\infty + \tfrac{1}{2}\rho u^2 + \rho_f g(z + h) \tag{3.3-141}$$

$$p_2 = p_\infty + \rho_f g z \tag{3.3-142}$$

Then

$$p_1 - p_2 = \tfrac{1}{2}\rho u^2 + \rho_f g h \tag{3.3-143}$$

But for the manometer fluid

$$p_1 - p_2 = \rho_m g h \tag{3.3-144}$$

so that

$$\tfrac{1}{2}\rho u^2 = (\rho_m - \rho_f) g h \tag{3.3-145}$$

The well manometer (Fig. 3.3-32), uses a small-bore glass tube for the low-pressure side and a much larger diameter well for the high-pressure side. Elevation changes of the well are very small so that only the height of fluid in the column needs to be measured. The manometer equation gives

$$p_H - p_L = \rho_m g h_1 \left(1 + \frac{d^2}{D^2}\right) \tag{3.3-146}$$

The diameter ratio is made to be very small so that error is minimal. Often the vertical scale is adjusted to make direct reading exact. The tube of a well-type manometer can be inclined at an angle α from horizontal to increase sensitivity by the factor $1/\sin \alpha$ (Fig. 3.3-33).

The dynamic response of manometer systems is generally poor and consideration of errors due to response should be considered. Although such systems are of second order, the pressures are often communicated through very small taps so that the damping is high and it becomes meaningful to consider response time. Benedict[14] considers a typical measuring system for air and shows that between 1 and 2 s are required to achieve 99% of asymptotic reading for various step input changes.

Pressure gauges usually employ a bourdon-tube-sensing element. This element has eliptical cross section and is bent in a circular shape. Under internal pressure the cross section tends to become more circular, causing motion of the free end (Fig. 3.3-34). The tube thus responds to the difference between internal and external (usually atmospheric) pressure.

The tip may be connected through a linkage and sector-pinion amplifier assembly to drive a pointer. Alternatively it may be connected to an electric transducer, such as variable resistor, to give electric output.

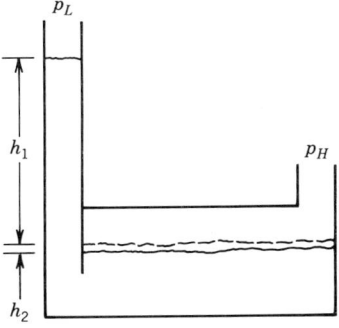

Fig. 3.3-32 Well-type manometer.

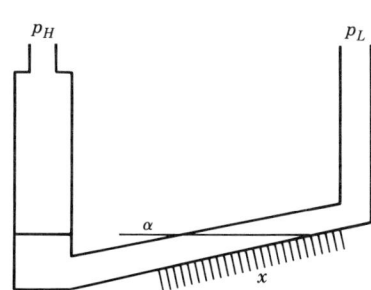

Fig. 3.3-33 Inclined manometer.

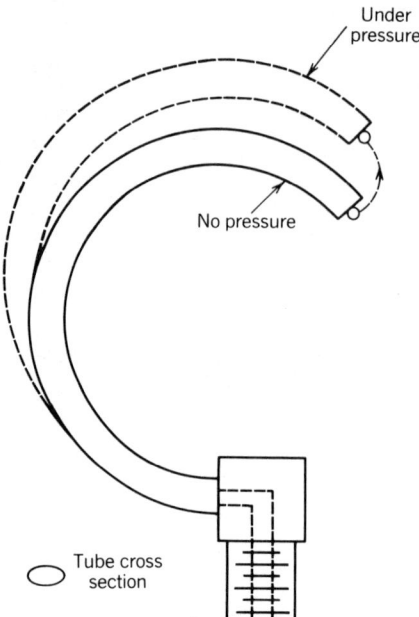

Fig. 3.3-34 Bourdon tube.

Diaphragms and bellows assemblies may also be used to convert pressure to mechanical motion. These can be used to generate motion of an observable device such as a pointer. More often, however, they are used to generate an electric signal for remote readout or for input into system controls. Electric transducers include bonded strain gauges, etched semiconductors, capacitive, variable reluctance, linear variable differential transformer (LVDT), and resistive. Figure 3.3-35 shows a bellows-type transducer coupled to a LDVT. This arrangement provides extension to low-pressure measurement but can be difficult to use with liquids unless great care is exercised in removing air bubbles. Figure 3.3-36 shows a dual-diaphragm arrangement to provide for differential pressure measurement with conductive liquids. The central chamber isolates the strain gauges from the measuring liquid.

Calibration of pressure transducers is usually carried out with deadweight testers. Particular attention should be paid to zero shift, hysteresis, and common-mode shift. In addition, pressure

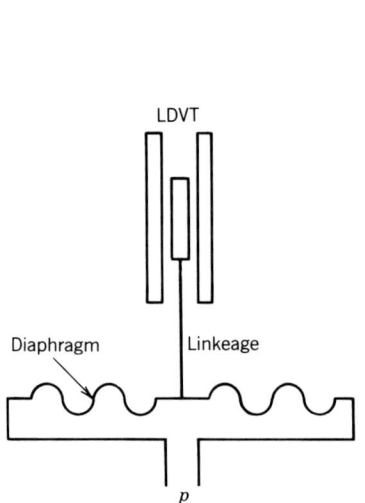

Fig. 3.3-35 Diaphragm transducer.

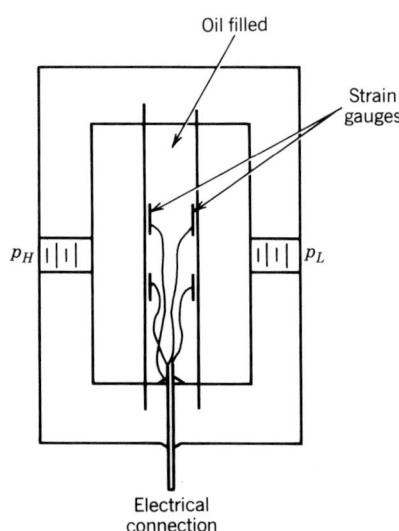

Fig. 3.3-36 DP cell.

measurement system response is of concern when the measured pressure(s) are not steady. Manufacturers usually quote diaphragm natural frequency in air at standard temperature and pressure. It can be considerably lower when liquids are being measured. In addition, because of the small volume changes, the measuring system can behave more like a transmission line than a second-order system[14] so that change of signal shape and hence averaging may be significant.

REFERENCES

3.3-1 F. M. White, *Viscous Fluid Flow*, McGraw-Hill, New York, 1974.

3.3-2 L. F. Moody, "Friction Factors for Pipe Flow," *Trans. ASME* **66** (1944).

3.3-3 R. P. Benedict, *Fundamentals of Pipe Flow*, Wiley, New York, 1980.

3.3-4 E. R. Van Driest and C. B. Blumer, *AIAAJ*. **1**, 1303–1306 (1963).

3.3-5 H. L. Dryden, *J. Aeronaut. Sci.* **20**, 477–482 (1953).

3.3-6 S. F. Hoerner and H. V. Borst, *Fluid Dynamic Lift*, Hoerner, 1975.

3.3-7 I. H. Abbott and A. E. von Doenhoff, *Theory of Wing Sections*, Dover, 1959.

3.3-8 L. M. Milne-Thompson, *Theoretical Hydrodynamics*, 4th Ed., Macmillan, New York, 1960.

3.3-9 A. M. Kuethe and J. D. Schetzer, *Foundations of Aerodynamics*, 2nd ed., Wiley, New York, 1963.

3.3-10 R. Kobori, S. Yokoyama, and H. Miyashiro, "Propagation Velocity of Pressure Wave in Pipe Line," *Hitachi Hyoron* **37**(10), October 1955.

3.3-11 R. D. Blevins, *Flow Induced Vibration*, van Nostrand-Reinhold, New York, 1977.

3.3-12 J. O. Hinze, *Turbulence*, 2nd ed., McGraw-Hill, New York, 1975.

3.3-13 G. Compte-Bellot, *Annual Review of Fluid Mechanics* **8** (1976).

3.3-14 R. P. Benedict, *Fundamentals of Temperature, Pressure, and Flow Measurements*, 2nd ed., Wiley, New York, 1977.

BIBLIOGRAPHY

Liepmann, H. W., and A. Roshko. *Elements of Gas Dynamics*. Wiley, New York, 1962.

Prandtl, L., and O. G. Tietjens. *Fundamentals of Hydro and Aeromechanics*. Dover, New York, 1957.

Schlichting, H. *Boundary Layer Theory*, trans. by J. Kestin, 6th ed., McGraw-Hill, New York, 1968.

Shames, I. M. *Mechanics of Fluids*, 2nd ed. McGraw-Hill, New York, 1982.

Shapiro, A. M. *The Dynamics and Thermodynamics of Compressible Fluid Flow*, Ronald Press, New York, 1958.

Vennard, J. K. and R. L. Street. *Elementary Fluid Mechanics*, 5th ed., Wiley, New York, 1975.

Wyle, B. E. and V. L. Streeter. *Fluid Transients*, McGraw-Hill, New York, 1978.

CHAPTER **4**

ENERGY SYSTEM TECHNOLOGY

RUDOLF FEHLAU

Borg-Warner Corporation
Long Beach, California

WARREN H. BROWN

Borg-Warner Corporation
Long Beach, California

J. B. THOMAS

Ingersoll-Rand Company
Allentown, Pennsylvania

JOHN A. MAYER Jr.

Worcester Polytechnic Institute
Worcester, Massachusetts

FRANK P. BLEIER

536 West Cornelia Ave.
Chicago, Illinois

JOHN M. SIMONSON

Valtek
Springville, Utah

**E. B. BRANCH AND
  C. A. PODCZERWINSKI**

Sargent and Lundy
Chicago, Illinois

RICHARD S. FEIN

Richard S. Fein Associates
Poughkeepsie, New York

CHESTER L. NACHTIGAL

Weyerhaeuser Company
Tacoma, Washington

A. W. LOOMIS

Stewartsville, New Jersey

DOMINIQUE N. BROCARD

Worcester Polytechnic Institute
Worcester, Massachusetts

JOHN M. BURNS

Stone & Webster Engineering Corporation
Boston, Massachusetts

JOSEPH H. DUFF

Burns and Roe, Inc.
Oradell, New Jersey

KENNETH J. BELL

Oklahoma State University
Stillwater, Oklahoma

4.1 PUMPS

4.1-1 Centrifugal Pumps: Design, Operation, Maintenance

Rudolf Fehlau and Warren H. Brown

Symbols and definitions commonly used in the pump industry are shown in Table 4.1-1 in customary U.S. and metric units. Useful formulas are listed in Table 4.1-2, expressed with constants to be utilized with customary U.S. units. (Tables and figures in Section 4.1-1 are provided courtesy of Byron Jackson Pumps Division, Borg-Warner Corporation, unless designated otherwise.)

Centrifugal pump performance is described in terms of head, brake horsepower, efficiency, and net positive suction head (NPSH) plotted as functions of capacity (discharge flow) at a constant rotational speed. Figure 4.1-1 illustrates a typical performance curve. The useful operating range from minimum to maximum flow is defined by the pump manufacturer. Operation at less than minimum flow or more than maximum flow can damage the pump.

Affinity laws relate to the effects of speed change and to the relationship of the performance of a model pump to the performance of a full-size pump. The mathematical relationships are contained in Table 4.1-2.

Figure 4.1-2 is an example of an isoefficiency performance curve showing the effect of speed variation on head and capacity. This curve assumes no reduction of efficiency when the speed is reduced. Small speed variations, such as those due to load changes of induction motors, do not affect pump efficiency. Much larger speed changes can be made without significant change in efficiency in large pumps with high Reynolds numbers. Large speed changes in small pumps require corrections for Reynolds number effects.

The parabola drawn through each efficiency point from zero to the 100% speed curve is a line of corresponding points along which the flow pattern within the pump is similar.

Scaling of a small model or an existing pump design is a cost-effective way to develop a large pump. An efficiency correction factor generally will be applicable because the efficiency of the full-size pump will be somewhat higher than that of the scale model.

Impeller trim formulas are of the same form as those for speed change, except that correction factors are included. These correction factors are a function of pump design and amount of trim and should be obtained from the pump manufacturer.

Figure 4.1-3 illustrates the isoefficiency performance curve for a pump with various impeller diameters from maximum to minimum recommended diameter. The efficiency for this particular pump increases and then decreases as the impeller is trimmed. Many pumps have their best efficiency point (BEP) at maximum diameter. Lines of constant horsepower are also plotted on this curve.

A *range chart* showing the performance for a line of pumps is shown in Fig. 4.1-4. A tentative pump selection can easily be made from this chart.

Viscosity effects—Most pump performance tests are run with fresh water at room temperature, and most performance curves are plotted for this fluid. Figure 4.1-5 illustrates the difference between the performance when pumping water and when pumping a viscous fluid.

Specific speed is a type number, calculated using the conditions at the BEP, which classifies centrifugal pumps according to their geometric proportions and performance characteristics. This is illustrated in Fig. 4.1-6, which shows the influence of specific speed on the BEP efficiency of single-suction centrifugal pumps. The values shown are considered to be typical. This figure also shows the specific speed ranges for radial, Francis, mixed-flow, and axial flow impeller types. Specific speed also determines to a large degree the typical shape of the performance curves as illustrated in Fig. 4.1-7a–c for pumps of 500, 2500, and 8000 specific speeds, respectively. The shape of the curves for pumps of any given specific speed can be varied to some degree by the hydraulic design.

Pump efficiency is maximized when losses are minimized. Pump losses can be categorized as mechanical, disk friction, leakage, and hydraulic losses.

Mechanical losses typically are very low. Hydraulic losses are relatively small in low-specific-speed pumps and increase with specific speed. The low efficiency of low-specific-speed pumps is the result of disk friction and leakage losses.

For a pump operating at other than BEP, the reduced efficiency is primarily caused by increased hydraulic losses, which are the result of recirculation in the impeller and the mismatch of the flow angles with the impeller and diffuser vane angles.

NPSH available is the net positive suction head above the vapor pressure of the pumped liquid, available at the impeller inlet. If the NPSH available is greater than the NPSH required by the pump, the pressure at the impeller inlet vane tips remains above the vapor pressure of the liquid, and cavitation does not occur.

NPSH required is shown on the pump performance curve by the pump manufacturer. Data for this curve are obtained from a laboratory suppression test and generally are plotted for a 3% head drop.

Other curves of NPSH required can be produced by selecting the point where the head drop is zero. In special cases the onset of cavitation is determined by visual observation or by other techniques.

TABLE 4.1-1 SYMBOLS AND DEFINITIONS

Symbol	Definition	U.S.		Metric	
Q	Capacity (discharge flow)	Gallons per minute	gpm	Cubic meters per hour	m^3/hr
H	Total head	Feet	ft	Meters	m
H_v	Velocity head	Feet	ft	Meters	m
H_{vd}	Discharge velocity head	Feet	ft	Meters	m
H_{vs}	Suction velocity head	Feet	ft	Meters	m
H_s	Total suction head	Feet	ft	Meters	m
H_{sv}	Net positive suction head (NPSH)	Feet	ft	Meters	m
P	Absolute pressure	Pounds per square inch	psia	Kilograms per square centimeter	kg/cm^2 abs
ΔP	Differential pressure	Pounds per square inch	psi	Kilograms per square centimeter	kg/cm^2
P_s	Suction pressure, absolute	Pounds per square inch	psia	Kilograms per square centimeter	kg/cm^2 abs
P_{vp}	Vapor pressure of fluid, absolute	Pounds per square inch	psia	Kilograms per square centimeter	kg/cm^2 abs
g	Acceleration due to gravity	Feet per second per second	ft/s^2	Meters per second per second	m/s^2
V	Average velocity of the fluid	Feet per second	ft/s	Meters per second	m/s
A	Area of the cross section of a fluid passage	Square inches	$in.^2$	Square centimeters	cm^2
bhp	Brake horsepower input to pump shaft	Horsepower	hp	Kilowatts	kW
η	Pump efficiency, brake horsepower/fluid horsepower	Expressed as decimal		Expressed as decimal	
sp.gr.	Specific gravity of the fluid compared to cold water	Expressed as decimal		Expressed as decimal	
f	Model scale factor	Expressed as decimal		Expressed as decimal	
n	Rotational speed	Revolutions per minute	rpm	Revolutions per minute	rpm
D	Impeller outside diameter	Inches	in.	Centimeters	cm
D_c	Impeller outside diameter, after trim (cut)	Inches	in.	Centimeters	cm
D_m	Impeller outside diameter, model	Inches	in.	Centimeters	cm
D_p	Impeller outside diameter, full size pump	Inches	in.	Centimeters	cm
K_h	Impeller trim correction factor, head	Expressed as decimal		Expressed as decimal	
K_{bhp}	Impeller trim correction factor, brake horsepower	Expressed as decimal		Expressed as decimal	
N_s	Specific speed	Used as type number		Used as type number	
S	Suction specific speed	Used as type number		Used as type number	

Subscripts

1	Initial operating condition or model pump condition
2	Final operating condition or full-size pump condition
m	Model
p	Pump, full size

TABLE 4.1-2   COMMONLY USED FORMULAS, U.S. UNITS

Total head	$H = 2.31(\Delta P)/(\text{sp.gr.}) + H_{vd} - H_{vs}$
Velocity head	$H_v = V^2/2g = V^2/64.4$
Fluid velocity	$V = 0.321Q/A$
Brake horsepower	$\text{bhp} = (Q)(H)(\text{sp.gr.})/3960(\eta)$
Efficiency	$\eta = (Q)(H)(\text{sp.gr.})/3960(\text{bhp})$
Affinity laws—speed change	$Q_2 = (Q_1)(n_2/n_1),$
	$H_2 = (H_1)(n_2/n_1)^2,$
	$\text{bhp}_2 = (\text{bhp}_1)(n_2/n_1)^3$
Model factor	$f = D_m/D_p$
Affinity laws—model pump	$Q_2 = (Q_1)(n_2/n_1)/(f)^3,$
	$H_2 = (H_1)(n_2/n_1)^2/(f)^2,$
	$\text{bhp}_2 = (\text{bhp}_1)(n_2/n_1)^3/(f)^5$
Impeller trim	$Q_2 = (Q_1)(D_c/D),$
	$H_2 = (K_h)(H_1)(D_c/D)^2,$
	$\text{bhp}_2 = (K_{\text{bhp}})(\text{bhp}_1)(D_c/D)^3$
Specific speed	$N_s = (n)(Q)^{0.5}/(H)^{0.75}$
NPSH available	$H_{sv} = H_s - H_{vp} = 2.31(P_s - P_{vp})/(\text{sp.gr.}) + (V_s)^2/2g$
Suction specific speed	$S = (n)(Q)^{0.5}/(H_{sv})^{0.75}$

At the point of 3% head drop, or even at the point where the head drop is zero, some cavitation exists in the eye of the impeller. Figure 4.1-8 illustrates the characteristic curves of NPSH required as a function of capacity for 3% head drop, 0% head drop, and for onset of cavitation.

Most pumps handling passive fluids can operate at 3% head drop without significant cavitation damage, and no NPSH margin is necessary. Passive fluids include propane, butane, and most hydrocarbon mixtures. On the other hand, water can cause cavitation damage under the same conditions.

NPSH margin is defined as the NPSH available minus the NPSH required and is often expressed as a percentage of NPSH required. The amount of NPSH margin that must be provided is a function of flow, speed, pump design, materials of construction, proximity of the operating point to the BEP, and the fluid pumped.

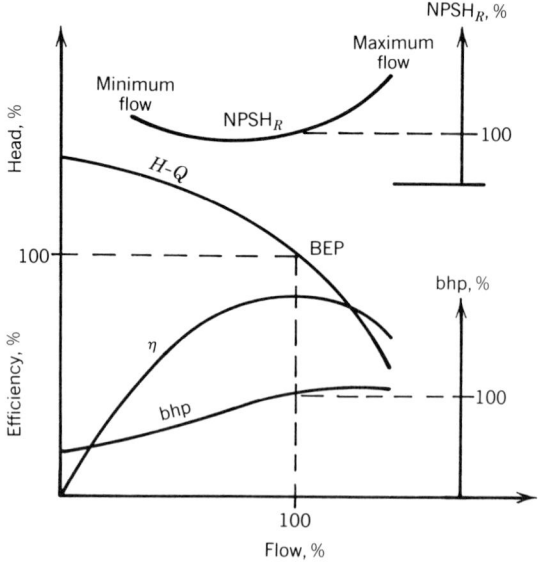

Fig. 4.1-1   Typical pump performance curves.

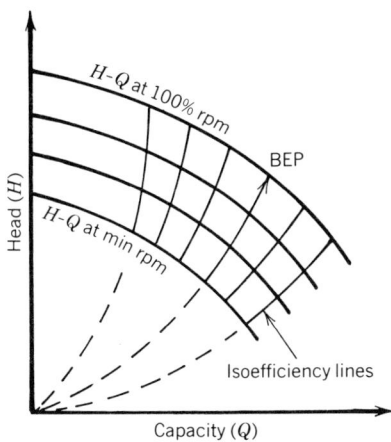

Fig. 4.1-2 H–Q as a function of pump speed.

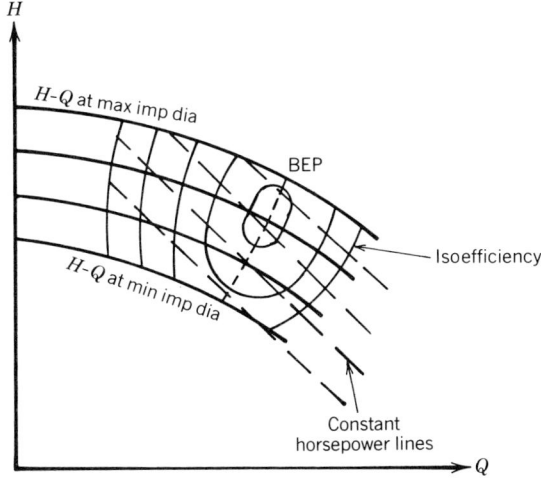

Fig. 4.1-3 H–Q as a function of impeller diameter.

Recommended values of NPSH available are given in *Hydraulic Institute Standards* and are considered to be conservative. This reference also contains submergence requirements to avoid suction vortices.

Suction-specific speed is a type number that classifies centrifugal pumps according to their NPSH requirements. Like specific speed, it is calculated using the conditions at the BEP of the pump.

The suction-specific speed of a double-suction pump is calculated using one-half of the total flow through the pump.

Upper limits of vibration, as recommended in *Hydraulic Institute Standards*,[4] are shown in Fig. 4.1-9. This graph applies to field operating conditions when handling clean liquids at flows at or near BEP without externally imposed vibration from improperly applied valves, piping supports, and so on.

A vibration identification chart, helpful in identifying the cause of vibration so that corrective' action can be taken, is shown in Fig. 4.1-10.[6]

Critical speeds are not normally found in centrifugal pumps because of the restoring forces created in annular clearance spaces such as wear rings, throttle bushings, and balance drums, which are subject to pressure differential.[2] These forces tend to raise the critical speed above the operating speed and promote stability. Such restoring forces are not present in vertical-pump lineshafts because the bearings are not subject to pressure differential.

Noise levels generated by centrifugal pumps are moderate when they are properly installed and operated. In high-horsepower installations the combined noise generated by the pump and related

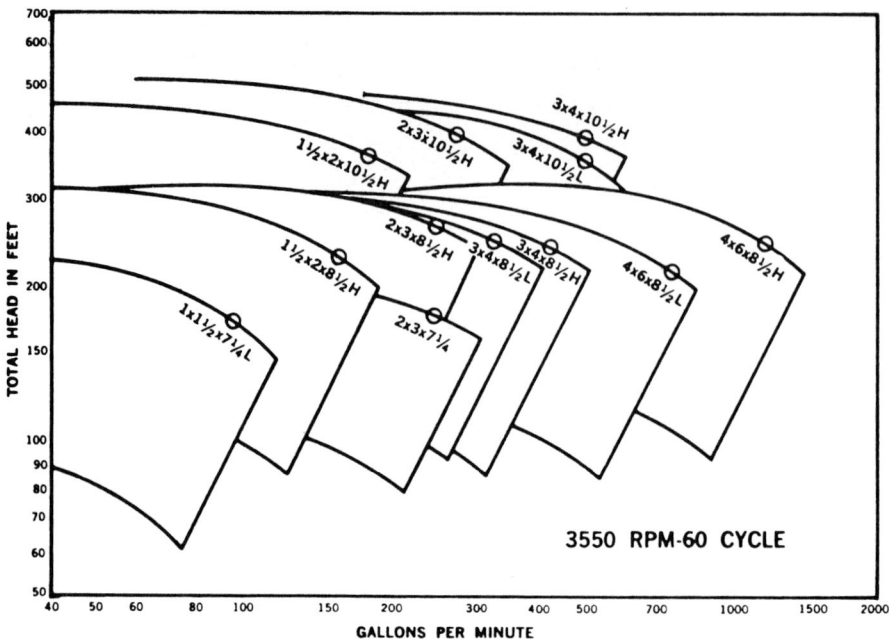

Fig. 4.1-4 Range chart.

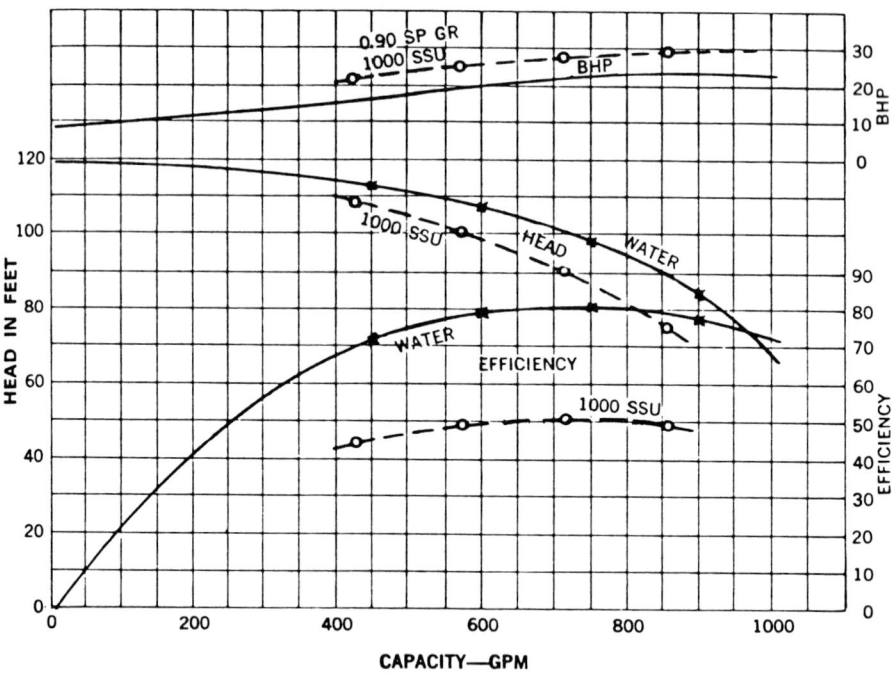

Fig. 4.1-5 Effects of viscosity on pump performance. (*Hydraulic Institute Standards*, 14th ed. 1983. Reprinted with permission of the Hydraulic Institute, Cleveland, OH.)

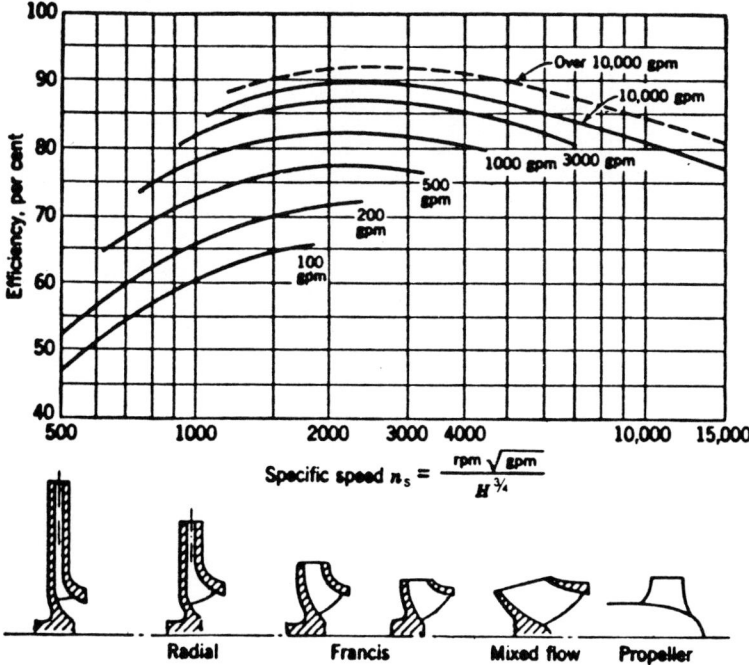

Pump efficiency versus specific speed and pump size

Fig. 4.1-6 Effects of specific speed on impeller profiles and pump efficiency. (From ref. 1; reprinted with permission.)

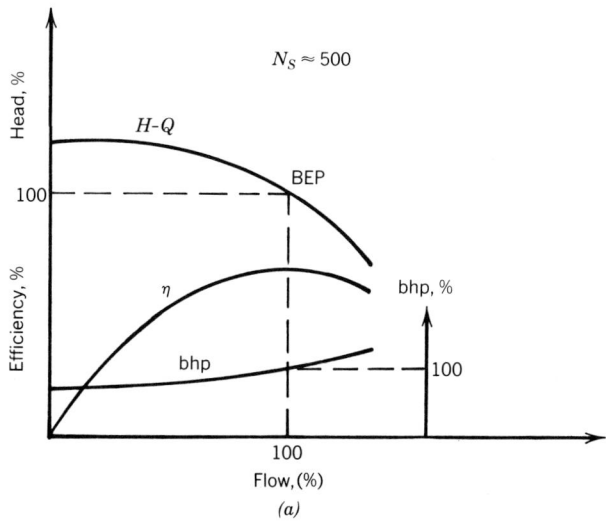

Fig. 4.1-7a Typical pump performance curve: low specific speed.

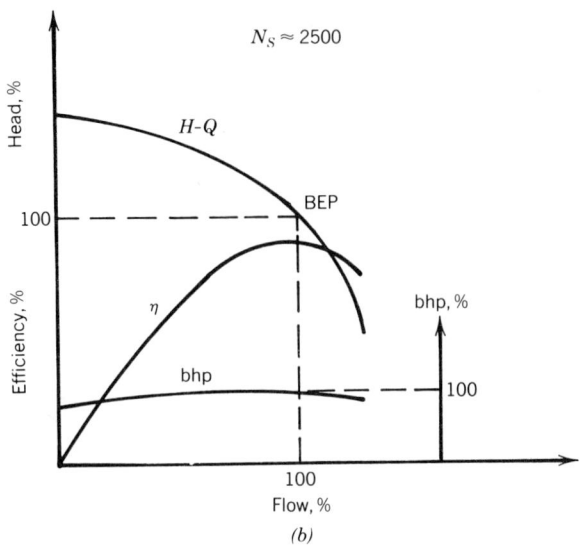

Fig. 4.1-7b Typical pump performance curve: medium specific speed.

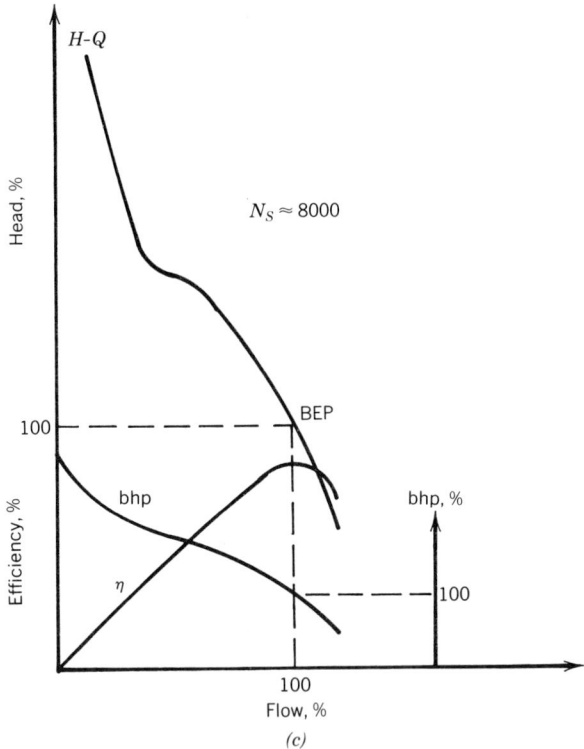

Fig. 4.1-7c Typical pump performance curve: high specific speed.

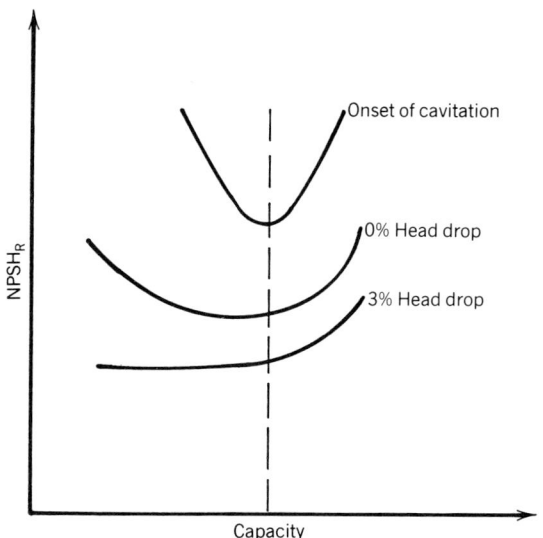

Fig. 4.1-8 Curves of net positive suction head required.

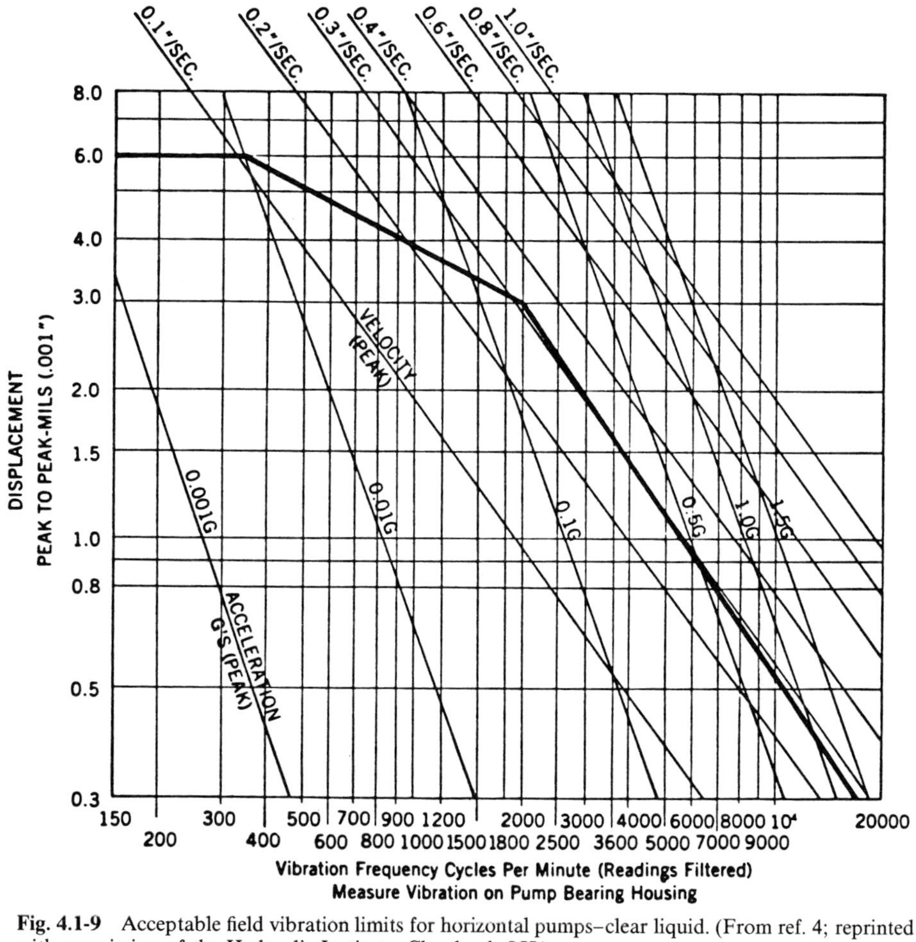

8.0

6.0

4.0

3.0

2.0

1.5

1.0

0.8

0.5

0.3

DISPLACEMENT
PEAK TO PEAK-MILS (.001")

0.1"/SEC.
0.2"/SEC.
0.3"/SEC.
0.4"/SEC.
0.6"/SEC.
0.8"/SEC.
1.0"/SEC.

VELOCITY (PEAK)

ACCELERATION G'S (PEAK)

0.001G
0.01G
0.1G
0.5G
1.0G
1.5G

150 200 300 400 500 600 700 800 900 1000 1200 1500 1800 2000 2500 3000 3600 4000 5000 6000 7000 8000 9000 10⁴ 20000

Vibration Frequency Cycles Per Minute (Readings Filtered)
Measure Vibration on Pump Bearing Housing

Fig. 4.1-9 Acceptable field vibration limits for horizontal pumps–clear liquid. (From ref. 4; reprinted with permission of the Hydraulic Institute, Cleveland, OH.)

VIBRATION IDENTIFICATION

CAUSE	AMPLITUDE	FREQUENCY	PHASE	REMARKS
Unbalance	Proportional to unbalance. Largest in radial direction.	1 x RPM	Single reference mark.	Most common cause of vibration.
Misalignment couplings or bearings and bent shaft	Large in axial direction 50% or more of radial vibration	1 x RPM usual 2 & 3 x RPM sometimes	Single double or triple	Best found by appearance of large axial vibration. Use dial indicators or other method for positive diagnosis. If sleeve bearing machine and no coupling misalignment balance the rotor.
Bad bearings anti-friction type	Unsteady - use velocity measurement if possible	Very high several times RPM	Erratic	Bearing responsible most likely the one nearest point of largest high-frequency vibration.
Eccentric journals	Usually not large	1 x RPM	Single mark	If on gears largest vibration in line with gear centers. If on motor or generator vibration disappears when power is turned off. If on pump or blower attempt to balance.
Bad gears or gear noise	Low - use velocity measure if possible	Very high gear teeth times RPM	Erratic	
Mechanical looseness		2 x RPM	Two reference marks. Slightly erratic.	Usually accompanied by unbalance and/or misalignment.
Bad drive belts	Erratic or pulsing	1, 2, 3 & 4 x RPM of belts	One or two depending on frequency. Usually unsteady.	Strob light best tool to freeze faulty belt.
Electrical	Disappears when power is turned off.	1 x RPM or 1 or 2 x synchronous frequency.	Single or rotating double mark.	If vibration amplitude drops off instantly when power is turned off cause is electrical.
Aerodynamic hydraulic forces		1 x RPM or number of blades on fan or impeller x RPM		Rare as a cause of trouble except in cases of resonance.
Reciprocating forces		1, 2 & higher orders x RPM		Inherent in reciprocating machines can only be reduced by design changes or isolation.

Fig. 4.1-10 Vibration identification chart; Form #393. (Reprinted with permission of IRD Mechanalysis, Inc., Columbus, OH. Copyright 1964.)

systems can be sufficient to cause concern and possibly exceed the levels allowed by regulatory agencies. In such cases acoustic insulation may be required.

A pump system consists of the pump and the connecting piping, valves, fittings, and vessels or tanks. The flow in the system is determined by the intersection of the pump head-capacity curve with the system head-capacity curve. The system head is the total static head plus the frictional resistance of the piping, fittings, valves, and other components of the system. The static head may be an elevation head or a pressure head such as the pressure in a boiler drum against which a boiler feed pump must operate. Figure 4.1-11 illustrates a typical system. The operating range is determined by the variations of static head and valve settings. If the pump has a variable-speed drive, several head–capacity curves are drawn from maximum to minimum speed to determine the full operating range.

Flow instability in a pump system is created by the interaction of a pump with an unstable head–capacity curve and a system with unstable characteristics. An unstable system has relatively high static head, relatively low friction head, and at least one free surface or other source of variations in static head.

Power plant feedwater systems, for example, have these inherent unstable characteristics, and a stable pump head–capacity curve (the pump head is continuously rising as the flow is reduced to minimum flow) is necessary for successful operation of such systems. On the other hand, pumps operating in pipelines in which the static differential head between pumping stations is small will not exhibit flow instability regardless of the shape of the pump head–capacity curve because the system is inherently stable. These two types of systems are illustrated in Fig. 4.1-12, which shows that with high static head unstable operation can result because the system curve intersects the pump curve at two different points. Pump specifications may unnecessarily require the pump to have a continuously rising curve. Such a requirement may compromise pump selection and thus increase both initial and operating costs. The necessity for specifying such a curve is determined by analysis of the pump system.

Parallel operation of two pumps is illustrated in Fig. 4.1-13. For any head, the flow of pump A is added to the flow of pump B to obtain the head–capacity point for the two pumps operating in parallel. The same principle applies when more than two pumps are operated in parallel. It is common, but not absolutely necessary, that the pumps be identical. If the pumps are not identical, their performance characteristics must be carefully matched so that each pump will share the flow and no pump will operate at less than minimum or more than maximum recommended flow. Boiler feed pump systems for central station power plants are good examples of parallel pump installations. Typically, two half-capacity feed pumps operate in parallel. A smaller or identical spare pump is generally provided.

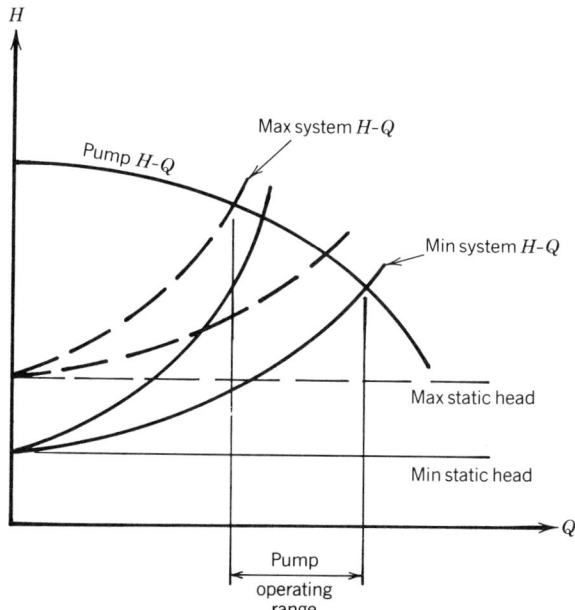

Fig. 4.1-11 Pump operating range as determined by pump and system curve.

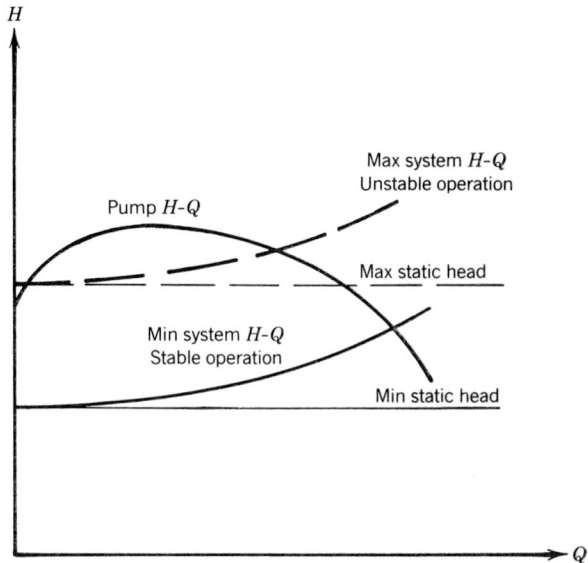

Fig. 4.1-12   Stable versus unstable operation.

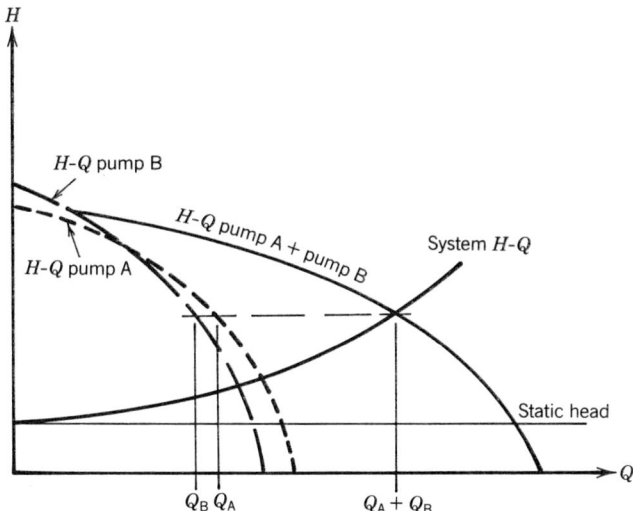

Fig. 4.1-13   Parallel pump operation.

Series operation of two pumps is shown in Fig. 4.1-14. In this case, for any flow the head of pump A is added to the head of pump B to obtain the head–capacity point for the two pumps operating in series. The same principle applies to more than two pumps operating in series and to multistage pumps, which can be considered as several pumps in series assembled on a common shaft. Typical examples of pumps operating in series include low-speed booster pumps feeding high-speed pumps in pipeline or feedwater systems.

Mechanical design features—Many varieties of shaft seal, bearing, coupling, impeller, casing, wear ring, and shaft sleeve designs are utilized in centrifugal pumps. The service for which the pump is intended dictates which mechanical design features are most desirable.

Shaft seals—The area where the shaft extends through the pump casing is called the stuffing box, because traditionally the shaft was sealed by rings of soft packing (Fig. 4.1-15). Modern pumps more often employ face-type mechanical seals for this purpose (Fig. 4.1-16).

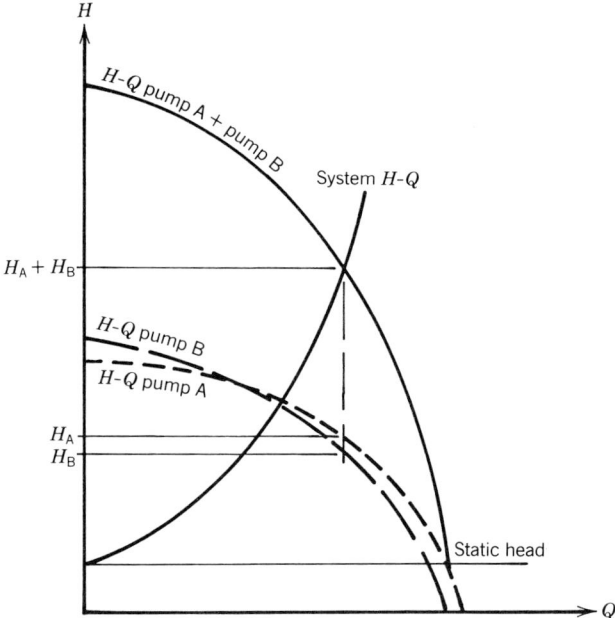

Fig. 4.1-14 Series pump operation.

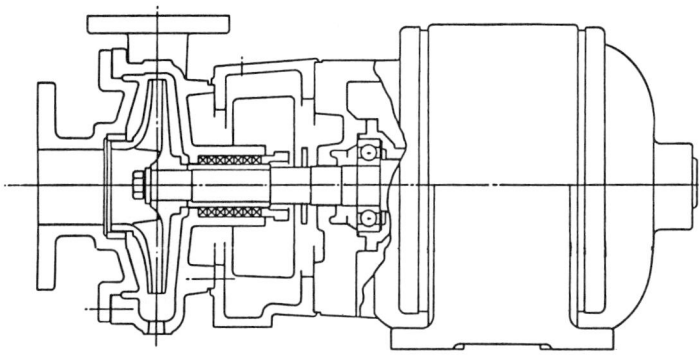

Fig. 4.1-15 Close-coupled or motor-mounted pump (typical).

Variations include double seals, with a buffer fluid between the seals so that the pumped liquid does not escape to the atmosphere, and tandem seals, in which two or more face seals are installed in series. Sometimes a throttling arrangement is used to divide the pressure equally among the seals (Fig. 4.1-27). On very large high-speed machines such as central station boiler feed pumps, throttle bushing arrangements are often used with controlled leakage of feedwater back to lower-pressure parts of the system (Fig. 4.1-20).

Bearings—In small or low-speed pumps grease-lubricated ball bearings are common. In larger pumps or with higher rotational speeds, and when greater reliability is required, oil-lubricated ball bearings (Fig. 4.1-17), sleeve and anti–oil-whip-type radial bearings, and pivot shoe thrust bearings are used (Fig. 4.1-18).

Vertical pumps employ lineshaft bearings that are often lubricated by the fluid being pumped (Fig. 4.1-24). If the liquid pumped contains abrasives, the lineshaft can be enclosed and the bearings lubricated with clean liquid or oil.

Couplings—Most vertical pumps utilize rigid couplings, which transmit axial forces from the pump to the motor thrust bearing. An adjusting nut or other device to position the pump rotor axially is generally provided (Fig. 4.1-23). Horizontal pumps utilize flexible couplings, which will tolerate a

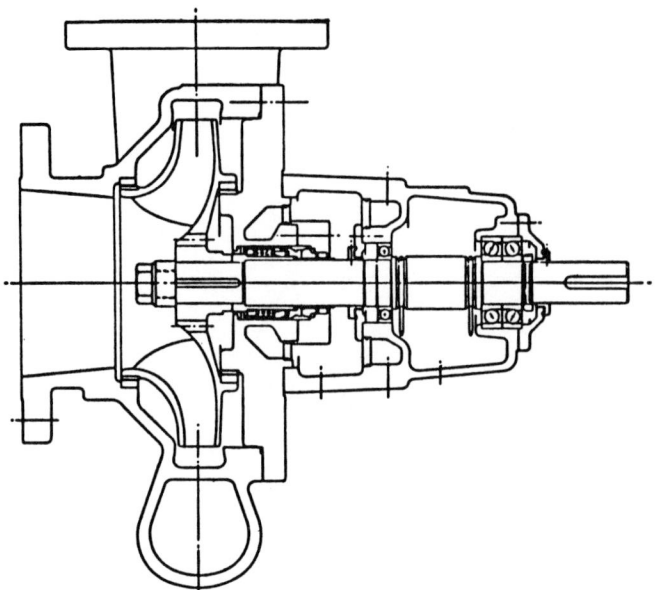

Fig. 4.1-16 Horizontal process pump. (Courtesy of Byron Jackson Pump Division, Borg-Warner Corporation.)

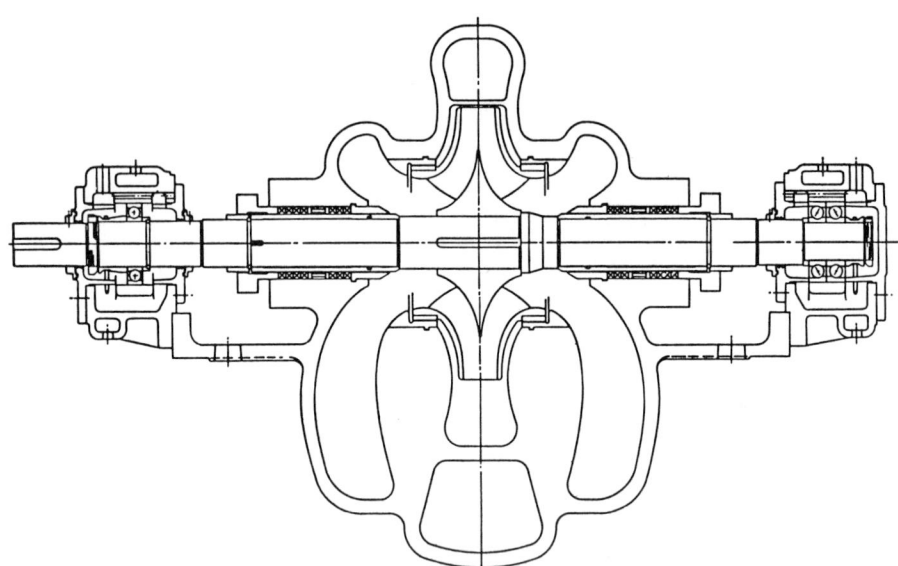

Fig. 4.1-17 Double-suction pump, axially split (typical).

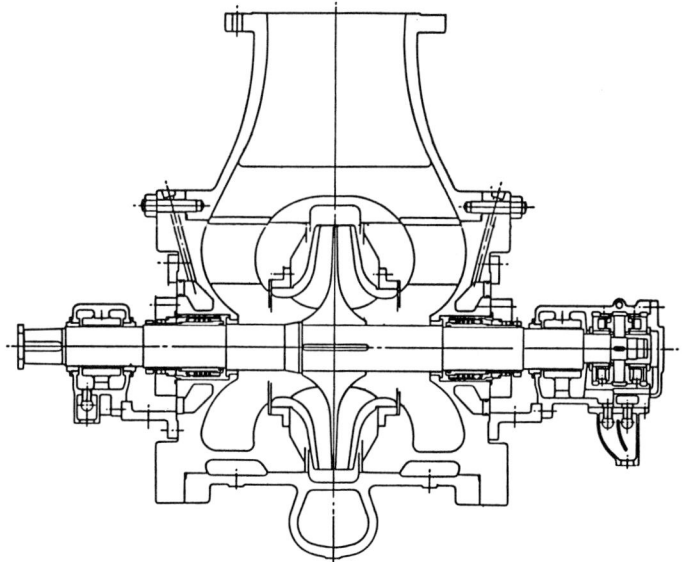

Fig. 4.1-18 Double-suction pump, radially split. (Courtesy of Byron Jackson Pump Division, Borg-Warner Corporation.)

slight misalignment between the pump and the driver. Large horizontal motors often do not have thrust bearings, and a limited end-float coupling is used to position the motor rotor axially.

Spacer-type couplings are commonly used in both horizontal and vertical pumps so that pump shaft seals and bearings can be serviced without disturbing the mounting of the pump or driver.

Impellers—Closed impellers with two shrouds and wear rings to minimize leakage losses are common in low to medium-specific-speed pumps (Fig. 4.1-16). Semiopen impellers are often used in the midrange of specific speed. The back or hub side of the impeller is shrouded, and the front or eye side is unshrouded and runs close to the pump casing or a contoured liner (Fig. 4.1-22). Adjustment to compensate for wear is therefore possible. Higher-specific-speed pumps, such as axial flow designs, utilize open (completely unshrouded) impellers. Open impellers also are utilized for special applications such as pumping abrasive slurry.

Pump casings can be axially split (Fig. 4.1-17), radially split (Fig. 4.1-18e), or of double-case design (Figs. 4.1-20 and 4.1-21). It is desirable that the pump be designed in such a fashion that the pump casing will remain in place without disturbing the mounting or the piping when the pump is serviced. For high pressures and very high or very low temperatures, radially split designs are required. Mounting of the pump casing at the horizontal centerline minimizes thermal distortions and misalignment caused by temperature changes. The stuffing box area of high-temperature pumps is often water jacketed to keep the temperature of the packing or seals within allowable limits (Figs. 4.1-16 and 4.1-18). Double volutes or diffusers are utilized in preference to single volutes to minimize radial thrust.

Wear rings and shaft sleeves—As wear occurs, leakage losses increase and pump performance deteriorates. This can be corrected by restoring the wear ring clearances.

Sometimes both case and impeller wear rings are replaceable, but there is an increasing tendency to eliminate rotating wear rings. Replaceable shaft sleeves are used to protect the shaft from wear.

Maintenance—Centrifugal pumps are inherently highly reliable, durable machines. Routine maintenance normally consists of renewal of shaft seals, bearings, wear rings, shaft sleeves, and gaskets. In most services maintenance requirements are minimal.

Pump types and classifications—For light duties and in small sizes, the pump can be mounted directly on the driver without a shaft coupling. These are known as close-coupled or motor-mounted pumps and are illustrated in Fig. 4.1-15. Special motors with mounting flanges and extended shafts are used.

Close-coupled construction is often used with vertical-shaft pumps, called in-line pumps (not illustrated). These pumps are often installed in and supported by the piping, so a pump foundation is not required. Heavy-duty in-line pumps are sometimes used for process pump applications.

A centerline-supported, overhung process pump is shown in Fig. 4.1-16. Variations of this design include models with two stages or with double-suction impellers. These pumps often are built to API

Standard 610 and are generally used for heavy-duty process applications, pumping corrosive, volatile, and hot liquids.[7]

Foot-mounted process pumps of overhung design (not illustrated) are used in corrosive and abrasive liquids in the chemical process industry. These pumps often are constructed to ANSI Specification B-73.1, which includes standardized mounting dimensions and performance parameters.

Figure 4.1-17 shows an axially split, double-suction pump. These are general-service pumps used in cooling tower, water supply, booster, fire pump, and transfer service.

For higher temperatures and heavy-duty service, horizontal double-suction pumps are constructed with radially split casings and centerline mounting, as illustrated in Fig. 4.1-18. A variation of this design incorporates two single-suction stages in series.

Axially split multistage pumps (Fig. 4.1-19) are widely used for boiler feed service in small power plants, as charge pumps in chemical and petrochemical processes, and as pipeline pumps.

For higher pressures and temperatures, the double-case or barrel pump (Fig. 4.1-20) is required. Central electric power generation plants use double-case pumps for boiler feed service. This construction is also required for hot-oil-charge pumps and for high-pressure pipeline and injection services. Figure 4.1-21 shows a double-case pump of diffuser-type construction.

Figure 4.1-22 shows a large vertical circulating pump of the kind used for cooling tower, dewatering, transfer, river intake, and other similar services.

A vertical pump with a suction barrel (Fig. 4.1-23) is often utilized in an installation where the NPSH available at the suction of the pump is insufficient or where floor space is limited. Additional

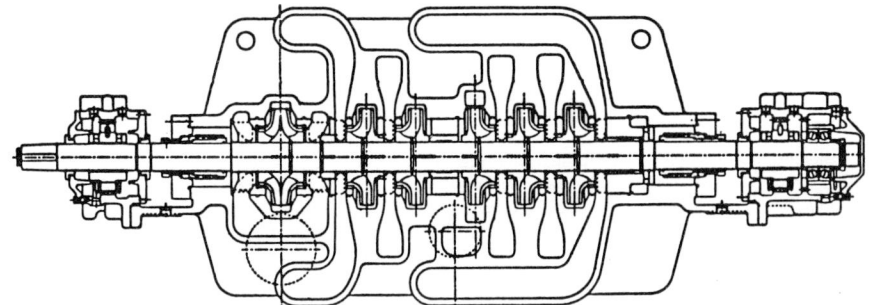

Fig. 4.1-19 Multistage pump, axially split. (Courtesy of Byron Jackson Pump Division, Borg-Warner Corporation.)

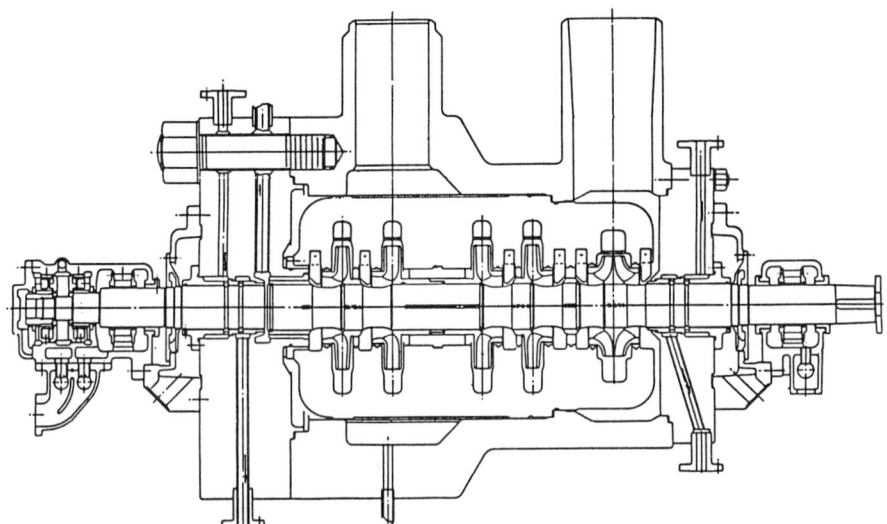

Fig. 4.1-20 Double-case or barrel pump, double volute type. (Courtesy of Byron Jackson Pump Division, Borg-Warner Corporation.)

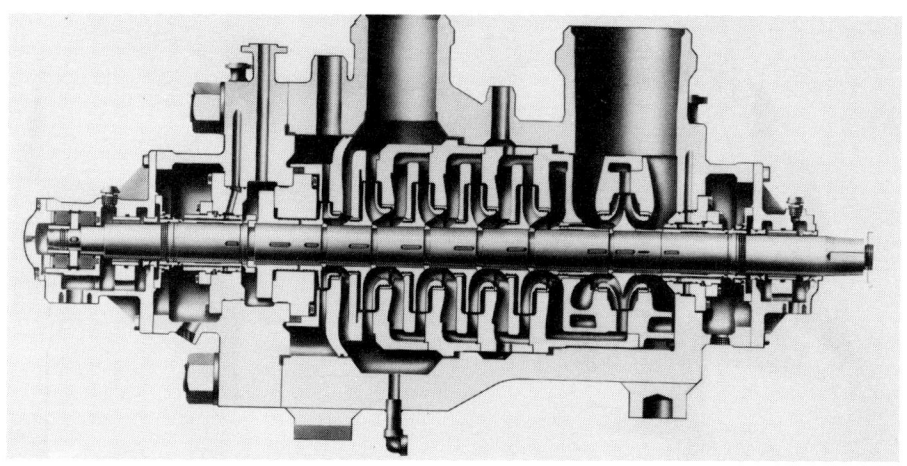

Fig. 4.1-21 Double-case or barrel pump, diffuser type. (CHTA Boiler Feed Pump, courtesy of Ingersoll-Rand Co., Phillipsburg, NJ.)

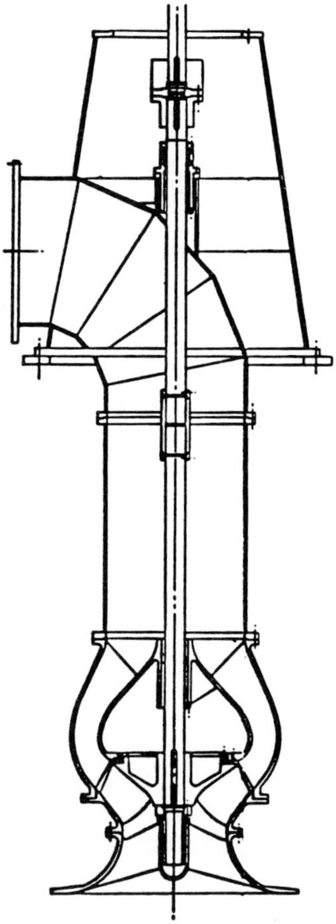

Fig. 4.1-22 Vertical circulating pump. (Courtesy of Byron Jackson Pump Division, Borg-Warner Corporation.)

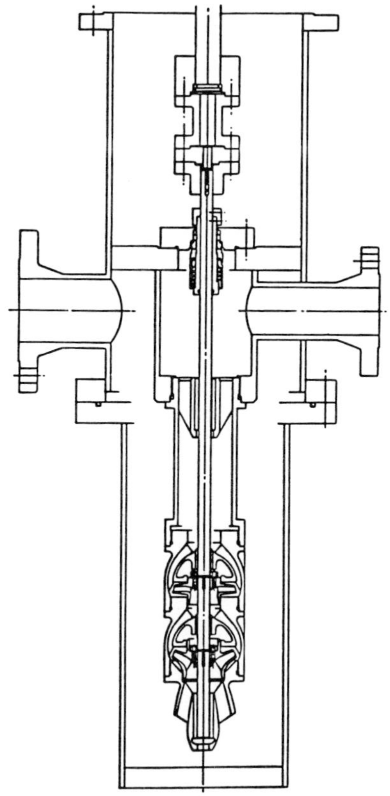

Fig. 4.1-23 Vertical process pump. (Courtesy of Byron Jackson Pump Division, Borg-Warner Corporation.)

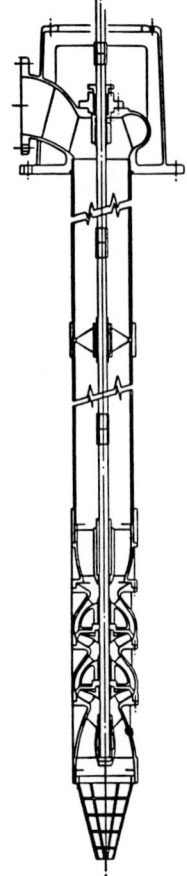

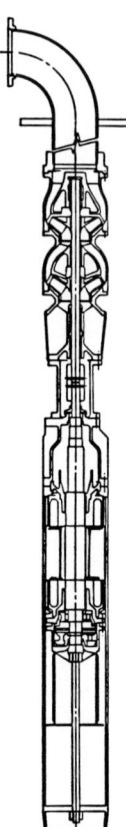

Fig. 4.1-24 Vertical turbine pump (typical). (Courtesy of Byron Jackson Pump Division, Borg-Warner Corporation.)

Fig. 4.1-25 Submersible-motor pump. (Courtesy of Byron Jackson Pump Division, Borg-Warner Corporation.)

NPSH is provided to the first-stage impeller by locating it below the suction nozzle. Such pumps are used for process and transfer service and as condensate pumps and boiler feed booster pumps.

Lineshaft vertical turbine pumps (Fig. 4.1-24) are used in agricultural, industrial, domestic, and municipal water supply service, pumping from wells up to 1000 ft deep. Figure 4.1-25 shows a deep-well pump driven by a submersible motor. This construction eliminates the need for the lineshaft and surface motor. The motor is of a special design, cooled by and sealed against the pumped fluid. Electric cables strapped to the discharge pipe supply power to the motor, which is rigidily coupled to the pump. With the motor below ground and only a discharge connection at the surface, submersible pump installations are very quiet and thus are suitable for use in residential areas. Often they are supplied for deep or hot-water wells, where lineshaft pumps are not practical.

Canned-motor units (Fig. 4.1-26) are specially designed for very high suction pressure applications and for processes where no leakage from the pump is allowed. The motor windings are protected by encapsulation or by thin stainless steel isolating liners. This type of pump is used for reactor coolant pumps on nuclear submarines as well as in high-pressure, high-temperature processes.

Figure 4.1-27 shows a large nuclear reactor coolant pump. This pump is designed to handle large flows at the high pressures and temperatures of nuclear systems.

Figure 4.1-28 shows the liquid sodium circulating pump design for the Clinch River Breeder Reactor Project. These pumps handle molten sodium at temperatures above 1000°F.

For pumping abrasive and corrosive slurries, a special pump construction is required, as illustrated in Fig. 4.1-29. Replaceable liners of hard metal, elastomer, or rubber are used to protect the case from abrasive wear. These pumps are used in mining and metal processing plants, coal washing, and coal slurry services.

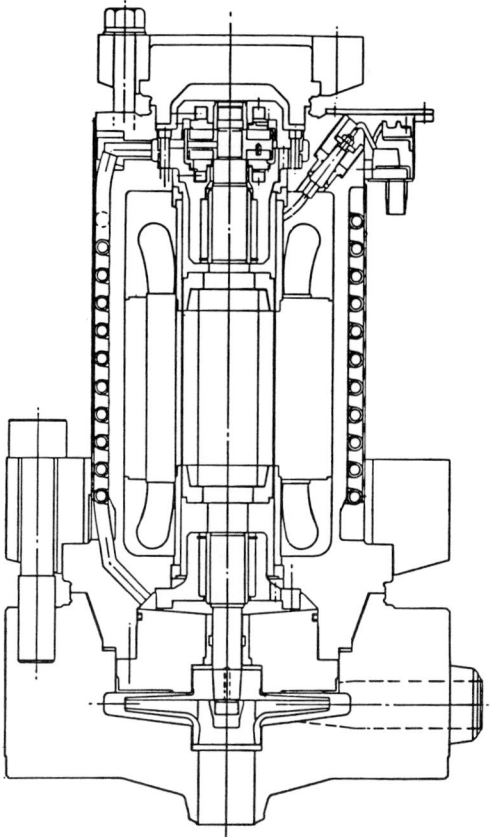

Fig. 4.1-26   Canned-motor pump. (Courtesy of Byron Jackson Pump Division, Borg-Warner Corporation.)

4.1-2   Reciprocating Pumps

J. B. Thomas

Steam

Steam pumps consist of a liquid and steam cylinder joined together by a spacer cradle. These pumps may be steam or air driven. The liquid end consists of liquid inlet and outlet ports, valves, and a piston or plunger. The steam end consists of a cylinder, inlet and outlet ports for steam or air, valve mechanism, and pistons.

Power and Metering

Power and metering pumps consist of a liquid end and a power end. These pumps are generally driven by electric motors or reciprocating engines or steam turbines. The liquid end consists of liquid inlet and outlet ports, valves, and pistons or plungers.

The power end consists of the frame, crankshaft, bearings, connecting rods, crossheads, and sometimes reduction gears.

Diaphragm

A class of pump consisting of a liquid end and a power end separated by a diaphragm that may be mechanically or fluid operated.

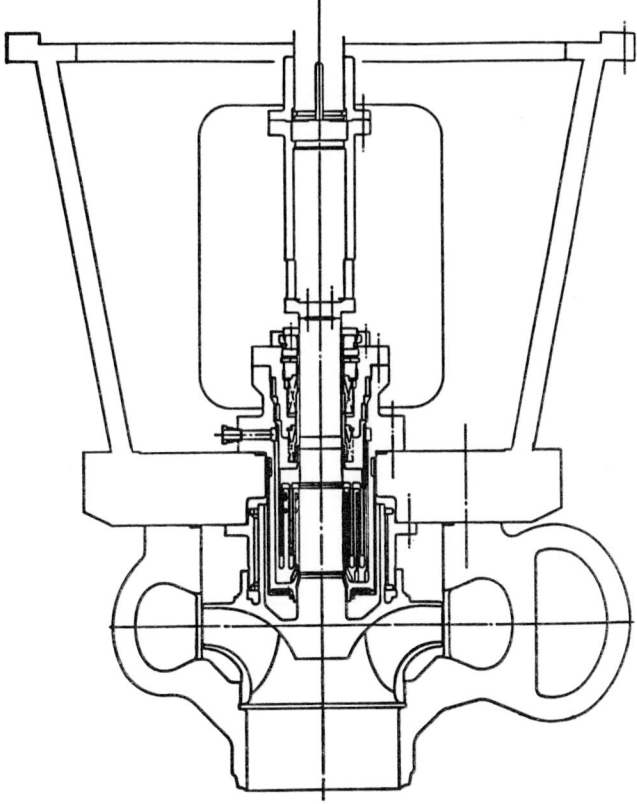

Fig. 4.1-27 Nuclear reactor coolant pump. (Courtesy of Byron Jackson Pump Division, Borg-Warner Corporation.)

For a diagram outlining classes of reciprocating pumps, see Fig. 4.1-30.
Some basic types of reciprocating pumps are as follows (see Figs. 4.1-31 to 4.1-36):

A simplex pump is a reciprocating pump having one piston or its equivalent, that is, a single- or double-acting plunger and/or diaphragm.

A duplex pump is a reciprocating pump having two pistons or their equivalent, that is, single- or double-acting plungers and/or diaphragms.

A triplex pump is a reciprocating pump having three pistons or their equivalent, that is, single- or double-acting plungers and/or diaphragms.

In a single-acting pump liquid is discharged only during the forward motion of the plunger or piston, that is, during one-half of the stroke or revolution.

In a double-acting pump liquid is discharged during both the forward and return motion of the plunger or piston, that is, discharge takes place during the entire stroke or revolution.

There are primarily two types of reciprocating power pumps that are common to industry:

1. Reciprocating piston design
2. Reciprocating plunger design

Piston pumps differ from plunger pumps insofar as the piston (displacement member) has a moving seal that operates through a fixed liner (see Fig. 4.1-37).

The plunger of the reciprocating plunger design operates through a stationary seal referred to as packing (see Fig. 4.1-38).

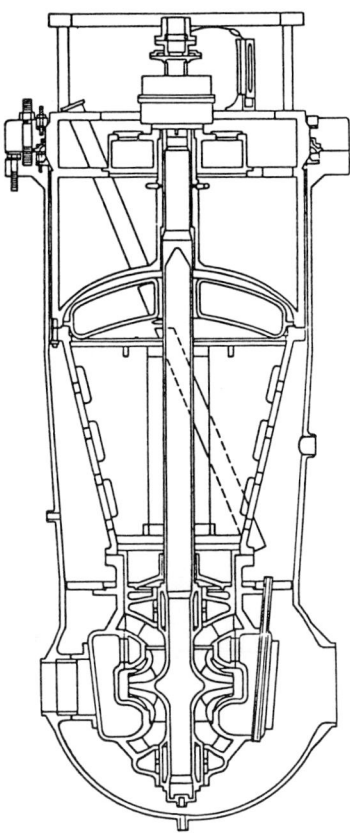

Fig. 4.1-28 Liquid sodium pump for Clinch River Breeder Reactor. (Courtesy of Byron Jackson Pump Division, Borg-Warner Corporation and General Electric Company.)

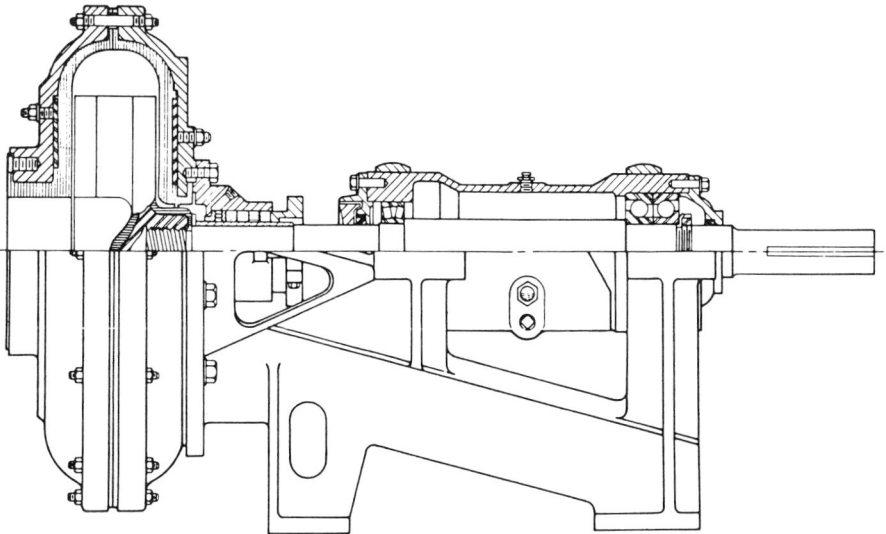

Fig. 4.1-29 Slurry pump rubber lined. (Courtesy of Galigher Ash Company, Salt Lake City, UT.)

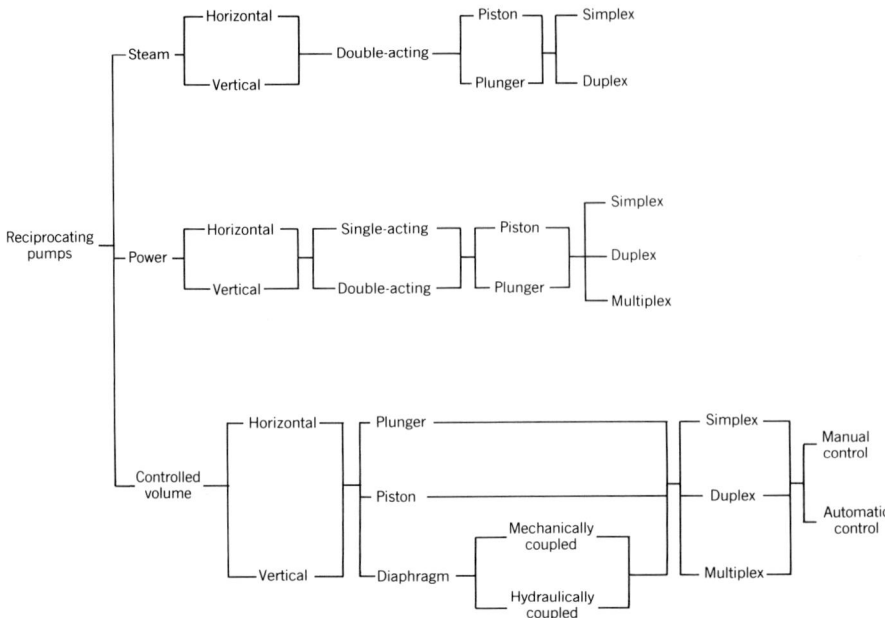

Fig. 4.1-30 Classes of reciprocating pumps. (Reprinted with the permission of the Hydraulic Institute, Cleveland, OH.)

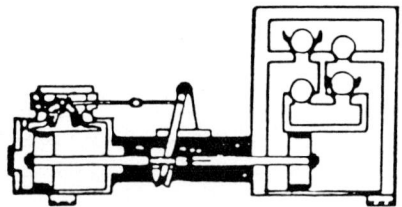

Fig. 4.1-31 Horizontal double-acting steam pump. (Reprinted with the permission of the Hydraulic Institute, Cleveland, OH.)

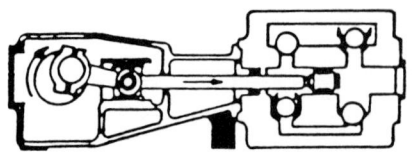

Fig. 4.1-32 Horizontal double-acting piston power pump. (Reprinted with the permission of the Hydraulic Institute, Cleveland, OH.)

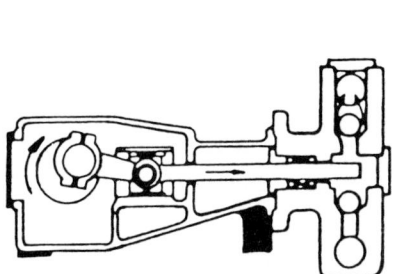

Fig. 4.1-33 Horizontal single-acting plunger power pump. (Reprinted with the permission of the Hydraulic Institute, Cleveland, OH.)

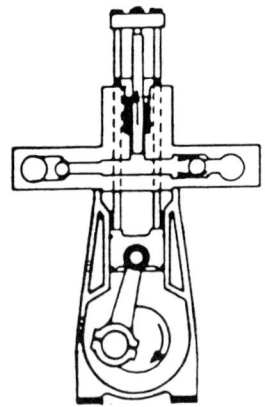

Fig. 4.1-34 Vertical single-acting plunger power pump. (Reprinted with the permission of the Hydraulic Institute, Cleveland, OH.)

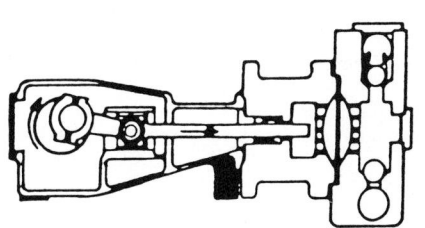

Fig. 4.1-35 Horizontal single-acting flat diaphragm pump. (Reprinted with the permission of the Hydraulic Institute, Cleveland, OH.)

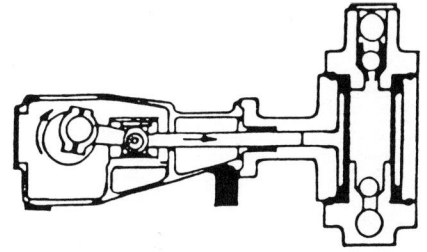

Fig. 4.1-36 Horizontal single-acting cylindrical diaphragm pump. (Reprinted with the permission of the Hydraulic Institute, Cleveland, OH.)

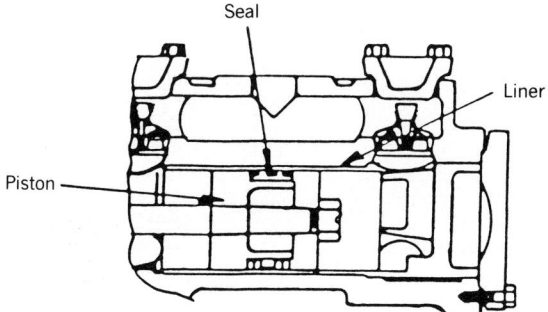

Fig. 4.1-37 Piston pump stuffing box. (Reprinted with the permission of the Hydraulic Institute, Cleveland, OH.)

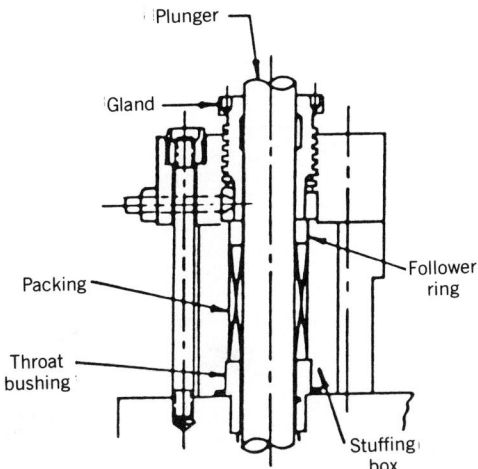

Fig. 4.1-38 Plunger pump stuffing box. (Reprinted with the permission of the Hydraulic Institute, Cleveland, OH.)

A power pump can be defined as a constant-speed, constant-torque and nearly a constant-capacity reciprocating machine whose plungers or pistons are driven through a crankshaft from an external source.

In cases where variable capacity is required, a variable-speed drive or variable-stroke design pump can be utilized.

The pump's capacity fluctuates with the number of plungers or pistons, In general, the higher the number, the less capacity variation at a given rpm. The pump is designed for a specific speed, pressure, capacity, and horsepower.

Pumps are built in both horizontal and vertical construction. The choice to construct vertical rather than horizontal is indicated by the following design features.

1. Vertically mounted and acting plunger results in compression force and plunger weight vectors in vertical direction, hence no eccentric wear of plungers, packing, or bushings as on a horizontal plunger pump.

2. Vertical design requires less floor space than comparable horsepower design horizontal pumps. This feature reduces foundation costs, and valuable floor space utilized for a vertical pump is usually one-third of the space of a comparable horizontal unit.

Piston pumps usually operate below 2000 psi (13,790 kPa) and the decision to go from a piston-type reciprocating pump to a plunger-type single-acting design becomes an economical factor.

The area of the piston will be large for low discharge pressure and small for high discharge pressure for a given rod load.

A reduction of piston diameter reduces the displacement volume, and the piston rod becomes an appreciable part of the crank-end volume, and the design starts to approach a plunger pump.

The reciprocating plunger pump can be supplied (vertical construction) up to 3000 hp. Plunger pumps are used for pressures ranging from 1000 psi (6895 kPa) to 30,000 psi (206,850 kPa).

Brake horsepower for the pump is

$$\text{bhp} = \frac{Q \times \text{Ptd}}{1714 \times \text{ME}} \tag{4.1-1}$$

where Q = delivered capacity, U.S. gpm
$\quad$ Ptd = developed pressure, psi
$\quad$ ME = mechanical efficiency, %

A reciprocating pump is the most widely used pump for medium- to high-pressure hydraulic services. It is better adapted to high-pressure service than the centrifugal pump because of its higher uniform efficiency. (Typical plunger pumps used in hydraulic services have a volumetric efficiency of 95% and a mechanical efficiency of 90%.)

Reciprocating pumps are the first choice for slurry pump application, especially for handling a medium- to high-abrasivity slurry.

Centrifugal pumps can and do pump slurries but are generally limited to low heads of 200 ft (86.7 psi or 597.79 kPa). This is due to the hydrokinetic principle of converting velocity to pressure, which involves velocity, and the magnitude of the velocity in slurries involves errosion of impeller and case. Thus, if a very high head is required, it becomes apparent that the centrifugal pump is not a practical consideration.

Practical Hydraulics for Positive-Displacement Pumps

Pressure surges are natural for all positive-displacement pumps, and if the system is designed properly, they should not be troublesome. However, if pressure surges are ignored in the system design, the result will be severe damage to the equipment and piping.

Pressure surges can be reduced by

1. properly designed piping systems,
2. consideration of net positive suction head,
3. streamlining the flow lines to and from the pump, and
4. use of suction stabilizers and discharge dampeners.

The design of the suction system should be simple. It should provide a positive suction head on each pump cylinder to assure that pulsations are kept to a minimum.

The NPSH for a reciprocating pump is simply that pressure required to "completely" fill the cylinder on each stroke. The cylinder on naturally fed pumps never fills 100%. A decrease in suction pressure on a reciprocating pump promotes a gradual decrease in volumetric efficiency. The greater the suction pressure, up to a limit, the better the hydraulic performance of the pump. Temperature of the incoming liquid has a decided effect on the suction pressure requirement of a pump. The high-pressure discharge system must be a safe and reliable system, made of components that should not have to be replaced or maintained during the life of the pump.

The positive-displacement plunger pump is a pulse generator. The discharge of the pump is an amplified version of the suction conditions. A pump operating with inadequate suction will have higher discharge pulsations than a pump with good suction conditions. Since pulsating flow cannot be totally eliminated; the engineer must be aware of its effects. Systems designed with only the static pressure taken into account may fail because of metal fatigue caued by pressure pulsation and mechanical vibration. Keeping the amplitude of pressure fluctuations and mechanical vibrations to a minimum reduces fluctuating stresses in the discharge piping.

The use of pulsation dampeners can be very effective in reducing pressure pulsations. Suction stabilizers offer stability to the all-too-vulnerable suction line. This relatively large-volume scientifically designed bottle isolates the effect of a suction line.

4.1-3   Jet Pumping Devices

John A. Mayer Jr.

Jet pumping devices are known by a variety of names: injectors, ejectors, syphons, water jet heat exchangers, eductors, and jet pumps. When compared with centrifugal pumps, these devices offer the advantage of simplicity, reliability, and low initial cost, at the expense of efficiency and high pressure gain per stage. The schematic representation of these devices is shown in Fig. 4.1-39.[9-12]

Some general observations about the relation of the three fluid streams are:

1. The pressure of the motive fluid supplied is greater than that of pumped fluid supply pressure.
2. The mass flow of the motive fluid is usually less than that of the pumped fluid.
3. The discharge pressure may range from that of the pumped fluid supply to well over that of the motive fluid (in some applications).

It is common to classify a particular device according to fluid phase and fluid type. As an example, note the classification of three common jet pumping devices that will be discussed:

Water jet pump—liquid phase, one fluid.
Steam air ejector—gaseous phase, two fluids.
Steam jet pump, or injector—two-phase, one fluid.

Of course, other combinations are possible, but these three and their closely related variations account for virtually all jet pump use today.

The Water Jet Pump

Water jet pumps are currently used in a variety of capacities and applications that require low head delivery. These include recirculation of high-pressure water inside the reactor vessel of boiling water reactors, sewerage pumping, slurry pumping, and sump service. They have been built in sizes that range from under 1 in. to sizes larger than 24 in.

The basic arrangement of a liquid jet pump is seen in Fig. 4.1-40. The driving head H_1 and capacity Q_1 are furnished externally. The pumped capacity Q_2 enters the pump suction under the supply head H_2. The discharge $Q = Q_1 + Q_2$ is delivered at head H_D.

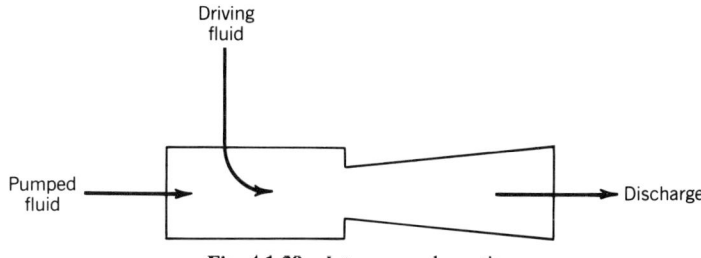

Fig. 4.1-39   Jet pump schematic.

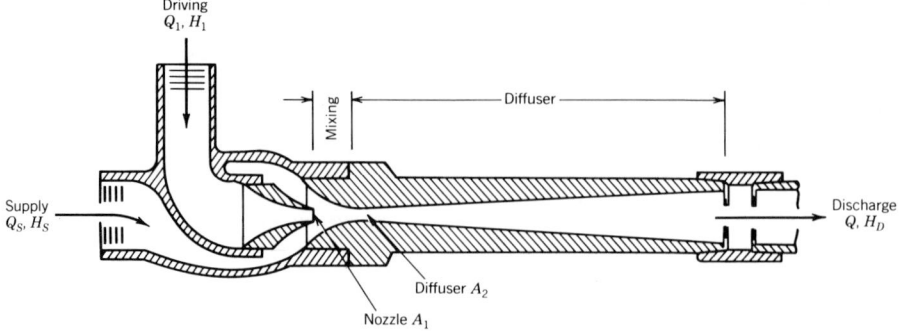

Fig. 4.1-40   Liquid jet pump.

In operation, the pump is started after making the water supply available at H_2, and the discharge line open. (A closed discharge will direct the driving stream out the supply line.) Driving stream Q_1 is accelerated through nozzle A_1 and into diffuser throat A_2, transferring energy to any fluid it entrains in the region from A_1 to A_2 (mixing region). This lowers the pressure in this region, stimulating the flow Q_2. Pressure recovery is accomplished in the diffuser section. Cavitation can occur if the pressure at the entrance of the diffuser drops below the saturation pressure corresponding to the liquid temperature at that location. Noise, flow limitation and erosion or pitting of the diffuser throat will occur under these conditions. Typical head–capacity curves for an ejector operating at various driving fluid conditions are shown in Fig. 4.1-41.

It is convenient to design a jet pump using a variation of the technique of Stepanoff.[1] He defines

$$N = \frac{H_D - H_S}{H_1 - H_D} = \text{head ratio} \tag{4.1-2}$$

$$M = \frac{Q_2}{Q_1} = \text{capacity ratio} \tag{4.1-3}$$

$$R = \frac{A_1}{A_2} = \text{nozzle/throat ratio} \tag{4.1-4}$$

The first of these, the head ratio N, is established by the service (head) intended for the pump and the driving head available. All heads are measured at the ports of the pump during operation. The most efficient nozzle–throat ratio R for this value of N can be found from Fig. 4.1-42, which represents Stepanoff's[1] survey of water jet pumps in the 1–3-in. size and application of a linear M-N relation to the data. M is then found by applying the law of conservation of momentum across the mixing region:

$$M = \frac{1}{\sqrt{R}} - 1 \tag{4.1-5}$$

The jet pump efficiency is defined as

$$e_j = MN \tag{4.1-6}$$

and the driving flow is

$$Q_1 = \frac{Q_2}{M} \tag{4.1-7}$$

Knowing the desired flow, the nozzle area required can be found:

$$A_1 = \frac{Q_1}{0.95\sqrt{2g(H_1 - H_2)}} \tag{4.1-8}$$

where the 0.95 represents the nozzle discharge coefficient and the small pressure drop of the supply from H_S to the mixing region is neglected.

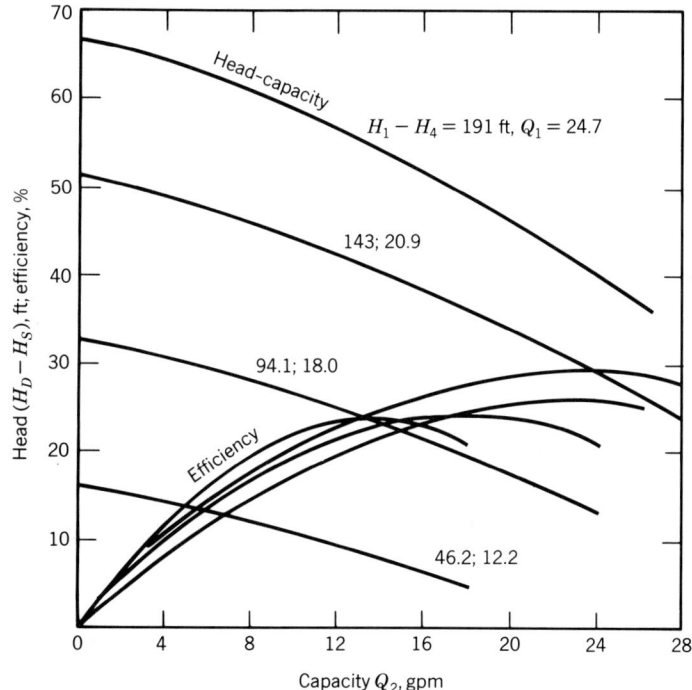

Fig. 4.1-41 Water jet pump performance. (From ref. 1; reprinted with permission.)

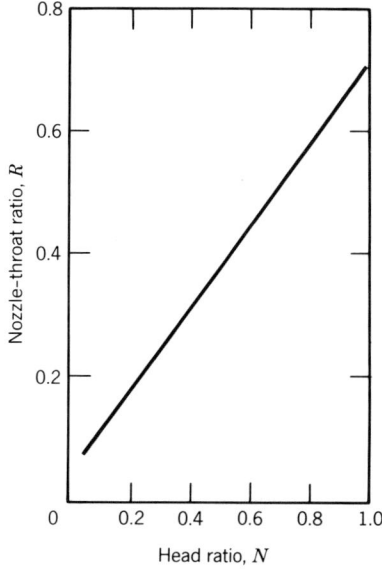

Fig. 4.1-42 Nozzle–throat area ratio for maximum efficiency.

The throat area is then computed:

$$A_2 = \frac{A_1}{R}$$

(4.1-9)

The nozzle-to-throat spacing should be set at one nozzle diameter and the diffuser length equal to six throat diameters. For higher velocities and flows, better performance may be realized if these dimensions are increased slightly.

The characteristic curve of the injector can be approximated by a straight line on an N-M plot, with intercepts of the M and N axes given by

$$M_0 = 2M$$

(4.1-10)

$$N_0 = 2N$$

(4.1-11)

Finally, if the suction conditions (temperature and pressures) are such as to cause cavitation, the nozzle–throat ratio R should be increased. That larger pump will run at reduced capacity, at less than maximum efficiency, to produce the required flow.

Example 4.1-1.   Consider the design of a water jet pump to be used to pump out a large basin. The injector will be submerged in a sump that assures a minimum head of 10 ft during the operation. A minimum flow of 200 gpm of cold water is to be removed by the pump and the friction requirements of the discharge piping and lift require a 45-ft head. Cold water at 80 psig is available.

Proceeding as above,

$$H_1 = (80 + 14.7)(2.31) = 218.8 \text{ ft}$$

$$H_S = 10 \text{ ft}$$

$$H_D = 45 \text{ ft}$$

$$N = \frac{45 - 10}{218.8 - 45} = 0.201$$

$$R = 0.18$$

$$M = \frac{1}{\sqrt{0.18}} - 1 = 1.36$$

$$e_j = 1.36(0.201) = 0.273$$

$$Q_1 = \frac{200}{1.36} = 147 \text{ gpm}$$

$$Q_1 = \frac{147}{448.8} = 0.328 \text{ ft}^3/\text{s}$$

$$A_1 = \frac{0.328}{0.95\sqrt{2g(218.8 - 10)}} = 0.00292 \text{ ft}^2$$

$$A_2 = \frac{0.00292}{0.18} = 0.0162 \text{ ft}^2$$

$$D_1 = 12\sqrt{\frac{4}{\pi}(0.00292)} = 0.732 \text{ in.}\quad \text{nozzle diameter}$$

$$D_2 = 12\sqrt{\frac{4}{\pi}(0.0162)} = 1.72 \text{ in.}\quad \text{throat diameter}$$

Set nozzle-to-throat dimension at $\frac{3}{4}$ in. and diffuser length at 10 in.

The expected performance can be plotted by calculating $M_0 = 2.72$ and $N_0 = 0.402$ and plotting the N-M straight line as shown in Fig. 4.1-43, then using this data to construct the characteristic curve of the unit as in Fig. 4.1-44. Note that this method does not show the slightly concave downward curvature of the head-flow curve seen in practice.

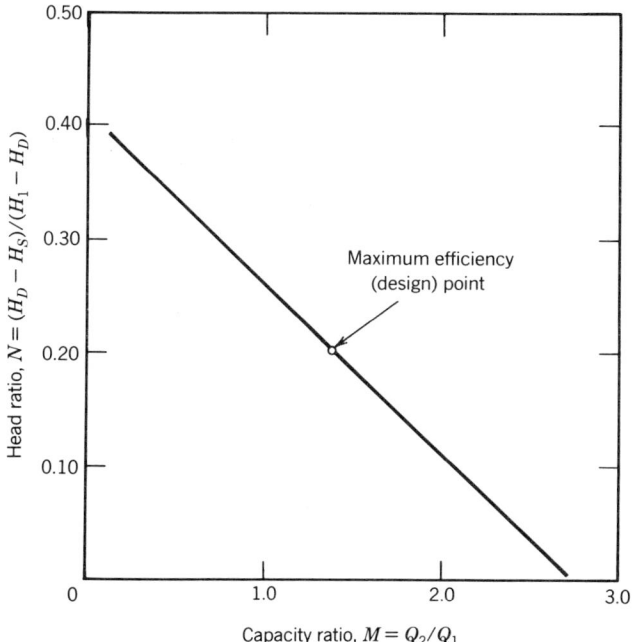

Fig. 4.1-43   Sample problem N–M curve.

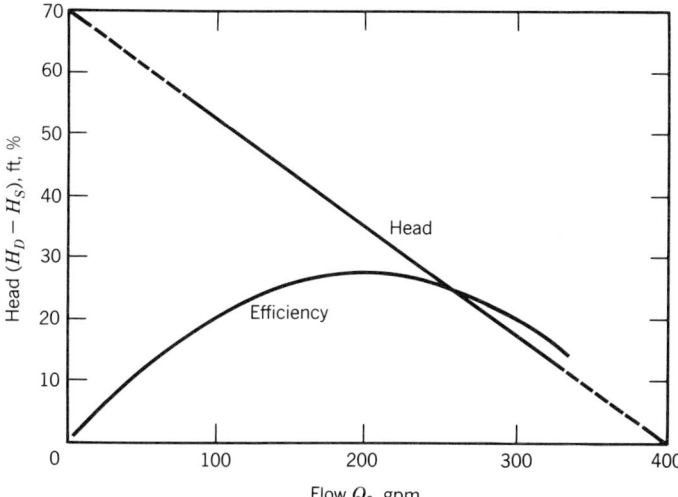

Fig. 4.1-44   Example problem characteristic curve.

The Steam Air Ejector

The steam air ejector (Fig. 4.1-45) is widely used as a vacuum pump, removing air or other gases from equipment. The principal application of the device is in steam power plants, where it is used to remove air from the condenser.

A single-stage ejector is able to produce suction pressures down to 3 in. Hg absolute when operated as a vacuum pump. Acting as a compressor, it can develop a compression ratio of 6 : 1.

In operation, the high-velocity steam jet issues from the nozzle into the suction volume of the ejector. As this jet moves into the throat of the diffuser, the shear stresses between the jet and the

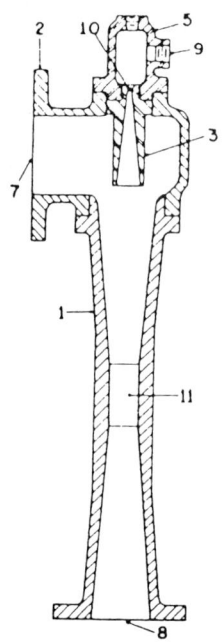

1. Diffuser   7. Suction
2. Suction Chamber   8. Discharge
3. Steam Nozzle   9. Steam Inlet
4. Nozzle Extensions (if used)   10. Nozzle Throat
5. Steam Chest   11. Diffuser Throat
6. Nozzle Plate (if used)

Fig. 4.1-45 Typical steam jet ejector stage assemblies. (From ref. 10; courtesy of Heat Exchange Institute, 1230 Keith Building, Cleveland, OH.)

surrounding gas cause kinetic energy to be transferred to the gas. Mixing occurs, and the pressure increases as the mixture moves through the diffuser to the discharge. When large gas capacities are desired, multiple-nozzle ejectors are used to increase the surface area available for energy transfer. Ejectors should be operated with dry, high-pressure steam. Moisture significantly reduces the energy available per point of steam, requiring greater mass flow. As a practical limit the highest pressures for injectors in condenser service is about 500 psi at the injector inlet.

Generally, injectors are classified according to arrangement, how staged, and exhaust conditions. Two-stage ejector installations can provide suction pressures of less than 1 in. Hg absolute, and multistage units of 3, 4, or 5 stages are capable of producing suction pressures well below 1 mm Hg absolute. The exhaust conditions may be either condensing or noncondensing. The condensing unit is more economical to operate but has a higher initial cost. The individual ejectors of a multistage unit may operate either condensing or noncondensing, according to the performance specifications for the unit.

During normal steam plant operation, noncondensable gases accumulate in the condenser as a result of leakage into the low-pressure portions of the system and due to production of some gas by feedwater treatment. If these gases are not removed, they will blanket condenser tubes and reduce the effective heat transfer surface available.

To remove air prior to operation of the turbine, a noncondensing ejector is used. This hogging jet provides suction pressures under 10 in. Hg absolute in the condenser prior to admission of steam to the turbine. Table 4.1-3 shows the capacities recommended by the Heat Exchange Institute (HEI) for this service.

During operation, condenser design pressures are in the order of 1 in. Hg absolute, and two-stage condensing ejectors are used. Figure 4.1-46 shows the arrangement of a two-stage, twin-element

TABLE 4.1-3   HOGGER CAPACITIES[a]

Total Steam Condensed (lb/h)	SCFM[b]—Dry Air at 10 in. Hg abs Design Suction Pressure
Up to 100,000	50
100,001 to 250,000	100
250,001 to 500,000	200
500,001 to 1,000,000	350
1,000,001 to 2,000,000	700
2,000,001 to 3,000,000	1050
3,000,001 to 4,000,000	1400
4,000,001 to 5,000,000	1750
5,000,001 to 6,000,000	2100
6,000,001 to 7,000,000	2450
7,000,001 to 8,000,000	2800
8,000,001 to 9,000,000	3150
9,000,001 to 10,000,000	3500

Source.   Courtesy Heat Exchange Institute, Cleveland, OH.

[a] In the range of 500,000 lb/h steam condensed and above, the above table provides evacuation of the air in the condenser and L.P. turbine from atmospheric pressure to 10 in. Hg abs in about 30 min if the volume of condenser and L.P. turbine is assumed to be 26 ft^3 per 1000 lb/h of steam condensed.

[b] SCFM—14.7 psia at 70°F.

condensing ejector. Normal service usually requires one of the injector sets; the second set is available if the first is inoperative or if excessive air leakage into the system occurs. Note that the primary ejectors and their intercondensers are significantly larger than the secondary ejectors and their aftercondenser: the second stage does not have to pump the steam supplied to the first stage or the moisture entrained in the gas from the condenser. This burden is removed in the intercondenser. Representative values for the conditions in a two-stage condensing ejector are shown in the flow diagram of Fig. 4.1-47. Air ejectors are used in a regenerative mode; that is, the heat removed in the inter- and aftercondensers is returned to the condensate system.

For systems that require condenser pressures below 1 in. Hg absolute, three-stage ejectors are used. Figure 4.1-48 shows the typical increase in performance offered by the use of the additional stage.

For virtually all power plant designs the Heat Exchange Institute recommends ejector design for 1 in. Hg absolute suction pressure and 71.5°F suction temperature. Table 4.1-4 shows the HEI recommended minimum ejector capacity requirements based on the steam flow and number of exhaust openings for a one-condenser shell. HEI provides recommendations for other conditions also. For this service, the steam consumption in pounds of steam per pound dry air may range from 13 to 14.5 for a two-stage ejector and from 9.5 to 10 for a three-stage ejector. The lower rates correspond to the operation of the ejector of higher-pressure steam.

Example 4.1-2.   A 425-MW double-casing, double-flow turbine operates at 2400 psig, 1000°/1000° throttle conditions. At full load, flow to the throttle is 2,884,000 lb/h and flow to the condenser is 776,000 lb/h. Estimate the steam requirements to the air ejector.

The double-flow low-pressure turbine has two exhaust trunks; therefore, the flow per main exhaust opening is

$$\frac{776,000}{2} = 388,000 \text{ lb/h}$$

From Table 4.1-4, at 388,000 lb/h and two exhaust openings, the ejector should handle (at a

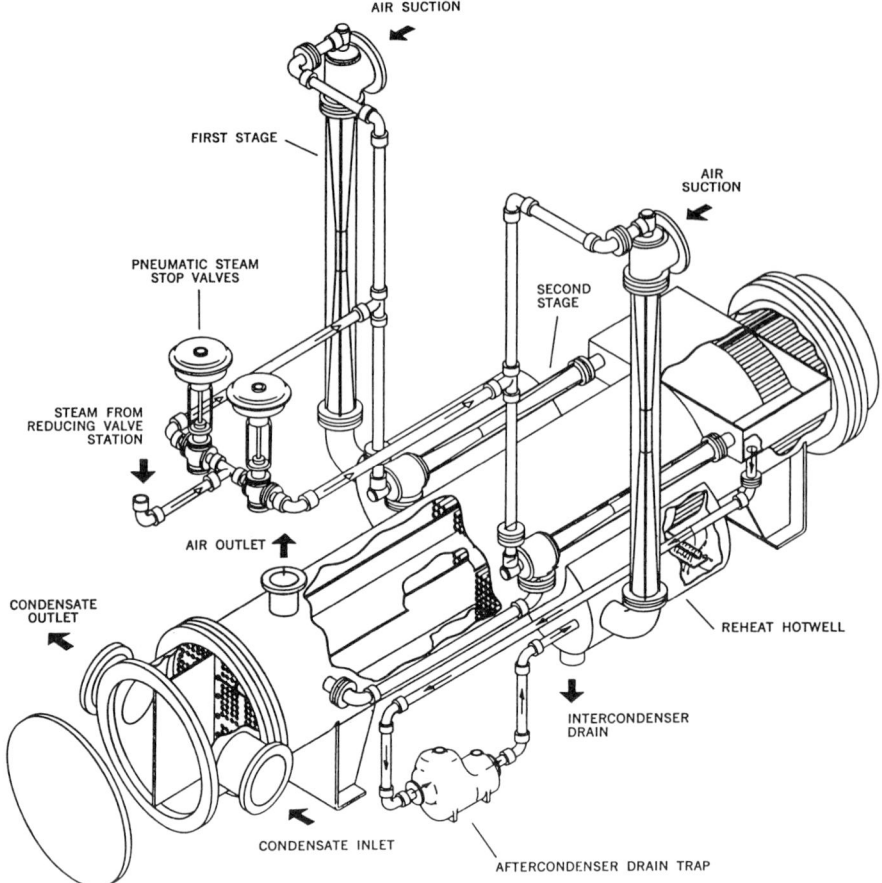

Fig. 4.1-46 Twin element two-stage ejector unit. Foster Wheeler steam jet air ejectors (SP-66-14). (Courtesy of Foster Wheeler Energy Corp.)

minimum)

    15.6 SCFM air (at 14.7 psia, 70°F)
    67.5 lb/h dry air
    148.5 lb/h water vapor
    216.0 lb/h total mixture

Main steam will be used with a regulator to reduce the pressure to the injector to 500 psi, so at 13 lb of steam per pound of dry air, a minimum of 13(67.5) = 878 lb/h of steam is required: specify slightly more, say 950 lb/h. The inter- and aftercondensers must provide sufficient heat transfer surface to condense 148.5 + 950 = 1099 lb/h of steam.

    If a separate noncondensing hogging ejector is provided for use at start-up, that ejector should be capable of removing 350 SCFM of dry air at 10 in. Hg absolute suction pressure (from Table 4.1-3).

The Steam Jet Pump

Since its introduction in the late 1800s the steam jet pump, or injector, has been widely used for steam locomotive boiler feed service. It offers simple construction, ease of operation, and reliability at the price of low efficiency. However, all losses appear as heat in the pumped stream; thus, the device functions as a combined boiler feed pump and feedwater heater.

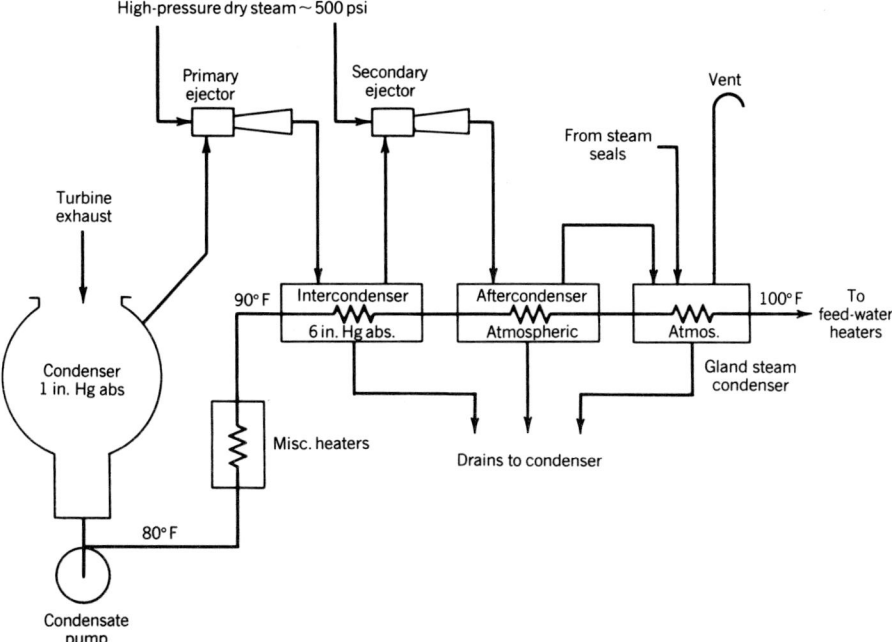

High-pressure dry steam ~ 500 psi

Primary ejector

Secondary ejector

Vent

From steam seals

Turbine exhaust

90° F | Intercondenser | Aftercondenser | | 100° F | To feed-water heaters
6 in. Hg abs. | Atmospheric | Atmos.

Gland steam condenser

Condenser 1 in. Hg abs

Misc. heaters

Drains to condenser

80° F

Condensate pump

Fig. 4.1-47 Flow diagram for two-stage condensing ejector system.

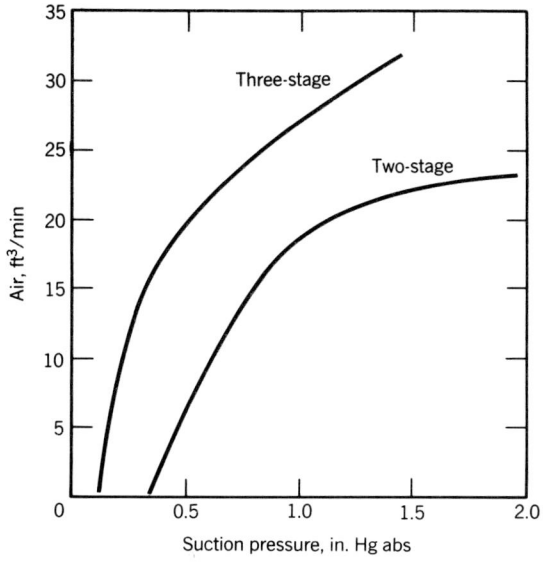

Three-stage

Two-stage

Air, ft³/min

Suction pressure, in. Hg abs

Fig. 4.1-48 Relative capacity of two-stage and three-stage ejectors for the same steam capacity.

TABLE 4.1-4   VENTING EQUIPMENT CAPACITIES[a]

A. One Condenser Shell

Effective Steam Flow Each Main Exhaust Opening (lb/h)		Total Number of Exhaust Openings								
		1	2	3	4	5	6	7	8	9
Up to 25,000	scfm[b]	3.0	4.0	5.0	5.0	7.5	7.5	7.5	10.0	10.0
	Dry Air lb/h	13.5	18.0	22.5	22.5	33.8	33.8	33.8	45.0	45.0
	Water Vapor lb/h	29.7	39.6	49.5	49.5	74.4	74.4	74.4	99.0	99.0
	Total Mixture lb/h	43.2	57.6	72.0	72.0	108.2	108.2	108.2	144.0	144.0
25,001 to 50,000	scfm[b]	4.0	5.0	7.5	7.5	10.0	10.0	10.0	12.5	12.5
	Dry Air lb/h	18.0	22.5	33.8	33.8	45.0	45.0	45.0	56.2	56.2
	Water Vapor lb/h	39.6	49.5	74.4	74.4	99.0	99.0	99.0	123.6	123.6
	Total Mixture lb/h	57.6	72.0	108.2	108.2	144.0	144.0	144.0	179.8	179.8
50,001 to 100,000	scfm[b]	5.0	7.5	10.0	10.0	12.5	12.5	15.0	15.0	15.0
	Dry Air lb/h	22.5	33.8	45.0	45.0	56.2	56.2	67.5	67.5	67.5
	Water Vapor lb/h	49.5	74.4	99.0	99.0	123.6	123.6	148.5	148.5	148.5
	Total Mixture lb/h	72.0	108.2	144.0	144.0	179.8	179.8	216.0	216.0	216.0
100,001 to 250,000	scfm[b]	7.5	12.5	12.5	15.0	17.5	20.0	20.0	25.0	25.0
	Dry Air lb/h	33.8	56.2	56.2	67.5	78.7	90.0	90.0	112.5	112.5
	Water Vapor lb/h	74.4	123.6	123.6	148.5	175.1	198.0	198.0	247.5	247.5
	Total Mixture lb/h	108.2	179.8	179.8	216.0	251.8	288.0	288.0	360.0	360.0
250,001 to 500,000	scfm[b]	10.0	15.0	17.5	20.0	25.0	25.0	30.0	30.0	35.0
	Dry Air lb/h	45.0	67.5	78.7	90.0	112.5	112.5	135.0	135.0	157.5
	Water Vapor lb/h	99.0	148.5	173.1	198.0	247.5	247.5	297.0	297.0	346.5
	Total Mixture lb/h	144.0	216.0	251.8	288.0	360.0	360.0	432.0	432.0	504.0
500,001 to 1,000,000	scfm[b]	12.5	20.0	20.0	25.0	30.0	30.0	35.0	40.0	40.0
	Dry Air lb/h	56.2	90.0	90.0	112.5	135.0	135.0	157.5	180.0	180.0
	Water Vapor lb/h	123.6	198.0	198.0	247.5	297.0	297.0	346.5	396.0	396.0
	Total Mixture lb/h	179.8	288.0	288.0	360.0	432.0	432.0	504.0	576.0	576.0
1,000,001 to 2,000,000	scfm[b]	15.0	25.0	25.0	30.0	35.0	40.0	40.0	45.0	50.0
	Dry Air lb/h	67.5	112.5	112.5	135.0	157.5	180.0	180.0	202.5	225.0
	Water Vapor lb/h	148.5	247.5	247.5	297.0	346.5	396.0	396.0	445.5	495.0
	Total Mixture lb/h	216.0	360.0	360.0	432.0	504.0	576.0	576.0	648.0	720.0
2,000,001 to 3,000,000	scfm[b]	17.5	25.0	30.0	35.0	40.0	45.0	50.0	55.0	60.0
	Dry Air lb/h	78.7	112.5	135.0	157.5	180.0	202.5	225.0	247.5	270.0
	Water Vapor lb/h	173.1	247.5	297.0	346.5	396.0	445.5	495.0	544.5	594.0
	Total Mixture lb/h	251.8	360.0	432.0	504.0	576.0	648.0	720.0	792.0	864.0
3,000,001 to 4,000,000	scfm[b]	20.0	30.0	35.0	40.0	45.0	50.0	55.0	60.0	65.0
	Dry Air lb/h	90.0	135.0	157.5	180.0	202.5	225.0	247.5	270.0	292.5
	Water Vapor lb/h	198.0	297.0	346.5	396.0	445.5	495.0	554.5	594.0	643.5
	Total Mixture lb/h	288.0	432.0	504.0	576.0	648.0	720.0	792.0	864.0	936.0

Source.   Courtesy of Heat Exchange Institute, Cleveland, OH.

[a] These tables are based on air leakage only and the air vapor mixture at 1 in. Hg abs and 71.5°F.
[b] 14.7 psia at 70°F.

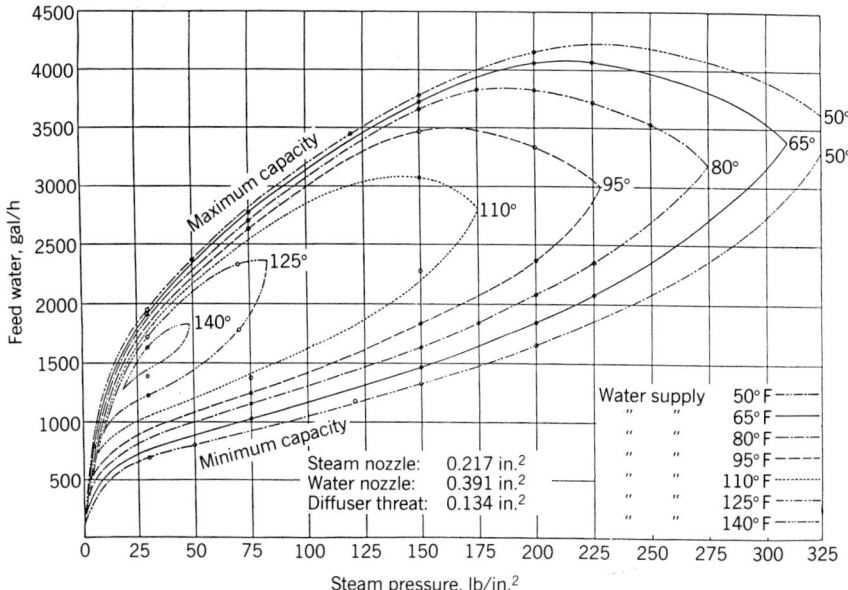

Fig. 4.1-49 Performance map of sellers No. $10\frac{1}{2}$ injector. (Source: *Locomotive Injectors and Boiler Attachments*, Sellers Injection Systems, 1928. Courtesy of Prosser-East Division of Purex.)

The performance map of a typical injector for this type of service is seen in Fig. 4.1-49, where delivery is into a boiler operating at the same pressure as the steam supplied to the injector. Units having capacities to 12,000 gal/h have been built.

Today injector applications are limited to a few special uses. One is seen in Fig. 4.1-50, which shows a third inlet port provided for the purpose of introducing detergent into the flow stream. Cold water, steam, and detergent are combined in the injector to produce a hot stream with effective cleansing properties. Other injectors have been designed and operated at pressures to several hundred psi.

In operation, high-velocity steam mixes with the pumped water in the combining tube (Fig. 4.1-51), transferring energy to the water. This is a region of two-phase flow at supersonic velocities. Heat transfer between the two components takes place across the large steam–water interface at temperatures only slightly above that of the inlet water. The pressure in this tube is therefore low at the saturation pressure corresponding to the temperature of the region. Vent holes in the side of the combining tube and the gap between combining tube and delivery tube serve to hold the combining tube at near constant pressure.

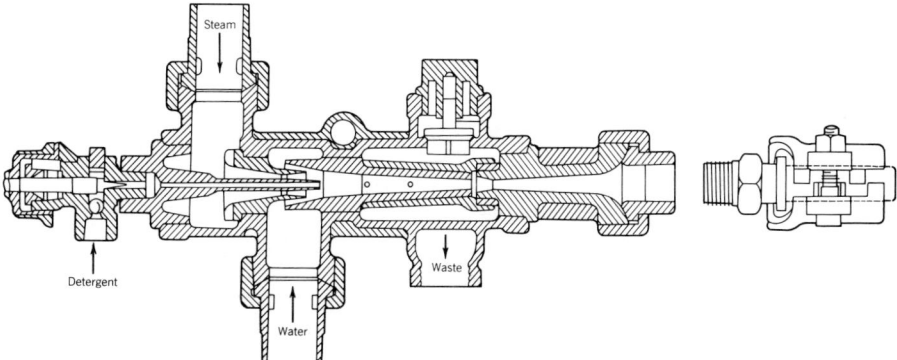

Fig. 4.1-50 Sellers jet cleaner. (Source: *Operating Guide*, Bulletin 420-E, Sellers Injector Systems. Courtesy of Prosser-East Division of Purex.)

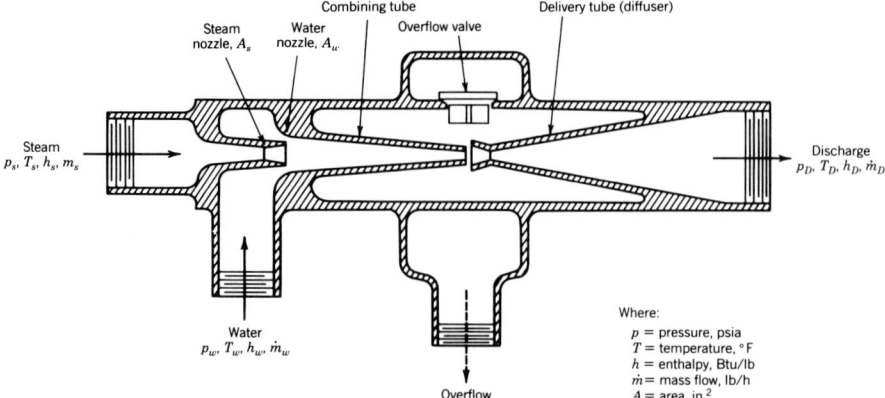

Fig. 4.1-51 The steam jet pump (injector).

Upon entry of the flow into the diffuser, a shock wave forms, the steam void collapses, and a sudden rise in pressure occurs. The shock moves in response to downstream conditions: toward the exit for low-discharge pressures, toward the inlet in response to higher pressures.

This device is unique in that the water flow through the unit is determined by the pressure drop across the inlet water nozzle, which is a function of water supply temperature and pressure. The discharge conditions do not influence the flow, save in the extreme! Flow control can only be achieved by throttling the inlet water stream.

Figure 4.1-52 shows the dependence of flow on inlet temperatures and Fig. 4.1-53 the constancy of flow with changing discharge pressure. The figure also shows the attainment of discharge pressures twice the inlet steam pressure and typical water-to-steam mass flow ratios under 10 : 1.

To accommodate shutoff of the discharge, the overflow valve and port are provided to maintain flow through the combining tube.

When the discharge is subsequently opened, the momentum of the fluid in the combining tube carries into the delivery tube, reestablishing flow.

The thermodynamic efficiency of the injector as a pump may be found by comparing the pump work actually achieved to that possible if the inlet steam were isentropically expanded to the saturation pressure corresponding to the temperature of the discharge, $_iP_D$. Thus, the pumping efficiency is

$$\eta_p = \frac{0.00297}{(h_s - {}_ih_D)}\left[\frac{\dot{m}_w}{\dot{m}_s}(P_D - P_w) + (P_D - {}_iP_D)\right] \tag{4.1-12}$$

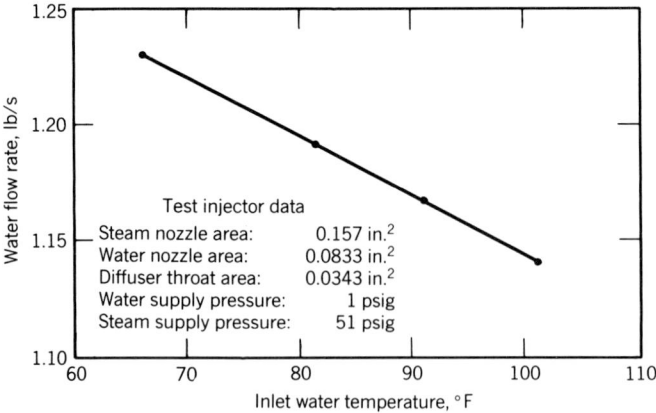

Fig. 4.1-52 Injector flow dependence on inlet temperature.

Steam nozzle area: 0.157 in.^{2}
Water nozzle area: 0.0833 in.^{2}
Diffuser throat area: 0.0343 in.^{2}
Water supply pressure: 1 psig
Water supply temperature: 60°F

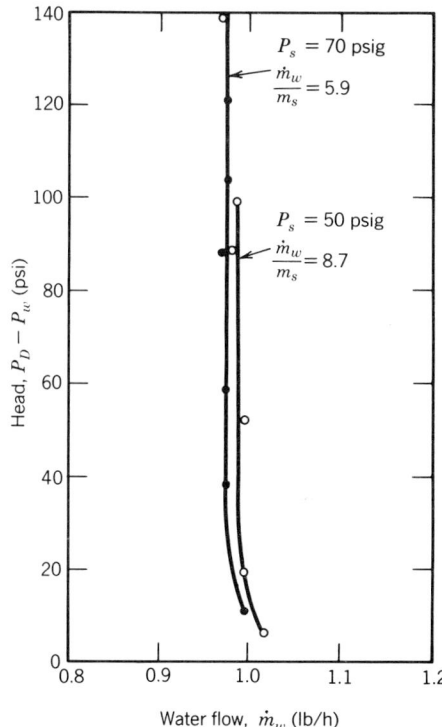

Water flow, $\dot{m}_w$ (lb/h)

Fig. 4.1-53 Injector characteristic curves.

where $\dot{m}_w/\dot{m}_s$ = water–steam mass flow ratio, lbm/s water/lbm/s steam
h_s = enthalpy of inlet steam, Btu/b
$_i h_D$ = enthalpy of steam if expanded isentropically from P_s to $_i P_D$, pressures in psia.

Typical values for maximum attainable pump efficiencies are on the order of 3%.
Since the losses appear as heat, the temperature relations for the injector are established by a heat balance:

$$T_D = \frac{(1 - \eta_p)\, h_s + \eta_{pi} h_D + 32 + (\dot{m}_w/\dot{m}_s)\, T_w}{1 + \dot{m}_w/\dot{m}_s} \tag{4.1-13}$$

where T_D = discharge temperature, °F
T_w = water supply temperature, °F

By treating the steam as a diatomic perfect gas flowing through a critical nozzle, the flow and throat diameter may be related to the steam inlet conditions by Napier's formula:

$$\dot{m}_s = 0.31 A_s \sqrt{\frac{P_s}{v_s}} \tag{4.1-14}$$

where v_s = specific volume of steam at the inlet,
A_s = steam nozzle throat area, in.^{2}

The size of the water nozzle is determined by the pressure difference across the nozzle. Condensation of the steam at the entrance to the combining tube will occur at the saturation pressure corresponding to the inlet water temperature, P_c.

$$A_w = \frac{1.52 \, \dot{m}_w}{0.95\sqrt{2g(P_w - P_c)}} \tag{4.1-15}$$

where $0.95 =$ nozzle discharge coefficient
       $P_c =$ combining tube pressure, psia

Example 4.1-3.   Estimate the steam flow requirements for an ejector to pump 14,400 lb of $70°F$ water per hour from 5 psig to 100 psig, with a discharge temperature of $175°F$. Saturated steam at 150 psig is available.

Since the discharge temperature is known, the saturation pressure and enthalpy of steam isentropically expanded to that pressure can be found from the steam tables:

$$h_s = 1196 \text{ Btu/lb}$$

$$T_D = 175°F$$

$$_iP_D = 6.7 \text{ psia}$$

$$_ih_D = 982 \text{ Btu/lb}$$

The mass flow ratio $\dot{m}_w/\dot{m}_s$ can be found by solving Eqs. (4.1-12) and (4.1-13).

$$\frac{\dot{m}_w}{\dot{m}_s} = \frac{h + 32 - T_D - 0.00297(P_D - {_i}P_D)}{T_D - T_w + 0.000297(P_D - P_w)}$$

Thus,

$$\frac{\dot{m}_w}{\dot{m}_s} = \frac{1196 + 32 - 175 - 0.00297(114.7 - 6.7)}{175 - 70 + 0.00297(114.7 - 19.7)} = 10.0$$

The efficiency is then found:

$$\eta_p = \frac{0.00297}{(1196 - 982)}[10.0(114.7 - 19.7) + (114.7 - 6.7)] = 0.014$$

This efficiency is realizable as it is well under 3%. This implies that if the discharge should be restricted for some reason, the unit will continue to operate at significantly higher heads. Proceeding to size the steam nozzle,

$$\dot{m}_s = \frac{\dot{m}_w}{10} = \frac{14,400}{10} = 1440 \text{ lb/h} = 0.4 \text{ lb/s}$$

$$P_s = 164.7 \text{ psia}$$

$$v_s = 2.75 \text{ ft}^3/\text{lb}$$

Then

$$A_s = \frac{0.4}{0.31 - \sqrt{164.7/2.75}} = 0.1667 \text{ in.}^2$$

and the steam nozzle throat diameter is

$$D_s = \sqrt{4/\pi(0.1667)} = 0.460 \text{ in.}$$

The saturation pressures corresponding to the inlet water ($70°F$) is 0.36 psia. Thus, the water nozzle size can be estimated by taking the estimated pressure differential:

$$A_{w(low)} = \frac{1.52(4)}{\sqrt{0.95[2g(19.7 - .36)]}} = 0.181 \text{ in.}^2$$

REFERENCES

4.1-1 A. J. Stepanoff, *Centrifugal and Axial Flow Pumps*, 2nd ed., Wiley, New York, 1957.

4.1-2 S. Gopalakrishnan, R. Fehlau, and J. Lorett, "Critical Speed in Centrifugal Pumps," ASME Paper 82-GT-277, American Society of Mechanical Engineers, New York, 1982.

4.1-3 I. J. Karassik and R. Carter, *Centrifugal Pumps—Selection, Operation, and Maintenance*, McGraw-Hill, New York, 1960.

4.1-4 Hydraulic Institute Standards, 14th ed., Hydraulic Institute, Cleveland, OH, 1983.

4.1-5 A. Agostinelli, D. Nobles, and C. R. Mockridge, "An Experimental Investigation of Radial Thrust in Centrifugal Pumps," ASME Paper 59-HYD-2, American Society of Mechanical Engineers, New York, 1959.

4.1-6 R. L. Baxter and D. L. Bernhard, "Vibration Tolerances for Industry," ASME Paper 67-PEM-14, American Society of Mechanical Engineers, New York, 1967.

4.1-7 "Centrifugal Pumps for General Refinery Services" API Standard 610, 6th ed., American Petroleum Institute, Washington, D.C., January 1981.

4.1-8 Hydraulic Institute Standards, 13th ed., Hydraulic Institute, Cleveland, OH, 1975.

4.1-9 *Standards For Steam Surface Condensers*, Heat Exchange Institute, Cleveland, OH, 1978.

4.1-10 Ronald P. Rose, *Steam Jet Pump Analysis and Experiments*, WAPD-TM-227, Bettis Atomic Power Laboratory, Pittsburgh, 1960

4.1-11 Philip J. Potter, *Power Plant Theory and Design*, Wiley, New York, 1959.

4.1-12 *Standards for Steam Jet Ejectors*, Heat Exchange Institute, Cleveland, OH, 1967.

4.2 FANS[†]

Frank P. Bleier, PE, Consulting Engineer for Fan Design

LIST OF SYMBOLS

AF	Airfoil
ahp	Air horsepower
AMCA	Air Movement and Control Association
BC	Backward-curved
bhp	Brake horsepower
BI	Backward-inclined
cfm	Cubic feet per minute
CCW	Counterclockwise
CW	Clockwise
DWDI	Double-width, double-inlet
EE	Electrical efficiency (sometimes called motor efficiency)
FC	Forward-curved
FD	Free delivery
fpm	Feet per minute
hp	Horsepower
ME	Mechanical efficiency (sometimes called total efficiency)
ND	No delivery
OV	Outlet Velocity
PF	Propeller fan
RB	Radial blade
RT	Radial tip
SE	Set efficiency (not static efficiency)
SP	Static pressure
SWSI	Single-width, single-inlet
TAF	Tubeaxial fan
TP	Total pressure
V	Velocity
VAF	Vaneaxial fan
VP	Velocity pressure

[†]All fans illustrated in this section were designed by the author.

4.2-1   Air Flow Fundamentals

Three types of air pressure are used in fan engineering: static pressure (SP), velocity pressure (VP), and total pressure (TP).

When the piston in Fig. 4.2-1 is lowered, the air volume below is compressed and the manometer will register a positive static pressure relative to the atmospheric pressure (which is considered zero pressure in fan engineering). This compressed air then has potential energy, that is, the potential to expand to its original volume. If, on the other hand, the piston is raised, the air volume is expanded and the manometer will register a negative static pressure relative to the atmospheric pressure. This expanded air also has potential energy, that is, the potential to contract to its original volume.

Positive and negative static pressure may exist in moving air as well as in stationary air. A fan blowing into a system (including such resistances as ducts, elbows, filters, heating coils, cooling coils) produces positive static pressure, which is used to overcome the various resistances. A fan exhausting from a duct system produces negative static pressure, which again is used to overcome the resistance of the system.

Air flowing through a smooth, round duct has a velocity distribution, as shown in Fig. 4.2-2. For duct diameters of 6–10 in. and for air velocities of 1000–3000 fpm, the average velocity V is approximately equal to 91% of the maximum velocity. To find the average velocity in larger ducts and for larger air velocities, a so-called Pitot tube traverse across the duct is taken (see Fig. 4.2-3). From the average velocity V (in feet per minute) and the air density d (in pounds per cubic foot), we can calculate the velocity pressure VP (in inches of water column, in. WC)

$$VP = d\left(\frac{V}{1096.2}\right)^2 \qquad (4.2\text{-}1)$$

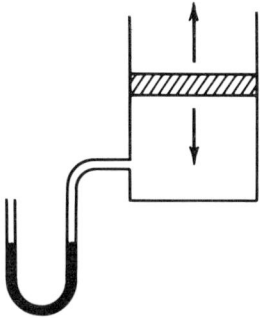

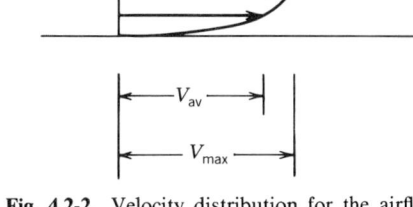

Fig. 4.2-1   Manometer. As the piston moves, the static pressure below it will become either positive or negative.

Fig. 4.2-2   Velocity distribution for the airflow through a round duct.

Fig. 4.2-3   27 in. diameter test duct with two supports for Pitot tube traverses and with throttling device at end of duct. (Courtesy of Circle Steel Corp, Taylorville, IL.)

or for standard air density of $d = 0.075$ lb/ft^3

$$VP = \left(\frac{V}{4005}\right)^2 \qquad (4.2\text{-}2)$$

Velocity pressure is the pressure we can feel when we hold our hand in the airstream. It represents kinetic energy.

Total pressure $TP = SP + VP$. In this equation VP is always positive. SP and TP may be positive or negative. For example,

SP = +2.2 in.	SP = −0.5 in.	SP = −1.4 in.
VP = 0.8 in.	VP = 0.8 in.	VP = 0.8 in.
TP = +3.0 in.	TP = +0.3 in.	TP = −0.6 in.

Let us consider a fan having an outlet area OA = 4.00 ft^2, blowing 16,000 cfm into a system and producing 3 in. SP in order to overcome the resistance of the system. This fan will have an average outlet velocity

$$V = \frac{\text{cfm}}{\text{OA}} = \frac{16{,}000}{4.00} = 4000 \text{ fpm}$$

and a velocity pressure

$$VP = \left(\frac{4000}{4005}\right)^2 = 1.00 \text{ in. WC}$$

The total pressure will be $TP = 3.00 + 1.00 = 4.00$ in. WC and the power output of the fan (called air horsepower, ahp) will be

$$\text{ahp} = \frac{\text{cfm} \times \text{TP}}{6356} = 10.07 \text{ hp}$$

If the motor output (= fan input) is 15 bhp, the fan efficiency at this point of operation will be the mechanical efficiency

$$\text{ME} = \frac{10.07}{15.0} = 0.67 = 67\%$$

If the motor input is 12.7 kW, the motor efficiency (or electrical efficiency) is

$$\text{EE} = \frac{0.746 \times \text{bhp}}{\text{kW}} = 0.88 = 88\%$$

The efficiency of the set (fan plus motor), called the set efficiency, is SE = ME × EE = 0.67 × 0.88 = 0.59 = 59%.

In selecting a fan for a certain application, the fan efficiency is of great importance, because with higher efficiency we can obtain the same air horsepower with less power input. This will not only

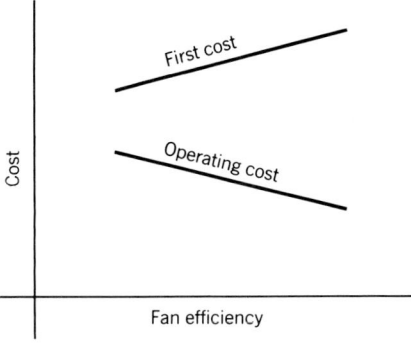

Fig. 4.2-4   Fan efficiency versus cost. Selecting a fan of higher efficiency normally results in a higher first cost but in a lower operating cost.

reduce the operating cost, but at the same time it will save energy. High-efficiency fans, on the other hand, are also more expensive (see Fig. 4.2-4). It should be attempted, therefore, to find a balance between first cost and operating cost, taking into consideration that the first cost of the fan unit itself often is only a small portion of the system's first cost.

4.2-2 Operating Principles of Fans

According to their operating principles, fans can be divided into two basic groups: axial flow fans and centrifugal fans.

Figure 4.2-5 shows a typical axial flow fan wheel. It can be assembled inside a cylindrical housing, possibly equipped with a venturi inlet and with guide vanes. The airstream passes through the unit essentially parallel to the axis of rotation and the flow is produced by deflection of the air by the fan blades.

Figures 4.2-6 and 4.2-7 show the general shape of a typical centrifugal fan. The airstream enters the unit parallel to the axis of rotation, then forms a funnel-shaped pattern, turning 90° radially outward, then is deflected by the rotating blades into several spiral pathways, and finally is collected again into a single airstream by means of a scroll housing. The flow is produced only in small part by deflection of the air by the fan blades; mainly it is produced by the centrifugal forces exerted on the rotating air.

As a result of this centrifugal force, centrifugal fans generally produce more SP than axial flow fans of the same wheel diameter and running speed. Conversely, axial flow fans usually produce more air volume (in cubic feet per minute) at low pressures; they also have the advantages of greater compactness, of easier installation, and usually of lower first cost.

4.2-3 Types of Fans

Axial flow fans are subdivided into four categories: propeller fans, tubeaxial fans, vaneaxial fans, and double-stage axial flow fans. Centrifugal fans are subdivided into six categories, according to their blade shapes: airfoil, backward-curved, backward-inclined, radial tip, forward-curved, and radial blade centrigual fans. All these are commonly used fan types.

In addition, there are three other fan types: mixed-flow fans, cross-flow fans, and vortex fans. Their operating principles are still deflection, centrifugal force, and combinations thereof. These three types are used for special applications, even though they have the disadvantage of lower efficiencies.

Finally, there is a very popular type of fan, the roof ventilator, which comes in a variety of configurations. It can use either an axial flow fan wheel or a centrifugal fan wheel, inside a special type of housing, designed either for radial discharge or for upblast.

Nomenclature and Symbols

For many years it has been customary to use the term *fans* for axial flow fans and the term *blowers* for centrifugal fans. Even today this terminology is often used. The terms *axial flow fans* and

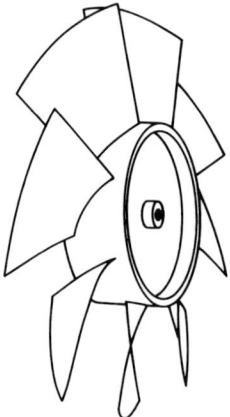

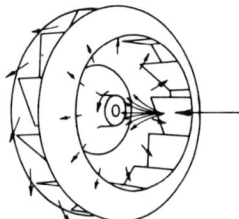

Fig. 4.2-5 Typical axial flow fan wheel with a hub tip ratio of about 0.4 and with 8 blades.

Fig. 4.2-6 Typical centrifugal fan wheel with backplate and shroud and with 12 backward-inclined, flat blades.

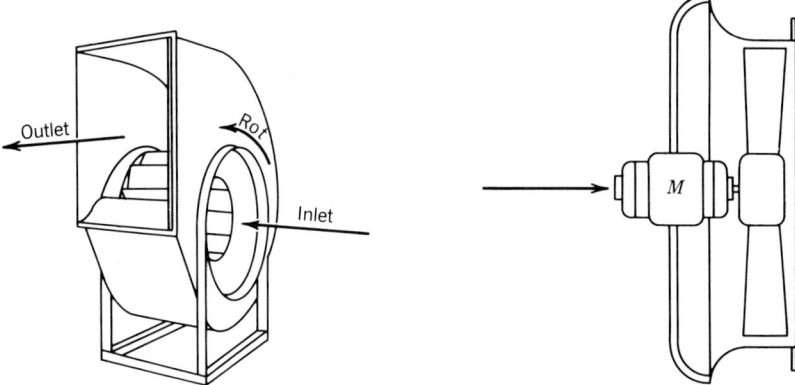

Fig. 4.2-7 Typical centrifugal fan unit with scroll housing, showing inlet, and outlet side.

Fig. 4.2-8 Schematic sketch of propeller fan with motor on inlet side.

centrifugal fans, however, are more descriptive and may avoid some misunderstandings, as either fan type may be used for blowing as well as for exhausting.

The rotating part of a fan unit is called the fan wheel, the stationary part the fan housing. The fan wheel has a number of blades. The fan housing may contain some stationary vanes. An axial flow fan wheel consists of a hub or a spider on which the blades are mounted. A centrifugal fan wheel may consist of a back plate, some blades, and a retaining ring or shroud. The streamline housing inlet (usually a spinning) of an axial flow fan housing is called the inlet bell or inlet venturi; in a centrifugal fan housing it is called the inlet cone because of its mainly conical shape.

4.2-4 Axial Flow Fans

The sequence, propeller–tubeaxial–vaneaxial, also indicates the general trend of increasing weight, price, hub diameter, static pressure, air horsepower, and efficiency.

The *propeller fan*, as shown in Fig. 4.2-8, is the lightest, least expensive, and most commonly used unit. The fan wheel runs inside a narrow mounting ring that is often shaped like an inlet bell and is sometimes extended to a square mounting panel. The mounting ring carries the motor support with the motor, which usually is located on the inlet side as shown in Fig. 4.2-8, but in special applications it is on the outlet side. The fan wheel has a small hub diameter, from 20 to 45% of the wheel diameter and is designed to operate in the range near free delivery. For lower cost the small hub is often replaced by a so-called spider, to which the blades are riveted.

The *tubeaxial fan*, as shown in Fig. 4.2-9, is a glorified propeller fan. Its outside appearance is that of a short cylinder about 1 diameter long. The cylindrical housing contains the motor support, the motor, and the fan wheel, which can be located upstream or downstream of the motor. The wheel has

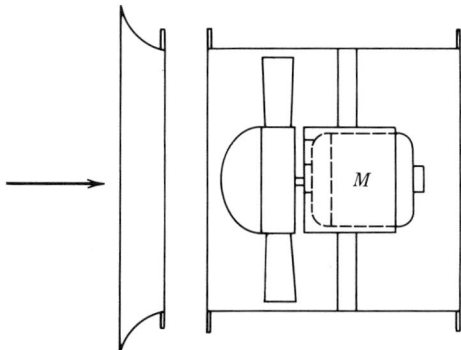

Fig. 4.2-9 Schematic sketch of tubeaxial fan with motor on outlet side and with separate inlet bell as an accessory.

a medium-size hub diameter, from 35 to 60% of the wheel diameter, and the unit is designed to operate in the range of moderate static pressures.

The *vaneaxial fan*, as shown in Fig. 4.2-10, is a more elaborate unit. It has the outside appearance of a cylinder, $1-1\frac{1}{2}$ diameters long, which contains the same parts as the tubeaxial fan and in addition a set of guide vanes and usually an inner shell. The guide vanes are usually downstream from the fan wheel, but occasionally they are upstream. They eliminate most of the air spin (which is an energy loss) past the fan and convert it into additional static pressure. The outlet vanes (see Fig. 4.2-11) do

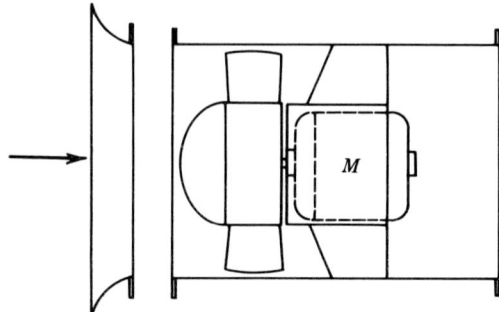

Fig. 4.2-10 Schematic sketch of single-stage vaneaxial fan with motor and guide vanes on outlet side and with separate inlet bell as an accessory.

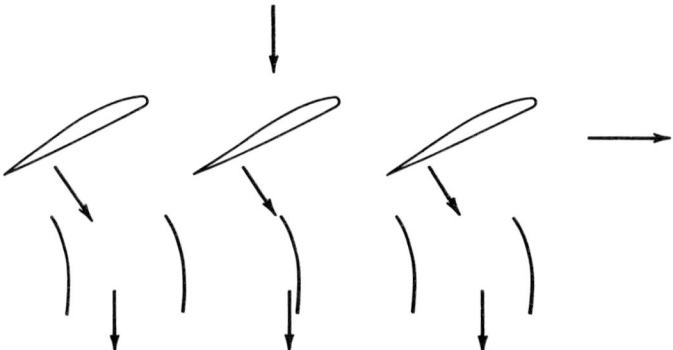

Fig. 4.2-11 Outlet vanes will guide the spiral air pattern produced by the blades, back into an axial direction.

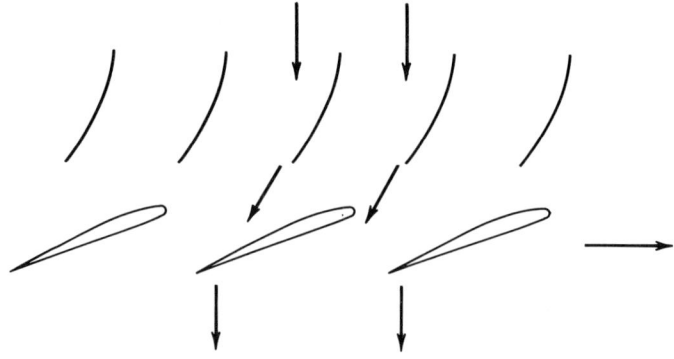

Fig. 4.2-12 Inlet vanes give the airstream ahead of the blades a spiral motion opposite to the fan rotation.

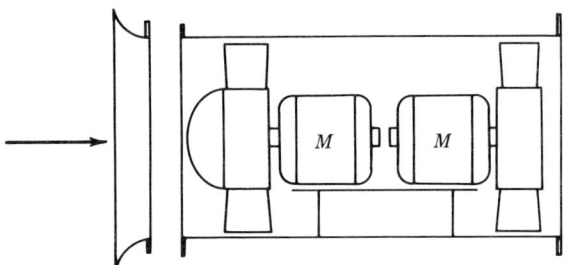

Fig. 4.2-13 Schematic of double-stage axial flow fan with counterrotating fan wheels (no guide vanes) and with separate inlet bell as an accessory.

this by guiding the spiral air motion past the blades back into an axial direction. The inlet vanes (see Fig. 4.2-12) give the airstream a spiral motion opposite to the fan rotation, and the subsequent blades neutralize the inlet spin. The inlet vanes, therefore, result in a condition similar to that of an increased fan speed, so that not only the ahp is increased but also the noise level. The fan wheel has a large hub diameter, from 50 to 80% of the wheel diameter, and the unit is designed to operate in the range of high static pressures.

The *double-stage axial flow fan*, as shown in Fig. 4.2-13, has the outside appearance of a long cylinder, $1\frac{1}{2}$–2 diameters long. It contains two fan wheels instead of one and usually two separate motors for counterrotation of the two fan wheels. This results in more than double the static pressure and eliminates the need for guide vanes, as the counterrotating second stage neutralizes the air spin produced by the first stage.

Features of Axial Flow Fans

The cylindrical shape of tubeaxial, vaneaxial, and double-stage fans makes them suitable for straight-line installation. A round duct can be connected to either end of the unit. A fan installed at the inlet of a system needs an inlet bell, as indicated in Figs. 4.2-9, 4.2-10, and 4.2-13. The inlet bell prevents the forming of a vena contracta (see Fig. 4.2-14) on the casing edge. The lack of an inlet bell would result in blade tips operating in turbulent air flow and being starved for air, particularly if the fan wheel is located near the housing inlet, as in Figs. 4.2-9 and 4.2-10. The use of an inlet bell, therefore, will increase the cfm and the fan efficiency and reduce the noise level.

Figure 4.2-15 shows the highlights of the above general description for the various types of axial flow fans. Each type can be built either with direct motor drive or with belt drive. Direct drive generally is preferable in small sizes so that obstructions to the air flow, belt losses, maintenance, and the extra expense for bearings, brackets, shaft, and belt are avoided. Belt drive is preferable in large sizes so that the running speed can be kept low without the use of expensive low-speed motors.

For good efficiency, the air flow of an axial flow fan should be evenly distributed over the working face of the fan wheel. The velocity of the rotating blade, on the other hand, is not evenly distributed: it is low near the center and increases toward the tip. This gradient should be compensated by a twist in the blade, resulting in larger blade angles near the center and smaller blade angles toward the tip. Low-cost fans sometimes do not have this variation of the blade angles from hub (or spider) to tip. This will result in a loss of fan efficiency because most of the air flow will be produced by the outer portion of the blades, even at low static pressures; at higher static pressures the blade twist is even

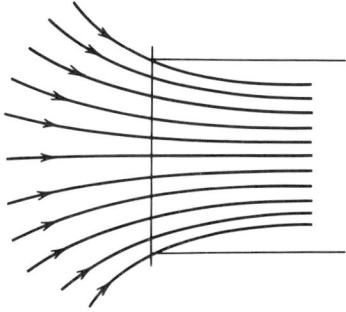

Fig. 4.2-14 A vena contracta (contracted vein) will form on the casing edge if no inlet bell is provided.

Type	Casing	Motor Support	Guide Vanes	Hub Diameter	Static Pressure	Max ME (%)
Propeller fan	Mounting ring or mounting panel	Inlet side of panel preferred	None	Small	Low	70
Tubeaxial fan	Short cylindrical housing	Inside housing, outlet side preferred	None	Medium	Moderate	75
Single-stage vaneaxial fan	Cylindrical housing	Inside housing, either side used	Around motor support	Large	High	90
Double-stage axial flow fan	Long cylindrical housing	Inside housing, between the 2 stages	None	Large	Very high	70

Fig. 4.2-15   Axial flow fans.

Fig. 4.2-16 Front view of 18-in. VAF wheel, 10 hp, 3450 rpm, aluminum casting, airfoil blades. (Courtesy of Circle Steel Corp., Taylorville, IL.)

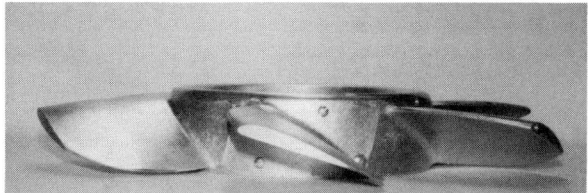

Fig. 4.2-17 Side view of 22-in. VAF wheel, 15 hp, 3450 rpm, aluminum casting, airfoil blades; note twist in the blade angles. (Courtesy of Circle Steel Corp., Taylorville, IL.)

Fig. 4.2-18 Inlet view of 18-in. VAF housing, 10 hp, 3450 rpm; note space provided for motor and for fan wheel. (Courtesy of Circle Steel Corp., Taylorville, IL.)

Fig. 4.2-19 Side view of 10-in. VAF wheel, $\frac{1}{2}$ hp, 3450 rpm, plastic, single-thickness blades; note twist in the blade angles. (Courtesy of Coppus Engineering Corp., Worcester, MA.)

more important because without it the inner portion of the blade will stall and will permit reverse air flow.

In some fans the static pressure requirement is so high that the larger blade angles near the center can no longer prevent the reverse air flow. Therefore, in vaneaxial fans the hub diameter is made larger so that reverse air flow near the center is prevented, even at high static pressures.

Best efficiencies are obtained by the use of airfoil shapes as cross sections of the blades, because airfoils have a large lift–drag ratio, resulting in more pressure with less losses. Airfoil blades are usually made as aluminum castings, incorporating both features—the twist in the blade angles and the airfoil shape in the cross sections. Figures 4.2-16 and 4.2-17 show such fan wheels. Another construction, used by a few manufacturers, is that of hollow-steel airfoil blades welded to a fabricated hub. Figure 4.2-18 shows the corresponding vaneaxial fan housing with the inlet bell attached.

Many fan wheels, however, use single-thickness blades, instead of airfoil blades, but even these blades should have the desired variation of the blade angles. Figure 4.2-19 shows such a fan wheel, featuring a large hub-tip ratio and an excellent blade twist. The cross sections of the blades are curves approximating a mean line of an airfoil shape. Oversimplified blades as are sometimes found in low-priced propeller fans will result in inefficient and noisy operation.

Performance of Axial Flow Fans

The performance of fans is determined by laboratory tests. The test methods are described in a manual prepared jointly by the ASHRAE (American Society of Heating, Refrigerating and Air Conditioning Engineers) and the AMCA (Air Movement and Control Association). Various standard test setups are specified, such as outlet ducts or inlet ducts (10 diameters long) and outlet chambers or inlet chambers.

Ducts with Pitot tube traverses (see Fig. 4.2-3) can be used only for testing fans that produce enough static pressure to overcome the resistance of the test duct plus accessories. For low-pressure fans such as propeller fans and some roof ventilators, more elaborate nozzle chambers are needed. These are equipped with calibrated nozzles and booster fans, which help overcome the resistance of the nozzles, screens, and so on.

When a propeller fan is tested on an inlet chamber, an outlet duct 2–3 diameters long may be added. This will result in a somewhat improved performance, due to static recovery. That is, some of the outlet velocity pressure is converted into static pressure as the air flow passes through the short outlet duct.

The principle used in fan performance tests is to set up the fan in a manner similar to that in which it presumably will operate when installed in the field. Then establish different operating conditions by gradually throttling the test system. For each condition, measure everything that is needed in order to determine by subsequent calculation the fan output (cfm, SP), the fan input (bhp), the rpm, and the fan efficiency so that characteristic curves can be drawn.

Figure 4.2-20 shows the general shape of the performance curves for an axial flow fan. Four curves are plotted against volume output, showing the variation of static pressure, brake horsepower, efficiency, and sound level for varying cfm. The following characteristics are to be noted in the shape of such performance curves.

The pressure curve, starting at the free delivery point, rises to a peak pressure where it starts to bend over and where the operating range ends. Here separation of the air flow from the trailing edge of the blade takes place and stalling occurs. The stalling range, that is, the range to the left of the stalling point, consists of a more or less pronounced stalling dip and a succeeding stalling rise in which the fan operates more like an inefficient mixed-flow fan.

The maximum efficiency and the minimum sound level occur at about 70% of the peak pressure prior to stalling. The entire operating range shows good efficiencies and low sound levels. At the stalling point, where the air flow becomes turbulent, the sound level increases abruptly and the quality of the sound changes noticeably from a predominantly musical note to a low rumbling noise. The frequency of the musical note is equal to the number of blades times the revolutions per second. Precaution must be taken in the selection of fans, so that the unit, when installed in the field, will not operate in any part of the inefficient and noisy stalling range. The bhp curve also has a stalling dip, and the bhp at no delivery may be higher or lower than the rest of the curve, depending on the design.

Figure 4.2-21 shows a comparison of the static pressure curves for the four types of axial flow fans, all having the same wheel diameter and the same running speed. While the general shape of the four curves is similar, the range of static pressures is quite different.

4.2-5 System Characteristics

Figure 4.2-21 also shows four system characteristics. These are parabolic curves characterizing the system in a similar way as the pressure curve characterizes the performance of a fan. Wherever the system characteristic intersects the fan characteristic, this will be the point of operation.

One point of the system characteristic defines its shape. The equation for these parabolic curves is simply $SP = K \, (cfm)^2$, where K determines the steepness of the parabola. This equation indicates that for a specific system the static pressure needed to overcome its resistance varies as the square of the cfm passing through the system. For example, if we want to increase the cfm in the ratio 1.2 (i.e., if we

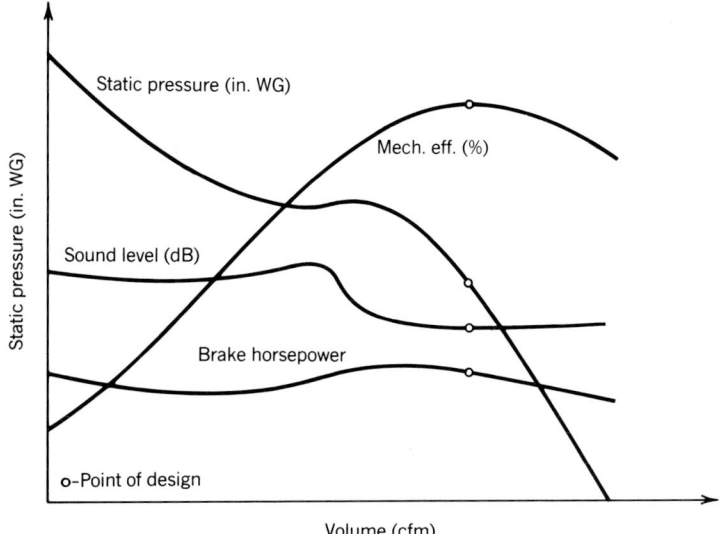

Fig. 4.2-20   General shape of axial flow fan performance curves.

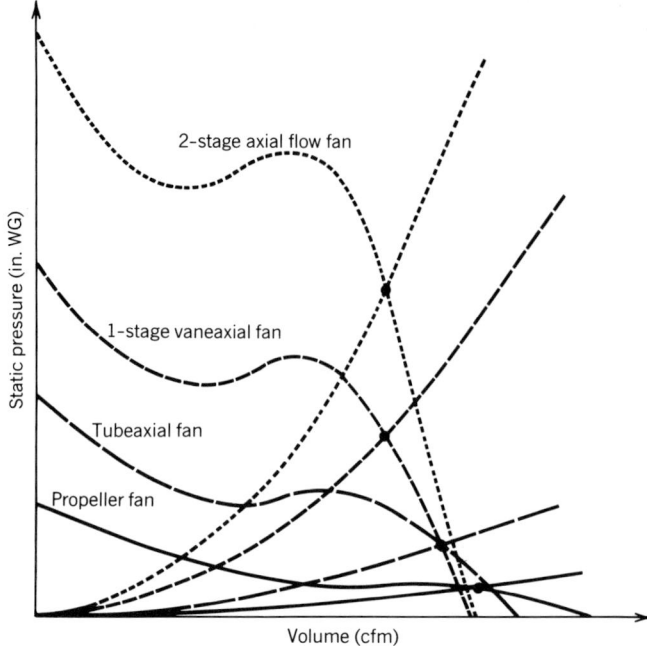

Fig. 4.2-21 Performance of different types of axial flow fans. Parabolic lines indicate system characteristics against which different fans operate.

want to force 20% more cfm through a system), the static pressure has to increase in the ratio $1.2^2 = 1.44$ (i.e., we need 44% more static pressure).

The above equation applies to duct systems, as used in ventilation. In other systems where the air velocities are much lower (such as agricultural bins for grain drying or bag houses for air pollution control), the exponent is 1.5 or 1 (instead of 2), which will result in a flatter parabola or even a straight line.

Figure 4.2-22 shows the performance of a typical 36-in. vaneaxial fan, running at 1750 rpm, with the blade angles at the tip varied from 13° (10 hp) to a maximum of 37° (60 hp). As the blade angles are increased, the cfm more than doubles, while the static pressure increases by about 50% and the stalling dip deepens. Some manufacturers make axial flow fan wheels with adjustable-pitch airfoil blades. Usually, this angular adjustment is done with the power shut off, but some designs permit automatic "in-flight" adjustment.

From the performance shown for this 36-in. vaneaxial fan at 1750 rpm, we could calculate the performance for other running speeds and for other geometrically similar sizes by using certain formulas known as fan laws.

4.2-6 Fan Laws

Fan laws are a set of formulas that can be used to convert the performance of a fan from one condition to another condition. They apply not only to axial flow fans but also to centrifugal fans, mixed-flow fans, or other types of fans, and they apply regardless of whether the fan is blowing or exhausting. Each point on the performance curve has to be converted separately. There are a number of fan laws, but the three most important ones are the fan laws for variation of rpm, for variation of fan size, and for variation of air density.

If the rpm is varied:

The cfm varies as the rpm.
The SP varies as the rpm^2.
The bhp varies as the rpm^3.

For example, a fan running at 2000 rpm delivers 16,000 cfm against 3 in. SP, thereby consuming 15

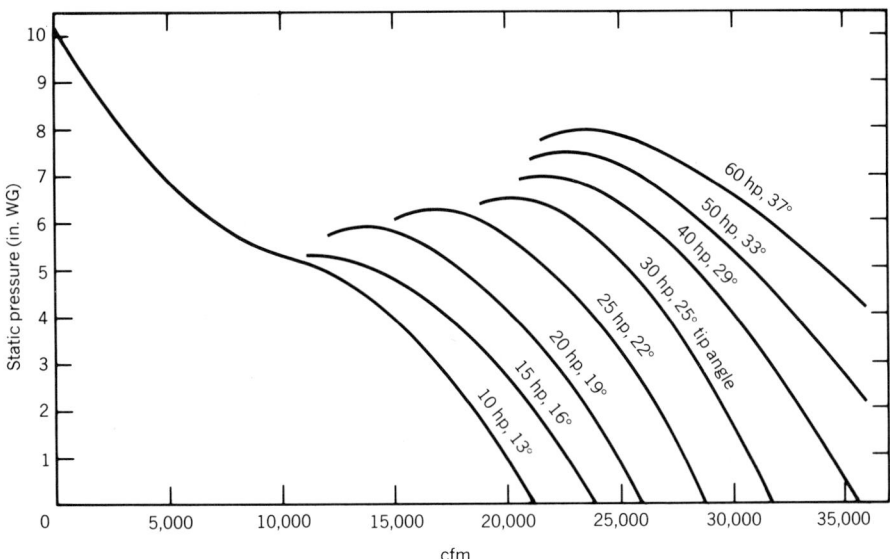

Fig. 4.2-22 Performance of a typical 36-in. vaneaxial fan at 1750 rpm, with tip angles varied from 13° (10 hp) to 37° (60 hp).

bhp. If the rpm is increased by 10%, to 2200 rpm, so that the rpm ratio is 1.1:

The cfm will increase from 16,000 to 1.1 × 16,000 = 17,600 cfm.
The SP will increase from 3 to 1.21 × 3 = 3.63 in. WC.
The bhp will increase from 15 to 1.331 × 15 = 19.97 hp.

From the above formulas, it appears that the SP also varies as the cfm². This happens to be the same relationship as for the system characteristic. That is, each point on the performance curve moves along its system characteristic when the rpm is varied. In other words, if the rpm of a fan is changed but the system remains unchanged, the point of operation (intersection of the two curves) will remain in the same relative location on the SP curve and the fan efficiency will remain the same. There is no risk, therefore, that speeding up a fan may shift the operating point from a good operating range into the stalling range. What a friendly gesture of cooperation by nature! On the other hand, suppose a fan was installed on a certain system (whose resistance had been determined incorrectly) and this fan–system combination resulted in the fan operating in the stalling range. Any attempt to rectify this condition by speeding up or slowing down this fan will be futile.

Fan manufacturers usually make a line of fans consisting of various sizes in geometric proportion. In that case they do not have to run performance tests on each size but can derive the performance of several sizes from one test size. This derivation is done by means of the fan laws for the variation of size. If the fan wheel diameter D is varied:

The cfm varies as the D^3.
The SP varies as the D^2.
The bhp varies as the D^5.

For example, a 27-in. vaneaxial fan running at 2000 rpm delivers 16,000 cfm against 3 in. SP, thereby consuming 15 bhp. If the fan wheel diameter D is increased from 27 to 30 in., so that the diameter ratio is 1.11:

The cfm will increase from 16,000 to 1.37 × 16,000 = 21,950 cfm.
The SP will increase from 3 to 1.23 × 3 = 3.70 in. WC.
The bhp will increase from 15 to 1.69 × 15 = 25.40 hp.

The two above fan laws (for rpm and for size) indicate that an increase in rpm boosts the SP more than the cfm. On the other hand, an increase in size boosts the cfm more than the SP.

An interesting conclusion can be reached by combining the two above fan laws for SP:

The SP varies as the rpm^2 ⎫   From these two relationships we can obtain another
the SP varies as the D^2 ⎭   relationship: SP varies as $(D \times \text{rpm})^2$.

Since $D \times$ rpm is proportional to the tip speed, we can write

$$\text{The SP varies as the (tip speed)}^2.$$

That is, two geometrically similar fans with the same tip speed will produce the same SP.
    If the air density d is varied,

The cfm does not vary.
The SP varies as the density d.
The bhp varies as the density d.

For example, at standard air density a fan delivers 16,000 cfm against 3 in. SP, thereby consuming 15 bhp. If the fan will operate in Denver where due to the high altitude the air density is only 82.4% of the standard air density:

The cfm will remain constant at 16,000 cfm.
The SP will decrease from 3 to $0.824 \times 3 = 2.47$ in. WC.
The bhp will decrease from 15 to $0.824 \times 15 = 12.4$ hp.

Sound Level

The sound level produced by well-designed axial flow fans is lower than that of centrifugal fans of the same tip speed but is more sensitive to the effect of turbulent air flow (which will raise the sound level considerably). A list of the factors (in the order of importance) that produce noise and therefore should be avoided when quiet operation is desired follow:

1. Operation in the stalling range.
2. High tip speed.
3. Lack of an inlet bell (if installed without an inlet duct).
4. Obstructions in the airstream ahead of and close to the blades (support arms, conduit pipes, etc.).
5. Sharp elbows in the ductwork ahead of and close to the fan inlet.
6. Inlet vanes.
7. Obstructions in the airstream past and close to the fan blades.
8. Vibration due to poor balance or due to a resonance condition.
9. Single-thickness blades produce slightly more noise than airfoil blades.
10. Many narrow blades produce more noise than fewer and wider blades because it is the blade edges (particularly the trailing edges) rather than the blade surfaces that produce turbulence and therefore noise.

4.2-7 Fan Selection

The problem of fan selection usually presents itself in the following manner: cfm and SP are specified; type, size, speed, and hp are to be selected. The first step is to make a tentative decision regarding the fan type and to study the rating tables, published in fan catalogs for this type of fan, in order to determine size, rpm, and hp. Usually, the same cfm and SP requirements can be met either by a larger fan at lower rpm or by a smaller fan at higher rpm. The larger fan at lower rpm has two advantages: It will have a lower tip speed and therefore a lower sound level; it also will usually have a lower power consumption and therefore a lower operating cost. The smaller fan, on the other hand, will have a lower first cost.
    In order to confirm that a good selection of the fan type has been made, use the fan laws for rpm and size and compare the data with the graph shown in Fig. 4.2-23. This graph shows the volume–pressure plane, divided into five ranges, as marked. The maximum load or the upper limit for good performance is indicated by the dotted curve. Most fans will stay well below this maximum load. The dividing lines are system characteristics similar to those shown in Fig. 4.2-21. They are here straight lines because of the logarithmic scales. The ranges are to be considered recommendations

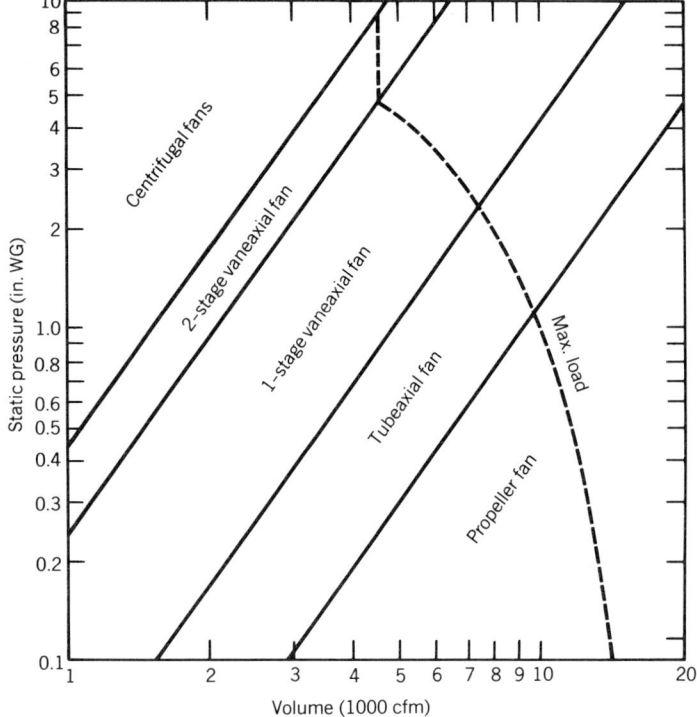

Fig. 4.2-23 Operating ranges of various types of fans. Classification sheet for 24-in. wheel diameter and 1750 rpm.

rather than strict rules, as the individual ranges are widely overlapping. In particular, centrifugal fans will be suited not only for the indicated range but will overlap into a good portion of the other ranges. Therefore, whenever the air flow requirements can be met by either an axial flow fan or a centrifugal fan, the decision between them will be based on other considerations, such as those listed below.

The centrifugal fan has the following advantages over the vaneaxial fan:

1. Natural adaptability to installations requiring a 90° turn of the airstream.
2. Better accessibility of the motor, compared with vaneaxial fans using direct motor drive.
3. Better protection of the motor against hot or contaminated gases than for a vaneaxial fan with direct motor drive.
4. Better adaptability to installations with fluctuating system characteristic.

The vaneaxial fan has the following advantages over the centrifugal fan:

1. Greater compactness.
2. Lower first cost.
3. Straight-line installation, resulting in lower installation cost.
4. Lower sound level at the same tip speed.

The above shows that each type has its definite place. It will also serve as a guide in the selection of fans, as illustrated by three examples:

1. A fan should be selected that will deliver 10,000 cfm against a static pressure of 0.1 in. WC and that will be quiet enough for installation as an exhaust fan in a record store. We select a 30-in. propeller fan at 855 rpm, having a tip speed of 6700 fpm. Converted to the standard size and speed of Fig. 4.2-23 (24 in. and 1750 rpm), this fan would deliver 10,500 cfm against 0.27 in. WC. Figure 4.2-23 confirms that this should be a propeller fan. If the fan were for installation in a factory, we could select a 24-in. propeller fan at 1750 rpm, having a tip speed of 11,000 fpm. This fan would be less

expensive due to the smaller size and the higher motor speed, but at the same time it would have a higher noise level due to the considerably higher tip speed.

2. An inexpensive fan should be selected that will deliver 23,000 cfm against 4 in. SP. We select a 36-in. tubeaxial fan at 1750 rpm. Converted to 24 in. and 1750 rpm, this fan would deliver 6800 cfm against 1.78 in. WC, which according to Fig. 4.2-23 still falls within the range of tubeaxial fans, but comes close to the vaneaxial fan range. If high efficiency would be more important than first cost, we would rather use a 36-in. vaneaxial fan at 1750 rpm.

3. A vaneaxial fan for straight-line installation should be selected that will deliver 2000 cfm against 5 in. SP. We decide on a 15-in. vaneaxial fan at 3450 rpm. Converted to 24 in. and 1750 rpm, this fan would deliver 4160 cfm against 3.3 in. SP, which according to Fig. 4.2-23 is still within the vaneaxial fan range. If good accessibility of the motor would be more important than straight-line installation, a centrifugal fan would be more suitable.

4.2-8 Centrifugal Fans

As previously mentioned, centrifugal fans generally are for higher SP. According to their blade shapes, they can be subdivided into the following six categories: AF (airfoil), BC (backward-curved), BI (backward-inclined), RT (radial tip), FC (forward-curved), and RB (radial blade). The sketch (Fig. 4.2-24) shows these six commonly used blade shapes. (Other variations are occasionally used.) Each of these blade shapes has its advantages and disadvantages and accordingly each is well suited for certain applications.

Figure 4.2-24 also shows the approximate maximum efficiencies that can be attained with these blade shapes. Many fans, however, are built for low cost and have efficiencies below those shown. The highest efficiencies can be obtained with AF blades, the lowest with radial blades.

AF blades are not only the most efficient but have the further advantage of high structural strength, so that they can run at high speeds and produce up to 20 in. SP. The blades usually are hollow airfoils of welded steel (see Fig. 4.2-25). Cast aluminum wheels are sometimes used, but only in small sizes.

BC and *BI blades* still have good efficiencies, but their maximum rpm and maximum SP are lower, unless the blade width is reduced (less cfm) or special reinforcements are provided which further impair the efficiency.

Figure 4.2-26 shows the performance curves for a typical BI centrifugal fan, $7\frac{1}{2}$ hp, 27-in. wheel diameter, 1140 rpm. For AF and BC units the curves are of similar shape. Figure 4.2-26 shows the following features:

1. Good efficiency.
2. The optimum operating range (maximum efficiency, lowest sound level) is between 50 and 75% of the FD cfm and this is the range in which these fans should be used.
3. Close to the ND point, there is sometimes a slight pressure drop (as indicated by the dashed line), but apart from this the SP curve is of steadily rising shape and free of dips such as usually occur in the SP curves of FC fans and of VAF units. This characteristic results in stability of performance, which makes AF, BC, and BI fans suitable for fluctuating systems and for parallel operation.
4. The bhp curve is of the so-called nonoverloading type, which again is a distinct advantage of AF, BC, and BI units only. The bhp curve reaches its peak in the optimum operating range. The motor horsepower then is selected equal to this peak value, and an overloading of the motor therefore is impossible.

Centrifugal fans with *RT blades* are used mainly in large sizes, in combination with bag houses, for air pollution control. The SP curve is still stable (no dip), but the efficiency is somewhat lower and the bhp curve is overloading at and near FD. The radial tip shape results in a self-cleaning action, an important advantage in the presence of dust and fly ash. DWDI and inlet boxes are popular for RT units. Renewable wearing plates, fastened to the pressure side of the blades, and interchangeable scrolls or scroll linings are sometimes provided. The bearings are often water cooled. Instead of RT blades, BI blades with steep blade angles are occasionally used to handle dust-laden air; they still have

AF	BC	BI	RT	FC	RB
92%	85%	78%	70%	65%	60%

Fig. 4.2-24 Six blade shapes commonly used in centrifugal fans. The approximate maximum efficiency attainable for each type is shown.

Fig. 4.2-25 Angular view of 16-in. centrifugal fan wheel, 5 hp, 3450 rpm, with eight airfoil blades, welded steel, with flat backplate and spun shroud for tapered blades. (Courtesy of Coppus Engineering Corp., Worcester, MA.)

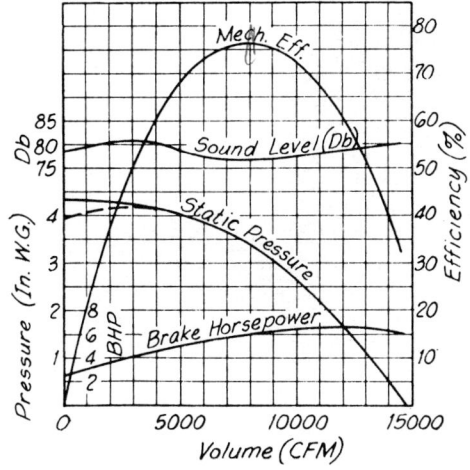

Fig. 4.2-26 Performance curves of a typical BI centrifugal fan, $7\frac{1}{2}$ hp, 27 in. wheel diameter, 1140 rpm.

the self-cleaning feature, but the efficiencies are slightly higher, the bhp curve is less overloading, and the manufacturing cost is slightly lower.

Centrifugal fans with *FC blades* are special in that they deliver considerably more cfm and produce higher SP than AF, BC, or BI blades of the same size and speed. Conversely, FC blades can run at about one-half the rpm to operate in a comparable range of cfm and SP. This is shown in Figs. 4.2-27 and 4.2-28. Note that the scales for cfm and SP, used in Fig. 4.2-27, are reduced to one-half the values of Fig. 4.2-26. In Fig. 4.2-28 the same scales are used as in Fig. 4.2-26, but the rpm was reduced to one-half.

The SP curve for FC units has a dip which in certain installations may cause unstable operation. Precaution, therefore, should be taken so that FC fans are not used for applications such as fluctuating

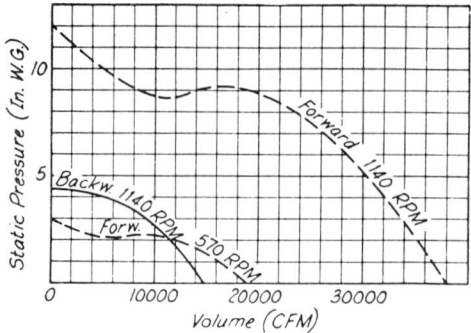

Fig. 4.2-27   Comparison of SP curves for BI and FC centrifugal fans, 27-in. wheel diameters.

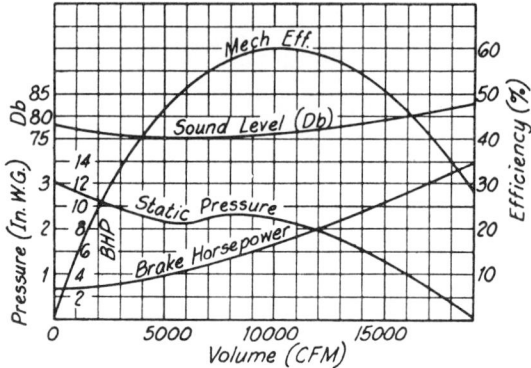

Fig. 4.2-28   Performance curves of a typical FC centrifugal fan. 10 hp, 27-in. wheel diameter, 570 rpm.

systems or parallel operation, and that even in other installations the unit will not operate in the unstable range of the SP curve.

The bhp curve of FC units is even more overloading in the low-pressure range than that of RT units. At FD, the bhp is more than twice the bhp in the range of maximum efficiency. Since centrifugal fans in general are not built for operation at or near FD (PFs or TAFs perform this function with greater efficiency), this shape of the bhp curve results in power requirements outside the operating range that are higher by far than those within the operating range. This overloading bhp curve is a serious disadvantage of FC fans.

FC fans are used mainly in small sizes where their lower efficiency is less objectionable. In these small sizes an oversize motor with an hp rating equal to the maximum bhp at FD is usually selected so that operation at any condition will be safe. For large units, however, the increased price of the oversize motor would be prohibitive, and the motor hp is selected only slightly larger than the bhp for the prospective operating condition. Precaution must then be taken so that the unit, when installed in the field, will not operate against too low an SP as this would result in an overload for the motor. The permissible operating range of the FC fan, being limited to the left by the dip in the SP curve and to the right by the rising bhp curve, therefore, is narrower than that of the AF, BC, or BI fan.

The sound level of the FC fan is comparable to that of the BC fan if the operating ranges are comparable, that is, if the FC fan runs at about one-half the rpm.

The considerably larger cfm of FC wheels is a result of the scooping action of the blades, producing high flow velocities as the airstream leaves the blade tips, higher than the tip speed (velocity of the blade tips itself). This high air velocity (kinetic energy) is then gradually slowed down in the scroll housing and partially converted into SP (potential energy).

Figure 4.2-29 shows a water jet hitting an egg cup (a forward-curved surface), which reverses its flow by almost 180°. A similar action can be observed when children play in a pool. By curving their fingers forward and scooping the water at the surface, they can produce water jets of high velocities, higher than the velocity of their moving hand.

Fig. 4.2-29 View of water jet being reversed when hitting a forward-curved surface.

To accommodate the large cfm potential of FC wheels, such fans have large inlets (resulting in shorter blades) and more blades. While the five other wheel types use from 6 to 16 blades, FC wheels use from 24 to 64 blades.

Summarizing, we can say that BC blades have the following advantages over FC blades:

1. Stable SP curve (no dip).
2. Higher efficiency, resulting in lower operating cost.
3. Nonoverloading bhp characteristic.
4. Higher operating speeds, which for direct drive may result in less expensive motors.

FC blades, on the other hand, have the following advantages over BC blades:

1. Compactness.
2. Lower running speed, resulting in easier balancing.
3. Lower first cost, particularly in small sizes.

Fig. 4.2-30 Angular view of a scroll housing showing drive side and outlet, for belt drive, CCW rotation, single-inlet and top horizontal discharge. (Courtesy of Ilg Industries, Inc., Chicago, IL.)

From the above, it appears that FC fans usually will be preferable in small sizes, while AF, BC, and BI fans will be preferable in large sizes.

Centrifugal fans with *radial blades* are used in pressure blowers, turbo blowers, and steel plate exhausters. Wheels for pressure blowers and turbo blowers usually have narrow blades, a backplate and sometimes a shroud ring, and these units can reach efficiencies of up to 60%. They are for low cfm and high SP. Steel plate exhausters have fairly wide blades and are available with and without backplate and shroud. The open steel plate fan wheels have blades mounted on a heavy spider so that backplate and shroud often are omitted. Their main advantages are ruggedness and ability to handle dust-laden air without plugging up. They can withstand severe service conditions. Because these units are often employed for conveying materials, such as grinding dust, sawdust, cotton, and shavings, the flat blades and the wide spacing of the blades are an advantage; curved blades and narrow passages would tend to become plugged up by deposits of dirt or of the material conveyed.

In addition to blade shape, there are several other parameters according to which centrifugal fans may be classified, such as drive arrangement, direction of rotation, type of inlet, and type of discharge. For example, the unit shown in Fig. 4.2-30 has belt drive, CCW rotation (viewed from the drive side), single inlet, and top-horizontal discharge.

Drive Arrangement

In the majority of cases, fan wheels are driven by electric motors because of the high efficiency and smooth operation of such motors. Other drives, such as engine drive or turbine drive, are occasionally used, especially in places where electric power is not available or in explosive atmospheres where electric sparks must be avoided. The fan wheel can be mounted directly on the motor shaft or the power can be transmitted by a rigid coupling, flexible coupling, or belt drive. Ten possible drive arrangements have been adopted as standards by AMCA. The decision on the drive arrangement to be used in a specific instance depends on the required running speed of the unit and the type of installation. Direct drive should be used wherever possible, but in many applications belt drive is indicated.

Belt drive has two advantages over direct drive:

1. Greater flexibility, since any running speed can be selected. Even after the unit has been installed, the rpm can still be changed within limits (as long as the motor will not be overloaded).

2. Lower first cost in cases of low-speed units, since high-speed electric motors can be used instead of expensive low-speed motors.

Direct drive, on the other hand, has two advantages over belt drive:

1. Lower maintenance cost and no expenses for shafts, bearings, pulleys, and belts.

2. Somewhat lower power consumption as belt and other losses are eliminated.

For these reasons, direct drive is prevalent in small sizes (high running speeds) and belt drive in large sizes (low running speeds).

Type of Inlet

The unit shown in Fig. 4.2-30 is of the SWSI type. By combining a CW and a CCW fan into one unit, a DWDI fan is obtained, having an inlet on each side and one common outlet. It delivers almost double the cfm against the same SP, while consuming about double the bhp. This construction is a practical and inexpensive solution for many applications where large air volumes are required, yet the size of the fan should not become excessive. The only disadvantage is the obstruction caused by the drive arrangement (base, bearing, and pulley) at the inlet on one side, resulting in a slight loss of cfm and efficiency.

For exhaust installations the fan inlet is sometimes equipped with inlet boxes attached to the scroll housing, especially on DWDI fans, but occasionally also on SWSI fans. The inlet boxes connect the duct work with the inlet openings of the fan. They usually are of simple shape, more or less rectangular or slightly tapered, and sometimes with turning vanes on the inside. Inlet boxes reduce the cfm somewhat, not only because of the friction loss of the additional obstruction, but also because they represent an elbow close to the fan inlet, thus producing an uneven velocity distribution as the air flow enters the fan.

In all fan installations great care must be taken to avoid inlet turbulence or an uneven inlet flow, which could be caused by obstructions or elbows near the fan inlet. Such conditions will result in reduced cfm, in lower efficiencies, and in increased noise levels. As previously mentioned, axial flow fans are more sensitive than centrifugal fans to such disturbances. The reason for this is that in axial flow fans the inlet flow normally approaches the blades in undisturbed straight lines, while in centrifugal fans the air flow approaching the blades already has passed through a 90° turn in the funnel-shaped flow pattern and therefore already has a certain amount of turbulence, which already has somewhat increased the noise level, so that additional turbulence is of less consequence.

Type of Discharge

In most centrifugal fans a scroll housing is provided, as shown in Figs. 4.2-7 and 4.2-30. Its main function is to collect the individual airstreams from the blades and to reunite them into the single airstream discharging through the outlet opening. The location and direction of the discharging airstream relative to the incoming airstream will vary according to the demands of the installation. Sixteen different discharge arrangements (eight for each rotation) have been adopted as standards by AMCA. Additional in-between positions are also used occasionally. Most centrifugal fans are designed in such a manner that the manufacturer can furnish them for any discharge arrangement required to suit the customer's need. Some designs are for universal discharge, meaning that even after complete assembly the discharge arrangement can still be changed by simply removing a few bolts and moving the scroll housing to a different angular position.

A different type of discharge is sometimes used for applications in which air is to be delivered into a large space rather than into a duct. In such cases (roof ventilators, plug fans) the scroll housing is completely omitted and the individual airstreams leaving the blade channels simply diffuse outward into space. This arrangement is called radial discharge or circumferential discharge. The term *radial discharge* is used more often, even though the term *circumferential discharge* comes closer to the truth because the direction of the flow velocity is closer to circumferential than to radial. This type of discharge has the advantages of lower cost, greatly increased cfm at low SP, and a considerable reduction of the sound level as the cutoff (a major source of noise) is eliminated. FC wheels cannot be used for this application as they require a scroll housing for the conversion of VP into SP, and without a scroll housing their performance would be poor.

AMCA Standards

AMCA has established standards for certain dimensions (such as maximum wheel diameters, maximum inlet diameters, and maximum outlet areas) for three commonly used lines of centrifugal fans. Most fan manufacturers conform to these standards. A short description of these three fan lines for which standards have been established follows:

1. Centrifugal fans that are in general use, particularly in heating, ventilating, and air conditioning, with either AF, BC, BI, or FC wheels. These units come in both SWSI and DWDI executions. They have large areas at the fan inlet and outlet, providing sufficient room for axially wide blades. However, reduced blade widths (for less cfm) also are available. The AF, BC, and BI wheels have small blade angles (usually between 40° and 45° at the tip) for nonoverloading bhp curves. The number of blades usually is between 8 and 12. The units are built for classes I, II, III, and IV; class I has low tip speeds and class IV (heavy duty) has the highest tip speeds, pressures, and prices. The AF units, as previously mentioned, can produce efficiencies of more than 90%. There are a total of 25 standard sizes, with wheel diameters ranging from $12\frac{1}{4}$ to $132\frac{1}{2}$ in. The wheel diameters form a geometric series with a progression ratio of about 1.1.

2. Industrial centrifugal fans, formerly called industrial exhausters, because they were predominantly used for exhausting sawdust, shavings, and granular material passing through the unit. Most fan manufacturers offer these units with a choice of four different wheels, sometimes called AH (air-handling wheel), MH (material-handling wheel), LR (long-shavings wheel with a backplate and usually with a shroud or rim), and LS (long-shavings open wheel, no backplate, no shroud). These units come only with SWSI. They have small areas at the fan inlet and outlet and the axial blade widths are somewhat smaller. The AH wheels have flat BI blades (flat for low cost, BI for good efficiencies), but the blade angles are steeper, resulting in a slightly overloading bhp curve. The three other wheels have approximately radial blades for self-cleaning action, resulting in lower efficiencies and in more overload at free delivery. There is a total of 16 standard sizes, with inlet diameters ranging from 11 to 60 in. and with wheel diameters ranging from $19\frac{1}{8}$ to $104\frac{1}{4}$ in. The progressions ratio is not exactly constant (for the sake of round numbers of the inlet diameters), but it is approximately 1.12.

Figure 4.2-31 shows a rating table for an air-handling unit with a $36\frac{1}{2}$-in.-diameter wheel and a 21-in.-diameter housing inlet. This table shows the usual arrangement in which the performance of belt-driven fans is presented in catalogs, for convenient selection by the customer. For example, an AH fan should be selected that can be connected to a 21-in. inlet duct and will deliver 7200 cfm against 5 in. SP. The table indicates that this fan will run at 888 rpm and will consume 8.07 bhp. A 10-hp, 1750-rpm fan will be used, with a pulley ratio of 0.507. The table shows that this fan will operate at 78% efficiency, which is good for a BI wheel with flat blades. The table also shows that for

AIR HANDLING WHEEL

21AH

WHEEL 36½ INCH DIAMETER 9.56 FT. CIRCUMFERENCE
INLET 21.0 IN. OUTSIDE DIAMETER 2.32 SQ. FT. AREA INSIDE
OUTLET 20⁵⁄₁₆ x 17¹¹⁄₁₆ INCH OUTSIDE 2.40 SQ. FT. AREA INSIDE

F.P.M. = OUTLET VELOCITY IN FEET PER MINUTE S.E. = STATIC EFFICIENCY M.E. = MECHANICAL EFFICIENCY

CFM	FPM	½" S.P. RPM	BHP	SE	ME	1" S.P. RPM	BHP	SE	ME	2" S.P. RPM	BHP	SE	ME	3" S.P. RPM	BHP	SE	ME	4" S.P. RPM	BHP	SE	ME
2402	1000	284	.27	.69	.78	383	.53	.71	.76	538	1.30	.70	.73	658	2.30	.69	.72	760	3.51	.69	.72
2882	1200	296	.34	.66	.78	390	.64	.69	.77	542	1.49	.71	.76	661	2.57	.71	.75	762	3.86	.71	.74
3363	1400	311	.43	.62	.77	400	.76	.69	.78	548	1.69	.71	.77	666	2.86	.71	.76	766	4.25	.71	.76
3843	1600	328	.53	.57	.75	412	.90	.67	.78	556	1.92	.71	.78	672	3.18	.71	.77	772	4.66	.71	.77
4324	1800	346	.65	.53	.74	427	1.06	.64	.77	567	2.17	.69	.78	681	3.52	.71	.78	780	5.10	.71	.77
4804	2000	364	.79	.49	.72	442	1.24	.61	.77	579	2.45	.68	.78	692	3.90	.70	.78	790	5.57	.70	.78
5284	2200	384	.94	.44	.71	459	1.44	.58	.76	593	2.75	.66	.78	703	4.30	.68	.78	800	6.07	.69	.78
5765	2400	403	1.12	.41	.70	476	1.66	.55	.75	607	3.08	.64	.77	716	4.73	.67	.78	812	6.61	.68	.78
6245	2600	424	1.32	.37	.68	494	1.91	.52	.74	623	3.44	.62	.77	730	5.20	.65	.77	825	7.19	.67	.78
6726	2800	446	1.56	.34	.67	513	2.18	.49	.73	639	3.83	.59	.76	745	5.70	.64	.77	839	7.80	.66	.78
7206	3000	468	1.82	.31	.66	532	2.48	.46	.72	656	4.25	.57	.75	761	6.24	.62	.77	853	8.46	.64	.77
7686	3200	490	2.11	.29	.65	551	2.81	.43	.71	674	4.71	.55	.75	777	6.81	.60	.76	868	9.15	.63	.77
8167	3400	512	2.43	.26	.65	571	3.18	.41	.70	691	5.20	.53	.74	793	7.43	.58	.76	884	9.88	.61	.77
8647	3600	535	2.79	.24	.64	591	3.58	.38	.69	710	5.72	.51	.73	811	8.08	.56	.75	900	10.7	.60	.76
9128	3800	558	3.18	.22	.63	613	4.02	.36	.68	729	6.29	.49	.72	828	8.77	.55	.75	917	11.5	.58	.76
9608	4000	581	3.61	.21	.63	635	4.50	.34	.67	748	6.89	.46	.72	846	9.51	.53	.74	934	12.4	.57	.75
10088	4200	604	4.08	.19	.62	657	5.02	.32	.66	767	7.53	.44	.71	864	10.3	.51	.73	952	13.3	.55	.75
10569	4400	627	4.59	.18	.62	679	5.59	.30	.66	787	8.22	.42	.70	883	11.1	.49	.73	970	14.2	.54	.74
11049	4600	651	5.14	.17	.62	701	6.19	.28	.65	807	8.96	.41	.70	902	12.0	.48	.72				
11530	4800	675	5.74	.15	.61	723	6.85	.27	.65	827	9.74	.39	.69								
12010	5000	699	6.39	.14	.61	746	7.55	.25	.64												

CFM	FPM	5" S.P. RPM	BHP	SE	ME	6" S.P. RPM	BHP	SE	ME	7" S.P. RPM	BHP	SE	ME	8" S.P. RPM	BHP	SE	ME	9" S.P. RPM	BHP	SE	ME
2882	1200	850	4.92	.60	.72																
3363	1400	852	5.37	.71	.74																
3843	1600	856	5.84	.71	.76	931	6.55	.70	.72												
4324	1800	862	6.35	.71	.77	934	7.09	.71	.74	1005	7.77	.66	.71								
4804	2000	869	6.89	.71	.77	937	7.65	.71	.75	1006	8.37	.70	.73	1075	9.79	.69	.71				
5284	2200	878	7.46	.71	.78	943	8.25	.71	.75	1009	9.01	.71	.74	1076	10.4	.70	.73	1140	11.8	.69	.72
5765	2400	888	8.07	.70	.78	950	8.90	.71	.77	1013	9.67	.71	.75	1079	11.1	.71	.74	1142	12.5	.70	.74
6245	2600					958	9.57	.71	.78	1018	10.4	.71	.76	1083	11.8	.71	.76	1145	13.5	.71	.75
6726	2800									1025	11.1	.71	.77	1089	12.7	.71	.76	1150	14.3	.71	.76
7206	3000																				

7686	3200	899	8.71	.69	.78	968	10.3	.70	.78	1033	11.9	.71	.77	1096	13.6	.71	.77	1155	15.2	.71	.76
8167	3400	911	9.40	.68	.78	979	11.0	.70	.78	1043	12.7	.70	.78	1104	14.4	.71	.77	1162	16.2	.71	.77
8647	3600	924	10.1	.67	.78	990	11.8	.69	.78	1053	13.6	.70	.78	1113	15.4	.71	.78	1170	17.2	.71	.77
9128	3800	938	10.9	.56	.78	1003	12.7	.68	.78	1064	14.5	.69	.78	1123	16.4	.70	.78	1180	18.3	.71	.78
9608	4000	952	11.7	.64	.77	1016	13.6	.67	.78	1076	15.5	.68	.78	1134	17.4	.69	.78	1190	19.4	.70	.78
10088	4200	967	12.6	.63	.77	1030	14.5	.65	.78	1089	16.5	.67	.78	1146	18.5	.68	.78	1200	20.5	.69	.78
10569	4400	983	13.5	.62	.77	1044	15.5	.64	.77	1103	17.5	.66	.78	1158	19.6	.68	.78	1212	21.7	.69	.78
11049	4600	999	14.4	.60	.76	1059	16.5	.63	.77	1117	18.6	.65	.77	1172	20.8	.67	.78	1224	23.0	.68	.78
11530	4800	1015	15.4	.59	.76	1075	17.6	.62	.77	1132	19.8	.64	.77	1186	22.0	.66	.78	1237	24.3	.67	.78
12010	5000	1032	16.5	.58	.75	1091	18.7	.61	.76	1147	21.0	.63	.77	1200	23.3	.65	.77	1251	25.6	.66	.78
12490	5200	1050	17.6	.56	.75	1107	19.9	.59	.76	1162	22.3	.62	.77	1215	24.7	.64	.77	1265	27.1	.65	.77
12971	5400	1067	18.8	.55	.75	1124	21.2	.58	.76	1178	23.6	.61	.76	1230	26.1	.63	.77	1280	28.5	.64	.77
13451	5600	1085	20.0	.53	.74	1141	22.5	.57	.75	1195	25.0	.59	.76	1246	27.5	.62	.77	1295	30.1	.63	.77
13932	5800	1103	21.2	.52	.74	1159	23.8	.56	.75	1212	26.4	.58	.76	1262	29.1	.60	.76	1310	31.7	.62	.77
14412	6000	1122	22.6	.51	.73	1176	25.3	.54	.74	1229	27.9	.57	.75	1278	30.6	.59	.76	1326	33.4	.61	.77
14892	6200	1140	24.0	.49	.73	1194	26.7	.53	.74	1245	29.5	.56	.75	1295	32.3	.58	.76	1342	35.1	.60	.76
15373	6400	1159	25.4	.48	.72	1213	28.3	.52	.74	1264	31.1	.55	.75	1312	34.0	.57	.75	1359	36.9	.59	.76
15853	6600	1178	26.9	.47	.72	1231	29.9	.51	.73	1281	32.8	.54	.74	1330	35.8	.56	.75	1376	38.8	.58	.76

CFM	FPM	10″ S.P.				12″ S.P.				14″ S.P.				16″ S.P.				18″ S.P.			
		RPM	BHP	SE	ME	RPM	BHP	SE	ME	RPM	BHP	SE	ME	RPM	BHP	SE	ME	RPM	BHP	SE	ME
3843	1600																				
4324	1800																				
4804	2000																				
5284	2200																				
5765	2400	1202	13.3	.68	.71																
6245	2600	1203	14.2	.70	.73	1316	17.4	.68	.71												
6726	2800	1205	15.1	.71	.74	1317	18.4	.69	.72												
7206	3000	1208	16.0	.71	.75	1319	19.4	.70	.74	1422	23.0	.69	.72	1520	26.7	.68	.71				
7686	3200	1213	17.0	.71	.76	1322	20.5	.71	.75	1424	24.2	.70	.73	1521	28.1	.69	.72	1612	32.0	.68	.71
8167	3400	1219	18.0	.71	.77	1326	21.7	.71	.76	1427	25.5	.71	.74	1522	29.5	.70	.73	1613	33.5	.69	.72
8647	3600	1226	19.1	.71	.77	1331	22.9	.71	.76	1431	26.8	.71	.75	1525	30.9	.71	.74	1614	35.1	.70	.73
9128	3800	1234	20.2	.71	.77	1338	24.1	.71	.77	1435	28.2	.71	.76	1528	32.4	.71	.75	1617	36.7	.71	.74
9608	4000	1243	21.3	.71	.78	1345	25.4	.71	.77	1441	29.6	.71	.76	1533	34.0	.71	.76	1620	38.4	.71	.75
10088	4200	1253	22.6	.70	.78	1353	26.8	.71	.77	1448	31.1	.71	.77	1538	35.6	.71	.76	1625	40.2	.71	.76
10569	4400	1264	23.8	.69	.78	1363	28.2	.71	.78	1456	32.7	.71	.77	1545	37.2	.71	.77	1630	42.0	.71	.76
11049	4600	1275	25.2	.69	.78	1373	29.7	.70	.78	1465	34.3	.71	.78	1552	39.0	.71	.77	1636	43.8	.71	.77
11530	4800	1288	26.5	.68	.78	1383	31.2	.70	.78	1474	35.9	.70	.78	1561	40.8	.71	.77	1644	45.7	.71	.77
12010	5000	1300	28.0	.67	.78	1395	32.8	.69	.78	1484	37.6	.70	.78	1570	42.6	.71	.78	1651	47.7	.71	.77
12490	5200	1314	29.5	.66	.78	1407	34.4	.68	.78	1495	39.4	.69	.78	1579	44.5	.70	.78	1660	49.8	.71	.78
12971	5400	1328	31.0	.66	.78	1420	36.1	.68	.78	1507	41.3	.69	.78	1590	46.5	.70	.78	1669	51.9	.71	.78
13451	5600	1342	32.7	.65	.77	1433	37.9	.67	.78	1519	43.2	.68	.78	1601	48.6	.69	.78	1679	54.1	.70	.78
13932	5800	1357	34.3	.64	.77	1446	39.7	.66	.78	1531	45.2	.68	.78	1612	50.7	.69	.78	1690	56.3	.70	.78
14412	6000	1372	36.1	.63	.77	1461	41.6	.65	.77	1544	47.2	.67	.78	1624	52.9	.68	.78	1701	58.7	.69	.78
14892	6200	1388	37.9	.62	.77	1475	43.6	.64	.77	1558	49.3	.66	.78	1637	55.2	.68	.78	1713	61.1	.69	.78
15373	6400	1404	39.8	.61	.76	1490	45.6	.64	.77	1572	51.5	.66	.78	1650	57.5	.67	.78	1725	63.6	.68	.78

MAXIMUM SPEED UNSHADED AREA 1570 RPM. SHADED AREA INDICATES EXTRA HEAVY CONSTRUCTION. MAXIMUM SPEED 1750 RPM.

Fig. 4.2-31 Rating tables for centrifugal fans, belt drive. (Courtesy of Ilg Industries, Inc., Chicago, IL.)

884 rpm and 4 in. SP (instead of 5 in. SP) the fan will consume 9.88 bhp, so the 10-hp motor will still be safe, even if the SP should be somewhat lower than anticipated. However, if the SP should drop to 3 in., this fan at 888 rpm would consume about 11.3 bhp, which might be too much overload for the motor. For larger pressures, on the other hand, the brake horsepower will be lower and the 10-hp motor will be safe.

3. Industrial centrifugal fans with cast housings are sometimes called pressure blowers or volume fans. Most of these units have not only cast housings but also cast fan wheels. These wheels usually have a backplate, but no shroud, and the blades often are flat and radial (resulting in an overloading bhp curve) although other blade shapes are occasionally used. These fans come only with SWSI. They are used for both exhausting and blowing. Typical applications are exhaust of welding fumes, toxic gases, and grinding dust and supply of combustion air. These units are usually directly driven by electric motors, running at 3450 rpm in small sizes and at 1750 rpm in larger sizes. There is a total of 11 standard sizes, with inlet diameters ranging from 4 to 19 in.; the wheel diameters are not standardized and are left to the discretion of the manufacturer.

Two centrifugal fans in series are occasionally used to produce higher static pressures, but this arrangement is not as simple and inexpensive as it is for axial flow fans. For very high pressures, such as are needed in vacuum cleaners, in gas boosters, for pneumatic conveying, and for some other applications, multistage turbo blowers are used more often. These have several rotating wheels mounted on a common shaft with stationary guide vanes between the stages. They produce considerable compression ratios and result in an airstream leaving the unit heated by adiabatic compression. Low-cost fabricated units are available as well as high-priced cast units, machined for small clearances and designed with long diffuser passages for good efficiencies. Figure 4.2-32 shows a three-stage turbo blower driven by a 135-hp engine, for pneumatic conveying of grain while unloading it from a ship and loading it onto trucks. The cyclone on the right side helps in by passing the fan so that the grain will not be damaged by the fan blades.

Two centrifugal fans in parallel can be used when the cfm required is larger than available from a certain fan size, but the SP produced is adequate. Using two fans in parallel rather than one larger fan can have the following advantages:

1. When air is blown into a large space, such as a bin for storing and drying grain, a more even distribution can be obtained by two (or more) fans operating in parallel. (This principle applies to axial flow fans as well.)

2. If one motor should fail, at least the other fan can still be used. In fact, a second fan is sometimes used as a stand-by, either just to be on the safe side or if later need for additional capacity, due to a change in the system (e.g., a mine), can be anticipated.

Fig. 4.2-32 Three-stage turboblower, 27-in. wheel diameter, 5000 rpm, 135 hp, for grain conveying. (Courtesy of Dunbar Kapple Inc., St. Charles, IL.)

3. Two small fans may fit into the space available while one larger fan may not. The conditions here are somewhat similar to those of a DWDI fan replacing a larger SWSI fan.

4. Two smaller fans and motors may be more economical, particularly if they operate in the efficient performance range while one large fan may not.

AF, BC, and BI fans are safe for operation in parallel. FC fans are not recommended, as there is a risk of instability (dip in the pressure curve) or of resonance conditions.

Volume Control

In some installations the fan is selected for maximum output requirements and means are provided so that the air flow can be reduced at times, either manually or automatically. A discussion of the three most common methods to accomplish this are follows.

   1. **Variable running speed.** This method can be applied to any type of fan. The effect of speed reduction on the pressure–volume curve of a BC fan is shown in Fig. 4.2-33. As previously mentioned, each point on the curve follows the fan laws and moves along a parabolic system characteristic. The fan efficiency remains unchanged, and there is no risk of any shifting into an unstable performance range (in case of an FC fan). Another advantage is that this method of flow reduction results in greatest power economy since the bhp of a fan varies as the third power of the rpm. A third advantage is that due to the reduced rpm the noise level is correspondingly reduced. The disadvantage of the method is that the first cost usually is high, especially if continuous variation is desired.

   2. **Adjustable outlet dampers.** A shutterlike mechanism is mounted on the fan outlet. The effect of such a throttle on the pressure–volume curve of an FC fan is shown in Fig. 4.2-34. Compared with

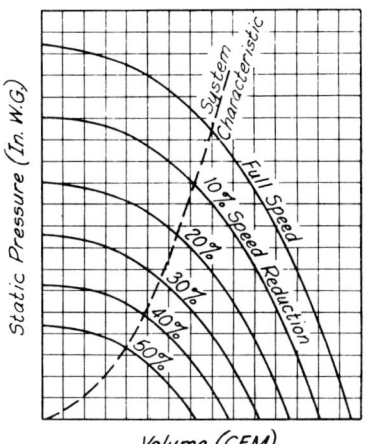

Fig. 4.2-33 Volume control by variable running speed.

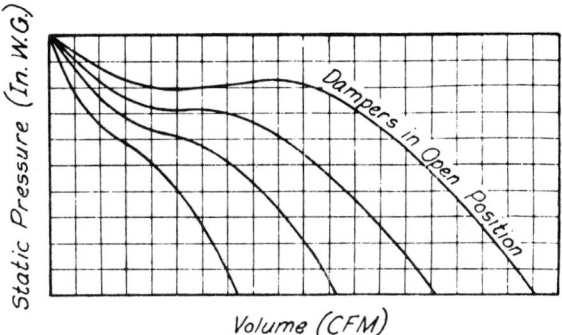

Fig. 4.2-34 Volume control by adjustable outlet dampers.

the two other methods, the savings in power consumption are somewhat smaller, but this method is the simplest in construction and the lowest in first cost. It therefore is the one generally used for flow reduction in small sizes, where the effected power savings are small at any rate. For medium and large sizes it shares the field with the two other methods.

   3. **Adjustable inlet vanes.** Not only do adjustable inlet vanes throttle the air flow they also impart to the entering airstream a spiral motion in the direction of the fan rotation that results in a reduction of cfm, SP, and bhp. As to both power savings and first cost, this method is somewhere between the two others. Inlet vanes have the additional advantage of acting as air guides and thereby creating predetermined inlet flow conditions so that disturbances due to inlet turbulence, sometimes caused by inlet boxes, elbows, or other obstructions at the inlet, are minimized.

   Inlet vanes can also be used for boosting instead of reducing the output of a centrifugal fan. In this case they have to produce an inlet spin opposite to the fan rotation. However, this will result in a larger motor horsepower and in a reduced fan efficiency. This method has been used occasionally on large units for mine ventilation if the running speed cannot be increased and for structural or other reasons.

Selection of Centrifugal Fans

For axial flow fans a classification sheet (Fig. 4.2-23) facilitates the selection of the proper type, size, and speed. For centrifugal fans, due to the various blade shapes available, a chart of this type would be too involved to be practical. Instead, we shall give a number of typical examples illustrating the application and selection of centrifugal fans.

   1. A centrifugal fan should be selected for ship ventilation to deliver 10,000 cfm against 4 in. WC and to be directly driven by a 10-hp, 1140-rpm motor. Under no condition must the fan overload the motor, and the sound level must be less than 83 dB. From the specifications we can calculate that an efficiency of 67% is required at the point of design; thus, a BI fan will meet this requirement; it will also have a nonoverloading bhp curve. From a rating table we find that a 30-in. wheel diameter will produce the specified cfm and SP. Converted (according to the fan laws) to a 27-in. wheel diameter, the specifications become 7290 cfm against 3.24 in. WC. By comparing these data with Fig. 4.2-26, we confirm that 30 in. and 1140 rpm was a good selection. The tip speed will be 8960 fpm, indicating that we can also meet the sound level requirement. If the fan were for installation in a noisy factory, where the sound level requirement could be left out, we would select a 24-in. BI fan at 1750 rpm. This fan would be less expensive due to the smaller size and higher motor speed, but it would require a 73% efficiency and the noise level would be higher, due to the higher tip speed.

   2. An inexpensive fan should be selected that will deliver 1000 cfm against 1.5 in. WC. This fan will be built into an industrial air heater and should be as compact as possible, as the space available is tight. Power consumption and noise level, on the other hand, are of secondary importance. We select an 8-in. FC fan, running at 1750 rpm and blowing into the heating coils. Converted to 27 in. and 570 rpm, the specifications become 12,520 cfm against 1.81 in. WC, which according to Fig. 4.2-28 falls well within the operating range of an FC fan. A 3/4-hp motor will be ample for the operating condition, but a 1-hp motor will have to be selected if the free-delivery bhp should also be covered.

   3. A DWDI fan for induced draft should be selected that will handle air at 610°F at a 700-ft altitude and under these conditions is to deliver 47,000 cfm against 3 in. WC. The correction factor for standard air density (70°F and sea level) is 2.07, and the converted specifications therefore read 47,000 cfm against 6.21 in. WC. We select a $44\frac{1}{2}$-in. fan with RT blades running at 860 rpm. With a fan efficiency of 70%, this fan would consume 78.8 bhp at standard air conditions. By dividing by the correction factor 2.07, we find that at the actual operating conditions 38.1 bhp will be sufficient. We therefore select a 40-hp motor.

   4. A pressure blower for exhausting the dust from a small grinding wheel should be selected. It should handle 350 cfm against 7 in. SP and have an outlet velocity of about 4000 fpm. We select a unit with a 4-in. diameter inlet and outlet and with an 11 in. diameter wheel at 3450 rpm, having radial blades. With 1 in. VP and 8 in. TP, the ahp will be 0.44, and with 55% efficiency at the operating point, the bhp will be 0.8. We select a 1-hp motor which will be safe even for lower SP, down to 4 in. SP, but would be overloaded at pressures below 4 in. SP.

   5. A belt-driven fan is to be selected that can be built into a small apparatus and will pull a cooling airstream through some narrow passages and discharge it into the atmosphere. Thus, 60 cfm should be delivered against 0.6 in. SP. Moreover, as a special requirement, the fan should be reversible and the same cfm should be exhausted regardless of the direction of rotation. In view of the small quantities involved, efficiency and power consumption are of minor importance. A good solution is found in a 5-in. diameter fan wheel with straight, radial blades and circumferential discharge, running at 2500 rpm. Quiet operation, extreme compactness, and draftless discharge are special features of this selection.

4.2-9 Other Fans

Mixed-flow fans, also called axial centrifugal fans or in-line fans, take a place between the axial flow fans and the centrifugal fans. Figure 4.2-35 shows a schematic of this arrangement, using direct drive and a barrel-shaped housing, resulting in smaller inlet and outlet diameters, for connection to smaller duct sizes. Some mixed-flow fans, however, use a simple cylindrical housing for lower first cost. In either case guide vanes past the fan wheel are used to eliminate the air spin as in the case of VAFs. Mixed-flow fans have the advantage of easy installation as part of the ductwork (like axial flow fans). At the same time, they produce more SP than axial flow fans of the same tip speed, although not as much as centrifugal fans. Unfortunately, their efficiency is lower than that of VAFs or of centrifugal fans. FC fan wheels cannot be used because, as previously mentioned, they require a scroll housing for proper performance.

Roof ventilators have become very popular and can be seen on many buildings. They come in various designs and can again be subdivided in several ways:

1. **Exhaust or supply units.** The great majority of all roof ventilators are for exhaust from buildings, but occasionally supply units are used for makeup air. Exhaust units can be for upblast or radial discharge.

2. **Upblast or radial discharge.** Radial discharge has a lower, more pleasing silhouette (Fig. 4.2-36 shows a typical contour), but upblast is required for the exhaust of grease-laden air from restaurant kitchens, as an accumulation of grease on the roof top would be a fire hazard.

3. **Axial flow or centrifugal fan wheels.** Axial flow fan wheels are less expensive and are used in most installations with little or no ductwork. Also, all supply units use axial flow fan wheels. Centrifugal fan wheels produce more SP and therefore are often used in connection with ductwork; most of them have BI blades, some have BC blades. FC blades, as previously mentioned, will not function without a scroll housing and therefore cannot be used in roof ventilators.

4. **Direct or belt drive.** Direct drive is simpler and requires less maintenance; it is generally used in small sizes. Belt drive is used in large sizes (to avoid expensive low-speed motors) and has more flexibility with regard to cfm and SP.

Cross-flow blowers are not a popular type of fan. They are unique in that the airstream passes through the fan blading twice. The fan wheel looks similar to an FC wheel, having many short FC blades, but there are two differences: The cross-flow wheels can be much wider in axial direction (in fact, there is no theoretical limit to the axial width) and the two ends of the wheel are closed by solid disks. The housing inlet has a special shape, directing the incoming air flow toward the blade tips at such an angle that the blades deflect it inward (against centrifugal force), helped along by the negative pressure produced in the center by the second passage of the air moving outward. There is a more scientific explanation of the cross-flow blower, but the above will do for a basic understanding.

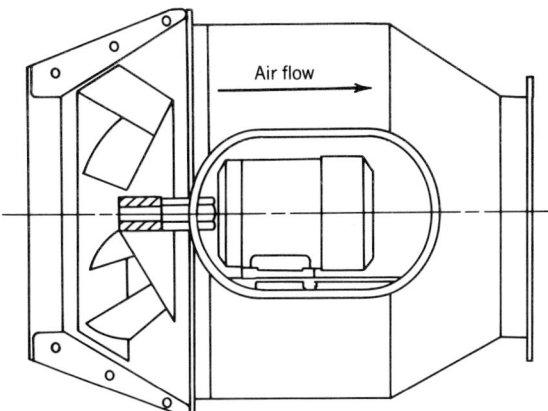

Fig. 4.2-35 Schematic sketch of a mixed-flow fan, consisting of a 26-in. diameter fan wheel with conical backplate and of a barrel-shaped housing with a 20-in. diameter inlet and outlet. Direct drive from a $7\frac{1}{2}$ hp, 1750 rpm motor. (Courtesy of DeBothezat Fan Division, Swartout Industries Inc., Sherman TX.)

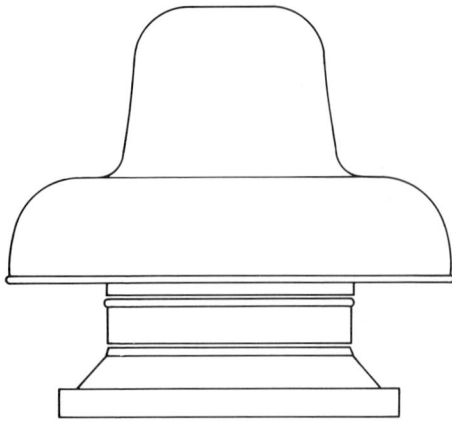

Fig. 4.2-36 Schematic sketch of a centrifugal roof exhauster, direct drive, radial discharge, 15-in. wheel diameter, 1 hp, 1725 rpm. (Courtesy of FloAire Inc., Cornwells Heights, PA.)

Cross-flow blowers are mostly built in small sizes where their comparatively low efficiencies are of less importance. Typical applications are in heaters and as air curtains where they produce a wide and thin layer of moving air, resulting from their geometric configuration and from their wide and thin outlet area.

A vortex fan, again a unique but not a popular type of fan, is basically a centrifugal fan of special design. They are for low cfm and very high SP, up to 100 in. WC. Efficiencies are very low and noise levels are so high that both the inlet and the outlet are equipped with sound-attenuating sleeves. The space in which the rotating blades and the stationary housing side come together has the shape of a toroid (doughnut). Within this toroid the air flow circles outward and inward, thereby passing through the blade section several times for each revolution. This equivalence to a multistage unit is responsible for the high SP. The main virtue is compactness, but at the expense of large motor horsepowers. Special applications include pneumatic conveying, suction to hold down sheets or other flat parts, supplying air or oxygen to submerged parts, and producing high velocity jets for aspiration.

4.3 VALVES

John M. Simonson

4.3-1 Introduction

Valves are devices that control fluid flow by varying the total piping system resistance. The range of resistance may be only from "on" to "off," or it may cover a wide range for modulating or throttling control. The resistance may even have directional characteristics that vary with flow direction.

4.3-2 System Analysis

The energy equation can be used to explain valve action in a piping system. This equation states that the difference between the kinetic energy, flow work, and potential energy entering and leaving a pipeline equals the sum of the work added, the internal energy changes, and the heat transfer. The last two items are generally referred to as a "head loss." The flow will accelerate or slow down to create a match. The head loss and energy-available terms can be plotted as a function of the flow rate as shown in Fig. 4.3-1. The intersection of the two curves gives the condition when the available energy equals the losses and indicates the operating point of the system. The flow is controlled by adjusting the energy loss curve, which is accomplished by adjusting the valve.

4.3-3 Valve Capacity

Valve resistance can be controlled from an infinite resistance when a valve is closed to the minimum resistance determined by the valve size and type. Rather than specifying the resistance, a valve's capacity is usually specified by a flow coefficient, C_v. An alternate specification may be an effective

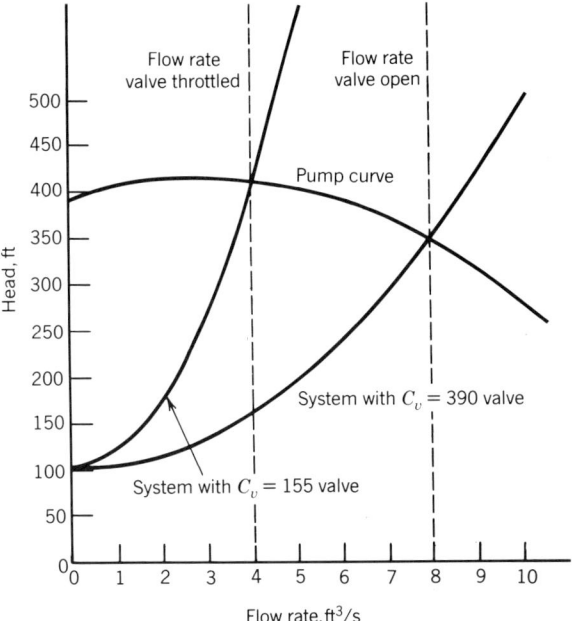

Fig. 4.3-1 Valve piping systems curve.

area and port area. C_v is defined by the equation

$$C_v = \frac{Q}{\sqrt{\Delta p / G_f}}$$

where C_v = flow coefficient
$\quad\quad\ Q$ = flow rate, in gpm
$\quad\quad\ p$ = pressure drop across valve
$\quad\quad\ G_f$ = specific gravity of flowing fluid

The flow coefficient C_v is a measured value. The method of measurement is specified by ISA.[1] Typical *CV* values are given in Table 4.3-1 for a number of valve types. Exact values must be measured as they depend upon the specific geometry. An approximation can be made for the C_v of globe valves by

TABLE 4.3-1 TYPICAL VALVE FLOW COEFFICIENT (gpm/$\sqrt{\text{psi}}$)

				Type				
Valve Size	Character Plug	Cage Guided	Gate/ Plug	Full Ball	Angle	Y Pattern	Segmented Ball	Disk
1	12.5	20.6	69.6	45	22	19.8		
1.5	33.4	39.2	144	125	53	43		
2	55.0	72.9	206	165	53.8	70		
3	123	148	608	350	155	183	220	220
4	199	236	895	775	208	326	530	420
6	438	433	1102	1000	350	666	1150	910
8	672	688	8710	2000		1040	2160	1720
10	1090		4920	3150		1710	3030	2780
12				5200				4000

the equation

$$\frac{C_v}{\text{Unit seat area}} = 40 - 19.5\frac{A_p}{A_0}$$

where A_p = port area
A_0 = valve outlet area

The definition of C_v may be extended to cover compressible flow including volume modification:

$$W = 19.3 p_1 Y \sqrt{\frac{xT}{MZ}}$$

where Y = compressibility factor, $\cong 1 - x/3F_k x_T$
p_1 = upstream pressure, psia
x = pressure drop ratio, $\Delta p/p_1$, values of x up to x_T affect flow (choking)
x_T = pressure drop to cause choking, dependent upon geometry (Typical values are given in Tables 4.3-2 and 4.3-3.)
Z = compressibility factor (accounts for real gas effects)
F_k = specific heat ratio factor = $k/1.4$
k = ratio of specific heats
T = absolute temperature, °R
M = molecular weight of gas

The compressibility factor is a thermodynamic function of the reduced temperature (T/T_c) and the reduced pressure (p/p_c) and becomes very significant for vapors flowing near saturation.

4.3-4 Check Valves

The simplest form of on/off valves is the check valve whose geometry may take a number of different forms. All forms permit flow with low resistance in one flow direction while preventing flow with shutoff in the other direction. A simple check valve is the globe lift check valve shown in Fig. 4.3-2. In this case, as the flow moves in one direction, the pressure difference across the seat lifts the plug, opening the valve-passing flow. If an attempt is made to reverse the flow, the pressure differential causes the plug to close. Check valves have C_v values similar to globe valves.

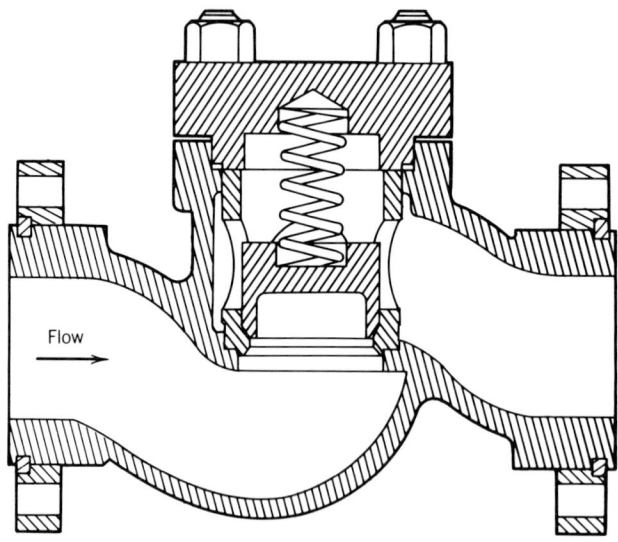

Fig. 4.3-2 Lift check valve. (Courtesy of Voltek, Inc., Springville, Utah.)

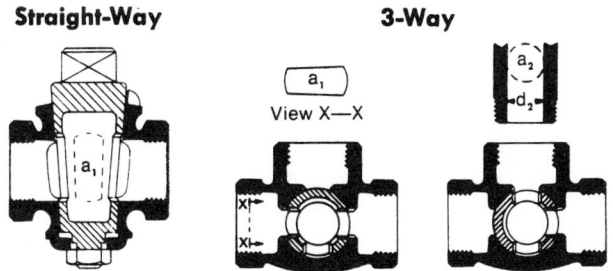

Straight-Way   **3-Way**

View X—X

Fig. 4.3-3   Typical plug valves or cocks. (Courtesy of Crane Co.)

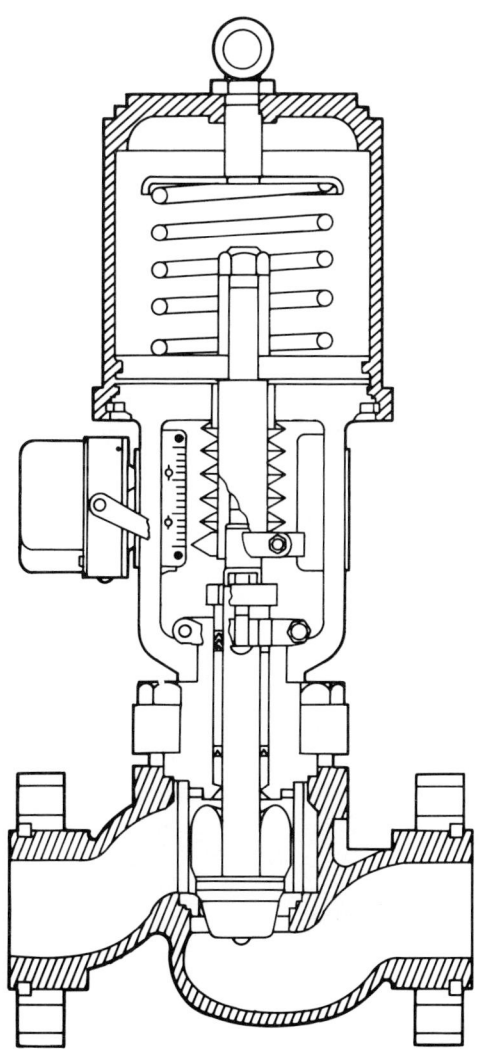

Fig. 4.3-4   Plug-characterized globe valve. (Courtesy of Valtek, Inc.)

4.3-5 Plug and Ball Valves

Typical plug valves or cocks are shown in Fig. 4.3-3. Simple plug valves are extensively used for on/off control. An extension of the plug valve that may be used for control is the ball valve, which is available both in the full-ball and the segmented-ball valve. The segmented ball is especially useful in slurries such as paper pulp, as it has a shearing action that clears the flow. C_v values obtainable from this type of a valve are higher for a given valve size than would be the case for a globe valve. Table 4.3-1 shows typical C_v values obtainable for plug, ball, and segmented-ball valves.

4.3-6 Globe Valves

Globe valves are the most common valves used for throttling control because of their good control characteristics. The maximum capacity obtainable from a globe is not as large as obtained in other types of valves. A typical automatic control valve with a globe pattern is shown in Fig. 4.3-4. The globe valve may also be manufactured in an angle pattern and a Y body. The angle pattern is especially well suited when there are particles entrained in the fluid.

4.3-7 Gate Valves

A gate valve has a sliding gate to control the flow resistance (Fig. 4.3-5). Like the globe valve the gate valve can be obtained in a number of geometries, especially when designed for throttling control. Hand-operated gate valves, because of the variation of C_v versus stroke, are a poor choice for modulation or throttling service.

Fig. 4.3-5   Gate valve. (Courtesy of DeZurik, a Unit of General Signal Corporation.)

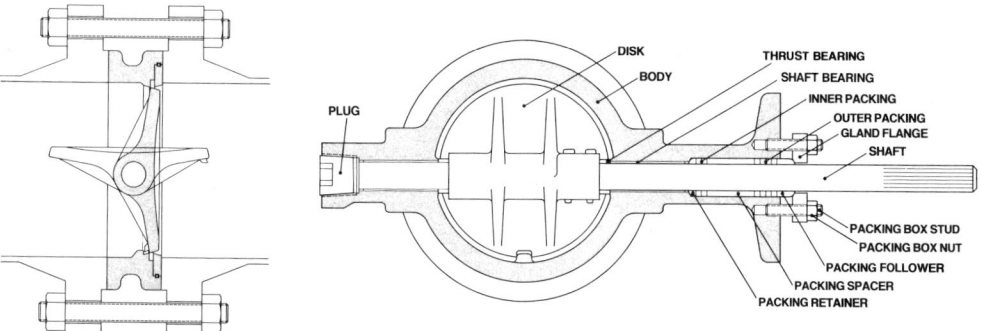

Fig. 4.3-6   High-performance disk valve.

4.3-8   Disk Valves

A butterfly, or disk, valve consists essentially of a disk that rotates about a shaft placed in the flow stream. When the disk is perpendicular to the flow channel, the minimum flow is established with the maximum resistance. When the shaft is turned so that the disk is parallel to the flow, the minimum resistance or maximum C_v is obtained. A simple disk valve, called a "swing-through" butterfly, may be used where tight shutoff is not required.

High-performance disk valves have been increasingly popular by offering the relative simplicity and light weight of a butterfly valve with much of the performance of other valves. Figure 4.3-6 illustrates a typical disk valve. This design can include an eccentrically cammed and offset shaft so that the center of rotation is not at the center of the disk. This permits tight shutoff when the valve is closed but at the same time reduces the wear on the seat during throttling. The design requires precision machining to work properly. Typical maximum C_v obtainable for disk valves are given in Table 4.3-1.

4.3-9   Valve Actuation

The method of valve opening or actuation is an important part of valve selection. The most common is the handwheel or a hand lever, but power actuating is often necessary for large valves and for automatic control. The relative size of the valve opening can be varied by the handwheel from open to closed by the turn of the handle, and hence the resistance in the flow system is controlled. Large valves use gear reduction to assist the manual operation.

Computers and automatic control systems and large flows have generated the need for automatic, remote-control valve actuators. Pneumatic, hydraulic, and electric actuator power units are widely used. Pneumatics have the widest use because of high power in a small physical package, being easily controlled, and creating few safety problems. Figures 4.3-4, and 4.3-6 show typical actuators. A simple spring-diaphragm is the most common type of actuator in use. A piston actuator with a bidirectional control and higher actuator pressures can be used to give a stiffer system. These actuators have a much higher frequency response with more precise control because a positioner with position feedback is used.

Actuators to actuate quarter-turn rotary valves, such as ball and high-performance butterfly valves, are often adaptations of linear actuators, although special designs have been used.

4.3-10   Valve Characteristics

The variation of capacity (C_v) as a function input, either handwheel or automatic control, is known as the valve characteristics. Some characteristics are inherent to the valve type or they may be controlled. The three most common controlled valve characteristics are "quick-open," "linear," and "equal-percent" (see Fig. 4.3-7).

Valve characteristics for throttling are usually chosen to result in a system response approximately linear with the input signal; however, quick-open characteristics are used primarily for on/off operation. The quick-open characteristic gives maximum flow rate with minimum plug stroke. The linear valve characteristics are used when the valve is the primary head loss in the system and other fittings and piping losses are relatively unimportant. Equal-percent characteristic is used when valves and piping both have significant flow resistance. The characteristic is designed for equal percentage change of C_v for equal changes of input signal.

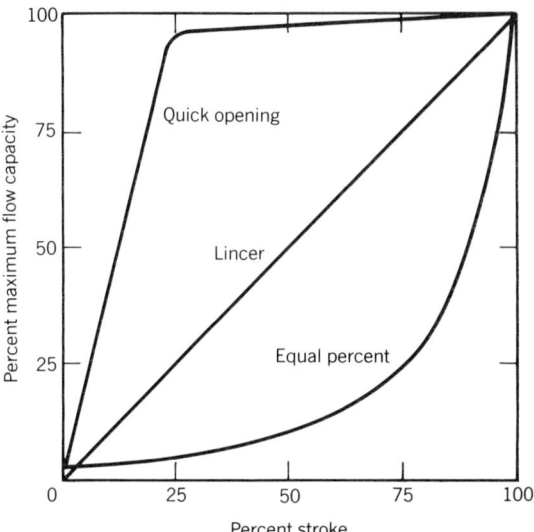

Fig. 4.3-7 Idealized valve characteristics.

Most computer-controlled systems use a transducer to convert the digital signal to an analog signal for the valve actuator; however, some direct digital valves are marketed.

4.3-11 Valve Pressure Ratings

ANSI standards and ASME codes have established pressure–temperature ratings of valves. The designation of an ANSI class specifies the basic physical dimensions and the pressure capabilities of a valve design as a function of temperature. One design can then be used for making a valve from a number of different materials with known limits and basic sizes. ANSI B16.34[2] for steel valves and ANSI B16.24 for bronze valves contain tables for pressure–temperature ratings for the various classes and materials used. Typical values are shown in Tables 4.3-2 and 4.3-3.

4.3-12 Valve Dimensions

Unfortunately, with the evolution of valve sizes and shapes has come a multitude of valve dimensions. ANSI B16.10[3] gives dimensions for valves in different pressure classes, different valve styles, and different materials. Tables 4.3-4 to 4.3-6 give the face-to-face dimensions for a number of commonly used valve types.

4.3-13 Noise Control

A common valve problem is the noise generation often present with pressure reduction. The primary mechanism of noise generation in liquids is cavitation. This occurs when the pressure locally drops below the vapor pressure and small bubbles form that collapse upon recompression. This is physically damaging and noisy. Special designs are available to reduce not only the damage but the noise generation. Manufacturers of proprietary designs should be contacted when noise is a problem.

The primary cause of noise in gaseous flow is turbulence such as that formed by shocks when supersonic flow is returned to subsonic speeds. Again, special propriety designs are available with up to 30 dBA noise reduction possible.

Most special designs work by dropping the pressure continuously or in a large number of small steps rather than in a single step, thereby avoiding localized supersonic flow or localized cavitation.

TABLE 4.3-2 TYPICAL CARBON STEEL PRESSURE–TEMPERATURE RATINGS

Applicable ASTM Specifications

Group 1 Materials

Material Group No.	Nominal Designation Steel	Forgings Spec.–Grade	Notes	Castings Spec.–Grade	Notes	Plates Spec.–Grade	Notes	Bars and Shapes Spec.–Grade	Notes	Tubular Products Spec.–Grade	Notes
1.1	Carbon	A 105 A 350-LF2	(1)(3)	A 216-WCB	(1)	A 515-70 A 516-70	{1} {1}	A 105 A 350-LF2 A 675-70 A 696 Gr.B	(1)(3) (1)(8)(5)	A 672-B70 A 672-C70	
	C–Mn–Si					A 537 Cl.1					

Permissible, but not recommended for prolonged usage above about 800° F.

STANDARD CLASS VALVES — FLANGED AND BUTTWELDING END

Working Pressure by Classes, psig

Temperature, °F	150	300	400	600	900	1500	2500	4500
−20 to 100	285	740	990	1480	2220	3705	6170	11110
200	260	675	900	1350	2025	3375	5625	10120
300	230	655	875	1315	1970	3280	5470	9845
400	200	635	845	1270	1900	3170	5280	9505
500	170	600	800	1200	1795	2995	4990	8980
600	140	550	730	1095	1640	2735	4560	8210
650	125	535	715	1075	1610	2685	4475	8055
700	110	535	710	1065	1600	2665	4440	7990
750	95	505	670	1010	1510	2520	4200	7560
800	80	410	550	825	1235	2060	3430	6170
850	65	270	355	535	805	1340	2230	4010
900	50	170	230	345	515	860	1430	2570
950	35	105	140	205	310	515	860	1545
1000	20	50	70	105	155	260	430	770

(Continued)

TABLE 4.3-2 *(Continued)*

Applicable ASTM Specifications

Group 2 Materials		Forgings		Castings		Plates		Bars and Shapes		Tubular Products	
Material Group No.	Nominal Designation Steel	Spec.–Grade	Notes	Spec.–Grade	Notes	Spec.–Grade	Notes	Spec.–Grade	Notes	Spec.–Grade	Notes
2.1	18 Cr-8Ni	A 182-F304 A 182-F304H	(6)			A 240-304 A 240-304H	(6)(7)	A 182-F304 A 182-F304H A 479-304 A 479-304H	(6) (6)(7)	A 312-TP304 A 312-TP304H A 358-304 A 376-TP304 A 376-TP304H A 430-TP304 A 430-TP304H	(6) (6) (6) (6)
	18 Cr-8Ni			A 351-CF3 A 351-CF8	(5)						

STANDARD CLASS VALVES – FLANGED AND BUTTWELDING END

Working Pressure by Classes, psig

Temperature, °F	150	300	400	600	900	1500	2500	4500
−20 to 100	275	720	960	1440	2160	3600	6000	10800
200	240	620	825	1240	1860	3095	5160	9290
300	215	560	745	1120	1680	2795	4660	8390
400	195	515	685	1030	1540	2570	4280	7705
500	170	480	635	955	1435	2390	3980	7165
600	140	450	600	905	1355	2255	3760	6770
650	125	445	590	890	1330	2220	3700	6660
700	110	430	575	865	1295	2160	3600	6480
750	95	425	565	845	1270	2110	3520	6335
800	80	415	555	830	1245	2075	3460	6230
850	65	405	540	810	1215	2030	3320	6085
900	50	395	525	790	1180	1970	3280	5905
950	35	385	515	775	1160	1930	3220	5795
1000	20	365	485	725	1090	1820	3030	5450
1050	20(1)	360	480	720	1080	1800	3000	5400
1100	20(1)	325	430	645	965	1610	2685	4835
1150	20(1)	275	365	550	825	1370	2285	4115
1200	20(1)	205	275	410	620	1030	1715	3085
1250	20(1)	180	245	365	545	910	1515	2725
1300	20(1)	140	185	275	410	685	1145	2060
1350	20(1)	105	140	205	310	515	860	1545
1400	20(1)	75	100	150	225	380	630	1130
1450	20(1)	60	80	115	175	290	485	875
1500	15(1)	40	55	85	125	205	345	620

NOTE:
(1) For welding end valves only. Flanged end ratings terminate at 1000° F.

Source. Ref. 4.3-2. Used by permission

388

TABLE 4.3-3 TYPICAL STAINLESS STEEL PRESSURE–TEMPERATURE RATINGS

Material Group No.	Nominal Designation Steel	Forgings Spec.—Grade	Notes	Castings Spec.—Grade	Notes	Plates Spec.—Grade	Notes	Bars and Shapes Spec.—Grade	Notes	Tubular Products Spec.—Grade	Notes
				Applicable ASTM Specifications							
2.1	18 Cr–8NI	A182-F304	(6)			A240-304	(6)(7)	A182-F304	(6)	A312-TP304	(6)
		A182-F304H				A240-304H		A182-F304H		A312-TP304H	
								A479-304	(6)(7)	A358-304	(6)
								A479-304H		A376-TP304	(6)
										A376-TP304H	
										A430-FP304	(6)
										A430-TP304H	
	18 Cr–8NI			A351-CF3	(5)						
				A351-CF8							

Working Pressure by Classes, psig

Standard Class Valves—Flanged and Buttwelding End

Temperature, °F	150	300	400	600	900	1500	2500	4500
−20 to 100	275	720	960	1440	2160	3600	6000	10800
200	240	620	825	1240	1860	3095	5160	9290
300	215	560	745	1120	1680	2795	4660	8390
400	195	515	685	1030	1540	2570	4280	7705
500	170	480	635	955	1435	2390	3980	7165
600	140	450	600	905	1355	2255	3760	6770
650	125	445	590	890	1330	2220	3700	6660
700	110	430	575	865	1295	2160	3600	6480
750	95	425	565	845	1270	2110	3520	6335
800	80	415	555	830	1245	2075	3460	6230

(Continued)

389

TABLE 4.3-3 (*Continued*)

Temperature, °F	\multicolumn{8}{c}{Working Pressure by Classes, psig}

Temperature, °F	150	300	400	600	900	1500	2500	4500
\multicolumn{9}{l}{Standard Class Valves—Flanged and Buttwelding End (Continued)}
850	65	405	540	810	1215	2030	3320	6085
900	50	395	525	790	1180	1970	3280	5905
950	35	385	515	775	1160	1930	3220	5795
1000	20	365	485	725	1090	1820	3030	5450
1050	20[a]	360	480	720	1080	1800	3000	5400
1100	20[a]	325	430	645	965	1610	2685	4835
1150	20[a]	275	365	550	825	1370	2285	4115
1200	20[a]	205	275	410	620	1030	1715	3085
1250	20[a]	180	245	365	545	910	1515	2725
1300	20[a]	140	185	275	410	685	1145	2060
1350	20[a]	105	140	205	310	515	860	1545
1400	20[a]	75	100	150	225	380	630	1130
1450	20[a]	60	80	115	175	290	485	875
1500	15[a]	40	55	85	125	205	345	620

[a] For welding end valves only. Flanged end ratings terminate at 1000°F.

Source.   Ref. 4.3-2. Used by permission.

TABLE 4.3-4 AMERICAN NATIONAL STANDARD FACE-TO-FACE AND END-TO-END DIMENSIONS OF FERROUS VALVES: CLASS 800 HYDRAULIC CAST IRON AND CLASS 600 STEEL VALVES

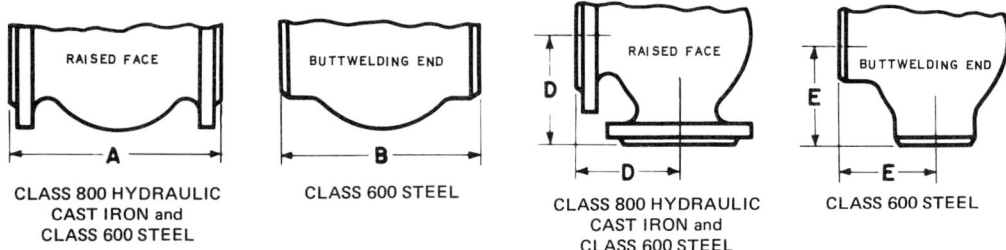

CLASS 800 HYDRAULIC CAST IRON and CLASS 600 STEEL — CLASS 600 STEEL — CLASS 800 HYDRAULIC CAST IRON and CLASS 600 STEEL — CLASS 600 STEEL

Valves with Class 800 Hydraulic or Class 600 End Flanges, or with Welding Ends
FACE-TO-FACE AND END-TO-END DIMENSIONS[1]

Dimensions shown are in inches.

1	2	3	4	5	6	7	8	9	10	11	12	13	14	15
	Class 800 Hyd. Cast Iron			Class 600 Steel										
	Flanged End (1/4 In. Raised Face)			Flanged End (1/4 In. Raised Face) and Welding End										
					Gate			Plug		Globe, Lift Check, and Swing Check	Globe, Lift Check, and Swing Check	Angle and Lift Check	Angle and Lift Check	
Nominal Valve Size	Gate, Solid Wedge, and Double Disc	Plug	Swing Check	Ball Long Pattern	Solid Wedge,[4] Double Disc,[5] and Conduit	Short Pattern[3]	Regular and Venturi Pattern	Round Bore Full Port	Round Bore Full Port	Regular Pattern	Short Pattern[3]	Regular Pattern	Short Pattern[3]	Control[2]
	A	A	A	A & B	A & B	B	A & B	A	B	A & B	B	D & E	E	A
1/2				6-1/2	6-1/2[6]					6-1/2		3-1/4		8
3/4				7-1/2	7-1/2[6]					7-1/2		3-3/4		8-1/8
1				8-1/2	8-1/2	5-1/4	8-1/2[8]	10		8-1/2	5-1/4	4-1/4		8-1/4
1-1/4				9	9	5-3/4	9[8]			9	5-3/4	4-1/2		
1-1/2				9-1/2	9-1/2	6	9-1/2	12-1/2		9-1/2	6	4-3/4		9-7/8
2	11-1/2	11-1/2	11-1/2	11-1/2	11-1/2	7	11-1/2	13		11-1/2	7	5-3/4	4-1/4	11-1/4
2-1/2	13	13	13	13	13	8-1/2	13	15		13	8-1/2	6-1/2	5	12-1/4
3	14	14	14	14	14	10	14	17-1/2		14	10	7	6	13-1/4
3-1/2														
4	17	17	17	17	17	12	17	20	22	17	12	8-1/2	7	15-1/2
5					20	15				20	15	10	8-1/2	
6	22	22	22	22	22	18	22	26	28	22	18	11	10	20
8	26	26	26	26	26	23	26	31-1/4	33-1/4	26	23	13		24
10	31	31	31	31	31	28	31	37	40	31	28	15-1/2		
12	33	33	33	33	33	32	33	42	42	33	32	16-1/2		
14				35	35	35	35			35[10]				
16				39	39	39	39			39[10]				
18				43	43	43	43[9]			43[10]				
20				47	47	47	47[9]			47[10]				
22				51	51		51[9]			51[10]				
24				55	55	55	55[9]			55[10]				
26				57	57		57[9]			57[10]				
28				61	61					63[10]				
30				65	65		65[9]			65[10]				
32				70	70[7]		70[9]							
34				76	76[7]		76[9]							
36				82	82[7]		82[9]			82[10]				

[1] API Standards 6A, 6D, 599, and 600 conform to the dimensions shown for corresponding sizes, valve type, and flange class or welding end.
[2] Instrument Society of America Recommended Practice ISA-RP4.1 face-to-face conform to dimensions shown in column 15.
[3] These dimensions apply to pressure seal or flangeless bonnet valves only. They may be applied at the manufacturer's option to valves with flanged bonnets.
[4] API Standard 597 Venturi Gate Valves have face-to-face and flange class or welding end dimensions sams as nominal size of the valve ends. A 4" x 3" x 4" face-to-face would be the same as a 4" gate valve.
[5] API Standard 6D: Flanged and Welding End Double Disc Gate Valves.
[6] Solid Wedge only.
[7] Double Disc and Conduit only.
[8] Regular Pattern only.
[9] Venturi Pattern only.
[10] Swing Check only.

Source: Ref. 4.3-3. Used by permission.

TABLE 4.3-5 AMERICAN NATIONAL STANDARD FACE-TO-FACE AND END-TO-END DIMENSIONS OF FERROUS VALVES: CLASS 1500 STEEL VALVES

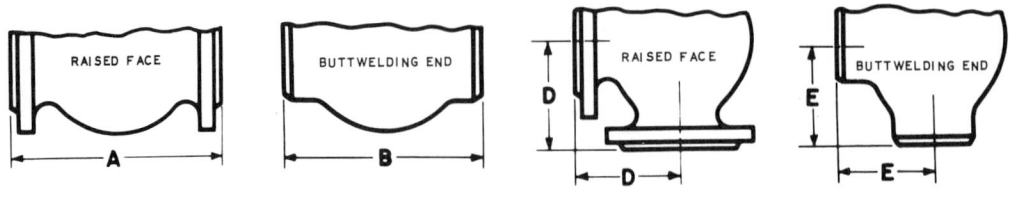

Valves with Class 1500 End Flanges or with Welding Ends
FACE-TO-FACE AND END-TO-END DIMENSIONS[1]

Dimensions shown are in inches.

1	2	3	4	5	6	7	8
	Class 1500 Steel						
	Flanged End (1/4 In. Raised Face) and Welding End						
	Gate		Plug		Globe, Lift Check, and Swing Check Regular Pattern A & B	Swing Check Short Pattern[3] B	Angle and Lift Check Regular Pattern D & E
Nominal Valve Size	Solid Wedge,[2] Double Disc,[4] and Conduit A & B	Short Pattern[3] B	Regular and Venturi Pattern A & B	Round Port Full Bore A			
1/2					8-1/2[8]		4-1/4
3/4					9		4-1/2
1	10[5]	5-1/2	10[6]		10		5
1-1/4	11[5]	6-1/2	11[6]		11		5-1/2
1-1/2	12[5]	7	12[6]		12		6
2	14-1/2	8-1/2	14-1/2[6]	15-3/8	14-1/2	8-1/2	7-1/4
2-1/2	16-1/2	10	16-1/2[6]	17-7/8	16-1/2	10	8-1/4
3	18-1/2	12	18-1/2[6]	20-5/8	18-1/2	12	9-1/4
3-1/2							
4	21-1/2	16	21-1/2[6]	24-5/8	21-1/2	16	10-3/4
5	26-1/2	19			26-1/2	19	13-1/4
6	27-3/4	22	27-3/4	31	27-3/4	22	13-7/8
8	32-3/4	28	32-3/4	35	32-3/4	28	16-3/8
10	39	34	39	42	39	34	19-1/2
12	44-1/2	39	44-1/2	48	44-1/2	39	22-1/4
14	49-1/2	42			49-1/2	42	24-3/4
16	54-1/2	47	54-1/2[7]		54-1/2[9]	47	
18	60-1/2	53			60-1/2[9]		
20	65-1/2	58			65-1/2[9]		
22							
24	76-1/2				76-1/2[9]		

[1] API Standards 6A, 6D, 599, and 600 conform to the dimensions shown for corresponding sizes, valve type, and flange class or welding end.
[2] API Standard 597 Venturi Gate Valves have face-to-face and flange class or welding end dimensions same as nominal size of the valve ends. A 4" × 3" × 4" face-to-face would be the same as a 4" gate valve.
[3] These dimensions apply to pressure seal or flangeless bonnet valves. They may be applied at the manufacturer's option to valves with flanged bonnets.
[4] API Standard 6D: Flanged and Welding End Double Disc Gate Valves.
[5] Solid Wedge only.
[6] Regular Pattern only.
[7] Venturi Pattern only.
[8] Globe and Lift Check only.
[9] Swing Check only.

Source: Ref. 4.3-3. Used by permission.

REFERENCES

4.3-1 "Control Valve Capacity Test Procedure," Instrument Society of America, ISA-S75.02, 1981.
4.3-2 "Valves—Flanged and Buttwelding End," The American Society of Mechanical Engineers, ANSI B16.34, 1981.
4.3-3 "Face-to-Face and End-to-End Dimension of Ferrous Valves," The American Society of Mechanical Engineers, ANSI B16.10, 1973.

TABLE 4.3-6   AMERICAN NATIONAL STANDARD FACE-TO-FACE AND END-TO-END DIMENSIONS OF FERROUS VALVES: CLASS 2500 STEEL VALVES

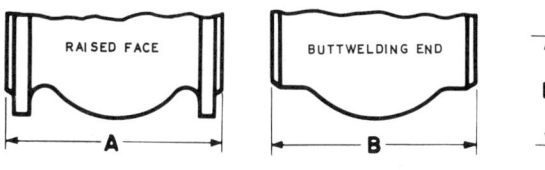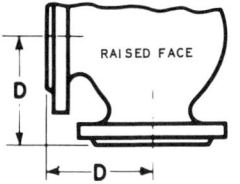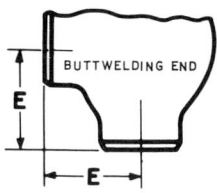

Valves with Class 2500 End Flanges or with Welding Ends
FACE-TO-FACE AND END-TO-END DIMENSIONS[1]

Dimensions shown are in inches.

1	2	3	4	5	6	7
	Class 2500 Steel					
	Flanged End (1/4 In. Raised Face) and Welding End					
Nominal Valve Size	Gate		Plug Regular Pattern A & B	Globe, Lift Check, and Swing Check Regular Pattern A & B	Swing Check Short Pattern[3] B	Angle and Lift Check Regular Pattern D & E
	Solid Wedge,[4] Double Disc[2] A & B	Short Pattern[3] B				
1/2	10-3/8[5]			10-3/8		5-3/16
3/4	10-3/4[5]			10-3/4		5-3/8
1	12-1/8[5]	7-5/16	12-1/8	12-1/8		6-1/16
1-1/4	13-3/4[5]	9-1/8		13-3/4		6-7/8
1-1/2	15-1/8[5]	9-1/8	15-1/8	15-1/8		7-9/16
2	17-3/4	11	17-3/4	17-3/4	11	8-7/8
2-1/2	20	13	20	20	13	10
3	22-3/4	14-1/2	22-3/4	22-3/4	14-1/2	11-3/8
3-1/2						
4	26-1/2	18	26-1/2	26-1/2	18	13-1/4
5	31-1/4	21	31-1/4	31-1/4	21	15-5/8
6	36	24	36	36	24	18
8	40-1/4	30	40-1/4	40-1/4	30	20-1/8
10	50	36	50	50	36	25
12	56	41	56	56	41	28
14		44				
16		49				
18		55				
20						
22						
24						

[1] API Standards 6D, 599, and 600 conform to the dimensions shown for corresponding sizes, valve type, and flange class or welding end.
[2] API Standard 6D: Flanged and Welding End Double Disc Gate Valves.
[3] These dimensions apply to pressure seal or flangeless bonnet valves. They may be applied at the manufacturer's option to valves with flanged bonnets.
[4] API Standard 597 Venturi Gate Valves have face-to-face and flange class or welding end dimensions same as nominal size of the valve ends. A 4″ x 3″ x 4″ face-to-face would be the same as a 4″ gate valve.
[5] Solid Wedge only.

Source: Ref. 4.3-3. Used by permission.

4.4   PIPING

E. B. Branch and C. A. Podczerwinski

4.4-1   Introduction

Piping is used in energy systems to transport fluid from one location in the system to another. Piping design is a major portion of the installation design for a system. Installation design is the process of converting the conceptual design of a fluid-handling circuit into detailed design drawings and purchase documents that are used for material procurement and installation of the system.

Within a system, pipelines can be categorized by the functions that they perform. Lines whose function is to carry process fluid from one piece of primary equipment to another are referred to as process lines. Lines in this category usually operate continuously. There are several categories of lines that operate intermittently. These are listed below.

Category	Function
Vent lines ⎫ Drain lines ⎭	Drainage for maintenance
Sensing lines	Pressure or flow measurement
Sampling lines	Taking process fluid samples
Test lines	Attachment of test equipment
Charging lines	Initial system pressurization
Discharge lines	Pressure relief
Bypass lines	To bypass equipment
Warmup lines	System warmup at low flow
Recirculation lines	Startup and testing

Design in each category requires different considerations, as service conditions and the vibration environment can vary considerably from one category to another.

Service conditions can range from near-atmospheric pressure and room temperature in simple service water systems to temperature and pressure extremes. An example is the main steam piping in modern fossil-fueled power stations, which operates in the neighborhood of 1000°F and 3500 psig. High temperatures are also involved in the reactor coolant systems for liquid-metal reactors.

When the service conditions or the function of the system is such that the safety of employees or the public would be compromised by a failure, piping design, fabrication, and construction activities are regulated by state and federal authorities. Control of these activities is accomplished by laws that require conformance with piping codes that are developed and maintained by ANSI (American National Standards Institute) and ASME (American Society of Mechanical Engineers). Codes corresponding to several quality levels have been established. Design complexity, quality control complexity, and cost vary by significant amounts depending on quality level. These codes control the design, fabrication, examination, and testing of a majority of the piping in energy systems.

Piping systems are assembled using pipe, fitting, and pipe support hardware products. The manufacture and rating of pipe and fitting products have been extensively standardized by ANSI and ASTM (American Society of Testing Materials). Preparing the installation design for a piping system consists of utilizing these standard products to establish a routing and support configuration that complies with piping code design rules for the required quality level. Drawings and purchase documents are then prepared to describe the design to the parties responsible for material procurement, fabrication, and installation. The remainder of this article will discuss the design process, piping products, and codes and will summarize current design practices.

4.4-2 Design Process

The installation design of a system is a two-step process. Figure 4.4-1 illustrates the process for a large-diameter, high-pressure system. Often, the two steps are carried out by two different parties. Communication between the parties is via a design specification and a piping and instrumentation diagram for the system.

The amount of work under the design activity heading varies a great deal, depending on the pipe size, service conditions, and quality level (piping code requirements). At one extreme is small-diameter, low-pressure, low-temperature piping. The design of these lines should be done by a tradesman as he installs the piping, working with simple span length and hardware selection rules from the specification. Design drawings, other than a piping and instrumentation diagram, are not involved. The other extreme is large-diameter, high-pressure, high-temperature piping. These lines require careful layout, detailed structural analysis, unique support designs, and the preparation of detailed design drawings for the piping and each support assembly.

The approach used must be carefully matched to the quality level to avoid unnecessary expense. The individual steps in the process are described below.

Piping Design Process

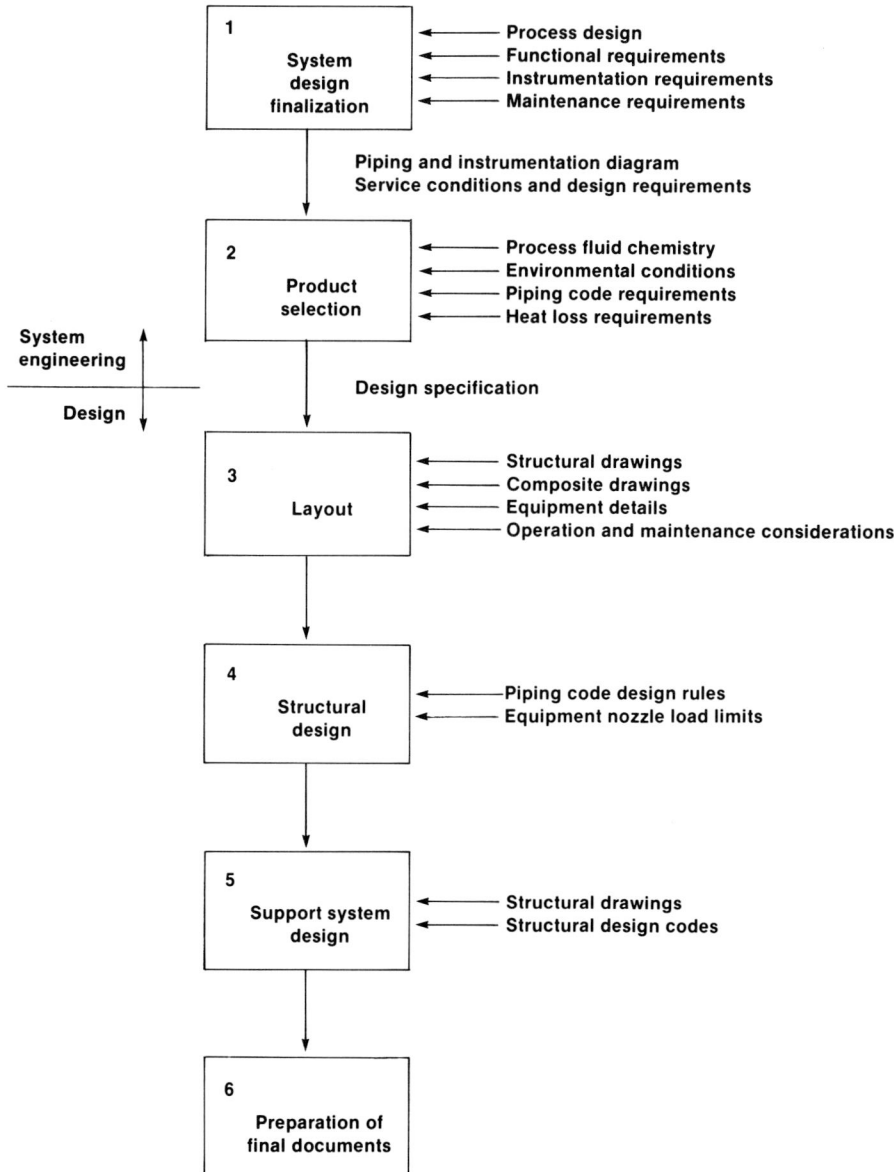

Fig. 4.4-1 Piping design process.

System Design Finalization

System design finalization is the process of converting the conceptual process design into a fluid-handling system whose operation can be controlled, monitored, and maintained during operation of the plant. The steps involved are:

1. Selection of primary equipment.
2. Pipe sizing for the primary process flow paths.

3. Location and sizing of bypass lines, warmup lines, and recirculation lines needed for starting up and testing the system.

4. Establishment of required operating modes and the determination of the service conditions during those modes.

5. Location and selection of control valves and the control system.

6. Location and selection of overpressure protection devices such as rupture discs and safety relief valves.

7. Location of vents, drains, and valves for maintenance of the system.

8. Location and selection of instruments and sampling stations for monitoring system operation.

The end product of this step is a detailed schematic diagram of the system [commonly referred to as a piping and instrumentation diagram (P & ID)], purchase documents for all equipment, and a detailed description of the service conditions required for piping product selection and design.

Product Selection

Once the design conditions and pipe sizes have been established, pipe, fitting, and insulation product selections are made. The following action is required for each pipeline in the system.

1. **Pipe product selection.** This is accomplished by a careful comparison of the service conditions, process fluid chemistry, and flow rate with the raw material and process used for manufacture of the piping being considered. Standard products manufactured in accordance with ASTM specifications are generally selected.

2. **Fitting product selection.** Raw material and the manufacturing process are selected for compatibility with the piping product selected in step 1. Standard products manufactured in accordance with ASTM specifications are generally selected. ASTM specifications exist for fittings that are compatible with each pipe product covered by an ASTM specification.

3. **Insulation product selection.** Insulation product selection is based on the service conditions and the function to be provided by the insulation material. Basic functions are

    (a) retention of heat to conserve energy,

    (b) prevention of sweating and heat absorption,

    (c) establishment of personnel protection,

    (d) prevention of freezing, and

    (e) maintenance of adjacent environmental temperatures at acceptable levels.

    Calcium silicate materials are typically used on hot piping systems. Fibrous glass or foam plastic material is typically used for chilled water piping. Details are provided in Sect. 4.4-9.

Layout Design

During layout design, the physical arrangement of the piping and its support system is established. The work is accomplished by a careful review of the physical surroundings; this is best done at the location that the piping is to be installed. When piping design is done prior to building construction, composite drawings or a design model are generally used. The product of the layout design step is a preliminary pipe routing and support arrangement that is sufficiently flexible to accommodate thermal expansion. The routing must be such that the system can be supported without building expensive auxiliary support structures. Equipment and valves must be accessible for operation and maintenance, and proper drainage must be assured.

Structural Design

The purpose of the structural design step is to verify the structural integrity of the design represented by the product selection and layout, and to determine design loadings for support system hardware. Two basic substeps are involved:

1. Pressure design.

2. Design analysis.

The structural design is accomplished by evaluating the design using the design rules and acceptability criteria provided by the piping code governing the design. The level of detail required depends on the service conditions and the size of the line being designed. The structural design of lower-temperature, smaller-diameter lines is usually done during the layout step by the utilization of simple span length and bending leg length sizing rules, developed from the piping code design rules. Larger-diameter lines are handled by employing a computer to perform a structural analysis of the entire configuration. The results of the structural analysis are then used to prepare piping code design calculations.

Support System Design

Once the configuration is finalized and design loadings for the support system are established, a detailed design of the support system is prepared. The level of detail required is dependent on the service conditions and the line size. The support design for low-temperature, smaller-diameter lines is generally done during the layout step by the selection of prequalified standard hardware assemblies. On large-diameter and high-temperature lines where load levels are high, each support assembly is a unique design that is released for construction on its own design drawings. The design of these supports requires hardware selection and design, structural steel and concrete design, and the preparation of detailed design drawings.

Preparation of Final Documents

A set of design documents is prepared to communicate a description of the configuration to the parties responsible for procurement of material, fabrication, and installation of the system. This set of documents consists of a specification that describes material, fabrication, and installation requirements, and a set of design drawings that describe the configuration. The level of detail required depends on the quality level, service conditions, and line size. Federal quality assurance requirements for nuclear power station construction requires extensive documentation of all design activities.

4.4-3 Commercial Piping Products and Their Application Pipe Products

The manufacture of commercial pipe and fitting products has been standardized by trade groups that produce and use the products. Pipe products are categorized into types based on application. Each type has been standardized differently by the trade group concerned with its specific application. Three types of pipe are commonly used in energy systems.

Type	Application
Standard pipe	Low-pressure mechanical service and structural service
Pressure pipe	Liquid, gas, or vapor service at elevated temperature and/or pressures
Line pipe	Water, oil, or gas line service

Nominal dimensions used for the manufacture of all three types have been consolidated into a single standard, ANSI B36.10. The manufacturing processes used for standard pipe and pressure pipe have been standardized by ASTM. The manufacture of line pipe has been standardized by API. Standard specifications prepared by these organizations are used extensively as the basis for purchase of piping.

A specific piping product is identified by four pieces of information:

1. Nominal pipe size.
2. Wall thickness designation.
3. Manufacturing standard.
4. Grade or type designation.

Nominal Pipe Size (NPS)

Nominal size is a nominal outside diameter. Pipe from $\frac{1}{8}$- to 12-in. NPS is manufactured with standardized outside diameters. These diameters were originally selected so that standard-weight pipe would have an inside diameter approximately equal to its NPS. Piping 14 in. and larger is manufactured with an outside diameter equal to its NPS.

Wall Thickness Designations

Wall thickness designations are unique to the pipe product type:

Type	Wall Thickness	Designation
Standard pipe and line pipe	Standard	(STD)
	Extra-Strong	(XS)
	Double extra-strong	(XXS)
Pressure pipe	Pipe schedule numbers	10, 20, 30, 40, 60, 80, 100, 120, 140, 160

Pipe schedule numbers are approximate values of the expression $1000 \times (P/S)$, where P is the design pressure and S is the allowable stress for the pipe material. This relationship is convenient to use for preliminary wall thickness selection. Table 4.4-1 lists the outside diameter and wall thickness of commonly used pipe products.

TABLE 4.4-1   OUTSIDE DIAMETER AND NOMINAL WALL THICKNESS OF STANDARD PIPE PRODUCTS

Nominal Size (in.)	Outside Diameter (in.)	Wall Thickness Designation		Nominal Wall Thickness (in.)
		STD-XS-XXS	Schedule No.	
$\frac{1}{8}$	0.405	STD	40	0.068
$\frac{1}{8}$	0.405	XS	80	0.095
$\frac{3}{8}$	0.675	STD	40	0.091
$\frac{3}{8}$	0.675	XS	80	0.126
$\frac{1}{2}$	0.840	STD	40	0.109
$\frac{1}{2}$	0.840	XS	80	0.147
$\frac{1}{2}$	0.840		160	0.188
$\frac{1}{2}$	0.840	XXS		0.294
$\frac{3}{4}$	1.050	STD	40	0.113
$\frac{3}{4}$	1.050	XS	80	0.154
$\frac{3}{4}$	1.050		160	0.219
$\frac{3}{4}$	1.050	XXS		0.808
1	1.315	STD	40	0.133
1	1.315	XS	80	0.179
1	1.315		160	0.250
1	1.315	XXS		0.358
$1\frac{1}{4}$	1.660	STD	40	0.140
$1\frac{1}{4}$	1.660	XS	80	0.191
$1\frac{1}{4}$	1.660		160	0.250
$1\frac{1}{4}$	1.660	XXS		0.382
$1\frac{1}{2}$	1.900	STD	40	0.145
$1\frac{1}{2}$	1.900	XS	80	0.200
$1\frac{1}{2}$	1.900		160	0.281
$1\frac{1}{2}$	1.900	XXS		0.400
2	2.375	STD	40	0.154
2	2.375	XS	80	0.218
2	2.375		160	0.344

TABLE 4.4-1 (*Continued*)

Nominal Size (in.)	Outside Diameter (in.)	Wall Thickness Designation STD-XS-XXS	Wall Thickness Designation Schedule No.	Nominal Wall Thickness (in.)
2	2.375	XXS		0.436
$2\frac{1}{2}$	2.875	STD	40	0.203
$2\frac{1}{2}$	2.875	XS	80	0.276
$2\frac{1}{2}$	2.875		160	0.375
$2\frac{1}{2}$	2.875	XXS		0.552
3	3.500	STD	40	0.216
3	3.500	XS	80	0.300
3	3.500		160	0.438
3	3.500	XXS		0.600
$3\frac{1}{2}$	4.000	STD	40	0.226
$3\frac{1}{2}$	4.000	XS	80	0.318
4	4.500	STD	40	0.237
4	4.500	XS	80	0.337
4	4.500		120	0.438
4	4.500		160	0.531
4	4.500	XXS		0.674
5	5.563	STD	40	0.258
5	5.563	XS	80	0.375
5	5.563		120	0.500
5	5.563		160	0.625
5	5.563	XXS		0.750
6	6.625	STD	40	0.280
6	6.625	XS	80	0.432
6	6.325		120	0.562
6	6.625		160	0.719
6	6.625	XXS		0.884
8	8.625		20	0.250
8	8.625		30	0.277
8	8.625	STD	40	0.322
8	8.625		60	0.406
8	8.625	XS	80	0.500
8	8.625		100	0.594
8	8.625		120	0.719
8	8.625		140	0.812
8	8.625	XXS		0.875
8	8.625		160	0.906
10	10.750		20	0.250
10	10.750		30	0.307
10	10.750	STD	40	0.365
10	10.750	XS	60	0.500
10	10.750		80	0.594
10	10.750		100	0.719
10	10.750		120	0.844
10	10.750	XXS	140	1.000
10	10.750		160	1.125
12	12.750		20	0.250
12	12.750		30	0.330

TABLE 4.4-1 (*Continued*)

Nominal Size (in.)	Outside Diameter (in.)	Wall Thickness Designation		Nominal Wall Thickness (in.)
		STD-XS-XXS	Schedule No.	
12	12.750	STD		0.375
12	12.750		40	0.406
12	12.750	XS		0.500
12	12.750		60	0.562
12	12.750		80	0.688
12	12.750		100	0.844
12	12.750	XXS	120	1.000
12	12.750		140	1.125
12	12.750		160	1.312
14	14.000		10	0.250
14	14.000		20	0.312
14	14.000	STD	30	0.735
14	14.000		40	0.438
14	14.000	XS		0.500
14	14.000		60	0.594
14	14.000		80	0.750
14	14.000		100	0.938
14	14.000		120	1.094
14	14.000		140	1.250
14	14.000		160	1.406
16	16.000		10	0.250
16	16.000		20	0.312
16	16.000	STD	30	0.375
16	16.000	XS	40	0.500
16	16.000		60	0.656
16	16.000		80	0.844
16	16.000		100	1.031
16	16.000		120	1.219
16	16.000		140	1.438
16	16.000		160	1.594
18	18.000		10	0.250
18	18.000		20	0.312
18	18.000	STD		0.375
18	18.000		30	0.438
18	18.000	XS		0.500
18	18.000		40	0.562
18	18.000		60	0.750
18	18.000		80	0.938
18	18.000		100	1.156
18	18.000		120	1.375
18	18.000		140	1.562
18	18.000		160	1.781
20	20.000		10	0.250
20	20.000	STD	20	0.375
20	20.000	XS	30	0.500
20	20.000		40	0.594
20	20.000		60	0.812
20	20.000		80	1.031

TABLE 4.4-1 (*Continued*)

Nominal Size (in.)	Outside Diameter (in.)	Wall Thickness Designation		Nominal Wall Thickness (in.)
		STD-XS-XXS	Schedule No.	
20	20.000		100	1.281
20	20.000		120	1.500
20	20.000		140	1.750
20	20.000		160	1.969
22	22.000		10	0.250
22	22.000	STD	20	0.375
22	22.000	XS	30	0.500
22	22.000		60	0.875
22	22.000		80	1.125
22	22.000		100	1.375
22	22.000		120	1.625
22	22.000		140	1.875
22	22.000		160	2.125
24	24.000		10	0.250
24	24.000	STD	20	0.375
24	24.000	XS		0.500
24	24.000		30	0.562
24	24.000		40	0.688
24	24.000		60	0.969
24	24.000		80	1.219
24	24.000		100	1.531
24	24.000		120	1.812
24	24.000		140	2.002
24	24.000		160	2.344
26	26.000		10	0.312
26	26.000	STD		0.375
26	26.000	XS	20	0.500
28	28.000		10	0.312
28	28.000	STD		0.375
28	28.000	XS	20	0.500
28	28.000		30	0.625
30	30.000		10	0.312
30	30.000	STD		0.375
30	30.000	XS	20	0.500
30	30.000		30	0.625
32	32.000	STD		0.375
32	32.000	XS	20	0.500
32	32.000		30	0.625
32	32.000		40	0.688
34	34.000		10	0.312
34	34.000	STD		0.375
34	34.000	XS	20	0.500
34	34.000		30	0.625
34	34.000		40	0.688
36	36.000		10	0.312
36	36.000	STD		0.375
36	36.000	XS	20	0.500
36	36.000		30	0.625

TABLE 4.4-1   (*Continued*)

Nominal Size (in.)	Outside Diameter (in.)	Wall Thickness Designation		Nominal Wall Thickness (in.)
		STD-XS-XXS	Schedule No.	
36	36.000		40	0.750
38	38.000	STD		0.375
38	38.000	XS		0.500
40	40.000	STD		0.375
40	40.000	XS		0.500
42	42.000	STD		0.375
42	42.000	XS		0.500
44	44.000	STD		0.375
44	44.000	XS		0.500
46	46.000	STD		0.375
46	46.000	XS		0.500
48	48.000	STD		0.375
48	48.000	XS		0.500

Manufacturing Standards

ASTM and API have prepared standard specifications for a broad range of products. These specifications are a precise statement of requirements that are to be satisfied by the finished pipe. Each specification will generally provide for two or more grades or types of the product. A product is identified using the name of the specification to which it is manufactured and the designation for the grade or type of pipe desired.

Example: A106   Grade B
                                             grade or type
                        Specification   designation

These specifications provide requirements for the following manufacturing and testing activities:

1. Manufacturing process.
2. Ordering information.
3. Heat treatment.
4. Raw material chemistry.
5. Product analysis.
6. Tensile strength test.
7. Ductility tests.
8. Nondestructive examination.
9. Hydrostatic tests.
10. Tolerances for weight and dimensions.
11. Finish.
12. Marking and packaging.
13. Inspection and certification.

Table 4.4-2 lists the range of steel pipe and tube products manufactured under ASTM or API specifications and their application. Table 4.4-3 is a comparison of requirements for the most commonly used products.

Fitting Products

The raw materials used to manufacture commercial piping products are also produced in the form of standard fittings. Manufacturing standards for the fittings are organized in a manner that provides a

TABLE 4.4-2   STEEL PIPE PRODUCTS AND THEIR APPLICATION

ASTM Specification	Product and Application
A430	Austenitic steel-forged and bored pipe for high-temperature service
A691	Carbon and alloy steel pipe, electric-fusion-welded for high-pressure service at high temperatures
A452	Centrifugally cast austenitic steel cold-wrought pipe for high-temperature service
A451	Centrifugally cast austenitic steel pipe for high-temperature service
A660	Centrifugally cast carbon steel pipe for high-temperature service
A426	Centrifugally cast ferritic alloy steel pipe for high-temperature service
A358	Electric-fusion-welded austenitic chromium–nickel alloy steel pipe for high-temperature service
A134	Electric-fusion (arc)-welded steel plate pipe (sizes 16 in. and over)
A139	Electric-fusion (arc)-welded steel pipe (Sizes 4 in. and over)
A672	Electric-fusion-welded steel pipe for high-pressure service at moderate temperatures
A671	Electric-fusion-welded steel pipe for atmospheric and lower temperatures
A135	Electric-resistance-welded steel pipe
A587	Electric-welded low-carbon steel pipe for the chemical industry
A369	Ferritic alloy-steel-forged and bored pipe for high-temperature service
A530	General requirements for specialized carbon and alloy steel pipe
A714	High-strength low-alloy welded and seamless steel pipe
A381	Metal-arc-welded steel pipe for use with high-pressure transmission systems
A523	Plain-end seamless and electric-resistance-welded steel pipe for high-pressure pipe-type cable circuits
A120	Pipe, steel black and hot-dipped, zinc-coated (galvanized) welded and seamless, for ordinary uses
A53	Pipe, steel, black and hot-dipped, zinc-coated welded and seamless
A312	Seamless and welded austentic stainless steel pipe
A589	Seamless and welded carbon steel water-well pipe
A731	Seamless and welded ferritic stainless steel pipe
A333	Seamless and welded steel pipe for low-temperature service
A376	Seamless austenitic steel pipe for high-temperature central-station service
A106	Seamless carbon steel pipe for high-temperature service
A524	Seamless carbon steel pipe for Atmospheric and Lower Temperatures
A335	Seamless ferritic alloy steel pipe for high-temperature service

TABLE 4.4-2   (*Continued*)

ASTM Specification	Product and Application
A405	Seamless ferritic alloy steel pipe specially heat treated for high-temperature service
A655	Special requirements for pipe and tubing for nuclear and other special applications
A211	Spiral-welded steel or iron pipe
A696	Steel bars, carbon, hot-rolled, and cold-finished, special quality, for pressure piping components and other pressure-containing parts
A733	Welded and seamless carbon steel and austenitic stainless steel pipe nipples
A252	Welded and seamless steel pipe piles
A409	Welded large-diameter austenitic steel pipe for corrosive or high-temperature service
API Specification	
5L	Seamless and welded steel line pipe
5LS	Spiral welded steel line pipe
5LX	High-test seamless and welded steel line pipe

compatible fitting for each pipe product. Three types of fittings are commonly used with energy system piping:

1. Butt-welded fittings.
2. Socket-welded fittings.
3. Threaded fittings.

Nominal dimensions and rules for establishing the pressure–temperature capacity ratings for standard fittings of primary interest for energy systems are contained in the standards listed below.

ANSI B16.5	Steel pipe flanges and flanged fittings
ANSI B16.9	Factory-made wrought-steel butt-welding fittings
ANSI B16.11	Forged-steel fittings, socket welded and threaded
ANSI B16.28	Wrought-steel butt-welding short-radius elbows and returns

ASTM has prepared specifications that standardize the manufacturing process for fittings. A specific pipe fitting is identified by five pieces of information:

1. Type of fitting.
2. Nominal pipe size.
3. Pressure–temperature rating designation.
4. Manufacturing standard.
5. Grade or type designation.

Commonly used fitting types are illustrated in Fig. 4.4-2. Nominal pipe size refers to the NPS of the pipe to which the fitting will be joined.

Pressure–Temperature Rating Class

The pressure-retaining strength of standard fittings is designed by hydrostatic testing. They are then assigned a rating which is a measure of their pressure-retaining ability. Two procedures are employed. Butt-welded fittings manufactured to ANSI B16.9 and ANSI B16.28 are designed by burst testing to verify that their bursting strength is equal to or greater than that of the pipe to which they are designated to be used. A specimen is constructed using the subject fitting and a basis pipe. During the testing, the test pressure is increased until the pipe fails. The fitting is then rated for use with the basis

TABLE 4.4-3 COMPARISON OF COMMONLY USED PIPE PRODUCTS

ASTM Spec.	Product	Pipe Sizes	Wall Thickness Tolerance	Outside Diameter Tolerance	Types and Grades	Tests	Finish
A53	Carbon steel pipe—general service	$\frac{1}{8}$–26 in.	Minimum Not less than $12\frac{1}{2}$% under nominal	$+\frac{1}{64}$ to $-\frac{1}{32}$ in. For $1\frac{1}{2}$ in. and below ±1% For 2 in. and Above	Type F—Furnace butt-welded seam Type E—Electric-resistance-welded seam Type S—Seamless Grades A and B	Tensile Bending Flattening Hydrostatic	No specific surface quality criteria
A106	Carbon steel pipe—high-temperature service	$\frac{1}{8}$–26 in.	Minimum Not less than $12\frac{1}{2}$% under nominal	$-\frac{1}{32}$ in. $+\frac{1}{64}$ to $\frac{1}{8}$ in. Depending on NPS	Grades A, B, C Tensile strength 48–70 ksi Yield strength 30–40 ksi	Tensile Bending Flattening Hydrostatic	Defects deeper than $\frac{1}{16}$ in. or 5% of nominal wall to be removed
A134	Plate pipe	16 in. and larger	Minimum Not less than $12\frac{1}{2}$% under nominal	Limit is on circumference +1% max. $\frac{3}{4}$ in. from nominal	Grades A, B, C, D Depends on raw material plate	Tensile[a] Hydrostatic[a]	No specific surface quality criteria
A672	Plate pipe for high-pressure and moderate-temperature service	16 in. and larger	Minimum Not less than $12\frac{1}{2}$% under nominal	+0.5% from nominal	See specification 28 grades available	Tensile Flattening Bending Hydrostatic Weed radiography on same grades	No specific surface quality criteria
A333	Carbon steel pipe—low-temperature service	Check with vendor	Minimum Not less than $12\frac{1}{2}$% under nominal	$-\frac{1}{32}$ in. $+\frac{1}{64}$ to $\frac{3}{16}$ in. Depending on NPS	Grades 1–9 Tensile strength 55–100 ksi Yield strength 30–75 ksi	Tensile Impact Flattening Hydrostatic	Defects deeper than $\frac{1}{16}$ in. or 5% of nominal wall to be removed
A335	Ferritic alloy steel—high-temperature service	All commercial sizes	Minimum Not less than $12\frac{1}{2}$% under nominal	$-\frac{1}{32}$ in. $+\frac{1}{64}$ to $\frac{3}{16}$ in. Depending on NPS	See specification 12 grades available	Tensile Flattening Bending Hydrostatic	Defects deeper than $\frac{1}{16}$ in. or 5% of nominal wall to be removed
A312	Austenitic stainless steel low-temperature service	All commercial sizes	Minimum Not less than $12\frac{1}{2}$% under nominal	$-\frac{1}{32}$ in. $+\frac{1}{64}$ to $\frac{3}{16}$ in. Depending on NPS	See specification 24 grades available	Tensile Flattening Hydrostatic	No specific surface quality criteria
A376	Austenitic stainless steel high-temperature service	Check with vendor	Minimum Not less than $12\frac{1}{2}$% under nominal	$-\frac{1}{32}$ in. $+\frac{1}{64}$ to $\frac{3}{16}$ in. Depending on NPS	See specification 14 grades available	Tensile Flattening Hydrostatic	Defects Deeper than $\frac{1}{16}$ in. or 5% of nominal wall to be removed

[a] Required on finished pipe. See ASTM specification for raw material plate to identify tests on raw material.

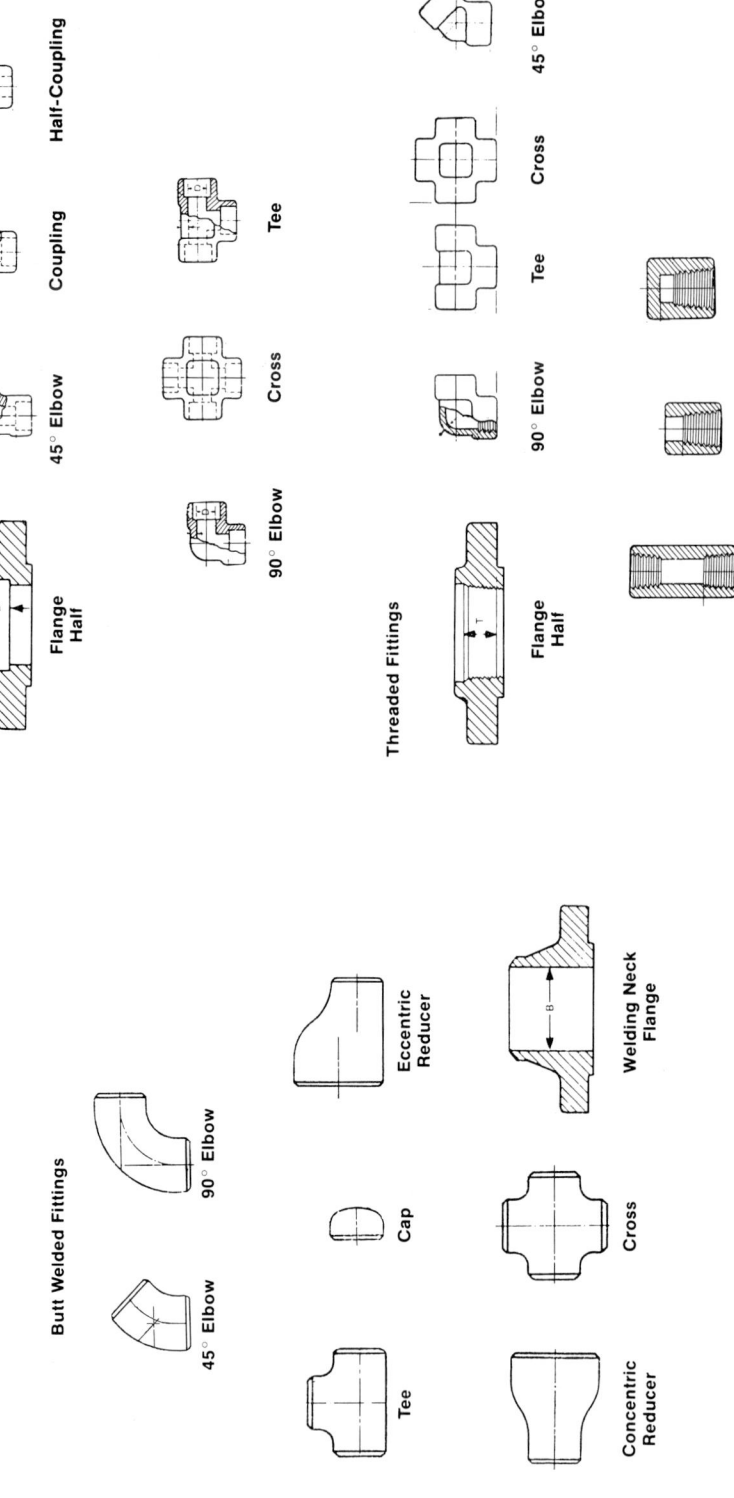

Fig. 4.4-2 Standard forged steel pipe fittings.

pipe. The pressure rating designation for these fittings is the wall thickness designation of the basis pipe.

ANSI B16.5 flanged joints and B16.11 fittings are organized into pressure–temperature rating classes. Each class is associated with a pressure-retaining capacity. The capacities vary with temperature. There are seven classes for flanges and three classes for B16.11 fittings:

	B16.11 Fitting Classes (lb)	
B16.5 Flanged Joint Classes	Socket Welded	Threaded
150	2000	3000
300	3000	6000
400	6000	9000
600		
900		
1500		
2500		

The pressure-retaining capacity of flanged joints manufactured to B16.5 is verified by hydrostatic test at 1.5 times their rated pressure at 100°F. ANSI B16.11 fittings are rated using the burst test procedure discussed above. Schedule 80, 160, and XXS pipe are used as the rating basis.

Manufacturing Standards

ASTM has prepared standard specifications for the manufacture of fittings for use with each standard piping product defined by their specifications. Each specification will generally provide for two or more grades of each fitting product. Table 4.4-4 lists commonly used products and their applications.

Insulation Products

Several types of thermal insulation products are used on piping in energy systems. The type of product required for a given line depends on the service conditions for the line and the function to be provided by the insulation. Three types of products are commonly used.

Material	Application
Calcium silicate	Hot piping
Metallic reflective insulation	Hot piping in radioactive areas of nuclear stations
Fibrous materials	Moderate temperature and chilled water piping
Form plastic materials	Chilled water or refrigerant piping

The manufacture of these materials has also been standardized. ASTM has prepared a set of standard specifications that are commonly used as basis for the purchase of these materials.

Calcium silicate is a hard molded material. It is manufactured in the form of flat blocks, curved blocks, or piping insulation. Calcium silicate piping insulation is available in the form of cylinders, half-cylinders, and curved segments that fit standard sizes of pipe and tube. It is manufactured in nominal thicknesses that range from 1 to 3 in. in $\frac{1}{2}$-in. increments. It is generally held in place by tie wires. When a thickness of 2 in. or greater is required, the insulation is applied in layers. Metal or canvas jacketing is used to protect the material from moisture. The manufacture of calcium silicate insulation products has been standardized. ASTM specification C-533 contains the requirements for these standard products.

Metallic reflective insulation is used in areas of nuclear power stations where radioactivity levels are high. It is a system of stainless steel or aluminum foils designed to reflect heat back onto the pipe wall. There are generally several layers of foils mounted within a heavy stainless steel jacket. Figure 4.4-3 illustrates this construction. Metal reflective insulation is generally custom designed to provide the same retention capacity as standardized mass insulation. It is used because it does not present a radioactive waste problem when it becomes contaminated. It can be reused after flooding and can be easily removed for in-service inspection of welds.

TABLE 4.4-4   FITTING AND BOLTING PRODUCTS

ASTM Specification	
Fitting Products	
A774	As-welded wrought austenitic stainless steel fittings for general corrosive service at low and moderate temperatures
A758	Butt welding, wrought carbon steel, piping fittings with improved notch toughness
A420	Piping fittings of wrought carbon steel and alloy steel for low-temperature service
A234	Piping fittings of wrought carbon steel and alloy steel for moderate and elevated temperatures
A652	Special requirements for wrought steel welding fittings for nuclear and other special applications
A403	Wrought austenitic stainless steel piping fittings
Bolting Products	
A193	Alloy steel and stainless steel bolting materials for high-temperature service
A320	Alloy steel bolting materials for low-temperature service
A540	Alloy steel bolting materials for special applications
A437	Alloy steel turbine-type bolting material specially heat treated for high-temperature service
A453	Bolting materials, high-temperature, 50–120 ksi yield strength, with expansion coefficients comparable to austenitic steel
A194	Carbon and alloy steel nuts for bolts for high-pressure and high-temperature service
A563	Carbon and alloy steel nuts
A307	Carbon steel externally and internally threaded standard fasteners
A354	Quenched and tempered alloy steel bolts, studs, and other externally threaded fasteners
A614	Special requirements for bolting material for nuclear and other special applications
A489	Carbon steel eyebolts
A449	Quenched and tempered steel bolts and studs

Fibrous materials are manufactured from mineral substances such as rock, slag, or glass that are processed into fibers. They are manufactured in the form of boards, blankets, or as preformed pipe insulation. Mineral fiber pipe insulation is available in the form of cylinders, half-cylinders, and segments. It is manufactured in nominal thicknesses ranging from $\frac{1}{2}$ to 4 in. Metal or canvas jacketing is used to protect the material from moisture. When used on low-temperature piping, it is important for the insulation to be carefully sealed against moisture to prevent sweating on the pipe wall and the subsequent collection of moisture inside the insulation. ASTM specification C-547 contains the requirements for mineral fiber pipe insulation.

Foam plastic products are used primarily on chilled water piping or lines that carry refrigerant. They are manufactured in the form of sheets or preformed pipe insulation. These coverings are available for pipe sizes up to 8-in. nominal size and 1 in. in thickness. ASTM specification C-534 contains requirements for these products.

Table 4.4-5 is a comparison of the physical properties of these materials over the ranges of their application. The thermal transmission properties of these materials vary with temperature, tempera-

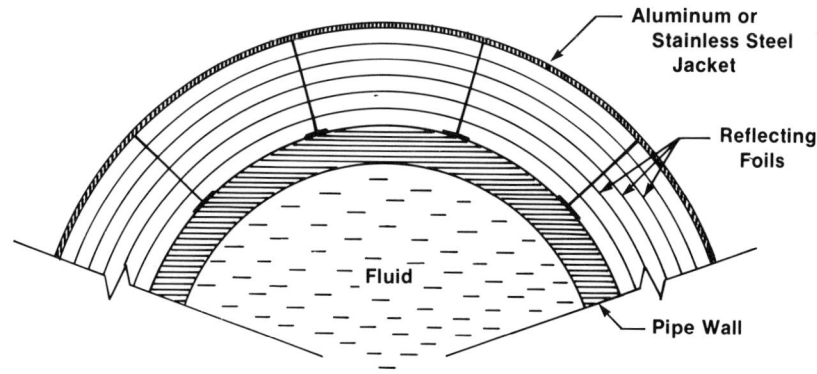

Fig. 4.4-3 Metal reflective insulation construction.

TABLE 4.4-5   PHYSICAL PROPERTIES OF COMMONLY USED PIPE INSULATION PRODUCTS

		ASTM C533 Calcium Silicate		ASTM C547 Mineral Fiber Products			ASTM C534 Foam Products
		Type 1	Type 2	Class 1	Class 2	Class 3	
Density (pcf)		15	15	10	12	18	4.5–8.5
	0°F						0.28
	25°F			0.24			0.29
	50°F			0.25			0.29
	75°F			0.26	0.31		0.30
Apparent	100°F			0.27	0.33	0.36	0.31
Thermal	200°F	0.45		0.32	0.39	0.42	
Conductivity	300°F	0.50			0.46	0.49	
$\dfrac{\text{Btu in.}}{\text{h}/\text{ft}^2 \,^\circ\text{F}}$	400°F	0.55	0.66		0.53	0.56	
	500°F	0.60	0.70			0.64	
	600°F	0.66	0.76				
	700°F	0.71	0.82				
	800°F		0.94				

ture gradient, thickness, and shape. The conductivity values in the table are based on samples tested using the methods specified in ASTM specification C-335. They may not represent the installed performance of the insulation under use conditions differing substantially from the test.

4.4-4   Government Regulation and Industry Codes

Design, construction, examination, and inspection of a majority of piping in energy systems is regulated by federal, state, and municipal authorities. Regulation is accomplished through the passage of boiler and pressure vessel laws. These laws establish requirements for initial design, and construction of boilers and pressure vessels and inspection of them during operation. The laws are administered by a department within the state and/or municipal government having jurisdiction at the installation site. Their approval of the design is required prior to installation. They inspect the completed installation prior to operation and conduct periodic inspections during operation. The federal government regulates the design and construction of safety-related piping in nuclear plants through 10 cfr 50.[1]

Each of these laws utilize piping codes that have been developed by ANSI and ASME as quality standards. Compliance with the requirements in the codes is mandatory if the fluid-handling system is subject to the boiler and pressure vessel laws or federal regulation. Piping codes have been developed to cover a broad range of applications. Each of the codes contain the following information:

1. Requirements for material specifications and product standards that are acceptable for usage.
2. The designation of dimensional standards for the pipe and fittings to be used.
3. Requirements for the design of component parts and assembled units, including pipe support and restraint components.
4. Requirements for the evaluation and limitation of stresses, reactions, and movements associated with pressure, temperature, and external forces.
5. Requirements for fabrication, assembly, and erection of piping systems.
6. Requirements for examination, testing, and inspection prior to assembly and erection and of completed systems after erection.
7. Prohibitions in areas where practices or designs are known to be unsafe.

Design complexity, quality control complexity, and cost vary by significant amounts depending on the code that is used. The boiler laws and regulations that require use of these codes are unique to each jurisdiction. To ensure that the appropriate requirements are being used for the design, the authorities having jurisdiction at the installation site should be contacted prior to the start of work. A summary of the requirements established by regulatory authorities throughout the United States and Canada is published by the Uniform Boiler and Pressure Vessel Laws Society.[2]

The codes listed below are used to regulate the design and construction of piping in energy systems.

ANSI Piping Codes

B31.1	Power piping
B31.2	Industrial gas and air piping
B31.3	Petroleum refinery piping
B31.4	Oil transportation piping
B31.5	Refrigeration piping
B31.6	Chemical industry process piping
B31.7	Nuclear piping
B31.8	Gas transmission and distribution piping systems

ASME Boiler and Pressure Vessel Code

Sections	
I	Power boilers
II	Material specifications
III	Nuclear power plant components
IV	Heating boilers
V	Nondestructive examination
VI	Recommended rules for care and operation of heating boilers
VII	Recommended rules for care of power boilers
VIII	Presure vessels
IX	Welding and brazing qualifications
X	Fiberglass—Reinforced plastic pressure vessels
XI	Rules for in-service inspection of nuclear power plant components

All the codes listed above are maintained by ASME code committees. The committees meet regularly to keep the codes up to date in context and in step with developments in materials and construction practices. Revisions to the codes are issued periodically in the form of addenda. New editions are published at 3-year intervals. Orderly procedures have been established to consider requests for interpretations of code requirements. When an approved reply to an inquiry is made, the interpretation is made public through the issuance of a code case. The cases are published in *Mechanical Engineering*, a magazine published by ASME.

4.4-5 Pipe Size Selection

The object of this step in the design process is to select the nominal pipe size that will result in a fluid velocity during operation that is appropriate for the service that the line provides. From the standpoint of material cost, it is desirable to keep the velocity high. High velocities, however, result in high pressure drops and energy losses, erosion problems, vibration problems, and water hammers. The appropriate service velocity for a line is the velocity that represents the lowest material cost (NPS), yet does not cause high velocity service problems. Sizing an individual line is a two-step process:

1. Establish what the maximum service velocity is to be.
2. Select the nominal size that meets the limit.

The maximum service velocity should be the lowest of the service velocities established using the three bases listed below.

1. The allowable pressure drop between point of supply and point of consumption.
2. Limits provided by equipment manufacturers for the purpose of precluding problems with their equipment.
3. Maximum limits known from experience to preclude erosion, vibration, and water hammer problems.

A comprehensive source of data and procedures for the determination of pressure drops in piping systems is available and is widely used.[3] The manufacturers of major equipment in the system should be requested to provide limits for the flow velocity at the inlets and outlets of their equipment and any specific pipe size requirements they may have. Table 4.4-6 lists maximum velocity values, which represent average practice for different types of service.

4.4-6 Pressure Design

The first step in the structural design process is to establish the pressure-retaining capacity for each line in the system. Developing the pressure design for a line consists of four steps:

1. Selection of pipe wall thickness designation.
2. Selection of the pressure–temperature rating for fittings and flanged joints that will be used.
3. Selection of the type of joints that will be used to join the pipe spools and fittings.
4. Design of fabricated or extruded branch connections and tee connections.

Each step is regulated in detail by the design and fabrication rules in the piping codes. The general philosophy behind the pressure design rules in all the codes is essentially the same. The current rules in ANSI B31.1 will be used as a basis for this discussion

TABLE 4.4-6   MAXIMUM VELOCITY VALUES FOR PIPE SIZING

Type of Service	Maximum Velocity (fpm)
Saturated steam greater than 25 psia and superheated steam with less than 25°F superheat	12500[a] 9000[b]
Superheated steam with more than 25°F superheat	20000[a] 9000[b]
Low-pressure wet steam including saturated steam up to 25 psia	7500
Feedwater piping	720
Condensate piping	720
Hot-suction piping	480–720
Feedwater heater drain piping	420
General-service piping	600
City water service	420

[a] For lines that do not contain 90° elbows.
[b] Lines that contain 90° elbows.

TABLE 4.4-7   VALUES[a] OF y

Temperature, °F	900[b] and below	950	1000	1050	1100	1150 and above
Temperature, °C	482 and below	510	538	566	593	621 and above
Ferritic Steels	0.4	0.5	0.7	0.7	0.7	0.7
Austenitic Steels	0.4	0.4	0.4	0.4	0.5	0.7

[a] The value of y may be interpolated between the 50°F (27.8°C) values shown. For nonferrous materials and cast iron, $y = 0.4$.
[b] For pipe with a D_o/t_m ratio of less than 6, the value of y for ferritic and austenitic steels designed for temperatures of 900°F (480°C) and below shall be taken as

$$y = \frac{d}{d + D_o}$$

Pipe Wall Selection

Pipe wall selection is done using the piping code requirement for minimum wall thickness (t_m), which is in the form of a simple equation. The current ANSI B31.1 requirement is illustrated below.

$$t_m = \frac{PD_o}{2(SE + Py)} + A$$

where P = internal design pressure
  D_o = outside diameter of the pipe
  SE = allowable stress specified by the piping code for the product that was selected
  A = additional thickness allowance to account for material removed in threading, grooving, corrosion, and erosion or to specify additional thickness that may be required for mechanical strength
  y = temperature coefficient having the values in Table 4.4-7

   The object of this requirement is to limit the proportion of the mechanical strength of the pipe that is utilized to contain its design pressure. The pipe selected must be manufactured with a minimum wall thickness that is greater than the t_m value. The relationship between the minimum wall thickness for a specific pipe and its nominal wall thickness can be found in the standard specification that was selected to control its manufacture.

Selection of Pressure–Temperature Ratings

The next step in the process is to select pressure–temperature ratings for the fittings and flanged joints that are to be used for fabrication of the line. For butt-welded fittings manufactured to ANSI B16.9 or B16.28, the pressure rating designation is the same as the wall thickness designation of the pipe that was selected, provided that the raw material for the pipe and fitting are identical. For ANSI B16.5 flanges and ANSI B16.11 fittings, the proper rating class can be selected using tables contained in those standards.

Selection of Joints

Each piping code contains unique requirements that govern the selection of joint type. General requirements are written into the design requirement sections of the codes. Detailed requirements for the joints themselves are contained in separate sections which specify fabrication requirements. Generally, socket-welded joints and threaded joints are used on smaller-diameter, or low-pressure piping. Full-penetration butt-welded joints are used for larger-diameter, high-pressure applications.
   Threaded joints and socket-welded joints are undesirable for applications that involve corrosive fluids or vibration. When they are used, specific action must be taken to minimize piping vibration.

Design of Fabricated Tee and Branch Connections

Fabricated tee and branch connections are made by joining a branch pipe to the run pipe without the use of a rated standard fitting. The pipe at the branch connection is weakened by the opening made in the run. Reinforcement of these connections is required when the pipe wall thickness is not sufficiently

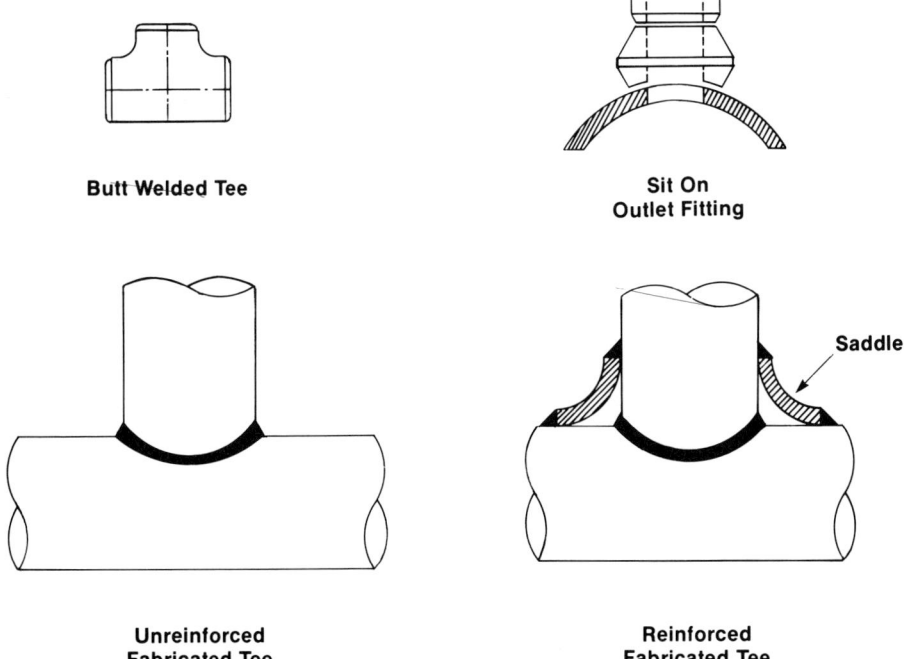

Butt Welded Tee

**Sit On
Outlet Fitting**

Saddle

**Unreinforced
Fabricated Tee**

**Reinforced
Fabricated Tee**

Fig. 4.4-4 Branch and tee connection construction.

in excess of the minimum required. Welded saddles, pads, or commercially available sit-on outlet fittings are used to reinforce the connections when required. Figure 4.4-4 illustrates the types of construction used.

Design of these connections consists of sizing the reinforcement. Each of the piping codes contains a detailed set of rules for the design. All rules are based on the idea of replacing the area of material removed by making the opening within the reinforcement metal area. The piping code that governs the selected design should be referred to for details.

4.4-7 Piping Layout and Structural Design

The object of these two steps in the design process is to establish the arrangement of the piping and its support system. They are closely related and in practice are often done in a single step. Their end result must be a piping and support arrangement that will

1. safely contain its fluid during all operating conditions,
2. provide an operable flow path during all operating conditions requiring that function,
3. provide for safe and efficient operation and maintenance of the system during its service life, and
4. allow installation at reasonable cost.

The level of design effort that is practical for an individual pipeline within a system is dependent on the pipe size, service conditions, and the function that the line is to provide under those conditions. Developing the structural design for a system is a three-step process:

1. Categorize the individual pipelines in the system by pipe size, service conditions, function, and performance requirements.
2. Establish a design approach for each category of line that is the most practical for that category of line.
3. Develop the design for each category of line using the appropriate approach.

Large-diameter, high-temperature, high-pressure process lines require careful attention during design. A bursting failure of one of these lines could result in the death of all nearby personnel and catastrophic damage to the plant. Thermal expansion of these lines generates large internal forces capable of causing major damage to equipment and building structures if not properly acknowledged. A detailed structural analysis is necessary for the design of these lines. Each support and restraint assembly in the suspension systems for these lines is a unique design that is released for construction on individual drawings. Detailed engineering evaluations are required when changes are made during installation.

The design of large-diameter lines with moderate service conditions such as test lines, charging lines, some warmup lines, and service water lines require a less detailed approach to obtain the same level of quality. These lines generate internal forces that are much lower and operate intermittently. High-quality designs for these lines can be developed using manual design procedures, which are developed using piping code design rules as a basis, and standardized support hardware assemblies. Layout design and structural design of these lines is done as a single work step. Piping and support configurations designed in this manner are simple and less costly to install. Detailed engineering evaluations of changes made during construction are required much less frequently because design margins inherent in these approaches are sufficient to provide for normally experienced deviations.

Low-temperature, small-diameter lines such as drains, vents, sensing lines, and sampling lines are done in a similar manner. It is most economical to design these lines and their supports in the plant after all process lines have been installed. This work is traditionally done by the piping erection contractor. The piping and supports for these lines are field fabricated using bulk-ordered pipe and hardware drawn from stock at the time it is installed. The layout and structural design is done in one work step using simple span length, bending leg sizing, and hardware selection rules which are provided in an erection specification.

Compliance with piping code requirements provides the design with the ability to safely contain its fluid and remain operable when necessary. Safe operation, maintainability, and reasonable cost are features that are built-in during layout design.

Layout Design

Layout design is accomplished by careful review of physical surroundings. It is best done at the actual location where the line will be installed. When piping design is done prior to building construction, composite drawings or a design model are used to prepare the layout. The installation cost for a line is extremely sensitive to the arrangement that is established. The following guidelines should be adhered to.

1. The line must be located directly under, over, or immediately adjacent to the building structure that it will be supported from or restrained to. Careful attention must be given to minimizing the size and number of auxiliary steel structures that are required to support the line. The most desirable design is one that can be directly connected to the existing building structure with standard hanger hardware components.

2. The lines should be arranged in a neat and orderly manner. They should be routed in groups and banks whenever possible.

3. High-pressure and high-temperature lines, or lines carrying toxic fluids, must not be located in areas where frequent personnel traffic is likely.

4. All in-line equipment requiring regular attention must be located such that it is readily accessible to maintenance personnel.

5. Clear working spaces must be maintained around and above equipment such as valves, pumps, heat exchangers, and strainers to permit maintenance and removal.

6. Valves should be located such that they can be operated, repacked, or replaced from floor level, permanent platforms, or portable platforms. Care should be taken to avoid locating heavy valves directly adjacent to equipment nozzles.

7. Oil or gas lines must not be located in the vicinity of or above electric switchgear or motors.

8. Care must be taken to avoid pockets (traps). These are particularly hazardous in steam lines and relief valve discharge piping. Water slugs accumulate in pockets. When the line is brought on stream, the slugs flash into steam or are accelerated to high velocities, causing severe vibration damage. Pockets can be inadvertently formed by concentric reducers, orifice plates, reducing valves, or excessive sag in the line.

9. Steam lines should be sloped to allow proper drainage.

10. Pump suction piping should be arranged so that the flow is as smooth as possible.

11. A support must be placed at the top of each riser pipe in the system to avoid instability.

12. A support should be located adjacent to each valve in the system.

13. All supports on plain straight pipe should be spaced as uniformly as possible.

14. Sample lines, instruments, vent lines, and drain lines must be arranged to minimize vibration, particularly if they are located on a process line near a piece of rotating equipment.

15. Suction and discharge piping associated with reciprocating pumps or compressors must be arranged and restrained to minimize vibration.

16. High-velocity bypass lines, and lines that discharge through a breakdown orifice, should be supported to minimize vibration.

17. Relief wave discharge piping and vent stack piping must be arranged so that they can be effectively restrained against fluid transient thrusts. It must be routed to discharge in a safe area and should be self-draining.

Structural Design

Once the piping and support arrangement has been established, it is necessary to verify that the design configuration represented by that arrangement satisfies the piping code analysis requirements applicable to it and to determine design loadings for all support system components. When moderate service conditions are involved, this is accomplished by using the piping code requirements to develop the design rules used during the layout process. Application of the rules yields a design that meets code requirements and a set of support capacity requirements that are then used for support design. When severe service conditions are involved, or the line connects to a fragile piece of equipment, the work is accomplished by doing a detailed structural analysis. The general philosophy behind the piping code requirements and practical procedures for implementing them are discussed below.

The analysis requirements in the piping codes were designed to prevent failure of piping by the mechanisms listed below.

1. Bursting.
2. Collapse
3. Fatigue fracture
4. Excessive distortion

They do so by limiting the internal bending moments in the line that are caused by the service conditions and the proportion of the pipe's pressure-retaining capacity that is utilized to contain its design pressure. The rules are based on well-established engineering principles. They acknowledge the significance of the service stress, the distribution of the stress through the pipe wall, and the loading mechanism to the load-resisting capacity of the pipe.

Protection against bursting failure and collapse is accomplished by limiting the membrane stress (average stress across the pipe wall), which is in equilibrium with the service pressure and externally applied primary loads. Primary loads are not self-limiting (relieved by plastic deformation) in nature. They are physically capable of causing uncontrolled plastic deformation if not adequately controlled. Deadweight loads, building vibration loads, relief valve thrusts, fluid transient thrusts, and liveweight loads are examples of primary loads.

Protection against fatigue failure is accomplished by limiting the range of peak stress that occurs at points of stress concentration in the system (fittings) as it is moved from one operating mode to another during plant operation. Control of these stress levels is accomplished through the use of stress intensification factors provided by the codes. The stress intensification factor for a fitting is established in a manner similar to that used to establish its pressure rating. A fatigue test is done using a specimen that is constructed using the subject fitting and a basis pipe. The stress intensification factor is selected such that when it is applied with the design rule it will provide the fitting with the same fatigue life as the plain straight pipe to which it is joined.

Piping code design rules allow designs that operate at stress levels above the yield strength of the piping material. Materials that operate at these stress levels are susceptible to low-cycle fatigue failures that result from strain concentrations and to incremental plastic distortion. These two failure mechanisms are active when the range of service stress that the material cycles through during operation is above twice its yield strength. Protection against failure by these mechanisms is provided by a design rule that limits the sum of the stress caused by sustained loads and the thermal expansion stress range.

The code version applicable to a specific design should be referred to for specific requirements.

Manual Design Approaches

A manual design approach is more practical to use for small-diameter lines and larger-diameter lines with moderate service conditions. Using a manual approach requires developing a layout design procedure from the piping code requirements. These procedures consist of the following:

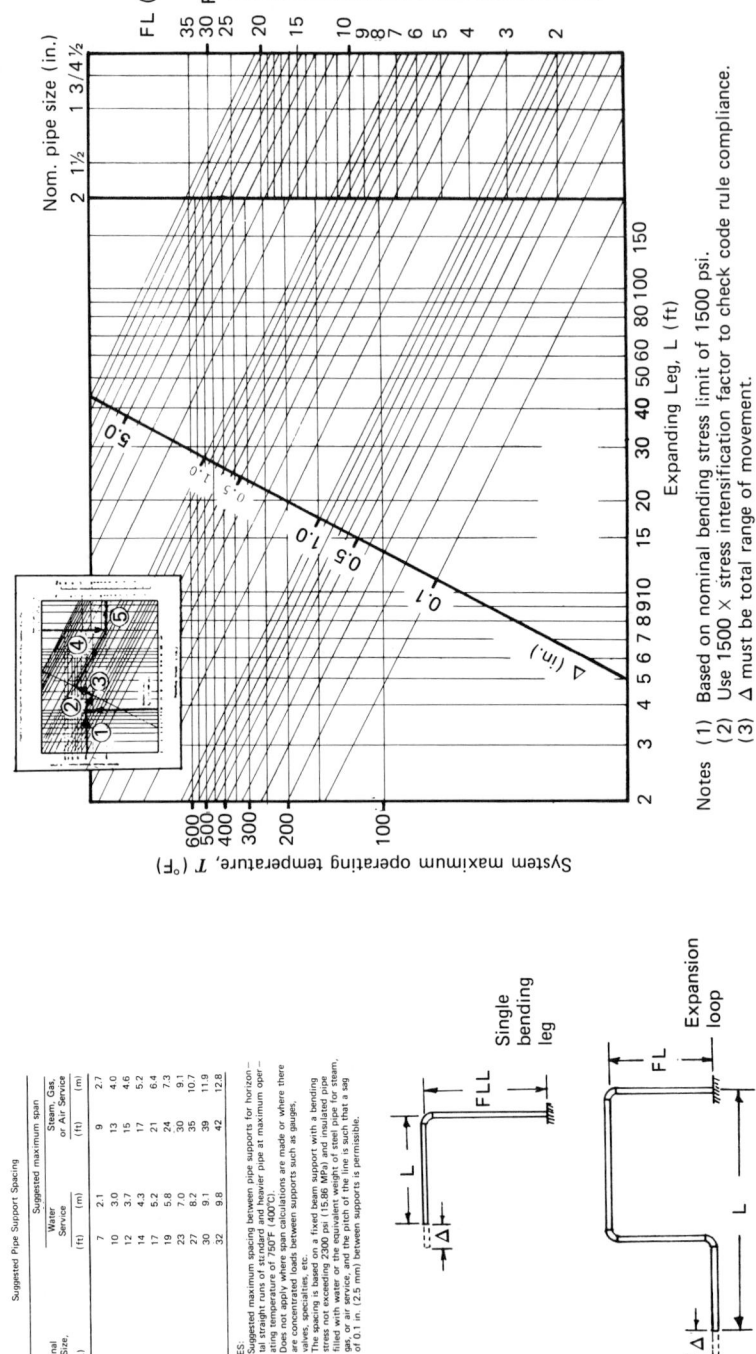

Suggested Pipe Support Spacing

Nominal Pipe Size (in.)	Water Service		Steam, Gas, or Air Service	
	(ft)	(m)	(ft)	(m)
1	7	2.1	9	2.7
2	10	3.0	13	4.0
3	12	3.7	15	4.6
4	14	4.3	17	5.2
6	17	5.2	21	6.4
8	19	5.8	24	7.3
12	23	7.0	30	9.1
16	27	8.2	35	10.7
20	30	9.1	39	11.9
24	32	9.8	42	12.8

NOTES:
(1) Suggested maximum spacing between pipe supports for horizontal straight runs of standard and heavier pipe at maximum operating temperature of 750°F (400°C).
(2) Does not apply where span calculations are made or where there are concentrated loads between supports such as gauges, valves, specialties, etc.
(3) The spacing is based on a fixed beam support with a bending stress not exceeding 2300 psi (15.86 MPa) and insulated pipe filled with water or the equivalent weight of steel pipe for steam, gas, or air service, and the pitch of the line is such that a sag of 0.1 in. (2.5 mm) between supports is permissible.

Single bending leg

Expansion loop

Notes (1) Based on nominal bending stress limit of 1500 psi.
(2) Use 1500 × stress intensification factor to check code rule compliance.
(3) Δ must be total range of movement.

Fig. 4.4-5 Typical piping arrangement rules.

1. A table of maximum support span lengths for plain straight pipe.
2. Practices for locating supports near in-line equipment, supporting riser pipes, and piping near tee and branch connections (see "Layout Design" above).
3. Rules for providing bending legs in the pipe routing to assure adequate flexibility.

Figure 4.4-5 contains span length and bending leg length requirements that are used for ANSI B31.1 piping design. A comprehensive source of data and design calculation formulas for organizing manual design procedures is available.[4]

Recent developments in computer-aided drafting and information management technology have simplified computer use to the extent that it is cost-effective to design shop-fabricated piping using a structural analysis when these types of facilities are available. At this point in time a manual design approach is still the most appropriate for small-diameter lines.

Computer-aided Structural Analysis

Lines with severe service conditions are designed using a computer-aided structural analysis. Establishing an arrangement for these types of lines is an iterative process. A preliminary arrangement is first established using average layout practice. The structural analysis is then prepared to verify that the arrangement satisfies piping code acceptability criteria. During the process, the piping and support arrangement is adjusted as necessary to meet the piping code requirements and limits on equipment nozzle reactions provided by the equipment vendor. When a suitable arrangement is established, final design loads are calculated for the support and restraint components on the system.

Several commercially available computer programs have been developed specifically for piping design analysis. The programs work with a geometric description of the piping and support arrangement and a description of the service conditions that are provided by the designer. Using this input, the programs carry out a detailed structural analysis for each loading case that the designer specifies. The analyses of the individual loading cases produce internal bending moments that are required for piping code design calculations and support reactions. The programs collect that information, organize and combine it as necessary, and then perform all the piping code design calculations required. The programs output a report containing the results of the design calculations and design loads for support and restraint components on the system. Piping analysis programs that cover a broad range of applications are available on a time-sharing basis through computer service bureaus.

4.4-8 Support and Restraint System Design

Once the piping and support arrangement has been finalized and design loadings for the support system are established, it is nececsary to specify or develop a detailed design for the support system. The support system for a line consists of a number of structural assemblies whose purpose is to support the line, brace it against the action of operating loads, and protect the equipment. A general support or restraint assembly has four basic parts.

Lower Attachment: The means by which the support is attached to the pipe.

Component: A hardware component that was designed to perform a specific function.

Auxiliary Steel: Steel framework required to connect the component to a primary building structure.

Upper attachment: The means by which the assembly is attached to the primary building structure.

Lower attachments, components, and a majority of upper attachments are selected or assembled from standard pipe support hardware. Welded attachments, made to the pipe's pressure boundary for the purpose of attaching support hardware, are designed to meet piping code requirements. Auxiliary steel framing, welded connections within the structures, and building connections are generally designed to requirements established by the American Institute of Steel Construction (AISC). Each piping code contains general requirements concerning the support system that must be met by the final design. Designing a general support assembly consists of selecting the appropriate lower attachment and component hardware and then designing the framing required to connect it to the building structure.

Standard Pipe Support Hardware

The design, manufacture, and application of commonly used pipe support hardware has been standardized by the Manufacturers Standardization Society of the Valve and Fitting Industry (MSS).

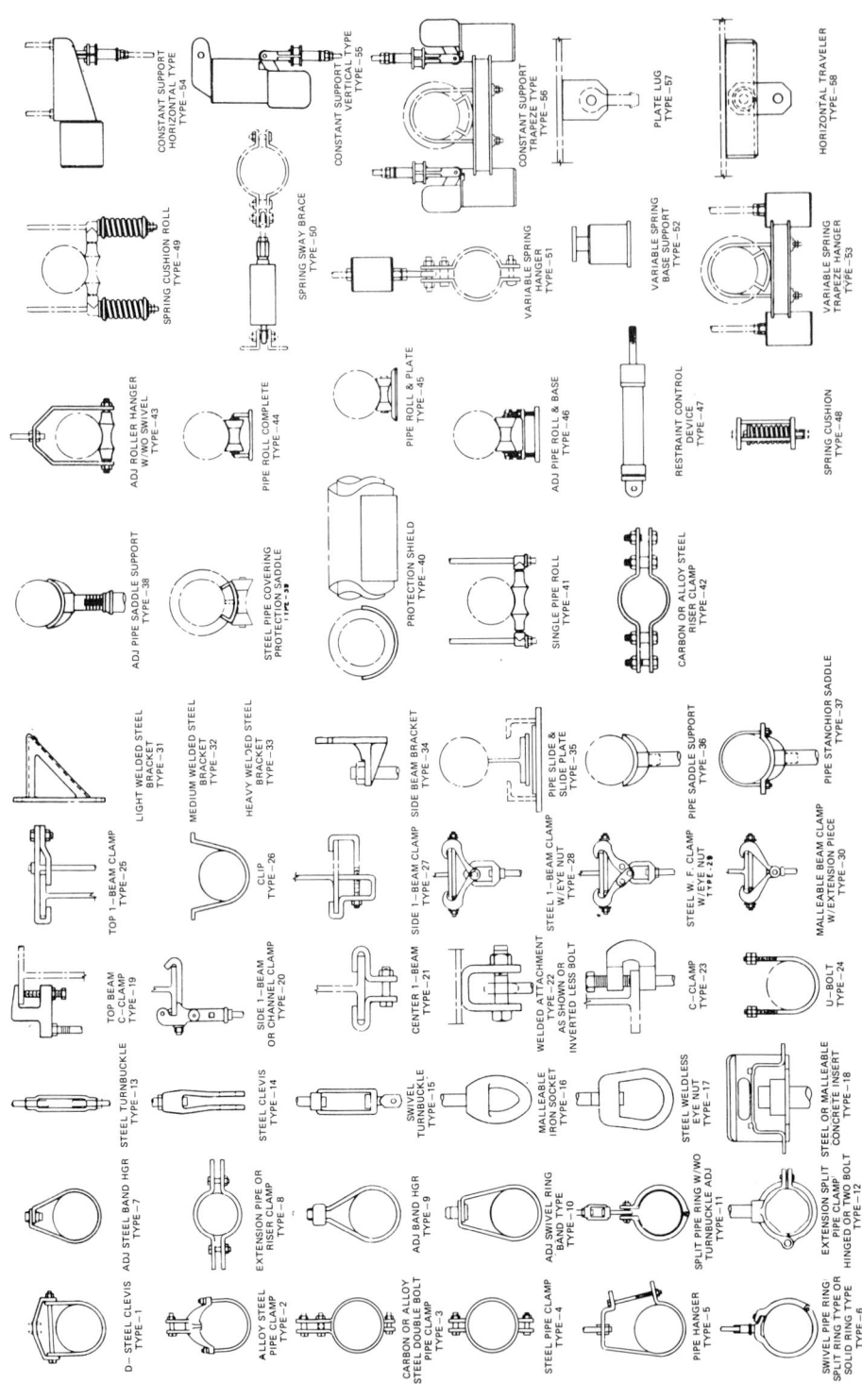

Fig. 4.4-6 MSS-SP-69 standard support hardware components.

CONSTANT SUPPORT HORIZONTAL TYPE—54

CONSTANT SUPPORT VERTICAL TYPE—55

CONSTANT SUPPORT TRAPEZE TYPE—56

PLATE LUG TYPE—57

HORIZONTAL TRAVELER TYPE—58

SPRING CUSHION ROLL TYPE—49

SPRING SWAY BRACE TYPE—50

VARIABLE SPRING HANGER TYPE—51

VARIABLE SPRING BASE SUPPORT TYPE—52

VARIABLE SPRING TRAPEZE HANGER TYPE—53

ADJ ROLLER HANGER W/WO SWIVEL TYPE—43

PIPE ROLL COMPLETE TYPE—44

PIPE ROLL & PLATE TYPE—45

ADJ PIPE ROLL & BASE TYPE—46

RESTRAINT CONTROL DEVICE TYPE—47

SPRING CUSHION TYPE—48

ADJ PIPE SADDLE SUPPORT TYPE—38

STEEL PIPE COVERING PROTECTION SADDLE TYPE—39

PROTECTION SHIELD TYPE—40

SINGLE PIPE ROLL TYPE—41

CARBON OR ALLOY STEEL RISER CLAMP TYPE—42

LIGHT WELDED STEEL BRACKET TYPE—31

MEDIUM WELDED STEEL BRACKET TYPE—32

HEAVY WELDED STEEL BRACKET TYPE—33

SIDE BEAM BRACKET TYPE—34

PIPE SLIDE & SLIDE PLATE TYPE—35

PIPE SADDLE SUPPORT TYPE—36

PIPE STANCHIOR SADDLE TYPE—37

TOP I—BEAM CLAMP TYPE—25

CLIP TYPE—26

SIDE I—BEAM CLAMP TYPE—27

STEEL I—BEAM CLAMP W/EYE NUT TYPE—28

STEEL W. F. CLAMP W/EYE NUT TYPE—29

MALLEABLE BEAM CLAMP W/EXTENSION PIECE TYPE—30

TOP BEAM C—CLAMP TYPE—19

SIDE I—BEAM OR CHANNEL CLAMP TYPE—20

CENTER I—BEAM TYPE—21

WELDED ATTACHMENT TYPE—22 AS SHOWN OR INVERTED LESS BOLT

C—CLAMP TYPE—23

U—BOLT TYPE—24

STEEL TURNBUCKLE TYPE—13

STEEL CLEVIS TYPE—14

SWIVEL TURNBUCKLE TYPE—15

MALLEABLE IRON SOCKET TYPE—16

STEEL WELDLESS EYE NUT TYPE—17

STEEL OR MALLEABLE CONCRETE INSERT TYPE—18

ADJ STEEL BAND HGR TYPE—7

EXTENSION PIPE OR RISER CLAMP TYPE—8

ADJ BAND HGR TYPE—9

ADJ SWIVEL RING BAND TYPE TYPE—10

SPLIT PIPE RING W/WO TURNBUCKLE ADJ TYPE—11

EXTENSION SPLIT PIPE CLAMP HINGED OR TWO BOLT TYPE—12

D—STEEL CLEVIS TYPE—1

ALLOY STEEL PIPE CLAMP TYPE—2

CARBON OR ALLOY STEEL DOUBLE BOLT PIPE CLAMP TYPE—3

STEEL PIPE CLAMP TYPE—4

PIPE HANGER TYPE—5

SWIVEL PIPE RING SPLIT RING TYPE OR SOLID RING TYPE TYPE—6

418

They have published three standards that apply to pipe support hardware.

SP-58	Pipe hangers and supports—Materials, Design, and manufacture
SP-69	Pipe hangers and supports—Selection and application
SP-77	Guidelines for pipe support contractual relationships

Most commercially available pipe support hardware is manufactured and assigned load ratings according to the requirements in these standards. Figure 4.4-6 illustrates the types of lower and upper attachment hardware most commonly used.

Support Components

Five types of components are used to support the weight of piping systems:

1. Rod hangers.
2. Resting supports.
3. Spring cushions.
4. Calibrated spring hangers/supports.
5. Constant-effort hangers/supports.

The appropriate component to use at a given location depends on the vertical thermal expansion movement of the pipe.

Rod hangers or rigid resting supports are the most desirable components to use from a cost standpoint. They should be used as often as possible. Rod hangers and resting supports are rigid and will only carry a load in one direction. They should not be used at locations that will move upward during thermal expansion of the system. The upward movement will cause the pipe to lift off the support. The lift-off results in a redistribution of the dead load to adjacent supports or equipment that may not be designed to accommodate it.

Spring cushions or calibrated spring hangers should be used when there is a vertical thermal movement that will cause a weight-shifting problem. Spring cushions should be used on manually designed lines where movements are small (less than $\frac{1}{4}$ in.). Calibrated spring hangers should be used when larger loadings and movements are involved. Settings for the calibrated spring hangers are determined during the structural analysis of the line.

Constant-effort hangers and support units are used at locations where the thermal expansion movement is large, typically more than $\frac{3}{4}$ in. These units are designed to provide a relatively constant supporting force as the pipe moves. Instructions for sizing spring and constant-support units are provided by the hardware vendor.

Restraints and Restraint Components

Six types of components are used to brace and restrain piping systems:

1. Structural anchors.
2. Guides.
3. Limit stops.
4. Rigid struts.
5. Sway suppressors.
6. Snubbers.

A structural anchor is a rigid restraint that prevents all pipe movement and rotation at the point where it is installed. The most desirable place to locate an anchor is at a wall or floor penetration. Smaller-diameter, low-temperature lines can be grouted in at wall or floor penetrations. Large-diameter or high-temperature lines are anchored using plate and sleeve assemblies (see Fig. 4.4-7). Anchor structure framework is required when anchors are placed at locations other than wall or floor penetrations.

Guides are rigid restraints that prevent lateral movement of the pipe. They allow movement along the centerline of the pipe and rotation. Guides are generally steel frame structures. Wherever guides or frame-type anchors are required, the pipe should be located as close to the building structure as possible to allow the structures to be sufficiently stiff and to minimize cost.

Limit stops are one-direction, rigid restraints designed to limit the displacements of the pipe to a specified value. They are generally used to reduce the loading on equipment nozzles. Generally, limit stops are frame structures as well.

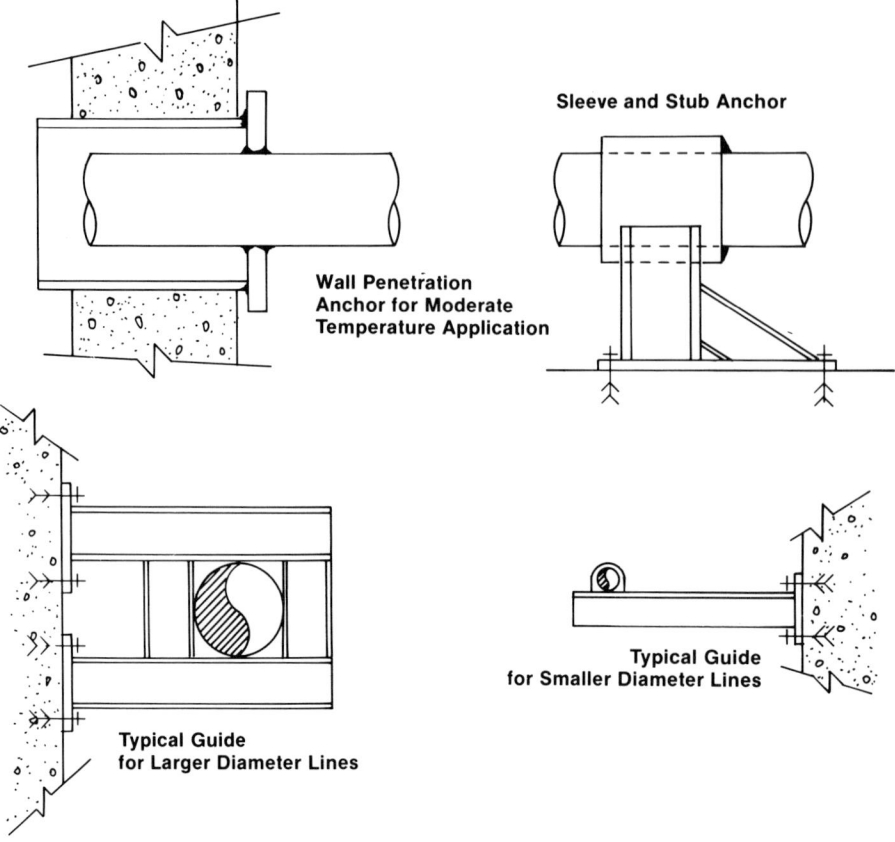

Fig. 4.4-7 Commonly used anchors and guides.

Rigid struts, sway suppressors, and snubbers are pin-connected restraint components. They are connected to the pipe with a clamp. They are used to limit vibration and to control the piping during fluid transients, wind loading, or seismic loading. The appropriate component to use at a given location depends on the thermal expansion displacement in the direction to be restrained and the stiffness required.

Rigid struts are most desirable to use from both cost and effectiveness points of view. They are simple rigid links that will sustain a load in tension or compression. Rigid struts are generally purchased as load-rated assemblies consisting of the strut, a clamp, and any brackets required for installation. Rigid struts should be used at all locations where their use does not cause an equipment overload or thermal expansion overstress problem.

Sway suppressors are used to reduce vibration and provide load control at locations on a line where limiting displacement to small values is not required. They allow thermal expansion of the line along the direction of restraint. A sway suppressor is a preloaded spring device. It will limit displacement to practically zero whenever the load applied to it is less than the preload. It will act like a spring under a load that is greater than the preload.

Snubbers (or shock arrestors) are used to sustain dynamic loadings such as shock or severe transient vibrations while allowing practically free thermal expansion movement of the pipe. Two types are available: hydraulic snubbers and mechanical shock arrestors. Hydraulic snubbers are designed to lock up and act as a solid restraint when the velocity of the pipe exceeds a small value. Mechanical shock arrestors are mechanical brakes. They limit the acceleration of the pipe to a small value. They should only be used to sustain oscillating loads that reverse in direction.

Both types of units deteriorate in service. They require maintenance and periodic functional testing during operation of the plant. They should only be used at locations with large thermal movements and only after it has been carefully established that they are necessary.

Welded Attachments

Two basic types of welded attachments are used to connect restraint and support components to pipe: lugs and pipe stubs. They are used with a variety of support and restraint hardware to form welded attachment assemblies. The most commonly used types are illustrated in Fig. 4.4-8.

Riser clamp-shear lug assemblies are used to support risers or to connect an axial restraint to a leg of pipe. Three-bolt clamp-shear lug assemblies are used with restraint components whose line of action is not perpendicular to the pipe centerline. Single lugs are used to attach supports to inclined pipe. They are only suitable for smaller-diameter, thick-walled pipe. Elbow lugs are used to hang risers from the top. Pipe stands are used to support pipe that is routed near the top surface of a slab. Trunnions are used to attach restraint components to pipe.

Welded attachments must be fabricated using bar stock, plate, or pipe that has the same nominal chemical composition as the pipe they are attached to. A full penetration weld must be used to join the attachment to the pipe wall on high-temperature lines. The weld bead should be ground concave to reduce the stress concentration. Welded attachments made to low-temperature lines may be made with fillet or partial penetration welds.

Welded attachments cause local bending of the pipe wall. The bending stresses are very high when thin-walled pipe is involved and must be carefully acknowledged during design. Thermal stress

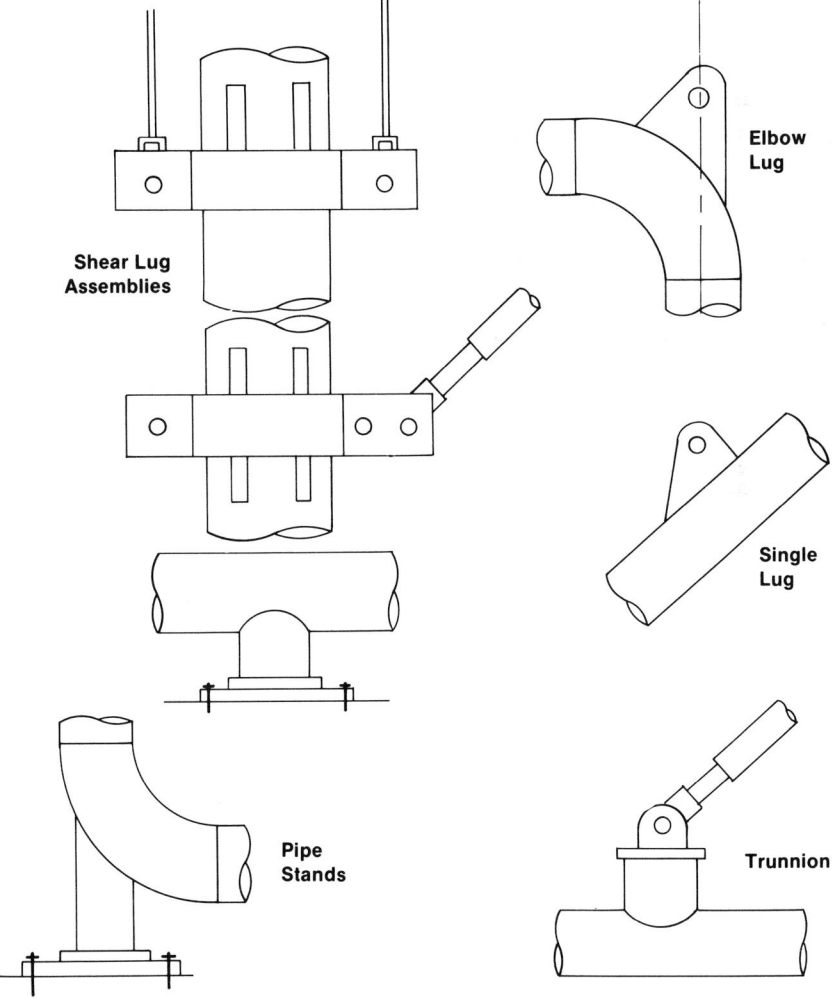

Fig. 4.4-8 Commonly used welded attachment assemblies.

problems can be caused on high-temperature lines if the attachments are not properly insulated. Shear lugs that can be completely covered by insulation are the only type of attachment that should be used on high-temperature lines.

Procedures that can be used to design welded attachments that meet the intent of piping code requirements have been published by the Welding Research Council.[5,6]

Auxiliary Steel Structures

Many anchors, guides, and limit stops are welded steel frame structures. Frames are also used to connect support and restraint components to building structures. They are generally built using general-purpose carbon steel angles, channels, smaller wide-flange sections, and tube manufactured to ASTM specification A36. Connections to building structures are made via embedded plates, bolted anchor plates, or welded connections to structural steel. The most commonly used types of frame are simple cantilevers, knee braces, simply supported spans, and box guides. Examples of these are illustrated in Fig. 4.4-9. If the piping arrangement calls for more complicated framework, then the pipe routing should be changed as necessary to minimize the extent of framework required.

It is necessary for the design of a pipe support frame to meet two general requirements:

1. Its load-resisting capacity must be adequate to safely control the load placed on it by pipe.
2. It must be stiff enough to perform its intended function.

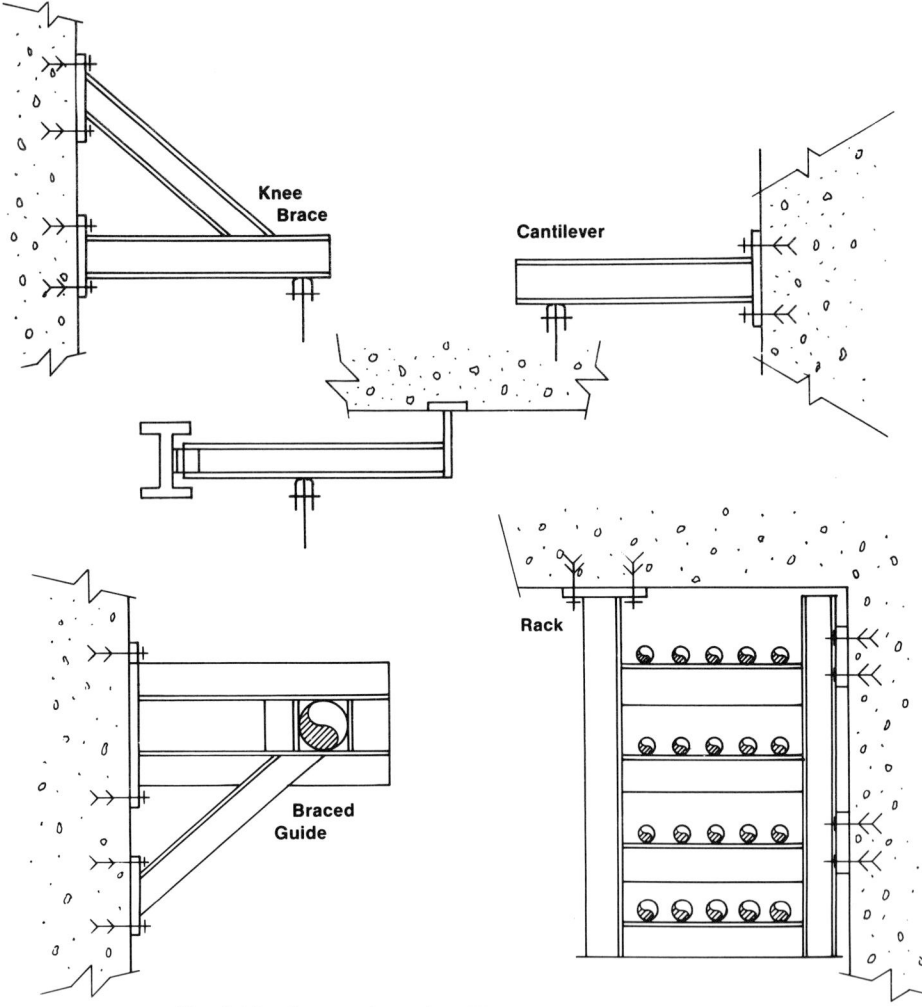

Fig. 4.4-9 Commonly used auxiliary steel framing arrangements.

Proper load-resisting capacity is built into the design by meeting the structural integrity criteria required by the piping code governing design. ANSI piping codes require that auxiliary steel design be done in accordance with standard design procedures and requirements that have been established by AISC. The requirements are published in their specification for design, fabrication, and erection of structural steel for buildings.[7] The document contains requirements for the following aspects of the design:

1. Material selection.
2. Types of construction.
3. Allowable stresses.
4. Stability.
5. Bolted connections and anchors.
6. Welded connections.
7. Fabrication.
8. Erection.
9. Quality control.

Section III of the ASME Boiler and Pressure Vessel Code contains its own set of support design requirements (subsection NF).

The following actions should be taken to assure that the frame is stiff enough to perform its intended function:

1. Determine what the deflection of pipe would be if the subject support was not acting.
2. Design the support such that its deflection under the piping load is a reasonable small fraction of that value.
3. Assure that the stiffness of adjacent restraints or supports, in the same direction as the one being worked on, are not drastically different.

Frames that are too flexible will not limit the deflection of the pipe to the extent required to control the loading or assure that the actual support reactions and equipment loads are distributed in a manner that is consistent with the design. This aspect of the design is particularly important when dealing with braces that are used for thermal expansion control or with restraints that are placed near equipment nozzles to control nozzle loads.

A comprehensive source of procedures and accepted practices that are used for the design of welded steel structures is available.[8]

4.4-9 Insulation Application

Insulation is applied to pipelines in a fluid-handling system on an individual basis. The type of insulation construction required for a line depends on the function of the line, its service conditions, and its location in the plant.

Lines that form the primary process flow path are insulated to conserve energy. The function of the insulation on these lines is to conserve energy by retaining or preventing the absorption of heat. Its design is based on the cost of energy. Application of insulation on these lines provides a substantial return.

Lines that operate a small percentage of the time, such as warmup lines, some bypass lines, test lines, recirculation lines, and drains, are insulated to provide the following:

Personnel Protection: In this capacity the function of the insulation is to limit the surface temperature of the line at locations where contact with it by people is likely.

Fire Protection: In this capacity its function is to limit the surface temperature of the line to prevent the ignition of materials that come in contact with the line.

Environmental Control: Insulation is applied for the purpose of maintaining the temperature of the surrounding environment at a level that permits the safe operation of nearby equipment and personnel access during operation. Its design is based on ventilation cost savings.

Instrument lines, sampling lines, charging lines, vents, and relief valve discharge lines are typically not insulated. Intermittently operating water lines that run outdoors are often insulated and heat traced (heated) to prevent freezing.

During operation of a line, the insulation constructed on it functions as part of a system, which consists of the following components:

1. Fluid flowing in the line.
2. Pipe wall.

3. Insulating material and its protective covering.
4. Insulation construction details.
5. Hardware protrusions.
6. Pipe orientation.
7. Surrounding air conditions and movement.

The performance of the insulation and its costs are dependent on the nature of each detail. They are sensitive to each detail to the extent that it is necessary to design the insulation as a system, using knowledge of the actual performance of that system in a service environment identical to the environment in which it will operate.

Selection of piping insulation for a system consists of four basic steps:

1. Identify the function of each line, its service conditions, the frequency with which it operates its location in the plant, surrounding equipment, and environmental conditions.
2. Based on that information, determine if it is necessary to insulate the line. If so, identify the function that the insulation is to provide (heat flow control or surface temperature control) and its performance requirements (heat flow and surface temperature limits).
3. Select the appropriate insulating material, protective covering, and construction details for each line based on actual performance data.
4. Prepare a specification and necessary drawings.

Performance data, testing procedures, design procedures, field fabrication procedures, material specifications, and guidelines for the preparation of installation specifications are available.[9-11]

REFERENCES

4.4-1 Title 10, Code of Federal Regulations, Part 50.
4.4-2 "Synopsis of Boiler and Pressure Vessel Laws and Regulations," Uniform Boiler and Pressure Vessel Laws Society.
4.4-3 "Flow of Fluids Through Valves, Fittings and Pipe," Technical Paper 410, The Crane Company, New York, 1980.
4.4-4 *Design of Piping Systems*, W. M. Kellog Company, Wiley, New York, 1956.
4.4-5 "Local Stresses in Spherical and Cylindrical Shells due to External Loadings," Welding Research Council Bulletin 107, Welding Research Council, New York, 1956.
4.4-6 "Stress Indices at Lug Supports on Piping Systems," Welding Research Council Bulletin 198, Welding Research Council, New York, 1974.
4.4-7 *Manual of Steel Construction*, American Institute of Steel Construction, Inc., New York, 1983.
4.4-8 *Design of Welded Structures*, The James F. Lincoln Arc Welding Foundation, Cleveland, 1966.
4.4-9 John F. Malloy, *Thermal Insulation*, Van Nostrand Reinhold, New York, 1969.
4.4-10 *Annual Book of ASTM Standards*, Part 18, "Thermal and Cryogenic Insulating Materials Building Seals and Sealants Fire Standards, Building Constructions Environmental Acoustics," American Society for Testing and Materials, Philadelphia, 1984.
4.4-11 *Thermal Insulation Performance*, ASTM Publication STP-718, American Society for Testing and Materials, Philadelphia, 1980.

4.5 LUBRICATION

R. S. Fein

4.5-1 Introduction

Energy systems all contain components that are loaded against each other and are required to move relative to each other. Lubrication controls the frictional resistance to motion between the "bearing" surfaces and reduces wear and other surface damage resulting from the motion under load. Lubrication is accomplished by separating the components with a material termed a "lubricant." Lubricants commonly also perform secondary functions including scavenging of heat, dirt, and wear debris, transfer of force and energy, and sealing.[2,11-13,18,20]

Wear is uniform attrition of bearing surface material which gradually changes dimensions.[2,16,20] Surface damage other than gradual attrition of surface material (wear) consists of a wide variety of phenomena which substantially modify surface topography. Terms such as scoring, scuffing, smearing, galling, frosting, and spalling are used to describe the damage.

A key feature of surface damage is the rapidity of the bearing surface change—often gross changes occur within a few seconds. A second key feature is that the bearing damage is often sufficiently severe that it can prevent normal functioning of the mechanism containing the bearing—in some cases it can cause the mechanism to self-destruct.

Sudden significant increases in friction, wear, and surface damage that occur for comparatively small changes in operating conditions are sometimes termed "transitions." In the cases of severe surface disruption, the conditions at which the transition occurs are variously called "load-carrying capacity," "scoring" (or "scuffing" or "galling"), or "smearing" limits.[14,15,17] The term that is used depends on the nature of the phenomenon, the operating conditions under which the transition is obtained, and historical usage in a particular industry and country. These types of surface damage usually involve transfer of comparatively large particles from one surface to another and/or give plastic deformation of one or both bearing surfaces.

Bearings support or guide the components moving relative to each other. The bearing materials generally are metals or polymeric materials.[2,4,10,12,19] Often the bearing materials are self-lubricating in that they internally contain friction and wear reducing constituents or are made with surficial films containing such constituents.

Lubricants that are not part of the bearings commonly are "mineral oil" based hydrocarbon liquids ("oils") or such liquids thickened to semisolid or solid consistency ("greases"). Water-based lubricants also are often used for their fire resistance, cooling capability, and environmental compatibility. In addition, other natural and synthetic liquid lubricants, dry solids, air, and process fluids may be used. Base lubricants often have their properties modified by "additive" materials.

Table 4.5-1 provides perspective on the sizes involved in separation of bearing surfaces by lubricants. Wear, high friction, and the type of surface damage termed scuffing, scoring, galling, and so on generally occur only when the bearing surfaces are sufficiently close together that some of the load is carried by the surface high points ("asperities") or by contaminant particles in the fluid film separating the surfaces. Under these "boundary-lubrication" circumstances, it is inferred that a boundary film of boundary lubricant is necessary to limit the bearing-material stress and strain at the locations of interference between the surface asperities or the contaminant particles and the confining surfaces.

The following subsections deal principally with hydrodynamic and boundary lubrication of bearings by liquid lubricants and with properties of liquid mineral-oil based lubricants. Emphasis is on lubrication and lubricant characteristics commonly important to engineering of bearing-containing devices. Comprehensive coverage of most of the characteristics may be found in the recent *CRC Handbook of Lubrication*[11,12] and the *ASME Wear Control Handbook*.[2]

Current developments concerning lubrication appear in *Lubrication Engineering* (American Society of Lubrication Engineers), *ASLE Transactions*, *Transactions of ASME* [especially *Journal of Tribology* (formerly *Journal of Lubrication Technology*)], and *Wear* (Elsevier Sequoia).

4.5-2 Fluid Film Lubrication

Bearing surfaces can be completely separated by the pressure in the fluid between the surfaces.[2,5,12,17,20] The pressure can be applied externally. This is a "hydrostatic" bearing. The pressure can also be developed by the movement of the bearing surfaces. This is a "hydrodynamic" bearing.

TABLE 4.5-1   SIZE SCALES FOR BEARINGS

	Approximate size Range (μm)
Monomolecular layer	0.0002–0.002
Sliding wear debris	0.002–0.1
Boundary film thickness	0.002–3
Elastohydrodynamic film thickness	0.01–5
Asperity height	0.01–25
Rolling wear debris	0.7–10
Asperity load-support width	0.7–10
Classical hydrodynamic film thickness	2–100
Asperity tip radius	10–1000
Concentrated contact width	30–500
Engineered counterformal effective radius	1000–100000
Counterformal load-support width	1000–700000
Engineered conformal effective radius	2000000–2.5×10^9

Hydrodynamic pressure development results from movement of the bearing surfaces toward each other. This tends to expel lubricant from between the surfaces. The pressure develops from the flow resistance of the lubricant. This flow resistance is characterized by the "dynamic viscosity" (units of stress × time) of the fluid which is defined as

$$\text{dynamic viscosity} = \frac{\text{shear stress}}{\text{shear rate}}$$

in which shear rate is the velocity gradient with distance normal to the velocity direction (units of time^{-1}).

Bearing surfaces move toward each other and develop hydrodynamic pressure because of motion normal to the surfaces ("squeeze" action) or because of tangential motion of nonparallel surfaces toward a region of closest approach ("convergent wedge" action). Most hydrodynamic bearings operate on the convergent wedge principle. However, some, such as automotive engine crankshaft bearings and many bearings subjected to vibration, operate with large squeeze components of hydrodynamic action.

Classical hydrodynamic bearings such as sleeve or "journal" bearings, tapered shoe thrust bearings, and so on have surfaces that nominally conform to each other. Such "conformal" bearings have small angles between the surfaces and develop load-supporting pressure over much of the apparent bearing area.

Figure 4.5-1 shows typical dependence of the minimum surface separation (minimum film thickness = h_m) and friction coefficient f on the dimensionless ratio of the viscosity and speed product to bearing load. The dimensionless ratio, when multiplied by the geometrical factor $(R/C)^2$ for a journal bearing is variously called the "bearing characteristic number," "Sommerfeld number," 1/"duty parameter," and so on.

Minimum film thickness and friction for common types of classical hydrodynamic bearings can be readily calculated from theoretically based curves available in handbooks.[2,12] Curves are also available for lubricant flow, temperature, stiffness, and damping characteristics of such bearings. These curves permit design and performance calculations for classical hydrodynamic bearings that have adequate accuracy for many purposes. Numerical methods are available for more exact solutions.[12]

"Elastohydrodynamic" (or EHD or EHL bearings such as gear teeth, ball and roller bearings, and cams and camfollowers) have nominally nonconforming "counterformal" surfaces which appear to concentrate the load on an infinitely small "line" or "point" area (concentrated contact). Pressures between the surfaces of such bearings usually cause elastic deformation.[2,5,12]

Counterformal bearing surfaces of high elastic moduli materials such as metals lubricated with an organic liquid lubricant develop sufficient pressure between the surfaces to increase lubricant viscosity many-fold in the region of closest surface approach. This EHD lubrication is variously termed high modulus, piezoviscous-elastic, or elastic-viscous.

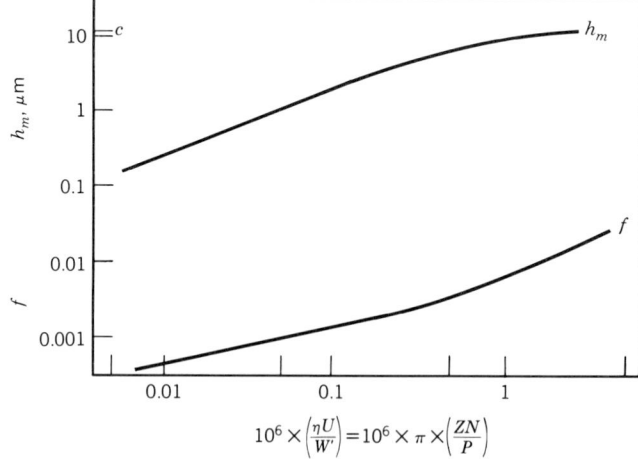

Fig. 4.5-1 Example of calculated classical hydrodynamic film thickness h_m and friction coefficient f. 25 mm full journal bearing, $R/C = 1000$, $L/D = 1$. Radial clearance = C, diameter = D, radius = R, $\eta = Z$ = dynamic viscosity, U = velocity of journal surface, N = rotational speed of shaft, W' = load/width, P = pressure.

For such high modulus EHD bearings, the ratio h_m/R of minimum film thickness to effective radius of curvature (R = product divided by sum of surface radii) may be estimated within about a factor of 2 (if viscosity is accurately known) with available theoretically based equations.[2,12] h_m/R increases strongly with the dimensionless ratio of the product of viscosity–pressure coefficient, dynamic viscosity, and mean surface velocity to effective radius. Figure 4.5-2 illustrates a typical EHD film thickness dependence on this viscosity–speed parameter. Effects of load and elastic modulus are comparatively weak.

Friction for high modulus elastohydrodynamic bearings depends strongly on the load and on the "sliding" or "slip" velocity (difference between the surface velocities)[12,17,18] Figure 4.5-3 schematically illustrates the typical dependence for friction or "traction" coefficient. Friction at "slide-to-roll ratio" (sliding velocity/mean velocity) over the order of 1% is generally best estimated from empirical equations for the particular component as installed in a particular device. In the absence of an

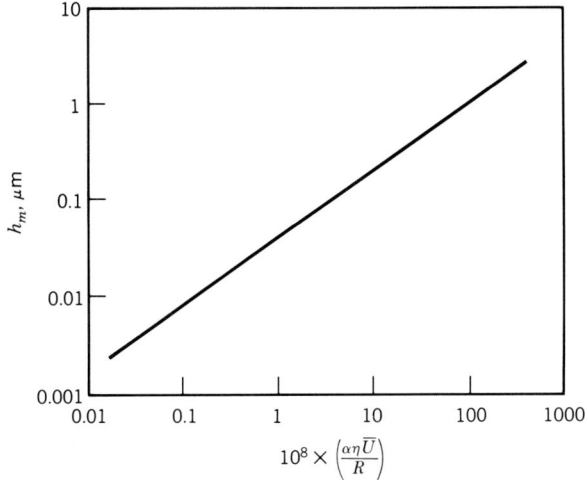

Fig. 4.5-2 Example of calculated EHD film thickness h_m. 82.5 mm center distance, 20° pressure angle spur gears with 15-tooth pinion and 16-tooth gear. Mineral oil lubricant. α = viscosity–pressure coefficient, η = dynamic viscosity, $\overline{U}$ = mean surface velocity, and R = effective radius.

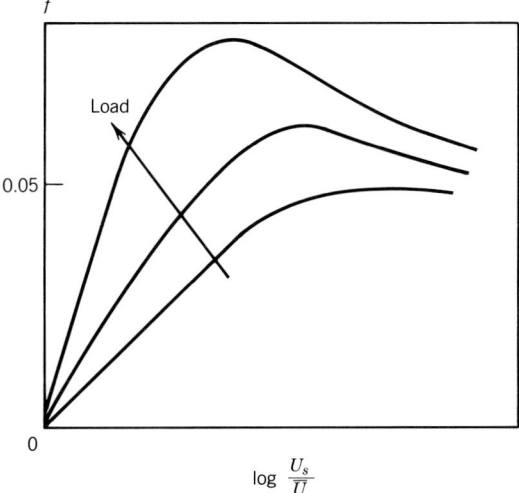

Fig. 4.5-3 Schematic piezoviscous–elastic EHD friction coefficient variation with slide-to-roll ratio and load. U_s = difference between surface velocities and $\overline{U}$ = mean surface velocity.

empirical equation, the friction coefficient at high slide-to-roll ratio can be roughly approximated as 0.06.

Low elastic modulus polymeric (including elastomeric) bearings and some bearings falling between counterformal and conformal in effective radius develop sufficient pressure to produce elastic bearing deformation but not enough to affect lubricant viscosity. These are termed "low modulus," "iso-viscous-elastic," and so on. EHD bearings. h_m/R can also be estimated within about a factor of 2 using available equations.[2] h_m/R increases strongly with the dimensionless "speed parameter" that is the ratio of the product of dynamic viscosity and mean surface velocity to the product of effective elastic modulus and effective radius. h_m/R values decrease comparatively weakly with the "load parameter" which is the dimensionless ratio of the load to the product of effective elastic modulus and square of the effective radius.

Note that film thickness increases with the product of dynamic viscosity and surface velocity for all hydrodynamic bearings.

4.5-3   Boundary Lubrication

"Boundary lubrication" comes into effect when separation of bearing surfaces by fluid films is reduced to the same magnitude as the largest asperity height and/or the size of the largest contaminant particles in the fluid. The asperities on surfaces begin to interact appreciably when the ratio of the hydrodynamic film thickness to the "composite roughness" (film thickness parameter) falls below about 2 to 3. The composite roughness $s_c = (s_1^2 + s_2^2)^{1/2}$, in which s_1 and s_2 are average roughness of bearing surfaces 1 and 2, respectively.

Under these circumstances control of friction and wear and prevention of surface damage must be at least partially exercised by the boundary lubricant. The role of bulk lubricant viscosity diminishes and friction, wear, and surface damage often become highly sensitive to combinations of chemical and physical conditions. The sensitivity depends on the compositions of the bearing materials, lubricant material, and atmosphere and also on configuration and operating conditions for the bearing system and on the history of operation of the system.[7,8,12,14,15-18]

The sensitivity is illustrated in Table 4.5-2 for sliding under conditions of negligible calculated hydrodynamic film thickness. The table shows variations in friction and wear coefficients with combinations of bearing, lubricant, and atmosphere materials.

Wear coefficient is a dimensionless number defined as the volume of material worn from a surface times the indentation hardness of the surface divided by the product of the load and the sliding distance.[2] It is also equal to the depth of wear times the hardness divided by the product of pressure and the sliding distance. (Reference 21 and some other sources define wear coefficient 3 times higher than the above definition from ref. 2.)

Formulas expressing relationships for wear volume and depth for various bearing situations are given in ref. 2. Reference 2 also contains tables of wear coefficients corresponding to results of common bench tests for the wear properties of lubricants. Both the formulas and tables are useful for bearing design or trouble-shooting purposes since wear coefficient tends to be independent of the bearing system for a given lubricant–metal–atmosphere combination.

Table 4.5-2 shows over a 100,000,000-fold range of wear coefficients under conditions of full load support by boundary lubrication but only about a 100-fold range of friction coefficient. Experience

TABLE 4.5-2   TYPICAL FRICTION AND WEAR COEFFICIENTS

Bearing	Lubricant/ Atmosphere	Friction Coefficient	Wear Coefficient
Mild steel/ mild steel	Air	0.6	7×10^{-2}
52100/52100	Dry argon	0.5	10^{-2}
52100/52100	Dry air	0.4	10^{-3}
Brass/tool steel	Air	0.24	6×10^{-4}
52100/52100	Isooctane/air	0.3	10^{-5}
Polyyethylene/ tool steel	Air	0.5	10^{-7}
52100/52100	Squalane/air	0.12	10^{-7}
52100/52100	Aromatic oil/air	0.06	10^{-9}
52100/52100	Engine oil/air	—	$< 2 \times 10^{-10}$
Copper/copper	Fatty acid/cetane/air	0.007	—

indicates that most high speed machines require wear coefficients below about 10^{-7} to achieve a useful life. Wear protection by boundary lubricants alone does not appear possible for machines requiring wear coefficients below about 10^{-10}. Such machines require hydrodynamic or hydrostatic surface separation.

Under conditions of essentially all load support by surface roughness ($h_m/s_c \ll 1$), relatively constant levels of wear and friction coefficients often occur between abrupt load-carrying transitions. Absolute values of wear and friction coefficients are often similar when measured with different tests and in service if the materials and operating conditions are not too divergent (bearing configurations often can be greatly different). Qualitative correlation of load-carrying limits among tests and with service usually require much closer correspondence of materials, bearing configuration, and operating conditions.

Boundary lubricants can be typed as gases, "indifferent" or "nonreactive," "oiliness" or "lubricity," "reactive" or "extreme pressure" (EP), and "solid." Gases (all the gases in the air, but especially oxygen, water, and organic materials) are usually inadvertent lubricants applied in light duty sliding bearings.[7,8,12]

Solid boundary lubricants can consist of the bearing materials with their naturally occurring surface films.[10] Soft metal films (e.g., indium, lead, and silver) on the bearing surfaces are also solid lubricants. Other solid lubricants are easily sheared solids such as graphite, molybdenum disulfide, and polytetrafluoroethylene (e.g., Teflon) applied as films or dry powders to bearing surfaces or as components of bearing materials. Solid lubricants are used to control wear and surface damage when liquid or grease lubricants are inconvenient, not tolerable, or inadequate. Solid lubricants are sometimes used as additives in liquid and grease lubricants.

In most cases where little wear and small friction are required, liquid or grease (thickened liquid) lubricants are used. Indifferent base oils (commonly straight mineral oils) are used under conditions of full hydrodynamic support of loads or little to modest roughness interaction.

Oiliness additives (at concentrations ranging from ppm to several tens of percent) are used in the base oils to reduce friction under conditions to modest to large roughness interaction. Oiliness additives particularly reduce friction at low sliding speeds. They contribute to smooth, quiet sliding by reducing or eliminating any friction-induced oscillations. Oiliness additives also often reduce wear if loads and sliding velocities are not too high and the film thickness parameter h_m/s_c is below the order of 2 to 3.

Reactive "antiwear" additives reduce wear at all severities of roughness interactions. Reactive additives also eliminate or otherwise control surface damage under conditions of extremely severe roughness interactions (i.e., high load and/or sliding velocity). Additives used for this purpose are commonly termed extreme pressure or EP additives. Typically reactive additives contain one or more of the elements sulfur, phosphorus, or chlorine in concentrations from ppm to several percent.

Physical and Chemical Effects

Low concentrations of additives can have large effects on wear, friction, and surface damage. Similarly, base-oil sulfur (which occurs naturally in mineral oils) and dissolved atmospheric oxygen can have important effects. Chemically active materials in process fluids can also be important (e.g., sulfur in gasoline or diesel fuel or methanol in a motor fuel).

Water can be an extremely important chemical component from the wear and load-carrying standpoints. It also generally will decrease pitting-limited life of bearing surfaces in concentrated conjunctions. Water gets into lubricants by condensation and dissolution from the air and from exposure to liquid water (e.g., rainwater and road splash in motor vehicles, coolants from metal working, etc).

Additive Interactions

EP and oiliness additives interact to give both synergistic and antagonistic effects. Optimum mixture concentrations exist and these optimums are different for wear than they are for surface damage.

4.5-4   Lubricant Properties

The lubricant properties providing the primary lubricant functions of wear, friction, and surface damage control often are achieved more easily than the properties providing secondary lubricant functions. Consequently, in many applications, the secondary properties predominate in importance for the design of a lubricated mechanism and lubricant selection for a mechanism. Properties of lubricants and their importance to service performance are discussed in refs. 1, 2, 6, 11, 12, 17–20, and 23.

Liquid and grease lubricant properties can be classified as bulk-material properties and surface-film related properties. The surface-film related properties often change with lubricant use due to bulk lubricant oxidation, suspension or solubilization of contaminants, or chemical reaction products formed on the bearing surfaces.

Lubricant properties may also be classified as basic properties and operational properties. Basic properties are physical or chemical characteristics that are essentially independent of the method by which they are measured. Operational properties involve the interaction of more than one basic property with the configuration and operating conditions of the device used for measurement.

Table 4.5-3 lists major properties of lubricants. The table indicates which properties are bulk (B) or surface-film related (S), which are independent (I) of measurement method and which are dependent (D). Properties peculiar to greases are indicated with the letter G.

Greases and grease properties are covered in refs. 12 and 18. Greases are thickened oils used in situations where the cooling by the lubricant is not important, where a liquid lubricant is inconvenient, or where sealing by excess solid or semisolid lubricant is desirable. The grease thickener holds the oil in place, but the oil (with comparatively little contribution from the thickener) keeps the bearing surfaces separated. Thickeners largely determine the unique properties of greases.

Viscosity characteristics of liquid lubricants and several of the most important other lubricant properties are discussed in the following.

TABLE 4.5-3   LUBRICANT PROPERTIES

Property	Bulk (B)/ Surface (S)	Independent (I)/ Dependent (D)	Grease
Density	B	I	
Viscosity	B	I	
Pourability	B	D	
Flammability	B	D	
Filterability	S	D	
Compatibility	S	D	
Stability, thermal	S	D	
Stability, oxidative	S	D	
Stability, hydrolytic	S	D	
Stability, shear	B	D	
Stability, oil separation	B	D	G
Bulk modulus	B	I	
Heat capacity	B	I	
Solvency	B	D	
Friction	S	D	
Wear	S	D	
Load-carrying	S	D	
Pitting fatigue	S	D	
Corrosion	S	D	
Wettability	S	D	
Thermal conductivity	B	I	
Electrical conductivity	B	I	
Penetration	S	D	G
Volatility	B	D	
Paintability	S	D	
Miscibility	S	D	
Emulsibility	S	D	
Surface tension	S	D	
Slumpability	B	D	G
Dropping point	B	D	G
Shear stability	B	D	
Viscosity loss, temporary	B	D	
Viscosity loss, permanent	B	D	
Foaming	S	D	
Air release	S	D	

4.5-5   Viscosity and Temperature Effects

Viscosity predominates as the most important property of a liquid lubricant. It determines the fluid film thickness for a given bearing configuration and operating conditions. It also often correlates with wear, friction, and surface damage even under conditions of negligible fluid film thickness ($h_m/s_c \ll 1$). Performance of many of the secondary functions of a lubricant also often correlates with viscosity.

Viscosity is defined by the OECD as "that bulk property of a fluid, semi-fluid, or semi-solid substance which causes it to resist flow."[2] Shear stress per unit change of shear rate is variously called "dynamic viscosity" or "absolute viscosity." It is commonly represented by the lower-case Greek symbol η or μ.

The cgs unit of viscosity is the poise with $1P = 1$ dyne $\cdot$ s/cm^2, with the centipoise (100 cP = 1 P) being commonly used. The SI unit of viscosity is N $\cdot$ s/m^2 or Pa $\cdot$ s. One Pa $\cdot$ s is sometimes called the "Poiseuille." Note that 1 cP = 1 mPa $\cdot$ s.

The ratio of dynamic viscosity to density is the "kinematic visocity" and is commonly represented by the lower-case Greek letter ν. Kinematic viscosity occasionally (in some older publications) is called static viscosity. The cgs unit of kinematic viscosity is the stoke with 1 St = 1 cm^2/s. Kinematic viscosity customarily is expressed in centistokes (100 cSt = 1 St). The SI unit is m^2/s. Hence, 1 cSt = 1 mm^2/s. For conversion between kinematic viscosity in cSt and dynamic viscosity in cP, the fluid density should be in g/mL.

Kinematic viscosity is sometimes measured and expressed as Saybolt Universal seconds (SUS) or Saybolt Furol seconds (SFS). Conversions to cSt are given in ASTM Standard D-2161.[3] Roughly, cSt = SUS/4.65 for viscosity over 140 SUS (30 cSt) and cSt = SFS/0.475 for viscosity over 30 SFS (60 cSt); the approximations improve as viscosity increases.

Viscosity of a given lubricant can vary over orders of magnitude with temperature changes. Viscosity at a fixed temperature can similarly change orders of magnitude among available lubricants. Solids such as some common engineering polymeric materials represent extremely high viscosity while air and gasoline represent extremely low viscosity. For hydrocarbon-based lubricants, the range of practical importance at atmospheric pressure is from about 1 to 10^8 mm^2/s or mPa $\cdot$ s.

The temperature variation of viscosity is commonly represented by the "Standard Viscosity–Temperature Charts for Liquid Petroleum Products" which is ASTM Standard D 341-77.[3] These charts, which can be purchased from ASTM (1916 Race Street, Philadelphia, PA 19103), plot kinematic viscosity versus temperature as a straight line over a considerable range of temperature. Thus, if viscosity is known at two temperatures, the charts enable determination of viscosity at other temperatures by interpolation or extrapolation. The charts can also be used for calculating the viscosity of blends of lubricants.[3]

For calculation purposes, it is often convenient to use the equations underlying the ASTM viscosity–temperature charts. In the range of viscosity from 2×10^7 to 2 cSt, the basic equation is

$$\log\log(\nu + 0.7) = A - B \log T$$

in which ν = kinematic viscosity in cSt (mm^2/s), T = absolute temperature in K, and A and B are constants. In much of the older U.S. literature, kinematic viscosity is reported at 100 and 210°F. Currently, the customary reporting temperatures are 40 and 100°C. The constant 0.7 changes with viscosity between 2 cSt and 0.21 cSt.

Viscosity temperature coefficients for typical mineral oils range from about 1%/°C (0.5%/°F) at a kinematic viscosity of 1 cSt to about 5–7%/°C (3–4%/°F) at a kinematic viscosity of 10^6 cSt. Since viscosity is such a sensitive function of temperature (and viscosity level), *hydrodynamic film thickness calculations require accurate estimation of the effective bearing temperature.*

The viscosity index or VI of a lubricant is an arbitrary measure of its viscosity–temperature sensitivity. VI dates back to the 1920s when lubricating oils were separated from crude oils by boiling point only (i.e., by distillation only). At that time oils from Pennsylvania crude showed small change in viscosity with temperature compared to oils from Gulf (of Mexico) coast crudes. VI was an empirical scheme which assigned a value of 100 to "paraffinic" Pennsylvania crude based lubricating oils and 0 to oils based on "naphthenic" Gulf coast crude.

Reference oils were selected from tables which had the same viscosity at the high reference temperature (210°F) as the unknown oil. The kinematic viscosities at the low reference temperature (100°F) of the 0 VI reference oil L, 100 VI reference oil H, and unknown oil U were then used to calculate VI. Thus, VI was defined as

$$VI = \left(\frac{L - U}{L - H} \right) \times 100$$

This scheme persists today and has been extended to VIs over 100. The scheme is embodied in ANSI/ASTM D-2270-77 which is also designated as IP 226/68 and U.S., British, and West German government standards.[3] D-2270-77 gives equations as well as tables that can be used to calculate VI from kinematic viscosities at 40 and 1000°C. Today, lubricating oils can be made from various crude oils to have approximately the same VIs. Also VIs can be increased by adding polymeric thickeners ("VI improvers") to lubricating oils.

At low temperatures, viscosity commonly increases more rapidly with decreasing temperature than predicted by extrapolation using the ASTM viscosity–temperature equation. In addition, the viscosity becomes time and shear dependent. Ultimately, the lubricant becomes a plastic solid which does not pour.

"Pour point" is measured by the ASTM D-97 procedure. It generally increases with viscosity grade and typically ranges from about −55°C (−65°F) to −10°C (15°F) for common commercial lubricants. Lower pour points in this range are often obtained by addition of polymeric additives termed "pour depressants."

Adequate lubrication requires the ability of a lubricant to flow by viscous drag or pumping to surfaces needing lubrication. These flow characteristics are not adequately measured by pour point. Hence, the low temperature flow characteristics of lubricants are measured by other viscous flow and performance type tests. Commonly used are the Cold Cranking Simulator (ASTM 2602), Brookfield Viscometer (ASTM D-2983), and Borderline Pumping Temperature of Engine Oil (ASTM D-3829).

Viscosity loss of thickened oils such as multigrade oils (e.g., SAE 10W-40) and greases occurs with increasing shear rate.[18,23] Some of the loss is temporary with viscosity increasing again when shear rate decreases. However, some of the loss may be permanent for oils. ASTM D-2603 and D-3945 are two laboratory methods of estimating permanent viscosity loss to be expected in service.

For hydrodynamic film thickness and other lubricant flow considerations, the viscosity of importance is that at the controlling shear rate in the system. Thus, consideration of shear rates is important for lubrication with a thickened oil. Greases and most multiviscosity graded oils (e.g., SAE 10W-40) have viscosities that depend on the rate of shear (the change in velocity with distance perpendicular to the velocity direction). For most elastohydrodynamic lubrication film thickness calculations, the base oil viscosity of thickened oils is a reasonable approximation. For classical hydrodynamic and hydrostatic bearings measurements on the particular oil at the shear rates of interest are needed.

4.5-6 Viscosity Pressure Effects

Viscosity increases with pressure. This is an appreciable effect only for pressures greater than a few million pascals (1 pascal = 1 N/m^2 = 1.45×10^{-4} psi). The Barus equation is a simple commonly used expression of the dependence of dynamic viscosity η at pressure P on the dynamic viscosity at atmospheric pressure η_0

$$\frac{\eta}{\eta_0} = \exp(\alpha P)$$

in which α is termed the "viscosity–pressure coefficient" or "pressure coefficient of viscosity."

The Roelands semiempirical equation more accurately represents viscosity versus pressure data on lubricants as

$$\log \eta + 1.200 = (\log \eta_0 + 1.200)\left(1 + \frac{P}{C}\right)^z$$

in which dynamic viscosity is in cP (mPa · s) and Z and C are constants characteristic of the fluid and of the gauge pressure units, respectively.[22] The constant C is 2000 kgf/cm^2, 1.961×10^8 Pa (0.1961 GPa), 1936 atm, 1961 bars, or 2.844×10^4 psi.

For mineral oils and hydrocarbons (including polymers) which do not differ too much from typical mineral oils in molecular structure, Z can be estimated from the dynamic viscosity at the conventional viscosity reference temperatures of 40 and 100°C. Letting

$$H_{t,0} = \log(\log \eta_{t,0} + 1.200)$$

in which the subscript indicates the temperature $t°C$ and atmospheric pressure (zero gauge pressure), then

$$Z = 7.8(H_{40,0} - H_{100,0})F_{t,0}$$

$F_{t,0} = (0.81 - 0.55H_{t,0})$ for mineral oils and related hydrocarbons, the order of 0.4 for polyglycols, and 0.95 for polymethylsiloxanes. Also, approximately, $Z = 0.10$ for water, 0.17 for glycerol, and 0.40 for polyglycols and fatty oils (castor oil, rapeseed oil, etc.) (derived from ref. 22).

Figure 4.5-4 shows effective viscosity–pressure coefficients for use in EHD calculations. The product of effective viscosity–pressure coefficient and viscosity has units of s^{-1}. It is the "lubricant parameter" or LP used in design calculations for rolling element bearings[2,12] and EHD calculations for such bearings, gears, and cams.[2]

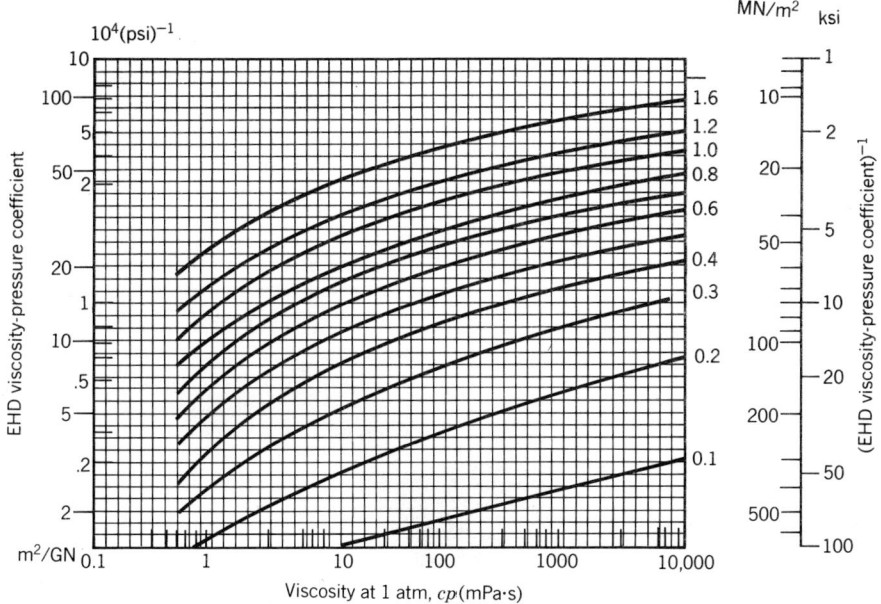

Fig. 4.5-4 EHD viscosity–pressure coefficient (α) dependence.

The atmospheric pressure viscosity and lubricant parameter used for EHD calculations should be at the temperature of the surfaces at the entrance to the load-carrying conjunction. The viscosity and lubricant parameter usually have appreciable uncertainty because of the difficulty of accurately measuring or calculating the appropriate temperature.

4.5-7 Other Properties

The density of lubricants required to convert between dynamic and kinematic viscosity depends only weakly on temperature and pressure.[6,12,22] Hence, for most hydrodynamic calculations with liquid lubricants, density variations with temperature and pressure can be neglected. Density is commonly expressed in units of g/mL, specific gravity (ratio of density to that of water), or API gravity [°API = 141.5/(sg 15.6/15.6°C) − 131.5]. Density for mineral oils can be approximated as about 875 kg/m³ (0.875 g/mL = 30.2°API) in the absence of data.

The most important stability property of lubricants is "oxidative stability." Oxidation of lubricants causes them to become acidic, corrode metals, increase in viscosity (thicken), and produce deposits such as sludges and varnishes.

Oxidative stability is sensitive to time at a given temperature, availability of oxygen, and to water, catalysts, and the degree of mixing and recycling of volatile oxidation products. Consequently, a variety of test methods is used to measure oxidation characteristics. These tests have been developed to correlate with the requirements of specific types of service. The tests are used to determine whether a lubricant qualifies for a specific service and whether a used lubricant is suitable for continued service.

Frequently used oxidation test procedures are ASTM D-943 (95°C), D-1313 (140°C), D-2272 (150°C), and D-2893 (95°C). Correlations of results by these and other test methods with each other and with service range from good to poor.

Oxidation stability becomes of concern when lubricants are exposed to oxygen at temperatures above about 65°C (150°F). At this temperature, oxidation appreciably degrades mineral oils without additives at exposure times of the order of 1000 h. Small quantities of "oxidation inhibitor" additives extend the times to years. With increased temperature to the order of 95°C (200°F), times for nonadditive mineral oils drop to several days. Times for oxidation inhibited oils may be of the order of a year or more at 95°C, depending on inhibitor additive type, other additives, and the nature of the operating conditions.

Additives used for other purposes also affect oxidation stability with extreme pressure additives commonly having a degrading effect.

Thermal stability is the resistance of a lubricant to breakdown in the absence of reactants such as oxygen or water. Gas formation, change in viscosity, and formation of gummy and carbonaceous deposits can occur when the lubricant molecules are broken.

Thermal breakdown of nonadditive mineral-oil lubricants becomes of concern with exposures to temperatures of the order of 410–435°C (770–815°F) occur for about an hour. It also becomes of concern with exposures to temperatures of the order of 330–350°C (626–662°F) for about a 1000 h. Additives in mineral oils generally do not increase thermal stability and often are themselves thermally unstable at substantially lower temperatures than the base oil.

Foaming and air entrainment are properties measuring the tendency of gases (particularly air) to form stable bubbles. Such bubble formation, if excessive, interferes with lubricant flow and affects heat transfer properties. It also may increase the volume of lubricant to the point where lubricant is expelled from the system. ASTM D-892 measures foaming properties.

Filterability of lubricants is often crucial to the prevention of excessive wear caused by abrasive contaminants (2) and to preventing sludge and other deposit-forming materials from preventing adequate lubricant flows or motions between parts.

Corrosion by lubricant oxidation products can damage bearings and accelerate (i.e., catalyze) the oxidation process. Corrosion characteristics of lubricants are often measured in performance type tests and in many of the standard oxidation bench tests. The acidity or acid neutralizing characteristics of lubricants (e.g., as measured by ASTM D-664) is often used as a crude measure of corrosion properties.

Aqueous corrosion of ferrous metals to give "rust" is also important because lubricants contain water and lubricant-wetted parts are exposed to humid air. ASTM D-665 and D-3603 measure rust-preventing characteristics of lubricating oils. Rust particles can damage bearings and prevent motion of close-clearance parts such as valves. Rusting can breach containment systems and weaken parts.

Corrosion inhibiting additives are very effective in preventing corrosion. The specific additive effects are sensitive to the service conditions and the presence of other additives.

TABLE 4.5-4   VISCOSITY GRADE EQUIVALENCE

ISO Viscosity Grade	AGMA Number	Approximately Equivalent SAE Class	
		Engine Oil	Gear Oil
2			
3			
5			
7			
10			
15			
22			
32		10W	
46	1		
68	2	10W-30, 20	
100	3	10W-40, 30	
150	4	20W-50, 40	80W-90
220	5	50	
320	6		
460	7		85W-140
680	8		
1000	8A		
1500			
ASTM D-2422[a]	AGMA 250.04	SAE J 300	SAE J 306
	R&O	Viscosity at 100°C	
	Compounded w/10% fat	determines number	
	EP	W suffix indicates additional low-temperature requirement	

[a] Grade determined by cSt @ 40°C

4.5-8 Lubricant Selection

Lubricant selection commonly involves a compromise among the properties required of a lubricant in a given system. Lubricant suppliers generally market lubricants representing various property compromises required by major applications of lubricants. Typically, large users of lubricants—governments and trade and technical organizations such as the American Gear Manufacturers' Association, American Petroleum Institute, American Society of Lubrication Engineers, and Society of Automotive Engineers—specify limits on properties. These specification limits determine the property compromises available in the marketplace.

Generally, lubricant selection should be from commercially available materials. Except in rare cases, these will be more economical and perform more reliably than specially prepared materials. Table 4.5-4 compares common viscosity grades of lubricants. References 11, 12, and 20 provide guidance on selection and utilization of lubricants for common classes of components and industrial applications. Additional guidance is available from the journal *Lubrication* (Texaco, Inc.) and from lubricant suppliers.

REFERENCES

4.5-1 American Society for Metals, *Technological Impact of Surfaces—Relationship to Forming, Welding, and Painting*, Proceedings of Conference in Dearborn, Mich., April 14–15, 1981 (1982).

4.5-2 ASME, *Wear Control Handbook*, M. B. Peterson and W. O. Winer (eds.), American Society of Mechanical Engineers, New York, 1980.

4.5-3 ASTM, *1983 Annual Book of ASTM Standards*, American Society for Testing Materials, 1916 Race St., Philadelphia, PA 19103.

4.5-4 G. M. Bartenev and V. V. Lavrentev, *Friction and Wear of Polymers*, Tribology Series 6, L. H. Lee and K. C. Ludema (eds.), Elsevier, Amsterdam, 1981.

4.5-5 F. T. Barwell, *Bearing Systems: Principles and Practice*, Oxford University Press, Oxford, 1979.

4.5-6 A. Bondi, *Physical Chemistry of Lubricating Oils*, Reinhold, New York, 1951.

4.5-7 F. P. Bowden and D. Tabor, *The Friction and Lubrication of Solids*, Part I, Clarendon Press, Oxford, 1954.

4.5-8 F. P. Bowden and D. Tabor, *The Friction and Lubrication of Solids*, Part II, Clarendon Press, Oxford, 1964.

4.5-9 E. R. Braithwaite, *Lubrication and Lubricants*, Elsevier, Amsterdam, 1967.

4.5-10 F. J. Clauss, *Solid Lubricants and Self-Lubricating Solids*, Academic Press, New York, 1972.

4.5-11 *CRC Handbook of Lubrication: Applications and Maintenance*, Volume I, E. R. Booser (ed.), CRC Press, Boca Raton, FL, 1983.

4.5-12 *CRC Handbook of Lubrication: Theory and Design*, Volume II, E. R. Booser (ed.), CRC Press, Boca Raton, FL, 1983.

4.5-13 H. Czichos, *Tribology: A Systems Approach to the Science and Technology of Friction, Lubrication and Wear*, Tribology Series 1, Elsevier, Amsterdam, 1978.

4.5-14 A. Dyson, Scuffing—a review, *Tribology International*, April, pp. 77–87 (1975).

4.5-15 A. Dyson, Scuffing—a review; Part 2, The Mechanism of Scuffing, *Tribology International*, June, pp. 117–122 (1975).

4.5-16 *Interdisciplinary Approach to Friction and Wear*, P. M. Ku (ed.), NASA SP-181, U.S. Government Printing Office, Washington, D.C., 1968.

4.5-17 *Interdisciplinary Approach to the Lubrication of Concentrated Contacts*, P. M. Ku (ed.), NASA SP-237, U.S. Government Printing Office, Washington, D.C., 1970.

4.5-18 *Interdisciplinary Approach to Liquid Lubricant Technology*, P. M. Ku (ed.), NASA SP-318, U.S. Government Printing Office, Washington, D.C., 1972.

4.5-19 D. F. Moore, *The Friction and Wear of Elastomers*, Pergamon, New York, 1972.

4.5-20 M. J. Neale, *Tribology Handbook*, Wiley, New York, 1973.

4.5-21 E. Rabinowicz, *Friction and Wear of Materials*, Wiley, New York, 1965.

4.5-22 C. J. A. Roelands, *Correlational Aspects of the Viscosity–Temperature–Pressure Relationship of Lubricating Oils*, Doctoral Thesis, Technische Hogeschool te Delft, Druk. V.R.B., Groningen, Netherlands, 1966.

4.5-23 SAE, *1983 SAE Handbook*, Society of Automotive Engineers.

4.6 INSTRUMENTATION AND CONTROLS

Chester L. Nachtigal

The fields of instrumentation and feedback control are interwoven in several ways:

1. Both are dynamic systems and therefore rely on such common analysis tools as characterization by transfer functions, frequency response, and time response.

2. Feedback control systems rely heavily on measurement transducers to carry out the task of information feedback.

3. Both fields are now heavily dependent on microprocessors.

4.6-1 Dynamic Systems Performance

Introduction

All engineering systems are dynamic as opposed to static systems, but we frequently restrict our attention to only some limited range of inputs and describe the system as dynamic or static for that particular range of inputs.[1] By *dynamic* we mean that the system's response is dependent upon the rate of change of the input. Suppose the dynamic system is a measurement transducer. If the input changes very rapidly, the transducer output will generally be different in relation to the input at any instant of time than if the input changes very slowly. Given that the response of a system is changing with time, it is not appropriate to label either the system or its response as dynamic. This term is reserved for the character of a system independent of its input and/or output.[2] As an illustration of the above, a mechanical spring is usually modeled as a static element if its deflection response is in phase with its input force. But in some range of frequencies the mass of the spring becomes significant and its output is not instantaneously proportional to the input. Therefore, the system must be labeled a dynamic system unless it is clearly understood that the range of input frequencies is well below that range where its mass becomes significant.

Instrumentation components and systems also exhibit dynamic behavior.[3,4] Frequently the primary sensing element takes some time, however small, to respond to its measurand. For example, a strain gauge pressure transducer has a diaphragm that may be modeled as a mechanical spring and a mass. This results in a lightly damped second-order response with a high natural frequency. Most temperature sensors such as thermocouples or thermisters have an associated mass that requires a small amount of time to equilibrate to a new temperature. This results in a first-order response with a time constant directly proportional to the thermal mass of the sensing element. Virtually all transducers convert the sensed signal to the electrical domain. Capacitive and inductive effects cause additional dynamic effects. With the advent of digital electronics and microcomputers in measurement systems, computation and/or conversion to the digital domain are additional reasons why the response is not instantaneously related to the input. In order to use instrumentation systems effectively, the reasons for their dynamic as opposed to static behavior must be understood and quantified. For example, if a control system has a frequency bandwidth of only 100 Hz, the fact that its feedback transducer exhibits dynamic behavior in the neighborhood of 1000 Hz does not rule out its use for the 100-Hz application. Two types of dynamic systems are of particular interest in this section:

1. Instrumentation systems, especially temperature, pressure, and flow.
2. Feedback control systems.

Transfer Function

The fundamental tool for characterizing a dynamic system is its transfer function. If this expression is known, its time domain response to standard inputs such as step functions, impulse functions, or sinusoids can be easily determined or predicted. Many design methodologies for improving system performance are based on knowledge of the system transfer function. The transfer function can be written by inspection if the differential equation relating input and output is known. Consider the following linear, constant coefficient differential equation, where $x(t)$ is the input and $y(t)$ is the response, as shown in Fig. 4.6-1:

$$a_n \frac{d^n y}{dt^n} + a_{n-1} \frac{d^{n-1} y}{dt^{n-1}} + \cdots + a_0 y = b_m \frac{d^m x}{dt^m} + \cdots + b_0 x \qquad (4.6\text{-}1)$$

This equation may be written in operator form by introducing the notation $D = d/dt$, $D^2 = d^2/dt^2, \ldots, D^n = d^n/dt^n$.

$$a_n D^n y + a_{n-1} D^{n-1} y + \cdots + a_0 D^0 y = b_m D^m x + \cdots + b_0 D^0 x \qquad (4.6\text{-}2)$$

Rewriting eq. (4.6-2),

$$\left(a_n D^n + \cdots + a_0 \right) y = \left(b_m D^m + \cdots + b_0 \right) x \qquad (4.6\text{-}3)$$

The transfer function in operator notation is shown in Fig. 4.6-2 and Eq. (4.6-4).

$$\frac{Y}{X}(D) = T(D) = \frac{b_m D^m + \cdots + b_0}{a_n D^n + \cdots + a_0} \qquad (4.6\text{-}4)$$

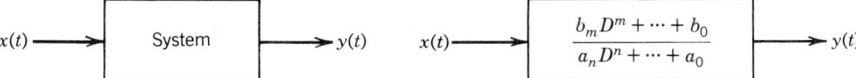

Fig. 4.6-1 Input–output block diagram.   **Fig. 4.6-2** Transfer function in operator notation.

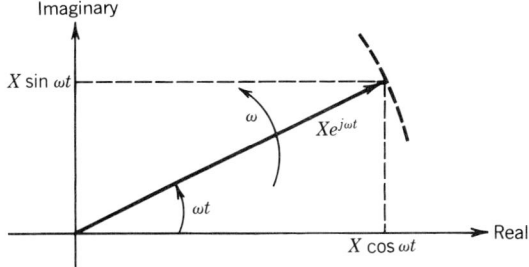

Fig. 4.6-3 Constant amplitude exponential forcing function.

A similar result is obtained when an input exponential forcing function of the form $x(t) = Xe^{j\omega t}$ is assumed. The rationale for using an exponential input is that it contains both the sine and cosine functions, as shown in Fig. 4.6-3. For the system differential equation of Eq. (4.6-1), let $x(t) = Xe^{j\omega t}$ and $y(t) = Ye^{j\omega t}$. Solving for the unknown Y,

$$Y = \frac{b_m(j\omega)^m + \cdots + b_0}{a_n(j\omega)^n + \cdots + a_0} X$$

$$= T(j\omega) \cdot X \tag{4.6-5}$$

Note that $T(j\omega)$ and $T(D)$ are of the same form. Using the assumed solution for $y(t)$ and the expression for Y, we have

$$y(t) = T(j\omega) \cdot Xe^{j\omega t}$$

$$= T(j\omega) \cdot x(t) \tag{4.6-6}$$

Figure 4.6-4 shows this result. Notice that the linear differential equation solution to an exponential forcing function may be written in multiplicative form. This occurs as a consequence of transforming to the frequency domain. This result will be analyzed in more detail in the section on frequency response.

One additional form of the transfer function is important and useful for time domain solutions to linear differential equations. Using the Laplace transform of time derivatives,

$$\mathcal{L}\left[\frac{d^n y}{dt^n}\right] = s^n Y(s) - s^{n-1} y(0) - s^{n-2} \frac{dy(0)}{dt} - \cdots - \frac{d^{n-1} y(0)}{dt^{n-1}} \tag{4.6-7}$$

Applying this transform to Eq. (4.6-1),

$$a_n\left[s^n Y(s) - \cdots - y^{(n-1)}(0)\right] + \cdots + a_0 Y(s) = b_m s^m X(s)$$

$$+ b_{m-1} s^{m-1} X(s) + \cdots + b_0 X(s) \tag{4.6-8}$$

$$Y(s) = \frac{b_m s^m + \cdots + b_0}{a_n s^n + \cdots + a_0} X(s) + \text{Initial conditions}\,(s) \tag{4.6-9}$$

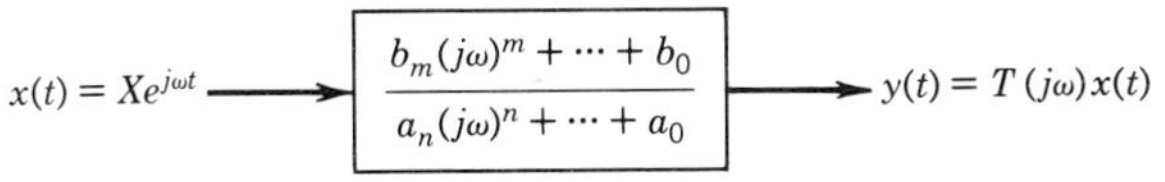

$$x(t) = Xe^{j\omega t} \longrightarrow \boxed{\frac{b_m(j\omega)^m + \cdots + b_0}{a_n(j\omega)^n + \cdots + a_0}} \longrightarrow y(t) = T(j\omega)x(t)$$

Fig. 4.6-4 Transfer function in the frequency domain.

The transfer function $T(s)$ is that portion of $Y(s)$ that relates the forcing function and the forced response.

$$T(s) = \frac{Y}{X}(s) = \frac{b_m s^m + \cdots + b_0}{a_n s^n + \cdots + a_0} \qquad (4.6\text{-}10)$$

Again the form of $T(s)$ is identical to that of $T(D)$ and $T(j\omega)$. Only the argument changes, depending on the type of solution. This last result will be examined in the next section.

Time Domain Response

The roots of the denominator polynomial of $T(s)$ are indicators of the form of the time domain transient response while the form of the steady-state response is the same as that of the forcing function. Thus, the denominator root locations in the s plane determine the rapidity with which the system completes its transition from one steady state to another steady state. Suppose the characteristic roots are known or are calculated, allowing us to write

$$T(s) = \frac{b_m s^m + \cdots + b_0}{(s + r_1)(s + r_2) \cdots (s + r_n)} \qquad (4.6\text{-}11)$$

If any roots are complex, they appear as complex conjugate pairs that are conveniently left as a second-order term.

$$T(s) = \frac{b_m s^m + \cdots + b_0}{\left(s^2 + 2\zeta\omega_n s + \omega_n^2\right) \cdots (s + r_n)} \qquad (4.6\text{-}12)$$

The roots of a second-order term are

$$s_{1,2} = -\zeta\omega_n \pm j\omega_n\sqrt{1 - \zeta^2}$$
$$= -\zeta\omega_n \pm j\omega_d \qquad (4.6\text{-}13)$$

where ζ is the damping ratio, ω_n is the natural frequency, and ω_d is the damped natural frequency. These roots may be placed on an s-plane root plot as shown in Fig. 4.6-5. The real root at $s = -r_n$ in Eq. (4.6-12) is also shown in Fig. 4.6-5. The angle β is related to the damping ratio by the expression

$$\cos \beta = \zeta \qquad (4.6\text{-}14)$$

Note that changing only the damping ratio to larger and larger values in Eqs. (4.6-12) and (4.6-13) moves the pole locations s_1 and s_2 along a circular path to the negative real axis. At $\zeta = 1$ the two poles merge at the point $s = -\omega_n$.

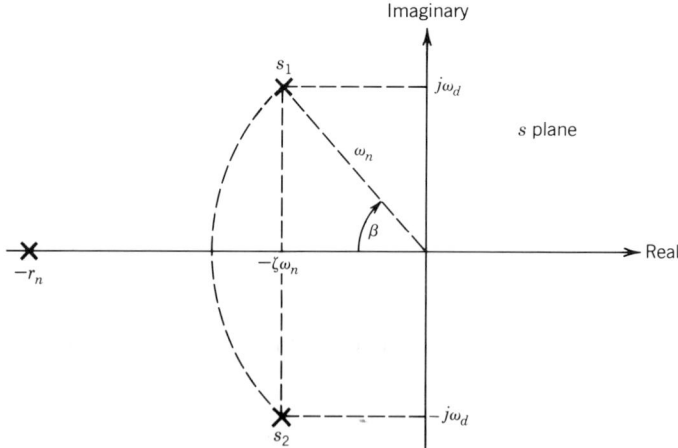

Fig. 4.6-5 System root locations in the s-plane.

The time response for some forcing function $x(t)$, $t > 0$, and some set of initial conditions $y_0, \dot{y}_0, \ddot{y}_0, \ldots, d^{(n-1)}y(0)/dt^{(n-1)}$ may be found by inverting the Laplace transform expression for $y(t)$.

Example 4.6-1. Let $a_1 \dot{y} + a_0 y = b_0 x(t)$; $x(t)$ is a step function of amplitude X, $x(t) = Xu_s(t)$.

$$a_1 s Y(s) - a_1 \dot{y}(0) + a_0 Y(s) = \frac{b_0 X}{s}$$

$$Y(s) = \frac{b_0}{a_1 s + a_0} \cdot \frac{X}{s} + \frac{a_1 y_0}{a_1 s + a_0}$$

The ratio a_1/a_0 is called the time constant τ. The system transfer function is

$$T(s) = \frac{b_0/a_0}{\tau s + 1}$$

For purpose of inverting a Laplace transform expression, it is necessary to set the coefficient of s in a first-order term to unity.

$$T(s) = \frac{b_0/a_0\tau}{s + 1/\tau}$$

$$Y(s) = \frac{Xb_0/a_0\tau}{s(s + 1/\tau)} + \frac{y_0}{s + 1/\tau}$$

Partial fraction expansion is used to reduce denominator product terms such as the first term of $Y(s)$ above to a sum of expressions involving only a single s term.

$$Y(s) = \frac{A_1}{s} + \frac{A_2}{s + 1/\tau} + \frac{y_0}{s + 1/\tau}$$

where $A_1 = Xb_0/a_0$; $A_2 = -Xb_0/a_0$. The final step is to simply look up the corresponding time function for each term in $Y(s)$.

$$y(t) = \frac{Xb_0}{a_0} u_s(t) - \frac{Xb_0}{a_0} e^{-t/\tau} y(t) = \frac{Xb_0}{a_0}(t) - \frac{Xb_0}{a_0} e^{-t/\tau} + y_0 e^{-t/\tau}; \qquad (t > 0)$$

Combine the first and second terms since they result from the forcing function $Xu_s(t)$.

$$y(t) = \underbrace{\frac{Xb_0}{a_0}[1 - e^{-t/\tau}]}_{\text{Forced response}} + \underbrace{y_0 e^{-t/\tau}}_{\text{Initial condition response}}$$

Example 4.6-2. $a_2 \ddot{y} + a_1 \dot{y} + a_0 y = b_0 x(t)$; $x(t)$ is a step function of amplitude X, $x(t) = Xu_s(t)$. Rewrite $\ddot{y} + 2\zeta\omega_n \dot{y} + \omega_n^2 y = (b_0/a_2)x(t)$, where $\omega_n^2 = a_0/a_2$, $2\zeta\omega_n = a_1/a_2$, and

$$\zeta = \frac{a_0}{2\sqrt{a_0 a_2}}$$

Rewrite the right-hand side:

$$\frac{b_0}{a_2} x(t) = \frac{b_0}{a_0} \omega_n^2 x(t)$$

$$\ddot{y} + 2\zeta\omega_n \dot{y} + \omega_n^2 y = \frac{b_0}{a_0} \omega_n^2 x(t)$$

Note that the steady-state solution may be written by inspection:

$$\omega_n^2 y_{ss}(t) = \frac{b_0}{a_0} \omega_n^2 x_{ss}(t)$$

$$y_{ss}(t) = \frac{b_0}{a_0} x_{ss}(t)$$

Notice that the ratio of the lowest-order coefficients is the static "gain" or static sensitivity of the system. Laplace transform the differential equation and the input:

$$s^2 Y(s) - s y_0 - \dot{y}_0 + 2\zeta \omega_n [sY(s) - y_0] + \omega_n^2 Y(s) = \frac{X}{s}$$

$$Y(s) = \frac{X b_0 \omega_n^2 / a_0}{s(s^2 + 2\zeta \omega_n s + \omega_n^2)} + \frac{y_0 s + \dot{y}_0 + 2\zeta \omega_n y_0}{s^2 + 2\zeta \omega_n s + \omega_n^2}$$

Rewrite the last term:

$$Y_{\text{I.C.}}(s) = \frac{y_0 (s + 2\zeta \omega_n)}{s^2 + 2\zeta \omega_n s + \omega_n^2} + \frac{\dot{y}_0}{s^2 + 2\zeta \omega_n s + \omega_n^2}$$

The three terms of $Y(s)$ are usually tabulated in Laplace transform tables:

$$y(t) = \frac{X b_0}{a_0}\left[1 - \frac{e^{-\zeta \omega_n t}}{\sqrt{1 - \zeta^2}} \sin(\omega_d t - \phi)\right] + \frac{y_0 e^{-\zeta \omega_n t}}{\sqrt{1 - \zeta^2}} \sin(\omega_d t + \phi) + \frac{\dot{y}_0 e^{-\zeta \omega_n t}}{\omega_d} \sin \omega_d t$$

where

$$\phi = \tan^{-1} \frac{\sqrt{1 - \zeta^2}}{\zeta} \qquad \omega_d = \omega_n \sqrt{1 - \zeta^2}$$

All three portions of the response have a transient response term which is a damped sinusoid that decays with a time constant $1/\zeta \omega_n$ and oscillates at the damped natural frequency ω_d. Step responses of a second-order system with various damping ratios are shown in Fig. 4.6-6. Figure 4.6-7 shows step responses for individual real-axis roots and complex conjugate root pairs as a function of s-plane location. When any of the roots are in the right half of the s plane, the system is said to be unstable. Any perturbation would cause the system response to grow without bound in theory. In practice, nonlinearities and/or fixed constraints would limit the output. Passive linear systems such as vibrational systems or other systems with no source of energy other than the input cannot exhibit unstable response. Feedback control systems, on the other hand, have energy sources, and consequently much design work is devoted to ensuring that these systems exhibit well-behaved stable responses.

Frequency Response

Equation (4.6-5) presents the ratio of the response amplitude to the input amplitude of a rotating phasor forcing function as the system transfer function:

$$T(j\omega) = \frac{Y}{X}(j\omega) = \frac{b_m (j\omega)^m + \cdots + b_0}{a_n (j\omega)^n + \cdots + a_0} \qquad (4.6\text{-}15)$$

The numerator and denominator polynomials of $T(j\omega)$ are themselves complex numbers whose real

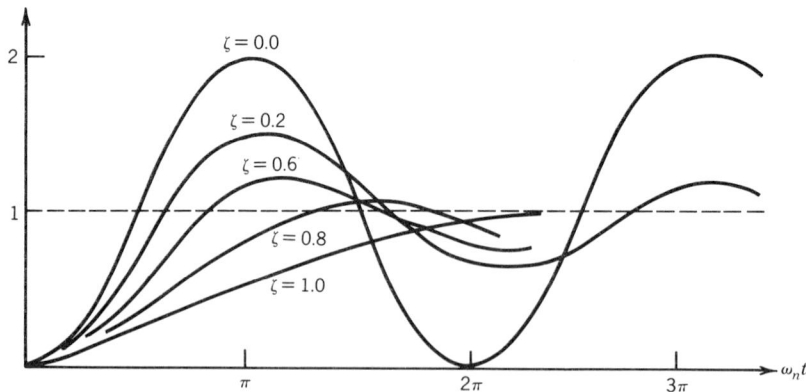

Fig. 4.6-6 Second-order system step response.

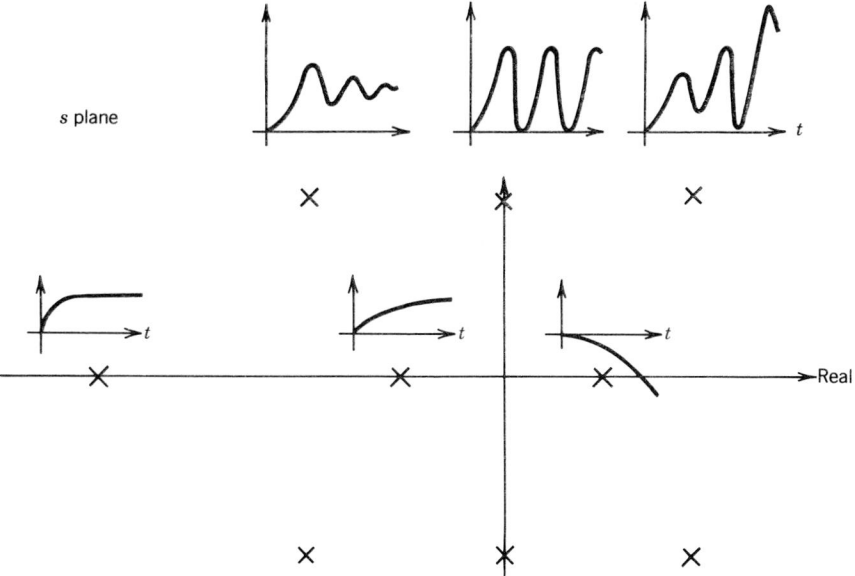

Fig. 4.6-7   Step responses for various pole locations in the s plane.

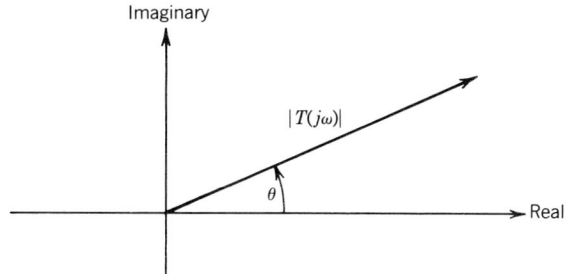

Fig. 4.6-8   Frequency domain transfer function shown as a magnitude and direction.

and imaginary components vary as the frequency is varied. The numerator and denominator complex numbers may be combined into a single complex number that may be expressed as a magnitude and an angular direction in the s plane, as shown in Fig. 4.6-8 and Eq. (4.6-16):

$$T(j\omega) = |T(j\omega)|e^{j\theta} \qquad (4.6\text{-}16)$$

The response of a system described by a linear differential equation and driven by an exponential forcing function may now be written by combining Eqs. (4.6-6) and (4.6-16):

$$y(t) = |T(j\omega)|e^{j\theta}Xe^{j\omega t}$$
$$= |T(j\omega)|Xe^{j(\omega t + \theta)} \qquad (4.6\text{-}17)$$

The exponential forcing function $Xe^{j\omega t}$ is merely a technique to derive the response to a sine or a cosine function. These responses are simply the imaginary and real components, respectively, of the response in Eq. (4.6-17).

For $x(t) = X\sin\omega t$,

$$y_I(t) = |T(j\omega)|X\sin(\omega t + \theta) \qquad (4.6\text{-}18)$$

For $x(t) = X \cos \omega t$,

$$y_R(t) = |T(j\omega)| X \cos(\omega t + \theta) \tag{4.6-19}$$

Note that in each case the response is simply the input with its amplitude magnified by the amplitude ratio of $T(j\omega)$ and its phase shifted by the phase angle associated with $T(j\omega)$. Therefore, the problem of sinusoidal (both sine and cosine) response reduces to the determination of $|T(j\omega)|$ and $\sphericalangle T(j\omega) = \theta$ versus ω. A graphical method of determining $|T(j\omega)|$ and θ is known as Bode plotting.

Bode plotting requires writing the transfer function in factored pole-zero form*:

$$T(s) = \frac{(s + z_1)(s + z_2) \cdots (s + z_m)}{(s + p_1)(s + p_2) \cdots (s + p_n)} \tag{4.6-20}$$

As was indicated by Eq. (4.6-12), denominator complex conjugate roots (system poles) may be left as second-order terms. Similarly, numerator roots (system zeros) may also appear as complex conjugate pairs left as second-order terms. Write eq. (4.6-20) in normalized form with $j\omega$ substituted for the s operator.

$$T(j\omega) = K \frac{(\tau_{z1} j\omega + 1)(\tau_{z2} j\omega + 1) \cdots (\tau_{zm} j\omega + 1)}{[1 - \omega^2/\omega_n^2 + j2\zeta(\omega/\omega_n)](\tau_{p3} j\omega + 1) \cdots (\tau_{pn} j\omega + 1)} \tag{4.6-21}$$

The magnitude and phase angle of the transfer function are

$$|T(j\omega)| = K \frac{|\tau_{z1} j\omega + 1||\tau_{z2} j\omega + 1| \cdots |\tau_{zm} j\omega + 1|}{|1 - \omega^2/\omega_n^2 + j2\zeta(\omega/\omega_n)||\tau_{p3} j\omega + 1| \cdots |\tau_{pn} j\omega + 1|} \tag{4.6-22}$$

$$\sphericalangle T(j\omega) = \sphericalangle(\tau_{z1} j\omega + 1) + \sphericalangle(\tau_{z2} j\omega + 1) + \cdots$$

$$- \sphericalangle \left(1 - \frac{\omega^2}{\omega_n^2} + j2\zeta \frac{\omega}{\omega_n}\right) - \sphericalangle(\tau_{p3} j\omega + 1) - \cdots \tag{4.6-23}$$

There are only four types of terms in the expression for $T(j\omega)$ of any linear system:

K	Constant gain or sensitivity (assume $K > 0$)
$\tau j\omega + 1$	Simple lead or lag (lead if in numerator, lag if in denominator)
$(j\omega)^n$	nth derivative or nth integral (if $(j\omega)^n$ is in the denominator)
$1 - \dfrac{\omega^2}{\omega_n^2} + j2\zeta \dfrac{\omega}{\omega_n}$	Second-order with $\zeta < 1.0$

(4.6-24)

If the phase shift and magnitude of these four terms can be constructed, the Bode plots of any arbitrary transfer function may also be constructed with very little additional effort. Further, if the resulting frequency response is unacceptable, design modifications may be made quickly. This last feature is especially useful in closed-loop control system design.

The multiplicative form of $|T(j\omega)|$ in Eq. (4.6-22) does not lend itself to combine two or more terms graphically. If a logarithmic transform is applied to Eq. (4.6-22), multiplication becomes addition and division becomes subtraction. Because logarithmic units are rather insensitive to small changes in $|T(j\omega)|$, a smaller unit, called the *decibel*, is employed. The resulting decibel, or dB, measure of the transfer function amplitude ratio is

$$20 \log_{10}|T(j\omega)| = 20 \log_{10} K + 20 \log_{10}|\tau_{z1} j\omega + 1| + \cdots + 20 \log_{10}|\tau_{zm} j\omega + 1|$$

$$- 20 \log_{10}\left|1 - \frac{\omega^2}{\omega^2} + j2\zeta \frac{\omega}{\omega_n}\right| - \cdots - 20 \log_{10}|\tau_{pn} j\omega + 1| \tag{4.6-25}$$

Now plot the decibel measure of each of the sample terms of Eq. (4.6-24) versus ω on a logarithmic coordinate. These results are shown in Figs. 4.6-9 to 4.6-12. Note the low- and high-frequency asymptotes in the first- and second-order lags of Figs. 4.6-10 and 4.6-12. Most of the frequency

---

*The modeling process normally results in a form that may readily be written in factored pole-zero form in the s domain.

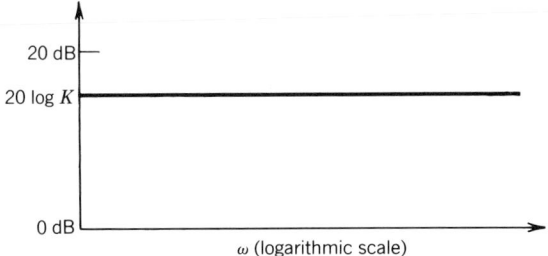

Fig. 4.6-9   Log magnitude plot of a constant gain.

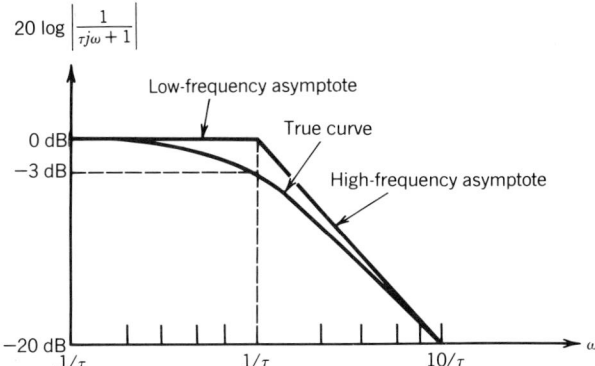

Fig. 4.6-10   Log magnitude plot of a simple lag.

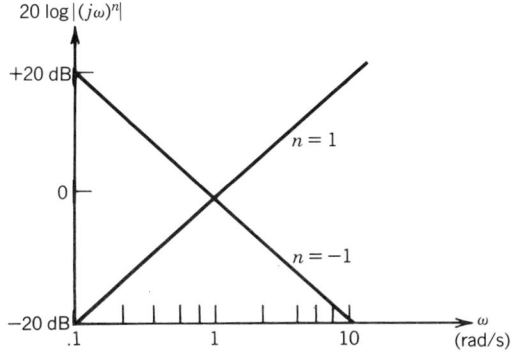

Fig. 4.6-11   Log magnitude plot of a differentiator and an integrator.

spectrum is adequately represented by the asymptotes themselves, except in the neighborhood of the break frequency, $1/\tau$, and ω_n, respectively. Thus, when combining various terms of a transfer function graphically, the low- and high-frequency asymptotes are combined first and then more detail is filled in in the neighborhood of the break frequencies.

Example 4.6-3.   Construct the magnitude Bode plot for

$$T(j\omega) = \frac{Kj\omega}{(\tau_1 j\omega + 1)(\tau_2 j\omega + 1)}$$

where $\tau_1 = 0.1$, $\tau_2 = 0.001$, and $K = 1$. If the gain is other than unity, the entire composite system curve is shifted vertically by $20 \log K$ units.

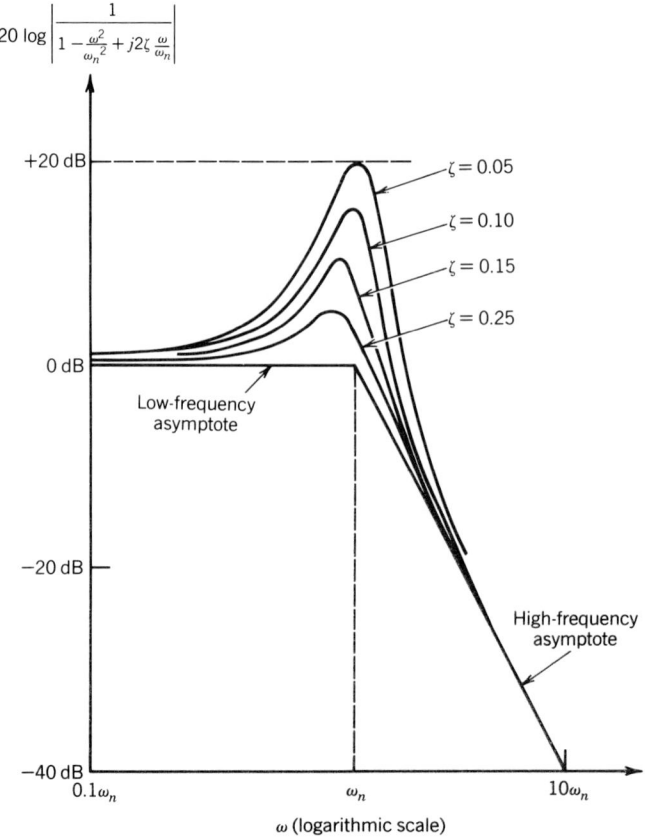

Fig. 4.6-12 Log magnitude plot of a second-order lag.

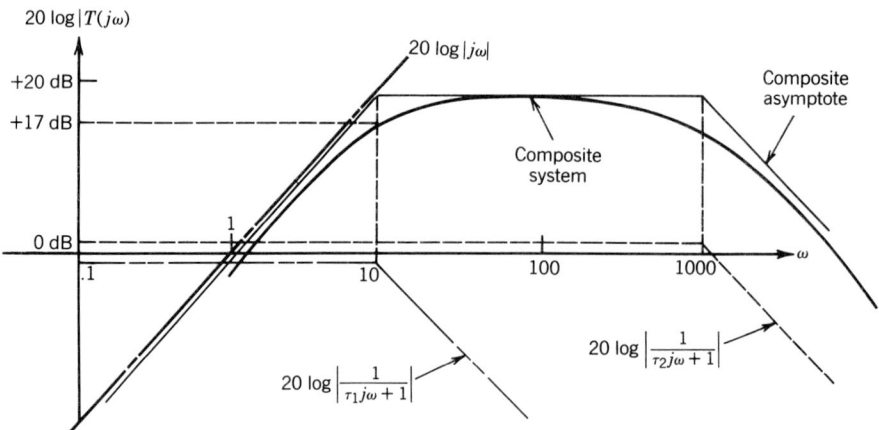

Fig. 4.6-13 Log magnitude plot of an overall transfer function.

444

If a first- or second-order term is in the numerator of a transfer function, the high-frequency asymptotes in Figs. 4.6-10 and 4.6-12 would slope up at the break frequency points. The entire Bode plot is simply reflected about the 0-dB horizontal line. Figure 4.6-13 shows the overall magnitude plot.

Graphical construction of the phase shift characteristics of the terms in Eq. (4.6-24) does not require logarithmic transformation since phase shifts of the individual dynamic terms of a transfer function are additive, as shown in Eq. (4.6-23). The expression for phase shift of a first-order lag is

$$\sphericalangle \frac{1}{\tau j\omega + 1} = -\tan^{-1}\frac{\text{imaginary}}{\text{real}}$$

$$= -\tan^{-1}\frac{\tau\omega}{1} \qquad (4.6\text{-}26)$$

The name *lag* stems from the fact that its phase shift is negative, or lagging, for all positive frequencies. If the term $\tau j\omega + 1$ is in the numerator, the term is called a simple lead. Its phase shift is identical to that given in Eq. (4.6-26) except that the sign is positive. Thus, the phase is positive, or leading, for all positive frequencies. These phase shifts are shown in Fig. 4.6-14. Note that the phase shift is exactly one-half of the overall phase shift at $\omega = 1/\tau$, the break frequency. The phase response of a second-order lag term is given by

$$\sphericalangle \frac{1}{1 - \omega^2/\omega_n^2 + j2\zeta(\omega/\omega_n)} = -\tan^{-1}\frac{\text{imaginary}}{\text{real}}$$

$$= -\tan^{-1}\frac{2\zeta\omega/\omega_n}{1 - \omega^2/\omega_n^2} \qquad (4.6\text{-}27)$$

Note that for frequencies below ω_n the phase shift is between $0°$ and $-90°$ (Fig. 4.6-15). Above ω_n the phase decreases from $-90°$ to $-180°$. If the damping ratio is zero, the phase angle changes abruptly from $0°$ to $-180°$ at $\omega = \omega_n$, as shown in Fig. 4.6-16. Again the phase shift is one-half the total at $\omega = \omega_n$, the break frequency. The procedure for constructing the phase response Bode plot of a composite transfer function is to construct the phase response of each term separately on linear versus $\log \omega$ coordinates. Then add the terms to obtain the composite phase response.

Example 4.6-4. Let

$$T(j\omega) = \frac{Kj\omega}{(\tau_1 j\omega + 1)(\tau_2 j\omega + 1)}$$

where $\tau_1 = 0.1$ and $\tau_2 = 0.001$.

The overall phase response is shown in Fig. 4.6-17.

It is common to construct the phase and magnitude Bode plots and gather all required information from these plots without actually writing the time response expression for a sinusoidal forcing

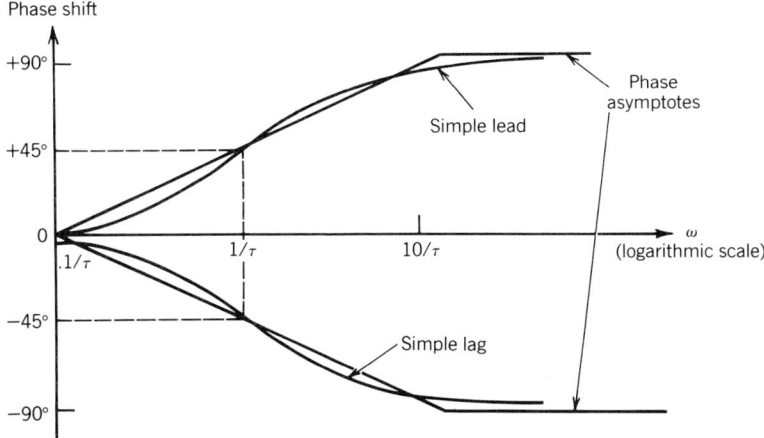

Fig. 4.6-14 Phase-shift plots of simple load and simple lag terms.

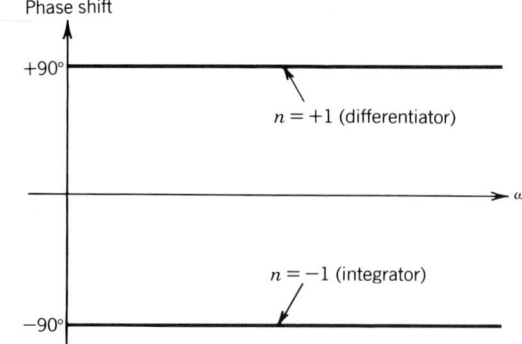

Fig. 4.6-15 Phase-shift plots of a differentiator and an integrator.

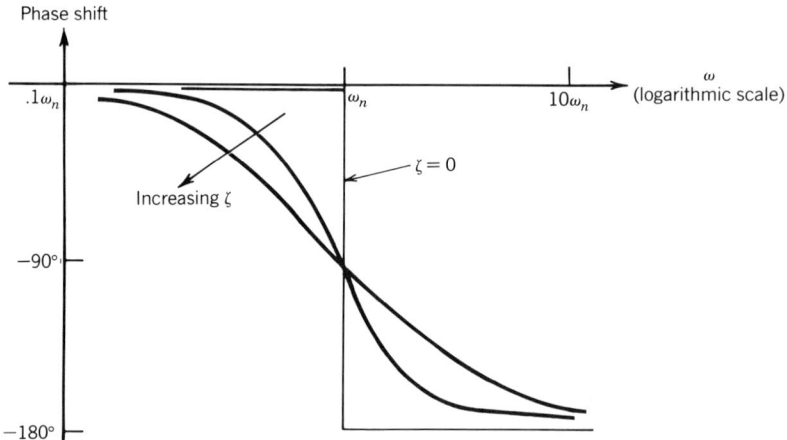

Fig. 4.6-16 Phase-shift plot of a second-order lag term.

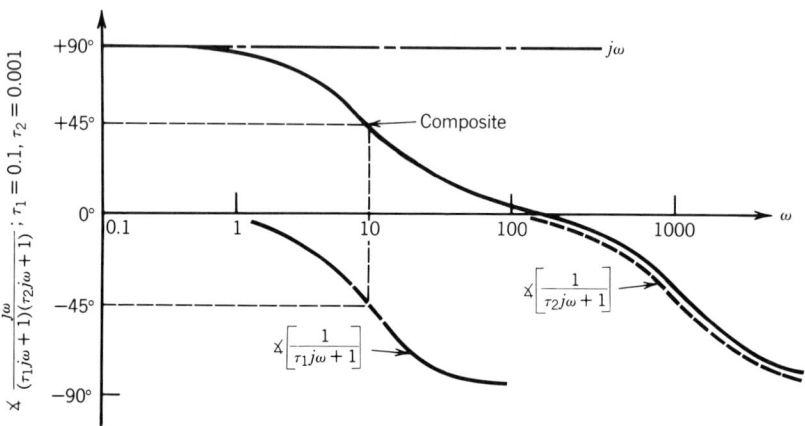

Fig. 4.6-17 Phase-shift plot of an overall transfer function.

function. Thus, the Bode plots are themselves the end-use tool for making design changes in the system. Put in another way, solution to the system differential equations is not necessarily required to complete the tasks of analysis and design.

4.6-2 Instrumentation for Energy Systems

Transducers for measuring three variables will be discussed in this section—temperature, pressure, and fluid flow. Only a relatively small number of transducers for each of these variables will be discussed. Additional transducers will be mentioned at the end of this section.

Temperature

It is difficult to define the variable we know as temperature. Suffice it to say that the temperature of a body is an indication of the level of molecular activity within that body. A more useful qualitative definition of temperature is that it represents the level of ability to transfer thermal energy to another body. The greater the temperature of the first body, the greater the rate of energy transfer to the second body.

Transducers may be categorized by their electrical principles of operation, such as variation of electric resistance, inductance, or capacitance as a function of the measured variable. Most temperature transducers are based on variable resistance. Included in this group are resistance thermometers (also known as resistance temperature detectors or simply RTDs) and thermistors. Still another transducer principle of operation is that of self-generation of a voltage or a current. Any self-generating transducer must obtain its output power from the input variable source. For example, a thermocouple, which is an example of a self-generating temperature transducer, converts some of the thermal energy in the surrounding medium to electric energy that flows out of the thermocouple circuitry.

The *thermocouple* is one of the most frequently used temperature transducers where conversion from the thermal to the electrical domain is required.[5] Its range of application is extremely wide, ranging from $+1650°C$ down to $-165°C$. In its simplest form a thermocouple is merely a pair of dissimilar wires joined at their two ends and a means of reading the generated voltage or current. This output signal is generated by a temperature difference across the two junctions where the dissimilar wires are joined together. Such a circuit junction is shown in Fig. 4.6-18. Junction p is called the measuring junction, and q is the reference junction. The latter is usually maintained at a specific constant temperature such as 0 or 20°C. A meter is shown here to symbolically represent some means of reading out the generated voltage or current.

The thermocouple dates back to 1821 when T. J. Seebeck discovered that current flows in a circuit when the measured and reference junction temperatures are not identical. Because the wires have internal resistance, some electric energy is converted to thermal energy. Since there is no electric power source in the simple thermocouple of Fig. 4.6-18, this power must be drawn from the measured temperature source or the reference temperature source, depending on which junction is the hotter of the two. The Peltier effect (1834) relates the absorption or evolution of heat at each junction of a thermocouple to the flow of current across the junction. It is strictly a local junction effect. The Thomson effect, predicted by Thomson (Lord Kelvin) in 1847, is a relation between the electromotive force (emf) generated in a single homogeneous wire and the temperature difference between the ends of the wire. The Thomson emf is proportional to both the temperature of the wire and the temperature difference in the wire. It is also a function of the type of material used. Again there is absorption or liberation of heat along the wire when a current flows through a homogeneous metal in which a temperature gradient exists. If the relative temperature across the two A-B junctions should change sign, the current direction would reverse itself and the sign of the heat flow relative to each homogeneous material and to each junction would switch. In most practical thermocouple measuring circuits, the Thomson emf is quite small and may be disregarded with the proper selection of materials. If both the measuring and reference junctions are at the same temperature, the voltage drop at each junction will be of the same magnitude and direction and therefore no current will flow. But if the temperatures are not the same, one voltage drop will be greater than the other, thus causing a

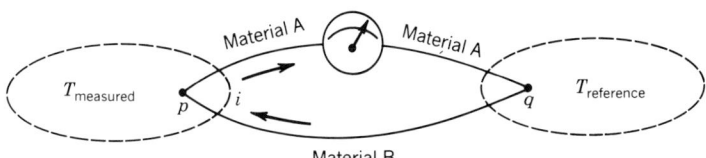

Fig. 4.6-18 Elementary thermocouple circuit.

current to flow. If the meter is a voltage-type device, it will measure the net potential around the circuit and will impede the flow of current.

Practical application of the thermoelectric effects is based on three fundamental thermocouple laws.

1. Law of homogeneous circuits. This law states that an electric current cannot be sustained in a circuit of a single homogeneous metal, however varying in section, by the application of heat alone. This law essentially states the requirement that a thermocouple circuit must have junctions between dissimilar metals.

2. Law of intermediate metals. This law states that the insertion of an intermediate metal into a thermocouple circuit such as the meter lead wires in Fig. 4.6-19 will not affect the net emf, provided that the two junctions introduced by the lead wire material C are at the same temperature. This law also permits the use of a third material such as silver solder to join the two original dissimilar metals, thereby forming a junction.

3. Law of intermediate temperatures. This law states that if a thermocouple circuit develops a net emf $e_{1\text{-}2}$ across the voltmeter junctions in Fig. 4.6-19 for measuring and reference junction temperatures of T_1 and T_2, respectively, and a net emf $e_{2\text{-}3}$ when its junction temperatures are T_2 and T_3, respectively, then it will develop a net emf of $e_{1\text{-}2} + e_{2\text{-}3}$ when its junctions are at temperatures T_1 and T_3.

The law of intermediate temperatures may be expressed by the following equations:

$$e_{1\text{-}2} = S_\theta (T_1 - T_2) \tag{4.6-28}$$

$$e_{2\text{-}3} = S_\theta (T_2 - T_3) \tag{4.6-29}$$

$$e_{1\text{-}2} + e_{2\text{-}3} = S_\theta [(T_1 - T_2) + (T_2 - T_3)]$$

$$= S_\theta (T_1 - T_3) = e_{1\text{-}3} \tag{4.6-30}$$

where S_θ is the thermocouple conversion coefficient or sensitivity. Figure 4.6-20 illustrates this law of

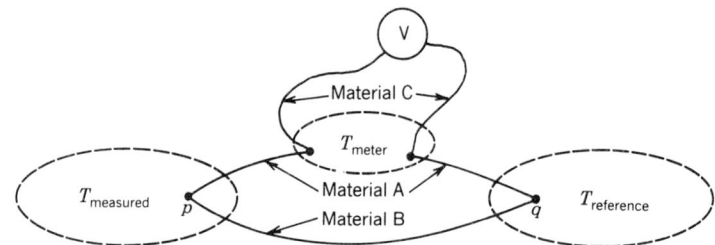

Fig. 4.6-19 Thermocouple circuit with additional junctions.

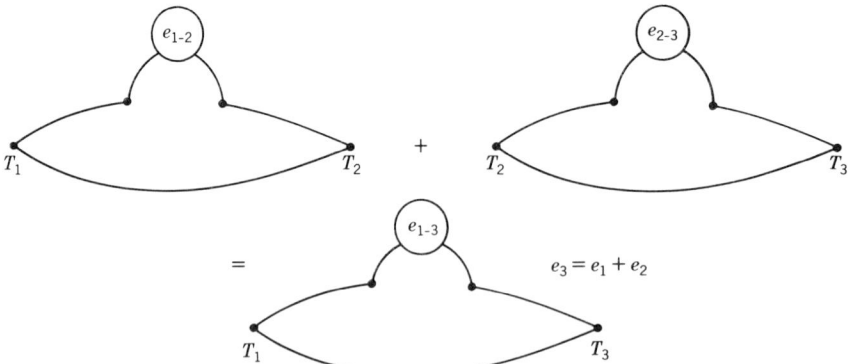

Fig. 4.6-20 Superposition of thermocouple input temperatures and responses.

intermediate temperatures still further in the form of superposition of temperatures and voltage responses.

Materials commonly used in the construction of thermocouples are as follows:

Copper	Rhodium	Platinum
Iron	Iridium	

Constantan: 60% copper, 40% nickel

Chromel: 10% chromium, 90% nickel

Alumel: 2% aluminum, 90% nickel, 8% silicon and manganese

Various combinations of these materials have practically become standards for certain temperature ranges, as shown in Table 4.6-1.

The value of the thermocouple sensitivity S_θ in Eqs. (4.6-28) to (4.6-30) is quite small. Typical emf versus temperature curves for the above frequently encountered thermocouples are shown in Fig. 4.6-21. Notice that the value of S_θ is approximately 0.085 mV/°C for the most sensitive thermocouple shown. Also note that the reference junction temperature is 0°C. This value for the reference temperature is almost universally the one on which thermocouple tables or curves of output versus measuring junction temperature are based. The reason is rather simple—it is relatively easy to maintain this temperature by using a mixture of ice and water. Let us assume that we are faced with a reference junction temperature other than 0°C but the only curves or tables available are referenced to the freezing point of water. The superposition principle illustrating the law of intermediate temperatures in Fig. 4.6-20 may be used for this situation. Given that an iron–constantan thermocouple with a reference junction temperature of 100°C produces a 16-mV response, what is the temperature T of the measuring junction? From Eq. (4.6-30) we have

$$e_{T-0°} = e_{T-100°} + e_{100°-0°} \tag{4.6-31}$$

From Fig. 4.6-21 we read $e_{100°-0°} = 6$ mV, and since $e_{T-100°}$ is 16 mV, $e_{T-0°} = 22$ mV. Again referring to Fig. 4.6-21, we note that $T = 400°C$.

Additional topics that the reader may wish to pursue are lead wire arrangements between the thermocouple and its signal-conditioning amplifier, design of thermocouples and thermocouple wells for fast dynamic response, reference junction compensation, and thermocouple signal conditioning.

A method of temperature measurement wherein the thermal sensor responds by a variation in its electric resistance is the use of *resistance temperature detectors* (RTDs). These devices are also referred to as resistance thermometers in the classic literature on temperature measurement. Several different materials have been used for RTDs, but four are most commonly used—platinum, nickel, Balco (a nickel–iron alloy), and tungsten. Of these metals, platinum is by far the most commonly used material in resistance thermometers. It is stable and exhibits a highly linear resistance characteristic versus

TABLE 4.6-1 FREQUENTLY ENCOUNTERED TYPES OF THERMOCOUPLES

Type		Temperature range
	Base Metal Thermocouples	
T	Copper–Constantan:	$-200°C$ to $+300°C$
J	Iron–Constantan:	$-200°C$ to $+1100°C$
K	Chromel–Alumel:	$-200°C$ to $+1200°C$
E	Chromel–Constantan:	$0°C$ to $+1100°C$
	Noble Metal Thermocouples	
R	Platinum, 13% Rhodium–Platinum:	$0°C$ to $+1450°C$
S	Platinum, 10% Rhodium–Platinum:	$0°C$ to $+1450°C$
B	Platinum, 30% Rhodium–Platinum, 6% Rhodium:	$900°C$ to $1700°C$

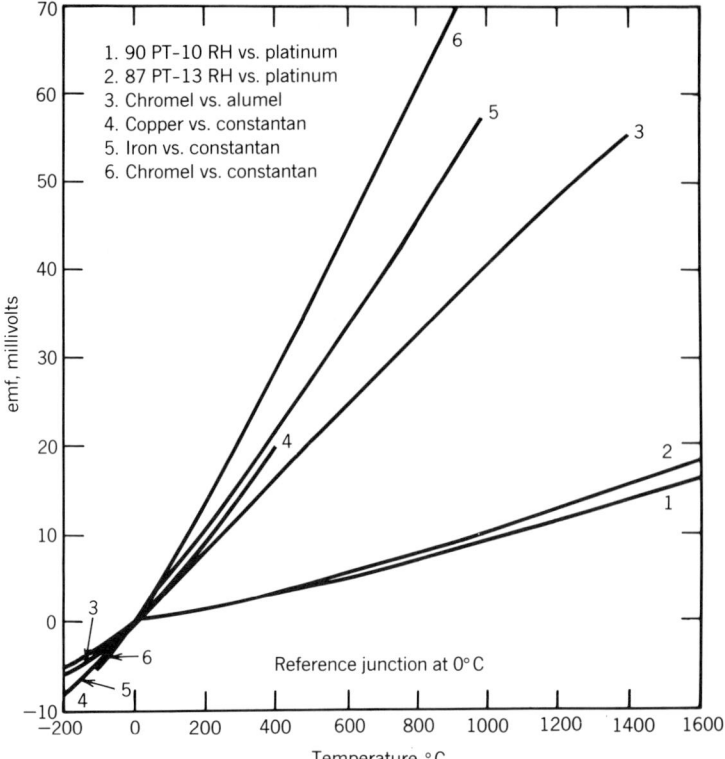

Fig. 4.6-21 Output voltage versus temperature for common types of thermocouples.

temperature, particularly in the range of 0–700°C. Figure 4.6-22 shows the relationship between resistance and temperature for the most common metals used in RTDs. Notice that each curve is presented in normalized form. To determine the actual resistance at any given temperature, simply multiply the resistance ratio at the temperature in question by the ice-point resistance of the particular metal. For platinum, the value of $R_{0°C}$ is 200 Ω.

In the temperature range from the ice point to the freezing point of antimony (575°C), the resistance versus temperature characteristic of platinum may be expressed as follows:

$$R = R_0(1 + AT + BT^2) \tag{4.6-32}$$

In the low-temperature range from the oxygen to the ice points (-177–0°C), this relationship may be further refined to read

$$R = R_0[1 + AT + BT^2 + C(T - 100)T^3] \tag{4.6-33}$$

where R_0 = resistance of platinum at the ice-point
       T = temperature,
    A, B = constants calculated from the resistance data obtained at the steam and sulfur points
       C = constant determined from data obtained at the oxygen point

A simplified relationship is often used for RTDs, which essentially is a linearized approximation to the full resistance–temperature characteristic.

$$R = R_0(1 + k\,\Delta T) \tag{4.6-34}$$

In this relation R_0 is the resistance of the sensor at a reference temperature such as 20°C. The k factor is known as the thermal coefficient of resistance, or simply the temperature sensitivity. As the purity of a metal increases, the thermal coefficient of resistance also increases. Except for its relatively

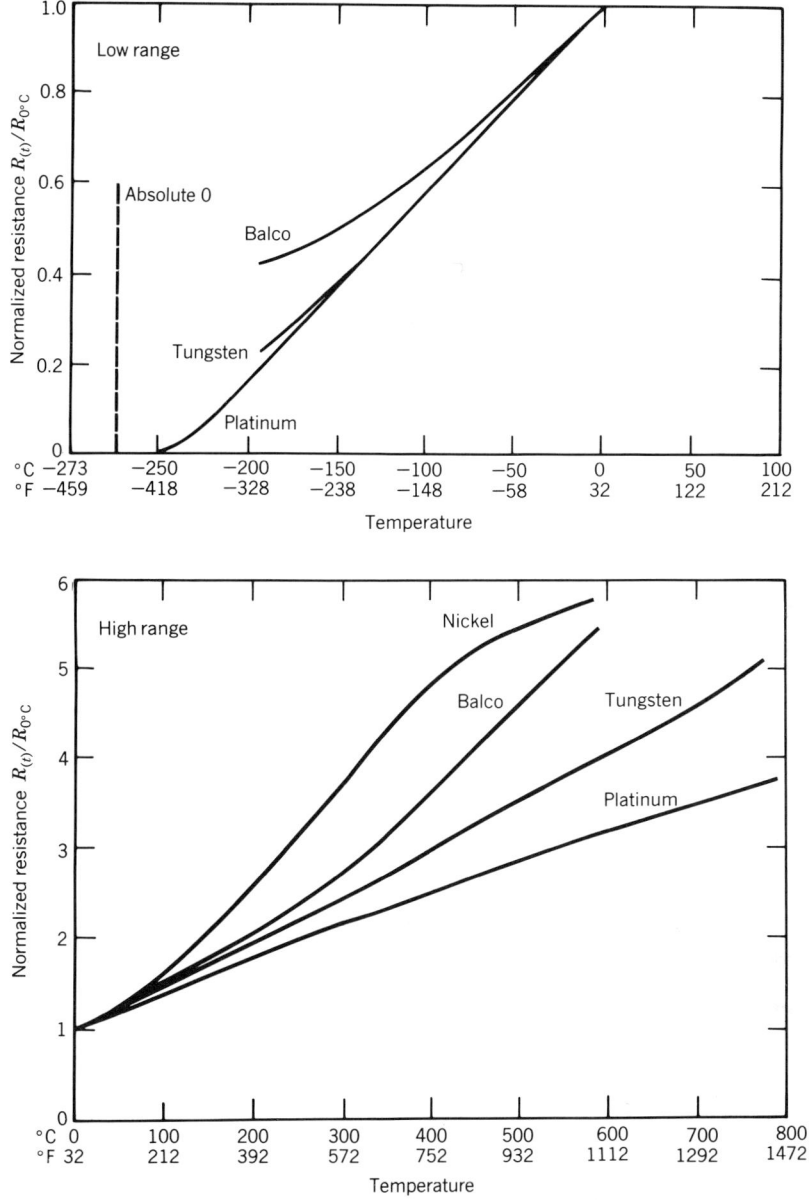

Fig. 4.6-22   Resistance versus temperature relationships of various temperature-sensing materials.

high cost, platinum is the most suitable of all metals for use in RTDs. Since it is available in a highly purified form, the values of the temperature coefficient are constant throughout its useful range. Therefore, platinum exhibits a highly linear resistance versus temperature characteristic.

Nickel is popular primarily because it is low in cost. The concave upward curvature of its characteristic is often utilized in bridge circuits that require temperature compensation and/or linearization. Balco, a 70% nickel and 30% iron alloy, is also quite inexpensive. In addition, it has a much larger sensitivity, in the 300–600°C range. This is a great advantage in electrically noisy environments because the signal is likely to be larger in amplitude, thus maintaining a better signal-to-noise ratio. Tungsten offers the most linear characteristic of the four metals used for RTDs. As in the case of platinum, tungsten has a relatively high melting point, making it attractive for high-temperature measurement.

Since it is not convenient to use varying resistance information in a temperature-measuring system, it is necessary to convert to a voltage or a current output. A bridge circuit is most frequently employed for this purpose. The simplest such configuration is shown in Fig. 4.6-23a. If the resistances of the two leads vary, the change will be interpreted as a change in R_s, the sensor resistance. The three-lead hookup shown in Fig. 4.6-23b compensates for most of the lead resistance change. Notice that the upper and lower lead resistances are effectively in separate but adjacent arms of the bridge and thus cancel each other. The center lead resistance is in series with the input impedance of the signal conditioner amplifier that receives e_0 as its input. If this input impedance is very large, any change in R_l will have a negligible effect on the output of the signal conditioner. Figure 4.6-23c illustrates a four-lead hookup that provides for full compensation of lead resistance changes so long as the sum of the lead resistances in the R_2 arm is equal to the sum of the lead resistances in the R_s arm. Since all four leads are in the same wire bundle and constructed from the same type of wire, it is reasonably certain that the lead resistances change by identical amounts.

The relationship between bridge output voltage and resistance change is given by the standard bridge equation:

$$e_0 = E_{ex} \frac{R_1 R_3 - R_2 R_4}{(R_1 + R_4)(R_2 + R_3)} \tag{4.6-35}$$

If we neglect the lead resistance in Fig. 4.6-23 and substitute $R_3 = R_s = R_0(1 + k\,\Delta T)$ from Eq. (4.6-34) and $R_1 = R_2 = R_4 = R_0$, Eq. (4.6-35) becomes

$$e_0 = E_{ex} \frac{kR_0\,\Delta T/R_0}{4 + 2kR_0\,\Delta T/R_0}$$

$$e_0 = E_{ex} \frac{k\,\Delta T}{4 + 2k\,\Delta T} \tag{4.6-36}$$

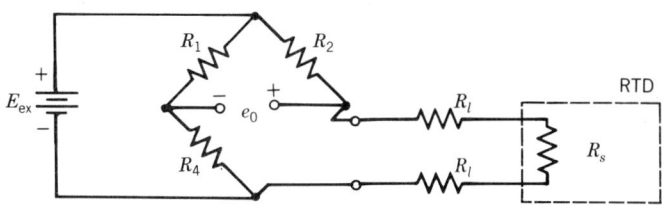

(a) Two-lead hook-up

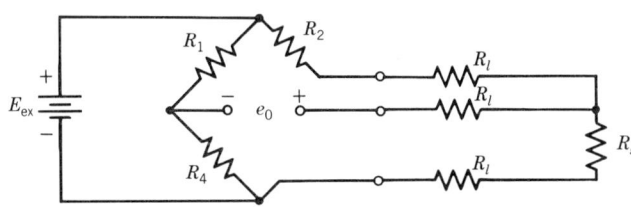

(b) Three-lead hook-up

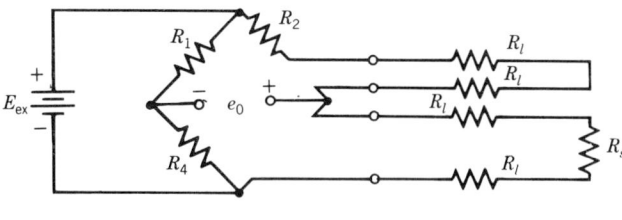

(c) Four-lead hook-up

Fig. 4.6-23 Various bridge circuits for resistance temperature detectors.

If $2k\,\Delta t$ is much smaller than 4, the relationship between output voltage and temperature change is reasonably linear. Let us examine the term $2k\,\Delta T$ for a platinum RTD and for a ΔT of 100°C. From Fig. 4.6-22b, $R_{100°C}/R_{0°C} \cong 1.4$. The output voltage is

$$e_0 = E_{ex}\frac{0.4}{4 + 0.8} \tag{4.6-37}$$

Since the term $2k\,\Delta T$ is nonnegligible compared with 4 in the denominator, the output voltage is nonlinear with temperature.

Clearly the linearity of the above relationship is sensitive to the magnitude of the temperature swing and the departure from linearity is appreciable for a 100°C temperature change. If the lead resistance $R_l + R_l$ is significant relative to the value of R_s, the overall nonlinearity effect will be reduced slightly but at the expense of overall bridge transducer sensitivity.

Additional topics include well design and signal conditioner design.

Thermistors are essentially semiconductors that behave as thermal resistors—that is, resistors with a high and usually negative temperature coefficient of resistance. As a brief comparison with resistance thermometers, thermistors are considered less stable versus time and have a narrower practical operating range ($-250°C$ to $+260°C$ versus $-250°C$ to $+800°C$ for resistance thermometers). On the other hand, a thermistor has a much higher temperature sensitivity. In some cases the resistance at room temperature may change by as much as 6% for each Celsius degree rise in ambient temperature. When compared with metallic conductors used for resistance thermometers, the thermal coefficient of resistance for a thermistor at room temperature is approximately 10 times greater.

The relationship between a thermistor's electric resistance versus the input temperature is of primary concern when using this transducer in a temperature measurement application. This relationship follows the form

$$R = R_0 e^{\beta(1/T - 1/T_0)} \tag{4.6-38}$$

where R = resistance at the absolute temperature T, in degrees Kelvin
 R_0 = resistance at the reference temperature T_0
 β = approximately constant over moderate temperature ranges and depends on composition and manufacturing process

It is useful to define a temperature coefficient of resistance, or a k factor, as in Eq. (4.6-34):

$$\frac{dR}{dT} = R_0 k \tag{4.6-39}$$

Differentiating Eq. (4.6-38) yields

$$\frac{dR}{dT} = -\frac{R_0\beta}{T^2}e^{\beta(1/T - 1/T_0)} \tag{4.6-40}$$

Therefore,

$$k = \frac{1}{R_0}\frac{dR}{dT} = -\frac{\beta}{T^2}e^{\beta(1/T - 1/T_0)} \tag{4.6-41}$$

At $T = T_0$, we have

$$k|_{T=T_0} = \frac{1}{R_0}\frac{dR}{dT}\bigg|_{T=T_0} = -\frac{\beta}{T_0^2} \tag{4.6-42}$$

Referring to Eq. (4.6-41), we see that k is the normalized change in resistance per unit change in temperature. This parameter expresses the percentage change in resistance in response to a temperature change of 1°C. Figure 4.6-24 shows plots of the resistance characteristics versus temperature for several negative temperature coefficient (NTC) materials, a positive temperature coefficient material, and copper. The k factor at a given temperature is simply the slope of the characteristic in question evaluated at the given temperature.

From the form of Eq. (4.6-41), we note that the temperature coefficient of resistance is a very strong function of the temperature. Ideally, we would like to have this temperature sensitivity be large but constant. To compensate for this drastic drop in k (see Fig. 4.6-24), what is needed are thermistors that have an increasing β with temperature to partially offset the plummeting temperature coefficient. Certain types of thermistors have been developed that in fact display an increase in β with

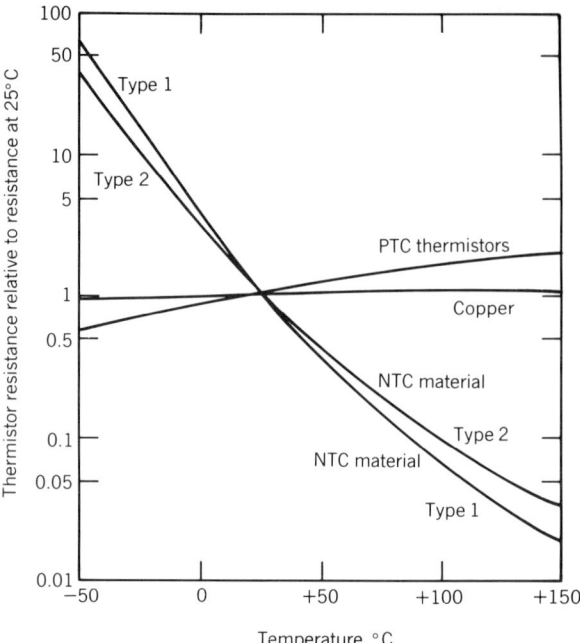

Fig. 4.6-24 Variation of thermistor resistance with temperature.

temperature. Some typical values for this parameter are 1000–1300 for measuring in the liquid oxygen–liquid hydrogen range. These values correspond to temperature coefficients between 13.0 and 17.5%/°C for liquid oxygen and 18–24%/°C for liquid nitrogen (−195.6°C). Thermistors for room temperature applications have been developed with β values from 1100 to 2800. For temperatures somewhat above ambient, thermistors with β values between 3400 and 4800 are available.

When a thermistor is used as a simple measurement transducer, it may be used much like a resistance thermometer—the active temperature-sensitive arm in a bridge circuit. Thus, the variable resistance is converted to a variable voltage output, as shown in Fig. 4.6-25. Figure 4.6-26 compares the linearity of the resistance ratio versus temperature with the linearity of the bridge output voltage ratio versus temperature. The reference temperature was taken to be 25°C with a β value of 4000 K and a reference temperature resistance value of 92,800 Ω.

The salient features of the three temperature transducers are compared in Table 4.6-2. Note that a thermistor has a relatively narrow temperature range in comparison with a thermocouple. If linearity and accuracy are of high priority, a resistance temperature detector is strongly advised.

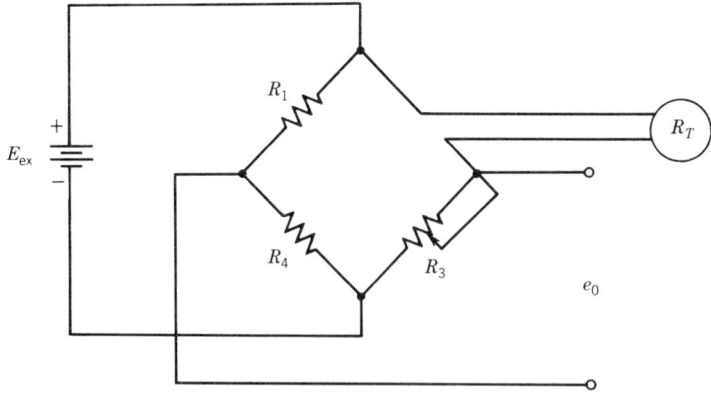

Fig. 4.6-25

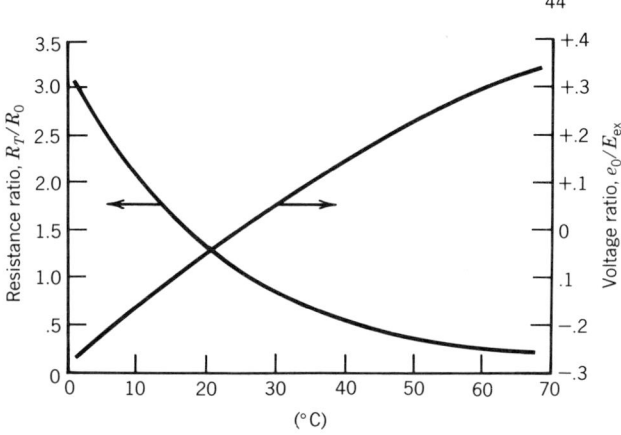

Fig. 4.6-26 Linearizing effect of a bridge network.

TABLE 4.6-2 COMPARISON OF THERMOCOUPLES, THERMISTORS, AND RESISTANCE TEMPERATURE DETECTORS

	Thermocouple	Thermistor	RTD
Temperature Range	$-200°C$ to $2500°C$ $(-320°F$ to $4500°F)$	Cryogenic to $300°C$ $(600°F)$	$-270°C$ to $900°C$ some to $1200°C$
Linearity	10% to 25%	10% to 25%	$0°C$ to $100°C$ $0°C$ to $400°C$
Signal Level	0.04 mV/°C and less	1 V to 2 V/°C with bridge	up to 360 mV/°C with proper bridge
Cold Junction	Required	N/A	N/A
Accuracy	0.1% to 5%	5% nominal	0.01% easy
Stability	Excellent	Poor	Excellent
Interchangeability	Good	Poor	Excellent
Sensing at a Point	Excellent	Excellent	Poor
Sensing Area	Can be very small	Very Small	Large

A frequently employed method used for measurement of high temperatures is by the use of noncontact *pyrometry*. Three different types of pyrometers may be identified:

1. Radiation pyrometer.
2. Photoelectric pyrometer.
3. Optical pyrometer.

In all cases the measurement principle involves use of radiation heat transfer. A *radiation pyrometer* focuses heat radiated from a source such as molten steel onto a small receiving disk. To the disk are attached the measuring junctions of a number of thermocouples. These thermocouples are connected in series in such a way that the outputs of the individual junctions all add. The resulting multiple-thermocouple circuit is called a thermopile. As radiative energy is received by the disk from the source, the receiving mass increases in temperature. Equilibrium is reached when the incoming radiative energy is just equal to the heat lost by convection. A *photoelectric pyrometer* is usually employed in the temperature range from approximately 800 to 2800°C. Its principle of operation is a photomultiplier tube that produces an electric current proportional to the amount of radiation received by it.

An *optical pyrometer* is not satisfactory for dynamic temperature measurement because of the need for a manual adjustment. A typical transducer of this type focuses radiation from the target surface onto a screen. The brightness of the image is then observed by a human operator. For comparison, a calibrated tungsten lamp provides a second source of radiation. The lamp filament radiation is superimposed on the radiation from the target source. The current though the standard lamp is then manually adjusted until the image of the filament just disappears against the background of the target surface. At that point the lamp current is a measure of the temperature of the external radiating surface. Temperatures of up to 4200°C are measurable with this type of pyrometer. Manual adjustment times are made in a matter of seconds.

Pressure

The general term for a pressure-measuring device capable of transduction to the electrical domain is *pressure transducer*.[6] All of the ones discussed in this section convert pressure to a proportional voltage and are able to follow a nonstationary pressure up to some upper value in frequency. This upper frequency may be as low as 5 Hz or as high as 100,000 Hz, depending on the nature of the transduction process. Pressure transducers are usually named by the manner in which the final conversion to a voltage is carried out; for example, a strain gauge pressure transducer uses strain gages to accomplish this task. A wide variety of such conversions are in use today, including:

Unbonded wire, bonded foil, and thin-film strain gauge.

Diffused and bonded bar semiconductor strain gauge.

Variable reluctance.

Piezoelectric.

Potentiometric.

Vibrating wire or tube.

Force balance.

Differential transformer.

Capacitative.

In each of the above cases a primary sensor interfaces with the pressurized medium and deflects in a proportionate manner. In some transducers the primary sensor is also the elastic member, which not only senses the pressure but also converts it. In others the primary sensor couples with an elastic secondary sensor to which is attached the conversion device.

The primary sensor is also called the pressure force summing device since it sums the incoming pressure force, the elastic force transmitted to the conversion device, and its own elastic force. A number of different primary sensors are incorporated in modern pressure transducers, largely dictated by the force displacement characteristics of the conversion element. For example, a linear variable differential transformer (LVDT) requires a greater input displacement per unit of response voltage than does a piezoelectric crystal. Figure 4.6-27 shows a number of commonly used pressure force summing devices. Pressure transducer manufacturers frequently offer a line of units for measuring differential pressure, absolute pressure, or gauge pressure. Usually only a slight change internal to the pressure transducer is required to go from one type of pressure measurement to another. For example, the P_2 chamber of the flat-diaphragm transducer illustrated in Fig. 4.6-28 would be evacuated and sealed if P_1 were to be an absolute pressure and P_2 open to the atmosphere for gauge pressure measurement at P_1 and connected to a second pressure source for a differential pressure measurement. An interesting spurious phenomenon occurs if the pressure transducer is installed in a vibratory medium—the diaphragm and associated elastic element unavoidably have mass, and thus an acceleration may cause a significant response unrelated to a pressure input.

Bonded strain gauge transducers have an especially rugged construction and withstand considerable abuse in handling and operation. In some instances overstressing due to accidental excess pressure will cause zero and sensitivity shifts while not destroying the transducer. For medium- and high-pressure units, the strain gauges are bonded directly onto the pressure-sensing diaphragm. On the other hand, low-pressure units may have a linkage to a bending beam to which the strain gauges are bonded.

Frequently, high frequency response pressure transducers may be ordered with their pressure input cavities filled with a silicon fluid to reduce the compliance between the pressure source and the sensing diaphragm. The compressibility of the intervening material between source and sensor is a major source of band-limiting frequency response.

Solid-state strain gauges produce a much larger sensitivity for a given strain. A gauge factor of 150 compared with 2 for that of a bonded strain gauge are typical values. Four strain gauges that form a four-active-arm bridge are diffused directly onto a silicon wafer that serves as the pressure-sensitive diaphragm. The process is very similar to that employed in the production of semiconductor electronics. Some transducers include integrated circuit signal conditioning to provide a 4–20 mA output signal, corresponding to zero to full-scale pressure input. Because of noisy environments and long signal transmission paths in many applications, it is preferable to transmit pressure information via a current signal rather than in the form of a voltage.

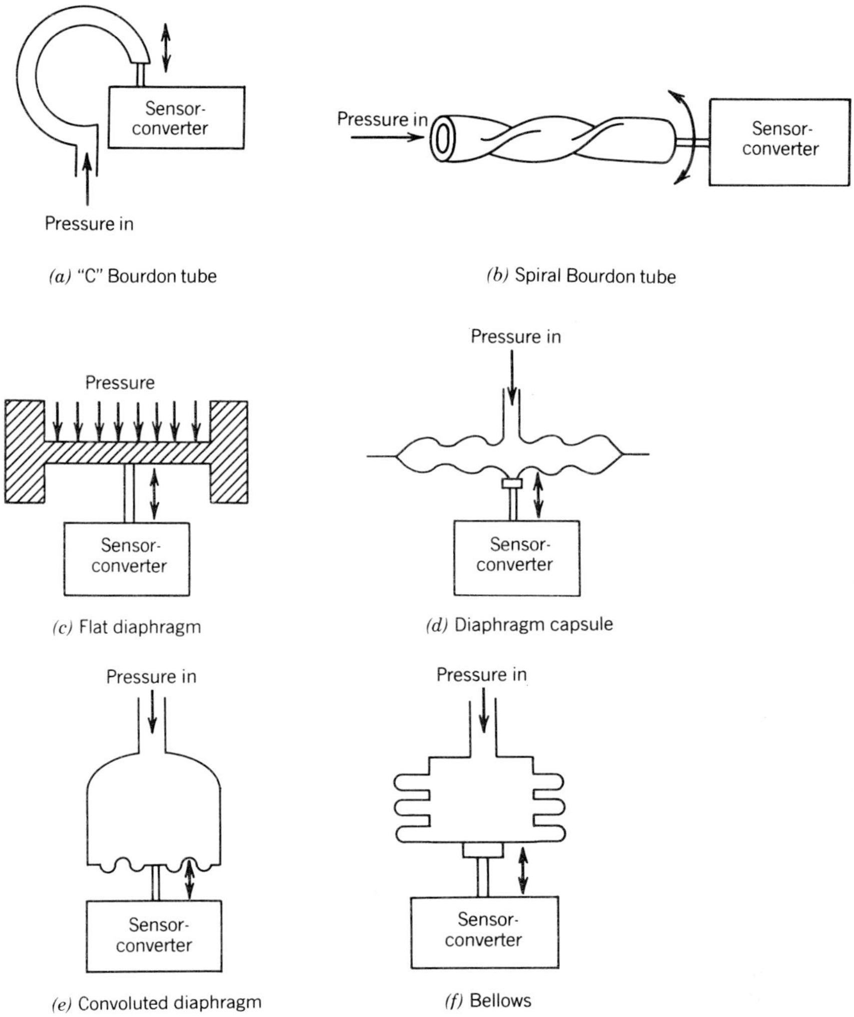

(a) "C" Bourdon tube

(b) Spiral Bourdon tube

Pressure in

Pressure

(c) Flat diaphragm

Pressure in

(d) Diaphragm capsule

Pressure in

Pressure in

(e) Convoluted diaphragm

(f) Bellows

Fig. 4.6-27 Pressure force summing devices.

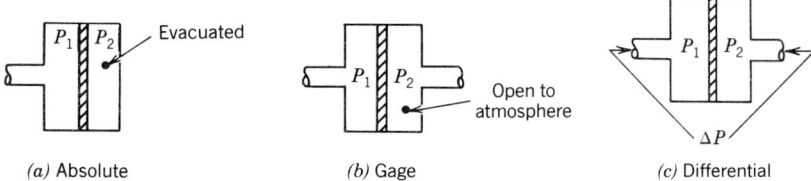

(a) Absolute

(b) Gage

(c) Differential

Fig. 4.6-28 Alternative designs for various types of pressure measurements.

Unbonded strain gauges have been used extensively in pressure transducers. The four-active-arm unbonded strain gauges serve as a spring-type sensing-conversion element connected to a diaphragm force summing area. This design assures overpressure protection up to 10 times the rated full-scale pressure without destruction of the transducer.

Piezoelectric pressure transducers are used where extremely high-frequency, short rise time pressures occur. A typical resonant frequency is 400,000 Hz. This type of transducer cannot read constant values of pressure.

The force summing device for an LVDT pressure transducer is frequently a diaphragm but may also be a bellows or a Bourdon tube. Because these latter designs have considerable mass associated with the movable portion, they are restricted to applications where a very low level of shock and vibration is encountered. A push rod links the pressure sensor to the core of the LVDT. The frequency response of this type of unit is considerably less than for other types. A representative differential transformer transducer has a natural frequency between 160 and 2000 Hz, depending on the selected pressure range. As opposed to the unbonded strain gauge pressure transducer, the LVDT transducer has a much less generous overload pressure—only 100–200% over the specified range.

Other less frequently used methods of converting pressure to an electric signal include *capacitive*, *variable reluctance*, *potentiometric*, and *vibrating element* with frequency-to-voltage conversion for analog output or frequency-to-count for digital output. There are also a number of nonelectrical methods of measuring pressure such as *U tube* and *inclined manometers* and *Bourdon tube* pressure gauges. These devices are generally not used for automatic control feedback because they do not convert their pressure reading to the electrical domain and their response times are relatively slow.

Flow

The measurement of fluid flow is nearly always carried out indirectly by measuring some other variable related to the flow rate, such as temperature, pressure, or velocity. Unfortunately, flow measurement transducers usually require insertion of some device into the flow conduit, which impedes the flow. A case in point is an orifice plate flow transducer wherein the flow rate is inferred by measuring the pressure drop across the orifice plate. Although most of the pressure drop is recovered in a downstream section of the conduit, some permanent pressure loss occurs. This drop represents an impedance to flow that would not be present if the measurement device were not present. A familiar example of this loading situation is the case of an ammeter inserted in series to measure current. The ammeter internal resistance can never be made zero except by the use of active components. This small resistance reduces the current in the conductor by a small but nonzero amount.

Direct measurement of flow rate is carried out by some sort of quantizing method such as weighing or measuring the accumulated volume in a tank and dividing by the time required to acquire the volume change. This method has several obvious shortcomings:

1. It must wait some period of time before delivering a new reading.
2. It can at best average the flow rate over the filling time interval.

Therefore, a direct measuring flowmeter is generally not applicable in the instantaneous determination of time-varying flow rates. All of the flow-measuring systems discussed in this section will be of the indirect type.

The class of *obstruction flowmeter* transducers includes all those where some type of restriction is placed in the pipe, which causes a pressure drop related to the flow rate through the reduced area. The flow rate may be inferred by measuring pressure drop across the restriction. The most common transducers that make use of this principle include the orifice plate, the flow nozzle, the venturi tube, the Dall flow tube, and the laminar flow element.

The laminar flow element is particularly interesting because the relation between flow and pressure drop is linear. For incompressible fluids, this is

$$Q = \frac{\pi D^4}{128 \mu L} \Delta P \tag{4.6-43}$$

where Q = volume flow rate, m^3/s
$\quad D$ = flow tube inside diameter, m
$\quad \mu$ = fluid viscosity, N·s/m^2
$\quad L$ = tube length across which the pressure drop is measured, m
$\quad \Delta P$ = pressure drop, N/m^2

This method is restricted, however, to flows with a Reynolds number of less than 2000, where

$$\text{Re} = \frac{VD\rho}{\mu} \tag{4.6-44}$$

For greater linearity of the pressure–flow characteristic, it is advisable to design for a Reynolds number of less than 1000. If a large flow rate is anticipated, a number of flow tubes may be arranged in parallel fashion. One of the problems associated with such a linear pressure drop flowmeter is that there is very little pressure drop across the flow tubes even for large changes in flow. If the pressure level is reasonably high, the pressure transducer must be capable of withstanding a large nominal pressure and still be sensitive enough to measure the variational component with sufficient accuracy. It would be best to use a differential pressure transducer to measure ΔP directly. In this way the full-scale range of the transducer can be selected to meet the full-scale range of the varying pressures rather than that of the nominal pressures, as would be the case if two individual transducers were used.

The *sharp-edge orifice plate* flow transducer is the most widely used flowmeter today. It is simple, low in cost, and has no moving parts. Its principal disadvantages are the square-law relationship between flow and pressure [Eq. (4.6-45)] and the erosion of the sharp edge of the orifice, which may cause flow system parameter variation. Figure 4.6-29 shows a simple schematic of an orifice plate flow transducer and the envelope of fluid flow lines. Clearly the minimum flow area is not the area of the orifice plate but is somewhat smaller and lies downstream of the orifice plate.

Ideally we need to know this minimum area and its location for each and every flow rate we wish to measure. Obviously it is not possible to know these values nor to adjust the pressure tap locations accordingly. In order to take into account these discrepancies, a calibration curve for a given installation is determined and a so-called coefficient of discharge is defined. With this adjustment factor to accommodate various flow rates, the volumetric flow rate for an incompressible frictionless fluid may be written.

$$Q = \frac{C_d A_2}{\sqrt{1 - (A_2/A_1)^2}} \sqrt{\frac{2}{\rho}(P_1 - P_2)} \qquad (4.6\text{-}45)$$

where A_1 = pipe cross-sectional area
  A_2 = orifice cross-sectional area
  ρ = fluid density in mass/volume
  P_1, P_2 = static pressures at distances l_1 and l_2, respectively, from the orifice plate
  C_d = coefficient of discharge

It has been found that the coefficient of discharge varies with the Reynolds number approximately as shown in Fig. 4.6-30. Fortunately, this coefficient stays constant above a Reynolds number of approximately 10^4 and remains roughly the same for a wide variation of orifice-to-pipe diameter ratios.

The process industry has made extensive use of the orifice plate differential pressure principle of flow measurement. An entire flow measurement system is usually offered as a package, which includes the orifice plate and body, a differential pressure transducer, and an electronic signal conditioning package. The signal conditioner converts the output of the differential pressure transducer to a

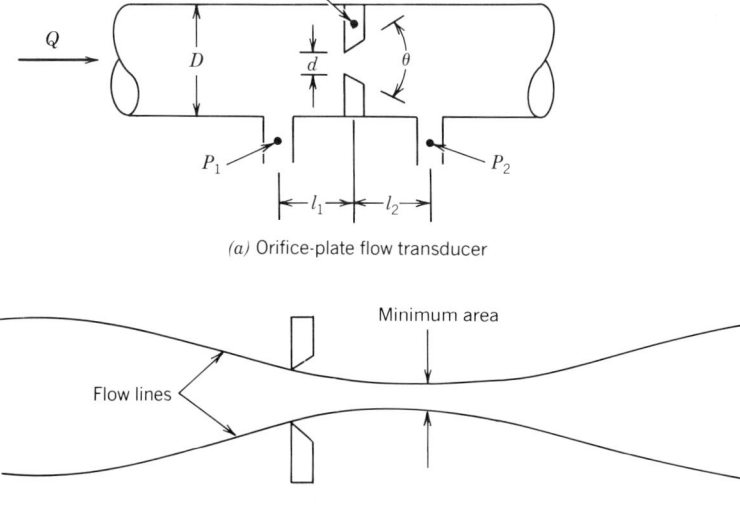

(a) Orifice-plate flow transducer

(b) Vena-contracta flow lines

Fig. 4.6-29 Sharp-edge orifice flow measurement.

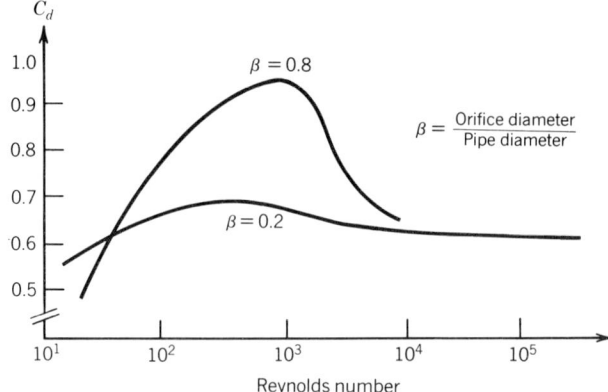

Fig. 4.6-30   Variation of the coefficient of discharge.

high-level voltage signal proportional to the pressure drop, extracts the square root of the differential pressure signal to convert to flow information, and delivers a 4- to 20-mA output current. This type of output is for transmitting the measured flow information to a remote location such as the control room of the processing plant. The overall flow measurement system is called a DP cell to designate its use of differential pressure to measure flow. The various types of DP cells are strain gauge, inductive, or capacitive, named after the method by which the differential pressure transducer converts pressure to an electric signal. Recent developments in this field are focused on replacing the analog output pressure transmitter by a microprocessor-based digital system. However, conversion to the digital domain is predicted to be slow in coming.

    There are many other methods of measuring fluid flow. The *target flowmeter* consists of a blunt body inserted in the flow stream and a means of measuring its deflection, such as by strain gauges attached to the target support arm. *Turbine flowmeters* are constructed so that a turbine inserted in the flow stream has a rotational velocity proportional to flow rate. This type of transducer may be used for either liquids or gases. *Hot-wire* and *hot-film* anemometers measure the convective heat loss from an electrically heated sensor inserted in the flowing fluid. Usually the velocity of the fluid is the measured variable, and flow rate is inferred.

    Several relatively recent developments in flow measurement include the *vortex* flowmeter, the *magnetic* flowmeter, and the *ultrasonic* flowmeter. Two types of vortex meters are available, one based on the vortex precession principle and the other based on the von Karman vortex-shedding principle. The vortices are sensed by changes in fluid temperature (e.g., measurement of these changes is by thermistors). A magnetic flowmeter has a chief advantage in that no flow obstructions are needed to measure the flow rate. Its principle of operation is based on Faraday's law of electromagnetic induction. A magnetic field is applied across the flow stream, and the velocity of the fluid causes an induced voltage across the flow tube. An ultrasonic flow-measuring system is based on the principle of sending a sonic pulse into the flowing medium and measuring the Doppler frequency shift.

4.6-3   Control Systems

Introduction

The word *control* has many different meanings and applications, even within the field of process control. For example, merely turning a switch on or off may be a form of control. A slightly more sophisticated form of control is to turn the switch on or off depending on whether a desired process variable is outside of or within a preestablished range. A home-heating system is an example of the latter system. When the temperature, as measured by the thermostat, is outside of a small preselected band, the switch closes, the heating system turns full on, and the temperature begins to rise. When the temperature reaches the set point value, the switch opens and the furnace turns off.

    The type of control discussed in this handbook includes those systems where an automatically measured value of the controlled variable is fed back and compared with its corresponding desired value. Thus, a difference signal is created that is positive if the feedback signal is less than the desired input signal.

    The error signal, in general, causes the actuator to produce a driving force to correct the mismatch between the measurement of the controlled variable and the desired or reference value of the controlled variable. Up to this point the preceding description could apply to the home-heating control system. The next delimiter will separate the home-heating system from the systems to be discussed here—the output of the actuator in most industrial control systems can be varied. On the other hand, the output of the home-heating system actuator, the furnace, cannot be varied. It is either full on or off. Furthermore, if the output of the actuator and every other output of each component in the system is linearly related to the input, the entire system is said to be a *linear* system. Stated very simply, a linear component has the property that doubling the input doubles the response.

The importance of the linearity property is that a number of mathematical techniques may be used to analyze the response of a system to a variety of inputs without having to build a prototype or even develop a computer simulation. Oftentimes a system may be analyzed directly and the quality of its response may be assessed without actually applying an input to the system or its simulation. A very useful property of a closed-loop system is that its overall transfer function characteristic (ratio of output to input) may be nearly linear in spite of one or more nonlinear characteristics or components in the system. The mathematical analysis tools that will be used here are the Laplace transform and the frequency response function (see Section 4.6-1). These tools are well developed in any text on classical automatic control systems principles and practices.

A rather recent development has had and is having a profound impact on automatic control systems—the microprocessor. In the early 1970s the proliferation of the minicomputer reduced the cost of digital computer hardware to the extent that the digital control system was emerging at a steady pace. Then along came the microprocessor in the early 1980s, and the era of the digital control system received a gigantic thrust. This new development made it economically possible to insert a truly powerful digital computer into many more single-loop control systems than was feasible with the earlier minicomputer. The significant improvements over formerly analog control systems are essentially threefold:

1. Convenient data storage or memory.
2. Vastly increased computational capability.
3. Changes in the control algorithm can be carried out much more simply by modifying the software on an off-line host computer without disturbing the closed-loop system.

As the computing rate of the new digital computer technology of the eighties continues to increase, the computational time delays inherent in discrete numerical computation are steadily diminishing relative to the speeds of response of the other components of the control system. This enables us to analyze and design a control system containing a microprocessor using linear system mathematics such as the Laplace transform without concern for the discrete behavior of its computational portion.

Block Diagrams

A feedback control system performs three tasks, as shown in Fig. 4.6-31:

1. Measure the variable to be controlled and thereby generate the feedback signal $v_f(t)$.
2. Subtract the feedback signal from the reference signal $r(t)$, generate an error signal $e(t)$, and compute an actuator drive signal $m(t)$.
3. Generate an actuator output to effect the desired change in the plant, thereby causing the controlled variable $c(t)$ to be restored to its correct value.

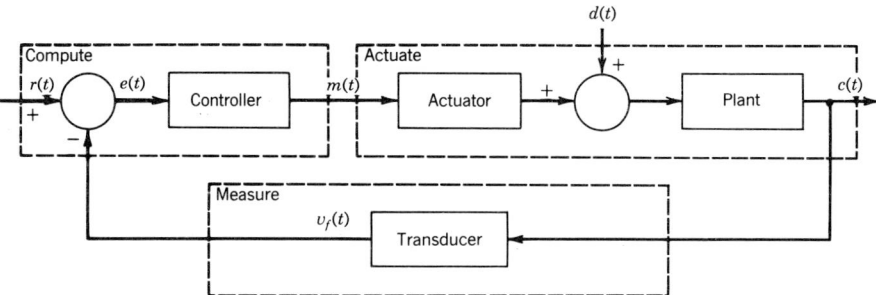

Fig. 4.6-31   Three tasks performed by a feedback control system.

There are four principal reasons for adding a feedback loop and its associated complexities:

1. Maintain accurate control when disturbances are present.
2. Convert a rate-type actuator to a servo having constant steady-state response.
3. Improve the overall performance.
4. Minimize the effects of changing component parameters.

Returning to the first reason listed, consider a system with only the feed-forward elements, a controller, an actuator, and a plant. With reasonable care, one could compute the time history of $m(t)$ necessary to cause a certain controlled variable $c(t)$ to occur over time. But any unpredicted and unmeasured disturbance $d(t)$ would cause errors in $c(t)$. The word *plant* refers to the portion of the system which is the useful end result, such as a thermal environment in a steel manufacturing reheat furnace or a machine tool cutting a particular profile on a work piece.

Now consider the second reason why feedback is used. Very often a control actuator produces a rate-of-change response for a constant input. For example, Fig. 4.6-32 shows two mechanical examples. In 4.6-32a a constant valve opening causes a constant flow rate, which in turn causes a steady-state velocity of the piston. The example in 4.6-32b shows a dc electric motor that produces a steady-state shaft speed. If the position of the plant is to be controlled, measuring and feeding back position changes the servo from an inherently rate-type system to a position control system. Similarly, a heat source actuator is another example of a rate-of-change output. A constant input causes a steady-state heating effect, which in turn causes a constant rate of change of the temperature of the medium where the heat is applied.

The third justification for using feedback is that it enables improvement of overall performance such as steady-state following accuracy, increased bandwidth, and reduced response time. The fourth need for feedback is that it minimizes the effects of changing component parameters, such as individual block gains. The overall system input–output characteristic is not as sensitive to a parameter change as is the individual block.

The block diagram shown in Fig. 4.6-33 is the most common representation of feedback control systems for both analysis and design. Each named function in the boxes is replaced by the transfer function of the mathematical operation performed by that function. Traditionally, forward path blocks are labeled $G_1(s)$, $G_2(s)$, $G_3(s)$, and so on, while feedback path blocks are designated $H_1(s)$, $H_2(s)$, and so forth. In most cases the feedback path contains only a single block, the feedback transducer. This component is typically approximated by a constant sensitivity over the range of frequencies that the control system is able to follow. The argument s is the Laplace transform

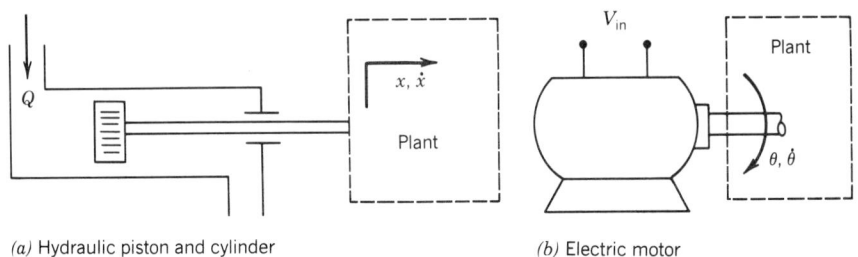

(a) Hydraulic piston and cylinder   *(b)* Electric motor

Fig. 4.6-32   Actuators which produce a rate of change.

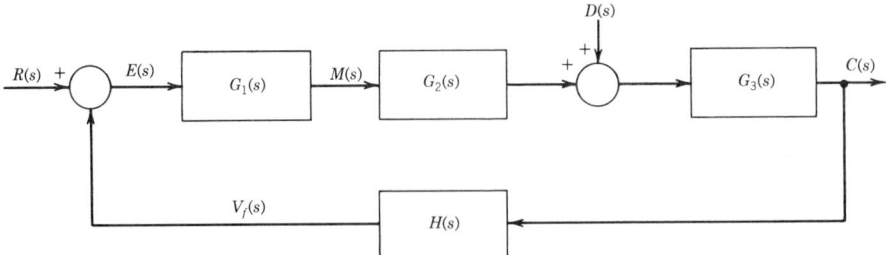

Fig. 4.6-33   Transfer functions used in representing a closed-log control system.

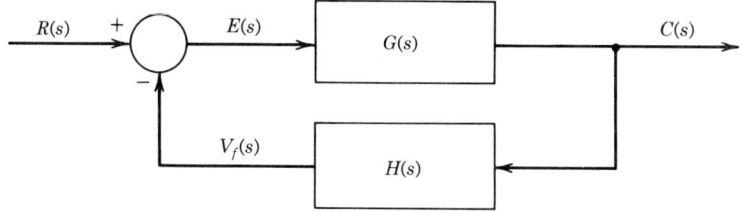

Fig. 4.6-34   Reduced block diagram representation.

variable. As discussed in Section 4.6-1, the variable s may be replaced by $j\omega$ for purposes of examining frequency response of the closed-loop system or any individual blocks thereof.

Figure 4.6-34 shows the closed-loop control system of Fig. 4.6-33, but with transfer functions instead of the various task names. Notice that the time domain variables are now represented by their Laplace transform symbols. Each individual transfer function is derived from the differential equation relating its input and output variables. We could also go back to the differential equation from knowledge of the transfer function alone. But the utility of the transfer function concept is that we can perform simple algebraic operations to convert the block diagram of Fig. 4.6-33 to the one shown in Fig. 4.6-34 and finally to the block diagram of Fig. 4.6-35. The forward-path transfer function $G(s)$ is simply the product of the individual transfer functions:

$$G(s) = G_1(s)G_2(s)G_3(s) \tag{4.6-46}$$

Reduction of the block diagram of Fig. 4.6-34 to that of Fig. 4.6-35 is as follows:

$$E(s) = R(s) - V_f(s) = R(s) - C(s)H(s) \tag{4.6-47}$$

The output is related to the error signal by

$$C(s) = E(s)G(s) \tag{4.6-48}$$

Substituting this result for $E(s)$ in Eq. (4.6-47),

$$\frac{C(s)}{G(s)} = R(s) - C(s)H(s)$$

$$C(s) + C(s)G(s)H(s) = R(s)G(s) \tag{4.6-49}$$

$$C(s) = R(s)\frac{G(s)}{1 + G(s)H(s)}$$

and

$$\frac{C}{R}(s) = \frac{G(s)}{1 + G(s)H(s)} \tag{4.6-50}$$

$$\frac{C}{R}(s) = \frac{G_1(s)G_2(s)G_3(s)}{1 + G_1(s)G_2(s)G_3(s)H(s)} \tag{4.6-51}$$

A similar derivation could be performed to yield the transfer function relationship between the disturbance $D(s)$ and its response at the output of the system:

$$\frac{C}{D}(s) = \frac{G_3(s)}{1 + G_1(s)G_2(s)G_3(s)H(s)} \tag{4.6-52}$$

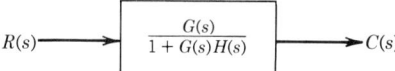

Fig. 4.6-35   Overall transfer function representation.

TABLE 4.6-3   BLOCK DIAGRAM MANIPULATION

Operation	Original Block Diagram	Final Diagram

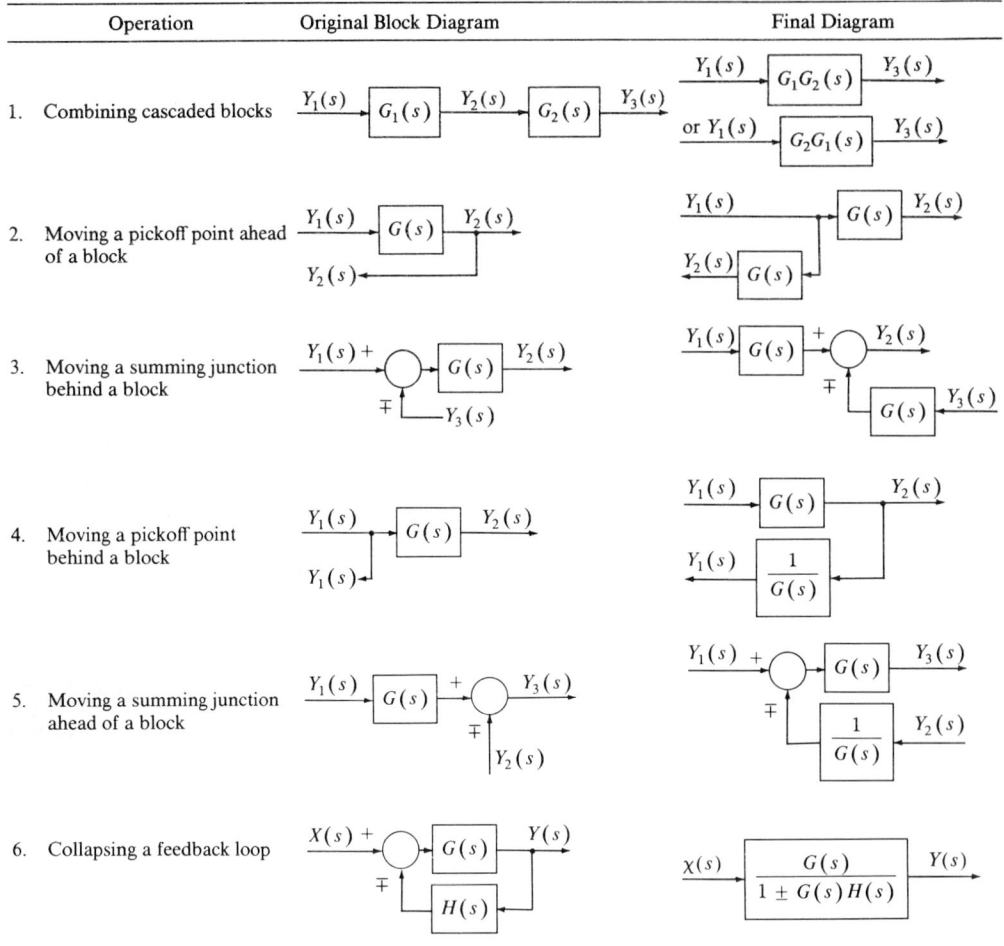

1. Combining cascaded blocks

2. Moving a pickoff point ahead of a block

3. Moving a summing junction behind a block

4. Moving a pickoff point behind a block

5. Moving a summing junction ahead of a block

6. Collapsing a feedback loop

Notice that the denominators of $C(s)/D(s)$ and $C(s)/R(s)$ are identical. Stated another way, all characteristic functions of each input–output pair of a linear system are identical. The manipulations commonly used to reduce a block diagram to the form of Fig. 4.6-35 are presented in Table 4.6-3.

Stability and Performance Analysis

Addition of a feedback loop to control a system has a serious liability—the system may go unstable. To understand why this is so, consider the block diagram of Fig. 4.6-36, which is the same as Fig. 4.6-33 with its feedback path opened and with s replaced by $j\omega$. If there is some frequency $\omega = \omega_c$ where the phase angle of all of the loop transfer blocks taken together is $-180°$ and such that the amplitude ratio of all the blocks is greater than unity, the closed-loop system is unstable. These two conditions are

$$\angle GH_{OL}(j\omega_c) = \angle G_1(j\omega_c) + \angle G_2(j\omega_c) + \angle G_3(j\omega_c) + \angle H(j\omega_c)$$

$$= -180 \tag{4.6-53}$$

$$\frac{V_f}{R}(j\omega_c) = |G_1(j\omega_c)||G_2(j\omega_c)| \cdot |G_3(j\omega_c)||H(j\omega_c)| > 1 \tag{4.6-54}$$

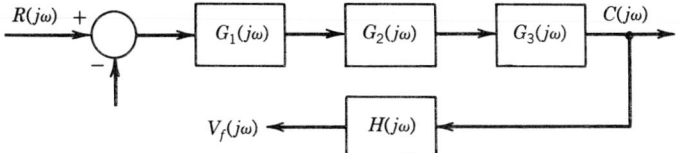

Fig. 4.6-36   Feedback signal disconnected.

The frequency $\omega = \omega_c$, which causes a $-180°$ phase shift, is called the *crossover frequency*. This term is also used in Nyquist stability analysis. But referring to Fig. 4.6-36, if there is a frequency ω_c such that Eq. (4.6-53) holds, the additional $-180°$ phase shift caused by the $(-)$ sign at the summing junction results in an overall phase shift of $360°$ in the ω_c component of $V_f(t)$. Furthermore, if Eq. (4.6-54) holds, the ω_c component is magnified as it traverses around the loop. Using this reasoning, it is clear that multiple recirculations of any signal or noise components at $\omega = \omega_c$, however small in amplitude, will grow and therefore the system is unstable.

A rigorous stability analysis for linear systems called the *Nyquist stability criterion* develops the same test conditions for stability as those expressed by Eqs. (4.6-53) and (4.6-54). Complex variable theory is used to determine whether the closed-loop system is stable by plotting the magnitude and phase characteristic of the open-loop system $GH(s)$. The procedure is as follows:

1.   Select a contour that completely encloses the right half of the s plane.
2.   Map the values of $GH(s)$ onto another complex plane called the GH plane as s takes on successive values around the s-plane contour.
3.   Count the number of encirclements of $GH(s)$ around the point $-1 + j \cdot 0$.
4.   Find the number of zeros of $1 + GH$ in the right-half s plane by a simple formula.

Figures 4.6-37 and 4.6-38 are examples of the contours chosen in step 1. Step 2 usually consists of plotting the $s = j\omega$, $0 \le \omega \le \infty$, portion of $GH(s)$ since the portion of the contour where $s = \infty$ (radius $= \infty$ in Figs. 4.6-37 and 4.6-38) causes $GH(s)$ to shrink to the origin of the GH plane. The portion of the contour $s = j\omega$, $-\infty \le \omega \le 0$, need not be plotted because it is a mirror image of the mapping for positive values of ω. Simply reflect the GH mapping about the real axis of the GH plane.

Consider the following open-loop transfer function and its mapping, shown in Fig. 4.6-39:

$$GH(s) = \frac{K}{(\tau s + 1)(s^2/\omega_n^2 + 2\zeta s/\omega_n + 1)} \tag{4.6-55}$$

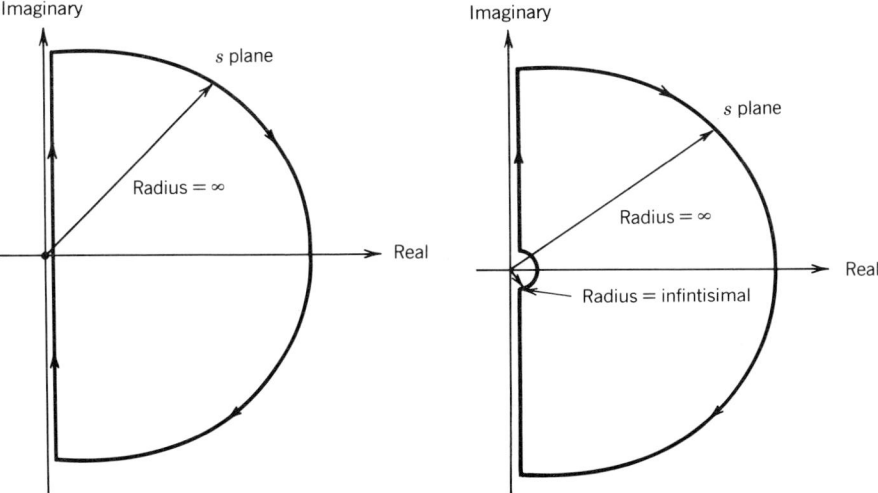

Fig. 4.6-37   Contour enclosing right-half s plane for open-loop system with no poles at the origin.   **Fig. 4.6-38**   Contour enclosing right-half s plane for open-loop system with poles at the origin.

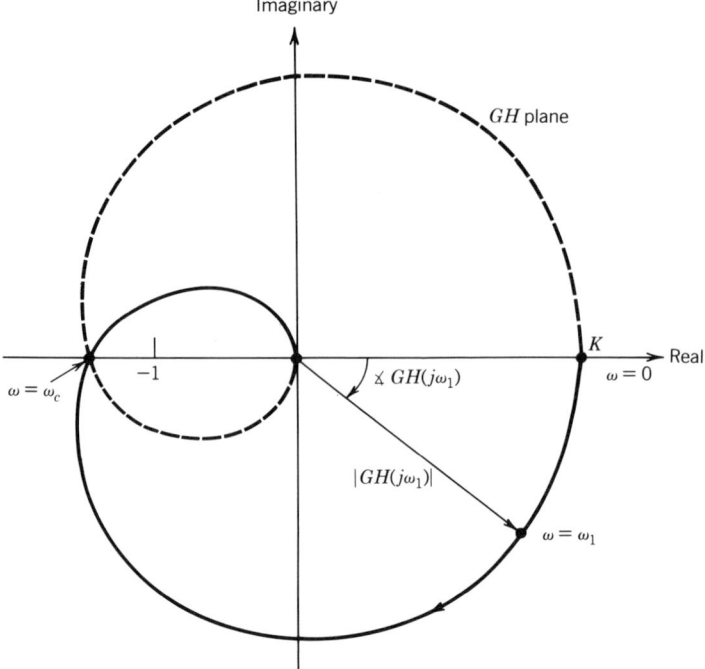

Fig. 4.6-39 Mapping of $GH(s)$ in Eq. (4.6-55) into the $GH(s)$ plane.

The dotted portion of Fig. 4.6-39 represents the range of values of ω from $-\infty$ to 0 and is obtained simply by reflecting the solid line portion about the real axis. Referring to step 3, the number of encirclements of -1 is 2. The formula for step 4 is

$$Z_{1+GH} = N_{GH,-1} + P_{GH} \qquad (4.6\text{-}56)$$

where Z_{1+GH} = number of zeros of $1 + GH(s)$ in the right-half s plane; $N_{GH,-1}$ = number of encirclements of $GH(s)$ about the -1 point; and P_{GH} = number of poles of the open-loop transfer function $GH(s)$ in the right-half s plane. Notice that stability of the closed-loop system can be determined from information about the open-loop system. This is an important point—modeling and analysis of a controlled system generally provides full details of the open-loop system. Because $P_{GH} = 0$ and $N_{GH,-1} = 2$, it is concluded that the closed-loop system resulting from the mapping in Fig. 4.6-39 is unstable, with two poles in the right-half s plane. Any input or disturbance will cause the system response to grow larger and larger. Figure 4.6-40 shows the polar plot* of the same transfer function as in Fig. 4.6-39 but with a smaller value of the loop gain K. For this value of gain there are no encirclements of -1, and thus the corresponding closed-loop system is stable. Thus, the Nyquist stability analysis can be used to determine the range of loop gain values for which the closed-loop system is stable. Figure 4.6-41 shows the polar plots of several typical transfer functions. The last transfer function in this figure has a free s in its denominator, which represents a pole at the origin of the s plane. Therefore, an s-plane contour of the type shown in Fig. 4.6-38 would be used to perform the $GH(s)$ mapping.

The Nyquist stability criterion may also be used to establish indices of performance for a stable closed-loop system. Extending the concept that the loop gain may be decreased to stabilize a system, it is also true that the nearer the gain is to its instability limit, the poorer the closed-loop system performance will be. *Gain margin*, or GM, is defined as that factor which, when multiplied by $|GH(j\omega_c)|$, results in a marginally stable closed-loop system.

$$GM \cdot |GH(j\omega_c)| = 1 \qquad (4.6\text{-}57)$$

$$GM = \frac{1}{|GH(j\omega_c)|} \qquad (4.6\text{-}58)$$

---

*Strictly speaking, a polar plot is the frequency response plotted in the form of magnitude and phase angle with frequency as a parameter. Its appearance is the same as plotting real and imaginary components of $GH(j\omega)$ with ω as a parameter.

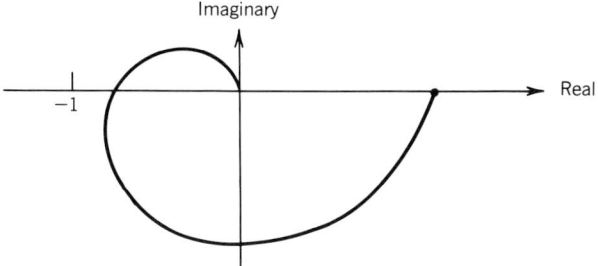

Fig. 4.6-40 Mapping of third-order $GH(j\omega)$ with stable loop gain

Referring to Fig. 4.6-41c, the crossover frequency ω_c is that frequency which causes a $-180°$ phase shift in the open-loop characteristic $GH(j\omega)$. For the transfer function of Fig. 4.6-41c, the crossover frequency is the system natural frequency, $\omega_c = \omega_n$. The gain margin may also be expressed in decibel measure.

$$\text{GM}_{\text{dB}} = 20\log_{10}\frac{1}{|GH(j\omega_c)|} \tag{4.6-59}$$

A gain margin of 2 generally yields a well-behaved closed-loop response.

Another index of performance derived from the Nyquist criterion is the *phase margin*. The phase margin is defined as the additional phase angle which, when subtracted from $\angle GH(j\omega)$ at $|GH(j\omega)| = 1$, yields a total of $-180°$:

$$\angle GH(j\omega_1) - \phi_m = -180° \tag{4.6-60}$$

Figure 4.6-41b, c illustrates the phase margin ϕ_m. If ϕ_m is negative, the open-loop transfer function at unity gain has a phase shift more negative than $-180°$. Therefore, the open-loop characteristic very likely encircles -1 and the closed-loop system is unstable. A phase margin of $+45°$ generally yields well-behaved closed-loop performance.

$$\text{GM} = 2 \qquad \text{Satisfactory closed-}$$

$$\phi_m = +45° \qquad \text{loop performance}$$

Example 4.6-5

$$GH(s) = \frac{K}{s\left(s^2/\omega_n^2 + 2\zeta s/\omega_n + 1\right)}$$

To find the gain margin, find $|GH(j\omega)|$ at $\omega = \omega_c$ (the $-180°$ phase shift frequency).

$$GH(j\omega) = \frac{K}{j\omega\left(1 - \omega^2/\omega_n^2 + j2\zeta\omega/\omega_n\right)} = \frac{K}{-2\zeta\omega^2/\omega_n + j\omega\left(1 - \omega^2/\omega_n^2\right)}$$

At the crossover frequency, the imaginary part of $GH(j\omega)$ vanishes. Setting the imaginary coefficient to zero at $\omega = \omega_c$,

$$\omega_c\left(1 - \frac{\omega_c^2}{\omega_n^2}\right) = 0 \qquad \omega_c = \omega_n$$

$$\text{GM} \cdot |GH(j\omega_c)| = 1$$

$$\text{GM} = \frac{1}{|GH(j\omega_c)|} = \frac{1}{\left|\dfrac{K}{-2\zeta\omega_c^2/\omega_n}\right|}$$

$$\text{GM} = \frac{2\zeta\omega_n}{K}$$

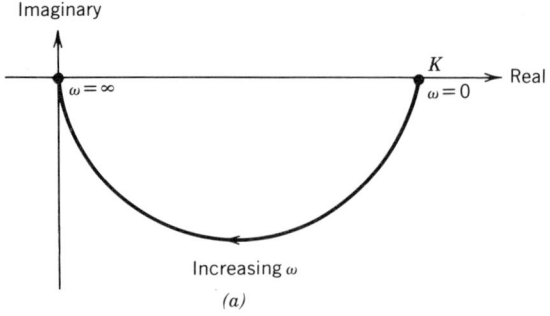

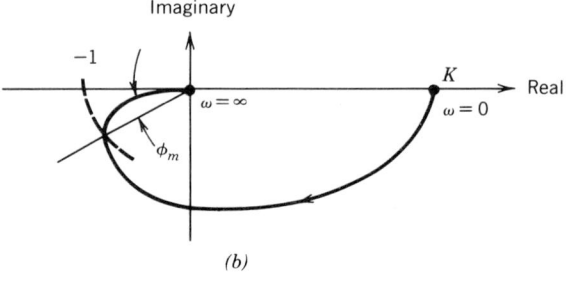

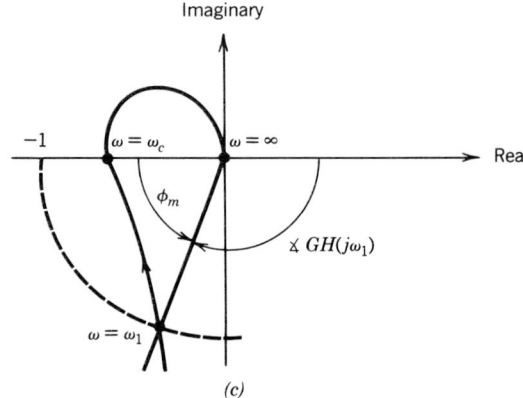

Fig. 4.6-41 Polar plots of some common open-loop transfer functions. (a) First-order system:

$$GH(s) = \frac{K}{\tau s + 1}$$

Closed-loop system will always be stable. (b) Second-order system:

$$GH(s) = \frac{K}{\dfrac{s^2}{\omega_n^2} + \dfrac{2\xi s}{\omega_n} + 1}$$

Closed-loop systems will always be stable but response will be less damped as K increases. (c) Third-order system having a free integration:

$$GH(s) = \frac{K}{s\left(\dfrac{s^2}{\omega_n^2} + \dfrac{2\xi s}{\omega_n} + 1\right)}$$

Closed-loop systems can be unstable, depending on how large K is.

Finding the phase margin is much more difficult. It is best to construct the Bode plots and read ϕ_m at the frequency where $|GH(j\omega)| = 1$ (0 dB). Figure 4.6-42 shows the amplitude ratio and phase shift Bode plots for $K = 50$, $\omega_n = 100 r/s$, $\zeta = 0.5$. When $GH(j\omega) = -180°$, $|GH(j\omega)| = 0.5$, or $20 \log_{10}|GH(j\omega)| = 6$ dB. Therefore, GM $= 2.0$, or $+6$ dB. When $|GH(j\omega)| = 1.0$, or $20 \log_{10}|GH(j\omega)| = 0$ dB, $\angle GH(j\omega) \cong -130°$. Therefore, $\phi_m = 180° + \angle GH = 180° - 130° = +50°$.

A problem that frequently arises in the control of systems with flowing fluids is the presence of a pure time delay. A person who prefers shower baths has experienced this phenomenon many times. If a hot or cold water pressure change suddenly occurs, the shower temperature changes a short time later. If the appropriate hot or cold water valve is adjusted, the temperature change is sensed yet another short time later. The tendency is to overadjust because there is a delay in the temperature response. In general, the time delay problem occurs in control systems where the sensor is located at a point some distance removed from the actuation location. In a heat exchanger the temperature sensor may require some "flow time" before it senses an upstream change in fluid temperature. Typically, the time delay can be approximated by a distance–velocity ratio, $T_d = d/v$. The block diagram of a

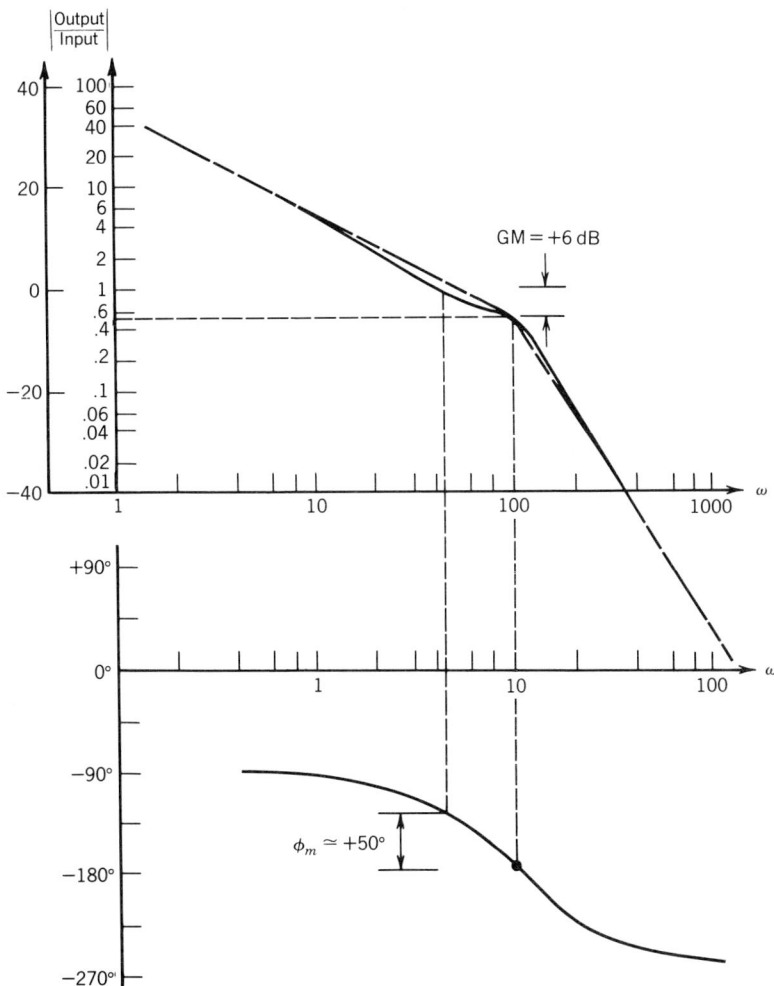

Fig. 4.6-42 Gain margin and phase margin. Read from Bode plots. Assume $K = 50$, $\omega_n = 100$, $\xi = 0.5$. At $\omega = \omega_c = \omega_n$

$$|GH(j\omega_n)| = \frac{50}{2 \times 0.5 \times 100} = 0.5$$

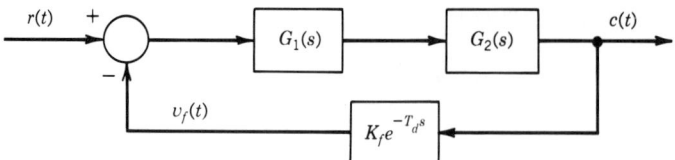

Fig. 4.6-43   Block diagram of a feedback system with time delays.

system with measurement delay is shown in Fig. 4.6-43. Stability of this system can be determined graphically by the Nyquist criterion. The first step is to plot the frequency response of the loop transfer function without the time delay term $e^{-T_d j\omega}$. Then rotate points along the resulting plot by the phase angle $-T_d\omega$. Finally, draw the overall open-loop plot by connecting the locus of phase-shifted points from the original curve as is done in Fig. 4.6-44. If the phase-shifted curve encircles the -1 point, the closed-loop system is unstable. Notice the multiple encirclements of the origin at higher frequencies. This behavior is due to the increasingly greater phase shift at high frequency.

A stability analysis tool that yields a clear picture of the closed-loop system time domain response to elementary time functions such as steps, ramps, and sinusoids is known as *root-locus* analysis. The name comes from the fact that it is a study of the loci of the roots of the closed-loop characteristic equation. The paths are traced out by the roots as they change position in the s plane due to changes in the loop gain. The root-locus analysis begins with placement of the open-loop poles (x) and open-loop zeros (0) of $GH(s)$ on the s plane. As the loop gain is increased, the closed-loop poles migrate from the open-loop pole locations to the open-loop zeros or to infinitely large complex numbers. The directions in which these roots migrate toward infinity dictate whether the system can become unstable. If the poles, or roots, enter or remain in the right half of the s plane, the closed-loop system is unstable for the corresponding range of loop gain values.

The root loci are derived from the characteristic equation (setting the denominator equal to zero) of the closed-loop transfer function. From Eq. (4.6-50), we have

$$G(s)H(s) + 1 = 0 \qquad\qquad (4.6\text{-}61)$$

Rewrite $G(s)$ and $H(s)$ to expose the open-loop poles and zeros and the loop gain.

$$K\frac{N_G(s)}{D_G(s)}\frac{N_H(s)}{D_H(s)} + 1 = 0 \qquad\qquad (4.6\text{-}62)$$

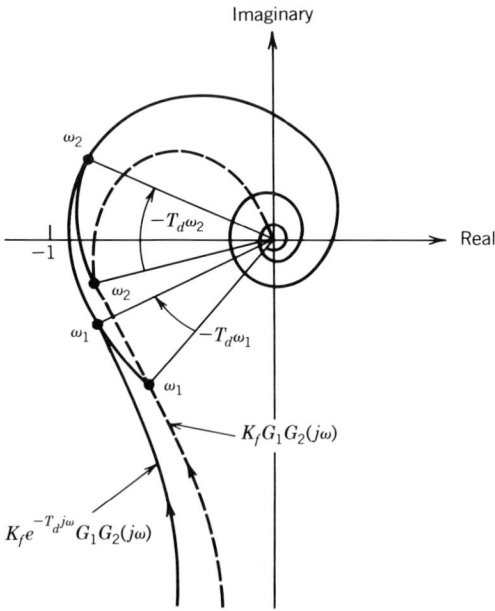

Fig. 4.6-44   Nyquist plot of an open-loop system with time delays.

where the terms N_G, N_H, D_G, and D_H are cascaded combinations of

$$s^n \qquad \tau s + 1 \qquad \text{and} \qquad \frac{s^2}{\omega_n^2} + \frac{2\zeta}{\omega_n}s + 1$$

From Eq. (4.6-62), we write

$$\measuredangle K\frac{N_{GH}(s)}{D_{GH}(s)} = 180 \pm k \cdot 360° \qquad (k = 0,1,2,3,\dots) \qquad (4.6\text{-}63)$$

$$\left| K\frac{N_{GH}(s)}{D_{GH}(s)} \right| = 1 \qquad\qquad (4.6\text{-}64)$$

Any complex number that satisfies the angle criterion is a point on a root locus. Any complex number that satisfies the magnitude condition for a given value of gain K and also satisfies the angle criterion is a pole of the closed-loop system.

Two additional rules must be invoked in order to draw a root-locus diagram:

$$\phi_{\text{asymp}} = \frac{180° \pm k \cdot 360°}{n - m} \qquad (4.6\text{-}65)$$

$$\text{Center of gravity} = \frac{\sum_{1}^{n}(-p_i) - \sum_{1}^{m}(-z_i)}{n - m} \qquad (4.6\text{-}66)$$

where n = number of poles of $GH(s)$ and m = number of zeros of $GH(s)$, and the numbers $-p_i$ and $-z_i$ represent s-plane locations of the poles and zeros of $GH(s)$. The numerator and denominator of $GH(s)$ when written in pole-zero form are

$$N_{GH}(s) = \frac{1}{z_1 z_2 \cdots z_m}(s + z_1)(s + z_2) \cdots (s + z_m) \qquad (4.6\text{-}67)$$

$$D_{GH}(s) = \frac{1}{p_1 p_2 \cdots p_n}(s + p_1)(s + p_2) \cdots (s + p_n) \qquad (4.6\text{-}68)$$

The values $-z_i$ are called zeros because $GH(s)$ goes to zero at those points in the s plane where s takes on a value $-z_i$. Similarly, when s takes on a value $-p_i$, the denominator of $GH(s)$ goes to zero, causing the magnitude of $GH(s)$ to become infinitely large, hence the name *pole*. Turning back to the task of constructing the closed-loop pole paths, or root loci, the center of gravity is calculated according to Eq. (4.6-66). This value represents the point along the real axis in the s plane from which the asymptotes of the loci are drawn. For Example,

$$GH(s) = \frac{K}{(\tau s + 1)\left[s^2/\omega_n^2 + (2\zeta/\omega_n)s + 1 \right]}$$

One pole at $s = -1/\tau$; two poles at $s = -\zeta\omega_n \pm j\omega_n\sqrt{1 - \zeta^2}$. The term $\omega_n\sqrt{1 - \zeta^2}$ is also called the damped natural frequency ω_d because the natural vibration of an underdamped system occurs at the frequency ω_d. Figure 4.6-45 shows the three open-loop pole placements, marked with an $\times$ at $-p_1$, $-p_2$, and $-p_3$. The center of gravity is (assume $1/\tau = 2\zeta\omega$ for convenience)

$$\text{CG} = \frac{(-\zeta\omega_n + j\omega_d) + (-\zeta\omega_n - j\omega_d) + (-1/\tau)}{3}$$

$$= \frac{2\zeta/\omega_n - 2\zeta/\omega_n}{3} = -\frac{4}{3}\frac{\zeta}{\omega_n}$$

$$\phi_{\text{asymp}} = \frac{180°}{3} ; \frac{180° + 360°}{3} ; \frac{180° - 360°}{3}$$

$$\phi_{\text{asymp}} = +60°, +180°, -60°$$

The point where the two complex conjugate poles cross the imaginary axis represents the value of K at which the system is marginally stable. This value of gain produces a real–imaginary Nyquist plot

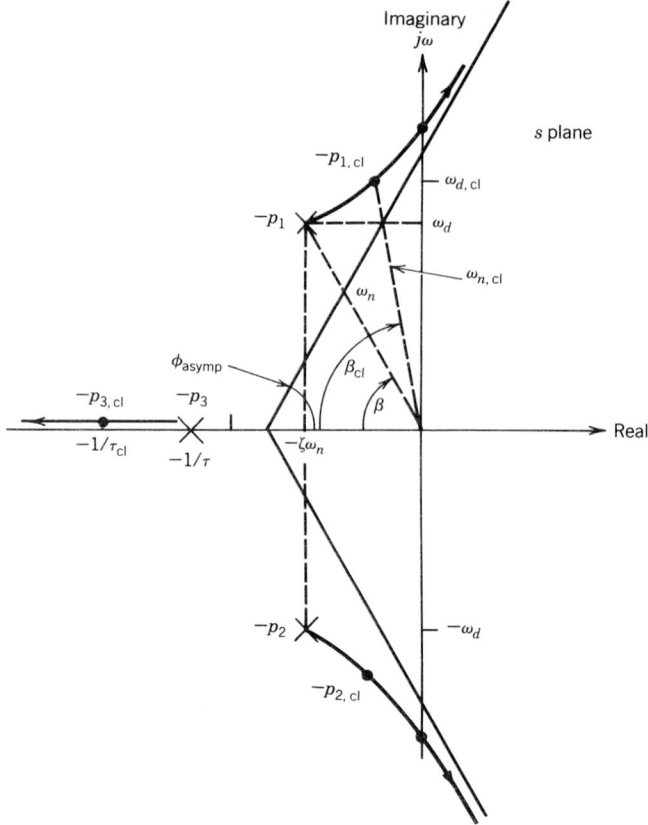

Fig. 4.6-45   Root loci for a third-order system.

(Fig. 4.6-39) that passes through the -1 point. The frequency of oscillation at this marginally stable condition is the crossover frequency. Two different methods may be used to calculate the marginally stable value of gain K:

1. Use the root-locus magnitude condition or Eq. (4.6-64) and solve for the gain graphically using the imaginary axis points of Fig. 4.6-45.
2. Use the Nyquist criterion by finding the crossover frequency ω_c and calculate the gain required to make $|GH(j\omega_c)| = 1$. A larger value of gain would cause the closed-loop poles to migrate into the right half of the s plane, thereby causing instability.

Now suppose the gain is adjusted to place the closed-loop complex conjugate poles at $-p_{1,\text{cl}}$ and $-p_{2,\text{cl}}$. Notice that the closed-loop damping ratio is smaller than the damping ratio for the open-loop complex conjugate poles. The value of the damping ratio may be calculated by the simple expression

$$\cos \beta = \zeta \qquad (4.6\text{-}69)$$

Note that the damped natural frequency of the closed-loop system is larger than for the open-loop system. Because of the latter, the rise time of the closed-loop system will be shorter; because of the former, the settling time for the closed-loop system may be greater, thereby resulting in an unacceptable response for certain applications. Figure 4.6-46 summarizes the character of the step responses for several different pole locations in the s plane. This figure illustrates that a pair of complex conjugate poles cause subcritically damped oscillatory response while individual real-axis poles cause exponential response.

There is another well-known stability analysis tool—the *Routh–Hurwitz* criterion, which indicates whether a system is stable or unstable by examining the closed-loop characteristic equation. Given a

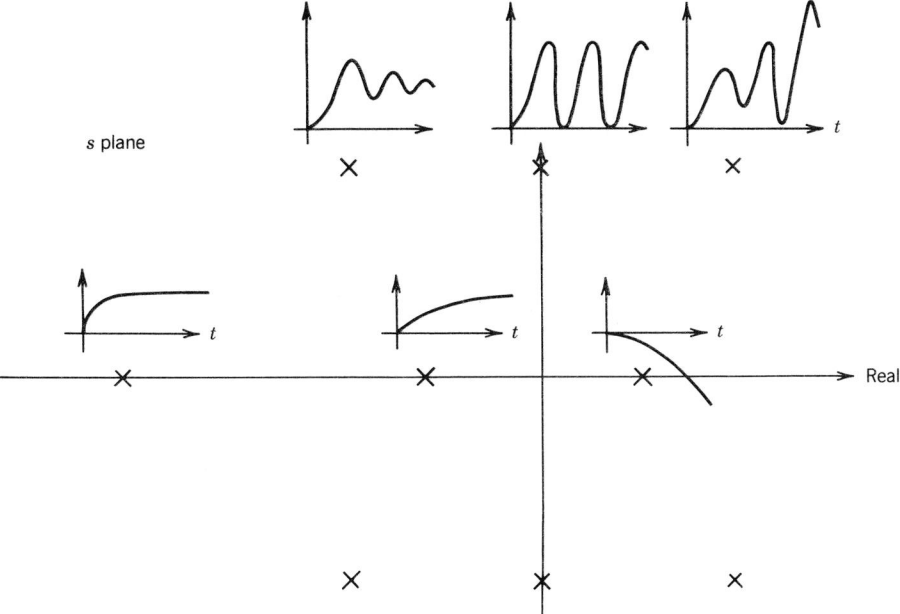

Fig. 4.6-46 Step responses for various pole locations in the *s* plane.

closed-loop characteristic equation of the form

$$a_n s^n + a_{n-1} s^{n-1} + \cdots + a_1 s + a_0 = 0 \qquad (4.6\text{-}70)$$

an array is formed as follows:

$$
\begin{array}{llll}
s^n: & a_n & a_{n-2} & a_{n-4} & \cdots \\
s^{n-1}: & a_{n-1} & a_{n-3} & a_{n-5} & \cdots \\
s^{n-2}: & b_{n-2} & b_{n-4} & \cdots \\
s^{n-3}: & c_{n-3} & c_{n-5} \\
& \vdots \\
s^0: & f_0
\end{array}
\qquad (4.6\text{-}71)
$$

where

$$b_{n-2} = \frac{a_{n-1} a_{n-2} - a_n a_{n-3}}{a_{n-1}}$$

$$b_{n-4} = \frac{a_{n-1} a_{n-4} - a_n a_{n-5}}{a_{n-1}} \qquad (4.6\text{-}72)$$

$$c_{n-3} = \frac{b_{n-2} a_{n-3} - a_{n-1} b_{n-4}}{b_{n-2}}$$

The Routh–Hurwitz criterion states that the number of closed-loop poles in the right-half *s* plane is equal to the number of sign changes in the first column of the array. It is not a very useful tool because it provides very little information about system performance as the stability limit is approached.

Controller Design

The preceding analysis tools do not address the question of what can be done to improve system performance if it is found unacceptable. In some instances the analysis may point up the elements or components of the system that are primarily responsible for the undesirable behavior. But if these components cannot be altered, some additional design information must be called forth to "fix up" the system. The portion of the closed-loop system most amenable to design modifications is the controller shown in Fig. 4.6-31. The intelligence of the system is concentrated in this block. In recent years and into the future, more and more controllers will be equipped with a microcomputer to replace the analog electronics of the recent past. This will make it increasingly easier to alter the system performance by changing only software, with minimal disturbance and downtime in the system.[11]

Typically, design of the controller element begins with overall system performance requirements. A frequent requirement is the allowable maximum error at steady state, such as for a step or ramp input.[7] From Fig. 4.6-34, we may write

$$E(s) = \frac{1}{1 + GH(s)} \cdot R(s) \qquad (4.6\text{-}73)$$

Using the final-value theorem, the steady-state error in the time domain is

$$e_{ss}(t) = \lim_{t \to \infty} e(t) = \lim_{s \to 0} sE(s)$$

$$= \lim_{s \to 0} \frac{s \cdot R(s)}{1 + GH(s)} \qquad (4.6\text{-}74)$$

For a step input of amplitude A,

$$e_{ss}(t) = \lim_{s \to 0} \frac{s \cdot A/s}{1 + GH(s)} = \frac{A}{1 + GH(0)} \qquad (4.6\text{-}75)$$

If the system is a type zero system (no free integrators), the term $GH(0)$ is simply the loop gain. It is also referred to as the position error constant K_p. The steady-state error is

$$e_{ss}(t) = \frac{A}{1 + K_p} \qquad (4.6\text{-}76)$$

If the system has one or more free integrators, $GH(0) = \infty$ and we have $e_{ss}(t) = 0$. For a ramp input (constant velocity of B length units/time unit), the steady-state error is

$$e_{ss}(t) = \lim_{s \to 0} \frac{s(B/s^2)}{1 + GH(s)} = \lim_{s \to 0} \frac{B}{sGH(s)} \qquad (4.6\text{-}77)$$

If $GH(s)$ is type zero, $e_{ss}(t) = B/0 \cdot K_p = \infty$. If $GH(s)$ is type one (one free integrator), $e_{ss}(t) = B/K_v$, where K_v is called the velocity error constant. If the system is type two (two free integrators),

$$\lim_{s \to 0} sGH(s) = \infty$$

and thus $e_{ss}(t) = 0$.

For a system with acceleration input $r(t) = Ct^2/2$, the steady-state error is

$$e_{ss}(t) = \lim_{s \to 0} \frac{s(C/s^3)}{1 + GH(s)} = \lim_{s \to 0} \frac{C}{s^2GH(s)} \qquad (4.6\text{-}78)$$

If the system has zero or one free integrator, $e_{ss}(t) = \infty$. If it is a type two system, $e_{ss}(t) = C/K_a$, where K_a is called the acceleration constant. If the system has more than two free integrators, $e_{ss}(t) = 0$ for an acceleration input. These cases are summarized in Table 4.6-4. It is important to note that in the preceding work, the steady-state error is given in units of the error signal, which is usually volts. To convert this error to units of the controlled variable, simply divide $e_{ss}(t)$ by the feedback transducer sensitivity, which is in units of volts per unit change of the controlled variable $c(t)$.

Now that a tool for predicting steady-state performance is in hand, the question of how to design the system to meet the steady-state requirements may be addressed. A well-known controller form is

TABLE 4.6-4 STEADY STATE ERRORS VERSUS TYPE NUMBER AND INPUT

Input System Type Number	Step: $r(t) = A$ $R(s) = A/s$	Ramp: $r(t) = Bt$ $R(s) = B/s^2$	Acceleration: $r(t) = Ct^2/2$ $R(s) = C/s^3$
0	$\dfrac{A}{1 + K_p}$	∞	∞
1	0	B/K_v	∞
2	0	0	C/K_a
3	0	0	0

called a proportional-integral-derivative controller (PID controller). Because differentiation tends to magnify noise components in the input and feedback signals, a PI controller is frequently used. These two are written as

$$m(t) = Ke(t) + K_I \int^t e(t)\, dt + K_D \frac{de(t)}{dt}$$

$$G_c(s) = K + \frac{K_I}{s} + K_D s$$

$$G_c(s) = \frac{K_D s^2 + Ks + K_I}{s} \tag{4.6-79}$$

and

$$G_c(s) = \frac{Ks + K_I}{s} \tag{4.6-80}$$

Very often, the system actuator has a free integrator. Thus, the addition of a PI or a PID controller results in a closed-loop control system which is type two. As shown in Table 4.6-4, this system can follow a ramp with zero steady-state error. In addition, the derivative action of the controller works to prevent large overshoot and unduly long transient oscillatory behavior in response to step inputs. Choice of the appropriate values for the controller gains K, K_I, and K_D may be determined by several different means:

1. Perform a root-locus analysis on the system and try several values of controller gains. Note that a PID controller added to a given open-loop system adds two zeros and a pole to those of $GH(s)$. The approach is generally one of setting the controller gains so that two complex conjugate zeros placed in the left half of the s plane attract two complex conjugate poles of $GH(s)$ that would otherwise migrate to the right half of the s plane.
2. Apply the Ziegler–Nichols method.[10]

A second method of modifying the intelligence of a system for improved closed-loop performance is adding in-the-loop cascaded compensation in the form of a lead-function or a lag-function network. Suppose the feedback control system is as shown in Fig. 4.6-47, where $G_p(s)$ represents the plant and $G_c(s)$ is the compensation network as well as an adjustable gain. The feedback element is a transducer with static gain S_{vc}. Transfer functions for lead-function and lag-function compensators are, respectively,

$$G_c(s) = K \frac{\alpha \tau_c s + 1}{\tau_c s + 1} \qquad \alpha > 1 \tag{4.6-81}$$

and

$$G_c(s) = K \frac{\tau_c s + 1}{\alpha \tau_c s + 1} \qquad \alpha > 1 \tag{4.6-82}$$

The Bode plots for these transfer functions are given in Figs. 4.6-48 and 4.6-49, respectively. Note that the phase shift of the lead-function compensator is leading at all frequencies and is lagging for the

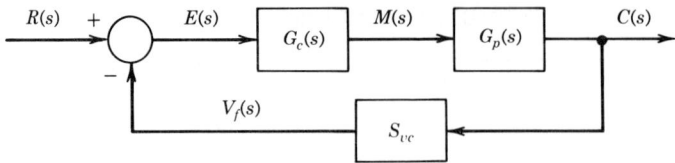

Fig. 4.6-47 In-the-loop cascaded compensation.

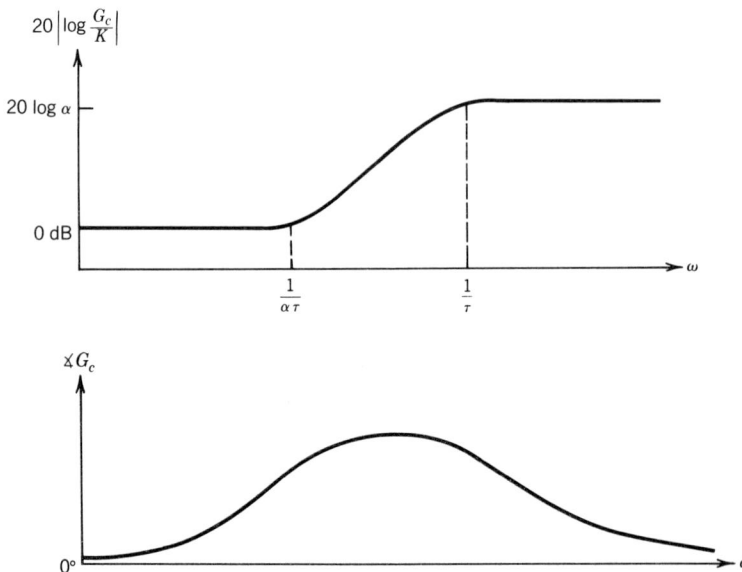

Fig. 4.6-48 Lead-function compensation.

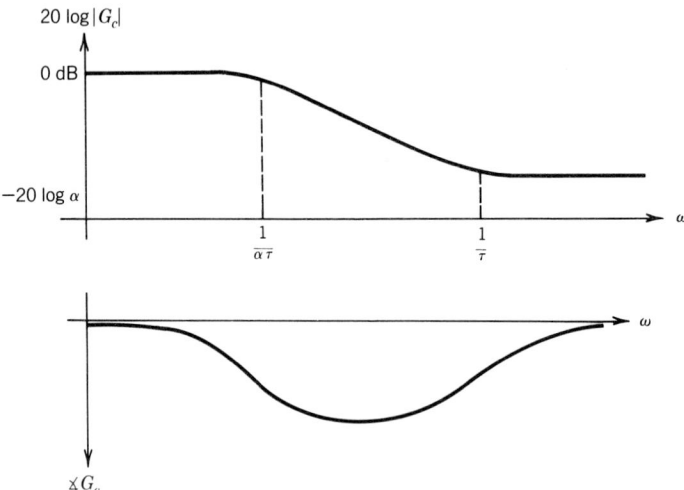

Fig. 4.6-49 Lag-function compensation.

476

lag-function network throughout the frequency spectrum, hence the names of the respective compensation networks.

The lead-function compensator is used to extend the bandwidth of a closed-loop system.[8] Assume that the plant transfer function is a first-order lag. If the numerator break frequency of the compensator, $1/\alpha\tau_c$, is approximately the same as the plant denominator break frequency, the bandwidth is extended by the bandwidth of the compensator. These statements are illustrated in Fig. 4.6-50. The parameter α of the lead-function network is normally in the range of 1–10, possibly as high as 20 for low-noise systems. Figure 4.6-50 does not reveal a great deal about the performance of the closed-loop compensated system. Consider the following examples.

Example 4.6-6. An electromechanical dc position servo is modeled as two first-order lags and a free integrator:

$$G_p(s) = \frac{K_p}{s(0.1s + 1)(0.01s + 1)}$$

It is desired to increase the bandwidth of the closed-loop servo system without decreasing the degree of stability. Figure 4.6-51 shows both the open-loop uncompensated and compensated systems. The design approach is to cancel the low break frequency pole with a compensator zero at the same location. Note that the gain margin of both the uncompensated and the compensated system is +40 dB = 100. But the crossover frequency has been raised from 30 to 100 r/s.

Example 4.6-7. The same dc position servo as in Example 4.6-6 is required to follow a ramp input. The uncompensated system exhibits a greater following error than is allowed. Use lag-function compensation to increase the gain margin. Thus, the static gain can be increased without decreasing the degree of stability. Figure 4.6-52 shows the result of choosing a lag-function compensator with its denominator break frequency set at one-tenth the lowest break frequency of the plant. Notice that the crossover frequency is essentially unaffected and that the gain margin has been increased by nearly 20 dB = 10.

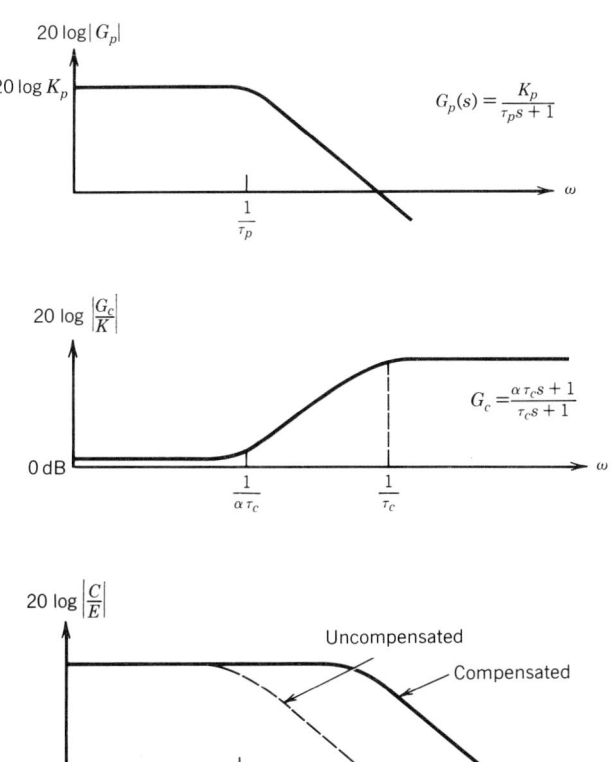

Fig. 4.6-50 Lead-function compensation of a first-order plant.

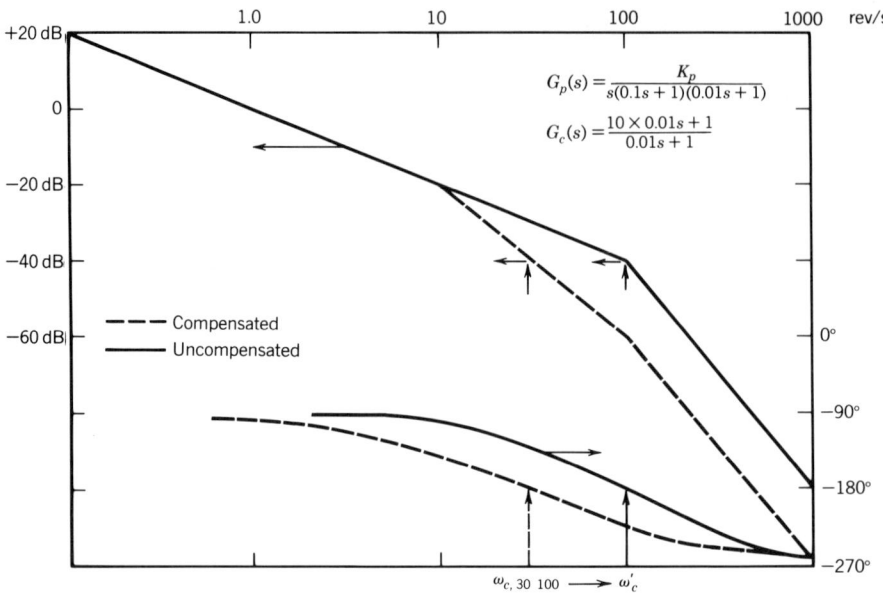

Fig. 4.6-51 Lead-function compensation of a dc Servo.

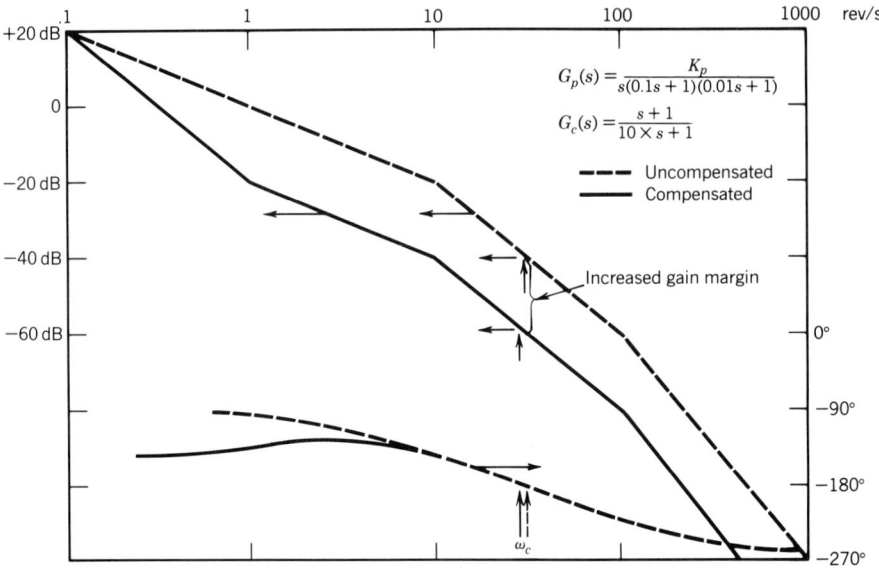

Fig. 4.6-52 Log-function compensation of a dc Servo.

Additional Topics

The feedback control topics presented here represent only a brief treatment of this rapidly expanding field. Only so-called classical control topics were treated. Additional topics under the heading of classical control include the following:

1. Linearizing nonlinear functional relationships.
2. Describing function analysis to predict limit cycle behavior.

3. Hall chart to predict closed-loop response from open-loop response.
4. Nichols chart to predict closed-loop response from open-loop response and to carry out the design of compensation networks.
5. Indices of performance for control systems such as rise time, settling time, peak overshoot, and resonant frequency.
6. Indices of performance such as integral-square error (ISE), integral-absolute error (IAE), integral-time absolute error (ITAE), and integral-time-square error (ITSE).[7]
7. Computer simulation of control systems, including analog, digital, and hybrid computers.

The recent and vast impact that the microcomputer is having and will continue to have is spawning numerous topics in control. Some of these are

1. sampled-data theory, including Z transforms;[11]
2. programming languages for real-time applications;
3. digital controller algorithms to replace the PID algorithm; and[12]
4. communication protocols for networks of digital computers.

Another group of topics that fall under the general heading of "modern control" have been available for the past 10–15 years but are still not widely used in industrial control. These include the following:

1. State-space time domain description of open- and closed-loop control systems (as opposed to Laplace transform description).[9]
2. Observability and controllability.[15]
3. Optimal control, using a quadratic index of performance and state variable feedback.[13,14,15,16]
4. System identification.[17]
5. Nonlinear stability analysis.[13]

REFERENCES

4.6-1 J. Lowen Shearer, Arthur T. Murphy, and Herbert H. Richardson, *Introduction to Systems Dynamics*, Addison-Wesley, Reading, MA, 1967.
4.6-2 Robert H. Cannon, *Dynamics of Physical Systems*, McGraw-Hill, New York, 1967.
4.6-3 N. H. Cook and E. Rabinowicz, *Physical Measurement and Analysis*, Addison-Wesley, Reading, MA, 1963.
4.6-4 Ernest O. Doebelin, *Measurement Systems: Application and Design*, McGraw-Hill, New York, 1976.
4.6-5 *Temperature Measurement Handbook and Encyclopedia*, Omega Engineering, Inc., Stamford, CT.
4.6-6 *Bell and Howell Pressure Transducer Handbook*, CEC/Instruments Division, Bell and Howell Company, Pasadena, CA, 1974.
4.6-7 Richard C. Dorf, *Modern Control Systems*, Addison-Wesley, Reading, MA, 1967.
4.6-8 Benjamin C. Kuo, *Automatic Control Systems*, Prentice-Hall, Englewood Cliffs, NJ, 1975.
4.6-9 William J. Palm, *Modeling, Analysis and Control of Dynamic Systems*, Wiley, New York, 1983.
4.6-10 J. G. Ziegler and N. B. Nichols, "Optimum Settings for Automatic Controllers," *Transactions of the ASME* **64**, November 1942.
4.6-11 Roberto Saucedo and Earl E. Schiring, *Introduction to Continuous and Digital Control Systems*, Macmillan, New York, 1968.
4.6-12 Gene F. Franklin and J. David Powell, *Digital Control of Dynamic Systems*, Addison-Wesley, Reading, MA, 1980.
4.6-13 Katsuhiko Ogata, *Modern Control Engineering*, Prentice-Hall, Englewood Cliffs, NJ, 1970.
4.6-14 Jay C. Hsu and Andrew U. Meyer, *Modern Control Principles and Applications*, McGraw-Hill, New York, 1968.
4.6-15 Yasundo Takahashi, Michael J. Rabins, and David M. Auslander, *Control and Dynamic Systems*, Addison-Wesley, Reading, MA, 1970.
4.6-16 Huibert Kwakernaak and Raphael Sivan, *Linear Optimal Control Systems*, Wiley, New York, 1972.
4.6-17 N. Sinha and B. Kuszta, *Modeling and Identification of Dynamic Systems*, Van Nostrand and Reinhold, New York, 1983.

4.7   GAS COMPRESSORS

A. W. Loomis

4.7-1   Introduction

Any device designed to compress a gas, including air, or gas mixtures to a pressure above the inlet pressure is a gas compressor. Most such devices are either positive-displacement compressors or dynamic compressors.

Positive-displacement units may be (1) reciprocating units in which the compressing element is a piston having a reciprocating motion in a cylinder; (2) vane-type rotaries having a vaned rotor eccentrically mounted in a cylindrical casing; (3) liquid piston rotaries, in which a liquid acts as a piston in an elliptical casing; (4) straight-lobe rotaries in which two mating lobes trap gas in a dual cylinder and carry it from inlet to discharge, compressing by back pressure from the discharge system; or (5) helical-lobe rotaries in which two intermeshing rotors in helical form compress and displace the gas.

Dynamic compressors are continuous-flow machines in which the compressing element rotates, rapidly accelerating the gas as it flows through it, creating velocity head, which is converted to pressure in stationary diffusors or blades.

Dynamic units may be (1) centrifugal units in which the compressing elements are impellers, (2) axial flow units in which the compressing elements are rotating blades followed by stationary blades, (3) mixed-flow units in which the impellers combine some characteristics of both centrifugal and axial types. Diagrams of the various types are shown in Figs. 4.7-1 and 4.7-2.

Fans, blowers, and thermal compressors for low-pressure use are covered in Section 4.2 as are ejectors for vacuum use. Major applications for compressors are shown in Table 4.7-1.

4.7-2   Compressor Arrangements and Characteristics

Reciprocating compressors are built in an almost innumerable number of different cylinder arrangements. Some of the more common are shown in Fig. 4.7-3.

Compressing a gas raises its temperature, so it is necessary to control cylinder temperatures by cooling.

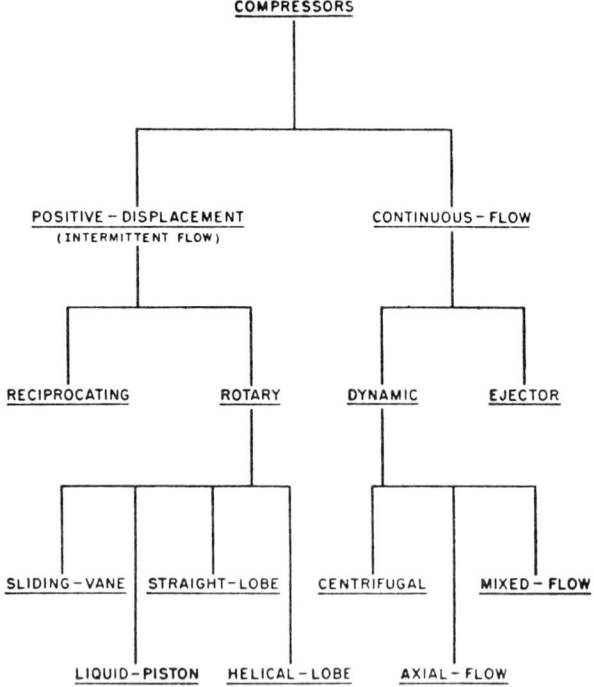

Fig. 4.7-1   Diagram of various compressor types. (courtesy of Ingersoll-Rand.)

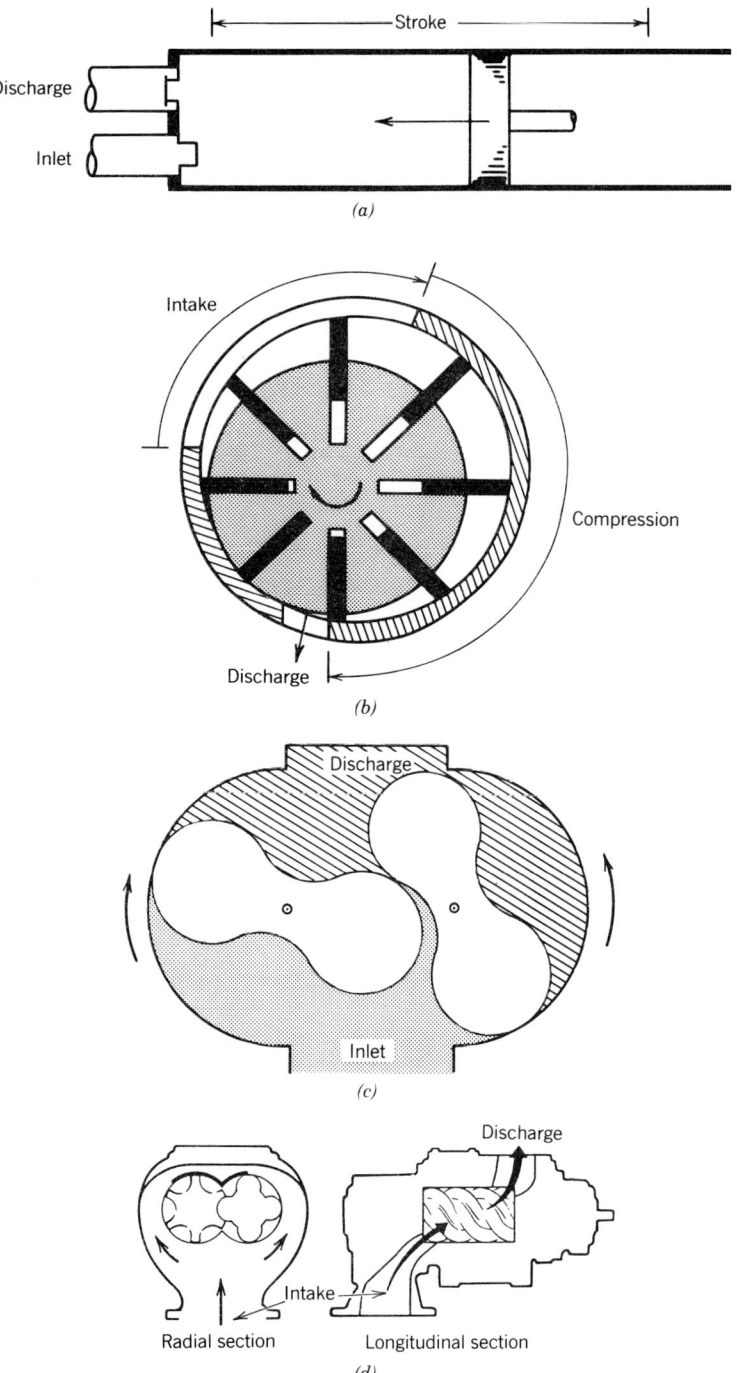

Fig. 4.7-2 Positive-displacement compressors: (*a*) Reciprocating (courtesy of Ingersoll-Rand); (*b*) vane-type rotary (courtesy of Ingersoll-Rand); (*c*) straight-lobe rotary (courtesy of Ingersoll-Rand); (*d*) helical-lobe rotary (courtesy of Ingersoll-Rand); (*e*) liquid-piston rotary (courtesy of Nash Engineering Company). Diagrams of a variety of compressor types: (*f*) Single-stage centrifugal (courtesy of Ingersoll-Rand); (*g*) multistage uncooled centrifugal (courtesy of Ingersoll-Rand); (*h*) axial flow (courtesy of Ingersoll-Rand); (*i*) single-stage mixed-flow compressor (courtesy of Ingersoll-Rand).

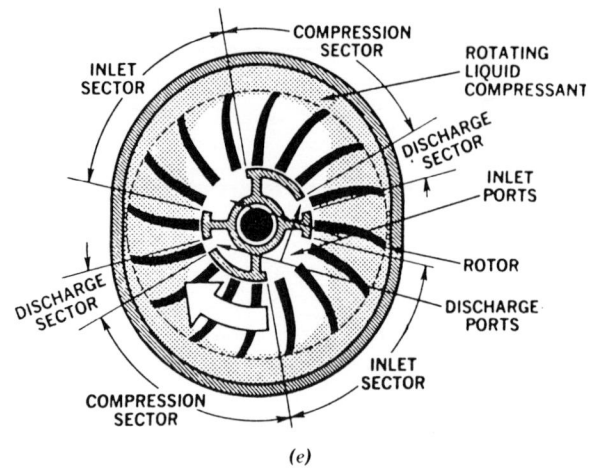

(e)

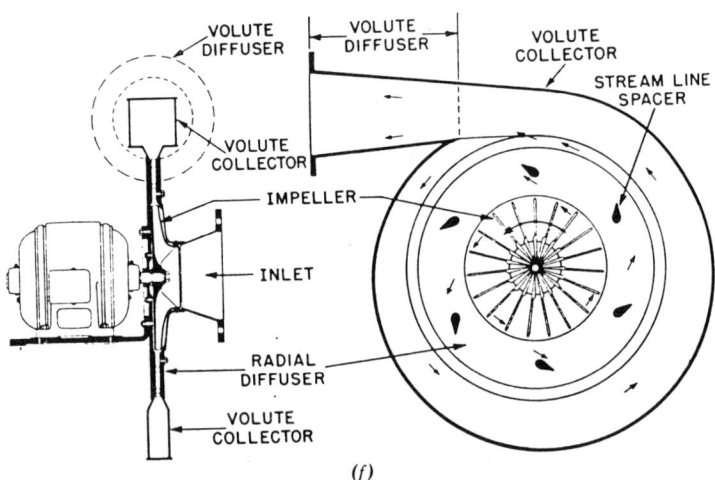

(f)

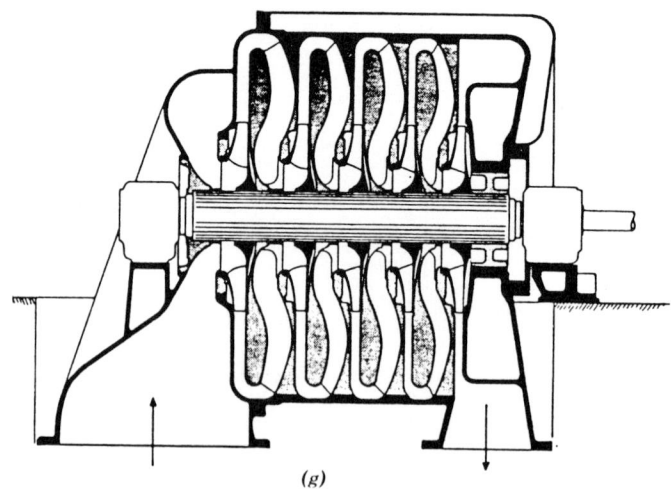

(g)

Fig. 4.7-2 (*Continued*)

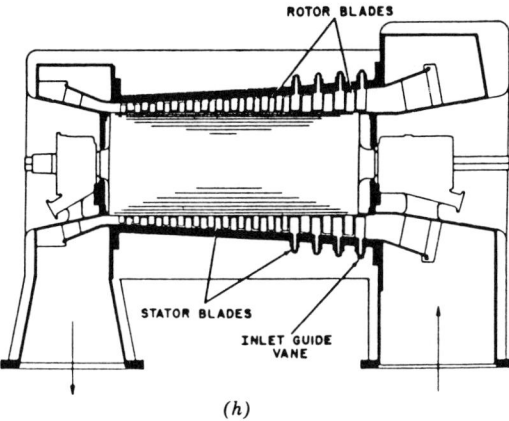

(h)

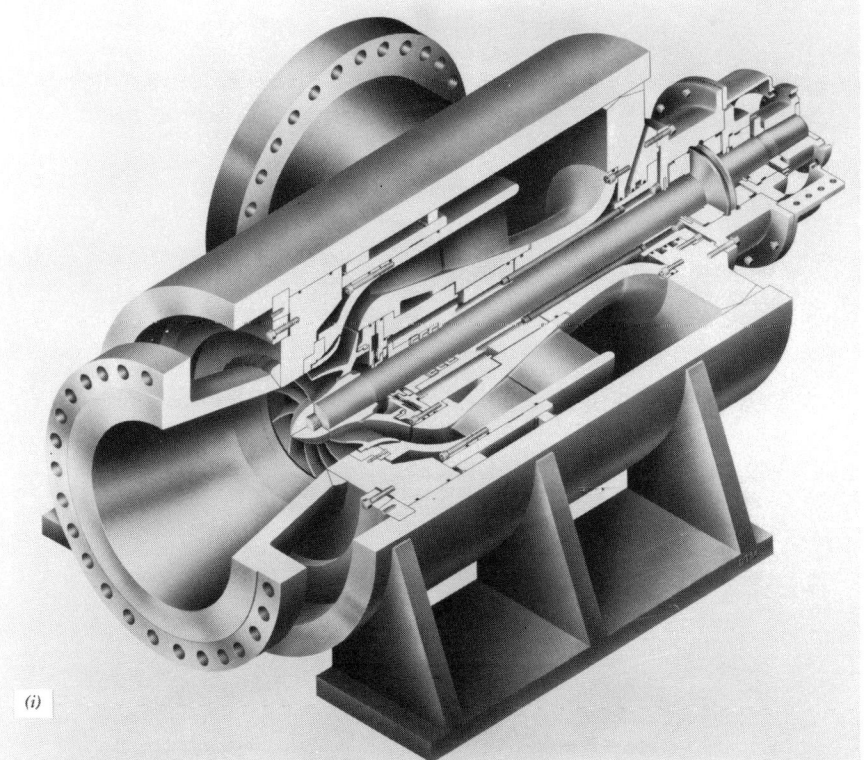

(i)

Fig. 4.7-2 (*Continued*)

In some of the smaller units the cylinders may be air-cooled by means of fins cast in the sides. In larger units they are usually water-cooled by jacketing the cylinders.

Since raising the temperature of a gas also increases its volume, most compressors utilize coolers (intercoolers) between stages to reduce gas temperature and volume. This saves power as well as limiting the final discharge temperature of the gas to a safer figure.

Reciprocating compressors have the following characteristics:

1. They are essentially constant-volume, variable-pressure units.
2. They deliver a pulsating flow of gas. Pulsation suppression devices are commonly used to minimize the effect of pulsation in piping.

TABLE 4.7-1 MAJOR APPLICATIONS FOR COMPRESSORS

Gas Handled	Industry or Application	Service
Air	Mining and quarrying	Rock drill operation (energy transfer)
		Tunnel ventilation
	Metal manufacturing	Tool operation (energy transfer)
	Contracting, construction	Rock drill or tool operation (energy transfer)
	Iron and steel	Combustion and oxidation, air
	Power plants	Soot blowing and tool operation, instruments
	Sewage treatment	Agitation
	Food and pharmaceutical processing	Agitation
	Material handling	Conveying
	Glass	Bottle molding
	Iron ore concentration	Pelletizing
	Engines	Supercharging and scavenging
	Plant power	Tool operation and instruments
	Nitric acid, ammonia, air separation	
Acetylene	Chemical and petrochemical	Gas recovery
Ammonia	Meat packaging—brewing	Refrigeration
Blast furnace gas	Iron and steel	
Carbon dioxide	Chemical, petrochemical, brewing, petroleum	Refrigeration, enhanced recovery of petroleum
Coke oven gas	Iron and steel	Compressing and exhausting, feed gas
Helium	Chemical	Recovery

Hydrocarbons
  Butadiene
  Butane
  Ethane
  Ethylene
  Hydrocarbon mixtures
  Isobutane
  Methane
  Propane
  Propylene
	Chemical and petroleum processing	Compression, feed gas, refrigeration, catalytic reforming, alkylation, catalytic cracking, raw gas supply, refrigeration

Natural gas	Petroleum and gas distribution Petroleum Natural gas	Repressuring oil wells, transmission, gasoline separation
Nitrogen	Air separation	Distribution, refrigeration, enhanced recovery
Oxygen	Air separation	Distribution
Syn-gas	Ammonia manufacture	Recirculation, fertilizer manufacture, synthesis

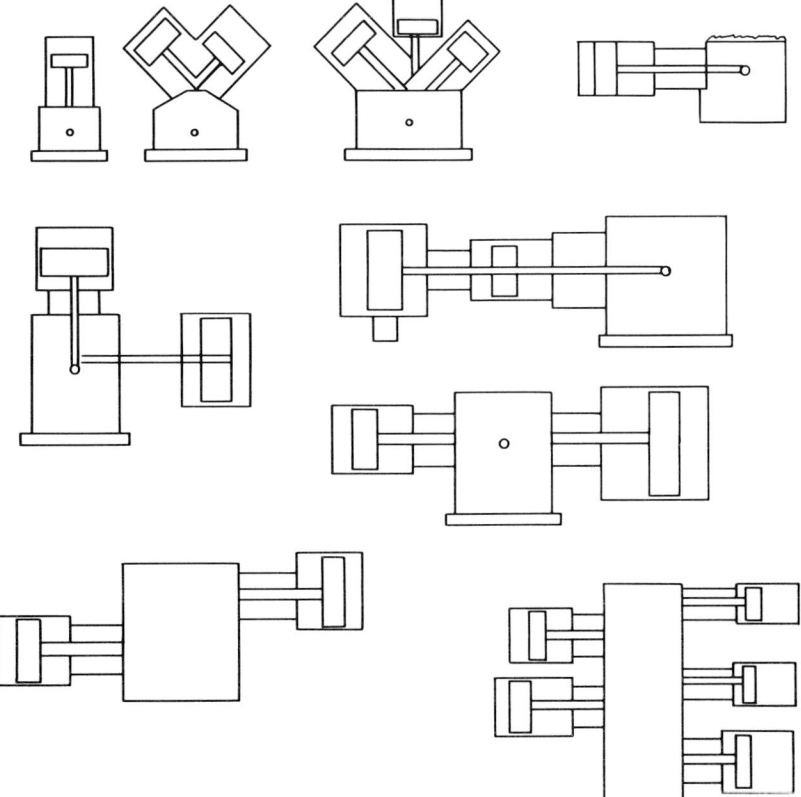

Fig. 4.7-3 A few of the possible cylinder arrangements for reciprocating compressors.

3. They cannot satisfactorily handle liquids but will handle vapors if condensation does not take place within the cylinders.

4. They can be fitted with capacity controls that will maintain their efficiencies at partial capacities.

5. They may have inertia forces that must be absorbed in the foundation. Sometimes this requires a larger foundation than might be required for a rotating-type compressor. Some are balanced-opposed with no unbalanced forces.

6. They require less power than any other equivalent unit.

Liquid piston rotary compressors are available in a limited range of sizes for a limited pressure range. They are cooled by feeding extra liquid into the casing. The extra liquid is separated after discharge by conventional baffle or centrifugal separators. They may be two-staged by connecting two machines in series.

They have the following characteristics.

1. Discharge is slightly pulsating.

2. Discharge gas is saturated at the discharge temperature of the compressing liquid.

3. Can handle saturated vapors, entrained liquid, and occasional foreign matter. Power required is considerably more than an equivalent reciprocating unit.

4. Only require bearing lubrication outside the compressing unit, and the delivered gas is therefore oil-free.

5. There are no unbalanced forces and foundation requirements are minimal.

Straight-lobe rotary compressors usually, but not necessarily, have two (sometimes three) symmetrical lobes of a figure-eight cross section. They are kept in phase by timing gears. Compression is by backflow from the discharge line. They may be two-staged by connecting two machines in series. They are generally air-cooled. Their characteristics are as follows.

1. Discharge is pulsating.
2. Some designs will handle a considerable amount of liquid carry-over in the intake gas—other designs should be protected from such carry-over.
3. Since there are few unbalanced forces, foundations are minimal in size.
4. Power requirements are high compared to reciprocating units.

Vane-type rotary compressors are built in two types—a conventional lubricating system with water-jacketed cylinders and water-cooled intercoolers and an oil-flooded type in which the machine operates in a flood of oil. The oil picks up the heat of compression, which is dissipated in a radiator or water-cooled oil cooler. Oil is separated from the gas before being cooled.
   Their characteristics are as follows:

1. High-frequency pulsations at inlet and discharge are not objectionable except as noise.
2. Require slightly more power than equivalent reciprocating units.
3. Have higher power requirement at partial loads than reciprocating units.
4. Clean gas is necessary to prevent undue vane wear.
5. There are no shaking forces so foundation requirements are minimal.

Helical-lobe rotary compressors are built in a dry or nonlubricated type and in an oil-flooded type. In the oil-flooded type oil separation devices separate the oil from the compressed gas. Two-stage units are available with both stages in a single casing with internal connections or two-staging may be accomplished by connecting two properly sized units together. Their characteristics are as follows:

1. Heavy-duty machines that can operate at full load for long periods.
2. Dry types deliver oil-free gas.
3. Require more power than equivalent reciprocating units at full and part load.
4. Foundations required are minimal due to lack of unbalanced forces.
5. Clean gas is necessary to prevent undue wear, loss of efficiency, and eventual failure.

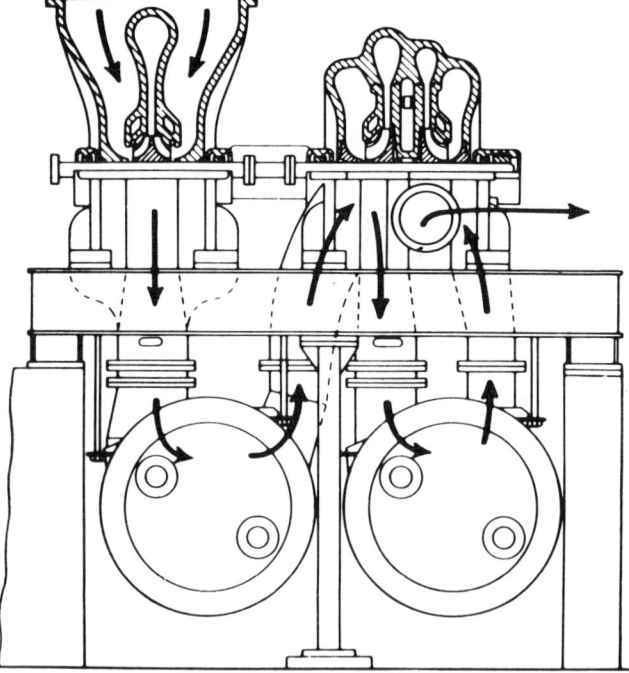

Fig. 4.7-4 A two-casing, three-stage centrifugal compressor with intercoolers between stages. (Courtesy of Ingersoll-Rand.)

Dynamic compressors are built in a variety of configurations for various uses with anywhere from 1 to 10 stages for centrifugal units and up to 15 stages in axial flow units in a single casing. More stages can be obtained by connecting two or more units in series (Figs. 4.7-4 to 4.7-6). Casings may be horizontally split, usually for moderate-pressure service, or vertically split for high-pressure service (Fig. 4.7-7). Low-pressure units are sometimes uncooled, but to save power and reduce temperatures, some interstage cooling is often used. This may be external cooling between stages or internal cooling in a package-type design. See Fig. 4.7-8.

Axial flow units are not usually cooled except after the final discharge stage (aftercoolers). Their characteristics are as follows:

1. They are basically large-capacity machines, with the trend toward the availability of smaller sizes; at present, down to 1200 cfm (34 m^3/m).

2. The operating speeds are high compared to other compressors.

3. They are well suited to steam or gas turbine drive.

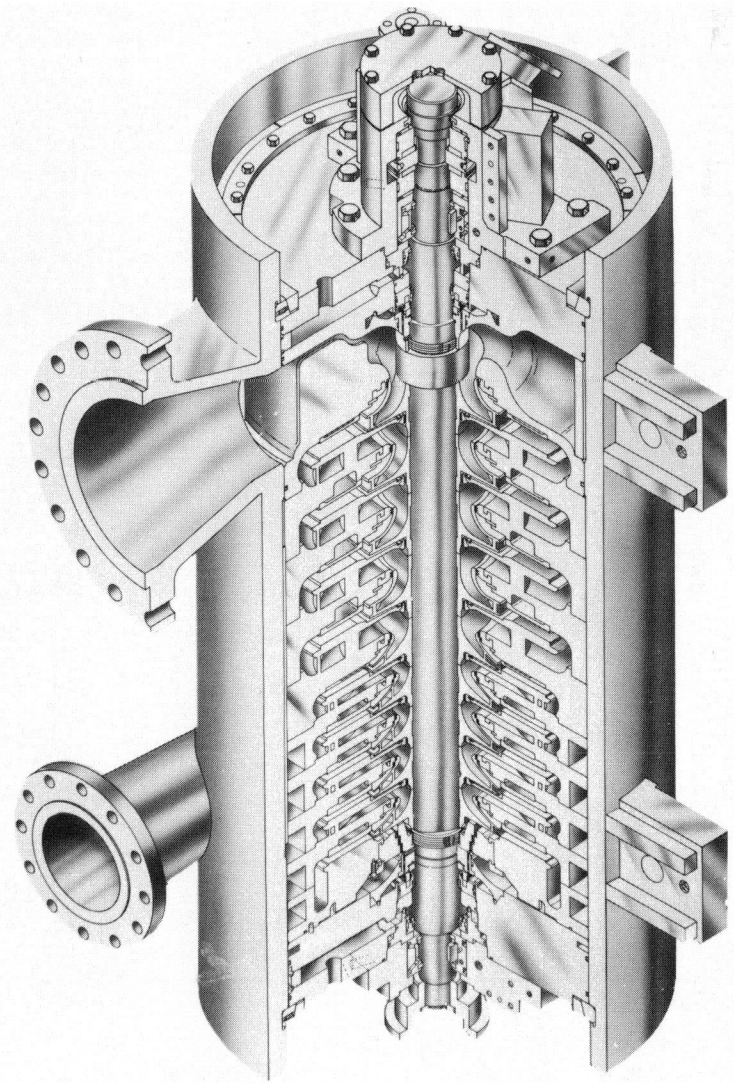

Fig. 4.7-5 Cross section of a 9-stage centrifugal compressor designed for comparatively low volume. (Courtesy of Ingersoll-Rand.)

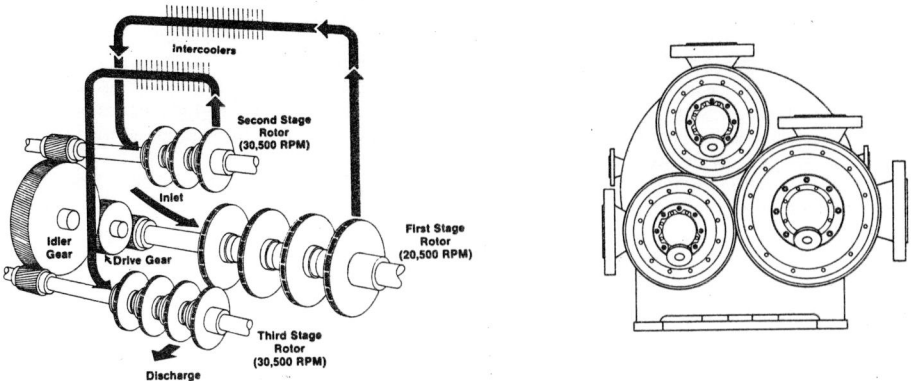

Fig. 4.7-6 A Multicasing centrifugal compressor with all casings in a single structure. Such units are built with one, two, or three casings depending on the pressure required. (Courtesy of Ingersoll-Rand.)

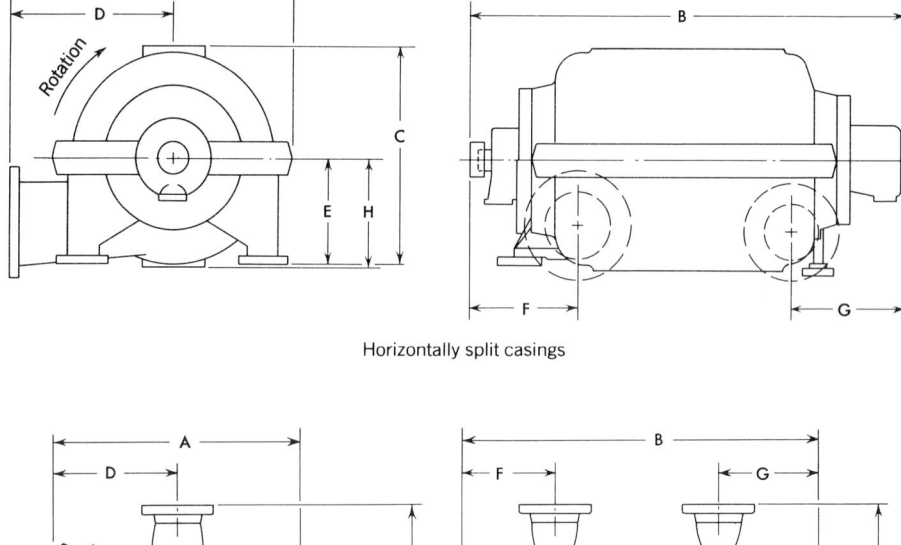

Horizontally split casings

Vertically split casings

Fig. 4.7-7 Diagrams of centrifugal compressors with horizontal and vertically split casings. (Courtesy of Ingersoll-Rand.)

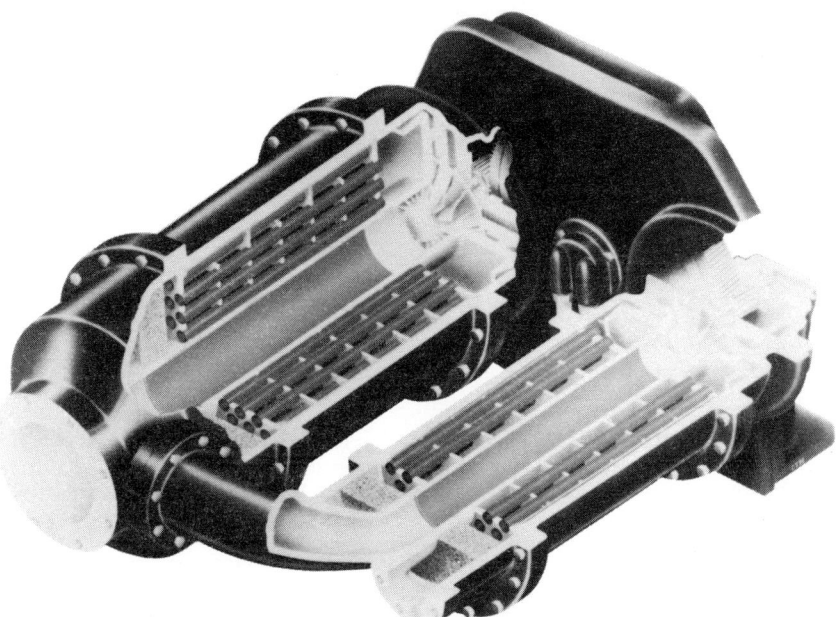

Fig. 4.7-8   Intercooling on two-stage package-type centrifugal compressor. (Courtesy of Ingersoll-Rand.)

4. Can be designed to handle higher-temperature gases than reciprocating compressors.
5. Have high availability—frequently operate without shutdown for 2 or 3 years.
6. Smooth nonpulsating flow within their stable operating range.
7. No inertia forces so foundations are comparatively small.

4.7-3   Selecting the Compressor

Almost any gas or vapor can be handled by a compressor. Some gases, however, may require compressors of special construction or materials to avoid corrosion or of nonlubricating type to avoid gas contamination or for safety or other reasons. A compressor manufacturer should be consulted who has experience with almost all gases and can produce a machine to suit your needs.

If your requirements are essentially similar to the requirements of other manufacturers or processors, the compressor builder will likely have a standard machine to fit your needs. He may even have it in stock ready to ship. If you have special needs, the builder can furnish a machine to meet those needs.

Determination of capacity and pressure requirements and possible variations in these factors during the process is important as is the possibility of introduced contaminants. Table 4.7-2 gives a rough idea of the sizes usually available in the different compressor types.

Many other considerations influence selection. In all cases the selection must be made on a basis of overall economy—considering both first cost and continuing costs. A few of the important considerations are discussed next.

1. **Floor space.**   Size and shape of available space will at times influence selection. The manufacturer will in many cases be able to offer designs to fit special needs.

2. **Foundations.**   Rotating units usually require smaller foundations because there are no unbalanced forces to be dampened. Reciprocating units usually require larger foundations for the same reason. Any of the rotating units (rotary or dynamic) may be supported on structural steel if necessary. Some reciprocating units may also be mounted on skids or other steel supports. All units must have sufficient support to handle the deadweight.

3. **Capacity control.**   The capacity of reciprocating air compressors can be controlled from zero to full capacity using several different types of control. Exact control needs should be discussed with

TABLE 4.7-2

Compressor Type	Approximate Maximum Power		Approximate Maximum Pressure (gauge)	
	(bhp)	(kW)	(psi)	(kPA)
Portable	500	375	350	2,415
Reciprocating	12,000 +	8,950	100,000	690,000
Helical-lobe rotary	6,000	4,475	350	2,415
Vane-type rotary	900	670	400	2,760
Centrifugal (dynamic)	35,000 +	26,100	10,000	69,000
Axial flow (dynamic)	100,000 +	74,600	100–150	690–1040

the manufacturer to select the most economical type. In process work involving gases other than air, it is usually not necessary to have as wide a range of capacity control.

Dynamic compressors can also be controlled as needed, but most cannot be operated at less than 60–80% of rated capacity because operating becomes unstable at low volumes.

Practically any type of control that is needed can be supplied from manual to semiautomatic to completely automatic. These can include automatic starting and protective devices for automatic warning or shutdown in case of unusual or unsafe operating conditions.

4. Volume. Process compressors are called upon to handle many different gases at varying inlet and temperature conditions—often in the same process. In some processes a large volume of gas is handled by dynamic compressors, and because the final discharge pressure is high, the final stages of compression are handled by reciprocating compressors. An example of this is a combination centrifugal and reciprocating unit, skid mounted, that can be installed in the field to produce nitrogen to inject for enhanced oil and gas recovery. It is normally more economical than carbon dioxide or natural gas for such services.

5. Temperature. There are practical limits to the maximum or minimum gas temperatures that each type of compressor can handle. In air service reciprocators are the least affected by temperature extremes. Centrifugal units are sensitive to temperature extremes. In process service dynamic compressors are less affected by temperature extremes than reciprocators. Centrifugal compressors have been used handling gas to 425°C (800°F) or even to 540°C (1000°F). Both dynamic and reciprocating compressors have been used with gas inlet temperatures below −75°C (−100°F).

6. Inlet and discharge pressures. Specifying the possible variations in inlet and discharge pressures when selecting a compressor is very important. If these changes are not foreseen, the design limits of the machine or its driver can be exceeded. Lowering the inlet pressure and maintaining the discharge pressure on a multistage reciprocating unit will lower the overall horsepower, lower the differential pressure on all stages except the last, and increase the differential pressure and temperature rise in the last stage. Raising the inlet pressure will have the opposite effect, raising the overall horsepower, increasing the differential pressures on all stages except the last, and lowering the temperature rise and pressure differential on the last stage (see Fig. 4.7-9).

In dynamic compressors, if inlet pressures are lowered, the machine will not compress to the desired discharge pressure. If inlet pressures are increased, the discharge pressure may increase depending on system resistance.

7. Cost. First cost is influenced by gas-specific gravity, foundation requirements, power source, and other factors. Power cost during the service life of a compressor is many times the first cost. Compressor efficiency is therefore of great importance and even though the most efficient machine may have a higher first cost, the power savings can quickly make up the difference.

8. Driver. Driver choice is usually dictated by the available source of power, use of waste gas, heat balance, or other factors in the process or plant. Compressors that best fit the power source should be considered. Steam or gas turbines are ideal drivers for many dynamic compressors. Electric motors fit the speed requirements of most reciprocating compressors and some dynamic units. If increasing or reducing gears are required, costs may increase.

Major types of drivers used for compressor drivers along with their characteristics are shown in Table 4.7-3. Not all possible combinations have been covered and larger sizes of some drivers can be obtained.

9. Driver-compressor assemblies. Usually the compressor manufacturer supplies the driver for a compressor. This is advantageous as it puts the responsibility for all engineering decisions on the

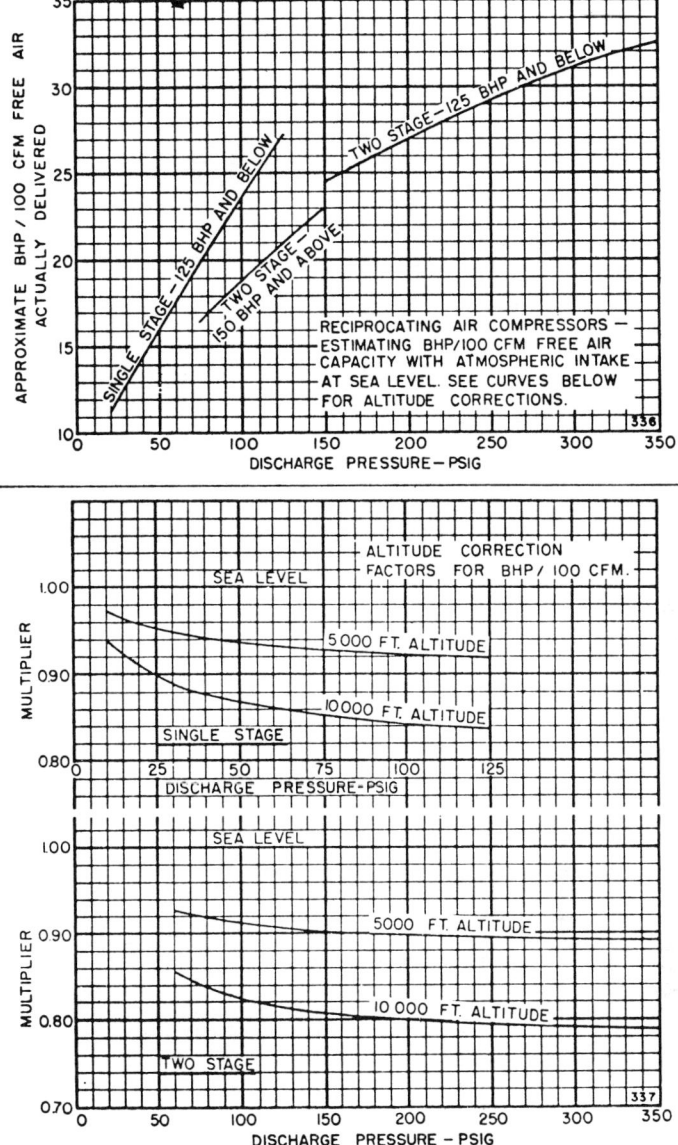

Fig. 4.7-9 Curves for estimating brake horsepower on reciprocating compressors, along with the necessary corrections for altitude. (Courtesy of Ingersoll-Rand.)

coupling of the two or more units in one place. The most common mounting of drivers for the various compressor types are shown in Table 4.7-4.

    10. Industrial engines. Industrial-type (as opposed to heavy-duty type) natural gas, diesel, and gasoline engines are used to drive compressors by direct connection or V belts. Such engines are available from 100 to 2000 hp and operate in a speed range of 600–2200 rpm. A service factor needs to be applied to the nominal rating for most compressor services otherwise there will be excessive maintenance and downtime.

       65%   For 24 h/day operation with minimum downtime and maintenance

       80%   For 24 h/day operation and reasonable time off for maintenance

TABLE 4.7-3

Driver type	Power Range (hp)	(kW)	Available Speed (rpm)	Possible Speed Variation	Efficiency	Starting Torque And amperes (% full load)	Stalling Torque (% full load)
Induction motor	1–15,000	0.75–12,000	440–3525 (with power at 60 Hz)	Constant speed	10 hp-83% 100 hp-91% 1000 hp-94%	60–100% torque and 550–650% amp	150% or more (min)
Synchronous motor	100–20,000	75–14,900	180–1800 (with power at 60 Hz)	Constant speed	93–99%	40% under 514 rpm 40–100% over 514 300 to 500% (amp)	150%
Steam engine	55–4000 +		140–400	100–20%	50–70%	About 120%	About 115%
Combustion gas turbine	To 50,000 +		1800–34,000	100–50%	16–38% thermal efficiency for simple open cycle 27–35% with regeneration[a]	Require a sizable starting motor or turbine. Single-shaft design has poor part load torque characteristics and requires larger starter. Two shaft turbines have good torque characteristics	
Steam turbine	50,000 +		1800–34,000	100–25%	35–82%	175–300%	Up to 300%
Integral gas engine	85 hp at 600 rpm, 6000 hp at 300 rpm			100–60%	Up to 40% overall thermal efficiency	None—started with compressed air	About 120%
Integral diesel engine	100 hp at 600 rpm ~ 1300 at lower rpm			100–60%	32% high-heat value	None—started with compressed air	About 120%
Coupled gas or diesel engine	100 hp at 600 rpm To 5000 hp at ~ 360 rpm			100–60%	41% on gas low-heat value 36.67 on oil high-heat value	None—started with compressed air	About 120%

[a]Regeneration is effective only with heavy industrial or low-pressure ratio gas turbines.

TABLE 4.7-4

Driver Type	Compressor Type			
	Reciprocating	Vane Rotary	Helical-Lobe Rotary	Dynamic
Induction motor	Belted Coupled Flange-mounted Geared	Belted Coupled	Belted Coupled Geared	Coupled Geared
Synchronous motor	Coupled Flange-mounted Geared Shaft-mounted	Seldom	Coupled Geared	Coupled Geared
Steam engine	Integral			
Steam turbine	Geared	Geared	Coupled Geared	Coupled
Gas turbine	Geared		Geared	Coupled Geared
Hydraulic turbine	Geared		Coupled Geared	
Expander	Integral Geared		Coupled Geared	Coupled
Gas engine or diesel engine (heavy duty)	Coupled Integral		Geared	Geared
Industrial engine	Geared			
	Belted Coupled	Belted Coupled	Coupled Geared	Belted Geared

90%	For maximum power and speed for intervals of up to 1 h followed by equal periods below maximum
100%	For maximum power and speed during short periods (5 min) followed by periods of reduced output

11. Expanders. Expanders are used for power recovery and to obtain refrigeration. As a power recovery unit, they can be used to drive compressors or generators. Essentially they expand a gas that exists at an elevated pressure and temperature down to a lower level of energy and extract useful work. They are also used for refrigeration by lowering the gas temperature through expansion. All expanders must be loaded to produce either power or refrigeration—the loading device is often a compressor, generator, or pump.

12. Hydraulic turbines. A hydraulic turbine is essentially a centrifugal pump operating in reverse. It is a suitable driver for a compressor when a liquid exists at a pressure level higher than is needed for its ultimate use. The turbine must be designed specially for the pressure conditions that exist in the plant.

BIBLIOGRAPHY

Compressed Air and Gas Data. Ingersoll-Rand, Washington, NJ, 1979.

Compressed Gas Handbook, NASA 3045, Supt. of Documents, U.S. Government Printing Office, Washington, D.C., 1969.

Cunningham, E. R. "Air Compressors," *Plant Engineering* **34**, 58–65, May 1, 1981.

Engineering Data Book. Gas Processors Suppliers Assn., Tulsa, OK.

"Equipment Guide: Compressors," *Process Engineering*, October 1981.

O'Keefe, W. "Compressed Air Systems," *Design* **37**, 48 (1981).

Popan, A. "How to Size Gas Compressors." *World Oil* **192**, 129–30, May 1981 and **192**, 225–226, June 1981.

"Quick Guide to Product Selection (Compressors, Vacuum Pumps, Blowers)," *Hydraulics and Pneumatics* **34**, 140, January 1981 and **35**, 136, January 1982.

Rex, M. J. "Compressor Packages." *Process Engineering*, June 1981.

Standards of the Compressed Air and Gas Institute. Keith Bldg., Cleveland, OH.

Thompson, Phillip A. *Compressible Fluid Dynamics*. McGraw-Hill, New York, 1972.

Vincenti and Krueger. *Physical Gas Dynamics*. Wiley, New York, 1965.

4.8 COOLING SYSTEMS

Dominique N. Brocard and John M. Burns

4.8-1 Introduction

Thermoelectric power production can only achieve partial transformation of heat to electric power, the remaining being released to the environment as waste heat. The amount of waste heat is now about 1.6 times the electric power output for fossil fuel plants and 2.1 times the electric output for nuclear plants.[1] From the point of view of the power production cycle, this heat loss occurs during cooling of the primary fluid in the condenser. The cooling medium is practically always water, which should have as low a temperature as possible to achieve low turbine back pressure and high cycle efficiency.

In once-through cooling systems cooling water is borrowed from a large body of water such as the ocean or a great lake or from a river and returned after passage through the plant. The water body size or flow insures constant supply of unheated water.

Closed-loop systems utilize a fixed amount of water (except for makeup and blowdown) which is cooled by passage through a cooling lake (natural) or pond (man-made), spray pond, or cooling tower. In all cases the ultimate heat sink is the atmosphere, but each system has individual characteristics, with advantages, disadvantages, and requirements, as summarized in Table 4.8-1. Waste heat utilization is still in its infancy, with attempts in the areas of residential or greenhouse heating, aquaculture, increased agricultural yields, or deicing of shipping waterways.

TABLE 4.8-1 CHARACTERISTICS OF COOLING SYSTEMS

Type of Cooling System	Advantages	Disadvantages	Requirements
Once-through			
Rivers and estuaries Lakes and oceans	Comparatively low cost Minimum water loss (through added) Lowest intake temperature Minimum visual impact	Environmental concerns Waste-heat-related mortality or migration Water quality impairment Fish containment or impingement at intakes	Adequate flow (rivers), flushing (estuaries) or net current (lakes or oceans) Discharge structure to produce rapid mixing of cooling water with ambient (may require extensive piping) Fish exclusion systems
Closed Loop Cooling ponds and lakes	Comparatively low cost when conditions are favorable Moderate water loss (through evaporation and seepage) Recreational value	Environmental concerns (for lakes) Waste-heat-related mortality or migration Water quality impairment through enhanced stratification	Suitable lake or large floodable area (0.5–2 acres/MW$_t$)
Spray ponds	High cooling per unit surface area	Operational problems High water loss (through evaporation)	Water treatment
Cooling towers	See Table 4.8-3		

4.8-2 Intake Structures

Once-through cooling systems and cooling ponds use cooling water intakes, which can be located on the shore or submerged offshore. Shore intakes involve a pump pit open on one side to the water body while offshore intakes are connected to the pit by a pipeline or tunnel. Several axial- or mixed-flow pumps are usually used with a wet-pit configuration. The design of the pit or sump is critical in determining the approach flow to the pumps and the occurrence of vortices, which can cause pump vibrations, reduced flow rate, and air entrainment. Recommended pit configurations and dimensions are given by the Hydraulic Institute[2] and the British Hydromechanic Research Association.[3] Dimensions from the latter are given in Fig. 4.8-1, but hydraulic scale model tests are often required to confirm or perfect the design, particularly for intakes from streams. Fish exclusion systems are often needed at cooling water intakes. Fine-mesh screens can be used with low throughflow velocities (0.2–0.5 ft/s) to avoid impingement. Angled screens with intake bypass (Fig. 4.8-2) have been shown to be very effective for shoreline intakes.[4] When the bypass cannot be easily connected back to the

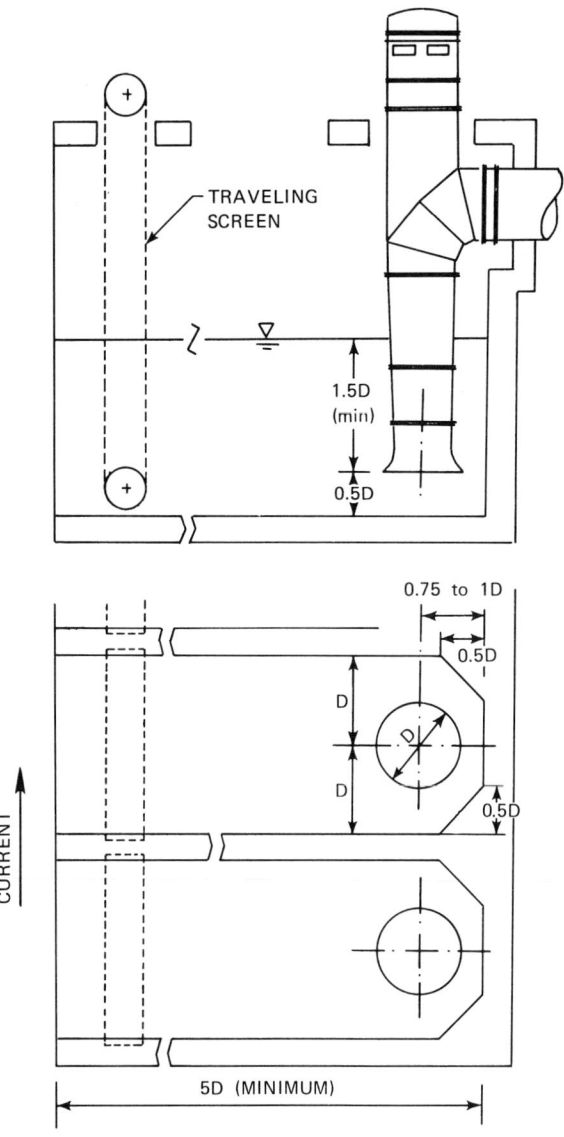

Fig. 4.8-1 Recommended pump pit dimensions. (Adapted from ref. 32.)

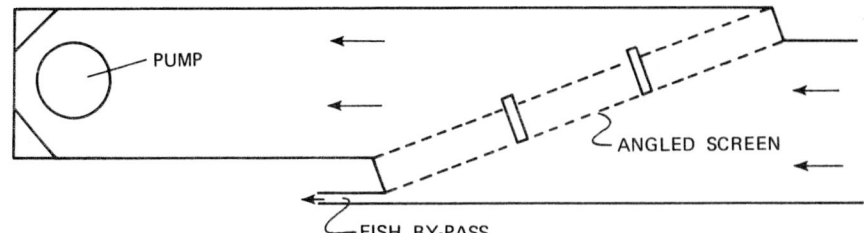

Fig. 4.8-2 Angled screen for fish exclusion.

ambient, fish transport through pipelines may be required and peripheral jet pumps can be used to drive the flow with minimal impact to the fish. Behavioral barriers, such as bubble screens, hanging chains, or water jet curtains, in the place of screens reduce but do not eliminate fish entrainment.

Recirculation (from discharge to intake) can usually be avoided by properly choosing the relative locations of the intake and discharge structures, for example, intake upstream of discharge. Since heated discharges are buoyant, they often result in stratification where the waste heat is confined to a surface layer, and in this case recirculation may be avoided through selective withdrawal.[5-8] At surface intakes this can be achieved with a skimmer wall, as shown in Fig. 4.8-3. For the two-dimensional channel case the critical withdrawal flow rate above which fluid from the upper layer becomes entrained is[9]

$$Q_c = B\sqrt{g'\left(\tfrac{2}{3}h_r\right)^3}$$

where the dimensions are as shown in Fig. 4.8-3 and g' = buoyancy of the upper layer = $g\Delta\rho/\rho$ = $g\beta\Delta T$ with g = acceleration of gravity, $\Delta\rho$ = layer density difference, ρ = ambient water density, β = coefficient of thermal expansion, and ΔT = layer temperature difference. The critical flow is independent of the opening height, h_s, but the latter should be less than the lower layer depth, h_2, in the channel. This depth is such that

$$h_r = h_2 + \frac{Q^2}{2B^2 g' h_2^2}$$

If the withdrawal flow rate Q is equal to Q_c, then $h_2 = \tfrac{2}{3}h_r$. These results can be applied to a radial skimmer wall by replacing B by the wall length $\theta_w r_w$. Submerged intakes can be simulated by the

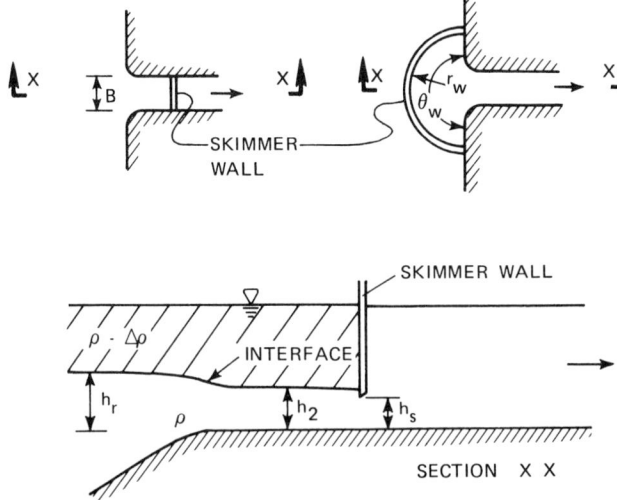

Fig. 4.8-3 Skimmer wall. (From ref. 9; used by permission.)

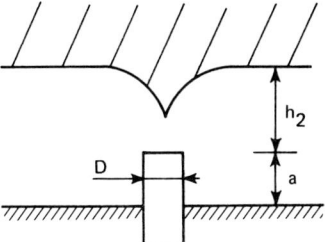

Fig. 4.8-4   Selective withdrawal for submerged intake. (From ref. 9; used by permission.)

configuration shown in Fig. 4.8-4. The critical flow rates are[6]

$$Q_c = 1.61\sqrt{g'h_2^5} \quad \text{for} \quad \begin{cases} a = 0 \\ 0.3 < D/h_2 < 1.1 \end{cases}$$

$$Q_c = 4.45\sqrt{\frac{D}{h_2}}\sqrt{g'h_2^5} \quad \text{for} \quad \begin{cases} a = \infty \\ 0.3 < D/h_2 < 14 \end{cases}$$

4.8-3   Discharge Structures—Nearfield Thermal Analyses

Waste heat discharges can be at the surface or submerged. Surface discharges are typically through a channel at an angle to the shoreline, producing a surface buoyant jet. Submerged outfalls may involve a single port discharging a submerged buoyant jet or a multiport diffuser discharging numerous buoyant jets that coalesce to form a more complex plume. In general, multiport diffusers, of which there are several types, achieve a more rapid dilution of the effluent than either surface or submerged single-port discharges.

Near the discharge, in a region called the *discharge nearfield*, the mixing of the effluent with ambient water is primarily due to turbulent entrainment into the discharge jet (or series of jets) and the thermal plume characteristics are very dependent on the discharge conditions. Because of the generally irregular boundary conditions, the number of physical phenomena to be considered, and the complexity of fluid turbulence, realistic plume analyses often require rather involved modeling efforts: Mathematical models have been developed using numerical methods to integrate simplified forms of the governing equations, but physical hydraulic models are still required in some cases. The following nearfield results, however, can be used for simple cases or for preliminary analyses.

Parameters common to all nearfield analyses are the discharge flow rate Q_0, its velocity u_0, and its buoyancy $g'_0 = g\Delta\rho_0/\rho = g\beta\Delta T_0$ with g = acceleration of gravity, $\Delta\rho_0$ = initial density difference, ρ = ambient density, and ΔT_0 = discharge temperature rise. In most cases an ambient current, V, must be taken into account, as well as the depth of the receiving water, H.

Surface buoyant jets result from surface discharges of cooling water, typically through a channel of depth h_0 and width $2b_0$ (see Fig. 4.8-5). Equivalent depths and widths must be defined for nonrectangular sections. The discharge densimetric Froude number, $F_0 = u_0/\sqrt{g'_0 h_0}$, characterizes the relative magnitude of the jet momentum and buoyancy. For $F_0 < 1$ (high buoyancy) an ambient

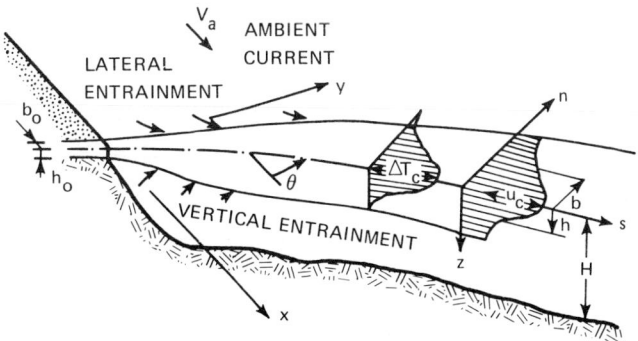

Fig. 4.8-5   Surface buoyant jet.

water wedge intrudes in the channel and the discharge depth h_0' is such that $u_0/\sqrt{g_0' h_0'} = 1$. These conditions may be approached in a cooling pond where entrance mixing should be minimized. For once-through cooling systems rapid mixing of the effluent with ambient water is desired and values of F_0 much larger than 1 are used (high momentum). When there is no current and the receiving water depth H is large, the longitudinal variations of the jet centerline temperature, width, and depth can be represented in condensed form,[10] as shown in Fig. 4.8-6, where $F_0' = u_0/\sqrt{g_0' l_0}$ and $l_0 = \sqrt{h_0 b_0}$. The half-width, b, and depth, h, given in Fig. 4.8-6 are actually length scales corresponding to assumed gaussian forms of the lateral and vertical profiles of excess velocity and temperature:

$$u = V\cos\theta + u_c \exp\left(-\frac{n^2}{b^2}\right)\exp\left(-\frac{z^2}{h^2}\right)$$

$$\Delta T = \Delta T_c \exp\left(-\frac{n^2}{b^2}\right)\exp\left(-\frac{z^2}{h^2}\right)$$

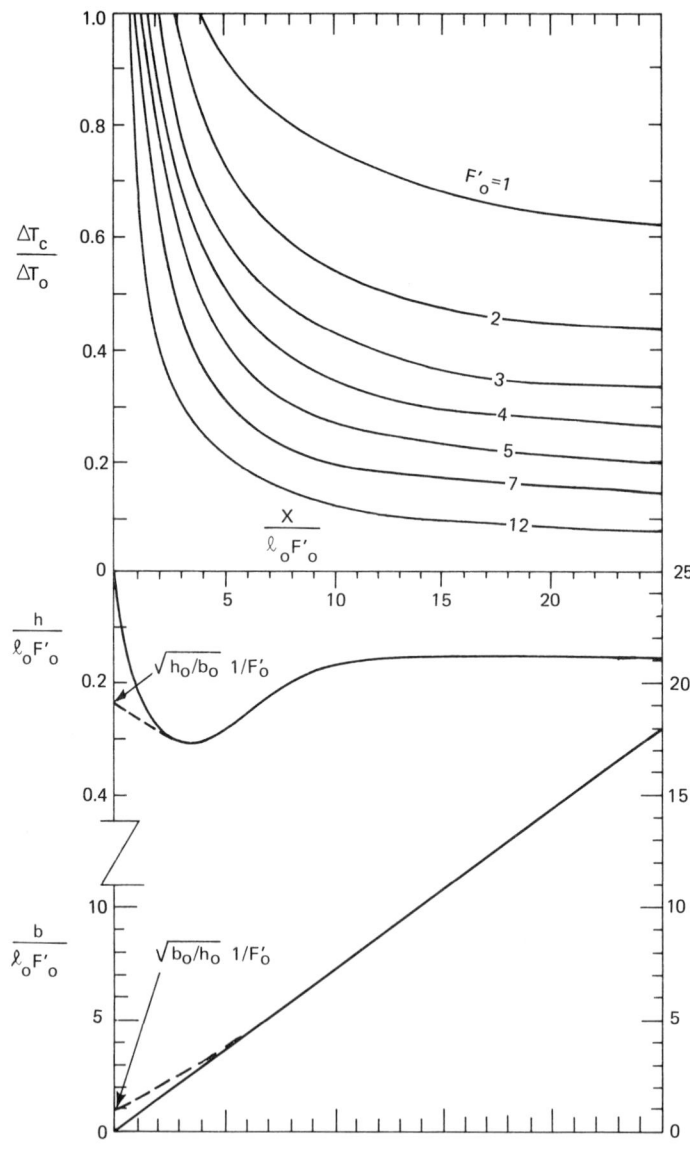

Fig. 4.8-6   Surface buoyant jet characteristics in deep stagnant water.

The subscript c stands for centerline, and the variables are shown in Fig. 4.8-5. Although the gaussian profiles extend to infinity, the total jet half-width and depth are often taken as $\sqrt{2}\,b$ and $\sqrt{2}\,h$. It can be seen that the jet depth first increases, but after some distance buoyancy accelerates the lateral spreading of the jet and h decreases to a plateau at about $0.17 l_0 F_0'$. For shallow receiving, water vertical entrainment through the bottom of the jet is hindered and the longitudinal temperature decay is slower than with large depths. Bottom hindrance begins for H less than about $0.56 l_0 F_0'$.[10]

The effect of a current is to deflect the jet and somewhat increase the rate of dilution. An example trajectory, temperature, width, and depth plot obtained with a mathematical model[11] is shown in Fig. 4.8-7 for $F_0 = 4$, $b_0/h_0 = 5$, and different cross-current velocities. For other values of these parameters one can use other such plots from the same source or a mathematical model, of which several are available.[12-14]

Submerged buoyant jets result from single-port submerged discharges in deep water. The jet characteristics are not very dependent on the shape of the discharge port, but rather on its area. The discharge port is, therefore, characterized by its diameter, or equivalent diameter, D_0, equal to that of a circle with equal area. The discharge densimetric Froude number is $F_0 = u_0/\sqrt{g_0' D_0}$. When the ambient is stratified, with a current, temperature predictions can be made with an integral-type mathematical model.[15-17] Plots of jet trajectory, centerline temperature, and width are also available for many sets of conditions.[18] Such results are given in Fig. 4.8-8 for a stagnant, unstratified ambient, with discharge angles $\alpha = 0°$ and $30°$ above horizontal. Jet width b corresponds to assumed gaussian profiles for excess velocity and temperature:

$$u = u_c\exp\left(-\frac{r^2}{b^2}\right) \qquad \Delta T = \Delta T_c\exp\left(-\frac{r^2}{\lambda^2 b^2}\right)$$

where r = radial distance from centerline and $\lambda = 1.16$ is the Schmidt number, introduced to account for the more rapid diffusion of heat than momentum. Centerline temperature and jet width are given in Fig. 4.8-9 for vertical discharges ($\alpha = 90°$). When the ambient is stratified, the jet trajectory may be limited vertically to a maximum height of rise because entrainment of deep, dense ambient water, together with the vertically decreasing ambient density lead to a situation where the jet is denser than its surroundings.

Multiport diffusers are designed to provide rapid dilution of the effluent with ambient water by increasing the total outside jet area through which entrainment takes place. The behavior of diffusers

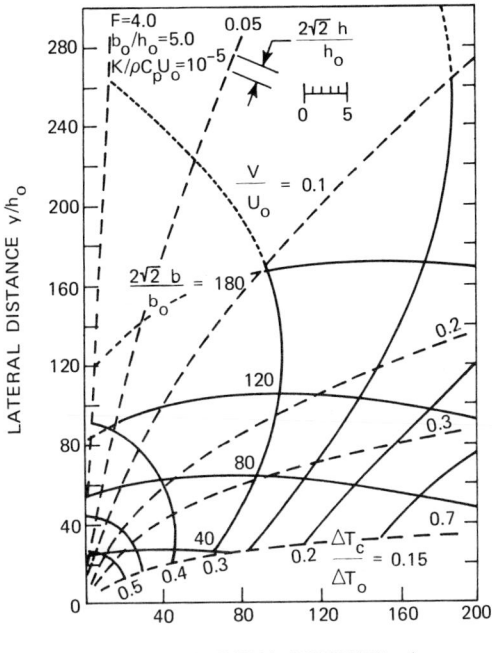

Fig. 4.8-7 Trajectory, temperature, width, and depth plot for surface buoyant jet showing the effect of current speed. (From ref. 11.)

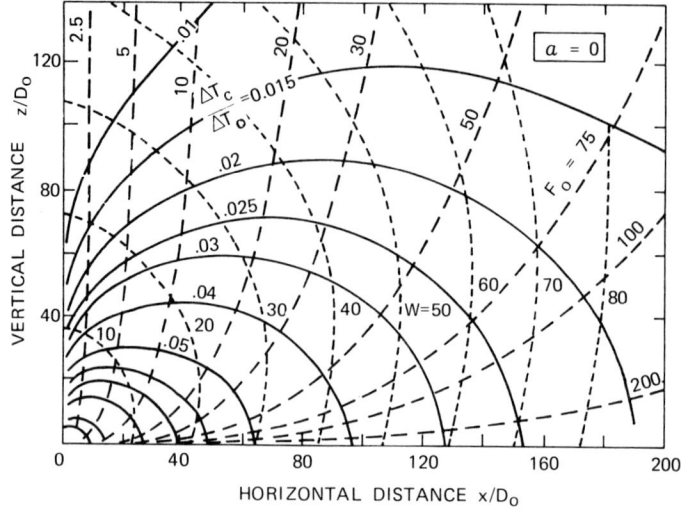

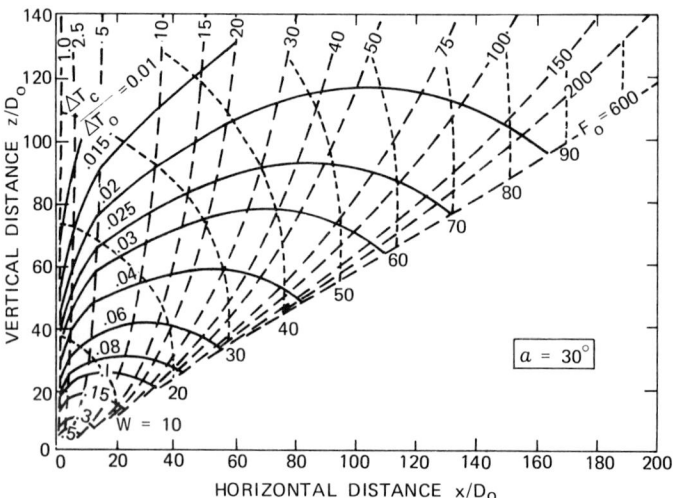

Fig. 4.8-8 Trajectory, centerline temperature, and width (W-$2\sqrt{2}\,\lambda b/D_0$) of submerged round buoyant jets discharged horizontally (top) and at 30° above horizontal. (From ref. 18.)

is markedly different in shallow water, where the plume occupies the full water depth above the diffuser, and in deep water. A complete thermal analysis of a diffuser plume requires sophisticated mathematical models or a physical hydraulic model when the topography is complex or when current reversals exist. Different diffuser types with different arrangement and orientation of nozzles have been developed. Figure 4.8-10 shows four basic types, with formulas for the maximum nearfield temperature rise in shallow water conditions. The nozzle spacing is then usually taken to be on the same order as the water depth and, with the jets rapidly merging, the diffuser can be simulated as a line source of mass and momentum with an equivalent width $B_0 = Q_0/u_0 L$. The coflowing diffuser is to be used when the ambient current is only in one direction, such as a river. The momentum of the discharge then causes the entrainment of water coming from an upstream region wider than the diffuser.[19] Both the tee and staged diffusers induce a net offshore momentum, directing the effluent away from the ecologically important nearshore, and both diffusers are indifferent to the current direction. The dilution of the tee diffuser, however, decreases with increasing ambient current, while that of the staged diffuser increases with current.[20-22] Staged diffusers are therefore preferred when currents are prevalent and the net offshore momentum reduces reentrainment upon changes of current direction.

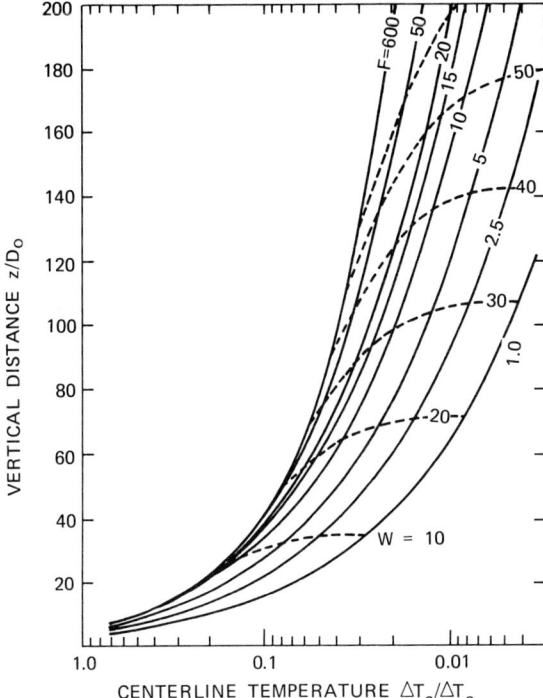

Fig. 4.8-9 Centerline temperature and jet width ($W = 2\sqrt{2}\,\lambda b/D_0$) for buoyant jet discharged vertically. (From ref. 18.)

The alternating diffuser relies on a buoyancy-induced countercurrent to obtain dilution water, with no net discharge momentum and small induced current velocities.[23]

In deep water a diffuser discharge can be modeled as a submerged two-dimensional buoyant jet.[24] When the ambient is stratified with a current, the plume can be analyzed with an integral mathematical model. Results for various discharge angles, ambient currents, and stratification are available.[18] Figure 4.8-11 gives plume trajectories, centerline temperature, and width for a horizontal discharge in a stagnant, uniform ambient.

4.8-4 Farfield Thermal Analyses

At large distances from the discharge point the mechanisms controlling the heated plume temperatures are diffusion by ambient turbulence and surface heat transfer to the atmosphere. The discharge configuration, its momentum, and buoyancy become irrelevant. The waste heat is often distributed over the total water depth or an upper layer depth, such as the epilimnion, and the vertically integrated conservation of heat equation can be used:

$$\frac{\partial \Delta T}{\partial t} + u\frac{\partial \Delta T}{\partial x} + v\frac{\partial \Delta T}{\partial y} = \frac{\partial}{\partial x}\left(E_x\frac{\partial \Delta T}{\partial x}\right) + \frac{\partial}{\partial y}\left(E_y\frac{\partial \Delta T}{\partial y}\right) + \frac{\phi_n}{\rho C_p H} \tag{4.8-1}$$

where t = time
$u, v,$ = velocity components in x and y directions
E_x, E_y = dispersion coefficients
ϕ_n = net surface heat flux (see Section 4.8-5)
C_p = specific heat of water
H = depth

Estimating the dispersion coefficients is a difficult problem; the following expression can be used for practical applications[4,25] (fortunately, temperature results are not very sensitive to the exact value used): $E_x = E_y \simeq 0.023 l^{1.15}$, where l is the width of the plume (3 standard deviations) in centimeters, giving E_x and E_y in square centimeters per second. For applications with complex topography and

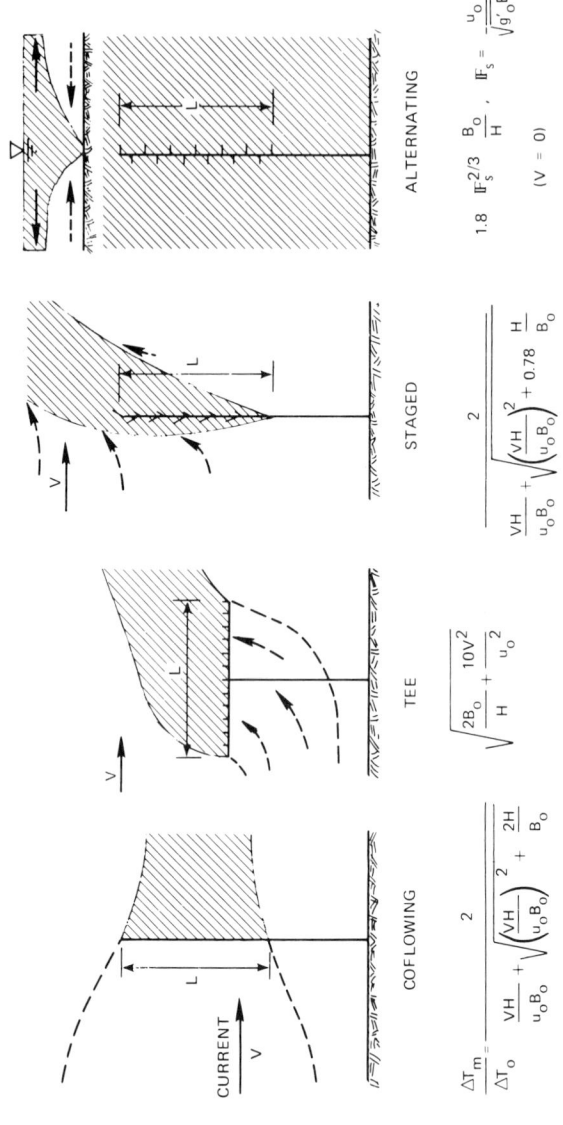

Fig. 4.8-10 Multiport diffuser types and nearfield performance in shallow water.

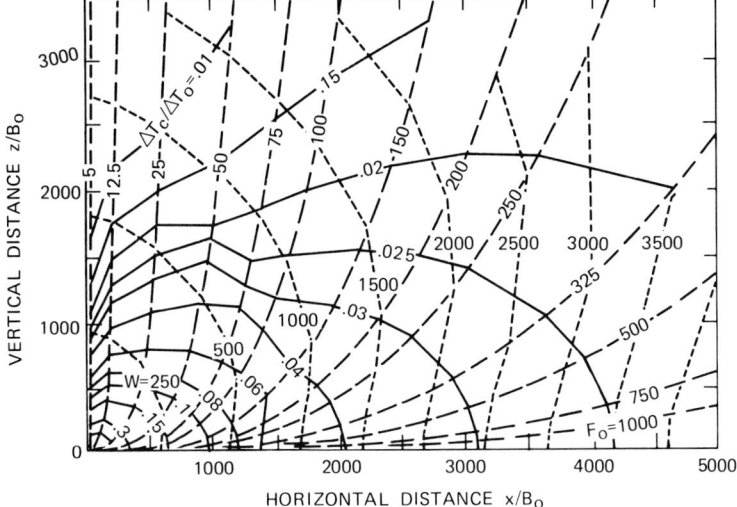

VERTICAL DISTANCE z/B_0

HORIZONTAL DISTANCE x/B_0

Fig. 4.8-11 Trajectory, centerline temperature, and width ($W = 2\sqrt{2}\,\lambda b/B_0$), for submerged slot buoyant jet discharged horizontally. (From ref. 18.)

current patterns Eq. (4.8-1) must be integrated numerically, with finite differences or finite elements techniques.[26,27] This often requires the prior integration of the continuity and momentum equations to obtain the flow field.[28,29] For conditions of steady uniform current and constant depth, closed-form solutions can be obtained. For a localized (point) source the temperature pattern is given by

$$\Delta T = \frac{Q_0\,\Delta T_0}{2\pi H\left(E_x E_y\right)^{1/2}}\, e^{xu/2E_x} K_0\left[\frac{\sqrt{\left(E_y x^2 + E_x y^2\right)\left(U^2 E_y + 4E_x E_y k\right)}}{2E_x E_y}\right]$$

in which K_0 is the modified Bessel function of the second kind of order zero and $k = K/\rho C_p H$, K being the surface heat flux coefficient (see Section 4.8-5). For a nearfield approximating a source distributed over a width $2b_s$ perpendicular to the current,[30]

$$\Delta T = \frac{Q_0\,\Delta T_0}{uHb_s}\, e^{-kx/u}\left[\text{erf}\left(\frac{y + b_s}{\sqrt{4E_y x/u}}\right) - \text{erf}\left(\frac{y - b_s}{\sqrt{4E_y x/u}}\right)\right]$$

in which erf = error function (see Section 17).

River or estuary discharges often become distributed over the entire river cross section within a short distance of the discharge point and they can be modeled one-dimensionally by solving the following cross-sectional integrated heat conservation equation[31]:

$$\frac{\partial}{\partial t}(A\,\Delta T) + \frac{\partial}{\partial x}(AU\Delta T) = \frac{\partial}{\partial x}\left(AE_L \frac{\partial\,\Delta T}{\partial x}\right) - \frac{B\phi_n}{\rho C_p} \tag{4.8-2}$$

where A = river cross-sectional area
B = river top width
U = cross-sectional averaged velocity
E_L = longitudinal dispersion coefficient

Discounting the effect of bends (which may increase or decrease dispersion) and dead zones (which increase dispersion), the dispersion coefficient for real streams is approximately given by[32]

$$E_L = 0.011\frac{U^2 B^2}{Hu^*} \tag{4.8-3}$$

in which $u^* = \sqrt{\tau_0/\rho}$ is the shear velocity with $\tau_0 = f\rho U^2/8$ is the bottom shear stress and f is the

504   **ENERGY SYSTEM TECHNOLOGY**

friction factor (see Figure 3.3-2). In estuaries, tidal flow reversals and salinity gradients lead to higher values of the dispersion coefficient.[33] For uniform-river geometry, a constant river flow, and a constant rate of heat discharge, the effect of longitudinal dispersion is negligible and the longitudinal temperature distribution is given by

$$\Delta T = \frac{Q_0 \Delta T_0}{A U} \exp\left(-\frac{kx}{U}\right)$$

4.8-5 Free-Surface Heat Transfer

The net heat flux across a water surface is the sum of individual fluxes due to different physical phenomena. These fluxes are listed in Fig. 4.8-12 and semiempirical formulas are given below for their calculation as a function of the water surface temperature and a limited number of meteorological parameters. Commonly used flux units are Btu/ft^2 · day and W/m^2.

The *solar radiation flux*, ϕ_s, depends on the latitude, day of the year, hour of the day, and cloud cover. The most reliable determination is by direct measurement using a radiometer. Useful estimates on a daily averaged basis, however, can be obtained by using the clear-sky radiation given in Fig. 4.8-13[34] multiplied by $1 - 0.65C^2$ to account for cloud effects.[35] The coefficient C is the cloudiness ratio (0 for clear sky to 1 for overcast). This incoming solar radiation must be further reduced by the percent reflected or albedo listed in Table 4.8-2.

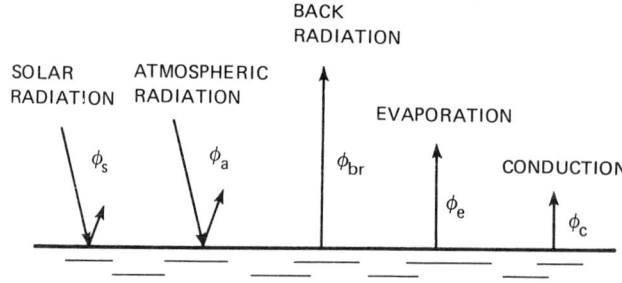

Fig. 4.8-12 Water surface heat fluxes.

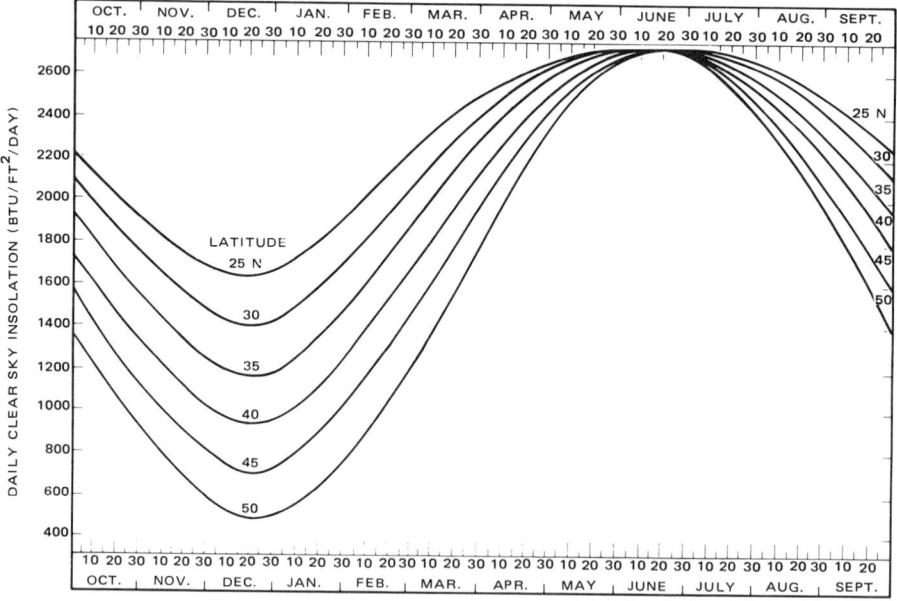

Fig. 4.8-13 Daily clear sky radiation (1 Btu/ft^2/day − 0.13 W/m^2). (From ref. 34.)

TABLE 4.8-2 PERCENT OF INCOMING SOLAR RADIATION REFLECTED BY THE WATER SURFACE

Month	Jan.	Feb.	Mar.	Apr.	May	June	July	Aug.	Sept.	Oct.	Nov.	Dec.
Percent reflected	9	7	7	6	6	6	6	6	7	7	9	10

The *atmospheric radiation*, ϕ_a, which is characterized by much longer wavelengths than the solar radiation, can be estimated with the following formula[36,37]: $\phi_a = c_1(1 + 0.17C^2)T_a^6$, with $c_1 = 1.16 \times 10^{-13}$ Btu/ft$^2 \cdot$ R^6 or 5.19×10^{-13} W/m$^2 \cdot$ K^6 and T_a = absolute temperature of the air, in degrees Rankine (R) or Kelvin (K). Note that clouds, which darken the atmosphere, tend to increase its radiation.

The *back radiation flux*, ϕ_{br}, from the water surface depends only on the absolute temperature of the water surface, T_s, and is given by $\phi_{br} = c_2 T_s^4$ with $c_2 = 4.0 \times 10^{-8}$ Btu/ft$^2 \cdot$ day $\cdot$ R^4 or 5.52×10^{-8} W/m$^2 \cdot$ K^4.

The *evaporative flux*, ϕ_e, can be estimated as follows, by extention of Dalton's law: $\phi_e = f(W)(e_s - e_a)$ in which $f(W)$ is a wind function, e_s is the saturation vapor pressure of water in air at the same temperature as the water surface, and e_a is the actual atmospheric vapor pressure. The latter is equal to $r \times e_{as}$, r being the relative humidity and e_{as} the saturation vapor pressure corresponding to the actual air temperature. Many empirical formulas have been proposed for the wind function $f(W)$. For natural water surface temperatures the evaporative mass transfer occurs exclusively by forced convection due to wind. At higher water surface temperatures free convection, due to the smaller density of the warm, moist air at the water surface than the air above, may also participate. The following two expressions represent these two regimes, and for practical applications, the larger of the two values should be used[35]:

$$f(W) = c_3 W_2 \quad \text{or} \quad c_4 W_2 + c_5 \left[\frac{T_s}{1 - 0.378 e_s/P} - \frac{T_a}{1 - 0.378 e_a/P} \right]^{1/3}$$

in which W_2 is the wind speed at a height of 2 m above the water surface and P is the atmospheric pressure. Measuring the vapor pressures in millimeters of mercury, the coefficients are $c_3 = 17$, $c_4 = 14$, and $c_5 = 22.4$ with W_2 in miles per hour and T_s and T_a in degrees Fahrenheit to obtain ϕ_e in Btu/ft$^2 \cdot$ day or $c_3 = 1.4$, $c_4 = 1.1$, and $c_5 = 3.58$ with W_2 in kilometers per hour, and T_s and T_a in degrees Celsius to obtain ϕ_e in watts per square meter. Wind speed data at a height z above the water surface can be converted to values at 2 m by the following formula, based on the logarithmic nature of the velocity profile in the atmospheric boundary layer:

$$W_2 = W_z \frac{\ln(2/z_0)}{\ln(z/z_0)}$$

where z_0 is the roughness height of the water surface, which, because of waves, increases with wind speed. Values back-fitted from measured wind profiles range from 0.001 to 0.005 m.

The *conductive heat flux*, ϕ_c, is approximately given by $\phi_c = c_6 f(W)(T_s - T_a)$ with $c_6 = 0.255$ for T_s and T_a in degrees Fahrenheit to get ϕ_c in Btu/ft$^2 \cdot$ day and $c_6 = 0.459$ with T_s and T_a in degrees Celsius to get ϕ_c in watts per square meters.

The *net surface heat flux*, ϕ_n, is then $\phi_n = \phi_s + \phi_a - \phi_{br} - \phi_e - \phi_c$. The meteorological parameters required to calculate ϕ_n are, thus, air temperature T_a, relative humidity r, cloudiness ratio C, and wind speed W_2. The net surface heat flux also depends on the water surface temperature. The water surface temperature for which, given a set of meteorologial conditions, the net surface heat flux is equal to zero is called the equilibrium temperature, T_E.

The form of the formulas given above for the individual fluxes does not allow a direct expression for T_E as a function of the meteorological parameters. An iterative approach such as Newton's method is therefore required to obtain T_E. For small ranges of the water surface temperature, the formula for the net surface heat flux may be linearized as follows: $\phi_n = -K(T_s - T_E)$, where K is the gross surface heat flux coefficient. This expression is useful when closed-form solutions are sought to the equations governing the temperature distribution in water bodies. Also, the surface heat flux coefficient gives a convenient estimate of the atmospheric cooling capacity, which is valuable for the comparison of sites, seasons, and the selection of critical conditions relative to waste heat disposal.

4.8-6 Cooling Lakes and Cooling Ponds

The term *cooling lake* is usually reserved to cooling impoundment of natural origin or obtained by damming a stream, while *cooling pond* refers to entirely artificial impoundments such as inside levees. Cooling lakes or ponds can be used for closed-loop cooling or for once-through precooling before discharge back to a river or other water body. For both cases the plant temperature rise ΔT_0 is fixed and the pond performance is measured by the intake temperature (T_i) from the pond, which should be as low and constant as possible. For the once-through precooling mode the discharge temperature into the pond is fixed at $T_0 = T_r + \Delta T_0$, T_r being the ambient river temperature. For the more prevalent closed-loop operation mode, the discharge temperature into the pond ($T_0 = T_i + \Delta T_0$) depends on the cooling capacity of the pond.

For a given surface area the most efficient cooling pond is deep, with minimal entrance mixing and submerged or skimmer wall intake.[38,39] The large pond depth allows the through-flow to only occupy a surface layer (see Fig. 4.8-14) with maximum utilization of density currents to spread the warm effluent over the entire pond area, including dead zones and side arms.[39,40] From this surface layer the flow downwells into the bottom layer whose size provides thermal inertia. The low entrance mixing ensures the warmest surface layer temperature possible for highest surface cooling (see Section 4.8-5) and the submerged intake is to withdraw the coolest lower-layer water. A cooling lake or pond is considered deep if [38]

$$P = \left[\frac{f_i}{4} - \frac{Q_0^2 L D_v^3}{g\beta\Delta T_0\, H^4 W^2} \right]^{1/4} < 0.3$$

where f_i is the interfacial friction factor [approximately half the Darcy Weisbach bottom friction factor for the same flow conditions], L is pond length, H is pond depth, W is pond width (or equivalent), and D_v is the vertical entrance dilution ratio. For a surface discharge, $D_v \approx 1.2 F_0' - 0.2$[38] with a minimum of 1.5; see Section 4.8-3 for the definition of F_0'. For steady-state load and meteorology the intake temperature from a deep cooling lake or pond is given by [41]

$$\frac{T_i - T_E}{\Delta T_0} = \frac{1}{D_v} \frac{e^{-r/D_v}}{1 - e^{-r/D_v}}$$

in which $r = KA_p/\rho C_p Q_0$ and A_p is the pond area. This relationship is plotted in Fig. 4.8-15, showing the improved performance for low D_v.

For $0.3 < P < 1.0$ the cooling pond is partially stratified, and for $P > 1.0$ it is vertically mixed. Such "shallow" conditions may be unavoidable for cooling ponds contained by levees. In this case recirculation zones or eddies should be avoided through the use of internal baffling or radial discharges and the pond can be simulated as one-dimensional with Eq. (4.8-2). The intake temperature is then given by [41]

$$\frac{T_i - T_E}{\Delta T_0} = \frac{4ae^{1/2E^*}}{(1 + a)^2 e^{a/2E^*} - (1 - a)^2 e^{-a/2E^*} - 4ae^{1/2E^*}}$$

in which $a = \sqrt{1 + 4rE^*}$ and $E^* = E_L/UL$, with E_L given by Eq. (4.8-3). This equation is plotted in Fig. 4.8-16. Shallow cooling ponds also tend to exhibit larger intake temperature fluctuations as a result of meteorological or plant load variations.

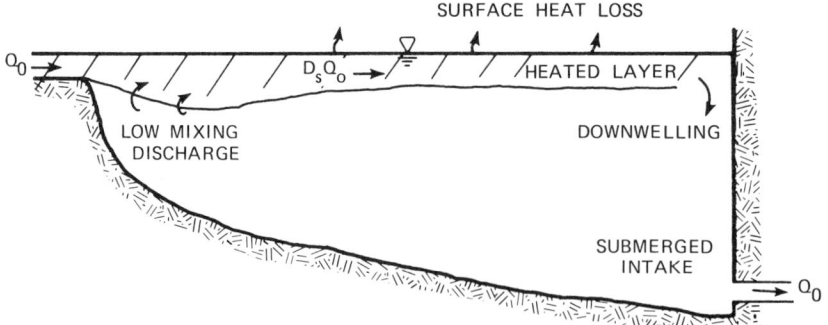

Fig. 4.8-14   Deep cooling lake.

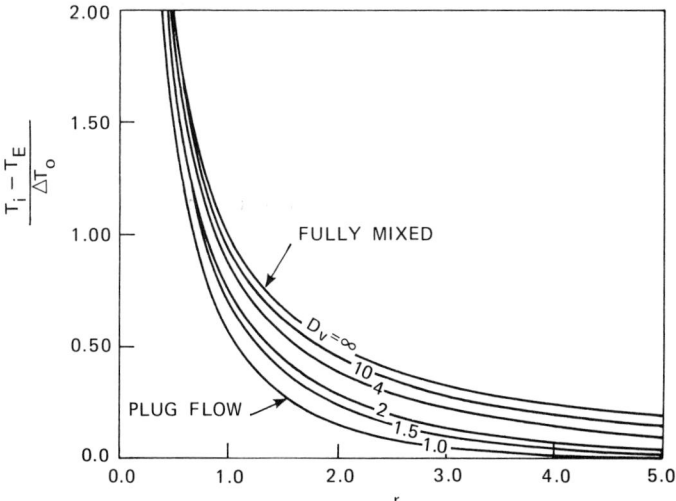

Fig. 4.8-15   Intake temperature of deep stratified cooling lake.

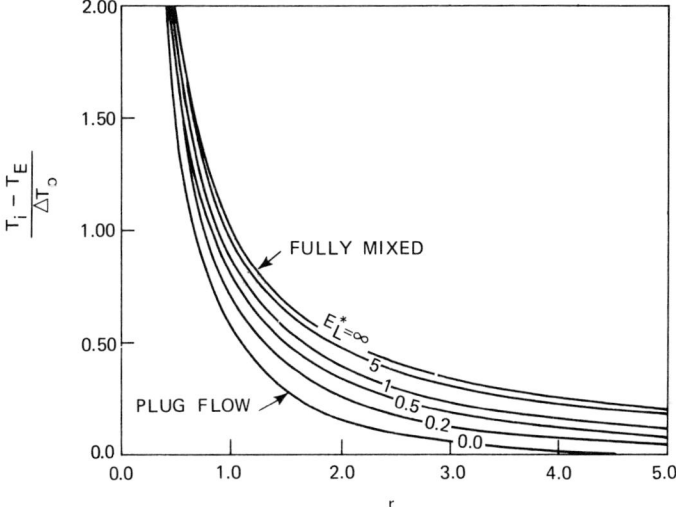

Fig. 4.8-16   Intake temperature of shallow dispersive cooling pond.

4.8-7   Cooling Towers

Formerly the term *cooling tower* referred to a device that conveyed its heat load to the atmosphere mainly by the mass transfer process of evaporation. The recent decade has, however, required more exact definitions and so this device has become known as a *wet cooling tower*. To differentiate between classes of towers, the term *dry cooling tower* was coined. It describes an air-cooling device that involves only sensible heat exchange and consists of fixed tubes that totally enclose the cooling medium so that no mass transfer is possible.

Wet towers are usually used if sufficient water to accommodate the evaporation and blowdown requirements is available. Dry towers are used when the cooling requirements are more extraordinary: if the water is not available, if the environmental siting constraints are stringent, or if the gas or liquid to be cooled cannot, for a variety of reasons, be exposed to the atmosphere. A wet cooling tower is generally the economic engineering choice when a closed-loop cooling system is required. Compared to a cooling pond or dry tower, it requires considerably less land area and will attain a better level of cooling at a considerably lower capital and/or operating cost.

Because wet cooling towers are more widely used, and because a dry tower is engineered in most respects like other air-to-water heat exchangers, the wet-tower technology will be emphasized throughout the remainder of this section. The information in this section is of two types. First, generalities and typical values are presented to develop an overall understanding and appreciation of the cooling tower. Second, several frequently encountered computational procedures are offered that allow a quantitative assessment of existing tower parameters or an estimate of future requirements.

Regardless of the kind of wet cooling system considered, the quantity of water it evaporates for a given heat load will be similar because the heat transfer mechanism is the same. As a result, the selection of a particular wet-tower system type generally depends on site environmental considerations. Further, the influence of the economics of the selection is usually slight because the environmental requirements of a specific site restrict the possible options.

With the exception of the natural draft tower, all types of cooling systems can be economically scaled up or down to fit the heat load being released. The tall shell of the natural draft tower, a feature necessitated by the relative humidity typically encountered in the United States, makes it uneconomical for cooling less than about 50,000 gpm through a 20°F range, that is, a heat load of under 5×10^8 Btu/h.

Table 4.8-3 extends the broad classification of cooling systems listed in Table 4.8-1 and presents the major characteristics of cooling towers. Desirable choices should be identified with the aid of Table 4.8-3 and the best system should then be established by an economic comparison of the candidate cooling tower systems. Figure 4.8-17 sketches the cross sections of typical types of cooling towers.

The thermal performance of wet-type cooling towers depends on the ambient wet-bulb temperature, and the performance of dry-type cooling towers depends on the dry-bulb temperature. The draft developed by natural draft towers and the fogging potential of wet/dry towers is significantly dependent on the ambient relative humidity. The latter parameter is a function of the wet- and dry-bulb temperatures. Use of summer weather data as a design basis and, if appropriate, addition of the effect of the warm exhaust that may recirculate back into the tower to establish the inlet wet-bulb temperature generally ensures that the tower will be sized to provide an adequate level of cooling throughout the year. Figure 4.8-18 shows statistical ambient wet- and dry-bulb temperatures within the United States that can be used for this design condition in the absence of more detailed site information.

Performance estimates of evaporation, draft, pressure losses, and so on, are accomplished with psychrometric chart data. For convenience, a psychrometric chart is presented in Fig. 4.8-19.

Approximate sizes of towers can be estimated from the characteristics of the towers defined later in this chapter.

Heat Load Estimate

The second law of thermodynamics defines the heat rejected (Q) by a power cycle producing W units of work which has a cycle thermal efficiency of η. That is, for W kilowatts, the heat rejected in Btu/h is

$$Q = \frac{\eta - 1}{n} \times 3413 \times W$$

Common values of η are 0.42 for fossil reheat-regenerative cycles, 0.32 for light-water reactor nuclear plants, and 0.40 for advanced-concept nuclear cycles. Similarly, heat rejection from air conditioning or refrigeration loads can be estimated per ton of refrigeration effect as $Q = 6000 \times$ tons where the coefficient of performance is taken as 2.5. Determining the heat load of other mechanical or chemical process cycles requires a knowledge of the process details.

Common Terms and Typical Values

Wet-bulb temperature is the lowest temperature at which evaporation can occur for the specific conditions of the atmosphere. It closely approximates the adiabatic saturation temperature and is measured by a psychrometer.

Dry-bulb temperature is the sensible temperature of the atmosphere as measured by a mercury-in-glass thermometer or other device.

Approach. For wet cooling towers, *approach* is the temperature difference between the liquid that was cooled and the wet bulb temperature. Figure 4.8-20 illustrates the temperature relationships. In a dry-tower system, generally the approach is the difference between the initial temperature of

TABLE 4.8-3   COOLING TOWER OVERVIEW

Cooling System Type	Description	Relative Merits	Relative Demerits
Wet mechanical draft rectangular	Generally cross flow and serviced by induced draft fans; multicell, multifan; concrete or wood and fiberglass siding construction.	Least expensive; quick shipment and erection; low drift; low profile; lightweight; occupied land area of single towers is small.	Will produce fog downwind; potential of high pockets of drift just beyond property line; placement sensitive to wind direction; multiple units require great spacing between each of the large towers to reduce recirculation; fans consume considerable horsepower; relatively noisy.
Wet mechanical draft round	Generally counterflow, large, are serviced by induced draft fans. Multifan; composed of same elements as rectangular, concrete, or wood and fiber glass siding construction.	Like rectangular but requires less space particularly on multiple towers for these can be spaced within 100 ft of one another without excessive recirculation; more buoyant plume hence less potential for ground fog; low pump head; not sensitive to wind direction.	Potential of high pockets of drift within a mile or so of tower; fans consume considerable horsepower; as noisy as rectangular.
Wet fan assisted	Counterflow; can have a bottle-shaped or short hyperbolic profile; height of stack is always over 100 ft; may have multiple peripheral vertical fans at base or a single immense fan at the top of fill inside; concrete.	Low pump head requirements; fan horsepower requirement reduced by 20–30% due to the natural draft; buoyant elevated plume discharge reduces fogging potential further; can space closely; no recirculation.	More expensive than round; high drift pockets potential; noisy; little current experience with hyperbolic profile type.
Wet natural draft	Mostly are counterflow now; hyperbolic stack profile of several hundred feet in height. Fill section may be wood if cross flow; counterflow towers no longer contain a wood fill support structure.	No expenditure of energy for moving the air; wide and low density dispersion of drift; elevated buoyant plume does not permit ground fogging; 5 dBA less noisy than mechanical draft; single units do not take much space; little maintenance.	Higher capital cost than round mechanical draft; often takes two construction seasons to erect; heavy, requires good foundations; wide and tall profile; moderate pumping horsepower expended particularly for cross flow; multiple units at same site are difficult to space; does not function in a desertlike low-humidity climate; performance influenced by inversions.

509

TABLE 4.8-3 (*Continued*)

Cooling System Type	Description	Relative Merits	Relative Demerits
Spray ponds/canals	Multitude of swirl spray nozzles spaced and connected to suitable piping in a large pond, hot water sprayed upward in a coarse spray to about 15–20 ft of height. Canals contain multiplicity of self-contained floating pump sprays spaced sequentially.	Hardware inexpensive; quickly installed; low profile; majority of drift remains on the property, relatively quiet.	Takes considerable land area for canals or pond; subject to performance dependency on wind; higher pumping head required; stringent performance unattainable particularly as capacity increases; subject to nozzle clogging/maintenance of pumps.
Evaporative coolers	Similar to a small mechanical draft rectangular tower with a cooling coil arranged as the fill. Within this coil the fluid or gas to be cooled circulates before returning to the heat source. The outside of the coil is deluged with recirculated cooled water that acts as the cooling medium.	Keeps the cooled fluid or gas separate from the cooling medium, thus not requiring a separate heat exchanger; comparative high system performance due to wet-bulb-temperature approach.	Sensitive to improper water treatment; requires more maintenance than cooling towers.
Dry	Utilizes forced draft fans; round or elliptical shaped finned tube; tube bundles in tent or lean-to shape over the frame; usually 4–6 tube rows deep; subject to recirculation effects. A condensing vapor rather than cooling water can be circulated inside the tubes.	Allows flexible plant siting because of very reduced water requirement of cooling systems; no fogging, icing, or drift; no affect on local aquatic environment.	Expensive; typically produces higher cooled water temperatures. Relatively noisy; requires more space than wet cooling towers. Pump head and fan horsepower requirements high; natural draft very large.
Wet–dry	A hybrid similar in geometry to wet cooling towers; generally serviced by induced draft fans; tube bundle above wet section provides parallel air flow path.	Reduced plume for many high-relative-humidity low-temperature conditions that normally cause fog in cooler seasons; reduced water consumption.	Lies between wet- and dry-tower types in all respects.

the process gas or fluid which required the cooling (for example, the exhaust temperature of the exhaust steam shown in Fig. 4.8-17) and the ambient dry-bulb temperature. The wet-tower approach limit, roughly proportional to the size and type of tower, can be as low as 7°F to the ambient-design wet-bulb temperature if mechanical draft and 13°F if natural draft. Note, however, that the approach of any wet tower, in contrast to dry towers, does vary with temperature significantly. The wet tower that has an approach of 7°F to an 80°F design wet-bulb temperature may have a 40°F approach at a 20°F wet bulb. A dry tower can approach the dry-bulb temperature as closely as 18°F.

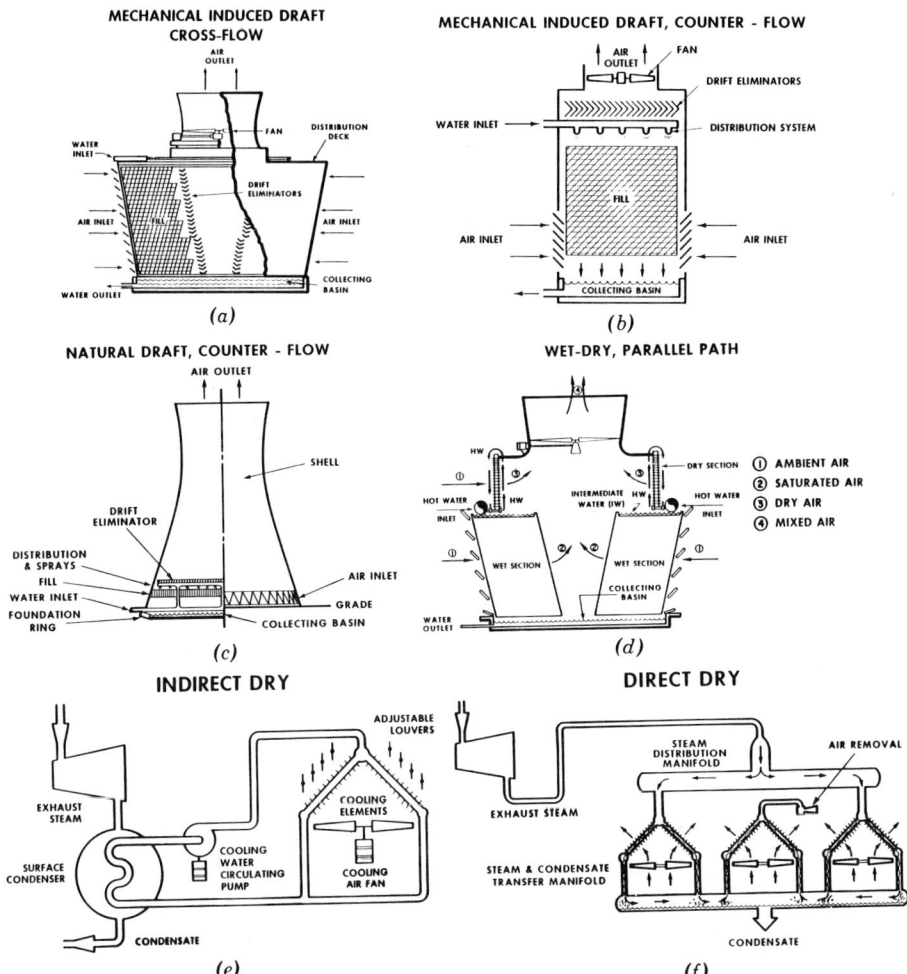

MECHANICAL INDUCED DRAFT
CROSS-FLOW

(a)

MECHANICAL INDUCED DRAFT, COUNTER - FLOW

(b)

NATURAL DRAFT, COUNTER - FLOW

(c)

WET-DRY, PARALLEL PATH

① AMBIENT AIR
② SATURATED AIR
③ DRY AIR
④ MIXED AIR

(d)

INDIRECT DRY

(e)

DIRECT DRY

(f)

Fig. 4.8-17 Types of cooling towers. (Courtesy of Marley Cooling Tower Co.)

Range refers to the temperature range between the initial and final cooled temperature, namely, the hot and cold temperatures. It is usually between 15 and 35°F.

Recirculation is that percentage of the exhaust of a cooling tower drawn back into its inlet. Natural draft towers generally do not recirculate. It is a function of the size of the tower and the magnitude and direction of the wind speed and causes an increase in the air temperatures at the inlet. Though is some cases it can be as high as 10%, it is usually 2–4% and results in a 1–2°F average inlet rise in the wet-bulb temperature.

Cross flow or **counterflow** describes the major direction of airflow compared to the liquid flow and characterizes the tower design.

Natural draft/hyperbolic.   Natural draft towers produce an airflow by a chimney effect, and the large concrete towers for utilities are generally hyperbolic in shape. Though this form accommodates the airflow within the tower, a hyperbolic shape was chosen primarily to facilitate the construction and improve the structural stiffness of the thin shell. Typically its height is about 1.3–1.5 times the base diameter.

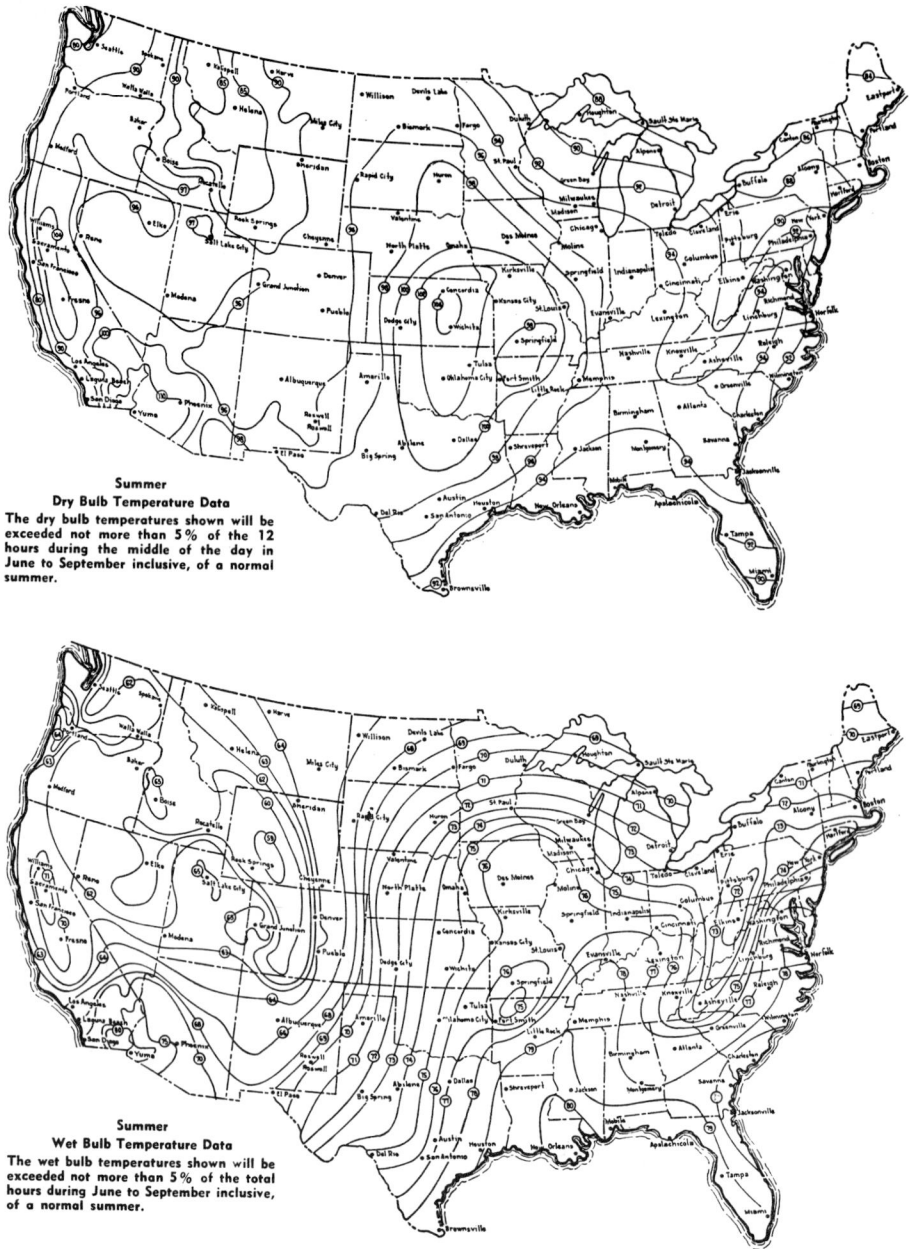

Summer
Dry Bulb Temperature Data
The dry bulb temperatures shown will be exceeded not more than 5% of the 12 hours during the middle of the day in June to September inclusive, of a normal summer.

Summer
Wet Bulb Temperature Data
The wet bulb temperatures shown will be exceeded not more than 5% of the total hours during June to September inclusive, of a normal summer.

Fig. 4.8-18 U.S. summer wet- and dry-bulb temperature data. Summer dry-bulb temperature data: The dry-bulb temperatures shown will be exceeded not more than 5% of the 12 h during the middle of the day in June to September inclusive, or a normal summer. Summer wet-bulb temperature data: The wet-bulb temperatures shown will be exceeded not more than 5% of the total hours during June to September inclusive, of a normal summer. (Reprinted by permission from *Engineering Data Book*, 8th ed., Gas Processors Suppliers Association.)

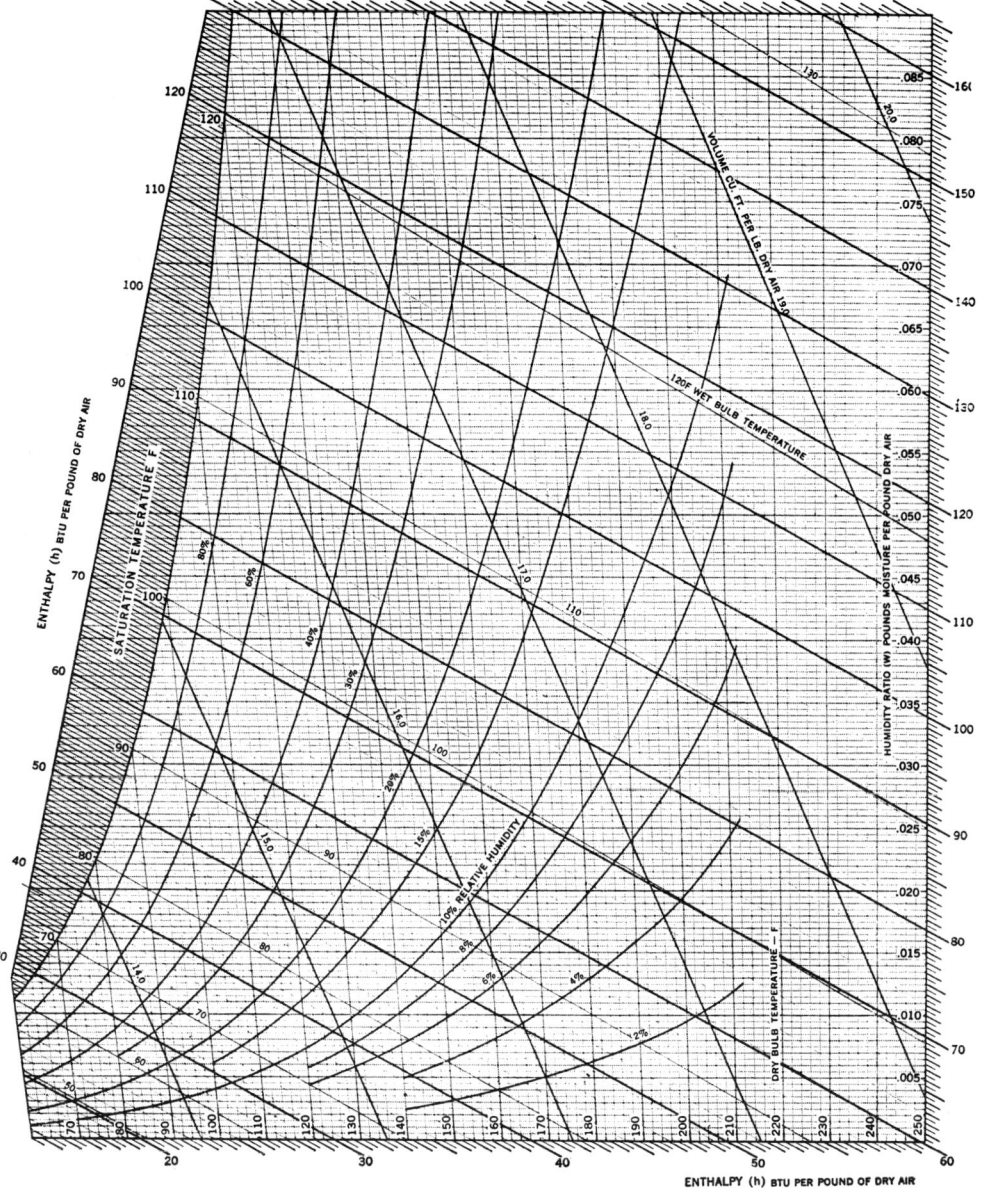

Fig. 4.8-19 Psychrometric chart (at sea level). [Reprinted by permission from American Society of Heating, Refrigerating and Air-Conditioning Engineers, Inc. (ASHRAE).]

Water load. The liquid cooled is usually given as a total quantity per unit of time; for some purposes it may be expressed as a flux, that is, a loading per unit of area and time. It is usually 3–5 gpm/ft² for a counterflow wet tower and 12–16 gpm/ft² for a typical cross-flow tower, whether mechanical or natural draft.

Air load. The gas or air cooling the liquid is expressed like the water load. Generally, it ranges from 1500 to 2500 lb/h-ft².

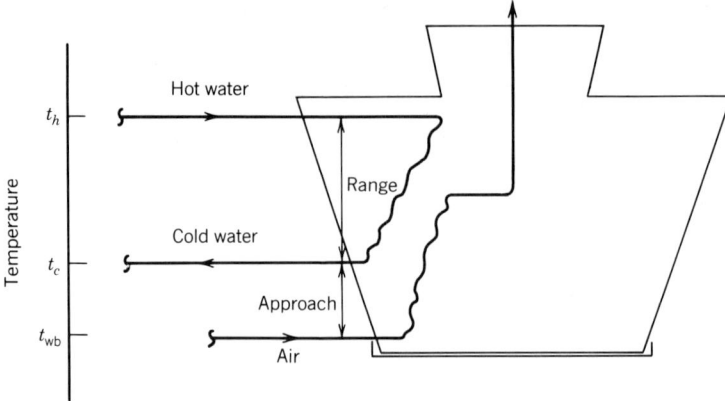

Fig. 4.8-20   Temperature definitions.

Drift is the quantity of liquid entrained in the exhaust of the tower. It can be expressed as a total quantity or a droplet spectrum. The total drift is typically less than 0.005% of the quantity of water cooled.

L/G ratio is the ratio of liquid cooled to (gas) air flow. It ranges from 1.3 to 1.8.

Nomenclature

A	Approach, °F
a	Surface area of liquid per unit volume of tower, ft^2/ft^3
acfm	Actual volumetric air flow, cubic feet per minute
CF	Fill thermal performance correction factors
CL	Head loss coefficient
G	Gas rate, lb/h
g	Gas flux, lb/h-ft^2
gpm	Liquid rate, gallons per minute
H	Shell height of natural draft tower measured from exit elevation of fill, ft
i	Enthalpy of air, Btu/lb
K	Mass transfer coefficient, lb/h-ft^2
L	Liquid rate, lb/h
l	Liquid flux, lb/h-ft^2
Q	Heat rejected, Btu/h
R	Range $(t_h - t_c)$, °F
tons	Tons of refrigeration (12,000 Btu/h)
t	Temperature, °F
U	Local air velocity, ft/s
V	Tower fill volume, ft^3
W	Work, kW
y	Height of fill, ft
Z	Total depth of air travel from basin to bottom of fill and from top of fill to top elevation of distribution sprays, ft
ΔP_T	Total pressure loss including one exit velocity head, inches of water
ΔP_S	Static pressure loss of air, inches of water
ρ	Density, lb/ft^3
η	Thermal efficiency

Subscripts

A	Ambient value
B	Bulk air surrounding liquid droplets at its wet-bulb temperature
base	Base test value of variable
c	Cold
DB	Dry bulb
exit	Exit value
h	Hot
i	Air at interface of liquid droplets (100% saturated at approximately the droplet temperature)
in	Inlet value
WB	Wet bulb
X	Actual value of variable

Materials

It is imperative that wet cooling tower components be compatible with the relatively severe corrosion environment existing: water that is aerated and typically contains minerals or solids concentrated through several cycles. It is also desirable that any material used be noncombustible. If this property is not present in the design, then fire protection sprays must be employed for periods when the tower is not in operation. Dry cooling tower components are not required to be selected as carefully unless operated in an extreme condition, such as a seacoast environment. Table 4.8-4 lists materials that have traditionally exhibited an acceptable service life.

Wet Cooling Tower Performance

Besides sizing estimates (refer to the section "Common Terms and Typical Values" for guidelines), engineering estimates of cooling towers most often involve some aspect of thermal performance. Assessing the adequacy of proposed designs to determine if the tower is large enough, predicting fan horsepower requirements, computing the evaporation associated with operation, reviewing test results, or approximating the performance at conditions other than design are examples of these topics and will be presented in this section. Since several evaluation techniques depend on quantitative estimates of the thermal performance itself, this important subject will be initially addressed and emphasized.

As shown in Fig. 4.8-17, the section of the tower that contains the heat transfer surfaces is denoted as the fill section. Originally filled with a large quantity of randomly stacked material, the surfaces are now usually precisely dispersed. The heat of the liquid is transferred to the cooling air. The air is drawn through this area by a fan or by a natural draft. Within the fill, the water is broken up, retarded from falling into the basin, and efficiently contacts the air. The major fill design objectives are that it be a durable, noncombustible, economical material that is easily installed, compact, and of uniform configuration and combines maximum thermal performance with a low air side pressure loss. In practice, the fill design is a compromise. A film fill consists of parallel layers of closely spaced thin materials vertically positioned so that the water to be cooled flows down the surfaces of the sheets into the basin. Splash fills are composed of horizontal bars of material spaced in a vertical array to enable the cooled water to splash on several bars as it falls down into the basin. High-efficiency fills provide the combined properties of both of the above. Typically, present-day high-efficiency counterflow fills are from 2 to 5 ft deep. Sheet film fills are twice this depth. Splash fills can be up to 25 ft deep. Cross-flow film fills are generally up to 15 ft deep; splash fills may be up to 50 ft tall.

The interaction of the fill thermal capability with the particular performance requirement is analogous to that of a pump curve and the system head curve. The tower operates at the intersection of the two curves as depicted by Fig. 4.8-21. The figure is called the tower characteristic and is plotted on log-graph paper because the fill capability is usually a simple power function between $(-)$ 0.6 and $(-)$ 0.8. Its abcissa is the mass ratio of liquid to gas flow (usually 1.3–1.8), and its ordinate is a dependent variable, called the *characteristic*, that is directly related to the thermal performance. Both terms are nondimensional. The capability of a specific fill design to meet a series of thermal performances can generally be determined only by full-scale testing of typical small modular sections of fill. Typically, the magnitude of the fill characteristic is directly proportional to the fill depth. Once established, the characteristic data has wide applicability to varying atmospheric and heat load conditions. It is computed for a particular fill from measurements of the air and water flows, the range, approach, and wet bulb.

TABLE 4.8-4   MATERIALS

Component	Materials
Wet Towers	
Structure (field erected) (factory assembled)	Exterior grade, pressure-treated Douglas fir or heartwood grade redwood
Fan deck	Stainless steel, Douglas fir plywood or galvanized carbon steel
Casing, stack, louvers, and fill support	Fiberglass
Fan blades	Fiberglass or high-strength, lightweight plastic
Film fill, eliminators	PVC
Splash fill	PVC; fiberglass
Wood connectors, washers	Fiberglass; PVC
Bolts, nuts, nails	Galvanized (minimum treatment); 300 series stainless steel; monel; silicon bronze
Column anchors	Cast iron
Basin	Concrete, T/G redwood, marine plywood
Handrails, stairs	Douglas fir, redwood
Water distribution system	Fiberglass, PVC (small-diameter headers); stainless steel valve stems, cast-iron valve bodies
Concrete for natural draft/mechanical draft	Type II concrete dense, aerated and well vibrated with $1\frac{1}{2}$-in. cover over rebar to prevent chloride intrusion
Dry Towers	
Headers, structure	Carbon steel
Tubes, fins	Carbon steel, aluminum

The fill will always operate thermally along the characteristic. To predict the performance at any other condition, determine first the effect in L/G of that new condition, then the modified operating point, and lastly the new approach and/or range and/or wet-bulb temperature that satisfies the new characteristic.

Both wet cooling tower capability and its requirement, the ordinate of Fig. 4.8-21, is expressed as a number of transfer units. The value is determined by the integrated form of the following equation for a minute volume within the fill:

$$\frac{KaV}{L} = \int_{t_c}^{t_h} \frac{dt}{i_i - i_B} \quad \text{(equivalent to } KaY/l \text{ on a unit plan area basis)}$$

Observe that all the parameters related directly to the thermal performance are on the right side of the equation, whereas the size of tower required is characterized by those terms on the left side. Thus, once the left side is defined for a set of conditions, it will be essentially constant provided the air or water flow do not change. This facilitates predictions of the performance (right side) under changing conditions.

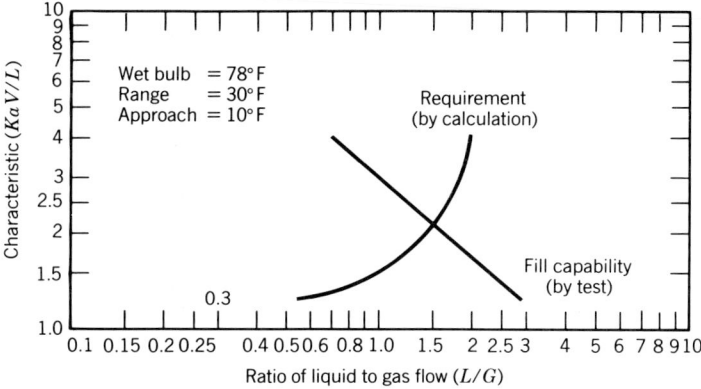

Fig. 4.8-21 Tower characteristic curve.

It has been indicated that the thermal performance data are established by experiment. As a result, the concept and quantifying equation is very versatile even though the theoretical derivation is specific only to a counterflow fill with water and air at moderate temperatures. It has been successfully applied to cross-flow fill by segmenting the volume into a two-dimensional array of elemental volumes of a unit depth. Spray canal and fill spray performance data can also be developed on this basis. Indeed, it can be used to correlate and predict the performance or change in performance of most cooling devices, whatever their type or condition of operation. Characteristic data for several different kinds of fill are available.[42-46]

For more exact calculations, the cross-flow or counterflow characteristic should be modified by several empirical correction factors (CF). The base characteristic, KaV/L, is multiplied by the corrections. For performance predictions at water temperatures warmer than those used in base tests, a reasonable correction from base data, t_{base} (°F), to any warmer water temperature t_X is estimated by using

$$CF = 1 + \frac{(t_{base} - t_X)}{10} \times 0.03$$

This is required because the performance of the fill diminishes as the water temperature increases when the fill performance is evaluated in accordance with the methods described here or has been determined by the typical test procedures currently in use. To account for a water loading that is different from that used in a base test, employ,

$$CF = \left(\frac{l_{base}}{l_X}\right)^{-0.2}$$

Altitude correction factors should be applied[47] to the required performance if the altitude is over 5000 ft.

The performance demand of tower-cooling water that contains over 15,000 ppm of salt is more difficult. The requirement can be determined exactly by accounting for the decreased vapor pressure of the water when computing the interfacial air enthalpy similar to the method utilized by Hamilton.[47] Generally, however, this effect is not more than 3%.

Due to its significance, it is necessary to be able to perform the numerical analysis that quantifies the thermal requirements. The integration to determine the characteristic cannot be performed directly because all mathematical approximations of the variation of enthalpy during the process lead to an unintegrable expression. Numerical integration is required. The Tchebycheff four-ordinate method provides sufficient accuracy and has been frequently used by cooling tower practitioners. Also, it should be noted that for difficult duty towers the value of characteristic determined by numerical integration will depend somewhat on the method employed, and so the assessment of the performance should be consistent with the manner by which the characteristic was originally established. Counterflow cooling tower computations using this technique have been developed for most conditions of interest by the Cooling Tower Institute.[48] Some of the manufacturers also have available data or curves similar to the counterflow characteristics.

A complete example of the Tchebycheff calculation is included in Table 4.8-5. To compute the characteristic, complete data for one tower- or fill-operating point must be known or closely estimated.

TABLE 4.8-5   TOWER CHARACTERISTIC ESTIMATE

Presume that it is required to calculate the characteristic of a 100,000-gpm counterflow tower near Dallas. Its range is 30°F and its approach is 10°F. The recirculation, based on its size, is approximated as 1°F. The L/G has been previously determined as 1.5 from an evaluation of the pressure losses through the tower and the fan capacity curve.

From Fig. 4.8-18 ambient bulb is 77°F, inlet wet bulb is $77 + 1 = 78°F$.

$$t_h = 78 + 10 + 30 = 118°F$$

$$t_c = 78 + 10 = 88°F$$

From the psychometric chart of Fig. 4.8-19 at 78°F,

$$i_{in} = 41.5 \text{ Btu/lb}, \qquad i_{exit} = i_{in} + \frac{L}{G}(t_h - t_c) = 41.5 + 1.5(118 - 88) = 86.5 \text{ Btu/lb}$$

The characteristic is

$$KaV/L = \int_{t_c}^{t_h} \frac{dt}{i_i - i_B}$$

The Tchebycheff approximation is

$$\frac{KaV}{L} = \int_{t_c}^{t_h} \frac{dt}{i_i - i_B} \simeq \frac{t_h - t_c}{4} \left[\frac{1}{(i_i - i_B)_1} + \frac{1}{(i_i - i_B)_2} + \frac{1}{(i_i - i_B)_3} + \frac{1}{(i_i - i_B)_4} \right]$$

where the enthalpy potential difference between the interfacial film and bulk air stream is evaluated at the following temperatures. Note that the bulk enthalpy is computed from an energy balance to that temperature.

$$(i_i - i_B)_1 \qquad \text{at } t_c + 0.1(t_h - t_c)$$

$$(i_i - i_B)_2 \qquad \text{at } t_c + 0.4(t_h - t_c)$$

$$(i_i - i_B)_3 \qquad \text{at } t_h - 0.4(t_h - t_c)$$

$$(i_i - i_B)_4 \qquad \text{at } t_h - 0.1(t_h - t_c)$$

Water Temperature (t_x)	Interface Enthalpy (i_i)	Bulk Air Enthalpy (i_B)	$(i_i - i_B)$	$\dfrac{1}{i_i - i_B}$
$t_c = 88$		$i_{in} = 41.5$		
$t_c + 0.1(t_h - t_c) = 91$	57	$i_{in} + 0.1L/G(t_h - t_c) = 46.0$	11.0	0.0909
$t_c + 0.4(t_h - t_c) = 100$	71.5	$i_{in} + 0.4L/G(t_h - t_c) = 59.5$	12.0	0.0833
$t_h - 0.4(t_h - t_c) = 106$	83.5	$i_{exit} - 0.4L/G(t_h - t_c) = 68.5$	15.0	0.0666
$t_h - 0.1(t_h - t_c) = 115$	105.0	$i_{exit} - 0.1L/G(t_h - t_c) = 82.0$	23.0	0.0435
$t_h = 118$		$i_{exit} = 86.5$		

From the Tchebycheff approximation equation

$$\frac{KaV}{L} = \frac{118 - 88}{4}(0.0909 + 0.0833 + 0.0666 + 0.0435) = 2.13 \text{ units}$$

The result may be presented graphically, as in Fig. 4.8-21. Refer to the text for the applicability of the computed value above.

Assume the data of Table 4.8-5 applies and the operating characteric is 2.13. To predict the approach of this tower at a 59°F ambient wet bulb, with everything else remaining the same, neglect changes in air flow to the tower. Add 1°F to account for the recirculation, bringing the 60°F wet-bulb temperature to 60°F. For several different approach values recompute the performance requirement at $L/G = 1.5$ until it approximates 2.13. The new approach will be determined as 19°F. Note that the characteristic has remained the same. The same technique would be followed if the prediction is required for a different heat load (range).

The water flow and air flow change if, for example, more water is put over the tower. This modifies the fan air flow because more water represents more air resistance. Due to the higher air pressure loss that must be overcome, the fan drops in capacity. The overall effect is an increase in the L/G. A modified value of the characteristic would be estimated from the slope of the fill curve. Say the L/G increased to 2.0, then from a chart similar to Table 4.8-5, the modified characteristic would be 1.67. The performance then would be predicted by computing the characteristic as outlined initially in this paragraph for the modified range and at various approaches until the value of 1.67 was achieved. Other variations in this theme are similarly executed.

Cross-flow characteristics are available in *Kelly's Handbook*.[49] Use of that text is recommended for those determinations.

The thermal test of a tower can be performed by reference to the Cooling Tower Institute Test Code, ATC 105, or the American Society of Mechanical Engineers Performance Test Code, PTC 23, Atmospheric Water Cooling Towers; spray ponds and canals can be tested by procedures in ASME PTC 23.1. Both codes are based on the concept that the resulting test capability of the tower is expressed as that quantity of water the tower can cool to the equivalent specified design range and approach.

The complete performance prediction of a tower also requires an assessment of the draft or fan motor kW (as applicable). The air pressure loss is first estimated. Static pressure loss of the design air flow necessitates a knowledge of the flow path and the loss coefficients. The initial step is to determine the mean velocity head at the inlet, through the fill, drift eliminators, and the exit of the tower or a cell for the actual or assumed air flow.

For detailed estimates, an accounting should be made of the changes in air density, moisture, and flow rate as the air heats and absorbs moisture. An energy balance across any wet tower, between the water and air, neglecting the quantity of water evaporated, yields

$$i_{\text{exit}} \sim \frac{L}{G}(t_h - t_c) + i_{\text{in}}$$

It is appropriate to assume that the exit air from the towers is saturated. Compute the exit enthalpy of the air from the above equation, and using a psychometric chart, determine the exit temperature, moisture, and specific volume. Following this, the mean and exit densities, unit change in moisture velocities, and air side pressure losses can be established for the tower.

Often an estimate of the evaporation is required. In summer months approximately 80% of the heat released by a wet cooling tower is evaporative with 20% by sensible heat transfer. During the winter 60% of the heat transfer is evaporative. Evaporation is computed more exactly by determining the increase in moisture per unit mass of air between the tower inlet and exit as described in the previous paragraph. To compute the total evaporation, multiply this by the expected air flow G.

The static pressure losses of the air as it flows through the tower are computed from the velocity head and loss coefficient related to that section by the equation

$$\Delta P_S = CL \times \rho \times U^2 = \frac{12}{2 \times 32.2 \times 62.4}$$

Loss coefficients in terms of local velocity heads are typically as follows: counterflow inlet, 2.7; counterflow inlet with louvers, 4.0; cross-flow inlet with louvers, 3.0; rain + distribution (counterflow only), $0.16Z(L/G)^{1.3}$, where Z is the total depth (ft) traveled in these areas; eliminators, 3.0–4.0; cross-flow turn into plenum, 0.5; fan entrance, 0.5 $(1 - \text{area at fan ring/area of plenum})$, where these areas are the cross sections normal to the air flow; fan stack discharge, 1.0; natural draft exit, 1.0. The static loss coefficients for the fill are more complex functions of the square of the fill air velocity and depend on the air and water flux, fill geometry, and length of air travel. The latter loss is usually from 0.1 to 0.4 in. of water and represents about three-fourths of the overall air pressure loss. Some typical fill pressure loss data are available.[42,43,46,49–51]

The total motor input power for the propeller fan is

$$\text{kW} = \frac{\text{acfm} \times (\Delta P_T) \times 0.745 \times 62.4}{33,000 \times 0.8 \times 0.96 \times 0.92 \times 12}$$

where total pressure efficiency is 0.8 and overall motor and gear efficiency is 0.92 and 0.96, respectively, acfm is the actual air flow at the fan, and pressure loss is in inches of water. The height of a natural draft tower to develop a draft that matches the losses is a modified from of the Bernoulli equation:

$$H = \frac{(\Delta P_T) \times 62.4/12}{(\rho_A - \rho_{\text{exit}})}$$

where H is defined from the elevation of the fill exit and densities are those of the ambient and exit air.

Spray Cooling

The performance of spray ponds and canals is generally not as good as that of wet cooling towers because ambient winds have difficulty fully penetrating the spray field, the spray droplets are usually very coarse, and the droplet residence time in the air is very short. The equivalent characteristic values of KaV/L at a wind speed of about 5 mph vary from 0.2 to 0.4 units. Ryan and Meyers[44] survey the analysis techniques and data available; Dutkiewicz[42] describes tests of a cooling tower fill composed entirely of sprays.

Dry Cooling Tower Performance

The thermal performance can be determined by utilizing the principles presented in the previous sections on wet-tower cooling. Specific heat transfer coefficients can be assessed with the aid of Section 3.2. Air properties are those of dry air. Tube bundle losses can be estimated from data presented by Kays and London.[52] The pressure losses, recirculation, and propeller fan characteristics of both wet and dry towers are similar. Note, however, that since the effective specific heat of dry air is only 0.24 Btu/lb · °F, the air flow requirements, everything else being equal, are approximately 3 to 4 times that of a wet cooling tower.

Dry-tower testing can be accomplished by referring to the American Society of Mechanical Engineers Performance Test Code 30, Air Cooled Exchangers, or a similar guideline developed by the American Society of Chemical Engineers.

Hybrid Tower Performance

The performance of wet/dry towers and evaporative coolers can be determined by separately analyzing the individual dry-section and wet-section performances and then combining the result in a manner compatible with the air flow, water flow, or heat flow paths. The wet side of the evaporative cooler performs like a wet cooling tower with a very small range; the tube inside heat transfer coefficients are typical of those of other heat exchangers. Wet/dry towers are sized with a dry section of adequate size to reduce the water evaporated or to produce an exit air condition sufficiently unsaturated to minimize the occurrence of fogging as the plume cools.

REFERENCES

4.8-1 F. L. Parker and P. A. Krenkel, *Physical and Engineering Aspects of Thermal Pollution*, CRC Press, Cleveland, OH, 1970.

4.8-2 *Hydraulic Institute Standards for Centrifugal, Rotary and Reciprocating Pumps*, 13th ed., Hydraulics Institute, Cleveland, OH, 1975.

4.8-3 M. J. Prosser, "The Hydraulic Design of Pump Sumps and Intakes," Report No. ISBN 0-86017-027-6, British Hydromechanics Research Association, Cranfield, England, July 1977.

4.8-4 Task Committee on Fish Handling Capability of Intake Structures, "Design of Water Intake Structures for Fish Protection," ASCE, 1982.

4.8-5 J. P. Bohan and J. L. Grace, Jr., "Selective Withdrawal From Man-Made Lakes," Technical Report H-73-4, U.S. Army Engineering Waterways Experiment Stations, Vicksburg, MS, March 1973.

4.8-6 N. H. Brooks and R. C. Y. Koh, "Selective Withdrawal From Density Stratified Reservoirs," *J. Hydraulics Division, ASCE* **95**(HY4), July 1969.

4.8-7 A. Craya, "Recherches Theoriques sur L'Ecoulement de Couches Superposees de Fluides de Densite Differentes," *La Houille Blanche*, January-February 1949.

4.8-8 J. Imberger, "Selective Withdrawal: A Review," Proceedings of the Second International Symposium on Stratified Flows, IAHR, Trondheim, Norway, June 1980.

4.8-9 D. R. F. Harleman and R. A. Elder, "Withdrawal from Two-Layer Stratified Flows," *J. Hydraulics Division, ASCE* **91**(HY4), July 1965.

4.8-10 G. H. Jirka, E. A. Adams, and K. D. Stolzenbach, "Buoyant Surface Jets," *J. Hydraulics Division, ASCE* **107**(HY11), November 1981.

4.8-11 M. A. Shirazi and L. R. Davis, "Workbook of Thermal Plume Prediction. Vol. 2. Surface Discharges," EPA Report R2-72-005b, May 1974.

4.8-12 W. E. Dunn, A. J. Policastio, and R. A. Paddock, "Surface Thermal Plumes: Evaluation of Mathematical Models for the Near and Complete Field," Report No. ANL/WR-75-3, Argonne National Laboratory, Argonne, IL, August 1975.

4.8-13 J. J. McGuirk and W. Rodi, "Mathematical Modeling of Three-Dimensional Heated Surface Jets," *J. Fluid Mechanics* **95**, Part 4, 1979.

4.8-14 K. D. Stolzenbach and D. R. F. Harleman, "Three-Dimensional Heated Surface Jets," *Water Resources Research*, **9**(1), February 1973.

4.8-15 L. N. Fan, "Turbulent Buoyant Jets into Stratified or Flowing Ambient Fluids," W. M. Keck Laboratory of Hydraulics and Water Resources, Report No. KH-R-15, California Institute of Technology, Pasadena, CA, 1967.

4.8-16 E. Hirst, "Buoyant Jets with Three-Dimensional Trajectories," *J. Hydraulics Division, ASCE* **98**(HY11), November 1972.

4.8-17 R. C. Y. Koh and L. N. Fan, "Mathematical Models for the Prediction of Temperature Distributions Resulting from the Discharge of Heated Water into Large Bodies of Water," Environmental Protection Agency, Water Pollution Control Research Series Report 16130 DW0 10/70, October 1970.

4.8-18 M. A. Shirazi and L. R. Davis, "Workbook of Thermal Plume Prediction. Vol. 1. Submerged Discharges," EPA Report R2-72-005a, August 1972.

4.8-19 J. H. Lee and G. H. Jirka, "Multiport Diffuser as Line Source of Momentum in Shallow Water," *Water Resources Research* **16**(4), August 1980.

4.8-20 C. W. Almquist and K. D. Stolzenbach, "Staged Multiport Diffusers," *J. Hydraulics Division, ASCE* **106**(HY2), February 1980.

4.8-21 D. N. Brocard, Discussion, "Staged Multiport Diffusers," *J. Hydraulics Division, ASCE* **106**(HY10), October 1980.

4.8-22 J. H. W. Lee, "Near Field Mixing of Staged Diffusers," *J. Hydraulics Division, ASCE* **106**(HY8), August 1980.

4.8-23 G. H. Jirka and D. R. F. Harleman, "Stability and Mixing of a Vertical Plane Buoyant Jet in Confined Depth," *J. Fluid Mechanics* **94** (Part 2) (1979).

4.8-24 P. J. W. Roberts, "Line Plume and Ocean Outfall Dispersion," *J. Hydraulics Division, ASCE*, **105**(HY4), April 1979.

4.8-25 A. Okubo, "Oceanic Diffusion Diagrams," *Deep Sea Research* **18** (1971).

4.8-26 F. Boulot, "Modeling of Heated Water Discharges on the French Coast of the English Channel," in *Transport Models for Inland and Coastal Waters*, H. B. Fischer (ed.), Academic Press, New York, 1981.

4.8-27 W. Leimkuhler, J. J. Connor, J. D. Wang, G. Christodoulou, and S. Sundgren, "Two-Dimensional Finite Elements Dispersion Model," Proceedings, Symposium on Modeling Techniques, ASCE, San Francisco, September 1975.

4.8-28 J. J. Leendertse, "Aspects of Computational Model for Long Period Water Wave Propagation," Memorandum RM-5294-PR, Rand Corporation, Santa Monica, CA, May 1967.

4.8-29 J. D. Wang, "Real Time Flow in Unstratified Shallow Water," *J. Waterways*, Port, Coastal and Ocean Division, ASCE, **104**(WW1), June 1977.

4.8-30 G. T. Csanady, *Turbulent Diffusion in the Environment*, D. Reidel, Boston, 1973.

4.8-31 D. N. Brocard and D. R. F. Harleman, "One-Dimensional Temperature Predictions in Unsteady Flows," *J. Hydraulics Division, ASCE* **102**(HY3), March 1976.

4.8-32 H. B. Fischer, E. J. List, R. C. Y. Koh, J. Imberger, and N. H. Brooks, *Mixing in Inland and Coastal Waters*, Academic Press, New York, 1979.

4.8-33 E. R. Holley, D. R. F. Harleman, and H. B. Fischer, "Dispersion in Homogeneous Estuary Flow," *J. Hydraulics Division ASCE* **98**(HY8), August 1970.

4.8-34 R. W. Hamon, L. L. Weiss, and W. T. Wilson, "Insolation as an Empirical Function of Daily Sunshine Duration," *Monthly Weather Review* **82**(6), June 1954.

4.8-35 P. J. Ryan, D. R. F. Harleman, and K. D. Stolzenbach, "Surface Heat Loss From Cooling Ponds," *Water Resources Research* **10**(5), October 1974.

4.8-36 W. C. Swinbank, "Long Wave Radiation from Clear Skies," *Quarterly J. Royal Meteorological Society London* **89**, July 1963.

4.8-37 W. O. Wunderlich, "Heat and Mass Transfer Between a Water Surface and the Atmosphere," Report No. 14, TVA Engineering Laboratory, Norris, TN, 1972.

4.8-38 G. H. Jirka and D. R. F. Harleman, "Cooling Impoundments: Classification and Analysis," *J. Energy Division, ASCE* **105**(EY2), August 1979.

4.8-39 P. J. Ryan and D. R. F. Harleman, "An Analytical and Experimental Study of Transient Cooling Pond Behavior," Technical Report No. 161, R. M. Parsons Laboratory for Water Resources and Hydrodynamics, MIT, Cambridge, MA, 1973.

4.8-40 D. N. Brocard and D. R. F. Harleman, "Two Layer Model for Shallow Horizontal Convective Circulation," *J. Fluid Mechanics* **100** (Part 1) (1980).

4.8-41 G. H. Jirka and M. Watanabe, "Steady State Estimation of Cooling Pond Performance," *J. Hydraulics Division, ASCE* **106**(HY6), June 1980.

4.8-42 R. Dutkiewicz, "Natural Draught Spray Cooling Tower," Institute of Mechanical Engineers, 1966.

4.8-43 H. Lowe and D. Christie, "Heat Transfer and Pressure Drop Data on Cooling Tower Packings and Model Studies of the Resistance of Natural Draught Towers to Airflow," American Society of Mechanical Engineers, 1962.

4.8-44 P. Ryan and D. Myers, "Spray Cooling: A Review of Thermal Performance Models," American Power Conference, 1976.

4.8-45 C. Savery and M. Hammill, "Evaporative Cooling Tower Performance Predictions," American Society of Mechanical Engineers, 1972.

4.8-46 J. Skold, "Energy Savings in Cooling Tower Packings," Chemical Engineering Progress, October 1981.

4.8-47 T. Hamilton "Effect of Altitude in Cooling Tower Rating and Performance," Cooling Tower Institute Bulletin TPR 125, July 1962.

4.8-48 *Cooling Tower Performance Curves*, Cooling Tower Institute, Houston, TX, 1967.

4.8-49 N. Kelly, *Kelly's Handbook of Crossflow Cooling Tower Performance*, Kelly and Associates, Kansas City, MO, 1976.

4.8-50 K. Keyes, *Methods of Calculation for Natural Draft Cooling Towers*, American Institute of Chemical Engineers, 1972.

4.8-51 N. Snyder, "Effect of Air Rate, Water Rate, Temperature and Packing Density in a Cross Flow Cooling Tower," *Chemical Engineering Progress* **52** (18), (1951).

4.8-52 W. Kays and A. London, *Compact Heat Exchangers*, National Press, Palo Alto, CA, 1955.

BIBLIOGRAPHY

Adams, E. E. "Dilution Analysis for Unidirectional Diffusers," *J. Hydraulics Division, ASCE* **108**(HY3), March 1982.

Billington J., and Billington, S. D. "Stability Analysis of Cooling Towers: A Review of Current Methods" IASS Conference on Shell Structures and Climatic Influences, 1972.

Capano, G., and Bradley, W. "Noise Prediction Techniques for Siting Large Natural Draft and Mechanical Draft Cooling Towers," American Power Conference, 1976.

Forman, G., and Kelly, N. "Cooling Tower Fan Performance," American Society of Mechanical Engineers, 1959.

Kosten, G., Morgan, J., Burns, J., and Curlett, P. "Operating Experience and Performance Testing of the World's Largest Air Cooled Condenser," American Power Conference, 1981.

Landon, R., and Houx, J. "Plume Abatement and Water Conservation with the Wet-Dry Cooling Tower," American Power Conference, 1973.

Lichtenstein, J. "Performance and Selection of Mechanical Draft Cooling Towers," American Society of Mechanical Engineers, Vol. 65, 1943.

Reisman, J. "A Study of Cooling Tower Recirculation," American Society of Mechanical Engineers, 1972.

Warner, M., and Lefevre, M. "Salt Water Natural Draft Cooling Tower Design Considerations," American Power Conference, 1974.

Wilbur, K., Shofner, E., and Margetts, M. "Measurement and Characterization of Cooling Tower Drift," American Society of Mechanical Engineers, 1974.

Wright, S. J. "Mean Behavior of Buoyant Jets in a Crossflow," *J. Hydraulics Division, ASCE* **103**(HY5), May 1977.

Zamuner, N. "Crossflow Cooling Tower Analysis and Design," *ASHRAE J.*, April 1962.

4.9 WATER TREATMENT SYSTEMS

Joseph H. Duff

4.9-1 Introduction

Water is a unique material of great value in energy systems. Favorable thermodynamic properties, relatively low boiling point, and a high latent heat, combined with low cost, general availability, and ease of handling, make water the favored heat transfer medium in power generation. Low viscosity and density favor low pumping costs, also making water useful in mass transport systems. The general use of water for cleanup operations and ash transport systems is widely recognized. Solid fuels such as coal may also be transported in water slurry form.

4.9-2　Solvent Properties

Water approaches being a universal solvent by dissolving at least traces of virtually everything with which it comes in contact. Therefore, there is no pure water found in nature. Small quantities of atmospheric gases, minerals from the earth, organic materials, plant and animal life, and man-made pollution may be found in all exposed water supplies. The water present on earth is continuously recycled from surface water bodies and plants to the atmosphere. Condensation, rain, and melted snow run off through streams and rivers to lakes and oceans. Water that percolates into underground strata dissolves mineral matter and may be recovered as well water or appear in springs.

Freshly fallen rainwater or snow will contain very small traces of other materials dispersed in the atmosphere along with dissolved nitrogen, oxygen, and carbon dioxide. Gases present from industrial pollution include sulfur dioxide, nitrogen oxides, and ammonia. Surface waters, rivers, streams, and lakes generally contain more dissolved materials from contact with the earth and some organic material, including industrial pollutants picked up by runoff.

Water that percolates deep into the ground generally contains larger quantities of inorganic mineral materials and smaller amounts of organic materials than surface supplies. The specific mineral content will vary with the underground strata contacted by the water. Generally, groundwater sources are stable in composition because the underground water approaches an equilibrium with its surroundings. Variation in composition or temperature does not occur unless the groundwater aquifer is rapidly depleted by excessive drawdown and water intrudes from other sources.

The definition of *contaminants* is arbitrary, depending on the intended use of the water supply. For example, potable water may contain significant amounts of mineral material but must be free of harmful bacteria. Water for high-pressure boilers must be essentially free of dissolved salts, but small numbers of bacteria are not significant. It is essential to adequately analyze available water supplies and carefully define all parameters requiring control for each water use.

4.9-3　Water Analysis

The very low chemical concentrations found in most fresh waters require the use of specialized analytical techniques. In all cases the analytical methods used should be identified. There are two sources of validated analytical methods suitable for water.[1,2] The USEPA also publishes a volume of water analysis methods found acceptable to that agency.[3] The major constituents dissolved in natural waters ionize into positively charged cations and negatively charged anions. In order to preserve the electroneutrality of the aqueous solution, the number of cation and anion charges must be equivalent. Recognition of this characteristic can be of value in checking the adequacy of analytical information provided.

Analyses may be reported in any one or combinations of units to confirm ionic balance. Balance may be made using equivalents per million (epm), milliequivalents per liter (meq/L) or the traditional equivalent unit, milligrams per liter as calcium carbonate (mg/L as $CaCO_3$). Conversion from concentration by weight as individual ions to calcium carbonate equivalents is done by multiplying the ion concentration by the factor

$$\frac{\text{Equivalent weight calcium carbonate}}{\text{Equivalent weight of ion}}$$

A list of the significant cations and anions in water supplies with conversion factors to calcium carbonate equivalents is given in Table 4.9-1.

TABLE 4.9-1　CONVERSION FACTORS[a]

Cations	Factor	Anions	Factor
Ca^{2+}	2.50	HCO_3^-	0.82
Mg^{2+}	4.10	CO_2	1.14[b]
Na^+	2.18	Cl^-	1.41
K^+	1.28	SO_4^{2-}	1.04
NH_4^+	2.78		

[a] Ion mg/L × factor = ion mg/L as $CaCO_3$.
[b] Factor shown is for ion exchange as a monovalent anion.

TABLE 4.9-2 EFFECTS OF WATER CONTAMINANTS

Constituent	Difficulties Caused	Means of Treatment
Turbidity	Deposits in water lines, process equipment, etc. Interferes with most process uses.	Coagulation, settling and filtration.
Color	May cause foaming in boilers. Hinders precipitation methods such as iron removal and softening.	Coagulation, settling and filtration. Adsorption on activated carbon.
Hardness, calcium and magnesium salts	Chief source of scale in heat exchange equipment, boilers, pipelines, etc.	Softening. Demineralization. Internal boiler water treatment. Surface-active agents.
Alkalinity, bicarbonate (HCO_3), carbonate (CO_3), and hydrate (OH)	Foaming and carry-over of solids with steam. Embrittlement of boiler steel. Bicarbonate and carbonate produce CO_2 in steam, a source of corrosion in condensate lines.	Lime or lime–soda softening. Acid treatment. Hydrogen cation softening. Demineralization. Dealkalization by anion exchange.
Free mineral acids, H_2SO_4, HCl, etc.	Corrosion.	Neutralization with alkali
Carbon dioxide (CO_2)	Corrosion in water lines and particularly steam and condensate lines.	Aeration. Deaeration. Neutralization with alkali
pH $pH = -\log\dfrac{1}{[H^+]}$	pH varies according to acidic or alkaline solids in water. Most natural waters have a pH of 6.0–8.0.	pH can be increased by alkalie or decreased by acid.
Sulfate (SO_4)	In itself is not usually significant. Combines with calcium to form calcium sulfate scale.	Demineralization.
Chloride (Cl^-)	Increases corrosive character of water.	Demineralization.
Silica (SiO_2)	Scale in boilers and cooling water systems. Insoluble turbine blade deposits due to silica vaporization.	Hot-process or cold lime softening with magnesium salts. Adsorption by highly basic anion exchange resins, in demineralization process.
Iron, Fe^{2+} (ferrous), Fe^{3+} (ferric) Manganese(Mn^{2+})	Source of deposits in water lines, boilers, etc.	Aeration. Coagulation and filtration. Lime softening. Cation exchange. Contact filtration. Surface-active agents.
Oxygen (O_2)	Corrosion of water lines, heat exchange equipment, boilers, return lines, etc.	Deaeration. Sodium sulfite or hydrazine. Corrosion inhibitors.
Hydrogen sulfide (H_2S)	Cause of "rotten egg" odor. Corrosion.	Aeration. Chlorination. Highly basic anion exchange.
Ammonia (NH_3)	Corrosion of copper and zinc alloys by formation of soluble complex ion.	Cation exchange with hydrogen zeolite. Chlorination. Deaeration.
Dissolved solids	High concentration of dissolved solids is a cause of foaming in boilers.	Various softening processes, such as lime softening and cation exchange by hydrogen zeolite, will reduce dissolved solids. Demineralization.
Suspended solids Gravimetric measure of undissolved matter.	Suspended solids cause deposits in heat exchange equipment, boilers, water lines, etc.	Sedimentation. Filtration, usually preceded by coagulation and settling.

Note that some constituents commonly found in water are not included in Table 4.9-1 because they are so weakly ionized that they need not be included when validating the cation–anion balance. Examples are iron, manganese, copper, silica, and free carbon dioxide.

4.9-4 Effects of Common Contaminants

The effects of a number of common contaminants found in water supplies are shown in Table 4.9-2. This is not a complete list but is illustrative of the problems caused by water contaminants.

4.9-5 Approaches to Contaminant Removal

Mechanisms for the removal of impurities from water include the following:

Precipitation—utilizing solubility-product-constant properties of contaminants for removal of sparingly soluble species such as calcium carbonate, magnesium hydroxide, iron, and other metal hydroxides.

Adsorption—the removal of soluble materials by physical interaction with a solid phase: silica adsorption on freshly precipitated iron or magnesium hydroxide, color and organics on alum floc, chlorine on activated carbon, oil on sorbent materials, and so on.

Coagulation—physical agglomeration of colloidal organic and inorganic materials, usually followed by flocculation and sedimentation or filtration for removal of suspended solids.

The above-listed methods will not reduce dissolved salt concentrations significantly, with the exception of those possible via lime-softening reactions:

$$Ca(OH)_2 + Ca(HCO_3)_2 \rightarrow 2CaCO_3 \downarrow + 2H_2O$$

$$2Ca(OH)_2 + Mg(HCO_3)_2 \rightarrow 2CaCO_3 \downarrow + Mg(OH)_2 + 2H_2O$$

When dissolved solids reduction is required, treatment may include the following:

Evaporation—the removal of water from the dissolved solids by adding heat to the water.

Electrodialysis—the removal of ionizable salts from water through ion-permselective membranes under the influence of electric potential difference.

Reverse Osmosis—the transport of water through semipermeable membranes from a region of high salt concentration to one of low salt concentration by application of a pressure in excess of the osmotic pressure of the salt solution.

Ion Exchange—the removal of strong or weak electrolytes from solution by interaction with ion exchange material, usually resin.

4.9-6 Water Quality Requirements

Specific water quality recommendations for industrial power applications may be obtained from trade associations, such as The American Pulp and Paper Institute (TAPPI) and the American Petroleum Institute (API). If the water or steam is to be used in food or medicinal products, water quality and treatment may be regulated by the FDA or other regulatory agencies. Regulations covering water for drinking or washing the person[4] are usually administered by state or local agencies.

4.9-7 Boiler Water

Water quality recommendations for power boiler applications[5] are available from the American Boiler Manufacturers Association (ABMA). Table 4.9-3 shows current ABMA recommendations for power boilers.

Industrial boilers, usually lower-pressure units, commonly utilize chemical treatment within the boiler itself in addition to treatment systems on the makeup and condensate systems. Internal treatment provides additional flexibility to protect the boiler against corrosion and deposits in the event of minor upsets in feedwater quality or boiler water quality. These internal treatment programs are usually of a proprietary nature and incorporate the use of phosphates to precipitate calcium, other alkaline materials for pH adjustment, reducing agents such as sodium sulfite for dissolved oxygen control, dispersing agents to prevent sludge accumulation, and sludge conditioning agents and other agents to modify scale characteristics and reduce adherence to heat transfer surfaces. Some proprietary systems include the use of chelating agents to permit carrying calcium, magnesium, and iron inventories in solution at elevated levels.

TABLE 4.9-3 RECOMMENDED BOILER WATER LIMITS AND ASSOCIATED STEAM PURITY AT STEADY-STATE FULL-LOAD OPERATION

Drum Pressure, psig	Range Total Dissolved Solids[a] Boiler Water, ppm (max)	Range Total Alkalinity[b] Boiler Water, ppm	Suspended Solids Boiler Water, ppm (max)	Range Total Dissolved Solids[b,c] Steam, ppm (max expected value)
Drum-Type Boilers				
0–300	700–3500	140–700	15	0.2–1.0
301–450	600–3000	120–600	10	0.2–1.0
451–600	500–2500	100–500	8	0.2–1.0
601–750	200–1000	40–200	3	0.1–0.5
751–900	150–750	30–150	2	0.1–0.5
901–1000	125–625	25–125	1	0.1–0.5
1001–1800	100	[d]	1	0.1
1801–2350	50		N/A	0.1
2351–2600	25		N/A	0.05
2601–2900	15		N/A	0.05
Once-Through Boilers				
1400 and above	0.05	N/A	N/A	0.05

Source. (Courtesy of American Boiler Manufacturers Association, copyrighted material.)

[a] Actual values within the range reflect the total dissolved solids in the feedwater. Higher values are for high solids, lower values are for low solids.
[b] Actual values within the range are directly proportional to the actual value of total dissolved solids in the boiler water. Higher values are for the high solids, lower values are for low solids.
[c] Exclusive of silica.
[d] Dictated by boiler water treatment.

4.9-8 Steam and Condensate

When high-pressure steam will be used to drive a power turbine, tighter limitations may be applicable.[6] Table 4.9-4 shows steam-purity recommendations given by some manufacturers of large utility turbines. It may be necessary to advise the boiler manufacturer of the turbine steam-purity requirements in order to obtain the boiler manufacturer's recommendation for suitable boiler water conditions to produce the necessary steam purity. This in turn dictates feedwater and makeup treatment system requirements.

TABLE 4.9-4 STEAM-PURITY RECOMMENDATIONS FOR HIGH-PRESSURE TURBINES

Parameter	Normal Operating Limits[a]	
	Westinghouse	General Electric
Cation Conductivity (μS/cm)	0.3	0.2
Sodium	5	3
Chlorides	5	—
Silica	10	[b]
Copper	2	[b]
Iron	20	[b]
Dissolved oxygen	10	[b]

[a] Units: Micrograms per liter in steam samples condensed and adjusted to 25°C, unless otherwise noted.
[b] Limits: Not required when manufacturer's boiler water limits are not exceeded.

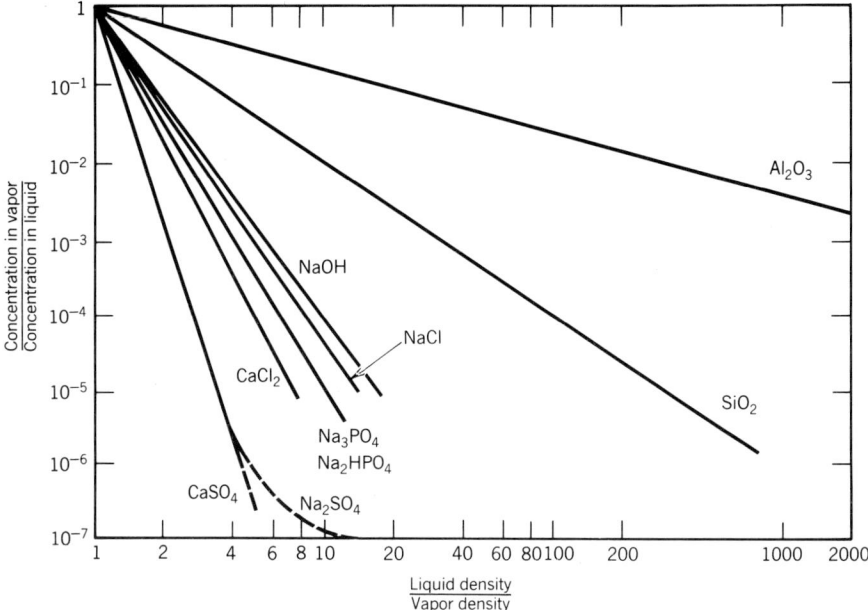

Fig. 4.9-1 Vapor phase: liquid-phase concentration ratios as a function of steam conditions. (Source: *Combustion*, March 1977. Courtesy of O.K. Jonas, Westinghouse Electric Co.)

Turbine steam impurities may originate in the boiler by two mechanisms. Materials present in the boiler water will be present in any mist carry-over to the steam. Another component of steam impurities will be due to solubility of materials in the steam phase. The solubility of a number of salts in steam as a function of steam conditions are shown in Fig. 4.9-1.

4.9-9 Cooling Water Systems

A major requirement of industrial operations and power generation is heat rejection, often requiring large amounts of cooling water. Recirculating cooling water systems significantly reduce the quantity of makeup water required by a facility but increase the treatment required to protect systems against corrosion and deposit damage. The circulating cooling water must be nonaggressive to the materials of construction within the cooling system. This requires not only control of cooling water chemistry but care in the design of the cooling water system and selection of the materials to avoid electrolytic corrosion due to incompatible metals in contact with the cooling water. Water chemistry must also be balanced to avoid deposits and corrosion from occurring at different points in the cycle, where there may be temperature differences of 20–40°C. Control may include the use of a Langelier saturation index,[7] which is based on control of the tendency of calcium carbonate to precipitate from the water. Under ideal conditions water with a very slight tendency to precipitate calcium carbonate is relatively noncorrosive to ferrous metal surfaces. Water with a strong tendency to precipitate calcium carbonate tends to form hard calcium carbonate deposits on heat transfer surfaces. Water with a tendency to dissolve calcium carbonate, that is, having a negative Langelier index, would be aggressive to steel, concrete, and asbestos–cement surfaces. Makeup and circulating water chemistry is controlled to avoid the accumulation of suspended solids within the system to minimize deposits and erosion. The suspended solids concentration in most circulating water systems is limited to about 100 mg/L. In smaller systems dispersing agents and crystal-modifying agents may be added to help control the formation of scales and deposits especially if there is not always adequate supervisory attention. These techniques may interfere with reuse of cooling water in other systems, however.

Biological fouling is also a concern to operators of cooling water systems. Bio-films often form on heat exchanger or condenser surfaces, interfering with heat transfer and impeding water flow. These bio-films tend to be sticky and trap suspended matter from the circulating water to worsen the problem. Biocides such as chlorine are effective in controlling bio-films if a rigorous controlled program is practiced. Biocides are a preventive measure and not generally completely effective at removing accumulated deposits. Once films become established, it is often necessary to mechanically clean condenser tubes in order to regain effective control. In most cases, by the time back-pressure and temperature differential measurements show the presence of a bio-film, it is too late to regain control by simple biocide feed.

4.9-10   Other Uses

Potable water requirements have been cited previously. Good practice separates completely the potable water—sanitary wastewater system from other plant water systems. The requirements for treatment and treatment equipment do not always dovetail with those of other plant water systems.

Wet flue gas scrubbing systems have makeup water requirements that vary with the specific chemistries involved. For example, the chloride ion may be limited in some scrubbers. When plant wastewater such as cooling tower blowdown is used for makeup, the operation of the upstream system is governed by the needs of the scrubber. Scrubber demister wash may require an independent supply of low dissolved solids water.

Water quality requirements for discharge of wastewaters from the plant site are governed by the Clean Water Act and regulated by the USEPA and state and local agencies. Systems for pH adjustment and removal of suspended solids, iron, copper, oil, and grease are typical requirements.

4.9-11   Removing Suspended Matter

Some suspended matter is present in virtually all surface water supplies and many well waters. The suspended matter may range from relatively large gritty materials that settle out in a few hours under quiescent conditions to submicron colloidal particles requiring years to settle. Composition varies from inorganic minerals carried in rivers and streams to organic material of biological origin, such as color bodies found in swamp waters. Deep well waters are usually free of color and other organics but may contain small quantities of colloidal silica or iron and manganese oxides. These metals may be in solution in water deep within the ground but oxidize and precipitate upon exposure to the air.

4.9-12   Clarification

Water that will be subjected to any ion exchange process must be free of suspended matter. Surface waters, or well waters containing colloidal materials or significant amounts of iron and manganese, are strong candidates for suspended solids removal even if appearing clear to the eye. A suspended solids removal system includes provision for feeding treatment chemicals, coagulation, flocculation, and sedimentation. Dosage, concentration, and order of chemical addition must be optimized in order to form reactive precipitates (flocs) that will adsorb and enmesh colloidal materials as the floc particles grow. This system will include a flash-mix section where chemicals are mixed with the incoming raw water. Flash-mix sections range in residence time from 30 s to 2 min depending on system design and treatment requirements.

After the addition of treatment chemicals, gentle agitation and flocculation are necessary to permit chemical reactions and adsorption reactions to proceed. Flocculation section residence may range from 10 to 30 min depending on design and treatment requirements. Warm waters being treated for suspended matter removal only require a relatively short residence time, while colored, cold waters may require significantly more time. After the completion of the flocculation operation, the water is passed to a separation section for the removal of the precipitates by gravity settling. Some industrial waste systems utilize air flotation for this separation step. The residence time in a separation zone may range from less than 30 min in systems of advanced design to as much as 8 h in conservatively designed conventional systems.

Some system designs recirculate freshly formed precipitates to the flash-mix or flocculation zone. The large surface area enhances solid-phase growth within the floc. Similar results are obtained passing the process flow through an upflow-suspended "sludge blanket." Solids contact can be of significant value in lime softening, reducing postprecipitation problems and the number of maintenance shutdowns for removal of hard lime scale. Deposits form on the recirculating precipitates rather than on component surfaces. Also, the recirculation of lime-softening precipitates reduces the residence time required for completion of chemical reactions.

System Arrangements

The components required for effective pretreatment may be assembled in a number of ways. Figure 4.9-2 shows a conventional arrangement having separate flocculation and sedimentation compartments. This type of arrangement is generally most suited to large supply systems and is primarily of concrete construction. The mechanical components are shop assembled prior to installation.

Figure 4.9-3 shows a solids contact clarification unit with variable-speed-drive recirculation impeller, separate flash-mixing, flocculation, and separation zones. Precipitates reaching the floor are maintained in a fluid condition by the mechanically driven scraper. Some unit designs of this type

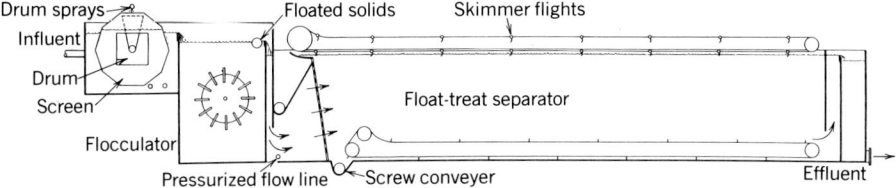

Fig. 4.9-2 Large conventional clarifier arrangement. (Courtesy of Envirex, a Rexnord Company.)

incorporate elevated lip concentrators that require that the precipitates be raised to a point above the floor for collection, concentration, and blowoff. The concrete pad beneath the unit usually includes a sloped floor section with one or more sludge hoppers and buried lines to an external sludge blowoff sump(s).

The sludge blowoff system should be automatic in proportion to flow. When the suspended solid loads are high or variable, and in lime softening, automatic sludge line backflush is a very valuable operating tool.

Figure 4.9-4 shows a modular system with modular tube settlers in the settling zone. Flash-mix and flocculation systems are similar to the conventional system shown in Fig. 4.9-2, but the residence time and the area requirements of the separation zone are significantly reduced by the use of tube settlers. The increased rate capability is due to a rather short separation distance for floc particles. In a conventional system floc particles may be required to settle 3–5 m whereas in tube settlers the vertical distance is a few centimeters. As the water flows up the inclined tube, precipitates slide back down the tube to the entrance end in a concentrated stream. Under some conditions, tube settler units may be sensitive to the accumulation of deposits and biological growths. Units are available equipped with a mechanical sludge concentrator or without it. Some tube settlers may be adapted to circular solids contact units to increase throughput capacity.

Figure 4.9-5 shows a system incorporating inclined plate separators. Though similar to tube units, flow paths in the separation units allow greater hydraulic capacity. Inclined plate separators offer the most compact arrangement available for industrial treatment systems and may be of significant value when retrofitting. These units may be sensitive to biological growths or deposit accumulation, particularly when lime softening without sludge recirculation.

Operating Conditions

All chemical coagulation, flocculation, and sedimentation systems operate best under continuous modulated-flow conditions with chemical addition in proportion to the flow rate. The units are sensitive to changes in chemical treatment demand. Demand is determined empirically by a jar test.[2] Clarifiers can only reflect and enhance the chemical treatment applied. That chemical treatment must be capable of producing the desired results. Units may sometimes be operated successfully on intermittent flow if conservatively sized initially. However, some units require 3–5 days of operation in order to achieve optimum results. In addition to variations in chemical treatment demand, excess air in the incoming water may form microscopic bubbles, causing floc particles to float and carry over into the separation zone. Temperature changes may cause thermal upset, short-circuiting across the separation zone. Sun shining on darkly painted units can cause heavy floc carry-over around the periphery.

Operating Results

Pretreatment units incorporating chemical coagulation, flocculation, and sedimentation are primary treatment units and do not remove all visible suspended material. Most manufacturers guarantee turbidity of less than 5 NTU; color, less than 5 PCU; and iron, less than 0.3 mg/L.

Units designed for lime-softening applications should carry calcium, magnesium, and alkalinity guarantees, silica guarantees are given only in special cases.

Treatment Chemicals

Clarification equipment, even the most sophisticated, cannot produce high-quality water unless a suitable chemical treatment is applied in a manner that forms precipitates and floc particles that will separate from the water. In the absence of definitive experience on the individual supply, the chemical feeds as given in Table 4.9-5 are suggested for jar tests.

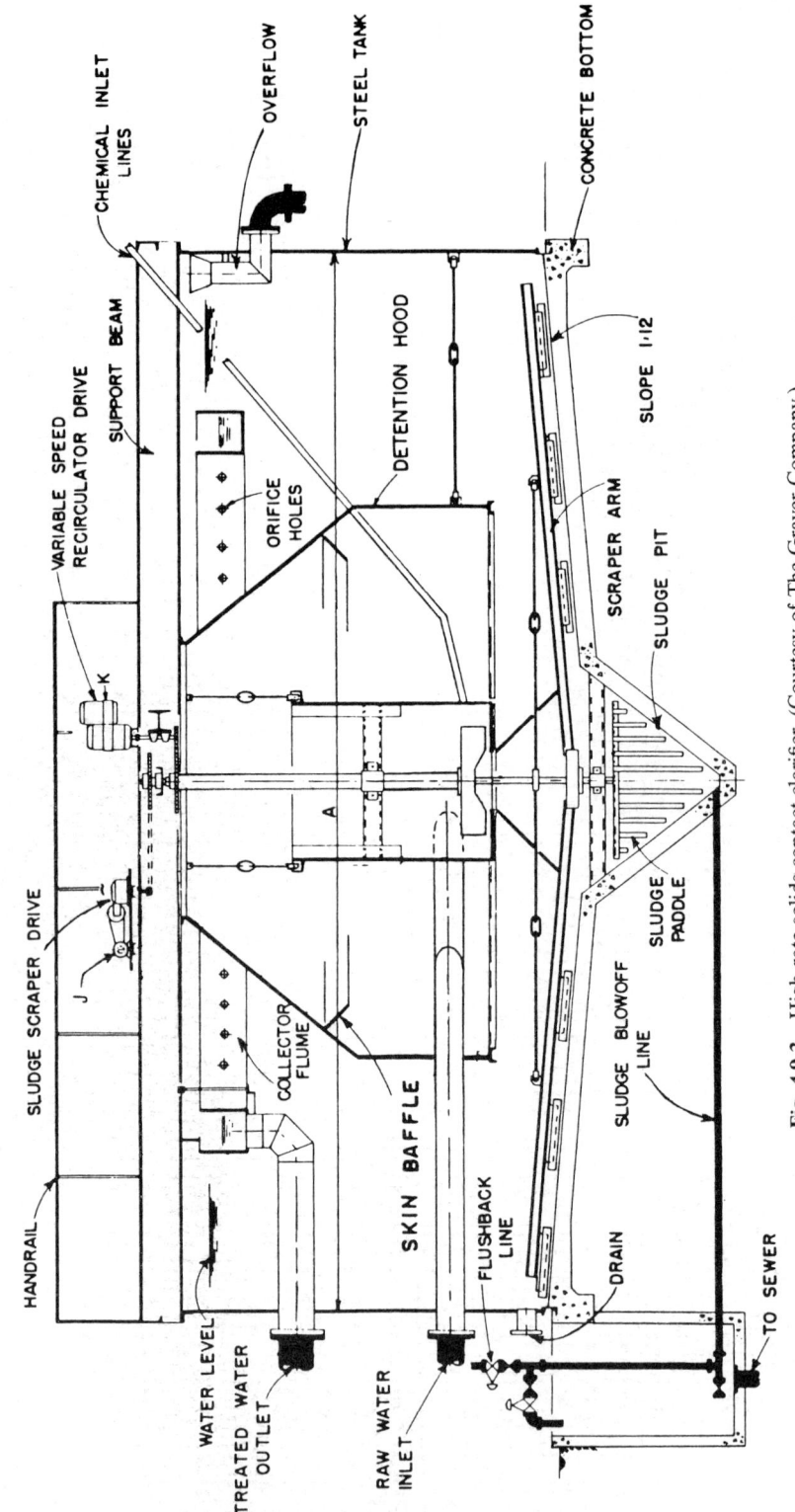

Fig. 4.9-3 High rate solids contact clarifier. (Courtesy of The Graver Company.)

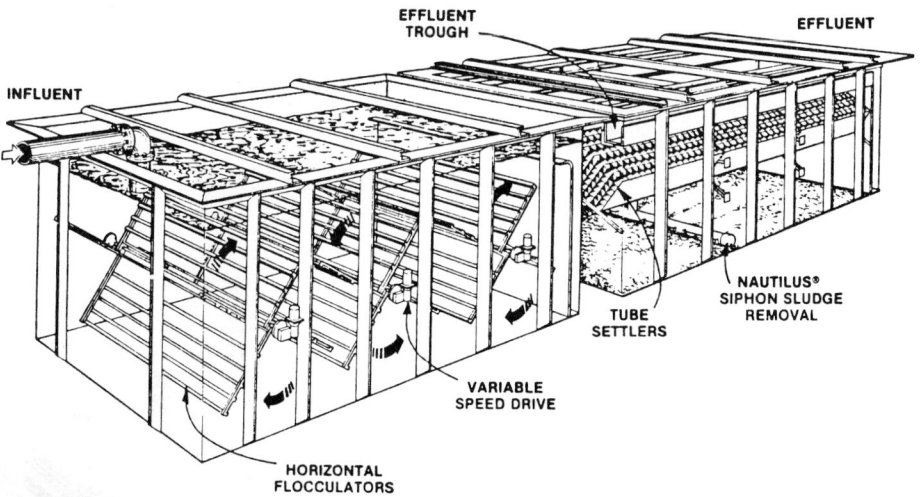

Fig. 4.9-4 Conventional clarifier with tube settlers for improved efficiency. (Courtesy of Neptune Microfloc, Inc.)

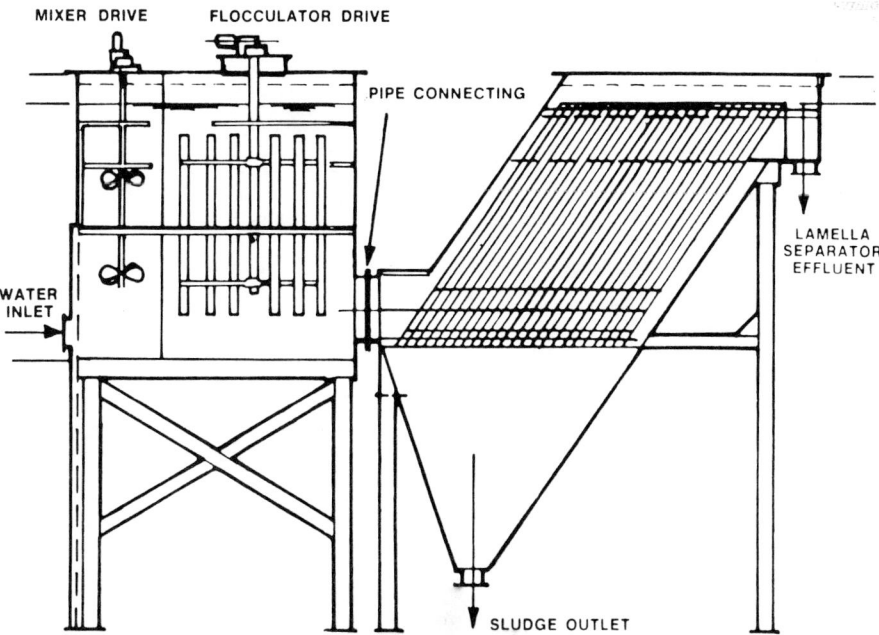

Fig. 4.9-5 Compact, preassembled inclined plate clarifier separator with sludge concentrator. (Courtesy of Passavant Corp.)

4.9-13 Filtration

Clarified water may be of suitable quality for a number of industrial applications without additional treatment. However, there will be periods of upset due to a change in treatment demand, shortage of chemical in a feed tank, chemical feedline plugging such as with lime, treated water demand surge, and mechanical component failure. During such periods relatively high turbidity may last from a few minutes to several hours depending on operator response. Systems providing water to downstream services sensitive to water quality usually require filters after the clarifier. Most industrial filters use

TABLE 4.9-5   Typical Clarifier Chemical Feeds[a]

Coagulants	Dosage, mg/L	Chemical Designation
Aluminum sulfate (filter alum) (alum)	10–40	$Al_2(SO_4)_3 \cdot 18H_2O$
Sodium aluminate	10–20	$Na_2Al_2O_4$
Ferrous sulfate (copperas)	20–40	$FeSO_4 \cdot 4H_2O$
Ferric sulfate (iron sulfate)	10–20	$Fe_2(SO_4)_3 \cdot H_2O$
Cationic polyelectrolyte	2–5	(Various)

Coagulant Aids	Dosage, mg/L	Notes
Cationic polyelectrolyte	$\frac{1}{2}$–3	
Nonionic polyelectrolyte	1–5	
Anionic polyelectrolyte	$\frac{1}{2}$–3	
Bentonite	2–10	Surface-cationic
Chlorine	1–5	Oxidant
Activated silica	5–10	Cationic

Softening	Dosage, mg/L	Chemical Designation
Calcium hydroxide (slaked lime)	[b]	$Ca(OH)_2$
Calcium oxide (unslaked, chemical, or quick lime)	[b]	CaO
Sodium carbonate (soda ash)	[b]	Na_2CO_3

[a]*Note:* Optimum coagulation treatment can require more than one coagulant, including chemicals listed under "Softening." Similarly, softening can require coagulant aid(s).
[b]Lime-softening chemical reactions are described in detail in several of the books listed in the Bibliography. Dosages depend upon the degree of treatment desired.

granular media such as silica sand (effective size 0.4–0.5 mm), anthracite (effective size 0.6–0.8 mm), or both. Dual-media filters have improved holding capacity. The coarser anthracite is placed on top of the finer, higher-density sand. Some systems employ a bottom layer of fine, high-density material (effective size ~ 0.2 mm). The exact mechanism of deep-bed media filtration is not clearly understood and probably includes a number of separate mechanisms such as mechanical straining, surface coagulation, interstitial sedimentation, adsorption, and electrokinetic effects.

Filter designs range from open concrete boxes to factory-built complete systems. The total system must include a container for the media and a freeboard to keep the media flooded, avoid air binding, and allow water level increase over a reasonable head loss range as foreign matter accumulates within the bed. The filtered water collector prevents the passage of filter media and may function as a part of the system for cleaning the media when the solids-holding capacity is exhausted.

Cleaning may be by simple flow reversal at a higher rate than the service run, with or without air assist. Mechanically or hydraulically driven sweeps or a system of fixed-orifice distributors in the upper portion of the bed for subsurface washing combined with backwash may assist in removal of tenacious matter. Some units are available with a partial bed removal and a cleaning and replacement cycle. Filtration rates range from 5 to 10 m/h varying with filter design, raw water characteristics, treated water demand, and so on. When estimating space requirements, allow at least 1 m along the operating face of each filter for the external pipe and valves needed. Water requirements for backwashing will range from 30 to 50 m/h.

Industrial filters in common use include those in Fig. 4.9-6, that is, gravity type with no prepumping. This avoids breakup of clarifier floc carried over and improves filtered water quality. A filter of this type stores its own backwash supply and may have single or dual media or even multimedia if desired. An appropriately designed underdrain system can very effectively clean dual media with air assist using as little as 20 L/dm^2 of backwash water.

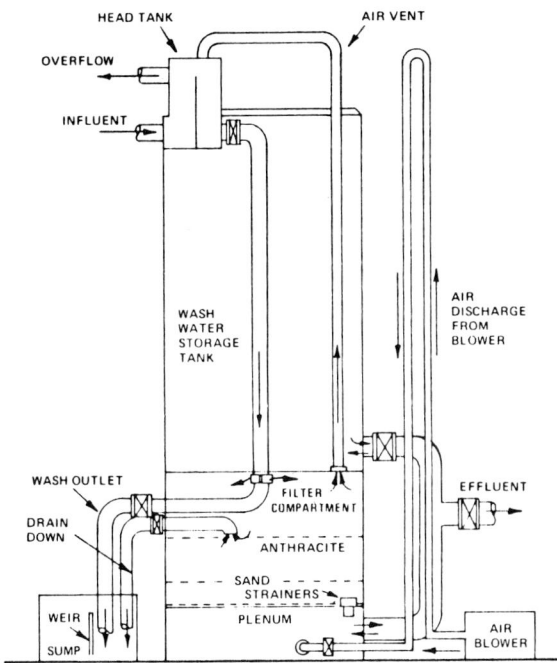

Fig. 4.9-6 Gravity filter with self-contained backwash supply. (Courtesy of the Permutit Company, Inc.)

Other filter types use single media with graded sand and gravel underdrains where extremely high quality water is not required and where the inlet water is available under pressure. Such filters may require 80 L/dm^2 filtered water supplied under pressure at 30–40 m/h (i.e., 30–40 m^3/h per m^2 filter area).

Filters must be protected against freezing and ingress of foreign material, including insects. Filtered water backwash storage should be covered to protect the water quality.

Wastewater is produced by filters during the backwashing operation in a flow rate range from 30 m/h to peak rates in excess of 60 m/h. Individual unit wash volume rates should be recognized in the design of the wastewater handling system. At times it may be necessary to wash two or more filters sequentially.

A direct-filtration process has been found applicable to a number of water supplies after long-term testing. For such waters a modified chemical coagulation and flocculation procedure precedes direct removal of precipitates by filtration in modified filter beds. Such systems must be provided with close control of chemical feed and frequent confirmation of chemical treatment demand to avoid turbidity breakthrough or excessive pressure drop buildup resulting from chemical overfeeding. Plants using direct turbidity removal by filtration operate with relatively short filter runs and increased wastewater production.

A modified process utilizes upflow filtration through graded media beds (Fig. 4.9-7) with the media ranging in size from about 20 mm in diameter at the bottom to about 0.5 mm at the top. Often, such beds are preceded by modified chemical coagulation and flocculation facilities to assure the removal of suspended and colloidal materials by the filter bed. Total media depths range from 1 to 3 m. Service rates are comparable to downflow systems. At the onset of limiting pressure drop or turbidity breakthrough, the bed is cleaned by diverting the flow to waste and increasing the upward velocity. This may be accomplished by supplemental air assist. At the completion of this forward-washing operation, a rinse step is required to free the upper portions of the filter media and the freeboard space of turbid water before returning to service.

Expendable precoat filters may be used in light-duty or specialized service. Permanent precoat retaining elements are fabricated in cylindrical or flat-plate form utilizing wound wire, woven screens, sintered metal, fabric, wound yarn, or the like, upon which the expendable precoat is placed. The actual filtering takes place upon the expendable precoat of diatomaceous earth, perlite, cellulose, or similar materials. At the pressure drop end point of the filtering cycle, the precoat media is removed either by backwashing with water or with a mixture of air and water. The filtration volume may be

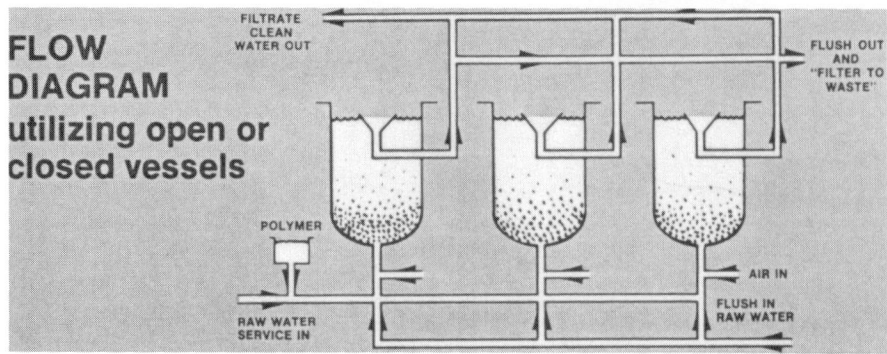

Fig. 4.9-7 Upflow graded media filter process. (Courtesy of L'eau Claire Systems, Inc., New Orleans, LA.)

increased in some cases by body feed of additional precoat material to increase the porosity of the surface film being formed on the elements. Body feed pumps must be selected to provide reasonable operating life with abrasive materials. Precoat filters are typically used in light-duty treatment service such as swimming pools and in specialized applications such as radioactive waste handling. Where liquid-waste volumes must be kept to an absolute minimum, the precoat may be removed using centrifugal force or air or a combination of the two.

Expendable cartridge filters are often used in polishing applications to protect final product purity. This type of filter may be selected with porosity from 200 μm to less than 1 μm, which removes even most bacteria. (Such waters are not actually sterile but contain few bacteria.) At the end of its service cycle the disposable filter cartridge is removed from the permanent holder and replaced.

4.9-14 Ultrafiltration

Ultrafiltration is a developing technique for removing suspended and colloidal material, including submicron particles, from water without chemical treatment. The membrane filter media is kept relatively free of surface fouling by controlling system pressure and maintaining high water velocity parallel to the membrane surface. About 75% of the water entering an ultrafiltration module will be discharged to waste or recycled. Ultrafiltration may permit treatment of surface waters without the addition of chemical coagulants or biocides and provide high-quality water suitable for ion exchange.

4.9-15 Removing Dissolved Matter

Many water applications require removal of some dissolved solids from the available water supply before use or reuse. Some examples are listed in Table 4.9-2. On occasion dissolved solids are too high to meet drinking water standards or low-pressure boiler makeup water requirements. Membrane, evaporation, and ion exchange processes can be used individually, in combination, and in conjunction with degasification to meet these needs.

4.9-16 Membrane Processes

Two membrane processes are available for dissolved solids reduction: electrodialysis and reverse osmosis.[8] In electrodialysis the salts dissolved in the water are caused to migrate, under the influence of applied potential difference, through cation-selective and anion-selective membranes. The ion concentration in the process stream is thus depleted. The processed water retains weakly ionized and suspended materials.

Reverse osmosis is important for industrial quality water production in energy systems. Reverse osmosis utilizes semipermeable membranes designed to permit the passage of water while preventing passage of most of the dissolved salts. The pressure differential across the membrane must significantly exceed the osmotic pressure of the salts in solution. Membranes are fabricated in sheet material or in the form of extremely small diameter fibers. Sheet membranes are generally rolled with appropriate spacer materials into sealed cylindrical form for insertion in cylindrical pressure vessels. Hollow fiber bundle assemblies usually resemble U-tube heat exchangers in FRP enclosures. Reverse-osmosis modules are available for treating waters of virtually any dissolved solids concentration ranging from seawater to relatively low concentration levels. The membranes are designed for specific service ranges. Membranes designed for seawater service would not be economical for low-solids water, for example.

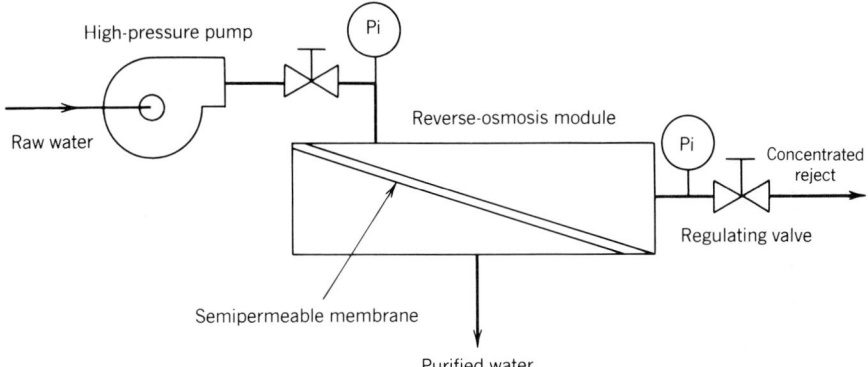

Fig. 4.9-8 A schematic representation of the reverse osmosis process. (Courtesy of Infilco Degremont, Inc.)

Fresh water reverse-osmosis systems are typically designed to remove 95% or more of the salts present in the incoming water (80–95% of dissolved silica) and to concentrate the salts in a waste stream, approximately 25% of the system feed stream. Most industrial systems operate at feed pressures of about 2400–4100 kPa (Fig. 4.9-8).

Feedwater to reverse osmosis units must be essentially free of suspended or colloidal materials to avoid plugging the membranes. Manufacturers often recommend a Silting Density Index[8] of less than 4 to provide reasonable periods of operation between chemical cleaning outages.

Concentrating salts and silica from the incoming feed stream to a relatively small volume waste stream requires that internal chemistry be very carefully controlled. It is often necessary to adjust the pH to an optimum range to avoid calcium carbonate deposition within the module or to feed dispersing agents (polyphosphates or other crystal modifiers) to avoid scale formation within the reverse-osmosis modules. Because the process operates at constant output, system suppliers usually package the entire system in modular fashion.

4.9-17 Evaporation

Thermal evaporation as a method of separating water from dissolved solids is not generally cost-effective, particularly for small energy systems. Multistage compression or flash evaporator systems are often used in wastewater recovery or radwaste processing systems. On seawater service, manufacturers typically guarantee product water containing about 10 mg/L of dissolved salts. This may be suited for some industrial applications and low-pressure boiler requirements. The need for large volumes of high-quality water for boiler feed usually dictates that one or more of the other processing systems be used for economy.

4.9-18 Ion Exchange

The ion exchange process removes ions from the process stream in exchange for ions of like charges released into the process stream from the ion exchange resin. A chemical regenerant is then applied to the resin in sufficient strength to reverse the process and again warehouse desirable ions on the resin. A familiar example of sodium cycle cation exchange is "water softening." Cation resin is treated with sodium chloride. During the softening cycle calcium, magnesium, and other multivalent hardness ions are exchanged for sodium ions. Where CR represents cation resin and Me^{2+} a metal ion,

$$2CR^-Na^+ + Me^{2+} = (CR)_2Me + 2Na^+$$

Most cation exchange resins are copolymers of styrene and divinylbenzene having strongly acidic sulfonic acid functional groups. Copolymers of acrylic acid and divinylbenzene having carboxylic weak acid functional groups are also important in special applications.

When strong acid cation resin is regenerated with a strong acid, virtually all detectable cations can be exchanged from the processed water which then contains only acids corresponding to the anions originally present. The treated water is low in pH, a property recognized in subsequent processing. Because the copolymer is subject to attack by oxidizing agents, especially at low pH, chlorine should be removed prior to hydrogen cycle cation exchange.

$$CR^-H^+ + Me^+ = CRMe + H^+$$

Anion exchange resins are similar to the cation materials except that the functional groups are weakly cationic tertiary amine or strongly cationic quarternary amines. Weakly basic anion exchange resins will only function at a low pH, removing strongly dissociated hydrochloric, sulfuric, and nitric acids. Strongly basic anion resins regenerated to the hydroxide form function over a wide pH range to remove both strongly and weakly dissociated anions, including silica and carbon dioxide. Where AR represents anion exchange resin and I^- an anion,

$$AR^+OH^- + I^- = ARI + OH^-$$

When a water is first treated by hydrogen cycle cation exchange and then hydroxide cycle anion exchange in a balanced process, the H^+ and OH^- exchanged from the resins combine to form water at neutral pH. Similar results are seen when cation and anion resins are used in a mixed-bed configuration. The quantity and type of any dissolved solids remaining in the water depend on the particulars of the selected process steps, the type of ion exchange resin, and the system configurations. Typical treated water characteristics are shown in Table 4.9-6.

TABLE 4.9-6   RESULTS OF ION EXCHANGE PROCESSES[a]

Process[b]	Softening	Soften and Dealkalize[c]	Soften and Dealkalize[c]	Demineralize[d]
Resin(s) employed	Strong acid cation	Strong acid cation Strong base anion	Weak acid cation and degasifier	Strong acid cation Strong base anion
Chemical regenerant(s)	Sodium chloride	Sodium chloride Sodium hydroxide[e]	Hydrochloric or sulfuric acid	Sodium hydroxide Hydrochloric or sulfuric acid
Calcium hardness	0.2	0.2	0–0.1	0
Magnesium hardness	0.1	0.1	nil	0
Alkalinity	Unchanged	4	0–0.1	0
Other cations	f	f	Unchanged	2–3
Other anions	Unchanged	g	Unchanged	2–3
Silica	Unchanged	Unchanged	Unchanged	0.05–0.3
Iron + manganese	0.07	0.07	0.07	0
pH	Unchanged	5–6	7–8	8–9.5
Wastewater volume (%)	3–5	6–10	5–8	8–12

[a] Units: hardness and alkalinity expressed in mg/L as $CaCO_3$, other parameters expressed in mg/L. Results are typical.

[b] Certain waters may prove (by history or pilot testing) to benefit by treatment of "organic scavenging" resins or activated carbon to remove organic molecules that can foul downstream ion exchange resins or escape treatment (see the Bibliography).

[c] Partial softening (hardness reduction) and alkalinity reduction may be achieved using parallel sodium cycle softening and acid-regenerated weak-acid cation units. Results vary in the blended outlet stream according to inlet water chemistry and may be acidic. See the Bibliography for design guidance.

[d] Results are for cation bed followed by anion bed, conventional regeneration. Use of countercurrent regeneration or addition of a polishing mixed-bed demineralizer can improve treated water quality as follows: cations 0–0.5, anions 0–0.5, silica 0–0.1 mg/L.

[e] Addition of sodium hydroxide to the salt regenerant (1 : 10) minimizes alkalinity leakage, but a filter in the regenerant line is recommended to remove calcium and magnesium precipitates. Treated water pH is ~ 8.3.

[f] An amount of sodium is added equivalent to calcium + magnesium removed.

[g] Chloride ion replaces other anions in the treated water.

Demineralizing Equipment

Ion exchange processes for boiler makeup water treatment typically utilize vertical cylindrical pressure vessels with the internals, as shown in Fig. 4.9-9. Internals include an inlet distributor, a separate regenerant chemical distributor, an ion exchange resin bed, and an underdrain collector. Sufficient freeboard must be maintained to prevent loss of the 0.5-mm-diameter resin beds during backwashing.

Units having a single type of resin may be designed for conventional downflow operation or for counterflow operation, which passes the regenerant through an undisturbed bed in the opposite direction to service flow (Fig. 4.9-10). Counterflow units may operate with upflow process and downflow regenerant or vice versa. Operating regenerant efficiency is much improved by this arrangement, provided continuous flow is maintained. Experience shows that any advantage may be lost when partially exhausted exchangers remain idle for a period.

Some ion exchange processes utilize mixed-bed configuration with two or more different types of resin within the same vessel. In such cases the vessels are often equipped with spent regenerant collectors in the interface zone between resin types to avoid contaminating large amounts of one ion exchange resin with the spent regenerant from another.

Ion exchange units are typically sized to operate in the cross-sectional area flow rate range of 15–30 m/h with mixed beds operating at rates of up to 50 m/h. The resin bed depth typically ranges from 0.8 to 2.5 m. The system capacity and arrangement must include treated water for plant needs plus system regeneration needs.

Operating Cycle

Systems are usually designed to minimize the amount of operator attention and required downtime. One unit of each type is removed from service, regenerated, rinsed, and returned to service in "train" control fashion. The outage time of a unit or train ranges from about 60 min for sodium cycle softening to about 4–6 h for a system having cation, anion, and mixed-bed exchangers such as used to produce high-quality boiler makeup water. Ion exchange units will produce between 11 and 14 L of wastewater for each liter of ion exchange resin regenerated. This water may contain suspended matter as well as acid and alkaline regenerant chemicals requiring neutralization before discharge to the facility drain system.

Regeneration Systems

Regenerant chemicals are introduced to ion exchange units in a relatively dilute form for optimum results. Salt-regenerated resins function best when contacted with 8–12% NaCl brine. Cation exchange resins regenerated by sulfuric acid may require several stages of acid concentration to avoid the potential formation of sparingly soluble calcium sulfate within the resin particles and in the resin bed. On the other hand, high acid concentration results in more effective regenerant utilization. Typical sulfuric acid concentrations range from 2 to 10% in contact with the bed. Hydrochloric acid (8–12%) may be used to regenerate resins to the hydrogen form without risk of in-bed precipitation. Operating cost may suffer. Where downstream components are sensitive to trace chloride ion, its use is not recommended.

Anion exchange resins regenerated with sodium chloride (such as for alkalinity reduction) typically function well with a regenerant concentration of about 5% NaCl. Small amounts of sodium hydroxide (0.5–1%) may be added to increase carbon dioxide removal and increase capacity. Weakly basic anion exchange resins may be regenerated with ammonia, sodium carbonate, or sodium hydroxide, 2–4% by weight. Strong-base anion exchange resins used for silica removal typically provide optimum results when regenerated with approximately 4% sodium hydroxide when the bed is at a temperature of about 50°C to aid in the removal of silica from the resin bed.

4.9-19   Condensate Polishing

Condensate in industrial and utility operations is often contaminated with at least traces of material that could be detrimental to the performance of the steam generator or boiler. For example, papermill dryers typically contribute significant quantities of iron to the returning condensate. Most industrial applications with long return lines have iron contamination of the condensate due to corrosion caused by traces of carbon dioxide and oxygen and unfavorable pH conditions. Small amounts of calcium, magnesium, and so on, may enter the condensate via cooling system leaks. Cycle corrosion, contamination from makeup, and cooling water leaks also affect even the tightest utility systems.

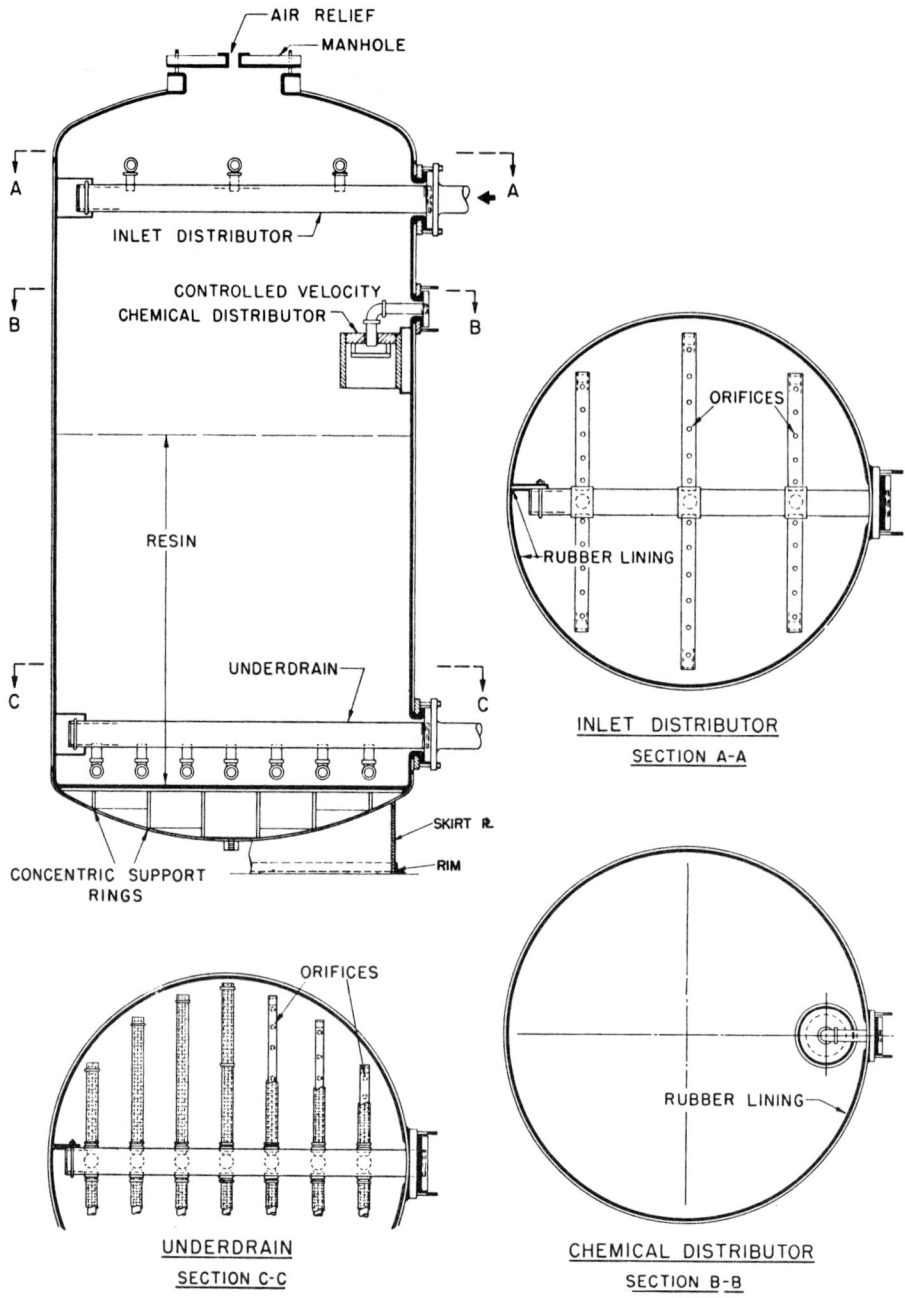

Fig. 4.9-9 Typical ion exchange vessel internals. (Courtesy of Infilco Degremont, Inc.)

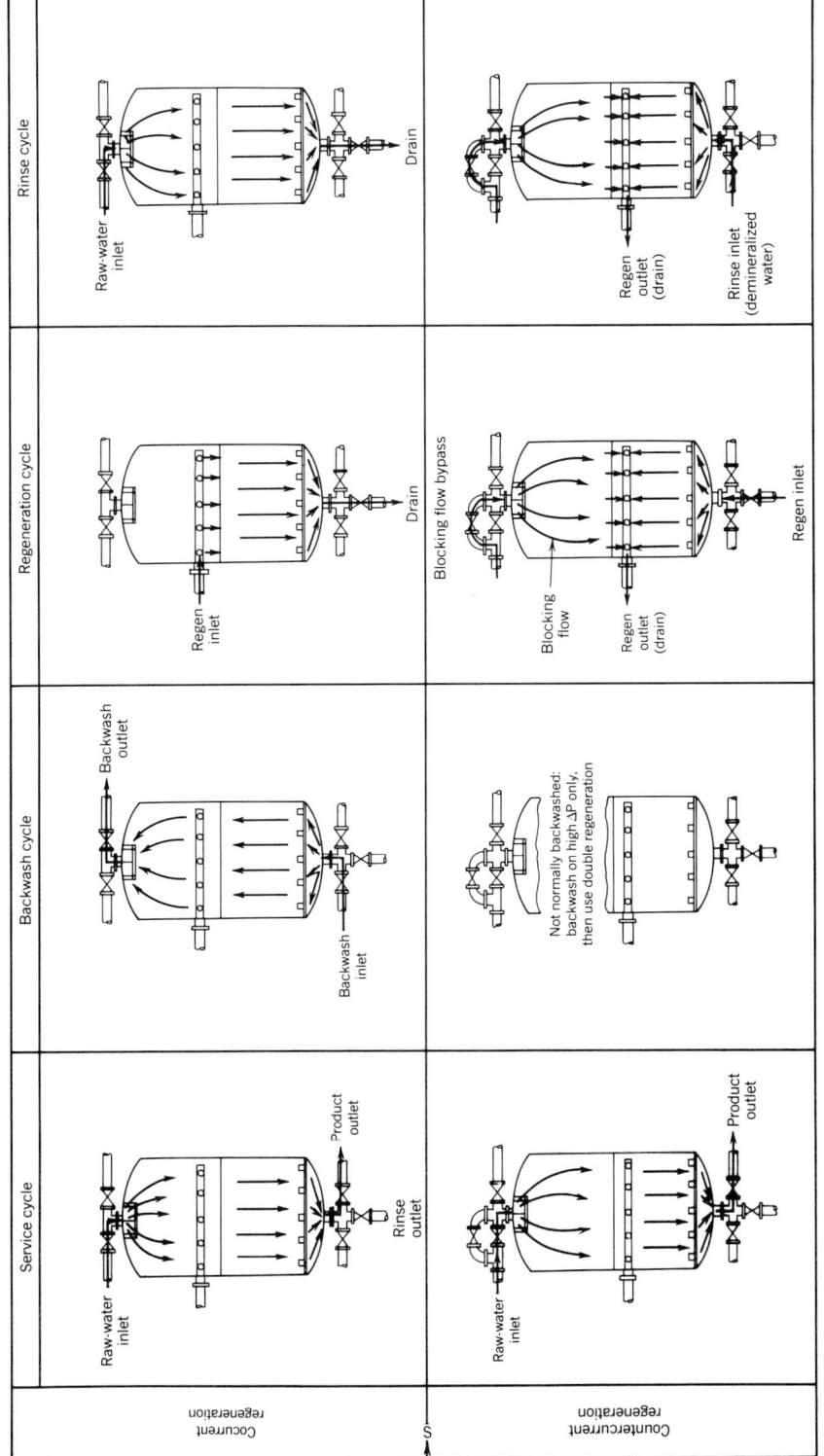

Fig. 4.9-10 A comparison of cocurrent and countercurrent regeneration designs for downflow ion exchange service. (Courtesy of Envirex, a Rexnord Company.)

539

Industrial Systems

Industrial condensate systems requiring removal of calcium, magnesium, and iron utilize polishing equipment very similar to salt-regenerated makeup units. The principal differences include higher pressure, flow rate, and ion exchange resin grade designed for the temperature and pressure differential. Units usually include resin bed depths of 0.8–1 m and operate at area flow rates of up to about 61 m/h. The units are often equipped with subsurface washers or rotary surface washers to break up surface accumulations or deposits in the upper 15–25 cm of the ion exchange resin bed for easier removal by backwashing.

Where higher-quality condensate is required and possible contamination of the feedwater with sodium chloride cannot be tolerated, units similar to makeup mixed-bed units are utilized, operated at approximately 61 m/h. This type of unit is falling into disfavor because of the possibility of the inadvertent contamination of the feedwater with regenerant chemicals.

Utility Systems

Condensate polishing systems (regenerable type) now almost exclusively sluice the ion exchange resin from the service vessel into a low-pressure regeneration facility for maximum protection against cycle contamination. The polishers are often arranged with two or more service vessels located near the hotwell condensate pumps to minimize the pressure drop. The service units are equipped with inlet water distributors and treated water collectors plus resin inlet distributor and resin transfer outlet. The units are often provided with recirculation pumps for rinse before the unit is placed into service because ion exchange resins do add very small amounts of contaminants to the water in the vessel during no-flow conditions. Such hydrolysis products can be removed by simply recirculating the water through the resin bed before returning to service. Air is used in the mixing of the ion exchange resins after regeneration and in some resin transfer systems. Some installations, therefore, recirculate the rinse back to the hotwell through a spray pipe in the air removal section of the condenser for oxygen removal.

Spent or exhausted ion exchange resins from the service units are transferred in slurry form from the service units to a regeneration facility equipped to carry out all regeneration operations: cleaning of the resins, separation of cation and anion exchange resins into separate regeneration vessels (so that each resin may be separately and thoroughly regenerated), and recombining, mixing, and holding resins ready for return to the service units. Although there are a few exceptions, the regeneration facility should be located no further than 150 m from the service units. Transfer lines must be accessible for disassembly and flushing in the event of a line blockage. Resin slurry transport is reliable as long as the flow is not interrupted during the transfer. If interrupted, line blockage is almost certain.

Service units are typically sized for area flow rates of 122 m/h, with bed depths of 0.8–1.5 m. The regeneration facility design is similar to makeup demineralizing systems with resin bed depths limited to about 2–2.5 m.

Some deficiencies of chemically regenerated condensate polishing systems are the remote risk of chemical contamination, the 12–24 h required for regeneration, the large quantities of condensate water required for regeneration, and the 16–17 L of wastewater for each liter of ion exchange resin processed. In boiling-water reactor facilities the large volume of radioactive regenerant waste for processing is a major disadvantage. Procedures to reduce the volume of wastewater include ultrasonic resin cleaning to remove iron more effectively than simple washing. This can reduce the frequency of chemical regeneration. There are techniques to permit operation beyond the normal ammonia exhaustion point and permit ammonia to pass through. In some applications it is more economical to utilize expendable ion exchange resin, eliminating the regeneration facility. Deep beds or a powdered ion exchange resin precoat on a specifically designed septum may be used.

Powdered Resin Systems

Powdered resin systems can efficiently produce polished condensate of quality comparable to chemically regenerated deep-bed systems (Fig. 4.9-11). There is substantially less ion exchange capacity in place with powdered resin because of the small quantity used. Typical precoat dosage levels provide only about one-sixth as much resin capacity as a deep-bed system might have. (A portion of this difference can be made up by providing a powdered resin body feed system.) Powdered resin systems have higher chemical costs, require only 30–60 min to precoat, need negligible amounts of condensate for cleaning and precoating operations, and produce very small volumes of wastewater for disposal and in radioactive service, significantly less radwaste than a chemically regenerable system of equivalent capacity. The space required for a typical powdered resin system would be about equivalent to the service units of a deep-bed system, since no space is required for external regeneration tanks, chemical storage tanks, and chemical handling systems.

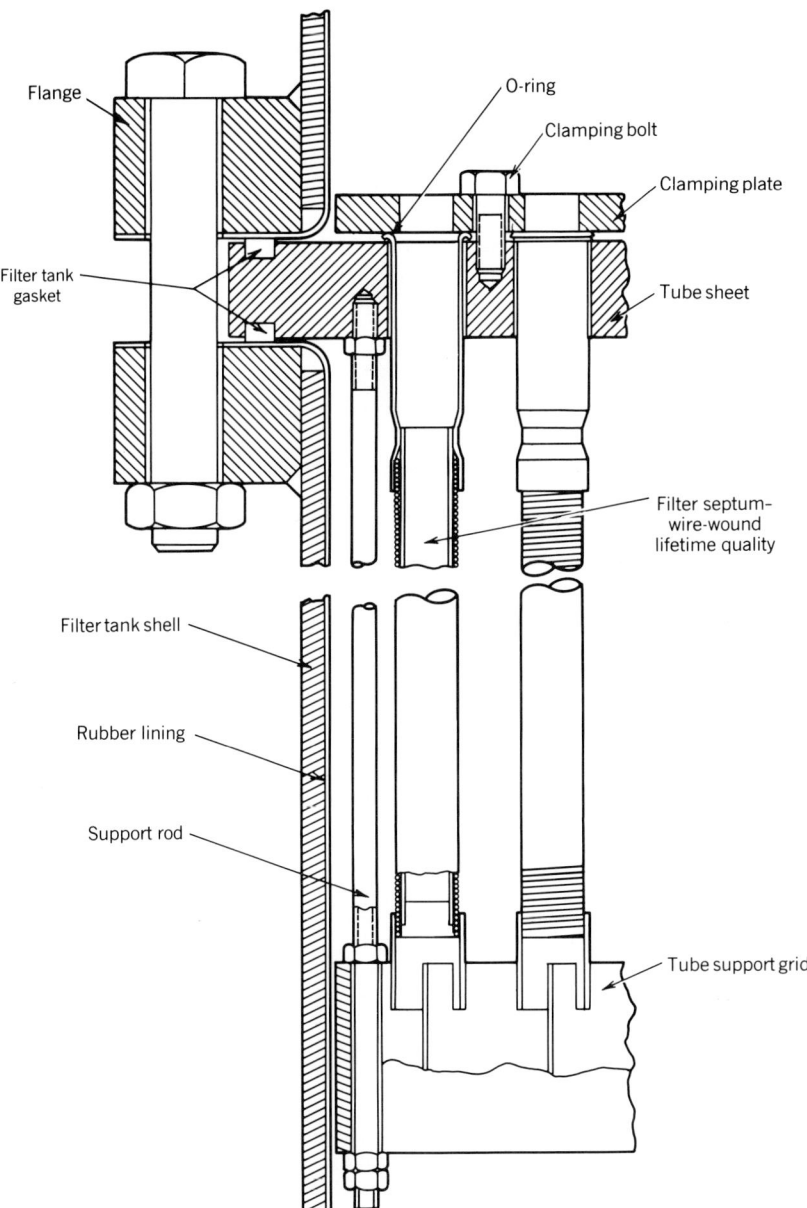

Fig. 4.9-11 Cutaway view of filter septum assembly. Powdered resin precoats are applied to the septa exterior. (Courtesy of Transamerica Delaval, Inc.)

Oil Removal

When returning condensates are heavily contaminated with oil, direct in-line treatment and reuse are not recommended. Makeup requirements of 100% may be offset by recovering condensate via one or more of the unit processes applicable to oily wastewaters. Lightly contaminated condensate may be successfully treated by means of an oil-adsorbent precoat applied to septum filters. The equipment is similar to that for the powdered ion exchange process. Direct-contact coalescing filters are also available, but temperature limitations apply to many available coalescing (oleophilic) materials. Since

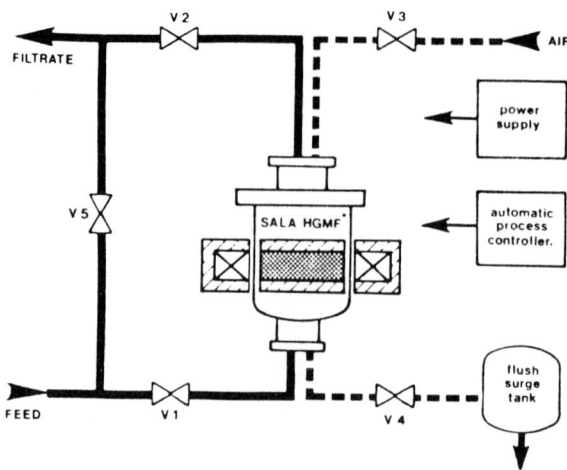

Fig. 4.9-12 Electromagnetic filter process components. (Courtesy of Sala Magnetics, Inc., Subs of Allis-Chalmers Corporation.)

oil renders condensate polishing resins inefficient to ineffective, oil removal precedes demineralizing. Powdered ion exchange resin condensate polishers may offer an advantage at this point, since the resins and the accumulation of trace oils are discarded following each cycle.

Electromagnetic Filters

Ion exchange systems, chemically regenerable or disposable resins, have an effective temperature limit of about 50°C if silica removal from the condensate is required and about 103°C for removing suspended matter only under normal operating conditions. Electromagnetic filters using steel media can be operated at any temperature up to the economizer inlet and effectively remove iron and other paramagnetic material. Such devices require little wastewater—washing at high velocities for only about 30 s—to effectively clean the media (Fig. 4.9-12).

Wastewater handling requires special consideration for magnetic filters operating at or near economizer inlet conditions. A primary advantage of this type of equipment is minimal space and power requirements. Typical units operate at 245 m/h.

4.9-20 Degasification and Deaeration

Dissolved carbon dioxide and oxygen may be removed from water by stripping processes that follow Henry's law.

Degasification is the process of passing atmospheric air upward through water splashing downward over a packing such as rings or saddles. The exposure of a large water surface area to the counterflowing air reduces the partial pressure of the gases being stripped, usually carbon dioxide, methane, and so on, so that the concentration of each gas in the liquid phase equilibrates at a low level.

Deaeration removes oxygen and other gases by reducing gas partial pressure. Surface exposure is used in combination with either vacuum or heating steam.

4.9-21 Degasifiers

Degasifiers are commonly vertical cylindrical vessels filled with packing such as polyethylene or polypropylene to avoid leaching of soluble species into the treated water.

Degasifiers using commercially available ring or saddle-type packing normally operate at cross-sectional area flow rates in the range of 50–75 $m^3/h \cdot m^2$. Air flow ranges between 5 and 12 $m^3/h \cdot m^2$ varying with the incoming water composition, packing depth, water temperature, and area flow rate.

Carbon dioxide is the most commonly stripped gas in industrial practice. Under normal operating conditions, carbon dioxide may be reduced to levels of about 5 mg/L. Degasifiers are commonly used in ion exchange systems. When sodium cycle exchangers operate in parallel with hydrogen cycle cation exchangers, the degasifier removes free carbon dioxide from the low-pH blend stream. A degasifier may also be used in demineralizing systems following a hydrogen cycle cation exchange unit to reduce the anion exchange load placed on strongly basic anion resin. Degasifiers may also be installed upstream of pretreatment clarifiers to reduce carbon dioxide or methane and other gases. By reducing carbon dioxide in lime-softening systems, the total lime demand may be reduced. Degasifiers may also remove sulfides from incoming water and transfer enough oxygen to the water to oxidize soluble iron.

4.9-22   Vacuum Deaerators

Vacuum deaerators (vacuum degasifiers) utilizing packed beds of rings or saddles similar to those in forced-draft degasifiers operate at cross-sectional flow rates of $50-100$ m^3/h $\cdot$ m^2 and at pressures within a few millimeters of the saturation pressure of the water being treated. Carbon dioxide is reduced to about the same levels as with forced-draft degasifiers, but dissolved oxygen may be reduced to a range of $0.01-0.3$ mg/L by this process. Low oxygen levels are most often obtained using two-stage deaerators to reduce the size and capacity of the vacuum pumping equipment required. Oil-sealed mechanical vacuum pumps are most commonly used for the vacuum service in industrial systems. Means must be provided to ensure against oil entering the treated water in the event of a power failure.

4.9-23   Deaerating Feedwater Heaters

A most effective means of removing dissolved oxygen, carbon dioxide, and other gases from feedwater is with deaerating feedwater heaters. These devices are direct-contact heat exchangers utilizing heating steam at pressures in the range of $0.2-3.4$ atm in industrial feedwater service and much higher pressures in utility cycles. A deaerated storage compartment is usually incorporated in the total system design. Deaerating feedwater heaters provide for the removal of dissolved oxygen and other gases to very low levels, direct-contact heating of feedwater to approximately heating steam saturation temperature, and the storage of deaerated, heated water above boiler feed pump suction (Fig. 4.9-13).

Utility deaerating heaters operate in an environment very low in dissolved oxygen and other gases. Therefore, they need not be designed to operate with high-inlet dissolved gas loads. Because of the high flow rates and the thermodynamic efficiency offered by tray deaerators, this type of equipment is preferred for utility service. Tray deaerators operate with virtually no pressure differential between the incoming heating steam and the operating environment of the deaerating heater itself. The deaerated heated water temperature leaving the deaerator is essentially at the saturation temperature of the incoming heating steam. Utility-tray-type deaerators are designed to operate with a very small quantity of makeup, virtually always less than 1% of feedwater flow, and are not expected to reduce the dissolved oxygen to normal operating levels under severe upset conditions. This type of deaerator is usually rated to produce deaerated water containing no more than $0.005-0.007$ mg/L of dissolved oxygen under normal operating conditions (Fig. 4.9-14).

Deaerating feedwater heaters used in industrial power service may be expected to deaerate a flow having zero returned condensate, that is, 100% makeup, with dissolved oxygen up to saturation. This type of service requires substantially more oxygen transfer capability than generating utility heaters. Industrial heaters typically have two sections. The incoming water is sprayed into a steam environment and then flows through a scrubbing section countercurrent to the incoming heating steam. Appropriately designed heaters of this type will produce deaerated water containing not more than $0.005-0.007$ mg/L of dissolved oxygen under essentially any condition of makeup and dissolved oxygen concentration. The major difference between the utility tray deaerating heater and the industrial spray-scrubber type is the apparent difference in thermodynamic efficiency. The lower apparent efficiency of the scrubber-type deaerator is due to the steam pressure drop through the scrubbing stage, which may amount to $0.5-1$ m of water. In most industrial cycles this minor loss is of no significance when compared to the capability of the deaerator to produce low-oxygen feedwater.

In some cases designers attempt to utilize the deaerating heater to remove significant quantities of carbon dioxide as well as oxygen and other gases. While this is possible, simple conventional deaerator designs must be substantially altered to ensure that the other gases are effectively removed and that local low-pH conditions within the deaerator do not cause excessive corrosion. This requires the extensive use of stainless steel within the deaerator, beyond that normally provided for wetted surface areas in the first stage.

Utility deaerating feedwater heaters are usually vented to the condenser. Provision must be made for atmospheric venting during startup and upset conditions to permit reduction of cycle oxygen concentration if the condenser is not operating under vacuum conditions. Low-pressure industrial-cycle deaerating heaters may be continuously vented to the atmosphere. Provision must be made to adjust vent rates to control the quantity of steam lost due to venting while providing sufficient venting for the applied oxygen load. Deaerating heater performance may be evaluated using ASME PTC 12.3.

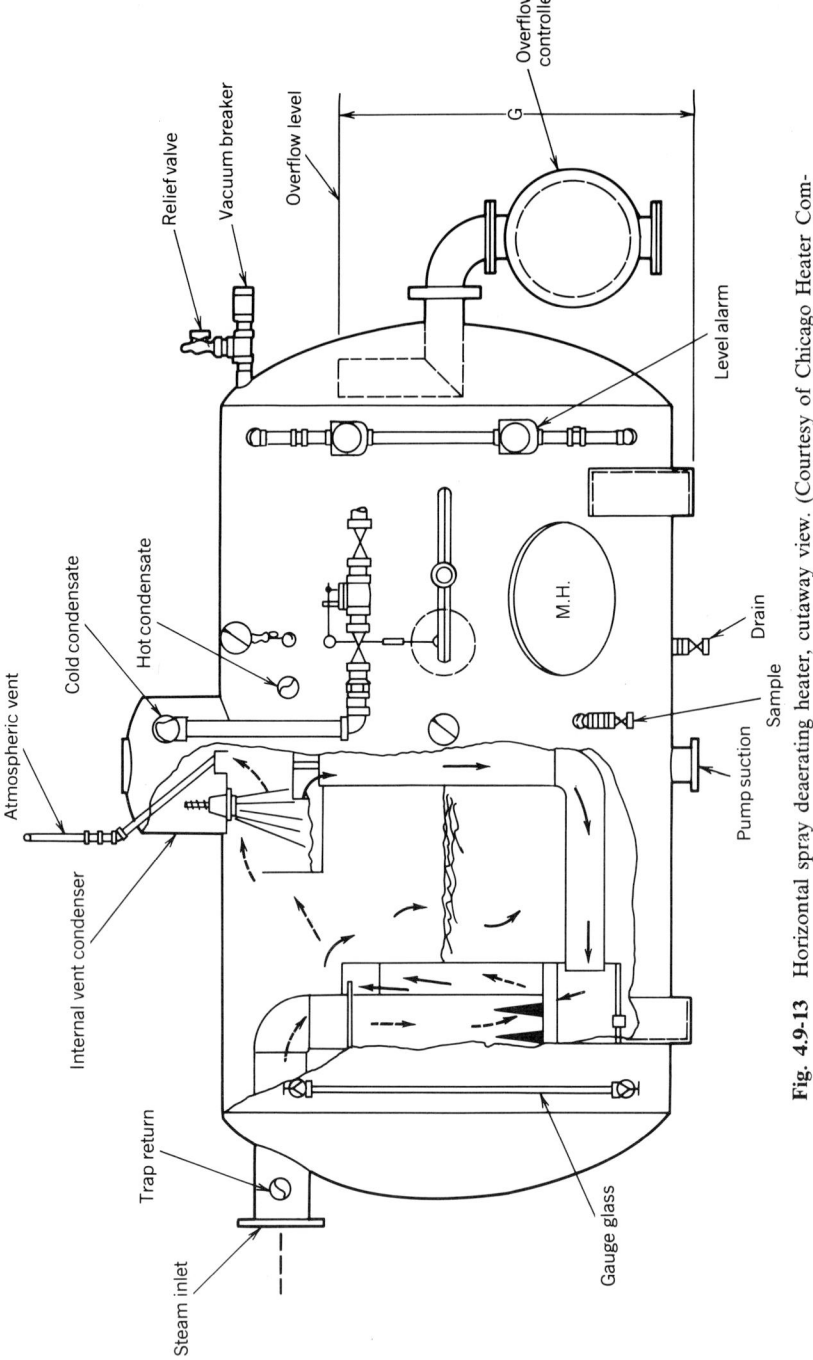

Fig. 4.9-13 Horizontal spray deaerating heater, cutaway view. (Courtesy of Chicago Heater Company, Inc.)

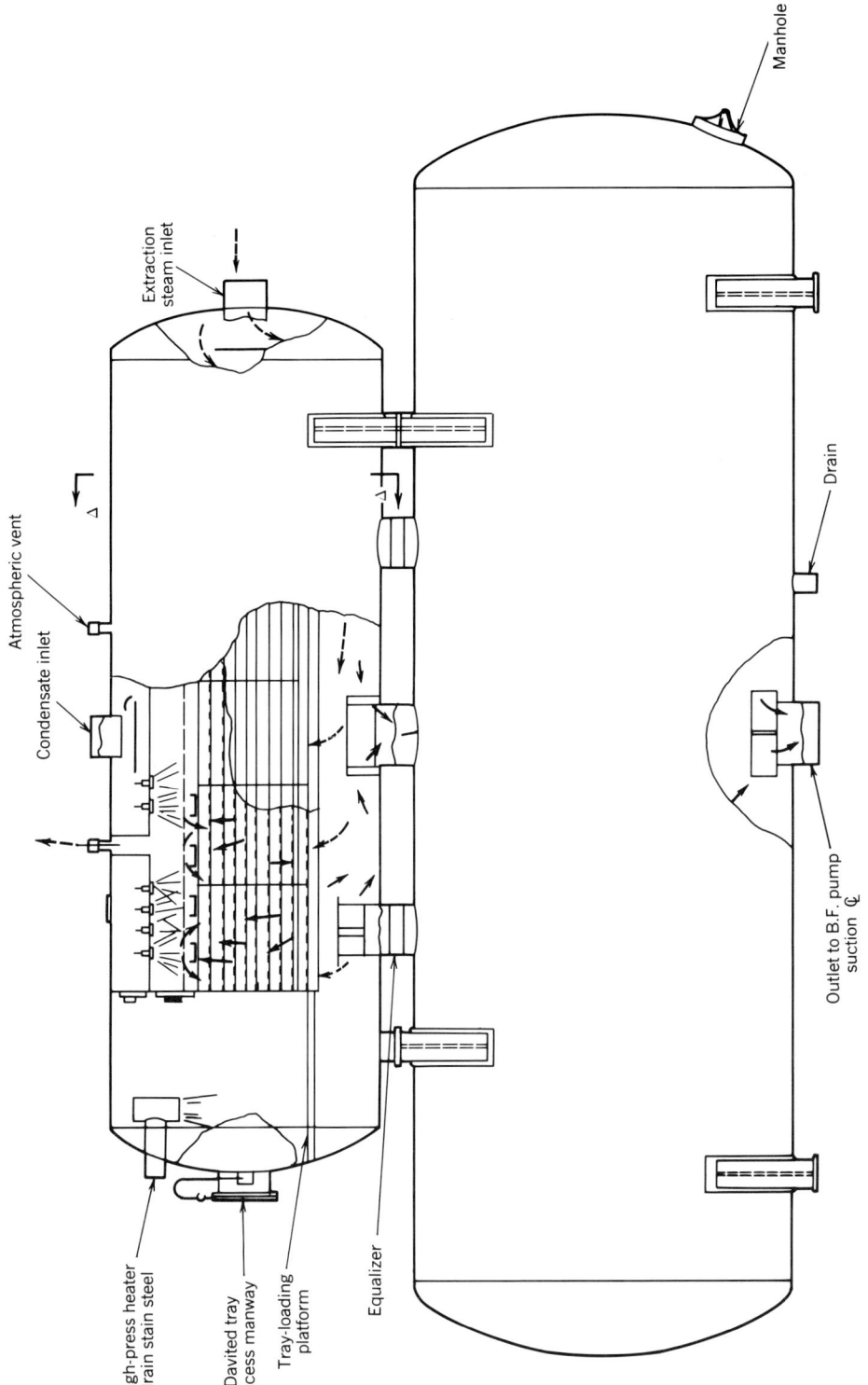

Fig. 4.9-14 The low head loss tray deaerating heater preferred for utilities operating at base load. (Courtesy of Chicago Heater Company, Inc., Subsidiary of the Marley Co.)

Manhole

Extraction steam inlet

Atmospheric vent

Condensate inlet

High-press heater drain stain steel

Davited tray access manway

Tray-loading platform

Equalizer

Drain

Outlet to B.F. pump suction ₵

4.9-24 Chemical Feed Systems

Chemical feed systems are used in pretreatment systems and ion exchange systems and to inject internal treatment chemicals into boilers and other systems. These systems may be shop assembled having metering pumps, a feed tank, a measuring tank, and all controls. Components should be constructed of durable, corrosion-resistant materials having long life and low maintenance requirements. Stainless steel (or other resistant metal) is often preferred.

When corrosive or hazardous chemicals are handled, a measuring tank should be included above the feed tank. This permits measurement by gauge glass of the chemical pumped from remote or drum storage. The feed tank should also be equipped with a gauge glass and provisions made to ensure against any possible backup of treatment chemicals into the dilution water supply line. Under no conditions should such tanks be directly connected to potable water or condensate systems. A calibration tube should be provided between the feed tank and the metering pumps.

Bulk Liquid Systems

Bulk liquid systems should be designed to contain an adequate residual and accept a convenient shipment of replacement chemicals. Typically, 23 m^3 has been found to be a suitable minimum. Tank materials should be selected for long-term maintenance free life. Steel, appropriately lined, (or FRP when suited to the chemical) is most often used.

Dry Feed Systems

Dry chemical feed systems should be designed to always recognize the hygroscopic nature of most chemicals used in water treatment. Chemical storage areas should be kept dry, isolated from wet areas. Dry chemical feeders are typically volumetric unless precision is required. In lime-softening service gravimetric feed is often preferable. Dry lime bulk silos should have air recirculation systems to exclude humid ambient air. Exhaust air from the bag filter may be recirculated back to the hopper bottom for fluidization.

ACKNOWLEDGMENT

I wish to sincerely thank M. H. Kerr for the technical and organizational expertise that made the completion of this section possible.

REFERENCES

4.9-1 American Public Health Association, *Standard Methods for the Examination of Water and Waste Water*, 15th ed., APHA, Washington, D.C., 1981.
4.9-2 American Society for Testing and Materials, *1982 Annual Book of ASTM Standards*, Part 11, *Water*, ASTM, Philadelphia, PA, 1982.
4.9-3 United States Environmental Protection Agency, *Methods for Chemical Analysis of Water and Wastes*, USEPA, Washington, D.C., 1974.
4.9-4 40CFR143, "Environmental Protection Agency National Secondary Drinking Water Regulations," 44FR42198, July 19, 1979.
4.9-5 American Boiler Manufacturers Association, *Boiler Water Limits and Steam Purity Recommendations for Water Tube Boilers*, 3rd ed., ABMA, Arlington, VA, 1982.
4.9-6 Electric Power Research Institute, *PWR Secondary Water Chemistry Guidelines*, EPRI NP-2704-SR, EPRI, Palo Alto, CA, 1982.
4.9-7 W. F. Langlier, *Journal of the American Water Works Association*, **8**, 1500–1521 (1936).
4.9-8 R. E. Lacey and S. Loeb, *Industrial Processing With Membranes*, Wiley, New York, 1972.

BIBLIOGRAPHY

BETZ Handbook of Industrial Water Conditioning, 7th ed. Betz Laboratories, Trevose, PA, 1976.

Corrosion, *The Journal of Science and Engineering*. The National Association of Corrosion Engineers, Houston (published monthly).

Technical Reports listed in "Electric Power Research Institute (EPRI) Guide," Electric Power Research Institute, Palo Alto, CA., updated periodically.

White, G. C. *Handbook of Chlorination*. Van Nostrand, New York, 1972.

The NALCO Water Handbook. McGraw-Hill, New York, 1979.

Proceedings of the American Power Conference. Water Technology Sessions. Illinois Institute of Technology, Chicago (published annually).

Proceedings of the International Water Conference. Engineers' Society of Western Pennsylvania, Pittsburgh, (published annually).

Proceedings of the International Conference on Water Chemistry of Nuclear Reactor Systems. Bournemouth, U.K. (published subsequent to meetings).

Water Conditioning Manual. Dow Chemical Co., Midland, MI, 1974.

Water Treatment Handbook, 5th ed. Halstead Press (Wiley), New York, 1979.

Water Treatment Handbook, Degrémont, Halstead Press, N.Y., 1979.

Water and Waste Treatment Data Book. The Permutit Co., Inc., Paramus, NJ, 1961.

Water Treatment for Industry, S. T. Powell, Inc., McGraw-Hill, New York 1954.

4.10   SHELL-AND-TUBE HEAT EXCHANGERS

Kenneth J. Bell

4.10-1   Areas of Application for Shell-and-Tube Heat Exchangers

The shell-and-tube heat exchanger provides reasonably compact and economical heat transfer surface in a readily constructed and mechanically strong structure. Provisions can be made for easy mechanical cleaning of the tube side and generally acceptable cleaning of the shell side. A wide variety of design features can be provided to meet special-service requirements, such as extreme pressures and temperatures, vibration, erosion, corrosion, venting, phase separation, and replacement of components. Available sizes range from 1 to over 1,000,000 ft^2 of heat transfer area in a single shell.

Therefore, the shell-and-tube heat exchanger as a generic type is widely used in the energy, environmental control, and process industries for a vast variety of services. Single-phase vaporization and condensing processes can be accommodated either in the tubes or on the shell as dictated by the rules of allocation (see Section 4.10-4). Multiple shells arranged in series or parallel are used to achieve heat transfer over a very large temperature change in each stream, or to reduce pressure drop, or to obtain a high degree of flexibility and control in operation.

The major drawback to a shell-and-tube exchanger is that it is not readily modified once it is constructed (unlike some other types of exchangers described in Section 4.11). Good design methods are available (for the best general reference see Vol. 3 of Schlunder[1]) and judicious overdesign will generally suffice for an exchanger in a well-defined service. However, significant changes in process conditions may well require complete redesign and reconstruction of an existing shell-and-tube exchanger.

4.10-2   Construction of Shell-and-Tube Exchangers

Basic Components

While there is an enormous variety of specific design features that can be used in shell-and-tube exchangers, the number of basic components is relatively small. These components are shown in Fig. 4.10-1 and are briefly described in the following paragraphs. More detail on all aspects of exchanger construction can be found (see Section 4.2 of Schlunder[1] and the TEMA Standards[2] for the general mechanical standard to which shell-and-tube exchangers are usually constructed).

Tubes.   The tube, labeled A, in Fig. 4.10-1, is the basic component of the shell-and-tube exchanger, providing the heat transfer surface between one fluid flowing inside the tube and the other fluid

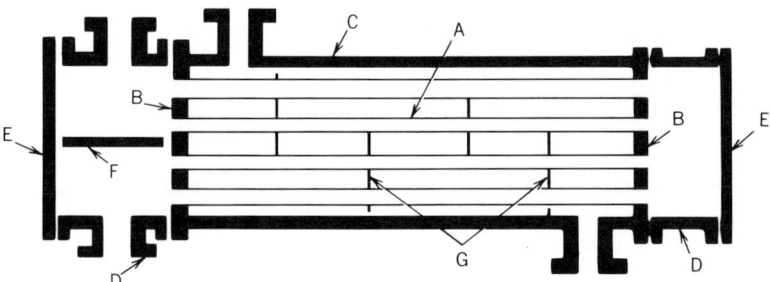

Fig. 4.10-1   Diagram of a typical fixed tube sheet shell and tube heat exchanger: A—Tube, B—tube sheet, C—shell, D—tubeside channel and nozzles, E—channel dover, F—pass divider, G—baffle.

flowing across the outside of the tubes. The tubes may be either drawn or extruded seamless metal or welded. The metal may be low-carbon steel, low-alloy steel, stainless steel, copper, admiralty, cupronickel, Monel, Inconel, aluminum (in the form of various alloys), or titanium, though many other materials may be specified for special applications.

The tubes may be either plain or with low fins on the outside. Low-fin tubes are used when the fluid on the shell side has a substantially lower heat transfer coefficient than the tube-side fluid. Low-fin surface provides $2\frac{1}{2}$–5 times as much heat transfer area on the outside as the corresponding bare tube, and this area ratio helps to offset the lower heat transfer coefficient. If low-fin tubes are used, it is necessary to consider the additional resistance to heat transfer caused by the longer conductive path; fin resistance or fin efficiency relationships have been developed for a number of important cases.[3] Fin efficiencies will typically be on the order of 0.9. Low-fin tubes have a fin outside diameter slightly less than the unfinned ends so that they may be inserted into the bundle through the tube sheet holes. Unfinned lands along the tube may be provided when the tubes must pass through the baffle holes and be supported.

The tubes are arranged in a regular geometric pattern, usually one of those shown in Fig. 4.10-2. A further characterizing parameter is the pitch ratio, the ratio of the center-to-center distance between adjacent tubes to the tube diameter. The smaller the pitch ratio, the greater the heat transfer area per unit volume. Practical mechanical considerations limit the pitch ratio to values above about 1.18; 1.25–1.50 is the normal range in application.

Tube outside diameters range from $\frac{1}{4}$ in. (6.35 mm) in the smallest exchangers to $\frac{3}{4}$ in. (19.0 mm) and 1 in. (25.4 mm) in the largest ones, though larger-diameter tubes are occasionally used.

Tube Sheets. A tube sheet, labeled B in Fig. 4.10-1, is usually a single round plate of metal that has been suitably drilled and grooved to take the tubes (in the desired pattern), the gaskets, and the spacer rods (not shown). It may be attached to the shell by flanging and bolting or by welding.

Where mixing between the two fluids (by means of tube hole leaks) must be avoided, a double tube sheet, as shown in Fig. 4.10-3, may be provided (at considerable increase in cost). The space between the tube sheets is open to the atmosphere so any leakage of either fluid should be quickly detected. Triple tube sheets (to allow each fluid to leak separately to the atmosphere without mixing) and designs with inert gas shrouds and/or leakage recycling systems are used in cases of extreme hazard or high value of the fluid.

The tubes are inserted into holes in the tube sheet, and there either expanded into radial grooves cut into the holes or welded to the tube sheet. In some low-stress situations (e.g., power plant condensers) the tubes can be simply expanded without grooving the tube sheet. Expansion into grooves provides a strong joint, and welding is resorted to primarily to provide an additional barrier to leakage through the tube sheet.

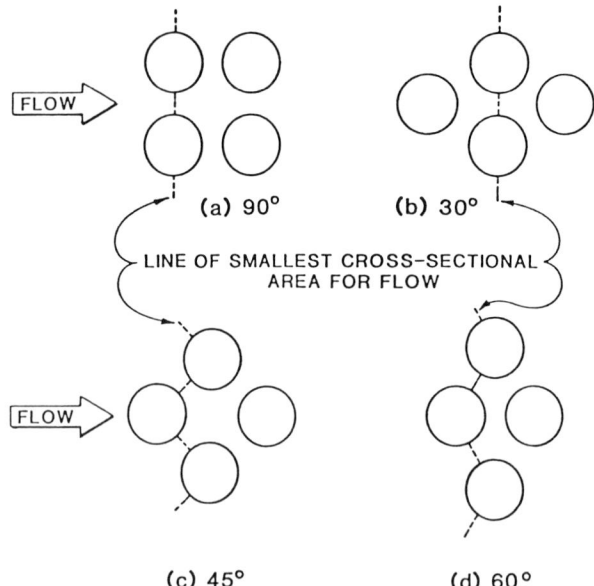

Fig. 4.10-2 Typical tube layouts used in shell-and-tube exchangers.

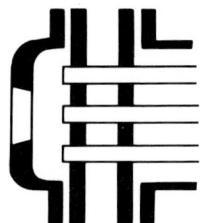

Fig. 4.10-3 Diagram of a double-tube sheet to avoid leakage of one fluid into the other.

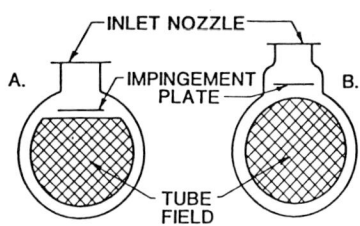

Fig. 4.10-4 Inlet nozzle arrangements with impingement plate: (A) Normal nozzle and (B) expanded nozzle.

In addition to its mechanical requirements, the tube sheet must withstand corrosive attack by both fluids in the heat exchanger and must be electrochemically compatible with the tubes and all tube-side material. Tube sheets are sometimes made from low-carbon steel with a thin layer of corrosion-resistant alloy metallurgically bonded to one side; the bonding may be achieved by various detonation-bonding processes or by laying down a continuous weld deposit of the cladding metal.

Shell and Shell-Side Nozzles. The shell (C in Fig. 4.10-1) is the container for the shell-side fluid, and the nozzles are the inlet and exit ports. The shell has a circular cross section and is commonly made by rolling a metal plate of the appropriate dimensions into a cylinder and welding the longitudinal joint ("rolled shells"). Small-diameter shells [up to ~ 24 in. (0.6 m) in diameter] can be made by cutting pipe of the desired diameter to the correct length ("pipe shells"). The roundness of the shell is important in fixing the maximum diameter of the baffles (see below) that can be inserted and therefore the effect of shell-to-baffle leakage. Pipe shells are more nearly round than rolled shells unless particular care is taken in rolling. In order to minimize out-of-roundness, small shells are occasionally expanded over a mandrel; in extreme cases the shell is cast and then bored out on a boring mill. In large exchangers the shell is made out of low-carbon steel wherever possible for reasons of economy, though other alloys are used when corrosion or high-temperature strength demands must be met.

The inlet nozzle often has an impingement plate (Fig. 4.10-4) set just inside the shell to prevent the incoming fluid jet from impacting directly at high velocity on the top row of tubes. Such impact can cause erosion, cavitation, and/or vibration. In order to leave enough flow area between the shell and impingement plate for the flow to discharge without excessive pressure loss and a high escape velocity, it is usually necessary to omit some tubes from the full-circle pattern. Alternatively, the nozzle may have an expanded section where it joins the shell (Fig. 4.10-4B).

Another very effective, if more expensive, solution to the fluid entrance problem is the vapor belt, or annular distributor (Fig. 4.10-5). In this design the inlet (and sometimes the outlet) nozzle is welded to an enlarged annular shell section, or knuckle, with the actual exchanger shell forming the inner surface of the annulus. This shell is slotted around the periphery to allow the fluid to flow in through a much larger total cross section than would be possible with any normal-sized nozzle.

Tube-Side Channels and Nozzles. Tube-side channels and nozzles (D in Fig. 4.10-1) control the flow of the tube-side fluid into and out of the tubes of the exchanger. Since the tube-side fluid is often the more corrosive, these channels and nozzles will frequently be made of alloy materials (compatible with the tubes and tube sheets, of course). They may be clad instead of solid alloy.

Channel Covers. The channel covers (E in Fig. 4.10-1) bolt to the channel flanges and can be removed for tube inspection without disturbing the tube-side piping. In smaller heat exchangers

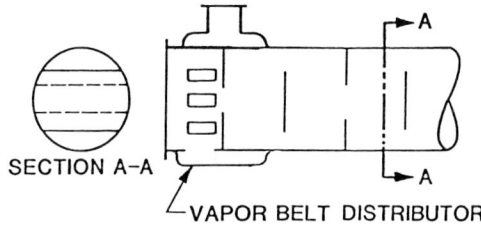

Fig. 4.10-5 A vapor belt, or annular distributor, shown in conjunction with double segmental baffles.

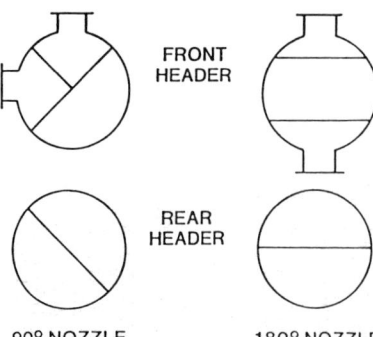

FRONT HEADER

REAR HEADER

90° NOZZLE   180° NOZZLE

Fig. 4.10-6 Alternative pass divider arrangements for a four-tube pass exchanger.

bonnets with flanged nozzles or threaded connections for the tube-side piping are often used instead of channels and channel covers.

Pass Divider. A pass divider (F in Fig. 4.10-1) is needed in one channel or bonnet for an exchanger having two tube-side passes, as the one illustrated in the figure, and they are needed in both channels or bonnets for an exchanger having more than two passes. If cast channels or bonnets are used, the dividers are integrally cast and then faced to give a smooth bearing surface on the gasket between the divider and the tube sheet. If the channels are rolled from plate or built up from pipe, the dividers are welded in place.

The pass dividers must fit tightly and cleanly into the grooves in the tube sheet and channel cover in order to minimize the possibility of leakage from one pass compartment to the next and a consequent serious deterioration of performance. The actual sealing is done by the gaskets, which must be well-set and periodically checked and replaced.

The arrangement of the dividers in multiple-pass exchangers is somewhat arbitrary, the usual intent being to provide nearly the same number of tubes in each pass, to minimize the pressure difference across any one pass divider (to minimize leakage), to provide adequate bearing pressure on all parts of the sealing gaskets, and of course to minimize fabrication complexity and cost. For four tube passes two arrangements are used (Fig. 4.10-6), mostly depending on piping convenience.

Baffles. Baffles (G in Fig. 4.10-1) serve two functions: first, and most importantly, they support the tubes in the proper position during assembly and operation and prevent vibration of the tubes caused by flow-induced eddies and, second, they guide the shell-side flow back and forth across the tube field, increasing the velocity and the heat transfer coefficient.

The most common baffle shape is the single segmental, shown in Fig. 4.10-7. The segment sheared off must be less than half of the diameter in order to ensure that adjacent baffles overlap at least one full tube row. For liquid flows on the shell side a baffle cut to 20–25% of the diameter is common; for low-pressure gas flows 40–45% (i.e., close to maximum allowable cut) is more common in order to minimize pressure drop. The baffle spacing should be chosen to make the free flow areas through the "window" (the area between baffle edge and shell) and across the tube bank roughly equal.

For many high-velocity gas flows the single segmental baffle configuration results in an undesirably high shell-side pressure drop. One way to retain the structural advantages of the segmental baffle and reduce the pressure drop (and, regrettably, to some extent the heat transfer coefficient) is to use the double segmental baffle shown in Fig. 4.10-8. Exact comparisons must be made on a case-to-case basis, but the rough effect is to halve the local velocity and therefore reduce the pressure drop by a factor of about 4 compared to a single segmental unit of the same dimensions.

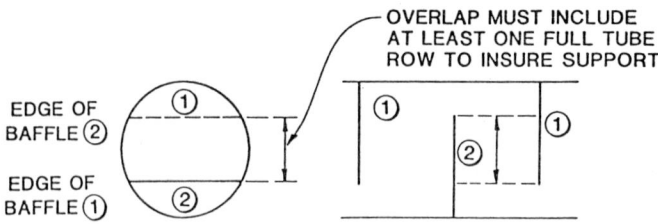

OVERLAP MUST INCLUDE AT LEAST ONE FULL TUBE ROW TO INSURE SUPPORT

EDGE OF BAFFLE ②

EDGE OF BAFFLE ①

Fig. 4.10-7 Sketch of typical segmental baffle arrangement.

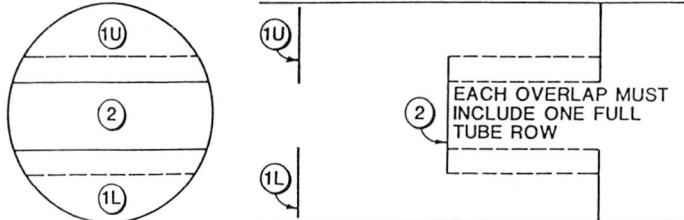

Fig. 4.10-8 Sketch of a double segmental baffle arrangement.

For sufficiently large units a triple segmental arrangement and ultimately strip and rod baffles can be used. The important point is always to ensure that every tube is positively constrained at periodic distances to prevent sagging and vibration.

Other baffle configurations such as disc-and-donut and orifice baffles have been used in the past but are seldom seen now.

A small clearance between tube outside diameter and baffle hole diameter is required to allow assembly of the tube bundle (and tube replacement, if desired). An excessive clearance provides too little tube support and possible vibration, as well as excessive leakage of fluid across the baffle. Too little clearance makes assembly and tube replacement difficult. Current TEMA standards[2] call for $\frac{1}{32}$ in. (0.8 mm) diametral clearance in most cases, with $\frac{1}{64}$ in. (0.4 mm) diametral clearance in special cases. In some air-conditioning applications the tubes are expanded into the baffle holes after assembly in order to eliminate the clearance altogether; this complicates tube replacement and possibly causes some thermal stress problems but may reduce potential tube vibration problems.

The outer diameter of the baffle must be less than the shell inside diameter to allow assembly, but the clearance should be as little as possible to minimize the shell-to-baffle leakage flow rate. (The shell-to-baffle leakage typically is the greatest penalty against the shell-side heat transfer coefficient.) TEMA[2] has set standards for this clearance, allowing an extra clearance for the out-of-roundness of rolled shells.

A modified baffle arrangement, particularly useful when vibration is expected, is the no-tubes-in-the-window (NTIW) design shown in Fig. 4.10-9. As the name implies, no tubes pass through the window regions of the heat exchanger, the baffle cuts being somewhat smaller than they would be in a typical fully tubed design. The advantage of the NTIW design is that now every tube passes through and is supported by every baffle guiding the flow. Additionally, tube support plates may be inserted as required between the baffles; these plates support the tubes against vibration rather than guiding the flow. The loss of tubes can be minimized by small baffle cuts and possibly by an increase in the shell-side velocity and hence heat transfer coefficient.

In cases where absolute protection against vibration is needed, or where pressure drop is to be minimized, the RODbaffle design may be utilized (Fig. 4.10-10; see also Small and Young[4]). Tube support is in the form of arrays of solid rods passing transversely between rows of tubes arranged in a square layout. The rods are the same diameter as the tube-to-tube spacing so that each rod holds two rows of tubes apart so they cannot vibrate or touch at that point. The rods are welded to an annular ring; each ring has a complete set of rods spaced two tube rows apart, with each successive ring having rods either filling the alternate spaces to the first ring or rotated to the other axis. The pattern is repeated for the length of the exchanger. Ring spacing is 6–12 in. (150–300 mm). Thus, each tube is supported on all four sides with each set of four rings. A vibration analysis has not been done on this method of support, but the structure is very rigid and no RODbaffle exchanger has ever been reported as vibrating.

While originally developed to resist severe vibration problems, the RODbaffle is also a very low pressure drop device. The low pressure drop feature has made it very interesting for condensers,

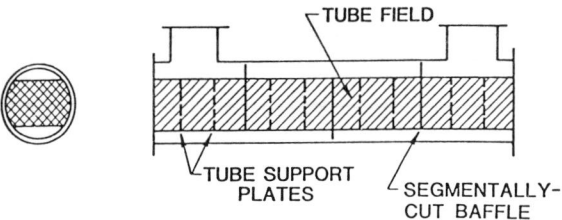

Fig. 4.10-9 Diagram of a shell-and-tube exchanger with no tubes in the window.

Fig. 4.10-10   Baffle and tie rod assembly for RODbaffle exchangers.

especially if noncondensable gases are present and one is designing to very close temperature differences between coolant and condensing vapor.

Provisions for Thermal Stress

The Thermal Stress Problem.   Since the shell of the heat exchanger will be at a different temperature than the tubes, differential expansion will result in stresses existing in both components and being transmitted through the tube sheets. Shells have been buckled or torn loose from supports, and tubes have been buckled or pulled out of the tube sheet or simply pulled apart. The fixed tube sheet exchanger shown in Fig. 4.10-1 is especially vulnerable to this kind of damage because there is no provision made for accommodating differential expansion. If the inlet temperatures of the two streams differ by more than perhaps 100°F (50–60°C), the thermal stress problem needs to be examined carefully, considering differences in materials and their properties, temperature level of operation, start-up and cycling operational procedures, and so on. Detailed treatments are given in Section 4.1 of Schlunder[1] and in the TEMA standards.[2] Various techniques for dealing with the thermal stress problem are given in the following paragraphs.

Expansion Joints.   The most obvious solution to the thermal expansion problem is to put an expansion roll or joint in the shell, as shown in Fig. 4.10-11. This becomes less attractive for large-diameter shells and/or increasing shell-side pressure. However, very-large-diameter, near-atmospheric pressure shells have been designed with a partial ball joint in the shell designed to allow the shell to partially "rotate" to accommodate stresses. Expansion bellows may also be used on the tube-side piping internal to a deep bonnet, as illustrated in Fig. 4.10-12. Bellows are available to at least 36 in. (0.9 m) in diameter at pressures and temperatures typical of process applications.

U-Tube Design.   The U-tube exchanger shown in Fig. 4.10-13 allows completely independent expansion of the tubes and hence solves the thermal stress problem perfectly. However, individual tubes cannot be replaced except in the outer row, the tube side cannot be mechanically cleaned, and erosion can occur inside the tubes in the U bend. In the large bundle diameters the long unsupported span of the outermost tubes in the U bend has caused many destructive vibration problems; vibration can be prevented by the use of an "egg-crate" of rods inserted between the tubes and fastened together outside the bundle. For easier comparison with other candidate bundle designs, the various features of this and the succeeding configurations are summarized in Table 4.10-1.

Floating-Head Design.   Several different designs of "floating-head" shell and tube exchangers are in common use. The goal in each case, of course, is to solve the thermal stress problem and each design does accomplish that goal. Inevitably, however, something must be given up, and each configuration

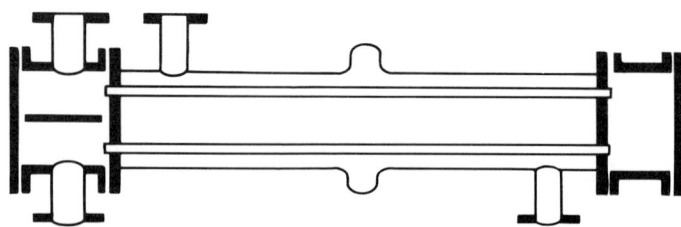

Fig. 4.10-11   Fixed tubesheet shell-and-tube exchanger incorporating a shell-side expansion joint (TEMA AEL).

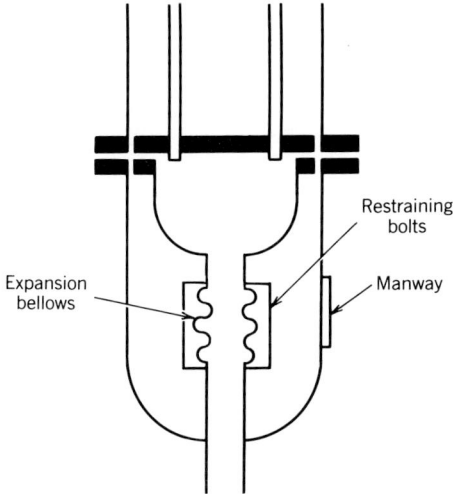

Fig. 4.10-12   Internal (tube-side) expansion bellows.

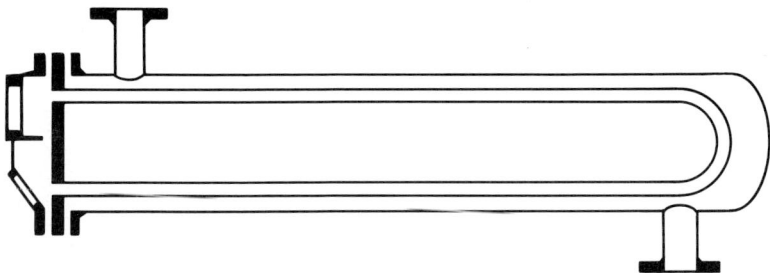

Fig. 4.10-13   Diagram of a U-tube exchanger (TEMA BEU). (Courtesy Patterson-Kelley Co., Inc., East Stroudsburg, PA.)

has a somewhat different set of drawbacks to be considered when choosing one. These are summarized in Table 4.10-1.

The simplest floating-head design is the "pull-through bundle" type shown in Fig. 4.10-14. One of the tube sheets is made small enough so that it and its gasketed bonnet may be pulled completely through the shell for shell-side inspection and cleaning. Unfortunately, many tubes must be omitted from the edge of the full bundle to allow for the bonnet flange and bolt circle. Also, the missing tubes around the periphery open up a low-resistance flow path for the shell-side fluid, the so-called bundle bypass stream. The reduced flow across the tube bundle has a significantly lower heat transfer coefficient than the ideal flow; additionally, the effective mean temperature difference is reduced. These problems can be reduced by the use of sealing strips put in longitudinally to partially block the bypass stream, as shown in Fig. 4.10-15. The sealing strips are always inserted in pairs and are placed symmetrically. They may be snapped into notches or welded to the baffles.

Another floating-head design that partially offsets the above disadvantages is the "split-ring floating-head" type (Fig. 4.10-16). Here the floating-head cover is bolted to a split backing ring rather than to the tube sheet, though there still must be a gasket-bearing surface. At some cost in added mechanical complexity, most of the tubes lost from the bundle in the pull-through design have been restored. The split-ring head is widely used to meet the severe demands of the petroleum industry, though the pull-through floating head is still used for the most severe services, particularly where positive sealing of the tube-side fluid is vital.

Two other types, the outside-packed lantern ring (Fig. 4.10-17) and the outside-packed stuffing box (Fig. 4.10-18), are less positively sealed against leakage to the atmosphere than the foregoing types but have the advantage of allowing single tube-side pass construction. (The pull-through and split-ring types can be modified for single tube pass design, but they then give up the advantage of positive sealing so important in high-pressure or hazardous fluid service.)

TABLE 4.10-1 COMPARATIVE FEATURES OF THE VARIOUS SHELL AND TUBE CONFIGURATIONS

Type of Design	U Tube	Fixed Tube Sheet	Floating-Head Pull-Through Bundle	Floating-Head Outside-Packed Lantern Ring	Floating-Head Split Backing Ring	Floating-Head Outside-Packed Stuffing Box
Relative cost increases from (A) Least expensive through (E) Most expensive	A	B	C	C	D	E
Provision for differential expansion	Individual tubes free to expand	Expansion joint in shell	Floating head	Floating head	Floating head	Floating head
Removable bundle	Yes	No	Yes	Yes	Yes	Yes
Replacement bundle possible	Yes	Not practical	Yes	Yes	Yes	Yes
Individual tubes replaceable	Only those in outside row	Yes	Yes	Yes	Yes	Yes
Tube interiors cleanable	Difficult to do mechanically, can do chemically	Yes, mechanically or chemically	Yes, mechanically or chemically	Yes, mechanically or chemically	Yes, mechanically or chemically	Yes, mechanically or chemically
Tube exteriors with triangular pitch cleanable	Chemically only	Chemically only	Chemically only	Chemically only	Chemically only	Chemically only
Tube exteriors with square pitch cleanable	Yes, mechanically or chemically	Chemically only	Yes, mechanically or chemically	Yes, mechanically or chemically	Yes, mechanically or chemically	Yes, mechanically or chemically
Double tube sheet feasible	Yes	Yes	No	No	No	Yes
Number of tube passes	Any practical even number possible	No practical limitations	No practical limitation (for single pass, floating head requires packed joint)	Limited to single or double pass	No practical limitation (for single pass, floating head requires packed joint)	No practical limitation
Internal gaskets eliminated	Yes	Yes	No	Yes	No	Yes

Source. Patterson-Kelley Co., *Heat Exchangers*, Manual No. 700-A, East Stroudsburg, PA, 1959.

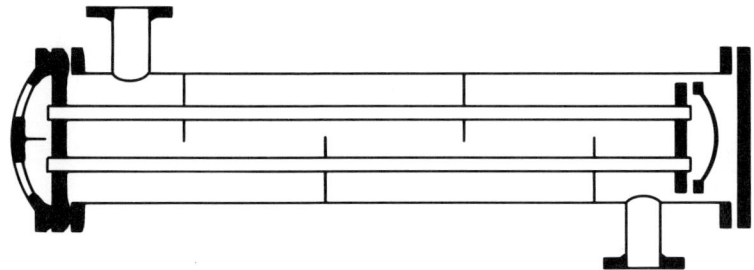

Fig. 4.10-14 Pull-through floating head exchanger (TEMA BET). (Courtesy Patterson-Kelley Co., Inc., East Stroudsburg, PA.)

SHELL

TUBE FIELD

BAFFLE CUTS

SEALING STRIPS

Fig. 4.10-15 Use of sealing strips to partially block bypass stream between tube field and shell.

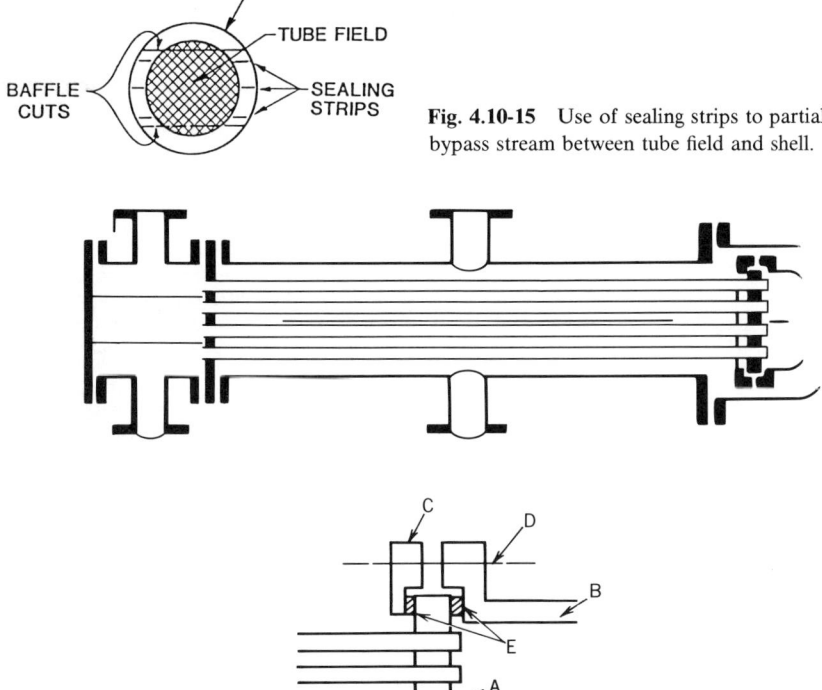

Fig. 4.10-16 Split ring floating head exchanger (TEMA AGS): (A) Floating tube sheet, (B) floating head cover, (C) split ring backing device, (D) bolt circle, and (E) gaskets. (Courtesy of Patterson-Kelley Co., East Stroudsburg, PA.)

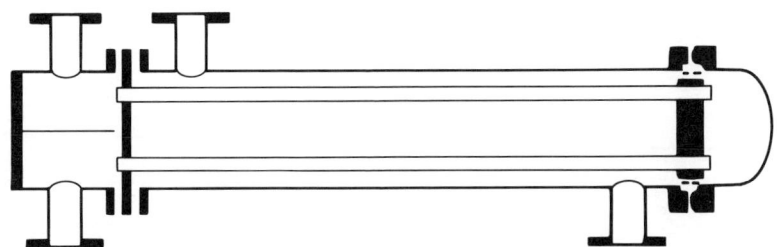

Fig. 4.10-17 Outside packed langern ring floating head exchanger (TEMA AEW). (Courtesy of Patterson-Kelley Co., East Stroudsburg, PA.)

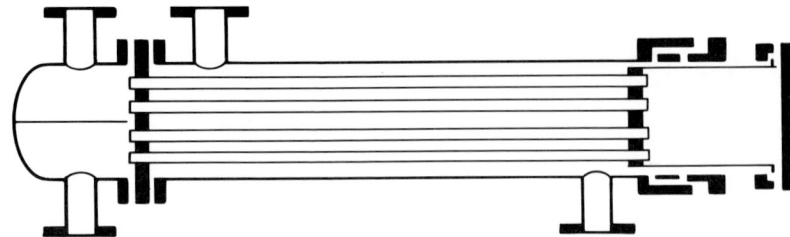

Fig. 4.10-18 Outside packed stuffing box floating head exchanger (TEMA BEP). (Courtesy of Patterson-Kelley Co., East Stroudsburg, PA.)

Other Shell Types

TEMA Notation. As indicated above and developed at greater length below, a wide variety of shell and front and rear head configurations is available in shell and tube designs. To simplify specification, TEMA[2] has developed a notation system for the major types by which a basic configuration can be identified by three letters, the first identifying the front head, the second the shell, and the third the rear head. Figure 4.10-19 gives the TEMA notations, and several of the exchangers in the figures are so identified. Most of the examples to this point have shown the E shell, with shell-side nozzles located at opposite ends of the shell. This is the most common configuration, but a variety of other designs (F, G, H, J, K, and X) are used for specific purposes. The following paragraphs will discuss these types, their applications, and some cautionary considerations.

The F Shell. The F shell has a longitudinal baffle in the shell that in principle provides for two shell-side passes in series. The main reason for doing this is to provide (at least on paper) a countercurrent flow arrangement when one has two tube-side passes and therefore to allow the use of a configuration correction factor (F) of 1.0 (see the Mean Temperature Difference concept in Section 4.10-3). If one traces through the flow paths, one sees that the two streams are always countercurrent to one another. The principle could be extended to multiple shell-side passes to match multiple tube-side passes, but this is seldom done in practice.

Even the provision of a single shell-side longitudinal baffle raises a number of fabrication, operation, and maintenance problems. Unless an insulated baffle is used, there will be heat transfer from the hot shell-side pass to the cold, which violates one of the assumptions of the mean temperature difference formulation (see Section 4.10-3). Furthermore, unless the baffle is welded longitudinally to the shell or specially gasketed (such as thin metal leaves mounted on the baffle edges along the entire length), the pressure difference across the baffle will cause physical leakage of fluid to occur. This completely violates the no-bypassing assumption built into the mean temperature difference derivation. Analyses have been made of the problem (Rozenman and Taborek[5]), which warn one when the penalty may become severe; but in general the use of multiple shell-side passes in heat exchangers is to be discouraged especially when there is sensible heat transfer occurring on the shell side.

The G and H Shells. The G and H shells, diagrammed in Fig. 4.10-19, are structurally related to the F shell, being essentially a doubling and quadrupling of the basic F-shell geometry. However, the flows are no longer purely countercurrent, and F is no longer even theoretically equal to unity in sensible heat transfer service (see the mean temperature difference concept in Section 4.10-3). In fact, these shells are seldom used for sensible service, for the good reason that they have no obvious advantages (except for a small improvement of F compared to an E shell) and several drawbacks. They are used in phase-change service, particularly as horizontal thermosiphon reboilers and partial and total condensers. In these services there is usually little penalty for thermal and physical leakage across the longitudinal baffle, whereas the baffle does help distribute the flow and flush-out concentrations of noncondensable or nonvolatile components that need to be continuously removed from the exchanger through vents or blowdown lines.

The J Shell. The J shell, or divided-flow shell, can be analyzed thermally as two E shells placed end to end. The inlet flow can be either to the single nozzle at the center and exiting out the double nozzles at the end or vice versa. In fact, it is common practice to use the J shells in stacked pairs, usually with the single nozzles as the inlet and outlet for the array.

The J shell's main advantage is the sharply reduced shell-side pressure drop compared to an E shell of comparable dimensions—on the order of one-fourth to one-eighth of the E-shell value, depending on flow regime. The J shell also gives a slightly improved F value compared to an E shell in the same service.

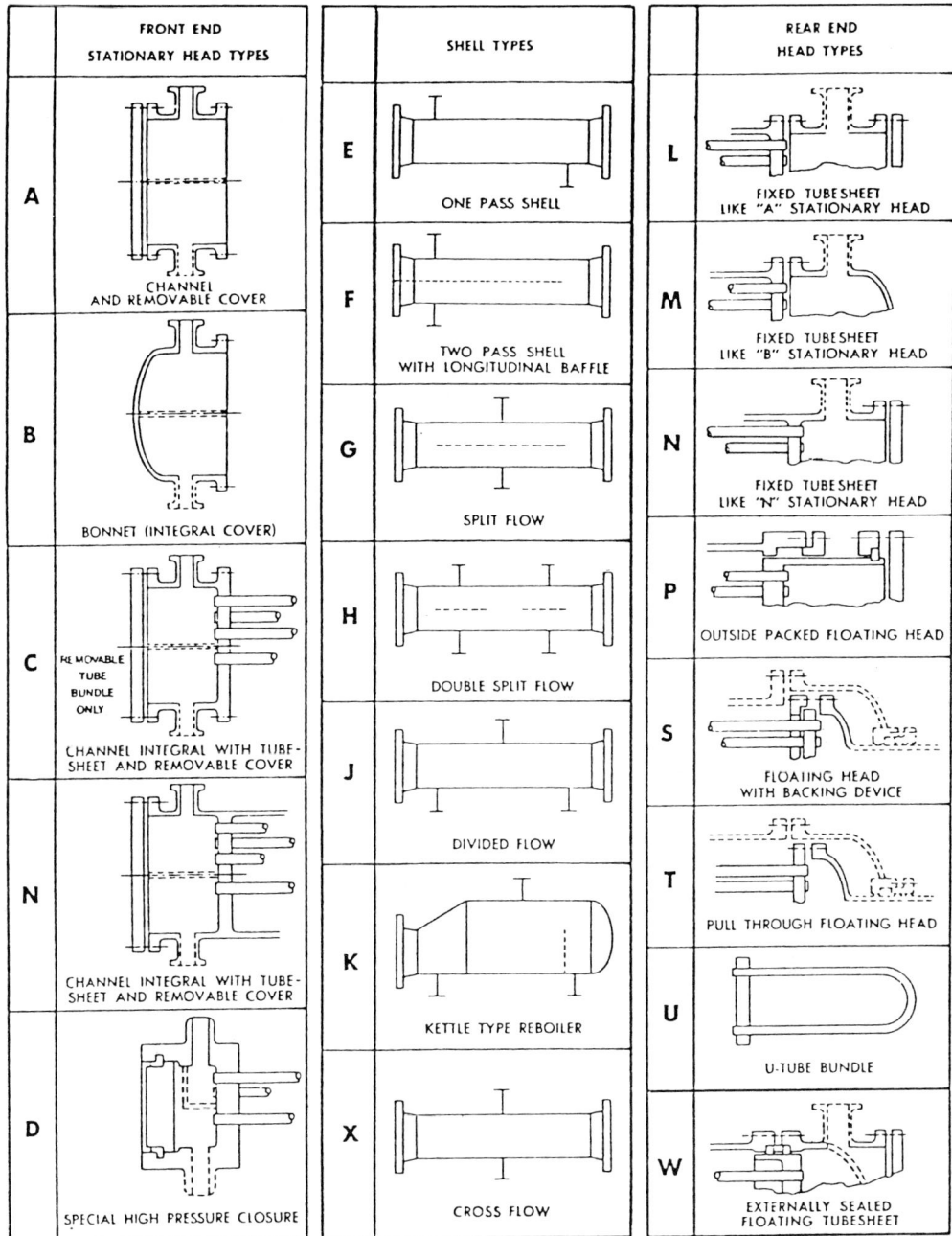

Fig. 4.10-19 Standard notation system for major types of heat exchangers. (From ref. 2; © 1978 by Tubular Exchanger Manufacturers Association.)

The K Shell. The K-shell design (also termed *kettle reboiler* or *flooded chiller* in the process and refrigeration industries, respectively) is employed when a portion of a stream needs to be vaporized, typically to a distillation column or an expansion device in a refrigeration cycle. The tube bundle has a diameter somewhat less than the smaller-diameter shell, and the liquid being vaporized ordinarily just covers the tube field. The large-diameter shell provides a vapor–liquid separation space. The feed liquid enters the shell at the nozzle near the tube sheet, the nearly dry vapor exits out the top nozzle, and the nonvaporized liquid overflows the end weir and exits through the right-hand nozzle. (It is essential that some liquid be removed from this nozzle continuously, or at least at frequent intervals, to limit the buildup of nonvolatile material.) Mesh demisters or centrifugal vane extractors are sometimes used at the vapor nozzle to minimize liquid carry-over.

The tube bundle is commonly a U-tube configuration, but a floating tube sheet may also be used. In a few applications a double fixed-tube sheet design has been used with a single tube pass; this requires very careful start-up and shutdown to minimize thermal stress. K shells are an expensive construction, especially for high-pressure vaporization, because of the large-diameter shell required. However, it is the only reboiler design that will produce dry or nearly dry vapor without an external vapor–liquid separator.

The X Shell.   The X, or cross-flow, shell is mostly used in condensing applications (including most central station steam power plant condensers), though it is occasionally employed in low-pressure gas heating or cooling. It is a very low pressure drop arrangement (on the shell side), which is vital in vacuum condensing applications. This, however, does complicate getting good distribution of the vapor or vapor–gas mixture throughout the bundle, and it is important to provide a clearway free of tubes at the top of the bundle and running the length of the bundle. Alternatively, multiple-inlet nozzles or a "bathtub" nozzle (a large inverted troughlike manifold mounted on top of and running most of the length of the shell) can be employed. Careful attention to removal of noncondensables is vital in X-shell condensers.

4.10-3   Basic Equations for Heat Exchanger Design

Film and Overall Heat Transfer Coefficients

In the heat exchangers considered here the heat transfer between the bulk fluid and the solid surface is by convection, which may be defined as transport of heat from one point to another in a flowing fluid, the heat being carried as internal energy. The rate of heat transferred, Q, is for most processes proportional to the temperature difference between the fluid and the wall $(T_f - T_w)$ and the wall surface area in contact with the fluid, A. Thus,

$$Q = hA(T_f - T_w) \tag{4.10-1}$$

where h is the film coefficient of heat transfer. Q is customarily in Btu per hour or watts, and h then has units of $Btu/h \cdot ft^2 \cdot °F$ or $W/m^2 \cdot K$. The value of h depends on the geometry of the system, the physical properties of the fluid, and the velocity of flow; the particular relationships for a large number of cases can be found in Section 3.2 of this handbook, volumes 2 and 3 of Schlunder,[1] Rohsenow and Hartnett,[6] and numerous other heat transfer textbooks. Specific equations for h will not be given here, though a table of typical values for common applications is given in Table 4.10-4. Heat transfer coefficients are always referred to a specific area, usually the area on which the convective heat transfer process is occurring. Therefore, properly one should say, for example, that h_i in Eq. (4.10-2) is the inside film heat transfer coefficient based on the inside heat transfer area of the tube. The importance of this becomes clearer when we define the overall heat transfer coefficient.

We sometimes employ a film coefficient in cases where the flux is not even approximately proportional to the temperature difference (e.g., nucleate boiling), contradictory to the implication of Eq. (4.10-1). This practice offers a useful basis of comparison of the relative resistances of the several heat transfer processes in a given problem but has no fundamental significance. However, when the individual heat transfer coefficients are reasonably constant for a given exchanger, the design problem is greatly simplified.

Figure 4.10-20 illustrates the heat transfer path between two fluids on opposite sides of a solid surface in a heat exchanger. Heat is being transferred from the fluid inside the tube (at a local bulk temperature of T), through a dirt or fouling film, through the tube wall, through another fouling film, to the outside fluid at a local bulk temperature of t. A_i and A_0 are, respectively, inside and outside surface areas for heat transfer for a given length of tube.

The heat transfer rate between the fluid inside the tube and the surface of the inside fouling film is by convection and is given by Eq. (4.10-2):

$$Q = h_i A_i (T - T_{f_i}) \tag{4.10-2}$$

where T_{f_i} is the surface temperature of the inside fouling layer.

Fouling layers exist on almost all heat transfer surfaces and constitute a resistance to heat transfer that must be considered in design. Fouling arises from a variety of sources, including sedimentation of suspended material on the surface, crystallization from a supersaturated solution, corrosion of the surface, polymerization or decomposition of the fluid, and biological growth. Often two or more of these processes occur simultaneously. The designer almost never knows either the thickness or the

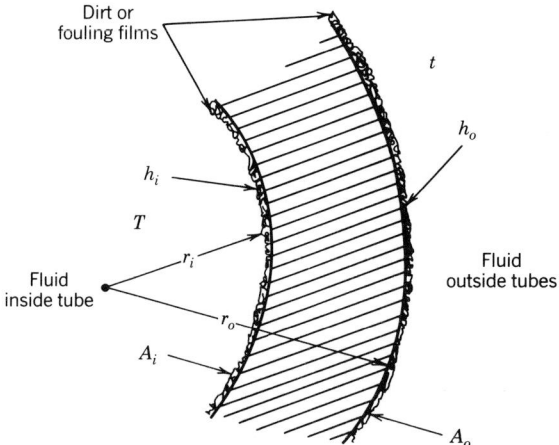

Fig. 4.10-20 Cross section of fluid-to-fluid heat transfer through a tube wall.

thermal conductivity of the fouling, which in any case usually change with time. Therefore, it is customary to treat the fouling as a simple resistance term, R_{f_i}, as defined by

$$R_{f_i} = \frac{A_i\left(T_{f_i} - T_{w_i}\right)}{Q} \qquad (4.10\text{-}3)$$

where T_{w_i} is the temperature of the tube wall in contact with the inside fouling layer, so $T_{f_i} - T_{w_i}$ is the temperature difference across the fouling layer. The corresponding units of R_f are $h \cdot ft^2 \cdot {}^\circ F/Btu$ or $m^2 \cdot K/W$. R_f is usually found in tables of values for various services and conditions; these tables (see Schlunder,[1] the TEMA standards,[2] and Kern[9] for example) are the consensus of experience and at best are rough approximations. Epstein[7] sets forth the current state of knowledge on fouling, still the greatest unsolved problem in heat exchanger design.

Heat transfer through the tube wall is by conduction and the exact equation is

$$Q = \frac{A_{w,lm} k_w \left(T_{w_i} - T_{w_o}\right)}{r_o - r_i} \qquad (4.10\text{-}4a)$$

$$A_{w,lm} = 2\pi L\left(\frac{r_o - r_i}{\ln(r_o/r_i)}\right) \qquad (4.10\text{-}4b)$$

In the above equations k_w is the thermal conductivity of the tube material; T_{w_i} and T_{w_o} are the surface temperatures of the inside and outside tube walls, respectively; r_i and r_o the inside and outside tube radii, respectively (so $r_o - r_i$ is the tube wall thickness); $A_{w,lm}$ the logarithmic mean wall heat transfer area; and L the length of the tube. For most practical cases the arithmetic mean wall area

$$A_{w,am} = \pi L\left(r_i + r_o\right) \qquad (4.10\text{-}4c)$$

can be used with sufficient accuracy instead of $A_{w,lm}$, and Eq. (4.10-4a) becomes

$$Q \approx \frac{\pi L\left(r_i + r_o\right) k_w \left(T_{w_i} - T_{w_o}\right)}{r_o - r_i} \qquad (4.10\text{-}4d)$$

Heat transfer through the outside fouling layer is

$$Q = \frac{A_o\left(T_{w_o} - T_{f_o}\right)}{R_{f_o}} \qquad (4.10\text{-}5)$$

where R_{f_o} is the fouling resistance for the fluid on the outside of the tube and T_{f_o} is the temperature of the outer surface of the fouling layer.

Heat transfer from the outside of the fouling to the outside fluid is given by

$$Q = h_o A_o \left(T_{f_o} - t \right) \tag{4.10-6}$$

where h_o is the outside heat transfer coefficient and t is the bulk temperature of the outside fluid.

Since only steady-state behavior is considered here, the same amount of heat must pass through each of the processes. Therefore, the equations may be combined to eliminate the unknown interface temperatures, and the relationship between the overall temperature difference $(T - t)$ and the heat transfer rate Q is found to be

$$Q = \frac{T - t}{\dfrac{1}{h_i A_i} + \dfrac{R_{f_i}}{A_i} + \dfrac{\ln(r_o/r_i)}{2\pi L k_w} + \dfrac{R_{f_o}}{A_o} + \dfrac{1}{h_o A_o}} \tag{4.10-7}$$

It is convenient to define an overall heat transfer coefficient U^* based on any convenient reference area A^*:

$$Q = U^* A^* (T - t) \tag{4.10-8}$$

Then, by comparison with Eq. (4.10-7),

$$U^* = \frac{1}{\dfrac{A^*}{h_i A_i} + \dfrac{R_{f_i} A^*}{A_i} + \dfrac{A^* \ln(r_o/r_i)}{2\pi L k_w} + \dfrac{R_f A^*}{A_o} + \dfrac{A^*}{h_o A_o}} \tag{4.10-9}$$

The reference area A^* may be any convenient definable area, but it must be clearly identified in each case. Often, the outside tube area A_o (or the total outside area of all of the tubes in an exchanger) is selected, and then Eqs. (4.10-8) and (4.10-9) may be written as

$$Q = U_o A_o (T - t) \tag{4.10-10a}$$

$$U_o = \frac{1}{\dfrac{A_o}{h_i A_i} + \dfrac{R_{f_i} A_o}{A_i} + \dfrac{A_o \ln(r_o/r_i)}{2\pi L k_w} + R_{f_o} + \dfrac{1}{h_o}} \tag{4.10-10b}$$

The third term in the denominator of Eq. (4.10-10b) may usually be written with negligible error as

$$\frac{A_o \ln(r_o/r_i)}{2\pi L k_w} \approx \frac{2 r_o (r_o - r_i)}{(r_o + r_i) k_w} \tag{4.10-10c}$$

Finally, an overall heat transfer coefficient may be converted to any area basis by the relationship

$$U^* A^* = U_o A_o = U_i A_i, \dots \tag{4.10-11}$$

The Basic Design Equation

Equation (4.10-8) and similar forms relate the rate of heat transfer (Q) to the temperature difference between the hot and cold streams $(T - t)$ at a given point. However, in almost all heat exchangers at least one and usually both stream temperatures are changing from point to point in the exchanger in response to heat being lost or gained by each stream.

In order to apply Eq. (4.10-8) to the design of a heat exchanger in which the temperature difference between the two streams is not constant, we first write Eq. (4.10-8) in differential form

$$dA^* = \frac{dQ}{U^* (T - t)} \tag{4.10-12}$$

and then formally integrate over the entire heat duty of the exchanger (Q_T):

$$A^* = \int_0^{Q_T} \frac{dQ}{U^* (T - t)} \tag{4.10-13}$$

This is the basic heat exchanger design equation. There are several assumptions implicit in this equation, some obvious (steady state) and some not so obvious (negligible longitudinal heat conduction in the wall). Detailed discussion of all of these assumptions is beyond the scope of this text, and this equation serves well enough for most design calculations. U^* may be, and in practice sometimes is, a function of the amount of heat exchanged. Additionally, for the hot stream the transfer of an amount of heat dQ will result in cooling the stream by an amount $-dT$:

$$-dT = \frac{dQ}{\dot{M}C_p} \qquad (4.10\text{-}14a)$$

where $\dot{M}$ is the mass flow rate and C_p is the specific heat of the hot stream. Correspondingly, the cold stream is heated by

$$dt = \frac{dQ}{\dot{m}c_p} \qquad (4.10\text{-}14b)$$

where $\dot{m}$ and c_p are the mass flow rate and specific heat for the cold stream.

If the paths of the hot and cold streams are specified relative to one another and the heat transfer surface, Eqs. (4.10-12), (4.10-14a), and (4.10-14b) can be written in finite-difference form and evaluated using numerical techniques, giving the heat exchanger area required for a given service. This is the way most computer-based design methods work and is essential if the heat transfer coefficients and/or the effective specific heats vary strongly with temperature or position or if the flow arrangement is complex.

The Mean Temperature Difference Concept

Fortunately, for many cases reasonable assumptions can be made that greatly simplify the calculations and lead to the mean temperature difference concept. These assumptions are as follows:

1.   All elements of a given stream have the same thermal history.
2.   The heat exchanger is at steady state.
3.   Each stream has a constant specific heat.
4.   The overall heat transfer coefficient is constant.
5.   The flow is either entirely cocurrent or entirely countercurrent.
6.   The heat exchanger does not exchange heat with the surroundings.

The first assumption is often overlooked; it means that all elements of a given stream that enter an exchanger follow paths through the exchanger that have the same heat transfer characteristics and have the same exposure to heat transfer surface. However, because of clearances between tubes and baffles, between baffles and shell, and between bundle and shell, there are alternative, or bypass, paths through the exchanger that may have lower resistance to flow than the prescribed path across the bundle and little or no heat transfer surface. Modern design methods make allowance for the reduction in performance, but the designer must take all reasonable care to minimize the possibilities of bypass in the exchanger.

The fifth assumption also requires some explanation in its application to shell-and-tube exchangers. Figure 4.10-21a shows the flow paths in a 1–1 exchanger having several baffles. The shell-side fluid in effect passes through a series of cross-flow sections exchanging heat with the tube-side fluid. The overall flow arrangement, however, is such that the shell-side fluid flows countercurrent to the tube-side fluid, as shown in Fig. 4.10-21b. If there are more than three baffles and if the outlet temperature of the hot stream is equal to or greater than that of the cold stream, the 1–1 exchanger can be treated as if the flow were purely countercurrent; that is, assumption 5 is valid. This question is carefully examined in Gardner and Taborek.[8]

If the assumptions given above are valid, Eq. (4.10-13) can be analytically integrated to the form

$$A^* = \frac{Q_T}{U^*(LMTD)} \qquad (4.10\text{-}15a)$$

where LMTD is the logarithmic mean temperature difference given by

$$LMTD = \frac{(T_i - t_o) - (T_o - t_i)}{\ln \dfrac{T_i - t_o}{T_o - t_i}} \qquad (4.10\text{-}15b)$$

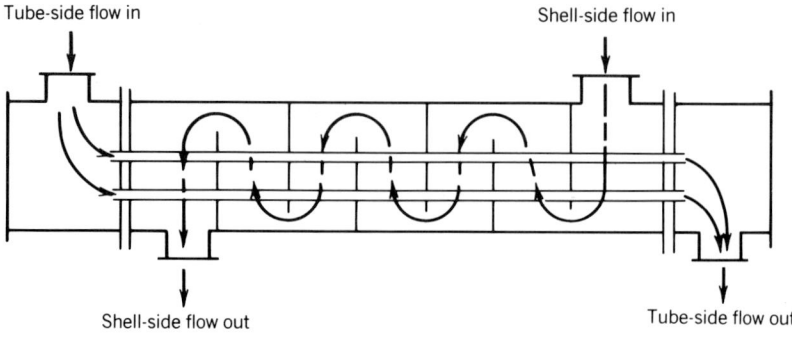

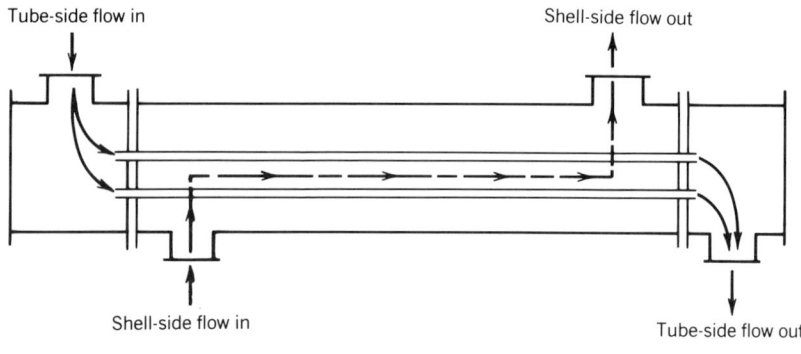

Fig. 4.10-21 Countercurrent and cocurrent flow arrangements in a 1-1 shell-and-tube exchanger: (*a*) Shell-side flow pattern with baffles, countercurrent; (*b*) idealized shell-side flow, countercurrent; and (*c*) idealized shell-side flow, cocurrent.

for countercurrent flow and by

$$\text{LMTD} = \frac{(T_i - t_i) - (T_o - t_o)}{\ln \dfrac{T_i - t_i}{T_o - t_o}} \tag{4.10-15c}$$

for cocurrent flow (as diagrammed in Fig. 4.10-21c for a shell-and-tube exchanger). In the above equations T_i and t_i are the inlet temperatures of the hot and cold streams, respectively, and T_o and t_o are the outlet temperatures.

If both streams change temperature (i.e., no isothermal phase transition), the LMTD for counter-current flow is always greater than that for cocurrent flow and therefore less area is required. Further, the outlet temperature of the hot stream can be less than the outlet temperature of the cold stream, which is not possible in the cocurrent exchanger. However, the tube wall temperature varies less with cocurrent flow, and this arrangement is occasionally used where an excessively high or low tube wall temperature must be avoided.

For countercurrent flow, if $T_o - t_i = T_i - t_o$, Eq. (4.10-15b) becomes indeterminate; observe that this corresponds to a uniform temperature difference through the exchanger, and LMTD $= T_o - t_i = T_i - t_o$.

One of the assumptions of the LMTD derivation was that the flow was purely countercurrent or cocurrent. However, in order to accommodate thermal stresses by U-tube or floating-heat construction or to maintain a high tube-side velocity, it may be necessary to go to multiple tube pass arrangements. The idealized flow patterns in a 1–2 exchanger are shown in Fig. 4.10-22a and a possible set of temperature profiles in Fig. 4.10-22b. Note that on the first tube-side pass the tube fluid is in countercurrent flow to the shell-side fluid, whereas on the second tube pass the tube fluid is in cocurrent flow with the shell-side fluid. It is possible for the outlet tube-side temperature to be somewhat greater than the outlet shell-side temperature, giving a "temperature cross." The maximum possible tube outlet temperature that can be achieved in this case, assuming constant overall heat transfer coefficient, is

$$t_{o,max} = 2T_o - t_i \tag{4.10-16}$$

One seldom designs for a temperature cross, but they sometimes occur operationally, especially during wintertime cooling service.

An alternative arrangement of a 1–2 exchanger is shown in Fig. 4.10-23a and corresponding temperature profiles in Fig. 4-10-23b. In this case t^* cannot exceed T_o. In spite of the very different appearance of these two cases, it turns out that they give identical values of the effective mean terminal temperature difference for identical terminal temperatures.

The problem of computing an effective mean temperature difference for the 1–2 case can be carried out similarly to the LMTD. The basic assumptions are the same (except for the pure cocurrent or countercurrent limitation), though in addition it is assumed that each pass has the same amount of heat transfer area. Kern[9] gives a detailed derivation for the equation for the mean temperature difference (MTD). Rather than calculate the MTD directly, however, it is preferable to compute a correction factor F on the LMTD calculated, assuming pure countercurrent flow; that is,

$$F = \frac{MTD}{LMTD} \tag{4.10-17}$$

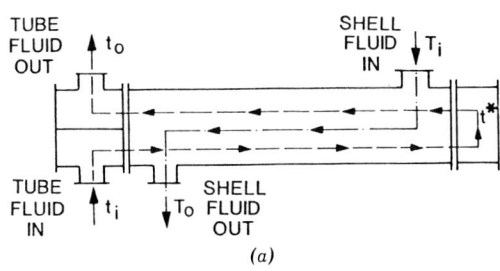

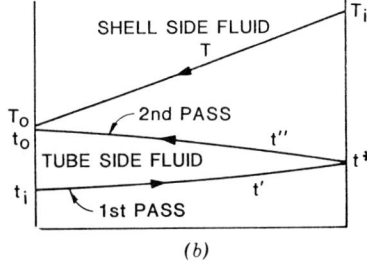

Fig. 4.10-22 (a) Diagram of a 1–2 exchanger. (b) Possible temperature profile for this exchanger.

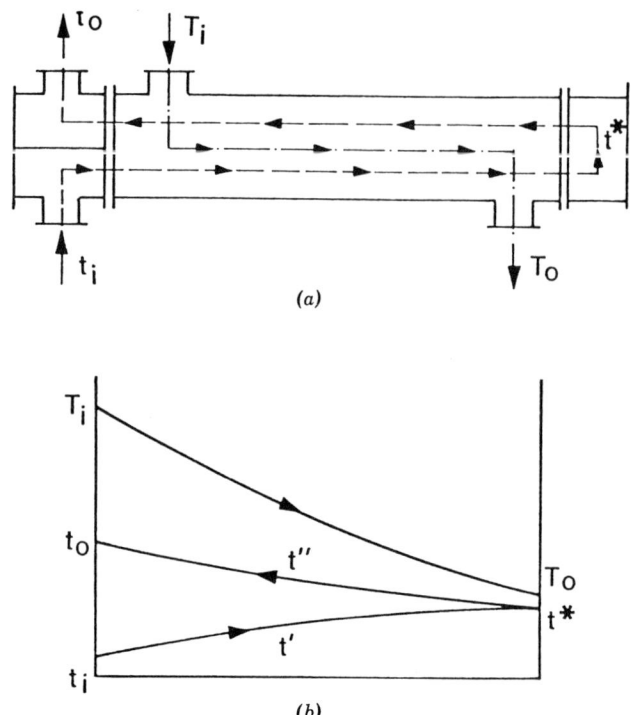

(a)

(b)

Fig. 4.10-23 (a) Diagram of a 1–2 exchanger with a different nozzle arrangement. (b) Possible temperature profiles for this case.

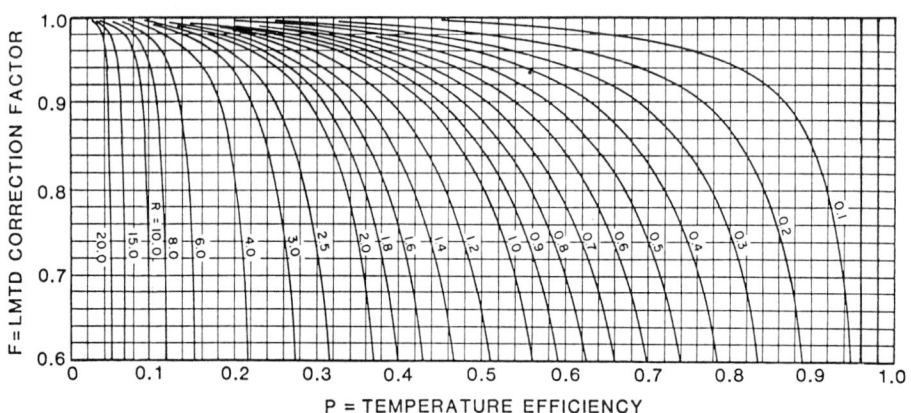

P = TEMPERATURE EFFICIENCY

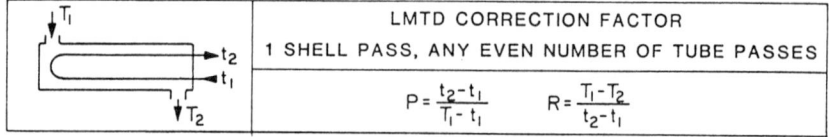

| LMTD CORRECTION FACTOR |
| 1 SHELL PASS, ANY EVEN NUMBER OF TUBE PASSES |
| $P = \dfrac{t_2 - t_1}{T_1 - t_1}$ $R = \dfrac{T_1 - T_2}{t_2 - t_1}$ |

Fig. 4.10-24 Configuration correction factor for the LMTD for one 1–2 shell. (© 1978 by Tubular Exchanger Manufacturers Association.)

where LMTD is given by Eq. (4.10-15b) and where $F = 1$ indicates the flow situation is equivalent to countercurrent flow. Lower values very clearly and directly show what penalty (ultimately expressed in area required) is being paid for the 1–2 configuration compared to the countercurrent case.

The correction factor F is shown in Fig. 4.10-24 as a function of two parameters R and P defined as (in terms of the nomenclature given on the chart)

$$R = \frac{T_1 - T_2}{t_2 - t_1} = \frac{\text{range of the shell-side fluid}}{\text{range of tube-side fluid}} \qquad (4.10\text{-}18a)$$

$$P = \frac{t_2 - t_1}{T_1 - t_1} = \frac{\text{range of tube-side fluid}}{\text{difference between inlet temperatures}} \qquad (4.10\text{-}18b)$$

The chart given here is adapted from the TEMA standards[2] and is almost identical to the one in Kern.[9] In the cases given here it is immaterial whether t refers to the tube-side fluid and T to the shell-side or whether t refers to the cold fluid and T the hot; see Singh[10] for a definitive discussion. However, there are other different (but finally equivalent) formulations and each one should be used carefully with its own definitions. Volume 1 of Schlunder[1] is the most comprehensive collection and discussion of the correction factor charts.

Examination of Fig. 4.10-24 reveals that for each value of R the curve becomes suddenly and increasingly steep at some value of P. This is due to the tube-side temperature approaching one of the thermodynamic limits discussed above. It is extremely dangerous to design an exchanger near this steep region, because even a small failure of one of the basic assumptions can easily render the exchanger thermodynamically incapable of rendering the specified performance no matter how much excess surface is provided; the first assumption is especially critical in this case. Therefore, there is a generally accepted rule of thumb that no exchanger will be designed to $F < 0.75$ or 0.8.

More than two tube-side passes are possible and frequently used. The correction factors for any even number of tube-side passes are within about 2% of those for two passes, so it is common practice to use Fig. 4.10-24 for all 1–n exchangers where n is any even number.

Since a multiple tube pass exchanger is effectively limited to applications in which the cold-stream outlet temperature is near or below the hot-stream outlet temperature, the designer needs to find some means to move toward a more nearly countercurrent flow arrangement. One possibility is the F shell described earlier, which for a double tube pass design apparently provides countercurrent flow. However, this advantage may be illusory because of the difficulty of avoiding thermal leakage and/or bypass flow around the longitudinal baffle and the corresponding construction limitations.

The usual solution to exchanging heat over a wide temperature range while using 1–$2n$ shells is to use multiple shells (usually identical) in series. This is illustrated in Fig. 4.10-25 for two 1–2 shells, but the use of six or even more shells in series is not uncommon. Qualitatively, the overall flow arrangement of the two streams is countercurrent, even though the flow within each shell is still mixed. Since, however, the temperature change of each stream in one shell is only a fraction of the total

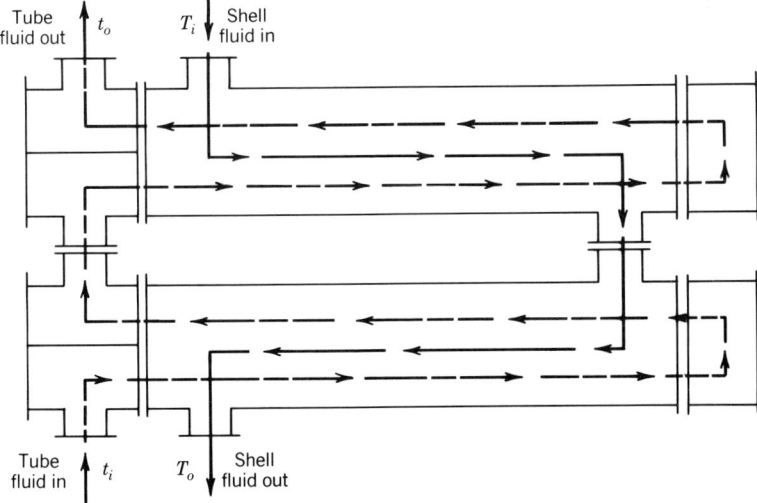

Fig. 4.10-25 Two 1–2 exchangers in series.

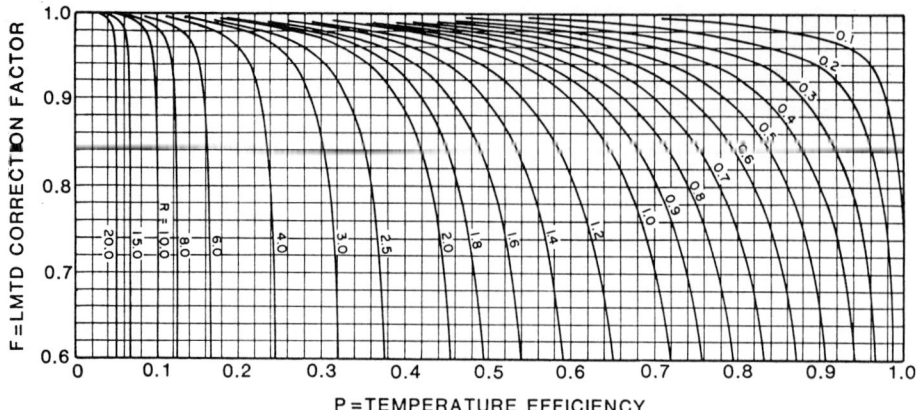

Fig. 4.10-26 Configuration correction factor for the LMTD for two shells in series. (From ref 3; © 1978 by Tubular Exchanger Manufacturers Association.)

change, the departure from true countercurrent flow is less. As the number of shells in series becomes infinite, the heat transfer process approaches true countercurrent flow and $F \to 1.0$.

Charts have been derived for the F values as functions of P and R for up to six identical shells in series (extensive sets are available in Schlunder,[1] the TEMA standards,[2] Rohsenow and Hartnett,[6] and Kern[9]). Only the one for two shells in series is included here (Fig. 4.10-26); as the reader can quickly verify, for a given set of values of P and R the value of F read from Fig. 4.10-26 is higher than that from Fig. 4.10-24. It must be remembered that all of the same assumptions underlying the validity of the MTD concept still apply (except assumption 5), and it is further assumed that all of the exchangers are identical.

The Number of Transfer Units (NTU)-Effectiveness Concept

To use the MTD approach directly, it is necessary to know at least three of the terminal temperatures; the fourth may then be calculated by a heat balance if not specified a priori. For a typical design problem in the process industries these conditions are often satisfied.

However, in heat recovery service it is sometimes necessary to calculate the outlet temperatures of both streams from an exchanger, given the inlet temperatures, flow rates, and exchanger design. To use the LMTD F method in this case involves trial-and-error calculations and an alternative but entirely equivalent procedure, the NTU-effectiveness method, is more convenient.

In this method one first calculates the thermal capacity of both streams. *Thermal capacity* is defined as the mass flow rate times the specific heat of a stream, $C = \dot{m} C_p$ (Btu/h · °F or W/K), and is the amount of heat that must be added to a stream per hour to change its temperature 1°F or 1 K. An isothermal phase change corresponds to a C_p and a C of ∞, but otherwise proceeds similarly. In general, one obtains two different thermal capacities; the smaller one is denoted by C_{min} and the larger by C_{max}. The ratio of the thermal capacities, C_{min}/C_{max}, can lie between 0 and 1.0.

Effectiveness, ε, is defined as

$$\varepsilon = \frac{\Delta T_{min}}{T_i - t_i} \qquad (4.10\text{-}19a)$$

where ΔT_{min} is the temperature change of the stream with the *smaller* thermal capacity, C_{min}. Note that by heat balance, $C_{min} \Delta T_{min} = C_{max} \Delta T_{max}$; ΔT_{min} is actually the *larger* of the two stream temperature changes. T_i and t_i are the inlet temperatures of the hot and cold streams, respectively; therefore, $T_i - t_i$ is the maximum temperature difference that can occur between any two points anywhere in the heat exchanger. Hence, ε is the ratio of the actual heat transferred to the maximum amount of heat that could be transferred under the most optimal conditions—infinite area and countercurrent flow. So the value of ε lies between 0 and 1.

The NTU is defined as

$$\text{NTU} = \frac{UA}{C_{\min}} \qquad (4.10\text{-}19b)$$

where A is the exchanger area and U the overall coefficient (on the same area basis as A).

The relationships between the three parameters ($C_{\min}/C_{\max}$, ε, and NTU) have been derived analytically and represented graphically for a number of different geometries. Two typical ones—for pure countercurrent flow in Fig. 4.10-27 and for a $1\text{--}2n$ exchanger in Fig. 4.10-28—are given here. Much more complete charts and equations are given in Kays and London,[11] and an extended discussion of the relationships between the various representations is given in Volume 1 of Schlunder.[1] It must be remembered that the relationships between the parameters $C_{\min}/C_{\max}$, ε, and NTU arise from exactly the same analyses and assumptions as the F, R, and P parameters for correcting the LMTD for various configurations, and the NTU-ε method has exactly the same limitations.

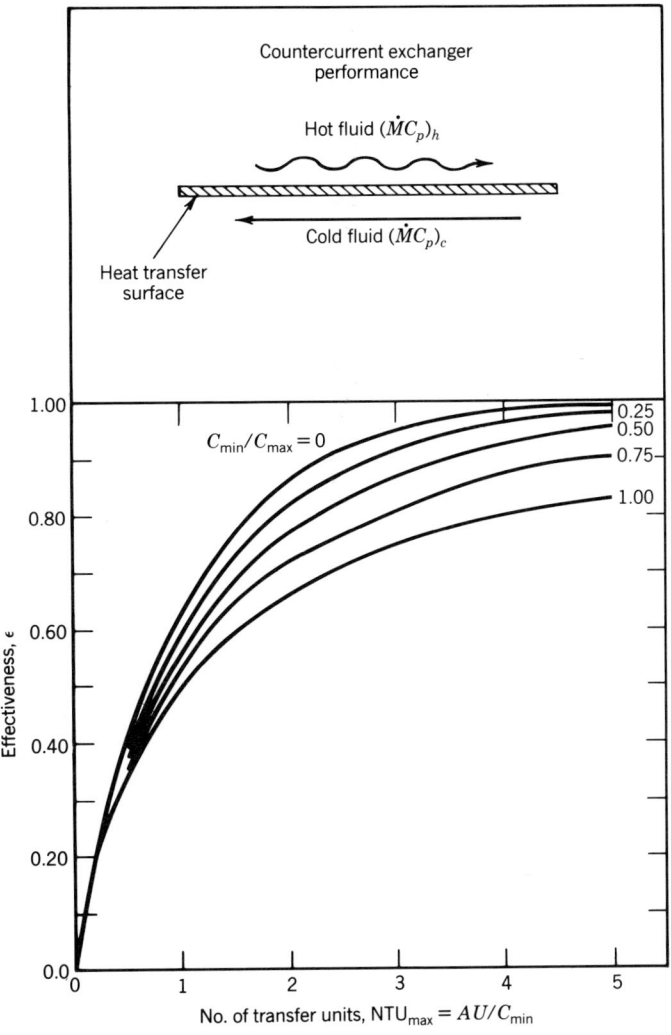

Fig. 4.10-27 ε-NTU chart for countercurrent exchangers. (Adapted from Kays and London, ref. 11.)

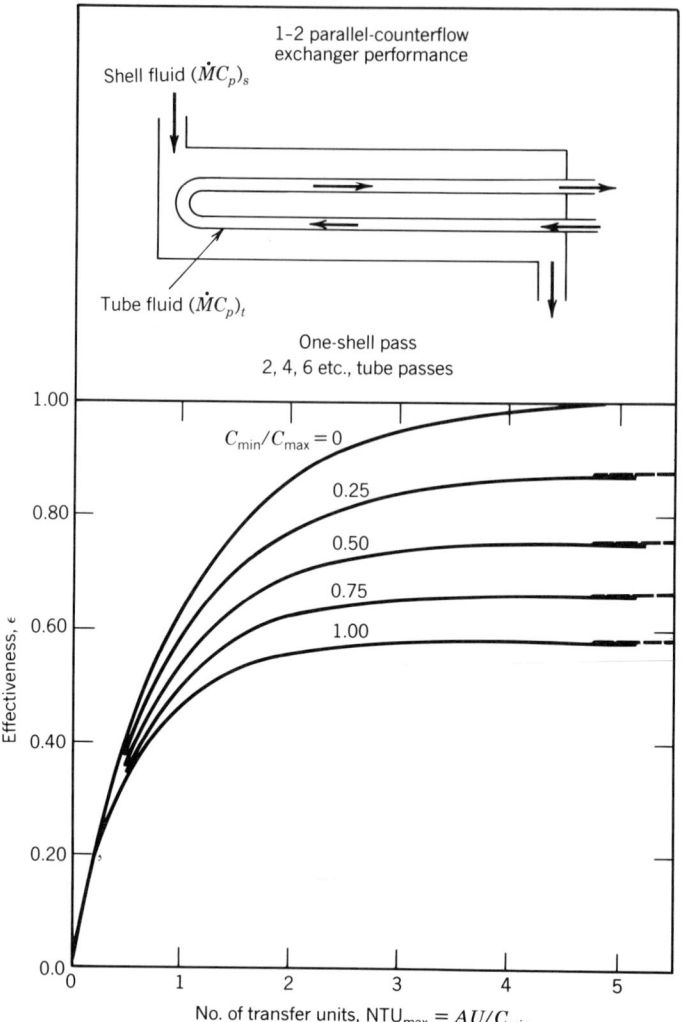

Fig. 4.10-28 ε-NTU chart for 1–2 and 1–n exchangers. (Adapted from Kays and London, ref. 11.)

4.10-4 Preliminary Design of Shell-and-Tube Exchangers in Sensible Heat Transfer Service

Criteria for Heat Exchanger Selection

The criteria that a heat exchanger must satisfy can be given in rather broad statements.

First, the heat exchanger must meet the process requirements. That is, it must effect the desired change in the thermal conditions of the streams within the allowable pressure drops, and it must continue to do this until the next scheduled shutdown of the plant for maintenance.

Second, the exchanger must withstand the service conditions of the plant environment. This includes the mechanical stresses of installation and the thermal stresses induced by the temperature differences. It must also resist corrosion by the process and service streams (as well as by the environment); this is usually mainly a matter of choice of materials of construction, but mechanical design does have some effect. Desirably, the exchanger should also resist fouling, but there is not much the designer can do with confidence in this regard except keep the velocities as high as pressure drop, erosion, and vibration limits permit.

Third, the exchanger must be maintainable, which usually implies choosing a configuration that permits cleaning—tube side and/or shell side, as may be indicated—and replacement of tubes,

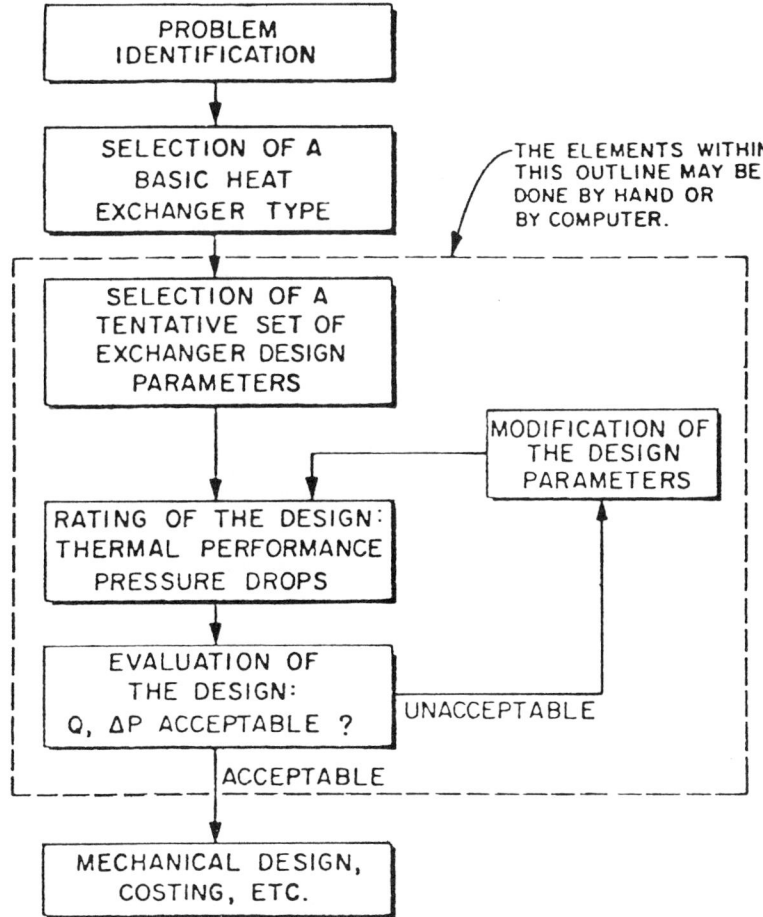

Fig. 4.10-29   Basic logical structure for process heat exchanger design.

gaskets, and any other components that may be especially vulnerable to corrosion, erosion, vibration, or aging. This requirement may also place limitations on positioning the exchanger and in providing clear space around it.

Fourth, the exchanger should cost as little as is consistent with the above requirements; this refers to first cost or installed cost, since operating cost and the cost of lost production due to exchanger unavailability should have already been considered in the earlier and more important criteria.

Finally, there may be limitations on exchanger diameter, length, weight, and/or tube specifications due to site requirements, lifting and servicing capabilities, or inventory considerations.

Logical Structure of The Design Process

The basic logical structure of the process heat exchanger design procedure is shown in Fig. 4.10-29. The basic structure is the same whether hand or computer-based design methods are used; all that is different is the replacement of the very subtle and complicated human thought processes by an algorithm suited to a fast but inflexible computer. In fact, all of the most important decisions are made outside the dashed line bounding the computer's contribution to the solution.

First, the problem must be identified as completely and unambiguously as possible, including stream compositions, flow rates and temperatures, and the likely ranges of variations in these parameters during operation. Any design problem will have certain contextual considerations the designer needs to know in order to arrive at a near-optimal design. A major judgment, usually made almost instinctively, is the level of engineering effort justified by the actual value of the exchanger in the process. This effort may be greater than the design organization can afford based on the actual cost of the exchanger, and it may be worth an additional investment by the customer to obtain the best engineering advice. Also, at this point the single most important decision is made (often by default): what basic configuration of exchanger is to be chosen and designed? While this section is devoted to shell-and-tube exchangers, it is always worth reviewing the other options as described in Section 4.11 to verify that in fact a shell and tube is the right choice.

The next decision is what design method to be used. Basically, these fall into two categories: hand design and computer design. Hand design methods in the most recent literature and applied by a competent designer are still valid for a large fraction of all heat exchanger problems. Volume 3 of Schlunder[1] is probably the best source of information on hand methods, especially for identifying their applicability. If one chooses to use a computer design method, one still has the task of selecting the level of the method. There are short-cut and detailed computer design methods available for most exchanger types.

The next step is to select a tentative set of exchanger geometrical parameters. The better the starting design, the sooner the designer will come to the final design, and this is very important for hand calculation methods. Such a method is offered below. On a computer, however, it is usually faster to give the computer a very conservative (oversized) starting point and use its enormous computational speed to move toward the desired design.

In either case the initial design will be "rated"; that is, the thermal performance and the pressure drops for both streams will be calculated for this design. The rating program is diagrammed in Fig. 4.10-30. In the rating program the problem specifications and the preliminary estimate of the exchanger configuration are used as input data; the exchanger configuration given is tested for its ability to effect the required temperature changes on the streams within the pressure drop limitations specified.

The rating process carries out basically three kinds of calculations: First, it computes a number of internal geometry parameters that are required as further input into the heat transfer and pressure drop correlations. Then the heat transfer coefficient and pressure drop are calculated for each stream in the configuration specified.

The results from the rating program are either the outlet temperatures of the streams if the length of the heat exchanger has been fixed or the length of the heat exchanger required to effect the necessary thermal change if the duty has been fixed. In either case the rating program will also calculate the pressure drops for both streams in the exchanger. Volume 3 of Schlunder[1] gives hand-rating methods for a number of heat exchanger types and applications.

If the calculation shows that the required amount of heat cannot be transferred or if one or both allowable pressure drops are exceeded, it is necessary to select a different, usually larger, heat exchanger and rerate. Alternatively, if one or both pressure drops are much smaller than allowable, a better selection of parameters may result in a smaller and less costly heat exchanger, while utilizing more of the available pressure drop.

The design modification program takes the output from the rating program and modifies the configuration in such a way that the new configuration will do a "better" job of solving the heat transfer problem.

A computer-based configuration modification program is a complex one logically because it must determine what limits the performance of the heat exchanger and what can be done to remove that limitation without adversely affecting either the cost of the exchanger or the operational characteristics of the exchanger, which are satisfactory. If, for example, it finds that the heat exchanger is limited by the amount of heat that it can transfer, the program will try either to increase the heat transfer coefficient or to increase the area of the heat exchanger, depending on whether or not pressure drop is available. To increase the tube-side coefficient, one can increase the number of tube passes, thereby increasing the tube-side velocity. If shell-side heat transfer is limiting, one can try decreasing baffle spacing or decreasing the baffle cut. To increase area, one can increase the length of the exchanger, or increase the shell diameter, or go to multiple shells in series or in parallel.

Clearly the possibilities are enormous, and the configuration modification program must be very tightly written to avoid wandering off into impossible designs or loops without an exit. A designer

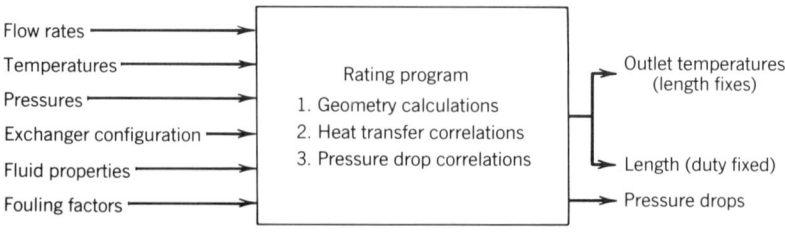

Fig. 4.10-30   The rating program.

using a hand method does not face those dangers but must very carefully identify the proper direction and distance to move in if an acceptable design is to be reached in a reasonable time.

If a computer-based design program is used, every key output item should be personally inspected to verify its basic rationality. This inspection should be done by a heat exchanger specialist who can very quickly determine if the design the computer has generated is a reasonable one. It is important to do this every time. While design programs are checked out by running a few test cases, there are typically 10^9–10^{15} distinct paths through the programs and any given case may have entered a faulty logical loop for the first time. Any questionable item in the output should be verified by analysis at whatever level of detail is required to establish its basic validity. It is not too much to say that no one should use a heat exchanger design program without knowing ahead of time pretty closely what the final design should look like.

Allocation of Streams

An early decision must be made as to which fluid will go in the tubes and which in the shell. The following guidelines apply:

1. The high-pressure fluid goes in the tubes, other things being equal. Because of their small diameter, normal thickness tubes are able to take quite high pressures and only the tube-side channels and nozzles need normally be designed to withstand high pressure. If it is necessary to put the high-pressure fluid in the shell, the most economical exchanger will normally tend to be a longer, smaller-diameter configuration.

2. The corrosive fluid goes in the tubes, other things being equal. Corrosion is usually controlled by using alloys, and it is much less expensive to avoid using alloy shells. The tube side of the tube sheet, the tube-side channels, and the channel covers can usually be made of low-carbon steel faced with the alloy.

3. The more seriously fouling fluid goes in the tubes, other things being equal. The tube side is easier to clean, especially if mechanical cleaning (brushing, high-velocity water jets, etc.) is required.

4. The stream with the lower allowable pressure drop usually goes on the shell side, other things being equal. This is a rather curious anomaly because the heat–momentum transfer analogies indicate that one can get more heat transfer per unit of pressure drop expended in a nonseparating geometry (e.g., inside tubes) than in a separating one (e.g., cross flow on the shell side). However, the range of design variables on the shell side is so great compared to the tube side that it is usually possible to come up with a better mechanical design with the low pressure drop fluid on the shell side.

5. The stream with the lower heat transfer coefficient goes on the shell side, other things being equal. This allows the use of low-finned tubing to partially offset the low coefficient with an increased area ratio. However, tube-side enhancement is now becoming available with acceptable engineering characteristics, and this rule is fading.

Problems arise when the above requirements are in conflict, for example, when one fluid is fouling and the other is at high pressure. Then the designer must estimate trade-offs and select the most economic choice, remembering that the heat exchanger must perform its thermal duty with a very high degree of reliability and process availability.

Approximate Size Estimation

Besides providing a starting point for detailed design by a hand method, an approximate sizing method for exchangers may provide a sufficient answer for preliminary estimates of equipment and plant size, cost, and layout. Such a method is presented in this section; the procedure is identical to that in Schlunder.[1]

Basic Design Equation. The equation used is

$$A_o = \frac{Q}{U_o(\text{MTD})} = \frac{Q}{U_o F(\text{LMTD})} \tag{4.10-20}$$

where A_o is the total heat transfer area required in the exchanger (in square feet or square meters) calculated based on the outside diameter of the tube; Q is the heat duty of the exchanger (Btu per hour or watts); U_o is the overall heat transfer coefficient (Btu/h · ft^2 · °F or W/m^2 · K (based on A_o)); LMTD is the logarithmic mean temperature difference calculated for countercurrent flow (in degrees Fahrenheit or Kelvin) [Eq. (4.10-15b)]; and F is the correction factor on the LMTD.

The validity of Eq. (4.10-20) is discussed in Section 4.10-3 (the MTD concept). The conditions for exact validity are often not met, and a more elaborate formulation is required in principle. However, that defeats the purpose of this procedure; moreover, most of these departures from ideality introduce errors smaller than the probable error in the other approximations made in the method.

Estimation of Heat Load. The heat load for sensible heat transfer is

$$Q = \dot{M}C_p\,(T_1 - T_2) = \dot{m}c_p\,(t_2 - t_1) \tag{4.10-21}$$

where $\dot{M}$ and $\dot{m}$ are the mass flow rates (in pounds per hour or kilograms per second) of the hot and cold fluids, respectively; C_p and c_p are the specific heats (Btu/lb · °F or kJ/kg · K); T_1 and T_2 are the inlet and outlet temperatures of the hot stream (degrees Fahrenheit or Kelvin); and t_1 and t_2 are the inlet and outlet temperatures of the cold stream. If one of the streams undergoes isothermal phase change (such as condensing steam),

$$Q = \dot{M}\lambda \tag{4.10-22}$$

where $\dot{M}$ is the mass of the stream changing phase per unit time (pounds per hour or kilogram per second and λ is the latent heat of the phase change (Btu per pound or kilojoule per kilogram).

Complex cases such as partial condensers require more elaborate analysis than space permits, though the present method can be applied (with care!) to obtain at least rough estimates even in these cases.

Estimation of the Mean Temperature Difference. The LMTD for countercurrent flow is given by Eq. (4.10-15b). For many purposes, the LMTD can be sufficiently approximated by the arithmetic mean temperature difference (AMTD):

$$\text{AMTD} = \tfrac{1}{2}\big[(T_1 - t_2) + (T_2 - t_1)\big] \tag{4.10-23}$$

LMTD is always equal to or less than the AMTD. The difference between the LMTD and AMTD increases with decreasing ratio of the smaller temperature difference to the larger. F should be found from the appropriate chart, but for well-designed $1{-}2n$ exchanger systems it may be estimated to be between 0.8 and 0.9.

There is a rapid graphical technique for estimating a sufficient number of shells in series. The procedure is illustrated in Fig. 4.10-31:

1. Plot the terminal temperatures of the two streams on linear graph paper, the hot-fluid inlet temperature and the cold-fluid outlet temperature on the left-hand ordinate, and the hot-fluid

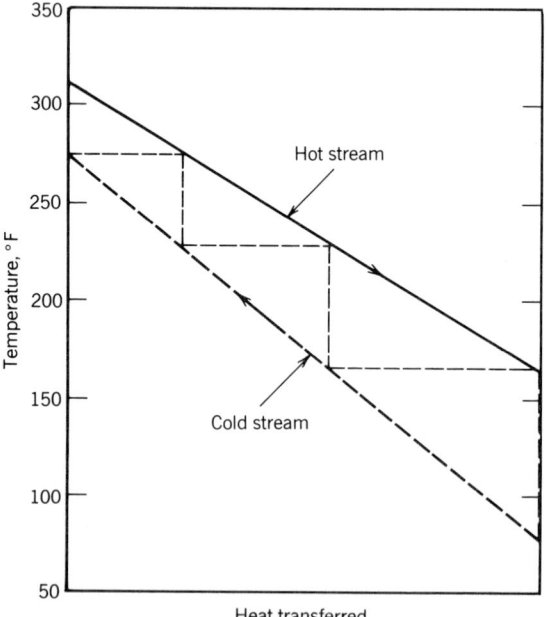

Fig. 4.10-31   Estimation of required number of shells in series.

outlet and cold-fluid inlet temperature on the right-hand ordinate. The distance between the ordinates is arbitrary.

2. If the specific heat of each stream is constant, straight lines are drawn from the inlet to the outlet temperature for each stream. If the specific heat of one or both streams varies or if there is a phase change, it is necessary to calculate the temperature of that stream as a function of the amount of heat transferred, resulting in one or both operating lines being curved. The procedure given here for finding a sufficient number of shells in series will still be valid. However, an additional error will be introduced into the area estimation procedure, the magnitude depending on the curvature of the lines and the variation in U_o.

3. Starting with the cold-fluid outlet temperature (275°F in Fig. 4.10-31), a horizontal line is laid off until it intercepts the hot-fluid line. From that point a vertical line is dropped to the cold-fluid line. This operation defines a heat exchanger in which the hot-fluid temperature is never less than any temperature reached by the cold so there is no temperature cross and no thermodynamic difficulty can arise if this operation is carried out in one shell.

4. The process is repeated until a vertical line intercepts the cold-fluid operating line at or below the cold-fluid inlet temperature or until a horizontal line crosses the right-hand ordinate.

5. The number of horizontal lines (including the one that intersects the right-hand ordinate) is equal to the number of shells in series that is clearly sufficient to perform the duty. In the case of the example problem this number is 3.

Following the above procedure will result in a number of shells usually having an overall F between 0.8 and 0.9.

Estimation of U_o. The step with the greatest uncertainty in preliminary calculations is estimating the overall heat transfer coefficient. The best procedure is to build up the value of U_o from the individual h values and the wall, fin, and fouling resistances using Eq. (4.10-10b). Typical values of coefficients and resistances are given in Table 4.10-2. In using such values, it is important to remember that individual cases may vary widely. However, it is often the case that one or at most two terms dominate the value of U_o and greater attention should be concentrated on these. A more extensive table and discussion are given in Vol. 3 of Schlunder[1].

Calculation of A_o. Once Q, MTD, and U_o are known, the total outside heat transfer area A_o is readily found from Eq. (4.10-20). What set of heat exchanger dimensions will accommodate that heat transfer area? The following section presents a technique for answering that question.

Estimation of Major Exchanger Parameters. Figure 4.10-32 gives A_o as a function of shell inside diameter and effective tube length for a fixed tubesheet 1–1 exchanger, fully tubed with $\frac{3}{4}$ in. outside diameter plain tubes on a $\frac{15}{16}$ in. triangular pitch. This figure shows immediately the combinations of tube length and shell diameter that will provide that area in a single shell for an exchanger of the given tube size and layout. A technique of applying Fig. 4.10-32 to other tube sizes, layouts, number of tube passes, and shell type will be given below.

The dashed lines in the parameter in Fig. 4.10-32 (marked 3 : 1, 6 : 1, etc.) show the approximate locus of shells having a given effective tube length to shell diameter ratio. Shells shorter than 3 : 1 may suffer from poor fluid distribution and excessive entry and exit losses and are likely to be more expensive than a longer, smaller-diameter unit of the same area, especially if the shell-side fluid is at high pressure. Shells longer than 15 : 1 are likely to be difficult to handle mechanically, require a large clearway for bundle removal or retubing, and show the effect of diminishing returns on cost. Many heat exchangers fall into the 6 : 1 to 8 : 1 range, with a pronounced trend toward the higher values as pressure drop prediction procedures have improved.

Figure 4.10-11 can be employed to estimate the required length and diameter of exchangers for other tube sizes and layouts, multiple tube-side passes, and bundle constructions. Define an "effective" area, A_o', by

$$A_o' = A_o F_1 F_2 F_3 \qquad (4.10\text{-}24)$$

where A_o' = area on ordinate of Fig. 4.10-32

A_o = required outside area of heat exchanger as calculated from Eq. (4.10-20)

F_1 = correction factor for unit cell tube array (= 1.00 for $\frac{3}{4}$-in. tubes on a $\frac{15}{16}$-in. triangular pitch)

F_2 = correction factor for number of tube passes (= 1.00 for one tube pass)

F_3 = correction factor for shell construction/tube bundle layout type (= 1.00 for fixed tube sheet)

Values of the correction factors are given in Tables 4.10-3 (F_1), 4.10-4 (F_2), and 4.10-5 (F_3).

TABLE 4.10-2 TYPICAL FILM HEAT TRANSFER COEFFICIENTS AND FOULING RESISTANCES[a]

Fluid	Heat Transfer Process[b,c,d]	h Btu/h · ft^2 · °F	R_f h · ft · °F/Btu
Water[e,f]	Sensible	1000–1200	0.0005–0.002
Water[f]	Condensing	1500–2500	0.0005
Water[f]	Vaporizing	1000–2500	0.0005–0.002
Gas, 10 psig[g]	Sensible	15–25	0
Gas, 100 psig[g]	Sensible	35–50	0
Gas, 1000 psig[g]	Sensible	70–125	0
Light organics[h]	Sensible	250–350	0.0005–0.001
Light organics[h]	Condensing	300–500	0.0005
Light organics[h]	Vaporizing	200–500	0.0005–0.001
Medium organics[i]	Sensible	150–250	0.001–0.002
Medium organics[i]	Condensing	200–400	0.0005–0.001
Medium organics[i]	Vaporizing	125–300	0.001–0.002
Heavy organics[j]	Sensible	30–100	0.002–0.005
Heavy organics[j]	Condensing	100–150	0.001–0.002
Heavy organics[j]	Vaporizing	75–200	0.002–0.010

Wall Resistances, $\Delta x/k_w$, h · ft^2 · °F/Btu

Material (BWG)	10	12	14	16	18	20
Low-carbon steel	0.0004	0.0003	0.00025	0.0002	0.00015	0.0001
304 stainless steel	0.0011	0.0009	0.0007	0.00055	0.0004	0.0003
90/10 cupronickel	0.00035	0.0003	0.0002	0.0002	0.00015	0.0001
70/30 cupronickel	0.0006	0.0005	0.00035	0.0003	0.0002	0.00015
Copper, deoxidized	0.00006	0.00005	0.00004	0.00003	0.00002	0.00002
Titanium	0.0010	0.0008	0.0006	0.0005	0.00035	0.00025
Admiralty	0.00015	0.00012	0.00009	0.00007	0.00005	0.00004
Aluminum 3004	0.00011	0.0009	0.00007	0.00006	0.00004	0.00003

[a] Heat transfer coefficients and fouling resistances are based on area in contact with the fluid. These coefficients assume that the equipment is designed for typical pressure drops, that is, about 10 psi for each stream, except for low-pressure systems where the pressure drop is limited to 5–10% of the absolute pressure, and very viscous fluids where the pressure drop may be 20 or 30 psi.

[b] "Sensible" means that there is no phase change during the heat transfer process.

[c] "Condensing" means that a saturated or superheated vapor with a narrow condensing range and negligible noncondensable gas content is condensing on a surface whose temperature is at or below the saturation temperature of the vapor at the existing pressure (see Section 4.10-5).

[d] "Vaporizing" means that a liquid, either pure or with a narrow boiling range, is being vaporized with a wall temperature low enough that nucleate boiling is occurring (see Section 4.10-6).

[e] Aqueous solutions give values similar to water as long as the solutions do not become saturated and deposit on the wall.

[f] Ammonia gives values similar to water.

[g] Hydrogen and helium give values up to several times those for other gases.

[h] "Light organics" include liquids with viscosities less than about 1 cP.

[i] "Medium organics" include liquids with viscosities between about 1 and 5 cP.

[j] "Heavy organics" include liquids with viscosities greater than about 5 cP. Care must be taken with fluids of viscosity greater than about 50 cP because of possible congelation on cold surfaces and excessive thermal decomposition and fouling on hot surfaces.

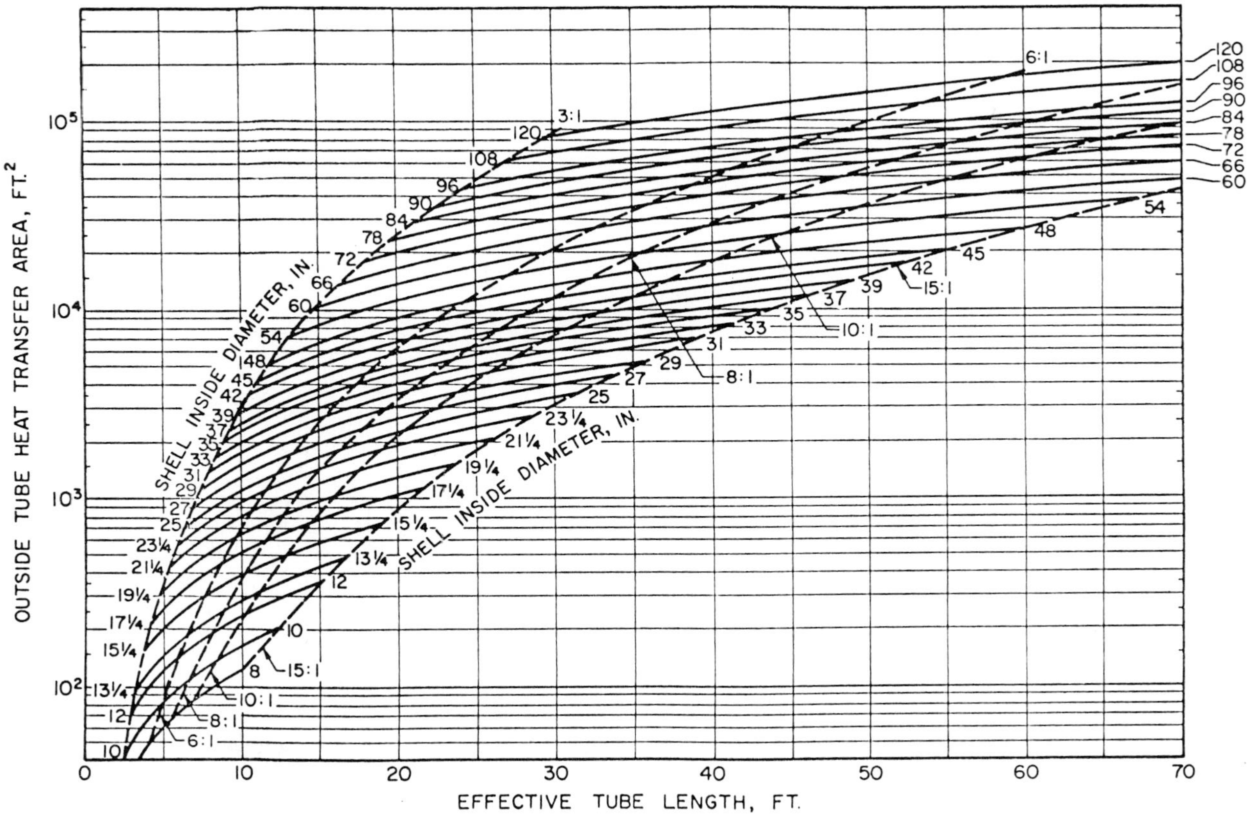

Fig. 4.10-32 Heat transfer area as a function of shell inside diameter and effective tube length for $\frac{3}{4}$-in. tube on a $\frac{15}{16}$-in. equilateral triangular tube layout, fixed tube sheet, one tubeside pass, fully tubed shell.

TABLE 4.10-3 VALUES OF F_1 FOR VARIOUS TUBE DIAMETERS AND LAYOUTS[a]

Tube Outside Diameter, in.	Tube Pitch, in.	Layout	F_1
$\frac{1}{2}$	$\frac{5}{8}$	→ ◁	0.67
$\frac{1}{2}$	$\frac{5}{8}$	→ ◇, □	0.77
$\frac{1}{2}$	$\frac{11}{16}$	→ ◁	0.81
$\frac{1}{2}$	$\frac{11}{16}$	→ ◇, □	0.93
$\frac{5}{8}$	$\frac{13}{16}$	→ ◁	0.90
$\frac{5}{8}$	$\frac{13}{16}$	→ ◇, □	1.04
$\frac{5}{8}$	$\frac{7}{8}$	→ ◁	1.05
$\frac{5}{8}$	$\frac{7}{8}$	→ ◇, □	1.21
$\frac{3}{4}$	$\frac{15}{16}$	→ ◁	1.00
$\frac{3}{4}$	$\frac{15}{16}$	→ ◇, □	1.16
$\frac{3}{4}$	1	→ ◁	1.14
$\frac{3}{4}$	1	→ ◇, □	1.31
1	$1\frac{1}{4}$	→ ◁	1.34
1	$1\frac{1}{4}$	→ ◇, □	1.54
1	$1\frac{3}{8}$	→ ◁	1.61
1	$1\frac{3}{8}$	→ ◇, □	1.86

[a]Note: $F_1 = \dfrac{\dfrac{(\text{heat transfer area})}{(\text{cross-sectional area of unit cell})_{\text{reference}}}}{\dfrac{(\text{heat transfer area})}{(\text{cross-sectional area of unit cell})_{\text{new case}}}}$

This table may also be used for low-finned tubing in the following way: The value estimated for U_o should be based on the total outside area (including fins) of the finned tube. This value will generally be somewhat less (10–30%) than the plain tube values given in Table 4.10-2. Then the required value of A_o is based on the finned tube area, and the above values of F_1 are *divided* by the ratio of the finned tube area to the plain tube area (per unit length). Typically this value will be from 2.5 to 4.

TABLE 4.10-4 VALUES OF F_2 FOR VARIOUS NUMBERS OF TUBE-SIDE PASSES[a]

Inside shell diameter, in.	Number of Tube-Side Passes			
	2	4	6	8
Up to 12	1.20	1.40	1.80	—
$13\frac{1}{4}$–$17\frac{1}{4}$	1.06	1.18	1.25	1.50
$19\frac{1}{4}$–$23\frac{1}{4}$	1.04	1.14	1.19	1.35
25–33	1.03	1.12	1.16	1.20
35–45	1.02	1.08	1.12	1.16
48–60	1.02	1.05	1.08	1.12
Above 60	1.01	1.03	1.04	1.06

[a]Since U-tube bundles must always have at least two passes, use of this table is essential for U-tube bundle estimation. Most floating-head bundles also require an even number of passes.

TABLE 4.10-5   VALUES OF F_3 FOR VARIOUS TUBE BUNDLE CONSTRUCTIONS

Type of Tube Bundle Construction	Inside Shell Diameter, in.				
	Up to 12	13–22	23–36	37–48	Above 48
Split-Backing Ring (TEMA S)	1.30	1.15	1.09	1.06	1.04
Outside-Packed Floating-Head (TEMA P)	1.30	1.15	1.09	1.06	1.04
U-Tube[a] (TEMA U)	1.12	1.08	1.03	1.01	1.01
Pull-Through Floating-Head (TEMA T)	—	1.40	1.25	1.18	1.15

[a]Since U-tube bundles must always have at least two tube-side passes, it is essential to also use Table 4.10-4 for this configuration.

4.10-5   Condensation in Shell-and-Tube Exchangers

Condensation Heat Transfer Mechanisms

There are four basic mechanisms of condensation generally recognized: dropwise, filmwise, direct contact, and homogeneous. In dropwise condensation drops of liquid form from the vapor at particular nucleation sites on the cold solid surface. The drops grow by further condensation and by agglomeration until they become large enough to be removed from the surface by gravity or vapor shear. When they depart, the surface is left dry and the process of nucleation, growth, and removal is repeated. In order for dropwise condensation to occur, the surface must be nonwetting. Dropwise condensation results in very high heat transfer coefficients, but no generally acceptable method exists for guaranteeing a nonwetting surface in industrial practice.

In filmwise condensation the drops initially formed quickly coalesce to produce a continuous liquid film on the surface, through which heat must be transferred to condense more vapor. The film flows from the surface by gravity and/or vapor shear, but a continuous liquid film always remains behind.

In direct-contact condensation vapor condenses directly on a subcooled liquid sprayed into the vapor space. The coolant may be the same as the vapor being condensed or a chemically different species. The barometric condenser is an example of this type of condensation.

In homogeneous condensation the liquid phase forms directly from a supersaturated vapor, away from any macroscopic surface. However, in practice, there are always sufficient numbers of dust or mist particles present to serve as nucleation sites. By whatever mechanism, the process results in fog formation, causing serious problems in loss of product and environmental contamination. Homogeneous condensation is never a design mode.

Therefore, essentially all condensation occurring in shell-and-tube heat exchangers is assumed to be by the filmwise mechanism. Condensation may occur either on the shell side or on the tube side. Specific correlations are available for a number of different combinations of geometry and condensate flow mechanisms, but it is beyond the scope of this section to discuss them in detail. In general, condensing heat transfer is usually an efficient heat process, especially since most condensing vapors carry little or no fouling material. The following sections concentrate on certain aspects of condensation as they apply in various shell-and-tube configurations. Detailed discussions of the condensation fluid flow and heat transfer mechanisms and correlations for these processes are given in Schlunder[1] and Rohsenow and Hartnett.[6]

Condensation of Pure Components

Condensing.   The condensation of a pure component is an isothermal process if the pressure drop is negligible. If there is significant pressure drop, the temperature of the condensing stream decreases and there are small sensible heat effects; however, the major effect is the reduction in the temperature driving force against the coolant and therefore a reduction in the heat transfer rate. Since condensation is generally an efficient heat transfer process even when gravity controls the condensate removal process, high vapor velocities and excessive pressure drop may result in a reduction rather than an increase in the condensing capacity of a piece of equipment. In general, condensation of a pure vapor poses few difficulties for the designer.

Desuperheating. Vapor admitted into a condenser may be substantially superheated above the saturation temperature. However, if the temperature of the condensing surface is below the saturation temperature at the pressure in the vapor space, condensation will occur directly from the superheated vapor. Further, it can be shown (Bell[12]) that this desuperheating process proceeds with heat fluxes equal to or greater than the subsequent condensation of the saturated vapor. Therefore, it is both simple and correct to design a pure component condenser with desuperheating duty as if the entire heat load were transferred by the condensing heat transfer mechanism and with a corresponding heat transfer coefficient. The only requirement is that the coolant be everywhere at a temperature below the saturation temperature of the condensing vapor. It is unnecessary and even self-defeating to provide separate desuperheating sections in condensers unless one is deliberately trying to heat the coolant above the vapor saturation temperature, a condition commercially employed only in high-pressure feedwater heaters in steam power plants (Fig. 4.10-33). This problem is discussed at greater length in Rubin.[13]

Subcooling. It is often desirable to subcool the condensate either to provide NPSH for a pump or to cool a product for safe storage. Where positive subcooling is desired, it is best to use a separate subcooler with its own coolant supply and controller. This provides not only for closer design but also for better control. The problem of design of condensers with integral subcooling is discussed in Vol. 3 of Schlunder.[1]

Condensation of Mixtures

Vapor with Noncondensable Gas. Many vapors to be condensed contain appreciable quantities of noncondensable gas. In this case the condensation process may start off rapidly at a condensing temperature corresponding to the partial pressure of the condensable vapor. However, further into the condensing process, appreciable quantities of noncondensable gas are swept to the interface, resulting in a mass transfer resistance to the transport of further vapor to the interface. As the concentration of condensable vapor in the bulk gas vapor flow decreases, the saturation temperature also decreases. The result of these mechanisms is that the heat transfer coefficient as well as the condensing temperature decreases sharply toward the end of the condensing process. Furthermore, it is never possible to condense all of the remaining condensable vapor in the noncondensable gas. The original analysis of this problem is given in Colburn and Hougan[14] and is discussed at greater length in Schlunder,[1] Rohsenow and Hartnett,[6] and Kern.[9]

Venting. All condensers must be vented. Certainly, those with appreciable quantities of noncondensable gas present must have a venting system designed to maintain a very low level of the

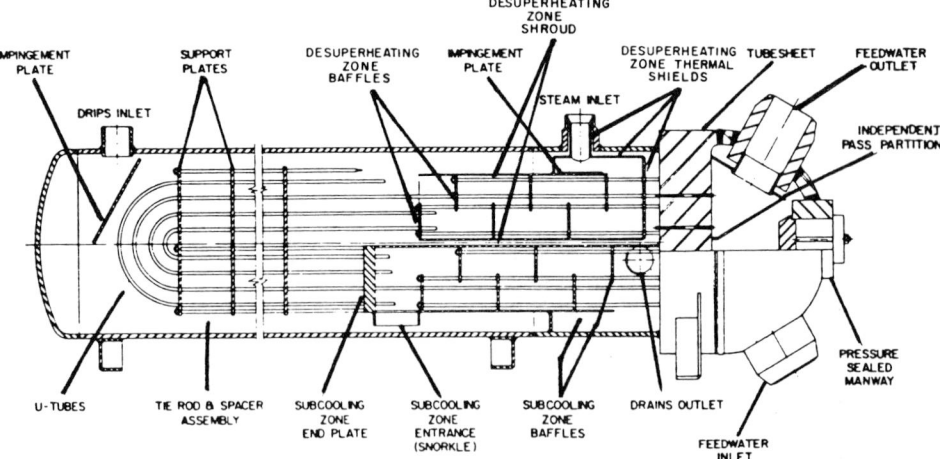

Fig. 4.10-33 Multizone condenser. Horizontal U-tube feedwater heater with desuperheating, condensing, and subcooling zones. (Courtesy of Foster Wheeler Energy Corporation and Hemisphere Publishing Corporation, New York.)

noncondensable gas at any point within the condenser. However, even nominally pure vapor or totally condensable mixtures also require vents on the condenser, if only to remove the inert gas present in the system before start-up. In fact, these vapors usually contain small amounts of noncondensable gas that will quickly accumulate if not vented. The flow within the condenser should be designed so that the noncondensable gases are swept to one or a few clearly identifiable points, with the vents located at those points. If the condenser is operating at subatmospheric pressure, the vents must be provided with pumps or ejectors to raise the pressure to that of the exhaust system. If the noncondensable gas carries sufficient condensable vapor to pose safety, environmental, or economic problems, a vent condenser may be provided with a refrigerated coolant to remove as much as possible of the remaining condensable.

Multicomponent Vapors. Many vapors to be condensed are composed of several components. It is first necessary to calculate the condensing curve, that is, the temperature and composition of the condensing mixture as a function of the amount of heat removed. These calculations are beyond the scope of this section, but any standard chemical engineering thermodynamics textbook may be consulted for the principles involved. From the condenser design standpoint, there are always sensible heat effects for both liquid and vapor phases as well as the heat of condensation. Additionally, there are diffusive mass transfer resistances in both phases. No complete and practical theoretical treatment is available for these effects, so approximation methods are used for design of equipment in this service. Bell and Ghaly[15] describe one of these methods.

Shell-and-Tube Configurations Especially for Condensing

While most of the shell styles can be used for condensing service, two in particular are especially recommended for severe pressure drop limitations. In addition, the unique requirements of the main condensers on steam power plants introduce additional considerations and special features. These cases are described in greater detail in the following sections.

Divided-Flow Shell (TEMA J Shell). The divided-flow, or TEMA J, shell is of interest when low pressure drop is a major consideration. Compared to an E shell of corresponding configuration, the pressure drop in a J shell will be about one-fourth to one-sixth of that in the E shell. Two J shells may be used in series with exit nozzles on one shell matching the inlet nozzles on the other. The baffles are present primarily to support the tubes and only incidentally to guide the flow. RODbaffle arrangements may be used to further reduce the pressure drop compared to a segmental plate baffle arrangement. Vents should be placed near or in the exit nozzles and in the ends of the shell on the single nozzle side.

Cross-Flow Shell (TEMA X Shell). The cross-flow shell gives minimum shell-side pressure drop, provided that there is adequate flow area across the top of the bundle for the vapor to distribute longitudinally. Alternatively, the vapor manifold may be connected to the shell by multiple nozzles. Sometimes a "bathtub" nozzle is used, an elongated nozzle of partial cylindrical cross section running nearly the full length of the shell above the tube field and open to it. It is important to provide correct venting in an X shell; in addition to venting at or near the exit nozzles (the hot wells), vents should be provided at any point that may not be well flushed out by the flow of vapor. The areas adjacent to the tube sheets are particularly susceptible to accumulations of noncondensables.

Power Plant Condensers. Power plant condensers are characterized primarily by their very large size, up to 1,250,000 ft^2 (120,000 m^2) of heat transfer surface. They always operate at pressures of at most a few inches of mercury absolute (< 20 kPa) and therefore must be configured so as to minimize pressure drop. Finally, there is always a significant amount of air introduced into the incoming steam through deaeration of the feedwater and leaks in the system, placing a premium on careful venting arrangements. They are usually mounted immediately under or adjacent to the lower-pressure stages of the turbine so that the kinetic energy of the high-velocity exhaust stream can be recovered by expansion in the connecting conduits.

Many different but closely related configurations have been proposed to solve these problems. The design in Fig. 4.10-34[16] is representative. The tube field, the shaded portion in the right-hand part of the figure, is so arranged that steam can penetrate into the interior of the tube nest without excessive pressure losses and so that the noncondensable gas (saturated with water vapor) remaining at the bottom of the tube field must pass over a number of coolant tubes before entering the air removal line. Other designs provide for air removal from near the bottom of the tube field near the hotwell. Modern design methods for power condensers are characterized by using two- and three-dimensional analyses of the flow field, together with the heat transfer correlations. Marto and Nunn[17] give a good overview of current technology in this field.

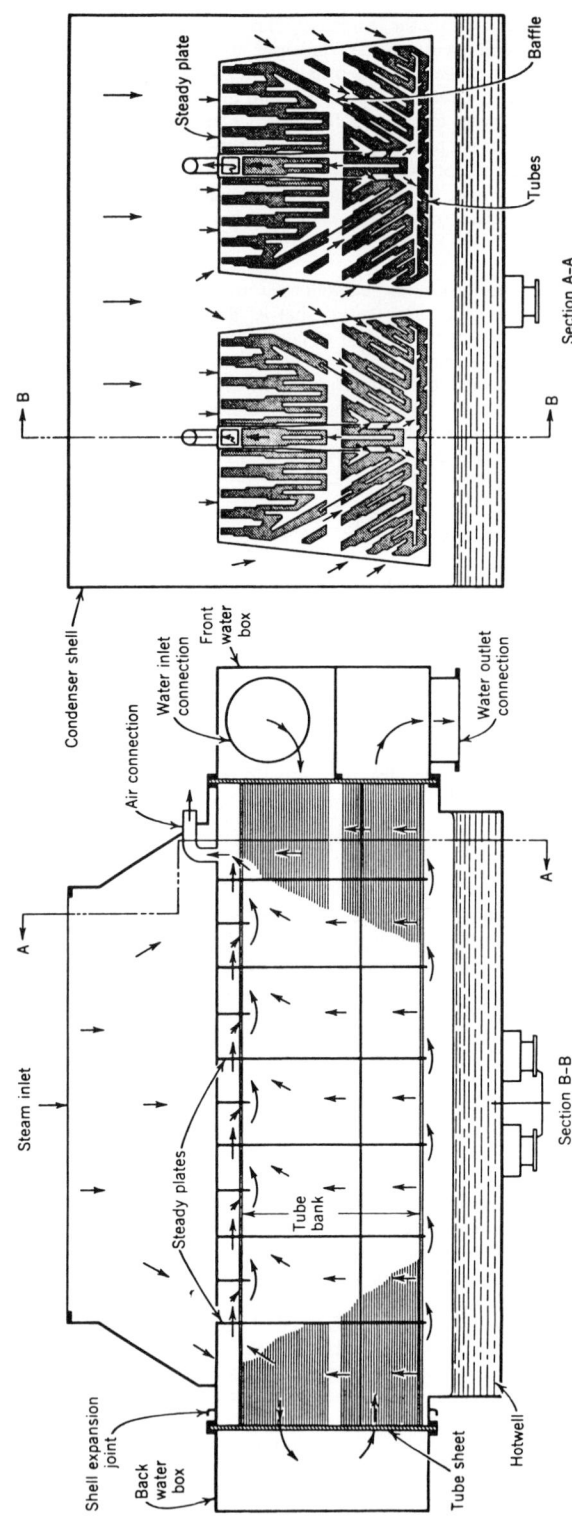

Fig. 4.10-34 Sections through a typical two-pass surface condenser for a large steam power plant. (Courtesy Allis–Chalmers Manufacturing Co.)

Baffle

Tubes

Section A–A

Steady plate

Condenser shell

B

B

Front water box

Water inlet connection

Air connection

Water outlet connection

A

A

Steam inlet

Section B–B

Steady plates

Tube bank

Shell expansion joint

Back water box

Tube sheet

Hotwell

4.10-6 Vaporization in Shell-and-Tube Exchangers

Vaporization Heat Transfer

A typical curve of heat flux versus temperature difference between surface and liquid saturation temperature for saturated pool boiling is shown in Fig. 4.10-35. Various boiling regimes indicated in the figure are as follows.

1. The *natural convection regime* is characterized by low ΔT. The liquid in contact with the hot surface is superheated and rises by natural convection to the vapor–liquid interface where the superheat is released by quiescent vaporization. There is no vapor bubble formation, and the heat transfer coefficients are characteristic of those of natural convection processes. For a smooth metal tube and water the upper limit on this region is a ΔT of about 10°F. However, the limit is lower for other fluids, and specially developed surfaces are commercially available that provide for nucleate boiling at ΔT of 1°F or less. Vapor generators are not designed to operate in the natural convection regime.

2. In the *nucleate boiling regime*, vapor bubbles are formed at preferred nucleation sites, typically small pits or scratches. Vapor bubbles form repeatedly at the sites and rise to the surface. As the temperature difference increases, more sites become active and the heat flux increases as ΔT^{2-4}. This is a very efficient heat transfer region, and vapor generators are designed to operate within it.

3. The point indicated by 3 in Fig. 4.10-35 is the *maximum, peak, burnout, or critical heat flux*, that is, the highest attainable heat flux for any reasonable surface temperature. Good correlations exist to predict this value for both single tubes and tube bundles (e.g., see Palen and Small[18]); typically the design heat flux is about 70% of the predicted maximum. ΔT at the peak heat flux can vary from about 25 to 50°F at normal pressures and temperatures. Attempts to increase the ΔT above this result in the bubbles leaving the surface preventing the inflow of liquid to the surface. The peak heat flux is essentially independent of the nature of the surface, so the special nucleating surfaces referred to above have about the same peak heat flux as ordinary tube surfaces.

4. In region 4 patches of the surface become periodically dry, resulting in a *lower average heat flux*. This is an unstable regime and prone to severe fouling, and it is important to ensure that equipment is neither designed nor operated in this region.

5. Region 5 is the *film boiling regime*, which exists at large temperature differences. A stable film of vapor exists between the surface and the liquid pool. Heat transfer is by conduction across the vapor to the liquid, resulting in low heat fluxes and correspondingly very low heat transfer coefficients. This surface may become hot enough to thermally degrade the substance being boiled; thus, fouling is a very probable consequence of operating in this region.

Convective Vaporization. The regimes described in the previous paragraph occur on a hot surface in a relatively quiescent pool of liquid. In most boiling apparatus the liquid or the vapor–liquid mixture flows past the surface at a significant velocity giving rise to convective two-phase vaporization processes. In fact, the presence of the vapor shear on the surface may suppress the pool boiling processes, but nucleate boiling is always essential at the beginning of the boiling process to initiate the vapor phase. While convective boiling processes result in very high heat transfer coefficients, they are still limited to a ΔT between surface and boiling fluid of less than about 50°F in order to minimize

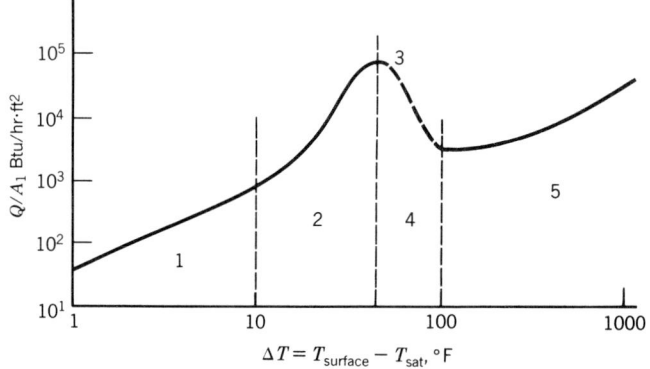

Fig. 4.10-35 Typical saturated pool boiling curve.

the possibility of drying all of the liquid off of the wall. Detailed discussion of correlations applicable to these processes is beyond the present scope, but Schlunder,[1] Fair,[19] and Collier[20] comprehensively survey the literature.

Design methods for vapor generation equipment have been developed for a majority of the types used in the process and energy industries. Inasmuch as these design methods must also consider the flow of the two-phase fluid and the associated pressure effects, the methods are very complex and the calculations lengthy and generally suited only for use with a computer. The foregoing references give a good review of the state of the art of design methods.

Vaporization of Mixtures

It is frequently necessary to vaporize a liquid containing a number of different components. The comments in Section 4.10-5 referring to multicomponent condensation are also generally applicable to this problem. For example, when a mixture is vaporized, the more volatile components near the hot surface will be preferentially vaporized, leading to a concentration of the less volatile components near the surface. This tends to suppress both the local heat transfer coefficient and the effective temperature difference, resulting in a significant reduction in heat flux.

Provision should be made in any vaporizer to ensure that a liquid phase is also withdrawn regularly, if not continuously, from the boiling space so as to prevent the accumulation of significant quantities of nonvolatile material. Fouling is a more serious problem in vaporizers than in condensers owing to the fact that the heavy fouling materials will tend to concentrate in the liquid phase. Therefore, reboilers and vaporizers should be selected with a view to minimizing the possibilities of fouling and making the cleaning process as simple as possible in those situations where fouling cannot be avoided.

Shell-and-Tube Configurations Used for Vaporization

While vaporization can be carried out in a wide variety of shell-and-tube heat exchangers (and other kinds of heat exchangers as well), there are certain configurations that have achieved particular application for the production of vapor both for feed to distillation columns and chemical processes and for waste heat recovery. Several of these specialized applications are given in the following sections.

1. Kettle or flooded bundle (TEMA K Shell). A typical configuration for the use of a kettle reboiler in a process application is shown in Fig. 4.10-36. While usually analyzed as a pool boiling device, the kettle reboiler actually has a well-defined circulation pattern of the vapor–liquid mixture up through the tube field, separating into vapor and liquid phases above the tube bundle, with the liquid recirculating down the sides of the bundle to the bottom. Nearly dry vapor is removed from the top of the large-diameter shell (demister pads or centrifugal separators are used to eliminate the remaining liquid carry-over, if necessary) and the bottom product is withdrawn from the end of the reboiler with the aid of a weir. U-tube bundles are commonly used in this configuration,

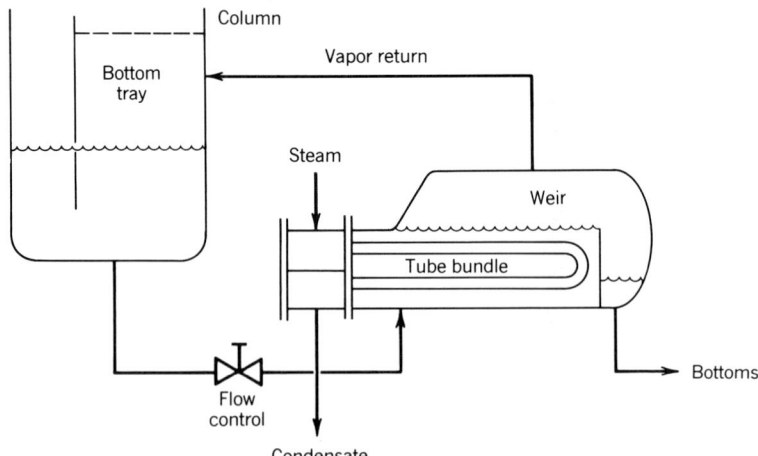

Fig. 4.10-36 Kettle or flooded bundle vaporizer.

allowing the bundle to be easily removed. If an open square pitch is used, cleaning is ordinarily not too difficult. The kettle reboiler gives heat transfer rates comparable to the other configurations, but it is rather easily fouled and has a high residence time in the heated zone. Construction is also relatively expensive due to the large-diameter shell.

   2. Vertical thermosiphon reboiler.   The vertical thermosiphon in-tube reboiler is shown in Fig. 4.10-37. It is arranged so that the boiling fluid flows up through the tubes, vaporizing as it rises and thereby giving a lower average density than in the all-liquid leg in the column and downcomer piping. The resulting hydrostatic pressure difference results in a natural circulation through the tubes. The liquid level in the column is usually at the level of the top tube sheet, though it may be dropped to about half the height of the tubes for vacuum operation.

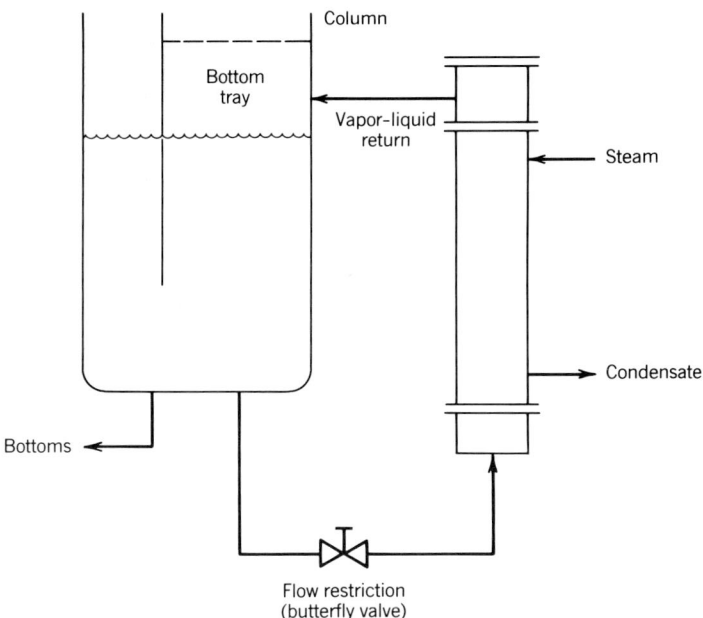

Fig. 4.10-37   Vertical thermosiphon reboiler.

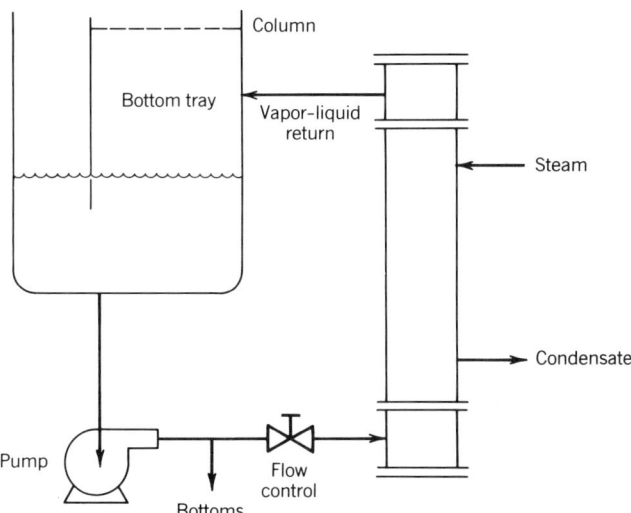

Fig. 4.10-38   Vertical forced circulation reboiler.

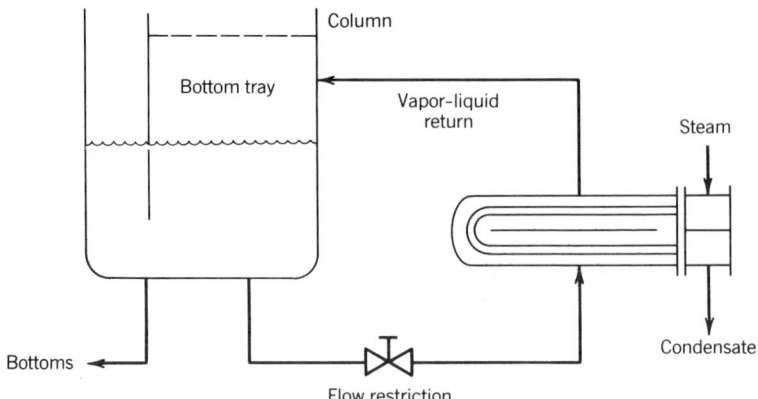

Fig. 4.10-39 Horizontal thermosiphon reboiler (using a TEMA G shell).

This configuration is comparatively inexpensive, gives high heat transfer rates and low residence times, and is relatively resistant to fouling. (However, some streams foul so badly that the mere selection of a particular configuration cannot eliminate the fouling.) The thermosiphon requires little ground space but does require a high skirt on the column to provide the liquid level; the vertical positioning also complicates the problem of cleaning and tube bundle maintenance. The vertical thermosiphon is subject to the phenomenon of instability, resulting in rapid and violent reversals of flow; the instability phenomenon has been studied and can be predicted and avoided with the aid of a computer program. Essential steps to control instability are to provide a variable-flow resistance in the inlet line (frequently a butterfly valve is used for this purpose) and to minimize any flow resistance in the exit two-phase piping.

3. **Forced-circulation reboiler.** A forced-circulation reboiler is shown in Fig. 4.10-38 in a vertical orientation with boiling in the tubes. However, it may also be used in a horizontal arrangement and/or with boiling on the shell side. A pump is used to force flow through the exchanger, thereby eliminating dependence on the natural circulation process and also eliminating the problem of instability. Otherwise, the basic features of the forced-convection units are similar to those of the corresponding thermosiphon design.

4. **Horizontal thermosiphon reboiler.** A horizontal thermosiphon reboiler is shown in Fig. 4.10-39. The operational principle is similar to that of the vertical in-tube thermosiphon but now the vaporization occurs on the shell side. The horizontal thermosiphon is more subject to fouling and has a relatively long residence time compared to the vertical thermosiphon, but it is less prone to instability. However, a flow control device should be provided in the liquid piping. A longitudinal baffle is provided to force the flow to the ends of the bundle. It is important to provide liquid withdrawal lines at the ends of the tube bundle to prevent the buildup of nonvolatile material.

REFERENCES

4.10-1 E. U. Schlunder (ed.), *Heat Exchanger Design Handbook*, Hemisphere, Washington D.C., 1982.

4.10-2 *Tubular Exchangers Manufacturers Association Standards*, 6th ed., New York (1978).

4.10-3 D. Q. Kern and A. D. Kraus, *Extended Surface Heat Transfer*, McGraw-Hill, New York, 1972.

4.10-4 W. M. Small and R. K. Young, "The RODbaffle Heat Exchanger," *Heat Transfer Engineering* **1**(1), 21–27 (1979).

4.10-5 T. Rozenman and J. Taborek, "The Effect of Leakage Through the Longitudinal Baffle on the Performance of Two-Pass Shell Exchangers," AIChE Symposium Series No. 118, "Heat Transfer—Tulsa," **68**, 12–20 (1974).

4.10-6 W. M. Rohsenow, and J. P. Hartnett, *Handbook of Heat Transfer*, 2nd ed., McGraw-Hill, New York, 1983.

4.10-7 N. Epstein, "Thinking About Heat Transfer Fouling: A 5 × 5 Matrix," *Heat Transfer Engineering* **4**(1), 43–56 (1983).

4.10-8 K. A. Gardner and J. Taborek, "Mean Temperature Difference: A Reappraisal," *AIChE J.* **23**(6), 777–786 (1977).

4.10-9 D. Q. Kern, *Process Heat Transfer*, McGraw-Hill, New York, 1950.

4.10-10 K. P. Singh, "On the Necessary Criteria for Stream—Symmetric Tubular Heat Exchanger Geometries," *Heat Transfer Engineering* **3**(1), 19–22 (1981).

4.10-11 W. M. Kays and A. L. London, *Compact Heat Exchangers*, McGraw-Hill, New York, 1960.

4.10-12 K. J. Bell, "Temperature Profiles in Condensers," *Chem. Eng. Prog.* **68**(7), 81–82 (1972).

4.10-13 F. L. Rubin, "Multizone Condensers: Desuperheating, Condensing, Subcooling," *Heat Transfer Engineering* **3**(1), 49–59 (1981).

4.10-14 A. P. Colburn and O. A. Hougen, "Design of Cooler-Condensers for Mixtures of Vapors with Noncondensing Gases," *Ind. Eng. Chem.* **26**(11), 1178–1182 (1934).

4.10-15 K. J. Bell and M. A. Ghaly, "An Approximate Generalized Design Method for Multicomponent/Partial Condensers," *AIChE Symposium Series*, No. 131, **69**, 72–79 (1972).

4.10-16 A. P. Fraas and M. N. Ozisik, *Heat Exchanger Design*, Wiley, New York, 1965.

4.10-17 P. J. Marto and R. H. Nunn (eds.), *Power Condenser Heat Transfer Technology: Computer Modeling, Design, Fouling*, Hemisphere, Washington, D.C., 1981.

4.10-18 J. W. Palen and W. M. Small, "A New Way to Design Kettle and Internal Reboilers," *Hydrocarbon Proc.* **43**(11), 199–208 (1964).

4.10-19 J. R. Fair, "What You Need to Design Thermosiphon Reboilers," *Pet. Ref.* **39**(2), 105–123 (1960).

4.10-20 J. G. Collier, *Convective Boiling and Condensation*, 2nd ed., McGraw-Hill, New York, 1981.

4.11 HEAT EXCHANGERS OTHER THAN SHELL-AND-TUBE TYPE

Kenneth J. Bell

4.11-1 General Considerations Concerning Other Types of Heat Exchangers

Many different configurations of heat exchangers other than shell and tube (Section 4.10) are commercially available, and each type has its area of application in which its performance and cost characteristics are superior or at least comparable to competing configurations. Some of the more common types are described in this section, but no list can include every device that is available and applicable as a heat exchanger. Many of the types described in this section are proprietary with respect to fabrication technology and design methods, including the basic thermal-hydraulic correlations. In fact, "design" for several types rests more on past experience and engineering judgment than it does on understanding the basic fluid mechanics and heat transfer mechanisms and equations derived therefrom.

Therefore, the emphasis in this section will be on a physical description of the equipment and the areas of application in which each type is especially useful and likely to be competitive. For many configurations in many services fundamental correlations of heat and mass transfer can be adapted with reasonable accuracy for most engineering purposes if the designer is careful to identify the controlling momentum, heat, and mass transfer mechanisms, and some specific suggestions are included here. Several of the heat exchanger configurations to be discussed in this section are considered in greater detail in Schlunder.[1] Volume 3 of this text is particularly helpful from the thermal-hydraulic standpoint, while the mechanical features are treated at greater length in Vol. 4. For rough estimating purposes, the typical heat transfer coefficients given in Table 4.10-2 may be used for most of these configurations.

4.11-2 Double-Pipe Heat Exchangers

Construction Elements

The basic configuration of the double-pipe heat exchanger is shown in Fig. 4.11-1. In principle, it is simply two concentric pipes with one fluid flowing inside the inner pipe or tube and the second flowing in the annulus between the inner and outer pipes, the heat transfer surface being the surface of the inner pipe. Appropriate fittings are attached to each pipe to allow the fluids to flow from one section

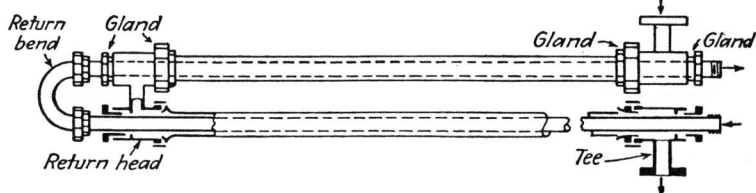

Fig. 4.11-1 Schematic of a basic double-pipe heat exchanger. (From ref. 2; courtesy of McGraw-Hill Book Co.)

to the next. Usually the flow arrangement is countercurrent, though other configurations may be used either to (1) accommodate two fluids of widely different flow rates, in which case some sort of series–parallel layout is commonly used, or (2) to maintain a more nearly constant wall temperature to minimize thermal degradation or solid deposition, in which case a cocurrent flow arrangement may be used.

For many applications a plain wall pipe or tube is used for the internal conduit. However, for other applications (especially if one fluid has a substantially lower heat transfer coefficient than the other), longitudinally finned tubes as diagrammed in Fig. 4.11-2 may be used. Type A would commonly be used with turbulent flow of the low-coefficient fluid (often a gas) on the finned side. If it is preferable to put the low-coefficient fluid inside the tube, internally finned tubes are also available.

For laminar flow, it is important to break up periodically the adverse temperature gradient that is formed in the fluid adjacent to the heat transfer surface, and type B tubes with interrupted fins are used; a different kind of interruption is shown in type C. Type D provides finned surfaces on both sides of the tube and is usually used when both fluids have a low heat transfer coefficient (e.g., gas to gas). When extended surface is used, a fin efficiency or fin resistance term must be used with the overall heat transfer coefficient. This concept is developed at length in Kern[2] and Kern and Kraus.[3]

A wide variety of fittings on the tube and annulus sides is available to meet the specific requirements in a given application. Specialized cases include the following:

1. High-pressure seals on either the tube or the annulus.
2. Double seals and drain lines for positive control of leakage of the annulus fluid.
3. Easily removable fittings for cleaning, either in the tube or in the annulus.

A closely related configuration is the multitube section, in which multiple longitudinal tubes with U bends are contained within a single "annulus" (Fig. 4.11-3). The tubes are commonly (though not necessarily) finned, resulting in a very large effective heat transfer area per single section.

If suitable design features are used, double-pipe heat exchangers can be designed for tube-side pressures to 5000 psi (35 MPa) and above, and annulus pressures of 1000 psi (7 MPa) and above. They may also be designed for temperatures well in excess of 1000°F (540°C), depending on material of construction, operating pressure, and particularly the choice of gasket and sealing material. In principle, double-pipe units can be manifolded in parallel to achieve any heat transfer area; however, most applications fall in the area below about 200 ft^2 (20 m^2) of bare tube area.

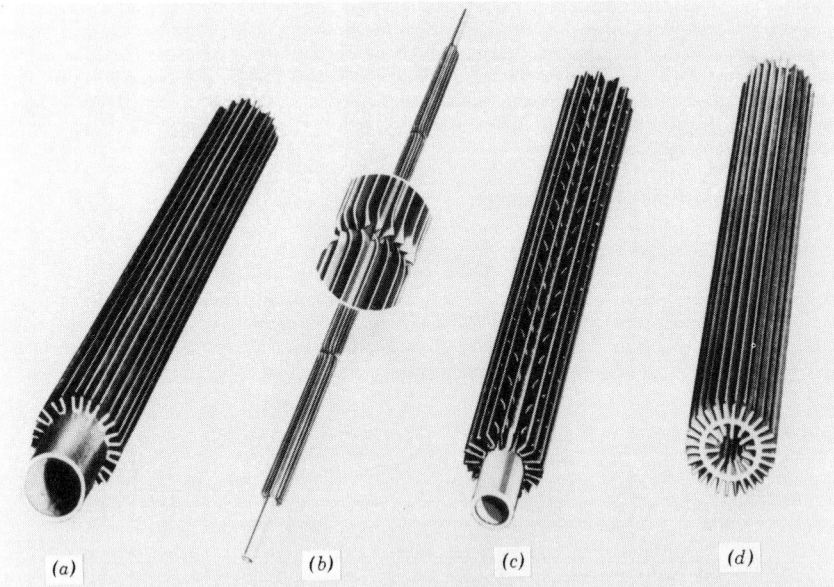

Fig. 4.11-2 Examples of longitudinally finned tubes for use in double-pipe exchangers. (Courtesy of Brown Fintube Co., Houston, TX.)

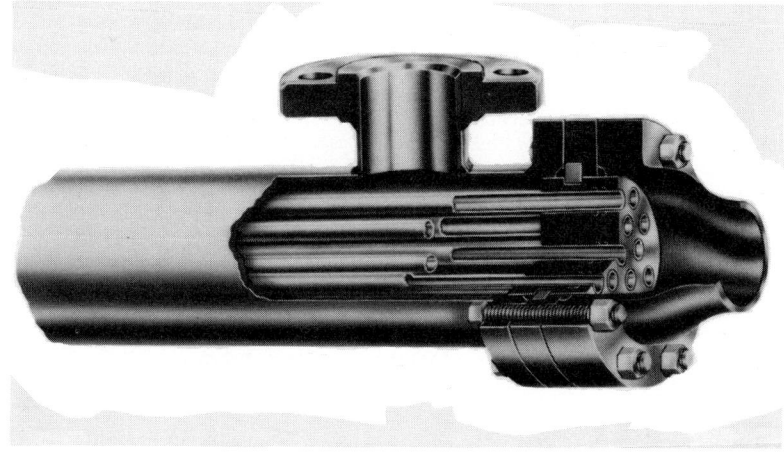

Fig. 4.11-3   Multitube exchanger: (Upper) Tubeside joint and closure; (lower) return bend construction. (Courtesy of Brown Fintube Co., Houston, TX.).

Areas of Application

Double-pipe heat exchangers can in principle be used for almost any heat transfer application, including vaporizing and condensing. The usual limitation on their use is the relatively large mass and volume of the heat exchanger per unit of heat transfer area; this results in correspondingly high surface cost per square foot and tends to limit double-pipe exchangers to relatively small surfaces and correspondingly low duties. While no hard and fast line can be drawn, any application requiring more than about 100–200 ft^2 of heat transfer surface in a double-pipe heat exchanger should be checked in order to determine if in fact a shell-and-tube heat exchanger (or perhaps another type) would not do the same job at significantly lower cost as well as weight and volume savings. However, applications in excess of 1000 ft^2 of surface are not unknown.

The greatest single advantage of the double-pipe heat exchanger is its ease of construction and resulting process flexibility. Double-pipe exchangers can be built up quickly from standard components carried in inventory, using regular maintenance crews. Thus, an existing installation can be literally modified overnight to meet new process conditions and/or remedy deficiencies in the existing capacity. Double-pipe exchangers can be applied over a wide range of temperatures and pressures using a variety of materials of construction. Flexible piping arrangements are possible to accommodate streams with poorly matched flow rates and thermal capacitances.

A double-pipe heat exchanger is often specified when it is necessary to heat or especially to cool a very viscous liquid. For this application a heat exchanger having multiple parallel channels (such as the tubes in a shell-and-tube exchanger) may exhibit a form of flow instability resulting in severe maldistribution of the viscous fluid among the channels and inadequate heat transfer performance. Such behavior is impossible in an arrangement of a single double-pipe exchanger fed by its own pump.

By proper selection of end-fittings, the tube side may be cleaned mechanically without removing the tube from the annulus. Alternatively, if proper fittings are selected, the tube can be easily removed and the annulus exposed for mechanical cleaning.

Generally very good design methods exist for double-pipe heat exchangers and the expertise and parts for their construction are widely available.

Design Basis

The design basis for double-pipe heat exchangers is relatively straightforward. Pressure drops in the tube sections are readily and fairly accurately calculated using standard in-tube correlations. Good correlations for the annulus also exist, and the equivalent diameter concept permits in-tube correlations to be used to an acceptable accuracy. However, a major contribution to the pressure drop is in the connecting fittings, and vendor's data are generally necessary to obtain accurate predictions.

Heat transfer coefficients for plain tubes and annuli may be calculated in the turbulent regime from in-tube correlations for both sensible and phase-change cases (using the equivalent diameter for the annulus). For finned tubes or other forms of enhancement or for laminar flow in the annulus, specialized (and often proprietary) correlations are needed.

The logarithmic mean temperature difference concept is applicable to double-pipe exchangers in either countercurrent or cocurrent flow arrangements. For series–parallel arrangements Kern[2] may be consulted.

4.11-3   Gasketed-Plate Heat Exchangers

Construction Elements

Figure 4.11-4 shows an exploded view of a typical gasketed-plate heat exchanger and Fig. 4.11-5 shows two typical plates. The heat transfer surface is provided by the stack of plates that are gasketed around the outer edge to prevent leakage of fluid to the surroundings and are held together by long bolts clamping two end plates together and compressing the gaskets to form a seal. The plates are pressed with a corrugated pattern, and alternate plates are arranged so that the corrugations contact or cross the corresponding corrugations on the adjacent plates. This arrangement provides a very large number of support points between adjacent plates to provide the necessary stiffening against pressure differentials across the plates and, somewhat incidentally, to provide a strongly eddied flow pattern resulting in high heat transfer coefficients (and correspondingly high pressure drops).

Flow between plates is controlled by the gasketing pattern around the ports; when flow is to be directed into a given plate, the gasketing is omitted between the inlet and exit ports for that fluid and the heat transfer surface of the plate. The two fluids flow between alternate plates, and by the use of blind or blanked ports, various combinations of parallel and series flow of a given fluid through a stack of plates are possible. Double gasketing is used between the plate and the ports for the fluid not admitted to that plate, and the dead space thus created is vented to the surroundings by a gap in the gasketing. This practically guarantees that the two fluids cannot leak into one another, though of course each may independently leak to the surroundings.

The plates may be pressed from a very wide variety of metals, including stainless steels, titanium, high-nickel alloys, and tantalum. It is feasible to press less expensive metals, but this tends to reduce the economic attractiveness of a plate heat exchanger compared to an equivalent shell-and-tube unit. The plates may be pressed with either an acute or an oblique chevron angle. Stacking obtuse angle plates adjacent to one another provides a very large number of contact points between the two plates in the direction of flow and correspondingly high heat transfer rates (or high number of transfer units) and high pressure drops. Stacking two plates with acute angles results in a smaller number of contact points in the direction of flow, lower heat transfer coefficients, and lower pressure drops per plate. Using a mixture of the two chevron angles in an exchanger allows the designer to very closely match the thermal requirements against the available pressure drop using a minimum number of plates.

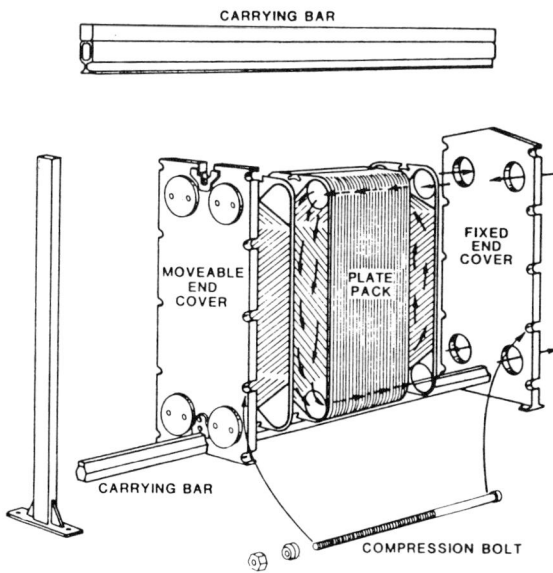

Fig. 4.11-4 Exploded view of a gasketed plate heat exchanger. (From ref. 1; courtesy of Hemisphere Publishing Corp., Washington, D.C.)

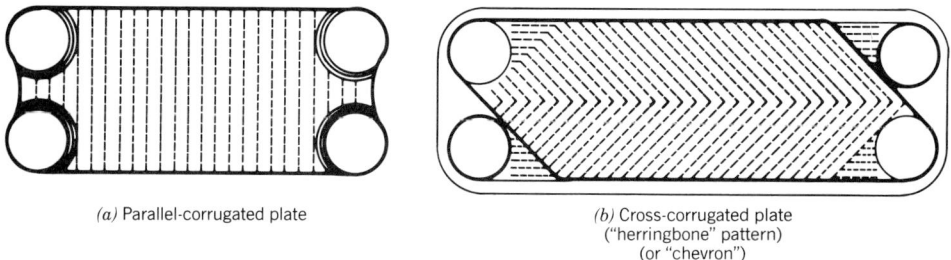

(a) Parallel-corrugated plate

(b) Cross-corrugated plate
("herringbone" pattern)
(or "chevron")

Fig. 4.11-5 Two different types of gasketed plate for the PHE. (Courtesy of Alfa Laval, Lund, Sweden.)

Gaskets are made from a variety of standard gasket materials including nitrile, butyl, and Viton elastomers, Teflon, and compressed asbestos. Since the gaskets represent the actual barrier between the heat exchanger fluids and the surroundings, great care must be taken to select a material which is both chemically compatible with the fluids and displays suitable strength and lifetime characteristics at the maximum temperatures and pressures to be used. Since plate heat exchangers are used in many services where frequent disassembly is required, gasket reusability is of prime concern. The compressed asbestos gaskets cannot be reused, so their use is limited to only the very highest temperature applications.

Effective heat transfer plate area ranges from about 0.34 ft^2 (0.032 m^2) up to about 40 ft^2 (3.7 m^2). Up to 600 plates may be put into a single frame, resulting in heat transfer areas of up to 24,000 ft^2 (2200 m^2) per unit. A wide variety of flow arrangements ranging from all-channels-in-series to all-channels-in-parallel flow is possible by suitable selection of plates with blind ports. One fluid may also be put in cross flow to the other by omitting the gaskets from the side of every adjacent pair of plates. This arrangement is especially useful when a very large volume of one fluid must be used with minimum pressure drop.

The main area of application of gasketed-plate heat exchangers is in the sensible heating or cooling of fluids at modest temperatures and pressures. Within the range of applicability, plate heat exchangers typically cost as little as half of that of the corresponding shell-and-tube exchanger where alloy construction is required. Because of the narrow clearance between the plates, gases are ordinarily not handled in plate exchangers, though vapors such as ammonia and propylene at several atmospheres have been successfully condensed. A few applications involving partial vaporization of one of the streams have also been reported. A major area of application is in heat rejection to brackish water or seawater using titanium plates. This is often combined with a chemically controlled water system on the other side, which is recirculated through equipment such as inter- and after-coolers and condensers.

The gaskets confine the role of the plate heat exchanger to relatively low temperatures and pressures. Most plate exchangers are capable of pressures of 150 psig (1 MPa) at fairly low temperatures, with a few designs being capable of pressures as high as 350 psig (2.5 MPa), also at relatively low temperatures. Conventional (reusable) gaskets are limited in temperature applications to below about 300°F (150°C), a value achieved only at relatively low operating pressures. Compressed asbestos gaskets have an upper operating range of about 480–500°F (250°C) and as low as about −40°F (−40°C), but they are not generally reusable.

Because of the difficulty in ensuring positive leak protection against the atmosphere, some users will employ plate heat exchangers only in applications where small amounts of leakage to the atmosphere are neither unsafe nor economically unacceptable. Other users have a less conservative philosophy and find that they are able to adequately control leakage problems by careful maintenance and constant inspection.

Plate heat exchangers, especially in the smaller sizes, are used in many applications in the food processing and pharmaceutical products industry, where frequent periodic disassembly for cleaning and sterilization is required. Ease of disassembly is lost quickly as plate sizes increase.

Design Basis

The basic design data for gasketed-plate heat exchangers consists of experimentally determined friction and j factors for the various plate configurations as a function of Prandtl and Reynolds numbers. Until very recently these data, obtained by the individual manufacturers, were considered highly proprietary and were very closely guarded. Under such conditions, a potential customer had

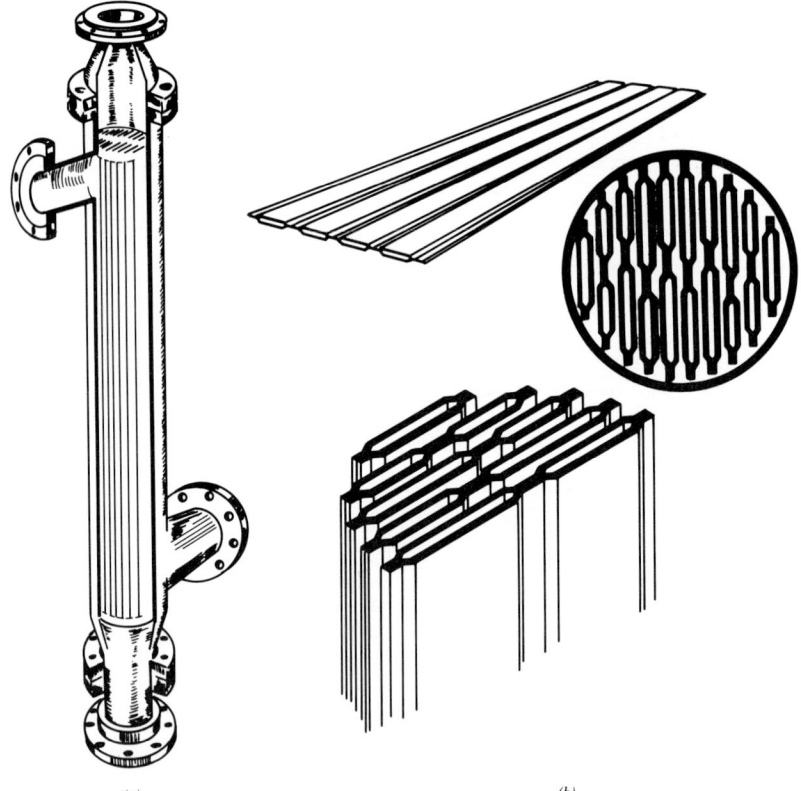

(a)   (b)

Fig. 4.11-6 Ramen welded plate heat exchanger: (a) Ramen LHE and (b) Lamella bundles for the LHE. (Courtesy of Alfa Laval, Lund, Sweden.)

little choice but to trust the quotation of the vendor. More recently, partly in response to the demands of the customers and partly in response to the growing competition in the field, this situation has started to change. Some basic data have been published both by the manufacturers and independent laboratories, and computer-based design procedures are now generally available from the vendors. This has made it possible to examine more closely the relative performance of several plate exchanger designs and also to compare them with competing configurations. However, no comprehensive set of correlations is generally available in the open literature.

As a general statement, plate exchangers result in very high heat transfer coefficients and correspondingly high pressure drops.

4.11-4 Welded-Plate Heat Exchangers

Recently a number of proprietary designs involving plate-type construction but with welded closures of the channels have become available. While it is impossible to describe all of these configurations, Figs. 4.11-6 and 4.11-7 show two types of this general class.

The common characteristic of this kind of heat exchanger is that the flow channels and their respective header devices are welded together, thus eliminating the need for internal gasketing to seal the fluids from each other and the surroundings and greatly increasing the temperatures and pressures at which these heat exchangers can be used.

In some types only one fluid may be at a high pressure owing to large flat containment surfaces on the other side, and it is always necessary to take into account the special problems posed by thermal stresses if these units are to be used at large temperature differentials.

These heat exchangers may be made out of any metals that may be pressed and welded; special welding techniques are often called for, and particular care must be taken to ensure the integrity of the welds. Because of their generally lower material requirements per unit of heat transfer area, these units are particularly attractive when alloy construction is required for corrosion or high-temperature strength.

Fig. 4.11-7 Bavex welded plate heat exchanger. (Courtesy of DVT-USA, Larkspur, CA.)

Areas of Application

Welded-plate exchangers can be employed in a wide variety of heat transfer services, especially including high temperatures and high pressure at least on one side. Some of them can be mechanically cleaned on one side, thus allowing their use in fouling service; however, others cannot be mechanically cleaned and must generally be considered only for use with clean streams or those that can be chemically cleaned. These exchangers generally offer more heat transfer surface per unit weight and volume than equivalent shell-and-tube heat exchangers and are thus attractive where space or weight limitations are significant.

Design Basis

No general rules can be provided for predicting heat transfer and pressure drop in these kinds of heat exchangers because of their widely variant surface configurations. In some cases (such as the Ramen exchanger) the channels on both sides have constant cross sections, and pressure drop and heat transfer can probably be adequately predicted by using the in-tube pressure drop and heat transfer correlations with the equivalent diameter concept. In other cases, such as represented by the Bavex exchanger, empirical correlations based on experimental information are essential because of the irregularly shaped flow paths. While some of the correlations are proprietary and closely held, most manufacturers will make their design basis available to bona fide customers.

4.11-5   Spiral-Plate Exchangers

Construction Elements

The spiral-plate exchanger is shown diagramatically in Fig. 4.11-8. In this case a long embossed plate is rolled using a special mandrel into a spiral configuration, so that the channel for one fluid is continuously wrapped between adjacent channels for the other fluid. Adjacent plates are then welded together on alternate sides of the exchanger, the spiral assembly enclosed in a cylindrical shell with gasketed face plates to seal the open channel, and nozzles welded at the appropriate points. The edge of each plate is gasketed against each face plate where they touch; however, since the channels for one fluid are open only on one side of the exchanger or the other, if an internal gasket fails, it only means that the fluid can short-circuit from one of its spiral passages to the next and cannot mix with the other fluid. The gasketing around the periphery of the outermost channel and against the face plate seals each fluid against the atmosphere. Once again, the gasket materials limit the temperatures and pressures that can be sustained in these units. In general, the ranges of temperature and pressure for which these units are applicable are comparable to those of the gasketed-plate exchanger, with of

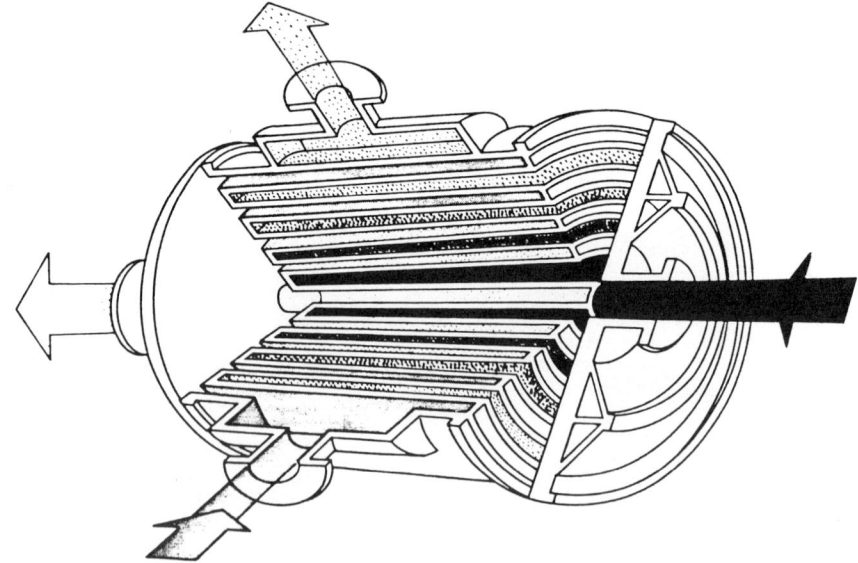

Fig. 4.11-8   Spiral plate exchanger. (Courtesy of Alfa-Laval, Lund, Sweden.)

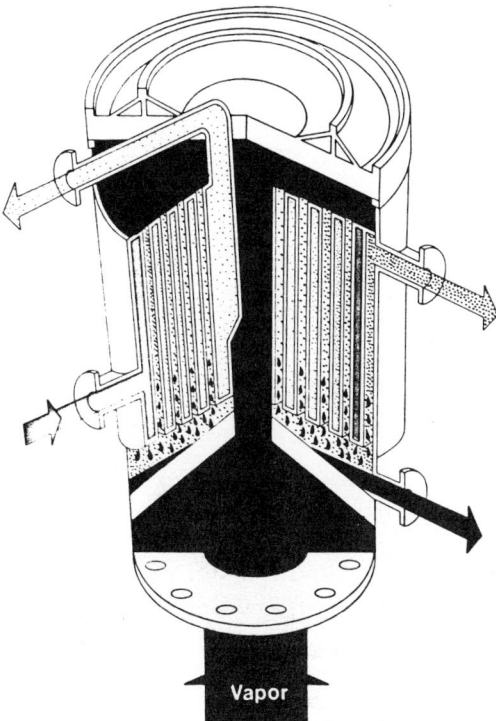

Fig. 4.11-9 Spiral plate exchanger used as an internal reflux condenser. (Courtesy of Alfa-Laval, Lund, Sweden.)

course the higher pressures being limited to the smaller sizes and lower-temperature levels of operation.

A modified configuration, indicated in Fig. 4.11-9, is often used as an internal reflux condenser in a distillation column. Because of the cylindrical symmetry, the unit fits compactly in the top of the column, eliminating the necessity to withdraw vapors from the column into a separate condenser. An annular trough may be arranged below the condenser in order to collect product condensate for withdrawal from the column.

Areas of Application

Spiral-plate exchangers do represent a relatively very compact heat transfer surface and have the advantage of allowing mechanical cleanability of both sides of the plate on removal of the corresponding face plates. Because of the spiral motion of the flow and the consequent secondary flows, spiral-plate exchangers are reported to show a lower tendency to foul in a given service than the corresponding shell-and-tube units. Again, the economic savings of this kind of exchanger become particularly manifest when alloy material construction is required.

Design Basis

Since this is a proprietary design, few design data have appeared in the literature. Hargis et al.[4] is probably the best available source. Estimates made in turbulent flow, using standard tube-side correlations and employing the equivalent diameter concept, will probably be on the conservative side for heat transfer and slightly nonconservative for pressure drop.

4.11-6 Plate Fin (Matrix) Heat Exchangers

Construction Features

The basic constructional elements of a plate fin heat exchanger are shown in Fig. 4.11-10. Basically the construction is a series of parallel plain sheets separated by layers of matrix material. The matrix material is formed from a sheet of aluminum, folded to the desired configuration, and perforated or

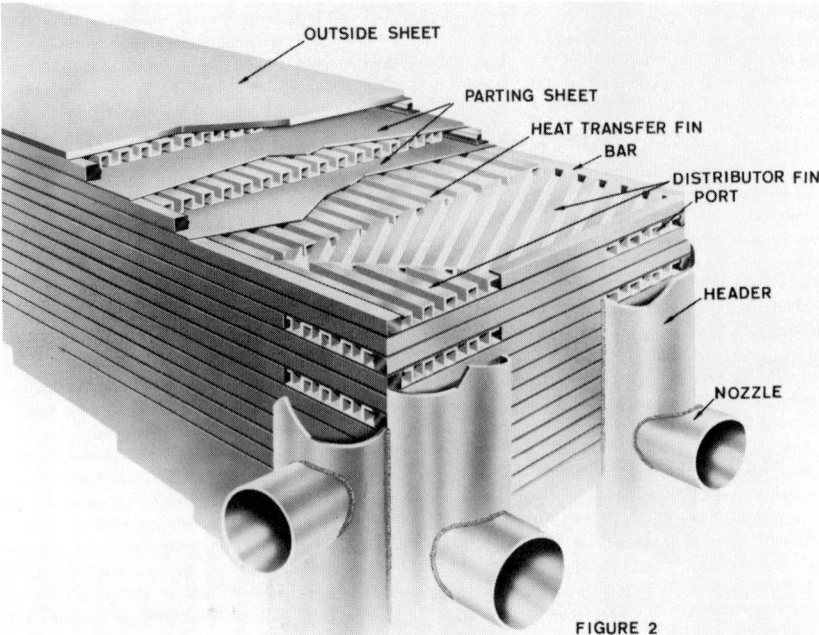

FIGURE 2

Fig. 4.11-10 Exploded view of a matrix heat exchanger. (Courtesy of The Trane Company, LaCrosse, WI.)

lanced on the sides of the folds in order to give flow enhancement elements on the surface. Alternate layers of parting sheets and matrix material are laid up, using side bars to seal the edges between parting sheets. The axis of the folding in a layer of the matrix material is in the direction of flow of that fluid. Once the desired configuration of the exchanger has been built up, the assembly is permanently brazed together, using either a salt bath or a furnace. The resulting block is fitted with flanges and manifolds to control the flow of fluids in and out of the various channels.

The matrix element has the dual purpose of supporting the parting plates and of providing heat transfer surface. By suitable manifolding, more than two streams can be and usually are incorporated in each matrix exchanger. The usual material of construction for cryogenic services is aluminum, but this basic configuration is available in a wide variety of materials using furnace brazing.

Because of the size of available salt baths, the largest salt-bath-brazed heat exchangers are approximately 40 in. on each side in cross section by approximately 20 ft long ($1 \times 1 \times 6$ m). The surface is extremely compact, totaling hundreds of square feet of heat transfer surface per cubic foot of volume. (It should be noted that because plate fin heat exchangers can handle more than two fluids, it is customary to include all heat transfer surface in contact with any fluid when referring to the heat transfer area of a given exchanger. This is in contrast to, e.g., a shell-and-tube heat exchanger, where it is customary to refer to the heat transfer area on the outside of all of the tubes in the exchanger as the total heat transfer area.)

Areas of Application

Because of their very high volumetric density of surface, plate fin heat exchangers are very desirable in critical heat transfer services where volume must be kept to a minimum. But their construction makes mechanical cleaning impossible, and they must ordinarily be used with nonfouling fluids. They can withstand quite high pressures, up to 1000 lb/in.^{2} (7 MPa) in the smaller sizes; those made of aluminum are suitable for use at very low temperatures. Accordingly, the major area of application of plate fin heat exchangers has been in cryogenic processes, where thermodynamic efficiency demands that very close temperature approaches be designed into the equipment. Examples of cryogenic processes in which they find application include liquified natural gas production, recovery of liquids from natural gas streams, recovery and purification of helium, and hydrogen purification. They are also used in gas-to-gas heat transfer.

Design Basis

The best generally available work on compact heat exchangers is Kays and London,[5] which includes some friction and *j*-factor curves for typical matrix configurations. Some extrapolation of these curves to other geometries using equivalent diameter concepts is possible. Limited data on two-phase flow and boiling heat transfer in these geometries is beginning to become available, primarily through the work of Robertson.[6]

A major practical problem in the design and application of plate fin heat exchangers is the need to secure nearly perfect distribution of the fluids among the hundreds, indeed thousands, of parallel-flow channels. This cannot be accomplished with a two-phase mixture, so conservative engineering design requires that only single-phase (gas/vapor or liquid) fluids be introduced into a given section of the matrix. If condensation or vaporization occurs in the matrix, it is necessary to separate the phases at the exit before once again introducing them into the channels of a plate fin exchanger. While this procedure imposes severe equipment limitations and thermodynamic penalties upon the process, it is the only basis upon which rational engineering design may be carried out.

Chemical cleaning of plate fin heat exchangers is possible but must be carried out with care. If salt-bath-brazed aluminum exchangers are used, they are in a dead soft condition and subject to attack by a wide variety of chemical agents, particularly alkaline solutions.

4.11-7   Tube-in-Plate Heat Exchangers

Construction Elements

A section of a typical tube-in-plate heat exchanger is shown in Fig. 4.11-11. It consists of a stack of plates, flat or molded, penetrated by a set of tubes or flattened channels. The tubes are attached to the plates by expansion, brazing, or galvanizing in order to assure mechanical rigidity and good thermal

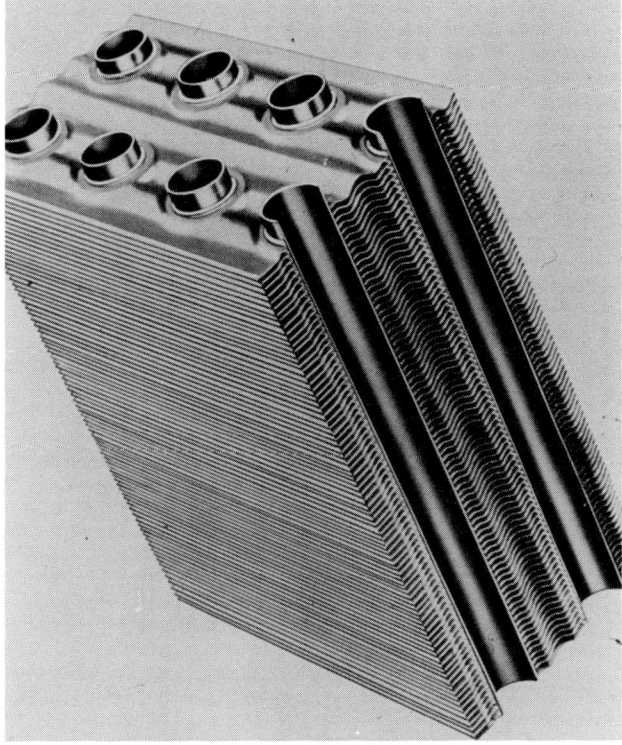

Fig. 4.11-11   Section of a typical tube-in-plate heat exchanger. (Courtesy of The Trane Company, LaCrosse, WI.)

contact. Tube-side headers or manifolds are attached to control the flow path through the tubes. These exchangers often have a fan to provide air flow across the fin surface. Various enhancements may be used inside the tubes to increase heat transfer effectiveness, and a variety of patterns of the plate fins are used for the same purpose.

Areas of Application

Tube-in-plate heat exchangers are used in many services where heat rejection to air is desired; such applications include (but are not limited to) automobile and truck radiators, refrigeration systems, and space heaters. They have application in any case in which the tube-side fluid has a significantly higher heat transfer coefficient than the plate-side fluid (which is usually air). Tube-side fluid is often jacket-cooling water from engines or condensing steam in the case of space heaters. Tube-side boiling or condensation of a refrigerant occurs in such applications as air-conditioning systems.

Design Basis

Analysis of the tube-side pressure drop and heat transfer performance is relatively straightforward, using correlations developed for in-tube flows. In many applications the heat transfer on the tube side is high enough that it constitutes a small part of the resistance to heat transfer and great accuracy is not required.

The flow and heat transfer characteristics on the plate are dependent on the configuration of the surface and are normally determined experimentally for each different configuration. Since many tube-in-plate heat exchangers are intended for mass production, the concern is less on a close a priori prediction of the performance than on optimizing the results by testing under operational conditions. Once an acceptable performance has been obtained, the task is to ensure through quality control techniques that each production unit has essentially the same performance characteristics as the prototype. Some estimate can be obtained of typical pressure drop and heat transfer characteristics using the correlations in Kays and London.[5]

4.11-8   Air-Cooled Heat Exchangers

Construction Features

The two basic configurations of air-cooled heat exchangers as used in the process and power industries are the forced-draft arrangement (Fig. 4.11-12) and the induced-draft arrangement (Fig. 4.11-13). (In the power industry these units are often called dry cooling towers.)

The essential heat transfer element is a shallow bank of high finned tubes (see below) connected to manifolds or headers at each end. Air is blown across the tube field in the forced-draft unit or drawn across in the induced-draft unit, using an axial flow fan in each case. Sheet metal plenums are provided to guide the air flow. A number of other construction elements, including louvers, preheat coils, fan guards, and so on, are provided to meet particular needs.

The dominant consideration in the construction of an air-cooled heat exchanger is the very low density of the air to be moved. In order to provide a sufficient mass flow rate to remove the heat, very large volumes of air must be moved, and the only economically feasible device to do this is the single-stage axial flow fan. These fans develop only very low pressure heads, so the flow path must be

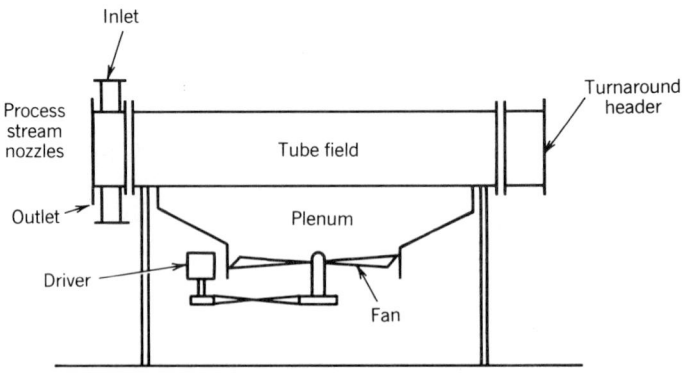

Fig. 4.11-12   Forced-draft air-cooled heat exchanger.

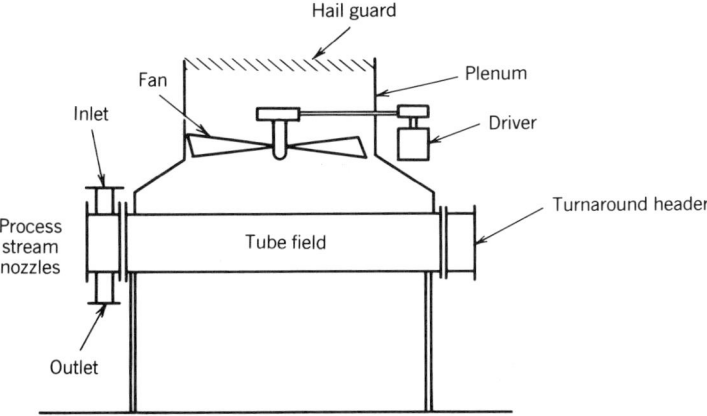

Fig. 4.11-13   Induced-draft air-cooled heat exchanger.

very short and the velocities must be low. Directly, this requires large cross-sectional face areas for flow and, indirectly, results in a very low heat transfer coefficient on the air side. Since the air-side heat transfer coefficient is usually one to two orders of magnitude lower than the tube-side coefficient, the use of extended surface—in this case high-finned tubes—on the air side is strongly indicated.

A number of ways of attaching fins to the outside of a tube are in common use and are illustrated in Fig. 4.11-14. The least expensive is the edge-wound or tension-wound fin in Fig. 4.11-14a. In this case the fins are wrapped under tension into a very shallow groove on the surface of the tube. This leaves the contact between fin and tube exposed to the atmosphere and subject to corrosion, which

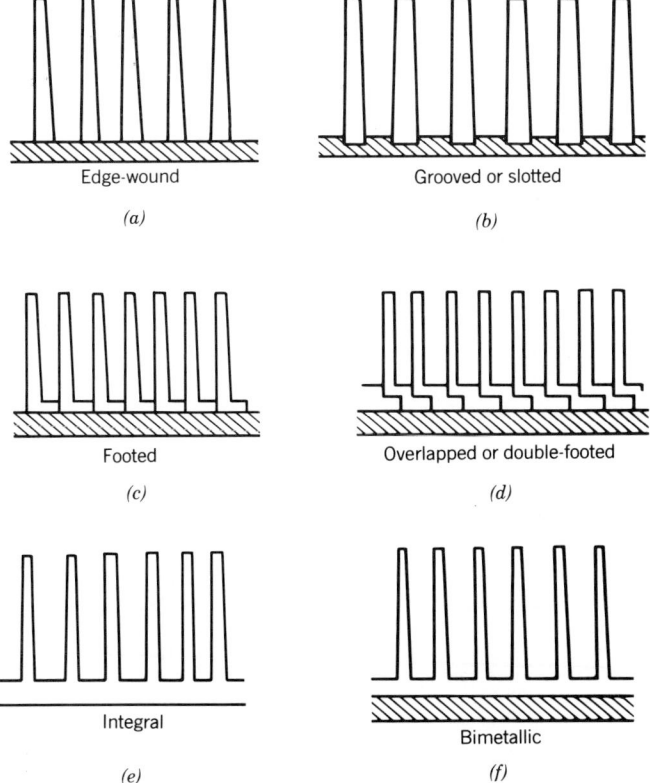

Fig. 4.11-14   Types of finning for tubes for air-cooled heat exchangers.

eventually results in deterioration of the thermal contact between the fin and the base tube. Depending on atmospheric conditions and the extent of thermal cycling, these tubes have relatively very short useful lifetimes.

Figure 4.11-14*b* shows a tube in which the fins have been inlaid into a fairly deep groove cut into the surface of the tube. This is a more expensive construction but results in the fin-to-tube contact area being better protected from atmospheric attack and subsequent deterioration. The L-footed tube shown in Fig. 4.11-14*c* is tension wound on the tube, and the foot provides a greater contact area for heat flow as well as protection against atmospheric attack. A variant of the L-footed tube is the knurled tube, in which the tube is longitudinally grooved before the fin is wrapped on; then a knurling tool forces the foot of the tube into the grooves. This tube is reported to have superior resistance to thermal cycling.

In Fig. 4.11-14*d* a double-footed or overlapped finned tube is shown. This provides further protection against penetration by the weather. An integral finned tube is shown in Fig. 4.11-14*e*. In this tube the fins are formed directly from the metal tube wall. This construction obviously eliminates the contact resistance problem, but it can only be produced from certain aluminum and copper alloys. If these materials are not suitable for corrosion resistance to the tube-side fluid, one may use the bimetallic tube shown in Fig. 4.11-14*f*. This tube is formed from a thick-walled muff tube and a liner tube that may be of any material required for corrosion resistance or strength. At the same time that the fins are raised from the muff tube, the tube is being swaged onto the liner tube with a substantial compressive stress. This construction eliminates atmospheric attack and meets the requirement for corrosion resistance on the tube side but introduces a possible substantial contact resistance at the bimetallic interface. Since the muff tube is of aluminum and has a thermal expansion coefficient greater than almost any other engineering metal, heating the tube results in a decrease in the compressive stress at the interface. At a certain temperature this stress is reduced to zero and the muff tube begins to grow away from the liner tube. This is the upper temperature limit of application of this tube. Repeated thermal cycling can have the same effect at even lower temperature levels.

While forced-flow and induced-flow units are used in about equal numbers, each has very specific advantages and disadvantages. The forced-flow arrangement has the following advantages:

1. The fan is in cool air, increasing the mass flow efficiency of the fan and possibly permitting the use of less expensive fan materials.
2. The fan and driver are more accessible for maintenance.

The induced-flow arrangement has the following advantages:

1. Exhaust air comes off of the heat exchanger at a much higher velocity (typically $2-2\frac{1}{2}$ times as high) and therefore penetrates more deeply into the surrounding atmosphere, reducing the possibility of significant recirculation of warm air.
2. The air flow across the tube field is more uniform.
3. The plenum protects the tube field against rain and hail, though a hail guard is needed to protect the fan. (If a rain storm does occur and the tube surfaces get wet, the effective sink temperature of the heat exchanger changes suddenly from the dry-bulb temperature of the air to near the wet-bulb temperature. In a dry climate this can amount to a sudden temperature change of 30–40°F and may cause a significant upset in plant process conditions.)
4. In the induced-draft arrangement trash is less likely to be ingested into the fan and tube field.

A hybrid configuration uses an induced-draft fan with the driver below the tube field. The driver is more readily accessible for maintenance, but the shaft penetrates the tube field; and unless this penetration is blocked, there will be a substantial internal bypass stream effect that can lead to deteriorated performance.

Areas of Application

Air-cooled heat exchangers are extensively used where air is the only available or economic medium for cooling. A variant on that theme is to heat air passing over the fin tubes with a hot fluid or condensing steam inside the tubes. This air is commonly used either for space heating or for drying processes.

The problem of freeze-up in cold climates is a very serious one, especially for steam condensers, because the ice phase formed can physically damage the tube. A variety of techniques is available to solve this problem, and the exact one to be used is dependent on the severity of the problem. For the least severe case variable-pitch or variable-speed fans may be used to reduce the air flow to a lower-than-designed value. The fans may be stopped altogether, leaving the heat exchanger to reject heat by natural convection. In multifan systems some of the fans may be reversed in order to deliberately recirculate air. Louvers are often used to control air flow with constant-speed fans. In sensible cooling services the flow may be reversed and directed through the heat exchanger cocurrent to the air flow. Heavy walled air preheat tubes may be employed underneath the bundle to preheat the incoming air. In extreme cases the air-cooled exchanger may be built into a houselike structure with controlled recirculation of the hot air tempered by a controlled amount of incoming cold air to give the desired air inlet temperature.

Basis of Design

The design and operational characteristics of an air-cooled heat exchanger are constrained by three limits.

    1. Fan limit. The single-stage axial flow fan is capable of producing pressure rises of only a few inches of water. Air-cooled exchangers are usually limited to an overall pressure drop of less than 1 in. of water (about 0.036 lb/in.^{2}, or about 250 Pa). A common design figure is $\frac{1}{2}$ in. of water. This pressure drop corresponds to face air velocities of about the magnitude shown in Table 4.11-1.

    2. Thermodynamic limit. A sufficient mass flow rate of air must be provided in order to remove the total amount of heat to be rejected. This mass flow rate of air is given by the equation

$$\dot{m} = \frac{q}{c_p(t_2 - t_1)} \tag{4.11-1}$$

where $\dot{m}$ = mass flow rate of air, lb/h

      q = total amount of heat to be removed, Btu/h

      c_p = specific heat of air, 0.24–0.25 Btu/lb·°F, depending on the temperature and moisture content of the air

    t_1, t_2 = inlet and outlet air temperatures respectively, °F.

    3. Rate limit. Because of the limited range of air velocities and tube geometries encountered in practice, and because the air side is usually the dominant resistance to heat transfer in air-cooled exchangers, overall heat transfer coefficients also tend to fall into a very narrow range of values. Typical values of the air-side transfer coefficient as a function of face velocity are given in Table 4.11-2.

    The overall coefficient is built up from the individual coefficients and resistances in Eq. (4.11-2):

$$U_o = \frac{1}{\dfrac{1}{h_0} + R_{f_o} + R_{\text{fin}} + \dfrac{\Delta x}{k_w}\dfrac{A_o}{A_m} + \left(R_{f_i} + \dfrac{1}{h_i}\right)\dfrac{A_o}{A_i}} \tag{4.11-2}$$

where R_{f_o} is the fouling resistance on the air side and is usually taken as zero and R_{fin} is the "fin resistance," which takes into account the fin geometry and efficiency but is essentially independent of the outside coefficient. For estimation purposes, R_{fin} is between about 2×10^{-3} h·ft^2·°F/Btu for low thick fins and 1×10^{-2} h·ft^2·°F/Btu for high thin fins (3.5×10^{-4} to 1.8×10^{-3} m^2·K/W). In any case, it usually constitutes no more than about 5% of the overall resistance. Δx and k_w are tube wall thickness and thermal conductivity and A_m is the logarithmic mean wall heat transfer area for the

TABLE 4.11-1 TYPICAL FACE VELOCITIES FOR AIR-COOLED EXCHANGERS

No. of Tube Rows	Face Velocity[a]	
	ft/min	m/s
3	750–900	3.8–4.5
4	650–800	3.3–4.0
5	550–700	2.8–3.5
6	500–600	2.5–3.0
8	425–500	2.1–2.5
10	370–450	1.8–2.2
12	330–380	1.6–1.9

[a]Average velocity of the air flowing across the entire face area of the heat exchanger.

TABLE 4.11-2   TYPICAL AIR-SIDE HEAT TRANSFER COEFFICIENTS

Face Velocity		$^a h_o$	
ft/min	m/s	Btu/h · ft² · °F	W/m² · K
900	4.6	10–12	57–68
800	4.1	9.5–11	54–62
700	3.6	9–10.5	51–60
600	3.0	8–9	45–51
500	2.5	7–8.5	40–48
400	2.0	6–7	34–40
300	1.5	5–6	38–34
250	1.3	4.5–5	26–28

$^a h_o$ is based on the total outside area of the tube including fins.

tube without fins:

$$A_m = \frac{\pi L (d_r - d_i)}{\ln(d_r/d_i)} \approx \tfrac{1}{2}\pi L (d_r + d_i) \tag{4.11-3}$$

Here d_r is the root diameter of the tube, that is, the outside diameter of the tube without fins, and d_i is the inside diameter. R_{f_i} and h_i are the inside fouling resistance and heat transfer coefficient, respectively.

The total heat transferred is then given by

$$q = U_o A_o F(\text{LMTD}) \tag{4.11-4}$$

where LMTD is calculated as if the fluids were in countercurrent flow:

$$\text{LMTD} = \frac{(T_1 - t_2) - (T_2 - t_1)}{\ln[(T_1 - t_2)/(T_2 - t_1)]} \tag{4.11-5}$$

where T_1 and T_2 are the inlet and outlet process fluid temperatures and F is the configuration correction factor for the cross-flow arrangement. Graphs of F for the various arrangements can be found in Schlunder[1] and Kern[2] but will usually be between 0.85 and 1.00 for most air-cooler applications.

In any given problem the approximate range of the design can be quickly calculated from the above equations (taking U_o between 60 and 80% of h_o for an estimate). (For more detailed calculation procedures see Schlunder.[1])

4.11-9   Heavy-Duty Extended Surface

Construction Features

Heavy-duty extended surface is used for heat transfer to and from gases at high temperatures and/or under corrosive conditions. The finned tubes described in the previous section cannot withstand these conditions (indeed many of the applications are above the melting point of aluminum) and a more rugged construction is required. Typical tubes for this service are shown in Fig. 4.11-15. The raised surface elements, whether they are fins or studs, are welded to the surface of the tube. Fin spacing varies from about 4 per inch to less than 1 per inch and the fins may be as much as $\frac{1}{8}$ or $\frac{3}{16}$ in. in thickness. These tubes are available in various steels and high-temperature alloys. The tubes are ordinarily plain on the inside, though certain types of enhancement can be used. The tubes can be connected by welded U bends or by welding into manifolds. Banks of these tubes are often mounted across flue gas ducts for heat recovery service.

Areas of Application

Heavy-duty enhanced surface is used wherever it is necessary to transfer heat to or from a high-temperature gas. Typical examples include air preheaters using flue gas as the heat source, waste heat coils in which the tube-side fluid is high-pressure water or boiling water, chemical reaction

(a)

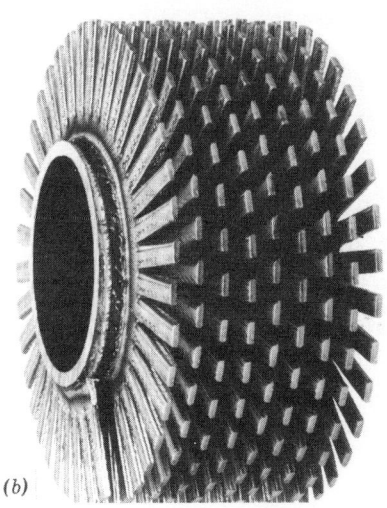

(b)

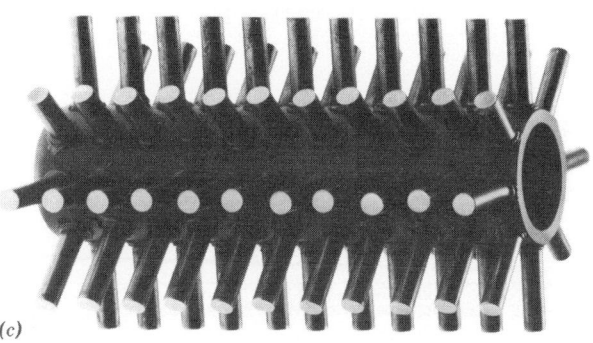

(c)

Fig. 4.11-15 Typical heavy-duty extended surfaces: (*a*) continuous fin, (*b*) cut or segmented fin, and (*c*) stud fin. (Courtesy of Escoa Fintube Corp., Pryor, OK.)

sections ("cracking sections") in process furnaces, and so on. In some applications the finned surface is exposed to flame as well as a flow of a hot gas; these are commonly called radiant sections and involve special heat transfer considerations on the flame side. Pressure drops during flow across banks of these tubes are almost always very limited; some applications are purely by natural convection on the gas side.

Design Considerations

Heat transfer and pressure drop in banks of heavy-duty finned tubes are calculated mainly through specialized and generally proprietary correlations. Because of the low pressure drops generally available, velocities and heat transfer coefficients are correspondingly low. Each installation is likely to be custom designed and constructed. Design in the presence of both convective and radiant contributions to the heat load is especially difficult and frequently requires the use of highly specialized computer programs. Vendors should be generally consulted for detailed design calculations.

4.11-10   Heat Pipes

Construction Features

The diagram of the heat pipe principle is shown in Fig. 4.11-16. It is basically a pipe with sealed ends containing a fluid that will vaporize in the desired temperature and pressure range, with provision made for the flow of the condensed liquid back to the heat addition zone of the pipe by surface tension or capillary action. The heat pipe operates in the following fashion. The liquid vaporizes at the heat addition end of the pipe, and the vapor flows to the heat removal or condensing section of the pipe. The wall is covered with a wick or grooved longitudinally, and the condensed liquid moves by capillary action through the wick or the grooves back to the heat addition end of the pipe. Depending on the temperature range to be covered, a variety of substances may be used as the working fluid; these include the common refrigerants at low temperatures, water and the organic transfer media in the middle range of temperatures, and liquid metals at very high temperatures. For a pure working substance in a heat pipe designed so that there is a very small pressure gradient required to cause a vapor flow, the two ends of the pipe are at very nearly the same temperature. This has the effect of causing the transport of large quantities of heat in the form of latent heat of vaporization of the working fluid under a very small temperature difference. If one defines the effective thermal conductivity of the heat pipe to be the quantity of heat transported divided by the cross-sectional area for flow and the temperature difference, the value is typically many times higher than even the best solid conductors of heat. Various surface treatments can be used in the vaporization and condensation sections of the heat pipe to give large heat transfer fluxes with very small temperature differences.

Area of Application

The heat pipe provides a means for transferring very large quantities of heat with a very small temperature difference. The distances between the two ends can be quite considerable, 5–10 m in some specialized applications. Therefore, heat pipes can be used wherever there is a desire to transfer heat from one medium to another some distance away.

One example of such an application is the transfer of heat from a hot flue gas into the incoming combustion air for a furnace. The heat pipe in this case would probably have finned surfaces on the outside because of the low heat transfer coefficient of the gases involved. In this case a bank of heat pipes would be used to provide the necessary heat transfer area between the two streams.

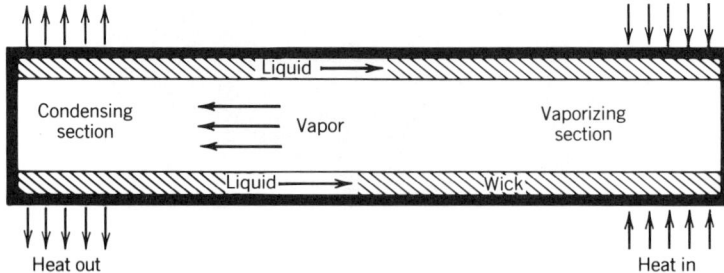

Fig. 4.11-16   Essential elements of a heat pipe.

The heat pipe is a relatively new invention and undoubtedly its greatest areas of usefulness are still to be developed.

Design Considerations

A large number of papers have been written exploring the basic flow and heat transfer mechanisms in a heat pipe. Because of the wide variety of geometries that can be used for the vaporizing, condensing, and wick sections, no general design procedure can be given at this time. However, Section 3.10 in Schlunder[1] contains an exhaustive summary of the available literature. Again, essentially every application is custom designed at this time.

4.11-11   Rotary Heat Recovery Devices

Construction Features

All of the types of heat exchangers discussed to this point have been stationary, at least as far as the heat transfer surface is concerned. However, there are two general types in which the surface is mechanically driven through the fluid. The first of these is usually termed *regenerative heat exchanger* and is typified by the Ljungstrom air preheater (see Fig. 4.11-17). The exchanger is placed so that the hot flue gas passes through one side of the heat transfer matrix and the incoming air to be preheated passes (ordinarily countercurrently) through the other side. The matrix revolves continuously, first receiving and storing heat from the hot gas and then, in the other part of the rotation, giving up the stored heat to the cold air.

No heat is transferred through the surface from one fluid to the other–only into and then out of the metal making up the heat transfer surface. The temperature of the metal matrix rises rapidly as it first contacts the hot flue gas and then more slowly as its own temperature rises and the heat diffuses into the interior of the metal. The process is reversed on the cold side. Therefore, the design of the matrix is concerned not only with optimizing the convection heat transfer between gas and surface but also in providing a corresponding heat storage capability. Both metal and ceramic matrices are used.

While these devices operate at pressures on both streams very close to atmospheric pressure, a certain amount of leakage between the two streams does occur and is potentially very detrimental to

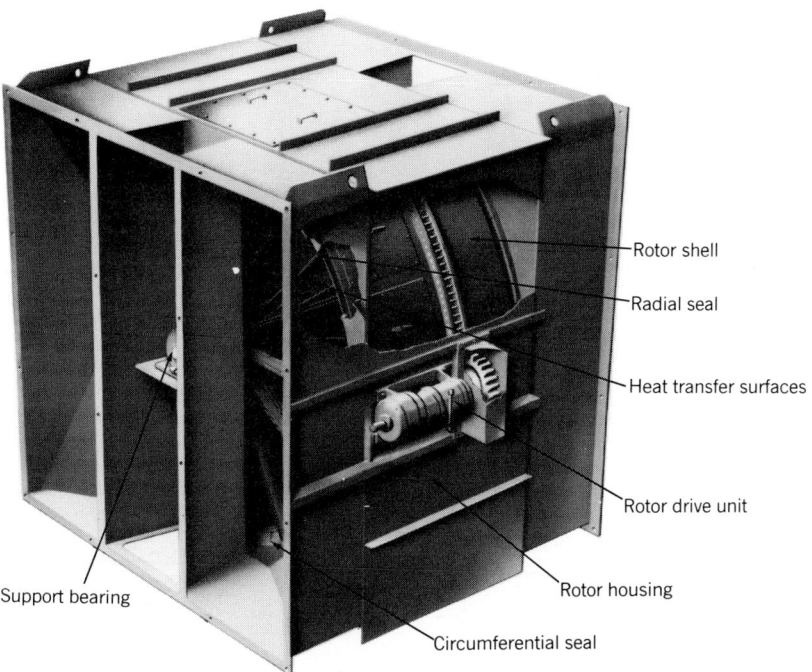

Fig. 4.11-17   Ljungstrom regenerative air preheater. (Courtesy of C-E Air Preheater, Wellsville, NY.)

the efficiency of the system. Therefore, great attention is paid to the proper design and subsequent maintenance of the radial and peripheral seals between the rotor and the housing. For the same reason, and for mechanical support, the rotor is divided into a number of radial sections. The rotor is usually driven by an electric motor and a circumferential gear drive. All of the components need to be designed to withstand high temperatures and severe temperature gradients.

Areas of Application

While the majority of applications of the rotary regenerative type of exchanger has been in conventional and gas turbine power plants, the principle is applicable any time heat needs to be transferred between two gas streams at essentially the same pressure. An attractive area of application is to high-temperature process furnaces (glass plants, foundries, cement kilns, etc.), but there have been difficulties with gas side fouling and development of matrix materials that withstand very high temperatures and temperature gradients and are not excessively brittle.

In a different temperature range the same principle has been used for heat recovery in building environmental systems.

Design Considerations

The regenerative heat exchanger poses design problems in addition to those considered previously. Under ideal steady-state plant operating conditions the temperature at any given fixed *spatial* point in the exchanger remains constant. However, any given element of the surface is cyclically heated and cooled, and the transient conduction problem in the matrix element needs to be analyzed. The problem of convection coupled with transient conduction, suitably integrated over the entire surface, is a formidable one mathematically. Schmidt and Willmott[7] treat the problem in detail as well as providing correlations for convective heat transfer and pressure drop.

4.11-12   Mechanically Aided Heat Exchangers

Agitators and Related Devices

There is a wide variety of heat exchangers that use external mechanical power to enhance the heat transfer process. Broadly they may be divided into two types: (1) tanks containing agitators such as propellors, which act on the bulk fluid away from the heat transfer surface and (2) vessels containing rotating blades that disrupt the fluid in the immediate vicinity of the surface. This section considers the first type, which is schematically depicted in Fig. 4.11-18a.

Stirred or agitated tanks are used in a variety of industries, and there is often a need to add or remove heat to the fluid contained. The heat transfer surface is most conveniently provided in the

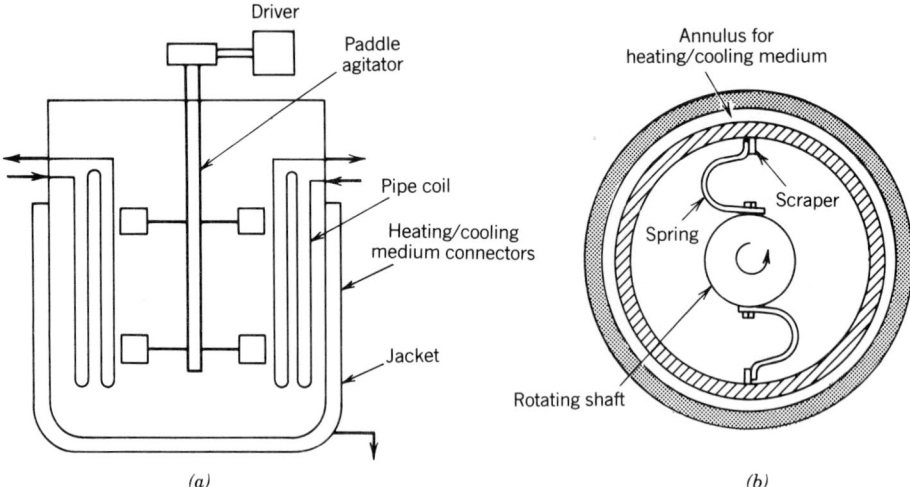

Fig. 4.11-18 Schematic diagrams of two types of mechanically aided heat transfer equipment: (*a*) stirred tank, with a paddle agitator and internal pipe coils and a jacket for heat transfer and (*b*) sectional view of a close-clearance heat exchanger, with spring-loaded blades.

form of a jacket on the outside of the tank with the heating or cooling service fluid circulated through it. Often, however, not enough heat can be transferred this way, and the internal heat transfer surface is provided in the form of pipe coils or plates with internal channels for the service fluid. Occasionally, the heat transfer surface may be incorporated in the agitator itself, though this poses obvious mechanical problems with rotating seals.

The agitator may assume various configurations: propellors, turbines, paddles, or helical flights. The purpose is to provide mixing of the fluid (particularly if a chemical reaction is taking place simultaneously) as well as enhanced turbulence or boundary layer disruption on the heat transfer surface. This mixing and heat transfer improvement is at the cost of increased power consumption, so some balance must be achieved between benefits and costs. Also, the energy expended in agitating the fluid shows up as heat added to the fluid; therefore, if the purpose is to cool the fluid, care must be taken not to overagitate.

Given the possible range of geometrical variables, fluid properties, and heat removal/addition requirements (which may vary with time in the case of a chemical reactor), it is hardly surprising that there is no universal correlation for heat transfer and power requirements in this kind of equipment. The more common cases have been studied experimentally and correlations of some generality developed for these. Schlunder[1] surveys the problem in his Section 3.14 while Uhl and Gray[8] and Nagata[9] provide extended discussions.

Close-Clearance Equipment

In close-clearance mechanically aided heat transfer equipment a rotating blade is in contact with, or very close to, the heat transfer surface, so that the fluid at the surface is very strongly sheared. By disrupting the temperature profile at the surface as rapidly as it can be built up, the heat transfer process is significantly improved compared to the fully developed laminar conduction–convection mechanism that would otherwise exist. This equipment is also used in cases where the fluid tends to deposit a solid phase, usually poorly conducting, on the surface; examples include wax in lubricating oil plants and crystals in fractional crystallization processes. In this case the blade physically limits the thickness of the solid layer.

Figure 4.11-18b illustrates a case in which the blade is held against the wall by a compressed spring. In such cases the blade hardness is chosen so that it is sacrificial to the wall, and the blade is regularly replaced. "Hydrodynamic" blades are also used; they pivot from the rotating arm, so that centrifugal force swings the blade outwards, balanced against the hydrodynamic forces of the fluid being sheared (similar to a slipper bearing). A properly designed blade never touches the wall, so blade wear is sharply reduced (but so is the shearing action of the blade on any deposit).

Close-clearance equipment is very expensive per square foot of heat transfer surface, and it is expensive to operate. Outages due to mechanical failure are frequent and maintenance charges high. However, it is the only equipment that will serve for certain applications. Custom design, with a generous contribution of experience, is the rule for close-clearance equipment (see Section 3.14 of Schlunder[1] for helpful information).

REFERENCES

4.11-1 E. U. Schlunder (ed.), *Heat Exchanger Design Handbook*, Hemisphere, Washington, D.C., 1983.

4.11-2 D. Q. Kern, *Process Heat Transfer*, McGraw-Hill, New York, 1950.

4.11-3 D. Q. Kern and A. D. Kraus, *Extended Surface Heat Transfer*, McGraw-Hill, New York, 1972.

4.11-4 A. M. Hargis, A. T. Beckmann, and J. J. Loiacono, "Applications of Spiral Plate Heat Exchangers," *Chem. Eng. Prog.* **63**(7), 62–67 (1967).

4.11-5 W. M. Kays and A. L. London, *Compact Heat Exchangers*, McGraw-Hill, New York, 1958.

4.11-6 J. M. Robertson, "Boiling Heat Transfer with Liquid Nitrogen in Brazed-Aluminum Plate-Fin Heat Exchangers", *AIChE Symposium Series* **75**(189), 151–159 (1979).

4.11-7 F. W. Schmidt and A. J. Willmott, *Thermal Energy Storage and Regeneration*, Hemisphere, Washington D.C., 1981.

4.11-8 V. W. Uhl and J. B. Gray, *Mixing Theory and Practice*, Academic Press, New York, 1967.

4.11-9 S. Nagata, *Mixing: Principles and Applications*, Wiley, New York, 1975.

CHAPTER 5
PRIME MOVERS

HARLAN T. HOLMES

Garver and Garver, Inc.
Little Rock, Arkansas

THOMAS E. STOTT

Thomas Stott and Associates
Cummaquid, Massachusetts

CHI CHENG

General Motors Corporation
La Grange, Illinois

GORDON C. OATES

University of Washington
Seattle, Washington

5.1 STEAM TURBINES

Harlan T. Holmes

5.1-1 Classification

A steam turbine is a heat engine that transforms heat into kinetic energy. It can be classified according to arrangement of steam flow (stage type, number of parallel paths, and flow related to plane of shaft), speed, inlet and exhaust steam conditions, and application. The steam turbine has been adapted commercially in many ways from mechanical drives to the major electric generation station prime movers. They are designed to accept steam up to supercritical (above 3200 psig) pressures heated to the 1000°F range, to allow steam to be bled from its internal passages for various process uses or for boiler recycling for reheating, and to exhaust into systems at various pressures from a near-perfect vacuum to a set or variable-pressure steam process requirement. Turbine design may fit specific or general operational conditions varying from steady state to load peaking including quick starts. Designs of turbine internal components are optimized to produce efficient conversion of heat energy to mechanical motion at desired operation ranges. Component shapes are sensitive to steam flow streamlining for high performance. This is accomplished by optimizing the relative wheel-rotating

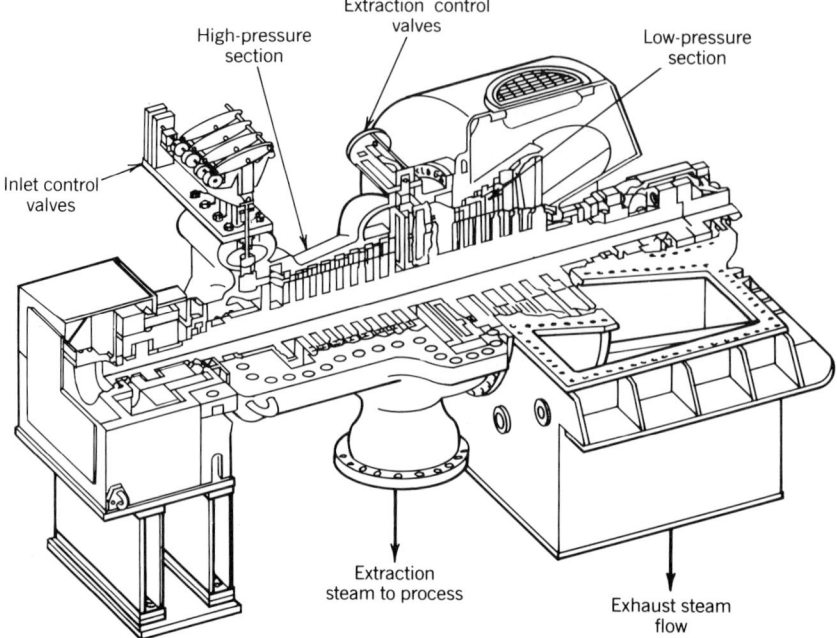

Fig. 5.1-1 Typical single automatic extraction steam turbine. (Courtesy of General Electric Company.)

velocities and steam flow velocities at various shaft speeds, loads, and steam flow path conditions (cleanliness, minimum leakages, and wear). In addition, designs must accommodate desired loading, thermal expansion, maintenance procedures, safety interlocks and trips, and operating conveniences (monitoring, automatic control, handling of limited emergencies). A typical steam turbine is shown in Fig. 5.1-1.

Energy Conversion

In the process of energy conversion steam exchanges pressure and temperature characteristics into velocity. Like the windmill, the moving steam stream acts on blades attached to a shaft, causing rotation. This results in a reduction of steam velocity, at which time the cycle repeats, starting with another exchange of pressure and temperature for velocity. The exchange has limits. Velocities must stay below critical speeds to provide control, to protect materials, and to set velocity ratios (rotating blade to steam) that maximize energy conversion efficiency.

In the larger turbines many of these cycles are in series. Each cycle stage is made to occur by routing the steam through physical structures referred to as nozzles (velocity increasers) and blades or buckets (energy form exchangers). Either the nozzles or blades can be the rotating element. If the nozzles move, the turbine stage is a reaction stage. If the blades move due to the impulsive forces of high-velocity steam, it is an impulse stage. A turbine shaft may combine the principles of reaction and impulse stages to provide the design that best optimizes steam conditions, performance, loading requirements, and turbine physical size. The reaction and impulse principles are illustrated in Fig. 5.1-2. The spherical vessel in the reaction principle generates steam under pressure, which exits the nozzle to a lower pressure and establishes a steam velocity. In reaction to this jet, motion occurs. Motion by impulse results when the jet strikes the blades on the rotating wheel. As motion occurs, the steam velocity reduces. If the steam could be gathered at the lower pressure into a similar vessel, it could exit through additional nozzles to an even lower pressure and cause another rotation force. By coupling the rotating forces to the same shaft, a multistage turbine is formed. Figure 5.1-3 illustrates this symbolically as it is done in an actual turbine. In Fig. 5.1-3a a stationary nozzle causes a velocity increase that impinges on the rotating blade. Figure 5.1-3b shows the moving part to be the nozzle moving in response to the reaction of the jet. A general steam pressure and velocity trace through the series of stages is shown, indicating points at which the pressure reduction causes a velocity increase and where work release causes a velocity reduction. With reduction in pressure and temperature,

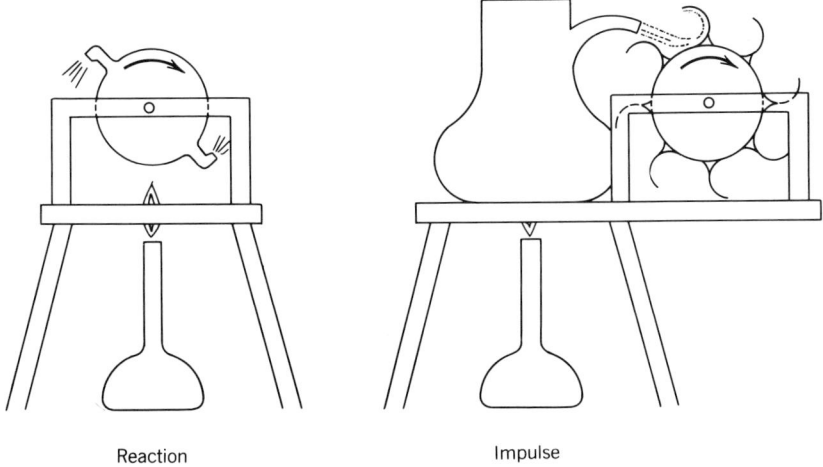

Reaction   Impulse

Fig. 5.1-2 Heat energy to kinetic energy.

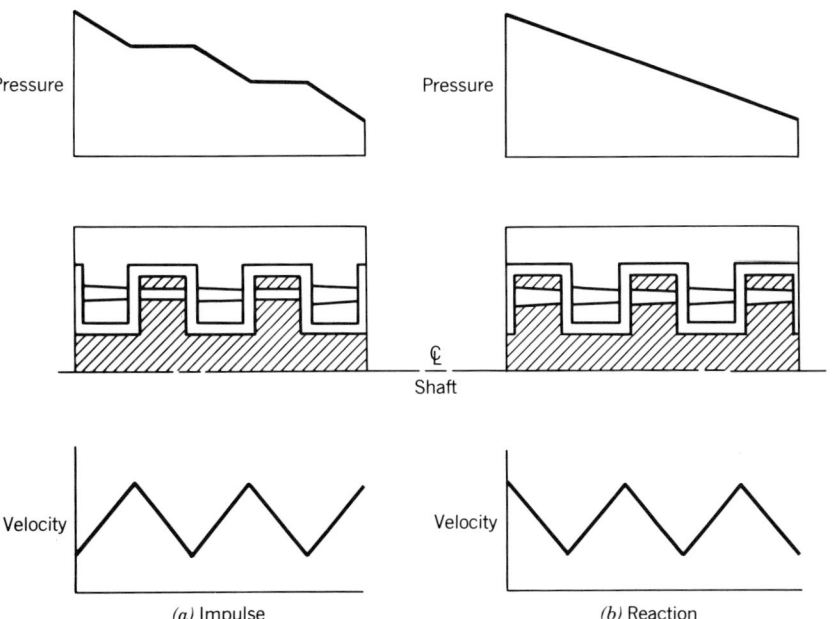

(a) Impulse   *(b)* Reaction

Fig. 5.1-3 Steam turbine; multistage steam conditions.

steam volume is increased, requiring an enlargement of path openings and turbine physical size. An end view of blading shapes (Fig. 5.1-4) shows the rotation forces from a different angle. Each stage consists of a stationary row followed by a moving row. The moving row is acting as nozzles in the reaction turbine. In the impulse turbine the stationary row acts as a nozzle and the rotating row (buckets) as a turning vane. In the impulse turbine, if the blades turn the steam jet a full 180°, about twice the force is exerted on the blade compared to the blade turning the steam only 30°. In one small, single-wheel mechanical drive turbine (Terry turbine), steam is directed toward reversing chambers carved into the turbine wheel periphery. Steam flows (helical pattern) through several reversing chambers as the wheel turns exhausting steam from a port several radial degrees away in the rotating directions.

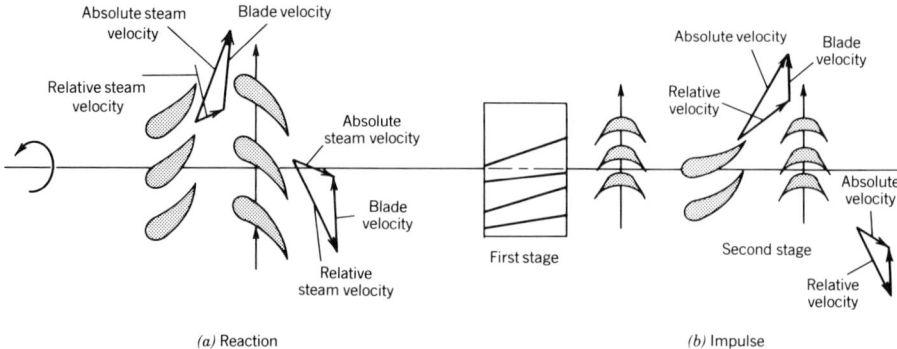

(a) Reaction   *(b)* Impulse

Fig. 5.1-4   Steam turbine-steam velocity vectors.

Stage Efficiency

Turbine stage heat conversion efficiency varies with the ratio of blade speed to theoretical steam velocity at the blade (Fig. 5.1-5). The reaction stage approaches 90% efficiency as blade speed approaches about 70% of theoretical steam velocity (223.8 times the square root isentropic enthalpy drop). The one-row impulse stage approaches a maximum efficiency of 86% at a ratio of 0.45. For equal shaft speeds the stage enthalpy drop or pressure drop must be lower for reaction blading than for impulse blading to maintain maximum efficiency. The impulse turbine, which can work with higher steam velocities and pressure drops, has an advantage in being able to utilize steam with small specific volumes (high pressure drops) in the high-pressure end or where high steam pressure drop per stage or stage grouping (multirow) is required.

Shaft Speed

In case of turbines driving electric generators, shaft (blade) speed is a constant—established by the electrical frequency requirements. Designers of turbines driving variable-speed equipment (pumps, fans, etc.) must deal with the varying speed of both steam and blades. In both types of turbines blade speed often establishes velocity of steam at the stage nozzle exit, which in turn establishes nozzle pressure drop and steam conditions. For a single stage the variables (with possible maximum limits) may be initial steam conditions, shaft rpm, or wheel diameter. The pressure diagrams (Fig. 5.1-3) show the pressure dropping gradually over several stages. In order to use high (supercritical) steam pressures, combinations of various types of stages are used. A few impulse stages preceding a series of reaction stages would allow for a large initial pressure reduction to help solve the turbine physical length problem caused by the need for many pressure-reducing stages.

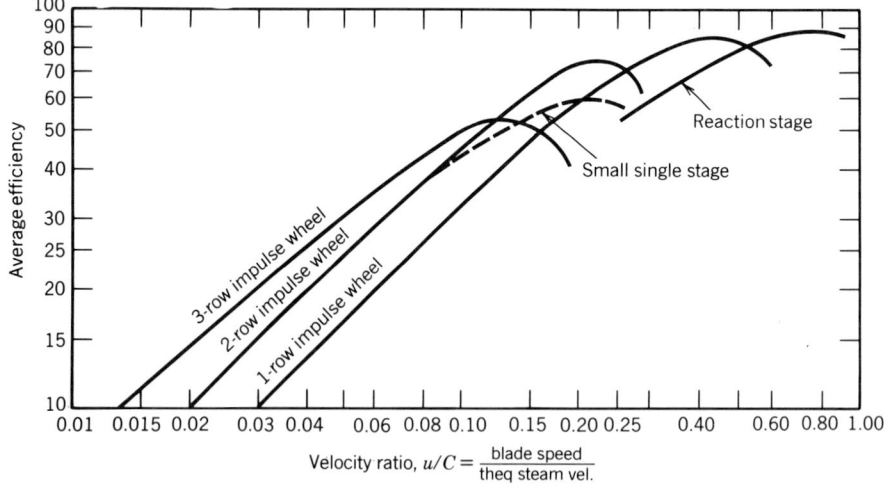

Fig. 5.1-5   Average efficiency of turbine stages. (Courtesy of Transamerica De Laval, Lawrenceville, NJ.)

Turbine Efficiency

Figure 5.1-4 indicates the velocity vector relationships of moving blades and steam flow. The ratios of mean blade speed to steam speed, along with the steam approach and exit angles to plane of blade rotation, are the bases of turbine blade design. The vectors show that the reaction turbine ratio might approach unity (blade velocity about equal to steam velocity). In practice, the ratio varies between 0.6 and 0.85 with the smaller ratio applied to smaller, less efficient turbines and the higher ratio to the large utility plant turbines. Impulse turbine vectors show a smaller blade velocity relative to steam velocity. The best values of speed ratio (impulse) are about one-half that of the reaction stage (about 0.3–0.45). This produces a stage efficiency of about 85% for the single-row impulse stage. Reference is made to Fig. 5.1-5 as an illustration of the variation of efficiency of different-type turbines.

In comparison, the ideal work per pound of steam flowing in the impulse stage is twice that of the reactive stage at the same blade speed. However, the enthalpy drops are about equal if the whole stage (nozzle and rotating blade) performance is considered. For this given overall enthalpy drop per stage, the reaction stage would have a speed equal to the square root of two times faster than the impulse stage. Under the equal speed and work (enthalpy drop) conditions, the ideal reaction turbine would have twice as many stages as the impulse turbine and a steam velocity lower by a ratio of the reciprocal of the square root of 2. Friction losses are less in the reaction stages due to the lower velocity. A significant design difference is the approximately zero pressure drop across the impulse turbine's moving blades as compared to the reaction blading. Sealing the impulse pressure chambers are by "stage" pressure drop rather than by separate nozzles and blades, thus making steam sealing easier in the impulse turbine.

During reduced load conditions, steam may flow over only part of the total periphery of the stage. In the idle sections moving blades churn stagnant steam, thus increasing windage loss. The reaction turbine needs a steam flow in all parts of its periphery, at all loads, while in the impulse turbine (moving bucket-zero pressure drop) only a part of the initial stage in operation is needed to move steam uniformly at the low loads. The impulse turbine, then, is more efficient at the small sizes with the differences in performance decreasing at larger sizes. Most small turbines are the impulse type.

In the large, multistage impulse turbine, partial-load performance is improved with partial admission (few nozzles keeping steam velocities high, velocity ratios lower, and efficiencies higher). An impulse turbine operating on high-pressure–high-temperature steam (about 450 Btu/lb available enthalpy differential) and exhausting to a vacuum would require about 20 single-row stages to hold high efficiency.

Up to a point, turbine efficiency improves with higher blade speeds and/or lower steam velocities. High blade speeds are obtained by increasing shaft rpm or blade row diameter. However, these increases quickly bring on strength problems and rotation loss (varies with fifth-power-of-diameter). To optimize a turbine to a low-speed pump, speed reduction gearing is used to ensure high blade velocities (speed) at a minimum wheel diameter.

To obtain maximum conversion to work, a streamlined (relative) steam flow must occur. At both entrance and exit the steam should move into blading parallel with the openings. This condition is influenced by turbine speed and loading and by condition of turbine (blade deposits or metal loss), clearances between moving and stationary parts with accompanying interference of leakages and leakage reentra, steam stagnation (windage), steam quality (water droplets), and the influences of any extractions of steam.

Expansion of steam occurs throughout the turbine. As pressure and temperature are exchanged for work, steam volume could increase by a factor of 1000 (vacuum exhaust). To allow for this and to hold reasonable steam velocities, passages are enlarged by lengthening last-stage blades to about 30 in. in height (condensing turbine). These blades, installed on a 4- to 5-ft shaft wheel would make the last-stage wheel diameter about 10 ft. When wheel size begins to cause peripheral speeds to approach the speed of sound, designs require dual expansion (steam exhaust through more than one port) using multiple low-pressure turbine sections.

In the reaction turbine pressure drop (causing expansion) occurs across moving blades but absolute velocity decreases (see Fig. 5.1-4). Expansion in moving blades does increase velocity *relative* to moving blades, but the movement of blades reduces the absolute velocity. In the impulse turbine pressure drop and expansion occurs in the stationary blades. The cresent-shaped rotating blades permit free entry and discharge at little steam pressure drop (friction only). Even though relative velocity changes little across impulse moving blades, the energy transfer causes a reduction in absolute velocity.

Blade Losses

All of the energy of the steam jet is not converted to useful work. Some of its energy goes to overcome the forces resisting steam passage through the blades. This resistance is due to friction, natural flow eddies, and turbulence caused by blade damage, deposits, or (if steam is in wet region) droplets of water moving through it. The amount of energy used to overcome this resistance is influenced by steam approach and exit patterns, blade design (shape, width, curvature, etc.), blade condition (smoothness at leading and trailing edges, cleanliness, and errosion), and pressure sealing features and turbine load (steam conditions). The longer blades (less rigidity) are subject to vibrations that add another factor influencing streamlined steam flow (conversion efficiency) and blade life. Since these losses (by design) are influenced by loading as well as by throttle and exhaust conditions, correction curves are available to predict output and efficiency at conditions other than design. Use of such curves allows for separation of effects of off-design operating conditions, allowing a more definitive effect of turbine internal conditions.

A factor of considerable importance in condensing turbines is called "exhaust loss." The high-velocity, exhaust steam contains an energy component equivalent to this velocity. With no turbine blading left to reclaim this energy, it is lost to (or reclaimed by) the exhaust heat exchangers. Careful design of exhaust passage (condenser and its internal equipment layout) can minimize the pressure loss needed to move steam to its condensing point. An additional low-load exhaust loss occurs when the latter turbine stages have insufficient pressure drop (windage loss) due to low flow volume.

Prior to the latter stages of the condensing turbine, sufficient energy has been removed from the steam to place it in a wet (water-bearing) condition. Due to this water (condensing steam), blade deposits and erosions are most severe in the last few stages, causing further reduction in efficiency and increased exhaust loss. Exhaust loss for the topping turbine (see 5.1-2) can be reclaimed by any process using the heat energy contained in the turbine exhaust.

Summary

Through the use of reaction or impulse (or combination) bladings, the steam turbine changes heat energy into kinetic energy. The blading stages can be used singularly or series-multiple to allow use of very-high-pressure steam. Performance depends on the relative blade-to-steam velocity ratio and on maintaining steam approach and exit angles to blading. Work conversion, which is maximized by the correct ratio and proper turbine application and conditions, varies from this maximum upon ratio changes [steam velocity changes due to load changes or turbine (blade) speed changes], steam path deterioration (cleanliness or metal surface changes), steam flow interferences (seal leakage or extraction changes), or steam impurity (water droplets and deposits). Losses include a turbine exhaust loss made up of the velocity energy of exhausting steam and energy needed to move steam into and through the exhaust chamber.

5.1-2 Steam Turbine Applications

General

The commercial steam turbine is a versatile prime mover. It can be adapted to use steam of almost any characteristic (up to supercritical pressure, 1100°F temperature, and 30% quality) and can be designed to provide shaft output ratings of up to about 1200 MW or to serve in cogenerating service as a drive (prime mover) for a mechanical or electric device while, at same time, being a steam-pressure-reducing device for furnishing steam to a process. The turbine steam path can be a simple straight-through pattern or a complex pattern allowing for reversing (series or multiple parallel) paths to facilitate turbine sizing and thrust control. The path can be arranged in series through separate (common or dual-shafted) turbines (rated as high pressure or low pressure) with piping systems allowing for extractions, mixing with other steams, or rerouting to the heat source (boiler) for steam reheating. The single-shaft (tandem) turbines have a common shaft speed, but the dual-shaft (cross-compound) turbines generally operate at different speeds. Steam turbines can exhaust into a pressure system from which steam for other process purposes is used or into a vacuum chamber where exhaust losses and heat of vaporization are removed by cooling water, thus changing steam into recoverable condensate. In the cogeneration mode the turbine may drive mechanical equipment (pumps, fans, compressors, refrigeration machinery, etc.) or electric generating equipment. It can have shaft extensions on both sides driving a mechanical system on one side and electric generator on other side.

The various types of turbines are identified relative to internal steam flow patterns (Fig. 5.1-6) and by general external steam circuits (Fig. 5.1-7). See Table 5.1-1 to further identify the various types by characteristics and applications. With regard to exhaust, turbines are divided into two classes—condensing and noncondensing. Both classes contain a variety of steam flow patterns—the full steam through-flow or the steam-extracting turbine that exhausts less steam than enters the throttle by the amount that is removed through extractions. This extraction may be of variable pressure (with load) or of constant pressure at a delivery point over a designated load range. The automatic feature is an internal regulating valve that varies steam flow to subsequent stages in order to maintain the set extraction pressure. Topping turbines (steam pressure reducers in addition to a mechanical power

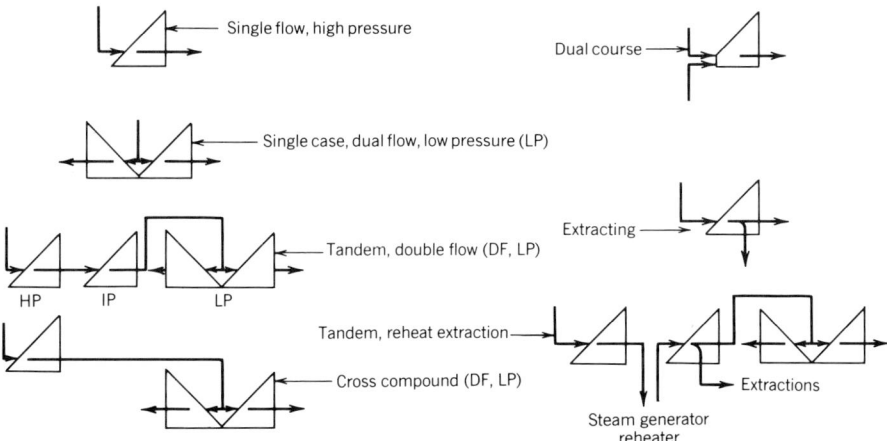

Fig. 5.1-6 Steam turbines, internal steam flow.

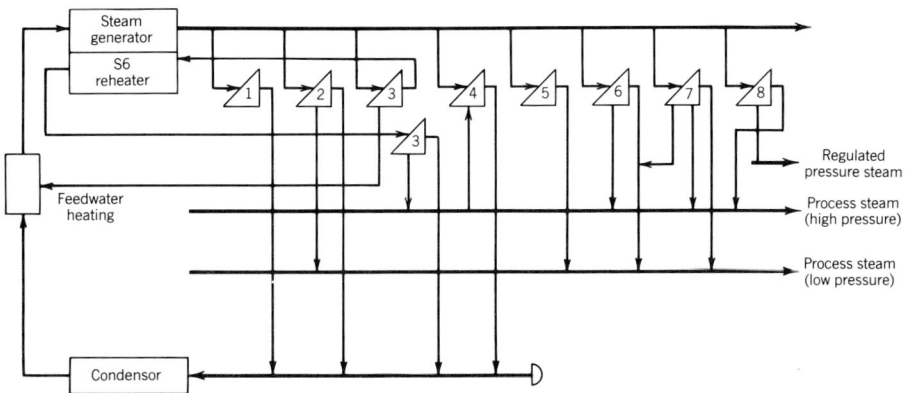

Fig. 5.1-7 Steam turbines, external steam flow. Type description: 1, straight condensing; 2, condensing, extraction (single); 3, condensing, extracting (multiple), reheat; 4, condensing, mixed pressure; 5, noncondensing (topping); 6, topping, extraction (NC); 7, topping, multiextracting (NC); 8, automatic extracting (NC).

source) may be required to exhaust at a controlled process system pressure. Most turbines require a regulated throttle pressure (by boiler) but can function on a limited reduced throttle pressure. Allowable throttle overpressure is limited to about 5%. If a regulated pressure (throttle, exhaust, or extraction) steam system is needed, it may also be supplied from the main steam line through a standard pressure-regulating valve rather than through the turbine. Under some conditions of system operation, it may be desirable to drive the steam turbine with an alternate steam supply. These mixed-pressure turbines are designed to receive energy from an alternative steam supply introduced into a divided steam chest (governing valves) or into the turbine casing several stages downstream of the throttle inlet.

During the expansion of steam through energy removal, considerable moisture appears as steam conditions cross the saturation line (Mollier chart) and enter the wet regions. In the region of condensation (turbine low pressure stages) water droplets cause an efficiency loss due to flow interferences and deposits or material damage. To offset this, the "reheat" turbine allows for exhaust or extraction steam from the slightly superheated range to be rerouted to the steam generator for superheating to initial temperature, allowing steam expansion in the dry range (above saturation) further into the remaining turbine stages. Steam reheating can be done in main boilers to reclaim heat in flue gas (boiler efficiency improvement), in a separately fired reheater, or to some extent by injection of high-temperature steam. In some turbines final-stage blades have built-in serations to facilitate water removal by slinging condensation into the casing opening from which it is removed.

TABLE 5.1-1 TURBINE CHARACTERISTICS

Classification	Type	Characteristics		Relative Advantages	Applications
		Physical	Operational		
Impulse	Simple impulse	One nozzle or set of nozzles. Single disc with one row of blades. One passage of steam across blades. Simple casing construction.	Little heating necessary— quick starting. Clearances usually ample. Simple governing systems.	Low first cost. Comparatively small rise. Heat drop limited to 100 Btu for high efficiency; larger heat drops lead to decreased efficiency. Small floor space.	Certain high- or low-pressure units with small heat drops. Auxiliary drives. Turbo-blowers and -compressors.
	Simple velocity stage or Curtis	One nozzle or set of nozzles. Single stage with two or more rows of revolving blades and necessary intermediate reversing blades. One passage of steam across each blade row.	Moderate wheel speeds. Large pressure drop in nozzle. Large end clearance for blades. Same pressure throughout stage. Simple governing systems. Can use comparatively large pressure drop.	Low first cost. Large power in small size. Moderate efficiency. Can utilize comparatively high heat drops. Strongly built. Small floor space.	Auxiliary drives, as stokers, pumps, exciters, small dc and ac generators, fans, blowers, etc. For emergency units requiring quick starting as reserve boiler feed pumps, house turbines, etc. Noncondensing units for industrial plants in connection with process.
	Reentry	One nozzle or set of nozzles. Single row of buckets or blades. Reversing chambers to redirect steam on buckets or blades one or more times. Simple casing construction.	Usually moderate wheel speeds. High wheel speeds used in some geared sets. Can employ comparatively large pressure drop. Simple governing devices.	Low first cost. May secure relatively good efficiency. Ruggedly built. Small floor space	Auxiliary drives of all kinds. Geared sets. Emergency reserve units for quick starting. Paper mill drives. Small marine-geared propelling units for tugs, etc.
	Helical flow				
	Multistage impulse or Rateau	Several simple impulse wheels in series, separated by diaphragms carrying nozzles or orifices. Usually many stages. Large clearances at ends of blades; small side clearance at nozzle.	Small heat drop per stage. Hence most efficient ratio of wheel speed to steam speed may be had. Fine clearances necessary at diaphragm glands. Throttling or nozzle governing, latter preferred.	High efficiencies possible. High blade speeds needed in last rows of large condensing units. Built in large as well as small sizes. Blade limitations in exhaust end of large units. Requires careful heating before starting.	Turbogenerator drive. Large auxiliary drive. Marine propulsion. Centrifugal pumping sets. Turbo-blowers and -compressors. Bleeder and extraction units. Mill drives.

	Velocity-compounded or multistage Curtis	Series of simple Curtis stages separated by diaphragms carrying nozzles. Large end clearances on blades; small side clearance at nozzle. Relatively few stages.	Relatively large pressure drop per stage. Fine clearances at diaphragm glands. Can be started quickly. Few operating troubles. Usually nozzle governing.	Small floor space. Moderate efficiency. Rugged construction fit for hard service.	Noncondensing service with large auxiliaries, house turbines, exciters, etc. Condensing turbogenerator sets. Marine reversing turbines. Bleeder and extraction units.
Impulse and reaction	Axial flow Parsons	Series of alternate rows of converging fixed and moving orifices. Both diameters and blade lengths steadily increase. Steam flows axially. Fine clearances at ends of some types of blading. End tightening used with certain blading. Usually many rows.	Small heat drop per row permits use of ratios of wheel speed to steam speed conducive to high efficiency. Usually throttle governing. Require extended warming-up period before starting.	Usually require large floor space. High efficiency possible in all sizes. Blade limitations in exhaust ends of large units. Easy to bleed for feedwater heating.	Turbogenerators. Marine propulsion. Geared drives. Extraction units. Mill drives.
	Radial flow Ljungstrom	Alternate series of radial rings carrying converging reaction blading ranged for radial steam flow. Low-pressure stages are axial flow on large units. Alternate rings revolve in opposite directions. Hence two generators tied together electrically. Small clearances over blades.	High relative velocities lead to high efficiency with few blade rows. Radial rings require nearly constant blade length. Small leakage losses. Can be started quickly. Can use high temperatures. No critical speed through which to pass.	Compact, relatively small. No foundation for unit which is supported by condenser. Quickly erected. High efficiency, particularly in high-pressure and in small-size condensing units. No outer heat insulation needed.	Turbogenerators. Extraction units.
Combination types	Velocity multistage impulse or Curtis–Ratcan	One Curtis stage followed by series of simple impulse stages with intermediate diaphragms carrying nozzles. Shorter than straight multistage impulse unit.	Only moderate pressures and temperatures in casing, due to large pressure drop in first nozzle. Favorable ratios of wheel speed to steam speed possible in both Curtis and other stages	Compact and short. Efficiency practically the same as multistage impulse up to 5000 kW and fair in larger sizes. Rugged construction in moderate sizes. Lower pressure on high-pressure gland packing box due to Curtis stage.	Turbogenerators. Auxiliary drives for larger units. Centrifugal pump drives. Turbo-blowers and -compressors. Mill drive through gearing. Bleeder and extraction units.
	Velocity reaction or Curtis–Parsons	One Curtis stage followed by series of Parsons stages. Frequently of disc-and-drum construction. Larger clearances than in Parsons.	Pressure and temperature in casing reduced by expansion in Curtis nozzle.	Shorter and more rugged than standard Parsons. Efficiency good in small sizes, and fair in large sizes.	Turbogenerators. Marine propulsion. High-efficiency marine-geared auxiliary generator sets. Bleeder and extraction units. Mill Drives

Prime Movers

In correct application and design the steam turbine will fit systems of almost any steam supply or load requirement. As an operating tool, it is flexible in its ability to respond to a wide variety of constant or varying loads. The absence of reciprocating parts and their stress limitations allow the turbine to be designed for high speeds and power outputs, making it a compact, portable, prime mover for marine and mobile service, and as a high-output, high-efficiency stationary drive for industry or electric generation stations. Metallurgy associated with turbine materials has provided metals compatible with steam conditions (Mollier chart), which allow for maximum available enthalpy differentials associated with the 1000°F range, as well as with strength requirements associated with high velocities of blades and steam. Factory construction, field installation, and operation/maintenance are proven activities allowing for up to several years operation between maintenance inspections [high mean time to failure (MTTF)] and for fast, routine maintenance [low mean time to repair (MTTR)]. The ability to control and use electronic control systems have made steam turbines standard industrial mechanical drives with ranges of up to 15,000 hp and utility electric generator drives with ratings of up to 1200 MW. In these applications the turbine is capable of variable-speed operation, of fitting to industrial processes, of performance under adverse loading (range, speed of response, speed regulation, etc.) and environmental conditions, and of being easy to install, start up, and operate. In the cogeneration cycle the turbine offers industry an electric or mechanical power source with beneficial energy economics. The economics improve if the industry has a need for process steam and a "waste" fuel source. Typical cogeneration service calls for small-size units (up to 3500 hp) using steam from an industrial boiler (up to 1400 psi) in a "topping" service. Extraction and/or exhaust steam (up to 150 psig) is used by the process. To be economical, process steam use should exceed 100,000 lb/h with maximum speed ranges of up to 5500 rpm.

Earlier discussions noted that the impulse stage had an efficiency advantage in small sizes. Many of the industrial prime movers are impulse turbines due to this design characteristic, but also due to low cost, ease of installation and operation, and small space requirements. A variation of the impulse turbine is the "reentry" (helical flow) turbine. The same advantages apply; in addition (through use of stationary, reversing reentry chambers and multiple rotating blades prior to exhaust) this type turbine can function with a comparatively larger pressure drop. For even larger initial-stage pressure drops (up to 500 psig), a series of simple velocity (one nozzle, one rotating impulse row) RATEAU stages may be used or, for supercritical throttle pressure with even larger pressure drops, velocity-staged (one nozzle followed by multiple rows of impulse blading separated by reversing blades) CURTIS wheels might be used. Turbines with higher ratings (output and pressure) utilize these quick pressure drop stages followed by series of reaction PARSON stages.

Electric Generation

The steam turbine as a prime mover for an electric generator furnishes the mechanical power needed to directly respond (with other generators in parallel) to variations in demand from the electric utility system. This is done at a speed (rpm) that sets the exacting electric power frequency (60 cycles per second). An elaborate control system varies turbine steam flow in quick response to supply energy for this varying mechanical power demand. To minimize control problems, other turbine systems (boiler for throttle steam conditions, condenser for stable exhaust pressure, electro-hydraulic governor for speed, etc.) contribute controlled features over a broad range of steam flow demands. Because utility electric loads are large, many turbines run in parallel to control an "area" electric frequency by matching supply to demand. This parallel operation requires a turbine governor control with a "drooping" characteristic speed (rpm) that appears to be slightly higher at no load versus full load. This characteristic is measured in terms of "percent speed regulation." To protect turbine and system equipment, design must include special safety devices and the turbine must be capable of dropping full load instantly upon a safety device function. On major utility systems containing dozens of turbo-generators operating in parallel, there is an economic load-dispatching control. This automatic control responds to load fluctuations by searching out the most economical multigenerator loading and automatically loading specific units. To accomplish this, controller input requires system load measurements and turbine efficiency performance curves. Generally, the dispatching center is centrally located and communicates through electronic networks with load-measuring stations and individual turbine governor controls to automatically establish system loading.

In an industrial mechanical or electric cogeneration system, operation may be simpler. In-house loads may require wider load range but allow a wider tolerance on control of exact load, speed, or frequency. Other industries, such as those with rolling mills, may need turbines matching processes of reasonably constant loads once in operation but require a steam turbine with greater speed control capabilities. Sizing of turbines (output rating and system application) on these systems is done on the basis of load characteristics and economics. Loading curves estimating future demand by projecting from past history establish existing load factors and future power (turbine) needs. The economics of energy cost related to output establish turbine application (size, type, and characteristics). Usable information includes plant energy billing and product output records; market analysis of energy cost, facility costs, and product demand; and engineering analysis of turbine application and process alternatives. Load demand curves aid in developing needed capability if adjustments are made for utilization of connected load and needed capacity reserves. For an industrial plant without records, sizing of turbine applications can be based on plant steam use or on connected load times a diversification factor that identifies maximum use of connected loads. The factor usually runs from 50 to 70% indicating a full load capacity reserves of 100% to about 40%. Design and construction periods (up to 10 years for a large turbine generator unit) include sufficient time to determine needs and provide the capability by the time of commercial operation data (COD). Factors to consider include expected growth, load demand and alternate power sources, needed capacity reserve to handle equipment outages, and economics. Economics include comparisons of the alternate choices on installed cost, energy efficiency, fuels cost, expected operation and maintenance cost, financing methods, and tax effects. Trends in turbine characteristics and application include technical advances [higher steam pressure (supercritical—3500 psig) and temperatures (up to 1100°F), vacuum exhaust, designs using steam reheat systems, improved metallurgy, etc.] for lower heat rates (heat input per unit output).

In addition to standard prime mover applications, turbines are being used in electric cogeneration or combined-cycle systems, especially where a combustible waste fuel is available and a process steam need exists. Figure 5.1-8 provides basic sketches on various turbine applications. New steam generators, initially designed to provide process steam, can be designed to economically add incremental heat energy (superheat) to the steam for use by a topping, cogeneration turbine. The turbine utilizes this added energy for its output while steam compatible (pressure and temperature) with process

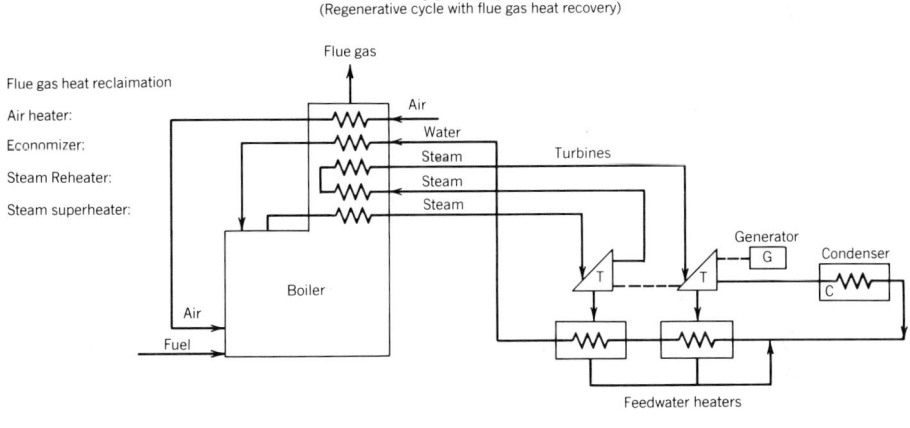

(a)

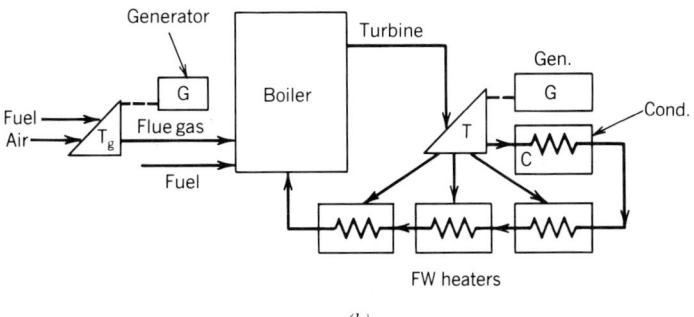

(b)

Fig. 5.1-8   Steam turbine applications.

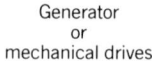

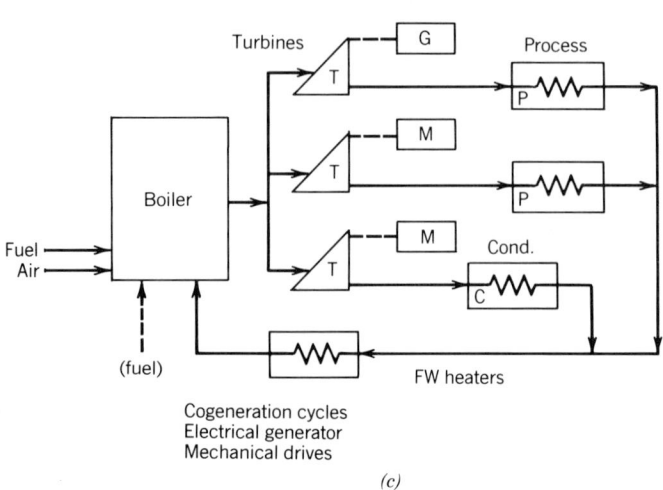

Cogeneration cycles
Electrical generator
Mechanical drives

(c)

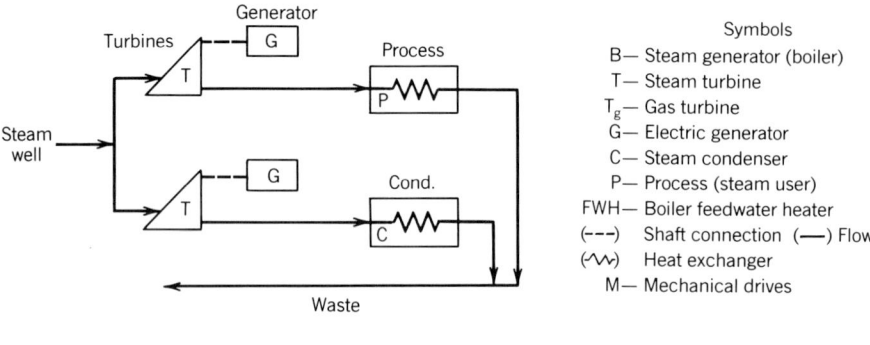

(d)

Fig. 5.1-8 (*Continued*)

needs. Thermal efficiency is maximized when the process uses the heat of vaporization rather than allowing this to go to the environment through a condenser. In a typical large utility turbine application (Fig. 5.1-8a) steam at about 1460 Btu/lb is available for work conversion. Of this initial heat, about 950 Btu/lb (exhaust enthalpy minus condensate enthalpy) is lost to the environment through the condenser. Energy conversions to work occur at about 38% overall thermal efficiency or about 8270 Btu per equivalent kilowatt-hour at turbine output. Any application that reduces the condenser duty (recovery of heat of vaporization) increases efficiency. The utility "regenerative" cycle does this by extracting (bleeding) steam from the turbine prior to the exhaust and reclaiming total heat content of this steam for heating boiler feedwater. An industry with a process steam need (Fig. 5.1-8c, d) does this by reclaiming heat of vaporization with the process. If combined with a low-cost energy source (Fig. 5.1-8c, d) or gas turbine (Fig. 5.1-8b), steam turbine applications for cogeneration are very feasible.

Peaking

Most mechanical or electric loads will, over the years, change character (size, extensive operation at extremes of load range, and cyclic changes) in magnitude and frequency. A typical nonreheat utility turbine annual load duration curve (Fig. 5.1-9) shows that peak loads may occur at short periods of

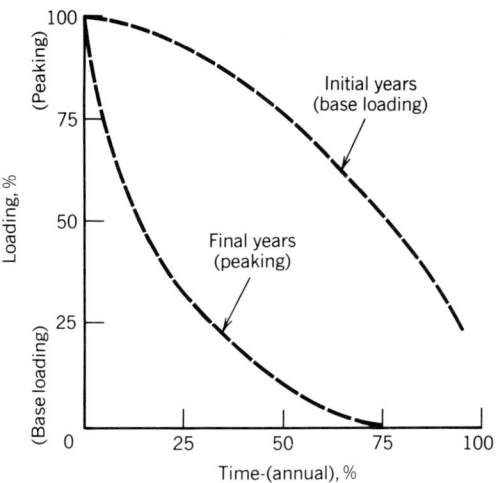

Fig. 5.1-9 Turbine load duration curve.

time and that there is a base load (about full time) carried by most efficient supply sources. A specific turbine enters into the base-load region in its initial service years, progressing upward into peaking service in final years as new turbine systems are added or process loads decrease. Changes in equipment condition and, more importantly, technical advancements and the need for production cost reductions bring this about. The steam turbine can be designed specifically for base loading, for peak service, or modified during operation for peaking service. Its contribution toward the production of an economic product will establish its "base-load" position. If several turbines are available to carry a system load, the base part of the load (longest time duration) is carried by the turbine best rated in fuel economy and reliance. Others would be added in this order as load increases to the peak demand. Turbine features influenced by design-loading criteria include steam conditions, materials and metallurgy, thermal expansion rates and allowable stresses, seals and clearances, internal steam paths (admission, bypasses, extractions, etc.), condensation removal, exhaust chamber cooling, turbine system cooling, instrumentation, and a governor system. Considerations for turbines operating at base load include establishing the highest efficiency point at expected loading, valve point operation (minimum throttling loss), small clearances, minimum drainage, long warm-up periods, best materials and metallurgy, and rigid steam-purity requirements.

The original turbine designs may be adjusted to accommodate conversion to peaking service. Full-steam arc (steam entering all quadrants of the first stage) admission equipment is installed; clearances are enlarged; greater condensate drainage are provided; new and faster operating procedures are established (low-pressure start-up and steam-to-metal temperature equalizing); additional automatic control (start-up, variable range, and rate of loading), operating mode conversion, and protection are added; and more frequent preventative maintenance is done to allow for the more flexible operation and frequent shutdowns. Industrial cogeneration turbines have similar life-cycles depending on technical progress (process or product) or business cycles.

Cogeneration

Basic flow diagrams of the electrical and mechanical cogeneration cycle are illustrated in Fig. 5.1-8. On a large scale a cogeneration system might include a major coal degassification plant providing coal derivative fuels (low heating value gas and industry raw materials—(coal oils, coke, sulfur, etc.) as well as process steam and electric power. In addition to fossil fuels, inputs to the energy center could include combustible waste (industries or municipal sources). The steam turbine, as a prime mover for the energy center, would drive electric generators with ratings up to 1000 MW and mechanical equipment rated up to thousands of horsepower.

On the smaller scale the cogeneration turbine may substitute for motor drives in a steam using process requiring mechanical or electric prime movers, and having available a raw material (fossil or waste) fuel of low economic value. The small condensing cogeneration turbine driving an electric generator does not compete economically with the electric utilities' large central-station units unless lower fuel cost offset the better central-station efficiency. To do this, fuel cycle cost would need to be less than 60% of the electric power cost to offset an expected 30% higher heat rate and 10% higher equipment capital investment.

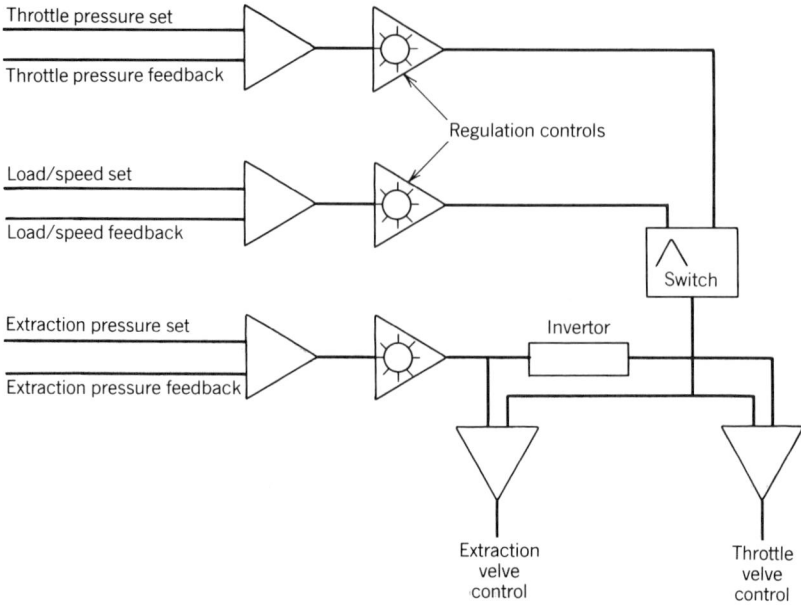

Fig. 5.1-10   Basic turbine control circuit.

Turbine selection depends on design criteria established by the application. A known process will establish prime mover design alternatives. Consideration is given to process growth, peak demand, load cycles, and redundancy. Alternatives are considered on the basis of economics, reliability, and a match to process requirements and operation policies. Should an industry have multiple energy sources (central-station cogeneration, utility companies, or in-house sources, including wastes), a computer-based power supply advisory and control system may be needed to select most economical combinations of source options and specific loadings. Computer output would develop best system energy cost economy, acceptable operating margins, and operations that minimize labor and maintenance cost.

Cogeneration turbines can be operated at constant speed, controlling a load by steam flow (turbines driving electric generators) or controlling driven equipment speed through a hydraulic coupling. They can be operated at constant pressure at either throttle, an extraction point, or at the exhaust point as is done for topping turbines exhausting steam into a constant-pressure process. Load control would then be by inlet pressure or speed. Figure 5.1-10 illustrates a simplified control system. Loading demand (feedback) signals enter to work against a "set" on throttle valve position to control steam flow to the turbine. If extraction pressure is regulated, variations from the set point cause adjustment of extraction and throttle valves in opposite directions to establish desired pressure. Load and extraction pressure will adjust until both are at desired points at speeds established by automatic control. The same basic control can be applied to other combinations such as throttle pressure and turbine exhaust pressure.

Electric Cogeneration

The electric cogenerating application may be a design capable of producing more electric power than is needed at a plant (full output or off-peak periods). Under this design the cogenerator sells power to the local electric utility. Normally, a power purchase contract (cogenerator from utility) can be made for power designated as "standby" (during cogenerators' equipment outages), "supplemental with demand clause" (needs beyond equipment capability), or "interruptible" (utility sales only on their off-peak periods). Power sales (cogenerator to utility) are similar—based on rates that adjust to utilities' load conditions. Federal regulations, intending to encourage cogeneration, compel the utility to purchase a cogenerator's excess electric power at a price determined by the utilities ("avoided cost") or cost of power displaced by the cogenerator's output. Sales value may have a "capability" value as well as an "energy" value depending on reliability to the utility of the cogenerators power source. Sales value also relates to delivery voltage (utility may need to adjust voltage for transmission) and delivery time. The cost (to utility) of power displaced represents the highest-cost power being generated at delivery time. This time may be on a peaking period (high load demand) during use of an older, low-efficiency turbine (hence a high value), or on off-peak resulting in a lower value.

If utility sales are not a consideration, sizing of the industrial electric cogeneration turbine relates totally to plant needs. Often, multiple units, rated at a percentage of plant peak loads, are chosen to increase reliability and provide backup. Sizing is often based on historic or expected peak loads adjusted to near-term future-load growth plus the starting requirements of the largest plant motor plus "backup" capability. The reserve capability may vary from 100% above peak loads in essential-service facilities to 0% in plants capable of withstanding emergency outages. Based on results of reliability studies, which indicate high outage risk components of systems or turbines, reserve capability can be minimized by providing redundancy for these specific components.

Application Summary

Steam turbine applications (Fig. 5.1-8) can be classified according to service. In general, the services are as listed below with designations identifying a condensing (C) or noncondensing (NC) service, a steam flow/constant-speed (SF-CS) control, a steam flow/variable speed (SF-VS) control, or a steam flow/back pressure (SF-BP) control.

Service

1. Utility electric generator drive: C and SF-CS.
2. Mechanical drive through hydraulic coupling: C and SF-CS.
3. Mechanical drive through gearing: C and SF-VS.
4. Marine drive through gearing: C and SF-VS.
5. Topping to process steam, electric generator drives: NC, SF-CS, and SF-BP.
6. Topping to process steam, mechanical drive: NC, SF-CS or VS, and SF-BP.
7. Electric generator drive: (C) and (SF-CS).
8. Standby, emergency, or peaking (electric or mechanical): NC and (SF-CS or VS).

Classification

1. Steam conditions: super- or subcritical pressure, superheated or saturated temperature.
2. Type blading: impulse, reaction, single stage, velocity stage, pressure stage, combination, reentry.
3. Steam flow pattern: single, double, multiexhaust flow, multicylinders, tandem, cross-compound, axial flow, radial flow, reheat.
4. Steam discharges: condensing, back pressure (also called topping and noncondensing), extraction, or bleed.
5. Speed: constant for 60-cycle electric generation, variable in response to loading of drive device, reversing through gearing.
6. Loading: base load, variable load, peaking and quick start.

Users

1. Electric utilities are interested in the best economics in higher capacities, in reliability (less backup) under rather adverse load conditions, and in technically advanced turbines. Technical advancements include turbine adaption to the regenerative cycle extracting steam (from up to 8 points) for boiler feedwater heating or steam reheating prior to entra into low-pressure turbine sections, utilization of highest possible throttle steam conditions and highest reheated steam temperatures, and use of large base-load design turbines for electric generation and small cogeneration turbines for pump or fan drives.
2. Marine (shipboard) designers use turbines for main propulsion and auxiliary equipment drives. They are especially developed for fuel economy, fast speed changes compatible with speed reduction and reversing gears, compactness and low weight, reliability, and adaptability to ship motions. In use are the reaction turbines with high efficiency at the lower marine steam pressures and temperature and, where space is limited, the combination pressure and velocity stage turbines with smaller physical size per unit of power and relatively good efficiency in small size ratings.
3. Industry is often near an economical fuel source (including waste) and needs a steam supply for a process. Under these circumstances the topping or extracting turbine is used. If sizes are

small and available steam heat (enthalpy) drop is 100 Btu/lb or less, the simple impulse turbine is applicable. This turbine is principally used as a drive for auxiliary equipment (water pumps, fans, compressors, refrigeration machines, or small generators) supporting a large process.

5.1-3 Components

General

Steam turbines are heat engines in which energy in steam is transformed into kinetic energy by directing steam through nozzles and using the resulting jet velocity to move blading attached to a rotating part (shaft). Four fundamental components are needed: the nozzle and steam flow passage, the rotating part carrying the blades or buckets, the enclosure (cylinder or casing) housing and supporting these components, and a system to control and protect the turbine. Supporting features for these components are built into the turbine casing in the form of bearings and their lubrication system, steam seals, valves and servomechanisms, shaft penetration and structural connection points to interface with piping, wiring, foundations, exhaust chamber, and driven equipment.

Nozzles

Once steam passes through control valves, it enters the nozzles. Under ideal conditions velocity is developed proportional to a constant times the square root of steam enthalpy drop across the nozzle. In the simplest, theoretical nozzle flow begins when the discharge pressure begins to be less than the inlet pressure. For superheated steam, velocity will increase with pressure differential up to a maximum, which is established when the discharge pressure drops to a point equaling 55% of the inlet pressure. This velocity is then about the speed of sound. Within the nozzle, steam flow will be converging as long as the rate of increase of velocity is faster than the rate of increase of steam volume. Discharge pressures below 55% of inlet pressure will allow the reverse, causing a divergence in the steam flow pattern. To do useful work the steam high-velocity jet must be directed toward the blades at a definite angle to the plane of the blade row. Actual nozzle performance varies from theoretical due to added resistances to steam flow, nozzle dimension changes, and other losses associated with steam conditions and variations in the approach and exit angle. Energy not converted to kinetic energy shows up as an increase in discharge steam enthalpy above that expected on an isentropic (constant-entropy) expansion. If the nozzle is operating in the superheat range (no moisture content), an increase in temperature will also occur. The degree of change in stage (or stage group) efficiency in an existing turbine can be detected by comparing the actual turbine expansion line to the manufacturer's predicted "new-turbine" expansion line.

In all but the first stage of a multistage turbine, steam enters the nozzles as discharge from a preceding stage at a varying velocity and direction. Designers improve initial efficiency by improving these flow patterns (over larger load range and full blade length) and improve sustaining efficiency by maintaining longer blade life (better materials and steam quality control). In turbines using a high first-stage pressure drop, nozzles are fitted into a massive structural nozzle block formed like a ring around the shaft. Steam is introduced (through governor valves) into separate block chambers from which steam exits through sections of nozzles. Chambers and nozzle spacing may allow steam entry into turbine a full circle around the shaft or only a portion of this arc. If nozzle use is not full-circle, steam entry is partial arc admission. By using select nozzle sections (control valve sequence), steam nozzle velocities can be held near design even at low-load operation. For uniform heating during start-up the control system may allow for temporarily supplying steam to all nozzles (full-arc admission) and restoration to normal control after warmup. At rated rotor speeds and initial loading, steam flow patterns soon return to an effective "full arc." As steam volume increases due to drop in pressure, nozzle diametrical pitch and heights increase. This enlargement in dimension has the effect of lowering design nozzle efficiency especially in the low-pressure stages of the turbines. Steam quality (percent moisture) affects efficiency as steam expansion causes entry into the wet regions. Droplets of water form and upset the streamline patterns. As condensation occurs, steam impurities also deposit on nozzles and blading to lower efficiency. These particles, coupled with the high velocity, deteriorate the smooth surfaces by erosion. With decreasing efficiency, the enthalpy of the steam increases as losses show up as heat. A comparison of the Mollier chart turbine expansion line with the ideal isentropic expansion line will show an increasing bending away to higher, equal-pressure enthalpy values in the low-pressure stages due to these effects.

Blades

Responding to the steam kinetic energy created by the nozzle, the blades force a rotation of the shaft upon which they are installed. In perfection, the blades offer no resistance to steam flow, allow streamline flow with no eddies, and convert all kinetic energy to useful work, exhausting zero-velocity steam. Blade shapes and material are selected to approach zero resistance and streamlined flow over a wide variation in output. Exhaust velocities are required to move steam to following stage or exhaust. Results under actual conditions approach 85–90% conversion of available kinetic energy. Nozzle jet velocity is proportional to the square root of its steam enthalpy drop, while the work the jet can do on moving blades varies with the square of the average steam velocity through blading. Actual work computations would consider velocity reductions caused by friction and actual steam entrance and exit angles. Detailed work equations will show the advantage of using small entrance and exit angles with plane of blade motion in maintaining maximum energy utilization.

If high-pressure steam is expanded to near-atmospheric pressure, the velocity generated is unmanageable unless the expansion is broken up into small steps. Early turbine designs provided the highest blade speed possible to minimize the number of steps but encountered mechanical problems. The methods of efficiently using high-pressure, high-temperature steam by small-"step" expansion include pressure and velocity stage compounding. In the pressure staging of the reaction turbines the total pressure is broken up into a series of small expansions throughout the rows of alternate stationary and moving blades. Velocities are kept low in each stage, allowing for reasonable mechanical designs and an efficient way of handling steam. In this type of multistage turbine the lower velocities cause less friction losses. Energy overcoming fluid flow friction (enthalpy increase) is partially recovered by the next stage. The leaving energy of velocity is also partly used by the next stage. The extent of use of these energies is identified by the carry-over ratio, which varies from zero (at discharge of final stage) to fairly close to unity. In impulse velocity staging the initial high kinetic energy of steam is converted to a high velocity, which is "step"-reduced through a series of alternate stationary and rotating buckets (blades). This combination of a nozzle row followed by multiple velocity-reducing rows may be repeated along the same shaft for additional large pressure drops. The use of over two moving velocity-reducing rows introduces considerable friction loss and thus lowers efficiency. Blading designs attempt to produce the desired low steam angles and to provide sufficient cross-sectional area between blades to pass the increasing volume of steam. In a supercritical, condensing turbine exhaust volume may be as high as 900 times as great as initial steam volume. To handle this volume of steam, large turbines are often designed with multiple, parallel-flow, low-pressure turbines, each handling its share of the total volume.

In a specific impulse row the moving blade spacing and space cross section is practically constant. Therefore, these blades allow steam passage with very little pressure drop and expansion. A row of reaction blading is more complex in design in that both stationary and moving blades function as a nozzle. Steam flow pattern may be converging or converging–diverging. In either case there is a pressure drop across both, as well as a volume increase. Specific geometric construction is not in accord with known mathematical theory but is experimentally developed on the basis that (1) entrance direction matches relative discharge direction from the previous row, (2) surfaces on both sides are rounded and smooth for friction control, (3) curvatures should tail out to straight-line trailing edges, (4) a compromise must be reached between the thin-streamlined blade forms and need for structural features, (5) total steam path cross section of the following blade row increases to allow expansion without undue flow turbulence or structural weaknesses, (6) the work done is equally shared (blade pitch optimization), and (7) steam sealing is effective despite the requirement to seal against pressure differential under relative motions caused by travel, force distortions, and thermal expansions.

Rotors

The turbine rotor (shaft) receives the torque forces exerted by individual rows of blades, totalizes these forces, and transmits them through a shaft coupling to the driven equipment. In addition to the rotational forces, the rotor operates under forces developed by blade thrust in axial direction and by internal stress forces resulting from thermal expansion, long-span support system, and transient forces caused by drastic load changes, out-of-balance vibrational forces, and normal load cycling—especially cycling in the lower 50% of rating. Rotors are forged alloy steel, thoroughly heat treated (stress relieved) and tested to ensure no internal defects. A rotor used for impulse turbines may be essentially constant diameter with disks carrying the blades shrunk or keyed on. In large turbines the disks may be machine cut from material of original rotor cresting. Reaction turbine rotors, on which blades are mounted directly, increase in diameter toward the low-pressure end. Steel alloys are selected for ease of forging and machining, strength at high temperatures, and resistance to corrosion and erosion. Blades or buckets are mounted in machine peripheral slots providing structural tie sufficient to handle rotational and centrifugal forces. Blade-fastening methods include inside and outside dovetails or tee-slots. Individual blades enter through a slot wall opening, are rotated around the wheel filling the slot, and pinned in with a dummy blade base to close up the slot opening. The blade tips are usually connected by a shroud ring to give greater rigidity and to prevent steam spillover across blades or between blades. Steam seals work in combination with this shroud either axially or radially. The most usual method of shroud fastening is to rivet shroud to the blade using an extension of blade material as the rivet.

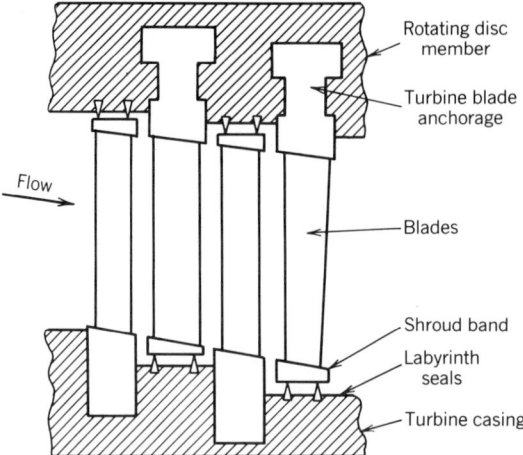

Fig. 5.1-11   Typical blading cross section.

Care must be taken in blade and shroud weight distribution to ensure very low dynamic shaft vibration. Final weight balance is usually adjusted during initial turbine testing and operation with small weights attached to the shaft (at points determined by testing equipment) until shaft vibration is less than 2 mils magnitude.

In Fig. 5.1-11 (typical blading cross section) features of the rotor are shown, including disc, blade anchorage, stage nozzle, shroud band, and seals. The turbine casing encloses the rotor and supports the bearings on which the rotor turns. Rotor speeds vary up to 5500 rpm for mechanical drive turbines. Turbines directly driving electric generators rotate at a speed (rpm) equal to electric cycles per second (cps) divided by the number of generator pole pairs times 60 to convert to time. A two-pole generator for 60 cps would rotate at 3600 rpm.

Care is taken in design to prevent operation at a natural frequency that, in resonance with others, would cause a vibration of excessive magnitude. The point at which this operational frequency occurs in the "critical speed" and is usually established at least 20% above or below design speed. Although mathematics can predict the needed rotor mass to control natural vibrations, rotors are tested under simulated working conditions, and mass adjustments (machining) are made to shift critical speeds the desired distance from operating speeds. In actual operation, if critical speed is below normal operating speed, start-up procedures require timing of speed increases from slow (to allow rotor warmup) to rapid (to get through the critical-speed range as fast as possible). Excessive periods and magnitude of rotor vibration decreases material life through fatigue and lowers efficiency through seal wear (if rubs against stationary parts occur). Rotor vibrations, reflected into the long low-pressure blading can be especially severe on the life expectancy of these blades.

The rotors for individual sections of large compound turbines are connected by shaft couplings. Accurate alignment of each shaft to others is essential to prevent vibrations, excessive bearing wear, and coupling failures. During manufacture and any following rotor machining, any stress concentration points (such as sharp edge cuts at shaft diameter changes) are eliminated.

Bearings

The turbine rotor develops thrust forces (a push in the direction of shaft axis) from several sources. Steam forces on turbine blading accounts for the principal thrust. Although the steam flow vectors of Fig. 5.1-4 show only steam and blade velocity, it can be seen that velocity vectors have a component force perpendicular to the blade velocity vector. This component force (in line with shaft) produces a shaft thrust that must be compensated. Thrust also comes from steam chamber pressure applied to unequal stationary and rotating chamber areas. The impulse turbine is less susceptible to this thrust from rotating wheels due to almost equal pressure and steam velocity blade differentials. Steam balance holes are usually cut through the impulse wheel to ensure equalized pressures. Thrusts may come from connected loads such as pumps, compressors, generators, and so on. These loads should have their own thrust control and be coupled to the turbine so that thrusts are not transferred. For the reaction turbine there is a pressure drop across rotating blades and thus a thrust from both velocity and pressure differentials. Attempts to carry reaction blading thrust totally with a thrust bearing have been unsuccessful. In addition to a thrust bearing, thrusts are handled by balancing thrust forces. Turbine sections (or a dual-flow section) are arranged with steam flow in opposite directions to produce a minimum net thrust and (for area equalization) dummy thrust compensators provide a balancing thrust. Remaining (not offset) thrust is transferred to stationary parts by the thrust bearing. This bearing positions the rotor relative to a point on the turbine casing. The thrust bearing adjustment establishes clearances which are set considering relative thermal expansion and steam sealing at stages and dummy piston (thrust compensator). Axial clearances are most affected by shaft position, including those preventing intrastage steam leakage as well as those spacing the passageway for steam flow. In the impulse turbine the rotor is set by adjusting the thrust bearing so that axial clearance is minimal (0.1–0.2 in.) between nozzle exit and the inlet of the rotating blade to ensure smooth steam flow patterns. In the reaction turbine using blade shrouds, the ruling clearance may be that associated with steam leakage, hence the primary purpose in thrust bearing setting. Axial clearances of the dummy piston located near the thrust bearing are set first, allowing for rotating blades to have equal clearances (upstream and downstream) within space between stationary blading.

Antifriction ball or roller bearings will handle a limited amount of thrust. Therefore, the turbine thrust bearing generally consists of a hardened, rotating collar bearing on stationary, oiled, babbitted plates on either side. A common type of thrust bearing is the pivoted, segmented-shoe, or Kingsbury, bearing. The dual-thrust Kingsbury bearing has shoes on both sides of the rotating collar. The turbine shaft may be radially supported by antifriction bearings or sleeve bearings either self-lubricated or pressure lubricated. For reliability, most turbine bearings are the babbitted journal type usually with features such as oil rings rotated by shaft for lubrication, sealing features to prevent oil leakage, cooling systems, and electrically insulated or shaft grounded to prevent induced shaft electric current. Larger turbines use a pressurized oil lubrication system from an external oil system containing storage reservoir, oil purification, and main and backup pumps. In large turbines this system may contain up to 10,000 gal of lubricating oil. A typical babbitted journal bearing has about 1 mil (0.001 in.) clearance per inch of shaft diameter under normal conditions and is in need of rebabbitting at about 2.5 times this clearance.

Seals

The steam turbine sealing system is designed to prevent steam leakage between stages and through casing penetrations, to hold vacuum (prevent air in leakage) if exhausting into a vacuum, and to aid in maintaining high turbine efficiency. Seals work by forming a high-resistant steam leakage flow path while at same time allowing adequate clearance between rotating and stationary parts. The exception to required clearances is the carbon-ring packing seal that rides directly on the rotating shaft. Garter coil springs encircling the carbon ring hold the carbon segments against the shaft to provide the seal. Another type of seal is the labyrinth seal, which offers a tortuous flow path to leakage steam. This seal consists of a number of alternate (rotating and stationary) disk-type rings encircling the shaft at the sealing point. Radially, the rings overlap, forcing leakage steam to take a turbulent out-and-back flow path. Rings are sharp edged and close fit (about 10 mils plus one mil per inch of shaft diameter at seal). Steam leaking through is withdrawing from the seal to a low-pressure use and often stopped from leakage to the environment by installing a water seal around the shaft at the low-pressure point. This consists of a chamber formed in the casing at a shaft penetration. Pressurized water fills the chamber to condense any steam escaping the seal. Some water leakage occurs in both directions, requiring a pure water source if pollution by in-leakage is unacceptable.

The stuffing-box seal consists of a chamber around the shaft into which a packing material is placed. An adjustable plug (split ring) enters the chamber axially, applying pressure to packing and forcing it against the shaft for a seal. Packing materials must have a lubricating, a lasting, and a high-temperature moisture resistance characteristic. Use of this type of packing is limited to small machines for shaft sealing at the low-pressure end. Application of these seals varies considerably in actual practice. Condensing turbines use water or externally supplied steam seals to prevent air in-leakage at shaft penetrations at exhaust. Between-stage seals are the labyrinth type. Valve stems seal with a flexible or close-fit metallic packing with leakage recapture. For turbines needing a vacuum established at the exhaust prior to starting, an external steam or water supply is used in existing seals to seal air in-leakage until sufficient turbine steam leakage is available for shaft sealing.

Oil System

All turbines have an oil system for lubrication, and many have a combined system that also performs control functions. The system contains temperature controls, oil purification and pumping systems, and instrumentation to monitor oil characteristics and provide indicators for abnormal conditions. The simplest form of lubrication, as used on small turbines, is the journal-sleeve bearing with its oil reservoir and self-lubrication features. Metal rings encompassing and resting on the shaft journal dip into the oil reservoir and, upon shaft and ring rotation, bring oil to the top of the shaft to lubricate the bearing. As the shaft rotates, oil is wedged between the journal and bearing to provide the oil film supporting and lubricating shaft. Oil drains to the reservoir for cooling and recycling. Large turbines have self-contained forced-feed or pressure systems consisting of pumps, coolers, piping, conditioners, and backup and monitoring features.

Journal bearing oil has the characteristic of wetting the metals it separates, high resistance to being squeezed out from between the metals, and no impurities to hinder its passage through the space between metals (usually about 0.005 in. for large high-speed turbine bearings) or to cause corrosion in the bearing. Bearing performance depends on design features (loading, clearance ratio, dimensions and position, unit pressure, and shaft speed) but even more so on oil characteristics such as coefficient for leakage and viscosity. For acceptable range of viscosity, oils should enter bearings between 105 and 140°F and exit between 130 and 160°F. As bearing wear occurs, oil flow or cooling can be adjusted to allow for operation in these ranges. Oil return temperatures from each bearing should be monitored to detect gradual or sudden changes. Degrading of initial oil characteristics could cause an increase in the bearing coefficient of friction causing increase in losses and rising temperature. Oil inhibitors added to new oils are adjusted throughout use to prevent oxidation, to adjust pH, to retard corrosion, and to stop the tendency to foam or emulsify in the presence of air or water. Large turbines, with up to 10,000-gal oil reservoirs, should have a continuous oil purification system (rated at about 10% of main flow) operating in parallel with main oil flow. It is used for oil filtration, water removal, and a receiver for normal oil makeup (a few gallons per day). Turbine oil life, properly treated, is many years. Where investment in an oil charge is not high, a spare batch is often maintained and rotated into operation at annual or biannual frequency. Piping and storage systems for lubricating oils are arranged to minimize fire hazards by routing away from hot areas and encasement in a drainable larger pipe so that any leaks will be contained and carried away from hot areas. Fire-extinguishing systems (CO_2 or water) are installed around the hot areas and oil storage parts of the system.

Oil systems are also used for control and emergency turbine protective systems. Some turbines may have a high-pressure oil system for introducing oil under bearing journals, thus providing an oil film support for initial shaft rotation from an at-rest position, The hydraulic control system uses oil servomotors to amplify small force signals from a governor into strong forces that operate turbine steam admission values. Turbine emergency trip systems operate to release control system oil pressure, thus closing steam admission values.

Governors

The steam turbine governor controls the quantity of steam entering the turbine to match energy input to required output, including losses. Governor types range from the simplest, direct-acting fly-ball governor (Fig. 5.1-12) to those controlling shaft speed through hydraulic (shaft impeller), magnetic, or electronic connections. A variation in shaft speed due to a load change develops a motion within the governor that repositions the turbine control or governor valves. The fly-ball governor shaft is directly connected to the turbine shaft, thus rotating the flyballs in direct relationship. A centrifugal force is produced, tending to move ball weights outward from the shaft with an increase in speed. This fly-ball action, which is opposed by a spring, causes a governor spindle movement, which is transmitted to the turbine steam valve. The resulting valve operation adjusts steam flow to the turbine. The governor spring tensions are made adjustable (manual or motor operated) for turbine speed (or load) changes. The relationship between the opposing forces from fly balls and spring as applied to the spindle establishes the turbine "speed regulations"—the change in speed from no load to full load divided by full-load speed. This characteristic affects the ability of the turbine to respond to load demands and to operate in parallel with other turbines. Zero percent regulations (equal fly-ball and spring forces) produce wild governor hunting. Providing a spring force that increases faster than the fly-ball force increases regulation but eliminates hunting. Figure 5.1-13 illustrates a governor of 4% regulation with a spring force greater than fly-ball forces. Two dashed lines are shown above and below the spring force curve. The vertical distance between the two lines measures the governors dead band or "sensitivity." This is the speed change needed to produce a corrective governor action. If it is desirable to keep shaft speed constant at all loads, an adjusting force (speed changer) is applied. It may be applied to directly adjust the spring or as a force applied to mechanical linkage, as shown in Fig. 5.1-13b.

In turbine parallel operation equal sharing between turbines of a load change may be desirable. This may be accomplished by using turbines of equal regulation. As an example, assume two turbines: A, with twice the rating of B but with a 4% regulation versus B's 2%. An additional load applied to these two units would be divided (directly with ratings and inversely with speed regulation drop) equally. If both units operate by remote-control speed changer, the system load control would adjust unit loading as desired. The speed changer is essential for a turbine driving an electric generator if a constant system frequency (generator speed) is to be maintained. An oncoming turbine generator with a speed regulator about equal to others and a rating small compared to the system ties in with no significant initial load pickup. To force the turbine to pick up the load, the speed changer is moved toward a faster turbine speed. Because the single turbine generator cannot influence the much larger system frequency (speed), the impulse to overspeed the system turns out to be a sharing of the system load at the set speed.

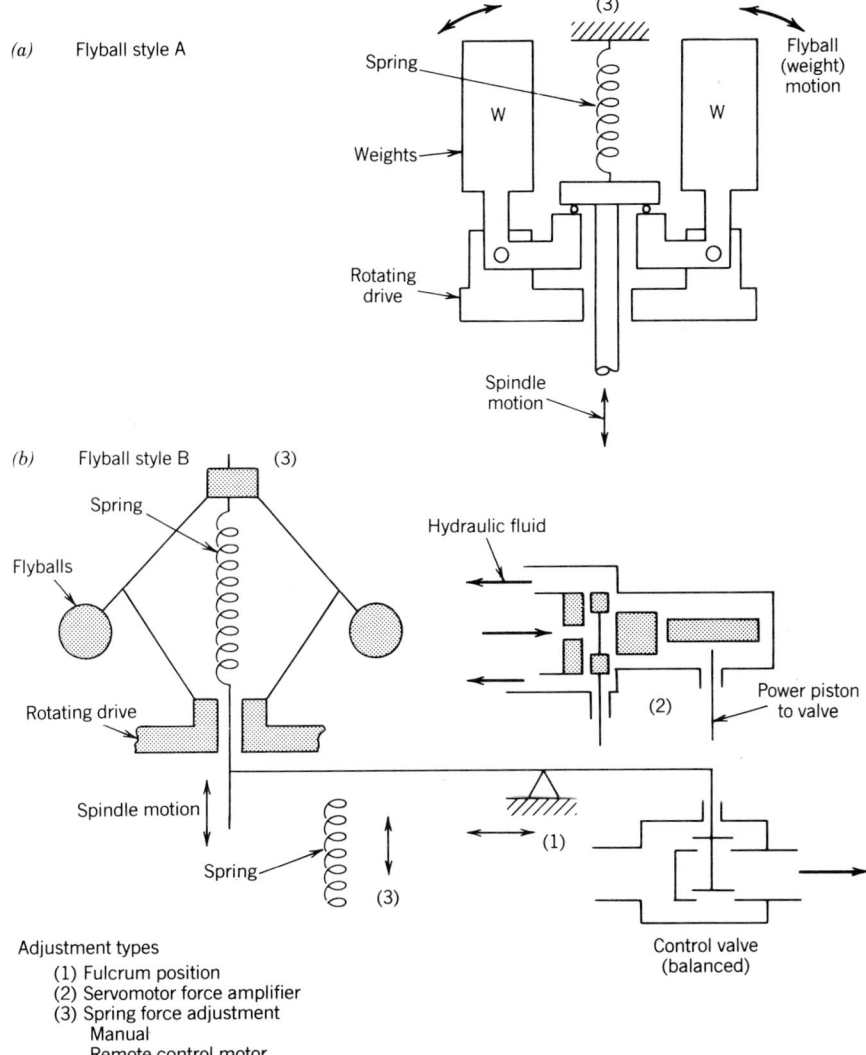

Adjustment types
(1) Fulcrum position
(2) Servomotor force amplifier
(3) Spring force adjustment
    Manual
    Remote control motor

Fig. 5.1-12 Governor fly ball and linkage.

The governor actions control the turbine steam inlet (control) valves. These valves are of the globe type with a linear steam flow characteristic through about the center 80% of disc travel. Outside the range, 10% off fully closed or fully open, regulation changes considerably, often in the direction of instability or hunting. In the case of multiple governor valves, setting of mechanical linkages can adjust sequence timing to allow for operation of the next valve prior to entering the nonlinear section of the valve in motion. This can move the insensitive regions away from the normal turbine operating range, thus improving select-range governor response.

A governor dead band may be caused by friction forces in the governor, slack in the connection linkage, or by the control valve characteristics. Friction and looseness must be minimized, especially for prompt response to small load change demands.

Figure 5.1-14 illustrates the more complex electric hydraulic governor. The system consists of magnetic or electronic turbine shaft speed indicators, a speed changer control station, a governor control system, a servomechanism, and an electric power and hydraulic oil supply. These systems are equipped to receive anticipatory signals from other systems for additional control and protection. A turbine protective system, which is automatically able to shut down the turbine in response to trouble signals, is built into the hydraulic controls. The action is to shut the steam valve (emergency stop and governor) with a dump of hydraulic oil pressure.

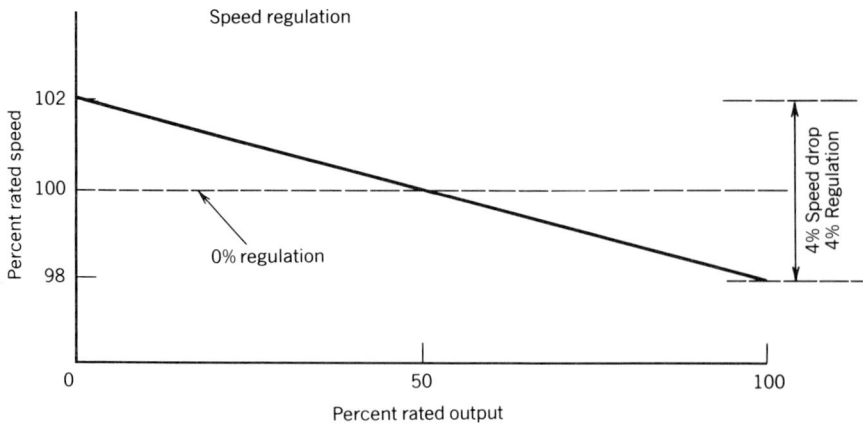

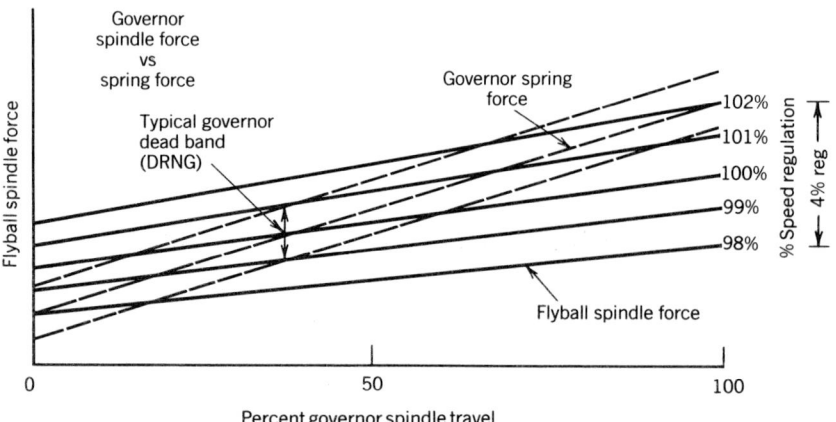

Fig. 5.1-13   Fly ball governor performance.

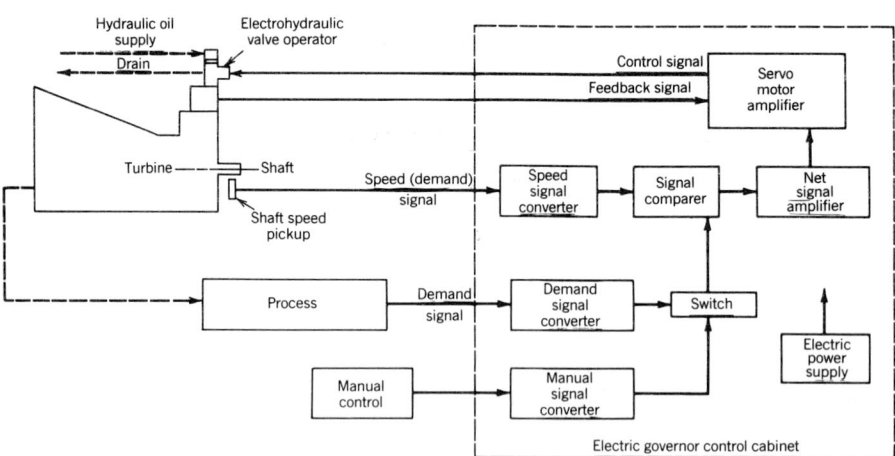

Fig. 5.1-14   Typical hydroelectric control system.

Valves

The large steam turbine systems contain valves for steam flow (load) control, steam pressure extraction or exhaust control, steam leak-off or seal control, steam antireverse flow valves for overspeed protection, quick closing stop and intercept valves for overspeed protection, condensed steam drain valves, steam warmup valves, vacuum breakers to aid shutdown, and many small valves associated with support systems such as lubrication, cooling, fire protection, water shaft seals, and instrumentation. Of these, the automatically operated and often combined throttle and stop valve and following governor valves are most significant. Valve characteristics vary with purpose. Most large valves are hydraulically operated. For the protective valves, opening and closing times of less than a few seconds and tight shutoff are important. The valves used in control functions have a designated flow characteristic—usually a wide region in which flow varies linearly with valve opening. In small turbines a single governor valve usually of the balanced flow type controls steam flow to first-stage nozzles. Larger turbines may have a steam chest that house several control valves opening in a set sequence to admit steam to selective sets of nozzles. The sequence may be controlled by a hydraulic-operated shaft upon which cams are placed and set for each valve. The setting and cam shape establish operating sequence and timing. Another arrangement often used is to lift the valve discs with a bar. Valve stems are attached to the bar through holes allowing a limited bar motion relative to the stem. The bar end positions are moved up and down by stems from two hydraulic pistons. Sequence of valve operation is set by varying stem lengths between bar and stem stop.

If load is suddenly removed from the steam turbine, shaft overspeed (5–8% of rated speed) is prevented by a quick shutoff of steam. In large reheat turbines pressurized steam exists in piping from steam chest to nozzles, in piping to and from boiler reheat section, and in crossover piping between turbines. Upon sudden release of load, shaft overspeed action causes a slam shut of the steam stop valve and governor valves. Steam in reheat piping is prevented from contributing to overspeed by slam shut operation of an intercept valve in the return line from boiler reheater section. Energy supply to the boiler shuts off, and boiler and reheater safety valves dump steam to the atmosphere to prevent overpressuring these systems. Thus, the whole energy flow stream, being suddenly interrupted, is provided a valve system to automatically shut off and release energy for equipment protection.

Valve components consist of a body housing the pressurized flowing medium, a combination seat-disc port through which the medium flows, a motion system to allow the disc to open or plug the port, a disc stem penetrating the body for external control, a medium-sealing system, and the external control, which may vary from a simple valve handle to a complex mechanical or electric stem drive system.

Piping

Steam turbine piping is used to transport steam from boiler to turbine, from entrance (throttle or stop) valves to governor valves, from these to first-stage nozzles, from extraction points to steam use points, from exhaust to a heat exchanger, between turbine sections, to and from boiler reheater sections, from steam leak-off ports to heat recovery devices, and to the sealing system. Other piping is used in connection with turbine lubrication, hydraulic operators, drains, cooling, instrumentation, and fire protection. Steam turbine piping is designed according to ANSI and ASME codes for designated service. Code procedures and equations are used to determine diameter, schedule number, and wall thickness. This determination is optimized to desired pressure drops across pipe and valve runs, space allocations, and costs. To protect turbines from water impingement damage, pipe drainage systems are installed and controlled to remove any accumulation of condensate. Pipe support designs must consider weight, fluid dynamic forces, expansion needs, and support attachments. And, for high-temperature piping, insulation is installed to minimize heat loss and provide personnel protection. Where instruments are required, piping is arranged to accommodate primary elements or attachments for pressure or temperature sensors.

Foundations

The support for steam turbines influences the turbine's ability to remain positioned without settlement, to operate without excessive vibrations, to thermally expand without stress damage, and to handle dynamic forces associated with shaft rotation, sudden load changes, and external loads from connecting pipe or driven equipment. Any foundation misalignment between turbine units or driven machines results in vibrations that can cause internal damage at close clearances. Foundation forms must allow space for turbine piping, driven equipment connections, condenser (if used), lubrication oil system including tankage, and access to machinery and valves for operations and maintenance. For the electric utility turbine the foundations are tall to provide needed clearances—often up to 40 ft. Foundations, or pedestals, for these turbines are either concrete or steel. The steel foundation allows greater clearance for piping and equipment, but carefully designed bracing and stiffening are needed.

To prevent transmission of vibrations, foundations are often built as an "island" isolated from other building floors. Generally, turbine manufacturers request a 50% overdesign in foundation for load allowance for the high-speed turbine, a 20-mil maximum deflection for main support beams, conservative bearing values to prevent foundation settlement, a space allowance for turbine sole plate shims for alignment and grouting, and a concrete mass about 2.5 times the weight of the machinery.

5.1-4 Auxiliaries

Instruments

Steam turbine reliability strongly relates to operation and maintenance procedures, and these relate to the turbine operating information provided. The turbine instrumentation, which provides this information, serves to maintain operating efficiency, provides operation procedure guidance, detects and alarms off-normal conditions, predicts needs for maintenance, provides diagnostic history, provides protection for equipment, and records performance (history, sequence of events, off normals, etc.). Its purpose is to monitor and protect the turbine, to reduce frequency and length of maintenance outages, and to maximize operating efficiency. Modern instrumentation attempts to increase monitoring response and to analyze two or more sensor responses through a computer application to provide trend-type analyses and computations.

In addition to the normal performance-indicating instruments, there are supervisory instrument systems providing data on vibration, thermal expansion, steam impurities, seal performance, bearing conditions, the relationship between actual operations and predetermined operating limits, and operating conditions for automatically performing start-up or shutdown procedures, controlling hydraulic systems, and providing for remote loading control.

The small mechanical drive turbine instrument needs are minimum: pressure and temperature gauges on throttle and exhaust, shaft seals, lubrication oil, and shaft vibration. The control system may require hydraulic oil pressure gauges and valve position indicators. Safety-related sensors include indicators for overspeed, load limiting, exhaust pressure, low oil pressure, high-thrust loading, and (on the extractions) indicators to detect any reverse steam flow. Instrumentation associated with automatic controls provide shaft speed for load control and emergency overspeed trip, initial throttle pressure to reduce load on dropping steam pressure, load indication to prevent exceeding a set load range or a set loading rate, and the various detectors for emergency turbine shutdown (shaft overspeed or vibration, thrust bearing overload, oil systems integrity, etc.).

The larger turbines possess these instrument packages plus more elaborate sensor systems for computerized analysis, control, and trending. During the turbine start-up period, spindle eccentricity is measured to indicate a shaft bow that limits the increasing of shaft speeds until rotation by turning gear reduces eccentricity. Relative side cylinder expansion is detected at points on opposite ends of the cylinder from the anchorage cylinder keys to ensure no cylinder bow, drain valve positions are detected in open position to ensure removal of condensate, and metal temperature indicators provide timed temperature differentials (about $200°F/h$) for limiting thermal stresses.

Computers that receive and interpret data from several sensors aid the operator in decisions critical to turbine health and performance. Through rapid surveillance of many inputs relative to each other, these systems are able to display or print out operating data, alarms for conditions near operating limits, analysis of systems or procedures for their cost-effectiveness or efficiency or operation, trends over any time period referenced to past data to aid in early discovery of specific turbine deterioration or impending major maintenance needs, and decisions for automatic implementation through control systems.

Pressure, temperature, and flow sensors provide steam state at various points on the turbine steam path, including inlet and exhaust, first stage, extractions, and leakages. If the turbine drives a generator, electric sensors (instrument transformers) and instruments such as wattmeters, voltmeters, ammeters, and relays provide output measurement, protective relaying, controlling data and operating history. The mechanical drive turbine output may be obtained from driven equipments' pumped fluid conditions (pressures, temperatures, and flow) and efficiency. Flowmeters indicating flow components of turbine may measure steam flow to turbine, feedwater flow to a dedicated steam generator at zero makeup, condensate from the exhaust condenser, and turbine extractions. A tachometer is required to detect turbine shaft speed providing operator guidance during start-up and normal (constant or variable) speed operation.

Although turbine efficiency varies with loading, departure from a specific-load "design efficiency" is detectable and may be due to (1) excessive interstage leakage due to metal erosion or seal wear, (2) blade-to-nozzle deposits or failure, (3) steam quality, or (4) increase in throttling or exhaust losses. In general, variance from design conditions will increase steam rate (pounds of steam per unit output) or heat rate (Btu per unit output). A forecast of this eventuality can be obtained from the instrument package through a comparison of specific existing and expected conditions.

Turbine manufacturers will supply design steam characteristics throughout the turbine (expansion line) showing expected conditions at various stages related to loading. Information from instruments showing actual operation comparisons, corrected to equal throttle and exhaust conditions, will show effects of turbine wear. Another effective test is repetitive "maximum load" obtained under (or corrected to) similar test conditions. The maximum load may be measured by "output" instruments on the mechanical or electric device being driven. Turbine manufacturers also provide the correction curves to standardize measured output to a set throttle and exhaust condition.

In addition to instruments indicating operating conditions, recorders or computers can provide records on turbine performance—trends, output for accounting, sequence of events for normal or emergency operation, or log sheets with built-in computations to provide efficiencies. Gradual changes of conditions may be detected and timed along with any sudden change. Such trends, as related to other events (emergency trips, intentional loading change, steam quality changes, feedwater treatment, planned maintenance, or operating procedures), provide operators with the guidance needed to maintain high plant efficiency, high mean time to failure (MTTF), and low mean time to repair (MTTR).

Lubrication and Hydraulic Systems

Basic hydraulic systems provide bearing lubrication, a cooling medium for removing the heat generated by bearing friction, and the hydraulics for turbine control and protective systems. The systems may share an oil reservoir containing oil-cooling facilities and multiple oil pumps for mechanical redundancy (up to 300%) and backup (d-c motor) for loss of a-c power supply. Oil pressure regulation, purification, and fire control equipment are also included. The bearing oil system usually operates up to 40 psi and the control system at about 100 psi. Hydraulic systems may be supplied by a single pump or, in large turbines, by a pump driven by high-pressure oil from a direct-connected turbine shaft oil pump. Either system is usually backed up by pumps taking power from a different energy source to ensure reliability. Hydraulic piping layout allows for system continuity during component maintenance and for oil conditioning.

Oil conditioning consists of filtration of particles down to 3–5 μm, water removal, and adjustment of oil inhibitors. An auxiliary, parallel-operated water removal and filtration system is often used for continuous cleaning, operating at a rate of up to 10% of main oil flow. Absorbent purifiers using filtration medium such as Fuller's earth are effective but often remove desirable oil inhibitors. Annual chemical analysis and adjustments of oil will ensure the desired level of inhibitors and neutralization number. These inhibitors are used to adjust pH, help prevent metal from rusting, prevent water from emulsifying, and in general improve lubricating performance and extend oil life.

The flash point of oils should be above 350°F. Viscosity should be a compromise between the upper viscosity oils (higher friction losses and oxidation associated with high temperatures) and the lower viscosity oils (lower lubricating qualities). Most satisfactory operation occurs using a lower viscosity oil (175–205 SUS at 130°F) operating at temperatures of 140°F (cooled) to 160°F (bearing return).

Cooling

Turbine losses include the energy needed to overcome bearing friction. This energy, reappearing as heat (temperature rise) in lubrication oils is removed by cooling through oil-to-air, or to-water, heat exchangers transferring heat to waste or a reclaim system. On the closed-cycle, condensing turbine the steam exhaust is condensed at a vacuum, and the resulting condensate is pumped to the steam generator as feedwater. Condensing occurs in a heat exchanger (condenser) usually cooled by water. Turbine bearing losses are small (about 0.5% turbine rating), but in condensing cycles the low-head exhaust steam heat loss (to environment) is large—amounting to 60–70% of the heat generated by the boiler. In the cogeneration application the higher-head steam exhaust is used as a heat source for process requirements. This application reclaims heat normally lost but severely reduces turbine capability compared to the condensing turbine design. Cooling to remove heat losses associated with turbine-driven equipment is required. Energy losses, which are required to overcome bearing friction, windage, electric current and magnetic field flow resistances, and equipment inefficiency, are removed by cooling water or air. Cooling water may be applied in a once-through circuit or in a closed circuit (recirculating) with water makeup to replace evaporation and to maintain water purity. Regardless of the type of cycle or cooling medium, heat pickup at equipment loss points must be given up or disbursed into another "heat sink." Once-through systems disburse heat into environmental air or water. Closed systems, by using an additional heat transfer device (direct- or indirect-contact cooling tower or spray pond), move heat to the environment indirectly. Cooling towers transfer heat from cooling water to air partly by sensible cooling of water but mostly by evaporation. Where rivers or lakes are available, cooling water for the once-through system is obtained from (and returned to) this source. Care is taken to prevent recirculation of discharging warm water back into the intake. In small reservoirs independent spray systems are often added to increase water-to-air contact, thus increasing cooling effect. In moderate-humidity climates reservoirs for natural cooling are estimated at about one acre of surface per megawatt rating of the typical utility system condensing turbine.

5.1-5 Theory

Steam Tables and Charts

Steam characteristics during the process of conversion of heat energy to mechanical motion can be traced on the Mollier chart of the steam tables. The chart is plotted with steam enthalpy (Btu/lb) as horizontal lines and entropy (Btu/lb-°F) as vertical lines. Curved lines sloping upward to the right represent steam pressure (psia). At about chart center, a single line (forming a dome) represents steam conditions at saturation. Lines originating on this saturation line and sloping upward to become almost horizontal represent temperature (°F). Above this saturation line steam is in the superheat region and below this line in the wet region. Additional lines indicate the degrees of superheat and the percent of moisture in these regions. Knowing steam conditions at the turbine throttle allows entering this chart at a point of intersection between pressure lines and temperature lines. From this plotted point, enthalpy (Btu/lb) and an entropy (Btu/lb-°F) can be determined. As the turbine converts the energy of steam into mechanical motion, steam pressure and temperature conditions drop. If this energy transfer had occurred at 100% efficiency, the plot of resulting steam conditions would follow the vertical entropy line down to the turbine exhaust pressure. Because steam turbine efficiency is not 100%, the expansion of the steam occurs down to the same pressure line but at a higher Btu per pound value. The difference between the two Btu per pound values on this exhaust pressure line represents available heat not converted to work. Upon request, the turbine manufacturers will provide design steam expansion lines or work output plots. Field comparisons of actual turbine performance with these design expectations will provide indicators of turbine performance.

Thermodynamics

The energy flowchart of a simple condensing, extracting turbine in a regenerative cycle is shown in Fig. 5.1-15. In addition to the single extraction shown, there may be other much smaller energy flows not shown. These are added energies from pumps and fans and lost energies, such as turbine mechanical losses to lube oil cooling or steam leakage losses. Turbine thermal losses (throttling and stage inefficiency) reappear in the steam partly to be recovered by downstream stages but mostly lost to the exhaust.

A plot of this energy pattern on a temperature–enthalpy plot of the steam tables is shown in Fig. 5.1-16. The area bounded by the numbers 1, 2, 6, and 7 represent heat available for work up to the first extraction. Area below this to the exhaust pressure (2, 3, 4, and 5) is heat available for work between the extraction point and the exhaust. Additional extraction would be represented by additional, but smaller, areas between throttle and exhaust. The area below line 3, 4′ represents heat unavailable for work (in turbine exhaust). This heat is a product of the exhaust flow and enthalpy difference between the exhaust and condensed exhaust steam. To reclaim a part of this loss, the exhaust heat content is reduced by the amount of totally recoverable extraction flow (lb/h) times its enthalpy (Btu/lb) difference as extraction steam and condensate. The transfer of energy from exhaust to extraction moves point 6 to point 5 to point 4 following a line parallel to line 6, 4′ to the exhaust pressure. A reduction of heat for work occurs, but a larger reduction of unavailable heat (4, 4′, *e*, *d*) occurs, thus increasing the thermal efficiency. The large utility turbine often has up to seven extractions in this regenerative cycle. This type of turbine may also use reheated steam (a turbine

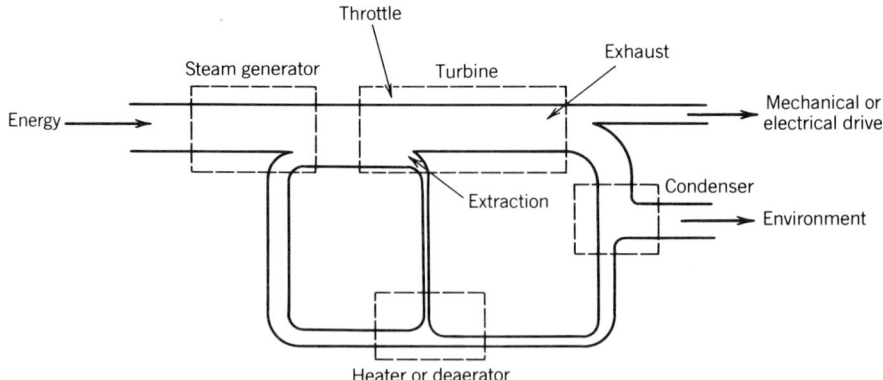

Fig. 5.1-15 Energy flow pattern.

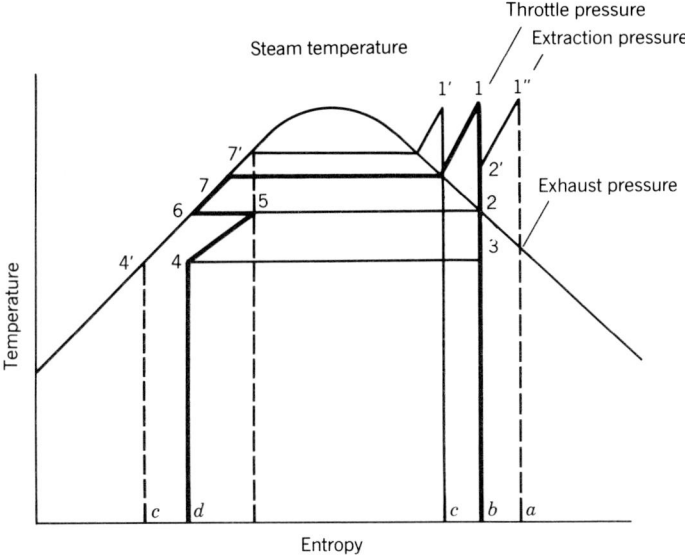

Fig. 5.1-16 Steam turbine energy diagram—T/S plot.

section exhaust reheated by a steam generator to a higher temperature). Reheating increases the heat available for work. This is illustrated by the area added by the dotted line ($2', 1'', a, b$). Steam, having produced some work, is reheated to the original temperature and expanded down a new constant entropy line to improve the ratio of available to nonavailable heat. In this process a pressure drop occurs (in piping systems between turbine and steam generator), which could be used for turbine output. The economic penalties of this pressure loss plus greater equipment cost and possibility of superheated exhaust tend to limit reheat stages to either single or double cycles.

Steam turbines are often operated up to 5% over pressure or at reduced pressure. The effect of increasing steam pressure is shown (line $7', 1'$) as an additional area representing available heat. Any increase in available heat area (above exhaust pressure line) relative to the area below (unavailable) increases thermal efficiency. Turbine manufacturers also provide curves on the expected "following-stage" pressure at various steam flows "to following stages." These are good references to detect (by operating data comparisons) internal troubles, such as intrastage leakages, erosions, deposits, worn seals, and so on. These deteriorating conditions cause a reduction in use of heat available for work that causes lines such as ($2, 5$) or ($3, 4$) to move higher transferring area (heat) to the unavailable zone.

For throttle-governed turbines (no separate valve–nozzle grouping), Fig. 5.1-17 indicates (a) theoretical throttling to a lower pressure (line $1, 2$) at no steam enthalpy change and (b) energy utilization at constant-entropy expansion (line $1, 7$; 100% conversion). Like most ideal cases, neither is

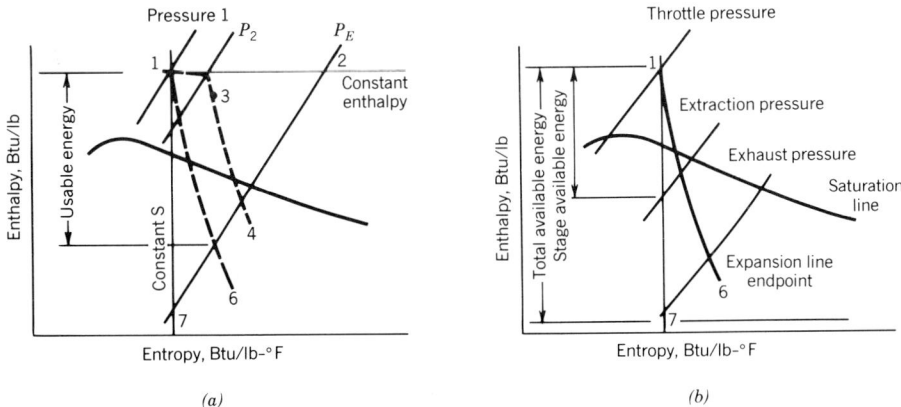

(a)   (b)

Fig. 5.1-17 Steam expansion plots.

possible. The steam turbine experiences throttling losses when governor valves are partially opened, causing a higher pressure drop across the valve to move steam through. In valve-throttling steam from P_1 to P_2 (Fig. 5.1-17a), expansion will tend to follow the constant-enthalpy line $(1, 2)$ to the valve discharge pressure. The actual expansion line $(1, 3, 4)$ takes a less efficient slope to the outlet of the first-stage pressure (point 3) from which the slope generally parallels the full-load expansion line $(1, 6)$. The overall resulting available enthalpy is less than that available on the full-load expansion line, and the utilization of enthalpy is less, resulting in lower output at a lower efficiency.

Figure 5.1-18 plots first-stage pressure versus load for the throttle-governed turbine. At most efficiency loads, inlet pressure is lessened by a moderate throttle valve pressure drop. At loads of less than this first-stage pressure is proportional to steam flow. At greater loads (steam flow) throttle valve losses (pressure drop) increase substantially, causing a change in the pressure versus load curve slope. A second curve (Willans line) relates steam flow versus load. The line may be assumed to be linear between zero and the most efficient load. Like the pressure line, the Willans line changes slope above most efficient loads. With more than one governor valve the line would be a series of straight lines with slightly changing slopes at the fully opened point of each valve. Steam turbine performance tests are most often made at designated loads (max, 100, 75, 50, and 25%) but can be made at points at which individual governor valves are fully opened. The latter test procedure will indicate a lower turbine heat rate curve (heat input per unit vs. output) as a result of minimizing throttling losses. Large utility turbines group a set number of first-stage nozzles to receive steam directly from specific governor valves. Nozzles being supplied steam by all fully opened governor valves will hold to the design ratio of blade/steam velocity to yield best efficiency.

Generally, a straight Willans line is accurate enough to be used as a reference and, if obtained under new-turbine conditions, serves as a convenient reference for later turbine performance comparisons. The vertical distance on the zero-load line (ordinate) between zero and intersection with the Willans line represents steam required to overcome losses at zero load and rated speed. This quantity of steam represents the energy needed to overcome the constant mechanical losses and variable full-speed losses. Similarly, pressure at the point of intersection of the first-stage pressure line and the ordinate represents the pressure needed to provide the steam flow overcoming these losses. A general method of establishing the Willans line is to determine (under standard turbine test procedures) first-stage pressure, steam flow, and load. For best results and later test duplication, test loading points are set at the point each control valve reaches wide-open position. A plot of recorded turbine characteristics against load will establish a present operating Willans line to compare (slope and zero-load readings) with the original lines. A variation from normal would point out a need for further turbine monitoring or performance investigation.

Data from the Willans line of a specific turbine can be rearranged to produce the turbine steam or heat rate curve and its following incremental heat rate curve (rate of change of steam flow or heat rate with rate of change of output). The steam rate curve is determined by dividing the steam flow from the Willans line by its corresponding output and plotting against the output. By multiplying values of steam rate (pounds of steam per unit output) by the corresponding load steam enthalpy (Btu/lb) drop across turbine cycle, the heat rate (Btu per unit output) is determined. These values plotted against output produce a heat rate curve. For the topping turbine with no extraction, a multiple of the steam rate at an output and the corresponding heat content (Btu/lb) differential between throttle and

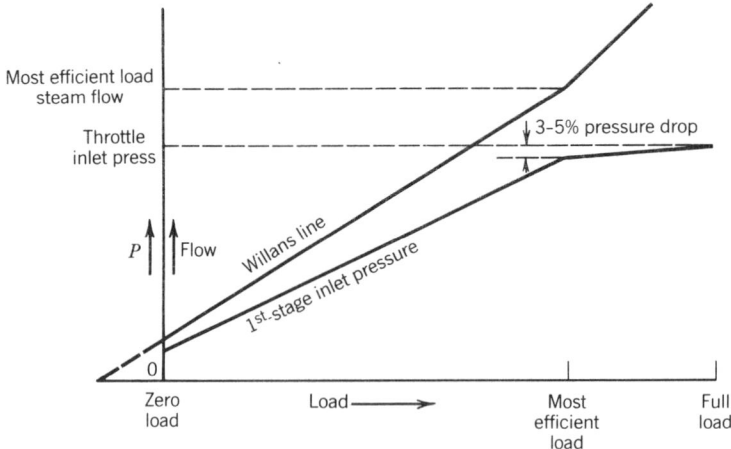

Fig. 5.1-18   Typical turbine steam flow and pressure lines.

exhaust conditions would be the corresponding heat rate. Larger electric utility turbines on a regenerative, condensing cycle (no process heat used) determine turbine heat rate as steam rate times the heat content (enthalpy) differential between throttle steam and feedwater at the steam generator inlet. By developing a formula (least-squares method) for the heat rate curve and taking the first derivative (calculus procedure), the resulting formula plotted against load will represent the incremental heat rate curve (change in heat rate with change in load). This type of curve is very useful in the economic load dispatch procedures of the major utility networks.

Theoretical Steam Rate (TSR) and Heat Rate (THR)

The Mollier steam chart plots of turbine throttle and exhaust steam conditions will provide data to determine the theoretical amount of energy available for conversion to work. Initial steam pressure and temperature locate a point on the Mollier chart from which the initial enthalpy (Btu/lb) and entropy (Btu/lb/degree) are obtained. Following the constant-entropy line (vertical downward) from the point established by the throttle conditions to the final pressure line establishes the theoretical "final" enthalpy of the steam. The difference between these two heat contents divided by output is the theoretical heat rate. This available heat (Btu/lb) divided into the Theoretical Heat Rate (THR) equivalent of a horsepower-hour (2545 Btu) or a kilowatt-hour (3413 Btu) provides the steam per unit output (TSR) as pounds of steam to produce the unit of energy of the ideal turbine at 100% efficiency.

Since actual turbines are not 100% efficient, all available heat is not converted to useful work. The exhaust contains more heat than theoretical exhaust conditions. A new, actual operation exhaust point is located at actual heat content and exhaust pressure. The operational expansion line then changes from the vertical (constant entropy) to a line originating at the same throttle steam condition and sloping downward to the right, terminating on the exhaust pressure line at a higher heat content.

As an illustration, assume a mechanical topping turbine with initial steam conditions at 900 psi and 840°F and exhausting into a 50-psia process system. The initial steam contains 1416 Btu/lb with an entropy of 1.6. Following the 1.6 entropy line downward on the Mollier chart to the 50-psia line gives a final enthalpy of 1130 Btu/lb. The difference in enthalpy (286 Btu/lb) is theoretical heat energy available for work. Assuming the turbine converts this at the rate of 70% efficiency, then the Btu/lb actually converted is 200 and exhaust enthalpy is $1416 - 200$ (1216 Btu/lb). The actual expansion line begins at throttle steam conditions, slopes downward to the right, and terminates at a point on the 50-psia line at 1216 Btu/lb. Corresponding theoretical and actual steam rates (pounds steam per horsepower hour) are 8.9 (2545/286) and 12.7 (2545/200). The theoretical heat rate (2545 Btu/hp · hr) then compares to an actual heat rate of 12.7×286 or 3630 Btu/hp · hr at the assumed 70% (2545/3630) efficiency of conversion.

In the event the steam turbine is a reheat unit (extracting steam for rerouting through reheat sections of the steam generator to increase temperature), two expansion lines exist—first, from the initial throttle conditions to the reheat extraction pressure and, second (assuming no piping pressure drop), from the intersection on this extraction pressure line and the new reheat steam temperature line to the new exhaust pressure line point. As in the single expansion line turbine, the TSR is obtained by comparing the summation of the constant-entropy line enthalpy differences of both lines with the output heat equivalent.

The methods illustrated are used to determine the TSR of a variety of turbine throttle/exhaust applications and the results are plotted in Fig. 5.1-19. It is noticed that the TSR is very dependent on the initial steam conditions and that gains in TSR are more easily made with increases in throttle steam conditions in the lower pressure and temperature ranges. For instance, at an exhaust pressure of 60 psig, a 2.7-lb/kW · h improvement in TSR is made by increasing steam pressures from about 600 psig to about 900 psig and temperatures from 750°F to 825°F. In the higher pressure and temperature regions an increase of 1800 psig/1000°F to 2400 psig/1000°F only produces a 0.3-lb/kw · h improvement in TSR. Cost of equipment increases with physical requirements of higher pressures and temperatures but will slightly decrease due to "economy of size." The net gain in capital cost for the advanced technical gains combined with decreasing gain in benefits (TSR) results in a harder economic justification of the technical progress.

To illustrate the basic economics, assume an energy cost of $3 per million Btu, a 10,000-kW topping turbine at 70% thermal efficiency, and a plant cost differential of $500/kW for the affected parts between the 600- and 900-psig cycles. Theoretical steam rate savings (2.7 lb/kW · h) using the higher pressure system would be 27,000 lb/h or 38,570 lb/h at 70% turbine thermal efficiency. At a regenerative cycle maximum expected feedwater temperature of 300°F, the steam generator adds about 1120 Btu/lb of steam produced. This quantity of steam represents about 52.7 million Btu/h heat input (at 82%) to the steam generator. At the above energy price the energy input is equivalent to about $158/h savings. Assuming operation at 8000 h/yr, this is $1.26 million energy savings annually. Differential costs of the systems ($500/kW) is $5 million, which is invested for the $1.26 million annual energy cost savings (25% return on investment). To justify the investment, savings must support a profit, funds to recover plant investment, the escalating cost of added operation and

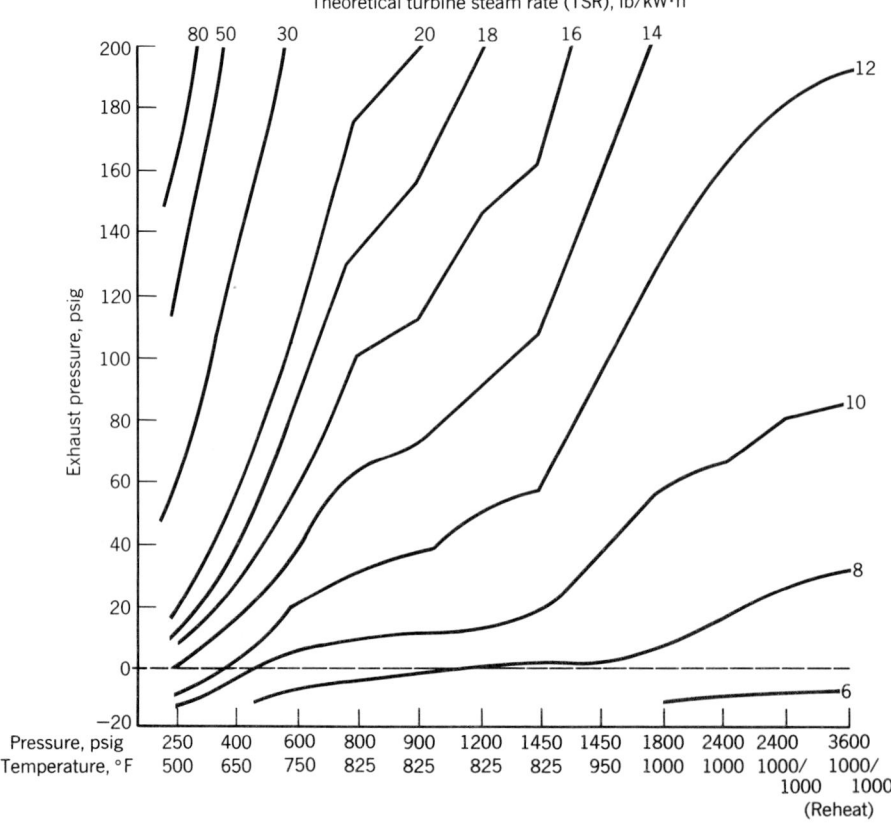

Fig. 5.1-19 Approximate turbine steam rates.

maintenance, and net tax influences (tax savings). In this illustration the return would be marginal depending on the industries' financial practices. Based on the 0.3 lb/kW · h TSR improvement of the 1800- to 2400-psig change, justification is even more difficult.

Heat Balance

The steam turbine operates in a system with a definite energy balance. Because all input energies can be accounted for as load or losses, it is possible (subject to adequate measurements) to compute this balance. Figure 5.1-20 represents a compact system in which energy balance is measurable with a minimum of instruments. The input is fossil fuel, which supplies energy for load and losses. Fuel energy (Btu/h) can be obtained from quantity (units per hour) and heating value (Btu per unit) measurements obtained from billing instruments. Combustion equations, along with a flue gas (Orsat) measurement of oxygen (excess air) and a flue gas temperature, will quantify one of the major losses (stack gas). Output is measured as a product of the pumped water quantity (flowmeter) and the pressure (gauges) rise from pump inlet to outlet (pounds per hour of flow times pressure gain as feet of water). The losses that appear in the system cooling water as heat are measured by obtaining a cooling water flow (orifice, weir, or Pitot tube) and temperatures (thermometers) into and from the system (pounds per hour times temperature rise). Minor losses in leaks (heat to ambient, water to storage level changes or blowdowns) may be measured or cut off for short test period durations. Within a few percentage points, this external energy accounting should balance (sum of outputs equal input). For the general heat balance, power inputs to motors driving auxiliary pumps or fans are not considered due to their small relative size. However, they are considered in reporting of *net* steam or heat rates, which are based on an output less power to auxiliary equipment. Figure 5.1-21 takes the energy balance from the system concept (external interfaces) to an internal cycle breakdown with emphasis on the steam turbine. Figure 5.1-21*a* represents the schematic layout and utilization of steam typical of either the utility (mechanical power to electric generator) or mechanical cogeneration application.

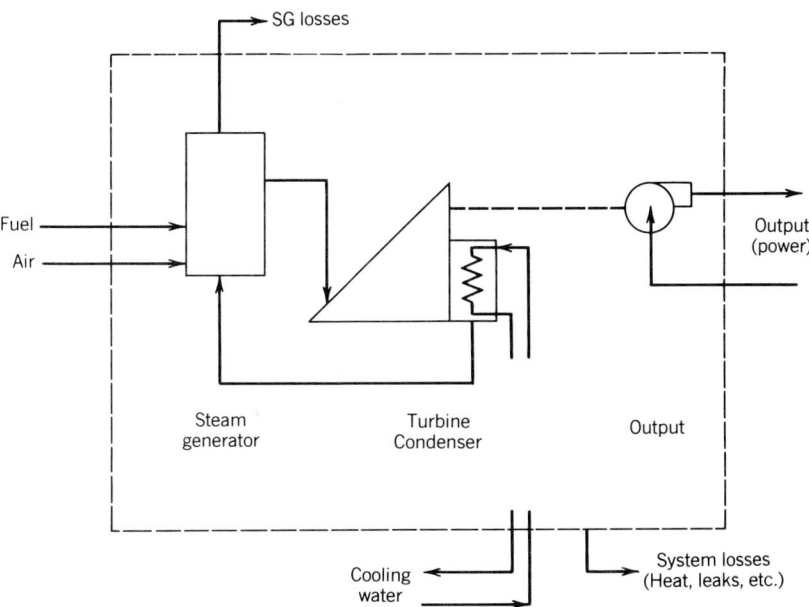

Fig. 5.1-20 Typical turbine system energy flow.

Figure 5.1-21*b* is the turbine steam expansion line (dashed) plotted on the Mollier chart beginning at turbine throttle conditions (psia and degrees Fahrenheit) and extending into the wet region to exhaust pressure. An enthalpy (Btu/lb) value can be taken from the chart at known throttle steam and extraction (steam bleed) conditions (P and T) if in the superheat region. However, instruments indicating steam conditions in the wet region indicate only pressure and saturation temperatures corresponding to pressure. These conditions in themselves do not reflect wet-region steam enthalpy. For the heat balance computations the exhaust point enthalpy must be estimated and the calculations performed using the estimate; if computed exhaust enthalpy differs, it must be reestimated using the new computed enthalpy. The computation is repeated until estimated and computed values match.

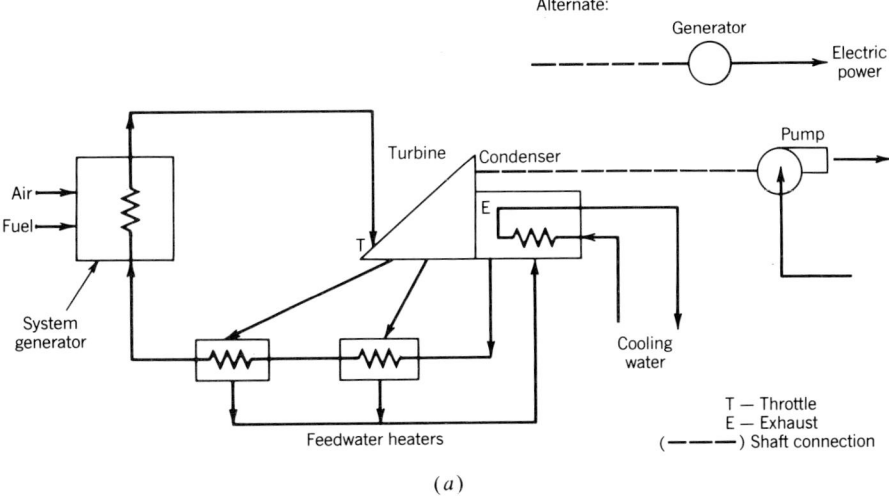

(*a*)

Fig. 5.1-21 Typical turbine cycle steam expansion.

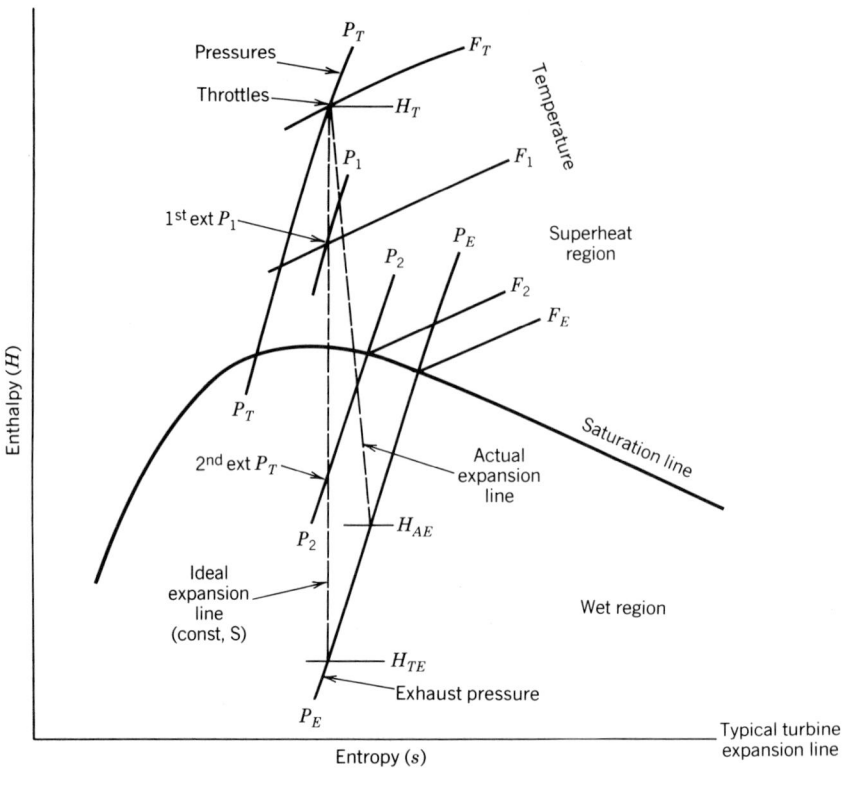

(b)

Fig. 5.1-21 (*Continued*)

Standard measurements include flows (at throttle, at condensate pump, and at boiler feed inlet) and pressures and temperatures [of steam at throttle, turbine extraction points, exhaust, feedwater heater shells, and of various liquids in the boiler feedwater circuit (condenser outlet, feedwater heater condensate in and out and shell drain, and feed to boiler)].

Assuming the single-feedwater circuit Rankine cycle illustrated in Fig. 5.1-22, the following heat balance equations are obtained:

1. Heat to steam.

$$H_s = Q_s (h_1 - h_5) \qquad \text{Btu/h}$$

where H_s = heat to steam, Btu/h
Q_s = steam flow, lb/h
$h_1 - h_5$ = enthalpy increase, Btu/lb
h_1 = throttle steam enthalpy, Btu/lb
h_5 = boiler feed enthalpy, Btu/lb

2. Heat to boiler.

$$H_b = H_s (1/U_B)$$

where H_b = heat to boiler, Btu/h
U_B = boiler efficiency, % $\times 10^{-2}$

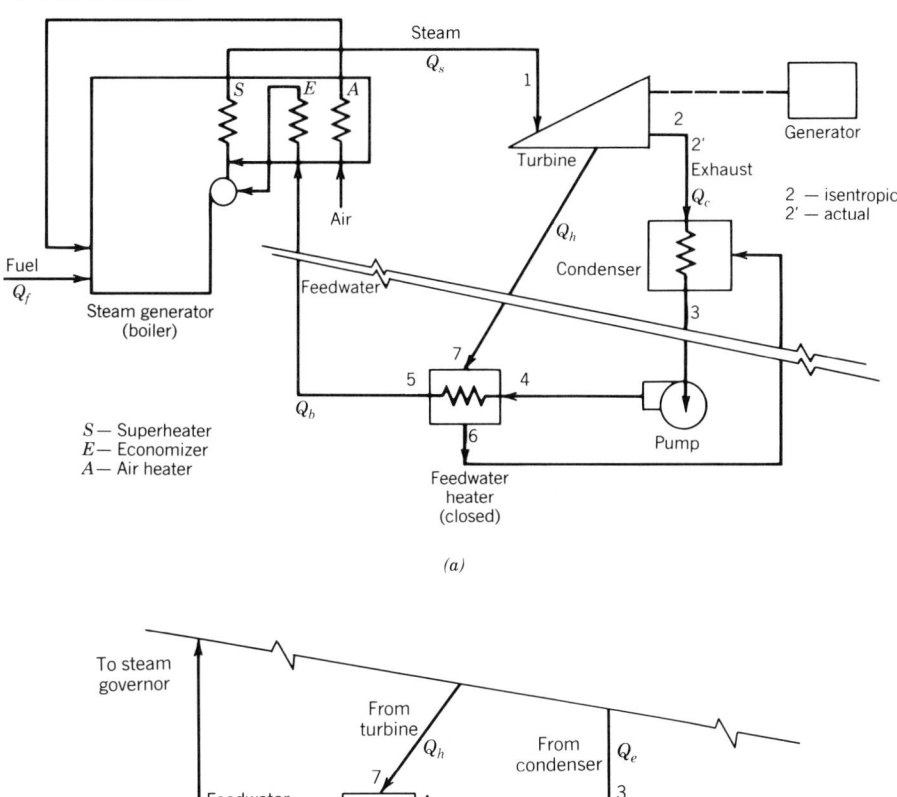

(b)

Fig. 5.1-22 Single-heater turbine cycle. (a) Closed FW heater cycle. (b) Open FW heater cycle.

3. **Heat balance across feedwater heater (FWH).**

$$\text{Closed FWH:} \qquad Q_b(h_5 - h_4) = Q_h(h_7 - h_6)$$

$$\text{Open FWH:} \qquad Q_b(h_5 - h_4) = Q_h(h_7 - h_4)$$

$$\text{Flows:} \qquad Q_s = Q_b = Q_h + Q_e$$

where Q_b = feedwater flow to boiler, lb/h (also $= Q_s$, lb/h)
Q_h = steam flow to FWH, lb/h
Q_e = steam flow to exhaust, lb/h
h_4 = enthalpy condensate to FWH, Btu/lb
h_7 = enthalpy steam to FWH, Btu/lb

4. **Extraction steam to heater (from turbine).**

$$\text{Closed FWH:} \qquad Q_h = Q_b(h_5 - h_4)/(h_7 - h_6) \qquad \text{lb/h}$$

$$\text{Open FWH:} \qquad Q_h = Q_b(h_5 - h_4)/(h_7 - h_4) \qquad \text{lb/h}$$

5. **Heat added to feedwater by pump.**

$$\text{Closed FWH:} \quad H_p = Q_b(h_4 - h_3) = (Q_e + Q_h)(h_4 - h_3)$$
$$\text{Open FWH:} \quad H_p = Q_e(h_4 - h_3)$$

where H_p = heat added by pump
h_3 = enthalpy condensate from condenser

6. **Heat rejected to condenser.**

$$\text{Closed FWH:} \quad H_c = Q_e(h'_2 - h_3) + Q_h(h_6 - h_3)$$

$$\text{Open FWH:} \quad H_c = Q_e(h'_2 - h_3)$$

where H_c = heat to condenser, Btu/h
$h'_2 - h_3$ = enthalpy loss to condenser from exhaust, Btu/lb
$h_6 - h_3$ = enthalpy loss to condenser from heater drain, Btu/lb
h'_2 = actual exhaust enthalpy (wet region), Btu/lb

7. **Work (heat equivalent) of turbine.**

$$\text{Extraction steam:} \quad Q_h(h_1 - h_7) \quad \text{Btu/h}$$

$$\text{Exhaust steam:} \quad Q_e(h_1 - h'_2) \quad \text{Btu/h}$$

$$W_t(\text{total}) = Q_h(h_1 - h_7) + Q_e(h_1 - h'_2), \text{Btu/h}$$

where h'_2 = actual enthalpy exhaust steam
W_t = heat equivalent of turbine work

8. **Steam turbine efficiency (thermal).**

$$u_t = (h_1 - h'_2)/(h_1 - h_2)$$

where h_2 = exhaust enthalpy at 100% efficiency (constant entropy from throttle conditions), Btu/lb
u_t = efficiency

9. **Generator output.**

$$\text{KW} = (W_t/3413)(u_g)$$

where KW = generator output, kW
u_g = generator efficiency
and 3413 is British thermal units per kilowatt.

10. **Overall cycle thermal efficiency.** Efficiency = $(\text{KW}(3413)/H_b)(100)$, %

11. **Heat Rate.**

$$\text{(Gross):} \quad \text{Btu/kW} \cdot \text{h} = H_b/\text{KW}$$

$$\text{or} \quad \text{(Net):} \quad \text{Btu/kW} \cdot \text{h} = H_b(\text{KW} - \text{KW to Auxiliary Systems})$$

5.1-6 Turbine Selection

The starting point for a steam turbine application is a description of the cycle (typical of Fig. 5.1-21) with performance intent and boundary energy balance. The heat balance (see discussion of heat balance above and Fig. 5.1-22) is extensive, optimizing (by trial and results) various steam and water conditions, flows, efficiencies, and costs throughout the proposed detail cycle. The basis of the computations is the required output in kilowatts if turbine drives on electric generator or in kilowatts or horsepower plus a process steam requirement for an electric or mechanical cogeneration cycle. Fuel sources (availability and cycle cost including waste disposal) and availability of cooling water establish other bases. From these basic criteria the heat balance computations will yield steam and water conditions and flows specifically leading to the turbine design duties. The turbine called for, now generally defined, must be characterized in its detail performance and matched to establish cycle duties. Turbine characteristics are described in general and specifically by manufacturer in forms of working curves. The general turbine curves include the following:

1. Theoretical steam rates.
2. Maximum feedwater enthalpy rise.
3. Stage efficiency related to ratio blade velocity to steam velocity.
4. Basic efficiency (mechanical drive turbine) versus output horsepower with correction curves for degrees superheat and speed.
5. Expected heat rate correction factors for change in throttle conditions.
6. Expected gains from feedwater heating.
7. Expected gains from reheat.

For a specific turbine, the manufacturer is able to provide the following:

1. Steam rate (SRC) and heat rate (HRC) curves.
2. Corrections to SRC and HRC for various throttle pressures and temperatures.
3. Corrections to SRC and HRC for various back pressures (exhaust with flow).
4. Exhaust loss variations with load.
5. Steam expansion line at various loadings.
6. Steam expansion line end point varying with back pressure and throttle and reheat flow.
7. Generator losses varying with output (in kilowatts).
8. Extraction pressures as a function of flow to following stage.
9. Willans line.
10. First-stage pressures for variations of load.
11. Packing leak enthalpy and quantity varying with throttle flow.

Turbine designers must consider ratios of blade to steam velocities by stage, number, and load balance of stages and steam path (areas throughout blade length, seals, moisture, etc.). Others who would basically size a turbine think in terms of theoretical steam rate, overall cycle efficiency, fit to cycle requirements (output, steam demand, space, loading characteristics, etc.), and cost of system as affected by turbine selection. Utility turbines are generally offered by frame size with a heat rate curve series as shown in Fig. 5.1-23. At any design load there are generally two frame-size choices, each with

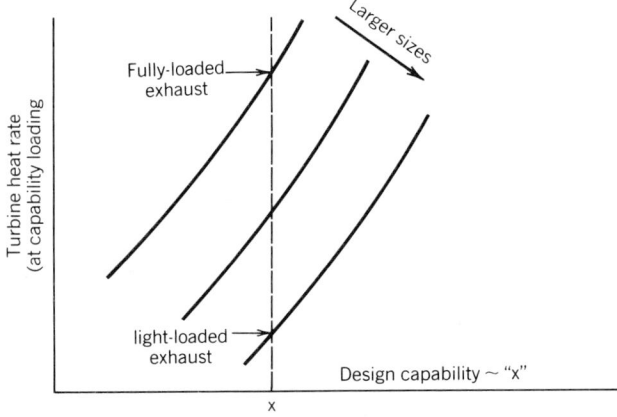

Fig. 5.1-23 Typical heat rate curves for turbine frame sizes. Heat rate curves established by (1) number of shafts, (2) number of exhaust flows, (3) size of the last-stage blades, and (4) throttle conditions (P/T).

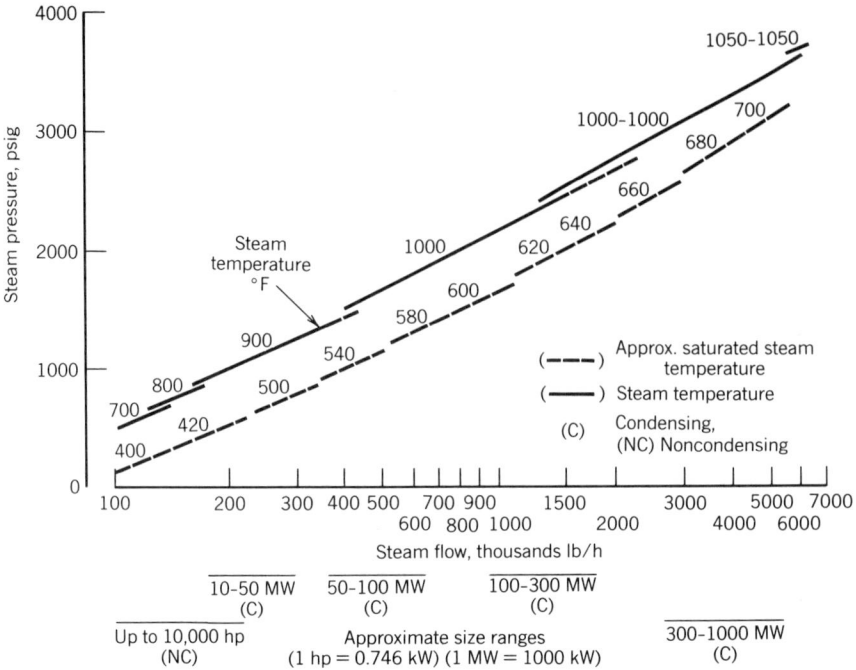

Fig. 5.1-24 Typical turbine steam throttle conditions.

a corresponding heat rate and capital cost. Those operators sizing and financing a utility turbine selection are faced with optimizing to expected load duration patterns, to heat rate (fuel economy), to turbine frame size (capital cost of turbine and its effect on balance of plant cost), and to expected operation and maintenance differentials.

The available energy expressed as theoretical steam rate may vary considerably with throttle steam conditions, back pressure, and extraction use. Once these cycle parameters are set, the detail turbine designs can establish specific load point, maximum turbine efficiency, and a load range over which this highest possible efficiency is maintained. Due to the ability to design high performance into turbines over a wide range of applications, the specific turbine selection may be done with some confidence based on a study of the economics of cycle alternatives as determined by accurate cycle factors. The overall economics may vary considerably (lower TSR resulting in lower fuel cost but more expensive equipment and operating cost), thus requiring the accuracy associated with accurately estimating a minimum number of cycle variables. In comparison with other estimates, an estimate of turbine efficiency at the expected maximum is less effective on results. Further, if steam conditions recognized as "standards" (Fig. 5.1-24) are utilized, the estimate of turbine efficiency is allowed an even wider tolerance without influencing results. Such an approach in turbine selection may be illustrated in the following illustrations:

Illustration 1.   Assume that an industry has a motor-driven pump requiring 1500 hp shaft input and 25,050 lb/h (40 psia saturated) steam demand. Fuel cost is $5 per million Btu and electric power (including demand charges) cost is $0.10/kW · h. A cogenerating steam turbine (mechanical) alternate for pump drive is to be considered. Assume the turbine is 74% efficient and steam conditions at throttle may vary between 400 psia/600°F and 1450 psia/950°F. The turbine exhausts into the steam process, which returns all condensate at 200°F. There are no extractions, boiler efficiency is assumed to be 84%, and motor efficiency is 93%.

The possibility of utilizing a cogeneration turbine exists because of the need for process steam and availability of an economical fuel.

The advantage of shifting electric power input to a fuel input shows up as a reduction in energy cost due to the lower cost of the fuel. Energy consumption will be slightly higher after the shift. This is due to the electric power entering at 93% efficiency as compared to fuel energy entering at a lower [84% (boiler) times 74% (turbine)] efficiency. A lower fuel input energy cost is needed to offset this lower-use efficiency in addition to the higher cost of cycle equipment. Fuel cost would need to be less than 70% of electric energy cost to approximately balance overall costs.

Assuming the cycle of Fig. 5.1-25, the heat balance computation begins with a selection of standardized throttle steam conditions that will allow sufficient enthalpy drop to produce required shaft power at the expected turbine thermal efficiency. Based on known exhaust pressure and the selected turbine throttle steam conditions and assumed efficiency, the actual exhaust enthalpy is computed [Eff $= (h_1 - h_2')/(h_1 - h_2)$]. Comparing the enthalpy drop with heat equivalent of required shaft output [(horsepower $\times$ 2545)/($h_1 - h_5$)] provides turbine steam flow (Q_s), which should equal (in heat equivalent) the heat requirement of process steam—as in the 400 psia/600°F illustration. If it does not (as in the other illustration), a restart with alternate expansion line assumptions is made. If the turbine needs less steam than process, the difference may be made up directly from the boiler (note circuit of Fig. 5.1-25 and line 8 of the following analysis). Once flows and enthalpies are determined, a total energy input comparison (line 10) can be made. By applying local energy costs (line 13), these inputs are converted to costs (line 14). A summation of computed data follows:

		Motor	Turbine	Turbine	Turbine
1.	Cycle				
2.	Throttle pressure/temperature (psia/°F)		400/600	800/800	1450/950
3.	Throttle enthalpy (h_1) (Btu/lb)		1,306	1,397	1,463
4.	Exhaust enthalpy (h_2): At 40 psia/const. S (Btu/lb)		1,106	1,110	1,106
	At 40 psia/74% TE (h_2') (Btu/lb)		1,158.0	1,184.6	1,198.8
5.	Heat balance: Shaft output at 1500 hp (Btu/h $\times 10^3$)		3,818	3,818	3,818
	Steam flow (lb/h) (Q_s)		25,800	17,980	14,451
6.	Process heat demand (Q_p) at 25,000 lb/h (Btu/h)		25,000	25,000	25,000
	Process heat demand equivalent at H_E (Btu/h)		25,300	24,640	24,301
7.	Steam flow ratio (%) (Q_s/Q_p)		102	73.0	59.5
8.	Boiler steam supplement (lb/h)		0	5,510	7,840
9.	Boiler steam flow (Q_b) (lb/h)	25,000	25,800	23,490	22,291
10.	Inputs				
	To boiler (Btu/h) $\times 10^3$	29,821	34,953	34,368	34,365
	To motor (Btu/h) $\times 10^3$	4,105	0	0	0
	Total (Btu/h) $\times 10^3$	33,986	34,953	34,368	34,365
11.	Turbine steam				
	(lb/hP $\cdot$ h)		(17.2)	(12.0)	(9.6)
	(lb/kW $\cdot$ h)		(23.1)	(16.1)	(12.9)

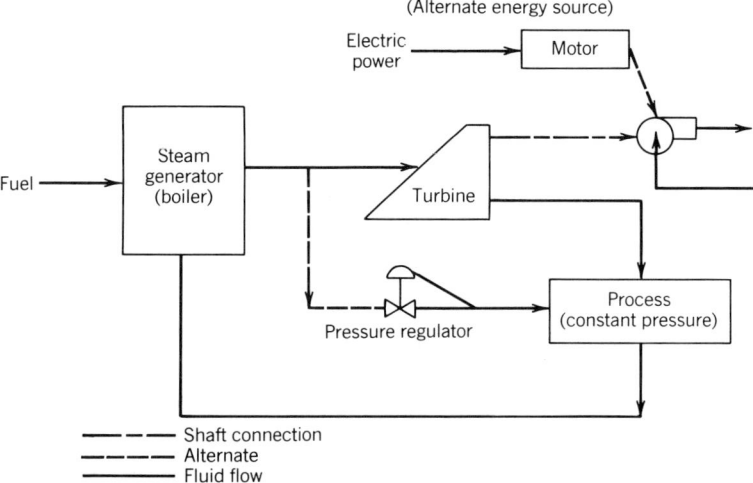

Fig. 5.1-25 Typical steam turbine cogeneration cycle.

12.	Turbine heat rate: At 84% BE (Btu/hp · h)		23,300	17,550	14,800
13.	Energy cost (unit): Fuels ($/MBtu)		5.00	5.00	5.00
	Electric power ($/kW · h)	0.10	—	—	—
	Electric power equivalent ($/MBtu)	29.3	—	—	—
14.	Boiler energy cost ($/h)	269.4	174.7	171.8	171.8
15.	Savings ($/h)	Base	(94.7)	(97.6)	(97.6)
	(% of base cost)		(35)	(36)	(36)

Other factors are considered prior to knowing actual cost advantages. These include the cost of financing new equipment installation, increased operation (including energy) and maintenance, and the effects of loading (load duration curve), taxes, and other owner costs. To continue the illustration, assume that new equipment costs $3 million, that the load duration curve indicated an average 80% loading over 8000 h/yr, that turbine efficiency is equal at 100 and 80% loading, that a 16%/yr rate will repay capital, that annual operating and maintenance cost is 5% of capital, and that annual tax savings consist of 50% of the 10% energy investment and 15% depreciation. Annual net income would be $756,800 less $480,000 (capital recovery) and $150,000 (operating and maintenance) plus tax savings of $150,000 (energy investment, first year only), $150,000 investment credit, and $255,000 (depreciation) for a net of about $510,000 or a simple payback period of about 6 years.

Illustration 2. Assume a 60-MW electric generator is driven by a condensing turbine. For comparison, steam conditions are chosen to be either 1200 psia/800°F or 2400 psia/1000°F and a single heater (either closed or open type) is applied to both cycles. Assumptions include a turbine thermal efficiency of 74%, a steam generator efficiency of 84%, and a feedwater heater heating condensate to 42% of maximum possible temperature use at a zero terminal difference. (Outlet temperature is equal to saturated steam temperature of extraction pressure.)

Utilizing the heat balance equation (see discussion on heat balance) previously discussed, the following results are obtained:

Throttle (psia/°F)		1200/800		2400/1000	
Generator output (MW)		60		60	
Turbine exhaust (psia)		2		2	
Heat contents (Btu/lb):					
At point (see Fig. 5.1-21)	1 (THR)	1380		1460	
	2 (EXH)	1020		1038	
	3 (C)	69		69	
	5 (FW)	267		308	
	7 (EXT)	1174		1217	
Type of FW Heater	Closed	Open	Closed	Open	
Throttle flow (lb/h)	660,168	647,437	575,225	559,988	
Extraction flow (lb/h)	143,918	115,244	151,285	116,985	
Exhaust flow (lb/h)	516,250	532,193	423,940	443,003	
Steam rate (lb/kW · h)	10.5	1.03	9.1	8.9	
Turbine heat rate (Btu/kW · h)	11,634	11,409	10,492	10,214	
Cycle heat rate (Btu/kW · h)	13,850	13,583	12,491	12,159	
Cycle thermal efficiency (%)	24.6	25.1	27.3	28.1	

Notes: Actual exhaust enthalpy from Eq. (8), Heat Balance.
Extraction enthalpy from plot of expansion line at pressure corresponding to temperature at 42% of max. rise.
Cycle heat rate based on steam generator input.
Heat rates are gross hour (cycle auxiliary power use credited).

The analysis of the two turbine cycles [different throttle conditions and type of feedwater (FW) heater] indicate an advantage in the higher throttle steam conditions (about 10% in heat rate) and in the open FW heater (about 2% in heat rate). This savings, expressed as dollars of annual fuel savings, is offset by added costs of the more expensive equipment and the increased operation and maintenance cost. A comparison of savings and added cost will indicate the net gain or simple payback period of investment. In general terms, if capital cost increases about 25%, annual financial charges (assumed at 16% annual financing rate) would increase about 4% and operating and maintenance (assume 10%) increases 3%. Other miscellaneous costs and tax benefits might improve on the energy savings up to 2% for a net gain of about 4.7%.

5.1-7 Turbine Performance from Test Data

Assume that the 100-MW utility turbine of Illustration 1 has been in operation for several years. A recent maximum-load test resulted in a 112.5-MW generator output at turbine throttle conditions of 1260 psia (5% over pressure) per 805°F and an exhaust pressure of 3 psia (6.1 in. Hg absolute). The extraction pressure is 28 psia, condensate flow from condenser is 950,000 lb/h, and boiler feedwater is 244.5°F. A heat balance is to be computed resulting in steam flows at throttle and to deaerating feedwater heater, steam rate, heat rate at an assumed 84% boiler efficiency, and the maximum load corrected to design throttle conditions (1200 psia/800°F) and 1 in. Hg absolute back pressure. A 5% pressure drop is assumed from turbine to deaerating heater.

On the Mollier chart, a plot of the expected turbine expansion line (1260 psia/805°F to the 3-psia exhaust pressure line at about 78% turbine efficiency) indicates both extraction and exhaust to be in the wet region. A steam calorimeter test of the pressurized extraction steam shows about 6% moisture. Because instrument readings (steam pressure and temperature) of wet steam locate points only on the saturation line, the exact points (and enthalpy) of the expansion line in the wet region are not known. One method of computing the expansion line end point (ELEP) is to estimate and reestimate the exhaust enthalpy until the equations show a flow and heat balance. In this case the estimation process will result in an expansion line reasonably in line with the design line as well as a balanced heat flow.

Flow and heat balance equations are set up and manipulated to an equation relating extraction enthalpy to exhaust enthalpy. Various estimates of exhaust enthalpy are inserted until a plot of the extraction and exhaust points result in a normal expansion line. The resulting enthalpies will allow flows to be computed; which, when inserted in the heat balance equations, show a balance.

The following equations apply to this cycle:

1. Turbine flow: $Q_T = Q_D + Q_E$

    where Q_T = throttle flow, lb/h
    $\quad\quad Q_D$ = extraction flow to DA heater, lb/h
    $\quad\quad Q_E$ = exhaust flow, lb/h

2. Turbine heat balance: $Q_T H_T = Q_D H_D + Q_E H_E + (\text{kW}/U) \times 3413$

    where H_T = throttle enthalpy, Btu/lb
    $\quad\quad H_D$ = extraction enthalpy, Btu/lb
    $\quad\quad H_E$ = exhaust enthalpy, Btu/lb
    $\quad\quad U$ = generator efficiency, %

3. Deaerating feedwater (FW) heat balance: $Q_D H_D + Q_E H_C = Q_T H_{FW}$

    where H_C = condensate enthalpy, Btu/lb
    $\quad\quad H_{FW}$ = boiler feedwater enthalpy, Btu/lb

    For the given data,

    $\quad Q_E = 950,000$ lb/h
    $\quad H_T = 1379.8$ Btu/lb (1260 psia/805°F)
    $\quad H_C = 109.4$ Btu/lb (3 psia)
    $\quad H_{FW} = 212.6$ Btu/lb (244.5°F)
    $\quad$ DA pressure $= 26.6$ (95% of 28 psia)

    The above procedure results in

    $\quad H_E = 985.0$ Btu/lb
    $\quad H_D = 1112.8$ Btu/lb
    $\quad Q_D = 108,970$ lb/h
    $\quad Q_T = 1,059,470$ lb/h
    $\quad$ Steam rate $= 10.6$ lb/kW
    $\quad$ Heat rate $= 12,366$ Btu/kW
    $\quad$ Turbine thermal efficiency $= 84.1\%$
    $\quad$ Rankine cycle efficiency $= 27.6\%$

To obtain corrected value of maximum load at design conditions, manufacturer's correction curves for this specific turbine are needed. Typical curves shown (Figure 5) would indicate corrections to load as follows:

From 1260 to 1200 psia—	×0.988
From 805 to 800°F—	×0.992
From 3 in. psia exhaust to 1 psia—	×1.005
Product	0.985

Correct load: $11.25 \times 0.985 = 110.8$ MW

5.1-8   Operation and Maintenance

General

Because the turbine is a heat-converting machine, it is subject to operating limits typical of metal performance under heat stress conditions. Since mechanical motion is the first step in the conversion, turbine-operating procedures must ensure that turbine operation is safely within any dynamic restraints imposed by the motion of fluids (steam) or rotating metals. With the most sensitive turbine parts enclosed and out of sight, operation is done utilizing an instrument system that informs the operators of conditions within the turbine. Using these indicators and routine inspections to discover wear, deposits, or damage, the operator is able to start, operate, hold in an extended shutdown state, and maintain the turbine to ensure minimized outages for maintenance and high efficiencies during operation. Modern large turbines are operated with automatic controls, capable of a programmed operation for the turbine to minimize thermal and rotating stresses, to bring to rated speed, to initially apply a load, to follow a load demand, to shut down the turbine under normal and emergency conditions, and to monitor, control, and protect the turbine and the auxiliary systems during these events.

Turbine loading

Factors affecting the ability to load a steam turbine range from those internal to the machine or its driven equipment to external (system) influences. Internal variance from design conditions may only limit loading to a fraction of the capability or be extensive enough to cause a forced outage. External factors include the influence of turbine mechanical support systems (steam supply, condenser, controls, etc.), as well as the economical conditions under which turbines operate (output needs, costs, economic dispatching of a utility turbine, etc.). The dispatching commands, as well as protective signals from driven equipment or system disturbances, are imposed on the turbine governor system to produce desired turbine loading. Internal influences, whose effects are also imposed through the governor system, may require manually established load limits or rate of loading and automatic protective actions (load reduction or dump, cooling or fire protection system implementation, etc.).

In the start-up mode the large turbine shaft is rolled at a low rate of speed (3–10 rpm) to work out any shaft bow occurring during shutdown. A manual or automatically connected turning gear performs this task. This may be done for several hours until the eccentricity meter indicates an acceptable shaft straightness. Upon admission of steam, speed increasing automatically releases the turning gear. At this point thermal saturation (stable metal temperatures), condensation (water accumulation), thermal expansion, critical speeds (resonant vibration), and (on condensing turbines) exhaust heating become concerns and are monitored carefully. To aid in evening metal temperatures, a full-arc admission start-up may be used. By manipulating all control valves to wide-open positions and admitting steam through a small main stop valve bypass to the valve chest, steam is forced to enter first stages in full circle for even metal heating. At a later point in start-up normal governor valve operation is restored. Start-up control systems often perform this valve manipulation automatically as well as perform checks on other start-up criteria. Continuation of start-up procedures is allowed upon compliance with start-up criteria, including assurance that turbine drain valves are open, that allowable metal temperature differentials exist, that normal expansion is occurring, and so forth.

Turbine component natural vibrations relate to shaft speed. If a number of component vibrations resonate at a specific speed, the vibration magnitude increases. In most turbines a resonant speed is encountered requiring that start-up procedures recognize this "critical" speed and pass through as rapidly as possible. The steam energy required to accelerate shaft speeds is momentarily reduced after normal speed is reached. Then, as the initial loading (up to 20%) is applied, additional turbine metal is exposed to the hotter steam, again requiring a monitoring of expansion, vibrations, and drainage.

The control system is usually hydraulic with pilot valves, sensitive to small forces, amplifying the small hydraulic signals to larger forces needed to manipulate large control valves. Very often it is necessary to limit the maximum load or loading range on a turbine due to some mechanical reason. The instrument and control system provides a way to do this through limits on hydraulic pressure, mechanical stop on valves, setting of limit switches on governor control motors, linkage adjustment in governor systems, or by set points in the electronics of the electro-hydraulic control systems.

Load control may extend beyond the specific turbine. In an industrial plant where many mechanical drive turbines operate in parallel, or in a major utility operating parallel turbine electric generators, a control system is utilized for economic assignment (dispatching) of system loads. Each turbine is tested to determine the cost of a unit output over its load range. The instrument package (along with the proper test procedure) produces an accurate accounting of inputs (steam conditions and flow) and outputs (horsepower, kilowatts, etc.) for the production of a heat rate curve (Btu per unit output vs. output), which can also be plotted as cost. The first derivative of this curve is replotted as an incremental heat rate curve to show incremental cost as a function of incremental load changes over the turbine load range. A central load control center able to analyze system needs automatically surveys individual turbine conditions and, through use of these curves, adjusts turbines to produce most economical sources of the system needs. Timing and size of incremental adjustments, which depend on number and rating of system turbines as well as the dispatching computer, are done in seconds at loading increments of less than 1% of system capability. Such an automatic, economic dispatch control system requires a system (plant) and turbine instrument and control system capable of providing accurate incremental loading and heat rate curves, of accepting remote control, and of providing remote feedback information confirming responses to loading demands.

Operations

With the introduction of hot steam, the turbine system is subjected to the hazards of thermal stresses, expansion, and water droplets caused by condensation. Thermal stresses are built up within the heavier sections by temperature differentials created during initial starting (warmup) or during significant load changes. These stresses, particularly if cycled frequently, may cause cracks in the heavier steel sections, such as in first-stage nozzle blocks. To minimize thermal stress effects, different operating procedures are established for cold or hot turbine starts, for turbines with (or without) a full arc steam admission start-up system, for turbines designed with smaller sections to allow quick starts (peaking or main service), or for turbines instrumented for temperature differentials. The readout of differences (usually held to about $200°F/h$) removes some uncertainties, thus allowing a faster start-up. Uniform heating (full arc admission) also allows faster start-up due to reduction of temperature differentials within heavier sections.

As hot steam enters the turbine and metal sections are heated, expansion occurs. The rotor, heating faster than the casing, will stretch out faster, thus affecting the clearance in seals. Turbines designed for high efficiency (close seal clearance) are allowed more thermal expansion time to ensure casing expansion along with rotor expansion. The peaking or quick-start features include large seal clearances, allowing for relative expansion at the expense of efficiency. The turbine designs include anchors for both casing (foundation keys at low pressure end) and rotor thrust bearing at high steam pressure end. As casing is heating, either side expansion is observed, and often instrumented and recorded, to ensure no casing bend due to uneven expansion on one side relative to other. To inform the operator of the position of the shaft, some turbines are equipped with a micrometer extending through the casing to contact a rotating part. A clearance reading is observed by screwing the micrometer shaft in until a light contact is made. This procedure is most useful for obtaining dummy piston clearances in reaction turbines.

As hot steam contacts the cooler turbine metal, condensation occurs. This results in formation of droplets of water, which cause metal erosion, rotor vibrations, and efficiency reduction. Formation of water occurs mostly during start-up (warmup period) or (during operation) at the turbine stages where steam characteristics enter the wet region (Mollier chart). During start-up, turbine casing drain line valves are opened, allowing removal of water. During initial operations or until thermal saturation (equal steam–metal temperatures) occurs, these drain valves remain open. During normal operation, transfer of heat energy from steam to shaft kinetic energy, wet steam forms in the final stages and deposits appear on these turbine blades. Operators can expect these deposits if main steam contains impurities and can detect significant deposits from instrumentation. Deposits will introduce flow resistance, causing a pressure increase about normal in turbine stages upstream from deposits and a decrease in maximum turbine output.

Steam purity (lack of impurities) is related to water chemistry of boiler water and boiler feedwater. Deposits often found in turbine include oxides of iron and copper and scales containing carbonates, phosphates, and sulfates. The oxides (appearing as a hard scale in high-pressure turbine sections) come from corrosion of metal parts of boiler, piping, or turbine components in contact with any oxygen in the steam. To minimize this formation, boiler feed and boiler water are treated for pH adjustments, scale control and with an oxygen-scavenging agent. Copper comes from copper tubes of heat exchangers in the water or steam cycles through the erosion or corrosion process. Other scales come from boiler water as carry-over in steam generation process. Excess amounts of oxygen scavenger (SO_3) or softeners (PO_4) as well as boiler makeup impurities (concentrated in boiler water) may be sources of turbine blade deposits.

In addition to minimizing thermal effects by uniform, slow heating and condensation control (high initial superheat and reheat cycles), turbine shells are well insulated, high-temperature valve stem leak-off are not reintroduced but used externally, and special turbine metallorgy is used.

Steam turbine rotors turn at speeds from about 1800 to 5500 rpm. These speeds, combined with close clearances at seals, blading, and bearings, require very low vibration levels (less than 5 mil shaft vibration). Vibrations are caused by rotating frequencies resonating with natural frequencies of structural parts (critical speeds), by unbalanced conditions of rotating shaft, by shaft bends, by poor shaft alignments, by contacts between rotating and stationary parts, by steam flow patterns that may pulsate due to condensation (water droplets) or deposits, by lubricating oil whip due to excessive bearing clearances or low oil temperature, or by a rotating part defect (crack, loose blade component, etc). Defects may be caused by internal stresses, excessive dynamic forces (overspeed), weakness due to wear, severe load cycling, or poor maintenance procedures.

The most common causes of vibrations are shaft out-of-balance or misalignment. Throughout the life of the turbine (especially after major overhauls), shaft balance is checked and adjusted as required. A static balance is initially done, but the most popular, dynamic balance check is field performed with the shaft being rotated in place by steam up to normal speeds. Special instrumentation measures speed and pickup from vibration meters. From this data the instrumentation computes size and location on wheel periphery of weights to be adjusted. The turbine is shut down, the weights adjusted, and the turbine restarted. By trial and error, each adjustment should improve vibration until full speed is reached at acceptable vibration. Usually good dynamic balance is obtained prior to encountering a critical speed (resonance), which probably occurs at about 75% of rated speed. Turbine start-up procedures require warmup periods (holding at low speeds for about an hour) at less than the critical speed and a fast increase through the critical range. Other critical speeds (at multiples of the first) occur at rpm above rated speed and so are not encountered. Alignment of individual turbine (HP, IP, and LP) shafts and with driven equipment is critical if no vibration effect is to be encountered. Large turbine shafts often take on a small bend (vibration cause) during extremely long outage periods. Prior to start-up (shaft roll on steam), those bends can be removed by turning the shaft at about 5 rpm with a motorized "turning gear" for several hours.

Turbine Operation

Prior to starting a turbine, operators should be knowledgeable about the characteristics of the boiler and its ability to provide a controlled pressure–temperature steam supply, the piping systems (including steam line and turbine warmup controls), all turbine auxiliary systems (including exhaust control and lubrication), computerized start-up programs and methods of clearing start-up restraints if applied, and turbine-operating limitations (heat-up rates, loading rates, mechanical limits, governor actions, and the characteristics of the driven equipment). Then the operator can begin general start-up:

1. For a small noncondensing turbine:
   (a) Turbine is isolated from steam source.
   (b) Lubrication and cooling systems are put in operation.
   (c) Turbine is placed on turning gear (if available).
   (d) Drains are opened.
   (e) Governor system is placed on manual control.
   (f) Steam is made available up to turbine main stop valve.
   (g) Exhaust steam use systems are lined up.
   (h) Turbine is rolled on steam using main valve bypass.
   (i) Follow predetermined warmup procedure to rated speed,
   (j) Safety checks (overspeed, emergency trip, etc.) are performed.
   (k) Steam flow control is placed on governor system.
   (l) Safety checks (bearings, expansion, vibrations, etc.) are performed.
   (m) Turbine is tied to load and small load is applied.
   (n) Proceed with checks and turbine warmup.
2. For a larger condensing turbine:
   (a) Lined up as with noncondensing turbine, making steam available to stop valve,
   (b) Shaft sealing system is placed in service.
   (c) Cooling water is provided to condenser.
   (d) Condenser system is started and vacuum (air evacuation) is established (turbine exhaust must be cooled and condensed to prevent overheating of exhaust and condenser chamber).
   (e) Continue as in noncondensing turbine following warmup procedure.
3. For loading a turbine from a starting load position:
   (a) Governor and load control system is set to automatic.
   (b) Slow load increase is continued up to about 20% of rated load. (*Note:* Because hotter steam enters latter turbine stages during this period, metal heating and condensation continue. Uniform heating and drainage continue.)

    (c)   If start is on full arc admission, operation is converted to normal governor.

    (d)   Drain valves closed.

    (e)   Performance checks made for lubrication, vibration, expansion, exhaust conditions, and cooling systems.

    (f)   Turbine set to automatic load control.

4. For normal turbine shutdown:

    (a)   Turbine set on manual control and load slowly dropped (in tandem with transfer of load).

    (b)   At about 5% (or 0% if valve zero-steam-flow is to be checked) load turbine is tripped with emergency trip.

    (c)   Exhaust is secured (if condensing, break vacuum).

    (d)   At zero turbine speed, turning gear is applied.

    (e)   Drain valves are opened.

    (f)   Exhaust condenser systems are secured.

    (g)   After turbine cools to ambient temperatures, turning gear, lubrication system, and cooling systems are secured.

The general turbine-operating procedures are set to minimize maintenance by attempting to eliminate the hazards of thermal stresses, expansions, and water formation and to provide the required lubrication and cooling at all times.

BIBLIOGRAPHY

"ASME Steam Tables," American Society Mechanical Engineers, 1967.

Bartlett, R. L. *Steam Turbine Performances and Economics*. McGraw-Hill, New York, 1958.

Baumeister, T. *Marks Standard Handbook for Mechanical Engineers*, 8th ed. McGraw-Hill, New York, 1978.

Church, Jr., E. F. *Steam Turbines*. McGraw-Hill, New York, 1950.

Croft, T. W. *Steam Turbine Principles and Practice*. McGraw-Hill, New York, 1940.

Gaffert, G. A. *Steam Power Stations*. McGraw-Hill, New York, 1946.

Gartmann, Hans. *DeLaval Engineering Handbook, 1970*. McGraw-Hill, New York, 1970.

Kent, R. T. *Kent's Mechanical Engineers Handbook—Power*. Wiley, New York, 1942.

Kovacik, J. M. "Cogeneration Considerations in the 1980s." General Electric

"Library of Practical Power Generation," Power Report, 1949.

Lindsey, J. R., M. R. Bishop, and R. L. Ullinger, "Improving Steam Turbine Reliability," Power Report, 1976.

Newman, L. E. *Modern Turbines*, General Electric Co. Wiley, New York, 1947.

Newman, L. E. *Modern Turbines*. Wiley, New York, 1944.

Nock, H. T., J. B. Wagner, and C. W. Medeiros, "Integrating Steam Turbine Control with Energy Management Systems," ASME Industrial Power Conference, 1981.

Salisbury, J. K. *Steam Turbines and Their Cycles*. Wiley, New York, 1950.

Schofield, P. "Maintaining Optimum Steam Turbine-Generator Thermal Performance," presented to Missouri Valley Electric Association, 1981.

"Steam Turbines," Power Special Report, 1962.

"Theoretical Steam Rate Tables," American Society Mechanical Engineers, 1969.

"Water Induction in Large Steam Turbines—Design Recommendations," General Electric Bulletin GEK25504E, 1981.

Woodruff, E. B. and H. B. Lammers. *Steam Plant Operations*, 3rd ed. McGraw-Hill, New York, 1967.

5.2   STEAM ENGINES

Harlan T. Holmes

5.2-1   History

The displacement-type steam engine appeared as a practical power source in the latter years of the eighteenth century and served for over a hundred years as the principal prime mover. It has been almost entirely replaced by the steam turbine. In small sizes the noncondensing engine may still be found where a combined steam and power need exists and where the power characteristics include a wide variation of speed with a high power output at the lower speeds. Ratings varied from a few horsepower for driving displacement-type water pumps to about 15,000 hp at 60–160 rpm for the massive marine engines. The steam engines were heavy, cast-iron structures (50–100 lb/hp) operating on low steam conditions (less than 300 psi with a relatively small degree of superheat). Steam engine losses and practical engine sizes limited steam expansion ratios (representing better economy), precluding the utilization of high vacuum exhaust conditions (due to high steam specific volumes).

The first steam engines (around 1705) served as both boiler and condenser with little heat being actually used. In 1769 James Watt brought about changes that led to higher efficiency by providing separate boiler and condenser and introducing full-pressure steam for only a part of the piston stroke. After steam cutoff "expansive working" continued to force piston movement until, at end of the stroke, steam pressure was reduced (maximum energy utilization). At this point connection was made to a condenser for expanded steam exhaust. Watt also devised an instrument (indicator) that when mechanically connected to piston travel and to piston steam pressure, recorded relative steam pressure throughout the stroke. This instrument contained a small drum that rotated a part revolution by being driven by piston movement. A paper was applied to the drum and a pen, moving up and down with cylinder steam pressure, recorded the pressure at every point of the stroke. The resulting diagram made it possible to study and improve the steam engine and also aided in determining the properties of steam and other gases. Watt's patent of 1782 shows the thermal efficiency improvement using the steam expansion principle. He indicated that thermal efficiency would be improved by a factor of 2.3.

Engine efficiency gradually increased as higher steam pressures were used and with more complete expansion. However, over the century, much additional information on steam and heat transfer was developed, leading to continuous improvements in engine design and efficiency. Mechanical improvements included making the engine drive a revolving shaft through use of crank and connecting rod (1781), development of a compound engine (two different-sized cylinders in series to allow further expansion—1781), and successive application of steam engine drive to railroad locomotives (1804), pumping stations (1850), and marine service (1854).

5.2-2 Classification

Steam engines are broadly classified by type of steam valve and corresponding steam flow pattern and by the process of steam expansion. Valves used include:

Slide (plain D-slide)
Slide (piston)
Disc/seat (poppet)
Cylinder ports (power piston operated)

These valve arrangements must admit steam to the cylinder at a beginning point (top dead center) where the piston is able to move down-cylinder as steam pressure is applied. The valve must shut steam off at some point (adjustable) prior to the end of the piston stroke. The exhaust valve opens, allowing steam to escape the cylinder just prior to piston return stroke and closes at about completion of the return stroke. To accomplish this automatically, valve control is through a mechanism connecting the valve to piston operation.

Slide Valves

The D-slide valve is a valve whose cross section resembles the letter D (see Fig. 5.2-1). The valve slides back and forth over ports connected to a cylinder at each end. It is positioned in a chamber containing full-pressure steam outside the D cup and an exhaust inside. Movement of the slide simultaneously opens the two ports, allowing steam to enter the cylinder and move the piston as well as allowing the piston to push out spent steam. The slide valve action may be timed for a variable "cutoff" to control steam flow during the stroke independently of exhaust valve action. At the end of the piston stroke the slide valve action continues, reopening both ports with a reversal in connection to the steam supply and exhaust. This allows pressure to build up behind the piston, forcing movement in the opposite direction.

In addition to the D-slide valve, but working on the same principle, is the piston valve (slide is in piston form) operating mostly on high-pressure steam engine sections. Variations include the balanced valve (differential pressure relief to reduce force and friction), multiport valve (several parallel passageways to reduce valve travel for less friction and quicker operation), and riding cutoff valve (separate control for main valve and cutoff valve for more feasible control). The advantages of the slide valve are its simplicity and low cost and maintenance. Disadvantages include the need for large size to handle steam volume (except for piston type, which are unsuitable for high-steam pressure–temperature); source of condensation; high pressure drops due to poor steam flow patterns; and, except for riding cutoff, poor independent adjustment of valves.

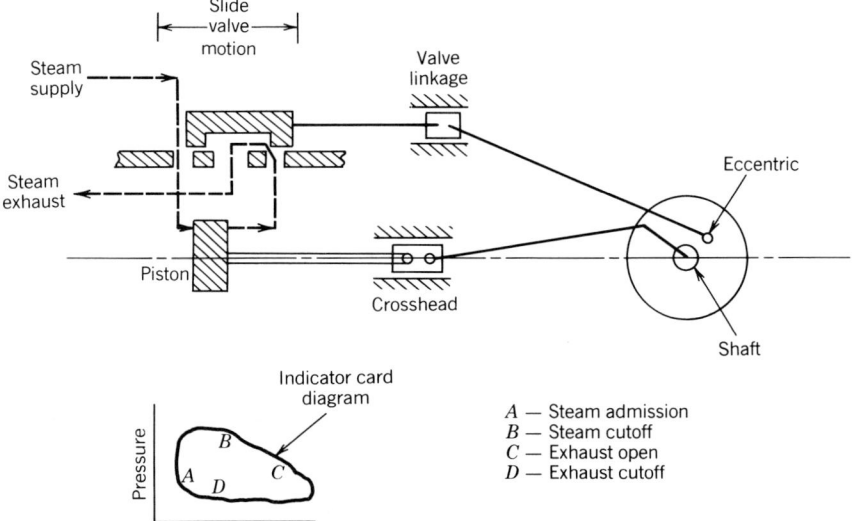

Fig. 5.2-1 Typical valve operation. Indicator card: simple slide valve engine.

In the Corliss engine an attempt was made to eliminate some of the disadvantages of the plain-slide valves. The cylinder contains four integral valves: two upper for steam admission and two lower for exhaust. These valves are long and cylindrical with multiple inlet edges (see Fig. 5.2-2). By exhausting the cooler expanded steam through the alternate bottom valve route, condensation is reduced. By shortening travel, quicker valve action is possible. The valve stem is given either an up–down (sliding) motion or a rotational motion (piston) by an external mechanism to provide the opening–closing of steam flow ports with a relatively low friction, quick-acting motion. The Corliss

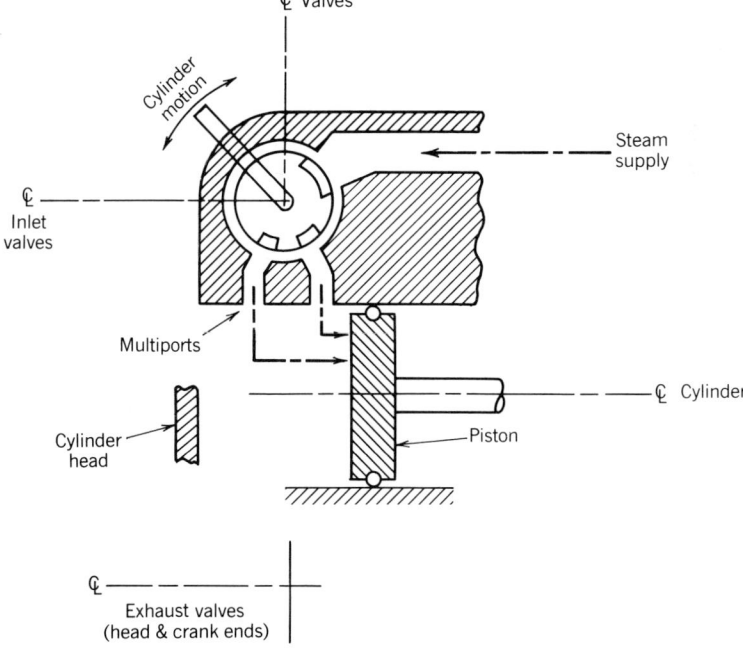

Fig. 5.2-2 Typical four-valve engine inlet valve.

engine is a slow-speed engine (less than 400 rpm) to allow for satisfactory mechanical action of the valve mechanisms. This action could be "releasing" or "nonreleasing," depending on the valve-closing operation. The releasing type allowed a dashpot or spring to close valves at a point of cutoff after mechanical release. Otherwise, the more positive action remained (directly mechanically connected.)

A "uniflow" principle was invented about 1885 and, during the next half century, predominated all other types of steam engines in stationary, marine, and locomotive applications. Modifications were chiefly concerned with improvements (exhaust systems, high-pressure designs, better valves and mechanisms, and elimination of initial condensation) of this uniflow engine. The conventional slide valve engine is classified as a counterflow, meaning that steam enters the cylinder, reverses at the end of expansion (stroke), and flows out of the same end. For the four-valve Corliss engine the same pattern occurs except steam exhausts out of the two bottom valves rather than the inlet passage. In the uniflow engine the exhaust ports are located in the cylinder midpoint. As the piston approaches the end of the stroke (about half cylinder length), it uncovers the exhaust ports, releasing steam. The piston performs its power function as well as acting as an exhaust valve. To accomplish this, piston length is about nine-tenths the length of the stroke. After exhaust ports are closed by piston return movement, remaining steam is compressed, requiring auxiliary exhaust ports if operating noncondensing. This possibility of excessive compression (to pressures above initial steam pressure) makes the uniflow engine primarily intended to operate condensing (vacuum exhaust). The use of valved auxiliary exhaust ports (as in the Corliss engine) on the noncondensing engine allows end clearances to be reduced from about 12–20% to about 2–4% of volume, thus restoring steam economy. The uniflow pattern significantly reduces initial condensation because, without expanded steam cooling, intake valves and head remain at relatively constant temperatures. In various designs this exhaust port may be located at the end of the cylinder or at a point on the cylinder where compression may be allowed to start without excessive pressure buildup, also aiding in maintaining constant head temperature.

In addition to the slide valve classification, there is a "poppet valve" engine. This valve consists of a disk fitted to a valve stem through which control functions are applied. The disk fits into a seat or port located as an integral part of the ends of cylinders. If it is balanced, the valve disk will fit into an upper and lower seat. Poppet valves are used equally well on all engines and are usually set directly over the end of the cylinders, allowing closer clearance. They can be used with high pressure–temperature steam. This type of valve, associated with proper operating mechanism, opens and closes quickly, allowing for more closely timed control. These features make the poppet valve almost universally used on uniflow engines.

TABLE 5.2-1   TYPICAL STEAM ENGINE DATA[a]

	Simple D-slide	Corliss 4-valve	Uniflow (condensing)
Pressure range, psig	50–150	75–200	100–250
Superheat range, °F	0	0–200	0–200
Size range, hp			
Stationary (C)	—	300–500	500–100
Stationary (NC)	10–100	100–500	—
Locomotive (NC)	500–6000	—	—
Marine (C)	500–9000	3000–12,000	5000–12,000
Unit weight range, lb/hp	50–100	110–140	140–160
Speed range, rpm	50–350	100–400	100–350
Volumetric clearance range, %	5–12	3–10	2–4
Steam rate range, lb/ihp			
Stationary (C)	—	16–19	11–15
Stationary (NC)	28–40	22–25	—
Typical MIP (150 P throttle), psig			
Stationary (C)	—	50	48
Stationary (NC)	64	60	—
Mean indicated pressure			
C	—	50	48
NC	64	60	—
Available heat conversion, %			
Stationary (C)	—	—	48–54
Stationary (NC)	—	57–67	—

[a] C, condensing; NC, noncondensing.

Classification includes the degree of "compounding," number of expansions, speed and size, pressure–temperature rating, and characteristics of exhaust. The tandem compound has identical, equal steam cycle cylinders operating pistons on the same piston rod, which in turn drives a single connecting rod and crank. Expansion compounding (cross-compound) places different steam cycle cylinders side by side, each with a separate connecting rod and crank but all applying power to a single shaft. The steam path may be through the high-pressure (HP) cylinder, the intermediate-pressure (IP) cylinder, and (due to volume increase at lower pressure) multiple low-pressure (LP) cylinders operating in parallel. Typical characteristics of speed, size, and throttle and exhaust conditions are shown in Table 5.2-1.

5.2-3 Selection of Engine

For small sizes where steam economy is less important than reliability, the simplest slide valve engine is selected. On the opposite end, for the economy of a mass power producer, in condensing service, a uniflow engine is applied. In applications where a process or heating steam is required, a noncondensing slide valve engine discharging into the process is used. A compound engine from which steam is bled from between IP and LP cylinders may also be applied. Uniflow engines, having a relatively wide-range, flat efficiency curve are best applied to widely varying loads, whereas the compound engine best fits a constant load. High back-pressure exhausts best fit the simple slide valve engines. Marine engines, usually high power rated and condensing, utilize the compound multiple-expansion principle. For locomotive service superheat in the simple engine is favored over the compound engine. Marine engines, large engines driving generators and pumping stations, are vertical engines to help utilize floor space. Corliss engines are not used with high pressure or superheated steam.

5.2-4 Components

Pistons and Cylinders

The power cylinder provides the pressure chamber into which steam is introduced. An internal piston, moving back and forth under the influence of this pressurized steam, converts heat energy into kinetic (motion) energy. Piston systems are designed to minimize motion friction (accurate machining), to prevent steam leakage around piston (piston rings), to have motion guidance (cross head), to prevent steam leakage to environment (piston rod packing gland), and to have maintenance access (cylinder head). Cylinders (cast iron) need a complicated casting process due to needs for dimension accuracy, internal steam passages, integral valve chambers, thermal expansion, and wear control. Small pistons are solid and cylindrical shaped with circumferential grooves for holding rings that press against the cylinder walls to provide steam seal. Large pistons are often made up of bolted disk and ring sections to facilitate installation of the large rings. Piston motion (a percentage of cylinder length) approaches cylinder ends with as small a linear "clearance" as possible—limited by physical requirements (head and valve features) and by steam cycle conditions. This clearance (volumetric) is expressed as a percentage of piston displacement volume and varies from about 9% for slide valve engines to 2% for poppet valve engines (better steam use economy).

Piston Rods, Crossheads, and Connecting Rods

Piston forces are transmitted through this linkage to the shaft crank arm. The rod is rigidly attached to the piston and (at the opposite end) to the crosshead. The whole unit moves in the same back-and-forth plane established by the cylinder and crosshead guides. Such rigid alignment allows for minimum friction movement, low wear, and rod sealing at the packing stuffing box. The connecting rod coupling the crosshead to the shaft crank arm has a two-plane motion (back and forth as well as up and down as the crank rotates) requiring a bearing coupling with crosshead and crank arm. These bearings require lubrication, cooling, and adjustments for wear. Because piston linear clearance is set by the length of the two rods, designs allow for a total length (rod plus arm) adjustment.

Crank, Flywheel, and Valve Mechanisms

The crank is an offset to the shaft and allows for a leverage arm upon which piston forces are applied. It forces the crank end of the connecting rod to take a circular path. The piston makes a stroke for each half shaft revolution. Steam force is applied alternately to each side of the piston for a power stroke both ways. These steam forces are delivered with a varying intensity (nonuniform throughout stroke) and are smoothed out by a shaft flywheel, which absorbs peak energy (inertia) and returns it to the shaft during minimum-force periods. The flywheel size must match engine size, speed, and service conditions. Flywheel (shaft) direction of rotation is set to keep the energy absorption (peak) portion of cycle such that (at this time) connecting rod forces push downward on crosshead guide, thus extending crosshead life and minimizing number of clearance adjustments. Engine valve systems (see Section 5.2-2) are automatically operated through linkages that time their actions to piston position. This operates from an "eccentric" (shaft tie offset by a "radius of throw" from the center of the shaft), which establishes and times the valve movement. More efficient valve operation requires two eccentrics for separate control of inlet and exhaust valve operations. This also allows the governor to control speed and reverse engine by adjusting steam inlet valve separately.

Governors

Variable load requires automatic governor operation. Safety systems (engine protection) work through a governor system for automatic emergency shutdown. Certain types of loads (electric generators, mills, etc.), requiring a "wide-load-range" constant speed, require a quick-action governor with a small speed regulation (change in speed from no load to full load expressed as a percentage of full-load speed). Regulation depends on the governor's sensitivity (small speed change detection), response (a steam flow correction based on a small speed change), and stability (rapid and significant load changes without speed ictuations/hunting). A common governor is one that responds to centrifugal forces created by rotation (speed). These apply the motion of a weight (fly balls) being moved by centrifugal forces to position a valve controlling steam flow to the engine. Increasing centrifugal force (higher speed) is compensated for by a force of gravity (fly ball) or spring tensions. When spring tensions are used, engine speed or constant-speed load can be adjusted by changing spring tension. Emergency engine shutdown is usually accomplished through the governor in response to shaft overspeed, mechanical failures, or emergency loading actions. In the case of governor malfunction design features force a safe shutdown (steam valve shut).

5.2-5   Engine Performance

Indicator

James Watt's invention and use of the steam engine indicator opened the door to determining the performance of steam engines. The indicator provided a diagram indicating cylinder steam pressure at any point of the piston stroke. Steam pressure operates a small piston that moves against a spring tension. The motion moves a pen (in direct ratio to pressure) up and down on the diagram, which is also moving horizontally (circular motion) on a small drum being rotationally oscillated by a ratioed connection to the crosshead. The combined pen–drum movement produced a record of performance. Figure 5-2.3 shows a typical diagram.

An ideal indicator card represents the greatest possible work obtained from a given operating condition. Work would be maximum when the area inside the pattern is maximum for the conditions.

Area would be maximum if inlet valves opened at the end of the stroke instantly, filling the cylinder with throttle-pressure steam until stroke completion and exhaust valves instantly opened,

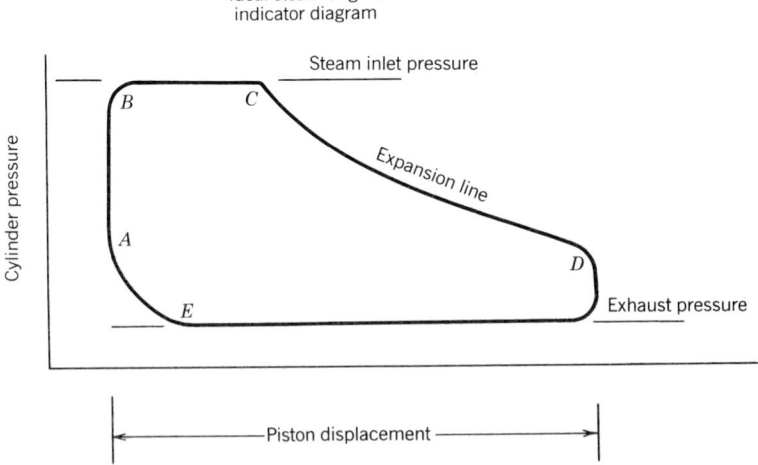

Fig. 5.2-3 Ideal steam engine indicator diagram: point A, steam inlet valve open; A to C, steam feed to cylinder; point B, steam full-pressure point; point C, steam cutoff point; C to D, steam expansion; point D, exhaust valve open; D to E, steam exhausting; point E, exhaust valve closed; E to A, steam compression.

emptying the cylinder of steam for the return stroke without compression. For best steam economy, steam expansion is utilized by stopping steam inlet at about one-third of the stroke, exhausting with the help of the piston action through about 85% of return stroke, and allowing a closely timed period for steam through-flow (minimum compression, full pressure of start of stroke without excessive entrance flow losses, cutoff for expansion to about exhaust pressure at end of stroke, and minimum exhaust flow losses). Should other conditions exist (poor valve and cutoff timing), the indicator diagram will show such effects as looped or abnormally sloping lines resulting in a less enclosed area (work).

Indicated Horsepower (ihp)

Horsepower (equivalent to 33,000 ft · lb/min) is a measure of the steam engine output. It is obtained by a measurement of piston motion (in feet) per unit time and the average force applied to the piston during this motion. Force is a multiple of piston area times the steam mean effective pressure (MEP). The MEP is obtained from the indicator diagram by obtaining the area (in.^{2}) enclosed by the diagram lines, dividing by the diagram length (in.), and multiplying by the indicator spring constant. Indicator card area is easily measured using a planimeter. These data are used in the formula ihp = PLAN/33,000 in which P is the mean effective pressure (psi), L is stroke length (ft), A is piston area (in.^{2}), and N is the number of power strokes per minute (usually twice the rpm of the shaft). Should the piston have a dual power stroke (steam to each end), indicator cards are taken on both the head and crank end of the cylinder, crank end piston area is reduced by the area of the piston rod, and ihp is computed for each side and added for the total. Not all of the horsepower is transmitted to the engine shaft due to engine losses. The output (brake horsepower—bhp) is the product of ihp and engine mechanical efficiency. Actual bhp may be measured using a friction brake (Proney brake) on the flywheel, weighing the reaction on a platform scale and applying scale reading (pounds F), length of leverage arm (feet L), and shaft speed (rpm) in the formula bhp = [(2π)FL(rpm)]/33,000. Engine mechanical efficiency at full load varies as follows:

Slide valve engine	88–90%
Corliss four-valve	90–92%
Uniflow	90–91%

Low-load operation is at an efficiency of about 60% of these values and one-half load at about 90%. Internal, or thermal, efficiency measures steam heat being converted to work relative to isotropic heat available. Figure 5.2-4 shows engine steam expansion on the Mollier chart. Theoretical expansion

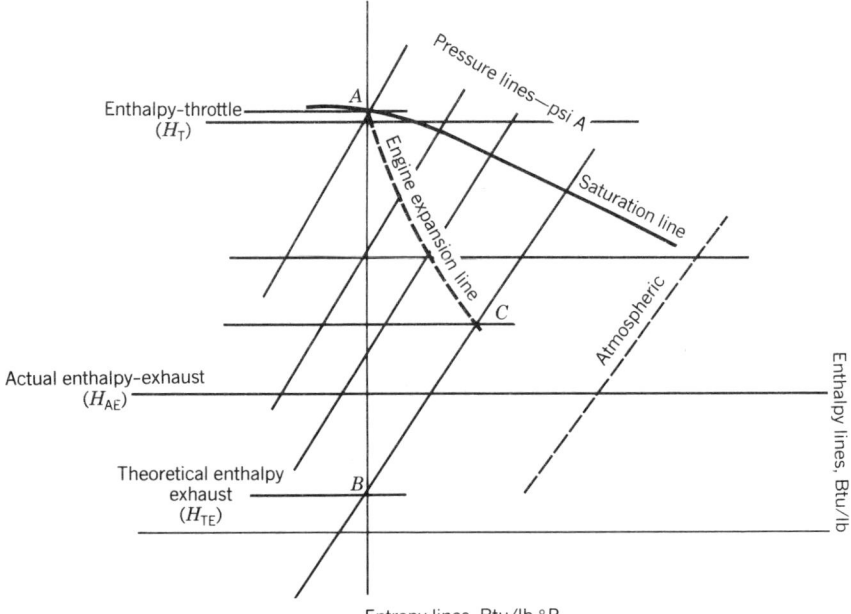

Fig. 5.2-4   Mollier chart. Steam expansion.

(100% efficiency) occurs on the constant-entropy line (AB), whereas actual expansion occurs along the sloping line (AC). Because $H_T - H_{AE}$ represents heat converted to work, engine thermal efficiency is the ratio of ($H_T - H_{AE}$)/($H_T - H_{TE}$) expressed as a percent.
Typical valves are:

	Noncondensing	Condensing
Corliss four-valve	62%	Not recommended
Uniflow	68%	54%

The large triple-expansion and compound Corliss engines exhibit the highest efficiency but do not compare in performance to the 80–85% thermal efficiencies obtained by the later developed condensing steam turbines. Engine theoretical steam rates (TSR) can be obtained by dividing Btu equivalent to horsepower (2546) by $H_T - H_{TE}$ and actual steam rate determined by dividing this by the thermal efficiency. Assuming an engine with throttle conditions of 250 psia, saturated ($H_T = 1201$ Btu/lb) and exhausting at 25 psia ($H_{TE} = 1026$ Btu/lb of constant entropy), there is 175 Btu/lb of heat available for conversion to work. TSR is 2546/175, or 14.5 lb of steam per ihp. At 60% conversion, the actual steam rate is 14.5/0.6, or 24 lb/ihp. Actual exhaust enthalpy (H_{AE}) is obtained from TE = ($H_T - H_{AE}$)/($H_T - H_{TE}$) in which 0.6 = (1202 − H_{AE})/(1201 − 1026). Actual exhaust enthalpy is 1096 Btu/lb.

5.2-6   Operation and Maintenance

Procedures

To ensure economical engine operation, all adjustments must be initially correct and rechecked frequently. The adjustments include:

1. Alignment of engine, including bearings, cylinders, crosshead, and rods.
2. Thrust bearing adjustment.
3. Piston clearance.
4. Valve settings and linkage adjustments.
5. Bearing clearances.
6. Steam-sealing adjustments.
7. Drive belt (if used) adjustment.
8. Lubrication feed rate adjustment.
9. Governor setting and adjustments.
10. Emergency trip settings and relief valve settings.

The operation of steam engines consists of starting the necessary auxiliary equipment such as air, cooling, and lubrication systems as well as establishing the exhaust-receiving system (process use or condenser). A steam supply at rated pressures and temperatures is made available. Variations in these procedures depend on engine size, method of lubrication, and power source for auxiliaries (engine or separately driven). An engine warmup cycle is necessary to minimize condensation and prevent the structural stresses or clearance problems associated with unequal expansion. The accumulation of condensate (water) can cause severe damage in the close piston clearance as a result of water hammer or corrosion. Steam lines and engine drain valves on jackets, cylinders, and valve chests are kept open throughout the warmup period. If a forced lubrication system is used, it is placed in service and checked for leaks and adequate bearing lubrication flow. Any oil purification system should be working. Engines should be tried slowly at first to ensure warmup and a motion response to a throttle operation. Reversing operations are checked if applicable. After a 30-min warmup period, drain valves are closed.

The simple, constant-speed engine may be started without changing the governor mechanism since governor control takes over only after controlled speed is reached. Cross-compound engines may be started by manually admitting high-pressure steam to a cylinder in which the piston is not on or near the top dead center. Tandem-compound engines may be started similarly, but all cylinders must be drained and lubricated.

Lubricants reduce friction and, as a result, reduce wear and heating of the bearings or cylinders. Two types are used: one capable of high-temperature operation of the steam valves and cylinders and another for bearings of shaft or rod pins. The lubricants must resist being worked away or emulsified by water, have lubricating qualities at temperature and clearances without corrosive influences, and must have a sealing quality at piston rings. Lubricators must store sufficient quantities of oil to last for a period of time at the desired feed rate and must pressurize feeds to overcome cylinder pressure.

Generally, 1 pint of cylinder oil per hour is fed to each million square feet of rubbing surface, meaning that a 125-rpm, 24-in. diameter by 24-in. stroke, dual-acting cylinder would require about 1 pint of oil every 5 h. Splash, gravity feed, or self-contained circulating oil systems are used on bearings with the operator observing oil through-flow and bearing temperatures to ensure continued lubrication.

Efficiency

Engine efficiency relates the amount of energy as productive work to the energy of input. The difference in these two energies is that energy needed to overcome mechanical losses (friction), thermal losses, and heat (steam) leakage. These losses may be classified as follows:

1. Friction losses (piston movement, bearings, crosshead, mechanical linkage, and stuffing boxes) should be less than 10% of ihp (minimized by correct alignment; bearing, linkage, and seal adjustments; and application of correct lubrication and cooling procedures).

2. Various thermal losses (throttle or cylinder valve throttling, moisture by admission or condensation, and incomplete steam expansion) should be less than 35% of ihp. Kept low by operating at loads corresponding to best design efficiency (75–100% of nameplate rating), proper setting of clearance volume and cylinder valves (inlet, cutoff, and exhaust), and correct cooling rates and upkeep of jacket insulation. Throttle steam moisture, which does no work, adds the possibility of physical damage, scale deposits from impurities (boiler carry-over), and utilization of energy if flashing occurs. Use of moisture separators or small levels of steam superheat at throttle helps solve these problems.

3. Energy leakage to the environment by conduction through insulation as well as by leakage through safety valves, rod seals, or steam traps should be kept to a minimum by frequent maintenance checks.

Unusual Operations

1. Priming. Priming is a dangerous condition that results from excessive moisture (water) in the cylinders. It is accompanied with a noise at the ends of the stroke. Anything that eliminates and removes accumulating moisture is done. This includes correcting boiler carry-over, prevention of steam use surges by slower load changes, and good drainage of condensate being formed.

2. Pounding. Pounding is a vibration or sound caused by poor distribution of power in multicylinder engines, by excessive bearing clearance, by unbalance, or by varying friction (grabbing of moving parts). Cures include adjustment of cutoff for more equal power distribution, use of a heavier oil for improved oil film, looser packing at rods; adjustment of controls (throttle and exhaust pressure or flow) to prevent any external fluctuations from reflecting into engine performance, or a check of foundation alignment or loose parts to ensure stability. For a cylinder of a multiple-expansion (single LP cylinder) engine to carry an equal share of total load, the product of that cylinder's pressure drop times its ratio to the HP cylinder should about equal HP cylinder pressure drop. Commercial cylinder ratios (triple expansion—200 psig) are about $1:3.7:9.0$ for HP, IP, and LP cylinders. Rotating parts are balanced by the use of counterweights, usually installed opposite the crank.

3. Hot-piston or rod. Temperature buildup of piston or rod is a serious matter due to potential damage to rings, rod metallic packing, or lubrication system associated with piston or crosshead. Low operating temperatures are maintained by proper adjustment of packing, adequate lubrication, alignment, and cleanliness of rod.

4. Abnormal exhaust. Engine capability is altered if total pressure drop (throttle pressure minus exhaust pressure) is decreased. Exhaust structures may overheat, especially if an exhaust vacuum is lost by failure of condenser cooling system. In this event engines should be shut down until restoration of normal exhaust conditions.

5. Oil contamination. Lubricating oils, in contact with water, air, and churning actions of connections rods, may emulsify or foam. Decomposed oil will plug passages, act as an insulation and increase temperatures, and cause rapid aging. Such oils must be removed and filtered, lubrication systems cleaned, and water leakage control implemented. All oil leakage into the condensate being returned to the boiler must be removed by oil separators.

BIBLIOGRAPHY

Encyclopaedia Britannica, 1960.

Gaffert, G. A. *Steam Power Stations*. McGraw-Hill, New York, 1946.

Kent, R. T. *Kent's Mechanical Engineers Handbook—Power*. Wiley, New York, 1942.

Osbourne, A. *Modern Marine Engineer's Manual*. Cornell Maritime Press, Centreville, MD., 1944.

Woodruff, E. B. and H. B. Lammers, *Steam Plant Operations*, 3rd ed. McGraw-Hill, New York, 1967.

5.3   GAS TURBINES

Thomas E. Stott

5.3-1   Introduction

The gas turbine is a relatively simple machine that converts thermal energy into mechanical work via the Brayton cycle (see Fig. 5.3-1). The closed-cycle gas turbine is a true Brayton cycle with compression, heating, expansion, and cooling where more work is produced during expansion than is required to compress the working fluid.

The open-cycle gas turbine used in most designs of these machines utilizes air as a working fluid, which enters from the atmosphere and discharges to it. The air is compressed, temperature is increased by combustion of fuel added to the air, and the products of fuel and air expand, developing more work than required for compression (net positive work output). Since the products of combustion are exhausted to the atmosphere after expanding in the turbines, the open cycle is actually not a true cycle in the purest sense.

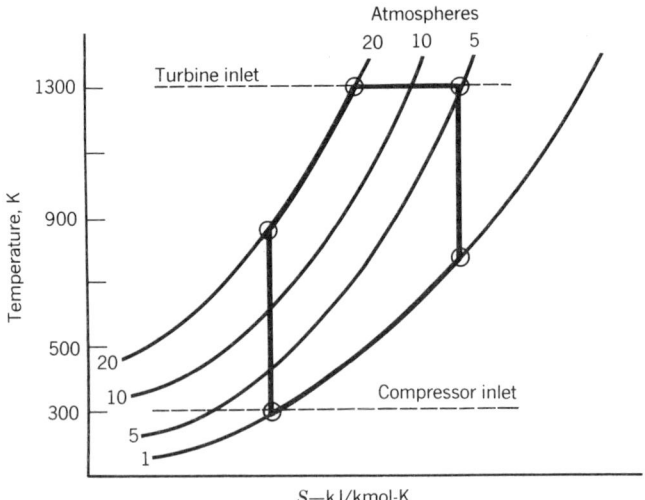

Fig. 5.3-1   Temperature–entropy diagram of Brayton closed-cycle gas turbine.

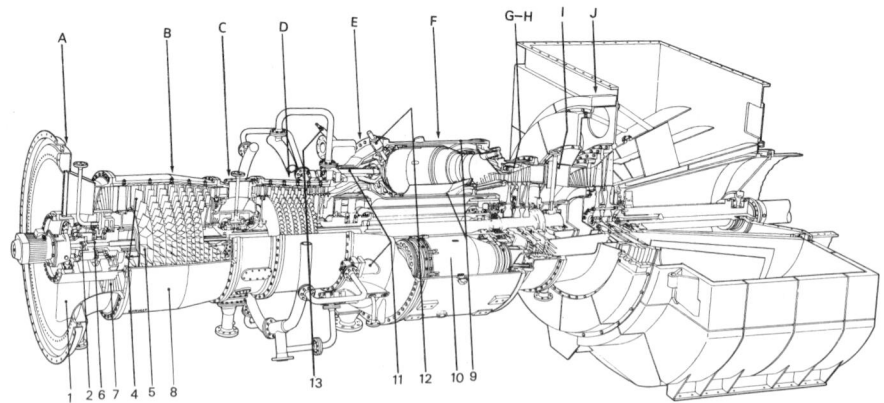

Fig. 5.3-2   Cross-section view of a gas turbine. (Courtesy ASEA Stal)

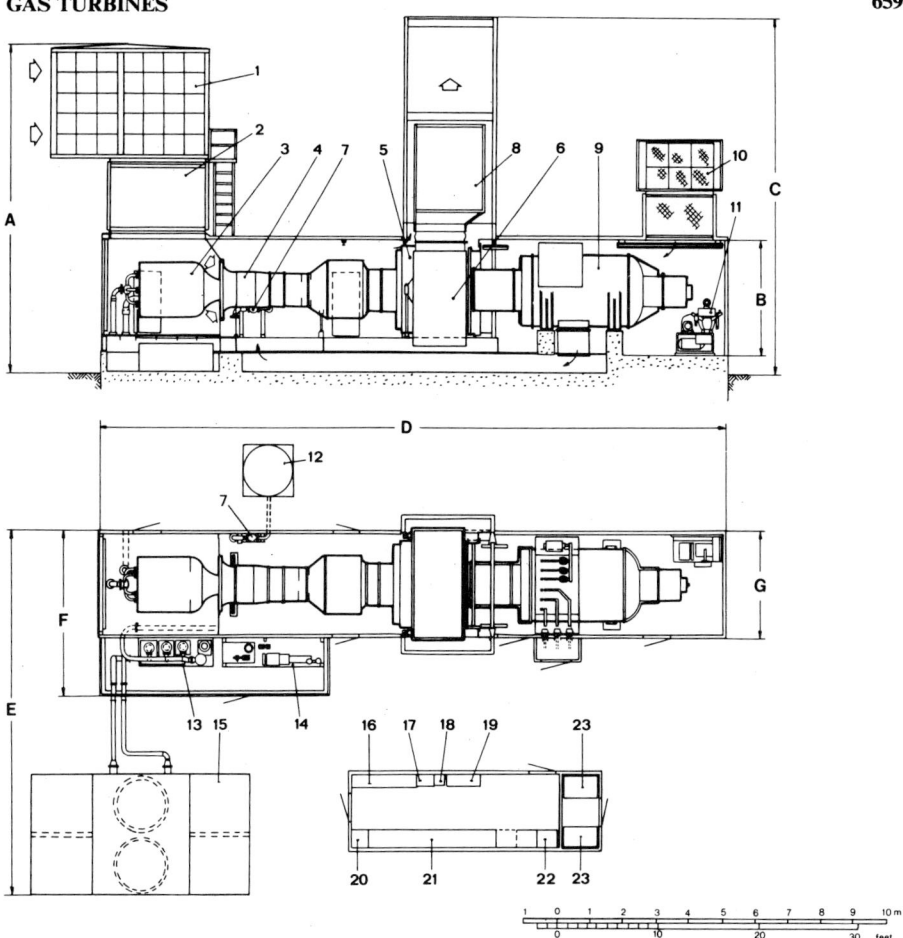

Fig. 5.3-3 Typical layout of a gas turbine power plant: (1) air intake unit, (2) inlet silencer, (3) starting device, (4) gas generator, (5) power turbine, (6) exhaust casing, (7) starting air unit, (8) exhaust silencer, (9) alternator, (10) cooling air intake for alternator, (11) air compressor unit, (12) air tank, (13) lubricating oil tank, (14) ancillary module, (15) oil cooler, (16) ac board, (17) inverter, (18) equipment for gas detector, (19) dc board, (20) automatic voltage, regulator (21) control panels, (22) rectifier 110 V, (23) batteries, 110 V (Courtesy ASEA Stal)

The gas turbine (see Fig. 5.3-2) is made up of compressors, combustion chambers, compressor turbines, a power output turbine, a control system, and auxiliary equipment, whereas the gas turbine power plant (see Fig. 5.3-3) consists of the gas turbine and essential equipment for the production of power in a useful form (i.e., electric generators, housing, fuel storage, auxiliaries not required in the main working fluid circuit, etc.).

5.3-2 Classification of Gas Turbine Power Plants

Gas turbine power plants can be used in many applications, from aircraft jet engines utilizing the hot pressurized working fluid to develop thrust to closed-cycle nuclear plants using an enclosed working fluid to transform the heat from the reactor to useful work.

Gas turbine power plants may be classified in the following ways:

Open Cycle. Working fluid enters gas turbine from atmosphere and discharges to it.

Closed Cycle. Working fluid is recycled independent of heat source.

Simple Cycle. Consists only of compression, combustion, and expansion.

Regenerative Cycle. Employs exhaust heat recovery and transfers it to working fluid before combustion chambers.

Intercooled Cycle.   Utilizes cooling of working fluid between stages of compression.

Single-Shaft Gas Turbine.   Compressor and turbine rotors are mechanically coupled with output shaft.

Multishaft Gas Turbine.   Includes two or three turbines running on independent shafts but working within same working fluid circuit.

Industrial-Type Gas Turbines.   Have relatively high weight-to-power ratios, derived from steam turbine and industrial compressor technology. Generally low turbine inlet temperatures.

Aircraft Derivative Gas Turbines.   Used as jet engines with low weight-to-power ratios; also aircraft units that have been modified for industrial use. Characteristically medium-to-high turbine inlet temperatures.

TABLE 5.3-1   DIMENSIONAL CONVERSION FACTORS: ENGLISH UNITS TO SI (METRIC) UNITS

Inches to millimeters	Inches $\times$ 25.4
Feet to meters	Feet $\times$ 0.305
Pounds to kilograms	Pounds $\times$ 0.454
Long tons to metric tonnes	Tons $\times$ 0.984
Btu to kilojoules	Btu $\times$ 1.055
Kilocalories to kilojoules	Kilocalories $\times$ 4.187
Horsepower to kilowatts	Horsepower $\times$ 0.746
psi to kilopascal	psi $\times$ 6.895
Degrees Fahrenheit to degrees Celsius	(Degrees Fahrenheit $-$ 32) $\times \frac{5}{9}$
Pounds per second to kilograms per second	Pounds per second $\times$ 0.454
Btu per pound to kilojoules per kilogram	Btu per pound $\times$ 0.479
Foot-pounds to newton-meters	Foot-pounds $\times$ 1.356

TABLE 5.3-2   VISCOSITY CONVERSION TABLE

Redwood No. 1 (s)	Saybolt Universal (s)	Engler Units	Kinematic Viscosity (cSt)
28	30	1.05	1.4
30	33	1.12	2.0
32	35	1.18	2.6
34	38.4	1.26	3.05
38	43.3	1.40	4.7
42	48.2	1.54	6.2
46	53.0	1.67	7.6
50	57.8	1.81	9.0
54	62.6	1.95	10.5
58	67.3	2.08	11.7
62	71.9	2.21	13.0
66	77.1	2.34	14.3
70	81.9	2.48	15.5
74	86.6	2.62	16.9
78	91.3	2.75	18.0
82	95.9	2.89	19.1
86	100.6	3.02	20.2
90	105.3	3.15	21.4
98	114.7	3.43	23.6
115	135.7	4.01	28.0

5.3-3 Standard Reference Conditions

Since most gas turbines are of the open-cycle type, utilizing air as a working fluid, they are more sensitive to changes in atmospheric conditions than other types of prime movers. Power output will vary depending on ambient air temperature, atmospheric pressure, relative humidity, and elevation.

In order that power, efficiency, and specific fuel consumption may be evaluated from the same base, the International Organization for Standardization (ISO) has established the following reference conditions for design and performance testing of gas turbines (ISO documents 3977 and 2314):

Atmospheric pressure	101.3 kPa (= 750.1 mm Hg)
Ambient temperature	15°C
Elevation	Sea level
Relative humidity	60% (may generally be ignored)
Cooling water	15°C (if used)

Manufacturers of gas turbines have correction curves that allow the user to estimate operation at site conditions. These correction data will also be used to evaluate tests of the machines for guaranteed performance of output power and fuel consumption.

Significant power reduction due to elevated air temperatures may be expected under tropical conditions. Under arctic conditions care must be taken that mechanical design limits are not exceeded due to excess power.

Conversion factors are given in Tables 5.3-1 and 5.3-2 as well as in Chapter 18.

5.3-4 Environmental Considerations

Site Conditions and Access

During the early planning stages of a gas turbine installation, thought must be given to standard conditions as previously given and what effect the site conditions will have on the installation, operation, and desired power. It should be determined what effects the site will have on the following:

Space available for installation.

Type of enclosure desired.

Maximum size and weight of largest pieces to be transported and handled at site.

Soil characteristics.

Expected wind loads, snow fall, rain fall, earthquake zone.

Detail of access, that is, waterway, railway or road, distance of unloading point to site.

Type, quantity, temperature, and chemical analysis of cooling water.

Details of electric, hydraulic, or pneumatic power required during construction.

Painting and any special protection required.

Nature of surroundings affecting purity of air and potential contamination by dirt, dust, or salt.

Recognized prevailing wind direction and possibility of stack gases being drawn into adjacent buildings.

Containment facilities for potential spillages from fuel storage and handling areas.

Sound

If an operator is to stand watch inside the gas turbine enclosure, appropriate acoustical treatment should be provided to reduce sound levels to those required by relevant standards. If no standard prevails, ISO document 1999 gives guidance in determining proper sound levels. Ear protectors may be used for entering enclosures that are normally unmanned.

The sound level outside the gas turbine plant must be limited to prevent damaging or disturbing noise. Normally, local authorities will have established limits. Guidance for acceptable sound levels outdoors are given in ISO document 1996 (see Fig. 5.3-4).

Site Measurements. Noise level as measured in the vicinity of a gas turbine installation will depend not only on the intensity of the emitted noise throughout the frequency range but also on the distance, angle, and elevation of the test point in relation to the noise source. Surrounding buildings, roads, trees, and so on, will also influence site measurements because of unpredictable reflection and absorption of various frequencies and because of their own emitted noise.

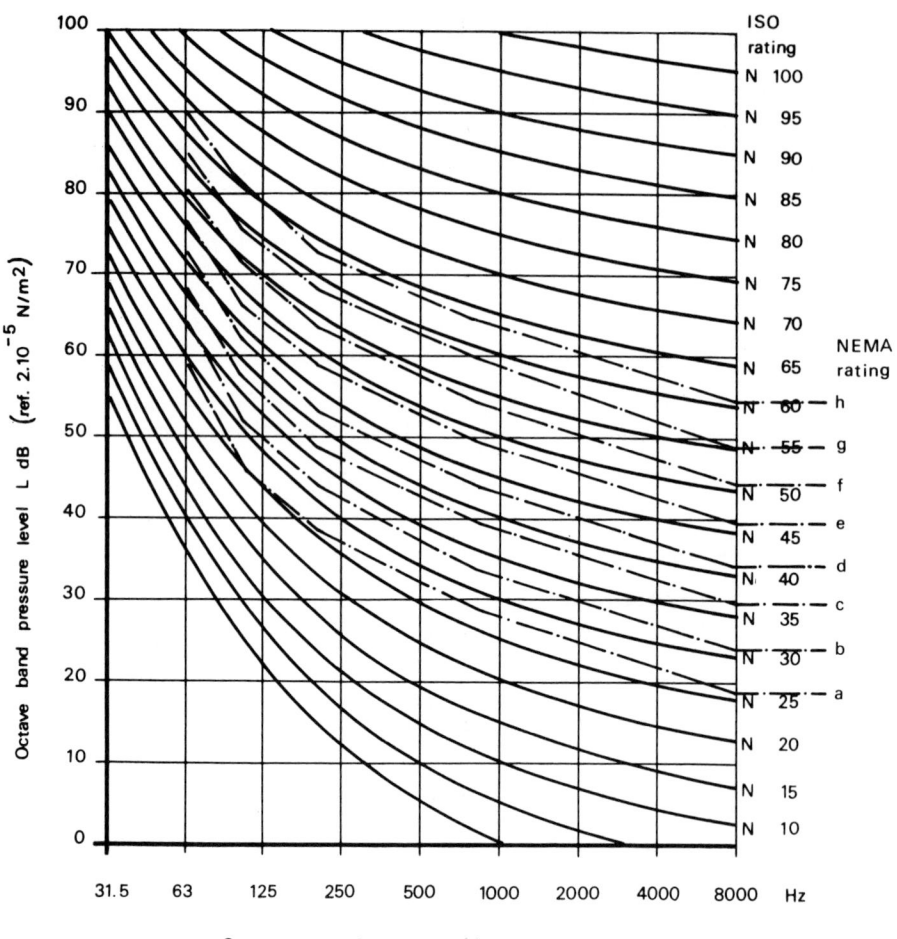

Center frequencies of octave bands

Fig. 5.3-4 Gas turbine noise rating curves. Noise rating: the lowest noise rating curve which is not exceeded at any point by the octave band sound pressure level of the measured noise. Curves are identified by N numbers on the ISO scale or by letters on the NEMA scale. Sound pressure level: the logarithmic ratio (L) between a measured sound pressure p and a reference sound pressure p_0 with the general relationship $L = 20 \log_{10} p/p_0$ dB where $p_0 = 2 \times 10^{-5}$ N/m².

Effect of Distance. Assuming hemispherical radiation, the sound intensity of a source reduces inversely as the square of the distance R. Thus, a practical expression for sound intensity as a function of distance includes a factor of $10 \log R_1^2/R_2$, which means a 6-dB reduction for each doubling of distance ($10 \log 2^2 = 6$). This rule is reliable only in a far field, that is, at a distance exceeding about 50 m for a single gas turbine plant and 75 m for two adjacent units.

The methods of calculation are:

$$55 \text{ dB at } 200 \text{ m} = 55 + 10 \log \frac{(200)^2}{(100)} \text{ at } 100 \text{ m}$$

$$= 61 \text{ dB at } 100 \text{ m}$$

$$50 \text{ dB at } 400 \text{ ft} = 50 + 10 \log \frac{(400 \times 0.305)^2}{(200)} \text{ at } 200 \text{ m}$$

$$= 45 \text{ dB at } 200 \text{ m}$$

Effect of Multiple Sources. Two or more noises can be summed only on the basis of the respective sound intensity levels of their sources. A noise of 50 dB has a source 10^5 times more powerful than the standard reference level of 10^{-12} W/m², 60 dB is 10^6 times more powerful, and so on (ISO R337). Therefore, the method of adding is

$$55 + 55 = 10\log(2 \times 10^{5.5}) = 58 \text{ dB}$$

$$60 + 50 = 10\log(10^6 + 10^5) = 60.4 \text{ dB}$$

This rule is not necessarily true when adding complex noise sources having a wide frequency spectrum. In general, however, two adjacent gas turbine plants in one station will have a combined rating 3 dB higher than a single unit ($10\log 2 = 3$), similary three units will be 5 dB higher than a single unit ($10\log 3 = 5$, approximate).

Smoke and Exhaust Emissions

Both visible smoke and nonvisible emissions may be objectional. Therefore, methods of measurement and limits of acceptability must be established. Local codes and standards should be consulted to determine compliance requirements. Most common elements to be investigated are unburned hydro-carbons, oxides of nitrogen, and sulfur. If a large gas turbine power plant or a number of smaller units are to be installed near an airport flight path, effects of turbulence induced by exhaust gases should be evaluated.

Smoke. Gas turbines utilizing gaseous fuels should operate with no visible smoke; however, plants operating with liquid fuels are highly suseptible to developing smoke during light off, idle, full load, or shutdown since only 0.1% of unburned carbon can result in dense exhaust smoke. Tests and procedures for defining public response are included in ISO 3977.

Oxides of Nitrogen. NO_x can contribute to atmospheric polution particularly in smog-prone areas. ISO 3977 gives methods of determining the amount in gas turbine exhausts, but it will be noted that it is very difficult to obtain a consistently accurate measurement. Water or steam injection into the exhaust gases has proved to be an effective way to reduce nitrogen oxides emissions.

5.3-5 Power Ratings and Determination of Output

All manufacturers of gas turbines can supply rating curves (see Figs. 5.3-5 and 5.3-6) for their machines, and most base these curves on the ISO standard conditions described in Section 5.3-3. In

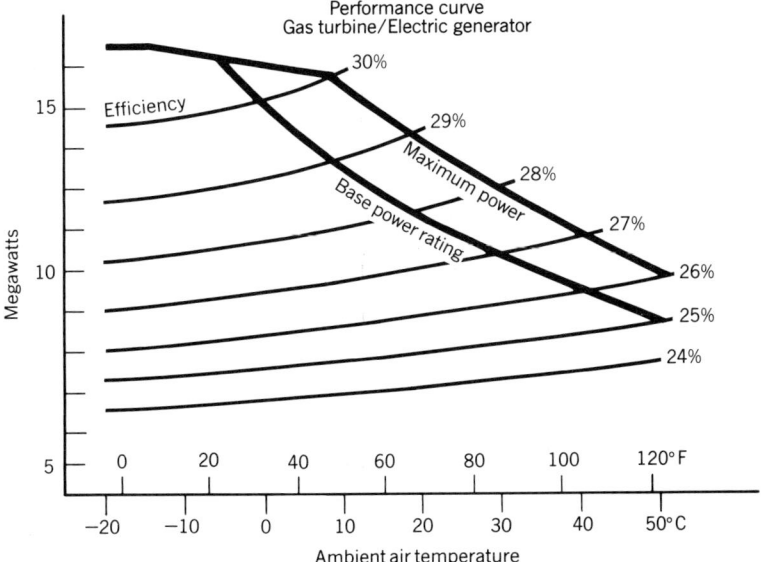

Fig. 5.3-5 Gas turbine performance curves—power. (Courtesy ASEA Stal.)

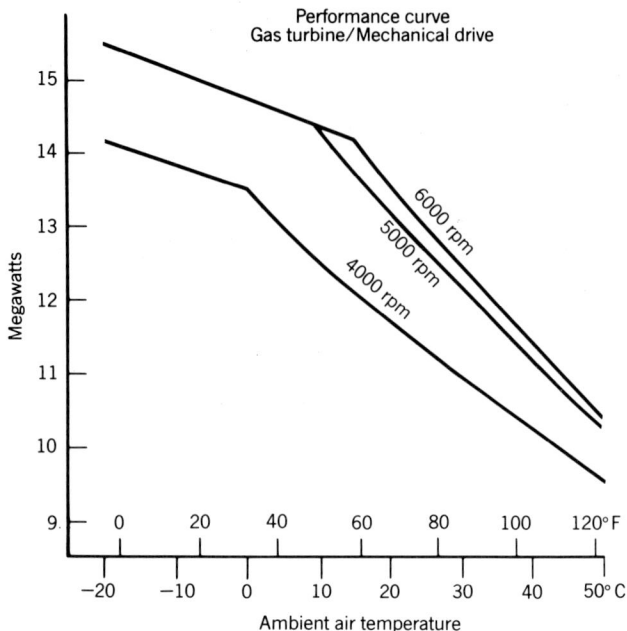

Fig. 5.3-6   Gas turbine performance curves—mechanical drive. (Courtesy ASEA Stal.)

addition to these conditions a standard fuel heat content must be established as well as the ancillary power required (see Table 5.3-9) when comparing gas turbines from different manufacturers. It is also necessary to calculate the power available, efficiency (fuel required), and available fuel (fuel characteristics) at actual site conditions to design an acceptable installation.

Calculation of Power Rating at Site Conditions

In order to calculate actual on-site gas turbine plant power, the following data must be available:

Atmospheric pressure.
Average, high, and low ambient-air temperatures and condition of ambient air (dust, salt, aerosol, etc.).
Elevation (altitude).
Hours per year of operation and number of starts per year.
Load characteristics (base, peak, etc.).
Type of fuel available and heat content.
Acceptable noise level.

With this information we can define the characteristics of the proposed gas turbine to be matched at these site conditions.
The following will be required from the turbine manufacturer:

Curves showing power versus ambient temperature.
Correction for elevation.
Load curves or operating profile, h/yr (to establish maintenance options).
Effect of number of starts.
Acceptability of fuel, specific fuel consumption, fuel treatment if required.
Inlet air filtration and effect on power rating.
Acoustic requirements and effect on power rating.
Ancillary loads to be deducted from rated power.

The following steps should be followed to determine site power, fuel consumption, and maintenance schedule:

Select power curve for base or peak load.

Determine power from curve at site average ambient temperature.

Determine if inlet air filtration requires extra stages (higher inlet pressure drop).

Determine if noise level requires extra silencing (higher exhaust loss).

Determine elevation correction (air density).

Determine fuel correction for heat content.

Determine standard "hot-section inspection" period and "time between overhauls" and correct for number of starts and operating profile of hours per year at base and peak loads.

The power at the selected ambient-air temperature is corrected for atmospheric pressures, pressure losses, elevation, and ancillary loads. The specific fuel consumption is corrected for heat content and total fuel is calculated. Maintenance time is based on the operator's start/operations profile and the manufacturer's time-limited components. The on-site calculated power should be checked at the lowest anticipated ambient-air temperature to ensure that the machine will not be overloaded and at the highest temperature to find the minimum power available.

Effect of Power Level on Material Creep Life

Base-load operation normally is limited by the creep life of the material used in time-limited discs and blades of the gas turbine (see Table 5.3-3). One common design criteria is 100,000 h of operation. In

TABLE 5.3-3 TYPICAL GAS TURBINE MATERIALS

Component	Material	Creep/Life-Hours
LP compressor blades	CR-NI-MO steel	100,000
Guide vanes	CR-NI-MO steel	100,000
Discs	Hardened/tempered steel	100,000
HP compressor blades	CR-MO-steel	100,000
Guide vanes	CR-MO-steel	100,000
Discs	Hardened/tempered steel	100,000
Combustion sec. flame tube	CR-NI-steel	100,000
Flame tube head	Nimonic 75	100,000
HP comp. turbine blades	Inconel 738	100,000 (30,000)[a]
Guide vanes	X-40	100,000
discs	CR-NI-MO-TI-V steel	100,000 (30,000)[a]
LP comp. turbine blades	Udimet 500	100,000 (30,000)[a]
Guide vanes	Udimet 500	100,000 (30,000)[a]
Discs	Cr-NI-MO-TI-V steel	100,000
Power turbine blades	Nimonic 90	100,000
Guide vanes	X-40	100,000
discs	CR-NI-MO-TI-V Steel	100,000

Source. Courtesy of ASEA STAL.

[a] Peak rating reduced life.

some cases these figures will be reduced to 30,000 h if the same machine is used extensively at the higher peak load (short duration). Another factor in the use of gas turbines for peak load is the high number of starts per stops per year. In this case some parts may be time limited by low-cycle fatigue.

5.3-6 Fuels

Gaseous Fuels

Gaseous fuels are often a mixture of several components. The flammable elements are "simple" hydrocarbons such as methane (CH_4), ethane (C_2H_6), and so on. Some gases also contain hydrogen (H_2) and carbon monoxide (CO). The nonflammable components are usually carbon dioxide (CO_2), nitrogen (N_2), and water vapor (H_2O) (see Table 5.3-4).

The most common gaseous fuel is natural gas, but certain blast furnace gases and gaseous products from oil refineries can also be used. Table 5.3-5 shows the analysis of typical gases. For simplicity, the various hydrocarbons are represented in the table simply by the number of carbon atoms they contain.

Natural gases are often regarded as ideal fuels. They are "clean" in that they burn without smoke and seldom cause carbonization or fouling of the turbine. Corrosive components such as sulfur dioxide (SO_2) and hydrogen sulfide (H_2S) can often be removed from the gas. However, small quantities of these impurities (1–2 vol. %) can be accepted without gas cleaning.

A natural gas has a characteristic temperature known as the hydrocarbon dewpoint. Heavier fractions such as C_5 and C_6 condense if the temperature is less than this value. If the fuel is then supplied to the combustion chambers as a mixture of the gas and liquid phases, it can cause overheating and considerable turbine damage.

Both fuel handling and the fuel system are extremely simple for a turbine using natural gas. The only complication normally occurring is if gas heating must be employed to maintain the gas above the hydrocarbon dewpoint.

Gases with low heat content contain large amounts of nonflammable fractions. These ballast gases are both advantageous and disadvantageous. Although they are not burned, the compressed gases represent a certain amount of energy released in the turbine. For example, 60% by weight of CO_2 in a gas increases the output of the gas turbine by about 8%. If, however, the gas is supplied at low pressure and has to be compressed before admission to the combustion chambers, the net output gain decreases considerably. This is particularly true with blast furnace gas, which is usually received at atmospheric pressure and high temperature and for which compression is so expensive that the ballast is a definite disadvantage.

Low heat content gases involve a large gas flow to the combustion chambers. The gas volume is about 3 times greater than that of natural gas. The greater volume necessitates larger sizes of pipes and fittings in the fuel system and also involves special design of combustion chambers. Blast furnace gas usually requires pilot combustion from liquid fuel supplied through separate injectors.

TABLE 5.3-4   GASEOUS FUELS

Component[a] (vol. %)	Natural Gas		Lean Natural Gas	Blast Furnace Gas	Refinery Gas
	Africa	Holland			
C_1	89.86	81.20	27.04	0.30	36.40
C_2	5.15	2.76	1.91	—	17.40
C_3	1.62	0.37	0.53	—	2.00
Flammable C_4	0.53	0.13	0.30	—	9.60
components C_5	0.04	0.40	0.07	—	1.50
C_6	—	0.05	—	—	—
C_7	—	—	—	—	—
H_2	—	—	—	2.30	22.90
CO	—	—	—	31.00	2.40
Nonflammable CO_2	2.62	0.93	66.89	9.00	0.20
components N_2	0.17	14.40	3.26	57.40	7.50
O_2	—	0.01	—	—	0.10
Density, kg/m³	0.82	0.84	1.60	1.28	0.99
Heat content, kcal/kg	11,050	9,085	1,780	800	10,320

[a]At 15°C.
(Courtesy ASEA Stal.)

TABLE 5.3-5 CALORIFIC VALUES OF GASEOUS FUELS[a]

Substance	Formula	Calorific Values (kJ/kg) Higher	Lower
Carbon	C	32,780	32,780
Hydrogen	H_2	142,120	120,075
Oxygen	O_2	—	—
Nitrogen	N_2	—	—
Carbon monoxide	CO	10,110	10,110
Carbon dioxide	CO_2	—	—
Methane	CH_4	55,545	50,000
Ethane	C_2H_6	51,920	47,525
Propane	C_3H_8	50,385	46,390
n-Butane	C_4H_{10}	49,565	45,775
Isobutane	C_4H_{10}	49,445	45,660
n-Pentane	C_5H_{12}	49,060	45,400
Isopentane	C_5H_{12}	48,970	45,305
Neopentane	C_5H_{12}	48,780	45,115
n-Hexane	C_6H_{14}	48,710	45,130
Ethylene	C_2H_4	50,345	47,205
Propylene	C_3H_6	49,940	45,800
n-Butane (butylene)	C_4H_8	48,475	45,350
Isobutene	C_4H_8	48,220	48,085
n-Pentane	C_5H_{10}	48,180	45,040
Benzene	C_6H_6	42,360	40,660
Toluene	C_7H_8	42,890	40,985
Xylene	C_8H_{10}	43,380	41,310
Acetylene	C_2H_2	50,010	48,325
Naphthalene	$C_{10}H_8$	40,235	38,865
Methanol	CH_3OH	23,865	21,115
Ethanol	C_2H_5OH	30,615	27,750
Ammonia	NH_3	22,490	18,610
Sulfur	S	9,265	9,270
Hydrogen sulfide	H_2S	16,515	15,225

[a]At 15°C.
(Courtesy ASEA Stal.)

Blast furnace gas contains corrosive components in both gaseous and solid form. Solid contaminants cause not only corrosion but also erosion and deposits in the turbines. The gas often has to be cleaned of solid particles before burning. Fuel specifications restrict both the size range and amount of solid particles.

Refinery gas is sometimes considered as a possible gas turbine fuel. The gas is obtained as a by-product of oil refining and usually contains up to 50% hydrogen. Hydrogen is known for its violent reaction with oxygen. It is not expected that this type of gas mixture would involve any special combustion problems, but until it reaches the combustion chambers, it must be handled with special care. The fuel system must be so designed that leakage and risk of explosion are avoided.

Liquid Fuels

The relatively simple chemical compositions that are a feature of gaseous fuels become considerably more complicated for liquid fuels. Oils contain a large number of hydrocarbons, usually with long chains of carbon and hydrogen. These chains can also contain other elements such as sulfur (S), nitrogen (N), and vanadium (V).

Liquid fuels have widely varying properties depending on their source and method of refining. Standard specifications (such as ASTM and DIN) classify fuels according to viscosity. Typical nomenclatures arising from this classification are kerosene, fuel oil, or heavy oil.

TABLE 5.3-6   CHARACTERISTICS OF LIQUID FUELS

Grade	0GT Naphtha	1GT Kerosene	2GT No. 1 Fuel Oil	3GT Heavy Distillate	4GT Residual Oil or Heavy Oil	
					Untreated	Washed/Dosed
Carbon/hydrogen	6.0	6.1	6.4	7.6	7.8	7.8
Viscosity, cSt,						
20°C	0.5	1.7	3.5	—	—	—
50°C	—	—	—	10	80	80
kg/m at 15°C	0.68	0.79	0.83	0.90	0.94	0.94
Pour point, °C	− 70	− 40	− 9	+ 20	+ 20	+ 20
Effective heat content, kJ/kg (kcal/kg)	44,370 (10,600)	43,530 (10,400)	42,700 (10,200)	41,860 (10,000)	41,000 (9,800)	41,000 (9,800)
Content, ppm						
Ash	1	1	2	10	250	900
Na			0.1	1	50	5
V			0.1	1	50	50
Mg					—	150
Ash sticking point, °C	> 900	> 900	> 900	> 900	600	> 900

(Courtesy ASEA Stal.)

When considering a fuel for use in a gas turbine, properties other than name and viscosity must be taken into account. Table 5.3-6 lists a number of examples of fuels used in gas turbines. Figure 5.3-7 shows temperature viscosity characteristics for the fuels.

Naphtha and kerosene, available in large quantities in certain areas, are excellent gas turbine fuels. Because naphtha and to a lesser extent kerosene have low viscosities, a lubricating additive is necessary to avoid pump seizures. The ash content of both naphtha and kerosene is very low, which means that there is little risk of high-temperature corrosion. Both fuels have good fat-dissolving properties so that a storage tank becomes effectively degreased. However, if there should be any water in the fuel, corrosion will very quickly arise on the tank walls, leading to deposits of rust in the fuel system.

Naphtha and kerosene are primarily used as fuels for gas turbines of the aircraft derivative type.

No. 1 fuel oil is freely available due to its common use for domestic heating. Modified aircraft jet engines can use this fuel. It is ideal for industrial gas turbines. Normally, the melting point of the ash in the fuel is above 900°C, but occasionally contamination arising in transport can give a lower melting point with subsequent risk for high-temperature corrosion.

Heavy distillate is something between No. 1 fuel oil and heavy oils. Its solidifying point is about 25°C but can be as high as 50°C. This low solidification temperature necessitates heating, both in storage and of all fuel-carrying pipes, as well as final heating before combustion.

Heavy distillates usually have low ash content with compatible ash components. There is, however, a risk that undesirable components can be present, and this risk is greater with heavy distillates than with fuels in the No. 1 fuel oil group. Heavy distillates are suitable as fuel for industrial gas turbines. The low sulfur content that is a characteristic of most fuels of this type makes it especially interesting. Development work is being carried out so that it can also be used in jet-type engines. A recently discovered factor is that fuel has a limited storage time due to the formation of unstable combinations in storage. However, the amount of these components can be kept low by the use of special absorption filters.

Heavy-oil residual oils are only used in larger industrial gas turbines. Because of their high viscosity and low solidifying temperature, these oils must be kept warm in storage and preheated before burning. Heavy oils usually have a high and damaging ash content that causes high-temperature corrosion in the combustion chambers and in the turbine. Preventive measures are necessary to combat this corrosion risk.

High-temperature corrosion involves the destruction of the blade metal through reaction with oxygen and sulfur. Molten products of sodium, vanadium, oxygen, and sulfur participate in this reaction as catalysts.

These low-melting-point components have their origin in the fuel's contaminants. Some manufacturers' fuel specifications require that the melting temperature of the ash be higher than 900°C. This is to eliminate or modify the dangerous low-melting-point ash components.

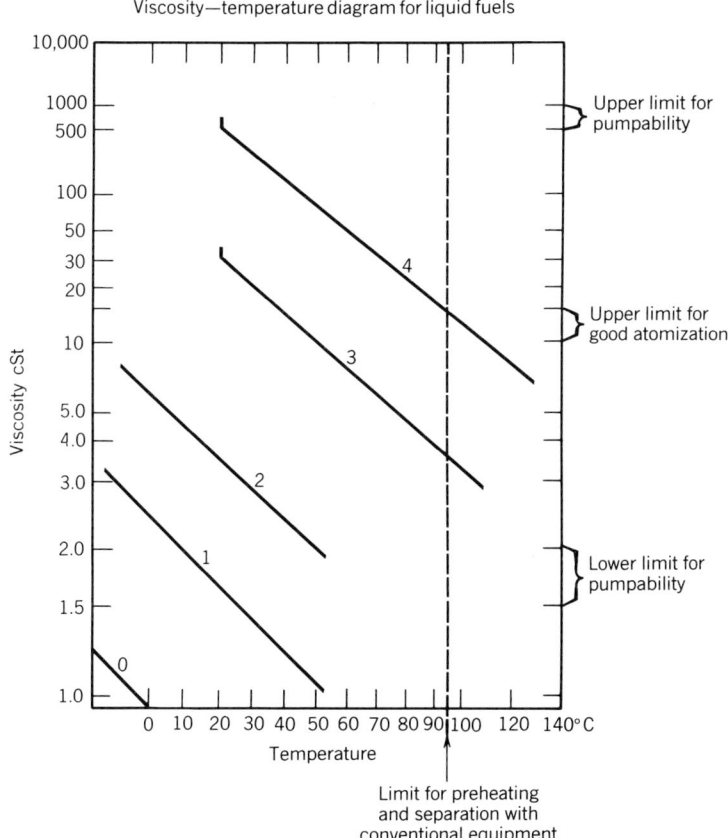

Fig. 5.3-7   Viscosity–temperature diagram—liquid fuels. (Courtesy ASEA Stal.)

Heavy-oil treatment consists of washing and dosing with an anticorrosion additive. Washing is intended to reduce the sodium content of the fuel. The fuel is mixed with water, which dissolves the sodium salts, and the water is then removed in a separator after a demulsifier has been added to the emulsion to make it so treatable. The reduction of the sodium content alone does not raise the melting point of the ash. An aqueous solution of magnesium sulfate may be dispersed in the oil. This means that after burning in the combustion chamber the vanadium is bound in the form of magnesium vanadate with a melting point above the turbine inlet temperature.

Experiments with the addition of a further additive, silicon dioxide (SiO_2), in different forms have given improvements. By using a combination of magnesium as an anticorrosion additive and silicon dioxide as an antifouling additive, long operating times can be attained without a stop.

Those deposits that do occur can be removed during full-load operation by the use of crushed nut shells blown into the turbine. The particles move through the turbine and "blast" away deposits.

Solid and Synthetic Fuels

Coal, peat, and coke are currently under investigation for potential fuels in open-cycle (direct-fired) gas turbines, and programs are in progress to use these fuels in gasified or liquified states as fuels.

5.3-7   Driven Equipment

Electric Generators

If aircraft use is ignored the most common use of gas turbines is for electric power generation. The following must be established when designing a gas turbine generating plant:

Base load or peak rating.
Operating profile (attended or unattended operation).

Voltage frequency and power factor.

Electric generator characteristics.

Heat rate and fuel consumption.

Fuel specifications, requirements, and storage.

Ambient and site conditions.

Codes, standards, and environmental requirements.

Cooling requirements and availability.

Fire protection requirement.

Controls, instrumentation, and alarms.

Load control.

Auxiliary systems and power requirements.

Test inspection and quality assurance requirements.

Black-start capabilities.

Mechanical Drive

Compressors and Pumps. The second major use of gas turbines is for gas and oil pipeline compressors and pump drivers. The following information is required in addition to that needed for an electric generator application:

Special control requirements (e.g., constant-delivery pressure).

Speed–torque characteristics required.

Turbine output speed profile (speed range).

Lubrication requirements of driven equipment.

Type of coupling required.

Geared or nongeared application.

Starting arrangements.

Critical-speed analysis (lateral and torsional).

Hot Gas–Mass Flow

The most common use of gas turbines is in aircraft jet engines. Fuel is added to the compressed air flow and combusted, hence flowing to the compressor turbines. After the compressor turbines, the heated air–fuel mass is exhausted through the jet exhaust, transforming the available energy into pounds of thrust instead of the torque in the power turbine of a gas turbine power plant.

Another example of this type of gas turbine in common use is the ground power unit (GPU) used to start aircraft jet engines.

5.3-8 Application and Type Selection

Section 5.3-2 listed the various types and configurations of gas turbine power plants. These machines are generally listed as "Industrial" (IND) or "Aircraft-derivative" (A/D) types. The following is a guide to the application and commonly used types of machines for the service:

Aircraft jet engines	A/D
Peaking electric plants	A/D
Base-load electric plants	IND
Oil pipelines	IND or A/D
Gas pipelines	IND or A/D
Offshore platforms	A/D
Cogeneration plants	IND
Naval ship propulsion	A/D
Oil refinery applications	IND
Combined gas–steam turbine cycles	IND
Vehicular	A/D or special

The above list is for guidance only. Applications of either type are found in all applications except aircraft.

5.3-9 Components of a Gas Turbine

All gas turbines have a compressor section, combustor(s), and compressor turbine(s). The gas turbine power plant converts the energy available into useful work by adding a power output turbine to the exhaust of the compressor turbine(s). A typical plant (see Fig. 5.3-2) with two compressors (low and high pressure) operating at different speeds and a power turbine is described as follows: Air is compressed in two multistage axial compressors, heated directly through combustion of fuel in the combustion chambers, and finally expanded through three turbine stages. The first turbine stage drives the high-pressure (HP) compressor and the two other stages drive the low-pressure (LP) compressor. The gas generator and power turbine consist of the following (Fig. 5.3-2):

Inlet section (A)
Low-pressure compressor (B)
Compressor intermediate section (C)
High-pressure compressor (D)
Diffuser section (E)
Combustion chambers (F)
High-pressure turbine (G)
Low-pressure turbine (H)
Intermediate section (I)
Two-stage power turbine (J)

The inlet section is of bell mouth form and has a number of profiled struts (1). An electric-motor-driven turning gear is located in the center of the inlet section (2). The motor rotates the LP compressor during the cooling-down period as well as during the initial part of the starting sequence.

The LP and HP compressors rotate in the same direction but at different speeds. At full load the HP compressor rotates at about 7000 rpm while the LP compressor rotates at about 5800 rpm. At part load the speeds are somewhat lower and the relationship between them changes. The LP compressor speed drops quicker than the HP compressor speed.

Each compressor has 10 stages. A rotor stage is made up of a disc (5) with blades (4) attached to a rim at the periphery of the disc. The 10 discs are bolted together to form a stiff drum by means of central tiebolts (6).

The stator blading (7) (guide vanes) is mounted in unsplit rings within the outer compressor housing (8), which itself has no horizontal flanges. The unsplit symmetrical design permits small clearances between the rotating and stationary parts.

The rotors are carried in journal and thrust bearings of the pad type. The LP rotor shaft rotates within the HP rotor and is carried in two relative bearings. The combustion chamber consists of two cylindrical housings (9), which together form an annular pressure chamber. Within this annular space there are seven flame tubes (10). In the forward part of each flame tube there is a fuel nozzle (11) through which the fuel is injected. Within one flame tube there is a flame detector (12) and within two others there are igniters (13) that are energized during start-up.

The HP and LP turbines are built up in the same way as the compressors. The HP turbine is the first stage of the compressor turbine, while the LP turbine is made up of the second and third stages.

There is no mechanical connection between the compressor turbines G and H. The hot gases pass through the intermediate section I into the power turbine inlet J.

The two-stage power turbine operates at a speed independent of the compressor turbines and can be optimally matched to the requirements of the driven equipment (electric generator, compressor, etc.).

5.3-10 Control and Protection

1. Starting may be manual, semiautomatic, or automatic.

Manual starting requires the operator to start all auxiliary systems, to initiate all steps in the starting sequence and to bring the plant up to the minimum governor setting.

Semiautomatic starting usually requires the operator to start all auxiliary systems and to make only one action to shift into an automatic mode sequencing from standby to minimum governor setting.

Automatic starting requires the operator (or automatic signal) to perform one action to bring auxiliary systems to operating mode and automatic sequencing to minimum governor setting.

2. Loading and shutdown may also be manual, semiautomatic, or automatic with possible periods of dwell at specific loads to provide for warming requirements.

TABLE 5.3-7 TYPICAL ALARM POINTS IN GAS TURBINE GENERATING PLANT

Low voltage—dc system.
Ground fault—dc system.
Low voltage—temperature supervision.
Ground fault—temperature supervision.
Inverter fault.
Ground fault—alarm circuit.
Fire fighting, gas detector trip.
Low voltage–115 VAC system.
Low voltage—480 VAC system.
Ignition—tripped.
Barring motor blocked.
Tripped motor protection; fuel/governing: lube; ventilation.
Low lube oil pressure.
Standby lube oil pump—in service.
Emergency lube oil pump—in service.
High-temperature lube oil system.
Low air pressure—operating air system.
Fuel supply failure.
High differential pressure intake air filter.
High average turbine inlet temperature.
High turbine seal air temperature.
Generator rotor ground fault protection.
Generator stator—high temperature.
High vibration—LP compressor, power turbine, exciter.

TABLE 5.3-8 TYPICAL AUTOMATIC SHUTDOWN POINTS IN GAS TURBINE GENERATING PLANT

Overcurrent protection.
Underexcitation protection.
Overvoltage protection.
Generator stator ground fault.
Low frequency.
High differential pressure turbine air intake filter.
Low lube oil pressure.
Lube oil pumps 1 and 2 out of service.
Low lube oil level.
Fuel control out of position.
Overspeed (mechanical trip).
Low governor pressure.
Air intake door open.
Too long for ignition to idle.
Too long for auxiliary preparation.
Emergency manual stop.
High lube oil temperature.
High generator speed.
Fire flaps closed.
Flame detector protection.
Undervoltage, 115 VAC.
High vibration—LP compressor, power turbine, exciter.
Fire.

Emergency shutdown devices shall be available for manual operation and shall also occur automatically as a result of plant protection devices. See Tables 5.3-7 and 5.3-8 for typical alarm and shutdown points in a gas turbine electric generating plant.

3. When using gaseous fuel, it is necessary to provide an automatic purging period in the starting sequence of sufficient duration to permit the gas turbine to displace at least three times the volume of the entire exhaust system at least three times before firing the unit. Special precautions must also be taken when using highly volatile liquid fuels.

4. Gas turbines operating at constant speed (i.e., electric generators) require a governor that senses output speed. The no-load speed is usually adjustable within the range of 95–105% of rated speed. Another important factor is the compatibility of speed control with other speed changers of units running in parallel.

Governors for mechanical drive applications generally limit the output speed to 105% of rated speed under steady-load conditions!

In all cases the governor systems should be capable of preventing the gas turbine from reaching turbine overspeed trip point with instantaneous loss of load.

5. In addition to the fuel control governor, a separate shutoff valve to stop all fuel flow should be installed. This valve should be deactivated (closed) in any shutdown mode and not allowed to open unless prescribed conditions are met.

6. An overspeed governor or trip should be installed on each shaft system. The setting should be at a speed that will not allow the system to exceed a maximum safe limit with sudden loss of load.

Accelerations on loss of load may cause the speed to continue rising after trip. The turbine should be designed to withstand all anticipated speed levels. Gas turbines with separate power turbines or heat exchangers may require additional overspeed protection from stored energy in the system.

Tables 5.3-7 and 5.3-8 list typical alarm and shutdown points. These listings are given for guidance only. The requirements for an actual installation may exceed the points listed. Recommendations for all alarm and protective devices should be solicited from the manufacturer or system designer.

5.3-11 Gas Turbine Auxiliary Systems

Starting

Gas turbines must have a separate source of power to bring them up to a speed where fuel may be introduced into the pressured air and fired to develop more power than that required to operate the compressors (self-sustaining). This may be accomplished by using declutchable electric motors, air–gas starting turbines, or pressurized air (from air receiver) blowing through the gas turbine using its own turbines as a starting device. Ground power units developing pressurized hot gases are used for most aircraft jet engines.

The starting sequence usually follows periods of cranking purging, firing, and acceleration to load desired.

Ignition

All gas turbines have some type of ignition system that "fires" the atomized or gaseous fuel in the combustion chamber at a predetermined point in the starting sequence.

A typical system has retractable high-energy spark plugs connected to transformers. When operating, the plugs are pneumatically lowered into the combustion chamber, the fuel ignited, flame spread via ignition flame tubes, and igniters retracted by spring tension.

Lubrication

Most gas turbine plants are fitted with a common lubrication system for the gas generator, power turbine, and driven equipment. The system supplies the bearings (and coupling in some cases) with oil for lubricating and cooling purposes. The industrial-type gas turbines generally use mineral turbine oils while aircraft derivative gas turbines use synthetic lubricants.

Most systems have an oil tank of sufficient volume for extended retention and air separation. Air or water cooling is required. Normal ac-driven pumps may be backed up by a dc-driven emergency pump. To avoid leakage and assist with air separation, an oil vapor fan may be fitted to the system to maintain a slight vacuum in the oil tank and the bearing housings.

Sealing Air

To prevent oil and air–gas leakage, units have labyrinth seals, which are sealed by air bled from the compressors. The turbine discs may be cooled by this leak-off air.

Filtration and Noise Abatement

The air (working fluid) usually requires some type of filtration and silencing prior to entering the LP compressor. A typical installation would include the following components in series:

Weather louvre inlet.
First-stage (high-efficiency) filter.
Second-stage filter.
Turning vanes.
Inertial separator.
Silencer.

A silencer is required due to reflective noise from the compressor. Note that right-angle bends, although increasing inlet pressure drop, give significant damping to the undesirable noise and also assist in the separation of unwanted airborne contaminants. A silencer is also usually designed into the exhaust stack. A blow-in door should be fitted into the inlet air plenum if rapid clogging or icing of the filter system can occur.

Compressor Blade Cleaning

It is impossible to avoid deposits on the compressor blades even with the most effective filtration, and to remove these deposits, most gas turbines are equipped with a blade-cleaning system. The cleaning media (carboblast or detergent) may be stored in a pressurized container and periodically injected into the airstream through nozzles at the inlet to the LP compressor.

Anti-icing

If a gas turbine is operating in a climate where icing may occur, provisions should be made to provide an anti-icing system. Generally, this is accomplished by bleeding pressurized hot air from the compressor section and reintroducing it through holes in the bell mouth and struts of the inlet section. Output power will be reduced during this operation in proportion to the amount of bled air.

Fire Fighting

Gas turbine power plants should be protected by automatic fire-fighting systems. A typical system would consist of bottles of inlet gases (CO_2 or Halon) placed in spaces to be protected and released automatically by temperature switches in the event of fire and/or manually by fire alarm boxes. When the system is activated, the fuel shutoff valves, fire valves, and fire damper should all be designed to close automatically. Some type of acoustic alarm should automatically be sounded and the turbine tripped.

5.3-12 Auxiliary Power Requirements

Table 5.3-9 shows the normal auxiliary power requirements for a typical installation. It also gives the rated connected load and operating loads for the intended operation (starting, continuous, or cooling down). Net operating loads are given for most operating modes.

5.3-13 Technical Information Required to Design a Turnkey Installation

In Sections 5.3-4, 5.3-5, and 5.3-7 we listed most data that must be considered. However, there are other factors required, including the following:

Specific location indicating type of adjacent area (industrial, residential, etc.) sea coast, desert.
Prevailing winds (speed and direction).
Snow loading.
Seismic requirements.
Soil-bearing capacity.
Ambient-air dust conditions and particle size.
Existing plant layout.
Local codes and standards.
Construction power availability.
Cooling media available.
Site storage conditions and time in storage.

TABLE 5.3-9   AUXILIARY ELECTRIC LOADS

	Rated Output (kW)	Continuous Load (kW)	Intermittent Load (kW)	Load Still	Load Start	Load Normal
Barring motor	4.0	—	1.7	—	1.7	—
Main lube	10.5	7.6	—	—	7.6	7.6
Emergency lube	4.0	—	3.0	—	3.0	—
Oil vapor fan	0.65	0.4	—	—	0.4	0.4
Oil tank heater 1	9.0	—	9.0	9.0	—	—
Governing oil pump	2.5	2.0	—	—	2.0	2.0
Oil tank heater 2	1.0	—	1.0	1.0		
Air compressor	18.5	—	13.5	13.5	13.5	13.5
Generator heaters	6.0	6.0	—	6.0	—	—
Direct current required	6.3	—	6.3	6.3	6.3	6.3
Controls	0.8	0.8	0.5	0.5	0.8	0.8
Ventilation	4.4	4.0	—	—	4.0	4.0
Heating	0.7	—	0.7	0.7	0.7	0.7
Lighting	0.4	—	0.4	0.4	0.4	0.4
Power consumption, kW				37.4	40.4	35.7

5.3-14   Shipping Preparations and Specifications

The following is a guide to shipping preparations:

> Thoroughly coat all exposed machine surfaces with a rust preventative.
> Paint all exterior parts except machined surfaces and instruments.
> Clean and coat with rust preventative all equipment surfaces contacted by the lube oil.

TABLE 5.3-10   TYPICAL SHIPPING WEIGHTS AND DIMENSIONS

Description	Quantity	Length (m)	Width (m)	Height (m)	Weight (tons)
Gas turbine main module (lifting tools included)	1	12.3	4.2	4.0	46.2
Generator (ac)	1	5.9	2.4	2.9	38.8
Generator enclosure (roof and wall sections)	1	7.2	3.6	1.3	6.2
	1	3.4	3.2	0.3	1.0
Air intake silencer	1	3.2	3.1	3.7	6.2
Air intake central unit	1	4.9	3.3	3.8	5.8
Air intake filters with hood	2	4.9	2.8	3.8	3.5
Exhaust stack with silencer (N55/100m)	1	4.9	4.2	3.2	15.5
Air intake silencer (generator)	1	2.1	1.4	1.1	0.3
Air intake filter unit (generator)	1	3.8	2.8	1.5	1.5
Lubricating oil tank	1	3.6	2.9	2.9	4.9
Ancillary module	1	3.4	3.3	1.2	3.9
Ancillary module enclosure (roof and wall sections)	1	7.2	2.0	1.0	2.0
Oil cooler with diffuser	1	3.9	2.4	1.4	3.2
	1	2.5	2.5	1.6	0.7
Air receiver	1	5.2	1.7	1.9	5.7
Control room	1	8.2	2.4	2.9	10.0

Spray all interior parts of the turbine with a rust preventative of a type that can be removed by solvent flushing.

Provide closures for all flanged openings and plugs or caps in or over threaded pipe connections.

Sections and modules should be securely packed and boxed as required for type and distance to be shipped to the site, with lifting weights and points clearly marked.

The shipping specification should contain a list of the major components and give the dimensions and weights of each. A typical list is shown in Table 5.3-10.

5.3-15   Erection on Site

When preparing for the erection on site, the first step is to outline the specific responsibilities of the turbine supplier, erection contractor, foundation contractor (if separate), and purchaser. Table 5.3-11 gives a typical erector's plan and crane schedule and the facilities and personnel required at site.

5.3-16   Acceptance Tests

A number of tests are generally carried out by the manufacturer. Some of these tests are performed so that the manufacturer is assured that the machine will meet its warrantees (first unit of a new design series, etc.); other tests should be considered as compulsory by the purchaser.
Compulsory tests should determine the following:

Sufficient power is delivered under the contract conditions.

Fuel consumption is equal to or less than that specified in the contract and under the contract conditions.

Protective devices and special instrumentation operate as specified.

Optional tests should be included if agreed to by the contracting parties during the contracting phase of the project and may include:

Operating characteristics (i.e., starting, loading, etc.).
Amplitude and frequency of vibration.
Stack emission.
Waste heat recovery.
Sound levels.
Governing system response.
Pressures and temperatures lubrication system.
Operation of auxiliary systems.

TABLE 5.3-11   TYPICAL ERECTORS PLAN

Skill	Duty	Men, No.	Weeks, No.
Mechanics	Erect mechanical	3	6
Mechanic helper	Erect mechanical	3	6
Electrician	Cabling and wiring	3	5
Electrician helper	Cabling and Wiring	2	5
Metal Worker	Assemble buildings	2	5
Welder	Foundation and piping	1	6
Painter	Painting	1	2

Cranes Required[a]			Time Required
Capacity, tons	Jib Radius, m	Hook Height, m	
15	6	15	6 weeks
25	8	18	1 day
50	7	10	2 days

[a] Including operator.

Operating and Standard Conditions

Reasonable efforts should be made to operate as close as possible to the contractual site conditions (see Section 5.3-4). In order that the plant power and fuel consumption may be proven, it is suggested that the purchaser use the ISO standard conditions (Section 5.3-3) and correct the acceptance test run to these design conditions. The fuel should be very close to that specified in the contract.

Instrumentation

The following instrumentation and measuring devices should be used to determine the results of the compulsory tests:

Power-measuring device (torque, speed, or electric).
Fuel-measuring device (calibrated meters or weighing).
Laboratory analysis of fuel-specific gravity and viscosity.
Calibrated pressure gauges and manometers at appropriate locations in gas turbine system.
Barometer and humidity measurements.
Thermocouples and thermometers at appropriate locations in the gas turbine system.
rpm indicators and revolution counters.
Master clock with synchronized signal.

Samples of fuel should be taken at the beginning, midpoint, and end of the test run. (Note that calculation of the liquid fuel heating value may be more accurate than using a calorimeter.)

Fuel meters may require calibration before and after test running.

Pressure measurements should be taken by manometers or deadweight gauges. If elastic-type gauges are used, they should be calibrated against deadweight gauges.

Calculation of Heat Consumption and Thermal Efficiency

The rate of heat consumption is determined by the equation

$$Q = M(\text{LHV} + H_1 - H_0)$$

where Q = rate of heat consumption, kW
M = rate of fuel consumption, kg/s
LHV = lower calorific value of fuel at 15°C, kJ/kg
H_1 = specific enthalpy of fuel at test temperature, kJ/kg
H_0 = specific enthalpy of fuel at 15°C, kJ/kg

Thermal efficiency is determined by the equation

$$R = \frac{P}{Q}$$

where R = thermal efficiency
P = net shaft power output, kW
Q = rate of heat consumption, kW

Correction of Test Results. As previously stated, it is desirable to operate the test as close as possible to design conditions. However, if this is not possible, it will be necessary to correct the test results for speed, power, or thermal efficiency.

5.3-17 Maintenance

Since gas turbines operate at high temperatures, high speeds, and generally in unattended locations and are designed with life-limited components and may operate with less than optimum fuel, it is imperative that a proper maintenance schedule be developed. This schedule should incorporate the training of operating and maintenance personnel.

Inspections

Inspections on an operating time basis should be adopted for the following:

> Hot section of burner elements and turbines.
> Turbine blading.
> Compressor blading.

It should be noted that a gas turbine operating on a cyclical basis (high number of starts and stops to total operating time), such as peak electric service, will have a marked reduction in time between inspections and reduction in life of some components.

The types of inspections can be classified as routine, hot section, and major. They may cover the following:

1. Routine inspection consists of the external visual inspection of a complete plant with a partial boroscope inspection of the hot section and a test run and should be supervised by qualified service engineer.
2. Hot-section inspection consists of the external visual inspection of a complete plant; boroscope inspection of the compressors and LP turbine; visual inspection of the combustors, HP turbine, and fuel nozzles; and functional checks of the control, governor, and protective devices. The test run should be supervised by one or more qualified service engineers.
3. Major inspection consists of hot-section inspection as described above plus additional replacement of turbine sections as indicated by running results and time on the life-limited parts. This should be done by a team of qualified service engineers and mechanics.

Site–Off-Site Repairs and Modular Replacement

Different types of gas turbines offer repair and replacement alternatives. In general, the large heavy-duty units must be repaired on site. The light aircraft types usually have the gas generator section overhauled in off-site maintenance shops equipped with special tools and fixtures. Some types of units take a third approach that allows for modular replacement of complete sections of turbines and compressors on-site and overhaul at an off-site repair facility.

5.3-18 Waste Heat Recovery and Cogeneration

An open-cycle gas turbine by itself generally has a thermal efficiency of 30–35%. The exhaust gas temperatures and mass flow are quite high and are suitable for waste heat recovery.

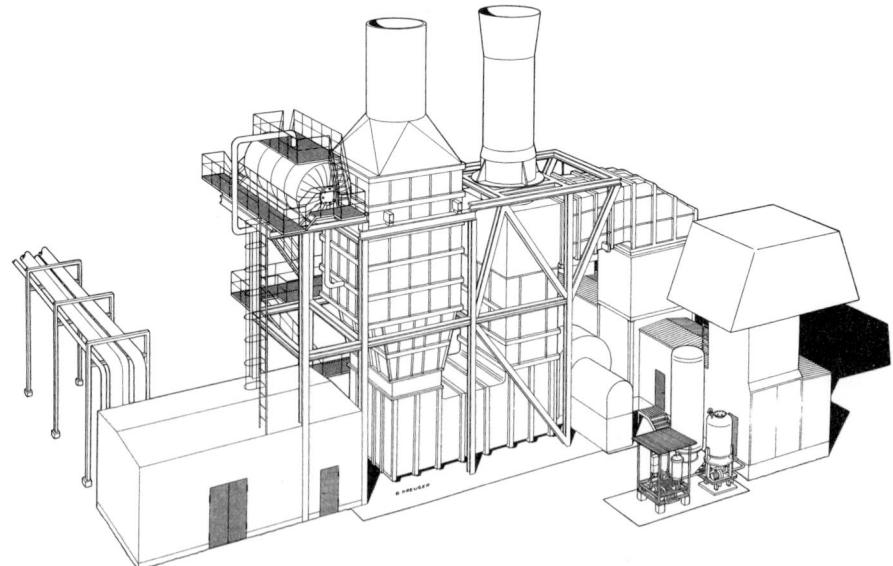

Fig. 5.3-8 Typical gas turbine plant with waste heat recovery—GT35. (Courtesy ASEA Stal)

Since each turbine manufacturer has his own design criteria for temperature and pressure distribution as well as mass flow within the cycle, it is necessary to work with him to develop a balanced installation. The potential gains are quite high and should be considered in each case.

If the exhaust gas is used to generate steam and the steam is used to gain additional power through the use of a condensing steam turbine, the thermal efficiency may be increased to approximately 40%. However, if some use is found for the steam or hot water for process purposes, then the thermal efficiency may be increased to 70–80%.

Various combinations of gas turbine cycles with waste heat recovery through the use of condensing steam turbines, back-pressure steam turbines, unfired waste heat boilers, and supplementary fired waste heat boilers offer a challenge to designers. A typical gas turbine plant with waste heat recovery is shown in Fig. 5.3-8.

REFERENCES

5.3-1   International Standard ISO 3977, "Gas Turbines—Procurement," No. ISO 3977.
5.3-2   International Standard ISO 2314, "Gas Turbines—Acceptance Tests," No. ISO 2314.
5.3-3   A. S. Campbell, *Thermodynamic Analysis of Combustion Engines*, Wiley, New York, 1979.
5.3-4   "Gas Turbine Performance Manual," ASEA STAL.
5.3-5   "Gas Turbine Fuels," ASEA STAL.
5.3-6   "Gas Turbine Procurement Standard," American National Standards Institute, B133.

5.4   INTERNAL COMBUSTION ENGINES

Chi Cheng

5.4-1   Basic Engines

Spark Ignition Engine

Spark ignition engines use gasoline, volatile liquids, or gas as fuel. They operate at a compression ratio of around 8 (or between 4 and 12) on the Otto cycle. Gasoline, kerosene, and gasohol are widely used in most engines. Commercial gas is used mostly in stationary engines. Load and speed are usually controlled by throttling the charge.

These engines have low first cost, low specific weights, low cranking effort required, and a wide variation in speed and load. See Table 5.4-1 for a comparison with compression ignition engines.

Compression Ignition Engines

Compression ignition engines use liquid fuel of low volatility varying from low-grade kerosene and distillates to crude oil such as diesel II, have compression ratios of 11.5 : 1 and 22 : 1, and operate on the diesel cycle. No ignition devices are used except for extreme cold start. Load and speed are controlled by varying the fuel quantity injected.

The dual-fuel or LCR (low compression ratio) diesel engines need an electric glow plug or a pilot injection of secondary fuel with good ignition quality to initiate the combustion process. Usually the intake and exhaust valves overlap in open at top dead center (TDC). Fuel is admitted after the exhaust valve is closed. The LCR has higher power output per weight, or BMEP (brake mean effective pressure).

The diesel engine is low in specific fuel consumption, economic in operation, thermal efficient at front loads, and low in fuel cost (see Table 5.4-1 for a comparison).

The four-stroke engines require four piston strokes or two crankshaft revolutions per cycle. This engine cycle is popular in automobiles, tractors, and aircraft engines with the exception of outboard engines. The two-stroke engines use only two piston strokes or one crankshaft revolution for each

TABLE 5.4-1

Engine	Cost Advantage	Specific Weight per Horsepower	Efficiency	Starting Characteristics
Spark ignition	Lower purchase price	Low	High mechanical efficiency	Low cranking effort required
Compression ignition	Lower fuel cost	High	High thermal efficiency	Need starting aid for cold start

TABLE 5.4-2

Engine	Power Output	Cost	Lubrication
Four-stroke	Increases with higher piston speed	Low specific fuel consumption	Secured, good lubrication
Two-stroke	150–80% higher at same speed	Low first cost for valveless design	Harder load on piston with less oil cooling and lubrication

cycle. Exhaust ports in the cylinder wall are uncovered by the piston or exhaust valves in the cylinder head are open near the end of the expansion stroke. Intake ports or valves are arranged similar to the exhaust. See Table 5.4-2 for a comparison of two- and four-stroke engines.

5.4-2 Fuel and Fuel Systems

Fuel for internal combustion engines are usually petroleum products, or natural and manufactured gases. Because most engines are designed to operate on specific types of fuels, the selection of a fuel injection system is dependent on the physical properties of the fuel.

Gasoline is the main fuel for spark ignition engines. Compression ignition engines use diesel fuel and dual-fuel engines use residual. The specification of these fuels should be maintained within the limits shown in Table 5.4-3.

The preferred physical properties of fuel for the present internal combustion engines are summarized as follows:

Viscosity should be 32 SSU minimum at 100°F (37.8°C) for diesel engines. Residual fuel should be heated to reduce the fuel to 150 SSU or less to enter the injection pump.

High gravity means heavier fuel with more Btu per gallon.

Cetane number indicates the ignition quality of the fuel. For smaller diesel engines, it should exceed 35.

For volatile fuels the Reid method is used to determine the vapor pressure of gasoline at 100°F in the air–fuel ratio of 4:1 by volume.

The effect of fuel volatility on engine performance is indicated by the 10 and 90% points of the ASTM distillation curve. The lowest temperature at which an engine can start is estimated at the 10% point in Fig. 5.4-1. A higher ASTM 90% temperature decreases the volumetric efficiency and power of the engine and increases the tendency to knock and to vaporlock.

TABLE 5.4-3

Fuel Designation Grade or Type	Gasoline, Automotive MIL-G-3056B		Diesel Fuel ASTM D975-53T		Residual, Bunker C, No. 6
	Type I	Type II	No. 1-D		
Gravity API	55–65	55–65	44–45		7–22
Specific gravity	0.72–0.76	0.72–0.76	0.805–0.815, 0.83		0.966 (A_v)
Viscosity					
CS @ 37.8°C	0.648	0.648	1.4 (min), 1.8–5.8		260–750
SSU @ 100°F, 37.8°C			30.5 (min), 32–45		123 SF-
Flash point, °F (min)			100 or legal	125 or legal	160
Distillation, 10%	158	122	420		
°F (max), 90%	356	302		675	
Sulfur, wt. %	0.25	0.25	0.50	1.0	3.0
Cetan no. (min)			40	40	12–30
Octane no. Research (min)	89	89			

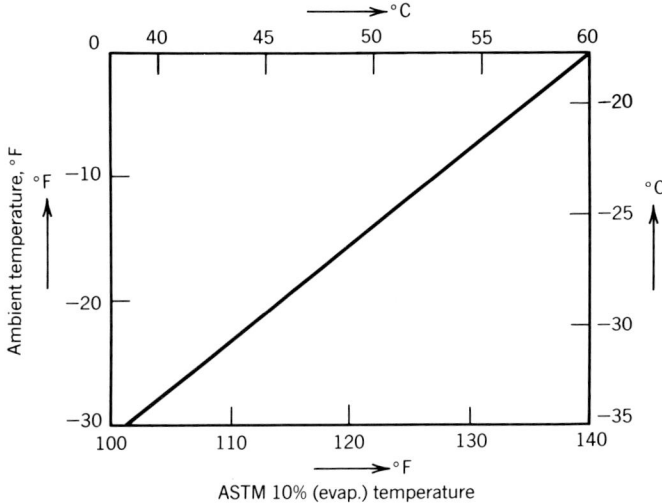

Fig. 5.4-1 Cold-start fuel volatility.

Fuel System

The supply of fuel to the injector of the fuel pump is driven by a transfer pump. The type used can be gear, vane and impeller, plunger, diaphragm, or hand primer.

The complete fuel supply system comprises the fuel tanks, transfer pumps, filters, and piping. The arrangement of these components differ to suit the application, for example, vehicle, industrial or marine, and type of fuel used.

Diesel Fuel Injection Systems

Three general systems of mechanical fuel injection have been developed: the constant pressure or common rail, the spring pressure or accumulator type, and the jerk pump.

The common-rail system maintains its fuel at a constant pressure in a manifold connected to cam-actuated nozzles or with a timing and distributor valve and pressure-operated nozzles. Constant injection pressures up to 12,000 psi are obtained by (1) large fuel manifold, (2) a pump of excess capacity, and (3) bypassing the excess fuel from the accumulator through a governor-controlled pressure-regulating valve.

The accumulator fuel system has its fuel quantity controlled by spring, cam, or accumulator injection. The spring injection uses an eccentric cam to load the spring to push the plunger against the trapped fuel. It may control the fuel pressure through a delivery groove. The fuel pump only delivers the fuel to the accumulator.

The jerk pump times, meters, and forces the fuel at high pressures through the spray nozzle. Plunger pumps are used exclusively and the plunger is actuated by a cam whose contour exerts considerable control on the injection characteristics. The spray duration in crank degrees increases with speed and fuel quantity, but not to the extent of the common-rail system. It has been widely adopted for high-speed engines as well as for those of low and medium speeds.

Table 5.4-4 lists the basic functions of these three mechanical fuel injection systems.

Modern Injection Pumps

This section will only cover some of the pumps and injectors being manufactured and used in the United States.

Single-Plunger Units. These pumps are flange mounted directly over the engine camshaft, one for each cylinder of the engine. They are of the port-controlled type. American Bosch and Bendix designed and manufactured these pumps.

Multiple-Plunger Units. These pumps have as many pumping elements as there are cylinders. They are generally of a bloc construction with self-contained camshaft and tappet mechanism. This simplifies the installation of injection equipment on the engine. It enables the pump to be calibrated

TABLE 5.4-4

Fuel System	Company	Fuel Pump To	Fuel Injection
Common rail			
Cam-actuated nozzle		Header and accumulator	Mechanically operated nozzles
Distributor	Cooper Bessemer	Distributor to nozzle actuated by plunger	Pressure-operated, differential-valve nozzles
Electrically operated nozzles	Atlas-Imperial	Accumulator	Electromagnetically controlled injection valves
Accumulator			Accumulator or spring
Spring injection	Ratellier	Cell between two plungers	Plunger through delivery groove indexing
Hydraulic		Accumulator volume through check valve	Expansion of fuel through nozzle
Jerk pump		Cam-actuated plunger	Plunger to spray nozzle
Variable stroke	Sheppard	Plunger chamber	Plunger with variable stroke
Throttled inlet	Demco	Plunger chamber	Plunger
Throttled bypass		Plunger chamber	Plunger
Port control		Plunger chamber	Plunger with helical groove

and serviced as a unit. American Bosch APE pumps, Simms' SPGE pumps, Robert Bosch model M pump, and other lines of multiple-plunger pumps are of this type.

Distributor. Lower cost by reduction of parts and inherent calibration are features of distributor pumps for small high-speed engines. One or two pumping elements are used to deliver the fuel to all engine cylinders. International Harvester once produced it. American Bosch, PSB, and Roosa-Master by Hartford Machine Screw Company are of this type.

Unit Injector. The unit injector combines the pump and spray nozzle in a single unit, which is mounted on the cylinder head. This design eliminates the problems of pressure waves and fuel compressibility in long-discharge tubings. It requires the means to actuate the injectors. General Motors, American Bosch, Murphy, and Cummins have their own version of the unit injector. The high fuel injection pressure is limited by the contact fatigue of such actuating means as cam lobes. See Table 5.4-5 for a comparison.

TABLE 5.4-5

Company	Fuel Injection	Fuel Quantity Control
General Motor, Series 71	Cam-actuated plunger	Rotating the plunger by the rack and gear
American Bosch	Spring-actuated plunger	Metering valve
Murphy	Cam-actuated plunger	Rotating the plunger by the rack and gear
Cummins PT system	Cam-actuated plunger	Pressure- and time-dependent orifice

Dual-Fuel Pumps. Dual-fuel engines are able to operate completely on fuel oil or predominantly on gaseous fuel with oil ignition and are fully convertible during operation from one to the other. The stacked plungers and telescopic plungers of American Bosch are of this type.

5.4-3 Mechanical Design

Introduction

The first step in determining the engine displacement and its drive train is simulated in a performance analysis with the applications where engine will be deployed. Next a representative engine BMEP curve is established based on the test data of other engines. Fuel and torque data for several engine displacements are then generated using the model BMEP curve. Packaging and drive train are finally used in the selection of basic engine design due to prior establishment of specific engine compartment parameters. The necessity of installation in the specific application is included in the packaging consideration.

Once the available compartment space is established, engine design can be started. Due to the recent developments in computers and such numerical analyses as finite-element methods, the stress and vibration of the whole engine block as well as its components can be accurately analyzed. Designs of engine block, pistons, connecting rods, crankshaft, camshaft, and valves have all been done by using finite-element methods. As an example, a design on the intake valve will be given next.

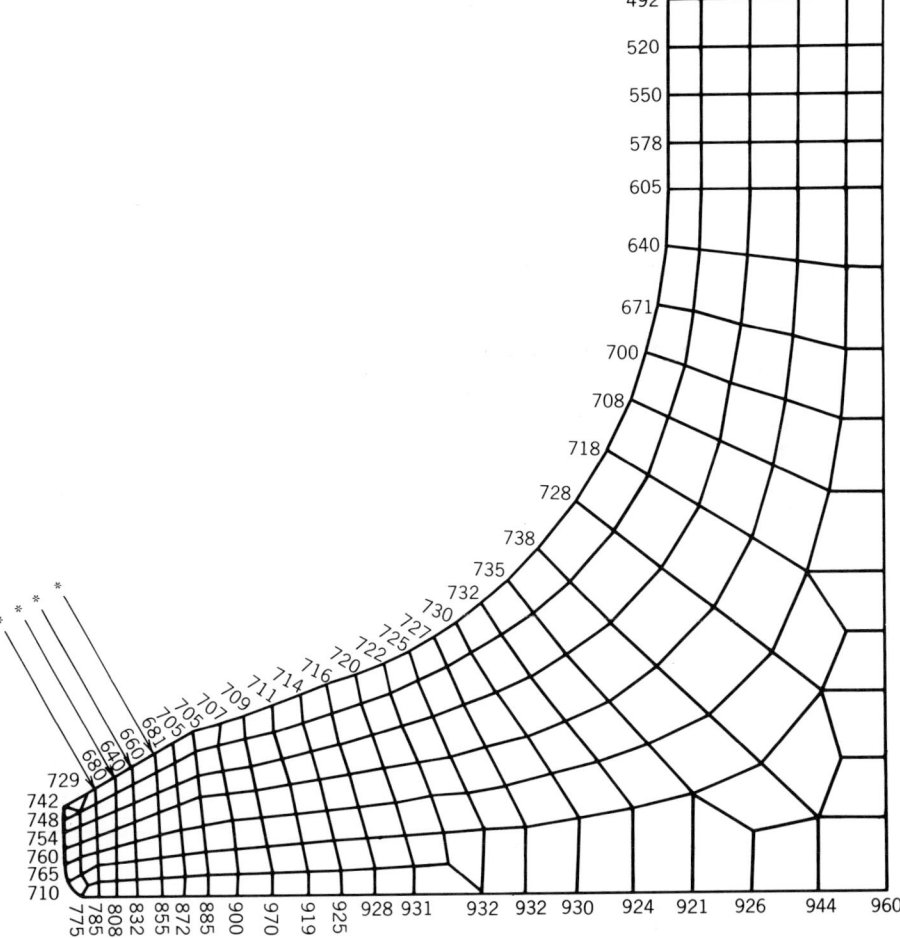

Fig. 5.4-2 Boundary temperature distribution of an intake valve. (Courtesy of International Harvester Co.)

PRIME MOVERS

Valves

The valve with $\theta \leq 30°$ valve seat inclined angle will have less wear than the valve with $\theta \geq 45°$. The valve stem and valve insert serve as the heat sink through conduction. The local temperature gradient on the valve seat generally imposes a high, tensile, thermal stress on the valve seating area. The combined cyclic thermal and mechanical stress should be considered in the Goodman diagram, which is based on the valve seat material at high operating temperatures.

An axisymmetric model for finite-element methods can be generated either manually, digitized, or automatically from computer graphics. At the critical valve seat location the mesh must be fine enough to pick up the local temperature gradient. The pilot valves have been actually tested for 1 h and had their temperature measured through the surface hardness conversion method—Tukon measurement. Using a heat equilibrium computer program, the temperature distribution over the entire valve body elements has been calculated and set up. The valve seat at the seating condition is simulated by an infinite stiff spring. The valve is simply supported. Only sliding along the seating surface is allowed. The peak firing pressure is distributed over the head face and the valve spring load over the stem. Analysis can be done for mechanical, thermal, or combined loading conditions.

Figure 5.4-2 shows the boundary temperature distribution and model of the finite-element method for International Harvesters' DT817C intake valve. For design and results, please contact author for reference.

Piston and Piston Ring

The piston bowl shape is determined by (1) the swirl rate requirement, (2) the combustion and fuel injection into the combustion chamber, and (3) the clearance with valve head, injector tip, or spark plug. Small engines rely on high swirl during intake, and bowl shape must be eccentric and accommodate the swirl rate requirement. A large engine's bowl shape is more symmetric for four valves and is offset for two valves by the off-center spark plug or fuel injector. They have low swirl rates and rely on high-pressure fuel injection for better combustion and engine performance. The depth of the piston bowl is determined by the bowl volume required for the designed compression ratio.

Because the piston cylinder liner has the largest share of frictional loss for the whole engine, the piston, piston rings, and fuel injection serve as the heart of the internal combustion engine. Piston

TABLE 5.4-6

Unit	Side Clearance	
	mm	in.
Petrol Engines		
Cylinder bore	60–120	2.36–4.72
Ring width tolerance	0.015	0.0006
Groove tolerance	0.025	0.001
Normal minimum side clearance	0.04	0.0016
Side-clearance range	0.04/0.08	0.0016/0.0032
Diesel Engines		
	millimeters (inches)	
Cylinder bore	up to 180(7.09)	over 180(7.09)
Ring width tolerance	0.015(0.0006)	0.025(0.001)
Groove tolerance	0.025(0.001)	0.025(0.001)
Normal minimum side clearance		
Top ring	0.06(0.0024)	0.08(0.0032)
Other ring	0.04(0.0016)	0.06(0.0024)
Side clearance range		
Top ring	0.06/0.10(0.0024/0.004)	0.08/0.13(0.0032/0.0051)
Other ring	0.04/0.08(0.0016/0.0032)	0.06/0.11(0.0024/0.0043)

rings must seat well on the piston ring grove, which should not distort too much due to thermal expansion. The rings should be coated to reduce the friction loss and wear rate. For these goals the side clearance of the ring in the groove must be kept to a minimum. The practical ring width tolerance is 0.015 mm and groove tolerance is 0.020 mm. Typical side clearances for spark ignition and compression ignition engines are listed in Table 5.4-6. Since early 1980 the preferred ring coating is an electrodeposited chromium plate for least friction and best durability if lubrication conditions are suitable. For sticking problems keystone (taper-sided) rings are used. For more detailed information, see Reference 2 at the end.

Connection Rod

As numerical analyses such as the finite-element method become widely used, design and stress analyses should be more precise to cover the L10 life at 10,000 h, or infinity. Some are designed for 10% overspeed and 10% overload depending on their application. For the loading condition the inertia load is calculated with the reciprocating mass of the piston, piston pin, and one-half to two-thirds of the connecting rod. Acceleration (a), of the reciprocating parts is given by $a = (V^2/R)(\cos\theta + r\cos2\theta/l)$, where R is the crank radius, V is the crank pin velocity, and θ is the crank angle when the centerline of the pistons and crankshaft lie in the same plane.

In the future, with faster computers and better models, transient models will replace the finite-element model with static loading case in the analysis. One way is that the reciprocating mass element of the connecting rod will impose the reciprocating force upon their own elements rather than considering the one-half to two-thirds of the connecting rod at the end.

However, the analysis should cover not only the stress and strength requirements but also the torsional natural frequency for the shank, the bolts, rod ends and cap, and the split-shell bearings.

5.4-4   Lubrication

Introduction

The lubrication system of an internal combustion engine consists mainly of a lubricant, a pump, the bearings, and the circulation piping. Most of them are using oil with additives. The base oils are usually mineral oils, but some are synthetic. The purpose of lubrication is (1) to reduce the wear rate, frictional resistance, and deposit accumulation; (2) to remove the deposit accumulation and to provide corrosion and oxidation protection; and (3) to cool the piston, piston ring, and other high-temperature components for endurance life. On hard-to-reach areas the lubricant might be grease, grease and oil, or dry lubricant instead of engine oil.

Lubricant

Consideration of which lubricant to use is mainly dependent on such properties as viscosity, rust and corrosion protection, and scavenging for contaminants. Petroleum oils with different additives serve the lubrication of internal combustion engine very well.

Figures 5.4-3 and 5.4-4 give the viscosity of SAE 10W–50W oil and synthetic oil, respectively. For cold start, low-weight oil with relatively low viscosity is vital. For high-temperature operation high weight oil with relatively high viscosity will ensure an adequate oil film for lubrication and prevent excess oil consumption. Hence, a 10W–40W or 20W–50W engine oil will be ideal for both cold-start and high-temperature operation requirements. For an engine with piston top ring oil temperature above 300°F (149°C), synthetic oil should be considered or engine output or design should be changed such that mineral engine oil temperature can be lowered below 250°F (121°C).

Oil Pump and Lubricating System

A small internal combustion engine can have its components lubricated with just an oil splash system. One way is to stir the oil in the oil pan and splash the oil through an end blade attached to the rotating connecting rod. For larger and high-horsepower engines, a gear or other type of oil pump is usually required to provide a pressurized oil to lubricate all the moving components and bearings. A duet or piping system is provided for the oil circulation and an aftercooler may be used to cool the hot-oil lubricant. For petro-derived oil the temperature must be limited to or under 250°F (121°C).

In the design of a set of spur gear oil pumps, the oil flow rate is estimated for all the engine components. They include crankshaft main and pin bearings, connecting rod bearings, cam follower bodies, cam bearings, piston ring and piston cooling jets, turbocharger, and so on. These individual flow rates are summed and magnified by the inverse of pump efficiency, say 90%. The speeds of oil pump drives and pinion gears are determined by the engine rated speed and gear teeth ratio.

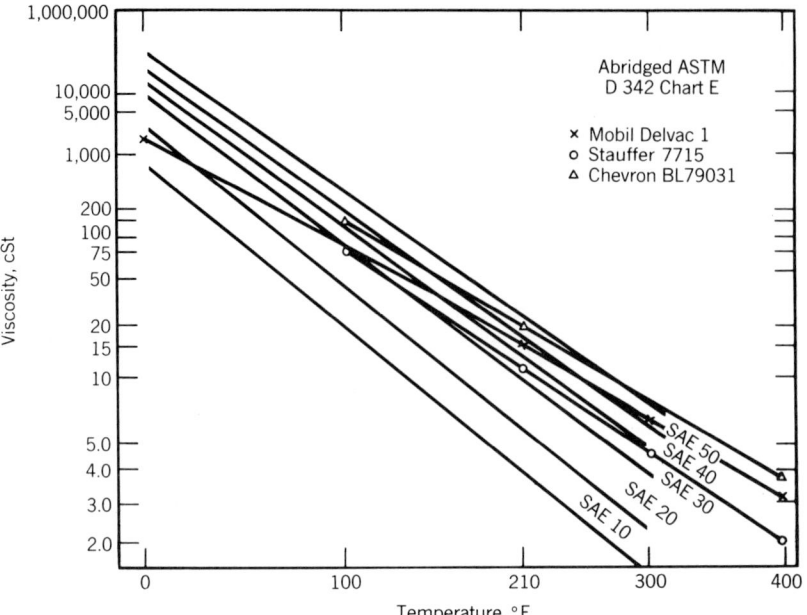

Fig. 5.4-3 Viscosity of synthetic and SAE oils. (Copyright, ASTM, 1916 Race Street, Philadelphia, PA 19103. Reprinted with permission.)

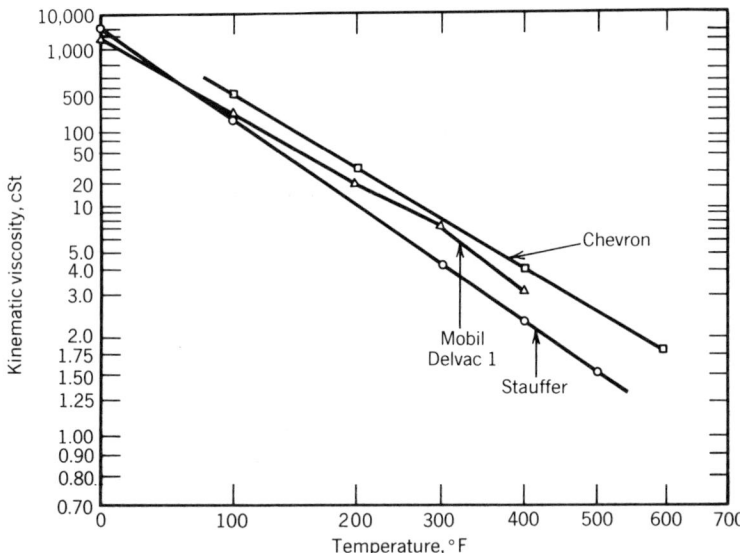

Fig. 5.4-4 Viscosity of synthetic oil. (Mobil data courtesy of Mobil Oil Corp. Stauffer data courtesy of Stauffer Chemical Co. Chevron data, for an experimental product, courtesy of Chevron Research Co.)

The minimum space or gear size of the spur gear oil pump can now be calculated from the overall oil flow rate relation

$$Q = \frac{dLn}{63p}$$

where Q = flow rate, L/min
d = pitch diameter of gear, cm
L = gear width, cm

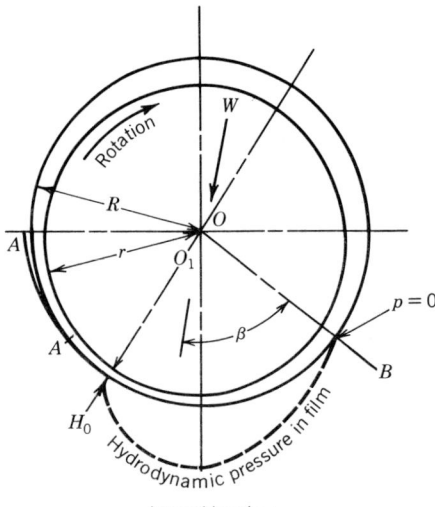

Journal bearing
with perfect lubrication

Fig. 5.4-5 Journal bearing with perfect lubrication.

$n =$ angular speed of gear, rpm
$p =$ diametral pitch, $1/$module

For a metric gear set a module of 5, 6, 7, and 8 and a gear teeth number of 9, 10, or 12 are combined and tried. The required flow rate must be satisfied and gear parameters are generated for the selection of oil pump gears.

Bearings

The bearings of an internal combustion engine include cylindrical journal bearings, thrust bearings, and antifriction ball and roller bearings.

The ball and roller bearings can follow the design manual set up by the bearing manufacturer. To name a few, SKF has a design manual for its ball and roller bearings. Timken is specialized in roller and tapered roller bearings. Torrington provides needle roller bearings. In these design manuals the bearing load (F) has to be the cube power weighted average $[(\Sigma F_i^3)^{1/3}]$ for ball bearings and the ten-thirds power weighted average $[(\Sigma F_i^{10/3})^{3/10}]$ for roller bearings.

The cylindrical journal bearing has a hydrodynamic oil film developed for perfect lubrication to carry the radial load. See Fig. 5.4-5 for the oil film pressure distribution. The reader is referred to Fuller or Forbes, Pope, and Everitt, or to the ASME conference listed in the bibliography. Similarly these references can be used for the design of thrust bearings. The oil supply hole and groove must be located at the low oil pressure location.

5.4-5   Cooling

Water Coolings

A noncirculating (open) cooling system is only suitable for application where cooling water is abundant. For a large engine the water would be wasted at a rate of 2–4 gal/bhp-h. Natural circulation for the engine requires large piping and connections. In this case hot water rises in the engine jacket, flows to the radiator, is cooled, descends, and flows back to the engine jacket.

In high-output engines forced circulation is achieved through a water pump with a capacity of from 10 to 15 gps. The coolant is directed at the hottest spot within the cylinder head, such as the exhaust valve seat area, with an inlet manifold for a common cylinder block jacket to avoid overheating.

Using the fully developed duct flow, the size of piping can be calculated from the required cooling load capacity, water flow rate, and a selected temperature rise for the application. The cooling load is equal to the difference of the energy input and the sum of brake horsepower and exhaust energy. It ranges from 15 to 35% for diesel and automotive engines. For heavy-duty truck engines it is 30%. The pressure head of the cooling system must provide water at 10–15 psi (0.07–0.10 Mpa) to enter different jackets (except for the piston).

A typical water pump is an external or internal gear pump. The water pump shaft should be corrosion resistant and can be inertia-welded with stainless steel at the end seal.

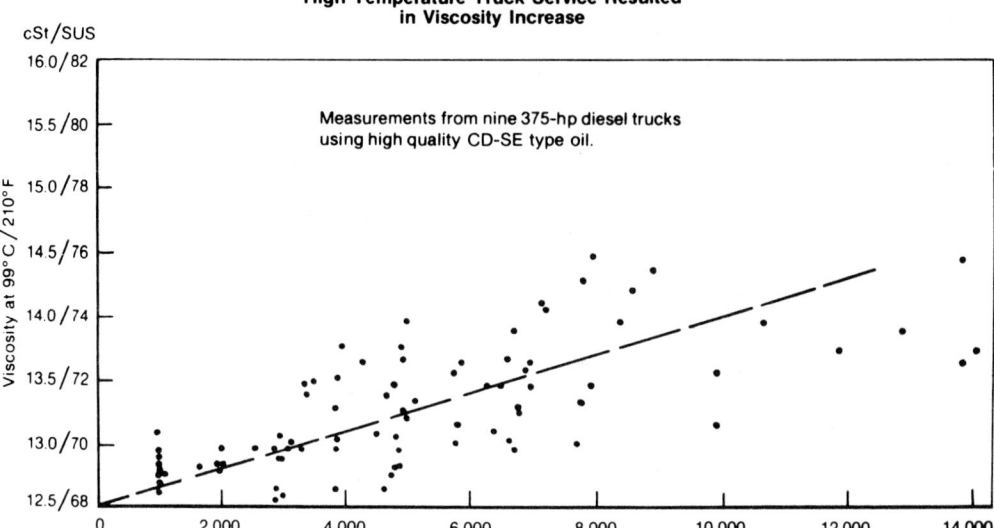

Fig. 5.4-6 High temperature truck service resulted in viscosity increase. (Courtesy Chevron Research Company.)

Oil Cooling

Heavy-duty engines usually use oil not only as a lubricant but also as a coolant to piston, cylinder wall, and bearing areas. The oil temperature should be maintained under 250°F (121°C) for conventional engine using SAE 10W–50W oil. Higher oil temperatures of 300°F (149°C) will lead to the oxidation and decomposition of the oil. Synthetic oils such as Mobile Delvac I, Stauffer 7715 and Chevron BL79031 can endure quite well up to 232°C (450°F). An oil–water or oil–air aftercooler should be used to lower the oil temperature to within limits.

Oil life and consumption are affected by engine speed, load, temperature, viscosity and volatility. A low viscosity base oil with shear and thermal stable, high viscosity additive improves the oil life. See Figure 5.4-6 for viscosity-life relationship.

Air Cooling

Many aircraft engines, most motorcycle engines, and some automotive and lawnmower engines are air cooled. The heat dissipation is usually achieved through the heat convection and radiation of the long fins on the cylinder head and cylinder block. High-output aircraft engines utilize the cooling from long fins at the expense of considerable air resistance or drag.

The heat dissipation for copper or aluminum fins in a parallel blast of air is

$$H = \left[0.0247 - 0.0054 \frac{l^{0.8}}{p^{0.4}}\right] v^{0.73} \qquad \text{Btu/ft}^2\text{-min-°F}$$

or

$$H = \left[0.178 - 0.0277 \frac{L^{0.8}}{p^{0.4}}\right] V^{0.73} \qquad \text{cal/m}^2\text{-min-°C}$$

and heat dissipation rate is

$$Q = HA \Delta T$$

where A is fin surface area, ΔT the temperature difference between the mean fin temperature and the incoming temperature; l is the length of the fins and p is the distance between fins (in inches) (or P and L in centimeters); and v is the air velocity in mph (or V in kmph). For steel fins it is 5–10% greater.*

5.4-6 Performance

The performance of internal combustion engines became of primary interest after the energy crisis of 1973. Turbochargers have become the main implement, and electronics and minicomputers will dictate future advances. For example, computer-controlled coil ignition has replaced distributors in GM's 2.5-L 4-cylinder engine and the three-way catalyst with underhood electronics has helped engines in emission control. Electronic-controlled fuel injection has become the main goal of research and development: the sequential fuel injection by Buick and AMC's sequential control on the spark timing with carburetors by adding a knock sensor. The hot-film sensor follows the MAP sensor to measure air flow for emission control. Different versions of electronic fuel injection and control will be developed by companies like Bosch and Bendix in joint venture with the auto manufacturers.

Theoretical Engine Cycle Performance

The theoretical indicated thermal efficiency, η_a, of the air standard Otto cycle depends only on the volumetric compression ratio γ_c:

$$\eta_a = 1 - \left(\frac{1}{\gamma_c}\right)^{0.4} \tag{5.4-1}$$

For the diesel cycle

$$\eta_a = 1 - \frac{\gamma_d^{1.4} - 1}{1.4\gamma_c^{0.4}(\gamma_d - 1)} \tag{5.4-2}$$

where γ_d is the volumetric expansion ratio during constant-pressure combustion.

The mean effective pressure is equal to the net work divided by displacement. Engine horsepower is

$$HP = (MEP)\, LA\frac{N}{K}$$

where HP is in kilowatts (hp) and

MEP = mean effective pressure, MPa (psi)
  L = stroke, cm (ft)
  A = net piston area, cm^2 (in.^{2})
  N = cycle per minute
  K = 32,630 (or 33,000 if English system is used)

Thermal efficiency with maximum power mixture (90% theoretical air) is

$$\eta_{mp} = 1 - \left(\frac{1}{\gamma_c}\right)^{0.22}$$

Increasing the compression ratio and the air–fuel ratio increases the thermal efficiency of both the Otto and Diesel cycles. The temperature and strength limitation of the material and the emission and combustion control of the engine put a limit to the compression ratio. Mean effective pressures depend on the fuel and air supply and the compression ratio. See Figs. 5.4-7 and 5.4-8 for mean effective pressures.

Turbocharging

Turbocharging as a means of performance improvement and emission control has been increasing at a rapid rate for diesel engines for years and for gasoline engines recently. In 1983 and 1984 turbochargers beef up the small automobile engines, gasoline and diesel, to replace larger automobiles. The horsepower is increased by 70–80% and peak torque stays high for most of the engine speed range.

*Gibson, Inst. Automo. (London), 1920.

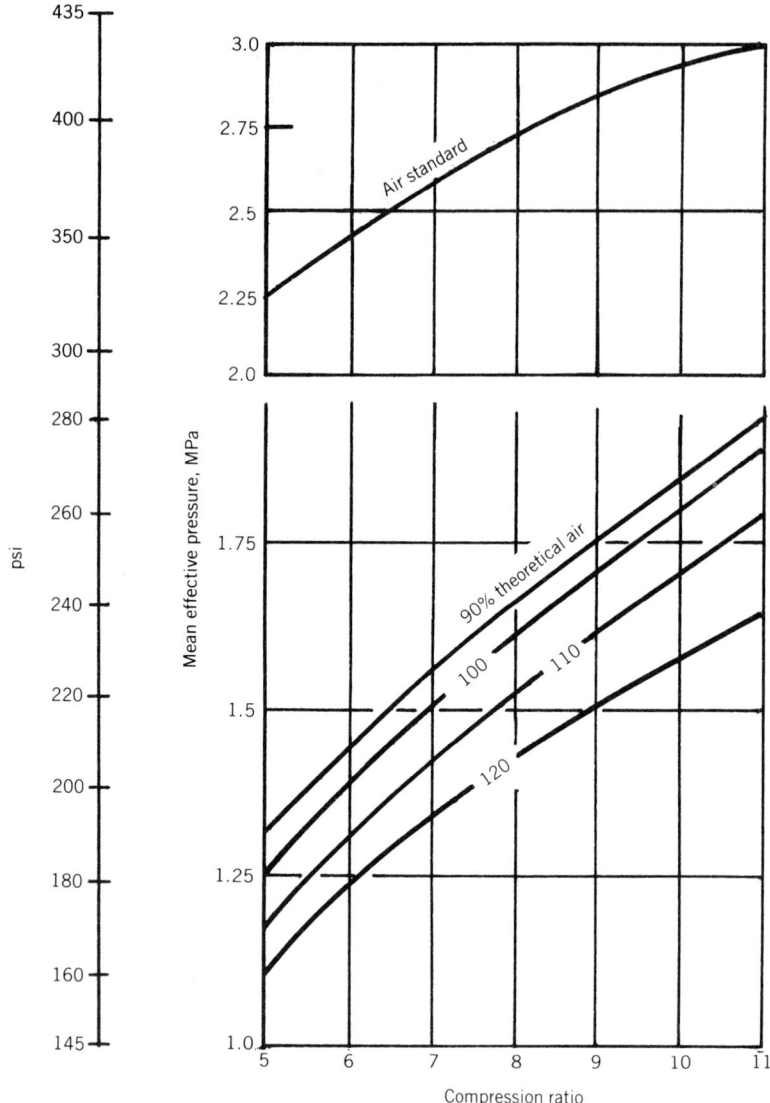

Fig. 5.4-7 MEP for Otto cycle engine. Air and gasoline at 15.6 °C.

For optimum fuel consumption and better engine performance, the compressor supplies the required air flow over a range in engine speeds at various loads and along the engine torque curve between torque peak and rated speeds. Figure 5.4-9 shows the compressor and engine match characteristics. The engine must operate within the range in mass flow capability of the compressor due to the surge. To have relatively high compressor efficiency at rated speed and load, and to avoid surge at torque peak speed, it is necessary to limit the ratio of torque peak engine speed to rated speed to approximately 0.6. To supply the same torque curve over a higher altitude, the turbocharger must be able to increase its rotative speeds to handle the overall increase in the required pressure ratio.

The turbine characteristics can be plotted as a comparison of the corrected mass flows versus corrected rotor speeds with lines sloping downwards to the right as shown in Fig. 5.4-10. The line in the central region of constant efficiency levels of the map represents approximately the series of the steady-state operating points for a given compressor.

With the firing pressure and an exhaust temperature limitation from the engine hardware, the turbine extracts the energy from the exhaust inlet gas with an overall efficiency of about 60–65%. It lowers the amount of available energy required for pumping air to approximately 40% of the air

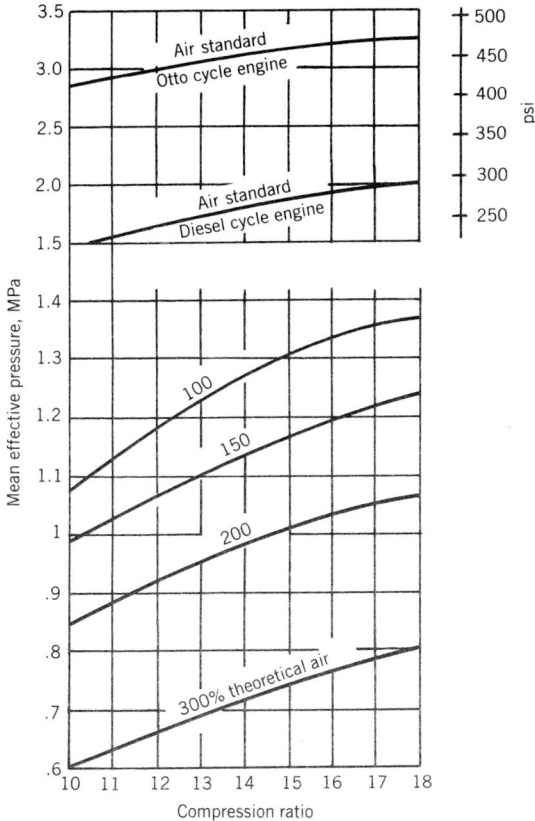

Fig. 5.4-8 MEP for diesel cycle engine.

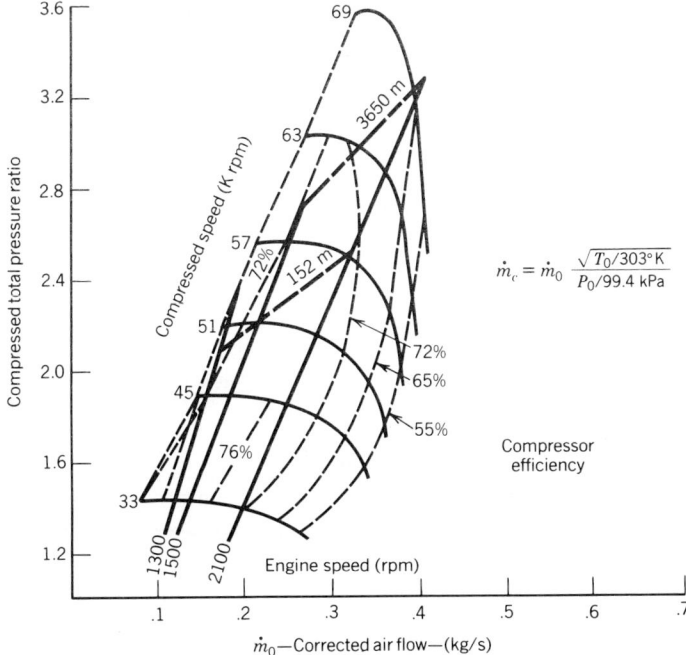

$$\dot{m}_c = \dot{m}_0 \frac{\sqrt{T_0/303°\text{K}}}{P_0/99.4\text{ kPa}}$$

Fig. 5.4-9 Compressor and engine match characteristics. Nomenclature: $\dot{m}_c$ = corrected mass flow rate; m_0 = observed flow rate; P_0 = machine inlet total pressure; T_0 = machine inlet temperature. (Reprinted with permission. © 1979 Society of Automotive Engineers, Inc.)

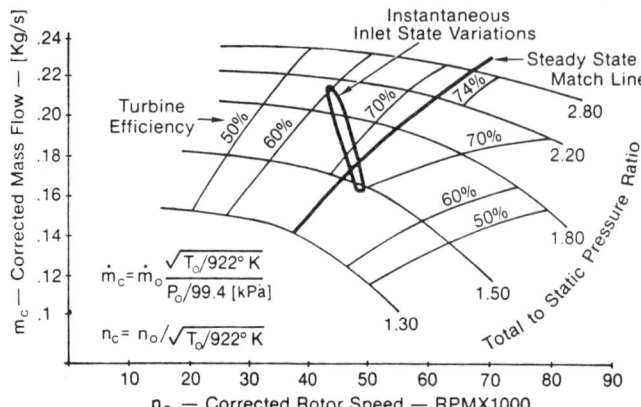

Fig. 5.4-10 Compressor and engine match characteristics. Nomenclature: $\dot{m}_c$ = corrected mass flow rate; m_0 = observed mass flow rate; n_c = corrected rotor speed; n_0 = observed rotor speed; P_0 = machine inlet total pressure; T_0 = machine inlet total temperature. (Reprinted with permission. © 1979 Society of Automotive Engineers, Inc.)

horsepower. When the in-cylinder closed-cycle improvement equals the rate of out-of-cylinder available energy consumption for a given "increment" of air flow, the optimum BSFC (brake-specific fuel consumption) occurs.

Turbine casing size is selected at a minimum to enhance the torque attainable under the constraints of firing pressure and turbine wheel speed. A variable-geometry turbine casing will meet the engine transient response to load and speed. It can provide a rapid transient torque rise and a nearly uniform fuel–air ratio supply during engine acceleration.

With efforts on turbomachinery design for efficient broad-range compressor configurations and responsive variable geometry turbine configurations, the overall engine performance will be further improved.

For more details refer to the listings for Watson and Janota in the bibliography.

5.4-7 Applications

Internal combustion engines are used in automobiles, trucks, buses, tractors, industries, aircraft stationary, locomotives, and marines.

Automobile Engines

Spark ignition gasoline engines are still the dominant power plant for automobile application, with small diesels getting an increasing share of attention. Fuel efficiency and engine performance are the focus in the eighties and the future. Due to the need for fuel efficiency, new engines have become smaller, have four cylinders only, have adopted turbochargers for higher engine horsepower, and use electronic fuel injection and control. Tables 5.4-7 and 5.4-8 list selected gas and diesel engines for automobile application.

Catalysts with underhood electronics commanded the power plant in early 1980. Since 1983, turbos have enhanced the power of small engines, which will replace the large gas-guzzler engines. General Motors replaced conventional distributors with computer-controlled coil ignition. Ford's Mustang adopted a 2.3-L turbocharged engine with an electronically controlled wastegate. Chrysler's 2.2-L engine is equipped with a Garrett turbocharger to boost its intake air and power. American Motors introduced an A.M.-Bendix electronically controlled fuel injection system in its 1.4-L engine through a throttle body assembly.

These turbos boost the horsepower by at least 40% to as much as 70%. The peak torque for Pontiac-Buick 1.8-L engine stays at 203 N · m (150 lb-ft) at 2800 rpm and flattens out until 4200 rpm.

Diesel Engines

Table 5.4-9 lists selected engines from each major engine manufacturing company. Long before the spark ignition engine the diesel engine had adopted the turbocharger or comprex. It even had used turbocompound machinery to extract energy from exhaust gas for better engine fuel efficiency. These high-output engines have produced almost 1 hp (0.7457 kW) out of 2 in.^{3} (0.016 L) engine displacement.

TABLE 5.4-7 SELECTED U.S. AUTOMOBILE ENGINES, 1983–1984[a]

	Cylinder No.	Bore mm	Stroke mm	Displacement (L)	Compression Ratio (to 1)	kW @ rpm	Torque N · m @ rpm
American Motors	4	98.4	81	2.5	9.2		132@2800
American Honda	4	77	94	1.8	8.8	56@4500	130@3000
Chrysler							
2.2LT	4	87.5	92	2.2	8.1	106@5600	217@3600
2.2L	4	87.5	92	2.2	9.0	74.6@5200	165@3200
Ford							
2.3LT	4	96	79.4	2.3	8.0	108@4600	244@3600
2.3L	4	96	79.4	2.3	9.0	67@4600	165@2600
G.M.							
2L	4	88.9	80	2.0	9.3	67@5000	149@3600
1.8LT	4	89	74	1.84	9.0	112@5500	203@2800
1.8L	4	89	74	1.84	9.0	66@5100	136@2800
V.W. of America	4	79.5	86.4	1.7	8.2	55@5000	122@3000

Source. Public Relation Dept. of American Motors, Chrysler, Ford, and General Motors; 1983–84; *Automotive Inds.* 13–14, 4 (1983).

[a] T—turbocharge.

TABLE 5.4-8 SELECTED FOREIGN CAR ENGINES, 1983[a]

	Cylinder No.	Bore (mm)	Stroke (mm)	Displacement (L)	Compression Ratio (to 1)	kW @ rpm	Torque, N · m @ rpm
Audi T	5	79.5	86.4	2.1	8.2	74.6@5100	152@3000
BMW	6	89	86	3.2	8.8	135@6000	264@4000
Datsun	6	83	73.7	2.39	8.9	89.5@5200	182@2800
Nissan	4	84.5	88	1.97	8.5	65.6@5200	152@2800
Fiat	4	84	90	1.9	8.1	76@5500	
Honda	4	74	86.5	1.5	9.3		
Isuzu Diesel	4	84	82	1.8	22	34@5000	97.6@2000
Mazda	4	86	86	1.99	8.6	62@4800	149@2500
Mercedes-Benz TD	5	90.9	92.4	3.0	21.5	89.5@4350	250@2400
Mitsubishi	4	91	98	2.6	8.2	74.6@5000	186@2500
Peugeot TD	4	94	82.8	2.3	21	59.7@4150	184@2000
Porsche	4	100	78.9	2.5	9.5	107@5500	186@3000
Renault	4	76	77	1.4	8.8	38@5000	96.3@3000
Rolls-Royce/Bentley	8	104	99	6.75	8.1		
Subaru	4	91.9	67	1.8	8.7	54.4@4400	127@2400
Toyota EGR	4	76	71.4	1.3	9.3	56@6000	
Volkswagen	4	79.5	86.4	1.7	8.2	55.2@5000	122@3000
Volvo	4	96	80	2.3	10.3	80@5400	172@3520

Source. *Auto. Inds.* 18–21, 4 (1983).

[a] T—Turbocharge; TD–turbodiesel; EGR—exhaust gas recirculation.

In some engines an overhead camshaft that drives the unit injector has simplified the required engine components. One-stage turbocharger becomes most efficient and reaches 65% overall efficiency. Top compression piston rings have been widely tried for least dead crevice volume and better fuel efficiency. The fuel injection pressure has been increased for the same size camshaft. The only difference between the engines is whether they solve the associated high contact stress, high temperature, and high wear problems.

TABLE 5.4-9 1983 DIESEL ENGINE FOR TRUCK, BUS, TRACTOR, INDUSTRIAL, AND LOCOMOTIVE

Make and Model	Designed For[a]	No. of Cylinders	Displacement (L)	Bore and Stroke (mm)	Bare Engine Maximum kW @ rpm	With Standard Accessories Maximum Intermittent kW @ rpm	Continuous Sustained kW @ rpm	Compression Ratio (to 1)	Max. Torque (N · m) @ rpm	Overall Dimension mm Length	Width	Height	Engine Weight (kg)
Alco 251F	I, MA	12	131			2297@1200	2088@1200			4570	1680	3000	15,377
Al.-Chalmers	TR, I	6	7.0	108 × 127	130@2400	119@2400	103@2200	15.5	621@1500	1118	584	1041	649
Caterpillar 3406	T, B OF	6	14.6	137.2 × 165.1			298@2100	14.5	1715@1300	1524	902	1298	1288
Continental TMD 2.7	I	4	2.7	91 × 103.2	49@3000	47	42	20.5	164@1800	709.3	558.5	677.9	253
Cummins L10-270	T, B	6	10	125 × 136		201@2100		16	1163@1400	1308	805	1101	885
NTA-400	OF, I	6	14	140 × 152	298@2100	328@2100	247@1800	14	1627@1400	1541	779	1292	1301
Deere & Co. 6466A	I	6	7.6	116 × 121		182@2100	135@1800	14.9		1194	559	1245	816
Detroit D.A. 3V-92T	I	8	12.1	122.9 × 127	321@2100		224@1800	17.0	1658@1400	1219	991	1321	1086
EMD-G.M. 16-645E3C	L, I	16	169	230 × 254	2610			16		5375	1719	2213	16727
Ford BSD-666T	I	6	6.6	111.8 × 111.8		119@2500	99@2200	16.5	565@1600	1156	637	917	
Int. Harv. DT-466	I, TR, T	6	7.6	109.2 × 135.9		139@2500	127 × 2400	16.3	650@1600	1118	914	1168	559
Mack E(C)6-350R	O–OH	6	11.0	124 × 152	261@1800			15.0	1534@1400				982
Onan L423	I	4	2.3	88.9 × 91.9		43@3600	39@3600	21.5	199@2200	579	490	632	224
Perkins T6.3544	T, TR, I	6	5.8	98.5 × 127	116@2600	105@2600	89@2250	16	498@1700		743	790	447
Waukesha VRD-220S	TR, I	4	3.6	98.3 × 118.3	67@2600			16	268@1800	813	576	1016	397

Source. Truck & Off-Highway Industries, **3 / 4** (1983).

[a]TR, Tractor; I, Industrial; T, Truck; O–OH, On-off Highway; B, Bus; OF, Off Highway; L, Locomotive.

5.4-8 Regulation

In natural aspirated four-stroke gasoline engines speed and load regulation are usually obtained by throttling the charge or by quantitative fuel governing. The air–fuel ratio may vary some with load for different engine requirements. Intake air throttling reduces the cylinder pressure at the end of the intake stroke and reduces the mass of charge air trapped per cycle. It increases the pumping work and consequently decreases the efficiency of the engine. It is the simplest regulation to provide a stable idling condition.

Due to the cylinder scavenging, carburetted two-stroke gasoline engines have difficulty in using throttling for the power control. Under the lean limit of flammability, qualitative fuel governing is used in some cases with constant-speed gasoline engines. However, the engine thermal efficiency will be sacrificed by lean burn unless the ignition timing is advanced. For each change in speed and load, ignition timing has to be changed for engine optimization. The variation in ignition timing will change the thermal efficiency and may overheat the engine.

Power control of a natural aspirated diesel engine is obtained by governing the amount of fuel injected per cycle. When engine load is increased, the fuel–air mixture becomes richer. With more gasoline and diesel engines being turbocharged, engine regulation has been greatly improved. See the discussion on turbocharging in Section 5.4-6.

A diesel engine has a speed governor that regulates the fuel amount and speed at all operating conditions. The governor can be mechanical, hydraulic, electronic, or a combination. In locomotive engines overspeed automatic shutdown control is provided. In automobile and truck engines one-bank operation fuel shutdown to some cylinders and an exhaust brake have been used to regulate the engine for part-load or no-load operation.

5.4-9 Advanced Concepts

The important considerations of internal combustion engines are fuel economy, exhaust emission, horsepower, performance, size, and weight. Alternative fuel must be considered in the near future.

In light of these issues two concepts of internal combustion engines will be discussed next: the adiabatic diesel engine and the Wankel rotary engine.

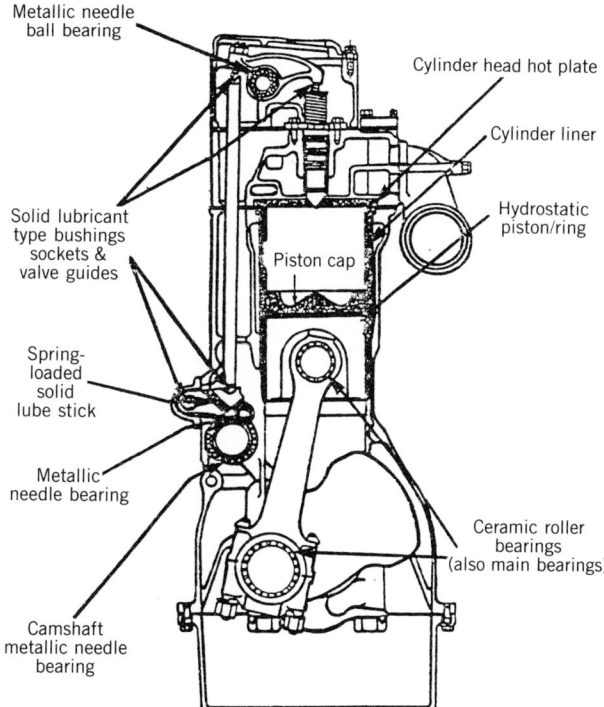

Fig. 5.4-11 Advanced adiabatic diesel engine. (Courtesy of Cummins Engine Co., Columbus, IN.)

Adiabatic Diesel Engine

An adiabatic diesel engine improves the performance of an internal combustion engine through insulation, which reduces the heat loss to coolant, and extracts the excess exhaust energy through secondary power plant such as turbocompounder or Rankine cycle. Figure 5.4-11 shows the schematic drawing of an adiabatic engine. Beginning in 1972 Cummins Engine Co. with funding from the U.S. Army Tank Research, has developed such an adiabatic diesel engine. The engine combustion chamber is insulated with ZrO_2 capped piston, ZrO_2 face plate cylinder head, and insulated cylinder liner. Also, the exhaust port is insulated, and water, water pump, and radiator are eliminated.

It has been demonstrated that the engine has better fuel economy with turbocompound machinery, can burn alternative fuel with relative ease due to the higher combustion temperature, and is less noisy due to secondary turbomachinery. It may improve the engine emission and run the engine much more efficiently if its gas-lubricated piston proves successful.

The difficulty of such an ambitious engine is the brittleness of the insulated ceramic material, its durability without water cooling, the high cost of lubricating synthetic oil, and the tight clearance requirement of a gas-lubricated piston. Further development is needed. Currently, Cummins, DDA, Caterpillar, and Kyoto-Ceramic are doing or watching the research.

Wankel Engine

Rotary internal combustion engines were developed in the twentieth century. Cooley designed a rotary piston steam engine (1901). Umpleby transformed it into an internal combustion engine (1908–1910). Wallinder and Skook (1923) and de Lavand tried to define a broad hypocycloid series of possible

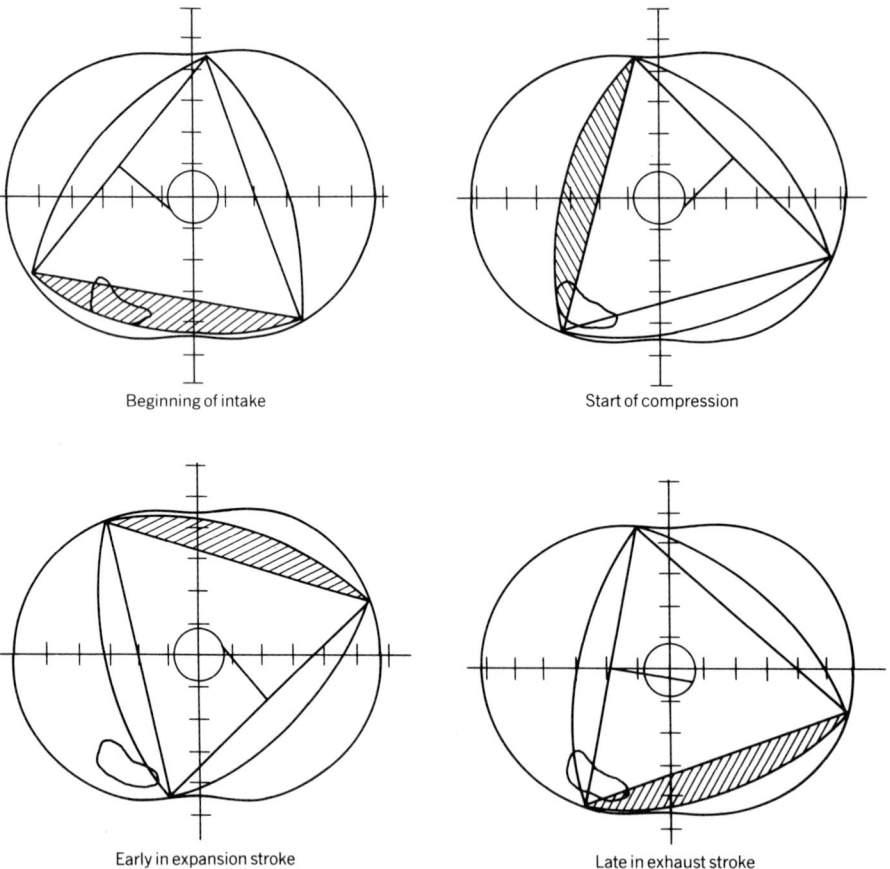

Beginning of intake

Start of compression

Early in expansion stroke

Late in exhaust stroke

Fig. 5.4-12 Four-cycle process of Wankel engine. (Adapted from SAE Paper 886D. Courtesy Curtiss-Wright Corp.)

engine configurations. In 1943 Maillard developed a three-sided rotor in an enveloping outer encounter. Felix Wankel modified the prior arrangement to eliminate radial motion of the rotor apex seal in their slots, essentially by geometric inversion, and devised a basic three-plane sealing configuration.

This rotary engine does not have intake and exhaust valves. Attractive features are 38% fewer parts, light specific weight, extreme high specific power output, natural complete inertia balance, freedom from vibration, and output smoothness. The operation of a Wankel engine and its four-cycle process is illustrated in Fig. 5.4-12. The housing is an epitrochoid plus an equidistance in construction. It contains an enveloped rotor.

The difficulty of the rotary combustion engine is in the development of an apex seal/trochoid sealing system for the high-output heavy-duty engine. Better and economical coating for the contact seal (which provides high compression pressure) has to be developed.

BIBLIOGRAPHY

Burman, P. G. and Deluca, F. *Fuel Injection and Control for Internal Combustion Engines.* American Bosch, Simmons-Boardman Publishing, New York, 1962.

Cook, R. D. *Concepts and Application of Finite Element Analysis.* Wiley, New York, 1974.

"Engineering Handbook of Piston Rings, Seal Rings, Mechanical Shaft Seals" Koppers Company, Inc., 1975.

Flynn, P. F. "Turbocharging Four-Cycle Diesel Engines," SAE 790314.

Forbes, W. G., Pope, C. L., Everitt, W. T. Lubrication of Industrial and Marine Machinery, Wiley, New York, 1954.

Fowle, T. I. "Lubricants for Fluid Film and Hertzian Contact Conditions," *Proc. Instr. Mech. Engrs.* 1967–1968.

Fuller, D. D. *Theory and Practice of Lubrication for Engineers*, Wiley, New York, 1956.

O'Connor, J. J. Boyd, J. and Avallone, E. A. *Standard Handbook of Lubrication Engineering.* McGraw-Hill, New York, 1968.

Sitkei, György. *Heat Transfer and Thermal Loading in Internal Combustion Engines.* Akadémiai Kiado, Budapest, 1974.

Taylor, C. F. and Taylor, E. S. *The Internal-Combustion Engine.* International Textbook, 1961.

Watson, N. and Janota, M. S. *Turbocharging the Internal Combustion Engine*, Wiley, New York, 1982.

5.5 JET AND ROCKET ENGINES

Gordon C. Oates

LIST OF SYMBOLS

A	Cross-sectional area
a	Speed of sound
C	Effective exhaust velocity
C^*	Reference velocity $= \left(\dfrac{\gamma + 1}{2}\right)^{(\gamma+1)/2(\gamma-1)} \sqrt{\dfrac{T_{tc}}{\gamma}\dfrac{R_u}{w}}$
C_F	Thrust coefficient
C_p	Specific heat at constant pressure
D_{add}	Additive drag
D_{ext}	External drag
e	Polytropic efficiency
F	Thrust
f	Fuel–air ratio
h	Heating value of fuel, altitude, enthalpy
H_{scl}	Scale height
I_{sp}	Specific impulse
M	Mach number
m	Mass
$\dot{m}$	Mass flow rate
P	Pressure

P_c	Chamber stagnation pressure
P_t	Stagnation pressure
R_u	Universal gas constant
S	Specific fuel consumption
T	Temperature
T_t	Stagnation temperature
U	Velocity
Δv	Velocity increment
w	Molecular weight
y	Defined in Eq. (5.5-5)
α	Bypass ratio, $= \dot{m}_F / \dot{m}_c$
γ	Ratio of specific heats
δ	Turning angle
η_p	Propulsive efficiency
η_{th}	Thermal efficiency
θ	Shock wave angle
π	Ratio of outgoing stagnation pressure to incoming stagnation pressure. Exception: $\pi_r = P_{t0}/P_0 = \tau_r^{\gamma/(\gamma-1)}$
Π	Product of pressure ratios $= \pi_r \pi_d \pi_c \pi_b \pi_n$
τ	Ratio of outgoing stagnation temperature to incoming stagnation temperature. Exceptions:

$$\tau_r = \frac{T_{t0}}{T_0} = 1 + \frac{\gamma - 1}{2} M_0^2$$

$$\tau_\lambda = \frac{C_{pt} T_{t4}}{C_{pc} T_0}$$

Subscripts

0–9	Station numbers
a	Ambient
b	Burner
c	Compressor (also core for $\dot{m}_c$), thermodynamic properties upstream of burner (γ_c, C_{pc})
d	Diffuser
e	Exiting
F	Fan stream ($\dot{m}_F$)
i	Installed
n	Nozzle
R	Reference conditions
s	Conditions preceding shock wave
SL	Sea level
t	Turbine, throat, thermodynamic properties downstream of the burner (γ_t, C_{pt})
u	Uninstalled

Superscripts

$'$	Refers to fan stream
$*$	Refers to optimal quantities

5.5-1 Principles of Jet Propulsion

The Thrust Equation

Airbreathing and rocket propulsion are basically similar in that thrust is obtained by generating rearward momentum in one or more streams of gas. In the case of a rocket the propulsive gas originates from on board the vehicle, whereas in the aircraft case (most of) the propellant gas originates from the free air surrounding the vehicle. The effect of the incoming momentum of air must

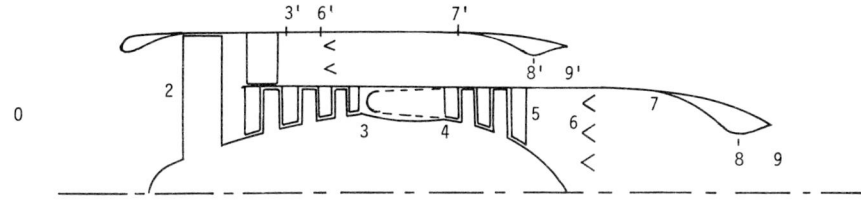

Turbofan Station Numbering

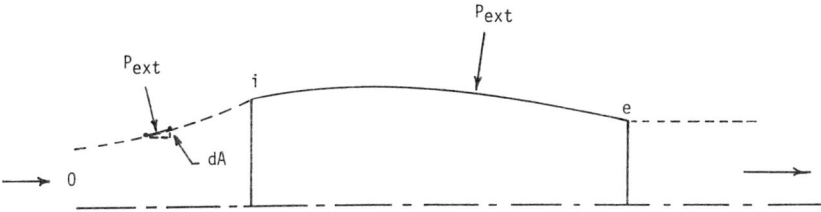

Fig. 5.5-1   Turbofan station numbering and control volume for thrust balance.

be accounted for in the airbreathing case, and it is to be noted that if, as in some concepts, air that has already been retarded by the aircraft surface is to be used for propulsion, then a significant effect upon the entering momentum contribution will accrue.

A careful application of Newton's laws for the case of thrust aligned in the flight direction, and for the case where the "free-stream" air is used for the propulsion, leads to the expression for the installed thrust of the engine, F_i:

$$F_i = F_u - D_{add} - D_{ext} \qquad (5.5-1)$$

where F_u = uninstalled thrust = $\dot{m}_9 U_9 - \dot{m}_0 U_0 + (P_9 - P_0) A_9$
$\quad D_{add}$ = additive drag = $\int_0^i (P_{ext} - P_0) \, dA$
$\quad D_{ext}$ = external drag = $\int_i^e (P_{ext} - P_0) \, dA$

See Fig. 5.5-1.

As written, the viscous contribution to the external drag has not been included. It is a matter of preference whether the viscous contribution is "book-kept" as airplane or engine drag, but it is slightly more usual to include it as an airplane drag. The presence of multiple streams can be easily included in the expression for F_u.

The terms D_{add} and D_{ext} arise because of the need to reference the incoming momentum to the far upstream state. It is to be noted that for a well-designed inlet the two terms very nearly cancel each other, the "external drag" becoming negative and appearing as a lip forward suction. Thus, the difference of the two terms is usually quite small, but the individual terms can be very large. Typical values for a 200,000-N thrust engine at start of roll are D_{add} = 50,000 N and $D_{add} + D_{ext}$ = 500 N. It is to be noted that the problem of recovering the additive drag through the lip suction can be very acute for an aircraft designed for supersonic flight and hence with a sharp-edged inlet. Thus, when such aircraft are flown subsonically, the lack of such recovery leads to large "spillage drag."

5.5-2   Performance Measures

The two most commonly used performance measures are specific thrust, $F/\dot{m}$, and specific fuel consumption, S. These performance measures are themselves related to the more fundamental efficiency measures, the thermal efficiency of the engine, η_{th}, and the propulsive efficiency, η_P. The thermal efficiency is the ratio of the rate of generation of kinetic energy and the rate of input of thermal (chemical) energy, whereas the propulsive efficiency is the ratio of power transmitted to the flight vehicle and rate of generation of kinetic energy. It follows simply that for an engine with single exhaust stream (ignoring the effect of fuel addition and assuming no exit pressure mismatch)

$$\frac{F}{\dot{m}} = U_9 - U_0 \qquad (5.5-2)$$

$$\eta_p = \frac{2}{1 + U_9/U_0} = \frac{1}{1 + F/(2U_0\dot{m})} \qquad (5.5-3)$$

Elementary manipulations show that the specific fuel consumption is proportional to the flight velocity and inversely proportional to the product $\eta_p \eta_{th}$. Thus, if follows that

$$S\alpha \frac{U_0}{\eta_p \eta_{th}} = \frac{1}{\eta_{th}}\left\{U_0 + \frac{1}{2}\frac{F}{\dot{m}}\right\} \tag{5.5-4}$$

The relationship between the specific fuel consumption and specific thrust shown in Eqs. (5.5-2) to (5.5-4) has important implications. Thus, it is of course desirable to have as low a specific fuel consumption as possible for a given flight speed, which itself requires high thermal and propulsive efficiencies. Obviously, however, a high propulsive efficiency can only be obtained when low specific thrusts are utilized, which presents the designer with a difficult optimization problem. When a low specific thrust is developed, it is necessary to utilize large mass flows, and as a consequence the engine installation problems increase. Use of a very large diameter fan with large bypass ratio, for example, might require the use of inordinately long landing gear as well as perhaps the use of a gear box to better match the fan tip speed with the tip speed of the turbine driving the fan.

Choice of the appropriate engine is also much affected by the possibly conflicting requirements of take-off thrust and cruise thrust. It is evident that for zero flight velocity the power required to supply a given thrust level is proportional to the exhaust velocity (inversely proportional to the mass flow). At high flight velocities large powers are required, and as a result engines with low specific thrusts sized for take-off thrust requirements are found to be underpowered at high forward speeds (as is the case for conventional turboprops). Conversely, very high specific thrust engines (turbojets) must be made oversize to satisfy the take-off requirement and hence tend to be too powerful for cruise flight at subsonic speeds. This latter condition results in throttled back operation at cruise with consequent loss in thermal efficiency because of the related reduction in compressor pressure ratio.

It is of interest to note the recent interest in very high powered turboprops, designed to fly at Mach numbers of up to 0.8. Such engines are so powerful that in the take-off condition they are operated at part throttle so as to allow the use of lighter gearboxes and to prevent propeller stalling.

It is worthy of note, also, that the turbofan, so popular in commercial airline use, provides an excellent balance between the take-off thrust and cruise thrust requirements.

5.5-3   Engine Types and Limitations

In Section 5.5-5 simple equations for estimating the design performance of jet engines will be summarized, but we first list example types of engines and their present approximate ranges of utility.

Turboprop

The turboprop engine enjoys popularity for use in a wide variety of aircraft with flight speeds up to a flight mach number (M_0) of approximately 0.6. Example aircraft types include personal and business passenger aircraft, short take-off and landing (STOL) aircraft, crop dusters, heavy lifter aircraft, short-range commuter aircraft, and so on. Some of the virtues that lead to the popularity of the turboprop are high take-off thrust, high reliability compared to piston aircraft, and low specific fuel consumption compared to turbofans or turbojets.

The flight speeds are limited to $M_0 \approx 0.6$ because of the onset of supersonic flow relative to the propeller blade tips, which in turn leads to the onset of shock losses and consequent drop-off in performance. The relatively heavy gearbox represents a disadvantage compared to a turbofan, and in addition, some reliability problems have been encountered.

High Disc Loading Turboprop

A new engine type, presently in its early stages of development, the high disc loading turboprop represents an attempt to circumvent the flight speed limitations of the turboprops while retaining some of the advantages of high take-off thrust and low specific fuel consumption (arising because of the high propulsive efficiency).

The concept involves the use of a relatively small diameter propeller with many (8 or 10) blades. It is hoped that the use of the smaller-diameter blades, with lower tip speed, will forestall the onset of serious shock losses for flight Mach numbers in excess of 0.8. The blades of the propeller will be severely swept to aid in forestalling the shock losses as well as to reduce the noiselevel (Fig. 5.5-2). A further advantage of the configuration results because the smaller blade diameter will allow a somewhat higher rotational speed, and as a result a lighter gearbox may be employed.

Turbofan

The turbofan engine has proven to be the most suitable choice of power plant for most of the world's military and commercial aircraft in the flight range $0.7 \leq M_0 \leq 2.0$. Appropriate choice of such design

Fig. 5.5-2 High disc loading turboprop. (Courtesy United Technologies/Hamilton Standard.)

parameters as bypass ratio (α), fan pressure ratio (π_c'), and compressor pressure ratio (π_c) is very much dependent on the aircraft mission requirements. A simple technique for estimating the (uninstalled) performance of a turbofan as a function of flight requirements, design limitations, and design choices is given in Section 5.5-5, but certain trends can be identified without calculation.

Low fuel consumption is a dominant concern when an engine is to be selected for use with a subsonic commercial transport aircraft. As could be inferred from the results of Section 5.5-2, we should then expect the optimal engine to have a low specific thrust and hence a high bypass ratio with a low bypass pressure ratio. In addition, in order to increase the thermal efficiency, such engines would be equipped with very high pressure ratio compressors. Figure 5.5-3 shows the Pratt & Whitney PW2037, a very recently introduced engine of 5.8 bypass ratio, 1.4 fan pressure ratio, and 32 compressor pressure ratio.

The extreme performance demands of the military environment lead to the selection of quite different design choices. Thus, high specific thrust is required for flight at high Mach numbers or for maneuvering flight at transonic Mach numbers. As a result lower fan bypass ratios and higher fan

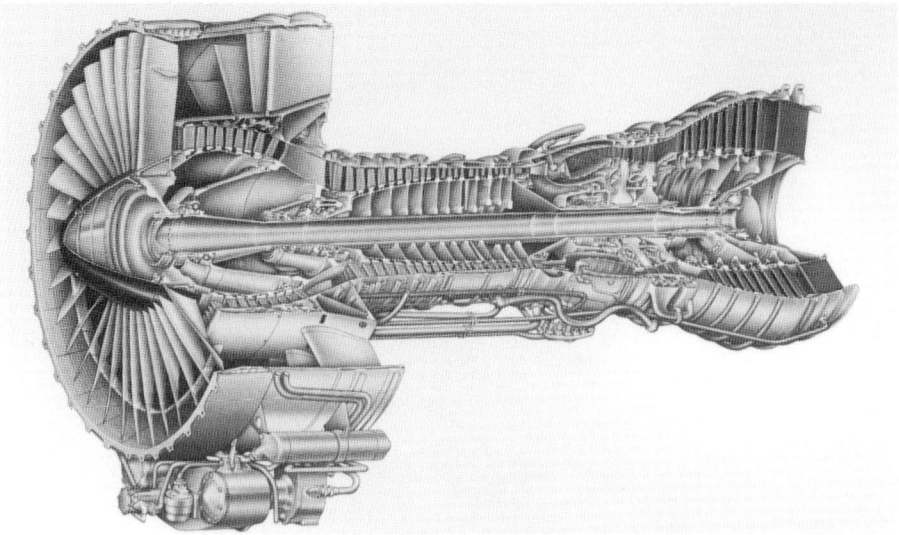

Fig. 5.5-3   Commercial turbofan engine. (Courtesy United Technologies/Hamilton Standard.)

Fig. 5.5-4   Afterburning turbofan engine F100-PW-100. (Courtesy United Technologies/Hamilton Standard.)

pressure ratios are found to be suitable. Even then, such aircraft must have acceptable subsonic cruise capability so that a compromise between high specific thrust and low specific fuel consumption is inevitable. The demand for compromise of such multimission aircraft is somewhat eased through incorporation of afterburning, which greatly increases the specific thrust at the high performance condition and hence allows use of a more fuel efficient system for subsonic cruise. Figure 5.5-4 shows the Pratt & Whitney F100 afterburning turbofan engine. It is to be noted that this engine has a three-stage fan with a pressure ratio of ~ 3 and a by-pass ratio of 0.78. The compressor pressure ratio is 25, which is relatively high for this class of engine, and is clearly incorporated to aid the subsonic fuel efficiency.

Turbojet

At very high Mach numbers the increased enthalpy of the incoming air leads to extreme power demands from the turbine to drive both the fan and compressor; hence, in the Mach number range of 1.5–3.0 the cruise condition requirements favor use of a turbojet, or "pure jet." (It is to be noted that in some cases for aircraft with such high speed capability, the optimum engine, selected on an overall mission requirement basis, will still utilize a fan stream.)

In the higher Mach number ranges the "ram pressure" becomes very large indeed (~ 36 for $M_0 = 3$), so that most of the air compression occurs in the inlet. Because of the high power demand for compression at such elevated enthalpies, and because such a high compression has already been achieved within the inlet, compressor pressure ratios for use in high Mach number engines are usually much lower than their subsonic counterparts (in the neighborhood of 3 for a flight-Mach-number-of-3 machine).

At even more elevated flight Mach numbers the compressed air departing even a low pressure ratio compressor reaches temperatures approaching the limiting temperature of the compressor blade materials. Further, the presence of the turbine in the gas stream limits the allowable fuel addition and hence the specific thrust level attainable by the engine. It is in this high flight Mach number regime that the ramjet becomes a viable concept.

Ramjets and SCRAMJETS

The ramjet engine accomplishes the required compression entirely within the inlet and expansion entirely within the nozzle. The top operating temperature can be substantially higher than that in the main burner of a turbojet because of the absence of a turbine within the flow field. At high Mach numbers the related high ram pressure ratios lead to surprisingly efficient operation for ramjets, but there are several severe design problems that limit ramjet performance.

It is apparent that at very low flight Mach numbers no useful cycle pressure ratio exists, so that no thrust can be produced. It is hence necessary that ramjets be boosted to usable (high) flight Mach numbers prior to engine light-up. A further problem arises in the difficulty of designing an inlet that will give efficient compression over a wide M_0 range; the inlet must be capable of substantial, and precise, geometrical variations.

The upper limit of the flight Mach number for conventional ramjet operation occurs when the stagnation enthalpy of the air entering the combustor is so extreme that when fuel addition occurs substantial dissociation of the products of combustion results. In this circumstance the chemical energy remains in the energy of dissociation, rather than appearing as translational energy of the combustion gases, and hence does not contribute to the momentum of the departing stream.

The supersonic combustion ramjet (SCRAMJET) is a device envisioned for use in the hypersonic regime of winged flight ($M_0 > 4$). In a SCRAMJET the incoming air would be only partially diffused —to supersonic Mach numbers—prior to the introduction of fuel (probably molecular hydrogen) into the air stream. As a result, the temperature of the air stream will remain lower than would be the case if full diffusion had been employed, and hence a much reduced level of dissociation should occur. A problem, however, is that burning at high Mach numbers carries with it unavoidable pressure losses. SCRAMJET engines are, at this time, only in the research phases, but some promising results have been attained.

Combined Cycles

The requirement for operation over a wide flight Mach number range often calls for extreme compromises when a single engine type is employed. It is obvious that a given vehicle could carry more than one engine type, but in most cases this requirement itself forces unacceptable compromises in vehicle design.

Engines with more than one mode of operation are often considered for aircraft with multiple-mission requirements. In a sense, engines employing afterburners (Fig. 5.5-4) can be considered "two-mode" engines. Several ingenious schemes[1] for varying the bypass ratio of an engine in flight have been suggested for use in advanced supersonic transport aircraft.

More extreme concepts are typified by the "turboramjet," which consists of a central turbojet surrounded by an annulus fitted with burners. The turbojet is used to raise the Mach number to where the surrounding ramjet can contribute useful thrust, at which time the ramjet portion is ignited. At higher Mach numbers either or both of the systems can be used. A very similar concept, the "ramrocket," utilizes a central rocket with surrounding ramjet.

5.5-4   Engine Components and Measures of Performance

Inlets

The flow within all inlets is virtually exactly adiabatic, so the stagnation enthalpy of the air does not change. Inlet performance is, hence, usually measured as the ratio of leaving to entering stagnation pressure, π_d. In an inlet with purely subsonic flow the stagnation pressure losses occur within the boundary layers unless, as can happen with the inlet at extreme angle of attack, separation occurs.

The design of subsonic inlets is dominated by the requirements to retard separation at extreme angle of attack and high air demand (as would occur in a two-engine aircraft with engine failure at take-off) and to retard the onset of both internal and external shock waves in transonic flight. These two requirements tend to be in conflict, because a somewhat "fat" lower inlet lip best suits the high angle of attack requirement, whereas a thin inlet lip best suits the high Mach number requirement. Modern development for the best compromise design is greatly aided by the advent of high-speed electronic computation, which allows analytical estimation of the complex flow fields and related losses.

Estimation of the losses within supersonic inlets is an easier task than for subsonic inlets for the simple reason that the major losses occur across shock waves and hence may be estimated using the relatively simple shock wave formulas. More exacting estimates require estimation of the boundary layer and separation losses.

Figure 5.5-5 illustrates the three principal classes of supersonic inlets: internal compression, external compression and mixed compression. External compression offers the advantages of relatively simple construction, short axial length, and good off-design performance. A disadvantage arises for use at high flight Mach numbers because the required flow turning is so great that the external cowl angle becomes excessive and strong external shocks are formed.

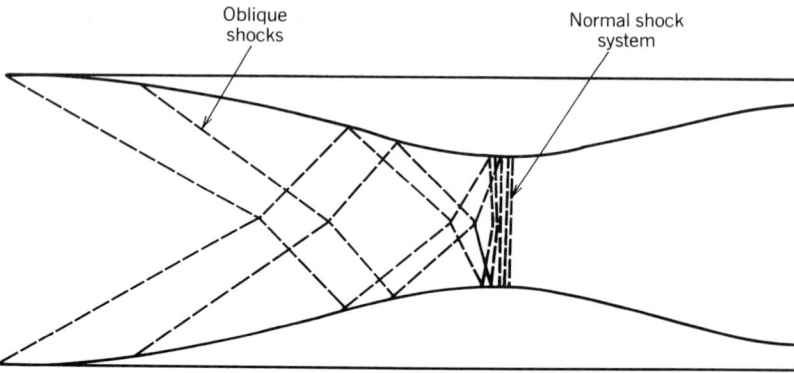

Internal compression inlet

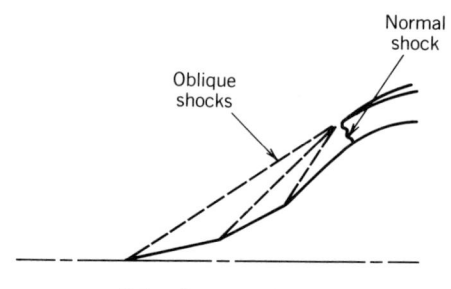

External compression inlet

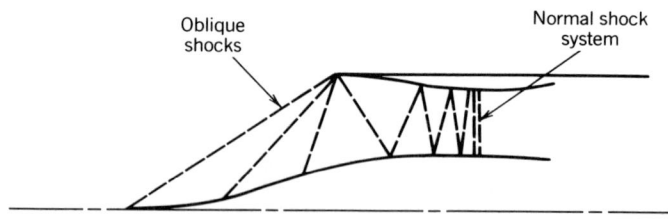

Mixed compression inlet

Fig. 5.5-5 Supersonic inlets.

The internal compression inlet does not suffer from the onset of excessive external drags at high design flight Mach numbers, but does have its own disadvantages. Thus, the geometry is such that excessive inlet lengths must be utilized, and the off-design characteristics can be unacceptable if sophisticated variable geometry is not employed. In order to "start" the inlet, the shock system must first be swallowed and then the geometry varied to locate the normal shock near the throat. Quick-acting throat and subsonic diffuser bleed systems must also be provided to prevent the sudden disgorging of the shock system (inlet "unstart") with change in engine air demand.

The mixed-flow inlet design[2] offers a useful compromise to these two designs for use at high flight Mach numbers. Though it suffers from some of the disadvantages of both, it offers the possibility of decreased inlet length and geometrical complication compared to the internal compression inlet.

Estimation of Losses in Supersonic Inlets

An upper limit to supersonic inlet performance can be estimated by assuming all the losses occur across the shock waves.

Figure 5.5-6 indicates the geometry and nomenclature for shock wave interaction and includes an example external compression inlet with three oblique shock waves followed by a normal shock wave. With the availability of desk top computers or calculators with branching and looping capabilities it is now more convenient to calculate desired quantities directly rather than to refer to tables. To this end, the following summary of equations is suggested for estimation of inlet shock losses.

Shock Losses

Oblique shocks

Input: M_i, δ, γ
Equations:

$$y = \frac{1}{M_i^2} + \left\{ \frac{1}{M_i^2} + \frac{\gamma + 1}{2} - y \right\} \left\{ \frac{y}{1 - y} \right\}^{1/2} \tan \delta \qquad (5.5\text{-}5)$$

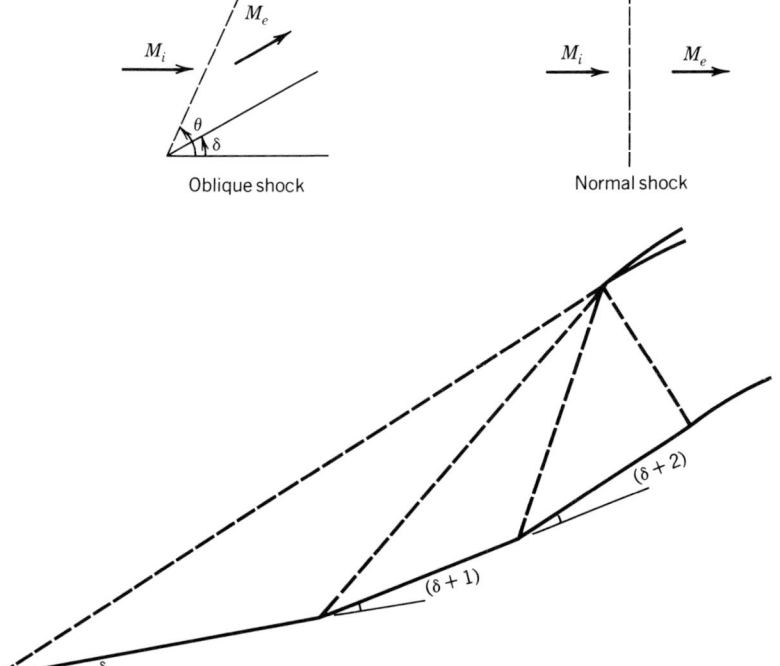

Fig. 5.5-6 Shock wave nomenclature and example external compression inlet.

To solve this equation, assume $y = 1/M_i^2$ in the right side and calculate a new value of y. Use this new value of y as an updated value on the right side. Continue until appropriate convergence. (*Note:* y is actually $\sin^2\theta$.)

$$\frac{P_{te}}{P_{ti}} = \left\{ \frac{\gamma + 1}{2\gamma M_i^2 - (\gamma - 1)} \right\}^{1/(\gamma - 1)} \left\{ \frac{(\gamma + 1) M_i^2 y}{2 + (\gamma - 1) M_i^2 y} \right\}^{\gamma/(\gamma - 1)} \tag{5.5-6}$$

$$M_e = \left[\frac{4 + 4(\gamma - 1) M_i^2 y + (\gamma + 1)^2 M_i^4 y - 4\gamma M_i^4 y^2}{\{2\gamma M_i^2 y - (\gamma - 1)\}\{2 + (\gamma - 1) M_i^2 y\}} \right]^{1/2} \tag{5.5-7}$$

Normal Shocks.   For a normal shock wave the above equations may be used with $y = 1$; hence

$$\frac{P_{te}}{P_{ti}} = \left\{ \frac{\gamma + 1}{2\gamma M_i^2 - (\gamma - 1)} \right\}^{1/(\gamma - 1)} \left\{ \frac{(\gamma + 1) M_i^2}{2 + (\gamma - 1) M_i^2} \right\}^{\gamma/(\gamma - 1)} \tag{5.5-8}$$

$$M_e = \left[\frac{2 + (\gamma - 1) M_i^2}{2\gamma M_i^2 - (\gamma - 1)} \right]^{1/2} \tag{5.5-9}$$

As an example use of these equations consider the two-dimensional external compression inlet of Fig. 5.5-6.

The inlet ramp consists of three ramps with angle of intersection increased by one degree at each succeeding ramp. The three oblique shock waves are followed by a normal shock wave. Successive application of Eqs. (5.5-5) to (5.5-9) for the case $\gamma = 1.4$, $M_0 = 2.5$, leads to the result

δ	8	9	10	11
π_d	0.8999	0.916	0.923	0.914

Compressor and Fans

There are two major classes of compressors used in aircraft gas turbines: the centrifugal compressor and the axial compressor. In the centrifugal compressor air is taken into the compressor near the axis and "centrifuged" to the outer radius. Subsequently, the swirl of the outlet air is removed and the air diffused prior to entry into another compressor stage or into the combustor. Centrifugal compressors have the advantage that they are rugged and deliver a high pressure ratio per stage. In addition, they are easily made in relatively small sizes. The disadvantages of the centrifugal compressor are that it is generally less efficient than an axial compressor and it has a large cross section compared to the cross section of the inlet flow. In modern usage centrifugal compressors are used with relatively small engines or as a final stage (following an axial compressor) in larger engines.

Axial compressors are used in the majority of the world's larger gas turbine engines. In such compressors enthalpy addition occurs in the rotating rows (rotors) in which, usually, both the kinetic energy and static pressure are increased. The stator rows remove some of the swirl velocity, thereby decreasing the kinetic energy and consequently increasing the static pressure. The limiting pressure rise through an axial compressor row occurs when the adverse pressure gradient on the blade suction surface becomes so severe that flow separation occurs. When substantial separation occurs, the entire compressor may surge (that is massive flow reversal will occur) or rotating stall may occur. Rotating stall is the condition where the flow in several blades stalls (becomes almost stagnant), and then the "package" of stalled fluid rotates around the blade row. The rotating stall condition is particularly dangerous because very large vibratory stresses can occur as the blades enter and depart the stall.

In order to achieve a high limiting pressure rise per stage, it is beneficial to design the stage so that the static pressure rise in each row is almost the same (so that one row will not stall prematurely). The degree of reaction is defined as the ratio of the static pressure rise in the rotor divided by the static pressure rise across the stage, and provides a measure of how well balanced the blade row loadings are. When detailed designs are investigated, however, it is found that, inevitably, the degree of reaction increases with increase of radius from hub to tip. A related result is that stator blades are limited in their performance at the hub, whereas rotor blades are usually limited at the tip. Further, the effect of variation of the degree of reaction with radius results in rows with large tip-to-hub ratios being more limited in attainable pressure rise than rows with small tip-to-hub ratios.

Compressors operate with very nearly adiabatic flow, and their "adiabatic efficiency," η_c, is defined as the ratio of the ideal work interaction required to achieve the desired pressure ratio.

Provided the compressed gas can be considered calorically perfect, this leads to the relationship

$$\eta_c = \frac{\pi_c^{(\gamma_c-1)/\gamma_c} - 1}{\tau_c - 1} \tag{5.5-10}$$

A closely related efficiency is the polytropic efficiency e_c, which is defined analogously to η_c except that the pressure rise is considered to be infinitesimal. The utility of the polytropic efficiency concept arises because e_c may be considered a constant over a substantial range of *design point* pressure ratios. Thus, in preliminary design estimates various trial values of compressor pressure ratio may be considered, without the requirement of adjusting the overall compressor efficiency, η_c. The relationships are true when e_c is constant:

$$\tau_c = \pi_c^{(\gamma_c-1)/\gamma_c e_c} \tag{5.5-11}$$

$$\eta_c = \frac{\pi_c^{(\gamma_c-1)/\gamma_c} - 1}{\pi_c^{(\gamma_c-1)/\gamma_c e_c} - 1} \tag{5.5-12}$$

Equation (5.5-12) gives an estimate of the design point compressor efficiency as a function of pressure ratio that could be expected to be achieved for a given level of "technology," e_c. Thus, for example, if $e_c = 0.92$, $\gamma_c = 1.4$, and $\pi_c = 32$, we find $\eta_c = 0.875$.

Combustors

Combustors are described by two performance measures: the combustion efficiency η_b, which is defined as the ratio of the actual stagnation enthalpy rise to the ideal stagnation enthalpy rise, and the ratio of outgoing stagnation pressure to incoming stagnation pressure, π_b.

Modern burners have very high combustion efficiencies ($\eta_b \geq 0.99$), so most present development is directed toward the removal of "hot spots" at exit from the burner and to the reduction of pollutant formation. Substantial efforts are also directed toward achieving stable operation in off-design operation and to efficient operation with alternative fuels.

Turbines

Virtually all turbines used in aircraft gas turbine engines are of the axial flow type and hence are superficially similar to an axial compressor operating in reverse. The engineering limitations on the performance of a turbine stage are, however, very different than the engineering limitations for a compressor stage. The large decrease in pressure found in turbines much decreases the tendency for suction surface separation, so turbine stages can be designed with very large pressure ratios. The gas enters the turbine at very high temperatures, however, and hence the initial turbine stages must be cooled by passing air from the outlet of the compressor through the blades.

In spite of the presence of substantial cooling in modern turbines, the concept of "adiabatic efficiency" is still employed. The efficiency definition is extended to include the effects of cooling by introducing a multiple-stream analysis.[3] For preliminary design purposes the effects of cooling may be ignored provided it is recognized that the predicted results will be slightly optimistic. It should be noted, also, that the predicted performance will continue to increase with increase in turbine inlet temperature, whereas if the required increase in cooling penalty is included, an optimal turbine inlet temperature would be found.[4]

The adiabatic turbine efficiency, η_t, and polytropic efficiency, e_t, are defined analogously to the compressor efficiencies to give

$$\eta_t = \frac{1 - \tau_t}{1 - \pi_t^{(\gamma_t-1)/\gamma_t}} \tag{5.5-13}$$

$$\pi_t = \tau_t^{\gamma_t/(\gamma_t-1)e_t}$$

$$\eta_t = \frac{1 - \pi_t^{(\gamma_t-1)e_t/\gamma_t}}{1 - \pi_t^{(\gamma_t-1)/\gamma_t}} \tag{5.5-14}$$

Nozzles

The final component of the aircraft gas turbine engine, the nozzle, accelerates the high-pressure exhaust gas to close to the ambient pressure. The primary design difficulties for nozzle design arise for nozzles intended for use in aircraft with wide Mach number capability. Flight over a wide Mach number range introduces a wide range of ram pressure ratios, with consequent wide range of nozzle pressure ratios. Optimum nozzle performance occurs when the nozzle exit pressure is not far from ambient, so for nozzles with a large operating range in pressure ratio, substantial geometrical variation must be possible.

As a result of the geometrical restraints required for good matching with the external flow field, the major effects of nozzle performance tend to be identified with the effect of exit pressure mismatch and the installation effects on installed thrust through the effect of "boat-tail" drag or exhaust plume back-pressuring.

The nozzle performance measure for use in estimation of the uninstalled performance is the stagnation pressure ratio, π_n. If cooling air is present (as can be the case if afterburning is employed), the effects of such cooling air are included through a multiple-stream analysis.

5.5-5   Estimation of Engine Performance—On-Design

In order to predict engine performance, three classes of "input" information are required:

1. The designer must know the intended design flight condition.
2. The designer must be aware of design limitations such as allowable turbine inlet enthalpy and possibly the allowable compressor outlet temperature and pressure; in addition, he must have estimates of the component performance measures and gas thermodynamic properties.
3. The designer has the flexibility to choose design variables to lead to the optimal performance.

As a relatively simple example, we consider estimation of the uninstalled performance of a turbofan engine. We consider the case with exit pressures matched ($P_9 = P_9' = P_0$) and for which no afterburning occurs. The gas is approximated as calorically perfect both before the burner (γ_c, C_{pc}) and after the burner (γ_t, C_{pt}), and the effects of turbine cooling and shaft auxiliary power take-off and inefficiency are not included. The analysis leads to estimation of the specific power and specific fuel consumption, and it remains to size the engine for the given aircraft requirement.

In terms of symbols, the three classes of required input listed above require specification of the following quantities (see List of Symbols):

1. Flight condition: T_0 (K), $\gamma_c = 1.4$, $C_{pc} = 1005$ (J/kg-K).
2. Limitations: τ_γ, γ_t, h (J/kg), $\pi_d, \pi_n, \pi_n', \pi_b, \eta_b, e_c, e_c', e_t$.
3. Design choices: π_c, π_c', α.

Prescription of these listed variables is sufficient to allow estimation of the uninstalled performance through the systematic process of cycle analysis.[3,5] Thus, the performance can be calculated by systematically relating conditions at exit (where $P_9 = P_9' = P_0$) to the ratios of stagnation properties across the internal components (thereby introducing the component performance measures). The system of equation is closed by invoking the power balance between fan, compressor, and turbine and by invoking the energy balance across the combustor.

The resulting equation set can be summarized as follows. Note that all calculations are sequential and do not require iteration. The equations can be entered into a computer or advanced calculator and allow calculation of a very wide variety of examples.

Estimation of Uninstalled Performance of a Turbofan Engine

Inputs: T_0 (K), $\tau_\lambda, \gamma_t, h$ (J/kg), $\pi_d, \pi_n, \pi_n', \eta_b, e_c, e_c', et, \pi_c, \pi_c', \alpha$.
Equations:

$$a_0 = 20\sqrt{T_0} \qquad \text{m/s} \tag{5.5-15}$$

$$\tau_r = 1 + \frac{M_0^2}{5} \tag{5.5-16}$$

$$\pi_r = \tau_r^{3.5} \tag{5.5-17}$$

$$\tau_c = \pi_c^{1/3.5e_c} \tag{5.5-18}$$

$$\tau_{c'} = \pi_{c'}^{1/3.5e_{c'}} \tag{5.5-19}$$

$$f = \frac{\tau_\lambda - \tau_r \tau_c}{\dfrac{h\eta_0}{1005 T_0} - \tau_\lambda} \tag{5.5-20}$$

$$\frac{P'_{t9}}{P'_9} = \pi_r \pi_d \pi'_c \pi'_n \tag{5.5-21}$$

$$M_0 \frac{U'_9}{U_0} = \left[5\tau_r \tau'_c \left\{ 1 - \left(\frac{P'_{t9}}{P'_9} \right)^{(-1/3.5)} \right\} \right]^{1/2} \tag{5.5-22}$$

$$\tau_t = 1 - \frac{1}{1 + f} \frac{\tau_r}{\tau_\lambda} \left\{ \tau_c - 1 + \alpha \left(\tau'_c - 1 \right) \right\} \tag{5.5-23}$$

$$\pi_t = \tau_t^{(\gamma_t/(\gamma_t - 1)e_t)} \tag{5.5-24}$$

$$\frac{P_{t9}}{P_9} = \pi_r \pi_d \pi_c \pi_b \pi_t \pi_n \tag{5.5-25}$$

$$M_0 \frac{U_9}{U_0} = \left[5\tau_\lambda \tau_t \left\{ 1 - \left(\frac{P_{t9}}{P_9} \right)^{-(\gamma_t - 1/\gamma_t)} \right\} \right]^{1/2} \tag{5.5-26}$$

$$\frac{F_u}{\dot{m}_F + \dot{m}_c} = \frac{a_0}{1 + \alpha} \left[(1 + f) M_0 \frac{U_9}{U_0} - M_0 + \alpha \left\{ M_0 \frac{U'_9}{U_0} - M_0 \right\} \right] \quad \frac{\text{N}}{\text{kg/s}} \tag{5.5-27}$$

$$S = \frac{f(10^6)}{(1 + \alpha) \frac{F_u}{\dot{m}_F + \dot{m}_c}} \quad \frac{\text{mg/s}}{\text{N}} \tag{5.5-28}$$

The set of equations given in the summary may be thought of as a functional relationship of the form $S = S(\pi_c, \pi'_c, \alpha)$. It is in fact possible to locate the minimum of S with both π'_c and α quite simply. The location of the minimum of S with α for prescribed π'_c and π_c is a particularly useful point, because it effectively locates the point of best core and bypass match. It should be noted that prescription of π'_c effectively determines the specific thrust level, and though the value of π'_c leading to minimum S can also be determined easily analytically, the related specific thrust levels are usually too low to be of interest.

When the optimum case (denoted by an asterisk) is desired, the related turbine temperature ratio is given by[5]

$$\tau_t^* = \left(\frac{1}{\Pi} \tau_t^* \right)^{-(1 - e_t)/e_t} + \frac{1}{0.8\tau_\lambda} \left[\frac{\tau_r \left(\tau'_c - 1 \right)}{M_0 \frac{U'_9}{U_0} - M_0} \left\{ 1 + \frac{1 - e_t}{e_t} \left(\frac{1}{\Pi} \tau_t^* \right)^{-1/e_t} \right\} \right]^2 \tag{5.5-29}$$

This equation may be solved by assuming an appropriate value for τ_t^*, substituting into the right side to obtain an updated value for τ_t^* and then continuing till convergence. An appropriate "start value," τ_{t0}^*, is

$$\tau_{t0}^* = \frac{1}{\Pi} + \frac{1}{0.8\tau_\lambda} \left[\frac{\tau_r \left(\tau'_c - 1 \right)}{\left(M_0 \frac{U'_9}{U_0} - M_0 \right)} \right]^2 \tag{5.5-30}$$

The related bypass ratio follows from

$$\alpha^* = \frac{1}{\tau'_c - 1} \left[(1 + f) \frac{\tau_\lambda}{\tau_r} \left(1 - \tau_t^* \right) - \tau_c + 1 \right] \tag{5.5-31}$$

The performance variables for the optimal case can hence be calculated with the preceding summary except that Eq. (5.5-30) must replace Eq. (5.5-23) and α^* must be calculated (not input) from Eq. (5.5-31).

Typical ranges of the limitation parameters given above that are found in modern subsonic civilian aircraft engines follow:

T_{t4}	γ_t	C_{pt}	h	π_d
1400	1.30	1245	$4.6(10^6)$	0.995
1200	1.35	1100	$4.4(10^6)$	0.980

π_n, π_n'	π_b	η_b	e_c	e_c'	e_t
0.995	0.98	0.995	0.93	0.92	0.91
0.99	0.94	0.980	0.90	0.90	0.88

As an example calculation, consider flight at $M_0 = 0.85$ with $T_{t4} = 1250$ K, $T_0 = 220$ K, $\gamma_t = 1.33$, $C_{pt} = 1157$, $h = 4.5(10^6)$, and all other values equal to the highest values listed above. We note immediately

$$\tau_\lambda = \frac{1157(1250)}{1005(220)} = 6.54$$

Two fan pressure ratios are considered, $\pi_c' = 1.149$ (which leads to the minimum possible specific fuel consumption) and $\pi_c' = 1.5$. The predicted performance variables are shown plotted in Fig. 5.5-7. The optimal case is considered, so the related bypass ratio is also displayed.

It is of interest to note that the "best" fan pressure ratio, in fact, leads to an unacceptably low specific thrust and an unacceptably high bypass ratio. Installation requirements would greatly favor the use of the higher fan pressure ratio ($\pi_c' = 1.5$). Note that even in the high fan pressure ratio case little penalty is incurred by reducing the compressor pressure ratio to, say, $\pi_c = 35$, from that value giving minimum specific fuel consumption. Further, an optimum design would probably favor use of a bypass ratio less than the "optimal" value of $\alpha^* = 9.2$ (corresponding to $\pi_c = 35$). Thus, the circled points indicate the performance to be obtained for the design choices $\pi_c = 35$, $\pi_c' = 1.5$, and $\alpha = 7$. Note that the specific thrust (as compared to the "optimal case") increases about 20%, whereas the

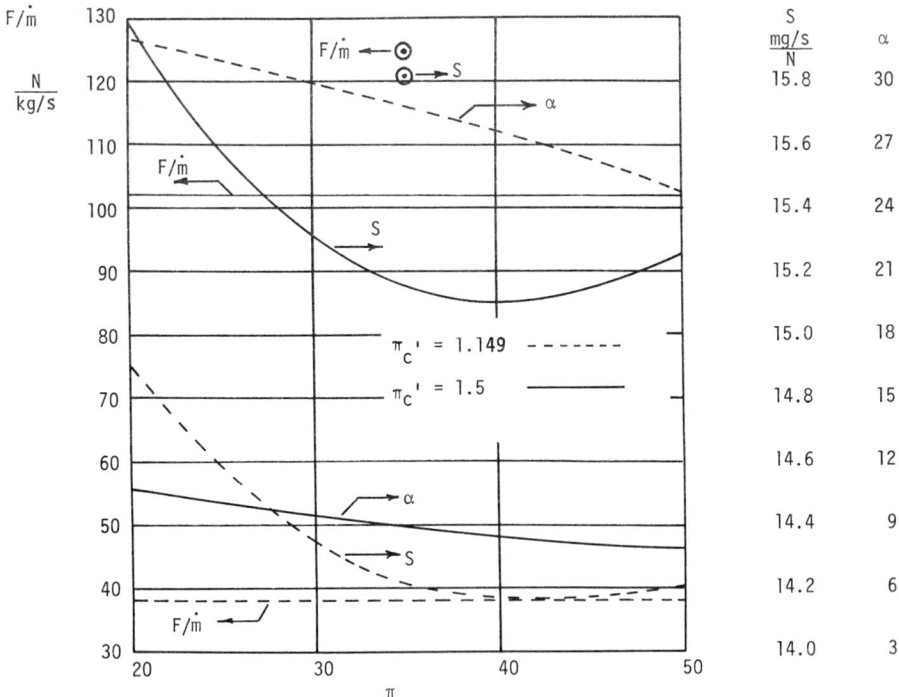

Fig. 5.5-7 Turbofan performance.

penalty in specific fuel consumption is approximately 4%. an additional advantage could accrue from the reduced turbine expansion required to drive the smaller fan in that fewer turbine stages might be required.

The very large number of input parameters to the equations listed in the summary allow investigation of many of the complex trends existing in engine design. Not only the effect of design choices but also the effect of technological advances (as found in increased turbine inlet temperatures and increased component efficiencies) may be easily investigated by use of such preliminary design analysis.

5.5-6 Estimation of Engine Performance—Off-Design

In Section 5.5-5 the prediction of engine performance at the design condition was presented. In the design case the designer is free to assign the variables τ_c, τ_c', and α. Once the design configuration has been determined, however, it remains to determine how the variables τ_c, τ_c', and α change with change in flight condition (M_0, T_0, etc.) or throttle setting (τ_λ). As a simple example we consider the turbojet (see refs. 3 and 5 to extend the analysis to more complicated configurations).

Modern turbojets have large pressure ratios across the engines, and as a result the flow remains choked at both the turbine inlet area and primary nozzle area over much of the operating regime of the engines. As a direct consequence, provided the flow areas at turbine inlet and primary nozzle remain fixed, the turbine expansion ratio remains very nearly constant. A further consequence, which follows from the power balance between turbine and compressor, is that the compressor stagnation temperature ratio τ_c is given in terms of τ_λ/τ_r and reference conditions by

$$\tau_c = 1 + (\tau_{CR} - 1)\frac{\tau_\lambda/\tau_r}{(\tau_\lambda/\tau_r)_R} \tag{5.5-32}$$

The compressor pressure ratio then follows from

$$\pi_c = \left[1 + \eta_c(\tau_c - 1)\right]^{\gamma_c/(\gamma_c - 1)} \tag{5.5-33}$$

The performance of the engine can then be estimated from the performance equations of the preceding section with the change, however, that the appropriate relationships utilizing the component efficiencies (η_c, η_t) rather than the polytropic efficiencies (e_c, e_t) be employed.

As an example, consider an engine with $\pi_{CR} = 9$, $\eta_{CR} = 0.90$ at $M_{OR} = 0$. The engine is flown at $M_0 = 3$ at an altitude where $T_0/T_{OR} = 0.759$ and $\eta_c = 0.88$. The turbine inlet temperature is held constant throughout. Straightforward calculation yields $\pi_c = 3.26$.

This very wide variation in operating compressor pressure ratio would indicate that the compressor would probably have to operate with substantial compressor bleed at the high Mach number condition. Note, however, that the overall cycle pressure ratio (including ram pressure ratio) has increased enormously. Because of the wide range in component operating conditions experienced in engines utilized for missions with large Mach number variation requirements, the restrictions imposed by the fixed geometries are presently being subjected to some scrutiny. Many ingenious geometrical variations are under active consideration for future engines.

5.5-7 Rocket Propulsion

The Thrust Equation

In the case of rocket propulsion (Fig. 5.5-8) the equation for the thrust is considerably simplified as compared to the airbreathing engine case because of the lack of incoming momentum to the vehicle. Thus, the expression may be written as

$$F = \dot{m}_e U_e + (P_e - P_a)A_e = \dot{m}_e C \tag{5.5-34}$$

The term C on the right-hand side is the "effective exhaust velocity." It is usual, also, to introduce the specific impulse I_{sp} defined by

$$I_{sp} = \frac{C}{g_0} \qquad \frac{m/s}{\dfrac{m}{s^2}} \tag{5.5-35}$$

Here g_0 is the acceleration of gravity at the standard reference condition.

The dimensions of specific impulse are usually in seconds, and the definition has the advantage that the numerical value of specific impulse is the same in SI units as in British units.

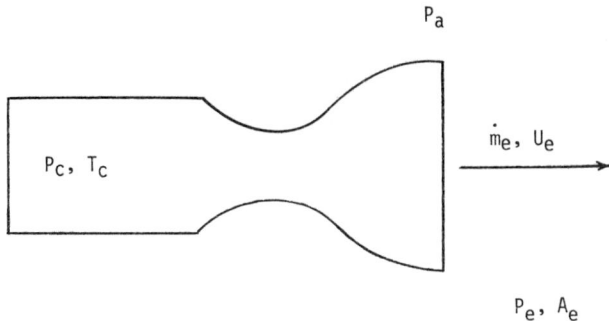

Fig. 5.5-8 Schematic of rocket engine with nomenclature.

A convenient formulation of the thrust equation follows if the exhaust gases may be considered (locally) calorically perfect. Thus,

$$C = C^* C_F \tag{5.5-36}$$

The convenience of this formulation arises because the term C^* is primarily a function of propellant characteristics (and a weak function of combustion chamber pressure), whereas C_F is primarily a function of the nozzle design and altitude.

An approximation of the thrust coefficient (C_F) may be obtained simply if it is assumed that the nozzle flow is isentropic and that the propellant gas is calorically perfect. In such a case it follows that

$$C_F = \gamma \sqrt{\frac{2}{\gamma - 1} \left(\frac{2}{\gamma + 1} \right)^{(\gamma+1)/2(\gamma-1)} \left\{ 1 - \left(\frac{P_e}{P_c} \right)^{(\gamma-1)/\gamma} \right\}^{1/2}}$$

$$\times \left[1 + \frac{\gamma - 1}{2\gamma} \left(\frac{P_c}{P_e} \right)^{1/\gamma} \frac{\dfrac{P_e}{P_c} - \dfrac{P_a}{P_c}}{\left\{ 1 - \left(\dfrac{P_e}{P_c} \right)^{\gamma-1/\gamma} \right\}} \right] \tag{5.5-37}$$

Nozzle Performance and Sizing

It is obvious from physical reasoning as well as analytically from Eq. (5.5-36) that the thrust (and hence C_F) will be a maximum for given chamber conditions and ambient pressure if the nozzle exit pressure is made to equal the ambient pressure. This is, of course, not a difficult task if the combustor and ambient pressures are held constant, but booster rockets traverse enormous altitude ranges and hence enormous ambient pressure ranges.

Provided that flow separation does not occur within the nozzle, the ratio of nozzle exit area to throat area required to provide the desired design pressure ratio, P_{ed}/P_c, is given by

$$\frac{A_e}{A_t} = \sqrt{\frac{\gamma - 1}{2} \left(\frac{2}{\gamma + 1} \right)^{(\gamma+1)/2(\gamma-1)} \left(\frac{P_c}{P_{ed}} \right)^{1/\gamma} \left\{ 1 - \left(\frac{P_{ed}}{P_c} \right)^{(\gamma-1)/\gamma} \right\}^{-1/2}} \tag{5.5-38}$$

Thus, for all altitudes above the "design altitude" (where $P_{ed} = P_{ad}$), the exit pressure is fixed at P_{ed} and is larger than the local ambient pressure P_a.

When a flight at altitudes below the design altitudes occur, the ambient pressure exceeds the nozzle exit pressure. When conditions are such that the ambient pressure greatly exceeds the design exit pressure, the strong oblique shock waves that are formed at the exit due to the overpressure move into the nozzle, thereby changing the effective exit pressure [for use in Eq. (5.5-37)] to that just preceding the shock wave. Under these circumstances the exit pressure is no longer determined by the chamber pressure and nozzle area ratio (that is, $= P_{ed}$) but by the ambient pressure.

A very simple approximate method of estimating the resulting effective exit pressure, $P_{e,\text{eff}} = P_s$, was suggested by Summerfield[6]:

$$\frac{P_s}{P_c} = \frac{1}{K}\frac{P_a}{P_c} \tag{5.5-39}$$

where K is a constant of value approximately 2.7–2.8.

The entire altitude performance of a given nozzle can now be calculated by utilizing Eq. (5.5-39) for altitudes below which the predicted value of P_s is less than P_{ed} and using $P_{ed} = P_e$ above this "separation" altitude. Particularly simple results can be obtained if the pressure variation with altitude may be approximated as exponential. Thus,

$$\frac{P_a}{P_{\text{SL}}} = e^{-h/H_{\text{scl}}} \tag{5.5-40}$$

In such a case it follows also that

$$h_{\text{sep}} = h_d - H_{\text{scl}}\ln K \tag{5.5-41}$$

Figure 5.5-9 indicates the resulting behavior of C_F versus altitude for a rocket designed with $\gamma = 1.28$, $P_c/P_{\text{SL}} = 50$, $K = 2.7$, $H_{\text{scl}} = 7000$ m, and $h_d = 10{,}000$ m. Hence

$$\frac{P_{ed}}{P_c} = \frac{P_{sl}}{P_c}e^{-h_d/H_{\text{scl}}} = 0.004793$$

As is illustrated in Fig. 5.5-9, the thrust coefficient of a fixed area ratio nozzle increases substantially with increases in altitude. It is evident, however, that a substantial penalty occurs when the fixed-nozzle performance is compared to that which could be obtained if ideal expansion was achieved at all altitudes (shown as the dotted line in Fig. 5.5-9). In an effort to alleviate the effects of this off-design penalty, considerable effort has led to the development of nozzles with adjustable area ratios (e.g., see Fig. 5.5-10). The illustrated nozzle not only has the capability of adjusting to "on-design" at three different altitudes, but it has the further advantage that the stage with which it is identified can be more compactly joined with the lower stage.

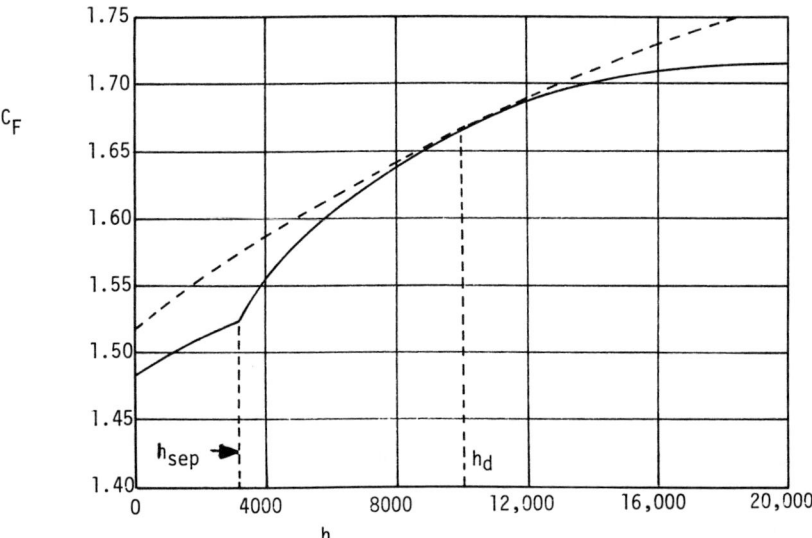

Fig. 5.5-9 Variation of thrust coefficient with altitude.

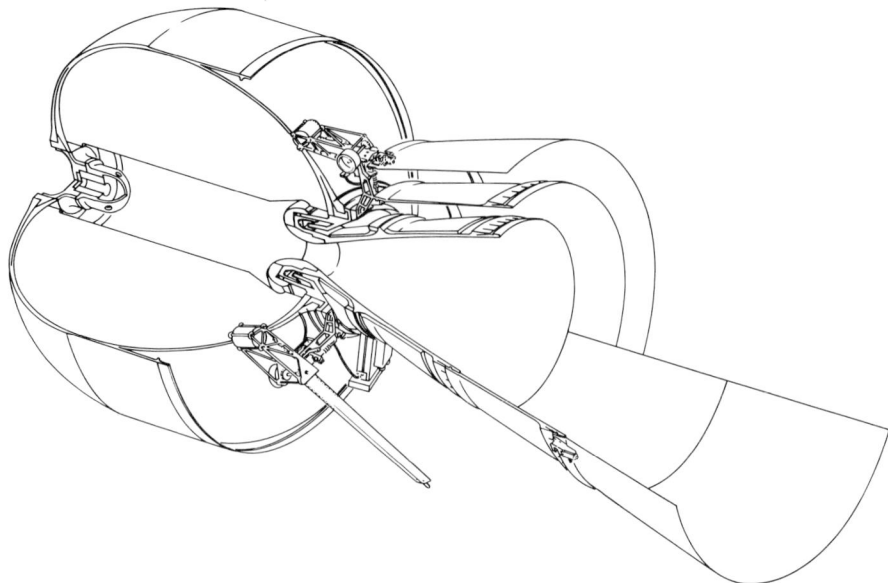

Fig. 5.5-10 Rocket engine with extendable exit cone (EEC). (Courtesy of Chemical Systems Division, United Technologies.)

Choice of Propellants

The function of the rocket nozzle is to convert the stagnation enthalpy of the combustion products into directed kinetic energy of the exhaust. It follows that the highest specific impulse propellants will be those corresponding to the products of combustion with highest stagnation enthalpy. At present the highest usable enthalpies are obtained with molecular hydrogen as fuel and molecular oxygen as oxidizer. (This leads to a "vacuum-specific impulse" of approximately 460 s.)

Aggressive development has led to the development of safe handling techniques for these very energetic propellants. The very low density of molecular hydrogen, however, leads to volumetric problems for use in the first stage of large booster rockets. As a result the optimal design solution favors use of hydrogen–oxygen upper stages and higher density propellant first stages.

5.5-8 Nonchemical Rockets

The mass ratio required to impart a given change in velocity to a vehicle, for the case where no drag forces are present, where no significant change in potential energy of the vehicle occurs, and for the case where the effective velocity C remains constant during firing, follows very simply from the equation of motion to give

$$\frac{m_f}{m_0} = e^{-\Delta v / C} \tag{5.5-42}$$

It is evident from this expression that if very energetic missions (large Δv) are to be carried out, it will be necessary to utilize propellants with very large C or the mass ratio m_f / m_0 will become unacceptably small.

The exhaust velocity that occurs in space is very nearly given by

$$C \approx \sqrt{2h_c} = \sqrt{\frac{2\gamma}{\gamma - 1} \frac{R_U}{w} T_c} \quad \text{(calorically perfect)} \tag{5.5-43}$$

Because the enthalpy per mass attainable through chemical reaction is limited to approximately that of the hydrogen–oxygen reaction, alternative methods of providing very high enthalpies have been investigated. Such methods involve the use of energy addition from sources other than a chemical reaction, so the choice of propellant becomes a separate issue from the source of energy.

Nuclear-Heated Rocket

A conceptually simple concept is that of the nuclear-heated rocket, which operates by having the propellant pass through heat exchange passages within a reactor and thence through the propelling nozzle. The most desirable propellant for such a system is that which gives the maximum possible enthalpy per mass for a given (limiting) stagnation temperature. Clearly, molecular hydrogen is the most suitable choice of propellant, so with $T_c = 2100$ K, the attainable specific impulse for a nuclear-heated rocket becomes

$$I_{sp} = \frac{C}{g_0} = \frac{1}{9.8} \left\{ \frac{2(1.4)}{0.4} \frac{8317}{2} (2100) \right\}^{1/2} \quad \text{or} \quad I_{sp} = 798 \qquad (5.5\text{-}44)$$

Thus, even though the attainable wall temperatures of the reactor restrict the gas temperature to be much less than that found in chemical rockets, the attainable specific impulse is much higher because of the use of the very low molecular weight propellant.

It is unfortunate that, to date, in spite of great success in the Rover and Nerva development programs, nuclear propulsion has not found an appropriate "mission." Thus, the great increase in specific impulse is not sufficient to justify use of the related massive reactor for Earth–lunar missions, and the attained specific impulses are still too small for use in manned missions to the planets.

Electrical Rockets

When very high energy missions are contemplated, missions analysis indicates that very high specific impulse thrustors will be required. In order to attain such huge specific impulses, very large energy addition to the propellants will be required. A variety of electrical concepts have been considered for this purpose, almost all of which require the addition of a radiator to the spacecraft for the rejection of waste heat. It becomes evident for such devices that the optimum choice of specific impulse is not a limitingly large value. This is because the mass of the power supply and radiator increase as the propulsive stream energy increases, so that the combination of propellant mass and power supply mass passes through a minimum at an intermediate specific impulse value. For manned solar missions the optimal range of specific impulse values has been found to be in the range 3000–5000 s.

Electrothermal Thrustors (Arc Jets)

A conceptually simple device for supplying high specific impulse is the electrothermal thrustor. In this device electric energy is added to a low molecular weight propellant to increase the stagnation enthalpy, after which the propellant is expanded through a conventional nozzle.

Electrothermal thrustors are limited in performance by the onset of high ionization (or dissociation) losses as well as high thermal losses to the containing walls. At present useful performance is limited to about $I_{sp} = 1700$ s, so that the devices are inappropriate for planetary missions. Their relative simplicity makes them viable candidates for use in orbit perturbation and station keeping.

Electrostatic Thrustors (Ion Rockets)

At the very high specific impulse end of the spectrum the exhaust energy of the propellant is so huge that the ionization energy of the propellant can be considered a very small loss. Thus, electrostatic thrustors operate by accelerating a stream of ions in an electrostatic field, and subsequently neutralizing the exhaust beam by injection of electrons.

With such very high energy devices, the performance limitation occurs because of the difficulty of creating sufficient thrust per area. (The beam becomes "space charge limited.") The beam thrust is proportional to the mass–charge ratio squared, to the fourth power of the specific impulse, and to the inverse square of the anode–cathode spacing. As a result, very small spacings and very high mass–charge propellants (cesium or mercury) are used.

To date, electrostatic thrustors have demonstrated successful performance in the range $I_{sp} > 5000$ s. The great remaining problem is the generation of the required power at acceptable power-to-mass ratios.

REFERENCES

5.5-1 G. W. Klees, and A. D. Welliver, "Variable Cycle Engines for the Second Generation Supersonic Transport," SAE paper no. 750630; presented at the Air Transportation Meeting, Hartford, CT, May, 1975.

5.5-2 E. Tjonneland, "The Design, Development and Testing of a Supersonic Transport Intake System," paper no. 18, AGARD-CP-91-71, Proceedings on Inlets and Nozzles for Aerospace Engines in Sandefjorde, Norway, September, 1971.

5.5-3 G. C. Oates, ed., "The Aerothermodynamics of Aircraft Gas Turbine Engines," AFAPL-TR-78-52, July, 1978.

5.5-4 J. W. Ramsay and G. C. Oates, "Potential Operating Advantages of a Variable Area Turbine Turbojet," ASME paper no. 72-WA/AERO-4; presented at the ASME Winter Annual Meeting, New York, November 26–30, 1972.

5.5-5 G. C. Oates, "Aerothermodynamics of Gas Turbine and Rocket Propulsion," AIAA Education Series, J. Przemieniecki, Series Editor, New York, July, 1984.

5.5-6 M. Summerfield, et al., "Flow Separation in Over-Expanded Supersonic Exhaust Nozzles," *Jet Propulsion* **24**, September–October (1954), pp. 319–321.

CHAPTER **6**

COAL TECHNOLOGY

ROBERT D. BESSETTE

Old Ben Coal Company
Lexington, Kentucky

ERIC H. SMITH

Consultant
Holden, Massachusetts

ALBERT H. RAWDON

Riley Stoker Corporation
Worcester, Massachusetts

ROBERT W. FITZMAURICE

Allen Sherman Hoff Company
Malvern, Pennsylvania

FRANCISCO PALACIOS

Riley Stoker Corporation
Worcester, Massachusetts

6.1 COAL FORMATION AND PROPERTIES

Robert D. Bessette

ACKNOWLEDGMENTS

There are many people who have provided input to the co-authors in the form of editing, typing, technical knowledge, and support. These we sincerely thank. However, the completion of this section must be credited to the efforts of one man, Eric H. Smith, who died prior to seeing it completed. Without Smith's dedication, knowledge, motivation, support and countless hours of work, this section could not have been completed—a memorium to a friend of the industry for over 60 years.

6.1-1 Introduction

Interdependencies of coal applications, production, and preparation systems will be presented as they relate to combustion, conversion, equipment and utilization systems—more simply stated, "coal from seam to stack." Discussions presented herein will provide the necessary information from which the engineer, student, or manager may learn the interrelations and considerations from which a detailed coal system may be selected and developed as the most practical and appropriate for the specific requirements of that industry under consideration.

The scope of the presentation will be limited to a brief discussion of the advantages, disadvantages, and total system interdependencies of individual items and considerations. No detail presentation will be made for any specific area of discussion; however, bibliographical reference will be provided for further detailed investigation. Individual sections have been designed for quick reference and understanding. A summary of each section will provide a brief statement of what the interdependencies may be.

In order to understand the coal technology system it is first necessary to know and understand the basic fuel characteristics. "What is it?" Coal is an inhomogeneous soft black rock containing energy that can be liberated through combustion or converted to a usable form via incomplete combustion, mechanical, or chemical processes. In actuality, under petrographic analysis (close physical analysis) in very thin layers coal is layered in dark red and gold bands that can correspond to the varied layers of vegetable matter and impurities deposited before, during, and after the coalification process.

6.1-2 Coal Formation

The process of coal formation is a continual one. Even now coal is being formed. Although coal was being formed 400 million years ago, most high-ranked bituminous and anthracite coal used today was formed during the carboniferous period some 200–300 million years ago. Lignite and subbituminous coals are young coals on the order of 1–70 million years old. During the carboniferous period the earth was wet and steamy. Plants grew rapidly, died, and fell into large areas of stagnant swamp and marshland.

When discussing coal formation it is necessary to think in terms of many thousands of years and vast areas of swamp spanning tens of thousands of square miles. In the span of these years rivers could move from one end of the swamp to the other, depositing stone, silt, and other impurities carried with them from the rains carrying volcanic dust and eroding mountains. During this time dust storms, geologic upheaval, and volcanos were also changing the landscape and structure of the coal beds. Major geological change following the coal formation process (e.g., the coming and going of mountains) also affected the coalification process. Impurities layered within the decaying vegetation are the source of ash and impurities in coal today. Figure 6.1-1 is an artist's rendition of what this time period might have been like.

The coal formation process, or coalification process, in itself was and is a long process requiring stagnant waters with low free oxygen, as was the case in the early times. Vegetable matter and trees die, fall into water, and decay with minimal oxidation to form peat, the first step in the coalification process. Variations in plant components collecting in segregated areas are responsible for different types of coal such as splint and cannel coals. The process of changing peat to coal requires the peat to be covered and sealed with rock and mud. Then, over time and with increasing pressure from the overlying layers increasing temperatures, the coalification process occurs and coal is formed. Classification of coal is by rank from metaanthracite to lignite. Table 6.1-1 shows the classification of coal by rank as developed by the American Society for Testing and Materials (ASTM).

There is still much research being directed into the actual composition and form of coal. It has been seen that the proportion of carbon ring structures increases with rank. The base composition of coal is affected by the coalification process. As the pressure of overlying layers of earth and the temperature within the peat or coal increase, the coalification process begins and continues. Peat is compressed and the wood, vegetable matter, contaminants (ash, rock, etc.), and moisture begin to reform into various carbon and carbon–hydrogen chain molecules. Moisture and oxygen trapped within the peat or coal is begun or continues to be forced out until coal is formed or the rank is changed.

Through the continuation of the coalification process coal density increases forcing more moisture and oxygen out in the form of CO, CO_2, and other gases or liquids. As a result, the inherent moisture decreases, the oxygen content decreases, and the density increases. One result of the increasing density is a change from a soft indefinitely defined lignite, to a harder and definitely banded bituminous, to a very hard lustrous and nonbanded anthracite. There are also internal changes in the volatile content

Fig. 6.1-1 Coal formation and coalification environment, volcanic dust, rain erosion, and sedimentation.

and composition of the coal. Volatile content decreases with degree of coalification and coal rank. A comparison of the changes in inherent moisture, oxygen, volatile content, and density can be traced using Table 6.1-2, showing general characteristics of the four major coal classes and peat.

6.1-3   Coal Composition

Coal composition must be discussed in terms of its chemical and physical properties or characteristics. Although the chronological age of coal plays an important part in the coalification process, the developmental age involving temperature and pressure considerations is a more accurate method of evaluating the actual chemical and physical properties as they affect the mining, preparation, and utilization of coal.

Chemical Properties: The Proximate Analysis

The basic components of coal are the proximate and ultimate analyses (see ASTM D3172). The proximate analysis consists of the four building blocks of coal: carbon, volatile matter, moisture, and ash.

A fairly good indication of relative performance can be achieved based on a knowledge of equipment design or past experience. As seen above, the type of vegetation and coalification processes determines the proportions of the various constituents. Table 6.1-2 shows the variation of moisture and volatile content relative to classification.

Moisture Content.   The reference to moisture content in the proximate analysis is to the total moisture within the coal as delivered to a using facility. Total moisture is a composite of inherent and

TABLE 6.1-1 CLASSIFICATION OF COALS BY RANK[a]

Class		Group	Fixed Carbon Limits, % (Dry, Mineral-Matter-Free Basis)		Volatile Matter Limits, % (Dry, Mineral-Matter-Free Basis)		Calorific Value Limits. Btu/lb (Moist,[b] Mineral-Matter-Free Basis)		Agglomerating Character
			Equal or Greater Than	Less Than	Greater Than	Equal or Less Than	Equal or Greater Than	Less Than	
I. Anthracitic	1.	Meta-anthracite	98	—	—	2	—	—	Nonagglomerating
	2.	Anthracite	92	98	2	8	—	—	
	3.	Semianthracite[c]	86	92	8	14	—	—	
II. Bituminous	1.	Low-volatile bituminous coal	78	86	14	22	—	—	Commonly agglomerating[e]
	2.	Medium-volatile bituminous coal	69	78	22	31	—	—	
	3.	High-volatile A bituminous coal	—	69	31	—	14,000[d]	—	
	4.	High volatile B bituminous coal	—	—	—	—	13,000[d]	14,000	
	5.	High volatile C bituminous coal	—	—	—	—	11,500	13,000	Agglomerating
							10,500	11,500	
III. Subbituminous	1.	Subbituminous A coal	—	—	—	—	10,500	11,500	Nonagglomerating
	2.	Subbituminous B coal	—	—	—	—	9,500	10,500	Nonagglomerating
	3.	Subbituminous C coal	—	—	—	—	8,300	9,500	
IV. Lignitic	1.	Lignite A	—	—	—	—	6,300	8,300	Nonagglomerating
	2.	Lignite B	—	—	—	—	—	6,300	

[a]This classification does not include a few coals, principally nonbanded varieties, which have unusual physical and chemical properties and which come within the limits of fixed carbon or calorific value of the high-volatile bituminous and subbituminous ranks. All of these coals either contain less than 48% dry, mineral-matter-free fixed carbon or have more than 15,500 moist, mineral-matter-free Btu/lb.
[b]Moist refers to coal containing its natural inherent moisture but not including visible water on the surface of the coal.
[c]If agglomerating, classify in low-volatile group of the bituminous class.
[d]Coals having 69% or more fixed carbon on the dry, mineral-matter-free basis shall be classified according to fixed carbon, regardless of calorific value.
[e]It is recognized that there may be nonagglomerating varieties in these groups of the bituminous class, and there are notable exceptions in high-volatile C bituminous group.

**TABLE 6.1-2   COALIFICATION EFFECTS ON MAJOR COAL
COMPONENTS RELATIVE TO RANK AND CLASSIFICATION**

Classification	Moisture[a]	Volatile[b]	Oxygen[c]	Density[d]
Anthracitic	1.0	5.0	3.0	99.0
Bituminous				
Low volatile	1.5	18.0	4.0	85.0
Medium volatile	2.0	25.0	5.5	82.0
High volatile				
	1.5	34.0	5.5	
	6.0	36.0	9.0	80.0
	9.0	39.0	13.0	
Subbituminous	17.0	40.0	15.0	72.0
Lignite	35.0	45.0	18.0	62.0
Peat	60.0	65.0	30.0	55.0

[a] Inherent or in seam moisture percent.
[b] Volatile content percent expressed as dry volatile.
[c] Oxygen as dry ultimate analysis percent.
[d] Density is given in in-seam pounds per cubic feet.

surface moisture. Surface moisture is that moisture added to the coal as a result of mining, preparation, or natural causes after the coalification process. Seepage, rain, high humidity, or dust control sprays can be causes of additional surface moisture.

Inherent Moisture.   Sometimes called equilibrium moisture, inherent moisture is contained as an intimate part of the coal as a result of the coalification process. An analytical method for determining inherent moisture is ASTM D1412. The surface moisture is removed by drying, leaving the inherent moisture. The inherent moisture determination can vary with coal particle size and relative humidity at the time of determination. Inherent moisture is an indicator of coal rank. It is also an important consideration in combination with volatile content, oxygen content, free swelling characteristics, coal structure, and sizing when evaluating coal reactivity as it could apply to the combustion process.

Volatile Content.   Another aspect used in determining coal rank is volatile content. Volatile content or matter is that portion of the total coal that will distill or volatilize at high temperatures prior to combustion or conversion. The individual elements of volatile matter vary greatly from lignite to anthracite. This is a result of the coalification process. Some elements of the volatile matter are thiophene, olefins, aromatics, tars, organic sulfides and oxygen, nitrogen compounds, and other carbon–hydrogen chains. Inorganic volatiles include organicly bound chlorides, mineral carbonates, and others.

During the combustion process the volatile content of a coal acts much like an ignitor to the fixed carbon. The heat produced from the combustion of the volatiles provides the necessary energy for the combustion of the remaining combustibles. Besides being an indicator of coal rank, volatile content is the major indicator of coal reactivity and combustibility. It is also that portion of the coal removed during the coking process of the iron and steel industry as well as that portion removed in the first step of a fixed-bed gasification system.

Fixed Carbon.   During the combustion and conversion process, after the volatile content of the coal is distilled following moisture evaporation, the remaining combustible matter is referred to as fixed carbon. This does not include the ash content, which is noncombustible. The composition of the fixed carbon is mostly carbon; however, small portions of hydrogen, oxygen, nitrogen, and sulfur remain. The fixed-carbon portion is normally the greatest portion of the total coal and is the single largest contributor to the coal Btu content.

Ash.   The ash is that portion of the coal remaining after the coal has been combusted. Ash as seen above is generally a contaminant found within coal resulting from the deposition of these contaminants before, during, and after the coalification process. The elemental content of ash varies greatly from one coal seam to another as well as within one coal seam and from one location to another.

The ash found in coal can be either as a parting, which may range from a millimeter thick in some cases to many feet thick in other places. This ash can be classified as extraneous or free ash. Other ash which is finely divided throughout the coal is considered to be inherent ash. This may be a substantial portion of the total ash depending on the circumstances at the time of initial formation. The distribution of ash throughout the coal bed is a very important consideration when coal preparation and beneficiation techniques are evaluated relative to the ash behavior during and after combustion or conversion.

The wide variation of ash content and characteristics has caused problems in predicting how a given coal or ash will behave in a specific application. Many tests have been developed by chemists, boiler manufacturers, steel producers, and others to determine these properties. However, most of these tests are statistically derived empirical determinations. These determinations will be covered in more detail when the physical properties of coal and equipment design considerations are discussed.

Heating Value.   The proximate analysis of coal also includes the representative Btu value of the coal —the heating value. There are a number of forms in which this can be expressed: as received, air dried, dry and moisture-ash-free (MAF). These are representations of gross Btu value, or the higher heating value of the coal (HHV). Another method of calculating the heating value of the coal is the lower heating value (LHV). The gross Btu is a function of the carbon, volatile, and sulfur percents in the specific coal as determined in a calorimeter per ASTM D2015 and adjusted for moisture content. This includes all Btu generated from the combustion of the coal. The specific elements from which Btu content is generated will be discussed later (see "The Ultimate Analysis").

The lower heating value is a somewhat vague representation depending on the value set for the latent heat of the moisture and moisture created by the combustion of fuel hydrogen to moisture. The latent heat is dependent on the temperature at which the gases from combustion are emitted to the atmosphere and lost. This value is normally considered to be between 1040 and 1050 Btu/lb of moisture per pound of fuel. At a boiler exit temperature of 350°F the equation for calculating LVH from HHV would be

$$LHV = HHV - [H_2(8.936) + M]1050$$

where H_2 = pounds of hydrogen per pound of fuel
M = pounds of moisture per pound of fuel
8.963 = pounds of H_2O per pound of hydrogen
1050 = latent heat of vaporization

A simple definition of the LHV may be stated as the amount of heat available from a given coal to do work—assuming no excess air or dry gas losses. These will be discussed in the section on combustion efficiency.

Most companies in the United States use the HHV as a means for comparing coals. However, in Europe and other parts of the world the lower heating value is the main consideration. It is important to know which is being used; if the LHV is used, it is further important to know the latent heat value to be used in the calculation.

When evaluating coals for application purposes, the as received Btu should be considered. This is the most representative evaluation of what a specific piece of equipment will be using. However, since moisture can vary, an evaluation of two or more coals may better be accomplished by using the dry Btu and considering the moisture separately. The dry basis will vary only with the base fixed carbon, volatile content, and ash.

For any specific seam of coal the moisture-ash-free Btu will vary very little and should be in the range of ±50 Btus. Variation outside this range could be an indicator of multiple seam blending or an outcrop or oxidized coal. The MAF Btu can be used as a tool for quality control when multiple seams are blended or to check compliance of coal received to that of coal purchased.

The air dry Btu is another consideration used by some companies when buying coal. Air-dried considerations are such that coal is allowed to dry for a certain period of time; then the Btu is obtained and calculated to the level with inherent and some surface moisture. There is not a general consensus as to the method of drying or the level of drying required. Like the LHV, the air-dried Btu method should be agreed to prior to sales and delivery of coal to the purchaser's facility.

Sulfur.   There are three forms of sulfur found in coal. The pyritic forms of iron pyrite and marcasite (specific gravity 5.0) are the most abundant in coal. Organic sulfur is also present in lesser quantities and is uniformly distributed throughout the coal with the coal structure, bound with carbon, nitrogen, oxygen, and hydrogen. Sulfur is also present in very small quantities as sulfate bound with calcium and/or iron.

The proportions of the types of sulfur, as well as the amount of sulfur, in the coal varies significantly from one seam to another and from one part of the country to another. The forms and proportions of sulfur are dependent on the formation and coalification process of the coal. Pyritic forms of sulfur are normally found in partings within the seams, in massive sulfur balls, or in finely distributed crystals throughout the seams. These were added as contaminants to the seam rather than as part of the vegetation.

Chemical Properties: The Ultimate Analysis

A representation of a coal's chemical content can be found in the ultimate analysis, a listing of the basic elements of composition that affect the chemical reactions for combustion or conversion. These are: carbon, hydrogen, nitrogen, chlorine, sulfur, ash, and oxygen. The methods by which these are determined is discussed briefly in Section 6.3-5.

Carbon. Carbon, like other elements to be discussed, is not found in its elemental form but in its organic and inorganic compounds. The total carbon will be discussed in more detail as a major contributor to the products of combustion, coal gasification, and coal conversion processes. The amount of carbon found in a given coal is dependent on the initial formation and the coalification process. The carbon content is needed to predict the products of combustion, conversion, and other energy technology systems, such as CO, CO_2, short- and long-chain hydrocarbons, and other hydrocarbon structures.

Hydrogen. Hydrogen is also found in its organic and inorganic compounds and as part of the moisture in the coal. Hydrogen is the second major contributor toward the formation of combustion and conversion products. It is also a prime contributor to the total volatile content of the coal as a function of the coalification process.

The combustion of hydrogen produces water vapor (H_2O) and in some cases in combination with chlorine or other elements in the coal can produce other compounds such as hydrochloric acid or hydrogen sulfide. During coal conversion processes and gasification the hydrogen content is an important consideration in determining feedstock requirements and final-product slate and quality. Both hydrogen and carbon are normally associated with the organic composition of coal.

Nitrogen. Nitrogen is the only element in coal to be found totally in the organic matter of coal. As such it is very stable and virtually impossible to separate from the coal. For many years nitrogen was thought to be an insignificant element since it added nothing to the combustion process as a noncombustible. However, in recent years, with the advent of emission considerations, it has been found that fuel nitrogen is the major contribution to the formation of the oxides of nitrogen. Special considerations or techniques usually have to be implemented to reduce the oxides of nitrogen produced from combustion.

Chlorine. The occurrence of chlorine in coal is in most cases in the form of inorganic salts of sodium, calcium, potassium, and, for some coals, magnesium and iron. The chlorine content is more a function of salt addition after or during the coalification process. However, there is some chlorine found in the organic structure of coal. Because chlorine reacts vigorously with hydrogen to form hydrochloric acid, which vaporizes, it is not found in the remaining ash after combustion. However, it is reported as part of the ultimate analysis.

The main concern regarding the total chlorine content of a coal is to determine the corrosive nature of the fuel. Some accepted rules follow when considering chlorine content to guard against high and low temperature corrosion:

1. Standard design procedures are sufficient to guard against corrosion when using coals of up to 0.2% chlorine.
2. Some precautions may be required for certain applications using coal below 0.3% chlorine.
3. For coal with a chlorine content of between 0.3 and 0.5% precautions must be considered to prevent corrosion.
4. The use of coal with greater than 0.5% chlorine is not recommended in conventional boiler or conversion systems.

Sulfur. The sulfur content of coal is found in both the organic and inorganic forms. The main concern regarding the total sulfur content of a coal is to predict the emission levels of sulfur oxides, oxygen requirements, and the propensity of the coal to form complex sulfate salts. Depending on the application, sulfur can be a most troublesome element. A more detailed discussion of sulfur emissions and control can be found in the section on emission control.

Ash. The ash content of a coal is that portion of the coal remaining after a coal has been combusted or converted to another form. The total ash is important; however, the specific elemental composition is more important and will be discussed in some detail in the section on the physical characteristics of coal. In order to accurately design the ash handling, emission control, and disposal equipment, the total coal ash content must be known.

Oxygen. The final consideration in the ultimate analysis is the oxygen content of the coal. The oxygen content is found in the organic and inorganic contents of the base coal in the forms of water, silicates, carbonates, and carboxyl groups. The presence of organic oxygen in the coal is a contributor to the reactivity of the coal and source of oxygen for the intimate combustion of coal carbon. The oxygen content must be known in order that the theoretical oxygen needed for complete combustion can be determined. The oxygen is also an indicator of coal rank and as such is a function of the coalification process.

Physical Characteristics of Coal

The physical characteristics of coal are those properties that affect its application or state before and during the chemical reaction of its conversion to a usable energy form. These characteristics are specific gravity, density, hardness, grindability, friability, size, free-swelling index, fluidity or plasticity, and reactivity. These are all necessary considerations for determining coal preparation, handling, storage and pulverization characteristics, as well as the degree to which the chemical reaction for combustion or conversion can take place.

Specific Gravity.   The specific gravity of coal is an important consideration for two reasons. The first and main reason is for conventional coal preparation. The second is with regard to coal density. Depending on the coalification process, coal will have a specific gravity of between 1.00 and 1.70, increasing with rank. Most bituminous coals have a specific gravity of approximately 1.3, which is a composite of the coal and some inherent or retained mineral matter. The specific gravity of the coal portion is determined by the compactness or porosity of the carbon structure—the coalification process. The mineral ash portion of the coal can have a specific gravity ranging from 1.8 to 5.0 (for pyrites). More specific gravity considerations will be discussed in the general category of coal preparation.

Density.   Coal density (in pounds per cubic feet) is a function of the specific gravity and degree of coal cleaning. If coal has a specific gravity of 1.3, the weight of a solid cubic foot would be about 80 lb. However, coal is not normally stored, used, or handled in solid cubic foot blocks. It is utilized as a sized product. Table 6.1-3 shows the approximate density of various grades or sizes of bituminous coal. The actual density of coal will vary with the rank, grade, and compaction of the specific sample considered. Compaction is a function of coal sizing and the techniques used to store and handle the coal. These will be discussed in more detail in the section on coal storage.

Hardness.   Hardness is a function of the vegetable matter composition and coalification process. Hardness is an indicator of the coal's structure—the form of the carbon molecules that resist breakage or shattering on impact. Hardness also varies with the age or rank of the coal. Lignite and subbituminous coals are relatively soft in structure, whereas high volatile bituminous coal and anthracite coals are very hard. During the coalification process a point is reached where devolatilization increases as the coal goes from high volatile to medium volatile to low volatile. This devolatilization causes a cracking and change in the coal structure, promoting an easy shattering or breaking. There are always variations such as cannel or splint coals which because of composition and the coalification process are extremely hard. These are the exceptions rather than the rule.

Grindability.   Grindability is not solely related to hardness but is also an indicator of a coal's pulverizability to a given percentage through a 200-mesh sieve. This is a function of the coal moisture and ash content, structure, and hardness and is necessary to determine when coal pulverization is

TABLE 6.1-3   APPROXIMATE COAL DENSITY BY SIZE AND GRADE[a]

Size	Uncompacted (lbm/ft^3)	Compacted (lbm/ft^3)
Run of mine, 6×0 in.^{2}	50	50–60
Crushed run of mine or blend, 2×0 in.^{2}	45	45–70
Stoker, $1\frac{1}{2} \times \frac{1}{4}$ in.^{2}	43	45–55
Modified stoker, 20% fines	44	45–60
Fines, $\frac{1}{4} \times 0$ in.^{2}	45	45–50

[a]Degree of compaction, specific gravity, and size distribution can affect actual densities.

desired. The main parameter for pulverizor sizing is the grindability relative to the percentage of coal required to pass through a 200-mesh sieve in order to obtain efficient combustion. The most widely accepted method to determine coal grindability is the Hardgrove grindability index (HGI). Refer to ASTM D409.

Friability. Many times friability is confused with grindability; however, they are not the same. Friability depends on the coal structure and hardness relative to how the coal will shatter, crack, or degrade on impact or through abrasion. There are specific tests to determine friability. However, these tests are not run routinely and coal is usually classified as either friable or nonfriable based on past experience. The friability of a coal affects the design of coal-handling and storage equipment. It also affects the size requirements for coal feed to spreader stoker fired boilers.

Weathering. Weathering, sometimes referred to as slacking, is a process that occurs in coal exposed to the atmosphere or an unstable environment. This characteristic is predominantly a function of the coal structure and inherent moisture content. A high inherent moisture content coal that tends to lose moisture too fast could degrade like a lignite and produce a greater number of fines. In the same way a coal exposed to freeze–thaw cycles will tend to degrade because of the expansion and contraction of the water. In both cases the severity of the weathering depends on the inherent moisture content, environmental conditions, and the time of exposure.

Coal Sizing. Coal sizing, although dependent on the above considerations, is more a function of the method of mining and preparation for use. Variation in mining techniques impact the resulting size consistency of a coal in different ways depending on the coal structure, hardness, friability, and weathering characteristics. Coal sizing considerations are important relative to handling, storage, and applications in stoker-fired boilers and coal gasification systems. Specific considerations regarding these systems will be found later.

Free-Swelling Index. The free-swelling index, sometimes known as the coke button, is an indicator of a coal's ability or resistance to give up its volatile content when heated. It is also an indicator of the structural changes and plasticity of a coal as the volatiles are distilled when heated. An ASTM standard (ASTM D720-67) has been developed to standardize this procedure. The expansion of a coal heated in such a way so as not to ignite the carbon or affect the surface characteristics of the coal measured against a standard chart is the indicator.

As the coal is heated, the coal structure becomes plastic. The degree to which this retards volatile distillation and swells the coal is the indicator from 1 to $9\frac{1}{2}$. The lower the index the easier the volatile distillation and the more free burning the coal is considered to be. The FSI is a function of the coal structure and surface characteristics such as total surface area and porosity. Lower-ranked coals such as lignite and subbituminous coals are extremely free burning because of a noncompressed structure and very porous surface that allows for easy volatile distillation. Coal oxidized by exposure to the atmosphere will have a change in the surface characteristics, increasing porosity and allowing for an easier distillation of the volatiles. It is not uncommon that a coal will lose as much as two FSI points over a 2- to 3-month period when exposed to the atmosphere. This is the reason outcrop coal tends to exhibit a lower FSI than the unexposed portion of the seam.

The FSI is of major importance when considering underfeed and mass-fed stokers, two-stage low-Btu coal gasification systems, and coal coking systems. It is also an important part of a coal reactivity evaluation.

Plasticity and Fluidity. As coal is heated, it has a tendency to become plastic or fluid. Research is ongoing to try to determine the cause of the alteration of the coal structure as it is heated. This aspect is beyond the scope of this work; however, it is important to understand that the plastic characteristics do impact on coal energy utilization systems. Plasticity and fluidity have been used primarily for selecting coal for blending to produce optimum coke qualities for the iron and steel industry. To date there has been little if any correlation of fluidity with conventional combustion and conversion systems. Like free swelling and grindability, fluidity is a relative indicator of a coal's characteristics. The coalification process tends to take coal through a stage in which it will become plastic or fluid when heated. This is evident primarily in the higher-ranked bituminous coals. Low volatile bituminous coals tend to have lower fluidities than do medium and high volatile bituminous coals. This could be an important consideration with regard to coal particle separation during and prior to pulverized coal combustion, clinkering in stoker-fired boilers, and agglomeration in fixed-bed low-Btu gasification systems. The relative fluidities are also important when blending two or more coals for application in systems where an intimate reaction of the coals takes place.

Agglomeration. Agglomeration is that characteristic of coal that indicates the relative ability of the particles to become plastic, swell, and stick to other coal particles during the heating and combustion or conversion processes. The FSI is the main determining factor for agglomerating tendency. An FSI rating between 0 and 3 is considered nonagglomerating. An FSI rating of more than 3 but less than $5\frac{1}{2}$ is considered to be moderately agglomerating. If the FSI is greater than $5\frac{1}{2}$, the coal is considered to be agglomerating. These coals and some moderately agglomerating coals have a tendency to stick together when heated and form clinkers or in some cases coke in underfeed boilers, mass-fed boilers, and coal gasification systems. Coal agglomeration is minimized if the volatiles are combusted as they are distilled.

Reactivity. The reactivity of a coal is at present a subjective evaluation of a coal's propensity to facilitate the chemical reactions of combustion or conversion. This may be the most important of all the physical considerations with regard to the total energy system. Reactivity is not yet a well-defined value and much difference of opinion exists as to its definition. The considerations that impact on the reactivity of a coal are surface and inherent moisture, volatile content, hydrogen content, oxygen content, surface area, porosity, FSI, plasticity, and the agglomerative characteristics. In most cases the reactivity is higher if (1) the surface moisture is lower; (2) the inherent moisture is higher; (3) the volatile, hydrogen, and/or oxygen contents are higher; (4) the total surface area or porosity is greater; (5) the FSI is lower; and (6) the plasticity and agglomerative characteristics are lower. In all cases it is important to consider the total interdependencies of all aspects of a coal.

Washability. Coal washability is of primary concern with regard to coal preparation and the determination of a coal's ability to be cleaned using conventional beneficiation techniques. Washability is determined by the washing of various sizes of coal at the same and different specific gravities. This produces a set of data showing the relative portions of ash and coal lost at the different specific gravities, allowing for the selection of the optimum level of beneficiation for a given coal. The washability of a given coal is dependent on the relative amounts of free and inherent ash, the specific gravity of the ash, and the coal structure as it relates to its ability to release or liberate the ash by crushing.

Coal Ash Characteristics

Ash composition is extremely variable, in many cases even more so than coal because of the methods by which it contaminates the coal seams. An understanding of coal ash is of critical importance when considering the utilization of coal. Ash is the resultant solid product of combustion or conversion. It must be handled and disposed of properly. The following discussion will consider the various aspects of coal ash to provide an integrated understanding for the total energy system.

Ash Composition. There are no specific formulas available to determine the ash composition. The major elements that must be determined individually are phosphorus pentoxide, silica, ferric oxide, alumina, titania, lime, magnesia, sulfur trioxide, potassium oxide, and sodium oxide. There are many other elements found in coal; however, these are found in such small portions that they are not normally reported. The above elements make up the basic coal ash and the relative proportions are dependent on the original coal formation environment and seam contamination before, during, and after the coalification process.

For coal applications, mineral ash considerations are normally confined to the above basic elements and the interrelationships between each metal, alkali, or fluxing agent in a reducing or oxidizing atmosphere. Heating caused by combustion, depending on whether rapid or slow, can have a varying effect on how the ash will react in a given environment associated with either combustion or conversion. Many articles and papers have been written on this subject, and some of these are listed in the bibliography of Section 6.11.

Ash Fusion Temperatures. The ash fusion temperatures as obtained by using the ASTM methods are probably the most widely used indicators of mineral ash plasticity of fluidity. These are performed in both oxidizing and reducing atmospheres. The fusion temperatures (ASTM D1857) given as initial deformation (I.D.), softening ($H = W$), fusion ($H = \frac{1}{2}W$), and fluid ($H = 1.6$ mm) are an indication of ash characteristics and the propensity of the ash to agglomerate, fuse, or melt at a given temperature in the boiler, on a stoker, or within a conversion process. It must be remembered that these temperatures are indicators and a complete analysis must be performed using the basic mineral ash composition in order to obtain a true understanding of the potential ash behavior. In most cases, if the flue gas or process temperature is less than the initial deformation temperature of the coal, and in some cases the softening temperature of the coal, there should be no problem with regard to slagging or fouling.

Slagging Considerations. If the mineral composition is such that at a given temperature the ash is plastic or fluid, it can build up on boiler or process equipment parts exposed to radiant heat. This is called slagging and is normally found in the furnace area of the boiler or in the combustion chamber of other process equipment. Higher-ranked coals with high iron, magnesium, and calcium contents relative to the silica, alumina, and titania have lower fusion characteristics and are potentially a problem, slagging coals. This is a major design consideration and will be discussed in more detail later.

Fouling Characteristics. The fouling characteristics of a coal ash are such that when the ash accumulates as a dust in areas without exposure to radiant heat but in areas where the temperature is above 300 or 400°F there is a tendency to sinter or fuse to a solid. These fouling deposits are for the most part brittle and easily removed. However, because fouling is a function of the ash composition, sintering time, temperature, and ash loading, fouling can cause severe problems with boiler and conversion equipment if proper precautions are not taken. Equipment design must be carefully evaluated relative to coal-fouling potential. The alkali content of the coal ash is the main indicator of the fouling potential of a given coal. Other elements such as sulfur, calcium, magnesium, and iron also have an affect on the fouling process. These considerations will be discussed in more detail later.

Ash Viscosity. In some applications, especially where ash is removed in the molten state, it is important to know the ash viscosity characteristics as well as the ash fusion temperatures. This is a more accurate method for determining the reaction between the various elements in an ash and correlating them to the operating temperatures within the coal system. The 250-P viscosity temperature (T, 250°F) is an indicator of the point at which the ash is a liquid capable of flowing on a vertical surface. Ash viscosity is usually a calculated value, but commercial laboratories are capable of testing for this value.

Resistivity Characteristics. Resistivity is that characteristic of coal ash that minimizes its ability to accept an electric charge. This is a very important consideration when an electrostatic precipitator is to be used or considered for particulate emission control. Resistivity is a function of the temperature, flue gas characteristics, electric field strength, precipitator design and velocity, ash carbon, sulfur, sodium, and certain trace elements. Trace element interactions are beginning to be investigated and are beyond the scope of this work. The most important consideration today is the sulfur in the coal. Coals with greater than 1.5% sulfur content generate the best collection efficiencies; and coals with less than 1.5% sulfur are usually very difficult to remove without special considerations.

There are two ways in which the resistivity of a coal ash can be affected. The first is to adjust the ash mineral composition, which can have side effects on the other ash characteristics. Second, the precipitator design can be adjusted to decrease velocities, increase field strength, and optimize performance. The latter is the most likely. These considerations will be discussed in more detail in the section on electrostatic precipitators.

Ash Abrasivity. In most cases ash is characterized as being highly abrasive, requiring that ash-handling systems and other systems subjected to the ash be designed to take this into consideration. In many instances this has been taken for granted. With the advent of fluidized bed boilers and the projected use of coal in oil-fired designed boilers, ash abrasivity is becoming a greater concern. The main consideration from which the abrasive characteristics of the ash can be determined are the silica, as quartz, and alumina content in relation to the calcium and alkali contents, which can produce a very hard, glassy, fused particle. Eastern-type coals tend to produce a more abrasive ash than the western-type coals. As a result, more consideration must be given to potential problems in high-velocity systems when using eastern in lieu of western coals. There is no current standardized test for ash abrasivity. This is still a subjective evaluation based on past experience with a given coal.

Corrosion. There have been many works dedicated to the subject of corrosion. Some of these are listed in the bibliography. The corrosion associated with flyash is that associated with the condensation of complex alkali salts in combination with ash buildup in the back passes of boilers or conversion systems. These elements, in combination with moisture and chlorine, tend to aggravate the problems associated with corrosion due to fly ash. The greater the fouling potential of a coal the greater the risk of corrosion.

6.1-4 Summary

Coal is a highly variable substance with many characteristics, depending on the original composition materials and the coalification process. The chemical characteristics describe and define the coal's combustibility or convertability to another form of energy. The physical characteristics of coal and its by-products are those that either facilitate or hinder the initiation or completion of the chemical reactions of combustion or conversion.

The synergistic effects of the various combinations of chemical and physical characteristics must be considered in total as the total system is developed. An understanding of the above is of the utmost importance when considering any coal energy system.

BIBLIOGRAPHY

Coal, A Concise Authoritative Survey in Usable Form. Encyclopedia of Chemical Technology, The Interscience Encyclopedia, New York, 1949, Vol. 4, pp. 86–134.

Arthur Raistrick, A. and Marshall, C. E. *The Nature and Origin of Coal and Coal Seams*. English University Press, London, 1939.

6.2 COAL MINING

Robert D. Bessette

Coal mining might be considered the extraction of a black rock through a hole in the ground. There is always an oversimplified estimation of what any given process might entail. In order to fully understand the potential impact of coal within the total energy system, it is necessary to understand the mining of and mining affects on coal and its costs. A true cost comparison of alternative mining methods is beyond the scope of this book, but the basics will be provided to permit full understanding of the general application impacts and uses of each method.

Two methods of mining will be considered, underground mining and surface mining. How and why they are used, the advantages and disadvantages of each method, and the general application interactions and requirements will be discussed.

6.2-1 Underground Mining

Before the development of large earth-moving equipment and technology for the surface mining of coal, underground mining was the backbone of the coal industry. There have now been many advances in both surface and underground mining with possibly the greatest advances in underground mining. It is now possible to remove more coal faster, from deeper seams, and from thinner or thicker seams than was ever thought possible 20 years ago. The mining industry has developed from the "mule-pulled" industry of yesterday to the modern efficient industry of today.

Today underground mining is done in basically three ways: conventional, continuous with the use of the room-and-pillar method, and longwall methods. However, prior to the actual extraction consideration the coal to be mined must be reached and made accessible to the miner and machines.

Underground mining has in many cases been classified by method of seam access, which includes shaft, slope, drift, punch, and box-cut. This section will place more emphasis on the extraction aspects of coal mining. However, mining costs are predicated on seam accessibility; and a discussion of the methods by which a seam is reached is important.

Shaft Mining

Shaft mining (see Fig. 6.2-1) is a technique by which the coal is reached via a vertical shaft and is usually used for coals where seams are excessively deep; that is, greater than 300 ft below the surface or where some obstacle prevents another method of access. The main shaft is used to deliver men and supplies to the coal seam. Separate shafts are used to provide ventilation and, in some cases, for the removal of coal. The depth of the seam can in most cases be an indicator of the cost of mining. The deeper the seam, the higher the initial capital investment required to open a new mine.

Slope Mining

Slope mining (Fig. 6.2-2) is a method of reaching a seam of coal that is not normally deeper than 600–700 ft. Miners, equipment, and coal can be taken in and out by cable cars, track vehicles, and/or conveyor belts. This eliminates the need for expensive elevators and facilitates the movements to and from the coal seam. Ventilation for slope mines is also provided through separate fan shafts.

Drift Mining

Drift mining (Fig. 6.2-3) is used when there is an outcrop of a seam available for access and other mining methods are not usable. In this case the mine is developed with horizontal shafts, which are also used for ventilation. Coal, men, and equipment are brought in and out by the use of conveyors, trolleys, or other vehicles. The energy to move the coal and equipment is minimized, allowing for lower costs of operation and a shorter development lead time.

Punch Mining

Punch mining is a derivative of drift mining and is sometimes used in conjunction with surface mining techniques where all of the coal cannot be removed, such as surface mining to the high-wall. The

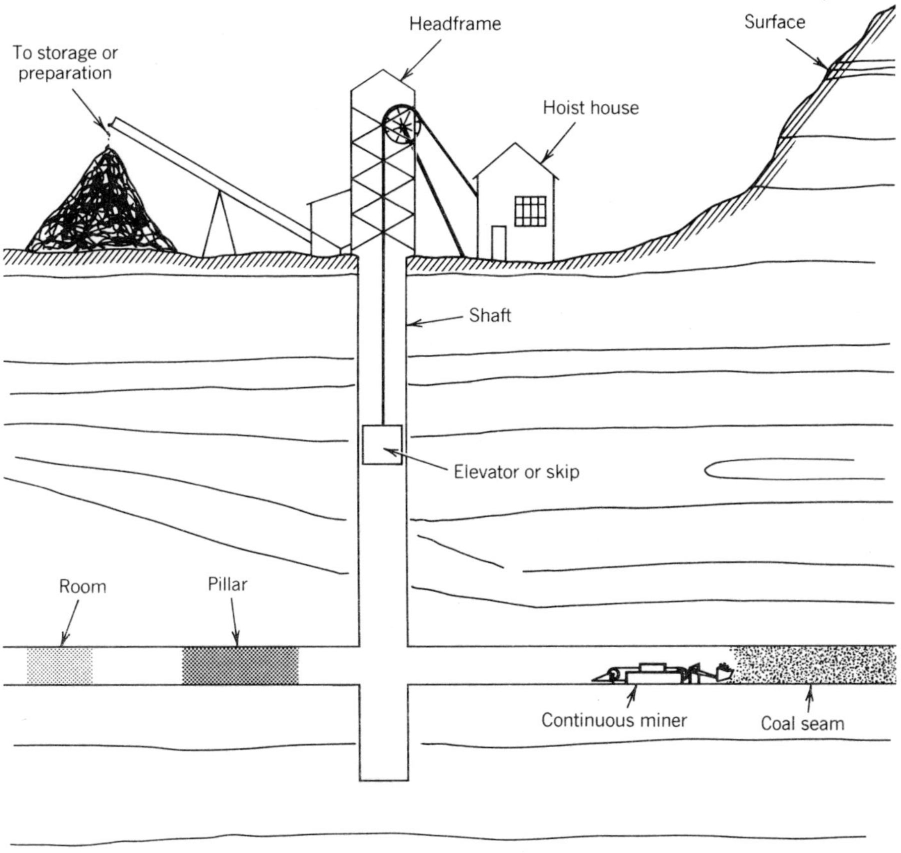

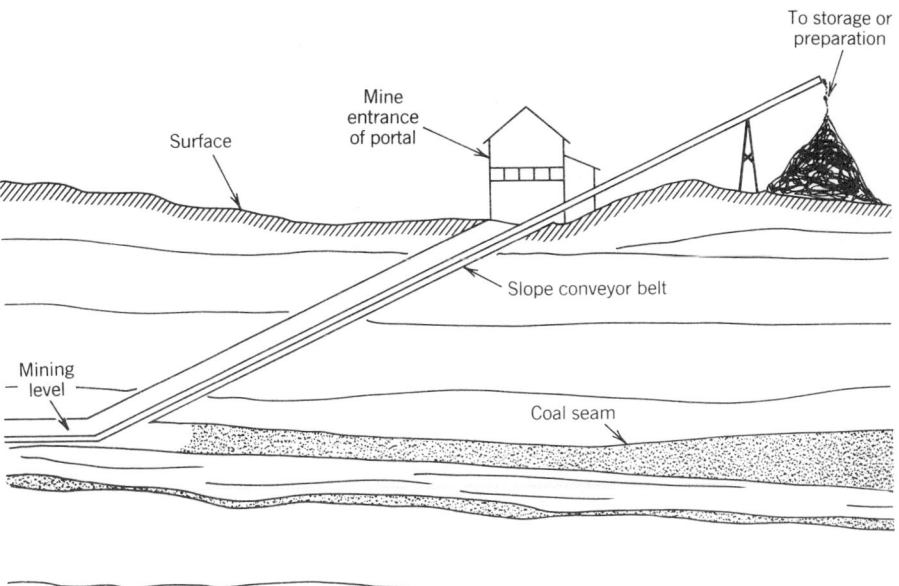

Fig. 6.2-1 Shaft mining.

To storage or preparation

Headframe

Surface

Hoist house

Shaft

Elevator or skip

Room

Pillar

Continuous miner

Coal seam

To storage or preparation

Mine entrance of portal

Surface

Slope conveyor belt

Mining level

Coal seam

Fig. 6.2-2 Slope mining.

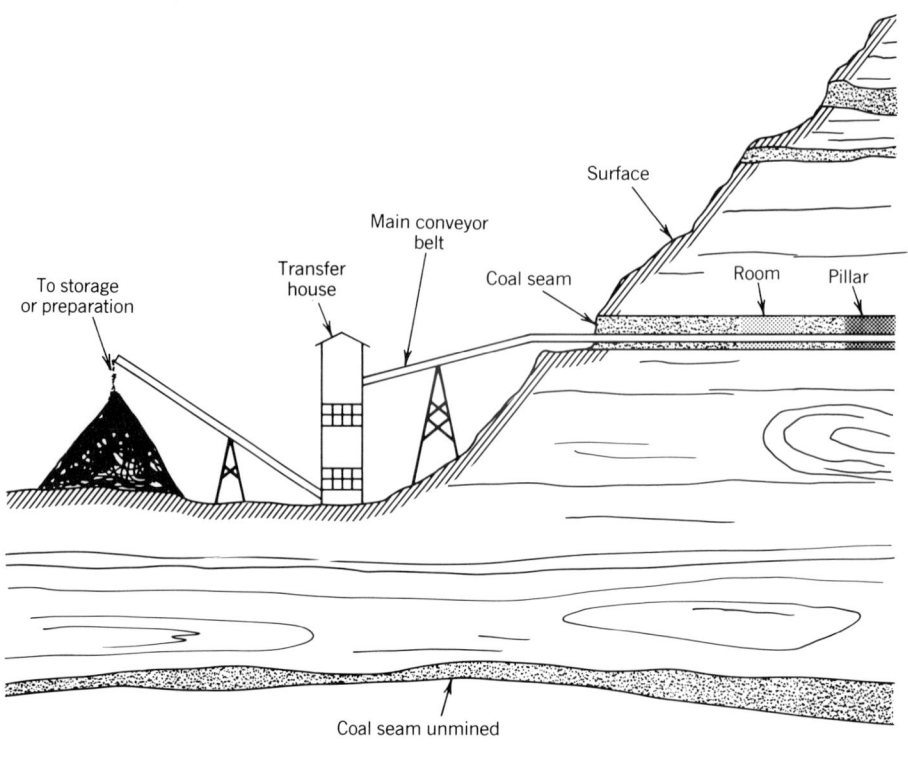

Fig. 6.2-3 Drift mining.

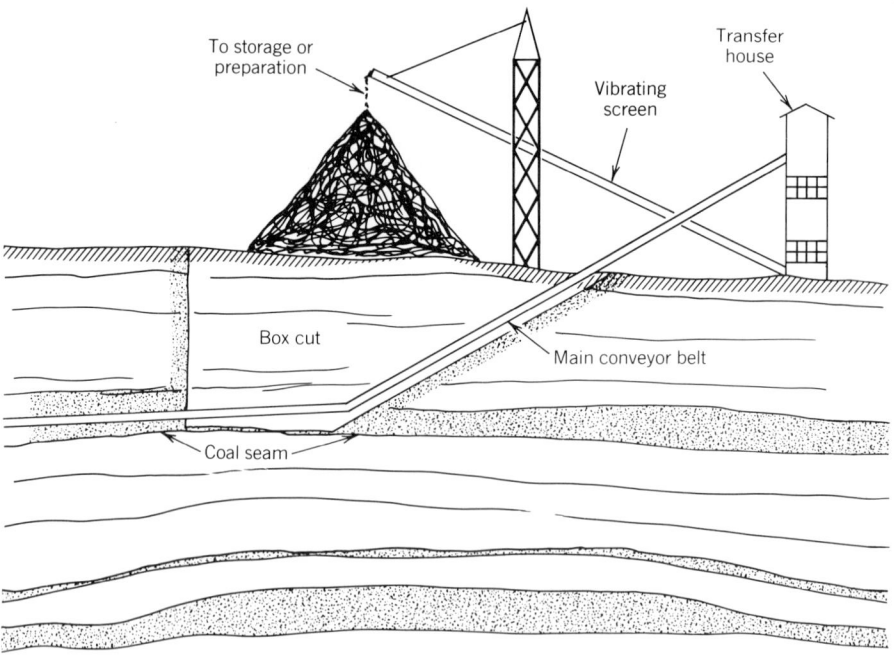

Fig. 6.2-4 Box cut mining.

characteristics of the drift mine apply; however, mining is restricted to a single shaft with the depth limited by ventilation requirements.

Box-Cut Mining

The box-cut mining method (Fig. 6.2-4) is another variation of drift mining. Here a box-type hole is dug to provide access to the seam. Drift mining techniques can then be employed. Ventilation can be provided for either by a separate shaft at the box-cut entry level or by vertical shafts at other locations. This type of mining is used when the seam has a high point near the surface that slopes off, making long-term surface mining impractical. The use of the box-cut minimizes the initial capital cost and provides operating flexibility similar to that of the drift mine.

6.2-2 Underground Mining Methods

Upon reaching the seam face, the method of coal extraction must be considered in light of the cost of mining in terms of labor, productivity, and capital costs as well as the effect on the physical and chemical characteristics of the coal being mined. The two main underground mining methods are the room-and-pillar method and the longwall method. These will be discussed with regard to the above considerations.

Room-and-Pillar Method

The room-and-pillar method of mining is such that upon reaching the coal seam a corridor or horizontal shaft of coal is extracted from which a series of rooms perpendicular to the corridor are extracted. These rooms are then connected by another corridor, leaving only a pillar for roof or top support. This method of mining can reclaim or extract 50–60% of the coal in a given seam depending on the size of the pillar that must be left for roof support.

Pillar size is a function of the top and bottom, the stratum immediately above and below the coal seam, the depth of the seam, and the height of the coal seam. In most cases these pillars do not provide all of the roof support. To provide the additional roof support necessary, beams and blocks are sometimes used. However, roof bolts and truss bolts are now used as standard support devices.

Depending on the overhead strata, a roof-bolting machine drills a hole 4–10 ft long into the upper strata. A resin is packed into the hole, which when compressed and heated as a result of the iron support bolt being screwed into the hole acts as a glue or cement. This acts to cement the upper strata together, thus supporting a larger area. A small beam at the end of the roof bolt assists in the support at the immediate surface of the roof. Roof bolt spacing, like roof bolt length, depends also on the upper strata characteristics.

During and upon completion of the corridor and rooms, fresh air for oxygen and circulation to remove any methane or other gases that might be hazardous to the miners must be provided. Health and safety standards must be maintained at all times. To do this, a series of curtains and walls are used to direct clean air past the point of mining and then return it via another route to the outside. The proper ventilation of an active mine is a complex function and is heavily regulated by government.

Another area that must be considered is rock dusting. Most coals, as a function of their coalification process, have some methane associated with them and within the seam. Methane is both explosive and toxic. Air circulation is used to maintain nonexplosive amounts of methane in the mine air. However, there are or can be pockets of methane, which are easily ignited, that can produce enough heat to ignite any coal dust that may have been shaken from the walls of the mine. Rock dust is used to minimize this potential. Limestone is pulverized and used as the rock dust. It works in two ways. First, during a puff caused by methane explosion limestone and coal dust are shaken from the walls. The limestone acts as a heat sink, lowering the temperature in the immediate area of the coal particle and preventing combustion. Second, the use of limestone rock dust on the walls of the mine helps reflect the light from the miners' lights and increases the overall brightness within the mine. As a result, a coal mine looks white underground rather than black as most people think.

Conventional Mining

Conventional mining is the process by which a special-design 12–17-ft cutting bar (chain saw) is used to undercut the coal seam with an approximately 4-in. high cut. The seam is then drilled at two levels and packed with water-based nonsparking explosives. The seam is then shot, allowing the first level to explode first and expand into the space generated by the cutter. The upper level of charge is then exploded. The intent of this method was to produce as few fines as possible, and it was capable of doing so until the spacing was regulated for safety purposes to a maximum of 3 ms. As a result, there is a high percentage of fines generated that are less than $\frac{1}{4}$ in., and of those there is a high percentage of less than 100-mesh particles. This is dependent on the coal characteristics and varies from one coal to another. The relatively high percentage of fines produced and the labor-intensive cutting, drilling, blasting, and removing steps tend to generate low productivity levels. These are the two main disadvantages of conventional mining.

Although there are a large amount of fines, this method produces the fewest number of total fines and results in the greatest amount of coarse coal for stoker applications and other size-sensitive applications. Another advantage of conventional mining is that only the coal is removed in the mining process. This includes any partings that may be present within the seam. However, this method does not take any of the top or bottom, which can be taken by other methods.

Shooting from the solid is a variation of the conventional mining technique. In this case the undercutting step is omitted. The depth of the coal to be shot is decreased and caution must be taken during the shooting process to minimize the potential of explosions because of pulverized coal dust. Because the coal does not have any space to expand into upon shooting, there are extremely large amounts of fines produced. This can be on the order of 50–70% of the total product.

Following the shooting of the coal, the coal is loaded onto shuttle cars and/or onto conveyors for removal from the mine. The coal is then delivered to storage, preparation plant, or loadout. In the mining process the roof bolters and rock dusters follow the miners.

Continuous Mining

Unlike conventional mining, the continuous mining method is an attempt to decrease the labor required per ton of coal and to increase productivity. In this case a machine similar to a movable rotating drum with teeth cuts into the seam, removing the coal without blasting. As the coal is dug from the seam, two gathering arms continuously move the coal to a conveyor to be loaded onto shuttle cars for removal from the mine. Roof bolting and rock dusting must follow the mining machines as they form a series of rooms and pillars for primary support. The only difference from conventional mining is the mining machine and coal removal equipment.

Continuous mining machines save time and increase productivity. This is the most important advantage, but another advantage is the lower percent of ultrafines (less than 100 mesh) produced as a result of a digging process in lieu of a blasting process, which causes very high percentages of the ultrafine coal to be produced. The digging process does produce a greater percent of fine coal, which decreases the amount of stoker or coarse coal available. This can be an advantage when optimum beneficiation is required.

The production of a high percentage of fines may be considered a disadvantage for the users of stoker-fired boilers or gasification equipment. The greatest disadvantage with continuous mining is the mining of top and bottom, which increases the total ash content of the raw coal. This increases the reject rate in the coal-cleaning cycle and the associated cost of handling, maintenance, and capacity. The quality variation of the coal is also increased. This can also cause severe problems with normal operation and availability of coal burning and conversion systems.

Longwall Mining

The longwall mining method is usually considered when increased productivity, lower reject rates, and increased recovery of reserves are required. With the longwall mining method approximately 70–90% of the reserves are recoverable. The need for roof bolting and other top support is limited to the two main access shafts along each side of the longwall section to be removed. The longwall is a series of hydraulic yielding jacks or self-advancing jack units that support the top above a moving conveyor and rotating cutting drum (shearer) or steel plow (planer), which remove the coal from the face and move it to the side where shuttle cars or conveyors collect it for extraction to the surface.

The longwall may be as much as 600 ft wide and may remove 100% of the coal from a 4000-ft run. The mining machine creeps forward as the coal is removed and the top is allowed to fall in behind the machine. Ventilation is simple, and the machine requires little operation labor. These, with 100% coal removal and minimum top and bottom mining, make up the main advantages of the longwall system. This method is not the answer to all mining problems. The application of a longwall is limited to very specific mining conditions.

The capital cost of a longwall mining system is significant and must be evaluated against productivity and recovery increases. The coal sizing produced with this mining method is similar to that produced with a continuous mining operation in the same type coal. Because the longwall mine runs parallel to the face of the seam, it can be designed to mine coal and only or very little top and bottom. However, because the top is allowed to fall in behind the miner, consideration must be given to assure adequate depth of mining to prevent surface subsidence. This usually limits the application of this method to mines that are deeper than 600 ft below the local surface. The shortwall mining system is a variation of the longwall system where shorter widths and runs are used and in some cases a continuous miner is used in lieu of a shearer or planer.

Dust Control

There is a large amount of dust generated from mining in a confined area. The effects on health and respiration and the explosive nature of the dust must be considered. These problems are handled in the same way, through the use of water and surfactant sprays. The surfactant is used to increase the wetting characteristics of the water and minimize the total water required to eliminate the dust problems. The amount of water used is insignificant when considering the amount of coal mined.

6.2-3 Surface Mining

Within the last 20 or 30 years, larger and larger pieces of earth-moving equipment have been produced, and as a result there has been an increase in the consideration of surface mining. Here coal seams close to the surface become economically accessible by removing the surface above the coal, exposing it for direct extraction with 100% reserve recovery. The general rule of thumb is for every foot of coal 10 ft of overburden can be removed. Depending on the quality, more or less overburden removal may actually be justified. In any case there seems to be a limit at about a 15 : 1 overburden-to-coal ratio where it is less expensive to surface mine than to deep mine.

Surface mining techniques recover 85–95% of the coal reserves and usually exhibit higher productivity than an equally sized deep mine. Items such as capital cost and reclamation become serious considerations when surface mining is selected.

The main objective of surface mining is to expose the coal seam and extract the coal with a scoop-type loader, shovel, rotary bucket loader, or coal auger. The most widely used method is the scoop- or bucket-type loader. During the surface mining operation, the coal may be exposed to heavy rain or adverse weather conditions. This can cause the retention of substantial amounts of moisture and a loss of Btu. However, the use of surface mining methods produces a larger proportion of coarse coal than does underground mining. This decreases the amount of surface moisture because of the lower total surface area and can be a benefit for systems that need low moisture content, such as pulverizers without preheated air for drying. Another benefit is the increased flexibility to produce larger quantities of sized coal for stoker-fired boilers and other specialty systems. The larger quantities of coarse coal require a substantial amount of crushing to liberate the inherent ash and small partings of ash trapped within the seam.

Surface-mined coal by its nature, being close to the surface, is visible in outcropings. As a result of this, some seams could exhibit a higher inherent moisture and increased level of oxidation, resulting in lower MAF Btu level and FSI. There is also the possibility of ash contamination from overburden not being completely removed or the extraction of some of the bottom with the coal. This can be a serious problem when the coal is used as raw coal.

Mountaintop Removal

In the East, where much of the high-quality coal has been mined through underground mining, lower-quality coals are becoming more attractive for combustion and other applications. One method of providing access to the coal is to remove the tops of the mountains and fill the valleys with the overburden, creating usable land in the process. The 10 : 1 overburden-to-coal ratio applies. Large quantities of overburden must be moved and surface mine reclamation standards must be met. The cost of reclamation can be the largest cost of surface mining in some areas.

Contour Mining

When a coal seam is accessible from the side of a mountain and the overburden is too great to warrant mountaintop removal, a method called contour mining is used to obtain the maximum reserve recovery without underground mining methods. Here the overburden is removed to a point where the economic advantage is reached for surface mining. The result is what is known as a High-Wall Reclamation must follow as for other surface mining methods in accord with current reclamation standards.

In some cases the seam quality may warrant the recovery of the coal under the mountain by punch mining, drift mining, or augering. Augering is a method where only a large diameter drill is used to auger a hole and remove the coal. This coal is much the consistency of continuously mined coal, having a very high percentage of fines. The use of auger-type mining procedures has been discouraged in many areas because of the safety problems associated with the holes remaining after the drilling.

Area Mining

When areas are relatively flat, with less than a 10% grade, and large areas of reserves are available to be recovered, area or dragline mining methods can be employed to remove overburden in a continuous, around-the-clock manner. The dragline in many cases is the most efficient method for mining coal. The size of the drag bucket can be larger than 60 to 70 cubic yards. The capital cost of such large pieces is very high, possibly the main disadvantage for the area mining method. Large shovels are sometimes used in lieu of draglines where the overburden is rocky and difficult to remove. In many cases there may be multiple layers of top soil, which must be removed and segregated for reclamation purposes. Reclamation to the original contours and conditions can become a severe problem, especially when acidity becomes a problem to contend with. Area mining methods are primarily used in the West and Midwest, although some may be found in other areas. Upon exposure of the coal, the coal is extracted like other surface-mined coal by scoop loaders and shovels.

6.2-4 Cost of Mining

The first considerations are always the seam thickness, depth, location, and top and bottom. These are usually evaluated through a core drilling and sampling program. From this information an analysis of alternative mining methods versus their relative costs can be completed. The coal seam location will determine the entry method, that is, shaft, slope, drift, or surface methods. The labor and productivity requirements will determine the type of mining equipment to be considered. The total reserves and recovery percent will determine the life expectancy and general parameters around which the total evaluation will be based.

Once the access and extraction methods are determined, the fixed costs for royalties, capital, overhead, union fees, taxes, and government regulation for running a mine can be added. The last area of concern is the yearly level of production, which is determined relative to the economic marketability of the resulting coal on a raw or washed basis. Upon completion of the above, the cost of that particular coal can be projected.

Three paragraphs are not sufficient to develop the total cost structure for coal mining. It must be noted that the cost for a similar operation, that is, seam thickness, depth, location, and capacity, is approximately the same and the resultant cost of the final raw coal is approximately the same and based upon these costs. The only areas where any appreciable savings can be found are in relation to the productivity of the labor force and the operation of the mine at full capacity.

6.2-5 Summary

Within the total energy system the coal mining aspect is very important from a cost-of-coal standpoint. However, it is also important from a coal sizing standpoint. The trend toward the improvement of productivity has increased the total percent fines content in the final raw coal product through the application of continuous and longwall mining equipment. Although this may be a problem with regard to the availability of sized coal for speciality applications, it could most definitely be an advantage with regard to the higher levels of beneficiation achievable with finer coal.

BIBLIOGRAPHY

Cassidy, Samuel (ed.) *Elements of Practical Coal Mining*. American Institute of Mining, Metallurgical, and Petroleum Engineers, Inc. (AIME), New York, 1973.

Stefanko, R. and Bisc, C. J. (eds), *Coal Mining Technology—Theory and Practice*, American Institute of Mining, Metallurgical and Petroleum Engineers, Inc. (AIME), New York, 1983.

6.3 COAL BENEFICIATION

Robert D. Bessette

Coal beneficiation, or preparation, as it is commonly referred to, is the maximization of coal quality by the use of mechanical, chemical, or other means of cleaning.

The United States' most abundant energy resource, coal, can be an available source of energy for the next two or three and possibly more centuries. Coal has the reputation of being a dirty solid fuel that requires a significantly greater amount of equipment and effort than gas or oil. An objective of coal beneficiation is to remove the contaminants deposited with the coal before, during, and after the coalification process.

An understanding of the ash content of coal is necessary both from a quality and distribution standpoint in order to obtain maximum economic beneficiation. In coal preparation the difference between the specific gravity of coal and that of ash is used to obtain a separation of coal and ash. Sulfur removal can be obtained as a by-product of the ash removal process because of the high specific gravity of pyritic sulfur. New technologies are being developed that are specifically designed to remove sulfur. In most cases these use chemical rather than physical methods of coal cleaning.

There is a limit to which a coal can be beneficiated. The rank of the coal and the carbon and volatile content of the base coal as laid down and coalified determines that limit. This can be expressed by the MAF Btu. From improved Btu and ash and moisture contents the cost of beneficiation can be obtained. Various types and methods of beneficiation will have different costs depending on the reject rate and carbon recovery of each. This can be predicted through washability testing.

6.3-1   Conventional Coal Cleaning

Conventional coal cleaning is that method of coal beneficiation that depends on the specific gravity of the coal and ash. Some advances in fine coal cleaning are being used based on the surface characteristics of the coal rather than the specific gravity of the coal. This is called Froth flotation.

Coal Sizing

Much rhetoric is used when discussing coal sizing and the meanings of the various terms have changed. For example, double-screened stoker coal originally meant a coal that had been screened once to produce a $1\frac{1}{2}$- or $1\frac{1}{4}$-in. top size and a $\frac{1}{4}$- or $\frac{3}{8}$-in. bottom size. Then the coal was screened a second time to remove the fines generated through degradation. Today double-screened stoker coal is a coal that is screened with two screens to produce the stoker product. No rescreening of the product is performed, and the resulting product is shipped with the fines caused by degradation.

It is important to understand what is meant by the definitions used regarding coal sizing. There is no single accepted list of terms used by all suppliers of coal. A list of specific terms used today is provided below. The top and bottom sizes are given such that not more than 5% of the coal, by weight, is above the top size and not more than 5% by weight of the coal is below the bottom size. The designations are as follows:

Furnace	6 in. $\times$ 4 in.
Egg	5 in. $\times$ 3 in.
Stove	4 in. $\times$ 3 in.
Nut	3 in. $\times$ 2 in.
Stoker	$1\frac{1}{2}$ in. $\times$ $\frac{3}{8}$ in. or $1\frac{1}{4}$ in. $\times$ $\frac{1}{4}$ in.
Modified stoker	Stoker with a percent fines
Modified	Maximum 20% fines
Controlled	Maximum 25–30% fines
Standard	Maximum 35–45% fines
Fines or carbon	$\frac{1}{4}$ or $\frac{3}{8}$ in. by 0 coal
Run-of-mine (ROM)	Total coal as mined
Crushed-run-of-mine (CROM)	ROM crushed to a top size specified
Blend	CROM minus stoker coal 50–70% fines
Resultant	CROM minus intermediate sizes 70–90% fines
Filter cake	28 mesh $\times$ 0 coal
Nut/slack	Any blend of coarse and fine coal

The size of coal specified should be as specific as possible relative to the top size required and the percent less than $\frac{1}{4}$ in. tolerable within the system. Flexibility will be achieved when the widest range of coal sizing is specified. The total system design should be structured with this in mind.

The Coal Preparation Plant

One way to understand the preparation plant is to look at it from a simple arrangement of crushing and screening to a complex arrangement of complete fine and coarse coal washing. Many old plants were designed to wash the coarse and fine coal and screen a complete slate of coal sizes for multiple, single-car loadout. Today these might be considered "dinosaur" preparation plants. The modern plant consists of advanced washing facilities with limited coal-sizing capabilities. Each of these will be discussed in some detail.

Crushing and Screening Plant

The coal crushing and screening facility has been the backbone of the coal preparation industry and is still being used today by many small producers. Figure 6.3-1 is a typical flow diagram for this type of facility. The objectives of this type of facility are improved handlability and providing the specific sizing required by some applications.

As can be seen in Fig. 6.3-1, the coal is received and unloaded into a receiving hopper and transported to a crusher designed to minimize fines production. The crusher crushes the coal to a preset top size, depending on the final product mix desired and the structure and hardness of the coal. In some cases the coal passes a manual picking table where pieces of slate and shale are removed by hand to improve the final product quality. A lump coal specialty product is also made this way. A rotary drum breaker has been used in place of a crusher where a degree of beneficiation can be achieved. Here the coal is fed into a rotating drum where it degrades and then passes through holes in the inner drum. The holes are preset to the top size required. The slate and other refuse contained with the coal are much harder and will not degrade in the time it passes through the breaker. The travel time through the breaker is dependent on the speed, the pitch, and the length of the breaker relative to the coal type. The refuse is removed from the end of the breaker and the coal (with some smaller pieces of refuse that fall through the holes) is passed on to the screens for screening or passed to the coal loadout for shipment to the end user.

This type of system has virtually 100% carbon recovery and very small reject rates when picking tables or rotary breakers are used. It is important to consider the type of mining being done and the relative coal sizing that will be delivered to the facility. These techniques of beneficiation could be advantageous if conventional and surface mining were used. However, if continuous or longwall mining are used, the high percentage of less than 2 in. of coal would tend to make the rotary breaker useless.

The benefits in this type of system lie in the low capital costs and its ability to produce a sized and, in some cases, limited beneficiated coal. The disadvantage of this type of system is that it does not reduce the amount of ash or moisture associated with the coal. There is also no provision for the

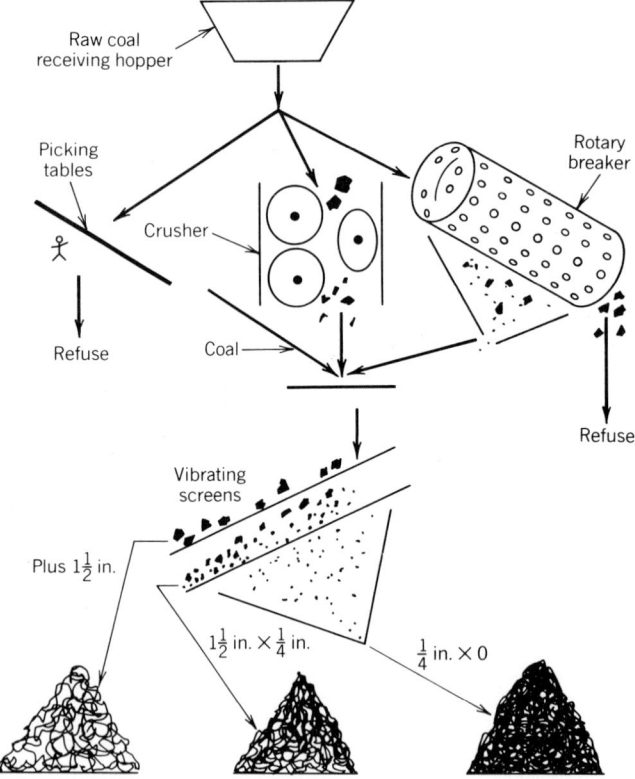

Fig. 6.3-1   Flow diagram for a simple crushing and screening operation.

minimization of coal variation and the coal as mined is that which is shipped to the user. Special consideration must be given to the coal selected for an application sensitive to variation in coal quality and ash characteristics when a raw product is used. These considerations evaluated with the costs of the product and its use within the system can be compared to the cost of washed and other products available for use at the facility.

The Older-Type Facility

The older-type facility, which is being replaced today, had the capability to wash and dry coal, producing a higher-quality product with a greater variety of product sizes. Figure 6.3-2 is a diagram of the typical old-type plant. An area of note with the older plant is the absence of a crusher. This allowed for the preparation of many size grades of coal. These plants included a washer, mainly a jig type washer or in some cases a heavy media bath for coarse coal, and concentrators for fine coal cleaning. Thermal dryers were used to reduce moisture increase due to washing, and to maximize Btu recovery in the final fine coal product. Coarse coal dewatering consisted mainly of vibrating and dewatering screens that were sufficient to lower the surface moisture content of the coal because of the relative lower total surface of the coarse coal. Differences in separation, dewatering, and sizing capabilities between the new and old cleaning facilities will be discussed. These plants are still used today, but they are too expensive to build and do not provide the level of beneficiation required for economic justification.

The Modern Cleaning Plant

As the cost of coal production rose, there was a need to increase carbon recovery, to maximize the recovery of coal, and to improve the overall quality. There have been many advances in coal cleaning with the advent of higher transport costs, production costs, and environmental regulations. These are

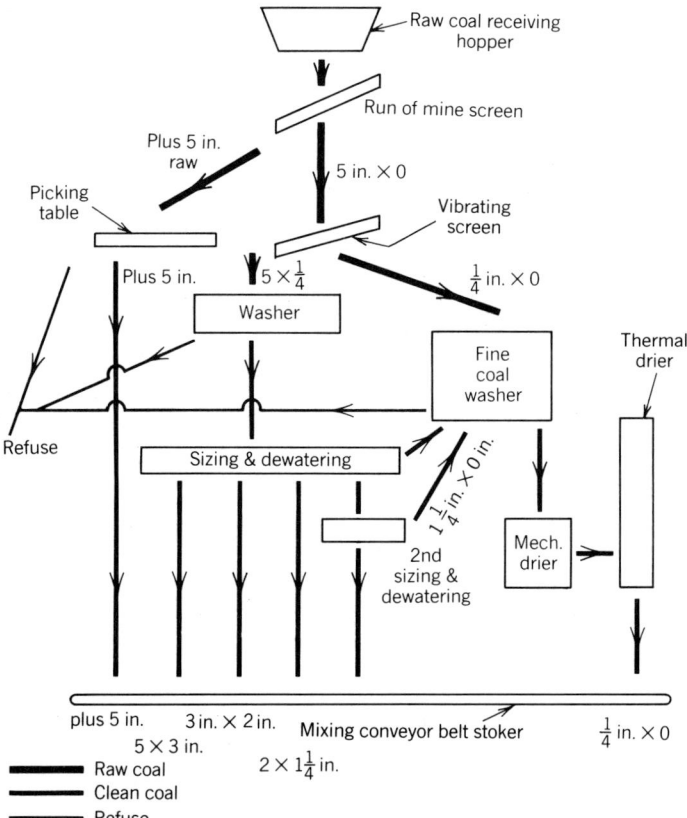

Fig. 6.3-2 Older type coal cleaning plant with multiple- or single-car loading and multisize capability.

driving the research further toward even better quality. The basic conventional coal-cleaning system consists of three separate sections: a coarse-coal section, a fine-coal section for $\frac{3}{8}$ in. by 28 mesh, and a fine-coal section for the 28 mesh by 0 coal. To achieve this maximum level of beneficiation, with maximum carbon recovery to minimize costs, the full coal washing system may be needed. Figure 6.3-3 is a flow diagram of what could be a state-of-the-art coal cleaning facility.

In this case the coarse coal is screened after primary crushing to a 6-in. top size. The resultant 6 in. $\times \frac{1}{4}$ in. coarse coal product is washed in a heavy media vessel to remove ash contaminants by gravity separation. The washed product is then passed over drying and dewatering screens where a natural stoker product is removed. No further drying of the stoker product is normally applied to minimize degradation. The coarse-coal fraction from the screening is passed through a secondary crusher to obtain a specific top size (2 in.), which may vary depending on the application. This coal can also be rescreened to obtain the maximum amount of stoker coal from a given run-of-mine coal.

The $\frac{1}{4}$- or $\frac{3}{8}$-in. screened coal is passed through an initial coal-cleaning device that removes the contamination from the $\frac{3}{8}$ in. by 28 mesh coal, and the 28 mesh by 0 coal is passed on for further cleaning. The cleaned portion of both steps is then dried through mechanical and possibly thermal dryers. The drying levels achievable for mechanical dryers of fine coal are from 6 to 10% and for thermal dryers from 3 to 5%.

The 28 mesh by 0 coal which is not effectively washed on concentrating tables is mixed with the 28 mesh by 0 screened portion and hydraulically transported to water cyclones where the ash fraction is rejected. The resultant cleaned coal is passed through another set of cyclones in which the plus 100 mesh coal is rejected to the clean coal side and the less than 100 mesh material is pumped to a thickener and then onto a settling pond. This product is passed through a centrifugal dryer, then

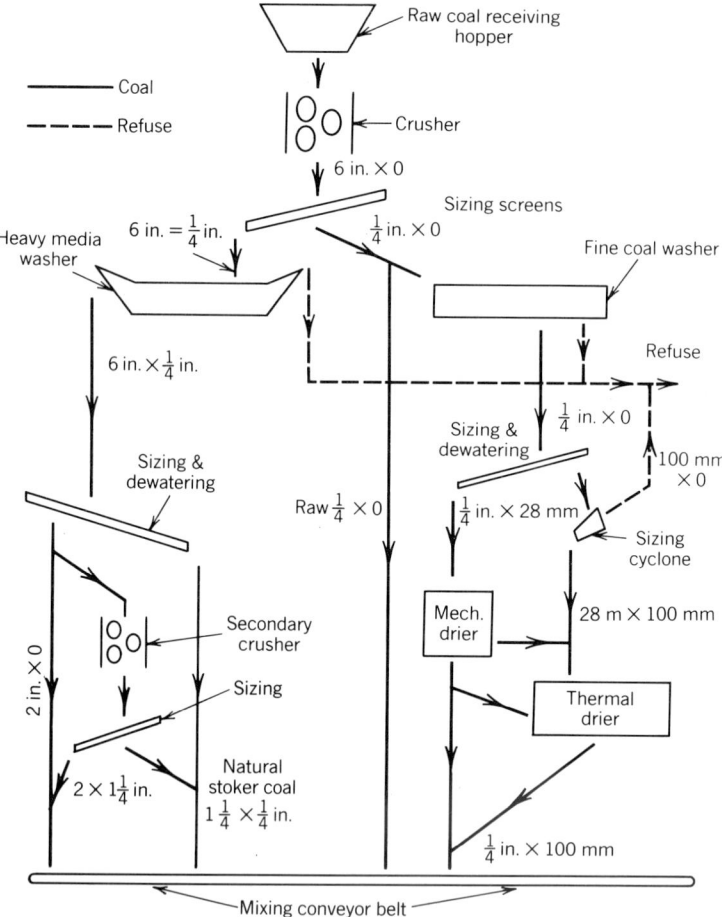

Fig. 6.3-3 Conventional coal cleaning plant with heavy media separation.

mixed with the $\frac{1}{4}$ in. by 28 mesh product entering the thermal dryer. This fine coal washed product is blended with the 2 in. by 0 fraction from the coarse coal washing system to produce the blend product. If the stoker coal is blended into this product, the resultant product is the total cleaned coal product or the crushed-run-of-mine (CROM) product. The various products produced are then transported to storage or loadout.

In this case there are no multiple sizing screens and a limited product slate is produced with the 100 mesh by 0 coal rejected. A modification of this system allows for the recovery of the 100 mesh by 0 coal. This is Froth flotation which can be applied to the 28 mesh by 0 product in lieu of cyclones. The use of flotation is expensive and uses the principle of surface chemistry difference between coal and ash to create a separation. A chemical agent is added to the system, which causes the coal to repel water and rise to the surface with air bubbles. These techniques will be discussed in more detail within the following sections.

Advanced Cleaning Systems

When the ash is distributed throughout a coal seam, it is difficult to remove most of the ash by coarse coal cleaning. In this case the coal must be crushed to a finer size to liberate the greatest amount of ash for removal. The finer the coal is crushed, the better the ash removal and the carbon recovery. Figure 6.3-4 is a flow diagram of a conventional system that addresses this consideration. All coal is crushed to a $1\frac{3}{4}$-in. maximum top size and passed through a heavy-media cyclone that produces a very good coal/ash separation. The resultant cleaned coal is rinsed to recover the specific gravity control agent. Advanced centrifugal dryers are used to decrease the moisture content to the 6–10% range. In this system, as with others, the 100 mesh by 0 coal is rejected, unless froth flotation systems are used to increase carbon recovery.

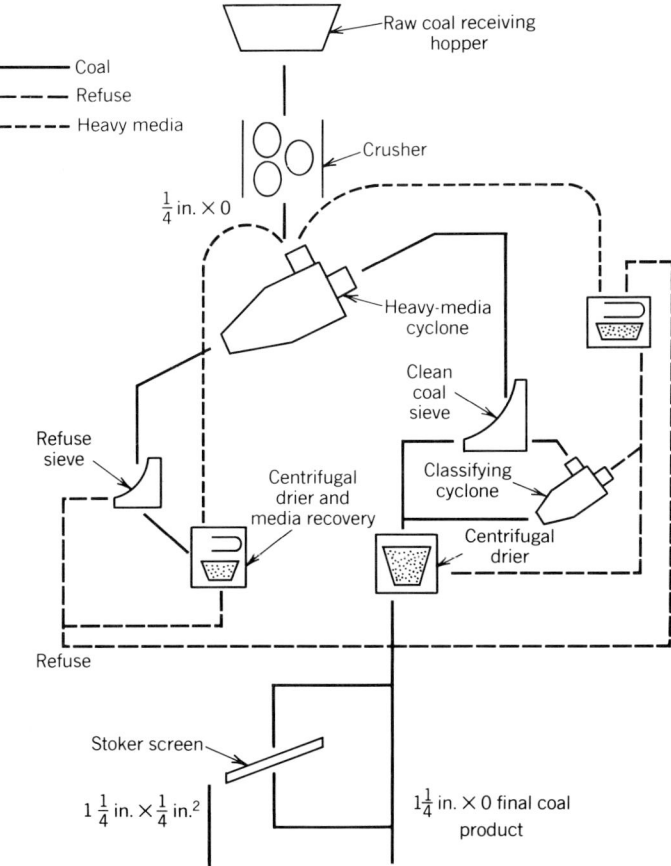

Fig. 6.3-4 Advanced conventional coal cleaning plant with maximum coal cleaning and carbon recovery.

The systems described above depend on coal washability and size distribution data for the coal as delivered to the preparation plant, which is dependent on the type of mining utilized at the given mine. The use of advanced cleaning technologies is becoming more widely accepted as the costs of transportation, mining, availability, maintenance and efficiency of utility, industrial systems, and conversion systems increase.

6.3-2   Conventional Beneficiation Equipment

The process by which conventional cleaning works is that of specific gravity separation. When coal with a specific gravity of 1.15–1.8, depending on rank and the coalification process, and ash with a specific gravity of 1.3–5.0 are washed together, a mechanical means can be used to achieve separation, thereby allowing the retention of the cleaned coal fraction and the rejection of the ash or refuse fraction of the coal. This discussion will consider bituminous coal with a specific gravity of approximately 1.3 and ash with a specific gravity of approximately 1.8. It must always be remembered that coal and ash are not homogeneous substances and that the ash distribution throughout the coal will have a serious impact on the ability to achieve finite separation. The following discussion of the types of equipment will be based on efficiency of separation and practical application to the various coal sizes that might be washed.

Jig Washers

If coal is allowed to flow over or through a chamber in a stream of water, it will have a tendency to settle out much like a rock in a deep pool of a river. The longer the path of travel over the chamber, the deeper the coal will settle, such that it could be deposited within the chamber and not follow the flow of water. The jig washer uses this principle to provide a means of separation. The heavier the particle, the faster the settling rate for that particle. Therefore, if a chamber is designed properly at a given flow rate, the heavier ash particles will settle faster than the coal and a separation can be made. Because this is a weight-dependent consideration, the size of the particles can have a serious effect on the coal/ash separation. A large piece of coal could in a low-flow situation settle and be rejected. The opposite is also true where a small piece of ash could be carried with the coal at a given flow rate. To maximize beneficiation and minimize ash carry-over, an increase in settling time must be achieved. In the jig this is accomplished through the use of a bellows system, which helps lift the flowing mass in steps to increase separation over distance and not decrease flow rate. The coal is in this way carried with the water and the ash is retained within the jig and discarded. Figure 6.3-5 is an illustration of this.

The jig-type washer is one of the oldest systems used today and can be the best system for specific coarse coal cleaning. Because of the flow to settling time ratio, it is not normally used when high percentages of fine coal are to be cleaned. It is very good when a narrow size range is to be cleaned and there is a sufficient specific gravity difference to achieve the necessary separation. This type of washer is not normally used when coal has a high inherent ash distribution or when there is the possibility of a clay buildup in the plant water system. The resulting increase in the specific gravity of the water will directly affect the settling rate and cleaning efficiency. A possible coating of a clay-type residue will also affect the combustion of the coal in stoker-fired boilers.

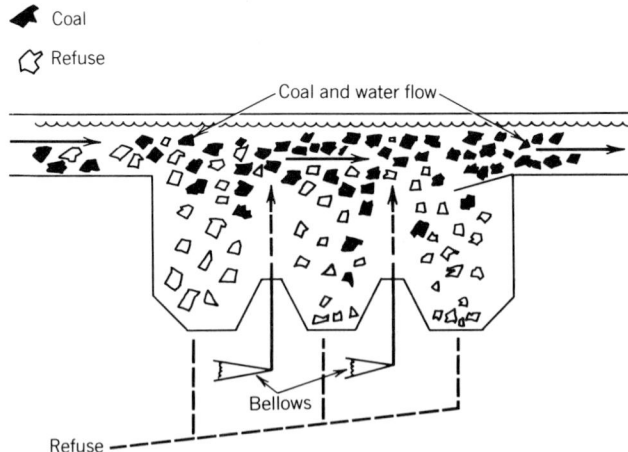

Fig. 6.3-5   Jig-type washer using coal and ash settling rate differences to obtain separation.

Heavy-Media Washers

Heavy-media or dense-media systems use magnetite or some other agent to increase the specific gravity of the coal/ash separating media. If the specific gravity of the separating media is greater than that of the coal, the coal will float and the heavier ash will sink, allowing for a fairly clean separation. Heavy-media washing is one of the simplest methods of washing, but it is susceptible to high fines content coals. High fines content can tend to increase the specific gravity of the media or be carried out with the cleaned coal, increasing the resultant ash content. The heavy-media washer is normally applied to the coarse-coal side of a preparation plant. The use of a two-stage washing system with one stage at a very low specific gravity and the second stage at a higher specific gravity can produce a very clean product and a middlings product which in total increases the total carbon recovery. Only systems capable of handling a high-ash coal, such as fluidized bed boilers or specifically designed conventional boilers, should consider the use of a middlings product. The ash distribution within the coal and the gravity difference will determine if this type of washing can be used.

Heavy-media washing is dependent on flow rate, the gravity control agent, and turbulence. Flow rate and turbulence must be balanced to allow the coal and ash to separate, yet not allow ash to be carried over with the cleaned coal. The gravity control agent must be controlled to maintain a given specific gravity level in the system, compensating for agent lost with the coal and refuse leaving the system. Agent selection is important with regard to cost and effect on the coal from either contamination or surface chemistry, which could be important for some coking or conversion systems.

The use of magnetite is generally accepted as the most advantageous because of its recoverability and inert nature. Magnetite losses of $\frac{1}{2}$ lb/ton of coal are acceptable in heavy-media systems. The magnetite is pulverized and controlled automatically to maintain the preset specific gravity needed to obtain optimum beneficiation. This is one of the important advantages of the heavy-media system. The other is in the ability of the system to generate a size-independent separation of a coal and an ash, minimizing carbon loss. Coal washability testing is most important when using these systems.

Hydro and Heavy-Media Cyclones

The use of a cyclone-type separator to separate particles of different specific gravities and weights is common and in recent years has been applied to coal-cleaning facilities. Figure 6.3-6 is an illustration of the process as applied to coal beneficiation. Cyclones achieve separation by dynamic means rather

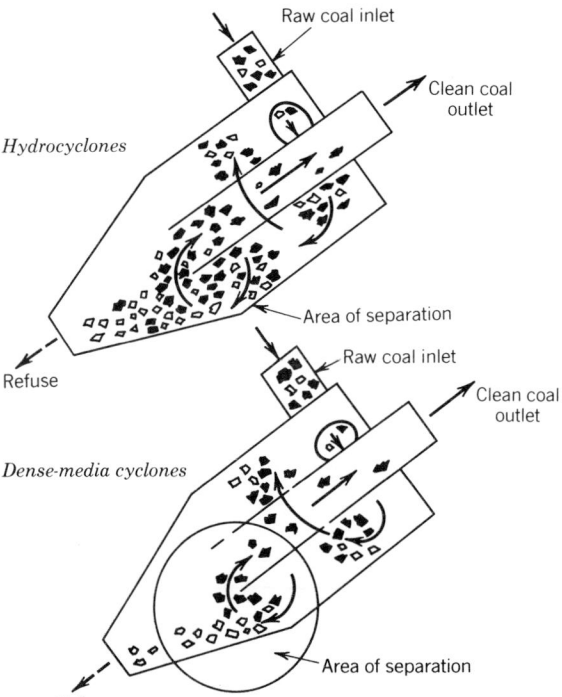

Fig. 6.3-6 Cyclone separation, the use of dynamic in lieu of static separation techniques.

than through a semistatic means like the above systems. In the water-only cyclone there is no clear-cut specific gravity separation. This system works much like the jig washer in which larger pieces of coal can be discarded as ash and the refuse is forced to the outside of the cyclone and out the bottom. Lighter ash particles can also be carried with the coal, thus increasing the ash content. The hydrocyclone is sensitive to particle size distribution and, as a result, is normally used on systems where the size of the raw coal to be washed is kept to within a narrow range. This system is also used when a definite specific gravity separation can be seen between the coal and the ash. The greater the separation, the more effective the cleaning.

Size control can be accomplished through the use of the hydrocyclone and in many cases the 100 mesh by 0 product is separated as overflow from a classifying cyclone. In all cases the coal feed to a cyclone must be maintained at a maximum of about $1\frac{3}{4}$ in. A smaller top size improves operation.

Heavy-media cyclones act as a hydrocyclone in principle, providing dynamic separation by weight and having the additional capability of gravity control within the system, making the coal lighter than the carrying media and assuring the best possible separation. Some small near-gravity material may be carried over with the coal; however, this is insignificant compared with other systems. The coal-cleaning cyclone principle is a simple technology to apply and there is very little problem in its application since the overflow is the clean coal product and the underflow is the refuse product. This simplifies the piping and handling systems within the preparation plant and is one of the main advantages of the cyclone systems.

The heavy-media cyclone has some of the same problems as the standard heavy-media bath system. However, the use of dynamic in lieu of static separation allows for better separation and better cleaning with higher carbon recovery and lower ash carry-over without regard to sizing. This is another and possibly the most important advantage of the heavy-media system. This method of heavy-media separation can be seen in Fig. 6.3-6 (circled section). Here is where the actual separation is achieved. With heavy media there should be a clear-cut separation and a distinct line of coal and ash. In reality there is a somewhat incongruous line of separation.

Concentrating Tables

An old and very effective method of washing coal is the concentrating table. This piece of equipment is primarily used for processing fine coal usually less than $\frac{3}{8}$ in. and greater than 28 mesh. The concentrating table acts to concentrate ash and coal along different sides of the table. Actual operation of the table is based on a specific gravity principle relative to flow and on the design of ribs that direct ash to one side of the table. The system consists of a table slightly inclined in the back with ribs built into the table such that the table vibrates in the direction of the ribs and a flow of water and coal is directed across the ribs. The ribs act to produce the separation, with the lighter coal being carried over the ribs to one side of the table while the ash that cannot be carried over the ribs is moved by the vibration to the other side of the table. Figure 6.3-7 is a simple sketch of this system.

The concentrating table is very susceptible to changes in flow rates and coal sizing relative to the degree of beneficiation required. For a specific coal with normal variation, the concentrating table is possibly the most trouble-free over the life of a preparation plant. This is possibly its major advantage. Larger coal sizes up to $1\frac{3}{4}$ in. top size coal has been used with limited success on some modified tables.

Froth Flotation

With the rising cost of mining and transporting it is becoming necessary to achieve maximum carbon recovery. The rejection of the 100 mesh by 0 product can prove to be costly if a substantial portion of

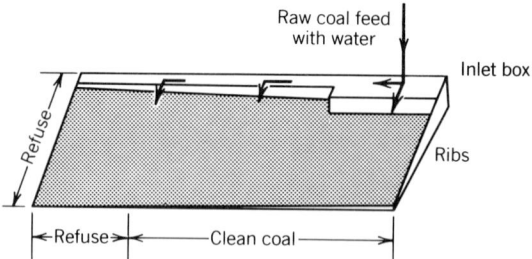

Fig. 6.3-7 The concentrating table is a table suspended and vibrating in such a way as to allow coal to pass over the ribs with the flowing water and the ash to follow the ribs for disposal.

the product is in this size range. As mentioned above, cleaning the 100 mesh by 0 coal is very difficult using size or weight techniques. Froth flotation was developed to provide a method of cleaning this product.

Froth flotation uses the surface characteristics of coal to promote separation. Chemicals are added to a water media to, first, increase the hydrophobic nature of coal and, second, create a foaming action on the water's surface. The addition of small air bubbles at the bottom of the tank causes the coal particles to collect these bubbles and rise to the surface as a foam. The coal cleaned by this method is usually no larger than $\frac{1}{2}$ mm or 28 mesh.

The ash, which is not as hydrophobic as coal, does not collect air bubbles and is removed with a liquid bleed from the flotation cell as makeup water and chemicals are added. These ash particles and water are transported to a thickener prior to being sent to large settling ponds for disposal. The foam containing the cleaned coal particles can be removed as it increases by using paddles or scrapers. The coal is defoamed and rinsed and then sent to vacuum filter for dewatering.

The equipment and chemical additives used for fine coal cleaning systems are relatively expensive and a complete evaluation of costs versus benefits is necessary. It should be noted that since the coal is very fine, most of the ash is available to be removed to produce an exceptionally clean product with high carbon recovery. One disadvantage of the froth flotation method is that the fine coal has very high surface area and a resulting very high moisture content associated with these products. This in on the order of 10–15% after the vacuum filter and must be accounted for in the final product for shipment. This higher moisture content and extremely good cleaning may be a direct benefit to coal water mixture preparation where the retained surface moisture becomes part of the 30% water in the mixture.

Dewatering Systems

One of the negative aspects of conventional coal cleaning is the water used in the washing cycle. Within the system 10% ash may be removed and 10% moisture may be added with a net gain in Btu of 0 in the final product. It is therefore imperative that some dewatering be applied to increase the resulting Btu after washing. Moisture is both surface and inherent. Only the surface moisture can be decreased effectively with mechanical dewatering systems. Inherent moisture can be decreased in some instances when the coal is of relatively low rank and thermal drying is employed.

Of all mechanical systems used for dewatering, vibrating screens are the most common. These screens vibrate and shake the water from the coal as it passes over the screen. They are fairly effective on coarse coal, however, because they remove only the excess surface moisture that drains from the coal. These screens are also used for sizing the coal for the different processes within the plant as well as for shipment. The sieve bend is a variation on the standard dewatering screen where the coal is forced to change directions on the screen. The centrifugal force exerted in the change of directions assists in the moisture removal and the 28 mesh by 0 coal and ash is also removed with the water as slurry.

The above might be considered a primary dewatering system for fine coal. Because of the high moisture content of fine coal a dynamic method of dewatering must be used to lower the surface moisture content. A centrifugal dryer is one of these methods. This dryer acts much like a spin dryer in a washing machine, forcing the water off the coal and through a screen-type basket. The centrifugal technique is very effective on $\frac{3}{8}$ in. by 28 mesh coal and with special design modifications, including fine mesh baskets, can reduce the moisture content of 28 mesh by 0 coal to acceptable levels. Depending on the hydrophobic characteristics of a given coal and the total surface area of the product, a 3–5% moisture level can be achieved. These systems are replacing the more expensive and more difficult to maintain vacuum filter.

Vacuum filters are mechanisms by which particle-laden slurries are dewatered using a vacuum to pull the water through a filter that will not allow the coal to pass. The coal is collected on the filter as a filter cake and removed as the filter rotates. These systems are capable of handling fine coal on the order of 28 mesh by 0 and are used in fine coal circuits, especially where froth flotation is included. They recover most of the coal and minimize the loss of coal to the rejects. The capital, operating, and maintenance costs of this type system are substantial and must be evaluated with the benefits derived.

Thermal drying uses an outside source of heat to dry the coal. It can decrease the surface moisture content of the coal to a level of from 1 to 3%. There are many types of dryers used today, all with different advantages and disadvantages. It is beyond the scope of this text to discuss these advantages and disadvantages in detail. In all cases, as long as the relative cost of the Btu input is less than the relative market value gain from the dryer coal product, there is a potential for using the thermal drying system.

Thermal dryers must meet environmental air emission regulations, which has caused costs to increase drastically. Thus, the high capital requirement has made evaluation of this application very difficult. However, as the cost of transportation increases, and the shipping of Btu in lieu of water becomes more valuable, the application of thermal dryers will be easier to justify.

Waste Disposal

Coal preparation plants are usually located at or near a mining site and have direct access to local disposal areas. It can be a direct benefit to a potential coal user to take advantage of this aspect of coal preparation. Assume 24,000 tons of 20% ash raw coal are delivered to a preparation plant and are washed with 30% rejects, leaving an 8% ash-washed coal. To burn the raw coal, a user would have to dispose of 4800 tons of ash in lieu of 1920 tons of ash from the washed coal. This is a savings of the total cost of disposal of 2880 tons of ash. Waste disposal at the mining site is eminently easier than the same disposal at a user's facility.

Coal Loadout

When considering the total energy system, major changes in attitude toward coal preparation and loadout have taken place in order to achieve the best possible delivered price at the user's facility. Many railroads have unit train (7000 or 10,000 tons) tariffs that can save substantial monies but require 4-h loading times. These are not possible with the older single-car loading systems and special systems have been designed to flood load continuously moving railcars. The emphasis has changed in such a way as to try to maximize loadout efficiency in lieu of multisize single-car flexibility. Track arrangements and storage capabilities are now designed to take advantage of the most favorable freight rates to the markets served. As freight and transportation costs continue to increase, more and more emphasis will be placed on the efficient use of this and other portions of the energy system sometimes taken for granted.

6.3-3   New Coal-Cleaning Systems

Many new physical coal-cleaning systems that are in the research and development stages are aimed at enhancing ash, sulfur, or moisture reduction as a result of coal crushing or using magnetic separation in lieu of dynamic liquid separation. Other research is being directed toward the improvement of separating media in lieu of magnetite to improve carbon recovery. Any discussion of these is beyond the scope of this work.

The area of major concern in technology development is that of chemical coal cleaning whereby not only pyritic sulfur is removed, but also a significant portion of the organically bound sulfur is removed. Many coals have a significant portion of the total sulfur bound in the organic state so many of these coals will not and cannot be beneficiated to a level that will allow their use in conventional combustion or conversion systems. In many instances the future of high-sulfur coals will lie with the potential of chemical cleaning.

Chemical Coal Cleaning

For most chemical coal cleaning, the coal is crushed to less than 14 mesh in top size and has been cleaned of a major portion of the pyritic sulfur by conventional methods or advanced conventional cleaning. Pyritic sulfur can hinder the chemical reactions needed to remove the organic sulfur. The coal is slurried with water or water/chemical additives and passed to a heated and/or pressurized sealed reactor where the sulfur is reacted with the chemical additives. The basic reaction is that of oxidizing the sulfur within the coal structure to form SO_x or a caustic replacement.

These systems are highly dependent on the coal structure and the time available for the sulfur to leach from the coal to the chemical water slurry. Research on these technologies is only in the infant stages of development and center around additives that only attack sulfur. Potential losses of chemical and costly sulfur recovery systems are of major concern. The resultant product is less than 14 mesh top size, which requires substantial dewatering and/or thermal drying to maximize Btu recovery or quality in the final coal.

The total sulfur removed with both conventional and chemical cleaning is on the order of 90–95% pyritic sulfur and anywhere from 15 to 60% of the organic sulfur depending on the process used. In general there is approximately 40% of the organic sulfur removed when using a leaching-type chemical process. When considering a carbon or Btu loss, there is anywhere from a negligible amount to a 10% loss as predicted by various technology developers.

The costs of chemical coal cleaning must be evaluated with its risks against the cost of flue gas desulfurization today. As these technologies develop in conjunction with physical coal cleaning, they may prove to be a better alternative than flue gas desulfurization with sludge disposal.

6.3-4   Cost of Coal Beneficiation

The cost of coal beneficiation is based on three basic elements: (1) process capital and operating costs per ton of raw coal input, (2) reject rate of refuse, and (3) coal or carbon recovery. These three basic elements determine the final cost of the coal at the mine. If there is no beneficiation, the cost of the coal is the cost of production plus an appropriate profit. The cost of prepared coal is somewhat more complex but can be understood by considering the interrelationships of the three basic cost elements of coal beneficiation.

As indicated above, the selection of a cleaning mode is dependent on the type of coal, its relative sizing, and its washability. Various capital and operation costs can be estimated for each system considered and added to the cost of the raw coal tonnage projected on a yearly basis. This may amount to a cost addition of $1 or $2 per ton of raw coal based on 1982 data. Once a preparation plant is constructed, the raw coal cost is increased to provide the proper payback on the facility. Hence, the true cost of raw versus washed coal at a facility with a cleaning plant lies in the cost of rejects and carbon loss as a function of the coal and washing process selected. An equation that describes this relationship is as follows:

$$C_w = \frac{C_r}{R_t \times R_c}$$

where C_w = cost of the washed coal in $/MMBtu
C_r = cost of the raw coal input (total mining, capital, and operating cost)
R_t = total product recovery percentage as clean coal
R_c = total carbon recovery index given by

$$R_c = \frac{\text{clean-coal Btu (as received)}}{\text{raw-coal Btu (as received)}}$$

Example 6.3-1. A raw coal with 12,000 Btu and 20% ash is to be processed yielding a 15,000-Btu cleaned coal product with 20% (100% ash) rejects. This is 100% carbon recovery with no change in moisture—a highly unlikely case. If the cost of the raw coal is given at $24 per ton or $1 per million Btu, then the above equation can be written as follows:

$$C_w = \frac{\$1/\text{MMBtu}}{0.8 \times 15,000\ \text{Btu}/12,000\ \text{Btu}}$$

or

$$C_w = \$1/\text{MMBtu}$$

At 15,000 Btu per pound there are 30 million Btu per ton of coal and the resultant clean coal cost is $30 per ton in lieu of the $24 per ton of raw coal. Although the cost per ton of cleaned coal is higher, the cost per Btu is equal. It can therefore be said that if there is 100% carbon or Btu recovery, the cost of beneficiation is limited to the capital and operating cost of the preparation plant.

When considering the reality of coal and coal cleaning, it is impossible to recover 100% of the coal carbon or Btu. Some carbon is always lost with the reject fraction of the coal as a result of the washing process. This is normally determined through detailed washability testing prior to the selection of the coal preparation plant design.

A second more realistic example may help clarify this consideration.

Example 6.3-2. The same raw coal is considered with 12,000 Btu/lb 20% ash at $24 per ton. The reject rate is 20%, leaving a 13,200 Btu/lb cleaned coal product. The above equation can then be written as follows:

$$C_w = \frac{\$1/\text{MMBtu}}{0.8 \times 13,200/12,000}$$

or

$$C_w = \frac{\$1/\text{MMBtu}}{0.8 \times 1.10} = \$1.136/\text{MMBtu}$$

For 26.4 million Btu per ton of coal the cost per ton at $1.136 per million Btu is $30.00 per ton. This is the same cost of washed coal on a per ton basis as in Example 6.3-1.

In both cases the cost per ton is dependent on the reject rate of the coal beneficiation plant. The reject rate affects the cost per ton of coal. However, the actual cost of preparation in cost per million Btu is dependent on the carbon loss as a result of washing, which is a function of the degree of separation achievable within the washing plant. The greater the degree of carbon recovery, the lower the cost of preparation. The cost of washing in Example 6.3-2 is $0.136 per million Btu or $3.59 per ton of washed coal. If a 4% moisture content is used in both cases, the MAF Btu is 15,789 and the resultant ash content in Example 6.3-2, will equal 12.4% in lieu of 20% for the raw coal, a decrease of 7.6% ash.

The justification for washing the coal to this level is determined, such that, only if the washed coal at 13,200 Btu and 12% ash can be sold for 13.6 cents per million Btu, more than the raw coal is it beneficial. The total cost for using washed coal must be considered relative to the total system cost effect.

6.3-5   Coal Sampling and Analysis

Coal preparation deals with the beneficiation of coal quality and a procedure for quality control must be used. One method for doing this is the ASTM procedure, which will be discussed in some detail. In all quality control measures the importance of a representative product sample cannot be overemphasized. This section will discuss the requirements of sampling and analysis to assure quality control and compliance with user specifications and other requirements. The following discussion will be considered from three points of view: that of a producer, that of a small industrial user, and that of a large utility or conversion facility.

Quality control is essential for optimum operation of a production and cleaning facility. A precise system of sampling, analysis, and reporting must be maintained. The small industrial user may not have the resources to develop, operate, and maintain an elaborate, high-precision system for sampling and may elect to use the producer's analysis as an indicator of coal quality. In most cases the industrial user is small and without ASTM facilities. Here the producer's analysis is usually best. The utility or conversion company using large quantities of coal may elect to develop its own ASTM quality control procedures. A 1% difference in coal quality can mean a $2 million penalty each year on a $200 million per year fuel bill. In these cases the utility sampling and analysis should be equal to or better than the coal producer's. The most important area of concern is the representativeness of the gross sample. Table 6.3-1 is a listing of ASTM procedures that can be followed. There are other methods used around the world; however, only ASTM will be discussed here.

Collection of the Gross Sample

ASTM D-2234 might be called the most important aspect of sampling and testing. It deals with the collection of the gross sample from which all tests and analyses are taken. The level of accuracy in the collection of the gross sample must be such that a $\pm 10\%$ variation of the ash content is achievable in 95 out of 100 samplings of a given coal shipment.

A sugar scoop of coal cannot be considered representative of a 100-ton coal shipment. Per D-2234 fifteen 6-lb increments of 2 in. $\times$ 0 washed coal must be taken and thirty-five 6-lb increments of 2 in. $\times$ 0 raw coal must be taken. This is because of the greater variability of the raw coal product. The weight of the increment is dependent on the size of the coal to be sampled. A $\frac{5}{8}$-in. product requires a 2-lb increment and a 6-in. product requires a 15-lb increment.

Each increment must be representative of the total coal produced and have an unbiased collection of the complete range of coal sizes, in the relative proportions, as found in the complete coal shipment —hence, the different increment weights required for the various coal sizes to be sampled. A sample is considered biased when a disproportionate amount of one size is present in the gross sample. This is evident when a high percentage of high-ash raw-coal fines are present in the gross sample, causing an unrealistically high reported ash content in the coal.

A general rule of thumb when evaluating the effectiveness of a coal-sampling system is to question the process and methodology: "Does every piece of coal have an equal chance of being in the final sample?" It if does not, the sample can be considered biased and not representative of the shipment.

Methods of Sampling

Although the number and size of the increments is important, the method by which they are taken is also important. There are four approved methods under ASTM D-2234. In order of degree of accuracy these are: (1) stop belt cut, (2) full stream cut, (3) part stream cut, and (4) stationary sampling. These can be taken either with or without human discretion and in a systematic or random manner. The ordering of these alternatives relative to accuracy is that a system that is without human discretion is better than one with human discretion and a systematic method is better than a random system.

When collecting coal samples, precautions must be taken to prevent the contamination of the coal from outside sources of ash. The same precautions must be taken with regard to moisture loss or gain due to rain, dust control sprays, or high humidity. Changes or contamination of the gross sample in any way will cause the sample to be nonrepresentative. This is of special concern when the as-received values of coal properties are to be used for quality guarantee purposes.

TABLE 6.3-1   ASTM TEST METHODS FOR COAL CHARACTERIZATION

Definition	Testing Method (ASTM)
Sampling	
Sampling	D 2234
Preparation for Analysis	D 2013
Coal Characteristics	
Total moisture, %	D 3302
Equilibrium moisture, %	D 1412
Btu per pound (dry)	D 2015
Moisture-ash-free Btu	D 3180
Hardgrove grindability index	D 409
Free-swelling index	D 720
Coal size (in.)	D 410
Fluidity or plastic properties	D 1812
Proximate Analysis (Dry), %	
Ash	D 3174
Volatile matter	D 3175
Fixed carbon	D 3172
Ultimate Analysis (Dry), %	
Carbon	D 3178
Hydrogen	D 3178
Nitrogen	D 3179
Chlorine	D 2361
Sulfur	D 3177
Ash	D 3174
Oxygen	By difference
Sulfur Forms (Dry), %	
Pyritic	D 2492
Organic	D 2492
Sulfate	D 2492
Ash Fusion[a]	
Initial Deformation (ID), °F	D 1857
Softening ($H = W$) °F	D 1857
Fusion ($H = \frac{1}{2}W$) °F	D 1858
Fluid °F	D 1857
Mineral Ash Analysis[b]	
Phosphorus entoxide, silica, ferric oxide, alumina, titania, lime, magnesia, sulfur trioxide, potassium oxide, sodium oxide, and undetermined	D 2795

[a] Reducing and oxidizing atmospheres.
[b] Wt. % ignited basis.

The stop belt cut is probably the most accurate method used to collect an increment. With this method the increment is usually much greater than the required increment, because when the belt is stopped, a portion $2\frac{1}{2}$ times the belt width is segregated and the coal within this area is removed to be incorporated with the gross sample. Care must be taken to collect every piece of coal and refuse within the area so as not to bias the increment. This method of sampling is very time consuming and requires a disruption of the loading process to collect the sample. However, the accuracy of the method provides for a method of bias testing other methods of collecting a sample.

The full stream cut is a method by which a sample of a flowing goal can be taken automatically to collect the proper increment. Requirements exist such that a cutter cannot cut a stream at a velocity greater than 18 in./s and the cutter opening must be at least $2\frac{1}{2}$ times the width of the largest piece of coal to be sampled. This method of sampling is the most widely used in industry for accurate sampling of coal. The only problem with the full stream cut is the potential for very large and unruly samples when large transfer belts and high-capacity loadouts are being sampled. To provide a sample that is handlable, secondary and in some cases tertiary sampling may be required prior to obtaining the final gross sample for analysis.

In cases requiring secondary or tertiary cutting, it must be remembered that there must be at least six secondary or tertiary increments for each primary or secondary cut. Each increment shall not be less than the weight described above for the size coal being sampled. In cases where tertiary sampling is needed, a crusher is normally added to decrease the size of the gross sample. When the crusher is used, the coal is crushed to a maximum of $\frac{5}{8}$ in. and not less than $\frac{1}{8}$ in. to prevent moisture loss or gain.

When a small belt with a relatively low flow rate is used, a normal cut may be taken, providing the cutter width meets the requirements and an increment of the appropriate weight is taken. This can be a very effective sampling method for the smaller industrial or mining facility using other systems.

The part stream cut is similar in most respects to the full stream cut in that it takes a sample of a moving stream of coal as the coal is being transferred from one point to another. The only difference is that it takes only one-half the stream assuming that the coal is distributed evenly across the belt. However, it is important to check the belt loading to assure an even distribution of coarse and fine coal on both halves of the belt. If this distribution does not exist, there can be a bias and the final sample will not be accurate.

The part stream cut is used when there are space limits that cannot be met. This method is not used when high precision is required or when there is a better sampler at some other point in the total system. Precautions must be taken to prevent the potential of a size distribution bias. In all other aspects the method is the same as a full stream cut.

Stationary sampling has been considered the backbone of industrial and small-facility sampling. This method of sampling, although not as accurate as the stop belt cut or the full stream cut, has the advantage of virtually no capital expense or elaborate arrangement. It is possible to collect a good sample provided some basic precautions are followed. The sample must be free of bias. The collecting shovel or scoop (see Fig. 6.3-8) must be at least $2\frac{1}{2}$ times the width of the largest piece of coal to be sampled and have sides equal to width of the largest piece of coal. The increment should be as specified in ASTM D-2234 for both number of increments and weight of each increment as part of the gross sample. An increment should *never* be taken from the surface of a coal to be sampled since segregation would most likely exist in this location. The basic increment should be taken at least 1 ft into the coal to be sampled. This type of sample does not collect a complete cross section and could generate a bias.

One method used to collect a better increment is the auger-type sampler (where a good cross section of coal pile is collected) railroad car, truck, or barge. These samplers collect a much larger gross sample and secondary splitting or cutting is necessary in order to obtain a handlable final gross sample.

Collection of the gross sample increments will vary for different sampling systems. It must always be remembered that coarse coal has a tendency to flow toward the outside of a pile, away from the point of impact, and fine coal has a tendency to stay at or near the point of impact. With this in mind, storage pile sampling should be done on the flat top and at various level on the sides of the pile to obtain the best gross sample. Figure 6.3-9 is an example of sampling for a conical pile without a flat top. The increment collection begins at about 3–4 ft above the base of the pile and circles the pile ending at about 3 or 4 ft below the top.

Car top sampling uses the same consideration relative to the center of the car. Figure 6.3-10 shows some alternatives.

Truck sampling is similar to car top sampling with a need for at least two increments per truck to obtain a fairly unbiased gross sample. These increments should be protected from moisture loss or gain until the total composites are collected for the gross sample. Since the minimum number of increments for a gross sample of washed coal is 15, two increments from eight trucks would have to be composited to obtain a standard gross sample. When large numbers of trucks are to be sampled, the sampling methodology should be agreed upon by all parties involved or affected by the outcome.

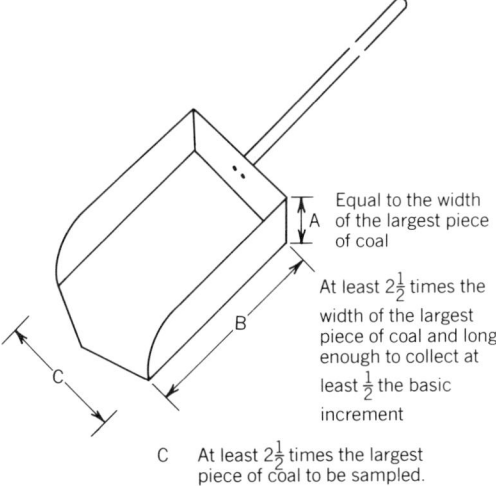

A — Equal to the width of the largest piece of coal

B — At least $2\frac{1}{2}$ times the width of the largest piece of coal and long enough to collect at least $\frac{1}{2}$ the basic increment

C — At least $2\frac{1}{2}$ times the largest piece of coal to be sampled.

Fig. 6.3-8 Sample shovel or scoop for stationary sampling of cars, trucks, stock piles, and so on.

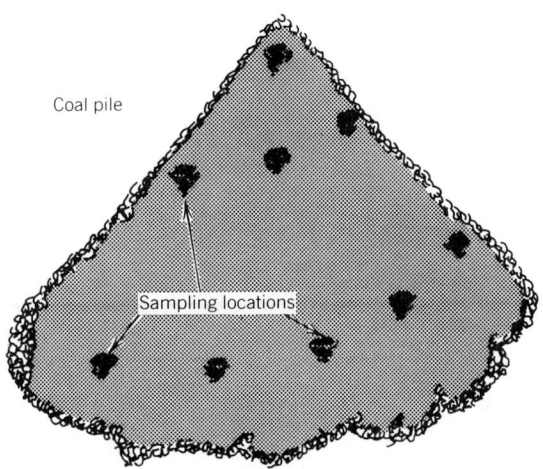

Coal pile

Sampling locations

Fig. 6.3-9 Coal pile sampling location for a conical pile. The total number of increments should equal that required per ASTM standards for the tons in the pile.

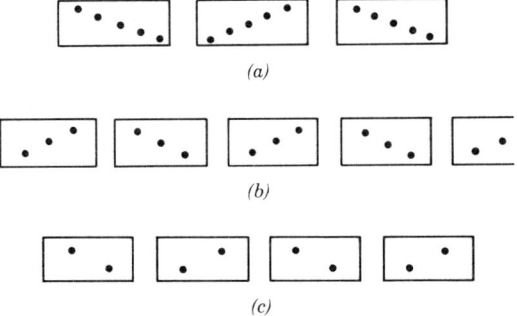

(a)

(b)

(c)

Fig. 6.3-10 Alternative sampling locations for car top sampling for different size shipments. (a) for three-car gross sample; (b) for five-car gross sample; (c) for 1000-ton or larger shipment. At least two increments per car should be taken and sampling methodology established prior to shipment. ●—sampling location.

Preparation of the Gross Sample for Analysis

ASTM D-2013 provides the basic guidelines for preparation of the gross sample for analysis. The basic concerns are similar to those relative to the collection of the gross sample:

1. The final sample must be representative and free from bias and contamination.
2. The moisture must be measured at each point where a change could occur.
3. The analysis must be repeatable.

In essence, any piece or piece of a piece of coal must have an equal opportunity to become part of the final sample without contamination or moisture change prior to weighing.

No coal sample may be split or reduced without primary crushing unless six increments of the proper weight for the size coal sampled are taken per ASTM D-2234. Following the primary crushing, unbiased sample splitting is required to obtain the final sample for analysis. Prior to any reduction of any gross sample, the sample should be weighed, air dried, and reweighed for moisture loss. Then and only then is reduction of the coal size allowed to be less then $\frac{1}{8}$ in. Prior to each crushing step the coal must be weighed, air dried, and reweighed for moisture loss. Upon reaching a coal particle size of 8 mesh the gross sample can then be split by riffling until a 500-g gross sample is available for pulverization and final analysis.

Analysis of Coal Characteristics

Many analyses must be performed on a coal to determine its applicability to various systems. The specific methodology for each of these analyses is beyond the scope of this work. ASTM Part 26, *Annual Book of Standards*, can provide all that is needed for each of these analyses. Table 6.3-1 is a listing of those analyses that should be incorporated into any request for coal.

Other analyses and calculable indexes can and are used for special consideration regarding critical areas of some combustion and conversion systems. A number of these areas will be discussed as part of the various combustion systems considered. Some of these are the silica/aluminia ratio, the dolomite ratio, and the T_{250} temperature.

6.3-6   Summary Coal Preparation and Analysis

Coal preparation is the art and science of cleaning coal. Unlike other production industries a coal company cannot make coal. Nature and the geologic coalification process have made coal and the most that a coal company can do is extract it from the ground, clean it of contaminants (ash, by mechanical means; sulfur, by mechanical and chemical means), and size it for various applications by the use of sizing screens and crushers.

Beneficiation is a function of the coal type and structure. The ash content and distribution within the seam can be determined through the use of washability testing and seam parting evaluation. A balance of the coal's washability relative to quality improvement and carbon loss through the washing process provides a method for the evaluation of coal preparation costs as they would compare with the benefits derived by using cleaned coal in steam generating, electric production, and conversion systems. As new technologies are developed and proven, carbon losses resulting from coal beneficiation will be minimized decreasing the cost of beneficiation on a dollars per million Btu bases.

If coal can be cleaned without carbon loss, there is only a minor cost of beneficiation due to the capital and operating costs of the preparation plant. In this case the cost of the prepared coal would be almost equal to the cost of the raw product. It is expected in the future that all coal will be prepared to the minimum ash and sulfur levels.

With the cost of coal, transportation, and operation increasing, it is becoming more and more necessary to assure quality control and have an accurate knowledge of the coal specifications prior to the utilization of the coal within the combustion or conversion system. These requirements have necessitated the initiation and use of high-accuracy sampling and testing procedures. There are many methods for doing this and many standardized practices have been developed around the world to assure some degree of uniformity. In the United States the standards developed by the American Society for Testing and Materials has become the guideline and standard for most work assuring the quality control needed to obtain the optimum results from the coal energy system.

BIBLIOGRAPHY

Annual Book of ASTM Standards, Part 26, Gaseous Fuels, Coal and Coke, Atmospheric Analysis. American Society for Testing and Materials, Philadelphia (revised annually).

Hutton C. A. and Gould R. N. *Cleaning Up Coal.* An Inform Book, Ballinger Publishing, Cambridge, MA, 1982.

Leonard J. W. and Mitchell D. R. (eds.) *Coal Preparation*, 3rd ed. The American Institute of Mining, Metallurgical, and Petroleum Engineers, Inc. (AIME), New York, 1968.

6.4   COAL TRANSPORTATION

Robert D. Bessette

Coal transportation is often thought of as the movement of coal from one point to another, without consideration for the interrelationships that exist between the coal, the producer, and the user. Unless there is direct involvement through the transportation company, a producer, or another company directly involved with the transportation process, the complete transportation system is taken for granted. The development of an integrated transportation system not only demands the location of two points but also a total logistical approach, including type of product to be transported, type of facilities at both receiving and shipping points, quantity of the product to be shipped, flexibility with regard to coal quality, and storage and system availability.

The evaluation of the total system begins with the consideration of the process or equipment. The second consideration is the coal requirements relative to the process. The third concern is the relative location of the using facility and the coal producer's location. Finally the transportation mode and model are considered. There is some evaluation relative to what has been done in the past at a given location; however, very little consideration has been given to the multitude of options available to the user for a given location. It is to the user's advantage to investigate the total range of options available to him.

With transportation costs rising as high as 40–60% of the delivered cost of coal, substantial savings can be obtained if the total logistics system is considered as an integral step in the total energy system. When an existing facility is considered, the location of that facility is fixed; the coal, with its alternative delivery modes and models, must be considered to achieve an optimum balance of cost and benefit relative to system requirements. The following section will evaluate various modes and models relative to their availability, equipment considerations, cost, and effect on coal characteristics.

6.4-1   Type of Coal Transportation

The types of coal transportation to be considered are truck, rail, barge, vessel, slurry, and miscellaneous modes.

Truck

The use of trucks has been increasing as a result of competition among other modes of transportation. The use of any method of transportation must be evaluated with regard to current and future cost and availability expectations. From a coal user's point of view, truck transportation could possibly be the most flexible mode of transportation. This is primarily a result of the simplification of the unloading—the dumping ability of the truck minimizes the need for elaborate handling and distribution systems to move coal to either active or long-term storage.

Trucks use the U.S. highway system to deliver coal to a user's facility and, as a result, are subject to the weight restrictions imposed as well as road-use taxes and ICC regulations for interstate movements. A trucker must have an ICC permit for the interstate movement of coal that is not owned by the trucker. In most cases truck weight limits are set at 80,000 lb, which usually equates to approximately a 25-ton coal load capacity. This is the main application limitation for truck transportation. A facility using 200,000 tons of coal per year over a 220-day delivery period would require approximately 8000 trucks over the course of the year. This would generate truck traffic of about 36 trucks per day and could cause a substantial amount of congestion if not accounted for in the initial consideration as a potential problem. Regulation with regard to truck weight restrictions can be a problem because of differing regulations from state to state, minimizing the total coal hauling capability of a truck over an interstate movement. One final consideration is the monitoring of truck weight and coal receipts at a given plant when truck deliveries may be made at odd hours or when plant personnel are not available to check in the trucks. Truck rates are dependent on the utilization of the truck—the greater the restriction placed on the trucker, the greater the rate.

The advantages of truck movement of coal are usually based on its flexibility, as mentioned above. The delivery time from mine to the user's facility is on the order of hours, instead of days as for other types of transportation. Hauling ranges vary from 1 to 2 mi to as many as 500–600 mi in the Western United States. Depending on competition from other modes, costs, and user requirements, it is not uncommon to see truck movements on the order of 200–400 mi. The short transportation time minimizes the problem of frozen coal during the winter months as well as minimizes the total amount of coal storage required at a given facility to maintain a secure level of operation. Because trucks are classified as self-unloaders, they, as such, do not require special unloading systems to unload the coal and transport it to the utilization equipment, active storage, or long-term storage. An added flexibility of truck movement is its ability to move over the U.S. highway system and normal roads, eliminating the need of rail or water access or multiple handling systems.

Truck movement of coal is possibly the most efficient mode of transportation for the small user and the least capital-intensive method of moving coal in all cases. For large users of coal where large quantities of coal must be moved, the higher operation cost of the truck movement, because of the small incremental tonnages moved per unit of power or energy, may outweigh the advantages of the lower capital costs and flexibilities. These must be considered on an individual plant basis.

When looking at competition among truckers, there are many independent truckers that can help to keep the operating costs down. However, union truckers may be required for a specific facility. This could affect the outcome of a critical evaluation of different transportation alternatives. In any case the availability of trucks for coal transportation would not be a problem since multiple trips are possible for short hauls on relatively short notice. New equipment, both tractors and trailers, are available on short notice from manufacturers. This may not be the case for other modes of transportation if a high coal demand market condition occurs.

The basic considerations for truck transportation application can be summarized as dependent on time, distance, quantity, flexibility, weather conditions, and whether an interstate or intrastate movement. Flexibility of unloading can be very valuable for a critical evaluation of various alternatives for a given plant.

Railroads

Rail transportation has been considered to be and is the the single largest transporter of coal within the United States and possibly the world. Coal production facilities are usually located in areas where access is limited to trucks or railroads. Because of the large tonnages produced at many facilities, rail is usually the primary means of transportation used. However, many producers have only one railroad serving a mine and as such are at the mercy of that railroad for railcars and power to move the coal produced. The railroads have maintained a monopoly structure with significant rate regulation. For many years there has been very little competition from other sources. With rail deregulation in 1980, and the resulting rapid escalation in prices, competition from trucks and barge modes has minimized the effect of deregulation in some cases. In other cases this is not so and can be considered the major disadvantage of the railroad system. The availability of competition must be considered.

When considering the use of rail transportation, transit times on the order of 4 to 20 days must be considered in lieu of hours for trucks. Rail transportation is capable of moving larger quantities of coal over distances ranging from 10 to 1500 mi. The basic equipment used for these movements is a 100-ton capacity bottom-dump hopper car. Other older 50- and 70-ton cars are also used. These cars can be shipped as single cars or as unit trains with up to 10,000 tons of total capacity. Many railroads offer a savings for unit train movements; however, the savings may vary with the size of the shipment, terrain to be crossed, and availability of power to pull the coal. An evaluation of a railroad and its rates is needed in all cases.

The major advantage of rail transportation is its ability to move large quantities of coal over long distances at economical prices. Other advantages are that railroads are available for direct shipment of coal to facilities with rail unloading equipment, the economies of scale obtainable for energy required per ton of coal shipped, control of coal weights, and management of shipments and receipts.

Disadvantages of rail transportation are mostly due to the monopolistic structure of a railroad, which has resulted in an unsympathetic attitude toward the needs of coal producers or users. This, associated with a very strong union, creates a number of disadvantages that must be evaluated against the advantages of the railroad mode of transportation. Strikes, bunching, demurrage, turn-around time, and weather are some of these disadvantages. Railroads are susceptible to strikes, and, as such, the user must be prepared with alternate modes for delivery or increased storage capability. In many cases coal unloading systems are limited to a number of cars per day. This can cause problems when a railroad bunches cars shipped and delivers a greater number than can be handled or when freight rates depend on car turn-around times, which must be met, and there is the possibility of delays because of severe weather and unloading problems. These can cause higher rates or the possibility of demurrage. Many railroads have provisions for bunching that can cause demurrage. These considerations must be taken as part of a critical evaluation for rail movements.

As coal utilization increases, the railroads will continue to be the backbone of the coal transportation system. Car and power supply needed to move the tonnages of coal produced should be adequate, barring any rapid surge in demand or major change in traffic requirements for a given area. Car shortages could develop in the short term; however, car availability for any long-term contract coal shipments should not be affected.

Railroads are more economical than trucks for long-distance transport of coal. In some cases they can be as competitive as trucks for the short haul when there are unit train rates developed for specific contract movements of coal to a user's facility. In all cases total cost must be considered based on capital equipment available for unloading and the distribution system peculiarities. These and the other system considerations must be evaluated as part of the total energy system.

River Barge

Many facilities designed to use coal have been or are being built either close to the coal reserves or on a river to use barge transportation as the primary transportation mode. Much like trucks, the barge is able to use an existing roadway of rivers with minimal capital investment per ton of coal moved. The barge is limited to the U.S. river system. This can be an advantage when the coal is moved downstream and the river is used for power, again decreasing the cost of transportation per ton of coal. The economies of barge movements are obtained when both shipper and receiver have direct access to the river for loading or unloading, respectively. Any combination movement to compensate for the user or producer being away from the river can affect the relative economic benefit for the barge movement.

Barge movements on the inland waterways are anywhere from 10 to 1000 mi long and can take from 1 to 15 days, much like rail. Barging is not normally used for facilities using less than 50,000 tons of coal per year because of the high cost of barge unloading. The barge sizes for river movements are from 600 to 3000 tons, with the normal being 1500 tons and moved in 15-barge tows (more than two unit trains). Three-thousand-ton barges are being developed to handle the larger tonnages of coal and to take advantage of the lower freight rates available.

The advantages of river movements are primarily those associated with economics, competition, and longer-term price stability. The ability to move larger quantities of coal with lower fuel and other operating costs is the main reason for this advantage. The longer the distance or the larger the quantity, the greater the advantage. For large users a barge-accessible river location has very attractive advantages.

The disadvantages of barge movements are mainly associated with river accessibility, loading, and unloading. Barges, unlike rail and truck, are not self-unloaders and must be manually or mechanically unloaded. Depending on the method used, there may be as much as 2–5% of the coal left in the barge. If a mine or user is not located on the river, coal must be delivered to the river through a transfer facility to the barge and unloaded through another transfer facility for movement to the user's plant. The longer the river portion of this transportation model, the more competitive the total cost will be relative to other forms of transportation. In winter and in times of low and high water on the river, system delays could occur or portions of the upper river system may be closed, hindering deliveries on a year-around schedule.

Barging of coal on the inland waterways is very economical. However, when considering the loading and unloading facilities, capital, and operating costs, only facilities that use in excess of 50,000 tons of coal per year can usually justify barge transportation of coal. Other areas of concern are the potential for an increase in the total moisture content of the coal from rain or river splashing. Barges do not drain as rail cars do, and this can cause problems. Covered barges can be used; however, these increase problems of availability, loading, and unloading.

In general, barge availability has not been a problem; however, as with rails, if demand for shipping coal by barge was to increase rapidly, there could be some delays and shortages. The resulting delay would be similar to those for rail with the major difference being that there is competition on the rivers, which allows for a greater flexibility. Unregulated competition should keep prices relative to demand.

Ocean Barge

Ocean barges are much like river barges except that they are larger (3000–30,000 net tons) and are usually covered to prevent the intake of water in the event of high waves. These barges are used in the intercoastal and coastal waters and not in the open ocean. The time requirements for ocean barge movements of coal will be on the order of from 1 to 3 weeks and, like a river barge, require approximately 50,000 tons of annual consumption for economic justification.

The advantages of the barge method are again associated with the economics of the single large shipment and economies of scale for large users. The disadvantages are associated with accessibility to the producing and user facilities, barge availability, and longer lead times. There are fewer barges available and a greater storage of coal is required to compensate for the transport time and possible delays. These problems are all exaggerated because of fewer ocean transloading facilities and competition with ocean vessels.

Lake and Ocean-Going Vessels

In the case of lake and ocean-going vessels, as conventional transportation models have been considered, the capacity or unit tonnage transport has increased and the resultant cost per net ton has decreased. Also, for each case, the respective coal loading and unloading system has increased in size, complexity, and capital cost. Lake vessels range from 12,000 to 50,000 net tons and ocean vessels from 15,000 to over 100,000 net tons. There are some vessels being built well in excess of the 100,000-ton size. Transportation time for lake vessels and ocean vessels are on the order of 2 to 4 weeks and 1 to 3 months, respectively. Some short dedicated movements involve much shorter time periods.

The advantages of ocean transportation lay mostly with the economies of size and the ability or necessity to use water as a means of transport. The disadvantages are mostly associated with the loading and unloading and must be considered in some detail in order to provide a basis for understanding the impact on the total system. When a vessel is to be considered for coal shipping, the draft must be known for the vessel and the port. All ports cannot serve all vessels. Port load-out capability must also be considered, as well as the number of other vessels to be loaded and possible delays to be expected. It must also be remembered that all coal cannot be delivered to all ports economically because of the railroad rate structure. It would require approximately 2400 trucks or 700 railcars to supply coal for one 60,000-ton vessel. There are very few transloading facilities that have storage to minimize the logistical problems associated with the movement of this quantity of coal from mine to port. Demurrage and port congestion are also areas of major concern.

The costs for this method of transportation can show significant savings over other methods when large quantities of coal must be moved. The advantages are found in the amount of power required to move a large quantity of coal in one increment. The economy of size is also evident in cost differences between vessels of different sizes. The disadvantages are found in the transit times and vessel availability for ports from which specific coals can be loaded. These must be evaluated within the total energy system.

Slurry Pipelines

The slurry pipeline is beginning to receive a considerable amount of attention as a potential alternative for all rail transportation. This system would be a capital-intensive method requiring very high capital investments but would provide for extremely low operating cost on a per ton of coal basis. These operating costs would be mainly for pumping and either dewatering of the coal at the site or for creating a mixture that could be burned without dewatering. Mixtures discussed to date for this mode of transportation range from 6-in. $\times$ 0 coal to 70% pulverized coal with 30% water. A slurry pipeline, in conjunction with an advanced conventional coal preparation plant, could minimize the total system costs. The best known pipeline in existence at this time is the Black Mesa pipeline where coal is slurried at Kayenta, Arizona, and delivered 273 mi to Mohave, Arizona, where it is dewatered and burned in a boiler. Pulverized coal slurries and coal–water mixtures will be discussed in greater detail in Section 6.10.

Problems associated with the use and implementation of this mode of transportation mainly deal with the very high capital cost, water availability, and government regulations favoring the railroad system within the United States. The capital costs, hindered by right-of-way considerations and a lack of government assistance, speak for themselves. However, water availability in the arid western areas could severely limit the applicability of slurry pipelines. Mixtures using methanol or other liquids are being developed to eliminate the need for water and increase the flexibility of application. Full application of pipeline transportation as a viable alternative to rail may not come until early in the 1990s.

Other modes of transportation have been used and considered in competition with truck and rail movements. Conveyors have been used for direct movements of coal from the mine to a power plant. The longest one in existence is on the order of 13 mi. The capital investment is very high and the maintenance of the system can be substantially more than a pipeline system. However, the operating costs for the continuous movement of coal is much less than that for other systems and can more than offset the disadvantages. Terrain, distance, and right-of-way must all be considered.

6.4-2   Transloading Facilities

Transportation is becoming more a part of the total energy system. As transportation alternatives are evaluated, it is evident that a multiple-mode system can be the most economical and provide the greatest flexibility with regard to storage, handling, and blending. The major aspect of the multimode transportation system is the transloading facility, with capabilities for one or more of the following: rail, truck or barge unloading and loading, coal screening, active and inactive storage, and multiple-coal blending. Many of these features can be found at various locations on the waterway systems, tide water, and gulf coastal areas.

The advantages of the transloading facilities are their flexibilities to store coal, blend coal to meet specifications, and to use the most effective combination of transportation alternatives, minimizing costs for the overall transportation portion of the total energy system. The ability to store coal has two direct benefits: (1) to the producer, which allows for storage away from the mine in the event of a strike and (2) for the end user, which can minimize on-site storage and the costs of maintaining high levels of inventory. The ability to blend coals from various producing facilities allows flexibility in the selection of coals to meet required specifications and quality while at the same time assuring the lowest possible costs.

The disadvantages of the transloading facility have to do with the effects on the physical characteristics of coal. The degradation due to handling is approximately 5% for each full handling

TABLE 6.4-1   TRANSPORTATION SUMMARY

Mode	Range (mi.)	Quantity (tons)	Transit Time[a]	Equipment Capacity	Benefits	Risks
Truck	5–400	25–200,000	1–10 h	23–35 tons	Lead time Self-unload Flexibility	Regulations No. of trucks Union
Rail	10–1500	50–4,000,000	4–30 days	50–100 tons	Tonnages Distance Cost	Monopoly Union Car availability Bunching Demurrage
Barge, river	10–1000	50,000 to unlimited	2–25 days	600–3000 tons	Tonnages Costs Stable prices	Limited access Weather Water level Availability
Barge, ocean	Intercoastal points	50,000 to unlimited	1–4 weeks	3000–30,000 tons	Tonnages Costs Stable prices	Port congestion Availability Access Power
Vessel, lake	Lake system	50,000–3,000,000	2–5 weeks	12,000–50,000 tons	Economics Stability	Access Lake season Availability
Vessel, ocean	Unlimited	100,000 to unlimited	1–3 months	15,000–100,000 tonnes	Economics	Congestion Demurrage Availability

[a] Time from mine to user's facility.

cycle from unloading to and out of storage to unloading. This holds true for coals with less than approximately 35% fines an will tend to decrease to about 3% per cycle as the fines increase and act as a buffer to further degradation. Where size control is of critical importance, the total number of handling cycles and the methods of loading at each point must be watched. The design of the facility will have an important bearing on the blending capabilities, through-put, and physical degradation within the system.

The number of transloading facilities are limited within the United States and the world. The scope of the equipment at these facilities may vary from a simple truck or barge load-out to a highly complex truck, rail (single or unit train), barge, or vessel center with storage and blending capabilities. The greater the flexibility, the greater the through-put charges and tonnages required to justify the higher capital cost for the facilities. In all cases transloading facilities provide an alternative means by which competition can be increased and economies of scale can be achieved when the total energy system is evaluated.

6.4-3   Transportation Summary

Coal transportation charges can amount to from 40 to 70% of the coal's delivered price at a using facility and must be taken as a very important consideration within the total energy system. The use of trucks, railroads, barges or vessels will be dependent on the distance, quantity, user and producer capabilities, and flexibilities required. Table 6.4-1 is a summary of various modes of transportation considerations and their advantages and disadvantages. It must always be kept in mind that the system selected should allow for maximum competition to minimize escalation in prices over the life of the using facility. The transportation system design should also take into consideration the most practical economies of scale, the lowest cost per ton mile, and provide benefits for the total system with regard to quality, storage, degradation, and transit times commensurate with the operation flexibilities needed.

BIBLIOGRAPHY

"National Energy Transportation Study, A Preliminary Report to the President by the Secretary of Transportation and the Secretary of Energy," Washington, D.C., 1980.

National Transportation Policies Through the Year 2000, *Final Report*. U.S. Government Printing Office, Washington, D.C., June 1979.

Proceedings of the Coal Technology Conference, Houston, TX, 1978, 1979, 1980, 1981 and 1982.

United Stated Code Annotated, *Title 49*, *Transportation* (partial revision). West Publishing Company, St. Paul, 1983.

6.5 COAL UNLOADING

Robert D. Bessette

Coal unloading is dependent on quantity, flowability, location, and two basic fundamental considerations of whether the coal is shipped by a self-unloading carrier or a carrier that must be manually unloaded, such as a barge. Concurrent with the transportation mode selection process, the method and cost of unloading must be considered. It is important that the unloading system cost and capabilities be economical and compatible with the method of transportation selected.

The flowability of coal is an important consideration when gravity unloading systems are used. Flowability depends on the coal moisture content, the fines content, and the type of contaminants such as ash, clay, and mud. The angle of repose of a coal pile is a good indicator of the flowability of a coal. The steeper the angle, the lower the probability that the coal will flow on its own. The higher the fines content, the greater the compaction, the greater the angle of repose, and the more difficult it is for the coal to flow on its own. Coal surface moisture tends to aggravate the above concerns, however, it must be remembered that coal will not stick to coal and that it is the contaminants that cause most of the problems associated with coal handling.

The following discussion will address the receipt and unloading of coal based on the mode of transportation selected, the quantity of coal, flowability, and location of the facility.

6.5-1 Gravity Discharge Systems

Truck Receiving and Unloading

A truck is possibly the easiest method for receiving coal. With a dumping bed truck, coal can be discharged either into ground storage or onto a 4-in. × 4-in. grizzly into the coal handling system. Truck traffic may be a problem when large tonnages of coal are delivered to one location on any one day. A traffic pattern should be developed to minimize truck dumping time by eliminating the need for trucks to back into restrictive areas or be detained by other plant road or rail traffic. Scales should be available for weighing trucks in and out as a check against load-out weigh bills. Provisions should also be available to collect a coal sample from each truck of coal received as a check against the producer analysis.

In order for a trucker to obtain the maximum economic advantage, the equipment must be used as much of the day as possible. The power house operator should be capable of handling one or two trucks on an evening shift; however, if many trucks are to be received, a full-time scale operator and yard foreman may be needed.

Environmental considerations for truck unloading are minimal, and in most cases dust suppression systems are not needed. If they are required, they are limited to a small enclosure to prevent the circulation of any dust. The road system for unloading the truck is an important part.

Rail Receiving and Unloading

Rail is the most widely used method of transportation and can be the most economical for larger quantities and longer distances. However, as the incremental tonnage received at any one time increases, the cost and complexity of the unloading system must also increase. Furthermore, the flexibility of dumping at either storage or into the handling system are very limited. Many variations can be applied for rail unloading, depending on whether the quantity being received is a single car or a unit train with set turn-around times.

The smaller facilities, which normally unload 1–8 cars per day, must consider the unloading time required and balance labor and equipment. A trestle or undertrack hopper can be the best methods for unloading in this case. Consideration must also be given to unloading rail car bottom hoppers with minimum repositioning of the car. This can cause substantial delays in the unloading process. A system designed to unload all three hoppers may be the most costly but will allow the complete unloading of a car with only one positioning.

The main constraints for the small facility are usually labor availability, capital, quantity to be unloaded per day, and the allowable time for unloading cars without demurrage. A balance must be achieved with consideration built into the system to account for coal flowability variations and other problems such as bunching and frozen coal.

The location of a facility can be an important factor when flexibility is being considered relative to the weather conditions to be encountered. Frozen coal and rain can create additional problems with unloading systems. When unloading coal from railroad cars, flowability is of prime concern, and compaction because of the shaking of the rail cars can hinder the ability and increase unloading time requirements. There are two basic methods used to assist in removing coal from railroad cars. One of these is the car shaker which when set on top or attached to the side of a car will vibrate and shake the car, loosening the coal and promoting flowability. The coal's physical characteristics must be considered when selecting a car shaker. Very high noise levels can be an environmental problem requiring noise suppression equipment or a building to enclose the unloading system. The other basic method is the hydraulic hoe, which can help push the coal out of the car and scrape the sides of any coal which would not flow on its own. This system could be used in conjunction with a shaker for frozen coal applications.

When unit trains are being used, the single-car unloading system will not be sufficient to meet the time and capacity handling requirements predicated by the rate tariff. If a 10,000-ton unit train is to be unloaded in 4 hours, about 120 cars have to be dumped, one every 2 minutes. Considering this, unloading coal via the above methods is impossible. A high-moisture 100% fines product may take a day to unload using this type system with a small shaker and single-hopper unloading point. With a heavy-duty shaker and a three-hopper unloading system, it could take as long as an hour or more.

Two methods for unloading unit trains are generally accepted as being the most economical: the rapid-discharge bottom dump and the rotary dumper. The rapid-discharge car is designed in such a way that the railcar bottom opens completely, allowing the coal to fall with very little hindrance. This is accomplished either by stopping the car over an unloading hopper or by using mechanical door openers to open the doors over the hopper while the train is moving. This latter case, using the continuously moving train, requires minimum trackage, a loop track, and a coal with a relatively high degree of flowability.

When the ability to fully empty a car is in question, the use of a rotary dumping device is usually preferred. The rotary dumper is designed in such a way that one or two cars are positioned in the dumper, clamped into position, and then rotated with the track 120–140°, emptying the coal from the top of the car. This dumping action allows most of the coal to be dumped by gravity. The cars and track are returned to their upright position and new cars are spotted. Using this method, approximately two cars can be unloaded every 2 minutes.

Positioning of the cars is accomplished by a pulley system or a dedicated engine or special car mover. When unit train systems are used, space must be provided to handle a maximum number of loaded cars, as well as the same number of empties. Track layout can become the limiting factor for unloading time. When short unloading times and large tonnages are involved, consideration must be given to the use of dedicated cars, swivel couplings, and loop tracks. Here the main engine pulls the cars and spots the cars in the rotary dumper. The cars are then rotated without the couplings being detached, saving time, and promoting a rapid turn-around time for the dedicated cars. These systems can become extremely complicated and costly and must be considered in light of the economic savings that can be acheived through a reduction in freight rates.

A rotary dumping system or a rapid unloading system are susceptible to environmental dust control measures because of the rapid dumping action. This usually requires an enclosure and water sprays to suppress the dust and noise created in the dumping process. if water sprays are used for dust control prior to the sampling system, consideration must be given to a possible moisture bias in the gross sample to be analyzed.

The cost associated with railcar unloading systems are mostly associated with increasing the capabilities to minimize unloading time, decreasing labor required per ton of coal, and minimizing the space required to accomplish the unloading of the coal. The larger the system, the greater the economies of scale. In any case a complete evaluation of the benefits on a cost per ton of coal must be completed with regard to rate savings, demurrage, labor savings, and capital costs.

Self-Unloading Barges or Vessels

The last in what is considered to be a gravity-discharge-type unloading system is the self-unloading barge or vessel. Here the carriers are equipped with a built-in reclamation system that allows coal to be unloaded directly onto a dock, eliminating the need for an external unloading system. This, plus the ability to use water as the main means of transport, are the main advantages.

There are a number of limitations that must be considered when a self-unloader is to be used. If no other forms of unloading are provided for, the receipt of coal to that facility will be limited to the availibility of the more costly, less available, self-unloading carriers. This could pose a difficult problem. Sufficient dock storage must be maintained to allow for the complete unloading of the carrier upon arrival. The interface with the coal handling system can require another coal handling cycle and increase degradation of the coal.

6.5-2   Manual Unloading Systems

Manual unloading systems require the coal to be extracted by shoveling or digging with external equipment. Here gravity must be overcome to remove the coal. Manual unloading is usually associated with either barges or vessels.

Barge Unloading Systems

The primary system for unloading the single barge is the clamshell-type shovel that will unload coal from the barge to a truck, storage pile, or coal handling system. The same considerations must be given with regard to evaluating the barge unloading system as those for evaluating the rail or truck systems. These are the quantity, the time allowed, and the space available to store one complete shipment as it is unloaded upon delivery. When large numbers of barges are received in 10- to 15-barge tows, there can be justification for considering the more expensive rotary-bucket-type unloaders. These operate as continuous unloading systems and can minimize the turn-around time required for unloading a shipment and minimizing any possible demurrage.

Systems designed for continuous unloading can in some cases minimize the problems with coal being left in the barges. It is very difficult to remove all the coal. This loss of coal can be substantial and in some cases justify a barge cleaning operation. When this is used, a small tractor is lowered into the barge and the coal is pushed into a pile for removal by the clamshell. Where time is critical, many companies specify clean barges and return the barges with some loss of coal. Services can be contracted to clean the barges. When a dedicated barge movement is considered, there is a one-time loss of coal, which stays with the barge until the barge is periodically cleaned.

Vessel Unloading Systems

In the future vessels having capacities of 100,000 metric tons or more will require large unloading facilities. There must also be sufficient space available at or near the dock to unload the complete vessel. The clamshell is the most widely used type of unloading equipment and possibly the most convenient because of the compartmentalized structure of the vessels. Access to the cargo hold is through openings in the deck, which are covered during transport. These covers must be removed during the unloading process and can create a problem in removing all the coal from the holds. Like barge unloading, a small tractor is lowered into the hold for cleaning up the coal prior to the vessel leaving the dock.

Constraints such as demurrage, port and dock draft, and vessel movement requirements can have an impact on the overall economics of the unloading system. The larger the vessels to be unloaded, the greater the capital cost of the systems, and the greater the operating costs associated with the system if the system is not used a large percentage of the time.

Combination Systems

In order to promote competition, which minimizes cost escalation because of a monopoly control over a single mode of transportation, coal unloading systems must be designed to allow for maximum flexibility to receive varying quantities of coal from varying modes of transportation. When railroads are the prime method of coal receipt, provisions should be made for trucks to be able to dump on the railcar unloading hopper at the small facility or into the coal handling system at another location. Where coal is received by water, it is essential to have a backup coal receiving system capable of taking coal by either railroad or truck movements. A simple three-car hopper unloading system with a shaker and a proper interface with the system's coal handling system is more than sufficient. This type of unloading system is also capable of handling trucks, although most trucks can unload directly into the long- or short-term storage. An assurance of coal supply must be maintained with coal unloading flexibility. The facility receiving coal by truck without access to railroads or water transportation must maintain a higher level of storage capacity to compensate for the possibility of strikes or work stoppages.

6.5-3   Frozen Coal

A frozen car of 2-in. $\times$ 0 coal could take five men or more one day to unload using salamander-type heaters with two or three drums of expensive fuel. In locations such as northern Michigan it could take even longer and delay the unloading of other cars received that same day, and thus begin accumulating demurrage after the second day of not being unloaded. This can amount to a substantial demurrage bill plus a loss of coal receipts and possible production downtime. With a high probability of freezing, seasonal storage increases should be considered.

Coal does not freeze—water does and freezes to the coal. The amount of surface moisture in a given coal shipment is dependent on the total surface area of the coal or the percent of coal fines, their type, and the weather encountered at loading and during transportation. The greater the moisture, the colder the temperature, the harder the freeze, and the more difficult it is to unload the coal regardless of the method of transportation used.

Car Shakers

For coals with low surface area and low surface moisture, like stoker coals or modified stoker coals (which are basically free flowing), a car shaker may be all that is needed. However, when there is a potential for snow and rain during transit time, plus severe cold, a car shaker may be ineffective.

The freezing action in a railroad car is such that the water tends to migrate to the sides and bottom of the car and freeze from the outside in, forming a shell which may or may not be broken by a shaker.

Warming Sheds

When transit times are minimum and rapid unloading systems are used, warming sheds may be all that is needed. In this case the outside of the car is thawed to allow for the coal to be dumped. The time required for thawing will depend on the frozen condition of the coal. In severe cold this may be insufficient to promote timely coal unloading.

The capital cost of the shaker can be amortized based on a full year's use. The capital cost of the warming shed must be written off only in the months it is used, and these capital costs are much higher than the shaker. There are also higher operating costs for electric radiant heaters or fuel for gas- or oil-fired heaters. The cost of operating this type of system must be evaluated relative to the severity of the freezing and the amount of coal to be unloaded. It is used mostly by utility systems.

Freeze Prevention Solutions

The above methods attack ice after it has formed. Salts and oils are used to minimize the freezing tendency of water or the water's ability to freeze to the coal. The addition of sodium or calcium chlorides to the coal to be shipped depresses the freezing temperature of water to approximately 20°F. However, these additives may cause fouling and/or corrosion during handling and combustion.

The addition of oil to the coal surface minimizes problems associated with water freezing and allows the unloading function to proceed smoothly using a car shaker. Oil begins to get viscous at about 20°F and can become ineffective at temperatures lower than this. The application of oil is usually in the 2-qt to 1-gal per ton range, depending on the fines content of the coal. Stoker coal, $1\frac{1}{4}$ in. $\times \frac{1}{4}$ in., with less than 10% fines is normally treated with 2 qt of oil per ton of coal. A CROM would be treated with about 3 qt per ton. At 150,000 Btu per gallon of oil and 25 million Btu per ton of coal, the Btu addition is substantially less than 1% and considered negligible.

Some chemical or conditioning agents have been developed that lower the oil's effective-use temperature to about −20°F. This extends the application range of oil; however, it is only marginally effective for high fines, high moisture, and very low temperatures. In order for the oil spray to be effective, the oil must be sprayed evenly over the surface of the coal. If not, patchy freezing could occur, eliminating the expected benefits of the oil.

Freeze Conditioning Agents

During the winter of 1979 and 1980, extreme low temperatures were recorded and severe problems with coal unloading were also experienced by all companies in the north using the types of handling and unloading systems described above. As a result, freeze conditioning agents were developed to affect the structure and hardness of ice. These conditioning agents work on the principle of an antifreeze additive, which prevents the ice from freezing as a solid by allowing the water to freeze, concentrating the antifreeze until a temperature is reached where the mixture will not freeze. In this way the water freezes in a crystalline structure, allowing it to be broken apart with a shaker or rotary dump mechanism.

The compounds are, in most cases, glycol-based compounds of either ethylene or proplyene glycol in a 70% glycol, 30% water mixture. This gives the highest viscosity at the lowest temperature for the mixtures, thereby improving mixture handleability and pumpability at low outside temperatures. These mixtures are all water soluble and to be effective must migrate through the total surface moisture of the coal. It is mandatory that the conditioning agent be sprayed evenly over the surface of the coal upon loading into the transit mode.

The last statement above cannot be overemphasized. Any of the pure glycol-based freeze conditioning agents can work if sprayed evenly throughout the coal. The amount sprayed is dependent

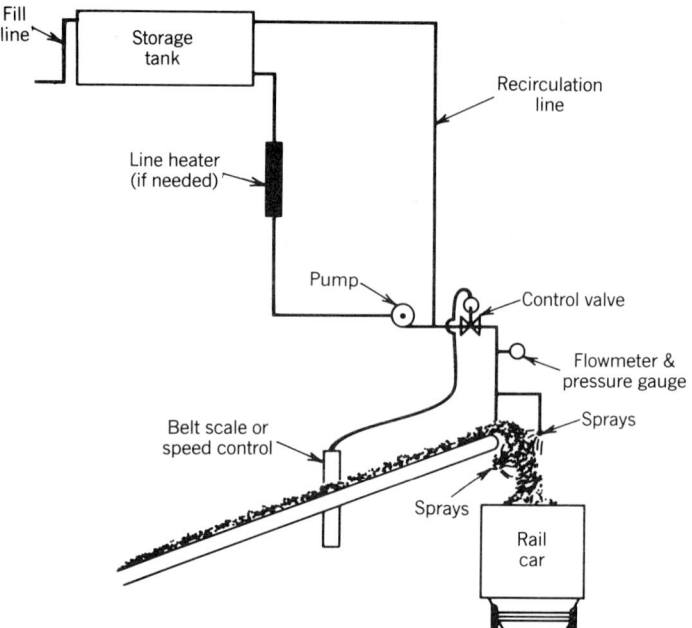

Fig. 6.5-1   Typical freeze conditioning spray system diagram.

on the surface moisture content, the percent fines, the temperature, and the time that the coal will be exposed at those temperatures. Two pints per ton of 10% moisture, 35% fines, CROM coal transported for 10 days to Detroit from Kentucky or West Virginia has been shown to be effective with most coal unloading down to temperatures as low as −20°F. Increasing the pints per ton can increase the effectiveness of the agent at lower temperatures and higher fines. Figure 6.5-1 is a typical spray system diagram.

There have been further developments within the field of freeze conditioning agents. These have been in the addition of triggering agents to promote migration throughout the coal surface moisture, and other chemicals that promote the conditioning agents' ability to adhere to the coal rather than be washed from the coal because of rain or melting snow. These could be the best alternatives to handle winter freezing.

Another development has been the side release agent, where the sides of the car are treated to form a barrier that causes the outer layer of ice, which forms because of moisture migration, to easily break away from the car when the car is dumped. This has been shown to be effective when rapid unloading systems are used and as much as 4–6 in. of coal are being left in the cars after dumping. This coal, which sticks to the sides of the cars, must bear the total delivered cost of the coal when lost and can reflect a fairly large loss. The use of these agents must be considered relative to the conditions encountered.

In all cases it must be remembered that these agents are water soluble. If treated coal is exposed to water and the agent is washed from the coal, a sudden drop in temperature could cause the coal to freeze as if there were no freeze conditioning agent applied. If the coal is sprayed with an insufficient amount of agent, or if the agent is not evenly applied, there could be problems with the coal freezing in pockets within the car. It must always be remembered that the effectiveness of the application is based on the distribution of the agent, the amount of the surface moisture, and the temperatures encountered.

Freeze conditioning agent costs are dependent on the quantity to be applied, the labor required, and the cost of the base product. The latter accounts for the greatest percentage of the cost. The capital equipment used in the application, as seen in Fig. 6.5-1, is insignificant when large tonnages of coal are to be treated. In most cases portable equipment is available from agent suppliers on a free or yearly rental basis. Labor is primarily associated with equipment maintenance and the cleaning and adjustment of sprays. The cost of the base agent is related to the concentration of the active material in the agent and the market value of that ingredient, be it ethylene or propylene glycol or a mineral salt. These costs can be further controlled by a monitoring of the expected weather patterns for the winter season and the adjustment of the need for spraying accordingly. On an evaluated basis these costs are usually lower than other methods used for unloading frozen coal.

6.5-4 Summary of Coal Unloading

Coal unloading is dependent on the transportation mode used, the quantity to be unloaded, the region where the coal is unloaded, and the flowability of the coal. Flowability is further dependent on the physical characteristics of the coal, that is, whether it is raw or washed, and how many fines are present in the coal as received at the unloading facility. These characteristics directly affect the ability to use a gravity or manual unloading system in their simplest design. A balance of coal type and system flexibility must be achieved.

In many cases economies of scale obtained for large tonnages of coal are dependent on the turn-around time for unloading the coal. The savings associated with this unloading, and the ability to receive unit trains or other large tonnage transporters, will offset capital costs for the equipment. Coal unloading systems must be designed to consider time, labor, capital equipment, space limitations, weather, and the ability of the system to unload a complete shipment.

System design must also take into consideration the flowability of the coal under normal and adverse weather conditions such as high humidity and rain and the potential for coal freezing en route to the facility. Each step of the unloading is dependent on the total system—from the coal to the ability to deliver the coal to the boiler or bunker.

BIBLIOGRAPHY

Annual Coal Handling & Storage Symposium, Sponsored by Coal Mining & Processing and Industrial Presentations, Inc., 1980, 1981, 1982 and 1983.

6.6   COAL STORAGE AND HANDLING

Robert D. Bessette

Following receipt and unloading of the coal, the coal must be delivered to either storage or active facility bunkers for transport to the boilers or for conversion. This section will discuss both the storage and handling aspects of the various systems as they affect the coal characteristics and delivery to the facility for utilization.

6.6-1   Coal Storage

The main objective of coal storage is to ensure the necessary fuel availability in the event problems occur within the energy system, causing a disruption of supply. It is also used to handle differing transportation times for the different coal carriers. A secondary requirement of the storage system is to maintain quality and consistency of the coal to ensure usability on demand. The problems that affect the ability to achieve the objectives are spontaneous combustion, slaking, FSI loss, and Btu loss. These can have an adverse effect on the coal quality and consistency for a given application.

There are five fundamental considerations regarding coal storage problems and their control or minimization:

1. Coal chemical and physical characteristics.
2. Quantity to be stored.
3. Time for which storage is needed.
4. Desired accessability of the pile.
5. Space limitations and access for storage.

An understanding of the interactions of coal and storage to a facility's handling system must be achieved prior to a more detailed handling system consideration.

Spontaneous Combustion

Spontaneous combustion and coal pile heating are possibly the largest problems associated with coal storage. Spontaneous combustion is caused by a buildup of heat within a combustible mixture because of slow oxidation or chemical reactions that produce heat, thereby accelerating the oxidation process, until self-sustained oxidation or ignition occurs. For spontaneous combustion to occur, fuel, oxygen, and the proper temperature must be present. In a coal pile these are associated with the coal sulfur content, the coal sizing and reactivity, the type of storage, possible contaminants, and the moisture content.

Coal sulfur content and reactivity are the key instigators. Pyritic forms of sulfur and iron sulfides react with sulfur and oxygen to produce heat. A typical equation for this is

$$2FeS_2 + 7O_2 + 2H_2O \rightarrow 2H_2SO_4 + 2FeSO_4 + 62{,}300 \text{ cal}$$

The production of this heat in a confined area, where the heat is not removed as fast as it is generated, can cause a rapid increase in the oxidation rate in that localized area until ignition is achieved and combustion begins. The greater the reactivity of the coal, the more rapid the increase and potential for ignition. Oxygen must be present for the reaction to happen. If contaminants are present such as paper, plastics, or wood (which have a lower ignition temperature), these can promote more rapid oxidation and spontaneous combustion. It has been seen that washed, prepared coal without contaminants and free of a large percentage of the pyrites has a lower propensity for spontaneous combustion.

In order to achieve rapid oxidation, oxygen must be present in sufficient quantities to sustain or increase that oxidation. If there is an air flow through a pile, depending on the fines content and the amount of flow, there can be an increase in the oxidation rate when channeling and low flow with high fines, or there can be a lowering of the oxidation rate when the coal is a $1\frac{1}{2}$-in. $\times$ $\frac{1}{4}$-in. stoker coal with low surface area and the flow is high, taking the heat away with the air, decreasing the temperature. This should be minimized.

The moisture content of coal can affect the initial reaction with the sulfur to promote oxidation; however, rain and high humidity can also increase this reaction and promote the beginning of the rapid oxidation cycle. Coal oxidation is an ongoing process that occurs very slowly at temperatures below 80°F and increases as the temperature rises to 150°F where rapid oxidation can begin. At these temperatures, spontaneous combustion can be reached in hours if precautions are not taken.

Some precautions should be taken prior to putting the first pound of coal into storage. The base for the pile should be crowned to allow for drainage, and all contaminants, such as wood, wires, refuse, and so on, should be removed. Coal laid on to the base should be layered over the entire base in 9- to 12-in. layers to prevent segregation of coal sizes, promoting air passage within the pile. Each layer should be compacted to further eliminate these air passages. Pile edges should also be compacted to prevent air infiltration because of wind-generated eddy currents. The prevailing wind direction should be noted to assist in pile layout to minimize these eddy currents. The pile should be shaped to promote water runoff during heavy rain and again minimize air entering the pile. When considering long-term storage, sealing the pile with an asphalt or fine coal coating maximizes protection.

When coal piles begin to get warm, the following steps can be taken as the temperature increases. At 100°F a temperature probe should be used to determine the location and temperature of the hot spots. A monitoring program should be instituted on a weekly basis. When the pile temperature reaches 140°F, a temperature reading should be taken on a two-day or daily basis. Action should be taken to reduce the pile temperature when the temperature reaches 160–175°F. Do not put water on a hot spot in an attempt to cool the spot and stop the oxidation. The water will fuel the sulfur reaction and increase the temperature, thereby increasing the rate of oxidation. The easiest method to decrease the temperature of a pile is to open the pile and expose the hot spot to the cool air, which will remove the heat as fast as it is generated. When large piles have heated, it may be necessary to move a complete pile by layering the coal, using 6- to 8-in. layers, and allowing time to cool between layers, and then compacting each layer to prevent oxygen access to the coal. Dousing the thin layers with water may be used to decrease the cooling time; however, it is necessary that for the heat generated by the sulfur reaction be allowed to dissipate. Opening a pile in order to cool a hot spot may necessitate moving the pile because oxygen now has access to the coal.

One problem area where not much can be done concerns coal moved from the north into a high-humidity area, such as New Orleans, where water will condense on the cooler coal. This is an exothermic reaction and will also happen to coal that has been dried below the natural inherent moisture of the coal. This will tend to add to the problems of putting water on a hot coal pile to cool it.

Free-Swelling Index and Fluidity Degradation

The free-swelling index (FSI) and fluidity are important when coal is to be used in the coking process. The surface characteristics, which hinder the devolatilization of the coal and promote the FSI and fluidity of the coal, are affected by oxidation. The more reactive a coal, the more susceptible it is to oxidation. It is also not normally considered a good coking coal. However, some of the good coking coals are very susceptible to these losses and can cause problems if this is not considered.

Btu Loss

When coal is stored for a year or more, and precautions are not taken to minimize oxidation, it is possible for the coal to lose anywhere from 100 to 500 Btu/lb depending on the coal's reactivity. A standard bituminous coal may lose only 200 Btu over a two-year period, and a lignite may lose as much as 700 Btu over the same period. Btu loss is a function of oxidation, and the same precautions taken for spontaneous combustion must be followed.

Slaking

When coal goes through freeze–thaw cycles, or when the coal is very porous and the inherent moisture is high, there is a possibility that the coal will degrade, or slake, during storage. This can be a significant problem when specific size is required for a process. The lower the rank and softer the structure, the greater the potential for slaking. Storage methods should be considered relative to the coal.

The above considerations are most important when long-term storage is needed. However, the general precautions should be considered whenever any storage is needed.

6.6-2 Types of Coal Storage

Long-Term Storage

Storage is considered long term when coal will be stored for periods longer than 1 year for bituminous coal and is measured in months for the lignitic, more reactive type coals. The considerations and precautions outlined above become most important for long-term storage of coal. Silos are one method of storage that can be used. The cost for a substantial amount of storage can be prohibitive, and precautions to allow for ventilation of methane gas generated in some coal can also promote oxidation.

Most long-term storage piles are well-compacted ground storage, sealed to prevent oxygen infiltration and sometimes covered to prevent water intake because of high rainfall. Access to these piles is usually limited to emergencies or extreme need. An example of this could be a low-sulfur coal storage pile needed for environmental upset. Because of the nature of the pile, minimum precautions are needed to meet dust and water pollution control requirements.

Short-Term Storage

Short-term storage is normally associated with seasonal coal demand requirements and transportation lead times. This can be a substantial amount of coal for large users where lake movements of coal are considered or when coal is received only a small number of times to take advantage of transportation economics.

The location of short-term storage must be such as to offer the degree of accessibility required when coal will be needed at the using facility. In many cases the short-term storage pile is adjacent to the active storage pile and provisions are made such that over the storage duration the full pile will be cycled through the system, minimizing the need for elaborate pile heating control. Only layering and simple compaction are required. This can be accomplished with a bulldozer with a coal blade on the front. Because short-term storage is a cyclical process of putting in and taking out, care must be taken to minimize coal degradation. A change in coal type with a given system may cause an increase in the degradation of the product.

To minimize the problem of coal degradation in ground-type storage piles because of moving the coal and running over it with wheeled or track vehicles, storage silos are sometimes used. These can be somewhat costly. However, when small tonnages are involved, there are benefits in lower degradation and control. This is especially true for stoker coal and when dust from the moving process may be a problem.

Because of concern with pile dust control, a spray system or enclosure must be considered. Water runoff must also be captured with any coal pile. State regulations for settling ponds usually prevail. Another alternative for dust control is a chemical coal pile sealant, which will minimize wind erosion and assist in preventing water penetration without affecting pile accessibility. Water runoff collection is still needed in this case.

Active Storage

Active storage is normally short-term (1 to 3 days) storage at or near the using facility. This storage is used as a surge for the daily needs of the facility and can be considered a continually cycling storage. Because it is for immediate use, storage silos and covered storage become first-line considerations when coal quality, sizing, and space limitations close to the boiler or conversion facility are a problem. For the small user the silo is used as the bunker and as such is an integral part of the coal handling system. When a facility is located in an area where there are many neighbors, the silo can solve many of the problems associated with esthetics. The use of silos as active storage is easier than as long-term storage since there is little problem with spontaneous combustion.

If a silo is not used, the problems associated with dust control and water runoff are similar to those of the long- and short-term storage piles. However, the activity is greater and the problems are closer to the plant. Spray towers and enclosures at transfer points are needed to minimize dust. Ponds or water treatment prior to discharge into sewers is needed for water disposal.

6.6-3   Summary of Coal Storage Problems

Coal storage is necessary to assure a continuous operation or compliance with regulations. The cost associated with coal storage is the land, the interest on the money tied up in the coal and transportation to that point, and the costs associated with the preservation of quality and consistency. The questions of storage quantity arise from the degree of perceived risk of supply stability and transportation assurance. These can only be evaluated based on the coal mining company selected, the method or methods of transportation selected, and the flexibility built into the coal unloading facility. The greater the tonnages required, the greater the problems associated with storage, reclamation, and spontaneous combustion. Coal storage requirements must be defined prior to selection of the overall handling system.

6.6-4   The Coal Handling System

The coal handling system is that system which can take coal from the coal unloading area to storage or to the active facility feed bunker or move coal from one storage area to another. The coal handling system discussed below is that system beginning at the coal unloading grizzly or at the handling system feed shoot and ending at the facility active bunker outlet. The discussion will center around the interface of various equipment pieces for various sizes of using facility. The main constraints to be discussed will be the type of coal that can be handled, the quantity for which the equipment is applicable, and the effects on the coal size consistency and segregation.

The coal handling system is dependent on four basic considerations:

1.   Type of storage.
2.   Quantity to be handled.
3.   Coal type and physical characteristics.
4.   Any space limitations.

The coal considerations are associated with the coal's flowability, degradation, and segregation. The prime objective of any coal handling system is to provide maximum flexibility in moving coal from storage or unloading to the using equipment as economically as possible.

Large utility and synfuel facilities using over 500,000 tons of coal per year must be able to handle the total coal received over an 8-hour day, 5-day week. This is about 250 tons per hour. If, in reality, it operated 50% of the time, this would mean about 500 tons per hour required or a car every 12 minutes. With facilities of this order of magnitude, as well as conversion facilities that expect to use up to 20,000 tons per day with 2 million tons of storage, large automatic handling and reclamation systems can and must be used. A discussion of the design details of these systems is beyond the scope of this discussion. It is enough to say that systems designed to handle large tonnages are not limited by coal that does not flow as are smaller systems. In many cases the problems associated with the smaller systems can provide guidelines with regard to larger system design.

The industrial system, in most cases, applies to an existing plant with virtually no space. Systems designed for industrial applications must be as simple as possible, while at the same time designed to assure quality and consistency. Tonnages handled within the normal industrial or commercial system may vary from 1500 to over 250,000 tons per year for a large industrial facility. There is no standard system that can be developed to meet the diversity of the industrial system requirements. Industrial systems will also vary in complexity as they vary in capacity. The following section will discuss the equipment used.

6.6-5   Coal Handling Equipment

There are two pieces of equipment necessary in any plant. These are an 8-lb sledge hammer and a 10-ft iron rod. These do wonders to improve coal flowability within a handling system.

The Vibrating Feeder

The vibrating feeder is used to move coal for short distances, on the order of 10–15 ft. The feeders are designed for a maximum feed rate by setting the angle of feed and speed of vibration.

Coal flowability is of concern since the vibrating feeder assists in the flowing characteristics of coal by vibrating an incline. This piece of equipment is most applicable to coal systems where the coal exhibits some degree of flowability. A 2-in. × 0 CROM with 35% fines product that needs to be moved 10 ft horizontally, with a 5-ft vertical space, would be a good example. Because by vibrating they improve coal flowability, the feeders are susceptible to plugging if the coal flowability changes. There can also be problems if large tonnages are required to be moved. The bed depth will control the feed rate at a given vibration rate and incline. Another factor that must be considered with regard to coal movement assisted by vibration is the potential for segregation when the feeder receives coal perpendicular to the direction of feed. This can be minimized if the coal is moved away from the feeder at a perpendicular direction. When considering degradation, this feeder may be the least prone to degrade the coal and produce additional fines.

The Screw Conveyor

Unlike the vibrating feeder, the screw conveyor not only assists the flow of coal but can move coal 40 or 50 ft at a 45° angle. The screw conveyor works on a volume basis, and the weight of the coal per cubic foot can be important when considering longer movements without a transfer point. The diameter of the conveyor and the distance limit the quantity of coal that can be moved.

The screw conveyor should be designed such that the inside diameter is at least four times the top size of the coal to be handled. Because of the closed nature of the screw conveyor, coals that have a tendency to plug with high percentages of fines and high moisture can cause serious problems, especially if there are contaminants such as wood, rocks, large bolts, or clay. Raw coal with high pyritic and moisture content can cause problems because of erosion and corrosion of screw internals, which could lead to major system upsets and high maintenance.

The Drag Chain

The drag chain is also a positive mover of coal. Here the coal is pulled along and within an open trough or chute by a chain and plate or bar arrangement. The flexibility of access for maintenance simplifies the application, and there are not as many fine coal or plugging problems. However, the chain and dragging motion of the conveyor can cause alignment and wear problems, more so than with the screw conveyor. There is more flexibility when using the drag chain. However, the space required is greater than that for a screw conveyor. The drag chain is used in many cases as a main transfer conveyor between major handling system interfaces of the larger or medium-size industrial user. Because of the open design and ease of access, there are usually less problems with corrosion and erosion.

Capacity is a function of the conveyor width, bar height, and chain speed, which can be a limiting factor depending on the coal size. A modification of the chain conveyor in a totally enclosed housing has been used to move coal and fine coal vertically in relatively confined areas. These systems have been proven effective for the smaller industrial system; however, they are not usually applied to larger systems.

The Belt Conveyor

The belt conveyor is the most efficient method for moving coal in the energy system. It is not limited to any type, quantity, or quality of coal. There is minimum wear and maintenance required. Coal is fed evenly on the moving belt through a transfer mechanism such as a vibrating feeder. The belt can carry the coal to another transfer point where it can be deposited in storage, to active bunkers, or to system coal feed. Because there is no dragging, there is virtually no degradation or segregation in the moving process; however, with the large quantity of coal capable of being handled per unit time, there is the possibility of coal-to-coal impact at transfer points. If there is any delay in the coal flow from one transfer point to another, degradation can increase sharply. If the coal stops at a transfer point, it can be considered a handling cycle. As a result, it is important to minimize the transfer points in this or any system.

In all systems it is important to look for the most direct route. The belt system is limited to a straight-line move from point to point or transfer point to transfer point with an angle of elevation of about 30° maximum because of coal flowability and inertia. Coal will have a tendency to roll back on a moving belt or incline if the incline of the belt is greater than 30°. The greater the flowability of the coal, the easier it is for the coal to roll back, and the lower the allowable angle of incline. Because of this, the space required to elevate coal by belt can be the limiting factor in the system.

The Bucket Elevator

The bucket elevator is the most direct way to move coal from one elevation to another. Coal is fed into the bucket elevator by a coal chute from a vibrating feeder or other feed mechanism. The coal is elevated to the desired level, and, as the buckets make a 180° turn over the top of the elevator, coal is emptied by gravity to a transfer chute for delivery to the active bunkers or storage. Bucket elevators are very effective for coals that exhibit very good flowability characteristics with low fines and low moisture. Raw coals that tend to stick, ball, and plug can remain in the buckets, causing serious problems and difficulties with operation and availability. The application of bucket elevators is primarily for the small industrial facility that uses up to 200,000 tons of stoker or modified stoker coal and the plant layout is not such that a belt conveyor can be used.

The Dense-Phase Pneumatic Conveyor

The dense-phase pneumatic conveyor is used to convey coal in a closed system using a pulse-type process. Coal is fed into a pressurized vessel and a pulse of high-pressure air forces the coal out of the vessel through a pipe to storage or the active bunkers. This system has the ability to move coal vertically and around corners by the use of wide-radius bends. System pipe design diameter should be approximately three times the top size of the largest piece of coal to be moved. High-moisture, high-fines raw coals can cause problems.

The cost of the dense-phase system is somewhat higher than a conventional system. However, the advantage of full automation and no moving parts, little or no segregation during movement, and very little degradation can produce benefits in labor savings and operation. Pneumatic coal handling systems are now beginning to appear and could be a viable consideration when the total system is considered for industrial users.

Coal Crushers

A coal crusher can be added to a system to provide for on-site coal sizing and flexibility in coal purchasing. Most washed coals are available with a 2-in. × 0 size consistency or as a $1\frac{1}{2}$-in. × $\frac{1}{4}$-in. stoker coal. For stoker-fired units a combination of a screen to remove the $1\frac{1}{2}$-in. × 0 coal and a crusher to crush the $+1\frac{1}{2}$-in. coal to the stoker size could allow for substantial savings in a normal market. A coal crusher can also help in winter with frozen coal lumps.

Coal Sampling

A simple coal sampling system station should be built into the coal handling system for collection of a coal sample as described in Section 6.3-5. This should be considered as part of the handling system for either a new or older plant.

Coal Transfer Equipment

Coal transfer equipment is used to transfer coal from one piece of equipment to another, that is, from unloading to belt, from bucket elevator to active bunker. These are important pieces of equipment and usually the least considered and most troublesome within the total handling system. In many cases the transfer mechanism is a chute, undersized and at an angle at which the coal will not flow. Chutes should be designed with at least a 60° angle and a size equal to or larger than the equipment feeding it. Stainless steel or some other abrasion-resistant material should be used to assure continuous flow. In long chutes, with a 2-in. × 0 coal, segregation can occur and cause problems if the receiving system requires an evenly distributed product. In this case special care must be taken when coal is being fed into the active bunker. These same precautions must be taken when coal is being delivered to storage and segregation can be a problem.

The design of coal transfer should consider the following:

1. Any decrease in flow rate into the device should be avoided.
2. The shortest transfer distance should be used.
3. The impact of coal on anything should be minimized or eliminated.
4. Segregation must be considered.

When coal is fed into the active storage or bunker, it is of prime importance that the coal be delivered to that piece of equipment free of segregation, especially when the ultimate user depends on a consistent size product for optimum operation and efficiency. The method of feed into the bunker should be such as to minimize segregation within the equipment. For a long bunker a continuous moving belt unloader can provide the best results. However, when multiple bunkers are used, it is important that the coal be fed into the bunker at least at two points 50% between the centerline and the outside of the bunker to prevent a high fines buildup over the bunker outlet, if it is a single-point bunker outlet.

The Coal Bunker

Bunker design can be an important consideration for two reasons. The first is coal flowability and the ability of the bunker to minimize plugging with side angle consideration and vibrators. The second is segregation. The system must be designed with consideration for the coal characteristics and flexibility desired to receive a specific range of coals. If the coal size range to be used is very narrow, $\frac{3}{4}$ in. $\times \frac{1}{4}$ in. for traveling grate stokers, there should be little problem. However, if a 2-in. $\times$ 0 coal is used, close attention must be paid.

6.6-6 Summary of Coal Handling Techniques

Coal handling is the art of moving coal from one place to another efficiently and as needed without changing its physical characteristics other than as designed. Coal handling systems must match the coal unloading and the boiler or coal converter feed systems. They must also provide adequate flexibility between coal receiving and storage reclamation. Coal handling systems can vary from small industrial/commercial systems to very large utility/synfuels systems handling 5–10 million tons of coal per year. Because of the diversity of these systems, no one system can meet the needs of each, and consideration must be given to the method of coal receipt, the coal flowability, the storage requirements, and the space limitations.

From information included above, a general industrial system concept for a plant should be able to be developed with the proper considerations. An evaluation of a system's performance relative to projected needs should also be possible. Utility systems, because of their magnitude, have similar problems magnified because of the scale of operation. These systems should be discussed with the major coal handling and storage equipment design companies to take advantage of the most appropriate total system methodologies.

BIBLIOGRAPHY

Harris, E. R., Connell, G. F., and Dengiz, F. *Coal Handling Equipment and Storage (For Industrial Plants)*. The American Society of Mechanical Engineers (ASME), Paper No. 75-IPWR-1, New York, 1975.

Stock, A. J. *Coal Segregation, Coal Flow and Bunker Fires*. Presented at the Industrial Coal Conference, Purdue University, Lafayette, IN, 1966.

Torma, B. and Weishar, A. R. *Key Considerations In Coal Handling System Design*. The American Society of Mechanical Engineers (ASME), Paper No. 82-JPGC-Pwr-39, New York, 1982.

"The Coal Storage Manual, A Power Engineering Special Feature," *Power Engineering Magazine*, S1–S24, March 1956.

6.7 COMBUSTION OF COAL

*Eric H. Smith**

6.7-1 Grate Burning

Effect of Heat

An examination of the reaction of coal under heat is the first step in considering the combustion process. Coal is so diverse in its properties that a complete study of the effect of heat on all coals is beyond the range of this handbook. More detail can be found in the bibliography.

Bituminous coal when first heated, gives off moisture. Then at about 700°F there is an increase in the gas evolved and a viscous oil makes its appearance. Coal particles with a high caking tendency will tend to stick together during combustion. If these coals are burned on underfeed stokers, or hopper-fed stokers, this solidified cake effectively blocks the air for combustion from flowing through the coal uniformly. Chunks of this solidified coal produce smokey fires, large unburned coke masses

*Deceased 1983. Text revisions by personnel of Riley Stoker Company, Worcester, Massachusetts.

and high excess air. This caking quality is measured in the laboratory by the *Free Swelling Index Test* (FSI ASTM Test D-720). The test consists of heating a 1 gram sample of pulverized coal to $820 \pm 5°C$ in a covered crucible and comparing the size and shape of the coke button obtained with the outlines of a set of standard profiles numbered in half units from 0 to 9. For underfeed stokers, a Coke Button FSI under 3 is recommended to secure a free burning bed.

Overfeed Burning. The fuel is fed by placing the raw coal on the surface of the burning fuel bed. The air for combustion flows upward in a direction opposite the flow of the fuel. The new fuel is heated by the hot combustion gases. The composition of the fuel bed can be divided into four reaction zones, the zones overlapping one with the other. Figure 6.7-1 shows these zones. One of the earliest experiments of the Bureau of Mines investigated the reactions taking place and analyzed the composition of the gases being evolved in the various reaction zones. Figure 6.7-2 shows the results of these investigations.

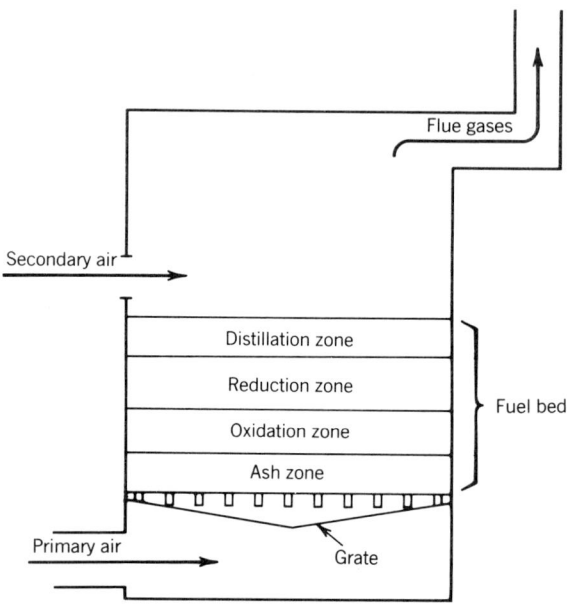

Fig. 6.7-1 Zones in the fuel bed of a furnace. Primary air enters *under* the grate, and secondary air is admitted *over* the fuel bed in order to complete combustion. U.S. Bureau of Mines study.

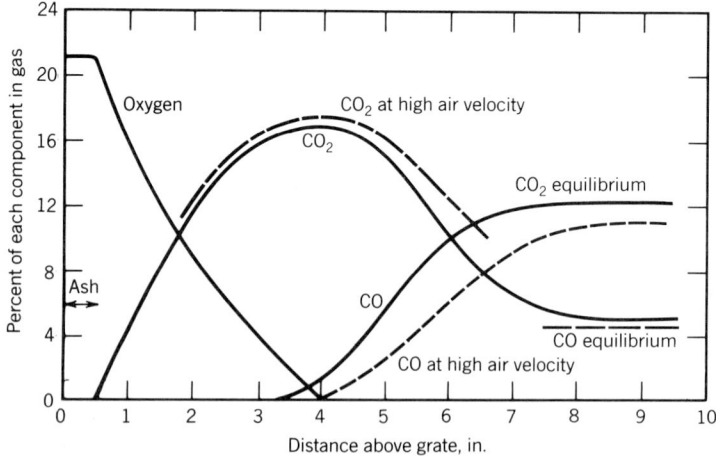

Fig. 6.7-2 Composition of the gases within the fuel bed of a coal-fired furnace. All the oxygen is consumed at about 4 in. above the grate. U.S. Bureau of Mines study.

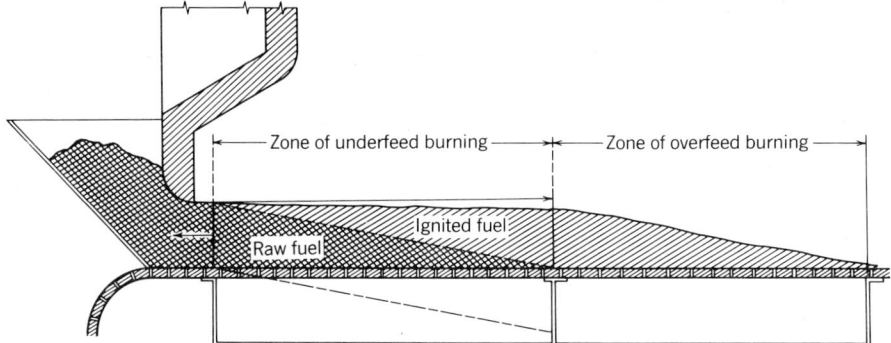

Fig. 6.7-3 Diagram of traveling-grate stoker. For convenience, horizontal scale is shorter than in conventional beds.

The oxygen in the air is rapidly consumed by the hot coke and forms CO_2. This process is essentially complete at a point 4 in. above the grate or 4 in. above the top of the bed of ashes. The ascending gas, now consisting of CO_2 and nitrogen, is reduced by the hot fuel bed. The CO_2 percentage drops to about 5% and the CO increases to 12% at a point 7 in. above the grate. The formation of CO is undesirable so stokers which work with a thin fuel bed, like the spreader stoker, have a definite advantage. The mixture of gases will leave the fuel bed at a temperature between 2200 and 2700°F. The higher temperature is attained at a burning rate of about 50 lb/ft^2 · h while firing low moisture Eastern bituminous coal.

Chain or Traveling Grate Burning. This combustion process is shown in Fig. 6.7-3. The grate moves continuously from left to right and as the fuel emerges from the hopper under the guillotine, the bed ignites from the top. The heat is supplied by radiation from the burning fuel and reflection from the front ignition arch.

To achieve the maximum rate of ignition, the air flow through the fuel bed at the front of the stoker must be limited or the cold air will cool the fuel bed faster than it can be heated. The rate of feed and the stoker grate speed are affected by the rate of ignition. Various characteristics of the coal, such as size, moisture content, volatile matter, free swelling index, and coal rank influence this rate of ignition. As the coal moves away from the ignition arch, the ignition front gradually moves downward leaving burning fuel above. The volatile is distilled, the coke burns to CO_2 as the fuel progresses, and finally the ash bed is discharged over the rear of the stoker.

Spreader Stoker Burning. The spreader stoker operates on the true overfeed principle. Raw fuel is fed over the plane of ignition using a thin fuel bed. In the path of the flame and hot combustion gases the small coal particles burn in suspension. The larger pieces are ignited as they land on the burning coal bed. Ignition is extremely rapid. As the pieces of coal burn away, the ash is immediately chilled by the combustion air and remains granular. The traveling grate then dumps this ash into a hopper.

Underfeed Burning. The term underfeed burning denotes that coal is fed from under the plane of ignition. As the raw coal reaches the plane of ignition it is heated and the moisture is driven off followed by the volatiles. These gases then pass through the burning fuel bed where the hydrocarbons mix with oxygen from the air. The compounds which are formed mix with more oxygen and while passing through the burning coke bed are so well stirred that they burn with a relatively small amount of overfire air at full load—none at all at light loads.

Underfeed stokers operate with thick fuel beds. They are essentially gasifiers and emit only a small amount of fly ash when properly operated.

To illustrate the principle, Fig. 6.7-4 shows the functioning of a very small underfeed stoker.

6.7-2 Suspension Burning

The larger industrial steam generating units and practically all utility steam generating units require that the fuel be burned in suspension.

There are serious problems in operating a stoker-fired boiler of 300,000 lb of steam per hour. Any breakdown of the grate or stoker mechanism requires that the whole unit be shut down. Pulverized coal steam generators have been made larger than 6 million pounds of steam per hour. Damage of burners or pulverizers only requires a fractional reduction in load. Hence suspension firing is the preferred method of firing the larger industrial boilers and practically all utility boilers.

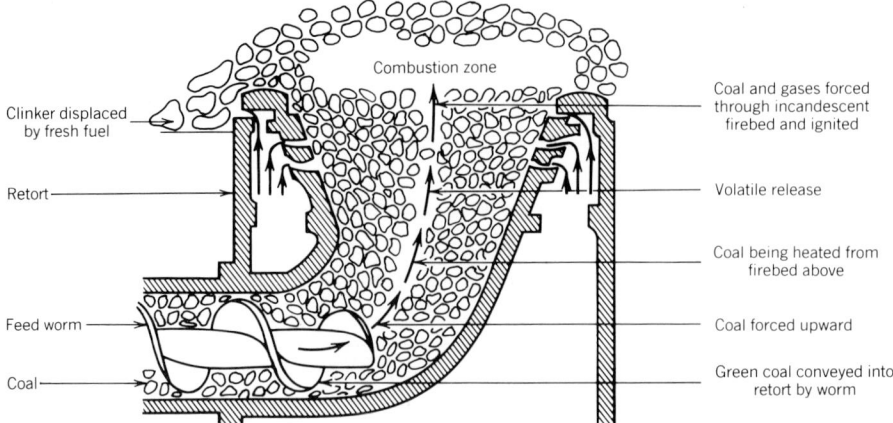

Fig. 6.7-4   Underfeed stoker.

The requirements for successful suspension burning are:

1.   The coal air mixture must ignite readily.
2.   The resulting flame must be stable.
3.   There must be a high degree of carbon burnout, at least 98%.
4.   The size of the flame must be reasonable, not less than 60,000 Btu/ft^3-h

To achieve these results the pulverizer and burner selection must produce:

1.   **Adequate fineness.**   A coal fineness size distribution ranging between 70 and 85% passing 200 mesh for all coals. Both pulverizer power and maintenance cost increase with fineness increase.
2.   **Primary air flow—drying.**   Enough primary air quantity and/or adequate temperature is needed to evaporate all the surface and some of the inherent moisture. The higher the heat content of the primary air, the faster the rate of ignition.
3.   **Flame ignition and stability.**   A stable flame requires that the primary air quantity be between 12 and 20% of the total air and the burner design must provide a means to recirculate hot combustion products back to the burner.
4.   **Primary air temperatures.**   Primary air temperatures exiting the pulverizer generally average between 140 and 180°F with a minimum air to coal ratio of 1.10.
5.   **Primary air velocity.**   Burner primary air nozzle velocities vary from 3000 to 6000 fpm at the coal–air mixture temperatures. The coal air mixture velocities inside the piping range from 4500 to 5400 fpm depending on the piping system design. These velocities have been established to prevent settling of pulverized coal in the piping, which is referred to as "coal pipe layout."

Burning Profile

The burning profile provides a measure of the oxidation intensity of a fuel. Weighing of the sample is performed simultaneously with the heating and subsequent burning of the sample. Plotting the rate of

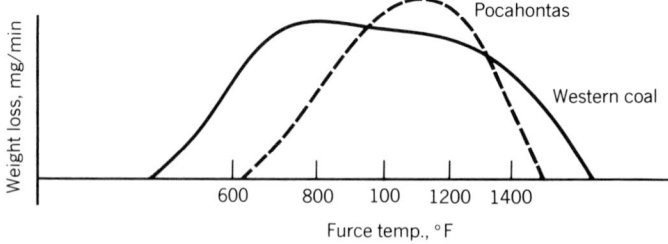

Fig. 6.7-5   Burning profile.

oxidation against furnace temperature gives a typical curve such as the one shown in Fig. 6.7-5. Curves from coals whose performance in given equipment is known can then be compared in the laboratory and office.

The engineer involved in the design of the equipment or in the performance of existing equipment can then make an evaluation before a commitment to buy a particular coal is made.

The Percentage of Volatile Matter

The proximate analysis has been the most common analysis used since well before the start of the twentieth century. Its use in predicting a coal's performance is and has been found very reliable. The percentage of volatile matter is one of the most important items in the proximate analysis.

The pulverized coal particle, almost always under 74 μm in diameter, is airborne from the pulverizer to the burner.

As the particle emerges from the burner nozzle or tube, it is exposed to the radiant heat from the flame front.

The moisture that has not been evaporated in the pulverizer or piping immediately flashes into steam. Much of the volatile matter then distills off as a gas and mixes with the combustion air.

Mixing, ignition, and burning are very rapid as the reactions are taking place between molecules. The greater the percentage of volatile matter, the faster the ignition as shown below.

Coal	% Volatile Matter	Ignition Time, ms
Illinois #6 Seam	33.9	27.5
Pittsburgh Seam	33.6	32.0
Pocahontas #3 Seam	15.3	38.0

Agglomerating Index

There is a significant difference between the actions of pulverized coal particles depending on where the coal originated. High-rank Eastern coals will soften before they burn. If two or more particles collide, a very frequent happenstance in such a rich mixture, they will stick together and agglomerate. This slows down ignition and the agglomerated particle will increase the carbon loss. This condition can be reduced by finer grinding.

Anthracite, sub-bituminous coals and lignite will burn without softening or changing shape. They are non-agglomerating.

Reactivity Index

This is a guide to the ease or difficulty with which a coal ignites and burns. The reactivity index is based on the rate of temperature rise of a sample at the instant when the furnace and the sample are at the same temperature.

Under these conditions the rate of rise becomes a measure of the reactivity of the sample.

The Pocahontas #3 seam with a volatile matter percentage of 17.3 has a reactivity of 239. The reactivities of American coals at T_{15} and T_{75} are relatively the same from 22% volatile to 35% volatile, where T_{15} and T_{75} are the reactivity indices based on rates of temperature rise in pure oxygen of 15 and 75°C/min, respectively.

6.7-3 The Combustion Process

Albert H. Rawdon, Jr.

The normal coal combustion process consists of the rapid and complete oxidation of the combustible components of the fuel with atmospheric air as the source of oxygen. This process is exothermic and releases a large amount of heat. The major combustible elements of coal are hydrocarbons and carbon. Minor combustible elements in fuel are sulfur and polycyclic nitrogen. The hydrocarbon in coal is associated with the volatile fraction and contains both hydrogen and carbon. Other materials present in coal are oxygen, water, and inert ash. The oxygen is available for burning of combustible elements. The water is inert and requires heat to change it to the vapor state. The ash is usually considered inert and only vaporizes to a negligible amount.

Generally speaking, the heat energy involved from the combining of a combustible element with oxygen (air) is dependent only on the ultimate end product and not upon the means or intermediary steps. That is, hydrocarbons burn as though they were broken down in carbon and hydrogen and combusted completely to carbon dioxide and water vapor. To ensure that this happens, combustion for the generation of power is done with an excess of oxygen (air). The following chemical equations represent the major coal combustion process:

$$C + O_2 \rightarrow CO_2 + 14,100 \text{ Btu/lb C}$$

$$2H_2 + O_2 \rightarrow 2H_2O + 61,100 \text{ Btu/lb } H_2$$

The largest minor constituent that burns is sulfur, which burns principally to sulfur dioxide. The following equation represents this:

$$S + O_2 \rightarrow SO_2 + 3983 \text{ Btu/lb S}$$

The principal constituents of air are nitrogen, oxygen, and water vapor. Other rare inert gases, such as argon, in the air are routinely lumped with nitrogen. Thus dry air is considered to be 76.85% nitrogen and 23.15% oxygen by weight. If the moisture is not known in the combustion air, it is assumed to be at 60% RH and 80°F db and the generally accepted value of 0.0132 lb of moisture/lb dry air is used.

Products of Combustion Calculations

Since the composition of all the compounds in coal are not known, the methods of calculating the products of combustion are not as precise as, for example, computing the products of combustion for methane gas. Calculation of products of combustion are usually done to size equipment, complete heat balances, and as part of the determination of SGU efficiency by the method of losses (ASME short form). The most precise method is by use of a table of combustion constants. Table 6.7-1 has been excerpted from a general table presented by the American Gas Association. To utilize this table, a sample problem has been prepared which requires an ultimate analysis of coal, the excess air required for the combustion system, and the humidity of the air to be known.

Example 6.7-1   Illinois Coal Fired at 20% Excess Air.

	Ultimate Analysis As Fired	Theoretical Oxygen	
		Table 6.7-1 Constants	O_2
C	0.632	2.66	1.681
H_2	0.046	7.94	0.365
O_2	0.081		(deduct) -0.081
N_2	0.011		
S	0.040	1.00	0.040
H_2O	0.100		
Ash	0.090		
Total	1.000		2.005 lb O_2/lb coal

Dry air required = 2.005/.2315 = 8.661.
Dry air supplied @ 20% excess = (1.20)(8.66) = 10.39 lb dry air/lb coal.
Oxygen supplied @ 20% excess = (1.20)(2.005) = 2.41 lb O_2/lb coal.
Excess oxygen = 2.41 − 2.005 = 0.405 lb O_2/lb coal.
lb N_2 from air/lb coal = (10.39)(.7685) = 7.985 lb N_2/lb coal.
If moisture in air is not known, assume 0.013 lb/lb dry air.
H_2O from air/lb coal = (.013)(10.39) = 0.1351 lb H_2O/lb coal.
Air supplied (lb) = $O_2 + N_2 + H_2O$ = 2.41 + 7.985 + 0.1351 = 10.53.

	Ultimate Analysis as Fired Coal	Products of Combustion	
		Table 6.7-1 Constants	lb Product/lb Coal
C	0.632	3.66	2.313 CO_2
H_2	0.046	8.94	0.411 H_2O
O_2	0.081	No value	0.000
N_2	0.011	pass thru	0.011 N_2
S	0.040	2.00	0.080 SO_2
H_2O	0.100	pass thru	0.100 H_2O
Ash	0.090	No value	0.000

TABLE 6.7-1

	Theoretical Oxygen		Products[a]		
Element	lb O_2/lb element	lb CO_2/lb element	lb H_2O/lb element	lb SO_2/lb element
Carbon	2.66	3.66	0	0
Hydrogen	7.94	0	8.94	0
Sulfur	1.00	0	0	2.00

[a]*Note*: (1) Fuel nitrogen and water pass through and do not engage in combustion except in trace quantities. (2) Ash and O_2 in fuel are not considered to be in the products of combustion.

Also the following appear in products of combustion:

Excess oxygen	0.405	lb O_2/lb coal
Nitrogen from Air	7.985	lb N_2/lb coal
H_2O from air	0.135	lb H_2O/lb coal
Total	11.440	lb products/lb coal

The products of combustion are then combined to show the makeup as follows for 1 lb of coal: 2.313 lb CO_2, 0.646 lb H_2O, 7.996 lb N_2, 0.405 lb O_2, and 0.080 lb SO_2.

Another, less precise, method of calculating the air required and products of combustion is from knowledge of the proximate analysis. For the coal in the previous example the following proximate analysis was given:

$$\text{Moisture} = 0.100$$
$$\text{Ash} = 0.090$$
$$\text{Volatile} = 0.381$$
$$\text{Fixed carbon} = 0.429$$
$$\text{HHV Btu/lb} = 11{,}800$$

Figure 6.7-6 is reproduced from *Steam* by Babcock and Wilcox with permission. It shows the air required to completely burn 10,000 Btu of the coal as a function of the volatile (V.M.) in the dry ash free (d.a.f.) coal. For the above proximate analysis:

$$\frac{\text{v.m.}}{\text{d.a.f.}} = \frac{0.381}{1 - 0.1 - 0.09} = 0.470$$

From the figure, theoretical air is then 7.56 lb dry air/10,000 Btu.
The air supplied per pound of fuel fired at 20% excess and with 0.013 lb moisture/lb dry air is

$$\frac{7.56}{10{,}000}(11{,}800)(1.2)(1.013) = 10.84 \text{ lb air/lb coal}$$

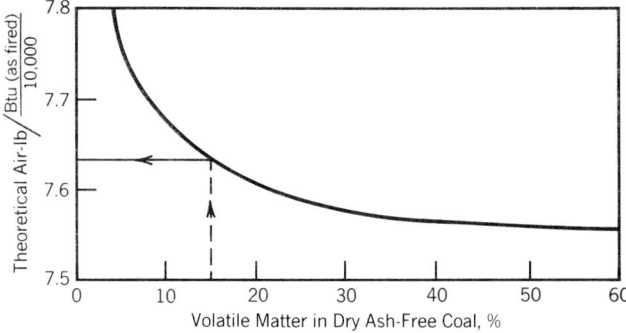

Fig. 6.7-6 Theoretical air in pounds per 10,000 Btu heat value of coal with a range of volatile.

The products of combustion are equal to the coal gasified plus the air supplied

$$\left(1.00 - 0.09\frac{\text{lb ash}}{\text{lb coal}}\right) + 10.84 = 11.75 \text{ lb products/lb coal}$$

This is less than 3% error from the more precise calculation method.

6.7-4 Fluidized-Bed Combustion

Robert D. Bessette

Fluidized-bed combustion has been around since the early 1920s and is now being applied to the combustion of coal. The objective of developing fluidized-bed combustion (FBC) for coal is twofold: first, to be able to remove sulfur during the combustion process in lieu of a wet scrubber; second, to be able to burn low-grade, nonconventional combustible coals within a system without elaborate modifications to be able to handle the slagging and fouling characteristics.

Unlike stoker firing, which can be 100% bed burning or partial bed and partial suspension burning of larger-sized particles, and unlike the rapid combustion of pulverized coal, FBC is 100% combustion of larger-sized particles suspended in a furnace with inert fluidized particles. The techniques employed with fluidized-bed combustion can have a definite advantage over conventional equipment for specific fuels when the total system is considered.

Work is under way on two different FBC models, an atmospheric model and a pressurized model. The atmospheric fluidized-bed combustor (AFBC) has been proven, and there are a number of these installations around the world that have been demonstrated. The following discussion will center on AFBC systems. Pressurized systems are in the research and development stages and are being designed for combined-cycle applications where hot flue gas under pressure will be expanded through a gas turbine, where the waste gases will then be used to produce steam. These PFBC systems will only be discussed briefly; however, the techniques applicable to PFBC are those that will be discussed in detail for AFBC.

The Fluidized Bed

The fluidized bed is a bed of coal or fuel and inert rock or limestone particles that are fluidized by an airstream. The degree of fluidization is a basis for classification by type. Figure 6.7-7 is a representation of the bed configurations used by various boiler manufacturers. Each has its advantages and disadvantages. It is important to understand what the bed is and how it works prior to discussing the other aspects of fluidized-bed combustion.

The bed itself consists of approximately 97% inert material and 3% carbon or combustible material at any one time. Because of the very high inert content, the bed acts as a heat sink, much like the large

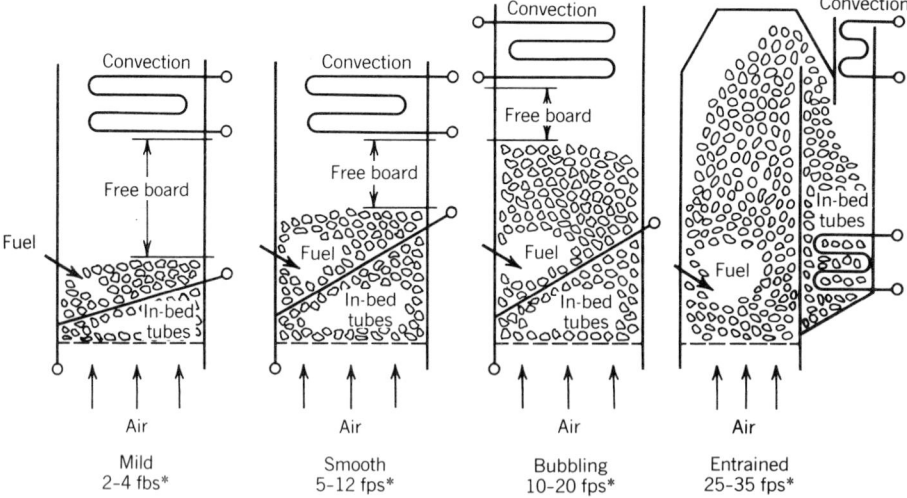

Fig. 6.7-7 Fluidized bed types. Asterisk indicates air velocity with coal top size at $\frac{1}{4}-\frac{1}{2}$ in.

area of a water tower where many gallons of water can be removed without appreciably lowering the level of the water. The inverse is also true. When a level of bed temperature is reached and sustained combustion is possible, the bed is able to maintain ignition. This normally happens at about 1000–1100°F. However, bed temperatures are normally elevated to the 1500–1700°F range to give better efficiency and have a reserve of energy and a greater temperature difference for heat transfer.

Bed composition is maintained either by the addition of coal ash, as a result of coal being added to the bed, or by the addition of coal and limestone, which may be added for sulfur removal, and a removal of excess bed material from the above feeds by bleeding off a portion of the bed. As a result of bed turbulence, some material will degrade and, when fine enough, elutriate with the flue gases. The amount will be dependent on the air and flue gas velocity, the size of the coal feed, and the strength and consistency of the ash or limestone used.

The four types of fluidized beds can be achieved in one of two ways, either by maintaining air velocity and decreasing coal size or by maintaining coal size and increasing the air velocity. This is a very important consideration with regard to the overall operation of the fluidized-bed combustor. Fan horsepower requirements will be dependent on the air velocity required for a given AFBC configuration and could impose a significant penalty for one system over another.

Coal top size requirements are dependent on the type of feed to the bed. The fines content must be controlled to a maximum of 5% less than 28 mesh, which could promote rapid elutriation and a high carbon loss. Here also the size distribution of the coal will have an effect on bed definition and elutriation. A narrow-size spread will allow better control of the bed and assist in combustion control optimization.

Combustion Within A Bed

Combustion within a bed is an interesting and unique situation. The interactions of the inert bed material with the combustion of individual particles of coal and the heat liberated by the coal and then absorbed by the bed becomes a much different environment than either stoker or pulverized coal firing. The principles of FBC can best be understood through a discussion of ignition within a heat sink.

If the ignition temperature of a given coal is 750°F, and the coal is a highly reactive coal giving up its total energy very quickly after ignition, the coal should ignite immediately upon entering a 750°F bed. This is not the case. An individual particle may ignite and release its energy to the bed particles in its immediate vicinity. However, since the bed is composed of 97% inert material acting as a heat sink, there is not enough heat in the bed to ignite a cold coal particle and sustain combustion. This is overcome when the bed temperature increases above the 1000–1100°F level.

The bed action as a heat sink and heat storage is such that once the ignition temperature is reached for a given coal, sustained combustion is a minor concern, unlike that for a stoker-fired boiler or a pulverized-coal-fired unit. Temperatures within the bed may vary, and it is important that air flow and turbulence be maintained as even as possible to assure even coal combustion and complete combustion of carbon.

When large units are considered, or turn-down capabilities desired, multibed configurations are sometimes used. In the start-up mode the beds are heated to approximately 1000°F by gas or oil igniters. Coal is then fed into the bed and upon sustained ignition the igniters are turned off and bed temperatures are raised to the operating level of from 1400 to 1800°F. In this mode only one bed need remain active. Other beds may be slumped with fluidizing air turned off. Increases in load are accomplished by fluidizing one bed at a time, elevating the temperature to a +1000°F and coal is then fed with or without limestone. This process is repeated until each bed is ignited and the unit reaches full capacity.

Heat Transfer

Heat transfer in a fluidized bed is a solid-to-solid transfer, which is much more efficient than a gas-to-solid heat transfer. The friction action of high-temperature particles imparts the particle's heat rapidly to the cooler surface. As a result, there can be greater heat absorption than for a conventional boiler. Radiant heat is still present at a somewhat reduced level because of the amount absorbed by the bed. Flue gas temperatures leaving the bed are at the bed temperature. At these temperatures there is a lower differential temperature and a need for more convective surface to achieve the same absorption. The total surface required for equal heat transfer with a conventional boiler is less if in-bed tubes are present.

Solids circulating within the bed provide an increased heat transfer, but it is a mistake to think the bed to be of uniform temperature in the turbulent state. Depending on air distribution patterns, and the solids' movement within the bed, temperature gradients may exist where higher velocities and bubbling may produce lower temperatures. Differences within the bed should not be more than 100 to 150°F if the fuel and air distribution are even.

Carbon Burnout

Carbon burnout is a function of the coal reactivity, temperature, and bed residence time at particle size. In order to achieve a fluidized bed, high velocities are required which, as coal mass is lost during combustion, can cause lighter coal not completely combusted to be carried away from the combustion zone. This can result in fairly high percentages of unburned carbon, that is, efficiency losses and a potential fire hazard in the ash removal system. Flyash reinjection and bed recirculation in the fast or entrained bed systems has improved combustion efficiency to equal or slightly better than that for a conventional stoker-fired boiler. The double-bed system with a higher-temperature second bed can accomplish the same results if an optimum coal is selected.

FBC *Fuel Specification*

To achieve optimum performance, the fuel characteristics should be closely evaluated. One of the main advantages of fluidized-bed combustion is the ability to use low-grade, poor-quality coal. The general specifications for this coal are as follows:

Moisture	10–30%
Ash	30% maximum
Top size	Per feed mechanism
Bottom size	Maximum 5% less then 28 mesh

Other than the above, any coals and most fuels can be combusted within a fluidized-bed boiler. Bed depth and free-board area above the bed can be designed to provide longer residence times relative to coal sizing and reactivity such that +95% of the carbon can be burned. Volatile, structure, friability, and reactivity are indicators for expected carbon loss. Reinjection can be used to minimize this.

Bed temperatures in the range of 1400–1800°F are substantially below the reducing atmosphere initial deformation temperature, eliminating the need for slagging and fouling control in the boilers. If an FBC system is operated using coal ash as the inert bed media in lieu of sand or limestone, special consideration should be given to the fouling or sintering characteristics of the ash. Special considerations must be given to the solid-to-fluid bed interface or, in the event of a multibed system, sintering in the slumped beds could be a problem. Inert bed composition may have to be modified in these cases by the addition of specialty sands or chemical additives. This can also be a problem in the ash systems.

Coal sizing has some effect, and the presence of high specific-gravity rock and material could cause settling problems that might cause upsets in air distribution. A washed or middlings-type coal could be recommended in lieu of a raw run of mine product.

Coal Feed Methods

As mentioned above the coal size selected is dependent on the type of coal feed method. The three methods most commonly used are the overfeed, underfeed, and in-bed feed. Each method is different in application and requirements, and their implications must be considered when evaluating an FBC system with a conventional system or other FBC design.

Overfeed

Coal is fed via gravity drop pipes, screw feeders, or spreader stoker-type feeders onto the top of the fluidized bed. Coal sizing must be such that the coal will submerge into the bed, ignite, and become fluidized. A standard stoker coal of $1\frac{1}{4}$ in. $\times \frac{1}{4}$ in. is normally specified with a maximum of 5–10% fines. Coal particles are degraded by the action of the bed, with larger pieces becoming part of the bed and the smaller particles being elutriated from the bed. Because of this, flyash reinjection must be provided.

Precautions must also be taken to protect the feeders from overheating when out of service. A sufficient pressure or air flow must be maintained in the feeder mechanism to offset furnace pressures when the unit is operating. If not, depending on the FSI and plastic propertities of the coal, coking or agglomeration could occur. This cooling air will prevent overheating, carbonization, and downtime.

Underfeed Systems

Although not as susceptible to coal sizing as the overfeed system, coal must be crushed to a top size that can be pneumatically transported to the bed and fluidized. Injection ports or nozzles are located at the bottom of the bed in this case and may vary in location and design from one manufacturer to another. The number of nozzles is such to give even distribution of the coal within the bed. The top size of this coal, as fed to the bed, is on the order of between $\frac{1}{8}$ and $\frac{1}{4}$ in., and the bottom size must be watched to provide that a minimum of coal is produced that is less than 28 mesh. Should a significant percentage of the coal be less than 28 mesh, a high percentage of these particles may elutriate from the bed as unburned carbon. It is also more difficult to reinject the ultrafine particles from this bed than the larger particles from the overfeed system. There is some degree of controversy with regard to which system generates the lowest carbon loss. This is a question of fuel characteristics.

Coal requirements for top size are dependent on what the crusher can handle. The lower the percentage of fines, which could increase the percent of 28 mesh, the better. High percents of 100 mesh or less coal could burn as a PC flame and create a wide range of temperature gradients in the bed, possibly causing stability problems. Crusher selection relative to coal friability and structure must be considered with regard to producing the optimum top size and distribution to the bed. As mentioned above, bottom size should also be considered.

In-Bed Feeding

In-bed feeding is a compromise between the overbed and underbed feed systems. Coal sizing can vary depending on the bed depth and location of the coal feed within the bed. In this case coal is crushed to approximately $\frac{1}{2}$ in. Materials of construction and bed pressure considerations tend to minimize the application of this system. An advantage of this method of feeding would be use of a simplified crusher and the ability to use a 2-in. $\times$ 0 product in lieu of a stoker coal.

Limestone Feed

When sulfur removal is required, limestone is fed into the bed by either mixing the limestone with the coal or by adding it separately into the bed through gravity drop pipes or pneumatic feeders. Limestone reactivity and strength must be taken into consideration to minimize degradation and rapid elutriation. The mechanism for sulfur removal will be discussed in Section 6.9-3.

Boiler Design

Based on the above considerations of bed characteristics and feeding method, there can be some variation in boiler design with regard to required surface and a balancing of in-bed to water wall and freeboard tube surface above the bed. Heat transfer coefficients (Btu/h-ft^2-$°$F) vary from 40 to 70 with 40–50 being normal for water wall and in-bed tube surface and from 5 to 20 with 10–15 being normal for freeboard surface. Except for the entrained-bed FBC, heat transfer in the convection section is the same as that for a conventional boiler, depending on the temperature difference between the flue gas and the tube temperatures. Heat transfer in the entrained-bed FBC is accomplished through freeboard and tubes in the sand recirculation section. Because the bed is recirculated, there is better carbon burnout and more time for heat transfer in the recirculation chamber. Total heat transfer surface is about the same.

Erosion

Erosion is the greatest problem when considering the turbulent fluidized bed of highly abrasive particles. The abrasiveness of the bed will be dependent on the composition of the bed (limestone, coal ash, sand, and combinations of these) and the angle at which the tubes are exposed to the bed particle flow, which is normally in the vertical plane. Perpendicular tubes in the bed would be subjected to the greatest erosion and those parallel to the flow would suffer the least erosion. Hence, in most cases an angled tube within the bed is used with the angle, depending on bed depth and the need for optimum heat transfer. The greater the particle-to-tube contact, the greater the heat transfer. Particle size and kinetics will provide the necessary information to determine erosion rate and the tube thickness required for a given life expectancy. Above the bed, erosion is similar to that in a conventional unit.

Air Distribution

The air-handling system for an FBC unit must be able to generate sufficient pressure and velocity to fluidize the bed. Air must be delivered evenly throughout the bed to minimize temperature and combustion differences. Two methods are currently in use. In the first method an air distribution plate distributes the air under the bed. Unfluidizable particles may settle onto the plate and cause an upset in the air distribution, causing the bed to be removed from service. Provisions for ash removal must be provided. In the second method nozzles are used to inject the air above the base plate, fluidizing the bed. This leaves a dead zone where heavier particles will fall and not affect the bed. Ash removal must be provided to maintain the level of the dead zone and minimize nozzle plugging.

Load Variations

Load fluctuations can be handled with a fluidized-bed boiler by the use of multiple beds, bed depth control, bed temperature control, in-bed tube surface control relative to bed depth, and rate of recirculation in the entrained bed. For the mild, smooth, bubbling beds a combination of the above and a multiple can be arranged to provide turndown equal to or better than the stoker-fired boiler. In the case of the entrained-bed boiler with primary heat transfer, depending on the bed recycling rate and bed temperature, the load fluctuations are controlled by direct control of these.

Ash Removal

Ash removal can be accomplished by continuously removing a portion of the bed with a bleed line, which will maintain bed density. Since the bed is only 3% unburned carbon, very little efficiency is lost in this process. The amount of the bleed required is dependent on the amount of inert material being fed into the bed as ash. Higher ash contents will require increased bleed rate and result in higher carbon losses. Large bleed requirements could result in bed temperature distortions.

Another method for ash removal involves the dead zone below the air distribution nozzles where heavier ash particles will fall into the dead zone for removal through ash ports at the bottom of the zone. A water-cooled conveyor is used to carry off the ash. In this case no bleed system is used. Whenever a dead zone is considered, the fouling characteristics of the coal must be evaluated to minimize sintering in the dead zones or in any beds that might be slumped. When sulfur removal is used and the ash is high in calcium sulfate and unreacted lime, care must be taken to minimize moisture addition, which could cause the ash to harden like concrete.

Emissions

Emissions from FBC systems are similar to those of a conventional unit with regard to particulates and NO_x. However, sulfur emission regulations can be met because of the ability to inject limestone and capture the sulfur. Particulate emissions will be greater than that for a stoker-fired boiler but less than that for a pulverized-coal-fired boiler. A baghouse or electrostatic precipitator must be used to meet current standards. NO_x emissions will be equal to or less than those for a similar-sized conventional boiler, because lower bed temperatures limit the formation of thermal NO_x but increase the time for the conversion of fuel nitrogen to NO_x. Thin-bed two-stage combustion, which minimizes the time of formation, has been used successfully to minimize the formation of NO_x.

SO_2 removal is accomplished in an in-bed gas-to-solid reaction. Bed residence time, accessibility of sulfur to limestone, and limestone reactivity are important considerations. Because the reaction is a surface reaction, it is important to have sufficient limestone surface for the reaction of the total SO_2. The larger the limestone, the smaller the surface. On the other hand, the bottom size of the limestone is limited by the rate of elutriation at the bed pressure and the air velocity. This limitation of the total surface of the limestone will increase the stoichiometric ratio of limestone required to obtain a given percentage of sulfur removal efficiency. Increasing the limestone reactivity can improve the stoichiometric ratio to some extent; however, care must be taken to use a limestone structured such that degradation within the bed is minimized. In some cases, if low-sulfur coal is used, the rate of reaction forming a $CaSO_4$ coating on the limestone is insufficient to prevent degradation, leading to rapid elutriation and very high stoichiometric ratios. The stoichiometric ratios normally associated with AFBC range from $2:1$ to $6:1$. In rare cases when a specific coal, limestone, and AFBC arrangement come together, a stoichiometric ratio of $1.2:1$ to $1.5:1$ may be obtained. The fast or entrained-bed system has some advantages because of the recycling of the bed. The two-stage fluidized bed with a high-temperature second stage can produce the same stoichiometric ratios if the proper coal is selected.

A short discussion is needed regarding the total system economics and design. Because of the crushing, air pressure and high-velocity air, ash disposal and limestone injection systems have a higher cost than that for a conventional boiler using low-sulfur coal. When the total system is considered at a given location and there is the ability to use low-price, low-quality locally available coal, the use of an FBC boiler with higher capital and operating costs can be justified. The evaluation of an FBC unit must be considered with that of a conventional boiler. The higher operating costs for using limestone at high stoichiometric ratios can be an important consideration for both handling and disposal systems.

Pressurized Fluidized Bed (PFBC)

PFBC systems have been designed to increase overall system efficiency by the combustion of coal in a combined-cycle system using an FBC boiler and a hot-gas turbine at a pressure of from 8 to 16 atm and a temperature of between 1400 and 1700°F. This temperature is limited by the coal ash characteristics, acidic flue gas emissions, and turbine design parameters. There are many questions regarding materials of construction that must be answered as they can be affected by a specific coal's ash characteristics. PFBC can be classified in a research and development stage.

Technological areas of concern range from the boiler system through the hot-gas cleanup and turbine system. In a PFBC the hot gas is expanded to produce electric power and the hot off gas is passed through a waste heat boiler or economizer to generate steam. This steam is combined with any steam produced in the pressurized fluidized bed and used to produce steam for increased electrical capacity at lower overall costs. Design for the pressure vessel to contain the combustion reaction at the high temperature and pressure and the fuel and sorbent feed systems and hot gas cleanup capability must be questioned. Many advances have been made; however, methods for breaking the pressure barrier for feeding fuel and sorbent and for waste removal seem to pose the greatest challenge to developers.

The high-temperature, high-pressure flue gas cleanup system must be capable of maintaining gas temperature and pressure and remove ash and corrosive alkali and chloride contaminants from coal combustion. The gas clean-up problem is made more difficult by a lack of knowledge about what conventional hot-gas turbine requirements for gas quality are or what modifications may be required to facilitate coal utilization. In any case, the chloride content of the coal and its effect on the total system remain to be addressed in detail. Following a more accurate definition of the problems, some answers may be developed by selective testing of new materials.

Control system and safety interlock with a hot-gas turbine are among other problems that must be addressed. The basic principles of FBC apply for startup and shutdown; however, consideration for load following, emergency shutdown, and process control instrumentation are only in the developmental stages.

Environmental compliance can be achieved with normal emission control technology; however, the cost of doing so may alter any economic advantage projected for the PFBC technology because of the greater temperature and pressure requirements. At this time there is insufficient data to determine the cost of a commercial system for either capital requirements or for operating costs.

FBC *Summary*

The fluidized-bed combustor or combustion boiler is a viable part of the total system evaluation and must be considered whenever there is a requirement for sulfur removal and a supply of very low price or low-quality coal locally available to the potential user's facility. Although the specifications of the coal are not as important as in the conventional boiler, the reactivity of the coal and its sizing can be important considerations depending on the method of feed used. Coal sulfur content and limestone reactivity must also be considered when strict SO_2 emissions are to be met or when there is a high cost for ash disposal and high stoichiometric ratios of limestone must be used.

Pressurized fluidized-bed combustion is in the developing stages, and for the most part the technology for AFBC can be used to understand the system. Problems with the high-temperature, high-pressure environment are still to be addressed.

BIBLIOGRAPHY

ASME Performance Test Codes PTC 4.1.

Breen, B. P. "A General Understanding of Low Nitric Oxide Operation of Stationary Combustors." Prepared for UCLS Seminar: Fundamentals of Pollutant Formation in Combustion Processes, August 1970.

Combustion: Fossil Power Systems. Combustion Engineering Inc., 1981.

Greenwald, R. F. and Redfield, R. S. "How to Choose the Right Fluidized Bed Boiler." *Power Magazine*, December, 1982.

Harris, M. E., Rowe, V. R., Cook, E. B., and Grumer, J. "Reduction of Air Pollutants from Gas Burner Flames." Bureau of Mines, Bulletin No. 653, 1970.

Miller, S. A., Vogel, G. J., Gehl, S. M., Hanway, J. E., Jr., Henery, R. F., Parker, K. M., Smyk, E. B., Swift, W. M., and Podolski, W. F. "Technical Evaluation: Pressurized Fluidized-Bed Combustion Technology." Argonne National Laboratory, Argonne, IL, April, 1982.

Proceedings of the Conference on Engineering Fluidized Bed Combustion for Industrial Use. Ohio Department of Energy and Battelle's Columbus Laboratories, Columbus, OH, 1977.

Proceedings of the 7th International Conference on Fluidized Bed Combustion. The U.S. Department of Energy, Electric Power Research Institute, and Tennessee Valley Authority, October, 1982.

Rawdon, A. H. and Johnson, S. A. "Control of NO_x Emissions from Power Boilers." Presented at the 35th American Power Conference, Chicago, IL, May 10, 1973.

Zeldovich, J. "The Oxidation of Nitrogen in Combustion and Explosives." *Acta Physiochemica, URSS,* **21**, 577 (1946).

6.8   STEAM GENERATING EQUIPMENT

*Eric H. Smith**

6.8-1   Stokers[†]

In the early days a "stoker" was someone employed to tend a furnace. To stoke a fire meant to do whatever was necessary to keep it burning properly, including poking, stirring, and supplying it with fuel. Eventually these stokers were replaced with mechanical devices that automatically fed fuel to the furnaces and in some cases automatically disposed of the ash. These were called mechanical stokers, and first appeared in the 1800s.

Today there are a variety of mechanical stokers in use. They differ in the way in which solid fuel is fed onto the grate, and they differ in the way in which ash is removed from the grate. Each method of firing has its own unique operating characteristics, and each has its own best applications. This section will discuss the types of mechanical stokers commonly used today for firing coal.

Today's stokers can be sorted into three categories based on the way in which fuel is fed onto the grate. These categories are: (1) underfeed stokers, (2) overfeed stokers, and (3) spreader stokers. Within each category the stokers differ according to the way in which the grate handles the ash. Each stoker category will be discussed, and the most common stoker types will be illustrated.

Underfeed Stokers

In an underfeed stoker, the fuel is introduced through long troughs, called "retorts," at a level below the location of air admission to the fuel bed. Thus the green coal (or raw coal) is at the bottom. The ash moves away from the retort and combustion takes place in between, constantly receiving a fresh supply of green coal from below and displacing the ash.

Underfeed stokers were developed in the 1800s, and were very popular before World War II. After World War II, they were gradually replaced by larger spreader stokers and overfeed stokers. However, they have many useful applications and are still being sold today.

The smallest of the underfeed stokers has a single or double retort into which the coal is fed. The coal feed is by either a screw or mechanical ram which forces the coal the length of the retort and upward. The ash on this type of stoker is normally discharged with side dumping grates. These stokers will fire boilers in the size range of up to 20,000 pounds of steam per hour.

The ash moves to the sides where it can be periodically dumped into ash pits and removed. The "tuyeres," through which air is admitted, are above the retort.

Most modern underfeed stokers are also equipped with overfire air jets to provide turbulence to mix the volatiles with the air for more complete combustion.

Multiple Retort

The larger underfeed stokers are of the multiple retort type having as many as 18 retorts inclined at an angle of 25–30° to aid the movement of coal and ash. An illustration of this type of stoker is given in Fig. 6.8-1. In this type of stoker the ash is discharged at the rear either intermittently with a dumping grate or continuously by means of a clinker grinder. The assembly illustrated in Fig. 6.8-2 has a rocker dump mechanism which ensures continuous and automatic ash discharge.

The multiple retort stoker employs large mechanical rams and pusher blocks to feed and distribute the coal. It also employs overfire air when necessary. These units are used in boilers which range in size from 20,000 to 50,000 pounds of steam per hour.

Underfeed stokers operate with very thick fuel beds. This results in a high thermal inertia, or a slow response time to changes in steam loading. They also have trouble burning certain grades of coal, such as high-coking coals, which form a coke that expands and arches off the grate, and free burning sub-bituminous coals whose loose ash leave sections of the grate bare causing overheating and grate damage. Low ash bituminous coals when burned may not generate sufficient ash to protect the grate surface.

On the positive side underfeed stokers have a clean smokeless combustion when fired with the proper coals and they have low flyash carryover. The smokeless combustion results from feeding the coal from underneath the combustion zone. As the coal is heated, the volatiles are driven off and pass upward through the incandescent burning coals where most are consumed before they pass completely through the fuel bed.

*Deceased 1983. Text revisions by personnel of Riley Stoker Co., Worcester, MA.
[†] Portions adapted by permission from *A Guide to Clean and Efficient Operation of Coal Stoker Fired Boilers*. Prepared under contract to USDOE and USEPA and published by the American Boiler Manufacturers Association, May 1981.

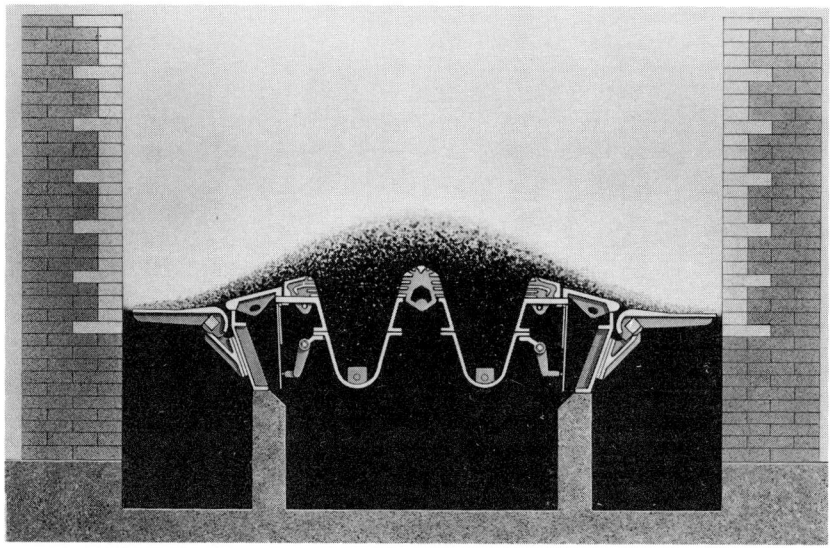

Fig. 6.8-1 Multiple retort stoker. (Courtesy of Riley Stoker Corp.)

Overfeed Stokers

In an overfeed stoker the coal is fed onto the grate above the point of air admission. The two basic types of overfeed stokers are the chain, or traveling grate stoker, and the water-cooled vibrating grate stokers. Technically, a spreader stoker is also a type of overfeed stokers. However, because it is commonly considered in a class by itself due to its unique features, it will be classified separately in this text.

Traveling Grate Stoker. A typical traveling grate overfeed stoker setting is illustrated in Fig. 6.8-3. It consists of a continuously moving grate. Coal is deposited on one end of the grate by gravity feed from

Fig. 6.8-2 Multiple retort stoker. (Courtesy of Riley Stoker Corp.)

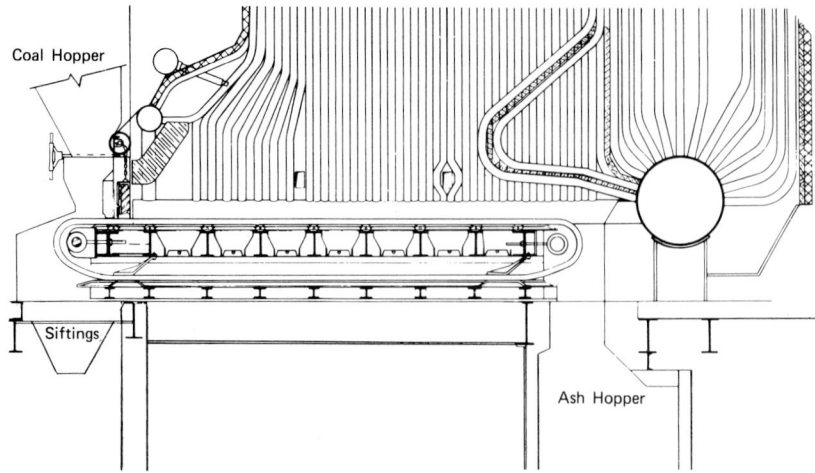

Fig. 6.8-3   Traveling grate stoker. (Courtesy of Riley Stoker Corp.)

a coal hopper. The coal depth is adjusted by a guillotine-like movable coal gate to a thickness of about 4–12 in. The coal is burned as it passes slowly through the furnace at grate speeds of less than 30 feet per hour. The ash is continuously discharged off the rear of the grate into an ash pit.

As noted, the two types of grate construction in use are classified as chain or traveling grates, although the chain grate also travels. A typical traveling grate as shown in Fig. 6.8-4 has a surface which consists of a series of grate clips mounted on lateral rack bars. The rack bars are attached to

Fig. 6.8-4   Traveling grate. (Courtesy of Riley Stoker Corp.)

rugged skid-mounted chain link assemblies. Other grate assemblies may employ roller chains. The undergrate air passes through openings in the grate clips which are designed to provide uniform air distribution, grate surface cooling, and minimum fuel ash siftings.

Undergrate air is controlled through individual air zones or compartments underneath the grate. The amount of air entering each zone is manually controlled by the operator. Many of the overfeed stoker-fired boilers have a rear arch, depending on coal type, which directs any remaining volatile gases and cinders from the burnout zone back toward the flame zone where they may be burned. Rows of high-pressure overfire air jets are used to mix the volatiles with the air for more complete combustion. Chain and traveling grate stokers have been built and are available for boilers as large as 200,000 pounds of steam per hour.

Water-Cooled Vibrating Grate Stoker. The water-cooled vibrating grate stoker uses vibration and gravity to move the coal. This type of stoker originated in Europe and was introduced to the United States in the mid-1950s.

It consists of a water-cooled grate supported by flexing plates and inclined at an angle of about 14°. Coal is gravity fed at the top of the grate and passes under a guillotine-type gate which controls the bed thickness. The grate is vibrated for about 5 seconds every 2 minutes. The interval and duration of the vibrations is tied to the automatic controls and determines the rate at which the coal is moved through the active burning zone. Ash is discharged at the rear of the stoker.

As with the chain and traveling grate stokers, the vibrating grate stoker has individually controlled air zones. The boiler often has a rear arch to direct any remaining volatile gases from the burnout zone back into the active combustion zone. It also uses high-pressure overfire air jets on the front wall to promote mixing of the volatile gases and the air for more complete combustion.

In general, overfeed stokers are characterized by low flyash carryover. They burn most coals although high-coking coals may be a problem. The overfeed stoker's response time to rapid changes in load is slower than that of the spreader stoker and such stokers require a larger grate size for a given heat input than a spreader stoker.

Spreader Stokers

In a spreader stoker coal is spread evenly over the entire grate surface by mechanical feeders located at the stoker front above the grate. Because the coal is thrown onto the grate, there is some suspension burning of the coal fines. This suspension burning coupled with a very thin fuel bed allows the spreader stoker to respond to rapid load changes.

There are many types of mechanical feeders in operation. They all utilize rotor speed and coal trajectory to adjust coal distribution on the grate. One type of mechanical feeder is illustrated in Fig. 6.8-5. In this typical arrangement coal feeds through a feeder drum which provides uniform fuel

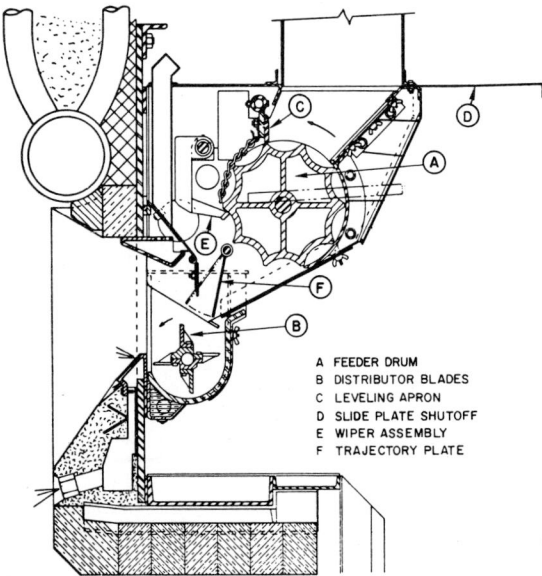

A FEEDER DRUM
B DISTRIBUTOR BLADES
C LEVELING APRON
D SLIDE PLATE SHUTOFF
E WIPER ASSEMBLY
F TRAJECTORY PLATE

Fig. 6.8-5 Mechanical feeder. (Courtesy of Riley Stoker Corp.)

RILEY TRAVELING GRATES

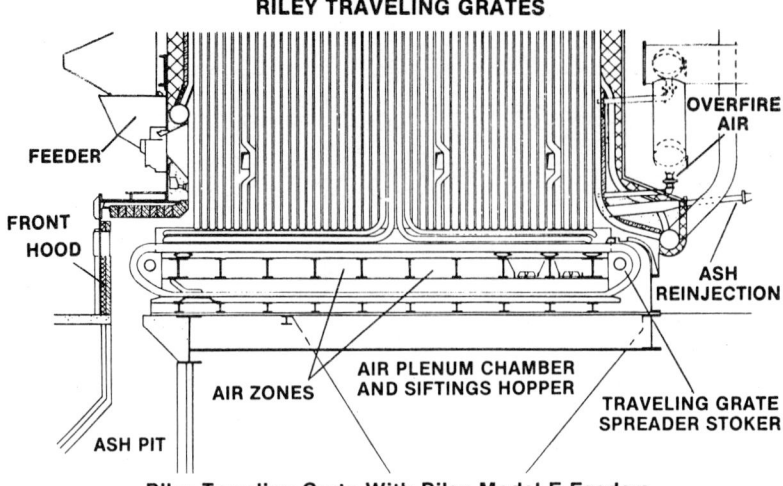

Riley Traveling Grate With Riley Model F Feeders

Fig. 6.8-6   Spreader stoker. (Courtesy of Riley Stoker Corp.)

metering onto a rotating "overthrow rotor," which is comprised of a series of distributor blades spinning at several hundred revolutions per minute. As the coal moves over the "trajectory plate" it is struck by the distributor blades and thrown into the furnace. The blades are designed to distribute the coal over a wide area. The rotor speed, trajectory plate, and distributor blade orientation can be adjusted to provide an even distribution of coal on the grate. Other coal feeders are available, mainly on larger installations, which have overthrow rotors and trajectory plates but employ gravity feed.

A typical arrangement of a spreader stoker with traveling grate is illustrated in Fig. 6.8-6. Note that the ash is discharged at the front of the stoker rather than at the rear. The spreader stoker is equipped with overfire air jets front and rear to provide turbulence for mixing the volatile gases with the air, and to hold the flames off the water walls and out of the feeder throat. Spreader stokers are usually not equipped with a complete series of separately controlled undergrate air zones because these have been found to be generally unnecessary. Traveling grate spreader stokers have been built and are available for boilers as large as 400,000 pounds of steam per hour.

The spreader stoker is characterized by a thin bed and partial suspension burning. As a result, it responds rapidly to changes in load. It is capable of firing a wide range of coal grades and types. The spreader stoker has high availability, simplicity of operation, and high operating efficiency, but it has high flyash carryover and a high flyash combustible heat loss. Cinder reinjection is used on spreader stokers to recover some of the carbon in the collected flyash.

Static Spreader Stoker Applications

In keeping with the principle of true coal spreader feeding are other grates that are comprised of a basic static design but include specific localized methods of both fuel and ash removal. Two such examples are termed the Dump Grate and Oscillating Grate Spreader Stokers.

Oscillating Grates.   The grate movement is obtained by small-amplitude motion of a series of flexure plates which span between the rigid base stoker frame and the interlocked grate bars. In some instances the grate is not always horizontal and has a small incline to take advantage of gravity assistance. A typical construction is illustrated in Fig. 6.8-7 which can attain boiler steaming capacities of up to 150,000 lb/h. The oscillating drive mechanism gives motion for approximately 5 s every 10 min, being governed by a timer control that relates to boiler load demands. The undergrate air control is by manually operated zone dampers which effectively distribute the air. The damper controls are conveniently located at the stoker side. The grate combustion efficiency, while not as efficient as the handling grate, is acceptable when considering the lower grate ratings used for static installations.

Dump Grate.   A typical design of a dump grate mechanism is shown in Fig. 6.8-8. The grate fuel ash discharge is accomplished by having the grate bars pivoted. By using a power cylinder, interconnecting rods operate the grate bar levers in specific zones using interconnecting levers. Manual operation is also possible. These grate designs have been successfully installed for boiler capacities of up to 50,000 lb/h when fired with low grade bituminous and sub-butuminous coals.

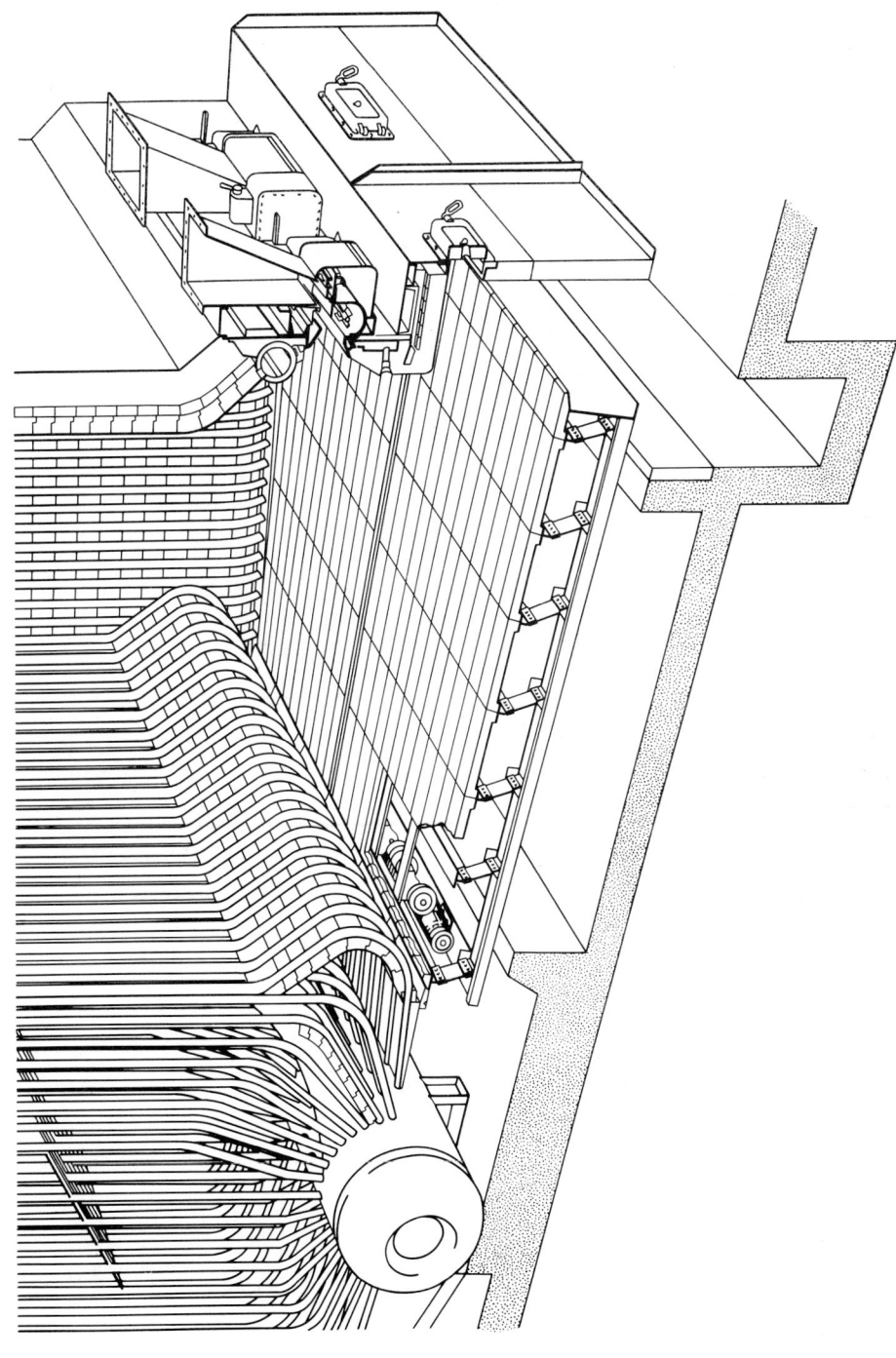

Fig. 6.8-7 Oscillating grate spreader stoker. (Courtesy of Riley Stoker Corp.)

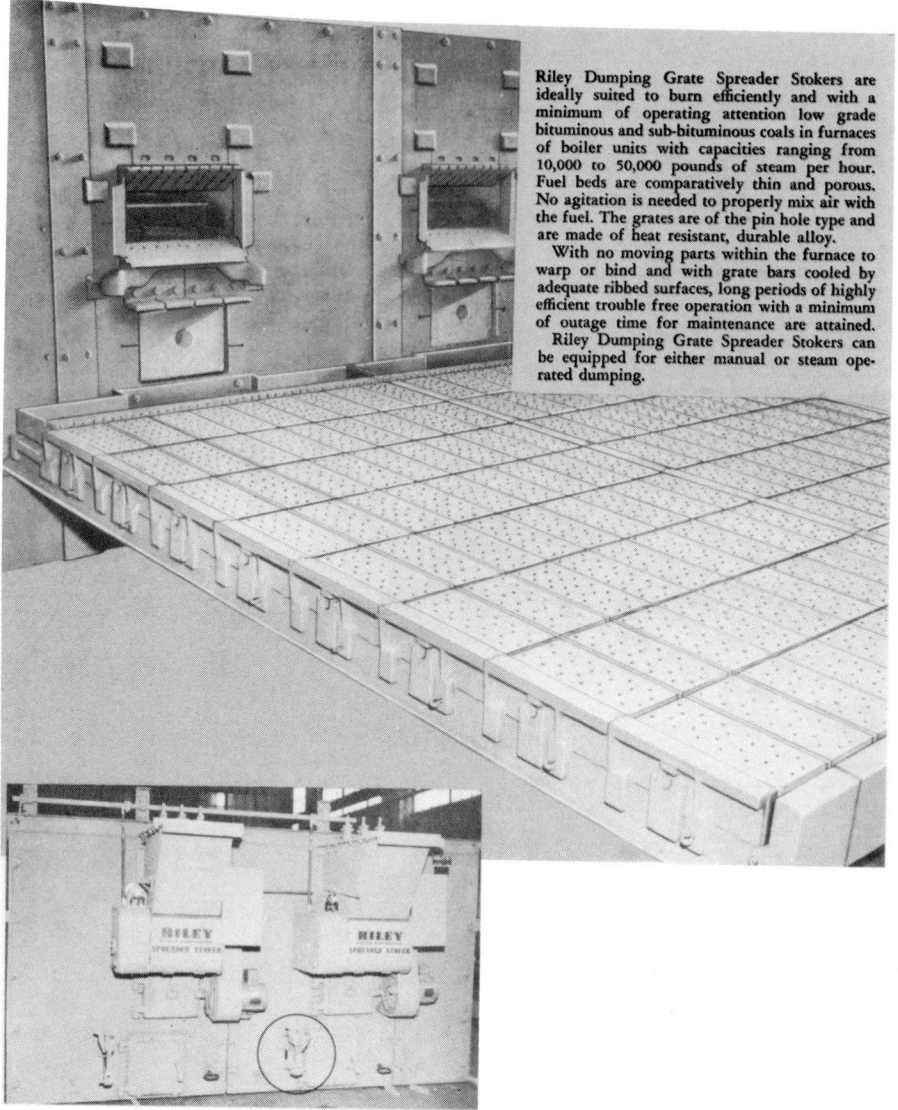

Riley Dumping Grate Spreader Stokers are ideally suited to burn efficiently and with a minimum of operating attention low grade bituminous and sub-bituminous coals in furnaces of boiler units with capacities ranging from 10,000 to 50,000 pounds of steam per hour. Fuel beds are comparatively thin and porous. No agitation is needed to properly mix air with the fuel. The grates are of the pin hole type and are made of heat resistant, durable alloy.

With no moving parts within the furnace to warp or bind and with grate bars cooled by adequate ribbed surfaces, long periods of highly efficient trouble free operation with a minimum of outage time for maintenance are attained.

Riley Dumping Grate Spreader Stokers can be equipped for either manual or steam operated dumping.

When desired the dump operation can be hand operated as shown in inset. The dump mechanism is conveniently located at the stoker front. A locking device on the dumping mechanism assures proper positioning of the grate surface.

Fig. 6.8-8 Dumping grate spreader stoker. (Courtesy of Riley Stoker Corp.)

The self-cleaning design of the pinhole grate bars permits easy ash dumping and quick positioning of the grate surface without binding or jamming. The grate efficiency or carbon loss is comparative to oscillating grates but not as effective as traveling grate stoker designs.

Boiler Efficiency

Most stoker fired boilers are between 65 and 85% efficient in converting the coal's energy into steam. This means that some of the coal's energy is wasted. Some of this waste is unavoidable, as we will see, but some of it can be avoided if the operator understands how it occurs.

Conservation of Energy.   One of the laws of nature states that energy is conserved. It may change form, as from chemical energy to thermal energy, but you always end up with as much as you started with. So in a coal-fired stoker boiler, if you add up the energy in the steam, the energy in the flue gas, the energy in the feed water, the energy radiated from the boiler, and all the other forms of energy leaving the boiler, they will exactly equal the energy in the coal burned. Another way of stating it is:

$$\text{Energy in Feedwater Energy in Coal} = \text{Energy in Steam}$$
$$+ \text{Energy in Heat Losses}$$

Boiler Efficiency. Boiler efficiency is the percentage of the coal's energy converted into steam energy. An accurate way of determining boiler efficiency acceptable in the industry is to use what is known as the heat loss method. In this method you simply measure the individual heat losses (expressed as percent of heat input) and subtract them from 100%, since energy is conserved.

$$\text{Boiler Efficiency} + 100\% - \text{Heat Losses}$$

Stack Gas Heat Loss. The largest loss of energy in most boiler plants is the energy in the flue gas going out the stack, which might account for as much as 30% of the fuel input in the worst cases. This is called the stack gas heat loss. When engineers measure this heat loss, it is broken down into three components: heat loss due to dry gas, heat loss due to moisture in the fuel, and heat loss due to H_2O from the combustion of H_2. As far as the operator is concerned, these three can all be lumped into one category—stack gas heat loss.

The biggest gains in boiler efficiency can be obtained by minimizing the stack gas loss. This loss depends on the temperature and the volume of gas leaving the boiler, so a reduction in either one of these will reduce the heat loss.

The practical minimum flue gas temperature limit is about 300°F, subject to an evaluation of the fuel. At lower temperatures, sulfuric acid vapor in the flue gas condenses on cold metal surfaces and causes severe corrosion. Therefore, some stack gas heat loss is unavoidable. To eliminate the stack gas heat loss altogether, it would be necessary to reduce the stack gas temperature to the temperature of the air around the boiler. Practically speaking, this is impossible.

There are three basic strategies for minimizing the stack gas heat loss:

1. Minimizing excess air.
2. Keeping heat transfer surfaces clean.
3. Adding flue gas heat recovery equipment where justified.

When excess air is reduced, it reduces the volume of flue gas leaving the boiler. It also reduces the temperature of the flue gas because the gas velocities are reduced and the flue gas spends more time in the boiler where heat can be absorbed. As a result of reducing the volume and temperature of the flue gas, the stack gas heat loss is reduced and the boiler efficiency is increased. As a rule of thumb, boiler efficiency can be increased one percent for each 15% reduction in excess air, 1.3% reduction in O_2, or 40°F reduction of stack gas temperature.

Keeping heat transfer surfaces clean must also be considered when operating the boiler. Ash deposits on furnace water walls (slagging) and on boiler connective tube surface (fouling), and scale deposits on the water/steam side tube surfaces act as insulation. This reduces the amount of heat the boiler water absorbs from the flue gas. As a result, the flue gas temperature is high and the boiler efficiency is low. Fouling and slagging are controlled by proper use of soot blowers, overfire air, and excess air. Internal tube scale deposits are controlled by feedwater treatment and proper use of blowdown.

The third strategy for minimizing stack gas heat loss is to install additional heat recovery equipment such as an air heater or economizer. If the boiler's flue gas temperature is greater than 450°F when tuned-up, it is worthwhile considering one of these devices. An engineering study will be required to determine if such a device will be cost-effective and that the coal quality being fired is acceptable for higher combustion air temperatures.

Combustible Heat Loss. The second largest heat loss in stokers is due to unburned fuel. This is called the combustible heat loss, and can be greater than 5% of the coal's energy. It occurs in three major ways:

1. Carbon in the bottom ash.
2. Carbon in the flyash.
3. Combustible gases in the flue gas.

Carbon in the bottom ash results when coal is dumped into the ash pit before it has been completely burned. The operator should take care to adjust the stoker so that the carbon is completely burned out. On traveling and chain grate stokers, this is done by properly adjusting the coal gate position, grate speed, and the undergrate air flow. On spreader stokers, the operator must adjust the grate speed, undergrate air flow, and the coal feeders if necessary. Experience will show which stoker adjustments work best.

Recent tests on 18 stokers gave the following range of heat losses due to combustibles in the bottom ash:

	Lowest	Highest	Average
Spreader stokers	0.0%	3.4%	0.9%
Overfeed stokers	0.4%	8.1%	2.4%
Underfeed stokers	1.2%	3.9%	3.2%

Carbon in the flyash is the result of small coal particles being blown off the grate or, in the case of spreader stokers, small particles being caught up (or entrained) in the gas flow before they land on the grate or have sufficient time to complete combustion in the furnace. This is also called flycarbon, carbon carryover, or combustibles in the flyash. To reduce this heat loss, and reduce the particulate loading, increasing the overfire air has been found to be effective in some cases. Reducing the excess air may also work in some cases, and using coal with fewer fines or less ash and fixed carbon may help.

Tests on 18 stokers gave the following range of heat losses due to combustibles in the flyash:

	Lowest	Highest	Average
Spreader stokers	0.5%	9.2%	4.4%
Overfeed stokers	0.3%	1.1%	0.5%
Underfeed stokers	0.1%	0.2%	0.1%

Combustible gases are volatiles which did not burn to completion. The most commonly measured combustible gas is carbon monoxide (CO). Carbon monoxide is formed when the excess air is too low in some area of the grate, when there is inadequate mixing, or when the flame impinges on a cold water wall. If the coal is evenly distributed on the grate, and if the undergrate air and overfire air are properly adjusted, this heat loss is minimized. On most stokers, the carbon monoxide concentration can be maintained below 400 ppm (0.4%). This is about a 0.2% heat loss.

Radiation Heat Loss. Some of the heat of combustion escapes through the walls of the furnace without being absorbed by the boiler water. This energy loss is called the radiation heat loss. Some radiation heat loss is unavoidable. If there were none, the outside surfaces of the stoker boiler would be the same temperature as the surrounding air. Most stoker-boilers are properly insulated when installed. It is necessary to maintain the insulation in good condition. This includes all hot surfaces—water walls, ducting upstream of heat recovery devices, and steam pipes.

6.8-2   Pulverizers and Pulverized Coal*

General

Pulverized coal utilized for the firing of steam-generating equipment is dried, ground, and classified to a fineness which may be transported and burned in suspension with air. The degree of fineness generally utilized ranges from 60 to 90% through a 200-mesh sieve. A typical sizing is represented by this screen analysis: 99.5% through 50 mesh; 96.5% through 100 mesh; and 80.0% through 200 mesh. The surface area of the particles of a pound of coal classified to this fineness is approximately 105,000 in.^{2} This area represents an increase of approximately 2300 times that of a single 1-lb lump of coal.

Energy Required

Energy required for preparing, grinding, and transporting coal ranges from 10 to 20 kW-h/ton of coal. The factors that materially influence the actual power required for a particular installation are type of equipment, grindability of coal, and degree of fineness. In general, the power is also influenced by the rating at which the equipment is operated with respect to its design capacity.

Coal Preparation

The necessity for supplying raw coal of uniform quality to the pulverizer at a metered rate requires cleaning, rough sizing, and bunkering of the coal. Cleaning necessitates the removal of foreign material, such as large pieces of wood, straw, rags, and iron. This foreign material, often present in the

*Revised and updated by permission from Kent, *Mechanical Engineers Handbook*, 12th ed., by Riley Stoker Co., Worcester, MA.

coal as delivered to the plant, is removed to facilitate uniform feeding and to protect equipment from damage. Iron may be removed by magnetic separators in the conveyor system; other foreign materials may be removed by screening, by manual removal, or by the equipment utilized for rough sizing or crushing.

Rough sizing or crushing eliminates oversize pieces of material which would jam the feeder or cause an irregular feed rate, and better distributes the moisture in the raw coal. The preferred sizing of the coal produced by the crusher is generally limited to a maximum of all through $\frac{3}{4}$ in. round screen. For small-capacity mills and to obtain a more uniform moisture distribution, the coal may be crushed down to all through a $\frac{1}{2}$ in. round screen. The bunker must be designed to give storage space for this prepared raw coal, and to provide a uniform supply of coal to the feeder. The method of filling the bunker must be considered as this is found to influence the degree of segregation and packing.

Coal Feeders

An uninterrupted uniform feed to the pulverizer is essential to successful operation of a pulverized fuel system. In many cases the feeder is used as a source of metering the fuel supply to the system. Performance of the feeder is thus an important design consideration. Two types of feeders are generally used, the belt type and the drum type. The belt type features a continuous belt wrapped around two rolls—one the driver and the other the idler or take-up roll. Coal from the bunker spout flows vertically onto the belt at the idler end and is horizontally translated to the driver roll end and vertically discharged to the mill. This feeder can be a volumetric or gravimetric device depending on whether or not weighing provisions are incorporated in the design. In the gravimetric mode, weight accuracies having less than 0.5% error are common over extended feed rate ranges.

The drum type feeder consists of a rotating spider with pockets which fill from the bunker coal spout and discharge to the mill. This is a volumetric feed device that finds extensive use in ball mill systems and in other applications where accurate fuel metering is not required. It is, however, more susceptible to pluggage and stoppages due to foreign matter.

Principles of Grinding / Theories of Comminution

The principles of grinding are impact, attrition, and crushing. The application of one or more of these principles is employed in the various types of mills.

Although known that the energy required for grinding is a function of feed size, product fineness, coal grindability, and, to a degree, moisture in the coal, there is no generally accepted theory of the relation of energy to fineness of grinding. Three such theories, however, have been developed and promulgated: *Rittingers Law*—"The work to produce material of a given size from a large size is proportional to the new surface produced"; *Kick's Law*—"The energy required to effect crushing or pulverizing is proportional to the volume reduction of the particle;" *Bond's Work Index* or *Third Theory of Comminution*—"The total work useful in breakage which has been applied to a stated weight of homogeneous broken material is inversely proportional to the square root of the diameter of the product particles."

Pulverizer Function

The function of a pulverizer is to grind, dry, and classify coal to a state in which it can be successfully transported and burned in mechanical suspension.

Predicted Pulverizer Performance

For lack of an exact law governing the energy required for grinding and a simple method for determining new surface, it has been desirable to predict mill grinding performance on the basis of grindability and 200-mesh sieve fineness rather than on surface area produced.

The Hardgrove method of grindability determination has been generally used because it is a direct measure of the 200-mesh fineness produced by a standard unit of energy input. Grindability is determined by placing a 50 g sample of air-dried coal, sized to minus 16 and plus 30 mesh, in the mortar of a specially designed test machine. After turning the machine through 60 revolutions at 20 rpm, while loading with 64 lb, the sample is removed and screened. The quantity passing a 200-mesh sieve is used to determine the Hardgrove grindability index by the following empirical formula:

$$G = 6.93W + 13$$

where W is the weight in grams of the sample that passes a 200-mesh sieve.

Alternatively, for material other than coal, and in most applications outside the United States, the Bond Work Index has found increasing acceptance for grindability determination. Bond has developed

TABLE 6.8-1   GRINDABILITY OF U.S. COALS

State	County	Coal Class	Grindability
Alabama	Jefferson	Bit	62–87
Arkansas	Franklin	Bit	99–102
Colorado	Las Animas	Bit	44–54
Illinois	Williamson	Bit	52–59
Indiana	Clay, Greene, Vigo	Bit	62–66
Iowa	Appanoose, Wayne	Bit	60–70
Kansas	Cherokee	Bit	61
Kentucky	Pike	Bit	50–60
Kentucky	Hopkins, Muhlenberg	Bit	60–65
Maryland	Alleghany	Bit	95–100
Missouri	Adair	Bit	72–75
Montana	Carbon, Musselshell	Sub-bit	47–56
New Mexico	McKinley	Bit	29–41
North Dakota	Mercer	Lig	45
Ohio	Belmont	Bit	50–60
Oklahoma	Pittsburgh	Bit	47–67
Pennsylvania	Schuylkill	Anth	35–45
Pennsylvania	Luzerne, Lackawanna	Anth	25–30
Pennsylvania	Cambria	Bit	85–107
Pennsylvania	Somerset	Bit	87–100
Pennsylvania	Westmoreland	Bit	60–70
Tennessee	Campbell	Bit	45–55
Texas	Bowie, LaSalle	Lig	53–79
Utah	Carbon	Bit	43–49
Virginia	Montgomery	Anth	83
Washington	Kittitas	Bit	49–52
West Virginia	Kanawha, Fayette	Bit	40–60
Wyoming	Campbell	Sub-bit	52

the following correlation of Work Index (W_i) to Hardgrove Grindability Index (HGI):

$$W_i = \frac{435}{\mathrm{HGI}^{0.91}}$$

where the work index is determined on an as-received sample. For air-dried samples, the work index is given as 1.3 times the above or

$$W_i = \frac{565.5}{\mathrm{HGI}^{0.91}}$$

The grindability index of various coals permit the manufacturer to predict the performance of a particular type mill with the coals to be used. Table 6.8-1 shows the Hardgrove Grindability of many U.S. coals.

Coal Drying in the Pulverizer

The ability of a particular mill to dry coal depends on the heat added in the form of hot air or gas, and the energy that is converted to heat due to grinding. To utilize properly the heat added by hot air or gas, there must be intimate mixing of the air and coal, and the incoming feed should be rapidly mixed with the dryer coal being pulverized or contained in the circulating load. Different types of pulverizers show considerable variation in drying performance. The ability to use high gas or air temperatures entering the mill rather than large quantities of gas or air also varies considerably with mills of different types. To enable high inlet drying temperatures there must be high velocities of gas or air. Dead pockets within the mill are a fire hazard. High inlet temperatures and consequent low gas or air

TABLE 6.8-2 STANDARD SCREEN SIZES

U.S. Standard Sieve			W. S. Tyler Sieve		
Mesh	Inches	Millimeters	Mesh	Inches	Millimeters
20	0.033	0.84	20	0.033	0.83
30	0.023	0.59	28	0.023	0.59
40	0.0165	0.42	35	0.016	0.42
50	0.0117	0.30	48	0.0116	0.30
60	0.0098	0.25	60	0.0097	0.25
100	0.0058	0.149	100	0.0058	0.15
140	0.0041	0.105	150	0.0041	0.10
200	0.0029	0.074	200	0.0029	0.074
325	0.0017	0.044	325	0.0017	0.043

quantities are desirable because of lower power requirements for handling a smaller weight of air, lower tempering air requirements (thus higher efficiency on direct-fired units supplied with gas-air heaters), and lower primary air quantities (thus better burner performance on direct-fired units).

The extent to which drying must be accomplished in a mill depends primarily on the type of coal. In general, it is necessary to remove all surface moisture, leaving only the inherent moisture in the pulverized coal. Failure to remove the surface moisture limits the capacity of the mill considerably beyond the reduction of capacity that occurs due to moisture in the raw coal, even when it is removed in the milling system.

Classification of Pulverized Coal

The classification of coal size is determined by screening. The screens generally used and their corresponding sizes are shown in Table 6.8-2.

Classification of coal in mills generally is accomplished by means of air separation. Oversize particles are separated by a change in direction and returned to the grinding chamber. Means for application and control of this air separation principle differ for various mills. The classifier performance has a direct bearing on the grinding power required to obtain satisfactory combustion results. Satisfactory combustion results require a minimum quantity of plus 50-mesh material and a large quantity of minus 200-mesh material. Good classification is thus measured by the retention of a minimum quantity of 50-mesh material with a given quantity of minus 200-mesh material.

Pulverizer Types

Pulverizers generally used are low speed ball mills, high speed impact mills and a variety of medium speed vertical spindle mills encompassing ring-roll, ball-race, and table-roll types.

Ball mills consist of a horizontally rotating cylinder less than half full of balls of various diameters. The speed of rotation is normally established at about 75% of critical speed—defined as the speed at which the balls would centrifuge. Balls carried up the periphery by lift liners in the direction of rotation continuously cascade toward the center. Coal mixed with the ball charge is pulverized by impact and attrition. Hot air passed through the mill dries the coal and removes the fines. A classifier is normally used to regulate the fineness of the finished product by returning the coarser particles. Low maintenance, high availability, and quick response to demand changes in output rate are characteristics of this mill. Specific power requirements, particularly at reduced capacity, are relatively high. Space requirements are relatively large for a given capacity, and with wet coal there is an extreme reduction in capacity. To offset this latter characteristic, crusher-dryers (high speed impactors) are frequently employed in tandem with ball mills. The large storage and heat capacity make this type of mill most suited for swing and base loaded applications and not suitable for intermittent operation requiring quick start-up and shutdown. Ball mills are the indicated choice for severely abrasive and hard-to-grind coals such as anthracites and high ash bituminous coals containing high levels of iron pyrites and silica in the form of quartz. A typical ball mill is shown in Fig. 6.8-9.

Impact mills consist of a series of hammers or lugs revolving at high speed in an enclosed chamber. Grinding is by impact and attrition. Air passing through the mill dries the coal and carries away the fines. A means of classification is generally provided for returning oversize particles. This type is compact, low in cost, and may be built in very small sizes. It is well adapted to drying, because there is intimate mixing of air and coal. Maintenance and power consumption are relatively high, and it is difficult to maintain uniform fineness over the life of the wearing parts. The small storage capacity

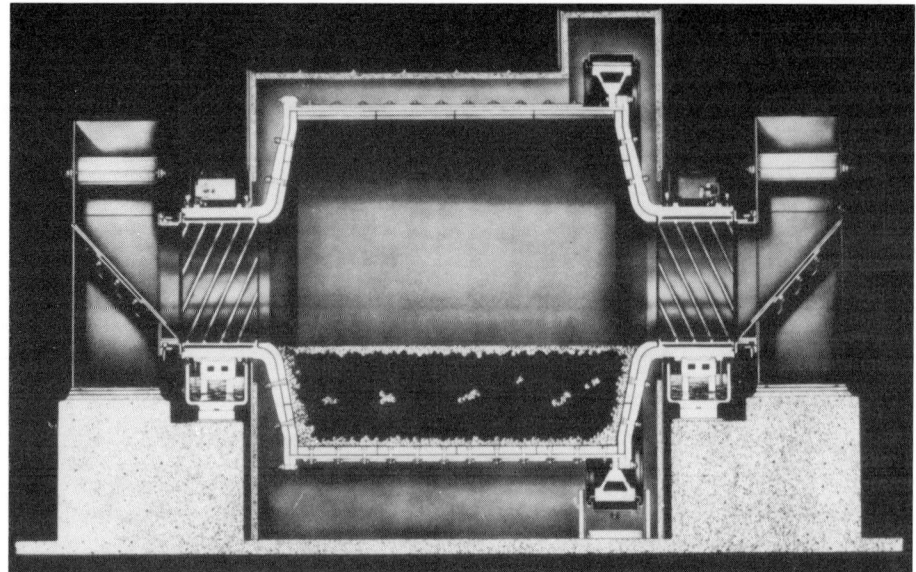

Fig. 6.8-9   Ball tube mill. (Courtesy of Riley Stoker Corp.)

makes this type well adapted to quick starting and intermittent operation. The less abrasive coals are best suited for these mills. A typical impact mill is shown in Fig. 6.8-10.

Vertical spindle mills pulverize by passing the coal between two surfaces, one rolling over the other. Grinding is accomplished by crushing and attrition. These mills have low power consumption, are compact, maintain reasonably well product fineness over the life of wearing parts, and handle high moisture coal with only a small reduction in capacity. Crushing force is provided either by hydraulic loading, mechanical springs, or centrifugal action. Characteristically, these mills are air swept and have integral, adjustable, static, or dynamic classifiers. Coal is fed from above and heated air enters from below. In the grinding zone, active mixing of the air and coal and recirculation of partially ground coal dries and reduces the coal particles enabling air conveyance to the classifier where larger coal particles are extracted and returned to the grinding zone for further grinding. These mills operate in a "medium" speed range and typically utilize grinding velocities of 400 to 800 fpm. Virtually the entire range of coals are suited for these mills with the possible exception of the most severely abrasive ones. A typical vertical spindle mill is shown in Fig. 6.8-11.

Pulverized Coal Systems

Pulverized coal may be utilized in a storage or in a direct-fired system. The latter system is characterized by immediate supply of coal to the burners and furnace as it is ground, with no part of it diverted to storage bins.

The bin or storage system has the advantage of operating flexibility, and permissible arrangement and location of equipment. This system permits preparation of coal during off-peak load hours at a constant (maximum) mill output rate. A few large-capacity pulverizers may be used without sacrificing flexibility. The ability to keep to a minimum the quantity of primary or carrier air use to convey the pulverized coal to the furnace is of distinct advantage in securing stability and range of operation of the fuel-burning equipment. These advantages are more than offset, in most cases, by the considerably higher equipment cost, the complication of venting the drying air or gas, transporting and storing the pulverized fuel, feeding the pulverized coal, and maintaining and operating the additional equipment. Justification of the storage system is difficult, except where low-volatile, hard-to-burn fuels, such as anthracite, must be utilized. A typical storage system is illustrated in Fig. 6.8-12.

The direct-fired system as shown in Fig. 6.8-13 is by far the most common arrangement. The advantages of this system are lower initial cost, simplicity of operation, and compactness of equipment. This system requires intelligent coordination in selecting milling, burning, and steam-generating equipment to obtain a reasonable degree of flexibility. The coordination of equipment must represent a satisfactory compromise between simplicity, operating range, types of coals (i.e., moisture,

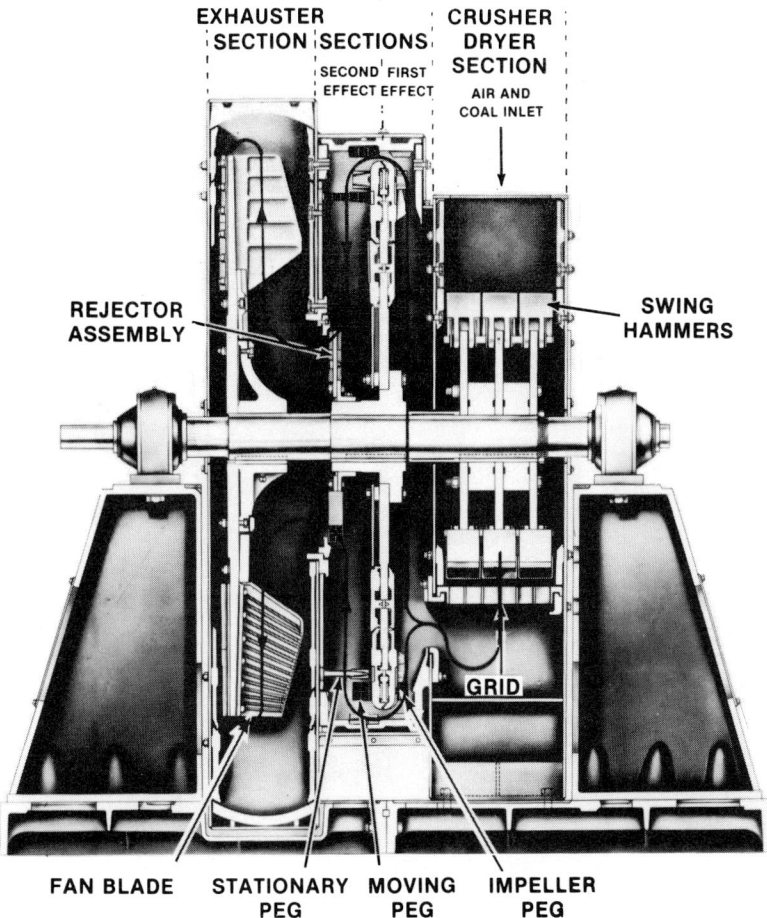

Fig. 6.8-10 Impact mill. (Courtesy of Riley Stoker Corp.)

grindability, and volatile matter) to be burned, reliability, overall cost, efficiency, and excess milling and drying capacity.

6.8-3 Pulverized Coal Burners

There are four principal types of pulverized coal burners. Each type is best suited for a particular boiler furnace design. The furnaces and the burners (Fig. 6.8-14) are:

Boiler Furnace	Outline	Burner Type
Straight wall	a, b	Horizontal circular
Tangential	c	Tangential
Turbo furnace	d	Directional flame
Vertically fired	e	Vertical or down shot

Horizontal Circular Burner

The horizontal circular burner, Fig. 6.8-15, consists of a center tube or nozzle through which the coal and primary air pass. The end of the coal nozzle is fitted with a component which is variously known as a rosette, a spreader, an impeller, or a diffuser. This serves to start to diffuse the primary coal and

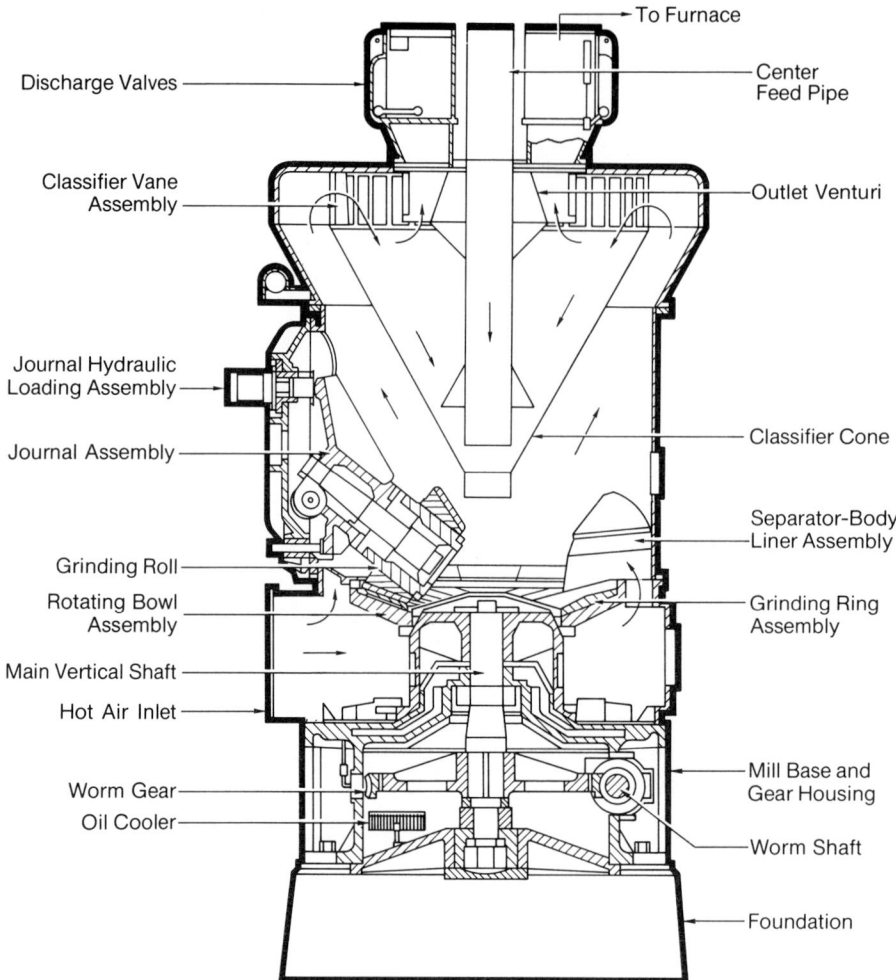

To Furnace

Discharge Valves

Center
Feed Pipe

Classifier Vane
Assembly

Outlet Venturi

Journal Hydraulic
Loading Assembly

Journal Assembly

Classifier Cone

Separator-Body
Liner Assembly

Grinding Roll

Rotating Bowl
Assembly

Grinding Ring
Assembly

Main Vertical Shaft

Hot Air Inlet

Worm Gear

Oil Cooler

Mill Base and
Gear Housing

Worm Shaft

Foundation

Fig. 6.8-11 Vertical spindle mill. (Reprinted with permission from *Combustion*: *Fossil Power Systems*, copyright © 1981, CE Inc.)

air into the hot secondary air. It also acts as a flame holder. The fuel load nozzle velocity may be as high as 8000 ft/min in a large burner.

The secondary air comes in around the coal nozzle, having been given a rotary motion by the secondary air vanes. The secondary air vanes are adjustable so that more or less spin can be induced into the secondary air stream. The horizontal velocity may be as high as 10,000 ft/min as it passes through the most constricted passage. The throat flares out as it approaches the inside of the furnace.

Both primary air and secondary air velocities have to be proportioned to the reactivity and flammability of the particular solid fuel being fired.

Because of the high turbulence and short flame of this type of burner, the NO_x emissions have usually exceeded the EPA requirements. All burner manufacturers have modified their designs in order to meet these requirements. Details of these modifications will be found in references.

Tangential Burners

Tangential burners, illustrated in Fig. 6.8-16, show a burner with coal and auxiliary oil capability. Each of the coal nozzles has a secondary air nozzle above and below it. Dampers control the air to each secondary air compartment.

A burner similar to the one illustrated is located at or near each of the four corners of the furnace. These burners are so angled that their center lines are tangent to an imaginary horizontal circle in the

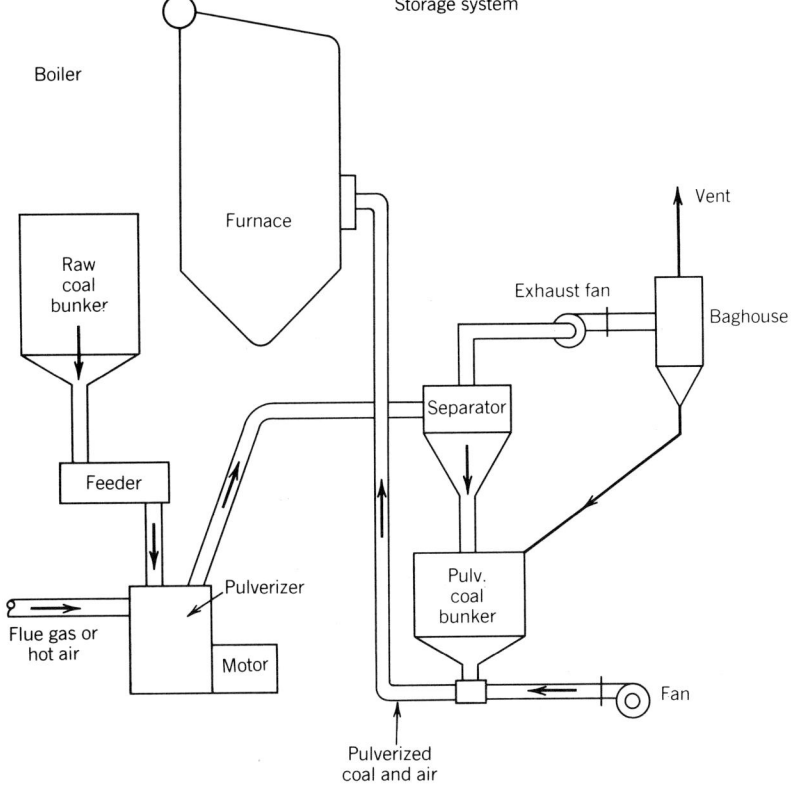

Fig. 6.8-12 Pulverized coal storage system.

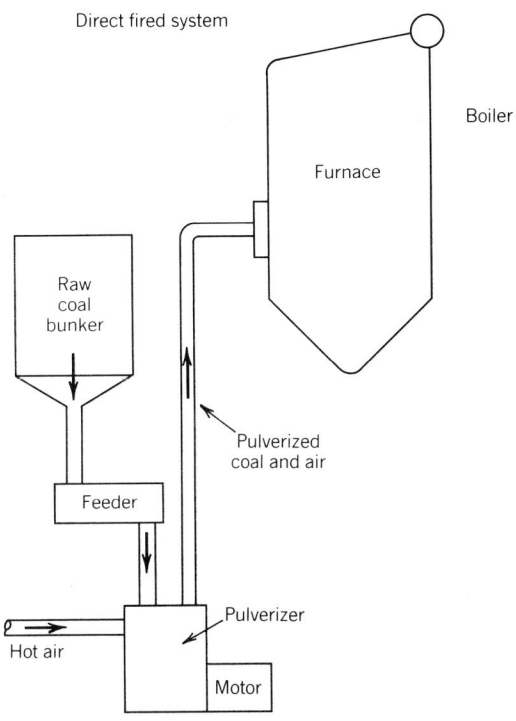

Fig. 6.8-13 Direct fired system.

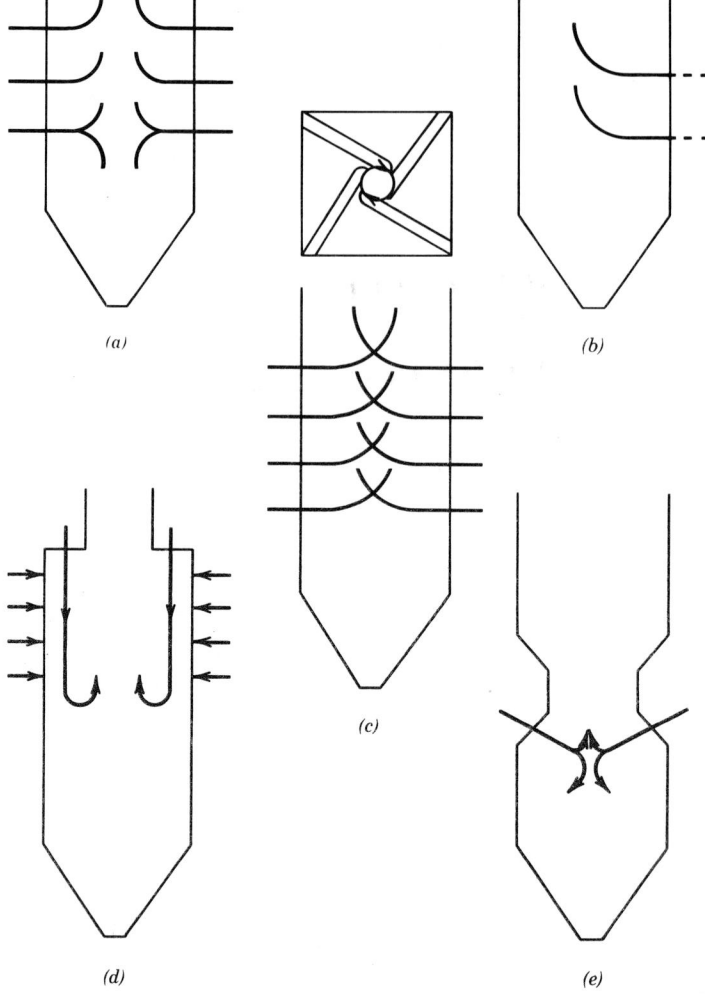

(a)   *(b)*

(c)

(d)   *(e)*

Fig. 6.8-14 Pulverized coal burner types: (*a*) one-side horizontal, (*b*) opposed horizontal, (*c*) tangential, (*d*) turbo furnace, (*e*) down shot.

center of the furnace as shown in Fig. 6.8-14c. The action of the streams is to produce a high degree of mixing in the furnace.

Each pulverizer is connected to a coal nozzle or nozzles in each of the four corner burners. The design permits the tilting of the burner components. This is done automatically to control furnace heat absorption and maintain superheat and reheat temperatures.

Directional Flame Burner

The directional flame burner is an integral part of the Riley Turbo Furnace. The burners are located below the characteristic venturi shape of the furnace. The front and rear furnace walls at this location are inclined downward at 25°.

Overfire air ports are installed at the neck of the venturi above the burners.

Figure 6.8-17 shows the arrangement of the burner. At the center of the two burner slots are the coal nozzles. In the secondary air slots, above and below the coal nozzles, are vanes capable of moving to deflect the air up or down. This action moves the fireball up or down in the furnace. Opposing burners are connected to the same pulverizer to keep the fire centered in the furnace.

This configuration produces a diffusion flame with a low NO_x emission.

Fig. 6.8-15 Horizontal circular burner. (Courtesy of the Babcock and Wilcox Co.)

Vertical or Down Shot Burner

This type of burner is used mainly for firing anthracite, coke, or similar low volatile fuel. The primary air and pulverized coal are discharged vertically downward surrounded by a small portion of the secondary air. The balance of the air for complete combustion is fed into the flame as it descends.

This is shown in Fig. 6.8-14e.

6.8-4 Combustion

The combustion system is the heart of the furnace wherein the oxidation of fuel and air converts the heat energy of the coal to heat in the product gas.

From the products of combustion, heat is extracted as explained in the next section. The combusting system attempts to minimize excess air, keep the local flame temperature at the desired level, and hold carbon loss to a minimum. Carbon loss and excess air are held to a minimum in order to maximize boiler efficiency. Local flame temperature must be maintained in order to control oxides of nitrogen formation and prevent large molten slag deposits. As will be explained under the section on environmental considerations, NO_x control means that local excess oxygen and local peak temperatures must be eliminated, by careful control of the developing flame. The burner controls the initial mixing and the flame shape. Furnace wall placement and furnace proportions must be such that flame impingement is avoided. In coal flames, impingement can cause slagging deposits, and local quenching of the flame. Slagging deposits are generally not acceptable in modern furnaces, dry ash or compacted ash are the acceptable forms of deposit. Local quenching leads to carbon monoxide formation and high unburned carbon losses. The above objectives are met by careful control of:

1. Mixing history.
2. Local available oxygen.
3. Temperature history.
4. Residence time.

Both the combustion system and the furnace contours can be adjusted to control these items. Residence time is the time from leaving the burner to flame extinction. This is related to the heat release rate per unit volume times the filling constant. The latter is the ratio of the actual burning

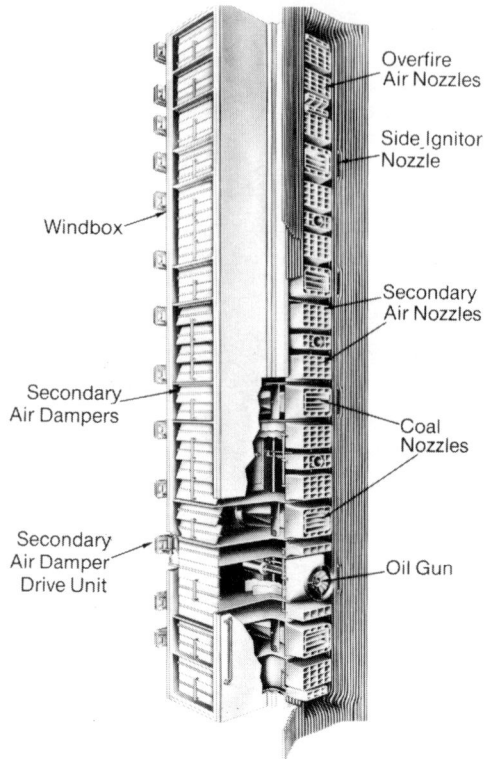

Fig. 6.8-16 Coal burner with auxiliary oil capability. (Reprinted with permission from *Combustion: Fossil Power System*, copyright © 1981, CE Inc.)

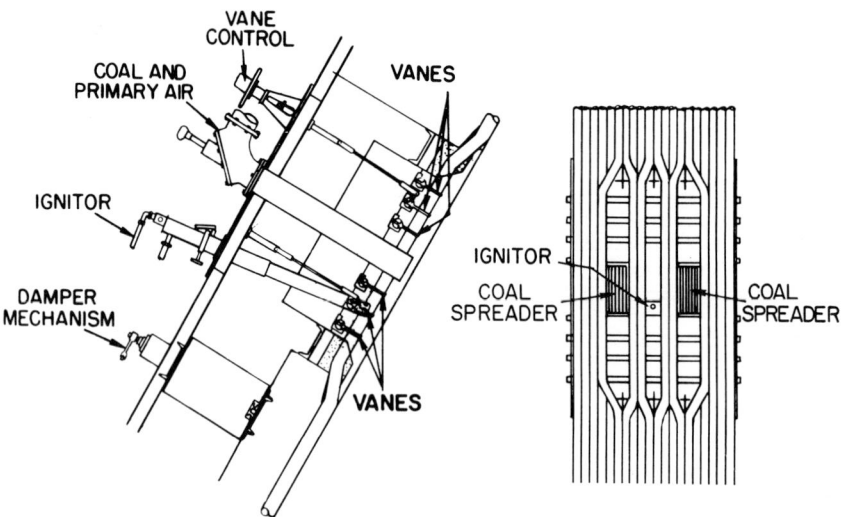

Fig. 6.8-17 Directional flame burner. (Courtesy of Riley Stoker Corp.)

volume to the furnace volume. In large utility furnaces, the burning volume may be less than 50% of the furnace volume.

Thermal Performance

The products of combustion, if not cooled, would produce an adiabatic flame temperature of about 3500°F for coal firing. Such a temperature is too high. Slagging to the point of blockage would occur if these hot gases were brought in close contact with the heat exchange tubing of the convection sections. Before entering the convection region the temperature must be reduced to approximately the initial deformation temperature of the coal ash. The implication of this is, that for all steam conditions, the same amount of heat must be extracted from each pound of flue gas. The necessary heat loss occurs mainly by radiation in the furnace. Thus, the furnace must remove the same amount of heat in a 900°F/900 psig non-reheat cycle as in a 1000°F/4700 psig supercritical cycle. The required gas cooling can be accomplished by means of an all water wall furnace design for 900°F/900 psig but such a simple design is no longer feasible as the pressure and temperature are raised to 1000°F/1800 psig. The latent (transition from water to steam) heat of the steam has substantially decreased at the higher pressure. If a furnace were constructed with all water walls for the foregoing conditions, it would result in underfiring, making sufficient steam but not much superheat and reheat. One method to overcome this is to convert a portion of the furnace walls from water cooling to steam cooling. As the gases cool to 2000°F by high temperature radiant energy exchange, some of the heat would then be absorbed by steam temperature rise and not by change of phase (boiling) at constant temperature. Other methods of achieving this are to use steam-cooled center walls, or draped pendants (widely spaced hanging curtains of steam-cooled tubing). The steam-cooled tubing in the furnace is much more difficult to design and maintain than water-cooled tubing. This is because the metal temperatures are much higher due to the lower film conductance of steam as compared to the film conductance of boiling water. With higher surface metal temperatures, the potential for liquid ash deposits and the attendant risk of liquid ash corrosion on the tubing is greatly increased.

Not only is the furnace arrangement a function of steam cycle conditions, but it is also a function of gross size. For example, up to furnace release rates of 500×10^6 Btu/h there is no need for draped superheat platens or water-cooled platens to reduce the gas temperature to 2000°F and still have the flame ending near the furnace exit. The furnace is one of the most costly items in the steam generator, so is kept as small as possible, consistent with good design.

Residence time has been shown to be related to heat release rate per unit of volume. Furnace temperature has been shown to be related to heat release rate per unit of wall area. The consequences of this is that in large sizes there is insufficient wall area to cool the flame, before the flame ends. Mathematically, the area of a furnace divided by the volume of the furnace is the surface to volume ratio which decreases with size. This, in combination with the usual practice of increasing the steam and reheat temperature with size, shows why the furnace design is so interrelated with steam conditions.

Mechanical

The furnace must be capable of enclosing the combustion process while absorbing heat and as such it acts as a duct to convey the gaseous products to the convection section. Since the combustion process can be irregular, the furnace walls must also be capable of withstanding several psi of internal pressure both in the form of an explosion or an implosion. Explosions (large increases in pressure) are contained in modern furnaces by membrane walls with buckstays at about 10-ft intervals. These buckstays are normally continuous bands or hoops of I beams that allow thermal expansion of the walls but retain the internal pressure. Implosions (large decreases in pressure) are caused by flame collapse and runaway ID fans and lead to buckling of the flanges of the I beams. One method of control is to place another set of beams normal to the first so that the outer flange is guided.

The furnace also acts as a support structure for the attached burners, radiant superheaters, slag accumulations, ducts, wall blowers, insulation, casing, and buckstays. All of these loads are carried through the tube system of the furnace to the rods and structural steel.

And finally the furnace acts as a dropout chamber for 10–30% of the coal ash and means for disposal of this must be made. The usual method is to use hoppers at the bottom of the furnace.

Environmental

In prior years, principally before passage of the Clean Air Act of 1970, the furnace was designed without much regard to environmental discharges. During those years, furnaces were optimized to produce the most economic design with regard to duty, slagging, and losses. In later years, furnaces have had to be enlarged to provide more cooling and more control of temperature of the developing flame. This enlargement of the furnace to control NO_x has led to dry ash firing as the only feasible means of compliance. Wet ash boilers, such as cyclones and slag tap bottoms, have not been designed since 1972, unless exempted for firing of mine waste material. Combustion systems have had to introduce mixing controls and local oxygen controls, which have taken the form of controlled mixing burners and the addition of overfire air. Both of these systems are used to reduce the formation of oxides of nitrogen (NO_x). These methods run the potential risk of local reducing atmospheres in the furnace, excessive carbon monoxide production, and high unburned carbon loss. These risks have been overcome by good design and today NO_x guarantees are being made by all the boiler manufacturers. Other environmental problems such as particulate control and sulfur dioxide control are usually done at the tail end of the heat exchange process.

Furnace Design

Furnace design is a part of the overall thermal considerations as such is linked into the superheat/ reheat control scheme and environmental control requirements, as well as the broad objectives stated in the first four paragraphs of this section. For example, at the control point (the lowest load at which superheat and reheat is maintained, usually 60% of full load) flue gas recirculation is often used to increase the gas mass flow over the superheater and reheater to enhance the heat recovery of these systems. Alternate methods are to tilt burners or fire burners higher in the furnace so as to void the lower sections at lower loads—thus raising the furnace exit gas temperature at the control point. In order to achieve a successful control of both superheat and reheat temperature the furnace exit gas temperature cannot depart more than about 100°F from the design value. Naturally, the more flexible the superheat/reheat control system, the less sensitive the superheat/reheat design is to a departure of the furnace exit gas temperature from the anticipated value. The furnace exit gas temperature is mainly a function of two factors: heat release rate per square foot of effective projected surface, and the wall deposits. The wall deposits are a function of the coal type, burner geometry, and the aforementioned heat release rate per unit of cooled area. Much has been written about the effect of the chemical constituents of the coal ash on slagging and fouling. The wall deposit is also affected by the physical properties of the ash, such as ash fusion temperatures, viscosity of the molten ash, the thermal conductivity, emissivity, and the sintering strength of ash deposits. The foregoing measurable or calculable factors are combined with experience factors to size the furnace. Digital computer programs have been written utilizing zonal models and can be used as an aid in designing the furnace.

The control of NO_x emission not only requires particular attention to the burner, but also to the furnace. For example, burners are now spaced much further apart, allowing more wall cooling of the flame from each burner. This prevents peak temperatures from developing. Also modern low NO_x burners have controlled mixing, which spreads out the burning volume. Some indices used to judge this are the heat released in the burner zone, usually called the burning area heat release which uses the heat release rate divided by the wall area in the vicinity of the burners. Various definitions are used to define this wall area, such as total burner height times the perimeter of walls, same as preceding but including the projected area of the hopper, same as preceding but including 10 ft of wall above and below the burners. These definitions are attempts to estimate the wall area where burning tempera-tures are high enough to produce thermal NO_x. In general, the lower the burning area release, the lower the NO_x emission, to a point where thermal NO_x is no longer of any consequence. At that point combustion dynamics govern NO_x production.

Another parameter of some importance in furnace design is the heat release rates per unit of plan area. This indication of the intensity of burning based on plan area is used as a guide to free flowing furnace ash hoppers and potential erosion problems in the superheater area, especially the platen superheater.

So far this design discussion has mentioned mainly the gross parameters used to size the furnace. In addition to these, several linear dimensions are important. The furnace width is chosen as a compromise between burner selection and other criteria beyond the furnace, such as steam drum release rates, retractable soot blower lengths, and gas path widths. The furnace depth must be adequate for the burner selected so as to avoid flame impingement. The distance from the lower burner to the beginning of the hopper slope must be adequate to prevent flame impingement on hopper slopes, which would then prevent free flow of fallen ash deposits to the ash removal system. The distance from the outside burner to the sidewall must be adequate to prevent flame impingement. And lastly, the distance from the top row of burners to the platen superheater must be adequate to prevent deposits from forming on platens and also to allow completion of combustion. All of these dimensions are a function of the burner type, size and arrangement.

6.8-5   Steam Generators

Babcock and Wilcox Carolina Radiant Boiler

Basic Design.   Figure 6.8-18 shows the side elevation of a Babcock and Wilcox Carolina-type radiant boiler for pulverized-coal firing 2875 psig with superheat and reheat at 1000°F and a capacity of 1,750,000 lb of steam per hour. The ash is removed in the dry state. Bleed feedwater heating lowers the

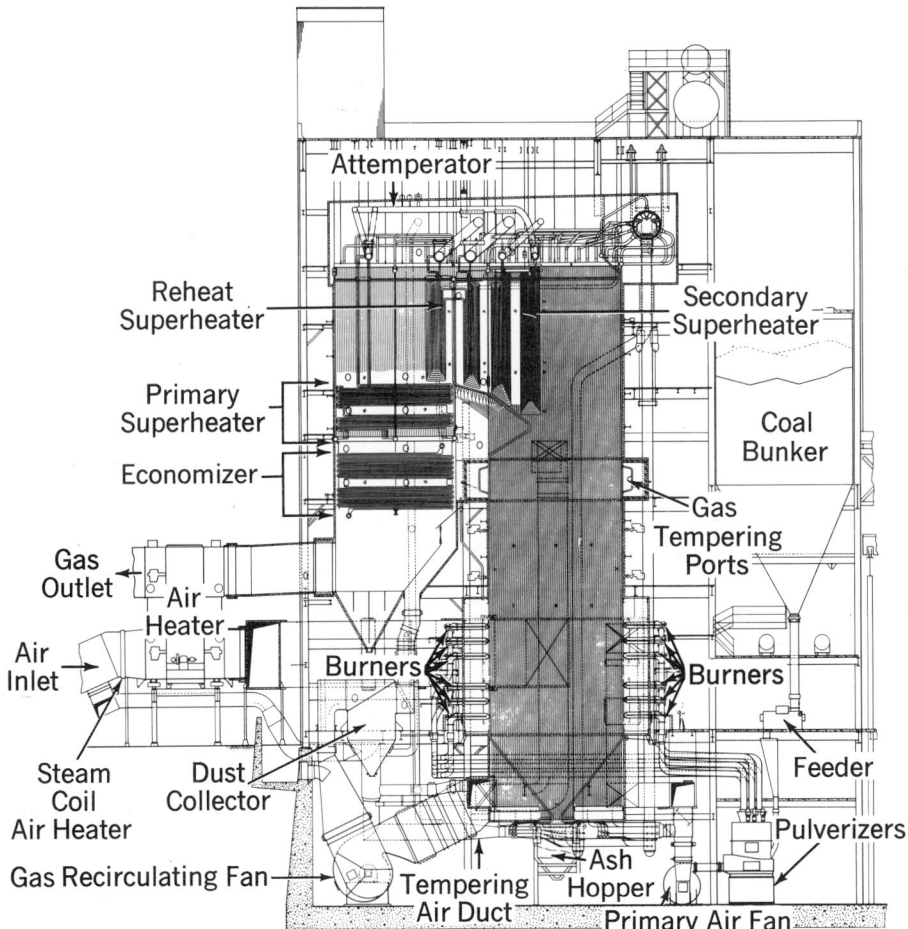

Fig. 6.8-18 Radiant boiler. (Courtesy of Babcock & Wilcox.)

temperature difference in an economizer and increases the temperature leaving the economizer. An air heater is used to lower this exit gas temperature and to supply hot air for combustion.

Foster Wheeler Boiler

Basic Design.   Figure 6.8-19 shows the side elevation of Central Louisiana Electric Co. Rodemacher No. 2 Steam Generator. This is the first of CLECO units to be designed to burn solid fuel. It has a capacity of 3,800,000 lb/h at 2620 psig with superheat and reheat at 1005°F. The turbine generator is 530 MW. The original design called for the boiler to be able to burn gas, oil, or western sub-bituminous coal.

During the design phase the decision was made to add future capability to burn Louisiana lignite. A comparison of these two solid fuels is shown in Table 6.8-3.

Primary Air Temperature.   When lignite, with its higher moisture content, was added to the Rodemacher unit's range of design fuels, it became necessary to increase the primary air temperature. This was achieved by substituting two primary and two secondary vertical shaft air heaters for the two trisector vertical shaft air heaters originally specified.

The major advantage of this configuration is that damper biasing can be used to distribute the gas flow between the primary and secondary air preheaters. Such proportional control of the gas flowing through the primary air preheater facilitates increasing or decreasing the preheater outlet air temperature to accommodate wide differences in fuel moisture. The conversion of the unit in the future to lignite firing will require the addition of 2 MBF pulverizers, each of 62 ton/h capacity and each serving four burners. Four extra burner ports have been provided in both front and rear walls.

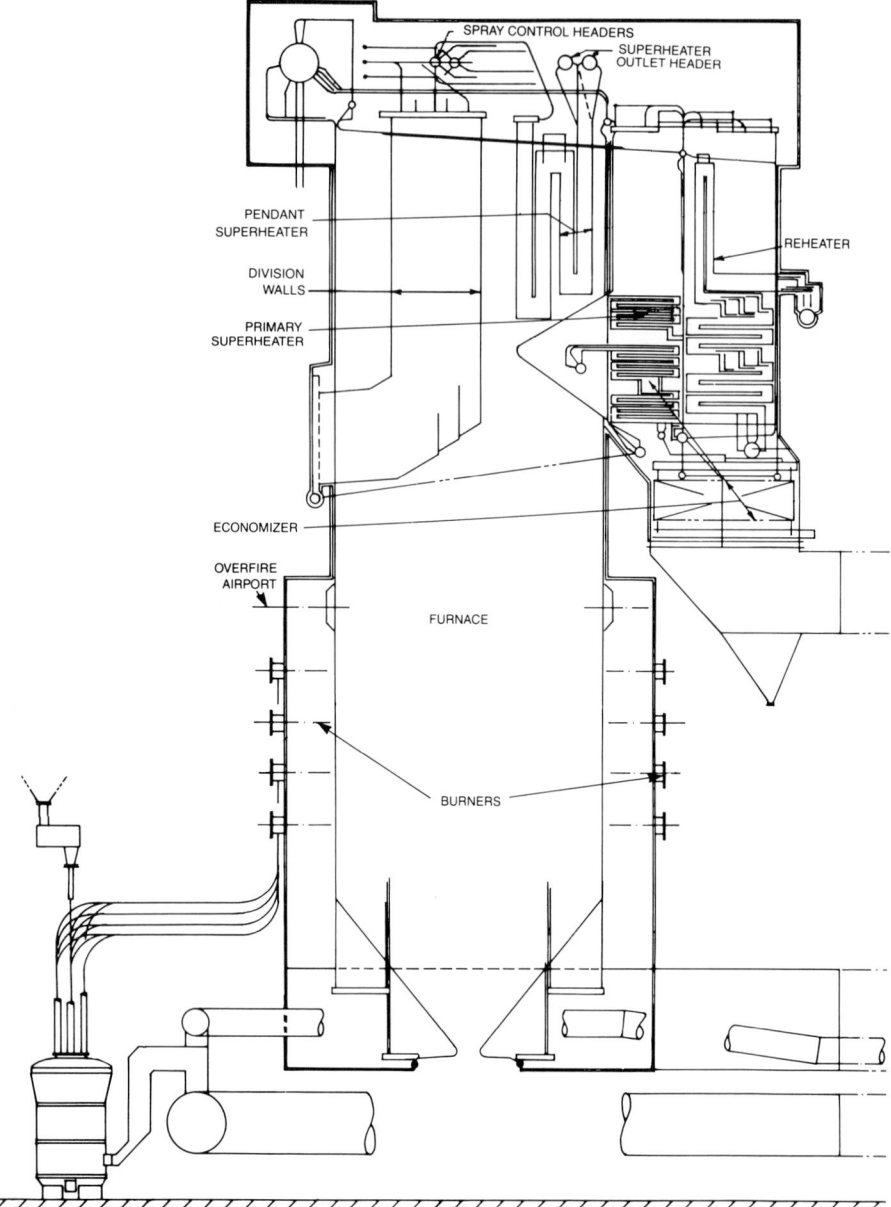

Fig. 6.8-19   Central Louisiana Electric Co. Rodemacher #2 Steam Generator. (Courtesy of Foster Wheeler Corp.)

Emission Control and Burner Selection.   The burners are designed for low NO_x emission using a double-register design. In addition, overfire air is provided through eight ports above the top rows of burners, four ports in front and four in the rear. Air from the windbox is introduced through slots in the hopper throat and along the sidewall slopes, as well as through ports located at the lower corners of the firing walls. This assures an oxidizing atmosphere in these areas.

Riley Turbo Furnace Boiler

Basic Design (Fig. 6.8-20).   The venturi extends from side-wall to side-wall with no reduction in furnace width. The venturi is situated somewhat below the midpoint in the side elevation.

TABLE 6.8-3   COMPARISON BETWEEN TYPICAL WESTERN COAL AND TYPICAL LIGNITE

	Western Coal	Lignite
Ultimate Analysis (as received)		
H_2O	28.95	32.59
Ash	5.79	12.28
Sulfur	0.48	0.84
Nitrogen	0.71	0.86
Carbon	48.61	36.68
Hydrogen	3.46	3.48
Oxygen	11.99	13.27
Chlorine	0.01	0.17
Heating value		
Higher heating value (Btu/lb)	8524	6956
Ash fusion temperature (for a reducing atmosphere)		
Hemispherical	2258°F	2300°F
Grindability		
Hardgrove Index	52.8	68.8
Mineral Analysis of Ash		
P_2O_5	0.49	0.25
SiO_2	29.14	48.96
Fe_2O_3	7.09	6.54
Al_2O_3	16.25	20.59
TiO_2	1.61	1.52
CaO	24.01	9.14
MgO	4.46	2.12
K_2O	0.38	0.73
Na_2O	1.54	2.09
SO_3	12.98	6.98
Undetermined	2.06	1.08

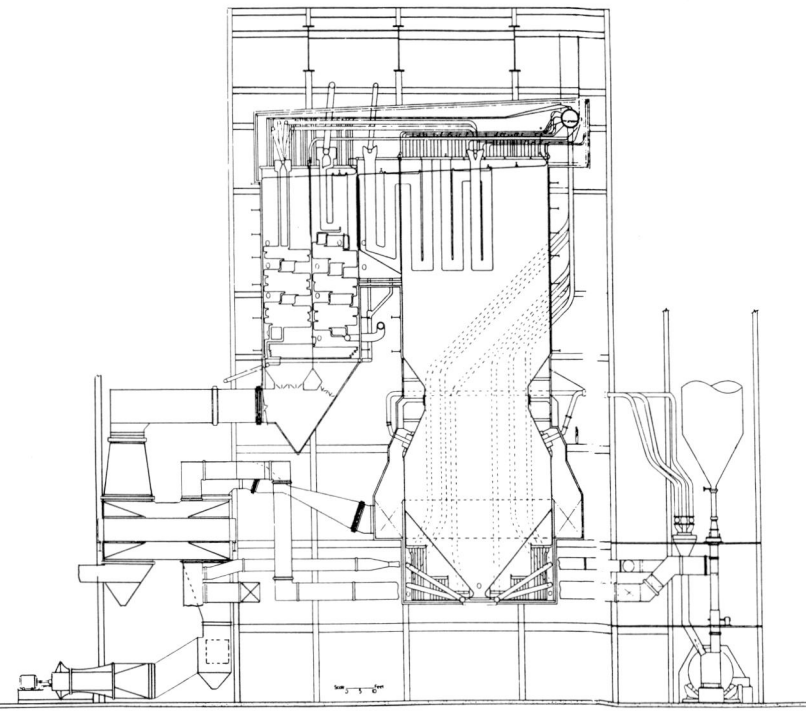

Fig. 6.8-20   Turbo furnace boiler. (Courtesy of Riley Stoker Corp.)

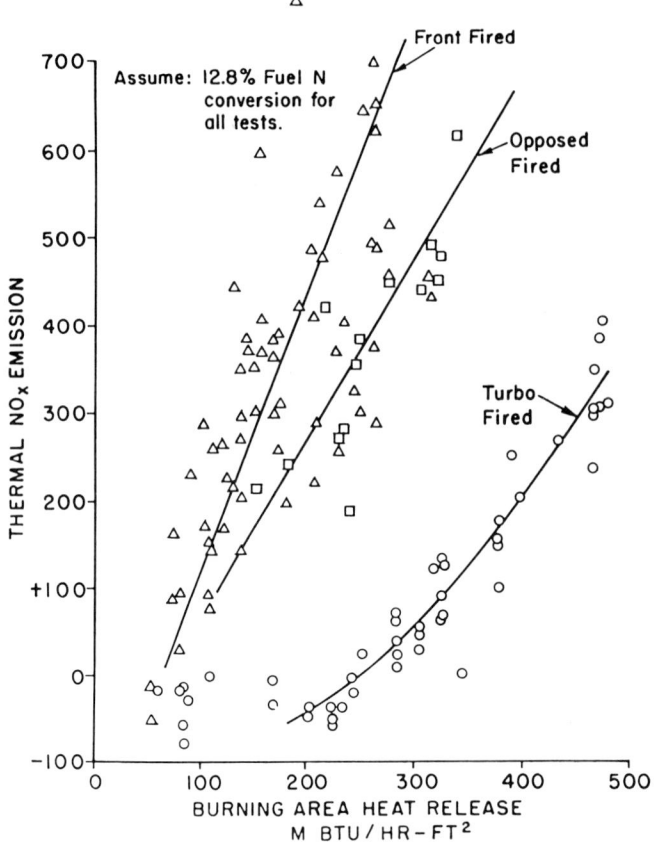

Fig. 6.8-21 Thermal NO_x emission. (Courtesy of Riley Stoker Corp.)

The slot type directional flame burners are inclined downward at an angle of about 25° and form the lower part of the venturi.

The Secondary Air slots lie immediately above and below the rectangular coal nozzles. The slots are equipped with movable vanes.

For burners firing large units, 400 kW and larger, the burners have two coal nozzles in each slot. Some air from the burner is piped directly into the midpoint of the venturi. This air is damper controlled. The nitrogen oxide production of this combination of furnace shape and burner design is in the lower range and is shown on the curve in Fig. 6.8-21.

6.8-6 The Effects of Changing Coal

It is often necessary for a variety of reasons to make a change to a different coal from that customarily used. The reasons for making the change may be new environmental regulations, cost of coal, availability, and suitability for the particular steam generating unit involved.

For each coal considered it is necessary to determine the advantages and disadvantages. The characteristics of the available coals need to be thoroughly investigated and compared with those of the coal customarily used. The study should include a determination of the necessary changes to equipment and operating procedure.

Data Needed

The "Recommended American Boiler Manufacturers Coal Guide Specification Form" can serve as a means of compiling much of the needed information.

Table 6.8-4 lists information from this sheet and some additional items. An explanation of some of the additional tests follows.

TABLE 6.8-4[a]

Coal rank

Handling considerations
    Coal size
    Coal friability
    % Surface moisture

Coal ignition and burnout characteristics
    % Volatile matter
    Heat value of volatile matter
    Burning profile
    Agglomerating value
    Free swelling index
    Flammability and reactivity
    Oxygen and flammability
    Rank and fineness

Environmental considerations
    Ash quantity
    % Nitrogen

Grinding considerations
    Hardgrove grindability
    Moisture

[a]Reprinted by permission from *Recommended American Boiler Manufacturers Coal Guide Specification Form.*

Coal Rank

United States coals decrease in rank from the eastern shore westward: Meta-anthracite found in Massachusetts and Rhode Island, anthracite in eastern Pennsylvania, followed by bituminous A in the Appalachian Mountains, bituminous B and C in Indiana, Illinois, and western Kentucky. Next in order are the large lignite deposits in Texas and the vast Northern Plains lignite fields of North and South Dakota and Montana. In Montana and in Wyoming there are the Powder River low-sulfur sub-bituminous deposits, which are surface mined from huge seams. The cost of mining is so low that this coal can be an economical fuel in the Midwest and as far south as northern Texas and Louisiana.

Many of the advantages and disadvantages of the coal properties listed under Table 6.8-2 and on the "Recommended ABMA Coal Guide Specification Form" will be obvious but others need further explanation.

Volatile Matter. A change in % volatile matter in the proximate analysis between 25 and 40% makes no difference in the reactivity.

Heat Value of Volatile Matter. The oxygen contained in the coal is driven off during the test to determine the % volatile matter. The oxygen adds nothing to the reactivity of the volatile matter. It should be subtracted from the % volatile matter, when comparing the ignitability of coals.

Burning Profile. This test employs derivative thermogravimetry in which a sample of the coal is oxidized under controlled conditions. The profile as plotted shows the rate of weight loss versus furnace temperature. One such plot at a furnace temperature of 842°F showed a weight loss as follows:

    Sub-bituminous—100%
    High-volatile bituminous—75%
    Low-volatile bituminous—28%

Agglomerating Value and Free Swelling Index. Pulverized coal from a seam with a high agglomerating (4 or 5) index will become plastic or semiliquid on heating. Particles will tend to stick together and, being larger, take longer to burn out. Such coals have to be ground to a smaller size. There is also a tendency on solidification to form a nonporous surface. There is a close resemblance between the agglomerating index and the Free Swelling Index (FSI).

The FSI Index is used to predict the behavior of coal when burned on a chain grate, traveling grate, or underfeed stoker. More difficulty is encountered in securing a low carbon loss when burning a high FSI on these stokers.

TABLE 6.8-5

	Passing 200 Mesh (74 U) Wt. %	Retained On 50 Mesh (297 U) Wt. %
Lignite and sub-bituminous C	60–70	2.0
Sub-bituminous B and high volatile C	65–72	2.0
Low volatile bituminous and high volatile C	70–75	2.0

TABLE 6.8-6   ASH DATA

Ash fusion temperature
   Initial deformation (ID)
   Ash softening temperature ($H = W$)
   Ash softening temperature ($H = \frac{1}{2}W$)
   Fluid temperature
   Spread between ID and fluid temperature
Base-acid ratio
   MgO content
   Na_2O content
   Fe : Ca ratio
   $Ash/1 \times 10^6$ Btu
   Ash friability

Grindability and Fineness

In order to achieve fast ignition and good burnout, experience has shown that a high Rank coal must be ground to a high fineness. The necessary fineness is given in Table 6.8-5.

Slagging and Fouling

The ability of a coal to develop full boiler capacity and maintain that capacity is very dependent on the composition of the ash.

Table 6.8-6 lists the information needed to determine the slagging and fouling potential of the ash.

A study of this information will determine the effect on the performance of the steam generating unit, what modifications have to be made to the boiler and/or what limitations have to be placed on the unit's output.

Ash Fusion Temperature

One criterion for furnace height is to design a furnace which will reduce the temperature of the combustion gases to 100°F (56°C) below the ash softening temperature before the gases enter the superheater. The *FT–ID Spread* indicates the range of temperatures over which the ash will remain sticky or viscous. The *base acid ratio* is

$$\frac{Fe_2O_3 + CaO + MgO + Na_2O + K_2O}{SiO_2 + AL_2O_3 + TiO_2}$$

This ratio multiplied by the dry sulfur percentage gives a factor, the magnitude of which gives the slagging factor. The tabulations that follow are only applicable to coals with an Eastern type ash.

Slagging Index (R_s)

Slagging Type	Index
Low	− 0.6
Medium	0.6–1.5
High	1.5–2.5
Severe	+ 2.5

Fouling Index (R_f = base acid $\times$ NO_2O)

Fouling Type	Index
Low	− 0.25
Medium	0.25–0.50
High	0.50–0.75
Severe	+ 0.75

For coals having a lignite type ash the following tabulation was developed by Bureau of Mines, Grand Forks, North Dakota.

Fouling Index

Fouling Type	Index
Low	% NO_2O in ash
Low	< 2.0%
Medium	2–6%
High	6–8%
Severe	> 8%

6.8-7   Ash Handling Systems

Robert W. Fitzmaurice

Hydraulic, pneumatic and mechanical systems are available for handling the products of fossil fuel combustion from steam generators. The refuse classed as ash which these systems must handle will include the following:

1. Bottom ash or slag dropped out of the main furnace.
2. Fly ash trapped by dust collectors.
3. Economizer and air heater ash (coarser than fly ash).
4. Mill rejects, or pyrites.

Applicable Systems

Each of the materials in these categories presents its own problems as far as the design of conveying systems is concerned. The quantity and quality (chemical composition) of the materials are factors to be considered in deciding what method of conveying is practical. Economic and environmental considerations cannot be overlooked.

Table 6.8-7 lists the available methods of handling each class of material and the choice which may be the best from an economic standpoint for either utility or industrial boilers.

Bottom Ash Systems (Including Slag and Mill Rejects)

The bottom ash systems operate on the following principles:

1. Sluicing by means of ejectors or centrifugal slurry pumps to a fill area, water not recovered.
2. Pumping or jetting to dewatering bins for removal by trucks or rail cars. (See Fig. 6.8-22)
3. Pumping or jetting to dewatering bins as in 2 with the addition of a complete water recovery system to permit recirculation and reuse of conveying water.
4. Moving by means of a mechanical drag chain conveyor to a belt conveyor for discharge directly to a disposal area, or to bins for removal by railroad cars or trucks, with or without a water recovery system.
5. Removal by a pneumatic (vacuum) system to a dry storage facility.

TABLE 6.8-7

	Practical Systems	Economic Preference	
		Utility	Industrial
Dry bottom ash	H-M-P	H-M	H-M-P
Slag (molten bottom ash)	H	H	H
Fly ash	P	P	P-M
Economizer and air heater ash	H-P-M	P-H	P-M
Mill rejects	H-P-M	H	H-M-P
	H = Hydraulic M = Mechanical P = Pneumatic		

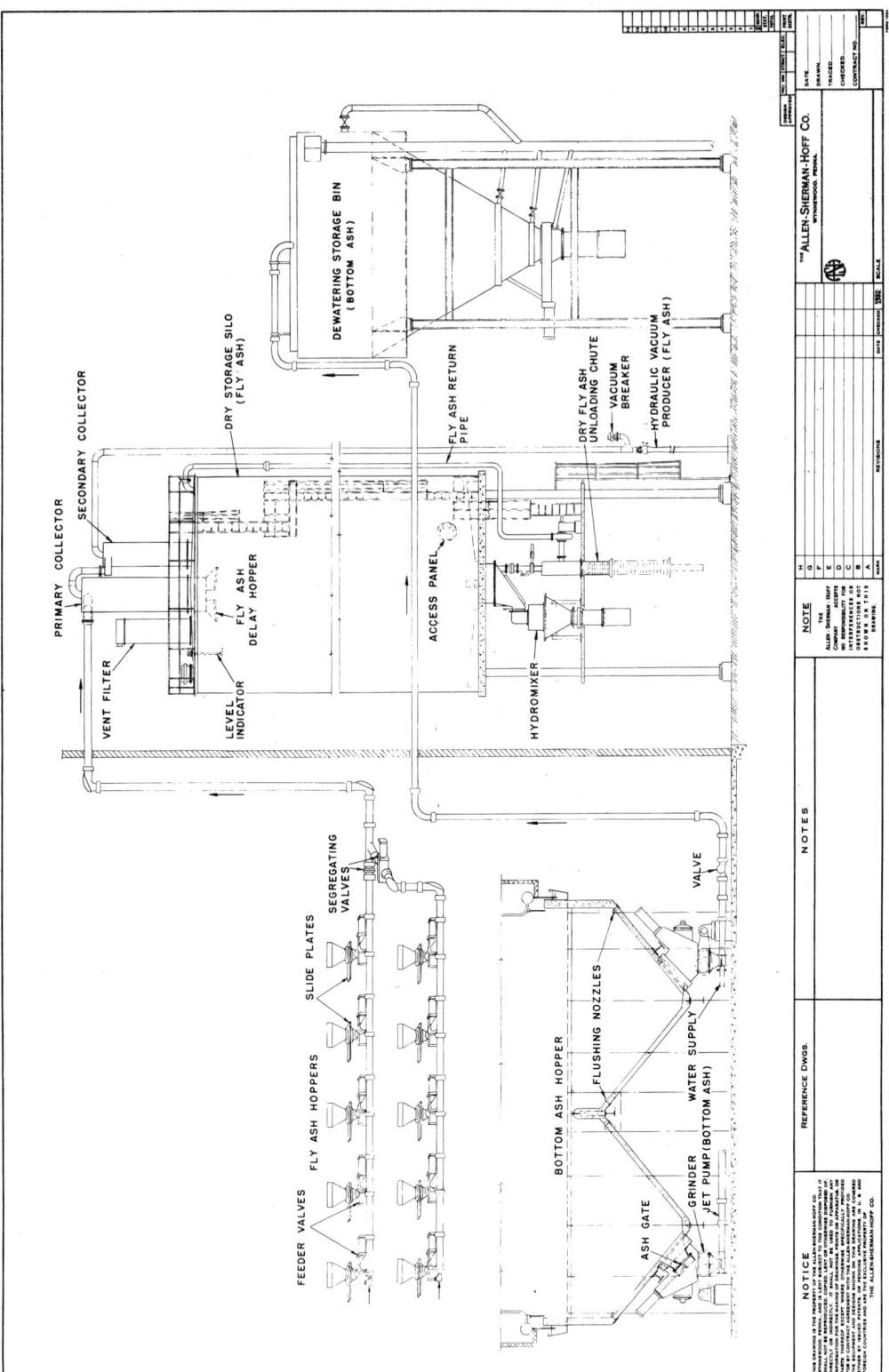

Fig. 6.8-22 Bottom ash system. (Courtesy of Allen–Sherman–Hoff Co.)

PRIMARY COLLECTOR

SECONDARY COLLECTOR

DRY STORAGE SILO (FLY ASH)

DEWATERING STORAGE BIN (BOTTOM ASH)

VENT FILTER

LEVEL INDICATOR

FLY ASH DELAY HOPPER

ACCESS PANEL

FLY ASH RETURN PIPE

DRY FLY ASH UNLOADING CHUTE

VACUUM BREAKER

HYDRAULIC VACUUM PRODUCER (FLY ASH)

HYDROMIXER

SLIDE PLATES

SEGREGATING VALVES

FLY ASH HOPPERS

FEEDER VALVES

BOTTOM ASH HOPPER

FLUSHING NOZZLES

VALVE

WATER SUPPLY

GRINDER

JET PUMP (BOTTOM ASH)

ASH GATE

REFERENCE DWGS.

NOTES

NOTE

THE ALLEN -SHERMAN -HOFF COMPANY ACCEPTS NO RESPONSIBILITY FOR INTERFERENCES OR OBSTRUCTIONS NOT SHOWN ON THIS DRAWING.

REVISIONS

ALLEN-SHERMAN-HOFF CO.
WYNNEWOOD, PENNA.

DATE.
DRAWN.
TRACED.
CHECKED.
CONTRACT NO.

SCALE.

TABLE 6.8-8

Pipe Diameter	P.C. Ash 7.5/9.0 fps		Stoker Ash 8.5/10.0 fps		Slag/Pyrites 10.0/12.0 fps	
	gpm[a]	tph	gpm[a]	tph	gpm[a]	tph
6	700–900	40	800–1000	40	900–1000	50
8	1200–1400	75	1350–1600	75	1600–1900	100
10	1900–2200	125	N.A.	—	2400–3000	150

[a]Range shown for gallons per minute (gpm) covers horizontal and vertical flow requirements. Where high vertical rises are expected use the higher amounts shown or reduce vertical pipe diameter to maintain proper minimum conveying velocity.

Hydraulic systems for bottom ash handling represent, by far, the greatest number of installations in operation in utility plants. These systems permit the temporary storage of ash in water-filled hoppers under boilers or stokers and periodic removal to a disposal point. The flexibility of design possible with closed pipe discharge lines is one of the attractive features of hydraulic bottom ash systems. Slagging boilers require the cooling and granulating effect made possible by hydraulic systems.

Table 6.8-8 lists water requirements and capacities for hydraulic systems of various sizes handling bottom ash and slag. Also shown are minimum conveying velocities in feet per second (fps) for horizontal and vertical lines.

A furnace bottom ash hopper should have a capacity equal to 12 hr of ash production. Operation should be based on 8 hr cycles. The difference of 4 hr will permit a down-time allowance for maintenance or a possible increase in coal ash content.

Whenever possible, ash hoppers should be designed with steep bottom slopes to permit gravity discharge of ash with minimum assistance from sluicing nozzles. See Fig. 6.8-23.

Manufacturers supply discharge gates and clinker grinders in various sizes compatible with ash hopper volumes and required discharge rates.

Where ashes are pumped to dewatering bins for removal by trucks or railcars sufficient capacity should be provided for weekend operations when equipment may not be available. Storage capacity equivalent to 72 hr is usually sufficient. In larger plants, this should be divided between two or more bins.

Table 6.8-9 lists the available dewatering storage bins for ash and slag based on bulk densities of 45 and 90 lb/ft^3, respectively.

Submerged mechanical drag chain conveyors can be applied to boilers burning pulverized coal and to stoker fired boilers. By incorporating an upward slope at the discharge end of the conveyor, bottom ash can be dewatered so that it can be deposited on other conveyors or into trucks or storage bins without further treatment. These systems are designed for continuous operation. Dimensions and speed of the conveyor flights are determined from the bottom ash production rate. Variable speed drives permit changes in flight speed to compensate for variations in boiler load or coal ash content.

Mill rejects can be moved from pulverizers to the bottom ash conveyor using similar drag chain conveyors.

The diagram in Fig. 6.8-24 shows the application of a mechanical drag chain conveyor to a pulverized coal fired boiler bottom.

Water is required for quenching the ash and if it is to be reused (recirculated) it must be cooled and clarified prior to return to the conveyor trough.

Pneumatic conveying of bottom ash is usually limited to industrial boilers of moderate size, particularly where headroom may be restricted or where water is scarce or difficult to dispose of. Capacity is low by comparison with hydraulic systems partially due to the fact that the feeding of ash into the conveyor must be manually assisted. Vacuum for these systems can be produced by various devices including steam-jet exhausters and mechanical vacuum pumps. Disposal is generally to dry storage silos from which the ash is fed to trucks or railcars after passing through wetting devices to make them acceptable for hauling over highways or railroads.

Mill reject systems can be of any of the three types shown in Table 6.8-6. Because of their weight, rejects (pyrites) must be conveyed at velocities considerably higher than required for pulverized coal bottom ash.

Most utility systems require individual mill storage hoppers from which the rejects can be discharged periodically to a holding bin having at least 8 hr storage capacity. Rejects can then be discharged as part of the bottom ash handling cycle to dewatering bins or to other disposal facilities. See Fig. 6.8-25.

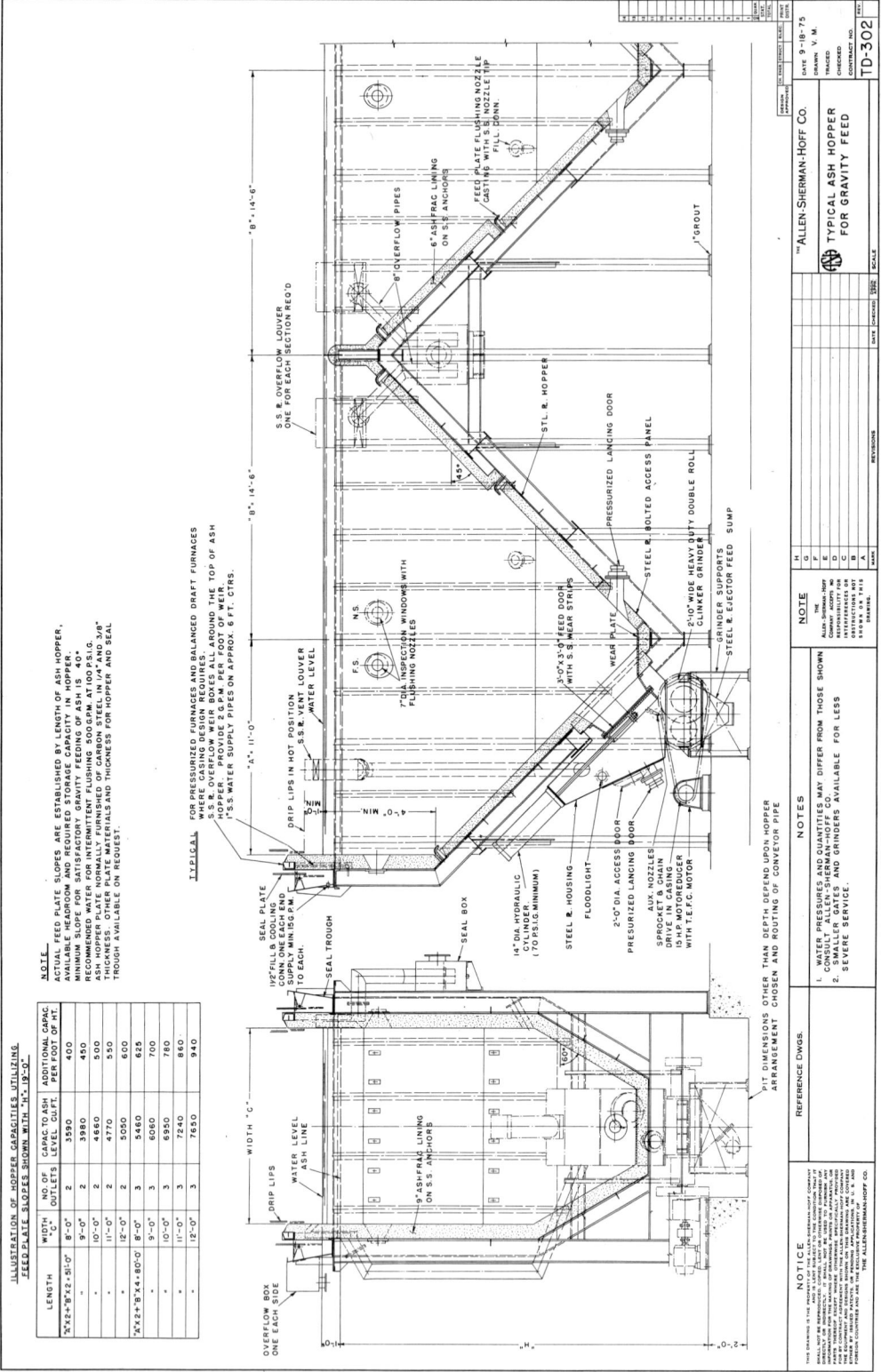

Fig. 6.8-23 Ash hoppers. (Courtesy of Allen–Sherman–Hoff Co.)

TABLE 6.8-9   DEWATERING BINS

Diameter	Volume Range (ft^3)	Storage Capacity (tons) P.C. Ash	Slag
16 ft-0 in.[a]	1110–4000	25–90	50–180
25 ft-0 in.	4500–13300	100–300	200–600
30 ft-0 in.	8888–22,222	200–500	—
30 ft-0 in.	8888–15,555	—	400–700
35 ft-0 in.	17,780–39,700	400–900	—
35 ft-0 in.	17,780–23,000	—	800–1035

[a]Limited to truck clearance.

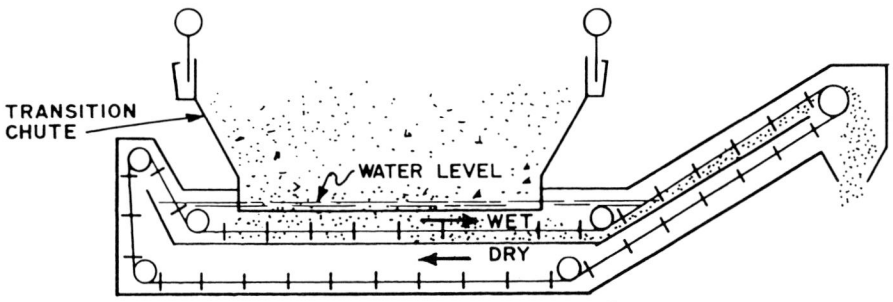

Fig. 6.8-24   Mechanical drag chain conveyor. (Courtesy of Allen–Sherman–Hoff Co.)

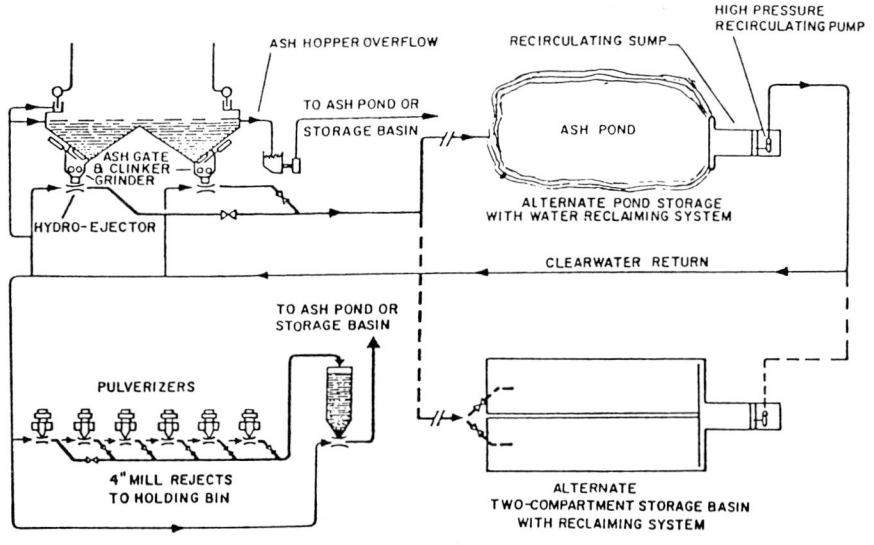

Fig. 6.8-25   Mill reject system. (Courtesy of Allen–Sherman–Hoff Co.)

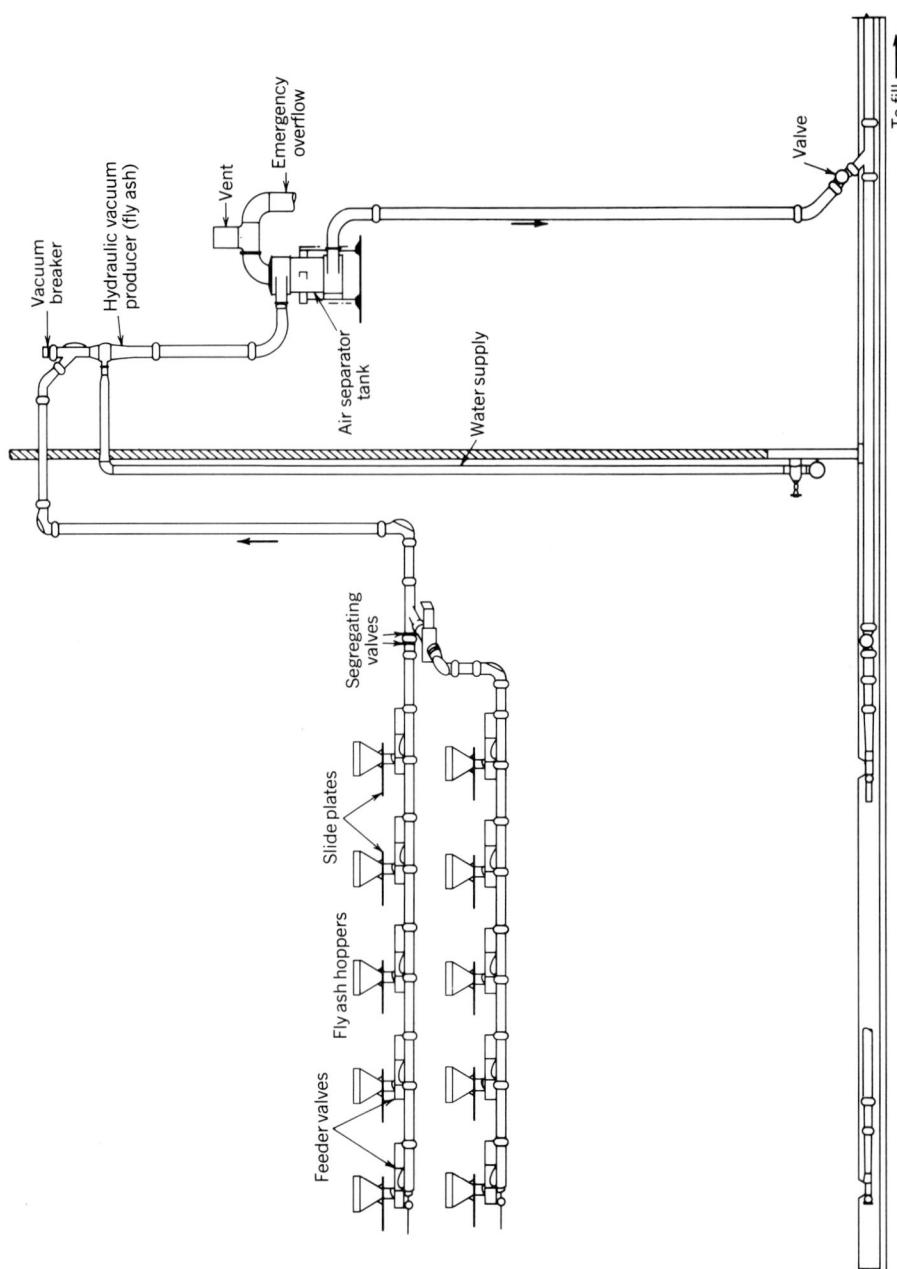

Fig. 6.8-26 Fly ash vacuum system. (Courtesy of Allen–Sherman–Hoff Co.)

Fly Ash Systems

The fly ash systems operate on the following principles:

1. Pneumatic *vacuum* conveying and wet disposal.
2. Pneumatic *vacuum* conveying to a dry storage silo.
3. Conveying by *pressure* from air-locks (dilute phase) or transporters (dense phase) to a dry storage silo.
4. Conveying pneumatically to a wetting device with the resulting slurry discharged by gravity through the bottom ash discharge line or through an independent discharge pipe.
5. Combining *vacuum and pressure* for economically conveying fly ash distances which are too great for vacuum systems.

Vacuum systems for fly ash handling have been widely used because of their relatively low cost compared to other methods, design flexibility, ease of control, and the fact that leakage at pipe joints will be inward preventing loss of fly ash from the conveying system into the plant or its surroundings.

Vacuum may be obtained by using hydraulic or mechanical vacuum producers. Hydraulic vacuum producers are specially suited to plants where the fly ash may be sluiced to a pond in which case it is pulled directly into the vacuum producer and mixed with water. The resulting slurry may then flow by gravity to the disposal area through the same discharge pipe serving the bottom ash system or through an independent line. See Fig. 6.8-26.

Fly ash may be stored in dry silos by interposing cyclone collectors and/or bag filters in the conveying line above the silo (see Fig. 6.8-22).

When mechanical vacuum producers are used, an efficient bag filter must be used in series with cyclone collectors to prevent carryover of fly ash into the vacuum pumps. Carryover is not as critical when hydraulic vacuum producers are employed if the particulates in the discharge water can be tolerated.

Curve A, Fig. 6.8-27, shows typical performance of a 6 in. vacuum system handling fly ash with a bulk density of 45 lb/ft^3 and a conveying velocity of 2400 ft/min at the point of fly ash pickup.

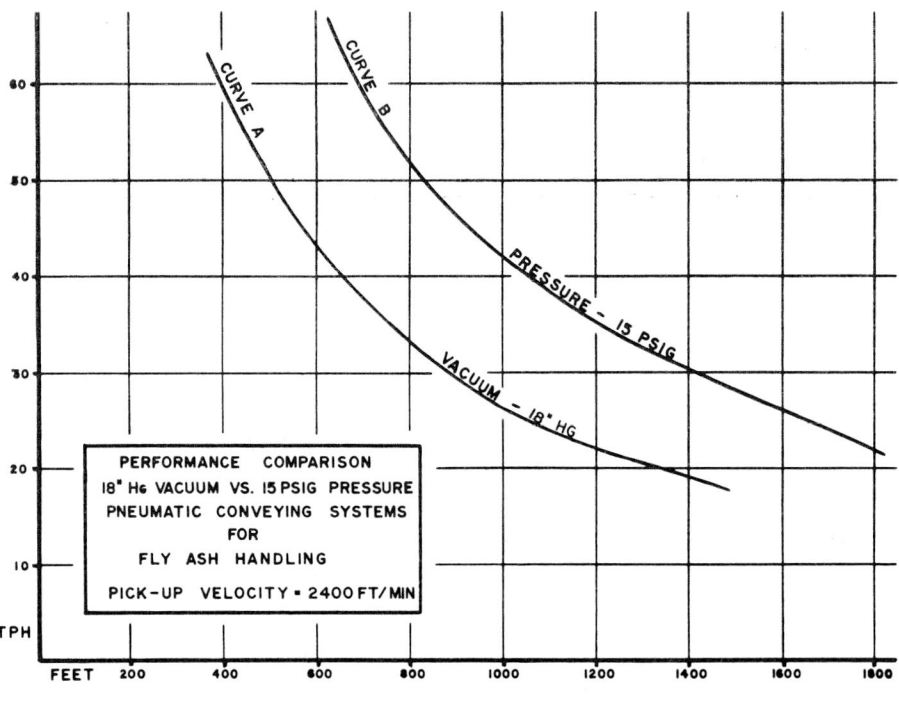

Fig. 6.8-27 Performance curves of fly ash vacuum and pressure systems. (Courtesy of Allen–Sherman–Hoff Co.)

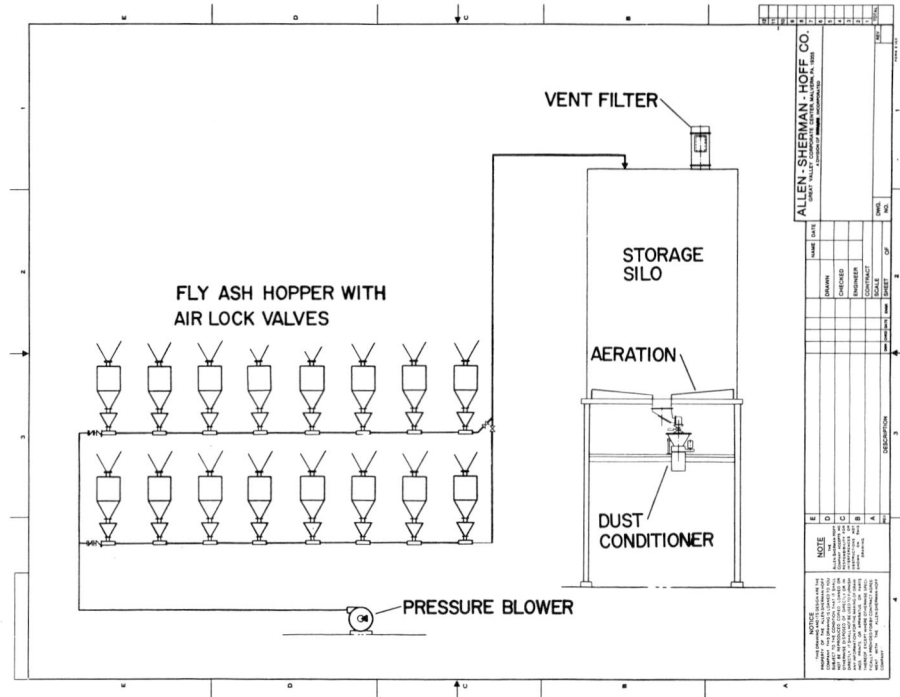

Fig. 6.8-28   Fly ash pressure system. Courtesy of Allen–Sherman–Hoff Co.)

Pressure systems may be of the *dilute* phase or *dense* phase type. Dilute phase systems handle fly ash directly as it flows from hoppers under precipitators or bag houses. Operating pressures in the range of 10–15 psig are common with conveying lines 4–10 in. diameter. See Fig. 6.8-28. Depending on system configuration, fly ash loading in dilute phase pressure systems will vary between 5 and 15 lb of fly ash per pound of conveying air.

Curve B, Fig. 6.8-27, shows performance of a pressurized dilute phase 6 in. system operating at 15 psig conveying fly ash having a bulk density of 45 lb/ft^3 and an air velocity of 2400 ft/min at system inlet.

Conveying velocities for vacuum systems as well as for dilute phase pressure systems should be carefully selected. Velocity requirements vary depending on material density as well as particle size and shape. Based on these factors minimum conveying velocities can range from a low of 1500 ft/min to a maximum of 4500 ft/min. Using conveying velocities higher than necessary for a given material will result in a conveying rate less than the optimum. For example, a fly ash which can be conveyed at a rate of 28 tons/hr (tph) using the correct pick-up velocity of 2400 ft/min will be conveyed through the same pipe line at only 14 tph if the pick-up velocity is increased to 4000 ft/min.

System capacity is also greatly affected by the number of 90° bends in the conveyor pipe. Designers should be aware of this and keep pneumatic conveyor lines as free from bends as possible. A 90° bend in a conveyor with a system inlet velocity of 2400 ft/min creates a loss equivalent to 200 ft of straight pipe. Losses in vertical risers are equal to 1.4 times losses in horizontal lines. Thus, at 2400 ft/min a system with a vertical lift of 100 ft, a horizontal length of 260 ft, and four 90° bends has an equivalent length of 1200 ft. From Curve A, Fig. 6.8-27, at a vacuum of 18 in. Hg, the conveying capacity in a 6 in. pipe will be 22.5 tons/hr.

At 4000 ft/min system inlet velocity, each 90° bend will cause a loss equivalent to 500 ft of straight pipe.

Vacuum-Pressure System

Vacuum systems are limited in the amount of energy available for conveying material and overcoming system friction losses. At high altitudes the energy that can be developed by vacuum is substantially less than at sea level. Therefore, vacuum systems, by themselves, may not be able to provide sufficient conveying capacities for long conveyor lines or for systems of moderate lengths at high altitude.

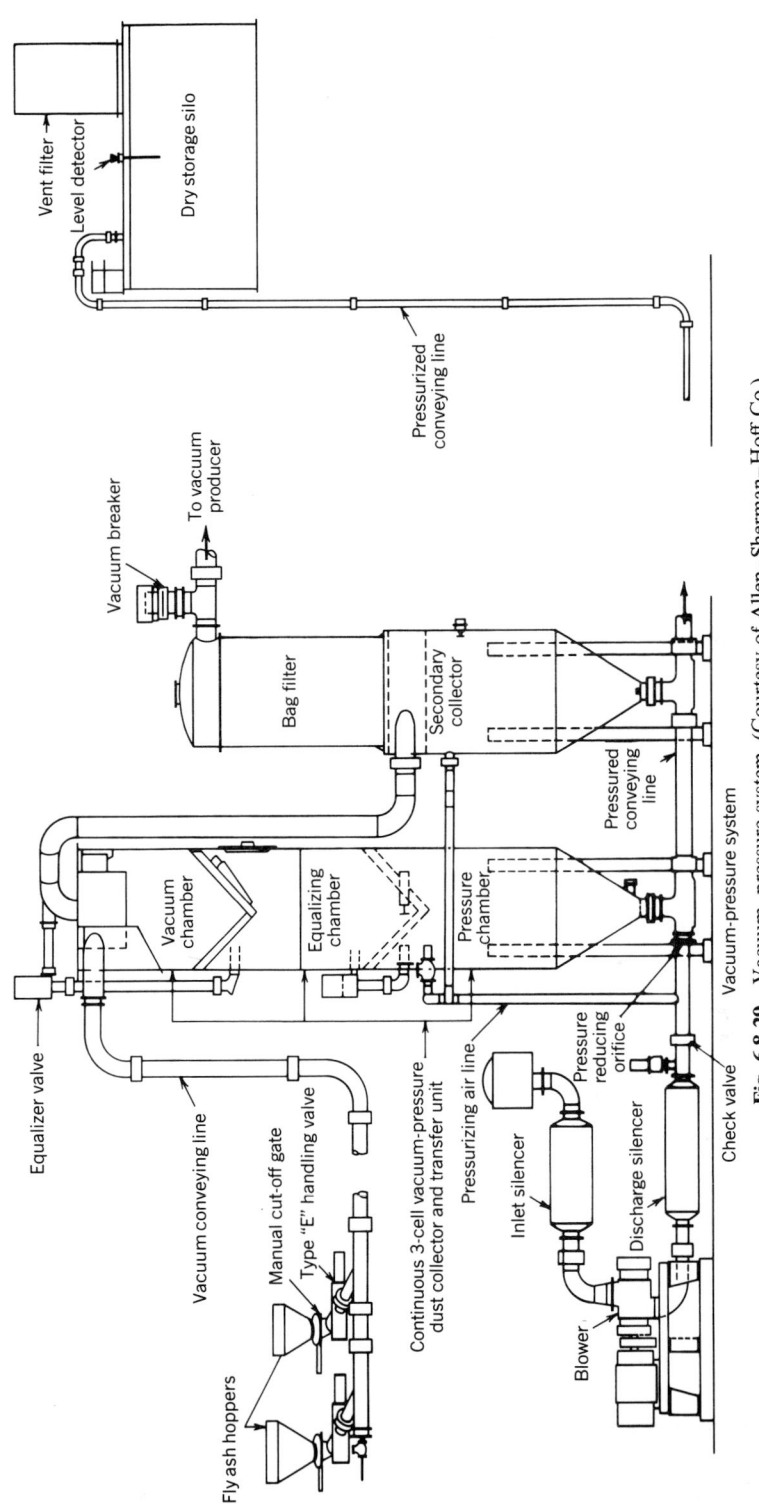

Vent filter

Level detector

Dry storage silo

Pressurized conveying line

Vacuum breaker

To vacuum producer

Bag filter

Secondary collector

Vacuum chamber

Equalizing chamber

Pressure chamber

Pressured conveying line

Vacuum-pressure system

Equalizer valve

Vacuum conveying line

Manual cut-off gate

Type "E" handling valve

Continuous 3-cell vacuum-pressure dust collector and transfer unit

Pressurizing air line

Pressure reducing orifice

Inlet silencer

Check valve

Discharge silencer

Blower

Fly ash hoppers

Fig. 6.8-29 Vacuum–pressure system. (Courtesy of Allen–Sherman–Hoff Co.)

TABLE 6.8-10   DENSE-PHASE SYSTEMS AT 25 psig[a]

Pipe Diameter (in.)	Horizontal (ft)	Vertical (ft)	Capacity (tph)
$2\frac{1}{2}$	400	50	10
3	500	50	15
4	1000	50	20
5	950	50	40
6	1100	100	50
8	2550	60	60

[a]All systems include six 90° bends. Fly ash bulk density 40–50 lb/ft³.

An economical solution of these situations is the use of combined vacuum and pressure conveyors. The advantages of a vacuum system can be applied at the precipitator or bag house outlets and the system kept as short and free of bends as possible. Fly ash is conveyed to a transfer point close to the fly ash outlets where a pressure system picks up the fly ash and conveys it the remaining distance. By keeping the vacuum system as short and compact as possible a high conveying rate can be obtained. The pressure system is then designed to match the rate of input from the vacuum system. The transfer of material from vacuum to pressure takes place in a three-cell chamber which permits continuous flow of fly ash from the vacuum state to the pressure state. See Fig. 6.8-29.

Dense phase systems make use of complex feeding devices at each fly ash pick up point. By carefully controlled fluidizing, fly ash is made to flow at a high concentration. As a result, conveyor pipe sizes for comparable system capacities are smaller than required by dilute phase operation. For the same capacities, power requirements for dense phase systems are usually less than for dilute phase systems of similar capacities because of lower air volume requirements. Dense phase conveying velocities are about 50% of those required for dilute phase systems.

Table 6.8-10 shows dense phase conveying capacities for a few representative systems operating at 25 psig. Depending on length of the system and the number of bends in the conveying line, fly ash concentrations of 30–50 lb/lb of air may be attained. Theoretically, much higher concentrations may be achieved in short conveyor lines without bends.

Comparative Costs

Material handling valves used to feed fly ash into a vacuum system are relatively simple and inexpensive compared to air-lock devices which control flow into dilute phase pressure systems or the pressure pots (transporters) required to feed dense phase systems. Consequently, vacuum systems usually are less expensive than other types of penumatic conveyors particularly where a large number of fly ash hoppers is served.

Offsetting the higher cost of feeding devices for dilute and dense phase pressure systems, in cases where fly ash is deposited in a silo, is the elimination of collectors which are required by vacuum systems.

Economizer and air heater ash can be handled in the same systems used for fly ash but certain modifications may be necessary because of temperature conditions.

Where coarse materials are formed, if ash is to be handled pneumatically, crushers should be provided at each outlet to reduce particles to $\frac{3}{8}$ in. diameter or smaller.

Alternately, small secondary hoppers can be installed under each economizer or air heater hopper to receive ash continuously from these areas. By removing ash from a hot gas stream the tendency to sinter will be reduced and pneumatic conveying can be applied in the same manner as for fly ash.

Economizer and air heater ash may, under suitable conditions, be deposited continuously in water filled tanks from which it can be pumped periodically. In this case, care must be taken to prevent vapors from rising in the spouts from the ash outlets which will cause buildup of ash in the spouts and eventual plugging. This may require the use of a fan to create a slight negative pressure within the storage tank. Hydraulic storage and removal is acceptable only when ash can be disposed of to a fill area. This ash cannot be dewatered satisfactorily when pumped into dewatering bins.

Caution: When ash is high in calcium and sulfates it must *not* come in contact with water and hydraulic handling cannot be attempted. Rapid set up in storage tanks will make removal impossible by ordinary means. Ash from most western coals falls into this category.

Conveyor Pipe

Most ash is highly abrasive and conveyor lines for wet or dry systems should be designed with that in mind. On a first cost basis, hard iron pipe with specially designed bend fittings is usually the least expensive and will have acceptable life in all but the most severe cases where corrosive attack accompanies the abrasive quality of ash or slag.

Many other types of pipe are available for ash handling service but most are significantly more expensive than conventional hard iron.

Pipe joints should permit easy removal and replacement of fittings or individual lengths of pipe and, in the case of hydraulic systems, should permit rotation of pipe in order to obtain maximum wear surface.

6.8-8   Combustion Control

Francisco Palacios

Combustion Control for Coal Pulverizers

Today, vertical spindle mills and ball tube mills are almost exclusively arranged for pressurized operation and supplied with the hot air necessary for drying and pulverized coal transport, called primary air, from a preheated source of pressurized air called forced draft or secondary air. The pressure of this secondary air is boosted by the action of primary air fans to the pressure required by the system.

On the firing systems for large utility boilers, two primary air fans handling cold air, discharging in a common manifold, and serving several mills is the most common arrangement. On smaller installations, one primary air fan per mill, handling hot air, is usually supplied.

High-speed attrition mills are commonly supplied with integral exhausters and therefore operate under suction at the pulverizer inlet.

The regulation of the coal flow in and out of the pulverizer and of the primary air flow follow specially designed control strategies adapted to the type of mill. However, there are certain control loops which are common, although with slight modifications to all types of pulverizers. The latter are: the primary air pressure control loop for cold air manifold systems, the control loop regulating the coal air mixture temperature, and the regulation of the seal air for systems designed for pressurized operation.

In the following, a particularized description of the controls associated with coal and air flow is presented for each pulverizer type, followed by descriptions of the above common control loops.

Vertical Spindle Pulverizer Control

Figure 6.8-30 shows a typical arrangement for a vertical spindle pulverizer system and its controls. Although there are several individual pulverizer designs, the controls for all models accomplish basically the same functions with slightly different control strategies depending on the manufacturer.

Coal flow to the burners is proportional to coal feed to the pulverizer after a relatively short dead time due to the grinding process and mixture transport through the coal pipes. Fast response is accomplished by increasing the primary air flow to the pulverizer simultaneously with feeder speed to pick up a quantity of pulverized coal already existing in the pulverizer.

There are two sets of dampers associated with a vertical spindle pulverizer. The hot air damper and the tempering air damper regulate the quantities of hot and cold air for temperature control purposes. A separate single air flow control damper regulates the air flow to achieve proper air to coal ratios.

The demand signal originating in the total fuel controller is applied directly to the feeder of each pulverizer. There is no need for a separate pulverizer controller because of the very good repeatability and stability of modern feeders and feeder drives.

The air flow is modulated by a ratio control loop which utilizes temperature compensated primary air flow measurement as feed back. A ratio station permits the operator to adjust the relationship between air flow and coal flow and a feed forward with derivative action accelerates the air flow modulation during fast load demand changes.

Signals from each coal feeder are summed in the total coal summator and their sum is used as the feedback signal for the total coal controller. See Fig. 6.8-31.

Ball Tube Mill Control

Figure 6.8-32 shows a typical arrangement of a ball tube mill system and its associated controls. There are two sets of air control dampers associated with a ball tube mill. The hot air damper and the

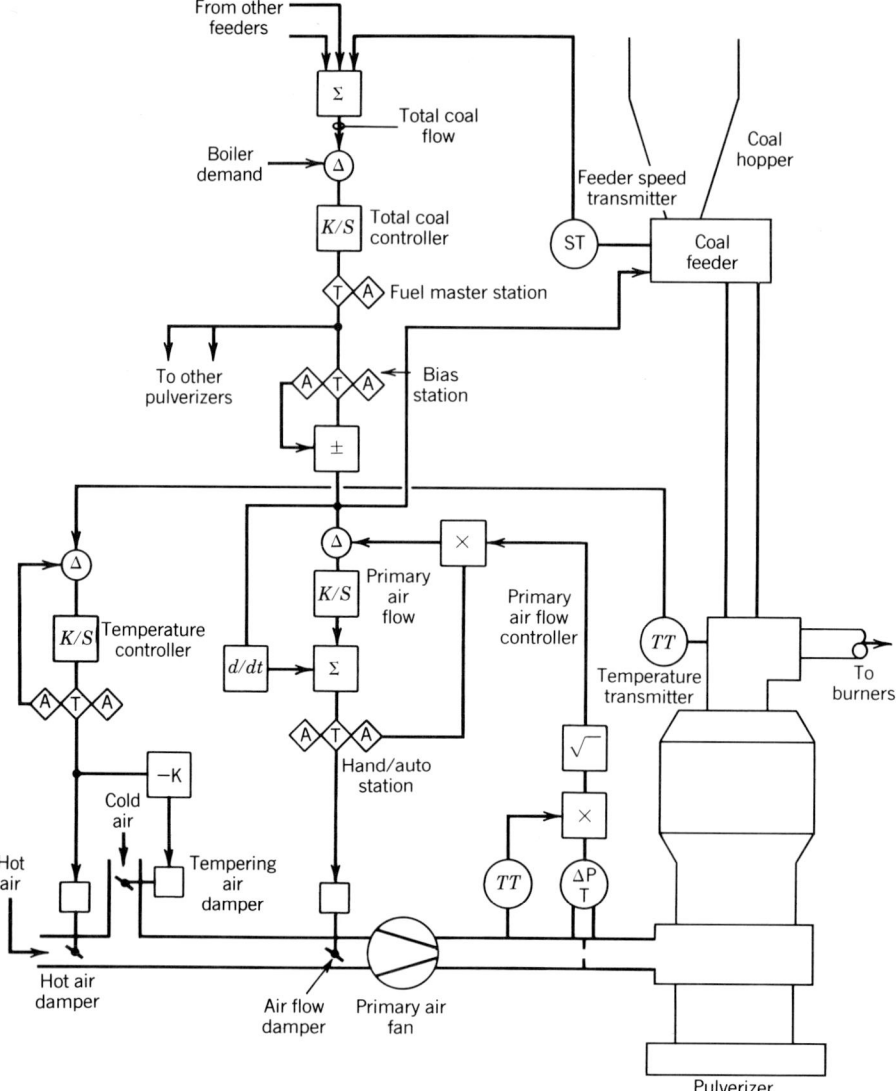

Fig. 6.8-30   Vertical spindle pulverizer with hot primary air fan.

tempering air damper regulate the quantities of hot and tempering air for temperature control purposes. The *rating* or capacity control damper and the bypass dampers regulate the quantity of air mass passing through the ball mill and thus the amount of coal delivered to the burners.

Although operation without bypass dampers is possible, the use of these dampers permits greater turndown of the mill system since, by opening them, it is possible to further reduce the air through the mill, and thus coal pick-up, without decreasing the coal–air mixture velocity in the coal pipes below the minimum allowable value.

There are five distinct control loops associated with ball tube mill systems: (1) for coal flow and air flow regulation, (2) for exit coal air mixture temperature control, (3) for mill coal charge control, (4) for seal air, and (5) for primary air duct pressure.

1.   Coal flow air flow regulation.   In a ball tube mill it is not possible to directly measure the amount of coal delivered to the burners. Signals of volume or weight of coal into the mill, readily available in the feeder hardware have no instantaneous relationship with pulverized coal delivery and therefore a more suitable coal flow signal is required for proper feed control. The pressure differential

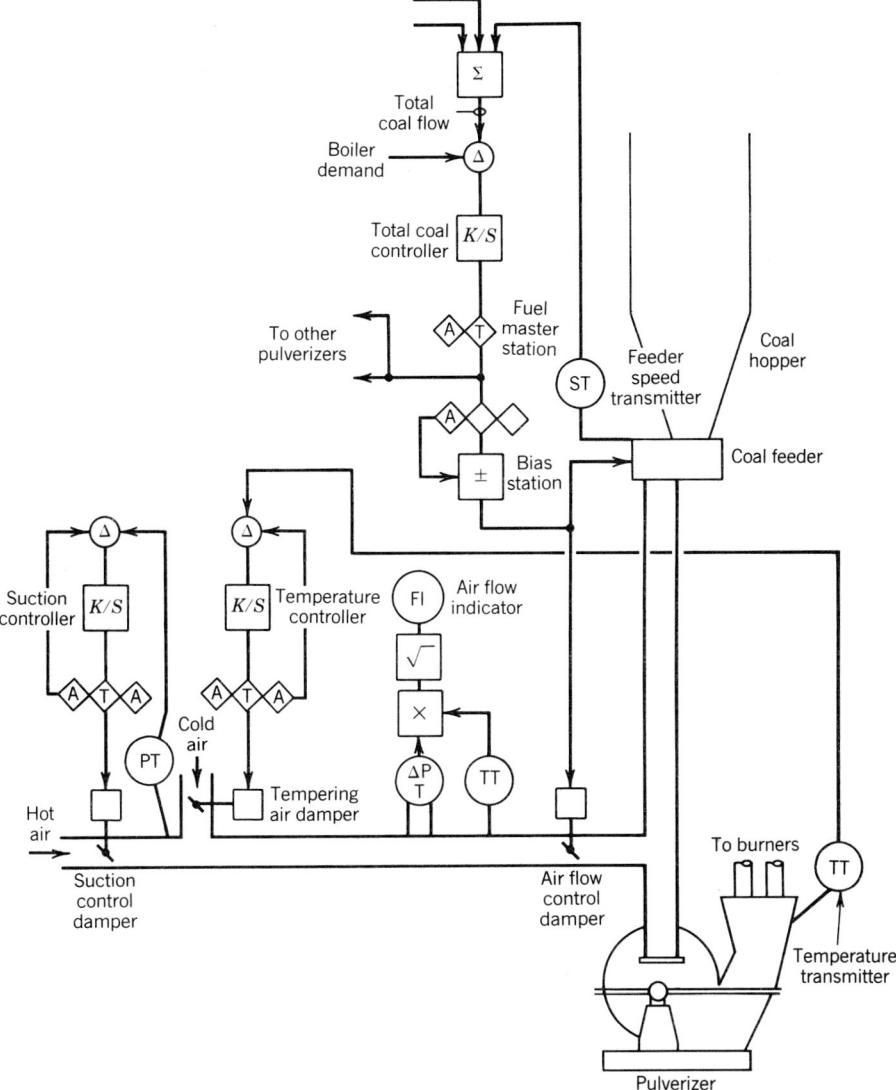

Fig. 6.8-31 High-speed attrition pulverizer suction primary air inlet.

between mill inlet and outlet or across the mill classifier on mills provided with external or no bypass dampers have been customarily used as an inferred coal flow signal for feedback purposes.

In the control loop, the demand signal originating in the total fuel controller is compared with the summation of the linearized mill end differential pressures to generate a control error signal which is amplified as required by the mill controller. A separate controller for each mill is recommended to ensure balanced operation among the mills. The output of this controller modulates both the rating damper and the bypass dampers. The coal flow is not exactly a square function of the mill end differential pressure and thus a function generator is preferred over a square root extractor to linearize the differential signal.

A filter located after the mill end differential transmitter is used to minimize the mechanical noise present in these signals.

Since there is no dead time associated with the grinding process—the mill maintains enough reserve capacity of pulverized coal—the response of coal flow to a change in air flow damper position is very fast, and after a few seconds, necessary for coal mixture transport, a change in the firing rate is achieved.

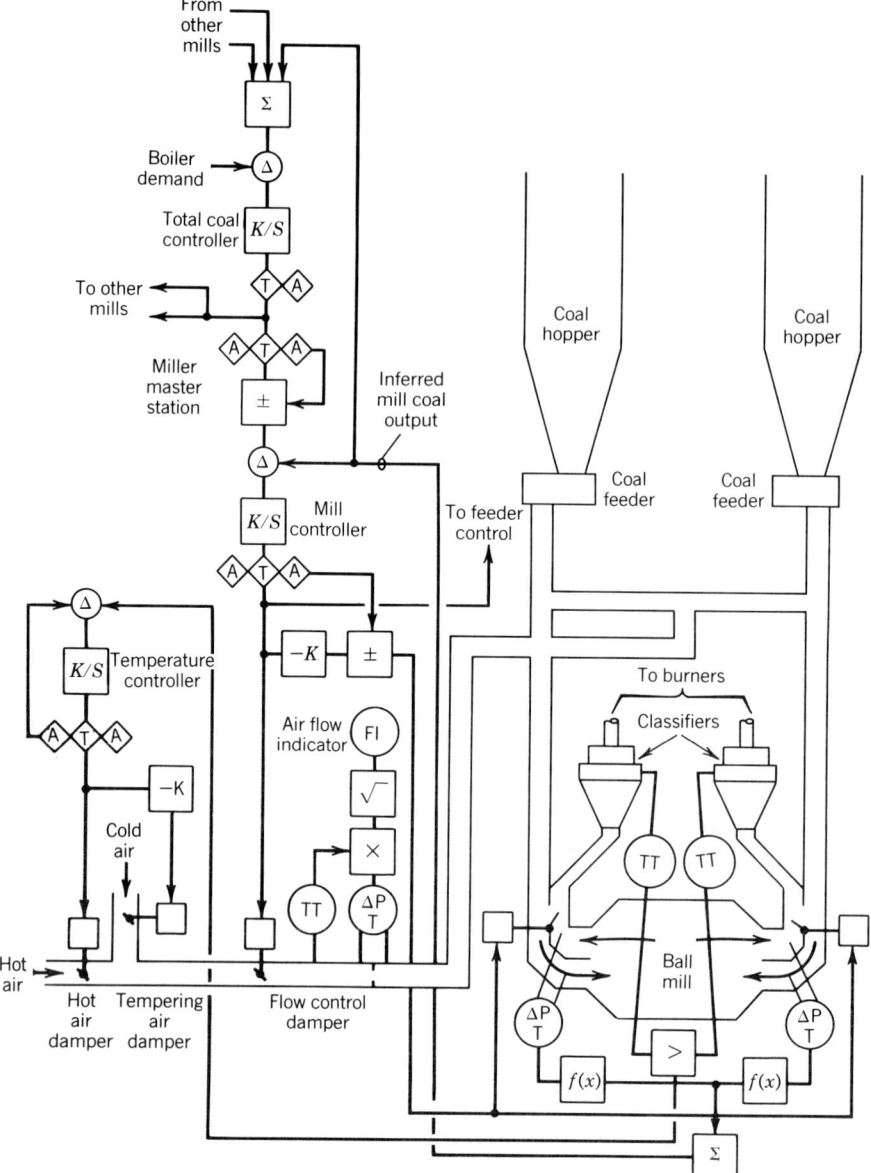

Fig. 6.8-32 Ball tube mill control.

For best control of the coal/air mixture the rating and bypass dampers modulate simultaneously and in opposite directions, their relative position determined by field tests to provide a linear variation of the coal air ratio through the load range.

2. Coal inventory control. To provide efficient grinding and rapid response to load change demands, a ready coal inventory is maintained in the mill. This control loop maintains the coal feed in balance with the coal delivered, under the condition that the mill must have at all times enough reserve to maintain its characteristic very fast response to boiler load changes.

Two systems have been successfully applied. The first system to evolve was the determination of the coal level by means of pneumatic probes located one at each end, inside the mill. The coal level probe consists of two sensing lines, one of them submerged inside the coal bed and the other above it. By introducing a continuous small amount of purge air through each tube, a back pressure differential proportional to the coal level is created and measured. The lowest of the two differentials is selected for control and compared to a manually selected set point. The output of the mill level controller regulates the output of the coal feeders. This arrangement is shown in Fig. 6.8-33.

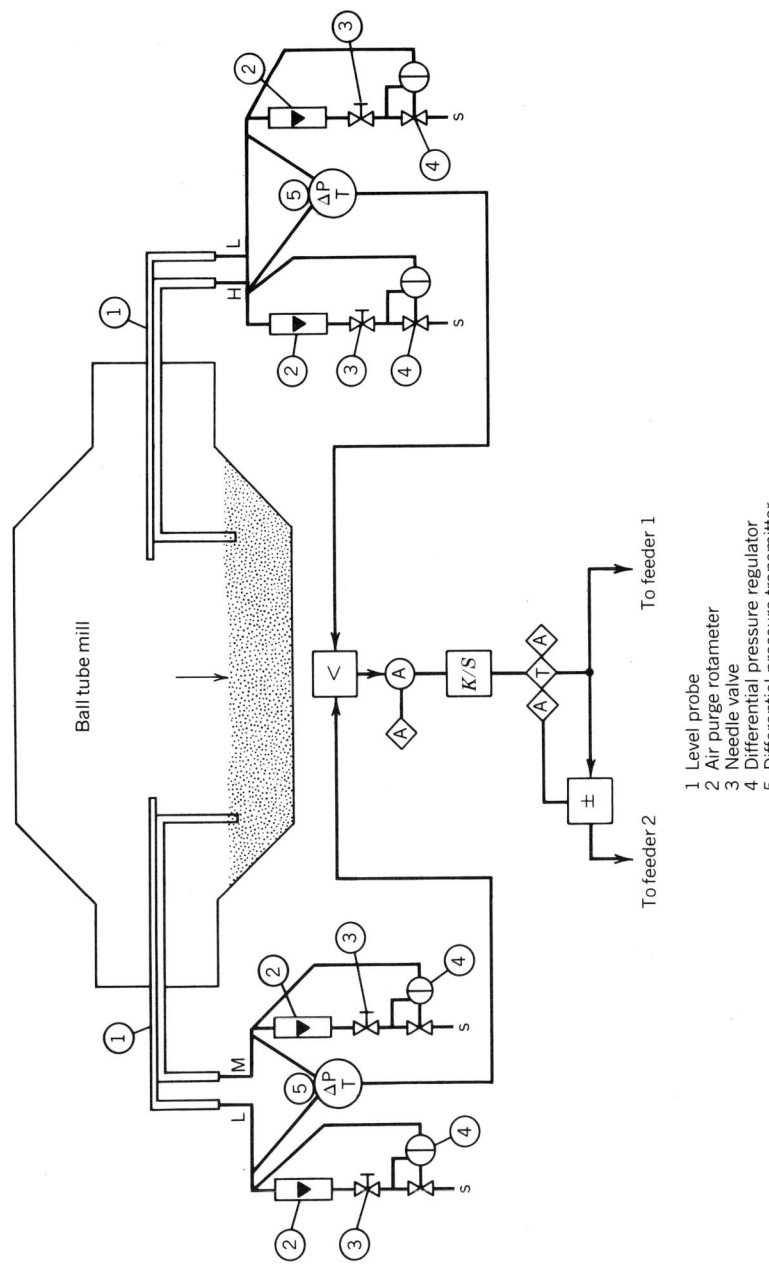

Ball tube mill

To feeder 1

To feeder 2

1 Level probe
2 Air purge rotameter
3 Needle valve
4 Differential pressure regulator
5 Differential pressure transmitter
s Instrument air supply

Fig. 6.8-33 Pneumatic probe type ball tube mill coal level control.

821

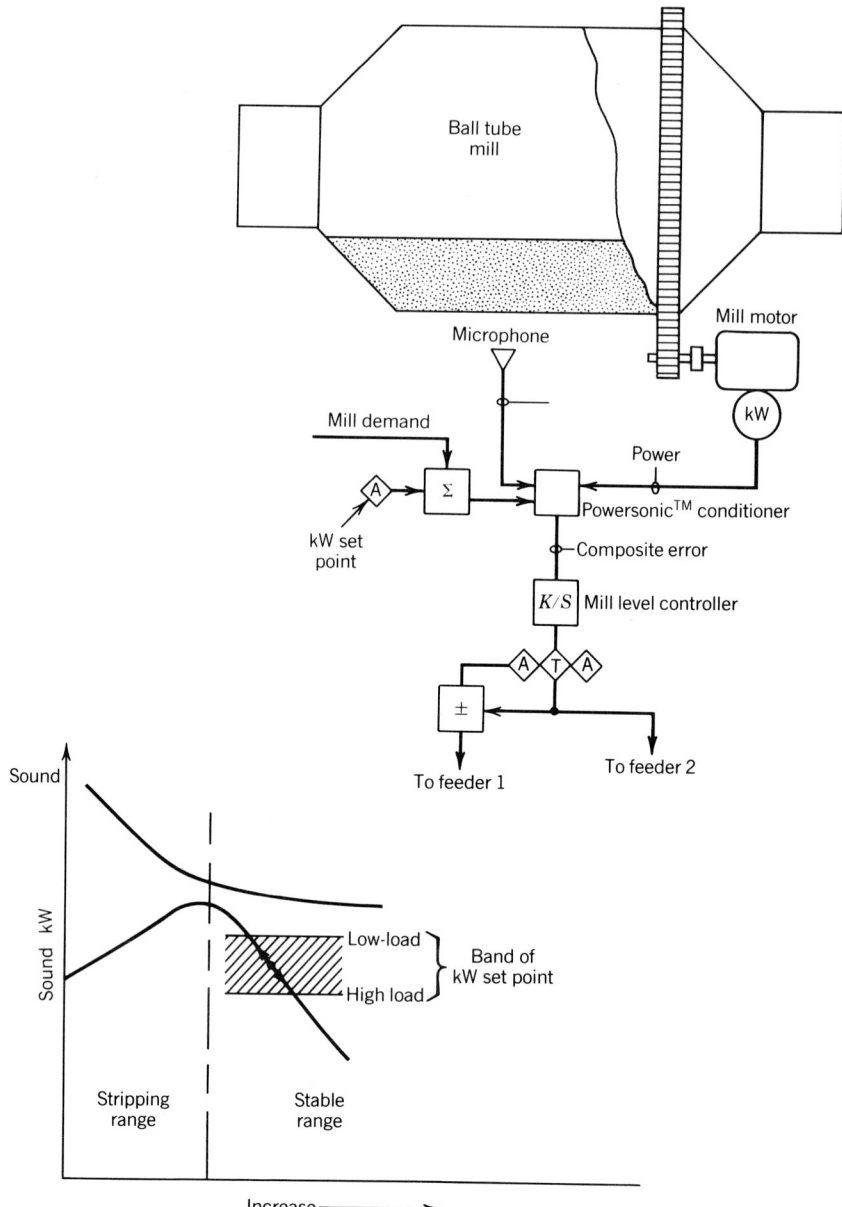

Fig. 6.8-34 Power-sound ball tube mill coal level control.

The second system takes advantage of the kilowatt consumption and sound characteristics of ball tube mills.

Starting with an empty mill, as coal is fed into the mill, the power input to the mill motor increases with coal charge. At the same time the sound level decreases. This trend continues until there is enough coal in the mill to fill all the cavities between the balls. At this point the power reaches a peak. At greater coal charges, the power consumption starts to decrease due to the effect of a displacement of the center of gravity of the mill closer to the axis of gyration, while the sound level decreases at a slower rate.

The region where power increases with charge is called the *stripping range* of the kilowatt (kW) curve, the region where power decreases with charge is called the *stable range* of the kW curve.

It is necessary to operate with a mill that has coal reserve charge in it for responding to quick demand increases. The ball tube mill is therefore operated in the stable range of the kW curve. A kW set point is selected in this range, and the resulting control error is brought into a conventional two-mode controller whose output repositions the feeders as required to satisfy the kW set point. In order to prevent operation in the stripping range and void of coal condition, the system takes advantage of the fact that the sound intensity increases in this range. If the sound level ever goes higher than a predetermined set point, the kW error is "swamped" and full signal is sent to the feeders in proportion to the deviation from the "sonic swamp" set point. Figure 6.8-34 shows the arrangement of this control loop.

Pulverizer Common Control Loops

1. **Seal air control.** Since many pulverizers and feeders operate under pressure, it is necessary to seal the areas between rotating and stationary members to prevent the escape of pulverized coal to the ambient. This is accomplished by air supply under pressure.

Most mill systems are provided with two seal air fans, one operating and the other stand-by. Seal air is supplied to the seals through a mill system seal air control damper. This damper modulates to control the differential pressure between the seal air entering the seals and the pressure inside the mill to a manually selected set point by the use of a single element control loop, as shown on Fig. 6.8-32.

2. **Primary air duct pressure control.** For cold primary air (PA) fan systems the PA fan inlet dampers are to be automatically operated to control the hot PA manifold pressure to a set point. The set point is constant during most of the boiler load range, but on certain conditions may have to be increased at the very high values of load. PA pressure is controlled by a single-loop arrangement as shown on Fig. 6.8-32.

3. **Temperature control.** The classifier exit temperature is controlled to a fixed selectable set point of approximately 140–160°F. On double-ended mills, the higher of the two end temperatures is selected and the controller modulates the hot and tempering air dampers in opposite directions to produce the required primary air temperature for proper drying.

Care should be exercised in determining the relative positions of these two dampers. Ideally, the dampers should be sized and field characterized to achieve an equal exchange of hot and cold air to minimize changes in the average mass flow which could be translated in unwanted variations in coal flow. Control loops for the pulverizer exit temperature control are shown on Figs. 6.8-30 and 6.8-32.

Combustion Controls for Stoker Fired Boilers

From the point of view of the control strategy coal stokers can be classified in two basic types: *thick bed* and *thin bed*.

Thick bed type stokers are commonly of the traveling grate type in which coal is fed by a single sliding coal gate mechanism which ensures a relatively constant bed height. The grate travels at a variable speed. To achieve acceptable load pick up capabilities, the load demand signal is applied to the air flow and to the grate speed rather than to the coal feed rate. In this manner an increase in air flow will stimulate combustion and Btu output faster than an increase in bed thickness.

Thin bed stokers are commonly of the spreader type in which coal or other fuels are projected into the furnace above a traveling or stationary grate. A high percentage of the fuel is burned in suspension and only a relatively thin bed of ashes and fuel is maintained on the grate.

Since the measurement of coal fuel flow is difficult with this type of stokers and, in addition, there is a relatively high dead time involved in the transportation and combustion processes, a parallel metering combustion loop is not normally employed and instead the combustion control strategy employs a simple positioning system to operate the air flow damper (in the case of traveling grate) or stoker feeders (in the case of spreader stokers).

Steam flow is a measure of the Btu input into the furnace and this signal is used to ratio the second combustion variable (grate speed in the case of traveling grate, air flow damper in the case of spreader stokers). In each case, an anticipation signal of boiler demand is imposed on the second combustion variable.

The ratio of steam flow to air flow can be set manually or automatically by means of a measurement of the oxygen controller. The set point of the oxygen controller can be made a variable function of the boiler load. Overfire air is generally set as a function of boiler load and therefore is directly positioned from boiler demand.

A stoker-fired boiler is often equipped with auxiliary gas and/or oil burners. A windbox with the required number of burners is placed above the grate and supplied with a source of hot secondary air from the air heater exit. For this arrangement a common control strategy is to control the secondary air pressure to a constant set point by manipulating the forced draft fan inlet damper. Air for the stoker undergrate, burner windbox and, if required, overfire air is drawn as required from this constant pressure source.

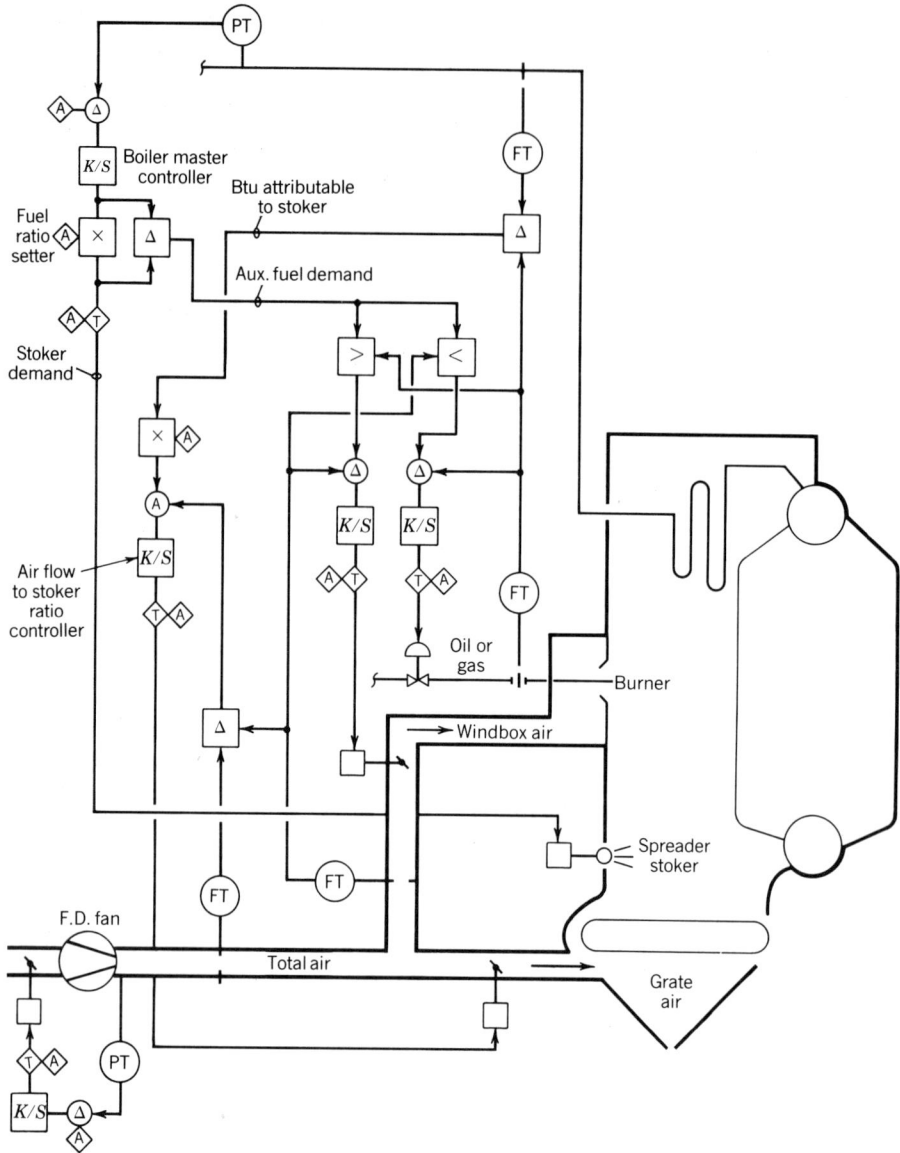

Fig. 6.8-35 Combustion control for simultaneous firing of stoker, spreader, and auxiliary (gas or oil). (Note: Steam temperature control, feed water control, and furnace pressure control not shown.)

The combustion of the auxiliary gaseous or liquid fuel is normally controlled by means of a separate parallel metering loop. The total steam flow is decreased by the amount of gas or oil burned and the resulting signal ratioed to position the undergrate air flow. Figure 6.8-35 shows a simple spreader stoker boiler and its controls.

BIBLIOGRAPHY

Axtman, W. and Eleniewski, M. A. "Field Test Results of Eighteen Industrial Coal Stoker Fired Boilers for Emission Control and Improved Efficiency." Paper 81-35-5, 74th Annual Meeting APCA, Philadelphia, June, 1981.

Carmen, E. P., Graf, E. G., and Corey, R. C. "Combustion of Solid Fuels in Thin Beds." Bureau of Mines Bulletin 565, 1957.

Ceely, F. J. and Wheater, R. I. "An Effort to Use a Laboratory Test as an Index of Combustion Performance." ASME paper 56-FU-6, ASME–AIME Joint Fuels Conference, Washington, D.C. October, 1956.

Ghosh, B. and Orning, A. A. "Influence of Physical Factors in Igniting Pulverized Coal," *Industrial Eng. Chem.* **47**(15), 117–121 (1955); *Combustion*, March, 1955, pp 57–61.

Hensel, R. P. "The Effects of Agglomerating Characteristics of Coals on Combustion in Pulverized Fuel Boilers," C. E. Publication TIS 4353.

Orning, A. A. "Reactivity of Solid Fuels." *Industrial Eng. Chem.* **36**, 813–816 (1944).

Mullen, J. F. and Gould, G. "The Influence of Volatile Matter on the Combustion of Pulverized Coal." ASME paper 60-FU-3, ASME–AIME Fuels Conference, October, 1980.

Wagoner, C. L. and Winegartner, E. C. "Further Development of the Burning Profile." Paper WA/FU-1, ASME Winter Annual Meeting, 1972.

6.9 EMISSION CONTROL SYSTEMS

Robert D. Bessette

This section discusses the various pollutants, their formation, and methods of control as they relate to the coal technology system. It will not discuss the regulations or their impact on the industrial, utility, or conversion systems. However, in all cases, an approach that is realistic, practical, and based on sound judgment and scientific fact must be developed to meet the needs of the community. This can be accomplished. The pollutants considered are nitrogen oxides (NO_x), sulfur oxides (SO_x), and total suspended particulate (TSP). There are different requirements for each of the major type systems, that is, industrial, utility, and conversion systems.

There are three areas that must be considered relative to pollution cleanup and/or minimization. The first is coal preparation in which, if all the ash and sulfur are removed, only nitrogen will remain as part of the coal. The second is combustion modification to prevent or minimize the formation of NO_x, SO_x, and TSP. The third is the use of flue gas cleanup before emission to the atmosphere. To provide the best, most economical, most practical solution to the problem of total system emissions, a combination of the three considerations must be investigated.

6.9-1 Nitrogen Oxide Emissions

Formation

Nitrogen oxides are formed during the combustion process and arise from two sources. The first is the fuel nitrogen, bound with the coal, and shown as nitrogen in the ultimate analysis. The total fuel nitrogen content in coal can vary from approximately 1% to over 2%, with the average nitrogen content being on the order of 1.5% or greater for most coals available in the United States. Fuel nitrogen is released during the combustion process in the forms of N, NH, or CN and can be oxidized to NO_x or recombine to form the very stable N_2 molecule. The availability of oxygen will dictate the direction the reactions will take.

The second method by which NO_x is formed is from thermal fixation of the atmospheric nitrogen (thermal NO_x). Thermal NO_x is dependent on the availability of oxygen, the furnace temperature, and residence time in the furnace. The formation of NO_x can best be described by the disassociation of O_2 to 2 oxygen molecules in the outer flame perimeter where the oxidation of carbon and hydrogen should take place. When the temperatures are 2800°F or greater, the following reaction can take place:

$$N_2 + O \rightleftarrows NO + N \text{ propagation} \qquad T > 2800°F$$

For all practical purposes the disassociation of N_2 is quenched at temperatures below 2800°F. However, fuel nitrogen is available for conversion at temperatures greater than 1500°F. In this case the following reaction can happen:

$$O_2 + N \rightleftarrows NO + O \text{ propagation} \qquad T > 1500°F$$

The complete cycle of formation is essentially stopped when the temperature is low enough that no free radicals are formed. Fuel nitrogen, to be available for these reactions, must be volatilized from the coal in the high-temperature region of the flame. It is also possible that in the following reaction the presence of NO in the flue gas will tend to minimize the formation of NO_x:

$$NO + (C \text{ or } H)N \rightleftarrows N_2 + O(C \text{ or } H)$$

With the basics of formation in mind it is now possible to consider the methods of control that could be applied for the coal combustion system.

Control Techniques

Fuel Selection.   Fuel nitrogen monitoring and selection is one method by which NO_x emissions can be minimized. A coal with a sufficiently low fuel nitrogen to meet the regulations may be the best method for solving the problem. Coal reactivity may also be decreased to slow down combustion and decrease temperatures in the furnace to minimize thermal NO_x production. Most industrial systems can operate without NO_x reduction techniques if the fuel nitrogen is maintained at a level less than 1.5% in the ultimate analysis of the coal. Utility systems with highly turbulent flames and high-temperature furnaces tend to require lower fuel nitrogen than is normally available in the quantities required. This necessitates the use of reduction techniques.

Furnace Sizing.   Increasing the furnace cooling surface attacks the high-temperature and time aspects of NO_x production. This is normally effective and can allow for some flexibility with regard to coal specifications and price. The advantages are that it is simple to accomplish in a new system design and requires no external power for reduction. The disadvantages are that it is difficult and expensive to apply this technique to an existing system. Any modification of pressure parts can cause serious problems with the specific unit operation.

Burner Selection.   The choice of a burner relative to the furnace design attacks the O_2 availability to form the NO_x while at the same time shaping the flame to minimize the 2800°F residence time. This is possibly the ideal method for using moderate- and low-nitrogen coal and meeting NO_x emission regulations. The process is simple, requires no power, and produces very good results.

Low Excess Air Combustion.   Low excess air combustion attacks the availability of oxygen and increases overall efficiency with the high-turbulence burner. The 2800°F residence time remains about the same, and only fair results can be achieved. When low excess air is used, a fairly complex system of controls is required to maintain optimum combustion. Coal ash behavior may also be a problem because of reduced O_2 levels in the furnace. Any low excess air or combustion air change must be considered relative to the coal reactivity and combustion characteristics.

Two-Stage Firing.   The two-stage firing system also attacks the oxygen availability by adding excess oxygen, needed for complete combustion, through overfire air ports. As a result, the rate of combustion is lowered and the 2800°F residence time is decreased. This produces good results with moderate-to-high fuel nitrogen coals. The disadvantages of this method are the possibility of decreased carbon burnout and possible increases in fireside deposits with potential corrosion and a decrease in furnace heat absorption.

Off-Stoichiometric Firing.   Similar to staged combustion, the off-stoichiometric system attacks O_2 availability and flame temperature, but, for different levels of burners on large units. This method is fair with regard to success and is easily applied to existing units. The same concerns with carbon loss and increased slagging exist as above.

Flue Gas Recirculation.   The flue gas recirculation technique takes advantage of the reaction that tends to drive the reaction of fuel nitrogen toward N_2 in the presence of NO. The presence of NO also has a tendency to minimize the formation of thermal NO_x, again by driving the reaction toward the more stable N_2 product. This is a last resort technique when regulations must be met with high-nitrogen coal. Although it is the most effective method, it is also the most difficult and expensive to install. Combustion control and operating requirements with fans, ducting, and air balancing are difficult and can be a problem.

Summary of NO_x Emissions Reduction

Combustion modifications, other than optimum design for efficiency and increased furnace cooling, tend to sacrifice combustion efficiency for NO_x reduction. Evaluation of fuel choice (nitrogen content and reactivity) is mandatory to obtain the most economical and practical system.

There is a limit to which NO_x can be lowered by the use of fuel selection and combustion modification techniques. As a result, NO_x scrubbers are being developed and could help reduce NO_x emissions from high fuel nitrogen fuels used in oil- and gas-designed boilers. These scrubbers are only in their early developmental stages and will not be discussed.

6.9-2 Particulate Emissions

Particulate emissions are those particles that result from ash contaminents, the noncombustible elements in the coal that leave the furnace and boiler area after combustion and can be emitted to the atmosphere. The amount and type of particulate are dependent on the type of coal and combustion equipment, that is, stoker, fluidized-bed, or pulverized-coal firing. The size of the ash particles and their plastic, electric, chemical, and mass properties are all important.

This section will discuss particulates in relation to boilers. From these considerations the basics of the control equipment selection can be developed for various types of equipment. The stoker-fired boiler produces a coarse ash with 75–90% of the total ash greater than 10 μm in size. Depending on the type of stoker, 20–60% of the total coal ash can be emitted as flyash. The pulverized-coal-fired boiler will generate 80% or more of the total coal ash as flyash with 80–90% of that ash being less than 10–15 μm in size. For FBC units, depending on the feed mechanism, flyash (including coal ash and bed material) will range from that of a stoker-type ash to that of a pulverized-coal-type ash. In most cases, with reinjection for carbon recovery, the ash will be on the order of 70–90% less than 15 μm. With this in mind, the appropriate control technology may be discussed.

Control Techniques

Mechanical Collection. The mechanical collector is the oldest of the particulate collection devices. Working on the principle of cyclonic or centrifugal separation, the collector causes heavier-than-air particles to be forced in a circular motion toward the outside of a cyclone and down until the air is caused to make a 180° turn upward, causing the effective separation of the ash and flue gas. Some of the less than 10-μm particles will be carried with the flue gas and not collected. Only about 50% of the less than 10-μm particles are collected.

Increases in collection efficiency or particle flue gas separation can be achieved by increasing flue gas velocity, causing an increase in the centrifugal force and promoting better separation. The increase of velocity will increase pressure drop requirements and necessitate an increase in power requirements. The normal expectation with regard to emission levels for a single mechanical collector is 0.6 lb/MMBtu for a spreader stoker-fired boiler. For the same boiler using two mechanical collectors in series, a 0.3 lb/MMBtu emission can be met. Firing methods that produce larger size particulates would be able to obtain better efficiencies than those listed above for the stoker-fired boiler.

Sidestream Separation. Sidestream separation or bottom hopper evacuation, as applied to mechanical collectors, was developed by General Motors and is a modification by which 10–20% of the flue gas is removed from the bottom hopper of the collector and cleaned in a small baghouse. This evacuation causes a slight negative pressure in the cyclone, promoting better separation as well as cleaning a high percentage of the very fine particulate and thus improving the overall collector performance. The flue gas from the baghouse is recombined with that from the collector.

In some cases on a spreader stoker-fired boiler there can be up to 35–50% efficiency improvement. This will allow the consuming facility the flexibility to use a higher-ash, lower-cost coal. For a spreader stoker-fired boiler with 15–25% of the flyash less than 10 μm, the sidestream arrangement can meet an emission level of 0.3 lb/MMBtu. In some cases it may be capable of meeting a level of 0.2 lb/MMBtu particulate. An emission level of 0.1 lb/MMBtu is not reachable on a consistent basis and any expected performance must be based on an evaluation of the flyash size distribution, the inlet loading of the flyash to the collector, and the velocity or pressure drop within the collector.

Electrostatic Precipitator. There are two types of electrostatic precipitators, the rigid-wire and the weighted-wire version, which operate in one of two modes. Either will work with hot (> 450°F) or cold (< 450°F) flue gas. The electrostatic precipitator operation is based on the ability of a coal ash to accept an electric charge. The lower the resistance of a coal ash to accept a charge, the better the collection capability.

The operation of an electrostatic precipitator is such that an electric field is set up between the wires and vertical plates. Coal ash passing between the wires and plates picks up a charge within the field and is attracted to the plates. A device called a rapper raps on the plates, causing the ash to slide down the plate into the collection hopper.

When designing for efficiency, areas of concern other than flyash specifications are velocity of the flue gas relative to the length of travel and the spacing of the plates and wires. The velocity and length of travel are important with regard to migration time for collection of the particles and the amount of reintrainment that might occur during rapping in a high-velocity flue gas stream. Wire and plate spacing relative to particle charge and electric field strength will determine the power requirements necessary to meet a collection efficiency. Specifications for design vary from manufacturer to manufacturer, however, the final design arrangement must be based on coal flyash specifications.

Flyash mineral characteristics, flue gas SO_3 content, and flyash carbon content, a function of coal reactivity and combustion efficiency, affect the resistivity of an ash at a given temperature. Flyash resistivity decreases with increasing temperatures to a point where the particles begin to lose the charge. This has a direct effect on efficiency.

General rules of application are if the coal sulfur content is greater than 1.5%, no special considerations will be required to meet a 0.1 lb/MMBtu particulate emission. In this case there is sufficient SO_3 present in the flue gas to minimize the impact of the ash minerals and carbon content of resistivity. In cases when low-sulfur coal is used, the resistivity of the ash becomes an important consideration. Here the use of lower velocities and stronger field strengths can help when high coal resistivities are encountered. When coal resistivity is such that the above methods are not enough to meet the needed collection efficiencies, the use of a hot electrostatic precipitator may be warranted. In this case the flue gas is cleaned before the air heater or final heat recovery step. The higher temperatures decrease the resistivity of the ash, and a stronger charge can be accepted, improving overall efficiency. It must be remembered that when cleaning hot flue gas the velocities must be maintained, which means a significantly larger piece of equipment for the same weight but increased volume of gas. The equipment arrangement and costs, because of size and difficult locations, is in many cases prohibitive. The precipitator is capable of meeting a 0.1 lb/MMBtu particulate emission rate.

Once a precipitator design has been selected and installed, it is very difficult to change coals and assure collection efficiency. Systems have been designed to add SO_3 or ammonia to the flue gas to improve the electric field strength, thereby increasing efficiency. Power pulsing has also been used where minor changes in flyash resistivity have been seen. However, these have been of only limited success on specific applications.

Baghouse Collector. The baghouse operates on the same principle as a bag-type vacuum cleaner. There are two types, each classified relative to the method of cleaning employed. The first is the reverse air system, which requires that a bag module be shut down and a reverse stream of air be blown through the bags, cleaning the flyash coating from the bags and into a hopper. The reverse air pressure is sufficient to remove the ash, however, but not to reentrain it into the flue gas. The second method is the pulse jet method where flyash is collected on the outside of the bag. A pulse of high-pressure air is forced down through the bag, flexing the bag, and shaking the ash from the bag into the ash hopper. This is accomplished during normal operation, and some ash is reentrained into the flue gas and collected on other bags. The operation of the pulse is cyclical and dependent on the pressure drop across the baghouse. The pressure drop across the baghouse must be kept to a minimum to keep power costs down and minimize potential boiler interruptions and implosion concerns.

Baghouse operation and efficiency has very little to do with coal or flyash characteristics other than ash mass flow to select the proper amount of bag surface to assure pressure drop limits over a reasonable cleaning frequency. This consideration is usually expressed in terms of a gas-to-cloth or air-to-cloth ratio—the ratio of flue gas (acfm) to effective cloth area in square feet. Ratios as high as 8 : 1 are used for some pulse jet systems because of more effective cleaning. Ratios of 4 : 1 or less are used when reverse air systems are considered. This also affects bag material considerations with regard to life expectancy.

Bag materials for this application range from cotton, wool, and synthetic soft fibers to Teflon, fiberglass, and Teflon-coated fiberglass in felted or woven arrangements. Within the baghouse environment the bags must be able to tolerate fairly high temperatures, an acid or possibly alkali condition, and exhibit good resistance to flexing caused by wear, especially in the pulse jet systems. The cost of replacement bags can be substantial and must be taken into consideration relative to bag life over the life of the equipment. Although not much evidence is available, bag life can vary from 3 years to a questionable 8–10 years.

Other areas of concern are the possibility of hot particles reaching the baghouse and burning or melting a hole in the bag. Blinding of the bags can occur if temperatures are allowed to fall below the dew point of the SO_3, around 280°F. Heaters and temperature monitors must be used to assure that maximum and minimum temperatures for the bag type are not exceeded. When stoker-fired boilers are considered, a mechanical collector or some other type of equipment should be considered to remove larger potentially hot particles.

Collection efficiency of a baghouse is very good and possibly better than all other types of equipment, with the ability to remove almost 100% of all the particulate greater than 0.4 μm. This provides flexibility to meet emission limits on the order of 0.03 lb/MMBtu particulates, independent of the coal characteristics.

Wet Scrubbing Systems. Wet scrubbing systems for particulate control can be operated at approximately the same efficiency levels as a baghouse. The basics of operation is a velocity difference between water droplets and ash particles. This is accomplished by the use of a flue gas venturi and water sprays. Flue gas is accelerated through the venturi, causing ash particles to reach high velocities and impact into larger slower moving water droplets. The water droplets are collected and removed in the downstream scrubber demister section.

In this case the main dependencies are particle size, velocity, and droplet size. The smaller the particle, the higher the velocity needed to assure impact with a droplet. Extremely fine particles greater than 50% less than 10 μm may require up to 30 in. of water pressure differential to acheive equivalent baghouse performance. An industrial spreader stoker-fired boiler could require a pressure drop of from 12 to 18 in. of water to meet a 0.1 lb/MMBtu particulate emission level.

There are two considerations that must be remembered when considering wet scrubbers for particulate removal. The first is water disposal. Because water is used to remove coal ash, the sulfur content and chlorine content of the coal will cause a lowering of the water's pH by the formation of sulfuric and hydrochloric acid. Provisions must be made with material selection and waste disposal to handle this. The second is related to the ability to remove the water from the flue gas. Adequate demisters must be provided to assure the complete removal of all the water from the flue gas at a temperature such that condensation does not take place on low-temperature days and cause localized acid rain. The main advantage of the wet scrubber is the potential for future SO_2 control or when an acidic stream is needed to neutralize a plant waste stream.

Particulate Removal Summary

Particulate removal costs can vary from one technology to another. If the mechanical collector is considered the base, the capital cost for particulate control may be described as follows: mechanical collector, 1.0; sidestream separator, 2.0; baghouse, 4.0; cold electrostatic precipitator, 4.0; hot electrostatic precipitator, 6.0; and wet scrubbers, 3.5. Operating costs depend on the power requirements to compensate for pressure drop or electric field generation, labor and maintenance for cleaning and replacement parts such as bags, and fuel cost premium if required to meet emission regulations.

The specific emission requirements for a given location must be developed prior to fuel selection in order that a true evaluation can be made over the total system. The effect of particulate emission requirements can in many cases determine the outcome of a coal system evaluation. The only more critical consideration within the environmental area is the sulfur dioxide emission rate allowed.

6.9-3 Sulfur Dioxide Emissions

Sulfur is part of the composition of every coal. There are no coals without it. Younger sub-bituminous and lignite have a lower percent than do those formed during the sulfur-rich environment of the carboniferous period. As a result, when coal is combusted in an oxygen-rich environment, sulfur is combusted to SO_2 or SO_3. If coal is combusted in an oxygen-poor or reducing atmosphere, H_2S is formed from the free hydrogen in the coal or from disassociated hydrogen from moisture, either inherent or surface. This latter is primarily associated with gasification or coking systems, and the former is associated with direct combustion systems.

The percentage of sulfur converted to SO_2 and SO_3 may vary depending on the coal ash characteristics and the combustion process. Concentrations of calcium and magnesium, the alkalis, and silica can form various combinations that tend to retain the SO_x as part of the ash rather than emitting it as a gaseous pollutant. The Environmental Protection Agency will tend to agree that approximately 5% of the total sulfur is retained in this way and that only about 95% is emitted as SO_x. For calculation and design purposes it is best to assume 100% sulfur conversion to SO_x. For application and coal purchase it is possible to consider the 95% conversion rate to set sulfur levels for the coal. However, this should be assured in writing by the appropriate regulatory body.

The following equation is used to calculate the SO_2 emissions at 100% conversion:

$$\text{lb } SO_2/\text{MMBtu} = \frac{1.0 \times 10^6 \text{ Btu}}{\text{Btu/lb coal}} \times (\%)(S) \times 1.998 \text{ lb } SO_2/\text{lb S}$$

To obtain the lb SO_2/MMBtu at the 95% conversion rate, the above equation would be multiplied by 0.95. Many companies have developed shortcut methods or alternative methods of calculating these values. It is possible and accepted practice to use 2.0 lb SO_x/MMBtu at the 100% conversion rate and 1.9 lb SO_x/MMBtu for the 95% conversion rate. The reason for this is that there is always a small portion of SO_3 formed at the relative weight of 2.497 lb SO_3/lb S.

In all sectors of industry there is a desire to minimize SO_x pollution through realistic control levels and regulations. The following discussion of SO_x control methods will be based on a 1.2 lb/SO_2/MMBtu emission rate.

Control Techniques

Coal Beneficiation. In many cases the use of washed coal is promoted as the best alternative for meeting sulfur regulations. It can be the most effective; however, transportation, availability, and price must be considered within the framework of the total system. The coal variability and ash content of a washed coal tend to favor the use of coal cleaning to meet sulfur regulations. In any case the cost of washing (the cost of the coal) to the desired sulfur level and the cost of transportation of the coal from a distant source may indicate that a locally available, lower-cost coal with a higher sulfur content may be the best choice when used with scrubbers. It is usually a combination of coal beneficiation and scrubbers that will meet the needs of the total system with locally available high-sulfur coal. Here the maximum level of benefit is generated by each method for an overall greater total system benefit.

Another method that should be considered is indirect beneficiation, which provides the ability to blend very low sulfur coal with high or moderately high sulfur to meet a given sulfur requirement at a location. Considerations of coal quality, reactivity, combustibility, and the resulting ash characteristics become a necessity, especially if coals of different rank and structure or reactivity are blended. Performance evaluations of coal blends is only now beginning, and sound performance projection information will come in the future.

Wet Nonregenerative Scrubbers.   The wet scrubber can be operated as either a "throw-away" system, where SO_2 is reacted with a sorbent such as limestone and the resultant product is disposed of, or a "recovery" system, where the SO_2 is reacted with a different sorbent and after some additional processing elemental sulfur is produced for sale. In the nonregenerative process the sorbent is lost in the cycle. The recovery system becomes an elaborate chemical plant and as such will not be discussed herein.

Wet scrubbers use the principle of a gas to liquid to solid phase chemical reaction to collect SO_2. SO_2 as a gas is scrubbed with a liquid–solid slurry of water and lime or limestone. The chemical reactions for this system are

Lime systems

$$SO_2 + CaO + \tfrac{1}{2}H_2O \rightarrow CaSO_3 \cdot \tfrac{1}{2}H_2O$$

and

$$SO_2 + \tfrac{1}{2}O_2 + CaO + 2H_2O \rightarrow CaSO_4 \cdot 2H_2O$$

Limestone systems

$$SO_2 + CaCO_3 + \tfrac{1}{2}H_2O \rightarrow CaSO_3 \cdot \tfrac{1}{2}H_2O + CO_2$$

and

$$SO_2 + \tfrac{1}{2}O_2 + CaCO_3 + 2H_2O \rightarrow CaSO_4 \cdot 2H_2O + CO_2$$

These reactions are dependent on inlet loadings of SO_2, sorbent concentrations, sorbent particle size, scrubber residence time, and collection efficiency required. The three most important of these are the particle size as it affects sorbent surface area, scrubber residence time for the reaction to take place, and the sulfur dioxide inlet loading dependent on the coal used in the process.

Systems designed for sulfur removal use many different arrangements from packed towers to multiple rod decks to agitated ball decks. Each design is provided to increase the residence time for the reaction to take place. The flue gas flow is normally counter to the sorbent flow, with the internals also acting to increase the gas-to-liquid contact by either providing a surface on which the contact can be made or a turbulence in which contact can be increased. Because the reactions are primarily a surface reaction on the surface of the reagent, the total surface is important. Greater surface can be provided by using slaked lime in lieu of pulverized limestone. This is similar to that surface required in a fluidized-bed boiler for sulfur removal. Consideration must be given to calcining and slaking costs for lime and its inherent pH and scaling problems. Pulverizing costs and efficiency must be considered for limestone.

The interdependencies of scrubber systems, as applied for SO_2 collection, are primarily concerned with SO_2 inlet loading from the coal sulfur content and the system capital cost. There are three levels of coal sulfur content at which the system capital costs increase sharply. The first is when the coal sulfur is above 1.7–1.8%; the second is when the coal sulfur content increases above 2.5%; and the third when coal increases above 3.2%. For each case total system size and complexity increase disproportionately. The nonregenerative system is the lowest-cost system to achieve a 90% SO_2 removal efficiency. The SO_2 is captured in the forms of calcium sulfite and some calcium sulfate with waters of hydration. The result is a heavy sludge-type product that cannot be fully dried to a solid product. This has been one of the main problems associated with wet scrubbing systems as applied to coal-fired systems. A 3% sulfur coal will generate about 1 ton of waste sludge for each 6.5 tons of coal consumed. The best stoichiometry of lime or limestone to sulfur have been on the order of 1.0–1.2 to 1. These are much better than the those commonly found for FBC systems.

Wet Regenerative Systems. The need for SO_2 removal has caused both utility and industrial companies to seek better, smaller, and more efficient methods to reach the SO_2 removal required when using high-sulfur coals. The reaction time in the lime/limestone system is long because of the gas to liquid to solid reaction. As such the equipment is large and costly. The dual-alkali system was developed to use a gas-to-liquid reaction that would speed up the collection process. Sodium hydroxide, NaOH, or sodium carbonate, both relatively expensive products, were used as reagents to provide for a reaction without solids buildup, scaling, or critical pH control. The second step, dual, had to be included to recover these products. Hence the reference of the dual-alkali system. The following are possible chemical reactions for this process:

Sodium carbonate reaction

$$Na_2CO_3 + SO_2 \rightarrow Na_2SO_3 + CO_2$$

Regeneration

$$Na_2SO_3 + CaCO_3 \rightarrow Na_2CO_3 + CaSO_3$$

Caustic reaction

$$2NaOH + SO_2 \rightarrow Na_2SO_3 + H_2O$$

$$Na_2SO_3 + SO_2 + 2H_2O \rightarrow 2NaHSO_3 + H_2O$$

$$2Na_2SO_3 + \tfrac{1}{2}O_2 \rightarrow 2Na_2SO_4$$

Regeneration

$$3Ca(OH)_2 + NaHSO_3 + Na_2SO_3 + Na_2SO_4 + 2\tfrac{1}{2}H_2O$$

$$\rightarrow 3NaOH + Na_2SO_3 + CaSO_3 \cdot \tfrac{1}{2}H_2O + CaSO_4 \cdot 2H_2O$$

$$+ H_2O + Ca(OH)_2$$

These reactions are much quicker in the scrubber, and a smaller scrubber can be used than for a lime or limestone system. These chemical reactions are more complicated, and the regeneration of the sorbent must be accomplished as quickly as possible without significant sorbent loss, which can increase operating costs well beyond those of the throw-away system. For these reactions to operate efficiently they need a fairly high flue gas SO_2 content. A $+2\%$ sulfur coal can help.

The dual-alkali system produces a solid calcium sulfite and calcium sulfate sludge, which is usually fixed by blending with the dry boiler ash and flyash prior to disposal. Extra regeneration equipment must be provided for those reactions, such as bleedlines and vacuum filters to recover as much of the high-cost sorbent as possible. There are always some losses, and these must be accounted for and minimized if the system is to be economical.

General Consideration for Wet Scrubbers. West systems, in general, operate in adverse atmospheres where low pH can cause corrosion. There is also what is known as a wet–dry interface between the flue gas and scrubber spray wet zone. Many problems have arisen in these areas. Spray nozzle erosion and piping wear can be substantial because of high-solids-slurry handling requirements. Because of this, materials of construction must be carefully selected to meet these requirements and those for materials corrosion due to possible chloride buildup within the systems. At temperatures below 300°F high chloride contents in the form of HCl can have serious repercussions within the system. Important consideration must be given to the chlorine content in the fuel selection process. Dual-alkali systems are extremely sensitive to the coal chlorine content. This is primarily a function of the need to minimize water-soluble sorbents. In most cases full particulate removal systems are required prior to the scrubbers, and provisions for handling low pH moisture-laden gases must be made in the ductwork and equipment downstream of the scrubber.

Dry Scrubbers. Wet–dry scrubbers or dry scrubbing, as it is sometimes referred to, have been developed to incorporate particulate removal to take advantage of any ash alkali content, eliminate problems with scaling, and eliminate the sludge handling system. This system was designed to use a sorbent mixed with water and mixed with the flue gas as an atomized spray. A gas to liquid to solid reaction takes place within the mixing chamber. At the same time the heat in the flue gas evaporates the slurry moisture, leaving a dry, solid product collectable in a baghouse collector. This system eliminates many of the problems associated with wet scrubber systems.

The chemical reaction for the dry scrubber system using slaked lime can be written as follows:

$$2SO_2 + 3CaO + 2\tfrac{1}{2}H_2O + \tfrac{1}{2}O_2 \rightarrow CaO + CaSO_3 \cdot \tfrac{1}{2}H_2O + CaSO_4 \cdot 2H_2O$$

This reaction is dependent on the flue gas inlet loading, the reactivity of the lime, the intimate and rapid mixing of sorbant and flue gas, the necessary residence time for the reaction to be completed, and the flue gas temperature that must be sufficient to evaporate 100% of the free moisture to produce a resultant dry, solid product. The inlet sulfur loading determines the amount of sorbant needed which, in turn, determines the total slurry moisture to be evaporated. If sulfur is high, slurry requirements and moisture may be such that flue gas temperature may not be enough to evaporate the total free moisture with a sufficient margin to assure efficient and effective baghouse operation.

Dry scrubbers are critically dependent on residence time and intimate mixing of the sorbant and flue gas than the wet scrubber systems with packed towers for multiple decks. In the dry scrubber a spray tower is used with atomized slurry and very low flue gas velocity to achieve a complete chemical reaction. Sorbant droplet size must be on the order of 10–15 μm to obtain best results. It has also been found that better evaporation can be achieved if the total flue gas is divided into small streams with individual atomizers much like oil atomizers in a multiple-burner throat arrangement leading into the evaporation chamber. Provisions such as tungsten carbide atomizer tips should be considered to prevent excessive tip erosion and degradation of the atomization.

Sorbants used must produce a solid and relatively dryable product. Sorbant reactivity must be considered relative to residence time to complete reaction within the drier. Some limestone has been tried on a limited basis with only limited success. Lime has proven to be the better sorbant. It is now being found in some instances that dolimitic lime may prove even better, with magnesium acting as a catalyst for the reaction. Sodium carbonate (Na_2CO_3) and soda ash ($Na_2CO_3 \cdot NaHCO_3H_2O$) have also been used with very good success on medium-sulfur coals. However, these are water soluble and can cause water pollution problems after product disposal.

In all cases the dry scrubbers perform better with lower-sulfur coals. With western coals a sulfur removal rate of about 90% can be expected. With eastern coals about 85% can be expected. There has not been much success with coals with greater than 2.0–2.5% sulfur. This is an important consideration when a dry scrubber is evaluated.

The simplicity of design and operation of the dry system tend to produce the lowest total investment. However, the higher cost for low-sulfur coal may tend to offset this if it is not locally available. The application of the dry scrubber may be better for industrial facilities because of the elimination of the need for sludge disposal required for other flue gas desulfurization systems. When western coals containing high concentrations of calcium and magnesium are used, the inherent sulfur removal characteristics of the ash can be taken advantage of, decreasing the total sorbant needed for sulfur removal.

A variation of the spray-tube dry scrubber, and possibly the forerunner to it, is dry injection of highly reactive nahcolite, which was sprayed into the flue gas stream dry and allowed to react with the flue gas in the baghouse. It can also be used in the dry scrubber but has the same problems as those associated with the disposal of any sodium-based resultant product. Nahcolite is a very inexpensive product mined as a raw material and, if these problems can be solved and the level of mining increased, it could become a very viable reagent source.

Other Sulfur Removal Systems. There are many other sulfur removal systems that have been developed for elemental sulfur production or to use a waste stream of some slurry by-product. Some of these are the use of magnesium oxide or ammonia. A simplified reaction for an ammonia system could be as follows:

$$NH_4OH + SO_2 \rightarrow NH_4HSO_3$$

$$(NH_4)_2SO_3 + SO_2 + H_2O \rightarrow 2(NH_4)HSO_3$$

$$2(NH_4)HSO_3 \xrightarrow{\Delta} (NH_4)SO_3 + SO_2 + H_2O$$

These systems can produce a concentrated stream of sulfur dioxide that can be used to produce elemental sulfur. If there is a market for sulfur, this may be a viable alternative.

There is one other method that should be mentioned, that is, the direct injection of lime or limestone into the furnace. This can be done as is the case for FBC boilers and is shown in Fig. 6.9-1 where the stability of Ca-O_2-S compounds can be seen relative to temperature and excess oxygen content. These reactions would be gas-to-solid reactions sensitive to temperature and O_2 environment. The gas to solid is the most difficult of the reactions to complete and requires a large total surface to react with and as such can require very high stoichiometries to obtain efficiencies required to meet emission limitations.

Summary SO₂ Removal

Wet scrubbing, dry scrubbing, or other methods of sulfur dioxide removal from the combustion process or from the flue gas are chemical reactions and as such are dependent on the reactivities of the

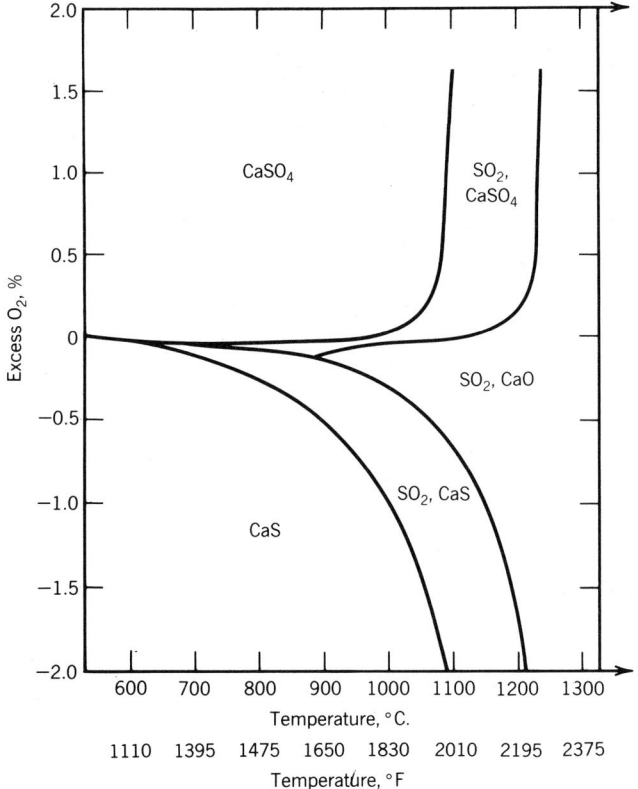

Fig. 6.9-1 High-temperature equilibrium diagram for Ca–O$_2$–S compounds. Adapted from Sillen and Anderson, Svensk Papperstidning, årg. 55, September 4, 1952, pp 622–631.

elements and the time and potential for contact of the elements. A complete evaluation of these systems, including fluidized-bed combustion, must be based on the total coal technology system discussed.

6.9-4   Emission Control Summary

The pollutants resulting from the combustion of coal are mainly nitrogen oxides (NO$_x$), sulfur oxides (SO$_x$) and total suspended particulates (flyash or particulate). Many systems have been designed that can be applied to coal combustion or conversion systems to meet realistic levels of emissions as promulgated by the regulating body, that is, city, state, or federal government. The ability to meet the limits imposed is an economical consideration and the fuel type, accessibility, and level of emissions are of primary concern when determining the type of equipment that must be used. This is a total system consideration. A combination of coal beneficiation and emission control equipment can provide the best results.

BIBLIOGRAPHY

Air Pollution Control, BNA Policy and Practice Series. The Bureau of National Affairs, Inc., Washington, D.C., 1983.

Baviello, M. A., Bowie, A. S., and Beerman, L. E. *The Scrubber Strategy*. An Inform Book, Ballinger Publishing, Cambridge, MA, 1982.

Cherry, S. S. "Effect of Fuel Nitrogen on Industrial Boiler NO$_x$ Emissions." U.S. EPA, Contract No. 68-02-3175 by KVB, Irvine, CA, 1980.

Oglesby, Jr., Sabert and Nichols, G. B. "A Manual of Electrostatic Precipitator Technology," Parts I and II, National Air Pollution Control Administration, Contract No. 22-69-73, Southern Research Institute, Birmingham, AL, 1970.

Rawdon, A. H. and Johnson, S. A. "Control of NO_x Emissions from Power Boilers." Presented at the Annual Meeting of the Institute of Fuel, Adelaide, Australia, November, 1974.

Sortor, C. J. "Successful Application of the 'Sidestream Separator'." Presented at the 1981 Industrial Fuel Conference, Purdue University, Lafayette, IN.

6.10   COAL CONVERSION AND ALTERNATE FUELS

Robert D. Bessette

6.10-1   Introduction

Since the early 1970s with the advent of the Clean Air Act and the first oil embargo, there has been an awareness that the availability of natural resources such as soil and gas is limited. A major effort was launched to develop coal-derived alternatives to the oil and gas so readily used as clean and inexpensive fuels. As the cost of oil and gas skyrocketed, the need for and development of various alternatives increased along with government funding. Within the U.S. economic structure the rate of alternative fuel development will always be tied to the cost of the other fuels. Only when the availability of the other fuels becomes questionable will coal-derived fuels have a market at an economic level that will justify commercial installations. There are only a small number of small installations that have been built or are being built.

Coal conversion may be considered the ultimate form of coal beneficiation, where coal is cleaned of ash and sulfur and converted to a form that can be used in existing oil- or gas-fired equipment. Here is the major consideration in the total system evaluation: the capital cost of replacing a given unit with a new coal-fired unit or the cost of retrofit for the existing unit to burn coal or the cost of converting the coal to a usable form for use in the existing unit. It is possible, in some cases, to mix coal with oil and/or water or water only and produce a mixture that will act like oil and not have the problems associated with direct coal combustion.

This section will discuss the technology of the various conversion and alternate fuel systems as they may be integrated into the total coal utilization system.

6.10-2   Coal Gasification

Basic Principles

The basic principles of coal gasification are relatively simple and amount to a distillation of the volatile content of the coal through a pyrolysis step, followed by the incomplete combustion of the remaining coal or char, which produces a gas high in carbon monoxide (CO) and hydrogen (H_2). Figure 6.10-1 gives a general description of the process and the fundamental reactions that must take place in order to gasify coal. In this way the ash is removed mainly as a reject, although some is carried out with the fuel gas and must be removed. As can be seen in the fuel gas makeup, not all systems will be able to use gas direct from the gasifier, and a certain amount of cleanup may be required to remove ash, tar, and sulfur from the fuel gas. The cleanup systems used for particulates and tar are similar to those used for conventional emission control. However, sulfur is produced in the form of H_2S and standard SO_2 scrubbers will not work. Special processes are available to reduce H_2S to elemental sulfur by the use of catalysis. The basic equation is as follows:

$$2H_2S + O_2 \rightarrow 2H_2O + 2S$$

There are many processes available to remove the sulfur, however, these will not be discussed here.

The air or oxygen feed noted at the bottom of Fig. 6.10-1 is important. The oxygen is needed to generate the reactions and produce CO gas. The steam is supplied for two purposes. The first, for all gasification techniques, to produce additional H_2 gas via the reduction of H_2O in the gasification process. The second, in the fixed-bed systems, to cool the ash and prevent agglomeration and upsets in the air flow through the gasifier. This steam is, in most cases, generated by waterwall cooling of the gasifier itself. If not, it must be supplied by another source.

It must be remembered that the objective of the gasification system is to produce a gas as similar to natural gas as possible for direct replacement of natural gas in existing systems where only gas will work or as an alternative to flue gas desulfurization. Gasification systems will be discussed relative to the type of gas to be produced, that is, low-, medium-, or high-Btu gas products.

Low-Btu Gasification

The purpose of coal gasification that produces a low Btu gas product is basically that process outlined in Fig. 6.10-1 where air is used in lieu of oxygen. As a result, for each pound of oxygen needed

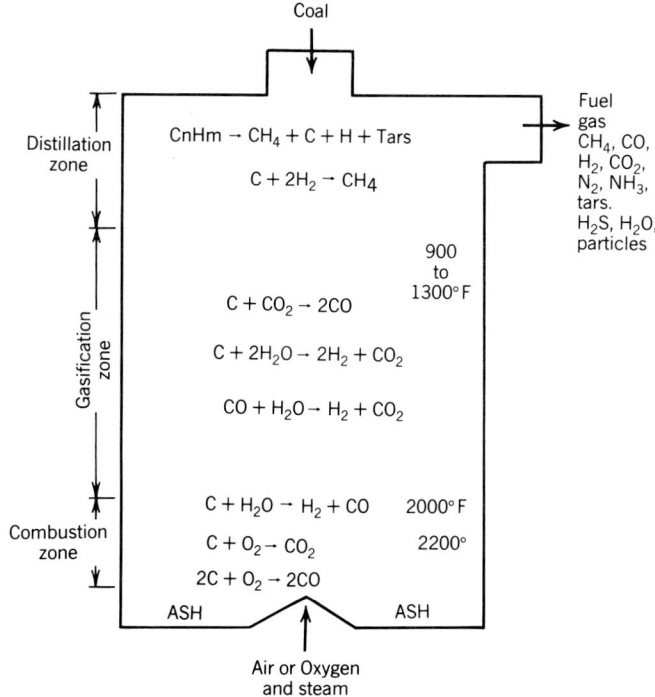

Coal

Distillation zone

$CnHm \rightarrow CH_4 + C + H + Tars$

$C + 2H_2 \rightarrow CH_4$

Fuel gas
CH_4, CO, H_2, CO_2, N_2, NH_3, tars. H_2S, H_2O, particles

900 to 1300°F

Gasification zone

$C + CO_2 \rightarrow 2CO$

$C + 2H_2O \rightarrow 2H_2 + CO_2$

$CO + H_2O \rightarrow H_2 + CO_2$

Combustion zone

$C + H_2O \rightarrow H_2 + CO$ 2000°F

$C + O_2 \rightarrow CO_2$ 2200°

$2C + O_2 \rightarrow 2CO$

ASH   ASH

Air or Oxygen and steam

Fig. 6.10-1 Coal gasification combustion sequence and process profile.

approximately 3 lb of nitrogen are also fed to the gasifier. This results in a dilution effect on the Btu content of the gas. Gas is considered to be low-Btu gas if the Btu is lower than 300 Btu per cubic foot. The actual Btu content will depend on the coal characteristics, the relative percentages of the lighter portions and tar portions of the volatile content, and the level of cleaning to which the gas is subjected. Because of the need to use the greatest percentage of the producer gas Btu, it is best to use a low-Btu producer directly connected to the equipment using the gas. In this case only a mechanical collector is used to collect particulate carried over with the gas. High gas volumes and tar content usually limit the transportation distance of the gas to about one-quarter mile.

If a cleaner gas is required and tar must be removed, the heating value of the tar is lost. If the gas is quenched when the tar is removed, the latent heat of the fuel gas is also lost, decreasing the delivered Btu of the gas further. This is diminished if the tar has a use in the total system or can be sold.

Because of the extra N_2 content of the gas, approximately six times more cubic feet of low-Btu producer gas are needed to equal a given amount of natural gas. The resultant products of combustion will be about two times greater than those of natural gas because of the lower pounds of air required per pound of fuel. When considering theoretical flame temperatures, low-Btu gas exhibits a flame temperature of about 3000°F. These become important considerations when an existing burner or heat recovery systems are used or if critical velocities are to be met.

Medium-Btu Gasification

For all practical purposes medium-Btu gasification is the same as low-Btu gasification, with the exception that O_2 is used in lieu of air and the dilution effect of the N_2 is eliminated. The Btu value of medium-Btu gas may range from 300 to 700 Btu per cubic foot depending on the coal type, volatile content breakdown, and the level of cleaning required. In this case only about twice as much gas is required to equal a natural gas supply, and the resultant products of combustion are approximately equal to those of natural gas because of the lower pounds of air required per pound of fuel. This gas can be used with virtually no modification. When considering theoretical flame temperatures, the flame temperature of medium-Btu gas is on the order of 3400°F.

Gas cleanup systems are less expensive for the medium-Btu systems than for the low-Btu system because of the lower volume that must be handled. There are also advantages in transportation of the gas over longer distances without significant cost compared to low-Btu gas. When any producer gas is transported for any distance, the tars should be removed to minimize the possibility that they will condense and settle in the transport piping, thus causing problems. Gas cleaning must be considered similar to that for the low-Btu gasification system, and the associated tar and latent heat losses must be considered.

There are a number of advantages for the medium-Btu gas; however, they require oxygen for the blast in lieu of air, and this can be very expensive. Most systems where medium-Btu gas has been considered have been large, where economies of scale have been able to justify an on-site oxygen plant. These units are usually considered where chemical plants require large amounts of petroleum feedstocks. There may be some application for boilers or process systems if natural gas becomes very expensive and the cost of an oxygen plant is less than the cost of gas piping and handling equipment for low-Btu gas.

High-Btu Gasification

High-Btu gasification is basically the same as medium-Btu gas with the exception that the gas is delivered to a catalytic process for conversion to primarily methane gas with a heating value (HHV) of between 920 and 1000 Btu per cubic foot. This is pipeline-quality gas and can be used as a substitute for natural gas without any differentiation. The cost of these systems is substantial and, as a result, they will possibly only be applied in areas where residential gas or chemical process feed stocks are needed in the future.

Gasification Equipment

There are a number of methods by which coal can be gasified. The main methods are the fixed-bed gasifier, the fluidized-bed gasifier, and the slagging gasifier. These and underground coal gasification will be discussed in some detail. There are other methods such as the molten salt bath and flash pyrolysis, which will not be discussed, but may also be available for consideration depending on the size of the total system and the overall goal to be achieved.

Fixed-Bed Gasification. There are many variations to the fixed-bed gasifier as they have been developed over the last century before the age of plentiful natural gas. The gas producer was the backbone of the lighting and heating industry before electricity. These technologies have been idle for many years and are now beginning to be redeveloped with modern technology to meet the new coal specifications, the operating regulations, and the environmental control regulations. The fixed-bed producer is primarily a large cylinder, similar to that shown Fig. 6.10-1, where coal is fed at the top, devolatilized, and gasified from heat produced by incomplete combustion in the lower area. Ash is removed from the bottom by a number of different techniques. Steam is used for ash cooling and to produce additional H_2 by the reduction of H_2O, which increases the total Btu content of the gas. The countercurrent coal and air and steam flow is most important and, as a result, coal sizing becomes very important. A 2 in. $\times$ 1-in. nut coal is ideal. However, a stoker coal $1\frac{1}{4}$ in. $\times \frac{1}{4}$ in. can be used if the FSI is very low (less than 4) and the plasticity of the coal and ash are such that they will not clinker or agglomerate.

Coals that tend to agglomerate can cause clinkers within the bed, and this will upset the air and coal flow, thus increasing bed pressure and decreasing gas production. Unless fixed-bed gasifiers are equipped with some method of poking (manual control of clinkers) or rotating arms or bars (automatic control of clinkers), they are usually limited to coal with a free-swelling index of less than 6 and a narrow size distribution such as 2 in. $\times$ 1 in. or 3 in. $\times$ 2 in. The original coal gasifiers were designed primarily for the conversion of anthracite coal.

In these fixed-bed systems the total gas must be cleaned when required. This is or can be a costly task. As a result, a double or two-stage gasifier was developed where two gas offtakes are provided. The top offtake is provided to remove the products of the devolatilization including tars, gasous hydrocarbons, and particulates. The bottom gas offtake was provided to remove gas produced from the gasification step consisting of primarily CO and H_2 gas. This allowed for the cleaning of only the top stream of gas and a significant total system cost savings. Many of these systems have been used in recent years and seem to be more accepted than the single stage.

Although the two-stage system is less expensive, there is a critical limitation. Because there is some restriction between the two chambers (distillation and gasification), the coal cannot have an FSI of greater than 3 or a plasticity such that any swelling or agglomeration could block the movement of coal and shut down the gasifier. As a result, it becomes of the upmost importance to check the availability of coal at a realistic price prior to the selection of the two-stage system in lieu of the single-stage system.

These systems are primarily used for industrial production, and an individual gas producer may be able to produce enough gas to feed a 60,000-pph boiler or its equivalent. This capacity would require a bank of gasifiers to produce very large quantities of gas, increasing the total cost of the system. It is possible to produce low- or medium-Btu gas directly or feed these gases to a methanization process to produce high-Btu gas. The number of gasifiers may be the determining factor.

Fluidized-Bed Gasification. The fluidized-bed gasifier was developed primarily for two reasons: the first to eliminate the consideration of low FSI and agglomeration; and the second is coal sizing so critical to the effective operation of the fixed-bed operation. In this case the FBC technology discussed in Section 6.7-4 is used with limited oxygen and a steam mixture to produce the gas. Here the steam is used for the production of additional H_2 and can have some negative effects on bed temperature and stability if proper control is not maintained. Bed temperature is maintained by the coal feed mechanism and the production of the steam required for gasification. Because of the use of this FBC technology, larger systems can be built and the need for a large number of producers may be eliminated for large users.

Unlike the case in the standard FBC where the bed is primarily inert, in the gasifier the bed is primarily carbon. As a result, the maintenance of bed density by a bleed or other method can cause significant carbon loss and reduced efficiency. There is also a need for a total gas cleanup system since the process of devolatilization and gasification happen together. The balancing of the air/O_2 feed throughout the bed from top to bottom is most important relative to the conversion efficiency of the system. For the additional information on coal requirements and operation see FBC combustion Section 6.7-4.

Slagging Gasification Systems

One of the main problems associated with the above gasifier systems is the ability to remove ash at an increased rate of gasification. As a solution for this problem, the slagging gasifier was developed where pulverized coal could be gasified rapidly at high temperatures and the ash could be removed in the molten liquid state. To maintain the rate of gasification, control and capacity oxygen must be used if optimum performance is to be achieved. Air blowing become a difficult problem with the excess nitrogen carried and the resultant higher velocities encountered. The devolatilization and gasification steps happen simultaneously, and the total gas produced must be cleaned. As a result of the rapid conversion, a substantial portion of the coal ash is carried out in a plastic state at about 2700°F and must be removed from the gas prior to use. The gas cleaning step becomes complicated in this regard, with multiple steps for particulate and tar removal. These systems are normally considered for larger facilities where the ultimate gas product is to be used for chemical feedstock or pipeline-quality gas. This system is normally considered for larger systems where one producer can take the place of a number of fixed-bed or fluidized-bed units.

Almost any coal can be gasified in this system; however, because of the high temperatures involved, chlorine content of the coal may be a problem with materials of construction in the downstream ash and tar removal systems where free hydrogen is available to form HCl acid if condensation occurs. There can also be problems handling large volumes of gas at these elevated temperatures which, if cooled, lose a substantial amount of the Btu due to the latent heat of the gas. Combined-cycle systems to capture this latent heat are being developed, such as using gas turbines. These could significantly help the overall economics of the gasification systems.

General Considerations

In all of the above cases coal handling and distribution systems are needed with lock hopper arrangements to prevent the producer gas from escaping into the surrounding environment. As the system progresses in complexity and capacity, more supporting systems are needed. One of the original reasons for the rebuilding of the gasifiers for modern utilization was an ability to use coal cleaned of sulfur and ash in a system that could not use coal directly. It is important to carefully evaluate the direct use of coal, if possible, as a reconversion or complete replacement with new equipment capable of directly using coal with the appropriate pollution control.

Underground Coal Gasification

When the total system is considered from seam to stack, the process of underground coal gasification may be a viable consideration. This is especially the case if the coal is relatively unminable or the coal location is immediately accessible to the user's facility. Provisions for the direct cleaning and conversion of this gas to high-Btu gas could also increase the attractiveness of this alternative if sufficient gas can be generated.

This process begins with two vertical shafts being drilled to the coal seam, and a horizontal shaft, or some other method by which the seam is fractured and a gas passage from one shaft to the other can be achieved. This is the main area of difficulty with this technology at this time. Once the coal is fractured, combustion is begun and a stream of air or oxygen and steam is passed from one of the shafts through the seam where gasification takes place, and the gas is removed through the other vertical shaft. The elimination of the coal mining and preparation steps in the total system can decrease the cost of the gas to the ultimate user.

There are still many questions that must be answered with regard to the technology of fracturing the coal and the amount of control that can be developed to generate gas over a long term from a given set of shafts. Until major systems are developed, such as those for the in situ oil shale retort, the availability and economics of underground coal gasification will not be on the commercial scale. This technology may be classified with the pressurized fluidized-bed boiler and will come as the cost of natural gas increases.

Summary of Coal Gasification

The market for producer gas continues to be tied to the cost and availability of natural gas, and seems to be one step behind, with the cost of technology increasing at a rate relative to the increasing cost of the natural gas. Systems such as lime kilns and gas turbines may be the first to use these technologies, which can produce a gaseous fuel cleaned of particulates and the contaminants of coal which can cause a wide range of products.

The types of gas—low, medium, or high Btu—and the number of the methods by which they can be produced and cause an extreme complication in the total energy system. The current level of knowledge with these technologies is such that they are normally considered only when a given system cannot directly use coal and will operate at a level substantially above 50% capacity. This is necessary to justify the relatively high level of capital need to produce a given quantity of gas.

6.10-3 Coal Liquefication Systems

Coal liquefication is a process by which coal is reformed into a liquid by adding hydrogen to create longer-chain molecules. There are four basic ways of doing this: (1) indirect liquefication, (2) pyrolysis, (3) hydrogenation, and (4) solvent extraction. To be economical these processes must be built on a very large scale, using on the order of 20,000 tons of coal per day and producing about 50,000 barrels of liquid fuel per day to achieve economies of scale. This is about 2–2.5 barrels of liquid fuel per ton of coal. Because of the magnitude of these processes, this discussion will be limited to a brief discussion of the main considerations and not a detailed discussion of each process. Substantial information on research is available in this regard and could be the subject of a handbook all in itself.

At a level of 20,000 tons of coal per day, suppliers of coal and the appropriate infrastructure for handling, transport, and storage must be sized to stand quantities on the order of 6–8 million tons of coal per year. In all cases the economics of the process are very dependent on the cost of the base feedstock coal and its inherent characteristics. Much research has been done in this regard and some common coal considerations must be met with regard to quality.

Sulfur content in most cases is not a problem as a sulfur (H_2S) removal system is an integral part of the system. However, the higher the sulfur, the greater the operating costs for the system. Oxygen, hydrogen, chlorine, and ash contents of the coal become key considerations with regard to the design and total system capacity. Low ash, oxygen, and chlorine are beneficial. High hydrogen content can save on hydrogen input requirements for the system. Because these are chemical processes, there is a potential for corrosion resulting from sulfur and chloride compounds generated during the liquefication process. Coal chlorine content is usually limited to a maximum of 0.2% of the coal, as reported in the ultimate analysis. A level of less than 0.1% is desired if a reasonable life expectancy for the system can be expected. Severe problems can be encountered when high-chlorine content coals are used, affecting the life and operation of the high-pressure and temperature system. Materials of construction selection becomes a critical task.

Variation in the systems under development are primarily associated with the ability to produce one or more of the following: light distillate, heavy distillate, gasoline, methanol, ammonia, and char. The processes are so varied that only an overview of the general topic will be discussed.

Indirect Liquefaction

The first step in indirect liquefaction is gasification. Producer gas is taken through a catalytic reactor for conversion to liquid products. The basic reaction is the conversion of producer gas to methanol, as in the following reaction:

$$CO + 2H_2 \rightarrow CH_3OH$$

This reaction takes place at pressures above 1450 psig and temperatures above 4800°F. These reactions are sensitive to the catalysis used and to the purity of the producer gas. Higher pressures and temperatures can minimize these problems. The methanol can be converted to formaldehyde, then to ethylene glycol, and on to ethylene, a basic building block for longer hydrocarbons. Another method (the Mobil process) is basically a dehydration of methanol to dimethylether (CH_2) via a catalytical reaction. This can then be converted to longer-chain hydrocarbons. The Fischer–Tropsch process is another method using catalysis to produce CH_2 from which liquids can be produced. This process dates back to the World War II era and is currently the heart of the Sasol facility in South Africa, currently the largest gasification and coal liquid producing facility in the world.

The production of liquids from a clean gas seems to be the most favored route taken in the world today and the basis from which development of various systems are being developed to improve conversion efficiency and product slate quantities for specific products such as light oil and gasoline. The greater the number of steps required to convert coal to a liquid the greater the cost and possibility for efficiency loss per barrel of high-grade fuel. The combination of facility costs and expected product market potential determine the relative economics of the system and the desired product slate. This liquefaction method can exhibit the lowest capital costs on a demonstrated basis for a limited product slate.

Rapid Pyrolysis

The process of rapid pyrolysis, like the Occidental flash pyrolysis process, is used to produce a gaseous, liquid, and solid fuel product slate. The solid fuel is a resultant char suitable for use as a utility boiler fuel. The faster the complete pyrolysis can be achieved, the greater the potential yield of liquid products due to a minimization of secondary decompositon of the volatile products bound in the coal. Some of the gas produced is used as a transport medium for the coal and the remaining is sold. Tars are hydrogenated either during the reaction or in a separate step following the reaction to increase the liquid yield. The resultant unreacted carbon or char by-product must be minimized and sold to make the process economical.

A major drawback to this process is the production of char, which is not readily accepted as a boiler fuel. However, the elimination of the complete gasification step could prove economical relative to the demand for the product slate components. Coal ash removal becomes simplified and does not play a significant role in the process development for liquid fuels since it is removed with the char.

Hydrogenation

Hydrogenation is the principle by which coal in oil is bombarded with hydrogen under high pressure and temperature to produce some gaseous but mostly liquid longer-chain liquid hydrocarbons. The hydrogen–coal process is a very good example of this type reaction. In this, a coal is pulverized and slurried with oil (natural or synthetic), heated, and pumped into a catalytic reactor at about 3000 psig and 850°F where it is bombarded with high-temperature hydrogen. Some gas is produced and cleaned of H_2S, and NH_3 is then available for sale. The liquids are distilled to produce a light, a heavy, and a bottom slurry. This last can be coked to remove all the liquids and produce a small portion of resultant char.

Unlike gasification and pyrolysis, the char and ash must be handled within the liquefaction process. This can cause serious problems and increases the importance of coal quality evaluation. As a result, high-level coal preparation is usually required to obtain minimum levels of ash within the system. Coal chlorine content can cause serious problems with corrosion in the liquid system materials of construction at temperatures in the range of 400–850°F. Depending on the catalysis, some mineral ash elements such as iron can affect catalysis reactivity and consumption. These systems eliminate the gasification step in favor of complete direct liquefaction. However, reactor vessels must be very large and capable of withstanding very high pressures and temperatures. Because of the high degree of coal beneficiation required, a middling product can be produced and used as fuel to produce steam for the hydrogen plant. Those plants that produce a char product can use it as a reducing agent in the hydrogen plant. The use of the greatest percent of the raw coal Btu can produce the best overall cost of the final product, as can be seen from Section 6.3-4. This technology is used to produce the greatest percent of light distillate oils and other high-quality refinery feedstocks.

Solvent Refining

In this method of liquefaction coal is slurried with a coal-derived solvent and is heated with a small amount of hydrogen to about 800°F at 1000 psig, substantially lower than that for hydrogenation, causing a complete solution of the organic matter. The gaseous portion is separated from the liquid portion and the liquid is filtered to remove ash and unreacted carbon. The resultant product is a liquid that will freeze at about 300–350°F. With a second step, including catalytic hydrogenation, the final product can be a suitable boiler or refinery feedstock. Here the ash is removed prior to the high-temperature high-pressure catalytic processing. The process limited to the solid-type product is less capital intensive than full liquefaction because of the lower temperatures and pressures; however, product yield is less and only about 60% of the organic sulfur is removed, which could cause emission problems with boiler applications. Problems with coal ash, chlorine, and hydrogen are similar to those with full hydrogenation. Important consideration must be given to these prior to total system development.

Summary Coal Liquefaction

The need and price of liquid fuel products will spur the development of these large systems to produce coal-derived liquids. The complexity of these systems and a detailed discussion is beyond the scope of this section. There is very little known regarding the application of these fuels, the precautions needed for handling, and the requirements for long-term storage. These questions are just now beginning to be addressed in the United States. The state of the art is changing rapidly and the above discussion should provide a foundation from which an understanding of a given process can be derived. Coal suppliers may find this most beneficial relative to coal specifications requested.

6.10-4   Coal–Oil Mixtures

There are a number of applications that must use or have been designed to use only oil or natural gas and others that have been converted from coal to oil or gas because of environmental regulations. When the price of oil and gas goes very high, it becomes necessary to consider the possible use of coal in one form or another. Coal–oil mixtures (COM) is one potential for using coal in an oil handling and combustion system with only minor modifications compared to that of converting to or back to coal. The objective of COM is to displace a maximum amount of the oil Btu, decrease the total system costs to the user, and obtain the needed level of operating performance. If the cost for production of the COM is less than the differential between the cost of oil and that of coal and a savings can be offered to the user that can provide a payback on the equipment modifications and a savings in total system fuel and operating costs, then COM can be a very viable alternative.

Production Considerations

The basic considerations that affect COM production are mainly associated with the coal Btu content and detailed chemical and physical characteristics, the oil type and viscosity, the coal concentration, its size, and the temperature. The combination of the above must be considered with regard to the handling characteristics of the mixture (viscosity) and the storage characteristics of the mixture (sedamentative and agglomerative stability). These will be discussed briefly below. The methods by which COM is produced can also have an affect on the characteristics of the final COM product and is the main difference between producers marketing the product. These will be handled separately.

Coal Characteristics.   The coal's surface characteristics (oilyophobic or oilyophilic) are its density or specific gravity, its ash content and characteristics, and the Btu content. Although the Btu content of coal is less than that of oil on a weight basis, it is possible that on a volume basis, such as a barrel, the total Btu is more because of increased weight. This is also dependent on the coal concentration. Coal reactivity and combustion consideration must be made for the coal within an oil flame. The surface character of the coal will determine the ease of suspension. If a coal is oilyophilic, it may be easier to produce a stable COM.

Coal Concentration.   Coal concentration is of primary concern relative to the overall economics of the system and oil Btu displacement. This is further dependent on the oil viscosity, coal fineness, and COM temperature. The coal concentration has a direct effect on the viscosity of the COM and as such can be considered a main limiting factor in COM production. Any COM product must be easy to transport. A good design point for COM is a maximum of 50,000 SSU at 140°F which can be handled with positive-displacement-type pumps. To obtain this viscosity level, coal concentrations are usually limited to a maximum of about 55% when using a high-quality No. 6 oil. The greater the coal concentration, the more resistance there is to settling, resulting in improved stability.

However, the increase in coal concentration can cause an increase in the agglomerative tendencies of the coal and a potential solid mass of coal at the bottom of the COM storage tank, barge, truck or railcar. The balance of these is very important to the applicability of COM for industrial or utility use.

Coal Fineness.   Coal fineness, as with pulverized-coal firing, is important relative to carbon burnout and available furnace residence time for combustion to be completed. The residence time in oil- and gas-fired equipment is much less than that for a standard coal-fired unit. In most cases the finer the coal, the more complete the combustion. Increasing the coal fineness, however, increases COM viscosity. This is the result of a greater number of particles being suspended in the same space, creating greater particle-to-particle interactions and hence higher viscosities. When this happens, the sedamentative stability increases such that almost indefinite stability can be achieved with coal pulverized to 100% less than 325 mesh and an average size of about 5 μm. When agglomerative stability is considered, the coal characteristics and sizing can play an important part. If the coal is more aphilic to itself than to oil, the greater number of particles can cause severe agglomeration, with suspended particles increasing the rate of sedimentation and strength of that sedimentation. The technologies of production are mainly concerned with these considerations.

Temperature. In all COMs the problem of balancing the viscosity, sedimentative stability, and agglomerative stability at maximum coal loading are of critical concern. Temperature affects each of these as it affects the viscosity of the oil. The thinner the oil, the less resistance there is to prevent settling and agglomeration. At temperatures above 180°F rapid settling occurs. At temperatures below 100°F the viscosity may be such that the COM cannot be handled.

Production Technologies

The technologies developed to manufacture COM have been geared to minimize and improve upon the above considerations at the lowest cost of production.

Continuous Agitation. When the proper combination of coal, oil, concentration, sizing, and temperature are found to give a usable viscosity, sedimentative and agglomerative stability considerations can be eliminated with continuous agitation.

Water Addition. It has been found that if a percentage of water is added to a COM in the mixing stage, a greater sedimentative stability can be achieved. The percent of water is dependent on the level of mixing and the coal and oil characteristics. Water acts to hinder sedimentation while at the same time minimizing viscosity. This allows for a greater concentration of coal with all other aspects being the same. The water can help atomization of COM; however, it does not help with agglomerative stability.

Chemical Addition. To minimize the problems associated with sedimentative and agglomerative stability, and increase the coal loading at usable viscosities, chemical dispersants and surfactants are used. These are sensitive to the type of coal and oil. An additive must be selected to be usable over a fairly wide range of coals and oils to be effective. COMs can be stored for 30 or more days with the use of additives. Agglomerative stability is also increased to allow for redispersion of any coal that might have settled out over that time. The length of time over which stability is required will determine the amount of additive needed, affecting the cost of the COM.

High-Energy Mixing. In an attempt to minimize the costs of production of COMs, yet provide a level of sedimentative stability, the use of high-energy mixers has been employed. This eliminates the need for a second mixing step and can save the cost of an additive. However, there is little or no effect on the agglomerative stability of the final product. A combination of this method with one or both of the above can be very effective. Low-level continuous agitation is often used as a backup, allowing for some flexibility in system design.

Ultrasonics Agitation. This is a more effective method of high-energy mixing when water is used as part of the COM. The ultrasonics cause the water to cavitate, producing a very intimate mix of the coal oil and water and a very stable nonsedimenting mixture. Long-term sedimentary stability is achievable provided that no coal is larger than 100 mesh. There are no provisions for agglomerative stability other than to prevent sedimentation. The cost of this process is about equal to that for chemical additives.

Ultrafine Grinding and Micronization. As mentioned above, if coal is pulverized fine enough, it can stay in suspension for long periods of time. There are two levels of fineness used today. The first is coal micronization, where coal is pulverized to about 100% less than 5 μm. The cost of this pulverization is substantial, and reliability of the pulverization systems is not fully accepted at this time. Large strides have been made within the last 2 years. The second method was developed to lower the cost of pulverization. In this method the coal is reduced to only 100% through a 325-mesh sieve. The stability of this mixture is good, and standard coal pulverizers can be used in lieu of the fluid-energy-type mill where steam is used to impact coal on coal to obtain micronized pulverization. There is one additional and important advantage of the fine coal systems. Better atomization of the mixture can be achieved and therefore coal combustion and carbon burnout is improved.

Combustion Application Considerations

There are four main considerations that should be mentioned: flame stability and ignition, carbon burnout, emission requirements, and process requirements. The first, flame stability, is dependent on the ability to effectively atomize the mixture. COM will exhibit a viscosity of about 1000–3000 SSU at 250°F. This may seem high, yet very good atomization can be achieved due to the shearing tendencies of the coal particles within the oil. The problem is not so much the ability to atomize but the ability to do it without excessive wear. Special nozzles have been used incorporating tungsten carbide or other highly abrasion-resistant parts, which have worked very effectively as long as the coal particle top size is maintained at 100% less than 50 mesh, preferably at 100% less than 100 mesh.

The second area of concern is that of carbon burnout and efficiency. Coal reactivity and sizing play an important part; however, the furnace residence time and atomization efficiency may be more important relative to the amount of excess air available for combustion. This could be the main limiting factor with regard to the application of COM to oil- and gas-fired boilers. The oil portion of the mixture tends to provide rapid ignition of the coal and as a result increases carbon burnout efficiency. Excess air requirements are comparable to those for coal rather than for oil or gas. It must be remembered that COM burns as a particle (solid) dispersed in a volatilized oil (gas) flame. We are only now beginning to research the kinetics of a pulverized-coal flame.

The third area of concern is that of the products of combustion: the NO_x, SO_x, and particulates. Consideration of nitrogen oxide is important. Anything done to improve carbon burnout will have the effect of increasing the NO_x emissions. Coal also has fuel nitrogen, which must be considered. For further information see Section 6.9-1. Sulfur is simply a combination of the oil sulfur and coal sulfurs on a weighted average basis. Particulate, on the other hand, is something that a gas- or oil-fired boiler has not seen. The ash content of oil is on the order of 0.05–0.1% where that for a COM may be as much as 2.5–5.0%. Emission control and furnace and boiler cleaning systems must be added or upgraded to handle this amount of ash. It is better to size equipment to handle 100% coal than to handle a 50% COM. Ash characteristics must be evaluated to minimize problems with slagging and fouling.

The final area to be covered with regard to COM is general process requirements to sustain combustion. The burner, piping train, pumping and heater set, and fan arrangement are major concerns. Only full-flow piping and wide-radius bends should be used to prevent erosion and minimize the potential for coal to settle out within any portion of the system. Continuous agitation and continuous circulation in the piping systems has been used to further prevent settling. Standard design positive-displacement pumps and heaters, which are easily accessible for cleaning, should be used. Oil and gas forced-draft-fired boilers should be converted to balanced-draft firing. Provisions for removing any ash buildup within the boiler should be considered. COM temperature control and monitoring are a must.

Summary of Coal Oil Mixtures

The COM market will only develop when the price differential between coal and oil generates adequate savings to pay back the conversion investment and provide the user with a financial incentive. Much is known about the production, storage handling, and burning of COM; however, there is still too little experience with the direct combustion of COM in boilers and process equipment. This will come when the price of oil increases or the availability decreases. It must be remembered that approximately 50% of the mixture is oil, so the economics will only pass about 50% of the savings of coal onto the user minus that cost of producing the COM and retrofitting the user's equipment.

6.10-5 Coal–Water Slurries

Coal–water slurries (CWS) or coal water mixtures is a sidestep development of the coal oil mixture development research. The object of its development was to eliminate the oil costs and associated problems from the coal–oil mixture, and it has been successful. CWS can be made with 70% coal and 30% water by weight with viscosity, sedimentative stability, and agglomerative stability to meet the needs of a utility or industrial consumer. The economics are better than those for COM since the technology takes advantage of the total cost differential between oil and coal within the total system evaluation. CWS, as COM, is a replacement fuel for oil or gas and must be considered for evaluation and future development.

Production Considerations

Slurry viscosity, sedimentative stability, and agglomerative stability are the important considerations for CWS as for COM, relative to handling and storage. Since there is no oil to act as an ignition fuel and there is a high percent water content to decrease flame temperature, the coal characteristics become very important relative to the combustion stability and efficiency. Because, unlike oil, water has no viscosity in itself capable of sustaining coal suspension, CSW must be produced using chemical additives in conjunction with very specific coal sizing requirements. Most of the companies involved with CWS production are in the research and development stages and not at the full-scale commercial production level. As a result, the technologies for production are classified as proprietary.

The additives used build chemical matrices or a lattice network with the coal maintaining the stability of the mixture. The chemicals are sensitive to the size of the coal particle and the relative distance between the coal surfaces. This has been accomplished by using a bimodal or controlled size unimodal distribution of the coal particle sizes. An example could be 75% standard pulverizer grind at 70% through a 200-mesh and 25% at 100% less than 325 mesh. Some research has been done where a coal can be pulverized to give a narrow range of size distribution with a top size on the order of 100 mesh or less. Once this particle distance is established for a certain coal concentration, the water content becomes fixed and must be monitored. The problems associated with agglomerative stability in COM are less of a problem with CWS because of the additives used. The reintroduction of water or additives can return the mixture to its proper viscosity and reentrain any coal that might have settled.

There is one aspect of CWS production that could solve a problem with the advanced coal beneficiation techniques, which require coal to be pulverized to very fine particles prior to cleaning. The same chemicals used to clean the coal are compatible with the chemicals used for CWS preparation. As a result, CWS made with superclean coal could be handled and combusted as oil. This could be the single most important advantage of CWS in the future. In any case this production technology is very dependent on the type of coal to be used, its surface characteristics, and the additives used. Minor changes in any of these can have negative results on the CWS.

Combustion Application Considerations

Like COM, there are four main areas of concern that must be addressed: flame ignition and stability, combustion and carbon burnout, emission requirements, and other process requirements. It must be remembered that in the case of CWS, 100% of the oil is replaced with coal and its associated ash. This can be of critical importance in oil or gas boilers.

The first area discussed is that of CWS ignition and stability. With a water content of about 30%, CWS might be considered similar to a lignitic-type fuel. However, the reactivity of the base fuel will be the key factor in determining the flame ignition and stability characteristics. Most coal considered for CWS applications are of the high-volatile (30% or more), low-ash (10% or less), high-Btu (+ 12,000 Btu), and the combination of other factors which indicate a highly reactive coal. Atomization is a greater problem than for COM, since a greater number of coal particles must be atomized to allow for water evaporization. The greater number of coal particles must also be considered with regard to atomizer erosion. Much progress has been made in these areas, and more continues to be made under current research projects with the Department of Energy, EPRI, and the boiler companies.

After atomization, combustion and the ability to achieve complete carbon burnout become the greatest concern. Because of the lower reactivities associated with the bituminous coal and the high percentage of surface moisture, longer residence time will be required for complete combustion. If furnace temperatures are hot enough and there is an initial reducing atmosphere, hydrogen disassociation can occur similar to a gasification reaction, which could balance the loss of ignition due to the surface moisture with the CWS. Because of the moisture content of the fuel, the lower heating value is also lower by approximately 3–4%, requiring an increased CWS feed to generate the same level of output. Actual flue gas volumes will be on the order of those generated when burning lignite and could be a significant factor in boiler derating. Slagging and fouling considerations and flue gas and ash particle velocity and mass flow must be considered relative to the boiler design.

Emission considerations with CWS is the same as that for the base coal, with the exception that there would be higher velocities due to the moisture content. Nitrogen oxide levels may be reduced with proper coal selection for reactivity and fuel nitrogen content. The flame-temperature-cooling effect of the water could assist in the reduction of NO_x formation.

Like COM there are a number of process requirements that must be considered with regard to the application of CWS for boilers or other processes. Piping considerations are the same as those for COM. However, storage is different in many ways. The two major concerns are associated with freezing and evaporation of the water. Provisions must be made for these concerns in the initial design of the CWS system. The control systems needed for combustion are similar to those for COM. These mixtures must be considered as non-Newtonian fluids. Special consideration must be given to the control system requirements to avoid the possibility of coal being forced out of suspension or settling in pipe restrictions.

Summary of Coal—Water Slurry

Coal–water slurries are in the pilot stages of development, with commercial applications expected in the near future. Atomization, combustion, and production techniques are still being improved, so the economics of these systems is still being developed as well. The same basic criteria of economic justification as applied to COM must hold for CWS in that the savings generated by the use of coal in lieu of oil must be sufficient to provide an acceptable level of return on investment for the retrofit equipment. Within the total system consideration all aspects of using coal such as transportation, handling, storage, and so on must be considered and paid for by someone.

6.10-6 Coal Briqueting

For a number of years some companies have been looking for effective ways to handle, transport, unload, and burn coal fines, especially clean fine coal that was at one time thrown away. Today, with the cleaning techniques available for the coal preparation plant operator, this is some of the cleanest coal available for sale and, in many cases, is used to blend with lower-quality coarse coal for indirect beneficiation. Coal briqueting can serve two main purposes: The first is to provide a relatively uniform size or multiple sizes of coal for a process that requires a coarse coal such as a fixed-bed gasification system. The second is to facilitate the use of very fine ultraclean coal through reforming the coal into larger pieces for transportation to the user. Within the total system consideration, as long as the cost of briqueting when added to the coal cost is less than the natural product, the technology could be a viable alternative.

The briqueting process may take many forms, from using cement-type binders and compressing coal to form the briquet to adding tars or using the coal volatiles in a reformation step under pressure and temperature and then extruding the briquet. This production is dependent on the coal type and its chemical and physical characteristics. It can be a very simple process. Limestone addition to the briquet for sulfur capture has been considered and may be possible if the reaction time for the gas-to-solid reaction to be completed and the temperatures are low enough (see Fig. 6.9-1).

The application for which the briquet is to be used must be considered. In the briqueting process a certain strength can be built into the briquet that will allow it to retain its shape or structure during transportation and handling. The briquet can also be produced with the proper strength to allow for degradation upon impact as would be required in a spreader stoker-fired boiler. When transport, handling, storage, and utilization are considered, it is possible that a coal briquet would be a viable alternative for a natural coal product. This may be an ideal solution for the needs of the low-Btu fixed-bed gasifier.

6.10-7 Summary

The conversion of coal to gaseous or liquid fuels or the production of fuels that contain some or all coal to replace oil or natural gas has been developing ever since the oil embargo of 1973 and 1974 when the price of oil and its availability became critical to an industry that had taken it for granted. Major efforts have been launched to create an energy independence. However, fluctuations in the world oil price and availability have created some very difficult short-term financial evaluations relative to inflation and quantities of funds needed to build the facilities. Coal is the most available energy source in the world today and can supply enough energy to meet the needs of the world for many generations to come if it has to. The need for these technologies to be developed any faster has not been exhibited in the relative economics of the competing systems and the overall total system. As these are evaluated within the total system, from seam to stack, relative to the needs of the market place, they will eventually find their time for commercial development.

BIBLIOGRAPHY

Herman, S. W., and Cannon, J. S. *Energy Futures*. An Inform Book, Inform, New York, 1976.

Proceedings of the 1981, 1982, and 1983 Coal Oil Mixture and Alternative Fuel Symposiums, U.S. Department of Energy, Pittsburgh Energy Technology Center, Pittsburgh.

6.11   SUMMARY

Robert D. Bessette

The overall topic of coal technology—the utilization of coal as an energy source to meet the needs of commercial, residential, industrial, utility, and process industries—must be addressed in a total systems approach considering the interrelations of the various aspects of the system from seam to stack. The system must be thought of as a whole, not merely the sum of the parts, as the synergistic effects of the integration of the parts. Possibly the best summary that could be provided would be a potential methodology by which the individual parts may be considered and integrated.

This section on coal technology, through its individual sections, has attempted to outline and somewhat detail the interrelations of the various subjects with the total system, as well as describe the important considerations that must be considered within the topic itself. It is believed that the foundation has been included from which the overall performance of a system may be considered and optimized. The following methodology could help the engineer, manager, consultant, and coal producer understand and begin the implementation process for the total systems approach.

6.11-1   Systems Integration Methodology

There are six important steps in the system integration process: (1) a true definition of need, (2) a fuel availability and characteristic evaluation, (3) technology review and selection, (4) primary fuel or feedstock selection, (5) fuel and design flexibility requirement, and (6) fuel purchasing and equipment specification. The intent of this method is to develop a performance-oriented specification to maximize the total system operation. The process must be built one step at a time with a concurrent understanding of any decision's impact on the total system. Each of these items will be discussed briefly below.

6.11-2   A True Definition of Need

Many projects are begun with concepts, and a direction is plotted. As the project develops, there are many and major changes that take place. One of these is the assumption that a fuel will be available by rail or truck. Another is the underestimation of environmental impact on the system. The main elements of concern at this step are the size of the equipment, pounds of steam per hour, pressure and temperature for boiler applications, the process capacities for conversion systems, the environmental specifics both at the plant location and with regard to regulations, space limitations, and transportation alternatives. With a true definition of the project needs, and a listing of all potential problem areas or interfaces, the foundation for the total system evaluation is complete.

6.11-3   Fuels Availability and Characteristic Evaluation

The logistical availability of coals to a plant site may be the most important initial consideration for the total project. The most effective way to do this is to go directly to the coal producers, requesting specific bids on the coal they would or could supply. A complete specification of the coal should be asked for on a raw or washed basis. This information should include the proximate analysis, ultimate analysis, sulfur forms, ash fusion temperatures, mineral ash analysis, grindibility, and free-swelling index. Also needed are physical characteristics such as size distribution, uncompacted density, and percent makeup by seam if it is not a single-seam product. When a blend of coals is considered to meet a specification, an average specification of the individual seams should be requested, including the long proximate analysis, ultimate analysis, mineral ash analysis, and FSI. Any major variation in the type of coal should be investigated relative to the type of equipment or process and should be specially noted prior to technology consideration.

Following the receipt of the information from the coal producers, an evaluation relative to available coal reserves of the various coals offered is made to assure a long-term supply for the facility. This should include market demand and transportation considerations. Further evaluations should include the amount of coal that would have to be unloaded, handled, and stored at the site, the expected ignition, reactivity, slagging, and fouling characteristics, the potential environmental emissions of NO_x, SO_x, and particulate, and finally the ash disposal requirements. When the evaluation of what is available is ready, and the relative costs are assigned, the review of available technology may proceed.

6.11-4   Technology Review and Selection

With the above in hand it is possible to evaluate the available technologies for optimum performance relative to the project needs and the logistically available fuels. These can be the differences between the stoker-fired boiler and the pulverized-coal-fired boiler or either of these with a scrubber versus a fluidized-bed boiler. A gasifier may be considered for a lime kiln process in lieu of natural gas or coal oil mixtures. Once this is done, it should be possible to predict an economic level of performance and prepare a preliminary budget estimate for the project and make a decision whether the project should be continued. If the project is continued, the primary fuel specification is selected.

6.11-5   Primary Fuel Selection

The fuel selected should assure adequate resources for the future and generate the level of performance expected over the long term. This specification will be the key to the success of the overall process.

Fuel and Design Flexibility

It is important that decisions be made on data that is as accurate as possible and not from assumptions or data files. The selection of coal specification range and reserve base considerations can be effectively made from the data received on the coal bids. The specifications of those coals that would be included within the range should be attached to the primary specification and used in the preparation of the equipment specifications. The selection of alternate coals should allow for increases competitive in all areas to minimize market price escalations and supply disruptions. At this point coal loading, unloading, and transportation compatibility must be considered for flexibility.

Fuel Purchasing and Equipment Specification

At this point the system design and performance requirements are virtually complete, and it is possible to begin coal supply negotiations and equipment specifications development. Both of these should be directed toward a performance-type structure. The coal supply agreement should promote quality control in the mining and preparation stage rather than after the fact with coal sampling and testing of shipments. In many cases standard ASTM testing procedures will be sufficient. Boiler and equipment specifications should be performance oriented to allow for maximum development of benefits to minimize power, operating, and maintenance costs at maximum efficiency and availability. Power costs, labor costs, and other cost factors should be included within the bid package to allow for this evaluation on the part of the equipment suppliers. Fuel costs should also be included in order to determine the true value of a percent efficiency. To optimize design, this must be used over the total system.

When a system is in existence, the inverse of the above process is followed. Equipment design and performance data are used to develop a set of performance specifications for coal. These performance specifications should include costs associated with coal handling, storage, combustion efficiency, maintenance availability, and ash disposal. With this level of information a coal producer should be able to determine which coal and at what level of beneficiation would produce the best total system economics. It could be possible to increase capacity and efficiency and decrease operation costs when using a higher-quality coal.

6.11-5 Conclusion

The coal technology environment has long been a segregated industry of equipment and fuel. When an industry, be it process, industrial or utility, looks at the total system, it must consider the effect on the final product cost. For the utility it is the bus bar energy costs. For a paper mill it is the energy costs per thousand pounds of steam needed per ton of paper produced. For the liquefaction facility it is the number of barrels of oil produced per ton of coal and their cost compared to the cost of the natural products. It is the synergistic effects of the total system interrelations that make up the total coal energy system.

BIBLIOGRAPHY

De Lorenzi, Otto (Ed.) *Combustion Engineering*. Combustion Engineering, New York, 1955.

Elloot, M. A. (Ed.) *Chemistry of Coal Utilization*. Second Supplementry Volume, Wiley, New York, 1981.

Lowry, H. H. (Ed.) *Chemistry of Coal Utilization*. New York, 1945.

Lowry, H. H. (Ed.) *Chemistry of Coal Utilization*. Supplementry Volume, Wiley, New York, 1963.

Reid W. T. *External Corrosion and Deposits*. American Elsevier, New York, 1971.

Singer, J. G. (Ed.) *Combustion, Fossil Power Systems*, 3rd ed. Combustion Engineering, Windsor, CT, 1981.

Steam, 39th ed. Babcock and Wilcox Company, New York, 1978.

CHAPTER 7

NUCLEAR TECHNOLOGY

MARC W. GOLDSMITH

Energy Research Group, Inc.
Waltham, Massachusetts

WARREN WITZIG
JOSEPH BENNER
JAMES SHILENN

Pennsylvania State University
University Park, Pennsylvania

CHONG CHIU

Combustion Engineering, Inc.
Windsor, Connecticut

DAVID D. LANNING

Massachusetts Institute of Technology
Cambridge, Massachusetts

ROBERT H. SIMON

GA Technologies, Inc.
San Diego, California

ROBERT A. EVANS

Helium Breeder Associates
La Jolla, California

FRANK J. RAHN

Electric Power Research Institute
Palo Alto, California

ARTHUR G. RANDOL III

Exxon Corporation USA
New Orleans, Louisiana

WALTON A. RODGER

OTHA, Inc.
Glen Echo, Maryland

NORMAN C. RASMUSSEN

Massachusetts Institute of Technology
Cambridge, Massachusetts

A. E. SCHERER and E. H. KENNEDY

Combustion Engineering, Inc.
Windsor, Connecticut

R. H. Simon is now with the R. H. Simon Company. Helium Breeder Associates merged with Gas-Cooled Reactor Associates in December 1982.

7.1 INTRODUCTION TO NUCLEAR TECHNOLOGY

Marc W. Goldsmith

Fission, the splitting of atoms to generate energy, was once only a dream of scientists. The task of engineers today is to assure that the atom is harnessed safely and efficiently. The technology to harness the atom requires a combination of all engineering disciplines working in an integrated manner to provide a design capable of producing electricity in a safe and reliable manner. Even today, in an era of skyrocketing construction and monetary costs, nuclear power, in most parts of the country, still provides the most economic alternative for new electric generating capacity.

In the late 1940s and early 1950s when nuclear technology emerged, there was no oil embargo or any obvious signs of an energy crisis. The driving forces for the rapid development of the atom were its fuel efficiency and its potential cost-effectiveness compared to its alternatives. Uranium was a cheap and abundant domestic fuel and the development of the technology provided new vistas and challenges for the engineering community. It was the goal of providing environmentally clean, abundant, and reasonably priced energy that motivated engineers then as now.

Nuclear technology developed under a mixture of government regulation and promotion and utility industry commercialization. The background that follows is to place in perspective the development and implementation of a technology largely resulting from the efforts of government to make the production of nuclear-powered electricity a commercial enterprise. This effort has largely succeeded, as greater than 10% of the electricity generated nationally is now provided by nuclear power.

7.1-1 Legislative History

Government involvement was an integral part of nuclear power technology development. In 1946, concurrent with the establishment of a joint congressional committee on atomic energy (JCAE) and a general advisory committee, Congress also created a civilian Atomic Energy Commission (AEC). These actions provided for private participation, but not private enterprise, in the development of nuclear-powered electric generation. The AEC's research and development efforts were concentrated in the national laboratories, each working on specific aspects of nuclear technology development. As part of this program, a national reactor testing site was established at Idaho Falls in the early 1950s. It was at this site that test facilities were constructed and the first safety research program initiated. In December 1951 the first electricity was generated by the AEC-operated experimental breeder reactor, EBR-1. Two years later, on December 8, 1953, President Eisenhower initiated the "atoms for peace" program. He declared that the development of the "peaceful atom" was essential to the future of both international and domestic affairs. In support of this program, in 1954, Congress passed a second atomic energy act in an effort to standardize regulations and licensing processes for the development of nuclear power by private industry. Private industry was specifically encouraged to participate in the development of nuclear energy. That same year, the federal government initiated a 5-year development program. Variations of fuels, coolants, and moderating materials were tested in pressurized water, boiling water, sodium-cooled, enriched-uranium, graphite-moderated, the breeder, and homogeneous reactor types. In addition, the AEC as part of its promotional function announced further research and development through a power reactor demonstration program. This program provided financial aid as an incentive to utilities to build and operate nuclear power plants. By 1957 there were 14 nuclear power plants in the planning or construction stages. This intense cooperation eventually led to an image of a close linkage between the regulators and the promoters of nuclear power. However, significant progress was made in the design, engineering, safety, and economics of nuclear-powered electric generation. By the end of the 1950s it was clear that private industry was capable of designing, constructing, and operating nuclear power reactors. The AEC felt that, having established engineering feasibility and safe operating experience, the next phase of nuclear power development was to establish the economics and reliability of this new generation of nuclear power stations, particularly in competition with the alternatives. Commercialization of nuclear power required that the developers not only establish the cost-competitiveness of nuclear power with fossil fuels, but also demonstrate an infrastructure capable of supporting the entire nuclear fuel cycle on a large scale. This required facilities to mine, process, enrich, and manufacture uranium fuels suitable for nuclear power plants and facilities for the disposal of radioactive waste products.

7.1-2 Technical Evolution

Once the scientific feasibility of extracting heat from nuclear fission had been established, practical application became the major challenge. In the early 1950s several industrial design teams began the process of developing nuclear steam supply systems using uranium fuel. They analyzed differences in

TABLE 7.1-1 U.S. NUCLEAR POWER PLANTS IN OPERATION, UNDER CONSTRUCTION, OR ON ORDER

Reactor Type	Operating License[a]		Under Construction		On Order	
	No.	MW(e)	No.	MW(e)	No.	MW(e)
Boiling water	29	21,197	19	20,094		
Pressurized water	52	43,316	40	44,880	4	4,790
High-temperature gas	1	330				
Graphite (steam)	1	860				
Fast breeder					1	350
Total	83	65,703	59	64,974	5	5,140

Source. Atomic Industrial Forum, Nuclear Power Plants in the United States, January 1983.

[a] Includes 73 reactors in commercial operation; 3 with full power licenses, 2 with low power licenses, 1 with suspended license, and 4 shut down indefinitely.

safety and efficiencies, options for fuels (e.g., natural uranium vs. enriched uranium and uranium metals vs. oxides and carbides), and the choices of moderators [e.g., graphite vs. heavy water (D_2O) vs. light water] and for coolants (organic liquids vs. water vs. liquid metals vs. gases). Choices were made to first narrow and focus on designs that offered good nuclear physics characteristics [stable operation, control, and fuel use (burnup)], adequate heat transfer under both normal and accident conditions, and efficient use of fuel. Safety was an all-pervading influence on the engineers and designers. Eleven small power reactor demonstrations were tried. These were all under 100 MW and provided the necessary trials of a variety of combinations of fuels, moderators, and coolants. These plants were built in the early 1960s and some operated into the mid-1970s before being decommissioned.

As a result of the research and operations conducted during the demonstration program and the naval propulsion program, concentration in the United States was placed on enriched-uranium, light-water-cooled, and moderated reactor designs. The so-called light-water reactor evolved into two major systems, the pressurized water reactor (PWR) (three major vendors) and the boiling water reactor (BWR) (one vendor). In Canada the natural-uranium, heavy-water reactor (CANDU) was selected, and in Europe the graphite-moderated gas-cooled reactor was selected.* The nuclear breeder that would provide a virtually inexhaustible uranium fuel supply has been pursued internationally on a parallel but slower track.

As electric demand grew rapidly in the 1960s and early 1970s and because of the belief (in the United States) in economies of scale, pressure was created to build larger and larger electric generating plants. As a result, nuclear plant size rapidly increased from the 60-MW electric [MW(e)] plant at Shippingport, Pa. (which achieved initial criticality December 1957) to a current maximum size plant allowed by regulation of 1300 MW(e). In just 15 years plant size had increased by a factor of around 20. The technology expanded as engineers designed, manufactured, and constructed even larger pipes, valves, pumps, and systems to control and operate increasingly larger and more complex systems. In addition, the industry expanded from an initial capability of designing, building, and constructing a single plant to one capable of producing tens of plants per year. This created problems as the scaling was occurring faster than the experience was being gained.

At the beginning of 1983, the twenty-fifth year of commercial nuclear power generation, the Department of Energy predicted that nuclear power would surpass hydroelectric as the second largest producer of electricity. Tables 7.1-1 and 7.1-2 show the current status and types of reactors with operating licenses, under construction and on order. Internationally, the types of reactors are more diverse and the plant size is typically smaller. Several countries (e.g., Belgium, France, Finland, and Switzerland) have a larger percentage of nuclear generation capacity than the United States.

The next section provides a practical view of the principles involved in reactor physics and heat removal; differentiates the engineering of reactor types; describes the supporting fuel cycle and waste management processes; considers the key elements in assuring nuclear safety; and provides an overview of the U.S. licensing process.

---

*In the United States two commercial high-temperature gas-cooled power plants were constructed. The Fort St. Vrain [330 MW(e)] power plant is the only high-temperature gas-cooled nuclear plant currently operating. France, one of the original developers of gas-graphite power reactors has now switched to the light water reactor. The other, the U.K., is in the process of switching.

TABLE 7.1-2   WORLD NUCLEAR POWER PLANTS IN OPERATION, UNDER CONSTRUCTION, OR ON ORDER

Reactor Type	Operating License		Under Construction		On Order	
	No.	MW(e)	No.	MW(e)	No.	MW(e)
Light-water-cooled, graphite-moderated	19	11,900	3	3,000		
Gas-cooled heavy-water-moderated	1	70				
Light-water-cooled, heavy-water-moderated	3	542^a	1	40		
High temperature			1	296		
Liquid metal fast breeder	4	1,433	2	1,495	1	300
Gas-cooled	36	8,260				
Advanced gas cooled	5	3,100	9	5,740		
Pressurized-heavy water-moderated and -cooled	14	6,436	21	12,720	4	1,840
Boiling water	36	23,199	19	18,422	4	4,051
Pressurized water	83	56,265	82	73,615	26	25,305
Total	201	111,205	138	115,328	35	31,496

Source.   Nuclear News, American Nuclear Society, February 1983.

^{a} Includes the 92-MW(e) steam generating HWR in the U.K.

7.2   NUCLEAR PHYSICS CONCEPTS

Warren F. Witzig, Joseph J. Benner and James K. Shillenn

7.2-1   Nature of the Nucleus

Nuclear Structure

In order to understand nuclear reactions or transformations it is often very helpful to use models that explain scientific observations. There are a number of models that can be used to describe the nature of the atom; however, for our purposes, a very simple model can be used.

An atom can be considered to be a dense core of particles called protons and neutrons forming a positively charged nucleus, surrounded by a cloud of orbiting electrons. Protons have a positive electric charge equal in magnitude to that of an electron. Neutrons have no charge and are approximately equal in mass to the protons. Collectively, the protons and neutrons are referred to as *nucleons*.

The number of protons in the nucleus is called the *atomic number* and is designated Z. The number of neutrons in the nucleus is designated N. The *mass number A* is the sum of $N + Z$.

One of two or more atoms with the same Z number but with a different A number are called *isotopes*. For example, hydrogen has three isotopes with $Z = 1$ and $A = 1, 2,$ or 3. Each isotope has the same chemical properties but differ in the number of neutrons. The nucleus of ordinary hydrogen is a positively charged proton; the deuteron (deuterium) consists of a proton plus a neutron, the triton (tritium) contains a proton plus two neutrons.

The symbols for a given nucleus are derived from the model previously discussed. That is, one indicates the number of protons and neutrons. For example, a nucleus might contain three protons and three neutrons—a total of six nucleons. This is represented by using the general designation

$$_Z^A X_N$$

where X is the chemical symbol. The model (lithium-6) can be represented by

$$_3^6 Li_3 \quad \text{or usually } ^6Li$$

A nuclide is a nucleus identified by its proton number Z and its neutron number N. All nuclides can be divided into three classes defined as follows:

Isotopes: $^A_Z X_N$ and $^{A^1}_{Z^1} X_{N^1}$, where $Z = Z^1$.
Isobars: $^A_Z X_N$ and $^{A^1}_{Z^1} X_{N^1}$, where $A = A^1$.
Isotones: $^A_Z X_N$ and $^{A^1}_{Z^1} X_{N^1}$, where $N = N^1$.

Size and Mass of a Nucleus

The nucleus of an atom is extremely small and dense compared to the whole atom. For example, the hydrogen atom has a radius of about 5×10^{-9} cm, while the nucleus has a radius of only 10^{-13} cm. A calculation for the radius R of the roughly spherical nucleus is

$$R \text{ (cm)} = 1.4 \times 10^{-13} A^{1/3}$$

where A is the mass number. The length of 10^{-13} is called a *fermi*.

The universal mass unit, abbreviated u (sometimes amu for atomic mass unit), is defined as one-twelfth of the mass of the ^{12}C atom which has been defined to be 12.000 00 u.

$$1 \text{ amu} = 1.66053 \times 10^{-24} \text{ g}$$

The rest masses of four familiar particles are presented below.

Particle	Symbol	Rest Mass, amu
Proton	p	1.007277
Electron	e^-	0.000548497
Hydrogen atom	^{1}H	1.007825
Neutron	n	1.008665

Mass–Energy Equivalence

The equivalence of mass and energy, or the conservation of mass–energy is sometimes called the *Einstein mass–energy equivalence*. It reads

$$E = mc^2$$

where E = energy in ergs
m = rest mass in grams
c = velocity of light = 3×10^{10} cm/sec.

A more frequently used energy unit is the electron volt or eV:

$$1 \text{ eV} \simeq 1.6 \times 10^{-12} \text{ ergs}$$

$$1 \text{ MeV (million electron volts)} \simeq 10^6 \text{ eV}$$

Thus, the energy equivalence of 1 amu using the mass–energy relationship is approximately 931 MeV.

Binding Energy

Binding energy can be defined as the energy difference between the nucleons in their free state and the energy of a given nucleus. Another way of saying this is the binding energy of a nuclide is the energy liberated on synthesizing a stationary, neutral atom in its ground state (unexcited) from neutron hydrogen atoms and neutrons initially at rest and completely separated from each other.

$$\text{Binding energy (MeV)} = \simeq 931 \left(Zm_\text{H} = Nm_n - M \right)$$

where

$$Zm_\text{H} = \text{atomic number } (Z) \times \text{mass of hydrogen atom (amu)}$$

$$Nm_n = \text{neutron number } (N) \times \text{mass of neutron (amu)}$$

$$M = \text{atomic mass of a neutral atom (amu)}.$$

Neglected in this relation is a small energy of atomic or chemical binding. The binding energy per nucleon, BE/A is a frequently used expression. Calculations such as these are required for several purposes—to compare the stability of one nucleus with that of another, to find the energy release in a nuclear reaction, and to predict the possibility for fission of a nucleus.

7.2-2   Types and Properties of Radiation Released

Radioactive Decay

Spontaneous reactions occur when nuclear instability exists. An unstable nucleus undergoes change by emitting different forms of radiation. These changes are known as *radioactive decay*, and the phenomenon is known as *radioactivity*.

Radiation that is emitted from radionuclides can be in the form of particles or waves. Since this radiation has the ability to eject electrons from atoms, forming charged atoms or ions, it is more specifically referred to as *ionizing radiation*.

In its simplest form radioactive decay can be represented by

$$X(\text{nuclide}) \xrightarrow{\text{decay}} Y(\text{nuclide}) + \gamma\,(\text{particle or wave})$$

The X represents the parent nucleus and Y represents the daughter. The daughter always has a lower ground-state energy (i.e., it has a higher binding energy) than the parent. The daughter may be stable or radioactive. If the daughter is radioactive, the series of decays that results is called a radioactive decay chain. Many nuclides decay by more than one mode. The most frequently observed is beta decay followed by the release of gamma radiation. The types of decay modes or γ in the above equation are alpha particle, beta particle, positron, gamma ray, neutron, proton, or electron capture.

Alpha Decay

Alpha particles are helium nuclei; that is, each particle consists of two protons and two neutrons giving it a $+2$ charge. The strong positive charge exerts a large attractive force on negatively charged electrons. Consequently, alpha particles can pull electrons from atoms with which they interact, causing ions to form. Alpha decay is observed for the elements heavier than lead and for a few nuclei as light as the lathanide elements. It can be written symbolically as

$$^{A}_{Z}X \dashrightarrow ^{A-4}_{Z-2}X + ^{4}_{2}He$$

An example of alpha decay is

$$^{210}_{84}Po \dashrightarrow ^{206}_{82}Pb + ^{4}_{2}He$$

Beta Decay

Beta radiation is another form of ionizing radiation which is particulate in nature. A beta particle is actually an electron moving at a very high velocity. The origin of a beta particle can be thought of as a neutron in a radionuclide changing into a proton and an electron. The proton stays in the nucleus, while the electron is emitted as a beta particle. It can be written symbolically as

$$^{A}_{Z}X \dashrightarrow ^{A}_{Z+1}X + \beta^{-}$$

For example,

$$^{60}_{27}Co \dashrightarrow ^{60}_{28}Ni + \beta^{-}$$

Beta decay is also accompanied by the release of an antineutrino which has a charge of 0 and a mass of 0. The antineutrino (and neutrino) does not react readily with matter. In fact, the interaction is so unlikely that a neutrino has a very high probability of passing through the entire earth without reacting. Because of this extremely low probability of reacting, neutrinos and antineutrinos are often omitted in writing beta decay reactions.

Positron Decay

Positron decay can be thought of as a proton in a radionuclide producing a neutron which remains in the nucleus and a positively charged electron (positron) which is ejected from the nucleus. Positron decay is accompanied by the production of a neutrino. A symbolic representation of positron decay is

$$_Z^A X \dashrightarrow _{Z-1}^A X + \beta^+$$

An example of positron decay is

$$_7^{13} N \dashrightarrow _6^{13} C + \beta^+$$

Electron Capture

The electron capture process (EC) can be written symbolically

$$_Z^A X \xrightarrow{EC} _{Z-1}^A X$$

In electron capture an electron is captured from one of the inner orbitals of the atom. This process is sometimes referred to as K-capture, L-capture, and so on, depending upon the electron shell from which the electron originates. The process can be thought of as the orbital electron combining with a proton in the nucleus and forming a neutron. Electron capture is the predominant mode of decay for a neutron deficient nucleus whose atomic number is greater than 80. An example of EC is

$$_{82}^{205} Pb \xrightarrow{EC} _{81}^{205} Tl$$

Gamma Emission and Internal Conversion

Alpha and beta particles which are emitted by the nucleus are in an excited state and this excitation is removed by a gamma ray or by a process called internal conversion. In most radioisotopes the emission of the gamma ray occurs immediately after the alpha or beta decay.

In some instances, however, the nucleus may remain in the higher energy state for a measurable length of time. The longer-lived excited nuclei are called *isomers*. An example is barium-137m (metastable) which decays with a half-life of 2.55 minutes to the ground state of barium-137. This decay is referred to as isomeric transition.

Internal conversion occurs when the excitation energy of the nucleus is transferred directly to an orbital electron. The electron is ejected and is referred to as the conversion electron. This process can be represented symbolically as

$$_Z^{Am} X \rightarrow _Z^A X^+ + _{-1}^0 e$$

Barium-137m decays by both the emission of a 0.66-MeV gamma ray and by the competitive process of internal conversion.

Delayed Proton and Neutron Emission

Radioactive decay rarely occurs with the emission of a proton or neutron. If a nucleus has sufficient energy to release a proton or a neutron, at least 8 MeV, the emission takes place in about 10^{-14} s. In this case the particle is so short-lived that it is not regarded as really existing. However, as a result of the fission process (see Section 7.2-6) or by bombardment of several nuclides with light particles, unstable species can be formed which emit neutrons or protons after the fission process or the bombardment is terminated. This is referred to as delayed proton or delayed neutron emission.

7.2-3 Radioactive Decay Laws

Radioactive decay is a spontaneous event that occurs in the life of a radionuclide. This means that radioactive decay is not influenced by temperature, pressure, or other external physical parameters. The likelihood or probability of the nucleus of a radioisotope decaying in a given time interval is a constant λ which is usually expressed in units of reciprocal seconds. If we have a specific number N of radioactive atoms then the decay rate or activity A is λN.

Since A and N are both functions of time they are indicated by writing $A(t)$ and $N(t)$. This can be expressed in decays or disintegrations per second (dps) or disintegrations per minute (dpm). Since $A(t)$ is often a large number, we often express activity in curies, Ci.

1 curie = 1 Ci = 3.7×10^{10} dis/s.
1 millicurie = 1 mCi = 10^{-3} curie.
1 microcurie = 1 μCi = 10^{-6} curie.
1 nanocurie = 1 nCi = 10^{-9} curie.
1 picocurie = 1 pCi = 10^{-12} curie.

The value of 3.7×10^{10} dps is close to the disintegration rate of 1 g of radium (radium-226), one of the earliest known radioisotopes.

SI Units

In the SI system (International System of Units) the radiation unit is the becquerel (Bq) and the activity is given in reciprocal seconds, s^{-1}. Although the curie is more commonly used, the Becquerel is the preferred unit of activity.

$$1 \text{ becquerel (Bq)} = 1 \text{ (disintegration) } s^{-1}$$

and

$$1 \text{ curie (Ci)} = 3.7 \times 10^{10} \text{ } s^{-1} \text{ (Bq)}$$

Decay Calculations

In order to calculate the activity of a radioisotope at time t the following equation can be used

$$A(t) = A_0 e^{-\lambda t}$$

where $A(t)$ = activity remaining after a time interval t
A_0 = activity of a sample at some original time
e = base of natural logarithms, 2.718
λ = decay constant for a particular radioactive element
t = time interval since the original time.

The same general expression can be used to calculate the number of unstable atoms remaining after a time interval t and count rate from a radiation detector after a time interval t. These expressions are

$$N(t) = N_0 e^{-\lambda t}$$

where $N(t)$ = number of unstable nuclei remaining after a time interval t,
N_0 = number of unstable nuclei at some original time,
and

$$CR(t) = CR_0 e^{-\lambda t}$$

where $CR(t)$ = count rate from a detector after a time interval t
CR_0 = count rate from a detector at some original time.

Average Lifetime

Instead of the decay constant λ, the average lifetime τ is sometimes used where

$$\tau = 1/\lambda$$

Half-life

Although the decay equations can be used to predict activities after some time interval (or original activity) a more common description of decay rate is half-life, $T_{1/2}$. Half-life is defined as the time it takes for the activity of a radioisotope to decrease by one-half. Another way of describing half-life is the time it takes for the number of radioactive atoms to decrease by one-half. Half-lives of radionuclides vary from a small fraction of a second to billions of years.
A useful equation for determining half life is

$$T_{1/2} = \ln(2/\lambda)$$

also

$$\lambda = \ln 2 / T_{1/2}$$

where $T_{1/2}$ = half-life
$\ln 2$ = natural logarithm of 2 (0.693)
λ = decay constant.

Radioisotope Data

The *Chart of the Nuclides* provides rapid information on half-lives and modes of decay. It also makes it very easy to determine if a nuclide is stable or radioactive and the decay mode of radioisotopes.

Successive Radioactive Decay

In many cases the decay of a radioactive nucleus results in the formation of a radioactive daughter. A general equation for the series of decays of this type is

$$A \xrightarrow{\lambda_1} B \xrightarrow{\lambda_2} C$$

where A decays to B which in turn undergoes decay to form C. λ_1 and λ_2 are the decay constants for the transitions of A to B and B to C, respectively. To solve for the number of atoms of radioisotope B (N_2) at time (t) the following equation is used

$$N_2 = \frac{\lambda_1}{\lambda_2 - \lambda_1} N_1^0 (e^{-\lambda_1 t} - e^{-\lambda_2 t}) + N_2^0 e^{-\lambda_2 t} \tag{7.2-1}$$

where N_1 = number of atoms of A at time interval t
N_2 = number of atoms of B at time interval t
N_1^0 = number of atoms of A at some original time
N_2^0 = number of atoms of B at some original time
λ_1 = decay constant of atom A
λ_2 = decay constant of atom B.

Secular Equilibrium

Secular equilibrium is a limiting case of radioactive equilibrium in which the half-life of the parent A is many times greater than the half-life of atom B. In order for secular equilibrium to occur the half-life of the parent must be at least 1000 times greater than the half-life of the daughter. In this case Eq. (7.2-1) can be simplified to

$$N_2 = \frac{\lambda_1}{\lambda_2} N_1^0 e^{-\lambda_1 t}$$

or

$$N_1 \lambda_1 = N_2 \lambda_2$$

Figure 7.2-1a demonstrates secular equilibrium.

Transient Equilibrium

Transient equilibrium is similar to secular equilibrium in that the half-life of the parent is greater than the half-life of the daughter. However, this type of equilibrium differs from secular equilibrium in that the half-lives differ only by a small factor (about 10 times). Accordingly, Eq. (7.2-1) can be simplified to

$$N_2 = \frac{\lambda_1}{\lambda_2 - \lambda_1} N_1^0 e^{-\lambda_1 t}$$

or

$$N_1 \lambda_1 = N_2 (\lambda_2 - \lambda_1)$$

As in the case of secular equilibrium when equilibrium is established both parent and daughter decrease at equal rates. The rate of the decrease is dependent on the half-life of the parent with a pronounced decrease in the total activity with time. Transient equilibrium is illustrated in Fig. 7.2-1b.

No equilibrium is established if the half-life of the parent is less than that of the daughter. This is illustrated in Fig. 7.2-1c. If one starts with pure parent, the activity of the daughter will increase, pass through a maximum, and then decrease. At the maximum of the daughter curve $dN_2/dt = 0$. The time required to reach this maximum can be calculated by differentiation of Eq. (7.2-1) setting the differential equal to zero and solving for t.

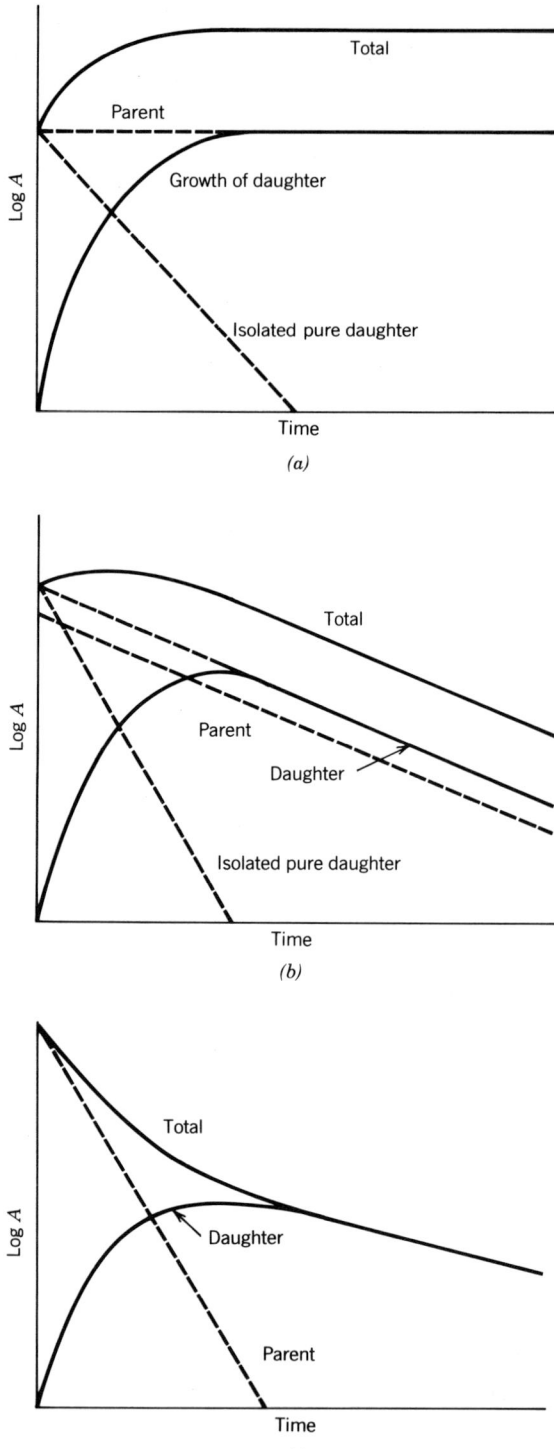

Fig. 7.2-1 (*a*) Secular equilibrium. (*b*) Transient equilibrium. (*c*) No equilibrium.

7.2-4 Shielding Concepts

In a modern 3000 MW(t) nuclear power reactor there occur approximately 9.4×10^{19} fissions per second which are accompanied by the release of fission neutrons, prompt gamma rays, and fission fragments with maximum energies of 5, 7, and 168 MeV, respectively. Additional radiations include beta particles, neutrinos, and capture gamma rays which are produced when various elements are activated by neutron capture. Subsequent to the fission event, many of the fission fragments are radioactive and decay by means of beta and gamma emission. It is obvious that adequate shielding must be used to protect personnel and electronic instrumentation from this hostile environment.

Particulate radiation, consisting of alpha radiation from the decay of uranium, beta radiation, and fission fragments, are easily stopped by a few millimeters of the reactor vessel; however, neutrons and gamma rays present difficult problems in shielding because they are very penetrating and react with the reactor vessel, coolant, and shielding material by producing secondary radiations.

Gamma rays interact with matter through the photoelectric effect, the Compton effect, and by pair production. See Fig. 7.2-2. In the photoelectric effect, a gamma collides with an orbital electron, is absorbed, and converted into kinetic energy of motion of the electron. As mentioned above, beta radiation is easily stopped; however, as the energy of the beta particle is removed, braking radiation or bremsstrahlung, which consists of continuous X-radiation emission, is produced.

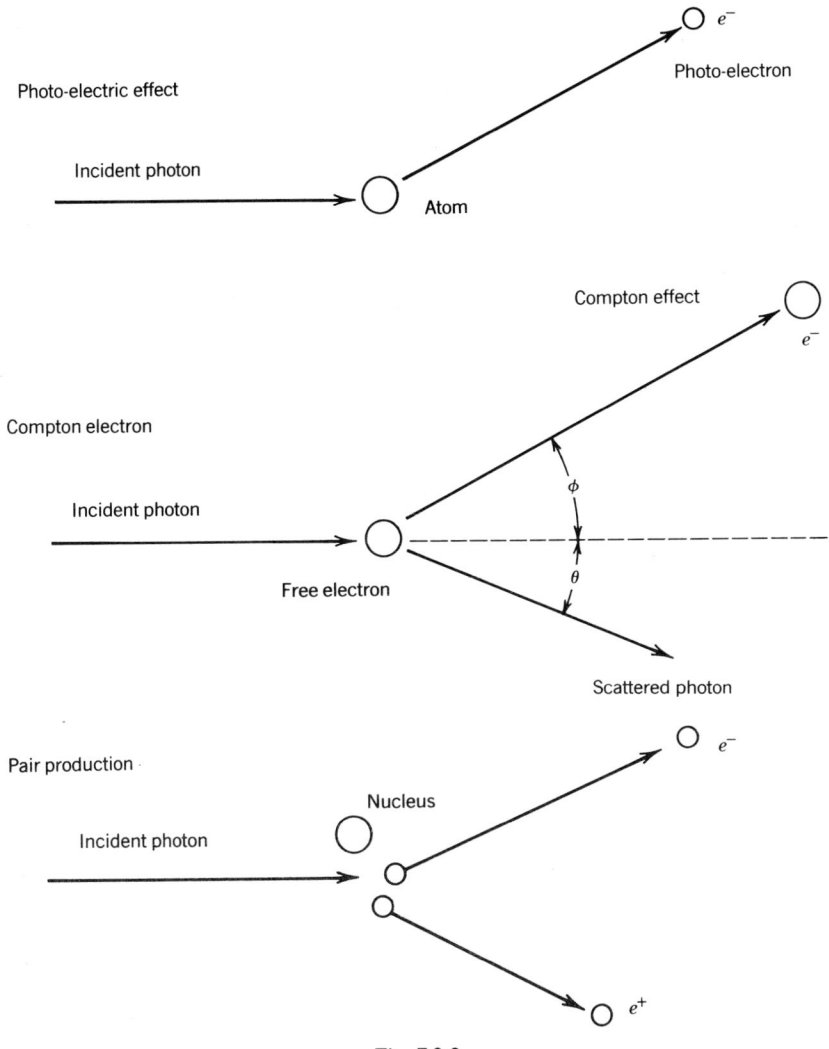

Fig. 7.2-2

When a Compton collision occurs, the initial gamma ray collides with a free electron in the shield in a classical billiard-ball-type collision. The electron moves off with a fraction of the initial gamma ray energy and a second gamma ray with decreased energy appears. The electron is easily stopped with the release of bremsstrahlung while the second gamma ray may undergo either a photoelectric or a Compton collision to produce further radiations.

Gamma rays with energies in excess of 1.022 MeV may interact with the nucleus of an atom in the shield to produce a positron/electron pair. The initial gamma ray is annihilated and its energy appears as rest mass energy of the pair as well as kinetic energy of motion. Again the particulate radiations are easily stopped with the production of bremsstrahlung. In this instance, the positron, being an antiparticle, will annihilate with an electron in the shield to produce two 0.511-MeV gamma rays which can interact through the photoelectric and Compton effects with the shield. The 0.511-MeV gamma rays are due to the conversion of the pair's rest mass into energy.

If the reactor contains either beryllium and/or deuterium in the moderator or the reflector, the gamma radiation may produce photoneutrons which are quite energetic and behave much like fission neutrons.

The uncollided flux as a function of distance can be shown to follow the simple relationship:

$$\phi(x) = \phi(0)e^{-\mu x}$$

where $\phi(0)$ is the initial gamma ray flux, e is the base of the natural log system, x is the distance, and μ is a quantity called the linear attenuation coefficient which is a function of both the gamma ray energy and the absorbing material. Specifically, the linear attenuation coefficient is the probability per unit length of travel that a gamma ray will interact with a given nuclei.

Because of the scattering that occurs when gamma rays traverse a relatively thick shield, secondary radiations of reduced energy are introduced into the system and the formula describing the uncollided flux is incorrect and actually underestimates the true gamma ray flux. Therefore, various schemes have been devised to generate correction factors, called buildup factors, to account for this secondary radiation.

Buildup factors have been produced by measuring the actual flux and tabulating the data and by calculating theoretically what the flux should be by means of the Monte Carlo method, which involves the use of probability theory.

The buildup factor may take several forms depending upon the degree of accuracy demanded. For example, the flux equation becomes

$$\phi(x) = B(\mu x)\phi(0)e^{-\mu x}$$

when $B(\mu x)$ is the buildup factor. See Table 7.2-1. Specifically, $B(\mu x)$ would be

$$B(\mu x) = 1 + b\mu x$$

where b is a number dependent on the gamma ray energy and the shield material. Another formula for a point source buildup factor is the *Taylor form* which is the sum of two exponential factors:

$$B(\mu x) = A_1 e^{-\alpha_1 \mu x} + A_2 e^{-\alpha_2 \mu x}$$

where A_1, A_2, α_1, and α_2 are again functions of the initial gamma ray energy and the absorbing material. See Table 7.2-2. Tables of buildup factors frequently are in units of μx, the number of mean free paths of shield.

Further complications arise in gamma ray shielding when the shield is composed of materials of different thickness and shielding strength. Also, it is important as to the order in which the different densities of shielding appear before the source. For example, a shield of water followed by lead is better than the reverse because the lead will absorb the buildup radiation produced in the water.

As with gamma rays, neutrons interact with shield material in a variety of ways. When neutrons are produced at fission, they possess enormous energies and as they traverse the reactor materials, they lose their energy by means of elastic and inelastic collisions. In an inelastic collision the energy loss appears as gamma radiation. Eventually the neutrons are slowed to the point at which they are either absorbed by the fuel and produce additional fissions or they are absorbed and activate the absorbing substance.

Fast neutrons can cause additional reactions. Interacting with the water in a reactor, neutrons cause the reaction $^{16}O(n, p)^{16}N$. The isotope ^{16}N has a half-life of 7.2 s with the emission of a 6-MeV gamma ray. If a reactor has dissolved air in the water, then the reaction $^{40}AR(n, \gamma)^{41}Ar$ occurs. The ^{41}Ar decays with a 1.82 h half-life and a 1.29-MeV gamma.

Thus, when shielding for neutrons, it is obvious that gamma radiation must also be considered. A formula similar to that for gamma radiation is applicable to neutron shielding:

$$\phi(x) = \phi(0)e^{-\Sigma x}$$

TABLE 7.2-1 DOSE BUILDUP FACTOR ($\mu_t x$) FOR A POINT ISOTROPIC SOURCE

Material	$\mu_t x$	Gamma-Ray Energy (MeV)							
		0.5	1.0	2.0	3.0	4.0	6.0	8.0	10.0
Water	1	2.52	2.13	1.83	1.69	1.58	1.46	1.38	1.33
	2	5.14	3.71	2.77	2.42	2.17	1.91	1.74	1.63
	4	14.3	7.68	4.88	3.91	3.34	2.76	2.40	2.19
	7	38.8	16.2	8.46	6.23	5.13	3.99	3.34	2.97
	10	77.6	27.1	12.4	8.63	6.94	5.18	4.25	3.72
	15	178	50.4	19.5	12.8	9.97	7.09	5.66	4.90
	20	334	82.2	27.7	17.0	12.9	8.85	6.95	5.98
Aluminum	1	2.37	2.02	1.75	1.64	1.53	1.42	1.34	1.28
	2	4.24	3.31	2.61	2.32	2.08	1.85	1.68	1.55
	4	9.47	6.57	4.62	3.78	3.22	2.70	2.37	2.12
	7	21.5	13.1	8.05	6.14	5.01	4.06	3.45	3.01
	10	38.9	21.2	11.9	8.65	6.88	5.49	4.58	3.96
	15	80.8	37.9	18.7	13.0	10.1	7.97	6.56	5.63
	20	141	58.5	26.3	17.7	13.4	10.4	8.52	7.32
Iron	1	1.98	1.87	1.76	1.55	1.45	1.34	1.27	1.20
	2	3.09	2.89	2.43	2.15	1.94	1.72	1.56	1.42
	4	5.98	5.39	4.13	3.51	3.03	2.58	2.23	1.95
	7	11.7	10.2	7.25	5.85	4.91	4.14	3.49	2.99
	10	19.2	16.2	10.9	8.51	7.11	6.02	5.07	4.35
	15	35.4	28.3	17.6	13.5	11.2	9.89	8.50	7.54
	20	55.6	42.7	25.1	19.1	16.0	14.7	13.0	12.4
Tungsten	1	1.28	1.44	1.42	1.36	1.29	1.20	1.14	1.11
	2	1.50	1.83	1.85	1.74	1.62	1.43	1.34	1.25
	4	1.84	2.57	2.72	2.59	2.41	2.07	1.81	1.64
	7	2.24	3.62	4.09	4.00	4.03	3.60	3.05	2.62
	10	2.61	4.64	5.27	5.92	6.27	6.29	5.40	4.65
	15	3.12	6.25	8.07	9.66	12.0	15.7	15.2	14.0
	20	—	7.35	10.6	14.1	20.9	36.3	41.9	39.3
Lead	1	1.24	1.37	1.39	1.34	1.27	1.18	1.14	1.11
	2	1.42	1.69	1.76	1.68	1.56	1.40	1.30	1.23
	4	1.69	2.26	2.51	2.43	2.25	1.97	1.74	1.58
	7	2.00	3.02	3.66	3.75	3.61	3.34	2.89	2.52
	10	2.27	3.74	4.84	5.30	5.44	5.69	5.07	4.34
	15	2.65	4.81	6.87	8.44	9.80	13.8	14.1	12.5
	20	2.73	5.86	9.00	12.3	16.3	32.7	44.6	39.2
Uranium	1	1.17	1.31	1.33	1.29	1.24	1.16	1.12	1.09
	2	1.30	1.56	1.64	1.58	1.50	1.36	1.27	1.20
	4	1.48	1.98	2.23	2.21	2.09	1.85	1.66	1.51
	7	1.67	2.50	3.09	3.27	3.21	2.96	2.61	2.26
	10	1.85	2.97	3.95	4.51	4.66	4.80	4.36	3.78
	15	2.08	3.67	5.36	6.97	8.01	10.8	11.2	10.5
	20	—	—	6.48	9.88	12.7	23.0	28.0	28.5

Source. Goldstein and Wilkins, "Calculations of the Penetrations of Gamma Rays—Final Report," N40-3075, June 30, 1984.

TABLE 7.2-2   COEFFICIENTS FOR AN ANALYTIC FIT TO BUILDUP FACTOR EQUATION

Material		Gamma-Ray Energy (MeV)							
		0.5	1	2	3	4	6	8	10
Water	A_1	—	13.5	8.1	5.6	4.5	3.4	2.8	2.5
	$-a_1$	—	0.100	0.068	0.059	0.0555	0.0525	0.05	0.0473
	$+a_2$	—	0.010	0.0405	0.073	0.11	0.156	0.17	0.1719
Aluminum	A_1	—	8.0	5.5	4.5	3.8	3.1	2.3	2.25
	$-a_1$	—	0.11	0.082	0.074	0.066	0.064	0.062	0.060
	$+a_2$	—	0.044	0.093	0.116	0.130	0.152	0.150	0.128
Iron	A_1	10.0	8.0	5.5	—	3.75	2.9	2.35	2.0
	$-a_1$	0.0948	0.0895	0.0788	—	0.075	0.0825	0.0833	0.095
	$+a_2$	0.012	0.04	0.07	—	0.082	0.075	0.0546	0.0116
Tungsten	A_1	—	3.3	2.9	2.7	2.05	1.2	0.7	0.6
	$-a_1$	—	0.043	0.069	0.086	0.118	0.171	0.205	0.212
	$+a_1$	—	0.148	0.188	0.134	0.070	0.00	0.052	0.144
Lead	A_1	1.65	2.45	2.60	2.15	1.65	0.96	0.67	0.50
	$-a_1$	0.032	0.045	0.071	0.097	0.123	0.175	0.204	0.214
	$+a_2$	0.296	0.178	0.103	0.077	0.064	0.059	0.067	0.08
Concrete	A_1	12.5	10.0	6.3	4.7	3.9	3.1	2.7	2.6
	$-a_1$	0.11	0.088	0.069	0.062	0.059	0.059	0.056	0.050
	$+a_2$	0.01	0.029	0.058	0.073	0.079	0.083	0.086	0.084

Source.   J. J. Taylor, "Application of Gamma-Ray Buildup Data to Shield Design," WAPD-RM-217, Jan. 28, 1954.

where Σ is the macroscopic cross-section and is the probability per unit length of travel that a neutron will interact with a given nuclei. It is a function of both neutron energy and absorbing material.

In the actual construction of shielding materials for a reactor, several layers of material are needed. To remove or attenuate the gamma radiation, materials rich in the heavy elements are needed. Concrete that has the mineral *barytes*, mainly barium sulfate, in place of the sand and gravel has been used extensively. To further increase the density of the shield, iron chips have been mixed in the concrete. In addition to attenuation of gamma rays, this form of concrete has the capability of slowing down the fast neutrons leaking out of the reactor core.

To absorb the neutrons, the procedure is to use a material rich in hydrogen to moderate or thermalize the fast neutrons that have passed through concrete shielding so that they may be more easily absorbed.

To absorb the thermal neutrons, materials rich in boron are used since these do not produce secondary high energy gamma rays upon neutron absorption.

7.2-5   Neutron Interactions

Since neutrons are electrically neutral they are neither attracted nor repelled by electrons or protons. As a result, neutrons pass through the electron cloud surrounding an atom and interact directly with the nucleus. This interaction with the nucleus can occur in a number of ways.

Elastic Scattering.   In this type of interaction the neutron strikes the nucleus of an atom in its ground state and results in the nucleus remaining in the ground state and the neutron reappears. There is an exchange of momentum between the neutron and the nucleus, with the result that the neutron's direction and speed are changed. Kinetic energy is conserved with the sum of the kinetic energy of the neutron and nucleus the same after the scattering as before.

Inelastic Scattering.   This process is similar to elastic scattering except that the interaction results in an excited nucleus along with the ejected neutron. The excited nucleus returns to the ground state by the emission of gamma rays.

Neutron Capture. A very common type of neutron reaction is neutron capture sometimes called radioactive capture because gamma radiation is almost always produced. In this case, the nucleus captures the neutron, forming a new isotope of the original nucleus with a mass number that is one higher than the original nucleus and one or more gamma rays—called capture gamma rays—are emitted. The resultant nucleus may be radioactive and undergo radioactive decay.

Charged Particle Reaction. Another neutron reaction results in the absorption of the neutron with the emission of either an alpha particle or proton.

Neutron Producing Reactions. With energetic neutrons striking a nucleus one or two neutrons can be extracted from the struck nucleus along with the original neutron. This is important in reactors containing heavy water (deuterium) or beryllium-9 since these nuclides have loosely bound neutrons that can easily be ejected.

Fission. The colliding of neutrons with certain nuclei may result in the splitting or fission of the nucleus. This fission process is the source of energy for present day nuclear power reactors and will be discussed in Section 7.1-6.

Cross Sections

The probability for a nuclear reaction is expressed in terms of the *reaction cross-section*. The unit of reaction probability is the *barn* when 1 barn (b) = 10^{-28} m^2 (or 10^{-24} cm^2). All neutron cross-sections are functions of the energy of the incident neutrons and the nature of the target nucleus. Neutron and charge-particle cross-section data are collected, evaluated, and distributed by the National Nuclear Data Center at Brookhaven National Laboratory, Upton, New York. Neutron cross-section data are published in the report *Neutron Cross Sections*, BNL-325, and other documents.

7.2-6 Nuclear Fission Process

Today the explanation seems obvious, but to two German radiochemists, O. Hahn and F. Strasman, the results of their experiments in 1938 were perplexing. When uranium was bombarded with neutrons, they identified barium, an atom with nearly half the mass of uranium, as one of the products. It was Lise Meitner and her nephew O. R. Frisch who suggested the correct interpretation of the results. In a letter dated January 16, 1939, published in the English scientific publication *Nature*, Meitner and Frisch wrote: "It seems possible that the uranium nucleus has only small stability of form and may, after neutron capture, divide itself into two nuclei of roughly equal size." American biologist William Archibold Arnold suggested that this splitting of the uranium nucleus into two halves be called *fission*, the term used for the dividing of living cells.

Thus, it was determined that when the atoms of certain heavy nuclides are bombarded by neutrons, some of the nuclei of these atoms will capture a neutron and become unstable. As a result of this instability, the atom splits or fissions into two smaller atoms. Together, the fission products weigh slightly less than the original atom and the bombarding neutron combined; this missing mass is converted into energy as described by Einstein's formula: $E = mc^2$.

Although a number of heavy nuclei will fission when bombarded with fast neutrons (neutrons with energies on the order of 1 MeV or more) only a few nuclides will fission with the absorption of a zero-energy neutron. These nuclei are said to be fissile and include nuclides such as uranium-235, plutonium-239, and uranium-233. Upon fissioning, these fissile nuclei release free neutrons which can produce additional fissions in neighboring fissile nuclei producing a *chain reaction*.

Fission fragments of various masses are formed as a result of the fission process. Chemical elements as light as zinc (atomic number 30) and as heavy as gadolinium (atomic number 64), with half-lives from fractions of seconds to millions of years are formed as a result of the fission process. Approximately 400 different nuclides have been identified as products in the fission of uranium-235 by neutrons.

The fission process for uranium-235 can be represented by

$$^{235}\text{U} + n \rightarrow \text{fission products} + \nu n$$

when ν is the number of free neutrons released as a result of uranium fissioning.

Most of the neutrons released in fission are emitted essentially at the instant of fission. These are called prompt neutrons to distinguish them from the delayed neutrons that are released comparatively long after the fission takes place. For each 0.0253-eV (thermal) neutron absorbed by the uranium nucleus about 2.5 (= ν) new neutrons are released. As the energy of the incident neutron increases the value of ν increases slightly. For every 6- to 7-MeV increase in neutron energy one additional neutron is emitted.

In a nuclear reactor the nuclear chain reactions are controlled so that an equilibrium state is reached, where for each fission only one of the new neutrons released results in a further fission. This will be discussed in greater detail in Section 7.1-8.

7.2-7   Energy Release from Fission and Decay

Before 1939 the largest known nuclear reaction energy was 22.2 MeV, associated with bombardment of lithium-6 with deuterons. In the fission process about 200 MeV is released. As in other nuclear reactions, the energy produced is equivalent to the difference in rest mass between the interacting particles and the final products.

The energy released in the fission process appears in several different forms. Most of the energy released (almost 85%) occurs as kinetic energy of the fission fragments. These fragments come to rest within about 10^{-3} cm of the fission site so that all of their energy is converted into heat. The remainder of the energy is released by radioactive decay energy of the fission products, gamma rays emitted at the instant of fission, and fission neutrons. Most of this energy does not escape from a nuclear power system. However, the energy of the neutrinos produced from fission product decay, which is about 12 MeV, will most likely escape from the system and cannot be recovered.

Neutrons released as a result of the fission process (approximately 2.42 neutrons per fission) can be absorbed by other nuclei in the system resulting in the production of one or more capture gamma rays, whose energies depend on the binding energy of the neutron to the new nucleus formed. This energy is recoverable in a nuclear power system. Table 7.2-3 gives an approximate distribution of emitted energy and recoverable energies from the fission of uranium-235.

Equivalent Energy of Coal and Oil

In a power reactor using uranium-235 as a fuel, the release of 1 MW of energy for 1 day requires approximately 1 g of uranium-235. This is equivalent to burning 3.17 short tons of 13,000 Btu/lb coal or 12.6 barrels of 6,500,000 Btu/barrel oil.

7.2-8   Reactor Control Concepts

In any system that produces power, there must be a method by which it is controlled. In a nuclear reactor this is accomplished by varying the reactivity of the core. By controlling the reactivity, the reactor can be kept in a shut down condition or can be made to start up and increase power at will. It can be made to maintain power for extended periods of time, decrease power, or shut down completely. During power operations, the core reactivity must also change to compensate for production of fission product poisons, temperature variations, and the depletion of fuel.

Because a nuclear reactor utilizes a chain reaction of neutrons inducing fissions, the power is effectively controlled by varying the neutron population in the core. If the number of neutrons increases, then the reactor power increases. Conversely, a decrease in the population of neutrons decreases the power.

TABLE 7.2-3   DISTRIBUTION OF EMITTED ENERGY AND RECOVERABLE ENERGY FROM ^{235}U

Form	Emitted Energy (MeV)	Recoverable Energy (MeV)
Fission fragments	168	168
Fission product decay		
Beta rays	8	8
Gamma rays	7	7
Neutrinos	12	—
Prompt gamma rays	7	7
Fission neutrons (kinetic energy)	5	5
Capture gamma rays	—	3–12
Total	207	198–207

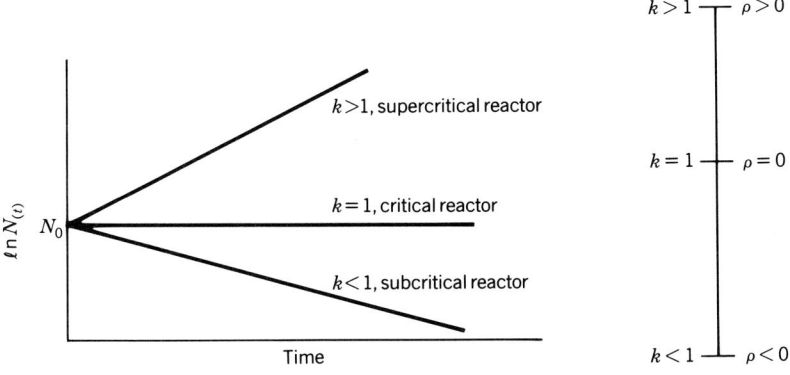

Fig. 7.2-3 Neutron population time dependence.

The ratio of the neutrons in the present generation as compared to those in the previous generation is known as the k (sometimes $k_{eff} = k$ effective) of the core.

$$k = \frac{\text{number of neutrons in present generation}}{\text{number of neutrons in previous generation}}$$

If $k = 1$ then the reactor is just critical and in a steady-state condition. If $k > 1$ or $k < 1$, the reactor is either supercritical and increasing in power or it is subcritical. See Fig. 7.2-3.

The reactivity of a core is defined as the deviation of the system from the just critical ($k = 1$) condition.

$$\rho = \frac{k-1}{k} = \frac{k_{ex}}{k}$$

where k_{ex} is the excess reactivity. It is obvious from the formula that $\rho = 0$ for $k = 1$. Similarly, $\rho > 0$ if the reactor is supercritical and $\rho < 0$ if subcritical. See Fig. 7.2-3.

Reactor control is achieved by manipulating the reactivity of the core and, consequently, the value of k. Reactivity is changed on demand through the use of control rods and chemical shims. As the reactor operates, intrinsic physical processes, such as fuel burn-up, fission produce inventories, and temperature changes in the core and moderator, also have an effect.

Reactivity is defined in units of dollars and cents. One dollar's worth of reactivity is that amount which will just cause the reactor to increase in power under the influence of only the prompt neutrons from fissions. Reactivity insertions of a dollar or more will cause the reactor to go prompt-critical with an extremely short period. Usually reactivity insertions are on the order of a few cents.

Reactor control must deal with events that occur in a time span as short as seconds and as long as months. Thus, a variety of control mechanisms are used. Sudden changes in fuel or moderator temperature and pressure, or void production (coolant bubbling) occur within seconds. The production of some fission product poisons usually requires hours in the cases of xenon-135 and days for samarium-149 while fuel burn-up and other fission product poisons may take place over the entire lifetime of the core.

Figures 7.2-4 illustrates the effects of xenon and samarium poisoning. The symbol ϕ indicates neutron flux and ψ the equilibrium poisoning that results. With xenon the time required to reach equilibrium poisoning is the same for different neutron fluxes; however, the equilibrium value of the poisoning increases with increasing flux. After shut down the poisoning peaks because the iodine-135 precursor of xenon-135 is decaying. Eventually the xenon poisoning decays away so that it is temporary. Samarium poisoning behaves differently. Regardless of the flux level, equilibrium poisoning from samarium is the same, but it takes longer to reach with decreasing flux levels. Samarium does not peak like xenon, but rather increases to a maximum level because of its precursor, promethium-149, decaying. Samarium is a permanent poison in the core and only a fraction of it can be removed through burn-up in later operation of the core.

A number of materials are useful as neutron absorbers (poisons). These include boron, cadmium, europium, hafnium, and samarium. Occasionally silver and indium which have large resonance absorption peaks are combined with cadmium. The essential characteristics of a control material are that it possesses a large absorption cross-section for thermal neutrons and is able to withstand the high temperatures and radiation fields present in the core.

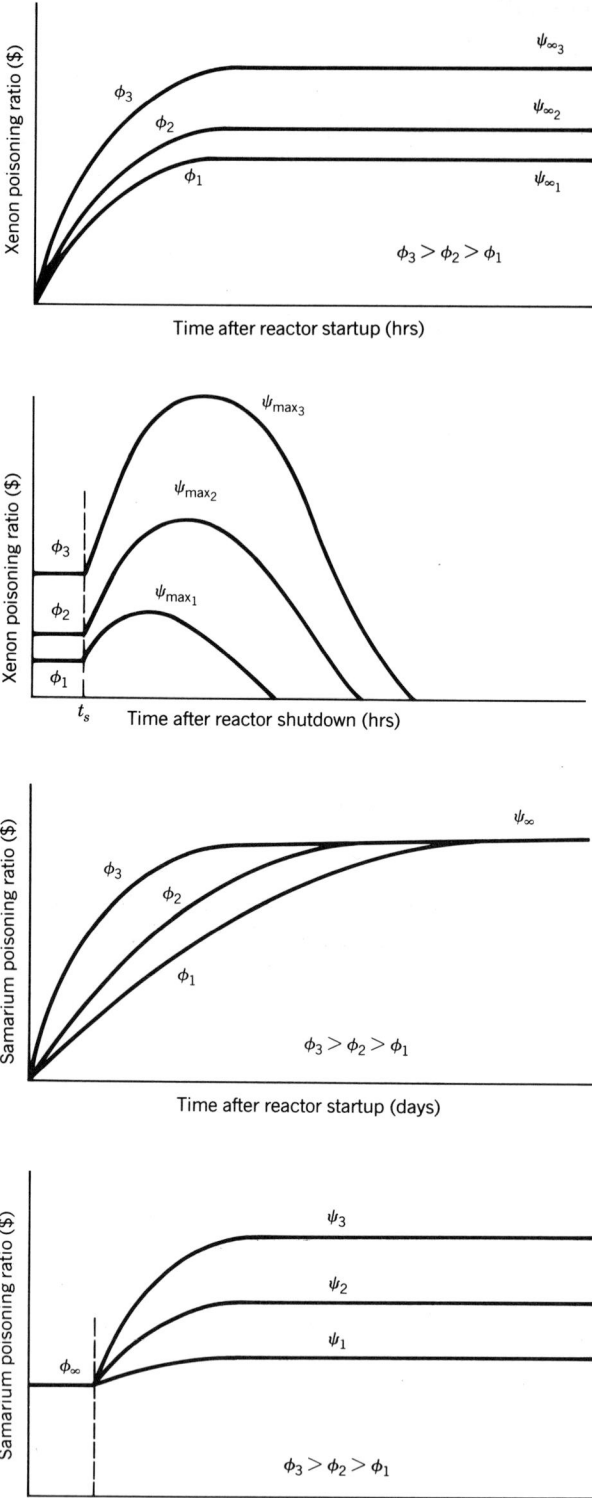

Fig. 7.2-4 (*a*) Xenon equilibrium for various flux levels. (*b*) Xenon peaking for various flux levels. (*c*) Samarium equilibrium for various flux levels. (*d*) Samarium buildup for various flux levels.

Reactor control poisons are used in three forms. They appear in control rods, as burnable poisons in fuel elements or separate rods, and as soluble poisons present in the moderator and/or coolant.

Because control rods can be manipulated easily and quickly, they are used to start up and shut down the reactor and to make changes in the power level in short periods of time. Control rods are available in a variety of forms. In a small research-type reactor, the control devices may be either plate type or cylindrical in shape. In a commercial power reactor the rods may be cruciform in shape or a spider assembly. Whatever the shape of the control rods, all must be capable of being inserted quickly into the core by means of gravity, spring, pneumatic, or hydraulic devices to produce what is called a reactor scram and shut down the reactor.

In boiling water reactors, the control rods enter the reactor core from the bottom to permit the placement of the steam dryer and separator assemblies at the top of the reactor assembly. In all other reactors the control rods enter from the top.

When a core is brand new, there is a need to have a large amount of negative reactivity available to control the reactor. However, as the reactor operates, the core experiences a depletion of the fuel and a buildup of its fission product (poison) inventory. Thus is needs a smaller amount of negative reactivity in the form of control poisons. It is possible to insert in the fresh core negative reactivity in place of some of the control rods. This accomplishes two things: First, it increases the negative reactivity needed for the fresh core and second, by reducing the number of control rods needed, more fuel can be loaded into the core to extend the core life. This negative reactivity is referred to as burnable poison and may be located in the fuel rods in the form of enriched boron or gadlinia.

Soluble poisons are used in pressurized water reactors in the form of boric acid. The amount of boric acid in solution can be controlled by the reactor operator through addition and/or removal of the acid and is commonly referred to as a chemical shim. The main use of a chemical shim is to make adjustments in the neutron population and the reactivity of the core uniformly throughout the core without changing the flux shape or the power distribution in the core. Using control rods to change the core reactivity changes the flux patterns and consequently the fuel burn-up in a nonuniform pattern. It is obvious that reactor control by chemical shiming cannot be used to control abrupt changes in reactivity. It is used in conjunction with control rods to account for reactivity changes over long periods of time.

It should be noted that chemical shims are not used in boiling water reactors because of the possible concentration of boric acid and negative reactivity that would result from the continual boiling of the water.

In addition to reactor control by the addition and subtraction of various types of poisons, there is yet another method present in the boiling water reactor (BWR) type and is due to the production of voids in the coolant/moderator of the reactor. If an increase of voids of bubbles occurs in the moderator—a substance used to moderate or slow the neutrons from the high energies with which they were born—then less moderation takes place, neutron leakage increases decreasing the rate of fissioning and, consequently, reducing the overall power being generated in the reactor. About 13% by weight of the fluid leaving the BWR core is steam. The balance, 87%, is recirculated through the core by a series of jet and recirculation pumps which surround the reactor vessel. Approximately one-third of the coolant water is taken through the recirculation pumps located outside the reactor vessel and jet pumps located between the core and the outer vessel wall. The flow of water is directed downward and causes an upward flow through the center of the core. If for some reason the recirculation loop flow is reduced, less cooling occurs, more water is converted to steam, and, consequently, the void content of the core is increased. This has the effect of moderating less effectively the neutrons in the core to introduce negative reactivity into the reactor. The overall effect is to decrease the core power.

7.2-9 Health Physics Concepts

Over the past 80 years, substantial knowledge and experience have been gained regarding the health effects of ionizing radiation. Although much still remains to be learned about the interaction of ionizing radiation and living organisms, more is known about the mechanisms of radiation damage on molecules, cells, and the whole living systems than about most other environmental agents.

Radiation Dose Units

Quantities and doses of ionizing radiation are measured in several units. An old unit of measurement that is rarely used is the roentgen, a measure of the quantity of ionization induced in air. The principal units used to express doses absorbed by living matter are the rad, an acronym for radiation absorbed dose (1 rad = 100 ergs per gram of tissue), and the SI unit the gray (1 gray = 1 joule per kilogram of tissue, or 100 rads). These units express the amount of energy deposited in a given mass of tissue.

Since for a given dose particulate radiation has the potential to do more radiation damage to living matter than X-rays or gamma rays, the dose equivalent units of the rem, an acronym for roentgen equivalent man, and the SI unit the sievert have been derived. These dose equivalent units are equal to the absorbed dose (rems, gray) times a quality factor times other modifying factors.

TABLE 7.2-4

Quality Factor Values Radiation	QF
Gamma rays from radium in equilibrium (0.5 platinum filter)	1
X-rays	1
Beta rays and electrons > 0.03 MeV	1
Beta rays and electrons < 0.03 MeV	1.7
Thermal neutrons	3
Fast neutrons	10
Protons	10
Alpha rays	10
Heavy ions	20

Quality Factor vs. Linear Energy Transfer Let $(keV/\mu m$ in water$)$	QF
3.5 or less	1
3.5–7.0	1–2
7.0–23	2–5
23–53	5–10
53–175	10–20

The quality factor is related to the linear energy transfer (LET) of the specific types of radiation being measured. LET is the rate of energy deposition along the path of ionizing radiation. Certain types of radiation cause more ionizations per micron in tissue. The greater the LET of a specific type of radiation, the greater the quality factor, and therefore, the greater the potential to cause biological damage (see Table 7.2-4).

Gamma rays, X-rays, and low-energy beta particles have a low LET and a quality factor equal to 1. Thus, a 1-rad dose of these types of radiation has an equivalent dose of 1 rem. Protons, fast neutrons, and alpha radiation have a relatively large LET and may have a quality factor as high as 10–20. Thus, a 1-rad dose of these types of radiations may have an equivalent dose of 10–20 rems.

There are other modifying factors that may contribute to the value of the dose equivalent. For example, when radioactive particles are ingested, they sometimes tend to concentrate in a particular organ or tissue of the body. This must be considered among the modifying factors when calculating dose equivalent.

Relative Biological Effectiveness

A ratio used in biological systems is the relative biological effectiveness (RBE), which takes into account that radiations of equal amount have different biological effects. RBE was developed in an attempt to standardize effects in living systems and can approximate the number of cells killed and amount of genetic damage in a particular organism under set conditions. RBE is defined as the ratio of the amount of energy of 200-keV X-rays required to produce a given effect to the energy required of any radiation to produce the same effects. Since the probability of injury depends on the concentration of molecular damage, ionizing particles generally have a higher RBE than do X-rays and gamma rays. At the same time, directly ionizing charged particles generally penetrate only a short distance in tissue, as compared with X-ray and gamma rays, so that they pose far less hazard if they originate outside the body and are not ingested or inhaled.

Population Dose

The collective dose to a population, which is used to assess the risks of radiation-induced cancer and genetic harm, is expressed in *person-rem*. This unit represents the produce of the average dose per person times the number of people exposed (e.g., 1 rem to each 1000 people = 1000 person-rem).

Radiation Effects

For purposes of radiological protection, radiation injuries are customarily divided into two categories —nonstochastic and stochastic effects.

Nonstochastic effects are observed only after exposure to very high doses of radiation resulting in damage to many cells. Such effects would only be experienced in a nuclear power facility as a result of a serious radiation accident leading to a large dose of radiation. The effects of this type of exposure include radiation sickness resulting from destruction of a large proportion of blood-forming cells in the bone marrow, the impairment of fertility, and cataract formation.

Stochastic effects are those that may conceivably result from damage to a single cell. These effects are assumed to have no threshold, although there is no evidence which excludes the possibility of a threshold. Such effects include damage to chromosomes and genes (mutagenic effects), disturbances in the growth and development of the embryo or fetus (teratogenic effects), and increase in the probability of certain types of cancer (carcinogenic effects). Such effects can not be distinguished as being caused by radiation as opposed to other causes. As a result, in the case of low-level radiation from nuclear power, which accounts for less than 0.1% of all natural and manmade radiation, the increase in frequency of effects is too small to be detected but is inferred statistically. Only a small percentage (0.5–3.0%) of genetically related diseases in the human population is attributable to natural background and manmade radiation.

Natural and manmade radiation are responsible for 1–3% of all cancers. This can be compared to the 30% or more due to smoking and dietary factors. The number of cancers attributable to occupational irradiation in radiation workers appears to be substantially less than 1% of the natural incidence in such persons. This increase in risk is no greater than, and in some cases less than, the risks assumed by workers in other occupations that are generally regarded as "safe."

Radiation Protection Standards

Very shortly after the discovery of X-rays in 1895, it was recognized that radiation could cause serious injury to the skin. As early as 1916 the Roentgen Society organized efforts to promote radiation safety standards.

In 1959 the U.S. Congress established the Federal Radiation Council (FRC) to advise the President on radiation matters affecting health and to recommend radiation standards for all federal agencies. The FRC relied heavily on the technical recommendations of the National Committee on

TABLE 7.2-5 DOSE-LIMITING RECOMMENDATIONS OF STANDARDS-SETTING BODIES; DOSES IN REMS[a]

Type of Exposure Group	FRC/EPA (1960)	EPA (1981)	NCRP (1971)
Occupational exposure			
Whole body			
prospective	12/yr; 3/qtr	5/yr	5/yr
retrospective	—	—	10–15 any year
to N years of age	5(N–18)	100 total career	5(N–18)
Skin	30/yr; 10/qtr		15/yr
Hands	75/yr; 25/qtr	50/yr	75/yr; 25/qtr
Forearms	75/yr; 25/qtr	—	30/yr; 10/qtr
Gonads	12/yr; 3/qtr	5/yr	5/yr
Lens of eye	12/yr; 3/qtr	5/yr	5/yr
Thyroid	30/yr; 10/qtr	—	15/yr
Any other organ	15/yr; 5/qtr	30/yr	15/yr; 5/qtr
Pregnant women	—	[b]	0.5 gestation
General population			
Individuals	0.5/yr	—	0.5/yr
Average	5/30 yr	—	5/30 yrs

[a] Based on FRC report 1, 1960; EPA notice in Federal Register, 46 FR 7836, 1981; NCRP report 39, 1971.
[b] Several alternative standards proposed.

Radiation Protection and Measurements (NCRP). The protection standards established by the FRC stipulated that occupational exposure of the whole body and the most radiosensitive organs of the body (bone marrow, eye lens, and gonads) should not deliver a cumulative dose (in rems) exceeding five times the age of the worker (in years) beyond 18. For example, a worker of age 28 would be allowed a cumulative dose of up to 50 rems. For the general public the standards were more conservative. An average dose of only 5 rems per generation (or 30 years) was specified as the maximum permissible limit (equivalent to 170 mrem/year).

In 1970 the functions and staff of FRC were transferred to the newly formed Environmental Protection Agency (EPA). Standards for exposure to radiation recommended by EPA, and earlier by the FRC, are called Radiation Protection Guides (RPGs). In 1981 EPA proposed new and somewhat more stringent RPGs for occupational exposure. For example, the 1981 proposed guides would prohibit annual doses of more than 5 rems and would limit total career dosage to 100 rems. These guidelines are still under consideration and the FRC RBGs are still in effect. A summary of recommended doses are presented in Table 7.2-5. Keep in mind that these doses are over and above those received from natural background radiation and doses from medical diagnosis and treatment.

In addition to the radiation protection guides there are two additional ground rules that are followed when exposure to an individual or a population is involved. The first assumes that any exposure to radiation may be potentially harmful and no deliberate exposure is justified unless there is some benefit. The ALARA principle is the second rule. ALARA is an acronym for "as low as reasonably achievable" which takes into account the state of technology for reducing exposure as well as the economics of reducing exposure relative to the benefits to be achieved. This ALARA concept is also used in connection with the emission of radioactive effluents from nuclear power plants. For example, the RPG standard for exposure to the general public is equivalent to 170 mrem/year while an average nuclear plant exposes the general public to less than 1 mrem/year through the use of technologies and procedures to reduce emissions.

Although there is no conclusive evidence from population studies that low levels of ionizing radiation produce any harmful effects, most experts in the area of radiation health effects make a number of arbitrary assumptions for setting radiation standards. These assumptions include:

1. Any dose of radiation produces an effect. There is no threshold dose of radiation below which there is no effect.
2. Doses of radiation are additive, no matter at what rate or at what intervals they may be delivered.

By using these assumptions it is generally felt that radiation protection standards are probably higher than necessary to ensure public and radiation workers' health and safety. However, in an area where there still remains much to be learned, this conservative approach is supported by most scientists in the radiation protection field.

BIBLIOGRAPHY

Bell, G. I. and S. Glasstone, *Nuclear Reactor Theory*, Van Nostrand Reinhold, New York, 1970.

Blizard, E. P. and L. S. Abbott (Eds.), *Reactor Handbook*, Vol. III, Part B: *Shielding*, 2nd ed., Interscience, New York, 1962.

Cember, H., *Introduction to Health Physics*, Pergamon Press, New York, 1969.

Chilton, A. B., J. K. Shultis, and R. E. Faw, *Principles of Radiation Shielding*, Prentice-Hall, Englewood Cliffs, N.J., 1969.

Connolly, T. J., *Foundations of Nuclear Engineering*, Wiley, New York, 1978.

Goldstein, H., *Fundamental Aspects of Reactor Shielding*, Addison-Wesley, Reading, Mass., 1959; reprinted by Johnson Reprint Corp., New York, 1971.

Jaeger, R. G. (Ed.), *Engineering Compendium on Radiation Shielding*, Vol. I: *Shielding Fundamentals and Methods*, Springer-Verlag, New York, 1968, Chap. 2.

Knoll, G. F., *Radiation Detection and Measurement*, Wiley, New York, 1979.

Lamarsh, J. R., *Introduction to Nuclear Engineering*, 2nd ed., Addison-Wesley, New York, 1983.

Lederer, C. M. and V. Shirley, *Table of Isotopes*, 7th ed., Wiley-Interscience, New York, 1978.

Murray, R. L., *Nuclear Energy*, 2nd ed., Pergamon Press, New York, 1980.

"Neutron Cross Sections," Report BNL-325 and Supplements, Brookhaven National Laboratory, Upton, N.Y. 1958, 1966.

"Radiation Quantities and Units," Report 33, International Commission on Radiation Units and Measurements, Washington, D.C., 1980.

Radiological Health Handbook, U.S. Government Printing Office, Washington, D.C., 1970.

"Reactor Physics Constants", ANL-5800, 2nd ed., U.S.A.E.C. (1963).

7.3   THERMAL-HYDRAULIC AND FUEL CONSIDERATIONS

Chong Chiu

7.3-1   LWR Design Evolution

The basic objective of this article is to present the principles underlying the thermal-hydraulic and fuel design of nuclear reactors. The emphasis is on light water reactors (LWRs), which are by far the most popular type of nuclear power generation reactors. Most of the design principles described here are also applicable to other types of nuclear reactors.

The rapid evolution of LWRs over the past 20 years can be characterized by continually increasing core power density and core inlet temperature. Table 7.3-1 lists for comparison the thermal-hydraulic and fuel design parameters of the early and the current generations of LWRs. The major driving force of this evolution is the reduction of capital investment and operating cost of a nuclear power plant. Under the current design philosophy, quantitative assessments of the safety thermal margin and operating thermal margin are needed, based on credible experimental evidence and analytical investigation. The safety margin assures that reactor safety and public health are not compromised when the reactor is operated within the acceptable design limits established and accepted by the reactor designers. The safety margin is the margin between the acceptable design limits and the "damaging conditions," such as loss of reactor core coolability during an accident, where risk to public health may occur.

The operating margin assures that the required operating flexibility and maneuverability exist. The operating margin is the margin between the steady-state operating conditions and the conditions at which the limiting set point of the reactor monitoring or protection system is reached.

Since the safety margin and the operating margin are of extreme importance in the design process, the significance of various thermal-hydraulic and fuel performance related phenomena are better understood in the context of these two thermal margins.

This article is divided into two major parts. The first part, consisting of Sections 7.3-2 to 7.3-6, describes the acceptable thermal-hydraulic and fuel design limits within which the reactor must be designed. The conservatism of these limits, that is, the safety margin, is discussed based on available experimental data. The second part, consisting of Sections 7.3-6 to 7.3-10, discusses the methodology of thermal margin design, which interrelates the thermal-hydraulic and fuel design limits, operating thermal margin, and the steady-state operating conditions.

7.3-2   Introduction to Thermal Hydraulics and Fuel Design Limits

One of the design requirements of nuclear plants is that release of radioactive materials during normal reactor operation and anticipated operational occurrences be kept as low as is reasonably possible.[5] In

TABLE 7.3-1   THERMAL HYDRAULICS AND FUEL DESIGN PARAMETERS[a]

	PWR Indian Pt.-I (1962)	PWR Palo Verde-I (1983)	BWR Dresden-I (1960)	BWR Grand Gulf-I (1982)
Plant net power (MW$_{th}$)	585	3800	627.3	3833
Reactor Pressure (MPa)	10.5	15.5	7.0	7.3
Reactor inlet temperature (K)	525	569	536	551
Core average inlet velocity (m/s)	3.9	4.7	1.4	2.2
Specific power (kW/kgU)	10.3	36.9	12.7	25.9
Power density (MW/m^3)	26.4	95.0	22.6	56.0
Plant efficiency	32.0%	33.4%	28.7%	32.6%
Fuel rod outside diameter (cm)	0.772	0.970	1.27	1.25
Cladding material	Stainless steel	Zr-4	Zr-2	Zr-2
Fuel material	UO$_2$–ThO$_2$	UO$_2$ (95% TD)	UO$_2$	UO$_2$ (94% TD)
Assembly configuration (number of rods)	16 × 16	16 × 16	5 × 5	8 × 8

[a]Data listed in the table are obtained from refs. 1, 2, 3, and 4.

addition, during any accidents, a person located at the plant site boundary for 2 h should not receive a radiation dose in excess of 25 rem whole-body radiation and 300 rem thyroid.[6] The source of the radioactive materials of a nuclear plant is the radioactive fission products, either in gaseous or solid form, generated via the fission process of uranium fuel. To minimize the release of radioactive materials, maintenance of the fuel cladding and reactor coolant boundary integrity is of primary importance. Fuel rod failure, generally defined as the loss of fuel rod hermeticity, is the breach of the first barrier of radioactive material containment.

Primary coolant boundary failure, generally defined as the abnormal leakage of the reactor coolant, is the breach of the second barrier of the radioactive material containment.

In order to preserve the integrity of the fuel cladding and reactor coolant boundary during normal operation and anticipated operational occurrences and to estimate the radiation dose release during an accident, various thermal-hydraulic and fuel design limits are developed by the designers. Operation within the limits avoids breach of the radioactive material containment barriers.

In the LWR industry it is conventional to categorize any event expected to occur one or more times during the life of the reactor as an anticipated operational occurrence (AOO).[7] Other events, including loss of core flow, that have lower frequency of occurrence but are specifically defined as anticipated operational occurrences[7] are also treated as anticipated operational occurrences by the designer. Any other event with a lower expected frequency is an accident. The permissible radioactive release is in general higher for an accident than for an anticipated operational occurrence because of its low probability of occurrence.

7.3-3   Design Limits for Normal Operation and Anticipated Operational Occurrences

During normal operation and anticipated operational occurrences, five phenomena[8] that could damage the fuel cladding have been identified. These damaging phenomena are fuel cladding overheating, pellet/cladding interaction (PCI), cladding hydriding, cladding collapse, and cladding fretting.

Cladding Overheating

The overheating of the cladding may be associated with the occurrence of the following conditions: departure from nucleate boiling (DNB) or dryout, fuel melting, and hydrodynamic flow instability. These three conditions are described below.

DNB and Dryout

Both departure from nucleate boiling and dryout phenomena, characterized by a sudden drop of the heat transfer rate, result in a temperature excursion of the fuel cladding. Departure from nucleate boiling[9] takes place when the bubble layer of bubbly flow adjacent to the fuel cladding becomes thick enough to impede the heat transfer process. Dryout[9] occurs when the liquid film of annular flow breaks down, resulting in dry patches. DNB will occur when the flow around the fuel pin is in the nucleate boiling heat transfer regime.[10-15] Dryout will occur when the flow is in the two-phase forced convection regime. PWRs tend to operate at low coolant quality, in single-phase forced convection and nucleate boiling, so DNB is the major concern. BWRs tend to operate at substantially higher quality, in two-phase forced convection, so dryout is the major concern.

It should be noted that the temperature excursion during DNB or dryout does not necessarily mean instant fuel cladding failure or even damage. For DNB, studies[16] suggest that during DNB the cladding will not be damaged as long as the cladding temperature does not exceed 870 K for more than 15 s.[17,18] Moreover, experimental data from the Power Burst Facility[19] reveal that it is possible to sustain post-DNB conditions for as long as 510 s without fuel failure, although actual behavior obviously depends on the specific operating conditions. For dryout, evidence from an in-pile experiment[20] demonstrated that sustaining postdryout conditions for less than 5 min did not cause any cladding damage for the operating conditions tested.

The margin to DNB or dryout can be expressed in terms of CHFR (critical heat flux ratio), which is defined as the ratio of critical heat flux (CHF) to local surface heat flux. CHF is the heat flux at which DNB or dryout occurs. Alternatively, the margin to dryout sometimes is expressed in terms of the critical quality beyond which dryout occurs. The behavior of both CHFR and critical quality have been extensively studied in the past 10 years with experimental data under simulated reactor operating conditions. Some good reviews are available on this subject.[9,21-23]

Several CHFR and critical quality correlations applicable to LWRs are summarized in Table 7.3-2. Some conservative limits of CHFR, at which there is at least a 95% probability at a 95% confidence level based on experimental data that the cladding will not experience DNB or dryout, are generally referred to as the specified acceptable fuel design limit (SAFDL) on CHF. The factors that affect CHF are listed in Table 7.3-3.

TABLE 7.3-2 CHF CORRELATIONS

DRY OUT	DNB
TYPICAL CORRELATIONS	TYPICAL CORRELATIONS

DRY OUT

- CISE[24] - BWR

$$\frac{\bar{q}_{bl}}{H_{fg}G} = \left(\frac{D_e}{4L_{bl}}\right) \frac{1-Pr}{(1.35G/10^6)^{1/3}}$$

$$x \frac{L_{bl}}{L_{bl} + \left[168\ (1/p_r)-1\right]^{0.4}(G/10^6)D_e^{1.4}}$$

where subscript (bl) means over boiling length and D_e and G are annular flow channels around the hot rod. All parameters in British units; L and D_e in feet.

- Hench- Gillis[25] - BWR

$$X_{crit} = \frac{0.50G^{-.43}L}{165 + 115G^{2.3} + L}\left[2-J_1 + \frac{0.19}{G}(J_1-1)^2 + J_3\right] + 0.006 -$$

$$0.0157\left(\frac{P-800}{1000}\right) - 0.0714\left(\frac{P-800}{1000}\right)^2$$

$$J_1 = -\frac{1}{32}\left[20fp + 2\ \Sigma f_i + \Sigma f_j\right]$$

$$J_3 = \left(\frac{0.14}{G + 0.25} - 0.10\right)$$ for central rods

$$L = \frac{n\ \pi\ D\ L_{bl}}{A}$$

f = peaking factors, n= number of rods in bundle

J_1, J_3 for corner and side rods are defined in (25)

- Bowring[26] - PWR & BWR

$$\frac{q_{CHF}}{10^6} = \frac{A-B_x\ H_{fg}}{C + Z\cdot Y + \left[1-4B\cdot10^6/(G\cdot D_h)\right]}$$

A, B, C & Y are defined in (26)

DNB

- W-3 [27][28] - PWR, SAFDL = 1.30

$$\frac{q''_{DNB,EU}}{10^6} = \left\{\left[2.02-0.430(p/10^3)\right] + \left[0.172-0.1(p/10^3)\right]\right.$$

$$x\ exp\left[18.2_x - 4.13_x(p/10^3)\right]\right\}\ (1.16-0.87_x)$$

$$x\left[(0.148 - 1.6_x + 0.173|_{x1})(G/10^6)+1.04\right]$$

$$x\left[0.266 + 0.836\ exp(-3.15D_e)\right]$$

$$x\left[0.826 + 0.0008(H_{sat} - H_{in})\right]$$

$$F = \frac{q''_{DNB,EU}}{q''_{DNB,U}}$$

$$= \frac{C\int_0^{1DNB} q''(z)\ exp\left[-C(1_{DNB} -z)dz\right]}{q''_{local}\ x\left[1-exp(-C1_{DNB,EU})\right]}$$

where $C = \left[0.15(1-x_{DNB})^{4.31}\right]/(G/10^6)^{0.478}\ in^{-1}$

- CE-1 [29] - PWR, SAFDL = 1.19

$$\frac{q''_{DNB,Uni}}{10^6} = 2.8922\ x\ 10^{-3}\left(\frac{D_h}{D_{hm}}\right)^{-.50749}\left\{(405.22-9.9290\ x\ 10^{-2}p)\right.$$

$$(G/10^6)\left[-0.67757 + 6.8235\ x\ 10^{-4}p\right] - G/10\ x\ H_{fg}\right\}/$$

$$(G/10^6)\ 3.1240\ x\ 10^{-4}-8.3245\cdot G/10^6)$$

where $C = \left[0.15(1-x_{DNB})^{4.31}\right]/(G/10^6)^{0.478}in^{-1}$ for F

- Babcock & Wilcox [30] [31] - PWR, SAFDL = 1.30

$$\frac{q''_{CHF,Uni}}{10^6} = (1.155 - 0.407D_e)$$

$$x\left\{0.37\ x\ 10^8(0.591\ x\ 10^{-6}G)\left[0.83+0.685(p/10^3 -2)\right]\right\}$$

$$-0.1521G_{xCHF}H_{fg}\right\}/\left\{12.71\right.$$

$$x\ 3.054G/10^6)\left[0.712+0.2073(p/10^3-2)\right]\right\}$$

where $C = 1.25\ x\ 0.249(1-x_{CHF})^{7.82}/(G/10^6)^{0.457}$ for F

Under pressure blowdown conditions with a rod bundle geometry, it was reported[51,52] that some steady-state CHF correlations predict well the available blowdown test data. Under high-pressure PWR blowdown conditions, the early CHF data, occurring less than 1 s into the transient, were well predicted by the CISE[24] and Biasi[53] correlations, and the delayed CHF data occurring under slow blowdown conditions were well predicted by the Griffith–Zuber[54] correlation. For lower-pressure BWR blowdown conditions the modified Biasi and the CISE-GE correlations were found adequate.[52]

TABLE 7.3-3 FACTORS AFFECTING CHF[a]

Factors	CHF (DNB)	CHF (Dryout)
Higher local coolant quality[32-37]	Decrease	Decrease
Higher local mass velocity[36,37]	Increase	Decrease
Higher system pressure[36,37,38]	Increase	Decrease
Higher lateral crossflow[39,40]	Increase	Increase
Lower upstream heat flux[36,41,42]	–	Increase
Adjacent fuel rods bow to near contact[43,44]	Decrease	Decrease
Greater axial spacing between grids[45-48]	Increase	Increase
Larger cold wall area[49-50]	Decrease	Decrease

[a]General references 32–37.

Under rapid power excursion conditions the data showed CHF to be much greater than that for steady state at the same local conditions.[9]

Fuel Melting

The melting point of UO_2, which is by far the most popular fuel material for LWRs, ranges from 3133 K as fresh fuel to 3055 K at around 50,000 MW(d)/tonne burnup.[55] The volume increase of the molten fuel, by as much as 7–9.5% relative to solid fuel,[56,57] may cause excessive cladding strain in both axial and radial directions. In addition, the axial and radial relocation of the molten fuel may result in direct contact between the molten fuel and the cladding. This contact could conceivably cause melting of the cladding. However, extensive in-pile operation of molten fuel at the center of solid fuel pellets at Halden Boiling Water Reactor showed no cladding failure.[58] Although some fuel melting is believed not deleterious to reactor operation, the avoidance of melting has been generally accepted as a design criterion.

To prevent fuel melting, a conservative limit of linear heat generation rate (LHGR) in terms of kilowatts per meter is established. The limit is generally referred to as the specified acceptable fuel design limit. Typical values of the limit for various LWR fuel designs are 62.3–68.9 kW/m for GE BWR-6,[3] 59 kW/m for Westinghouse 17 × 17 fuel,[59] and 68.9 kW/m for C-E 16 × 16 fuel.[4] In general, either DNB or dryout will occur before the LHGR reaches the specified acceptable fuel design limit on fuel melting.

Hydrodynamic Flow Instability

Hydrodynamic flow instability is defined as flow oscillations in a heated boiling channel. This phenomenon is of interest to the nuclear reactor designer due to its relationship with premature boiling crisis. The critical heat flux (CHF) at which DNB or dryout develops under flow oscillation conditions has been reported to be much lower than that under steady-flow conditions.[60,61] A number of types of hydrodynamic instabilities have been either observed or postulated. These are Ledinegg instability,[62] flow regime (e.g., bubbly/annular) instability,[63] system instability, density wave oscillations,[64] and void reactivity feedback oscillations.[65] Only the density wave and the void reactivity feedback oscillations have been found to be of concern.[3]

Void reactivity feedback instability is caused by the coupling effect between voids in the core and the reactivity. A temporary power increase will increase the core void fraction. A larger core void fraction will shift the neutron spectrum away from the thermal energy region and result in a reactivity (power) decrease. The cycle is repeated. Typical natural frequency for this oscillation is about 0.4 Hz.[3] Density wave oscillations occur when the total pressure drop across a boiling channel with subcooled inlet flow is imposed externally by the feeding system. The oscillation frequency ranges from 0.1 to 0.8 Hz and is proportional to the product raised to the 0.25 power of heat flux and mass flux.[60] When a temporary decrease of inlet flow rate occurs in the channel, the length of the single-phase region decreases. Sustained oscillations are possible under certain conditions.

Under PWR operating conditions of high-pressure, low-void fraction, and an open lattice array in the core, no hydrodynamic instability is expected.[66,67] Under BWR operating conditions, however, both the density wave and void reactivity oscillation are of concern. For a typical BWR-6[3] the decay ratios X_2/X_0 for the density wave and void reactivity instabilities are 0.57 and 0.98, respectively, at the 50% power natural circulation condition. The decay ratio X_2/X_0 is conventionally defined as the ratio of the magnitudes of the second overshoot to the first overshoot of a continuous oscillation. If the decay ratio reaches 1.0, a sustained oscillation will occur.

The characteristics of density wave and void reactivity oscillations are summarized as follows[34]:

1. Decreased reactor system pressure or core-coolant mass flow decreases the stability.

2. Decreased channel heat flux increases stability.

3. Increased inlet subcooling up to a critical value reduces stability. Beyond that value, a further increase in subcooling can reverse the trend and stabilize the flow.

4. Decreased delayed neutron fraction, such as at the end-of-fuel cycle, reduces the void reactivity feedback stability.

Pellet Cladding Interaction

Damage to Zircaloy fuel cladding during power ramps has been reported since 1966 for both types of LWR reactors.[68,69] Some of these cladding failures are attributed to the effects of pellet-cladding interaction (PCI). The failure scenarios of PCI are related to the cladding strain and stress imposed by the differential thermal expansion between the cladding inner surface and the fuel pellet outer surface during power ramps. At the same time the embrittlement-inducing agents, such as iodine[70,71] or cadmium,[72] are released to the gap between the fuel pellet and the cladding. Cracks develop at or near the ridges where the fuel pellet and cladding interacts under the combined adverse effects of stress (or strain) and embrittlement-inducing chemical agents. This type of failure has been attributed to either fission-product-induced stress corrosion cracking (SCC) or liquid metal embrittlement (LME).[72] Experiments performed to date have indicated that SCC is sensitive to neutron fast fluence, strain rate, stress, strain, gap conductance, fuel temperature, accumulated burnup before ramp, and fuel pellet and cladding configuration.

Several experiments[73,74] suggest that there exists a critical threshold stress below which cladding failure from SCC is not likely to occur. The critical stress can be regarded as being the crack formation stress above which failure may ensue. A typical value for critical stress of Zircaloy in an iodine environment for small cracks (1.27×10^{-4} m displacement) is 280 MN/m^2.[74] Similarly, there exists a critical strain above which the strain energy absorption will exceed the failure limit. It is observed that the strain at failure for cold-worked Zr-2 is 2% in an iodine environment at 623 K and 3×10^{-7}/s strain rate and is 3% in a cadmium environment at 573 K and 8×10^{-6}/s strain rate.[75]

Factors Affecting PCI During Operation

Gap conductance is related to PCI in that the lower the value of gap conductance, either due to low fill gas pressure or large gap size after fuel densification, the higher the fuel pellet average temperature. A higher fuel pellet average temperature is more likely to cause the embrittlement agents, such as iodine and cadmium, to migrate to the pellet-cladding interaction region via the thermally activated migration process and result in stress corrosion cracking.

The local fuel temperature after power ramps largely controls the release rate of fission products, including iodine and cadmium, from the fuel pellet. The total fission product release increases as the fuel pellet average temperature increases. The amount of the fission gas, including corrosive species, generated in the fuel is proportional to accumulated fuel burnup. Fuel pellet and cladding configuration affects susceptibility to PCI during power ramps. Short fuel pellet length, thick cladding, dished and chamferred fuel pellet ends, longitudinal undulations on the surface of the fuel pellet,[76] CANLUB[77] (graphite-and-siloxane), and zirconium[78] coatings between fuel pellet and cladding tend to reduce the susceptibility of fuel cladding to PCI failure.

Due to the complexity of the PCI phenomenon, a complete understanding requires a level of knowledge much beyond what our current technology can provide. As an intermediate approach to reduce the propensity of PCI failure, limitations on the rate of power ramps are imposed by several reactor vendors. For example, a 5% rated power per hour ramp for KWU's PWRs[79] and a 0.197–0.328 KW/m per hour power ramp for local power higher than 26 kW/m for BWRs[80,81] are used as operating guidelines for unconditioned fuels to avoid PCI failure.

Two interim criteria to prevent PCI in LWR fuels have been provided by the U.S. Nuclear Regulatory Commission[8]: (1) the uniform strain (elastic and inelastic) of the cladding is not to exceed 1% and (2) fuel melting is to be avoided.

Hydriding, Collapse, and Fretting

Hydriding and collapse of the clad can cause fuel failure early in the fuel life. Hydriding of the Zircaloy cladding of LWRs, the formation of brittle ZrH_2 at the inner side of the cladding, is prevented by keeping the level of moisture and other hydrogenous impurities very low during fabrication. The acceptable level of moisture in fuel pellets is 20 ppm and the corresponding limit for hydrogen from all sources is 2 ppm.[82] A moisture level of 2 mg H_2O per cm^3 in the voids, such as in fission gas plena, has been shown to be insufficient for primary hydride formation.[83]

Fuel collapse occurs when a large axial gap between fuel pellets and a high differential pressure across the cladding coexist. Large axial gaps develop when fuel pellets densify by the sintering effect and the clad grows due to irradiation. A large differential pressure across the cladding develops in the early stage of fuel life before a large amount of fission gas has been released from the fuel pellets.

Fretting, or wear, occurs on the fuel rod surfaces in contact with the spacer grid. This wear is due to flow-induced vibration and may be aggravated by a reduction in the spring loads as a result of the effects of irradiation-enhanced stress relaxation. For the spacer grids and fuels of current design, no fuel failures have been attributed to either fuel collapse or clad fretting.[69]

Design Criteria for Maintenance of Functional Capability

The design criteria by the reactor designer[3,4,8,59] to ensure that the design functions of the fuel and reactor systems are maintained are discussed here. Violation of these criteria does not mean a failure of the fuel system or the reactor pressure boundary. These design criteria are summarized below.

1. For all Zircaloy fuel components the cumulative number of strain fatigue cycles should be less than the limit established in O'Donnel and Langer[84] with a safety factor of 20, and the stress intensity should be less than the limit specified in ref. 85.

2. The primary coolant pressure should be less than 110% of the design pressure.[86]

3. The gas pressure within the cladding of uranium and burnable poison rods should be less than the system pressure during normal operation and anticipated operational occurrences,[8] unless otherwise justified, to ensure that the cladding does not expand away from the fuel pellet.

4. The cladding inelastic circumferential strain averaged over the cladding thickness and circumference should be less than 1% over the cladding lifetime.[8]

5. The effects of cladding oxidation, cladding crudding, rod bow, and fuel pellet densification should be considered conservatively and quantitatively in the design of the fuel system.

7.3-4 Design Limits for Accidents

During an accident the reactor coolant pressure may drop much below the fuel rod gas pressure or the fuel pellet temperature may increase drastically. The former condition happens during a loss of coolant accident (LOCA) and the latter condition happens during a reactivity insertion accident (RIA) when a control rod is accidentally removed with high speed from the core. Under these conditions the fuel cladding may perforate due to high-strain-rate fracture. For severe cases the cladding will fragment and lose its design geometry because of a combination of the adverse effects resulting from cladding melting, pellet-cladding mechanical interaction, cladding embrittlement, violent expulsion of fuel, extreme coplanar fuel rod ballooning,[85] or thermal shock during sudden quenching. Loss of fuel rod design geometry could reduce the long- and short-term coolability of fuels and pose problems in returning the reactor to a safe and controllable condition after the accident. Because of these concerns, the reactor designer has established several design limits to prevent the loss of coolability of the fuel rods. In addition to these limits on fuel coolability, the accumulated radiation dose released during an accident has to be kept within acceptable limits.[5]

The conservative design limits that assure the coolability of the core, that is, the coolability criteria, during LOCA and RIA are listed as follows:

LOCA: Local cladding oxidation is less than 17% equivalent-cladding-reacted.

RIA: Maximum local cladding temperature is less than 1477 K. The radial average fuel enthalpy of fuel pellets is less than 1048 J/g.

The first design limit assures that cladding fragmentation does not occur during reflood quenching after a LOCA. Before reflood quenching the steam oxidizes the surface of the Zircaloy cladding to ZrO_2 and to oxygen-enriched and stabilized alpha-phase Zircaloy. Both ZrO_2 and alpha-phase Zircaloy are much more brittle than the original beta-phase Zircaloy.[87] Upon reflood quenching, the stresses produced by thermal shock of a large temperature reduction may fragment the cladding if the thickness of the brittle ZrO_2 and alpha-phase zirconium exceeds a certain limit. This limit is a function of the cladding temperature at which the isothermal oxidation process takes place.[87]

A 17% limit on equivalent-cladding-reacted (ECR), defined as the equivalent thickness of ZrO_2 accounting for the oxygen concentration in both transition phase and alpha-phase Zircaloy, and a 1477 K limit on maximum cladding temperature have been accepted by the designer as a conservative design criterion to prevent cladding fragmentation during reflood quenching. A recent experiment at ANL[87] on Zr-4 cladding under typical LWR LOCA heating and cooling situations has shown that the cladding can survive the thermal shock loading as long as the ECR is less than 28% of the total cladding thickness at the isothermal oxidation temperature of 1775 K. The ECR limit increases to about 40% when the oxidation temperature decreases to 1400 K. No fragmentation occurs when the oxidation temperature is approximately less than 1300 K.

For a severe reactivity insertion accident disintegration of the cladding and fuel pellet into large fragments on sudden quenching after the power burst has been observed[88] for fuel pellets experiencing radial average enthalpy above 1048 J/g. During a RIA event high fuel enthalpy during the power burst is accompanied by high cladding surface heat flux, which will in turn induce film boiling and high cladding temperature. Extensive oxidation at high cladding temperature during the power burst contributes to the final disintegration of fuel rods when the coolant recontacts the cladding at the end of the power burst transient.

The above-discussed criteria are set up to ensure core coolable geometry during LOCA and RIA. Other coolability criteria may exist for other types of accidents. Those additional criteria are not standardized among the various fuel suppliers and NSSS vendors.

Another design limit is radiation dose. In order to quantify the radiation dose during an accident, cladding failure criteria during an accident are established[8] as follows:

1. Departure from nucleate boiling or dryout (cladding overheating)

2. UO_2 fuel pellet radial average enthalpy exceeding 587 J/g[88] (pellet-cladding mechanical interaction)

3. Cladding temperature exceeding the rupture temperature limit[89] (differential-pressure-induced rupture)

7.3-5 Introduction to Thermal Margin Design

Three of the major thermal-hydraulic and fuel design objectives are to provide adequate safety margins, adequate operating thermal margins, and to optimize the core steady-state thermal-hydraulic conditions. This section deals with the thermal margin design of the nuclear reactor core. First, the relationship between safety margin, adequate operating margin, and the steady-state core operating conditions is briefly described through the thermal margin diagram. Next the factors affecting the magnitude of these parameters are discussed. Finally, optimization of the steady-state thermal-hydraulic operating conditions is discussed.

The safety thermal margin is defined as the margin between the condition at which the reactor reaches the acceptable fuel design limits for the coolability criteria and the condition at which the cladding fails or the core loses its coolable geometry. The current generation of nuclear reactors is so designed that under any condition of expected transients and postulated accidents, the fuel design limits and the coolability criteria are not violated, respectively.

The operating margin provides plant operation flexibility. Tow types of constraints limit the plant operation flexibility. One type is the limiting conditions for operation (LCOs). The other type of constraint is the limiting safety system settings (LSSSs) of the reactor protection system. LCOs are monitored on-line routinely and are not to be exceeded during normal operation. Reactor operation within the LCOs assures that the design limits are not violated for any AOO. The LSSSs are the settings related to those variables monitored by the reactor protection system. When the settings are exceeded, an automatic reactor trip will occur.

7.3-6 Thermal Margin Diagram

The relationship between safety margin, operating margin, and steady-state core operation conditions can be best understood by the thermal margin diagram, as shown in Fig. 7.3-1. Similar diagrams have been used for thermal margin discussions.[90,91] In Fig. 7.3-1 two types of thermal margin diagrams are

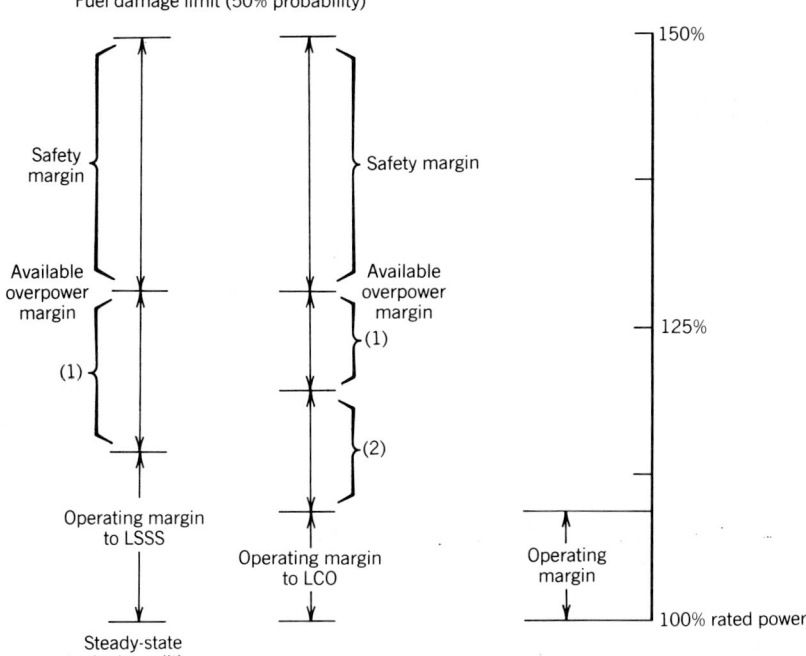

Fig. 7.3-1 Thermal margin diagram. (1) On-line system and measurement uncertainties; (2) margin set aside to accommodate the limiting anticipated operational occurrence or accident without violating acceptance criteria.

illustrated. The diagram on the right-hand side is called the LCO thermal margin diagram. The other diagram is the LSSS thermal margin diagram. The vertical scale of these two margin diagrams is in terms of percentage of rated power. The percentage of rated power is also called overpower and has been accepted by the majority of reactor designers as a convenient means to measure the magnitude of various types of margin. The magnitude of the thermal margin to LCO, LSSS, SAFDL, and fuel damage condition can be determined by continuously increasing the core power, while keeping the core inlet thermal-hydraulic conditions constant, until the LCO, LSSS, SAFDL, and fuel damage limit are respectively reached. The overpower at which the SAFDL is reached is called the available overpower margin. The overpower at which the LCO is reached is called the operating margin to the LCO. That at which the LSSS is reached is the operating margin to the LSSS. The operating margin of the core is the more limiting of the LCO and LSSS operating margins.

7.3-7 Available Overpower Margin

The magnitude of available overpower margin is important in optimization of the core operating margin and core steady-state operating conditions. A greater available overpower margin can result in steady-state operating conditions with greater plant electricity output for a given operating margin.

The available overpower margin is the overpower at which a SAFDL is violated. The available overpower margin to the SAFDL on fuel melting is generally much greater than that for the SAFDL on CHF and therefore has less impact on determining the core operating margin. Because of this, only the available overpower margin to the CHF SAFDL is discussed here.

The available overpower margin to the CHF SAFDL is computed by continuously increasing the core power, while holding the steady-state operating conditions constant, until the CHF SAFDL is reached. The factors affecting the available overpower margin to the CHF SAFDL,[91] in approximate order of relative importance, are the core-average fuel surface heat flux, core maximum integrated radial peaking factor, core-average inlet mass flux, core inlet temperature, ratio of rod diameter to subchannel flow area, normalized axial power shape, system pressure, and spacer grid design. A lower limiting value of SAFDL on CHF can also result in a greater available overpower margin.

The available overpower margin can be increased by 1% if the surface heat flux is decreased by about 1%, if the radial peaking factor is decreased by about 1%, if the core mass flux is increased by about 1%, if the core inlet temperature is decreased by about 1.0 K, if the ratio of the rod diameter to subchannel flow area is decreased by about 1%, if the axial shape index, defined as the normalized bottom-half core power minus the top-half core power, shifts to a more positive value (more bottom peaked) up to a threshold value by about 0.02, or if the system pressure is increased by about 0.14 MPa. These sensitivity parameters are derived in Lepine and Trapp[91] based on the W-3 CHF correlation at PWR nominal operating conditions. For different CHF correlations and operating conditions the values of these sensitivity parameters will deviate from the values reported here.

7.3-8 Limiting Anticipated Operational Occurrence

PWR

The limiting anticipated operational occurrence with the greatest thermal margin degradation for PWRs is in general the loss of flow event. The event sequence starts with the loss of electric power to all reactor coolant pumps and all coolant pumps instantly coast down. The core coolant flow continues to decrease to the final natural circulation flow rate. Figure 7.3-2 illustrates the characteristics of thermal margin degradation as a function of time during the initial seconds of a loss of flow event. The operating thermal margin to the LSSS trip set point in terms of overpower at the initiation of the loss of flow event in this figure is assumed to be 103%. In other words, the overpower to the LSSS set point is 103% of rated power. The thermal margin continues to degrade after the event initiation. When the operating thermal margin reaches 100%, the trip set point is exceeded. The time delay between the time of event initiation and the time when the set point conditions are reached is called trip activation delay.

At the time that the set point is reached, the reactor protection system (RPS) is activated. The RPS then generates a signal to open up the control rod trip breaker and cut off the electricity to the control rod magnetic holding coil. The lapse between the time that the set point is exceeded and the time that the trip breaker is open is called the RPS delay. After the breaker is open, it takes approximately 0.5 s[91] for the decay of the magnetic flux before the control assemblies drop into the core. This delay is called the holding coil delay. After about 1 s from the time that the control assemblies start to insert into the core, core power starts to decrease.

The total thermal margin degradation during the event, that is, the required overpower margin, is proportional to the time to reach minimum CHFR and the rate of flow coastdown. The time to reach minimum CHFR consists of trip activation, RPS, holding coil, and CEA worth delays. The rate of flow coastdown for the first few seconds of the event is proportional to the ratio of total pump torque to pump inertia (T/I), that is, $T/I = dM/dt$, as described by Newton's second law applied to pumps.

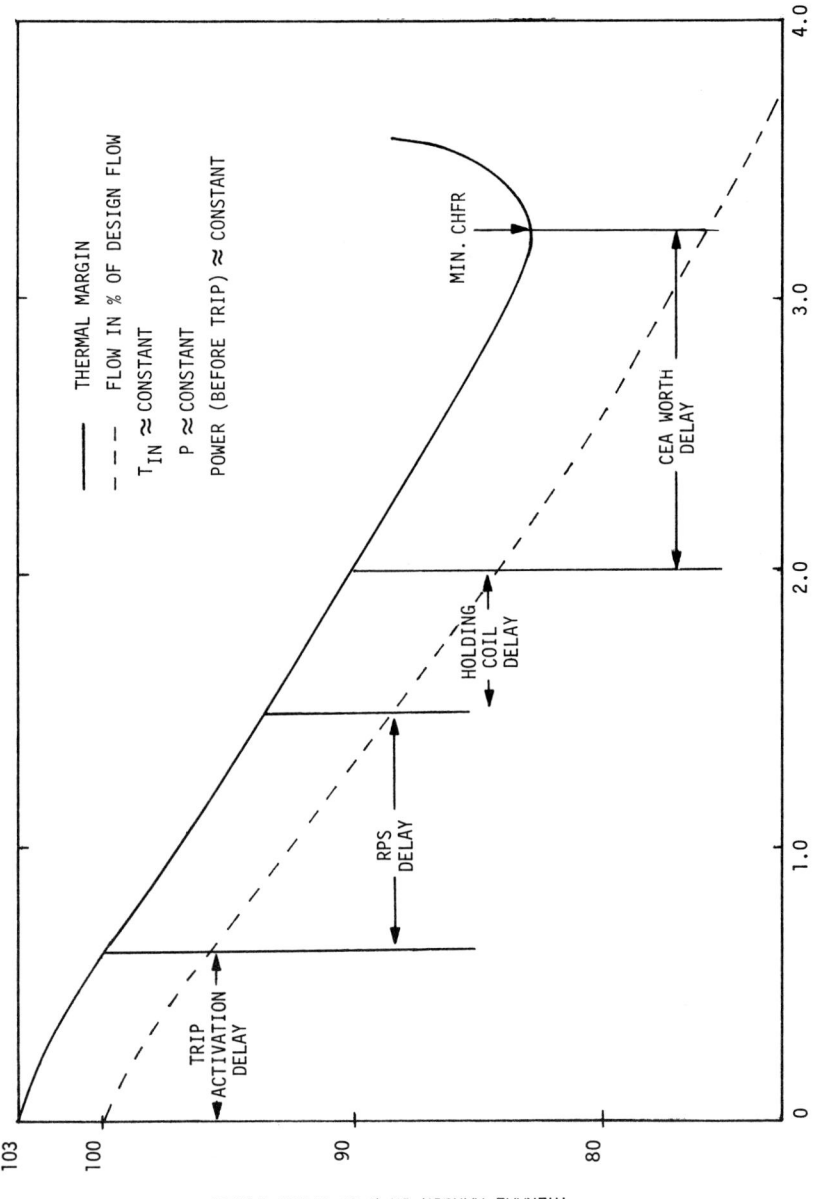

Fig. 7.3-2 Thermal margin degradation versus time for a typical PWR loss of flow event.

For a plant that has a low T/I ratio pump, high low-flow-trip setpoint, fast-response RPS, fast-decay magnetic holding coil, high control rod worth, and high control rod drop speed, the margin degradation during a loss of flow event will be small.

BWR

The limiting anticipated operational occurrence for BWRs is in general the loss of feedwater heater event. The loss of flow event for BWRs is not limiting because the void generated by low core flow automatically shuts down the reactor much sooner than the CEA insertion in PWRs. The loss of feedwater heater event results either from the closure of the steam extraction line to feedwater heaters or from steam bypass around the heaters. The margin degradation in terms of overpower for a typical 3800-MW$_{th}$ BWR-6[3] is 14%. The analyzed limiting event starts with highly subcooled coolant, approximately 55 K less than that for nominal operation, flowing into the core. The subcooled coolant collapses voids in the core and results in a positive reactivity insertion via negative void reactivity coefficient. The core power continuously increases until it reaches the set point of the high thermal power trip, typically at 120% of rated power and about 40 s after the event initiation.

Figure 7.3-3 illustrates the characteristics of the loss of feedwater heater event. Δt_1 is the sum of the RPS delay, control rod activation delay, and CEA worth delay. Δt_2 is the time delay for the surface heat flux to follow the neutron flux. To minimize the thermal margin degradation in this event, several approaches exist. One is to lower the neutron flux set point such that the conditions exceeding the set point will occur sooner. Caution has to be exercised for this approach because a low set point may cause unnecessary reactor trips during plant maneuvering, load following, or minor transients that plants are designed to accommodate. A second approach is the reduction of the RPS delay time, CEA worth delay time, and control rod activation delay time.

7.3-9   Loss of Coolant Accident

The loss of coolant accident in LWRs is the hypothetical design basis accident and is the most limiting accident for which the margin is set aside to avoid exceeding the coolability criteria. It is highly unlikely to happen but its consequences are most severe. The emergency core cooling system (ECCS) is designed to mitigate the consequences of the accident to an acceptable level, that is, the core coolable geometry is maintained. We will first discuss the characteristics of the LOCA accident for PWRs and BWRs, respectively. Then the factors affecting thermal margin degradation during a LOCA will be discussed.

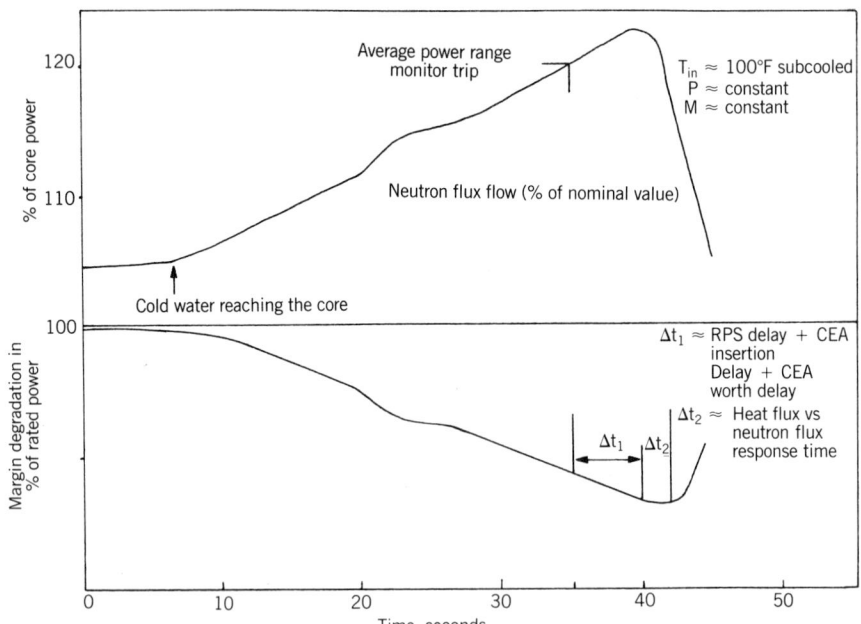

Fig. 7.3-3   Thermal margin degradation versus time for a typical BWR loss of feedwater heater event.

Characteristics of LOCA

The characteristics of a LOCA accident for a given plant depend strongly on the features of the ECCS and location of the break. The PWR ECCS generally consists of three subsystems: the high-pressure safety injection (HPSI), the low-pressure safety injection (LPSI), and the accummulators. The BWR ECCS consists of four subsystems: HPSI, LPSI, the high- and low-pressure core spray systems (HPCS and LPCS), and the automatic depressurization system (ADS).

The consequences of a LOCA, conventionally expressed in terms of peak cladding temperature, depend on break location. For PWRs the LOCA resulting from a double-ended, guillotine-type, cold-leg break located between the coolant pump and the pressure vessel is most limiting.[92] A similar break at the hot leg will result in a peak cladding temperature much lower than that during a cold-leg LOCA.[93] This low temperature results from a higher mass flow through the core during a hot-leg break LOCA[93] than during a cold-leg break. For BWRs, a double-ended break in the pump suction line of one of the two recirculation loops represents the most limiting case. A similar break of the steam line will result in a lower peak cladding temperature. BWR steam line break is usually followed by steam break flow rather than by the liquid–vapor mixture break flow after the recirculation line break. The steam break flow tends to maintain a high residual coolant inventory throughout the transient. As a result, the maximum cladding temperature during the steam line break accident is lower than that during the recirculation line break accident.

The characteristics of typical PWR and BWR limiting LOCAs will be discussed next. They are calculated with the conservative licensing models required by the Code of Federal Regulations, Title 10, Section 50, Appendix K.[94]

PWR Limiting LOCA Characteristics

Figure 7.3-4 illustrates the accident characteristics of a typical PWR limiting LOCA. These characteristics can be divided into four sequential phases: blowdown, refill, reflood, and long-term cooling. The detailed characteristics shown in Fig. 7.3-4 will change from plant to plant, depending on the plant-specific characteristics of the ECCS and plant layout.

The *blowdown*, or pressure-release phase, initiates immediately following the double-ended, cold-leg break. Primary system water in liquid form is expelled through the break into the containment. The system pressure drops to the saturation pressure of the local coolant temperature in less than 0.1 s. Below the saturation pressure local boiling and flashing take place and the reactor goes subcritical via the negative moderator reactivity feedback. The blowdown flow turns into a water–vapor mixture. The depressurization rate is reduced when core pressure falls below the saturation pressure after about 0.1 s, as can be seen in Fig. 7.3-4. At this time core stagnation flow, followed by a reverse flow (downward direction), develops because of a fast depressurization of the reactor lower plenum. The core flow reverses again to the upward direction after about 0.5–1.0 s following the accident because of a positive pump rotation inertia in the intact loops. DNB occurs at a high-void fraction[95] about 0.1–0.8 s into the LOCA transient.

Once DNB is reached, the rate of energy removal from the cladding is greatly reduced. Cladding temperature steadily increases at a rate up to 400 K/s after DNB due to redistribution of stored energy, that is, equalization of temperature of fuel pellet and cladding. The cladding temperature reaches a peak at about 1000 K when the cladding and fuel pellet temperatures are approximately equal. Around this time the water–vapor mixture flowing through the core is able to carry away more energy than that generated by the decay heat, the system pressure drops below the HPSI set point, and HPSI is activated. However, the HPSI coolant injected into the cold leg is entrained by the escaping steam. The water level continues to decrease until a large amount of water from the accummulators reaches the lower plenum. The accummulators are pressurized tanks filled with borated water. They release the contained water through check valves when the primary system pressure has dropped below the nitrogen pressure within the accummulators. After the actuation of the accummulator injection there is a 20–25-s delay before the injected water reaches the lower plenum. The delay is primarily because of the combined effects of the steam water countercurrent flow in the downcomer and the downcomer bypass. The countercurrent escaping steam prevents the accummulator water from reaching the lower plenum. The downcomer bypass means that the accummulator water injected into the intact loop passes through the upper downcomer region around the core and escapes from the break. As the system pressure decreases to the containment pressure, the break flow is reduced and the accummulator water begins to fill the lower plenum.

The *refill phase* starts when the accummulator water begins to fill the lower plenum. At this time the core is uncovered by water. The fuel rods are cooled primarily by thermal radiation and by steam natural convection. The temperature rises at a rate of 8–12 K/s due to an imbalance between energy generated by the decay heat and energy removed by radiation and steam natural convection. Typical temperature rise during the refill phase is about 50–250 K, depending on the core and ECCS designs and the thermal-hydraulic models used in evaluation.

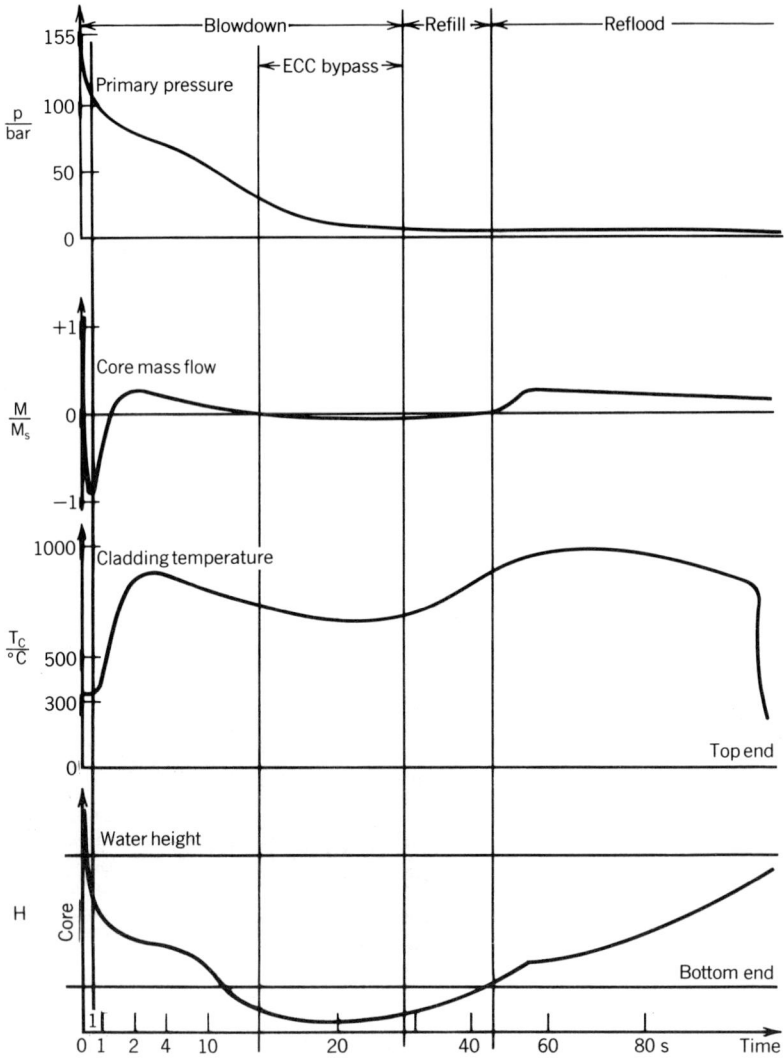

Fig. 7.3-4   Generalized loss of coolant behavior for large pipe breaks in a PWR.

The *reflood phase* begins when the water level reaches the bottom of the core. The water boils off and increases the upward steam flow. This steam flow entrains a substantial amount of water in the form of droplets, which provides an effective cooling mechanism when they deposit on the rod surface. The amount of steam flow increases as the water level rises. Thus, the steam cooling becomes increasingly effective. Finally, the steam cooling is capable of removing the total energy generated by the decay heat. The cladding temperature reaches its peak and starts to decrease.

The reflooding rate, the rate at which the water front rises, is largely determined by a pressure-driving force. This driving force is the difference in hydraulic static pressure between the downcomer water and the core water. More specifically, the driving force is equal to the pressure difference caused by the water level difference between the downcomer and the core minus the pressure drop of the steam flow from the core exit, through the steam generator, to the cold-leg break location. Therefore, a larger steam flow pressure drop will reduce the reflooding rate and result in a higher peak cladding temperature. This adverse effect on reflooding rate is generally referred to as the steam binding effect. The steam binding effect is more pronounced when the entrained water droplets evaporate by the stored energy released from the system structure or when the escaping steam has low density under low containment pressure conditions. The evaporation will increase the steam flow and result in a high pressure drop. The flow oscillation between the downcomer and the core, like the oscillation in a manometer tube, will also reduce the reflooding rate.[93] The oscillation, however, will increase the heat transfer coefficient in the lower part of the rod by as much as 20–30%. The total temperature rise during this phase is from 400–600 K. Typical reflood velocity ranges 2 to 5 cm/s.

The above reflood characteristics are associated with PWRs without vent valves on the core barrel, through which the core pressure is reduced during the accident by steam venting through the vent valve to the break location. With vent valves[96] the peak cladding temperature occurs at an early stage of the reflood phase, with magnitude much less than that without vent valves, everything else being equal.

The *long-term cooling phase* starts after the core has quenched to below about 900 K, at which point the Zircaloy–water reaction is suppressed, or the water level covers the active fuel. During this phase the water inventory is controlled by the HPSI pumps. The continuous operation of the HPSI is to ensure the long-term dissipation of the decay heat. One month after the event there is still about 0.13% of rated thermal power that has to be removed from the core.

BWR Limiting LOCA Characteristics

Figures 7.3-5 and 7.3-6 illustrate the accident characteristics of a typical BWR limiting LOCA. The course of this accident can be divided into four phases: coastdown phase, window phase, lower-plenum flashing phase, and the post lower-plenum flashing phase.

The *coastdown phase* starts when a double-ended break of a recirculation line develops. The power supplied to the recirculation pumps and steam to the feedwater pump turbine are assumed to fail at the event initiation. The recirculation pumps of the intact loop provide a decreasing flow to the core based on pump coastdown. At this time a strong natural circulation flow develops and causes a slight increase in the upward core flow. The break discharge fluid during this phase is in liquid form. Depressurization is relatively slow in this liquid-discharge phase.

The *window phase* begins when the liquid level has dropped below the jet pump inlet at approximately 7–8 s following the accident. The jet pump cavitates and the core flow drops to nearly zero. A short period of flow stagnation in the core occurs before more water escapes from the pressure vessel. At this time dryout occurs and the heat removal capability of the core flow is greatly reduced. The cladding temperature starts rising at a rate of up to 120 K/s due to the redistribution of stored energy between the fuel pellet and the cladding. As the water level in the downcomer continues to drop, the break at the recirculation line becomes uncovered by water. The break flow turns into steam and results in fast depressurization.

The *lower-plenum flashing phase* is characterized by a strong steam flow through the core caused by flashing of the water inventory in the lower plenum. At this time the cladding temperature decreases slightly. The film boiling heat transfer mechanism during this phase is responsible for removing the stored and decay energy from the fuel rods. When the lower plenum inventory depletes, the steam–water flow into the core diminishes. At about 50 s the depressurization process and lower-plenum flashing phase come to an end. Then the core heats up for a short period of time. No effective steam cooling takes place until the ECCS comes into action.

The *post lower-plenum flashing phase* starts around 50 s after the accident. At this time the spray flow rate reaches its rated value. The spray water flows down to the lower plenum and starts to refill and reflood at a typical rate of 7–10 cm/s. The temperature rises steadily at a rate of about 5–8 K/s until the steam cooling generated by the reflood water is able to remove the decay heat of the fuel rods. At this time the peak cladding temperature is reached. In general, steam binding is less of a problem for BWRs because of their relatively low loop hydraulic resistance.

The ECCS high-pressure core spray begins to deliver flow to the vessel within 30 s after the accident. A delay in the core reflooding process may occur in BWRs due to the effect of countercurrent steam and water flow. This countercurrent flow limited (CCFL)[97] condition is characterized by the phenomenon that the spray water injected into the top of the core is prevented from running down by the upward steam flow. The delay ends when the water upheld in the upper plenum finally penetrates into some of the low-power bundles and reaches the lower plenum.

Factors Affecting Limiting LOCA Thermal Margin Degradation

To determine the operating margin to the LCO on LOCA coolability criteria, the limit on the local LHGR beyond which a LOCA will result in a violation of the coolability criteria is first determined. Then the maximum LHGR expected to occur during the fuel cycle of interest is evaluated. After accounting for the monitoring system uncertainty, the ratio of the LHGR LOCA limit to the maximum LHGR expected during the fuel cycle of interest is the minimum expected operating margin to the LOCA LCO.

The LHGR LOCA limit reflects the amount of margin degradation during the limiting LOCA. Another indicator for margin degradation is the peak cladding temperature (PCT). For a given initial local LHGR the greater the margin degradation, the higher the peak cladding temperature. Approxi-

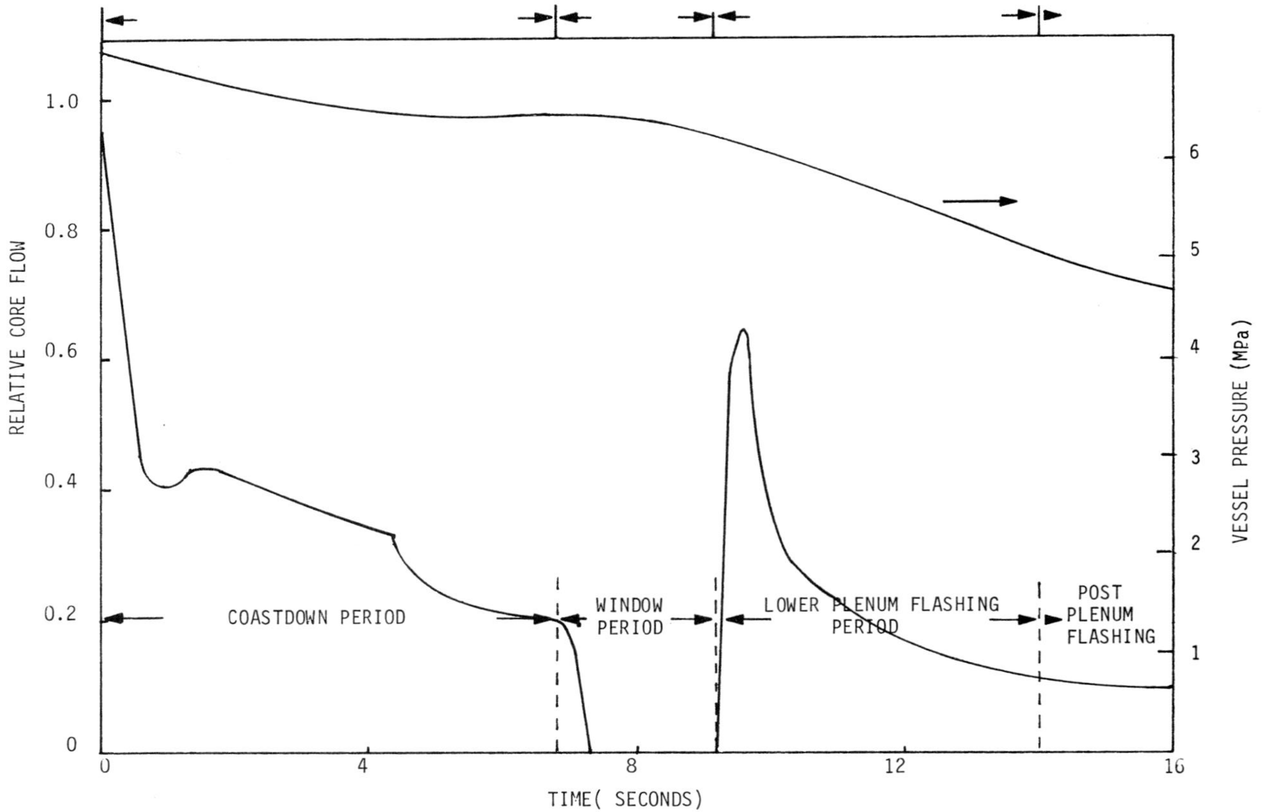

Fig. 7.3-5   Core flow and vessel pressure following a BWR limiting loca.

882

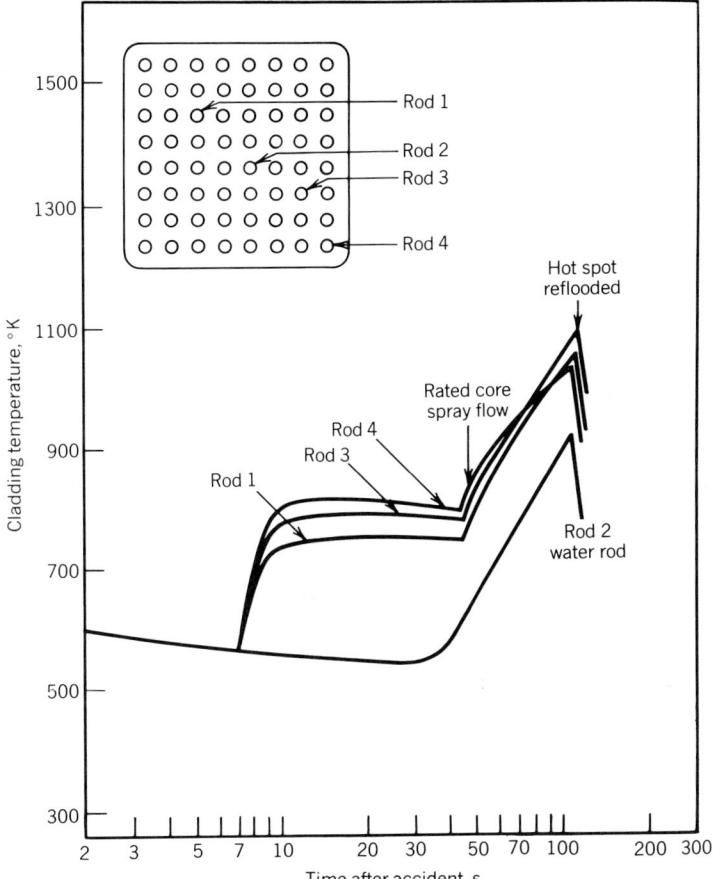

Fig. 7.3-6   Cladding temperature versus time for a BWR limiting loca.

mately every degree Kelvin increase of the peak cladding temperature, at a given initial local LHGR close to the LHGR LOCA limit, will decrease the LHGR LOCA limit by about 0.05 kW/m for PWRs without vent values. This direct conversion exists between the LHGR LOCA limit and the peak cladding temperature because the margin to the 1498 K temperature limit during the accident is more limiting than the margins to other coolability criteria.

The LHGR LOCA limit, ranging from 41 kW/m for BWR-6 to 59.4 kW/m[98] for some PWRs, is often calculated with the conservative thermal-hydraulic models.[94,99] The magnitude of the conservatism is hard to quantify because the best-estimate thermal hydraulic models are still under development. Nevertheless, replacing some of the conservative assumptions and models with more reasonable ones reduces the peak cladding temperature from 1498 to 866 K for a 15 × 15 fuel assembly PWR.[100]

As described in the previous section, the cladding temperature during the refill/reflood phase is usually much greater than the temperature peak resulting from stored energy redistribution. Reduction of the peak temperature caused by stored energy redistribution results in only a minor reduction of the maximum peak cladding temperature during reflood. This is because a higher temperature peak caused by stored energy redistribution will result in a higher heat transfer due to a greater temperature differential and a higher heat transfer coefficient during reflood.[101]

The temperature peak caused by stored energy redistribution is approximately proportional to the amount of the stored energy at the time of CHF. Before CHF is reached, part of the stored energy is removed by the nucleate boiling heat transfer process. The amount of stored energy at the time of CHF is strongly dependent on the initial stored energy during steady-state operation and the time to CHF after the accident. A high initial stored energy can result from a low gap conductance, a low conductivity of fuel material, and a high linear heat generation rate. The time to CHF decreases as the linear heat generation rate increases.[102] A calculation performed in Smith and Griffith[103] shows that a

0.1-s reduction of time to CHF can increase the peak temperature caused by stored energy redistribution by as much as 110 K.

Factors that significantly change the peak cladding temperature are those that affect the refill/reflood heat transfer process. This is because the temperature rise during the refill/reflood phase constitutes a significant portion of the total temperature rise. This temperature rise can amount to as much as 450–850 K for both PWRs and BWRs. From past experimental investigation some key factors that can reduce the PCT during refill and reflood phases are listed as follows:

1. High reflooding and refilling velocity.[101,104–109]
2. High containment pressure.[101,108,110]
3. Combined cold-leg and hot-leg injection versus cold-leg injection.[109,111,112]
4. Low fuel rod power density during steady-state operation and low fission product decay heat.[105,108]
5. High initial rod surface temperature at the time of reflood.[107,108,113,114]
6. High degree of subcooling of ECCS water.[115]

A high reflood velocity provides a high heat transfer coefficient via large amounts of entrained water droplets. It has been found[92] that the beneficial effect on PCT diminishes as the reflood velocity increases beyond 10 cm/s. When reflood velocity is below 2 cm/s, the effectiveness of the reflooding cooling process is drastically reduced. A higher containment pressure will equalize the reactor and containment pressure and start the refill process sooner. In addition, the steam density and the amount of entrained water droplets are greater with high containment pressure, resulting in a higher heat transfer coefficient.[101] Experimental data has shown that the total temperature rise during the reflood phase decreases from 100 to 60 K as the pressure increases from 1 to 6 bars. A similar trend has also been observed in the BWR-FLECHT test, but the effects on temperature rise are less pronounced.

For U.S.-built PWRs the low-pressure ECCS water from accummulators is injected into the cold legs. However, the accummulator water is injected into both cold legs and hot legs for PWRs manufactured in Germany. The cold water injected into the hot legs flows into the heated assemblies from the top. This cooling process will reduce the PCT as well as the quench time. Experimental data from the PKL tests[109] show that the PCT is reduced by about 80 K when combined injection is used instead of cold-leg injection alone.

An increase in fuel rod power density increases the temperature rise during refill and reflood. The temperature rise during refill is governed by the specific power of decay heat generated in the fuel rod, the energy removed by radiation heat transfer, and the time span of the refill phase. During reflooding a high fuel rod power density will result in a high steam flow and a slow-moving quench front. As a result, the reflood temperature rise will increase slightly. A high initial rod temperature during reflooding decreases slightly the reflood temperature rise but increases the quench time.[101,108] The main reason for the peak cladding temperature decrease with a high initial cladding temperature is that a large amount of steam can be produced at the lower part of the fuel rods, thus increasing the effectiveness of cooling the hot spot.

A high degree of ECCS subcooling will cause an early drop in heat transfer during reflooding but improve cooling effectiveness later. These effects slightly reduce the quench time. Experimental data[113] show that the quench time increases from about 400 to 700 s as the ECCS subcooling decreases from 115 to 3 K.

The experimental evidence described above gives the reactor designer some guidance in minimizing the thermal margin degradation of a limiting LOCA. As discussed earlier, the impact on PCT on refilling/reflooding velocity, containment pressure after the accident, ECCS design, and fuel rod power density are much more significant than other effects. Therefore, any design that can increase refilling/reflooding velocity, improve ECCS effectiveness, and reduce fuel rod specific power is favorably considered by the reactor designer. In general, a high containment pressure is not desirable from the standpoint of containment structural integrity and cost.

7.3-10   Protection and Monitoring System Uncertainties

The uncertainties shown in the thermal margin diagram (Fig. 7.3-1), account for statistical variations associated with various parameters that affect the on-line determination of the LCO or LSSS. With these uncertainties incorporated, the LCO and LSSS can be conservatively determined by the monitoring and protection systems, respectively. Combination of all the component uncertainties into the overall uncertainties for LCO and LSSS calculations can be achieved by a deterministic analysis, a statistical analysis, or a combination of the two types. A deterministic analysis combines the component uncertainties assuming that the worst variations occur simultaneously. Use of the deterministic analysis can lead to an overly conservative LCO and LSSS. As a result, the operating margin will be reduced. To combine these component uncertainties realistically, the probability distribution of each uncertainty must be taken into account and combined statistically. As the more detailed information needed has become available, the statistical uncertainty analysis method has been increasingly accepted in the nuclear industry. In view of this trend, the overall uncertainties discussed are assumed to be derived by statistical means.

First, a brief, generic description of the statistical uncertainty method is given. Second, the components of the overall uncertainties are discussed. Finally, the factors that impact the magnitude of the overall uncertainties are identified and discussed.

One-Sided Tolerance Limit

It is generally accepted by the nuclear reactor designer that the overall uncertainties used in the determination of the LCO and LSSS should provide 95% probability, at the 95% confidence level, that SAFDLs will not be violated. Figure 7.3-7a illustrates the probability density function (p.d.f) of the overall uncertainty. The 95% limit shown in this diagram bounds 95% of the population. In this diagram the probability density function is assumed known. In reality, a finite sample size is usually used to estimate the probability density function. Figure 7.3-7b shows the probability density function in histogram form with limited samples randomly selected from the true density function. To infer the 95% limit from this limited sample size, one can construct a one-sided tolerance limit that will bound 95% of the population at a specified confidence level. A 95% confidence level is generally regarded as an appropriate level of conservatism. This confidence level means that 95 out of 100 times the one-sided tolerance limit derived from a finite sample size will bound a given portion of the assumed

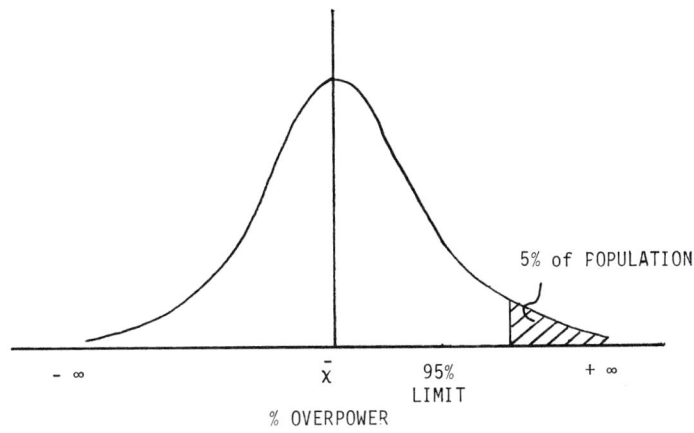

PROBABILITY DENSITY FUNCTION

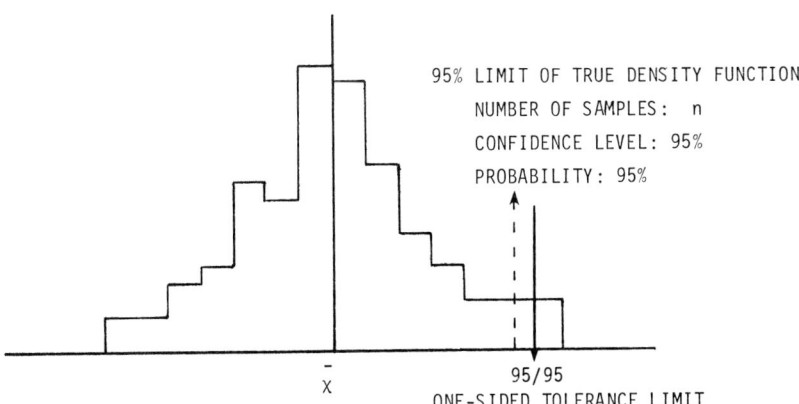

Fig. 7.3-7 (a) Probability density function; (b) observed probability density function.

population. If the probability density function is normally distributed, the one-sided tolerance limit of 95% probability at a 95% confidence level can be evaluated by the following formula[116]:

$$T_{95/95} \ (95/95 \text{ one-sided tolerance limit}) = X_m + kS$$

where S^2 = unbiased estimate of the variance for the overall uncertainty p.d.f.
  x_m = mean of p.d.f.
  k = a function of sample size, confidence level, and probability limit.

For more detailed information about the value of the k factor as a function of confidence level, probability limit, and sample size, Owen[116] and Hahn and Shapiro[116] can be consulted.

The conservative operating margin to the LSSS or LCO is determined by dividing the operating margin calculated without considering the uncertainties by this 95/95 one-sided tolerance limit.

Statistical Uncertainty Analysis

The overall uncertainty is affected by uncertainties of three types: system parameter uncertainties, state parameter uncertainties, and modeling uncertainties. The system parameters describe the physical system and its boundary conditions under consideration and are not monitored on-line during reactor operation. System parameter uncertainties account for the variation in fuel rod diameter, fuel pellet diameter, subchannel area, and so on. State parameters are those parameters monitored or measured on-line. The modeling uncertainties account for the differences between the on-line algorithms and the more detailed best-estimate codes, empirical correlation uncertainties, and the physics model uncertainties in the best-estimate codes. The on-line LCO and LSSS algorithms or correlations are simplified versions of the best-estimate codes, which are usually used off-line in the detailed thermal margin calculation with a much longer computer running time.

The purpose of the statistical uncertainty analysis is to determine the p.d.f. and the 95/95 one-sided tolerance limit for the overall uncertainty. A simplified flowchart of a generic CHF LCO and LSSS uncertainty analysis is illustrated in. Fig. 7.3-8. A similar flowchart can also be drawn for the determination of the LHGR LCO and LSSS uncertainties, but with a different set of component uncertainties. The statistical uncertainty method requires knowledge of the p.d.f. for each component uncertainty. The p.d.f. of each component uncertainty is stochastically sampled in the overall simulation process. This process is repeated many times to build up the p.d.f. of the overall uncertainty.

As shown in Fig. 7.3-8, the statistical uncertainty analysis starts with a random selection of a set of state and system parameters. The values of the state parameter should be within the reactor operating space over which the LCO or LSSS are expected to be valid. Once the values of the system parameters and state parameters are selected, the analysis proceeds in three paths. One path is to evaluate the LCO and LSSS overpower (X_i) which corresponds to the selected value of state and system parameters without any variations. The core simulator used in the LCO and LSSS overpower analysis is a detailed best-estimate computer code that simulates the detailed thermal-hydraulic and nuclear characteristics in the core. Another path is to calculate the LCO or LSSS overpower (Y_i) considering the variations of system parameters. The SAFDL on DNB in this case is determined by stochastic sampling of the p.d.f. for the independent variable CHFR (actual)/CHFR (predicted). The third path is aimed at the LCO or LSSS overpower (Z_i) considering the measurement uncertainties of the state parameters and the uncertainty for the difference between the simplified on-line algorithm and the detailed, best-estimate core simulator.

Components of Overall Uncertainty

Typical component uncertainties involved in the determination of CHF LCO and LSSS operating margins are summarized in Table 7.3-4 which lists the uncertainties of system parameters, the modeling uncertainties, and the uncertainties of state parameters. Typical values of the uncertainties are obtained from refs. 118 to 124 covering the PWRs and BWRs manufactured in the United States. Unless otherwise noted, the unit of the typical values is in percentage of rated power.

Factors Affecting Overall Uncertainty

As can be seen in the thermal margin diagram, the smaller the uncertainty, the greater the operating margin. Many methods exist to reduce the uncertainty, and some are more economical that others. The reactor designer often does a cost–benefit study before choosing the final methods to reduce the uncertainties. A summary of the methods that can reduce the uncertainties follows.

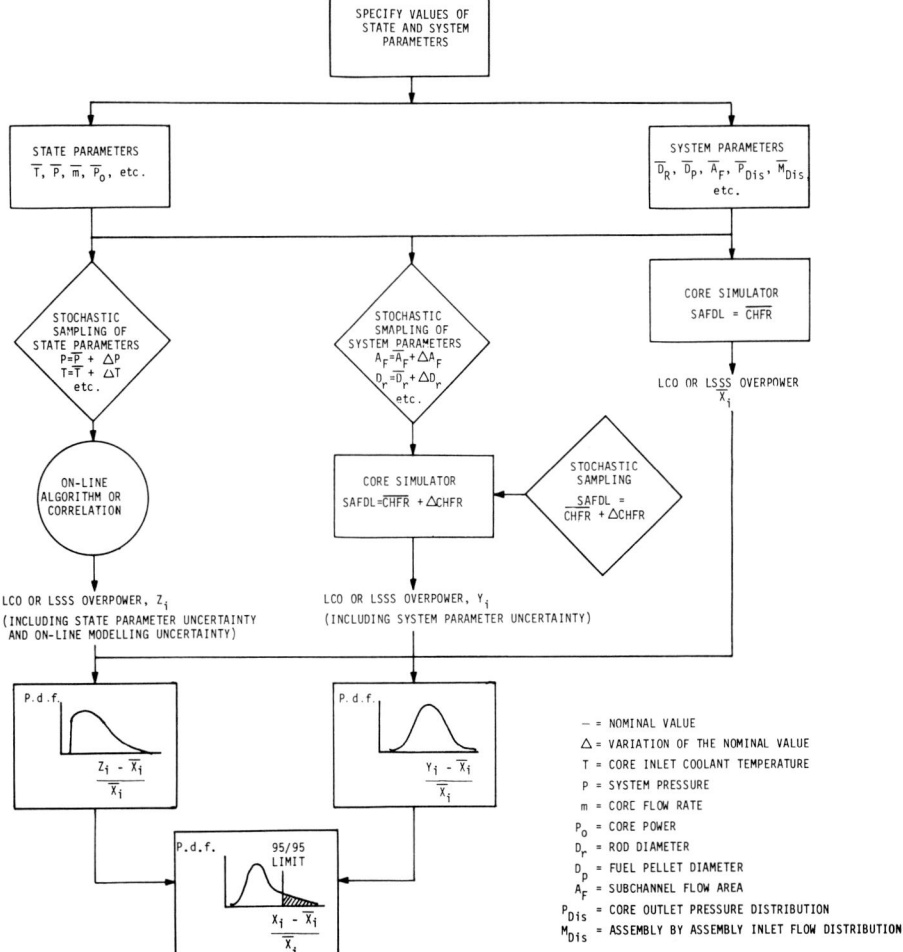

Fig. 7.3-8 DNB uncertainty analysis using stochastic simulation methodology.

1. Development of a CHF correlation that, in conjunction with a core simulator, can model the experimental CHF behavior more accurately, that is, reduce the uncertainty of the CHF correlation.

2. Development of on-line algorithms that model the core simulator more accurately.

3. Development of instruments or measurement methodology with greater accuracy.

4. Development of data reduction methodology with greater accuracy to obtain the "measured" radial peaking factor from in-core neutron detector signals.

5. Reduction of system parameter uncertainties through improved quality control.

7.3-11 Operating Thermal Margin and Steady-State Operating Conditions

A minimum operating margin is in general required to accommodate margin variations during steady-state operation. The following factors should be considered in the determination of this minimum required operating margin.

1. **Maneuvering capability.** If the plant is designed for rapid load-follow, some margin allowance is needed to accommodate the increase of local radial peaking factor due to flux depression effects of control rod insertion.[125]

TABLE 7.3-4 COMPONENT UNCERTAINTIES OF DNB LCO AND LSSS[a]

System Parameter Uncertainty	Typical Value 95/95	Modelling Uncertainty	Typical Value 95/95	Typical Value Uncertainty	State Parameter 95/95
Subchannel area	~ ±2.0%	CHF correlation (Including some of the modelling uncertainties in the T-H simulator)	17% ~ 30%	Inlet core flow	±4% ~ ±6%
Heat flux factor	~ ±2.5%				
Enthalpy rise factor	~ ±1.4%		—	Core inlet temperature	±2°F ~ ±4°F
Assembly-by-assembly inlet mass flux distribution	-	Core by-pass flow	~ ±1.3% of rated flow	System pressure	±30 F ~ ±45 F
Exit pressure distribution (153)	-	Radial peaking factor[b] Planar 1 (1) one pin/assembly (2) assembly (1) + (2) Integrated	~ ±2.1% ~ ±3.5% ~ ±5% ~ 7.5%	Core power Calorimetric power Neutron power (ex-core detectors)	±1.76% ~ ±2% ~ ±4%
				Axial power shape Synthesis from ex-core detectors Synthesis from in-core detectors Axial shape offset, or axial shape index[c]	~ 5.9% on F_q ~ 2.3% on F_q ~ ±.03 ASIU

[a]Typical values of the subuncertainties are quoted from references 118–124.
[b]The planar radial peaking factor is defined as the ratio of the maximum pin power to the average pin power ar a given axial plane. The integrated radial peaking factor, also called the $F_{\Delta H}$ hot channel factor, is the axial shape weighted planar peaking factor.
[c]Including both processing uncertainty and excore detector measurement uncertainty.

2. **Likelihood of design change.** Due to licensing or economic considerations, the design of the reactor may be altered after it begins commercial operation. One example is the extension of fuel cycle length at many operating plants.

3. **Flexibility of operation.** Current reactors are in general operating within a tight operating band. For example, the constant axial offset control strategy[125,126] maintains the axial shape index near a target value with a deviation of no more than ± 0.05 at 100% rated power. As the operating band increases, the operating margin allowance set aside to allow the core to operate at any point within the band will increase.

Once the minimum required operating margin is determined considering the above three factors, the "design" steady-state operating conditions can be optimized to result in the lowest power cost, considering both the capital cost and the operating cost of the plant. The constraint of optimization is that the resulting operating margin should be greater than the minimum required operating margin. Many optimization techniques, such as linear and nonlinear programming,[127–129] have been used to determine the optimum design operating conditions. However, for the most part the optimization process takes place with a fixed design of the steam generator and primary coolant pumps. This approach is encountered when standardized components are considered in design or when the reactor designer intends to increase the power level of an operating plant. In these circumstances the optimization process is greatly simplified with only two parameters, the core power and the coolant inlet temperature.

The simplified optimization process balances two effects associated with the change of core coolant temperature. One is that the plant efficiency increases as the core inlet temperature increases.[130] The other effect is that the available overpower margin decreases as the core temperature increases. When maintaining a constant margin degradation for the limiting AOO, constant monitoring and protection system uncertainties, and a constant minimum required operating margin, this decrease in available overpower margin will result in a decrease in the allowable core power. In Fig. 7.3-9 the plant efficiency and the allowable core power are plotted as a function of core inlet temperature. The product of the core power and the plant efficiency is electric output. As can be seen at the top of Fig. 7.3-9, the electricity output reaches a maximum at a certain inlet temperature. The assumption in this simplified optimization is that the turbine-generator is slightly oversized and is able to handle a range of electric output.

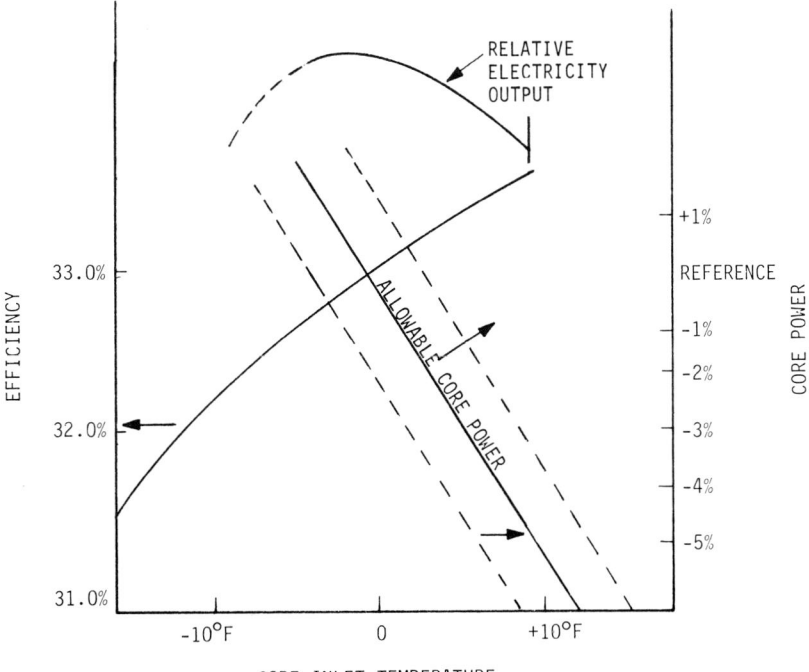

Fig. 7.3-9 Effects of core coolant inlet temperature on efficiency and allowable core power.

7.3-12   Other Critical Thermal-Hydraulic Phenomena—Small-Break LOCA

After the Three Mile Island accident, thermal-hydraulic phenomena during a small-break LOCA accident have been paid special attention. A small-break LOCA can be distinguished from an intermediate-break and a large-break LOCA by the small size (typically less than 80 cm^2) of the break.[93,131] Other characteristics of a small-break LOCA include that the core gradually uncovers with a phase separation when the coolant boils off and that the heat removal through the steam generators and condensers by natural circulation and reflex flows[132] are required to remove the decay heat., The rates of pressure decrease and loss of water inventory of a small-break LOCA are much lower than that for a limiting LOCA such that the operator's actions to mitigate the consequences are possible and needed. The consequences of a small-break LOCA are in general less severe than a large-break LOCA.[133,134] The consequences are sensitive to the design of the high-pressure safety injection systems and the emergency procedures that are used by the operators during the accident to remove the decay heat. Increasing the high-pressure makeup water capacity, enhancing the reactor coolant system depressurizing capability, and limiting the core power or top peaked axial power shapes during normal operation will mitigate the consequences of a small-break LOCA.

7.3-13   Concluding Remarks

This article puts various critical thermal-hydraulic and fuel-related phenomena into the perspective of reactor safety and operation. Relevant references are provided in the article such that more detailed information on subjects of interest can be researched.

Improving the operating thermal margin has been, and will continue to be, the focus of many thermal-hydraulic and fuel-related research activities. With a large operating margin available, many options exist to improve the performance of a nuclear plant. These options include stretch of the licensed power level, extension of the fuel cycle length, increase of the plant efficiency by increasing the core inlet temperature, improvement of operating flexibility with a wider operating band of axial power shape and other parameters, and converting a base-load nuclear plant into an intermediate peaking capacity generation station.[135]

Assessing the true safety margin between acceptable design limits and the fuel-damaging conditions will continue to be another subject of main interest. The incentive behind this kind of research program is to assure the public health and safety with an improved knowledge of how conservative these "acceptable" design limits are. If the safety margin can be demonstrated to be overly conservative, relaxation of the acceptable design limits may be possible. The benefit of the relaxation of design limits is an increase of the operating margin.

REFERENCES

7.3-1   W. C. Gumprich, "The Consolidated Edison Reactor," Annual Meeting of Atomic Industrial Forum, Washington, D.C., 1955.

7.3-2   L. E. Foster, "Technical Design of the Dresden Nuclear Power Station," Annual Meeting of Americal Society of Mechanical Engineers, 1958.

7.3-3   "General Electric Standards Safety Analysis Report," GESSAR-251, Docket-STN-50331-22, 1975

7.3-4   "System 80 CESSAR FSAR," Amendment No. 5, Combustion Engineering, Inc., 1981.

7.3-5   "Numerical Guides for Design Objections and Limiting Conditions for Opertion to Meet the Criteria," Code of Federal Regulations, 10CFR50, Appendix I, U.S. Government Printing Office, Washington, D.C., 1975.

7.3-6   "Reactor Site Criteria," Code of Federal Regulations, 10CFR100, U.S. Government Printing Office, Washington, D.C., 1975.

7.3-7   "General Design Criteria for Nuclear Power Plants," 10CFR50, Appendix A, U.S. Government Printing Office, Washington, D.C., 1975.

7.3-8   Standard Review Plan, NUREG-0800, Revision 2, 1980.

7.3-9   L. S. Tong, *Boiling Crisis and Critical Heat Flux*, ANS Critical Review Series, Americal Nuclear Society, LaGrange Park, IL, 1972.

7.3-10   A. E. Bergles and W. M. Rohsenow, "The Determination of Forced Convection Surface Boiling Heat Transfer," Paper 63-HT-22, 6th National Heat Transfer Conference, ASME-AIChE, Boston, 1963.

7.3-11   E. J. Davis and G. H. Anderson, "The Incipience of Nucleate Boiling in Forced Convection Flow," *AIChE J.* **12**, 774–780 (1966).

7.3-12   E. R. Hosler, "Flow Pattern in High Pressure Two-Phase (Steam Water) Flow with Heat Addition," WAPD-TM-658, Westinghouse Electric Corporation, Pittsburgh, PA, 1976.

7.3-13 A. E. Bergles, R. F. Lopina, and M. P. Fiori, "Critical Heat Flux and Flow Pattern Observations for Low Pressure Water Flow in Tubes," *J. Heat Trans.* **89C**(1), 69–74 (1976).

7.3-14 N. A. Radovich and R. Moissis, "The Transition from Two-Phase Bubble Flow to Slug Flow," Report No. 7-7673-22, Department of Mechanical Engineering, M.I.T., 1962.

7.3-15 A. Porteous, "Prediction of the Upper Limit of the Slug Flow Regime," *Brit. Chem. Engineering* **14**(9), 117–119 (1969).

7.3-16 S. E. Ritterbusch et al., "Factors Affecting Post-DNB Operation for LWR's–Volume 1," EPRI-NP-1999, *EPRI* (1981).

7.3-17 R. H. Stoudt et al., Factors Affecting Post-DNB Operation for LWR's—Volume 2," EPRI-NP-1999, *EPRI* (1981).

7.3-18 A. A. Bauer and L. M. Lowry, "Tensile Properties and Annealing Characteristics of H. B. Robinson Spent Fuel Cladding," *Nuclear Technology* **41**, 359–372 (1978).

7.3-19 A. A. Bauer et al., Effects of Annealing on Irradiated Spent Fuel Cladding," *Trans American Nuclear Society* **30**, 202–203 (1978).

7.3-20 S. Levy et al., "Experience with BWR Fuel Rods Operating Above Critical Flux," *Nucleonics*, April 1965.

7.3-21 G. F. Hewitt, "Critical Heat Flux in Flow Boiling," 6th International Heat Transfer Conference, Toronto, Canada, 1978.

7.3-22 V. Marinelli, "Critical Heat Flux: A Review of Recent Publications," *Nuclear Technology* **34**(2), 135–171 (1977).

7.3-23 E. D. Hughes, "A Compilation of Rod Array Critical Heat Flux Data Sources and Information," *Nucl. Eng. Des.* **30**(1) 20–35 (1974).

7.3-24 G. P. Gaspari et al., Italian Report CISE-R-276, 1968.

7.3-25 J. W. Hench and J. C. Gillis, "Correlation of Critical Heat Flux Data for Application to Boiling Water Reactor Conditions," EPRI-NP-1898, *EPRI* (1981).

7.3-26 R. W. Bowring, "A New Mixed Flow Cluster Dryout Correlation for Pressures in the Ranges 0.6–15.5 MN/m^2 (90–2250 psia)—For Use in a Transient Blowdown Code," Paper no. C217/77, *Mech IE* (1977).

7.3-27 L. S. Tong et al., *J. Nuclear Energy* **21** 241 (1967).

7.3-28 L. S. Tong et al., *Chem. Eng. Progr. Synp. Ser.* **62**(64), 35 (1965).

7.3-29 F. D. Lawrence, B. W. Woodman, C. F. Fighetti, and W. E. Hovemeyer, "Critical Heat Flow in PWR Fuel Assemblies," 77-HI-92, ASME, New York, 1978.

7.3-30 J. S. Gellerstedt, et al., *Two-Phase Flow and Heat Transfer in Rod Bundles*, ASME, New York, 1969, pp. 63–71.

7.3-31 R. H. Wilson, et al., *Two-Phase Flow and Heat Transfer in Rod Bundles*, ASME, New York, 1969, pp. 56–62.

7.3-32 R. J. Wetherhead, "Nuclear Boiling Characteristics and the Critical Heat Flux Occurrence in Sub-Cooled Axial Flow Water Systems," ANL 6675, 1973.

7.3-33 J. M. Healzer, J. E. Hench, E. Jannsen, and S. Levy, "Design Basis for Critical Heat Flux Condition in Boiling Water Reactors," APED-5286, GE, 1966.

7.3-34 E. Janssen and S. Levy, "Burnout Limit Curves for BWR's," APED-3892, GE, 1962.

7.3-35 L. S. Tong and G. F. Hewitt, "Overall Viewpoint of Flow Boiling CHF Mechanisms," 72-HT-545, ASME, New York, 1972.

7.3-36 J. Weisman and B. S. Pei, "Prediction of Critical Heat Flux in Flow Boiling at Low Qualities," Int. J. Heat and Mass Transfer, Vol. 26(1), pp. 1463–1472 (1983).

7.3-37 S. Levy et al., "Prediction of Critical Heat Flux for Annular Flow in Vertical Tubes," EPRI NP-1619 (1980).

7.3-38 P. B. Whalley, P. Hutchinson, and G. F. Hewitt, "The Calculation of Critical Heat Flux in Forced Convection Boiling," Paper B6.11, Fifth Int. Heat Transfer Conf., Tokyo, Japan, 1974.

7.3-39 R. D. Coffield, Jr., W. M. Rohrer, and L. S. Tong, "An Investigation of the Departure from Nucleate Boiling in a Crossed Rod Matrix with Normal Flow of Freon 113 Coolant," *Nucl. Eng. Design* **6**, 147 (1967).

7.3-40 S. P. Kezios, T. S. Kim, and F. M. Rafchiek, "Burnout in Crossed-Rod Matrices and Forced Convection Flow of Water," International Developments in Heat Transfer, Part II, p. 262, ASME, New York, 1961.

7.3-41 D. C. Groeneveld, "The Effect of Short Flux Spikes on the Dryout Power," AECL-4927 (1975).

7.3-42 K. W. Hill, F. E. Motley, and F. F. Cadek, "Effects on Critical Heat Flux of Local Heat Flux Spikes or Local Flow Blockage in Pressurized Water Reactor Rod Bundles," ASME paper WA/HT54 no. 54, 1974.

7.3-43 K. W. Hill, F. E. Motley, F. F. Cadek, and J. E. Casterline, "Effect of a Rod Bowed to Contact on CHF in PWR Rod Bundles," 75-WA/HT-77, ASME, New York, 1972.

7.3-44 E. R. Rosal et al., "High Pressure Rod Bundle DNB Data with Axially Non-uniform Heat Flux," *Nuclear Engineering Design* **31**, 1–20 (1974).

7.3-45 R. T. Lahey, Jr., E. E. Polomik, and G. E. Dix, "The Effect of Reduced Clearance and Row Bow on Critical Power in Simulated Nuclear Reactor Rod Bundles," Proceedings of an International Meeting on Reactor Heat Transfer, pp. 520–537, ANS, 1973.

7.3-46 T. Kobori, "Critical Heat Flux Measurements in a Full Scale Rod Cluster," *Bulletin of the Japan Society of Mechanical Engineers* **19**(131), 340–346 (1976).

7.3-47 L. A. Zielke and R. H. Wilson, "Transient Critical Heat Flux and Spacer Grid Studies," *Nuclear Technology* **24**, 13–19 (1977).

7.3-48 V. N. Smolin and V. K. Polyakov, "Coolant Boiling Crisis in Rod Assemblies," Sixth International Heat Transfer Conference, Vol. 5, pp. 47–52, 1978.

7.3-49 L. S. Tong, "Prediction of Departures from Nuclear Boiling for an Axially Non-Uniform Heat Flux Distribution," *J. Nucl. Energy* **21**, 241–248 (1967).

7.3-50 A. N. Ryabov et al., "Boiling Crisis and Pressure Drop in Rod Bundles with Heat Transfer Enhancement Device," *Heat Transfer—Soviet Research* **9**(1), 112 (1977).

7.3-51 J. C. M. Leung and K. A. Gallivan, "Analysis of Blow Down Heat Transfer Experiments and Critical Heat Flux," NUREG/CP-0014, Vol. 2, p. 1143, 1980.

7.3-52 R. E. Philips, R. W. Shumway, and K. H. Chu, "Improvements to the Prediction of Boiling Transition During Boiling Water Reactor Transients," in *Thermal-Hydraulics in Nuclear Power Technology*, 20th National Heat Transfer Conference, ASME, 1981.

7.3-53 L. Biasi et al., "Studies on Burnout: Pt. 3," *Energia Nucleare* **14**, 530 (1967).

7.3-54 P. Griffith, et al., "Counter-Current Flow Critical Heat Flux," presented at National Heat Transfer Conference, San Francisco, ASME, 1975.

7.3-55 J. A. Chistens, "Irradiation Effects on Uranium Dioxide Melting," HW-69234, Hanford Works, Richland, WA, 1962.

7.3-56 J. A. Christensen, "Thermal Expansion of UO_2," U.S. AEC Report HW-75148, 1962.

7.3-57 M. F. Lyons, R. C. Nelson, T. J. Peshos, and B. Weidenbaum, "Uranium Dioxide Fuel Rod Operation with Gross Central Melting," *Trans. Am. Nucl. Soc.* **6**, 155 (1963).

7.3-58 C. Lepescky, G. M. Testa, H. Houggard, and K. W. Jones, Experimental Investigation of In-Reactor Molten Fuel Performance," *Nuclear Technology* **16**, 367 (1972).

7.3-59 "Westinghouse Reference Safety Analysis Report, RESAR-414," Docket-STN-50572-1, Westinghouse Nuclear Systems, 1976.

7.3-60 D. B. Collins, M. Gacesa, and C. B. Parsons, "Study of the Onset of Premature Heat Transfer Crisis During Hydrodynamic Instability in a Full Scale Reactor Channel," ASME-AIChE Heat Transfer Conference, 71-HT-11, ASME, 1971.

7.3-61 F. Mayinger, O. Shud, and E. Weiss, "Research on the Critical Heat Flux (Burnout) in Boiling Water," Final Report, EURAEC-1811, EURATOM, 1973.

7.3-62 M. Ledinegg, "Instability of Flow During Natural and Force Circulation," *Die Warme* **61**, 8 (1938).

7.3-63 G. B. Wallis, "Some Hydrodynamic Aspects of Two-Phase Flow and Boiling," *Proc. Int. Dev. Heat Transfer*, Part II, ASME, New York, 1961.

7.3-64 L. E. Weaver, *A System Analysis of Nuclear Reactor Dynamics*, American Nuclear Society Monograph, LaGrange, IL, 1963.

7.3-65 A. W. Cramer, *Boiling Water Reactors*, Addison-Wesley, New York, 1958.

7.3-66 H. Kao, C. D. Morgon, and M. B. Parker, "Prediction of Flow Oscillation in Reactor Core Channel," *Trans. Am. Nucl. Soc.* **16**, 212 (1973).

7.3-67 T. N. Veeziroglu and S. S. Lee, "Boiling-Flow Instability in a Cross-Connected Parrellel-Channel Up Flow System," ASME Paper No. 71-HT-12, ASME, New York, 1971.

7.3-68 M. T. Pitek, "Utility Experience in LWR Fuel Performance," NUSCO-115, Joint ANS/CNA Meeting, Toronto, Canada, 1976.

7.3-69 B. L. Siegel and H. H. Hagen, "Fuel Failure Detection in Operating Reactors," NUREG-0401, U.S. NRC, 1977.

7.3-70 A. Garlick and P. D. Wolfenden, "Fracture of Zirconim Alloys in Iodine Vapor," *J. Nuclear Materials* **41**, 274–292 (1971).

7.3-71 J. C. Wood, "Factors Affecting Stress Corrosion Cracking of Zircaloy in Iodine Vapor," *J. Nuclear Materials* **45**, 105–122 (1973).

7.3-72 W. T. Grubb and M. H. Morgan III, "Cadmium Embrittlement of Zircaloy-2," pp. 295–304, Proceedings of ANS Topical Meeting on Water Reactor Fuel Performance, ANS, 1977.

7.3-73 R. L. Jones, "EPRI–NASA Cooperative Project on Stress Corrosion Cracking of Zircaloys," EPRI-RP-455, *EPRI* (1977).

7.3-74 F. Wunderlich, R. Holzer, H. P. Fuchs, and W. Hering, "The KWU PCI Modeling Approach," *Nuclear Engineering and Design* **56**, 3–9 (1980).

7.3-75 L. F. Coffin and R. P. Gangloff, "Integrated Laboratory Methods for Evaluating PCI in Zircaloy-2 Fuel Cladding," Proceedings of ANS Topical Meeting on Water Reactor Fuel Performance, ANS, 1977.

7.3-76 H. Magard, "Suppression of PCI Induced Defects by Lightly Undulating the Bore Surface of the Fuel Cladding," ANS Topical Meeting on Water Reactor Fuel Performance, ANS, 1977.

7.3-77 D. G. Hardy, A. S. Bain, and R. R. Meadowcroft, "Performance of CANDU Development Fuel in the NRU Reactor Loops," ANS Topical Meeting on Water Reactor Fuel Performance, pp. 198–206, ANS, 1977.

7.3-78 D. G. Franklin, "Improved LWR Fuel Design," *Trans. Am. Nucl. Soc.* **43**, 150 (1982).

7.3-79 R. Holzer, D. Knoder, and H. Stahler, "Pellet Clad Interaction: Experience, Testing, and Evaluation, A KWU Review," Proceedings of ANS Water Reactor Fuel Performance Topical Meeting, ANS, 1977.

7.3-80 Yasunori Bessho, Hiroshi Motoda, and Mitsutaka Watanable, "Principles for Control Rod Withdrawal Strategy During the Startup of Boiling Water Reactors," *Nuclear Technology* **58**, 113–119 (1982).

7.3-81 Thompson et al., "Conditioning of Nuclear Reactor Fuels," United States Patent. No. 4, 057, 466, 1977.

7.3-82 "Standard Specification for Sintered Uranium Dioxide Pellets," ASTM Standard C776-76, Part 45, 1977.

7.3-83 K. Joon, "Primary Hydride Failure of Zircaloy-Clad Fuel Rods," *Trans. Am. Nucl. Soc.* **15**, 186 (1972).

7.3-84 W. J. O'Donnel and B. F. Langer, "Fatigue Design Basis for Zircaloy Components," *Nuclear Science Engineering* **20**, 1–12 (1964).

7.3-85 ASME Boiler and Pressure Vessel Code and an American National Standard, Section III, Rules for Construction of Nuclear Power Plant Components, ASME, New York, 1980.

7.3-86 "Overpressure Protection," NB-7000, ASME Boiler and Pressure Vessel Code, Section III, ASME, New York, 1980.

7.3-87 H. M. Chung, A. M. Garde, and T. F. Kassner, "Development of an Oxygen Embrittlement Criterion for Zircaloy Cladding Application to Loss-of-Coolant Accident Conditions in Light Water Reactors," Zirconium in the Nuclear Industry (Fourth Conference), ASTM STP 681, pp. 600–621, 1979.

7.3-88 P. E. MacDonald et al., "Assessment of Light-Water Reactor Fuel Damage During a Reactivity-Initiated Accident," *Nuclear Safety* **21**(5) (1980).

7.3-89 D. A. Powers and R. D. Meyer, "Cladding Swelling and Rupture Models for LOCA Analysis," NUREG-0630, 1980.

7.3-90 R. T. Lahey, Jr., "Transient Response of Light Water Reactors," in *Nuclear Reactor Safety Heat Transfer*, (O. C. Jones, Jr., ed.), Hemisphere Publishing, New York, 1981.

7.3-91 C. W. Lepine and E. L. Trapp, "Nuclear Plant Operating Margin Optimization," C-E, TIS-5808, 1978.

7.3-92 Report to the American Physical Society by the Study Group on LWR Safety, *Rev. Modern Phys.* **27**(suppl. No. 1) (1975).

7.3-93 W. L. Riebold, "Loss-of-Coolant Accident," in *Nuclear Reactor Safety Heat Transfer*, (O. C. Jones, Jr., ed.). Hemisphere Publishing, New York, 1981.

7.3-94 "ECCS Evaluation Models," Code of Federal Regulations, 10 CFR 50, Appendix K, U.S. Government Printing Office, Washington, D.C., 1975.

7.3-95 L. S. Tong, "Loss-of-Coolant Accident," Thermal Analysis of Pressurized Water Reactors, Chapter 5.4.5, American Nuclear Society, 1979.

7.3-96 "Multinode Analysis of Core Flooding Line Break for B & W's 2568 MWt Internals Vent Valve Plants," BAW-10064, 1973.

7.3-97 D. D. Jones, "Subcooled Countercurrent Flow Limiting Characteristics of the Upper Region of a BWR Fuel Bundle," NEDG-23549, GE, 1977.

7.3-98 G. E. Hanson, "Normal Operation Controls," BAW-10122A, 1979.

7.3-99 "Acceptance Criteria for Emergency Core Coolant Systems for Light Water Cooled Nuclear Power Reactors," U.S. AEC, Docket No. RM-50-1, 1973.

7.3-100 C. L. Caso, J. Gallaghar, and M. M. J. Plumier, "Fuel Rod Performance During a Loss-of-Coolant Accident in Westinghouse Pressurized Water Reactors," BNES, Nuclear Fuel Performance International Conference, London, 1973.

7.3-101 K. Riedle and F. Winkler, "ECC-Reflooding Experiment with a 340 Rod Bundle," Proceedings of the CREST Specialist Meeting on Emergency Cooling for LWRs, Casching/Munchen, 1972.

7.3-102 A. H. Hsia, F. R. Hubbard, R. E. Schneider, and S. C. Rose, "Rod Bundle Blowdown Heat Transfer Tests Simulating PWR LOCA Conditions," EPRI NP-113, *EPRI* (1977).

7.3-103 R. A. Smith and P. Griffith, "Critical Heat Flux in Flow Reversed. Transients," EPRI-NP-151, *EPRI* (1979).

7.3-104 J. D. Duncan and J. E. Leonard, "Response of a Simulated BWR Fuel Bundle Cooled by Flooding Under Loss-of-Coolant Conditions," GEAP-10117, 1969.

7.3-105 J. A. Blaisdell, L. E. Hochreiter, and J. P. Waring, "PWR FLECHT-SET Phase A Report," WCAP-8238, 1973.

7.3-106 J. P. Waring et al., "PWR FLECHT-SET Phase B1 Data Report," WCAP-8431, 1974.

7.3-107 R. W. Griebe and J. W. McConnel, "Boiling Water Reactor—Full Length Emergency Core Cooling Heat Transfer (BWR-FLECHT) Test Projects Final Report on Atmospheric Pressure Stainless Steel Experiments," ANCR-1115, 1973.

7.3-108 F. F. Cadek et al., "PWR FLECHT Group II Test Report," WCAP-7544, 1970.

7.3-109 D. Hein, F. Mayinger, and F. Winkler, "The Influence of Loop Resistance on Refilling and Reflooding in the PKL-Tests," 5th Water Reactor Safety Research Information Meeting, U.S. NRC, 1977.

7.3-110 J. D. Duncan and J. E. Leonard, "BWR Standby Cooling Heat Transfer Performance Under Simulated Loss-of-Coolant Conditions Between 15 and 300 psia," GEAP-13190, 1971.

7.3-111 Schneider and Thomas, "Further Development of the Technology of Light Water-Cooled Reactors—Project 2b—Cooling Conditions of the Reactor Core in the Case of the Maximum Credible Accident During Refilling by the Safety Feed System—Final Report," AEC-TR-7396, 1973.

7.3-112 C. M. Moses and R. W. Griebe, "SECTH-III—An Experimental Investigation of Top and Bottom Flooding of a Nuclear Bundle Simulator," IV-1535, 1970.

7.3-113 W. J. Penn, R. K. Lo, and J. C. Wood, "CANDU Fuel–Power Ramp Performance Criteria," *Nuclear Technology* **24**, 249–268 (1977).

7.3-114 H. Ogasawan, J. Kashiwai, and Y. Takashima, "Coolant Mechanism of the Low Pressure Coolant Injection System of BWRs and Other Studies on the Loss-of-Coolant Accident Phenomena," CONF-730304, 1973.

7.3-115 D. Andreoni, M. Courtand, and R. Deruaz, "Heat Transfer During the Reflooding of a Tubular Test Section," European Two-Phase Flow Meeting, Harwell, 1974.

7.3-116 D. B. Owen, "Factors for On-Sided Tolerance Limits and for Variables Sampling Plans," Sandia Corporation Monograph, SCR-607, 1963.

7.3-117 G. J. Hahn and S. S. Shapiro, *Statistical Models in Engineering*, Wiley, New York, 1967.

7.3-118 "Statistical Combination of Uncertainties," CEN-124-(B), Parts 1, 2 and 3, Combustion Engineering, Inc., 1980.

7.3-119 E. Oelkers, A. S. Heller, D. A. Farnworth, and K. J. Kearfott, "Statistical Core Design," BAW-10139, Babcock & Wilcox, 1979.

7.3-120 J. F. Carew, "Process Computer Performance Evaluation Accuracy," NEDO-20340, General Electric Company, 1974.

7.3-121 H. E. Neuschaefer and J. R. Humpheries, "Synthesis of In-Core Power Distribution from Tri-Level Excore Detectors," TIS-5361, Combustion Engineering, Inc., 1977

7.3-122 H. Chelemer, L. H. Boman, and D. R. Sharp, "Improved Thermal Design Procedure," WCAP-8569, Westinghouse Electric Corporation, 1975.

7.3-123 W. B. Terney, G. H. Marks, E. A. Williamson, Jr., T. G. Ober, "Axial Power Distributions from Fourier Fitting of Fixed In-Core Detector Powers," TIS-4720, Combustion Engineering, Inc., 1975.

7.3-124 R. L. Williams, "RPS Limits and Setpoints," BAW-0121, Babcock & Wilcox Co., 1978.

7.3-125 T. Morita et al., "Power Distribution Control and Load Following Procedures," WCAP-8403, Westinghouse Electric Corporation, 1974.

7.3-126 "Power Distribution Control of Westinghouse Pressurized Water Reactors," WCAP-7811, 1971.

7.3-127 G. Hadley, *Non-Linear and Dynamic Programming*, Addison-Wesley, Reading, MA, 1964.

7.3-128 B. Gottfried and J. Weisman, *Introduction to Optimization Theory*, Prentice-Hall, Englewood Cliffs, N.J., 1973.

7.3-129 J. Weisman and A. G. Holzman, "Optimal Process System Design Under Risk," *Ind. Chem. Process Des. Dev.* **11**, 386 (1972).

7.3-130 L. S. Tong, "Heat Transfer in Water-Cooled Nuclear Reactors," *Nucl. Eng. Des.* **6**, 301 (1967).

7.3-131 "Safety Design Rationale for Nuclear Power Plants," Appendix IX of "Reactor Safety Study," USAEC-WASH-1400, 1975.

7.3-132 H. Weisshaupl and B. Brand, "PKL Small Break Tests and Energy Transport Mechanisms," in Proceedings of ANS Specialists Meeting on Small Break Loss-of-Coolant Accident Analyses in LWR's, ANS, 1981.

7.3-133 P. M. Gururaj, S. S. Dua, and A. S. Rao, "Review of Boiling Water Reactor Small Break Loss of Coolant Accidents," Proceedings of ANS Specialist Meeting on Small Break LOCA Analysis in LWRS, pp. 252–257, 1981.

7.3-134 S. E. Jensen, "Overview of Exxon Nuclear Small Break LOCA Activities," Proceedings of ANS Specialist Meeting on Small Break LOCA Analysis in LWR's, pp. 237–249, 1981.

7.3-135 C. R. Musick and R. W. Knapp, "C-E Meets Utility Load Change Capacity Needs With System 80 NSSS," TIS-6215, Combustion Engineering Inc., 1979.

7.4 METHODS AND COMPUTER CODES

Chong Chiu

7.4-1 Introduction

This section discusses the current thermal-hydraulic and fuel performance analysis methods for analyzing light water reactor (LWR) performance during steady-state operation and transients. Emphasis is given to methods and computer codes with which favorable experience has been developed and in which confidence has been established. The discussion is restricted to the methods and computer codes accessible to the general public or the nuclear industry.

This section is divided into three major parts, each devoted to a different aspect of nuclear power plant performance analysis. First, the methods and codes for analyzing the detailed thermal-hydraulic behavior within the reactor core are presented. These codes calculate the thermal-hydraulic conditions within flow subchannels and are usually referred to as subchannel codes. Second, the models and codes for analyzing the fuel performance, such as fission gas release, fuel temperature distribution, pellet-cladding gap conductance, and dimensional changes of the cladding, are discussed. These are called fuel performance codes. Last, the models and codes for analyzing the nuclear plant overall system behavior are discussed. The codes in this category are called system codes.

7.4-2 Subchannel Codes

Subchannel codes calculate detailed thermal-hydraulic conditions and the minimum critical heat flux ratio (CHFR) within the subchannels during steady-state operation and transients. A subchannel is conventionally defined by the solid surface of four adjacent fuel rods and the boundaries joining rod centers. A channel comprising more than one subchannel is conventionally called a *lumped channel*. The purpose of a subchannel calculation is to determine the available overpower margin and the required overpower margin needed for thermal margin and set point analyses, to predict the number of failed fuel pins during transients, or to interpret and correlate the experimental data on critical heat flux (CHF), friction factor, mixing, and other thermal-hydraulic parameters.

In general, input to subchannel codes can be categorized as follows:

1. **Fuel system geometry.** Rod diameter, rod pitch, axial location of spacer grids, and so on.
2. **Core thermal-hydraulic boundary conditions.** Core inlet coolant temperature or enthalpy, core inlet flow rate, assembly-by-assembly inlet flow or pressure distribution, assembly-by-assembly exit pressure distribution, and so on.
3. **Heat flux boundary conditions.** Core heat flux on fuel rod surface, heat generated in the coolant, normalized axial heat flux distribution, radial peaking factors, and so on.
4. **Coefficients for physical models or correlations.** Coefficients for fuel surface friction factors, spacer grid pressure loss, coolant turbulent mixing, and so on.
5. **Coefficients for numerical modeling.** Axial node length, parameters to define lumped channels and subchannels, numerical convergence criteria, and so on.
6. **Engineering factors.** Enthalpy rise engineering factor, heat flux engineering factor, flow area engineering factor, and so on.

For transients, the core thermal-hydraulic and heat transfer boundary conditions input to the subchannel codes are specified as a function of time. Assembly-by-assembly inlet flow and pressure distributions, which are determined from either analyses or experiments, are usually expressed in dimensionless form and are usually assumed to remain constant during the transient. The core thermal-hydraulic boundary conditions are typically calculated by a system code. The heat flux boundary conditions during transients are typically calculated by a one-, two-, or three-dimensional neutronics computer code.

The output of a subchannel code includes four fundamental thermal-hydraulic parameters: coolant enthalpy, mass flux, cross flow between subchannels, and pressure at various axial locations. Other parameters derived from or related to these four fundamental thermal-hydraulic parameters include coolant temperature, quality, void fraction, and critical heat flux ratio.

A number of subchannel codes, for example, COBRA,[1-4] HAMBO,[5] TORC,[6] THINC,[7] and LYNX,[8,9] have been written for use in core analysis. Although they were initially developed more or less independently, their basic thermal-hydraulic models are similar. Comparison of characteristics of some subchannel codes currently available to the public is provided in Table 7.4-1. Among these codes, the COBRA-IIIC code is by far the most popular in the nuclear industry. This is primarily because it has been extensively qualified with available experimental data and is used by the Nuclear

TABLE 7.4-1 DESCRIPTION OF SUBCHANNEL CODES

CODE CAPABILITY	COBRA-II (1)	COBRA-IIIC (2)	COBRA-IIIC/MIT (3)	COBRA-IV-I (4)
A. SUBCHANNEL REPRESENTATION				
BWR Assembly/subchannel	No	No	Yes	No
PWR assembly/subchannel	Yes	Yes	Yes	Yes
Variably sized channels	No	No	Yes (Chiu/Weisman)	No
One stage calculation	Yes	Yes	Yes	Yes
Multi-stage calculation	No	No	No	No
B. HYDRODYNAMIC CAPABILITY				
Transient	No	Yes	Yes	Yes
Homogeneous/Non-homogeneous	Homogeneous	Homogeneous	Homogeneous	Homogeneous
2-phase thermal equilibrium/Non-equilibrium	Equilibrium	Equilibrium	Equilibrium	Equilibrium
Lateral pressure difference	$\Delta P = C\|w\|w$	$\Delta P=C\ W\ W+\delta(UW)/\delta x+\frac{\delta W}{\delta t}$	Same as COBRA-IIIC	COBRA-IIIC equation $+\ \delta(VW)/\delta x$
Flow reversal	No	No	No	Yes
Natural recirculation	No	No	No	Yes
Flow blockage	No	Partial blockage	Partial blockage	Total channel blockage
C. FUEL/THERMAL HYDRAULIC MODELS				
Single/two-phase mixing	input/input	input/input	input/Beus/input	input/input
Coolant void fraction	homo./Mod. Armand/slip/Levy	homo./Mod. Armand/slip	homo/Levy/Mod. Armand/Smith	homo./Levy/Mod. Armand
Two phase friction factor	homo./Armand	homo./Armand	homo/Armand/Baroczy	homo./Armand
Gap conductance	input	input	detailed MATRO package	input
Fuel conductivity	temperature independent	temperature indep.	temperature dependent	temperature dependent
Critical Heat Flux	No	BAW-2/W-3	BAW-2/W-3/CISE-4/Hench Levy	BAW-2/W-3
Axial conduction	No	No	No	Yes
Wall conduction	No	No	No	both in fuel & coolant
Heat transfer correlation	input	input	BEEST package	RELAP-4 pkg.
D. USER CONVENIENCE				
Dynamic computer storage	No	No	Yes (IBM only)	No
Plotting	None	None	None	P, m, h, W_{ij}

References for models and correlations: homo.(10), Mod. Armand(11), slip(10), Levy(13), BAW-2(14), W-3(15), CISE(16), Hench&Levy(17), MATRO(18), BEEST(19), RELAP-4(20), Baroczy(21), Smith(22), Chiu(31), Weisman(3).

W: cross-flow
U: axial velocity
V: transverse velocity

Regulatory Commission (NRC) of the United States to evaluate proprietary subchannel codes developed by various organizations.

COBRA-II, a steady-state code and the predecessor of COBRA-IIIC, is based on a simplified transverse momentum equation. The convective terms $\delta(UW)/\delta x$ and $\delta(VW)/\delta y$, which are important for modeling cases with substantial cross flows, are neglected. COBRA-IIIC extends the capability of COBRA-II to modeling transient subchannel thermal-hydraulic phenomena and includes the convective term $\delta(UW)/\delta x$ in its transverse momentum equation. With the improved modeling of the transverse momentum transport process, COBRA-IIIC is capable of solving partial flow blockage problems. COBRA-IIIC/MIT was developed from COBRA-IIIC and these two codes have similar conservation equations. COBRA-IIIC/MIT has improved capability for modeling BWR assembly-to-assembly thermal-hydraulic analysis and incorporates some improved numerical techniques to reduce the computer running time. The computer running time of COBRA-IIIC/MIT is less than that of COBRA-IIIC by a factor of about 25 for a case of 32 subchannels.[23] This benefit in computation time increases as the number of subchannels increases.

COBRA-IV-I extends the transverse momentum equation in COBRA-IIIC with an additional convective term, $\delta(VW)/\delta y$. Because of this addition, COBRA-IV-I is capable of analyzing complete subchannel flow blockage problems. It has the added option of an explicit solution scheme, in addition to the implicit solution scheme employed in COBRA-IIIC, to solve the conservation equations. This numerical method enables COBRA-IV-I to calculate reverse flows. The other three codes discussed in Table 7.4-1 do not have this capability.

Application

A subchannel code is often used in thermal margin analysis to calculate the available overpower margin, the margin degradation during a transient, or the uncertainties associated with the on-line determination of operating margin. To calculate these parameters, the CHFR within the entire reactor core has to be determined. A typical LWR core contains thousands of subchannels. In order to identify the fuel pin with the minimum CHFR economically, two methods are used to reduce the number of subchannels that must be analyzed. These methods are briefly discussed below.

Identification of the minimum CHFR subchannel, or the limiting channel, can be carried out by two kinds of analysis: single-stage analysis[24] and multistage analysis.[7] In a single-stage analysis the limiting channel is first identified. The thermal-hydraulic conditions of the limiting channel and of a few subchannels or lumped channels close to the limiting channel are then analyzed to obtain the minimum CHFR. In a multistage analysis the process of identification of the limiting channel may proceed in stages. For example, in the first stage of analysis coarse lumped channels are established and analyzed to identify the most limiting coarse lumped channel. Then, in the second-stage analysis, the limiting coarse lumped channel is broken down into many smaller lumped channels. As a result of the second-stage analysis, the most limiting lumped channel is identified and chosen to be further broken down into subchannels. In the last stage of the analysis each subchannel within the limiting lumped channel is analyzed and the minimum CHFR is determined. The advantage of a single-stage analysis over the multistage analysis is simplicity and less computer cost. The disadvantage is the need of a prespecified criterion to identify the hot channel before the analysis. In theory a simple selection criterion can be used to identify the hot channel in a single-stage analysis. The criterion repeatedly compares any two-candidate hot channels according to the following formula until the hot channel is identified:

$$\Delta\text{CHFR} = \sum_i \left(\frac{\delta\,\text{CHFR}}{\delta f_i} \right)(f_{i,1} - f_{i,2}) \tag{7.3-1}$$

where ΔCHFR = CHFR difference between candidate hot channels 1 and 2. If CHFR ≥ 0, subchannel 2 is more limiting; otherwise, subchannel 1 is more limiting.

f_i = parameters that affect the CHFR value within a subchannel. Typical parameters are the radial peaking factor among the fuel pins surrounding the hot channel, core average surface heat flux, subchannel inlet mass flow, coolant inlet temperature, radial peaking factors of fuel pins close to the hot subchannel, etc.

$\dfrac{\delta\,\text{CHFR}}{\delta f_i}$ = sensitivity factor of CHFR for f_i.

The sensitivity factors are frequently evaluated at nominal operating conditions. Their exact values depend on the CHFR correlation and fuel system design.

The transport process of energy, momentum, and mass between two lumped channels of different sizes, or even between a lumped channel and a subchannel, is of special concern to the reactor thermal-hydraulic analyst. The transport process between two lumped channels has not been experimentally studied as extensively as that between two regular subchannels.[25-30] With the knowledge of the transport process between two lumped channels of arbitrary size, the thermal margin analysis can be performed with many fewer lumped channels or subchannels. In Chiu[31] the transport process between two different-sized lumped channels was related to the known transport process between two subchannels. This relationship is expressed by "transport coefficients" whose sensitivities relative to various operating conditions are briefly described in Chiu[31] and Chiu et al.[32]

7.4-3   Fuel Performance Codes

In general, fuel performance codes calculate fuel performance parameters, such as fuel cladding internal gas pressure, cladding strain and stress, fuel pellet stored energy, cladding oxidation, mechanical pellet-cladding interaction, and so on. The purpose of the calculation is to verify the integrity of fuel rods during steady-state operation and anticipated operational occurrences, to determine the specified acceptable fuel design limit on fuel centerline melting, and to calculate the maximum cladding temperature and the total energy deposited in the fuel during an accident.

Some of the fuel performance codes developed to date analyze the fuel performance of only a single fuel rod during a loss of coolant accident. These codes include DINO,[33] ECCSA4,[34] THETA-1B,[35] RELUX,[36] and STRIKIN.[37] Some codes, such as FRAPCON-1,[38] FRAP-S3,[39] BEHAVE-4,[40] GAPCON-THERMAL-2,[41] LIFE-THERMAL-1,[42] COMETHE-IIIJ,[43] TAFY,[44] CYGRO,[45] FMODEL,[46] and FATES,[47] have been developed to analyze steady-state fuel performance. Among the codes with multirod and transient fuel performance capabilities are SCORCH-B2,[48] PARCH,[49] TRAC,[50,51] FRAP-T5,[52] and RELAP-5.[53] In addition, HOTPIN[54] analyzes the molten fuel motion within the fuel cladding, GLUB[55] analyzes the water-logging transient caused by the existence of fuel cladding defects during an increase in power, and SPEAR-BETA[56] predicts the cladding failure probability.

Among the above codes, FRAPCON-1, BEHAVE-4, GAPCON-THERMAL-2, COMETHE-IIIJ, LIFE-THERMAL-1, and FRAP-T5 have been well studied and qualified with available experimental

TABLE 7.4-2   DESCRIPTION OF FUEL PERFORMANCE CODES

CODE CAPABILITIES	FRAPCON-1 & FRAP-S3	BEHAVE-4	GAPCON-T2	LIFE-T1	COMETHE-IIIJ	FRAP-T5
A. FUEL PELLET ANALYSIS						
• Pellet to pellet interface	YES	NO	NO	NO	(YES)	YES
• Plasticity	NO	(YES)	NO	NO	(YES)	NO
• Primary/secondary creep	NO	NO	NO	YES	NO	NO
• Stored energy	YES	NO	YES	NO	NO	YES
• Densification/restructuring	YES	YES	YES	YES	YES	YES
• Fission gas release to plenum	YES	YES	YES	YES	YES	YES
• Indigenous gas release	NO	NO	YES	NO	NO	NO
• Gas release other than H_e, X_e, K_r	YES	YES	YES	YES	YES	YES
• Irradation swelling/thermal expansion	YES	YES	YES	YES	YES	YES
• Hot pressing	NO	YES	YES	NO	YES	NO
• Gap conductance	YES	YES	YES	YES	YES	YES
B. CLADDING ANALYSIS						
• Elasticity/plasticity	YES	YES	YES/NO	YES	YES/NO	YES
• Primary creep/secondary creep	YES	YES	YES	(YES)	YES	YES
• Material texture effects on σ-ϵ and Fast flux irradiation growth	NO	NO/YES	NO	NO/YES	YES	NO
• Strain hardening/time hardening	YES/NO	YES	NO	NO/YES	YES	YES/NO
• Heat generation in cladding	NO	(YES)	(YES)	(YES)	(YES)	NO
C. OTHERS						
• Transient	NO	NO	NO	NO	NO	YES
• Cladding corosion	YES	NO	NO	YES	NO	YES
• CHFR, oxidation, & N_t-Z_r eutectic failures	YES	NO	NO	NO	NO	YES
• Stress corrosion cracking failure	YES	NO	NO	NO	NO	YES
• Pellet-cladding axial interaction friction force (limited slippage)	NO	YES	NO	YES	NO	NO
• Varied power history	NO	(YES)	(YES)	(YES)	(YES)	YES
• Cladding rupture (overstress failure)	NO	NO	NO	NO	NO	YES
• Cladding ballooning (overstrain failure)	NO	NO	NO	NO	NO	YES
• Channel thermal hydraulic analysis	YES	NO	YES	YES	YES	YES
D. USER CONVENIENCE						
• Link with other types of codes	FRAP-T5	NO	NO	NO	NO	FRAP-T5 RELAP-4
• Computer running time	FAIR	GOOD	GOOD	GOOD	GOOD	FAIR

YES :   TREATED EXPLICITLY BY MODELS OR CORRELATIONS

(YES):   TREATED INDIRECTLY (WITH OR WITHOUT INPUT SUPPORT)

data. FRAP-S3, the predecessor of FRAPCON-1, and GAPCON-THERMAL-2 are used for steady-state fuel performance analyses and FRAP-T5 is used for transient fuel performance analysis by the NRC. BEHAVE-4, GAPCON-THERMAL-2, LIFE-THERMAL-1, and COMETHE-IIIJ were extensively compared with experimental data in Freeburn et al.[57] For these reasons, the capabilities of only these six fuel performance codes are presented in detail here. Table 7.4-2 summarizes the capabilities of these six codes.

COMETHE-IIIJ was judged the most versatile among the four when applied to both thermal and structural analyses.[57] For fuel temperature predictions COMETHE-IIIJ and LIFE-THERMAL-1 provided better agreement with data. For fission gas release predictions LIFE-THERMAL-1 was considered to be the best. For rod structure analysis BEHAVE-4 was considered comparable to COMETHE-IIIJ and better than LIFE-THERMAL-1. The performance of GAPCON-THERMAL-2 was judged worse than the other codes, largely due to its lack of inelastic cladding models and overpredictions of fuel temperature and fission gas release. These deficiencies, however, have been improved in its more recent version, GAPCON-THERMAL-3.[58]

In general, the input to a fuel performance code can be categorized as follows:

1. **Fuel system geometry.** Rod diameter, cladding thickness, fuel and cladding surface roughness, fuel pellet geometry, initial gap size, fuel plenum length, fuel spring geometry, crud thickness, and so on.
2. **Subchannel coolant conditions.** Subchannel flow rate, inlet coolant temperature, system pressure, and so on.
3. **Material description.** Fuel enrichment, initial fill gas composition and pressure, fuel initial grain size, fuel initial water content, fuel initial nitrogen content, degree of cold-work of cladding, and so on.
4. **Operating conditions.** Axial power shape, fast neutron flux, initial fuel burnup, and so on.

The output of interest to the reactor designer includes the internal gas pressure, fuel centerline and cladding surface temperatures, cladding oxidation thickness, pellet-cladding interaction, fission gas release, and fuel pellet stored energy. All parameters are used to predict fuel pin performance and to determine how close the operation is to the design limits.

7.4-4 System Codes

The system codes calculate the thermal-hydraulic behavior of the reactor primary loop of both PWRs and BWRs or of both primary and secondary loops of PWRs. These codes analyze the system thermal-hydraulic performance during a system transient or an accident.

This type of code has been extensively used in conjunction with the subchannel codes and fuel performance codes to estimate the consequences of a system transient or an accident, to design the power plant components, to optimize the plant overall performance, to design various control systems, and to establish the set points of various protection systems. In general, the input to a system code can be categorized as follows:

1. **Plant components description.** Component geometry, layout of various components, location of valves, reactor core geometry, fuel system geometry, and so on.
2. **Characteristics of components.** Reactor coolant pump characteristics, safety injection pump characteristics, let-down and charging flow rates, feedwater pump characteristics, steam generator recirculation ratio, carry-over and carry-under characteristics, friction factors of various flow paths, and so on.
3. **Control and safety system characteristics.** Control logic, safety system set points, safety system delay time, and so on.
4. **Coefficients for physics models and correlations.** Selection of models and correlations, pipe break pressure, pipe-wall heat transfer coefficient, emissivity of fuel rods and flow channel walls, structure material conductivity, and so on.
5. **Coefficients for numerical modeling.** Numerical noding scheme, convergence criterion, allowable number of iterations, and so on.
6. **Operating conditions.** Initial operating conditions in power, flow, temperature, and so on.

The output of a system code that is of special interest to the system designer includes the system pressure, peak cladding temperature, core thermal-hydraulic boundary conditions (if a subchannel code is separately used to calculate core conditions), maximum cladding oxidation, and amount of reactor coolant leakage. In some cases a fuel performance code, such as FRAP-T5, is linked to a system code to perform a more detailed cladding oxidation and temperature calculation. A subchannel code can also be linked to a system code in the determination of the number of fuel rods predicted to

TABLE 7.4-3 SYSTEM CODES ANALYSIS CAPABILITIES

CODE CAPABILITIES	RELAP4/Mod7	RELAP5	TRAC/(PD2/BD1)	RETRAN
• REACTOR SYSTEM REPRESENTATION				
• BHR	Yes	No	Yes	Yes
• PWR	Yes	Yes	Yes	Yes
• No. of components connecting the 3-D Vessel	One	One	More than one*	One
• ECCS - LPI	P dep. fill	Fill table	P dep. fill	P. dep. fill
- HPI	P dep. fill	Fill table	P dep. fill	P. dep. fill
- accumulation	Active control vol.	Active control vol.	Active control vol.	Fill junction table
• Containment	Yes	Yes	Yes	No
• Trip logic	Yes (but simple)	Yes	Yes	Yes*
• Control systems model	No	No	No	Yes*
• Secondary system	Steam generator	Steam generator	Steam generator	Steam generator/Turbine-generator*
• Valves (check/motor driven)	Yes	Yes	Yes	Yes
• HYDRODYNAMIC MODELS				
• Homogeneous/non-homogeneous	Homogeneous	Non-homogeneous*	Non-homogeneous*	Non-Homogeneous (slip)
• 2-phase thermal equilbrium/non-equilibrium	Equilibrium	Non-equilibrium (2 fluid, 5 equ.)	Non-equilibrium (2 fluid, 6 equ.)	Non-Equilibrium (separate volume)
• Metal-water reaction	yes	No	Yes	Yes
• Phase separation	Empirical model (bubble rise)	Implicit calc.	Implicit calc.	Empirical model (bubble rise)
• Chocked flow model	Empirical models	2 fluid bound.* cond. nodal cal.	One-component, 2-Φ Equilibrium model	Empirical models
• Kinetics Model	Point	Point	Point	1-D, point
• Heat transfer models				
• 2-Φ interfacial heat transfer	NA	Yes	Yes	NA
• Rod-to-rod radiation heat transfer	Yes	Yes	Yes	No
• Blowdown/reflood (wall-to-fluid)	2 sets for blowdown reflood separately	One set for both blowdown/reflood	Same as RELAP4	Input
• Pressurizer model	2Φ, complete separation (constructed by user)	2Φ, complete separation (constructed by user)	2Φ, complete separation	2Φ, complete separation
• USER CONVENIENCE				
• Noding for reflood/refill after blowdown	Automatic	Automatic	Automatic	Input
• Dynamic computer storage	Yes	Yes	Yes	Yes

* Significant advance in modelling methodology

fail. The system pressure during an accident should be less than 110% of design pressure to avoid abnormal leakage. The maximum peak cladding temperature and cladding oxidation thickness should be maintained less than the design limits for adequate core coolability. The amount of reactor coolant leakage and the number of fuel rods predicted to fail during an accident are input to determine the amount of radiation dose released to the environment.

The system codes evolved quickly after 1966 when an Atomic Energy Commission (AEC) task force[59] was established to evaluate the adequacy of various safety systems. Since then, the main driving forces for system code development have been the emergency core cooling hearings in the early 1970s,[60] the increasing concern about the consequences of a hypothetical accident, and the increasing licensing requirements from the NRC. In recent years the major evolution of system codes has been in improvement of running time, versatility, and thermal-hydraulic modeling. The new generation of system codes with increased versatility include RELAP4/Mod 7,[20] TRAC-PD2 and BD1,[50,51] RELAP5,[53] and RETRAN.[61] RELAP4 is currently being used by the NRC as an evaluation code. RETRAN has been developed based on the concept of RELAP4 and has been employed extensively in the utility industry to analyze plant transients. TRAC-PD2 (PWR), TRAC-BD1 (BWR), and RELAP5 have made significant advances in technical capability, flexibility of use, and level of experimental assessment. They have been developed to provide best estimates of the system response during transients and accidents. One of the major advances in technical capability of the RELAP5 and TRAC codes is the two-phase nonhomogeneous and thermal nonequilibrium modelings. The two-fluid, nonhomogeneous hydrodynamic approach, describing the characteristics of steam and water separately, allows such important phenomena as countercurrent flow to be treated explicitly. The thermal nonequilibrium approach between steam and water permits more accurate modeling of the fluid dynamics through a direct treatment of flashing and condensation effects. The capabilities of RELAP4/Mod 7, TRAC-PD2/BD1, RELAP5, and RETRAN are described in Table 7.4-3.

Each reactor vendor has its own proprietary system codes, such as CESEC,[62] COAST,[63] CRAFT,[64] LOFTRAN,[65] MARVEL,[66] ODYN,[67] PUMP,[68] REFLOOD,[69] and SATAN.[70] In general, these codes are fast running due to the need of repeated use. The technical capabilities of these codes, however, are not as advanced as the best-estimate codes, for example, TRAC and RELAP5. In addition to the codes described previously, several system codes have been developed and released to the public by various research organizations. These codes[71-80] are described in Table 7.4-4.

The containment thermal-hydraulic behavior is also of interest to the reactor designer. The containment pressure during LOCA directly impacts the reactor core heat transfer coefficient and, in

TABLE 7.4-4 LIST OF AVAILABLE SYSTEM CODES

Code Name	Development Organization	Year Released	BWR/PWR	Local/ Transient	Calculation/Description
ASCOT-1	JAERI	1978	PWR	L	Thermal hydraulic conditions during blowdown
COMPARE	LANL	1977	PWR	Both	Primary-loop subcompartment analysis with homogeneous two-phase model
DYNAM	GA	1970	PWR	T	Flow oscillation study for PWR primary loop
FLASH(4,6)	BAPL	1976	PWR	L	Large-break LOCA analysis
MARCH	BCL	1980	Both	L	Core meltdown (after LOCA) analysis
NAIAD	AAEC (Australia)	1978	Both	Both	Thermal hydraulic conditions during transients and LOCA
NATRAN	ANL	1975	Both	T	Pressure pulse transient analysis
PARK-1	BAPL	1977	PWR	T	Simulation of primary and secondary loops; plant performance and transient calculation
TILT	(ITALY)	1973	BWR	Both	Thermal hydraulic conditions during transients and LOCA
WRAP-EM	SRL	1977	Both	L	LOCA analysis, developed by the NRC as an evaluation code

turn, the peak cladding temperature during the refill and reflood phases of the accident. The thermal-hydraulic behavior, including compartment pressure, temperature, mass and energy inventory, and structural temperature distribution of a BWR containment can be analyzed by CONTEMPT-LT.[81] For PWR containment analysis CONTEMPT-4[82] or BECON[83] can be used. CONTEMPT-LT and CONTEMPT-4 are lumped parameter codes, whereas BECON is based on a three-dimensional, two-fluid, thermal nonequilibrium modeling. This detailed modeling enables BECON to analyze the hydraulic loads on the intercompartment barriers and jet loads during the blowdown phase of a LOCA. For subcompartment thermal-hydraulic behavior, RELAP, TRAC, or COMPARE can be used.

REFERENCES

7.4-1 D. S. Rowe, "COBRA-II: A Digital Computer Program for Thermal-Hydraulic Subchannel Analysis of Rod Bundle Nuclear Fuel Elements," BNWL-1229, Pacific Northwest Laboratory, 1970.

7.4-2 D. S. Rowe, "COBRA-IIIC: A Digital Computer Program for Steady State and Transient Thermal Analysis of Rod Bundle Nuclear Fuel Elements," BNWL-1695, Pacific Northwest Laboratory, 1973.

7.4-3 N. E. Todreas et al., "COBRA-IIIC/MIT-2: A Digital Computer Program for Steady State and Transient Thermal-Hydraulic Analysis of Rod Bundle Nuclear Fuel Elements," MIT-EL-81-018, M.I.T., Cambridge, 1981.

7.4-4 C. L. Wheeler, C. W. Stewart, R. J. Cena, D. S. Rowe, and A. M. Sutey, "COBRA-IV-I: An Interim Version of COBRA for Thermal-Hydraulic Analysis of Rod Bundle Nuclear Fuel Elements and Cores," BNWL-1962, Pacific Northwest Laboratory, 1976.

7.4-5 R. W. Bowing, "HAMBO: A Computer Program for the Subchannel Analysis of the Hydraulic and Burnout Characteristics of Rod Clusters," Part 1: General Description, AEEW-R524 (1967), Part 2: The Equations, AEEW-R582, U.K. Atomic Energy Authority, Winfrith, 1968.

7.4-6 "TORC Code, A Computer Code for Determining the Thermal Margin of a Reactor Core," CENPD-161, 1975.

7.4-7 H. Chelemer, P. T. Chu, and L. E. Hochreiter, "THINC-IV—An Improved Program for Thermal-Hydraulic Analysis of Rod Bundle Cores," WCAP-7956, Westinghouse Electric Corporation, 1973.

7.4-8 "LYNX2—Subchannel Thermal Hydraulic Analysis Program," BAW-1013, Babcock & Wilcox, 1976.

7.4-9 "LYNX1—Reactor Fuel Assembly Thermal-Hydraulic Analysis Code," BAW-10129, Babcock & Wilcox, 1976.

7.4-10 J. G. Collier, Convective Boiling and Condensation, McGraw-Hill, England, 1972.

7.4-11 W. A. Messena, "Steam-Water Pressure Drop and Critical Discharge Flow—A Digital Computer Program," HW-65076, Hanford, Washington, 1960.

7.4-12 A. A. Armand, "The Resistance During the Movement of a Two-Phase System in Horizontal Pipes," Translated by V. Beak, AERE Trans., p. 828 (1946).

7.4-13 S. Levy, "Forced Convection Subcooled Boiling—Prediction of Vapor Volumetric Fraction," CEAP-5157, GE, California, 1966.

7.4-14 J. S. Gellerstedt et al., Two Phase Flow and Heat Transfer in Rod Bundles, ASME, New York, 1969, pp. 63–71.

7.4-15 L. S. Tong et al., J. Nuclear Energy 21, 241 (1968).

7.4-16 G. P. Gaspari et al., Italian Report CISE-R-276, 1968.

7.4-17 J. M. Healzer, J. E. Hench, E. Jannsen, and S. Levy, "Design Basis for Critical Heat Flux Condition in Boiling Water Reactors," APED-5286, General Electric Company, 1966.

7.4-18 Hench Levy, "MATRO-Version 09—A Handbook of Material Properties for Use in LWR's," TREE-NUREG 1005, 1976.

7.4-19 T. A. Bjornard and P. Griffith, "PWR Blowdown Heat Transfer," ASME Reactor Safety Symposium, New York, 1977.

7.4-20 S. R. Behling et al., "RELAP4/Mod 7—A Best Estimate Computer Program to Calculate Thermal and Hydraulic Phenomena in Nuclear Reactor or Related Systems," NUREG/CR-1998, 1981.

7.4-21 C. J. Baroczy, "A Systematic Correlation for Two-Phase Pressure Drop," AIChE-ASME Heat Transfer Conference, Los Angeles, 1965.

7.4-22 S. L. Smith, "Void Fractions in Two-Phase Flow: A Correlation Based Upon an Equal Velocity Head Model," Proc. I.M.E. 1(38), 649 (1969).

7.4-23 R. W. Bowring, "Timing Runs for the Original and Modified COBRA-IIIC," in Program Development Notes of RFP-227, EPRI, 560–583 (1975).

7.4-24 P. Moreno, C. Chiu, R. Bowring, E. Khan, J. Liu, and N. E. Todreas, MIT-EL-006, M.I.T., 1977.

7.4-25 J. J. Rogers and R. G. Rosehart, "Turbulent Interchange Mixing in Fuel Bundles," Trans. Conf. on Applied Mech., University of Waterloo, Canada, 1969.

7.4-26 K. Petrunik, "Turbulent Interchange in Simulated Rod Bundles," Ph.D. thesis, University of Windsor, Canada, 1973.

7.4-27 K. Singh, "Air Water Turbulent Mixing in Simplified Rod Bundle Geometries," Ph.D. thesis, University of Windsor, Canada, 1972.

7.4-28 D. S. Rowe and C. W. Angle, "Crossflow Mixing Between Parallel Flow Channels During Boiling, Part III: Effect of Spacer on Mixing Between Channels," BNWL-371, Pt. 3, Pacific Northwest Laboratory, 1969.

7.4-29 K. F. Rudzinski, K. Singh, and C. C. St. Pierre, *Can. J. Chem. Eng.* **50**, 297 (1972).

7.4-30 S. G. Beus, "A Two-Phase Turbulent Mixing Model for Flow in Rod Bundles," WAPD-T-2438, 1971.

7.4-31 C. Chiu, "Three-Dimensional Transport Coefficient Model and Prediction-Correction Numerical Method for Thermal Margin Analysis of PWR Cores," *Nuclear Engineering Design* **64**(1), 103–115 (1981).

7.4-32 C. Chiu, P. Moreno, N. E. Todreas, and R. W. Bowring, "Enthalpy Transfer Between PWR Fuel Assemblies in Analysis by the Lumped Subchannel Model," *Nuclear Engineering Design* **53**, 163–189 (1979).

7.4-33 H. Abel-Larson and L. Larson, "Heating in a Reactor Fuel Element Rod Under Transient Conditions, Part II, DINO-Temperature During a Loss-of-Coolant Accident," Danish Atomic Energy Agency, 80-3, 1972.

7.4-34 R. A. Cudnik, "ECCSA and MUCHA, Computer Codes for the Analysis of Emergency Core Cooling Systems," BMI 1916, 1977.

7.4-35 C. T. Hocevar and T. W. Wineinger, "THETA-1B, A Computer Code for Nuclear Reactor Core Analysis," IN-1145, 1971.

7.4-36 W. L. Kirchner, Reflood Heat Transfer in a Light Water Reactor," NUREG-0106, 1976.

7.4-37 "STRIKIN-II—A Cylindrical Geometry Fuel Rod Heat Transfer Program," CENPD-135, Combustion Engineering, Inc., 1974.

7.4-38 Gary A. Berna and M. P. Bohn, "FRAPCON-1: A Computer for the Steady State Analysis of Oxide Fuel Rods," NUREG/CR-1463, 1981.

7.4-39 J. A. Dearien et al., "FRAP-S2: A Computer Code for the Steady-State Analysis of Oxide Fuel Rods," TREE-NUREG-1107, 1977.

7.4-40 E. Y. Lim, E. T. Rumble, and R. G. Stuart, "BEHAVE-4 and GAPCON-THERMAL-2 Evaluations-Phase III Topical Report," SAI-030-76-PA, 1976.

7.4-41 C. E. Beyer, C. R. Hann, D. D. Lanning, F. E. Panisko, and L. J. Parchen, "GAPCON-THERMAL-2: A Computer Program for Calculating the Thermal Behavior of Oxide Fuel Rods," BNWL-1897 and BNWL-1898, Northwest Pacific Laboratory, 1975.

7.4-42 H. B. Meieran and E. L. Westermann, "Summary of Phase III of the EPRI Fuel Rod Modeling Code Evaluation Project Using the LIFE-THERMAL-1 Computer Program," ODAI-RP-397-2(51), 1976.

7.4-43 P. Verbeck and N. Hoppe, "COMETHE-IIIJ, A Computer Code for Predicting Mechanical and Thermal Behavior of a Fuel Pin, Part 1: General Description," Belgonucleaire S. A., 25 Rue du Champ De Mars, 1050 Bruxelles, 1976.

7.4-44 C. D. Morgan and H. S. Kao, "TAFY—Fuel Pin Temperature and Gas Pressure Analysis," BAW-10044, 1972.

7.4-45 J. B. Newman, J. F. Giovengo, and L. P. Conden, "The CYGRO-4 Fuel Rod Analysis Computer Program," WAPD-TM-1300, 1977.

7.4-46 F. J. Homan, W. J. Lackery, and C. M. Cox, "FMODEL—A FORTRAN IV Computer Code to Predict In-Reactor Behavior of LMFBR Fuel Pins," ORNL-4825, 1973.

7.4-47 "Fuel Evaluation Model," CENPD-139, Combustion Engineering, Inc., 1974.

7.4-48 K. Able and K. Sateo, "SCORCH-B2-Simulation Code of Reactor Core Heatup During LOCA," JAERI-M6678, 1976.

7.4-49 "PARCH—A FORTRAN IV Digital Program to Evaluate Pool Boiling, Axial Rod and Coolant Heatup," CENPD-138, Combustion Engineering, Inc., 1974.

7.4-50 "TRAC-PA1: An Advanced Best-Estimate Computer Program for PWR LOCA Analysis," NUREG/CR-0665, 1979.

7.4-51 J. W. Spore et al., "TRAC-BD1: An Advanced Best Estimate Computer Program for Boiling Water Reactor Loss-of-Coolant Accident Analysis," NUREG/CR-2178, 1981.

7.4-52 L. J. Seifken et al., "FRAP-T5: A Computer Code for the Transient Analysis of Oxide Fuel Rods," NUREG/CR-084D, TREE-1281, 1979.

7.4-53 V. H. Ransom et al., "RELAP5 Code Development," CDAP-TR-057, Vol. 1, 1979.

7.4-54 H. J. Willenber, "Improvements in HOTPIN Internal Fuel Motion Code," HEDL-TIME 76-56, 1977.

7.4-55 F. T. Dunchhorst et al., "GLUB-1-A FORTRAN IV Digital Computer Program for Water-Logged Fuel Element Analysis," WAPD-TM-569, 1966.

7.4-56 R. Christensen, "SPEAR-BETA Fuel Performance Code System, Volume 1: General Description," EPRI-NP-2291, *EPRI* (1983).

7.4-57 H. R. Freeburn et al., "Light Water Reactor Fuel Rod Modeling Code Evaluation," EPRI NP-369, 1977.

7.4-58 D. D. Lanning, F. E. Panisko, and C. L. Mohr, "GAPCON-THERMAL-3, Verification and Comparison to In-Reactor Data." PNL-2435, NUREG/CR-0218, 1978

7.4-59 Emergency Core Cooling, Report of Advisory Task Force on Power Reactor Emergency Cooling, USAEC, W. K. Ergen, Chairman (1967).

7.4-60 "Acceptance Criteria for ECCS for Light Water Cooled Nuclear Power Reactors Pursuant to the AEC's Notice of Rule-Making Hearing 50-1," Docket No. RM-501-1, 1973.

7.4-61 K. V. Moore et al., "RETRAN—A Program for One-Dimensional Transient Thermal-Hydraulic Analysis of Complex Fluid Systems," EPRI CCM-5, Vols. 1, 2, 3 and 4, *EPRI* (1978).

7.4-62 "CESEC-Digital Simulation of a Combustion Engineering Nuclear Steam Supply System," CENPD-107, Combustion Engineering, Inc., 1974.

7.4-63 R. S. Daleas, "COAST Code Description," CENPD-98, 1973.

7.4-64 R. A. Hendrick, J. J. Cudlin, and R. C. Foltz, "CRAFT-2-FORTRAN Program for Digital Simulation of a Multinode Reactor Plant During Loss-of-Coolant," BAW-10092, Rev. 1 and Rev. 2, 1974.

7.4-65 T. W. T. Burnett, C. J. McIntyre, and J. C. Buker, "LOFTRAN Code Description," WCAP-7907, 1972.

7.4-66 R. C. Krise and S. Miranda, "MARVEL-A Digital Computer Code for Transient Analysis of a Multiloop PWR System," WCAP-8844, 1977.

7.4-67 "Qualifying of the One Dimensional Core Transient Model for Boiling Water Reactors, Volumes 1 and 2," NEDO-24154, 1978.

7.4-68 M. R. Grandia, "PUMP-Analog-Hybrid Reactor Coolant Hydraulic Transient Model," BAW-10073, 1973.

7.4-69 B. E. Bingham and K. C. Shieh, "REFLOOD—Description of Model for Multinode Core Reflood Analysis," BAW-10093, 1974.

7.4-70 F. M. Bordelon et al., "SATAN VI Program—A Comprehensive Space-Time Dependent Analysis of Loss-of-Coolant," WCAP-8306, 1974.

7.4-71 K. Kobayaski and K. Sato, "ASCOT-1—A Computer Program for Analyzing the Thermal-Hydraulic Behavior in a PWR Core During a LOCA," JEARI-M-7917, 1978.

7.4-72 R. G. Gido, R. G. Lawton, C. I. Grimes, and J. A. Kudrick, "COMPARE: A Computer Program for the Transient Calculation of a System of Volumes Connected by Flowing Vents," LA-NUREG-6488-MS, 1976.

7.4-73 L. E. Efferding, "DYNAM, A Digital Computer Program for Study of the Dynamic Stability of Once-Through Boiling Flow with Steam Superheat," GAMD-8656, 1968.

7.4-74 J. J. Beyer, W. D. Peterson, D. A. Prelreivx, and G. W. Swartele, "FLASH-6: A FORTRAN-IV Computer Program for Reactor Plant Loss-of-Coolant Accident Analysis," WAPD-TM-1249, 1976.

7.4-75 R. O. Wooten, R. S. Denning, and P. Cybulakis, "Analysis of the Three Mile Island Accident and Alternative Sequences," NUREG/CR-1219, 1980.

7.4-76 G. D. Trimble and W. J. Turner, "NAIAD—A Computer Program for Calculation of the Steady State and Transient Behavior (Including LOCA) of Compressible Two-Phase Coolant in Networks," AAEC/E378, 1976.

7.4-77 Y. W. Shim and R. A. Valentin, "Two-Dimensional Fluid Hammer Analysis by the Method of Characteristics in a Closed Axisymmetric Cylindrical Domain," ANL-8090, 1974.

7.4-78 W. G. Clark et al., "PARK-1, A FORTRAN-IV Computer Program for Nuclear Reactor Power Plants Analysis," WAPD-TM-447(L), 1975.

7.4-79 A. Magni, "Tilt: A Digital Simulator Program for the Study of Hydrodynamic Processes and Core Heat of Boiling Water Pressure Tube Reactor During Transient Conditions," *Energia Nuclear* **20**, B5, Maggio (1973).

7.4-80 M. M. Anderson, "WRAP—A Water Reactor Analysis Package," DPST-NUREG-77-1, Savannah River Laboratory, 1977.

7.4-81 L. Wheat, R. J. Wagner, G. F. Niederauer, and C. F. Obenchain, "CONTEMPT-LT-A Computer Program for Prediction of Containment Pressure-Temperature Response to a Loss-of-Coolant," ANCR-1219, Version 26, SRD-83-76, 1976.

7.4-82 L. J. Metcalfe, W. J. Mings, J. E. Hartman, and A. C. Crail, "CONTEMPT4/MOD2, A Multicompartment Containment System Analysis Program," TREE-NUREG-1202, 1978.

7.4-83 R. A. Walls, "BECON/MOD2A—A Computer Program for Sub-Compartment Analysis of Nuclear Reactor Containment—A User's Manual," CDAP-TR0051, 1979.

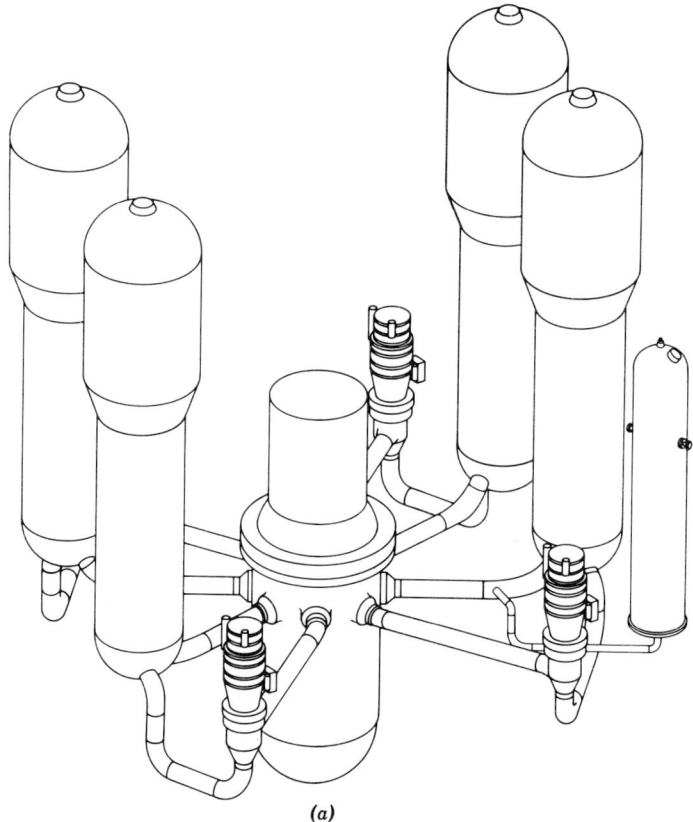

(a)

Fig. 7.5-1a Westinghouse nuclear steam supply system four-loop design. (Courtesy Westinghouse Electric Corporation.)

7.5 NUCLEAR POWER PLANT TYPES

David D. Lanning, Robert H. Simon, Robert A. Evans, and Frank J. Rahn*

7.5-1 Pressurized Water Reactor (PWR)

The pressurized water reactor (PWR) is the major commercial nuclear power plant that is now being utilized throughout the world. In the United States there are three different commercial PWR vendors and other vendors build this type of reactor in countries such as France and Germany. Although the reactor type is the same, there are detailed design differences provided by each PWR vendor. Arrangements of the primary system components are shown in Figs. 7.5-1a to c. Notice that all three designs incorporate the same basic components, namely a reactor vessel and a primary coolant circulating loops consisting of pumps, steam generator, and pressurizer. The specially treated ordinary water in these loops and the vessel is kept at a pressure of about 15.5 MPa (2250 psi) by adjustment of a steam bubble in the upper part of the pressurizer.

A more complete simplified diagram of the PWR nuclear plant is given in Fig. 7.5-2. As can be seen, the heat generated in the reactor is circulated to the steam generators where it is transferred to a secondary system. The ordinary water in the secondary side of the steam generator is boiled into steam at a pressure of about 6.9 MPa (1000 psi) and outlet temperature of about 285°C (545°F). This steam is circulated to the turbine, where it does the work of turning the turbine much like a windmill. The turbine is connected to drive an electric generator that produces the electricity. The steam then passes to a large condenser where a third circulating cooling water system takes the rejected heat and transports it to the environment, either to the atmosphere by a cooling tower or to a river, lake, or

*The material on gas-cooled reactors was contributed by R. H. Simon and R. A. Evans; that on liquid metal fast breeder reactors by F. J. Rahn.

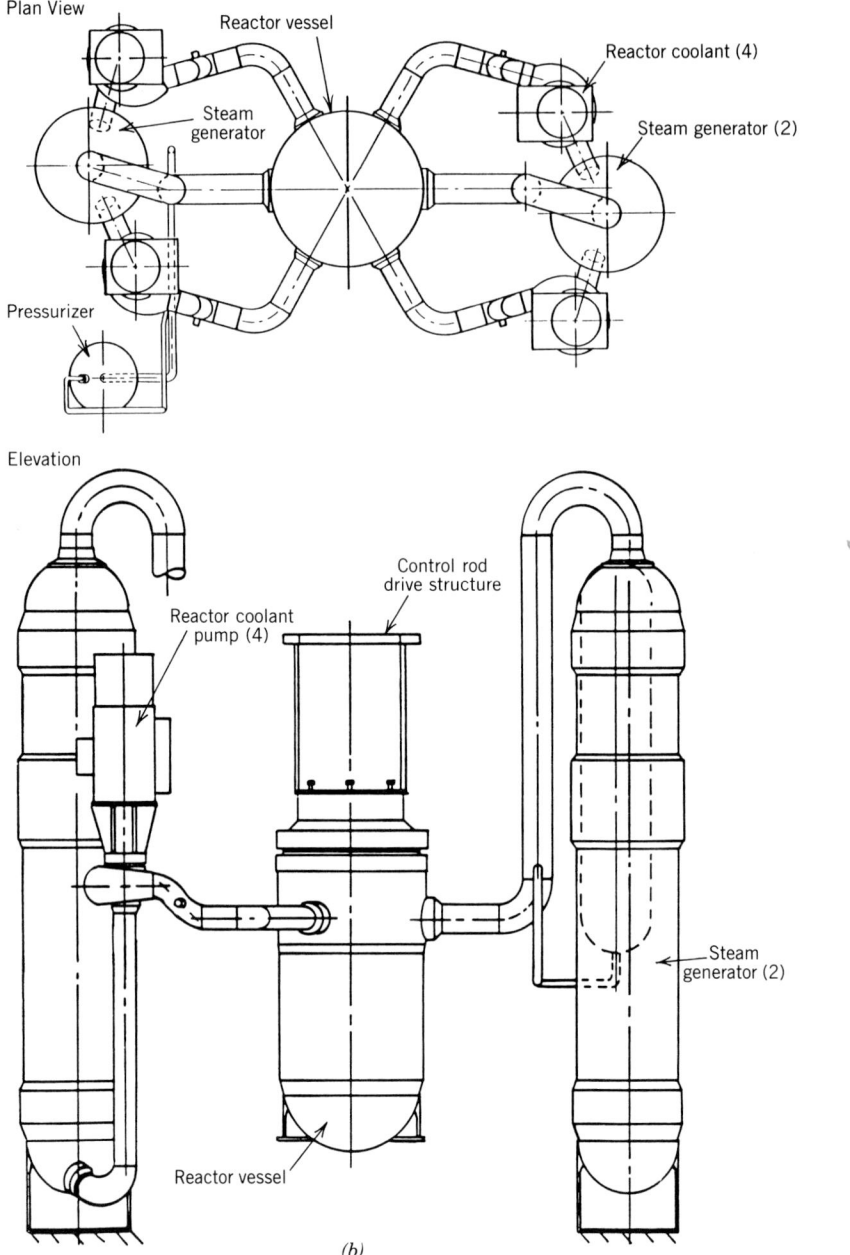

Plan View

Reactor vessel

Reactor coolant (4)

Steam
generator

Steam generator (2)

Pressurizer

Elevation

Control rod
drive structure

Reactor coolant
pump (4)

Steam
generator (2)

Reactor vessel

(b)

Fig. 7.5-1b Babcock & Wilcox reactor coolant system arrangement. (Courtesy of Babcock & Wilcox, Lynchburg, VA.)

ocean. The net efficiency of a typical PWR is about 32%, and hence one-third of the energy from the reactor is changed into electric energy and two-thirds is released as the rejected heat. More details of the PWR design are given in Section 7.6.

7.5-2 Boiling Water Reactor (BWR)

The boiling water reactor (BWR) electric generating power plant differs from the PWR system in that it affords a direct-cycle concept of producing steam. The steam is generated within the reactor and

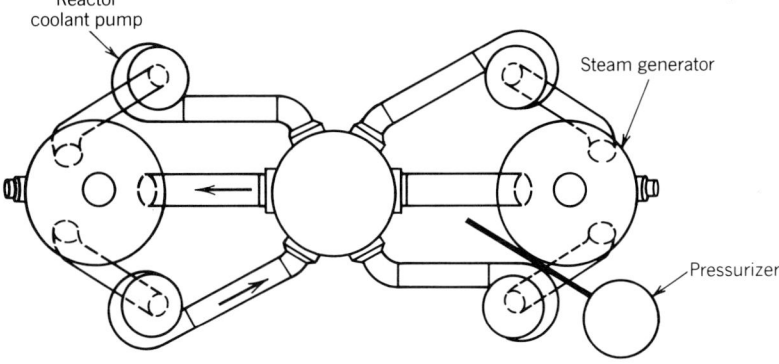

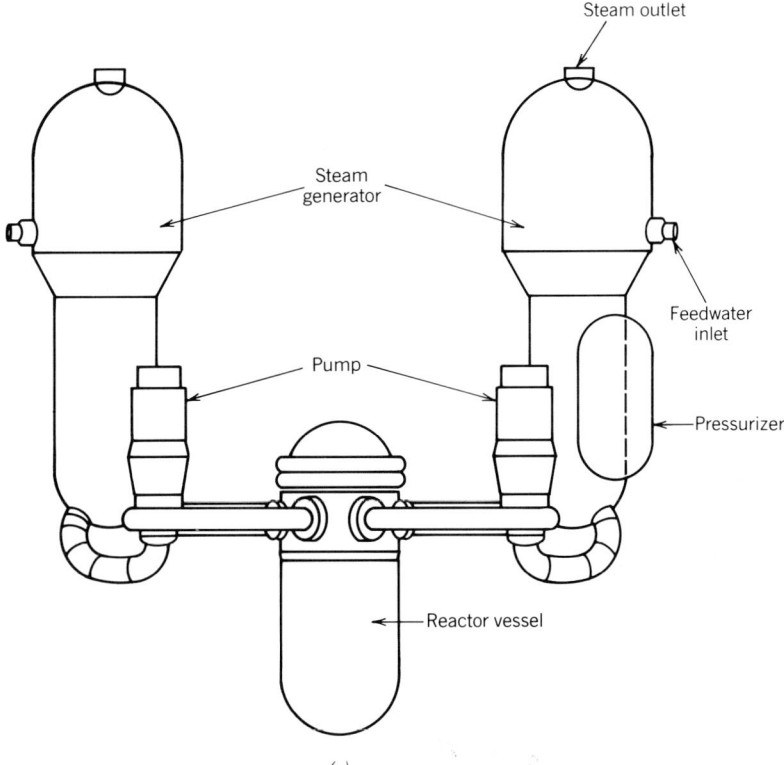

(c)

Fig. 7.5-1c Forked River reactor coolant system arrangement. (Courtesy Combustion Engineering Corporation.)

sent directly from the reactor to the steam-driven turbine. The concept of a BWR is shown in Fig. 7.5-3. The primary coolant water is heated and boiled in the core operating at about 290°C (550°F) and 7.0 MPa (1040 psi). As shown in Fig. 7.5-3, the steam drives the turbine generator to produce electricity and then the condensed steam becomes the feedwater and is pumped back into the reactor vessel for reheating and boiling.

The BWR is manufactured by one vendor in the United States. Several vendors in other countries, notably Sweden and Germany, produce similar versions of the BWR.

The primary system of the BWR is shown in Fig. 7.5-4. The containment system for the BWR includes redundant isolation valves on each steam line; these assure the capability for isolation of the

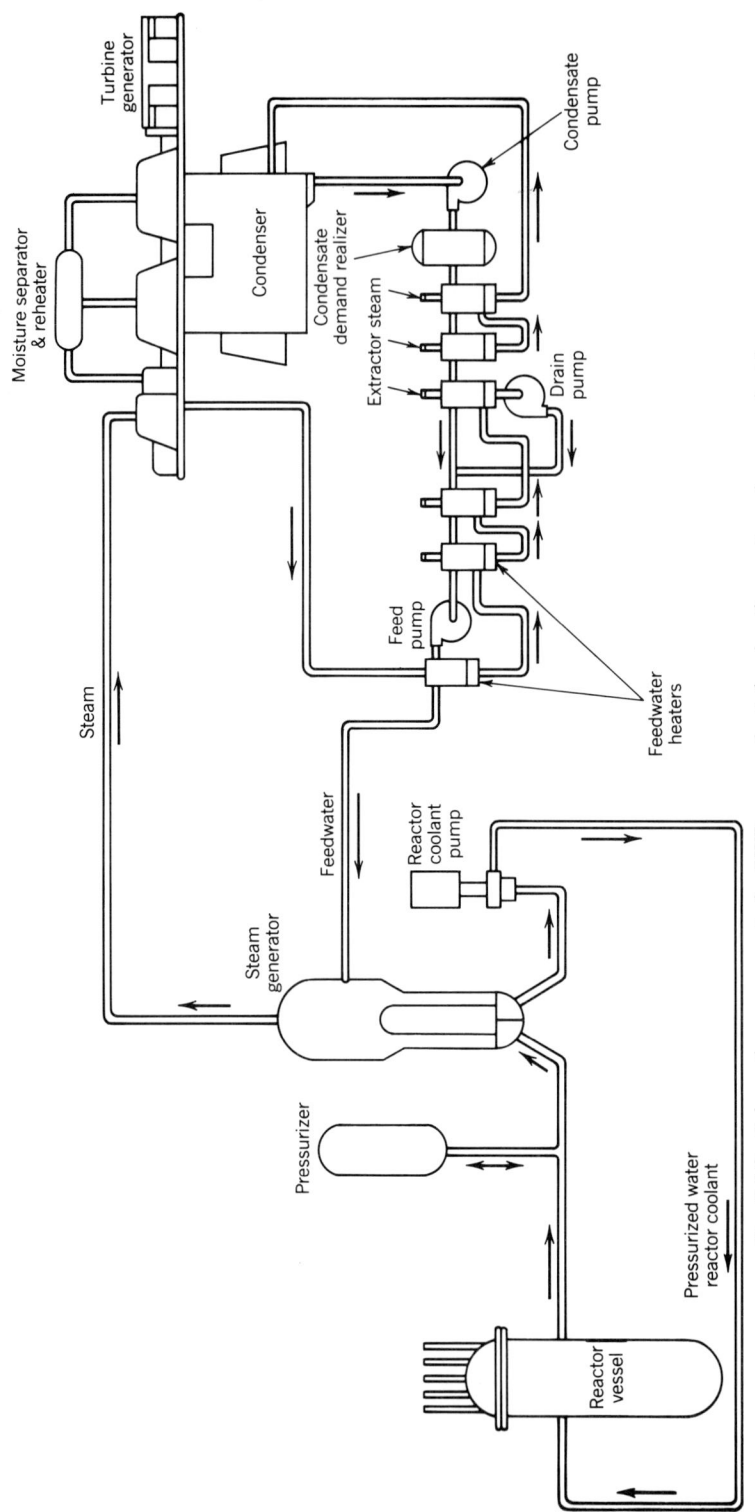

Fig. 7.5-2 Schematic arrangement of a PWR power plant showing the reactor primary coolant loop and the turbine, power train. Reprinted with permission from K. C. Lish, *Nuclear Power Plant Systems and Equipment*, Industrial Press, New York, 1972.

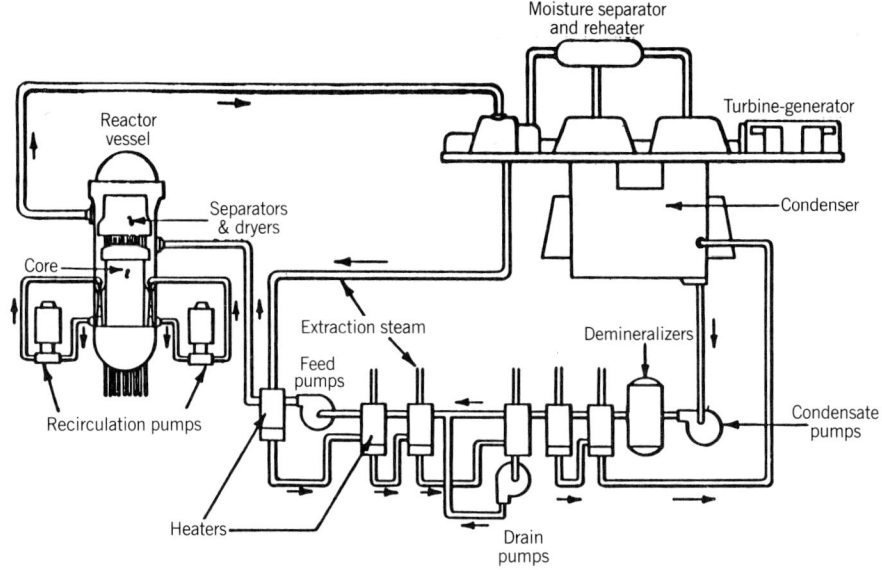

Fig. 7.5-3 Boiling water reactor system. (Courtesy General Electric Company.)

Moisture separator
and reheater

Turbine-generator

Reactor
vessel

Separators
& dryers

Condenser

Core

Extraction steam

Demineralizers

Feed
pumps

Recirculation pumps

Condensate
pumps

Heaters

Drain
pumps

Steam dryers

Main steam flow
to turbine

Steam
separators

Main feed flow
from turbine

Driving flow

Jet pump

Core

Recirculation
pump

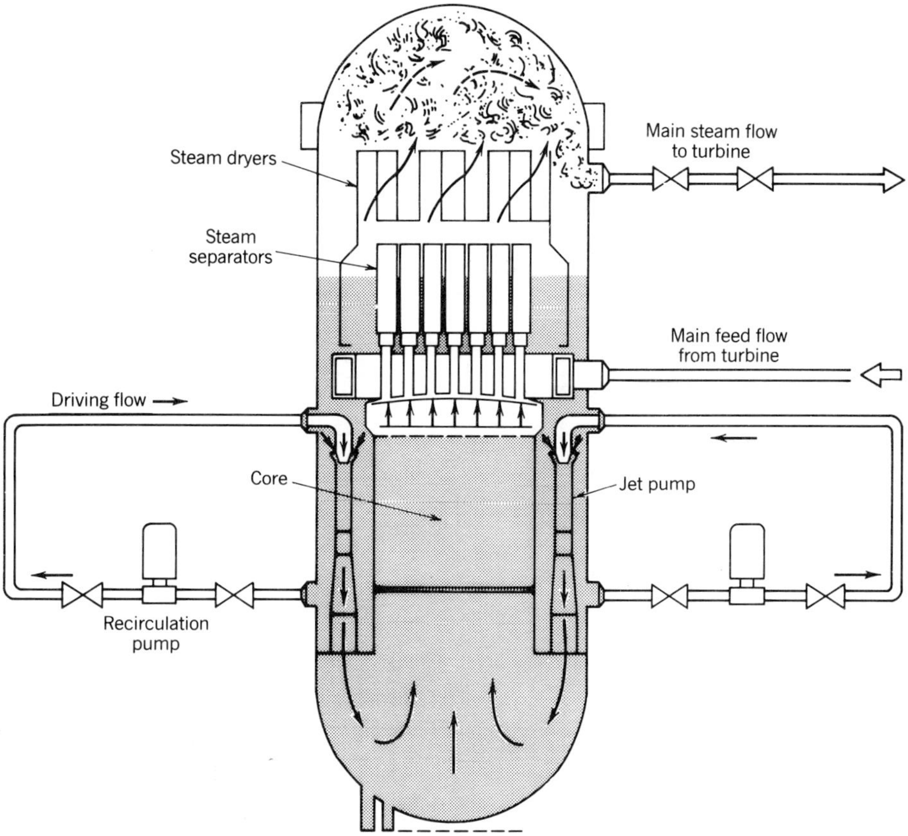

Fig. 7.5-4 Boiling water reactor primary system. (Courtesy General Electric Company.)

primary system within the protected region of the containment building. Since boiling of the primary coolant occurs in the core, the steam–water mixture at the core exit (15% steam by weight) is separated and the steam is further dried in the upper structures of the pressure vessel before exiting to the steam lines. The remaining water coolant is recirculated back through the core. Forced recirculation, rather than natural convection, is used to enhance the heat transfer and allows higher power densities to be used. The net efficiency for production of electricity is about 33%, and the remaining two-thirds of the total thermal power is reduced from the condenser cooling system. Further details of the BWR components are given in Section 7.6-2.

7.5-3   High-Temperature Gas-Cooled Reactor (HTGR) Plants

The HTGR comprises a family of advanced converter reactor systems using ceramic fuel, a graphite moderator, and helium as the coolant. These features provide the ability to operate at high temperatures, which results in high efficiency for electrical generation and offers the potential for unique applications of nuclear power to meet broad energy requirements, including cogeneration and process heat. The inherent HTGR design features, as discussed in Section 7.6-3, result in large safety margins, low environmental impacts, and low personnel radiation exposure.

The first HTGR to produce electricity was the Philadelphia Electric Company's 40-MW(e) Peach Bottom 1 developmental power plant, which began commercial operation in June 1967 and was shut down in October 1974 after a total of 1349 equivalent full-power days and production of more than 1.4 billion kilowatt-hours. This reactor served as proof of the HTGR concept. The net thermal efficiency was 35%, and the nuclear system availability was 88% over its $7\frac{1}{2}$-year life.

The first U.S. HTGR demonstration plant is the 330-MW(e) Fort St. Vrain Nuclear Generating Station, owned and operated since 1974 by the Public Service Company of Colorado. Fort St. Vrain (FSV) began commercial operation in 1976 and was taken to full power in 1981. Design innovations include use of silicon-carbide-coated particles, once-through modular steam generators with integral superheaters and reheaters, steam-driven helium circulators and a prestressed concrete reactor vessel (PCRV).

FSV has demonstrated high fuel integrity, low radiation exposure of plant personnel, and ease of plant control. The high core thermal capacity has resulted in slow, predictable and easily controlled temperature transients; that is, the HTGR is a forgiving reactor and allows the operators time to react to problems. Operation to date has achieved 1420°F (771°C) primary loop helium temperatures and 1000°F (538°C) steam temperatures. Figure 7.5-5 is a simplified schematic of the FSV heat cycle.

Steam Cycle/Cogeneration (HTGR-SC/C) Plant

In the HTGR-SC/C design, extending FSV technology, reactor thermal energy is used to generate high-temperature, high-pressure steam for either high-efficiency electricity production or for the cogeneration of steam and electricity for process applications.

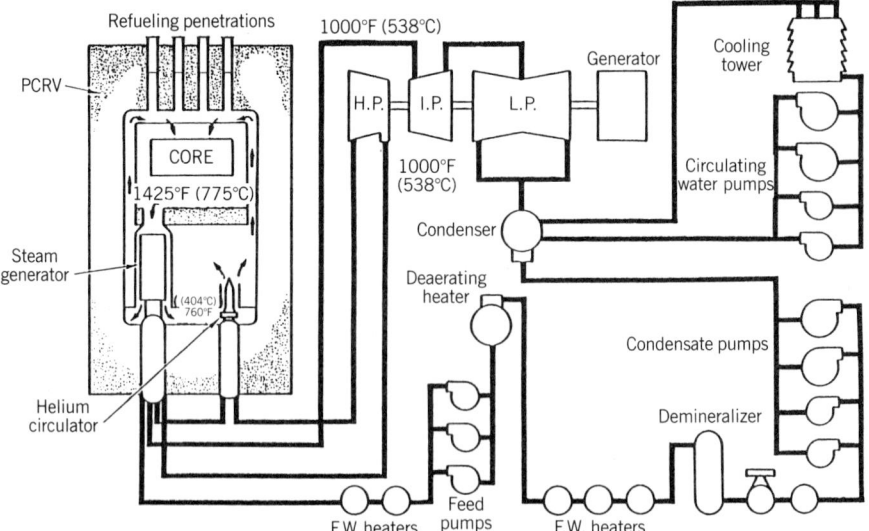

Fig. 7.5-5   Fort St. Vrain flow diagram. (Courtesy General Atomic Company, San Diego, CA.)

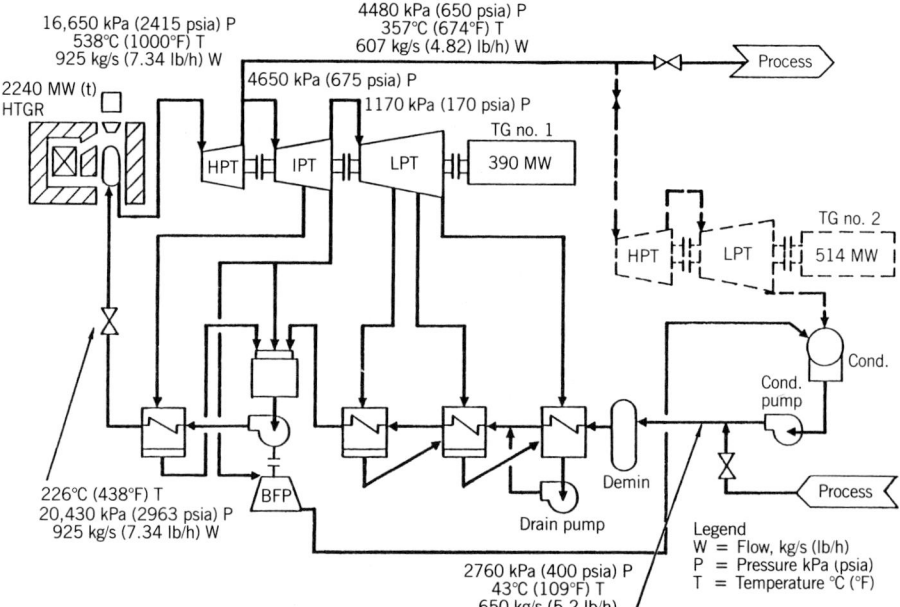

Fig. 7.5-6 HTGR-SC/S flow diagram. (Courtesy General Atomic Company, San Diego, CA.)

Figure 7.5-6 shows a simplified schematic for a typical cogeneration application with a 2240-MW(t) HTGR-SC/C system. In this cycle the output of the condensing turbine (turbine 2) would be varied depending on the demand for process steam. In the maximum steam generation mode the condensing turbine would be shut down and the steam delivered to industrial users.

Advanced Concepts

Two advanced configurations of the HTGR have been the subject of considerable design, analysis, and experimental work. One is the HTGR–gas turbine in which helium is used as the working fluid in a closed Brayton cycle gas turbine. The other is the HTGR–process heat with helium temperatures up to 1740°F (950°C). This opens up a greater range of process heat applications, for example, steam-methane reforming.

Foreign Developments

In 1955 the first gas-cooled reactor power station was started. This station was of the Magnox series at Calder Hall in the U.K. These plants use carbon dioxide as the coolant and natural uranium as the fuel. Thirty-six Magnox systems with a combined electrical rating of 8200 MW had accumulated over 600 reactor-years of operation by 1982. The advanced gas-cooled reactor (AGR) evolved from the Magnox reactor. A total of 15 AGR systems with a combined electrical rating of 8872 MW are scheduled for the U.K., including the four now operating.

The first gas-cooled reactor in Germany is the 15-MW(e) AVR "pebble bed" reactor (Fig. 7.5-7), which has been operating since 1967. This reactor, with on-line refueling of spherical fuel elements, has achieved burnups to 180 MWd/kg heavy metal and reached 1742°F (950°C) core outlet temperature in 1974. This success led to the THTR-300 pebble bed reactor demonstration plant, which is expected to be in operation in 1985.

7.5-4 Heavy-Water-Moderated Reactors (HWR)

The use of heavy water (D_2O) as a neutron moderator allows improved neutron utilization and hence better fuel utilization compared to a light water reactor; that is, lower enrichment or even natural uranium fuel can be used. In fact, many countries have adopted the heavy water reactor (HWR) for commercial power production because it can be operated with natural uranium fuel and thus does not require any dependency on uranium enrichment facilities. However, since D_2O is only about 1 part in

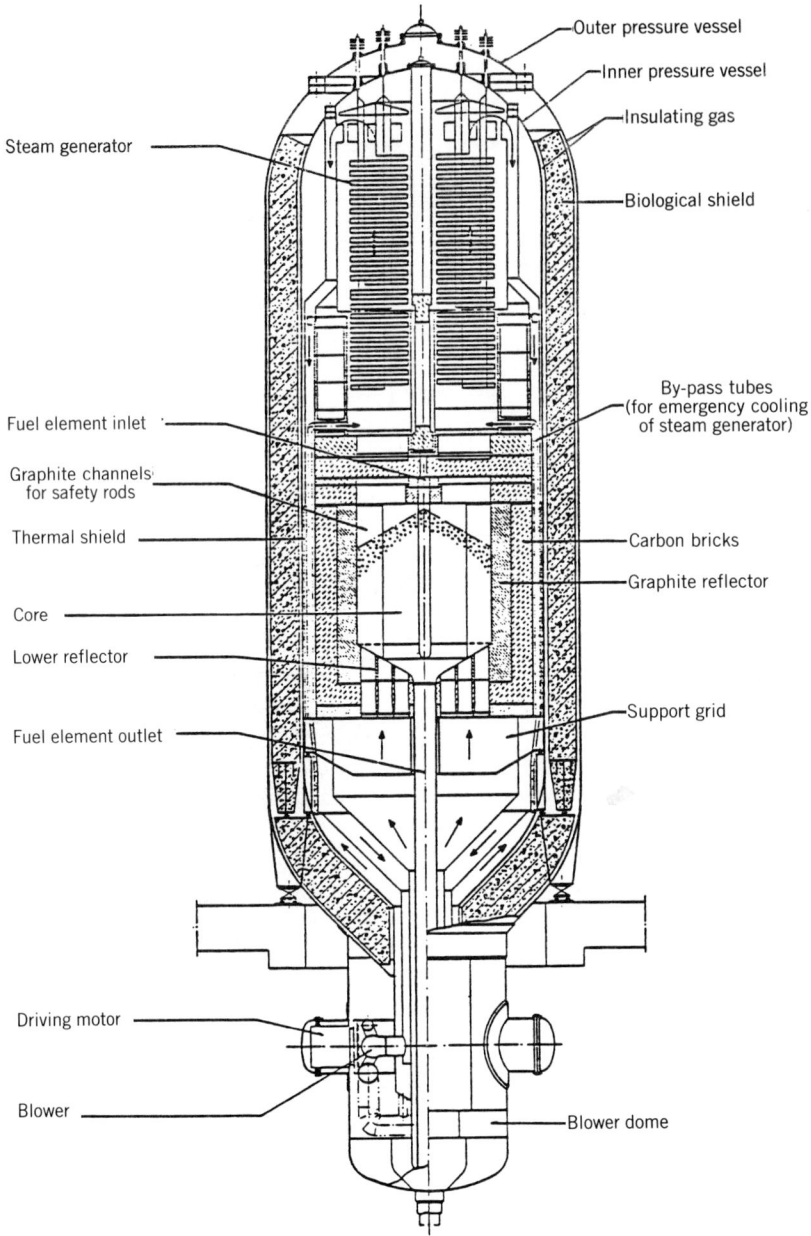

Fig. 7.5-7 Vertical section of AVR "pebble bed" reactor. (Courtesy Arbeitsgemeinschaft Versuchs-Reaktor Dusseldorf, Germany.)

6400 in ordinary water, the use of D_2O for the reactor requires the availability of heavy-water separation facilities.

The major commercial HWR in the world today is the Canadian design called the CANDU (*Can*adian *D*euterium *U*ranium) utilizing D_2O as a moderator and a separate circuit of pressurized D_2O as the coolant. The reactor arrangement involves a large low-pressure reactor vessel, called a calandria, filled with D_2O and penetrated by pressurized tubes containing the fuel and D_2O coolant. These tubes pass horizontally through the calandria as shown in the cutaway drawing (Fig. 7.5-8). The D_2O in the calandria at low (near-atmospheric) pressure is the neutron moderator and the pressurized

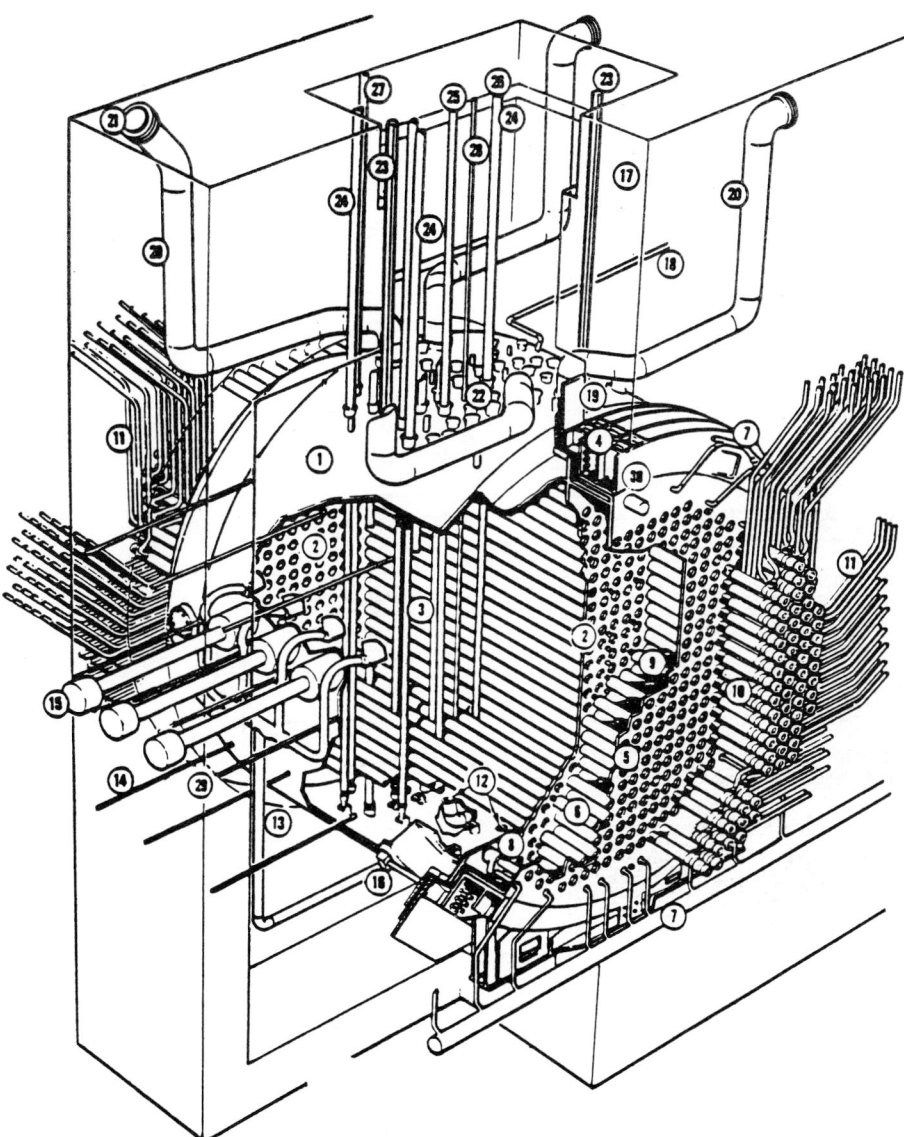

Fig. 7.5-8 CANDU reactor assembly: (1) calandria, (2) calandria—side tubesheet, (3) calandria tubes, (4) embedment ring, (5) fueling machine—side tubesheet, (6) end shield lattice tubes, (7) end shield cooling pipes, (8) inlet–outlet strainer, (9) steel ball shielding, (10) end fittings, (11) feeder pipes, (12) moderator outlet, (13) moderator inlet, (14) horizontal flux detector unit, (15) ion chamber, (16) earthquake restraint, (17) calandria vault wall, (18) moderator expansion to head tank, (19) curtain shielding slabs, (20) pressure relief pipes, (21) rupture disc, (22) reactivity control unit nozzles, (23) viewing port, (24) shutoff unit, (25) adjuster unit, (26) control absorber unit, (27) zone control unit, (28) vertical flux detector unit, (29) liquid injection shutdown nozzle, (30) ball filling pipe. (Courtesy Atomic Energy of Canada, Ltd.)

($\sim$ 1500 psi or 10 MPa) D_2O in the tubes removes the heat from the fuel and transfers it to the secondary light water system outside the tubes in the steam generators. The boiling light water in the steam generator provides the steam for driving the turbine system, as shown in Fig. 7.5-9. The secondary steam supply system is cooled in the condenser at the turbine outlet and a third circulating cooling water system, also light water, carries the rejected heat away from the condenser. In addition, some heat appears in the D_2O moderator due to gamma and neutron effects as well as some heat

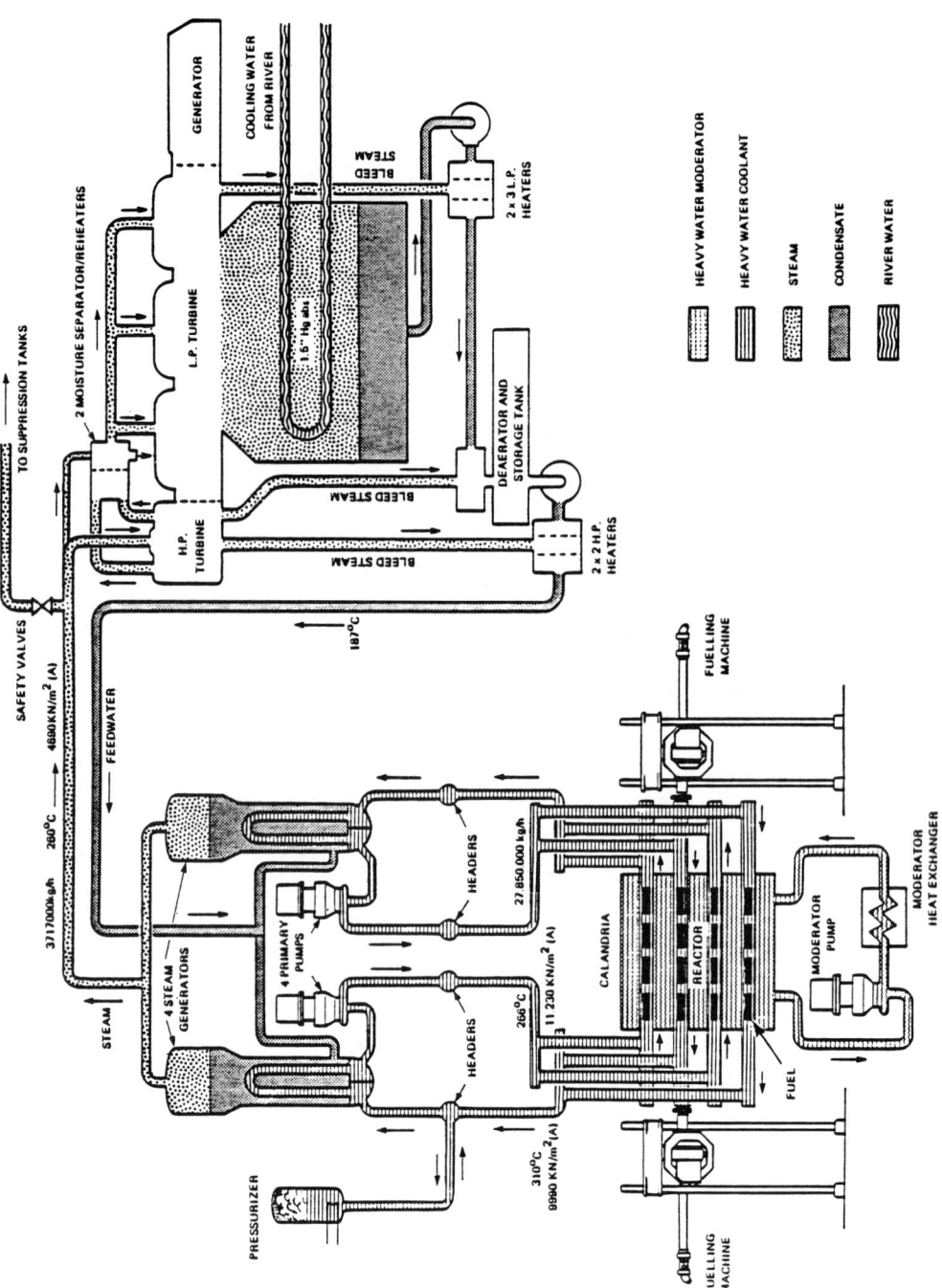

Fig. 7.5-9 CANDU nuclear power system. (Courtesy Atomic Energy of Canada, Ltd.)

transfer. This low-temperature unpressurized D_2O is cooled by a separate circulation system with a heat exchanger to remove the heat and transfer it to the cooling water system.

The use of natural uranium requires frequent refueling, and to avoid frequent reactor shutdowns, a system for on-line refueling has been developed. Fresh fuel bundles can be inserted at one end while spent fuel is removed from the other end with continued pressure and coolant flow provided by machines on each end of the reactor, without shutting the reactor down. The successful development of the machine has made these HWR plants capable of demonstrated availabilities notably higher than other commercial nuclear power plants. On the other hand, the use of lower primary system pressure, to reduce D_2O leakage possibilities, gives a typical thermal efficiency for the plants of 28–29%, slightly lower than the PWR and BWR plants. More design details are given in Section 7.6-4.

Although the CANDU HWR is the primary D_2O-moderated commercial reactor, there are other designs, notably the ATUCHA reactors in Argentina, designed by KWU in Germany. The ATUCHA I and II involve pressurized vessels containing the D_2O moderator at the same pressure as the D_2O coolant that passes through vertical tubes in the vessel. On-line refueling is also designed into the system by a machine that can be connected to adapters on the top head of the pressure vessels. Again, a high capacity factor has been demonstrated for ATUCHA I and higher plant thermal efficiency (32%) is expected for ATUCHA II by transferring heat from the moderator to the secondary system through feedwater heaters.

There have been other types of heavy-water-moderated reactors demonstrated and utilized. The French have used a heavy-water-moderated gas-cooled pressure tube design. The Canadians have demonstrated a heavy-water-moderated boiling light water pressure tube cooled design with natural uranium, and the English have demonstrated a similar system called the steam-generating heavy water reactor (SGHWR) utilizing slightly enriched uranium. In Japan a special type of HWR has been designed and demonstrated by the FUGEN-ATR (Advanced Test) reactor. The FUGEN reactor is designed to burn plutonium that has been produced in light-water reactors. The design consists of a low-pressure calandria with vertical tubes containing the PUO_2-UO_2 mixed-oxide fuel and pressurized boiling light water coolant, hence the remaining plant is similar to a BWR.

More details of a typical heavy water reactor (CANDU type) is given in Section 7.6-4.

7.5-5 Liquid Metal Fast Breeder Reactor (LMFBR)

A liquid metal fast breeder reactor (LMFBR) is a nuclear reactor that produces more fissile fuel than it consumes, using the liquid metal sodium to transport the heat in the core to the steam generator. An LMFBR is usually fueled with Pu-239. Excess neutrons can be captured by U-238 placed in or around the core creating additional Pu-239 by the reaction

$$U\text{-}238 + n \rightarrow Pu\text{-}239 + 2\beta$$

The net rate of Pu-239 production depends on the core design, although fuel doubling times of 15 years are typical.

The first electricity from a nuclear plant was produced by the EBR-I breeder reactor on December 20, 1951. Although the United States was the first to develop a breeder, other countries have since taken the lead. France, Britain, and the USSR have all proceeded with their own, original plans. A listing of the worldwide LMFBR projects is given in Table 7.5-1, while Table 7.5-2 shows some key differences for the major projects now under construction. LMFBRs have characteristics that are basically different than LWRs. The principal ones are: the radioactivity of the primary coolant systems; the flammability of sodium; the low vapor pressure of the coolant; and the vulnerability of thin-walled, hot pipes to seismic loads.

A basic difference between the various breeders is the design of their heat transport system. There are two possibilities. The first is a loop design, where the sodium coolant, after exiting the core, flows through an external piping loop, which contains a heat exchanger and pump. The second type is a pool design, where the primary sodium coolant never leaves the pressure vessel. In this variation the sodium, after exiting the core, circulates through heat exchangers and pumps that are placed inside the vessel itself. This is accomplished by means of flow baffles and structures that channel its flow. An example of the former design is the U.S. Clinch River reactor; while the French Super-Phénix employs the pool arrangement. Both plants will be described in Section 7.6.

A pool-type pressure vessel is much larger than that required for a loop design. It may be nearly 74 ft in diameter, compared to the 44 ft required for the loop-type pressure vessel. Pressure vessels this large are possible because sodium has a low vapor pressure at operating temperatures, so that high pressures, which would require heavy thick-walled vessels, are not attained. However, the pool design avoids the external intermediate heat transport piping. This in turn avoids special shielded compartments for the coolant pumps and intermediate heat exchangers, required in loop design because the primary sodium coolant is highly radioactive. Such compartments greatly complicate the design and increase the size of the reactor containment building. The relative merits of the pool and loop LMFBR designs are given in Table 7.5-3.

TABLE 7.5-1   LMFBR PROJECTS THROUGHOUT THE WORLD

Name	Country	Type	Power (MWt)	Power (MWe)	Initial Operation
In Operation					
BR-10	USSR	Loop	10	—	1959
EBR-II	United States	Pool	62	16	1963
Rapsodie	France	Loop	40	—	1967
BOR-60	USSR	Loop	60	12	1969
BN-350	USSR	Loop	1000	350	1972
Phenix	France	Pool	563	250	1973
PFR	United Kingdom	Pool	600	250	1974
JOYO	Japan	Loop	75	—	1977
KNK-2	Germany	Loop	58	20	1977
FFTF	United States	Loop	400	—	1980
BN-600	USSR	Pool	1470	600	1980
In Procurement/Construction					
Super-Phénix	France	Pool	3000	1200	1985
SNR-300	Germany	Loop	762	327	1987
MONJU	Japan	Loop	714	300	1987
CRBR	United States	Loop	975	350	Cancelled
In Planning/Design					
Super-Phénix 2 & 3	France	Pool	3750	1500	1992
BN-1600	USSR	Pool	4200	1600	1990
CDFR-1	United Kingdom	Pool	3250	1300	1992
DFBR	Japan	—	2800	1000	1993

Source.   From EPRI Report NP-1972.

7.5-6   Gas-Cooled Fast Breeder Reactor (GCFR)

The gas-cooled fast breeder reactor, in common with the liquid metal fast breeder reactor (LMFBR), generates more fissionable material than it consumes and produces electricity as well. The coolant is helium at a pressure of about 10.5 MPa (1520 psig). The use of helium has several advantages: (1) helium has a low absorption cross section to neutrons, which results in a hard neutron spectrum leading to a very high breeding ratio and low induced radioactivity; (2) helium is chemically inert; (3) helium is a gas under all reactor conditions, and therefore the problems associated with a coolant phase change are eliminated; and (4) helium is visually transparent, which along with the low induced radioactivity eases some of the operating and maintenance problems.

Fast reactors do not have a moderator and hence have a high volumetric core power density. As a result, liquid metal, because of its excellent heat transfer properties, was the initial choice of coolant and continues to be the coolant in all fast reactors in operation or under construction. Starting in the early 1960s, however, the possibility of using high-pressure helium as a coolant was investigated. Interest in the GCFR continues not only because of the advantages of helium mentioned above, but also because it increases the probability that a breeder which is technically, economically, and institutionally acceptable will be developed.

Work on GCFR design and development has proceeded in parallel over the past 20 years. The former has led to the conceptual design of the nuclear steam supply (NSS) system for a 360-MW(e) demonstration plant. This design combines the helium technology, systems, and components of the high-temperature gas-cooled reactor (HTGR) and the fuel and physics technology of the LMFBR. It features an upflow core, pressure-equalized and vented fuel rods, a multicavity prestressed concrete reactor vessel (PCRV), electric motor drives for the main helium circulators, and extensive provision for residual heat removal. There are three main and three core auxiliary cooling loops of diverse design, and a pony motor on the main loop provides a backup safety class system. In addition, the core auxiliary cooling system can operate with natural circulation. The GCFR is described in more detail in Section 7.6-6.

TABLE 7.5-2   DESIGN FEATURE COMPARISON

Design Feature	U.S. Clinch River	Japan Monju	France Super-Phenix	U.K. CDFR-1	West Germany SNR-300	USSR SN-600
Pool design			•	•		•
Loop design	•	•			•	
Heterogeneous core	•					
Human factors control room	•	•		•		
Containment/confinement with filtered venting	•	•	•		•	
All reactor buildings on a common mat	•	•		•	•	
Independent, redundant, and diverse shutdown systems	•	•	•	•	•	•
Short-shaft coolant pump	•					
Hot-leg primary coolant pump	•				•	
Wire replacer multiplexing	•					
Independent decay heat removal system	•	•	•	•	•	•
HCDA qualified reactor head	•	•	•	•	•	•
Natural circulation capability	•	•	•	•	•	•
Turbine-driven auxiliary feedwater	•	•	•	•	•	•
Steam generator qualified tube guillotine rupture	•	•	•	•	•	•
Sodium/water reaction accommodation of intermediate piping	•	•	•	•		•
Valveless intermediate loop	•					•
Primary and secondary systems cell liners	•	•	N/A	N/A	•	N/A
Bent-tube steam generator	•			•		
On-line accessibility of head access area	•	•	•	•	•	•
Superheated steam cycle	•	•	•	•	•	•
Inconel clad upper internals	•		N/A	N/A		N/A
Sodium/water reaction sensor at steam generator outlets	•	•	•	•	•	•
Straight-pull refueling machine	•	•	•	•	•	•
Limited free bow core restraint	•					

Source.   Energy Daily, Dec. 8, 1982.

TABLE 7.5-3   COMPARISON OF LOOP AND POOL DESIGNS

Advantages of Pool Design

Simpler, no need for external primary sodium piping.

Smaller reactor containment building required.

Neutron irradiation of pressure vessel is low, therefore, no neutron embrittlement and activation. As a result of the latter, the pressure vessel is accessible.

High Na inventory is an advantage in the case of an accident.

The pressure vessel sees low temperatures of cold sodium ($\sim 400°C$) uniformly. (Loop plants usually have cold sodium at bottom of pressure vessel, hot sodium ($\sim 510°C$) at top.)

Advantages of Loop Design

Smaller reactor pressure vessel.

Seismic design is easier.

Pressure vessel does not have to support the pumps and IHXs.

Thermal and hydraulic design easier.

7.6 NUCLEAR REACTOR SYSTEMS AND COMPONENTS

David D. Lanning, Robert H. Simon, Robert A. Evans, and Frank J. Rahn*

7.6-1 Pressurized Water Reactor (PWR)

The brief description in Section 7.5-1 shows the arrangements of the primary components of a PWR. This section provides a more detailed description of those components.

Reactor Core and Pressure Vessel

The core of the reactor is installed within the pressure vessel, as shown in Fig. 7.6-1 cutaway. This core consists of fuel assemblies and control rod mechanisms. Each fuel assembly is a square array of fuel rods with some lattice positions left unfueled so that control rods can be inserted into selected assemblies. Cooling water is circulated upward through the core within the spaces between the fuel rods. The fuel rods are typically long, thin (~ 1.0 cm diameter) zirconium alloy (Zircaloy) tubes containing uranium dioxide (UO_2) fuel pellets. In the fresh fuel the uranium is slightly enriched (approximately 3% U-235 with 97% U-238). Figure 7.6-2 is a diagram of a fuel rod and a fuel assembly. Typical dimensions and characteristics of the operating core and pressure vessel are given in Table 7.6-1.

Control Mechanisms

The reactor power level is controlled by adjustment of the neutron-absorbing control rods (typically, containing boron as the neutron absorber) or by adjustment of the neutron absorption by changes in boron concentration in the water coolant. Rapid adjustment of the power and power distribution can be made by moving the control rods, and the reactor can be quickly shut down by dropping the control rods to their fully inserted position. Control rods are grouped to move in and out of selected fuel assemblies, as shown in Fig. 7.6-3. Longer-term reactivity adjustments for fuel burnup and fission product effects can be compensated by changing the concentration of boron in the primary coolant. Thus, at full power most of the control rods are pulled up to their full-out position. Control characteristics are stable due to the negative (power-reducing) feedback effects when the fuel and coolant temperatures are raised (see Section 7.2). Also, it should be remembered that since the water in the core acts as a neutron moderator as well as the coolant, when the water density is reduced, such as by boiling, the reactivity is reduced and the reactor is inherently caused to reduce power or shut down.

Primary Coolant System

The power is removed from the core by circulating the pressurized and heated water from the core through a steam generator. These components, consisting of pipes, pumps and steam generators, are shown in Fig. 7.5-1. On one loop a pressurizer is attached and is used to control the pressure of the entire primary system. The pressurizer contains heaters in the lower section and sprays in the upper section. To increase pressure, the heaters are turned on, creating a larger steam bubble pressure in the upper section; to decrease the pressure, the sprays are turned on in order to condense some steam and reduce the steam bubble pressure.

The pressurized water (15.5 MPa) can be heated by the core to 325°C with no significant boiling. This heated water is circulated to the steam generator, where it passes through a large bundle of tubes. Outside the tubes, the secondary water is heated at a lower pressure (7 MPa) and boiled (at 288°C) to produce the desired steam to drive the turbine. Figure 7.6-4 gives a cutaway view of a typical steam generator.

Decay Heat Removal and Emergency Core Cooling System

Even after the reactor has been shut down by insertion of the control systems, there is a residual heat remaining in the fuel elements and a continued low power (less than 7% of full power) due to the fission product decay heat. Continued core cooling is required to remove this heat. Under normal shutdown conditions, initial cooling is provided by the steam rejection through turbine bypass valves to the condenser. After an initial cooling period the decay heat reduces to 4% of full power. After 1 hr

---

*The material on gas-cooled fast reactors was contributed by Robert H. Simon and Robert A. Evans; that on liquid metal fast breeder reactors by Frank J. Rahn.

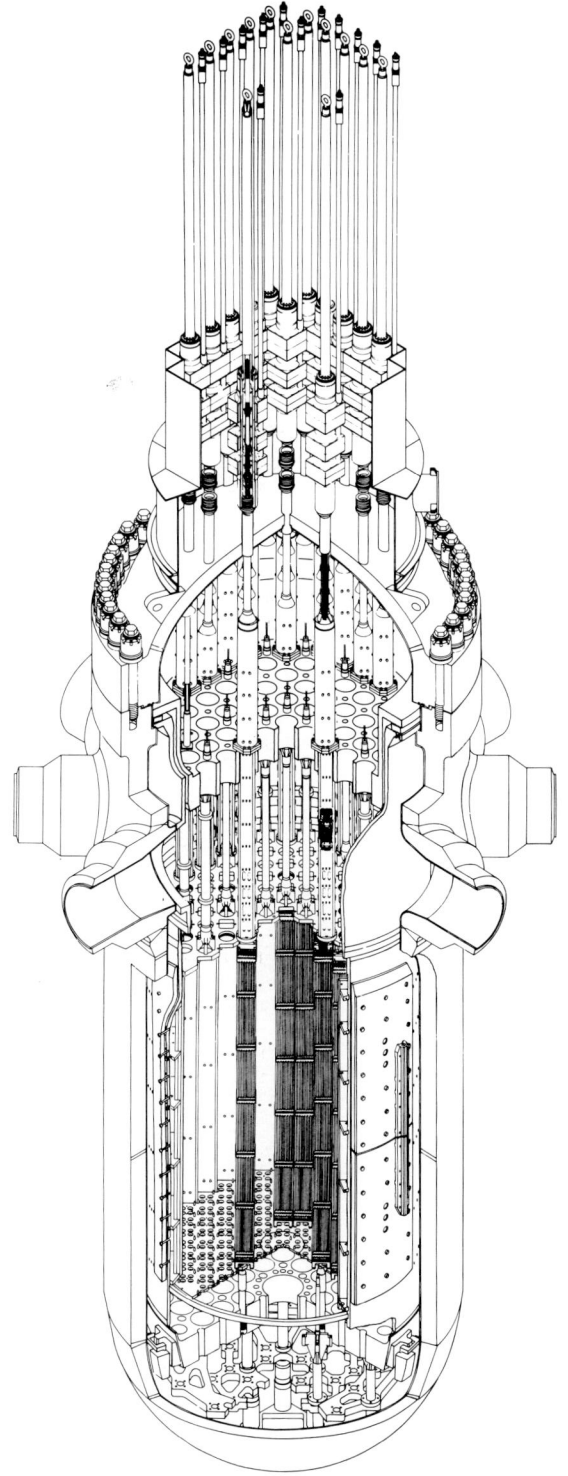

Fig. 7.6-1 Westinghouse pressurized water reactor. (Courtesy Westinghouse Electric Corporation.)

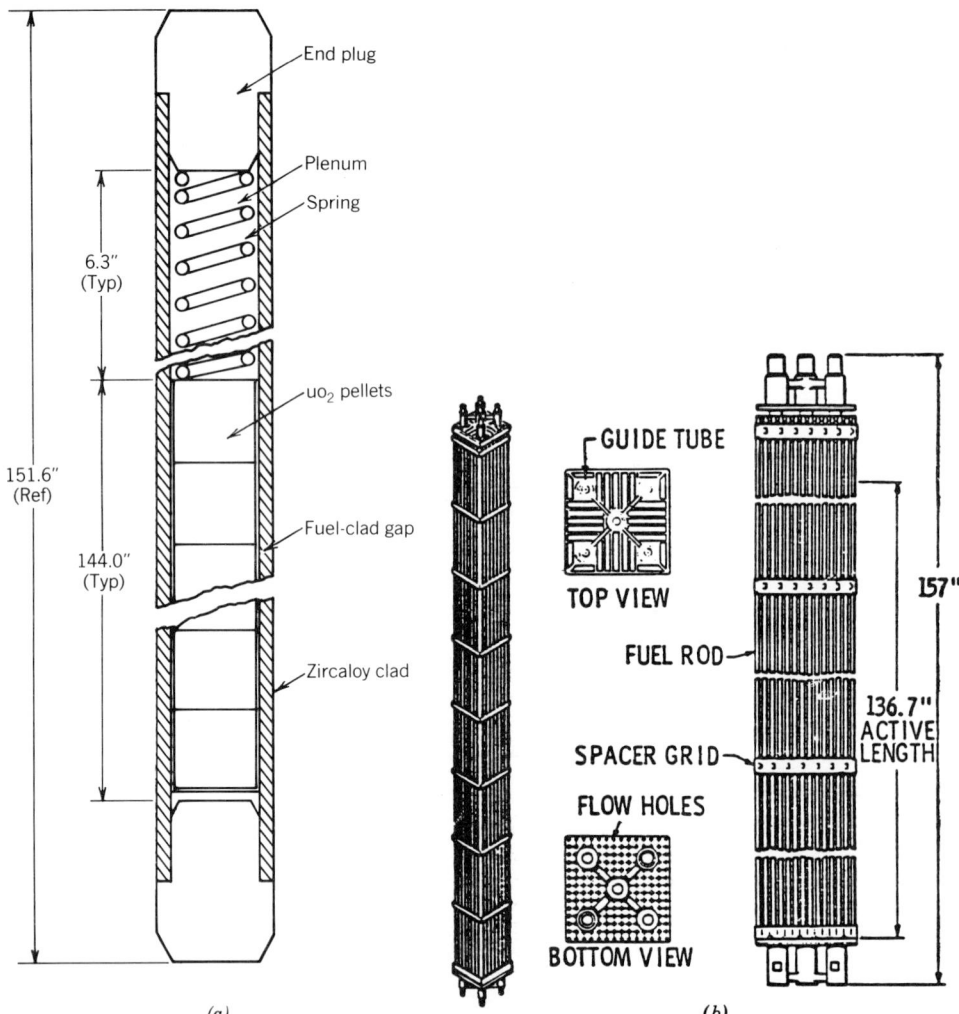

Fig. 7.6-2a   Fuel rod schematic. Specified dimensions depend on design variables such as prepressurization, power history, and a discharge burnup. (Courtesy Combustion Engineering.)
Fig. 7.6-2b   Fuel rod assembly. (Courtesy Combustion Engineering.)

the heat can be removed by a much lower flow rate provided by an auxiliary system of pumps and heat exchangers.

Thus, if there is a leak or a break in the primary coolant system, it is necessary to provide emergency coolant injection capability to assure that the core can always be properly cooled. Redundant emergency core cooling systems are provided for the PWR, as shown in Fig. 7.6-5. If the break is small, the system will remain pressurized and a high-pressure injection system is needed. If the break is large, a rapid injection of lower-pressure water is needed and is provided by the standby pressurized tanks and then the low-pressure core injection system. In all cases the coolant has a high boron concentration to assure that the reactor remains shut down.

Notice that the primary coolant pipes all enter the vessel at a vertical height above the core. Therefore, if water is returned to the core, the core will remain flooded and cooled, even if a pipe break remains open. The remaining heat is then removed by circulating the water through heat exchangers or through the steam generators.

TABLE 7.6-1 REPRESENTATIVE CHARACTERISTICS OF PRESSURIZED WATER REACTORS

Core thermal power	3411 MW(th)
Plant efficiency	32%
Plant electrical output	1100 MW(e)
Core diameter	134 in. (3.4 m)
Core (or fuel rod) active length	144 in. (3.7 m)
Core weight (mass)	276,000 lb (125 Mg)
Core power density	98 kW/L
Cladding material	Zircaloy-4
Cladding diameter (OD)	0.422 in. (1.07 cm)
Cladding thickness	0.024 in. (0.06 cm)
Fuel material	UO_2
Pellet diameter	0.37 in. (0.9 cm)
Pellet height	0.6 in. (1.5 cm)
Assembly array	15 × 15, open structure[a]
Number of assemblies	193
Total number of fuel rods	39,372[a]
Control rod type	B_4C or Ag-In-Cd in cylindrical rod
Number of control rod assemblies	60 (may vary considerably)
Number of control rods per control assembly	20 (may vary considerably)
Total amount of fuel (UO_2)	217,000 lb (98 Mg)
Fuel power density	38 MW/Te
Fuel/coolant ratio	1/4.1
Coolant	Water (liquid phase)
Total coolant flow rate	136×10^6 lb/h (17 Mg/s)
Core coolant velocity	15.5 ft/s (4.7 m/s)
Coolant pressure	2250 psi (15.5 MPa)
Coolant temperature (inlet at full power)	552°F (289°C)
Coolant temperature (outlet at full power)	617°F (325°C)
Nominal clad temperature	657°F (347°C)
Nominal fuel central temperature	4140°F (2282°C)
Radial peaking factor (variation in power density)	1.5
Axial peaking factor	1.7
Design fuel burnup	32,000 MWd/Te (heavy metal); varies
Fresh fuel assay	3.2% ^{235}U (less in initial load)
Spent fuel assay (design)	0.9% ^{235}U, 0.6% $^{239,241}Pu$
Refueling sequence	One-third of the fuel per year
Refueling time	17 days (minimum)

Source. Taken primarily from Westinghouse Electric Corp. specifications.

[a] PWRs now being licensed have a 17 × 17 assembly array, with thinner rods totaling 50,952. Other specifications may be slightly changed.

Containment Building

As a part of the defense in depth to prevent uncontrolled releases of radioactive fission products, the reactor designs include a containment building surrounding the primary reactor components. This containment building and components can be used to isolate the reactor such that if an unexpected event were to occur that caused damage to the reactor fuel, thus releasing some fission products into the primary coolant system, they would still be contained within the containment building. This building is designed to stay sealed and to withstand the pressures that might occur if the hot primary

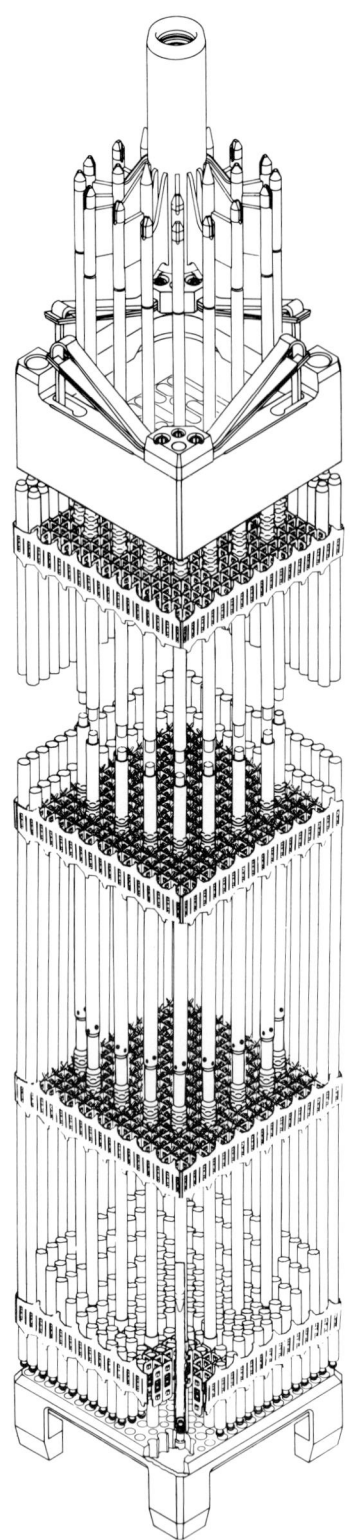

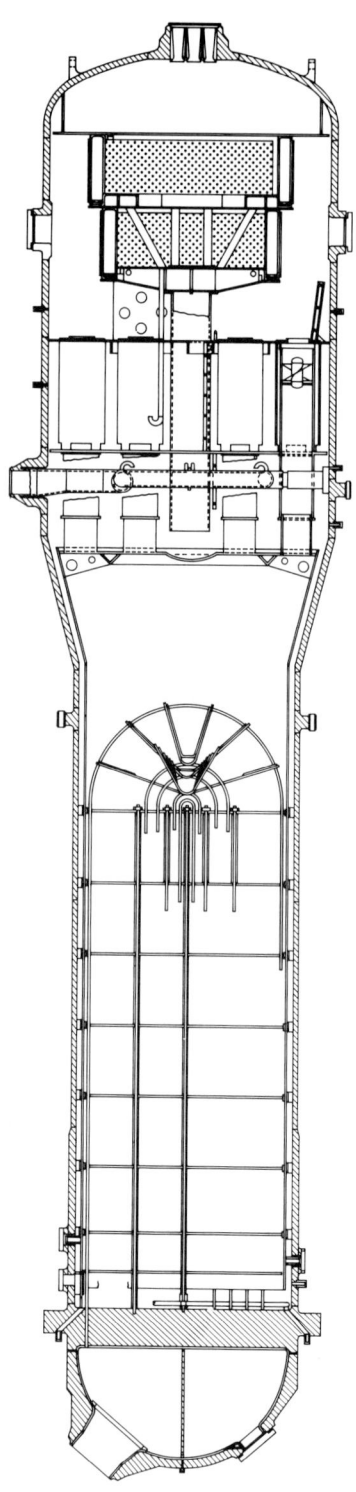

Fig. 7.6-3 Westinghouse 17×17 fuel assembly. (Courtesy Westinghouse Electric Corporation.)

Fig. 7.6-4 Westinghouse model F steam generator. (Courtesy Westinghouse Electric Corporation.)

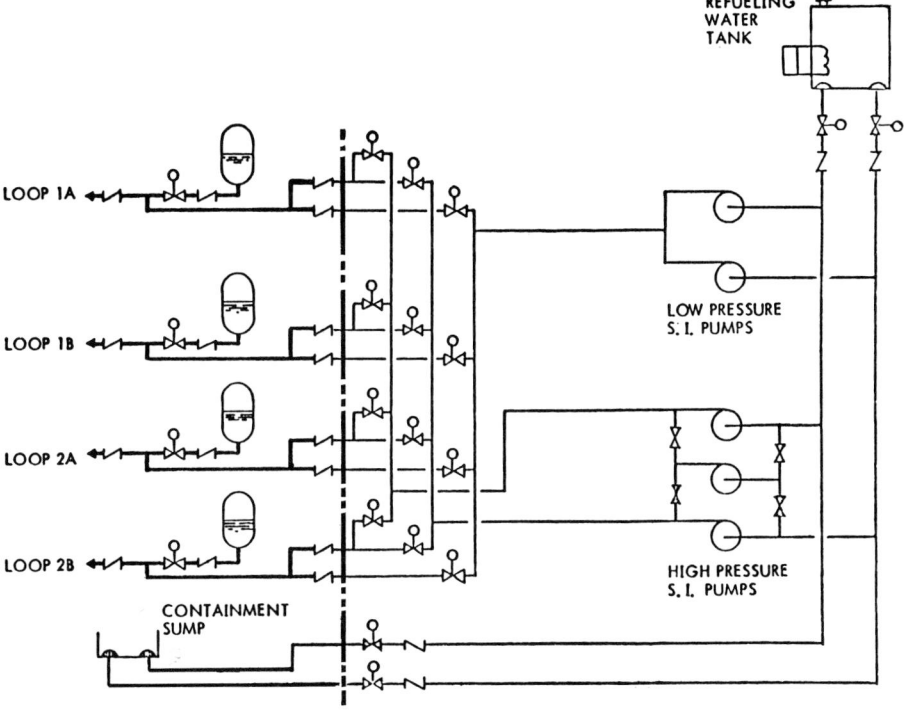

Fig. 7.6-5 PWR emergency core cooling system. (Courtesy Combustion Engineering Corp.)

coolant is released into the building atmosphere and flashes into steam. Special cooling sprays and auxiliary equipment such as hydrogen recombiners are included in the building. In addition, the containment includes thick concrete walls and top to the building, providing shielding of any radioactive material in the building and also affording some protection of the reactor system from external abnormal occurrences (i.e., windblown or falling objects, etc.). An example of a PWR containment building is shown in Fig. 7.6-6.

7.6-2   Boiling Water Reactor (BWR)

The brief description of a boiling water reactor given in Section 7.5-2 is an introduction to this section. In this section a more detailed description of the components is presented, including data on typical design dimensions and characteristics.

Reactor Core and Pressure Vessel

The core of the reactor is installed within the pressure vessel, as shown in Fig. 7.6-7 cutaway view. The core consists of fuel assemblies and control mechanisms. Each fuel assembly is a square array of fuel rods (typically 8×8 array). The fuel assemblies are enclosed on the sides within individual Zircaloy containers that channel the primary water upward through the spaces between the fuel rods for each assembly, removing the heat from the rods by heating and boiling the circulating water. The fuel rods are typically long, thin Zircaloy tubes (~ 1.2 cm in diameter, slightly larger than PWR fuel rods) containing uranium dioxide (UO_2) fuel pellets. In the fresh fuel the uranium is slightly enriched (enrichment varies from about 2.0 to 3.0% U-235 with 98–97% U-238) and some fuel pellets also contain gadolinia as a burnable neutron absorber, described in the section on control mechanisms. Figure 7.6-8 is a diagram of a fuel rod and fuel assembly. The different enrichments are used to give a uniform power production by using lower enrichment in fuel rods on the sides of the assembly where the neutron flux is peaked due to water-filled gaps between the fuel assemblies. The water-filled gaps provide the space for insertion of the control rods; these cruciform-shaped control rods contain boron (B_4C) in tubes, as shown in Fig. 7.6-9. Typical dimensions and characteristics of the operating core and pressure vessel are given in Table 7.6-2. Note that the control rod drives, which couple to the

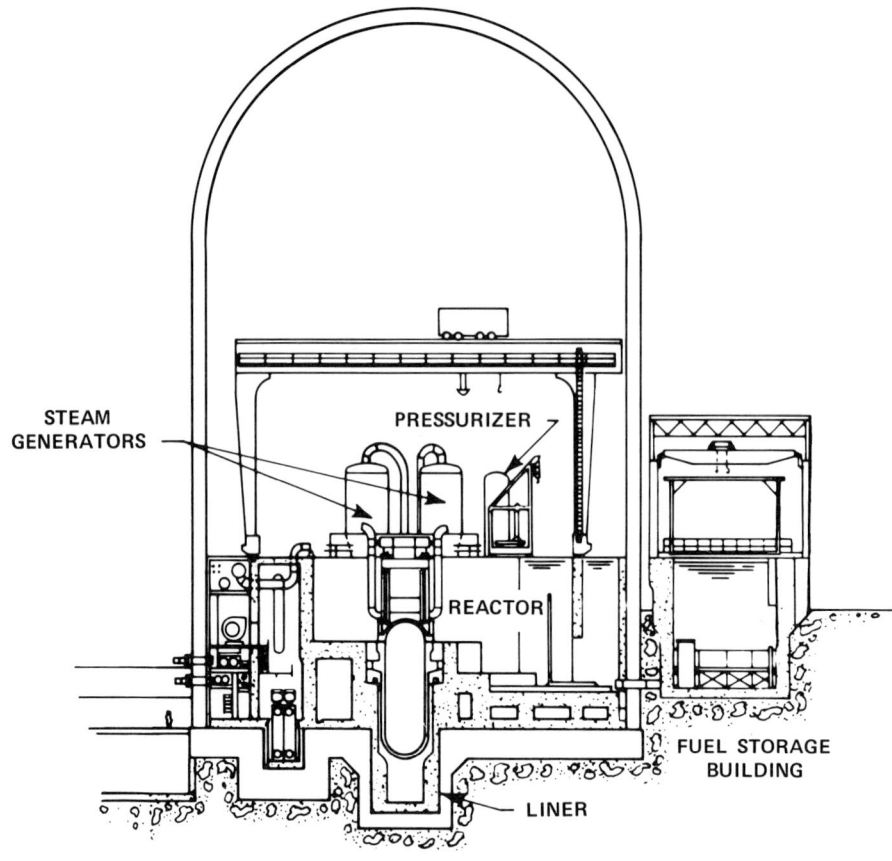

Fig. 7.6-6 Cross section of a typical Westinghouse pressurized water reactor plant. A four-loop system includes the following major components: (1) one reactor vessel, (2) four steam generators, (3) one pressurizer, (4) four reactor coolant pumps, (5) approximately 100 valves. (Courtesy Westinghouse Electric Corporation.)

control rods, enter through the bottom of the reactor. The upper section of the reactor vessel contains the steam separators and driers to remove the moisture from the steam before it leaves the vessel. The separated water is recirculated into the core via the jet pumps, (see the discussion of primary coolants below).

Control Mechanisms

The reactor power level is controlled by adjustment of both the neutron-absorbing control rods and by adjustment of the recirculation flow rate. The control rod drives are actuated from beneath the reactor and involve hydraulically driven pistons. The actuating controls are interlocked to allow only single selected rod stepwise insertion or withdrawal for control, but also to allow rapid insertion of all rods for reactor shutdown capability. Operation of the reactor typically involves start-up by removal of control rods in a uniform pattern across the core. Some control rods initially remain in the core and are used for long-term fuel burnup compensation. In addition to the control rod compensation, the freshly loaded fuel also contains some neutron-absorbing gadolinia (Gd_2O_3) dispersed in selected fuel rods in such a manner that the absorption reactivity effect of the gadolinium is "burned out" at about the same rate as the reactivity losses occur from the fuel U-235 burnout and fission products buildup. Hence, the gadolinium burnout compensates for much of the fuel burnup during a fuel cycle.

Power level control of a BWR involves one additional unique feature, namely the relation between power level and recirculation flow rate. As described in the next section, the recirculation pumps drive the flow of coolant through the jet pumps down to the lower plenum where the flow is distributed to

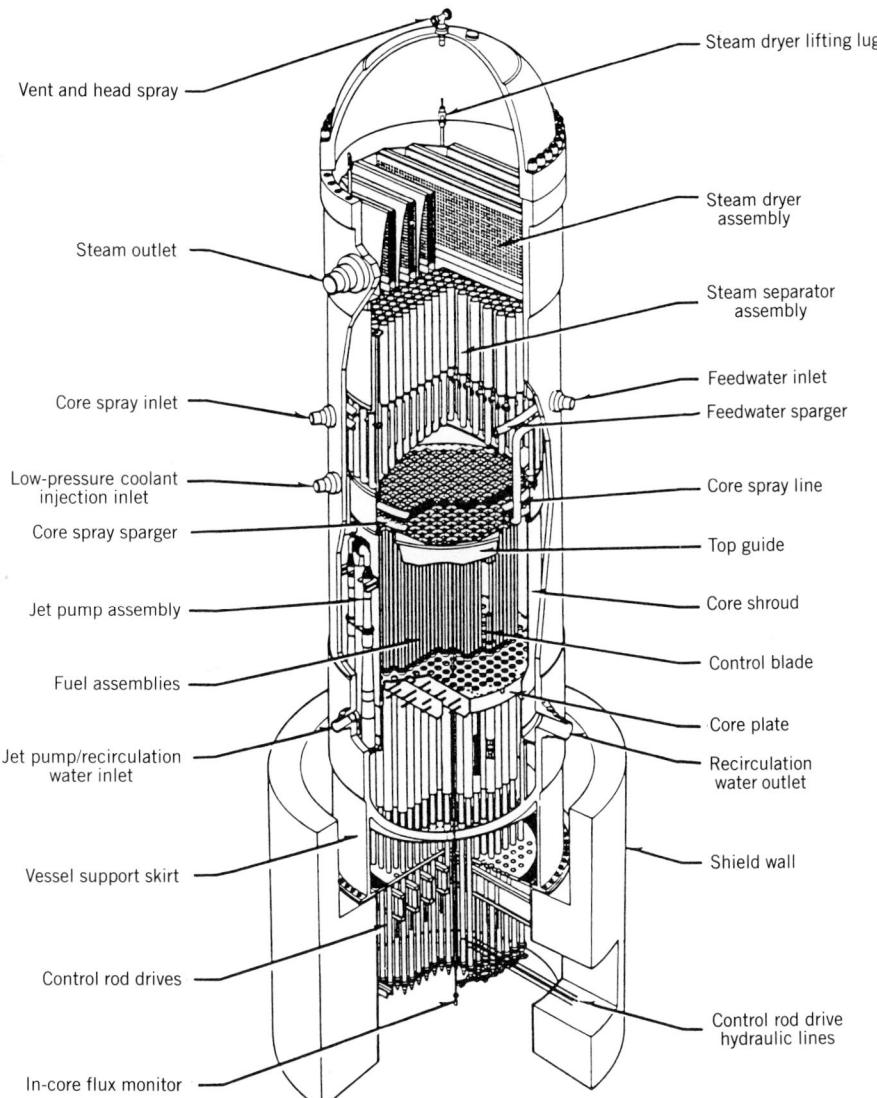

Vent and head spray

Steam dryer lifting lug

Steam dryer assembly

Steam outlet

Steam separator assembly

Core spray inlet

Feedwater inlet

Feedwater sparger

Low-pressure coolant injection inlet

Core spray line

Core spray sparger

Top guide

Jet pump assembly

Core shroud

Fuel assemblies

Control blade

Jet pump/recirculation water inlet

Core plate

Recirculation water outlet

Vessel support skirt

Shield wall

Control rod drives

Control rod drive hydraulic lines

In-core flux monitor

Fig. 7.6-7 Reactor assembly of a BWR power plant. (Courtesy General Electric Co.)

each fuel assembly. The water flows up through the fuel elements and is heated to boiling, and this causes a large reduction in the average water density, especially in the upper section of the core. It must be recalled that a reduction in water density reduces the neutron moderation and causes a reduction in the reactivity. Thus, more boiling will cause the reactivity to go down, causing a power reduction—when the power is reduced, there will be less boiling and the reactivity will increase. This feedback from water density to power is a stabilizing effect and the power level comes to a steady state for a given recirculation flow rate. Increasing the flow rate will remove heat from the core at a faster rate (increasing the average water density) and the power level will thus rise until the reactivity is again balanced. (The average water density is returned to nearly the previous value.)

This control of the power level by changing the recirculation flow is very responsive and affects the core uniformly; hence, it does not perturb the overall power distribution or local power peaking. Typically, the power level adjustments from 50% power-up to full power are made by using the recirculation flow control. In passing, it should be noted that the effect of water density on reactivity also makes the BWR sensitive to pressure. For example, if there is a sudden decrease in pressure (a pipe break, etc.), the water density is reduced by flashing to steam and the reactivity decreased, thus

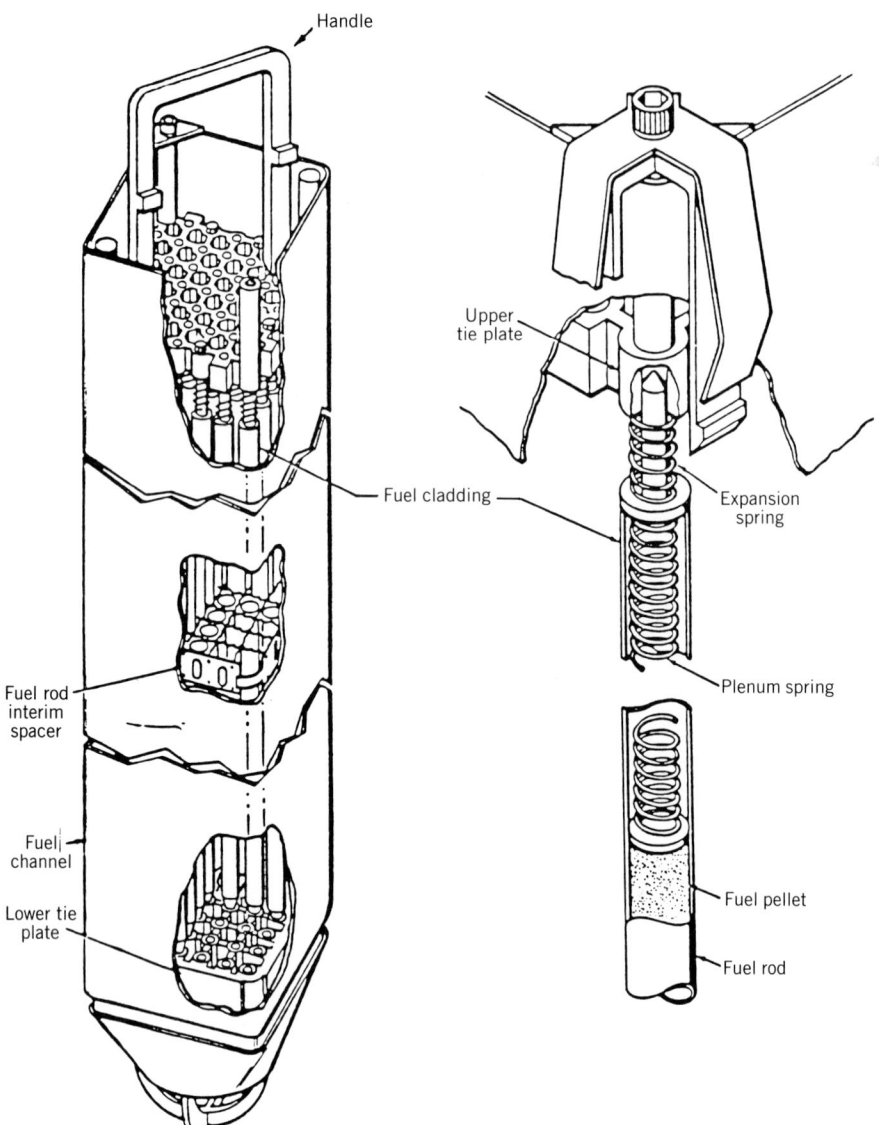

Fig. 7.6-8 BWR fuel assembly. (Courtesy General Electric Co.)

causing the power to be reduced or even completely shut down, depending on the size of the pressure reduction. Finally, a backup shutdown manually controlled mechanism is provided to inject a neutron-absorbing boron solution into the primary water.

In order for the operator to check if all parameters are within the safe operating limits, there are neutron detectors that can be positioned at various points throughout the core giving information that is related to the power production distribution. The operator also has indicators for the pressure, temperature, and flow measurements.

Primary Coolant

The power produced in the core is removed by the steam flowing out of the vessel to the turbine and condenser system. The water from the condenser is then pumped through water cleanup systems, fuel water heaters, and finally back into the upper part of a reactor vessel to mix with water being

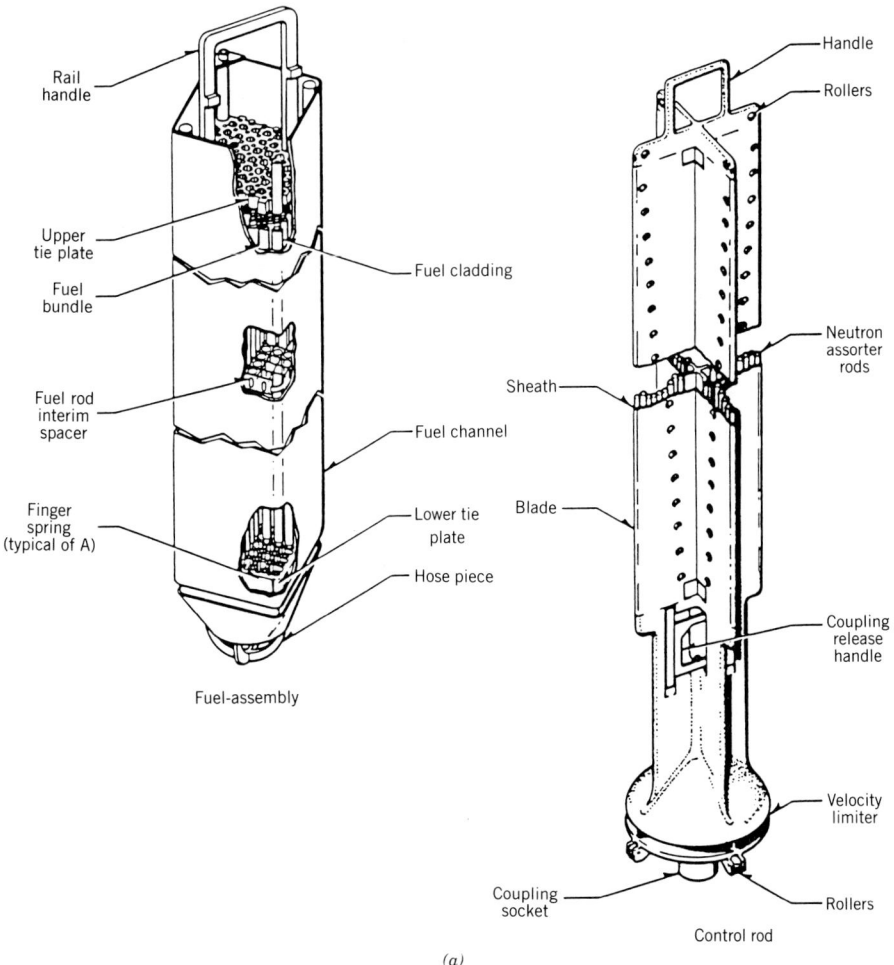

Fuel-assembly

Control rod

(a)

Fig. 7.6-9a Boiling water reactor, General Electric Co. fuel assembly and control rod. (Source: A. V. Nero, Jr., *Guidebook to Nuclear Reactors*, University of California Press, used by permission.)

recirculated back into the core. This is shown in Fig. 7.5-3. The rate at which the heat can be removed from the core depends on the boiling heat transfer that, in turn, depends on the recirculation flow rate. The recirculation pumps (typically two pumps) take the water from the region outside of the core in the downcomer region (see Fig. 7.6-7) and pump it back to the jet pump nozzles (typically 20 jet pumps). These pumps, located in the vessel downcomer, enhance the total flow through the core by sweeping more of the upper downcomer region water along with the high energy flow from the recirculation pumps. Typically, this enhancement is a factor of 2 to 3 times the recirculation pump flow rate. The total flow then enters the bottom plenum and passes through the bottom grid into the fuel assemblies. The speed of the recirculation pumps can be controlled by the operator. The reactor pressure is held consistent as MPa (1000 psia) by pressure regulation, for example, with bypass steam, and the steam demand for the turbine is then balanced by adjusting the recirculation flow rate to give the desired steady-state power. (See also the discussion of control mechanisms above.)

Emergency Core Cooling System

As described in the PWR section on emergency cooling, the residual heat and fission product decay heat must be removed even after the reactor is shut down. In normal shutdown operation the heat is removed by bypassing the turbine and condensing the steam in the condenser until the heat rate has

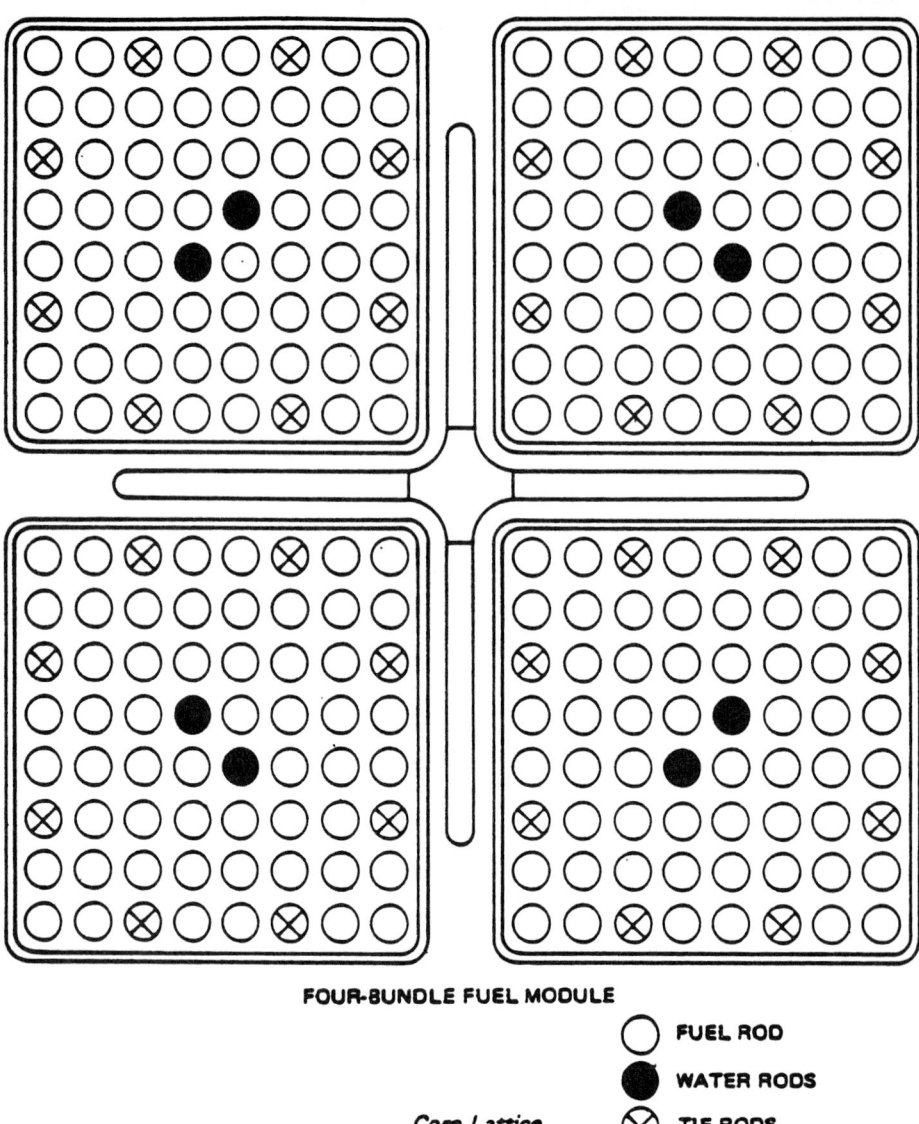

FOUR-BUNDLE FUEL MODULE

○ **FUEL ROD**

● **WATER RODS**

⊗ **TIE RODS**

Core Lattice

CORE LATTICE

The basic module of a BWR core is a set of four fuel bundles, with a control assembly at the point where they meet. Note that some of the positions in the fuel assemblies are taken by tie rods and others are occupied by water rods, which serve to flatten the power distribution in the assembly.

FUEL ASSEMBLY

A BWR fuel assembly consists of a square array of fuel rods, held together by upper and lower tie plates and interim spacers, and surrounded by a fuel channel. The bottom of the assembly serves to regulate the flow through the assembly.

CONTROL ROD

The BWR control rod is a four-bladed assembly containing neutron absorber rods. This assembly is driven from the bottom of the reactor vessel. (Figure courtesy of General Electric Co.)

(b)

Fig. 7.6-9b Boiling water reactor core components. (Source: A. V. Nero, Jr., *Guidebook to Nuclear Reactors*, University of California Press, used by permission.)

TABLE 7.6-2   REPRESENTATIVE CHARACTERISTICS OF BOILING WATER REACTORS

Core thermal power	3579 MW(th)
Plant efficiency	34%
Plant electrical output (nominal)	1220 MW(e)
Core diameter	193 in. (4.9 m)
Core (or fuel rod) active length	150 in. (3.8 m)
Core weight (fuel assemblies)	524,000 lb (238 Mg)
Core power density	54 kW/L
Cladding material	Zircaloy-2
Cladding diameter (OD)	0.483 in. (1.23 cm)
Cladding thickness	0.032 in. (0.81 mm)
Fuel material	UO_2
Pellet diameter	0.410 in. (1.04 cm)
Pellet height	0.41 in. (1.04 cm)
Assembly array	8 × 8, with fuel channel enclosing array
Number of assemblies	748
Total number of fuel rods	46,376
Control rod type	"Cruciform" control rods inserted from the bottom between sets of four assemblies
Number of control rods	177
Total amount of fuel (UO_2)	342,000 lb (155 Mg)
Fuel/coolant ratio	1/2.7, blades out; 1/2.5, blades in (cold)
Coolant	Water (two phase)
Total coolant flow rate	104×10^6 lb/h (13 Mg/s)
Coolant pressure	1,040 psia (7.0 MPa)
Coolant temperature (steam system design)	551°F (288°C)
Feedwater temperature	420°F (216°C)
Average coolant exit quality (percent steam weight)	14.7%
Average clad temperature	479°F (304°C)
Maximum fuel central temperature	3330°F (1832°C)
Average volumetric fuel temperature	1130°F (610°C)
Axial peaking factor	1.4 approx.
Design fuel burnup	28,400 MWd/Te
Fresh fuel assay	Average 2.8% ^{235}U (initial core: 1.7–2.1% ave.)
Spent fuel assay	0.8% ^{235}U, 0.6% $^{239,241}Pu$
Refueling sequence	Approximately one-fourth of the fuel per year to one-third per 18 months
Refueling time	188 h @ 100% efficiency
Vessel wall thickness min/max	5.7 in./6.46 in. (14.5 cm/16.4 cm)
Vessel material	Manganese-molybdenum-nickel steel internally clad with $\frac{1}{8}$ in. austenitic stainless steel
Vessel diameter (ID)	19 ft 10 in. (6.0 m)
Vessel height	71 ft (22 m)
Vessel weight (including head)	1,950,000 lb (884,500 kg)

Source.   General Electric Co. specifications.

decayed to a lower value and then the residual heat removal system of pumps and heat exchangers are used to continue removing the remaining heat production.

In the event of a leak or break in the primary coolant system, it is necessary to inject makeup water into the reactor vessel to assure that the fuel is properly cooled. If a small break were to occur, there are high-pressure injection systems and for protection in the event of large breaks, redundant low-pressure injection systems can be used. These systems include redundant spray injectors that are designed to spray water over the top region of all fuel assemblies, condensing the steam and cooling the elements until the core has been reflooded. Keeping the fuel surrounded by water will provide adequate cooling for the fuel assemblies and the heat can be removed by the normal decay heat removal circulation system.

Containment Building

As discussed for the PWR, the defense in depth for the BWR design also includes a containment building surrounding the reactor vessel. There are redundant isolation valves on the primary system pressurized lines such as the steam lines to allow complete isolation of the main reactor and coolant system within the containment building. BWR containment buildings typically include a "suppression" pool connected within the containment so that, in the event of a pipe break with the reactor coolant at elevated temperature and pressure, the steam from the primary coolant release will be conducted to the suppression pool where it is condensed, thus suppressing the pressure buildup in the containment building. The suppression pool also is used to condense steam from the safety relief valves that can be operated to prevent a pressure buildup in the reactor vessel when the coolant system is isolated from the main turbine condenser. In all other respects the containment building is similar to the PWR containment design.

7.6-3   High Temperature Gas-Cooled Reactor

The brief description in Section 7.5-3 gives an introduction to the types and general system arrangement for high-temperature gas-cooled reactors (HTGR). In this section a more detailed HTGR design description for one typical type is given.

Reactor Core and Primary System

The HTGR core is cooled with pressurized helium, moderated and reflected with graphite, and fueled with a mixture of uranium and thorium fuel particles with individual ceramic coatings. The reactor core is constructed of prismatic hexagonal graphite blocks with vertical holes for coolant channels, fuel rods, and control rods, as shown in Fig. 7.6-10.

As shown in Fig. 7.6-11, the entire reactor core, together with other major primary system components, is contained in a multicavity, prestressed concrete reactor vessel (PCRV).

In addition to the primary coolant loop, auxiliary cooling loops each consisting of a helium/water heat exchanger and electric-motor-driven circulator are provided. The PCRV and ancillary systems are housed inside a reactor containment building, a conventional steel-lined reinforced concrete structure.

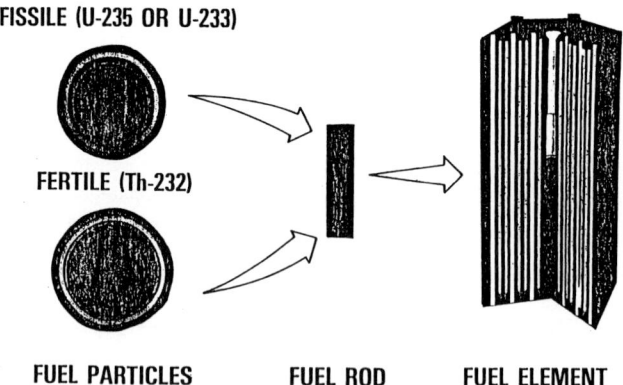

FISSILE (U-235 OR U-233)

FERTILE (Th-232)

FUEL PARTICLES   FUEL ROD   FUEL ELEMENT

Fig. 7.6-10   HTGR fuel components. (Courtesy General Atomic Co., San Diego, CA.)

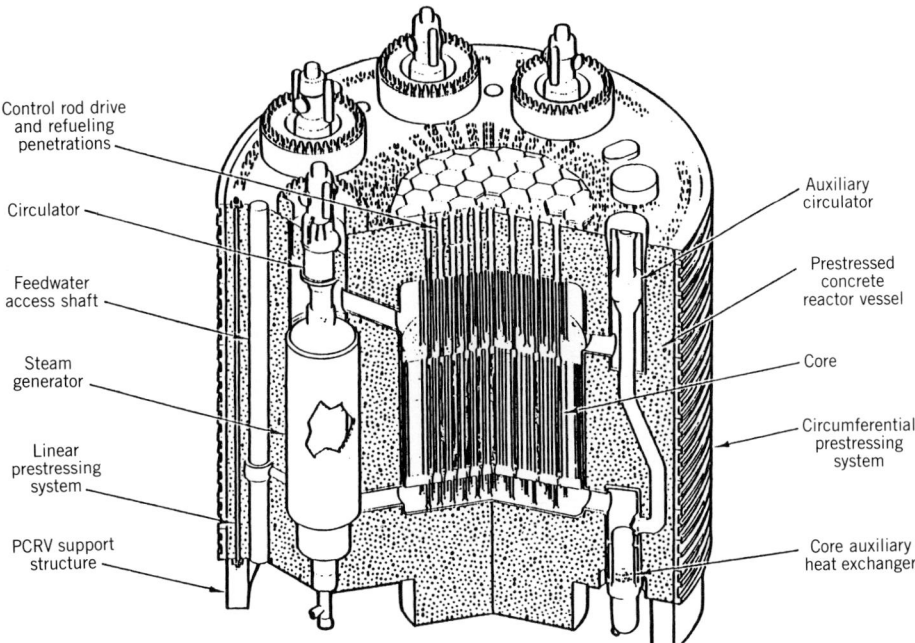

Control rod drive
and refueling
penetrations

Circulator

Feedwater
access shaft

Steam
generator

Linear
prestressing
system

PCRV support
structure

Auxiliary
circulator

Prestressed
concrete
reactor vessel

Core

Circumferential
prestressing
system

Core auxiliary
heat exchanger

Fig. 7.6-11 HTGR-SC nuclear steam system. (Courtesy General Atomic Co., San Diego, CA.)

Prestressed Concrete Reactor Vessel (PCRV)

The PCRV shown in Fig. 7.6-11 is prestressed circumferentially by steel strand cables wound under tension and vertically by large linear tendons. This prestress is designed to induce sufficient precompression in the concrete to counteract the primary and secondary stresses caused by cavity pressures and operating temperatures. Steel liners and the penetrations with their closures form the continuous gas-tight boundary of the PCRV.

Prestressed concrete reactor vessels have accumulated over 180 reactor-years of successful operating experience in gas-cooled nuclear power plants in Europe and the United States (see Fig. 7.6-12). The PCRV design is based on the ASME Boiler and Pressure Vessel Code, Section III, Division 2, for concrete reactor vessels.

Steam Generator

The HTGR-SC/C steam generator is similar to that developed for Fort St. Vrain (FSV) except that the separate reheater assembly has been deleted. The design of the economizer/evaporator/superheater (EES) heat transfer section of the HTGR-SC/C is a direct scale-up of FSV helically coiled EES bundles. The heat transfer section is a once-through counterflow heat exchanger with a helical-coil, $2\frac{1}{4}$ CR-1Mo, EES tube bundle followed by an Alloy 800 straight-tube finishing superheater located in the center of the helical bundle. The EES heat transfer section is 14 ft (4.07 m) in diameter and 48.3 ft (14.72 m) long. The outer envelope dimensions of the overall assembly including shrouds, tube sheets, and bundle support are 14.8 ft (4.51 m) in diameter by 71.3 ft (21.73 m) long. A diagram of the steam generator module is shown in Fig. 7.6-13. A comparison of the current HTGR design and other gas-cooled reactor steam generator data is given in Table 7.6-3.

Main Helium Circulator

The circulators are mounted vertically above the steam generators and share common PCRV cavities. The circulator impeller is of the centrifugal type and is removable through an opening in the primary closure without disturbing the concrete plug closing the steam generator cavity. Variable rotating speed is utilized for primary coolant flow control. A solid-state synchronous motor is combined with a variable-frequency power supply to provide the driving power to the circulator. The motor is fully enclosed and is air cooled with air-to-water heat exchangers mounted in the side cavities of the motor

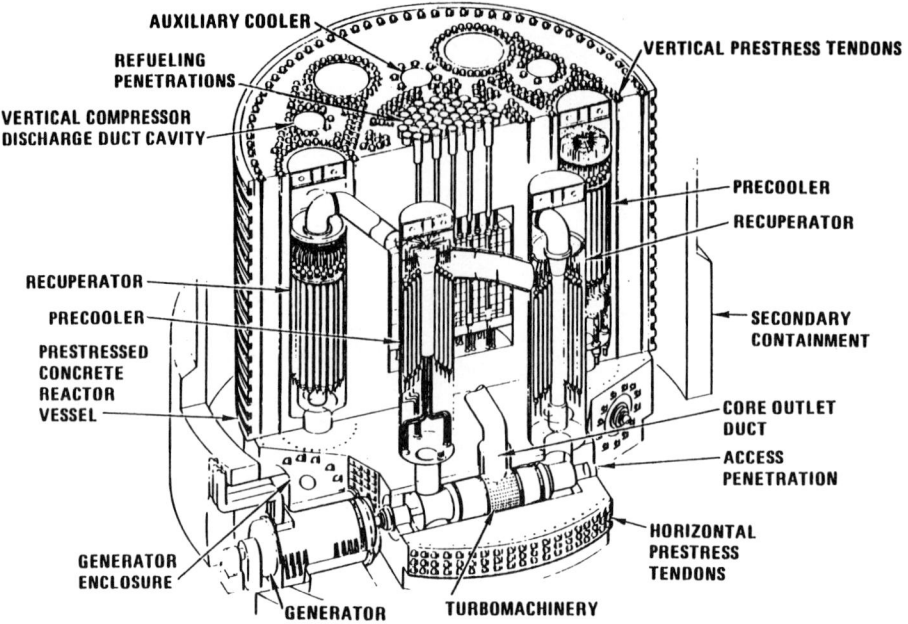

AUXILIARY COOLER

REFUELING
PENETRATIONS

VERTICAL COMPRESSOR
DISCHARGE DUCT CAVITY

VERTICAL PRESTRESS TENDONS

PRECOOLER

RECUPERATOR

RECUPERATOR

PRECOOLER

PRESTRESSED
CONCRETE
REACTOR
VESSEL

SECONDARY
CONTAINMENT

CORE OUTLET
DUCT

ACCESS
PENETRATION

HORIZONTAL
PRESTRESS
TENDONS

GENERATOR
ENCLOSURE

GENERATOR

TURBOMACHINERY

Fig. 7.6-12   HTGR gas turbine plant. (Courtesy General Atomic Co., San Diego, CA.)

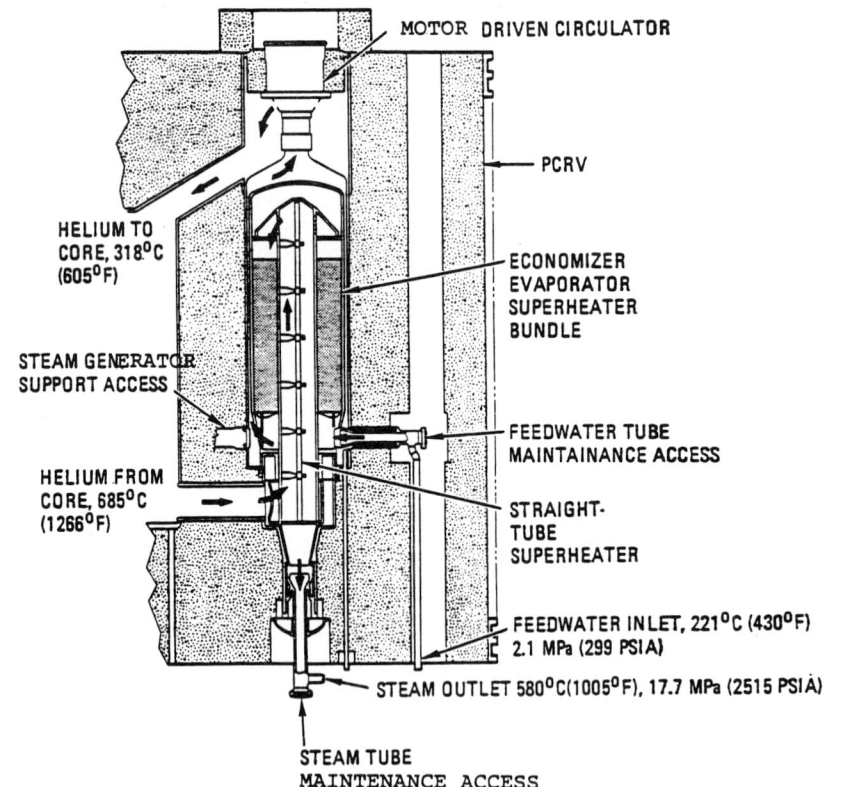

MOTOR DRIVEN CIRCULATOR

PCRV

HELIUM TO
CORE, 318°C
(605°F)

ECONOMIZER
EVAPORATOR
SUPERHEATER
BUNDLE

STEAM GENERATOR
SUPPORT ACCESS

FEEDWATER TUBE
MAINTAINANCE ACCESS

HELIUM FROM
CORE, 685°C
(1266°F)

STRAIGHT-
TUBE
SUPERHEATER

FEEDWATER INLET, 221°C (430°F)
2.1 MPa (299 PSIA)

STEAM OUTLET 580°C(1005°F), 17.7 MPa (2515 PSIA)

STEAM TUBE
MAINTENANCE ACCESS

Fig. 7.6-13   Steam generator arrangement HTGR. (Courtesy General Atomic Co., San Diego, CA.)

TABLE 7.6-3   COMPARISON OF GCR STEAM GENERATORS

	Peach Bottom HTGR		Fort St. Vrain HTGR		Hartlepool AGR		Large HTGR	
Bundle type	U tube		Helical		Helical		Helical	
Materials								
Economizer	Low carbon		$\frac{1}{2}$ Cr-$\frac{1}{2}$ Mo		Carbon		$2\frac{1}{4}$ Cr-1 Mo	
Evaporator	Carbon		$2\frac{1}{4}$ Cr-1 Mo		9 Cr-1 Mo		$2\frac{1}{4}$ Cr-1 Mo	
Presuperheater	NA		$2\frac{1}{4}$ Cr-1 Mo		9 Cr-1 Mo		$2\frac{1}{4}$ Cr-1 Mo	
Superheater	Alloy 800		Alloy 800		316 SS		Alloy 800	
Reheater	NA		Alloy 800		316 SS		NA	
Temperatures, °F (°C)								
Design inlet gas	1367	(742)	1427	(775)	1193	(645)	1266	(686)
Main steam	1005	(540)	1000	(538)	1009	(543)	1005	(540)
Reheat steam	NA		1000	(538)	1002	(539)	NA	
Outlet gas	628	(331)	700	(371)	550	(293)	596	(313)
Operating pressures, psig (kPa)								
Gas	329	(2,270)	688	(4,740)	570	(3,930)	1041	(7,180)
Main steam	1550	(10,690)	2512	(17,320)	2500	(17,240)	2515	(17,340)
Reheat steam	NA		600	(4,140)	550	(3,790)	NA	
Power (each generator), MW(th)	63		70		203		560	
Envelope								
Diameter, ft-in. (m)	7-6	(2.29)	5-7.25	(1.71)	10-0	(3.05)	14-10	(4.53)
Length (max, shipping), ft-in.	30-2	(9.19)	25-7	(7.80)	60-0	(18.29)	71-4	(21.74)
Shipping weight, tons (tonnes)	63	(56)	30	(27)			325	(290)

Courtesy General Atomic Co., San Diego, CA.

housing. The two radial bearings and the one thrust bearing in the motor are of the tilting-pad oil-lubricated type. Figure 7.6-14 shows the basic circulator and motor installation configuration. The basic operating parameters are given in Table 7.6-4. The circulator radial bearings are water lubricated, part hydrostatic-part hydrodynamic, with the bearing lubricant pump integrally mounted on the shaft. The circulator seals are designed to contain the primary coolant within the reactor and the lubricating water inside the circulator.

Core Auxiliary Cooling System (CACS)

The location of the CACS is shown within the PCRV in Fig. 7.6-11.

The CACS consists of three separate, independent, and functionally identical cooling loops. Each loop has circuits in two major systems, one helium to water in the nuclear steam supply (NSS) portion and the other water to air in the balance of plant (BOP) portion. The latter is generally referred to as the core auxiliary cooling water system (CACWS). The first (inner) circuit within the PCRV circulates primary helium coolant from the reactor core outlet to the inlet plenum passing through the straight-tube helium-to-water heat exchanger (core auxiliary heat exchanger—CAHE) via a flow-actuated split check valve (auxiliary shutoff valve) to the axial flow compressor (auxiliary circulator) which provides the driving force for the primary coolant flow.

Within the CACWS circuit pressurized water transfers heat in a closed system to an air-blast finned-tube-type heat exchanger. Water is circulated by conventional electric motor pumps and system pressure is established by the thermal expansion of the water inventory against a volume of inert gas in a pressurizer. Air is circulated by electric-motor-driven fans over the finned tubes of the heat exchanger. Thus, the heat taken from the reactor core is rejected to the atmosphere surrounding the plant, which is the ultimate heat sink for the CACS.

7.6-4   Heavy Water Reactor; CANDU Type

As introduced in Section 7.5-4, there are several heavy water reactor designs. Since the major commercial type is the Canadian design (CANDU), this section provides a more detailed description of the CANDU system.

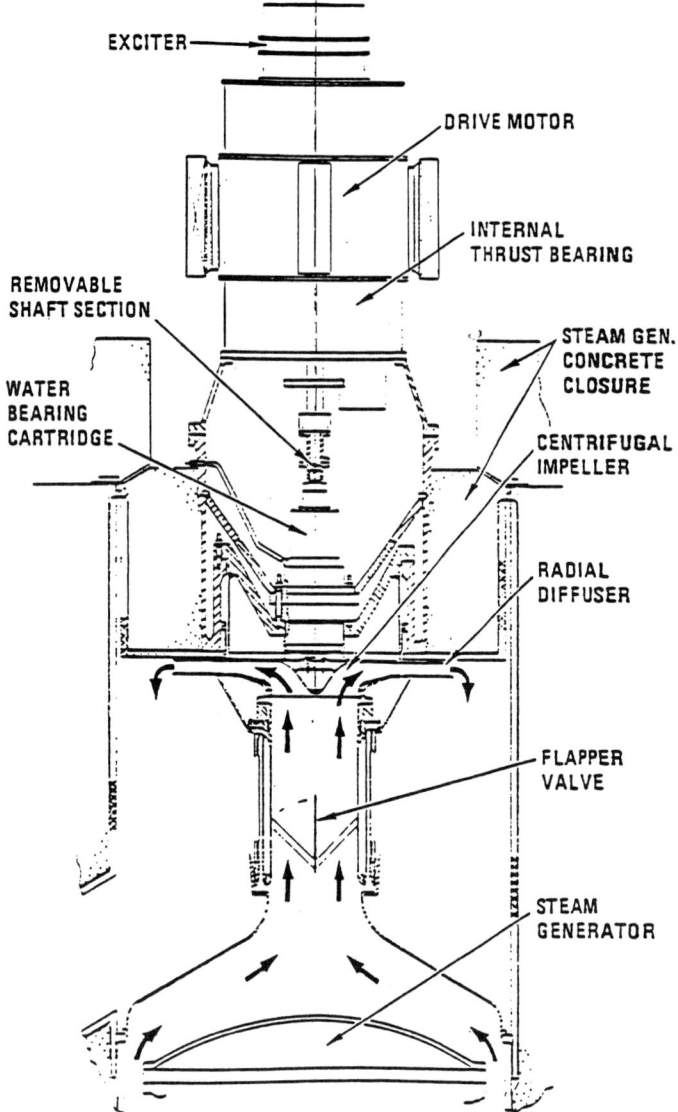

EXCITER

DRIVE MOTOR

INTERNAL
THRUST BEARING

REMOVABLE
SHAFT SECTION

STEAM GEN.
CONCRETE
CLOSURE

WATER
BEARING
CARTRIDGE

CENTRIFUGAL
IMPELLER

RADIAL
DIFFUSER

FLAPPER
VALVE

STEAM
GENERATOR

Fig. 7.6-14   Main helium circulator. (Courtesy General Atomic Co., San Diego, CA.)

Reactor Core and Calandria Vessel

The core of the reactor is installed within the unpressurized "calandria" vessels, as shown in Fig. 7.5-1. For the typical 600 MW(e) plant, this core consists of an array of 380 horizontal fuel channels surrounded by the D_2O moderator and interspersed by vertical and lateral control and monitoring mechanisms. Each fuel channel consists of a calandria tube (Zircaloy-2) surrounding a zirconium–niobium pressure tube with a small insulating gas space that reduces heat transfer to the D_2O moderator. Inside the pressure tube fuel bundles are cooled by pressurized (~ 10 MPa) D_2O. Each fuel bundle typically consists of an array of 37 fuel tubes each 13.1 mm diameter and 500 mm long; each tube contains natural uranium dioxide fuel pellets. The fuel bundle is pictured in Fig. 7.6-15 and sits in the fuel channel, as shown in Fig. 7.6-16. Table 7.6-5 gives further details of the dimensions.

TABLE 7.6-4 MAIN HELIUM CIRCULATOR OPERATING PARAMETERS

Inlet pressure	1028.3 psia (7090 kPa)
Pressure rise	21.7 psi (149.6 kPa)
Inlet temperature	596°F (313°C)
Mass flow rate	646 lb/s (293 kg/s)
Volumetric flow	1778 ft^3/s (50.35 m/s)
Adiabatic head	8600 ft (2621 m)
Shaft power	10,070 kW
Rotating speed	2200 rpm
Impeller type/diameter	Radial/72 in. (1.829 m)
Bearings	Water lubricated, hybrid, ultra-high stiffness
First critical speed	2700 rpm
Drive motor	
Type	Solid rotor, synchronous
Speed control	Solid state, self-commutated converter
Bearings	Oil lubricated, tilting pad, two radial plus single thrust

Courtesy General Atomic Co., San Diego, CA.

Control Mechanisms

The CANDU reactor is controlled by a variety of adjustments in neutron absorption and fuel reactivity. The following list is for a typical CANDU system.

1. **28 Shutoff rods.** Highly neutron-absorbing rods are held above the core in vertical tubes and drop into the reactor core to shut the reactor down when a trip is actuated.
2. **21 Adjuster rods.** Less absorbing rods are used for neutron flux distribution and power adjustments.
3. **6 Liquid-zone control compartments.** Tubular compartments are symmetrically located throughout the core and can be filled with light water to a desired level. The light water gives a small neutron-absorbing effect in the D$_2$O moderator and thus a power distribution control capability.
4. **6 Poison injector nozzle assemblies.** Rapid injection of a liquid neutron absorber (gadolinium nitrate) can be used to poison the moderator as a backup shutdown mechanism when a trip signal is activated.
5. **On-line refueling.** Natural uranium reactors do not carry a large excess of reactivity and part of the reactivity control is afforded by adding fresh fuel; this is done in a uniform manner by fueling every other tube in the opposite direction.
6. **Boron solution in the moderator.** Some slow transient control, such as compensation during burnout of an excess of fresh fuel, is provided by adding or diluting the boron concentration of the D$_2$O moderator.

The arrangement of these control mechanisms is shown in Fig. 7.6-17, and the on-line refueling mechanisms are depicted in Fig. 7.6-18. Note that these refueling machines are identical on each side and can load fresh fuel or receive spent fuel with proper cooling and storage.

Primary Coolant System

The power is removed from the core by circulating the pressurized D$_2$O through the tubes to headers where the combined D$_2$O flows through steam generators and is pumped back to headers for distribution to the pressure tubes in the core. Two loops each service half the core at every other tube. These components are shown in Fig. 7.5-2. On one leg of each of the two independent loops, a pressurizer is attached to control the pressure similar to a PWR (Section 7.6-1).

The pressurized D$_2$O can be heated by the core to 310°C with a small amount of boiling at the tube outlets. This heated water is circulated to the steam generator where it passes through a large

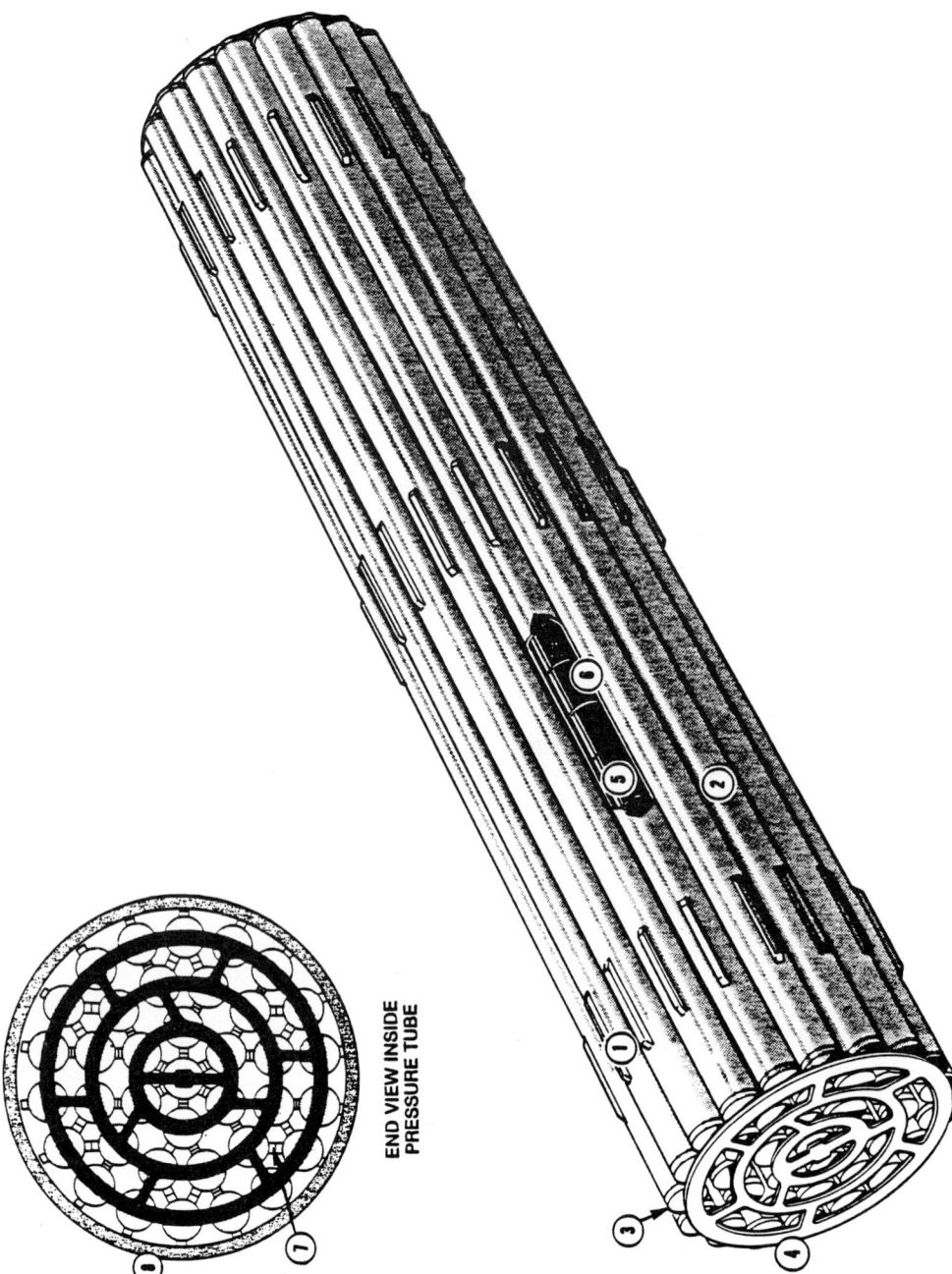

END VIEW INSIDE PRESSURE TUBE

Fig. 7.6-15 (1) Zircaloy bearing pads, (2) Zircaloy fuel sheath, (3) Zircaloy end cap, (4) Zircaloy end support plate, (6) Canlub graphite interlayer, (7) interelement spacers, (8) pressure tube. 37 element fuel bundle for CANDU reactor. (Courtesy Atomic Energy of Canada, Ltd.)

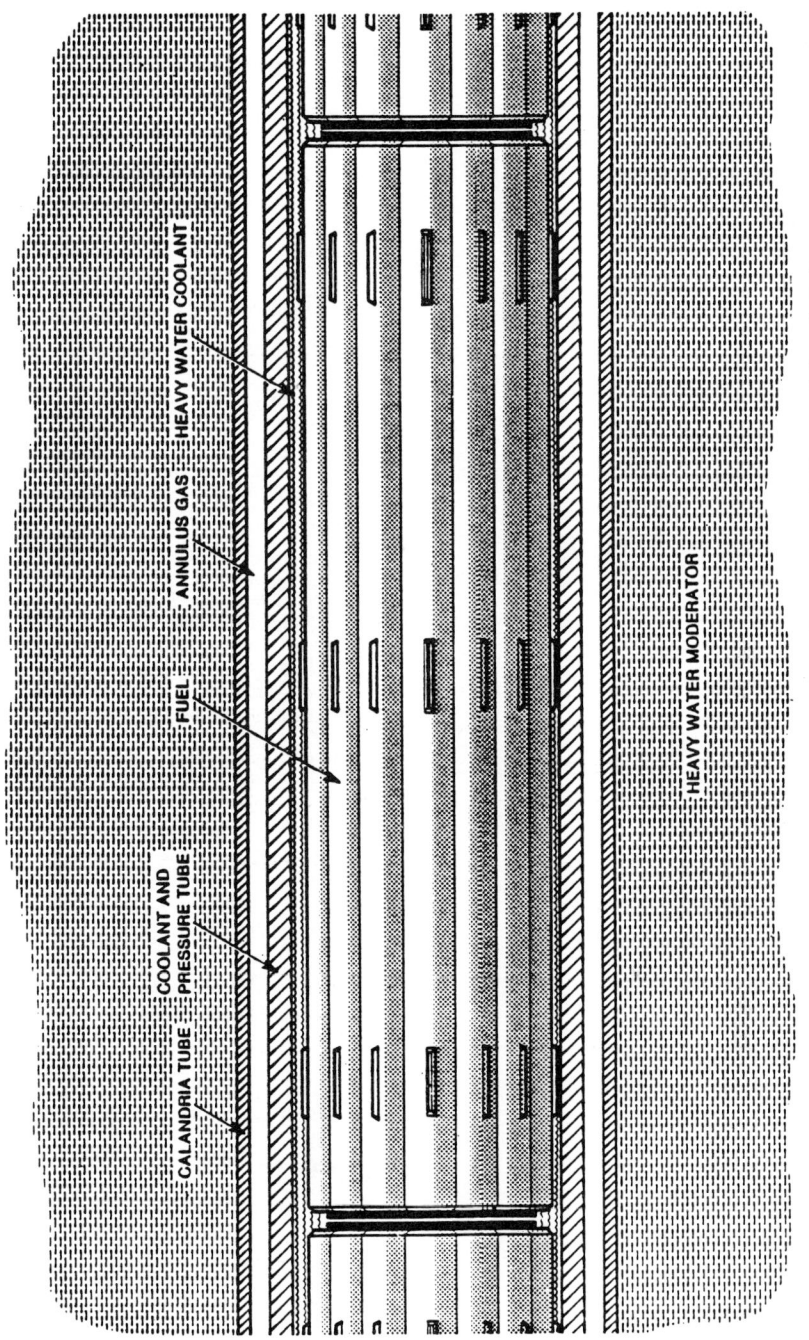

CALANDRIA TUBE

COOLANT AND PRESSURE TUBE

FUEL

ANNULUS GAS

HEAVY WATER COOLANT

HEAVY WATER MODERATOR

Fig. 7.6-16 CANDU fuel channel arrangement. (Courtesy Atomic Energy of Canada Ltd.)

937

TABLE 7.6-5   REPRESENTATIVE CHARACTERISTICS OF A CANDU 600 REACTOR

Core thermal power (total fission power)	2156 MW(th)
Plant efficiency	29.6%
Plant electrical output	638 MW (e) (net)
Core diameter, effective	628.6 cm
Core length	594.4 cm
Core weight (fuel bundles)	108 Mg
Core power density (core average within calandria)	12 kW/L
Calandria material	Zircaloy
Cladding diameter (OD)	13.08 mm
Cladding thickness	0.42 mm (collapsible under external coolant pressure
Fuel material	UO_2 (natural U)
Fuel bundle assembly	37 rods, arranged in concentric circles
Bundle length	495 mm
Bundle diameter (OD)	102.4 mm
Total number of bundles in core	4560
Total number of fuel rods	168720
Total amount of fuel (UO_2)	97 Mg
Control rod types	Variable neutron absorbers (e.g. light water components), adjustable absorber (such as stainless steel); shutdown by either absorbing rods or by poison injection
Number of control rods or compartments	28 shut-off rods 6 poison injection nozzles 6 light water zone control compartments 4 mechanical control absorber rods 21 adjuster units
Coolant	Pressurized heavy water
Total coolant flow rate	7.6 Mg/s
Coolant pressure (channel inlet)	11.04 MPa
Coolant pressure (channel outlet)	10.3 MPa
Coolant temperature (channel inlet)	267°C
Coolant temperature (channel outlet)	312°C
Average coolant exit quality (channel outlet)	3%
Moderator	Heavy water (99.85 wt% D_2O)
Moderator pressure	Approximately atmospheric
Moderator temperature (entrance)	49°C
Moderator temperature (exit)	77°C
Total heavy water inventory	470 Mg (195 Mg coolant, 275 Mg moderator)
Maximum fuel (UO_2) temperature (nominal design)	1570°C
Maximum clad temperature (nominal design)	336°C
Axial form factor (average bundle power/ max. bundle power)	0.553
Radial form factor (average channel power/ max. channel power)	0.823
Fuel residence time (average)	~ 280 full power days
Design natural uranium utilization	176 Mg (U)/GW (e) year
Fresh fuel array	0.71% ^{235}U (natural)
Spent fuel array	0.24% ^{235}U, 0.26% $^{239,241}Pu$
Refuelling sequence	On power, essentially continuous refuelling
Number of pressure tubes (Zirconium/Niobium alloy)	380
Pressure tube inside diameter	103.88 mm (minimum)
Pressure tube wall thickness	4.19 mm (minimum)
Number of calandria tubes (Zircaloy)	380
Lattice array	Square, 28.575 mm pitch

Courtesy Atomic Energy of Canada, Ltd., Mississanga, Ontario, Canada.

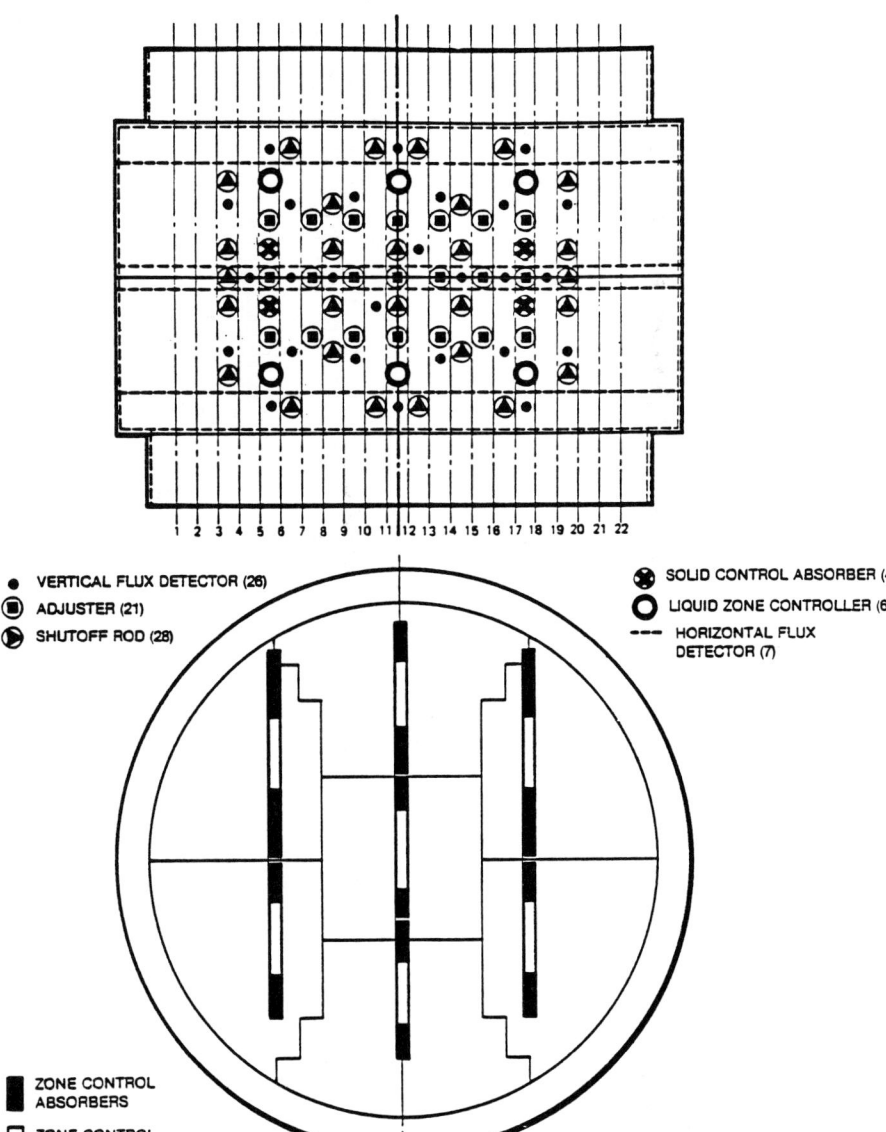

Fig. 7.6-17 CANDU reactivity mechanism layout. (Courtesy Atomic Energy of Canada, LTD.)

bundle of tubes. Outside the tubes the secondary light water at a lower pressure (4.55 MPa) is heated and boiled (at 260°C) to produce the desired steam to drive the turbine. These steam generators are similar to the PWR cutaway shown in Fig. 7.6-1 (PWR).

Decay Heat Removal and Emerging Core Cooling

As described in Section 7.6-1, the residual heat and fission product decay heat must be removed even after the reactor trip has inserted the neutron absorbers to shut the reactor down. (Loss of pressurized D_2O coolant does not inherently shut the CANDU reactor system down, as compared to the LWR, where the moderator is also the coolant; however, the redundant shutdown mechanisms and low reactivity control requirements provide the CANDU system with a highly reliable shutdown probability.) In normal operation the residual heat can be removed by continuing to cool through the steam generators and dumping steam through bypass lines to the condenser.

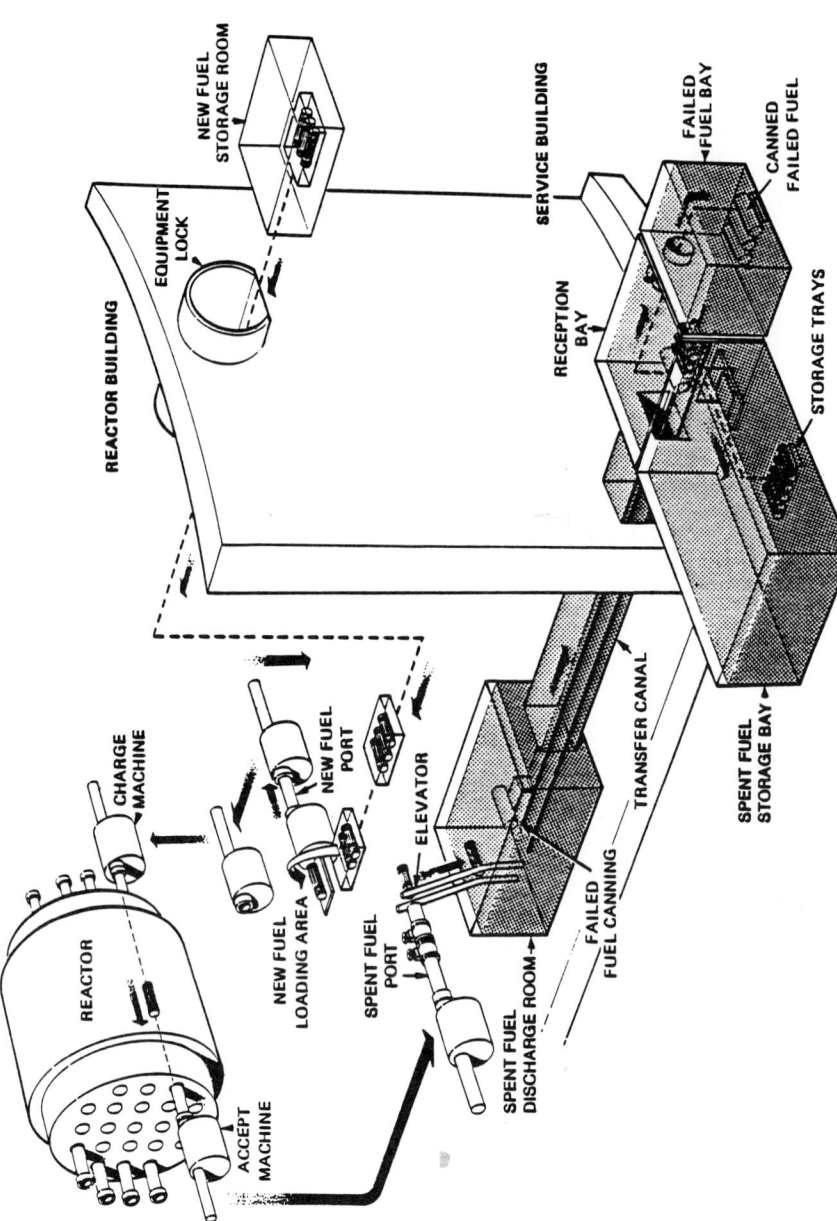

Fig. 7.6-18 CANDU fuel-handling schematic. (Courtesy Atomic Energy of Canada, Ltd.)

In the event of a leak or a break in the primary coolant system, it is necessary to inject makeup water into the pressure tubes to assure that the fuel is properly cooled. A large tank of light water is located in the ceiling of the containment system, called a dousing tank. Valve arrangements are such that this water can be directed to a broken low-pressure part of the primary system. For small leaks a pressurized D_2O water addition is provided. If the dousing tank is emptied, water from the reactor building sumps can be circulated by pumps to provide the needed cooling.

Containment Building

The containment building consists of a large prestressed concrete building surrounding the reactor system. For the CANDU systems the large tank of water (dousing tank) is constructed within the ceiling to provide emergency cooling and also to provide containment system sprays. These sprays will condense steam released in the building and reduce the pressure in the event of a loss of coolant accident. The containment system is designed to isolate and enclose any pressurized primary system break and contains the energy so that no significant radioactivity can be released. The defense-in-depth system is similar to the PWR and BWR containment buildings.

7.6-5   Liquid Metal Fast Breeder Reactor (LMFBR)

An introduction to the LMFBR systems is given in Section 7.5-5. The efficiency of any thermal plant, nuclear or otherwise, is determined by its peak operating temperature. The temperatures of an LMFBR are typically 200°C higher than in a LWR; thermal efficiency is correspondingly higher, 40 versus 32%. The design temperatures of an LMFBR are limited by the fuel element cladding, the thermal stress limits for the pressure vessel and outlet piping, and the behavior of the steam generator tubing under high steam pressures. Table 7.6-6 compares operating conditions of LMFBR and LWR reactors.

Sodium as a Coolant

In many ways, sodium is an ideal coolant. It has a high thermal conductivity and low vapor pressure. These properties allow high power density in the core and system pressures much lower than for a LWR. Sodium freezes, however, at 208°F (97.8°C) so that piping and components must be heated to prevent its solidification when the plant cools down. Sodium is very active chemically and it can burn in air and other oxidizing agents. This must be taken into account while designing an LMFBR, but if done properly, the plant can be safer than a LWR and easier to operate. To reduce the possibility of a sodium spill, LMFBRs have outer guard vessels around the main pressure vessel and piping to catch any leaking sodium and to maintain the sodium inventory in the core.

LMFBR Reactor Stability

Uncontrolled power excursions are potentially dangerous in any nuclear reactor core. Like all other reactors, LMFBRs have inherent physical properties that limit the power level any core can attain. These properties are the temperature reactivity coefficients of the core, which define the change in reactivity as the temperature in the core changes. The stability of an LMFBR core is dependent on the following parameters:

1. **Sodium void coefficient.**   An increase in reactivity occurs when the sodium coolant decreases density or voids the core.

TABLE 7.6-6   TYPICAL TEMPERATURE AND PRESSURE CONDITIONS LMFBR VERSUS PWR

	LMFBR	PWR
Reactor outlet temperature, °F	1020	597
Reactor inlet temperature, °F	740	553
Turbine inlet pressure, psia	2200–3000	710
Steam saturation temperature, °F	700	505
Turbine inlet temperature, °F	910	500
Steam reheat temperature, °F	455	453

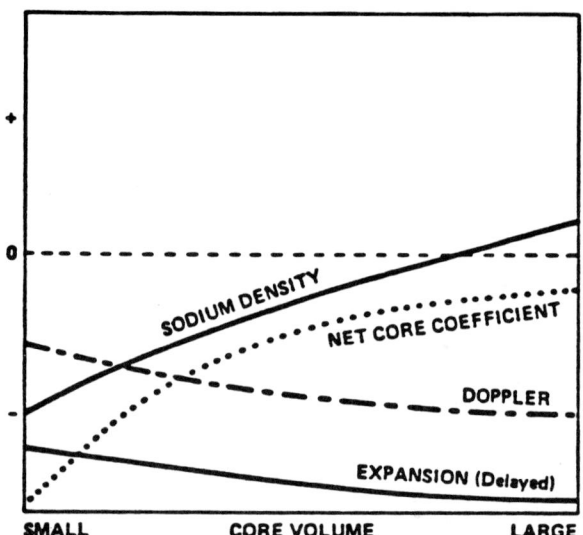

Fig. 7.6-19   General characteristics of power or temperature reacting coefficients for LMFBRs.

2. **Doppler coefficient.** The ability of the fuel to capture neutrons without fission increases rapidly with the temperature due to the thermal motion of the atoms. This mechanism, called the Doppler effect, is a prompt negative-feedback effect.
3. **Mechanical expansion.** As the power level in a core increases, thermal expansion of the fuel elements occur. This effectively increases the size of the core, thereby decreasing its reactivity. Three types of expansion occur: axial fuel expansion, fuel element bowing, and radial core expansion. The latter two phenomena, however, are somewhat delayed in time since they are related to the structural material.

Generally, it is not desirable to have any positive reactivity coefficients. This is not always possible, particularly in the case of the sodium void coefficient. What is important is that the overall reactivity coefficients be negative as the power increases and that any positive coefficient be made as small as possible. The characteristics of the reactivity coefficients as a function of core size are shown in Fig. 7.6-19.

LMFBR Plants

Two specific designs will be described as examples of the current state of the art in LMFBR design. The first, a loop plant, is the Clinch River breeder reactor (CRBR) under development in the United States. The CRBR was recently cancelled but is an example of the state of the art in loop designs. The second, a pool design, is the French Super-Phénix reactor under construction at Creys-Malville on the Rhone River. While the CRBR has not yet started construction, the Super-Phénix is expected to be operational in 1985. Both designs incorporate advanced technological features that are expected to be significant advances over previous LMFBR plants.

An LMFBR core is typically filled with hexagonal arrays of fuel assemblies. The fuel pins in the assemblies are about 6 mm in diameter stainless steel tubes containing mixed plutonium and uranium dioxides. These hexagonal pin arrays are contained in stainless steel ducts to direct the sodium coolant flow and pins are spaced by wire wraps. Control of the core is similar LWR control with rods containing neutron absorbers such as boron. The core of the reactor is surrounded by similar assemblies called blanket assemblies to absorb leakage neutrons and produce plutonium. The configuration of the fuel and core is similar to that shown for the GCFR in Figs. 7.6-23 and 7.6-25.

The Clinch River Breeder Reactor

The layout of the Clinch River breeder (Fig. 7.6-20) included two structural design categories: hardened and nonhardened. The hardened buildings contained those systems or equipment that had to remain functional under earthquake, tornado, or maximum possible flood-level conditions. The major hardened buildings that compose the "nuclear island"—that is, the nuclear steam supply system and

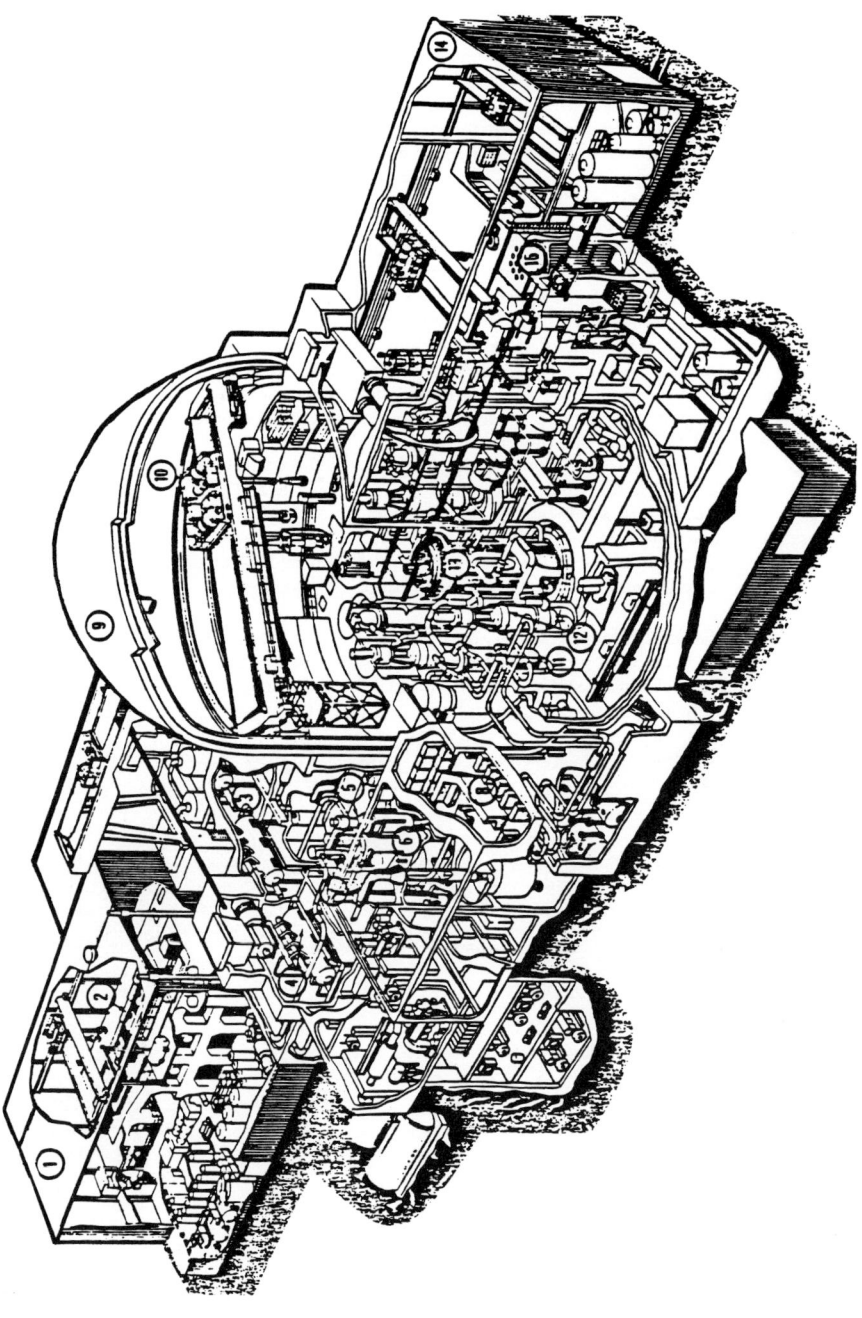

Fig. 7.6-20 Clinch River breeder reactor plant: (1) turbine building, (2) turbines, (3) steam generator building, (4) steam drum, (5) intermediate pump, (6) evaporator, (7) superheater, (8) control room, (9) containment confinement building, (10) containment building, (11) intermediate heat exchanger, (12) primary pump, (13) reactor, (14) reactor service building, (15) ex-vessel (fuel) storage tank. (Courtesy of the Breeder Reactor Corporation.)

other radioactive systems—were to be of reinforced concrete resting on a simple monolithic base mat of reinforced concrete 18 ft thick.

The reactor containment building enclosing the reactor primary heat transport systems and their support systems consisted of a building within a building. The inner containment was to be a steel shell, $1\frac{3}{8}$–$1\frac{3}{4}$ in. thick, above grade and a combination of the steel shell and concrete below grade. The outside confinement was of reinforced concrete totally enclosing the inner containment, both above and below grade. An annular space between the two structures provided containment for any possible leakage from the inner containment.

The reactor containment building was divided in half horizontally by an operating floor. The lower level consisted of inerted, steel-lined concrete vaults housing the reactor, radioactive sodium piping, primary pumps, the intermediate heat exchangers, and the sodium overflow system. The upper level was housed refueling equipment, reactor servicing equipment, and other primary components. A $44\frac{1}{2}$-ft diameter hatch between the reactor service building and containment building allowed for movement of equipment between the two.

A steam generator building housed the intermediate sodium systems, parts of the steam generator systems, and support systems. Two other hardened buildings were the control room building containing central plant control, protection, and instrumentation systems and the diesel generator building containing two emergency diesels to provide the plant with independent power for instrumentation and control in case of loss of off-site power.

There were to be three primary sodium loops carrying heat from the reactor to the intermediate heat exchangers (IHXs). Sodium enterd the reactor vessel at 730°F and left at 995°F. The primary pumps moved it to the three IHXs at a rate of 33,700 gal/min and 175 psi pressure. The sodium then flowed back through a check valve and flowmeter to the reactor. Primary piping is $\frac{1}{2}$-in.-thick welded stainless steel, 36 in. in diameter between the reactor and the pump and 24-in. elsewhere. Steam conditions at the turbine throttle were 906°F at 1550 psi pressure.

Super-Phénix (Crays-Malville) Reactor

The main feature of the Super-Phénix reactor is its pool design for the primary coolant circuit. The core [rated at 3000 MW(th)], four primary coolant pumps and eight IHXs, are placed in the main

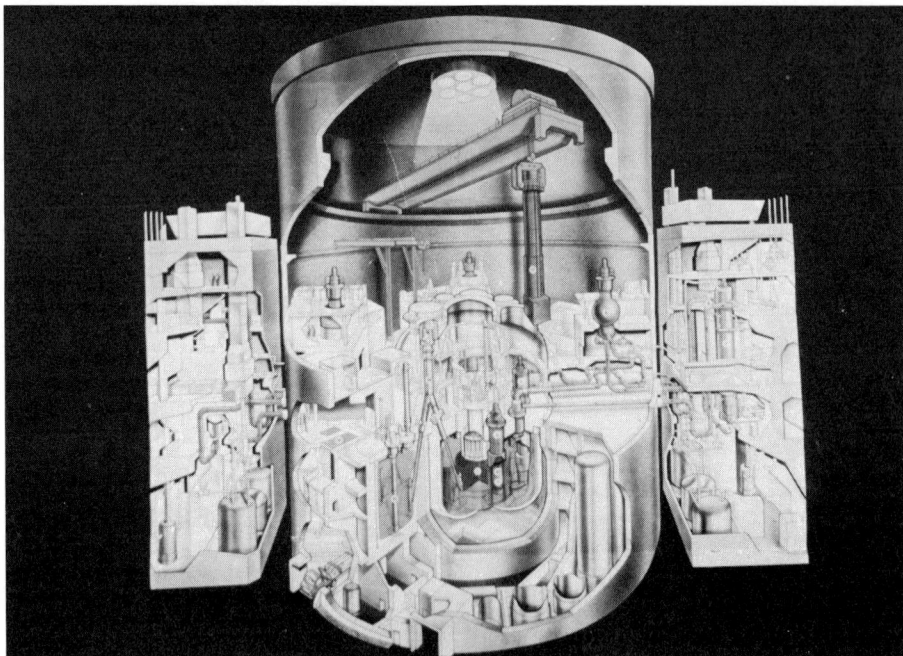

Fig. 7.6-21 Super-Phenix power station: (1) core, (2) primary pump, (3) immediate exchanger, (4) secondary pump, (5) steam generator, (6) steam generator, (7) steam output—490°C—180 bar, (8) secondary sodium input, (9) secondary sodium output, (10) primary sodium input, (11) primary sodium output, (12) control rods, (13) fuel transfer machine, (14) loading and unloading machine, (15) irradiated fuel storage, (16) fresh fuel storage, (17) large component handling device. (Courtesy of Novatome.)

TABLE 7.6-7 POWER CIRCUIT CHARACTERISTICS

Super-Phénix (Creys-Malville) Reactor

Primary circuits

Total mass of sodium in primary circuits	3200 Te
Nominal flow rate	4×4.10 t/s
IHX outlet temperature	392°C
Core inlet temperature	395°C
Core outlet temperature	545°C
IHX inlet temperature	542°C

Secondary circuits

Total mass of sodium in secondary circuits	1500 t
Nominal flow rate	4×3.30 t/s
Steam generator outlet temperature	345°C
IHX inlet temperature	345°C
IHX outlet temperature	525°C
Steam generator inlet temperature	525°C

Water–steam circuits

Water temperature at steam generator inlet	235°C
Water pressure at steam generator inlet	210 bars
Steam temperature at turbine stop valves	487°C
Steam pressure at turbine stop valves	177 bars
Nominal flow rate	4×340 kg/s

Source. Nuclear Engineering International, June 1978.

vessel (Fig. 7.6-21). The reactor vessel is 21 m in diameter, containing 3200 tons of primary sodium. Four independent loops provide, by means of intermediate sodium circulation, the transport of heat to the steam generators. Each steam generator (one per loop) is of the once-through type, with an economizer, evaporator, and superheat section. The secondary loops are at 90° to each other in order to provide maximum separation between them. The main operating characteristics of the plant are given in Table 7.6-7.

The primary and secondary sodium pumps used at Super-Phénix are motor-driven, centrifugal, free-level pumps. The sodium enters at the top of the pump, is driven by the impeller through the diffuser, and then to the discharge chamber. The pump shaft is supported at the top by a ball bearing and at the bottom by a sodium hydrostatic bearing.

The steam generators combine the evaporator and superheater into a single component for reasons of economy. No sodium reheat is used. Each steam generator is a vertical once-through unit, with helically wound tubes. The sodium enters the top of the unit and enters the bottom, while the water/steam flows in the opposite direction. This design does not have a tube sheet, a thick plate that supports the heat exchanger tubes that has often given problems in the past. Rather, the tubes are wound around a cylindrical support. An outer cylindrical shell is penetrated by each tube. This arrangement mitigates thermal shocks during power transients.

First started in 1977, commercial operation of the Super-Phénix reactor is expected in 1985; only 7 years from the start of construction.

Availability of Breeder Reactor Fuel

Using reasonable estimates of LWR-installed capacity by 1985, the amount of reactor-grade plutonium being made yearly in the United States by that time will be in the range of 25–30 tons per year.

TABLE 7.6-8   KEY DIFFERENCES BETWEEN GCFR AND HTGR

Difference	GCFR	HTGR	Reason and Effect
Fuel	UO_2/PuO_2 pellets inserted into cylindrical metal rods (cladding)	UO and ThO_2 kernels with pryolytic and silicon carbide coatings. The fuel particles are embedded in graphite rods, which are, in turn, placed in axial holes in graphite blocks.	Plutonium is the best fissile fuel in a fast reactor, uranium is the diluent and fertile material; U-235 and U-233 are the best fissile fuels in thermal reactors, thorium is the fertile material and transmutes to U-233.
Core	Assemblies of fuel rods	Stacked graphite blocks with axial coolant holes.	Graphite is both moderator and structural material in HTGR.
Core outlet temperature, °F (°C)	(970–1060) 520–570	(1260–1830) 680–1000	HTGR has all ceramic core; GCFR has metal-clad fuel.
Core power density, MW/m^3	120–320	6–8	No moderator in GCFR; therefore, higher pumping power required.
Core flow direction	Up	Down	Lack of significant heat capacity in GCFR core (no moderator) justifies design for high reliability of residual heat removal (after scram), e. g., addition of natural circulation capability.

Reprocessing of spent LWR fuel, however, will be required to separate the plutonium from the uranium and fission products. Against an initial plutonium requirement of 4–5 tons for a large breeder reactor, there appears to be no problem in attaining available plutonium from LWR spent fuel provided the reprocessing services are available. There is currently no U.S. commercial reprocessing of spent fuel, although a reprocessing plant is now operating in Cap La Hague, France.

7.6-6   Gas-cooled Fast Breeder Reactor (GCFR)

An introduction to the GCFR system is given in Section 7.5-6. The design of the GCFR is similar to that of the HTGR because the helium-cooled technology is common to both. The basis for this technology is the lessons learned from the design, construction, and operation of the Peach Bottom Unit 1 and the Fort St. Vrain Nuclear Generating Station described briefly in Section 7.5-3 and the continuing work on the HTGR. The significant differences between the fast and thermal concepts are listed in Table 7.6-8.

The similarities include the multicavity PCRV, the helically coiled steam generators, electric-motor-driven helium circulators with water-lubricated bearings, helium handling and purification systems, core auxiliary cooling systems, and most of the structural materials.

The design of the nuclear steam supply system for a 360-MW(e) GCFR demonstration plant[5] is shown in Fig. 7.6-22 and the plant parameters are given in Table 7.6-9.

The reactor coolant system consists of three main loops, each with an independent steam generator, a horizontally mounted electric-motor-driven circulator, and a gravity-closing isolation valve. Three core auxiliary cooling system (CACS) loops are also provided, each having a vertically mounted electrically driven circulator, a heat exchanger, and a gravity-opening isolation valve.

Prestressed Concrete Reactor Vessel (PCRV)

The entire primary coolant system is contained within the PCRV, which is a multicavity pressure vessel reinforced with steel rods and prestressed by a system of longitudinal tendons and circumferential wire wrappings. All PCRV interior surfaces are covered with a leak-tight steel liner that contains the primary coolant. This, in turn, is lined with thermal insulation to protect the PCRV from the high temperatures of the helium coolant.

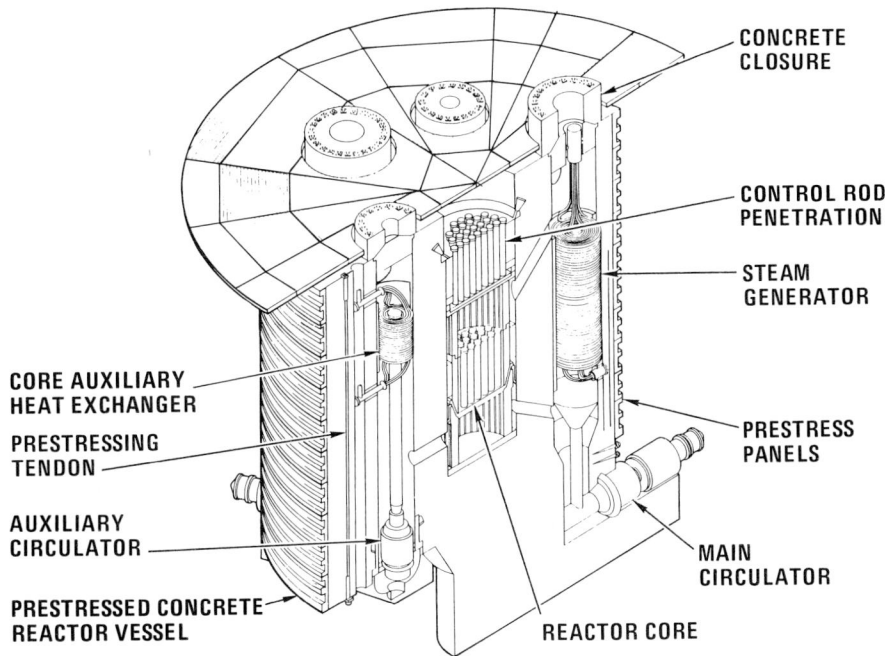

Fig. 7.6-22 GCFR demonstration plant NSS system. (Courtesy General Atomic, San Diego, CA.)

TABLE 7.6-9 GCFR DEMONSTRATION PLANT PARAMETERS*

Overall plant

Thermal power	1,090 MW(th)
Net electric power	360 MW(e)
Net plant efficiency	33%
Number of main loops	3
Circulator power requirement	(14,300 BHP/loop) 10.7 MW/loop

Reactor

Fuel material	$(Pu, U)O_2$
Fuel rod diameter	(0.31 in.) 8.0 mm
Blanket material	Depleted UO_2
Maximum cladding temperature	(1,382°F) 750°C
Reactor inlet temperature	(568°F) 298°C
Reactor outlet temperature	(975°F) 524°C
Helium pressure	(1,520 psig) 10.5 MPa
Reload interval	$\frac{1}{3}$ of core/year
Breeding ratio	1.35

Power Conversion System

Steam Turbine Inlet Pressure	(1,450 psig) 10.0 MPa
Steam Turbine Inlet Temperature	(900°F) 482°C

*Courtesy General Atomic, San Diego, CA.

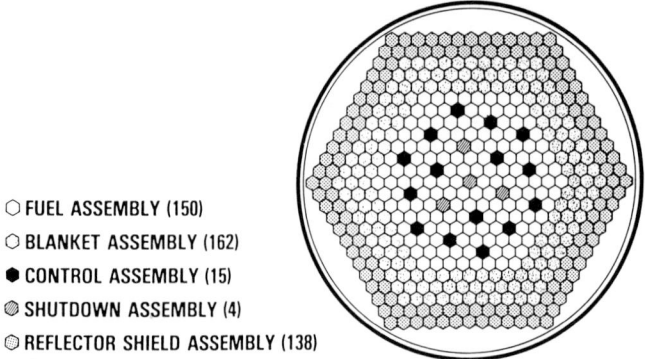

○ FUEL ASSEMBLY (150)
○ BLANKET ASSEMBLY (162)
● CONTROL ASSEMBLY (15)
◉ SHUTDOWN ASSEMBLY (4)
◎ REFLECTOR SHIELD ASSEMBLY (138)

Fig. 7.6-23   GCFR demonstration plant core plan view. (Courtesy General Atomic, San Diego, CA.)

Reactor Core

The general configuration of the demonstration plant reactor core is illustrated in Fig. 7.6-23. The active core region consists of 169 hexagonal assemblies; 150 of these are fuel assemblies, 15 are control assemblies, and 4 are secondary shutdown assemblies. The central core region is surrounded by a radial blanket region containing fertile material. Fertile blanket material is also included in the fuel assemblies above and below the active core (upper and lower axial blankets). The radial blanket is surrounded by 138 hexagonal reflector/shield assemblies to protect the core restraint.

The core assemblies are axially supported at their lower ends by the bottom-mounted grid support plate (Fig. 7.6-24) and laterally by a core restraint that consists of a cylindrical support barrel and core formers surrounding the core, mounted on the grid plate.

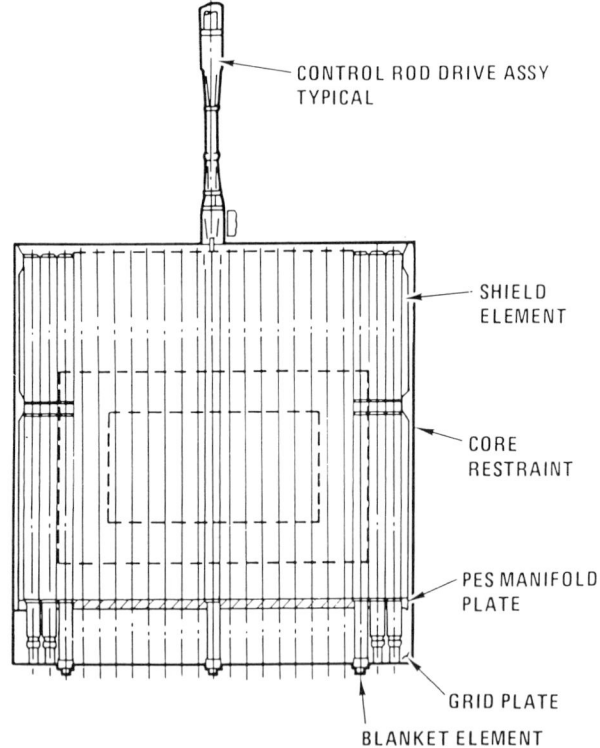

CONTROL ROD DRIVE ASSY
TYPICAL

SHIELD
ELEMENT

CORE
RESTRAINT

PES MANIFOLD
PLATE

GRID PLATE

BLANKET ELEMENT

Fig. 7.6-24   GCFR demonstration plant core elevation view. (Courtesy General Atomic, San Diego, CA.)

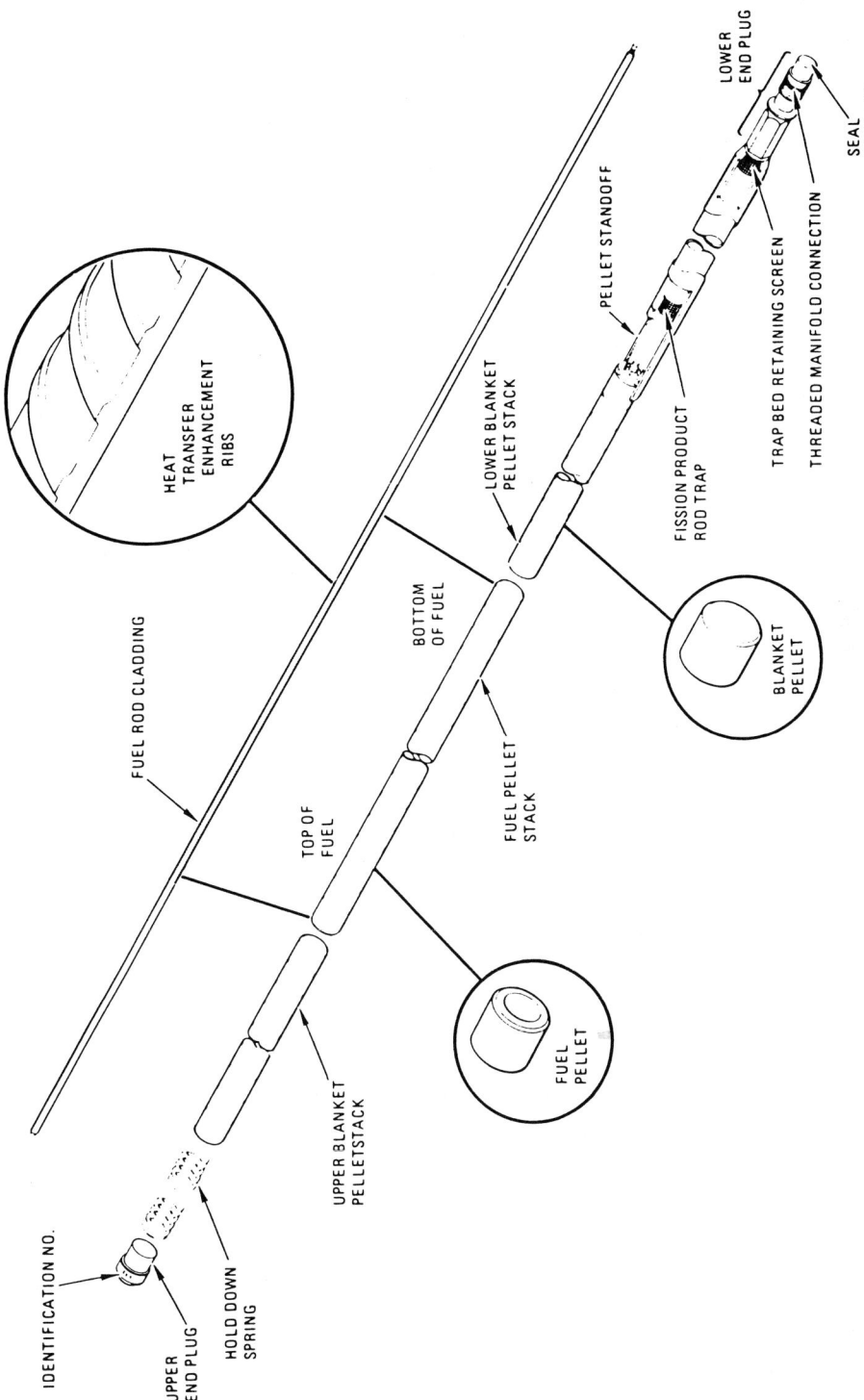

HEAT TRANSFER ENHANCEMENT RIBS

FUEL ROD CLADDING

TOP OF FUEL

BOTTOM OF FUEL

FUEL PELLET STACK

LOWER BLANKET PELLET STACK

PELLET STANDOFF

LOWER END PLUG

SEAL SURFACE

FISSION PRODUCT ROD TRAP

TRAP BED RETAINING SCREEN

THREADED MANIFOLD CONNECTION

BLANKET PELLET

FUEL PELLET

UPPER BLANKET PELLET STACK

IDENTIFICATION NO.

UPPER END PLUG

HOLD DOWN SPRING

Fig. 7.6-25 GCFR fuel rod assembly. (Courtesy General Atomic, San Diego, CA.)

The CGFR fuel assembly design is similar to designs employed in LMFBR programs, particularly the fuel rod design, which uses the same materials and has similar geometry and operating conditions. The unique characteristics incorporated in the GCFR fuel assembly design include (1) ribbed fuel rod cladding to enhance heat transfer to the helium coolant; (2) pressure-equalized and vented fuel rods, that essentially eliminate any pressure-induced stresses on the cladding from either the primary coolant system pressure or from fission gases generated within the rods; and (3) a large fuel rod pitch-to-diameter ratio relative to that commonly employed in the LMFBR designs.

The fuel assembly is hexagonally shaped, 201 mm across flats, and 5 m long, consisting of 265 fuel rods. A hexagonal flow duct is provided around the rod bundle for channeling the coolant flow. A schematic of the fuel rods is shown in Fig. 7.6-25.

The blanket assembly design is essentially the same as the fuel assembly, except that the blanket rod bundle consists of a smaller number of larger rods. Use of either depleted UO_2 or ThO_2 as blanket material is foreseen.

Steam Generators

The steam generators in the CGFR demonstration plant are based on design features similar to those already operating in the Fort St. Vrain plant and are expected to provide highly reliable operation. They are located in cavities in the PCRV. Hot helium enters at the top of the steam generator, flows downward on the shell side of the helically coiled tube bundle, gives up heat to water and steam, and exits at the bottom. The water enters through a feedwater tube sheet located at the bottom, flows upward inside the tubes, and exits at the top through a superheater tubesheet.

Main Helium Circulators

Each of the three main helium circulators is driven by a 13.9-MW (18,600-hp) variable-speed 0–3100-rpm synchronous, horizontal shaft electric motor located outside the PCRV. Speed is controlled by a variable-frequency solid-state electronic controller. Water-lubricated journal bearings and controlled leakage seals similar to those in the HTGR are used in the main circulators. Water is used for the lubricant in preference to oil because of its simple shaft-sealing systems. Any leakage would not contaminate the primary cooling system. Independently powered pony motors provide circulation for decay heat removal.

Residual Heat Removal

When the reactor is shut down, safety grade residual heat removal (RHR) systems are available to back up the normal power conversion system. Normally, core cooling is provided by the main loop cooling system (MLCS) (Fig. 7.6-26) composed of the main circulator, steam generator, turbine, condenser, boiler feed pumps, and feedwater heaters (not shown).

If for any reason the MLCS is not available, the cooling function automatically transfers to the shutdown cooling system (SCS). The SCS uses the main circulator driven by a pony motor, the steam generator, and a separate heat pump system composed of an atmospheric water storage tank with integral heat exchanger/condenser and a makeup water supply. Coastdown of the main circulator provides time for activating the emergency power supply for the pony motors for cases where both the turbine and off-site line power are lost. Water flow from the SCS starts immediately by natural circulation upon opening of one valve. The atmospheric pressure water storage tanks have 20–30 min of inventory before the makeup water supply must be established.

If for any reason MLCS/SCS cooling is lost, then the cooling function automatically transfers to the CACS in the forced circulation mode. The auxiliary loop isolation valves and auxiliary loop cooler louvers open and the auxiliary circulator provides helium flow.

If for any reason forced circulation on the CACS is lost, natural circulation provides adequate passive core cooling even using a single loop. The CACS gives up its heat in the CAHE to a pressurized water system that can also operate under natural convection. The heat is dumped to the atmosphere through a natural draft cooling tower.

Figure 7.6-27 shows the results of a calculation of the maximum fuel cladding temperature that would occur when the core, after scram and 90 s of main loop coastdown, is cooled by natural circulation using only one of the three CACS loops.

Fuel Cycles

The fissile fuel for the GCFR is plutonium and breeding is done in the fertile material U-238 which is transmuted to plutonium. An alternative fuel cycle for the GCFR is the use of thorium in the fertile blankets, which generates U-233. The U-233 is an excellent fuel for thermal spectrum reactors, particularly advanced converters (ACRs) such as the HTGR.[2,3] This symbiosis between ACRs and fast breeder reactors (FBRs) could lead to a scenario in which FBRs that are self-sufficient in

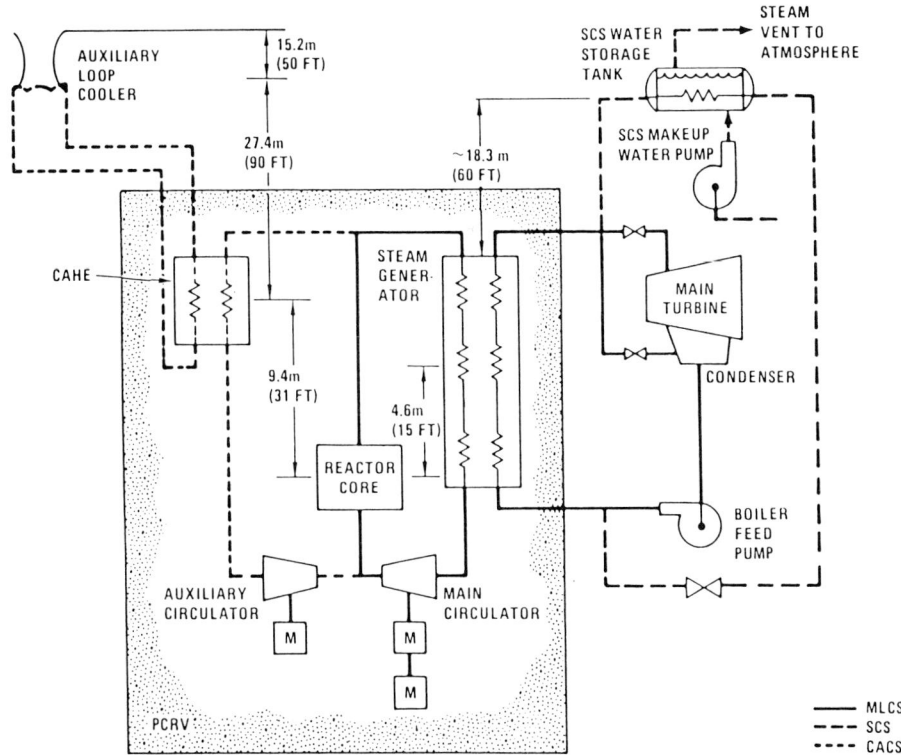

Fig. 7.6-26 Schematic of GCFR core cooling system. (Courtesy General Atomic, San Diego, CA.)

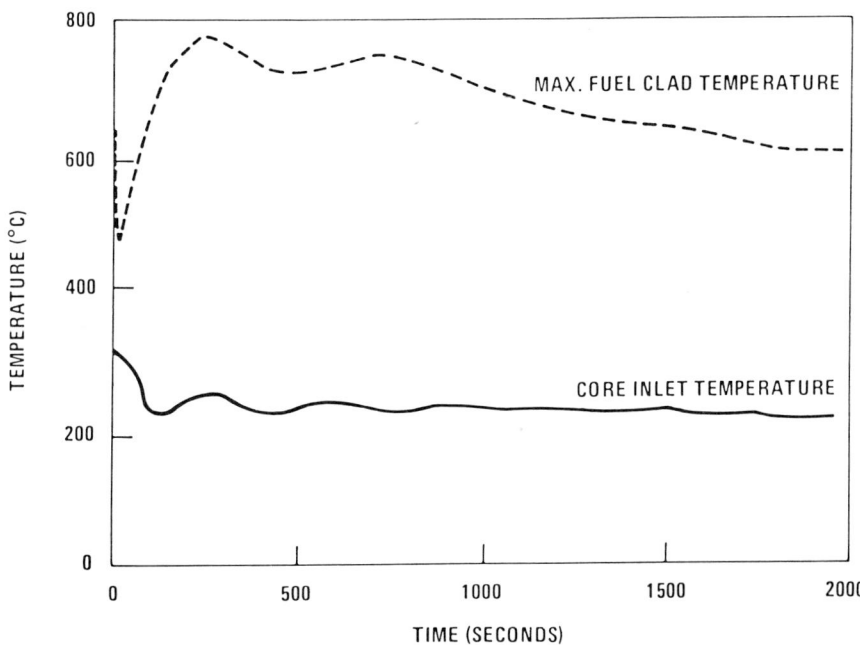

Fig. 7.6-27 GCFR core cooling by natural circulation only. (Courtesy General Atomic, San Diego, CA.)

plutonium and ACRs operating on the Th/U-233 cycle in a ratio of about 1 : 3 are the only nuclear power plants built.[4] This scenario could continue for many centuries.

Using advanced oxide fuel, a 1200-MW(e) GCFR with thorium blankets would have a breeding ratio of 1.54 (vs. 1.57 for U-238) and a compound system doubling time of 9.0 years (vs. 8.3 years for U-238). With advanced carbide fuel even better performance would be achieved with compound system doubling times of 6.2 years.[5]

REFERENCES

7.6-1 R. H. Simon, "Status of the Gas-Cooled Fast Breeder Reactor Program in the United States from the Industrial Viewpoint," IAEA Technical Committee Meeting on Gas-Cooled Reactors, October 1981, Minsk, USSR.

7.6-2 P. Fortescue, "Association of Breeder and Converter Reactors: A General Picture," General Atomic Rep. GA-A14025, July 1976.

7.6-3 C. L. Rickard and R. C. Dahlberg, "Nuclear Power: A Balanced Approach," *Science* **202**(4368) (Nov. 10, 1978).

7.6-4 R. F. Turner and D. M. Ligon, "The Role of Advanced Converters," ASME Paper 80-WA/NE-6, November 1980.

7.6-5 R. H. Simon, C. J. Hamilton, and R. S. Hunter, "Proceedings of the Conference on Gas-Cooled Reactors Today," Bristol, England, Sept. 1982, pp. 293-298, British Nuclear Energy Society, London.

BIBLIOGRAPHY

Acherbach, E. "Pebble Gas-Cooled Reactor," Part 2, in *Heat Transfer and Fluid Flow in Nuclear Systems* (H. Fenech, Ed.), Pergamon Press, New York, 1981.

"Development of Gas-Cooled Nuclear Reactors in the U.S.A.,"*Nuclear Engineering and Design* **26**, 1–186 (1974).

Fischer, P. U. and H. Oeme. "High-Temperature Gas-Cooled Reactors—Experience, Development, Status, and Outlook." GA Report GA-A15031, September 1978.

Fisher, C. R. And D. D. Orvis. "Licensing of HTGRs in the United States." GA-A15954, September 1978.

"Gas-Cooled Fast Breeder Reactor Engineering and Design," Nuclear Engineering and Design **40** (1) 1–233 (1977).

"GCFR Status Report," Organization for Economic Cooperation and Development, Nuclear Energy Agency, 1 and 3 supplements, OECD, Paris, 1980.

Gas-Cooled Reactor Associates. "High Temperature Gas-Cooled Reactor Gas Turbine Application Study." December 1980.

Gas-Cooled Reactor Associates. "High Temperature Gas-Cooled Reactor Reformer Application Study." August 1980.

Goodjohn, A. J. "HTGR Applications Development in the United States." *Nuclear Engineering International*, **24**, pp. 83–86 May 1979.

Kantor, M. E., et al. "Fort St. Vrain Reaches Full-Power." *Nuclear Engineering International* **27**, pp. 14–15 1982.

Lotts, A. L., et al. "HTGR Fuel and Fuel Cycle Technology." Nuclear Power and its Fuel Cycle, Vol. 3, IAEA-CN-36/578, 1977.

McDonald, C. F. and C. O. Peinado. "The Nuclear Gas Turbine—A Perspective on a Long Term Advanced Technology HTGR Plant Option." GA-A16461, November 1981.

McMain, Jr., A. T., Jr., et al. "Industrial Cogeneration Applications of the Steam Cycle High-Temperature Gas-Cooled Reactor." *Proceedings of the American Power Conference* **43**, 1981.

Peinado, C. O., et al. "Design of the 800 MW(e) HTGR Gas Turbine Power Plant." *Proceedings of the American Power Conference* **42**, pp. 130–135, 1980.

Quade, R. N. and C. F. McDonald. "HTGR for Process Heat Applications." ASME No. 81-WA/NE-13, July 1981.

Quade, R. N. and D. L. Vrable. "High Temperature Process Heat Applications with an HTGR." GA-A15868, April 1980.

Rahn, F. J. et al. A Guide to Nuclear Power Technology, J. Wiley and Sons, New York, 1984.

Shenoy, A. S. and B. I. Shamasunder. "Helium Cooled Systems, Part I, High Temperature Gas-Cooled Reactors," in *Heat Transfer and Fluid Flow in Nuclear Systems* (H. Fenech, Ed., Pergamon Press, New York, 1981.

"Status of and Prospects for Gas Cooled Reactors," Technical Reports Series No. 235, IAEA, Vienna, 1984.

Turner, R. F. "Role for Thorium in the Nuclear World." *Power* February 1979.

Vrable, D. L. and R. N. Quade. "Design of the HTGR for Process Heat Applications." GA Report, GA-A15875, May 1980.

Walker, R. F., et al. "Fort St. Vrain Nuclear Power Station." *Nuclear Engineering International*, **14**, pp. 1069–1093, 1969.

7.7 NUCLEAR FUEL PRODUCTION

Arthur G. Randol III

The production of new fuel for a power plant reactor and its disposition following discharge from the power plant is usually referred to as the "nuclear fuel cycle." The processing of fuel is cyclic in nature since sometime during a power plant's operation old or "depleted" fuel must be removed and new fuel inserted. For light water reactors this step typically occurs once every 12–18 months. Since the time required for mining of the raw ore to recovery of reusable fuel materials from discharged materials can span up to 8 years, the management of fuel to assure continuous power plant operation requires simultaneous handling of various aspects of several fuel cycles, for example, material is being mined for fuel to be inserted in a power plant 2 years into the future at the same time fuel is being reprocessed from a discharge 5 years prior.

The various fuel production steps in the light water reactor uranium fuel cycle include: mining of uranium ore, milling of uranium ore to produce natural uranium oxide concentrates (U_3O_8) or extraction of uranium concentrates by-products, conversion of uranium oxide concentrates to uranium hexafluoride (UF_6), enrichment of uranium hexafluoride, conversion of enriched uranium hexafluoride to uranium dioxide (UO_2), pelletizing uranium dioxide and fabrication of fuel assemblies, reprocessing of fuel discharged from a power plant for recovery of uranium and plutonium, recycling of recovered uranium as enrichment feed (UF_6) or as direct fuel blending material (UO_2), and recycling of recovered plutonium (PuO_2) as fissile material for fuel fabrication. These steps constitute the "closed" fuel cycle, which minimizes the amount of uranium required to sustain a given level of power production. The "open," or "once-through," fuel cycle includes all steps up to shipment of discharged fuel to a reprocessing plant. In this latter approach spent fuel, with its residual uranium and plutonium, is stored pending a decision on disposition of the various components of the fuel. In 1980 the total fuel cost for U.S. conventional light water reactors represented approximately 20% of the power generation cost.

Important aspects of each step in the fuel production process are discussed below. Section 7.5 provides specific details of typical fuel for various power plant types.

7.7-1 Worldwide Occurrences of Uranium

Uranium ore bodies with commercial potential occur in a large number of nations dominated by Australia, Canada, South Africa, and the United States. Table 7.7-1 lists estimates of uranium resources (uranium as U_3O_8) in selected countries in which there are currently or planned substantial levels of production. The two resource categories shown are Reasonably Assured Resources (RAR), which corresponds to uranium that occurs in known mineral deposits of such size, grade, and configuration that it could be recovered within the given production cost ranges with currently proven mining and processing technology, and Estimated Additional Resources (EAR), which corresponds to uranium expected to occur in extensions of well-explored deposits, little explored deposits, and undiscovered deposits believed to exist in a well-defined geological trend with known deposits. RAR uranium is assigned a high probability of recoverability while EAR uranium has a lower probability. The RAR category corresponds to the reserves and the EAR category corresponds to the probable potential resources classifications of the U.S. DOE. In addition to these two categories another resource category, speculative resources, has been established for uranium thought to exist, based on indirect evidence and geological extrapolations, in deposits discoverable with existing exploration techniques.

As with any minable resource, an economic "grade cutoff" is necessary to define the extent of the resource since uranium ore deposits are not discrete individual units that have one specific ore grade and tonnage. Typical average ore grades for the $80/kg RAR category are 0.10% (1 kg U_3O_8 per tonne of ore) and for the $130/kg RAR category are 0.07%. These cost bases include all forward costs of production including the operating and capital costs of mining, processing, and haulage. The production cost categories do not reflect market prices, which must account for other business-related

TABLE 7.7-1   SELECTED URANIUM RESOURCES AND PRODUCTION CAPABILITY DATA

| | Uranium Production (tonnes) | | Resources (1000 tonnes)[a] | | | |
| | | | Reasonably Assured[b] | | Estimated Additional[b] | |
Country	1980 Actual	1990 Planned	$80/kg U	$130/kg U	$80/kg U	$130/kg U
Australia	1,561	4,700	294	317	264	285
Canada	7,150	10,500	230	258	358	760
Central African Republic	—	1,000	18	18	0	0
France	2,634	4,050	59	75	28	47
Gabon	1,033	1,500	19	22	0	10
Namibia	4,042	4,150	19	135	30	53
Niger	4,100	12,000	160	160	53	53
South Africa	6,146	7,600	247	356	84	175
Spain	190	1,050	13	16	9	9
United States	16,800	21,800	362	605	681	1,097

Source. OECD Nuclear Energy Agency and the International Atomic Energy Agency, "Uranium Resources, Production and Demand," February 1982.

[a] Data available January 1, 1981.
[b] The production cost categories correspond to the mined uranium oxide as follows: $80/kg U = $30/lb U_3O_8; $130/kg U = $50/lb U_3O_8.

costs such as the cost of carrying capital, past exploration costs, return on investment, and so on. Actual prices may range from $1\frac{1}{2}$ to $2\frac{1}{2}$ times the estimated forward production cost.

Uranium occurs in a variety of geological settings. The dominant deposit in the RAR category worldwide is sandstone (40%). This is also the principal deposit source for U.S.-produced uranium. Other significant deposits occur in quartz-pebble conglomerates, Proterozoic unconformity related deposits, and disseminated deposits (15% each in the RAR category). Although uranium ore grades for conventionally mined deposits are typically 0.05–0.3%, occasionally ore bodies of significantly higher grade (e.g., 2.4% at Nabarlek in Australia) have been discovered.

7.7-2   Mining / Extraction Processes

The mining of uranium ore is conducted in much the same manner as other distributed resources are mined. Open-pit or surface mining is widely used. In the United States, for example, 45% of the ore mined in 1980 was from open-pit. Wyoming, Texas, and Washington state uranium extraction is dominated by open-pit activity. Underground mines are also widely used. Major underground mines in New Mexico, Colorado, and Utah contributed to 41% of the ore mined in the United States in 1980. Decisions to mine a deposit as either open-pit or underground are based on a number of factors, including proximity to surface, extent of deposit, and richness of grade. In the United States most of the total resource base is within 2000 ft of the surface with a significant fraction within 500 ft of the surface.

Some deposits, although substantial, preclude conventional mining options because of excessive economic extraction costs. In these cases *in situ* leaching or solution mining techniques are used. This scheme utilizes an injection well/extraction well system, which injects fluids into the ore-bearing strata and extracts dissolved uranium oxide from the withdrawn fluids. This technique is used extensively in Texas and, to a limited extent, in Wyoming. In 1980 this process provided 8% of total U.S. production. As in all cases where chemicals are used, strict regulations are imposed to assure acceptable levels of environmental impact for leaching agents injected into the ground.

The extensive natural occurrence of uranium in the earth's crust also permits economically attractive recovery from a variety of other mining activities. Uranium by-products from copper (Arizona, Utah), beryllium (Utah), and phosphoric acid (Florida, Louisiana) production operations provide a small fraction of U.S. uranium production (i.e., approximately 5%). In South Africa, however, by-product uranium from gold mining operations is a major component of total uranium production (i.e., approximately 70% in 1980). Some limited amounts of uranium is also recoverable by leaching from tailings piles stored above ground.

7.7-3 Milling

Uranium ore from conventional mining operations are transported by truck to nearby milling facilities which use a multistep chemical process to extract and concentrate the U_3O_8. A typical process would include either an acid or a carbonate leaching a solvent extraction or a column ion exchange. Mill recovery of uranium oxide from the ore can achieve approximately 90% at ore grades of 0.1–0.2%. However, as lower grades of ore are processed, the recovery rate is expected to be lower. Since the economics of ore transportation are a significant component in the cost of uranium, mills are constructed in the proximity of mining activities, typically within 50 mi.

The by-product from conventional mills is dominated by the residual ore, which still contains approximately 10% of the uranium oxide. After diversion to a tailings pond for removal of residual water by evaporation, the tailings are reclaimed. Federal and state license obligations of mill operators usually require a variety of precautions to be taken to secure tailings, including an impervious lining in the pond, several feet of soil cover, and surface vegetation to minimize erosion. In some instances, where practical, the tailings are returned to the mine site as land or cavity fill prior to reclamation.

Commercial uranium ore processing mills are located in Colorado, New Mexico, Texas, Utah, Washington, and Wyoming. Transportation of the natural uranium oxide concentrate, sometime referred to as "yellowcake" because of its initial postprocess coloration, is by truck or rail in 55-gal steel drums with gasketed lids. Although shipped as radioactive material, no shielding or overpack is required. At a price of $30/lb U_3O_8 uranium concentrates represents approximately 35% of the total fuel cost.

7.7-4 Conversion

The next step in the fuel cycle is a change in chemical form. Natural uranium concentrates are converted to uranium hexafluoride (UF_6). This transformation is critical since it provides the gaseous working material for the enrichment process. Uranium hexafluoride is a solid at room temperature, with a triple point of 64.8°C. In the United States commercial conversion plants are located in Oklahoma and Illinois. The natural uranium hexafluoride is shipped as a solid in metal cylinders designed for the handling of gases. The cost of conversion represents approximately 2% of the total fuel cost.

7.7-5 Enrichment

With the exception of the CANDU-type reactor fuel, fuel for commercial power plants must be upgraded from the naturally occurring 0.711% U-235 fissionable material content to some higher level, typically between 2 and 4%. Processes for enrichment of uranium rely on some mechanism for isotopic separation of the U-235 from the U-238 in natural uranium and concentration or "enriching" with U-235 isotopes. The uranium enrichment process usually takes as feed material uranium hexafluoride and produces both an enriched product and a product depleted in U-235 content. A typical process balance would involve delivery of 5.5 kg of uranium in the form of natural UF_6 and the production of 1 kg of 4.3% enriched UF_6 for shipment to a fabrication facility and 4.5 kg of 0.2% depleted UF_6 for storage at the enrichment plant. The depleted uranium, or "tails," is useful for blending with plutonium fuels to be used in light water reactors as a fertile fuel base for fast breeder reactors or as feed material for an atomic vapor laser isotope separation enrichment process.

Uranium fuel enrichment has been dominated by the process known as gaseous diffusion. This technology has been under development since the 1940s and has been deployed in the United States, United Kingdom, France, and the USSR for over 30 years. The process depends on the small difference in molecular mobility between gas molecules of different isotopic species (e.g., U-235 F_6 and U-238 F_6). Each stage in a gaseous diffusion separation plant provides only a small degree of enrichment so many stages are needed in sequence or "cascaded" to provide the capability to produce low enriched uranium. The separation factor, defined as the ratio of the product to the tails enrichment, is 1.004 for a single gaseous diffusion stage. Substantial amounts of electric energy are required to operate these plants because of the large pumping requirements. Government-owned gaseous diffusion plants are operated in Paducah, Kentucky; Portsmouth, Ohio; and Oak Ridge, Tennessee.

Stimulated in large part by the improved economics of operation, substantial international development has proceeded in gas centrifuge technology. This process relies upon the mass difference of the isotopes being separated. Centrifugal force causes the heavier U-238 F_6 molecules to move to the outside wall of the rotor. As in the gaseous diffusion plant, a single stage or rotor provides only a small amount of isotopic separation so many stages must be connected in series to provide the desired level of enrichment. The separation factor for a typical centrifuge is 1.3. An advanced technology centrifuge enrichment plant is estimated to require about 5% of the power necessary to run a diffusion plant. Production of low enriched uranium with centrifuge technology has been accomplished successfully in both a pilot plant in the United States (Oak Ridge, Tennessee) and small-scale commercial facilities in Europe. A major government-owned commercial facility utilizing this technology is planned for Portsmouth, Ohio. At a price of $100 per enrichment unit using conventional technology at an operating tail or 0.2%, the enrichment component represents approximately 40% of the total fuel cost.

Studies of more advanced enrichment processes were undertaken in the late 1960s and early 1970s. The U.S. government has narrowed its principal advanced isotope separation (AIS) program to the atomic vapor laser isotope separation (AVLIS) scheme. This process was originally invented by private U.S. industry, but development was suspended when commercial enrichment opportunities failed to materialize. Outside the United States considerable interest has been shown in an aerodynamic nozzle process (Germany) and a chemical exchange process (France).

The principle of AVLIS is based on the selective excitation and ionization of U-235 atoms in a uranium metal vapor by lasers. The U-235 ions are separated from neutral U-238 atoms by use of an electric field collection system. Other approaches studied in the United States include the molecular laser isotope separation (MLIS) process, which relies on selective excitation and decomposition of UF_6 gas using lasers. Yet another process is the plasma separation process (PSP), which relies on the selective excitation of a single isotopic constituent in a stable plasma through application of electric and magnetic fields. The incentive for development of AIS technology for commercial enrichment application includes both substantial operating and capital cost economies over both gaseous centrifuge and diffusion technology. Typical separation factors are estimated to be 10.0. In addition, these systems appear to be well suited for economic recovery of U-235 from conventional enrichment plant depleted "tails." Pilot scale facilities are not expected to be in operation in the United States until the 1990s.

Enriched uranium hexafluoride is reloaded in steel cylinders for shipment to fuel fabrication plants. A typical cylinder contains up to 2.5 tons of low enriched UF_6 and is placed in steel protective overpacks. No special shielding is required.

Strict inventory controls on both feed material and product streams are maintained.

7.7-6 Fabrication of Uranium Fuel Assemblies for LWRs

The initial process at the fabrication plant involves a chemical conversion of the enriched UF_6 to uranium dioxide (UO_2), a ceramic powder. After chemical conversion the uranium dioxide is carefully pressed into small pellets and sintered. Stacks of pellets are loaded into metal tubes, usually of a zirconium alloy, approximately 10–15 ft long and $\frac{1}{3} - \frac{1}{2}$ in. in diameter depending on the power plant features. The tubes are filled with an inert atmosphere such as helium to avoid inclusion of corrosion-inducing moisture and sealed.

A number of these "fuel rods" are then assembled into bundles with the use of mechanical spacer grids at various intervals along the length of the rods. These fuel assemblies may include an array of from 7×7 to 9×9 rods for a boiling water reactor (BWR) and from 14×14 to 17×17 rods for a pressurized water reactor (PWR). Not all positions in the fuel assembly are occupied by fuel rods—some positions in PWR assemblies provide guide tubes for insertion of control rods—and not all fuel rods contain fuel materials—some positions in BWR assemblies include materials such as gadolinium oxide, which are used as depletable neutron control materials.

Fuel assemblies are individually packaged in mechanical frames for land and sea transportation to power plant sites. Major U.S. commercial fabrication facilities are located in Connecticut, North Carolina, South Carolina, Virginia, and Washington. The cost of uranium fuel fabrication is approximately 15% of the total fuel cost.

Accountability for materials used during fuel fabrication is required to meet quality control, commercial, and licensing requirements. Both destructive and nondestructive techniques are used to determine uranium quantities and enrichments in the incoming uranium hexafluoride cylinders, the process lines, and the finished fuel assemblies.

7.7-7 Recovery of Reusable Uranium and Plutonium

In the "closed" fuel cycle fuel discharged from power plants is transported in heavily shielded casks to a reprocessing facility that chemically separates the residual uranium, the newly generated plutonium, and fission product wastes into three streams. Key process steps are discussed in Section 7.8. The residual uranium has an enrichment roughly comparable to that of natural uranium following an energy generation period of approximately 3 years. This uranium can be converted to a fluorine compound making it suitable for reintroduction into an enrichment plant for upgrading, it can be blended directly with the plutonium nitrate by-product and converted directly to a mixed-oxide (UO_2-PuO_2) fuel (coconversion), or it can be converted to uranium oxide for mechanical blending as a mixed-oxide or fertile fuel constituent.

Plutonium recovered from power plant fuel includes the fissionable isotopes of Pu-239 and Pu-241 as well as the fertile Pu-240 and parasitic Pu-242. Approximately 60–70% of the plutonium produced in LWR fuel at the time of discharge is fissionable. The fissile isotopes are essentially directly substitutable for the U-235 atoms loaded into LWR fuel. The plutonium product from reprocessing can be chemically blended with uranium product to yield mixtures for coconversion to usable mixed-oxide powder or it can be converted directly to oxide powder for either mechanical blending with uranium dioxide or storage. The blend for recycle fuel use in LWRs would require about 5–7% PuO_2 compared to 15–25% PuO_2 for fast breeder reactors.

Only a small fraction (approximately 3% by mass) of the fuel material is diverted to the fission product waste stream. The waste can be readily solidified (e.g., calcined) and rendered suitable for blending with a stabilizing medium such as glass frit to permit vitrification. Several nations have identified this process as suitable for producing a high-level waste form for disposal. Section 7.8 discusses this waste management activity in detail.

Technical measures such as coprocessing, reprocessing, in which the product is a mixture of uranium and plutonium, and coconversion are available safeguards against material diversion. Other key elements of safeguards include materials accountability on input and output, minimal physical access to the process, and plant physical security. Plutonium oxide and mixed oxide can be shipped in an insulated double-steel packaging system that provides for both heat dissipation and shielding. The overpack has a weight of approximately 1 ton and requires special equipment for handling.

7.7-8 Fabrication of Mixed-Oxide Fuel Assemblies

Fabrication of fuel assemblies using a mixed-oxide fuel base proceeds in much the same manner as fabrication of uranium-oxide-only fuel rods. Rather than enriched UF_6 feed, the starting material for the process is either a mixed-oxide powder from a reprocessing plant or a combination of plutonium dioxide powder and a uranium blending material, possibly depleted or natural uranium dioxide. Fuel powder with approximately 4–5% total fissile material is pressed into pellet form and sintered prior to loading in fuel tubes. Until the fuel tube is sealed, additional ventilation control is required to contain any possible dispersion of plutonium material. For some process feed material limited shielding may be required for personnel on the fabrication line. However, once encapsulated, mixed-oxide fuel rods can be handled under essentially the same procedures as uranium oxide rods.

Stringent procedures are used to account for all plutonium bearing material from the initial feed to the final finished fuel assembly and any associated process waste streams. Small-scale commercial production facilities for mixed-oxide fuel assemblies exist in the United States. Fabrication of fuel for the U.S. breeder demonstration program dominates the current demand for this capability although some LWR fuel has been fabricated for in-plant demonstration purposes.

BIBLIOGRAPHY

Atomic Industrial Forum. *Survey of Electric Utility Economics.* Washington, D.C., December 18, 1981.

International Atomic Energy Agency. *Advanced Fuel Cycle and Reactor Concepts: Report of Working Group 8.* Vienna, Austria, January 1980.

International Atomic Energy Agency. *Enrichment Availability: Report of Working Group 2.* Vienna, Austria, January 1980.

International Atomic Energy Agency. *Fuel and Heavy Water Availability: Report of Working Group 1.* Vienna, Austria, January 1980.

International Atomic Energy Agency. *Reprocessing, Plutonium Handling, Recycle: Report of Working Group 4.* Vienna, Austria, January 1980.

National Academy of Sciences. *Workshop on Concepts of Uranium Resources and Producibility.* Washington, D.C., 1978.

OECD Nuclear Energy Agency & International Atomic Energy Agency. *Uranium Resources, Production and Demand.* Paris, France, February 1982.

U.S. Department of Energy. *Nuclear Proliferation and Civilian Nuclear Power, Report of the Nonproliferation Alternative Systems Assessment Program: Volume II: Proliferation Resistance.* Washington, D.C., June 1980, DOE/NE-0001/2.

U.S. Department of Energy. *Nuclear Proliferation and Civilian Nuclear Power, Report of the Nonproliferation Alternative Systems Assessment Program: Volume IX: Reactor and Fuel Cycle Description.* Washington, D.C., June 1980, DOE/NE-0001/9.

U.S. Department of Energy. *Nuclear Proliferation and Civilian Nuclear Power, Report of the Nonproliferation Alternative Systems Assessment Program: Volume III: Resources and Fuel Cycle Facilities.* Washington, D.C., June 1980, DOE/NE-0001/3.

U.S. Department of Energy. *Statistical Data of the Uranium Industry.* Grand Junction, Colorado, January 1, 1981, GJO-100, 1981.

7.8   NUCLEAR WASTE MANAGEMENT

Walton A. Rodger

7.8-1   Sources and Characteristics of Wastes

Most of our activities have always produced waste products of one sort or another. Huxley[35] gives a humorous account of wastes throughout antiquity. So it should come as no surprise that some radioactive materials end up as waste products requiring management and disposal.

Public perception of nuclear waste hazards places them much higher on the "worry scale" than is justified by the actual hazard involved. While the public perception of these hazards appears to revolve mostly around high-level wastes, there are several other categories of wastes that must also be controlled and managed. The major sources of radioactive wastes are discussed below.

Reactors

During the operation of a nuclear reactor radioactive materials are formed in the fuel, coolant, and structural components. Any reactor that has operated for any length of time at a high power level represents a large inventory of radioactive material.

Under ordinary operating conditions essentially all these materials are maintained within the reactor. It is not possible to operate a nuclear reactor, however, without permitting some small quantities of radioactive materials to escape into the environment. It is the function of the various radwaste systems (see Section 7.6) to keep the amount that does escape to a very low level.

In an accident, such as that experienced at TMI-2, it is possible that a sizable portion of this inventory can be released from the core and the reactor. It is the function of the emergency core cooling systems and of containment, not of the reactor radwaste systems, to maintain control of such accidental releases.

Radioactive materials are produced in the fuel, coolants, and structural materials of all types of reactors. Fission products escape only from defective or failed metal clad elements. Most of the radioactive materials normally found in the coolant are formed by neutron activation of the fuel clad, coolant, and structural materials and of impurities carried by the coolant.

During operation the major radioactive isotopes in the coolant are due to the activation of the water by neutrons and protons (see Table 7.8-1). All of the water activation products are gases, most of them short-lived. The most significant activity is N-16, which reaches a sufficient concentration that some of the piping external to the reactor must be shielded. Upon reactor shutdown this short-lived activity decays within a few minutes. Some of the principal activation isotopes are also shown in Table 7.8-1.

Some fission products may be introduced into the coolant from "tramp" uranium on the surface of fuel elements and considerably more from a failure of fuel element cladding. Normally the fission product content of the waste will predominate over that of corrosion products. There are more than 100 isotopes formed in fission. Details on the characteristics of these fission products may be found in Etherington,[28] Meek and Rider,[47] and Strominger et al.[70]

From a practical standpoint only the gaseous isotopes and those with relatively long half-lives are important in considering waste management. A representative set of reactor coolant concentrations for BWR and PWR, operating at essentially equivalent failed fuel percentage, is given in Table 7.8-2. Leakage of reactor coolant from the primary system is the major source of radioactive waste at a reactor. Concentrations of radionuclides in PWR coolant tend to be higher than those in BWRs. This is because BWRs have three internal cleanup systems that operate on an effective turnover time of an hour or two, whereas PWRs have only a single cleanup system with a turnover time of about a day. On the other hand, leakage from a BWR tends to be higher than that from a PWR so that environmental releases from the two reactor types are similar.*

An estimate of waste volumes produced at reactors is shown in Table 7.8-3.[57] These numbers were compiled from a survey of 18 PWRs and 12 BWRs. The data represent a total of 63 years of operating experience for BWRs and 112 years for PWRs. Activity levels of the various waste forms are shown in Table 7.8-4.

Front-End Wastes

The "front end" of the fuel cycle consists of (1) mining and milling, (2) conversion to UF_6, (3) enrichment, and (4) fuel fabrication. Each step produces its own waste products. The radioactivity associated with them is naturally occurring.

---

*As long as the major BWR gas stream, the steam jet air ejector off-gas, receives additional treatment—see Section 7.6.

TABLE 7.8-1 PRINCIPAL ACTIVATION PRODUCTS IN LWR COOLANT

Isotope	Reaction	Half-Life	Total Decay Energy (MeV)
	Water Activation Products		
H-3[a]	H-2 (n, γ)	12.26 yr	0.0186
	Fission product		
	B-10 (n, α) Li-6 (n, α)		
N-13	O-16 (p, α)	10.0 min	2.21
N-16	O-16 (n, p)	7.4 s	10.4
	N-15 (n, γ)		
N-17	O-17 (n, p)	4.14 s	8.8
O-19	O-18 (n, γ)	29.0 s	4.80
F-18	O-18 (p, n)	1.87 h	1.67
Ar-37[b]	Ar-36 (n, γ)	35.0 day	0.82
Ar-41[b]	Ar-40 (n, γ)	1.83 h	2.49
	Impurity Activation Products		
Na-24	Al-27 (n, α)	15.0 h	5.51
	Na-23 (n, γ)		
Al-28	Al-27 (n, γ)	2.30 min	4.65
Si-31	P-31 (n, p)	2.62 h	1.48
Cl-38	Cl-37 (n, γ)	37.3 min	4.8
Ar-37	Ar-36 (n, γ)	35.0 day	0.82
Ar-41	Ar-40 (n, γ)	1.83 h	2.49
Ca-45	Ca-44 (n, γ)	160.0 day	0.25
Cr-51	Cr-50 (n, γ)	27.8 day	0.75
Mn-54	Fe-54 (n, p)	280.0 day	1.38
Fe-55	Fe-54 (n, γ)	2.6 yr	0.22
Mn-56	Fe-56 (n, p)	2.58 h	3.70
	Mn-55 (n, γ)		
Co-58	Ni-58 (n, p)	71.0 day	2.31
Fe-59	Co-59 (n, p)	45.0 day	1.56
Co-60	Co-59 (n, γ)	5.27 yr	2.81
Cu-64	Cu-63 (n, γ)	12.9 h	c
Ni-65	Ni-64 (n, γ)	2.56 h	2.10
Zn-65	Zn-64 (n, γ)	245.0 day	1.35
Cu-66	Cu-65 (n, γ)	5.1 min	2.63
Sr-89	Sr-88 (n, γ)	51.0 day	1.46
Zr-95	Zr-94 (n, γ)	65.0 day	1.12
Hf-181	Hf-180 (n, γ)	45.0 day	1.02
Ta-182	Ta-181 (n, γ)	115.0 day	1.73
Ta-183	Ta-182 (n, γ)	5.0 day	1.07
W-187	W-186 (n, γ)	24.0 h	1.3

[a] Tritium. While tritium is a gas, it usually is only present in readily detectable quantities as the oxide in liquid wastes discharged from light-water-cooled reactors and aqueous fuel reprocessing plants. It may be present in hazardous amounts in the coolant of heavy-water reactors.
[b] Ar-37 and Ar-41 may build up in small amounts if the water is not deaerated.
[c] EC and positron, 1.68; beta, 0.57.

TABLE 7.8-2   APPROXIMATE CONCENTRATION OF REACTOR COOLANT

Isotope[a]	Half-Life	BWR[b] Concentration (μCi/cm^3)	PWR[c] Concentration (μCi/cm^3)
Noble Gases			
Kr-83m	1.86 h	0.006	0.04
Kr-85m	4.4 h	0.004	0.2
Kr-85	10.8 yr	0.00001	0.02
Kr-87	76.0 min	0.01	0.1
Kr-88	2.8 h	0.01	0.4
Xe-133	5.27 day	0.005	6.5
Xe-135m	16.0 min	0.015	0.05
Xe-135	9.2 h	0.015	2.0
Xe-137	3.8 min	0.08	0.2
Xe-138	14.0 min	0.05	0.2
Iodines			
I-131	8.05 day	0.005	0.4
I-132	2.3 h	0.04	0.1
I-133	21.0 h	0.03	0.5
I-135	67.0 h	0.04	0.25
H-3 (tritium)		0.005	0.1
Fission Products			
Sr-90	28.8 yr	0.0001	0.01
Zr-95	65.0 day	0.00003	0.00006
Mo-99	66.0 h	0.008	0.4
Ru-103	39.6 day	0.00002	0.00004
Ru-106	1.0 yr	0.000006	0.00001
Te-131	25.0 min	0.001	0.005
Cs-134	2.07 yr	0.0003	0.02
Cs-136	13.0 day	0.0001	0.01
Cs-137	30.2 yr	0.0002	0.02
Ba-140	12.8 day	0.005	0.01
Ce-144	290.0 day	0.00002	0.0002
Activation Products			
Cr-51	28.0 day	0.0005	0.0008
Fe-55	2.6 yr	0.0015	0.0015
Co-58	71.0 day	0.003	0.006
Co-60	5.2 yr	0.0003	0.001

[a] Representative major components.
[b] At 100,000 μCi/s after 30 min, equivalent to 0.07% failed fuel.
[c] At 0.1% failed fuel.

Uranium is itself radioactive. Its two principal isotopes have the following properties:

Isotope	Half-life, yr	Natural Abundance, %
U-238	4.5E + 09	99.3
U-235	7.1E + 08	0.7

Naturally occurring uranium is in "secular equilibrium," that is, there is 1 Ci of each of the daughter products for each curie of the parent. U-238 has 13 daughters and U-235 has 12. In the course of carrying out the front-end operations, this material is moved and in some cases concentrated. Although no radioactive material is produced that did not already exist, the potential effects of that material upon man may have been enhanced and therefore these wastes are subject to regulation.

TABLE 7.8-3   AVERAGE PLANT UNTREATED WASTE VOLUMES

	Waste Volumes, ft^3/MW(e)-yr			
	BWR		PWR	
Waste Type	Deep Bed CPS^a	Precoat CPS	Without CPS	With CPS
Deep-bed resin	4.6	0.23	0.94	0.32
Concentrated liquids	12.7	0.6	3.9	4.8
Filter sludge	5.4	7.7	—	0.015
Cartridge filters	—	—	0.39	0.39
Trash				
Total	11.5	11.5	11.5	11.5
Compactible	7.8	7.8	7.6	7.6
Noncompactible	3.7	3.7	3.9	3.9
Total	34.2	20.0	16.7	17.2
Annual volume (ft^3/yr) for a 1000-MW(e) plant	34,200	20,000	16,700	17,200

Source.   "A Waste Inventory Report for Reactor and Fuel Fabrication Wastes," ONWI-20, March 1979.

^{a} Condensate polishing system.

TABLE 7.8-4   AVERAGE PLANT WASTE ACTIVITY

	Waste Activity, Ci/MW(e)-yr			
	BWR		PWR	
Waste Type	Deep-Bed CPS^a	Precoat CPS	Without CPS	With CPS
Deep-bed resin	1.9	0.0014	0.61	0.2
Concentrated liquids	0.58	0.016	0.20	0.024
Filter sludge	2.0	0.5	—	0.012
Cartridge filters	—	—	0.12	0.12
Trash				
Total	0.402	0.402	0.063	0.063
Compactible	0.0052	0.0052	0.0049	0.0049
Noncompactible	0.397	0.397	0.058	0.058
Total	4.88	0.92	1.00	0.42
Annual Activity (Ci/yr) for a 1000-MW(e) plant	4880	920	1000	420

Source.   "A Waste Inventory Report for Reactor and Fuel Fabrication Wastes," ONWI-20, March 1979.

^{a} Condensate polishing system.

Mining and Milling.   In the mining of uranium ore, typically containing 0.15% uranium, radioactivity is released from the mines in small amounts in the form of radon gas. The ore is hauled to a mill where it is crushed and the uranium extracted by chemical processing. The uranium is recovered as uranium oxide. The solid-waste residues, which contain nearly all of the radium, are called tailings. These are pumped to a waste collection area where the waste slurry is discharged and the solids settle out to form a tailings pile.

Radon gas is released from the uranium ore during the crushing process and also from the tailings pile. Particles of uranium-containing dust are also discharged during the crushing of the ore and calcination of the oxide.

Conversion.   Uranium concentrate must be converted to uranium hexafluoride (UF_6) in order to be enriched by the gaseous diffusion process. At the conversion facilities, the radionuclides present are uranium, thorium, radium, and their respective decay products. Some of the decay products are impurities in the mill concentrate and others reoccur there by the continuing radioactive decay of the uranium. Uranium is essentially the only source of radioactivity in the gaseous effluents. The radium, thorium, and decay products are separated from the uranium in the conversion process and appear in the liquid- and solid-waste streams.

Enrichment.   Natural uranium contains only 0.7% of fissionable U-235. LWRs utilize uranium that is enriched in U-235 to the range of 2–4%. Gaseous diffusion is the technology that has been developed in this country for performing the enrichment operation. Wastes released from this process are small and consist largely of uranium.

Fuel Fabrication.   LWR fuel is fabricated from uranium hexafluoride enriched to 2–4% in U-235. This slightly enriched uranium hexafluoride is shipped to a fuel fabrication facility where it is hydrolyzed to uranyl fluoride, converted to ammonium diuranate, and calcined to uranium dioxide. The dioxide powder is pelletized, sintered, and loaded into Zircaloy tubing, which is then capped and welded. Due to leakage, spillage, and breakage during the fuel fabrication steps some of the enriched uranium is released to the various waste streams.

Further information on front-end wastes and processes of the uranium fuel cycle may be found.[20,52]

Reprocessing

In the course of the operation of a reactor two things take place within the fuel assemblies: (1) neutrons combine with the U-235 atoms causing them to become so energetic that they break apart (undergo fission). In the course of doing this, they form fission products (the wastes of this process). (2) Neutrons also combine with the more abundant U-238 atoms. The U-238 atoms which so react do not undergo fission but rather produce (after two short intermediate steps) an isotope of a totally new element, plutonium (Pu-239). Thus there are three main components in spent fuel: (1) unburned U-235 and unconverted U-238; (2) Pu-239, which has been bred into fuel by conversion of U-238 and which was not consumed while the fuel was in the reactor; and (3) radioactive fission products from the fissioning of U-235 and Pu-239.

Reprocessing in simplest terms is the separation of these three components from one another. There are several steps to the process which is outlined schematically in Fig. 7.8-1.[51] This figure also shows the major waste streams from the process. These include: (1) hulls, (2) high-level waste (HLW), (3) gaseous effluents, (4) low-level waste (LLW), and (5) "TRU" waste. The quantities and characteristics of these waste streams and their dispositions are shown in Fig. 7.8-2. The most important of these waste streams is the HLW. Its approximate composition is shown in Table 7.8-5.

Research, Medical, and Industrial Uses

Hospitals, clinics, private physicians and laboratories, medical schools, universities, and colleges are sources of institutional radioactive waste.[7,8,34] Sources of industrial LLW are radiochemical manufacturers, pharmaceutical companies, and manufacturers of smoke detectors and luminous dials.[1,6,8] All of this waste may be classified as low level. Based on reports for the years 1972–1977 it is estimated that at least 200,000 m^3 of institutional and industrial waste had been shipped to commercial burial by the end of 1980. For the LLW shipped to commercial burial sites in 1979, the proportions of reactor, institutional and industrial, and government LLW were 50, 41, and 9 vol. %, respectively.[34] The total institutional waste volume shipped to burial in 1979 was estimated to be 15,000 m^3 and the industrial waste volume was estimated to be 17,000 m^3 (Table 7.8-6). The accumulated radioactivity (uncorrected for decay) contained in this waste was about 300,000 Ci, accounting for about 70% of the radioactivity shipped to commercial burial sites.

7.8-2   Waste Management Methods

General Principles

Since there is no practicable way in which the radioactive waste can be made to decay any faster than the fundamental half-life of its constituents, there are intrinsically only two basic ways to treat the material: (1) concentrate and contain and (2) dilute and disperse. Many of the waste sources (see Section 7.8-1) are sufficiently large that they must be concentrated and contained. On the other hand, there are some waste streams so low in radioactivity content that they may, and indeed must, be diluted and dispersed. A major objective of waste management is to see that the dilution/dispersion

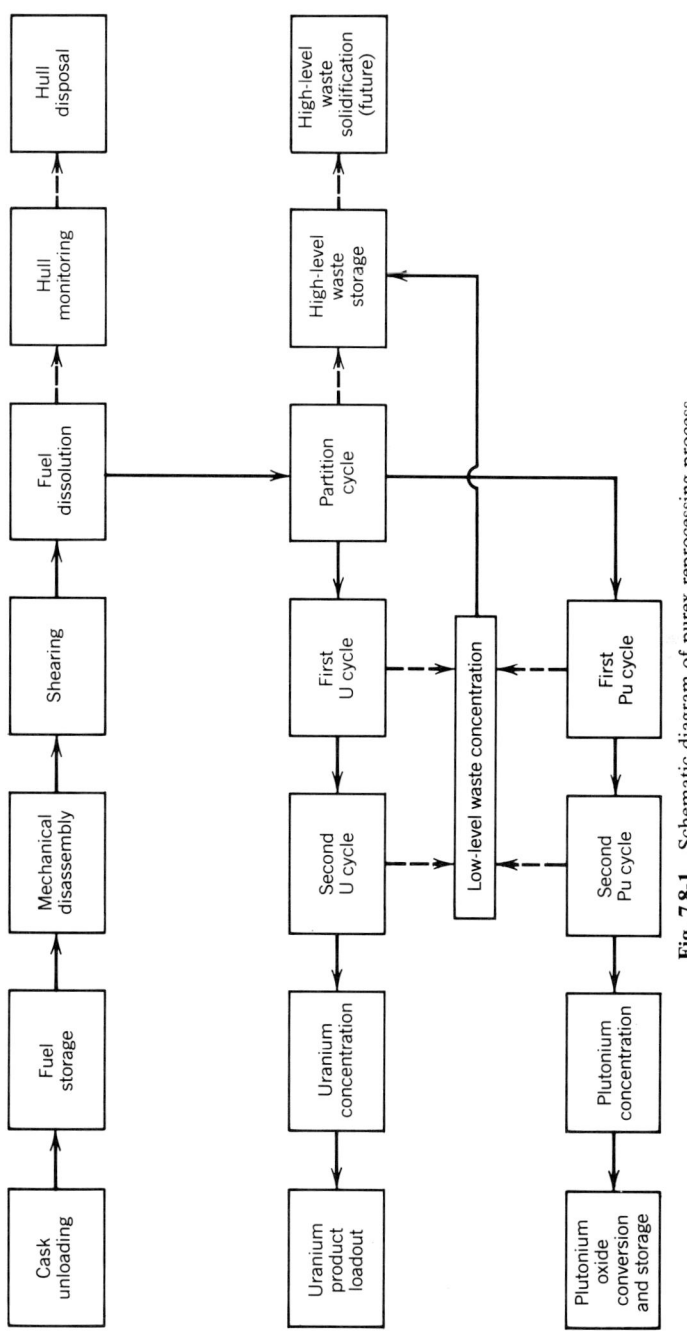

Fig. 7.8-1 Schematic diagram of purex reprocessing process.

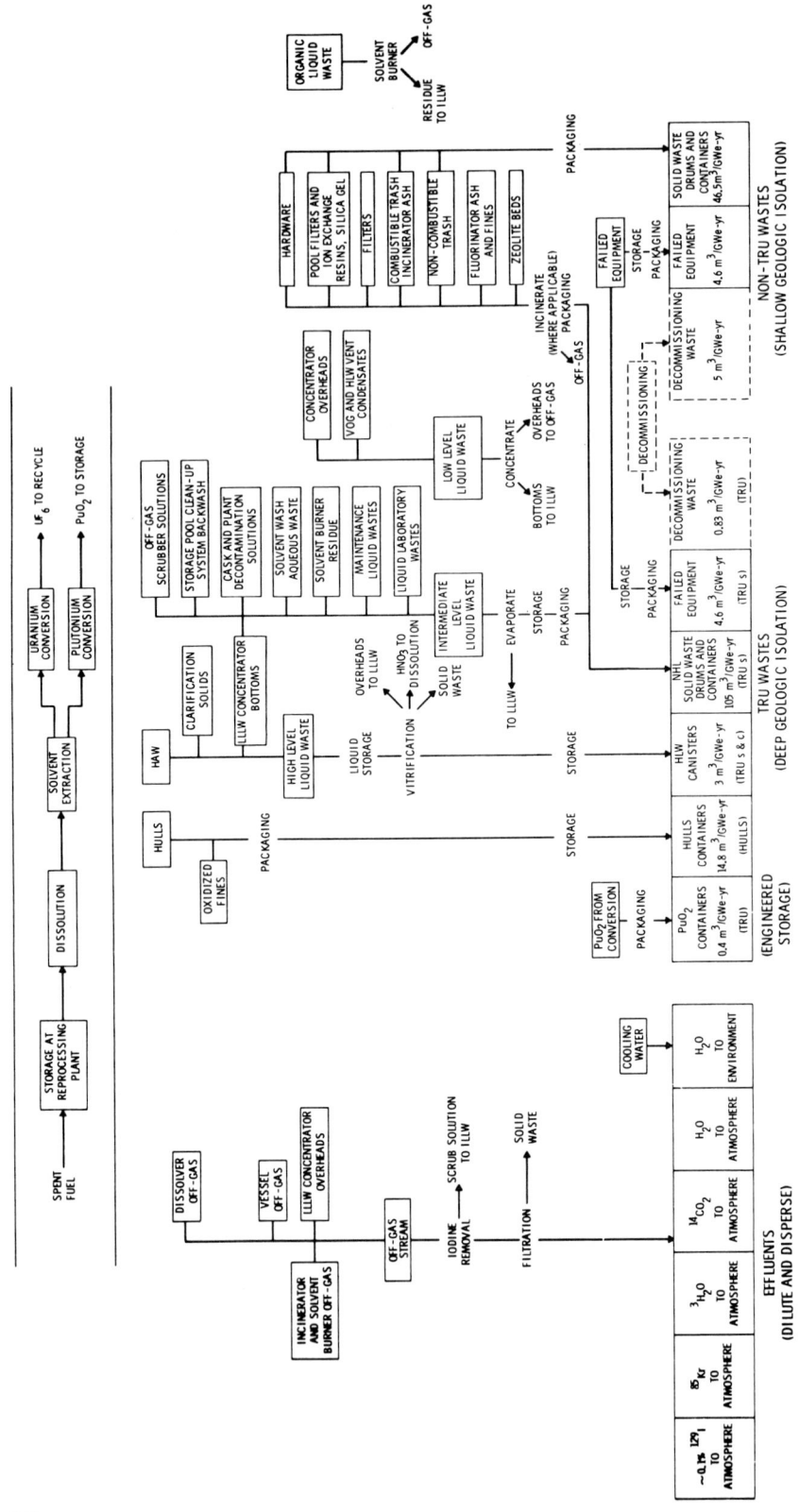

Fig. 7.8-2 Origin, treatment, and disposition of radioactive wastes from a reprocessing plant (mode II). (From "Alternatives from Managing Wastes from Reactors and Post-Fission Operations in the LWR Fuel Cycle," ERDA-76-43, **1**, May, 1976.)

TABLE 7.8-5 COMPOSITION OF HLW[a]

		Concentration, mol/L @ 378 L/MTU		Kilograms Nonvolatile Oxide/MTU	
	Constituent	Possible Range	Reference Values[a,d]	Constituent	Reference Values
Inerts (reprocessing chemicals)	H^+	1–7	2.0	—	—
	Na	0–3.0	0.01	Na_2O	0.12
	Fe	0.05–1.4	0.054	Fe_2O_3	1.6
	Cr	0.01–0.04	0.0096	Cr_2O_3	0.28
	Ni	0.005–0.02	0.0034	NiO	0.19
	PO_4^{3-}	0.025–0.30	0.042	P_2O_5	1.1
	SO_4^{2-}	0–0.90	0	SO_4^{2-}	0
	NO_3^{4-}	2.7–20	3.6	—	—
	F^-	0–0.25	0	F^-	0
	Gd^b	0–0.2	0.150	GD_2O_3	10.0
	B^b	0–5.0	0	B_2O_3	0
	Cd^b	0–1.5	0	CdO^3	0
Subtotal					13.0
Fission products	Rb	[c]	0.0095	Rb_2O	0.34
	Sr	[c]	0.017	SrO	0.68
	Y	[c]	0.0095	Y_2O_3	0.41
	Zr	[c]	0.074	Z_rO_2	3.5
	Mo	[c]	0.071	MoO_3	3.9
	Tc	[c]	0.017	Tc_2O_7	1.0
	Ru	[c]	0.044	RuO_2	2.2
	Rh	[c]	0.011	Rh_2O_3	0.53
	Pd	[c]	0.030	PdO	1.4
	Ag	[c]	0.0015	Ag_2O	0.067
	Cd	[c]	0.015	CdO	0.078
	Sn	[c]	0.0009	SnO_2	0.052
	Sb	[c]	0.0002	Sb_2O_2	0.013
	Te	[c]	0.0078	TeO_2	0.47
	Cs	[c]	0.039	Cs_2O	2.2
	Ba	[c]	0.023	BaO	1.3
	La	[c]	0.018	La_2O_3	1.1
	Ce	[c]	0.034	CeO_2	2.2
	Pr	[c]	0.016	Pr_5O_{22}	1.1
	Nd	[c]	0.055	Nd_2O_3	3.5
	Pm	[c]	0.0005	Pm_2O_3	0.035
	Sm	[c]	0.012	Sm_2O_3	0.80
	Eu	[c]	0.002	Eu_2O_3	0.13
	Gd	[c]	0.001	Gd_2O_3	0.076
Subtotal					27.0
Actinides	U	0.011–0.22	0.053	U_3O_8	5.7
	Np	[c]	0.003	NpO_2	0.31
	Pu	0.0001–0.006	0.002	PuO_2	0.17
	Am	[c]	0.009	Am_2O_3	0.96
	Cm	[c]	0.003	Cm_2O_3	0.25
Subtotal					7.4
Total					47.0

Source. *Alternatives for Managing Wastes from Reactors and Post-Fission Operations in the LWR Fuel Cycle*, ERDA-76-43, **2**, May, 1976.

[a] The reference HLW is representative of a large state-of-the-art commercial reprocessing plant such as BNFP.
[b] Potential soluble poisons that may be used during fuel dissolution.
[c] Depends on burnup of the fuel being reprocessed, should not vary over a factor of 2 from the reference waste composition (burnup = 25,000 MW(d)/MTU).

TABLE 7.8-6   VOLUME OF INSTITUTIONAL AND INDUSTRIAL LLW GENERATED IN 1978

Form	Source and Volume (m^3)			
	Bioresearch	Nonbioresearch	Medical	Total
Dry solids	3,900	1,600	920	6,400
Scintillation vials	5,700	21	42	5,800
Absorbed liquids	1,100	440	21	1,500
Biological	1,200	20	63	1,300
Subtotal	11,900	2,100	1,000	15,000
Miscellaneous industrial				18,000
Total				33,000

capacity of the environment is used judiciously, that releases are kept within statutory limits and as low as reasonably achievable (ALARA). A second objective of waste management practices is to convert as much of this waste as practicable into a volume and form that can be contained. Some of the methods used to do so are discussed in this section. Ultimate disposal of the wastes is discussed in Section 7.8-3.

There are two general principles that apply to all radioactive waste treatment.

1. Segregate at the source. Smaller volume streams tend to contain the bulk of the radioactivity in concentrated form while the larger volume streams have sufficiently low radioactivity that they can frequently be discarded after a minimum treatment without exceeding maximum permissible limits. The concentrated streams must be reduced in volume so that they can be economically stored until they decay to a very low level, or until they can be discarded in some particular circumstance known to be nonhazardous. If these two types of waste streams are mixed, it generally results in having to use low-volume methods on much larger volumes.

2. Don't let material become radioactive in the first place. That is, don't take anything into a radioactive area if this can be avoided.

In this section a number of methods for treating various types of wastes are described. It is axiomatic that it is uneconomical to choose methods that produce a higher degree of separation than is needed.

Reactors

Briefly described herein are "typical" radwaste systems for BWRs and PWRs even though there really is no such thing as a "typical" system. It is necessary to begin with a description of the primary systems, since the entire primary system is in reality part of the radwaste system of a reactor.

BWR.   Figure 7.8-3 is a simplified flow diagram of a typical BWR. The basic characteristic of a BWR is that the water in the pressure vessel is permitted to boil in contact with the core and the heat is removed by the resulting steam (103), most of which goes to the turbine.* The exhaust steam from the turbine is condensed (106) and pumped through two condensers and then through full-flow condensate demineralizers. The cleaned condensate (107) is pumped through feedwater heaters and then back into the reactor. There are two side streams taken off the primary steam (103): Stream 105 is used for the turbine gland seals; the other (104) is used to drive two stages of steam ejectors which are used to remove the noncondensibles from the main condensers. This stream is condensed. The noncondensibles used to be put through a holdup line, which provided about 30 min of holdup prior to discharge through a high stack. This stream, discharged about 0.1 Ci/s of noble gas activity to the environment, which led to the pressure to reduce emissions from LWRs and resulted in the promulgation by NRC of 10CFR50 (Appendix I), the effect of which is to limit doses from LWRs to individuals in unrestricted areas to about 5 mrem/yr. All BWRs today provide considerably further treatment for this air ejector off-gas. A number of systems are used. One common solution is to put the off-gas through a catalytic recombiner (to convert the contained hydrogen to water), condenser, drier, and then through beds of charcoal, which serve to hold up the krypton and xenon long enough to permit the shorter-lived noble gases to decay prior to release to the atmosphere. This also effectively removes the iodine isotopes from this stream.

*Numbers in parentheses are stream identification numbers from Fig. 7.8-3.

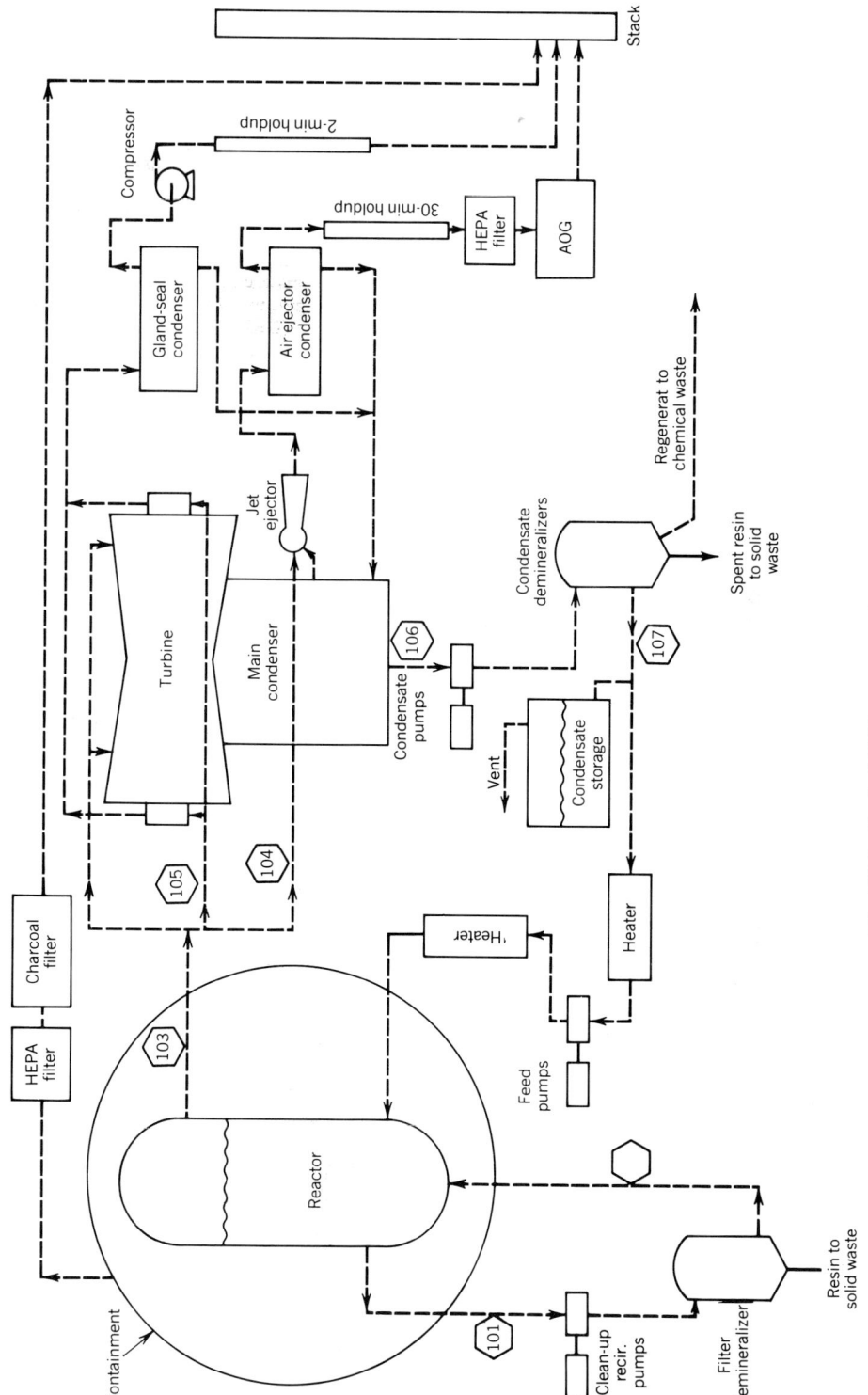

Fig. 7.8-3 Typical BWR gaseous radwaste system.

In order to keep the radioactivity content of the primary reactor water low, a sidestream of primary coolant (101) is drawn off, cooled, and put through a filter-demineralizer. The cleaned coolant (102) is returned to the reactor vessel. The flow rate of this side stream is such as to provide a total turnover of the reactor water about once every 2 h. This filter-demineralizer is precoated with a powdered filter aid. This is the major source of filter sludge wastes.

A second cleanup mechanism for BWR coolant is the boiling action itself followed by the full-flow demineralization of the condensate. The boiling results in continuous degassing of the noble gases which, by the action of the steam jet air ejector, are sent to the augmented off-gas cleanup system. In addition to all the noble gases the steam also carries over a small fraction of the fission and activation products in the coolant. In the case of iodines, which are partially volatile, a somewhat higher fraction is carried over. All of these end up on the condensate demineralizers. These units are periodically backwashed to the resin regeneration tank, where the mixed resins are mechanically separated and regenerated. The regenerant solutions are concentrated in evaporators to about 20–25% sodium sulfate. Some of the newer plants also use ultrasonic resin cleaning (URC). Since resins pick up larger quantities of "crud" than radioactivity, they generally must be regenerated due to increased pressure drop, rather than ionic breakthrough. The URC thus provides a way to reduce the frequency of regeneration.

The result of the multiple cleanup mechanisms is to produce a somewhat lower concentration of radioactivity in the primary coolant than would be found in a PWR operating with a similar level of failed fuel (see Section 7.8.1 and Table 7.8-2).

In addition to the steam jet air ejector gaseous waste stream, other minor losses of gaseous wastes can occur from BWR in the following ways: (1) with the discharge from the turbine gland seal system, (2) from leaks of steam or water into the various building ventilation systems, and (3) with the discharge of the mechanical vacuum pump system used during startups.

The liquid wastes of BWRs occur from the leakage of primary coolant or of steam followed by the condensation thereof. BWR liquid wastes are usually classified into four categories: (1) high purity, (2) low purity, (3) chemical, and (4) detergent.

High-purity wastes generally have low solids content, low conductivity, and variable radioactivity. They come from equipment drain sumps and from the backwash of resin transfer water used following the regeneration of the condensate demineralizers. Liquid wastes collected in the turbine building equipment drains may sometimes be included with the high-purity waste stream; more frequently they are returned directly to the main condenser hotwell. Reuse of processed high-purity waste is highly desirable.

Low-purity wastes have moderate conductivity and solids content. They come from building floor sumps and are generally high-purity wastes that have become contaminated by dirt, grease, and so on.

The chemical wastes that come from resin regeneration and laboratory drains contain high solids and relatively higher radioactivity and are usually segregated from the other waste streams.

Detergent wastes come from the laundry (if provided) and from personnel and equipment decontamination. They are very low in radioactivity content and the detergent content makes their processing difficult. They should be discarded without treatment if at all possible.

The basic tools at hand for processing radioactive liquid waste include (1) holdup for decay before release, (2) filtration, (3) ion exchange (regenerable or nonregenerable), (4) reverse osmosis, and (5) evaporation.

With four or five basic categories of waste for each type of reactor, which may be combined in a number of ways, with five basic tools with which to work which in turn may be combined in a large number of ways, it is clear that there is no such thing as a "standard" liquid radioactive waste disposal system. Figure 7.8-4 is a simplified flow diagram of one possible combination of treatments that may be considered "typical" of current BWR practice. This particular combination makes use of full segregation of waste by category and includes only those treatments for each stream that can be justified on a "cost–benefit" basis; that is, the estimated annual cost is less than the estimated reduction in annual population detriment* that results from the use of the system in question.

In the system shown in Fig. 7.8-4 the high-purity waste is collected in receiving tanks, filtered, and put through ion exchange and the treated product is collected in storage tanks. On a batch basis the product is sampled, assayed, and either discarded into the main cooling water discharge or reused in the plant. Because this waste is quite low in TDS to start with, ion exchange beds can be expected to last a fairly long time and it may be cheaper to discard the resins as solid waste than to regenerate them. The filters used in this system have been of various designs. Early systems tended to use removable cartridge filters, but these have not been too satisfactory. Some newer units are using a traveling bed or "flatbed" filter with considerable success.

Common alternatives to the system depicted for the high-purity waste are to add a second stage of ion exchange in series or to use a powdered resin demineralizer in place of the deep-bed ion

---

Detriment is a term used by the International Commission on Radiation Protection to refer to the total impact, somatic plus genetic, that a given amount of radioactivity discharged into the environment will visit upon the entire exposed population.

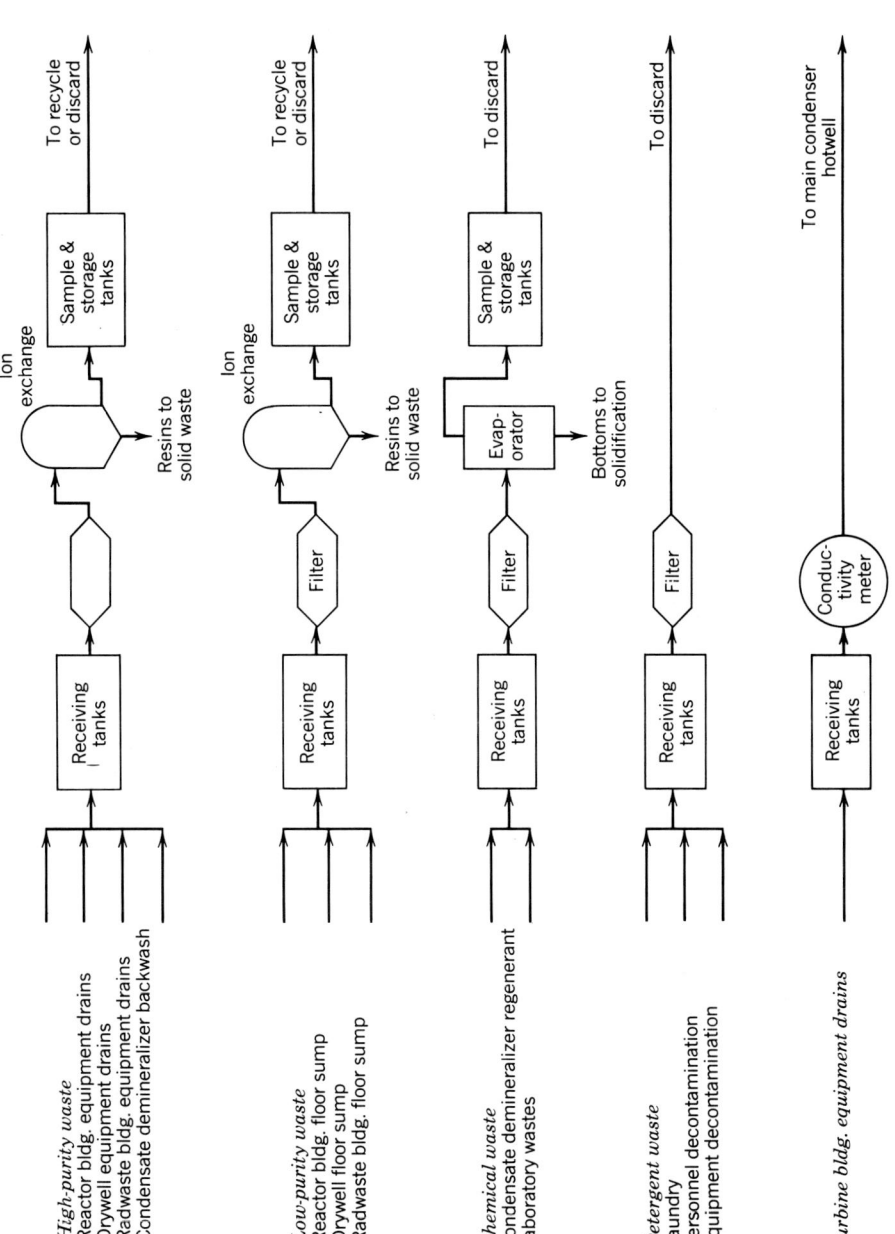

Fig. 7.8-4 Typical BWR liquid radwaste system.

969

exchangers. Off-specification turbine building drain waste is sometimes added to this stream for processing. There are installations in which evaporators have been used or considered for this stream, but the characteristics of the waste and the fact that reuse of most of the product is desirable make the use of evaporation uneconomical.

The low-purity waste is treated in the same manner as the high-purity waste in a parallel but separate set of equipment. Here too the resin beds are usually discarded and replaced rather than regenerated. Common alternatives to the system outlined are to replace the deep-bed ion exchangers with the powdered resin system and the exchangers to be replaced entirely with evaporators. The overhead from the evaporator may then be further treated with ion exchange. Bottoms from the evaporator must be solidified before shipment as solid waste to a licensed burial ground. Off-specification turbine building drain wastes are sometimes added to this stream for treatment.

The chemical waste is high enough in total solids and contains so much radioactivity that an evaporative step is essential in order that the product be sufficiently decontaminated to allow its disposal. This waste is collected in receiver tanks, filtered, and evaporated. The overhead is generally discarded. The bottoms, which contain the bulk of the radioactivity from this source, are solidified and discarded as solid waste in a licensed burial ground. Detergent wastes are sometimes added to this stream for treatment, but this usually causes problems with foaming and carry-over in the evaporator.

Detergent wastes contain so little activity that doing anything at all with them is hard to justify. Regulatory and environmental pressures have sometimes forced a utility to try to treat them. At one or two places reverse-osmosis treatment of this waste is being tested on an experimental basis.

The turbine building equipment drain waste is normally put directly back into the main condenser hotwell. When treatment is required, the stream is added to either the high-purity or low-purity stream for processing.

PWR. Figure 7.8-5 is a simplified flow diagram of a typical PWR. The basic characteristic of a PWR is that the primary coolant is maintained under sufficient pressure to prevent boiling of the coolant. Heat is removed from the core by the pressurized coolant stream (101),* which is conducted to a secondary steam generator where the heat is used to convert a secondary water stream to steam (L2). This steam goes to the turbine; the exhausted steam is condensed and the condensate collected in a hotwell. The condensate (105) is returned to the steam generator by pumps. Some blowdown (107) is taken from the steam generator to maintain the required water purity.

After the primary coolant has lost some of its heat to the steam in the generator, it is pumped back (102) into the reactor. A side stream from this flow (103) is cooled down, taken outside of containment, and put through an ion exchange system to remove some of the contained radioactivity. The rate at which this done is such that the full primary coolant volume is treated about once every 24 h. Unlike the BWR, which has three cleanup mechanisms at work, this is essentially the only cleanup mechanism provided for PWR coolant. Consequently, for a given level of fuel failure the isotopic concentrations in PWR coolant will be higher than those for BWR.

Another characteristic of PWR is that a soluble poison, boric acid, is added to the reactor coolant in concentrations approximating 2000 ppm at the beginning of the core cycle to help control the reactor. As the fuel in the core burns out, so does the boric acid and the reactivity of the system remains about constant.† The rate of burnout of the boric acid is not quite fast enough, however, so that some of it must be removed by taking a side stream (106) from the cleanup flow (103,104) at a rate approximating 2 gpm (about 5% of the cleanup flow) and sending this to an evaporator where the boric acid is concentrated to from 4 to 12%. The bottoms are saved and reused whenever boric acid is needed. Some of the overhead from the evaporator should be discarded so as to maintain the tritium concentration of the primary coolant acceptably low; most of it, however, will be recycled to the primary system.

There are a few places in the primary system of a PWR where small gas bleeds may be taken. These are collected and compressed and held in pressurized storage tanks for 30 to 60 days to allow decay of the short-lived noble gases before discharge, normally through a vent at the top of the building. Other sources of gaseous wastes from PWR are (1) steam generator condenser air ejector, (2) steam generator blowdown tank exhaust, (3) purging of the containment shell, (4) leakage into the auxiliary building ventilation, and (5) leakage into the turbine building ventilation.‡

The liquid wastes of PWRs occur in general from the leakage of primary coolant. PWR liquid wastes are usually classified into five categories: (1) clean waste, (2) dirty waste, (3) steam generator blowdown, (4) turbine building drains, and (5) detergent.§

Clean wastes are made up primarily of the "shim bleed" drawn off from the cleanup sidestream for control of the boric acid concentration. Added to this are some equipment, valve, and pump leakoffs

*Numbers in parentheses refer to stream identification numbers in Fig. 7.8-5.
† The burnout of boron is the major source of tritium in a PWR.
‡ The sources in (1), (2), and (5) will contain radioactivity only if there is steam generator tube leakage.
§ The streams in (3) and (4) would not contain appreciable radioactivity if there were no steam generator tube leakage.

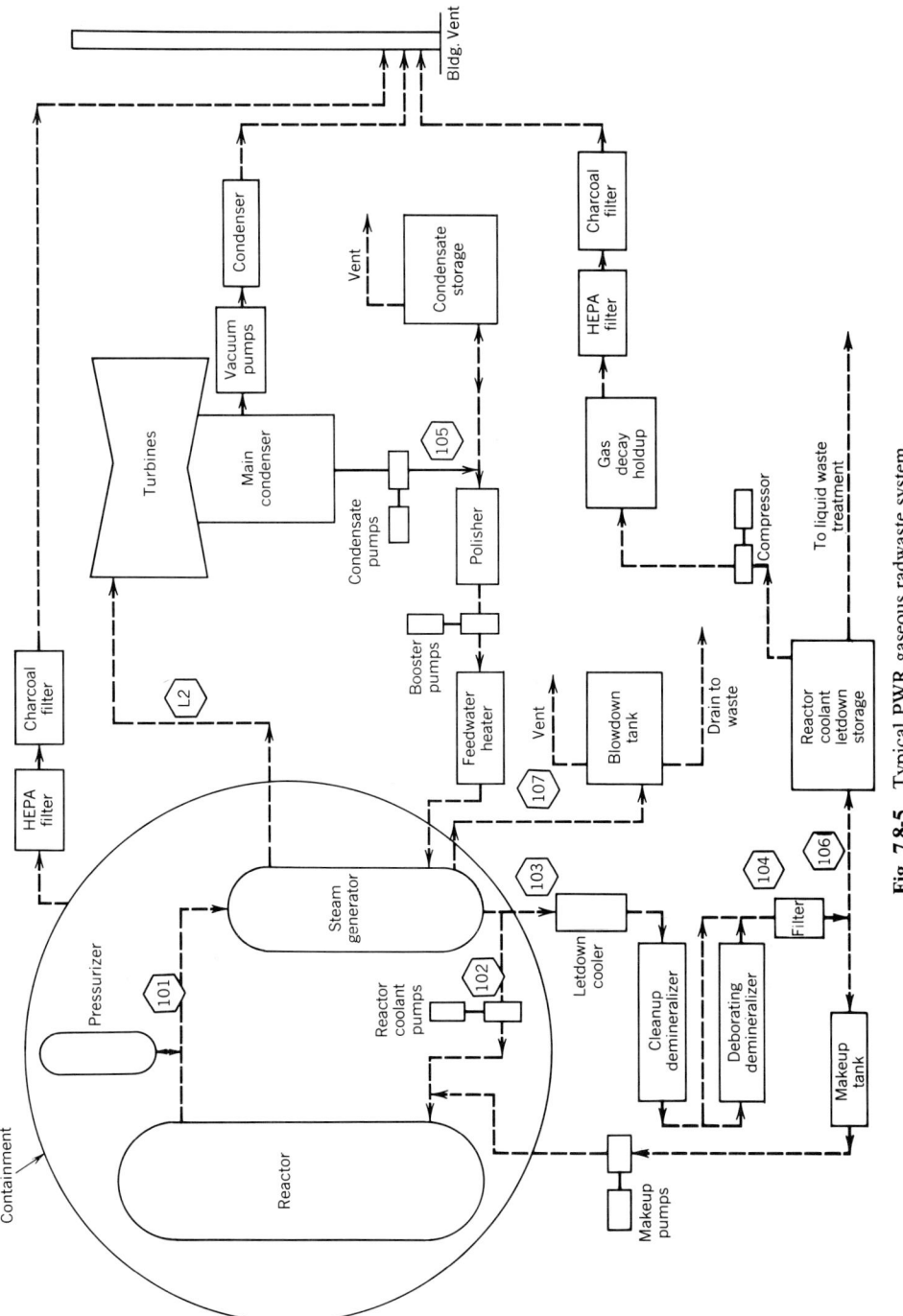

Fig. 7.8-5 Typical PWR gaseous radwaste system.

and seals. All of this waste is essentially reactor grade water and in a sense this is not a waste system at all. The concentrated boric acid is almost always reused and as much of the overhead products as possible also are reused.

Dirty wastes are of moderate conductivity and variable radioactivity content. They come from building sumps, laboratory drains, sample stations, and miscellaneous floor drains. The treated product may be discarded or reused.

Steam generator blowdown is required to maintain the desired water chemistry in the secondary system. The turbine building floor drains will not contain any radioactivity unless there is steam generator tube leakage.

As with BWR, detergent wastes come from the laundry (if provided) and from personnel and equipment decontamination. They are very low in radioactivity content and the detergent content makes their processing very difficult.

The basic tools available for managing PWR wastes are the same as those described for BWR. And again there are many possible combinations. Figure 7.8-6 is a simplified flow diagram of one possible combination of treatments that may be considered "typical" of current PWR practice. As with the BWR, this particular combination makes use of full segregation of waste by category and includes only those treatments for each stream that can be justified on a cost–benefit basis.

In the system shown in Fig. 7.8-6 the clean waste is collected in receiving tanks, filtered, evaporated, and collected in sample tanks. On a batch basis the product is sampled, assayed, and generally is returned to the primary water system. On occasion, some of this stream may be discarded into the main cooling water discharge. Similarly the bottoms from the evaporator are generally reused as boric acid makeup for the primary system. Some of the bottoms may be discarded by solidification and burial in a licensed burial site. A spare evaporator is sometimes provided, frequently serving both systems. The filters used have generally been of the replaceable cartridge type.

The dirty waste is treated essentially in the same manner as the clean waste in a parallel but separate set of equipment. The bottoms from this evaporator are generally discarded by solidification and burial in a licensed burial ground.

Originally no provision was made in PWRs to treat the steam generator blowdown. Since Appendix I limits were set, there has been pressure to provide treatment for this stream to handle the occasions when there is significant steam generator tube leakage and to provide better control of secondary side chemistry. Alternatives that have been suggested for the treatment of this stream include the addition of a second stage of ion exchange, the replacement of ion exchange with evaporation, and the addition of a final stage of ion exchange to the evaporator overhead. Reverse osmosis is also being considered.

The radioactivity content of the turbine building drains will be low. When this stream is treated, it is generally done by combining this source of waste into the input to the dirty waste system so that it will then get whatever treatment has been provided for the dirty wastes.

The situation with the detergent wastes is precisely the same as that described for BWR.

Front-End Wastes

The front-end segment of the uranium fuel cycle consists of uranium mining, milling, conversion, enrichment, and fuel fabrication. Wastes containing naturally occurring radioactivity are released during the processing.

The release of small amounts of radon gas is associated with mining. The concentration of the radon is reduced by dilution with large quantities of ventilation air in the underground mines. Atmospheric dilution suffices for the open-pit mines.

Waste associated with milling consists of radon gas and uranium-containing dust particles. At the mill the radon dilutes into ventilation air streams, which are equipped with dust collection devices. Radon gas and windblown dust from the tailings piles are minimized by keeping them wet during the time of the mill operation. After the mill is shut down, the tailings pile is covered with a thick cover of earth, which attentuates the radon emissions and prevents the tailings from becoming windblown.

Waste produced during conversion, enrichment, and fuel fabrication consists of gaseous releases and solid residues. The gaseous releases are treated by chemical scrubbing and fume washing. Any process dust generated is captured by dust collection devices. Solid wastes are either buried on-site or are shipped to a licensed waste burial site.

Reprocessing

Of the reprocessing waste streams discussed in Section 7.8-1, the one that has always received the most attention is that which is usually referred to as "high-level waste" (HLW) so the handling of this waste stream will be treated in the most detail. Management of the other major waste streams will be briefly discussed in this section.

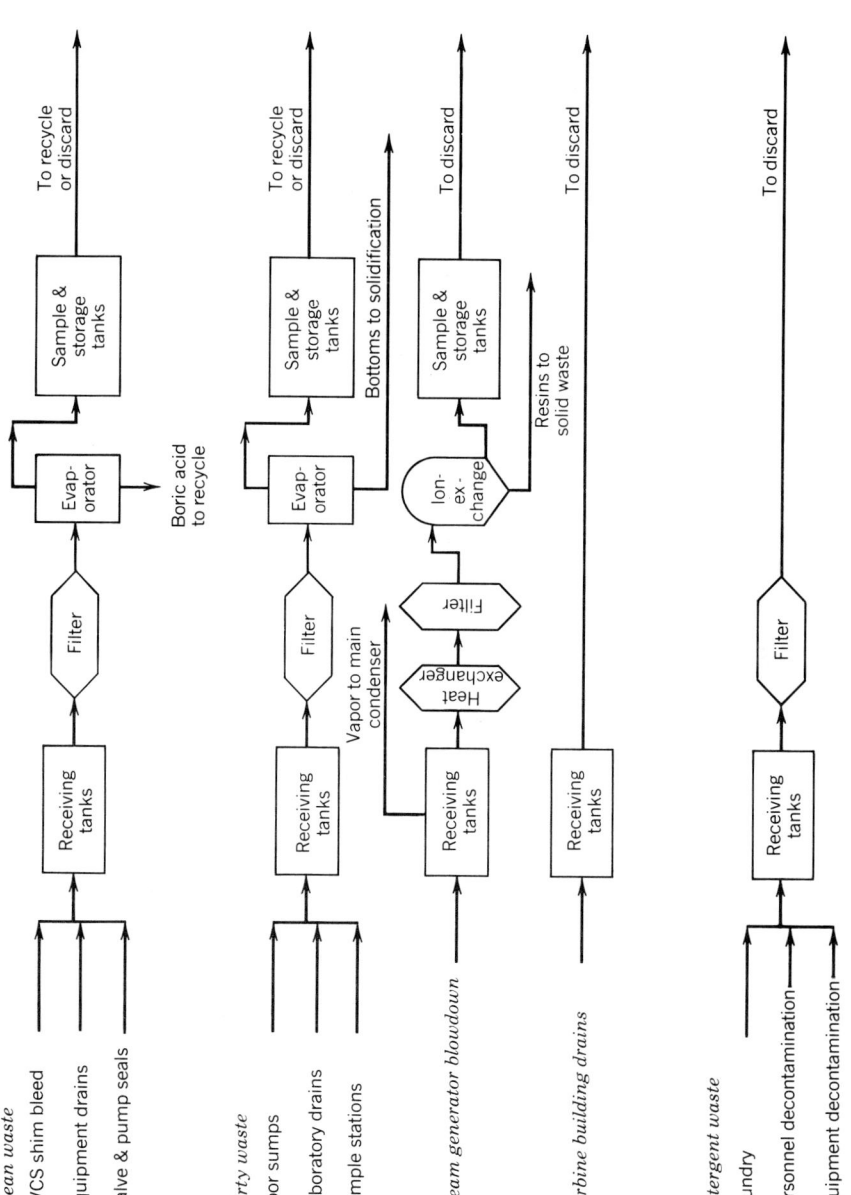

Fig. 7.8-6 Typical PWR liquid radwaste system.

973

HLW. The generally accepted definition of high-level wastes are those liquid wastes, containing over 99.9% of the nonvolatile fission products and more than 50% of the transplutonium actinides, generated in the solvent extraction cycles of nuclear fuel reprocessing facilities.[22]

Current regulations[10] require that liquid HLW be solidified within 5 years and shipped to a federal repository within 10 years from the time the spent fuel is reprocessed. However, since neither a repository nor a commercial reprocessing plant is now in operation, this regulation is currently moot. Instead discharged fuel is presently being held in storage pools. It is possible that the ultimate governmental decision will be to treat this spent fuel as the waste form to be disposed (see Section 7.8-3). If this comes to pass, none of the reprocessing wastes discussed in this section will exist.

A unique characteristic of HLWs is the fact that the radioactive decay produces heat. The amount of heat is sufficient to be a nuisance but not enough to be useful. It is large enough, however, that any considerable quantity stored as liquid will self-boil. Therefore liquid HLW interim storage tanks must be equipped with cooling systems. The fission product decay heating persists, of course, after the HLW is solidified and must be taken into account in connection with the heat transfer considerations of the packaged, solidified waste.

In terms of volume HLW currently amounts to about 200 gal/tonne of fuel processed.[25] After the waste is solidified, the volume will be about 3 ft^3/tonne.[26] If the spent fuel were to be considered as the HLW, the volume of spent fuel elements would be about 20 ft^3/tonne of fuel processed.[11]

The first production of nuclear wastes in any quantity occurred in 1944 at Oak Ridge.[67] Even in the earliest days time and effort were taken to assure that these waste products were kept under control. The fission product content and volume of these early wastes were relatively low, so the wastes did not self-boil. They were neutralized and collected in six 40,000-gal carbon-steel-lined Gunite tanks.

Beginning in early 1945 similar wastes higher in fission product content began to be produced at Hanford. These did self-boil. The method of handling these wastes was similar to that at Oak Ridge. The place chosen to build them was selected very carefully. They were located in a plain underlain by thick strata of basalt over which are thick strata of sedimentary soil materials. The soil material has high natural ion exchange properties and therefore a high capability for absorbing radioactive materials. The depth to the water table is 250–300 ft and the water is not replenished by direct infiltration of precipitation. The location is many miles from the nearest point where the underlying aquifer might be used as a potable water source. Thus, it was expected that, if there were any leakage from these tanks, there would be several defenses to assure that the activity did not get back to man in any concentration of consequence. Nor have they in the intervening three and a half decades.

Tanks constructed at Hanford before 1952 were vertical concrete cylinders about 75 ft in diameter, with a flat bottom and a self-supporting domed roof covered with 6–9 ft of earth. The tanks were lined with carbon steel on the bottom and sides. The waste was neutralized and had low heat evolution rates (less than 0.15 Btu/gal-h). The tanks were equipped with air-cooled condensers venting to the atmosphere.

Beginning in 1952 similar but larger tanks were built for storage of waste with higher heat-generating capacity. These tanks were equipped with airlift circulators and were used for wastes generating heat at rates from 1 to 10 Btu/gal-h, a rate great enough to cause the waste to boil for 1–10 years. The tanks were equipped with water-cooled condensers and had provisions for either returning condensate to the tank or sending it to a disposal site.

In 1958 the first of 20 confirmed cases of leaking tanks occurred at Hanford. In each case the remaining contents of the leaking tanks were pumped into another tank. In no case has any known effect been produced off-site.

Because of this experience, tanks constructed at Hanford since 1968 have been of double-wall design. Double-wall tanks intended for HLW contain air-lift circulators to stir the waste. An annulus between the primary and secondary tanks collects any leakage from the primary tank. Equipment is installed to detect and pump out liquid from the annular region and to pump liquids and slurries between tanks, tank farms, evaporators, and production plants.

Waste tanks at the Savannah River Plant (SRP) are built from carbon steel to three designs for self-heating waste and to another for low-heat waste that does not require cooling. Type I tanks were the first constructed at SRP (1951–1953). They are vertical tanks set in a carbon steel pan. The roof is supported by internal columns and the entire assembly is in a reinforced concrete vault. The tank is equipped with cooling coils and leak-detecting equipment.

Type II tanks were constructed in the period 1955–1956. The primary container of these tanks consisted of two steel cylinders, a small inner and a large outer, with a flat bottom and flat top. The inner cylinder houses a single supporting column for the vault roof. The primary tank sits in a carbon-steel pan equipped with leak-detecting facilities. Cooling coils are provided in the tank.

Type III tanks were the last constructed (1967–1972). This design was developed after investigation of the leaks that had occurred in Type I and Type II tanks. Design changes were made in the roof support column to minimize stresses in that area. These tanks are set in a concrete vault lined with carbon steel. Original Type III tanks are equipped with removable cooling coils. However, future tanks will have permanent cooling coils.

TABLE 7.8-7 PROJECTED INVENTORIES OF COMMERCIAL WASTES

High-Level Cumulative Waste Volumes

Year	Fuel Discharged (tonnes) Annual	Cumulative	Liquids[a] (gal)	Solids[b] (ft^3)	Spent Fuel as waste[c] (ft^3)	Cs-137 + Sr-90 Content[d] (Ci)	Plutonium Content High-Level waste[e] (kg)	Spent Fuel[f] (kg)
Thru 1975	2,120	2,120	4.2E05	6.4E03	4.0E04	4.0E08	6.4E01	1.3E04
1976	982	3,100	6.2E05	9.3E03	5.9E04	6.2E08	9.3E01	1.9E04
1977	1,236	4,340	8.7E05	1.3E04	8.3E04	8.6E08	1.3E02	2.6E04
1978	1,487	5,830	1.2E06	1.8E04	1.1E05	1.2E09	1.8E02	3.5E04
1979	1,668	7,500	1.5E06	2.3E04	1.4E05	1.5E09	2.3E02	4.5E04
1980	1,963	9,460	1.9E06	2.8E04	1.8E05	1.9E09	2.8E02	5.7E04
1981	2,208	11,700	2.3E06	3.5E04	2.2E05	2.3E09	3.5E02	7.0E04
1982	2,646	14,300	2.9E06	4.3E04	2.7E05	2.9E09	4.3E02	8.6E04
1983	3,305	17,600	3.5E06	5.3E04	3.3E05	3.5E09	5.3E02	1.1E05
1984	3,813	21,400	4.3E06	6.4E04	4.1E05	4.3E09	6.4E02	1.3E05
1985	4,499	25,900	5.2E06	7.8E04	4.9E05	5.2E09	7.8E02	1.6E05
1986	4,920	30,800	6.2E06	9.2E04	5.9E05	6.2E09	9.2E02	1.9E05
1987	5,305	36,200	7.2E06	1.1E05	6.9E05	7.2E09	1.1E03	2.2E05
1988	5,475	41,600	8.3E06	1.3E05	7.9E05	8.3E09	1.3E03	2.5E05
1989	5,641	47,300	9.5E06	1.4E05	9.0E05	9.4E09	1.4E03	2.8E05
1990	5,675	52,900	1.1E07	1.6E05	1.0E06	1.1E10	1.6E03	3.2E05
Fraction of present military inventory			0.15	0.02	0.15	18	1.9[g]	388[g]

[a]At 200 gal/tonne fuel processed.
[b]At 3 ft^3/tonne.
[c]At 19 ft^3/tonne.
[d]At 2E05 Ci/tonne total Cs + Sr, decay neglected.
[e]At 0.5% loss and 6 kg/tonne = 3E-2 kg/tonne.
[f]At 6 kg/tonne.
[g]Military waste taken as high level (625 kg) + liquids to ground (200 kg).

At SRP eight tanks have leaked very small amounts of radioactive waste from the primary tank into the annulus or into the ground. Extensive study of the single release to ground indicates that no more than a few tens of gallons of waste reached the soil and that this waste migrated only a few feet from the vault.

HLW generated at the Idaho Chemical Processing Plant (ICPP) is stored as an acid solution in stainless steel tanks. The tanks are contained in reinforced concrete vaults, the tops of which are about 10 ft below ground. The vaults have sumps to contain any leakage from the tank. A few leaks have occurred in transfer lines. No leaks have occurred in any of the stainless steel waste tanks at ICPP since startup in 1951.

Drawings of all the designs discussed above may be found in a NRC report.[53]

The early Hanford waste has been processed to remove the contained uranium. Some of the later Purex wastes have been treated to remove much of the Sr-90 and Cs-137. The remainder has then been solidified in some of the older tanks.

In the late 1950s Idaho elected to make use of a fluid-bed calcination process developed at Argonne National Laboratory to solidify some of its older waste, thus freeing up tankage for newer waste. The solid product of fluid-bed calcination is a free-flowing granular solid. This product is at present being stored at Idaho in stainless steel air-cooled bins. This may not be the final disposition of this waste. At none of the three sites has there ever been a case of exposure of a member of the public from the waste disposal operations.

It still remains to be seen what will be the ultimate disposition of these defense wastes. There is a strong presumption that criteria similar to those applied to the commercial wastes should also be applied to the wastes generated in the military program. A comparison of the projected inventory of various types of commercial waste with the 1975 inventory of military wastes is shown in Table 7.8-7.

Historically the generation of HLW from the reprocessing of spent commercial (power reactor) fuels has been limited to the Nuclear Fuel Services (NFS) West Valley plant where some 600,000 gal are currently in tank storage. This plant is no longer is operation and the permanent disposition of these wastes has not yet been decided. The 600,000 gal of neutralized high-level Purex waste, including an estimated 30,000 gal of sludge, are currently in storage in a single 750,000-gal carbon steel tank. (There is an empty similar tank as a spare.) There are also 12,000 gal of acid thorium waste in storage in a 15,000-gal stainless steel tank (which also has a spare).[50] A number of alternative approaches for management of NFS wastes have been considered.[18]

U.S. policy is to convert the HLW generated in the reprocessing of spent power reactor fuel to a "dry solid."[29]

Twenty years of research and development in the United States and abroad and large-scale production operations at ICPP have provided several solidification processing options ready for commercial plant implementation.[31,32,36,37,43,46,56,64-66,71,72] In particular, the conversion of HLW to a glass form has had the most effort and is expected to be the form of the waste in the first-generation operating commercial HLW management system. Glass is produced from HLW in a two-step process —*calcination* (conversion of the waste to a solid oxide particle form) and *vitrification* (mixture of the calcine with glass-forming materials at high temperature to form glass).

Waste solidification technology is well developed, with a number of processes having been tested and demonstrated at full radioactivity levels and some at production throughput. The major calcining methods are spray, fluid-bed, and rotary kiln. The major vitrification (glassmaking) methods are in-can, metallic, and ceramic smelters. Detailed descriptions of all of these processes can be found.[22]

The French have been developing vitrification processes since 1959. In 1971 work was begun on a continuous process using a rotary calciner coupled with a continuous melter. Liquid HLW is fed to a four-zone, electronically heated, rotary calciner.[25] Waste is evaporated in the first two zones and is calcined in the last two. The calcine is discharged directly to an Inconel melting pot. Glass beads are continuously added to the pot at a controlled rate to produce a glass melt. Molten glass is periodically drawn off to a canister, which is then cooled, sealed, decontaminated, and sent to storage. At the present time the cooled canisters are stored in air-cooled retrievable storage pending a decision as to final disposal. A similar unit with some improvements is presently scheduled for construction at the La Hague reprocessing plant.

Both the British and the Germans are also developing vitrification processes, mostly producing borosilicate glasses.

The waste, once solidified, must be packaged, stored, and transported to a repository. None of these operations require any technological breakthrough to accomplish. They all are either identical or similar to operations that have been conducted for a quarter of a century or more.

Ultimate disposal of the solidified waste is discussed in Section 7.8-3.

Hulls. As indicated in Table 7.8-7, about 3 ft^3/tonne of hulls are produced as waste from the mechanical head end operations and the fuel dissolution. The hulls are the non-fuel-bearing portions of the sheared elements that do not dissolve in nitric acid. They contain sufficient activation products within the metal pieces themselves to cause them to read several thousand roentgen/hour (R/h) on contact. In addition they contain perhaps 0.1% of the fuel that remained undissolved. This quantity of fuel contains enough transuranic (TRU) material that this waste stream may not meet the proposed 10CFR61 (vide post) TRU limit to qualify for disposal by shallow land burial. Therefore, this waste stream will have to be packaged remotely into shielded containers for handling and transport.

Once a waste repository is in operation, it is likely that hulls will be sent there for ultimate disposal. Hull packages could be used as part of the backfill of disposal drifts.

Trash. It is also expected that a reprocessing plant will produce about 50 ft^3/tonne fuel processed of trash. At nearly all nuclear installations trash qualifies easily for disposal by shallow land burial. At a reprocessing plant, however, some fraction of the trash may be high enough in TRU content that it may have to go to the repository and like hulls be used as part of the drift backfill.

Any volume reduction of the trash is likely to be done by compaction. Incineration may be required for wastes sent to the repository in order to meet waste form stability requirements.

Research and Development and Medical Uses

A wide variety of radioactive isotopes are used in academic, medical, and industrial research. The isotopes become associated with body wastes (*in vivo* application), animal wastes and carcasses (*in vitro* application), and with a wide variety of equipment and assorted trash.

It is frequently possible for a research institution to store its shorter-lived wastes until they decay to insignificance, after which they can be disposed of with the regular trash of the institution. All other research wastes generally qualify easily for shallow land burial at licensed burial sites.

7.8-3   Ultimate Disposal of Waste

Some of the components of nuclear waste have very long half-lives—many millions of years. Therefore, some residuum of that waste will be around essentially forever. This is not unique to nuclear waste. Arsenic, for example, has an infinite half-life; it *will* be around forever.*

---

*Arsenic is an interesting parallel. It is largely imported into this country and thus is specifically introduced into our environment even as nuclear waste is. The 1979 imports amounted to 12,325 short tons equivalent to

General Principles

In disposing of nuclear waste safely the goals are to

1. assure that the public is adequately protected in relation to its supplies of air and water and/or
2. protect individuals from harm to themselves due to intruding into the waste.

These are two very different goals, and in considering these goals, let us first define two terms:

Containment. Keeping the waste within the confines of its place of interment to the degree necessary to prevent significant leakage to the biosphere which results in harm to the general public;

Isolation. Emplacing the waste in a place or manner that humans are not likely to intrude and come into contact with the concentrated waste form.

The first goal, the protection of large populations by protecting their air and water supplies, depends on containment; isolation may be desirable but is not required. It does not, however, require "absolute" containment. What it does require is a degree of containment such that any movement from the point of interment will be small enough and slow enough that, when and if the material finally does get back into air or a public water supply, the concentration of all radioisotopes will clearly be within acceptable limits.

The second goal, that of preventing intrusion into the waste, either by individuals or by nature, is a very different matter. Here, if we assume that intrusion can be prevented during some period of institutional control, the demands upon the system will be dictated only by the long-lived isotopes. Furthermore, the basic requirement is isolation.

The first goal, protection of the public, leads to inventory limits that can be shown to be adequately contained by the defenses of the site in question. The second leads to concentration limits such that intrusion scenarios can be accepted without unduly exposing the intruder.

Low-Level Waste

As the name itself implies, low-level waste has the characteristics of low concentration and, despite the large volumes involved, low inventory. This type of waste has generally been handled by shallow land burial or disposal at sea.

Shallow Land Burial. Except for a small amount of waste disposed at sea, low-level waste in the United States has traditionally been disposed of by shallow land burial. Shallow land burial is very much akin to a sanitary landfill operation except that the sites are much more carefully selected, wastes are placed in discrete trenches and carefully filled and covered, and much better records are kept. Monitoring requirements are also much more stringent.

At one time six commercial burial sites were in operation in the United States: Beatty, Nevada; Maxey Flats, Kentucky; Richland, Washington; Sheffield, Illinois; West Valley, New York; and Barnwell, South Carolina.

Only three of these are still open.* The Sheffield site is full and those at West Valley and Maxey Flats have been closed. This has left Barnwell, South Carolina, as the only site east of the Mississippi. For a time nearly 90% of all the waste was going to Barnwell upon which the state imposed drastic volume limitations on the import of out-of-state waste. In 1980 Congress passed the Low-Level Waste Policy Act, which in effect legislates that regional burial sites shall be in operation by 1986 and that after that date host states may refuse to accept out-of-region waste.

The NRC has (July 1981) issued for public comment the proposed 10CFR61, a new set of regulations covering all aspects of shallow land burial. This proposed part includes tables of allowable isotopic concentrations for shallow land burial (see Table 7.8-8).

It must be considered that materials buried near the surface have a rather low degree of isolation, and that, with a few possible exceptions, any plausible scenario of intrusion that can occur will occur after termination of institutional control. Therefore, the only defense that has meaning is to establish concentration limits for all isotopes of importance such that the intrusion scenarios can be accepted without unduly exposing the intruder or the public if the intrusion results in potential public exposure. Note that it is the concentration, not the quantity, of the isotope that is important in this case, since the intruder comes in contact with discrete portions of the waste.

Two independent studies[42,54] have shown that the intrusion scenarios dominate the calculation of risk from shallow land burial. All of the limits in Table 7.8-8 are based on intrusion scenarios.[55]

---

about 5E10 lethal doses, enough to kill every man, woman, and child in the United States 200 times. However, we don't put this material into a deep geologic repository—we put it on our food crops and eat at least a portion of it.
*Beatty may be closed by the time this reaches print.

TABLE 7.8-8   PROPOSED 10 CFR PART 61

Part 1

Radionuclide[a]	Concentration, Ci/m³
C-14	8
C-14 in activated metal	80
Ni-59 in activated metal	220
Nb-94 in activated metal	0.2
Tc-99	3
I-129	0.08
Alpha-emitting transuranic nuclides with half-life greater than 5 yr[b]	100
Pu-241[b]	3,500
Cm-242[b]	20,000

Part 2

	Concentration, Ci/m³		
Radionuclide[c]	Column 1	Column 2	Column 3
Total nuclides with half-life of less than 5 yr	700	[b]	[b]
H-3	40	[b]	[b]
Co-60	700	[b]	[b]
Ni-63	3.5	70	700
Ni-63 in activated metal	35	700	7000
Sr-90	0.04	150	7000
Cs-137	1	44	4600

[a] If the waste contains only the radionuclides listed in Part 1, classification shall be determined as follows:

1. If the concentration does not exceed 0.1 times the value in Part 1, the waste is Class A.
2. If the concentration exceeds 0.1 times the value in Part 1, the waste is Class C.
3. If the concentration exceeds the value in Part 1, the waste is not generally acceptable for near-surface burial.
4. If the waste does not contain any of the radionuclides listed in Part 1, classification shall be determined based on the concentrations shown in Part 2.

[b] Units are nCi/g.

[c] There are no limits established for these radionuclides in Class B or C wastes. Practical considerations such as the effects of external radiation and internal heat generation on transportation, handling, and disposal will limit the concentrations for these wastes. These wastes shall be Class B unless the concentrations of other nuclides in Part 2 determine the waste to the Class C independent of these nuclides.

1. If the waste concentration does not exceed the value in column 1, the waste is Class A.
2. If the concentration exceeds the value in column 1, but does not exceed the value in column 2, the waste is Class B.
3. If the concentration exceeds the value in column 2, but does not exceed the value in column 3, the waste is Class C.
4. If the concentration exceeds the value in column 3, the waste is not generally acceptabie for near-surface burial.
5. If the waste contains radionuclides from both tables, the classification shall be determined as follows:
    (a) If the concentration of a nuclide listed in the first part of the table is less than 0.1 times the value, the class shall be that determined by the concentration of nuclides listed in the second part of the table.
    (b) If the concentration of a nuclide listed in the first part of the table exceeds 0.1 times the value, the waste shall be Class C, provided the concentration of nuclides listed in the second part does not exceed the value shown in column 3 of Part 2.

Sea Disposal. Prior to 20 years ago some U.S. low-level waste was disposed of at sea. The earliest stirrings of the antinuclear movement were directed against this practice. With the advent of commercially available shallow land burial (Beatty, Nevada, 1962) the then cost of land burial was found to be significantly lower than disposal at sea. Sea burial is still practiced extensively by the Western European nations[30] and by Japan.[73] Sea disposal makes eminent sense. The intrusion pathways that dictate the limits for shallow land burial are totally eliminated. The vast dilution power of the ocean provides very large inventory limits. In fact, it has been repeatedly suggested that high-level waste should be disposed into the ocean. Pressure to use this method of disposal may well increase in the years immediately ahead.

Other. The only other method of disposal that merits consideration is land burial at intermediate depth.* Intermediate burial ranges from placing selected wastes at the bottom of present-day shallow land burial trenches to the use of an abandoned salt mine at Asse by the Germans.[33] The advantage of deeper burial is that the intrusional scenarios are diminished and the allowable concentrations are increased.

High-Level Waste

HLW, either in the form of spent fuel elements or of solidified first-cycle reprocessing waste, is the major waste stream of the nuclear power industry. It contains over 99.9% of the fission products and a large fraction of the transuranic materials discarded from the system. As such, this stream has received the bulk of attention but there still does not yet exist an in-place repository for this waste. Under the best of circumstance an operating repository is still between 10 and 20 years away.

Most of the effort to date has been on disposal in deep geologic formations. Other methods will be discussed briefly such as subseabed disposal, hydrofracturing, and, simply since it keeps coming up, disposal into space.

Deep Geologic Disposal. The first comprehensive discussion of the possibilities of disposal of HLW in geologic formations was at a conference held nearly 30 years ago (1955) at Princeton University.[48] Although a number of different approaches for disposal were considered at this conference, the consensus of the 65 experts in the earth sciences, biology and medicine, chemistry, physics, and engineering was that disposal of solidified waste forms into salt formations provided the best possibility for technical feasibility, safety, and environmental protection.

Following the committee's recommendation, a detailed laboratory and field investigative program was initiated. This program culminated in the successful conclusion of Project Salt Vault, a test and demonstration program carried out in the Carey Salt Co. mine at Lyons, Kansas, in 1968. The results of this program have been extensively reported[9] and described.[44] As stated in ORNL-4555, "With the completion of this experiment, it can be concluded that most of the major technical problems pertinent to the disposal of highly radioactive waste in salt have been resolved. Project Salt Vault successfully demonstrated the feasibility and safety of handling highly radioactive materials in an underground environment. The stability of the salt under the effects of heat and radiation has been shown, as well as the capability of solving minor structural problems by standard mining techniques. The data obtained on the deformational characteristics of salt have made it possible to arrive at a suitable design for a mine disposal facility."

A conceptual design for Lyons was developed by ORNL in 1969, and in 1970 the scope of the design was expanded to include a capability for disposal of TRU wastes.[58] The NAS/NRC committee continued its oversight and endorsement of the program.[49]

Subsequently, more detailed investigation of the proposed repository location environs identified a number of exploratory oil and gas wells and a solution mining operation in the vicinity. This identification coupled with politico-institutional issues led to the conclusion that the Lyons site was not suitable for a federal waste repository. The negative conclusion regarding the proposed site was in no way associated with any lack of viability of the deep geologic repository concept, nor to any basic questions regarding the feasibility of salt formations as a host formation for HLW disposal.[45]

A study was then conducted of alternative locations for repository development. Salt formations in the Delaware subbasin of the Permian Basin near Carlsbad, New Mexico, was found to offer potential for a suitable repository site.[59] Around mid-1974 the federal program for a brief period was focused on the development of a retrievable surface storage facility (RSSF) rather than an ultimate disposal geologic repository. In April 1975 the newly established Energy Research and Development Administration (ERDA), which had assumed the AEC's waste management responsibilities, withdrew the RSSF program.[21]

Following this withdrawal, the Waste Isolation Pilot Plant (WIPP) project was reinitiated in the Delaware subbasin. It was stated by ERDA officials that the objective of providing facilities to

---

*Disposal into space crops up from time to time. But the costs are so vast that use of this method is highly unlikely.

permanently dispose of commercial and ERDA TRU waste was achievable with proven, existing analytical capabilities and technology.[60] Since that time work has progressed on the WIPP project through the issuance of a Geological Characterization Report[61] and a Draft Environmental Impact Statement[16] and the initiation of repository design.

In February 1980 President Carter[62] called for cancelation of the WIPP project but stated that the site will continue to be evaluated along with other sites and, if qualified, will be reserved as one of several candidate sites for possible use as a repository for defense and commercial HLW. WIPP has not been canceled but has been reserved for defense TRU waste and the experimental disposal of defense fission product waste.

With respect to the management of commercial reactor HLW, in 1976 ERDA launched what was characterized as a major, expanded national waste terminal storage (NWTS) program.[23,24] This stressed President Ford's emphasis of October 28, 1976,[75] "to speed up the program to demonstrate all components of waste management technology by 1978 and to demonstrate a complete repository for commercial high-level nuclear wastes by 1985." A simultaneously released Fact Sheet[27] provided additional details on the program implementation plans. This program continued to be based on the viability of the deep geologic repository and recognized the potential suitability of geologic formations in addition to rock salt, that is, crystalline rocks, including basalt and granite, and argillaceous formations. It also recognized the possibility that the waste form might be spent fuel.

With the advent of a new administration in January 1977 ERDA was replaced by the DOE. A task force published a draft report on waste management in February 1978.[15] The report concluded that "there appears to be a substantive consensus and valid technical basis for the view that present plans and actions should rely on geologic containment of waste which can be achieved in a safe and environmentally acceptable manner." The report also noted that this view has been promulgated over a period of time by several independent assessments ranging from that of the National Academy of Sciences in 1957 (and subsequent reaffirmations by that body) to that of the American Physical Society[69] in 1977 and further pointed out that similar findings have been expressed in government-supported reviews in other countries (e.g., West Germany, Sweden, and Canada).

Following the issuance of the Task Force report, a governmental review of overall nuclear waste management was directed by President Carter on March 13, 1978, to be carried out by an Interagency Review Group on Nuclear Waste Management. After receiving public comments on a draft report, the group issued its final report in March 1979.[38] Its principal conclusion was that "...no scientific or technical reason is known that would prevent identifying a site that is suitable for a repository provided that the systems view is utilized rigorously to evaluate the suitability of sites and designs. ... A suitable site is one at which a repository would...provide a high degree of assurance that radioactive waste can be successfully isolated from the biosphere for periods of thousands of years."[39] The report later on states, "Disposing of nuclear wastes in mined repositories is a highly promising approach to long-term isolation."[40]

Subsequently, in the DEIS issued in April 1979, DOE once again arrived at positive conclusions regarding the viability and validity of the deep geologic repository concept.[17]

More recently still another independent, authoritative group, under the aegis of the National Academy of Sciences, addressed the subject of nuclear wastes in connection with its comprehensive, 4-year energy study. It concluded that "No insurmountable technical obstacles are foreseen to preclude safe disposal of nuclear wastes in geologic formations."[12]

This positive status of waste disposal technology has also been recognized by the federal government. The president has emphasized to the Congress the importance of the timely passage of legislation supporting a waste management program.[63,19] The Congress has responded, at least in part, through overwhelming Senate passage of a comprehensive nuclear waste bill. Action in the House in early July 1982 had progressed to the point where legislative proposals had been reported out by three separate committees and the prospects of long-awaited legislation being passed by the 97th Congress appeared reasonably favorable.

Thus, successive federal agencies, groups operating within the most prestigious scientific organizations in the land, independent scientific and technical groups, foreign governments, and others, using repeated analyses and evaluations and results of research, development, and field tests, have all, over the past generation, generally concluded that mined repositories in deep geologic formations are capable of containing and isolating high-level nuclear wastes, including spent fuel, in a safe and environmentally acceptable manner. Clearly the barriers to an operating repository are not technical; they are political and will not be resolved without resolute and continuing political action.

A deep historical footnote serves to emphasize the fact that technology is not the problem. The discovery of six natural fission reactors in the Oklo mine in Gabon, West Africa, has demonstrated that fission products (including plutonium) can be immobilized by the soil's geocharacteristics over geologic periods of time.[3,13,14,74]

Oklo is the name of a uranium mine in the southeastern part of the Republic of Gabon in West Africa. The discovery that a natural fission reaction took place 2 billion years ago was made in 1972 by the French. It is now known that about 15,000 megawatt-years of fission energy was involved, that about 6 tons of fission products and $2\frac{1}{2}$ tons of plutonium were formed, and that these products remained undisturbed in the immediate area of the reactions for more than a billion years. Measurements have shown that those isotopes of most concern have stayed in place or decayed more rapidly than they migrated since criticality took place.

Of most significance is that the Oklo fission events revealed that the most publicly discussed product of nuclear fission, plutonium, had not migrated from the area even though it was in a water-containing environment over the hundreds of thousands of years duration of the nuclear reactions. Now, after almost 2 billion years, it is clear that plutonium never did migrate from the area.

Subseabed Disposal. The subseabed disposal concept has been under active investigation since 1973.[2] The concept involves the emplacement of waste into the red clay sediments in the middle of a tectonic plate, e.g., the North Pacific Plate, under the center of a surface circular water mass called a gyre. The major potential advantages cited are as follows:

Ability to make long-term predictions of stability and uniformity.

Lack of resources in the areas of interest, reducing likelihood of human intrusion.

Plastic nature of sediments, which will allow closure of any openings, man-made or natural.

Low-permeability and high-sorption qualities of the sediments.

Continuously depositional nature of the sediments, eliminating the risk of erosion down to or including buried waste.

Large size of the areas of interest.

Remoteness of the areas of interest from normal human activities.

Lack of need for mining activities or waste-handling facilities at the site.

Lack of need to resolve questions about federal–state relations and authority over disposal sites.

The geologic history of these regions as obtained from cores shows them to be remarkably uniform in extent (over areas of 6000 square nautical miles) with no indications of damaging perturbations for tens of millions of years. The formations are highly predictable, a highly advantageous attribute. The red clays also have very high sorption coefficients and therefore represent a major barrier to radionuclide migration.

Emplacement technology is considered to be essentially available state of the art.[4] However, engineering of the transport and emplacement systems needs to be undertaken on a specific basis. The major technical issue in the subseabed concept appears to be in the area of nearfield interactions between the sediments and the heat and, to some extent, the radiation from the waste. *In situ* field tests to investigate these interactions are planned for the early 1980s.

A program activity chart for the subseabed program indicates completion of detailed site characterization of an initial site in the northern oceans by 1985–1986, completion of detailed system design and issuance of construction permit by 1994, and acquisition of an operating license in 1998–1999.[68]

However, it should be recognized that from an institutional and public acceptance standpoint one can anticipate major international opposition to the concept. It has been suggested that implementation of the subseabed disposal concept might be in violation of existing U.S. law and inconsistent with international laws. Accordingly, it cannot be relied upon at this time as a usable system for waste disposal.

Other Methods

Other methods of disposal that were considered[17] are: very deep hole, rock melting, reverse well, space, ice sheet, and island. A few brief comments follow about each of these alternatives.

Very Deep Hole. A potential alternative is to drill or sink a shaft to isolate high-level wastes in a very deep hole. This concept relies on using surrounding rock to contain the wastes and on the great depths to delay the release and reentry of nuclear material to the biosphere. The utility of the deep-hole concept is affected by three principal factors.

First, the geologic characteristics of the site, including hydrologic conditions, rock strength, and rock/waste interactions at great depths, are not well known. Very deep holes located in strong, unfractured rock, such as crystalline rock (which typically has low water content) or some deep sedimentary basins, would be a good selection for a deep-hole site.

Second, the current capability to excavate a very deep hole has been established. Presently it is possible to drill a narrow deep hole to 10 km (35,000 ft) or to sink a wide shaft to about 4 km (15,000 ft).

Third, the safe emplacement of wastes in this concept may present severe engineering problems. Lowering waste canisters 10–12 km on a wire through high-density muds could significantly increase short-term risks.

The deep-hole concept cannot be evaluated as a disposal alternative without more information on the deep-groundwater system, rock strength under increased temperatures and pressure due to decay of wastes, and the sealing of the holes over long periods of time.

Rock Melting.   The rock-melting concept calls for the direct emplacement of liquid waste in a deep underground hole or cavity. Radioactive decay heat causes melting of the surrounding rock, which in turn dissolves the waste. In time the waste-rock solution refreezes, trapping the radioactive material in a relatively insoluble matrix deep underground. Presumed advantages of the method include ability to directly emplace HLW without the need for solidification or transportation, although presolidified waste or spent fuel can also be directly emplaced.

Reverse Well.   Reverse-well, or deep-well disposal of radioactive wastes encompasses two distinctly different techniques:

1. Injection of the waste in an acidic liquid form.
2. Injection of the waste in a slurry composed of neutralized liquid waste mixed with cement, clay, and other additives designed to ensure ultimate solidification after injection.

The deep-well liquid injection concept draws largely upon existing disposal practices in the oil and gas industry. Shale-grout injections of intermediate-level wastes have been undertaken by Oak Ridge National Laboratory at Oak Ridge, Tennessee.

Extensive technology development would be required in order to determine the viability of either alternative.

Space.   The dominant attraction of disposal of nuclear waste in space is the promise of permanent separation of selected wastes from the human environment. Some of the space-unique systems are in development (space shuttle and facilities) or are being planned for other requirements (orbital transfer vehicle). The major area of development is associated with the safety/environmental concerns and with cost. The practicable application of the use of space for disposal seems remote indeed.

Ice Sheet.   Disposal of nuclear waste in the Antarctic or Greenland ice sheets could offer the advantages of remoteness from human activities and potential isolation from the biosphere. If conditions prevalent in these areas for the past 2 or 3 million years persist, the ice sheets could isolate the waste for many thousands of years. The disadvantages of ice sheet disposal are long transport distances, high costs and difficulties of operations, and uncertainties in the long-range interactions between the ice sheets and the waste. Furthermore, ice sheet waste disposal would require new international initiatives, an amendment of the Antarctic Treaty, or a new treaty with Denmark. An extensive research program would be required to develop this concept.

Island Disposal.   This concept involves emplacement of solidified wastes within deep stable geologic formations beneath an island. Salt deposits are unlikely to be available at island sites; the most probable disposal formation is crystalline rock. Conventional geologic disposal concepts for crystalline rock would therefore be directly applicable to the island disposal concept.

The major difference between mainland and island disposal sites lies in the geohydrologic regime and the requirement of sea transportation.

7.8-4   Institutional Factors

Government Roles

The litany of government action (and inaction) given in Section 7.8-3 on the history of the waste repository program is clear evidence that institutional coordination is necessary if we are ever to implement a national waste management program. Successful and timely accomplishment of the goals of the nuclear waste management program requires both federal and state coordination and coordination and dedication among the federal agencies involved in the program. The administration is giving clear support to an accelerated program.[19] The strongest bill we have ever had supporting the effort has passed the Senate (S.1662). Companion legislation in the House is still uncertain. Given a strong and sustained commitment from Congress and the executive branch, with clearly delineated rules for state and local participation, a repository can become a reality in 10–20 years, because the technology is available. Lacking such commitment, no amount of additional technical research will lead to a successful conclusion of the program.

Perception of Risk.   At the heart of the problem of getting this governmental commitment is the sensitivity of public officials to public perception. And the public perception of the risks associated

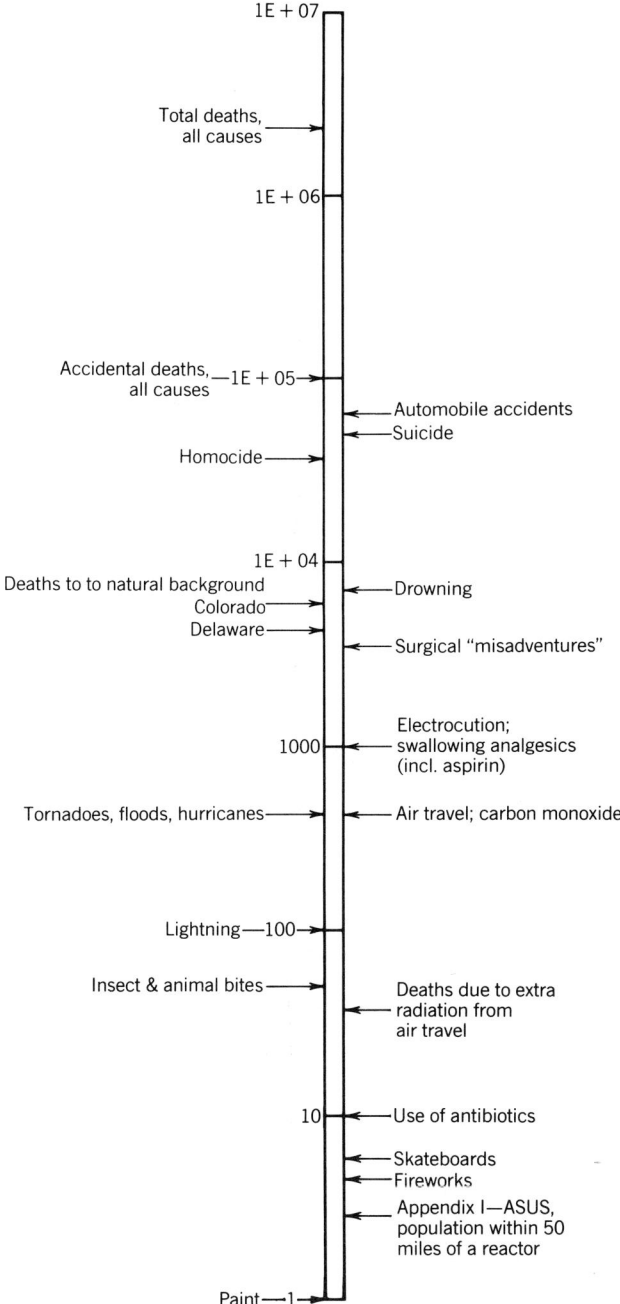

Fig. 7.8-7 Comparison of estimates of population deaths.

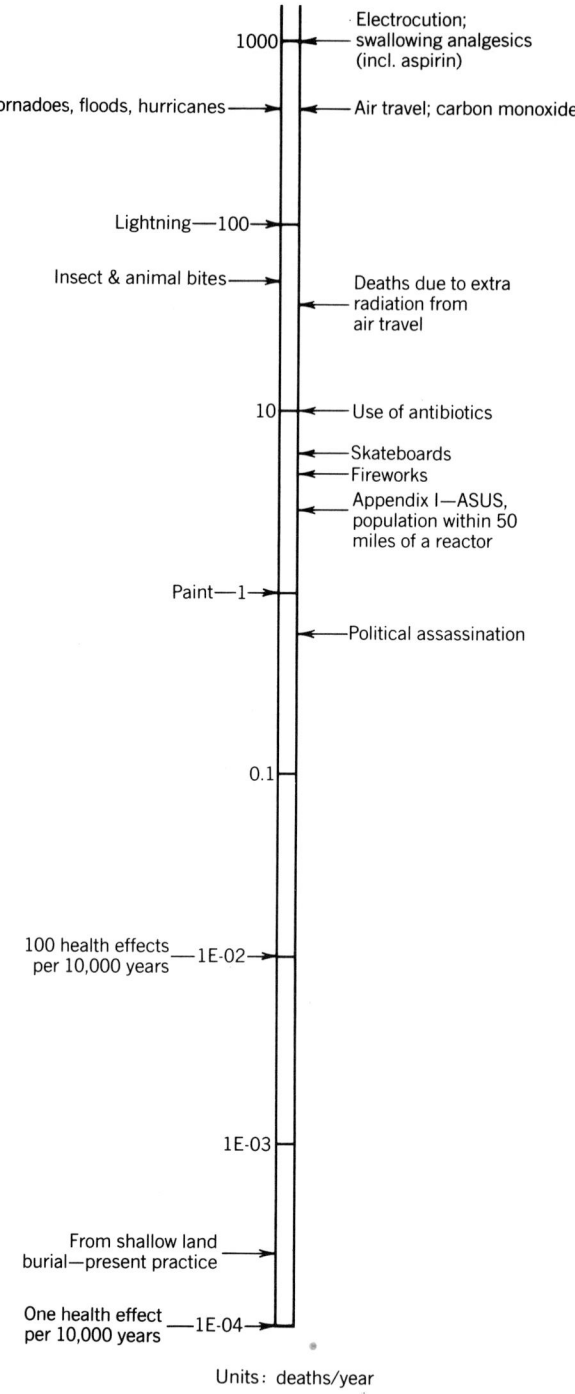

Units: deaths/year

Fig. 7.8-7 (*Continued*)

with nuclear power in general, and with waste disposal in particular, is routinely very much higher than the actual magnitude of those risks based on any rational standard. This was shown quite conclusively in a recent *Scientific American* article.[5] It is also illustrated in Fig. 7.8-7 wherein the real risks of low-level and high-level waste disposal are compared to those of other hazards we readily accept in our everyday lives. It can be seen that the risks* from waste disposal are orders of magnitude lower than, for instance, from political assassination or the use of skateboards.

Community Action. In the final analysis some communities are going to have to be persuaded to accept disposal sites in their areas. In such matters community action is generally driven by the NIMBY[†] principle. This is true for anything deemed to be undesirable, be it a prison, an airport, a sewage disposal plant, a sludge disposal area, a sanitary landfill, an industrial complex, low-cost housing, or a nuclear waste repository. Yet our civilization demands that each of these be built and they all have to go somewhere. The solution to the nuclear disposal problem may lie in a few communities coming to realize just how benign an entity a nuclear disposal site is and volunteering to accept it in lieu of other less desirable things.

REFERENCES

7.8-1 R. L. Andersen et al. *Institutional Radioactive Wastes*, NUREG/CR-0028, March 1978.

7.8-2 D. R. Anderson, *Nuclear Waste Disposal in Subseabed Geologic Formations: The Seabed Disposal Program*, SAND 78-221, Sandia Laboratories, May 1979.

7.8-3 K. E. Apt, compiler, "Investigation of the Oklo Natural Fission Reactors, July 1975–June 1976, LA6576-PR, November 1976.

7.8-4 Personal communication, R. Bauer, Global Marine, Inc, April 1980.

7.8-5 W. P. Bebbington, "The Reprocessing of Nuclear Fuels," *Scientific American* **735**(6), 30–41 (1976).

7.8-6 T. J. Beck et al. *Institutional Radioactive Wastes—1977*, NUREG/CR-1137, October 1979.

7.8-7 M. J. Bell, "Sources of Reactor Wastes, Their Characterization and Amounts," presented at *Radwaste Management Workshop*, sponsored by Oak Ridge National Laboratory, New Orleans, LA, January 12–14, 1977.

7.8-8 G. A. Benda, "Non-fuel Cycle Waste," *Proceedings of the Symposium on Waste Management* (*Waste Management '79*), Tucson, AR, February 26–March 1, 1979 (R. G. Post, ed.), University of Arizona, 1979.

7.8-9 R. Bradshaw and W. McClain, *Project Salt Vault: A Demonstration of the Disposal of High-Activity Solidified Wastes in Underground Salt Mines*, ORNL-4555, Oak Ridge National Laboratory, April 1971.

7.8-10 Code of Federal Regulations, Title 10, Part 50, Appendix F.

7.8-11 R. J. Cholister et al. *Nuclear Fuel Cycle Closure Alternatives*, Allied-General Nuclear Services, March 1976.

7.8-12 Committee on Nuclear and Alternative Energy Systems, National Academy of Sciences, *Energy in Transition, 1985–2010*, January 1980.

7.8-13 G. A. Cowan, "Oklo—A Prehistoric Natural Fission Reactor," Californium-252 Utilization Mtg, San Diego, CA, November 4–6, 1975.

7.8-14 G. A. Cowan, "A Natural Fission Reactor," *Scientific American*, July 1976.

7.8-15 Directorate of Energy Research, U.S. Department of Energy, *Draft Report of Task Force for Review of Nuclear Waste Management*, February 1978.

7.8-16 U.S. Department of Energy, *Draft Environmental Impact Statement on the Waste Isolation Pilot Project*, DOE/EIS-0076-D, 1979.

7.8-17 U.S. Department of Energy, *Draft Environment Impact Statement on the Management of Commercially Generated Radioactive Waste*, DOE/EIS-0046-D, April 1979.

7.8-18 U.S. Department of Energy, *Western New York Nuclear Service Center Study*, TID-28905-2, 1979.

7.8-19 Testimony of James B Edwards, Secretary of Energy, to House Committee on Energy and Commerce, June 8, 1982.

7.8-20 Environmental Protection Agency, *Environmental Analysis of the Uranium Fuel Cycle, Part 1*, Fuel Supply, EPA Office of Radiation Programs, PB-235 804, October 1973.

7.8-21 ERDA Press Release No. 75-51, *ERDA to Issue Expanded Environmental Impact Statement on Management of Commercial High-Level and Transuranium-Contaminated Wastes*, April 11, 1975.

---

*It should be emphasized that these "risks" from waste disposal are statistical risks based on the assumption of the linear hypothesis applied to very low levels of radiation. By contrast most of the other risks shown in Fig. 7.8-7 are real, identifiable deaths.

[†] Not In My Back Yard!

7.8-22 U.S. Energy Research and Development Administration, *Alternatives for Managing Wastes from Reactors and Post-Fission Operations in the LWR Fuel Cycle*, ERDA-76-43, **1**, May 1976.

7.8-23 U.S. Energy Research and Development Administration, *Overview: National Waste Terminal Storage Program of the United States Energy Research and Development Administration*, Press Conference, December 1976.

7.8-24 U.S. Energy Research and Development Administration, Press Conference, Press Release No. 76-355, December 2, 1976.

7.8-25 op. cit. ERD76, **2**.

7.8-26 Ibid.

7.8-27 U.S. Energy Research and Development Administration, *Fact Sheet: States Placed in Three Groups for Phased Approach in Establishing Commercial Nuclear Waste Sites*, December 1976.

7.8-28 Harold Etherington (ed.) *Nuclear Engineering Handbook*, Chapter 11, McGraw-Hill, New York, 1958.

7.8-29 *Federal Register*, **35**, No. 222, p. 17533, 1970.

7.8-30 U.S. General Accounting Office, "Hazards of Past Low-level Radioactive Waste Ocean Dumping Have Been Overemphasized," EMD-82-9, October 21, 1981.

7.8-31 J. P. Girand and G. LeBlaye, "Design of an Industrial Facility for the Incorporation into Glass of Fission Products and the Storage of Highly Radioactive Glass," *Symposium on the Management of Radioactive Wastes from Fuel Reprocessing*, Paris, France, November 27–December 1, 1982.

7.8-32 E. Gleuckauf, *Atomic Energy Waste, Its Nature, Use, and Disposal*, Interscience Publishers, 1961.

7.8-33 Gorleben Hearing ("Rede und Gegenrede," March 28–April 3, 1979), Hanover, Germany.

7.8-34 B. D. Guilbeault, *The 1979 State-by-State Assessment of Low-level Radioactive Wastes Shipped to Commercial Burial Grounds*, NUS-3440 (revision), November 1980.

7.8-35 Aldous Huxley, *Tomorrow, Tomorrow, and Tomorrow, and Other Essays*, Harper, New York, 1956.

7.8-36 "Disposal of Radioactive Wastes," in *Proceedings of IAEA Conference November 16–21, 1959*, International Atomic Energy Agency, 1960.

7.8-37 "Treatment and Storage of High-level Radioactive Wastes," in *Proceedings of IAEA Symposium in Vienna, October 8–12, 1962*, International Atomic Energy Agency, 1965.

7.8-38 Interagency Review Group, *Report to the President by the Interagency Review Group on Nuclear Waste Management*, TID-29442, March 1979.

7.8-39 Ibid., p 42.

7.8-40 Ibid., p. 66.

7.8-41 D. C. Kocher, "Nuclear Decay Data for Radionuclides Occurring in Routine Releases from Nuclear Fuel Cycle Facilities," ORNL/NUREG/TM-102, Oak Ridge National Laboratory, August 1977.

7.8-42 G. W. Leddicotte et al. "Suggested Concentration Limits for Shallow Land Burial of Radionuclides," paper presented at *Symposium, Waste Management '79*, Tucson, Arizona, March 7, 1978.

7.8-43 C. A. Mawson, *Management of Radioactive Wastes*, Van Nostrand, New York, 1965.

7.8-44 W. McClain and R. Bradshaw, *Status of Investigation of Salt Formations for Disposal of Highly Radioactive Power Reactor Wastes*, *Nuclear Safety*, **11**, No. 2, March–April 1970.

7.8-45 W. McClain, *Background and Review of Investigations in the Salina Salt Basin*, Oak Ridge National Laboratory presentation to Michigan Environmental Review Board, 1976.

7.8-46 J. McElroy et al. *Waste Solidification Program Summary Report, 11, Evaluation of WSEP High-level Waste Solidification Processes*, BNWL-1667, Battelle Northwest Pacific Laboratories, July 1972.

7.8-47 M. E. Meek and B. F. Rider, "Summary of Fission Products Yields for U-235, U-238, Pu-239, and Pu-241 at Thermal, Fission Spectrum and 14 Mev Neutron Energies," APED-5398-A, Class 1, revised, General Electric Co, Nucleonics Laboratory, Vallecitos Nuclear Center, Pleasanton, CA, October 1968.

7.8-48 National Academy of Sciences/National Research Council, *The Disposal of Radioactive Waste on Land*, Publication No. 519, April 1957.

7.8-49 National Academy of Sciences/Nuclear Regulatory Commission, *Disposal of Solid Radioactive Wastes in Bedded Salt Deposits*, 1970.

7.8-50 *Safety Analysis Report, NFS Reprocessing Plant*, Nuclear Fuel Services, Inc, Docket 50-201, p. V-6-1, 1973.

7.8-51 Nuclear Fuel Services, Inc, *Safety Analysis Report, NFS Reprocessing Plant*, West Valley, NY, Docket 50-201, *1*, Fig IV-1-1, p IV-1-2, Rev 1, January 24, 1975.

7.8-52 U.S. Nuclear Regulatory Commission, *Draft Generic Environmental Impact Statement on Uranium Milling*, NUREG-0511, April 1979.

7.8-53 U.S. Nuclear Regulatory Commission, *Environmental Survey of the Reprocessing and Waste Management Portions of the LWR Fuel Cycle*, NUREG-0116 (supplement 1 to WASH-1248), October 1976.

7.8-54 U.S. Nuclear Regulatory Commission, "A Classification System for Radioactive Waste Disposal—What Waste Goes Where?" June 1978.

7.8-55 U.S. Nuclear Regulatory Commission, "Licensing Requirements for Land Disposal of Radioactive Waste," Draft Environment Impact Statement on 10 CFR Part 61, September 1981.

7.8-56 "Management of Radioactive Waste from Fuel Reprocessing," in *Proceedings of Symposium, Paris, November 27–December 1, 1972*, OECD Publication Center.

7.8-57 Office of Nuclear Waste Isolation, "A Waste Inventory Report for Reactor and Fuel-Fabrication Facility Wastes," ONWI-20, UC-70, March 1979.

7.8-58 *Radioactive Waste Repository Project Technical Status Report*, ORNL-4751, Oak Ridge National Laboratory, 1971.

7.8-59 FY 1974, *Authorization Hearings before Joint Congressional Committee on Atomic Energy*, statement: F. K. Pittman, March 22, 1973.

7.8-60 Hearings on U.S. Energy Research and Development Administration FY 1976 Authorization before the Joint Committee on Atomic Energy, 94th Cong, 1st Sess, statement: F. K. Pittman, February 1975.

7.8-61 D. W. Powers et al. *Geological Characterization Report—Waste Isolation Pilot Project*, SAND 78-1596, Sandia Laboratories, 1978.

7.8-62 *President's Message to Congress on Radioactive Waste Management*, Weekly Comp of Pres Doc 303, February 12, 1980.

7.8-63 Letter from President Reagan to the Congress dated April 28, 1982.

7.8-64 *Proceedings of the Second United Nations International Conference on the Peaceful Uses on Atomic Energy, 18*, Sessions C-21 and C-22 on "The Treatment of Radioactive Wastes," September 1958.

7.8-65 "Radioactive Waste Management," in *Proceedings of the Third International Conference of the Peaceful Uses of Atomic Energy at Geneva, August 31–September 9, 1964, 14*, Session 3.11.

7.8-66 W. M. Regan (ed.), *Proceedings of the Symposium on the Solidification and Long-term Storage on Highly Radioactive Wastes, February 14–18, 1966*, Richland, Washington, CONF-660208, November 1966.

7.8-67 W. A. Rodger, "Reprocessing of Spent Nuclear Fuel," in *Hearings on Implementation of Nuclear Reprocessing and Waste Disposal Legislation before the California State Energy Resources Conservation and Development Commission*, March 7, 1977.

7.8-68 Sandia Laboratories, *Subseabed Disposal Program Plan, Vol II—Overview*, SAND 80-007/1, January 1980.

7.8-69 "Report to the American Physical Society by the Study Group on Nuclear Fuel Cycles and Waste Management, *Review of Modern Physics* **50**(1), January 1978.

7.8-70 D. Strominger et al., "Table of Isotopes," *Reviews of Modern Physics* **30**(2), Part 2, April 1958

7.8-71 *Fixation of Radioactivity in Stable, Solid Media*, report on Working Meeting at Johns Hopkins University, June 19–21, 1957, TID-7550, March 1958.

7.8-72 *Report on the Second Working Meeting on Fixation of Radioactivity in Stable, Solid Media, September 27–29, 1960*, TID-7613, February 1961.

7.8-73 M. E. Wacks, coord, "The State of Waste Disposal Technology, Mill Tailings, and Risk Analysis Models," *Proc Symp Waste Management, 1*, p 243, Roy Post, ed, Tucson, Arizona, March 10–14, 1980.

7.8-74 R. D. Walton, Jr., "Oklo, the World's First Nuclear Fission Reactor," *Health Physics Society 10th Mid Year Symposium*, October 12, 1976.

7.8-75 White House Release, *President's Nuclear Waste Management Plan*, Weekly Comp of Pres Doc, October 28, 1976.

7.9 NUCLEAR SAFETY CONSIDERATIONS

Norman C. Rasmussen

7.9-1 Non–Reactor Safety Issues

Nuclear safety as discussed here will be limited to a review of the risks associated with the radioactivity present in various parts of the fuel cycle. There are, of course, the usual occupational and public risks such as fires, explosions, falls, and so on, that are associated with industrial activities of this size even if no radioactivity is present. However, it is customary to limit discussions of nuclear safety to those risks created by the radioactivity that is produced and handled in the various activities associated with nuclear power production. The major steps in the nuclear fuel cycle are mining and milling, enrichment, fuel fabrication, power production, reprocessing, and waste disposal. Although radioactivity is present in each of these steps, the amounts present are much greater in the last three, and thus they are the main focus of those concerned about safety. The last two steps have been discussed in Section 7.4. Most of this section will be devoted to safety issues associated with the operation of the nuclear power plant, but a brief discussion of the other steps is included here.

In mining and milling the uranium ore is recovered from the deposit and then processed to produce yellow cake (U_3O_8), which is then sent to the enrichment plant where it is converted to a gas, UF_6, which is the form used in either the gaseous diffusion enrichment process or the centrifuge process. In underground mining the principal hazard is associated with the noble gas radon, one of the daughter products of the uranium decay series. This gas is capable of diffusing out of the ore deposit and reaching unsafe levels in the mine atmosphere. Its principal effect is to cause α radiation exposure to the bronchial tubes when inhaled. If high enough, this dose can increase the likelihood of cancer in that part of the respiratory system. Thus, it is an occupational risk to miners. The dose can be kept to safe levels by proper mine ventilation. In addition, sometimes mines are pressurized to reduce the rate of diffusion out of the deposit. Normal dust control procedures must also be employed to prevent inhalation of radioactive dust particles.

The mining and milling step also creates as a by-product substantial mine tailings piles. These piles contain small amounts of uranium and its daughter products, especially radon, which, if not controlled, can be an increased source of exposure to the general public in the area. Thus, it is important that the accumulation of tailings be stabilized in a manner so that potentially dangerous amounts of radioactivity are not distributed into the biosphere by wind, rain, or other means.

In the enrichment phase the uranium has been separated from its daughter product so that it is very weakly radioactive, and radon is no longer a serious problem. However, as the enrichment process proceeds, the possibility of accidental criticality increases. More than two dozen accidental criticalities outside of reactors have occurred.[1] Experience has shown that they have been limited to less than 10^{19} fissions by the release of enough energy to change the original critical geometry to one that is not critical. The burst of radiation released in such an event is usually enough to produce fatal doses to anyone who is unshielded in the immediate area, but the energy release is far short of enough to destroy buildings or other structures. Thus, such events are an occupational risk, although public health risks associated with them are minimal. The control of these risks is accomplished by using, where possible, tanks and vessels that have geometries that are always subcritical no matter what enrichment they contain, by the use of neutron absorbers and by administrative controls. Over the years this combination of criticality controls has kept this risk small compared to the other occupational risks of the process.

7.9-2   Reactor Safety

A typical large nuclear power station running at a power of about 1000 MW(e) will generate a large inventory of radioactivity in its core. The amount will depend on how long the plant has operated, but after even a few weeks will be in excess of 3×10^{10} Ci. Table 7.9-1 gives the quantities of some 50 of the key isotopes and their amounts for an equilibrium fuel cycle in a 1000-MW(e) light water reactor. If any substantial fraction of this inventory were accidentally released, the potential exists for an accident that could produce serious public health consequences and property damage. The safety design philosophy of reactors is to make the likelihood of such events acceptably small.

The basic safety philosophy of all reactors is essentially the same. All reactors include at least three separate barriers to prevent the release of fission products. The safety systems of the plant are designed to assure that at least one of these barriers remains effective under a variety of postulated system failures called design basis accidents (DBA). These barriers are the cladding of the fuel, the primary system boundary, and the containment building. The safety analysis must postulate mechanisms that might fail these barriers, and the probability of such events must be acceptably low. Most of the radioactivity (except for the noble gases) is effectively trapped in the uranium dioxide fuel at normal temperatures. Large releases can occur only if the fuel is heated to temperatures at or near the melting point of about 2800°C. Thus, the safety engineer must consider those failures that can lead to serious overheating of the fuel.

To provide protection for the barriers, a "defense-in-depth" philosophy is used. This approach has three stages: reliable design, protection systems, and engineered safeguards. Reliable design requires that very strict codes and standards be followed in the design, fabrication, inspection, and testing of the system and its components. The reactor must include protection systems that monitor plant conditions (temperatures, flow rates, pressure, etc.) and respond by alerting the operator, and, in the case of serious conditions, automatically tripping (stopping the chain reaction) the plant. If these two levels of defense fail, public protection is achieved through the operation of the engineered safeguards systems. These are systems designed to assure that there is no undue risk to the public even if a DBA occurs. These accidents include such internal failures as rupture of the largest primary system pipe, small pipe rupture, steam line break, loss of off-site power, and a number of others. In addition to

TABLE 7.9-1   INITIAL ACTIVITY OF RADIONUCLIDES IN NUCLEAR REACTOR CORE AT TIME OF HYPOTHETICAL ACCIDENT (REF. 2)

No.	Radionuclide	Radioactive Inventory Source (Ci $\times 10^{-8}$)	Half-Life (days)
1	Co-58	0.0078	71.0
2	Co-60	0.0029	1,920
3	Kr-85	0.0056	3,950
4	Kr-85m	0.24	0.183
5	Kr-87	0.47	0.0528
6	Kr-88	0.68	0.117
7	Rb-86	0.00026	18.7
8	Sr-89	0.94	52.1
9	Sr-90	0.037	11,030
10	Sr-91	1.1	0.403
11	Y-90	0.039	2.67
12	Y-91	1.2	59.0
13	Zr-95	1.5	65.2
14	Zr-97	1.5	0.71
15	Nb-95	1.5	35.0
16	Mo-99	1.6	2.8
17	Tc-99m	1.4	0.25
18	Ru-103	1.1	39.5
19	Ru-105	0.72	0.185
20	Ru-106	0.25	366
21	Rh-105	0.49	1.50
22	Te-127	0.059	0.391
23	Te-127m	0.011	109
24	Te-129	0.31	00.048
25	Te-129m	0.053	0.340
26	Te-131m	0.13	1.25
27	Te-132	1.2	3.25
28	Sb-127	0.061	3.88
29	Sb-129	0.33	0.179
30	I-131	0.85	8.05
31	I-132	1.2	0.0958
32	I-133	1.7	0.875
33	I-134	1.9	0.0366
34	I-135	1.5	0.280
35	Xe-133	1.7	5.28
36	Xe-135	0.34	0.384
37	Cs-134	0.075	750
38	Cs-136	0.030	13.0
39	Cs-137	0.047	11,000
40	Ba-140	1.6	12.8
41	La-140	1.6	1.67
42	Ce-141	1.5	32.3
43	Ce-143	1.3	1.38
44	Ce-144	0.85	284
45	Pr-143	1.3	13.7
46	Nd-147	0.60	11.1
47	Np-239	16.4	2.35
48	Pu-238	0.00057	32,500
49	Pu-239	0.00021	8.9×10^6
50	Pu-240	0.00021	2.4×10^6
51	Pu-241	0.034	5,350
52	Am-241	0.000017	1.5×10^5
53	Cm-242	0.0050	163
54	Cm-244	0.00023	6,630

these internal failures, the plant must be designed to withstand such external events as severe earthquakes, tornados, extreme floods, and certain others, such as tidal waves and airplane crashes, if appropriate.

All thermal reactors are designed to have negative temperature coefficients and rather long neutron lifetimes, which limit the size of a nuclear excursion to energy releases far below those of nuclear explosions. The worst postulated reactivity accident (control rod ejection) might damage some fuel pins, but energy releases are far below those needed to break the primary system. In fast reactors somewhat more energetic releases are possible, but even pessimistic calculations lead to releases equivalent to pounds of high explosive. The system is designed to handle such releases. For these reasons, the release of the core inventory of radioactivity through an uncontrolled nuclear excursion is not a realistic concern.

The major concern is that the coolant system will fail to remove heat at the required rate. This problem is complicated by the fact that even after the chain reaction is stopped, the large radioactive inventory continues to generate heat at significant rates. Figure 7.9-1 shows the rate of the decay heat generation as a fraction of the full power heat rate versus time for a reactor that has been at power for a substantial time (a few weeks or more). Right after shutdown this rate is about 7% of the full power rate. This value decreases with time as the shorter-lived isotopes decay away. For a 1000-MW(e) reactor, which typically operates at about 3200 MW(th), this would be a power level of about 225 MW at shutdown. It can be seen from Fig. 7.9-1 that significant cooling is required for many weeks after shutdown if fuel overheating is to be prevented.

There are two types of heat imbalances that can produce overheating of the fuel: over-power conditions, in which the heat generation rate exceeds the heat removal capability, and undercooling conditions, in which the cooling system fails to remove the expected amount of heat. The overpower conditions are associated with reactivity insertions that cause beyond-design increases in the power generation rate. A postulated cause of this is rapid removal of neutron poisons by sudden control rod withdrawal. A second cause could be the sudden addition of cold water. Since the reactor is designed to have a negative temperature coefficient, this cold water has the effect of adding positive reactivity to the system. Such events are considered only in a PWR where it is possible to postulate operating the plant with one of the primary loops shut down. The sudden startup of this cold loop could be a mechanism for suddenly introducing a large amount of cold water to the system. Such events, although analyzed and protected against, are not considered a major part of the risk.

There are two general classes of undercooling considered; loss of coolant accidents (LOCA) in which a rupture in the system causes the cooling fluid to be lost, and transients where, for planned or unplanned reasons, the reactor is asked to shut down. In these cases failure to stop the chain reaction or to establish decay heat removal may lead to an accident condition. Large LOCAs are events in

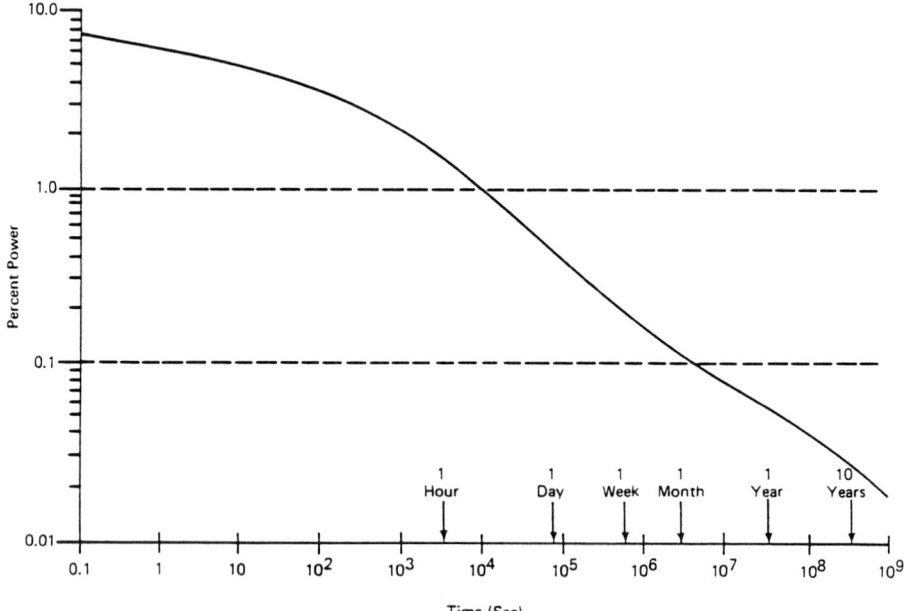

Fig. 7.9-1   Heat production of decay of fission products (ref. 2).

which the rupture is large enough to rapidly depressurize the primary system. Core overheating is prevented in this case by the addition of a large volume of water delivered against a low head. The system that accomplishes this is called the low-pressure emergency core cooling system (ECCS). In a small LOCA coolant inventory is lost, but at a rate that does not depressurize the system. A high-pressure ECCS system that delivers a small volume of fluid against a high head is required for this event. All water reactors must have ECCSs capable of preventing major cladding failure for pipe breaks up to the double-ended rupture of the largest system pipe. Clearly, the massive rupture of the pressure vessel would result in a much more serious LOCA that would be beyond the capability of the ECCS. Analysis of failures in pressure vessels, built to strict nuclear codes, has led to the conclusion that the failure rate (about 10^{-7} per year) is small enough so that no emergency safety feature is required to cope with this event. In all reactors there are connections between the high-pressure primary system and certain low-pressure systems (e.g., low-pressure ECCS). Typically, the low-pressure system is protected by check valves. A failure in this isolation system may lead to a LOCA that fails the low-pressure ECCS and violates the containment. This event, known as the interfacing LOCA, is uniquely different from other LOCAs and requires special consideration. In all LOCAs, the ECCS must have a water supply sufficient for several hours of operation. After that, the water that has leaked out of the break and collected in the reactor sump is recirculated.

Transients occur in plants at a rate of 5–10 per year. Most of these are what are termed *anticipated events* in that experience has shown they can be expected during normal plant operation at a frequency of a few per year. Certain more serious events, called unlikely transients, are expected at a frequency of between 10^{-1} and 10^{-2} per year. A list of such events for both the PWR and the BWR[2] are given in Tables 7.9-2 and 7.9-3. Successful protection for transient events requires that the plant be made subcritical and that the decay heat be successfully removed. In many of these events the heat sink is lost (e.g., turbine trip), thus making it very important to stop the chain reaction. Failure to do this leads to a major heat imbalance that will overpressurize the primary system. This class of event, called anticipated transients without scram (ATWS), are judged to be important contributors to the risk in many plants and must be carefully considered in the design process. The importance of successfully stopping the chain reaction in most transients and LOCAs requires exceptionally high reliability of the shutdown system. It is usually the most redundant and reliable of all the safety systems and typically has an estimated failure rate of 10^{-5} per demand or less. In many transients the main cooling system is lost, and a backup system called the auxiliary feedwater system (AFWS) is required to remove the decay heat. Because it is called upon a number of times each year, it must also be a very reliable system.

The major safety systems described above are all intended to prevent serious damage to the fuel cladding. However, two systems are designed to mitigate the consequences of a release of radioactivity from the primary system. They are the fission product removal system and the containment. The fission product removal system in a PWR consists of a set of pumps and water supplies that can spray the inside of the containment with a heavy rain. Since a large fraction of any postulated released fission products will be in the form of fine aerosols, this is a very effective way of removing them from the atmosphere and trapping them in water in the reactor basement, where they are much less mobile. In a BWR this same function is accomplished by forcing gases released from the primary system to pass through a pool of water (called the suppression pool) which provides a similar scrubbing action. The effect of either of these two systems is to greatly reduce the amount of airborne radioactivity that could be potentially released if the containment were to fail.

The containment is an airtight structure whose job it is to prevent the release of radioactivity should all the previously described systems fail. It is typically designed to withstand an internal pressure of 3–4 atm that might be created if the entire primary coolant inventory were released as steam. This value is somewhat lower in a BWR because the suppression pool is designed to condense steam and thus reduce pressure. When a containment isolation is called for, all major pipes penetrating the containment must be closed by isolation valves that are automatically actuated.

Regulations require that all of the safety systems described above must meet a single failure criterion. This rule states that if a failure occurs that requires one of these systems to operate, it must perform its design function with any single active component failed. In most cases off-site power is also assumed to be lost. In essence this means that all such systems, including their actuation circuits and the on-site power supplies, must be fully redundant.

In addition to the failures of the types discussed, the safety philosophy requires the designer to consider a series of possible external events that might affect plant operation. These include earthquakes, wind, floods, fires and, in some cases, airplane crashes into the plant. Of these, seismic considerations usually impact the plant design the most. In most regions of the United States plants must be able to be safely shut down following seismic accelerations of $0.2–0.3\,g$. In earthquake-prone regions the value will be even higher. This requirement leads to stiff structures and the need for a considerable number of snubbers on safety related pipe. In addition, all operating equipment needed for safe shutdown must be qualified to the earthquake loads. The seismic analysis is done with the aid of large and complicated computer codes. For most plants the design must safely survive 360-mph wind loads from a tornado. Usually, the limiting load is created by wind-driven missiles such as

TABLE 7.9-2 PWR TRANSIENTS

Likely Initiating Events	Likely Initiating Events
1. Turbine trip	15. Boron dilution by malfunctions in chemical volume and control system
2. Spurious signals from SICS	
3. Loss of condenser vacuum	16. Startup of inactive reactor coolant loop (in PWR with No RCS loop isolation valves)
4. Inadvertent closure of main steam line isolation valves	
5. Loss of main station generator with failure to relay auxiliary loads (e.g., main feedwater pumps, condensate pumps) to ac power incoming from off-site network	17. Accidental opening of pressurizer safety or relief valves
	18. Loss of RCS coolant flow (main RCS circulating pump malfunctions)
6. Loss of main circulating water pumps for condenser cooling	**Unlikely Initiating Events**
7. Loss of main feedwater pumps	1. Rupture of high-energy piping in secondary coolant system: (a) main feedwater lines and (b) rupture of lines in main steam system[a]
8. Loss of condensate pumps	
9. Loss of ac power incoming from off-site network	
10. Inadvertent opening of steam generator power-operated Relief Valves ($\sim$ 10% sudden load demand)	2. Rupture of steam generator (see Section 4.1-5)
	3. Rupture of control rod mechanism housing on reactor vessel leading to small LOCA and control rod ejection (see Sections 4.1-3 and 4.1-4)
11. Increase in main feedwater flow; malfunctions in feedwater flow control	
12. Malfunctions of Control Resulting in inadvertent opening of all turbine steam bypass valves ($\approx$ 40% sudden load demand)	4. Abrupt seizure of all main RCS recirculation pumps
	5. Startup of inactive reactor coolant loop with abrupt opening of both isolation of valves in one RCS loop in PWR plants employing RCS loop isolation valves
13. Uncontrolled rod withdrawal at (a) full power and (b) start-up	
14. Control rod assembly drop	

Source. From ref. 2.

[a] These ruptures are included somewhat arbitrarily within the unlikely event category. However, failures of lines in the PWR secondary coolant systems have occurred principally during plant testing and startup periods. These types of failures have included inadequate initial design of relief valve headers in the steam supply lines, discharge of secondary coolant from leaking feedwater valves, discharge of secondary coolant from cracks in main feedwater lines, etc. The RCS cooldown transients stemming from these failures would be less severe than those included under No. 12 of the likely event category. The potential impact of such high-energy line failures in specific locations of the plant, since they might commonly interact with and affect availability of the plant ESFs, was considered as part of this study.

telephone poles. If flooding is a problem, a hypothetical 10,000-year flood is used to determine the elevation of the plant or the height of surrounding dikes. If major dams are upstream, their failure must be considered. The defense against fires is provided by the use of fireproof materials, where possible; by requiring fire-fighting equipment be available; and by separating redundant safety equipment by fireproof barriers.

7.9-3 Probabilistic Risk Analysis (PRA)

PRA is an analytical method for assessing the risk to the public from potential accidents in the plant. The first major study of this type for reactors[2] was completed in 1975. Since that time many other such studies have been carried out. Although these more recent studies have included improvements over the 1975 study, they nonetheless have followed very nearly the same methodology and obtained very similar results for the risks. What follows is a brief summary of the approach and results of the Reactor Safety study.[2]

TABLE 7.9-3   BWR TRANSIENTS[a]

Likely Initiating Events	Likely Initiating Events
1. Rod withdrawal at power	12. Recirculation flow control failure (decreasing flow)
2. Feedwater controller failure, max. demand	13. Recirculation pump trip (one pump)
3. Recirculation flow control failure (increasing flow)	14. Recirculation pump seizure
4. Start-up of idle recirculation pump	15. T-G pressure regulator faiiure—rapid opening
5. Loss of feedwater heating	**Unlikely Initiating Events**
6. Inadvertent HPCI pump start	1. Rod ejection accident[a]
7. Loss of auxiliary power	2. Rod drop accident[a]
8. Loss of feedwater flow	3. Compound initiating events such as (a) seizure of two recirculation pumps, (b) startup of idle recirculation pump simultaneously with turbine trip, (c) rod withdrawal and simultaneous startup of idle recirculation loop
9. Electric load rejection (turbine valve closure)	
10. Turbine trip (stop valve closure)	
11. Main steam line isolation valve closure	

Source. From ref. 2.

[a]BWR plants have design features provided that make the probabilities for occurrence negligibly small.

Typically, such an analysis has six major steps, as illustrated in Fig. 7.9-2.

1. Identify the various accident sequences (combinations of failures) that can lead to a significant release of radioactivity.

2. Estimate the probability of occurrence of these accident sequences.

3. Calculate the probability and magnitude of the radioactive release resulting from the accident sequences.

4. Calculate how the released radioactivity will be dispersed in the environment under the influence of the prevailing weather conditions.

5. Calculate the health effects to the population produced by radiation exposure. Some studies also calculate the economic impact of damage to public property.

6. Sum the consequences of the individual accident sequences to obtain the overall risk.

To identify the accident sequences in an organized, logical way, a technique called event tree analysis is used for the initial analysis. Figure 7.9-3 shows a simplified version of an event tree to illustrate the method. The event tree starts with an initiating event, in this case a pipe break, which is a form of LOCA. The analyst must then, through his knowledge of the system, identify various plant functions that will affect the outcome of this initiating event. In this case the functions identified are availability of electric power, the availability of ECCS, the fission product removal system, and finally the integrity of the containment itself. In Fig. 7.9-3, at each branch point the path that goes up represents failure of this system to perform its design function. Thus, there are a variety of possible

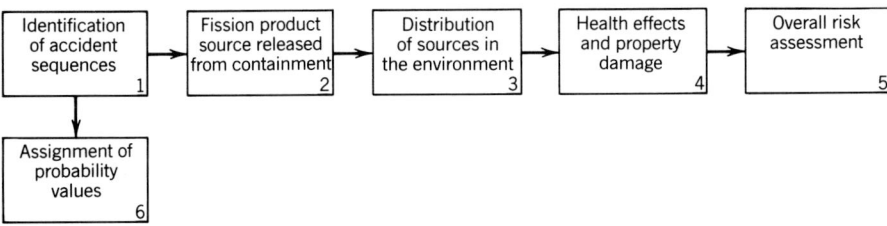

Fig. 7.9-2 Basic seven tasks in reactor safety study (ref. 2).

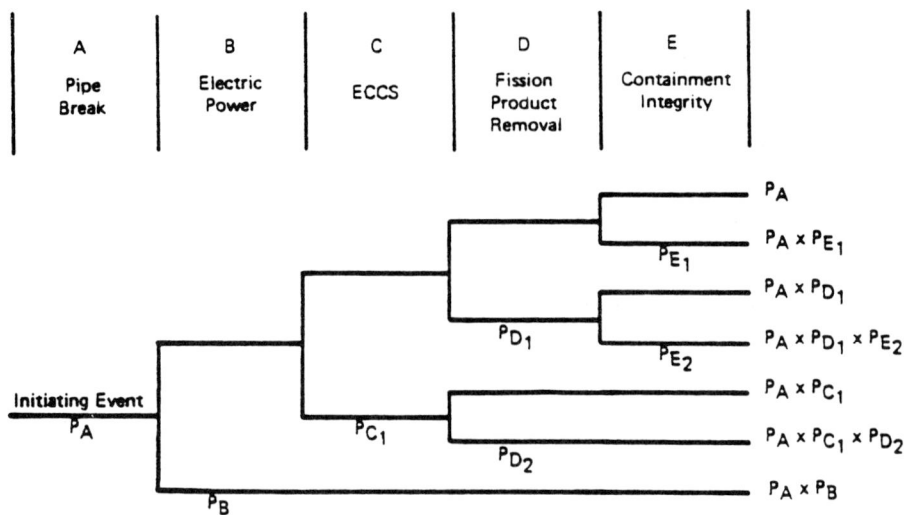

Fig. 7.9-3 Simplified event trees for a large loca (ref. 2). Since the probability of failure, P, is generally less than 0.1, the probability of success $(1 - P)$ is always close to 1. Thus, the probability associated with the upper (success) branches in the tree is assumed to be 1.

paths through the tree, each of which represents a possible state of the plant following the initial event, pipe break. Not all possible branches are shown because, for example, if electric power fails, the ECCS cannot operate, so there is no choice shown on that line. In addition, the fission product removal system cannot operate. If these two systems fail, the core will melt and rupture the containment, so containment integrity is not available either. Thus, the bottom line on this tree has one step, failure of electric power, and it leads directly to a very large release of radioactivity. In the figure P indicates the probability of that particular path at any particular branch point. Thus, any one path's total probability can be obtained by multiplying the probabilities of the various branches together. The final column on the chart shows this. It should be noted that, in general, these probabilities, the values of P, are quite small, so that a good approximation is to assume that $1 - P = 1$. That approximation has been made in determining the probability shown in the final column. Clearly, there may be dependences between the various probabilities. That is, the value of the probability of ECCS function failure P_{cl} may depend on, in some way, the conditions created by the pipe break itself P_A. Thus, in the analysis one must be careful to assess any dependences that may exist between the probabilities. This analysis is called the common-cause or common-mode failure analysis of the system. Thus, if done with care, the event tree allows the analyst to define all the possible final states that might result from the initial event, in this case a pipe break, and indicates what the probability of the various outcomes might be if a method is available to determine the values of P in the chart. In this study six different event trees are developed to cover the spectrum of initiating events that could lead to core melt. They are vessel rupture, large LOCA, medium LOCA, small LOCA, interfacing LOCA, and transients.

To determine the values of the probability, another logic method is used. This method is called fault tree analysis and is indicated in a very simplified way in Fig. 7.9-4. Shown in this figure is a simplified fault tree for the failure of the electric power system to provide power to the engineered safety features (ESF) of the plant. Note that the logic is somewhat the reverse of the event tree logic. The fault tree starts with some undesired outcome and asks how it might have happened. The event tree starts with some initial event and asks what are all the possible outcomes. Figure 7.9-4 shows that loss of power to the emergency safety features can be caused by loss of dc power or loss of ac power. This is because there is a direct current system that activates the controllers to the various electric equipment, and there is an ac system that provides the power to operate the equipment. Thus, loss of either means loss of ability for those systems to function properly, and on the chart it is indicated by the symbol called an OR gate, which simply means that the top event will occur if either one or the other of the two lower events happens.

To illustrate another type of logic that occurs in the tree, consider the loss of ac power. Alternating current power is provided by two different sources, each of which is capable of operating the engineered safety features. These are the off-site power, that is, the connection of the utility grid, and the on-site power provided by the station diesel generators. To lose all ac power, both sources must be lost, so they are related to the box loss of ac power through a symbol called an AND gate, which

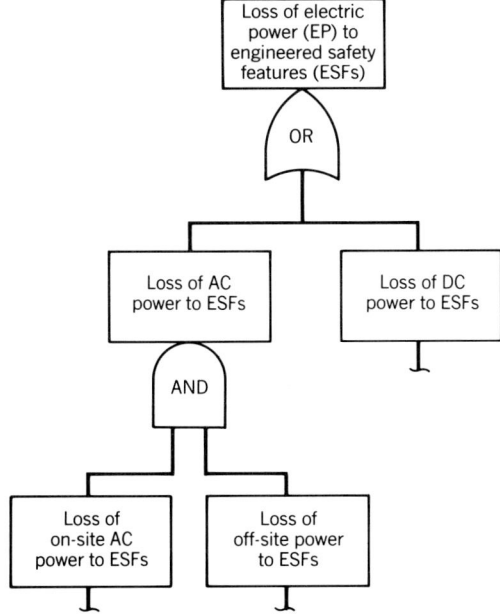

Fig. 7.9-4   Illustration of fault tree development (ref. 2).

indicates that both must occur for the event above to happen. The analyst knows that the probability of the top event in this case is the product of the probability of the two lower events, whereas in the case of the OR gate the probability of the event above is, to a first approximation, the sum of the probabilities of the two lower events. If the tree is developed to lower and lower levels, that is, to smaller and smaller parts of the plant, one finally obtains a series of failures, such as failure of the operator to throw the switch, failure of the switch to work properly, the solenoid sticks, the valve sticks, the pump fails, and so on. From experience with equipment of this type in a variety of operations under similar conditions, one can obtain an estimate of the probability of occurrence of these events. With the probabilities of these bottom events, called primary events, and the relationships that indicate how their probabilities are related to each other, one can propagate the probabilities upward through the tree and determine the probability of the top event.

Fault trees based on these principles are drawn for some 20 different plant systems. These are the systems identified by the event trees as affecting the outcome of various initiating events. Thus, the fault trees are used to obtain the probabilities, and the probabilities are combined, as indicated, by the event trees to determine the probability of various failure states of the plant. Careful analysis of these conditions is required to determine whether serious fuel overheating would result or not. The application of fault trees and event trees thus provide a method for carrying out tasks 1 and 2 in Fig. 7.9-2.

Task 3 is carried out by an analysis of the conditions created by the failures identified in task 1. These analyses determine whether fuel melting has occurred or not. And, given that fuel melting has occurred, further analysis is done to estimate how much radioactivity would be released. The release of radioactivity could come about by various kinds of failures of the containment itself. It is usually assumed that, if the core melts, the containment would surely fail. However, there are different ways in which the containment might fail, and these different ways would release different amounts of radioactivity. The result of step 3 is to generate a histogram of the magnitude of release of radioactivity versus the likelihood of release of radioactivity.

The final result of this analysis is a histogram of release probability versus release magnitude. Figure 7.9-5 is the histogram for a PWR[2] and Fig. 7.9-6 is for the BWR. The ordinate gives the probability per year of a release. The abscissa gives the release magnitude as one of nine different categories (PWR) and six different categories (BWR). Category 9 (PWR) and category 6 (BWR) represent quite small releases and category 1 is the largest release. In Table 7.9-4 the fraction of the core inventory of various fission products released in each category is tabulated.

In the analysis all identified accidents are assigned to one of the nine categories according to the calculated release of radioactivity. The probabilities of all accident sequences assigned to that category are then summed to get the final histograms shown.

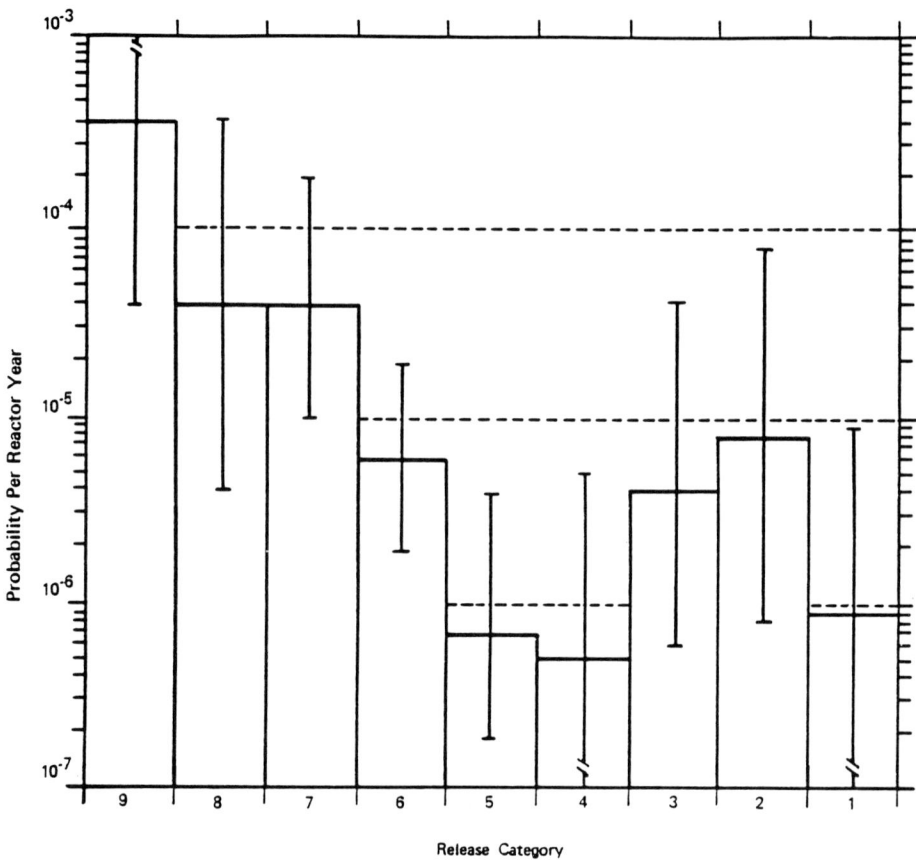

Release Category

Fig. 7.9-5   Histogram of PWR radioactive release probabilities (ref. 2).

The next step is to develop a model for calculating the consequences of the release of radioactivity. This computer code calculates five different health effects, plus the economic loss in dollars due to property damage, cleanup costs, and cost of relocating people. The five health effects considered are early fatalities (fatalities within 1 year); early injuries (nonfatal injuries requiring medical care); cancer fatalities (cancer deaths that might be expected in a 10- to 40-year period after the accident); thyroid injury (latent nonfatal effects of the thyroid gland that would require medical care); and genetic effects.

The consequence code, CRAC, uses a gaussian Plume model to predict how radioactivity is dispersed under prevailing weather conditions. The input data required is the release histogram, demographic data from the Census Bureau out to 500 mi from the site, and a year's worth of weather data from the site. Using a Monte Carlo method, the code calculates the magnitude of a large number of possible accident consequences and the probability of these various consequences. The reader interested in more details about the model is referred to Appendix 6 of *Reactor Safety Study.*[2]

The results of the analysis are usually presented in three different ways often used for expressing risk. The societal risk is defined as the average annual impact. The individual risk is the average probability that any particular individual would suffer a given health effect as a result of a nuclear plant accident. The results for these two types of risks are given in Table 7.9-5[2] for a 100-reactor industry.

The third method of presentation is complementary cumulative distribution functions that express the probability of an accident of any given size or larger. Such plots are prepared for each of the consequences. Figure 7.9-7 gives such a plot for the consequence of early fatalities. The results of the curves for the three latent health effects are summarized in Table 7.9-6. The numbers in the table are the expected annual rate of incidence of these three effects over a 30-year period, starting about 10 years after the accident. For comparison, the normal incidence rate experienced by the exposed population is also given. From Table 7.9-5 it can be seen that the latent cancer fatalities are on average about 700 times greater than the early fatalities. This is because, in a large fraction of the accidents,

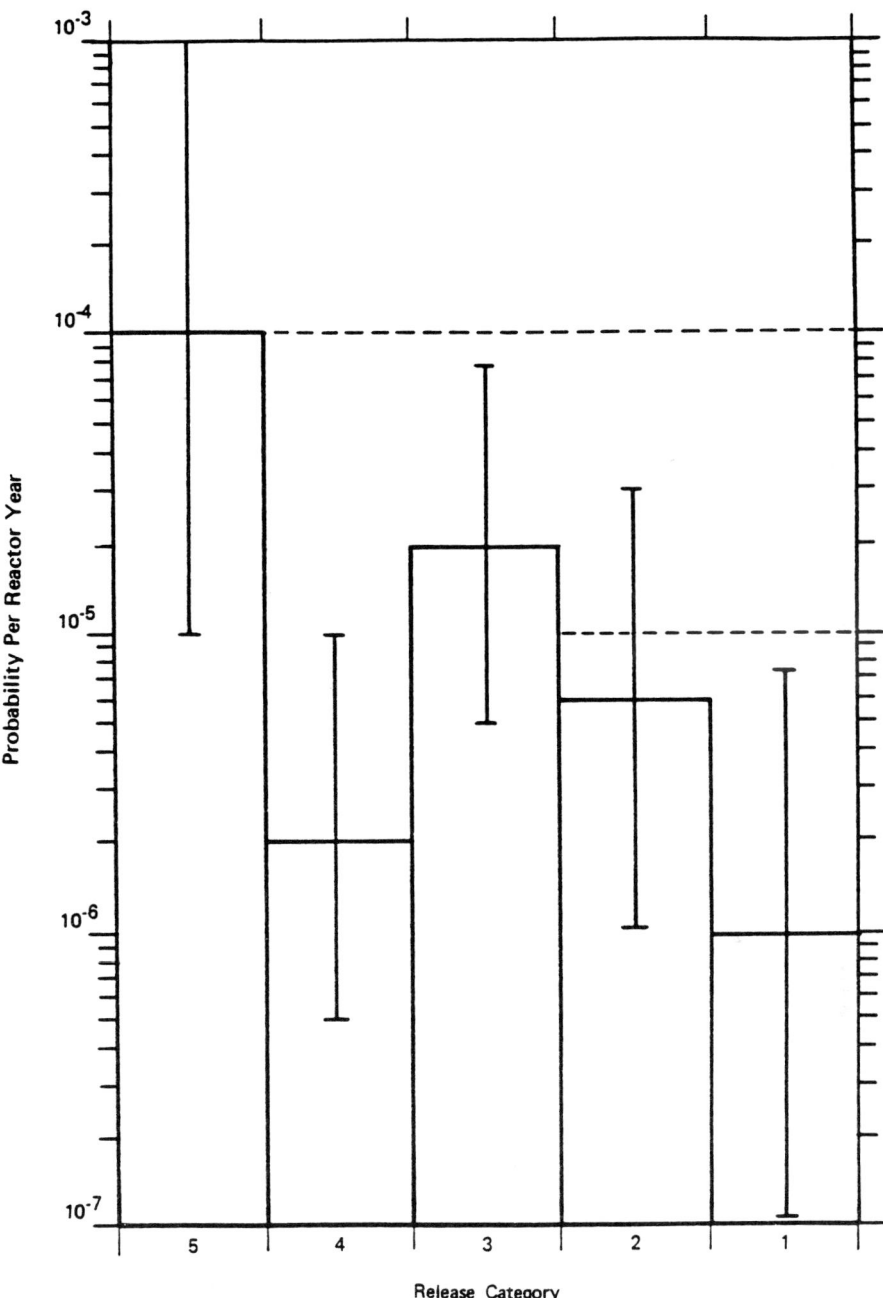

Fig. 7.9-6 Histogram of BWR radioactive release probabilities (ref. 2).

exposure levels are so low that no early fatalities are expected, but the relatively low doses to a fairly large population do lead to some expected latent cancer fatalities. In the largest accident, identified at a probability of 1 in 10^9 per plant per year, the ratio of latent fatalities to early fatalities is $45,000/3300 = 13.6$.

In the reactor safety study[2] a frequency of core melt of 5×10^{-5} per reactor-year was obtained. More recent studies on a number of different plants have obtained values between 10^{-5} and a few times 10^{-4} per reactor-year. The dominant contributors to this probability are transients and small

TABLE 7.9-4 SUMMARY OF ACCIDENTS INVOLVING CORE

RELEASE CATEGORY	PROBABILITY per Reactor-Yr	TIME OF RELEASE (Hr)	DURATION OF RELEASE (Hr)	WARNING TIME FOR EVACUATION (Hr)	ELEVATION OF RELEASE (Meters)	CONTAINMENT ENERGY RELEASE (10^6 Btu/Hr)	FRACTION OF CORE INVENTORY RELEASED[a]							
							Xe-Kr	Org. I	I	Cs-Rb	Te-Sb	Ba-Sr	Ru[b]	La[c]
PWR 1	9×10^{-7}	2.5	0.5	1.0	25	520[d]	0.9	6×10^{-3}	0.7	0.4	0.4	0.05	0.4	3×10^{-3}
PWR 2	8×10^{-6}	2.5	0.5	1.0	0	170	0.9	7×10^{-3}	0.7	0.5	0.3	0.06	0.02	4×10^{-3}
PWR 3	4×10^{-6}	5.0	1.5	2.0	0	6	0.8	6×10^{-3}	0.2	0.2	0.3	0.02	0.03	3×10^{-3}
PWR 4	5×10^{-7}	2.0	3.0	2.0	0	1	0.6	2×10^{-3}	0.09	0.04	0.03	5×10^{-3}	3×10^{-3}	4×10^{-4}
PWR 5	7×10^{-7}	2.0	4.0	1.0	0	0.3	0.3	2×10^{-3}	0.03	9×10^{-3}	5×10^{-3}	1×10^{-3}	6×10^{-4}	7×10^{-5}
PWR 6	6×10^{-6}	12.0	10.0	1.0	0	N/A	0.3	2×10^{-3}	8×10^{-4}	8×10^{-4}	1×10^{-3}	9×10^{-5}	7×10^{-5}	1×10^{-5}
PWR 7	4×10^{-5}	10.0	10.0	1.0	0	N/A	6×10^{-3}	2×10^{-5}	2×10^{-5}	1×10^{-5}	2×10^{-5}	1×10^{-6}	1×10^{-6}	2×10^{-7}
PWR 8	4×10^{-5}	0.5	0.5	N/A	0	N/A	2×10^{-3}	5×10^{-6}	1×10^{-4}	5×10^{-4}	1×10^{-6}	1×10^{-8}	0	0
PWR 9	4×10^{-4}	0.5	0.5	N/A	0	N/A	3×10^{-6}	7×10^{-9}	1×10^{-7}	6×10^{-7}	1×10^{-9}	1×10^{-11}	0	0
BWR 1	1×10^{-6}	2.0	2.0	1.5	25	130	1.0	7×10^{-3}	0.40	0.40	0.70	0.05	0.5	5×10^{-3}
BWR 2	6×10^{-6}	30.0	3.0	2.0	0	30	1.0	7×10^{-3}	0.90	0.50	0.30	0.10	0.03	4×10^{-3}
BWR 3	2×10^{-5}	30.0	3.0	2.0	25	20	1.0	7×10^{-3}	0.10	0.10	0.30	0.01	0.02	3×10^{-3}
BWR 4	2×10^{-6}	5.0	2.0	2.0	25	N/A	0.6	7×10^{-4}	8×10^{-4}	5×10^{-3}	4×10^{-3}	6×10^{-4}	6×10^{-4}	1×10^{-4}
BWR 5	1×10^{-4}	3.5	5.0	N/A	150	N/A	5×10^{-4}	2×10^{-9}	6×10^{-11}	4×10^{-9}	8×10^{-12}	8×10^{-14}	0	0

(a) A discussion of the isotopes used in the study is found in Appendix VI. Background on the isotope groups and release mechanisms is found in Appendix VII.

(b) Includes Mo, Rh, Tc, Co.

(c) Includes Nd, Y, Ce, Pr, La, Nb, Am, Cm, Pu, Np, Zr.

(d) A lower energy release rate than this value applies to part of the period over which the radioactivity is being released. The effect of lower energy release rates on consequences is found in Appendix VI.

TABLE 7-9-5 APPROXIMATE AVERAGE SOCIETAL AND INDIVIDUAL RISK PROBABILITIES PER YEAR FROM POTENTIAL NUCLEAR PLANT ACCIDENTS[a]

Consequence	Societal	Individual
Early fatalities[b]	3×10^{-3}	2×10^{-10}
Early illness[b]	2×10^{-1}	1×10^{-8}
Latent cancer fatalities[c]	7×10^{-2}/yr	3×10^{-10}/yr
Thyroid nodules[c]	7×10^{-1}/yr	3×10^{-9}/yr
Genetic effects[d]	1×10^{-2}/yr	7×10^{-11}/yr
Property damage ($)	2×10^6	—

Source. From ref. 2.

[a] Based on 100 reactors at 68 current sites.

[b] The individual risk value is based on the 15-million people living in the general vicinity of the first 100 nuclear power plants.

[c] This value is the rate of occurrence per year for about a 30-yr period following a potential accident. The individual rate is based on the total U.S. population.

[d] This value is the rate of occurrence per year for the first generation born after a potential accident; subsequent generations would experience effects at a lower rate. The individual rate is based on the total U.S. population.

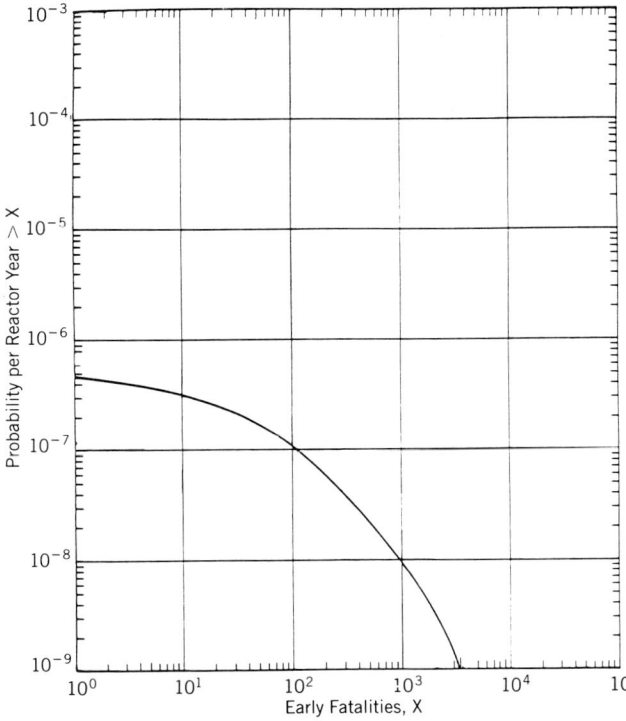

Fig. 7.9-7 Probability distribution for early fatalities per reactor year (ref. 2). Approximate uncertainties are estimated to be represented by factors of $\frac{1}{4}$ and 4 on consequence magnitudes and by factors of $\frac{1}{5}$ and 5 on probabilities.

TABLE 7.9-6 CONSEQUENCES OF REACTOR ACCIDENTS FOR VARIOUS PROBABILITIES FOR ONE REACTOR

Chance per Reactor-Year	Consequences		
	Latent Cancer[a] Fatalities (per year)	Thyroid Nodules[a] (per year)	Genetic Effects[b] (per year)
One in 20,000[c]	< 1.0	< 1.0	< 1.0
One in 1,000,000	170	1,400	25
One in 10,000,000	460	3,500	60
One in 100,000,000	860	6,000	110
One in 1,000,000,000	1,500	8,000	170
Normal incidence	17,000	8,000	8,000

Source. From ref. 2.

[a]Occurs approximately in the 10–40-yr range following a potential accident.
[b]Applies to the first generation born after a potential accident. Subsequent generations would experience effects at a lower rate.
[c]Predicted chance of core melt per reactor year.

LOCAs as a result of internal plant failures; earthquakes are the dominant external cause. Although the best estimate of the probability of an earthquake-caused core melt is usually smaller than the best estimate of an internally caused core melt, because of a much larger uncertainty in the seismic analysis seismic events tend to dominate the overall risk.

In the course of carrying out such an analysis, there are many difficult-to-calculate phenomena and many difficult-to-estimate probabilities. This leads to significant uncertainty in the results. The Reactor Safety Study[2] estimated the uncertainty in the probabilities as a factor of 5 ($\frac{1}{5}$–5 times) and the uncertainty in the consequence magnitude as a factor of 4. Recent work suggests that these error bounds may be unrealistically narrow and a better estimate may be a factor of 10 on probability and a factor of 5 on consequence magnitude. A very good critical review[3] of the methods of the reactor safety study[2] has been prepared by the Risk Assessment Review Group (RARG) for the U.S. Nuclear Regulatory Commission (NRC).

Although the goal of a PRA is to make a realistic analysis of the risk, there is a tendency to be conservative in cases where uncertainty exists. There is evidence accumulating to suggest that the release fractions (especially categories 1 and 2) of Table 7.9-4 is one area where this may be true. If these studies are correct, the release fractions in categories 1 and 2 may be a factor of 10 or more too high. If true, this would mean that the risk of early fatalities would go essentially to 0 and the delayed risks would be significantly reduced. The NRC has recently reviewed this issue and feels that the current evidence is not conclusive enough to warrant a reduction in the release fractions at this time. Many feel that experiments now under way may provide conclusive evidence within a few years.

7.9-4 Major Safety Issues

During the last decade there have been a number of safety issues raised that have required special analysis or in some cases major research programs to resolve them. Many of them have been resolved and others are in various stages of resolution. For the most part these have not been so serious that they required plant shutdown. However, shutdown was required for some TMI-type plants after that accident until the accident cause was understood and dealt with. In another case temporary shutdown was required until the seismic analysis of the pipe was redone. Below are listed some of the key safety issues that are still listed as unresolved. This list does not attempt to include all the current issues. The U.S. NRC keeps such a list and publishes an updated version regularly. This publication is known as the "Aqua Book" because of its color.

Pressurized Thermal Shock

Pressurized thermal shock is a concern that the pressure vessel might fail if it is filled with cold water and pressurized after an accident. For this to be a possibility the nil ductility temperature must be increased significantly due to neutron irradiation. In addition, there must be a flaw in the vessel larger than the critical crack size to permit a brittle fracture to occur. A postulated scenario for such an event is a small LOCA in which the reactor is tripped and the high-pressure ECCS is actuated. If these pumps are not throttled once the level is returned to normal, the system may be overpressurized. This could produce high pressure at cold temperatures, and if a critical-sized flaw were present, the vessel might fail. Conservative analysis shows that in a few of the older vessels having high copper content in the welds, the value of the nil ductility temperature may be in a range of concern within a few years. A major program is underway to better understand the issue and develop solutions. For the most part this is not an issue in BWRs since the jet pumps surrounding the core reduces the neutron fluence on the vessel. The problem does not appear to be severe in recent PWRs since the copper content in the welds is much less than in older plants. This leads to a much slower increase in nil ductility temperature with irradiation.

Water Hammer

The concern in this case is that damaging shock waves may be created in the water piping. This can happen because of the sudden closing of a valve or by the collapse of a steam bubble. Although this is a well-recognized problem in water systems, it has occurred enough times in reactors to remain an issue of concern.

Steam Generator Tube Integrity

The sudden rupture of a steam generator tube leads to a small LOCA that discharges into the secondary system and possibly through the steam dump valve to the atmosphere. Such an event would not lead to a significant radioactive release unless serious fuel damage also occurred. However, there has been a history of degradation of the steam generator tubes so that the event seems more likely than once believed. For this reason it has warranted more detailed study to assure that such events are recognized by operators and dealt with properly.

Pipe Cracking

Pipe cracking has been a particular problem in the older BWRs. The steels used in most BWR internals has been found to be very sensitive to stress corrosion cracking especially in the heat-affected zone of the welds. The testing of newly developed alloys suggests that the problem may have been solved for future plants but it remains a problem in the earlier BWRs.

Systems Interactions

Experience has shown some unexpected dependencies between certain reactor systems. For example, in one case a rather subtle failure led to the loss of a number of instruments in one reactor. This has led to some careful analysis to see if other such dependencies may exist between what were thought to be independent systems.

Hydrogen Control

It has long been recognized that if Zircaloy is heated above about 1100°C in the presence of steam, it will be oxidized and produce hydrogen. Prior to the TMI accident the design basis accident considered only very modest amounts of hydrogen. Subsequent to the TMI accident, however, the threats to the containment from much larger amounts of hydrogen had to be considered. Most large containments can withstand the burning of these larger amounts of hydrogen, but there is a concern about the possible detonation of it. In BWRs where the containment volume is rather small inerting with nitrogen is required. In a recent operating license of a PWR igniters were required to be sure the hydrogen was burned before it reached levels high enough to detonate. This, however, was an ice-condenser-type containment of rather small volume.

In addition to the above issues considerable work continues on improved methods for seismic design. Work also continues on the forces generated by reactor blowdown either by a large pipe break or by the opening of relief valves. These have been a particular problem for the older BWRs in which blowdowns are anticipated events. Both the shutdown and the electric power systems are very important safety systems and work continues to improve their reliability. Finally, all analyses show that the single biggest contributor to the risk is failure of humans to perform properly. For this reason research on how to design systems that will reduce the rate of operator failure is very active.

REFERENCES

7.9-1 T. J. Thompson and J. G. Beckerley, *The Technology of Nuclear Reactor Safety, Volume I,* Technology Press, Cambridge, MA, 1964, chapter 11.

7.9-2 U.S. Nuclear Regulatory Commission, *Reactor Safety Study: An Assessment of Accident Risks in U.S. Commercial Nuclear Power Plants,* WASH-1400 (NUREG-75/014), October 1975.

7.9-3 U.S. Nuclear Regulatory Commission, *Risk Assessment Review Group Report to the U.S. Nuclear Regulatory Commission*, (NUREG/CR-0400), September 1978.

BIBLIOGRAPHY

Nuclear Safety: A bimonthly Technical Progress Review prepared for the Nuclear Regulatory Commission by the Nuclear Safety Information Center at Oak Ridge National Laboratory.

U.S. Nuclear Regulatory Commission NUREG series of technical reports.

7.10   LICENSING AND REGULATION PROCEDURES

A. E. Scherer and E. H. Kennedy

The use of nuclear material in the United States—whether for medical uses, industrial radiography, or the production of electric power—is subject to extensive regulation by the federal government, the individual states, and, in some instances, even counties and municipalities. Since, however, federal preemption over nuclear issues generally prevails, this section focuses on the federal laws and regulations that define and govern the process by which an electric utility company receives authorization to build and operate a nuclear reactor for electric power generation. Particular emphasis is placed on the primary federal agency, the Nuclear Regulatory Commission (NRC), and on the major steps in the regulatory process. This discussion cannot, therefore, provide a comprehensive list or reference of all potentially applicable regulatory requirements.

7.10-1   Overview of the Licensing Process

Major Milestones

Figure 7.10-1 outlines the major milestones in the licensing of a nuclear power plant. The process has two major steps. First, the applicant (the utility) must apply for and receive a construction permit (CP) authorizing them to build the plant. After constructing the plant, the applicant must then obtain an operating license (OL) before they can operate the plant. Although the sequence of events in each of these major steps is similar, they can be separated quite distantly in time. (It takes 10–12 years or more to license and construct a nuclear power plant in the United States today.) After the plant has received its OL, it remains subject to federal regulatory authority throughout its lifetime.

The information submitted by the applicant and the review by the NRC are concentrated in three major areas: a review of the impact of the plant on public health and safety, a review of the environmental impact of the plant at its particular site, and a review of the antitrust implications of plant ownership. The review of plant safety and the environmental impact of the plant will be discussed in more detail below. The review of antitrust aspects, however, need not be treated further here since it is primarily a legal and economic issue.

Public Participation

With few exceptions, all documents submitted to the NRC as part of a licensing review and all NRC documents concerning that review are available to the public. Public document rooms are maintained at NRC headquarters in Washington, D.C., and at locations near each plant construction site. In addition, public notice of the receipt of a utility application and other milestones in the licensing process are routinely published in the Federal Register. The availability of these documents makes it possible for the public to monitor each of the elements of the NRC's review of a utility application. The primary forum for actual public participation in the licensing process, however, is in the public hearings before the Atomic Safety and Licensing Boards. To assure that each concern is fully ventilated, the hearing boards have in the past exercised wide latitude as to the admissibility of issues that are raised by the public. The result has been that the public hearing portion has, over the years, become an increasingly dominant and unpredictable part of the overall plant schedule. (The Atomic Safety and Licensing Boards are described further in Section 7.10-5.)

7.10-2   The Nuclear Regulatory Commission—Organization and Structure

The organization of the NRC is shown in Fig. 7.10-2. The commission itself consists of five members who are appointed by the president (and confirmed by the Senate) for nonconcurrent terms of 5 years.

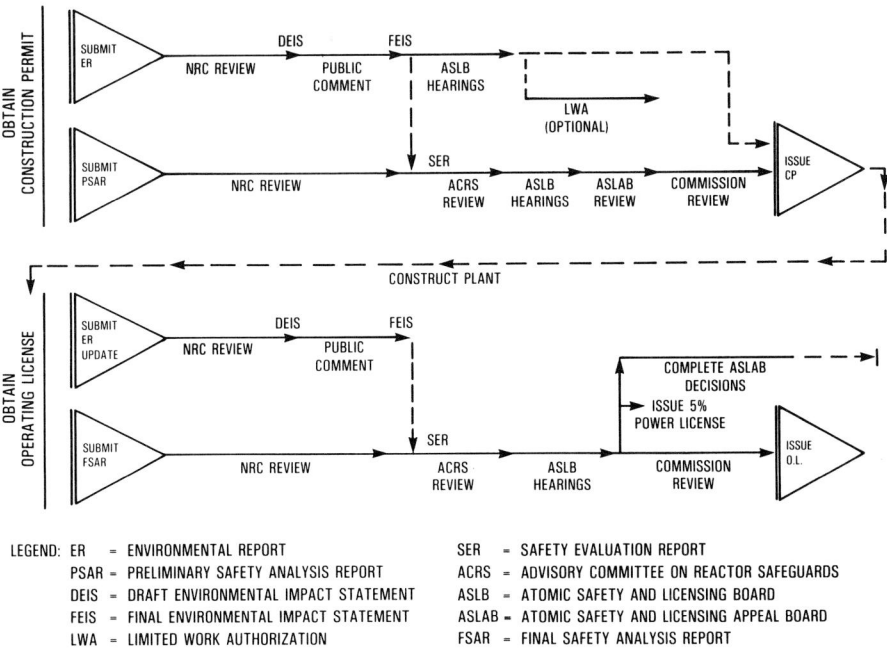

LEGEND: ER   = ENVIRONMENTAL REPORT   SER = SAFETY EVALUATION REPORT
        PSAR = PRELIMINARY SAFETY ANALYSIS REPORT   ACRS = ADVISORY COMMITTEE ON REACTOR SAFEGUARDS
        DEIS = DRAFT ENVIRONMENTAL IMPACT STATEMENT   ASLB = ATOMIC SAFETY AND LICENSING BOARD
        FEIS = FINAL ENVIRONMENTAL IMPACT STATEMENT   ASLAB = ATOMIC SAFETY AND LICENSING APPEAL BOARD
        LWA = LIMITED WORK AUTHORIZATION   FSAR = FINAL SAFETY ANALYSIS REPORT

Fig. 7.10-1 Major milestones in the licensing of a commercial nuclear power plant (not to scale).

The chairman of the NRC is appointed by the president and is the principal executive officer and the official spokesman of the commission. The commission, however, is a "collegial body" with each commissioner having one vote. Approval of any commission action requires the approval of a majority of the commissioners present.

The NRC staff numbers over 3000 people who are divided into two major categories: those reporting to the commission and those reporting solely to the chairman. The commission staff consists of Licensing and Appeal Board Panels and the offices of the General Counsel, Policy and Evaluation, and others. The chairman's staff consists of four major program offices, five regional offices, and a large support staff, all of which report through the Executive Director of Operations (EDO) who reports directly to the chairman. The EDO also coordinates the development of policy options for commission consideration. The functions of the more important offices within the NRC are outlined below.

Office of Nuclear Reactor Regulation

The Office of Nuclear Reactor Regulation licenses nuclear power, test, and research reactors. The Office of Nuclear Reactor Regulation reviews license applications to assure that each proposed facility can be built and operated without undue risk to the health and safety of the public and with minimal impact on the environment. The office licenses facility operators and continues to monitor operating reactor facilities throughout their lifetime up to and including decommissioning.

Office of Nuclear Material Safety and Safeguards

The Office of Nuclear Material Safety and Safeguards is responsible for the licensing and regulation of facilities and materials associated with the processing, transport, and handling of nuclear materials. This involves the disposal of nuclear waste as well as the regulation of uranium recovery facilities. The Office of Nuclear Material Safety and Safeguards reviews and assesses safeguards against potential threats, thefts, and sabotage for licensed facilities (including reactors). The office works with other NRC offices in coordinating safety and safeguards programs and in recommending research, standards, and policy options.

Nuclear Regulatory Commission

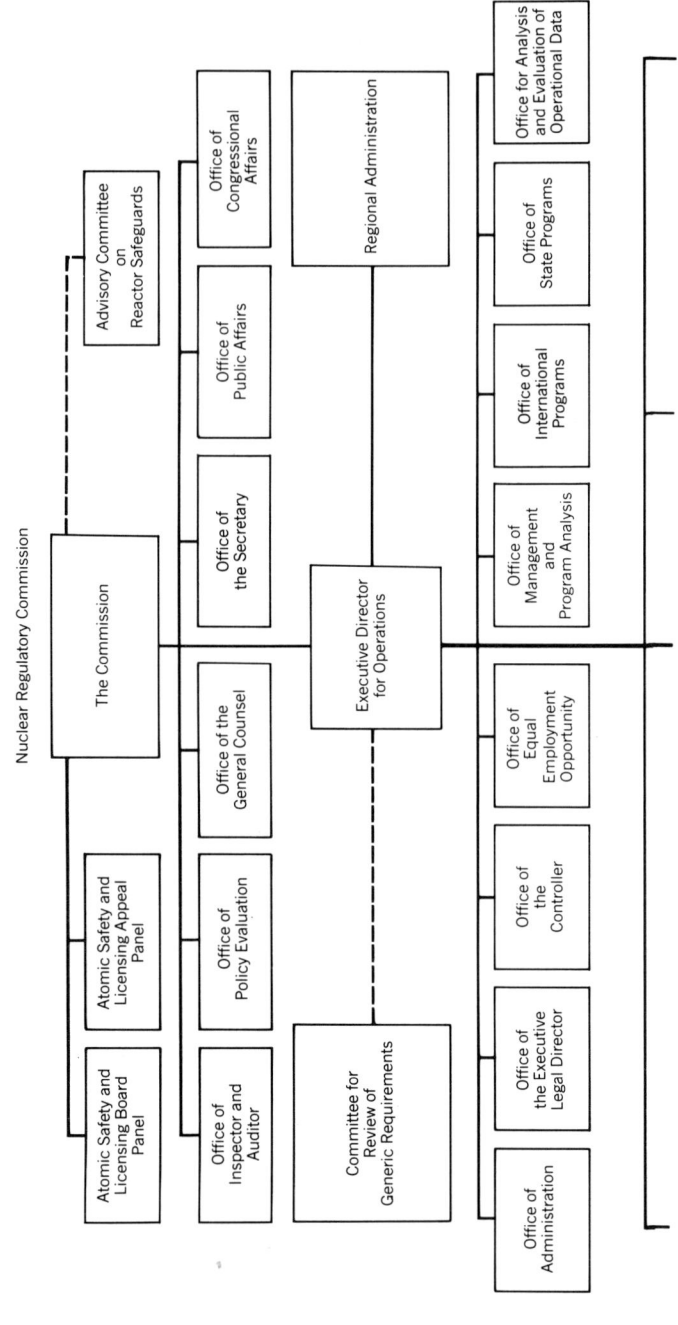

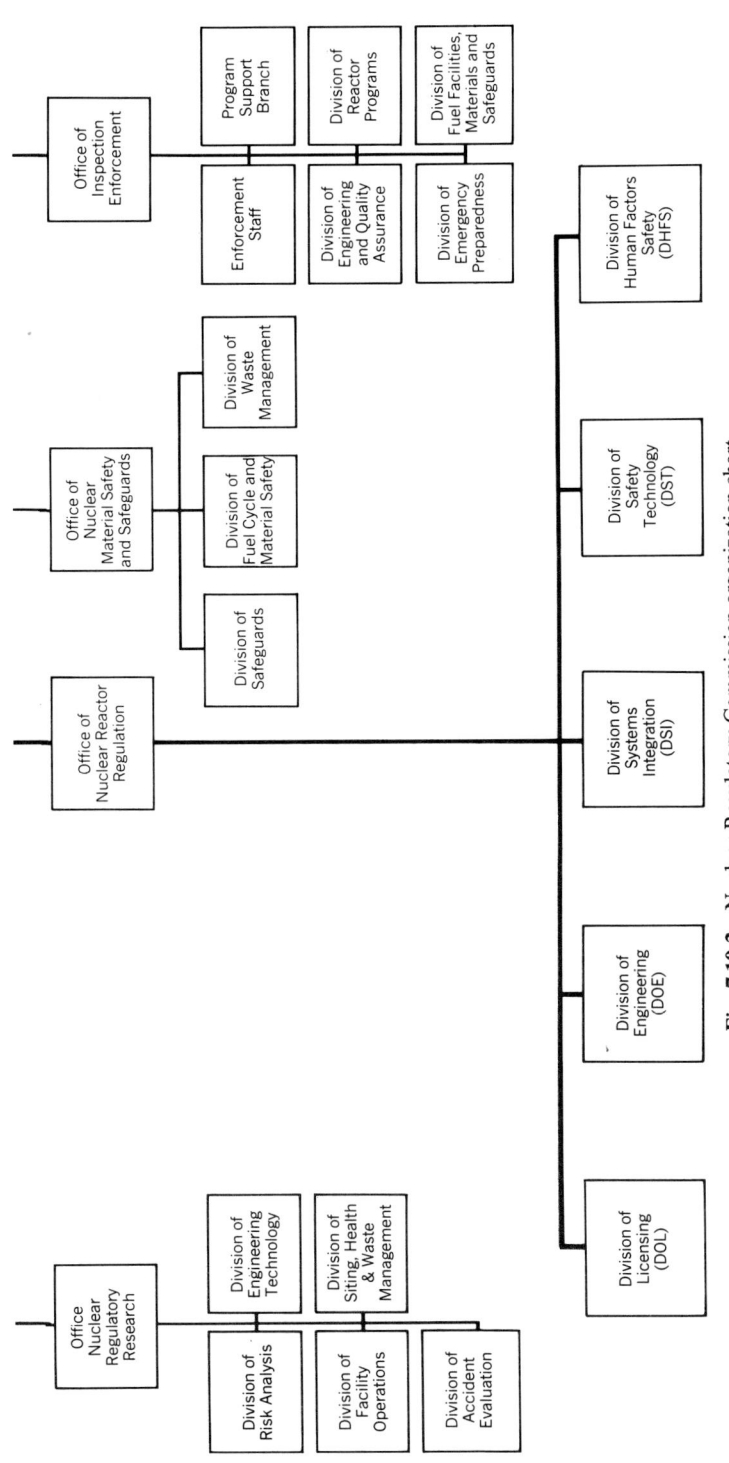

Fig. 7.10-2 Nuclear Regulatory Commission organization chart.

Office of Nuclear Regulatory Research

The Office of Nuclear Regulatory Research plans and conducts the comprehensive research and standards program deemed necessary for the performance of the commission's licensing and regulatory functions. The program covers areas such as facility operation, engineering technology, accident evaluation, probabilistic risk analysis, and siting, health, and waste management.

Office of Inspection and Enforcement

The Office of Inspection and Enforcement develops and monitors programs for the inspection of nuclear facilities and materials licensees to determine whether facilities are constructed, and operations are conducted, in compliance with license provisions and commission regulations. They also identify conditions that may adversely affect the protection of nuclear materials and facilities, the environment, or the health and safety of the public and provide a basis for recommending issuance or denial of licenses. The Office of Inspection and Enforcement enforces NRC regulations and license provisions as well as manages and directs all NRC actions related to emergency preparedness, including evaluation of state and local emergency plans performed by the Federal Emergency Management Agency. It also provides guidance to regional offices on program matters and appraises program performance in terms of effectiveness and uniformity.

Office of Investigation

The Office of Investigation (OI) is responsible for the investigation into alleged wrongdoing by non-NRC employees. OI investigators operate out of several field offices around the country.

Regional Offices

Regional offices execute established NRC policies and assigned programs relating to inspection, licensing, investigation, and enforcement within their regional boundaries. The NRC is in the process of shifting many of the licensing responsibilities of the Offices of Nuclear Reactor Regulation and Nuclear Materials Safety and Safeguards to the regional offices.

Office of General Counsel

The Office of General Counsel assists the commission in the review of Appeal Board decisions, petitions seeking direct commission relief, and rulemaking proceedings, and drafts legal documents necessary to carry out the commission's decisions. The General Counsel provides a legal analysis of proposed legislation affecting the commission's functions and assists in drafting legislation and preparing testimony. The General Counsel also represents the commission in court proceedings, frequently in conjunction with the Department of Justice.

Office of Executive Legal Director

The Office of Executive Legal Director provides legal advice and services to the Executive Director for Operations and staff, including representation in administrative proceedings involving the licensing of nuclear facilities and materials and the enforcement of license conditions and regulations; counseling with respect to safeguards matters, contracts, security, patents, administration, research, personnel; and the development of regulations to implement applicable Federal statutes.

Office of International Programs

The Office of International Programs plans and implements programs of international nuclear safety cooperation, creating and maintaining relationships with foreign regulatory agencies and international organizations; coordinates NRC export–import and international safeguards policies; issues export and import licenses; and coordinates responses by NRC to other agencies related to export–import actions and issues.

Office of State Programs

The Office of State Programs directs programs relating to regulatory relationships with state governments and organizations and interstate bodies, manages the NRC state agreements program, administers the indemnification program and performs financial qualification reviews of applicants and licensees. The Office of State Programs also verifies that applicants are not in violation of the antitrust laws.

Office for Analysis and Evaluation of Operational Data

This office (which was created in the aftermath of the accident at Three Mile Island) provides agency coordination for the collection, storage, and retrieval of operational data associated with licensed activities, analyzes and evaluates such operational experience, and feeds back the lessons of that experience to NRC licensing, standards, and inspection activities. The office oversees the overall effectiveness of the agencywide operational safety data program.

7.10-3   The Legal Framework of Nuclear Regulation

The NRC derives its authority over the licensing of domestic nuclear reactors from three congressional statutes: The Atomic Energy Act of 1954, The Energy Reorganization Act of 1974, and the National Environmental Policy Act.

The Atomic Energy Act of 1954 provides the fundamental legal framework for the regulation of atomic energy in the United States. Although it has been amended many times, the basic process has not changed significantly. The Atomic Energy Act contains broad descriptions and general authority for federal agencies in regulating nuclear power. Supreme Court rulings have continued to hold that it gives the primary federal agency (currently the NRC) wide latitude in implementing the general requirements of the act.

The Energy Reorganization Act of 1974 created the present Nuclear Regulatory Commission out of part of the original Atomic Energy Commission. Those functions of the original Atomic Energy Commission that dealt with promotion or development of nuclear power were transferred to other government agencies—most recently the Department of Energy. The NRC is clearly responsible for the regulation of nuclear power without any requirement that they ensure its promotion or development.

The National Environmental Policy Act (NEPA) is significant because it has been found to require the NRC and its predecessors to fully consider and document the potential environmental impact of the construction and operation of nuclear power plants. As a result of NEPA and, more specifically, the courts' interpretation of NEPA,* the NRC is required to consider not only potential radiological impacts, but nonradiological effects as well. As a result of a Circuit Court of Appeals decision in 1971 there was a virtual halt in the entire licensing process for well over a year. Since that time, however, the NRC has in general altered its review requirements to encompass NEPA reviews without major impact on licensing schedules.

These laws form the fundamental legal basis of nuclear power regulation. Because their requirements are stated broadly, however, they are seldom used by the designer or the constructor of a plant as a primary reference source. Instead, a substantial body of detailed federal regulations and other guidance documents have been developed and implemented over the years.

7.10-4   NRC Regulations and Guidance

Title 10, Chapter 1, of the Code of Federal Regulations (10CFR) contains the detailed regulatory requirements affecting nuclear-related materials and activities in the United States (Table 7.10-1). The basic requirements for licensing a commercial nuclear power plant in the United States are concentrated in Title 10, Part 50, of the Code of Federal Regulations (10CFR50). The Atomic Energy Act, the Energy Reorganization Act, and the National Environmental Protection Act are law. Title 10 of the Code of Federal Regulations can be viewed as the commission's interpretation of the law. It is considered binding on the NRC staff and on the applicant unless legally challenged. For example, a regulation could theoretically be challenged as inconsistent with the Atomic Energy Act. Such challenges, however, have seldom been ultimately successful.

The rules and regulations of the NRC are subject to revision through a process known as rulemaking (10CFR2.800). Rule changes may be initiated by the commission, NRC staff, Advisory Committee on Reactor Safeguards, or by members of the public via petitions. Generally, the NRC solicits public comments on rule change proposals and attempts to accommodate or rebut each comment prior to forwarding the final proposal to the commission. A majority vote of the commission is necessary to impose a rule change which, when effective, applies to all licensees unless specifically exempted.

The regulations in Title 10 vary from statements of general requirements to very detailed requirements concerning various design features or analytical techniques. Of particular interest to the designer of a nuclear power plant are the General Design Criteria contained in Appendix B of 10CFR50 and summarized here in Table 7.10-2. Title 10 of the Code of Federal Regulations, in particular Part 50, is the source that the designer or the constructor of a nuclear power plant, in fact,

---

*Calvert Cliffs Co-ordinating Committee v. United States Atomic Energy Commission, 449 F.2d 1109 (D.C. Cir. 1971).

TABLE 7.10-1   TITLE 10, CODE OF FEDERAL REGULATIONS, CHAPTER 1

Part	
0	Conduct of employees
1	Statement of organization and general information
2	Rules of practice for domestic licensing
4	Nondiscrimination in federally assisted commission programs
7	Advisory committees
8	Interpretations
9	Public records
10	Criteria and procedures for determining eligibility for access to Restricted Data/National Security Information
11	Criteria and procedures for determining eligibility for access or control over special nuclear material
14	Administrative claims under Federal Tort Claims Act
19	Notices, instructions, reports to workers; inspections
20	Standards for protection against radiation
21	Reporting defects, noncompliance
25	Access authorization, licensee personnel
30	Rules of general applicability to domestic licensing of by-product material
31	General domestic licenses
32	Specific domestic licenses manufacture/transfer items containing by-product material
33	Specific domestic licenses of broad scope
34	Licenses for radiography, radiation safety requirements for operations
35	Human uses of by-product material
40	Domestic licensing of source material
50	Domestic licensing of production and utilization facilities
51	Licensing regulatory policy and procedures for environmental protection
55	Operators' licenses
60	Disposal of high-level radioactive wastes in geologic repositories
70	Domestic licensing of special nuclear material
71	Packaging of radioactive material for transport under certain conditions
72	Licensing requirements for storage of spent fuel in ISFSI
73	Physical protection of plants and materials
75	Safeguard nuclear material
81	Standard specifications granting patent licenses
95	Security facility approval safeguarding national information and restricted data
100	Reactor site criteria
110	Export and import nuclear equipment, materials
140	Financial protection requirements, indemnity agreements,
150	Exemptions, continued regulatory authority in agreements, States and offshore waters under section 274
160	Trespassing on commission property
170	Fees for facilities and materials licenses
171–199	Reserved

TABLE 7.10-2 GENERAL DESIGN CRITERIA FOR NUCLEAR POWER PLANTS—10CFR50, APPENDIX A

Criteria	GDC Number	Criteria	GDC Number
I. Overall requirements		Fracture prevention of reactor coolant	
Quality standards and records	1	pressure boundary	31
Design bases for protection against natural		Inspection of reactor coolant pressure	32
phenomena	2	boundary reactor coolant makeup	33
Fire protection	3	Residual heat removal	34
Environmental and missile design bases	4	Emergency core cooling	35
Sharing of structures, systems,		Inspection of emergency core cooling system	36
and components	5	Testing of emergency core cooling system	37
		Containment heat removal	38
II. Protection by multiple-fission product		Inspection of containment heat removal system	39
barriers		Testing of containment heat removal system	40
Reactor design	10	Containment atmosphere cleanup	41
Reactor inherent protection	11	Inspection of containment atmosphere	
Suppression of reactor power oscillations	12	cleanup systems	42
Instrumentation and control	13	Testing of containment atmosphere	
Reactor coolant pressure boundary	14	cleanup systems	43
Reactor coolant system design	15	Cooling water	44
Containment design	16	Inspection of cooling water system	45
Electric power systems	17	Testing of cooling water system	46
Inspection and testing of electric			
power systems	18	V. Reactor containment	
Control room	19	Containment design basis	50
		Fracture prevention of containment pressure	
III. Protection and reactivity control systems		boundary	51
Protection system functions	20	Capability for containment leakage rate testing	52
Protection system reliability and testability	21	Provisions for containment testing and	
Protection system independence	22	inspection	53
Protection system failure modes	23	Systems penetrating containment	54
Separation of protection and control		Reactor coolant pressure boundary	
systems	24	penetrating containment	55
Protection system requirements for		Primary containment isolation	56
reactivity control malfunctions	25	Closed-system isolation valves	57
Reactivity control system redundancy			
and capability	26	VI. Fuel and radioactivity control	
Combined reactivity control systems	27	Control of releases of radioactive materials	
capability reactivity limits	28	to the environment	60
Protection against anticipated operational		Fuel storage and handling and radioactivity	
occurrences	29	control	61
		Prevention of criticality in fuel storage and	
IV. Fluid systems		handling	62
Quality of reactor coolant pressure		Monitoring fuel and waste storage	63
boundary	30	Monitoring radioactivity releases	64

uses to define the regulatory requirements for the design and licensing process. It is also one of the principal bases for enforcement of regulations by the NRC.

In addition to the legal requirements of federal legislation and the requirements of the Code of Federal Regulations, there is a significant body of more detailed guidance and interpretative information. Principal examples of these documents are regulatory guides and standard review plans. Figure 7.10-3 shows the relationship among these various documents.

There is a clear distinction between laws or regulations and the next level of documentation depicted in Fig. 7.10-3. Regulatory guides, standard review plans, and additional documentation are "guidance." They are considered by the NRC staff to be at least *one* acceptable way of meeting the NRC requirements embodied in Title 10 of the Code of Federal Regulations and can be thus viewed as an interpretation of the regulations by the NRC staff. Compliance is not required and alternatives should be acceptable if they can be shown to meet the basic federal regulation. In practice, however, approval of significant deviations from positions that have been taken by the NRC in Regulatory Guides and other similar documents are usually very difficult to obtain and result in more detailed and prolonged review by the NRC staff.

In addition to the regulatory guidance issued by the NRC, a comprehensive set of voluntary standards have been developed by the nuclear industry. Some of these standards, such as the

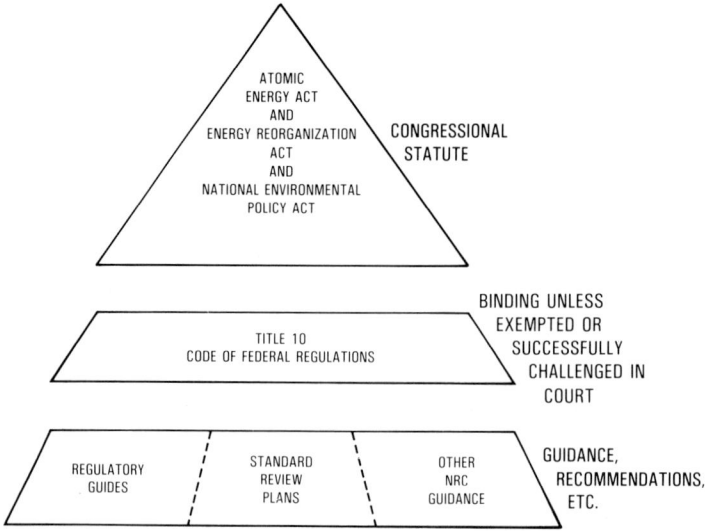

Fig. 7.10-3   Hierarchy of federal regulatory documents.

American Society of Mechanical Engineers Boiler and Pressure Vessel Codes, have achieved such wide acceptance that they are actually referenced in 10CFR50 as NRC requirements (e.g., in 10CFR50.55a). Other industry standards may be cited in Regulatory Guides as one acceptable method of meeting an NRC requirement (sometimes with additional requirements added by the NRC staff). A wide variety of Institute of Electrical and Electronics Engineers standards fall into this category. In general, voluntary compliance with recognized and accepted industry codes and standards promotes more rapid NRC review and acceptance.

7.10-5   Obtaining a Construction Permit

Determinations

For the commission to issue a construction permit to a utility applicant, the following determinations must be made (10CFR50.35 and 10CFR2.104)*:

1. The applicant has described the proposed design of the facility including the "principal architectural and engineering criteria" for the design and has "identified the major features or components incorporated therein for the protection of the health and safety of the public."
2. Technical information necessary to complete the safety analysis that is not available at the beginning of construction and that can reasonably be left for later consideration will be supplied in the Final Safety Analysis Report.
3. Safety features or components, if any, requiring research and development have been identified and a research and development program will be conducted to resolve any safety questions.
4. On the basis of determinations 1, 2, and 3 there is "reasonable assurance" that such safety questions will be satisfactorily resolved prior to the completion of construction and that "the proposed facility can be constructed and operated at the proposed location without undue risk to the health and safety of the public."
5. The applicant is "technically qualified to design and construct the proposed facility."
6. The issuance of a construction permit is "not inimical to the common defense and security or to the health and safety of the public."
7. The facility can be constructed in accordance with environmental protection policies in 10CFR51 and the requirements of the National Environmental Policy Act.
8. Appropriate action is determined with respect to conflicting factors contained in the record of the licensing proceeding.

*Parenthetical references to the appropriate regulations are provided for those readers who require a more precise statement.

In order to reach those conclusions, the NRC conducts a review of the license application as outlined below.

Application Documents

Once a utility has decided to construct a nuclear power reactor, it must submit an application for a construction permit to the NRC (10CFR50.30). Two primary documents submitted are the Preliminary Safety Analysis Report (PSAR) and the Environmental Report (ER). (As discussed in Section 7.10-1, there is additional information submitted on antitrust considerations, which is not discussed here.) The PSAR (10CFR50.34) describes the design bases and design features of the proposed plant. It also includes an extensive discussion of the predicted plant response to hypothetical transients and accidents, preliminary plans for the applicant's organization and training program, quality assurance plans, and other technical information. A current PSAR consists of approximately 20 volumes or more of written material. The ER (10CFR50.51), which may be 2 to 4 additional volumes, describes the basis for the site selection of the plant, the evaluation of alternatives to the proposed nuclear power plant, and the projected environmental impact of siting of that particular plant at that particular site.

These documents, together with other information required by the NRC staff, are first reviewed for completeness. (It should be noted that while in an ideal sense NRC licensing requirements could be satisfied by a description of appropriate design bases, methods, and acceptance criteria, it has become standard practice for the NRC to request a large amount of detailed, plant-specific design data.) If the NRC determines that sufficient information is presented to begin their detailed technical review, then the application is "docketed." The docket date of an application is important in that it later becomes the basis for determining the applicability of certain codes, standards, or regulatory requirements (10CFR50.55a).

Once the application documents have been docketed, the NRC staff begins their detailed technical review. This review has taken from as little as 6 months to significantly more than 2 years. There will likely be several hundred official questions from the NRC staff with responses from the applicant, and many additional discussions and meetings. One of the primary documents used internally by the NRC staff in their review of the PSAR is the Standard Review Plan (SRP). For each chapter of the PSAR there is a corresponding SRP chapter that describes the criteria that the NRC reviewer should use to evaluate the acceptability of the information provided in that chapter. As discussed in Section 7.10-4, the SRPs are not "requirements" in the legal sense; however, since they are relied upon by the NRC staff in their review, the designer and the applicant attempt to keep them in mind in preparing the licensing documents.

The result of the NRC staff's review is documented in the Safety Evaluation Report (SER). The SER describes the basis upon which the NRC concludes that a construction permit can (or cannot) be issued in conformance with the requirements of The Atomic Energy Act and the Code of Federal Regulations. Once the SER is issued, the recommendation that a construction permit should be issued (or denied) is referred to the Advisory Committee on Reactor Safeguards.

The review of the ER is conducted in parallel with the safety review. The NRC staff prepares a Draft Environmental Impact Statement based on their review of the ER. After receiving public comment, the staff produces a Final Environmental Impact Statement that documents their conclusion that the environmental impacts are acceptable. Although the ER requires a substantial commitment of resources by the utility and is reviewed in detail by the NRC, it is generally not the dominant factor in determining the overall length of the licensing review.

Advisory Committee on Reactor Safeguards (ACRS) Reviews

The Atomic Energy Act requires an independent review by the ACRS (10CFR50.58). The ACRS consists of 15 members with wide and varied technical backgrounds appointed by, and reporting to, the commission. Their review is not intended to be a detailed review, but rather an audit of the plant design, the site, and the applicant's qualifications. An ACRS subcommittee, aided by numerous consultants, may meet several times before referring the application to the full committee for final consideration. The ACRS is not limited in their selection of topics or area of interest but, in general, they select specific technical or management issues of interest. Upon completion of their review, the ACRS recommends to the commission, in writing, their opinion as to whether or not the construction permit should be granted together with any comments or conditions that they feel are appropriate.

Atomic Safety and Licensing Board (ASLB) Hearings

The hearings conducted by the ASLB provide the forum for public participation in the licensing process. A public hearing is required on each application for a construction permit (10CFR50.58).

An ASLB is established for each application, consisting of three members. One member is qualified in the conduct of administrative proceedings (usually with a legal background), while the remaining two members have technical or other appropriate qualifications (10CFR2.721). As soon as practicable after the application for a construction permit a notice of hearing is published in the Federal Register. Separate hearings may be held on antitrust issues, environmental issues, and radiological health and safety issues.

In response to the notice of hearing, any person "whose interest may be affected by a proceeding and who desires to participate" as a party in the hearing may file a petition to intervene. The petitioners must state their interests in the proceeding and the specific aspects in which they desire to intervene. In subsequent prehearing conferences the petitioner must specifically identify each issue they wish to challenge and the basis of their contentions (10CFR2.714).

After evaluating the petition to intervene, the ASLB may admit any petitioner as a full party to the hearing based on one or more admitted contentions. If, however, the ASLB determines that a petitioner's interest is limited to one or more issues, they may be allowed to participate only in those portions of the hearing. In addition to full or limited participation as a party to the hearing, the ASLB may permit other interested parties to make "limited appearances." Such parties may make oral or written statements on the issues at prehearing conferences or at the hearings themselves. They do not otherwise participate in the proceedings (10CFR2.715).

In addition to issues identified by the petitioners, the ASLB itself may put other matters into controversy, but only where they first determine that a serious safety, environmental, or common defense and security matter exists (10CRF2.760a). Such issues are said to be raised "*sua sponte.*" The proceedings before the ASLB are conducted under formal rules of practice (10CFR2). Although drafting of contentions, decisions on admissibility, and other such issues can take place over an extended period of time, the actual hearings are normally scheduled after the NRC staff has completed its review and issues the SER and Environmental Impact Statement (EIS). At the completion of hearings, the ASLB issues its findings and decisions.

Atomic Safety and Licensing Appeal Board (ASLAB) Appeals

Each decision of the ASLB is subject to automatic review by an Atomic Safety and Licensing Appeal Board. Decisions of the ASLB may be allowed to go into effect pending the completion of ASLAB review, or the effectiveness of the ASLB decisions can be "stayed" during the Appeal Board's review.

Issuance of the Construction Permit (CP)

Since the accident at Three Mile Island, ASLB decisions authorizing issuance of a construction permit do not take effect until both the ASLAB and the commission itself take further action. The ASLAB must review the record, rule on any motions to stay the ASLB's decision, and determine whether issuance of the construction permit may create novel safety or environmental issues or prejudice review of significant environmental or safety issues. Even upon a favorable finding by the ASLB and ASLAB, the commission must still finally approve the issuance of the CP (10CFR2.764). As of this writing no new CPs have been issued since the accident at TMI.

Limited Work Authorization

In addition to the process described above, the applicant may request a limited work authorization (LWA) in order to begin certain preliminary site construction activities (excluding safety-related structures) prior to receiving the full construction permit (10CFR50.10). As a condition of granting the LWA, the Final Environmental Impact Statement must have been issued and the ASLB must have made the applicable findings regarding site suitability. Additional construction activities may be pursued (such as the installation of structural foundations and certain safety-related structures) if specific additional authorization has separately been requested and granted.

7.10-6   Obtaining an Operating License

Having received the construction permit and substantially completed construction of the plant, the applicant must file an application for permission to operate the plant.

In order to issue the operating license, the commission must make the following findings (10CFR50.57 and 10CFR2.104):

1.   "Construction of the facility has been substantially completed, in conformity with the construction permit," the provisions of the Atomic Energy Act and the regulations of the commission.

2.   The "facility will operate in conformance with the application," the provisions of the Atomic Energy Act, and the regulations of the commission.

3. There is "reasonable assurance" that the activities authorized by the license can be conducted "without endangering the health and safety of the public."

4. The applicant is "technically qualified" to engage in the authorized activities.

5. The applicant has met the necessary financial protection and indemnity requirements (10CFR140).

6. "The issuance of the license will not be inimical to the common defense and security or to the health and safety of the public."

7. The impact on the environment, when evaluated against 10CFR51 requirements, is acceptable. (This finding is required for contested proceedings only.)

An additional requirement for plant operation is that the plant personnel who will actually operate the plant be individually licensed by the NRC to be reactor operators (10CFR55).

The process involved in obtaining an operating license is essentially the same as that involved in obtaining a construction permit. A Final Safety Analysis Report (FSAR) and an update to the environmental report must be submitted to the NRC. Certain information not available at the PSAR stage must be submitted in the FSAR (10CFR50.34). Of particular importance is "as built" information and a discussion of the utility's management organization and resources. (Since the accident at TMI, the NRC has directed increasing attention to assuring that the utility has obtained adequate numbers of qualified personnel and provided them with appropriate management authority.) An extended technical review of those documents, similar to the review of the PSAR and original ER, is conducted. At the conclusion of that review another SER and EIS is issued by the NRC staff, and additional ACRS meetings are held. Atomic Safety and Licensing Board hearings are not mandatory at this stage if no petition for intervention is received. Unlike the constraints at the CP stage, an ASLB decision authorizing the issuance of an operating license may become immediately effective insofar as it authorizes operation of the plant up to 5% power. Under current rules the commission itself must still approve the license for operations at greater than 5% power (10CFR2.764).

7.10-7 Standardization

As a supplement to the licensing process described above, the NRC also permits the limited use of standard designs. The purpose of employing standard designs is to simplify the licensing process, avoid duplicate reviews, and in general to make the design and construction process more efficient and economical.

NRC policy regarding standard designs is contained in a series of policy statements, culminating in the policy statement of August 31, 1978.* Under this policy, the NRC recognizes four standardization options.

Reference System

The reference system involves NRC preapproval of a standard design for a major portion of a nuclear power plant outside the context of a CP or OL application. Approval by the NRC is granted to a designer in the form of a preliminary design approval (PDA) or a final design approval (FDA), which are valid for a specified period of time. A utility applying for a CP may, in lieu of a detailed submittal, simply reference the standard design that has received the PDA or FDA. An applicant for an OL may reference an FDA in a similar manner for that portion of the plant design. The availability of an approved FDA would, therefore, permit what is in effect "one-step" licensing for the portion of the plant approved in the FDA, since both a CP and an OL could be approved if based on the same FDA.

Use of the reference system concept for standardization has been the most popular among vendors of the nuclear steam supply systems and balance of plants. Requirements for reference system approval applications are contained in 10CFR50, Appendix O.

Duplicate Plant

The duplicate plant concept involves a number of applications for construction and operation of nuclear power plants of essentially the same design to be located at different sites by one or more utility applicants. Requests for duplication must be made prior to tendering an application for a CP to the NRC. Several attempts have been made to use the duplicate plant option—some in conjunction with reference designs—with varying success to date. Requirements for application to construct duplicate plants are contained in 10CFR50, Appendix N.

---

*As of September 1982 the NRC was reviewing its position on standardization policy. Several proposed legislative changes were also being discussed.

Manufacturing License

The manufacturing license concept involves the submittal of an application for a number of identical plants that would be manufactured at one location and moved to a different location for operation. Only one application for a manufacturing license has been made to date. Requirements for a manufacturing license application are contained in 10CFR50, Appendix M.

Replicate Plant

The replicate plant concept involves the application to construct a nuclear power plant of essentially the same design as one recently reviewed and approved by the NRC. When an applicant proposes to replicate a previously approved plant, the NRC will perform a qualification review to determine the acceptability of the base plant for replication.

The replicate plant concept was intended by the NRC to be used for a relatively short time while the industry and the NRC fully implement the other standardization options; it has enjoyed mixed success to date. Requirements for an application to construct a replicate plant are contained in 10CFR50, Appendix O.

7.10-8   Regulation of Operating Plants

Operating plants continue to be subject to a wide variety of regulatory requirements. In response to new issues or to events at other operating plants, the NRC may request additional information, make recommendations, or even issue orders to the utility to implement changes. Some of the more significant regulatory areas for operating plants are discussed below.

Technical Specifications

As a part of the operating license, the NRC issues to the utility a set of technical specifications (10CFR50.36). This document defines certain operational restraints on the plant. For example, certain minimum safety equipment availabilities are established, certain instrument accuracies may be defined, allowable leakage limits may be established, and other such considerations. Each technical specification contains appropriate action statements that must be implemented if the specification is violated. These range from a specific time allowance in which the utility must fix the malfunctioning equipment to a requirement that the plant be shut down. NRC approval is required before any change may be made to the plant's technical specifications since many of the technical specifications are derived from the analyses done as part of the FSAR to assure safe response of the plant under abnormal or accident conditions. Violations of technical specifications must be reported to the NRC.

Changes to the Facility

Once the plant has received an operating license, proposed changes to the facility must be evaluated by the licensee under the provisions of 10CFR50.59. The holder of the license may make certain changes to the facility if changes to the technical specifications are not required or if an "unreviewed safety question" is not created. If a technical specification change is required or if an unreviewed safety question is created, then prior NRC approval is required before implementing the change. Summaries of evaluations pursuant to 10CFR50.59, even if negative, must be submitted to the NRC at least annually.

Refueling

Light water reactors are typically refueled at approximately 12- to 18-month intervals. Refuelings are considered to be changes to the facility, as described above, and are subject to the provisions of 10CFR50.59. In the event that no changes to the technical specifications are required as a result of the changes to the reactor core (or due to other planned changes in the plant during the refueling shutdown), and if no unreviewed safety question is created, then startup following refueling may proceed without prior NRC approval. If, as is more frequently the case, some technical specification changes are required or some additional testing or analysis must be conducted as a result of plant changes, then an amendment to the operating license must be sought from the NRC prior to return to operation. The NRC staff will conduct a technical review of the application for the amendment (similar to a review of the safety analysis report), issue a safety evaluation of the changes, and issue the amendment to the license.

TABLE 7.10-3   APPLICABILITY OF REPORTING REQUIREMENTS

Part 21	NRC licensee organizations (all)
	Nonlicensee organizations supplying nuclear components and services
	Individuals (director or responsible officer)
Construction deficiencies	Holder of construction permit
Reportable occurrences	Licensee of a production or utilization facility (Part 50)
Abnormal occurrences	NRC-regulated facilities and activities

Source.   Adapted from NUREG-0302, Rev. 1, October 1977.

Backfitting

Under the provisions of 10CFR50.109, the commission may, at any time after the issuance of the construction permit, require backfitting of components or structures but only if it finds that such action is required to provide "substantial additional protection" for "the public health and safety or the common defense and security."

As a practical matter most commission-generated changes to operating facilities have been imposed without resort to the provisions of 10CFR50.109. They usually occur following an extensive joint staff–industry evaluation of a safety concern that had arisen through operating experience or had been postulated by a member of the staff, licensee, the industry, or the public.

Public Hearings

On each occasion when an amendment to the existing operating license is requested or required, the NRC must determine if the amendment presents a "significant hazards consideration." If not, the commission may issue the amendment.* If it is determined that a significant hazards consideration is present, a notice of the proposed amendment will be published in the Federal Register, and any person whose interest may be affected is provided with an opportunity to request a public hearing on the amendment.

7.10-9   Reporting Requirements

Once the construction permit is issued for construction of a nuclear power plant, the applicant becomes subject to a variety of reporting requirements for any deficiency that may be discovered. During the construction phase of the plant the most common reporting requirement is defined by 10CFR50.55(e); once the plant has been granted an operating license, the reporting requirements included in the technical specifications apply (generally referred to as licensee event reports or LERs).

Another significant reporting regulation is 10CFR21, which is unique in several respects. Tables 7.10-3 and 7.10-4 summarize the applicability of the various reporting requirements and the conditions that must be reported. It should be noted that 10CFR21 is unique in that it applies not only to the holder of the license, but also to a large part of the nuclear supply chain. It also differs from other regulations in that it can result in a fine to individuals (i.e., "directors" or "responsible officers").

Those regulations that define these reporting requirements are treated quite seriously by the NRC and failure to observe them scrupulously may result in substantial penalties or fines. Because the regulations are sometimes quite difficult to interpret, a careful reading of the precise wording of the regulations is often necessary.

7.10-10   NRC Inspections

As discussed in Section 7.10-2, the activities of the regional offices and the NRC Office of Inspection and Enforcement continue after the plant has received its OL. The NRC has assigned at least one resident inspector at each operating facility to monitor the utility's operations and compliance to regulations. The inspectors report to the regional administrators. Other inspectors, the regional offices, and the Office of Inspection and Enforcement conduct both routine audits and special inspections

---

*As of this writing legal challenges are in progress on the NRC's method of issuing a license amendment based on a "no significant hazards" finding.

TABLE 7.10-4   SUBJECT(S) REQUIRING REPORTING

Part 21	Facility, activity, or basic component supplied that (1) fails to comply with Atomic Energy Act of 1954 or any applicable NRC requirement or (2) contains a defect that could create a substantial safety hazard.
Construction deficiencies	Breakdown in QA.
	Significant design deficiency.
	Significant construction deficiency or damage.
	Significant deviation from performance specifications.
Reportable occurrences	Operational events [9 classifications (Licensee Event Reports) identified in Regulatory Guide 1.16].
Abnormal occurrence	An unscheduled incident or event associated with any licensed facility or activity that the commission determines is significant from the standpoint of public health and safety.

Source.   Adapted from NUREG-0302, Rev. 1, October 1977.

throughout the operating life of a nuclear power plant. In addition, audits of some nonlicensees (such as major systems designers and component manufacturers) are conducted as part of the NRC-approved quality assurance programs.

7.10-11   The Economic Impact of Regulation

It has been estimated that the licensing process in the United States today adds several years to the design and construction schedules for nuclear power plants.* Considering the multibillion dollar investments necessary for plant construction and the effects of high interest rates, the financial impacts are considerable. More importantly, however, the current licensing process injects considerable uncertainty into the process. It is difficult to predict with any certainty when, or if, plant construction will be allowed. NRC regulations have also been likely to be revised (or reinterpreted) during the licensing and construction process, which may then lead to a cascade of changes.

   Furthermore, the costs of regulation do not cease once the plant has received its operating license. As noted in Section 7.10-8, operating plants continue to be bound by a wide variety of NRC rules and regulations. NRC-ordered plant shutdowns, affecting multiple plants and costing millions of dollars, are not unknown.

   In response to this perception of licensing instability, several positive changes have been occurring. Standardized designs (Section 7.10-7) have been developed to reduce (or eliminate) the imposition of new design requirements and in an effort to make the NRC review simpler and more predictable. Reforms of the laws and regulations that govern nuclear plant licensing have been proposed by Congress, by industry, and by the NRC itself (Section 7.10-12). In 1981 the NRC created the Committee to Review Generic Requirements in an effort to coordinate and control the proliferation of new NRC requirements. Even if only partly successful, such initiatives will substantially reduce the economic burden and uncertainty of federal regulation.

7.10-12   Licensing Reform

As noted above, the rules and regulations that govern the licensing of nuclear power plants in the United States are subject to continual change. The Atomic Energy Act gives the NRC great latitude in developing and enforcing a regulatory process adequate to protect the health and safety of the public. There has been a variety of proposed changes aimed at simplifying or streamlining the licensing process, either through new laws or by regulatory changes within the current statutory framework, which have been heralded as "licensing reform." Most of these proposals share two common themes: reform of the public hearing process and "one-step" licensing and a combined construction permit and operating license. The intent is clear: to stabilize the licensing process and insofar as possible to remove it from the critical path of nuclear plant design and construction without compromising the excellent overall safety record of commercial nuclear power plants.

---

*Licensing, Design, and Construction Problems: Priorities for Solutions, Atomic Industrial Forum, January, 1978.

CHAPTER **8**
PETROLEUM TECHNOLOGY

OWEN D. THOMAS

Owen Thomas & Associates, Inc.
Salt Lake City, Utah

HAROLD B. BOYD

The M. W. Kellogg Company
Houston, Texas

V. E. GRIMSHAW, ROBERT A. McELROY, JR.

Union Oil Company of California
Los Angeles, California

CYRUS CLARK, IVAN V. LA FAVE

CBI Industries, Inc.
Oak Brook, Illinois

CHARLES W. RUOFF, JOONG MIN SOH

ARCO Chemical Company
Newtown Square, Pennsylvania

CHAKRA J. SANTHANAM, R. PETER STICKLES

Arthur D. Little, Inc.
Cambridge, Massachusetts

RAVINDRA M. NADKARNI

Leach & Garner
Attleboro, Massachusetts

8.1   EXPLORATION AND DISCOVERY

Owen D. Thomas

8.1-1   Introduction

While oil was first discovered in the United States by drilling at Titusville, Pennsylvania, in 1859 by Colonel Edwin L. Drake, use of oil for medicinal purposes, boat caulking, lamps, and other uses dates back to biblical or earlier times.

Development of the internal combustion engine provided the principal market for petroleum and the economic incentive for the search for this valuable resource that fueled the economic expansion of the last 100 years. Logically the first wells were drilled near seeps where oil could be seen oozing from the earth. The origin of petroleum and the reasons why it accumulates in underground reservoirs in certain areas was poorly understood. There was little application of science to the search. As more wells were drilled, it was observed that gas was frequently encountered above the oil in some reservoirs and water was present beneath the oil. It was also observed that the rocks that make up the earth's crust are nearly always saturated with fluids, generally water, from the near surface water table downward. Early geologists gradually came to realize that oil was frequently found where the rock strata were bowed upward into domes or anticlines. The logic of these occurrences became apparent with the realization that in saturated reservoirs the most buoyant fluid, natural gas, would rise to the top, followed by oil with the ever-present water forming the underlying support. Once this was established, the search for petroleum shifted from drilling solely near seeps to drilling domes or anticlines in the earth's sedimentary strata. The presence of seeps, of course, was still considered a favorable sign because it indicated that petroleum had been generated in the earth in the vicinity.

Primarily, however, from the day that the anticlinal theory of oil accumulation was accepted until the present, the quest for oil has been focused on the search for domes or anticlines. Only recently has a full-fledged effort been made to search for non-anticlinal traps. Early explorers and geologists debated whether oil originated from igneous or sedimentary rock sources, but the solution of this problem did not concern the practical oil finder who concentrated his efforts on finding and mapping anticlinal structures. This was done originally by observing and recording the inclination of sedimentary layers or beds on a map. The strike, or bearing of a horizontal line in the plane of the bed, and the dip, or inclination of the bed perpendicular to strike, were plotted on maps. Ideally geologists preferred to record this information on a key bed or layer, which could be identified by its lithology or other characteristics over the area being mapped. Lacking key beds, observations on other beds were also recorded. When this information was plotted on maps, it frequently showed areas from which the beds dipped away in all directions, identifying a dome or closed anticline. Since these features generally have been eroded to some degree, it was observed that on an anticline older beds were found in the center of the structure and successively younger beds on its flanks. In areas where vegetation, soil, and debris have not concealed the outcrops, as in many of the Rocky Mountain states, a geologist could "walk the outcrop" of a particular bed completely around the structure to confirm that the beds were dipping away from the crest of the structure in all directions. Wells were drilled on the crests of such features. Some people confused the structure of the bed rock layers with the surface topography, which usually do not coincide inasmuch as these structures have generally undergone erosion over the millions of years since they were uplifted.

Surface geological studies still play an important part in the search for petroleum, augmented in more recent times by aerial photographs, stereoscopic pairs of aerial photos for better depth perception, and more recently space satellite or Landsat images of the earth.

8.1-2   The Seismograph

However, in the 1920s first thought was given in employing the seismograph, which had been used to record earthquake waves, in the search for petroleum. A series of miniature seismographs called geophones were connected by a cable along an accurately surveyed line. A charge of dynamite was detonated at a particular point adjacent to the line and the time required for the shock waves or seismic waves penetrating the earth to be "echoed" or reflected back to a recording center on the surface was precisely recorded on a strip of paper. This formed a wiggly line record of the various waves or echoes as they arrived at the recording station, revealing the depth of a particular sedimentary layer when an appropriate velocity of the waves was multiplied by the elapsed time from when the shot was fired until the wave or "echo" returned to the recording instrument. The reflection seismograph soon became the principal oil-finding tool of the industry and remains so to this day. It enabled the explorationist in effect to "see" the structure of the earth beneath the surface. By patiently assembling these data, together with what surface observations could be made, he was able to map the "buried" anticlines, which could not be seen on the surface, as well as to establish the configuration with depth of surface structures. This method has been perfected over the years to a fine tool, which has made it possible to map even very complex structures in places like the Rocky Mountain Overthrust Belt.

Recent improvements have been based on computer technology which facilitated improved processing of the seismic data to bring out not only details of the structure but also important stratigraphic information that has led to the discovery of many non-anticlinal types of oil and gas fields. The reflection seismic method was so successful and the rate of finding oil was so improved that, combined with the effects of the Great Depression of the 1930s, oil sold for as little as 10 cents per barrel. Oil companies established geophysical departments to work with their geological departments in locating prospects. This effort was so successful that most of the giant oil fields of the land areas of the lower 48 United States were discovered in the 1930–1960 era.

8.1-3 Rotary Drilling

At about the same time the reflection seismograph was being perfected as the most useful tool of the explorationist, the rotary drilling method was devised. Previously drilling had been done by the "churn" drill or cable tool type of rigs in which a bit was connected to a string of tools at the end of a cable. The cable raised and lowered the tools causing the bit to impact the bottom of the hole, chipping and cutting the formation. This method was relatively slow and not very well adapted to deep holes. The rotary method involved attaching a bit with cutting teeth on roller cones on the end of a drill pipe through which mud was circulated to pick up the cuttings and return them to the surface through a mud-filled hole. The bit could be rotated at optimum revolutions and with optimum weight, provided by a string of heavy drill collars above the bit, to cut the formation in the most efficient manner. The hydrostatic head of the column of mud in the hole prevented blowouts of the type encountered in the dry hole of the cable tool drilling method. When shows of oil were observed by the well site geologist, a formation test could be performed by running a test tool at the bottom of the drill string which effectively "packed off" the space between the drill string and the wall of the hole, relieving the formation of the hydrostatic head of the mud and allowing formation fluids to flow through a perforated pipe beneath the packer to the surface. The rotary drilling method is used to this day and it, together with the reflection seismic method, enabled the industry to find, drill, and produce oil-bearing structures at a rate that would have been impossible in earlier days.

8.1-4 Exploration Trends

The economic growth of the United States, its insatiable demands for petroleum, and the growth of petroleum technology caused the United States to be more thoroughly explored than any other part of the world. More than 75% of the world's oil and gas wells have been drilled in the United States.

It became obvious in the Gulf Coast region that fields found onshore continued on trend into the offshore. During the late 1930s the first offshore wells were drilled in pursuit of these objectives. The United States has led the world in development of offshore technology, including barge-mounted rigs, jackup platforms, drillships, and semi-submersible floating rigs. This effort has spread throughout the world. Wells have been drilled for petroleum in water as deep as 4800 ft and new depth records are certain to be recorded as prospects are found in deeper and deeper water, the price of oil increases to provide an economic incentive, and technology is devised to produce these wells.

Approximately 1 trillion barrels of oil have been found in the past by conventional drilling. The world has consumed about one-third of this amount, leaving about two-thirds of 1 trillion barrels as present recoverable reserves. These figures do not include oil from oil sands and oil shale and other nonconventional sources. The consensus of opinion of explorationists is that there probably is at least another trillion barrels to be found by drilling.

Geochemical studies indicate with certainty that oil is generated from the remains of animal and plant forms that died and were buried in a reducing environment in fine sediments on the bottom of the sea and sometimes in large lakes. This organic material is converted to petroleum when depth of burial is such (usually more than 5000 ft) that temperatures increased to a level where chemical changes convert the organic material to droplets of petroleum. Compaction of clays to shale expels the oil droplets and the formation water into porous reservoir rocks. The bouyancy of the oil and gas relative to water causes them to migrate updip until they are trapped in anticlines or other traps or escape to the surface. Past efforts have concentrated on the search for anticlines, but now with sophisticated seismic data processed in modern computer facilities, a much clearer picture of the stratigraphy can be obtained. These seismic sections show a great variety of non-anticlinal traps where oil is prevented from moving updip to surface outcrops and escaping to the atmosphere. A great deal of effort is now being devoted to the search for non-anticlinal traps that have been overlooked by past efforts. These traps plus the search for anticlines in the relatively unexplored areas of the Arctic, deep sea along the margins of the continents, and geologic complexes such as the Rocky Mountain Overthrust Belt, which could not be satisfactorily mapped until recently, constitute the future exploration provinces of the petroleum industry. How many areas are explored and how much oil is found in the future depends upon economic incentives.

BIBLIOGRAPHY

Halbouty, Michel T. and Moody, John D. *Exploration and Demand: Tenth World Petroleum Congress*, Vol. 2, 1979, pp. 291–301. The American Association of Petroleum Geologists, Tulsa, Oklahoma.

Owen, Edgar Wesley, Trek of the Oil Finders: *A History of Exploration for Petroleum:* The American Association of Petroleum Geologists, Memoir 6, 1975. Tulsa, Oklahoma.

8.2   CRUDE OIL PROCESSING

Harold B. Boyd

8.2-1   Characteristics and Properties of Crude Oil

Petroleum or crude oil consists of a wide range of hydrocarbon mixtures, along with lesser amounts of other components containing sulfur, nitrogen, oxygen, vanadium, and nickel. In naturally occurring crude oil, there are three basic hydrocarbon types present—paraffins, naphthenes (cycloparaffins), and aromatics. Heavy oils can be upgraded to produce a synthetic crude oil that resembles a naturally occurring crude oil. Depending on the processing employed, synthetic crude oils can also contain olefinic hydrocarbons.

The *paraffin* series of hydrocarbons have the general formula $C_n H_{n+2}$. Paraffins are characterized by carbon atoms being connected with a single bond and other bonds saturated with hydrogen atoms. Compounds with the same number of carbon and hydrogen atoms but differing structure are called isomers. These paraffins include straight-chain or normal paraffins and also branched chains or isoparaffins. Isomers, while having the same chemical composition, have different physical properties.

The *olefins* series of hydrocarbons are normally not found in naturally occurring crude oil. These compounds are formed in numerous refining processes. Olefinic hydrocarbons are similar to the paraffin series except that two of the carbon atoms are connected by double bonds. Olefins have the general formula $C_n H_{2n}$.

Naphthenic (cycloparaffinic) compounds are a saturated ring compound with all carbon atoms being connected with a single bond. Naphthenes have the general formula $C_n H_{2n}$.

The *aromatic* series of compounds all contain a benzene ring. The benzene ring consists of a six-carbon ring (hexagonal) in which carbon atoms are joined together alternately with single and double bonds. The benzene ring is very symmetrical and in general a very stable compound.

Examples of typical compounds and their physical properties from the various classes, all with six carbon atoms, are shown in Table 8.2-1.

Since chemical composition cannot be used to effectively characterize the wide variations in the properties of petroleum, various physical and chemical tests have been adopted by the industry to characterize crude oil and petroleum fractions. Some of the more common tests are described below.

ASTM *Distillation*

ASTM Distillation D-86 is a nonfractionating-type atmospheric distillation for use with gasoline, kerosene, and other light hydrocarbon fractions. This method should only be used up to about 500°F since cracking may occur at higher temperatures. Distillations above 500°F are performed at reduced pressure, 400 mm Hg total pressure and corrected to atmospheric pressure.

True Boiling Point Distillation

A true boiling point (TBP) distillation is performed in an apparatus with fractionation. The distillation curve (temperature vs. percent distilled) obtained from a TBP distillation is not materially altered by a greater degree of fractionation.

API *Gravity*

The density, or specific gravity, of a crude oil or petroleum fraction is expressed in terms of "degrees API." Specific gravity is related to °API by the following formula:

$$°API = \frac{141.5}{sg} - 131.5$$

where sg is the specific gravity at 60°F. Therefore, a crude oil of 25°API is heavier than a crude oil of 35°API.

TABLE 8.2-1   EXAMPLES OF C₆ HYDROCARBONS

Hydrocarbon Class	Name	Formula	Structure	Specific Gravity (60°F/60°F)	Boiling Point (°F)
Paraffin	*n*-Hexane	C_6H_{14}		0.664	155.7
	Isohexane	C_6H_{14}		0.658	140.5
Olefin	1-Hexene	C_6H_{12}		0.678	146.3
Naphthene	Cyclohexane	C_6H_{12}		0.783	177.3
	Methyl cyclopentane	C_6H_{12}		0.754	161.3
Aromatic	Benzene	C_6H_6		0.885	176.2

Characterization Factor

The Watson or UOP characterization factor (K) indicates the aromaticity or paraffinicity of a petroleum stock.

$$K = (\text{Tmabp})^{1/3}/\text{sg}$$

where Tmabp is the mean average boiling point in °R (460 + °F) and sg is the specific gravity at 60°F. The higher the K factor value, the more paraffinic a petroleum stock; conversely lower values indicate a greater amount of aromaticity. Practically, the Watson characterization factor only varies over a narrow range, about 10–14. Crude oil with a K of 10.5 would be aromatic while a value of 12.75 would indicate a paraffinic stock.

Bureau of Mines Correlation Index

Another correlation that indicates the aromaticity or paraffinicity of a petroleum fraction is the Bureau of Mines Correlation Index, or BMCI. The BMCI is also based on the boiling point and specific gravity of the petroleum fraction.

$$\text{BMCI} = \frac{87,552}{\text{Tmabp}} + 473.7(\text{sg}) - 456.8$$

The BMCI scale is based on n-paraffin (i.e., n-hexane) having a value of 0 and benzene a value of 100.

Carbon Residue

The carbon residue test is a measure of the carbon-forming tendency of a petroleum stock. There are two tests in general use in the industry—Conradson Carbon Residue Test ASTM-D-189-52 and Ramsbottom Carbon Residue Test ASTM D-524-52T. Both tests refer to heating a sample of oil, driving off the volatile matter, and continuing to heat until a dry hard coke deposit remains. Carbon residue is expressed as the weight percent of residual deposit compared to the sample of oil.

Pour Point

The pour point of a petroleum oil is the lowest temperature at which the oil will "pour" or flow when it is chilled without disturbance under prescribed conditions (ASTM D-97-47). Pour point is a rough indication of paraffinicity and aromaticity of petroleum stock. The lower the pour point, the lower the paraffin content of the material.

Properties of crude oils and their various fractions are available from the producer or marketer and the Bureau of Mines. An example of the properties of several crude oils is given in Table 8.2-2.

8.2-2   Refinery Processing Units

There are many processing units that are used in any given refinery. The type and sequence of processing units depend on crude type, product distribution, product specifications, overall plant economics, flexibility to handle heavier crudes, and so on. Generally, the processing units employed in

TABLE 8.2-2   PROPERTIES OF CRUDE OILS

Properties	Arabian Heavy (Saudi Arabia)	Bachequero (Venezuela)	Minas (Indonesia)	North Slope (United States)	Romashkinskaya (USSR)	Taching (China)
Gravity, °API	28.2	16.8	35.2	26.8	32.4	33.0
TBP Distillation, °F						
10%	190	380	290	250	195	
50%	695	855	730	690	605	
70%	—	—	920	900	—	
90%	—	—	—	—	—	
Vol. % at 1000°F	70.5	63.5	—	78.5	—	
Sulfur, wt. %	2.84	2.4	0.09	1.04	1.61	0.04
K Factor	11.9	11.5	12.4	11.7	11.9	12.5
Carbon residue, wt. %	7.8	10.5	3.0	5.0	4.5	2.7
Pour point, °F	−30	−10	90	−5	−20	30
Viscosity						
SUS at 100°F	93	1,360		83	50	138
SUS at 122°F			64[a]			

[a] Too waxy to measure at 100°F.

a refinery can be categorized as follows:

Separation processes.
Product upgrading.
Conversion processes.
Resid processing or bottom-of-barrel upgrading.

This section discusses the processing objectives of the major refinery units. For additional processes and more detailed information on the described processes, the licensors of the respective technologies should be contacted.

Crude Oil Desalting

Crude oil, as it is received at a refinery, may contain appreciable quantities of inorganic salts. The salts, if not removed from the crude oil, may cause plugging of heat exchangers, coking of process furnaces, and corrosion problems in downstream processing units. The inorganic salts are removed by either chemical or electrostatic desalting techniques. The treated crude oil from a desalter will contain 5 lb or less of salts per thousand barrels of oil.

Crude Oil Distillation

Raw crude oil or treated crude oil from a desalting unit is separated into various petroleum cuts, usually identified as a boiling range fraction. The distillation step may be a single fractionating column or multiple columns. The first fractionating column is operated at atmospheric pressure and is termed an atmospheric distillation unit.

The bottoms of the atmospheric column, reduced crude, may then be processed in a second fractionating column that is operated under vacuum conditions. This subatmospheric pressure unit is called a vacuum distillation unit. Vacuum conditions are required to maximize recovery of gas oils by vaporization without cracking the residue. The distillate product from this subatmospheric operation is vacuum (or heavy) gas oil and the bottom product is termed vacuum bottoms or residue.

The normal boiling range fractions recovered from crude oil distillation and associated boiling ranges are typically as follows:

Petroleum Fraction	Boiling Range, °F
Gases	Less than 80
Light naphtha	80–200
Heavy naphtha	200–400
Kerosene	350–500
Diesel	400–650
Light gas oil	450–675
Atmospheric bottoms	650 and heavier
Vacuum gas oil	650–1050
Vacuum residue	1050 and heavier

Alkylation

Alkylation processes combine isobutane with an olefin (propylene, butylene, or pentene) in the presence of either sulfuric or hydrofluoric acid to produce alkylate. The alkylate product is a high-quality blending component for gasoline production. As an example, when isobutane is alkylated with butylene, the major component in the alkylate product is isooctane (2-2,4-trimethyl pentane). Isooctane has an octane rating of 100.

The chemical reaction can be described as follows:

Isobutane	Butylene	Isooctane
C_4H_{10}	C_4H_8	C_8H_{18}

While alkylation can be an attractive process, its use in a refinery scheme is usually limited by the availability of isobutane.

Dimersol

The dimersol process catalytically dimerizes light olefins such as propylene and *n*-butylene, producing either a light, high-octane gasoline called Dimate or a mixture of C_6–C_8 olefins suitable for use in plasticizer synthesis. The propylene dimerization is illustrated as follows:

Propylene	Propylene	Hexene
C_3H_6	C_3H_6	C_6H_{12}

Dimersol has become an especially useful process because of the increasing demand for high-octane, no-lead gasoline components.

Isomerization

Isomerization technology has two basic applications in refinery operations, namely, conversion of n-butane to isobutane for alkylation or upgrading of C_5/C_6 feedstock streams.

The hydrocarbon feedstocks are combined with hydrogen and passed over a catalyst in a fixed-bed reactor. Typical operating conditions vary from 300–550°F and 300–1000 psig depending on the proprietary process and application. Yields are about 98–100% on feedstocks.

Isomerization is used as a means of upgrading light streams for gasoline blending since iso- or branched paraffins have a higher octane rating than the corresponding normal paraffins. For example:

Compound	Octane Rating (Research Method)
n-pentane	62
i-pentane	92
n-hexane	25
i-hexane	73

Catalytic Reforming

Catalytic reforming is a process that is used to upgrade low-octane naphthas to produce a high-octane reformate. Feedstocks to reforming units can consist of any low-quality naphthas such as straight-run material, hydrocracker naphthas, hydrotreated cracked, and coker naphthas. These feedstocks are usually in the C_6–400°F boiling range. The high-quality product reformate can be used as a component for gasoline blending or as a source of aromatics such as benzene, toluene, or xylenes.

The modern reforming processes use a bimetallic catalyst, usually platinum and another promoter such as rhenium. The use of bimetallic catalysts allows operation at lower pressures (100–150 psig) and higher severities without reducing cycle time than the older-type catalysts. Typical operating conditions range from about 900–1000°F and 100–250 psig depending on catalyst, feedstocks, and severity of operation.

The major chemical reactions that are involved in reforming are:

1. Isomerization of naphthenes:

Methyl Cyclopentane	Cyclohexane
C_6H_{12}	C_6H_{12}

2. Dehydrogenation to aromatics

$$
\underset{\substack{\text{Cyclohexane}\\ C_6H_{12}}}{
\begin{array}{c}
H_2 \\
C \\
H_2C \qquad CH_2 \\
| \qquad\qquad | \\
H_2C \qquad CH_2 \\
C \\
H_2
\end{array}}
\longrightarrow
\underset{\substack{\text{Benzene + Hydrogen}\\ C_6H_6 + 3H_2}}{
\begin{array}{c}
H \\
C \\
HC \qquad CH \\
\| \qquad\qquad | \\
HC \qquad CH \\
C \\
H
\end{array}}
+ 3H_2
$$

3. Cyclization of paraffins

$$
\underset{\substack{n\text{-Hexane}\\ C_6H_{14}}}{
\begin{array}{c}
H \;\; H \;\; H \;\; H \;\; H \;\; H \\
| \;\; | \;\; | \;\; | \;\; | \;\; | \\
H-C-C-C-C-C-C-H \\
| \;\; | \;\; | \;\; | \;\; | \;\; | \\
H \;\; H \;\; H \;\; H \;\; H \;\; H
\end{array}}
\longrightarrow
\underset{\substack{\text{Benzene}\\ C_6H_6}}{
\begin{array}{c}
H \\
C \\
HC \qquad CH \\
\| \qquad\qquad | \\
HC \qquad CH \\
C \\
H
\end{array}}
\quad \text{Hydrogen} + 4H_2
$$

Most feedstocks contain sulfur, arsenic, and nitrogen compounds; a hydrotreating step is usually used to remove these compounds before reforming. This is a necessary step since reforming catalyst will be poisoned by these compounds.

Reforming reactions take place in a reducing or hydrogen atmosphere. Therefore, a portion of the hydrogen evolved in the reforming reaction is recycled to the reactors. The remaining by-product hydrogen is available as a source of hydrogen in other refinery hydrogen-consuming processes such as hydrotreating, desulfurization, hydrocracking, and other petrochemical units.

Distillate Hydrotreating

Distillate hydrotreating is used to upgrade the quality of petroleum fractions to meet desired product specifications or as a feed preparation step. Hydrotreating removes sulfur and nitrogen compounds and can saturate olefins.

The feedstock to be treated is combined with hydrogen and heated to reactor conditions. The reaction takes place in the presence of a catalyst, usually cobalt-molybdenum or cobalt-nickel, in a fixed-bed reactor. The process conditions and hydrogen consumption used vary considerably with feedstock, type of operation, and catalyst. However, pressures in the range of 300–500 psig and temperature of about 700°F are used. Since hydrotreating is to upgrade the quality of a feedstock, not conversion, liquid yields are usually 99% or greater. Sulfur removal of 90% is common.

Fluid Catalytic Cracking

The most efficient conversion process in a modern refinery is fluid catalytic cracking. A fluid catalytic cracking unit (FCCU) selectively converts heavy gas oils into gasoline and lighter components. Earlier, heavy gas oil feedstocks were thermally cracked to produce gasoline, but today the catalytic process has replaced the thermal process.

Today's zeolite cracking catalysts provide superior characteristics to the older catalysts. Zeolite catalysts are (1) more stable, (2) able to withstand higher temperatures, and (3) less sensitive to metals poisoning.

With the high-activity zeolite cracking catalyst, the cracking reactions take place in a riser. The hydrocarbons are disengaged from the catalyst by means of cyclones. The hydrocarbons are recovered in the vapor recovery unit. In the cracking reactions coke is deposited on the catalyst. The coke is recovered by combustion with air in the regenerator. The hot catalyst is then returned to the riser reactor, thus supplying the heat required for the cracking reactions. Typical FCCU reactor temperatures are about 870–950°F, and regenerator temperatures are in the range of 1100–1250°F.

Typical FCCU yields when cracking a 650°F plus boiling range light Mideast gas oil are shown below:

	Volume (%)
Conversion	81.0
Yields	
C_2 & lighter (wt. %)	4.4
Propane/propylene	10.7
Butane/butylene	18.4
C_5–400°F EP gasoline	64.1
Light-cycle oil	14.0
Decant oil	5.0

Hydrocracking

The various hydrocracking processes combine the features of both hydrogenation and cracking. Hydrocracking is a relatively new advancement of refinery technology, with the process being introduced about 20 years ago. The major chemical reactions that take place in the hydrocracker reactor are hydrogenation, cracking, and isomerization.

Typically, hydrocrackers are operated at 600–2000 psig and 550–800°F. Hydrocracking is a very flexible process as to feedstock and product distribution. Hydrocrackers can efficiently process both virgin gas oil feedstocks and cracked material from fluid catalytic cracking and coking units. By adjusting the severity of operation, the product distribution can be varied from maximum gasoline to maximum middle distillate operation.

Since the reactions take place in a hydrogen atmosphere, contaminants, that is, sulfur and nitrogen, are minimized in the hydrocracked products. However, the gasoline fraction is of relatively low-octane value and usually requires treatment in a catalytic reforming unit before blending into the gasoline pool.

Operating costs are high for hydrocrackers due to the hydrogen consumed and the high pressure required. Hydrogen requirements are typically met by the use of reformer hydrogen and augmented by hydrogen produced in a steam reforming unit.

Coking

Coking is a severe thermal cracking process. Residual feedstocks are processed to produce coke, gas oil, gasoline, and gas. The coke produced, while a salable end product, is sometimes considered a by-product. The quantity of coke produced is strongly a function of the Conradson carbon content of the feedstock and to a lesser extent, the process variables of temperature and pressure.

The coke produced can be used as a fuel, electrode manufacture, metallurgical coke, and chemical production. The end use or value of petroleum coke is dependent on its properties, including volatile material, metals, and sulfur content.

There are several variations of the coking process, namely, delayed coking, fluid coking, and flexicoking.

1. **Delayed coking.** The residual feedstock is heated to about 900–950°F in a fixed tubular heater. Heater design is important to limit the extent of cracking in the tubes. The heater effluent then enters the coke drum where cracking and polymerization take place at long residence time at about 20–60 psig. Delayed coking units consist of at least two coke drums. One coke drum is in coking operation while the other is being decoked by use of high-pressure water jets.
2. **Fluid coking.** Residual feedstock is injected into a reactor drum where it is thermally cracked at about 950°F. Vapor products are quenched and recovered in a conventional manner. Coke is heated to yield light hydrocarbons and the fluid coke product.
3. **Flexicoking.** The flexicoking process integrates conventional fluid coking with coke gasification. The fluid coke produced is gasified at high temperatures in the presence of air and steam to produce a low-Btu gas.

Visbreaking

Visbreaking is a mild thermal cracking process in which atmospheric or vacuum resids are processed to reduce their viscosity. Conversion is often measured by the yield of gasoline. Visbreaking reactions are usually considered first order. A given conversion can be achieved at either high temperature and short reaction time or lower temperature and higher reaction time. Two types of furnace designs have evolved. One is a coil-only design that operates at a maximum temperature of about 900°F at low reaction time. The other design heats the feedstock to about 850°F and soaks in a reaction chamber at longer reaction time, achieving the same conversion. Maximum conversion of a given feedstock is limited by the stability of the fuel oil produced.

In refineries where residual material is blended with light distillate cutter stocks to produce residual fuel, the reduction of the residual portion requires less cutter stock to meet a given residual fuel oil viscosity. The total amount of residual fuel oil is also reduced, as seen in the following example:

	Vacuum Residue Blending	Visbreaking
Vacuum resid	100	(100)
Visbreaker resid	—	91
Distillate cutter stock	45	18
Total residual fuel oil	145	109

Visbreaking is a relatively low capital cost process to upgrade residual feedstocks. Visbreaking is also finding application in the synthetic fuels industry to process tar sands bitumen.

Resid HDS

With the need for refiners to handle heavier and higher-sulfur content crude oils, resid hydrodesulfurization (resid HDS) has become an important refining process.

The main reactions are hydrogenation and removal of sulfur, nitrogen, oxygenated compounds, and metals and the saturation of heavy hydrocarbons. Hydrodesulfurization reactions take place in a multireactor, fixed-bed system at about 2000–3000 psi and temperatures in the range of 750°F. Resid HDS can achieve conversion to lighter material of 20% with 70–90% sulfur removal and 90–95% removal of metals, depending on severity of operation. Hydrogen consumption is relatively high, typically in the order of 800–1200 scf/barrel.

Due to the pressure on refiners to better utilize the entire crude, resid HDS is usually used as a feed pretreatment step to other processing units such as delayed coking, fluid catalytic cracking, and hydrocracking. Considerable research is underway in the area of catalyst development to improve the economics of handling high-sulfur and metals resids.

Heavy Oil Cracking

The cracking of atmospheric and vacuum resid in a catalytic system is termed heavy oil cracking. This process is usually practical in a fluidized mode. These feedstocks have higher Conradson carbon value than conventional FCCU feedstocks and are higher in metals content. The higher Conradson carbon results in significantly higher coke yield. Excess heat evolved in the regenerator is used to produce high-pressure steam.

The higher metals content, especially vanadium and nickel, adversely affect the activity and selectivity of the zeolite catalysts. These effects can be minimized by the use of specially formulated catalysts and the use of metal passivators.

Hydrotreating the atmospheric residue improves the cracking characteristics of the feedstock. The following example shows typical yields when cracking a virgin and hydrotreated light Arabian atmospheric residue:

	Virgin	Hydrotreated*
Light Arabian Residue	← wt. % →	
Conversion	80.8	83.8
Yields		
C_5–400°F EP gasoline	43.6	51.2
Butane/butylene	10.6	12.4
Propane/propylene	5.4	6.4
Light-cycle oil	13.1	11.5
Heavy oil	6.1	4.7
Coke	16.1	10.2
C_2 and lighter	5.1	3.6

*Hydrotreated to 0.3 wt. % S.

Solvent Deasphalting

Vacuum residue is extracted with a light hydrocarbon solvent, propane to hexane, to produce a "deasphalted oil" (DAO) and a residual product, asphalt. Solvent type and solvent-to-feed ratio are selected to achieve the maximum yield of oil with a given quality (metals and carbon residues). For a given DAO yield, the lighter solvent is more selective (less metals in DAO product). Increasing the solvent to feed ratio when using a given solvent also increases selectivity.

Solvent deasphalting has several applications in a refinery: (1) to increase the yield of heavy gas oil, (2) to produce high-quality lube oil stock, and (3) to produce asphalt. The gas oil extracted from a deasphalting unit provides a good feedstock for further processing in a fluid catalytic cracking or hydrocracking unit.

H-Oil

H-oil is a catalytic process for hydrogenation of vacuum residue to produce upgraded distillate products. Liquid feed and hydrogen are fed to an ebullating bed reactor containing small catalyst particles, usually cobalt or nickel–molybdenum. The reactor operates as essentially isothermal conditions. Conversion of vacuum residue of 90–95% to lighter products is achievable.

This process is flexible as to feedstock characteristics and has been extended to shale oil, bitumen from tar sands, coal tars, and coal extracts. Metal and sulfur content of the feedstock is not a limitation with this process.

Recent economic and political pressures have been placed on the refining industry to minimize the consumption of crude oil. Considerable effort has been made to increase the selectivity of established technology and in the area of energy conservation. New process concepts have also been developed to meet these challenges. The following processes are typical of the types of developments that are ready to be commercialized.

ART

Engelhard has recently developed the asphalt residual treatment (ART) process. This process was developed to remove contaminants, carbon, and metals from heavy crudes and residual oils. In the ART process, ARTCAT* fluidizable contact material is employed to treat the heavy hydrocarbon feed. In the contactor the hydrocarbon components are selectively vaporized and the unvaporized contaminants, that is, coke precursors, metals, nitrogen, and sulfur, are deposited on the contacting medium. The deposited carbonaceous material is burned off in the burner section, providing the heat required in the contactor.

The ART process is applicable to (1) the treatment of reduced crudes prior to hydrotreating, fluid catalytic cracking, or hydrocracking and (2) the upgrading of heavy crude oils at their source.

The ART concept has been tested in both bench- and pilot-plant-scale equipment, in a 200-barrels per day (bpd) demonstration unit, and in a 10,000-bpd commercial unit over a wide range of feedstocks. The M.W. Kellogg Company is the exclusive worldwide licensing agent for this technology.

CANMET

The CANMET hydrocracking process was originally developed by the Department of Energy, Mines and Resources, Government of Canada. CANMET is a resid or heavy crude hydrocracking process that utilizes a unique additive to suppress coke deposition. This allows for high conversion of heavy materials to lighter distillates, while concentrating almost all metals contained in the feed into a pitch. The additive used is coal impregnated with an iron compound. CANMET technology is applicable to treating heavy crudes to produce a lighter syncrude or to provide feedstocks for further upgrading. A wide range of feedstocks, including resids, heavy crude oils, and tar sands bitumen, is possible.

This technology was developed in a pilot-plant-scale unit. A 5000-bpd commercial demonstration unit is planned for operation in 1984. PetroCanada is responsible for the commercialization of the CANMET process, with Lavalin providing further engineering development and marketing.

MTBE Production

Methyl tertiary butyl ether (MTBE) has become an important high-octane gasoline blending stock due to legislation limiting the lead content of gasoline. MTBE is produced by catalytically reacting

*ARTCAT is a trademark of Engelhard Corporation.

isobutylene with methanol according to the following equation:

$$\underset{\substack{\text{Isobutylene}\\C_4H_8}}{\overset{H}{\underset{H}{\diagdown}}C=\overset{CH_3}{\underset{|}{C}}-\overset{H}{\underset{|}{C}}-H} + \underset{\substack{\text{Methanol}\\CH_3OH}}{H-\overset{H}{\underset{|}{C}}-O-H} \rightarrow \underset{\substack{\text{MTBE}\\C_5H_{12}O}}{H-\overset{H}{\underset{H}{\overset{|}{C}}}-\overset{CH_3}{\underset{CH_3}{\overset{|}{C}}}-O-\overset{H}{\underset{H}{\overset{|}{C}}}-H}$$

Isobutylene is derived from FCC C_4's or dehydrogenation of isobutane. Typical conversion of isobutylene to MTBE is about 96% with selectivity to MTBE of 99%. Typical properties of MTBE are as follows:

Chemical Composition	
MTBE, wt. %	98.6
Isobutane, wt. %	Nil
Other C_4's, wt. %	0.02
Methanol, ppm	< 5
Tert-butyl alcohol, wt. %	1.0
Isobutene Dimer, wt. %	0.2
Water, ppm	< 5

Properties	
Density, lb/gal	6.18
Boiling point, °F	131
Freezing point, °F	−164
Blending octane numbers	
Research	121
Motor	105

8.2-3 Petroleum Products Specifications

The properties of the major classes of petroleum products (gasoline, kerosene, etc.) are, for the most part, governed by state and federal specifications. These specifications "define" the product and ensure that it will perform adequately in its intended use.

Incorporated into these specifications are a large number of specialized tests that are used primarily by the petroleum industry. The more important tests are listed below.

Flash Point. Flash point is the lowest temperature at which the vapor from an oil sample will ignite in the presence of a flame. Flash points are specified for jet fuels and heavier products to ensure that they do not contain any unexpected light components that could explode without special handling precautions.

Distillation. Distillation is a measure of the amount of oil vaporized as a function of increasing temperature. This test provides a practical means of characterizing the overall composition of the oil.

Sulfur Content. Sulfur content is the total amount of elemental and compounded sulfur present in an oil sample on a weight basis. Sulfur content must be limited in final products to ensure that the fuel will burn cleanly and not produce pollutant sulfur by-products.

Copper Strip Test. The copper strip test is a qualitative measurement of the degree of corrosion exhibited by a polished strip of copper immersed for 3 h in an oil sample maintained at 122°F (50°C). This test provides a relative indication of the corrosivity of the oil.

Reid Vapor Pressure. Reid vapor pressure is the vapor pressure of an oil sample measured at 100°F (38°C) in a bomb having a 4 : 1 ratio of air to liquid. This measurement is crucial in gasoline products to determine whether or not gasoline may cause "vapor lock" (premature vaporization between fuel tank and carburetor) in internal combustion engines.

TABLE 8.2-3 TYPICAL PETROLEUM PRODUCTS SPECIFICATIONS[a]

Properties	LPG		Gasoline (Motor Unleaded)	Jet Fuel	Kerosene	Diesel	Residual Fuel Oil
	Commercial Propane	Commercial Butane					
Gravity (°API)	(sg = 0.509)	(sg = 0.582)		45 (Min) 54 (Max)			25 (Max)
Flash point (°F)				110 (Min) 150 (Max)	100 (Min)	150 (Min)	150 (Min)
Distillation							
1BP							
10%			158 (Max)				
50%			257 (Max)				
90%			365 (Max)	370 (Max)		675 (Max)	
95%	−37 (Max)	36 (Max)		470 (Max)		725 (Max)	
EP			410 (Max)		540 (Max)		
Sulfur (wt. %)	[b]	[b]	0.1 (Max)	0.2 (Max)	0.15 (Max)	1.0 (Max)	3.0 (Max)
Corrosion, copper strip (no.)	1 (Max)	1 (Max)	1 (Max)	1 (Max)	1 (Max)	1 (Max)	
Reid vapor pressure (psig)			8.5 (Max)				
Viscosity (CS)				15 (Max) at −30°F		2.0 (Min) at 100°F	
						5.5 (Min) at 100°F	65 (Max) at 100°F
Octane no. (research clear)			90 (Min)				
Cetane no.						40 (Min)	
Heat of Combustion (Btu/lb)				18,400 (Min)	18,400 (Min)	18,400 (Min)	18,300 (Min)
Gum (mg/100 mL)			4 (Max)	7 (Max)			
Ash (wt. %)						0.02 (Max)	0.1 (Max)

[a]From refs. 1 and 2.
[b]Maximum sulfur = 15 g/100 ft³ of gas.

Viscosity.   Viscosity is a measurement of a fluid's resistance to flow. Viscosity is a crucial heating oil characteristic since it indicates the rate at which the oil will flow through the fuel system and the ease with which it can be atomized.

Octane Number.   Octane number is the percentage of isooctane in the particular blend of isooctane (Octane No. = 100) and *n*-heptane (Octane No. = 0) that produces the same knock intensity as a sample fuel in a standard laboratory engine. This number is the prime indicator of the ignition quality of a gasoline.

Cetane Number.   Cetane number is similar to octane number but using cetane (100) and alpha methyl naphthalene (0) for the reference blend. Used to indicate ignition quality of diesels.

Heat of Combustion.   Heat of combustion is the amount of heat produced when a fuel is burned completely. Particularly important in jet fuels where maximum power must be derived from a minimum amount of fuel.

Gum Content.   Gum content is the amount of insoluble deposits formed during the deterioration of petroleum products. Gum content must be minimized in combustion fuels to prevent excessive depositing on engine surfaces.

Ash Content.   Ash content is the amount of residue remaining after combustion of an oil. Important for heavy fuel oils since ash formation causes corrosion and fouling on boiler tubes.
   Specifications for six major classes of petroleum products are listed in Table 8.2-3 to illustrate how the above tests are used in actual practice.

8.2-4   Typical Refinery Configuration

The optimum refinery configuration is a function of crude to be processed, products to be produced or maximized, operating costs, capital requirements or limitations, crude and product values, and desired flexibility. Several refinery configurations are presented in this section to illustrate complexity. Figure 8.2-1 presents a simple topping refinery where the only processing units employed are those required to upgrade the various petroleum fractions to specification grade products.
   A more complex refinery scheme is shown in Fig. 8.2-2. This scheme utilizes a fluid catalytic cracking unit to convert vacuum gas oil to gasoline and an alkylation unit to increase the quantity and quality of the gasoline product. A visbreaking unit is employed to reduce the quantity of cutter stock required to meet heavy fuel oil viscosity specifications.
   Figure 8.2-3 presents a "modern" refining scheme to process heavier crudes while maximizing gasoline production. In this scheme the reduced crude is processed in a resid hydrotreating unit to reduce sulfur and metals content while converting a portion to kerosene and diesel blend stocks. The

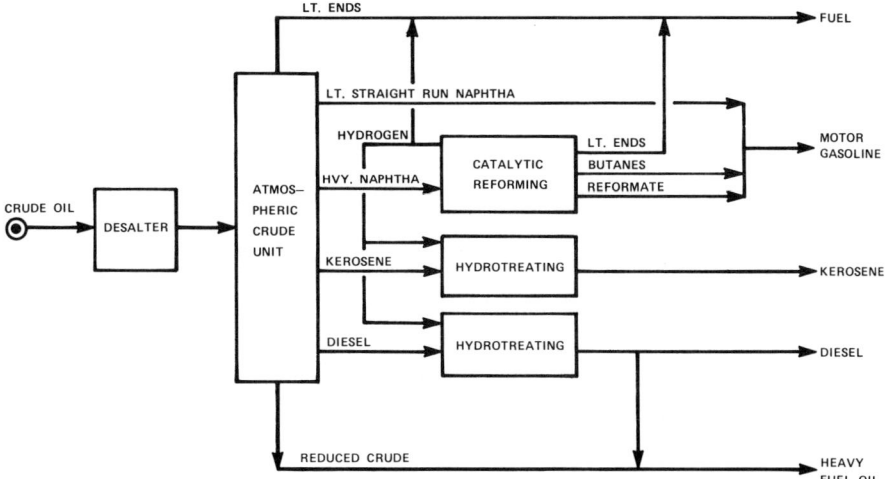

Fig. 8.2-1   Schematic flow diagram of simple topping refinery.

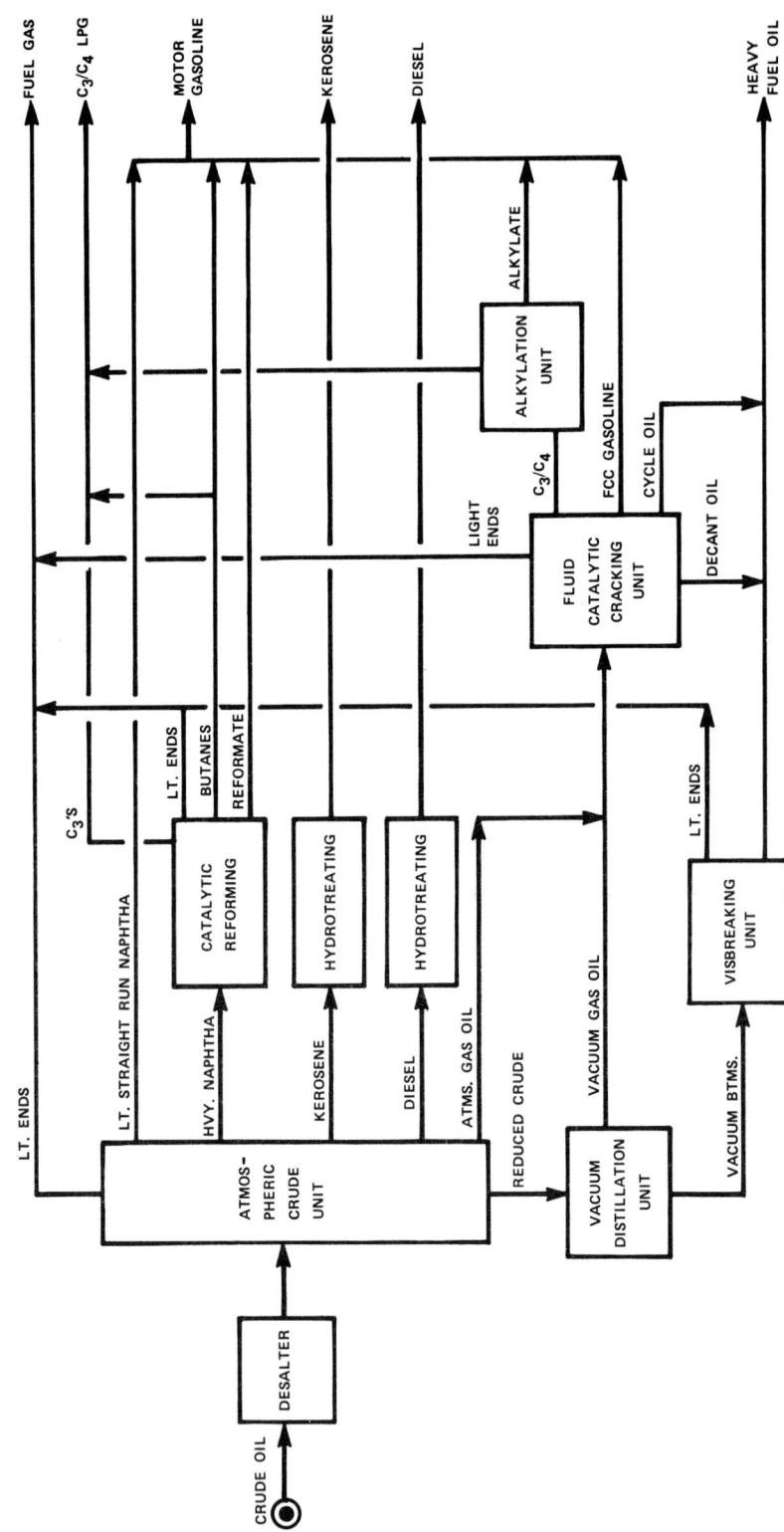

Fig. 8.2-2 Schematic flow diagram of typical conversion refinery (maximum gasoline production).

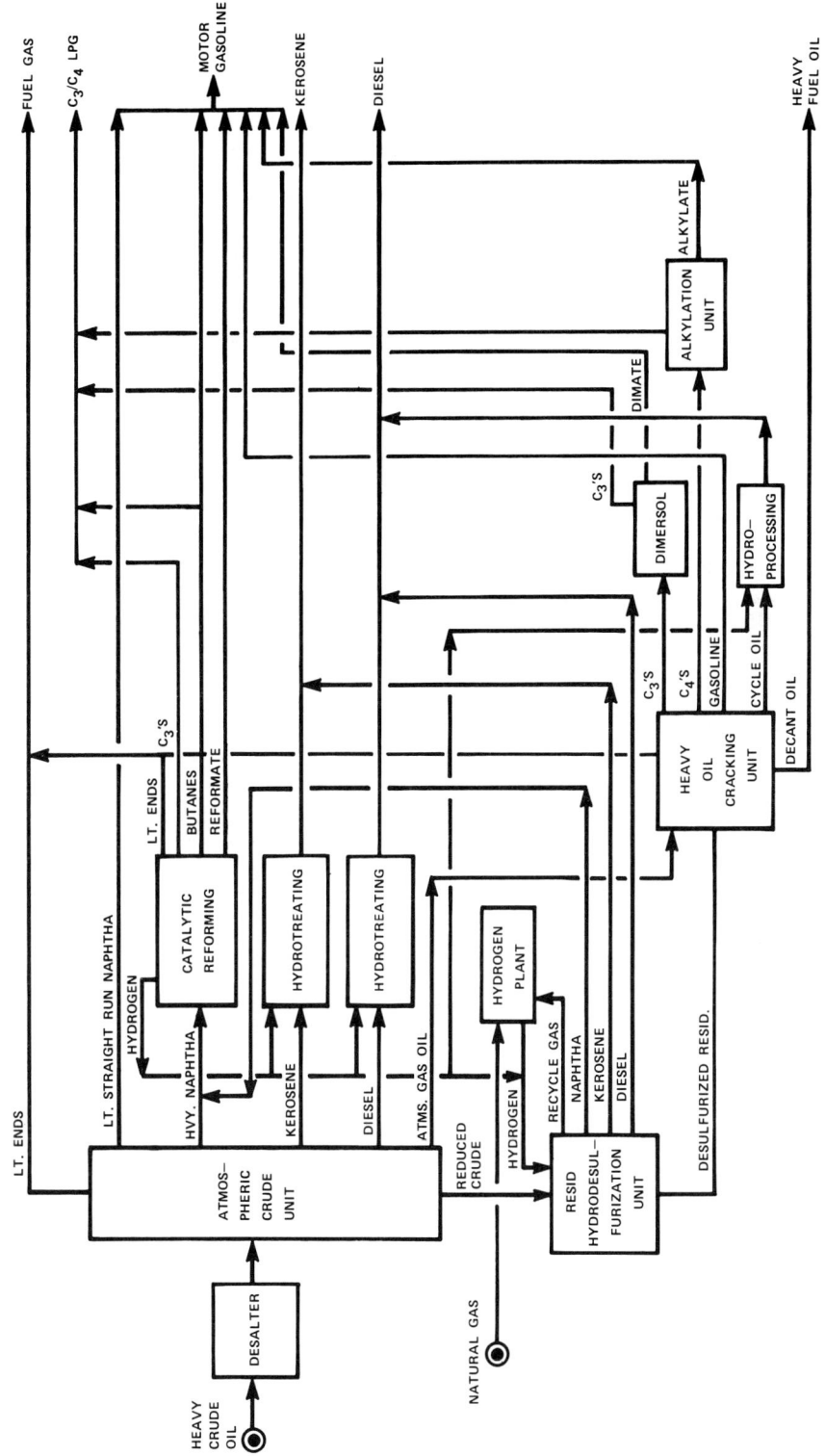

Fig. 8.2-3 Schematic flow diagram of modern refinery processing heavy crudes (maximum gasoline operation).

1033

desulfurized resid is then processed in a heavy oil cracking (HOC) unit to maximize gasoline production. Propylene produced in the HOC unit is converted to a gasoline blending stock in the dimersol unit. Butylenes and isobutane are reacted in an alkylation unit to produce alkylate for gasoline blending. The resid processing scheme shown was selected as the design basis for the Saber Energy refinery at Corpus Christi, Texas.

REFERENCES

7.2-1 W. F. Bland and R. L. Davidson, *Petroleum Processing Handbook*. McGraw-Hill, New York, 1967.
7.2-2 W. L. Nelson, *Petroleum Refinery Engineering*, 4th ed. McGraw-Hill, New York, 1969.

BIBLIOGRAPHY

ASTM Standards on Petroleum Products and Lubricants. American Society for Testing Materials, Philadelphia, 1958.

Bartholic, D. B. and Haseltine, R. P. "New Process Cleans Up FCC Feed." *Oil and Gas Journal*, June 1, 1981, p. 89.

Benedek, W. J. and Mauleon, J. L. "How First Dimersol is Working," *Hydrocarbon Processing*, May, 1980, pp. 143–149.

Boldt, K. and Hall, B. R. *Significance of Tests for Petroleum Products*. American Society for Testing and Materials, Philadelphia, 1977.

Buether, H., et al. "Recent Developments in the Technology of Residue Processing." *Chemical Engineering Progress Symposium Series*, No. 57, 1961, pp. 20–28.

Chambers, L. W., et al. "The Canmet Residuum Hydrocracking Process: An Update." Paper presented at the Second International Conference on Heavy Crude and Tar Sands, Caracas, Venezuela, February 1982.

Gary, J. H. and Handwerk, G. E. *Petroleum Refining*. Marcel Dekker, New York, 1975.

A Guide to World Export Crudes. The Petroleum Publishing Company, Tulsa, 1976.

Guthrie, V. B. *Petroleum Products Handbook*. McGraw-Hill, New York, 1960.

Hatch, L. F. "A Chemical View of Refining." *Hydrocarbon Processing*, February 1969, pp. 78–88.

Kirk-Othmer, *Encyclopedia of Chemical Technology*, Vol. 17. Wiley, New York, 1982.

"1980 Refining Process Handbook." *Hydrocarbon Processing*, September 1980, pp. 93–220.

Speight, J. G. *The Chemistry and Technology of Petroleum*. Marcel Dekker, New York, 1980.

Waddams, A. L. *Chemicals from Petroleum*, 4th ed. Gulf Publishing, Houston, 1980.

Whittington, E. L., et al. "Refinery Decisions for the 80's." *Chemical Engineering Progress*, February 1981, pp. 45–50.

8.3 TRANSPORTATION OF PETROLEUM

V. E. Grimshaw and R. A. McElroy, Jr.

8.3-1 Modes of Transportation

Crude oil and refined petroleum products are shipped by one or more modes of transportation, depending on several economic and technical factors. Domestically produced crude moves to refineries predominately by pipelines. Crude from foreign sources moves by ocean-going tankers to coastal ports, and some domestic crude moves short distances by barge, rail car, or truck; but virtually all of it moves by pipeline at some point between the producing well and the refinery.

Refined petroleum products are less pipeline dependent. The location of the refining center and its proximity to the ultimate market dictates the optimum modes of transportation for refined products.

Economic Considerations

Except for very large crude tankers (VLCCs), pipelines are typically the most economical transportation mode when significant volumes are to be moved over long periods of time. Pipelines require large capital investments and, once installed, are not readily moved and not usually adaptable for other uses. More flexible modes (barge, truck, rail car) are more suitable for low-volume movements and when a long-term need cannot be assured. The relative cost of various modes of transportation can be approximated using Table 8.3-1.

Table 8.3-1 indicates that the longer the haul and the higher the volume to be moved, the more attractive it becomes to make the large investments required for limited-use facilities such as large

TABLE 8.3-1 RELATIVE TRANSPORTATION COSTS

	Units per Barrel	
	Short Haul, Low Volume	Long Haul, High Volume
Pipelines	4–14	1–2
Tankers	—	2–4
Large tankers (VLCCs)	—	0.5–1
Rail car	11–13	6–8
Barges	6–8	3–5
Tank trucks	6–10	8–12

Source. Data derived from relative cost information contained in ref. 1.

tankers and pipelines. More versatile facilities are preferred to short-haul, low-volume situations. Pipelines may be preferable in some short-haul, low-volume situations, but generally such investments are not attractive.

Determining the most cost-effective mode or modes of transportation requires an assessment of (1) the daily average and peak volumes to be moved, (2) the time period over which the volumes are reasonably assured, (3) the capital cost of all facilities required, (4) the operating expenses associated with each alternative, and (5) the risks associated with the above assumptions.

Example 8.3-1. Determine the most economical transportation mode for 2000 barrels per day (bpd) of gasoline from a refinery to a distribution terminal 10 mi away if volume is assured for 5 years; annual capital and operating costs of a pipeline system are $500,000 and $50,000, respectively; and the trucking alternative would cost $0.50 per barrel.

Solution: Using rate of return (ROR) analysis*, the pipeline investment would generate an 18% annual ROR compared to the trucking alternative (differing methods of calculating ROR and variations in tax treatment will alter the results). An 18% ROR is usually adequate to encourage investment, so in this instance the pipeline appears to be the best transportation mode. Under the given conditions a prudent investor would elect to build the pipeline.

In the above example, if the volume can be assured for only 3 years, the ROR would be substantially less. Potential risks such as uncertainty of an assured volume over an adequate time period may make the investment unattractive. Also the trucking alternative would require approximately 12 truck-trailer loads daily over the proposed route. In some locales this may be undesirable, or even prohibited. Thus, while trucking may be an economic alternative, it may not be a realistic option. Environmental impacts need to be recognized when evaluating alternatives.

Symbols and Units

Symbols and units used in this section are shown in Table 8.3-2. The units used are those common to the U.S. petroleum industry. Conversion factors from these customary units to SI units are provided in Table 8.3-3.

8.3-2 Pipelines

Pipelines handle nearly 50% of the crude and refined products moved in the United States. Most U.S. pipelines operate as common carriers, transporting for all shippers of petroleum. Such lines are regulated by state or federal agencies. Some lines, however, operate strictly in proprietary service of the owners. Pipelines are divided into three general categories: crude gathering systems, crude trunk systems, and refined products systems.

Crude Gathering Systems

As crude enters a gathering system it is measured and sampled. The usual method is by transfer through a lease automatic custody transfer (LACT) unit where the crude is automatically metered and

---

*There are many different approaches to determining rate of return, one of which is detailed in ref. 2. This simplified example ignores the impact of inflation and the residual value of the investment.

TABLE 8.3-2   SYMBOLS

bph	Barrels per hour
C	Roughness factor
D	Outside diameter (in.)
d	Inside diameter (in.)
DWT	Deadweight tonnage
f	Friction factor
gpm	Gallons per minute
h_f	Frictional head loss (ft.)
hp	Horsepower
h_t	Total head loss (ft.)
L	Length of line (ft.)
P	Pounds per square inch (psi)
Q	Flow rate (bph)
R	Reynolds number
S	Fiber stress (psi)
shp	Shaft horsepower
SUS	Saybolt universal seconds
t	Pipe wall thickness (in.)
V	Flow velocity (ft./s)
v	Kinematic viscosity (cSt)
W	Liquid density (lb/ft^3)

TABLE 8.3-3   CONVERSION TO SI UNITS

	To Get SI Units	Multiply:	By:
Length	Millimeter, mm	Inches	2.5400E + 01
Area	Square meter, m^2	Square feet	9.2900E − 02
Volume	Cubic meter, m^3	Gallon	3.7854E − 03
		Barrel	1.5899E − 01
Mass	Kilogram, kg	Pound	0.4536
	Megagram, Mg	Long ton	1.0160
Pressure	Kilopascals, kPa	lb/in.^{2}	6.8948
Capacity	m^3/s	gal/min (gpm)	6.3090E − 05
		bbl/h (bph)	9.0129E − 05
Viscosity	m^2/s	ft^2/s	9.2903E − 02
Temperature	Kelvin, K	°F	$(t + 459.67)/1.8$
Energy	joule, J	Btu	1.0551E + 03
Power	Watt, W	Horsepower (hp)	7.457E + 02

sampled as it is pumped into the system. As the oil wells produce into field tanks the LACT, responding automatically to tank level, pumps the oil through a calibrated meter. A meter ticket is printed showing the measured volume, usually corrected to the standard temperature of 60°F. A sample is collected by means of a proportioning sampler, which draws a small increment of the passing stream. The aggregate sample, from 1 to 5 gal collected over a period of a few days is stored in a sealed container. Periodically, the meter ticket is changed, and the sample withdrawn and tested for quality. The quantity and value of the shipment is based on the meter data and the quality test results.

Where the volume produced from a lease is low, LACT equipment may not be economically justified. Such leases will produce into small calibrated field tanks where the crude is manually measured and sampled prior to shipment.

Gathering lines are sized to accommodate a number of lease facilities from a field or from adjacent fields for delivery into major trunk systems. Line size is typically from 3 to 8 in. nominal diameter.

Pressure requirements are usually low, from approximately 50 to 300 psi, if the system discharges into trunk line accumulation tanks. In some cases gathering lines deliver directly into the trunk line and then must be designed to match trunk line pressures, perhaps up to 1500 psi.

Crude Trunk Systems

Trunk lines are typically 8 in. diameter and up. They accumulate crude from several producing fields and move it to the refining centers. A trunk line will have at least an origin pump station and may have one or several intermediate pump stations (booster stations) along its length. Trunk lines operate at high pressures, limited by strength of the pipe and the capability of the pumping equipment.

Crude trunk lines are usually operated as single-commodity systems, with all gathering streams being blended together into one common stream. In some cases it is desirable to segregate crudes of different characteristics, either because one crude might have a detrimental effect on the common stream or because a crude has special refining qualities that the shipper wants to protect. When this happens, streams can be batched through the system on a segregated basis. Intermixing of two dissimilar adjacent batches in a pipeline is minimal if flow is maintained in the turbulent range and operating and valve-switching procedures are properly planned. Many pipelines can accommodate separating plugs, called "batching pigs," but normal practice is to batch without them. The commodity interface is commingled with one or both of the adjacent batches.

Batching two or more different crude streams through a system also requires more tankage at the origin and destination locations. The volume of the required tankage depends on the size of the batches being moved and on the amount of any one batch needed to ensure an adequate supply for refining operations while other commodities occupy the pipeline system.

Refined Products Systems

Product pipelines are similar in size and equipment to crude lines, but their operations are very different. Product pipelines typically handle several distinct grades of gasoline, diesel, jet fuel, and light fuel oil from refineries to distribution points. Tenders of these several commodities are handled for several different refiners and may have several destinations. Each tender must be carefully scheduled to match each shipper's needs and must be coordinated with all other tenders to ensure timely movement of all commodities with a minimum of product contamination. Most product pipelines are operated by computer-oriented control equipment programmed to recognize and react to the many parameters encountered in dispatching of multiproduct systems.

Some interfacial mixing of adjacent commodities in a pipeline is unavoidable. With large batch sizes the few barrels of intermixing that takes place is usually such a small percentage of the batch that product quality is not degraded. If intermixing does exceed allowable levels, "transmix" facilities are provided to receive the off-grade material, and it is returned to the refineries for reprocessing or is otherwise disposed of. Batching pigs are used to separate adjacent commodities in some systems, but usually satisfactory separation is maintained without them.

Economic Pipeline Sizing

An ideally sized pipeline is one that can accommodate the required volume with a proper balance of initial capital investment and annual operating costs. An undersized line, while requiring minimum initial capital, will result in excessive pumping power requirements and can severely restrict operating flexibility and opportunities for expansion. Conversely, an oversized line will result in minimal power costs, but will waste initial capital and may cause unacceptable operating problems such as excessive interfacial mixing of adjacent commodities due to low flow velocities.

Table 8.3-4 provides a rule-of-the-thumb guide to optimum line sizes and relative pumping costs for various flow rates.

The operating costs in Table 8.3-4 are the relative pumping costs through various size lines, expressed in terms of units per barrel-mile. The table shows, for example, 1 unit per barrel-mile cost to move 600 bph of products in an 8-in. pipeline, while 1600 bph could be moved in a 12-in. pipeline at a cost of 0.70 units per barrel-mile. Larger line sizes are more economical, providing there is sufficient volume to operate a system near optimum capacity.

8.3-3 Design of Pipelines

Prior to making a detailed engineering design of a pipeline system, a knowledge of the basic design parameters is required. These include the type of product to be transported, properties of the product (density, viscosity, temperature, pour point, etc.), volume to be moved in terms of average daily throughput and peak hourly pumping rates, and the desired origin and destination points of the commodity. Also an understanding of the physical properties of the pipe material is required. This section provides the basic formulas for evaluating these design parameters.

TABLE 8.3-4   OPTIMUM LINE SIZES

Flow Rate, bph	Optimum Line Size, in.		Relative Operating Costs
	Heavy Crude	Products	
200	6	4	1.70
400	8	6	1.25
600	10	8	1.00[a]
800	10–12	10	0.85
1200	12	10–12	0.75
1600	12–16	12	0.70
2000	16	12–16	0.65

[a]Arbitrarily assigned a relative operating cost of 1.00.

The American National Standards (ANSI) Code B31.4, Liquid Petroleum Transportation Piping Systems, is the basic standard for petroleum pipelines. This code, published by the American Society of Mechanical Engineers, covers the design, construction, inspection, testing, operation, and maintenance of pipeline systems. Also the United States Department of Transportation (DOT) issues a regulation that governs petroleum pipelines in interstate commerce. The DOT regulation, Minimum Federal Safety Standards For Liquid Pipelines, is published as part 195, title 49, Code of Federal Regulations.

Physical Properties of Line Pipe

The physical requirements of pipe used in pipeline systems are detailed in specifications published by the American Petroleum Institute (API). These specifications cover processes of manufacture, chemical properties of the pipe material, physical properties such as tensile and yield strengths, testing requirements, and standards governing dimensions, weights, and other design parameters.

API specification 5L covers standard grades of steel pipe having specified minimum yield strengths (SMYS) in the range of 30,000–35,000 psi. Specification 5LX covers higher-strength line pipe, with SMYS values of 42,000 psi and above. Higher-strength line pipe is generally used because such pipe will accommodate significantly higher working pressures for a given pipe wall thickness; however, many factors affect the final decision on the appropriate grade of material. A thorough understanding of the ANSI B31.4 code and the API line pipe specification is required.

Internal Design Pressure

The internal design pressure, or maximum allowable operating pressure, is determined by the equation $P = 2St/D$.[3] The allowable fiber stress, S, can in most cases be taken as 72% of the SMYS of the pipe. The DOT regulations provide for a lower percent for some specific installations. The above equation assumes pipe manufactured in accordance with specifications API 5L or API 5LX having a joint factor of 1.00. API 5L allows welding processes that result in reduced joint factors, but these are not recommended for line service. Many other specifications and grades of pipe are available. Designers must select line pipe on the basis of pressure requirement, pricing, and availability.

Example 8.3-2. Determine the maximum allowable working pressure of $10\frac{3}{4}$-in. outside diameter, 0.25-in. thick 5LX-52 pipe.

Solution: 5LX-52 pipe has a specified minimum yield strength of 52,000 psi. Thus, the 72% allowable fiber stress is 37,440 psi and

$$P = \frac{2(37,440)(0.25)}{10.75} \quad \text{or} \quad 1741 \text{ psi}$$

Hydraulic Surge

Hydraulic surge is a change of pressure in a pipeline caused by a sudden change in the velocity of the moving stream. This change in velocity may be the result of shutting down a pump, closing a valve, or any blockage of the moving stream. The magnitude of the hydraulic surge decreases as it moves away from the point of origin. In the design of pipelines the ANSI B31.4 code requires that "surge calculations shall be made, and adequate controls and protective equipment shall be provided so that the level of pressure rise due to surges and other variations from normal operation shall not exceed the internal design pressure at any point in the piping system and associated equipment by more than 10%."

Rapid closing of a main-line valve will cause surge pressures. The surge pressure above operating pressure, at the point of closure, can be approximated by the rule-of-thumb equation[4] $P = 0.8WV$. For a detailed calculation of surge pressure, see ref. 5.

In addition to increase in pressure due to surge upstream of the blockage, a pressure decrease of the same magnitude is experienced downstream of the blockage. This negative surge wave can cause downstream booster stations to experience low suction conditions.

Flow Formulas

Two basic formulas are used in the design of petroleum pipelines. Selection of the appropriate formula depends on the type of product to be transported The Darcy–Weisbach equation is appropriate for crude oil systems, and the Hazen and Williams formula is appropriate for refined products systems.

In pipeline flow calculations it is convenient to express head in feet of liquid. To convert between feet of head and psi:

$$P = \frac{\text{head in feet} \times \text{specific gravity}}{2.31}$$

Crude Oil Flow Calculations

A useful form of the Darcy–Weisbach equation is expressed as

$$H_f = \frac{0.0153 f L Q^2}{d^5} \tag{8.3-1}$$

Friction factor f is a dimensionless number that is a function of internal pipe roughness and liquid turbulence. Turbulence is expressed in the form of the Reynolds number, R. R is also dimensionless, and is determined as follows[6]:

$$R = \frac{2214Q}{dv} \tag{8.3-2}$$

For R below 2000 the stream is in laminar flow. For R above 3000 the stream is in turbulent flow. The transition range between 2000 and 3000 is a zone of instability where f could be as low as the laminar flow calculation or as high as that indicated in the turbulent range. A conservative assumption for pressure drop calculations is that turbulence is maintained throughout the unstable range.

Friction factor f in the laminar flow range is linear with respect to R, and is equal to $64/R$. In the turbulent range, f cannot be expressed as a simple equation, but Heltzel[3] developed an empirical formula based on experimental data that is reasonable for Reynolds numbers up to 57,600. Heltzel's formula is

$$f = \frac{0.364}{R^{0.265}} \tag{8.3-3}$$

Beyond R = 57,600, a reasonable approximation is

$$f = \frac{0.116}{R^{0.160}}$$

Refined Products Flow Calculations

For refined products systems, where viscosities are low (less than 31 SUS) and the liquid is flowing under turbulent conditions, the Hazen and Williams formula may be used. A useful form of this equation is

$$h_f = 0.001077L \left(\frac{100}{C}\right)^{1.85} \times \frac{Q^{1.85}}{d^{4.8655}}$$

The roughness factor, C, is 140 for new steel pipe. It may decrease to 100 or less over time as the pipe wall roughness increases. For design purposes a value of 120 is recommended. With a C of 120,

TABLE 8.3-5   VISCOSITY CONVERSION[a]

SUS	cS	SUS	cS
31	1.0	80	15.7
35	2.6	90	18.2
40	4.3	100	20.6
50	7.4	150	32.1
60	10.3	200	43.2
70	13.1	250	54.0

Source. from viscosity conversion tables in ref. 7.
[a]Above 250 SUS: SUS = 4.62 cS

the formula becomes

$$h_f = \frac{0.0007687 L Q^{1.85}}{d^{4.8655}} \tag{8.3-4}$$

Crude oil viscosity is generally measured and reported in SUS units. Table 8.3-5 converts SUS to kinematic viscosity in centistokes (cS).

Example 8.3-3.   Determine the head loss in feet of liquid for an 8-in. ($8\frac{5}{8}$-in. outside diameter, 0.312-in. wall thickness) 10-mi long pipeline, flowing at a rate of 1000 bph for (1) crude oil with a viscosity of 1500 SUS, (2) crude oil with a viscosity of 150 SUS, and (3) gasoline.

Solution:

1. For 1500 SUS crude oil (325 cS from Table 8.3-5), R [using eq. (8.3-2)] is 852, thus flow is laminar and $f = 0.075$. From Eq. (8.3-1) $h_f = 1849$ ft.
2. For 150 SUS crude oil (32.1 cS), R = 8625; thus flow is turbulent. From Eq. (8.3-3), f is found to be 0.033 and h_f is calculated to be 813 ft.
3. For gasoline Eq. (8.3-4) yields a head loss of 581 ft.

Thus, from the above example, the pressure drop ranges from 581 ft for gasoline to 1849 ft for a viscous fluid under otherwise equivalent pumping conditions.

Equations (8.3-1) and (8.3-4) can be solved for flow rate or line size. In these cases the total required head (H_t) is used in place of H_f. H_t for the pipeline system includes frictional head loss (H_f), differential elevation head, and back-pressure head.

Published graphs are available from several sources that translate the previously discussed flow formulas to graphic solutions. One source is the Hydraulic Institute.[7] Using these graphs, head loss per unit length can be directly determined.

Hydraulic Gradient

The hydraulic gradient provides a graphical representation of the head loss in a pipeline system. The gradient can be used to determine theoretical pump station spacing and allows an evaluation of the effect that differing operating parameters have on the head loss. It also identifies areas where the maximum allowable operating pressure of the pipe might be exceeded due to ground profile changes.

A typical gradient is shown in Fig. 8.3-1. The horizontal axis of the gradient is pipeline length in miles; the vertical axis is head in feet. The lower irregular line is a plot of the ground elevation along the proposed route. The upper irregular line is identical in shape to the ground profile. Its spacing, f, above the ground profile represents the maximum allowable head as dictated by pipe design limitations. At all points along the pipeline, the hydraulic gradient must lay within the bounds of the two irregular lines.

At the origin of the pipeline the maximum desired operating pressure is plotted. Using the appropriate flow formula, the head loss per unit of length is computed to determine the slope of the gradient. The point at which the gradient intersects the ground profile represents the maximum distance the stream can be pumped at the given flow conditions. If it falls short of the desired destination, a booster station is required or flow conditions must be modified (larger diameter pipe, lower flow rate, heating, etc.). Allowances must be made at intermediate booster stations for station

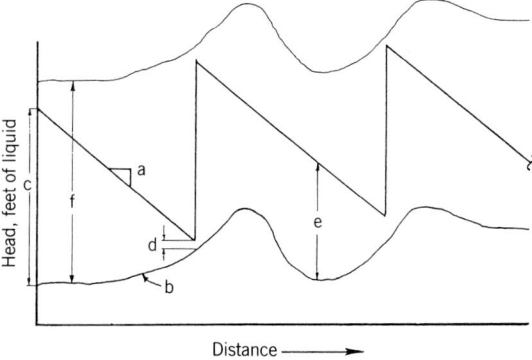

Fig. 8.3-1 Typical hydraulic gradient: (a) gradient equal to frictional head loss per unit of length, (b) ground profile along pipeline route, (c) initial head in feet of liquid being pumped, (d) required suction pressure at intermediate booster stations or back pressure at terminus point, (e) head in feet of liquid at any point along pipeline, (f) maximum allowable system operating pressure at any point along the pipeline.

piping head losses and required pump suction pressures (usually 25–50 psi) and at the line terminus for the elevation head of the receiving tanks. The head at any point along the pipeline is equal to the difference between ground elevation and the gradient.

Example 8.3-4. Determine the approximate spacing of booster stations required for 6-, 8-, and 10-in. pipelines if the maximum desired working pressure is 1000 psi, flow rate is 1000 bph of crude oil having a specific gravity of 0.90 and a viscosity of 500 SUS, and the pipeline route is essentially level.

Solution: The initial head is equivalent to 2567 ft. The head loss for 6-in. pipe [using Eqs. (8.3-1) to (8.3-3)] is equal to 438 ft/mi. This head loss, plotted as a gradient in Fig. 8.3-2 shows that the gradient approaches ground elevation at approximate 6-mi intervals. Thus a booster station would be required at least every 6 miles if a 6-in. line is operated under the given conditions.

The head loss for an 8-in. line (again using the flow equations) is equal to 112 ft/mi. Plotting this on Fig. 8.3-2 shows a spacing requirement of approximately 23 mi. Note that in this calculation the Reynolds number is 2560, thus flow is in the transition range between laminar and turbulent conditions. The higher turbulent flow friction factor is used.

The head loss for a 10-in. line (stream in laminar flow) is 23 ft/mi, thus no boosting would be required for 112 mi.

If, under the conditions given in Example 8.3-4 the total required pipeline distance was 40 mi, a 10-in. line would be larger than needed, and the 6-in. line would require too many booster stations. It would appear that an 8-in. line, requiring one booster station, might be the proper choice; however, a

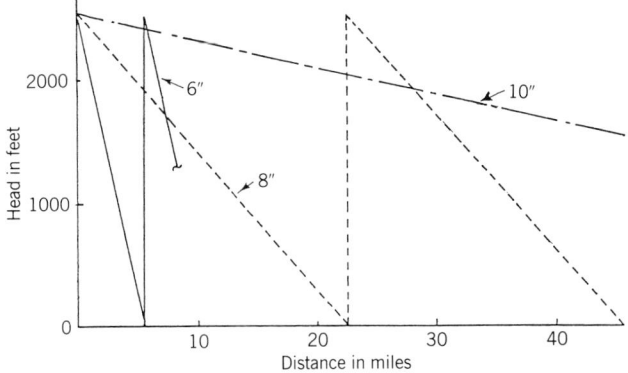

Fig. 8.3-2 Hydraulic gradients for 6-, 8-, and 10-in. pipelines.

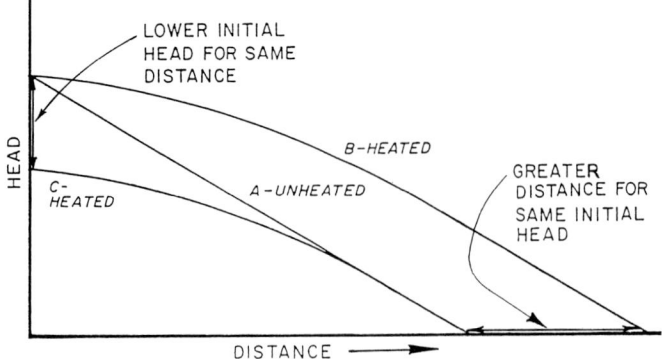

Fig. 8.3-3 Effect of heating on viscous crude oil.

close look at a 10-in. line is suggested. A 10-in. line, even though larger than needed for the given operating conditions, would avoid the added capital and operating costs of one intermediate booster station and would require considerably less power. The initial operating pressure of a 10-in. system would be approximately 285 psi compared to an operating pressure of approximately 920 psi at each of the two stations in an 8-in. system. Before deciding between an 8-in. and a 10-in. system, a detailed economic evaluation is needed to compare capital investment costs and operating costs over the projected life of the system. Another advantage of the 10-in. system is that it has substantial spare capacity to accommodate increased future volumes. A 40-mi 10-in. line could handle 1400 bph, 40% more than initial design requirements.

For safety and economy, it is desirable to operate a system at the lowest possible working pressures. For this reason, if an 8-in. line was used in the above example, the intermediate pump station would be located somewhere near the midpoint of the system, rather than having one station operating at the 1000 psi maximum and the other at about 820 psi.

Heated Oil Pipelines

Since the head loss in a pipeline is significantly affected by viscosity, and viscosity is a function of temperature, viscous crude and products are often heated to improve pumping characteristics. The hydraulic gradient in Fig. 8.3-3 shows the advantage of heating viscous liquids.

Gradient A represents the movement of an unheated viscous crude. Gradient B represents the movement after heating the crude, pumping at the same flow rate and initial discharge pressure. The initial slope of this gradient is less because the friction loss of the heated crude is lower, and the slope increases as cooling takes place down the line. The gradient slope will continue to increase until the oil temperature has dropped to the ambient level. Beyond that point, the slope will be the same as the unheated gradient. Note that for the same initial pumping head, the heated crude can be pumped much further. Gradient C represents the same crude heated to the same temperature as B and pumped at the same flow rate for the same distance as A. Note here that the required pumping head is much less.

The heat loss from a pipeline is a function of:

Pipeline length	Outside diameter
Initial oil temperature	Atmospheric temperature
Flow rate	Liquid density
Liquid specific heat	Soil conditions
Pipe coating	Pipe insulation

A coefficient, K (in units of $Btu/ft^2 \cdot h \cdot °F$) is used to express the effect of soil conditions, pipe coating, and insulation on the heat loss in a pipeline system. Table 8.3-6 gives approximate values of K for various conditions.

The basic procedure for calculating head loss in heated oil systems is as follows:

1. Determine the required maximum hourly pumping rate.
2. Select appropriate K values as shown in Table 8.3-6.
3. Select what appears to be an appropriate pipe diameter.

TABLE 8.3-6   HEAT LOSS COEFFICIENTS

Pipeline Condition	K Value Btu/ft^2-h-°F
Dry soil with 24-in. cover	
Uninsulated	0.25–0.40
2-in. thick insulation	0.05–0.15
Moist soil with 24-in. cover	
Uninsulated	0.50–0.60
2-in. thick insulation	0.10–0.15
Wet soil with 24-in. cover	
Uninsulated	1.1–1.3
2-in. thick insulation	0.15–0.20
Exposed pipeline	
Uninsulated	1.5–2.0
2-in. thick insulation	0.1–0.2
Offshore, unburied pipeline lying on bottom, exposed to water currents	
Uninsulated	8–12
2-in. thick insulation	0.1–0.2

Source.   Extracted from refs. 8 and 9.

4. Divide long pipelines into equal length sections. For most purposes, sections 2–4 mi long may be used. If the line is to pass through a variety of soil conditions, use appropriate K values for the various sections.

5. Determine the temperature drop in each section.

6. Calculate the head loss for each section using the viscosity corresponding to the mean temperature of the oil in that section.

7. Add the head losses for each section of the pipeline to determine the total head loss in the pipeline.

If the resulting total head loss is not within reason (above maximum allowable pressure or substantially less than a reasonable operating pressure), the process can be repeated using a different line size or changing the inlet temperature. It may be necessary to analyze two close alternatives from an economic standpoint for optimum selection of line diameter, pumping temperature, and pumping pressures.

It should be noted that the flow in most heated oil pipelines will pass through the transition zone between turbulent and laminar flow as the stream cools. This can introduce errors in the design calculations. Karge[8] provides detailed calculations that evaluate the magnitude of this potential error under various flow conditions. Assuming the flow to remain turbulent produces a conservative design.

Karge developed a technique that simplifies the determination of temperature drop of heated oil in pipelines. This technique is explained in Fig. 8.3-4. Using this technique, the mean temperature in each section of line is determined, then the fluid viscosity can be found for that temperature. Having determined viscosity, the hydraulic gradient can be plotted for the entire pipeline length using the flow formulas or published friction loss curves. Numerous computer programs are available to calculate the frictional head loss in heated oil pipeline systems.

Corrosion Protection

Pipeline coating is the primary defense against external corrosion. Many coating systems are currently available, and selection of the proper coating requires careful analysis.[10] The National Association of Corrosion Engineers defines qualities that a pipe coating should possess in NACE Standard RP-01-69 (Section 5). Desirable characteristics of pipeline coatings include ease of application and repair, ability to withstand normal handling, storage and installation, good adhesion to pipe surface, ability to resist development of holidays (openings to bare metal) with time, effective electrical insulator, ability to resist disbonding when under cathodic protection, and ability to maintain constant electrical resistivity with time.

Numerous types of pipe coatings are available for use such as bituminous enamels, asphalt mastics, extruded plastics, powder resins, liquid epoxy and phenolics, and tapes.

In addition to external coating, pipeline systems must be cathodically protected to ensure against corrosion that can occur where coating may become ineffective. Regardless of the best efforts toward

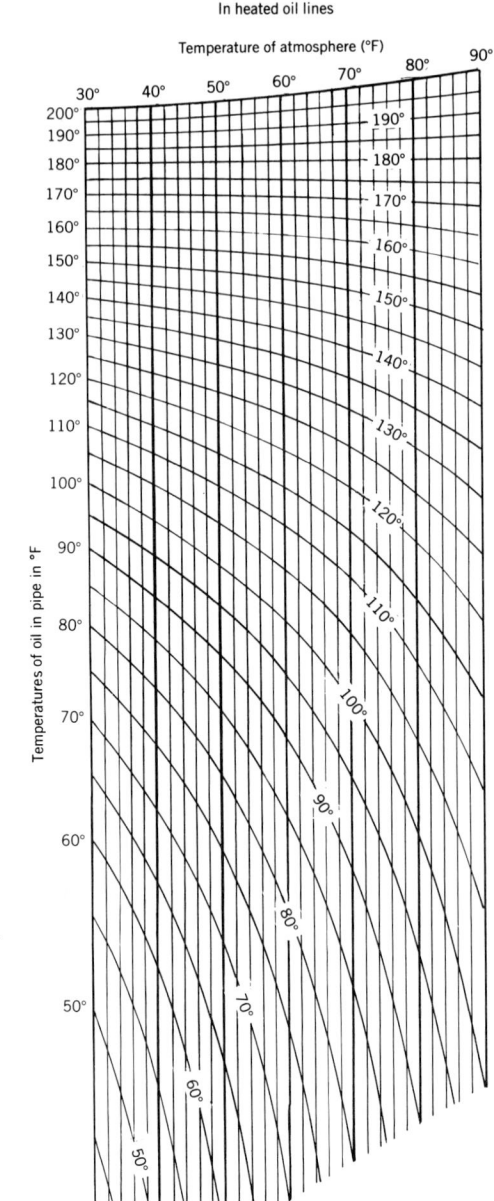

Fig. 8.3-4 Temperature drop in heated oil lines. (Reprinted by permission from Petroleum Engineer International.)

perfecting long-lasting coatings, problem spots will eventually develop and corrosion will begin, unless preventative steps are taken.

External corrosion of buried pipelines is primarily caused by electrical current flow, similar to the flow of current in an electrolytic cell. Conditions can exist where the pipe acts as the anode within the cell, and the anodic activity will deplete the pipe metal. To prevent this, cathodic protection equipment is connected in such a way that the pipe then acts as the cathode, and metal depletion ceases.

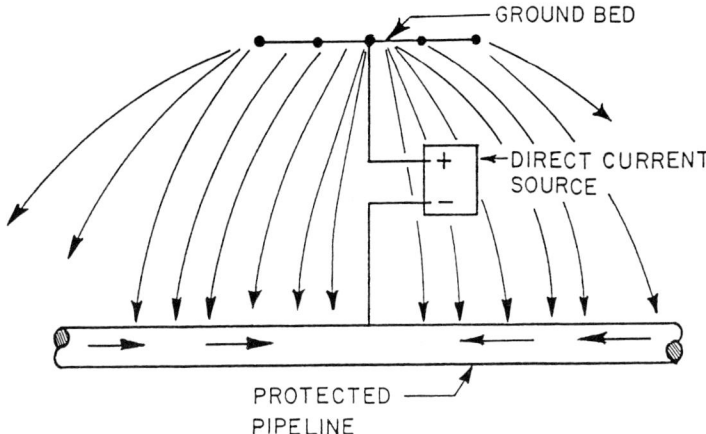

Fig. 8.3-5 Impressed current cathodic protection.

Sacrificial Anodes

The use of sacrificial anodes for cathodic protection is the simplest method. Galvanic anodes, usually of zinc or magnesium because these metals are strongly anodic compared to steel, are installed in the common electrolyte (the earth) adjacent to the line, and connected to the line to complete the cell. The zinc or magnesium then becomes the anode and is corroded, discharging current in the process. The current flows from the anode through the earth to the pipeline (now the cathode), then down the pipeline back to the anode. In the process the anode is consumed instead of the previously anodic pipeline.

The current available from sacrificial anodes is very low. Thus, they are usually used where currents required for protection are small or in locations where anodic activity is confined to short isolated sections of line.

Numerous factors determine the current output of sacrificial anodes. These factors include the depth at which an anode is buried, the number of anodes and their spacing, pipe-to-soil potential of the pipeline, resistivity of the soil surrounding the anode, and several other factors. Anode suppliers can provide approximate current output values of their products, knowing the particular application under consideration.

Impressed Current Anodes

Typically, the cathodic protection needs of a pipeline system exceed the low-current capabilities of sacrificial anodes. Where this is the case, voltage from an external power source is impressed on the circuit. This is done by means of variable-voltage rectifiers connected to the pipeline and anodic ground beds, as shown in Fig. 8.3-5. Direct current is forced from the source, through the ground bed, and onto exposed surfaces of the pipeline. This current, when properly regulated, will overcome corrosion currents from the pipeline, and there will be a net current flow onto the pipeline.

The design of the ground bed that serves as the anode is complicated and must recognize many factors, including anode material, adjacent soil resistivity, space available, proximity of other underground facilities, and accessibility for maintenance and servicing. It can be installed horizontally near the ground surface if space is available and other factors permit. A common installation where area adjacent to the pipeline is limited is in a deep well located on the pipeline right-of-way.

Typically, anodes are made of graphite installed in a bed of carbonaceous material, although cast iron and steel pipe or structural material can be used. The National Association of Corrosion Engineers publishes much detailed information on this subject.

A major concern that must be recognized when evaluating cathodic protection needs for a pipeline system is its potential adverse impact on other buried structures. Because the "electrolytic cell" that a pipeline passes through may contain other pipelines and other metallic structures, such objects may be endangered by cathodic protection installations that do not recognize their presence. It is often necessary to interconnect adjacent pipeline systems for mutual cathodic protection. Constant testing and inspection is required of all adjacent buried structures to be sure they are adequately protected and that their protective equipment is not causing damage to foreign structures.

8.3-4 Pipeline Operation

Scheduling

Scheduling of pipeline shipments is a complex job. Most pipeline operators use computer programs to develop and constantly update both long- and short-range operating schedules.

Typically, shippers are required to notify the pipeline operator in advance of the type, volume, source, and delivery point of the shipments required. This information, when received from all shippers, becomes the basis for the pipeline operator's schedule. Changes caused by pipeline operating problems, unscheduled refinery shutdowns, erratic tanker arrivals, or other unplanned circumstances necessitate frequent schedule revisions, and both shipper and pipeline operator need to be somewhat flexible.

Dispatching

Dispatching translates commodity movement requirements into the pipeline operating schedule and involves continuous surveillance and control of the hour-to-hour operation. Dispatching personnel are commonly situated at one central location. They carefully coordinate the operation with field personnel at various pipeline facilities. Dispatchers sometimes have direct control of the operations, using remote control supervisory equipment.

Batching

Some crude lines handle only one commodity and make no attempt to segregate differing grades of crude oil. Other crude systems, and most products pipelines, handle more than one commodity segregation by batching. Intermixing of adjacent batches is usually minimal, particularly if flow rates are uninterrupted and high enough to maintain turbulent flow conditions. Some intermixing of various grades of crude is usually tolerable. Intermixing of similar products is also tolerable to a degree. Intermixing can be held to a very small percentage by scheduling appropriate batch sizes. Generally, the pipeline operator will schedule batches within similar product cycles to minimize degradation of dissimilar products. To reduce contamination where it cannot be tolerated, some systems are equipped to insert commodity batching pigs in the line between adjacent commodities. However, this expensive operation can usually be avoided by proper sequencing of products and careful system operation.

Leak Detection

Computers are used on most modern pipeline systems to continuously monitor the system for leakage or other operating problems. The computer receives signals from all inlet and outlet meters, adjusts these signals for various operating conditions, and alarms the dispatcher if the system balance is outside of allowable tolerances. Line pressures along the systems are also constantly monitored.

8.3-5 Tanker Transportation

The principal parameters that influence selection of a tanker are the type and quantity of cargo to be transported; the location of, and distance between, loading and discharge ports; and operating restrictions such as depth of water, canal sizes, bridge clearances, turning basins, and terminal arrangements.

Vessel size is designated by deadweight tonnage, or DWT (1 DWT = 1 long ton, or 2240 lb). DWT is the total weight that can be loaded into a vessel, including the weight of cargo, fuel, water, lubricants, crew, and stores. Crude tankers range from 20,000 to 550,000 DWT and product tankers range from 12,000 to 45,000 DWT. Vessels between 200,000 and 300,000 DWT are classed as "very large crude carriers" (VLCCs) and those above 300,000 DWT are "ultra large crude carriers" (ULCCs). Because of their deep draft (60–80 ft) VLCCs and ULCCs cannot enter most ports and usually trade between deep-water offshore moorings.

Optimum size is influenced significantly by economies of scale since the size of a ship's crew is relatively unaffected by the size of the ship and capital costs per DWT decline as size increases. Generally then, the larger the ship, the more economical the transportation, providing proper consideration has been given to all components of the transportation system.

Tanker Decision

The physical design of tankers is a complex process that takes into account size, speed, optimum ship dimensions, type of power plant, and all applicable regulatory requirements.

Tanker dimensions vary with vessel size but conform to certain optimum relationships. These relationships are the length-to-beam ratio ranging from 5.5 to 6.5, the beam-to-draft ratio ranging from 2.0 to 2.8, and the draft-to-depth ratio ranging from 0.75 to 0.80. Minimal values of the length-to-beam ratio are optimum since ships with greater beam have a greater cargo capacity for a specified draft. However, operating restrictions may dictate less than optimum ratios. For example, vessels that must transit the Panama Canal are limited to a beam of 105 ft, so vessels designed for Panama Canal transit may have a higher length-to-beam ratio.

Ship propulsion plants include medium- and slow-speed diesel engines, steam turbines, and aircraft or heavy-duty gas turbines. All of these, except slow-speed diesel engines, operate at a higher rpm than the most efficient propeller speeds (80–90 rpm) and must be coupled to propulsion shafting through reduction gearing.

The primary objective in selecting the power plant is to minimize capital and operating costs. The installed cost for a typical 20,000 shp marine power plant, including auxiliaries, will range from $10 to $15 million. The major variable in operating costs is the cost of fuel. Most marine propulsion plants consume Bunker C fuel, although many medium speed diesel engines and gas turbines require diesel fuel.

Ballast Requirements

Tankers must have pumping equipment and storage capacity for ballast to maintain ship stability when the cargo tanks are empty. Pollution regulations require that clean ballast storage capacity, independent of cargo tanks, be provided. The segregated ballast capacity may be augmented by ballasting cargo spaces in severe weather conditions.

Crude carrier voyage patterns often result in the ballast operation up to 50% of the time in service. Product tanker voyages, however, are generally of shorter duration and involve multiple discharge ports with partial loads. In these instances ballast is utilized for list and trim control.

Safety and Environmental Considerations

Safety and pollution regulations for tankers are promulgated and enforced by many international, national, and local organizations. Several international treaties specify minimum standards for fire-fighting equipment, fire prevention, lifesaving appliances, vessel construction, pollution prevention, and other design and operating requirements.

Pollution regulations govern all types of potential tanker emissions such as stack exhausts, sewage effluent, bilge and ballast discharge, cargo and slop handling, and hydrocarbon emissions resulting from loading and gas freeing.

Vessels must have equipment fill the cargo tanks with inert gas as cargo is discharged. Vessels with steam propulsion or diesel ships with large auxiliary boilers utilize scrubbed flue gas for this purpose. Other vessels have independent inert-gas generators. These systems are designed to deliver inert gas having an oxygen content of less than 5% at a rate equal to about 125% of the ship's cargo pumping capacity.

Most tankers are equipped with cleaning equipment to control the accumulation of cargo residues, to prepare tanks for a change in product, and to clean tanks to permit access for repairs and inspections. Crude tankers generally use a crude oil washing system (COWS) that directs a high-velocity stream of crude oil, under pressure from the cargo pumps, against the interior structure of the tanks during discharge. The cleaning system is usually also arranged to provide seawater for rinsing the cargo tanks if required for gas freeing.

Product tankers are also fitted with fixed tank cleaning equipment usually consisting of heat exchangers to heat seawater to 150–200°F for cleaning the cargo tanks. The cleaning residue is transferred to shore facilities for disposal.

Cargo Handling

Tanker cargo handling systems are designed specifically to suit the particular cargoes to be transported and are influenced by cargo density, viscosity, corrosive properties, and safety hazards such as inflammability, toxicity, and pollution potential.

A crude cargo pumping system includes from two to six centrifugal cargo pumps mounted in a pump room located aft of the cargo tanks. Individual pump capacity may range from 2000 to 10,000 gpm. Total pump capacity is sized to permit complete discharge of cargo in 18–30 h. Cargo pump head is usually 250–400 ft depending on location of shore-side facilities. The arrangement of transfer connections must be carefully designed to ensure compatibility with shore-side facilities, which may vary from port to port. Crude cargo piping is usually arranged to isolate up to four different cargo segregations.

Product tankers are designed to carry from as few as 2 to as many as 25 different segregations of refined products and petrochemicals. Pumping schemes vary depending on the degree of segregation required. The pump room may be similar to that described for crude carriers, while some designs incorporate the pumps within the cargo tanks. Cargo pump capacities range from 500 to 3000 gpm with heads of 200–350 ft. Cargo discharge time may range from 10 to 20 h.

Some commodities must be heated to facilitate pumping. Tankers handling such cargoes are fitted with steam heating coils. Heating temperatures range from 120°F for certain fuel oils up to 300°F for asphalt.

8.3-6 Small-Volume Movements

Where the volume to be transported is inadequate to justify major investments in pipeline facilities, barges, rail cars, and tank trucks are utilized. Typically, this equipment is owned and operated by common carrier transport companies. Regulations governing design and operation of this equipment are promulgated and enforced by various federal, state, and local agencies.

Barges are used to advantage along the coastal and inland waterways, particularly for product movements to distribution and marketing facilities. In some instances large barges are competitive with small tankers for some intercoastal movements.

Rail movement of petroleum is generally limited to refined products and petrochemicals, although some crude also moves by rail. Rail transportation offers good flexibility because of the vast U.S. railroad network, provided adequate rail and storage facilities are available at the loading and discharge locations. "Unit trains," a complete train of interconnected cars capable of being loaded and unloaded from a single point, may be utilized where the volume to be moved will justify investment in the unique equipment and related facilities.

Tank trucks offer the most flexibility because they are essentially unrestricted in routes and because they require only minimum investment in ancillary facilities.

REFERENCES

8.3-1 Association of Oil Pipelines, "Pipeline Transportation, a Review of the Oil Pipeline Industry," Washington, D.C., June 4, 1975.

8.3-2 C. T. Horngren, *Introduction to Management Accounting*, Prentice-Hall, Englewood Cliffs, NJ, 1978.

8.3-3 *Line Pipe*, National Tube Division, United States Steel, Pittsburgh, Pa., 1954.

8.3-4 *Pipe Line Rules of Thumb Notebook*, "Crude and Products Line Surges," Pipeline Industry, Houston, TX, 1962, p. 32.

8.3-5 E. Cykowski, T. Connor, and M. Spratt, "Designing Liquids Pipelines for Surge Control in Lines," *Oil & Gas Journal*, June 21, 1982, pp. 183–188.

8.3-6 *Cameron Hydraulic Data*, Ingersoll-Rand, Woodcliff, NJ, 1977.

8.3-7 *Standards of the Hydraulic Institute*, Hydraulic Institute, New York, 1973.

8.3-8 Fritz Karge, "Design of Oil Pipelines," *The Petroleum Engineer*, March, April, May, 1945.

8.3-9 W. A. Smith, "Guidelines Set Out for Pumping Heavy Crudes," *The Oil & Gas Journal*, May, 1979, pp. 111–114.

8.3-10 R. N. Sloan, "How to Select Mill Applied Coatings," *Pipeline & Gas Journal*, February, 1982, pp. 43–46.

BIBLIOGRAPHY

American Petroleum Institute. "Manual of Petroleum Measurement Standards." Washington, D.C., 1981.

Arnold, C. L. "Temperature Effects of Hydraulics and Fluid Mechanics." Paper presented at the 1981 Annual Pipeline Design and Construction Symposium, Dallas, TX, April 30, 1981.

Cohn, A. R. and Nalley, R. R. "Using Regulators for Line Pressure Relief." *Pipeline Industry*, August, 1980, pp. 45–47.

Defense Fuel Supply Center. "Defense Petroleum Course." Alexandria, VA, November, 1976.

Fekete, L. A. "Structural Design of Pipelines Subject to Temperature Change." *Pipeline Industry*, August, 1974.

Hangs, F. E. "More Insulated Heated Oil Pipelines are in Prospect." *Oil and Gas Journal*, October 3, 1966, pp. 117–122.

Hooker, John N. "Oil-Pipeline Energy Efficiency, Studied for U.S." *Oil and Gas Journal*, February 15, 1982, pp. 114–116.

Huber, D. W. "Real-Time Transient Model for Batch Tracking, Line Balance and Leak Detection." Paper presented at the API Annual Pipeline Conference, Dallas, TX, April 28–29, 1981.

Leuba, H. R. "Inventory Management in the Petroleum Industry," Prepared for the Department of Energy by Evaluation Research Corporation, Falls Church, VA, 1978.

Military Traffic Management Command. "Pipelines for National Defense." Vol. I, Newport News, VA, 1982.

National Petroleum Council. "Petroleum Storage and Transportation Capacities." Washington, D.C., 1979.

O'Brien, H. H. "*Petroleum Tankage and Transmission*," Graver Tank and Mfg. Co., Inc. East Chicago, IN, 1951.

Packard, W. V. *Voyage Estimating*. Fairplay Publications Ltd, London, 1978.

Peabody, A. W. *Control of Pipeline Corrosion*. National Association of Corrosion Engineers, Houston, TX, 1967.

Smith, S. S. and Schulze, R. K. "Interfacial Mixing Characteristics for Products in Product Pipe Lines." *The Petroleum Engineer*, September–October, 1948.

Thompson, J. C. "New Development of High Temperature Pipeline Coatings." Paper presented at the Third International Conference of the Internal and External Protection of Pipes, London, 1979.

Wolbert, G. S., Jr. *U.S. Oil Pipelines*. American Petroleum Institute, Washington, DC, 1979.

Yarborough, V. A. "Causes of Pressure Surges in Pipelines." Paper presented at the 1981 Annual Pipeline Design and Construction Symposium, Dallas, TX, April 30, 1981.

8.4 STORAGE OF HYDROCARBON LIQUIDS

This section presents general information on storage of flammable and combustible liquids derived from crude oil (petroleum) and petroleum-related gases. The products considered in Section 8.4-1 are those stored at or near atmospheric pressures. This includes most hydrocarbons, both pure products and blended products. The products considered in Section 8.4-2 are liquid petroleum gases stored under pressure at ambient temperature or refrigerated temperature. This includes propane, butane, and propylene. Codes and standards applicable to storage are referenced in this section. Some important safety aspects are included.

8.4-1 Petroleum Base Products

Cyrus Clark

The petroleum products included in this broad base vary from very low vapor pressure fuel oils to high vapor pressure gasolines. Figure 8.4-1 includes examples of some products, both pure and blended hydrocarbons. Vapor pressures of these products are generally less than atmospheric pressure at ambient storage temperatures.

Tank Descriptions

Petroleum products in bulk quantities are stored in two basic types of tanks: the closed tank or floating-roof tank. Each type has specific application, which is discussed in general terms later in this section. Applications depend on factors such as product vapor pressure, tank capacity, and local or state regulations.

The closed tank (Fig. 8.4-2), whether horizontal or vertical in position, can be designed for atmospheric pressure upward to desired operating pressure. Products generally stored in this type tank are fuel oils, diesel fuels, gasolines (small quantities), and sometimes more volatile products associated with vapor recovery systems. The vertical open-vented closed tank (Fig. 8.4-3) with an internal floating roof (IFR) and the vertical open-top tank (Figs. 8.4-4 and 8.4-5) with an external floating roof (EFR) operate at atmospheric pressure. The IFR and EFR tanks are used to store volatile products of a wide range including gasolines, aviation fuels, crude oils, napthas, and many other intermediate products generated in the crude oil refining process.

The closed tank may range from a small vertical or horizontal aboveground or horizontal underground tank storing fuel oil for a small boiler installation to a very large diameter vertical cylindrical flat-bottom fixed-roof tank storing bulk quantities of heavy fuel oil for a power plant. The open-top or open-vented closed tank with EFR or IFR may range from a small diameter vertical cylindrical flat-bottom tank in a small gasoline distribution terminal to a very large tank in a high-volume crude oil shipping terminal. Some standard tank sizes are given in the standards discussed later in this section.

In recent years federal, state, and local environmental regulations have been established that influence the type of tank to be selected for storage of petroleum products. Two major factors will

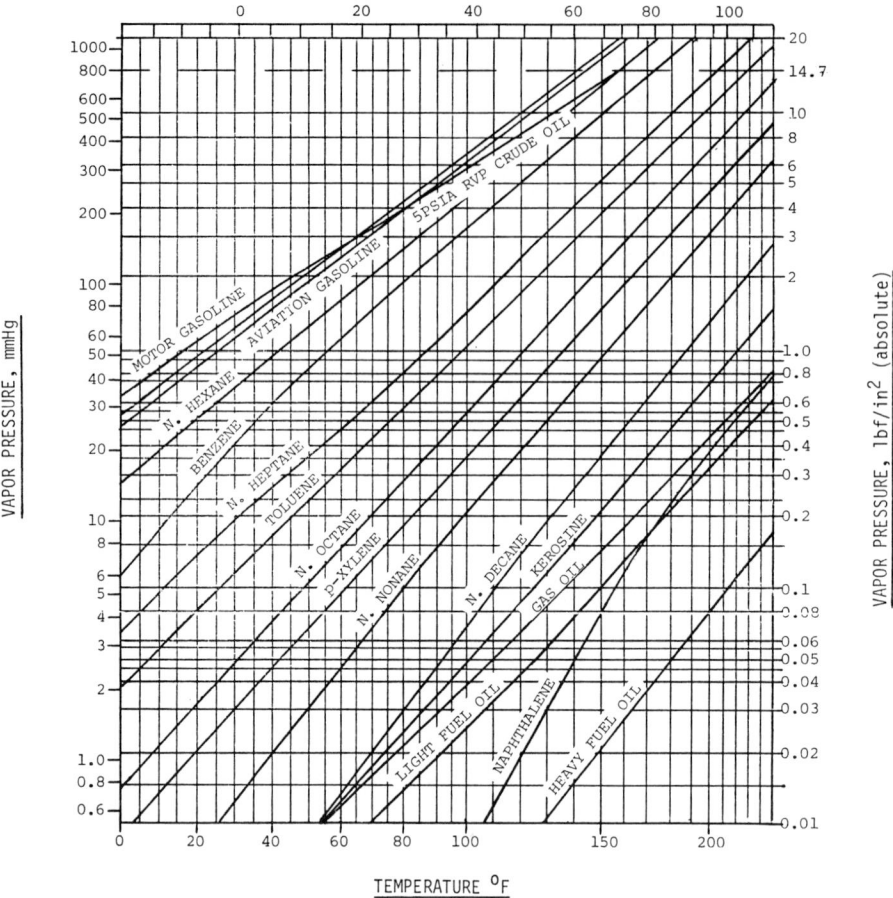

Fig. 8.4-1   Vapor pressures for some hydrocarbons. (Courtesy of CBI Industries.)

determine the type of tank required: quantity to be stored and true vapor pressure of the product. Most regulations allow quantities of any product less than 151,412 L (40,000 gal) to be stored in closed tanks with normal venting to the atmosphere. Quantities of product over 151,412 L (40,000 gal) and having a true vapor pressure 78 mm Hg (1.5 psia) or less may be stored in closed tanks with normal venting to the atmosphere. Quantities of product over 151,412 L (40,000 gal) and having a true vapor pressure over 78 mm Hg (1.5 psia), but not greater than 570 mm Hg (11.1 psia), are stored in an open-top tank with an EFR or an open-vented closed tank with an IFR. There may be slight variations of these rules enforced by environmental authorities in each state or district (usually county). Regulations should be determined clearly before a storage facility is designed. The federal rules are published[1] and are adopted by many states.

Tank Standards and Codes

Several storage tank standards or codes are available, and the application of each depends on the tank size, materials to be considered, exposure (whether underground or aboveground), and ultimate use. Each tank standard or code covers all aspects of design, materials, fabrication, construction, and testing. In each installation state and local jurisdictional requirements should be determined for the acceptance of an individual code or standard.

Standards for small shop-produced underground or aboveground closed tanks are Underwriters' Laboratories standards.[2,3] Both of these standards cover capacities up to 189,265 L (50,000 gal) and are used primarily for storage of small quantities of motor gasoline, diesel fuel, and a range of fuel

Fig. 8.4-2 Aboveground closed tanks. (Courtesy of CBI Industries.)

Fig. 8.4-3 Vertical open-vented closed tank with internal floating roof (KFR). Vents at top of tank wall and center roof. (Courtesy of CBI Industries.)

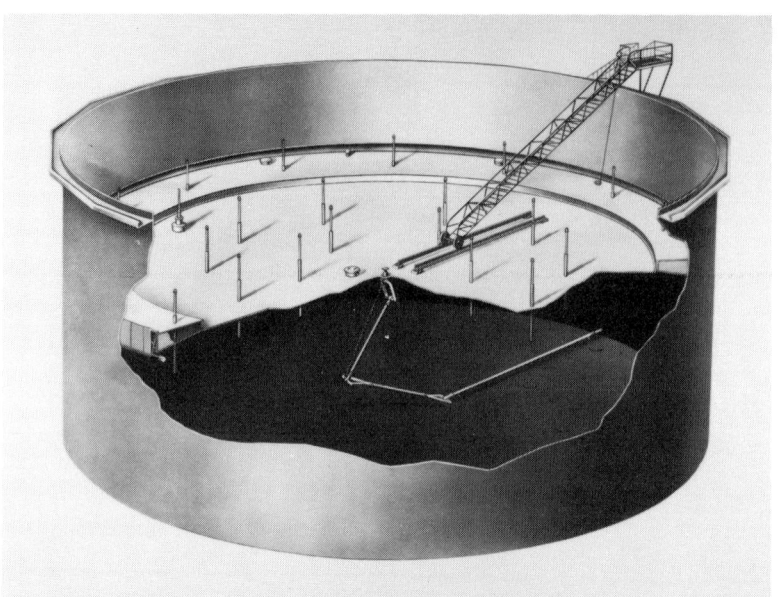

Fig. 8.4-4 Vertical open-top tank with external floating roof (EFR), pontoon type. (Courtesy of CBI Industries.)

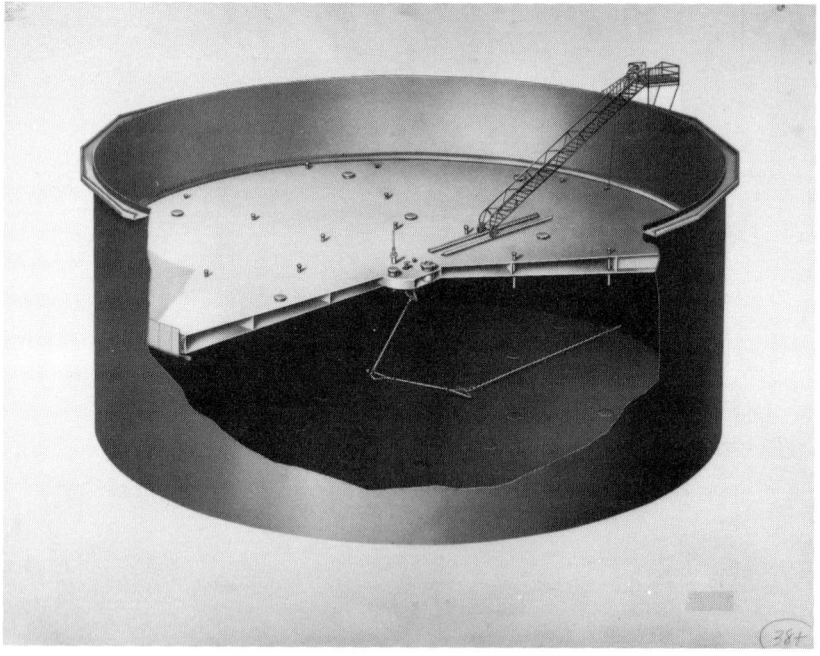

Fig. 8.4-5 Vertical open-top tank with external floating roof (EFR), double deck type. (Courtesy of CBI Industries.)

oils. This type tank is usually limited to maximum internal pressure of 6.89 kPa (1 psig) by the rules of the National Fire Prevention Association.[4] The tanks are shop assembled by welding preformed sections of relatively thin sheet or plate steel.

A standard for small closed tanks to be used inside a building up to 2500 L (660 gal) is published by Underwriters' Laboratories.[5] This type tank is used for storage of low vapor pressure fuel oils such as diesel fuel or light heating oil. The tanks are shop assembled by welding preformed sections of light gauge steel.

API Standard 650[6] is the most used standard for storage tanks for larger quantities. This standard provides requirements for vertical, cylindrical, aboveground, closed, or open-top welded storage tanks. The material used is a variety of carbon and carbon–manganese steel plate. The quality of plate required will depend on the design metal temperature conditions at the tank location. The size of the tank designed to this standard will usually vary from about 160 m^3 (1000, 42-gal barrels) to maximum diameters allowed by the maximum thickness limitations. Capacities up to 238,480 m^3 (1.5 million 42-gal. barrels) have been constructed. The standard covers both shop-assembled tanks and site-assembled tanks. It also covers design requirements for closed tanks with internal gas pressure up to 17.24 kPa (2.5 psig).

API Standard 620[7] provides rules for design of tanks to store products requiring internal pressures up to 103.42 kPa (15 psig). For the normal storage of petroleum products, tanks designed to this standard are uncommon unless an internal pressure is required for an inert-gas blanket on the product or if the tank is linked to a vapor recovery system to control emissions. Appendices Q and R in this standard provide rules for tank designs storing refrigerated hydrocarbon liquid gases such as butane, propane, ethylene, and methane. API Standard 620[7] is often specified for storage of petrochemicals and chemicals where low-pressure storage may be required because of the storage system employed.

The American Petroleum Institute (API) also publishes three other standards providing rules for tanks to store production liquids (oil field service). These standards are issued by the production department of API for tanks located at oil production sites. These standards cover bolted and welded vertical, cylindrical aboveground closed design, both shop assembled and site assembled. These standards are API Standard 12B[8], API Standard 12D[9], and API Standard 12F.[10]

Three other codes often referenced for design of a petroleum storage facility are NFPA 30,[4] NFPA 11,[11] and NFPA 78.[12] Chapter 2 in NFPA 30[4] discusses aspects of safety concerning tank storage. Appendix A in NFPA 11[11] describes requirements for foam extinguishing equipment. Chapter 6 in NFPA 78[12] describes minimum requirements for lightning protection for storage tanks. Similar information is also contained in API RP 2003.[13]

Tank Location and Spill Protection

Recommendations for locating storage tanks with respect to property lines, public ways, buildings, and to each other are given in NFPA 30.[4] This code is generally accepted throughout the United States for minimum requirements of tank spacing. Local authorities, tank owners, or insurance requirements may implement more restrictive requirements depending on such factors as terrain conditions, personnel concentrations, fire protection facilities, and amount of land available.

Accidental spill protection around tanks will vary depending on a number of factors, including land use, location or property, slope of property, tank spacing, and sometimes type of soil. Two types of impounding are generally considered: remote impounding or impounding by diking. Remote impounding is a concept more often used for larger-volume facilities and requires a sloping terrain. It is a relatively new concept, providing some advantages over the usual concept of diking, primarily providing a pathway for flow of liquid to a remote area away from the tanks. Impounding by diking may involve separate dikes around each tank or groups of tanks. Dikes usually consist of earth, concrete, or steel. The height of the dike will depend on the area allowed for protection around the tank or tanks. NFPA 30[4] provides recommendations for volume design of impounding provisions.

Foundations

Proper foundation design for storage tanks is extremely important to prevent nonuniform movement of the relatively thin cylindrical tank walls. Nonuniform movement can result in severe distortion of the relatively thin tank wall and tank bottom material, preventing proper operation of floating roofs or resulting in strains in the tank that may ultimately cause some degree of failure. The amount of allowable differential settlement of a storage tank will depend on a number of factors, such as tank diameter, tank height, thickness of material, internal details, tank connections, and piping arrangements.

The type of foundation selected to support a storage tank will be influenced by the site soil conditions and the size of tank. Concrete slab foundations placed on suitable soil may be economical for small shop-built tanks and some small site-constructed tanks. Larger site-erected tanks are generally supported on a combination concrete ringwall–grade foundation or a suitable grade-type foundation. Unusual conditions of very low bearing value soil may require pile-supported concrete slabs, especially if settlement must be severely restricted. Some factors to be considered for foundation design for site-constructed tanks are included in API Standard 620[7] and API Standard 650.[6]

Tank Venting

Excessive internal or external pressure is a common cause of storage tank failure. All storage tanks should have a dedicated connection for the purpose of venting. Minimum venting requirements for tanks constructed to the various API tank standards are given in API Standard 2000.[14] The Underwriters' Laboratories standards for shop-fabricated tanks contain minimum venting requirements by selection of the connection size from tables. All tanks storing petroleum products must have venting provisions that consider both normal operating venting conditions and fire exposure venting conditions.

Proper vent sizing is extremely important for protection for the tank integrity. This is especially important for a closed-type tank with the roof-to-tank wall joint not designed as a frangible joint applicable to and as described in API Standard 650.[6] Special in-line venting arrangements combined with inert-gas blanket systems or liquid traps should be avoided as a primary venting system unless carefully arranged for positive protection of the tank. With these types of primary venting arrangements, it is recommended that a separate dedicated vent to atmosphere be provided with a discharge pressure setting slightly higher than the primary system.

For closed-type tanks storing hot products such as heavy fuel oils, residual fuels, and asphalts, venting requirements should be carefully considered if the roof or shell of the tank is not insulated. Rapid cooling of a steel roof by a cold rain can result in a severe external pressure if the vent sizing is inadequate.

Evaporation Loss or Emission Control

Standards of performance for emission control are discussed in the beginning of this section and in Section 8.6. Many regulations require that best available control technology (BACT) be applied to new, modified, or additions to petroleum storage facilities. Emissions also means product loss by evaporation and, therefore, is an important factor for selections of tank and detail.

It is clear that closed tanks designed for approximate atmospheric pressure should generally be limited to very low vapor pressure product and to small quantities of product storage with higher vapor pressures so that evaporation losses are minimum.

External floating-roof (EFR) open-top tanks and internal floating-roof (IFR) open-vented closed tanks should be used for all products with vapor pressures of 78 mm Hg (1.5 psia) or higher as recommended by 40 CFR 60[1] for maximum evaporation loss control.

Any EFR should be designed for full contact with the liquid surface when fully buoyant in the liquid. The two basic EFRs are "pontoon" and "double deck." General design requirements are given in Appendix C, API Standard 650.[6] Some variations will be evident for each manufacturer's design and detail. Selection of EFR type may be influenced by loading conditions, weather conditions, drainage requirements, and economy. The pontoon roof is the most used EFR and when properly designed can contain large amounts of water if the primary roof drain is inoperable. The double-deck roof provides a positive water drainage to the roof sump, but the designs usually limit the amount of water containment capacity with the primary drain inoperable.

IFR generic designs are described in Appendix H, API Standard 650.[6] Also a description of the fundamental evaporation loss factors are given in API Standard 2519.[15] These factors should be evaluated for each design considered.

Selection of the peripheral seal system between the floating roof and tank wall rim space is a major factor for evaporation loss control whether an EFR or an IFR is selected. Loss factors are given in API Standard 2517[16] for each generic type of seal system. Figures 8.4-6 and 8.4-7 show two examples of seal systems for the floating-roof rim space. Evaluation of loss predictions can be made using the API Standard 2517[16] calculation procedures for each seal system. Results can be used to evaluate economic considerations and specific emission requirements.

Tank Maintenance

Regular inspection and maintenance is an important factor in extending tank life, maintaining safety conditions, and preventing product loss. The American Petroleum Institute regularly reviews and publishes a series of "Guide for Inspection of Refinery Equipment." Chapter XIII, "Atmospheric and Low Pressure Storage Tanks," provides guidelines for storage tank inspection.

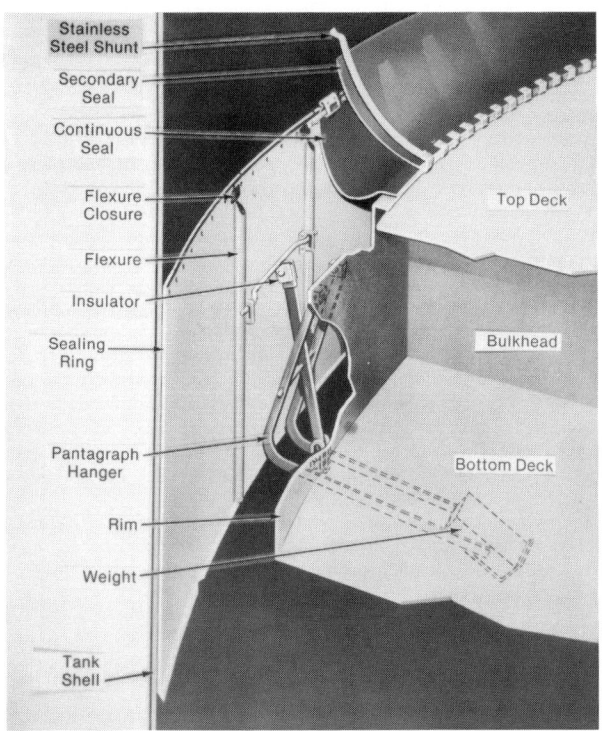

Fig. 8.4-6 Mechanical show primary seal with secondary seal. (Courtesy of CBI Industries.)

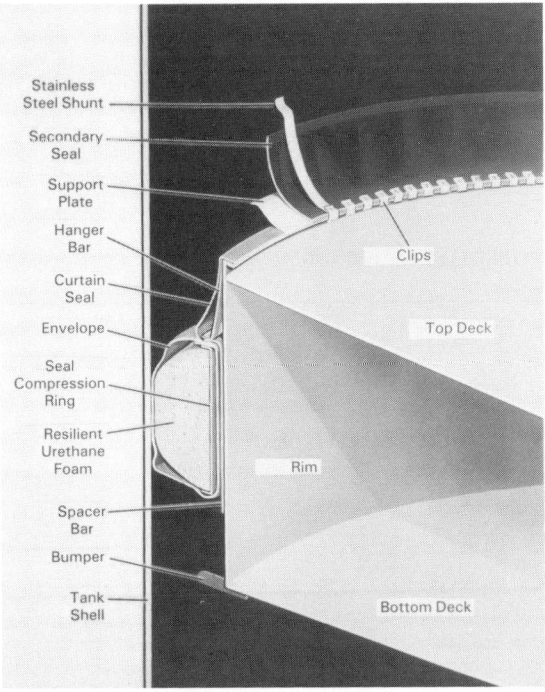

Fig. 8.4-7 Liquid-mounted resilient primary seal with secondary seal. (Courtesy of CBI Industries.)

8.4-2 Liquefied Petroleum Gas

Ivan V. LaFave

Liquefied petroleum gas is also referred to synonymously as LP-gas or LPG and sometimes as bottled gas. It is a very versatile energy source and is widely used in many parts of the world. It is usually a mixture of several hydrocarbon gases but can also be predominantly one of them. The principal constituents are propane, propylene, normal butane, isobutane, and the butylenes. In some mixtures there may also be small quantities of ethane, ethylene, isopentane, and normal pentane.

The LPG industry[21] has developed definitions and specifications for several different LP-gas mixtures used in commerce and industry. Included are specifications for commercial propane, commercial butane, butane–propane mixtures, and propane for motor fuel.

Liquefied petroleum gas is produced primarily from two sources. One source is the natural gas liquids (NGL), which are taken from the gas streams at natural gas processing plants. The other source is liquefied refinery gases (LRG), which are derived from the refinery cracking of petroleum.

When the source is natural gas liquids, the principal components are propane and the butanes. When the source is refinery gas, there is likely to be butylenes and propylene in the mixture.

Each of the principal hydrocarbons in LPG is a gas at normal ambient temperature. The critical temperature is above the ambient temperature range and each or any mixture of them can be liquefied under moderate pressure at normal ambient temperatures. Table 8.4-1 presents the properties of interest of the principal hydrocarbons found in LPG.

Practically all energy applications use the product in its gaseous state. When stored as a liquid under pressure, the vapor space above the liquid is filled with the product vapor at the storage pressure. The vapor pressure is dependent on the component hydrocarbon gases in the mixture and the temperature at the liquid surface.

LPG is readily storable and transportable in the liquid state, making it a convenient concentrated energy source. It is used in a variety of applications in food processing, industrial and agricultural processing, power generation, residential cooking, space heating and cooling, and in transportation as a fuel for motor vehicles.

LPG has utility because in the gaseous state it will burn in air, producing heat. It is highly flammable and can be hazardous when it is not handled properly. Odorization is used for the detection of leaks. There have been many serious accidents resulting in personal injuries and property damage as the result of improper procedures and failure to understand the fundamentals of liquefied petroleum gas handling. All who are concerned with LP-gas should have a full knowledge of its properties and should understand all the requirements for proper handling.[23]

The LP-gas liquid state can be achieved by refrigeration as well as compression of the gas. There are some cases when partial refrigeration is employed to reduce the required storage pressure. It is transported by ships, railroad cars, trucks, pipelines, and in portable containers. Full pressure is most often used to maintain the gas mixtures in the liquid state, but some is transported in special ships in the fully refrigerated condition.

LPG is stored at producing facilities, at bulk distribution plants, at utility gas plants, and at dealer and user facilities. The type of storage depends on the volume to be stored, the condition of the

TABLE 8.4-1[a] PROPERTIES OF PRINCIPAL HYDROCARBON GASES IN LIQUEFIED PETROLEUM GAS

Hydrocarbon Gas	Normal Boiling Point	Critical Temperature	Vapor Pressure (100°F psia)	Vapor Pressure (40°C kPa, abs)
Propane,	−42.07°C	96.67°C	188.0	1341
C_3H_8	−43.73°F	206.01°F		
n-Butane,	−0.49°C	152.01°C	51.54	377
C_4H_{10}	31.12°F	305.62°F		
Isobutane,	−11.81–C	134.98°C	72.39	528
C_4H_{10}	10.74°F	274.96°F		
1-Butene	−6.23°C	146.38°C	62.10	451.9
(butylene),	20.79°F	295.48°F		
C_4H_8				
Propylene,	−47.72°C	91.7°C	227.6	1596
C_3H_6	−53.90°F	197.06°F		

[a] References 21 and 22 give extensive property data for gases in LPG.

product when received, the rate at which it is received, and the use of the product or the method of transporting it from the facility. The choice between full pressure, full refrigeration, or partial refrigeration with reduced pressure is generally made based on an economic analysis of the costs for the alternative methods.

Producing facilities handle large volumes, and the storage is either in refrigerated low-pressure insulated tanks or large spherical or cylindrical pressure vessels. Some producing facilities have both types of storage.

At bulk distribution plants cylindrical pressure vessels are commonly used. These are either horizontal or vertical. A utility gas plant will utilize refrigerated storage or pressure storage and in some instances both. A dealer will usually have full-pressure storage in the form of horizontal cylindrical pressure vessels.

The user or consumer may have bulk storage, may receive and use it directly from a pipeline, or may receive it in portable containers or cylinders, which are transported to the point of use by motor carrier.

Most LPG installations will fall under the jurisdiction of some authority associated with the federal, state, or local government. This "authority having jurisdiction" will have promulgated or adopted a set of regulations that form the basis for the rules governing the siting, storage, handling, and safety for the installation. It is necessary to become knowledgeable regarding the authorities that have jurisdiction and the rules that apply to a specific installation.

The National Fire Protection Association (NFPA) publishes standards which are frequently adopted or referenced by regulatory authorities. Two NFPA standards specifically cover LPG.[18,19]

LPG storage containers are of two general types. For full-pressure storage and partially refrigerated storage, pressure vessels are used. Design pressure will range from 345 kPa (50 psi) to 2155 kPa (312.5 psi). Fully refrigerated LPG is stored in flat-bottom low-pressure cylindrical tanks.

The smallest tanks used for storage are the portable DOT cylinders, which are built and marked in accordance with the regulations of the U.S. Department of Transportation, CFR49.[17] They range in size up to 454 L (120 gal) or 453.6 kg (1000 lb) full water capacity. The largest cylinder will contain approximately 190.5 kg (420 lb) of propane. Filling densities (ratio of product weight to full water weight) are specified by the DOT regulations and NFPA 58. For residential, commercial, and small industrial loads, multiple-cylinder installations are frequently used.

Fig. 8.4-8 Cylindrical ASME pressure vessels for full-pressure LPG storage. (Courtesy of CBI Industries.)

Fig. 8.4-9 Spherical ASME pressure vessels for full-pressure LPG storage. (Courtesy of CBI Industries.)

Fig. 8.4-10 Insulated ASME pressure vessels for partially refrigerated LPG storage, capacity 50,000 barrels. (Courtesy of CBI Industries.)

Fig. 8.4-11 Insulated flat bottom API 620 tank for fully refrigerated LPG storage, capacity 600,000 barrels. (Courtesy of CBI Industries.)

For larger volumes storage is in ASME[20] cylindrical pressure vessels. Portable ASME tanks, which usually have skids attached, are built in capacities up to about 7571 L (2000 gal). These "skid tanks" must meet the requirements of DOT.[17] ASME cylindrical pressure vessels for fixed installations will range in size up to 454,250 L (120,000 gal). They are usually supported on two saddles that rest on concrete foundations (Fig. 8.4-8).

For storage volumes in excess of the 120,000-gal size, either multiple cylindrical ASME tanks or spherical pressure vessels are used (Figs. 8.4-9 and 8.4-10). The spheres are also designed and built in accordance with the ASME code and are usually of a size that requires construction at the LPG storage site. Spheres are built in capacities ranging from 159 m^3 (1000 barrels) (42,000 gal) to 7949 m^3 (50,000 barrels) (2. million gals).

Spherical pressure vessels are normally supported by cylindrical columns attached to the shell near the equator of the sphere. The number of columns varies with the size of the structure. The supports are designed to carry the gross weight of the vessel full of water. Usually each column rests on an individual concrete pier. Local conditions such as low soil bearing capacity or the possibility of earthquakes may necessitate the use of piling or the design of special foundations.

For still larger volumes of LPG the product is fully refrigerated to a temperature near the normal boiling point of the product and stored in field-constructed flat-bottom insulated tanks. This type of storage is frequently used at producing facilities, marine receiving terminals, and utility gas plants. The design temperature varies with the product and may be as low as $-50°C$ ($-58.0°F$). The design pressure of the tank usually will be between 6.89 kPa (1 psi) and 20.7 kPa (3 psi). Tank capacities range from 3975 m^3 (25,000 barrels) to 143, 090 m^3 (900,000 barrels) (Fig. 8.4-11).

The tank walls are cylindrical. The dome roofs are either spherical or ellipsoidal. The tank rests upon some form of load-bearing insulation, which transmits the weight of the contents to the foundation and the ground beneath the tank. A source of heat is required in the foundation to prevent the freezing isotherm from penetrating the soil beneath the tank.

Anchorage for flat-bottom tanks normally is required to hold the shell down against uplift forces from wind, earthquakes, and the internal pressure acting under the roof. Anchorage must also permit the vessel to move radially in response to temperature changes. A system of anchor straps or bolts attached to the tank shell embedded in a concrete foundation is the usual method of providing anchorage.

REFERENCES*

8.4-1 "Code of Federal Regulations, Title 40, Part 60, Subparts K and Ka," Office of the Federal Register National Archives and Records Service, General Services Administration, Washington, DC, 1980.

8.4-2 UL58, "Standard for Steel Underground Tanks for Flammable and Combustible Liquids," Underwriters' Laboratories, Inc., 1976.

8.4-3 UL142, "Standard for Steel Aboveground Tanks for Flammable and Combustible Liquids," Underwriters' Laboratories, Inc., 1981.

8.4-4 NFPA 30, "Flammable and Combustible Liquids Code," National Fire Protection Association, Quincy, MA, 1981.

8.4-5 UL80, "Standard for Steel Inside Tanks for Oil-Burner Fuel," Underwriters' Laboratories, Inc., 1980.

8.4-6 API Standard 650, "Welded Steel Tanks for Oil Storage," American Petroleum Institute, Washington, DC, 1980.

8.4-7 API Standard 620, "Recommended Rules for Design and Construction of Large, Welded, Low Pressure Storage Tanks," American Petroleum Institute, Washington, DC, 1982.

8.4-8 API Standard 12B, "Specification for Bolted Tanks for Storage of Production Liquids," American Petroleum Institute, Washington, DC, 1977.

8.4-9 API Standard 12D, "Specifications for Field Welded Tanks for Storage of Production Liquids," American Petroleum Institute, Washington, DC, 1982.

8.4-10 API Standard 12F, "Specifications for Shop Welded Tanks for Storage of Production Liquids," American Petroleum Institute, Washington, DC, 1982.

8.4-11 NFPA 11, "Foam Extinguishing Systems," National Fire Protection Association, Quincy, MA, 1983.

8.4-12 NFPA 78, "Lightning Protection Code," National Fire Protection Association, Quincy, MA, 1983.

8.4-13 API RP 2003, "Recommended Practice for Protection Against Ignitions Arising out of Static Lightning and Stray Currents," American Petroleum Institute, Washington, DC, 1982.

8.4-14 API Standard 2000, 2nd ed., "Venting Atmospheric and Low-Pressure Storage Tanks (*Nonrefrigerated and Refrigerated*)," American Petroleum Institute, Washington, DC, 1982.

8.4-15 API Standard 2519, 3rd ed., "Evaporation Loss from Internal Floating Roof Tanks," American Petroleum Institute, Washington, DC, 1983.

8.4-16 API Standard 2517, 2nd ed., "Evaporation Loss from External Floating Roof Tanks," American Petroleum Institute, Washington, DC, 1980.

8.4-17 U.S. Department of Transportation (CFR 49) Code of Federal Regulations Title 49 Part 178, Office of the Federal Register National Archives and Records Service, General Services Administration, Washington, DC, 1981.

8.4-18 NFPA 58, "Storage and Handling Liquefied Petroleum Gases," National Fire Protection Association, Quincy, MA, 1983.

8.4-19 NFPA 59, "LP-Gases At Utility Gas Plants," National Fire Protection Association, Quincy, MA, 1984.

8.4-20 *ASME Boiler and Pressure Vessel Code Section VIII*, American Society of Mechanical Engineers, New York, 1983.

8.4-21 *Engineering Data Book* and LPG product technical information, Gas Processors Association and Gas Processors Suppliers Association, Tulsa, OK, 1979.

8.4-22 *Matheson Gas Data Book*, Matheson Publications, Hasbrouck Heights, NJ, 1981.

8.4-23 Consumer and Safety Information, National LP-Gas Association, Oak Brook, IL.

8.5 FUEL OILS: HANDLING AND BURNING

Charles W. Ruoff and Joong Min Soh

8.5-1 Fuel Oils and Handling

Fuel Classification and Properties

There are many varying classifications, nomenclature, and properties of fuel oils. The broadest classification is light oils versus heavy oils. The chemical composition, specifically the hydrogen–carbon ratio plays the largest role in the fuel's characteristics. Table 8.5-1 shows a typical listing of fuel oils and their various characteristics.

*References 8.4-2 through 8.4-20 are on-going publications. The reader is advised to seek the latest edition. Reference 8.4-23 refers to a series of information bulletins issued at random intervals. They are undated.

TABLE 8.5-1

	Gasoline	No. 1 Oil	No. 2 Oil	No. 4 Oil	No. 5 Oil	No. 6 Oil
Flash point, °F	−45°F	100–165°F	100°F	130°F	over 130°F	above 150°F
Fire point, °F		152°F	200°F	230°F	260°F	325°F
Autoignition temperature, °F	495°F	444°F	494°F	505°F	650°F	765°F
Explosion limits of vapor in air, %	1.4–7.6	0.7–5.0	—	—	—	—
Average boiling range	90–363°F	345–510°F	93–365°F	—	—	—
Conradson carbon residue, wt. %		0.15	0.35	4	7	10–12
Specific heat at 40°F	0.492	0.450	0.436	0.428	0.423	0.415
Vapor pressure, mm Hg at 60°F	7	2	1	12	0.2	2.8
Ash, wt. % (max)	—	—	—	0.1	0.1	0.1

Light oils or distillates contain a larger portion of hydrogen than the heavier oils or residuals. Lighter oils ignite more easily and therefore have low flash and fire points. The flash point of an oil is the lowest temperature at which gases given off will give a flash when ignited. It is the lower temperature limit of flammability and serves as a criterion of fire hazards in storage and handling. The fire point is the temperature at which the gases given off may be ignited and will continue to burn. An oil that has a low flash point will burn more readily than one with a high flash point.

Pour point is the lowest temperature at which the oil flows under standard conditions. Cloud point is defined as the temperature at which a cloud or haze appears when the oil is cooled under specified conditions. Carbon residue is a measure of the solid material left from an oil sample subjected to evaporation and pyrolysis. It also provides a rough approximation of the tendency for the fuel to produce smoke in the flue gas and contribute to carbon deposition in certain types of burners. The common carbon residue tests in use are the Conradson and Ramsbottom tests.*

Stability is the resistance of oil to breakdown, forming sludge. Autoignition temperature is defined as the minimum temperature to which fuel must be heated to produce self-ignition without an external flame source. Ash content in fuel is the percentage of inorganic residue, with its value varying from a trace in gas oil to 0.1% in residue fuels.

Fuel oils can be classified by source such as residuals, crudes, distillates, and blends.[1] Residuals or heavy fuels are the remaining products left after removing the more volatile hydrocarbons. By removing these higher volatiles, the fuel is safer to handle and burn. Crude is what comes from the oil well without processing.

Distillates are derived by fractional distillation of the crudes and produce the lighter more volatile grades of oil. Blended oils are mixtures of two or more of the above.

Heavy fuels cannot be handled easily without first heating to lower the viscosity. Viscosity, the fluid's resistance to flow, is usually expressed in the fuels industry as Saybolt Universal Seconds (SUS) or centistokes. Table 8.5-2 gives standard viscosity conversions. All liquid fuels in order to be burned must be of a viscosity that will promote good atomization. For practical purposes this is usually in the range of 150–200 SUS for heavier fuels. Lighter fuels with lower viscosities do not need to be heated.

Specific gravity of fuel oil is the primary index of its classification and is measured by a hydrometer in degrees Baume or API (American Petroleum Institute). Knowing the specific gravity of a fuel, it may be converted to degrees API by the following equation:

$$°\text{API} = \frac{141.5}{\text{sgr at } 60°/60°\text{F}} - 131.5 \tag{8.5-1}$$

where sgr at 60°/60°F represents the ratio of oil density at 60°F to water density also at 60°F. Table 8.5-3 gives a range of analysis of fuel oils. Typically most fuel oils have an API range of 10–40.

---

*The Conradson residue test (ASTM D189-81) is commonly used for heavy residual fuels or coker feedstocks, which cannot be easily loaded into a Ramsbottom glass bulb. Direct flame heating is used in the Conradson test while molten bath heating is used in the Ramsbottom test (ASTM D524-81).

TABLE 8.5-2

Kinematic viscosity, centistokes	Equivalent Redwood No. 1 viscosity, seconds, at			Equivalent Saybolt Universal viscosity, seconds, at			Equivalent Engler viscosity, degrees — All temperatures	Kinematic viscosity, centistokes	Equivalent Saybolt Furol viscosity, at 122°F, seconds
	70°F	100°F	200°F	100°F	122°F	210°F			
2	30·22	30·65	31·22	32·62	32·67	32·85	1·141	50	26·1
3	32·72	33·15	33·72	36·03	36·08	36·28	1·225	55	28·3
4	35·33	35·65	36·33	39·14	39·20	39·41	1·309	60	30·6
								65	32·8
5	37·94	38·25	38·94	42·35	42·41	42·65	1·401	70	35·1
6	40·55	40·85	41·55	45·56	45·62	45·88	1·482		
7	43·26	43·55	44·21	48·77	48·84	49·11	1·565	75	37·4
8	46·07	46·25	46·97	52·09	52·16	52·45	1·655	80	39·6
9	48·93	49·05	49·73	55·50	55·58	55·89	1·749	85	41·9
								90	44·1
10	51·79	51·95	52·64	58·91	58·99	59·32	1·840	95	46·4
12	58·01	58·07	58·86	66·04	66·13	66·50	2·024		
14	64·49	64·54	65·39	73·57	73·68	74·09	2·224	100	48·6
16	71·32	71·39	72·37	81·30	81·41	81·87	2·439	110	53·2
18	78·29	78·40	79·54	89·44	89·57	90·06	2·650	120	57·8
								130	62·4
20	85·63	85·75	87·13	97·77	97·91	98·45	2·877	140	67·0
22	93·15	93·28	94·75	106·4	106·6	107·1	3·108		
24	100·7	100·8	102·5	115·0	115·2	115·8	3·344	150	71·7
26	108·4	108·6	110·3	123·7	123·9	124·5	3·584	160	76·3
28	116·1	116·3	118·3	132·5	132·7	133·4	3·830	170	81·0
								180	85·6
30	123·9	124·1	126·4	141·3	141·5	142·3	4·081	190	90·3
35	143·7	144·0	140·6	163·7	163·9	164·9	4·708		
40	163·7	164·1	167·2	186·3	186·6	187·6	5·350	200	95·0
45	183·8	184·3	188·1	209·1	209·4	210·5	5·993	220	104·3
50	203·9	204·4	208·9	232·1	232·4	233·8	6·650	240	113·7
								260	123·0
55	224·0	224·6	229·8	255·2	255·6	257·0	7·260	280	132·4
60	244·2	244·8	250·8	278·3	278·7	280·2	7·920		
65	264·4	265·0	271·6	301·4	301·8	303·5	8·580	300	141·8
70	284·8	285·4	292·6	324·4	324·9	326·7	9·240	320	151·2
75	304·7	305·4	313·4	347·6	348·1	350·0	9·900	340	160·6
								360	170·0
79·9	324·5	325·3	333·9	370·3	370·8	372·9	10·547	380	179·4
								399	188·3

Source: Technical Data on Fuel, The British National Committee, World Power Conference, London, 1962. Used by permission.

TABLE 8.5-3

Grade of Fuel Oil	No. 1	No. 2	No. 4	No. 5	No. 6
Weight, percent					
Sulfur	0.01-0.5	0.05-1.0	0.2-2.0	0.5-3.0	0.7-3.5
Hydrogen	13.3-14.1	11.8-13.9	(10.6-13.0)°	(10.5-12.0)°	(9.5-12.0)°
Carbon	85.9-86.7	86.1-88.2	(86.5-89.2)°	(86.5-89.2)°	(86.5-90.2)°
Nitrogen	Nil-0.1	Nil-0.1	—	—	—
Oxygen	—	—	—	—	—
Ash	—	—	0-0.1	0-0.1	0.01-0.5
Gravity					
Deg API	40-44	28-40	15-30	14-22	7-22
Specific	0.825-0.806	0.887-0.825	0.966-0.876	0.972-0.922	1.022-0.922
Lb per gal	6.87-6.71	7.39-6.87	8.04-7.30	8.10-7.68	8.51-7.68
Pour point, F	0 to −50	0 to −40	−10 to +50	−10 to +80	+15 to +85
Viscosity					
Centistokes @ 100F	1.4-2.2	1.9-3.0	10.5-65	65-200	260-750
SUS @ 100F	—	32-38	60-300	—	—
SSF @ 122F	—	—	—	20-40	45-300
Water & sediment, vol %	—	0-0.1	tr to 1.0	0.05-1.0	0.05-2.0
Heating value					
Btu per lb, gross (calculated)	19,670-19,860	19,170-19,750	18,280-19,400	18,100-19,020	17,410-18,990

° Estimated.

Courtesy of Babcock & Wilcox.

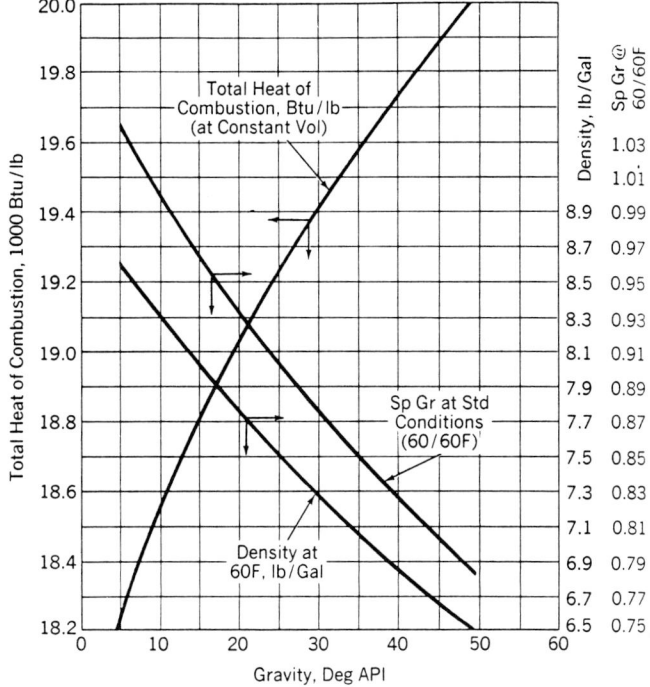

Fig. 8.5-1 Heating value, weight (lb/gal), and specific gravity of fuel oil for a range of API gravities. (Courtesy of Babcock & Wilcox.)

Heating Values

The heating value of a fuel is usually expressed as Btu per pound or Btu per gallon. The gross or higher heating value (HHV) of a fuel is the total heat that can be released through the heat of combustion per pound or gallon of fuel. Knowing the API gravity, the heating value of a fuel can be estimated using the following equation:

$$\text{Btu/lb of oil (HHV)} = 17{,}687 + 57.7 \times {}^\circ\text{API at } 60^\circ\text{F} \qquad (8.5\text{-}2)$$

This value (HHV) assumes that the water produced through combustion is condensed to liquid (which never happens in practice). The lower heating value (LHV) is based on the water remaining in the vapor phase. The relationship between HHV and LHV can be expressed as follows:

$$\text{LHV} = \text{HHV} - 1040\,W \qquad (8.5\text{-}3)$$

where 1040 is the heat of vaporization of water in Btu/lb and W is pound of water produced/pound of fuel.

Figure 8.5-1 shows the relationship of heating values of fuel oil for a range of API gravities.

Fuel Grades

Fuel oils are also classified for commercial purposes into grades ranging from No. 1 to No. 6, according to their gravity. Table 8.5-4 lists the six grades and their respective uses. References of Bunker C fuel mainly used in marine terminology refers to a commercial grade No. 6.

Storage and Handling

Requirements for safe storage and handling of fuel oils is usually explained and governed by national* and local codes and insurance company standards.

*For example, The National Fire Protection Association (NFPA).

TABLE 8.5-4

Detailed Requirements for Fuel Oils[A]

Grade of Fuel Oil	Flash Point, °C (°F) Min	Pour Point, °C (°F) Max	Water and Sediment, vol % Max	Carbon Residue on 10% Bottoms, % Max	Ash, weight % Max	Distillation Temperatures, °C (°F)			Saybolt Viscosity, s[D]				Kinematic Viscosity, cSt[D]						Specific Gravity 60/60°F (deg API) Max	Copper Strip Corrosion Max	Sulfur, % Max
						10% Point Max	90% Point Min	90% Point Max	Universal at 38°C (100°F) Min	Max	Furol at 50°C (122°F) Min	Max	At 38°C (100°F) Min	Max	At 40°C (104°F) Min	Max	At 50°C (122°F) Min	Max			
No. 1 A distillate oil intended for vaporizing pot-type burners and other burners requiring this grade of fuel	38 (100)	−18[C] (0)	0.05	0.15	...	215 (420)	...	288 (550)	...	...	...	...	1.4	2.2	1.3	2.1	...	...	0.8499 (35 min)	No. 3	0.5
No. 2 A distillate oil for general purpose heating for use in burners not requiring No. 1 fuel oil	38 (100)	−6[C] (20)	0.05	0.35	...	...	282[C] (540)	338 (640)	(32.6)	(37.9)	...	...	2.0[C]	3.6	1.9[C]	3.4	...	...	0.8762 (30 min)	No. 3	0.5[B]
No. 4 (Light) Preheating not usually required for handling or burning	38 (100)	−6[C] (20)	0.50	...	0.05	...	...	...	(32.6)	(45)	...	...	2.0	5.8	...	...	...	...	0.8762[G] (30 max)	...	...
No. 4 Preheating not usually required for handling or burning	55 (130)	−6[C] (20)	0.50	...	0.10	...	...	...	(45)	(125)	...	...	5.8	26.4[F]	5.5	24.0[F]	...	...	...	...	...
No. 5 (Light) Preheating may be required depending on climate and equipment	55 (130)	...	1.00	...	0.10	...	...	...	(>125)	(300)	...	...	>26.4	65[F]	>24.0	58[F]	...	...	...	...	...

Grade																			
No. 5 (Heavy) Preheating may be required for burning and, in cold climates, may be required for handling	55 (130)	1.00	...	0.10	...	...	...	(>300)	(900)	(23)	(40)	>65	194[k]	>58	168[k]	(42)	(81)	...	...
No. 6 Preheating required for burning and handling	60 (140)	2.00[e]	''	...	...	...	...	(>900)	(9000)	(>45)	(300)	...	...	...	...	>92	638[k]	...	...

[a] It is the intent of these classifications that failure to meet any requirement of a given grade does not automatically place an oil in the next lower grade unless in fact it meets all requirements of the lower grade.

[b] In countries outside the United States other sulfur limits may apply.

[c] Lower or higher pour points may be specified whenever required by conditions of storage or use. When pour point less than −18°C (0°F) is specified, the minimum viscosity for grade No. 2 shall be 1.7 cSt (31.5 SUS) and the minimum 90% point shall be waived.

[d] Viscosity values in parentheses are for information only and not necessarily limiting.

[e] The amount of water by distillation plus the sediment by extraction shall not exceed 2.00%. The amount of sediment by extraction shall not exceed 0.50%. A deduction in quantity shall be made for all water and sediment in excess of 1.0%.

[k] Where low sulfur fuel oil is required, fuel oil falling in the viscosity range of a lower numbered grade down to and including No. 4 may be supplied by agreement between purchaser and supplier. The viscosity range of the initial shipment shall be identified and advance notice shall be required when changing from one viscosity range to another. This notice shall be in sufficient time to permit the user to make the necessary adjustments.

[g] This limit guarantees a minimum heating value and also prevents misrepresentation and misapplication of this product as Grade No. 2.

[h] Where low sulfur fuel oil is required, Grade 6 fuel oil will be classified as low pour +15°C (60°F) max or high pour (no max). Low pour fuel oil should be used unless all tanks and lines are heated.

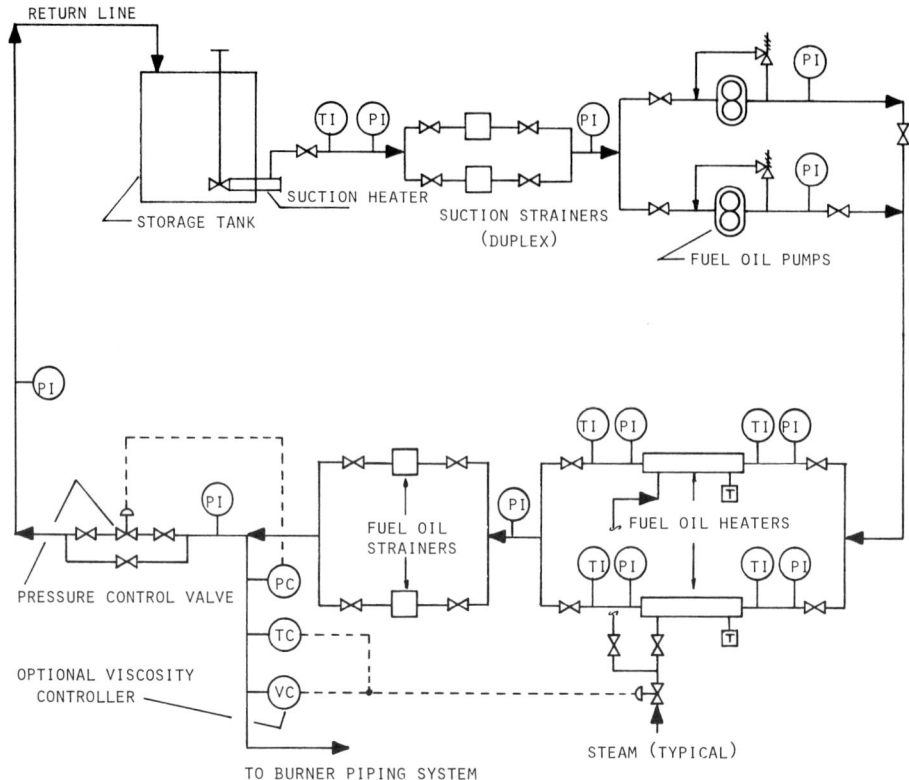

Fig. 8.5-2 Schematic of a typical heavy fuel oil handling system.

Proper attention must be given to the selection and arrangement of fuel handling equipment. The equipment must be able to function in extreme weather conditions handling the heaviest oil. A schematic of a typical fuel oil handling system is shown in Fig. 8.5-2.

All fuels, except carbon, burn as a gas. Liquid fuels will not burn in a liquid state and must first be converted into a gaseous state, or vaporized, and intimately mixed with air to form a combustible mixture. Vaporization of the fuel is accomplished through atomization by injecting the liquid fuel under high pressure through a small orifice or jet and/or by directing high-pressure steam or other atomizing medium into the fuel stream. This resulting pressure drop and shearing action break the fuel into tiny fine droplets. This atomization process is performed in the oil burner. The ability of the burner to atomize properly is primarily based on the fuel's viscosity. Heavier fuels cannot be pumped or atomized properly until the viscosity has been lowered by heating.

If heavy oils are being burned, a suction heater at the storage tank is required to lower the oil viscosity for pumping. The pumps are usually of a positive-displacement type and are sized to handle a specific flow and pressure at a given viscosity and speed. For liquid fuels such as naphtha and solvents, fuel lubricity needs to be considered in the selection of the pump. It is preferable to locate the circulating pumps as close as possible to the storage tank to prevent suction line vaporization. This will usually occur with about 8 in. Hg vacuum. Fuel oil strainers are provided at the pump suction and discharge. Usually a coarse screen is used on the suction (16–20 mesh) and a finer screen (40–60 mesh) is used on the discharge line. Strainers should be of a duplex design to facilitate cleaning one element while oil is passing through the other.

Heating and Control

The fuel oil heaters heat the oil to the proper temperature for atomization and burning. These heaters are usually supplied by steam, but can be electric. Typical fuel oil temperatures that will provide the

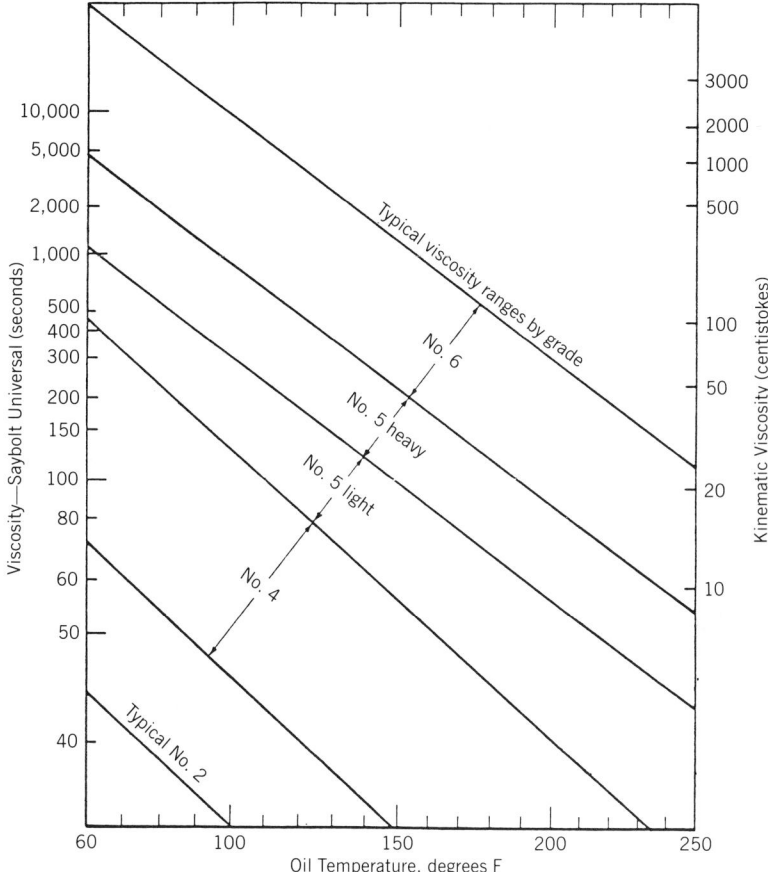

Fig. 8.5-3 Typical viscosity—relationships for various grades of fuel oil. (Copyright, ASTM, 1916 Race Street, Philadelphia, PA. 19103. Reprinted, with permission.)

proper viscosity for atomization are as follows:

Fuel Oil No.	Temperature at Burner (°F)
4	125–150
5	150–170
6	225–250

Lighter oils do not need to be heated. If the viscosity of the oil varies at a given temperature, then a viscosity control system, which varies oil temperature to maintain a specific viscosity, should be considered.

The maximum viscosity that oil burners will handle and still provide good atomization is about 150–200 SUS. A correlation of viscosity to temperature for various grades of fuel oils is shown in Fig. 8.5-3.

The fuel oil pressure controller is usually positioned in the oil circulating loop at a point as close to the burners as possible (Fig. 8.5-2). Constant fuel pressure is important to whichever fuel volume control is utilized. In order to assure good atomization when the burner is first started, the circulating loop should be designed so that heated oil is circulated as close to the burner as possible. This will reduce the amount of cold oil injected into the furnace before it ignites. Figures 8.5-4 and 8.5-5 show typical burner piping systems with the appropriate controls and safety devices. For heavy fuel oils such as No. 6 and tars, it is necessary to install a flushing connection in the burner fuel oil piping. After burner shutdown, the loops are purged with steam or lighter oil to prevent gumming.

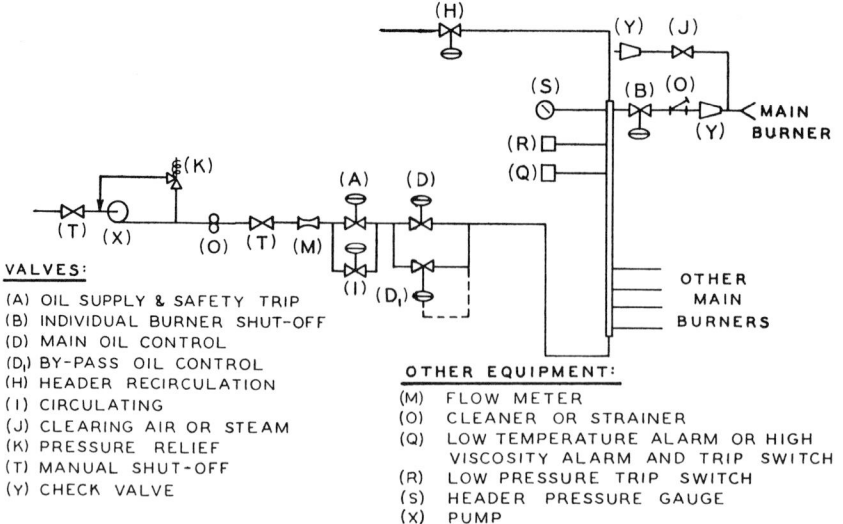

Fig. 8.5-4 Typical main oil burner (schematic)—mechanical atomizing. Valves: (A) oil supply and safety trip, (B) individual burner shutoff, (D) main oil control, (D₁) bypass oil control, (H) header recirculation, (I) circulating, (J) clearing air or steam, (K) pressure relief, (T) manual shutoff, (Y) check valve. Other Equipment: (M) flow meter, (O) cleaner or strainer, (Q) low-temperature alarm or high-viscosity alarm and trip switch, (R) low-pressure trip switch, (S) header pressure gauge, (X) pump. (Reprinted with permission from NFPA 85D-1978, Standard for Prevention of Furnace Explosions in Fuel Oil-Fired Multiple Burner Boiler-Furnaces, Copyright© 1978, National Fire Protection Association, Quincy, Massachusetts 02269. This reprinted material is not the complete and official position of the NFPA on the referenced subject, which is represented only by the standard in its entirety.)

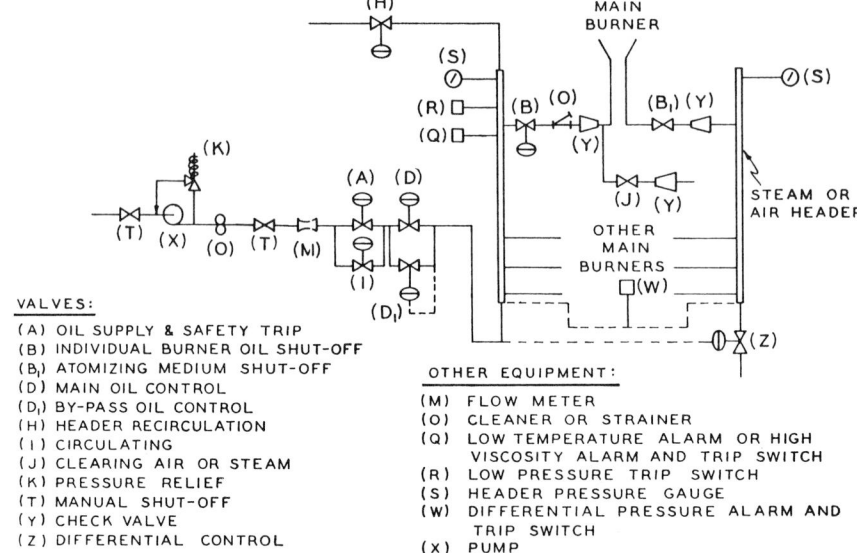

Fig. 8.5-5 Typical main oil burner (schematic)—steam or air atomizing. Valves: (A) oil supply and safety trip, (B) individual burner oil shutoff, (B₁) atomizing medium shutoff, (D) main oil control, (D₁) bypass oil control, (H) header recirculation, (I) circulating, (J) clearing air or steam, (K) pressure relief, (T) manual shutoff, (Y) check valve, (Z) differential control. Other Equipment: (M) flow meter. (O) cleaner or strainer, (Q) low-temperature alarm or high-viscosity alarm and trip switch, (R) low-pressure trip switch, (S) header pressure gauge, (W) differential pressure alarm and trip switch, (X) pump. (Reprinted with permission from NFPA 85D-1978, Standard for Prevention of Furnace Explosions in Fuel Oil-Fired Multiple Burner Boiler-Furnaces, Copyright© 1978, National Fire Protection Association, Quincy, Massachusetts 02269. This reprinted material is not the complete and official position of the NFPA on the referenced subject, which is represented only by the standard in its entirety.)

TABLE 8.5-5

Source of Crude Oil	Vanadium	Nickel	Sodium
Africa			
1	5.5	5	22
2	1	5	—
Middle East			
3	7	—	1
4	173	51	—
5	47	10	8
United States			
6	13	—	350
7	6	2.5	120
8	11	—	84
Venezuela			
9	—	6	480
10	57	13	72
11	380	60	70
12	113	21	49
13	93	—	38

Courtesy of Babcock & Wilcox.

Contaminants

The fundamental difference in fuel oils is their chemical composition. As can be seen in Table 8.5-3, fuels are composed of carbon, hydrogen, sulfur, nitrogen, and oxygen and, in the case of heavier fuels, ash. The carbon and hydrogen ratios affect combustibility, while the sulfur and nitrogen levels affect pollution generation. Ash in the fuel oil will tend to foul convection transfer surfaces. Oils with an ash content greater than 100 ppm require the use of soot blowers.

In the combustion process these fuel-bound elements of sulfur and nitrogen will oxidize and form various oxides of sulfur and nitrogen. In addition to being undesirable pollutants, they can be a source for both high- and low-temperature corrosion in the furnace.

The ash in heavier fuels usually contains elements of sodium, vanadium, nickel, and other trace metals. Table 8.5-5 lists the amounts of these elements found in various crude sources.

These metals, particularly vanadium and sodium, when involved in the combustion process, can form complex compounds having low melting points as indicated below[2]:

Sodium sulfate	1625°F
Sodium pyrosulate	1260°F
Vanadium pentoxide	1275°F
Sodium metavanadate	1165°F
Sodium pyrovanadate	1120–1210°F
Sodium orthovanadate	1500–1590°F

These compounds in the flue gas stream tend to become sticky and semimolten and will readily adhere to tube surfaces. Once this initial deposit is formed, continued fouling will occur until the tube deposit temperatures reach a point where they will no longer be sticky. Rate of buildup has been determined to be greatest when the sodium–vandium ratio in the fuel oil was 1 to 6.[3]

In addition to increased fouling, these sodium–vanadium complexes are corrosive when molten. This molten ash dissolves refractories and combines with protective metal oxide films to accelerate corrosion. As the gas and tube metal temperatures increase, corrosion increases. It is therefore important, when evaluating fuels, to pay particular attention to the ash content and the amounts of sodium and vanadium in the ash. Since sodium and vanadium attack has been known to cause catastrophic failures of tubes of high chrome alloy, some operations limit the sodium and vanadium content to 10 ppm in reforming catalyst tube applications.[4]

When burning high-vanadium fuels is a necessity, several common methods of dealing with the problem are as follows:

Furnace design.
Flue gas recirculation.
Excess air control.
More frequent tube cleaning.
Use of additives.

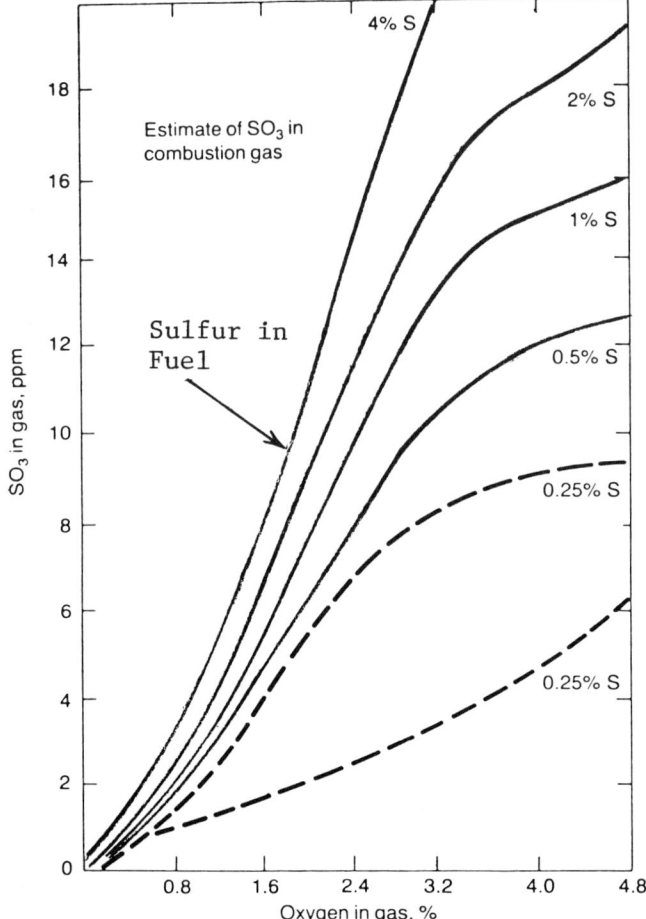

Fig. 8.5-6 Relationship between SO_3 and oxygen content. (Reproduced with permission from HYDROCARBON PROCESSING, May 1981.)

Design and operation changes can be made to keep this problem of molten ash off surfaces where it can do damage. Either molten ash above its fluid temperature or solid ash considerably below the critical deformation temperature poses little problem to refractories or metals. Changes in furnace geometry to allow the ash to cool, lowering of internal boiler temperatures, and installation of retractable soot blowers have been commonly used to deal with the problem.

Flue gas recirculation, which recycles comparatively cool flue gases to the fire box, can lower the furnace temperature, consequently changing the ash characteristics.

The reduction of excess air provides less of an opportunity for complete oxidation of vanadium and sulfur. High excess air increases the formation of V_2O_5, a corrosive low-melting-point vanadium salt instead of the higher-melting-point salts of V_2O_3 and V_2O_4. Reduction of excess air from 7% to 1–2% has been demonstrated to reduce boiler fouling and corrosion by 75% or more.[5]

Low-temperature sulfur corrosion is a particularly prevalent problem in oil-fired furnaces.[3] The sulfur in the oil is oxidized during combustion to SO_2 and SO_3. Usually 1–3% of the SO_2 converts to SO_3, depending on the excess oxygen level in the combustion process. Figure 8.5-6 shows SO_3 generation and oxygen content in the flue gas for different fuel sulfur contents. Vanadium in the fuel acts as a catalyst in converting SO_2 to SO_3 from the available excess oxygen in the flue gas stream.[3] When these flue gases come in contact with a surface that is at or below the sulfur dew point, the gases condense and form sulfuric acid (H_2SO_4).

The amount of SO_3 and subsequent H_2SO_4 formed can be controlled through fuel oil selection, treatment, furnace design, and furnace operation.

Figure 8.5-7 shows a correlation of sulfur dew point versus percent sulfur in the fuel, and Fig. 8.5-8 shows sulfur dew point versus the amount of SO_3 in ppm in the dry flue gas.

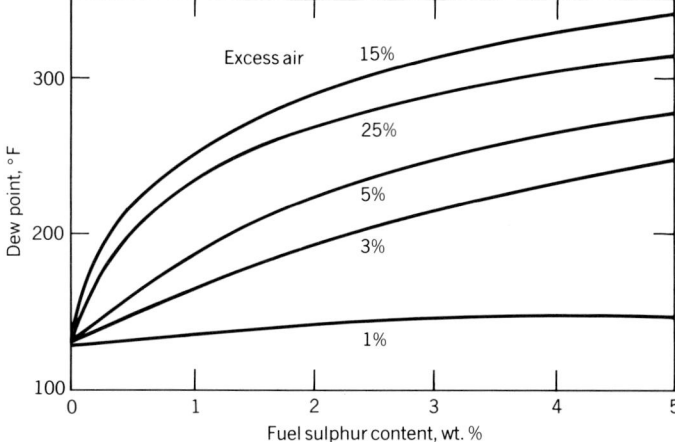

Fig. 8.5-7 Relationship between acid dew point and sulfur content of fuel. (Courtesy of Combustion Engineering, Inc.)

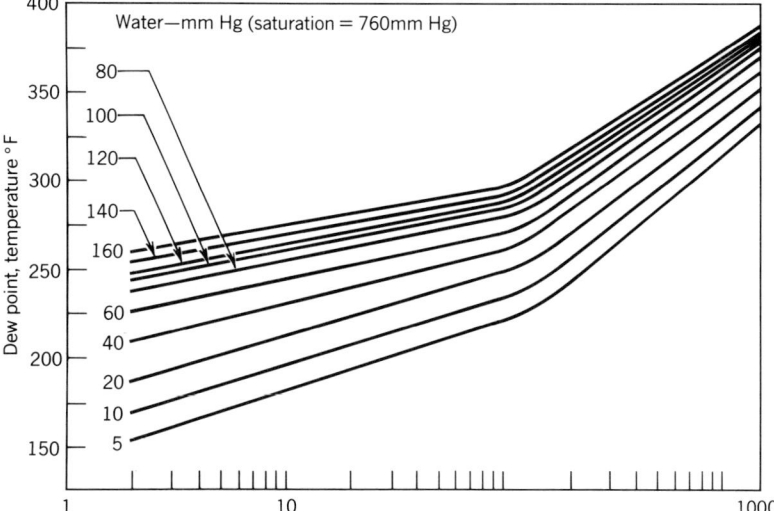

Fig. 8.5-8 SO_3 ppm in dry flue products. (Reproduced with permission from HYDROCARBON PROCESSING, June, 1974.)

Additives[6]

One of the methods for treating high- and low-temperature corrosion problems when burning vanadium and sodium-bearing fuels is the use of fuel additives. Additives change the characteristics of the boiler tube deposits in such a way as to permit their easy removal by soot blowers or air lances.

Most additives today are formulated as magnesium oxides (MgO), which are designed to chemically combine with the low-melting-point compounds. These new resulting compounds have higher melting points. The key is to get the V_2O_5 to react and be completely converted to magnesium orthovandate ($3MgO \cdot V_2O_5$). Since the dispersion and mixing of the MgO in the combustion process is not ideal, larger quantities of MgO than chemically required are added. The MgO can be either added to the fuel or dispersed into the combustion air.

Treatment of low-temperature corrosion, resulting from the formation of sulfuric acid (H_2SO_4), can also be handled with the same MgO formulations. The work done in treating high-temperature corrosion problems, such as converting V_2O_5 to $3MgO \cdot V_2O_5$, accomplishes several things. First, it ties up the vanadium, which is a catalyst in converting SO_2 to SO_3. Second, the magnesium orthovandate compound is a refractory material that will coat metal surfaces and reduce metal oxide catalyzing of SO_2 to SO_3.

Any excess MgO that does not combine with the V_2O_5 can combine with the SO_2 to form sulfates, thereby reducing the amount of SO_3 formation in the flue gas.

Injection of the additives can be in two steps. In the fuel or combustion air sufficient MgO is added to deal with the vanadium content of the fuel. This will deal primarily with the high-temperature corrosion of the vanadium compounds. Low-temperature corrosion protection can be accomplished by injecting the MgO compound directly into the flue gas stream ahead of the economizer or air heater. If good distribution can be accomplished, this has a distinct advantage over adding all the MgO in the front end in that it significantly reduces ash formation from the additives.

8.5-2 Burners

Function

The primary function of a liquid fuel burner is to provide a mechanism for atomization and intimate mixing of the fuel and combustion air. It acts as the control point for the proper amount of fuel and air for combustion. This chemical ratio of fuel to air for theoretical combustion is called the stoichiometric ratio. A mixture containing more fuel than stoichiometry is called fuel rich; likewise a mixture containing more air than stoichiometry is called fuel lean. The amount of air that exceeds this stoichiometric ratio is referred to as excess air. Excess air is usually measured as excess oxygen in the

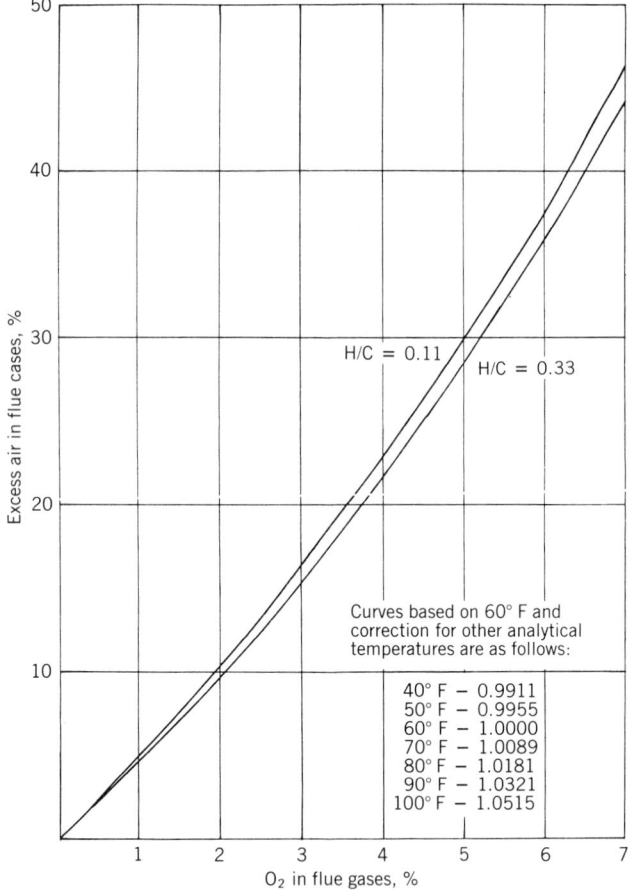

Fig. 8.5-9 Excess air indication by oxygen content of flue gases as based on the H:C ratio by weight of the fuel. H:C of 0.33 = methane; H:C of 0.11 = #6 (bunker C) oil. (From *Furnace Operations*, Third Edition, by Robert D. Reed. Copyright© 1981 by Gulf Publishing Company, Houston, Texas. Used with permission. All rights reserved.)

flue gas and can be approximated as excess air using the following equation[7]:

$$\text{Percent excess air} = \frac{100 \times O_2 - CO/2}{0.264 N_2 - (O_2 - CO/2)} \tag{8.5-4}$$

where O_2, CO, and N_2 are in vol. %. The amount of O_2, and CO if present, in the flue gas can be obtained with a suitable analyzer or through an Orsat analysis. Figure 8.5-9 is a graphical representation of the relationship expressed in Eq. (8.5-4).

The burner is designed to retain the flame produced by the combustion process in a fixed position and provide a particular flame pattern or geometry to match the furnace requirements. Consequently, a burner needs to be designed to the particular application, but in general should provide good flame stability over the specified operating range, that is, turndown range, with as little excess air as possible.

Types

Oil burners are usually classified by a variety of their characteristics, such as shape, type of atomizer, and whether forced or natural draft. The type of furnace and heat release required will dictate the type of burner used.

There are commonly two configurations or shapes of burners in practice today: the circular register type and the block type. Figure 8.5-10 shows a single circular register burner for oil firing and Fig. 8.5-11 shows a block burner for oil firing. In both configurations the air is introduced in such a fashion as to shape the flame and provide sufficient turbulence to mix the air with the atomized fuel.

Both designs are suitable for a wide range of applications, orientations, and furnace pressure conditions. Typically, however, the circular burner produces a circular flame pattern and is suited for boilers and cylindrical heaters. The block burner produces a flat or fan-shape flame and is typically suited for rectangular furnaces where burners must be placed closely to the furnace wall for maximum radiation effect.

The arrangement and orientation of the oil burners are designed so as not to directly impinge upon any tube or refractory surface, with the exception of radiant wall burners. Such flame impingement will cause tube failure or premature refractory failure. Also, the burners are placed so that adjacent flames do not overlap, which could cause fuel-rich zones.

As a result, burner shape and configuration are governed by the furnace geometry and the required flame pattern. If the furnace operates under negative draft conditions, the burners can either be of the natural draft, induced draft, or forced draft design. If the furnace operates under positive draft conditions, the burners must be of the forced draft design.

Atomization

In order for oil to burn, it must first be atomized into tiny fine droplets resembling a fine mist. These droplets evaporate in the presence of the airstream and form a combustible vapor that can be ignited.

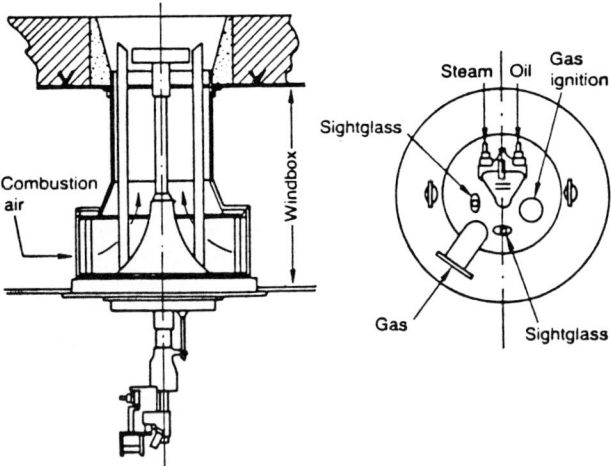

Fig. 8.5-10 Circular-type burner. (Reproduced with permission from HYDROCARBON PROCESSING, May, 1981.)

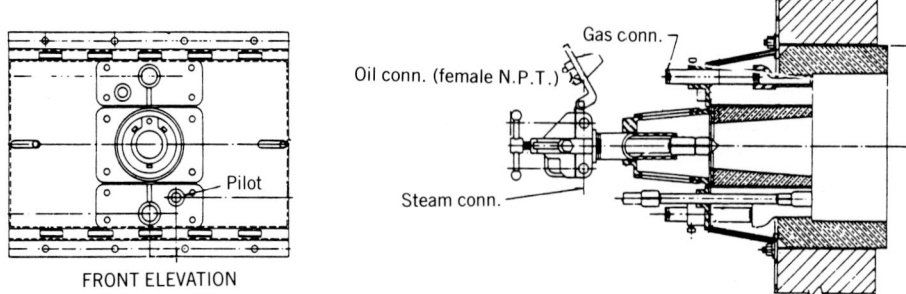

Fig. 8.5-11 Block burner. (Courtesy John Zink Co., Tulsa, OK.)

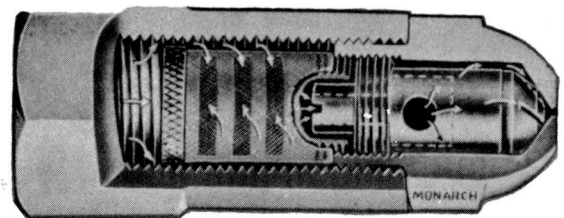

Fig. 8.5-12 Typical mechanical atomizer. (Courtesy Monarch Mfg. Works, Inc., Philadelphia.)

There are many ways of atomizing the fuel oil, but principally the most common methods are mechanical and steam or air atomization.

Mechanical or pressure atomization is mostly used where the fuels to be burned are lighter oils and in smaller heat release rates. The pressure atomizer depends on the shearing action caused by passing the fuel oil under high pressure through a small drilled orifice. No steam or atomizing medium is required. These tiny orifices or jets are very susceptible to pluggage and erosion and, consequently, high maintenance. Generally the fuel atomization is not as efficient as steam atomization, therefore, a higher degree of turbulence is required in the fuel–air mixing zone. The required oil pressure is generally several times higher than that of the steam atomizer for a given heat release rate. Since its capacity is controlled by pressure, it is usually small compared with the steam atomizer. At reduced flow rates atomization is poor, therefore limiting the ratio of maximum to minimum output (termed "turndown" ratio). Figure 8.5-12 shows a typical mechanical atomizer.

The most widely used type of atomizer is the steam or air-assisted atomizer (Figs. 8.5-13 and 8.5-14). The atomizing media (steam, gas, or compressed air) is used as an alternate energy source for fuel atomization. Oil pressure is much lower than for mechanical atomizers and is usually limited to 100 psi. Steam pressure is usually required at 20–40 psi above oil pressure.

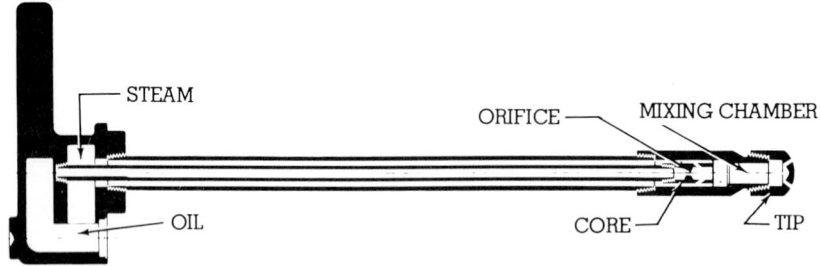

Fig. 8.5-13 Typical steam atomizer. (Courtesy National Airoil Burner Co., Philadelphia.)

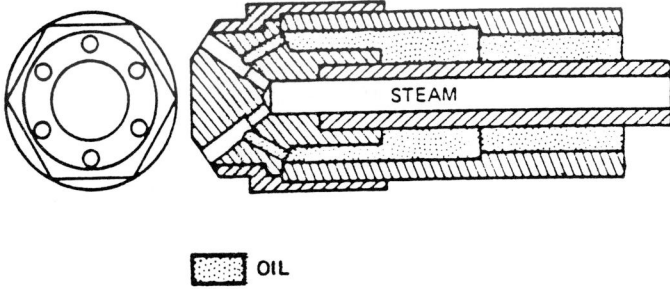

OIL

Fig. 8.5-14   Typical steam atomizing nozzle. (Courtesy AIChE.)

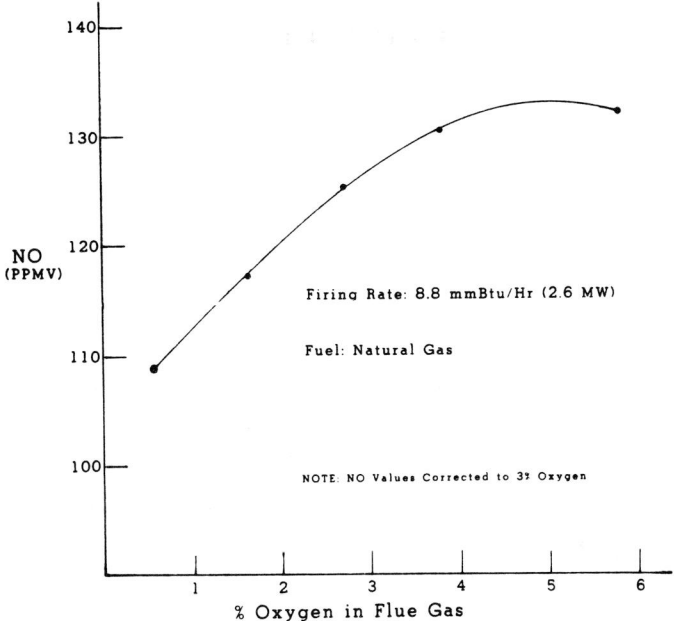

Firing Rate: 8.8 mmBtu/Hr (2.6 MW)

Fuel: Natural Gas

NOTE: NO Values Corrected to 3% Oxygen

% Oxygen in Flue Gas

Fig. 8.5-15   Influence of oxygen on NO_x generation. (Courtesy of G. M. Bitterlich, ASME 38th Petroleum Mechanical Engineering Workshop and Conference, September, 1982.)

Steam (or air) atomizers perform very efficiently and can be designed for turndowns of 5 : 1 or more. The atomizing steam must be dry since entrained moisture can cause flame quenching and poor vaporization. Properly designed and trapped atomizing steam piping is essential to good operation.

The amount of atomizing steam a burner requires depends on the type of fuel, its viscosity, and the burner design. Typically, however, this quantity can be as high as 0.5 lb steam per pound of fuel and for a good atomizer as low as 0.1 lb steam per pound of fuel at maximum capacity. All of the steam energy is lost in the combustion process and, depending on the consumption, can amount to a substantial amount of steam over a period of time.

Operation

The main objective in a successful burner operation is to operate a burner at as low an excess air level as possible, achieving complete combustion within the specified flame patterns.

Since air and fuel do not mix ideally, a minimum amount of excess air is required for any combustion process. Any excess air above the minimum required amount wastes fuel through additional stack gas losses. Therefore, reduction of excess air increases furnace efficiency, consequently reducing fuel input. Other benefits of low excess air operation are:

1. The reduction of flue gas SO_3 content and the corresponding dew point, resulting in the reduced incidence of low-temperature corrosion.
2. The reduction of NO_x generation. Figure 8.5-15 shows the reduction of NO_x as the oxygen content in the flue gas decreases.
3. The reduction of superheater deposits and its corrosion.

Low excess air operation will tend to increase the amount of particulate matter, but these particulates have a lesser fouling tendency.

It has been demonstrated in tests with an air atomizing burner that the critical level of excess air below which the benefits of reduced superheater slagging and corrosion are achieved is approximately 3%.[8]

For the measurement of excess air in the flue gas, oxygen offers an advantage over CO_2 measurement for liquid and gaseous fuels because it is relatively independent of fuel composition, while CO_2 on the other hand is very fuel dependent, such as in hydrogen-rich fuels where little or no CO_2 is produced.

Operational Problem Areas

Higher excess air operation than required is a prevalent burner operational problem, but usually not identified as such. Not only is it uneconomical, but can contribute to increased pollutant generation and its side effects. This problem can usually be solved by good housekeeping techniques and stack gas monitoring. High excess air operation is common in a "loose" furnace (experiencing many air leaks) and with those having loosely fitting air registers. In units with high heat liberation an automatic air register(s) or a stack damper control system to maintain low excess air can usually be justified on economic benefits.

Incomplete combustion is another common problem that is not easily detected without a combustible gas detection instrument. Combustibles in the 1–2% ranges can be present in the flue gas without generating visible smoke from the stack and, therefore, are not easily detected, This can happen even above the stoichiometric ratio of a fuel in a furnace with much air infiltration. Excess air in the flue gas, measured with an oxygen or CO_2 analyzer, would be assumed to be what is remaining after complete combustion. When, in fact, the air–fuel mixing zone may actually be fuel rich, with the excess air entering the furnace downstream of the combustion process before the analyzer. Incomplete combustion is undesirable, not only for its waste of fuel but also for its fouling of heat transfer surface in the convection section and the increased hazard of explosion.

Other common burner operational problems are irregular flame patterns and coke buildup in the burner block area, especially in the case of heavy fuel firing. Coke buildup and burner flame problems can be caused by the following situations:

Burner tip fouling.
Burner tip erosion due to particulate matter in the fuel.
Inadequate fuel pressure.
Inadequate fuel temperature for atomization.
Inadequate atomizing medium temperature.
Inadequate atomizing medium pressure.
Wet steam.
Inadequate positioning of the oil gun.
Misalignment of burner parts.
Insufficient draft.
Inadequate air supply.
Fuel viscosity fluctuation.

Burner Selection

After the process requirements for a furnace are fixed, burner selections will be made concerning their number and capacities. The turndown capacity of a burner will largely affect the selection of the number of burners in a given flame pattern.

In order to achieve a uniform heat flux along the tubes, obviously an infinite number of burners are preferable. However, practical and capital considerations will limit the actual number.

A common error in burner selection is oversizing, which will result in inefficient operation. This oversizing usually results by specifying more capacity than needed in anticipation of future needs. The burner designer in turn adds another 20–25% additional capacity as a safety margin. Consequently, the burner is sized well in excess of its immediate needs, has a reduced meaningful turndown capability, and because of reduced air flow pressure drop has less turbulence and lower combustion efficiency. Burner sizing needs to be closely coordinated with the manufacturer specifically specifying the normal, maximum, and minimum operating conditions and required fuel pressure.

A fuel analysis is required to check for possible contamination of water, PCBs, or halogen compounds as well as viscosity relationships. Viscosity changes over temperature ranges are required to define the fuel preheating requirement for atomization. Other parameters required in the fuel analysis are ash content, heating values, specific gravity, flash point, sulfur content, and vapor pressure. The ash content will identify the soot blower requirements.

Other important factors in burner selection to be considered are local regulations for noise and environmental emissions such as NO_x, SO_2, CO, and particulates.

In a stringent NO_x-regulated area, such as Los Angeles, most new installations require low NO_x burners, unless combustion modifications are employed to reduce the amount of NO_x finally being discharged. There exist several combustion modification methods such as ammonia or water injection, low excess air operation, flue gas recirculation, and staged combustion.

Insurance company requirements should also be considered in the combustion control and fuel piping train design.

Most burner users today require a burner performance test at the manufacturer's shop prior to actual burner installation. This is a good time to check ignition reliability, flame stability over its operating range, and the operating pressure requirements of both the fuel and atomizing medium.

REFERENCES

8.5-1 E. B. Woodruff and H. B. Lammers, *Steam-Plant Operation*, McGraw-Hill, New York, 1977, pp. 126–130.

8.5-2 J. A. Gray and J. Stramba, "The Use and Misuse of Residual Fuel Oil Inhibitors in Industrial Boilers," *Combuston*, March 1967, p. 21.

8.5-3 *Steam, Its Generation and Use*, Babcock & Wilcox, New York, 1972, pp. 15–24 to 15–25.

8.5-4 K. Manuel, "Steam-Hydrocarbon Reformer Furnace Design," ASME paper 69-Pet-18, 1969.

8.5-5 M. Bautoletti et al., "Recent Experiences with Organic Additives to Prevent Corrosion and Fouling of High and Low Temperature Surfaces in High Pressure Steam Generators," *Combustion*, May 1967, p. 32.

8.5-6 "Additives for Boiler and Gas Turbine Fuels," (a compilation of articles), *Power*, McGraw-Hill, New York, 1981.

8.5-7 *Steam, Its Generation and Use*, 38th ed., Babock & Wilcox, New York, 1972, pp. 6–19.

8.5-8 M. Chaikivsky and C. W. Siegmund, "Low-Excess-Air Combustion of Heavy Fuel: High-Temperature Deposits and Corrosion," ASME paper No. 64-WA/CD-2, 1964.

BIBLIOGRAPHY

Anson, D. et al. "Carbon Monoxide as a Combustion Control Parameter." *Combustion*, March 1972.

Baur, P. A. "Combustion Control and Burner Management." *Power*, September 1982.

Coe, W. W. "Combustion: Efficiency vs. NO_x." *Hydrocarbon Processing*, May 1980.

Coe, W. W. "How Burners Influence Combustion." *Hydrocarbon Processing*, May 1981.

Combustion Handbook, 2nd ed. North American Mfg. Co., Cleveland, OH, 1978.

Fletcher, R. "Maintain Combustion Systems." *Hydrocarbon Processing*, January 1979.

Gilbert, L. F. "Precise Combustion-Control Saves Fuel and Power." *Chemical Engineering*, June 21, 1976.

Goodger, E. M. *Hydrocarbon Fuels*. Wiley, New York, 1975.

Heenan, W. and Dalton, C. "Set Burner Air by Stack Analysis." *Hydrocarbon Processing*, February 1977.

Hoffman, H. "Better Furnace Operations." *Hydrocarbon Processing*, February 1977.

Johnson, R. K. "The Modern Approach to Process Heater Control Systems." *Combustion*, November 1978.

Kane, L. A. "Combustion Control Analyzers: What's Really Needed?" *Hydrocarbon Processing*, June 1980.

Leonard, D. J. and Kehoe, T. J. "Automatic Control Ups Heater Combustion Efficiency." *Oil & Gas Journal*, September 21, 1981.

Martin, R. R. et al. "Watch for Elevated Dew Points in SO_3 Bearing Stack Gases." *Hydrocarbon Processing*, June 1974.

Maxwell, J. B. *Data Book on Hydrocarbon*. Van Nostrand Reinhold, New York, 1950.

May, D. L. "Cutting Boiler Fuel Costs with Combustion Controls." *Chemical Engineering*, December 12, 1975.

McMullen, John J. "Boiler Problems Associated With Use of Bunker C Fuel." *Combustion* April 1960.

NFPA 85D, *Explosion Prevention Multiple Burner Boiler-Furnaces, Fuel Oil-Fired*. National Fire Protection Association, Boston, MA, 1978.

Reed, R. D. *Furnace Operations*. Gulf Publishing, Houston, 1976.

Reed, R. D. "Recover Energy from Furnace Stacks." *Hydrocarbon Processing*, May 1975.

Reed, R. D. "Residual O_2 in Flue Gas Requires Interpretation." *Oil & Gas Journal*, December 25, 1978.

Seebold, J. G. "How to Conserve Fuel in Furnaces and Flares." *Hydrocarbon Processing*, November 1981.

Seebold, J. G. "Reduce Heater NO_x in the Burner." *Hydrocarbon Processing*, November 1982.

Shore, D. E. and McElroy, M. W. "Tuning Industrial Boilers for High Efficiency, Low Emissions," Parts I, II, & III. *Power*, April, May, June 1977.

Spiers, H. M. *Technical Data on Fuel*. The British National Committee, World Power Conference, London, 1962.

Stoa, T. A. "Calculating Boiler Efficiency and Economics." *Chemical Engineering*, July 16, 1979.

Trinks, W. and Mawhinney, M. H. *Industrial Furnaces*, 5th ed. Wiley, New York, 1961.

U.S. Environmental Protection Agency. *Guidelines for Burner Adjustments of Commercial Oil-Fired Boilers*. EPA-600/2-76-08, March 1976.

Wilcox Jr., J. C. "Improving Boiler Efficiency." *Chemical Engineering*, October 9, 1978.

Wilfred, F. *Fuels and Fuel Technology*. Pergamon Press, London, 1965.

Williams, D. A. and Jones, G. *Liquid Fuels*, Macmillan, New York, 1963.

Woodard, A. M. "Improve Heater Efficiency." *Hydrocarbon Processing*, May 1975.

8.6   POLLUTON REDUCTION AND WASTE REMOVAL SYSTEMS

Ravindra M. Nadkarni, Chakra J. Santhanam and R. Peter Stickles

8.6-1   Introduction

The world oil industry produces thousands of varieties of crude oils. Each crude oil is different from any other in that it has different concentrations of major components such as aliphatics, aromatics, waxes, asphalt, and so on; different compositions of impurities such as sulfur compounds, particulates, heavy metal compounds, and so on; different distillation properties; different yields of products; and products with different blending properties. Even within a given oil field, the properties of a crude will vary with time and location in the field. In older fields, when secondary or tertiary oil recovery techniques are used, the chemicals used in the recovery can have a dramatic effect on the properties of the crude.

The petroleum industry can logically be divided into three major divisions: production, refining, and marketing. Production includes the operations involved in locating and drilling oil fields, removing oil from the ground, pretreatment at the well site, and transporting the crude to the refinery. Refining is limited to operations necessary to convert the crude into salable products. Marketing involves the distribution and sale of finished petroleum products. The regulatory framework for environmental control is different for each. The primary emphasis in this section will be the regulations pertaining to refining and storage of petroleum products. While the focus will be on the regulatory framework in the United States, analogous frameworks exist or are emerging throughout the industrial world.

As noted elsewhere in this handbook, the configuration of an individual petroleum refinery depends a great deal on the range of feedstock the refinery was designed for, the desired product slate, and the utility fuel(s) available to the refinery. A refinery will use some or most of the following processes:

Crude distillation.
Vacuum distillation.
Catalytic cracking.
Reforming.
Hydrocracking.
Hydrorefining.

Hydrotreating.

Alkylation.

Aromatic/isomerization.

Thermal processes ranging from low-severity processes such as gas oil cracking to high-severity processes such as fluid coking and delayed coking.

Lube oil production.

Auxiliary units such as crude desalting and sour water stripping.

By-product recovery facilities such as Claus sulfur and light gas purification plants.

Pollution control facilities for controlling air and water emissions.

Water treatment facilities.

Cooling towers.

Boilers for steam and/or electric power generation.

Facilities for hydrogen production.

It is important to note that the refinery fuel system is an integral part of the refinery operation. The source of purchased refinery fuel and the allocation of intermediate product or by-product streams to the utility system has a major effect on the overall refinery operation. For example, switching large process heaters and boilers from refinery gas to liquid fuels is technically feasible, and the displaced gas can be exported from the utility system to product upgrading, which is a higher-priority use. Similarly, the increasing cost of natural gas to petroleum refiners has improved the management of hydrogen-containing streams within the refinery and has increased the incentives for producing hydrogen from petroleum residues through partial oxidation, although current production is limited. The increasing cost of sweet crudes and the increasing demand for lighter- and lower-sulfur products has resulted in increased hydrotreating capability and increased hydrogen consumption within petroleum refineries.

8.6-2 Pollutant Generation by Petroleum Processing

Many of the unit operations within a petroleum refinery can be emitters of air pollutants, wastewater, and solid waste.

Air Pollution

The major sources of airborne emissions are:

1. Combustion sources such as process heaters and boilers that are a part of a refinery's fuel system.
2. Combustion sources within process units such as fluid cracking catalyst regenerators and fluid cokers where combustion is used to regenerate the catalyst and/or to generate the heat necessary for the endothermic cracking reaction.
3. Other process units, such as vacuum ejectors used in vacuum distillation, which are a source of hydrocarbon emissions.
4. Effluent control systems, principally the tail gas from Claus plants.
5. Miscellaneous hydrocarbon emissions from a variety of process equipment, including process drains, blowdown systems, valves, flanges, pressure relief valves (usually discharge to flare system), pump and compressor seals, vacuum jets, and so on.

Combustion Sources. Combustion sources emit particulates (usually small amounts), sulfur oxides, nitrogen oxides, and hydrocarbons. The emissions of sulfur oxides are a function of the sulfur content of the fuels burned. Traditionally, sulfur oxide emissions have been controlled at refinery combustors by reducing the sulfur content of the feedstock. Flue gas desulfurization technology has been proven in industrial boiler and electric utility boiler applications for a wide variety of high-sulfur fuels including coals. In the future this technology is likely to prove more cost-effective in many instances in controlling emissions of sulfur oxides. The emission of other pollutants is controlled through appropriate control of combustion conditions (for minimizing NO_x formation or for maximizing a hydrocarbon burnout) and by adding particulate control equipment, when necessary.

Control of particulates, SO_x and NO_x from utility systems is a major factor in coal-fired boilers discussed in Section 6.1. The reader is referred to that section for a perspective on combustion control.

FCC Regenerator. The regenerator of a fluid catalytic cracker (FCC) regenerates the catalyst by burning off carbon deposited on catalyst surfaces. The effluent from a traditional regenerator is rich in carbon monoxide. Many of the larger regenerator units typically burn the carbon monoxide and recover the energy in a carbon monoxide boiler. A new process called Ultracat Regeneration has been developed by Amoco, which allows superior regeneration of catalyst through more complete burnout and complete combustion of carbon monoxide within the regenerator. The process also increases the yield of gasoline during riser cracking.

The FCC regeneration vent is one of the major sources of SO_x emissions in a refinery. Both post combustion and in situ methods have been applied to reduce sulfur emissions. One oil company has installed eductive scrubbers to control such emissions. FCC catalyst additives have been employed to promote H_2S formation in the riser and reduced sulfur levels on the coke prior to regeneration.

Claus Plant Tail Gas. The tail gas from a Claus plant contains hydrogen sulfide and is typically incinerated before being vented to the atmosphere. The imposition of new sulfur emission standards for Claus plant tail gas by the EPA has required the industry to install additional Claus and/or tail gas cleanup system stages to reduce the emissions of oxidized or reduced sulfur compounds.

Hydrocarbons. Most continuous hydrocarbon emissions and intermittent reliefs from process units are collected and incinerated in ground flares or elevated flare stacks. Large, fixed-roof storage facilities require vapor recovery systems. Storage tanks for light products (gasoline, jet fuel) have to be of the floating-roof type to minimize such emissions from smaller tanks where a vapor recovery system is not installed.

Water Pollution

These sources are described below.

Wastewater. The major sources of wastewater in petroleum refining are:

1. Wastewater associated with incoming feedstock, such as ballast water from incoming tankers.
2. Wastewater generated through direct contact by water or steam with petroleum products.
3. Wastewater from the blowdown of the noncontact cooling water and steam circuits.
4. Wastewater associated with leaks and spills, including storm water runoff from process areas that eventually drains into the refinery's central sewer system.

It is typical in most refineries to collect all contaminated process wastewater streams and to combine them into a single stream for treatment in a central facility.

Many attempts have been made to correlate water emissions from a refinery with the technical characteristics of the refinery operation. The best equation derived in this fashion is as follows[1]:

$$\text{Flow} = 0.568 + 0.048A + 0.004\,(C - 59.2) + 0.046\,(K - 7.2) + 0.048L$$

where A = sum of asphalt treating process capacities in thousands of barrels per day throughput
K = sum of cracking processes in thousands of barrels per day throughput
C = sum of crude processes in thousands of barrels per day throughput
L = sum of lube processes in thousands of barrels per day throughput

The flow predicted by this equation is in units of million gallons per day. This equation was derived by linear regression analysis of 227 plants that provided the necessary data. The regression equation has a correlation coefficient (R^2) of 0.749, which indicates that the equation explains only about 75% of the variability in the data.

Solid Wastes

A petroleum refinery generates a variety of solid waste streams on a continuous and an intermittent basis. Much of the intermittent waste is generated during plant maintenance/plant turnaround, including process vessel sludges, vessel scale, storage tank sediments, and so on. Similarly, during plant turnaround, spent catalyst is removed from certain processing units and replaced with active catalyst. Thus, the annual production of intermittent waste is a function of the refinery processes and their waste management/housekeeping practices. Continuously generated solid wastes are produced by units such as delayed or fluidized cokers in the form of coke fines and spilled coke from coke handling facilities, spent catalysts and catalyst fines from fluid catalytic cracking units, spent treating clays from lube oil processing plants, and sludges from wastewater treatment facilities, including hydrocarbon-rich sludges as well as biological sludges from activated sludge units. Finally, if the plant utility system or Claus plant uses scrubbers to control emissions from combustion, flue gas cleaning sludges would also be included in this category.

The major constituents of these various solid wastes are listed below[2]:

Solid Waste	Major Constituents
Once-through cooling sludge	Arsenic
Cooling tower sludge	Benzo-*a*-pyrene
Alkylation sludge	Cadmium
Waste FCC catalyst	Chromium (total)
Treating clay	Chromium (Hexavalent)
Tank bottoms	Copper
Storm water silt	Cyanide
API separator bottoms	Fluoride
Air flotation float	Lead
Biological sludge	Mercury
Boiler treating sludges	Nickel/phenol/selenium/vanadium/zinc
Boiler treatment sludges	

The American Petroleum Institute conducted an industry survey for the year 1976.[2] This survey covered 57% of the U.S. crude capacity from which the results were extrapolated to the total industry. The results showed that the industry produced oily solid wastes in the range of 40,000–72,000 metric tons per year, nonoily solid wastes from 200,000 to 285,000 per year, and mixed oily solids in the range of 240,000–357,000 per year.

The major techniques used by the refineries for waste disposal include land farming, in which the wastes are spread on land and cultivated, landfilling in secure landfills for hazardous wastes, and in sanitary landfills for nonhazardous wastes, or lagooning (i.e., storage at site), and the selling of certain wastes for beneficial use. Only certain wastes such as spent caustic, spent catalysts, or high oil content sludges can be sold to companies in the reclamation business.

8.6-3   Regulatory Framework

Particularly in the United States and broadly throughout the industrial world, environmental regulations and technology have become closely intertwined. The formation of the U.S. Environmental Protection Agency marked a new era of national scope and responsibility in the development and dissemination of new technology for environmental control, including those relative to petroleum processing. Thus, a significant part of new technology development for pollution reduction has been under EPA sponsorship or funding. More than that, in the United States the close involvement of the EPA in technology development requires that in the future the planning and design of process facilities be based on appropriate interaction with the EPA at a rather early stage. This is a substantial change from the practice of a few years ago. Regulatory framework cannot be divorced from planning and design of process operations.

The environmental regulatory framework for refining and storage facilities for petroleum products can be categorized into four groups:

1. Air pollution control.
2. Water pollution control.
3. Solid waste management.
4. Emerging response measures including spill prevention.

Air Pollution Control

The regulatory framework includes both air quality standards and emission standards.

Air Quality Standards.   Under the Clean Air Act the air quality standards in the United States are presented in Table 8.6-1. Some modifications have been proposed, including inhalable particulate standards.[4] The regulatory framework also has provisions for prevention of significant deterioration (PSD) of air quality. EPA regulations[5] established three area classifications with corresponding maximum increments in PSD concentrations of SO_2 and particulates. All PSD areas were initially designated as class II, and individual states could redesignate these areas as class I or class III. Under the EPA regulations, PSD review is required for construction of new or modified facilities in 19 industrial categories including petroleum refining.

TABLE 8.6-1   U.S. AMBIENT AIR QUALITY STANDARDS

Contaminant		Secondary	Primary
Sulfur dioxide	(Annual) (24 h) (1 h)	 1300 μg/m^3 (0.5 ppm)^a	80 μg/m^3 (0.03 ppm) 365 μg/m^3 (0.14 ppm)^b
Suspended particulate matter	(Annual) (24 h)	60 μg/m^3 150 μg/m^{3,b}	75 μg/m^3 260 μg/m^{3,b}
Carbon monoxide	(8 h) (1 h)	10 mg/m^3 (9 ppm) 40 mg/m^3 (35 ppm)^b	
Oxidants (ozone)	(Annual) (24 h) (1 h)	 200 μg/m^3 (0.10 ppm)^b	
Nitrogen dioxide	(Annual) (24 h) (1 h)	100 μg/m^3 (0.05 ppm)	
Hydrocarbons	(3 h)^c	160 μg/m^3	160 μg/m^3

^{a}3-h concentration; not to be exceeded more than once per year.
^{b}Not to be exceeded more than once per year.
^{c}6–9 A.M. (max).

TABLE 8.6-2   PERMITTED INCREMENTS^a UNDER PSD^3

	Class I	Class II	Class III	NAAQS
SO$_2$				
Annual	2	20	40	80
24-h	5	91	182	365
3-h	25	512	700	1300 (s)
TSP				
Annual	5	19	37	75 60 (s)
24-h	10	37	75	260 150 (s)

^{a}In micrograms per cubic meter. All 24- and 3-h values may be exceeded once per year.

Permitted increases over baseline levels in particulates and sulfur oxides are shown in Table 8.6-2. The 1977 amendments define "baseline air quality" as the concentration levels existing at the time of the first application for a permit in an area subject to the new PSD rules. Determination is to be based on air quality data available from EPA or a state air pollution control agency and on monitoring data that the permit applicant is required to submit.

The baseline air quality concentrations must include all projected emissions from any major emitting facility that commenced construction before January 6, 1975, but that has not begun operation by the date of the baseline determination. SO$_2$ and particulate emissions from any major emitting facility that commenced construction after January 6, 1975, cannot be included in the baseline but must be counted against the maximum allowable increment limitation for the applicable PSD area. This latter clause may mean that in some PSD areas the increment limitation is already used up and no emissions are permissible.

Emission Standards: Refineries.   Emission standards[6] were promulgated for the following facilities in petroleum processing: fluid catalytic cracking (FCC) unit catalyst regenerators, fuel gas combustion devices, and all Claus sulfur recovery plants except Claus plants of 20 long tons per day (ltd) or less. The Claus sulfur recovery plant need not be physically located within the boundaries of a petroleum refinery to be an affected facility, provided it processes gases produced within a petroleum refinery. The standards are presented in Table 8.6-3.

Emission Standards: Storage Vessels.   Standards are in effect[8] on storage vessels greater than 151,412 L (40,000 gal) in volume. For vessels constructed before May 19, 1978, two categories of vessels are

TABLE 8.6-3   EMISSION STANDARDS PETROLEUM REFINING

No.	Detail	Standards	Comment
1	TSP	a. 1.0 kg/1000 kg of coke burnoff in the catalyst regenerator.	
		b. Opacity less than 30% except one 6-min average opacity in 1 h.	Ref. 7
		c. Of gases from fluid catalyst cracking (FCC) unit passes through incinerator or waste heat boiler, incremental emission from the latter shall not exceed 43.0 J/MJ (0.1 W/MMBtu).	Ref. 8
2	CO	Maximum of 0.050% by volume from FCC discharges.	
3	SO_2	a. Maximum H_2S in fuel gas of 230 mg/dscm (0.10 g/dscf) unless flue gas is treated.	Ref. 8
		b. Discharge from Claus plants in excess of	
		(1) 0.025% by volume SO_2 at 0% oxygen on a dry basis if emissions are controlled or	
		(2) 0.030 vol % of reduced sulfur compounds and 0.0010 vol % H_2S calculated as SO_2 at 0% oxygen on a dry basis if emissions are controlled by reduction control systems.	
4	Emission monitoring	a. Continuous monitoring for opacity on FCC unit catalyst requirement emissions.	
		a. CO monitoring on FCC unit catalyst requirement emissions.	Ref. 8
		c. SO_2 monitoring for emissions from fuel combustion units.	Ref. 8
		d. H_2S monitoring in fuel gases.	
		e. SO_2 monitoring for emissions from Claus units.	
5	Others	Test methods and procedures are defined.	Ref. 9

defined on which standards are to be met:

1. Has a capacity greater than 151,416 L (40,000 gal), but not exceeding 246,052 L (65,000 gal), and commences construction or modification after March 8, 1974, and prior to May 19, 1978.

2. Has a capacity greater than 246,052 L (65,000 gal) and commences construction or modification after June 11, 1973, and prior to May 19, 1978.

Basic standards are:

1. If the true vapor pressure of the petroleum liquid as stored is equal to or greater than 78 mm Hg (1.5 psia) but not greater than 570 mm Hg (11.1 psia), the storage vessel shall be equipped with a floating roof, a vapor recovery system, or their equivalents.

2. If the true vapor pressure of the petroleum liquid as stored is greater than 570 mm Hg (11.1 psia), the storage vessel shall be equipped with a vapor recovery system or its equivalent.

For vessels constructed after May 18, 1978, two vessel categories are defined:

1. Except as provided in paragraph (2) of this section, the affected facility to which this subpart applies is each storage vessel for petroleum liquids which has a storage capacity greater than 151,416 L (40,000 gal) and for which construction is commenced after May 18, 1978.

2. Each petroleum liquid storage vessel with a capacity of less than 1,589,873 L (420,000 gal) used for petroleum or condensate stored, processed, or treated prior to custody transfer is not an affected facility and, therefore, is exempt from the requirements of this standard.

The standards in effect apply to storage vessels handling petroleum liquid with a true vapor pressure equal to or greater than 10.3 kPa (1.5 psia) but not greater than 76.6 kPa (11.1 psia). One of the following will be required:

1. An external floating roof, consisting of a pontoon-type or double-deck-type cover that rests on the surface of the liquid contents and is equipped with a closure device between the tank wall and the roof edge. The closure device is to consist of two seals one above the other. Further details on the type of seal is also specified.[9]
2. A fixed roof with an internal floating-type cover equipped with a continuous closure device between the tank wall and the cover edge. The cover is to be floating at all times (i.e., off the leg supports) except during initial fill and when the tank is completely emptied and subsequently refilled.
3. A vapor recovery system, which collects all vapors and gases discharged from the storage vessel, and a vapor return or disposal system, which is designed to process such vapors and gases so as to reduce their emission to the atmosphere by at least 95% by weight.
4. A system equivalent to those mentioned above.

Water Pollution Control

Water Quality Standards. At present, there are no generally applicable water quality standards in the United States. The regulations of water pollution control activities is implemented by the National Pollutant Discharge Elimination System (NPDES) under the Clean Water Act. In 1979[10] EPA published proposed consolidation of NPDES permit regulations taking into account the following regulations:

1. NPDES under Clean Water Act.
2. Hazardous wastes under the Resource Conservation and Recovery Act (RCRA).
3. Underground injection control (UIC) under the Safe Drinking Water Act (SDWA).
4. Prevention of significant deterioration (PSD) under the Clean Air Act (CAA).
5. Section 404, Dredge or Fill Programs.

Recently EPA had proposed rules[11] for setting water quality standards. However, these are not likely to result in water quality standards except over the longer term.

Effluent Guidelines. Effluent guidelines have been established for the following categories of petroleum refinery facilities:

1. Topping subcategory.
2. Cracking subcategory.
3. Petrochemical subcategory.
4. Lube subcategory.
5. Integrated Facility subcategory.

A brief review of the standard in each of these is given below. For each of the above categories of petroleum processing facilities, effluent guidelines have been promulgated or are planned for the following cases:

1. Effluent limitations guidelines representing the degree of effluent reduction attainable by the application of the best practicable control technology currently available (BPT).
2. Effluent limitations guidelines representing the degree of effluent reduction attainable by the application of the best available technology economically achievable (BAT).
3. Effluent limitations guidelines representing the degree of effluent reduction attainable by the application of the best conventional pollutant control technology (BCT).
4. Pretreatment standards for existing sources (PSES).
5. Standards of performance for new sources (NSPS).
6. Pretreatment standards for new sources (PSNS).

Since the effluent guidelines for each subcategory are lengthy, it is proposed to describe the focus of each subcategory and outline the guidelines in full for the integrated category. For details on effluent guidelines in each of the other categories the reader may consult the literature.[12]

Topping Subcategory.   The provisions of this subpart apply to discharges from any facility that produces petroleum products by the use of topping and catalytic reforming, whether or not the facility includes any other process in addition to topping and catalytic reforming. The provisions of this subpart do not apply to facilities that include thermal processes (coking, visbreaking, etc.) or catalytic cracking.

Cracking Subcategory.   The provisions of this subpart are applicable to all discharges from any facility that produces petroleum products by the use of topping and cracking, whether or not the facility includes any process in addition to topping and cracking. The provisions of this subpart are not applicable, however, to facilities that include the processes specified other subcategories described below (petrochemical, lube or integrated).

Petrochemical Subcategory.   The provisions of this subpart are applicable to all discharges from any facility that produces petroleum products by the use of topping, cracking, and petrochemical operations whether or not the facility includes any process in addition to topping, cracking, and petrochemical operations. The provisions of this subpart shall not be applicable, however, to facilities that include the processes specified in lube integrated subcategories.

Lube Subcategory.   The provisions of this subpart are applicable to all discharges from any facility that produces petroleum products by the use of topping, cracking, and lube oil manufacturing processes, whether or not the facility includes any process in addition to topping, cracking, and lube oil manufacturing processes. The provisions of this subpart are not applicable, however, to facilities that include the processes specified in petrochemical or integrated subcategories.

Integrated Subcategory.   The provisions of this subpart are applicable to all discharges resulting from any facility that produces petroleum products by the use of topping, cracking, lube oil manufacturing processes, and petrochemical operations, whether or not the facility includes any process in addition to topping, cracking, lube oil manufacturing processes, and petrochemical operations.
For the integrated subcategory, the following tables outline the effluent guidelines:

1. Table 8.6-4 defines effluent limitation guidelines under the best practical control technology (BPT) standards.
2. Table 8.6-5 defines effluent limitation guidelines under best available technology (BAT) economically achievable.
3. Table 8.6-6 defines pretreatment standards for existing sources (PSES).
4. Table 8.6-7 defines standards of performance for new sources (NSPS).
5. Table 8.6-8 defines pretreatment standards for new sources (PSNS).

Solid Waste Disposal

The principal regulatory framework concerning solid waste management in the United States is the Resource Conservation and Recovery Act (RCRA). Figures 8.6-1 and 8.6-2 describe the process of determining if a certain waste is subject to RCRA regulations[13] and whether it is hazardous. Wastes to be regulated under RCRA are broadly classified as hazardous (Subtitle C) or nonhazardous (Subtitle D). The distinction is made on the basis of specified test protocol on the wastes from four perspectives: corrosivity, toxicity, reactivity, and ignitibility. Hazardous wastes are subject to EPA promulgated standards for waste management[13]; as of late 1982, these standards are still evolving. Nonhazardous wastes are also subject to regulations or guidelines, but these are less stringent. Mechanism exists under RCRA for nonhazardous wastes to be subject to regulations by individual states.
A special case of waste management in petroleum technology is drilling mud. In 1980[14] the U.S. Congress exempted drilling fluids, produced waters, and other wastes associated with the exploration, development, or production of crude oil, natural gas, or geothermal energy from the requirements of RCRA until further studies are completed.

Spill Prevention

The Environmental Protection Agency has promulgated regulations concerning oil pollution prevention at nontransportation-related onshore and offshore facilities. The Department of Transportation

TABLE 8.6-4   EFFLUENT GUIDELINES INTEGRATED REFINING—BPT[a]

a.   Effluent Limitation

Pollutant or Pollutant Property	BPT Effluent Limitations	
	Maximum for any one day	Average of daily values for 30 consecutive days shall not exceed
	Metric units (kg/1000 m³ of feedstock)	
BOD$_5$	54.4	28.9
TSS	37.3	23.7
COD	388.0	196.0
Oil and grease	17.1	9.1
Phenolic compounds	0.40	0.192
Ammonia as N	23.4	10.6
Sulfide	0.35	0.158
Total chromium	0.82	0.48
Hexavalent chromium	0.088	0.032
pH	6 to 9	6 to 9
	English units (lb/1000 bbl of feedstock)	
BOD	19.2	10.2
TSS	13.2	8.4
COD	136.0	70.0
Oil and grease	6.0	3.2
Phenolic compounds	0.14	0.068
Ammonia as N	8.3	3.8
Sulfide	0.124	0.056
Total chromium	0.29	0.17
Hexavalent chromium	0.025	0.011
pH	6 to 9	6 to 9

b.   The limits set forth above are to be multiplied by the following factors to calculate the maximum for any 1 day and maximum average of daily values for 30 consecutive days.

1.   Size Factor

1000 bbl of Feedstock per Stream Day	Size Factor
Less than 124.9	0.73
125–144.9	0.76
150–174.9	0.83
175.0–199.9	0.91
200–244.9	0.99
250 or greater	1.04

2.   Process Factor

Process Configuration	Process Factor
Less than 6.49	0.75
6.5–7.49	0.82
7.5–7.99	0.92
8.0–8.49	1.00
8.5–8.99	1.10
9.0–9.49	1.20
9.5–9.99	1.30
10.0–10.49	1.42
10.5–10.99	1.54
11.0–11.49	1.68
11.5–11.99	1.83
12.0–12.49	1.99
12.5–12.99	2.17
13.0 or greater	2.26

[a]Basis: Effluent limitations guidelines representing the degree of effluent reduction attainable by the application of the best practicable control technology currently available (BPT).

TABLE 8.6-5   EFFLUENT GUIDELINES INTEGRATED REFINING-BAT[a]

a.   Effluent Limitations

Pollutant or Pollutant Property	BAT Effluent Limitations	
	Maximum for any one day	Average of daily values for 30 consecutive days shall not exceed
	Metric units (kg/1000 m³ of feedstock)	
COD	388.0	198.0
Phenolic compounds	0.40	0.192
Ammonia as N	23.4	10.6
Sulfide	0.35	0.158
Total chromium	0.068	0.032
Hexavalent chromium	0.068	0.032
	English units (lb/1000 bbl of feedstock)	
COD	136.0	70.0
Phenolic compounds	0.14	0.068
Ammonia as N	8.3	3.8
Sulfide	0.124	0.056
Total chromium	0.29	0.17
Hexavalent chromium	0.025	0.011

b.   The limits set forth in section (a) above are to be multiplied by the following factors to calculate the maximum for any 1 day and maximum average of daily values for 30 consecutive days.

1.   Size Factor

1000 Barrels of Feedstock per Stream Day	Size Factor
Less than 124.9	0.73
125.0–149.9	0.76
150.0–174.9	0.83
175.0–199.9	0.91
200.0–224.9	0.99
225 or greater	1.04

2.   Process Factor

Process Configuration	Process Factor
Less than 6.49	0.76
6.5–7.49	0.82
7.5–7.99	0.92
8.0–8.49	1.00
8.5–8.99	1.10
9.0–9.49	1.20
9.5–9.99	1.30
10.0–10.49	1.42
10.5–10.99	1.54
11.0–11.49	1.68
11.5–11.99	1.83
12.0–12.49	1.99
12.5–12.99	2.17
13.0 or greater	2.26

[a]Basis: Effluent limitations guidelines representing the degree of effluent reduction attainable by the application of the best available technology economically achievable (BAT).

TABLE 8.6-6   EFFLUENT GUIDELINES INTEGRATED REFINERY—PSES[a]

Pollutant or Pollutant Property	Pretreatment Standards for Existing Sources [maximum for any 1 day (mg/L)][a]
Oil and grease	100
Ammonia (as N)	100[b]

[a] Basis: Pretreatment standards for existing sources (PSES). The following standards apply to the total refinery flow contribution to publicly owned treatment works (POTW).
[b] Where the discharge to the POTW consists solely of sour waters, the owner or operator has the option of complying with this limit or the daily maximum mass limitation for ammonia.

has promulgated standards for pipelines and other means to transport petroleum products. The latter is outside the scope of this section.

Under the EPA regulations[15] nontransportation-related onshore and offshore facilities are required to have spill prevention control and countermeasure (SPCC) plans. Guidelines for preparing SPCC plans have been published. Some of the guidelines are given below:

1. Onshore facilities. Appropriate containment and/or diversionary structures or equipment to prevent discharged oil from reaching a navigable water course should be provided. One of the following preventive systems or its equivalent should be used as a minimum:

(a) Dikes, berms, or retaining walls sufficiently impervious to contain spilled oil.

(b) Curbing.

(c) Culverting, gutters, or other drainage systems.

(d) Weirs, booms, or other barriers.

(e) Spill diversion ponds.

(f) Retention ponds.

(g) Sorbent materials.

(h) Offshore facilities.

(i) Curbing, drip pans.

(j) Sumps and collection systems.

When it is determined that the installation of structures or equipment to prevent discharged oil from reaching the navigable waters is not practicable from any onshore or offshore facility, the owner or operator should clearly demonstrate such impracticability and provide the following:

2. Bulk storage tanks (onshore), excluding production facilities).

(a) No tank should be used for the storage of oil unless its material and construction are compatible with the material stored and conditions of storage such as pressure, temperature, and so forth.

(b) All bulk storage tank installations should be constructed so that a secondary means of containment is provided for the entire contents of the largest single tank, plus sufficient freeboard to allow for precipitation. Diked areas should be sufficiently impervious to contain spilled oil. Dikes, containment curbs, and pits are commonly employed for this purpose, but they may not always be appropriate. An alternative system could consist of a complete drainage trench enclosure arranged so that a spill could terminate and be safely confined in an in-plant catch basin or holding pond.

(c) Drainage of rainwater from the diked area into a storm drain or an effluent discharge that empties into an open water course, lake or pond, and bypassing the in-plant treatment system may be acceptable under some conditions.

3. Facility transfer operations, pumping and in-plant process (onshore, excluding production facilities).

(a) Buried piping installations should have a protective wrapping and coating and should be cathodically protected if soil conditions warrant. If a section of buried line is exposed for any reason, it should be carefully examined for deterioration. If corrosion damage is found,

TABLE 8.6-7   EFFLUENT GUIDELINES INTEGRATED REFINING—NSPS[a]

a. Any new source subject to this subpart must achieve the following new source performance standards (NSPS).

Pollutant or Pollutant Property	NSPS Effluent Limitation	
	Maximum for any one day	Average of daily values for 30 consecutive days shall not exceed
	Metric units (kg/1000 m³ of feedstock)	
BOD$_5$	41.6	22.1
TSS	28.1	17.9
COD	295.0	152.0
Oil and grease	12.6	6.7
Phenolic compounds	0.30	0.14
Ammonia as N	23.4	10.7
Sulfide	0.26	0.12
Total chromium	0.64	0.37
Hexavalent chromium	0.052	0.024
pH	6 to 9	6 to 9
	English units (lb/1000 bbl of feedstock)	
BOD$_5$	14.7	7.8
TSS	9.9	6.0
COD	104.0	54.0
Oil and grease	4.5	2.4
Phenolic compounds	0.105	0.051
Ammonia as N	8.3	3.8
Sulfide	0.093	0.042
Total chromium	0.220	0.13
vHexavalent chromium	0.019	0.0084
pH	6 to 9	6 to 9

b. The limits set forth in section (a) above are to be multiplied by the following factors to calculate the maximum for any one day and maximum average of daily values for 30 consecutive days.

1. Size Factor

1000 bbl of Feedstock per Stream Day	Size Factor
Less than 124.9	0.73
125.0–149.9	0.76
150.0–174.9	0.83
175–199.9	0.91
200–224.9	0.99
225 or greater	1.04

2. Process Factor

Process Configuration	Process Factor
Less than 6.49	0.75
6.50–7.49	0.82
7.50–7.99	0.92
8.00–8.49	1.00
8.50–8.99	1.10
9.00–9.49	1.20
9.50–9.99	1.30
10.0–10.49	1.42
10.5–10.99	1.54
11.0–11.49	1.68
11.5–11.99	1.83
12.0–12.49	1.99
12.5–12.99	2.17
13.0 or greater	2.26

[a]Basis: Standards of performance for new sources (NSPS).

TABLE 8.6-8 EFFLUENT GUIDELINES INTEGRATED REFINING—PSNS[a]

a. The following standards apply to the total refinery flow contribution to the POTW.

Pollutant or Pollutant Property	Pretreatment Standards for Newer Sources Maximum for any One Day (mg/L)
Oil and grease	100
Ammonia (as N)	100[b]

b. The following standard is applied to the cooling tower discharge part of the total refinery flow to the POTW by multiplying (1) the standards, (2) by the total refinery flow to the POTW, and (3) by the ratio of the cooling tower discharge flow to the total refinery flow.

Pollutant or Pollutant Property	Pretreatment Standards for Newer Sources Maximum for any One Day (mg/L)
Total chromium	1

[a] Basis: Pretreatment standards for new sources (PSNS). Any new source subject to this subpart which introduces pollutants into a publicly owned treatment works must achieve the pretreatment standards specified for new sources (PSNS).
[b] Where the discharge to the POTW consists solely of sour waters, the owner or operator has the option of complying with this limit or the daily maximum mass limitation for ammonia.

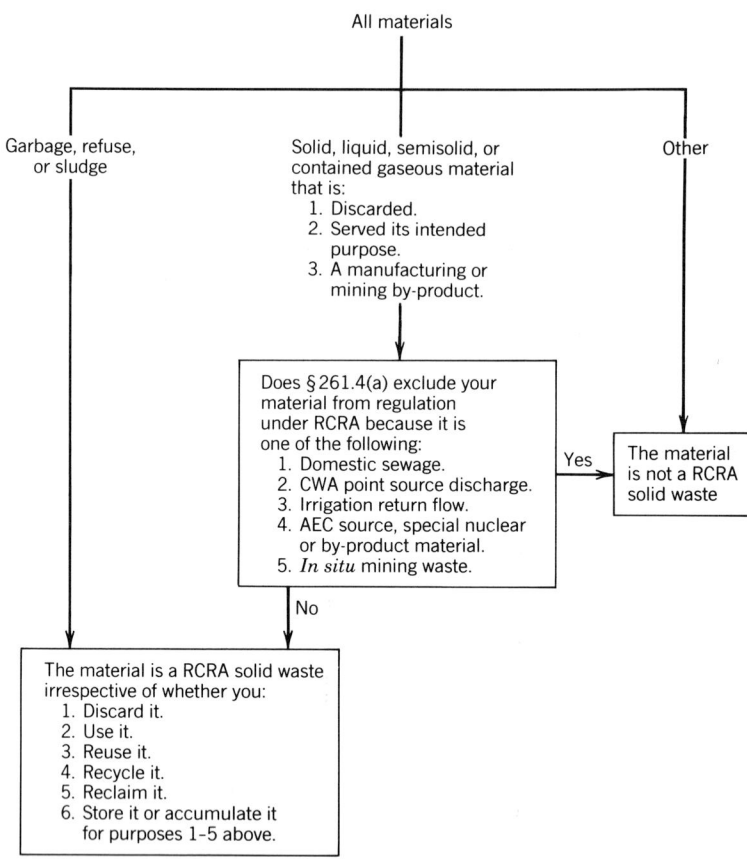

Fig. 8.6-1 Definition of a solid waste (from ref. 11, used by permission.)

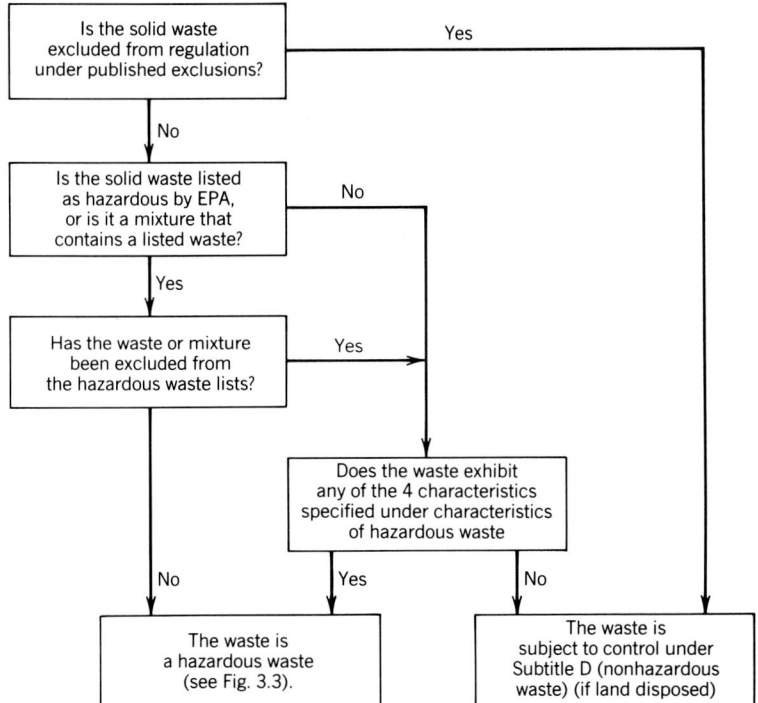

Fig. 8.6-2 Definition of a hazardous waste (from ref. 11; used by permission).

additional examination and corrective action should be taken as indicated by the magnitude of the damage. An alternative would be the more frequent use of exposed pipe corridors or galleries.

(b) When a pipeline is not in service or in standby service for an extended time, the terminal connection at the transfer point should be capped or blank-flanged and marked as to origin.

(c) Pipe supports should be properly designed to minimize abrasion and corrosion and allow for expansion and contraction.

(d) All aboveground valves and pipelines should be subjected to regular examinations by operating personnel, at which time the general condition of items such as flange joints, expansion joints, valve glands and bodies, catch pans, pipeline supports, locking of valves, and metal surfaces should be assessed. In addition, periodic pressure testing may be warranted for piping in areas where facility drainage is such that a failure might lead to a spill event.

(e) Vehicular traffic granted entry into the facility should be warned verbally or by appropriate signs to be sure that the vehicle, because of its size, will not endanger aboveground piping.

4. Facility tank car and tank truck loading/unloading rack (onshore).

(a) Tank car and tank truck loading/unloading procedures should meet the minimum requirements and regulations established by the Department of Transportation.

(b) Where rack area drainage does not flow into a catch basin or treatment facility designed to handle spills, a quick drainage system should be used for tank truck loading and unloading areas. The containment system should be designed to hold at least maximum capacity of any single compartment of a tank car or tank truck loaded or unloaded in the plant.

(c) An interlocked warning light or physical barrier system or warning signs should be provided in loading/unloading areas to prevent vehicular departure before complete disconnect of flexible or fixed transfer lines.

(d) Prior to filling and departure of any tank car or tank truck, the lowermost drain and all outlets of such vehicles should be closely examined for leakage, and if necessary, tightened, adjusted, or replaced to prevent liquid leakage while in transit.

Additional regulations exist for (1) oil production facilities (onshore), (2) facility transfer operations (onshore), and (3) oil drilling, production or workover facilities (offshore).

For details on these basically engineering standards, the reader is referred to ref. 16.

8.6-4 Environmental Control Practices

Air Pollution Control

Typical sources of emissions from petroleum facilities (refineries, storage tanks, etc.) can be categorized as that from production, refining, and marketing.

Crude Oil Production. The air contaminants emitted from crude oil production consist chiefly of the lighter saturated hydrocarbons. The main sources are process equipment and storage vessels. Hydrogen sulfide gas may be an additional contaminant in some production areas. Internal combustion equipment, such as natural-gas-fired compressors, contributes relatively small quantities of sulfur dioxide, nitrogen oxides, and particulate matter. Potential individual sources of air contaminants are shown in Table 8.6-9.

Air pollution control in crude oil production facilities vary with location. Environmental regulations and economic incentives (rising price of oil) have combined to result in increasing sophisticated levels of control; the saving from recovered or conserved hydrocarbons is considerable.

Refining. Air pollution control is necessary to comply with air pollution laws and also to try to improve community relations and prevent fire hazards. The air contaminants emitted from equipment associated with oil refining include hydrocarbons, carbon monoxide, sulfur and nitrogen compounds, malodorous materials, particulate matter, aldehydes, organic acids, and ammonia. Table 8.6-10 provides a typical list. Some of the typical control methods are listed in Table 8.6-11. A brief review of the major units for pollution control are given below. For more detailed assessment, the reader is referred to Danielson.[17]

Storage Vessels. Tanks used to store crude oil and volatile petroleum distillates are a significant potential source of hydrocarbon emissions. Hydrocarbons can be discharged to the atmosphere from a storage tank as a result of diurnal temperature changes, filling operations, and violation. Control efficiencies of 90–100% can be realized by using properly designed vapor recovery or disposal systems, floating-roof tanks, or pressure tanks.

Bulk Loading Facilities. The filling of vessels used for transport of petroleum products is potentially a source of hydrocarbon emissions. As the produce is loaded, it displaces gases containing hydrocarbons to the atmosphere. A method of preventing these emissions consists of collecting the vapors by enclosing the filling hatch and piping the captured vapors to recovery equipment. Submerged filling and bottom loading also reduce the amount of displaced hydrocarbon vapors.

Catalyst Regeneration. Refining processes use many solid-type catalysts. Cracking over a catalyst (typically alumina-silica formulations) is accomplished at slightly greater than atmospheric pressures and about 900°F. It gives a higher gasoline yield and better quality gasoline than thermal cracking. The charge stock is usually gas oil, a distillate intermediate between kerosene and fuel oil. Catalytic cracking yields a "synthetic crude" that is separated into gaseous hydrocarbons, gasoline, gas oil, and fuel oil. During the cracking process, which is usually continuous, coke deposits on the catalyst and is burned off in separate regenerating vessels.

Catalytic cracking units may be classified according to the method used for catalyst transfer. There are two methods used: moving bed and fluidized bed. The moving-bed system, typified by the thermofor catalytic cracking (TCC) units and the Houdriflow units, and the fluidized-bed or fluid catalytic cracking (FCC) units, are dominant in refinery operations. Details on these are discussed earlier in Section 8.2. The coke buildup on the catalyst requires regeneration and is usually done by burning off the coke under controlled combustion conditions. The resulting flue gases may contain catalyst dust, hydrocarbons, and other impurities originating in the charging stock, as well as the products of combustion.

Air pollution control associated with catalyst regeneration units may include the following:

1. The particulate emission problem encountered in regeneration of moving-bed-type catalysts requires control by water scrubber cyclones, precipitators, or other particulate control units, depending on the type of catalyst, the process, and the regenerator conditions.

2. Hydrocarbons, carbon monoxide, ammonia, and organic acids can be controlled effectively by incineration in carbon monoxide waste heat boilers. The waste heat boiler offers a secondary

TABLE 8.6-9 SOURCES AND CONTROL OF AIR CONTAMINANTS FROM CRUDE OIL PRODUCTION FACILITIES

Phase of Operation	Source	Contaminants	Typical Control Method
Well drilling, pumping	Gas venting for production rate test	Methane	Flares, wet-gas-gathering system
Production	Oil well pumping	Light hydrocarbon vapors	Proper maintenance
Operation	Effluent sumps	Hydrocarbon vapors, H_2S	Replacement with closed vessels connected to vapor recovery
Storage, shipment	Gas–oil separators	Light hydrocarbon vapors	Relief to wet-gathering system
	Storage tanks	Light hydrocarbon vapors, H_2S	Vapor recovery floating roofs, pressure tanks, white paint
	Dehydrating tanks	Hydrocarbon vapors, H_2S	Closed vessels connected to vapor recovery
	Tank truck loading	Hydrocarbon vapors	Vapor return, vapor recovery, vapor incineration bottom loading

Phase of Operation	Heaters, Boilers	H_2S, HC, SO_2, NO_x, particulate matter	Use of Gas Fuel
Compression, absorption dehydrating, water treating	Compressors, pumps	Hydrocarbon vapors, H_2S	Mechanical seals, packing glands vented to vapors recovery
	Scrubbers, KO pots	Hydrocarbon vapors, H_2	Relief to flare or vapor recovery
	Absorbers, fractionators, strippers	Hydrocarbon vapors	Relief to flare or vapor recovery
	Tank truck loading	Hydrocarbon vapors, H_2S	Vapor return, vapor recovery, vapor incineration, bottom loading
	Gas odorizing	H_2S mercaptans	Positive pumping, adsorption
	Waste-effluent treating	Hydrocarbon vapors	Enclosed separator vapor recovery or incineration
	Storage vessels	Hydrocarbon vapors, H_2S	Vapor recovery, vapor balance, floating roofs
	Heaters, boilers	Hydrocarbon, SO_2, NO_x, particulate matter	Proper operation of fuel control

TABLE 8.6-10 TYPICAL SOURCES OF SPECIFIC EMISSIONS FROM OIL REFINING

Oxides of sulfur	Boilers, process heater, catalytic cracking unit regenerators, treating units, H_2S flares, decoking operations
Hydrocarbons	Loading facilities, turnaround sampling, storage tanks, wastewater separators, blowdown systems, catalyst regenerators, pumps, valves, blind changing, cooling towers, vacuum jets, barometric condensers, air blowing, high-pressure equipment handling volatile hydrocarbons, process heaters, boilers, compressor engines
Oxides of nitrogen	Process heaters, boilers, compressor engines, catalyst regenerators, flares
Particulate matter	Catalyst regenerators, boilers, process heater, decoking operations incinerators
Aldehydes	Catalyst regenerators
Ammonia	Catalyst regenerators
Odors	Treating units (air blowing, steam blowing), drains, tank vents, barometric condenser sumps, wastewater separators
Carbon monoxide	Catalyst regeneration, coking operations, compressor engines, incinerators

control feature for plumes emitted from fluid catalytic cracking units. This type of visible plume, whose degree of opacity is dependent on atmospheric humidity, can be reduced or eliminated by using the carbon monoxide waste heat boiler.

Other parts of a refinery use solid and liquid catalysts; some of these can be regenerated wherever regeneration is practiced; minimization of air pollution by use of closed systems is recommended.

Particulate Emission Control

Cyclones. Cyclone separators are widely used for catalyst dust collection systems in refineries. The cyclones are employed as a single unit or in multiple two- or three-stage series systems. Large FCC unit regenerators may have as many as 10 or more three-stage cyclones, while smaller units may have only 1 two-stage cyclone. In general, high-efficiency cyclones have dust collection efficiencies of over 80% for particle sizes of more than 15 μm. The efficiency drops off rapidly for particles for sizes below this level.

Multiple cyclones are also used for catalyst fines collection. Dust collection efficiencies are in the same range as those for high-efficiency cyclones. Wet-type, centrifugal collectors or scrubbers adequately clean the gas streams from the catalyst elevators, and part of the regeneration gases from the kilns. Untreated water in the wet collectors can cause carbonate deposits.

Electrostatic Precipitators. Electrostatic precipitators are capable of high particulate collection efficiency and are widely used. Electrical precipitators are normally installed in parallel systems because of the large volume of regeneration gases involved in FCC unit regenerators. Power requirements for these precipitators may range from 30 kV A for small FCC units to 150 kV A for the larger installations. The hot gases from the regenerator must be cooled from approximately 1100°F to below 500°F before entering the precipitator. This is usually accomplished in a waste heat boiler. Ammonia injection upsteam of the precipitator can be employed, if needed, for conditioning the gas stream (i.e., to control resistivity). The precipitators usually employ either a continuous-type electrode-rapping and plate-vibrating sequence or an intermittent hourly rapping cycle.

Waste Heat Boilers

Large amounts of carbon monoxide are discharged to the atmosphere with the regeneration flue gases of an FCC unit. The carbon monoxide waste heat boiler is a means of using the heat of combustion of carbon monoxide and other combustible and the sensible heat of the regeneration gases. The CO boiler oxidizes the carbon monoxide and other combustibles, mainly hydrocarbons, to carbon dioxide and water. Usually, auxiliary fuel is required in addition to the carbon monoxide and may be either fuel oil, refinery process gas, or natural gas. The CO boiler may be a vertical structure with either a

TABLE 8.6-11   TYPICAL CONTROL METHODS—OIL REFINING

Storage vessels	Vapor recovery systems; floating-roof tanks; pressure banks; vapor balance; painting banks white
Catalyst regenerators	Cyclones, precipitator, CO boiler
Accumulator vents	Vapor recovery; vapor incineration
Blowdown systems	Smokeless flares—gas recovery
Pumps and compressors	Mechanical seals; vapor recovery; sealing glands by oil pressure; maintenance
Vacuum jets	Vapor incineration
Equipment valves	Inspection and maintenance
Pressure relief valves	Vapor recovery; vapor incineration; rupture discs; inspection and maintenance
Effluent waste disposal	Enclosing separators; covering sewer boxes and using liquid seal; liquid seals on drains
Bulk-loading facilities	Vapor collection with recovery or incineration; submerged or bottom loading
Acid treating	Replacement with catalytic hydrogeneration units; incinerate all vented cases; elimination of sludge burning
Acid sludge storage and shipping	Caustic scrubbing, incineration vapor return system; disposal at sea
Spent-caustic handling	Incineration; scrubbing
Doctor treating	Stream strip spend doctor solution to hydrocarbon recovery before air regeneration; replace treating unit with others, less objectionable units (Merox)
Sour-water treating	Use sour-water oxidizers and gas incineration; conversion to ammonium sulfate
Mercaptan disposal	Conversion to disulfides; adding to catalytic cracking charge stock; incineration; using material in organic synthesis
Asphalt blowing	Incineration; water scrubbing (nonrecirculating type)
Shutdowns, turnarounds	Depressure and purge to vapor recovery

rectangular or circular cross section with water-cooled walls. The boiler is usually equipped with a forced-draft fan and fixed, tangential-type burners.

Waste Gas Disposal System

Petroleum refineries are faced with the problem of safe disposal of volatile liquids and gases resulting from scheduled shutdowns and unexpected upsets in process units. Emergencies that can cause the sudden venting of excessive amounts of gases and vapors include fires, compressor failures, overpressures in process vessels, leaks, power failures, and so on. Uncontrolled releases of large volumes of gases also constitute a safety hazard to personnel and equipment. Hence, a system for safe handling and disposal of waste gases is an important part of pollution control of a refinery.

A typical system for disposal of emergency and waste refinery gases consists of a manifolded pressure-relieving or blowdown system, and a blowdown recovery system and a system of flares for the combustion of the excess gases. In addition to diposing of emergency and excess gas flows, these systems are used in the evacuation of units during shutdowns and turnarounds. Normally a unit is shut down by depressuring into a fuel gas or vapor recovery system with further depressuring to essentially atmospheric pressure by venting to a low-pressure flare system. Thus, overall emissions of refinery hydrocarbons are substantially reduced.

Typically, a blowdown (or pressure-relieving) system consists of relief valves, safety valves, manual bypass valves, blowdown headers, knockout vessels, and holding tanks. A blowdown recovery system also includes compressors and vapor surge vessels such as gas holders or vapor spheres. Flares are usually considered as part of the blowdown system in a refinery. For details on the design of blowdown systems, the reader is referred to the literature.[17-19]

The preferred control method for excess gases and vapors is to recover them in a blowdown recovery system and/or to incinerate them in an elevated-type flare. Flares have been further developed to ensure that this combustion is smokeless and in some cases nonluminous. Noise can also result in a nuisance problem if the refinery is located in an area with residences in close proximity.

Flares. There are, in general, three types of flares for the disposal of waste gases: elevated flares, ground-level flares, and burning pits.

Elevated Flares. Smokeless combustion can be obtained in an elevated flare by the injection of an inert gas to the combustion zone to provide turbulence and inspirate air. The most common inspirating medium is steam. Steam injection is generally believed to result in the following benefits:

1. Energy available at relatively low cost can be used to inspirate air and provide turbulence within the flame.
2. Steam reacts with the fuel to form compounds that burn readily at relatively low temperatures.
3. Water–gas reactions also occur with this same end result.
4. Steam reduces the partial pressure of the fuel and retards polymerization.

Ground-Level Flares. There are four basic types in use: horizontal venturi, water injection, multijet, and vertical venturi. Horizontal venturi systems use a knockout drum to remove liquids prior to sending the gas to the burners. The water injection systems use water spray to aspirate air; this type requires an adequate water supply and open space. The multijet flare uses two sets of burners: the smaller group handles normal gas leakage while both sets are used at higher rates of flaring. Vertical venturi type uses a blower and uses venturi-type burners. For details on these, the reader is referred to the literature.[17,18]

Burning Pits. The burning pits are reserved for extremely large gas flows caused by catastrophic emergencies in which the capacity of the primary smokeless flares is exceeded. Ordinarily, the main gas header to the flare system has a water seal bypass to a burning pit. Excessive pressure in the header blows the water seal and permits the vapors and gases to vent to burning pit where combustion occurs.

The essential parts of a flare are the burner, stack, seal, liquid trap, controls, and ignition system. In some cases vented gases flow through chemical solutions to receive treatment before combustion. As an example, gases vented from an isomerization unit that may contain small amounts of hydrochloric acid are usually scrubbed with caustic before being sent to the flare.

Water Recycle Treatment and Reuse

A modern refinery has a number of use points for water and steam. As such, the number of points producing wastewater are several. Proper management of the overall water balance around a refinery is a mix of several approaches.

1. Measures to reduce the amount of wastewater at various use points.
2. A properly designed sewer system that permits maximum water treatment recycle and reuse.
3. Conventional biological treatment systems for removal of BOD.
4. Advanced treatment processes for further reduction of pollutants in the effluent to be discharged. A logical extension of some of the above steps can potentially lead to cases where in some of the arid areas total recycle of the water, in a refinery, may be appropriate. A brief review of the major steps comprising water management in a modern refinery are briefly presented below.

Reduction of the Final Wastewater. The primary focus should be on the reduction or elimination of wastewater by a mix of the following:

1. Concentration of the contaminants.
2. Process substitution to eliminate waste streams.
3. Substitution of air cooling for water cooling.
4. More efficient heat exchange. The most significant constituents to water pollution in oil refineries are acids, caustics, sulfur, phenols, dissolved solids, free oil and dissolved oil. Sulfur is usually recovered as a product. Acid and caustic wastes from the refinery require special handling and can be regenerated for some reuse. The primary wastewaters containing phenols, dissolved solids, free oil, dissolved oil, and small quantities of other materials require several levels of treatment.

A number of methods are available for the reduction if not elimination of the refinery wastewater. The following examples illustrate some of the possibilities.

1. The classical barometric condenser can be substituted by an indirect condenser or by appropriate type of air cooling.
2. The trading of contaminated process water for steam condensate. A typical example is dilutions steam in cracking furnaces.
3. Concentration of the contaminant in a wastewater near the saturation point by recirculation of the bottom section of a water-wash tower.
4. Elimination of a water seal in a flash stack by other types of seals such as a molecular seal.
5. More judicious maintenance and housekeeping operations, which minimize the use of water for washing down the plant area.

The above are some typical possible methods for a reduction of wastewater. The proper attention to entering detail both in design and operation can reduce the total amount of wastewater for subsequent treatment by a very significant margin.

Sewer System. A properly designed sewer system that permits at least some level of segregation of wastewaters with relatively small amounts of contaminants permits selective recycle. It is recognized that such systems would be very difficult to achieve in some of the older refineries. For typical approaches in terms of overall piping design, including the sewer systems, the reader may refer to ref. 18.

Conventional Treatment Methods. In addition to proper minimization of wastewater and judicious layout of piping, a number of treatment methods are conventionally employed for treating refinery wastewaters.

Physical Treatment Methods. Physical treatment methods are still the most widely employed primary treatment methods for refinery wastes. Typical sequence involves one or more of the following:

1. Gravity separation using API or other type of separators is one of the more popular methods of treatment. While this is normally always employed, there are cases where other methods may also be necessary.
2. Another widely employed method for the disposal of wastewaters is through steam or flue gas stripping. Phenols, mercaptans, H_2S, and other compounds have been removed by this method. Waste gases from stripping operations are usually burned to reduce air pollution.
3. Absorption and extraction. Many compounds can be effectively removed from wastewaters from absorption or extraction techniques. A typical example is phenols that can be extracted by ethers.
4. Sedimentation, filtration, flotation, evaporation, and submerged combustion processes. These have been widely employed for the treatment of refinery wastewaters. For details on typical design practices in each of these, the reader may refer to refs. 20 and 21.
5. In addition to the above conventional and physical treatment methods, equalization basins are usually employed to provide a steady stream of wastewater flow to subsequent chemical and biological treatment operations. As such, equalization may be considered an extra physical step.

Chemical Treatment Methods. In addition to the physical treatment methods outlined above, a number of common chemical treatment methods are widely employed.

1. pH adjustment. Adjustment of pH for the addition of an appropriate acid or base is necessary both for proper functioning of biological treatment systems and for meeting effluent requirements.
2. Flocculation and chemical precipitation are often employed where emulsification is a potential problem.

Biological Treatment Processes. A number of variations of biological treatment processes are broadly employed in the petroleum industry. Typical variations are as follows:

1. **Activator sludge.** The activator sludge process is one of the more popular biological processes. Complicated petroleum industry wastes have been successfully handled by this and other biological treatment methods. For a more detailed review of the biological treatment processes, the reader is referred to ref. 21.
2. **Lagoons.** Simple lagoons without the sophistication of activator sludge processes is applicable if significant amounts of land area is available.
3. **Trickling filters.** Trickling filters again have been very widely used for removal of BOD.

Advanced Treatment Processes. In addition to the above, a number of more advanced physical treatment methods are also employed to supplement other methods. Table 8.6-12 provides one typical list of conventional and advanced.

1. **Filtration.** The use of filtration as a final treatment system is employed when it is desirable to obtain very high-quality effluent or recycle stream. Typically, dual-media filters are employed: anthracite and sand, activated carbon and sand, or resin beds and sand. Typical filtration rates are reported to be in the 5–6 gpm/ft^2 range. Whenever filtration is employed, some pilot studies are necessary for design.

2. **Carbon absorption:** There has been a major increase in the use of activated carbon for the removal of dissolved organic materials from wastewater. Carbon absorption can be the polishing step for treatment of an effluent from biological treatment processes. However, its use is applicable only if a high-quality effluent is required. Typically effluents from activated sludge processes can be treated in Haz-Ops units to reduce BOD to 3–10 mg/L, the oil content to less than 1 mg/L and phenol to almost zero. For most commercial-scale Haz-Ops units it is economical to regenerate the spent carbon for reuse. API Design Manual[22] provides information of basic design parameters.

Chemical Oxidation. The primary application of chemical oxidation is for the reduction of phenols and cyanides in waste streams by oxidizing agents as chlorine or ozone. These are only applicable to small concentrated streams where conventional biological oxidation is not applicable.

Factors Affecting Overall Treatment Systems. The complexity of the wastewater management system, of course, varies with the complexity of a refinery. In addition to the specific operations

TABLE 8.6-12   TYPICAL REFINERY WASTEWATER TREATMENT PROCESSES

Primary	Intermediate	Secondary/Tertiary
Sulfide/ammonia stripping	Flotation	Activated sludge
API separators	Coagulation	Trickling
Filters	Precipitation	
Tilted plate separators	Equalization	Waste stabilization ponds
Liquid–liquid extraction		Aerated lagoons
Filtration for oil removal		Filtration
pH control		Carbon adsorption
		Chemical oxidation

mentioned above, other factors that need to be taken into account in the design and implementation of a wastewater management scheme include factors such as town water retention pond and its tie-in to the separator, internal recycle of water after some treatment, and so on. A zero-discharge refinery has been proposed.[23] The tendency in a modern refinery is to work minimization of net effluent discharge and maximization of internal water recycle treatment and reuse.

Solid Waste Disposal

The sources of sludge from a petroleum refinery may be listed as follows:

1. Bottoms off API separators.
2. Dissolved air flotation scums.
3. Biooxidation sludge.
4. Steam plant blowdown sludge.

Depending on the overall effluent treatment scheme adopted, these sludges may be treated separately or combined.

API Separator Bottoms. A common method of oil-water separation in wastewater from a refinery is by means of an API gravity separator.[22] API separators are sized to permit most of the free oil to float to the surface and the heavier solids to sink to the bottoms. The oil is recovered. Where wastewaters have a high amount of free oil, they are segregated and treated in an API separator before mixing with nonoily water streams. Solids from API separators may equal 40–80 ppm of the weight of water processed. Although they may settle to the bottom 15–20% of the separator, they are removed as a 3–5% sludge. The principal factors that affect the design of oil separators and hence the characteristics of resultant sludge[24,25] are:

1. Specific gravity of the oil.
2. Specific gravity of the wastewater.
3. Temperature of the wastewater stream.
4. Presence or absence of emulsions.
5. Suspended solids concentration.

In addition to the API separator, other special-purpose separators based on analogous principles of operation are often employed for streams containing oil and suspended solids.

Dissolved Air Flotation Scums. Dissolved air flotation is usually considered a secondary or intermediate treatment process for reducing the quantity of oil and suspended solids present in API separator effluents.[25] Literature indicates that dissolved air flotation can remove 75–85% of the oil present and 30–50% of the suspended solids present. The oil and suspended solids float to the surface and are skimmed from the surface to accumulate as a sludge of about 3–6% dry substance (DS) by weight.

Bioxidation and Other Operation. The decoiled main effluent stream is then usually treated by other physical methods and biochemical treatment systems. These will produce a sludge that can be thickened to 2–4% by settling. In some cases the sludge has been thickened to 3–5% by flotation. For a 250,000-barrel refinery, the total weight of sludge dry substance would be between 5 and 10 tons, of which about 15–25% would be accounted for by the API separator bottoms.[21] Again using broad averages, this sludge will be about 70% volatile organics and 30% inert.

Steam Plant Blowdown. The petroleum industry requires large quantities of steam. As a part of the water management program in order to keep salts in balance, a part of the water is blowdown. A rough average is that the blowdown will equal 5% of the steam produced. Thus, a plant utilizing 120 tons/h steam will blowdown 6 tons/h. Compared to the flows mentioned earlier, this is a relatively small amount, but it does contribute to the overall load.

Treatment of Petroleum Sludges

Conventional treatment of sludges from petroleum refining (or marketing operations) may follow two practices: dewatering operations and ultimate disposal.

Dewatering. Table 8.6-13 gives generally accepted average performances for different dewatering equipment. A more detailed description of such equipment is contained in Section 6.3.

TABLE 8.6-13   DEWATERING PETROLEUM INDUSTRY SLUDGES

Dewatering Method	Typical Product Sludge (% dissolved solids in cake)
Centrifuge	10–15
Rotary vacuum filter	15–20
Gravity belt filter	15–20
Plate press	35–50
Pressure tank filter	35–50

Ultimate Disposal.   Ultimate treatment of sludges from petroleum refining and petrochemical sludges may include the following:

1. Wet impoundments (or lagooning).
2. Pitting.
3. Landfill.
4. Sludge farming.
5. Incineration.

Wet Impoundments (Lagooning).   This method of disposal is still practiced but its use is decreasing. Increasing value of land and laws for protection of groundwater are likely to make lagooning as a disposal option not very attractive in the future.

Pitting.   Sludges with a relatively high oil content such as dissolved air flotation scums are often sent directly to pits where normal biodegradation can take place. Then it must be prepared in such a way as to avoid difficulties from leaching and impact on groundwater. Usually several pits are used sequentially in time; periodically the supernatant water is drawn off and sent to the main bioxidation system. Pitting is also decreasing in popularity.

Landfilling.   Landfill of semisolid and solid wastes has been practiced for many years. Licensing of landfills now requires stricter controls to prevent leaching of potentially toxic elements to the subsurface water. Further, the Resource Conservation and Recovery Act of 1976 is leading to a further tightening of regulations for land disposal. As the controls become more stringent, the cost of landfilling will increase. Landfilling of oily waste tend to leave the oily waste unchanged for many years. In some cases, to make sludges amenable for landfill, chemical treatment processes to minimize leaching are employed.[26] Landfilling will probably continue to be used as a method of disposal of the ash product of incineration.

Sludge Farming.   Unlike landfilling, sludge farming biologically degrade the sludge and further has minimum air pollution problems (unlike incineration discussed below). On sludge farming degradation of hydrocarbons by living matter is carefully promoted by thoroughly bringing the sludge in contact with the soil that contains the bacteria. Oxygen is provided by tilling the soil nitrogen by fertilization.
Sludge farming consists of the following steps:[27]

Selection of a site with suitable drainage.
Preparation of the site as though for sowing of an agricultural crop (including addition of fertilizer, as necessary).
Spreading on the surface of the soil of the refinery sludge at an appropriate concentration.
Mixing the sludge with the top soil using conventional agricultural equipment.
Repeating the mixing at suitable intervals, ensuring adequate aeration of the soil/sludge layer.
Monitoring of the site by suitable sampling and analysis of soil samples.

It appears that:

1. Oil degradation rate is directly related to the percentage of oil in the soil.
2. Fertilization improves degradation rate.
3. Aeration (tilling) frequencies vary (from 1 week to 2 or 3 months).
4. Between 350 and 400 m^3 (2000–2500 barrels) of oil per hectare should be degraded in a 8- or 9-month growing season.
5. Sludge farming is about one-fifth as expensive as is incineration.

Incineration. Incineration involves oxidation of the carbonaceous material to CO_2 water and ash. Incineration system design should include flue gas cooling, and flyash fluidized-bed incinerators have been employed. Depending upon the type of incinerator used, the optimum moisture content of the sludge feed will vary. Typically, stationary incinerators handle 40–70% by weight sludge feed; fluid beds handle 2–10%. Incineration is an expensive process and should be a supplement to other methods of sludge management.

REFERENCES

8.6-1   "Development Document for Effluent Limitation Guidelines and Standards for the Petroleum Refining Point Source Category," EPA 440/1-79/014-b, December, 1979.

8.6-2   "The 1976 API Refinery Solid Waste Survey," Engineering Science Inc., Austin, TX, Report to the American Petroleum Institute, 1978.

8.6-3   Federal Register, 39, FR 42510 December 5, 1974.

8.6-4   "Particulate Matter: Inhalable Variety," *Env. Sci. Tech.* **15**(9), September 1981, pp. 983–986.

8.6-5   *Environment Reporter*, Bureau of National Affairs, Washington, D.C. pp. 121:1530–121:1533, (a continuously updated compendium).

8.6-6   Federal Register, 44. FR 61543, October 25, 1979.

8.6-7   Federal Register, 42, FR 32426, June 24, 1977.

8.6-8   Federal Register, 43, FR 10869, March 15, 1978.

8.6-9   Federal Register, 45, FR 23378, April 4, 1980.

8.6-10 Federal Register, 44 FR 34244, June 14, 1979.

8.6-11 *Inside EPA*, June 5, 1981, p. 14.

8.6-12 *Environment Reporter*, Bureau of National Affairs, Washington, D.C., pp. 135:0461–135:0471.

8.6-13 Federal Register, 45, FR 33063, May 19, 1980.

8.6-14 *Congressional Record*, H10174 to H10187, October 1, 1980.

8.6-15 Federal Register, 41, FR 12657, March 26, 1976.

8.6-16 *Environment Reporter*, Bureau of National Affairs, WAshington, D.C. pp. 131:0931–131:0936 and 131:0951.

8.6-17 J. A. Danielson, *Air Pollution Engineering Manual*, Environmental Protection Agency, Office of Air Quality Planning and Standards, Research Triangle Park, NC, May 1973.

8.6-18 F. L. Evans, Jr., *Equipment Design Handbook for Refineries and Chemical Plants*, Vols. 1 and 2, Gulf Publishing, Houston, TX, 1971.

8.6-19 *Recommended Practice for the Design and Installation of Pressure Relieving Systems in Refineries, Part I—Design and Part II Installation*, American Petroleum Institute, Washington, D.C., December 1976.

8.6-20 B. Lipsak (ed.), *Environmental Engineering Handbook*, Vols. 1, 2, and 3, Chilton, Radnor, PA, 1974.

8.6-21 W. W. Eckenfelder and C. J. Santhanam (eds.). *Sludge Treatment*, Marcel Dekker, New York, 1981.

8.6-22 "API Manual on Disposal of Refinery Wastes," American Petroleum Institute, Washington, D.C., 1976.

8.6-23 J. W. Porter, J. H. Blake, and R. T. Mulligan, "Zero Discharge of Wastewater from Petroleum Refineries," National Conference on Complete Water Reuse, Washington, D.C., April 1973.

8.6-24 B. B. Carnes, D. L. Ford, and S. Brady, "Treatment of Refinery Wastewater for Reuse," National Conference on Complete Water Reuse, Washington, D.C., April 1973.

8.6-25 D. L. Ford and L. F. Tismiller, "Meeting BPT Standards for Refining Wastewater Treatment," *Ind. Wastes*, 20–25, July/August 1977.

8.6-26 R. Wisniewski, "Process Converts Sludge to Landfill," *Oil & Gas Journal*, **73**, 133, 1975.

8.6-27 "Sludge Farming: A Technique for the Disposal of Oil Refinery Wastes," Report No. 3-80, CONCAWE, The Hague, Netherlands, March 1980.

CHAPTER 9

GAS TECHNOLOGY

WILLIAM W. BATHIE

Iowa State University
Ames, Iowa

9.1 INTRODUCTION

Gaseous fuels (natural and synthetic), when available, are ideal for combustion for the following reasons:

1. Storage at the point of use is either minimized or eliminated since the gas is either manufactured close to the point of use or is piped directly to the consumer.
2. It is easy to regulate and/or change the rate at which the gas is supplied to the combustion chamber.
3. The gas mixes readily with combustion air resulting in complete combustion with low amounts of excess air.
4. Combustion produces little or no ash residue thereby eliminating any ash disposal problems.
5. The fuel is low in sulfur and ash thereby minimizing the pollution control mechanisms needed. This also leads to a low level of maintenance on the combustion equipment.
6. The fuel is extremely low in noncombustible constituents.

Natural gas is the volatile portion of crude petroleum and is composed primarily of methane. Other constituents usually found in natural gas samples are small amounts of other hydrocarbons such as ethane, propane, butane, and pentane, and nonhydrocarbon constituents usually including, in varying amounts, nitrogen, carbon dioxide, helium, hydrogen sulfide, argon, and water vapor.

The first useful discovery of natural gas was reported in 1821 but its use did not rapidly expand until the late 1920s when improved methods to transport the gas via pipelines from wells to points of consumption were developed. Today, natural gas pipelines serve many parts of the world. Natural gas transported via pipelines is liquefied during periods of reduced demand so it can be stored near the point of consumption for use during high demand periods. It is also being liquefied and transported by ship from available sources to points of high consumption.

The supply and disposition of natural gas in the United States for the last 18 years is shown in Table 9.1-1. It should be noted that dry production of natural gas from wells in the United States peaked in 1973 and has declined approximately 18% since then. Consumption peaked in 1972 and has declined since that year.

TABLE 9.1-1   SUPPLY AND DISPOSITION OF NATURAL GAS IN THE UNITED STATES[a]

	Supply				Disposition				
Year	Dry Production	Withdrawals From Storage	Imports	Total	Additions to Storage	Exports	Unaccounted For	Consumption	Total
1965	15,286,280	959,865	456,394	16,702,539	1,077,980	26,132	318,711	15,279,716	16,702,539
1966	16,467,320	1,141,614	479,780	18,088,714	1,210,469	24,639	401,203	16,452,403	18,088,714
1967	17,386,791	1,132,534	564,226	19,083,551	1,317,363	81,614	296,214	17,388,360	19,083,551
1968	18,494,523	1,329,536	651,885	20,475,944	1,425,075	93,745	325,062	18,632,062	20,475,944
1969	19,831,680	1,379,488	726,951	21,938,119	1,498,988	51,304	331,587	20,056,240	21,938,119
1970	21,014,229	1,458,607	820,780	23,293,616	1,856,767	69,813	227,650	21,139,386	23,293,616
1971	21,609,885	1,507,630	934,548	24,052,063	1,839,398	80,212	338,999	21,793,454	24,052,063
1972	21,623,705	1,757,218	1,019,496	24,400,419	1,892,952	78,013	328,002	22,101,452	24,400,419
1973	21,730,998	1,532,820	1,032,901	24,296,719	1,974,324	77,169	195,863	22,049,363	24,296,179
1974	20,713,032	1,700,546	959,284	23,372,862	1,784,209	76,789	288,731	21,223,133	23,372,862
1975	19,236,379	1,759,565	953,008	21,948,952	2,103,619	72,675	235,065	19,537,593	21,948,952
1976	19,098,352	1,921,017	963,768	21,983,137	1,755,690	64,711	216,240	19,946,496	21,983,137
1977	19,162,900	1,749,884	1,011,001	21,923,785	2,306,515	55,626	41,063	19,520,581	21,923,785
1978	19,121,903	2,157,765	965,545	22,245,213	2,278,002	52,532	287,201	19,627,478	22,245,213
1979	19,663,415	2,047,000	1,253,383	22,963,798	2,295,034	55,673	372,330	20,240,761	22,963,798
1980	19,557,709	1,972,333	984,767	22,514,809	1,949,064	48,731	639,559	19,877,293	22,514,647
1981	19,356,963	1,930,092	903,949	22,191,004	2,227,522	59,372	500,501	19,403,858	22,191,253
1982	17,902,544	2,164,184	933,336	21,000,064	2,472,383	51,728	474,731	18,001,055	20,999,897

[a] In million cubic feet. Data from ref. 1.

Supplemental sources of natural gas from the North American continent are from Canada and Mexico. Declining proven resources in the lower 48 states and the rapid increase in the price of natural gas have lead to the transportation of gas by ship as liquefied natural gas (LNG) and consideration of the manufacture of synthetic gases.

World resources of natural gas are quite large. Natural gas proved reserves, as reported in the *U.S. Crude Oil, Natural Gas, and Natural Gas Liquids Reserves 1982 Annual Report*[1] ranged from 202 to 213 trillion cubic feet in the United States, 92 to 97 trillion cubic feet in Canada, 75 to 76 trillion cubic feet in Mexico, and 3024 to 3323 trillion cubic feet in the world. This shows that the proved reserves of the United States are about 6% of the total for the world and about 11% of the proved reserves are in North America.

Synthetic (manufactured) gases have been in existence for a long period of time. Most of the synthetic gases are from the conversion of coal or oil into gas and have low Btu contents (100–200 Btu/ft^3) or medium Btu contents (250–350 Btu/ft^3) as opposed to the traditional Btu levels of approximately 1000 Btu/ft^3 for natural gas.

Low Btu gas is produced when coal is gasified using air. It is the cheapest and easiest gaseous fuel to produce from coal. Because of its low Btu content, it usually is not transported (piped) over long distances and can lead to expensive burner modifications for units designed for natural gas.

Medium Btu synthetic gas is made by gasifying coal with oxygen instead of air. Once again, it usually is not piped over long distances because of its Btu content.

Other methods of manufacturing synthetic gas include the pyrolysis of refuse (solid waste) with oxygen and the generation of methane from manure, sewer gas, and gas from biomass.

The following sections discuss discovery, transportation, and use of natural and synthetic gases.

REFERENCE

9.1-1   *U.S. Crude Oil, Natural Gas and Natural Gas Liquids Reserves, 1982 Annual Report*, Energy Information Administration, Office of Oil and Gas, U.S. Department of Energy, Washington, D.C., DOE/EIA-0216 (82), September 1983.

9.2   EXPLORATION AND DISCOVERY

Natural gas was known to ancient civilization and was used in Europe and the United States, on a small scale, prior to the discovery of oil in Pennsylvania in 1859. Even when oil was discovered in

1859, the gas which was present with the oil was usually burned by flaring. It was necessary to develop the natural gas pipeline industry before the natural gas industry could be developed.

Medici[2] states that the first useful discovery in the United States was reported to have been at Fredonia in 1821 and that the first attempt to transport natural gas via a pipeline was made in 1870 in the United States at Bloomfield. This attempt had to be cancelled since the pipeline was made of white pine logs and developed many gas leaks.

Today, many large-diameter long-distance pipeline systems exist in the United States and Canada with large volumes of gas being piped from wells to consumer areas where it is used as an industrial and residential fuel and as a raw material for its chemical properties in producing end products such as fertilizer.

Natural gas is always present with crude oil and is also found in significant amounts in a reservoir without being in contact with a significant quantity of crude oil.

Associated natural gas is defined as natural gas occurring as a free gas in a crude oil reservoir. *Dissolved natural gas* is defined as natural gas in solution with crude oil. *Nonassociated natural gas* is defined as natural gas not in contact with an appreciable quantity of crude oil.

It is reported[1] that the estimated natural gas reserves in the United States were 62,082 billion cubic feet of associated-dissolved gas on December 31, 1982, 147,172 billion cubic feet of nonassociated natural gas on this date for a total proved reserve of 209,254 billion cubic feet. All volumes are for a pressure of 14.73 psia and are at 60°F.

Natural gas, as it is removed from the ground, may be either wet or dry.

A *gas condensate well* is defined as one producing from a gas reservoir that contains considerable quantities of liquid hydrocarbons in the pentane and heavier groups. *Natural gas liquids* are those hydrocarbons in natural gas which are separated from the gas through the process of absorption, condensation, or other methods. These liquids usually consist of propane and the heavier hydrocarbons. A *lease separator* is a facility located at the wellhead for the purpose of either separating the gases from the crude oil or for separating those constituents that liquefy at the separator temperature and pressure from the gases.

Natural gas usually contains nonhydrocarbon constituents such as hydrogen sulfide, nitrogen, carbon dioxide, helium, and water vapor. All except hydrogen sulfide are noncombustibles and reduce the heating value of the gas mixture. Carbon dioxide and helium are removed from the gas mixture at the wellhead site for commercial sale if they are present in sufficient amounts. Water is removed from the gas either by absorption or adsorption. Hydrogen sulfide is very corrosive and, if it is present in an appreciable amount, must be removed before the gas is transported via a pipeline or liquefied from transportation by ship.

Dry natural gas is defined as the volume of gas remaining after the natural gas liquids and nonhydrocarbon constituents have been removed in lease, field, or plant separators.

Prospecting for natural gas and liquid deposits is a long-term activity. It can involve many years between the time when prospecting is initiated and the point at which natural gas is produced and sent to the consumer. Large financial investments are required and the risks are high. Early exploration was over land, but recently exploration, prospecting, and production have occurred from offshore wells with offshore wells being more expensive than those on land.

There are many methods that can be used when prospecting for natural gas and/or crude petroleum. Several of the methods are discussed below.

Early techniques for discovering oil and/or natural gas consisted mainly of hunting for seepages of oil or gas. This was successful in many cases but soon gave way to using geology and related earth sciences to discover new fields.

Today, trying to accurately determine natural gas and liquid deposits deep in the earth involves determining the geological structure from readings taken at the earth's surface.

The methods described below are based on the variation in physical properties between reservoir rocks and other rocks in the vicinity. Prospecting techniques are based on variations in the earth's acoustical conductivity, magnetism, and gravity. All methods used involve trying to locate rock formations that are favorable to the containment of oil and/or gas.

The most widely used method today is the *seismic reflection* method. This method involves transmitting energy waves deep into the earth by means of an explosion and then interpreting, at the surface, the reflected waves. Wave velocities vary with rock type and depth. This technique provides only depth information.

Another method used is the *seismic refraction* method. This method provides information about depth and the refracting medium.

The *magnetic method* depends upon variations in the magnetic susceptibility of different rock types; the *gravimetric method* uses the principle of variation in the gravitational acceleration due to mass, size, and shape of the reservoir rock. The *electric resistivity method*, one of the earliest methods used, depends on the conductivity of the earth's structure.

Once a prospecting technique indicates the location of a natural gas and/or oil reservoir, it is necessary to drill one or more exploratory wells to confirm the find and to estimate the volume of proved reserves. The initial estimate is based upon the initial flow data and includes the flow rate,

TABLE 9.2-1   ESTIMATED TOTAL U.S. PROVED RESERVES AND PRODUCTION OF NATURAL GAS AND NATURAL GAS LIQUIDS, 1982[a]

Location	Dry Natural Gas[b]		Natural Gas Liquids[c]	
	Proved Reserves	Production	Proved Reserves	Production
Alaska				
Onshore and offshore	34,990	261	9	0
Lower 48 states				
Onshore	127,568	11,854	6,411	619
Offshore	38,954	5,391	801	102
Subtotal	166,522	17,245	7,212	721
U.S. Total	201,512	17,506	7,221	721

[a] Data from ref. 2.
[b] Billion cubic feet, 14.73 psia, 60°F.
[c] Million barrels of 42 U.S. gallons.

TABLE 9.2-2   TOP FIVE STATES OF TOTAL ESTIMATED PROVED RESERVES, 1982[a]

Total Dry Natural Gas

Rank	State	Reserves[b]
1	Texas	49,757
2	Louisiana[c]	44,916
3	Alaska	34,990
4	Oklahoma	16,207
5	New Mexico	12,418

Natural Gas Liquids

Rank	State	Reserves[d]
1	Texas	2,771
2	Louisiana	1,295
3	Oklahoma	745
4	Utah/Wyoming	681
5	New Mexico	531

[a] Data from ref. 2.
[b] Billion cubic feet, dry, 14.73 psia, 60°F.
[c] Includes offshore Alabama.
[d] Million barrels of 42 U.S. gallons.

thickness of the reservoir, and electrical and other measurements taken within the well hole. As additional wells are drilled and placed into production, additional data (such as thickness and extend) are used to update the proved reserve estimate.

Table 9.2-1 lists the estimated total United States proved reserves and production of natural gas and natural gas liquids as of 1982. One should observe that in the lower 48 states, 23% of the proved reserves and 31% of the production of dry natural gas was from offshore wells, that 17% of the proved reserves but only 1.5% of the production of dry natural gas came from Alaska.

The top five states for estimated proved natural gas and natural gas liquids at the end of 1982 are shown in Table 9.2-2.

Drilling is the only method available to determine if a location actually contains gas and/or oil. The rig and drilling tools used in sinking a well depend upon expected rock formations and depth.

Rotary drilling is at present the most common method used for both offshore and land drilling. This method was first used in Texas in 1901.[3]

Equipment selection and derrick height depend upon the following:

1. Type of rock expected.
2. Length of drill pipe sections.
3. Depth.
4. Platform structure for offshore drilling.

Another method is *cable drilling*. The reader is referred to ref. 1 or 3 for a detailed description of these methods.

Once a well has been drilled, it is necessary to complete the well. This involves installing the necessary casings, determining the production capability, and being certain a minimum amount of contaminants enter the well.

Gas withdrawn from an associated or nonassociated well is used in many ways. These include:

1. *Reservoir repressuring* which is gas returned to the gas formation to maintain well pressure and for cycling purposes.
2. *Extraction* of natural gas liquid constituents.
3. *Removal* (reduction) of nonhydrocarbon gases in treating or processing operations.
4. *Venting* or flaring at the gas well site.

The *marketed production* is defined as the amount of dry natural gas remaining after that used for repressuring, quantities vented or flared, and nonhydrocarbon constituents have been removed.

REFERENCES

9.2-1 M. Medici, *The Natural Gas Industry*, Newnes-Butterworths, London, 1974.
9.2-2 *U.S. Crude Oil, Natural Gas and Natural Gas Liquids Reserves, 1982 Annual Report*, Energy Information Administration, Office of Oil and Gas, U.S. Department of Energy, Washington, D.C., DOE/EIA-0216 (82), September 1983.
9.2-3 *Gas Engineers Handbook*, The Industrial Press, New York, 1965.

9.3 NATURAL GAS

Natural gas, as defined earlier, is the volatile portion of crude petroleum (associated gas) or a nonassociated gas which means it is gas not in contact with an appreciable quantity of crude oil. It consists of hydrocarbons primarily in the paraffin family along with nonhydrocarbon constituents which usually include hydrogen sulfide, nitrogen, carbon dioxide, helium, and water vapor.

The hydrocarbons usually found in natural gas are methane (CH_4), ethane (C_2H_6), and propane (C_3H_8) along with small amounts of the butanes, petanes, and heavier hydrocarbons. Methane is the most abundant hydrocarbon found in natural gas. The fraction of each hydrocarbon found in natural gas will vary with the source of the natural gas and has a great influence on the heating value of the natural gas.

Table 9.3-1 lists the physical properties of the hydrocarbons commonly found in natural gas. Physical properties listed include:

1. Molecular (formula) weight which is the sum of the atomic weights of the atoms in the molecule.
2. Specific gravity which is the density of the hydrocarbon gas divided by the density of dry air at the same temperature and pressure. Densities are determined at 60°F and 1 atm.
3. Boiling point (vaporization temperature), which is the temperature at which the hydrocarbon will change from a liquid to a vapor at 1 atm.
4. Flammability limits that represent the lower and upper limits where combustion will occur. These limits are a ratio of the volume of hydrocarbon gas to the total volume of hydrocarbon plus air and are expressed as a volume percent of hydrocarbon in the mixture.
   (a) Lower limit represents the minimum volume percent below which combustion will not occur and is commonly referred to as the *lean mixture limit*.
   (b) Upper limit represents the maximum volume percent above which combustion will not occur and is commonly referred to as the *rich mixture limit*.
5. Critical pressure and temperature. The *critical pressure* is the highest pressure at which one can distinguish between the liquid and vapor phases. The *critical temperature* is the highest temperature at which it is possible to distinguish between the liquid and vapor phase at

TABLE 9.3-1 PHYSICAL PROPERTIES OF SELECTED HYDROCARBONS[a]

Hydrocarbon	Chemical Formula	Molecular Weight	Specific Gravity of Real Gas at 60°F and 1 atm	Boiling Point in Air at 1 atm		Flammability Limits Volume Percent in Air Mixture		Critical Constants				Constant Pressure Heat of Combustion (Btu/ft³)	
								Pressure		Temperature			
				°F	°C	Lower	Upper	psi	atm	°F	°C	H₂O (l)	H₂O (g)
Methane	CH_4	16.043	0.55469	−258.69	−161.495	5.0	15	667.8	45.44	−116.63	−82.57	1011.5	910.73
Ethane	C_2H_6	30.070	1.0465	−127.48	−88.60	2.9	13	707.8	48.16	90.09	32.27	1783.5	1631.4
Propane	C_3H_8	44.097	1.5496	−43.67	−42.045	2.1	9.5	616.3	41.94	206.01	96.67	2563.1	2358.0
n-Butane	C_4H_{10}	58.124	2.0749	31.10	−0.50	1.8	8.4	550.7	37.47	305.65	152.03	3373.7	3113.8
Isobutane	C_4H_{10}	58.124	2.0687	10.90	−11.72	1.8	8.4	529.1	36.00	274.98	134.99	3354.1	3094.6
n-Pentane	C_5H_{12}	72.151	2.6073	96.92	36.064	1.4	8.3	488.6	33.25	385.7	196.5	3913.8	3624.3
Isopentane	C_5H_{12}	72.151	2.6091	82.12	27.843	1.4	8.3	490.4	33.37	369.10	187.28	4015.8	3715.9
Neopentane	C_5H_{12}	72.151	2.6181	49.10	9.499	1.4	8.3	464.0	31.57	321.13	160.63	4188.9	3871.8

[a]Data from ref. 1.

TABLE 9.3-2 PHYSICAL PROPERTIES OF NONHYDROCARBON CONSTITUENTS FOUND IN NATURAL GAS

Constituent	Chemical Formula	Formula Weight	Constant Pressure Heat of Combustion	
			Btu/lb-mole	kcal/g-mole
Hydrogen sulfide	H_2S	34.0799	−222,709	−123.81
Carbon dioxide	CO_2	44.0100	0	0
Nitrogen	N_2	28.0134	0	0
Helium	He	4.0024	0	0
Water vapor	H_2O	18.0153	0	0

equilibrium. The state where it is not possible to have a distinct transformation between the liquid and vapor phase is referred to as the *critical point*.

6. Constant pressure heat of combustion. This value represents the amount of energy released when a unit quantity of the hydrocarbon is burned completely and the products of combustion are cooled to the temperature at which the hydrocarbon and air (or oxygen) were supplied. The *higher heating value* is the absolute value of the constant pressure heat of combustion when all water formed from combustion is a liquid; the *lower heating value* is the absolute value of the constant pressure heat of combustion when all water formed from combustion is in the vapor phase. All values tabulated in Table 9.3-1 are expressed in Btu/ft³.

It should be observed, from the data in Table 9.3-1, that the boiling point (vaporization temperature) for methane is −258.69°F (−161.495°C), and that the boiling point for the heavier hydrocarbons are higher. The boiling point for *n*-butane and heavier hydrocarbons at 1 atm. is at the freezing point for water or higher. Therefore, when the natural gas is brought to the surface and reduced to atmospheric pressure and temperature, most of the heavier hydrocarbons will condense.

Also all hydrocarbons, except methane, have a specific gravity of 1 or higher. This means that methane is lighter than air, ethane is approximately the same as air, all other hydrocarbons are heavier than air.

The higher heating value, per cubic foot, increases as the molecular weight of the hydrocarbon increases. Therefore, natural gases with higher fractions of these heavier hydrocarbons will have higher heating values.

Other constituents normally found in natural gas include hydrogen sulfide, carbon dioxide, helium, nitrogen, and water vapor. The chemical formula, formula (molecular) weight, and constant pressure heat of combustion values for these constituents are listed in Table 9.3-2. It should be observed that other than for hydrogen sulfide, these constituents are noncombustibles; that is, their heat of combustion is zero and they decrease the higher heating value of a gas mixture.

The amount of air theoretically required for complete combustion depends on the composition of the fuel. When carbon, hydrogen, and sulfur are completely burned with oxygen, the resulting products are, respectively, carbon dioxide, water, and sulfur dioxide, or, for each of these constituents, the amount of oxygen theoretically required is

$$C + O_2 \rightarrow CO_2 \tag{9.3-1}$$

$$H_2 + \tfrac{1}{2}O_2 \rightarrow H_2O \tag{9.3-2}$$

$$S + O_2 \rightarrow SO_2 \tag{9.3-3}$$

The general reaction for the complete combustion of a hydrocarbon with the stoichiometric amount of oxygen is

$$C_xH_y + \left(x + \frac{y}{4}\right)O_2 \rightarrow xCO_2 + \frac{y}{2}H_2O \tag{9.3-4}$$

The composition of dry air is tabulated in U.S. Standard Atmosphere, 1962.[2] The main constituents from this analysis are listed in Table 9.3-3.

The amount of dry air theoretically required for the complete combustion of a hydrocarbon with chemical formula C_xH_y is

$$\frac{\text{moles dry air theor. req'd}}{\text{mole } C_xH_y} = \frac{x + y/4}{0.2095} \tag{9.3-5}$$

TABLE 9.3-3   DRY AIR VOLUMETRIC ANALYSIS[a]

Constituents	Chemical Formula	Molecular Weight	% Volume
Nitrogen	N_2	28.0134	78.09
Oxygen	O_2	31.9988	20.95
Argon	Ar	39.948	0.93
Carbon dioxide	CO_2	44.0100	0.03
			100.00

[a] The apparent molecular weight for the dry air analysis is 28.965.

TABLE 9.3-4   VOLUME OF DRY AIR THEORETICALLY REQUIRED TO COMPLETELY BURN SELECTED HYDROCARBONS

Hydrocarbon	Chemical Formula	Volume Dry Air/ Unit Volume Hydrocarbon
Methane	CH_4	9.55
Ethane	C_2H_6	16.7
Propane	C_3H_8	23.9
Butane	C_4H_{10}	31.0
Pentane	C_5H_{12}	38.2

The volume of dry air theoretically required to burn a unit volume of each of the hydrocarbons normally found in natural gas are given in Table 9.3-4. The values listed in Table 9.3-4 assume the hydrocarbon and dry air are gases supplied at the same pressure and temperature and dry air has the analysis given in Table 9.3-3.

The composition of natural gas, and therefore its heating value and the quantity of air theoretically required for complete combustion, varies considerably from source to source. Compositions, real heating values, the Wobbe number, and the air theoretically required for complete combustion for five wellhead natural gases from U.S. sources are tabulated in Table 9.3-5. These data are from ref. 3 which lists 320 wellhead gases and 18 liquefied and synthetic natural gases.

The heating values of the natural gases tabulated in ref. 3 vary from under 920 Btu/scf to over 1300 Btu/scf with the median value of approximately 1040 Btu/scf. The compositions of these natural gases vary considerably.

The five gases listed in Table 9.3-5 have heating values that vary from 991 to 1156 Btu/scf. The amount of carbon dioxide and nitrogen, the two noncombustibles in these natural gases, varies considerably. The amount of methane varies from approximately 80% to over 96% and the amount of air required varies from 9.3 to 10.9, it being very dependent on the composition of the natural gas.

The Wobbe number, which is equal to the heating value divided by the square root of the specific gravity, is indicative of the gas flow through an orifice.

Sour natural gas is natural gas with an excessive amount of sulfur compounds, namely, hydrogen sulfide, which is highly corrosive. This sulfur has to be removed before the gas enters the pipeline. Hydrogen sulfide can be removed as sulfur by combining H_2S with SO_2 or SO to obtain liquid sulfur.

Sweet natural gas has a low concentration of sulfur compounds.

The quantity used, average price, and use of natural gas have changed considerably over the last 40 years. The quantity and average price of natural gas used in the United States during the last 13 years is tabulated in Table 9.3-6. One should observe the following:

1. The average wellhead price, quite uniform until 1973, is now 13 times what it was 10 years ago.
2. Gross withdrawals peaked in 1971.
3. Approximately 6% of the gross withdrawals are used for repressuring.
4. The quantity of natural gas vented and flared has decreased drastically since 1970. The amount vented and flared was 17.5% of the gross withdrawals in 1945, 2.1% in 1970, and only 0.5% in 1982.

TABLE 9.3-5 SELECTED VOLUMETRIC ANALYSES OF WELLHEAD NATURAL GAS FROM U.S. FIELDS[a]

Constituents, % by Volume	Samples				
	1	2	3	4	5
CO_2	0.430%	0.490%	0.653%	3.850%	0.320%
N_2	6.730	1.930	0.289	3.890	4.320
CH_4	87.930	93.350	96.461	83.260	80.470
C_2H_6	3.860	3.130	2.084	4.200	9.200
C_3H_8	0.710	0.100	0.294	1.870	2.750
i-C_4H_{10}	0.050	—	0.065	0.710	0.200
n-C_4H_{10}	0.140	—	0.059	0.870	1.440
i-C_5H_{12}	0.030	—	0.026	0.390	—
n-C_5H_{12}	0.040	—	0.018	0.320	1.300
C_6H_{14}	0.080	—	0.051	0.640	—
	100.000%	100.000%	100.00%	100.000%	100.000%
Heat value, Btu/Std ft^3	991.0	1014.9	1030.9	1077.6	1155.9
Real specific gravity	0.619	0.584	0.579	0.701	0.698
Air required	9.312	9.535	9.687	10.128	10.860
Wobbe No.	1260	1328	1354	1287	1383

[a] Data from ref. 3.

TABLE 9.3-6 QUANTITY AND AVERAGE PRICE OF NATURAL GAS PRODUCTION[a]

Year	Gross Withdrawals	Used for Repressing	Vented and Flared	Marketed Production	Extraction Loss	Dry Production	Average Wellhead Price of Marketed Production
1970	23,786,453	1,376,351	489,460	21,920,642	906,413	21,014,229	0.17
1971	24,088,031	1,310,458	284,561	22,493,012	883,127	21,609,885	0.18
1972	24,016,109	1,236,292	248,119	22,531,698	907,993	21,623,705	0.19
1973	24,067,202	1,171,361	248,292	22,647,549	916,551	21,730,998	0.22
1974	22,849,793	1,079,890	169,381	21,600,522	887,490	20,713,032	0.30
1975	21,103,530	860,956	133,913	20,108,661	872,282	19,236,379	0.45
1976	20,943,778	859,410	131,930	19,952,438	854,086	19,098,352	0.58
1977	21,097,071	934,801	136,807	20,025,463	862,563	19,162,900	0.79
1978	21,308,815	1,181,432	153,350	19,974,033	852,130	19,121,903	0.91
1979	21,883,353	1,245,074	167,019	20,471,260	807,845	19,663,415	1.18
1980	21,869,692	1,365,454	125,451	20,179,724	776,605	19,403,119	1.59
1981	21,587,453	1,311,735	98,017	19,955,823	774,562	19,181,261	1.98
1982	20,209,928	1,388,392	93,365	18,519,675	761,942	17,757,733	2.46

[a] Volumes in million cubic feet; value in dollars per thousand cubic feet. Data from ref. 4.

Natural gas consumption for the last 13 years is tabulated in Table 9.3-7. Once again, one should observe the following:

1. The quantity of natural gas delivered to consumers peaked in 1972.
2. Approximately 10% of the amount consumed was used either as lease and plant fuel or pipeline fuel.
3. The percentage consumed by each of the groups has changed considerably in the last 10 years. This change is illustrated in Fig. 9.3-1 which shows that the percentage consumed by industrial consumers has decreased from 41.1% in 1972 to 35.8% in 1982, while commercial consumption has increased from 13.1% in 1972 to 16.0% in 1982.

TABLE 9.3-7   NATURAL GAS CONSUMPTION IN THE UNITED STATES[a]

	Lease and Plant Fuel	Pipeline Fuel	Delivered to Consumers				
Year			Residential	Commercial	Industrial	Electric Utilities	Total
1970	1,398,758	722,166	4,837,432	2,398,510	7,850,660	3,931,860	19,018,462
1971	1,413,650	742,592	4,971,690	2,508,977	8,180,527	3,976,018	19,637,212
1972	1,455,563	766,156	5,125,982	2,607,982	8,168,856	3,976,913	19,879,733
1973	1,495,915	728,177	4,879,387	2,597,037	8,688,675	3,660,172	19,825,271
1974	1,477,386	668,792	4,786,128	2,555,617	8,291,782	3,443,428	19,076,955
1975	1,396,277	582,963	4,924,124	2,508,293	6,968,267	3,157,669	17,558,353
1976	1,634,355	548,323	5,051,360	2,667,740	6,963,850	3,080,868	17,763,818
1977	1,659,145	532,669	4,821,485	2,500,793	6,815,289	3,191,200	17,328,767
1978	1,647,911	530,451	4,903,006	2,601,106	6,756,641	3,188,363	17,449,116
1979	1,498,530	600,964	4,965,365	2,785,961	6,899,418	3,490,523	18,141,267
1980	1,026,438	634,622	4,752,082	2,610,895	7,171,661	3,681,595	18,216,233
1981	927,591	642,325	4,546,450	2,519,791	7,127,548	3,640,153	17,833,942
1982	1,109,398	596,411	4,633,035	2,605,523	5,831,170	3,225,518	16,295,245

[a] In million cubic feet. Data from ref. 4.

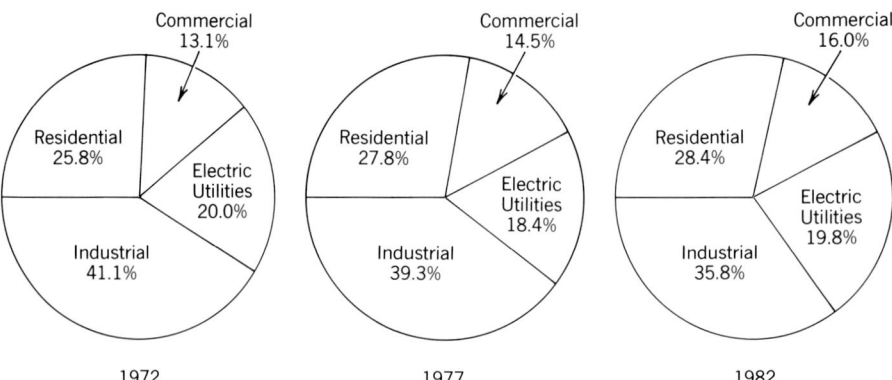

Fig. 9.3-1 Natural gas consumption in the United States by residential, commercial, industrial, and electric utilities.

REFERENCES

9.3-1 *Physical Constants of Hydrocarbons C_1 to C_{10}*, ASTM Data Series Publication DS 4A, American Society for Testing and Materials, Philadelphia, PA, 1971.
9.3-2 *U.S. Standard Atmosphere, 1962*, U.S. Government Printing Office, Washington, D.C., 1962.
9.3-3 J. C. Griffins, S. M. Connelly, and R. B. DeRemer, *Effect of Fuel Gas Composition on Appliance Performance*, GRI-82/0037, American Gas Association Laboratories, Cleveland, OH, December 1982.
9.3-4 *Natural Gas Annual, 1982*, Energy Information Administration, Office of Oil and Gas, U.S. Department of Energy, Washington, D.C., E3.11/2.2:982, October 1983.

9.4   LIQUEFIED NATURAL GAS AND LIQUEFIED PETROLEUM GAS

Data in Table 9.2-2 illustrate that the top five states with estimated total proved reserves of dry natural gas in the United States are in areas that are quite far from where most of the natural gas is consumed. It was also noted in Section 9.1 that the estimated proved reserves of natural gas in the United States are approximately 6% of the total proved reserves for the world. This means that a large percent of the natural gas is produced and available in areas where the market for natural gas is very limited. Therefore, it is necessary to transport natural gas from the gas well to the regions of high gas demand. This has resulted in the movement of natural gas over long distances via pipelines from the gas fields to the major centers where the gas is consumed.

Extensive pipelines exist in the United States, Canada, Europe, and other parts of the world, with large quantities of natural gas being transported within the United States and from Canada into the United States daily. Transporting natural gas by pipelines has a number of drawbacks including:

1. The pipeline capacity is fixed by the pipe diameter, length, maximum pressure, and location of compressor stations.
2. A pipeline involves a large capital investment and is a permanent installation from source to area of consumption. Both the source of natural gas and the major areas of consumption are subject to change.
3. It is impossible to transport natural gas as a gas between Europe and the United States or between any continents not connected by land because of its volume.

The consumption of natural gas in many of the consuming markets varies with the season, time of day, and industrial activity. This means that if pipelines are to be used to their capacity all year, some means must be used to store the natural gas near the point of consumption. One method is to store the gas in depleted oil or gas fields or in aquifers which are gas-tight formations from which the water has been removed. Another storage method is to liquefy the natural gas and store it in vessels either above or below ground.

The only economical way to transport natural gas by ocean vessels is as a liquid.

This section will discuss (1) the properties of liquefied gases, (2) natural gas preparation before it can be liquefied, and (3) methods of liquefying natural gas. Transport, storage, and safety standards associated with liquefied natural gas (LNG) are discussed in Section 9.6.

The volumetric analysis of a typical LNG would be 90–95% methane (CH_4), 5–7% ethane (C_2H_6), and the remainder propane (C_3H_8) and the heavier hydrocarbons. Its composition depends on the source.

Selected properties of methane, the principal LNG constituent, are listed in Table 9.4-1. One cubic foot of liquid methane is equivalent to approximately 625 ft^3 of gaseous methane at the same pressure. Selected properties of the other constituents in LNG are listed in Table 9.4-2.

Impurities such as water, carbon dioxide, nitrogen, and hydrogen sulfide must be removed before the start of the liquefaction process.

Hydrogen sulfide must be removed because it is very corrosive and has an offensive odor, the others because they will solidify, possibly causing a blockage in the equipment and are noncombustibles. Therefore, if they are not removed, they would add to the transportation cost and lower the heating value of the fuel.

Natural gas, once the impurities have been removed, must be liquefied. This is accomplished in peak-load shaving and base-load facilities.

Peak-load shaving is one means of meeting the daily, weekly, and/or seasonal variations in natural gas consumption without resorting to building an oversize, thereby inefficient, pipeline. Natural gas is liquefied and stored during the summer (lower demand period), vaporized and supplied to the consumer during the high demand days of the winter.

Peak-load shaving LNG plants are simpler than base-load plants which are used to prepare LNG for ocean vessel shipment. Since the natural gas supplied to base-load plants via a pipeline was cleaned at the wellhead, impurities such as hydrogen sulfide, water vapor, and possibly carbon dioxide and nitrogen were removed before the gas was supplied to the pipeline.

The first peak-load shaving LNG plant was built in 1940 but because of a disastrous fire at the Cleveland, Ohio facility in 1944, extensive development of LNG peak-load shaving facilities did not

TABLE 9.4-1 PROPERTIES OF METHANE[a]

Molecular weight	16.043	
Boiling point at 1 atm	−258.69°F	−161.495°C
Freezing point in air at 1 atm	−296.46°F	−182.48°C
Heat of vaporization at the normal boiling point, 1 atm	219.22 Btu/lb	1.955 kcal/mole

[a] Data from ref. 1.

TABLE 9.4-2 PROPERTIES OF TYPICAL LPG HYDROCARBONS[a]

Hydrocarbon	Chemical Formula	Molecular Weight	Boiling point at 1 atm (°F)	Freezing Point in Air at 1 atm (°F)	Density of Liquid at 60°F (lb/ft³)	Heat of Combustion of the Liquid at 60°F and Constant Pressure (Btu/lb)	
						H_2O (l)	H_2O (g)
Propane	C_3H_8	44.097	−43.67	−305.84	31.59	21,498	19,767
n-Butane	C_4H_{10}	58.124	31.10	−217.05	36.39	21,136	19,493
n-Pentane	C_5H_{12}	96.92	96.92	−201.51	39.28	20,926	19,339
Cyclopropane	C_3H_6	42.081	−27.0	−197.36	35.08	—	—
Cyclobutane	C_4H_8	56.108	54.52	−131.26	43.57	20,858	19,497
Cyclopentane	C_5H_{10}	70.135	120.65	−136.91	46.73	20,188	18,827

[a] Data from ref. 1; English units.

TABLE 9.4-3 PROPERTIES OF TYPICAL LPG HYDROCARBONS[a]

Hydrocarbon	Chemical Formula	Molecular Weight	Boiling Point at 1 atm (°C)	Freezing Point in Air at 1 atm (°C)	Density of Liquid at 15°C (g/cm³)	Heat of Combustion of the Liquid at 25°C and Constant Pressure (cal/g)	
						H_2O (l)	H_2O (g)
Propane	C_3H_8	44.097	−42.045	−187.69	0.5076	11,943	10,989
n-Butane	C_4H_{10}	58.124	−0.50	−138.362	0.5847	11,742	10,837
n-Pentane	C_5H_{12}	96.92	36.064	−129.730	0.63087	11,626	10,751
Cyclopropane	C_3H_6	42.081	−32.80	−127.42	0.563	—	—
Cyclobutane	C_4H_8	56.108	12.51	−90.70	0.6997	11,589	10,839
Cyclopentane	C_5H_{10}	70.135	49.252	−93.839	0.75016	11,215	10,465

[a] Data from ref. 1; metric units.

occur until the 1960s. International and U.S. peak-shaving plants in existence in 1979 are listed in Tiratsoo[1] and Lom.[2]

Base-load plants are designed to prepare LNG for export, usually by specially designed LNG ocean vessels. Tiratsoo[1] states that ten base-load LNG facilities were in operation around the world by mid-1978. The international base-load LNG plants in operation or under construction by 1979 are listed in Tiratsoo.

Terminal or receiving LNG plants must be able to transfer the LNG from the ocean vessel, store it, and, as needed, vaporize the gas for delivery to the consumer. These plants are relatively simple when compared to a base-load plant. Tiratsoo[1] lists the terminal (receiving) plants in operation and under construction in 1979.

Liquefied petroleum gas (LPG) consists of the hydrocarbons that are extracted at the wellhead from wet natural gas or are removed as liquids during the refining of crude oil. The main constituents in LPG are propane (C_3H_8) and butane (C_4H_{10}) with small amounts of propene (C_3H_6), butane (C_4H_8), and other C_2 through C_5 hydrocarbons. Properties of these hydrocarbons are listed in Tables 9.4-2 and 9.4-3.

The properties of LPG, as one observes from the data in Tables 9.4-2 and 9.4-3, depend on the concentrations of the different constituents present in the mixture. It is usually delivered in cylinders and is a convenient source of energy for both domestic and commercial uses in regions that are not accessible to natural gas pipelines or where a portable source is needed such as with camping trailers, portable driers, and domestic gas grills.

It is necessary to cool a gas below its critical temperature before it may be liquefied since only below the critical temperature is one able to distinguish between the liquid and gas phases. This means that natural gas must be cooled to extremely low temperatures since the critical temperature of methane, the main constituent in liquefied natural gas, is $-116.63°F$ ($-82.57°C$) and its boiling point is $-258.69°F$ ($-161.495°C$). Liquefaction of natural gas involves the removal of the sensible and latent heats so as to lower its temperature below the temperature at which it changes from a gas to a liquid.

Two well established principles, which have been used for years in the liquefaction of air, are used in the liquefaction of natural gas. These two principles are the Joule–Thompson effect and the fact that when a gas expands adiabatically and work is done by the gas, the temperature of the gas decreases.

The Joule–Thompson coefficient

$$\mu = \left.\frac{\partial T}{\partial p}\right|_h \tag{9.4-1}$$

is of extreme importance when a gas is throttled from a high to a low pressure. If $\mu = 0$, the temperature of the gas will remain constant as the pressure is decreased. If $\mu < 0$, the temperature of the gas will increase as the pressure is decreased. If $\mu > 0$, the temperature of the gas will decrease as the pressure is decreased.

This means that the only time a cooling effect will occur is when $\mu > 0$. The larger the Joule–Thompson coefficient, the greater the cooling effect. All liquefaction processes depend on this throttling process to achieve at least a portion of the necessary cooling effect.

Figure 9.4-1 illustrates one cycle that could be used to liquefy natural gas. In this cycle, natural gas enters at state 1 where it is mixed with the gases being recycled from the gas–liquid separator.

The gas is first compressed, then enters a heat exchanger where its temperature is decreased with a minimum decrease in the pressure. It then enters a throttling valve where, if the pressure change is sufficient, a portion of the gas will change to a liquid. It next enters a separator where the liquid is separated from the gas, the liquid being taken to storage and the gas recycled and mixed with makeup gas. This cycle is shown on a temperature–entropy diagram in Fig. 9.4-2. One should note that the compression process is shown in Fig. 9.4-2 as an isentropic process and that the heat transfer process is shown as a constant pressure process. These, of course, are idealizations of an actual cycle. In fact, it is desirable for the compression process to be as close as possible to an isothermal process as illustrated by the dashed line on the temperature–entropy diagram of Fig. 9.4-2. This reduces the maximum temperature that will occur during the compression process.

The type of cycle illustrated in Fig. 9.4-1 is not used because the maximum temperature and pressure necessary with a single-stage would be prohibitive. It would be necessary to compress methane to over 1000 atm to liquefy it if heat was rejected at room temperature.

A modification of the cycle illustrated in Fig. 9.4-1 is shown in Fig. 9.4-3. In this cycle, the high pressure gas, upon leaving the heat exchanger which uses cooling water, passes through a counterflow heat exchanger. Energy is transferred from the high-temperature, high-pressure gas to the low-temperature, low-pressure gas being returned. This cycle produces a lower-temperature gas at the inlet of the throttling valve without appreciably lowering the pressure at the inlet to the throttling valve, thereby resulting in a larger fraction of the fluid being a liquid in the flash tank (separator).

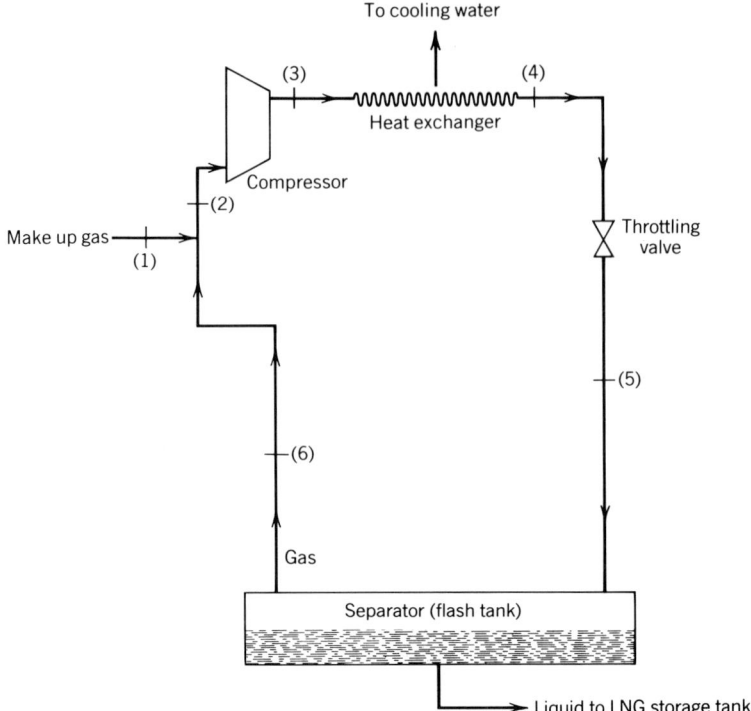

Fig. 9.4-1 Basic liquefaction cycle.

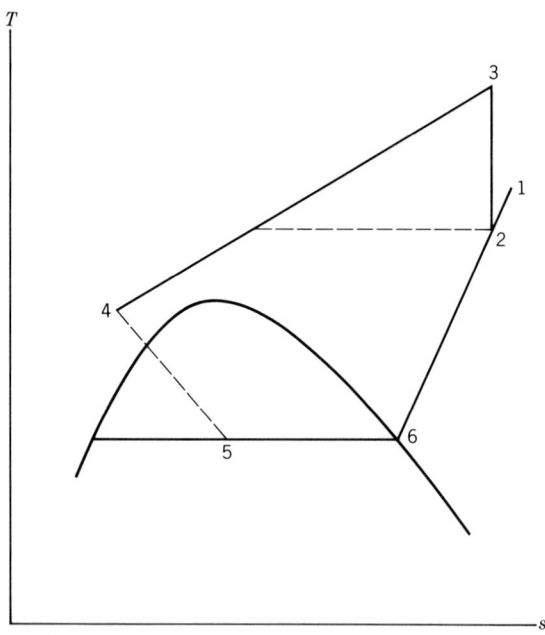

Fig. 9.4-2 Temperature–entropy diagram for basic liquefaction cycle.

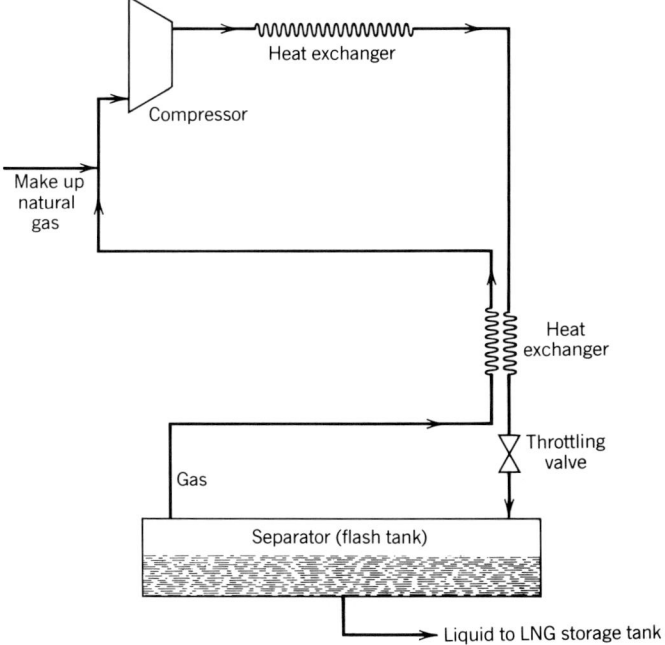

Fig. 9.4-3   Modified basic liquefaction cycle.

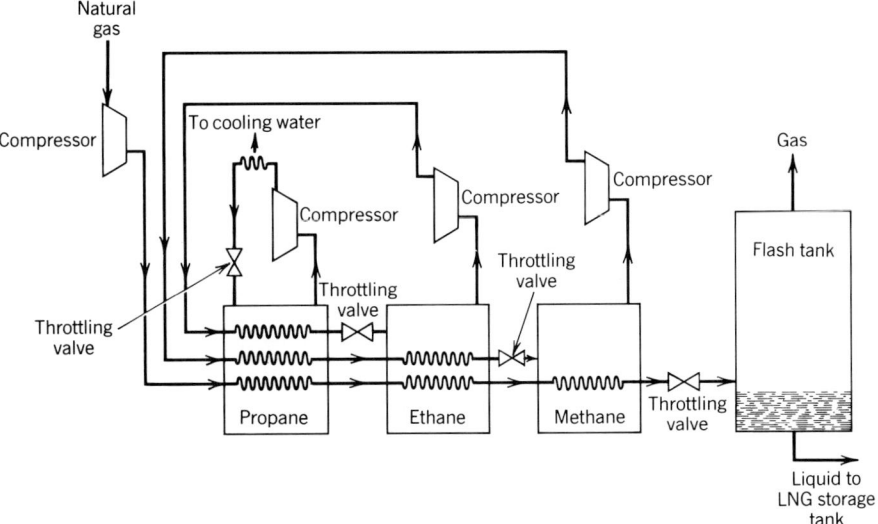

Fig. 9.4-4   Cascade liquefaction cycle.

The two principal liquefaction processes are the cascade cycle and the expander cycle.

The cascade cycle consists of a series of cycles as previously described. Each cycle rejects energy to the next higher-temperature cycle. One possible cascade cycle for liquefying natural gas is illustrated in Fig. 9.4-4. Conventional liquefied natural gas cycles usually use propane, ethane, and methane as the intermediate refrigerants.

Referring to Fig. 9.4-4, we see that natural gas enters a compressor where its pressure is increased. It then passes through three heat exchanger stages: the first one uses propane as the refrigerant, the second one uses ethane as the refrigerant, and the third one uses methane as the refrigerant. After

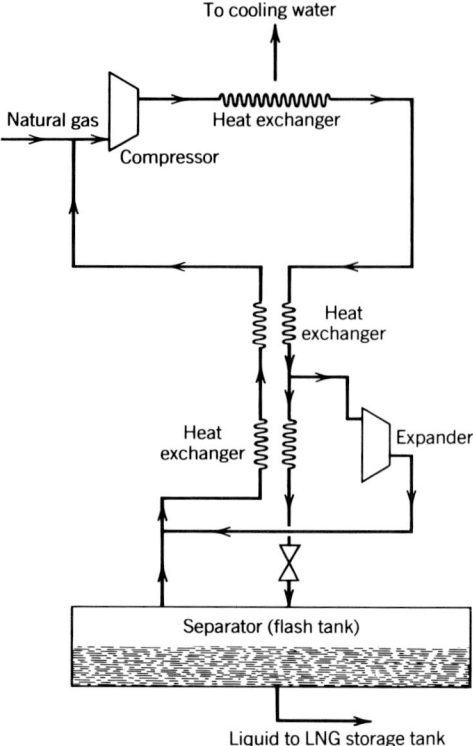

Fig. 9.4-5   Claude (expansion) cycle.

exiting the third heat exchanger, it is throttled to the storage pressure. In the flash tank, a portion of the fluid will be a liquid (LNG), a portion will be a gas.

The liquid in the flash tank is delivered to a storage tank while the gas from the flash tank is either recycled or used as a fuel.

The methane vapor from the refrigeration cycle that uses methane as the refrigerant is compressed, cooled in stages using propane and ethane as the refrigerants, then throttled for use as a refrigerant.

The ethane vapor from the refrigeration cycle that uses ethane as the refrigerant is compressed, cooled in the stage using propane as the refrigerant, then throttled for use as a refrigerant.

The propane vapor from the refrigeration cycle that uses propane as the refrigerant is compressed, passed through a heat exchanger that usually uses cooling water as the fluid to which heat is rejected, then throttled for use as a refrigerant.

The cycle illustrated in Fig. 9.4-4 and described above is one form of a cascade cycle. It is possible to use refrigerants other than propane, ethane, and methane. Also, one should observe that four compressors are required in this cycle. There are other cascade cycles that require only one or two compressors and have separators.

One major disadvantage of the cascade cycle is the fact that the throttling process is highly irreversible—that is, no work is realized.

A cycle that partially overcomes this drawback is the expander cycle. In the expander cycle, gas is compressed and passed through a heat exchanger which rejects heat to cooling water. The high-pressure gas then passes through a counterflow heat exchanger. At the exit, a portion of the gas is extracted and passes through a turboexpander where its pressure is reduced to that of the flash tank. This low-pressure, low-temperature gas is used to cool the high-pressure gas before it enters the throttling valve. This cycle, commonly called the Claude cycle, is illustrated in Fig. 9.4-5.

REFERENCES

9.4-1   E. N. Tiratsoo, *Natural Gas*, Gulf Publishing Company, Houston, TX, 1979.
9.4-2   W. L. Lom, *Liquefied Natural Gas*, Applied Science Publishers Ltd., London, 1974.

9.5   SYNTHETIC AND SUBSTITUTE NATURAL GASES

Many sources of synthetic and substitute natural gases have been investigated and used. These include gaseous fuels that resulted as by-products from coal conversion, coal gasification, and fuels from biomass and wastes.

Gasification of coal has been used for over a century to produce a gaseous fuel with a low to medium heating value. This gaseous fuel, because of its heating value, has usually been used at the point where it has been produced. It has been used as an industrial fuel and as a fuel for power generation. Recent studies have investigated ways to produce a high heating value fuel that can be economically transported from the point of production to point of consumption.

Early synthetic natural gases include coke-oven gas, water gas, producer gas, and refinery gas. These are defined below.

Coke-oven gas is the noncondensible gas that results during the production of coke. The composition of coke-oven gas depends on the composition of the coal and the coking process. It usually contains hydrogen sulfide which must be removed by scrubbing before the gas can be burned. This gas has a heating value of 500–600 Btu/ft^3.[1]

Water gas is a gas produced by passing steam through hot coke. The resulting reaction is the carbon in the coke combining with the steam to produce hydrogen and carbon monoxide or, in equation form,

$$C + H_2O \rightarrow CO + H_2 \qquad (9.5\text{-}1)$$

and has an estimated heating value of 500–550 Btu/ft^3.[1]

Producer gas is a gas that results when either coal or coke are burned with a deficiency of air. It has a relatively low heating value.

Refinery gases are the noncondensible gases that result during the distillation of a crude oil or that result from the cracking of oil in a refinery. Refinery gases consist mainly of methane, ethane, and hydrogen.

Biomass, which is defined as all plant and animal wastes including trees, shrubs, grasses, agricultural residues including manure, solid wastes from domestic and industrial sources, and human wastes, has been investigated for many years as an energy source. Conversion processes usually involve biological conversion which produces methane and carbon dioxide, pyrolysis, the thermal decomposition of organic materials in the absence of oxygen, and gasification which is the thermal decomposition of organic matter using oxygen, air, or steam.

Anaerobic digestion is a biological process in which organic material is decomposed by bacteria in the absence of air. This results in the production of methane and carbon dioxide and has been used for over 50 years in sewage treatment plants. This results in a gas consisting of 50–75% methane.

Both batch and continuous systems have been designed and built, at least as pilot plants. Batch or plug-flow systems usually are smaller systems built in areas such as farms or in remote areas. They have the advantage in that they may be loaded then ignored until they need to be recharged. The amount of gas produced is a function of time. It is necessary to design a storage system to be used with this type of system if it is necessary to have gas supplied at a fixed (steady) rate.

Continuous systems are usually used in industrial applications. They may be instrumented and produce gas at a near steady rate.

Two systems that have been built to convert refuse to a combustible gas are the Destrugas System and the Landguard System. These are described in detail in Porteous.[2]

The Destrugas System is a continuous flow system. Refuse is delivered to a receiving pit where it is stored until entering a grinder. After being shredded, it is conveyed to a retort maintained at 1000°C. This system was operational in Kolding, Denmark in 1967 and yields a gaseous fuel that is 54% hydrogen, 10% carbon monoxide, 10% methane, 2% other hydrocarbons, 23% carbon dioxide, and 1% nitrogen with an estimated heating value of 17 MJ/m^3 (455 Btu/ft^3) according to Porteous.[2]

The Landguard System was developed by Monsanto. The refuse is pulverized before being conveyed to a rotary kiln. The resulting gaseous fuel contains, according to Porteous,[2] 69.3% nitrogen and 11.4% carbon dioxide with the two main combustibles being hydrogen and carbon monoxide, each being 6.6% of the total. The gaseous fuel has an estimated heating value of 4.46 MJ/m^3 (120 Btu/ft^3).

Coal gasification, the conversion of coal to gas, may be classified by the gasification system used or the energy content of the resulting gas. It is necessary to transfer large quantities of heat to the coal in order to achieve complete gasification.

Coal is the most plentiful fuel in the United States but does contain ash and water, two noncombustibles, and sulfur, a combustible but one which is desirable to remove before burning. Coal gasification is one means by which coal can be converted at its source, cleaned, then transported to its point of consumption.

Gaseous fuels produced by coal gasification result in:

1. Low heating value fuels, usually considered to be those with a heating value of 200 Btu/ft^3 or lower. The combustibles in these fuels are usually carbon monoxide and hydrogen and may be quite high in nitrogen, a noncombustible. These fuels must be consumed at the point of generation since they cannot be economically transported. Air is usually used to produce gaseous fuels with low heating values.

2. Medium heating value fuels with heating values up to 500 Btu/ft^3. They, like low Btu fuels, consist mainly of carbon monoxide and hydrogen but have a higher percent of combustibles. Oxygen usually is used to produce gaseous fuels with medium heating values.

3. High heating value fuels which usually have heating values of approximately the same magnitude as natural gas; that is, of approximately 1000 Btu/ft^3. This type of fuel can be economically transported and consists mainly of methane.

The main methods used in converting coal to a gaseous fuel are fluidized bed systems, fixed bed systems, entrained bed systems, and molten bath bed systems.

Fixed bed systems, according to Norwacki[3] usually are counterflow systems with coal entering at the top and air or oxygen along with steam entering at the bottom. The maximum temperature, approximately 1700–2600°F, occurs at the bottom of the system. The residence time for coal in this system is between 1 and 3 hours. The main combustibles produced by a fixed bed system are carbon monoxide and hydrogen.

Three systems which have been built using fixed bed gasifiers are the British Gas/Lurgi Slagging Process, Wellman–Galusha Process, and Woodall–Duckham/Gas Integrate Process. These three processes are briefly described in Tables 9.5-1, 9.5-2, and 9.5-3. The reader is referred to Norwacki[3] for additional details on these cycles and for additional coal gasification systems using the fixed bed process.

Fluidized bed systems, as reported in Norwacki,[3] require coal of approximately 10–100 mesh size. The maximum bed temperature is fixed by the melting point of the ash. This means that bed temperatures will be in the 1500–1900°F range.

One fluidized bed system which is commercially in use is the Winkler process. Initial development began in the 1920s with the first commercial plant becoming operational in 1926 at Leuna, Germany. Norwacki[3] lists 22 plants with their locations, capacity, and operational dates. Data for this cycle are summarized in Table 9.5-4. The reader is referred to Norwacki[3] for additional details on this cycle and for other fluidized bed systems.

Entrained bed systems are finely sized coal which has been pulverized so that 70% will pass through a 200 mesh. The coal particles are conveyed by the gas (usually oxygen and steam) and carried into the reactor where the chemical reaction is completed. The gas is then separated from the ash.

One entrained bed system currently in commercial operation is the Koppers–Totzek process. The first demonstration unit, according to Nowacki,[3] was built in 1948 with the first commercial unit installed in Finland in 1952. Nowacki lists nine locations where Koppers–Totzek systems have been installed. Data for this system are summarized in Table 9.5-5.

The *molten bath system* is one in which crushed coal along with oxygen (or air) and steam are injected into a hot liquid bath which provides the heat transfer, heat storage, and separation of the fixed carbon from the waste. The fixed carbon then combines with oxygen to produce carbon monoxide. These systems use molten salt, iron, or coal ash as the molten bath.

TABLE 9.5-1   BRITISH GAS/LURGI SLAGGING PROCESS[a]

History
    1955—Ministry of Power purchased Lurgi's pilot gasifier
    1960—Wilson Committee recommended development of Lurgi gasifier
    1974—British Gas Corporation modified Lurgi gasifiers at Westfield, Scotland
    1978—All runs at Westfield were completed

Operating pressure—75–385 psia

Operating temperature—2300–2500°F in combustion zone
                             700–800°F exit temperature

Coal preparation—Crushed and screened to remove fines smaller than $\frac{1}{8}$ in. diameter

Coal residence time—10–15 minutes

Oxidant—oxygen

Heating value of gas—381 Btu/ft^3

[a] Data from ref. 3.

TABLE 9.5-2   WELLMAN—GALUSHA PROCESS[a]

History—Has been commercial for over 35 years with over 150 of the most recent gasifiers installed worldwide in industrial applications

Operating pressure—300 psig

Operating temperature—Approximately 2400°F in combustion zone
600–1000°F at exit for anthracite coals
1100–1200°F at exit for bituminous coals

Coal preparation—Coal crushed to $\frac{3}{16}$ to $\frac{9}{16}$ in. size for anthracite coals, 1–2 in. for bituminous coals

Residence time—Approximately 4 h

Oxidant—Air

Heating value of gas—168 Btu/ft^3 for bituminous coals
146 Btu/ft^3 for anthracite coals

[a] Data from ref. 3.

TABLE 9.5-3   WOODALL—DUCKHAM/GAS INTEGRATE PROCESS[a]

History—Developed by ‖ Gas Integrate, Milan, Italy. Has been used for over 30 years to produce industrial fuel gases

Operating pressure—Atmospheric

Operating temperature—2200°F in gasification zone

Coal preparation—Double-screened: $\frac{1}{4}$–1 in. and $\frac{1}{2}$–$1\frac{1}{2}$ inch sizes are typical

Coal residence time—Several hours

Oxidant—Air or oxygen

Heating value of gas—280 Btu/ft^3 (dry), oxygen blown
175 Btu/ft^3 (dry), air blown

[a] Data from ref. 3.

TABLE 9.5-4   WINKLER PROCESS[a]

Operating temperature—Approximately 1800°F fluidized bed maximum temperature using lignite as the fuel, 2100°F for less reactive coals

Operating pressure—Original systems operated at 1 atm; proposed systems have operating pressures of 1.5–15 atm

Residence time—20 to 30 min

Oxidants—Oxygen

Heating value of gas—62 Btu/ft^3 (dry) for oxygen blown lignite
108 Btu/ft^3 (dry) for air blown western U.S. subbituminous coal

[a] Data from ref. 3.

TABLE 9.5-5   KOPPERS–TOTZEK PROCESS[a]

Operating temperature—3500°F in combustion zone
2700°F exit temperature prior to water spray

Operating pressure—slightly above atmospheric

Residence time—1 s or less

Oxidant—Oxygen

Heating value of gas—286 Btu/ft^3 (dry)

[a] Data from ref. 3.

All coal gasification systems require some gas treatment to remove the sulfur and particulate from the gas before it is used as a fuel or supplied to a pipeline. The gas treatment required depends on the coal that is used and the process selected for producing the gaseous fuel.

REFERENCES

9.5-1 *Steam/Its Generation and Use*, Babcock & Wilcox, New York, 1972.
9.5-2 A. Porteous, *Refuse Derived Fuels*, Applied Science Publishers Ltd., London, 1981.
9.5-3 Perry Nowacki, *Coal Gasification Processes*, Noyes Data Corporation, Park Ridge, N.J., 1981.

9.6   TRANSPORTATION, STORAGE, AND SAFETY

Pipelines are an economical means for transporting natural gas from its source to its points of consumption. Many factors must be considered in the design of a long-distance trunk transmission pipeline since the initial capital investment is very large. Factors which must be considered are:

1. **Present and future market demands.** This will determine the volume of gas that must be transported through the pipeline.
2. **Length of the pipeline.** This will allow one to determine the estimated pressure drop and the location of compressor stations where the pressure is restored to its maximum value.
3. **Terrain and elevation.** This allows the designer of a pipeline to determine obstructions in a proposed pipeline path such as rivers, towns, forests, and other areas through which a pipeline cannot be built.

Pipelines usually are identified as gathering pipelines, trunk transmission pipelines, distribution lines, and service lines.

The gas must first be transported from the well (source) to extraction units where impurities such as water and sulfur are removed and to processing plants where the heavier, condensible hydrocarbons are removed. These lines are commonly called *gathering lines*.

Once the field processing and cleaning has been completed, the natural gas is ready for distribution. It is supplied to a *trunk transmission line* which consists of underground, high-pressure lines where the gas pressure after a compressor station may be as high as 1600 psia.

The pressure in a trunk transmission line will decrease with distance due to friction. The rate of pressure drop is a function of pressure, gas velocity, pipe diameter, and pipe roughness. Therefore, it is necessary to install compressor stations at regular intervals along the trunk transmission line. These compressor stations are used to increase the pressure to overcome the losses due to friction.

It has been estimated that it requires 150 Btu of energy to transport 1 ft^3 of natural gas 1000 mi. This illustrates the fact that natural gas with a heating value of approximately 1000 Btu/ft^3 may be economically transported but that manufactured (synthetic) gaseous fuels with low to medium heating values cannot be economically transported over long distances but must be consumed at or near the point of production.

Distribution lines are pipelines located within cities and towns that distribute the gas to the vicinity of the consumer. Service lines are pipelines servicing the individual consumer.

Projected shortages of natural gas and the need to transport natural gas between continents have led to the development of a large number of special designed tankers that are used to transport liquefied natural gas (LNG).

The first commercial transportation of LNG occurred in 1964 when LNG was shipped from Algeria to the United Kingdom.[1] These were the British vessels *Methane Progress* and *Methane Princess* each with a capacity of 27,400 m^3.

Today, an extensive fleet of LNG tankers exist for transporting LNG from Algeria, Alaska, Libya, and other regions of surplus natural gas to Japan, Spain, the United States, and the United Kingdom. All current tankers consist of a primary container which is in contact with the LNG, an insulation layer whose function is to minimize heat transfer, and an outer barrier.

The primary (inner) container makes it necessary to use aluminum, stainless steel, or steel containing at least 5% nickel with 9% nickel steel commonly used. Some tankers are designed so that the container is self-supporting and not a part of the ship's hull. Others are designed so the container is a part of the ship's hull. In all cases, the heat transfer must be minimized. The portion of LNG that is vaporized usually is used as fuel to propel the ship.

It is necessary to operate the long-distance trunk transmission pipelines at a high load factor, that is, at near capacity at all times. This allows the rate at which gas is withdrawn from the ground to be at a near constant rate. Pipelines that were designed to meet the peak demand which occurs on cold days with high industrial activity would be oversized and uneconomical because of the high initial capital investment required.

This means that some method must be available to store the gas at or near the point of consumption. Storing natural gas at or near the point of consumption allows the trunk transmission line to operate at a constant rate yet meet demands which are varying due to weather conditions, industrial activity, and other variables such as time of day, or the day of the week (weekday or weekend).

Many ways have been used to store natural gas either as a gas or as a liquid. These methods are usually identified as short-term versus long-term, those that store it as a gas versus LNG storage methods and above-ground versus below-ground storage methods.

Short-term variations in natural gas demand usually result because of variations due to time of day or industrial activity. Storage methods used to meet this type of variation include:

1. Large water-sealed above-ground storage tanks of conventional design.
2. Storage in transmission lines by allowing the pressure in the pipeline to increase as demand decreases.

Long-term variations in demand result because of seasonal variations in demand. Storage methods used to meet long-term variations in demand include:

1. Using depleted gas or oil wells if they are located in or near the vicinity of the consumer.
2. Storage as liquefied natural gas either above or below-ground.

The original LNG storage tanks were spherical metal tanks located above ground. Their design was similar to those of tanks used in ships. The inner vessel, which is in contact with the liquefied natural gas, was made of aluminum with cork used as the insulating material. The outer wall was made of carbon steel.

Later designs were cylindrical in shape and were located partially or almost entirely below ground. Some designs incorporated a concrete outer shell which was intended to retain the tank's contents in the event of an outer wall failure.

Safety has been of concern to those involved with the transportation, storage, and use of natural gases, synthetic (manufactured) gases, and liquefied natural gas. Below are listed a few of the many standards which have been adopted. The reader should determine which national, state, and/or local codes or standards apply to a given application. Many of the standards have been developed by the American Gas Association (AGA), American Society of Mechanical Engineers (ASME), American National Standards Institute, Inc. (NASI), and the National Fire Protection Association (NFPA).

One standard that applies to the transmission and distribution of natural gas is ANSI B31.8.[2] To meet a need for a national code for pressure piping, the American Engineering Standards Committee initiated Project B31 in March 1926 with the first edition being published in 1935. Over the years, this pressure code has been refined incorporating new materials standards, severity of service conditions, and progress in instrumentation, controls, and testing. It is now formally designated as an American National Standard.

ANSI B31.8 covers the design, fabrication, installation, inspection testing, and safety of gas transmission and distribution systems. It includes (1) materials and equipment, (2) welding, (3) piping system components and fabrication, (4) design, installation, and testing, (5) operation and maintenance, and (6) corrosion control. This code specifies that a distinctive odor must be added to any gas distributed to customers through gas mains or service lines if it does not have its own distinctive odor. The odor level must be detectable at gas concentrations equal to one-fifth the lower explosive limit and above.

ANSI B31.8 applies to (1) gas pipelines, (2) gas compressor stations, (3) gas main, and (4) service lines up to the customer's meter.

Two standards which apply to the installation of gas piping after the customer's meter set assembly are ANSI Z21.30, *Installation of Gas Appliances and Gas Piping*, and ANSI Z83.1, *Installation of Gas Piping and Gas Equipment on Industrial Premises and Certain Other Premises*.

ANSI Z21.30[3] has been developed and revised over the years and is intended to protect the public in the utilization of gas. A subcommittee was appointed by the National Fire Protection Association's Committee on Explosions and Combustibles in 1913–1914 to prepare a fire protection code for gas installations in cities. The first NFPA code was adopted in 1920. Over the years this code has been revised to take into account expanding uses of gas, new types of gas appliances, materials, testing, fabrication, and instrumentation. It applies to low-pressure domestic and commercial piping systems after the customer's meter set assembly. Low pressure is defined in this standard as pressures not to exceed 0.5 lb/in.^2 or 14 in. water column. Topics included in this standard are:

1. Gas piping installation including piping plan, meter location, size of piping, acceptable piping materials, gas shutoff valves, and leakage check requirements.
2. Appliance installation.
3. Venting of appliances.

Work was begun on the development of a standard for industrial gas piping and industrial gas utilization equipment in 1952. The need for such a standard resulted because ANSI Z21.30 specifically excluded industrial applications. The first standard for installation of gas piping was approved in 1966 and resulted in ANSI Z83.1.[4]

ANSI Z83.1 was patterned after ANSI Z21.30 and included:

1. General safety precautions.
2. Sizing and installation of gas piping systems
3. Installation of customer-owned gas meters, regulation, and gas equipment.
4. Maintenance of gas equipment.

In 1972, supervision of ANSI Z21.30 and Z83.1 was transferred to ANSI Z223 committee.

Storage and handling of liquefied petroleum gases and liquefied natural gas are covered by separate standards. Storage and handling of liquefied petroleum gases is covered by NFPA 58,[5] which is ANSI Z106.1 storage and handling of liquefied natural gas by NFPA 59A.[6]

The first LPG standard was adopted in 1932. NFPA 58 resulted in 1940 when several standards were combined. It, like all other standards, has been reviewed and revised, as necessary, with changing materials, equipment, instrumentation, applications, and so on. LPG is unique in that it is used in portable containers which must be considered in any standard. NFPA 58 includes standards on:

1. LPG properties and required odorization.
2. LPG equipment and appliances including container design, container markings, and safety devices.
3. Installation of LPG systems.
4. Liquid transfer.
5. Storage of containers.
6. Truck transportation of LPG.

Liquefied natural gas was recognized as requiring its own standard. Work on the initial standard began in 1960 with the first standard being tentatively adopted in 1966 with the first official edition adopted in 1967. Today, NFPA 59A is the basic LNG code.

Material contained in NFPA 59A includes:

1. Plant considerations including site, spill, and leak control, buildings and structures, and protection for cryogenic equipment.
2. Vaporization facilities.
3. Piping systems and components.
4. Transfer of LNG.
5. Fire protection, safety, and security.

There are many other standards which apply to the transportation, handling, and use of gaseous fuels. It is recommended that the most recent code be consulted to determine its applicability to a given situation.

REFERENCES

9.6-1 G. G. Haselden, *The Challenge of LNG in 1980*, in Cryogenic Progress and Equipment in Energy Systems, ASME, August 19–21, 1980.

9.6-2 ASNI B31.8-1975, *Gas Transmission and Distribution Piping Systems*, American Society of Mechanical Engineers, New York, 1975.

9.6-3 ANSI Z21.30, *Installation of Gas Appliances and Gas Piping*, American Gas Association, New York, 1964.

9.6-4 ANSI Z83.1, *Installation of Gas Piping and Gas Equipment on Industrial Premises and Certain Other Premises*, American Gas Association, Arlington, VA, 1974.

9.6-5 NFPA 58, *Storage and Handling of Liquefied Petroleum Gases*, National Fire Protection Association, Inc., Boston, MA, 1979.

9.6-6 NFPA 59A, *Storage and Handling of Liquefied Natural Gas*, National Fire Protection Association, Inc., Boston, MA, 1979.

9.7 BURNERS, EMISSION CONTROL, AND REGULATIONS

The function of the burner is to provide an efficient device in which the fuel and combustion air are mixed then ignited in a reliable manner with a minimum level of emissions. The burner must have stable and efficient operation from the minimum to the maximum firing rate.

A burner designed for use with gaseous fuels is much simpler in design than one designed for liquid or solid fuels since the fuel is already in the gaseous phase. Mixing of the fuel and combustion air is easily accomplished, the only problem being to supply the fuel and air in the proper amounts, that is, about 15% excess air. Burners using gaseous fuels are able to operate efficiently at a much lower percent excess air than can coal and oil burning equipment.

Burners used in steam generators usually are circular in design and located in the wall of the steam generator. Combustion air is supplied by fans.

Both atmospheric pressure and pressurized burners are used in domestic heating units. Burner design may be of the ribbon design or, like steam generators, circular in design.

A typical burner has two zones. These are the primary and secondary zones. The primary zone is the region where the fuel is injected and ignition occurs. The fuel is mixed with a portion of the combustion air. The velocity in this zone must be below the flame velocity so that the flame is not carried (blown) out of the the primary zone. The fuel and air must be evenly distributed.

The secondary zone is the region where combustion is completed and the remaining air is mixed with the combustion gases leaving the primary zone.

Pollutants that are emitted directly from combustion equipment are called *primary pollutants*. Pollutants that are formed by the chemical interaction between the primary pollutants, normal atmospheric constituents, and sunlight are referred to as *secondary pollutants*. Only primary pollutants will be discussed.

The primary pollutants are:

1. Particulate matter which is solid airborne particles and liquid particles larger than 0.0002 μm in diameter.
2. Oxides of sulfur which usually consist of sulfur dioxide, SO_2, sulfur trioxide, SO_3, and hydrogen sulfide, H_2S.
3. Oxides of nitrogen which usually consist of nitric oxide, NO, and nitrogen dioxide, NO_2.
4. Hydrocarbons which consist of the C_1 through C_5 hydrocarbons.
5. Oxides of carbon which consist of carbon monoxide, CO, and carbon dioxide, CO_2.

Over the past 25 years, considerable work has been done on controlling emissions from combustion processes. The main pollutants emitted from equipment burning gaseous fuels are the oxides of nitrogen. This is the only pollutant that will be discussed.

The oxides of nitrogen are produced in all fossil-fuel combustion processes that use air as the oxidant. There are several known oxides of nitrogen but the two that are important in the atmosphere are nitric oxide, NO, and nitrogen dioxide, NO_2. These are generally combined and expressed as the oxides of nitrogen, NO_x.

Nitric oxide is a colorless, odorless gas that is slightly soluble in water. It is the primary product formed during the combustion process when oxygen and nitrogen are both present. These two constituents combine according to the following endothermic reaction

$$N_2 + O_2 \rightarrow 2NO \qquad (9.7\text{-}1)$$

The sources of nitrogen for this reaction are the nitrogen in the fuel and the nitrogen that is present in the air (oxidant).

The amount of nitric oxide formed during a combustion process is influenced by thermodynamic and kinetic rate factors. Factors which influence the amount of nitric oxide formed and therefore emitted are:

1. Fuel type and composition.
2. Heat and mass transport rates.
3. Operating and design variables.
4. The temperature–time history of the combustion process.

Theory predicts that the operating variables which influence the amount of nitric oxide formed and emitted are (1) flame temperature, (2) excess air, and (3) residence time at a given temperature.

The constant-pressure adiabatic flame temperature is the temperature that results when a fuel is burned adiabatically during a constant-pressure process. The maximum temperature occurs when a

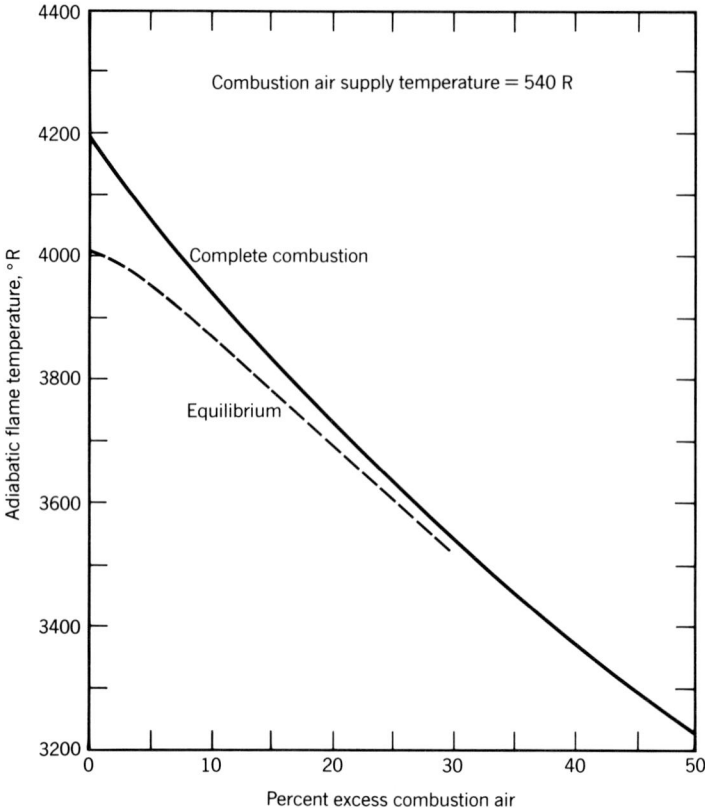

Fig. 9.7-1 Variation of adiabatic flame temperature with percent excess combustion air for air supply temperature of 540 R.

fuel is burned completely, that is, all carbon forms carbon dioxide and all hydrogen forms water. The general chemical reaction for the complete combustion of a hydrocarbon with oxygen is given by Eq. (9.3-4).

The variation of the maximum flame temperature with percent excess air is shown in Figs. 9.7-1 and 9.7-2. Both plots are for methane as the fuel. The dry air for the combustion process illustrated in Fig. 9.7-1 is supplied at 540 R, in Fig. 9.7-2 at 1100 R. One should observe the maximum theoretical adiabatic flame temperature, the difference between the maximum temperature in the two figures, and the rate at which the theoretical temperature decreases with increasing percent excess combustion air.

The actual adiabatic flame temperature is lower than the theoretical adiabatic flame temperature because of dissociation. Dissociation has an effect similar to incomplete combustion. The degree of dissociation increases with increasing temperature and is based on the second law of thermodynamics.

The equilibrium adiabatic flame temperature as a function of percent excess combustion air for two different combustion air supply temperatures is shown as the dashed lines in Figs. 9.7-1 and 9.7-2. One should observe the large difference in flame temperatures (theoretical versus equilibrium) with no excess combustion air, this difference decreasing as the percent excess combustion air increases.

The equilibrium concentration of nitric oxide is dependent on the temperature and the amount of nitrogen and oxygen present. The equilibrium composition, for a system at constant pressure and temperature, occurs when the Gibbs function is a minimum. The reaction which is of concern is Eq. (9.7-1). The equilibirium constant, K_p, based on partial pressures is, for the reaction in Eq. (9.7-1)

$$K_p = \frac{\left(p_{NO}\right)^2}{p_{N_2}\,p_{O_2}} \qquad (9.7-2)$$

Numerical values for the equilibrium constant for the formation of nitric oxide from molecular oxygen and molecular nitrogen are listed in Table 9.7-1.

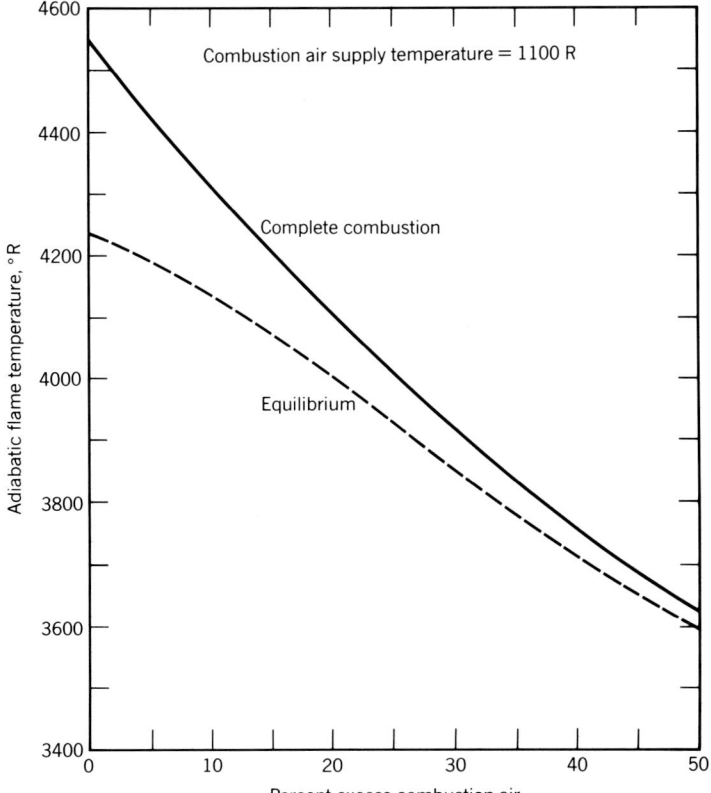

Fig. 9.7-2 Variation of adiabatic flame temperature with percent excess combustion air for air supply temperature of 1100 R.

TABLE 9.7-1 EQUILIBRIUM CONSTANT FOR THE FORMATION OF NITRIC OXIDE FROM MOLECULAR NITROGEN AND MOLECULAR OXYGEN

Temperature			Temperature		
R	K	K_p	R	K	K_p
3060	1700	5.7756×10^{-5}	3780	2100	6.6359×10^{-4}
3240	1800	1.1773×10^{-4}	3960	2200	1.0629×10^{-3}
3420	1900	2.2264×10^{-4}	4140	2300	1.6346×10^{-3}
3600	2000	3.9500×10^{-4}	4320	2400	2.4238×10^{-3}

The values tabulated in Table 9.7-1 illustrate that as the temperature increases, the amount of nitric oxide formed for fixed values of nitrogen and oxygen will increase. One must remember that as the percent excess combustion air increases, the equilibrium adiabatic flame temperature decreases but the amount of oxygen and nitrogen in the products increases. A decreasing flame temperature lowers the amount of nitric oxide formed while an increasing availability of oxygen and nitrogen increases the amount of nitric oxide formed.

The equilibrium concentration of nitric oxide in the products when methane is burned with dry air is shown in Fig. 9.7-3. The solid line is for combustion air supplied at 1100 R, the dashed line for combustion air supplied at 540 R. One should observe that the nitric oxide concentration increases with increasing percent excess combustion air reaching its maximum value between 15 and 25% excess combustion air. The maximum nitric oxide concentration and the percent excess air where it occurs depend on the fuel and the combustion air supply temperature.

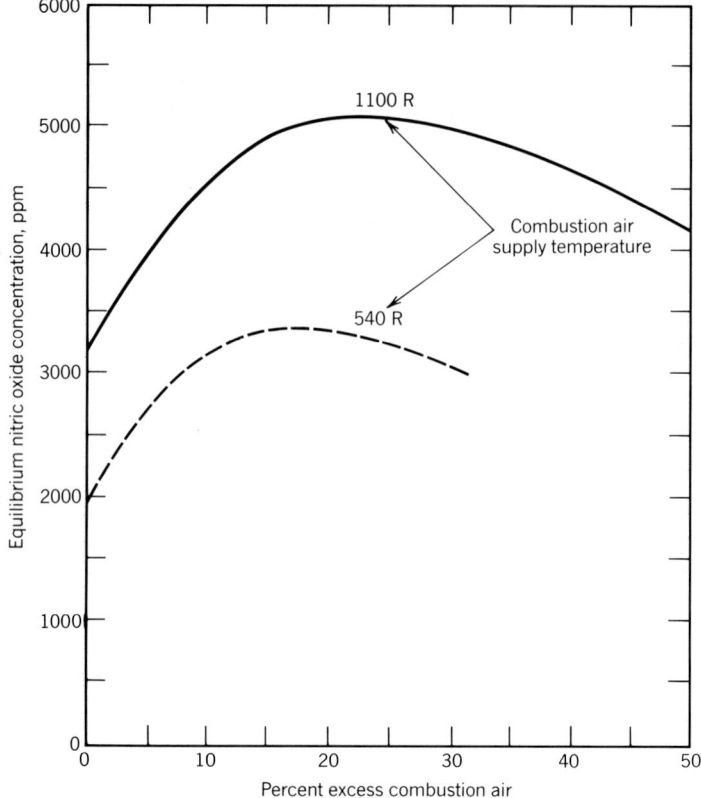

Fig. 9.7-3 Variation of equilibrium nitric oxide concentration with percent excess combustion air for two air supply temperatures.

The preceding discussion assumed that the equilibrium state is reached. This usually is not true, the amount of nitric oxide formed being dependent on the length of time the gases are at a given temperature. It is next to impossible to accurately predict the time–temperature history for a combustion process. It has been found that emissions will vary extensively between similar units which are operating under near-identical conditions. The reason for these variations probably are because of variations in the time–temperature history of the units.

The following theory will illustrate the influence of temperature and time on the amount of nitric oxide formed.

In 1946, Zeldovich proposed the following mechanism for the formation of nitric oxide:

$$O_2 + M \rightarrow 2O + M \qquad\qquad (9.7\text{-}3)$$

$$O + N_2 \rightarrow NO + N \qquad\qquad (9.7\text{-}4)$$

$$N + O_2 \rightarrow NO + O \qquad\qquad (9.7\text{-}5)$$

A simple rate expression may be derived for nitric oxide formation based on the Zeldovich mechanism. It is necessary to assume that the atomic oxygen concentration is in equilibrium for reaction (9.7-3) and that there is excess air. Danielson[1] estimated the theoretical time required to form 500 ppm nitric oxide for a mixture containing 3% oxygen and 75% nitrogen. The results for four temperatures are tabulated in Table 9.7-2.

Examination of the values in Table 9.7-2 illustrates the rapid change in the time required to form 500 ppm of nitric oxide with changing temperature. When the flame temperature is in the 3400–3800°F range (typical maximum flame temperatures), it is apparent that a fraction of a second residence time can have an appreciable effect on the amount of nitric oxide formed and emitted.

The above illustrates the importance of flame temperature, excess air, and residence time. One of the difficulties associated with theoretical calculations is accurately knowing the flame temperature and cooling rate within a furnace, that is, the time–temperature history of the combustion process.

TABLE 9.7-2   THEORETICAL TIME REQUIRED TO FORM 500 ppm OF NITRIC OXIDE[a]

Temperature (°F)	Time (s)	Temperature (°F)	Time (s)
2400	1370	3200	1.1
2800	16.2	3600	0.117

[a]Data from ref. 1.

It is apparent from the preceding discussion that there should be several techniques which can be used to control nitric oxide emissions. Possible techniques may be separated into the following categories:

1. Modification of operating conditions.
2. Furnace design modifications.
3. Flue gas cleaning.

Modifications to operating conditions involve ways to reduce the maximum temperature and/or minimize the availability of oxygen. They include:

1. **Low excess air combustion.** This reduces the amount of oxygen available. It does require careful furnace supervision to avoid the formation of smoke, carbon monoxide, unburned hydrocarbons, and other undesirable operating conditions.
2. **Flue gas recirculation.** The main effect from this technique is to lower the maximum temperature since the recirculated gases will absorb a considerable amount of energy. The quantity of flue gases recirculated to achieve the desired effect can vary greatly and will influence the feasibility of this technique. The location at which the recirculated flue gases are added back into the combustion system influences the effectiveness of this technique.
3. **Two-stage combustion.** In this technique, the fuel is burned with about 95% of the stoichiometric amount of air. After the resulting flue gases have been partially cooled, the remaining combustion air is supplied and the combustion process completed.
4. **Steam or water injection.** This technique has been found to be very effective. Injecting water reduces the maximum combustion temperature in the primary zone of the combustion chamber, thereby reducing the amount of nitric oxide formed and emitted.

Furnace design modifications which have been investigated include burner configuration and location. It has been found that a tangentially-fired unit has lower nitric oxide emissions than do units where the burners are being fired horizontally. This reduction in emissions probably occurs because of enhanced heat transfer, thereby reducing the time the combustion gases are at or near their peak temperature.

Several flue gas cleaning techniques have been investigated but none has proved to be as effective as modification of operating conditions.

Little is known about the pollution of our atmosphere prior to the 1950s. Some measurements were taken prior to the 1950s but most were very crude, taken at irregular intervals and the accuracy of the measurements were subject to question.

In 1953, sampling for suspended particulate matter was begun in 17 cities. In 1957 the National Air Sampling Network was established to routinely monitor suspended particulate matter. By 1966, this network had been expanded to include about 150 urban stations and 30 nonurban stations.

During 1959, sampling for the gases NO_2 and SO_2 was begun. During 1961, the Continuous Air Monitoring Project (CAMP) was begun. This program created six stations equipped to monitor and record, automatically, 24 hours a day, the ambient concentrations of carbon monoxide, nitric oxide, nitrogen dioxide, sulfur dioxide, total hydrocarbons, and total oxidants.

Air pollution episodes occurred and problems existed long before any measurements on ambient levels were made. Air pollution episodes, such as those in London in 1873, 1952, and 1956 or in 1930 in the Meuse Valley of Belgium or in 1948 in Donora, Pennsylvania, resulted from particulate and sulfur dioxide emissions and occurred in valleys or in highly industrialized areas during an atmospheric inversion. These episodes have led to an extensive ambient air monitoring system and to federal legislation.

The first federal legislation enacted was the Air Pollution Control Act of 1955 (Public Law 84-159, July 14, 1955). It recognized the growth in air pollution and that the public was endangered. The law was narrow in scope and considered the prevention and control of air pollution to be primarily the responsibility of state and local governments.

Two acts which amended the act of 1955 were the Air Pollution Control Act Amendments of 1960 (Public Law 86-493, June 6, 1960) and Amendments of 1962 (Public Law 87-761, October 9, 1962).

The Clean Air Act of 1963 (Public Law 88-206, December, 1963) provided matching grants to state and local agencies for air pollution control programs; initiated efforts to control emissions from federal facilities; provided for the development of air quality criteria; and encouraged automotive companies and fuel industries to prevent air pollution.

The Air Quality Act of 1967 (Public Law 90-148, November 21, 1967)

1. Established eight specific areas in the United States on the basis of common meteorology, climate, and topography.

2. Provided for the development and issuance of air quality criteria for specific pollutants that have been identified as having adverse effects on publich health and welfare.

This provision led to a series of documents for common air pollutants such as particulate matter, carbon monoxide, sulfur oxides, nitrogen oxides, hydrocarbons, oxidants, and so on. One document for each pollutant summarized what science at that time was able to measure of the effects of that pollutant on humans and the environment, a second document summarized information on techniques for controlling that pollutant.

The Clean Air Amendments of 1970 (Public Law 91-604, December 31, 1970) included the following major provisions:

1. National ambient air quality standards (both primary and secondary) were to be established by the Environmental Protection Agency (EPA).

2. Each state had the primary responsibility for assuring air quality within the entire geographic areas comprising such state.

3. Industry was to monitor and maintain emission records and make them available to EPA officials.

4. Standards of performance for new stationary sources were to be established and implemented.

The Clean Air Amendments of 1970 required that national ambient air quality standards, both primary and secondary, were to be established by EPA. These standards were published in the *Federal Register* in 1971[2] and set maximum concentration limits for carbon monoxide, hydrocarbons, nitrogen dioxide, sulfur dioxide, particulates, and oxidants.

The Clean Air Amendments of 1970 also required the establishment of emission standards from new stationary sources. These standards were published in the *Federal Register* in 1971.[3] These standards have undergone modifications with the latest standards being published in the *Code of Federal Regulations*.[4] These are discussed below.

The current Standards of Performance for New Stationary Sources as published in the *Code of Federal Regulations*[4] contain the following:

1. Standards have been established for fossil-fuel-fired steam generators, incinerators, portland cement plants, nitric acid plants, sulfuric acid plants, petroleum refineries, and storage vessels for petroleum liquids, along with smelters and sewage treatment plants.

2. Test methods and procedures.

3. Different standards of performance for fossil-fuel-fired steam generators depending on the date when construction commenced.

The standards of performance for fossil-fuel-fired steam generators for which construction is commenced after August 17, 1971 contained the following standard for nitrogen oxides:

no owner or operator subject to the provisions of this subpart shall cause to be discharged into the atmosphere from any affected facility any gases which contain nitrogen oxides, expressed as NO₂ in excess of:

    1. 36 nanograms per joule heat input (0.20 lb per million Btu) derived from gaseous fossil fuel or gaseous fossil fuel and wood residue.

    2. 130 nanograms per joule heat input (0.30 lb per million Btu) derived from liquid fossil fuel or liquid fossil fuel and wood residue.

    3. 300 nanograms per joule heat input (0.70 lb per million Btu) derived from solid fossil fuel or solid fossil fuel and wood residue (except lignite or a solid fossil fuel containing 25 percent by weight, or more of coal refuse).

The standards of performance for electric utility steam generating units for which construction is commenced after September 18, 1978 contained the following standard for nitrogen oxides:

no owner or operator subject to the provisions of this subpart shall cause to be discharged into the atmosphere from any affected facility, except as provided by paragraph (b) of this section, any gases which contain nitrogen oxides in excess of the following limits, based on a 30-day rolling average.

1. NO_x emission limits.

Fuel type	Emission Limit for Heat Input	
	ng/J	(lb/million Btu)
Gaseous fuels:		
Coal-derived fuels	210	0.50
All other fuels	86	0.20
Liquid fuels:		
Coal-derived fuels	210	0.50
Shale oil	210	0.50
All other fuels	130	0.30
Solid fuels		
Coal-derived fuels	210	0.50

2. NO_x reduction requirements.

Fuel type	Percent Reduction of Potential Combustion Concentration
Gaseous fuels	25
Liquid fuels	30
Solid fuels	65

The standards quoted above are just a small portion of a very lengthy standard. The reader is advised to consult the current *Code of Federal Regulations* to determine the current standards and how (or if) they apply to a given application.

REFERENCES

9.7-1 J. A. Danielson (ed.), *Air Pollution Engineering Manual*, 2nd ed., Environmental Protection Agency, May 1973.

9.7-2 Environmental Protection Agency, *National Primary and Secondary Ambient Air Quality Standards*, *Federal Register*, Vol. 36, No. 84, April 30, 1971.

9.7-3 Environmental Protection Agency, *Standards of Performance for New Stationary Sources*, *Federal Register*, Vol. 36, No. 247, December 23, 1971.

9.7-4 *Code of Federal Regulations*, Title 40-Part 60, U.S.

9.8 FUEL CELLS

To this point, the conversion devices discussed involve a burner which converts the chemical energy to a high-temperature gas before it is converted to electrical or mechanical energy. These devices usually require cooling water or heat rejection to the atmosphere and include steam power plants and internal combustion engines.

A device that converts chemical energy directly to electrical energy is the fuel cell. The essential components of a fuel cell are shown in Fig. 9.8-1 and include a fuel electrode, an oxidant electrode, and an electrolyte. The fuel electrode is the anode and the oxidant electrode the cathode.

The fuel cell illustrated in Fig. 9.8-1 is for hydrogen as the fuel and oxygen as the oxidant. Hydrogen is fed to the porous anode where it is ionized as follows:

$$2H_2 \rightarrow 4H^+ + 4e^- \tag{9.8-1}$$

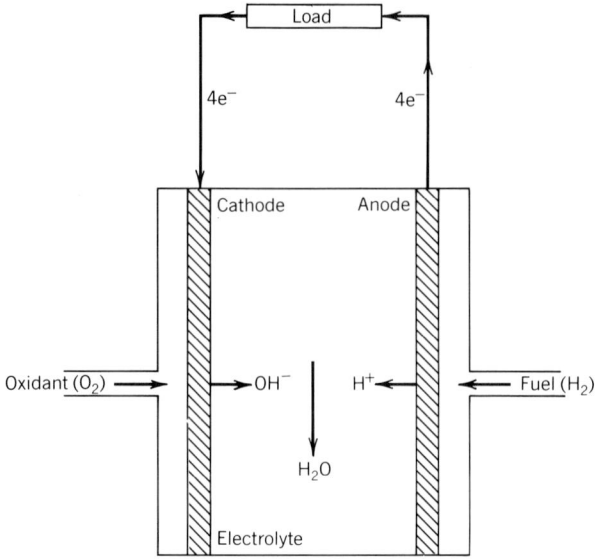

Fig. 9.8-1   Schematic diagram of a hydrogen–oxygen fuel cell.

The electrons pass through the external circuit (load) and the H^+ ions pass into the electrolyte. At the porous cathode, the oxygen acquires electrons and reacts with water from the electrolyte to form hydroxyl ions, OH^- or

$$O_2 + 2H_2O + 4e^- \rightarrow 4OH^- \qquad (9.8\text{-}2)$$

The electrolyte, potassium hydroxide in this example, has H^+ and OH^- ions in equilibrium with water. The overall (global) reaction for this type of fuel cell is

$$2H_2 + O_2 \rightarrow 2H_2O \qquad (9.8\text{-}3)$$

The electrical output from an individual cell of the type just described is quite low making it necessary to combine cells in series and/or parallel to obtain the desired voltage and current.

The first fuel cell was demonstrated by Sir William Grove in 1839 according to Crowe.[1] The first practical cell was a hydrogen–oxygen fuel cell developed by Francis T. Bacon who embarked on his project in 1932. The early purpose for development of a fuel cell was to have a device that would run on conventional fuels and air for electrical energy generation. Recent interest has been for developing a device for use on spacecraft. It has been used extensively on the Gemini and Apollo manned spacecraft.

The advantages of the fuel cell are:

1. Potential for very high efficiencies since a fuel cell is not limited by the Carnot cycle efficiency.
2. Minimal air and water pollution.
3. Few moving parts, therefore low vibration.
4. High reliability.

Noyes[2] classifies fuel cells as direct, indirect and regenerative. The direct fuel cell is one in which the fuel is fed directly to the anode. The indirect fuel cell is one in which the fuel is converted to hydrogen or a hydrogen-rich fuel before it is supplied to the anode. A regenerative fuel cell is one in which the products are converted back into reactants and recycled. The direct fuel cell is the most highly developed.

Another way to classify fuel cells is based on the fuel used. The following have been used and/or considered as fuel cell fuel-oxidant combinations:

1. Hydrogen–oxygen.
2. Natural gas–air or oxygen.
3. Synthetic gas–air or oxygen.

4. Ammonia–oxygen.
5. Hydrazine–oxygen.

The ideal fuel would have the following characteristics:

1. Low in cost.
2. Readily available.
3. Easy to store; minimum volume with no safety problems.
4. Easy to transport.

Hydrogen as a fuel and oxygen as the oxidant have been studied the most and, therefore, is the most highly developed fuel–oxidant combination for use in fuel cells.

Many companies did fuel cell research in the 1960s but, because of many problems, canceled their research in this area by the early 1970s.

One project where over $50 million has been invested is in a project involving 43 U.S. gas utilities and The Pratt & Whitney Aircraft Division of United Technologies Corporation.[2] This project was initiated in 1967 and was called the Team to Advance Research for Gas Energy Transformation (TARGET) and led to the development of a fuel cell using natural gas as the fuel which would be a total-energy alternative for use in a residential home, commercial building and/or industrial plant.

REFERENCES

9.8-1 B. J. Crow, *Fuel Cells—A Survey*, NASA SP-5115, National Aeronautics and Space Administration, Washington, D.C., 1973.
9.8-2 R. Noyes, (ed.), Fuel Cells for Public Utility and Industrial Power, Noyes Data Corporation, Park Ridge, NJ, 1977.

CHAPTER 10

HYDROELECTRIC POWER

JAMES J. GARRITY, PAUL F. SHIERS, FRED R. HARTY, JR., THOMAS J. LAMB

Stone & Webster Engineering Corporation
Boston, Massachusetts

LIST OF SYMBOLS

Symbol	Item	Units U.S.	SI
A	Flow area	ft^2	m^2
C	Head loss coefficient	—	—
D	Diameter of water conduit section or discharge diameter of turbine runner	ft	m
D_1	Runner discharge diameter of smaller turbine	ft	m
D_2	Runner discharge diameter of larger turbine	ft	m
E	Effective voltage at generator terminals	V	V
f	Friction factor for pipes (from Moody diagram)	—	—
f_{Hz}	Electrical frequency of transmission system	Hz	Hz
g	Gravitational acceleration	ft/s^2	m/s^2
H_G	Gross head	ft	m
H_n	Net head	ft	m
H_1	Net head on smaller turbine	ft	m
H_2	Net head on larger turbine	ft	m
H_{loss}	Head loss	ft	m
h_a	Absolute pressure head of the atmosphere at the turbine outlet	ft	m
h_P	Pressure head	ft	m
h_s	Geodetic suction head	ft	m
h_t	Total head	ft	m
h_v	Velocity head	ft	m
h_{va}	Vapor pressure head	ft	m
I	Effective current at generator terminals	A	A
K	Head loss coefficient	—	—
L	Length of water conduit	ft	m
N	Exponent for use in Moody formula	—	—
N_s	Specific speed of turbine	rpm	rpm

LIST OF SYMBOLS (*Continued*)

Symbol	Item	Units U.S.	SI
n_g	Rotational speed of generator	rpm	rpm
n_t	Rotational speed of turbine	rpm	rpm
n_1	Rotational speed of smaller turbine	rpm	rpm
n_2	Rotational speed of larger turbine	rpm	rpm
n_{11}	Reduced speed (SI units)	NA	rpm
P_G	Generator power output	kW	kW
p_g	Gauge pressure	psig	kPa, gauge
P_t	Turbine power output	kW	kW
P_1	Power output of smaller turbine	kW	kW
P_2	Power output of larger turbine	kW	kW
Q_p	Turbine discharge (total plant)	ft^3/s	m^3/s
Q	Turbine discharge (one unit)	ft^3/s	m^3/s
Q_1	Discharge of smaller turbine	ft^3/s	m^3/s
Q_2	Discharge of larger turbine	ft^3/s	m^3/s
v	Mean velocity	ft/s	m/s
z	Potential head (geodetic head)	ft	m
γ	Specific weight of water used in turbine	lb/ft^3	N/m^3
θ	Phase angle between E and I	degrees, radians	degrees, radians
ϕ	Velocity ratio (U.S. units)	—	—
η_g	Generator efficiency	%	%
η_p	Plant efficiency	%	%
η_s	Gearbox/flywheel efficiency	%	%
η_t	Turbine efficiency	%	%
η_1	Efficiency of smaller turbine	%	%
η_2	Efficiency of larger turbine	%	%
σ	Cavitation coefficient	—	—
ν_1	Kinematic viscosity of water used in smaller turbine	ft^2/s	m^2/s
ν_2	Kinematic viscosity of water used in larger turbine	ft^2/s	m^2/s

Extensive lists of definition are also given in the various codes and publications referenced herein. However, it is important to note that for many of these terms, there are a number of different interpretations in common usage, as discussed in the other sections. It is particularly important, therefore, that the terms be clearly defined in any computations.

Figures in Chapter 10 not otherwise designated are courtesy of Stone and Webster Engineering Corporation, Boston, Massachusetts.

10.1 THEORY AND SITE REQUIREMENTS

Paul F. Shiers

10.1-1 Basic Concepts and Associated Terms

Water Power Equation

Power is produced at a hydroelectric station when water from an elevated reservoir flows through a hydraulic turbine. This turbine is connected to an electric generator that converts the power to electricity. The power inherent in the water flow at any given instant is proportional to the product of the available head, the rate of water flow, and the plant efficiency.

The available head at a site involves two terms, *gross head* and *net head*. The gross or static head, H_G, is the total elevation difference between the river water surface at the inlet diversion and the water surface in the river where the water is returned and is defined for the conditions when no water is flowing. The net or effective head, H_n, is the head available to produce energy. It is defined as the gross head minus all hydraulic losses not attributed to the turbine, such as form and friction losses in the waterways and entrance and exit losses. The losses through the turbine, including the draft tube, are accounted for in the turbine efficiency curves. Also, for impulse turbines the net head is measured to the average elevation of the turbine buckets since the design of these machines prevents the use of any static head downstream of the turbine discharge. These terms are discussed in more detail in Section 10.2-2.

In considering a hydroelectric project, gross head is usually referenced since it takes into account all of the available water elevation difference at the site. For a given gross head H_G and total plant turbine discharge Q_p, the plant power output is given by the following water power equation:

$$\text{Total power output (kW)} = Q_p H_G \frac{0.7457}{550} \eta_p \quad \text{(U.S. units)}$$

$$\text{Total power output (kW)} = \frac{Q_p H_G \eta_p}{1000} \quad \text{(SI Units)}$$

Two terms are involved in the definition of efficiency: hydraulic efficiency and plant efficiency η_p. The hydraulic efficiency, which is the efficiency of the water conduit system exclusive of the turbine, is the ratio of the net head to gross head. The plant efficiency is the hydraulic efficiency multiplied by the combined efficiencies of the turbines, generators, flywheels, speed increasers, and transformers. The plant efficiency also incorporates a suitable reduction in plant output for station service power consumption.

The water power equation provides the instantaneous power output in kilowatts from the hydroelectric plant, given the instantaneous gross head and total turbine discharge. The energy in kilowatt-hours produced during any time period is equal to the power output multiplied by the number of hours of operation.

The hydraulic turbines and generators in present-day design have efficiencies of approximately 90 and 97%, respectively. This results in a combined efficiency of 87%. In the initial planning phases of a project, average efficiencies may be assumed as 85 and 95%, respectively, giving a conservative combined average efficiency of approximately 80%. Using this figure for efficiency and the net head, the average water power equation may be derived as follows (assuming fresh water at 62.4 lb/ft^3 or 9806 N/m^3):

$$\text{Average power (kW)} = \frac{0.8}{11.8} Q_p H_n \approx 0.07 Q_p H_n \quad \text{(U.S. units)}$$

$$\text{Average power (kW)} \approx 8.1 Q_p H_n \quad \text{(SI units)}$$

The power is linearly related to both the net head and flow variables, and thus a change in either would have an equal effect if the assumed average efficiency remains constant.

Other Power Production Terms

A hydroelectric plant can benefit a power system either by generating energy or by having the capability to do so at short notice. The following terms are often used to describe the general characteristics of hydropower.

Capacity and energy are two parameters used to measure and evaluate hydropower. Capacity is the maximum power that the turbines and generators can develop at full flow and rated head. Its basic unit is the kilowatt (kW). Energy, which is the measure of work performed, has a basic unit of kilowatt-hour (kW · h). Hydropower is evaluated in relation to both its energy value and its capacity value. Its economic feasibility is judged on its cost relative to alternative sources of equivalent capacity and energy. Capacity is a function of the flow rate available at rated head. Energy is a function of the flow over a time period, usually a year, and of the available head during each incremental time period.

The maximum power the plant can develop at full discharge and rated head is called the rated plant capacity. Since full discharge may not be available at all times and the head may vary, a number of related terms can be used to describe a plant's energy production and capacity.

Dependable or firm capacity is the power that a hydroelectric plant can supply even during a critical hydrologic period such as a drought. Similarly, firm or primary energy is the energy a hydroelectric plant will supply during a critical period. If the capacity or energy is available during peak periods, it can sometimes be considered firm, even if it is available only during a few specified hours in a given period of time. A plant's annual energy production will usually be greater than its potential firm energy output. The excess is known as secondary energy.

Firm or primary energy generally has more value than secondary energy. Similarly, available capacity, even if it is not actually generating, has an economic value if it is dependable or firm whereas nondependable capacity generally has no value.

Plant factor or capacity factor is the actual energy produced divided by the energy that could have been produced if the plant had operated at rated capacity over the entire period.

For additional definitions consult ref. 1.

Concepts for Development of Hydropower

There are several concepts for the development of hydropower, depending on site conditions. A run-of-river plant has no reservoir storage capability and must either use or spill whatever water is flowing in the river at any given time. This flow can vary considerably over a given period of time. A run-of-river plant often has little, if any, firm capacity. Its principal value is associated with its ability to displace other, more expensive, sources of energy from thermal-electric generating stations whenever sufficient water is available.

If a plant has storage capability, it can store water when the river inflow is high and release it later during low-flow periods, thereby achieving a reduction in spillage and increased energy output, when compared to a run-of-river plant. This type of plant is of high value since it maximizes the amount of firm capacity, primary energy, and secondary energy that can be developed. If the flow is released at a relatively constant rate, the plant can be used as a base-load plant, producing a relatively constant power output. Alternatively, the natural river flow can be stored during periods of low demand on the electrical system, typically nights and weekends, and subsequently released at a higher-than-average flow rate during periods of high power demand. This type of plant is called a peaking plant. A peaking plant will be credited with having firm capacity even when no energy is being generated, provided this capacity is quickly available when needed. Hydroelectric units can be started and brought up to full load within several minutes, which is an advantage over most other forms of generation. Hydroelectric plants may also be designed for startup without the use of an external power source.

Peaking power also can be provided by pumped storage plants. Unlike the previously noted plants that contain only hydraulic turbines, pumped storage plants may have either pumps and turbines or reversible pump–turbines. During off-peak hours, a utility's base-load thermal plants can be used as an economical power supply to pump water to an upper reservoir. During periods of peak electrical demand, the water is allowed to flow back to the lower reservoir through the turbines to generate electricity. Although some energy is lost in the pumping–generating process, there is a net benefit if the value of energy generated during peak periods is sufficiently greater than the cost of pumping energy during off-peak periods. This is normally the situation that exists in a typical system. Pumped storage plants are advantageous for instant power supply since they are often designed for rapid starting in both directions several times a day.

Tidal power can also be used to provide electrical generation. It can be harnessed either in one direction, usually with the outgoing tide, or in both directions. A further design refinement is the possible operation of a plant in a pumping mode in one, or both, directions. Water is pumped under low heads at one end of a tidal cycle and reused for generation under somewhat higher heads at the other end of the tidal cycle. This type of operation results in a net gain in energy production, and the benefits derived usually exceed the associated costs. Tidal plants using only one pool (basin) are not able to generate power on a continuous basis due to the lack of sufficient head and/or water at the end of the filling and emptying cycles. This feature, combined with the fact that the lunar cycle causes the tides to advance about 50 min/day, makes it very difficult to meet system peak power demands, which tend to occur at about the same time each day. While a two-pool scheme makes it possible to generate continuously, the level of power output will still vary on a daily and hourly basis due to the ever-changing magnitude of the tides and the mismatch between the lunar cycle and the solar day. In order to effectively use tidal power, some type of backup generation or reregulating facility is generally needed. A conventional pumped storage project is particularly well suited for such a use. However, the probable distance between these two types of plants could involve significant energy losses in the transmission lines.

10.1-2 Structures and Equipment

Figure 10.1-1 shows the more important elements of a hydroelectric plant. A hydroelectric project normally consists of a diversion structure; headworks; water conduits to convey water to and from the turbines; a powerhouse containing the hydraulic turbines, governing mechanisms, generators, controls, and switchgear; and a switchyard and transformers to connect the generators to the transmission system. There may also be environmental structures included in the project, such as fish ladders.

Diversion Structures

The diversion structures include a dam to maintain headwater, to divert water to the power plant, and in some cases, to provide water storage. A spillway is also included to pass excess flows through, over, or around the dam without jeopardizing the integrity of the structure. The spillway can be included as part of the dam, or as a separate structure. Bypass works with gates and conduits to divert water when the turbine is not operating may also be included in these structures. Dams are discussed in more detail in Section 10.3.

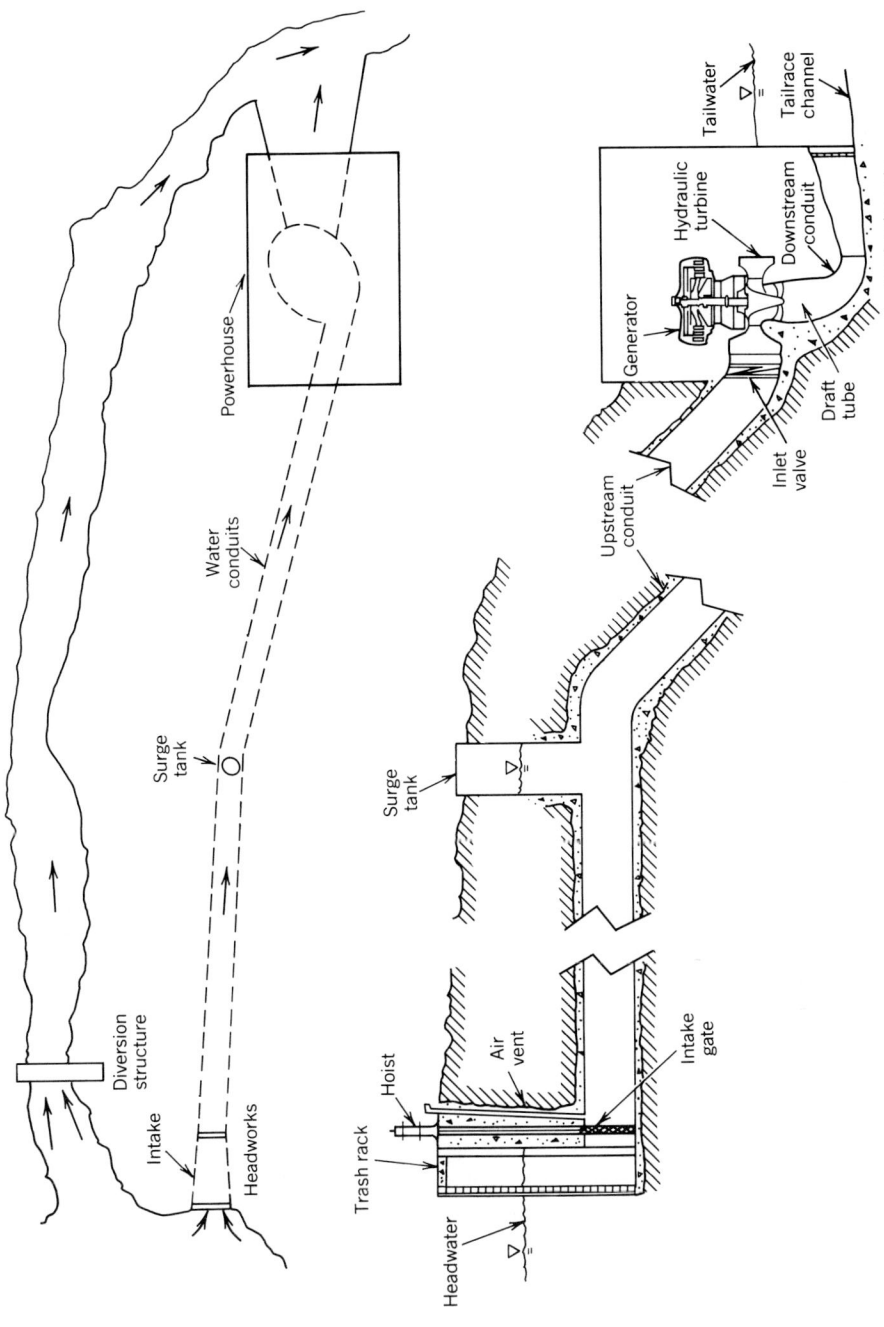

Fig. 10.1-1 Typical hydroelectric plant structures and equipment (not to scale).

Conveyance Works

If the powerhouse is located some distance from the diversion structure, a canal and forebay may be provided. The forebay, a small reservoir located upstream of the waterways, may be provided to supply additional water needed during the period immediately following startup and to contain or store water that continues to flow from the canal during the period immediately following shutdown.

The water flow approach to the headworks structure should be carefully arranged to provide uniform flow to the intake to avoid turbulence and vortices. The flow area should be dimensioned so that normal approach velocities are low to minimize head losses at the trashracks.

Headworks

The headworks element of the development is the intake section that connects the reservoir or entrance canal with the water conduits. The intake is generally equipped with trashracks and in some cases trashrakes. Various types of intakes are used depending on the type of hydro project. These include an intake tower or submerged intake in the reservoir, an integral intake in the dam or powerhouse structure, and a separate intake structure.[2]

Intakes for hydroelectric plants generally exist as part of the dam structure. The intake should be located high enough to minimize the gate lift distance if gates are provided. The entrance should be designed to prevent the formation of vortices, which could carry air into the water conduits and result in a lower hydraulic turbine efficiency and pressure surges in the system. Hydraulic model studies are frequently performed to confirm the adequacy of the intake design relative to potential vortex activity. The entrance sill should be located an adequate distance above the reservoir bottom to minimize the entrainment of stones, silt, and other bottom debris.

Trashracks are normally provided at the intake and are installed vertically or at a slight incline. The spacing of rack bars depends on the size and type of hydraulic turbine selected and is influenced by the size of trash expected. Other design considerations include the head losses due to the contraction and sudden expansion of the flow at the bars, and the allowable length of laterally unsupported rack bars to avoid vibration problems. The design of trashracks is discussed in further detail in ref. 2.

Upstream Water Conduits

Water is conducted from the headworks to the hydraulic turbines through an upstream conduit system. These conduits can be aboveground or belowground. Belowground conduits can be lined tunnels through dams, lined or unlined tunnels in abutments that extend downstream to the vicinity of the powerhouse, or closed cut-and-cover conduits. Aboveground conduits can be made of steel, reinforced concrete, wood, or fiberglass pipe.

Water conduits should have smooth linings and be as short as possible to minimize head losses. Dynamic forces must be resisted by pipe anchors or thrust blocks. This feature, plus the associated head losses, make the use of sharp bends undesirable. If bends are necessary, they should be located sufficiently upstream of the turbine to allow streamlined flow at the entrance to the turbine. Intakes, transitions, and branches are sometimes model tested at a hydraulic laboratory to determine the velocity variations, turbulence, and head losses associated with the expected flow conditions.

Penstocks are used to supply water to the hydraulic turbines. For sites with a short intake-to-powerhouse distance, individual penstocks are often provided for each hydraulic turbine. For longer distances a single penstock or tunnel may branch into separate penstocks at the lower end. Where a single penstock carries water to two turbines, or where a bypass is provided off the main conduit, a bifurcation is used to split the flow from one pipe to two. At large sites trifurcations or multiple bifurcations are frequently used.

Structural design considerations of the water conduit system are similar to other large pipes. Water hammer design considerations are essential due to the possibility of sudden changes in load on the turbine, which can result in sudden flow changes. The most significant flow variation occurs when the turbine–generator unit is shut down quickly following a sudden disconnection from the system. Operating stresses due to internal pressure and external pressure and temperature must be considered. The static head included in the internal pressure is the difference in elevation between the headwater at the dam and the elevation of the conduit. Water hammer pressure must be added to static head to establish the design pressure of the conduit for the selected wicket gate closure speed. A long penstock may be provided with a surge tank to limit water hammer pressures and to provide a temporary supply of water to meet a sudden load increase.[3]

The stresses caused during the handling and erection of the penstock must also be considered, and they are frequently the controlling design factor. To accommodate this, a minimum handling thickness, which is a function of penstock diameter, is generally required.[3]

Uncovered penstocks have the advantage of being accessible for inspection and repairs but require expansion joints, supports, anchors, and other appurtenances. They are usually supported on concrete cradles or steel ring girders. It is sometimes advisable to bury penstocks in severe climates to reduce the number of expansion joints required and to prevent the formation of ice. In some cases the most economical solution is to use a lined or unlined tunnel through rock.[4]

Head losses in the conduit include form losses (losses due to geometry of the conduit) and friction losses. Form losses are usually expressed as a head loss coefficient times the velocity head, $v^2/2g$ (see Section 10.2 for further discussion), of the flow at the point in question. This can also be expressed as a function of the discharge Q and the area A.

$$H_{\text{loss}} = \frac{Kv^2}{2g} = \frac{KQ^2}{2gA^2}$$

The form loss coefficients are fixed primarily by the geometry of the conduit, with only minor variations due to the scale. Thus the value of K for a given shape can be determined by model test or approximated from textbook values generally without reference to the size of the structure.

Similarly, the friction losses can be estimated from the Darcy–Weisbach equation, by incorporating the friction factor f, the length of conduit L, and the diameter of conduit D.

$$H_{\text{loss}} = \frac{fLv^2}{2gD} = \frac{fLQ^2}{2gDA^2}$$

For sections of a conduit of nonuniform diameter where no branching of flows occurs, and where the conduit includes several types of form losses in addition to the friction losses, it is convenient to combine all of the loss factors into a single term for the complete conduit:

$$H_{\text{loss}} = CQ^2$$

The coefficient C is useful in evaluating variations in discharge or alternative conduit arrangements since it represents the entire conduit system, exclusive of the turbine, with a single number.

Powerhouse

The powerhouse contains the hydraulic turbines, generators, and associated equipment. Conventional hydroelectric plants can be divided into four general categories: (1) an integral dam and powerhouse, (2) a separate powerhouse at the toe of the dam, (3) a separate powerhouse downstream of the diversion structure connected by a long upstream water conduit, and (4) an underground powerhouse as discussed in ref. 2.

The integral powerhouse is a mass concrete structure that includes a reinforced concrete intake, water passages, trashracks, intake and draft tube gates, hydraulic turbines, generators, and other supporting equipment.

The separate powerhouse at the toe of the dam may be added to an existing dam or constructed concurrently with a new dam. A short penstock connects the intake with the hydraulic turbines in the powerhouse.

The separate powerhouse connected by a long, upstream water conduit to the diversion structure is used where added head is available downstream of the dam site, which can be economically developed. In this type of project, a minimum flow may be required in the bypassed section of river between the dam and the powerhouse.

The underground powerhouse is built in locations where the excavation of the required caverns is more economical than the construction of another powerhouse type. This type of installation is also used where site space is unavailable for a conventional structure, and the site conditions make it desirable for environmental considerations. An underground powerhouse allows for deep setting of the generating units to take advantage of higher speed and smaller dimensioned turbine–generator units. An underground powerhouse might also be located to take advantage of additional head downstream of the dam by means of long tailrace tunnels.

Powerhouses generally include three main sections: a turbine–generator area, a station service area, and a maintenance/erection area. Access for maintenance should be provided between the turbine–generator units and the walls and sides of the powerhouse and between the individual turbine–generator units. Vertical access must also be provided for removal or replacement of any part of the main generating unit and supporting equipment. The erection area normally includes space for each individual part that may be removed during overhaul work.

Generator housings can be either indoor, semioutdoor, or outdoor types. The indoor type is fully enclosed by a building of adequate height to permit handling and transfer of the generator by an indoor crane. The semioutdoor type consists of a fully enclosed generator room with an external gantry crane and roof hatches used for equipment removal. The outdoor type has a weatherproof generator housing in lieu of walls and a roof. A portable crane is used to handle the heavy generator parts in this case.[2]

Hydraulic turbines, generators, and associated equipment are discussed in Section 10.2. Additional powerhouse equipment would include transformers, switchgear, circuit breakers, cranes, sump pumps, air compressors, and heating, ventilating, and air-conditioning equipment. Intake and draft tube gates are also usually provided to enable the turbine to be dewatered.

Downstream Water Conduits

After passing the hydraulic turbine, the water flow passes through a conduit system to the tailwater. For reaction turbines (refer to Section 10.2-1), this conduit system includes a draft tube that conveys water from the turbine runner and which can either discharge directly to tailwater or through a longer tailrace tunnel into a tailrace canal or outlet structure.

The primary purpose of the draft tube is to convert the high-velocity, low-pressure flow at the turbine runner exit to a lower-velocity, higher-pressure flow at the outlet structure. This is accomplished by means of a diverging flow section that minimizes the hydraulic head loss at the exit. If the turbine runner is set above tailwater level, the draft tube also extends the water passage below the tailwater level. This permits the use of the full gross head on the plant for power production by recovering a static suction head downstream of the turbine. The elevation of the turbine runner with respect to tailwater is set by requirements for cavitation protection (for a discussion of cavitation see Section 10.2-2). Also, since this elevation will vary with the specific speed of the turbine, a balance must be achieved between turbine–generator costs and excavation costs. The design aspects of the draft tube are discussed in further detail in ref. 2.

The shape and dimensions of the draft tube are usually developed by the turbine manufacturer, and the performance of the draft tube, including the head loss at the exit, must be included in the overall power and energy computations, as described in Section 10.2-2. The draft tube is not included in an installation with an impulse turbine (refer to Section 10.2-1) due to the free discharge of water to atmospheric pressure.

The tailrace is the downstream channel or tunnel that carries the water from the draft tube back to the river. Topography and economics will dictate whether a channel or a tunnel, or a combination of the two, will be used. When the bottom of the draft tube exit or tailrace tunnel exit is lower than the riverbed, the tailrace channel must slope upward. This slope is generally between $3H:1V$ and $6H:1V$, with the milder slopes being generally preferred for hydraulic reasons.

At the point where the upward-sloping part of the tailrace meets the riverbed, normal velocities should be reduced to 0.5–1.0 m/s. This can be accomplished by a gradual widening of the channel with a divergence angle of 15° or less.

Scouring and/or deposition of materials are sometimes encountered in the tailrace. An unlined tailrace may experience scouring of the channel bottom and a lowering of the tailwater elevation. A net removal of material may also occur in some cases where an upstream dam blocks the resupply of material deposits to the streambed. When not allowed for in the original design, this problem has resulted in the tailwater level dropping below the draft tube outlet and a consequent reduction in hydraulic turbine capacity due to a loss of static head on the downstream side. In this case either the draft tube must be lowered or an overflow weir must be constructed in the tailrace channel to raise the tailwater back to its normal elevation. The tailrace channel may also accumulate material and cause the tailwater elevation to rise. This may occur where the spillway is close to the tailrace and no training wall is provided. This results in a decrease in head at the plant, and remedial work of dredging or otherwise removing the material may be required.[4]

Surge Tanks

Surge tanks are sometimes provided in both the upstream and downstream conduit systems. They provide a supply of water to control changes in pressure, thereby protecting the conduits and equipment. For a load reduction the wicket gates on the turbine close to control the speed, resulting in a storage requirement for water on the way to the turbine. Similarly, the wicket gates open on a load increase that creates a demand for added water to maintain speed. The surge tank provides space for this water and thus improves the ability of the turbine–generator to regulate its load.

For a surge tank upstream from the turbine, a sudden reduction in turbine discharge causes a rise in water level in the surge tank, which is followed by the water surface cycling up and down until the surges are dampened out by fluid friction. Similarly, if the surge tank is located in the downstream side of the system, the reduction in turbine discharge causes a drop in water level in the surge tank. An equation, derived by Linsley and Franzini,[4] expresses a relationship between water surface level in the tank and velocity in the conduit during the period from the time of a complete and sudden shutdown until the top of the first surge is reached. This equation can be used to find the maximum surge height. The results provide a conservative estimate since friction, velocity head, and entrance losses are neglected. Also, the equation assumes that the surge results from instantaneous shutoff of the turbine discharge.

There are several types of surge tanks: simple, restricted orifice, differential, and closed. A simple vertical surge tank is usually designed to be open at the top and to be of sufficient height that overflow will not occur. The water flows in and out of this tank with minimal head losses. However, in some cases, where the water can be disposed of without damage, overflow is permitted. The simple tank design is frequently adequate for small tanks.

The other types of surge tanks can be used instead of larger tanks to reduce the total volume and cost of the surge tank while improving dampening action. One type of tank provides a restrictive orifice at the base of a simple surge tank and separates the storage function from the accelerating/decelerating function. This provides a head loss in the water flowing into and out of the tank, thereby restricting the total range of the surges and providing improved damping action. However, the rapid creation of accelerating or decelerating heads during load acceptance or rejection develops fluctuating heads on the turbine that complicate the governor mechanism. An alternative design is the differential surge tank, which combines the simple and restricted orifice tanks. In this tank a vertical riser of about the same diameter as the conduit is installed inside of a larger-diameter main tank. Orifice-type openings in the riser or around the base of the riser limit the flow of water into the main tank. The water level fluctuates more rapidly in the riser than in the tank. In this case the fluctuations are out of phase and can be dampened out more quickly so that the outer diameter need be only 70% of that of a simple surge tank. Another method of restricting the limits of the surge would be to seal the tank at the top, thereby providing an air cushion that can be compressed and expanded. The bottom of the surge tank must be set low enough so that water will remain in the tank at all times to prevent air from entering the conduit.[4]

The surge tank is intended primarily to protect the section of conduit between the surge tank and the reservoir. If the shutoff of the turbine discharge is so fast that the initial pressure waves can travel to the surge tank and return to the turbine after the wicket gates have closed completely, the closure is considered instantaneous. The presence of the surge tank will therefore not affect the maximum pressure experienced in the conduit just upstream from the turbine. If, however, the surge tank is relatively close to the turbine and the initial pressure wave can reach the surge tank and return to the turbine before the wicket gates are completely closed, the closure is not considered instantaneous, and the presence of the surge tank will reduce the maximum pressure upstream from the turbine. Therefore, the surge tank is generally located as close as possible to the turbine, both to minimize the length of pipe subjected to full pressure change and to limit the amount of pressure change to some extent.

Valves and Gates

Throughout the water conduit system, from headwater to tailwater, there is usually a series of valves, gates, or stop logs used for flow regulation, emergency shutoff, and dewatering and inspection.

Head gates or stop logs are usually provided at the intake to allow for dewatering of the conduits for inspection and repair. Slide gates are the preferred choice for large units. An air vent should be installed in the closed conduit downstream of the gate. This limits negative pressures that could otherwise result in a pipe collapse following closure of the gate.

An inlet valve may also be provided upstream of each hydraulic turbine. The inlet valve serves the following functions: emergency shutoff, dewatering of the unit for inspection or maintenance, and prevention of inadvertent rotation of the unit and loss of water due to excessive leakage through the turbine wicket gates. These wicket gates are the movable guide vanes that control the flow into the turbine runner and are normally used to shut off the flow through the unit.

High-head projects are usually equipped with large spherical (plug) valves located just upstream from the turbine either in the powerhouse itself or in a special gallery upstream of the powerhouse. A spherical valve consists of a large valve body housing and a plug with an opening diameter matching that of the penstock. This configuration allows full flow with negligible head loss. The valve is normally opened or closed only when water is not flowing in the penstock and pressure is balanced on both sides of the valve. However, these valves are usually designed to close under full-flow conditions in an emergency.

On low-head installations butterfly valves are more economical and are normally used as inlet valves. However, since this type of valve presents some obstruction to the flow, head losses are incurred.

Normally the inlet valves are used only in the fully opened or fully closed positions since flow control is usually provided by the turbine itself. However, in low-head, small-scale installations, the turbines may not have any flow control devices. In such cases the valves or gates are partially opened during unit startup and when full speed is reached, the generator is connected to the transmission systems. The valves or gates can then be fully opened.

Gates or stop logs are usually provided at the draft tube exit to permit dewatering of the turbine water passages. In some cases they may also be used as flow control devices. For example, on some low-head projects the turbines have been used as substitute spillways where appropriate equipment provisions are made. Thus upon a sudden disconnection of the plant from the transmission system, hydraulically operated roller gates located in the draft tubes will partially close and the normal turbine flow control devices will completely open. In this mode of operation large turbines can pass up to 70% of their normal flow until the transmission problem has been corrected or other measures have been taken to pass the water.

If there is a tailrace tunnel, stop logs or gates can be provided at the downstream end to permit dewatering and maintenance of the tunnel.

10.1-3 Development Alternatives

Conventional Reservoir and River Sites

A proposed hydropower development at reservoir and river sites may take any of the following forms:

1. A complete project at a new site.
2. Addition of new generating units at an existing power station.
3. Rehabilitation of an existing power station by either partial or complete replacement of the old turbines and generators with more efficient equipment of modern design.
4. Installation of generating units at existing nonpower facilities.

New Sites. For a new site a number of options are open with respect to the type of dam and equipment. The new site often allows development with fewer technical restrictions than would exist with the redevelopment of existing facilities. Normally, the development of a new site will require considerable technical investigation and environmental studies. Regulatory requirements are also more extensive for the development of a new site. Considerations for dam siting and selection of dam type are discussed Section 10.3. Associated equipment considerations are discussed in Section 10.2.

Rehabilitation and Expansion of Existing Hydropower Sites. Many hydroelectric plants have been in continuous service for over 50 years, and the effects of this service, plus repeated repairs, may have considerably reduced the efficiency of the unit's operation. Replacement or rehabilitation of worn-out equipment may therefore improve the energy production of an old plant. Furthermore, the use of new turbine design techniques can often result in higher capacities since modern turbines operating at higher speeds may produce more power output for a given runner diameter. The limiting factor in this upgrading can be the potential for cavitation, as discussed in Section 10.2.

The new units generally may be installed with a minimum of changes to existing civil work, although some minor powerhouse modifications will be required due to certain physical differences in configuration between the old and new units. The new, more efficient units would then operate nearly continuously at a high-capacity factor. The older, less efficient units would operate only during periods of high or low river flows beyond the flow capacities of the new units.

In cases where the existing powerhouse is still in operation, the project economic analysis should consider the energy lost during the period required to modify the powerhouse and install the new units. Also, if the existing installation has a small number of years remaining in its operating life, only the incremental energy produced by the rehabilitated plant during this period should be considered when evaluating the economic feasibility of modifications.

Rehabilitation of existing sites often involves the expansion of powerhouse facilities to add generating equipment. Construction procedures used with this type of development will usually permit the existing units to remain on-line during construction with the possible exception of the unit adjacent to the new construction. However, in some cases, the addition of units may require modifications to the existing plant conduits and intake structures. For example, the old conduits may have to be replaced with larger ones, or a new conduit added in parallel to the existing ones.[5]

Adding Generation At Existing Nonpower Facilities. Hydropower generation may also be added to a number of existing nongenerating facilities, including flood control dams, water storage and irrigation dams, and water supply facilities such as canals or conduits as summarized in this section and discussed in detail in ref. 5.

Existing water conduits associated with these facilities are critical in evaluating the possibility of adding generation to this type of facility. Existing conduits may be adequate for their intended use but not readily adaptable to power generation. If the existing conduits are inadequate, they must be modified or new conduits must be installed. Existing conduits buried under earth- and rock-fill dams usually cannot be modified, and tunneling through an abutment may be necessary. For a concrete dam, cutting a water passage is technically possible with suitable analysis and construction principles being employed, however, this may not be economical. In certain cases a siphon penstock over the dam may be a more suitable alternative to utilize a site potential.[3]

One type of existing facility that might be considered for the addition of hydropower generation is a flood control dam. At these dams water is usually released through a low-level outlet constructed in the base of the dam. An intake tower, slide gates, conduit, and outlet channels are often provided. The conduit usually has a large diameter to accommodate a relatively large flow. This conduit is generally concrete or concrete lined.

Locating a powerhouse close to the downstream end of the conduit, while using the existing dam outlet works as intake and penstock for conveying water to the new power plant, is a possible design approach for such a facility. A bypass would be constructed upstream of the new valve to carry water to the powerhouse. The conduit would be provided with a valve at the downstream end to maintain the availability of the outlet for flood releases. However, this arrangement would place the conduit under pressure for plant operation. Since this pressure may exceed the design pressure of the pipe, it could require the installation of a steel liner.

A more economical development scheme for a flood control dam may be achieved by not using the existing outlet facilities and constructing a new intake with a lined tunnel to convey water to the turbines. The turbines would usually discharge directly into the river below the dam, and no special outlet works would be needed other than a short excavated tailrace channel.

Water storage and irrigation dams are constructed similar to flood control dams except that the outlet works are sized for much smaller flows. These dams usually have control valves or gates at the downstream end, and consequently the conduits are designed to operate under pressure over their entire length. These characteristics make the installation of power generation equipment easier at these facilities than at flood control dams since they are designed to operate under pressure. However, this is balanced off by the reduced available capacity due to lower flows.

Major irrigation systems, which include flumes or canals, also provide development opportunities.[6] Power generating facilities can often be added in an irrigation canal to bypass a drop structure located either on the main canal or at a distribution point between two canals. The power generation facilities then perform the same function for the irrigation system as the drop structure flow-regulating equipment. With this type of project, power is not available on demand but only when and to the extent that water flow is required in the water distribution system.

The same principle can be applied to water supply pipeline networks. These systems typically include energy-dissipating valves at points of distribution to the water users. Power generation facilities may be installed on a bypass around these energy-dissipating structures. These pipelines are apt to be long and often are not designed to resist significant water hammer pressures. Therefore, water hammer pressures must be taken into account when designing the power generation facilities either in the selection of the hydraulic turbines or in the design of surge supression facilities.

Pumped Storage Developments

Hydroelectric power generation by conventional methods involves water flowing in one direction only. In pumped storage developments, water flows in both directions. During off-peak hours, water is pumped to an upper reservoir. During periods of peak power demand, this water is then released to generate power. Pumped storage hydropower is a form of economically creating a reserve store of energy in an elevated water storage reservoir. This reserve energy can be instantaneously called upon to satisfy peak-load requirements in the utility system.

For a pump storage plant to be attractive, the following things are necessary:

1. An existing system of thermal base-load generation capacity that can provide pumping power during off-peak hours.
2. High-peak energy demands.
3. Favorable site conditions (refer to Section 10.3).

A significant part of the capacity of a utility system is in the form of nuclear and fossil-fired base-load units. During off-peak periods, the inexpensive energy supplied by these sources can be used to pump water to an elevated storage reservoir, enabling the utility to store this energy for use in high-demand peak periods. This higher-valued energy can be called on as required to meet peak-load periods in the utility system load cycle.

The attractiveness of using pumped storage hydropower as a means of meeting fluctuations and peaks in system load demands include (1) improved total system reliability through plant "instant-on" capabilities (dead start), (2) more efficient use of available base-load units, and (3) control of system frequency through the start/stop capability of the units/plant and the ability for rapid change in load. Figure 10.1-2 shows the basic arrangement of a pumped storage plant, which in this case consists of an upper reservoir, a pressure tunnel or penstock, powerhouse, surge tank, tailrace tunnel, and lower reservoir.

Since the flow of water is in two directions, there are several plant feature and design considerations that differ from conventional hydropower developments.

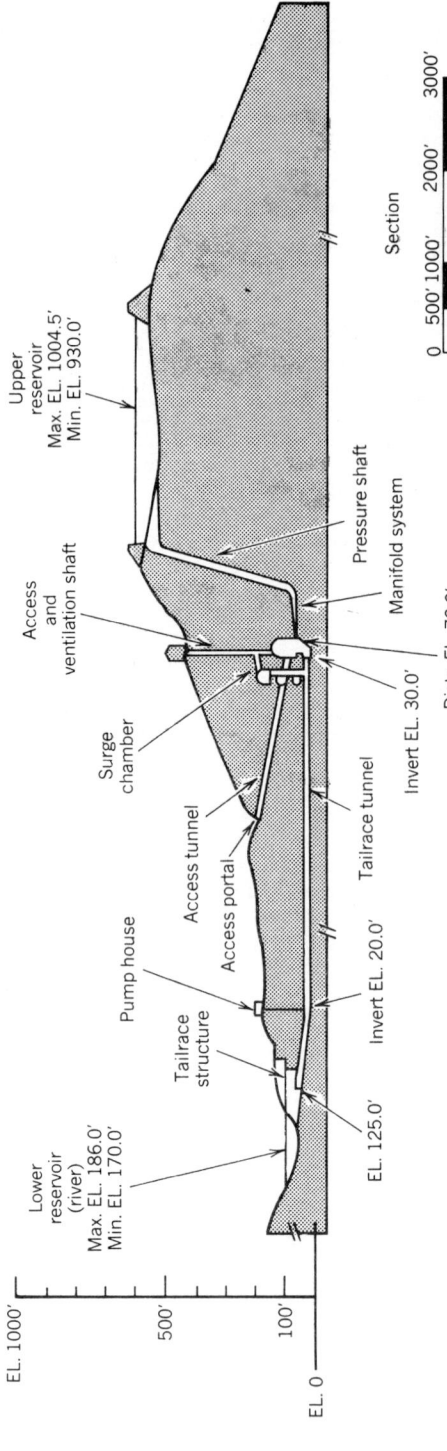

Fig. 10.1-2 Northfield Mountain pumped storage project.

Upper
reservoir
Max. EL. 1004.5'
Min. EL. 930.0'

Access
and
ventilation shaft

Pressure shaft

Manifold system

Surge
chamber

Distr. EL. 72.0'

Invert EL. 30.0'

Access tunnel

Access portal

Tailrace tunnel

Pump house

Invert EL. 20.0'

Tailrace
structure

EL. 125.0'

Lower reservoir
(river)
Max. EL. 186.0'
Min. EL. 170.0'

Section

0 500 1000' 2000' 3000'

EL. 1000'

500'

100'

EL. 0

Inlet–Outlet Structures. Inlet–outlet structures must have a design that considers flow in either direction. A well-designed inlet must avoid localized high-velocity jets when used as an outlet structure. Similarly, a well-designed outlet structure must also avoid vortices when used as an inlet. Because of these potential problems, as with the conventional sites, it is desirable to conduct hydraulic model tests of any proposed inlet–outlet structure to assure satisfactory performance during both modes of operation. Trashracks may also have to be supplied at both structures depending on the potential sources of trash.

Generating Equipment. The earliest pumped storage developments made use of separate pumps and turbines connected to a common motor/generator. This type of unit allows extremely rapid change-over from the pump to turbine mode because the machinery rotates in the same direction for both modes. Such a capability is very desirable for accommodating short-term changes in power demand on the electric system. It also offers the added advantage of optimum efficiencies for both modes of operation. However, because of higher equipment costs, this arrangement is now limited to very high head installations beyond the present state of the art for reversible pump–turbines. For such applications, impulse turbines can be combined effectively with multistaged pump units capable of very high head delivery and having relatively modest submergence requirements.[7]

Most of the recent pump storage developments have used reversible pump–turbines that have developed rapidly since the 1950s with significant improvements in terms of permissible head, flow capability, and overall efficiency. Reversible Francis pump–turbines have been utilized for projects with heads of over 600 m, and with outputs on the order of 400 MW.[8] These reversible pump–turbines have reasonable efficiencies in both directions and have a lower installed cost than do separate pumps and turbines, as well as lower maintenance costs due to the elimination of a major piece of machinery for each unit.

The major differences between turbines and pump–turbines with respect to the powerhouse structure and water conduit design are as follows:

1. Head losses and fluid transient effects must be evaluated for flow in both directions.
2. The flow throttling effect of a pump–turbine must be considered in the fluid transient analyses (during overspeed following load rejection, a pump–turbine can cause its own water hammer effect in addition to that caused by the closing of the wicket gates).
3. Submergence requirements are generally governed by the pumping mode.

Site Characteristics. Site features that have a significant impact on the development include the effective head, topography, length of flow line, existing facilities, source of reservoir fill and makeup water, and dam siting requirements[9] (discussed in Section 10.3). The available head is particularly important since it will influence the size and hence the cost of all the major features of the project, including power and tailrace tunnels and/or penstocks, pump–turbines, powerhouse, intake structures, and upper and lower reservoirs.

The manner in which the plant facility will operate and its total capacity will dictate reservoir size requirements. A plant designed to operate on daily peaking and storage cycles can normally be supported by relatively small upper and lower reservoirs. In utility systems where efforts have been undertaken to shift residential customer load demand to off-peak hours, the use of weekly or even monthly storage–release cycles may be indicated, requiring substantially larger and more costly reservoirs.[10]

A pumped storage project requires water for the initial filling, whether it be for the upper or lower reservoirs. Thereafter, it requires only enough water to replace evaporation and seepage losses. A large-capacity pumped storage project can, therefore, be built even in locations with small stream flow, if the permissible reservoir-filling period is sufficiently long.

Normally, the upper reservoir for a pumped storage project is formed by utilizing (1) a normal stream valley reservoir, (2) a high valley reservoir, or (3) a hilltop reservoir. Topography, size, soils/geologic considerations, and the type of dam structure are all important factors to be considered in the siting, technical design, and economic aspects of the reservoir location. Watertightness of the reservoir and dam structure behavior under large rapid draw-down conditions are special factors to be considered in both design and operating criteria for a pumped storage project since this type of project may have an essentially complete draw down and refill on a daily basis. The influence of the hydraulic aspects of the inlet–outlet structures is also an important factor to be considered in the design of any adjacent earth embankment sections since unfavorable flow patterns can be detrimental to these structures. A detailed discussion of reservoir selection and design considerations is available.[11,12]

To minimize project costs, it is advantageous to incorporate an existing reservoir impoundment into the project scheme as one of the reservoirs. However, new site developments that require the creation of both reservoir pools should not be excluded from consideration.

Lower reservoirs are commonly constructed on natural streams or rivers, often forming part of a conventional hydroelectric, flood control, or multiple-use water project. Existing dams creating these reservoirs can require extensive modifications to accommodate increased river flow and water level fluctuations caused by generating and pumping operations. New dams and reservoirs on rivers will constitute a major construction cost for the pumped storage project as it would for any project requiring a new dam on a river (see Section 10.3).

Pumped storage reservoirs are normally created by the construction of dams. Another pumped storage concept that has not yet been used (1982) but which is being considered is the use of a conventional surface reservoir and an underground lower reservoir, either excavated for this purpose or adapted from an abandoned mine. By this method, extremely high heads can be made available to develop a large capacity with a relatively small volume of water.

This can be accomplished in one of two ways. A single powerhouse can be located at the lower reservoir (single-drop arrangement). Alternatively, this reservoir can be of nominal size and can serve as the upper reservoir for another powerhouse and larger reservoir located at an even deeper level (multidrop arrangement). With the multidrop arrangement the height of the individual drops can be established to accommodate existing technology for reversible pump–turbines. Furthermore, while the entire gross head is developed by the combined power generating facilities, the maximum pressure in the system is that corresponding to the individual drops.

Although the underground storage concept has some potential advantages, the high cost of removing material to create the lower reservoir(s) is a significant factor affecting project feasibility. This factor would make those sites with existing underground caverns more favorable for project development. Increased depth provides increased head and a reduction in reservoir size. Since underground rock stresses tend to increase dramatically with depth, the quality of rock becomes quite important with regard to the planned excavation methods, stabilization procedures, and general safety requirements.[13]

10.1-4   Site Evaluation and Conceptual Design

A proposed development for a given site should balance the benefits of the power and energy produced against the associated cost of development. This is an iterative process where the tentative generating capacity, dam height, storage provisions, and project arrangement are all adjusted and evaluated to arrive at the optimum balance. The development of individual hydropower sites may also include an evaluation of the existing and potential development of the river basin under consideration.

Hydrological Analysis

The first step in evaluating a hydropower site is to perform a hydrologic analysis to establish the stream flow data. This is followed by a determination of the available head that can be developed through different arrangements of dam heights and water conduit lengths. Using a number of combinations of water flow and available head, it is then possible to evaluate alternatives with different installed capacities for the power generation facilities.

The characteristics of rivers used for hydropower developments are variable. Most rivers are subject to both extended periods of low flow caused by drought and periods of flood flows due to heavy rains. Periods of drought may reduce the supply of water to unacceptable levels. Heavy rains and the resulting flood runoffs may cause a substantial decrease in the available head due to the increase in tailwater level. These natural hydrologic characteristics are thus major considerations in sizing site facilities, selecting applicable hydroelectric equipment, and establishing plant operating conditions.[3]

Variations in natural stream flow require a detailed review of many years of hydrologic records to analyze the site's potential energy output. For most cases a period of 20–25 years, including samples of both high- and low-flow years and covering any critical dry periods, is considered adequate. For some analyses, due to a limitation of available data, shorter periods of 8–10 years may be used.[3]

The hydrologic data can be presented in two forms: (1) a continuous hydrograph that shows the sequential flow rates over the entire period of record or (2) a flow duration curve that indicates the frequency at which flow rates of a given magnitude have occurred over the period of record. If the project under study is to include storage provisions, the hydrograph can be used in simulating the operation of the plant with various volumes of storage plus either base-load or peaking operation. If the project under study is to be a run-of-river project with no storage, then the sequence of the natural flow is of less importance and a flow duration curve can be used.

A flow duration curve is a graphical plot of the range of historical river flows at a given location versus the percent of time these flows are exceeded. It provides a graphical illustration of flow distribution, regardless of the sequence of occurrence during a given period. The curve is developed by grouping all the hourly, daily, or monthly flow values into groups within set ranges of discharge. The number of classes selected must be adequate to reasonably define the curve (generally 10–30 classes). Beginning with the highest discharge class, the number of times when the limit is exceeded is summed

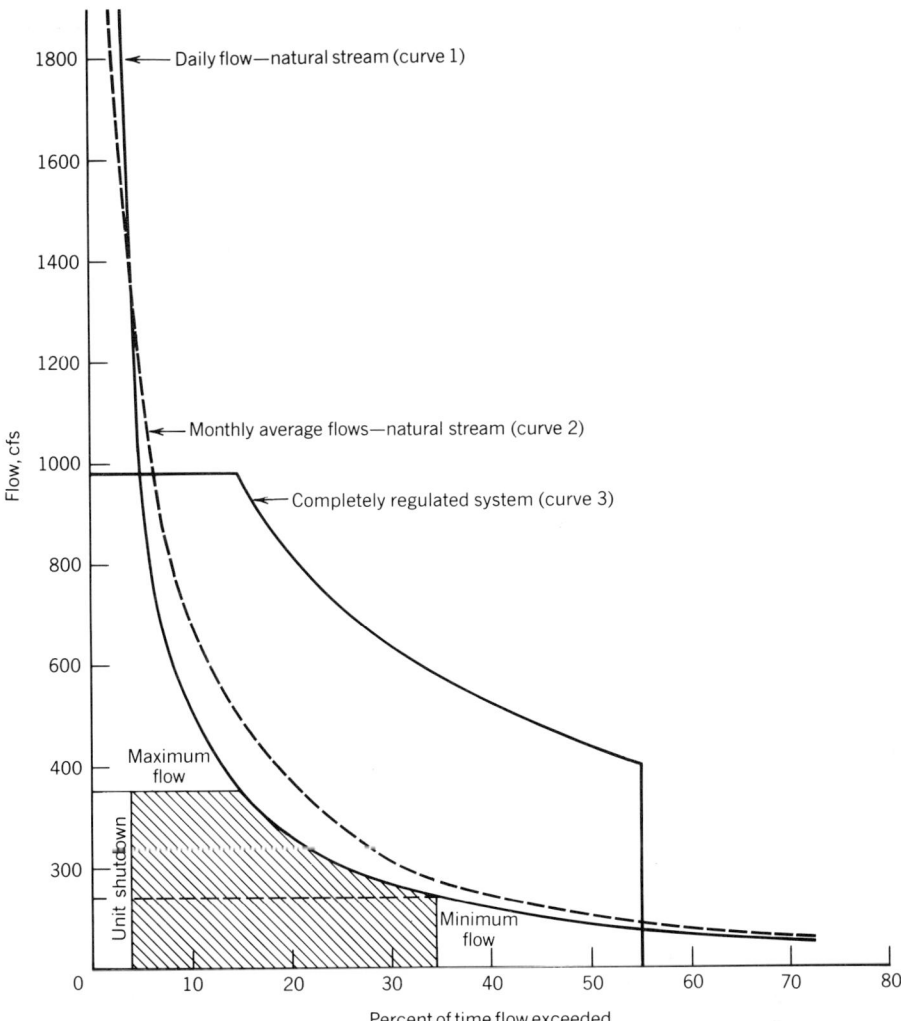

Fig. 10.1-3 Typical flow duration curves.

for successive classes. This summation is expressed as a percentage of the total number of recorded hours, days, or months. The curve is then plotted using the lower range limit of each class as the ordinate and the cumulative percentage of total events as the abscissa.[3]

Figure 10.1-3 shows a number of typical flow duration curves. The y axis indicates the values of flow; the x axis indicates the percent of time that any given flow is equalled or exceeded. The area under the curve represents flow volume. These curves can be developed for hourly, daily, or monthly average flow rates. A single curve may be developed for the entire period of record at sites where available head remains relatively constant throughout the year. Twelve separate curves (one for each month of the year) would be developed where the head variation from month to month is significant.

In Fig. 10.1-3, curve 1 is a flow duration curve for a natural stream based upon average daily flows. In this case it can be seen that 100 cfs is exceeded 45% of the time, while 500 cfs is exceeded only 10% of the time.

Curve 2 is computed on a monthly average flow basis for the same stream. Comparison of curves 1 and 2 indicates that the use of the monthly curve would result in an overestimate of available flow volume of 17%. This overestimate generally varies from 15–50% of the flow rates depending on the day-to-day variations of the flow at a given site. However, where daily flow data are not readily available, monthly flow records may be used for preliminary evaluation purposes.[3]

Curve 3 is a flow duration curve for a water supply system. This curve demonstrates flow pattern differences between a regulated system and a natural stream.[3]

Stream flow data are generally available from a variety of sources. The U.S. Geological Survey (USGS) maintains the most comprehensive records of stream flow data for the U.S.A. For each gaging station, actual daily flows, average annual flow, and flood-frequency analyses are generally available. The USGS has flow duration curves available that are derived from daily flow analyses. The WATSTORE water data storage and retrieval system, maintained by the USGS, is also a source of data where direct computer access is available. Hydrologic data are also maintained by a number of other agencies including state departments responsible for water resources, the U.S. Army Corps of Engineers, the U.S. Soil Conservation Service, the U.S. Bureau of Reclamation, local irrigation districts, water companies, local utilities, and private industries. Similar agencies provide data in other countries.[3,14]

Long-term stream flow data are seldom available at a proposed new hydropower site. In this case records for the nearest gaging stations with similar drainage basin areas and hydrologic characteristics can be used, preferably records for locations immediately upstream or downstream of the proposed site. These flows then may be proportioned by the ratio of their drainage areas above or below the gaging station to the site drainage area in order to transpose the data. Transposition may also be made between adjacent watersheds with similar characteristics. Data on drainage areas may be available from federal or state water resource agencies or may be determined by direct measurement of USGS topographic maps. To obtain more precise analyses, adjustments can be made in the transposed data to correct for significant evaporation losses, diversions, and seepage losses.[3]

Where the period of stream flow data is limited or where no adjacent gaging station appears suitable, it may be possible to develop a synthetic record using other similar areas or precipitation stations. The use of short-term records to generate long-term data is possible through stochastic or synthetic hydrology.[3]

The stream flow data will also include water level at the gaging station. Using this information and the channel characteristics, it is possible to determine a relationship between water level and rate of flow at a location just downstream of the site. This water level is to be the tailwater level of the potential hydroelectric plant and the relationship is known as the tailwater rating curve. The tailwater rating curve is used in conjunction with the headwater elevation to determine the head on the site under any given operating conditions.

Reservoir area and volume curves for varying headwater elevations may be prepared using available topographic maps for the project area. The reservoir volume curve is used in the simulation of plant operations involving the use of storage for flow regulation.

Conceptual Design

Once the hydrologic data have been compiled, alternatives can be considered. The steps in evaluating an alternative are:

1. Select a hydraulic turbine capacity, a dam height, and a general arrangement for the project and estimate the cost of the development.
2. Determine the power generation (capacity and energy potential) by one of three methods: approximation method, flow duration curve method, or sequential analysis method.

Hydraulic Turbine Capacity. Based on the available head, a preliminary selection of a suitable type (or types) of turbine can be made (refer to Section 10.2-1). The physical dimensions of the hydraulic turbines and generators normally control the size of the powerhouse. A trade-off must be made between the specific speed of the turbine and the amount of excavation required. A high specific speed results in a higher output for a given size of turbine, and thus results in a smaller powerhouse. However, a higher specific speed also requires a deeper setting to prevent cavitation (for discussion of cavitation see Section 10.2-2). This results in greater excavation costs.

The number of units the plant will have must also be chosen. A multiple-unit plant will generally cost more than a single-unit plant but only a portion of the capacity is lost if one unit in a multiple-unit plant is out of service. The high cost of replacement energy may thus make several smaller units preferable to one large unit. In addition, in a multiunit plant, the units can be of different types or sizes to accommodate the variations in flow. Thus, they can use a greater percentage of available flow. Furthermore, multiple-unit plants generally have a greater plant efficiency because each unit can be operated at or near its best efficiency point.

For preliminary studies an overall equipment efficiency of 80–85% is a reasonable assumption. This number assumes efficiencies in the high 90% range for the generators, transformers, and gear boxes, and somewhat lower peak efficiencies for the hydraulic turbines. It also reflects the fact that turbines will operate at less than peak efficiencies with typical variations in head and flow. To obtain the overall plant efficiency, variations in tailwater level must be accounted for and head losses in the water conduits must be established.

Water Conduits.   Trade-offs must be made in designing the plant's water conduit system. Larger conduits result in smaller energy losses but in most cases are more expensive to build than smaller conduits. An equation has been developed for the preliminary selection of an economical penstock diameter,[15,16] which can be used for preliminary studies.

Dam Height.   As part of the site evaluation, a tentative dam height must be selected based upon the site conditions. The greater the dam height, the higher the available head for power generation. A higher dam also increases the storage and, therefore, reduces the spillage to provide more available flow and a greater energy output. However, the cost of the dam increases significantly as height increases, as discussed in Section 10.3.

Civil Features.   The site access for construction and future maintenance, the channel conditions downstream of the dam, the foundation conditions, and the location of the dam spillway must be given consideration. The major civil features of the power plant include the excavation and construction of a powerhouse and the temporary works, including cofferdams, required during construction.[3]

Power Generation.   After the general arrangement, including turbine type and dam height, has been tentatively decided upon, the benefits of developing the site must be calculated. This can be done in one of three ways, depending upon the type of study being performed. Reconnaissance or appraisal studies generally use an approximation method. If the reconnaissance or appraisal study indicates that the site has a reasonable possibility for development, a feasibility study would then be performed. For run-of-river plants the flow duration curve method would be used since the plant must use or spill whatever water is flowing in the river at any given time. For plants with storage capability the sequential analysis method would be used since the daily, monthly, and seasonal sequences of river flow would be important.

Approximation Methods.   For reconnaissance or appraisal studies, a quick estimate of the power potential, as well as a starting point for sizing equipment, is obtained by solving the average water power equation using average annual flow and average available head.

Another approximation method is the use of an approximate flow duration curve. The shape of such a curve might be available for a region or for a similar site within the region. The flow rates would then be established by using the ratio of drainage basin areas. Caution must be exercised in using this method because storage and local features of the river might distort the flow duration curve of the site. Therefore, the approximate flow duration curve may be reasonable only if developed for a location in close proximity to the site or for determining long-term variations.

Flow Duration Curve Method.   For a feasibility study the use of a flow duration curve from the actual site is more appropriate than use of the average flow or a curve for the entire region. Evaluation of a run-of-river site using the flow duration curve method is carried out as follows:

1. A turbine–generator capacity is selected to accommodate a specified water flow. For a first trial, this flow is typically one which is exceeded 10–25% of the time (see Fig. 10.1-3).
2. The above selection gives the maximum flow that can be put through the hydraulic turbines. A minimum flow, which is some fraction of the maximum flow for acceptable operation, must also be considered. This minimum flow can be as high as 50% of the maximum flow or as low as 20%. A figure of 40% is reasonable for feasibility studies, and this establishes the lower limit of operation on the flow duration curve (see Fig. 10.1-3).
3. For any given flow on the flow duration curve, the available head on the turbines is determined by considering the tailwater effect and the head losses in the conduits. If high tailwater reduces the available head below acceptable limits for operation, the unit may be shut down during high flows, as shown in Fig. 10.1-3.
4. Power output for any point on the flow duration curve can be calculated using the water power equation with the appropriate values of head and flow, together with efficiencies obtained from typical turbine efficiency curves.
5. The total energy production is then determined by dividing the flow duration curve into representative sections based upon percent of time. The energy for each section is the power output of the section multiplied by the time period. The total energy is the sum of these individual energy increments.

The above process is repeated with a different turbine–generator capacity, number of units, dam height, water conduit size, or other combination of applicable parameters. The value of the energy produced is then compared with the associated cost.

Sequential Analysis Method. When storage capability is included, the power production analysis of a site or combination of sites involves simulated plant operation over a selected sequence of loading and hydrologic conditions. Evaluation of a site using the sequential analysis method is as follows:

1. A turbine–generator capacity is selected based on the anticipated manner of operation—base load or peaking.

2. An initial water level in the reservoir is selected. For each time increment, the natural inflows are added to the reservoir storage and the outflows, including turbine discharge, spillage, and minimum flow releases are subtracted. The net volume change is used to compute the new initial water level for the next time increment.

3. Energy production is determined using the head and flow through the turbines as discussed in relation to the flow duration curve method.

4. Dependability of power generation is established by identifying periods when power production must be reduced due to lack of water in the reservoir or excessively low tailwater level, or other restrictions.

The value of the energy produced is then compared with the associated costs. The above process is repeated with different parameters until an optimum is achieved.

The appropriate time interval for a sequential study varies according to project type and system. Long-term analysis of storage projects may be performed using monthly data, whereas 1- or 2-h time increments may be appropriate for weekly studies of projects with daily storage. The travel time of the water between reservoirs may be an important consideration in selecting the time interval when available storage capacities are significantly different.

In performing the mathematical simulation of operation, it must be recognized that some of the water in storage is not available for power generation. This water, known as dead storage, is located below the lowest reservoir outlet. Sometimes, water above this level cannot be released due to regulatory requirements. This is called inactive storage. Additionally, there may be conservation storage and flood control storage that might not be usable should there be a demand for power generation. A further discussion of reservoir hydraulics is provided by Koelzer.[12]

A project sequential analysis, whether for a critical week or a monthly analysis, can be performed by basic numerical calculations using storage–elevation–area curves, tailwater rating curves, and storage–evaporation–month of year tables. These curves can be combined to develop a kW/cfs nomograph that can then be used in sequential routing calculations.[17]

There are also numerous computer programs available for use in performing sequential analyses. The U.S. Army Corps of Engineers (COE) has two models available for use in hydropower studies: HEC-3 and HEC-5C. HEC-3 does monthly routing and investigates system power operation, yield maximization, peaking capability, evaporation, and power benefits. It is particularly applicable for determination of maximum firm yield in an interrelated system of reservoirs. HEC-5C has, in addition to the HEC-3 capabilities, six routing methods and can handle flood routing and pumped storage. This program is more applicable to system analyses requiring river routing capability using a shorter time interval. Engineers also have developed and documented similar computer programs for their respective uses. A more detailed discussion of the COE computer programs is contained in ref. 17.

Regulatory Considerations

The licensing process has a major impact on the development of a hydropower project in the United States. In addition to several federal agencies involved, many state and local agencies are also involved. Their requirements are subject to continuous updating and revision. This section briefly describes regulations in effect as of 1982. Further discussion on this subject is available in ref. 18. Requirements for a specific project should be determined by a review of the most current regulations.

FERC Licensing Requirements. The primary federal agency involved in the regulation of hydroelectric projects including dams, powerhouses, water conduits, reservoirs, and transmission lines in the United States is the Federal Energy Regulatory Commission (FERC). The predecessor agency (until 1977) was the Federal Power Commission (FPC). This agency regulates the construction and operation of hydroelectric projects under Part I of the Federal Power Act (FPA) and the sale of electricity in interstate commerce under Part II of the act. FERC has jurisdiction over hydropower sites where the project is owned and operated by a nonfederal entity: is located on a navigable waterway or the water discharged from the project affects a navigable waterway; affects interstate or foreign commerce; uses federal lands; or uses surplus water from a federally owned dam. FERC also has jurisdiction if power output is marketed to or interconnected with an interstate transmission system. The issue of jurisdiction may be resolved by an informal letter of opinion or the filing of a declaration of intent to construct a dam or other project works.[18]

Generally, a large hydropower development under FERC jurisdiction will require a license. However, Congress has authorized FERC to exempt some projects from the licensing process under certain conditions. These exemptions generally involve small-capacity projects at existing facilities and at facilities that are to be constructed for purposes other than power production, such as water distribution systems. The basic intent of these exemptions is to facilitate the installation of small-capacity electric generating equipment, where such installation would not materially change the site character from that which would have existed without the generating facilities.

For projects requiring FERC licenses, two categories of licenses are available, major licenses and minor licenses, depending upon the generating capacity. A license application for a large project requires some detailed engineering and environmental studies. Smaller projects, which in some cases can file a short-form application, still require extensive detail to describe the project. For this reason there exists an optional intermediate step known as the preliminary permit. The permit allows a potential developer to study the project, as well as to determine the environmental and economic feasibility, while maintaining priority over other developers for consideration in licensing the particular site.

The preliminary permit application documentation includes a project description, a study plan and schedule, estimated energy production, confirmation of financial ability to perform the study, and maps identifying the project boundaries. The preliminary permit review process usually takes 3–6 months. The permit term is generally 2 years at existing sites and 3 years at new sites. In order to retain priority status, the developer must apply for a license prior to expiration of the permit.

The detailed requirements for exemption, permit, and license applications are described in FERC regulations and FERC has published a detailed guide to the licensing process.[19]

At several points in the licensing process, the applications are open for public comment or formal intervention at FERC proceedings. Anyone interested in a site may comment on a permit or license application for a project during the comment period. Comments and formal interventions by persons interested in a site or affected in any way by issuance of a FERC license for a site are reviewed by the FERC staff and commission. FERC makes the final decision to hold or not to hold a hearing, and the results are published in the Federal Register, along with FERC's review of the application and final articles of license. A FERC license, when granted, can be in effect for up to 50 years.[18]

Fish and Wildlife Requirements.   The Fish and Wildlife Coordination Act requires a fish-and-wildlife plan be prepared for each license application. FERC coordinates the efforts of the Department of Interior's Fish and Wildlife Service with other appropriate state agencies. On exemption applications, the U.S. Fish and Wildlife Service may comment on the proposed exemption, and comments made with respect to requirements for minimum flows and fish passage must be included in the exemption certificate.

Army Corps of Engineers Requirements.   The Army Corps of Engineers (COE) responsibilities for hydroelectric projects are defined in regulations prepared pursuant to the Rivers and Harbors Act of 1889. Section 9 of this act requires COE approval prior to construction of any dam or dike across any navigable waterway. The approval may be in the form of a permit or, if the project is under FERC jurisdiction, the COE would review the plans for compliance with COE rules as part of the licensing process.

Dredge and fill operations associated with a hydro project require a permit from the COE. A permit is necessary to locate a structure or to excavate or discharge fill material in waters of the United States. The permit investigation process requires data on the location and nature of the project and details on the dredged material or any other feature of the project that is pertinent to COE's rules and regulations. The COE issues permits, subject to Environmental Protection Agency (EPA) approval, for the discharge of fill material and the use of disposal sites for dredged material.[17]

State and Local Requirements.   Throughout the United States, the review of licenses and permits differs from state to state. Resource agencies, which coordinate most of the departments that need to be contacted, exist in some states, facilitating a statewide review of a proposed facility. Other states have separate agencies that are contacted individually to ensure compliance with state regulations. The state agencies that must be contacted for hydro licensing include those with responsibility for dam safety; energy development, planning, and research; historical preservation; fish and wildlife preservation; flood control; water quality control; land use; water rights and resources; and Public Utilities Commissions.

Local agencies are principally involved in taxation, planning, property aquisition, and zoning. Local ordinances also affect such areas as construction, employment, and traffic.

REFERENCES

10.1-1   Inter-Agency Committee on Water Resources, *Glossary of Important Power and Rate Terms, Abbreviations, and Units of Measure*, U.S. Government Printing Office, Washington, D.C., 1965.

10.1-2 J. C. Stevens and C. V. Davis, "Hydroelectric Plants," *Handbook of Applied Hydraulics*, 3rd ed., McGraw-Hill, New York, 1970.

10.1-3 L. M. Pruce, "Power from Water," *Power Magazine* **124**(4), April 1980.

10.1-4 R. Linsley and J. Franzini, *Water Resources Engineering*, 3rd ed., McGraw-Hill, New York, 1979.

10.1-5 M. Vasilescu, *Physical Layout and Hydraulic Design*, 1980 Lecture Series on Small Scale Hydropower, BSCES/ASCE, Boston, 1981.

10.1-6 K. G. Laurence and W. E. Shaughnessy, *Case History—1.4 MW South Consolidated Hydro Plant*, Stone & Webster Engineering Corporation, Denver, 1982.

10.1-7 E. Mosonyi, *Water Power Development, Volume Two: High-Head Power Plants*, Hungarian Academy of Sciences, Budapest, 1965.

10.1-8 T. Yoshigawa, "Trends in Hydroelectric Generating Equipment Technology," *Hitachi Review* **28**(4), 1979.

10.1-9 R. H. Resch and D. Predpall, "Pumped Storage Site Selection: Engineering and Environmental Considerations," Engineering Foundation Conference on Converting Existing Hydro-electric Dams and Reservoirs into Pumped Storage Facilities, Rindge, NH, 1974.

10.1-10 Bureau of Reclamation, Combined Federal–Private Sector Study Tour Team, *Report on Pumped Storage Technology in Europe*, U.S. Department of the Interior, 1974.

10.1-11 R. D. Harza, "Pumped Storage," *Handbook of Applied Hydraulics*, 3rd ed., McGraw-Hill, New York, 1970.

10.1-12 V. A. Koelzer, "Reservoir Hydraulics," *Handbook of Applied Hydraulics*, 3rd ed., McGraw-Hill, New York, 1970.

10.1-13 F. L. Brennan and R. J. Conlon, *Hydroelectric Pumped Storage*, IEEE, New York, 1976.

10.1-14 P. Manley, *Hydrologic Analysis for Power Potential*, 1980 Lecture Series on Small Scale Hydropower, BSCES/ASCE, Boston, 1981.

10.1-15 G. S. Sakaria, "Economic Penstock Diameters: a 20 year Review," *Water Power and Dam Construction*, **33**(11), November 1979.

10.1-16 G. S. Sakaria, "Economical Diameter for Penstocks," *Water Power*, September 1958.

10.1-17 U.S. Army Corp of Engineers, *Feasibility Studies for Small Scale Hydropower Addition—A Guide Manual*, U.S. Army Corp of Engineers, 1979.

10.1-18 P. W. Brown, *Federal and State Licensing Process 1980 Lecture Series on Small Scale Hydropower*, BSCES/ASCE, Boston, 1981.

10.1-19 Federal Energy Regulatory Commission, Office of Electric Power Regulation, *Procedures to Apply for Hydropower Licenses and Permits*, Washington, D.C., March 1981.

10.2 HYDRAULIC TURBINES AND ASSOCIATED EQUIPMENT

Fred R. Harty, Jr.

10.2-1 Types of Turbines

There are several types of hydraulic turbines for generating electric power, each particularly suited to certain combinations of head, output, and site arrangement. In some cases the requirements for a specific application can be met by two or more types of turbines. Furthermore, within each type there are several possible variations in rotational speed, dimensions, and water passage configuration, which need to be considered for a given application.

Large hydroelectric developments can justify the cost of the custom designs that are needed to obtain maximum power production from the available water. For smaller developments the cost of custom design may not be justified by the slightly increased output relative to that obtained from standardized equipment. However, the user should still compare prices and performance of many different custom and standard turbines to determine which one most closely fits the requirements of the site.

Turbines are generally separated into two broad categories—impulse turbines and reaction turbines. Reaction turbines are further subdivided into Francis, Deriaz, cross-flow, fixed-blade propeller, and Kaplan propeller turbines. Within each category there are many variations in design and arrangement, depending upon the manufacturer and the desired shaft orientation and flow path.

The specific speed of a turbine describes the rotational speed with respect to the power output and head, as discussed in Section 10.2-2. In general, the higher the head, the lower the acceptable specific speed. Francis turbines have lower specific speeds than do propeller turbines and are used over the intermediate head range—from heads of less than 30 m to over 300 m. For heads in the upper range of Francis turbines and on into heads of several thousand meters, the impulse design has its primary range of application, although there is essentially no technical lower head limit for this type of turbine. For the lower head ranges, fixed-blade and Kaplan propeller turbines predominate, although Deriaz and cross-flow designs have also been used at medium and low heads. A representative chart of typical

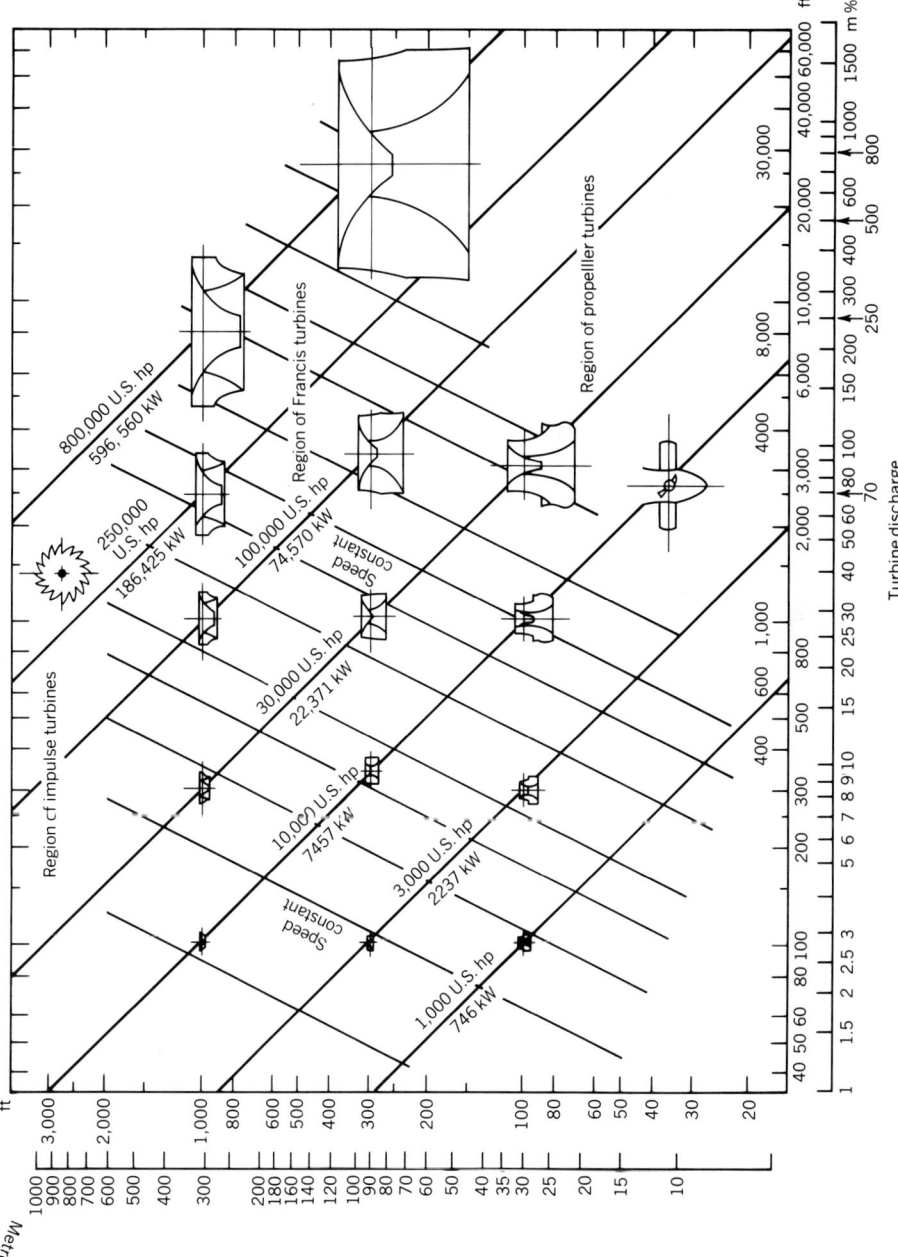

Fig. 10.2-1 Application diagram for types of hydraulic turbines. (Courtesy of Bureau of Reclamation.)

Region of impulse turbines

Region of Francis turbines

Region of propeller turbines

Turbine discharge

800,000 U.S. hp
596,560 kW

250,000 U.S. hp
186,425 kW

100,000 U.S. hp
74,570 kW

30,000 U.S. hp
22,371 kW

10,000 U.S. hp
7457 kW

3,000 U.S. hp
2237 kW

1,000 U.S. hp
746 kW

Speed constant

Speed constant

1155

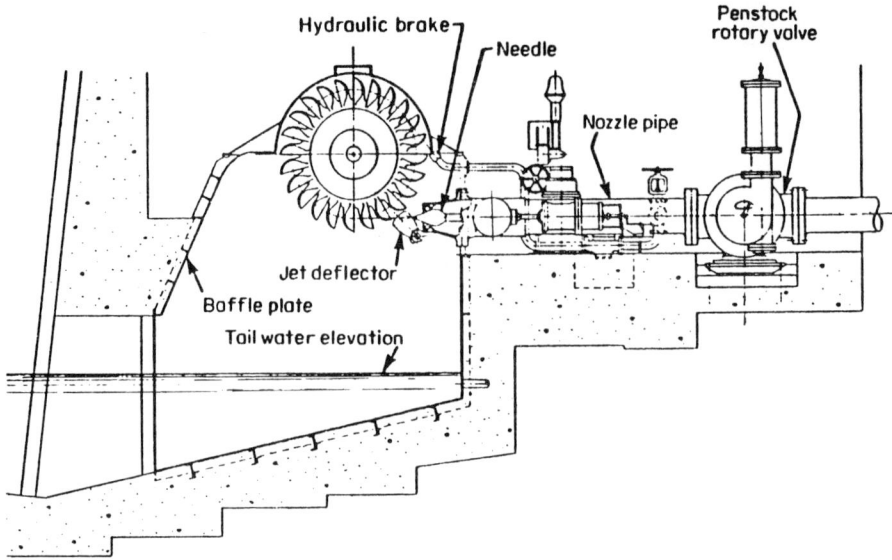

Fig. 10.2-2 Section through a horizontal impulse turbine.[1] (Courtesy of Allis Chalmers and McGraw-Hill Book Company.)

ranges of use for these turbine types is given in Fig. 10.2-1. There can, however, be considerable variation from the ranges indicated in the figure, and such charts should therefore be considered as approximate only.

Impulse Turbines

With an impulse turbine the water pressure in the conduit system is discharged as a high-velocity jet directed at the buckets on the outer periphery of the rotating turbine runner. The jets discharge into the atmosphere (Figs. 10.2-2 and 10.2-3).[1]

There can be one or more nozzles operating simultaneously on the turbine runner, with all jets controlled by synchronized needle valves. It is also possible to have backward-pointing nozzles for use in fast braking upon load rejection.

Water flow is controlled primarily by means of the needle valves. In the event of a load rejection or a sudden and large reduction in load, a set of jet deflectors can divert the water from the runner, thereby minimizing the overspeed. The needle valves would then close at a relatively slow rate, thereby limiting the overpressure in the water conduit.

A variation of the impulse turbine is the cross-flow turbine. The runner of a cross-flow turbine is like a cage with the blades fixed on the outer periphery like the bars of the cage. Water is directed to the upper portion of the runner, acts on the uppermost blades, falls through the center, and acts on the lowermost blades as it exits the runner. Water flow is controlled by adjustable guide vanes. These guide vanes can be divided into two sections if operating conditions require, and for most installations the ratio of the lengths of the two guide vane sections would be 1 : 2. This allows the smaller section to be used for low discharge, the larger section for average discharge, and both sections for high discharge. The divided guide vane arrangement makes it possible to obtain a flat efficiency curve over a wide range of head and discharge.

Reaction Turbines

In a reaction turbine the turbine runner is installed in a closed conduit system and the power is transmitted through a combination of velocity change and pressure effect on the runner blades.

Reaction turbines are divided into two broad catagories—Francis turbines and propeller turbines—with additional variations and combinations of the two.

Francis Turbines. A Francis turbine (Figs. 10.2-4 and 10.2-5) resembles a centrifugal pump, with an added feature of a series of overlapping wicket gates to control the water flow. In modern applications the water flows inward toward the shaft axis and exits axially from the center of the runner.

The Francis pump–turbine is by definition a combination pump and turbine, usually accomplishing the two modes of operation by reversing the direction of rotation. The pump–turbine design is a

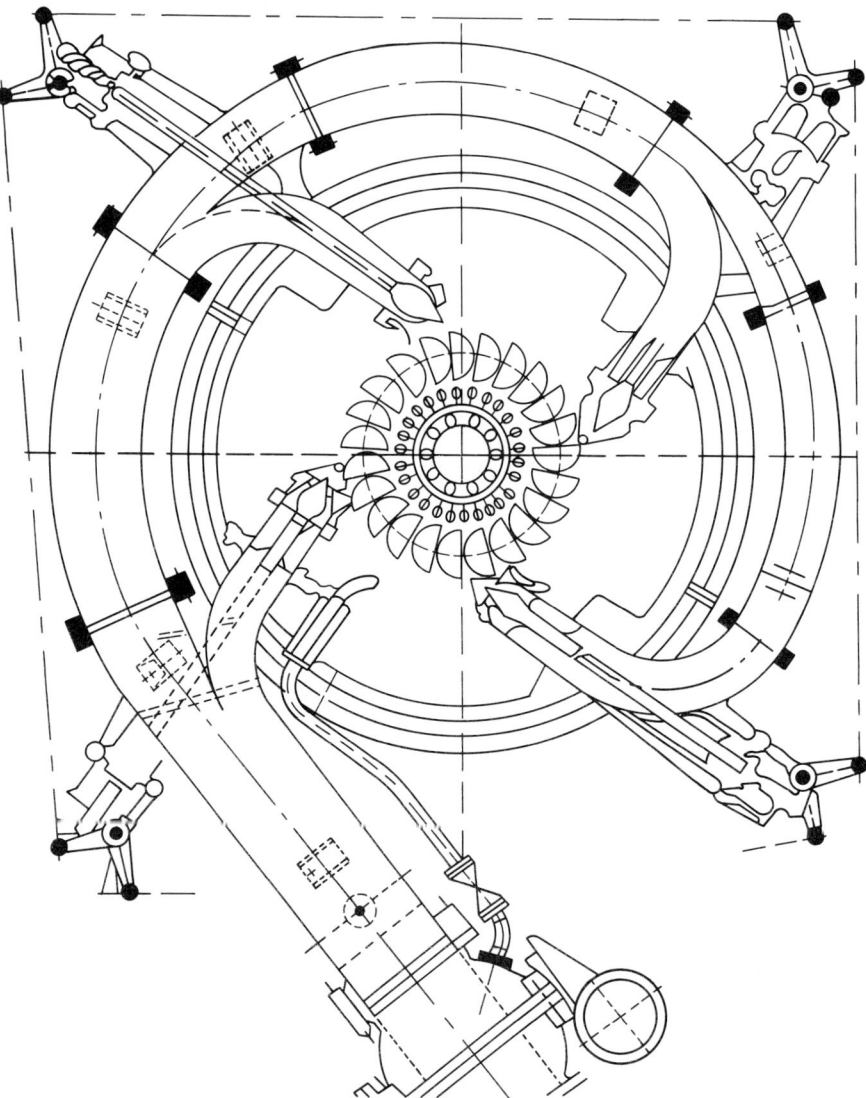

Fig. 10.2-3 Section through a vertical shaft impulse turbine. (Courtesy of Dominion Engineering Works, Ltd.)

compromise between the two functions and provides somewhat lower efficiency than either a standard pump or a standard turbine. It does, however, provide an attractive cost saving over the purchase of two separate machines.

A Deriaz turbine (Fig. 10.2-6)[1] has a flow arrangement that is halfway between the Francis and propeller arrangements since the flow enters the runner at an angle that is part way between radial flow and axial flow. Deriaz turbines can also have adjustable runner blades. Furthermore they can be designed as pump–turbines.

Propeller Turbines. The runner of a propeller turbine is similar to a ship's propeller, where the water flow is axial through the runner section (Fig. 10.2-7).

Propeller turbines are further identified as fixed blade (the blade angles are permanently fixed) and Kaplan (the blades are mounted on pivots and are adjustable). In a Kaplan turbine the runner blades can be continuously adjusted to approach optimum efficiency under varying conditions of head and

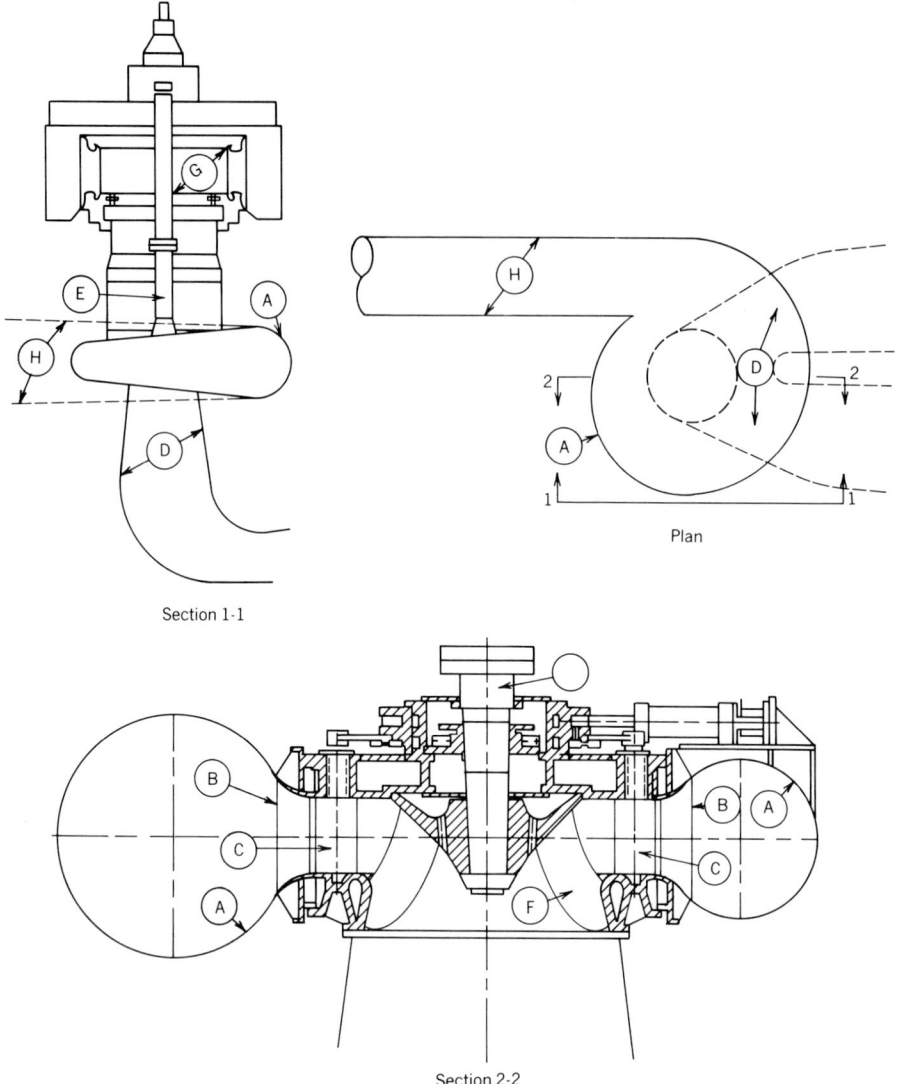

Fig. 10.2-4 Vertical-shaft Francis turbine: (A) spiral case, (B) stay vane, (C) wicket gate, (D) draft tube, (E) main shaft, (F) runner, (G) generator, (H) penstock. (Courtesy of James Leffel & Co.)

discharge. The use of adjustable blades provides for a relatively high efficiency over a wide range of outputs rather than the sharp peak at one output which exists with a fixed-blade propeller turbine.

One variation of the propeller turbine is the bulb turbine. In a bulb turbine (Figs. 10.2-8 and 10.2-9), the water flow is essentially axial from intake to draft tube exit and the generator is enclosed in a watertight "bulb" suspended in the water passage. Bulb turbines, as compared to conventional vertical-shaft propeller turbines, generally have a more constant efficiency over a wider range of operating conditions. They can also have higher specific speeds for a given set of operating conditions. The bulb turbine can be equipped with any combination of fixed and adjustable wicket gates and runner blades. If adjustable wicket gates are not provided, the startup and shutdown could be controlled by a separate gate or valve. A more precise degree of speed regulation would be available from the more conventional governor and wicket gate arrangement. However, precise speed control is often not required during startup of small-capacity units of this type because of the relatively small inertia of the generator. Furthermore, these units are usually connected to large systems where frequency control is established by other units. Therefore, speed control is not required when the unit is under load.

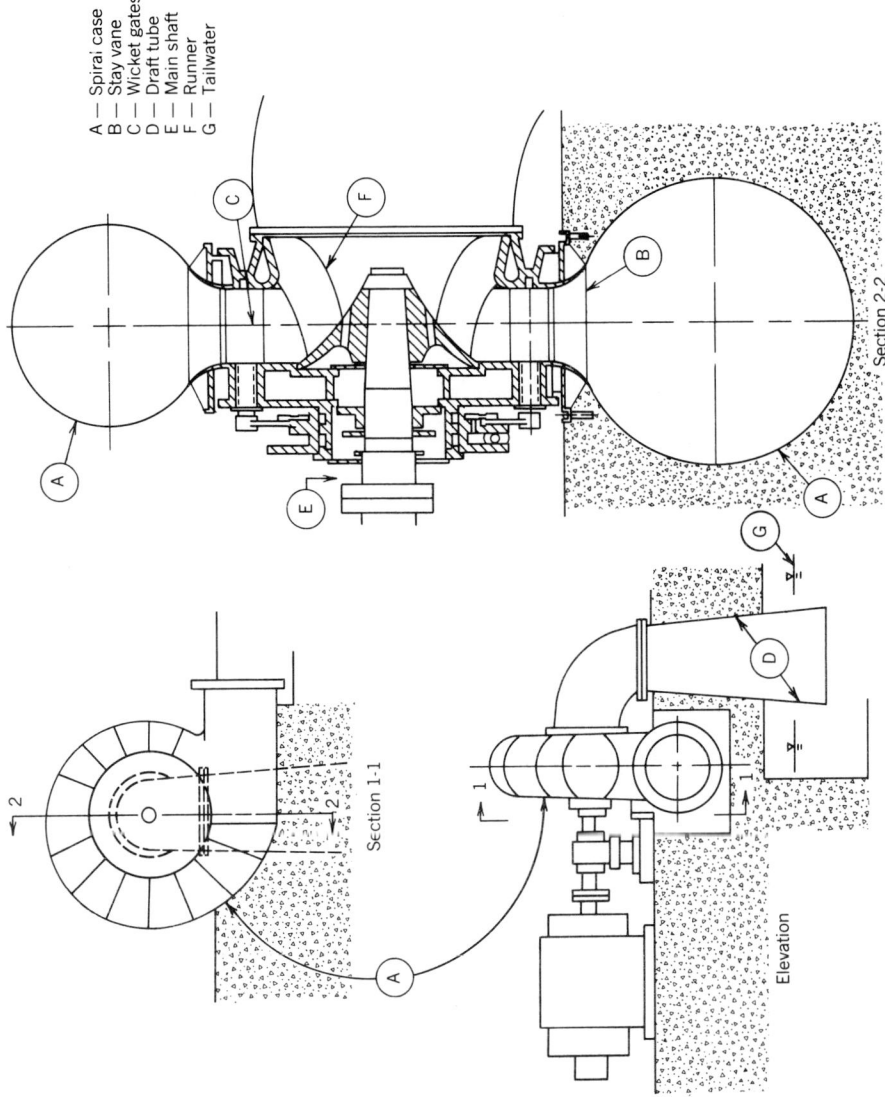

A — Spiral case
B — Stay vane
C — Wicket gates
D — Draft tube
E — Main shaft
F — Runner
G — Tailwater

Section 2-2

Section 1-1

Elevation

Fig. 10.2-5 Horizontal-shaft Francis turbine: (A) spiral case, (B) stay vane, (C) wicket gates, (D) draft tube, (E) main shaft, (F) runner, (G) tailwater. (Courtesy of James Leffel & Co.)

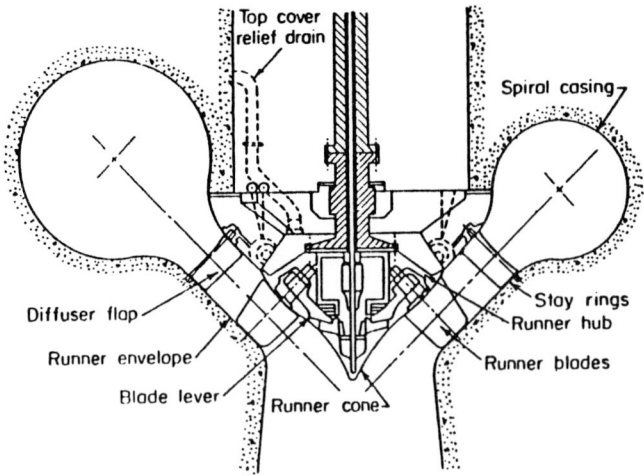

Fig. 10.2-6 Sectional elevation of a diagonal-flow (Deriaz) turbine.[1] (Courtesy of Allis Chalmers and McGraw-Hill Book Company.)

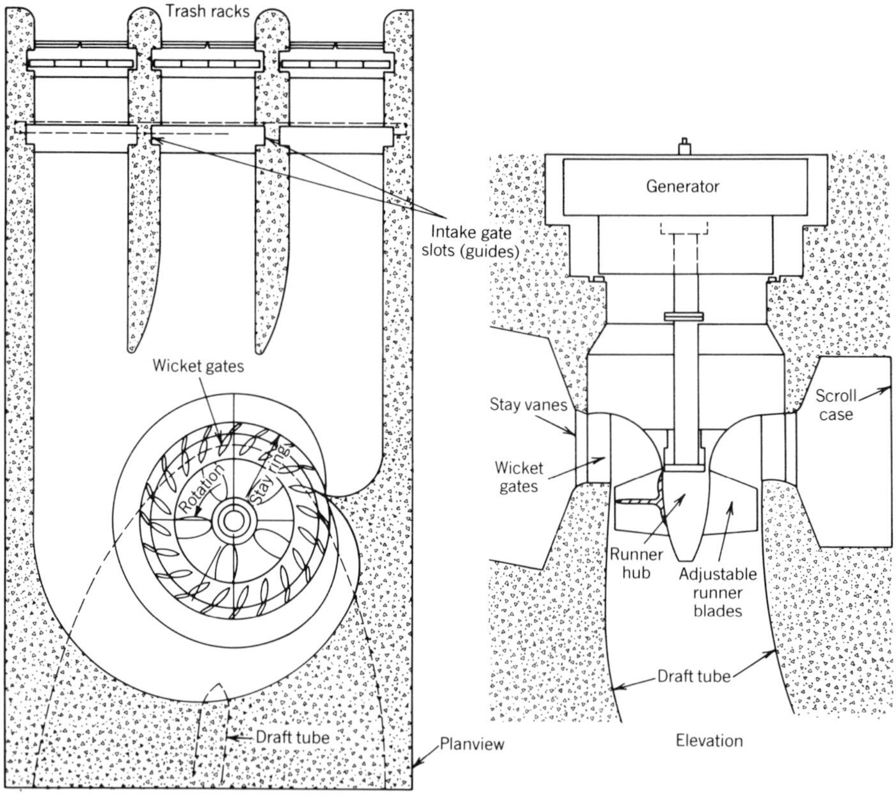

Fig. 10.2-7 Vertical-shaft Kaplan turbine.

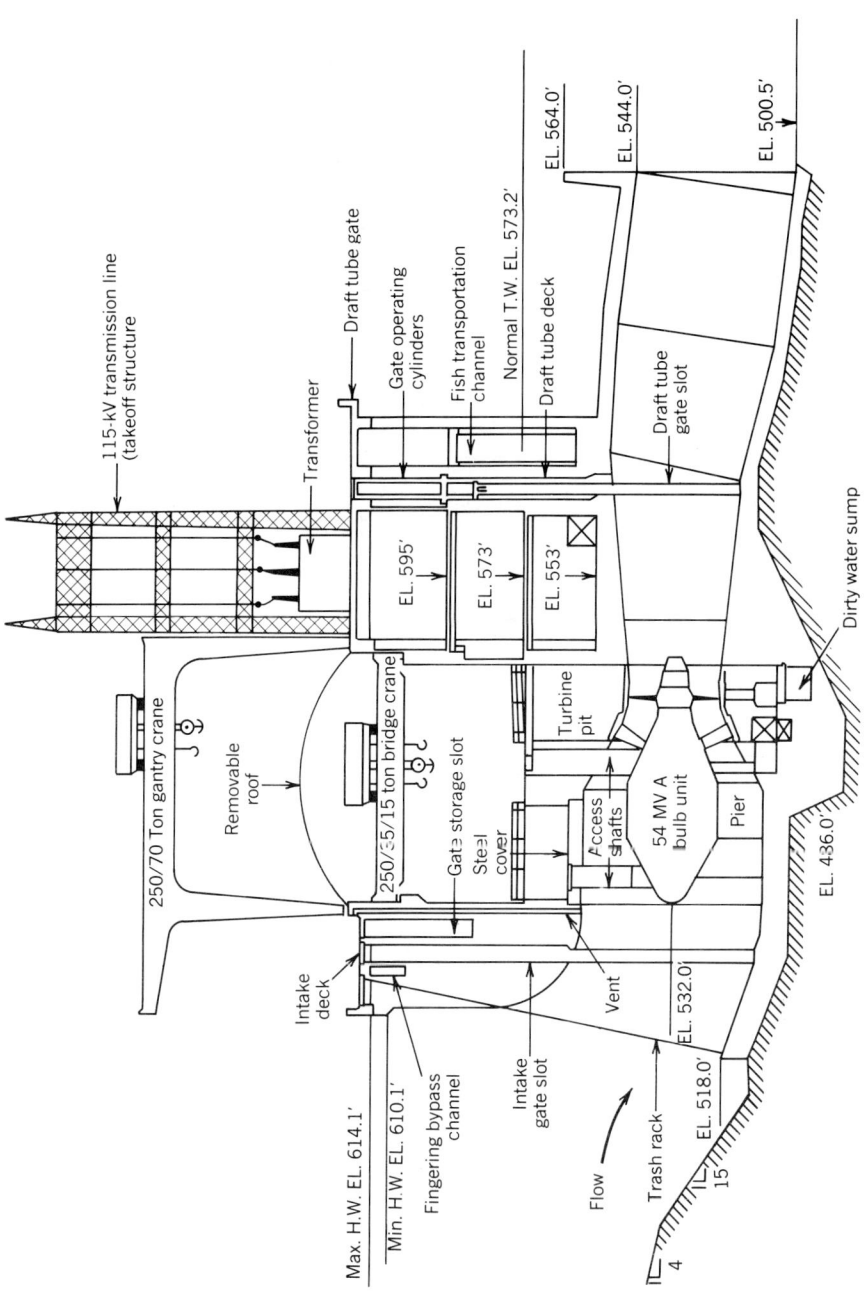

115-kV transmission line (takeoff structure

Draft tube gate

Gate operating cylinders

Fish transportation channel

Normal T.W. EL. 573.2'

Draft tube deck

EL. 564.0'

EL. 544.0'

EL. 500.5'

Transformer

Draft tube gate slot

250/70 Ton gantry crane

EL. 595'

EL. 573'

EL. 553'

Removable roof

250/55/15 ton bridge crane

Gate storage slot

Steel cover

Turbine pit

Draft tube deck

Dirty water sump

Intake deck

Access shafts

54 MV A bulb unit

Pier

EL. 436.0'

Vent

Max. H.W. EL. 614.1'

Min. H.W. EL. 610.1'

Fingering bypass channel

Intake gate slot

EL. 532.0'

Flow

Trash rack

EL. 518.0'

15

4

Fig. 10.2-8 Section through a bulb turbine powerhouse (Rock Island).

1161

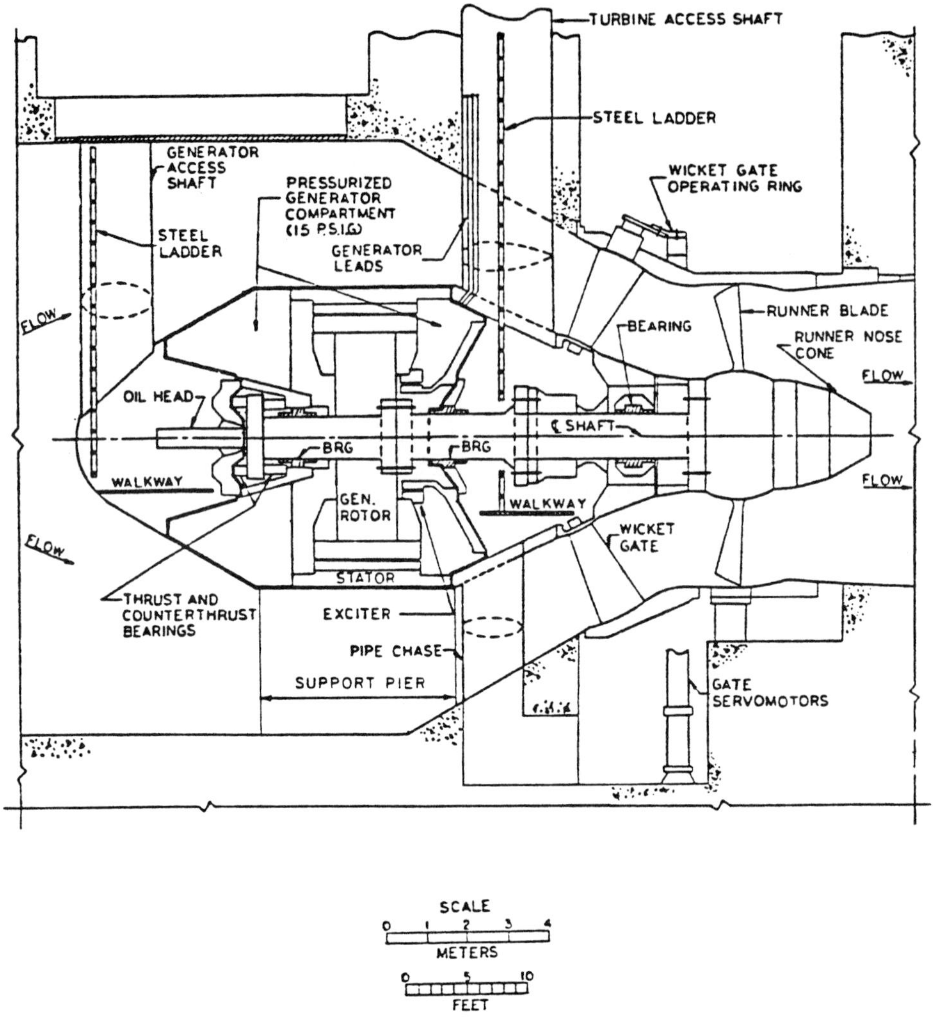

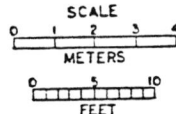

Fig. 10.2-9 Bulb turbine unit.

Another variation of a propeller turbine is the tubular turbine, which is often used for small-capacity low-head applications. In a tubular turbine the water passage makes a downward bend to allow the turbine shaft to extend through the boundary of the water passage, either horizontally or at a slant. This allows the generator to be located outside the water passage. For very small units a right-angle gear is sometimes used to allow the generator to be located outside of a straight water passage. Combinations of adjustable and fixed wicket gates and runner blades are available for tubular turbines. As with the bulb turbine, this allows matching of efficiency characteristics with the available discharge and head for applications having a wide variation in these parameters. The combination of adjustable runner blades and adjustable wicket gates will give the best efficiency over varying conditions of head and flow. Where simplified operating conditions are available, either the wicket gates or the runner blades can be fixed as required. Adjustable runner blades with fixed wicket gates are appropriate for constant heads with variable discharges. Fixed runner blades with adjustable wicket gates can be used for constant discharges with variable head. Another combination includes fixed runner blades and fixed wicket gates. This requires a nearly constant water discharge and head, and the flow cannot be regulated. Operation is possible on a continuous or intermittent basis, but only at full load. Figure 10.2-10 shows a representative efficiency variation as a function of output for these four combinations of wicket gates and runner blades for a propeller turbine.

The rim-generator/turbine is similar to the bulb turbine except that the generator rotor is mounted on the outer tips of the turbine runner blades. The turbine runner forms the generator spider. A

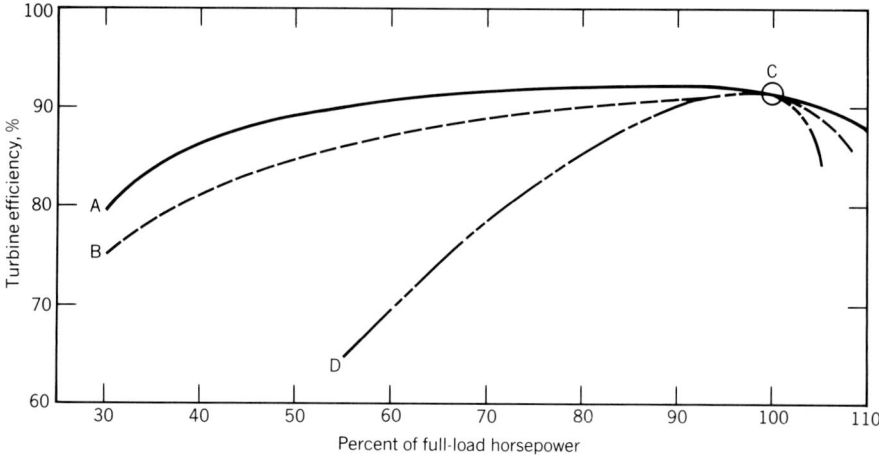

Fig. 10.2-10 Representative efficiency curves propeller-type turbine: (A) adjustable blades and adjustable gates, (B) adjustable blades and fixed gates, (C) fixed blades and fixed gates, (D) fixed blades and adjustable gates.

rotating seal is provided on the water passage side of the rotor to prevent entry of water into the generator compartment.

Pumps as Turbines. A pump is in principle a reaction turbine running in reverse, and for some applications the use of standard pumps rotating in reverse as turbines can offer some economic advantages when compared to equipment specifically designed for turbine application. Since pumps are manufactured in larger quantities than turbines, they might be obtained off-the-shelf and readily modified for operation as turbines. While high efficiency can be obtained, this efficiency is generally not as high as that obtainable from a conventional turbine. Further information is available in refs. 2 and 3.

10.2-2 Specifying and Evaluating the Hydraulic Turbines

Design Conditions Imposed by the Site

The turbines will be designed to fit the design condition for the site such as

> Dimensional constraints of the proposed civil works.
> Number of units.
> Water temperatures and chemistry.
> Turbine discharge.
> Net head.
> Tailwater level.

Dimensional Constraints of Proposed Civil Works. The topography of the site or the type of powerhouse may favor a particular balance of horizontal and vertical dimensions. This balance and any other physical site constraints must be established before the turbines are specified.

Number of Units. The number of units will be selected as discussed in Section 10.1-4.

Water Temperature and Chemistry. The specific weight of the water used in the turbine will affect the power output for a given volume rate of water flowing. The specific weight, γ, is the weight in air of a unit volume of water. Variations of specific weight of fresh water with temperature, latitude, and elevation above sea level are given in standard references such as the IEC publications noted herein. Additional variations will occur as a result of chemical composition. Water chemistry and temperature are also important for a determination of the materials to be used in fabrication of the turbine parts.

Turbine Discharge. The turbine discharge, Q, is the volume rate of water flow through the turbine. As defined in the International Electrotechnical Commission (IEC) Turbine Field Acceptance Test

Code,[4] the turbine discharge includes leakage water through stuffing boxes and any thrust relief pipes but excludes water required for operation of the generator or auxiliary equipment or for cooling of bearings.

Net Head.   Net head H_n is the head available for doing work on the turbine, and is therefore the gross head minus all those head losses not attributed to the turbine. Excluded from the net head may be, for example, head losses in the intake, trashracks, and penstocks.

The terms *gross head* and *net head* both refer to difference in total head h_t, which is defined at any point as the sum of potential head, pressure head, and velocity head. The measuring points for the net head should be selected at points where the streamlines can be assumed to be parallel so that the average values of pressure and velocity can be used in the computation of total heads.

The potential head, z, is defined as the elevation of the specified measuring point with respect to some common datum.

The pressure head, h_p, is related to the gauge pressure p_g by the following equation:

$$h_p = \frac{p_g}{\gamma}$$

All pressure heads for turbines are expressed as a column height of the water used in the turbine, even though the individual pressures may not be dependent on this specific weight. For example, the atmospheric pressure head, h_a, is defined as the atmospheric pressure divided by the specific weight of the water in the turbine, even though the absolute atmospheric pressure has nothing to do with the water in the turbine. This does, however, make h_a compatible with other pressure heads used in the computations.

The flow area A is defined as the cross-sectional area normal to the general direction of flow[4] and the mean velocity v is related to this area by $v = Q/A$. The velocity head h_v is the water column equivalent of the pressure represented by the mean velocity, defined as follows:

$$h_v = \frac{v^2}{2g}$$

This results in a total head of

$$h_t = z + h_p + h_v$$

As stated above, the net head on the turbine is the difference between the total head at the turbine inlet and the total head at the turbine outlet. For reaction turbines part of the total head at the turbine outlet is therefore in the form of a velocity head, which will become a head loss if the draft tube exits directly into a reservoir. This is an item for which the accounting practice may vary. The IEC codes[5,6] define the turbine net head with respect to a point upstream from the draft tube exit, while the ASME Test Code for Hydraulic Prime Movers, PTC 18-1949,[6] uses a point downstream from the exit, thereby including the exit loss in the turbine net head.

If the draft tube exit loss is excluded from the turbine net head, the shape, outlet dimensions, and overall length of the draft tube must be evaluated in order to separately account for the exit loss in comparing different equipment options. For impulse turbines the net head is defined as the total head upstream from the turbine minus the weighted average elevation of the turbine bucket/jet interfaces as described in ref. 4.

Tailwater Level.   A given reaction turbine operating under a specified set of conditions will require a certain "back pressure" to prevent cavitation. Cavitation occurs when vapor pockets or cavities form due to low pressure in the water passages. The turbine must therefore be set at a suitable level with respect to tailwater in order to provide this back pressure.

Hydraulic Specification.   Since turbine discharge, net head, and tailwater level can vary independently, the specification could include a matrix of these operating parameters. The turbine will then be selected to best fit the overall matrix, with due consideration given to the expected frequency of occurrence for the various operating conditions.

Physical Characteristics of The Turbine

The turbine will be selected to best fit the design conditions imposed by the site. This selection will yield the type of turbine, power output, runner diameter, rotational speed, and highest allowable elevation of the turbine runner.

The types of hydraulic turbines are discussed in Section 10.2-1. For the preliminary project layout or study, an initial turbine selection can be made using guideline charts such as those prepared by the U.S. Bureau of Reclamation.[7] This publication also provides preliminary dimensions. Turbine manufacturers can also provide similar and more detailed information. Having selected a suitable type, there remains the selection of the runner diameter and rotational speed. For a given application, there can be a spectrum of combinations of these two parameters. A generic term for describing these combinations is the specific speed of the turbine.

Specific Speed. Specific speed, N_s, is defined as a function of the turbine rotational speed, power output, and the net head as follows:

$$N_s = \frac{n_t\sqrt{P_t/0.7457}}{H_n^{5/4}} \quad \text{(U.S. units)}$$

$$N_s = \frac{n_t\sqrt{P_t}}{H_n^{5/4}} \quad \text{(SI units)}$$

Traditionally, specific speed has been calculated using U.S. horsepower in the U.S. system of units. Metric practice has included the use of both metric horsepower and kilowatts. Any value of specific speed must therefore take note of the units which were used since this may not be obvious from the magnitude of the quoted number.

The above equations define specific speed on the basis of U.S. horsepower (hence the 0.7457 conversion factor) for U.S. units. In SI units specific speed is defined above based on kilowatts. Another convention in the computation of specific speed is that for reaction turbines the entire power output of the turbine runner is used, and for impulse turbines the power output from a single jet is used.

Rotational Speed. The rotational speed of an ac generator must be a speed that is synchronous with the frequency of the transmission system to which the generator is connected. Synchronous speeds are computed as follows:

$$n_g = \frac{120 f_{Hz}}{\text{number of poles}}$$

The above equation yields the synchronous speed of the generator, given the frequency in hertz and the number of generator field poles. The number of poles must be an even integer, and, for larger machines, a multiple of 4 is preferred.

If the turbine is to be directly connected to the generator, then the turbine speed will be given by the above equation, with the number of poles being selected to achieve the best possible design of the turbine, consistent with the requirements of the generator.

Most hydroelectric generators are directly driven by the turbine shaft. However, small units often have gearboxes between the turbine shaft and generator shaft. If a gearbox or other type of speed changer is used, the generator will operate at a higher speed than the turbine. The economic trade-off is then the cost of the speed increaser versus the reduced cost of the higher speed (and physically smaller) generator and associated civil works with due allowance for the speed changer efficiency.

Performance Characteristics of The Turbine

The performance of the selected turbine can be expressed in terms of its power output, efficiency, and freedom from cavitation.

Power Output and Efficiency. The power output, P_t, (in kW) of a turbine at any given operating point is computed from the net head H_n, turbine discharge Q, the specific weight of water γ, and the turbine efficiency η_t as follows:

$$P_t = Q\gamma H_n \left(\frac{0.7457}{550}\right)\eta_t \quad \text{(U.S. units)}$$

$$P_t = \frac{Q\gamma H_n \eta_t}{1000} \quad \text{(SI units)}$$

The generator output P_G is usually measured at the generator terminals and is computed from the effective voltage E, effective current I, and power factor $\cos\theta$ by:

$$P_G = \frac{EI}{1000} \quad \text{(dc generator)}$$

$$P_G = \frac{\sqrt{3}\, EI \cos\theta}{1000} \quad \text{(three-phase, ac generator)}$$

The generator output is related to the turbine output by applying the generator efficiency, η_g, and the gearbox/flywheel efficiency, η_s, as follows:

$$P_G = P_t \eta_s \eta_g$$

In computing the generator efficiency η_g, the following losses are determined separately and added to the generator output to arrive at the generator input:

I^2R losses.
Core losses.
Stray load losses.
Windage losses.
Bearing friction losses.
Excitation losses.
Energy consumption in external cooling devices.

IEC Publication 34-2[8] and IEEE standard No. 115[9] differ slightly in the definition of standard temperature for which the I^2R losses are to be computed for reporting purposes. Furthermore, these standard temperatures may be different from the temperatures that are expected to occur in actual operation. The reported efficiency can therefore vary somewhat depending on the method of computation used. Generally, these differences would be on the order of a few tenths of a percent. However, even such small differences may be significant in a large power development. All computations for a given project should, therefore, be based on the same method.

Freedom from Cavitation. Cavitation in a hydraulic turbine can cause erosion of metal from the turbine runner and associated casings. It can also result in rough operation, excessive noise, and severe vibrations. This erosion and vibration can result in costly repairs and lost operating time for the unit.

To visualize the development of cavitation, consider first a hydraulic turbine on which the wicket gates are closed and water is not flowing. The pressures in the water passages of the runner will be the static pressures associated with the tailwater level and can be either below or above atmospheric pressure, depending on whether the runner is located above or below the tailwater level.

As the wicket gates are opened and water begins to flow, the pressure will drop according to Bernoulli's principle. If the local pressure at any point drops to the vapor pressure, the water will boil and form cavities filled with water vapor. As these cavities are then swept into zones of lower velocity (higher pressure), the cavities implode. This can sound like rocks passing through and hitting the sides of the water passages.

One measure of safety against cavitation is the geodetic suction head, h_s, which is the elevation of the turbine runner above the tailwater level and can be either positive (runner above tailwater level) or negative (runner submerged below tailwater.) A way of avoiding cavitation is to set the centerline of the runner at a very low elevation with respect to tailwater (a large negative value of h_s), thereby providing considerable static head in all parts of the water passage.

The geodetic suction head can be combined with the atmospheric pressure head, h_a, and the vapor pressure head, h_{va}, to form the quantity $h_a - h_s - h_{va}$. If this quantity is less than zero, then vapor cavities will be present. If this quantity is greater than zero, then vapor cavities may or may not form, depending on the value of the above quantity combined with the possible existence of flow separation and high local velocities (and corresponding low local pressures) within the water passages of the runner.

The cavitation coefficient sigma (σ) is the following dimensionless number, which is used in describing potential for cavitation:

$$\sigma = \frac{h_a - h_s - h_{va}}{H_n}$$

When the information for the power station is inserted into the above equation, the result is called plant sigma. This value of sigma must be greater than critical sigma (the value of sigma at which cavitation effects on hydraulic performance would be unacceptable) and a suitable margin over this value must be provided.

Plant sigma is fixed by factors external to the turbine, while critical sigma for a given net head is a function of turbine design, wicket gate position, and runner blade position. Critical sigma is generally determined during hydraulic model tests as described in Section 10.2-3. Generally, freedom from cavitation is best at the point of best efficiency since the water will at this point have the smoothest entrance to the runner blades, and the exit flow will be essentially in the axial direction without appreciable swirling action. For higher outputs there is generally increased cavitation activity due to the increased water flow and also due to local separation of flow because of the less than optimum approach and exit angles. There is also generally a zone of increased cavitation activity at small outputs. At these lower loads the local flow separation due to inefficient approach and exit angles may more than offset the reduction in average velocity head.

Setting the Turbine with Respect to Tailwater

The elevation of turbine runner must be set so that plant sigma is greater than critical sigma by a suitable margin. This margin is a matter of judgment since the determination of critical sigma is based on hydraulic factors and not upon a test specifically intended to predict a rate of metal removal. Examples of this type of margin are given in the literature.[10-12]

For preliminary studies and for comparison with the final results, a tentative value of plant sigma is often selected based on the specific speed of the turbine. Examples of these guidelines are available.[7] A high specific speed turbine will be physically smaller than a low specific speed turbine but will require a deeper setting. A trade-off must therefore be made.

10.2-3 Hydraulic Model Testing

Tests on a scale model, under laboratory conditions, can be more closely controlled than full-scale tests on the prototype where the quantities may be more difficult to measure. Such models are extensively used to develop the final dimensions and shapes and to predict performance characteristics of full-size turbines. The model tests are generally performed in the manufacturer's own laboratories, and further demonstration tests of the final designs may also be performed in independent laboratories.

Data recorded on a model can be stepped-up to predict full-scale turbine performance at conditions that are dynamically similar to those at which the model test was performed. Dynamic similarity exists when the velocity patterns are similar. The velocity ratio can be used to represent this similarity.

Velocity Ratio (Reduced Speed)

The velocity ratio ϕ is the ratio of runner tip speed ($\pi D n_t / 60$) to the theoretical spouting velocity of a water jet operating under the specified net head ($\sqrt{2gH_n}$).

$$\text{Velocity ratio} = \phi = \frac{\pi n_t D}{60 - \sqrt{2gH_n}} \quad \text{(U.S. units)}$$

This parameter is representative of the velocity triangle for flow at the runner discharge section. If the turbines are operating at equal values of ϕ, the velocity triangles will be similar and the conditions are said to be homologous.

The velocity ratio is used with the U.S. system of units. In SI units the parameter reduced speed, n_{11}, is used for the same purpose.

$$n_{11} = \frac{n_t D}{\sqrt{H_n}} \quad \text{(SI units)}$$

Equations of Homology

The following are the theoretical equations for stepping up the model data to full scale:

$$\left(\frac{P_2}{P_1}\right) = \left(\frac{n_2}{n_1}\right)^3 \left(\frac{D_2}{D_1}\right)^5$$

$$\left(\frac{Q_2}{Q_1}\right) = \left(\frac{n_2}{n_1}\right) \left(\frac{D_2}{D_1}\right)^3$$

$$\left(\frac{H_2}{H_1}\right) = \left(\frac{n_2}{n_1}\right)^2 \left(\frac{D_2}{D_1}\right)^2$$

$$\left(\frac{P_2}{P_1}\right) = \left(\frac{H_2}{H_1}\right)^{3/2}\left(\frac{D_2}{D_1}\right)^2$$

$$\left(\frac{Q_2}{Q_1}\right) = \left(\frac{H_2}{H_1}\right)^{1/2}\left(\frac{D_2}{D_1}\right)^2$$

The above equations apply to the performance step-up of two different size turbines when the following restrictions and caveats are observed:

1. The two turbines must be geometrically similar.
2. The operating points must be homologous. That is, the flow patterns, wicket gate angles, runner blade angles, and velocity ratio (or reduced speed), must be the same.
3. Cavitation must not occur.
4. Scale effects on turbine efficiency must be considered separately.

Turbine Efficiency Scale Effects

The equations of homology assume that a large turbine will have the same efficiency as a geometrically similar small turbine when the turbines are operated under homologous conditions. This is generally assumed to be correct for impulse turbines. Experience has shown, however, that in the case of reaction turbines, the larger turbine will have a higher efficiency. Two commonly used efficiency step-up formulas are the Moody formula and the Hutton formula. In the IEC Model Test Code,[5] the Moody formula is suggested for Francis turbines and the Hutton formula is suggested for Kaplan turbines and other propeller turbines.

$$\frac{1 - \eta_2}{1 - \eta_1} = \left(\frac{D_1}{D_2}\right)^N \quad \text{(Moody formula)}$$

Values of N have ranged from $\frac{1}{4}$ to $\frac{1}{5}$, and the value of $\frac{1}{5}$ is given in the IEC code.

$$\frac{1 - \eta_2}{1 - \eta_1} = 0.3 + 0.7\left[\frac{D_1 \nu_2}{D_2 \nu_1}\left(\frac{H_1}{H_2}\right)^{1/2}\right]^{1/5} \quad \text{(Hutton formula)}$$

Having established the quantity $(1 - \eta_2)/(1 - \eta_1)$ by either formula, the efficiency increment $\Delta\eta$ to be added to η_1 is computed as follows:

$$\Delta\eta = (1 - \eta_1)\left(1 - \frac{1 - \eta_2}{1 - \eta_1}\right)$$

The Moody and Hutton formulas are the two best-known formulas for predicting the expected efficiency after scale up. However, both formulas have been subject to some discussion with respect to accuracy, range of applicability, and method of applying the efficiency step-up to the measured model data.

In addition to defining the efficiency step-up formula, it is necessary to define the points at which the increment will be computed. According to the IEC Model Test Code, this increment is computed for each guaranteed point and is then added to the model efficiency at that point to obtain the predicted efficiency of the full-size turbine.

Another common practice is to compute the efficiency increment only at the point of best efficiency on the model and to then add this constant increment to the measured model efficiencies at all points to obtain the predicted efficiency of the full-size turbine.

The intended use of the efficiency step-up is for comparison of energy production between turbines whose models are of different size. That comparison may be made without defining the distribution of the efficiency increment between discharge and power. In practice, however, it is best to define this distribution in a consistent manner so that all computations on a given project will be based on identical numbers.

Cavitation Effects

A hydraulic model can be used to estimate the value of critical sigma for a particular turbine operating under specific conditions of net head, runner blade angle, and wicket gate opening. In this type of test the model operation is first stabilized at the specified operating conditions. A vacuum pump is then

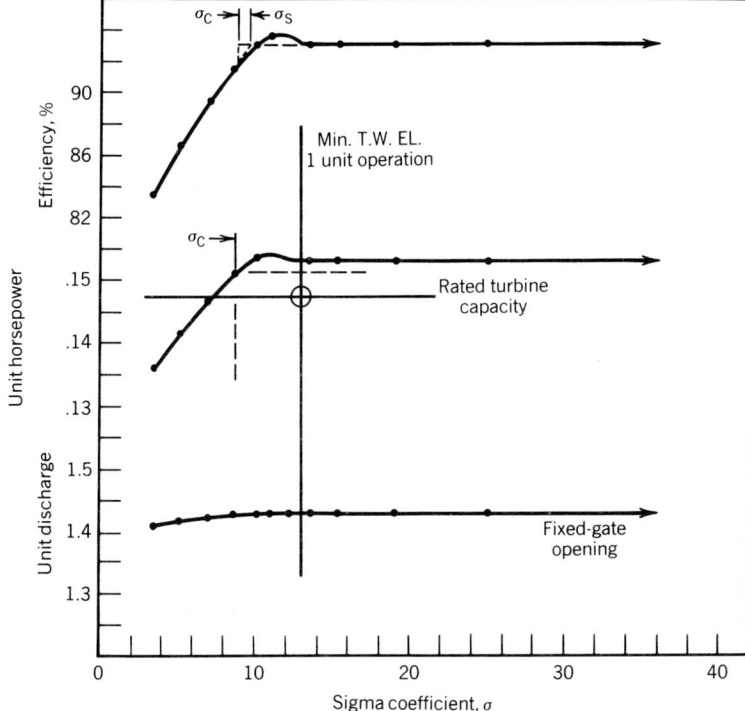

Fig. 10.2-11 Representative test data showing turbine performance versus sigma. (Courtesy of Dominion Engineering Works, Ltd.)

used to simultaneously reduce the pressure (h_a) in the closed tanks, which contain the headwater and the tailwater surfaces. A typical model will show a relatively high value of sigma at normal atmospheric pressure, but as the vacuum pump is operated, the value of the sigma is reduced. Since the net head remains the same, only the value of sigma will change as the pressure is reduced. The reduction in pressure continues until the efficiency drops off as a result of the cavitation effects. Critical sigma is then determined from the graph of efficiency versus sigma (Fig. 10.2-11). In practice the acceptable value of sigma will be determined by judgment in observing bubble formation and in selecting a suitable margin of safety.

10.2-4 Generators and Other Equipment

Generators

Generators can be either synchronous type or induction type. Large-capacity turbines are generally connected to synchronous generators, which have provisions for voltage control and power factor control. For small units it may be more economical to use a standard induction generator. However, if the induction generator is to supply power to an independent system, a capacitor bank or some other form of external excitation must be provided. Even if the induction generator is to be connected to a multiunit system, the system requirements may dictate the use of such external excitation.

Other Equipment

Governors. For a unit that will be used to control the frequency of an electric system, the governor will adjust the water flow to maintain the required speed (frequency) as the load changes. For a unit that is connected in parallel with other units and will not be used to control frequency, the governor will control the turbine–generator speed when the unit is off-line and the turbine–generator output when the unit is on-line. This is accomplished by comparing the speed or output signal to the set points and converting any deviation to a command for opening or closing the wicket gates or adjusting the runner blade angles. Wicket gate and blade adjustments are generally made through oil-pressure-operated servomotors.

For small units intended to operate only when connected to a large system, a conventional governor may not be necessary. Load control may be either by independent manual or programmed adjustments or by automatic control from reservoir water level.

Oil Systems. The oil systems generally serve three functions: operation of needle valves, wicket gates, and runner blades; lubrication of main shaft guide and thrust bearings; and lubrication of runner hub trunnions on adjustable-blade units.

Greasing Systems. An automatic greasing system is often provided to periodically inject grease at the turbine wicket gate, operating ring, and inlet valve bearings. Alternatively, some or all of these bearings could be provided with nonlubricated Teflon bearings.

Cooling Water. Filtered cooling water is generally required for the bearing oil heat exchangers, the generator coolers, the turbine shaft packing or seal, and, if the turbine is ever to be operated in air, the runner seals.

Tailwater Depression System. Many turbines and pump–turbines include provisions for operating the turbine in air for one or more of the following purposes:

1. To allow the generator to operate as a "synchronous condenser" to improve the transmission system power factor.
2. To maintain the unit on-line and quickly available without using any water from the reservoir.
3. To allow starting of a pump–turbine in the pumping mode under a relatively small load.

If the turbine runner is set above tailwater elevation, the runner can be operated in air by opening a vacuum breaker valve. If the runner is set below tailwater level, compressed air can be admitted to depress and hold the water surface below the turbine runner.

REFERENCES

10.2-1 D. G. Fink and H. W. Beaty, *Standard Handbook for Electrical Engineers*, 11th ed., McGraw-Hill, New York, 1978.

10.2-2 P. Cooper and R. Worthen, "Feasibility of Using Large Vertical Pumps as Turbines for Small Scale Hydropower," Final Technical Report to U.S. Dept. of Energy, Ingersoll-Rand Research Inc., Princeton, N.J. Oct. 1981.

10.2-3 H. Mayo, "Small Scale Hydro/Centrifugal Pumps as Turbines," *Waterpower '81 Proceedings*, Vol. I, U.S. Army Corp. of Engineers, 1981.

10.2-4 International Electrotechnical Commission, *International Code for the Field Acceptance Tests of Hydraulic Turbines*, IEC Publication 41, IEC, 1963.

10.2-5 International Electrotechnical Commission, *International Code for Model Acceptance Tests of Hydraulic Turbines*, IEC Publication 193, IEC, 1965 and Amendment No. 1, 1977.

10.2-6 American Society of Mechanical Engineers, *Hydraulic Prime Movers*, ASME PTC 18-1949, ASME, New York, 1949.

10.2-7 Bureau of Reclamation, *Selecting Hydraulic Reaction Turbines*, U.S. Department of the Interior, Washington, D.C. 1976.

10.2-8 International Electrotechnical Commission, *Rotating Electrical Machines*, IEC Publication 34-2A, IEC, 1972; IEC Publication 34-2A, 1974.

10.2-9 Institute of Electrical and Electronics Engineers, *Test Procedures for Synchronous Machines*, No. 115, IEEE, New York, 1965.

10.2-10 U.S. Army Corps of Engineers, *Engineering & Design Manual—Selecting Reaction Type Hydroelectric Generators and Generator Motors*, U.S. Army Corps of Engineers, to be published.

10.2-11 J. P. Davies, *Hydraulic Turbine Cavitation Pitting Damage*, Canadian Electrical Association, Montreal, 1981.

10.2-12 C. V. Davis and K. E. Sorensen, *Handbook of Applied Hydraulics*, 3rd ed., McGraw-Hill, New York, 1970.

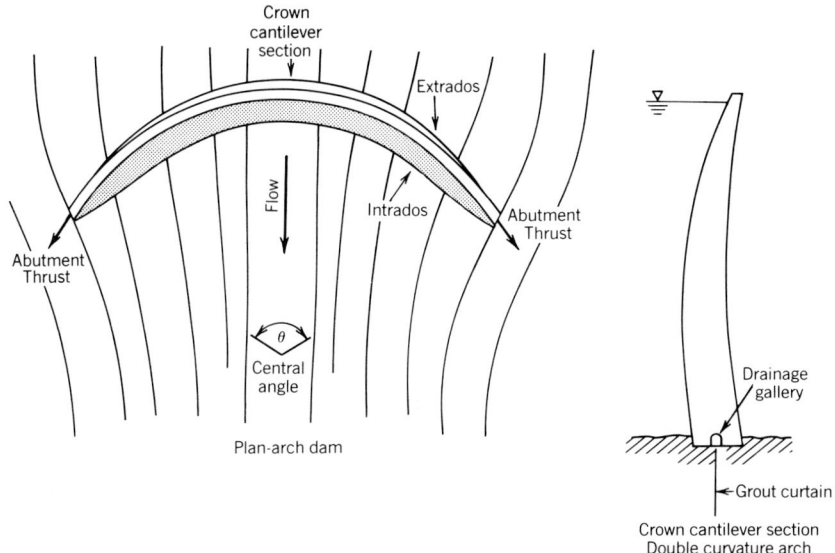

Fig. 10.3-1   Arch dam.

10.3   DAMS

Thomas J. Lamb

10.3-1   Dam Types*

Dams can be grouped into three broad categories: arch dams, gravity dams, and embankment dams. The design principles are different for each type. The site conditions, such as topography, geology, and availability of materials, will dictate which type is most suitable.

Arch Dams

Arch dams are unreinforced concrete structures. They can be of the single-curvature and double-curvature types (Fig. 10.3-1). The older, more common type, is the single-curvature arch, which is curved only in the horizontal plane and thus is essentially a segment of a right cylinder. The arch acts in compression, delivering the thrust to the rock abutments in the valley sides. The double-curvature arch, which is curved in both the horizontal and vertical planes, permits greater economy in the use of materials since the concrete sections can be made thinner in some areas.

Gravity Dams

Gravity dams, as the name implies, resist water loads through friction developed at the base due to the weight of the dam. The most familiar type is composed of unreinforced concrete monoliths with a near-vertical upstream face and a downstream face sloping at about 0.75 horizontal to 1 vertical (Fig. 10.3-2). Buttress dams, such as flat slab, buttress head, or multiple arch, also act as gravity structures, although in these the water load contributes to the downward force, improving the stability (Fig. 10.3-2).

The loading imposed on the foundation by a concrete gravity dam is not a significant problem unless the rock is very weak. The major concerns are the establishment of a good bond to transfer shear forces to the foundation, detection of any weak planes within the foundation upon which sliding could occur, prevention of excessive seepage through the foundation, and provision for drainage at the dam–foundation contact.

The choice between a conventional gravity dam section and a buttress-type dam is principally a matter of economics. Conventional sections require large amounts of concrete but very little rein-

*Portions of this section were inspired by a series of lectures given by W. F. Swiger in 1981.

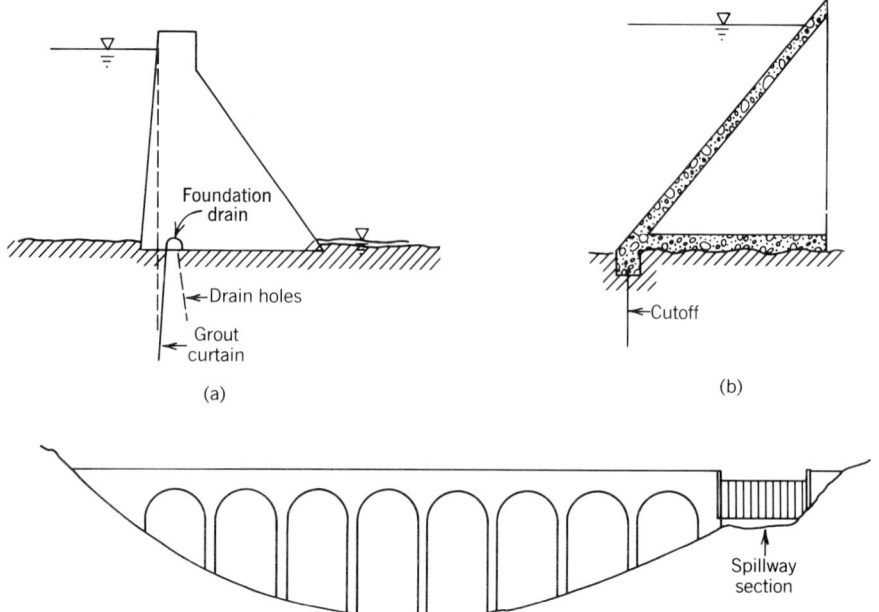

Fig. 10.3-2   Gravity dam. (*a*) Concrete gravity dam. (*b*) Buttress dam (multiple arch similar). (*c*) Multiple arch dam (buttress dam similar).

forcing and simple, reusable forms. Slab and buttress dams are quite economical in their use of concrete, but the shears and bending moments in the slabs can be quite large in high dams, requiring heavy reinforcing. Multiple-arch dams are economical in terms of the amount of concrete used, but the form work required is costly. In areas with severe winters, the underside of the relatively thin slabs or arches may require frequent maintenance to offset serious deterioration caused by temperature changes and freeze–thaw action.

Embankment Dams

An embankment dam is composed of impervious soil or of an impervious core or membrane with pervious soil or rock-fill shells. Embankment dams are adaptable to a wide variety of site conditions with the dam cross section being selected to make the most economical use of available materials (Figs. 10.3-3 and 10.3-4). Embankment dams are frequently the most economical type of dam and account for roughly 75% of the large dams in the United States.

10.3-2   Site Selection

Preliminary Selection of Potential Dam Sites

Preliminary selection of potential dam sites is based on topographic and geologic data. Published data are used to select a reasonable number of sites that appear to have the most favorable characteristics.

The cost of the dam itself is roughly proportional to its volume, provided that foundation and abutment conditions do not vary drastically. The preliminary identification of a site will therefore focus on locations where dams with relatively small volume and height can be constructed. Increases in dam height increase the head available for power generation. However, dam volume and, hence, cost roughly increase as the square of the height for dams in broad valleys with steep sides and as the cube of the height in narrow, V-shaped valleys. Therefore, a trade-off must be made between increased height and dam cost.

Geologic conditions at the dam site can have a very significant influence on project cost. The foundation and abutment treatment required for a site with unfavorable conditions often equals or exceeds the cost of the dam itself, and it may be more economical to construct a larger dam where conditions are more favorable. Geologic features of concern are summarized in Table 10.3-1.

Potential for leakage both under the dam and from other areas in the reservoir must also be carefully evaluated, both to provide for dam safety and because water lost through leakage is not

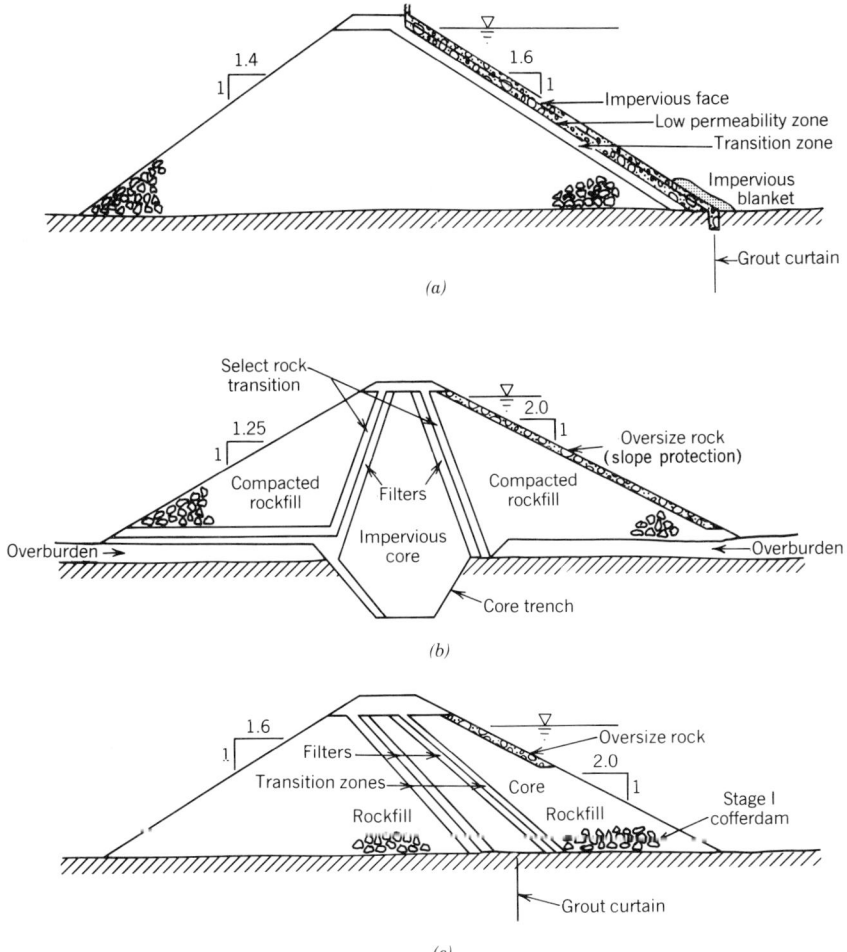

Fig. 10.3-3 Embankment dam. (*a*) Impervious face rockfill dam. (*b*) Central core rockfill dam. (*c*) Sloping core rockfill dam.

available for power generation. The latter consideration is particularly important for the upper reservoirs of pumped storage projects because additional pumping is required to replace the leakage. Measures to reduce leakage at unfavorable sites are generally very costly and are not always successful. Several dams have been constructed that never impounded water because leakage exceeded inflow.

After preliminary investigations have identified potential hydropower sites, the type of dam appropriate to each site must be determined. The type of dam to be considered is dictated by site conditions.

Arch Dams. High concrete arch dams are very efficient in the use of materials compared to other types and are well adapted to symmetrical, steep-sided, relatively narrow valleys (width-to-height ratio usually less than about 5) with strong, high-quality rock in the abutments and foundation. Long arches transmit considerable load to the foundation through cantilever action, so for low dams or broad valleys concrete gravity dams or embankment dams are generally more economical. Very large arch thrusts must be accommodated by the rock abutments, and their strength and deformability characteristics are of fundamental importance.

Gravity Dams. Concrete gravity dams are well suited for large rivers in relatively broad valleys where substantial spillway capacity is needed since it is relatively simple to incorporate a large spillway into the dam. Gravity dams are suitable for a wide variety of foundation conditions, including overburden for low dams.

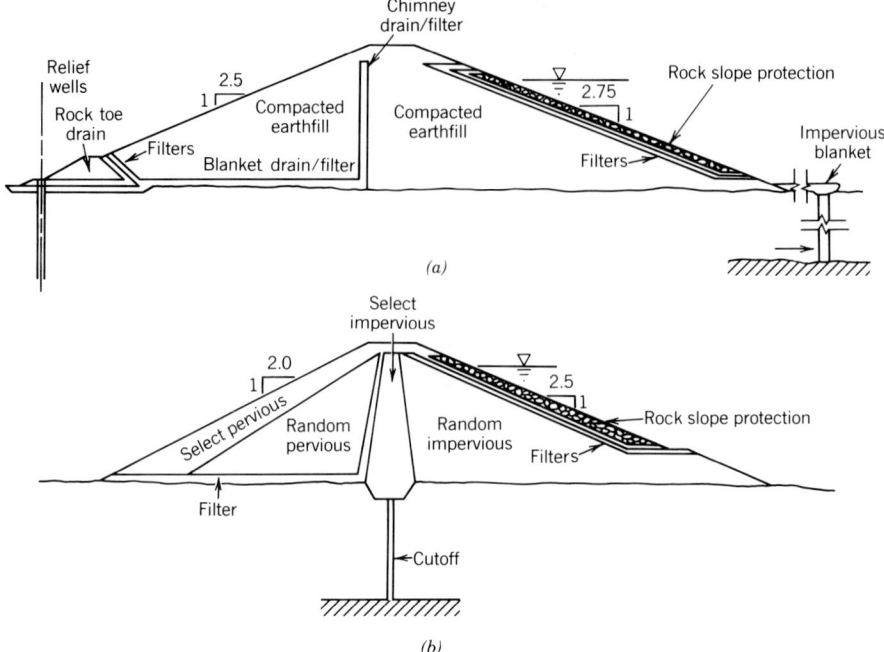

Fig. 10.3-4 Embankment dam. (*a*) Homogeneous earthfill dam. (*b*) Zoned earthfill dam.

TABLE 10.3-1 GEOLOGIC FEATURES OF CONCERN IN DAM SITING

Thickness and properties of overburden
Rock properties (particularly joints and faults)
Availability of concrete aggregate or embankment materials
Solution channels (in limestone, anhydrite, gypsum, or salt)
Porous rocks (lava or lava interbeds especially)
Buried stream channels
Lava pipes
Open joints or faults
Pervious soil strata (such as openwork gravel)
Active faults (or potential for fault reactivation)
Volcanic activity
Landslides, both ancient and active
Potential for slope failures in the abutments, foundation, or
    reservoir perimeter

Embankment Dams. A suitable embankment section can be designed for sites with widely varying geologic and topographic conditions. Embankment dams, if properly designed, can tolerate differential settlement and even modest displacement on faults in the foundation. Rock-fill dams can tolerate seepage, but earth-fill sections may be subject to failure by piping (internal erosion). Properly designed rock-fill dams can resist modest overtopping in an emergency, but a separate spillway or discharge structure is required for normal operation.

The large abutment thrusts associated with arch dams are not present with embankment dams. However, impervious core embankment dams in steep or irregular valleys are susceptible to cracking, and properly designed filters are essential. Care must also be taken to prevent seepage through the foundation or abutment from eroding the core.

TABLE 10.3-2   SITE CONDITIONS TO BE CONSIDERED DURING DAM SITING

Condition	Potential Effect on Dam	Sources of Data	Remarks/Remedial Measures
River flow	Required spillway capacity	See Section 10.1	Average and minimum flow determine power production, while maximum anticipated flows determine required spillway and diversion capacity.
Topography	Dam location, type, and size	Published maps (USGS), aerial surveys	See text
Geology	Dam type, cost, and safety	Published open file data (USGS and state), site investigations (borings, mapping, air photos, satellite photos, geophysical surveys, test audits, etc.)	Sites with unfavorable conditions in the foundation or abutments may require very elaborate and costly treatment
Seismicity	Dam type, cost, and safety	Historical earthquake records, geologic studies	Presence of active faulting can preclude siting a dam, particularly a concrete dam
Access to site and availability of material and labor for construction	Dam cost and schedule	Topographic maps, socioeconomic data	Remote sites can require extensive infrastructure improvements before construction can begin
Environmental impact	Unacceptable impacts on local inhabitants, fish, endangered species, or historic sites can prevent construction	State, local, and federal agencies, field studies	Many impacts can be mitigated at a cost. Examples include fish ladders, navigation locks, relocation of roads, railroads, or industries, and establishment of recreational areas

Final Selection

The preliminary investigations based on published data generally identify several potential dam sites. Further investigations must be carried out to determine which of these sites is the best one. Each potential site is examined by a team of experts in the various technical disciplines to confirm or revise the conclusions drawn from published data and to identify any problems not evident from the published data. Preliminary designs and cost estimates are developed for the more favorable sites. Sites are then ranked on the basis of engineering, economic, and environmental factors to identify those most favorable for further study. More detailed investigations, which may include geologic field work, limited borings, and environmental baseline studies, are performed at higher-ranking sites. The necessary studies vary from site to site and may require a substantial investment. The factors to be considered are listed in Table 10.3-2. All factors influencing the selection must be considered in order to select the dam site that most economically satisfies the competing technical and environmental requirements.

10.3-3   Dam Design and Construction

Arch Dams

The required thickness of the dam for a given span increases as the central angle decreases. The change in volume due to decreasing the arc length is overshadowed by the thickness change. However, as the central angle increases, the line of thrust becomes more nearly tangent to the valley wall, increasing the shear forces in the abutment. A balance between these is necessary to obtain the optimum line of thrust at the abutments.

Concrete arch dams tend to shorten after construction, due to temperature changes and drying shrinkage in the concrete. The hydrostatic water load behind the dam places the structure in compression. The modulus of the abutment rock is often comparable to that of concrete, considering joints and other rock defects; therefore, some yielding occurs under the arch thrust. Since the dam radius cannot be altered at the abutments, shear and bending stresses are set up in the dam. These effects can be compensated for to some extent by changing the radius near the abutments and thickening the abutment concrete to maintain the resultant thrust reasonably close to the middle of the section. To minimize temperature and shrinkage effects, arch dams are usually placed as a series of individual blocks, and the concrete is cooled during curing. The joints between the blocks are then pressure grouted. On some dams the arch is prestressed by placing flatjacks in the joints. This technique minimizes net movement when the reservoir is filled.

The classic method of analyzing arch dams is the trial load method. Another method is the use of model tests constructed in accordance with appropriate scaling laws and loaded by pistons. This method can be used for both preliminary and final design. However, the most common method of analysis at present involves the use of three-dimensional finite-element computer programs.

The behavior of the dam abutments under load has a large influence on the stresses within the dam. For this reason detailed investigations of the abutments, often employing test audits, are needed to discover any potentially weak or compressible zones. Large-scale field tests are used to accurately estimate the modulus of the abutment rock. Where unfavorable or nonuniform conditions are encountered in the abutment, extensive remedial measures may be required. Drainage of the abutment regions is of critical importance.

Gravity Dams

The stresses within concrete gravity dams are calculated using standard structural analysis procedures. Two-dimensional and three-dimensional finite-element programs are used where appropriate.

The overall stability of a gravity dam is calculated by comparing the applied horizontal loads, primarily hydrostatic water forces plus earthquake and ice loads, with the available sliding resistance due to the weight of the dam times an appropriate coefficient of friction plus a shear strength value due to the bond between the dam and the foundation rock. Uplift pressure beneath the dam reduces the normal force on the failure plane, the uplift pressure distribution being dependent on drainage conditions assumed. Modern practice generally assumes that the uplift acts over 100% of the base. Buttress dams are analyzed similarly.

Normal practice calls for keeping the resultant force within the middle third of the base under normal loading, which ensures compression everywhere beneath the dam. A zone of zero compression at the heel may be allowed for some extreme loading conditions, but full uplift pressure is assumed to exist as if there were a gap throughout this zone.

Several gravity dams have failed due to weak planes within the foundation, and such a mechanism must also be considered. Similarly, poor bonding between horizontal construction joints can allow the development of failure planes within the dam. A system of water stops and drains is installed in both horizontal and vertical construction joints to prevent the development of hydrostatic pressures. Stability should be checked at a series of elevations within the dam.

Concrete gravity dam monoliths are cast as a series of individual blocks, usually about 15 m wide, to limit shrinkage. The monoliths may be independent, with only water stops between them. Keys may be constructed to transfer load between blocks, or in some cases, the joints are grouted so that the structure behaves monolithically. The choice depends largely on the designer's evaluation of the effects of topography and rock conditions. Concrete lift height is limited to reduce shrinkage and cracking. Cooling of the concrete during cure may be required if schedule considerations dictate rapid completion of the monoliths.

A recent development that promises considerable cost savings is the use of "rollcrete."[1] Stiff concrete is batched, placed in lifts, and then compacted using earth-moving equipment. Experience to date with rollcrete has been favorable, but some difficulties have been encountered with bonding and leakage between lifts.

Embankment Dams

Embankment dams are constructed using earth and rock materials found within a short distance of the dam site. The use of processed or imported materials greatly increases costs, but is sometimes necessary, especially for filters, whose gradation must be tightly controlled.

Rock-fill Dams. Rock-fill dams can be divided into impervious face and impervious core types (Fig. 10.3-3). The former consist of an embankment of rock or pervious gravels with a graded low-permeability zone and an impervious membrane on the upstream face. Various materials are used for the face, including asphalt, steel, and wood, but reinforced concrete is the most popular in the United States. The concrete face is usually about 300 mm thick at the crest and increases in thickness by about 0.002–0.003 times the depth. The low-permeability zone serves a dual function as a base upon which to construct the slab and as a second line of defense in case leaks develop in the face. Attention to design details will minimize cracking of the slab, and the face is accessible for repairs if the reservoir can be drawn down. Some cracking and leakage should always be anticipated, particularly at the upstream toe. Modern designs often place impervious material on top of the face at the upstream toe to limit this leakage. Concrete-faced rock-fill dams over 150 m high have been constructed, and such dams are probably the safest type of dam against earthquake forces and are also one of the safest types for frequent rapid drawdowns. This type of dam can tolerate brief overtopping without failure and is a good choice for use as a cofferdam.

Because of the tendency of asphalt to creep, flatter upstream slopes are required for asphalt-faced dams than for concrete-faced dams ($1.8H:1V$ vs. $1.5H:1V$ or steeper). Since the face is placed with no provision for movement, temperature changes and settlement will result in cracking and leakage. Consequently, asphalt faces generally have built-in provisions for drainage, which permits the use of materials in the embankment that would not be suitable for a concrete-faced dam. Quality control of asphalt materials and the lack of adherence of the asphalt to the aggregate can be significant problems.

The required upstream and downstream slopes of impervious-faced rock fills are a function of the frictional characteristics of the rock fill, the strength of the foundation, and the loading on the dam. For example, weak foundation materials or large seismic loads require flatter slopes. The usual procedure is to assume a slope based on experience, develop a cross section, perform an analysis, and revise the slope if necessary for stability. This process is repeated until the desired factor of safety is obtained. The most economical embankment usually has minimum volume and the steepest slopes. However, if excess excavated material from other project areas must be disposed of, it may be more economical for the overall project to construct a dam with flatter slopes.

Impervious core rock-fill dams utilize a low-permeability zone (usually soil, but other materials, especially concrete and asphalt, are also used) contained within the pervious shells to control seepage. The core may be vertical or may be inclined upstream (Fig. 10.3-3). Both designs have advantages. Sloping-core designs undergo less overall deformation under reservoir load and provide better downstream slope stability, while vertical-core designs result in a more stable upstream slope. Vertical-core designs are popular for dams subjected to frequent draw down for this reason. Sloping cores are pressed against the abutments by the water load and are less susceptible to cracking than a thin vertical central core. The dimensions of the core are a function of the relative cost of core material versus shell material. The minimum core thickness is generally 0.3 times the hydrostatic head at any point, although dams with thinner cores have been constructed. Very thick cores are common where core material is plentiful. Minimum thickness for low dams may be dictated by construction procedures.

Well-graded glacial tills with plastic fines and clays make excellent core material, although clays are somewhat difficult to place. Silts are susceptible to piping and difficult to protect with filters; they are therefore less desirable.

Filters are required to protect the core. Properly designed filters are sufficiently pervious to collect seepage through the core, but their gradation prevents core material from migrating into the downstream shells. Upstream filters are used on dams subject to draw down and for protection against cracking. In rock-fill dams there is often a large discrepancy in size between the core and the shells, and two-stage filters may be necessary. Widely graded filters arc subject to segregation during placement and should be avoided.

The upstream and downstream slopes on impervious core rock-fill dams are usually flatter than for impervious-faced dams because of the relatively lower strength for the core material. Slopes generally range from $1.6H:1V$ to $2.5H:1V$, depending on the core material and geometry.

Zoned and Homogeneous Dams. Zoned dams and homogeneous earth-fill dams have wide variations in geometry depending upon the relative abundance of natural materials (Fig. 10.3-4). Zoned dams may have pervious shells and be essentially the same as a central core rock fill. Homogeneous dams are usually constructed from fine-grained soils, which are not free draining, and thus very careful attention must be paid to drains and filters to protect the embankment and ensure safety. Chimney drains, toe drains, blanket drains, and combinations of these are used to provide the necessary drainage (Fig. 10.3-4). Drains and filters require care during construction to assure that permeability is maintained and that the drain is not contaminated by embankment material. Placement procedures for the embankment generally result in horizontal permeabilities greater than vertical permeabilities; this should be considered in the design of drains.

The geometry of zoned and homogeneous embankments is more complex than the geometry of rock-fill dams. Slopes are flatter, and changes in slope or benches are commonly used. Berms are sometimes employed to improve stability of the dam toe, especially where foundation soils are weak.

Embankment Dam Stability. Analysis of an embankment dam requires the calculation of seepage patterns, deformations, and stability. Flow nets, constructed manually or with finite-element computer programs, provide seepage information. Similarly, deformations can be computed manually or using computer programs, depending on the complexity of the problem. Stability of the embankment is assessed by analyzing slip circles or sliding wedges, approximating the anticipated geometry of a failure surface. Computer programs are available that allow the rapid and economical calculation of factors of safety on many assumed failure surfaces. Sliding of the entire embankment is rarely a problem unless very weak soils or high pore pressures exist in the foundation.

Seismic Design of Embankment Dams.   Much recent progress has been made in designing dams to resist seismic forces. The traditional method of analysis—application of a pseudostatic horizontal inertial force—is still used and is adequately conservative in many cases where earthquake shaking does not reduce the soil strength. Many dams designed using this procedure have withstood earthquakes without damage. However the analysis of large dams and dams subjected to large earthquakes requires a complete dynamic analysis. Dynamic finite-element programs have been developed to do this.

Deformations of embankments subjected to dynamic loads should also be calculated. If calculated deformations are small, the dam may be perfectly safe, even though the calculated factor of safety under peak load may be less than one.

Earthquake loading has caused liquefaction (complete loss of strength) of loose sands and silts within the embankment or the foundation at several dams. Hydraulic fill dams, where the embankment is placed by hydraulic dredge, are especially susceptible to this type of failure, and such structures have performed poorly when subjected to seismic loads.

Slope Protection.   Embankment dams must be protected from wave and ice forces in the reservoir. Impervious faces are designed to resist wave action and ice loads. Riprap, concrete armor, or soil cement is placed on the upstream face of other embankments. Careful bedding of the slope protection is required to prevent wave action from washing embankment material out through the armor.

The embankment must also be protected downstream where tailwater flows might cause erosion. The entire downstream slope of homogeneous and zoned dams is generally seeded, stepped, or paved to prevent rainfall from causing erosion.

Embankment Dam Construction.   The usual method of construction employed in embankment dams is the placement and compaction of the various materials in thin lifts, with careful control of lift thickness and compactive effort. Core zones and homogeneous dams are placed in 200–250-mm lifts and heavily compacted using sheepsfoot or tamping foot rollers for plastic materials and rubber-tired rollers for tills. A minimum density of 95% of standard proctor is usually specified; the lower portions of high dams are often compacted to 95% of modified proctor to reduce settlement and cracking. Water content of plastic materials is generally maintained slightly wetter than optimum to achieve increased plasticity and lower permeability in the dam. Filters and drainage zones are lightly compacted to ensure they remain pervious. Rock fill is generally placed in 300–1200-mm lifts and compacted with heavy (10–15 ton) vibratory rollers.

Placement of impervious material is very sensitive to weather conditions, while pervious material can be placed in a wide variety of conditions. Sites with wet and dry seasons often find it expedient to maintain the various zones within the embankment at different levels, working on core placement during the dry season and placing pervious material during the wet season.

Careful inspection during construction is extremely important for embankment dams. It is essential that qualified personnel thoroughly familiar with the design basis be involved in the construction phase, since unforeseen changes in materials or foundation conditions can nullify design assumptions and, in extreme cases, require substantial design changes. During reservoir filling careful monitoring of the dam is required to assure that it functions as intended.

Foundation Preparation.   Foundation and abutment preparation is critical for embankment dams. Overhangs and significant irregularities under the core and filters must be filled in to prevent formation of voids or loose zones in the embankment. Care must be taken to prevent formation of a seepage path at the dam–rock contact. Equally important is the treatment of joints in the foundation to prevent embankment materials from entering them and being eroded away. Blanket grouting and careful dental concrete work are used beneath the core and filter zones to prevent piping. Soil foundations must also be suitably protected with filters to prevent piping. Grouted cutoffs are generally used to limit seepage under the dam, but impervious upstream blankets can be an effective alternative. Differential settlement can cause problems when rigid cutoffs such as concrete slurry trenches or ICOS-type walls are used. Foundation soils are often weaker than the embankment, and may be subjected to high pore pressures when the reservoir is full, so failure modes including foundation materials must be analyzed. Loose sands and silts are not only weak, they may liquefy in an earthquake and lose essentially all strength. Improvement of the strength of foundation soils is possible, but costs are generally prohibitive.

Diversion

Diversion of river flow during dam construction must be considered early in the planning of the project since diversion requirements can profoundly influence the dam design and the construction schedule. Diversion schemes include cofferdams to protect the construction area and a waterway to carry the diverted flow past the site. The waterway may be a portion of the existing channel, tunnels through the abutment, or sluiceways within the partially completed dam. Economies can be achieved at some sites by early construction of power tunnels or gated spillway structures and their subsequent use for passing diversion flows.

Cofferdams are temporary structures, but considerable care is needed in their design. The most economical design is dictated by the diversion scheme and available materials. Dumped rock fill with an impervious face (frequently dumped soil) is popular because the rock fill can be placed in water flowing with substantial velocity, an important consideration for closure sections. Steel sheetpile cells are also an economical and popular choice where pervious fill materials are available. Cells require a smaller area than a rock-fill embankment and are useful adjacent to an existing structure. However, cells are difficult to construct in flowing water. A major project will frequently use more than one type of cofferdam.

The design flood for the diversion facilities represents a balancing of costs and risks. Cofferdams are typically designed for floods with return periods between 10 and 500 years, but a greater or lesser flood is appropriate in some cases.

10.3-4 Dam Safety

Many dam failures have resulted in loss of life and extensive damage to property. The potential for loss of life and property after failure of a dam (hazard potential) depends on dam height, reservoir volume, and conditions downstream. Guidelines for evaluating hazard potential have been developed by the Corps of Engineers.[2] Even low dams can result in severe destruction should they fail. The hazard potential of a dam will influence design criteria.

Causes of Dam Failure

The U.S. Commission on Large Dams (USCOLD) investigated the causes of dam failures in the United States[3] and reached the following conclusions:

1. Most, but not all, failures involved older dams.
2. Nearly one-quarter of the failures were due to overtopping. Dams of recent design have not failed by overtopping.
3. Seepage through the enbankment or the foundation accounted for 45% of the failures.
4. Five percent of the failures involved slides on the downstream slope of embankment dams and an additional 5% involved sliding of concrete dams. Several near failures involved slides on the upstream slopes of embankment dams.

As noted above, failures by seepage and piping accounted for nearly half of the failures reported, and several modern dams were involved. Some seepage will occur through, beneath, or around any dam no matter how carefully grout curtains or seepage cutoffs are constructed. The dam must be designed to collect and control this seepage, and to prevent it from causing piping.

Spillways

Overtopping of a dam can cause erosion of earth embankments, overstressing, or sliding of concrete dams and higher water levels upstream. A spillway is used to pass flood flows without exceeding the maximum design water level upstream. Concrete dams can have the spillway incorporated into the structure, while embankment dams require a separate spillway, usually a concrete structure or a side channel excavated into the abutment. The type and size of the spillway depends largely on dam operation. Most major projects incorporate both a gated spillway to regulate reservoir elevation and maintain minimum flows downstream and an uncontrolled spillway at a relatively high elevation to pass large floods. Some small embankments have been constructed with conduits penetrating them to allow discharge, but great care is needed to prevent piping from developing along the conduit. The settlement and lateral spreading associated with large embankment dams may rupture a conduit installed through them, and such installations should be avoided.

Existing Dams

Many existing dams were designed for floods of lesser magnitude than are postulated in current design practice. Reanalysis based on current practice may result in postulated overtopping that was not anticipated in the original design. However, if a concrete dam is stable under the higher loads, the overtopping can be accommodated. If it is not stable, anchors might be installed to increase stability. An alternative, applicable to both concrete and embankment dams, is to provide additional spillway capacity; however, this may be technically unfeasible or quite costly at many sites.

Maintenance practices at existing dams vary from dam to dam. The repairs needed because of deterioration vary from major reconstruction to ensure safety down to minor and essentially cosmetic work. In some cases, particularly when the dams are of timber crib or masonry construction, deterioration may be so severe that economic and safety concerns dictate that the dam be breached.

Maintenance

Dam owners are responsible for the safety and adequacy of their dams. Most nonfederal dams associated with power generation are licensed by the Federal Energy Regulatory Commission (FERC). The licensee is required, under FERC Order 122,[4] to "use sound and prudent engineering practices," report safety-related incidents, report modifications to the project, maintain records, develop and file an emergency action plan, and, every 5 years, have a complete inspection made by an independent consultant. Low dams with small reservoirs are exempt form the latter requirement, unless they have a high hazard potential. Good maintenance and monitoring practices are necessary to meet FERC requirements; prudence dictates that they also be employed on exempt or unlicensed projects.

Maintenance of dams requires periodic inspections of the entire project and the repair of deteriorated structures and equipment. The items included in the maintenance program vary from project to project, but usually include the following items:

1. Embankments—check for seepage, settlement or deformation, movement or breakage of riprap, erosion, animal burrows, and growth of trees and shrubs.
2. Concrete sections—check for deterioration of concrete (spalling, erosion, exposed reinforcing, etc), opening of cracks, movement at joints, differential settlement or movement, and leakage.
3. Control structures—check proper functioning of gates, hoists, valves, and so on. Inspect various components for corrosion, wear, or signs of overstressing.
4. Drains—check proper functioning. Note long-term changes in flow.
5. Abutments and reservoir perimeter—check for seepage and development of instabilities.
6. Instrumentation—check that monitoring instruments are functioning properly and compare readings with those taken previously.

REFERENCES

10.3-1 American Concrete Institute, "Roller Compacted Concrete," Report No. ACI 207.SR-JO, *ACI Journal*, July/August 1980.

10.3-2 33 CFR Part 222.

10.3-3 U.S. Committee on Large Dams, *Lessons from Dam Incidents, USA*, American Society of Civil Engineers, New York, 1975.

10.3-4 18 CFR Part 12.

BIBLIOGRAPHY

American Society of Civil Engineers. *Foundations of Dams*, ASCE, New York, 1974.

American Society of Civil Engineers. *Inspection, Maintenance, and Rehabilitation of Old Dams*, ASCE, New York, 1974.

American Society of Civil Engineers. *Pumped Storage*. ASCE, New York, 1975.

Bachmann, J. "Standardizing Small Turbines," *Water Power and Dam Construction*, **32**, 7, July 1980.

Baumeister, T., Avallone, E. A., and Baumeister, T. III. *Marks' Standard Handbook for Mechanical Engineers*, 8th ed., chap. 9, McGraw-Hill, New York, 1979.

Bureau of Reclamation. *Design of Arch Dams*. U.S. Department of the Interior, 1977.

Bureau of Reclamation. *Design of Gravity Dams*. U.S. Department of the Interior, 1976.

Bureau of Reclamation. *Design of Small Dams*, 2nd ed., U.S. Department of the Interior, 1973.

Creager, W. P., Justin, J. D., and Hinds, J. *Engineering for Dams* (3 vols). Wiley, New York, 1975.

Critikos, T., Field, J. S., Rutherford, P. H., and Miller, A. V. *Powerhouse Design and Construction for Bulb Units*. Stone & Webster Engineering Corp., Boston, 1981.

Critikos, T., Makarechian, A. H., Garrity, J. J., and Elmore H. C. *The Rock Island Project—First on the Columbia River, First U.S. Bulb Turbine Installation*. Stone & Webster Engineering Corp., 1981.

de Moraes, J., Rodriquez, J., Gummer, J. H., and del Brenna, F. "Turbines for Itaipu," *Water Power and Dam Construction*, Part One, Vol. 33, No. 12, December 1982; Part Two, Vol. 34, No. 1, January 1981.

de Siervo, F. and de Leva, F. "Modern Trends in Selecting and Designing Francis Turbines." *Water Power and Dam Construction*, August 1976.

de Siervo, F. and de Leva, F. "Modern Trends in Selecting and Designing Kaplan Turbines." *Water Power and Dam Construction*, December 1977 and January 1978.

Doland, J. J. *Hydro Power Engineering*. Roland Press, New York, 1954.

Golze, A. R. *Handbook of Dam Engineering*. Van Nostrand Reinhold, New York, 1977.

Hainson, B. C. "Design of Underground Powerhouses and the Importance of Pre-excavation Stress Measurements." *Design Methods in Rock Mechanics*, Sixteenth Symposium on Rock Mechanics, University of Minnesota, 1975.

Harty, F. R. Jr. *Hydraulic Effects Associated with Preliminary Operation of the Northfield Pumped Storage Project*. Stone & Webster Engineering Corp, Boston, 1974.

Hirshfeld, R. C. and Poulos, S. J. *Embankment Dam Engineering*. Wiley, New York, 1973.

Hokenson, R. and Cunningham, C. "Selection of Installed Capacity," *Waterpower '81 Proceedings*, Vol. I, U.S. Army Corps of Engineers, 1981.

International Electrotechnical Commission. *International Code for Model Acceptance Tests of Hydraulic Turbines First Supplement*. IEC Publication 193A, IEC, 1972.

International Electrotechnical Commission. *International Code for the Field Acceptance Tests of Storage Pumps*. IEC Publication 198, IEC, 1966.

Makdesi, F. I. and Seed, H. B. "Simplified Procedure for Estimating Dam and Embankment Earthquake Induced Deformations." *Journal of the Geotechnical Engineering Division*, American Society of Civil Engineers **104**, (GT7), July 1978.

New England River Basins Commission. *Water, Watts, and Wilds: Hydropower and Competing Uses in New England*. New England River Basins Commission, Boston, 1981.

Nohab, Tampella and Leffel. *Small Hydro Turbines*. Nohab Tampella Hydro Energy & Leffel Hydro Energy, Finland, 1980.

St. Onge, G. A., Harty, F. R. Jr., and Click, J. E. *Start-up, Testing, and Performance of First Bulb-type Hydroelectric Project in the USA*. Stone & Webster Engineering Corp., Boston, 1981.

Schafer, L. and Agostinelli, A. "Using Pumps as Small Turbines." *Water Power and Dam Construction* **33**, (11), November 1981.

Seed, H. B., Makdesi, F. I., and DeAlba, P. "Performance of Earth Dams During Earthquakes." *Journal of the Geotechnical Engineering Division*, ASCE **104** (677), July 1978.

Sherad, J. L., Woodward, R. J., Gizienski, S. F., and Clevenger, W. A. *Earth and Earth Rock Dams*. Wiley, New York, 1963.

Strohmer, F. and Walsh, E. "Appropriate Technology for Small Turbines," *Water Power and Dam Construction*, **33**, (11), November 1981.

Wayne, W. W., Jr. *Hydroelectric Equipment Selection*. 1980 Lecture Series on Small Scale Hydropower, BSCES/ASCE, Boston, 1981.

Wayne, W. W. Jr. *Small Scale Hydropower Equipment*. Stone & Webster Engineering Corp., Boston, 1980.

Yang, K. H. "Design Trends of Hydro Plants in the USA." *Water Power and Dam Construction*, **32**, (7), July 1980.

CHAPTER 11
SOLAR-DERIVED POWER

AMIYA BASU, THOMAS LEPLEY, ERIC WEBER

Arizona Public Service Company
Phoenix, Arizona

ROBERT L. SCHEFFLER, MICHEL C. WEHREY

Southern California Edison Company
Rosemead, California

EDWARD S. LIPINSKY

Battelle Columbus Laboratories
Columbus, Ohio

DAVID A. TILLMAN

Envirosphere Company

DONALD L. KLASS

Institute of Gas Technology
Chicago, Illinois

PAUL C. YUEN

University of Hawaii at Manoa
Honolulu, Hawaii

11.1 DIRECT SOLAR ENERGY

11.1-1 Energy from the Sun

Amiya Basu

The creation of energy at the source is generally accepted as a hydrogen-to-helium thermonuclear reaction in the sun. This energy is produced in the interior of the solar sphere at approximately $20 \times 10^{6}\,°K$ and is radiated into the solar system at an equivalent blackbody temperature of $5760\,°K$

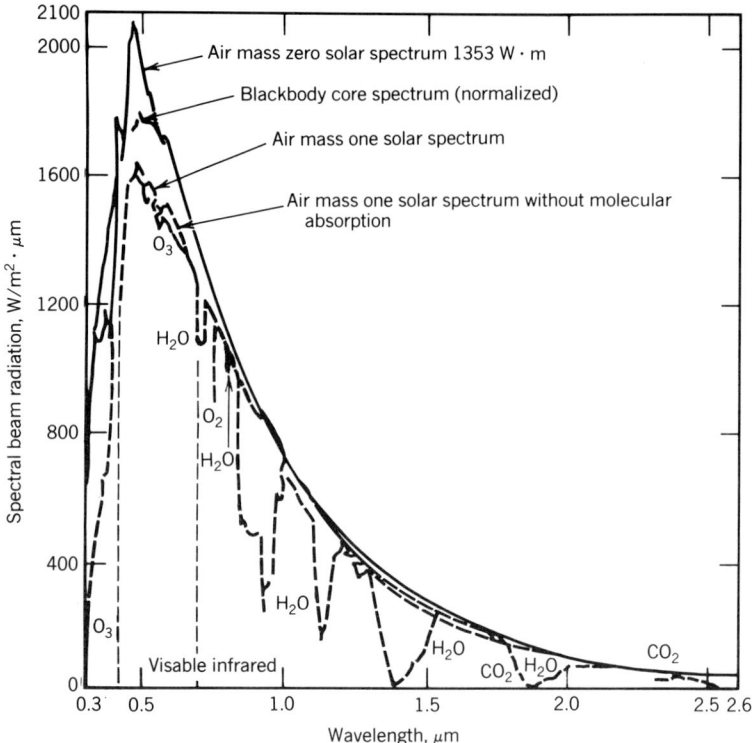

Fig. 11.1-1 Solar spectral intensity for air mass zero and one. (From ref. 13); reprinted with permission, Institute of Environmental Sciences, Mt. Prospect IL., originally published in "The Energy Crisis and Energy from the Sun," author, M. P. Thekaekara.)

from the outermost layer. Of the total power emitted from this layer, only a tiny fraction, 1.7×10^{14} kW is intercepted by the earth.[4]

The average amount of solar radiation in near-earth space is called the solar constant. The standard value for this constant is 1353 W/m². The characteristics of the sun and its special relationship to the earth result in a nearly fixed intensity of extraterrestrial solar radiation.[4]

Solar radiation received at the surface of the earth is dependent on the components of the atmosphere and variations in the earth/sun distance. Solar radiation of wavelengths less than 0.29 μm is absorbed high in the ionosphere by nitrogen, oxygen, and other atmospheric components; ozone absorbs most of the ultraviolet. Terrestrial applications, therefore, are limited to radiation of wavelengths between 0.29 and 2.5 μm. Figure 11.1-1 shows the atmospheric effect on the solar spectrum. Scattering from air molecules, water vapor, and dust result in further attenuation of the solar radiation.

Absorption of certain wavelengths limits terrestrial sunlight to 0.29–2.5 μm essentially across the globe. This component is termed *direct*, or *beam*, *radiation*. For the most part, scattering this component is dependent on weather and climatic conditions. Direct radiation that has changed direction by reflection or scattering by the atmosphere is diffuse radiation.

The incident solar radiation reaching the earth's surface is dilute and intermittent in nature. An important limitation of solar energy is its intermittency. Any development of solar energy as a large-scale source of power and heat would be dependent on daily, seasonal, and annual variations of irradiance. Furthermore, solar energy is unavailable during periods of inclement weather. All these variations would place an important design factor on the conversion, storage, and distribution of this energy.

Energy Conversion Modes

Solar energy is captured by being converted to other forms of energy. This can be done by three methods: (1) chemical reaction, (2) thermal excitation, and (3) photovoltaic effect. Sunlight is chemically converted into energy through the photosynthesis process. This directly produces food and wood and offers the possibility of cultivation of appropriate plants exclusively for the purpose of generating power. This could be realized through direct bioconversion or by pyrolytic conversion into liquid or gaseous fuel.

Simple thermal conversion devices, such as flat plate collectors, are suitable mainly for providing low-energy, high-entropy heat to systems of the same nature. These include hot water, space heat, and absorption cooling systems. The flat plate technology collectors can deliver temperatures up to approximately 100°C.

Medium to high temperatures can be achieved by a number of advanced technology devices by concentrating the incident sunlight. These devices provide heat for high-energy, low-entropy applications, such as generation of steam to produce electric power.

The direct conversion of sunlight to electricity by the means of solar cells is the photovoltaic effect. The solar cells utilize energetic photons of the incident solar radiation, converting sunlight into dc electricity. The theoretical efficiency of the photovoltaic process is dependent on the solar cell material and the fraction of the solar spectrum that specific material converts.

11.1-2   Solar Thermal Collectors and Conversion Methods

Amiya Basu

Solar thermal collectors are designed to absorb the radiant energy received from the sun and in turn heat up the working fluids flowing through them. Depending on the type of application and the solar energy available, the collectors could be flat plates, evacuated tube types, or concentrating collectors. Concentration of the sun's rays or evacuation of collectors are required in places where flat plate collectors are not suitable to provide the energy input at higher fluid temperature.

Flat Plate Collectors

The flat plate collector is most widely used among solar collectors in use today. It is relatively inexpensive and easy to construct. These collectors typically can deliver energy at temperatures of 100°C above ambient. The flat plate collector utilizes incident solar radiation to heat up the fluid flowing through the collector and generally does not require tracking. Flat plate collectors are commonly used in low-temperature applications such as domestic water heating, swimming pool heating, and so on.

A typical flat plate collector is made up of a black surface (absorber plate), thermal insulation on the back and edges of the black surface, and one or more transparent covers on the top of the surface. The solar energy absorbed by the black surface is transferred to the fluid flowing through the collector tube (liquid collector) or to the air passing through the duct (air collector). The cross-sectional views of an air and a liquid collector are shown in Figs. 11.1-2 and 11.1-3, respectively. One or more glass or plastic covers are used to minimize the heat loss from the black surface to the surrounding air. The covers help to reduce the convection and radiation losses from the surface as well as create a greenhouse effect (covers are opaque to the long-wave radiation).

The performance of a flat plate collector is dependent on factors such as energy absorbed by the absorber plate, energy losses to the surroundings, local meteorological conditions, and so on. In order to determine the amount of net energy collected by the absorber, it is necessary to determine the amount of heat absorbed, heat losses to the environment, and the optical losses. Although flat plate collector heat balance is fairly complex, the basic analysis proposed by Hottel and Woertz[1] and further derived by Duffie and Beckmen,[4] is a good approximation under steady-state conditions.

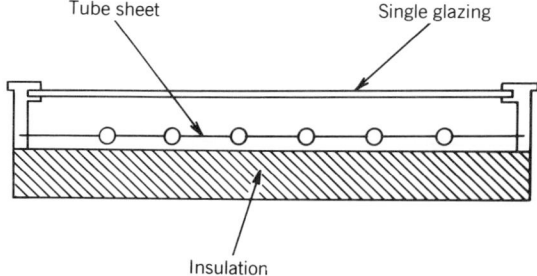

Fig. 11.1-2   Flat plate collector (liquid) cross section.

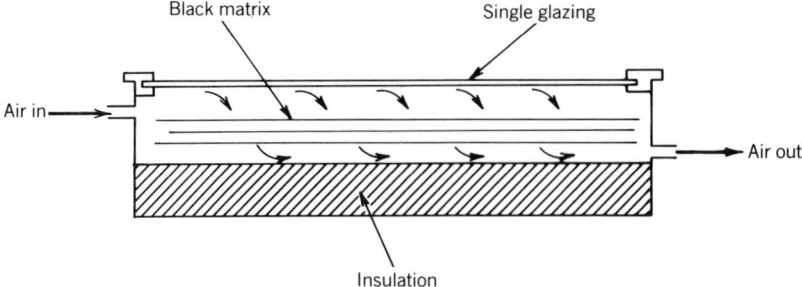

Fig. 11.1-3   Flat plate collector (air) cross section.

Evacuated Collectors

The performance of a flat plate collector can be considerably enhanced if the absorber heat losses can be reduced. In evacuated tube collectors this is done by surrounding the receiver with a vacuum tube. This results in minimal or no heat loss from the receiver surface to the outside air, thereby improving the thermal efficiency of the collector. The efficiency of these collectors can be improved further by adding concentrators around the evacuated tube. These collectors, although more expensive than the flat plate collectors, do not require tracking, but at the same time produce heat at relatively high temperatures.

The evacuated tube collector designs commercially sold today include either the collector fluid flow directly through the tube or through a sleeve inserted inside the vacuum tube. In the second case the collector fluid does not come in direct contact with the tube surface. The cross sections of these two types of tube configurations are shown in Figs. 11.1-4 and 11.1-5. A slight concentration, between 1.3 and 2.0, can also be achieved by coating the internal glass surface with a mirrored film.

The energy balance equation for an evacuated tube collector is similar to the flat plate collector analysis. The losses through the absorber plate and back and edges of a flat plate collector do not exist in the evacuated tube design. The analysis originally performed by Owens-Illinois Company[2] and subsequently developed by Kerith and Kreider[3] are referred here to approximate the thermal performance of an evacuated tube collector.

Concentrating Collectors

Concentrating collectors offer the distinct advantage of increasing the fluid temperature by optically concentrating the sun's ray. The concentration is achieved by using otherwise dilute solar energy to focus either at a point, called point focusing collector, or on a straight line, called line focusing collector. Concentrating on a smaller surface area as compared to a flat plate collector also helps to reduce the heat losses associated with the large absorber surface.

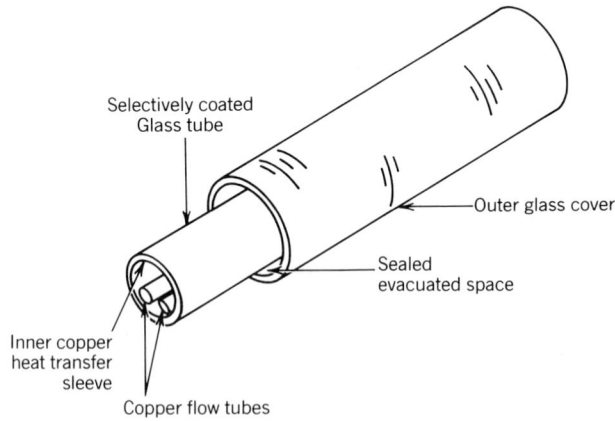

Fig. 11.1-4   Evacuated tube design. (Source: SOLERAS Program—Engineering Field Test of a Solar Cooling System. DOE/ET/20645-1 May 1983.)

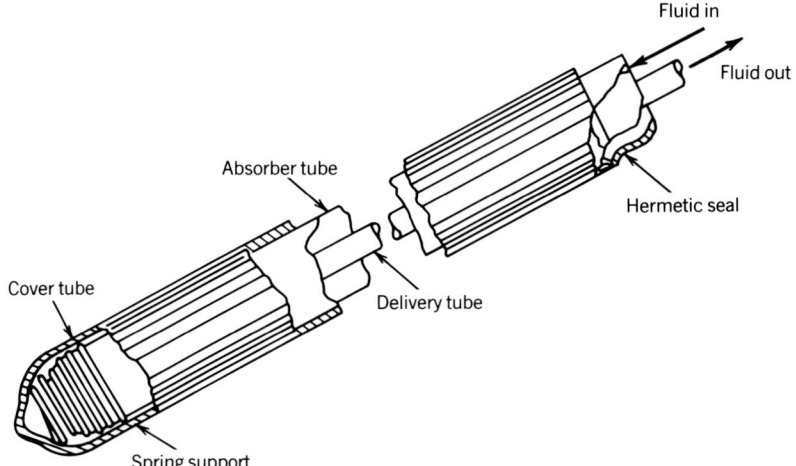

Fig. 11.1-5 Owens-Illinois Company evacuated tube design (from ref. 3).

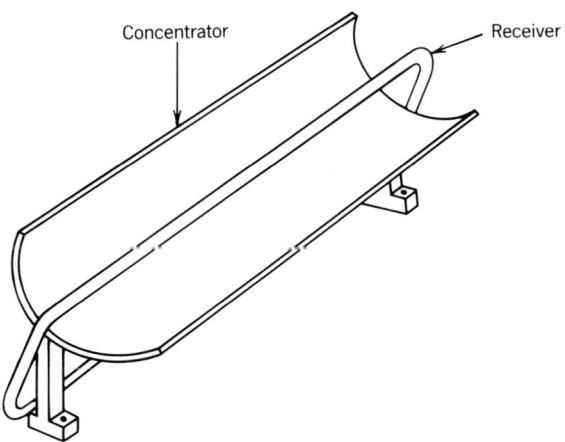

Fig. 11.1-6 Periodic concentrator.

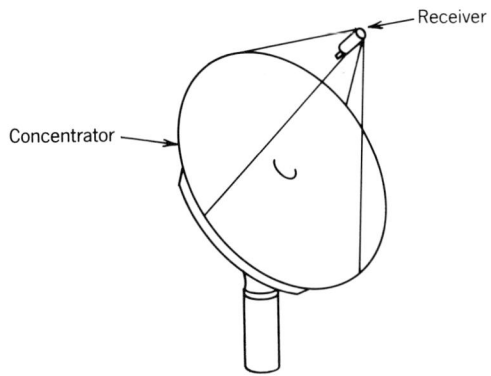

Fig. 11.1-7 Dish-type concentrator.

Rays from the sun are focused either at a point or in a straight line in imaging collectors. The sun's image is formed on the receivers by tracking the sun. The trough collectors are single-axis tracking types (two-dimensional) and the dish/bowl concentrators are driven by biaxial tracking mechanisms (three-dimensional). Examples of two- and three-dimensional tracking collectors are shown in Figs. 11.1-6 and 11.1-7.

These collectors can be used to provide thermal energy to a variety of solar energy systems. They can be designed to operate a solar absorption cooling system at 90°C (low temperature) as well as supply solar process heating requirements at 250–300°C (mid temperature). Concentrating collectors can also generate heat at very high temperatures (above 550°C).

11.1-3   Active Solar Heating and Cooling Systems

Amiya Basi

Solar Hot-Water Heating System

The main components in a solar water heater are flat plate collectors and the storage tank. Solar hot-water heaters could be classified as natural circulation or forced circulation systems. In a natural circulation system (Fig. 11.1-8) the storage tank is placed at a higher level than the collector. Once the water is heated up by the collector, it becomes less dense and rises to the higher elevation. Thus water starts to circulate by the natural convection (thermosiphon) process. In order to prevent reverse natural convection, the storage tank should be located above the collector elevation level.

A forced circulation solar hot-water system is shown schematically in Fig. 11.1-9. This system uses a pump to circulate the working fluid throughout the system. As shown in Fig. 11.1-9, check valves, controls, and auxiliary heaters are required to ensure efficient system operation. The check valve prevents the natural circulation at night while the controller would provide freeze protection as well as start or stop the system at design set points. Normally, the sensor placed on the collector absorber plate activates the pump in case the absorber plate approaches freezing condition.

Swimming Pool Heating System

The use of solar energy to heat swimming pools is technically feasible and in many cases economically viable. The design temperature requirement of 30°C for swimming pool heating suitably matches with the flat plate collector performance. At that low temperature the collector can perform efficiently and the pool itself serves as the storage medium. A typical swimming pool heating system is shown in Fig. 11.1-10. The swimming pool circulation pump is normally used to pump the water from the pool into the collector through the pool filter. Back flows are prevented by using on-line check valves. A differential temperature controller helps to maintain the preset pool temperature by mixing flows through the control valve.

The performance of a solar swimming pool heating system can be improved markedly by utilizing a pool cover. A pool cover helps to reduce convection and radiation, as well as evaporative losses of the pool.

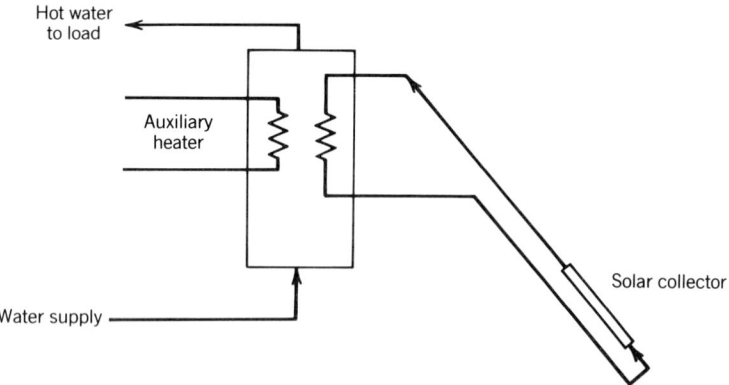

Fig. 11.1-8   Solar hot water heating system (natural circulation).

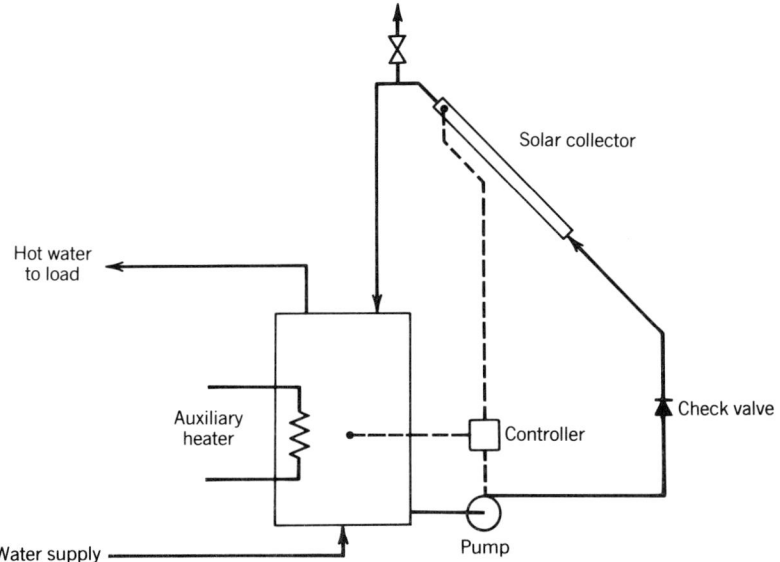

Fig. 11.1-9 Solar hot water heating system (forced circulation).

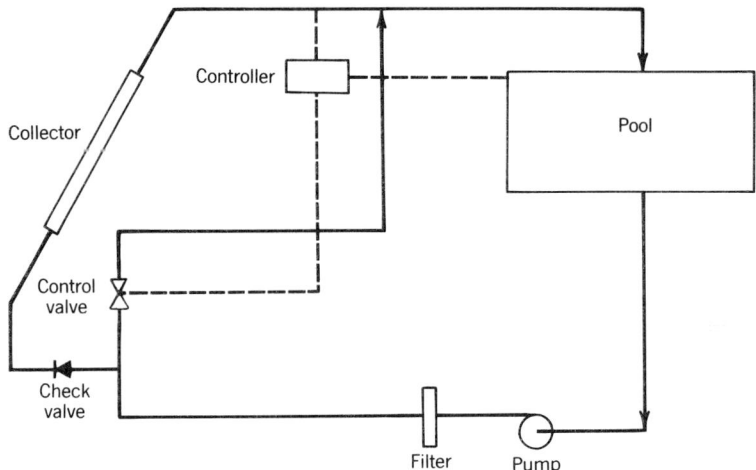

Fig. 11.1-10 Solar swimming pool heating system.

Air-Heating System

Depending on the thermal requirements, air-heating systems might be preferable to the liquid systems. Air is noncorrosive, does not freeze, is nontoxic, and free. There is no stagnation or vapor-lock problem and the design of the system is simple. Like any other system the air system also has its limitations. Due to relatively poorer heat transfer characteristics, air systems require larger parasitic power. Since the energy density of an air system is lower than a liquid system, energy storage requires more volume and, therefore, can be more expensive than the liquid system.

In air-heating systems flat plate collectors are used to heat the air. An air-heating system schematic is shown in Fig. 11.1-11. The operating modes are very similar to a liquid system, possibly with the exception of a rock-bed storage system. The efficiency of an air-heating system is severely reduced if there are leaks in the system. Air blowers tend to draw in cold air from outside through the leaks, drastically reducing the system performance.

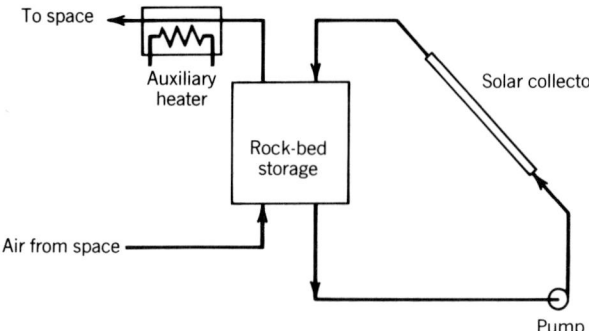

Fig. 11.1-11 Solar air heating system.

Space-Heating System

The heating of space using solar energy has been successfully tested in this century. Starting in 1938 a number of solar houses were built successively at MIT. The data obtained from the solar space-heating system in these houses provided the basis for subsequent solar space-heating technology development in the post–oil embargo era.

The schematic of the MIT house IV is given in Fig. 11.1-12. Above (60 m^2) of the collector area was installed at an angle of 60°. In the northern hemisphere fixed collectors are installed due south at an angle of about 15° greater than the local latitude. The system worked on the draindown principle (explained in the hot water section) and utilized a 946 liter draindown tank. The main storage tank was of 5678 liters storage capacity. An oil-fired 379 liter auxiliary tank was used as a backup source. The heat was transferred to the house through a water-to-air heat exchanger.

The collector pump operation was similar to the hot water system. Sensors located in the absorber plate and the storage tank controlled this loop. When the house thermostat called for heating, the hot water pump was turned on to circulate the hot water from the storage tank. If solar heat was not available, control was switched to the auxiliary operating mode.

Space-Cooling System

Among other direct uses of solar energy, the cooling of spaces by using the energy from the sun has been greeted with great enthusiasm. In tropical countries there is a special interest because everyone

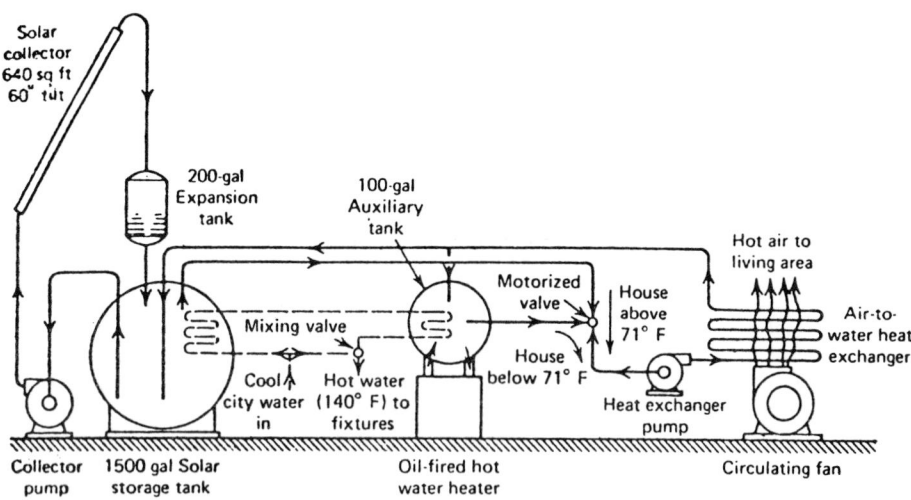

Schematic of MIT Solar House IV.

Fig. 11.1-12 Schematic of MIT solar house IV (from ref. 5).

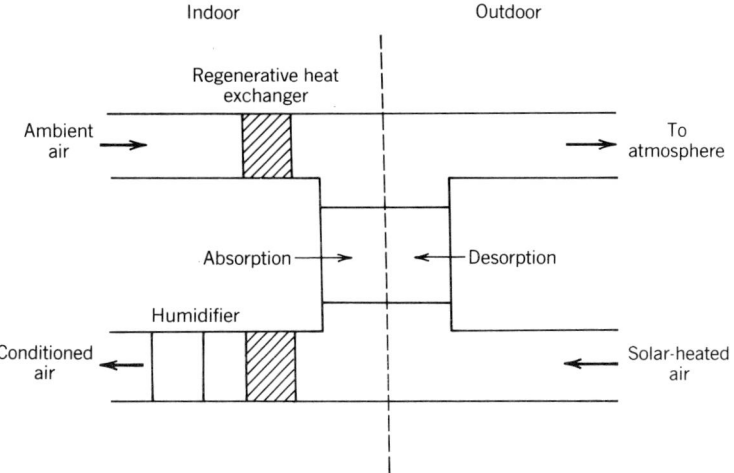

Fig. 11.1-13   Desiccant system (from ref. 11).

desires cooling, but the systems now available are expensive. Almost all the solar cooling systems being considered so far have used the absorption refrigeration system. Very recently, systems like the mechanical vapor compression cooling system driven by a Rankine cycle engine and desiccant dehumidification systems, using solar energy as the source of heat, have gained some popularity.

Desiccant Cooling System.   Solid and liquid desiccant materials have long been known and put to use in various industries. Desiccant materials are used extensively for processes such as dehydrating compressed air, recovering hydrocarbons from natural gas, and drying unsaturated hydrocarbons (ethylene, propylene, butadiene). Dehumidification in moist, hot climates is almost as good as cooling.

A desiccant cooling system is considerably different from other conventional cooling systems. Basically, the air from the room or the cooling space is passed through a desiccant material. The dehumidification, which is an exothermic process, increases the temperature of the air undergoing circulation. The heat is transferred to the outside.

The dried or dehumidified air is humidified again depending on the required inlet condition for the space to be conditioned, thereby cooling the ambient air to achieve air conditioning. The climatic conditions play a very important role in the use of desiccant dehumidification. This system is especially useful in a region characterized by high relative humidity and low day-to-night temperature swings. A schematic diagram of a desiccant system is shown in Fig. 11.1-13.

Rankine Cycle—Mechanical Vapor Compression Cooling System.   Another alternative method for producing air conditioning involves a mechanical vapor compression cooling cycle powered by a Rankine cycle engine. Almost all solar-powered air-conditioning systems built so far have used the absorption cycle. But recent expansive research on turbomachinery and Rankine cycle engines, using organic working fluids such as Freon, have made solar-powered compression refrigeration a technically feasible system.

The Rankine cycle–mechanical vapor compression cooling system is shown schematically in Fig. 11.1-14. This system is split into two loops, one a conventional Rankine power cycle with a turbine that drives a compressor and a conventional vapor refrigeration cycle. Basically, the solar-powered, Rankine cycle–mechanical vapor compression cooling system is composed of three different sections: a Rankine cycle, a vapor compression cooling loop, and a solar collector array.

In the Rankine cycle working fluid (in liquid state) is pumped into the generator and heated by the solar energy. The liquid vaporized at the generator expands through the turbine producing the shaft work. The heat is rejected to the atmosphere by the condenser, and the turbine exhaust is converted back to the liquid state. The liquid is then pumped back to the generator.

The shaft power available from the turbine is used to drive the compressor, which in turn compresses the vapor received from the evaporator. The compressed vapor is condensed to liquid before it passes through the expansion valve. The working fluid at high pressure is expanded by throttling through the expansion valve. Cooling is accomplished by evaporating the liquid (refrigerant) inside the evaporator. The vaporized refrigerant is then compressed again before condensation.

Absorption Refrigeration System.   Vapor compression is no doubt an efficient method to achieve refrigeration. But relative term system requires shaft work as energy input, which is a high-grade

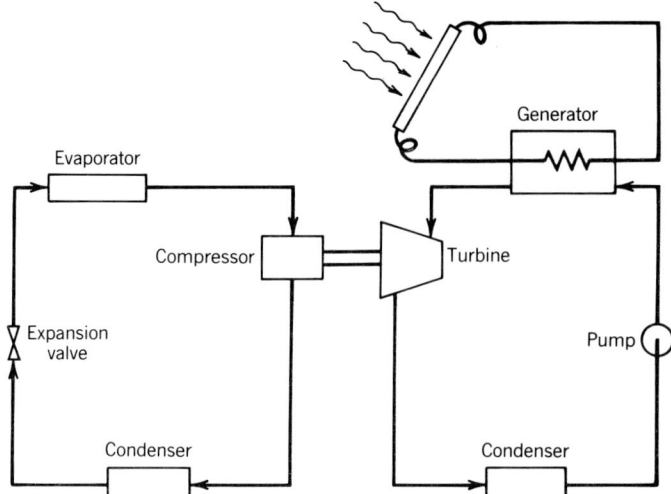

Fig. 11.1-14 Rankine cycle–mechanical vapor compression cooling system (from ref. 8).

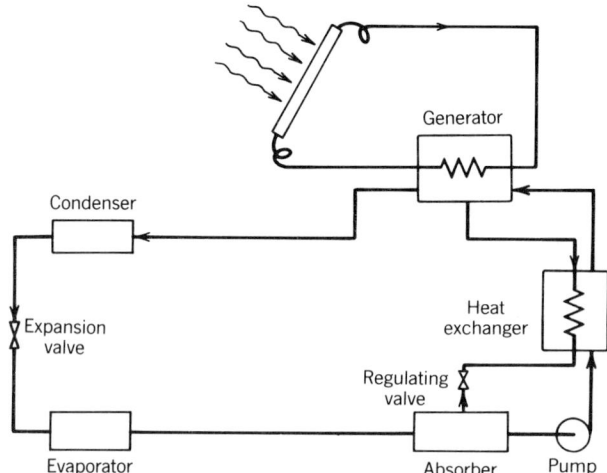

Fig. 11.1-15 Absorption refrigeration system (from ref. 8).

energy. In an absorption refrigeration system this work is reduced to a great extent by the binary mixture absorption process, which raises the pressure of the refrigerant appreciably.

The schematic of an absorption refrigeration system is given in Fig. 11.1-15. Solar heat from a flat plate collector is added to the generator that contains the refrigerant-absorbent solution. The addition of heat vaporizes the refrigerant. This increases the concentration of the refrigerant, which is desirable. This concentration is maintained by allowing the weak solution to pass through a valve and recirculate through the absorber.

Vaporized refrigerant is passed through the condenser to change the vapor state of the refrigerant to the liquid state. The liquid refrigerant is then passed through a throttle valve for expansion before it enters the evaporator. The low-temperature vapor refrigerant coming out of the evaporator is absorbed into the solution in the absorber. This increases the concentration of the refrigerant in the solution contained in the absorber, thereby making it a strong solution. The weak solution from the generator replaces the strong solution, leaving the absorber by setting up weak and strong solution circulation between generator and absorber. Thus, the cycle is continued.

11.1-4   Passive System

Amiya Basu

The solar systems described in this section so far are termed *active systems*. Passive systems employ nonmechanical means. In these systems heat transfer generally takes place by means of free convection and radiation.

In a solar passive heating system south-facing glass or plastic glazings are used as collectors and a thermal mass, a wall for example, is used for heat absorption, storage, and distribution. If the sunshine is allowed directly into the living space, heating up the walls, floor, and so on, the concept is known as direct-gain passive heating system. Another passive heating concept called the Trombe wall, was first designed by Felix Trombe of France. A large concrete or adobe wall is constructed and painted black. The Trombe wall is exposed to the sunlight and the heat absorbed by the wall during sunshine hours is transferred to the space by conduction, convection, and radiation. At night or during cloudy days the south wall glazing is covered to minimize the heat loss. During summer vents (provided on top and bottom) are opened on the top of the wall, to allow the hot air to escape outside. Vents at the bottom are opened up during heating season to set up natural convection current. The vents are closed at night to prevent reverse natural circulation.

A passive heating and cooling concept, first designed by Yellot and Hey,[6] also known as Sky-Therm, includes roof-supported water bags and a movable insulation as the major components. The water bags work as the collector and the storage. During the heating season the insulation is removed for the sun to heat the water contained in the bags. At night the insulation is moved back and the heat is transferred into the space. During summer the movable insulation is opened at night, and night sky convection and radiation help to cool the water. At daytime hours the insulation is pulled back and the space is kept cool by the water bags on the roof.

11.1-5   Solar Cells

Thomas Lepley

Background

The photovoltaic (PV) effect was first discovered in 1839 when Becquerel found that a photovoltage was developed when light was absorbed by an electrode immersed in an electrolyte. The effect was observed in a solid (selenium) about 40 years later by W. G. Adams and R. E. Day. Photovoltaics was used in photoelectric exposure meters for photography, but it was not until 1954 that it was considered as a possible source of electric energy. In that year a group of RCA scientists demonstrated that practical efficiencies could be achieved.

Solar cells were used for power production in the late 1950s and the 1960s in space applications, and in the 1970s work was begun to develop photovoltaics for terrestrial applications.

Some advantages of photovoltaic devices and systems are:

1. There is no inherent lifetime limit.[13]
2. Their efficiencies are independent of size.[13]
3. They are modular.
4. Once manufactured, they are quiet, benign, and compatible with almost all environments.[13]
5. They have a fairly constant voltage output independent of sunlight intensity.[13]
6. They respond instantaneously with illumination.[13]
7. They can operate with relatively little maintenance.
8. Operation and maintenance costs are expected to be low in most applications.
9. Cells have inherently high reliability and systems can be easily designed to have high reliability.
10. Simplicity (fewer moving parts).
11. Unattended operation has been demonstrated.
12. No cooling water requirements.

Some disadvantages of photovoltaic devices and systems are:

1. Their limited theoretical efficiency of about 25% combined with the low energy intensity of sunlight requires a relatively large collector (this is an inherent problem with all solar systems).[13]

2. Present capital costs make them economically uncompetitive with a majority of other energy sources and converters for most terrestrial applications today.[13]

3. They require storage or backup to supply any load requirements when there is insufficient sunlight.

4. They require dc to ac inversion equipment to supply ac loads.

5. Their manufacture may produce environmental hazards.

11.1-6   Operation of a Photovoltaic Cell

Thomas Lepley

The *photovoltaic effect* can be defined as the generation of a potential when radiation ionizes the region in or near the built-in potential barrier of a semiconductor. It is characterized by a self-generated emf and the ability to deliver power to a load, the primary power coming from the ionizing radiation.[22]

Effect of Light on a Semiconductor[22]

The photovoltaic potential developed when light strikes a semiconductor *pn* junction can be explained with the help of Fig. 11.1-16. The left side of Fig. 11.1-16 shows how carriers are generated when light photons enter a semiconductor crystal. The circles represent atoms in this two-dimensional crystal lattice and the double lines represent the two electrons in the covalent bonds, as in the case of silicon. Photons with energy above that required to break these bonds (1.1 eV for silicon) will create electron–hole pairs on a unit quantum efficiency basis as shown on the diagram. The right side of the figure shows the same process using the energy-band model for a semiconductor. Both an *n*- and *p*-type semiconductor are shown. In the case of *n*-type, a column V element contributes extra electrons, and in the case of a *p*-type, a column III element *does not* contribute enough electrons, and so contributes holes.

The Fermi level (denoted by *f* in Fig. 11.1-16 of the *n*-type material is near the top of the forbidden gap with many electrons, *n*, in the conduction band and few holes, *p*, in the valence band. The opposite is true in the *p*-type material. At a given temperature the product *np* is a constant ($np \sim 10^{21}$ for silicon at room temperature). For *n*-type highly conducting silicon, *n* can be $10^{17}/cm^3$ and *p* would therefore be $10^4/cm^3$. Thus, the concept of majority and minority carriers (electrons and holes, respectively, in the cited cases) is developed. When sunlight strikes this semiconductor, those photons having energy greater than the forbidden gap energy produce both types of carriers in equal numbers. The net effect is that an intense light source can increase the minority carrier density by many orders of magnitude while the effect on majority carrier density is negligible. Those photo-induced carriers are in excess of the thermal equilibrium number, and they will diffuse randomly about the semiconductor and recombine in times on the order of tenths of microseconds.

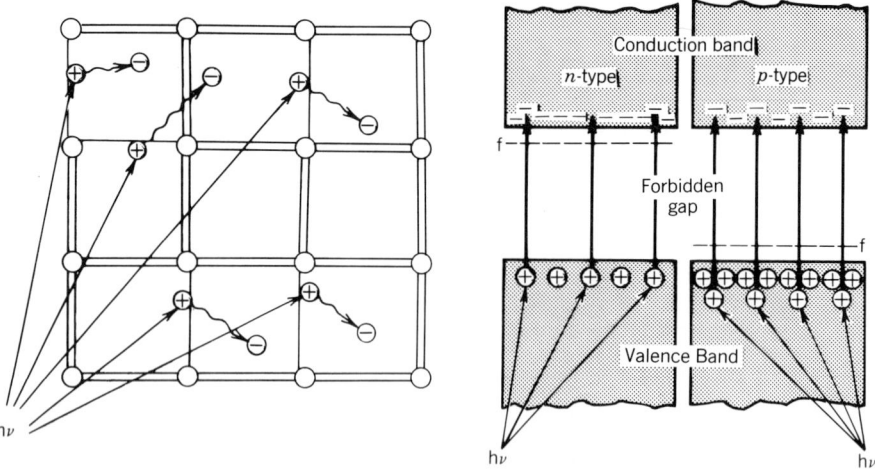

Fig. 11.1-16   Electron hole production by photons (from ref. 22).

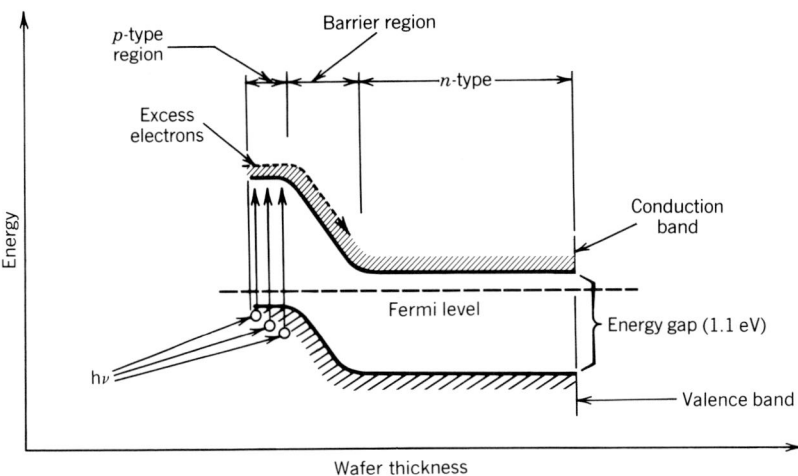

Energy level diagram of cell

Fig. 11.1-17 Energy level diagram of a cell (from ref. 22).

The p–n junction[22]

If the *n*- and *p*-type semiconductors are brought together, a *p–n* junction having a potential barrier is formed, because thermodynamics requires that the average energy of the carriers (the *Fermi level*) be the same in the two materials. Now, excess carriers that are within a *diffusion length* (the average distance minority carriers diffuse before they recombine) of the potential barrier will be "trapped" by the barrier and caused to flow across it in an attempt to reduce their energy. This is shown by the energy level diagram of a *p–n* junction in Fig. 11.1-17. The excess electrons flow to the right and the excess holes to the left. This flow constitutes an electric current, and when suitable electrical connections are made, such a device converts radiation into electricity. The current produced is proportional to the number of photons absorbed, and the voltage depends on the height of the barrier, which is always less than the width of the energy gap (E_g) in the semiconductor depending on how heavily the *p* and *n* regions are doped.

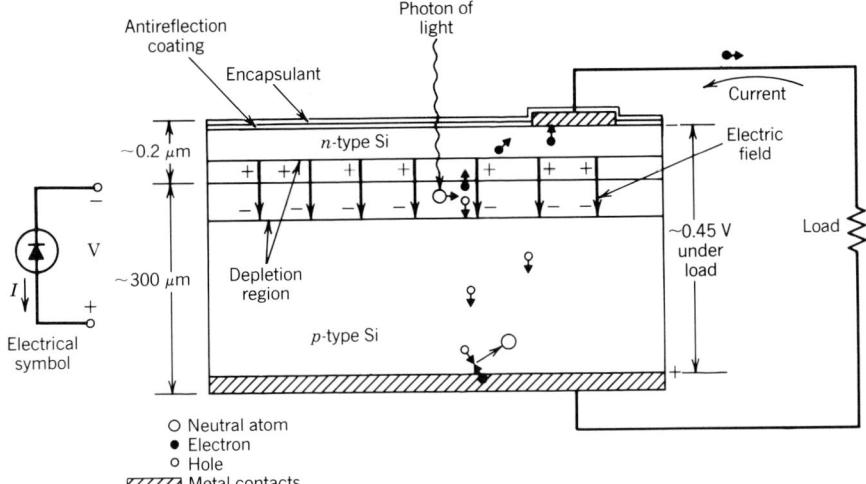

Fig. 11.1-18 Typical solar cell cross section. Note that current flow is, by convention, in the opposite direction to electron flow: ○ neutral atom, ○ hole, ● electron, ⬛⬛⬛ metal contacts.

The Solar Cell

The key factor in the operation of a solar cell is that the cell is designed in such a way that the photo-generated electrons and holes are separated before they can recombine. The electrons are forced to travel through an external circuit before recombining. Several structures are possible, including a semiconductor–metal junction (Schottky barrier), metal–insulator–semiconductor (MIS) junctions, and several types of heterojunctions (junctions consisting of two different materials). However, at present the most common structure is the silicon np homojunction.

A cross section of a typical cell is shown in Fig. 11.1-18. Light passes through a thin layer of n-type silicon and produces a hole–electron pair in the vicinity of the pn junction. Holes and electrons are separated by an internal electric field. The electrons are collected by a set of metal fingers or a grid on the top of the cell and travel through an external circuit. They recombine with holes at the bottom contact.

The voltage produced by the cell is a characteristic of the material used. For crystalline silicon the maximum voltage is about 0.6 V but may typically be about 0.45 V under load.

Efficiency

Solar cells cannot convert all of the incident sunlight into electricity. The majority of loss occurs because of the broad range of frequencies in the solar spectrum. Photons with energy less than the forbidden gap pass through the cell without being absorbed. Photons with energy greater than E_g are absorbed, but the excess energy is lost as heat. Figure 11.1-19 shows typical losses for a silicon cell. The maximum theoretical efficiency of several different semiconductors is shown in Fig. 11.1-20.

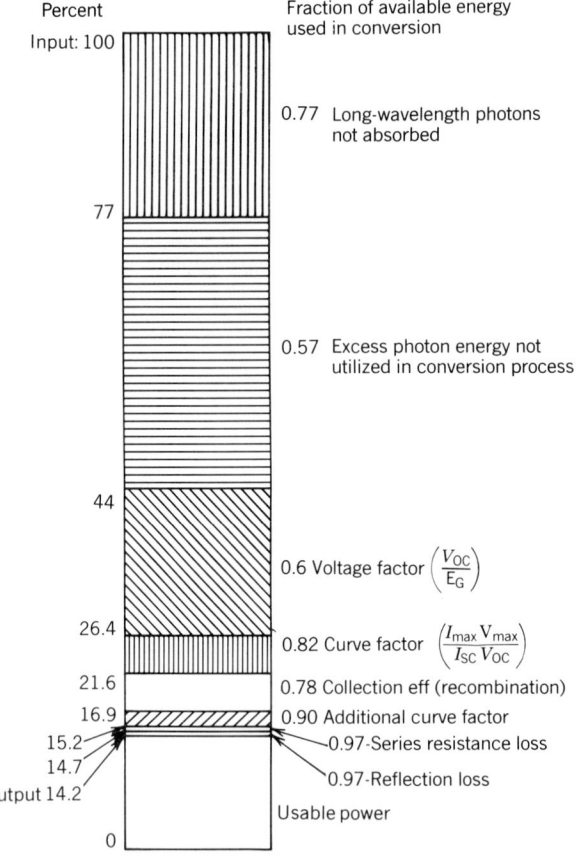

Fig. 11.1-19 Bar chart of distribution of energy losses in a typical silicon solar cell under air mass 1 (adapted from ref. 23; reprinted with permission.)

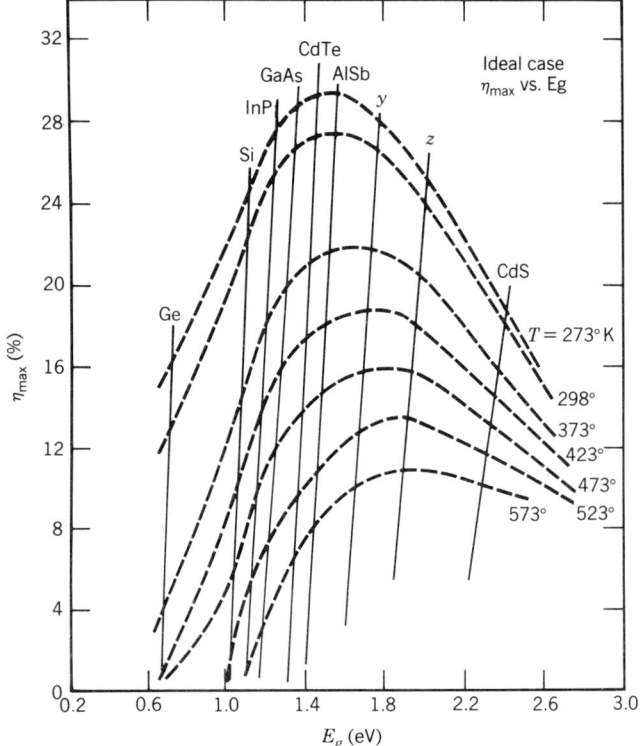

Fig. 11.1-20 Efficiency versus energy gap for various temperatures (from ref. 22).

Cell Voltage/Current Characteristics

The *pn* junction solar cell is simply a specialized type of diode. Figure 11.1-21 shows on *I–V* plot for a solar cell with and without illumination. It is customary to invert the waveform about the voltage axis as shown in Fig. 11.1-22. Figure 11.1-22 also shows the maximum power point, and defines the *short-circuit current* (I_{SC}), the *open circuit voltage* (V_{OC}), and the *fill factor*.

Figure 11.1-23 shows how the *I–V* characteristic varies with changing cell temperature and irradiance level. In general the short-circuit current of the cell is directly proportional to the irradiance level, and the voltage at the maximum power point is linearly dependent on cell temperature, decreasing about 0.5% of its 25°C value for each 1°C of increasing cell temperature.[25]

11.1-7 Conventional Cell Production Methods

Thomas Lepley

Polycrystalline Silicon Production

The first step in the production of silicon cells is purification of the silicon. Metallurgical (MG) grade silicon is generally produced by electrochemical reduction of silica from high-purity sandstone or quartz. The basic reaction is

$$SiO_2 + 2C \rightarrow Si + 2CO$$

Carbon acts as the reducing electrode in an electric arc furnace. The liquid silicon slag is 98–99% pure. It is further purified by reaction with hydrochloric acid to form trichlorosilane ($SiHCl_3$). This material is then cracked thermally to yield polycrystalline Si (*poly*) which is of semiconductor grade (99.999% pure).

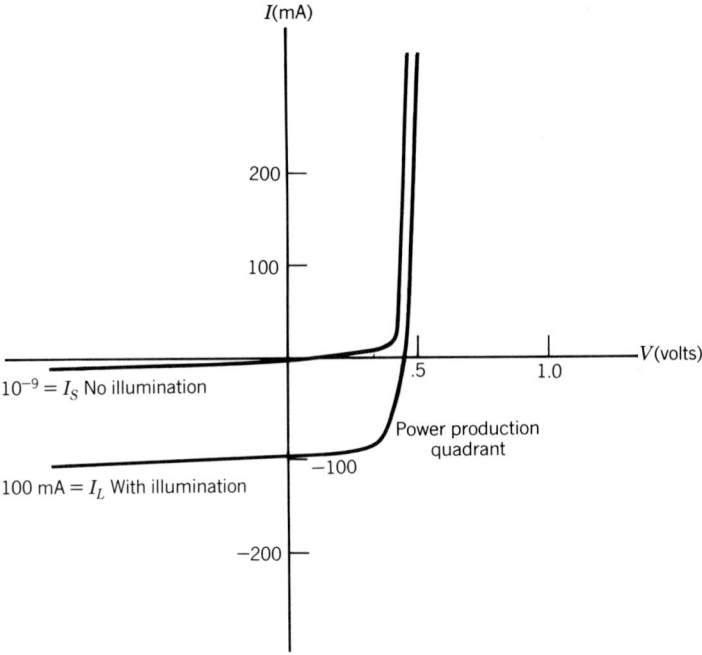

Fig. 11.1-21 Typical I–V plot of a solar cell: $I = I_s(e^{qV/kT} - 1) - I_2$, where $I_2 =$ strength of constant current source due to the incident light; $q =$ electron charge; $k =$ Boltzmann's constant = $8.62 + ID^{-5}$ eV/K; $T =$ temperature in K; $kT/q = 0.0259$ V at 300 K.

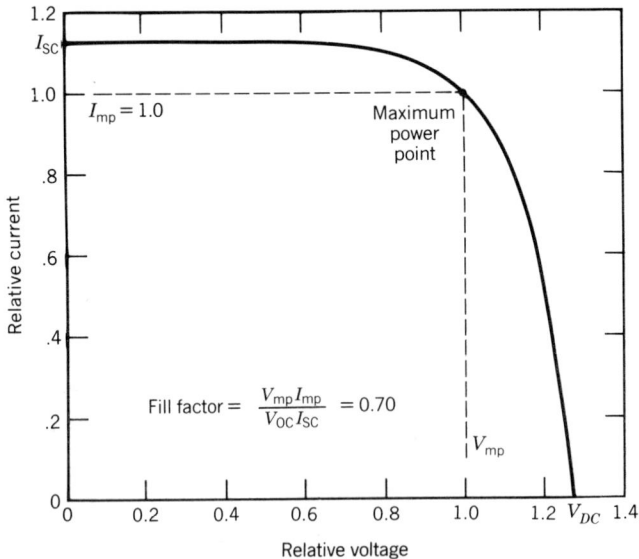

Fig. 11.1-22 Typical photovoltaic I–V curve at 100 mW/cm², 25°C cell temperature (from ref. 25).

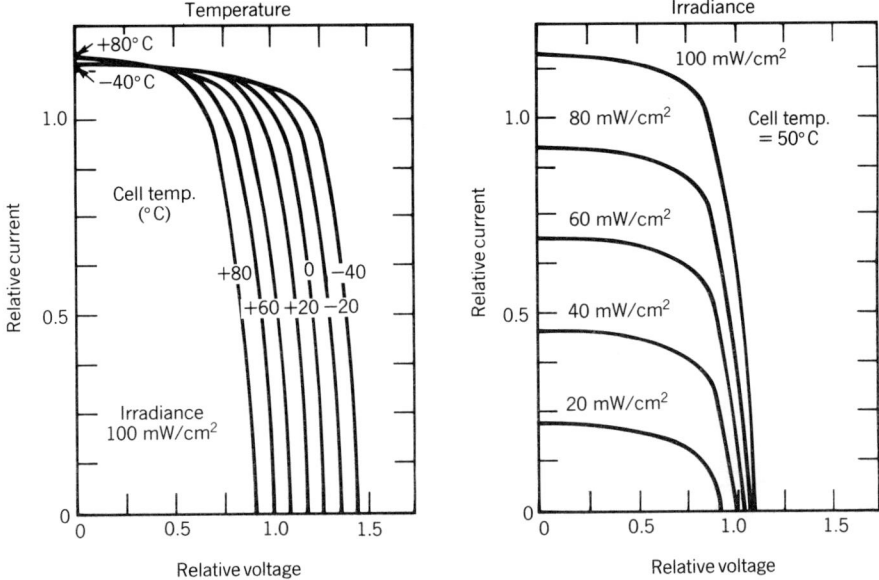

Fig. 11.1-23 Influence of irradiance level and cell temperature on array $I-V$ curve (from ref. 25).

Several low-cost polysilicon production processes are being developed. Two of these produce semiconductor grade poly (Battelle uses zinc reduction of silicon tetrachloride in a fluidized-bed reactor, and Union Carbide produces silane, SiH_4, from MG silicon and then deposits silicon from it). Two other approaches (by Westinghouse and SRI International) produce "solar grade" polysilicon, which has more impurities than semiconductor grade but less than MG, and may allow production of cheaper cells with acceptable performance. All of the approaches have the potential to reduce the price of poly from $60–75/kg to less than $10/kg.

Czochralski Method

The Czochralski method is presently the most common used for the production of commercial Si solar cells. Polysilicon and a dopant are melted in a quartz crucible inside a vacuum furnace. A seed crystal is dipped into the melt and slowly withdrawn while the seed and crucible are rotated in opposite directions. The molten Si solidifies at the interface with the seed. Gradually a single crystal ingot, or boule, is grown, typically at a rate of 10^{-4} to 10^{-2} cm/s. The boule may be 5–15 cm (2–6 in.) in diameter and up to 91 cm (36 in.) long. Thin circular wafers (250–400 μm thick) are sliced from the ingot. Approximately 50% of the crystal is lost as sawdust.

The wafer (solar cell "blank") is subsequently doped to produce the $p–n$ junction by surface diffusion or by ion implantation.

11.1-8 Advanced Cell Fabrication Technologies

Thomas Lepley

Single-Crystal Silicon–Bulk Crystal Methods

1. **Advanced Czochralski (Fig. 11.1-24).** This process is basically the same as the conventional Czochralski method, except that the crystal-growing furnaces are recharged with liquid or solid polysilicon. More than one ingot can be pulled from a single crucible before it must be discarded.
2. **Heat exchanger method (HEM) (Fig. 11.1-25).** Molten silicon is placed in a square crucible and allowed to solidify in a controlled fashion from a single crystal seed. A square ingot of silicon is produced, which is essentially a single crystal, except for the material near the edges. The ingots are normally 30–35 cm (12–14 in.) on a side.[13]

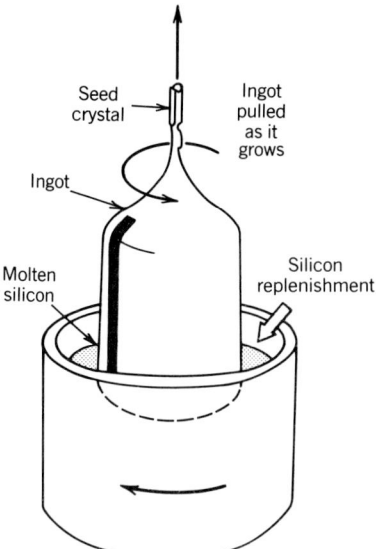

Fig. 11.1-24   Advanced Czochralski (from ref. 17).

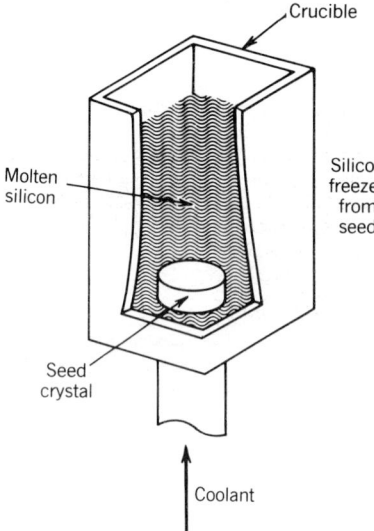

Fig. 11.1-25   New heat exchanger method (from ref. 17).

Single-Crystal Silicon—Ribbon Methods

Ribbon technologies involve the production of a thin ribbon of silicon, thereby avoiding the loss of single-crystal material as sawdust. Ribbon technologies also result in a rectangular cell that can be packed with high density into a module; and they are more feasible for continuous production.

1. **Die-shaping processes (Fig. 11.1-26).** The most developed die approach is *edge-defined film-fed growth* (*EFG*). A wetted die (usually graphite) is partially immersed in molten silicon. The silicon rises by capillary action and feeds the growing crystal at the top of the die. Since silicon is a very good solvent, it is difficult to keep it from picking up impurities from the die. However, cells with competitive efficiencies have been produced with EFG. The IBM *capillary action shaping technology* (CAST) operates very similarly to EFG. The *Stepanov* and *inverted Stepanov* techniques use nonwetted dies and have resultant lower contamination of the silicon.

2. **The dendritic web process (WEB).** This method involves the pulling of ribbon material from a molten silicon pool without the use of a die (Fig. 11.1-27). Two silicon dendrites are lowered

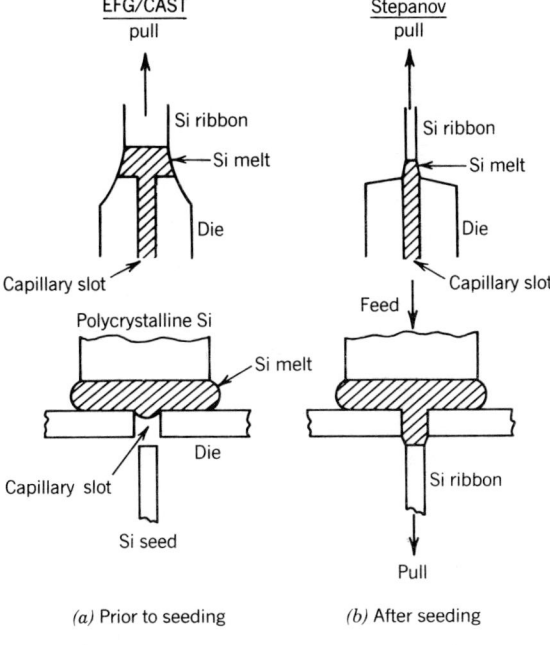

EFG/CAST pull

Si ribbon
Si melt
Die
Capillary slot

Stepanov pull

Si ribbon
Si melt
Die
Capillary slot

Polycrystalline Si
Si melt
Die
Capillary slot
Si seed

Feed
Si ribbon
Pull

(a) Prior to seeding   (b) After seeding

Inverted Stepanov

Fig. 11.1-26 Die shaping fabrication processes (from ref. 12).

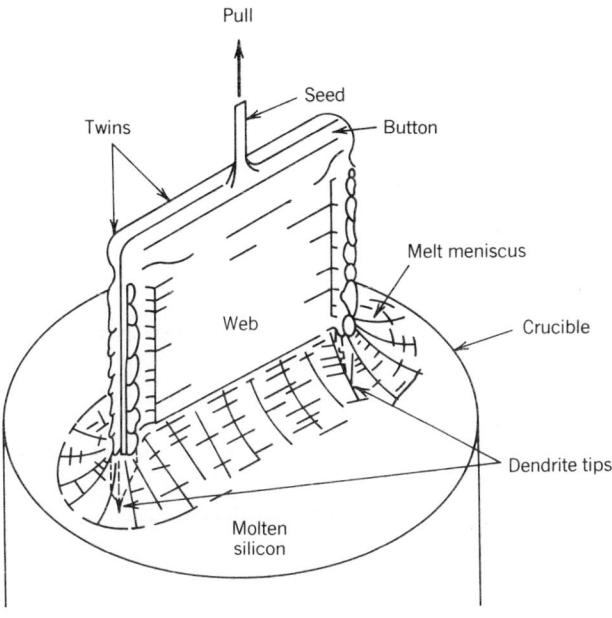

Pull
Seed
Twins
Button
Melt meniscus
Web
Crucible
Dendrite tips
Molten silicon

Fig. 11.1-27 Dendritic web process (from ref. 12).

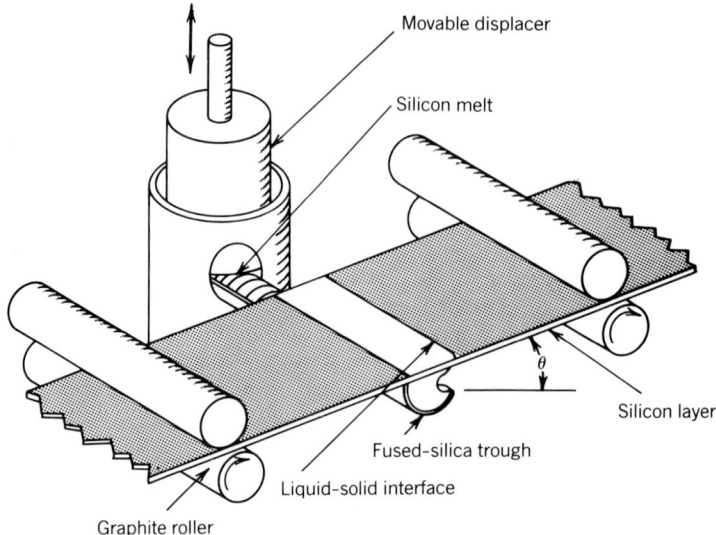

Fig. 11.1-28 Silicon on ceramic (from ref. 13).

into the silicon melt and slowly withdrawn. A web forms by capillary action and a single crystal ribbon solidifies. The cross section resembles a dumbbell.[13]

Multicrystalline Silicon

Techniques are being developed to make cells from polycrystalline (a relatively large number of small crystal grains) and semicrystalline (a relatively small number of large crystal grains) silicon. Crystal grain boundaries cause separated charge carriers to recombine without providing useful output. However, if the grains are large enough (greater than ~ 100 μm on a side), the multicrystalline cells may provide acceptable performance.

1. **Semicrystalline.** Solarex (through its subsidiary Semix) and Wacker Chemotronic, in West Germany, have been developing the multicrystalline approach. Semix casts silicon ingots in a low-cost container. The melt is held at a very uniform temperature, and large crystal grains grow as the material is slowly allowed to cool.
2. **Silicon on ceramic (SOC).** A mullite-based ceramic substrate is coated with a 250-μm thick layer of molten silicon (Fig. 11.1-28). The silicon solidifies into a large-grained, polycrystalline sheet.
3. **Ribbon to ribbon (RTR).** See Fig. 11.1-29.
4. **Chemical vapor deposition (CVD).** Epitaxial growth of silicon on a suitable substrate occurs at 300–1200°C in an atmosphere of gaseous silane (SiH_4) or chlorosilanes.

Thin-Film Materials and Methods

Many materials have higher absorbing probabilities than crystalline silicon. Sunlight can be absorbed in a film only 1 μm thick. Since less material is used and the crystallization step is generally avoided, the cost is much lower. However, so far efficiency has also been lower.

1. Hydrogenated amorphous silicon (a-Si : H) has shown promise as a low-cost cell. A thin film of silicon (~ 1 μm) is deposited on a substrate with the silicon atoms randomly distributed. Hydrogen bonds to the silicon and compensates for the large number of broken bonds (defect states) normally present in amorphous silicon. a-Si : H has a band gap of 1.6 eV, which is closer to the optimum than the 1.1 eV of crystalline Si. The theoretical efficiency limit is 15–20%[20] and 10% has been demonstrated.[32]
2. Cadmium sulfide/copper sulfide (Cu_2S/CdS) cells have demonstrated efficiencies in the laboratory of greater than 10% with a theoretical limit of 16%.[15] Photon Power is producing cells commercially with 4–6% efficiency.[19]

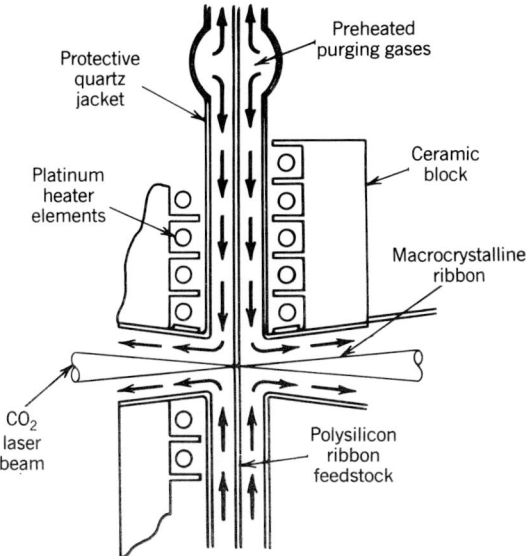

Fig. 11.1-29 Ribbon to ribbon. Motorola converts fine-grained polysilicon feedstock into large-grained macrocrystalline ribbons by heating and then recrystallizing them with a gas laser. This process requires no crucible or die and has a high throughput rate of 55 cm^2/min (from ref. 15). (Reprinted from *Electronics*, July 19, 1979. Copyright © McGraw-Hill Inc. 1979. All rights reserved.)

3. Numerous other compounds are receiving attention: gallium arsenide, zinc sulfide, cadmium telluride, and so on.

High-Efficiency Approaches

A relatively expensive solar cell could be economically justified if its efficiency were high enough. Several concepts are under development, with most of them attempting to use a low-cost optical concentration system to focus sunlight onto a high-efficiency cell.

1. Standard silicon and gallium arsenide cells can be operated at up to 1000 times concentration. If the cell is kept cool, its efficiency is greater at higher concentration levels than at 1 sun illumination. Efficiencies of 22–24% may be possible for silicon[19] and 26% for gallium arsenide.[15]
2. The split-beam approach is being developed by Varian. Silicon and gallium arsenide respond differently to different frequencies of light. Varian uses a spectral (dichroic) filter to split a 165 sun beam of light into high- and low-energy bands. The high-energy beam is directed to a 20% efficient GaAlAs cell and the low-energy beam to a 16% efficient silicon cell. The combined efficiency is 31.4% for the cells, with a net system efficiency of 28.5%. Varian is projecting 35% system efficiency at a 500 : 1 concentration ratio.[15]
3. Tandem or cascade (multiple-junction) cells consist of two or more cells with different band gaps stacked in optical series. Higher-energy photons are absorbed by the top cell; the next cell converts lower-energy photons, and so on. Efficiencies of 35–40% have been achieved, with 60% possible.[19]
4. Thermophotovoltaics (TPV) involves concentrating sunlight onto a tungsten radiator, which is heated to about 1800°C. The radiator emits energy with longer wavelengths than natural light, which are more usable by solar cells. The combination of this spectral shifting and recycling of the energy that is not converted while passing through the solar cell to heat the radiator may permit overall conversion efficiencies of 30–35%.[19,21] Twenty-six percent has been achieved.[15]

Energy Payback

Present-day cells require an estimated *energy payback period* of about 4 years. Using 1986 production processes, the period is expected to be reduced to as low as 2–4 months.[13]

TABLE 11.1-1   PHOTOVOLTAIC SYSTEM TERMINOLOGY[27]

Photovoltaic (PV) system.   The total components and subsystems that combine to convert solar energy into electric energy suitable for connection to an application load.

The major subsystems and their interfaces are the array, power-conditioning, monitor and control, storage, cabling, and power distribution units. The auxiliary energy subsystem is not part of the PV system but is included in the total energy system. For a PV–thermal system, a thermal subsystem also is included.

Array field.   The aggregate of all solar photvoltaic arrays generating power within a given system.

Array subfield.   A group of solar photovoltaic arrays associated by a distinguishing feature such as field geometry, electrical interconnection, or power conditioning.

Branch Circuit.   A number of modules or paralleled modules connected in series to provide dc power at the system voltage.

Array.   A mechanically integrated assembly of modules or panels with a support structure and foundation, tracking unit, thermal control, and other components, as required, to form a dc power producing unit.

Panel.   A collection of modules fastened together, preassembled and wired, and designed to provide a field-installable unit.

Module.   The smallest complete, environmentally protected assembly of solar cells, optics, and other components (exclusive of tracking) designed to generate dc power under unconcentrated terrestrial sunlight.

Receiver.   The component designed to operate under concentrated sunlight, incorporating the concentrator cell assembly and providing thermal energy removal.

Concentrator optics.   The optical concentrating portion of a module designed to operate with concentrated sunlight on the receiver.

11.1-9   Photovoltaic Systems

Thomas Lepley

Photovoltaic cells, along with other components, are assembled into systems. The specific design used will vary depending on the application. Photovoltaic system terminology is defined in Table 11.1-1 and illustrated in Fig. 11.1-30.

Modules

One of the first levels of assembly is the module. The primary functions of the module are to provide protection and support for the cells and to connect the cells together to provide a specific voltage and current output. Voltage adds for cells (or groups of cells) connected in series and current adds for cells (or groups of cells) connected in parallel. An optical system and cooling system may also be included in some types of concentrator modules. Figures 11.1-31 and 11.1-32 illustrate the makeup of typical flat plate and concentrator modules, respectively.

Collectors

At present photovoltaic solar collector systems fall into two major categories: flat plate and concentrating. In *flat plate arrays* solar cells cover almost the entire area exposed to the sun. The modules are typically positioned in a southerly direction with a fixed tilt angle, although they may be mounted on a tracking structure to increase the amount of energy they receive. Since flat plate modules operate on both direct and diffuse sunlight, they have almost universal geographical application.

In *concentrator arrays* some form of optics such as a Fresnel lens or a concentrating mirror is used to focus sunlight onto a concentrator-type solar cell. Current types of concentrator systems use only the direct component of sunlight and require tracking equipment. They will probably have widest application in sunny areas, such as the southwest United States. In the concentrator approach a relatively inexpensive lens is used, allowing more expensive, higher-efficiency solar cells to be utilized.

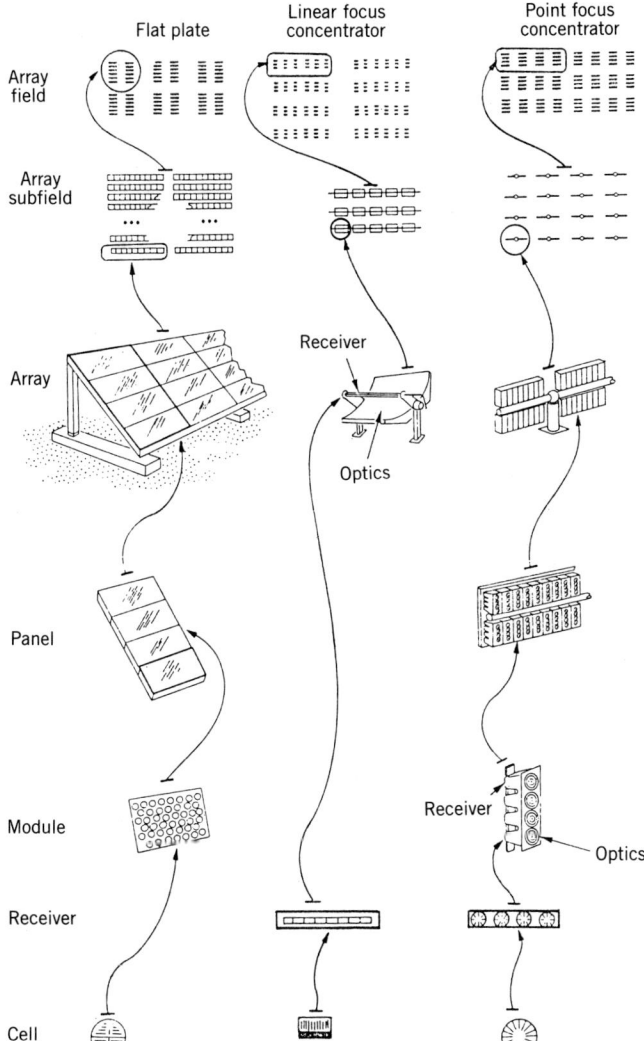

Fig. 11.1-30 Photovoltaic array hierarchy (from ref. 27).

System Design Approaches

The two basic approaches to PV systems are on-site (PV system located near the load, as for example a rooftop residential system), and central-station, where plants of up to 100–200 MW may be used to generate power directly into a utility grid. (The central-station plant may be owned by the utility or by third parties.) A third classification is the intermediate size (10 kW to 10 MW), which may have characteristics of either on-site or central station.

Stand-Alone On-Site Systems. On-site systems may be either stand-alone or utility interactive. Stand-alone systems may be economic in cases where the load is at such a distance from the utility lines that the expense of bringing in utility power is prohibitive. They are designed so that they do not require an external source of power. Storage or a backup generator (e.g., diesel or propane) would be used to supply any demand that occurs when the incident solar energy is too low to supply the load directly. Remote microwave repeater stations, water pumping, and cathodic protection are examples of present-day on-site applications. The on-site system may supply a dc load or an inverter may be used to convert the power for ac loads. Figure 11.1-33 is a general block diagram showing the components that may be used in a stand-alone system.

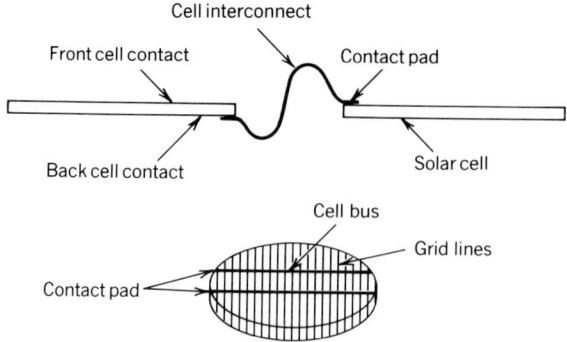

Fig. 11.1-31a   Typical flat-plate module cell contact elements (from ref. 27).

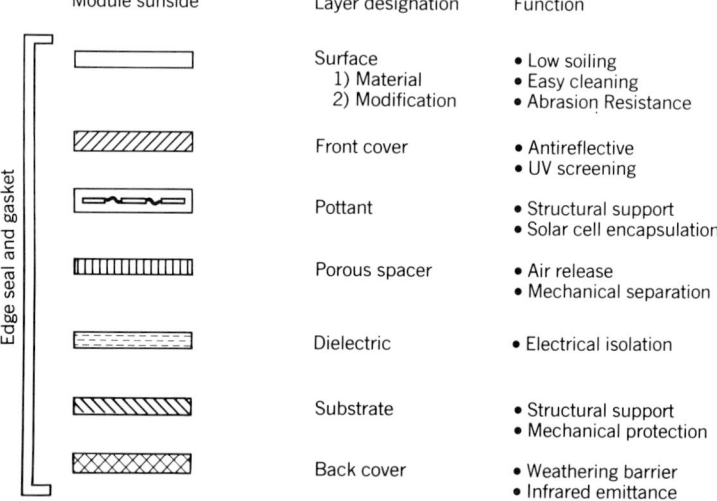

Fig. 11.1-31b   Typical flat-plate module encapsulation system. Structuring, designations, and functions of various elements of one type of flat-plate module encapsulation system; any necessary primers and adhesives are not shown (from ref. 27).

A tradeoff that needs to be made with stand-alone systems involves their sizing. A system may be sized to supply any load requirements for a 24-h period, for example. However, under conditions of extended cloudy weather the PV system would not be able to supply the load. Alternatively, the system can be oversized to meet the load for most or all of the conceivable weather conditions. However, this approach will result in a more expensive system and will mean that energy is wasted during much of the year.

The design of a cost-effective, stand-alone PV system involves the following steps[28]:

1.   Determination of the load.
2.   Computation of the insolation.
3.   Selection of the array and storage system size.
4.   Computation of the loss of load probability.
5.   Computation of the life-cycle costs.

A detailed discussion is beyond the scope of this handbook but is available in the literature.[28]

Utility Interactive On-Site Systems.   The most likely early approach to on-site applications will be utility interconnected systems. A typical residential installation is shown in Fig. 11.1-34. The utility will act as a backup during periods of low insolation and will purchase any excess electricity when the

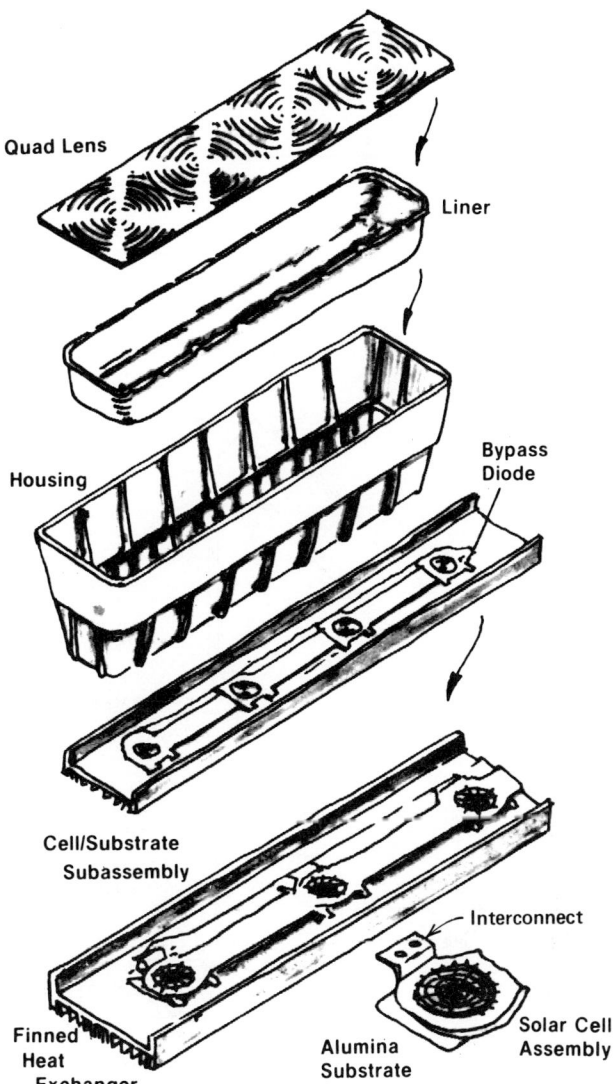

Quad Lens

Liner

Housing

Bypass
Diode

Cell/Substrate
Subassembly

Interconnect

Finned
Heat
Exchanger

Alumina
Substrate

Solar Cell
Assembly

Fig. 11.1-32 Typical module/heat exchanger point-focus.

PV output exceeds the local load. The utility acts as "storage" by storing coal, oil, or other fuel that would have been burned if the PV system were not there.

The DOE has targeted residential systems as the earliest applications of PV since they are perceived to have several advantages over central-station installations:

1. Residential tax credits are greater than tax credits for other groups.
2. In residential applications the panels can be mounted directly on the roof and a separate support structure may not be needed.
3. The residential PV system energy may compete against utility energy at retail rates.
4. Electrical losses due to transmission and distribution are minimized since the generation is located close to the load.

At present there is not total agreement on requirements for connecting small dispersed PV power producers with the electric utility system. However, a list of general *utility connection guidelines*

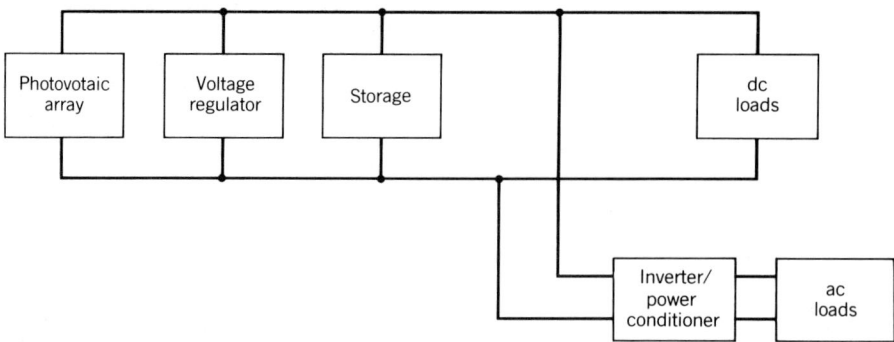

Fig. 11.1-33 Stand-alone photovoltaic system components. This is a general block diagram. Designs for a specific application may omit one or more of the blocks.

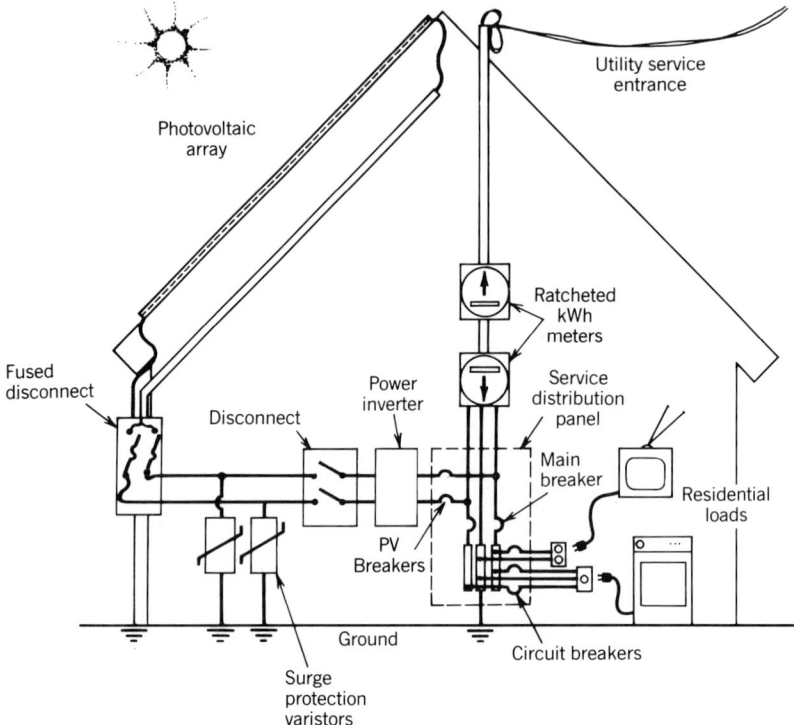

Fig. 11.1-34 A generic residential photovoltaic system. (From ref. 29. This article first appeared in the July issue of *SOLAR AGE*, © 1981. SolarVision, Inc., Harrisville, NH 03450 USA. All rights reserved. Reprinted and published by permission.)

follows.[30] (Note that these are only *guidelines*. If a utility interconnection is being considered, the local utility should be contacted for its specific requirements.)

1. An *isolation transformer* may be required to allow for grounding of one side of the PV array and to protect the distribution transformer and other user loads from dc current.

2. A *dedicated distribution transformer* may be required if an isolation transformer is not used.

3. A lockable (open position) manual load break *disconnect switch* should be provided to allow utility personnel to disconnect the PV system.

4. *Grounding* must be designed to meet codes.

5. A *ground fault interrupter* (GFI) should be used to protect the residential and distribution systems from ground faults in the array.

6. The PV system should include proper *synchronization* with the utility waveform.

7. Automatic system *startup* may be desirable.

8. The system should produce the proper ac *voltage* to match the utility service voltage.

9. The PV system should be designed to disconnect automatically when the utility supply is lost. An *over/under frequency* relay with 57/63 Hz as acceptable boundaries may be used in combination with an *over/under voltage* relay, with ANSI standard voltage range B (110–127 and 220–254 Vac) used for the settings.

10. After a loss-of-utility event the PV system should allow a modest time to elapse following *utility recovery* before commencing operation.

11. The PV system should avoid causing *voltage flicker* on the utility system.

12. The PV system should not exceed *harmonic distortion* limits. The utility may limit THD (total harmonic distortion) in the current waveform to a maximum of 5%, with no more than 3% in any single harmonic. The maximum voltage harmonic may be 2%, with no more than 1% in any single harmonic.

13. The PV system should not require excessive *reactive power* from the utility. Utilities may require that the power factor range should be 0.95 lagging to 0.95 leading from one-eighth to full power output.

14. The utility should attempt to distribute PV systems uniformly on the three phases of the feeder to avoid *feeder imbalance*.

15. The PV system should not produce excessive levels of conducted or radiated *electromagnetic interference* (EMI).

16. The PV system should include all NEC, building code, and UL requirements for *safety*.

Central-Station PV. Central-station plants have the same general layout as utility interactive on-site systems. The primary difference is in size. For example, a 100-MW plant may have twenty 5-MW inverters producing 34.5 kV (ac) output power. The 34.5-kV power would be collected at a central point and stepped up to transmission voltage. Central-station plants may make use of either stationary flat plate, tracking flat plate, or tracking concentrator systems. Advantages of central-station over on-site systems are:

1. Lower cost due to economies of scale in the power conditioner.

2. Relatively complex tracking systems are more feasible.

3. The plant will be under the control of the utility for construction, dispatching, and so on, so any potential utility distribution system problems and/or coordination problems would not occur.

4. Economy of scale in marketing may apply (since the builder of a large system can bypass much of the distribution chain for PV equipment).

5. Operation and maintenance will be easier to perform at one central location.

PV System Design

Array Field. Modules are connected in series and parallel to produce the desired voltage and current levels. The choice of specific modules may depend on how well they can be connected to meet the requirements of the load.

Example 11.1-1. A 4.4-kW (peak) flat plate array field is needed for a residential installation. The power will be fed into a 4-kW (ac) inverter with an input voltage rating of 180–220 V (dc).

Case A—modules that produce 50 W (5 V and 10 A).

Case B—modules that produce 33 W (16 V, 2.06 A).

Solution A. It is desirable to operate the inverter at the high end of its voltage rating to maximize efficiency. Thus, 220 V divided by 5 V per module = 44 modules in series. The 44 modules produce 44×50 W = 2.2 kW. Two of these branch circuits in parallel provide the required 4.4 kW.

Solution B. Find the number of modules in series. 220 V divided by 16 V per module = 13.75 modules. Use 13 modules in series to provide 208 V, 429 W (14 modules would exceed the input voltage rating of the inverter). Next find the number of branch circuits to provide the desired power. 4.4 kW divided by 0.429 kW per branch circuit = 10.26 branch circuits.

If 10 branch circuits are used, the total array field power is 4.29 kW. If 11 branch circuits are used, the total array field power is 4.719 kW.

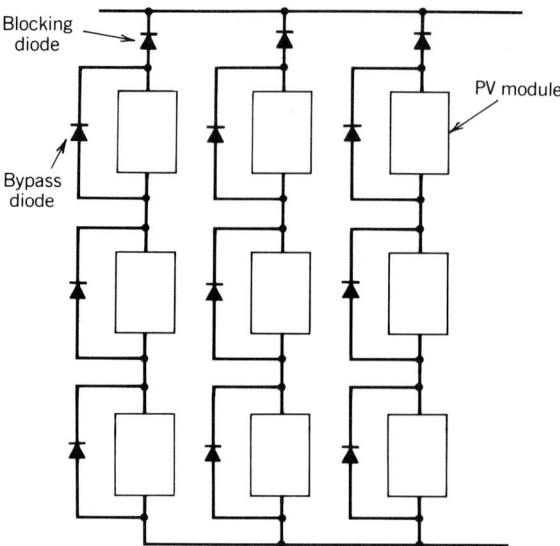

Fig. 11.1-35   Array diode placement.

If it is important to have as close to 4.4 kW as possible, then an alternate would be to use 12 modules in series and 11 branch circuits, which would give 4.36 kW. However, the output voltage would only be 192 V.

Diodes.   *Blocking diodes* are placed in series with each branch circuit, as shown in Fig. 11.1-35. They allow current to flow from the solar cells/modules to the bus but block reverse power flow when the branch circuit voltage is lower than the bus voltage for some reason (for example, shading or faults). *Bypass* diodes (or shunt diodes) are placed in parallel with a module or a string of several modules. They are normally reverse biased and do not conduct current. If the current flow through the module becomes limited due to shadowing or cell fracture, the diode becomes forward biased and shunts the string current around the module. This allows the branch circuit to continue operation in the event of a permanent open circuit in the module. In addition, it helps to prevent permanent failures from developing due to hot-spot heating, which occurs if a cell is temporarily reverse biased (due to shading, etc.).

Array Current.   The array short circuit or *fault current* may be only 15% greater than normal operating current. Use of overcurrent relays to detect faults is not possible since the fault current at reduced insolation levels is actually less than normal current at full insolation. Also, a PV system without storage cannot directly supply the startup currents required by some motors.

Array Voltage.   The maximum *open circuit voltage* on sunny, cold days may be as great as 175–180% of normal operating voltage. It is important to consider the maximum voltage withstand capability of the load or inverter and avoid connecting the array at times when it may cause damage.

Array voltage may be adjusted to provide maximum output from the array under varying conditions. This function, called *maximum power tracking*, is a common control circuit in many commercially available inverters. In the United States energy loss for fixed-voltage operation ranges from 0.7 to 2.5% depending on location.[25]

Module Orientation.   The tilt angle and east–west orientation must be considered in the design of flat plate, nontracking systems. If adjustable tilt is not provided, the maximum energy production on an annual basis occurs with the panels tilted at the latitude angle and facing directly south (or facing directly north in the southern hemisphere). In general, only minor loss in energy produced will occur if tilt is within ±15° latitude and orientation is within 15% east or west of due south. Table 11.1-2 and Figs. 11.1-36 and 11.1-37 provide a more quantitative measure of energy loss for nonoptimum tilt and azimuth orientation.

Characteristics of the application must be considered in the design of the module orientation. If, for example, the load is higher in the summer, then the tilt angle should be less than the latitude angle in order to increase the summer output.

TABLE 11.1-2 TILT ANGLE FOR OPTIMUM AVAILABLE SUNLIGHT

January	LAT + 29	July	LAT − 24
February	LAT + 18	August	LAT − 10
March	LAT + 3	September	LAT − 2
April	LAT − 10	October	LAT + 10
May	LAT − 22	November	LAT + 23
June	LAT − 25	December	LAT + 30

Source. "Designing Small Photovoltaic Power Systems," Monegon, Ltd., Gaithersburg, MD; used by permission.

LAT stands for Latitude angle. Numbers are in degrees.

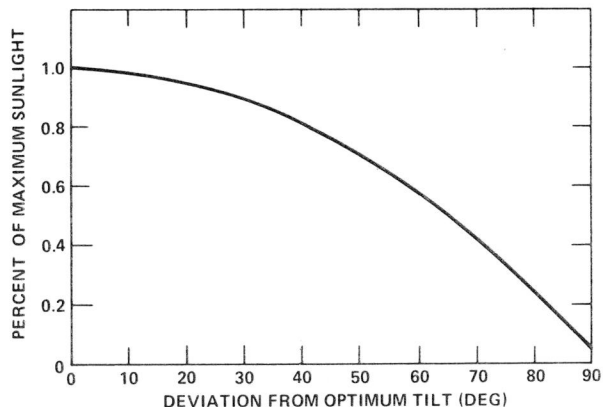

Fig. 11.1-36 Loss of available sunlight with nonoptimum tilt. (From "Designing Small Photovoltaic Power Systems," Monegon, Ltd., Gaithersburg, MD. Used by permission.)

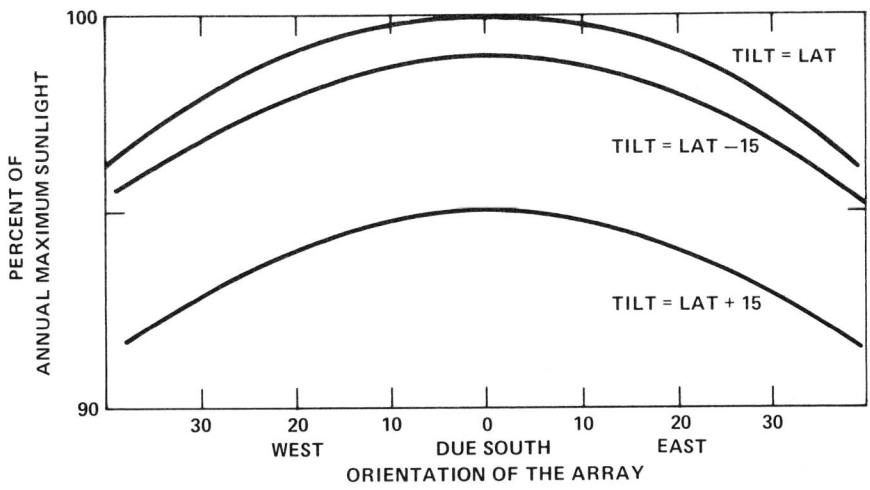

Fig. 11.1-37 Effect of variation of due south orientation on array power output. (From "Designing Small Photovoltaic Power Systems," Monegon, Ltd., Gaithersburg, MD. Used by permission.)

1211

Inverters

Inverters convert the dc power produced by PV systems into ac power. The two main types of inverters used in PV systems are *line commutated* and *self-commutated* (or force commutated). Line-commutated inverters generally require the utility voltage to provide the commutation or switching energy. It has been found, however, that existing designs may become self-excited without the utility voltage being present under some conditions when a source of reactive power, such as capacitors, is provided. Self-commutated inverters make use of energy storage elements within the inverter (capacitors and/or inductors) to provide the commutation energy. They can be designed to operate either in a utility interactive or stand-alone mode.

Ratings

A PV system cannot be rated in the same manner as a conventional power plant because the output varies greatly as a function of weather conditions, type of cells, location, module-mounting approach, and so on. To give some degree of consistency *standard reporting conditions* (SRC) or *peak reporting conditions* or *standard test conditions* (STC) have been adopted by the international photovoltaic community. The conditions are 100 mW/cm^2 irradiance level (1 sun), air mass 1.5 spectrum, and 25°C cell temperature. SRC is also convenient for laboratory measurements.

A second set of reference conditions are *standard operating conditions* (SOC): 100 mW/cm^2 irradiance and NOCT. NOCT (*Nominal operating cell temperature*) is defined as the operating temperature of the cells in the intended mounting configuration with an incident irradiance level of 80 mW/cm^2, an air temperature of 20°C, a wind velocity of 1 m/s, and the array open circuited. This set of conditions yields a temperature that accurately represents the average cell temperature in the field during periods of significant energy production. Roughly 50% of the energy will be produced above and below this temperature. Typical values of NOCT for ground-mounted arrays range from 45 to 50°C, and for roof-mounted arrays from 60 to 70°C.[25]

11.1-10 Economics

Thomas Lepley

The overall economic value of photovoltaics depends not only on the cost of the systems, but also on the cost of the alternatives and the amount of sunlight available. For example, as the cost of photovoltaic systems is reduced, one of the first areas where they will be cost-effective is the southwest United States because of the high insolation levels. However, the northeast United States will also be an early market. Even though the insolation is lower, the cost of electricity from alternate sources is higher.

Photovoltaic System Economics

The Department of Energy (DOE) has established price goals for PV modules and systems. The goals in 1980 dollars per peak watt for systems without storage are[31]:

	Collector Price (FOB Factory) $/Wp	Total Installed System Price $/Wp
1982	$2.80	$6.00–$13.00
1986	$0.70	$1.60–$2.20
1990	$0.15–$0.40	$1.10–$1.80

At present the cost of PV systems is being reduced, but not fast enough to meet the goals. There is optimism that the cost goals can be achieved, but not by the dates indicated.

Utility System Considerations

For photovoltaics to become a significant energy supplier, PV systems will have to be connected to the utility grid. PV generation without storage has different characteristics than conventional generation, which is traditionally base load, intermediate (or load following), or peaking. A better characterization of PV generation is "negative load." PV generation, whether central station or distributed, is not completely under the control of the utility dispatcher since its output is dependent on the amount of insolation, temperature, and so on.

The value of PV generation can be expressed in terms of the utility's *avoided cost*. Avoided costs are the amounts that a utility avoids spending due to the existence of the PV system. It has two components: variable O & M (operation and maintenance) costs (which are mostly fuel savings) and capacity savings. These values are highly dependent on the specific utility and location. For example, if the PV generation displaces oil generation, there would be a relatively high savings in O & M; if it displaces nuclear generation, the avoided O & M is low.

A PV system may statistically allow a utility to eliminate or delay new generating capacity. The value of the PV system depends on the type of unit that can be deferred. However, it can be a disadvantage to take a capacity credit in cases where the deferral of a new conventional generating station results in the postponed retirement of an existing oil- or gas-fired station.[34]

11.1-11   Solar Thermal Electric Systems

Eric Weber

The process of converting sunlight into thermal energy and its subsequent conversion into electricity utilizes heat transmission mechanisms of collection, reflection, absorption, transfer, and transport. Four basic solar systems have been developed that configurate the heat transmission mechanisms hardware in the most efficient manner. Each system is associated with a specific collection concept and operating temperature range dependent on material and collection concentrator limitations. The method of conversion of thermal to electric energy, the thermodynamic cycle, is dependent on working fluid and the operating temperature range. The systems that have been developed are the results of selective elimination through exhaustive system studies and hardware development.

11.1-12   Point-Focusing Central Receivers

Eric Weber

The point-focusing central receiver generic concept employs a large number of computer-guided tracking mirrors, called heliostats, to reflect the sun's energy to a receiver mounted on top of a tower. At the receiver the reflected energy is absorbed by the receiver surface and transferred as thermal energy to the circulating heat transfer fluid. This system, as utilized in a power plant, is now commonly referred to as the solar thermal central receiver power system.

A solar central receiver system concentrates the low-density [approximately 320 Btu/h ft^2 (1 kW(t)/m^2)] sunshine falling on the earth by using these many computer-guided tracking mirrors to redirect the sun's energy. The receiver on the tower absorbs and transfers 80–95% of the reflected energy to the heat transport fluid. The energy may be used with a turbine–generator to produce electricity or be stored as thermal energy for later use. Power levels from 3.4 to 5100 million Btu/h [1–1500 MW(t)] can be achieved by using from 30 to 50,000 large (540 ft^2, 50 m^2, reflective area) heliostats that pivot about two axes to track the sun. Using available technology, achievable, nominal working fluid conditions are 2400 psi, 1000°F (16.5 MPa, 540°C) for steam and 1500°F (840°C) for air.

System Description

There are two generic configurations of central receiver power plants (Fig. 11.1-38): stand-alone (solar power only) and hybrid (solar plus an auxiliary power source). In the hybrid concept the central receiver system is typically combined with a conventional fossil fuel energy source so that the plant can operate using either solar energy, fossil fuel, or a combination of the two. The hybrid configuration is also used in the retrofitting of existing fossil-fired power plants and has been termed *solar repowering*.

A typical central receiver system may include up to seven subsystems; collector, receiver, storage, electric power generating, master control, heat transport and exchanger, and nonsolar energy source. There are many options available for each of these subsystems, as indicated in Table 11.1-3. Selection of thermodynamic cycle and heat transport fluid normally determines the system configuration and component sizing requirement.

Collector Subsystem.   The main element of the collector subsystem is the heliostat (a typical example is shown in Fig. 11.1-39) arranged in either surrounding or north-field configuration, with the tower-receiver as the focal point (Fig. 11.1-40). Selection of collector field configuration, surrounding or north, is a function of site location (latitude), thermal power level, receiver configuration, and tower height. Southern latitudes, approaching the equator, high power levels, and an external surface receiver favor a surrounding field. More northerly latitudes, low to medium power levels, and cavity receivers favor a north field.

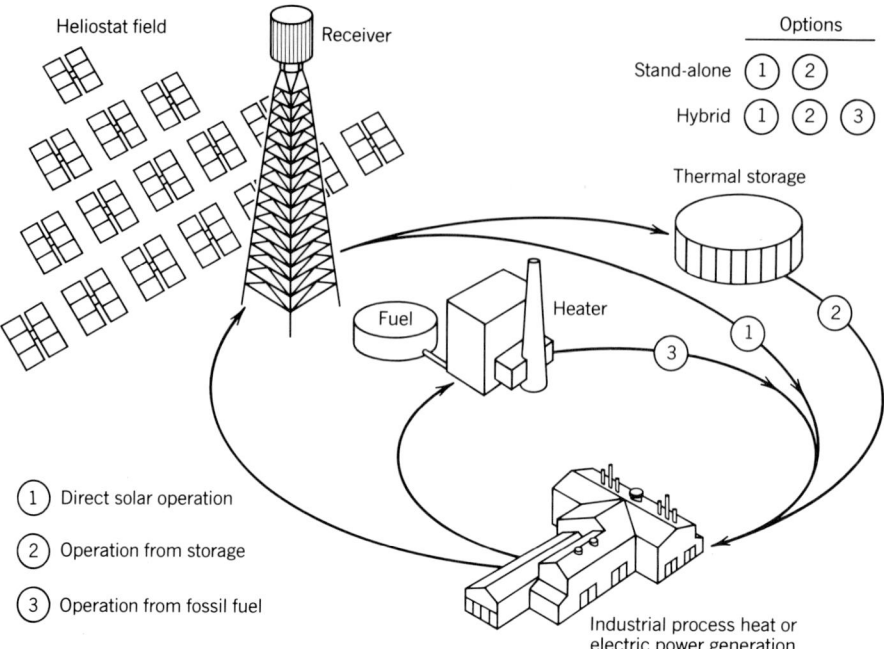

Fig. 11.1-38 Central receiver system configurations.

In typical southwestern regions of good insolation, a 100-MW(e) solar central receiver power plant with a peak thermal power of 1300 million Btu/h or 380 MW(t) would require 12,500 heliostats each of 540 ft (50 m²) reflective area.

Receiver Subsystem. There are two basic configurations of receivers: external and cavity (Fig. 11.1-41). For a given collector field the configuration selected depends on the working fluid, the desired inlet and outlet fluid temperatures, the incident solar flux (power per unit area) capability of the heat transfer surface, the thermal efficiency of the design, the heliostat field layout, and the cost. In general, the external receiver is smaller, lighter, and less costly but suffers greater thermal losses than a cavity receiver. For comparison purposes, several 100-MW(e) receivers are given in Table 11.1-4, which illustrates respective types, sizes, working fluid, and maximum incident solar fluxes.

The tower that supports the receiver can be constructed of either steel similar to office building framing or concrete similar to that of smoke stacks and cooling towers. Steel towers are favored for low power [1–10 MW(e)] and reinforced concrete for high-power systems on the basis of construction costs.

Heat Transport and Exchange Subsystem. The central receiver system is distinguished by the type of heat transport fluid used to deliver the thermal energy from the receiver to the electric power generation subsystem. Five heat transport fluids have been studied in considerable detail: water/steam, molten nitrate salt,[3] liquid sodium, oil, and gas (air or helium). Some fluid characteristics are listed in Table 11.1-3. Selection of fluid is dominated by the temperature requirements. In general, for current technology, the feasible temperature ranges are up to 550–800°F (290–425°C) for oil heat transport fluids (hydrocarbons and synthetic), up to 1000°F (540°C) for water/steam, 500–1050°F (260–565°C) for molten nitrate salt, 300–1100°F (150–590°C) for liquid sodium, and up to 1550°F (840°C) for gas.

Heat exchangers for any of the anticipated heat transport and working fluids are based on current technology. Nitrate salt experience exists in the chemical process industry although at lower temperatures than 1050°F (565°C). Sodium experience comes from the nuclear reactor industry (CRBR). Water/steam experience is modern power plant practice and gas experience is similarly available from gas turbine technology.

TABLE 11.1-3 PROPERTIES OF HEAT TRANSPORT FLUIDS (ENGLISH UNITS)

Working Fluid	TEMP (°F)	Density (lb/ft³)	Specific Heat (Btu/lb·°F)	Energy Density (Btu/ft³·°F)	Thermal Conductivity (Btu/h·ft·°F)	Viscosity (lb/ft·h)	Typical Forced-Convect Film Coefficient (Btu/h·ft²·°F)	Typical Peak Design Solar Flux on Tubes (Btu/h·ft²)
Nitrate Salt ($0.15/lb)	600	117.9	0.41	48.0	0.29	6.768	1000–2000	220,000
	800	113.6	0.38	43.0	0.31	3.996		
	1000	108.7	0.34	37.0	0.32	2.426		
Sodium ($0.40/lb)	700	53.7	0.31	16.7	41.8	0.684	4000–8000	500,000
	1000	51.2	0.30	15.4	37.8	0.504		
Water (sat. liquid)	400	53.6	1.08	57.9	0.381	0.33	1300–12,000	220,000
	600	42.4	1.51	64.0	0.292	0.21		
Steam (1500 psia)	600	3.55	1.52	5.4	0.040	0.08	1500	160,000
	800	2.30	0.70	1.6	0.040	0.083		
	1000	1.86	0.60	1.1	0.048	0.088		
Air (atm.)	600	0.0374	0.250	0.009	0.0271	0.072	120–250	70,000
	1000	0.0272	0.263	0.007	0.0362	0.089		
	1400	0.0213	0.274	0.006	0.0442	0.104		
	1800	0.0175	0.282	0.005	0.0512	0.117		
Oil (Caloria HT-43) ($0.35/lb)	200	50.3	0.50	25.2	0.075	12.0	100–1000	
	400	45.6	0.60	27.4	0.070	2.5		
	600	40.3	0.70	28.2	0.065	1.03		

Source. K. W. Battleson, "Solar Power Tower Design: Solar Thermal Central Receiver Power Systems, a Source of Electricity and/or Process Heat," SAND81-8005, Sandia National Laboratories, April 1981.

1215

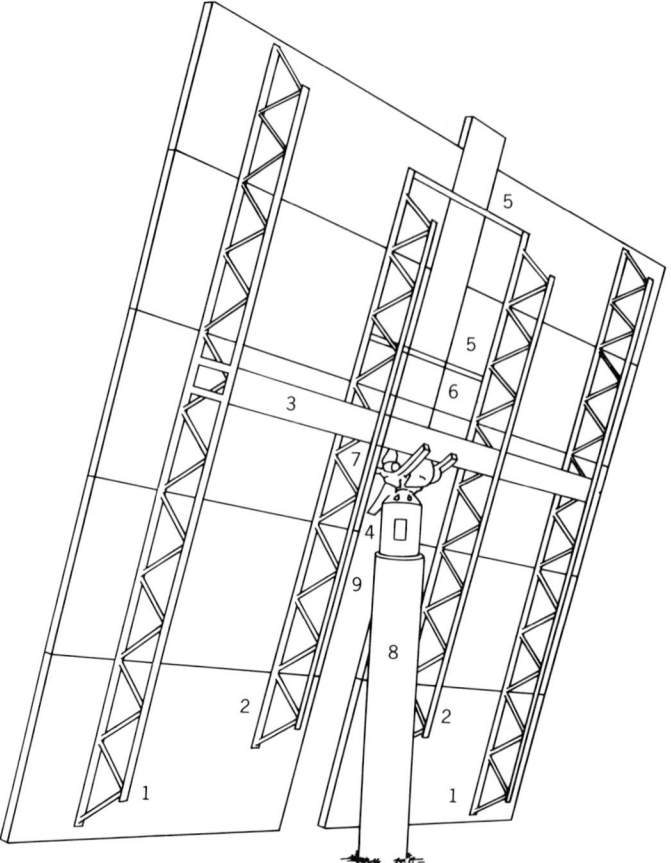

Fig. 11.1-39 Heliostat assembly—Martin Marietta Heliostat (from ref. 39). Heliostat assembly; reflective assembly: mirror assembly (11 total), mirror mounting studs. Rack assembly: (1) Long bar joist with mirror support tabs, (2) short bar joist with mirror support tabs, (3) elevation beam, (4) control arms, (5) mirror support stringer(s). Drive mechanism assembly: (6) drive mechanism; (7) drive motors, encoders, encoder couplers, encoder and limit switch brackets. Pedestal/foundation: (8) Pedestal/foundation, (9) pedestal interface tube (electronics access cover).

Thermal Storage Subsystem. Major considerations impacting the design of solar thermal power systems for commercial electric generation are needed to (1) provide continuous operation during periods of variable insolation, (2) extend operating periods into nonsolar hours, (3) buffer potentially harmful transients induced into systems by abrupt insolation changes, (4) ensure the availability of productive capacity in periods of inclement weather, and (5) optimize dispatch energy to the system load. Investigation of a broad spectrum of energy storage mechanisms has narrowed to sensible heat storage as the most convenient and cost-effective method to satisfy the stated requirements.

There are two types of energy storage systems that can be used: separate hot and cold tanks or single-tank thermocline. Molten nitrate salt and sodium are preferred fluids for high-temperature storage (1050–1100°F, 565–590°C), whereas oil/rock are suitable for temperatures up to 800°F (425°C). The volumetric energy storage density, which affects the cost of tankage or other containment materials, ranges from 7950 Btu/ft^3 (0.082 MW · h(t)/m^3) for sodium to 14,200 Btu/ft^3 (0.15 MW · h(t)/m^3) for rock (void fraction of 25%) and 21,300 Btu/ft^3 (0.22 MW · h(t)/m^3) for nitrate salt, for a temperature range of 550–1050°F (290–565°C). Molten nitrate salt storage is a preferred system for commercial solar power plants.

Master Control Subsystem. Central receiver power plants would use control systems somewhat more complex than those used in existing fossil power plants to integrate the independent controls of the subsystem (collector, receiver, storage, and electric power generation). The need for the greater sophistication is caused by the more dynamic nature of the solar power source, in particular the

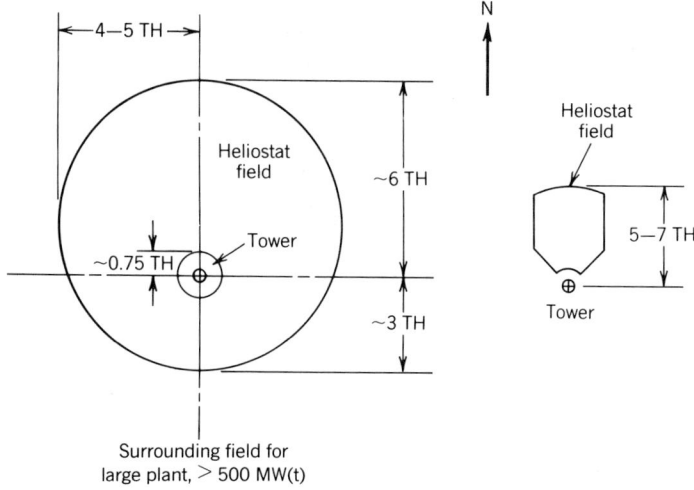

Surrounding field for
large plant, > 500 MW(t)

Fig. 11.1-40 Optimum heliostat field shape (from ref. 35), TH = tower height.

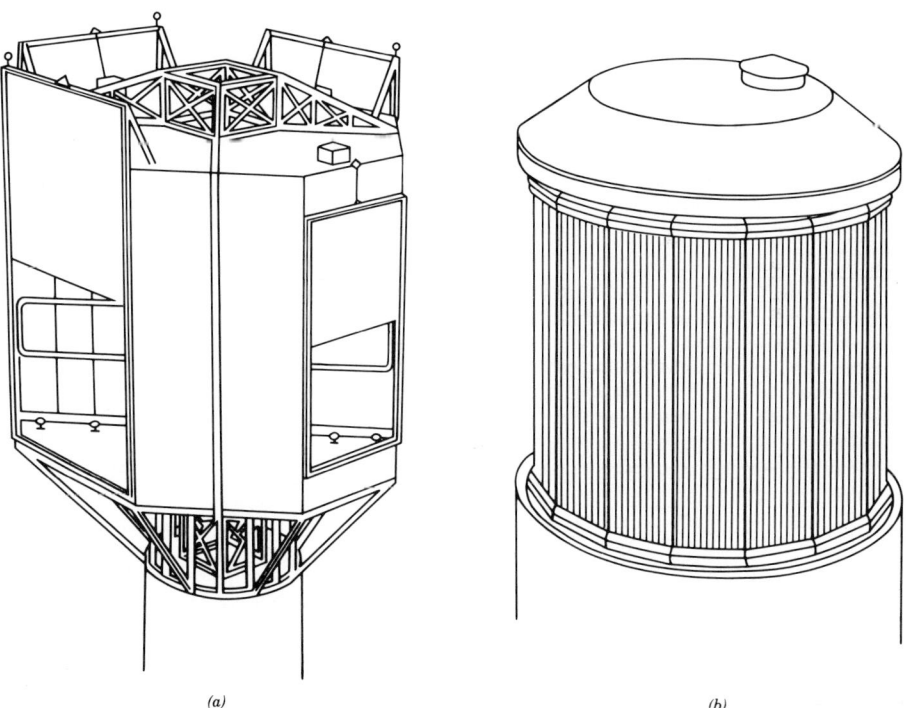

(a)   (b)

Fig. 11.1-41 Basic receiver configuration (figures courtesy of Sandia National Laboratories). (*a*) Example of a four-aperture cavity-type receiver design. It could be built as a natural circulation steam generator producing steam at 1800 psi (12.4 MPa) and 950°F (510°C), or as a molten salt or sodium heater producing 1050°F (565°C) fluids. (*b*) external-type receiver design. It could provide steam, molten salt, or sodium at conditions equivalent to those for a cavity receiver.

TABLE 11.1-4 CAVITY AND EXTERNAL RECEIVERS COMPARISON

Parameter	Quad Cavity (One-Side Heating)	Single Cavity	External
Design power output	320 MW(t)	326 MW(t)	390 MW(t)
Working fluid	Molten nitrate salt	Molten nitrate salt	Sodium
Receiver envelope (including steel frame)	106 × 103 × 175 ft high	134 × 100 × 196 ft high	5.28 ft high × 52.8 ft diameter
Aperture area–total	7268 ft^2	5250 ft^2	None (external)
Receiver panels	98	28	24
Panel sizes	31	1	1
Receiver weight	2600 tons	3200 tons	247 tons
Receiver door(s)	4	1	None
Tower height	508 ft	650 ft	544
Heliostats	8610	8184	14,100
Total panel area	22,300 ft^2	17,300 ft^2	7753 ft^2

requirements to respond to solar transients. For hybrid systems there is the additional requirement to coordinate the solar and fossil controls.

11.1-13 Point-Focusing Distributed Receivers

Eric Weber

The point-focusing distributed generic concept employs a multiplicity of either fixed or computer-guided tracking paraboloid of revolution (parabolic dish) concentrator mirrors to reflect the sun's energy to a receiver mounted at the focus point. At the receiver the reflected energy is absorbed by the receiver surface and transferred as thermal energy to the circulating heat transfer fluid. These distributed receiver systems are distinguished by either the receiver tracking the locus of the focal point of a fixed concentrator (bowl) or the receiver and mirror (dish) as a module tracking the sun in unison. These systems, as utilized in a power plant [10 kW to 10 MW(e)], are now commonly referred to as solar thermal distributed receiver power systems.

The parabolic dish or bowl solar collector can achieve relatively high concentration ratios (500–2000) and, as a result, rather high potential temperatures; values in the range of 500–1400°C are achievable. The thermal energy collection efficiency is also quite high; for example, the demonstrated efficiency is in the range of 70%.

These characteristics indicate both the strength and weakness of the parabolic solar collector. The high-concentration ratio and high fluxes at the focal zone make it difficult to design a receiver with low-temperature differences between tube wall and fluid. Heat pipes, spiral fluid tubes in insulated cavities, and porous ceramics are examples of design approaches for a suitable receiver. A three-dimensional surface with sufficient optical surface accuracy and the structural rigidity needed to withstand high wind loadings on the dish is a major design concern along with accurate two-axes tracking. A fixed bowl mitigates these design problems to some extent. The high potential temperatures and efficiencies are extremely attractive for driving a heat engine, but the high temperature makes it difficult to collect and transfer heat either because of material or transport limitations to bring the heat to the power conversion system.

Two approaches have been considered in the application of the parabolic dish for electric power production. The first is central generation where heat is collected from a field of dishes and transported to a central site by thermal transport for energy conversion. Large steam Rankine engines are efficient even at moderate temperatures (500°C) and are relatively inexpensive. A substantial heat transport subsystem is required for central electric generation. The extra costs and losses of this subsystem must be traded off against the lower costs of the central Rankine plant.

In the second approach, distributed generation, the heat engine is located near the dish. Electric power is generated at each dish and collected at a central site for external transmission. This second approach can use much higher temperatures since there is a very short heat transport path to the heat engine. Energy conversion devices that are suitable at high temperatures (greater than 500°C) are gas Brayton engines, Stirling engines, and advanced liquid metal Rankine systems. The Brayton and Stirling engine configurations are the favored systems. Small closed-cycle Brayton or Stirling engines are placed at each parabolic dish to minimize the distance required for high-temperature heat transport. A schematic of one module is shown in Fig. 11.1-42. Usually air cooling is used to reject

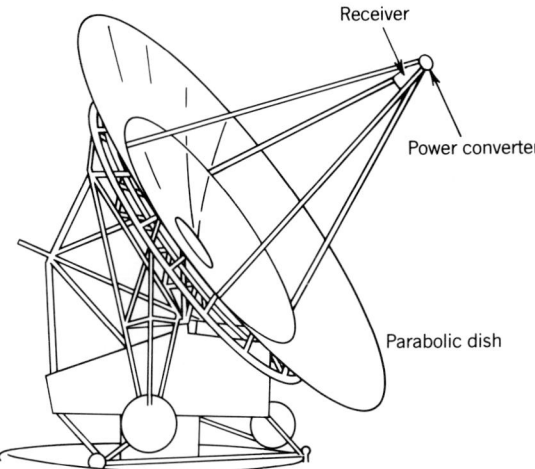

Fig. 11.1-42 Dish concentrator with power converter at the focus (from ref. 37; courtesy Jet Propulsion Laboratories).

heat to the environment to eliminate dependence on water cooling. Normally ambient air will be heated from 20°C to about 120°C to cool this engine. The amount of power generated by economically attractive designs of a dish coupled to a Brayton engine is small by central-plant standards. A dish that is 36 ft (11 m) in diameter can collect about 52 kW per day of thermal power out of 74 kW per day of incident direct solar power. At a peak insolation level of 1 kW/m² the thermal power collected would be 65 kW(t). The use of a small closed-cycle Brayton engine and generator at each collector to avoid long-distance transport of high-temperature heat limits the engine energy that would be collected from each engine and brought to the transmission terminal for power conditioning.

Thermal Performance

Thermal performance of the three usable cycles, Rankine, Stirling, and Brayton, represent respective successively higher operating temperatures and system efficiencies.

The Brayton cycle operating at the current state-of-the-art temperature of 815°C is a relatively high efficiency (30–36%) device. An air heat exchanger, which is equivalent to a dry tower, has a conversion efficiency of about 33%. Thus, an average of about 17 kW(e) is generated at each dish collector during the day.

These combinations of equipment have few advantages compared to central generation systems: very short startup time and low initial investment. Engine startup is accomplished in several minutes, rather than tens of minutes or even an hour, after achieving full temperatures in the working fluid.

Fixed-mirror, distributed-focus (FMDF) solar energy technology uses large, hemispherical, fixed-aperture collectors to concentrate solar energy on tracking linear receivers. FMDF systems operate at temperatures up to 750°C. However, because of their fixed apertures, their annual energy production on a unit basis is less than that for a two-axes tracking system. Using an FMDF, therefore, is economical only if systems can be produced at low cost so that performance–cost ratios are competitive with other solar thermal technologies.

To keep the cost of FMDF systems low, elastically formed, commercial-grade float glass is used as the reflective materials for FMDF mirrors, Some technical questions about bowl technology remain unanswered; the largest uncertainty in the technology's commercial future is its cost.

Applications

Versatility is a key attribute of solar thermal systems as illustrated in terms of the end product produced: electricity, process heat, steam, chemicals, and fuels. Parabolic dish power plants, although highly versatile in terms of application, are limited commercially by the need to compete with conventional resources. This restriction regulates the application to areas that have very high energy costs or to remote service areas that cannot be serviced by electric distribution lines because of installation cost.

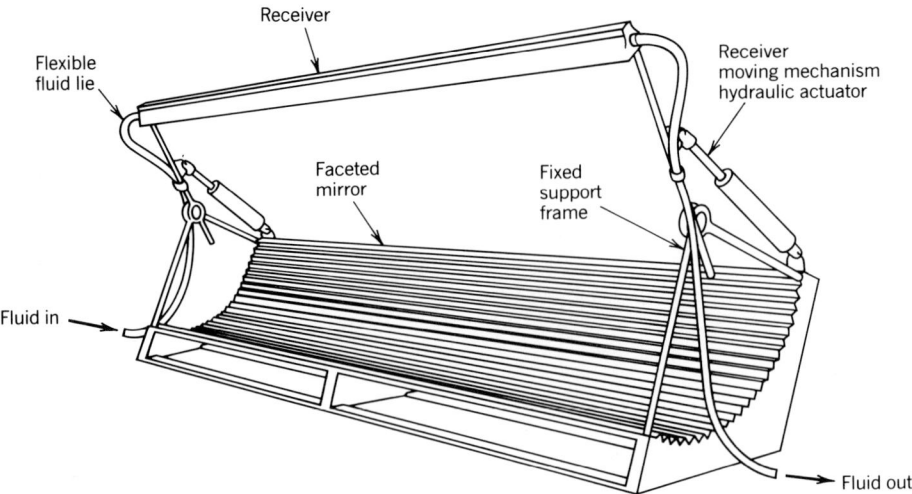

Fig. 11.1-43 Line-focusing distributed receiver concept using fixed mirrors and movable receiver (LFDR-TR) (from ref. 37; courtesy Jet Propulsion Laboratories).

11.1-14 Linear-Focusing Distributed Receiver

Eric Weber

This generic plant is commonly referred to as a parabolic trough, depending on whether the concentrator or the receiver is tracking (Fig. 11.1-43). In the first case the tracking troughs are placed in an east–west orientation with an achievable ground cover ratio of 0.35. Heat is transported from the collector to the central energy conversion unit by using an organic-based fluid, such as Dowtherm, which flows through an insulated piping grid. A heat exchanger utilizes the heat supplied by the transport fluid to vaporize toluene or trichlorinetrifluorethane. The organic Rankine engine/generator produces electricity, and any excess heat is routed to thermal storage. The storage medium is oil in combination with rock and sand. A wet cooling tower rejects excess condenser heat. A second variation of LFDR using a tracking receiver employs fixed mirrors. In this system the receiver tracks the linear image produced by the fixed-mirror parabolic trough. The concentrator flux intensity is increased by a factor of 2 by using a CPC as a secondary concentrator.

The heat transport fluid is Caloria HT43 or equivalent petroleum-base oils. As in the LFDR-TC designs this fluid exchanges heat with toluene to generate power. The transport fluid can also serve as the storage fluid in combination with sand and rock. A third version of the line-focus system is a one-axis tracking version of the point focus central-receiver (PFCR) system. This system has an array of long single-axis tracking heliostats that reflect the incident direct solar flux onto a cavity-type linear receiver. In addition to an elevation tracking mechanism, the heliostats use a mechanism to flex the reflective surface that changes the focal length. This is necessary because the illumination is generally off-axis during the early morning and late afternoon hours, introducing off-axis astigmatism. With an adjustable radius of curvature, a line focus can be maintained for off-axis illumination. The LFCR schematic diagram is very similar to the PFCR system and differs only in power plant layout.

Thermal Performance

Linear focus distributed systems are typically characterized by low-temperature, low-pressure cyclic conditions, 350–500°F and 50–100 psig.

Three types of energy transport are used in the systems described: electric, thermal, and optical. The product efficiencies of all three transports yield the total plant efficiency. The collector field efficiency, using as a measure the time in hours from solar noon, exhibits a decrease that follows a cosine curve. This effect is more noticeable with line focus systems than with point focus. Typically, the collector field efficiencies range from 42 to 62% (Fig. 11.1-44).

Electric and thermal transport can be related more directly to overall cycle efficiencies, which for steam Rankine and organic Rankine typically vary from 24 to 26%.

In order to compete with oil and ultimately coal plants for the intermediate-load segment of the large, firm-capacity utility market, PFDR technology will have to develop either a hybrid capability or

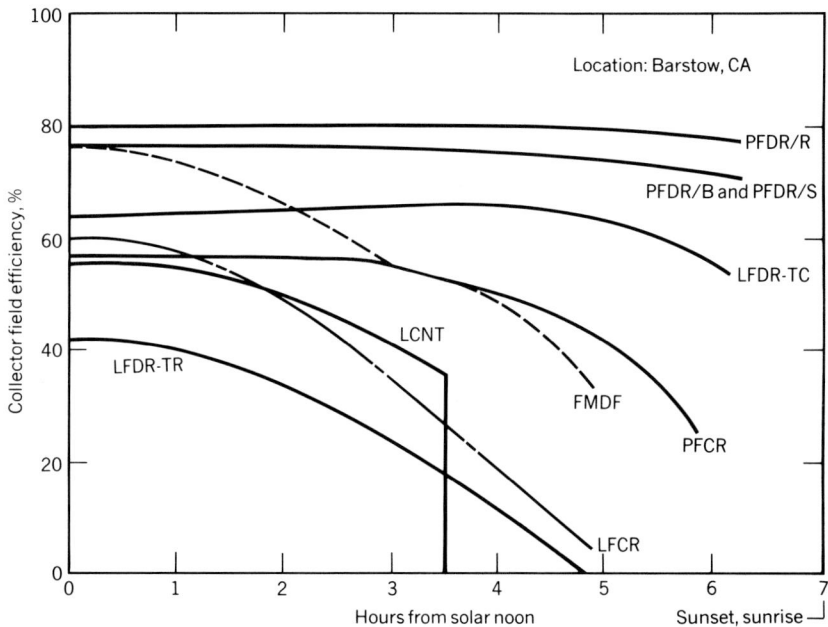

Fig. 11.1-44 Collector field efficiency as a function of time at summer solstice (transport losses are not included.) (Source: Center Receiver Solar Thermal Power Systems, SAN/1108-8/1-6 October 1977.)

low-cost electrical storage. The extra cost and complexity of the hybrid feature must be justified by a high cost of energy from conventional sources. Typically, this translates to high oil and gas costs in the greater than $35/barrel or $7.5/MBtu. High fuel costs are definitely in the future, the only question is when or how soon.

Industrial applications have to be mentioned in concert with utility applications, since the requirement to provide thermal source will precede the generation of electricity. This is true only because of a simpler and less costly system. By first penetrating the industrial market, a production base can be established to reduce the cost of the collector system. Figure 11.1-45 illustrates various types of solar power plants versus capacity factor to attain levelized energy costs.

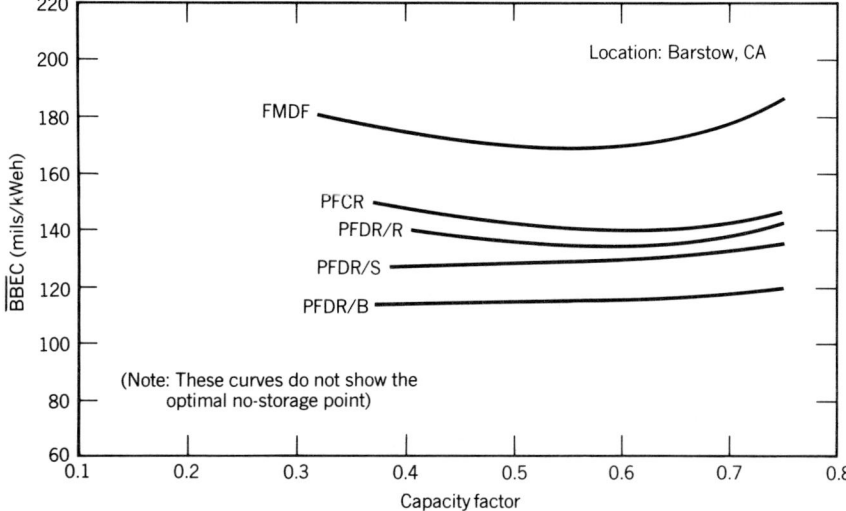

Fig. 11.1-45 Energy cost as a function of capacity factor for selected 1-MWe power plants (1978 dollars.) (Source: Center Receiver Solar Thermal Power Systems, SAN/1108-8/1-6 October 1977.)

REFERENCES

11.1-1 H. C. Hottel and B. B. Woertz, "The Performance of Flat Plate Solar-Heat Collectors," *Trans. ASME*, February 1942, pp. 91–104.

11.1-2 D. C. Beekly and G. R. Mather, "Analysis and Experimental Tests of a High Performance Evacuated Tube Collector," Owens-Illinois, Toledo, 1975.

11.1-3 F. Kreith and J. F. Kreigh, "Principles of Solar Engineering," Hemisphere Publishing Company, Washington, D.C. 1978.

11.1-4 J. A. Duffie and W. A. Beckman, *Solar Energy Thermal Processes*, Wiley, New York, 1974.

11.1-5 P. J. Lunde, *Solar Thermal Engineering*, Wiley, New York, 1980.

11.1-6 J. I. Yellot and H. R. Hey, "Thermal Analysis of a Building with Natural Air Conditioning," *Trans. ASHRAE* **75**, 78 (1969).

11.1-7 ASHRAE, *Applications Handbook*, Chapter 58, 1978.

11.1-8 A. Basu, "A Comparative Study of Desiccant Cooling, Rankine Cycle-Mechanical Vapor Compression Cooling, and Absorption Cooling Systems Using Solar Energy," thesis, New Mexico State University, 1975.

11.1-9 SERI Report TR-632-385, June 1983.

11.1-10 Southwest Project, DOE Contract No. EM-77-C-01-8720, Phoenix, 1979.

11.1-11 Peter Lunde, "Solar Desiccant System Analysis and Materials," NSF Solar Cooling for Building Workshop Proceedings, February 6–8, 1974.

11.1-12 Milton C. Krupka, "Decentralized Solar Photovoltaic Energy Systems," DOE/EV-0101, September 1980.

11.1-13 Charles E. Backus, "Photovoltaic Technology Assessment," Solar Technology Assessment Conference, Orlando, FL, January 29–30, 1981.

11.1-14 J. W. Harrison, "Photovoltaic Cell and Module Status Assessment," Vols 1 and 2, Report by Research Triangle Institute for Electric Power Research Institute, EPRI AP-2473, October 1982.

11.1-15 John Javetski, "A Burst of Energy in Photovoltaics," *Electronics*, July 19, 1979.

11.1-16 Jim Canavan, "Photovoltaics: The Drive to Reduce Costs," *Circuits Manufacturing*, August 1980.

11.1-17 Elmer Christensen, "Electricity from Photovoltaic Solar Cells," *JPL*, February 1980.

11.1-18 Aden B. Meinel and Marjorie P. Meinel, *Applied Solar Energy—An Introduction*, Addison-Wesley, Boston, 1976.

11.1-19 Paul D. Maycock and Edmond N. Stirewalt, *Photovoltaics—Sunlight to Electricity in One Step*, Brick House Publishing, 1981.

11.1-20 D. E. Carlson, "Photovoltaics V: Amorphous Silicon Cells," *IEEE Spectrum*, February 1980.

11.1-21 *EPRI Journal*, March 1978.

11.1-22 Paul Rappaport, "The Photovoltaic Effect and Its Utilization," reprinted in *Solar Cells*, Charles E. Backus (ed.), IEEE Press, New York, 1976.

11.1-23 M. Wolf, "A New Look at Silicon Solar Cell Performance," reprinted in *Solar Cells*, Charles E. Backus (ed.), IEEE Press, New York, 1976.

11.1-24 S. M. Sze, *Physics of Semiconductor Devices*, Wiley-Interscience, New York, 1969.

11.1-25 C. C. Gonzalez, G. M. Hill, and R. G. Ross, Jr., Draft "Baseline Specification for Power Conditioning Hardware in a Utility-Interactive Residential Photovoltaic System—Appendix A, Power Conditioning Input Characteristics for Photovoltaic Flat-Plate Array Applications," *JPL*, February 1982.

11.1-26 W. J. Stolte, EPRI Report AP-2474: "Photovoltaic Balance of System Assessment," June 1982.

11.1-27 Solar Energy Research Institute, "Performance Criteria for Photovoltaic Energy Systems," Vols. I and II SERI/TR-214-1567, July 1982.

11.1-28 H. L. Macomber, et al., "Photovoltaic Stand-Alone Systems, Preliminary Engineering Design Handbook," DOE/NASA/0195-1, August 1981.

11.1-29 Miles C. Russell, "An Apprentice's Guide to Photovoltaics," *Solar Age*, July 1981.

11.1-30 Draft of "Interim Working Guidelines and Discussion Concerning the Interface of Small Dispersed Photovoltaic Power Producers with Electric Utility Systems," October 30, 1981.

11.1-31 "National Photovoltaics Program, Multi-Year Program Plan," draft report by DOE, September 1980.

11.1-32 Molly McCormick, "Solar Cells: Thinking Thin," *Electro-Optical Systems Design*, November 1982.

11.1-33 Monegon, Ltd., "Designing Small Photovoltaic Power Systems," Monegon Publication No. Mlll, May 1981.

11.1-34 JBF Scientific Corporation, "Assessment of Distributed Photovoltaic Electric Power Systems," EPRI Report AP-2687-SY, October 1982.

11.1-35 K. W. Battleson, "Solar Power Tower Design Guide: Solar Thermal Central Receiver Power Systems, a Source of Electricity and/or Process Heat," SAND81-8005, Sandia National

Laboratories, April 1981.

11.1-36 "Conceptual Design of Advanced Central Receiver Power Systems Sodium-Cooled Receiver Concept," SAN-1483-1/2, Rockwell International Energy Systems Group, March 1979.

11.1-37 R. S. Cuputo, "An Initial Study of Solar Power Plants Using a Distributed Network of Point Focusing Collectors," EM 342-308, Jet Propulsion Laboratory, July 1975.

11.1-38 L. S. Rosenberg and W. R. Revere, "A Comparative Assessment of Solar Thermal Electric Power Plants in the 1–10 MW$_e$ Range," DOE/JPL-1060-21, Jet Propulsion Laboratory, June 1981.

11.1-39 E. R. Weber, "Advanced Conceptual Design for Solar Repowering of the Saguaro Power Plant," DOE/SF-M570/2, Arizona Public Service Company, April 1982.

11.2   WIND POWER

Robert L. Scheffler and Michel C. Wehrey

11.2-1   Energy Utilization Parameters

Wind Energy Conversion Equations

The energy of an airstream (wind) is kinetic and is proportional to

$$\tfrac{1}{2}\left(\text{mass} \times \text{velocity}^2\right)$$

The mass of air passing through an area A per unit time is $\rho A V$, and the total power available from the airstream is

$$P = \tfrac{1}{2}\rho A V (V)^2 = \tfrac{1}{2}\rho A V^3 \tag{11.2-1}$$

where ρ is the mass density of air, $\rho A V$ is the air mass flow, and V is the airstream velocity.

Consider a wind turbine with a rotor area A placed into a stream of air with a velocity V (Fig. 11.2-1). The airstream velocity at, and downstream of, the rotor disk can be expressed as a function of an interference factor a. It can be shown from momentum and energy equations that the airstream velocity could be reduced by a maximum factor of $2a$.

Wind turbines extract energy from the wind by absorbing some of the airstream kinetic energy. Assuming no rotational motion in the stream and no drag losses in the rotor, the energy per unit time, or power, P, extracted is

$$P = 2A\rho V^3 a(1 - a)^2 \tag{11.2-2}$$

Equation (11.2-2) has a maximum when $a = \tfrac{1}{3}$, and dividing Eq. (11.2-1) by Eq. (11.2-2) yields

$$\frac{P}{P_A} = 0.592$$

The 0.592 factor, or 59.2%, is sometimes called the Betz limit and is commonly considered to be the maximum energy conversion efficiency achievable by wind turbine rotors.

The axial thrust exerted by the airstream on a rotor of area A is

$$T = 2A\rho V^2 a(1 - a) \tag{11.2-3}$$

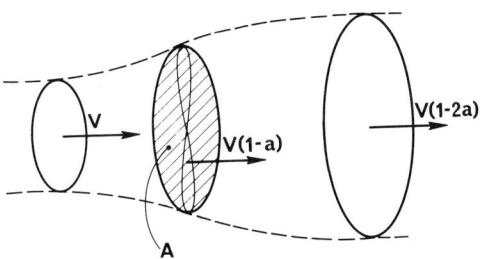

Fig. 11.2-1   Air velocity variation through a rotor of area A.

The power available from the wind to a rotor of area A, using Eq. (11.2-1) and the Betz limit, becomes

$$P = 0.592\left(\tfrac{1}{2}\rho A V^3\right) \tag{11.2-4}$$

The air density, ρ, varies according to atmospheric pressure and ambient temperature and can be expressed in relation to the standard density ρ_0 at 60°F (15.6°C) and sea level:

$$\sigma = \frac{\rho}{\rho_0} = \frac{p}{p_0}\frac{t_0}{t} = 17.336\frac{p}{t} \tag{11.2-5}$$

Where t is the temperature in degrees Rankine, p is the pressure in inches of mercury, and σ is the air density ratio.

Using a standard air density of 0.0763 lb/ft³ (12.0 N/m³), Eq. (11.2-4) can be rewritten as

$$P = 0.592 K \sigma A V^3 \tag{11.2-6}$$

where K has been calculated for various engineering units as follows:

Power (P)	Area (A)	Velocity (V)	K
W	ft²	mph	0.00508
W	ft²	m/s	0.0569
W	m²	mph	0.0547
W	m²	m/s	0.6125
W	ft²	knot	0.00776
hp	ft²	mph	0.00000681
hp	ft²	m/s	0.0000763

thereby expressing the maximum power a rotor can extract from the wind under any atmospheric pressure and ambient temperature conditions.

Wind Energy Resource Parameters

The basic source of winds is the uneven solar heating of the earth's surface resulting in convective air currents. These air currents are influenced by the earth's rotation and by local terrain features such as mountains and valleys. Wind is an intermittent source of energy, with even the windiest areas experiencing periods of low or calm wind conditions. Wind cannot be transported and, therefore, wind turbines must be sited at locations where the wind resource is present. The energy content of the wind, being related to the cube of the wind speed, varies significantly with only small changes in wind speed. This fact points out the importance of having accurate wind speed data when the wind energy resource is being evaluated.

Mean Wind Speed and Energy Resource.   The annual mean wind speed at a location is useful as an initial indicator of the value of the wind energy resource. The annual mean wind speed is commonly available from any wind speed monitoring site, such as weather stations and airports. Table 11.2-1 provides a general indication of the relationship between annual mean wind speed and the potential value of the wind energy resource.

In locations where existing data are not available, a qualitative indication of a high annual mean wind speed can be inferred from geographical location, topographical features, wind-induced soil erosion, and deformation of vegetation. However, accurate determination of the mean annual wind speed requires anemometer data for at least a 1-year period.

The mean wind speed for a given year varies from year to year but is generally within $\pm 10\%$ of the long-term mean. A $\pm 10\%$ variation in mean wind speed would indicate a variation in energy content of $\pm 30\%$ or more.

Some manufacturers provide projections of a wind turbine's annual energy output as a function of mean annual wind speed. This data can be useful but can also be somewhat misleading if the actual wind speed frequency distribution at a location is different from the distribution assumed by the manufacturer.

Wind Speed Frequency Distribution.   Because the energy in the wind is proportional to the cube of wind speed, wind speeds above the annual mean will have a disproportionate amount of the wind energy. As a result, two locations with identical mean wind speeds that have differing wind speed

TABLE 11.2-1   QUALITY OF WIND RESOURCE

Annual Mean Wind Speed at 33 ft (10 m) Height	Indicated Level of Wind Resource
Below 10 mph (4.5 m/s)	Poor
10 mph (4.5 m/s) to 12 mph (5.4 m/s)	Marginal
12 mph (5.4 m/s) to 15 mph (6.7 m/s)	Good to very good
Over 15 mph (6.7 m/s)	Exceptional

distributions could have quite different levels of wind energy resource. Furthermore, the frequency of occurrence of various wind speeds, including the extremes, must be known to determine how well the operating range of a particular wind turbine under consideration matches the wind speed distribution. Obtaining measured data on the wind speed frequency distribution requires use of an anemometer, which can provide a record of the wind speed as a function of time or duration of occurrence.

The wind speed frequency distribution at a given location is either tabulated from wind speed data measured as a function of time or is approximated by a probability distribution function based on measured data or assumed wind resource characteristics. The Weibull probability distribution function is often used to represent the wind speed frequency distribution, $F(V)$, and in this case

$$F(V) = \left(\frac{k}{c}\right)\left(\frac{V}{c}\right)^{k-1} \exp\left[-\left(\frac{V}{c}\right)^{k}\right]$$
(11.2-7)

where k is a dimensionless shape factor, c is a scale factor with units of speed, and $F(V)$ is the probability, or percentage, of occurrence per unit speed of the wind speed V. The factors c and k vary considerably, and to determine them, some data must be available for a location or assumptions must be made. Although generally applicable, for some sites the Weibull distribution does not always accurately represent the wind speed frequency distribution. Assumptions made for c and k should be verified upon receipt of measured data, and in those locations where the Weibull function is not an accurate representation, it would be necessary to use a curve-fitting technique or a tabulated wind speed frequency distribution based on measured data.

Figure 11.2-2 presents a wind speed frequency distribution curve used by the Department of Energy to represent a typical location with a 14-mph (6.3 m/s) annual mean wind speed at a height of 30 ft (9.1 m). This curve is based on a Weibull distribution with $c = 15.8$ mph (7.05 m/s) and $k = 2.25$. This curve is representative of a large number of sites in the United States. However, wind characteristics vary considerably, and, as an example of the potential variation, a second curve in Fig. 11.2-2 presents data measured at a Department of Energy meteorological tower at a desert site (San Gorgonio) near Palm Springs, California.[1,2] In both cases the annual mean wind speeds are similar, but the wind speed frequency distribution at the San Gorgonio site differs significantly from the Department of Energy curve and cannot be accurately represented by a Weibull function. Due to the potentially large variability in mean wind speeds and in wind speed frequency distributions between locations, it is important to obtain accurate site-specific wind resource data at any location being considered for installation of a large number of wind turbines.

The cumulative frequency distribution, or duration curve, is often a useful way to present the wind speed distribution. It represents the fraction of time or number of hours that the wind exceeds or is equal to a particular speed V. Figure 11.2-3 presents the cumulative frequency distribution curves corresponding to the frequency distributions presented in Fig. 11.2-2. The cumulative frequency distribution, if a Weibull distribution is used, can be expressed as

$$D(V) = 8760 \exp\left[-\left(\frac{V}{c}\right)^{k}\right]$$
(11.2-8)

where D is the number of hours in a year that the wind speed exceeds or is equal to V. The c and k factors are the same Weibull factors discussed previously.

Diurnal and seasonal wind speed variations should also be considered. The wind resource is more useful if peak winds occur at times when demand is high.

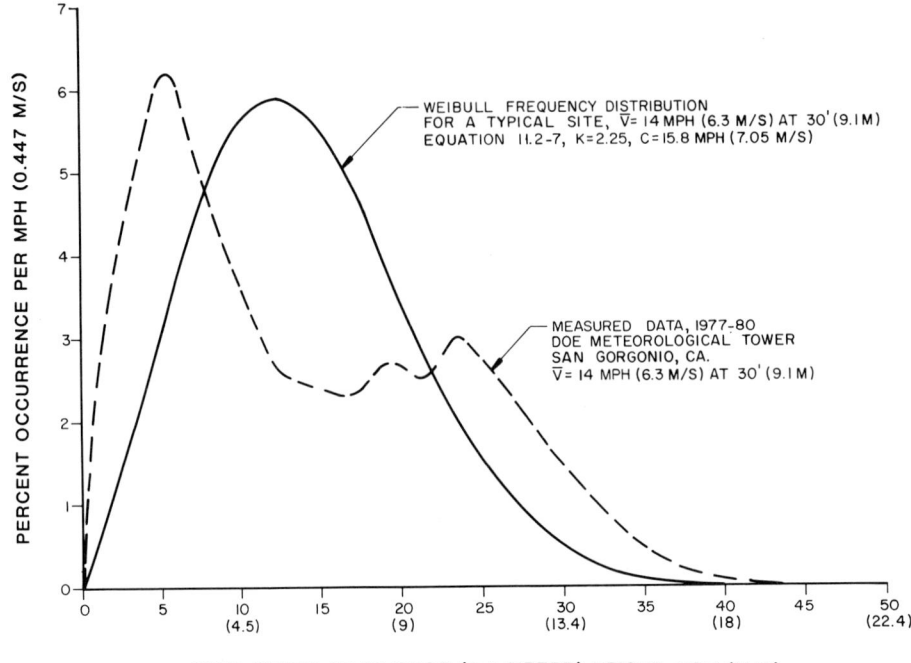

Fig. 11.2-2 Potential variation of wind speed frequency distributions.

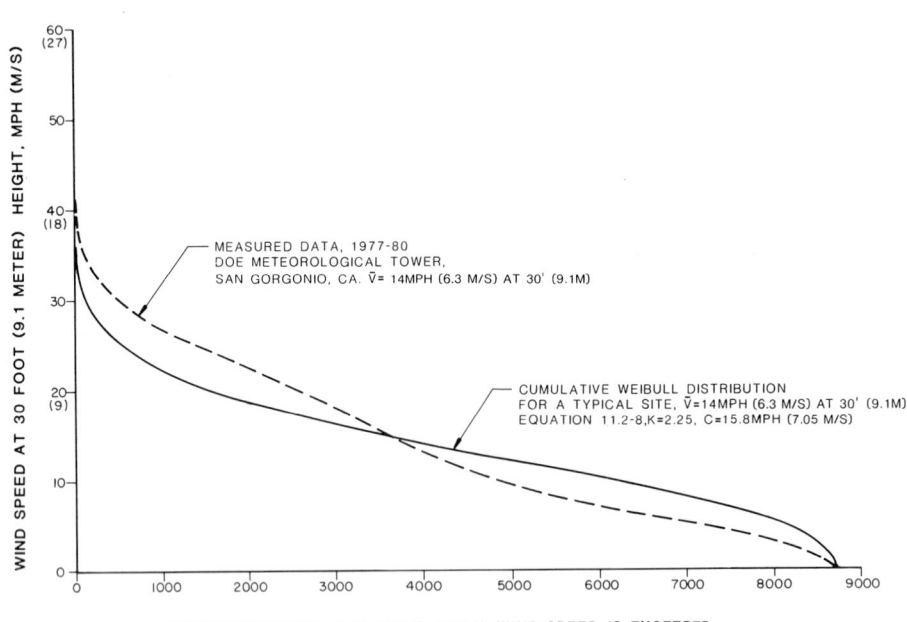

Fig. 11.2-3 Potential variation of cumulative wind speed frequency distributions.

Wind Direction Frequency Distribution. The directional characteristics of the winds at a location can be important. If winds are equally likely from all directions, then a wind turbine must be spaced an adequate distance in all directions from other wind turbines or obstacles that could disrupt the wind from an upstream direction. If winds are predominantly unidirectional, it may be permissible to locate wind turbines closer to other wind turbines or obstacles in the crosswind direction if adequate spacing is preserved in the direction of the prevailing wind. To obtain a wind direction frequency distribution, data must be recorded from a wind direction sensor. Prevailing wind direction may be indicated by the presence of directional deformation of vegetation.

Vertical Wind Shear. Due to friction with the surface of the earth, the wind speed is generally lowest at ground level and increases with height. The portion of the atmosphere affected by surface friction is called the boundary layer and can extend several hundred feet above ground level. The variation of wind speeds with height is called the boundary layer velocity profile or vertical wind shear. Although various equations have been presented in the literature, the following equation is frequently used to relate the wind speed, V_1, at a height, H_1, to the wind speed, V_2, at a corresponding height, H_2:

$$\frac{V_2}{V_1} = \left[\frac{H_2}{H_1}\right]^n \qquad (11.2\text{-}9)$$

A value of $\frac{1}{7}$(0.14) has been developed for n (the "one-seventh power law") for certain classes of turbulent flat plate boundary layer flow and is often used as a typical value for wind flow over flat terrain. The value of n can vary with terrain roughness, turbulence, speed, and height, but is usually in the range of 0.1–0.2 for most conditions in relatively flat terrain. Values of n can be outside this range and, in some cases, the value of n may even be negative, indicating a decrease in speed with height. This condition may result under certain conditions when winds at lower levels are locally accelerated near ridges or hills.

The wind speed variation with height can be very significant in terms of potential wind energy output. For example, using Eq. (11.2-9), the winds at 200 ft (61 m) would be 21% greater than those at 50 ft (15.2 m) for $n = 0.14$. Due to the cubic relationship between wind speed and energy, a wind turbine with a 200-ft (61-m) hub height would have access to an almost 80% greater energy resource than one with a 50-ft (15.2-m) hub height under these conditions. Similarly, it can be seen that a wind turbine with a large blade span may encounter substantially different wind speeds and loads between the upper and lower portions of the rotor disk and must be designed to adequately withstand this type of condition.

General values of n can be used for initial evaluation, but for site-specific data a tower of appropriate height with multiple levels of anemometers is required to accurately determine the wind shear. A less precise, but much less expensive, approach would be to use kite anemometers to take a series of wind speed measurements at various heights to obtain an initial indication of the vertical speed variation.

Turbulence. Turbulence is the rapid fluctuation in the speed and direction of the wind. The presence of turbulence results in additional vibrations and loads on a wind turbine. A higher turbulence level also has the effect of more rapidly reenergizing the wake behind a wind turbine and would allow closer windwise spacing of wind turbine arrays. Measurement of turbulence intensity requires sensitive instruments to measure deviations in speed or direction from the mean speed or direction, respectively. Turbulence can vary substantially with height, wind speed, wind direction, and distance from other wind turbines or obstacles.

Wind Turbine Performance Parameters

A wind turbine is employed to capture kinetic energy from the wind and convert it to a usable form, which may be mechanical (e.g., pumping water) or electrical (e.g., producing alternating or direct current for remote applications; producing alternating current for use in connection with a utility system). Because most modern applications involve generation of electric energy, emphasis will be given to electric wind turbine systems. However, the basic wind turbine performance characteristics are valid for mechanical as well as electric end-use applications.

Performance Characteristics. The shaft power produced by a wind turbine rotor depends on the power in the wind and the rotor aerodynamic efficiency C_p. The power content of the wind [Eq. (11.2-1)] intercepted by the rotor disk area is proportional to the air density, the square of the rotor diameter, and the cube of the wind speed.

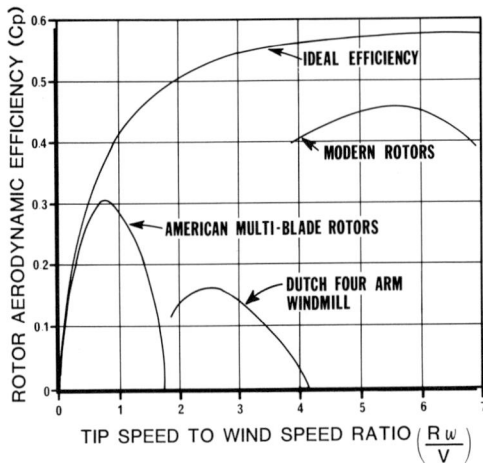

Fig. 11.2-4 Typical performance of wind turbines.

The rotor aerodynamic efficiency is the ratio of rotor shaft power, P_r, to the power content of the wind, P_w, and can be expressed as a function of the blade tip speed to wind speed ratio.

$$C_p = \frac{P_r}{P_w} = f\left(\frac{R\omega}{V}\right) \tag{11.2-10}$$

where R is the rotor radius, ω is rotor rotational speed, and V is the wind speed. Figure 11.2-4 presents typical aerodynamic efficiency values for various wind turbine designs. The performance of the overall wind turbine system depends on the rotor aerodynamic efficiency and the efficiency of the remainder of the system. For a wind turbine producing electric energy, the efficiency of the remainder of the system, which includes the drive train and electric generator, is generally in the 90% range.

The system power output of a wind turbine producing electric energy can be characterized by a curve where net electric power output is presented as a function of wind speed. This power curve is usually based on sea level conditions. To apply the curve at a specific site, the power values given by the curve must be multiplied by the air density ratio σ [Eq. (11.2-5)].* Due to the variation of wind speed with height, the wind speeds on which the power curve is based must be referenced to a particular height. Wind speed at hub height is commonly used, but wind speed at a 30-ft (9.1-m) height is also often used, especially for smaller wind turbines. Use of a 30-ft (9.1-m) reference height requires an assumption of a wind shear profile between the wind speed at the rotor height where the power is being captured and the wind speed present at 30 ft (9.1 m). Due to the variability of the vertical wind shear, the use of a power curve based on the 30-ft (9.1-m) reference height may not always be accurate for larger wind turbines at some sites having unusual wind shear conditions. Therefore, use of a power curve reference to hub height is recommended for larger wind turbines.

Figure 11.2-5 presents a power curve for a variable-pitch rotor, typical of larger horizontal axis wind turbines. As wind speed increases above the cut-in speed (the minimum wind speed at which power is produced), the power output increases from zero to the rated power, generally the maximum, at the "rated wind speed." As the wind speed increases further, the power output of the wind turbine is held constant by adjusting blade pitch to spill some of the available wind energy. At the "cut-out" wind speed the blades are feathered (blade pitch is adjusted for zero energy capture) and the unit is shut down to prevent damage, which might occur by operating the wind turbine in extreme winds.

Most vertical axis and small horizontal axis wind turbines operate at constant rotational speed and fixed blade pitch. The power curve for these types of wind turbines is represented by the one illustrated in Fig. 11.2-6.[3] The wind turbine power output increases from zero at the "cut-in" wind speed to a maximum and then begins to decrease due to the increasing presence of aerodynamic stall as wind speed continues to increase. At the cut-out wind speed the wind turbine is shut down.

The power curve is the primary tool used to describe the performance characteristics of a wind turbine design, but to evaluate its performance in terms of energy production, other considerations, primarily the wind resource level in which the unit will operate, must be taken into account.

---

*Adjustment of the power curve by multiplying it by the air density ratio is an approximation. A more detailed adjustment of the power curve to more accurately present power values at a specific site can generally be developed by the wind turbine manufacturer.

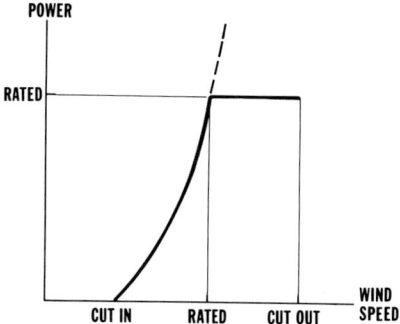

Fig. 11.2-5 Power versus wind speed curve of a wind turbine with a variable pitch rotor.

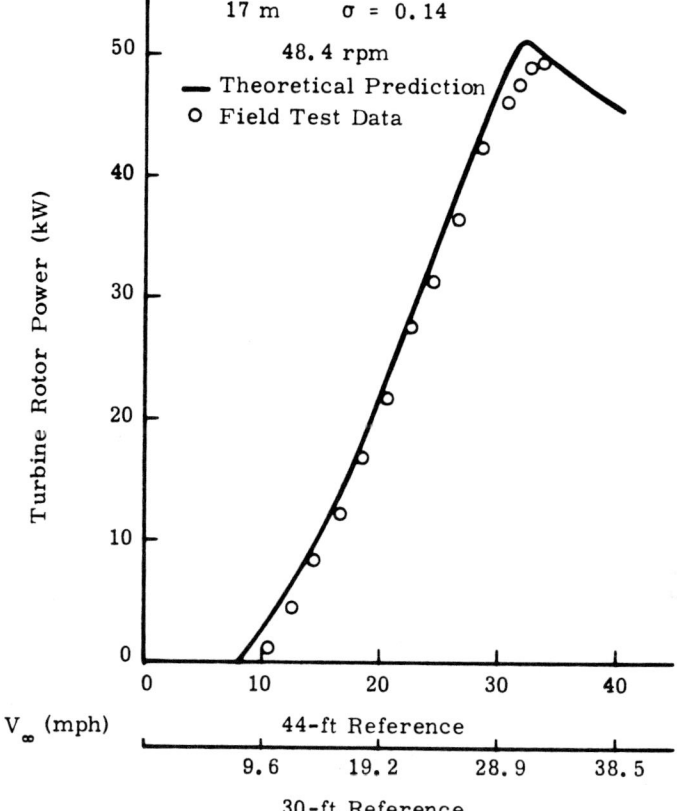

Fig. 11.2-6 Calculated and measured power of a vertical axis rotor as a function of wind speed (from ref. 3).

Estimating Annual Energy Production. The annual energy production expected from a wind turbine depends on the wind resource, the performance characteristics of the wind turbine, and the maintenance downtime projected for the wind turbine. Annual energy production from a wind turbine can be calculated from the equation

$$\text{AEP} = 8760\sigma A_f \int_{V_{ci}}^{V_{co}} P(V) F(V) \, dV \tag{11.2-11}$$

where AEP is the annual energy production (kW · h), $P(V)$ is the power produced by the wind turbine at wind speed V, $F(V)$ is the wind speed frequency distribution in percent per unit speed on

TABLE 11.2-2   ANNUAL ENERGY PRODUCTION USING HYPOTHETICAL TABULAR DATA

Wind Speed @ Hub Height, mph (m/s)	Annual Hours of Occurrence,[a] $8760 \times F(V_i)\Delta V_i$	Wind Turbine Power Output,[b] kW, $P(V_i)$[c]	Wind Turbine Energy Output, kW · h
0–10 (0–4.5)	3,000	0	0
11 (4.9)	200	15	3,000
12 (5.4)	200	30	6,000
13 (5.8)	200	45	9,000
14 (6.3)	200	60	12,000
15 (6.7)	200	80	16,000
16–25 (7.2–11.2)	2,000	100	200,000
over 25 (11.2)	2,760	0	0
Total			246,000
Assuming 90% wind turbine mechanical availability factor:			0.9
Estimated annual energy production:			221,400

[a]An approximation; winds ± 0.5 mph (0.22 m/s) of tabulated value are assumed to occur at the tabulated value; for example, winds occurring between 10.5 mph (4.7 m/s) and 11.5 mph (5.1 m/s) are assumed to occur at 11 mph (4.9 m/s).
[b]Equivalent to a hypothetical wind turbine power curve, similar to Fig. 11.2-5, for a case where cut-in, rated, and cut-out wind speeds at hub height are 10.5 (4.7), 15.5 (6.9), and 25.5 mph (11.4 m/s), respectively, and rated power is 100 kW. The air density ratio is assumed to be unity.
[c]Increments in which $P(V_i)$ is constant are grouped together in the table to simplify presentation.

an annual (8760-h) basis, V_{ci} is the cut-in wind speed, V_{co} is the cut-out wind speed of the wind turbine, σ is the air density ratio, and A_f is the wind turbine mechanical availability factor for the wind turbine accounting for downtime for maintenance. Various researchers have projected values for A_f in the range of 0.85–0.97 as reasonable for commercial wind turbines.

Due to variations in wind speed with height, the wind turbine power curve $P(V)$ and the wind speed frequency distribution $F(V)$ must be based on the same height (rotor hub height, for example). If tabular data based on the frequency distribution of increments of velocity, ΔV, is preferred, the equation becomes

$$\text{AEP} = 8760 \sigma A_f \sum_{i=1}^{n} P(V_i) F(V_i) \Delta V_i \tag{11.2-12}$$

where V_i is the median wind speed in increment i; $P(V_i)$ is the power output at V_i; ΔV_i is the change in wind speed in the increment i (width of the increment), typically taken to be one unit of speed; $F(V_i)$ is the percent occurrence of wind per unit of wind speed in increment i; n is the number of wind speed increments; σ is the air density ratio; and A_f is the wind turbine mechanical availability factor.

To illustrate the use of Eq. (11.2-12), Table 11.2-2 presents an example of a calculation of annual energy production for a hypothetical wind turbine and wind speed frequency distribution.

There is a wide range of wind turbine designs with a variety of rated, cut-in, and cut-out wind speeds. It is important to match the wind turbine design to the wind regime under consideration. If a wind turbine designed for a high-wind regime (high rated and cut-out speeds) were used in a low-wind regime, the rated capacity might seldom be reached and the ability (achieved at additional cost) to operate at high-wind speeds at or above rated capacity might seldom be utilized. Alternatively, if a wind turbine designed for low-wind conditions (low rated and cut-out speeds) were operated in a high-wind regime, the rated speed and cut-out speed would be exceeded for a significant amount of time and, therefore, the energy capture would be significantly lower than desired. In this case the time at rated power would be relatively high, but the actual energy capture would be lower than could be achieved by a machine designed for the high-wind regime.

In the application of the aforestated equations to project annual energy production at a particular site, two approaches can be taken depending on available wind data and required accuracy. The first approach uses an energy output versus annual mean wind speed curve, which can be calculated from the power curve of the wind turbine for an assumed wind speed distribution function and availability factor. If the annual mean wind speed for a site is known, the projected annual energy output can be easily read from the energy curve. Manufacturers often provide these types of energy output curves, which are convenient but may introduce errors if used for a specific site with a wind speed frequency distribution different from the assumed distribution. Using the second approach, annual energy production can be more accurately calculated for a specific site if the actual wind speed frequency distribution at that site is used along with the manufacturer's power curve and an estimate of the availability factor. This approach requires measuring or estimating the wind speed frequency distribution for each specific site under consideration.

In addition, it may be appropriate to take into account energy lost due to operational considerations including startup time (more significant for large WTGs—several minutes) and losses due to yaw errors as a result of fluctuating wind direction changes. This refinement requires minute-to-minute wind speed and direction data as well as detailed knowledge of the wind turbine operational characteristics. In many cases this level of analysis is not practical, but the effect on output should be recognized and a loss factor included if increased accuracy is desired. For example, this factor has been projected to be as high as 10–25% for some large megawatt size wind turbines.

Wind Turbine Array Spacing. The energy extracted from the wind by the wind turbine results in a downstream wake in which the energy content is reduced. The energy in the wake is gradually restored by turbulent mixing with adjacent wind flow. When siting an array of wind turbines, the performance of a unit in terms of energy output may be reduced because of interference from the wakes created by upstream wind turbines. The spacing between wind turbines must be adequate to minimize performance reduction, yet take into account the increased land requirements and related site development costs that result from increased spacing distances.

The performance of an array of wind turbines can be expressed in terms of array efficiency, or the energy output of an average wind turbine in an array compared to the energy output of an individual wind turbine of like design under the same wind conditions. Array efficiency would be 100% for an infinite separation for which there would be no wake interference from the other units. As spacing decreases, the efficiency decreases due to increasing array interference. Due to cumulative interference effects, a larger array of wind turbines would require a greater spacing distance than a smaller array to

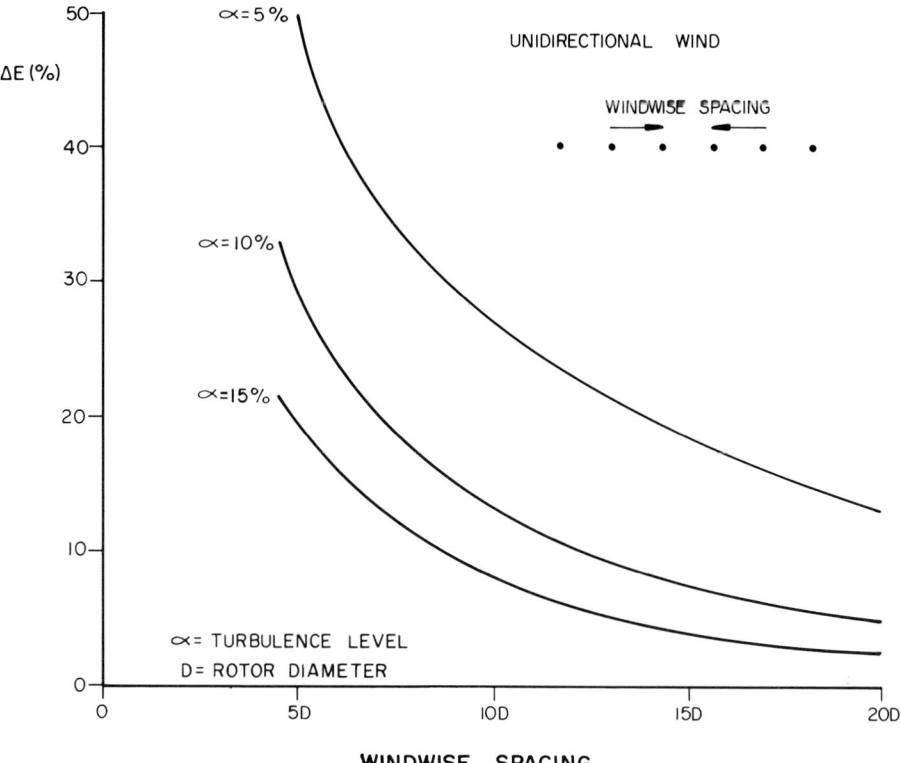

WINDWISE   SPACING

Fig. 11.2-7 Effect on array efficiency of windwise spacing, single-line array (AeroVironment Inc.) (from ref. 5).

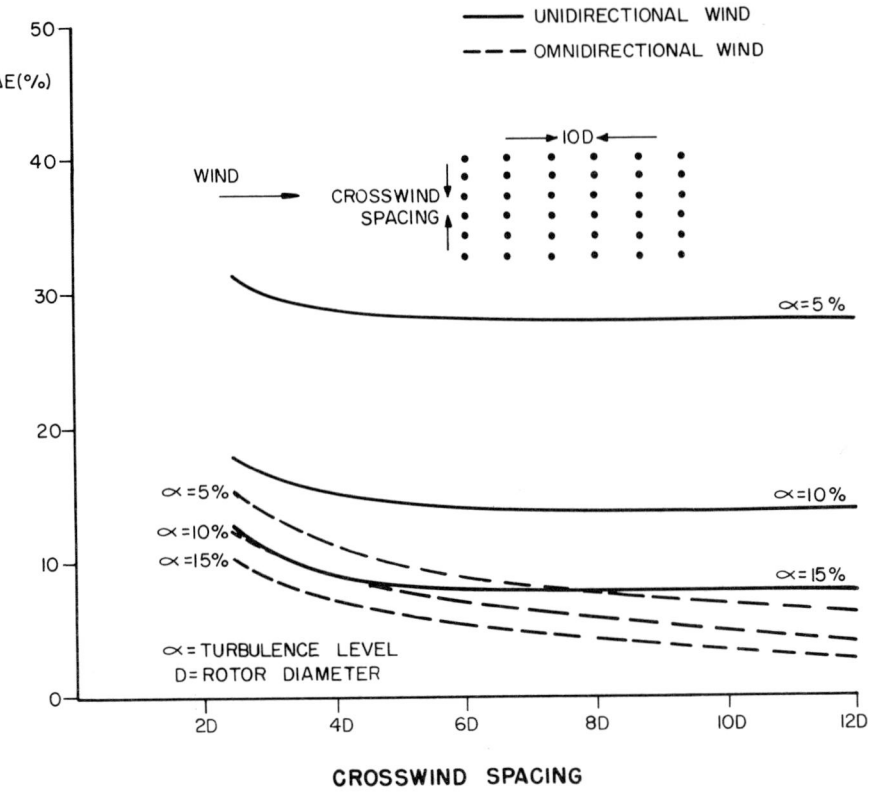

Fig. 11.2-8 Effect on array efficiency of crosswind spacing, rectangular array (Aerovironment Inc.) (from ref. 5).

achieve the same array efficiency. Analytical studies and wind tunnel tests on array spacing have been performed with widely varying results depending on the techniques and assumptions used.[4] A rigorous comparison of analytical results to actual operating performance of a wind turbine array is not yet available.

Although future refinements may be anticipated as more test data become available, some spacing guidelines can be inferred from representative results of a recent analytical array spacing study[5] illustrated in Figs. 11.2-7 and 11.2-8. Figure 11.2-7 represents the variation of array efficiency with windwise spacing (in terms of ΔE, decrease in energy output) for a single line array and a unidirectional wind for three levels of turbulence. Similarly, Fig. 11.2-8 presents the variation in array efficiency for a rectangular array in which windwise spacing is 10 rotor diameters and the crosswind spacing and turbulence levels vary. Efficiency is considerably different for omnidirectional and unidirectional wind speed distributions. In a unidirectional wind for a given windwise spacing, the array efficiency is a weak function of crosswind spacing. In this case it would be possible to decrease the crosswind spacing to a 3–5 rotor diameter value and reduce array land requirements without a significant decrease in array performance. Small additional improvements would be expected by staggering the rows of wind turbines. Other site-specific factors such as trees, terrain features, and structures may require alteration of the spacing layout of an array.

The optimum array configuration will be that which maximizes the overall cost effectiveness of the wind turbine array, taking into account the tradeoff between array efficiency and land cost and related site development costs.

11.2-2   Vertical Axis Wind Turbines

Characteristics

Vertical axis wind turbines have a rotor that receives the wind from any direction. They are relatively simple, and their rotor is not subjected to the gyroscopic forces that occur with horizontal axis rotors

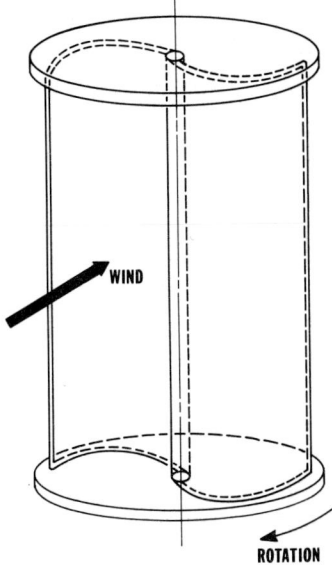

WIND

ROTATION   **Fig. 11.2-9**   Savonius rotor.

during a change of orientation. The end-use energy conversion device (electric generator, pump, etc.) is usually placed at the base of the rotor, thereby making the design and construction of the supporting structure easier. The rotor of these machines need not be installed on a tall tower.

Vertical axis wind turbines can be classified in two broad categories: drag types (or panemones) and lift types.

Drag types use plates, cups, or S-shaped cross-sectional rotors. Since they produce little or no lift forces, they operate at low blade tip to wind speed ratios and produce a relatively low power output for a given rotor size and weight. A typical example of the drag-type vertical axis concept is the Savonius rotor illustrated in Fig. 11.2-9.

Savonius rotors are self-starting and have been built with various numbers of blades (two or more). The height over diameter ratio, as well as the shape and separation of the vanes, vary with the designs. The rotor can be freestanding or kept in a vertical position by a set of guy cables.

Lift-type vertical axis turbines are generally based on a design invented by Georges Darrieus in the 1920s, which uses blades with an airfoil section moving in a circular fashion around a fixed axis of rotation. The blades can be curved in the shape of a turning rope (troposkien), straight and mounted parallel to the axis of rotation, or be of any other combination of orientation and shape. They are usually made of extruded aluminum and generally operate in a fixed position with respect to each other.

Mode of Operation

Darrieus-type rotors have little or no starting torque and therefore need an aerodynamic or mechanical device to set them in motion. This is due to the fact that at low or no rotational speed, the blade is stalled and does not produce adequate lift.

Figure 11.2.10 shows the relationship of absolute wind and blade velocities with the relative wind velocity to the blade, as well as the lift and drag forces on the blade. The tangential component of lift exceeds drag during most of the rotation and creates the rotor driving torque.

Darrieus wind turbines have been built with two-bladed or tri-bladed rotors. Rotors with one blade have also been proposed. Since the blades are generally mounted in a fixed pitch position on the rotor shaft, some means other than pitch change have to be provided to slow the rotor when the load is lost. Aerodynamic spoilers mounted at the midpoint of the blades have been tried but have been found unreliable and difficult to maintain. The method currently preferred is to use a brake mounted directly on the rotor shaft that is capable of absorbing the kinetic energy stored in the rotor and the full torque produced by the maximum operating (cut-out) wind speed. Care has to be exercised by the designer to ensure that the rotor's natural modes of vibration are not energized during the starting and particularly during the stopping of the unit. Minimum braking torque requirements may also be dictated by the need to prevent a situation where the rate of rotor speed decay is insufficient to prevent the excitation of resonant modes.

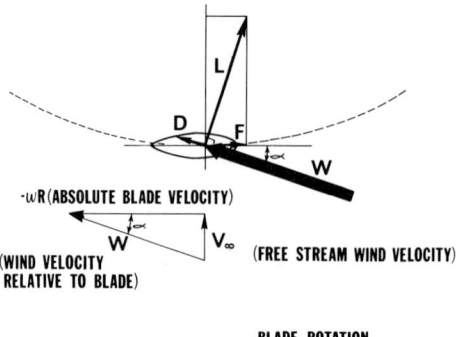

Fig. 11.2-10 The wind velocity W relative to the blade is the vectorial sum of the free stream wind velocity V and the absolute blade velocity $-\omega R$.

Vertical axis wind turbines usually operate at near-constant rotational speeds and, as a result, are most efficient in a narrow zone of wind speeds. To provide for the proper range of tip speed–wind speed ratios, the designer has to know the most frequently occurring wind speeds at the contemplated site and has to identify the optimum rotational speed for the rotor. This is done by choosing the speed increaser (gearbox) ratio that will match the rotor to its load and to the wind resource. Variable-ratio gearboxes have been considered but not used due to the economic penalty involved.

The aerodynamic performance of the wind turbine can be expressed by

$$P = \tfrac{1}{2}\rho A V^3 C_p \qquad\qquad (11.2\text{-}13)$$

where P is the rotor shaft power, ρ is the air density, A is the rotor swept area, V is the wind speed, and C_p is the aerodynamic efficiency expressed as a function of the blade tip speed–wind speed ratio $(R\omega/V)$ where R is the rotor radius and ω is the rotor rotational speed.

The aerodynamic efficiency C_p can be calculated or measured experimentally. A typical curve of C_p as a function of $R\omega/V$ is shown in Fig. 11.2-11.[3]

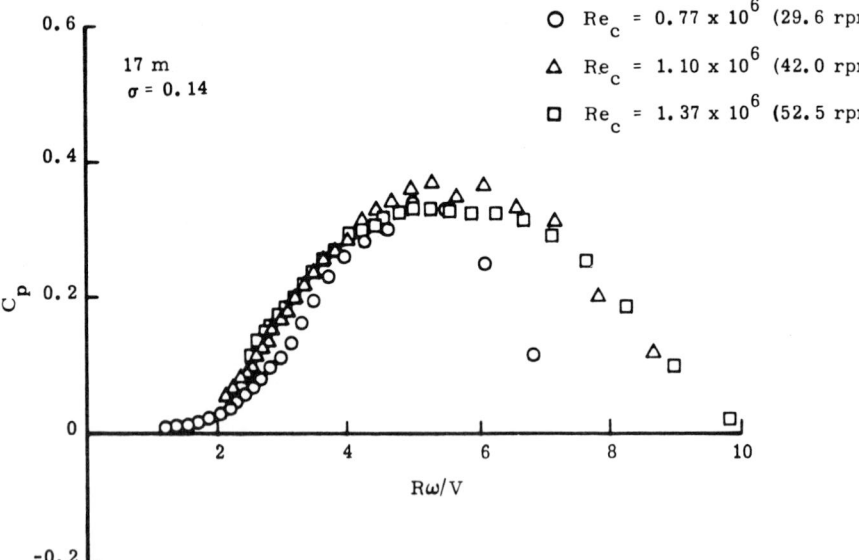

Fig. 11.2-11 Measured aerodynamic efficiency of a vertical axis rotor as a function of tipspeed–wind speed ratio (from ref. 3).

Design Considerations

Matching of Rotor to End-use Device. Since $R\omega$ is constant or near constant in most vertical axis wind turbine applications, the power P produced by the rotor can be directly expressed as a function of the wind speed. As shown by the power curve for a representative vertical axis unit[3] presented in Fig. 11.2-6, the shaft power is theoretically always less than some maximum value for all wind speeds. If the end-use device is rated above the maximum value, there is no need for the turbine to have a torque regulation mechanism. However, economics may dictate that the drive train and end-use device be sized for a lower power level since the utilization factor of a fully sized system may be low. If this is the case, the wind turbine has to be shut down when the wind speed exceeds a predetermined cut-out value.

Vertical axis wind turbines used for generation of electric energy rely on the "pullout" torque of the generator to achieve near-constant speed operation. Induction and synchronous generators, once tied to a relatively "stiff" utility grid (where the grid frequency is set by conventional generators several orders of magnitude larger than the wind turbines), have steep torque versus rpm characteristics when operated near their synchronous speed. However, the cutout wind speed has to be set low enough to prevent any pullout from occurring, since this would relieve the rotor from much of its load and could lead to a potentially disastrous overspeed condition.

Torque Ripple. An intrinsic mechanical phenomenon associated with the operation of any type of wind turbine, and particularly vertical axis machines, is called "torque ripple." This is the time variation of the torque supplied by the rotor to the drive train and to the end-use device. In a vertical axis turbine this cyclic torque variation is caused by the continuously changing orientation and magnitude of the wind speed relative to the blade. This causes continuous changes of lift on the blade and creates a torque ripple with a frequency which is established by the rotor rpm and its number of blades. For example, a two-bladed rotor will experience a "two per revolution" torque ripple as shown in Fig. 11.2-12.[6] Experiments and analyses have shown that three methods of reducing torque ripple are available: a reduction of the torsional rigidity of the drive train, a reduction of its power losses, and the addition of rotational inertia along the drive train.

Rotor Support. Lateral stability of the rotor of vertical axis turbines is generally achieved by guy cables anchored on the ground and connected at the top of the rotor through a bearing assembly. Steady and vibratory loads must be resisted by the guy cables, which are also subjected to tension and thermal loads. The number of cables used varies from three to six. Three may be a preferred number since it simplifies construction and provides for a better control of tension in each cable. Optimization studies have shown that the best elevation angle is 35°. The combined stiffness of the rotor shaft and

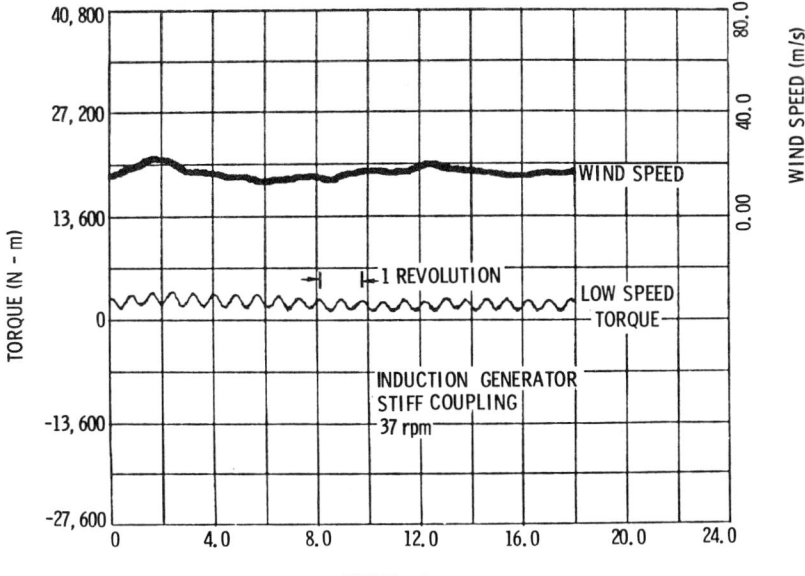

Fig. 11.2-12   Recorded torque ripple of a two-bladed vertical axis rotor (from ref. 6).

TABLE 11.2-3 MAIN SPECIFICATIONS OF REPRESENTATIVE VERTICAL AXIS WTG DESIGNS

Specification	Alcoa "100"	Alcoa "500"	Flow Wind "170"	McDonnell Douglas "Gyromill"	DAF Indal Ltd. "50"	DAF Indal Ltd. "500"	Vawt Power "185"
Rated power, kW	100	500	170	40	50	500	185
Rated windspeed,[a] mph (m/s)	37 (16.5)	35 (15.6)	44	20 (9.0)	45 (20.1)	43 (19.2)	37 (16.5)
Cut-in windspeed, mph (m/s)	13 (5.8)	14 (6.3)	11 (4.9)	10 (4.5)	14 (6.3)	15 (6.7)	13 (5.8)
Cut-out windspeed, mph (m/s)	45 (20.1)	60 (26.9)	60 (26.9)	40 (17.9)	NA[b]	55 (24.6)	50 (122.4)
Total height, ft (m)	95 (28.9)	137 (41.9)	92 (28)	96 (29.3)	66 (20.1)	132 (40.2)	99.8 (30.4)
Rotor height, ft (m)	83 (25.3)	123 (37.5)	75 (22.9)	42 (12.8)	56.5 (17.2)	122 (37.2)	79.5 (24.2)
Rotor diameter, ft (m)	55 (16.8)	82 (25)	57 (17)	58 (17.7)	38 (11.6)	80 (24.4)	57.5 (17.5)
Rotor swept area, ft^2 (m^2)	2915 (270)	6400 (595)	2600 (242)	2436 (226)	1300 (120)	6400 (595)	3000 (278.8)
Number of blades	2	3	2	3	2	2	2
Blade chord, in. (m)	24 (0.61)	29 (0.74)	24 (0.60)	27 (0.68)	14 (0.36)	29 (0.74)	29 (0.74)
Airfoil section	NACA 0015	NACA 0015	NACA 0015	NA	NACA 0015	NACA 0015	NACA 0015
Blade position	Fixed	Fixed	Fixed	Variable	Fixed	Fixed	Fixed
Rotor rpm	48	40	53	NA	80	45	48
Generator rpm	1200	1800	1800	NA	1200	720	1800
Generator type	Induction	Induction	Induction	NA	Induction	Induction	Induction
Generator number	1	1	1	1	1	2	1

[a] Windspeed at rotor midheight except for Alcoa and Flow Wind models where windspeeds are 30 ft (9.1 m) above ground level.
[b] Not available.

guy cable system is not appreciably increased by mounting the rotor rigidly to the base of the turbine. Most vertical axis turbine rotors are therefore designed with a capability to tolerate small out-of-plumb conditions through the use of a coupling device between the rotor and the speed increaser.

Cable tension is controlled by dynamic considerations, and the requirement to design a tiedown system that will not be excited by prevalent forcing functions, such as the two per revolution function of two-bladed rotors. The natural frequency of taut, heavy cables is given by

$$f_m = \frac{m\pi}{L}\left(\frac{Tg}{W}\right)^{1/2} \tag{11.2-14}$$

where m is the vibration mode number, T is the cable tension, g is the acceleration of gravity, W is the cable linear weight, and L is the cable length.

Thermal effects will cause significant changes in the cable tension and, therefore, must be taken into account. The total change in cable length due to thermal and mechanical strain is given by

$$\Delta L = \eta L \Delta T + \frac{LP}{EA} \tag{11.2-15}$$

where η is the thermal expansion coefficient; A is the cross-sectional area; E is Young's modulus; L and ΔL are length and length change, respectively; P is the load; and ΔT is the temperature change.

Representative Designs

Vertical axis wind turbines have been built in a wide variety of sizes and power ratings. Darrieus designs with ratings up to 500 kW have been built, and designs with megawatt ratings have been contemplated.

Table 11.2-3 summarizes the characteristics of representative turbines. Figure 11.2-13 shows the outline and dimensions of a two-bladed unit rated for 500 kW.

11.2-3 Horizontal Axis Wind Turbines

Characteristics

Horizontal axis wind turbines have a rotor (propeller) that faces the wind and is mounted on a tower. If the end-use energy conversion device is an electric generator, it is generally also mounted on the tower. Pumps are generally mounted at ground level and driven by a long shaft or a connecting rod. Horizontal axis turbines can be classified in two broad categories: upwind and downwind types.

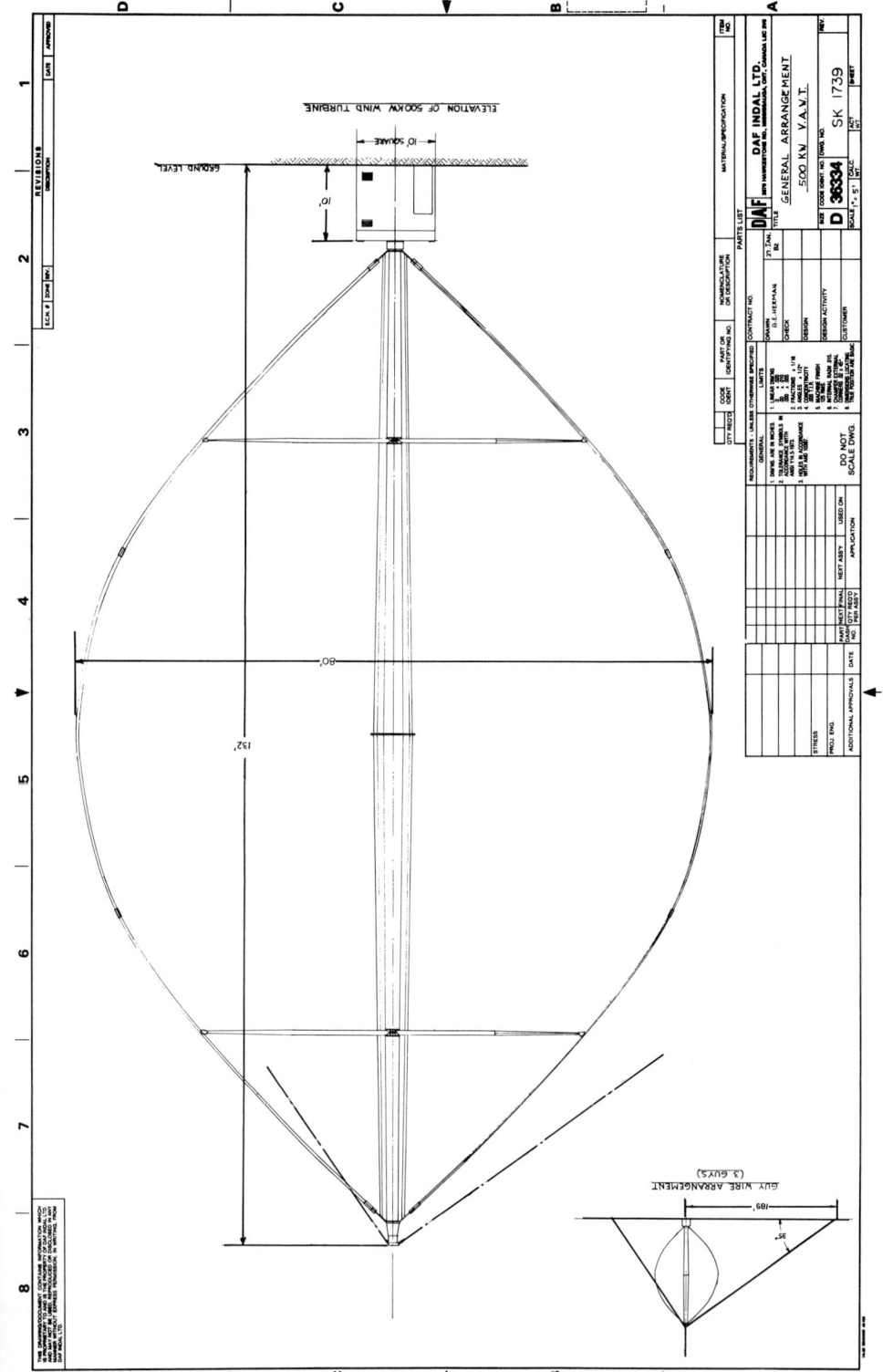

Fig. 11.2-13 Outline of the DAF Indal Ltd. two-bladed 500 kW vertical axis wind turbine. (Courtesy DAF Indol Ltd., Mississauga, Ontario, Canada.)

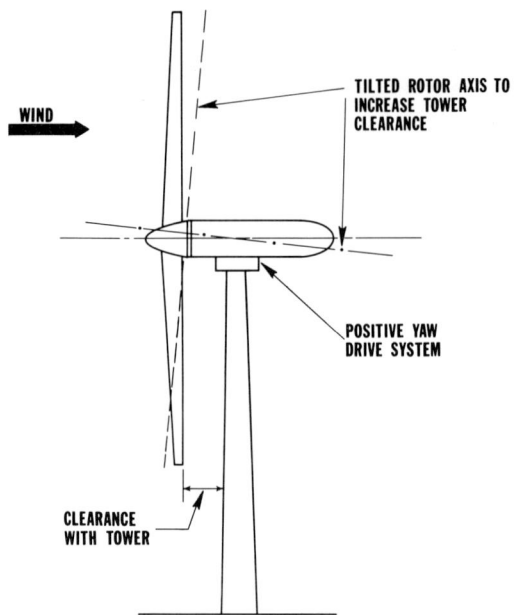

Fig. 11.2-14　Upwind horizontal axis wind turbines may have a tilted rotor to increase tip clearance with tower.

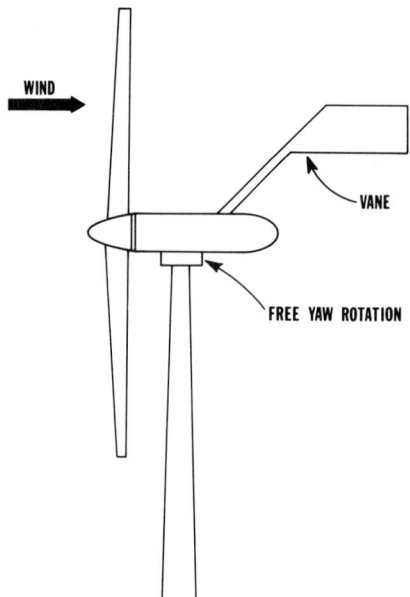

Fig. 11.2-15　Orientation of rotor achieved through wind vane.

Upwind turbines have their rotor mounted in front of the tower. Large units need a positive-yaw drive system with associated controls to achieve orientation in the wind. Clearance between the rotor blades may necessitate the rotor axis to be tilted upwards (Fig. 11.2-14). Smaller units (less than 150 kW) may achieve yaw control through the use of a vane mounted on the nacelle (Fig. 11.2-15). A benefit of the upwind rotor location for both large and small units is that the tower produces a much smaller wind "shadow" on the rotor.

Downwind turbines have their rotor mounted behind the tower and may not need a positive-yaw control system. The rotor acts as a weather vane and tends to orient itself into the wind. However, yaw

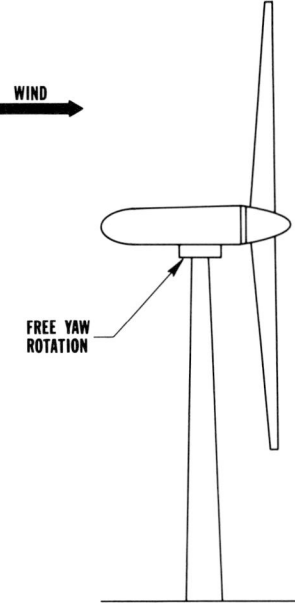

WIND

FREE YAW
ROTATION

Fig. 11.2-16   Passive yaw control on a downwind wind turbine.

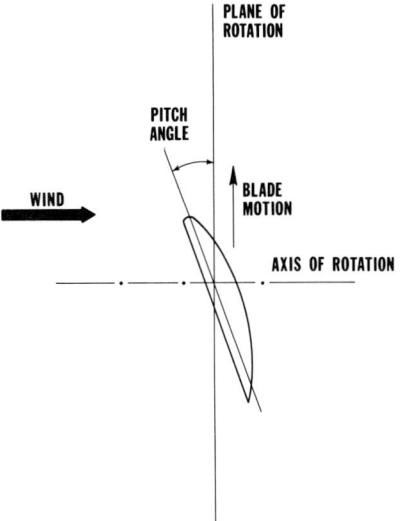

PLANE OF
ROTATION

PITCH
ANGLE

WIND

BLADE
MOTION

AXIS OF ROTATION

Fig. 11.2-17   Definition of pitch angle.

instabilities have been observed in small units. Large downwind units may need positive-yaw control if the yaw error cannot be kept within acceptable limits (Fig. 11.2-16).

Rotors of horizontal axis turbines have been built with various numbers of blades. The high-torque, low-rpm rotors of water-pumping windmills may have more than 30 blades whereas large 1-blade rotors have also been proposed. The number and size of blades determine the "solidity" of a rotor, defined as

$$\text{Solidity} = \frac{(\text{blade area}) \times (\text{number of blades})}{(\text{area swept by the rotor})}$$

Since the energy capture gains decrease rapidly as the number of blades increases, large modern wind turbines tend to have low solidity values to decrease rotor complexity and minimize the cost of energy.

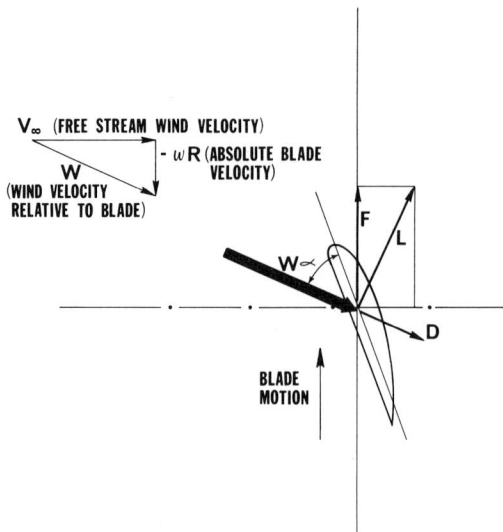

Fig. 11.2-18   The lift, L, is created by the wind speed, W, if the angle of attack, α, is not too large.

The pitch of the rotor blades is defined as the angle between the blades and their plane of rotation (Fig. 11.2-17). If the blade can be positioned parallel to the wind (near 90° pitch), it will generate no torque and is in a "feathered" position.

Horizontal axis wind turbines are generally self-starting if the blade pitch is large enough (or can be adjusted) to generate a torque sufficient to overcome the friction and inertia forces present in the drive train and end-use device.

Mode of Operation

The rotating blade of the horizontal axis rotor is subjected to lift and drag forces similar to those applied on the wing of an airplane. As opposed to drag devices, the blade operates under relative wind velocities in excess of the free stream wind velocity (Fig. 11.2-18). The lift generated by the blade creates the torque applied to the rotor.

As in the case of vertical axis turbines, the performance of a horizontal axis rotor can be expressed by

$$P = \tfrac{1}{2}\rho A V^3 C_p \qquad\qquad (11.2\text{-}16)$$

The aerodynamic efficiency of the rotor can also be expressed as a function of the tip speed–wind speed ratio $R\omega/V$. Figure 11.2-4 shows the typical aerodynamic efficiency values of various wind turbines and Fig. 11.2-19 shows the calculated aerodynamic efficiency of the 200-ft (61-m), two-bladed rotor of a megawatt size machine for various pitch angles.

Figure 11.2-19 also shows that maximum rotor efficiency is achieved for a narrow range of tip speed–wind speed ratios. Wind turbines operating at constant rotor rpm cannot maintain this condition at all wind speeds and therefore lose some efficiency. Variable rotor rpm operation can increase the energy capture of a wind turbine if the energy losses in the variable-speed transmission or variable-speed end-use device do not exceed the gains obtained from the increased rotor efficiency.

Figure 11.2-5 illustrates how the rated power output is maintained at a constant level as the wind speed increases from V_r (rated) to V_{co} (cut-out). The blade pitch is progressively increased to decrease the rotor efficiency, and the rotor "spills" some of the wind energy.

Design Considerations

Rotor.   The number of blades is dictated by the selected solidity factor and the blade area. The driving factors involved in selecting the number of blades are blade costs, hub complexity, and cost of the pitch control system. Dynamic interaction of the rotor with the tower may also influence the number of blades since the forcing function originating from the rotor is dependent on the rotor speed and the number of blades.

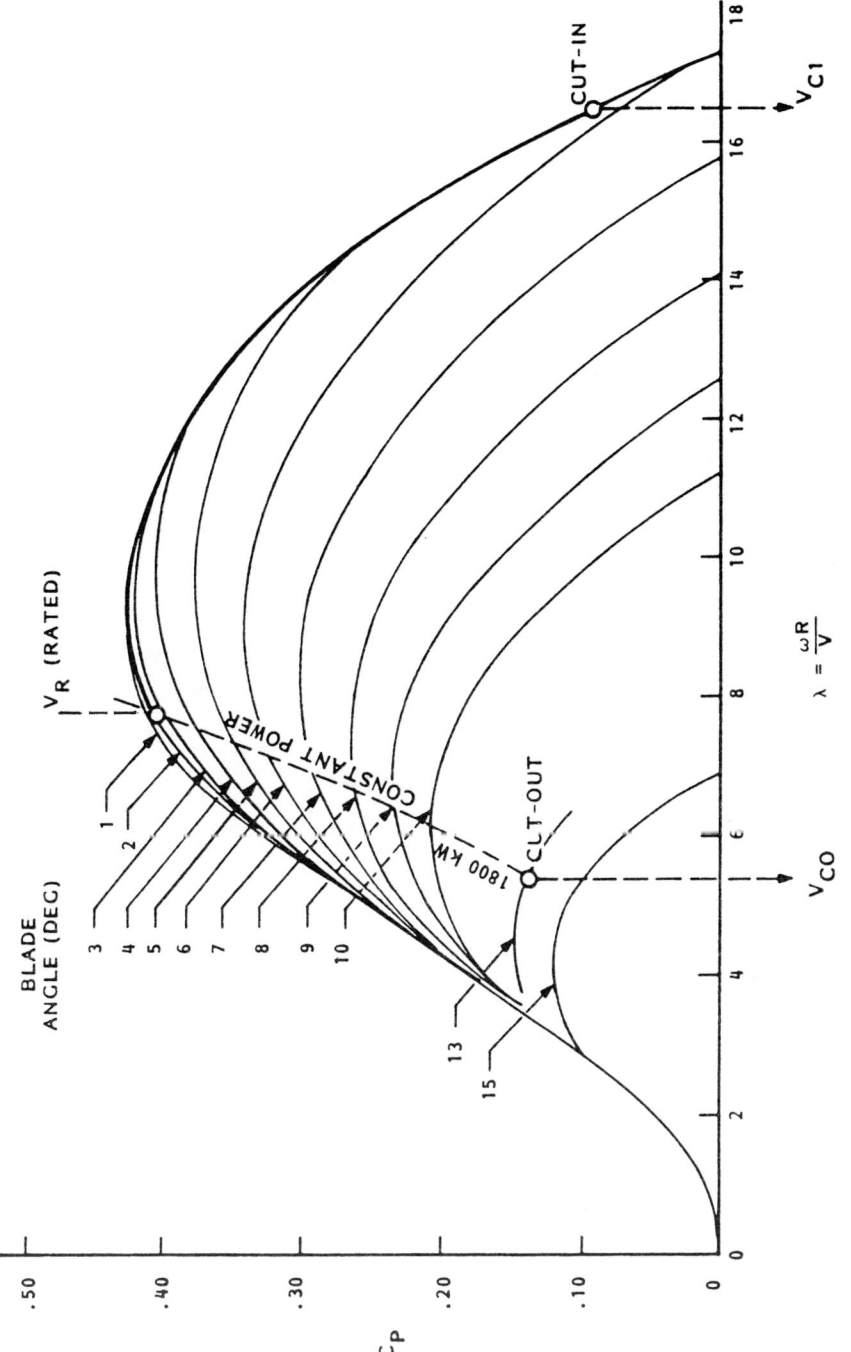

Fig. 11.2-19 Calculated performance of DOE's MOD-1 wind turbine.

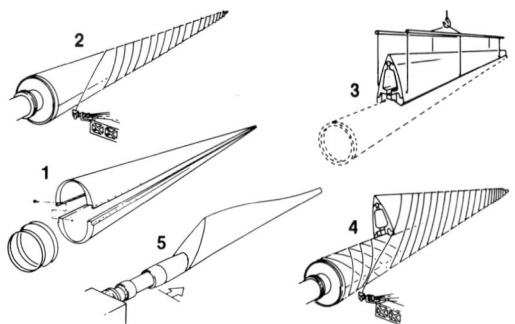

Fig. 11.2-20 Filament-wound construction method for 120-ft (36.6-m) composite rotor blades. (Courtesy of Hamilton Standard, Windsor Locks, CT.)

The airfoil section of the blades can be symmetrical, cambered, or flat bottomed. Airfoil shapes are selected to maximize lift, minimize drag, and provide a moment of inertia compatible with structural strength requirements. The blade can be designed with some or no twist, depending on tradeoffs between performance and manufacturing costs. A certain amount of twist can be selected to optimize the angle of attack of the airfoil along the length of the blade. The planform (outline) of the blade can be rectangular or tapered with the largest sections being located near the hub of the rotor. Planform shapes are selected to distribute the aerodynamic loads as evenly as possible.

Blade construction techniques include laminated wood, riveted stainless steel or aluminum, welded steel, filament-wound composites, and extruded aluminum. Solid laminated wood can be used for small blades, whereas large blades can be built with a laminated wood spar and plywood ribs. Layers of wood veneer shaped on a mandrel can also be used to construct hollow blades with a process similar to that used to construct boat hulls. Riveted construction and aluminum skin blades have not given encouraging results. Extruded aluminum has only been used for small turbines (100 kW or less) due to manufacturing and structural limitations. Construction methods using composite materials and filament-wound techniques are favored by some designers due to an absence of welded or riveted joints and due to their resistance to fatigue and corrosion. Composite blades also resist damage from foreign objects, are easier to repair, and have less impact on electromagnetic signals. Figure 11.2-20 illustrates a construction method using filament-wound techniques developed for 120-ft (36.6-m) blades.

Blade materials should be selected for:

1. Low material and fabrication costs.
2. The existence of prior experience on their use and the availability of design data.
3. Adequate mechanical properties:
   (a) High elastic modulus to resist buckling
   (b) High fatigue strength to resist initiation and propagation of cracks under cyclic loading
   (c) High ductility to avoid rapid crack propagation.
4. Resistance to degradation under adverse environment.
5. Repairability.
6. The availability of reliable inspection techniques.

Hub. The hub function is to retain the blades, provide for pitch angle changes, and allow for teetering or coning as selected by the designer. The hub retains the bearings mounted at the root of the blades and has to withstand the weight, centrifugal, and aerodynamic loads transmitted by the blades. Casting, forging, and welding are common techniques used to manufacture the hubs of modern wind turbines. Teetering is provided with two-bladed rotors to relieve the hub and blades from dynamic loads created by wind gusts and wind shear. To achieve teetering, the blades are allowed to pivot around an axis perpendicular to the axis of rotation, as shown in Fig. 11.2-21. Coning, as shown in Fig. 11.2-22, consists of tilting the blades downwind from a plane perpendicular to the axis of rotation to reduce the bending moments in the root sections. Coning can be built-in or variable through the use of hinges. When free to cone, the blades subjected to wind pressure and centrifugal forces will seek an equilibrium position. The angle selected for built-in coning is selected to match this equilibrium position under operating conditions.

Pitch Angle Control. Small wind turbines can operate with fixed-pitch blades and rely on aerodynamic or mechanical braking to limit the output of the rotor or stop it under shutdown conditions. Moderate

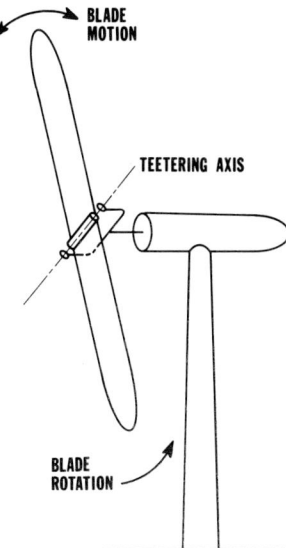

Fig. 11.2-21 Rotor teetering can reduce the dynamic loads of two-bladed horizontal axis rotors.

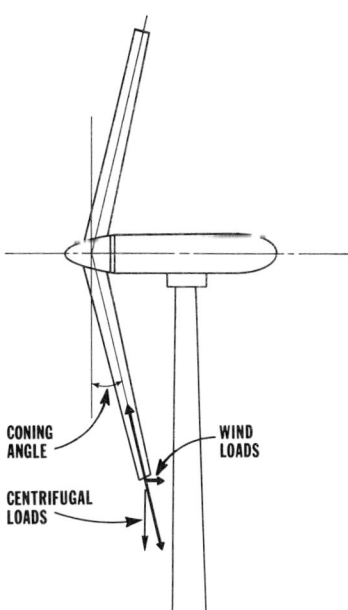

Fig. 11.2-22 Coning uses the centrifugal forces on the blades to reduce the bending loads near the rotor hub.

and large size machines need a pitch control mechanism to start the rotor, control its output at wind speeds above rated, dampen its dynamic oscillations, and slow it down to rotational speeds compatible with the energy-absorbing capability of the brake. Economic considerations may prevent the selection of brakes sized to absorb the full rated output of the rotor.

Full-span pitch control is achieved by rotating the entire blade in the rotor hub.

A typical full-pitch change mechanism, as shown in Fig. 11.2-23, consists of a hydraulic supply, a rack and pinion actuator, and gears to rotate the blades in the hub. The pair of racks is moved linearly back and forth by hydraulic pressure. The racks rotate a master gear that in turn rotates the blades through bevel gears mounted at the root of the blades. Most pitch change mechanisms include means of returning the blades to a fully feathered position in case of a loss of control power.

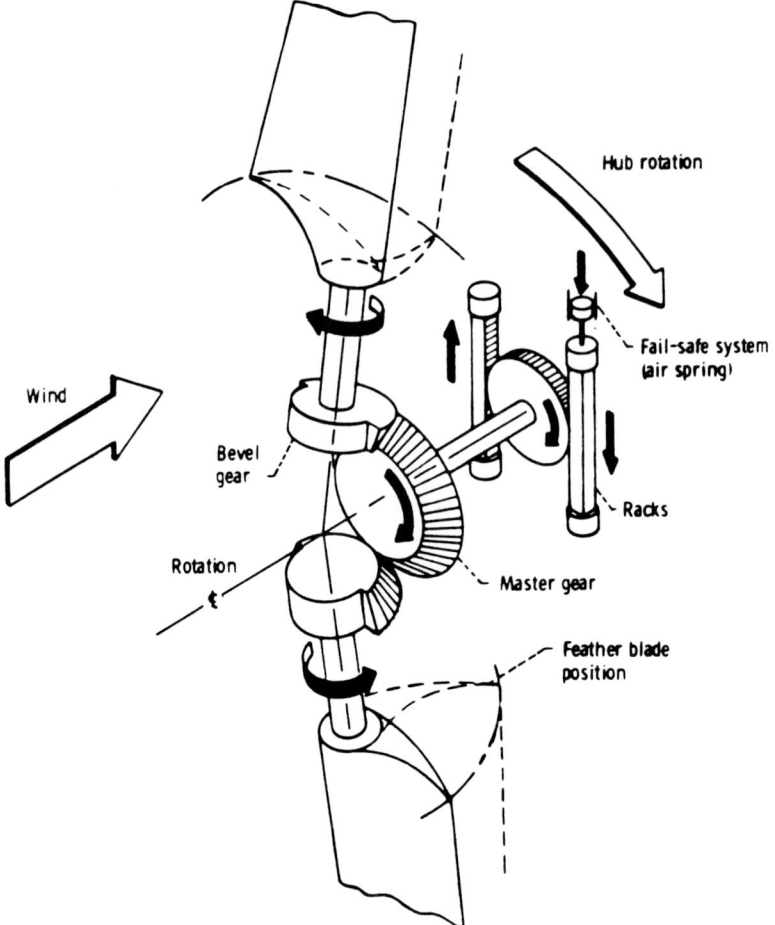

Fig. 11.2-23   Pitch control mechanism of DOE's MOD-0 wind turbine.

Partial span pitch control can be selected to simplify the construction of the hub and decrease the torque required to change the pitch angle. In this configuration the central portion of the blades is constructed with a fixed pitch, and only the tip of the blades is hinged and moved by actuators (Fig. 11.2-24).

Drive Train.   The drive train transmits the mechanical power of the rotor to the end-use device. Matching of rotational speeds is achieved through a gearbox. A brake mounted either on the low-speed (rotor) shaft or the high-speed (end-use device) shaft stops the rotor during normal and emergency shutdowns. Planetary gearboxes are generally preferred for large wind turbines because of their compactness, high efficiency, and lower weight (Fig. 11.2-25). Wind turbines producing electricity through the use of a synchronous generator require mechanical damping between the rotor and the generator. This damping can be achieved through the use of a fluid coupling device, a "quill" (soft) shaft, or a shock-mounted gearbox.

Yaw Control.   Orientation of the rotor into the wind is achieved by allowing the nacelle to rotate on a turntable bearing. A bull gear mounted on the base of the nacelle or on the top of the tower is usually driven by one or two pinion gears connected to hydraulic or electric motors. A locking mechanism is used to achieve the required torsional stiffness. Small, downwind wind turbines may not have a yaw drive system, but a locking mechanism is still required for maintenance and safety purposes.

Figure 11.2-26 shows the nacelle of a 200-kW two-bladed, downwind, wind turbine generator.

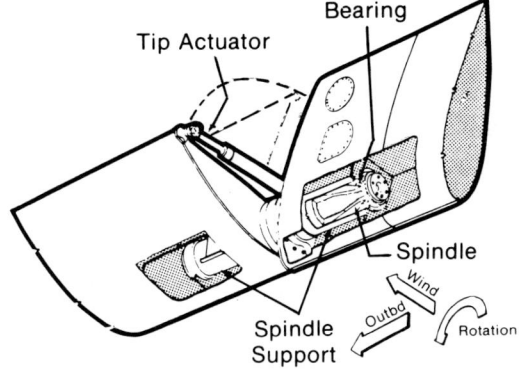

Fig. 11.2-24 Partial span pitch control of DOE's MOD-2 wind turbine.

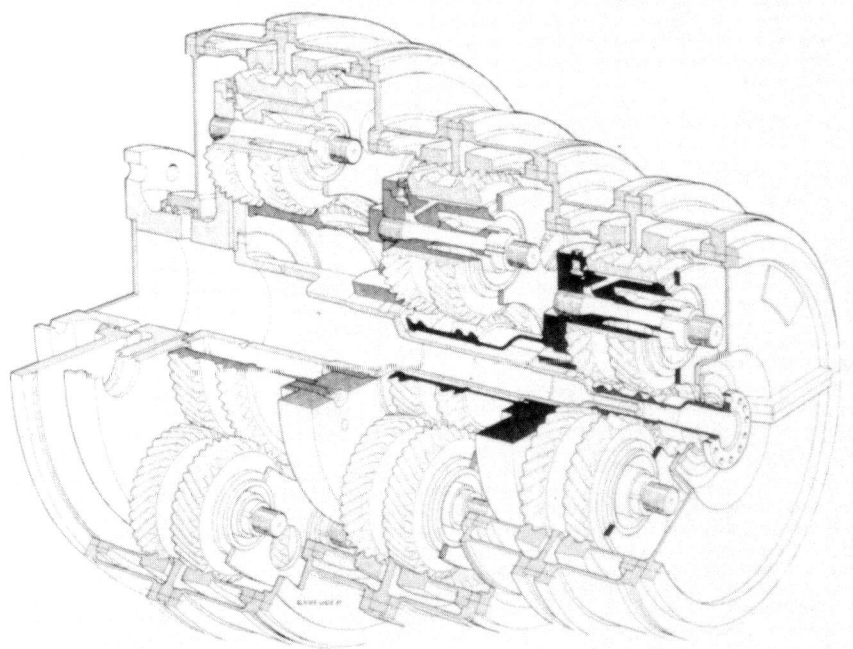

Fig. 11.2-25 Planetary gearboxes are compact, efficient, and light.

Representative Designs

The power rating of horizontal axis wind turbines built in the United States and Europe ranges from a few kilowatts to several megawatts. Rotor diameters in excess of 400 ft (122 m) have been considered. Tubular and truss-type steel construction are most often selected for the tower supporting the nacelle. Concrete towers have also been selected for their low cost and long life advantages. Table 11.2-4 summarizes the main characteristics of representative wind turbines in operation or planned. Figure 11.2-27 illustrates the evolution of large machines designed in the United States.

11.2-4 Application Potential

Theoretical Potential

It has been estimated[7] that up to 1.3×10^8 MW of power could theoretically be extracted from wind on a global basis and up to 2×10^6 MW from the continental United States. The potential is large;

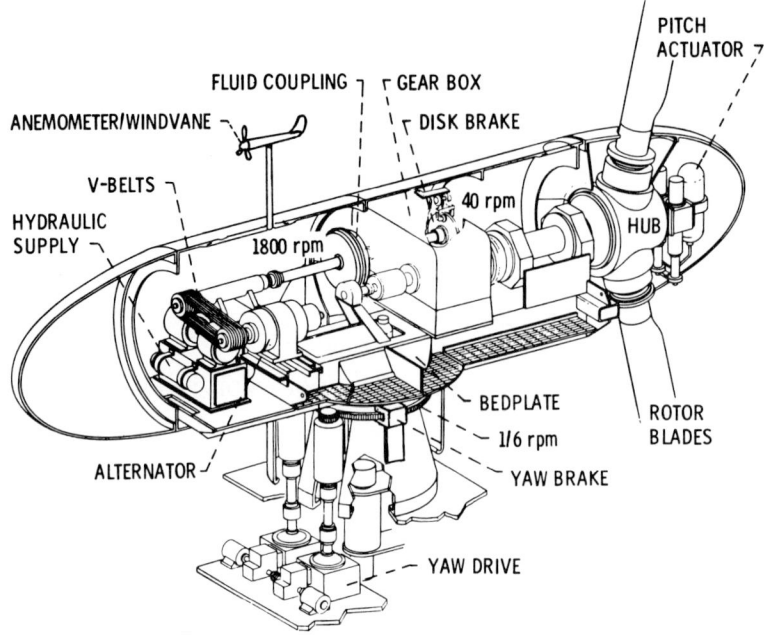

Fig. 11.2-26   Nacelle of DOE's MOD-0A wind turbine.

however, the use of this energy source can be influenced by a number of practical factors. Factors that affect the application of wind turbines as an energy source may include siting considerations, environmental impacts, and economics.

Siting Considerations

The initial consideration in siting wind turbines is the availability of the wind energy resource. One proposed methodology for selection and evaluation of sites with good wind resources, adapted from work by Zambrano and others,[8] consists of the following steps:

   1.   **Meteorological data assessment.**   Existing meteorological data bases (such as those from the U.S. National Weather Service, airports, U.S. Forest Service and state forest agencies, utility companies, and interviews with local residents) are reviewed, taking into account local site topography, to identify areas that may have a good wind energy resource.
   2.   **Field assessment.**   Promising regions identified in step 1 are investigated by field personnel. Measurements are taken of surface winds and winds aloft, ideally when the wind flow is in the prevailing direction. Presence of ecological indicators, such as vegetation flagging and tree deformation, are noted.
   3.   **Regional investigation/verification.**   Areas identified in steps 1 and 2 can be investigated by placing a network of anemometers, generally at standard heights of 30 ft (9.1 m) or 33 ft (10 m), to obtain longer-term (usually 1 year or more) data. This provides information on general power-producing capabilities and determines specific meteorological properties such as the variation of winds with location, general flow patterns, wind speed and direction frequency distributions, maximum wind speed, and diurnal and seasonal variations in wind speeds. This information can then be used to identify promising wind turbine siting areas.
   4.   **Site verification.**   Certain sites identified in step 3 would be equipped with a meteorological tower, as much as 150 (45.7) to 350 ft (107 m) in height depending on the height of the wind turbines under consideration. These towers would measure wind data at several levels to provide detailed wind characteristics, such as vertical wind shear, turbulence, wind speed and direction frequency distributions, wind variability, and maximum wind speeds as a function of height above ground level.
   5.   **Wind turbine selection/design and performance evaluation.**   Based on site-specific measured data obtained in step 4, wind turbine design specifications can be developed, or by assessing the performance characteristics of available wind turbines, a wind turbine design could be selected that is well matched to the wind regime at the site. After a design has been selected, the number of wind turbines in an array, and spacing parameters of a wind turbine array, can be established. Then the

TABLE 11.2-4 MAIN SPECIFICATIONS OF REPRESENTATIVE HORIZONTAL AXIS WTG DESIGNS

Specifications	Manufacturer							
	J. Carter CWG-25	WECS Tech WT-605	Westinghouse WWG-0500	Boeing MOD-2	Hamilton Standard WTS-4	Boeing MOD-5B	Nibe "B"	KaMeWa WTS-75
Rated power, kW	25	100	500	2,500	4,000	3000	630	2,000
Rated windspeed,[a] mph (m/s)	26 (11.6)	28 (12.5)	27.5 (12.3)	27.3 (12.2)	33.9 (15.1)	28 (12.5)	29.1 (13)	28 (12.5)
Cut-in windspeed, mph (m/s)	7.5 (3.4)	12 (5.4)	14.3 (6.4)	14 (6.3)	15.4 (6.9)	11 (4.9)	13.4 (6)	13.4 (6)
Cut-out windspeed, mph (m/s)	100 (44.7)	NA[b]	50 (22.4)	60 (26.8)	60.4 (27)	60 (26.8)	55.9 (25)	47 (21)
Nacelle height, ft (m)	60 (18.3)	75 (22.9)	93 (28.3)	200 (61)	262 (79.9)	200 (61)	148 (45)	262 (80)
Rotor diameter, ft (m)	32 (9.8)	58 (17.7)	125 (38.1)	300 (91.4)	256 (78)	320 (97.5)	131 (40)	246 (75)
Rotor swept area, ft² (m²)	804 (75.4)	2,640 (246)	12,270 (1,140)	70,690 (6,561)	51,472 (4,778)	80,425 (7,472)	13,478 (1,257)	47,529 (4,418)
Number of blades	2	3	2	2	2	2	3	2
Blade chord—root, in. (m)	42 (1.07)	76.7 (1.95)	54 (1.37)	136 (3.45)	177.6 (4.51)	NA	NA	NA
Blade chord—tip, in. (m)	13 (0.33)	38.4 (0.98)	18.75 (0.48)	57 (1.45)	39.6 (1.01)	NA	NA	NA
Airfoil section	23,012/23,021	NA	LS(1) – 04XX	230XX	28,000	NACA 230XX	4,412/4,434	NACA 64-4XX
Blade construction	Composite	Dacron-Aluminum	Wood	Steel	Fiberglass	Steel	Steel/fiberglass	Steel/composite
Pitch control	Passive	Passive	Full span	30% tip	Full span	35% tip	Full span	Full span
Rotor position	Downwind	Downwind	Downwind	Upwind	Downwind	Upwind	Upwind	Upwind
Blade twist, degrees	NA	20	25.6	NA	13	NA	11	15
Yaw control	Free	Free	Driven	Driven	Free	Driven	Driven	Driven
Tower type	Steel Column	Steel Column	Steel truss	Cylindrical steel	Cylindrical steel	Cylindrical steel	Concrete	Concrete
Rotor rpm	120	67.8	42	17.5	30	14.9–20.1	34	25
Generator rpm	1,800	1,800	1,800	1,800	1,800	1,530–2,070	1,530	1,512
Generator type	Induction	Induction	Induction or synchronous	Synchronous	Synchronous	Variable speed induction	Induction	Induction

[a]Windspeed at rotor hub height except for J. Carter model where windspeecs are 30 ft (9.1 m) above ground level.
[b]Not available.

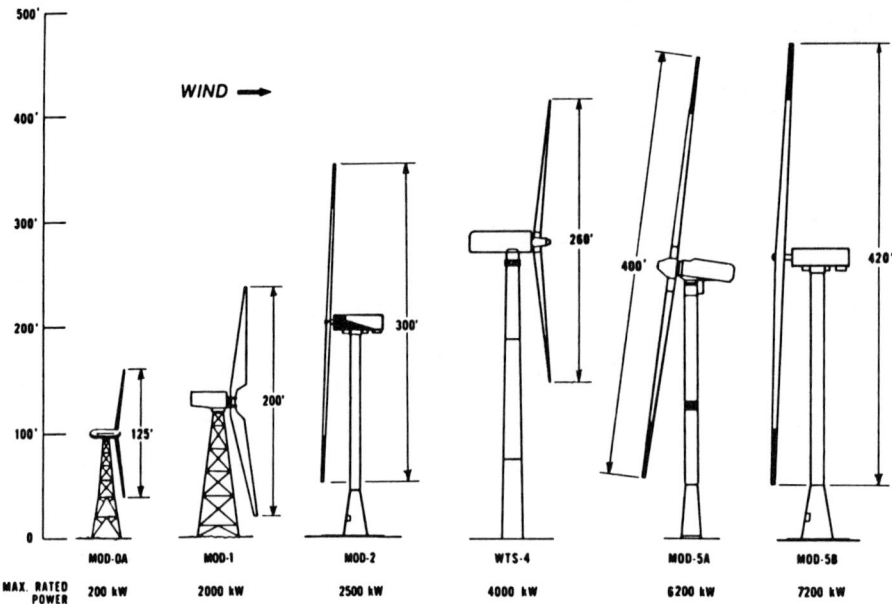

Fig. 11.2-27 Evolution of large wind turbines designed in the United States (DOE).

array energy output can be determined for use in economic analyses of the potential wind turbine arrays at the chosen sites.

Other siting considerations must be addressed, such as procuring land for siting a wind turbine array. Rights to land upwind from wind turbines or between adjacent wind turbines may have to be acquired to ensure access to the wind. Consequently, high land values could impact the desirability of a potential site. Also, local zoning ordinances, building codes, and other regulatory constraints could affect the ability to develop a potential site.

Promising sites in remote locations may require the construction or upgrading of access roads, transmission lines, and substation facilities, the costs of which must be taken into account in the site evaluation. In some remote sites increased maintenance time and expense should be anticipated due to the additional time required to transport maintenance personnel and parts to the site. Potential sites in rugged terrain may involve increased construction costs, which may also have an impact on the viability of a site.

Environmental Factors

Although wind turbines are generally considered to have relatively minor environmental impacts, some potential issues have been identified.[9] Most concerns center around noise, electromagnetic interference, migratory birds, safety, and visual aesthetics.

Noise produced by a wind turbine is due to the blade motion through the air and to the power train components, such as the gearbox and generator. Under most conditions noise from a wind turbine is not considered objectionable and is masked by the ambient noise of the wind itself. One notable exception has been the operation of the MOD-1 wind turbine in Boone, North Carolina, where nearby residents were conscious of a low-frequency infra-sound, presumably related to the passing of the downwind rotor blades behind the tower. It is believed that modification of rotor speed and other design changes (including use of upwind rotors) can mitigate this type of problem.

Electromagnetic interference can occur in areas near a wind turbine when the rotating blades reflect electromagnetic signals, causing interference between the transmitted and reflected signals. Some localized interference with television reception has been encountered at the MOD-OA installation at Block Island, Rhode Island, and at the MOD-1 installation at Boone, North Carolina. Interference with television and other electromagnetic signals can generally be mitigated by proper siting of the wind turbines. Other measures can include cable transmission, directional antennas, and use of nonmetallic wind turbine blades.

There is concern that migrating birds might fly into a wind turbine blade, tower, or guy cable supports. Although experience to date is limited, there has been little evidence to suggest that significant bird mortality would result from wind turbine placement or operation. If future experience indicates a problem, siting of wind turbines in migratory flyways may require consideration of mitigative measures.

Concerns about the safety of wind turbines generally involve ice throw, blade throw, or tower collapse. Adequate design criteria for icing shutdown, tower strength, blade strength, and overspeed control systems must be employed to allow reasonably small exclusion zones surrounding wind turbine installations. More definitive criteria for exclusion zones are expected as further experience is gained.

Aesthetics relate to the public's perception of the visual impact of wind turbines. Although public reaction has been generally positive,[10] some objections have been raised. Proper choice of colors, wind turbine design, and site location relative to surrounding landscape may be important in those cases where wind resources are present in visually sensitive areas.

Economics of Wind Energy

Cost of Energy.   A basic element of any application of wind turbines for energy conversion is the cost of the output energy, that is, electricity. The cost of energy produced by a wind turbine is a function of the capital cost of the wind turbine and all related facilities, the operating and maintenance costs, and the level of wind resource available, which dictates the level of energy production.

In the case of electric generation the cost of energy is generally calculated on a levelized life-cycle basis. This method results in a constant annual cost for each year of the economic life of the facility. The present value of this annual cost series over the facility life is equal to the present value of the actual facility costs including both fixed initial capital costs and projected variable expenses (operation and maintenance costs) over the facility life. The value of annual cost is divided by the projected annual energy production to obtain the cost of energy. The formula used to calculate the cost of energy is

$$\text{COE} = \frac{(\text{IC})(\text{FCR}) + (\text{AOM})(\text{LF})}{\text{AEP}} \tag{11.2-17}$$

The term COE (in dollars per kilowatt-hours) is the levelized cost of energy over the life of the facility; IC (in dollars) is the total initial capital cost of the facility and includes the installed wind turbine, land, electrical transmission lines, and all other related costs; FCR (in percent per year) is the fixed charge rate which amortizes the initial capital cost over the life of the project on an annual cost basis and also accounts for income taxes, property taxes, insurance, and other applicable overheads[*]; AOM (in dollars) is the projected first-year annual operating and maintenance cost, and because these costs will escalate during the life of the facility, the AOM is multiplied by a levelizing factor, LF, to convert the AOM to an equivalent constant, or levelized, annual cost[†]; and AEP (in kilowatt-hours per year) is the projected annual energy production for the facility. The AEP is calculated from the wind turbine performance characteristics and the wind resource available at the site with allowances made for maintenance outages and possible array wake interference effects.[‡]

Example 11.2-1.   Calculate the 30-year levelized life-cycle cost of energy for a wind turbine array project with the following assumed characteristics:

Initial cost, IC	= $10 million
Fixed charge rate, FCR	= 20%/yr
Annual operation and maintenance, AOM	= $100,000
Levelizing factor, LF	= 2.0
Annual energy production, AEP	= 20 million kW · h/yr net

$$\text{COE} = \frac{\$10,000,000 \times 0.2 + \$100,000 \times 2.0}{20,000,000 \text{ kW} \cdot \text{h}}$$

$$= \$0.11 \quad \text{or} \quad 11 \text{¢/kW} \cdot \text{h}$$

---

[*]For example, the FCR for private utilities is commonly in the range of 20–25% for a 30-year facility lifetime. For public utilities and government agencies with reduced cost of capital and tax obligations the FCR may be considerably less.
[†]The levelizing factor varies with the projected escalation rate and present-value interest rate and is obtained from commonly available tables of economic factors for converting geometrically increasing series to uniform (levelized) series.
[‡]See Section 11.2-1 for methods of calculating annual energy production with factors for maintenance downtime and array efficiency.

Value of Wind Energy. The electric energy produced by wind turbines on a utility system would allow fuel to be saved that would otherwise have been burned in a conventional generation facility. The value of the energy produced from the wind would be related to the fuel displaced. As a result, the value could vary considerably depending on the energy resources that a particular utility would be using.

Even though wind is an intermittent energy source, the presence of wind turbines on a utility system could reduce the need for a certain amount of other generation capacity. For example, one study[11] indicates that a wind turbine could be credited with a statistically firm capacity of 10–30% of its rated value, depending on the design characteristics of the wind turbine, a utility's existing generation mix, and the statistical relationship of wind occurrence and power generation demand. The value of wind energy as a fuel saver and for capacity displacement is highly variable and must be evaluated on a case-by-case basis.

Current Research and Development Issues

The technology for wind energy conversion is not yet mature, and there are a number of generally acknowledged areas that currently require continuing research and development efforts:

1. Additional actual operating experience demonstrating reliable and routine operation of individual wind turbines and wind turbine arrays integrated with utility system electrical networks is necessary. Projections of long-term energy production, reliability, and system life times have not yet been adequately verified, especially for the larger megawatt size wind turbines.

2. Predictions of lifetimes for critical wind turbine components (rotor, pitch control system, gearbox, etc.) need further verification.

3. Operational aspects, such as controlling wind turbine output to follow loads under certain conditions, increasing energy capture during stop–start conditions, and other control system and electrical interface optimization areas need further investigation.

4. Costs of wind turbines need to be further reduced through value engineering and appropriate manufacturing methods. Achievement of reasonably low levels of long-term operation and maintenance costs need to be demonstrated.

5. Further work in the area of environmental issues is required. The possible concerns related to visual impacts, noise, electromagnetic interference, safety, and effects on migratory birds should be evaluated further, along with investigation of cost-effective mitigation measures if needed.

6. Wind turbine array spacing criteria require further refinement and verification.

Efforts are under way in all of the above areas, and these efforts, along with the progress that has previously been made in the wind energy field, offer the promise of the utilization of wind energy as a significant contributor to the needs for energy in the future.

REFERENCES

11.2-1 "Candidate Wind Turbine Generator Site Meteorological Monitoring Program Monthly Reports," Western Scientific Services, Inc., Fort Collins, CO, NASA Lewis Research Center Contract NAS3-20452, January 1977–September 1978.

11.2-2 "Candidate Wind Turbine Generator Site Monthly Data Reports," Battelle-Pacific Northwest Laboratories, Richland, WA, DOE Contract DE-AC06-76RLO-1830, September 1978–December 1980.

11.2-3 M. H. Worstell, "Aerodynamic Performance of the 17-Meter Diameter Darrieus Wind Turbine," Sandia Laboratories Energy Report SAND 78-1737, January 1979.

11.2-4 J. J. Riley, E. W. Geller, M. D. Coon, and J. C. Schedvin, "A Review of Wind Turbine Wake Effects," Flow Industries, Kent, WA, DOE Contract DE-AC06-79ET23160, (DOE/ET/ 23160-80/1), January 1980.

11.2-5 P. B. S. Lissaman, A. D. Zalay, and G. W. Gyatt, "Design of Optimal Wind Turbine Arrays," Aerovironment Inc., Pasadena, CA, AV-TP-82/528, sponsored by Battelle-Pacific Northwest Laboratories under DOE Contract DE-AC06-76RLO-1830, Report No. PNL-4183, UC-60, Proceedings of the 1982 Wind/Solar Energy Conference, Kansas City, MO, April 1982.

11.2-6 R. C. Reuter, Jr. and M. H. Worstell, "Torque Ripple in a Vertical Axis Wind Turbine," Sandia Laboratories Energy Report SAND 78-0577, April 1978.

11.2-7 M. R. Gustavson, "Limits to Wind Power Utilization," *Science* **204**, 13–17, April, 1979.

11.2-8 T. G. Zambrano, S. N. Walker, and R. W. Baker, "Wind Energy Assessment of the Palm Springs–Whitewater Region," Prepared for Southern California Edison Company and California Energy Commission, Aerovironment Inc., Pasadena, CA, AV-R-9547, February 1980.

11.2-9 "San Gorgonio Wind Resource Study–Environmental Impact Report/Environmental Impact Statement," Prepared for Riverside County and the United States Bureau of Land Management, Wagstaff and Brady Inc., Berkeley, CA, August 1982.

11.2-10 "Public Attitudes Toward the Development of Wind Energy in the San Gorgonio Pass Area," Prepared for Southern California Edison Company, Harbricht Research Inc., Arcadia, CA, September 1981.

11.2-11 "Electric Utility Value Analysis for Wind Energy Conversion Systems," Aerospace Corporation, El Segundo, CA, ATR-81 (7869-01)-3, SERI Contract XH-9-8336-1, May 1981.

BIBLIOGRAPHY

Berke, B. L. and Meroney, R. N. "Energy from the Wind: Annotated Bibliography," Vols. I–IV. Colorado State University, Fort Collins, CO, 1975–1982.

Eldridge, F. R. *Wind Machines*, 2nd ed. Van Nostrand Reinhold, New York, 1980.

Golding, E. W. *The Generation of Electricity by Wind Power*. Philosophical Library, New York, 1956.

Hunt, V. D. *Windpower, A Handbook on Wind Energy Conversion Systems*. Van Nostrand Reinhold, New York, 1981.

Johnson, G. L., *Wind Energy Systems*, Prentice-Hall, Inc., Englewood Cliffs, NJ, 1985.

Kansas Wind Energy Handbook. Kansas Energy Office, Topeka, KS, 1981.

Koeppl, G. W. *Putnam's Power from the Wind*, 2nd ed. Van Nostrand Reinhold, New York, 1982.

"Proceedings of the Vertical Axis Wind Turbine (VAWT) Design Technology Seminar for Industry." Sandia Laboratories, Albuquerque, NM, SAND-80-0984, August 1980.

"Proceedings of the Fifth Biennial Wind Energy Conference and Workshop." Washington, D.C., October 1981, Vols. 1–3, SERI, Golden, CO, SERI/CP-6351340, July 1982.

Vachon, W. A. "Large Wind Turbine Generator Performance Assessment, Technology Status," Report, Nos. 1–5, Arthur D. Little, Inc., Cambridge, MA, Project 1996-1. Electric Power Research Institute, Palo Alto, CA, 1980–1982.

Wind Energy Report. P.O. Box 14, Rockville Centre, NY, 11571 (periodical).

Windletter. American Wind Energy Association, 1516 King St., Alexandria, VA 22314 (periodical).

Wind Industry News Digest. 107 S. Central Avenue, Milaca, MN, 56353 (periodical).

11.3 BIOMASS CONVERSION

Edward S. Lipinsky, David A. Tillman, and Donald L. Klass

Biomass is a term that designates materials originating recently from plant, microbial, or animal sources. Such fossil resources as coal, peat, or petroleum originated from the same sources but not recently. There is a complex chain of consumption of biomass in that a forage crop may be converted into animal manure or trees may be converted into paper products that become part of municipal solid waste. Thus, the scope of biomass resources is quite broad and includes *some* of the constituents of municipal solid wastes, landfills, and sewage sludge.

Biomass may be used directly as a fuel by direct combustion or converted into a wide array of biomass-derived fuels. However, solid biomass is bulky and some types are perishable. Such biomass-derived fuels as methane, methanol, and ethanol offer the advantages of clean burning, transportability, and fit with existing end-use devices that petroleum-derived fuels have offered. In addition, biomass is described as a renewable resource, at least compared with petroleum or peat. Actually, biomass is not strictly a renewable resource because it does consume vital soil nutrients and has an impact on land quality over the long term.

The major biomass energy applications are direct combustion, anaerobic digestion, thermochemical gasification, methanol production, and ethanol production. The state of the art varies widely for these applications, with wood combustion at the multiquad level of acceptance and others at the laboratory research stage. Each fuel conversion route draws upon a different set of biomass compositional considerations, leading to natural niches for biomass products. Thus, biomass products with simple carbohydrate constituents have their highest value in production of ethanol. Highly lignified biomass (e.g., wood, stumps, and bark) has advantages for direct combustion. High-moisture, low-lignin materials are attractive for anaerobic digestion.

11.3-1   Biomass for Generation of Steam and Electricity Generation

David A. Tillman

Introduction

Wood fuel use, which declined during the 100 years from 1870 to 1970, has reversed its downward trend and is now supplying the U.S. economy with 2.5 exajoules (EJ) or quadrillion Btu (quads) per year.[1]

Similarities Between Biomass- and Fossil-Fired Systems

Like coal, raw biomass is solid. (The use of liquid and gaseous fuels that can be derived from solid biomass is described in subsequent sections.) In practice, biomass may be used as a virtual substitute for coal and, to a lesser extent, oil and gas. The Rankine cycle is applicable equally to fossil fuels and to biomass. Thus, biomass or coal can be combusted to raise high-pressure steam (e.g., 58–238 atm), which is then expanded through a turbine to produce power. The steam may be condensed, as is done in a stand-alone power plant. Alternatively, the high-pressure steam may be expanded in the turbine, exhausted at pressures of 1–20 atm, and sent to manufacturing process or district heating, an alternative called cogeneration.

Like coal, biomass fuels can be adapted to the Brayton thermodynamic cycle (the gas turbine system), but only with considerable difficulty.[2] Consequently, the Rankine cycle remains the preferred power generation approach.

Biomass fuels, like fossil fuels, may be burned under boilers to raise process steam; they may be burned in kilns to supply direct process heat.

Differences Between Biomass- and Fossil-Based Systems

The differences between biomass- and fossil-fired systems result from the economic and technical characteristics of the fuels. Biomass fuels are typically by-products and residues resulting from the production of materials (e.g., lumber, paper) and foodstuffs. They are frequently geographically dispersed (e.g., logging residues). They are solid, low in bulk density, and modest in heating value, frequently wet, and sometimes dirty. They tend to deteriorate on storage and frequently require special handling to avoid spontaneous combustion.

As a consequence of the dispersed nature of biomass fuels, it is difficult to accumulate substantial quantities in a single location. The chemical compositions of biomass fuels are different from coals, petroleum fuels, and natural gas. Thus, biomass fuels exhibit unique combustion characteristics. These differences between biomass and fossil fuels result in substantial differences in boiler hardware design requirements.

Because the power generation cycle that is appropriate for biomass fuels is identical to those associated with coal, and because differences in the power plants are a consequence of fuel characteristics, this analysis will focus on the biomass fuels themselves. Biomass properties and the mechanisms associated with biomass combustion affect the selection and optimization of energy systems that use this resource.

Characteristics of Biomass Fuels

Biomass fuels result from the growth and/or processing of plant materials. The dominant fuel is wood; other fuels include crop and food-processing wastes and animal fecal matter. Typically, the economically available fuels are by-products of other processes. Hogged wood fuel, which consists of hammer-milled bark and sawdust, results from processing logs into lumber and pulp. Spent pulping liquor results from the delignification of woody tissues by chemical means for the production of paper and chemicals. Logging residuals are by-products of timber harvesting. Thinnings and release cuttings, which are generated in the forest, are by-products of silvicultural processes designed to enhance the growth of commercial timber.

Locational Characteristics of Biomass Fuels

The consequence of the dispersed nature of biomass is that these fuels are not inexpensive. Their costs are driven by biomass piece size, terrain where located, distance from the proposed point of utilization, and opportunities for alternative uses (e.g., as furnished for pulp, particle-board). The cost of fuel increases quickly as fuel requirements at a facility become larger. Illustrating this point are the marginal cost curves for wood fuels shown in Figs. 11.3-1 and 11.3-2. The Wyoming site has a smaller resource base, and marginal costs rise more rapidly than at the Oklahoma site.

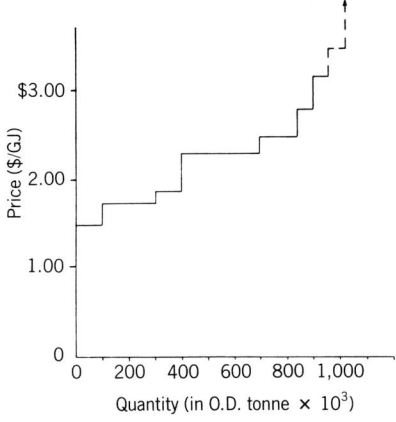

Fig. 11.3-1 Marginal cost curve for wood fuel in the Oklahoma area (from ref. 18).*

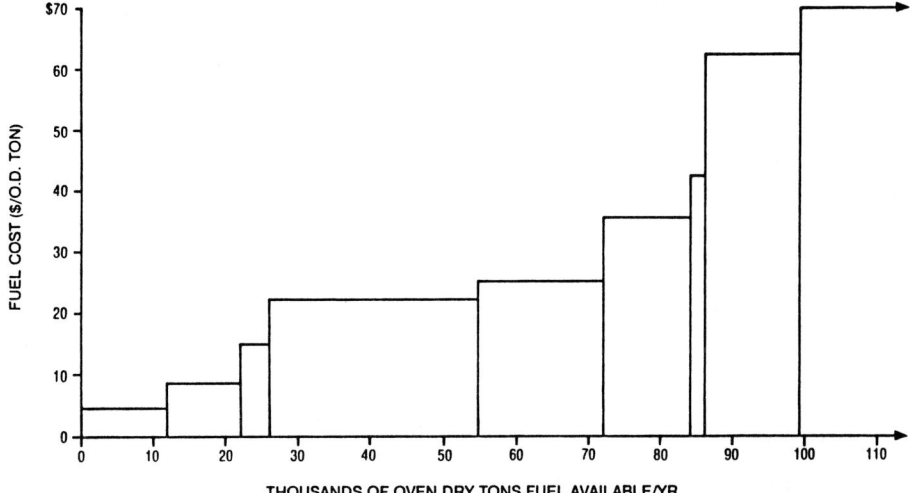

THOUSANDS OF OVEN DRY TONS FUEL AVAILABLE/YR

Fig. 11.3-2 Marginal cost curve of fuel delivered to a sawmill (from ref. 19).

Because the cost of biomass fuels rises as increasing quantities are demanded, practical limitations exist on the scale of operations for any given location. These limitations result from the fact that rising unit prices for biomass fuels can overwhelm economies of scale in capital investment and operating cost of the power plant. Each location has its own marginal cost curve, on which such tradeoffs must be analyzed.

Physical Characteristics of Biomass Fuels

Physical characteristics of biomass fuels include specific gravity and bulk density, moisture content, and particle size distribution. Again, these values are determined by the nature of biomass.

Specific gravity for biomass fuels ranges from about 0.2–0.3 for some crop wastes[3] to 0.35–0.65 for most wood species.[4] These values are on an oven-dry (O.D.) basis. For the most common commercial wood species (e.g., Douglas fir, loblolly pine) values range from 0.45 to 0.55 depending on age of tree, part of tree (e.g., bolewood or branch wood), and management regime.[5] These values contrast with coal, where specific gravities range from 1.2 to 1.4.[6]

Bulk densities are, perhaps, even more important, because they strongly affect storage and materials handling system requirements. Some representative bulk densities of various biomass fuels are presented in Table 11.3-1. They are presented on an "as-received," or green, basis. Clearly, these are low when compared to coal, which has bulk densities of 1040–1304 kg/m³.

*All figures and tables from Academic Press are used by permission.

TABLE 11.3-1 REPRESENTATIVE
AS-RECEIVED BULK DENSITIES
FOR SOME BIOMASS FUELS

Fuel	Bulk Density (kg/m^3)
Hogged bark and wood	200–450
Rice hulls	75–150
Straws (unprocessed)	45–60
Cotton gin trash	35–75
Bagasse	60–120
Peach pits	275–450

TABLE 11.3-2 SOME PHYSICAL PROPERTIES OF WOOD
MANUFACTURING RESIDUALS USED AS FUEL[a]

Residual	Size Range (cm)	Relative Surface/Volume Ratio	Moisture Content (wt. %)
Bark	0.08–10.2	1	25–75
Coarse wood Residuals	0.08–10.2	1	30–60
Planer shavings	0.08–1.27	5	16–40
Sawdust	0.08–0.95	6	25–40
Sanderdust	0.08	10	2–8

[a] From: ref. 14.

Table 11.3-2 gives moisture content and particle size data for a variety of wood fuels produced by lumber and plywood manufacturing processes.

Moisture content is an especially significant issue because it directly affects the energy that can be derived from fuels. Typical values range from 35 to 55% (as-received, or green, basis) for hogged wood and bark, with an average of 50%. A few crop wastes and food-processing wastes, however, may be much drier, with values typically in the 10–20% range. These latter wastes include straws, rice hulls, corn cobs, cotton gin trash, and nut shells. Some wood fuels such as planer shavings and (plywood) sanderdust also may be dry (moisture contents ranging from 8 to 15%). These moisture contents are a result of specific lumber and plywood processing activities.

There is a wide variation of particle sizes possible in a single biomass fuel pile, as illustrated by Fig. 11.3-3, which is a representative particle size distribution for hogged bark obtained at a Wyoming sawmill. As the figure illustrates, the designer must contend with wide variations in particle size and with a high percentage of fine (< 0.64 cm) and very fine (< 0.32 cm) particles.

The heterogeneity of specific gravity, moisture content, and particle size of wood size of wood fuels also occurs in many biomass products (e.g., sugarcane bagasse). The presence of a high percentage of fines is also common to virtually all biomass fuels.

Chemical Properties of Biomass Fuels

Combustion is a chemical process involving oxidation of reduced forms of carbon and hydrogen by free-radical processes. Chemical properties of the biomass fuels determine the higher heating value of the fuel and the pathways of combustion. Critical parameters are the proximate and ultimate analyses, the higher heating value (HHV), and some fundamental structural considerations.

Proximate and ultimate analyses for a variety of biomass fuels are presented in Table 11.3-3 along with HHV values for those fuels. Pittsburgh seam bituminous coal is also presented in this table for comparison purposes. All values for such fuels presented in Table 11.3-3 are on an O.D. basis. What becomes clear is the following: biomass fuels are highly reactive, volatile, oxygenated fuels of moderate heating value. The reactivity is reflected in the hydrogen/carbon (H/C) and oxygen/carbon (O/C) molar ratios of these fuels, and the volatile/fixed carbon ratio of these fuels. These measures of

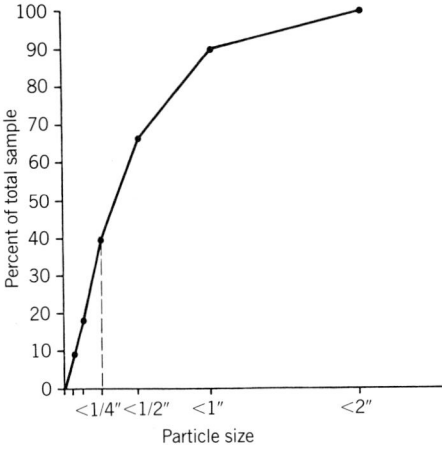

Fig. 11.3-3 Cumulative particle size distribution of bark (from ref. 19).

TABLE 11.3-3   PROXIMATE AND ULTIMATE ANALYSES OF SOME BIOMASS FUELS[a, b]

Analysis	Douglas-fir Wood	Western Hemlock	Pine Bark	Bagasse	Rice Hulls	Cotton Gin Trash	Pittsburgh Bituminous Coal[c]
Proximate, wt. %							
Volatiles	85.8	83.8	72.9	83.8	64.5	81.1	33.9
Fixed carbon	13.4	14.0	24.2	12.7	12.9	8.9	55.8
Ash	0.8	2.2	2.9	3.5	22.6	10.0	10.3
Ultimate, wt. %							
Hydrogen	6.3	5.8	5.6	5.8	4.4	5.2	5.0
Carbon	52.3	50.4	53.4	48.8	38.3	43.2	75.5
Oxygen	40.5	41.4	37.9	41.7	33.9	40.1	4.9
Nitrogen	0.1	0.1	0.1	0.2	0.8	1.5	1.2
Sulfur	—	0.1	0.1	—	—	—	3.1
Ash	0.8	2.2	2.9	3.5	22.6	10.0	10.3
Higher Heating Value, MJ/kg	21.05	20.05	21.0	19.37	13.81	16.51	31.74

[a] All values, oven dry basis.
[b] From refs. 5, 6, 14, 20, and 21.
[c] 3% moisture.

reactivity are presented in Table 11.3-4. Again, Pittsburgh seam bituminous coal is presented for comparison.

The vast difference in reactivity between biomass and coal is clearly shown in Tables 11.3-3 and 11.3-4. The difference is directly related to the aromaticity of the fuels. Biomass fuels are lignocellulosic materials. They are approximately 25–33% lignin and 67–75% holocellulose (cellulose and the hemicelluloses) on an extractive-free basis. The only aromatic structures occur in the lignin complex, and these structures occur in single units.

The various coals, however, contain more aromatic structures. Further, as the rank of coal increases, the number of rings per cluster increases. Bituminous coals, for example, may have 2–4 rings in a single cluster[7] as compared to the isolated rings in lignin. Aromaticity is one of the major determining factors in the volatiles/fixed carbon ratio. It aids in understanding why the biomass fuels are far more reactive than their solid fossil fuel counterparts.

Beyond reactivity, the chemical analyses demonstrate that biomass fuels contain little sulfur, nitrogen, and ash-forming constituents. As a consequence, clean biomass is relatively clean burning. Although a few biomass species contain significant quantities of ash within their cells (e.g., rice hulls), most high-ash biomass has adhering dirt as the source of the ash. The alkali metal salt content of biomass ash has significant combustor design implications because fuel ash is so corrosive.

TABLE 11.3-4   MEASURES OF REACTIVITY FOR BIOMASS FUELS[a]

	Measure of Reactivity		
Fuel	H/C Ratio (molar)	O/C Ratio (molar)	Volatile/Fixed Carbon Ratio (wt. %)
Douglas fir	1.44	0.58	6.40
Western hemlock	1.38	0.62	5.99
Pine bark	1.26	0.53	3.01
Bagasse	1.43	0.64	6.60
Rice hulls	1.38	0.67	5.00
Cotton gin trash	1.44	0.70	9.11
Pittsburgh seam Bituminous coal	0.79	0.05	0.61

[a] Complied from Table 11.3-3.

Biomass Fuels Combustion

As have been shown, biomass fuels enter the combustor typically in a wet (e.g., 50% moisture), dirty condition. Further, the fuel particles are heterogeneous with respect to moisture content and particle size. Yet they are light in weight and quite reactive. It is important, therefore, to consider how these fuels burn, particularly as the pathways of biomass oxidation influence system design.

Mechanisms of Biomass Combustion.   Figure 11.3-4 is a conceptual diagram of solid fuels combustion adapted from Edwards.[8] It shows solid fuels undergoing the following reactions: heating and drying; solid-particle pyrolysis, where volatiles and char are evolved; precombustion gas phase reactions; and char oxidation reactions.

The detailed chemistry associated with this model is beyond the scope of this treatment and can be found in the following references: Shafizadeh,[9] Tillman,[10] and Tillman, Rossi, and Kitto.[5] Certain critical points remain, however.

Pyrolysis is the initial chemical process associated with combustion. As can be seen from the proximate analyses, biomass pyrolysis results in the release of substantial quantities of volatiles. These include carbon dioxide, carbon monoxide, hydrogen, methane, ethane, ethylene, acetylene, methanol, phenols, and other combustibles. A far greater percentage of volatiles is released from biomass than from coal. As a consequence, gas phase combustion reactions play a more significant role in the use of biomass fuels.

Char production is increased if reactions take place at lower temperatures. As a result, increasing fuel particle sizes and quantities of moisture and ash may alter the volatiles/fixed carbon ratio in favor of char formation and subsequent char oxidation reactions.

The process of biomass combustion is highly complex, dominated by both physical and chemical reactions. Time, temperature, and turbulence govern the completeness, efficiency, and rate of biomass

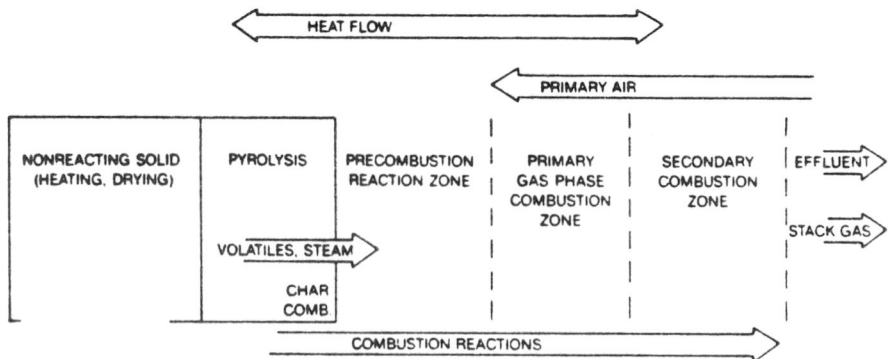

Fig. 11.3-4   The conceptual model of solid-fuels combustion. Adapted from Ref. 8.

combustion. These issues can be translated into requirements for combustion air, final stack temperatures, thermal efficiencies, flame temperatures achieved, and emissions resulting from biomass oxidation.

Calculations of Biomass Combustion. Calculations of air requirements, efficiencies, and flame temperatures are essential to the design and operation of biomass combustion. Consequently, techniques
for their calculation are presented below.

Air requirements for biomass fuels are typically higher than those associated with fossil fuels.
There are several reasons for this requirement, including the moisture content and insulative
properties of the fuel. For grate-fired systems the following general expression, postulated by G.
Villesvik, holds:

$$\text{Percent excess air} = 40 \frac{MC_g}{1 - MC_g} \qquad (11.3\text{-}1)$$

where moisture content (MC_g) is expressed on a total weight or green basis and is shown as a decimal
(fraction) rather than as a percentage.[5]

For typical biomass fuels at 50% moisture, excess air requirements are about 40% when grate-fired
systems are employed. For dry biomass fuels, a minimum level of excess air is about 25% in such
systems.

For suspension-fired systems and fluidized-bed combustors, the combustion air requirements are
much higher, typically in excess of 100%. They are high because the air must keep particles in
suspension or fluidize the bed medium.

Calculating the total air requirement is not the only basic issue associated with combustion air in
biomass-fired systems. The distribution and heating of that air is also of importance. Junge[11] has
determined that the optimum distribution of air is 43% undergrate and 57% as overfire air. Further, he
has determined that only the overfire air should be preheated. Undergrate air is held at ambient
conditions to protect grate life, while overfire air is heated to increase turbulence and mixing in the gas
phase reaction zone.

System efficiencies are determined by several factors, including fuel composition, moisture content,
and higher heating value; excess air employed; completeness of combustion; and final stack temperatures. The general formula for calculating system efficiency is as follows:

$$\frac{(1 - HL)}{HHV} \times 100 = \eta \qquad (11.3\text{-}2)$$

where HL represents all heat losses, HHV represents the higher heating value of the fuel, and η
represents percentage.

The specific heat losses that are added to achieve the general term HL include unburned carbon in
the ash, sensible heat in the ash, sensible heat in the dry stack gas, sensible heat in moisture contained
in the stack gas, the latent heat of vaporization of the moisture in the stack gas, and radiation-related
losses.

The critical losses are associated with the stack gas due to the moisture content of the fuel and the
consequent requirements for excess air. As a result, significant efforts are now under way to reduce the
final stack temperature by using air heaters and, periodically, economizers. In 1983, final stack
temperatures of approximately 450 K are considered reasonably economically achievable.

Moisture content (MC) influences efficiency more than any variable. As a result, heat balances
have been calculated for fuels at 50% MC and 17% MC (Fig. 11.3-5). A relatively efficient biomass
system firing 50% MC fuel will have a thermal efficiency of 65–68%. An efficient unit firing biomass
fuel at 15% MC or less will have a thermal efficiency approaching 80%.[5] The techniques for calculating
heat balances can be found in *Steam*[6] or in Hougen, Watson, and Ragatz's monograph.[12]

Thermal efficiency calculations show that biomass fuels burn less efficiently than do fossil fuels.
Efficiency leads to questions of flame temperature. Flame temperature is associated with such
efficiency concerns as estimation of the heat transfer surface area required in the boiler.

The critical variables for calculating adiabatic flame temperature are enthalpy of the fuel, moisture
content of the fuel, percent excess air, and level of preheat in the combustion air. If excess air is
measured as excess O_2 in total stack gas (as opposed to excess O_2 in dry stack gas), then the following
expression can be used to estimate adiabatic flame temperature[13]:

$$FT_a = 2405 - 8.7MC - 75.5EO_2 + 0.61(A\text{-}298) \qquad (11.3\text{-}3)$$

where FT_a is adiabatic flame temperature (K), MC is percentage moisture content of the fuel on a
green basis, EO_2 is the percent O_2 measured in the total stack gas, and A is the temperature of the
combustion air (K). The relationship of excess air and excess O_2 in total stack gas is shown in Table
11.3-5.

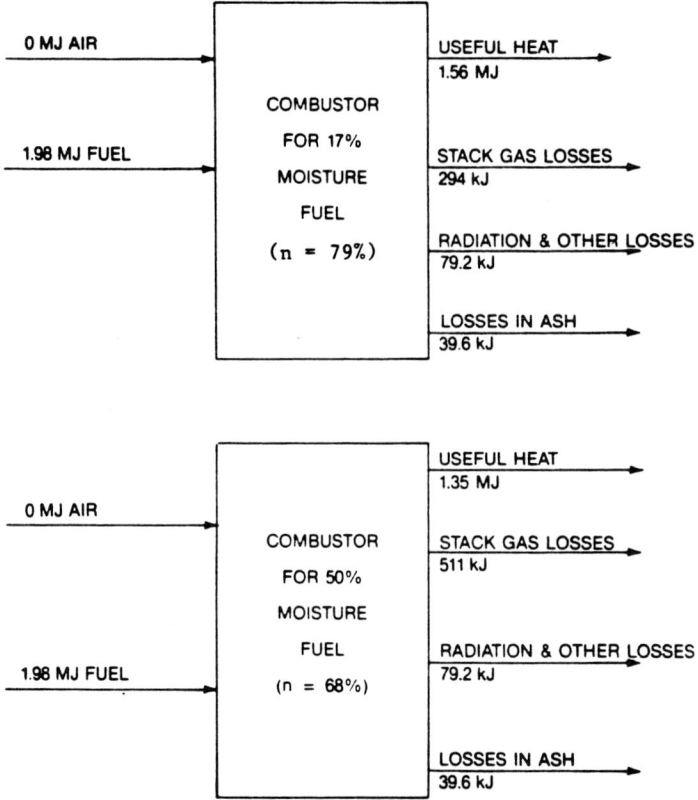

Fig. 11.3-5 Calculated heat balances for combustors using 17% and 50% MC fuel, holding energy input to the combustor constant (from ref. 5).

TABLE 11.3-5 RELATIONSHIP OF EXCESS AIR TO EXCESS O_2 IN TOTAL STACK GAS

Excess Air (%)	Excess O_2 in Total Stack Gas Composition (%)[a]
100	11–12
43	7–8
33	6–7
25	5–6
18	3–4
11	2–3

[a] Variation caused by percent H_2O in stack gas.

It is clear that at low levels of excess air adiabatic flame temperatures in excess of 1900 K are achievable. Actual flame temperatures will be lower, depending on the speed of combustion, whether furnace walls are refractory lined or comprised of boiler tubes, and other similar variables. Manufacturers, however, can guarantee greater than 1370 K across the superheater and temperatures of 1640 K using wood have been measured on the grate. These temperatures indicate that reasonably efficient power generation cycles can be designed based on biomass fuels.

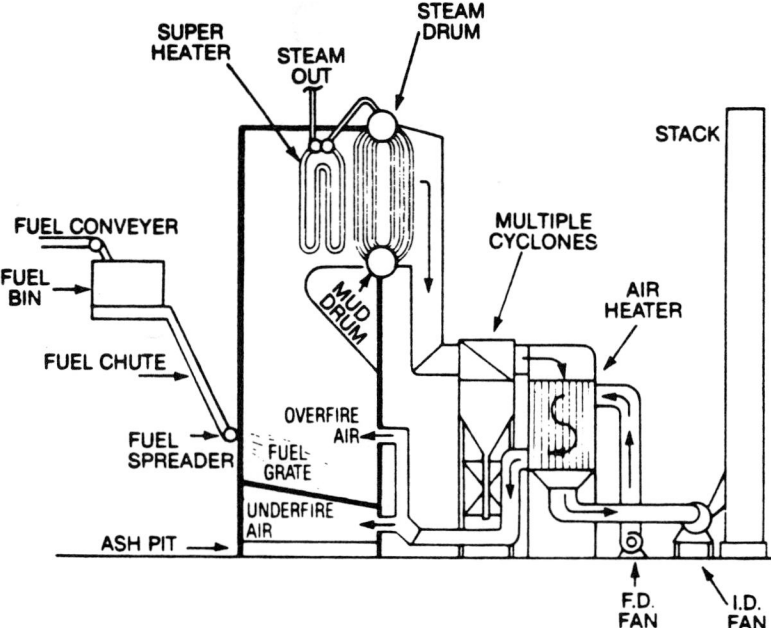

Fig. 11.3-6 The spreader-stoker system including steam drum, superheater, air pollution control equipment, and air heater (from ref. 14).

Hardware for Biomass Combustion

The design of hardware for biomass combustion involves conceiving of systems compatible with the combustion mechanism illustrated in Fig. 11.3-4. These systems must be capable of handling the low specific gravity, high moisture content, and high reactivity of biomass fuels.

Systems Descriptions.   Traditionally, wood has been burned in pile-burning systems such as Dutch ovens. Since 1950 the spreader stoker (with pneumatic, rather than mechanical, stoking) has become the dominant combustion system. The spreader stoker shown in Fig. 11.3-6 became dominant as the need for larger (e.g., greater than 100 GJ/h) and more efficient systems emerged. Since 1970 cyclonic and suspension-fired units have become more prevalent, and fluidized-bed combustion systems have been commercialized.

The pile-burning systems and spreader stoker both provide for staged combustion, separating the zones of reaction presented in Fig. 11.3-4. Suspension-fired units and fluidized-bed combustors merge reaction zones into one single space due to their design, the need to support the fuel particle with air, and the consequent high levels of primary air employed (frequently in excess of 100% excess air).

Pile-burning and spreader stoker systems have been employed in a wide variety of applications. Suspension firing, however, is used in specialized situations where dry duel (e.g., less than 15% MC) is available and where direct heat is required. Typical applications may be in pulp mill dry kilns, cement kilns, and other direct systems.

Fluidized beds are somewhat less efficient because of the heat loss associated with the bed media and because of their high excess air requirements. They are used most commonly for very wet fuels (greater than 50% MC) or very heterogeneous fuels.

There are certain design distinctions associated with systems that burn biomass. These include maximum practical size, moisture content limits, rates of heat release, and rates of airborne emissions. Size and moisture content factors, which have been previously discussed, are summarized in Table 11.3-6.

Rates of heat release are related to flame temperature, to heat content of the fuel, and to the fundamental expression developed by Shafizadeh and Degroot.[15]

$$I = \frac{dw(h)}{dt} \qquad (11.3\text{-}4)$$

TABLE 11.3-6   SIZE AND MOISTURE CONTENT LIMITATIONS AS A FUNCTION OF COMBUSTOR TYPE

Combustor Type	Maximum Size (GJ/h)	Maximum Moisture Content of Fuel (green or total wt. %)
Pile burning	100	65
Spreader stoker	600	57
Suspension burner	100	15
Fluidized bed	100	65

TABLE 11.3-7   HEAT RELEASE RATES FOR BIOMASS COMBUSTORS[a, b]

Combustor Type	Heat Release GJ/m^2 of Grate	Rate (MJ/m^3) Furnace Volume
Pile burner	8–9	470–590
Spreader stoker	10–11	500–780
Suspension burner	NA	470–1550

[a] Basis = 1 h.
[b] From ref. 16.

where I = flame intensity
dw = change in weight
h = heat of combustion of the fuel
dt = change in time

Heat release rate factors are also determined by the volatile–fixed carbon ratios as presented in Table 11.3-4. These heat release factors are presented in Table 11.3-7. For spreader stoker coal systems, the heat release rates are 8–8.5 GJ/m^2 of grate and 940 MJ-1.96 GJ/m^3 of furnace volume. Consequently, biomass-fired units have a smaller cross-sectional area and a much taller firebox than coal-fired units.

Pollution control is largely an issue of airborne emission management. It is influenced by the nitrogen, sulfur, and ash contents of the fuel (see Table 11.3-3) and by the combustor and the air quality control system.

Although traditional biomass combustors are less efficient than are fossil-fired units, they have broad fuel tolerances, reasonable rates of heat release, and low airborne emissions (particularly SO_2 and NO_x) when compared to coal-fired units.

New Combustor Designs.   In addition to the designs discussed above there have been major improvements in biomass combustion, particularly in pile burning. Two such system improvements are the inclined grate, depicted in Fig. 11.3-7, and the Lamb-Cargate pyrolytic wet-cell burner shown in Fig. 11.3-8.

The inclined grate is offered by such vendors as Solid Fuels and Foster-Wheeler Ltd. It provides such complete separation of the reaction zones that it is commonly referred to as a gasification-combustor. It permits pile burning of relatively wet biomass to be performed in combustor sizes up to 600 GJ/h. At the same time it minimizes airborne emissions by strict staging of combustion.

The Lamb-Cargate unit (Fig. 11.3-8) also strictly stages combustion, and separates homogeneous gas phase oxidation from hetergenous gas–solids oxidation reactions.[17] While it may be more limited in size than is the inclined grate, it offers advantages in pollution control. Like the inclined grate, it can handle feedstocks with moisture in excess of 60%.

These two designs will permit a wider choice of combustion systems, particularly when large units are needed or when pollution control is difficult to achieve. They are, consequently, highly useful innovations in biomass utilization.

Electricity Generation Systems for Biomass Fuels.   Biomass-fired units typically are limited to boilers of 227–273 tonnes steam per hour. This limitation is due to fuel availability/cost considerations and to materials-handling difficulties associated with low specific gravity fuels.

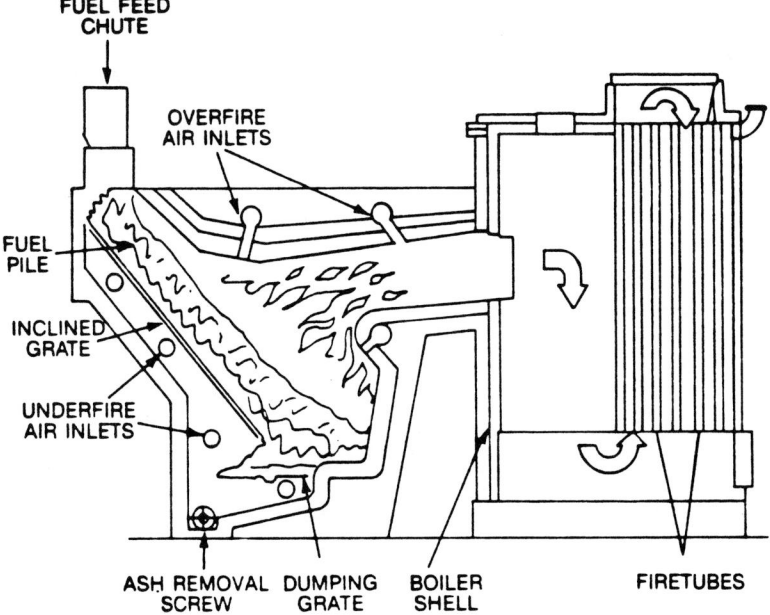

Fig. 11.3-7   An inclined-grate combustor of wood fuel (from ref. 5).

Because boilers are limited in size, the maximum economical pressure in the boiler is currently approximately 102 atm. This is substantially lower than the 163–238 atm pressures associated with coal firing. This maximum pressure penalty reduces the ΔP across the turbine and increases the steam rate (kg steam/kW · h) of the unit.

The size limitation has other serious implications, decreasing the efficiency of biomass-fired units. The economically justifiable number of feedwater heaters is between one and four. This contrasts with the eight feedwater heaters commonly associated with fossil-fired units. Further, reheat cycles are not economic due to the relatively small size of biomass systems.

The consequence of these size limitations is an efficiency penalty and a power plant size penalty. These are illustrated in Table 11.3-8, which is a compilation of factors associated with biomass to electricity systems. As a practical matter, cogeneration systems are typically 5–25 MW(e) in size, and condensing power systems designs are usually in the range of 10–50 MW(e). Throttle steam pressures are typically 58–85 psig; and heat rates are typically approximately 5.7 MJ/kW · h for cogeneration and approximately 14.8 MJ/kW · h for condensing power systems.

Figure 11.3-9 is a heat balance of a typical biomass cogeneration system and Fig. 11.3-10 is a heat balance of a typical condensing power system. These illustrate the scale and efficiency concerns associated with electricity generation from biomass fuels.

Despite these inefficiencies, biomass fuels are used by some industrial companies and electric utilities. They do so for the following reasons:

Local (including in-plant) availability of biomass at relatively low cost.

Ability to avoid costly pollution control (e.g., SO_2 removal) systems.

Opportunity to exert greater control over fuel supply and price.

Because numerous firms, largely in the forest products industry, make extensive use of wood fuel, special systems have been developed to obtain, process, and combust biomass. At the same time, however, the inherent difficulties associated with biomass handling and the efficiency and size limitations associated with biomass combustors restrict biomass fuels to site-specific situations.

11.3-2   Residential Wood Burning

Edward S. Lipinsky

Of the 2.5 EJ or quadrillion Btu (quads) per year of biomass energy in the United States, approximately 0.3–0.8 EJ are consumed in residential wood burning.[22] Some residential wood burning

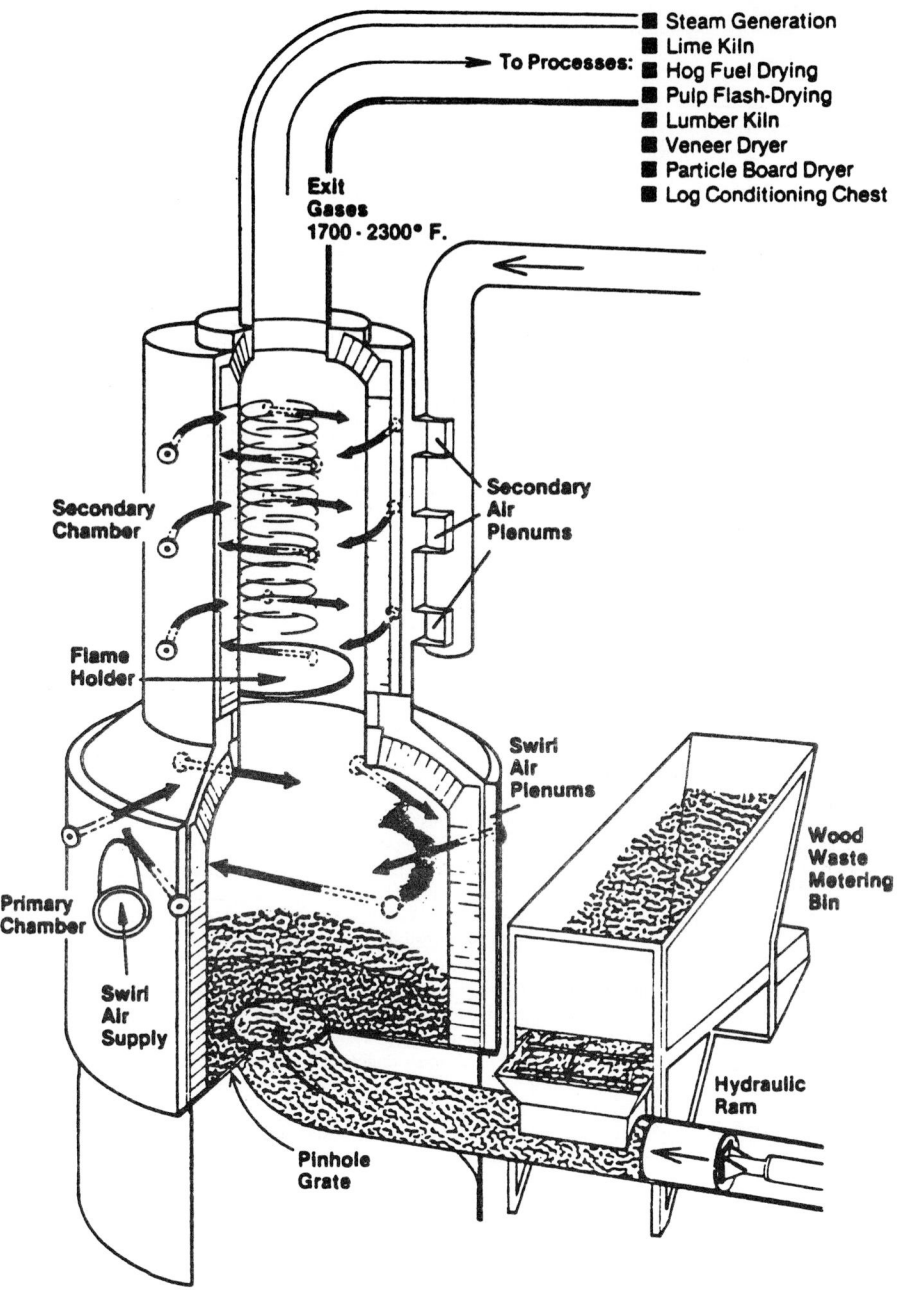

To Processes:
■ Steam Generation
■ Lime Kiln
■ Hog Fuel Drying
■ Pulp Flash-Drying
■ Lumber Kiln
■ Veneer Dryer
■ Particle Board Dryer
■ Log Conditioning Chest

Exit Gases 1700 - 2300° F.

Secondary Chamber

Secondary Air Plenums

Flame Holder

Swirl Air Plenums

Wood Waste Metering Bin

Primary Chamber

Swirl Air Supply

Hydraulic Ram

Pinhole Grate

Fig. 11.3-8 Lamb–Cargate combustion.

is an esthetic ritual, but most wood is burned in wood stoves or closed fireplaces where the objective is to substitute for or to supplement central heating.[23]

Wood as a Residential Fuel

Cordwood, which consists of split logs and/or sticks, is the most popular wood for residential use. The major factors in cordwood quality are its moisture content and density.[24] The moisture content, which is determined by whether the wood is freshly cut or air dried, determines the available energy

TABLE 11.3-8   BIOMASS POWER PLANT CHARACTERISTICS[a]

| | Generation Mode[b] | |
Characteristic	Cogeneration	Condensing Power
Size, MW		
Minimum	1	10
Maximum	35	50
Typical throttle		
Steam pressure, atm		
Minimum	30	40
Maximum	100	100
Typical steam rate, kg/kW · h		
Minimum	7.7	3.6
Maximum	13.6	5.4
Typical heat rates chargeable to Power, (MJ/kW · h)		
Minimum[c]	4.9	13.2
Maximum[d]	6.3	21.1

[a] From ref. 2.
[b] Back-pressure turbine.
[c] Based on large systems and 15% MC fuel.
[d] Based on small systems and 50% MC fuel.

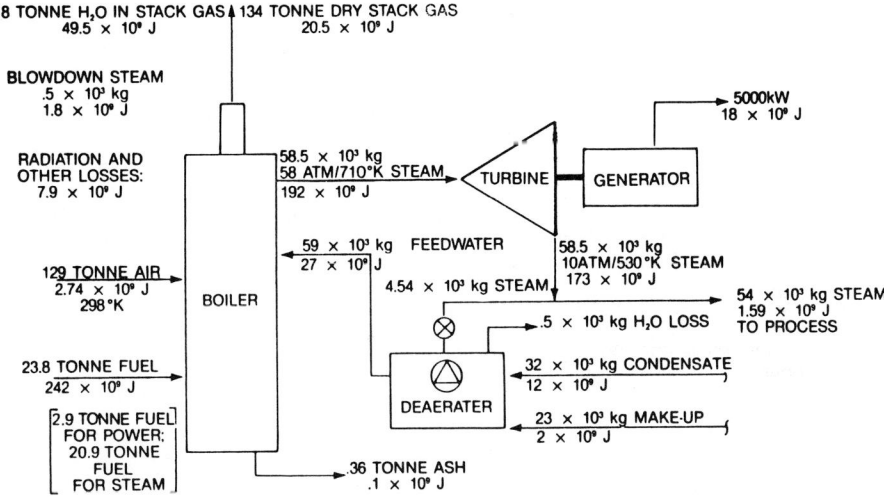

Fig. 11.3-9 Simplified material and energy balance for a 58.5 metric ton steam, 5 MW cogeneration system (from ref. 5).

content of the wood and affects its burning characteristics. Other factors that are important in selection of the fuel include the size and shape of the wood pieces (thick logs vs. thin sticks or chips) and the storage and the extent to which it adheres to the wood.

Residential Wood Fuel Combustion Characteristics

Residential wood burning needs to be contrasted with combustion conducted in electric utility or industrial boiler operations. Residential wood generally is burned in chunks that are massive compared with the chips, sawdust, bark, or sugarcane bagasse employed to produce steam and/or electricity on a commercial scale. The combustion process in residences is much more likely to release

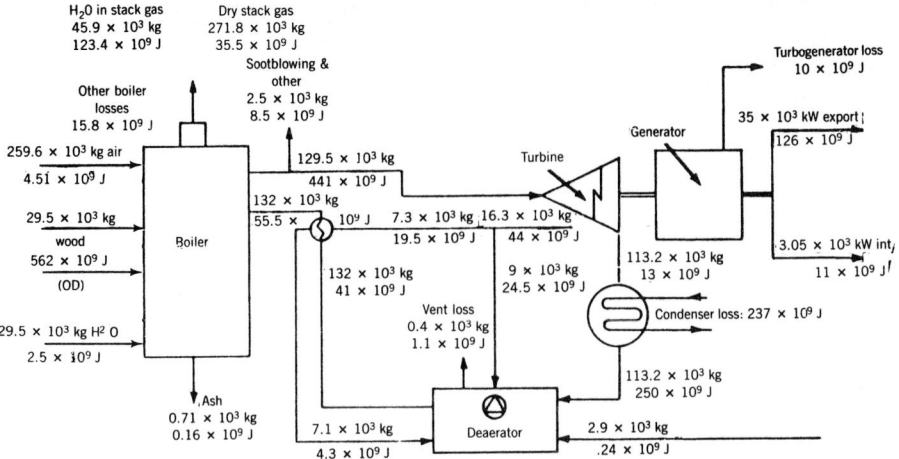

Fig. 11.3-10 Simplified material and energy balance for a 35 MW wood-fired power plant (basis of calculation 1; 1 h). The heat rate for this power plant is 14.9 MJ/kW · h (from ref. 5).

emissions containing polycyclic aromatic hydrocarbons (PAH). Burning rates are slow in residential wood burners, and the wood is burned with a deficiency of air to provide burning rate control. These factors tend to result in a greater variety of chemical reactions that would occur in a typical boiler, resulting in objectionable air pollutant emissions.

The processes that occur in wood log burning are summarized in Table 11.3-9.[25] With a massive chunk, a considerable time elapses during which the chunk absorbs the heat (especially radiant heat). Drying and surface charring occur initially, followed by pyrolysis, luminous flame activity, charring, and finally burnout.

Wood Stove Design and Performance

The four major generic wood stove systems are shown in Fig. 11.3-11.[25] The underlying strategy for each type of wood stove is to obtain as much heat as possible from pyrolysis gases, not only for reasons of economy but also because these pyrolysis gases contain carbon monoxide and frequently such pollutants as PAH, aldehydes, and ketones.

Updraft. Many simple box stoves use an updraft design. The primary air flow passes upward through the burning wood. A secondary air supply to burn the initial pyrolysis products is often provided. However, the primary combustion products may be cooled by wood that has not yet reached ignition temperature. Therefore, combustion may not always be attained in the secondary combustion space.

Downdraft. The primary air flow is downward through the wood, with combustion taking place on the grate that supports the wood. The major advantage of the downdraft burner is that the pyrolysis gases are drawn into the active burning region so that there is greater assurance of complete combustion.

Crossdraft. The primary air enters at one side of the base of the burning wood charge and leaves the primary burning area also at the base of the wood charge. The pyrolysis gases are released into a stagnant upper region of the firebox that serves as a fuel magazine. The off-gas and partially burned products do not escape directly into the stack because they are directed through an active burning region and burned there.

S-Flow. Many Scandinavian designs for wood stoves employ the S-flow air flow pattern in which the wood sticks burn slowly from one end to the other of their length. The baffle controls the pathway and turbulence of the gas flow.

Secondary Combustion

The most important single factor to consider in the selection and/or design of wood stoves is prevention of unburned organics and carbon monoxide leaving the stove.[26] The most important

TABLE 11.3-9 TYPICAL PROGRESSION OF LOCAL PHENOMENA IN WOOD LOG BURNING[a]

Location of Phenomenon	Log Introduction	Time Progression							Burnout
Distance from surface	Radiation influx	Radiation influx	Radiation influx	Radiation influx	Radiation flux near balance	High radiation outflow	High radiation outflow	High radiation outflow	Radiation ceases
Near to surface	Convection influx	Convection influx reduced by steam efflux	Convection influx is reduced by efflux of pyrolysis products and steam	Convection influx / Unattached luminous flame	Intense luminous flame	Decreasing luminous flame	Reduced flames less luminous	Diffusion control of surface burning	—
Surface phenomenon	Heat influx / Steam efflux	Heat influx / Steam and pyrolysis products efflux	Increased heat influx / Steam, pyrolysis gas, and liquids efflux	Radiation much increased / Large gas efflux	Highly radiant surface / Gas efflux reduced	Highly radiant surface / Attached blue flame; Highly radiant surface	Char burning / No flame; Highly radiant surface	Char burning	—
Surface material	Sensible heating	Moisture evaporation	Low-temperature pyrolysis	Exothermic pyrolysis	Char formation	Pyrolysis products cracking and steam dissociation within char layer	Surface lost	Surface lost	Surface
Subsurface material	No change	Sensible heating	Moisture evaporation	Low-temperature pyrolysis	Exothermic pyrolysis	Pyrolysis products	Char burning	Char burning	Surface lost
Interior material	No change	No change	Sensible heating	Moisture evaporation	Low-temperature pyrolysis	Exothermic pyrolysis	Pyrolysis products cracking	Char burning	Surface lost
Core, central material	No change	No change	No change	Sensible heating	Evaporation	Low-temperature pyrolysis	Exothermic pyrolysis	Char burning	Char burnout

[a] From ref. 25.

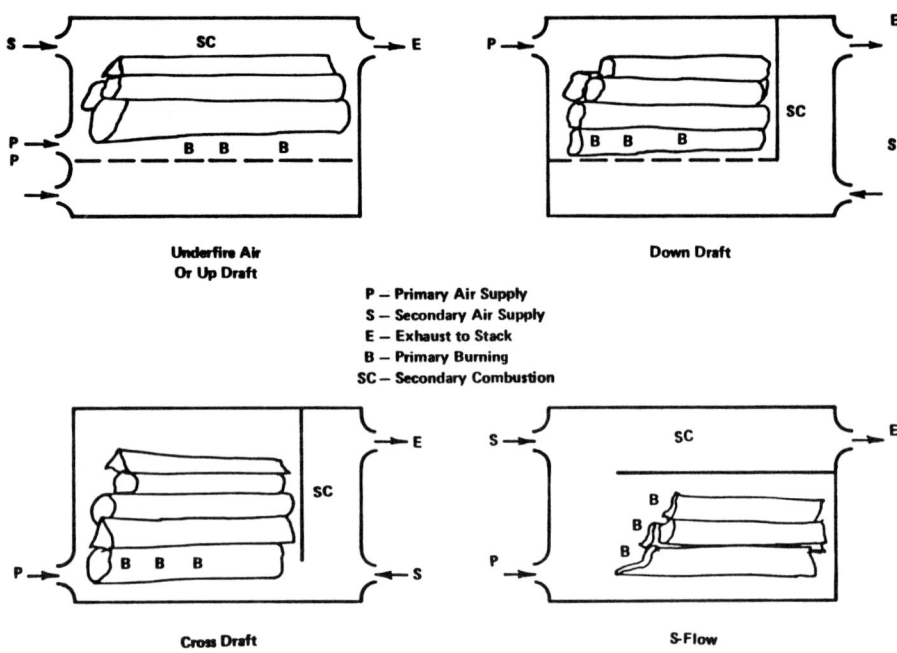

Underfire Air
Or Up Draft

Down Draft

P – Primary Air Supply
S – Secondary Air Supply
E – Exhaust to Stack
B – Primary Burning
SC – Secondary Combustion

Cross Draft

S-Flow

Fig. 11.3-11 Generic designs of wood stoves based on flow paths (from ref. 25).

phenomena to consider are the following:

1. Complete secondary combustion is impeded by insufficient turbulence and mixing rate in natural draft stoves.
2. The temperature in the secondary combustion area may be below that necessary for ignition and burning, especially at low rates of burning and in locations in the stove where relatively cold walls quench the flames from the wood.
3. A design that is sufficient to burn the average amount of pyrolysis gases generated by the stove will be insufficient during the peak combustion rates. Otherwise, excess air will be available when it is not needed.

Environmentally Responsible Design

Because many wood stoves are and will be used in urban areas, design of stoves to keep adverse emissions under control is especially important. The following are typical guidelines developed by Allen and Cooke (those interested in details should consult ref. 25).

1. All of the wood inventory within the stove should be kept at a low temperature until active local burning is started. This will preclude excessive preburning pyrolysis.
2. After combustion has been established, only large pieces of wood should be fired, consistent with desired burning rate maintenance. This is a fuel characteristic that may require more frequent attention or demanding operator techniques. By this means, pyrolysis can be kept to a minimum in existing stove designs.
3. A high turbulence level should be maintained in the active burning area by using stove design techniques. The use of an air blower with tailored air ducting can achieve this objective, but this design characteristic is not presently widely used.
4. High temperatures should be maintained in the active burning area to promote rapid and complete burning of fuel. This design characteristic should minimize the incomplete combustion associated with quenching of partially burned fuel and can be improved by use of insulating refractory materials enclosing the combustion chamber.
5. An abrupt reduction in the rate of primary air supply to an operating stove is to be avoided because it does not simultaneously reduce the rate of gas evolution or pyrolysis of the wood in the stove. Coupled with the reduced supply of air, the gas contains considerable unburned or partially oxidized fuels.

6. Increased turbulence in the primary burning zone will promote the complete burning of pyrolysis products released directly into the flame region, thus eliminating some or all of the products of incomplete burning.

7. Pyrolysis products from the magazine region should be ducted only into an active burning region where adequate air supply and temperatures will promote complete combustion.

8. The temperature in the secondary combustion chamber should be kept high to broaden the range of gas mixtures that will permit combustion. This is a design characteristic of the stove using insulation and proximity of the secondary air inlet to the primary combustion chamber to improve chamber heat retention.

9. The secondary air should be heated to avoid quenching secondary combustion that might otherwise occur. This design factor can be accomplished by within-the-stove heating of secondary air prior to mixing with primary combustion products.

10. Catalytic afterburners have been developed and applied to several wood stoves currently being marketed. By reducing the temperature at which combustion can be initiated, and by conserving the heat released to maintain the combustion, the catalytic unit might significantly reduce emissions under many, but not all, conditions of burning.

Additional information on wood burning may be found in section 13.2-8.

11.3-3 Anaerobic Digestion

Donald L. Klass

An overview of anaerobic digestion technology is shown in Fig. 11.3-12.[27] Anaerobic microorganisms use their enzyme systems as catalysts to convert organic matter into methane and carbon dioxide.

Prior to about 1970 anaerobic digestion was considered to be useful in the United States primarily as a waste stabilization process, particularly for sewage sludge. Digester gas by-product was flared in most cases or used as a supplemental fuel for the plant. In the last decade extensive work has been done to maximize the yield of this gas, which is composed of methane and CO_2. When organic wastes

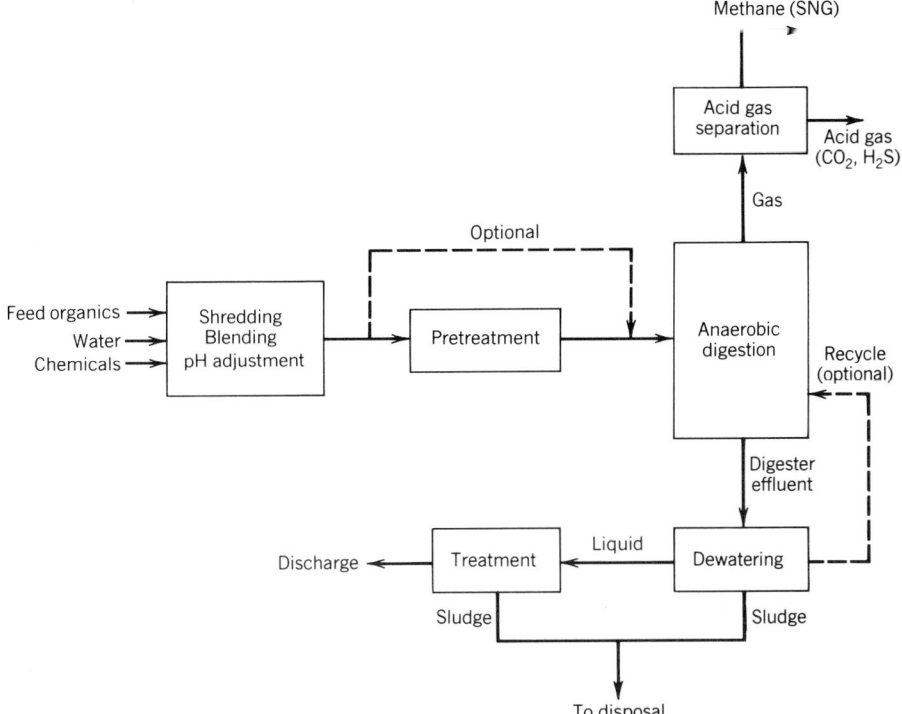

Fig. 11.3-12 Methane production by anaerobic digestion.

are used as feedstocks, waste stabilization and energy recovery can be achieved simultaneously. In addition, when wastes such as municipal solid waste (MSW) are used as feeds, recyclable commodities such as ferrous and aluminum metals and glass can be separated as salable products. As anaerobic digestion technology improves and the cost of anaerobic digestion decreases, the use of dedicated energy crops as resources for the production of methane is expected to become commercially more attractive.

Feedstock Evaluation and Characteristics

Many waste and biomass species as well as blends have been evaluated as feedstocks for anaerobic digestion. Although there are major differences in energy, moisture, volatile solids, and ash contents among the various raw materials, when balanced digestion can be attained with a natural biodegradable material, the gas production parameters, volatile solids reductions, and energy recovery efficiencies as methane in the product gas appear to span a relatively narrow range. This is illustrated by the data in Table 11.3-10, which shows digestion performance under similar high-rate conditions for several different feeds. Table 11.3-11 shows some of the typical analyses for the materials in Table 11.3-10. Major differences are apparent in the components and their concentrations. These compositional differences do not lead to very dissimilar gas production rates and methane yields per unit of organics (volatile solids, VS) added to the digester. One of the methods[28] of predicting biodegradability based on substrate composition is

$$B = 0.830 - 0.28X \tag{11.3-5}$$

where B = biodegradable fraction of volatile solids (VS)
X = VS lignin fraction determined by 72% sulfuric method

This estimation method, which is based on the concept that lignin controls the extent of substrate biodegradation, is not the most accurate but has the virtue of simplicity. The amount of the biodegradable fraction of the volatile solids alone generally will not permit accurate predictions to be made of methane yield. The methane yield per unit mass destroyed must also be known in making an accurate prediction of the maximum possible methane yield per mass of substrate added to the

TABLE 11.3-10   COMPARISON OF DIGESTION PERFORMANCE UNDER HIGH-RATE MESOPHILIC CONDITIONS[a, b]

Food	Gas Production[c] Rate, vol./vol-day	Methane in Gas, mol %	Methane Yield,[d] SCF/lb VS Added	Volatile Solids Reduction, %	Feed Energy Recovery as CH_4, %
Primary sewage sludge	0.78	68.5	5.30	41.5	46.2
Primary-activated sewage sludge[e]	0.89	65.5	5.53	49.0	54.4
RDF[f]-sewage sludge[g]	0.62	60.0	3.54	36.7	39.7
Biomass-waste[f, h]	0.55	62.0	2.40	33.3	38.3
Coastal Bermuda grass[i]	0.59	55.9	3.51	37.5	41.2
Kentucky blue grass[j]	0.55	60.4	2.54	25.1	27.6
Giant brown kelp[f]	0.66	58.4	3.87	43.7	49.1
Water hyacinth[f, k]	0.50	62.8	3.13	29.8	35.7

[a]Conditions were 35°C, daily feeding, continuous mixing, pH 6.7–7.2, 12-day detention time, 0.10 lb VS/ft³-day loading rate except for keep which was 0.13 lb VS/ft³-day.
[b]Adapted from refs. 29 (water hyacinth), 30 (giant brown kelp), 31 (Kentucky blue grass), 32 (coastal Bermuda grass), 33 (biomass waste), 34 (RDF-sewage sludge, sewage sludges).
[c]Standard volumes (60°F, 30.00 in. Hg, dry) of gas produced per culture volume per day.
[d]Calculated using the equation

$$\frac{\text{scf of gas}}{\text{lb vs. added}} = \frac{16.12 \times \text{experimental volume in liters of } CH_4}{\text{Sample wt.} \times \dfrac{\text{% dry solids of sample}}{100} \times \dfrac{\text{% Volatile solids}}{100}} \tag{11.3-6}$$

[e]Blend of 50 wt. % primary-activated sludge on a total solids basis.
[f]Ground with an Urschel Laboratory Grinder (Comitrol 3600) to give particle sizes of 1 mm or less.
[g]Blend of 80 wt. % RDF and 20 wt. % 50 : 50 primary-activated sludge on a total solids basis.
[h]Blend of 32.3 wt. % water hyacinth, 32.3 wt. % coastal Bermuda grass, 32.3 wt. % RDF, and 3.1 wt. % primary-activated sewage sludge (30 wt. % and 70 wt. % on a total solids basis) on a volatile solids basis.
[i]Feed slurry supplemented with NH_4Cl to overcome nitrogen deficiency.
[j]Passed through a 2-mm sieve.
[k]Harvested from a sewage-fed lagoon.

TABLE 11.3-11   COMPARISON OF COMPOSITIONS FOR BIOMASS AND WASTES EVALUATED UNDER HIGH-RATE CONDITIONS[a]

	Primary Sewage Sludge[b]	Primary- Activated Sludge	RDF- Sludge Blend	Biomass- Waste Blend	Coastal Bermuda Grass	Kentucky Blue Grass	Giant Brown Kelp	Water Hyacinth
Carbon, wt. % (dry)	43.7	41.8	42.10	43.1	47.1	46.2	26.0	41.0
Nitrogen, wt. % (dry)	(4.02)	4.32	1.91	1.64	1.96	4.3	2.55	1.96
Phosphorus, wt. % (dry)	0.59	1.30	0.81	0.43	0.24	—	0.48	0.46
Ash, wt. % of TS	26.5	23.5	8.4	17.2	5.05	10.5	45.8	22.7
Volatile matter,								
wt. % of TS	73.5	76.5	91.6	82.8	95.0	89.8	54.2	77.3
Crude Protein[c]	(25.1)	27.0	11.9	10.1	12.3	26.9	15.9	12.3
Cellulose	(11.5)	—	—	37.5	31.7	—	4.8	15.9
Hemicellulose	(23.2)	—	—	31.8	40.2	—	—	45.9
Lignin	(1.8)	—	—	4.6	4.1	—	—	6.8
Mannitol	—	—	—	—	—	—	18.7	—
Algin	—	—	—	—	—	—	14.2	—
Laminarin	—	—	—	—	—	—	0.7	—
Fucoidin	—	—	—	—	—	—	0.2	—
Heating value								
Btu/lb (dry)	8,537	7,874	7,396	8,994	8,185	8,250	4,409	6,886
Btu/lb (MAF)	12,573	10,293	8,074	10,862	8.616	9,187	8,135	8,908
Btu/lb C	19,535	18,837	17,568	20,868	17,378	17,857	16,958	16,795
C/N Ratio	—	9.7	22.0	26.3	24.0	10.7	10.2	20.9
C/P Ratio	74.1	32.2	52.0	100	196	—	54.2	89.1

[a]Except as indicated in b below, all data are for experimental runs in Table 11.3-10.
[b]Figures in parentheses on lot of primary sewage sludge other than that used for runs in Table 11.3-10.
[c]6.25 × total N.

TABLE 11.3-12   ORGANIC COMPONENT CONVERSION UNDER HIGH-RATE DIGESTION CONDITIONS[a]

	Coastal Bermuda Grass[b]		Giant Brown Kelp[c]		Biomass-Waste Blend[d]	
	wt. % of VS	wt. % Converted	wt. % of VS	wt. % Converted	wt. % of VS	wt. % Converted
Crude protein	12.3	—	29.3	8	12.0	24
Cellulose	31.7	65	8.9	8	44.6	32
Hemicellulose	40.2	67	—	—	37.8	86
Lignin	4.1	9	—	—	5.5	0
Mannitol	—	—	34.5	71	—	—
Algin	—	—	26.2	85	—	—

[a]Conditions were 35°C, daily feeding, continuous mixing, pH 6.7–7.2, 12-day detention time, 0.10 lb VS/ft^3-day, except for kelp which was 0.13 lb VS/ft^3-day.
[b]Ref. 32.
[c]Ref. 33.
[d]Ref. 34.

fermentor. However, the maximum methane yield possible with a given substrate can also be rapidly estimated, along with the heat of reaction and the gas composition, from stoichiometric relationships, provided a few analytical values are available.[34]

The main organic fractions in most wastes and biomass species are protein, cellulose, hemicellulose, and lignin, but there are exceptions as shown by the data in Table 11.3-11. The differences between terrestrial and marine biomass are particularly noteworthy. It is of interest to compare the amounts of these organic fractions converted under anaerobic digestion, as illustrated by the data in Table 11.3-12. Lignin conversion is small, if any, but the percentage of lignin present in the volatile solids is usually much less than that of the other components. The general order of decreasing

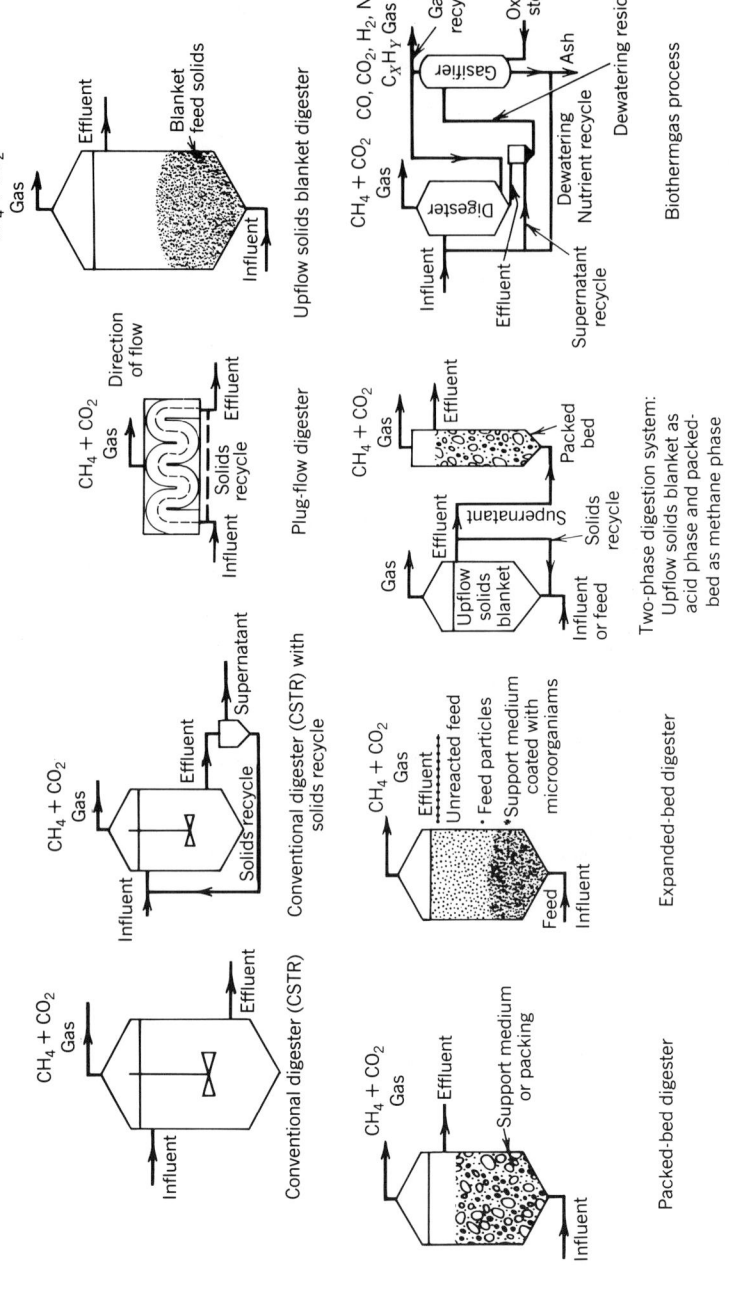

Fig. 11.3-13 Typical anaerobic digestors (from ref. 35).

CH₄ + CO₂ Gas — Effluent — Blanket feed solids — Influent — Upflow solids blanket digester

CH₄ + CO₂ Gas — Effluent — Direction of flow — Solids recycle — Influent — Plug-flow digester

CH₄ + CO₂ Gas — Effluent — Supernatant — Solids recycle — Influent — Conventional digester (CSTR) with solids recycle

CH₄ + CO₂ Gas — Effluent — Conventional digester (CSTR) — Influent

CO, CO₂, H₂, NH₃, C_xH_y Gas — Gas recycle — Gasifier — Oxygen steam — Ash — Dewatering residue — CH₄ + CO₂ Gas — Digester — Dewatering Nutrient recycle — Influent — Effluent — Supernatant recycle — Biothermgas process

CH₄ + CO₂ Gas — Effluent — Packed bed — Gas — Effluent — Supernatant — Solids recycle — Upflow solids blanket — Influent or feed — Two-phase digestion system: Upflow solids blanket as acid phase and packed-bed as methane phase

CH₄ + CO₂ Gas — Effluent — Unreacted feed — Feed particles — Support medium coated with microorganisms — Feed Influent — Expanded-bed digester

CH₄ + CO₂ Gas — Effluent — Support medium or packing — Influent — Packed-bed digester

anaerobic biodegradability is monosaccharides, hemicelluloses, cellulose, protein, and lignin. The same fraction in two substrates may have different biodegradabilities. This is supported by the data for cellulose conversion in Table 11.3-12, which indicates that the cellulose fraction in each substrate had different characteristics.

One way to increase methane yields in an anaerobic digestion process is to pretreat the feed so that the complex organic structures, especially the polymers, are broken down to lower molecular weight species, which then become more susceptible to microbial degradation. Acid hydrolysis of cellulose, for example, affords glucose, which has good biodegradability, and alkaline treatment can be used to break lignocellulose complexes to make the degradable components more accessible to attack.

Digester and Systems Designs

Stirred tank reactors, usually in the complete-mix mode, are still the mainstay of most anaerobic digestion systems, but significant advances are being made to develop other digester designs for methane production. Examples are plug-flow, two-phase, and fixed-film digesters (Fig. 11.3-13).[35]

Plug-flow digesters are reported to be quite suitable for feedstocks such as dairy cattle manure.[36] Table 11.3-13 shows a direct comparison of plug-flow and complete-mix digestion of dairy cattle manure under mesophilic conditions in farm-scale digesters (10,000 gal). The plug-flow unit afforded higher methane yields and volatile solids destruction efficiencies than the complete-mix unit did. The plug-flow digester was an in-ground, excavated channel covered with a plastic bag to maintain anaerobic conditions and to collect the gas. The cost was $10,000 compared with $30,000 for the equivalent capacity rigid-tank, complete-mix digester. Similar plug-flow digestion systems have been in operation on Michigan and Wisconsin farms since about 1978.[37]

Two-phase digestion in which the acid and methane formation phases are physically separated by control of the detention time and volatile solids loading rate has been reported for insoluble substrates.[38-40] Table 11.3-14 presents mesophilic digestion results for both phases in complete-mix laboratory digesters, and Table 11.3-15 presents data reported for a Phase-I complete-mix and a Phase-II anaerobic filter digestion system. Two-phase digestion permits the operating conditions to be optimized for each microbial population and thereby facilitates shorter residence times and more efficient conversion to methane. The benefits of two-phase digestion of sewage sludge in municipal treatment systems are more apparent by comparing the costs of high-rate and two-phase digestion plants of equivalent capacity, as shown in Table 11.3-16. The smaller vessels of the two-phase design are projected to reduce capital cost about 60% for a plant that has a volatile solids loading of about 250 tons/day. The high efficiency of two-phase digestion is clearly illustrated by the data shown in Table 11.3-17 for glucose.[41] Essentially complete conversion of glucose was obtained, about 11% of cellular biomass and 89% to gas. Also, about 81% of the feed energy was converted to gaseous energy —13.9% in the form of hydrogen from the first phase and 66.95% as methane from the second phase. Even higher methane yields could be obtained by cycling the Phase-I gas forward to the Phase-II reactor via biomethanation.

TABLE 11.3-13 DIRECT COMPARISON OF FARM-SCALE DIGESTION OF DAIRY CATTLE MANURE IN COMPLETE-MIX AND PLUG-FLOW DIGESTERS 65 HEAD CATTLE / DIGESTER[a]

	Complete Mix	Plug Flow
Temperature, °C (°F)	35 (95)	35 (95)
Culture volume, gal (m³)	9,247 (262)	10,040 (284)
Feed concentration, lb TS/gal	1.08	1.08
15-day detention time		
Feed rate, gal/day	616	669
Loading rate, lb VS/ft³-day	0.40	0.40
Gas production rate, vol/ vol-day	2.13	2.33
Gas composition, mol % CH_4	55	55
Methane yield, ft³/lb VS added	2.5	3.0
Volatile solids reduction, %	27.8	34.1

[a]Adapted from ref. 36.

TABLE 11.3-14 TWO-PHASE DIGESTION OF SEWAGE SLUDGE[a]

	Acid Phase	Methane Phase
Temperature, °C	37	37
pH controlled	No	No
pH	5.7–5.9	7.0–7.4
Detention time, days	0.5–1	6.5
Loading rate, lb VS/ft^3-day	1.5–2.7	—
Methane production rate, vol/vol-day	0.006–0.6	4.6–8.9
Methane concentration, mol %	19–44	61–78
Methane yield, scf/lb VS destroyed	0.1–1.2	8.5–12.8[b]
Effluent volatile acids, mg/L as acetic acid	3700–5100	100–150

[a] From refs. 38 and 39; feed was 90% activated sludge and 10% primary sludge obtained from the Metropolitan Sanitary District of Greater Chicago: the vessel sequence was a complete-mix Phase I digester (10 L), effluent storage vessel, and a complete-mix Phase II digester (10 L) with no sludge settling or recycling.
[b] High because sludge VS had HHV at 11,200–12,000 Btu/dry lb.

TABLE 11.3-15 TWO-PHASE LABORATORY DIGESTION OF DOG FOOD[a]

	Hydraulic Retention Time, days		
	9.4	4.7	3.1
Temperature, °C	33	33	33
Total reactor volume, gal (L)	5.28 (40)	5.28 (40)	5.28 (40)
Feed concentration, mg COD/L[b]	3800–6300	7700–9400	5400–9500
Gas yield, L/kg COD added	280	455	290
Methane content, mol %	80	65	—
Methane yield			
L/kg COD added	224	296	—
scf/lb VS added[b]	3.56	4.11	—
COD reduction, %[c]	93–98	93–98	93–97

[a] Adapted from ref. 40. Vessel sequence consisted of complete-mix Phase I reactor, sludge separator, and an upflow Phase II anaerobic filter filled with crushed stones.
[b] COD = chemical oxygen demand.
[c] Calculated from L/kg COD by assuming all of COD is $C_6H_{12}O_6$.

TABLE 11.3-16 COMPARISON OF COMMERCIAL HIGH-RATE AND TWO-PHASE DIGESTION OF SEWAGE SLUDGE

	High Rate[a]	Two Phase[b]
Volatile solids, dry ton/day	254.4	254.4
Methane production, 10^6 scf/day	1.15	1.69
Volatile solids destruction, %	34	43
Digester volume, 10^6 ft^3 (number digesters)	2.7	0.136 (4 Acid)
		0.818 (4 Methane)
Estimated ratio of installed plant costs	3	1

[a] Information provided by Metropolitan Sanitary District of Greater Chicago.
[b] Projected based on experimental data.

TABLE 11.3-17   TWO-PHASE LABORATORY DIGESTION OF GLUCOSE[a]

	Acid Phase	Methane Phase
Temperature, °C	30	30
pH controlled	Yes	No
pH	6.0	—
Detention time, h	10	100
Loading rate		
g Organic C/L-day	9.6	0.6
g VS/L-day	24	1.32[b]
lb VS/ft^3-day	1.5	0.083
Gas yield		
L/L feed	3.1	3.3
scf/lb VS added	4.96	5.28[c]
Gas composition,		
mol % CH_4	0	84.3
mol % H_2	58.1	0
mol % CO_2	41.9	15.7
H_2/CH_4 yield, scf/lb VS added	2.88 (H_2)	4.45 (CH_4)
Feed energy in gas, %	13.9	66.9
Volatile solids reduction, %	12	89[d]

[a] Adapted from ref. 41; vessel sequence consisted of Phase I complete-mix reactor (0.4 L), storage vessel at 8°C (1 L), and upflow Phase II reactor (4 L); medium contained inorganic salts plus 0.1425 g/L sodium citrate. Feed: 1.0 wt. % glucose in salt medium.
[b] Assumes VS is 45.5 wt. % C.
[c] Calculated from influent glucose concentration.
[d] Overall; 11% VS to cells.

Digesters containing solid supports for anaerobic organisms have shown promise for both efficient waste stabilization and methane production.[42-44] A 5000-gal/day fixed-film upflow bioreactor (ANFLOW) containing a packed bed of ceramic Raschig rings has high COD removals with raw sewage feeds at ambient temperatures and short residence times. Similar results have been observed with settled sewage sludge and upflow bioreactors containing small particles that behave like fluid beds. Anaerobic attached-film, expanded-bed bioreactors (AAFEB) are reported to be effective for treatment of primary settled sewage with a total COD as low as 80 mg/L as well as animal waste slurries with CODs of 20,000 mg/L. These units operate at low energy inputs and give a good quality effluent.

One of the difficulties of state-of-the art anaerobic digestion systems is that the process does not completely gasify the organics; residuals are always left in the digester effluent. An innovative approach to the solution of this problem has been to combine anaerobic digestion and thermal gasification.[45,46] IGT's BIOTHERMGAS process is exemplary.[45] In this process anaerobic digestion, dewatering, steam–oxygen gasification, and catalytic methanation or biomethanation of the hydrogen-rich gas from the thermal gasifier are combined in that order to produce methane. The refractory organics from the digester are thus completely gasified and waste heat as well as ammonia and inorganic nutrients can be recycled from the thermal gasifier to the digesters. Analysis of two designs (Table 11.3-18) indicates unexpectedly high operating efficiencies with Bermuda grass and refuse-sludge feeds.[45] The key to achieving practical utility of this process will be the development of economical processes to dewater effluent to 50% or less moisture content.

Operating Parameters

Of the many independent operating parameters that control anaerobic digestion, loading rate, detention time, and temperature probably have the most influence on the performance of processes designed to produce methane. Ideally, it is desirable to load the digesters as high as possible for the shortest period of time consistent with high methane yield and volatile solids destruction. Numerous projects have been directed to these goals. Operation at high temperatures should also improve methane production rate and possibly the biodegradability of refractory organics. A tradeoff must be made between the additional energy required for heating the digesters to a high temperature and the incremental increase in gas production.

A farm-scale digester was operated at thermophilic temperatures, as shown in Table 11.3-19.[47] The optimum conditions for methane production that resulted from this work, however, required about 45% of the methane output for process heat. In another large-scale test in which there was a direct comparison between mesophilic and thermophilic digestion of feedlot manure at the same time and with the same feed slurry (Table 11.3-20), the methane yields and volatile solids reduced were slightly lower for the thermophilic digester.[48] Based on these limited reports, it appears that unless the energy requirements for thermophilic digestion can be substantially reduced, mesophilic digestion may be preferable to maximize net energy output.

Anaerobic digesters operating in the semicontinuous and continuous modes are normally fed with slurries containing up to 10–12 wt. % solids. Batch systems have been operated at higher loadings. Landfills, for example, which are analogs of large-scale batch digesters operating on municipal and industrial wastes, have been reported to generate methane at solids contents as high as 70–80 wt. %. The rate of gas production increases as the solids content is reduced (and moisture is increased) from this level.[49]

TABLE 11.3-18 TECHNICAL ANALYSIS OF BIOTHERMGAS PROCESSING EXAMPLES[a]

	Example 1	Example 2
Feed	Bermuda grass	MSW/Sewage Sludge
Feed quantity, dry ton/day	10,000	1,000/112
Processing sequence[b]	Digestion	Digestion
	Dewatering	Dewatering
	Two-Stage Steam/ O_2 Gasification	Single-stage steam/ O_2 gasification
	Catalytic methanation	Biomethanation
Approximate SNG output, 10^9 Btu/day	120	7
Cold gas efficiency, %[c]	74	60
SNG recovery efficiency, %[d]	82	72

[a]Adapted from ref. 45.

[b]Front-end, gas purification, ammonia recovery, air separation, and sulfur recovery operations not listed.

[c]Energy inputs divided by energy output with gas at ambient temperature.

[d]SNG energy content divided by feed energy content.

TABLE 11.3-19 THERMOPHILIC DIGESTION OF CATTLE FEEDLOT MANURE IN PILOT-SCALE COMPLETE-MIX DIGESTER[a]

	Hydraulic Retention Time, days	
	12	4
Temperature, °C	55	55
Culture volume, gal (m^3)	1427	1347
	(5.4)	(5.1)
Feed concentration, wt. % VS	6.18	5.95
Loading rate, lb VS/ft^3-day	0.32	0.93
Gas production rate, vol/ vol-day	3.05	6.65
Methane concentration, mol %	55	50
Methane yield, scf/lb VS added	5.25	3.56
Volatile solids reduction, %	52.8	39.8

[a]Reference 47; Feed steers housed in partially roofed, concrete floor pens; steers fed 80% corn, 18% silage, 2% soybean meal, and no antibiotics.

**TABLE 11.3-20   DIRECT COMPARISON OF DIGESTION OF
CATTLE FEEDLOT MANURE AT MESOPHILIC AND
THERMOPHILIC CONDITIONS**[a]

	Mesophilic	Thermophilic
Temperature, °C (°F)	37.8 (100)	55.6 (132)
Culture volume, gal (m^3)	17,100 (484)	11,968 (339)
Feed concentration		
lb TS/gal[b]	0.57	0.57
lb VS/gal[b]	0.40	0.40
Detention time, days	20	12
Loading rate, lb VS/ft^3-day	0.15	0.25
Methane production rate, vol/vol-day	0.358	0.520
Methane yield, ft^3/lb VS added	3.12	3.05
Volatile solids reduction, %	25.7	17.7

[a]Adapted from ref. 48; mixing by effluent recirculation; solar augmented
heating.
[b]Estimated average values from data in ref. 48.

An interesting concept where high-solids digestion is conducted under so-called dry fermentation conditions, which are defined as fermentation of organic solids at concentrations higher than that at which water will drain from the substrate, is under development.[50] These high-solids systems might be considered to be the analog of *in-situ* digestion that takes place in municipal landfills containing refuse at high solids levels. Small-scale laboratory and pilot studies of dry fermentations have shown that 90% destruction of the biodegradable volatile solids can be achieved in 60–90 days at 55°C or in 120–200 days at 35°C with such substrates as wheat straw. Gas production in a 110-m^3 dry fermentor operating at mesophilic temperatures with wheat straw averaged 0.25 vol/vol of fermentor-day.

Commercial Methane Production Systems in the United States

Sewage Sludge. Many wastewater treatment plants throughout the United States are now making maximum use of digester gas for process heat, steam production, or electric power generation. An example is shown in Table 11.3-21. This table summarizes the gas utilization program underway at the West-Southwest Treatment Plant of the Metropolitan Sanitary District of Greater Chicago (MSDGC).[51] The economics of low-pressure steam production are quite favorable based on MSDGC's current gas price, but note that about 63% of the product gas is used for digester heating. MSDGC's digesters are operated in the mesophilic temperature range.

Manure. The only large-scale commercial plant in the United States for SNG production from cattle feedlot manure (Guymon, Oklahoma) was operated by Calorific Recovery Anaerobic Processes; it had a 20-year contract with Natural Gas Pipeline Co. to supply SNG at a rate of about 600 million cubic feet per year. It was shut down due to front-end materials-handling problems.[52] After extensive renovation the plant has reopened under the name CalFeed Corp., making dried animal feed from manure. All of the gas produced is used in drying the product.

A full-scale, feedlot-manure digestion plant for Lamar, Colorado, was designed by Bio-gas of Colorado.[53] This plant was to have converted manure from 50,000 head of cattle and the gas was to have been used for electric power generation. Cattle feeds were also to have been produced. Unfortunately, the gas costs were found to be too high for this plant in comparison with the cost of locally available natural gas so the project has been temporarily shelved. Many other small-scale anerobic digestion systems are in operation on farms and feedlots for waste disposal and methane production.

Biomass. No small- or large-scale biomass-fed anaerobic digestion plants have been commercialized in the United States (March 1983). Several integrated biomass production–digestion projects have been proposed for demonstration, and it is expected that if these projects are successfully completed, commercial ventures will be started.

Municipal Solid Waste. The 100-ton/day proof-of-concept facility in Pompano Beach, Florida, is still the only operating anaerobic digestion plant using MSW in the United States.[54] In late 1981 and 1982 the plant was operated in the thermophilic mode at 10-day retention times to provide gas yields

TABLE 11.3-21 DIGESTER GAS UTILIZATION AT WEST-SOUTHWEST SEWAGE TREATMENT PLANT OF METROPOLITAN SANITARY DISTRICT OF GREATER CHICAGO[a]

Digestion Plant Summary

Digester size	2.5×10^6 gal
Number of digesters	12
Operating conditions	High rate (95°F)
Design capacity	25 TPD TS/digester
Waste feed	3.3% TS activated sludge
Average gas production	2.7×10^6 ft^3/day
Average methane production	1.76×10^6 ft^3/day

Gas Utilization

For digester heating	1.7×10^6 ft^3/day
For low-pressure steam	1.0×10^6 ft^3/day
Steam rate	22,000 lb/h (210°F, 100 psig)

Gas Utilization Economics
  For Steam Production

Capital cost	$830,000
Annual operating cost	$150,000
MSDGC natural gas cost	$2.40/10^6 Btu
Annual gas cost for steam	$569,400
Approximate pay out from digester gas use	2 years

[a] From ref. 51.

of 6–8.5 ft^3/lb VS added and methane yields of 3–4.4 ft^3/lb VS added.[54] These results appear to be among the best obtained on this project to date. However, the unit has only been operated at a feed rate of a few tons per day because of operating and startup problems. It has never operated at its design capacity.

Landfill Gas. Landfills containing wastes such as municipal refuse correspond to batch analogs of anaerobic digestion. Methane can be recovered by drilling relatively shallow wells if the conditions within the landfill, particularly temperature and moisture content, are conducive to gas production rates that make it economic to recover methane. Currently, commercial methane recovery for utility and industrial use is taking place at 15–20 landfills across the United States, and many other sites are under test. The technology appears to be well established.[55]

11.3-4 THERMOCHEMICAL GASIFICATION

Edward S. Lipinsky

Thermochemical gasification of biomass can yield gaseous fuels that can operate turbines, internal combustion engines, and boilers originally designed for natural gas and petroleum liquids. Thus, equipment that was purchased when petroleum and natural gas were inexpensive need not be scrapped and replaced with solid biomass boilers. In addition, the quick turndown and pollution-free characteristics of natural gas can be obtained from a renewable resource.

Both anaerobic digestion and thermochemical gasification make available gaseous fuels from biomass. Each has its role to play; thermochemical gasification is especially suitable for biomass that contains considerable lignin and extractives, which makes it resistant to microorganisms as well as high in energy content. Anaerobic digestion is especially suitable for wet or even dilute biomass feedstocks that are susceptible to biological degradation. The two gasification methods complement each other more than they compete.

TABLE 11.3-22   CHEMICAL COMPOSITION OF BIOMASS RELEVANT TO GASIFICATION PERCENT (MOISTURE-FREE BASIS)[a]

Property	Lobolly Pine	Western White Pine	Tan Bark Oak	Fir Bark	Yellow Birch Bark	Sugarcane Bagasse
Cellulose and hemicellulose	62.0	59.7	58.0	27.4	40.7	71.0
Lignin	28.0	26.4	24.9	39.2	36.5	17.8
Extractives	9.0	13.7	16.3	30.4	19.9	7.8
Ash	0.4	0.2	0.8	3.1	2.9	3.4

[a] From refs. 56 and 57.

Biomass Properties Relevant to Gasification

Thermochemical gasification converts biomass into a mixture of gases that contain many or all of the following: hydrogen, carbon monoxide, methane, carbon dioxide, water, and simple olefins. The yield of these products, the rate of reaction, and the energy content of the gas depend on the chemical and physical properties of the biomass starting material. The terrestrial biomass species that are candidates for thermochemical gasification consist primarily of cellulose, hemicellulose, lignin, various extractives, and ash (Table 11.3-22). The detailed gasification mechanism differs substantially for each biomass constituent. The differences in the total content of cellulose and hemicellulose for softwoods and hardwoods are relatively small. However, tree bark has significantly less of these constituents and more lignin and extractives. Conifers have high-energy extractives (terpenoids) while hardwoods tend to have low-energy extractives (simple carbohydrates). Nonwoody biomass tends to be high in carbohydrates and low in both lignins and high-energy extractives.

Much of the engineering research and design activities relating to thermochemical gasification of biomass is focused not on the detailed polymer composition of the biomass but on proximate and ultimate analyses that are useful in estimating energy production and mass transfer needs. The volatile matter and fixed carbon for several major biomass species are shown in Table 11.3-3. High percentages of volatile matter are favorable because they represent readily gasified materials; high percentages of fixed carbon are undesirable because they represent resistant (but ultimately gasifiable) char. All of the biomass raw materials are substantially more reactive than the representative coal shown for comparison. However, bark is not as desirable as is true wood.

Comparison of ultimate analyses of various types of biomass with each other and with a typical coal yields significant insights. Regardless of source, biomass tends to be about $40 \pm 10\%$ oxygen whereas coals are well under 10% oxygen. Furthermore, the oxygen in biomass provides "weak links" that render gasification easy. Except for a few species (e.g., rice hulls), a clean biomass tends to be very low in ash compared with coal. Failure to appreciate these ultimate compositional differences between coal and biomass led to many abortive designs for gasifiers that were based on an overly literal transfer of coal gasification technology to biomass gasification technology.

The physical properties of biomass are of considerable value in determining the success of gasifier design and operation. As shown in Table 11.3-3, there is little difference among woody biomass species with respect to their higher heating value. Woody biomass tends to be perhaps 20% higher in heating value than nonwoody biomass; however, the presence or absence of triglycerides or terpenoids affects this generalization.

The key factor is not the heating value, which shows small differences, but the moisture content, which varies greatly by species of biomass resource and especially with its exposure to water. Lignocellulosic biomass from industrial processes can have relatively constant high moisture content (e.g., bagasse, 45–55% moisture) or low moisture content (sanderdust, 2–8% moisture). Forestry residues or agricultural residues that are exposed to the weather can vary tremendously in moisture content. For example, bark can vary from 25 to 75% moisture. Moisture can act as a diluent, as a heat absorber on vaporization, and as a reactant in thermochemical gasification reactions.

Although permeability, heat capacity, and thermal conductivity are very important in determining the kinetics of gasification reactions, little is known about these physical properties under conditions that pertain to gasification reactions. The values presented in Table 11.3-23 are typical of those available, but some were determined at temperatures that are not necessarily relevant to gasification.

Characteristics of Product Gases

Thermochemical gasification of biomass can be conducted to yield gaseous fuels that are suitable for combustion. Depending on whether or not the product contains nitrogen from air, the product is

TABLE 11.3-23   PHYSICAL PROPERTIES OF BIOMASS RELEVANT TO GASIFICATION[a]

Physical Property	Units	Typical Values
Bulk specific gravity	Dimensionless	0.38 (green ponderosa pine); 0.42 (oven dry ponderosa pine); 0.35–0.38 (quaking aspen)
Longitudinal permeability	$\dfrac{cm^3\ (air)}{cm\ atm}$	10,000 (red oak); 1 (Douglas fir heartwood)
Transverse permeability	$\dfrac{cm^3\ (air)}{cm\ atm}$	0.01–0.0001 (all species)
Heat capacity	Btu/lb · °F	0.57 (Oak); 0.67 (yellow pine) at 75°F
Thermal conductivity	Btu/(ft/h · °F)	0.09 (white pine) at 86°F
Higher heating value	MMBtu/ton	17.9 (poplar); 18.0 (Douglas fir); 18.0 (loblolly pine stemwood); 19.0 (Douglas fir bark); 13.6 (sugarcane bagasse)

[a] Based on data from ref. 56.

designated low Btu gas (generally in the 100–200 Btu/scf range) or medium Btu gas (generally in the 250–500 Btu/scf range), depending on process details.

Thermochemical gasification of biomass may be conducted to achieve synthesis gas, a chemical intermediate for the production of such liquid fuels as methanol or Fischer–Tropsch fuels.[58] Synthesis gas also can be employed to manufacture gases rich in methane, known as high-Btu gas or SNG.[59] Thus, thermochemical gasification is broadly useful as a source of relatively inexpensive fuel gases to be used as-is and as a source of high-quality gaseous and liquid fuels.

Gasification Principles

Reed, Milne, and Graboski have summarized gasification principles in terms that are useful to both the engineer and the chemist and the reader is referred to their treatise for additional information.[56]

Figure 11.3-14 places the gasification of carbon-containing materials in perspective. Not only can every material containing carbon, hydrogen, and oxygen be displayed on this triangular diagram but also the path from the reactants to products for every possible thermochemical gasification process can be displayed. The extreme right-hand portion of the triangle consists of compositions that are so high in oxygen that they have no fuel value. The line between the vertices that represent carbon and hydrogen display all possible hydrocarbons. The gaseous fuels occupy a limited area that includes hydrogen, a few low molecular weight hydrocarbons, and carbon monoxide. Gasification strategy is to move effectively from a specific starting material (e.g., pine bark) into the gaseous fuel area.

Gasification Strategies

There are four broad thermochemical gasification strategies that are sources of the hundreds of tactical alternatives available.

Pyrolysis.   The biomass is raised to a high temperature, leading to disproportionation into energy-rich gases, water, and a solid char. The char may be recovered to supply part of the heat required for this process.

Hydrogenation.   By introducing hydrogen into the system, hydrocarbons are formed instead of char. At the temperatures used in the gasification of biomass, it is not certain that hydrogenation is one of the primary steps. An alternative to hydrogen gas addition is the use of carbon monoxide, which reacts with water to give carbon dioxide and hydrogen.

Oxygen Gasification.   Introducing either air or pure oxygen, a small amount of the biomass can be combusted in the reactor to achieve the high temperature required for gasification. Oxygen gasification can be combined with steam gasification.

Steam Gasification.   Water at high temperatures can be used to transfer heat to biomass. In a high-temperature system such as a coal gasifier water reacts with carbon materials to yield carbon monoxide and hydrogen. This is an endothermic reaction requiring supplementary heat. Water vapor will also react with carbon monoxide to yield carbon dioxide and hydrogen.

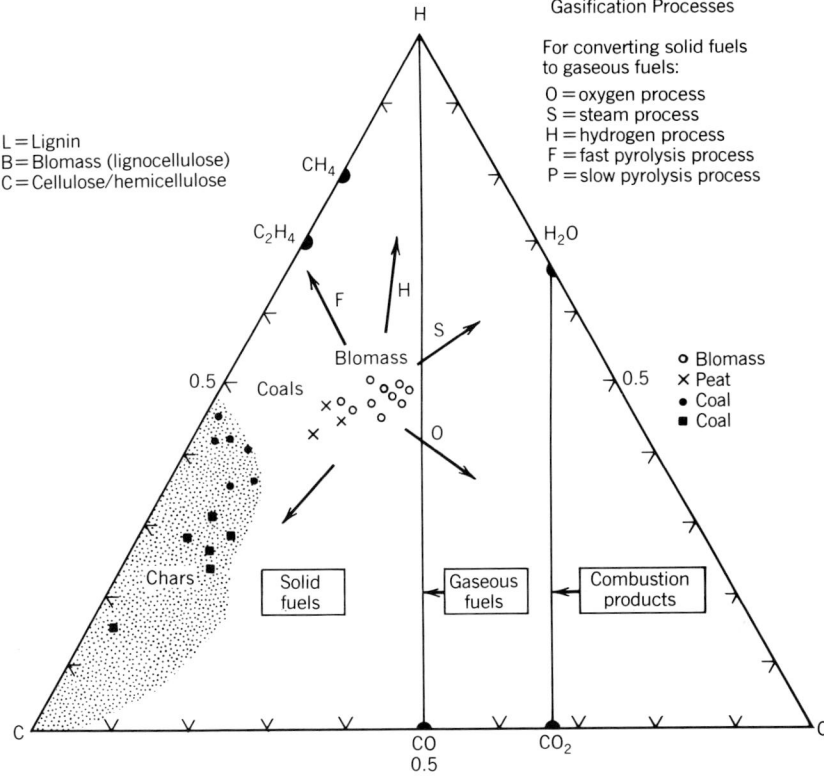

Fig. 11.3-14   Chemical changes during biomass gasification.

Action of High Temperatures on Biomass

Within an individual particle of biomass, especially a moist particle, exposure to a high thermal gradient is associated with complex phenomena (Fig. 11.3-15). Just how complex the situation is depends in part also on how large the particle size is. In the section on residential wood burning, the situation was discussed for biomass that is burning very slowly. Residence times in gasifiers range from under 1 s to less than 10 min. Most gasifiers operate with particles that are small compared with cordwood. Nevertheless, there are stagnant gas films, surface char that temporarily prevents the escape of liquids and gases, and similar phenomena.

As shown in Fig. 11.3-16, the phenomena in the system are more complex than those that pertain to a single particle. The primary gas, oil, and combustion gases from particles already undergoing gasification supply heat and chemical reactants to the newly introduced particles that are beginning gasification.

Engineering Aspects

The range of gasifiers that could be built to embody the various strategies of converting solids to gases of suitable composition is extremely large, when the need to accommodate the heat and mass transfer requirements is also taken into account. They are illustrated in Fig. 11.3-17, which makes the principles of updraft, downdraft, fluidized bed, suspended bed, multisolid fluidized bed, recirculating liquids, and recirculating gases more explicit.[60] The gasifier system of choice for a specific application will depend on the biomass feedstock available, the idiosyncrasies of the end use, and the state of the art with respect to commercialization of the numerous competitive systems. The attributes and selection criteria are illustrated for four typical gasifier systems in the next section.

Typical Gasifier Systems

Upflow Gasifiers.   Upflow gasifiers frequently are used to retrofit boiler systems that were originally designed to use fuel oil or natural gas. The design shown schematically in Fig. 11.3-18 has been

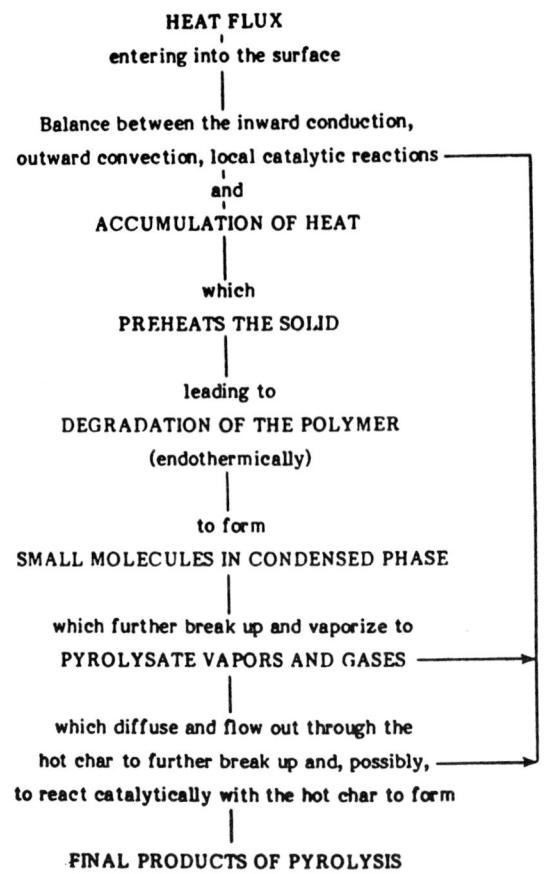

HEAT FLUX

entering into the surface

Balance between the inward conduction,
outward convection, local catalytic reactions ─────

and

ACCUMULATION OF HEAT

which

PREHEATS THE SOLID

leading to

DEGRADATION OF THE POLYMER

(endothermically)

to form

SMALL MOLECULES IN CONDENSED PHASE

which further break up and vaporize to

PYROLYSATE VAPORS AND GASES ─────→

which diffuse and flow out through the

hot char to further break up and, possibly, ─────→

to react catalytically with the hot char to form

FINAL PRODUCTS OF PYROLYSIS

Fig. 11.3-15   Heat flux (from ref. 56).

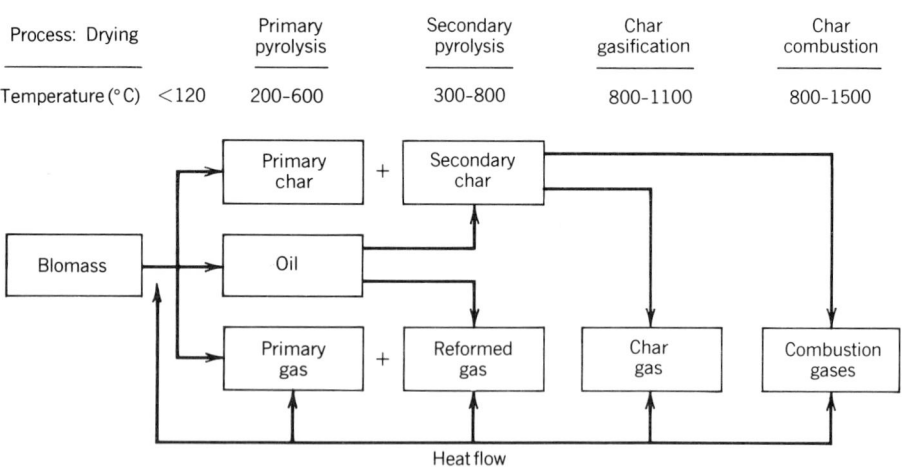

Fig. 11.3-16   Heat and mass flows in pyrolysis and gasification processes (from ref. 56).

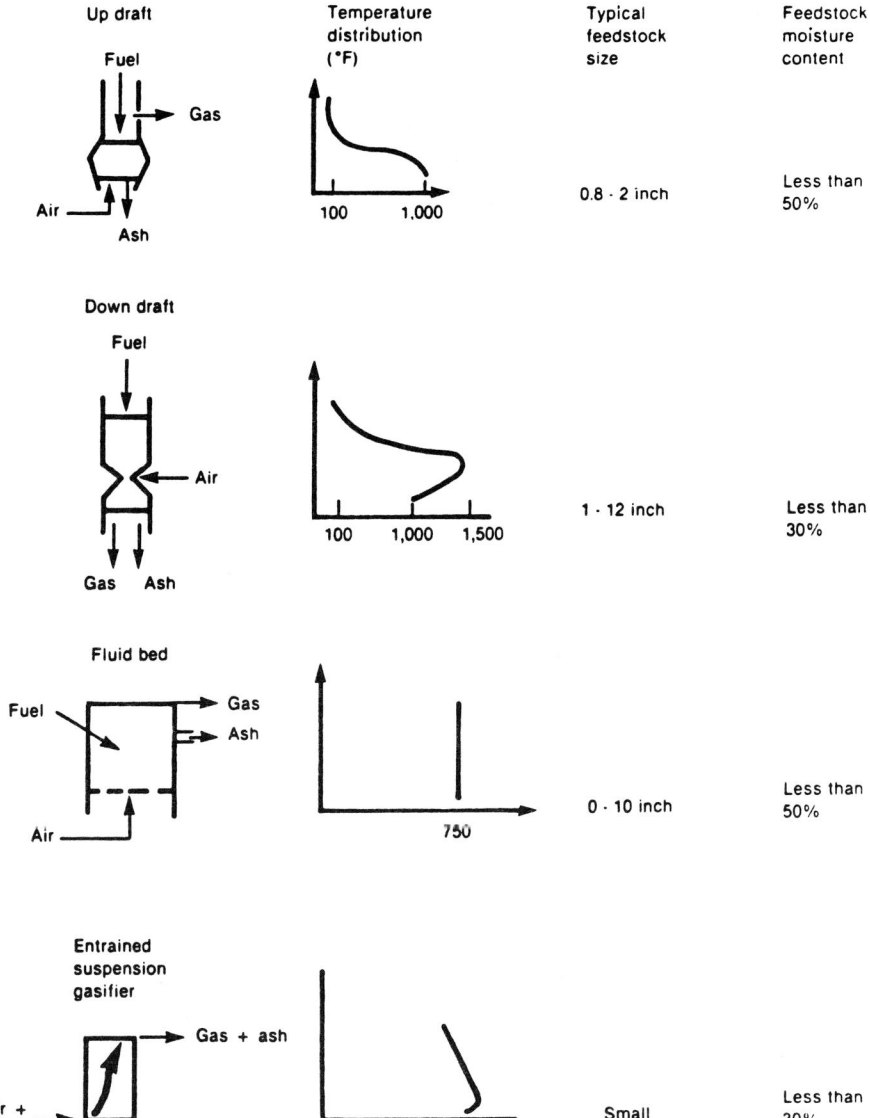

Fig. 11.3-17 Principles of various gasifier types (from ref. 60).

installed in a hospital steam plant and as a partial replacement for natural gas in a relatively large utility boiler, for example.[61] Wood chips containing approximately 50% moisture are introduced through the airlock and encounter increasingly high temperatures as the fixed bed moves downward in a countercurrent relationship to the hot gases. The biomass is first thoroughly dried by hot gases and then pyrolyzed into a mixture of low-Btu gas, oils, and char. Most of the oil and char move downward into a reduction zone in which they are converted to carbon. This carbon reacts with steam to form a lower Btu gas. The excess char encounters the process air, which burns part of it to generate the heat needed to operate the entire system. The initial carbon dioxide and water generated by this combustion is reduced by part of the char to carbon monoxide and hydrogen. The flows of biomass and process air are adjusted to achieve the desired product gas chemical composition and temperature. The low-Btu gas has a temperature of approximately 200°F and can be burned in installation designed for No. 6 fuel oil or natural gas at excess air ratios comparable to those used with conventional fossil fuels. Because such gasifiers do not have a capability for instant shutoff as does a natural gas burner, a flare system is employed for boiler shutdown.

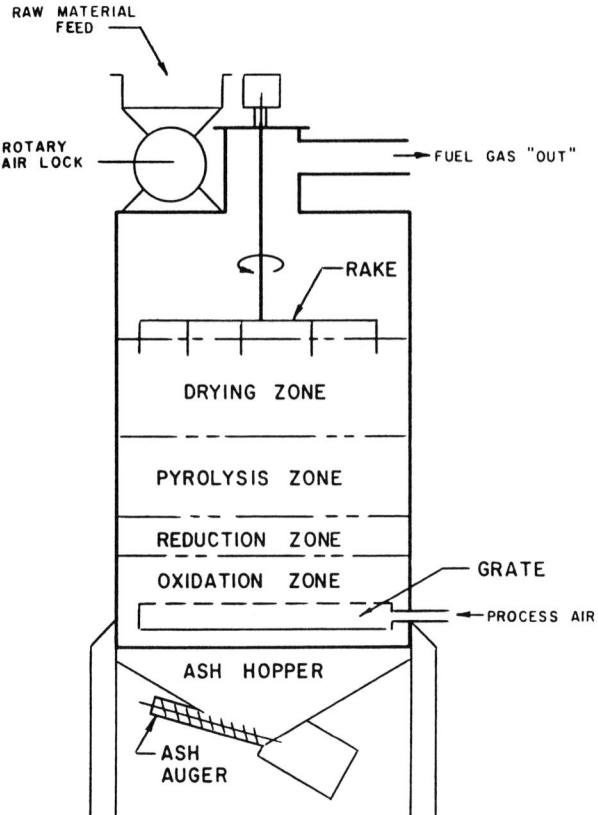

Fig. 11.3-18   Updraft gasifier (fixed bed) (from ref. 61).

TABLE 11.3-24   TYPICAL DRY BIOMASS-DERIVED FUEL GAS COMPOSITIONS

Composition	Low Btu	Medium	Btu
		G	
CO	28 ± 2^a	46.5^b	36.7^c
H_2	11 ± 1	14.9	19.4
CO_2	58	14.6	
N_2			
CH_4	2	17.8	12.2
C_2H_4		4.9	5.0
Btu/scf	157 ± 8	480–500	384

[a] Fixed-bed updraft air gasifier (from ref. 61).
[b] Dual fluidized-bed gasifier with no air or oxygen. Source: Reference 62.
[c] Fluidized-bed gasifier with oxygen (from ref. 63).

The composition of this low-Btu gas is compared with other gases in Table 11.3-24. Considerable energy losses would occur if the product were allowed to cool before use because sensible heat would be lost as would the heat contained in any entrained tars and oils.

Downdraft Gasifiers.   Although the fixed-bed updraft gasifier is quite suitable for relatively small installations in which the gas is burned directly, the generation of pyrolysis tars and oils is a serious drawback for gas that is to be used for the production of methanol, methane, or fuel applications calling for medium-Btu gas. As shown in Fig. 11.3-19, the downdraft gasifier forces the tars into a very

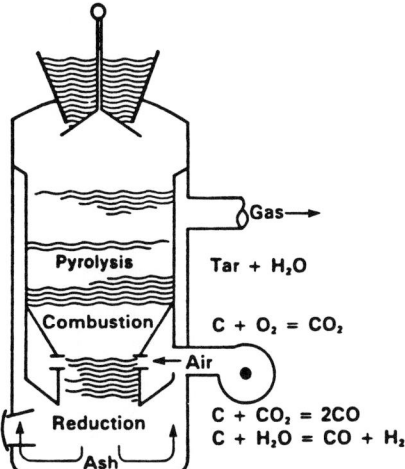

Fig. 11.3-19 Schematic diagram of downdraft gasifier.

hot zone where they either burn or are pyrolyzed to fuel gas products. This type of low-tar gas can be employed to operate internal combustion engines when operated with air.

The Solar Energy Research Institute (SERI) high-pressure oxygen downdraft gasifier has been chosen for presentation in Section 11.3-5 because it is designed to produce chemical synthesis gas.

Fluidized Bed Gasifiers. Both updraft and downdraft gasifiers are fixed beds through which gas must move. Their performance can suffer through plugging, as occurs when too much sawdust is fed with wood chips to a typical downdraft gasifier.

Fluidized-bed gasifiers provide efficient heat transfer between hot gases and solids of varying particle size and moisture content. A commercial example of a fluidized-bed gasifier is that of Omnifuel Gasification Systems, Ltd.[64]

A schematic of the gasifier (Fig. 11.3-20) shows how the air used for gasification is employed to agitate sand, which acts as the dense phase. Sawdust, bark, and wood chips are introduced into the fluidized bed and decomposed due to the high temperature that arises from combustion of a small portion of the biomass in the fluidized bed. The low-Btu gas is generated by reaction of the rest of the air with the char that is produced in the thermal decomposition of the wood. The gas is cleaned by means of a cyclone and appropriate filters so that it can be piped within an industrial facility to

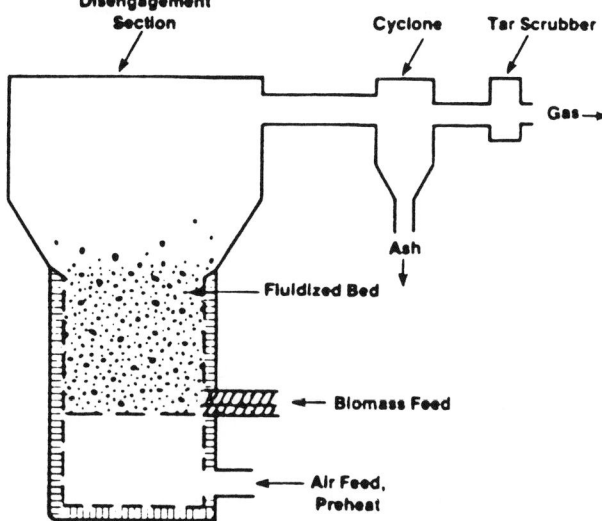

Fig. 11.3-20 Schematic diagram of fluidized-bed gasifier.

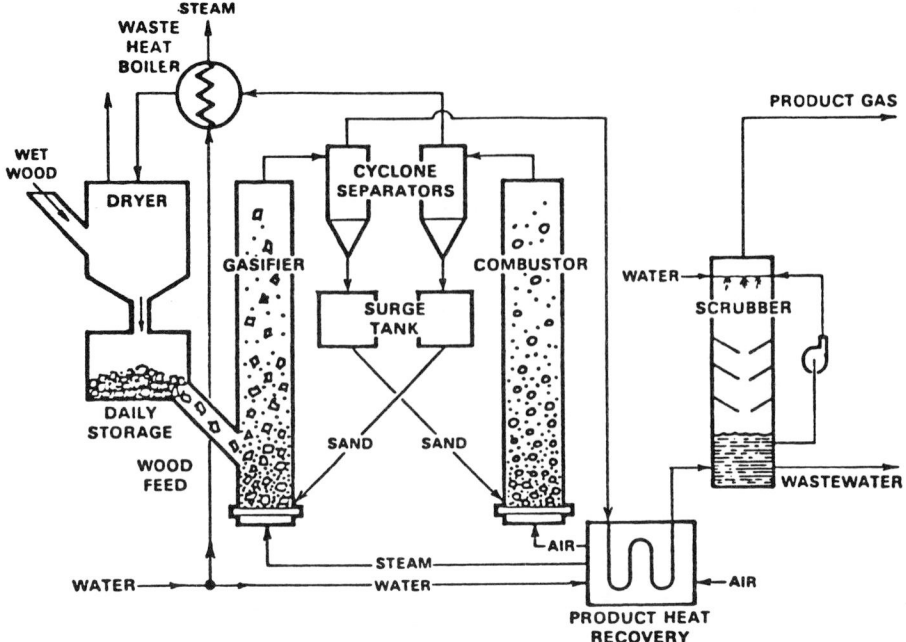

Fig. 11.3-21 Simplified flow diagram of Battelle biomass gasification system.

supplement or replace fossil fuels required in heaters and boilers. The air that is used for fluidization and gasification is preheated during the cooling and cleaning of the product gas.

The major advantage of a fluidized-bed gasification system is its high capacity per unit of reactor volume. By using oxygen instead of air, a medium-Btu gas could be obtained using this type of fluidized-bed system. It is also possible to operate fluidized-bed gasifiers in a pressure mode.

Therefore, fluidized-bed systems and downdraft pressure gasifier systems both should be considered for the entire range of gasification applications.

Dual Fluidized-Bed Systems. Updraft, downdraft, and conventional fluidized-bed gasifiers generate the heat to perform the gasification reaction within the gasification vessel. They generally make use of some of the combustion products as reactants in the gasification process. Although this strategy appears logical, efficient, and economical, it imposes constraints on the system with respect to temperature profiles and gas composition. For example, air gasification generates low-Btu gas because of the high nitrogen content of the air used for heating.

Bailie[65] and Feldmann[62] have pioneered in the development of dual fluidized-bed gasifiers in which the attainment of the high temperature required for the gasification is physically separated from the gasification reaction. A typical gasifier of the dual fluidized-bed type is shown in Fig. 11.3-21. The heat transfer agent is sand or other solid particles chosen for their physical and possible catalytic properties. The sand is heated in a fluidized-bed combustor and transferred to the gasifying fluidized bed where it encounters wood chips or other biomass feedstocks. The biomass has not been pretreated except for partial drying, which can be accomplished with flue gas from the combustor. The biomass is gasified with steam to produce product gas and char. Only enough char is produced to supply the fuel for the combustor. Char production is regulated by controlling the gasifier conditions. The light char is carried out of the gasifier, separated, and transferred to the combustor where it is combusted to reheat sand that is recirculated. The product gas is cooled in a heat exchanger that generates steam for the gasifier. Heat recovery from the flue gas is employed to partially dry the incoming wood.

Use of this dual fluidized-bed concept eliminates the need for an oxygen plant in the production of medium-Btu gas (Table 11.3-24). Although conceived as a dual fluidized-bed process, it is practiced as an entrained-bed gasifier linked to a fluidized-bed combustor. As long as the heat transfer medium (e.g., sand) is heated in a separate unit, it is not essential that the combustor be of a fluidized-bed design.

The high-Btu content of the gas from the dual fluidized-bed unit (475–500 Btu/scf) renders it an excellent fuel gas; however, the high energy content arises from a high methane content. Therefore, the gas is not attractive for production of methanol and related chemicals, which require pure hydrogen and carbon monoxide at a specific ratio.

11.3-5 METHANOL

Edward S. Lipinsky

Both solid and gaseous fuels have disadvantages with respect to end-use applications in which intermittent use is an important feature. Transportation applications are especially well suited to the use of liquid fuels. Although more complex liquid fuels than methanol could be made from biomass, methanol appears especially attractive. Experiments with internal combustion engines, electric utility gas turbines, and liquid fuel-fired boilers have demonstrated the clean burning characteristics of methanol, along with its ability to be stored and pumped.[66-68]

Methanol can be manufactured from virtually any carbon-containing material.[69] The synthesis is simple in principle in that the carbonaceous material is gasified to a mixture of hydrogen and carbon monoxide and the composition of the gas is "shifted" to a ratio that permits hydrogenation of carbon monoxide by hydrogen to yield methanol. The various reaction and side reactions taking place in methanol manufacture are shown in Table 13.3-25.

Relation of Methanol to Thermochemical Gasification

All of the processes for the manufacture of methanol use thermochemical gasification to achieve the gaseous raw material for conversion to methanol. The difference between thermochemical gasification for manufacture of direct gaseous fuels and methanol synthesis are primarily the following:

1. Gas used to synthesize methanol must be extremely pure with respect to trace quantities of compounds of sulfur and various metals.
2. Methanol is produced from a specific ratio of hydrogen to carbon monoxide and/or carbon dioxide—energy-rich methane is inert in this system.
3. Many biomass raw material supply systems function best with the delivery of less than 200 tons per day of wood whereas most (but not all) economical methanol systems require 1000–2000 tons per day.

Thus, biomass does have prospects for use in the production of methanol, where it is sufficiently plentiful and inexpensive.

Chemistry of Methanol Production

The equations shown in Table 11.3-25 summarize the reaction sequence by which methanol is synthesized, including synthesis gas production, shifting of the ratio of carbon monoxide to hydrogen to favorable ratios and hydrogenation of the hydrogen/CO mixture. Because of the need for chemically pure methanol for the production of industrial chemicals, considerable catalyst development has been done in the last few decades to reduce the production of alcohols that are heavier than methanol. For fuel methanol, higher alcohols would be desirable rather than detrimental. However, the hydrogenation of carbon monoxide to methane and other saturated hydrocarbons is detrimental

---

TABLE 11.3-25   CHEMISTRY OF METHANOL PRODUCTION

Synthesis Gas Production

$$Biomass \rightarrow CO + H_2 + CO_2 + H_2O + C + C_nH_{2n}$$
$$Biomass + O_2 \rightarrow CO + H_2 + CO_2 H_2O$$

Shift Reactions

$$H_2O + CO \rightleftarrows H_2 + CO_2$$

Reforming Reactions

$$CH_4 + H_2O \rightarrow 4H_2 + CO_2$$

Hydrogenation Reactions

$$2H_2 + CO \rightleftarrows CH_3OH$$
$$3H_2 + CO_2 \rightleftarrows CH_3OH + H_2O$$
$$(2n + 1)H_2 + nCO \rightleftarrows C_nH_{2n+2} + nH_2O$$
$$(n + 1)CO + 2(n + 1)H_2 \rightleftarrows CH_3(CH_2)_nOH + nH_2O$$
$$(2n + 1)CO + (n + 2)H_2 \rightleftarrows CH_3(CH_2)_nOH + CO_2$$

because the methane builds up and must be purged from the reactor. Similarly, synthesis gas production methods that yield methane, ethylene, ethane, and other hydrocarbons are not desirable because these products also are not converted to methanol in the conventional process.

There are methane-reforming reactions utilizing special catalysts. However, for economic reasons, the conventional approaches to methanol from biomass do not make use of this reaction.

Alternative Processes

The emphasis in the development of biomass-derived methanol is on the development of thermochemical gasification technology that meets the needs of the downstream shift and hydrogenation reactions. Thus, the gas that is transferred to the shift reactor is to be low in methane and other hydrocarbons, preferably under considerable pressure, and low in any trace impurities that might poison either the shift catalyst or the hydrogenation catalyst. The closer that the ratio of hydrogen to carbon monoxide is to that required for methanol production (2 : 1) the better for methanol production. Such a synthesis gas may be much less desirable as a fuel than gases rich in methane would be.

The basis for the emphasis on the thermochemical gasification aspect of methanol production by process developers is the belief that the technology for the shift reaction and the hydrogenation already are highly developed and that the need of biomass-based processes are approximately the same as coal-based processes. Since biomass facilities need to be substantially smaller than coal facilities, if the economics are to be comparable, there appears to be a need for development of shift reactors and hydrogenators that can be shop fabricated. In this way mass production could be employed to reduce costs instead of relying on economies of scale in field-erected units.

The major process options are:

1. Fixed-bed versus fluidized-bed designs.
2. Oxygen gasification to achieve high temperatures versus indirect methods of introducing heat (e.g., hot sand).
3. High pressure gasification versus atmospheric pressure gasification.

In general, avoidance of the use of oxygen has the potential to reduce costs but complicates the system. Although pressure gasification is especially valuable for methanol production that occurs at more than 50 atm and has the potential to reduce capital investment by achieving more throughput per unit volume, a number of engineering problems are introduced.

Rather than conduct the biomass gasification and shift reactions in separate vessels, it is possible to employ a catalyst to accomplish the shift reaction during gasification.[70] The H_2/CO ratio that is suitable for methanol production has been demonstrated, but the catalyst lifetimes need to be improved.

The concept of employing biomass as a portion of the feedstock for a larger methane to methanol facility can be justified on several grounds:[71]

1. Methane has too much hydrogen and biomass has too little hydrogen; therefore, a facility that utilizes both simultaneously needs to do less shifting and reforming.
2. Some otherwise desirable gasification processes yield methane that could be reformed along with a separately derived methane feedstock (e.g., natural gas).
3. A large-scale plant could be constructed that would provide the economies of scale that a biomass-only facility could not achieve.
4. Many biomass resources are seasonal in their availability and/or have considerable price fluctuations; these drawbacks could be overcome in a hybrid facility.

Typical Gasifier Options

The SERI gasifier (Fig. 11.3-22) operates with oxygen under pressure so that savings can be made in conversion of the gas into upgraded fuel products requiring high-pressure reaction conditions.[72] It is designed to be shop fabricated on a mass production basis. It will operate in a continuous mode. The gasification reactions occur in four zones. Incoming oxygen encounters tar vapors that were released by biomass pyrolysis and burns them. These gases and most of the oxygen move through a second zone in which pyrolysis is completed. In the third zone, these pyrolysis products are cracked and gasify.

The design goal for the SERI downdraft gasifier is to operate at higher pressures so that economies can be realized in methanol production. The low bulk density of biomass renders the biomass pressure feeding system critical in importance. The SERI gasifier uses a lock hopper system. A star valve delivers wood chips to the gasifier from a high-pressure hopper. The high-pressure hopper receives fuel from an intermediate-pressure hopper that in turn is fed by a hopper at atmospheric pressure. When level indicators determine that the pressurized hopper is low on wood chips, a slide valve opens and

SERI Oxygen Biomass Gasifier

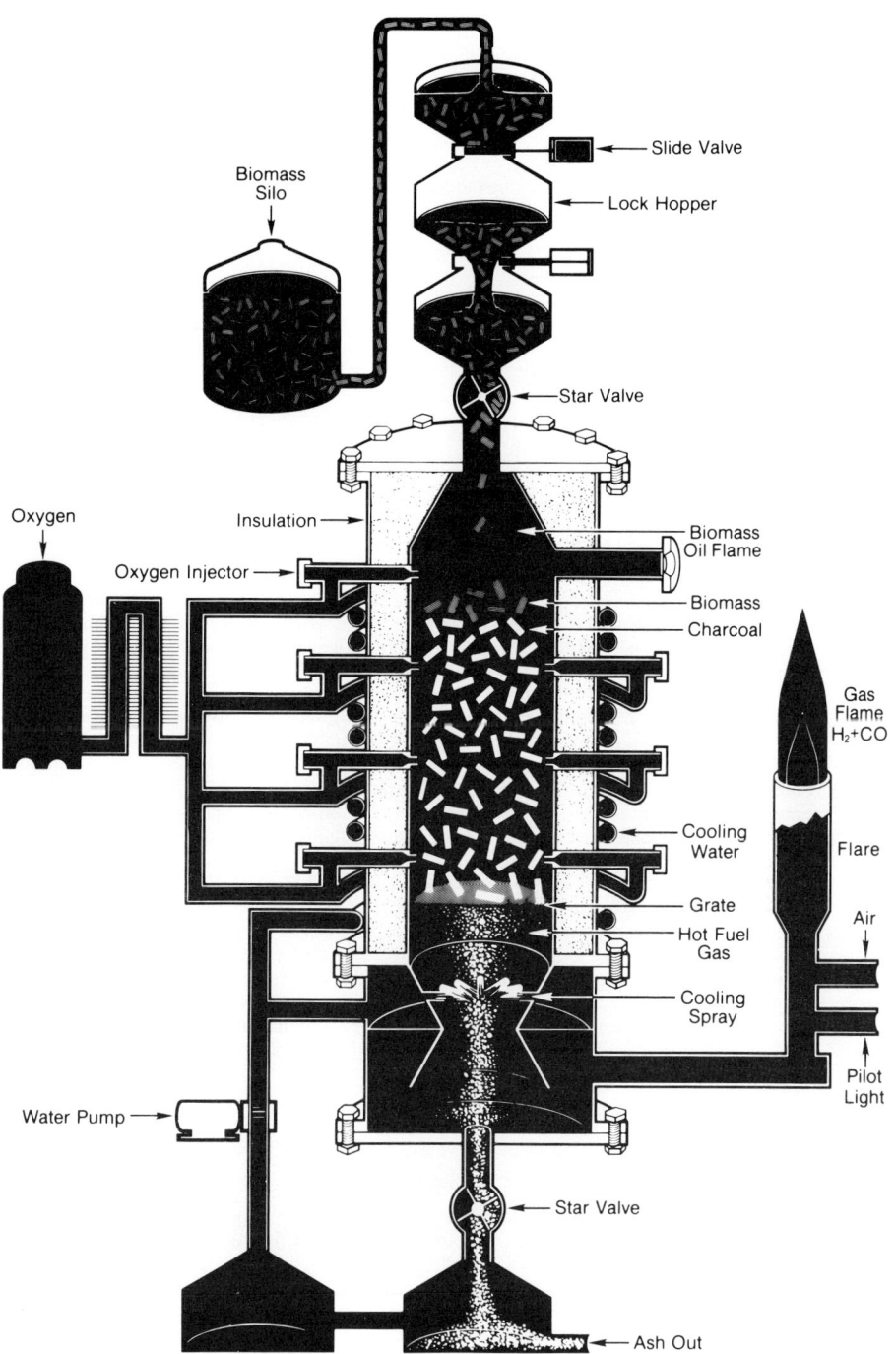

Fig. 11.3-22   SERI downdraft gasifier.

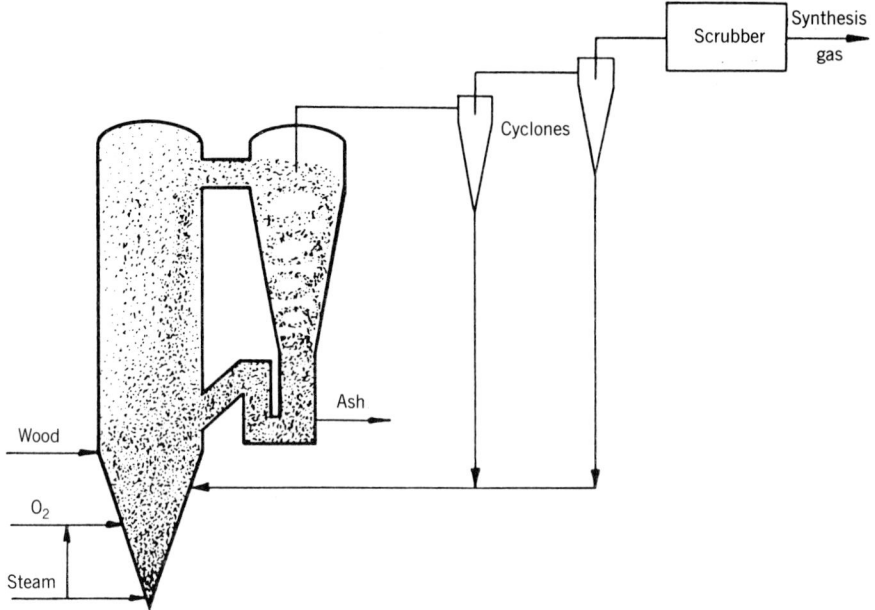

Fig. 11.3-23 Oxygen gasification in a circulating fluidized bed, the Lurgi process.

releases a fresh charge to the pressure hopper. After the slide valve closes, the pressure in the middle hopper is released to the atmosphere and the upper valve opens, releasing the wood chips into the fuel hopper. Numerous alternative pressure feeding systems are under consideration for biomass processing because biomass feeding problems are substantially more difficult than those of coal.

Both oxygen consumption and the ratio of hydrogen to carbon monoxide are affected adversely by moisture in the wood chips. Economical operation probably requires investment in wood-drying equipment that could make use of relatively low grade heat.

Crusot–Loire Process.[73] A single fluidized bed is employed in an oxygen gasification mode at approximately 15 atm pressure to achieve a synthetic gas suitable for methanol production. This process design is an extrapolation of an atmospheric pressure fluidized-bed oxygen gasification system that has been operated on wood for approximately 2 years.

Lurgi Process. A circulating fluidized bed (Fig. 11.3-23) is employed to achieve high throughputs. This type of fluidized bed has been used by Lurgi for coal combustion.[74]

Catalytic Generation of Methanol Synthesis Gas. The U.S. Department of Energy is developing a process in which hydrocracking catalysts are employed to attain specified ratios of hydrogen to carbon monoxide in biomass gasification reactions.[70] A pilot demonstration unit (Fig. 11.3-24) has demonstrated the technical feasibility of the concept, but catalyst lifetimes still are insufficient (1983).

11.3-6 ETHANOL

Edward S. Lipinsky

Conversion of carbohydrates into ethanol is a well-established technology that is evolving rapidly to achieve greater energy efficiency and to make possible the use of less expensive feedstocks. By-products credits and seasonality considerations play important roles in determining which technology is viable in specific circumstances.

Biomass Properties Required for Ethanol Production

The key reaction in the production of ethanol from biomass is shown below.

$$(C_6H_{12}O_6) \rightarrow 2C_2H_5OH + 2CO_2 \qquad (11.3\text{-}7)$$
$$\text{Hexose} \qquad\quad \text{Ethanol}$$

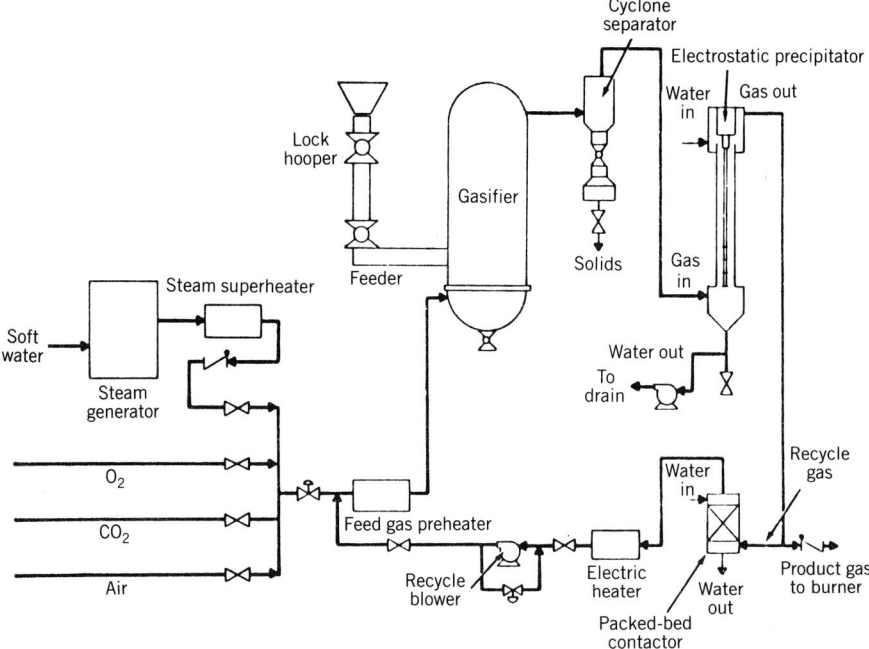

Fig. 11.3-24 Pacific northwest methanol reactor design (from ref. 70).

The overall reaction appears simple but more than 20 distinct enzymatic reactions occur during the transformation. Five-carbon sugars also are fermented to ethanol but by a more indirect pathway. Frequently the simple sugar is not a native constituent of the biomass resource but must be made by depolymerization of cellulose, starch, or inulin. Sucrose and lactose are examples of disaccharides that also are fermentable to ethanol. Lignin, triglycerides, and terpenoids are not convertible into ethanol by fermentation.

Elements of the Technology

As shown in Fig. 11.3-25, glucose or other simple sugars are generated either by enzymatic or acid-catalyzed reactions of the naturally occurring polymers. The simple sugars then are converted to ethanol and carbon dioxide by an appropriate microbial catalyst that may be either a yeast or a bacterium. This reaction occurs in relatively dilute solutions and at temperatures that are confined to rather narrow ranges in which the microorganism functions best. Ethanol is very soluble in water and

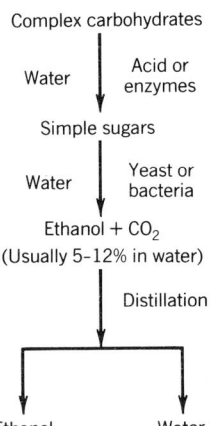

Fig. 11.3-25 Elements of ethanol production.

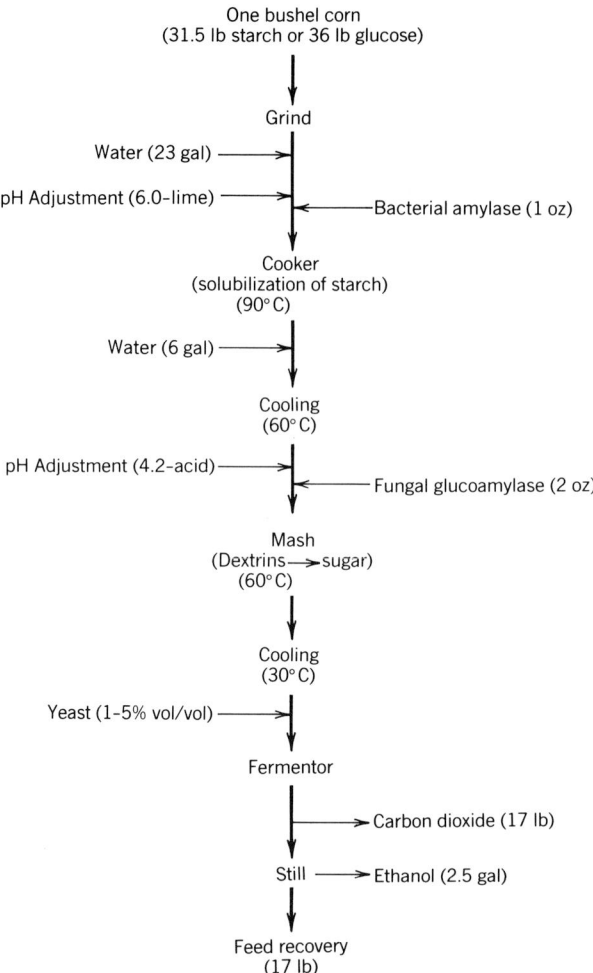

Fig. 11.3-26   Ethanol process flow diagram (from ref. 75).

separation is accomplished by distillation, although alternative methods that use solvent extraction or membrane technology are under development.

Conversion of Corn to Ethanol

Although the elements of the technology are simple to describe, the practical processes are quite sophisticated because of the need to conserve energy and to obtain useful by-products in high yields. Figure 11.3-26 illustrates the application of these principles to fermentation of dry-milled corn.[75] As the materials balance shows, roughly equal amounts of ethanol, carbon dioxide, and distillers dried grains and solubles (DDGS) are formed.

Because yeasts are not equipped with enzyme systems to convert starch to glucose, the ground corn is treated with enzymes from bacterial and/or fungal sources to convert starch to glucose. Improvements in enzyme technology have been directed toward reducing the time required for this conversion.[76] Because the fermentation time is related to the slow rate at which the disaccharides of glucose are converted to ethanol by yeast, improvements in enzyme systems that reduce disaccharide concentration have been important in fuel ethanol production.

When corn is wet milled instead of dry milled, a starch fraction is obtained that has been separated from the protein and corn oil fractions that normally appear in DDGS (Fig. 11.3-27).[77] Enzymatic or acid hydrolysis of this starch slurry yields a pumpable liquid that is well suited for use with packed-bed fermentors or other continuous fermentation systems. The fermentors can be linked in

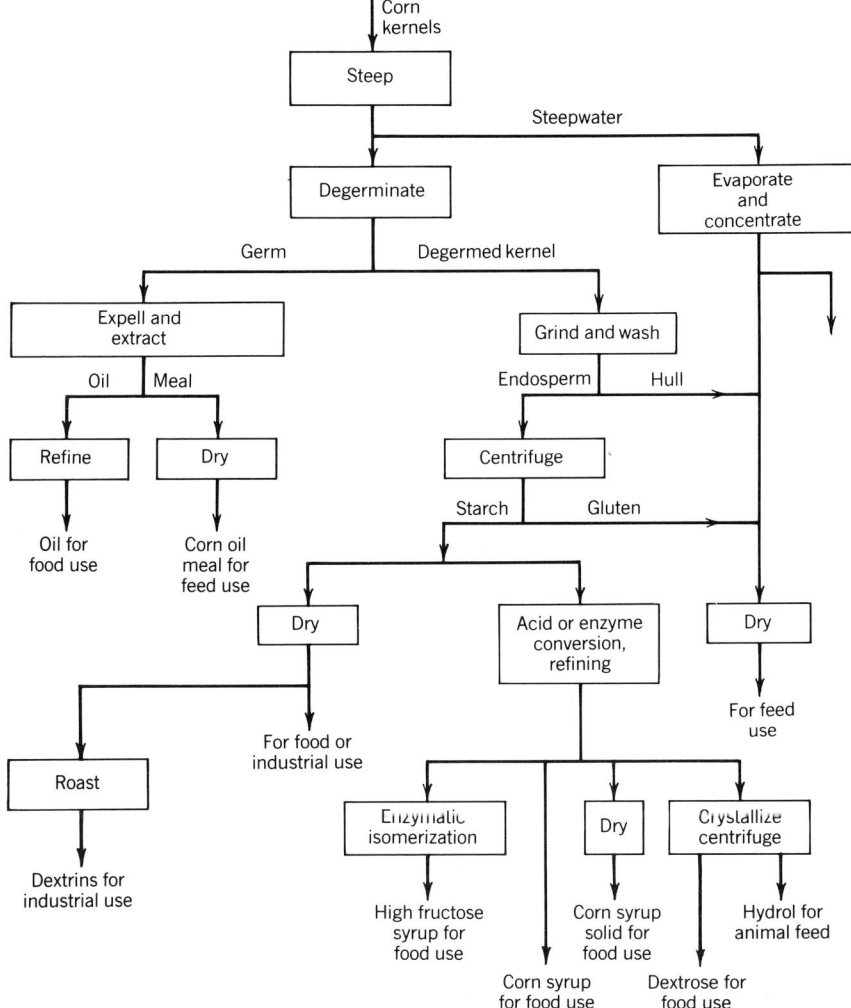

Fig. 11.3-27 Corn wet milling (from ref. 77).

series so that the yeast in the successive vats are acclimated to specific sugar/ethanol concentrations. This cascade approach can be used either with dry or wet milling of grain.[78]

The choice of a fuel ethanol plant employing dry-milled versus wet-milled corn is a complex one that concerns the overall business strategy of the company, not simply a choice of ethanol production technology. Wet milling can provide a product line that includes starch, glucose, syrups, and high-fructose corn syrup. Addition of an ethanol plant can help to sustain production of the overall facility during cycles in prices of raw materials and end products.

The recovery of anhydrous ethanol requires that the azeotrope of ethanol with water be broken; otherwise, approximately 5% water would remain in the product. Distillation of the initial ethanol product with benzene,[79] diethylether,[80] or gasoline[81] can break the azeotrope. To reduce steam consumption in a modern facility, use is made of recovered low-pressure steam wherever possible. For example, process steam requirements are estimated to be about 30 lb steam per gallon of anhydrous ethanol for a facility that uses dry milling, enzyme conversion, and the ACR gasoline azeotrope distillation process (Fig. 11.3-28).[82]

Conversion of Sugar Crops to Ethanol

Ethanol facilities that use sugar crops (primarily sugarcane, sugar beets, or sweet sorghum) as raw materials are simpler than those that use corn because the sugar juice is directly fermentable by

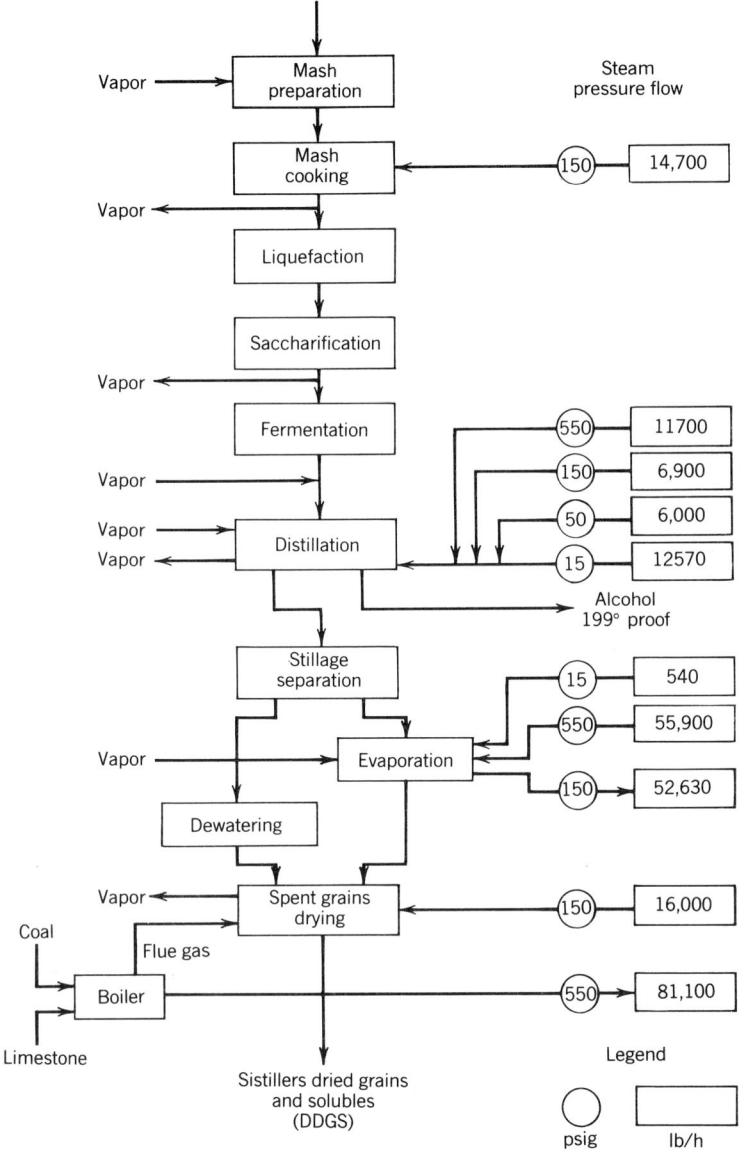

Fig. 11.3-28 Block flow diagram—steam usage in ethanol production by dry milling process.

yeasts.[83] Therefore, cooking and enzymatic hydrolysis are not needed. This technology is practiced with sugarcane on a wide range of scales of production in Brazil with considerable success. Ethanol production can be integrated with raw sugar production so that the mother liquor from the first crystallization can be converted into ethanol instead of undergoing recrystallization to achieve high raw sugar yields.[84] By-product molasses can also be converted to ethanol.

Because the fibrous portion of sugarcane (bagasse) is employed as fuel in sugarcane-based installations, the design of the distillation and stillage treatment operations is optimized to stay within the energy available from this indigenous resource. This design requirement contrasts with that of a corn conversion unit that uses purchased fuel. In the corn-based plant reduction of fuel consumption is a major design goal.

The stillage from sugarcane does not have high levels of protein, nutritious vegetable oil, or vitamins that stillage from corn processing has. Sugar crop stillage can be used as a fertilizer, low-quality fuel, or relatively low value additives for cattle rations.[85] It is also possible to anaerobically digest this stillage for production of methane as a supplementary fuel.

Conversion of Lignocellulose to Ethanol

Corn and sugarcane ethanol technologies are at the commercial scale with many plants with capacities of more than 20 million gallons per year, but conversion of lignocellulose to ethanol remains at the research and development stage in 1983. There are ethanol-from-wood facilities in the USSR, but these operate only because of special economic circumstances. Successful commercialization of technology to convert wood to ethanol would make available quantities of fuel ethanol that are orders of magnitude beyond what can be made available from grains and sugar crops. Many approaches are being explored by hundreds of research organizations throughout the world.

Relevant Properties of Wood

As shown previously in Table 11.3-22, woody biomass consists primarily of cellulose, hemicellulose, and lignin. Lignin cannot be converted to ethanol; therefore, its presence in wood reduces the potential yield of ethanol. However, the lignin can supply energy because of its high energy content or it can be a by-product that increases the revenue of the venture.

The three polymers that comprise lignocellulose form an interpenetrating network that is difficult to handle. Lignin protects the carbohydrates by shielding them from enzymes and providing a hydrophobic surface to keep out aqueous acids. There are chemical bonds between hemicellulose and lignin. For these reasons, it is especially dangerous to extrapolate from experiments done on purified cellulose or hemicellulose to processes that use native lignocellulose.

Cellulose, like starch, is a polymer of glucose. However, cellulose is much more refractory in its hydrolysis reactions to glucose than is starch. Cellulose is a partly crystalline polymer that resists enzymatic attack because the linkages that need to be broken are relatively inaccessible to the bulky enzymes that perform the hydrolysis.[86] When attacked by hot acid, the amorphous portions of the cellulose molecule are rapidly degraded to glucose, but the rate is slow for the crystalline portions. By the time the crystalline portions have been depolymerized to glucose, the glucose formed from the amorphous regions has had time to undergo side reactions.[86] Some of these side reactions create chemicals that are toxic to the microorganisms employed in making ethanol from glucose.

The chemical composition of hemicellulose depends on whether it is derived from hardwoods (deciduous trees) or softwoods (conifers). Hardwood hemicellulose primarily is composed of polymers of xylose, a 5-carbon sugar. Softwood hemicellulose consists primarily of 6-carbon sugars. Both types have uronic acid monomer units as well.

Hemicelluloses are much more readily depolymerized than is cellulose. However, there are no commercially viable processes for production of ethanol from 5-carbon sugars. Intensive research and development is under way to overcome the several drawbacks of fermentation systems that use 5-carbon sugars as raw materials (see below).

Conversion of Cellulose to Glucose

There are many variations of acid hydrolysis and enzymatic hydrolysis. Only the major aspects are outlined here. The reader is referred especially to Goldstein's[87] and Bungay's[88] monographs for more information.

Acid Hydrolysis. When cellulose or wood is treated with dilute acid (e.g., 0.5% sulfuric acid) at high temperatures in an autoclave, the cellulose is converted to glucose. It is probably best to remove the hemicellulose first so that degradation of the 5-carbon sugars does not interfere with the hydrolysis of cellulose. Development efforts on this approach are aimed at increasing yields that are hampered by the refractory nature of the crystalline cellulose. Another important objective is reduction of materials toxic to the yeast to be employed in ethanol production.

Concentrated acid processes make use of a combination of hydrolytic and solvent properties of concentrated hydrochloric acid.[87] Glucose is not formed under these conditions; a low molecular weight polymer of glucose is formed that can be converted to glucose in a separate facile reaction. Glucose yields by this method are high because the solvent power of hydrochloric acid overcomes the crystallinity barrier of cellulose. Ingenuity is required to overcome the cost of hydrochloric acid and its corrosive properties.

Enzymatic Hydrolysis. The cellulase enzyme complex produced by *Trichoderma reesii* employs several enzymes that cooperate to convert crystalline cellulose into amorphous cellulose, amorphous cellulose to cellobiose, and cellobiose to glucose. Frequently cellobiose builds up because of insufficient quantities of one of the enzymes (cellobiase) in the mixture. Intensive research and development efforts are under way to produce more powerful enzyme complexes and reduce the cost of enzyme production.[86] The enzyme complex clings to the woody fibers, and engineering approaches to recycling enzymes play an important role in commercialization efforts.

Enzymatic conversion is slow, usually involving more than 24 h of reaction time. In addition to the steric problems of using large enzymes to conduct the hydrolysis, the glucose end product tends to stick to the final enzyme inhibiting the hydrolysis.

Conversion of Glucose to Ethanol

The major differences between fermentation of wood-derived glucose and fermentation of sugarcane or corn usually center on the following problems:

1. Enzymatic hydrolysis usually yields quite dilute solutions.
2. Glucose obtained by acid hydrolysis contains other small molecules, (e.g, furfural).

A more dilute solution implies a high requirement for steam for distillation. The presence of some furfural or similar materials in the fermentation medium can be overcome by selecting (or genetically engineering) yeasts or bacteria that are not so sensitive to these impurities. This alternative may prove commercially more viable than achieving development of an acid-based process that yields none of these by-products.

SSF Process. The end product inhibition of glucose that retards enzymatic hydrolysis of cellulose could be overcome by an innovative process that is in the early states of commercialization. The SSF process (simultaneous saccharification and fermentation)[89] places yeast in the vessel in which cellulase enzymes are hydrolyzing cellulose. As the glucose is formed, it is promptly converted to ethanol, removing the cause of the end product inhibition.

Direct Fermentation. There are bacteria of the genus *Clostridia* that are capable of producing ethanol directly from cellulose, hemicellulose, or a mixture of both carbohydrates.[90] The yields and rates appear acceptable, but the ethanol concentration is rather low. Although this process circumvents many of the problems of lignocellulose hydrolysis and fermentation, the *Clostridia* mutants that have been examined thus far appear to be highly inhibited by the presence of lignin. Therefore, this process appears to require a more thorough-going lignin removal than the simple disruptions required for many alternative processes.

Conversion of Hemicellulose to Ethanol

Large-scale commercialization of fuel ethanol from lignocellulose is greatly enhanced in attractiveness if both the cellulose and the hemicellulose can be converted to ethanol. Except for the direct fermentation process referred to above, the hemicellulose fraction has proved to be even less tractible than the cellulose. When *P. tannophilus* is employed as the microbial catalyst for ethanol formation from xylose, the following drawbacks are observed:[91]

1. The xylose must be present in very low concentrations.
2. The microorganism is sensitive to dilute ethanol solutions.
3. The yield of ethanol from xylose is far below the yield from glucose.

A combination of enzymatic treatment and fermentation can greatly improve the rate of xylose fermentation and the concentration of ethanol achieved.[92] The superior results need to be considered in the context of the added cost of enzyme production and use. However, the enzyme that has been suggested is the same that is used on a very large scale to convert glucose to fructose.

REFERENCES

11.3-1 C. H. Norwood and W. L. Warnick, "Consumption of Wood Fuels in the United States 1971–1980," in *Progress in Biomass Conversion*, Vol. 3, Academic Press, New York, 1982, pp. 129–182.

11.3-2 D. A. Tillman, R. F. Schnorr, and J. W. Sale, "Alternative Cogeneration Systems Employing Biomass as a Fuel: an Incremental Analysis of Heat Rates," in *Progress in Biomass Conversion*, Vol. 3, Academic Press, New York, 1982, pp. 105–127.

11.3-3 A. J. Rossi, Personal communication, November 1982.

11.3-4 Forest Products Laboratory, USDA, *Wood Handbook: Wood as an Engineering Material*, U.S. Government Printing Office, Washington, D.C., 1974.

11.3-5 D. A. Tillman, A. J. Rossi, and W. D. Kitto, *Wood Combustion: Principles, Processes, and Economics*, Academic Press, New York, 1981.

11.3-6 Babcock and Wilcox, *Steam: Its Generation and Use*, 39th ed. Babcock and Wilcox, New York, 1978.

11.3-7 A. Sliepcevich, et al., *Assessment of Technology for the Liquefaction of Coal*, National Research Council, National Academy of Sciences, Washington, D.C., 1977.

11.3-8 J. Edwards, *Combustion: Formation and Emission of Trace Species*, Ann Arbor Science Publishers, Ann Arbor, MI, 1977.

11.3-9 F. Shafizadeh, "Chemistry of Pyrolysis and Combustion of Wood," in *Progress in Biomass Conversion*, Vol. 3, Academic Press, New York, 1982, pp. 51–76.

11.3-10 D. A. Tillman, "Review of Mechanisms Associated with Wood Combustion," *Wood Science* **13**(4), 177–184 (1981).

11.3-11 D. C. Junge, *Design Guideline Handbook for Industrial Spreader Stoker Boilers Fired with Wood and Bark Residue Fuels*, Oregon State University, Corvallis, OR, 1979.

11.3-12 O. A. Hougen, K. M. Watson, and R. A. Ragatz, *Chemical Process Principles*, part 1, 2nd ed., Wiley, New York, 1954.

11.3-13 D. A. Tillman and L. L. Anderson, "Computer Modelling of Wood Combustion with Emphasis on Adiabatic Flame Temperature." Presented at the Ninth Cellulose Conference, Syracuse, NY, May 26, 1982.

11.3-14 D. C. Junge, *Boilers Fired with Wood and Bark Residues*, Forest Products Laboratory, Oregon State University, Corvallis, OR, 1975.

11.3-15 F. Shafizadeh and W. F. Degroot, "Thermal Analysis of Forest Fuels," in *Fuels and Energy from Renewable Resources*, Academic Press, New York, 1977, pp. 93–114.

11.3-16 Envirosphere Co., "Program Negative Declaration Concerning the Biomass Demonstration Program," California Energy Commission, Ebasco Services, Inc., Newport Beach, CA, 1980.

11.3-17 B. Schweiger, "Power from Wood," *Power* **124**(2), S.1–S.32 (1980).

11.3-18 R. L. Jamison, "Wood Fuel Use in the Forest Products Industry," in *Progress in Biomass Conversion*, Vol. 1, Academic Press, New York, 1979, pp. 27–52.

11.3-19 Envirosphere Co., "The Feasibility of Wood Waste Fired Cogeneration in Hulett, Wyo." A report to Wyoming State Forestry Division, Cheyenne, WY, Ebasco Services, Inc., Bellevue, WA, 1981.

11.3-20 L. L. Anderson, "A Comparison of Biomass and Coal as Feedstocks for Synthetic Fuels," in *Progress in Biomass Conversion*, Vol. 3, Academic Press, New York, 1982, pp. 215–240.

11.3-21 W. R. Smith, Personal communication of laboratory analysis of rice hulls and cotton gin trash, February 27, 1981.

11.3-22 W. F. Lipfert and J. L. Dungan, "Residential Firewood Use in the United States," *Science* **219**, March 25, 1983.

11.3-23 U.S. Department of Energy, "Estimates of U.S. Wood Energy Consumption from 1949 to 1981," Washington, D.C. August 1982.

11.3-24 J. Shelton, *The Woodburners Encyclopedia*, Section I. Vermont Crossroads Press, Waitsfield, VT, 1976.

11.3-25 J. H. Allen and W. M. Cooke, United States Environmental Protection Agency, Final Report on "Control of Emissions from Residential Wood Burning by Combustion Modification," Battelle Columbus Laboratories, May 1981.

11.3-26 E. R. Frederick (ed.), Air Pollution Control Association, Proceedings on Residential Wood and Coal Combustion, Specialty Conference March 1 and 2, 1982, Louisville, KY, Published by Air Pollution Control Association, Pittsburgh.

11.3-27 D. L. Klass, "Fuels from Biomass," Kirk-Othmer *Encyclopedia of Chemical Technology*, 3rd ed., Vol. 11, Wiley, New York, 1980.

11.3-28 J. A. Chandler, et al., "Predicting Methane Fermentation Biodegradability," paper presented at the Second Symposium on Biotechnology in Energy Production and Conservation, October 3, 1979. Gatlinburg, TN.

11.3-29 D. L. Klass and S. Ghosh, "Methane Production by Anaerobic Digestion of Water Hyacinth (Eichhornia Crassipes) in *Fuels from Biomass and Wastes*, D. L. Klass and G. H. Emert (eds.), Ann Arbor Science Publishers, Ann Arbor, MI, 1981, pp. 129–149.

11.3-30 D. L. Klass, S. Ghosh, and D. P. Chynoweth, "Methane Production from Aquatic Biomass by Anaerobic Digestion of Giant Brown Kelp," *Process Biochem.* **14**(4) 18–22 (1979).

11.3-31 D. L. Klass, S. Ghosh, and J. R. Conrad, "The Conversion of Grass to Fuel Gas for Captive Use," Symposium Papers, Clean Fuels from Biomass, Sewage, Urban Refuse, Agricultural Wastes, pp. 229–52, Institute of Gas Technology, Orlando, FL, January 27–30, 1976.

11.3-32 D. L. Klass and S. Ghosh, "Methane Production by Anaerobic Digestion of Bermuda Grass," *Biomass as a Nonfossil Fuel Source*, D. L. Klass (ed.), ACS, Washington, D.C., 1981, pp. 229–249.

11.3-33 S. Ghosh, M. P. Henry, and D. L. Klass, "Bioconversion of Water Hyacinth-Coastal Bermuda Grass—MSW Sludge Blends to Methane," *Biotech and Bioeng.*, Symposium No. 10, 163–187 (1980).

11.3-34 D. L. Klass et al., "Brazilian BIOGAS®," Final Report, Project 8896, Institute of Gas Technology, Chicago, IL, October 1978.

11.3-35 D. P. Chynoweth, "Microbial Conversion of Biomass to Methane," Proceedings of the Eighth Annual Energy Technology Conference, Washington, D.C., March 9–11, 1981.

11.3-36 W. J. Jewell et al., "Low Cost Methane Generation on Small-Farms," Third Annual Biomass Energy Systems Conference Proceedings, sponsored by The U.S. Department of Energy, June 5–7, 1979. Golden, CO, pp. 547–80.

11.3-37 T. P. Abeles, D. F. Freedman, L. A. DeBaere, and D. A. Ellsworth, "Energy and Economic Assessment of Anaerobic Digesters and Biofuels for Rural Waste Management," report prepared for the Environmental Protection Agency, Contract R-804-457-010, June 1978.

11.3-38 S. Ghosh, J. R. Conrad, and D. L. Klass, "Anaerobic Acidogenesis of Wastewater Sludge," *J. Water Pollut. Cont. Fed.* **47**(1), 30–45, January 1975.

11.3-39 S. Ghosh and D. L. Klass, "Two-Phase Anaerobic Digestion," *Proc. Biochem.* **13**, 15–24, April 1978.

11.3-40 J. Norrman and B. Frostell, "Anaerobic Waste Water Treatment in a Two-Stage Reactor of New Design," paper presented at the 1977 Industrial Waste Conference, Purdue University, May 10, 1977, Lafayette, IN.

11.3-41 A. Cohen et al., "Anaerobic Digestion of Glucose with Separated Acid Production and Methane Formation," *Water Res.* **13**, 571–80 (1979).

11.3-42 W. W. Pitt and R. K. Genung, "Energy Conservation and Production in a Packed-Bed Anaerobic Bioreactor," in Symposium Papers, Clean Fuels from Wastes and Biomass, Institute of Gas Technology, Chicago, IL, January 21–25, 1980, Lake Buena Vista, FL.

11.3-43 W. J. Jewell, "Development of the Attached Microbial Film Expanded Bed Process for Aerobic and Anaerobic Waste Treatment," paper presented at the Conference on Biological Fluidized Bed Treatment of Water and Wastewater, April 14–17, 1980, Hertfordshire SGL-1TH, England.

11.3-44 W. J. Jewell, M. S. Switzenbaum, and J. W. Morris, "Sewage Treatment with the Anaerobic Attached Microbial Film Expanded Bed Process," paper presented at 52nd WPCF Conference, October 11, 1979, Houston.

11.3-45 D. P. Chynoweth, V. J. Srivastava, M. P. Henry, and P. B. Tarman, "Biothermal Gasification of Biomass," Energy from Biomass and Wastes, Lake Buena Vista, FL, January 21–25, 1980, Institute of Gas Technology, Chicago.

11.3-46 D. L. Wise, C. L. Cooney, and D. C. Augenstein, "Biomethanation: Anaerobic Fermentation of CO_2, H_2, and CO to Methane," *Biotechnology and Bioengineering* **20**, 1153–1172 (1978).

11.3-47 A. G. Hashimoto, Y. R. Chen, and R. P. Prior, "Thermophilic Anaerobic Fermentation of Beef Cattle Residue," Energy from Biomass and Wastes, August 14–18, 1978, Washington, D.C., Institute of Gas Technology, Chicago, pp. 379–402.

11.3-48 S. Schellenbach, "Imperial Valley Bio-Gas Project Operations and Methane Production from Cattle Manure, October 1978 to November 1979," Energy from Biomass and Wastes, January 21–25, 1980, Lake Buena Vista, FL, Institute of Gas Technology, Chicago.

11.3-49 M. L. Wilkey, R. E. Zimmerman, and H. R. Isaacson, "Methane from Landfills," Preliminary Assessment Workbook Argonne National Laboratory prepared for the U.S. Department of Energy, June 1982.

11.3-50 W. J. Jewell et al., "Dry Fermentation of Agricultural Residues," Final Report, SERI/STR-231-1892, December 1982.

11.3-51 R. R. Rimkus, J. M. Ryan, and A. Michuda, "Digester Gas Utilization Program," ASCE Convention, October 22–26, 1979, Atlanta.

11.3-52 D. L. Klass, "Energy from Biomass and Wastes: 1979 Update," Energy from Biomass and Wastes, January 21–25, 1980, Lake Buena Vista, FL, Institute of Gas Technology, Chicago.

11.3-53 *Bio-Energy Directory*, P. F. Bente (ed.), p. 506, April 1980.

11.3-54 D. K. Walter, "Anaerobic Digestion of Municipal Solid Waste to Produce Methane," Energy from Biomass and Wastes VII. January 24–28, 1983, Lake Buena Vista, FL, Institute of Gas Technology, Chicago.

11.3-55 M. L. Wilkey and J. J. Walsh, "Methane Recovery from Sanitary Landfills," Proceedings of 10th Energy Technology Conference, Washington, D.C., 1983.

11.3-56 T. B. Reed (ed.), *Biomass Gasification, Principles and Technology*, Noyes Data Corp., Park Ridge, NJ, 1981.

11.3-57 E. S. Lipinsky and S. Kresovich, "Sugar Stalk Crops for Fuels and Chemicals," in *Progress in Biomass Conversion*, Vol. 2, Sarkanen and Tillman (eds.), 1980.

11.3-58 Proceedings of Specialists' Workshop on Fast Pyrolysis of Biomass, Solar Energy Research Institute, SERI/CP622-1096, NTIS, Springfield, VA, 1981.

11.3-59 J. L. Jones and S. B. Radding, *Thermal Conversion of Solid Wastes and Biomass*, ACS Symposium Series 130. ACS, Washington, D.C., 1980.

11.3-60 Office of Technology Assessment, "Energy from Biological Processes, Vol. II—Technical and Environmental Analyses," U.S. Government Printing Office, Washington, D.C., September 1980.

11.3-61 J. F. Jackson, "Design, Commercial Operation, and Costs of Two 25 Million/Btu hr

Air-Blown Wood Gasifiers," Energy from Biomass and Wastes VI, January 25–29, 1983, Lake Buena Vista, FL, Institute of Gas Technology, Chicago.

11.3-62 H. F. Feldmann, M. A. Paisley, H. R. Appelbaum, and B. C. Kim. "Steam Gasification of Forest Residues in a High-Throughput Gasifier," The American Society of Mechanical Engineers, New York, 1982.

11.3-63 J. W. Black, "The Pressurized Fluidized Bed Gasifier in the Synthesis of Methanol from Wood," from Proceedings of Biomass-To-Methanol Specialists' Workshop, T. B. Reed and M. Graboski (eds.), 1982.

11.3-64 K. G. Bircher, "Economics of an 80 MM Btu/Hr Wood Gasifier Installation," Energy from Biomass and Wastes VI, January 25–29, 1982, Lake Buena Vista, FL, Institute of Gas Technology, Chicago, pp. 721–735.

11.3-65 R. C. Bailie, "Results from Commercial-Demonstration Pyrolysis Facilities (35–45 tons/day refuse) Extended to Producing Synfuels from Biomass," Energy from Biomass and Wastes V, January 26–30, 1981, Lake Buena Vista, FL, Institute of Gas Technology, Chicago, pp. 549–569.

11.3-66 U.S. Department of Energy, "Proceedings of Third International Symposium on Alcohol Fuels Technology," May 29–31, 1979, Asilomar, CA, 1980.

11.3-67 Proceedings of the IV International Symposium on Alcohol Fuels Technology, October 5–8, IPT, Sao Paulo, Brazil, 1980.

11.3-68 E. S. Lipinsky, D. A. Ball, and D. Anson, "Evaluation of Biomass Systems for Electricity Generation," prepared for Electric Power Research Institute (EPRI AP-2265) Palo Alto, CA, February 1982.

11.3-69 Proceedings Biomass-to-Methanol Specialists, Workshop, Tamarron Durango, CO, March 3–5, 1982, T. B. Reed and M. Graboski (eds.), SERI Report No. SERI/CP 234-1590.

11.3-70 L. K. Mudge et al., "Catalytic Generation of Methanol Synthesis Gas from Wood," Proceedings Biomass-to-Methanol Specialists' Workshop, Tamarron Durango, CO, March 3–5, 1982, J. B. Reed and M. Graboski (eds.), SERI Report—SERI/CP 234-1590.

11.3-71 K. L. Rock, "Production of Methanol from Mixed Synthesis Gas Derived from Wood and Natural Gas," Energy from Biomass and Wastes VI, January 25–29, 1982, Lake Buena Vista, FL, Institute of Gas Technology, Chicago, pp. 737–761.

11.3-72 T. B. Reed and M. Markson, "The SERI High Pressure Oxygen Gasifier," Proceedings Biomass-to-Methanol Specialists' Workshop, Tamarron Durango, CO, March 3–5, 1982, J. B. Reed and M. Graboski (eds.), SERI Report—SERI/CP 234-1590, pp. 151–178.

11.3-73 G. Chrysostome and J. M. LeMasle, "Pressurized Oxygen Blown Fluidized Bed Wood Gasifier," in *Energy from Biomass*, 2nd E. C. Conference, A. Strub, P. Chartier, and G. Schleser, (eds.), Berlin, September 1982.

11.3-74 A. Beenackers, "The Methanol from Wood Pilot Plant Program of the EC," Energy from Biomass and Wastes VII, January 24–28, 1982, Institute of Gas Technology, Chicago.

11.3-75 R. J. Bothast and R. W. Detroy, "What is Alcohol? How is it Made?" Northern Agricultural Energy Center, in *Alcohol and Vegetable Oil as Alternative Fuels Proceedings of Regional Workshops*, April 1981, pp. 31–37, USDA, Peoria, IL.

11.3-76 B. E. Norman and N. W. Lutzen, "Process Considerations for the Production of Ethanol from Cereals," in *Cereals: a Renewable Resource*, Y. Pomeranz and L. Munck, (eds.), 1981, pp. 651–665, NOVO Research Institute, Denmark.

11.3-77 E. S. Lipinsky, W. J. Sheppard, J. L. Otis, E. W. Helper, T. A. McClure, and D. A. Scantland, "Systems Study for Fuels for Sugarcane, Sweet Sorghum, Sugar Beets, and Corn, Vol. V: Comprehensive Evaluation of Corn," Battelle-Columbus Laboratories for DOE under Contract No. W-7405-eng-92, NTIS Report No. BMI-1975A-V5/LL, 1977.

11.3-78 A. S. Allen, "Full-Scale Continuous Fermentation for the Commercial Production of Fuel Ethanol," Energy from Biomass and Wastes VI, January 25–29, 1982, Buena Vista, FL, Institute of Gas Technology, Chicago, pp. 957–869.

11.3-79 O. Leppanen et al., "Energy Consumption in the Distillation of Fuel Alcohol," Proceedings of the IV International Symposium on Alcohol Fuels Technology, October 1980, IPT, São Paulo, Brazil, pp. 113–117.

11.3-80 D. R. Miller, "Development of Advanced, Commercial Cellulose-to-Ethanol Continuous Pretreatment, Fermentation, and Distillation Systems," Energy from Biomass and Wastes VI, January 25–29, 1982, Buena Vista, FL, Institute of Gas Technology, Chicago, pp. 833–856.

11.3-81 R. S. Chambers, R. A. Herendeen, J. J. Joyce, and P. S. Penner, "'Gasohol' Does it or Doesn't It Produce Positive Net Energy?" *Science*, **206**, p. 789. November 16, 1979.

11.3-82 J. N. Wade, "Kentucky's New Generation Power Alcohol Distillery," Energy from Biomass and Wastes VII, January 24–28, 1983, Buena Vista, FL. Institute of Gas Technology, Chicago.

11.3-83 E. S. Lipinsky and S. Kresovich, "Sugar Crops as a Solar Energy Converter," in *Experientia*, Supplement, Vol. 43, New Trends in Research and Utilization of Solar Energy Through Biological Systems, H. Mislin and R. Bachofen (eds.), 1982, pp. 19–23.

11.3-84 P. M. A. M. Chenu, "Alcohol Manufacture in a Sugar Factory," Proceedings of the International Society of Sugar Cane Technologists, XVI Congress, ISSCT, São Paulo, Brazil, 1978, pp. 3241–3251.

11.3-85 J. M. Paturau, *By-Products of the Cane Sugar Industry, An Introduction to Their Industrial Utilization*, Elsevier Publishing, New York, 1982.

11.3-86 R. D. Brown, Jr., and L. Jurasek, *Hydrolysis of Cellulose; Mechanisms of Enzymatic and Acid Catalysis*, Advances in Chemistry Series 181, ACS, 1979.

11.3-87 I. S. Goldstein (ed.), *Organic Chemicals from Biomass*. CRC Press, Boca Raton, FL, 1981.

11.3-88 H. R. Bungay, *Energy, The Biomass Options*, Wiley, New York, 1981.

11.3-89 G. H. Emert and R. Katzen, "Gulf's Cellulose-to-Ethanol Process," *Chemtech* **10**(10) 610–614 (1980).

11.3-90 G. C. Avgerinos et al., "A Novel, Single-Step Microbial Conversion of Cellulosic Biomass to Ethanol," *Advances in Biotechnology*, Vol. II, M. Moo-Young (G. ed.), C. W. Robinson, (ed.), Pergamon Press, Canada, 1981, pp. 119–124.

11.3-91 R. W. Detroy, R. L. Cunningham, R. J. Bothast, M. O. Bagby, and A. Herman, "Bioconversion of Wheat Straw Cellulose/Hemicellulose to Ethanol by *Saccharomyces warum* and *pachysolen tannophilus*," *Biotechnology and Bioengineering*, Vol. 24, Wiley, New York, 1982, pp. 1105–1113.

11.3-92 C. S. Gong, L. F. Chen, M. C. Flickinger, L. C. Chiang, and G. T. Tsao, *Appl. Environ. Microbiol.*, **41**, 430 (1981).

11.4 OCEAN ENERGY

Paul C. Yuen

Useful energy can be obtained from several properties of the ocean: tides, currents, waves, winds, salinity gradients, and thermal gradients. The use of tides, currents, waves, and winds has been studied for a long time and a number of ideas have been proposed and tried. The use of salinity gradients is in an early stage of research.

The most promising method for commercial electric power generation is the utilization of the temperature difference that exists between the warm surface waters of the temperate and tropic oceans and their deep cold waters. Large research and development efforts are being carried out by several countries, and base-load power from this source could be available before the end of the century.

11.4-1 Energy from Tides

Tidal power plants are similar to conventional hydroelectric power plants except that they must operate with lower pressure head and with water flowing in two opposite directions. The energy in the tides can be captured by forming one or more basins and using the ebb and flood of the tides to drive conventional or reversible electric generating units. Propeller design for the turbines is generally conventional except that they work with lower head. A major problem with tidal power systems is that the extractable power varies with time, so even though it is predictable, it essentially falls to zero at certain times of the day.

Estimates of the costs of tidal power plants show that they are not competitive with fossil fuel plants at today's fuel prices. Capital cost is estimated at $2500–3500/kW of generating capacity and the cost of electricity is estimated at 56–106 mills/kW · h. Another problem is that suitable tidal changes are found in only a few locations.

11.4-2 Energy from Ocean Current

The extraction of energy on a large scale from ocean currents is still in an experimental state. The principle basically is similar to that of extracting energy from the movement of tides; both seek to convert the kinetic energy of the ocean's movements into mechanical energy to drive an electric generator.

While propeller-driven wheels have been used for tidal power generation, various forms of water wheels have been proposed for ocean current energy extraction. Structures will be large; it has been estimated that a power plant extracting 1 MW from the Florida current would have a 30-m-diameter rotor and blades 25-m long. The mooring of such a massive unit in a fast-flowing current in deep water is technologically feasible, but its costs may be prohibitive.

11.4-3 Ocean Wave Energy

The possibility of converting the potential and kinetic energy in ocean waves has led to proposals for a large number of different energy conversion systems. An attractive concept is the use of a system that

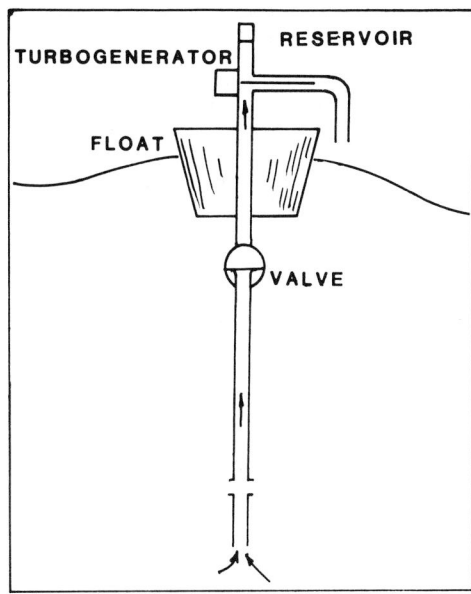

Fig. 11.4-1   Issacs' wave-powered generator.

can also function as a breakwater to absorb wave energy and to provide calm water in a harbor. No system, however, has proven to be economically feasible to date.

One design that has been used is the Issacs' generator (Fig. 11.4-1). The action of the waves forces water up through the pipe and into a reservoir. The check valve allows water only to move up the pipe. The water in the reservoir is allowed to flow through a turbogenerator, providing electricity. A pipe 90 m long and 0.9 m in diameter can produce up to 40 kW of power.

11.4-4   Energy from Ocean Winds

Once the primary source of energy for ocean-going vessels, ocean winds are a well-known resource. The advent of cheap petroleum and steam-driven engines marked the end of dependence on intermittent wind power. Recently the higher price of petroleum has led to a resurgence of interest in wind as an energy source. Sail-assisted ships (Fig. 11.4-2), ships with canvas or metal folding sails to

Fig. 11.4-2   Sail-assisted ship.

utilize available winds, could reduce fuel consumption and lower operating costs and are therefore gaining attention.

11.4-5 Energy from Salinity Gradients

Another method of extracting energy from the ocean is the use of salinity gradients such as occur near the mouths of rivers. When two solutions of different salinities are separated by a semipermeable membrane, the liquid with the lower salinity will flow across the membrane into the solution with the higher salinity, causing an increase in pressure on the side of the solution with the higher salinity. The pressure difference can then be used to obtain mechanical work and hence electric energy. This system is still in the early research stages, and commercial application is distant.

11.4-6 Ocean Thermal Energy Conversion (OTEC)

General Principle

The most promising method for commercial electric power generation utilizes the temperature differences between warm surface water and deep cold water. This is called ocean thermal energy conversion (OTEC). OTEC is a solar technology that produces electricity using a heat engine. The heat source is the warm surface ocean waters (27°C), which absorb the sun's radiation, and the heat sink is the approximately 1-km deep cold waters (4°C), which do not. The temperature difference, ΔT, is critical to the amount of power such an OTEC system is capable of generating. With a temperature difference of 22°C the maximum theoretical efficiency achievable is only 7%, and practical efficiencies may be as low as 2%. This low efficiency means large structures and capital costs; and the harsh ocean environment causes major problems in platform mooring, pipe design, and cable design and deployment.

OTEC is an attractive energy resource because it is a solar option capable of producing base-load power due to built-in self-storage. It is an enormous renewable resource and incurs no fuel costs or fuel dependency. OTEC also has a highly favorable net energy ratio, is probably environmentally benign, and requires no waste disposal, fresh water, or land area for offshore systems. Varied energy products are possible.

Resource Estimates

The ocean thermal resource for OTEC is enormous, even if it is limited to areas with a ΔT of at least 19°C. Because the highest incident solar radiation falls at the equator, most of the best OTEC areas are tropical. Figures 11.4-3 and 11.4-4 show the global annual average ΔT contours for 1-km depths. There may be 50 million square kilometers of ocean with suitable ΔT for OTEC, and with an assumed 0.8 MW/km² production rate, approximately 40 million megawatts can be produced.

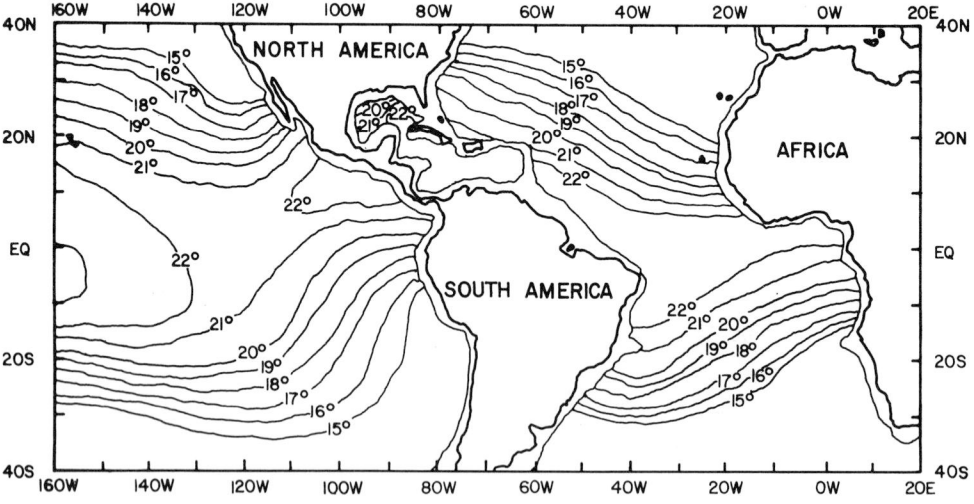

Fig. 11.4-3 OTEC thermal resource—ΔT (°C) between surface and 1000-m depth.

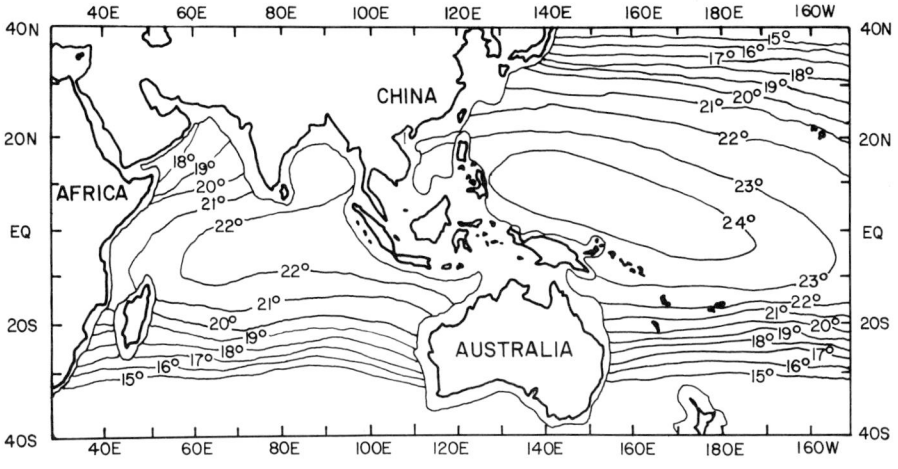

Fig. 11.4-4 OTEC thermal resource—ΔT (°C) between surface and 1000-m depth.

Types of OTEC Cycles

Three basic types of OTEC heat engines have been studied (see Table 11.4-1). The closed-cycle system (Fig. 11.4-5) pumps warm surface seawater into an evaporator where a subcooled working fluid such as ammonia under high pressure absorbs heat and is turned into vapor. The vapor drives a turbine-generator, producing electricity. Then the vapor is condensed in a condenser by the deep ocean water from the cold water pipe. The condensed fluid is pressurized again before being fed into the evaporator, thus closing the cycle.

A second, less-tested type known as an open-cycle OTEC (Fig. 11.4-6) uses seawater as the working fluid. Warm surface water is degasified and then evaporated in a vacuum in a boiler, producing low-pressure steam which is expanded in a very large diameter turbine-generator, producing electricity. Deep cold seawater then condenses the steam in a condenser.

A third cycle still in the laboratory research stage is the lift cycle (Fig. 11.4-7), which evaporates warm surface seawater into low-pressure steam in a way that traps substantial volumes of water with it. The condenser, which is physically located above the evaporator, condenses the vapor. Potential

TABLE 11.4-1 COMPARISON OF OTEC CYCLES

Cycle	Advantages	Disadvantages
Closed	1. Small turbine and vapor passage size 2. Experimental systems operated	1. Complexity cost and power to handle working fluid 2. Requires large, costly, high-pressure heat exchangers 3. Biofouling counter measures required
Open	1. Can produce fresh water as well as electricity 2. No large heat exchangers required 3. Higher thermodynamic efficiency 4. Simpler system 5. Less biofouling 6. Higher heat transfer coefficients 7. No potentially hazardous working fluid	1. Large turbine and steam duct size 2. Deaeration of seawater needed 3. Need to maintain vacuum in huge volumes 4. Increased parasitic power requirements
Lift	1. Can use lower-cost hydraulic turbines 2. No heat exchangers needed	1. Very early stage of development

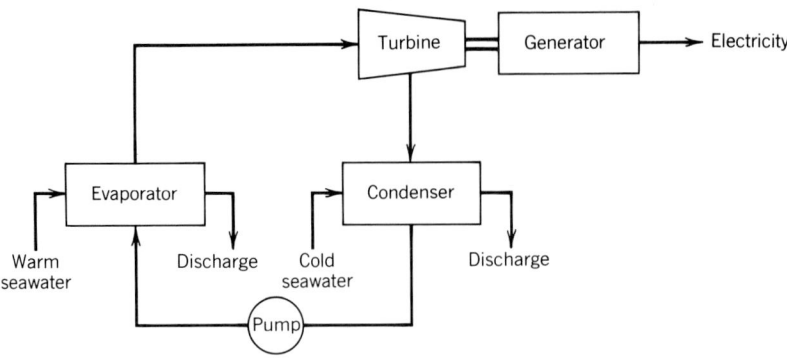

Fig. 11.4-5   Closed-cycle OTEC system.

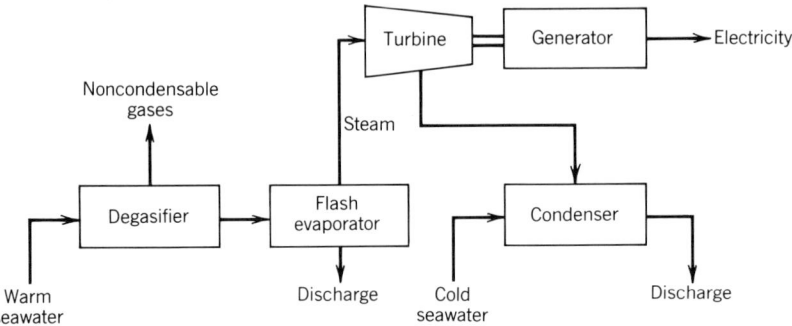

Fig. 11.4-6   Open-cycle OTEC system.

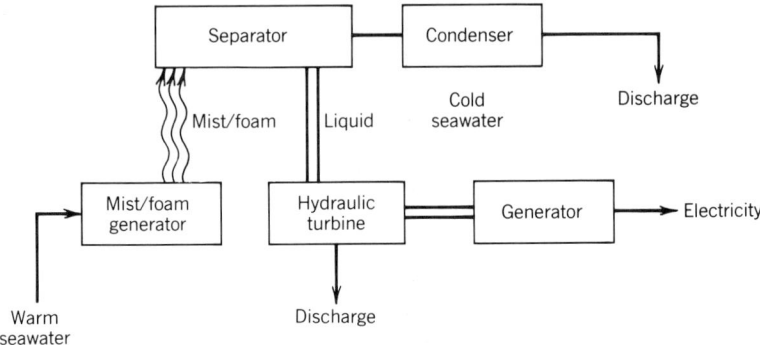

Fig. 11.4-7   Lift-cycle OTEC system.

energy of the resulting liquid, which has been raised to the condenser level, is then converted into rotational energy through the use of a hydraulic turbine-generator to produce electricity. Two approaches to the lift cycle are being pursued. The first requires the addition of a surfactant to the warm water to cause a foam. In the second the warm seawater is turned into a mist.

Types of Systems

Depending on their location, OTEC facilities are of several types: shore based, floating offshore, tower-mounted offshore, or grazing. Shore-based plants are easier to build and operate and can transmit the electric energy using existing grids. Many oceanic islands rise steeply from the ocean floor, have steep offshore slopes, and are prime shore-based OTEC sites.

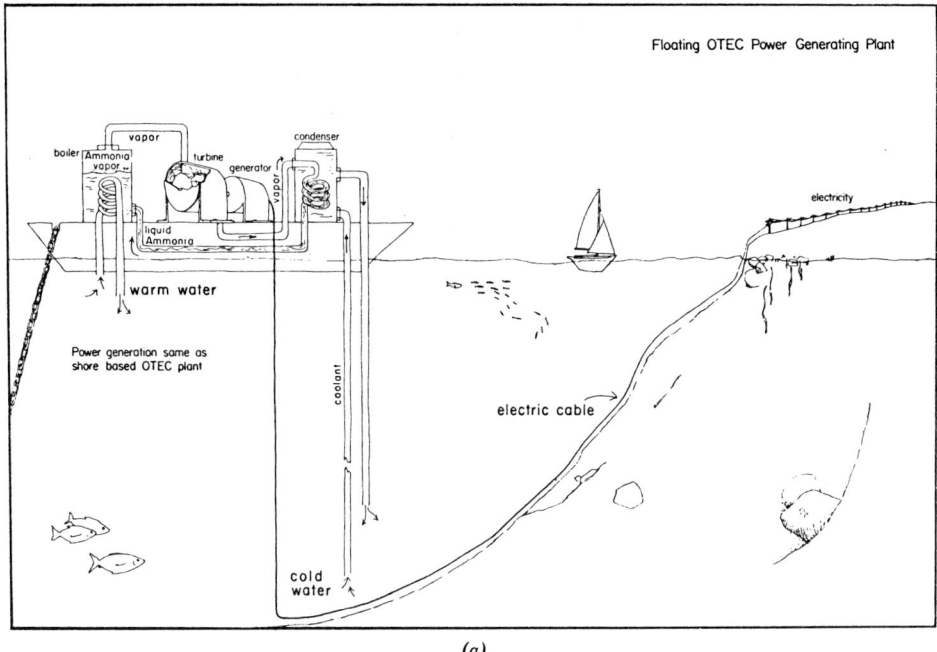

(a)

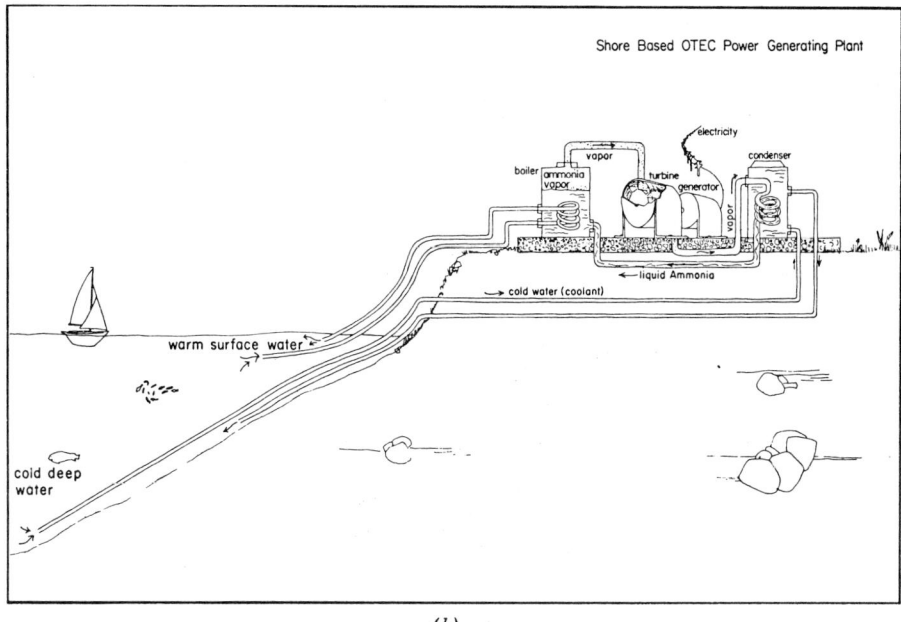

(b)

Fig. 11.4-8   OTEC power generating plants.

Offshore floating platforms (Fig. 11.4-8) are anchored or dynamically positioned at specific locations by propellers or water jets. The electricity produced is transmitted to shore via a riser cable, followed by a bottom cable leading to a land-based station. The cold water pipe is vertical and attached to the bottom of the platform. Protection needed for the pipe is simpler since it does not cross the surf zone; the electric cable will, however, need protection.

Tower-mounted structures are built and permanently located offshore in up to 90 m of water. An important advantage is that because they will not cross the surf zone, pipes bringing in the warm and cold water do not have to be specifically protected. Also, stationkeeping subsystems are eliminated, but the bottom electric cable to shore will have to be protected.

Grazing plantships are self-propelled platforms that move freely in the open ocean seeking the maximum ΔT to optimize their operating efficiency. Since they must be thousands of miles from shore, their output is an energy-intensive product such as ammonia, hydrogen, or metals from ores brought to the plantship. A cold water pipe suspended in the open ocean and the transportation costs that may increase product costs are the potential disadvantages of grazing OTECs.

OTEC Products

A variety of energy products and alternative applications are possible. The production of electric energy is the primary application of OTECs. Commercial-size plants are estimated to be in the 200–400-MW range.

Valuable fresh water can be derived from open-cycle OTEC plants. As much as 2.8 million liters per day per megawatt of power (based on a plant capacity factor of 90%) can be produced.

OTEC power plants are of value to aquaculture because the cold, deep, nutrient-rich water they use for cooling contains 60–200 times the nitrate-nitrogen and up to 20 times the phosphate-phosphorus of warmer surface waters. Phytoplankton convert these nutrients to organic material; and filter feeders such as oysters, clams, and other shellfish convert this organic material to more valuable meat protein.

Hydrogen, which can be produced by electrolyzing water, can store OTEC-produced energy and be transported as a gas via a pipeline to shore or be shipped as a liquid or a metal hydride. It can also be used on board OTEC for industrial purposes. Production of ammonia is one of the most attractive uses of hydrogen.

Many synthetic fuels could be made on site by transporting some source of carbon (coal, limestone, etc.) to the OTEC platform. The use of carbon dioxide from seawater as the carbon source to make methanol, for example, is possible.

The OTEC concept can be applied to land-based facilities with rejected heat sources: thermal plants, nuclear facilities, solar ponds, and freshwater thermoclines. The application of bottoming cycles to large-scale power plants allows the recovery of much of the large amount of waste heat produced, resulting in improved efficiency.

Mini-OTEC. On August 3, 1979, Mini-OTEC became the first modern OTEC plant to generate power. A U.S. Navy open hopper dump scow (37 × 11 m) from the Navy's mothball fleet was used for the Mini-OTEC platform. Offsite construction of heat exchangers and other plant components was begun in the latter part of 1978, and shipyard modifications to the test platform were begun in January and completed in June of 1979. Mini-OTEC is composed of four major subsystems: platform, power plant, cold water pipe, and mooring (Fig. 11.4-9). The plant design was not optimized but used stock components.

OTEC Power Systems. Shell-and-tube and plate-fin heat exchangers can be used in OTEC systems. Other shell-less designs, including trombone and tube-in-tube types, have also received attention. In all of these designs the working fluid and the seawater are separated by a barrier that transmits heat. The direct-contact exchanger is experimental only but holds promise.

Shell-and-Tube Heat Exchangers

For shell-and-tube heat exchangers the flow of shell-side fluid through the tube bundle creates unavoidable areas of high and low fluid velocities, so the use of seawater as the shell-side fluid could lead to deposition of debris, biofouling, and corrosion, Thus, the seawater should be placed inside the tube to allow close control of seawater velocities, use of conventional cleaning techniques to remove biofouling deposits, and easier access to heat exchanger components for repair. Conventional flooded-bundle shell-and-tube evaporators can be used. The bundle diameter should be made as large as possible in order to minimize the cost required to manifold a large number of parallel units and to decrease the total volume required for the system of evaporator units. Heat transfer enhancements can also be added to improve the performance of the evaporators and condensers. Conventional cleaning techniques such as the Amertap abrasive ball system can be used to remove biofouling deposits in smooth-tube seawater passages.

Compact Plate-Fin Heat Exchangers

Plate-fin heat exchangers consist of large banks of tall vertical channels arranged in parallel rows that guide the working fluid across the seawater flow. Because of their higher efficiency, the plate-fin heat exchangers are considerably more compact than conventional shell-and-tube configurations and have

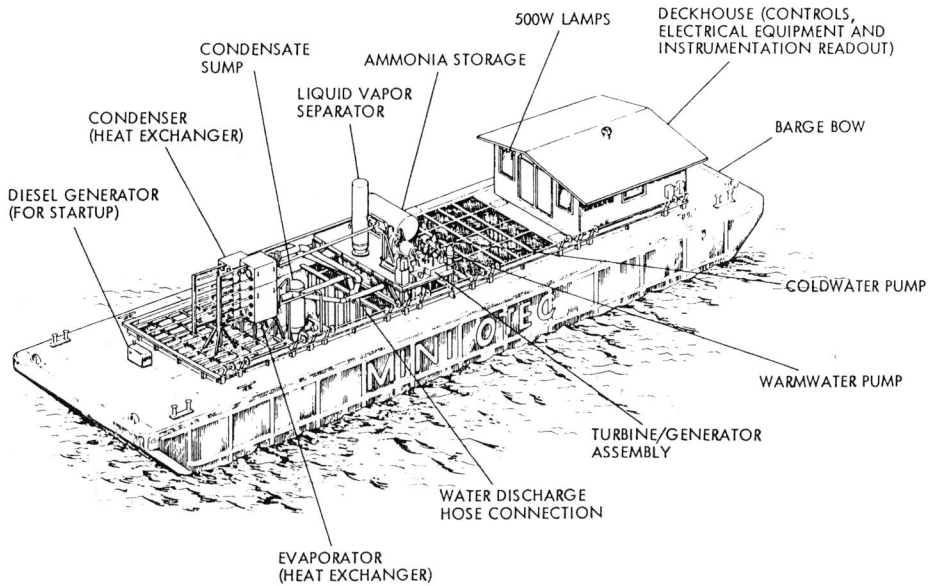

Fig. 11.4-9 Solar ocean energy liaison.

the potential of reducing construction material requirements per unit power delivered by 20–30% of that required by shell-and-tube arrangements. This decrease in material requirements could also lead to substantial reductions in structural requirements for the power plant containment.

The following mechanical characteristics are recommended for compact heat exchanger designs:

Fins protruding into the working fluid passages to increase the working fluid side heat transfer coefficient.

Seawater cross flow of the working fluid.

Small working fluid passages that optimize the tradeoff between working fluid pumping power requirements and enhanced heat transfer rates.

Vertical working fluid passages in both the evaporator and the condenser.

Thinnest possible walls surrounding all working fluid passages, which still ensures the integrity of the plate-fin unit.

Table 11.4-2 compares shell-and-tube and plate-fin designs for OTEC applications. Although the shell-and-tube concept remains closer to the state of the art, using conventional shell-and-tube heat exchangers for OTEC applications will lead to low heat transfer coefficients on the order of 1700–2800 $W/m^2 \cdot °C$. The compact plate-fin designs offer the possibility of reducing construction material requirements, but they need more development. Nonetheless, using improved plate-fin exchangers or modified shell-and-tube designs with surface enhancements may increase overall heat transfer rates by 50–100% and result in lower heat exchanger costs.

Heat Exchanger Materials

Metallic alloys being considered for the heat exchanger include titanium (grade 2), AL-6X stainless steel, CA-706 copper-nickel, and the aluminum alloys 5052 and Alcald 3003.

Titanium essentially will be inert under OTEC conditions, is hard enough to resist any mechanical cleaning techniques likely to be used, and can be fabricated into a wide variety of shapes. It is, however, expensive and its availability in quantities required for the OTEC program is doubtful. There is little doubt, though, that titanium will provide a 30-year lifetime for OTEC heat exchangers. The high-alloy stainless steels typified by AL-6X are readily available but their costs are comparable to titanium. Copper-nickel alloys have been used extensively in marine heat exchangers and should be suitable for OTEC.

The aluminum alloys appear to have the potential for substantially reducing the overall costs of heat exchangers. Aluminum's most serious drawback is its tendency to "pit" in the presence of heavy

TABLE 11.4-2   COMPARISON OF HEAT EXCHANGER TYPES

Advantages	Disadvantages
Shell and Tube Heat Exchangers	
Construction technology is available. Easy access for repairs. Conventional mechanical cleaning techniques can be used. Variety of internal and external tube enhancements are possible to improve heat exchanger performance. Heat transfer and pressure drop performance characteristics are verified.	Large quantities of material per unit power output are required, leading to high heat exchanger costs. High volume per unit surface area may lead to large OTEC containment structure requirements. Additional research into two-phase thermal fluid flow is needed especially in modified shell and tube designs.
Plate-Fin Heat Exchangers	
Small quantities of material per unit power output may reduce heat exchanger costs. Compact, large surface-to-volume ratios may lead to reductions in OTEC platform containment structure requirements. In mechanical construction designs units can be easily disassembled for repair and can utilize a wide selection of materials. Variety of surface enhancements, finned surfaces, and plate surface geometries are possible to improve heat exchanger performance. Heat transfer and pressure drop performance characteristics are verified.	Mechanical cleaning is difficult, especially in thin-plate mechanical construction designs. In brazed, welded, or adhesive joined units, removal of heat exchanger components for repair is difficult. In addition, selection of materials is limited to those which are suitable with these joining techniques. Assembly methods for large-scale units to meet OTEC power plant requirements are not yet developed. Additional research into two-phase thermal fluid flow is needed.

metal ions such as copper or iron. This could lead to premature failure of the thin walls of the heat exchanger.

Biofouling and Corrosion Countermeasures

The wide variety of countermeasures to control fouling in OTEC heat exchangers is shown in Table 11.4-3. All of these can be used on shell-and-tube and plate-fin designs, but only chemical methods (because they do not require heat exchanger disassembly) are practical for flat plate designs.

Working Fluids

Working fluids being considered for closed-cycle OTEC include sulfur dioxide, ammonia, propane, isobutane, methyl compounds, and refrigerants such as the Freon compounds. A good working fluid should have a high heat of vaporization, a high thermal conductivity, be relatively safe to handle, and be compatible with materials in the heat exchangers, pipes, pumps, valves, and turbines.

Open-Cycle Power System Technology

Water is the working fluid in open-cycle systems. But any large fluid system operating below atmospheric pressure is likely to leak, and leakage of air and oxygen into a system almost invariably

TABLE 11.4-3   POTENTIAL FOULING COUNTERMEASURES

Mechanical Methods	Chemical Methods
M.A.N. brushes	Halogens, including chlorine and bromine
Amertap balls	Chloride and chlorine dioxide
Slurries	Ozone, peroxide, and permanganate
Ultrasonics	Organic biocides
Increased fluid velocity	Antifouling coatings
Ultraviolet radiation	Periodic outage chemical treatments
Twisted tape inserts	Periodic outage detergent treatments
Deaeration	
Water jets	

causes corrosion and maintenance problems. In addition to this, the partial pressure of air in a boiling and condensing fluid system seriously impairs the operating performance. Very large and expensive turbines and additional equipment to remove noncondensable gases and maintain system vacuum are required for open-cycle plants. Operating this equipment entails increased parasitic power requirements.

Using water as a working fluid, however, does have some advantages. Of all working fluids considered, its specific latent heat of vaporization is the greatest. In an open-cycle plant without freshwater recovery, no warm side (evaporation) heat exchanger is required. Even in an open-cycle plant with freshwater recovery, improved heat transfer coefficients and reduced biofouling and corrosion are possible.

A major disadvantage of the open-cycle system is that because of the high specific volume of water vapor it requires a turbine of very large size. A single turbine yielding 100 MW could be as much as 60 m in diameter.

Condensers

For open-cycle with freshwater recovery a conventional shell-and-tube condenser similar to the kind used in closed-cycle plants can be used. When fresh water is not recovered, the direct-contact condenser for the open-cycle systems may be a jet type, in which the condensing water is delivered in jets through nozzles and directed into a discharge pipe at the lower end of the condenser. Steam entering the condenser from the turbine exhaust comes into direct contact with the water jets and is condensed.

Ocean Engineering

The three most important environmental parameters to be considered are winds, currents, and waves. Ocean environmental data are insufficient at present for ocean engineering design purposes, and forecasting wave models, based on the significant wave concept, have important uncertainties.

Platform

The selection of an optimum platform for OTEC plants is dependent on several considerations: site-specific environmental forces, types of cold water pipe, heat exchangers, and other equipment; relative ease of construction, deployment, operation, and maintenance; and life-cycle costs. Platform technology, however, does not appear to be a limiting factor in the construction of an OTEC plant.

Various platform configurations considered for commercial OTEC power plants include ship, spar, semi-submersible, submersible, disc, and tuned-sphere types.

Cold Water Pipe

The most critical component of the seawater system is the cold water pipe. Transferring the cold seawater from 1-km depths to the OTEC platform requires a pipe perhaps 20–30 m in diameter that must survive movement in all three axial directions as well as rotation about any of the axes. Pipe materials under consideration include steel, polyethylene, fiber-reinforced plastic, and elastomer; concrete was considered but rejected.

Stationkeeping

The stationkeeping subsystem must be designed to survive the 100-year storm and a 30-year service life. A mooring system that anchors the platform to the seafloor is favored for OTEC. This requires mooring lines such as wire rope in the 13- to 15-cm diameter range and chain in the 10- to 13-cm diameter range with 13,610- to 27,216-kg drag embedment type anchors. With this system the platform can maintain its position during normal operational conditions; but under extreme environmental conditions (i.e., hurricanes) electric cables will be slacked off or released; power generation may be shut off; and the cold water pipe may be uncoupled, if time permits, while the platform and mooring system remains in place.

The transmission of electric energy to shore is accomplished with a submarine cable consisting of two segments: a bottom cable extending from a land-based power subsystem to the bottom of the ocean floor and a riser cable extending from the termination of the bottom cable to the OTEC plant. These cables must operate and survive in 1–2 km of water rather than the 500-m maximum depth of present-day cables.

Bottom cables have to be embedded 2.5 m below the seabed whenever water depth is less than 90 m in order to prevent mechanical damage from anchors, trawlers, and so on. The cost of embedding could be enormous for such sites as Florida and New Orleans where the cable would have to be laid across 160 km of shallow shelf. Because buried cables are more susceptible to heat buildup, they will be larger and thus more expensive than cables laid on top of the seafloor.

Because of the action of waves, wind, and currents, the riser cable will experience forces along its entire unsupported length. Also vortex shedding and macrofouling will result in more stresses. The platform motion must be reduced as much as possible, therefore, to minimize additional stresses.

OTEC Economics

There are wide variations in estimates of capital costs and electricity costs from OTEC plants. Reliable cost estimates probably cannot be obtained without operating experience with OTEC plants. Present cost ranges are given:

Site	Range of Capital Cost Estimates, 1980 $/kW	Range of Cost of Electricity Estimates mills/kW · h
Gulf of Mexico	1980–5300	42–79
U.S. islands	1583–4300	31–96
Grazing	1584–2300	34–70

Environmental and Social Considerations of OTEC Development

Preliminary work indicates that OTEC is a relatively clean energy source with almost no wastes and little or no pollution, especially when it is compared to oil-fired, coal-fired, and nuclear-powered plants. Because no commercially sized OTEC plants have ever been in operation, however, the overall and long-term effects on the environment of such OTEC factors as displacement of seawater, both through warm and cold water intake and plume dispersal, biofouling and corrosion of heat exchangers and their countermeasures, leaking of working fluids, discharges of waste from the OTEC crew and spillage of OTEC products, and the artificial reef impact of an OTEC ship are not known.

Legal, Political, and Institutional Issues: Barriers and Incentives

A review of both jurisdictional and regulatory laws is generally favorable and indicates no legal obstacles to OTEC development: Nationally, most issues surrounding OTEC development have been settled by enactment of the Ocean Thermal Energy Conversion Act of 1980; internationally, neither current conventions nor the United Nations Draft Convention on the Law of the Sea is likely to affect OTEC adversely. In addition, legislative incentives make investment in OTEC more attractive. Nonetheless, caution by investors, utilities, and industries is still to be overcome.

CHAPTER 12

GEOTHERMAL
ENERGY SOURCES

HEMENDRA K. ACHARYA and **JOHN BRIEDIS**

Stone & Webster Engineering Corporation
Boston, Massachusetts

12.1 INTRODUCTION

Hemendra K. Acharya

It is well known from drilling, observation in mines, and various geophysical studies that temperature in the earth increases with depth. This increase in temperature (T) with depth (Z) is termed *geothermal gradient* (T/Z). At depths accessible to drilling (currently 6 mi or 10 km) the geothermal gradient is usually in the range 5–25°F/1000 ft (9–45°C/km), but in many areas it exceeds 25°F/1000 ft. Geothermal energy is the utilization of these steep thermal gradients at shallow depths for conversion to useful work.

Several factors presently limit the exploitability and potential uses of a geothermal resource. These include temperature, depth to the resource, the content of water and its long-term availability, the form of the water, and its pressure and geographical location. The most desirable characteristics with respect to these factors are rather obvious, namely hot, shallow, high-pressure steam in large volume near an energy load center. The search for these desirable characteristics has placed an emphasis on electric power production. Power generation is usually sought because electricity can be transported easier than heating fluid, it is thermodynamically worth more, and it is in steady demand.

Although many types of geothermal energy systems exist, economically significant geothermal anomalies occur in shallow (depth < 2 km), local hot spots where high temperatures (70–340°C) are found in porous rock containing water. Heat flow from depth is transported to the surface by the convective circulation of water. These geothermal systems are usually referred to as hydrothermal convection systems and may be either vapor dominated or liquid dominated.

In order to utilize geothermal energy for power production, it is necessary to have the resource at a depth of less than 2–3 km and of an areal extent greater than 2 km².[2] Most importantly, the temperature of the system should be greater than 180°C. In addition to power production there are many other areas for the utilization of steam and hot water produced from the deeply located geothermal reservoir if the temperature of the system is lower. Figure 12.1-1 shows the utilization of geothermal energy for various purposes depending on temperature.[3]

For any heat-intensive process such as space heating, power production, distillation, hot water supply, refrigeration, air conditioning, and certain other manufacturing processes, it is very important that inexpensive heat is available. It has been shown in Hungary, Iceland, New Zealand, and the Soviet Union that direct utilization of geothermal energy in industry, agriculture, and space heating is appreciably less expensive than the use of crude oil, gasoline, or diesel fuel for the same purposes. Natural gas, where obtainable, and coal are more nearly competitive, though still more expensive than

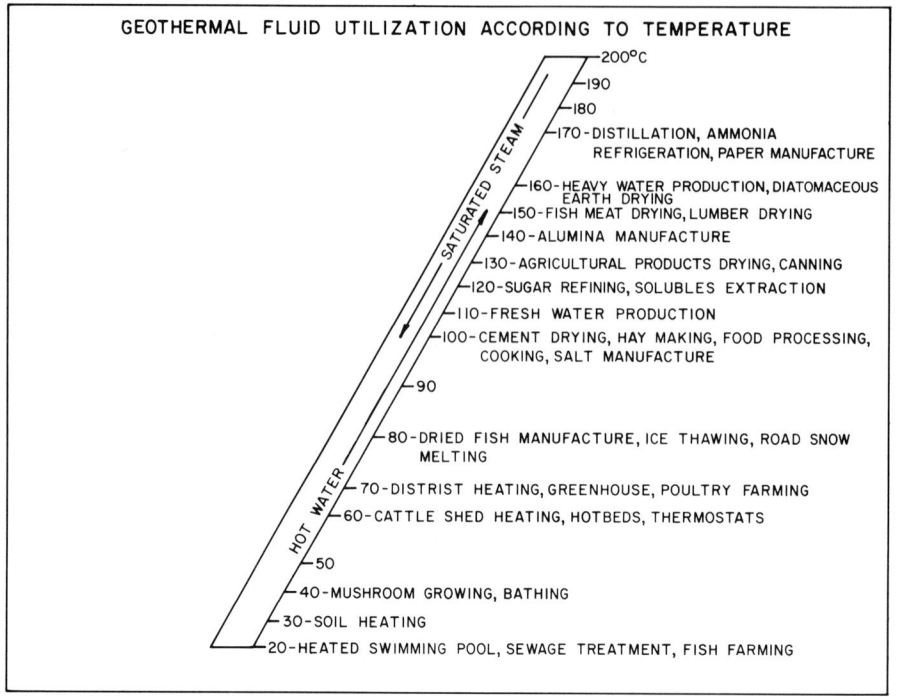

Fig. 12.1-1 Potential utilization of geothermal energy based on resource temperature (from ref. 3).

hot water for heating purposes. In the generation of electricity only hydroelectric power has been found to be cheaper and only in certain situations. In Iceland, for example, hydroelectric power was shown to be less expensive than geothermal power in most circumstances. However, the direct utilization of hot water for municipal heating is far cheaper than heating by hydroelectricity. At the Geysers in the United States geothermal electric power proved to be cheaper than power from other fuel sources, regardless of plant size.[4] For the operating year 1976 the cost per kilowatt-hour at the Geysers was 18 mills for geothermal as compared to 24 mills for nuclear, 26 mills for coal-fired plants, and 36 mills for oil-fired plants. In addition, geothermal plants were the least expensive to construct, being 20% cheaper than oil-fired plants, about half as expensive as coal-fired plants, and costing only 38% of a typical nuclear plant. The ability of geothermal generating systems to be developed economically in relatively small power units, 25–50 MW, is a major consideration for the developing countries where the load and load growth are commonly small. However, even in developed countries, geothermal power compares favorably with power from very large generating stations that have the advantage of economy of scale.

A geothermal power station utilizes a natural resource—a flow of more or less contaminated water at elevated pressure and temperature, which either flashes into steam or vaporizes an organic working fluid to drive a turbine. The system is capable of producing useful work in the form of an electric output because the raw material is found existing naturally in a state of nonequilibrium with the surroundings. The process of allowing this material to tend toward equilibrium with the atmosphere is responsible for the production of useful work. The process of this conversion is discussed in detail in Section 12.3.

12.1-1 Nature of Geothermal Energy

Muffler[1] has identified two types of geothermal energy:

1. Geothermal systems related to young igneous intrusions in the upper crust such as
   (a) convective hydrothermal systems,
   (b) hot dry rock, and
   (c) magma.

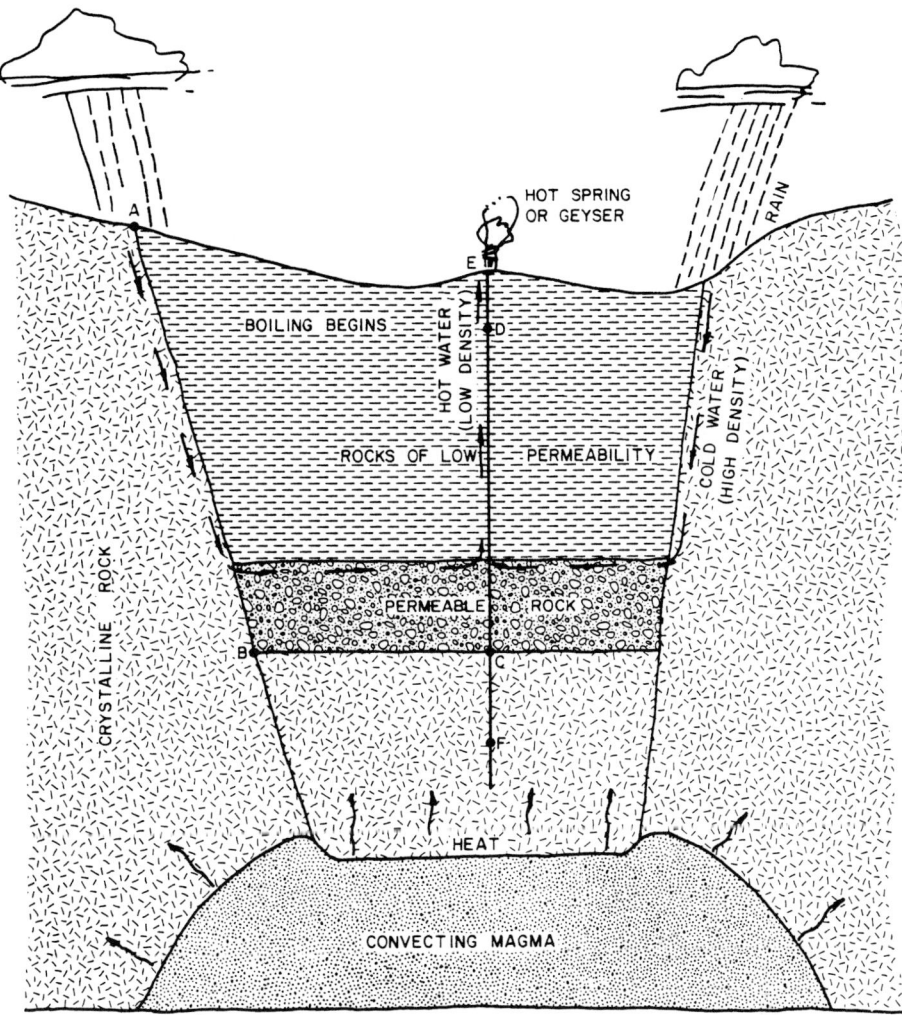

Fig. 12.1-2 A model of geothermal system (from ref. 5; used by permission).

2. Geothermal systems not related to young igneous intrusions in the upper crust, such as

(a) systems in low-porosity conductive environments,
(b) systems in low-porosity conductive environments modified by circulation of meteoric water,
(c) systems in high-porosity environments at hydrostatic pressure, and
(d) systems in high-porosity environments at pressures greatly in excess of hydrostatic (the geopressure systems).

Systems Associated With Young Volcanic Rock

It has long been recognized that many geothermal systems have a close spatial and generic relation to young volcanic centers. Intrusion of magma into the upper crust can generate one or all of the three types of geothermal systems: (1) hydrothermal convection, (2) magma, and (3) hot dry rock.

Igneous intrusions into permeable water-bearing rocks of the upper crust commonly set up overlying hydrothermal convection systems. Fractures and faults around the intrusion (Fig. 12.1-2) allow circulating groundwater to reach down to or near the cooling intrusion. Water absorbs some of the heat, and the density difference between hot and cold water[5] causes the heated water to rise, thus concentrating thermal energy in a near-surface geothermal reservoir. Geothermal reservoirs related to young intrusions can have very high temperatures (300–600°F/150–350°C) and have been developed for electric generation and direct uses in a number of localities.

Two major types of *hydrothermal convection systems* are recognized, which differ in the physical state of the dominant pressure-controlled state:

1. Vapor-dominated systems yield dry or superheated steam with no associated liquid. The world's two largest geothermal power producers, the Geysers in California and Larderello in Italy, are of this type. Such reservoirs are especially easy to exploit, but they are extremely rare.

2. Liquid-dominated systems, which contain superheated water, are more abundant but are harder to exploit.

Geologically, wet-steam and dry-steam systems are generally similar, emphasized by the fact that in some cases wells have produced wet steam initially and dry steam later.

Hot dry rock systems have considerable heat stored in reservoir rocks, and this heat exists even if no fluids are present. In many places hot rocks exist at shallow depths, but the methods of providing sufficient heat transfer surface in these rocks by fracturing and circulating water to extract heat are only in the initial stages of experimentation and development.

Magma chambers and lava pools are known to be associated with volcanic systems. Access to these regions and injection of a practical heat exchanger system into them are major problems. Yet these extremely large reservoirs of high-temperature heat are an attractive target and deserve serious research attention.

Geothermal Systems With No Associated Volcanism

Many geothermal systems derive their heat from large volumes of rock by deep circulation of water along permeable zones, which may be either stratigraphic beds or networks of faults and fractures. The temperature attained by water is primarily dependent on the magnitude of the regional heat flow and the depth to which water circulates. If such systems occur within the economic reach of drill holes, they can be exploited. If the artesian pressures are sufficiently great, thermal waters may flow at land surface.

Geopressure geothermal reservoirs are aquifers with pore pressure considerably greater than hydrostatic pressure and approaching lithostatic pressure. Less porous sediments that lie on top of geopressured reservoirs prevent upward passage of water that would ordinarily transport and lose the heat to the surface. Water in geopressured sediments thus contains an anomalous amount of heat as well as substantial amounts of dissolved methane. The technology for producing geothermal energy from such reservoirs is still being perfected, although basically it involves the use of the same tools and techniques required in very deep drilling and therefore is a costly undertaking.

12.1-2 Distribution of Geothermal Systems

The distribution of geothermal anomalies on the earth's surface is not random but is rather governed by global and local geologic processes. Figure 12.1-3 shows the principal areas of known geothermal occurrences on a world tectonic map.[1] It shows that many of the geothermal systems associated with young volcanism are located near active plate margins be they spreading ridges, subduction zones, or transform margins.[1,5] Geothermal fields explored or exploited in the vicinity of spreading ridges are located in Ethiopia, Kenya, Uganda, Iceland, Salton Sea–Cerro Prieto area of southern California, and northern Mexico. Geothermal fields associated with subduction zones are currently being exploited or explored in Central and South America, Japan, Philippines, New Zealand, Indonesia, China, India, and western North America. Some activity seemingly remote from plate margins, such as Yellowstone geysers and Hawaii, have been attributed to hot spots.[6]

Acharya[7] observed that the locations of geothermal fields in the circum-Pacific is influenced by plate boundary geometry. Geothermal fields, such as the Salton Sea and Cerro Prieto (Fig. 12.1-3), are situated in the environment of diverging plates. They are located near one end of the spreading systems which are bounded by transform faults. In other areas around the Pacific two plates converge, and one plate underthrusts beneath the other plate; Acharya[7] observed that many geothermal fields are situated in volcanic areas along the extension of transverse zones that subdivide the underthrusting zone into different segments.

Away from zones of young tectonism, deep regional aquifers have been developed extensively for direct use in France, Hungary, and the USSR. In the United States deep regional aquifers having geothermal potential are known to occur in the Williston Basin in the northern Great Plains and may also occur in the other large basins of the north, central, and western United States. The eastern and midwestern United States areas such as the Allegheny, Michigan, and Illinois basins, appear to offer similar opportunities for geothermal exploration. Geopressure systems with potential for exploitation are situated in the Gulf Coast region of the United States.

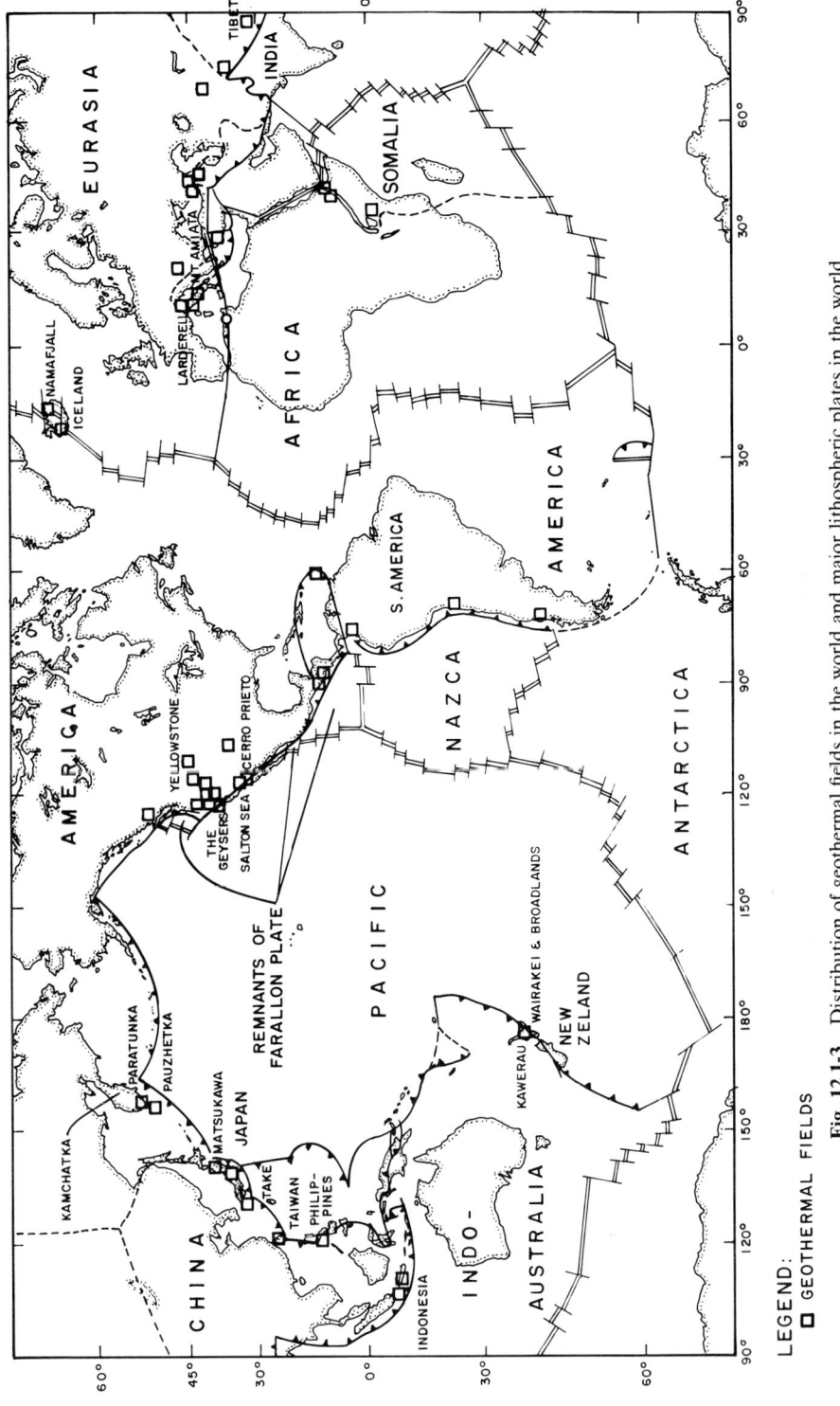

Fig. 12.1-3 Distribution of geothermal fields in the world and major lithospheric plates in the world (modified from ref. 1). Spreading ridges are shown by double lines, subduction zones by barbed lines, and transform faults by single solid lines. Plate boundaries of uncertain nature are shown by dashed lines. Geothermal fields are shown by open squares.

LEGEND:

□ GEOTHERMAL FIELDS

12.1-3 Exploration, Confirmation (Well Drilling), and Evaluation of the Resource

The objective of geothermal exploration is to search for positive geothermal anomalies—locations where relatively high temperatures will be encountered at the shallowest possible depth—to estimate the volume, temperature, chemical composition, and permeability at depth, to predict whether wells will produce dry steam or a mixture of water and steam in quantities to make it commercially feasible. Some exploration strategies have been offered.[8-10] Meidav and Tonani[11] have criticized certain aspects of these strategies. Clearly the exploration strategy for a recent volcanic caldera is different than that for a geopressured zone deep in a sedimentary basin. Because each prospect represents a unique combination of geological, hydrological, geochemical, geophysical, technical, and financial characteristics, no one exploration technique suffices for all situations. The exploration, confirmation, and exploitation of geothermal energy consists of five phases.[9]

Phase I: Reconnaissance Survey

The objective of the reconnaissance survey is to identify specific prospect areas. Having located the thermal manifestations, the following information should be collected by geological, hydrological, and geochemical surveys:

1. Type: Fumeroles, steaming ground, spring, well seepage, and so on.
2. Temperature.
3. Flow rate.
4. Local geologic control.
5. Chemistry.

The aims of the *geological survey* are to delimit the thermal area geographically; to understand the tectonic and stratigraphic setting of the area; and to locate recent faulting, the distribution, and age of young volcanic rocks, and the location and character of thermal manifestations including hydrothermally altered rock. The *hydrological survey* includes temperature and discharge measurements of hot and cold springs, chemical analysis of water from springs, determination of water table in available wells, and evaluation of surface and subsurface movements of water. Basic climate data (temperature, humidity, precipitation, etc.) help in the interpretation process.

The *geochemical survey* involves sampling and analysis of water and gases from hot springs and fumeroles. The objectives of this survey are to determine whether the system is hot water or vapor dominated and to estimate the minimum temperature expected at depth, the homogeneity of water supply, and chemical characteristics of water at depth, as well as the source of recharge waters. Although many chemical geothermometers have been considered and are in various stages of development,[12,13,14] the SIO_2[15] and Na-K-Ca[16] geothermometers have been most widely used.

Phase II: Geophysical Surveys

Geophysical surveys are carried out to indirectly measure temperature at depth and to investigate subsurface structural complexities and thereby define target areas for drilling.[2,8] The most useful techniques for understanding subsurface thermal characteristics are temperature probes or geothermal gradient surveys, heat flow determination, and electric resistivity surveys. Structural characteristics are explored by seismic reflection surveys, gravity studies, and microearthquake surveys.[17] Seismic noise surveys have also been used to examine thermal characteristics.[18]

Phase III: Drilling

Drilling is the only technique to determine actual geothermal reservoir characteristics such as temperature depth distribution, pressure depth distribution, porosity, permeability, lithology, stratigraphy, fluid composition, and fluid flow rate.[9] Prior to drilling full size holes, it is necessary to decide on the size and depth of the holes, casing program to be used, circulation system, frequency of collection and location of core samples, and the frequency and types of test measurements.

Drilling for geothermal resources uses standard oil well equipment and methods. The unique problems of geothermal drilling have to do with the elevated temperatures and corrosive fluids and with the hard, abrasive rock and fractured formations typical of geothermal reservoirs. Penetration rates are low, bit life is short, and there are frequent drill pipe failures and lost circulation problems.[19] The lack of adequate high-temperature drilling fluids is also a major problem. The upper temperature working limit for drilling muds is about 150–175°C (300–350°F). Brines are frequently used, and foam drilling has been gaining attention, although environmental cleanup problems are serious considerations with foams, particularly at high temperatures. Air drilling is commonly used to protect the formation from being clogged with drilling fluid. In the oil field, the drilling fluid is used to control the well by balancing the mud with the formation pressure. Most of the hydrothermal zones are underpressured and are controlled by other means.

Drilling and completion costs at a typical steam well at the Geysers have been estimated at about $1.1 million. Ryley and Di Pippo[19] have computed the following approximate relationship between the cost of drilling geothermal wells and depth of the well for both granite and sediments:

$$\text{Cost (\$)} = 5.5 \times 10^5 \exp{(0.327Z)} \quad \text{(granite)}$$

or

$$\text{Cost (\$)} = 1.9 \times 10^5 \exp{(0.327Z)} \quad \text{(sediments)}$$

where Z is the depth of the well in kilometers. In general, geothermal wells may be expected to cost 2–4 times as much as oil and gas wells. Ryley and Di Pippo[19] discuss advanced drilling methods that may eliminate some of the problems associated with this phase.

The term *well logging* refers to the measurements, observation, and recording of all items of interest associated with the drilling, completion, and testing of a well. The log will record the mechanics of the drilling procedure, the bit life, the loss in circulation of drilling fluid, the rate of penetration, the total drilling time, the collapse of a local section of the bore wall, and many other noteworthy occurrences. It is possible to make a number of geophysical measurements within the bore. These measurements are of value because they may confirm predictions made earlier by surface-based geophysical methods. Measurements include temperature, pressure, elastic wave velocity, electric resistivity, laterolog, induction log, and so on.

High temperatures affect all downhole tools required for testing the well. Geophysical logging tools are limited and those that are available often are not able to complete the run; the heat affects the wire cables and the connections. For this reason, the hole is logged on the trip into the hole rather than coming out of the hole. Downhole packers also have limited life and sometimes last only a few hours. Advances have been made in materials; however, formation testing is still limited. The high temperatures also affect the blow out preventers.

Phase IV: Evaluation

The exploration phase of the project does not end with proving the existence of a high-temperature geothermal reservoir having sufficient permeability for production. The next and probably the most difficult question to answer is what is the capacity of the field. The basic measurement for this evaluation is the variation of steam output with well head pressure and with time. Unfortunately, the structural complexity and the importance of fracture permeability in most geothermal fields make capacity projections based on only a few exploration wells highly unreliable. For this reason, there is no sharp division between exploration (drilling) and development drilling in geothermal projects as in the oil industry.[9] The problem of predicting production capacities of geothermal fields is circumvented by having sufficient exploration capital available to prove at the well head a sufficient amount of steam to operate a minimum size utilization plant economically. For a power plant this size could be in the range of 20–50 MW. Assuming an average production of 5 MW per well, there should be sufficient capital available, after a discovery is made, to drill 5–10 additional offset wells. McNitt[9] suggests that the offsets should not be drilled more than 200–300 m from the discovery well. Figure 12.1-4 shows the layout of wells for a 55-MW power plant at Ahuachapan, El Salvador. McNitt[9] also discusses problems for decision making if the first exploration well has marginal characteristics.

Phase V: Economic Feasibility

A feasibility study to determine the capital and operating costs of a power plant is carried out as the final step in the exploration and confirmation program.[9] This evaluation includes a review of well test data to verify production capacity and evaluate evidence for drawdown. The environmental impact of development is also assessed at this stage. Whether a particular field, properly delineated, can be characterized as renewable or of limited life cannot be decided by presently available techniques. Practice, however, suggests that a lifetime of 30–50 years of profitable exploitation may be expected.

12.1-4   Model of Geothermal Systems

The basic features of a geothermal field, wet or dry, are shown in Fig. 12.1-2. These features include a source of natural heat of great output, an adequate water supply, an aquifer or permeable rock, and a cap rock. Geothermal models are developed to better understand specific geothermal fields. A model of a geothermal system consists of a set of equations that describe the transport processes active within the system and the solution to these equations subject to conditions that prevail at a particular site.[20] This set of equations is general and may be used to consider two major subdivisions of

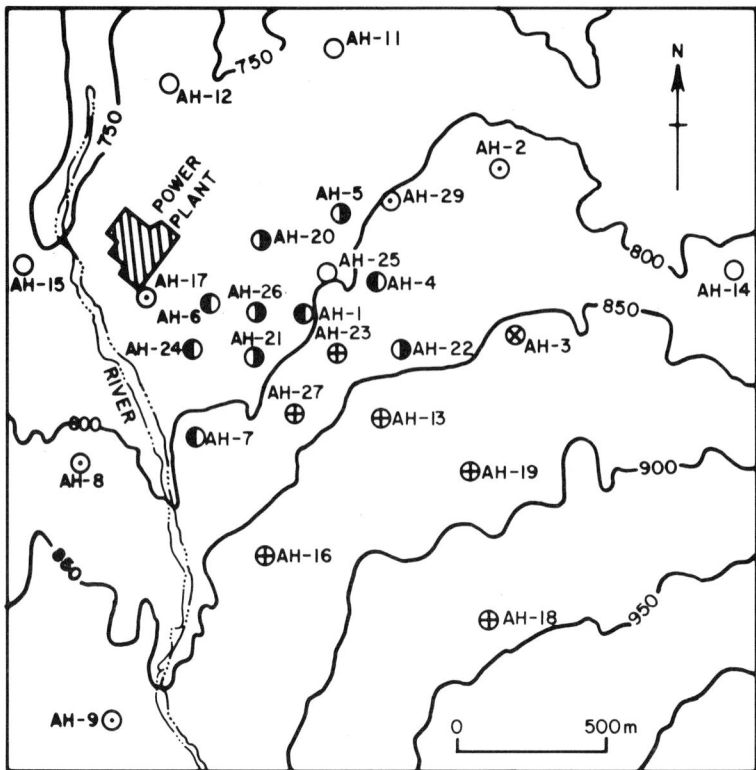

Fig. 12.1-4 Layout of production and reinjection wells at Ahuachapan, El Salvador (Hartley and DiPippo, 1980).

geothermal modeling: (1) modeling the geothermal systems under natural conditions in an effort to better understand how it forms and persists within the earth's crust and (2) modeling the system during exploitation in order to predict its behavior subject to man-made stresses.

A complete review of free convection models is included in Witherspoon et al.[21] The main thrust of these studies has been to attempt to understand how geothermal systems can form and prevail under various geological constraints known to exist in the earth's crust. Mathematical modeling of the natural conditions of heat and fluid flow in a geothermal region can be used to describe certain basic phenomena such as fluid convection induced by density differences and the associated heat transfer.

Data obtained from exploration studies (geological, hydrological, geochemical, and geophysical) are used to develop models of geothermal systems during the exploitation phase. These data provide equation parameters, boundary conditions, and initial conditions and are used in simulation runs from which crude initial estimates of recoverable energy may be calculated. As development of the reservoir continues, new data produced by additional drilling and production records can be used to refine the accuracy of the model, yielding more dependable predictions as well as a better reproduction of observed reservoir behavior. Additionally, the model may serve as a management tool to optimize engineering decisions such as producing schemes and well locations. Two different types of reservoir models have been developed: (1) lumped parameter and (2) distributed parameter.

1. *Lumped-parameter models* are the simplest models to describe the behavior of the reservoir during exploitation. Only the total amount of mass and energy within the system and crossing the boundaries is considered. Since time is the only independent variable, the system can be characterized mathematically by a set of ordinary differential equations or an equivalent set of algebraic expressions representing total mass and energy.[21]

2. In the *distributed-parameter model* the properties of rock and fluid are allowed to vary in space. The model is complex and can be solved analytically only under certain restrictive assumptions. Therefore, the alternative approach of solving the equations numerically is preferred by most investigators. Mercer and co-workers applied this model to the Wairekei field in New Zealand using a Galerkin finite-element method. They were able to reproduce historical temperature data for up to

1962, when large quantities of steam had formed in the reservoir. Mercer and Faust[20] discuss these and other models in detail.

Although numerical techniques are sufficiently powerful to solve the most difficult problems, the conceptual basis of these models needs reinforcement.[20] The geologic materials in which heat transport processes occur are not easily idealized. Constitutive relationships for thermal dispersion and relative permeability and the description of flow and transport in fractured rock are areas that need further study. Finally, additional field verification of the complicated mathematical models is needed.

REFERENCES

12.1-1 L. J. P. Muffler, "Tectonic and Hydrolic Control of the Nature and Distribution of Geothermal Resources," Proceedings of the Second U.N. Symposium on the Development and Use of Geothermal Resources, San Francisco, Vol. 1, 1976, pp. 400–507.

12.1-2 C. J. Banwell, "Geophysical Methods in Geothermal Exploration," Geothermal Energy (Earth Sciences, 12), Review of Research and Development, UNESCO Monograph, 1973, pp. 41–48.

12.1-3 "Harnessing the Earth's Thermal Energy," Geothermal Energy Research and Development Company Ltd., Tokyo, 1981.

12.1-4 D. J. Mahoney and A. C. Bangert, "Economic Impact of Geothermal Development, Sonoma and Lake Counties, California," Pacific Gas & Electric Company, San Francisco, 1977.

12.1-5 D. E. White, "Characteristics of Geothermal Resources," *Geothermal Energy*, P. Kruger and C. Otte, (eds.), Stanford University Press, Stanford, 1973, pp. 69–94.

12.1-6 H. R. Shaw and E. D. Jackson, "Linear Island Chains in the Pacific—Result of Thermal Plumes or Gravitational Anchors," *Journal Geophysical Research* **78**, 8634–8652 (1973).

12.1-7 H. K. Acharya, "Influence of Plate Tectonics on Location of Geothermal Fields," Proc. 1982 Geothermal Conference and 4th New Zealand Workshop, V.1, 145–150 (1982).

12.1-8 J. Coombs and L. J. P. Muffler, "Exploration for Geothermal Resources," *Geothermal Energy*, P. Kruger and C. Otte (eds.), Stanford University Press, Stanford, 1973, pp. 95–128.

12.1-9 J. R. McNitt, "Summary of United Nations Geothermal Exploration Experience, 1965–1975," in Proceedings of the Second U.N. Symposium on the Development and Use of Geothermal Resources, San Francisco, Vol. 2, 1976, pp. 1127–1134.

12.1-10 S. H. Ward, W. J. Perry, W. P. Nash, W. R. Sill, K. L. Cook, R. B. Smith, D. S. Chapman, F. H. Brown, J. A. Whelan, and J. R. Bowman, "A Summary of the Geology, Geochemistry, and Geophysics of the Roosevelt Hot Springs Thermal Area, Utah," *Geophysics* **43**, 1515–1542 (1978).

12.1-11 T. Meidav and F. Tonani, "A Critique of Geothermal Exploration Techniques," Proceedings of the Second U.N. Symposium on the Development and Use of Geothermal Resources, San Francisco, Vol. 2, 1976, pp. 1143–1154.

12.1-12 A. H. Truesdell, "Summary of Section III Geochemical Techniques in Exploration," Proceedings of the Second U.N. Symposium on the Development and Use of Geothermal Resources, San Francisco, Vol. 1, 1975, pp. 1iii–pxxix.

12.1-13 A. J. Ellis and W. A. J. Mahon, *Chemistry and Geothermal System*, Academic Press, New York, 1977.

12.1-14 W. F. McKenzie and A. G. Truesdell, "Geothermal Reservoir Temperatures Estimated from the Oxygen Isotope Composition of Dissolved Sulfate and Water from Hot Springs and Shallow Drill Holes," *Geothermics* **5**, 51–61 (1977).

12.1-15 R. O. Fournier and J. J. Rowe, "Estimation of Underground Temperatures from the Silica Content of Water from Hot Springs and Wet Steam Wells," *American Journal of Science* **264**, 685–697 (1966).

12.1-16 R. O. Fournier and A. H. Truesdell, "An Empirical Na-K-Ca Geothermometer for Natural Waters," *Geochemical et Cosmochimica* **37**, 1255–1275 (1973).

12.1-17 G. Palmason, "Geophysical Methods in Geothermal Exploration," Proceedings of the Second U.N. Symposium on the Development and Use of Geothermal Resources, San Francisco, Vol. 2, 1976, pp. 1175–1184.

12.1-18 H. M. Iyer and T. Hitchcock, "Seismic Noise in Long Valley, California," *Journal Geophysical Research* **81**, 821–840 (1976).

12.1-19 D. J. Ryley and R. Di Pippo, "Drilling for Geothermal Resources," in *Sourcebook on the Production of Electricity from Geothermal Energy*, J. Kestin (ed.), U.S. Department of Energy, Supt. of Documents, Washington, D.C., 1980, pp. 136–155.

12.1-20 J. W. Mercer and C. R. Faust, "Physics of Fluid Flow and Heat Transport in Geothermal Systems," in *Sourcebook on the Production of Electricity from Geothermal Energy*, J. Kestin (ed.), U.S. Department of Energy, Supt. of Documents, Washington, D.C., 1980, pp. 121–135.

12.1-21 P. A. Witherspoon, S. P. Neumann, J. L. Sorey, and M. J. Lippmann, "Modeling Geothermal Systems," paper presented at the International Meeting on Geothermal Phenomena and Its

Applications, Academia Nationale dei Lincei, Rome, 1975.

12.1-22 R. P. Hartley and R. Di Pippo, *Environmental Considerations: Source Book on the Production of Electricity from Geothermal Energy*, J. Kestin (ed.), U.S. Department of Energy, Supt. of Documents, Washington, D.C., 1980.

12.2 DIRECT-USE GEOTHERMAL ENERGY

John Briedis

Direct-use of geothermal energy is gaining acceptance throughout the world. Its principal applications are for space heating, agriculture, and low-temperature industrial processes. Tables 12.2-1 to 12.2-3 present a listing of the major direct-use projects around the world and illustrate the wide diversity, distribution, and ingenuity of applications. Also shown in Table 12.2-4 is a summary of the status of direct-use applications in the United States.

The major components of a direct-use geothermal system are the resource, the distribution system, and the user application. Each of these will be discussed here. With each discussion some applied examples are given for quick and simplified rule-of-thumb calculations. The examples have been adopted from a DOE publication.[13] It is assumed here that the user may be a lay person interested in residential, small business, or agricultural applications and may not be as adept in engineering skills as potential users of other segments of this handbook.

12.2-1 Resource

The main factors for consideration are its location, temperature, quality depth, and recharge. The lower-temperature direct-use geothermal resources are much more abundant globally than the high-temperature resources for power production. As discussed in previous sections, geothermal resources are normally found in volcanic and tectonically active areas of the world or areas that were active in the geologic past. Generally, 50°C is considered as the minimum temperature for practical applications, and resources of temperatures above 160–180°C, although available for direct-use applications, are generally considered for electric power development. The low-temperature geothermal fluids are usually less corrosive and contain less dissolved mineral matter for causing scaling than the high-temperature deposits. This advantage allows for considerable savings and less complexity in development, distribution, and design of user systems. Drilling is normally done by regular drilling equipment with no need for high-temperature bits and drill stem tools. In resource areas below 100°C downhole pumps are necessary to raise the geothermal fluid out of the well. For temperatures above 100°C the geothermal fluid will rise to the surface if pressure is allowed to drop below hydrostatic in the well, permitting flashing to occur. If flashing is undesirable for user application, pressure must be maintained by downhole pumping. The importance of the subsurface permeability of the reservoir rocks must not be understated. The natural permeability of the reservoir dictates the amount of fluid flow to the well and acts as a throttle and must be considered in well and distribution system design.[1]

12.2-2 Distribution System

Considerable experience exists worldwide with distribution systems. The materials and technology have been developed and are operating successfully. The longest distribution system to date from source to user is 60 km. It is in operation in Iceland.[2] It is important to note here that quite often direct-use applications are seasonal, and this fact must be considered in cost analysis and feasibility of the project. Given below are some rules of thumb for distribution system design adopted from the Department of Energy.[13]

The distribution system links the user system with the resource. Both the user system and the resource for the purposes of distribution design are fixed. The user system normally dictates the energy delivery rate it requires for successful operation. The system's demands are normally expressed as heat transfer rate (Q), temperature drop (Δt), and operating head (H). The resource normally dictates the well depth, fluid, quality, and temperature. Engineering factors to be developed include those affecting the system fluid flow rate and normally would include flow rate ($\dot{W}$), pipe size, and pump horsepower (hp).

Given the example:

Geothermal energy in the form of hot water is to be pumped from a 400-ft (121.9-m) depth, through 800 ft (243.8 m) of level transmission line, and applied to a user system that requires a 50-ft (15.2-m) operating head. The system requires 6×10^6 Btu/h (1.76×10^6 W) and causes a 40°F (22.2°C) temperature drop.

TABLE 12.2-1   RESIDENTIAL AND COMMERCIAL APPLICATIONS OF GEOTHERMAL ENERGY[a]

Country	Localities	Description of Application	Production, Steam Flow Rate or Water Flow Rate	Associated Power (MW)	Comments
Iceland	Reykjavik	District heating	490 GW · h in 1961 1500 GW · h in 1973	171 (ave.) 325 (peak)	The system's geothermal resources come from the Reykir area (3600 m³/h at 80°C 170 MW) and the Reykjavik area (1700 m³/h at 119°C, 155 MW).
	Olafsfjordur	District heating system serves the housing of 1000 inhabitants.	Utilizes 48°C and 56°C water		
	Selfoss	District heating system serves 154,000 m³ of housing and 75,000 m³ of public commercial and industrial buildings.	80 kg/s at 80°C	15.1 +	Boreholes are located 1.5 km away from the city. System started in 1948.
	Hveragerdi	District heating system serves housing of entire city (820) and a balneo-therapeutic institution for 140 patients, while also supplying heating for 30,000 m² of hot houses.	Utilizes 180°C water		The system was built in 1953.
	Saudarkroukur	District heating			This system serving the city of 2000 utilizes 70°C water.
	Akureyri	District heating	Utilizes 80–96°C water at 150 L/s		Serves 85% of population (12,000).
	Akranes and Borgarnes	District heating	Utilizes 95°C water at 60 L/s		Serves population of 6500.

TABLE 12.2-1 (*Continued*)

Country	Localities	Description of Application	Production, Steam Flow Rate or Water Flow Rate	Associated Power (MW)	Comments
United States	Boise, ID	District heating system serves about 200 houses and 10–12 businesses.	53 kg/s at 77°C (maximum capacity)	9.2 +	Built in 1970, one of the oldest district heating systems in the United States
	Klamath Falls, OR	Individual space heating of homes and businesses—presently 468 residences are heated geothermally and some commercial installations.	49 GWh	5.6	Space heating of home is generally accomplished with several home-owners sharing a well and using downhole heat exchange systems.
	El Centro, CA	Heating and hot water to community center	730 gpm at 250°F		Pilot project for district heating plan.
	Susanville, CA	District heating	175°F, 700 gpm		
	Diamond Ring Ranch, SD	Space heating	67°C, 10.7 L/s		
	Elko, NV	District heating			
	Philip, SD, Haakon School	District and space heating of schools and businesses	300 gpm, 157°F		
United States	Rexburg, ID	District heating			
	Monroe City, UT	District heating	600 gpm, 165°F		
	Pagosa Springs, CO	District heating	900 gpm, 140°F		
	St. Marys Hospital, Pierre, SD	Space heating	375 gpm, 106°F		
	Utah State Prison, Draper, UT	Space heating			

Country	Location	Use	Flow/Temperature	Value	Remarks
USSR	Warm Springs Hospital, Deer Lodge County, WY	Space heating	160°F, 250 gpm		
	Makhach-Kala		23 kg/s at 63°C plus 70 kg/s plus others	12.6 +	Several districts are supplied, one of which has 15,000 inhabitants.
	Zgoudidi town, Georgia	District heating	50 Gcal/h	58.1	
	Mendji, Georgia	Heating of meteorologic station and agricultural uses	2.0 Gcal/h	2.3	
	Zaichi, Georgia	Heating of meteorologic station, hot houses, and baths	2.1 Gcal/h	2.4	
	Iserback town, Daghestan	District heating	6 Gcal/h	7.0	Heating for 7500 inhabitants and industrial uses
	Caspillsk town, Daghestan	District heating and hot water supply	5 Gcal/h	5.8	
	Paratounka, Kamchatka	Heating of apartments	0.55 Gcal/h	0.64	Three apartment buildings of 48 apartments each
	Cherkesk, Stavropol	District heating, industrial uses, and hothouses	22 Gcal/h	25.6	Heating for 18,200 inhabitants, plus industrial uses and hothouses
New Zealand	Rotorua	Individual space heating of homes plus space cooling of a business			Over 700 geothermal bores serving many individual applications

TABLE 12.2-1 (*Continued*)

Country	Localities	Description of Application	Production, Steam Flow Rate or Water Flow Rate	Associated Power (MW)	Comments
Japan	Towada	District heating	14 kg/s at 70°C	2.1 +	System constructed in 1963 with 11.5 km transmission line from Sarukuro springs
	Okawa	District heating	22 kg/s at 70°C	3.2 +	System provides heating for 3000 houses from a 12-km transmission line
	Ukiyama	District heating	12 kg/s at 40°C	0.25 +	Water is heated to 55°C in fossil-fueled boiler
	Aomori	District heating	22 kg/s at 60°C	2.3 +	Provides heating for 34 hotels and 140 houses, with water from the Asamushi hot spring area
	Iwate, Kazuno	District heating	94–115°C, 400–1000 t/h		
Hungary	Szeged	District heating			University clinics and 1200 flats—226,000 m³
	Hodmezovasarhely	Individual space heating			Factory and hospital—172,000 m³
	Mako	Individual space heating			Hospital—80,000 m³
	Cherkesk Stavropol	District heating, industrial uses, and hothouses	22 Gcal/h	25.6	Heating for 18,200 inhabitants, plus industrial uses and hothouses
France	Melun	District heating	28 kg/s at 70°C	4.1 +	Heating and hot water for 3000 housing units

[a]Compiled from refs. 1–6.

TABLE 12.2-2 AGRICULTURAL APPLICATIONS OF GEOTHERMAL ENERGY[a]

Application	Country	Localities	Description of Application	Power Output	Associated Power MW	Comments
Greenhouses	Iceland	Various localities	Glass greenhouses heated by natural steam and/or hot water	120 Tcal/yr	15.9	Heat use either direct or heat exchangers
	USSR	Makhach-Kala and other localities		25,000,000 m², 56,000 m² of greenhouse	5011.2	1,002,240 tons/yr of tomatoes, cucumbers, and other vegetables
	Italy	Castelnuovo		3000 m² of greenhouse	0.6	Mild climate in Italy is responsible for low interest in heated greenhouses
	Hungary	Szentes and various localities	Typical horticulture	800,000 m² of greenhouse	160	Typical greenhouse vegetables plus paprika
	Japan	Various localities	Horticulture—various species of vegetables	15,528 m² of greenhouse	3.1	
	United States	Oregon	Greenhouse	26,000 ft² of greenhouse	0.48	70°F year-round-automatic environmental control system; heat exchanger
		Sandy, UT	Flowers	120°F, 300 gpm	1.02	
		Cotton City, NM	Plants		Small	
	New Zealand	Various localities	Mushrooms, tree nursery seedings, tomatoes	Not mentioned		Soil is sterilized and heated by using geothermal fluids directly
	India	Chumathong	Experimental		Small	
	Bulgaria	SW Bulgaria	Heating greenhouse		Small	
Animal husbandry	USSR	Lorinsk	Fowl runs	Small	Small	Part of the Chukotsk collective farm
		Various localities	Needs of cattle breeding	Small	Small	
	Hungary	Various localities	Heating and cleaning animal shelters	25 wells at 7 × 203 10⁶ kcal/h		
	Japan	Minamitzu	Heating poultry houses, drying droppings	115°C water at 300 L/min	1.6	Yoshisawa poultry yard, heating below floor with pipes; 8000 chickens
		Ueda, Beppu, Oita	Heating poultry houses, drying droppings	Small		Nakamura poultry yard, 1600 chickens
		Higashusa, Shizuoka	Alligator and crocodile breeding	105°C water at 2000 L/min	9.7	Geothermal water mixed with cold to attain 28–32°C
	New Zealand	Taupo	Pig farm heating, sterilizing	Unknown but small		Geothermal steam. Cook and sterilize garbage feed; warm piggeries floors at 85°F
			Dry sheep crutchings	Unknown but small		Drying of sheep crutchings
			Dry wool cuttings	Unknown but small		Boiling of sheep cuttings
Aquaculture	Japan	Hokaibo and Kocoskina prefectures	Eel breeding	Unknown but small		Utilizing water from hot springs
		Shikabe, Makkaido	Experimental breeding station	70 L/s at 70°C	10.2	Hot water Hokkaido hatching center; eels and carp
	Iceland	Various localities	Experimental salmon breeding station	7 L/s at 70°C	1	Kollajord experimental fish farm; rearing young salmon to the smolt stage
	Bulgaria	SW Bulgaria	Trout rearing		Small	
	United States	Coachella Valley, CA	Freshwater prawns	84–87°F water at 300 gpm		
Related	Japan	Kannawa, Bappu, Oita	Drying rice	Unknown		Daily rice processing capacity = 180 kg
	Iceland	Reykholar	Drying seaweed	80 L/s at 100°C	21.8	Drying seaweed for export
	Bulgaria	Sofia	Algae production for antibiotics	Unknown		
	United States	Pierre, SD	Grain drying	67°C, 655 L/min		

[a]Compiled from refs. 1, 5–9.

TABLE 12.2-3 INDUSTRIAL APPLICATIONS OF GEOTHERMAL ENERGY[a]

Application	Country	Localities	Description of Application	Production, Steam Flow Rate, or Water Flow Rate	Associated Power (MW)	Comments
Wood and paper industry						
Pulp and paper	New Zealand	Kawerau	Processing and a small amount of electric power generation. Kraft process used	400,000 lb/h of steam	100 to 125	Geothermal energy delivered to mills by 80,000 lb/h of 200-psig steam and 320,000 lb/h of 100-psig steam obtained by flashing wet steam at the wellbore
Veneer factory	New Zealand	Rotorua				
Timber drying	New Zealand	Rotorua				
Washing and drying of wood	Iceland	Hveragerdi, (Hengill area)	Steam drying			Reported to occur in other places
Mining						
Diatomaceous earth plant	Iceland	Manafjall	Production of dried diatomaceous earth recovered by wet mining techniques	Up to 50 tons/h of steam at 183°C/10 atg. Total steam consumption 40–50 tons/h according to the season. Wellbore flow—184.8 Gcal/h Utilized—30.5 Gcal/h	35	Dredging in the lake is done only in the summer while the plant runs throughout the year. The reported 30.5 Gcal/h appears to be high or assumes superheated steam at 10 atg
Salt plant	Japan	Shikabe, Hokkaido	Production of salt from seawater	150 tons salt/h		No longer in operation
Salt plant	Philippines	Tiwi, Albay	Production of salt from seawater		2.5	Seawater brought 3 km to plant; three grades of salt produced
Sulfur mining	Japan		Sulfur extraction from the gases issuing from a volcano			Unsophisticated operation that has become uneconomic

Process	Country	Location	Description	Capacity/Conditions	Value	Remarks
Calcium chloride	United States	Imperial Valley, CA	Recovery of potassium chloride from the geothermal brine	Uncertain but small		
Boric acid	Italy	Larderello	Geothermal steam used for processing imported ores	30 tons steam/h	15–19	
Boric acid, ammonium bicarbonate, ammonium sulfate, sulfur	Italy	Larderello	Recovery of substances from volatile components that accompany the geothermal steam	No longer in operation. Large production before 1966		
Dry ice	United States	Imperial Valley, CA	Production of dry ice from CO_2 in the Salton Sea geothermal area			
Miscellaneous						
Confectionary industry	Japan	Kannawa, Beppu, Oita		Daily rice processing capacity 180 kg		Few details given, uses 98°C water, spring source
Grain drying	Philippines	Kiwi, Albay	Geothermal steam heats rotary kiln dryer		2.5	Palay drying time cut to 10 min from 4–8 h; model under test
Dehydration of onions	United States, United States	Pierre, SD; Fernley, NV	Grain drying	67°C, 665 L/min	Small	
Brewing and distillation	Japan	Ibusuki		Uncertain		
Stock fish drying	Iceland	Reykjavik	Fish drying in shelf dryers			Uses excess water from commercial heating system in Reykjavik during summer in local stock fish processing center

TABLE 12.2-3 (*Continued*)

Application	Country	Localities	Description of Application	Production, Steam Flow Rate, or Water Flow Rate	Associated Power (MW)	Comments
Curing cement building slabs	Iceland		Curing of light aggregate cement building slabs			No details given
Washing and drying wool						Reported to occur in two or more countries
Linen processing	Bulgaria	Velingrad area	—	—	Unknown	
Seaweed	Iceland	Reykholar	Drying seaweed for export	80 L/s at 100°C. Production of 3600 tons of dry seaweed per year. Each ton requires $3.40 worth of energy per ton of seaweed if the energy costs $0.45 per Gcal	3–4	Description of proposed system given; only word of mouth indicated that system is presently in operation

[a]Compiled from refs. 1, 8, 10, and 11.

TABLE 12.2-4 STATUS OF U.S. DIRECT-USE GEOTHERMAL APPLICATIONS[a]

	Residential		Commercial		Industrial		Agriculture and Aquaculture	
	10^9 Btu	No.	10^9 Btu	No.	10^9 Btu	No.	10^9 Btu	No.
Alaska	2	2	—	—	8	1	28	3
California	59	2	2	2	110	2	535	6
Colorado	4	5	11	6	20	2	8	4
Idaho	44	7	3	3	—	—	909	14
Montana	3	3	5	6	—	—	102	2
Nevada	9	7	16	2	289	4	2	2
New Mexico	4	5	9	3	1	1	116	6
North Dakota	—	—	1	1	—	—	2	2
Oregon	39	11	21	4	277	4	128	9
South Dakota	0.5	1	32	5	100	1	79	2
Utah	1	1	21	3	—	—	23	4
Washington	—	—	—	—	10	1	2	1
Wyoming	2	3	15	2	10,000	3	1	1
Total	167.5	47	136	37	10,815	19	1,933	56

[a] Modified from ref. 13.

Determine:

1. Flow rate ($\dot{W}$).
2. Transmission pipe size.
3. Pump horsepower (hp).

Flow rate can be calculated from the following equation, assuming the geofluid has the characteristics of water:

$$\dot{W} = \frac{Q}{500t} \quad \text{or} \quad \dot{W} = \frac{Q}{(4.2 \times 10^3)}(\Delta t)$$

$$\dot{W} = \frac{6 \times 10^6}{500(40)} \qquad = \frac{1.76 \times 10^6}{(4.2 \times 10^3)} \quad (22.2)$$

$$\dot{W} = 300 \text{ gpm} \qquad = 18.9 \text{ L/s}$$

Pipe Size can be determined from Fig. 12.2-1 for the fluid velocities. Generally, to minimize friction loss and pumping cost, flow velocities below 10 ft/s (3.0 m/s) are used. In this example a 6-ft/s (1.8-m/s) velocity is used requiring 4.5-in. (11.4-cm) diameter transmission.

To estimate the required pump horsepower, total head (H), in feet (meters) of water, must be determined.

The head loss due to friction is derived from Fig.12.2-2 and is approximately 2.75 ft/100 ft of line. For a transmission distance of 800 ft (243.8 m), the friction loss is 22 ft (6.7 m). The total head for the system is calculated by summing heads due to

Pump depth	400 ft or	121.9 m
System depth	50 ft	15.2 m
Friction loss	$\frac{22 \text{ ft}}{472 \text{ ft}}$	$\frac{6.7 \text{ m}}{143.8 \text{ m}}$

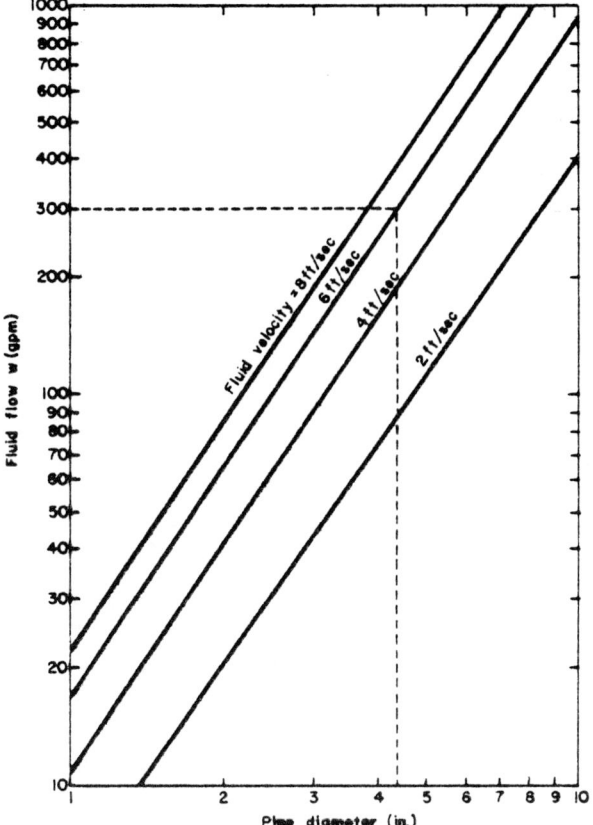

Fig. 12.2-1   Fluid flow versus pipe diameter (modified from ref. 14).

Assuming 100% pump efficiency, required horsepower can be computed by the following equation:

$$\text{hp} = \frac{8.331(\dot{W})(H)}{33,000} \quad \text{or} \quad \text{kW} = \frac{9.8(W)(H)}{1000}$$

$$= \frac{(8.331)(300)(472)}{33,000} \quad \text{or} \quad = \frac{9.8(18.9)(143.8)}{1000}$$

$$= 35.75 \qquad\qquad\qquad = 26.6$$

Assuming 80% pump efficiency, then required horsepower = 44.7 hp or 33.3 kW

12.2-3   User Applications

Space Heating

The most successful large-scale application of low-temperature geothermal resources for space heating is that of Iceland. It began with a pilot project in 1933 in the municipality of Reykjavik,[2] where today the entire city and 24 public district heating services are in operation serving 100 rural localities. Bjornsson estimates that at present about 70% of the population enjoy geothermal district heating. This figure will be increased to 80% in the next few years when the expansion projects already initiated are completed. As can be seen from Table 12.2-1, low-temperature geothermal energy is being applied successfully in many other parts of the world. In the United States district space heating applications are being successfully applied at Klamath Falls, Oregon, and Boise, Idaho. Its potential is basically

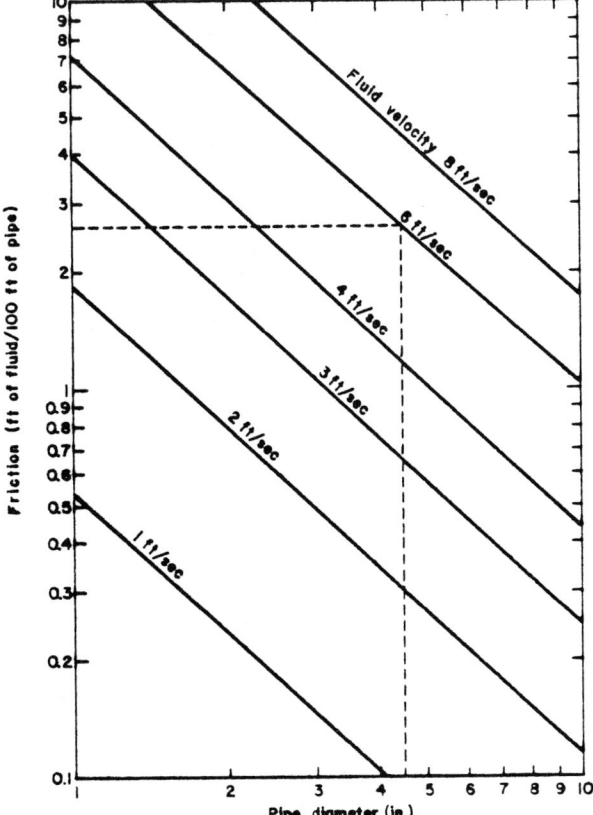

Fig. 12.2-2   Friction versus pipe diameter (modified from ref. 14).

applicable to parts of all the western states, Hawaii, and Alaska, and some parts east of the Mississippi.

Given below are some rules of thumb and an example developed for space heating applications published by the DOE.[13]

Factors normally to be determined for space heating applications include the maximum heat load (HL_{max}), annual heat load (HL_a), heat delivered to the system (Q), and the required geothermal fluid flow rate ($\dot{W}$). Factors normally dictated by the environment are the minimum outside design temperature (t_o) and the annual Fahrenheit degree days (DD). The factor dictated by the user is the desired inside temperature (t_i).

For example, determine the maximum and annual heat loads, the delivered heat requirement, assuming a 25% overdesign, and the required geothermal fluid flow rate.

Given is a modern well-insulated 2100-ft^2 (195-m^2) home requiring about 14,000 Btu/DD (1.477×10^7 J/DD) per year. Its outside design temperature (t_o) is $-15°F$ ($-26.1°C$) and desired inside temperature (t_i) is 70°F (21.1°C). The degree days per year for a given location can be obtained from a publication on "Climatological Data" available from the National Oceanic and Atmospheric Administration. The value assumed for this example is 4200 DD. A fluid temperature drop of 15°F (Δt) or 8.3°C is assumed.

Maximum heat load:

$$HL_{max} = 500(t_i - t_o) \qquad \text{or} \quad HL_{max} = 9.5 \times 10^5(t_i - t_0)$$

$$= 500[70 - (-15)] \qquad\qquad = 9.5 \times 10^5[21.1 - (26.1)]$$

$$= 42,500 \text{ Btu/h} \qquad\qquad = 4.48 \times 10^7 \text{ J/h}$$

$$= 1.25 \times 10^4 \text{ W}$$

Annual heat load:

$$HL_a = (14{,}000 \text{ Btu/DD/yr})(DD) \quad \text{or} \quad HL_a = (1.477 \times 10^7 \text{ J/DD/yr})(DD)$$

$$= (14{,}000)(4200) \qquad\qquad\qquad = (1.477 \times 10^7)(4200)$$

$$= 5.88 \times 10^7 \text{ Btu/yr} \qquad\qquad = 1.56 \times 10^4 \text{ W}$$

$$\qquad\qquad\qquad\qquad\qquad\qquad = 6.20 \times 10^{10} \text{ J/yr}$$

Delivered heat + 25% overdesign:

$$Q = 1.25(HL_{max}) \quad \text{or} \quad Q = 1.25(HL_{max})$$

$$= 1.25(42{,}500) \qquad\qquad = (1.25)(1.25 \times 10^4)$$

$$= 53{,}125 \text{ Btu/h} \qquad\quad = 1.56 \times 10^4 \text{ W}$$

Fluid flow rate:

$$W = \frac{Q}{500}(\Delta t) \quad \text{or} \quad W = \frac{Q}{4.2 \times 10^3}(\Delta t)$$

$$= \frac{53{,}125}{500}(15) \qquad\quad = \frac{1.56 \times 10^4}{4.2 \times 10^3}(8.3)$$

$$= 7.08 \text{ gpm} \qquad\qquad = 4.48 \times 10^{-1} \text{ L/s}$$

Agricultural and Industrial Applications

As can be seen from Table 12.2-2, a considerable amount of geothermal power is applied for agricultural use and a smaller yet significant amount of power is applied for industrial processes. The agricultural uses fall into four major categories[1]:

1. Greenhouses.
2. Animal husbandry.
3. Aquaculture.
4. Protein production.

By far, the greatest agricultural use is for greenhouses.

Industrial applications are much more diverse and range from different types of drying applications to chemical recovery and processing.[1] The two largest industrial applications are a diatomaceous earth plant in Iceland and a pulp, paper, and wood processing plant in New Zealand.[1]

Given below are some rules of thumb and an example of industrial process application derived from material published by the DOE.[13]

Industrial processes often entail the use of steam heat. For such processes geothermal fluids can often be flashed to provide the steam supply. Factors that need to be considered in such processes are (1) the percentage yield of saturated steam from the geothermal fluid and (2) the heat delivery rate.

Steam Yield

Figure 12.2-3 gives percentage steam yield. The dashed lines show that 400-degree (205°C) fluid flashed to 300°F (150°C) produces an approximate 11% steam yield. Assuming a 12,000-lb/h (1.51 kg/s) geofluid flow, actual steam yield is 1320 lb/h (1.7×10^{-1} kg/s.

Heat Delivery Rate

Figure 12.2-4a and 12.2-4b (metric) can be used to determine the heat delivery rate in Btu/h (watts). The 300°F (150°C) saturated steam is shown to contain approximately 910 Btu/lb (2.1×10^6 J/kg), resulting in a delivery rate of (1320 lb/h)(910 Btu/lb) equal to 1,201,200 Btu/h or (1.7×10^{-1} kg/s) (2.1×10^6 J/kg) equal to 3.6×10^5 W.

Other factors often addressed in applying geothermal fluid to an industrial process are the required heat exchange area (A) and the minimum economical geothermal resource fluid temperature (t_r). To

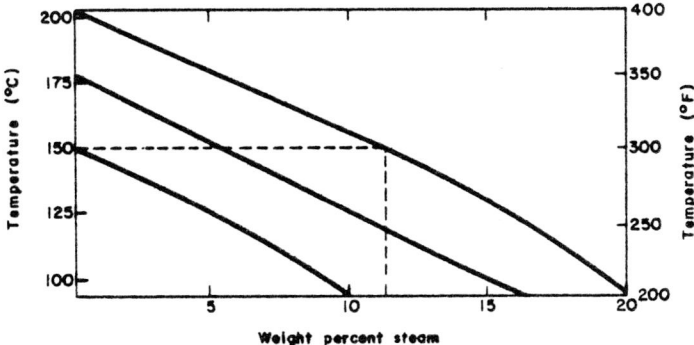

Fig. 12.2-3 Temperature versus weight percent steam (modified from ref. 13).

address these problems, the relationship for heat transfer rate (Q) is used,

$$Q = UA(\Delta t)$$

where Q = heat transfer rate, Btu/h (W)
     U = heat transfer coefficient
     A = heat transfer area, ft^2 (m^2)
     Δt = temperature difference, °F (°C)

and the relationship between heat exchange (T_{hx}) and economically usable resource temperature (t_r) is used and expressed as

$$T_{hx} = 0.6t_r - 70 \quad \text{or} \quad T_{hx} = 0.4t_r - 21 \quad \text{(metric)}$$

To find the required heat exchange area, for example, a water-filled cooking vessel, heated with geothermal fluid (water) at a rate (Q) of 175,000 Btu/h (51,275 W) and a temperature drop (Δt) of 80°F (44.4°C) across the vessel, the above equation is solved for A. The coefficient (U) for a typical

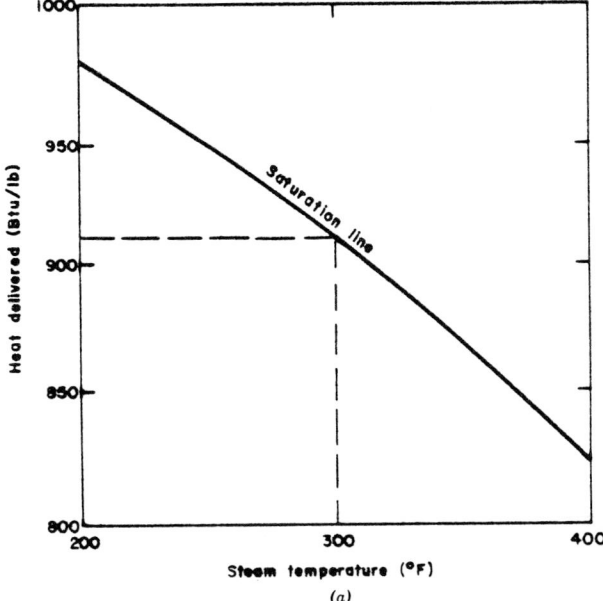

Fig. 12.2-4 (*a*) Heat delivered versus steam temperature (from ref. 13). (*b*) Heat delivered versus steam temperature (metric).

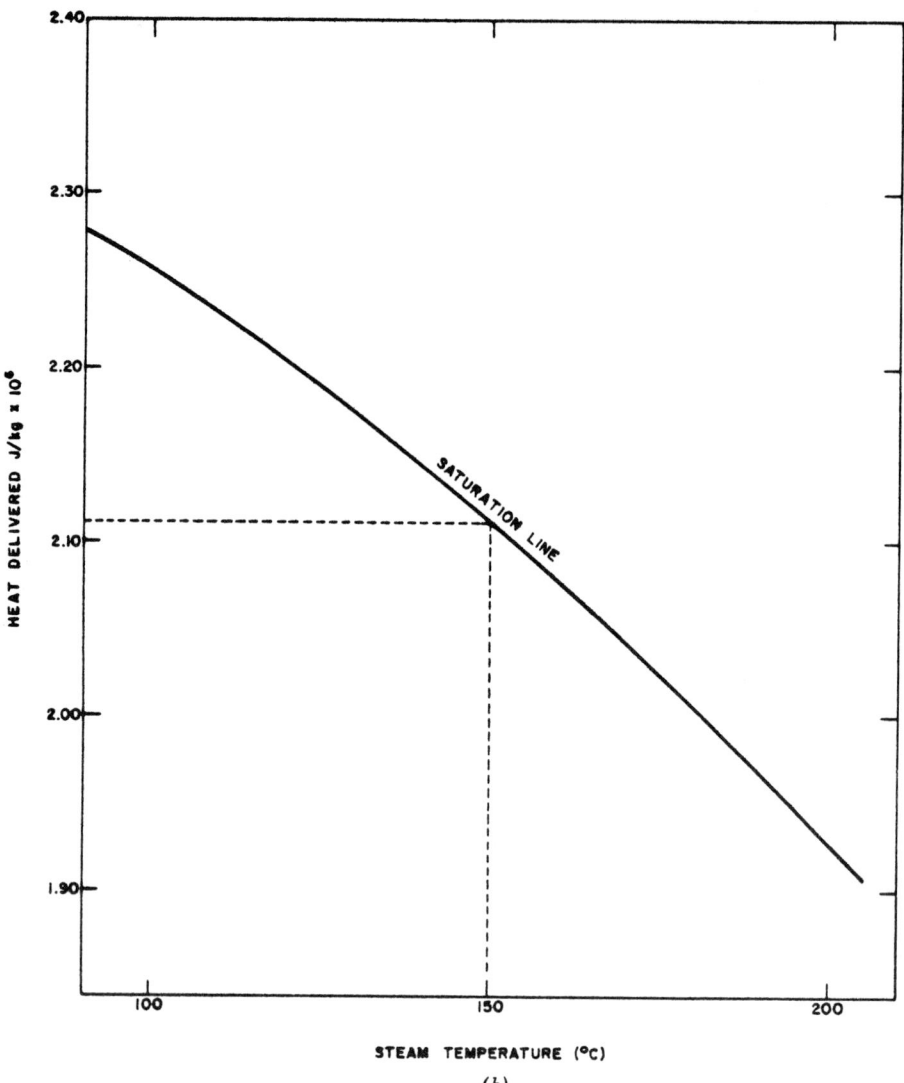

STEAM TEMPERATURE (°C)

(b)

Fig. 12.2-4 *(Continued)*

geothermal fluid and water is 200–250 Btu/h · ft² · °F or (1136–1420 W/m² · k).

$$A = \frac{Q}{U\Delta t} \qquad \text{or} \quad A = \frac{Q}{U\Delta t} \quad \text{(metric)}$$

$$= \frac{175,000}{(200)(80)} \qquad = \frac{(51,275)}{(1136)(44.4)}$$

$$= 10.9 \text{ ft}^2 \qquad = 1.0 \text{ m}^2$$

To find the economical resource fluid temperature, the previous equation is solved for t_r and applied to the example given:

$$t_r = \frac{T_{hx} + 70}{0.6} \qquad \text{or} \quad = \frac{T_{hx} + 21}{0.4}$$

$$= \frac{80 + 70}{0.6} \qquad = \frac{27 + 21}{0.4}$$

$$= 250°F \qquad = 120°C$$

REFERENCES

12.2-1 J. H. Howard, "Principal Conclusions of the Committee on the Challenges of Modern Society Nonelectrical Applications Project," Second United Nations Symposium on the Development and Use of Geothermal Resources, Vol. 3, May 1975, pp. 2127–2140.

12.2-2 Axel Bjornsson, "Exploration and Exploitation of Low-Temperature Geothermal Fields for District Heating in Akureyri, North Iceland, Geothermal Resources Council, *Transactions* 5, pp. 478–495, October 1981.

12.2-3 L. S. Georgsson, H. Johannesson, and E. Gunnlaugsson, "The Baer Thermal Area in Western Iceland: Exploration and Exploitation," Geothermal Resources Council, *Transactions*, 5, pp. 511–514, October 1981.

12.2-4 Sekioka, Mitsuru, Fujitomi, and Masaharu, "National Projects on Direct Utilization of Geothermal Resources in Japan," Geothermal Resources Council, *Transactions* 5, pp. 567–569, October 1981.

12.2-5 F. W. Childs, K. W. Jones, L. B. Nelson, J. A. Strawn, and M. K. Tucker, "Progress in District Heat Applications Projects," Geothermal Resources Council, *Transactions* 4, pp. 549–552, September 1980.

12.2-6 K. S. Robinson, "Status of Direct Heat Application Projects," Geothermal Resources Council, *Trans.* 5, pp. 563–566, October 1981.

12.2-7 Robert R. Lansford, L. N. Chaturvedi, G. H. Abernathy, B. J. Creel, D. C. Nelson, D. J. Cotler, N. R. Gollehon, T. S. Clevenger, and R. C. Patterson, "Utilization of Geothermal Energy for Agribusiness Development in Southwestern New Mexico," Geothermal Resources Council, *Transactions* 4, pp. 581–584, September 1980.

12.2-8 Robert B. McEuen, "Thermal Waters of Bulgaria: Present Use and Probable Origin," Geothermal Resources Council, *Transactions* 2, pp. 423–426, July 1978.

12.2-9 S. A. Subramaniam, "Present Status of Geothermal Resources Development in India, "Proceedings Second United Nations Symposium in the Development and Use of Geothermal Resources, Vol. 1, 1975, pp. 269–271.

12.2-10 T. J. Zeller, W. H. Grams, and S. M. Howard, "Direct Utilization of a Moderate Temperature Geothermal Resource in Agribusiness, "Geothermal Resources Council, *Transactions* 4, pp. 633–636, September 1980.

12.2-11 Paul A. Rodzianko, "Case History of Direct Use Geothermal Applications," Geothermal Resources Council, *Transactions* 3, p. 589, September 1979.

12.2-12 R. A. Walker and D. J. Entigh "Status of U.S. Direct Utilization of Geothermal Energy," Geothermal Resources Council, *Transactions* 5, October 1981.

12.2-13 Department of Energy, "Rules of Thumb for Geothermal Direct Applications," U.S. Department of Energy, Idaho Operations Office, 1978.

12.3 GEOTHERMAL ENERGY FOR ELECTRICITY GENERATION

John Briedis

The world generating capacity from geothermal sources stands at about 2500 MW as of 1981 and is expected to continue to grow at about 15% per year worldwide and at about 16% in the United States for the near future.[1] Table 12.2-1 shows the distribution by country of geothermal electric generating capacity worldwide. There are about 115 operating units in 14 countries.

12.3-1 Geothermal Design

The production of geothermal electricity and the design of geothermal plants are greatly dependent on the characteristics of the natural resource. The enthalpy, chemistry, pressure, and flow rate are all site, time, and resource dependent. The importance of a thorough resource characterization and assessment to the success of a geothermal project cannot be overemphasized. Only after such a thorough site exploration, testing, and resource analysis phase should the project proceed to initiate a final design phase and make decisions on the purchase of generating equipment. In general, the factors that govern geothermal plant design are the mastery of systems utilizing low-to-moderate temperature design, systems minimizing scaling and corrosion, systems designed to anticipate changes in the geothermal resource with time and to be readily modified to cope with possible changes, and systems having provisions for future expansion.

The actual generation of power from a large geothermal power plant is quite similar in many aspects to that of producing power from fossil fuel plants. Many of the design concerns and factors discussed on the design of fossil plants in other sections of this handbook also apply to geothermal plant power design and will not be repeated here. Instead, emphasis will be placed on some of the

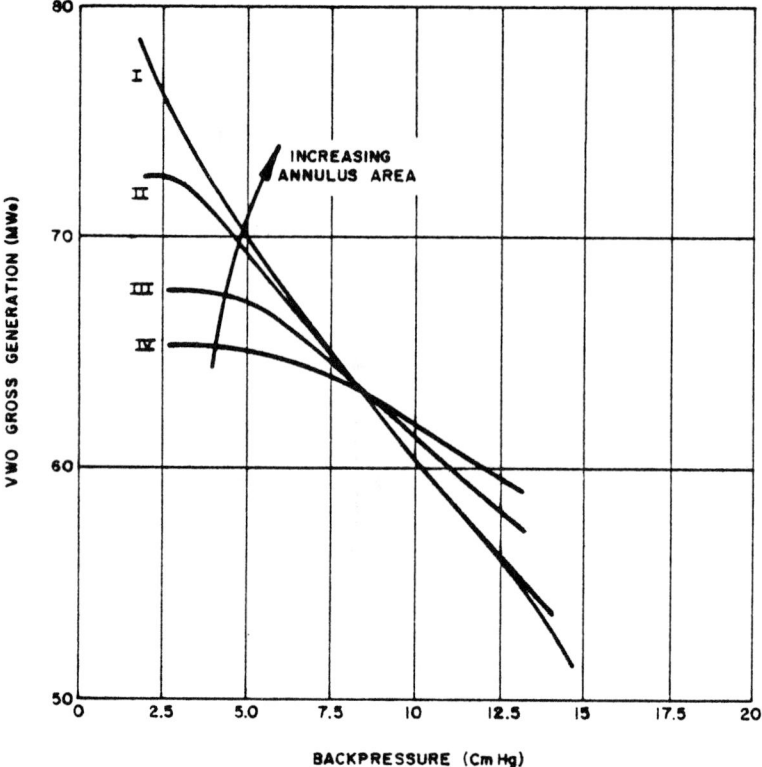

MODIFIED FROM:
    TUCKER & KLEINHANS (1981)

Fig. 12.3-1   Generation versus back pressure—geothermal turbines (modified from ref. 2).

principal differences and unique features of geothermal design. These factors are discussed below:

    1. Boiler.   The boiler is replaced by a complex system composed of the geothermal reservoir, production wells, distribution system, and steam separators.

    2. Use of low-pressure steam.   The energy available from the low-pressure saturated geothermal steam is considerably less than from the higher-pressure superheated steam generated in conventional fossil plants. A typical 55-MW geothermal unit requires approximately 140 kg/s (1,100,000 lb/h) of steam exhausted at 10 cm (4 in.) Hg A. This is about $2\frac{1}{2}$ times the rate of flow of a comparable size conventional fossil-fueled unit.[2] To describe geothermal turbine cycles, it is customary to use the term *steam rate* expressed in kg/kW · h (lb/kW · h) instead of the conventional plant usage of turbine *heat rate* usually expressed in J/kW · h (Btu/kW · h) which does not adequately describe the system.[2] The heat rate is somewhat ambiguous since the energy is produced by a complex natural system and not by the combustion of fuel.

    3. Turbines.   Many of the early power plants purchased steam on the basis of "mills per net kilowatt-hours." As a result, capital investments were of prime consideration, plant efficiency was not a contributing factor, and plant auxiliary power had little or no associated cost penalty.

    More recently steam contracts call for the purchase of steam on a "per pound" basis. Thus, all steam used (in either the generation of power or for auxiliary equipment) carries an associated cost per pound. For this reason, recent geothermal power plants are being designed more efficiently.

    One of the prime factors to consider in efficient geothermal turbine design is the length of the last-stage blades. In general, longer lasting stage blades reduce exhaust losses and prevent choking at low operating back pressures. The turbine performance compared to annulus is shown in Fig. 12.3-1. The largest annulus area, a result of longer last-stage blades, increases the efficiency while increasing the price of the turbine and pedestal. When steam is purchased on a per pound basis, the added performance generally justifies the increased cost.[2]

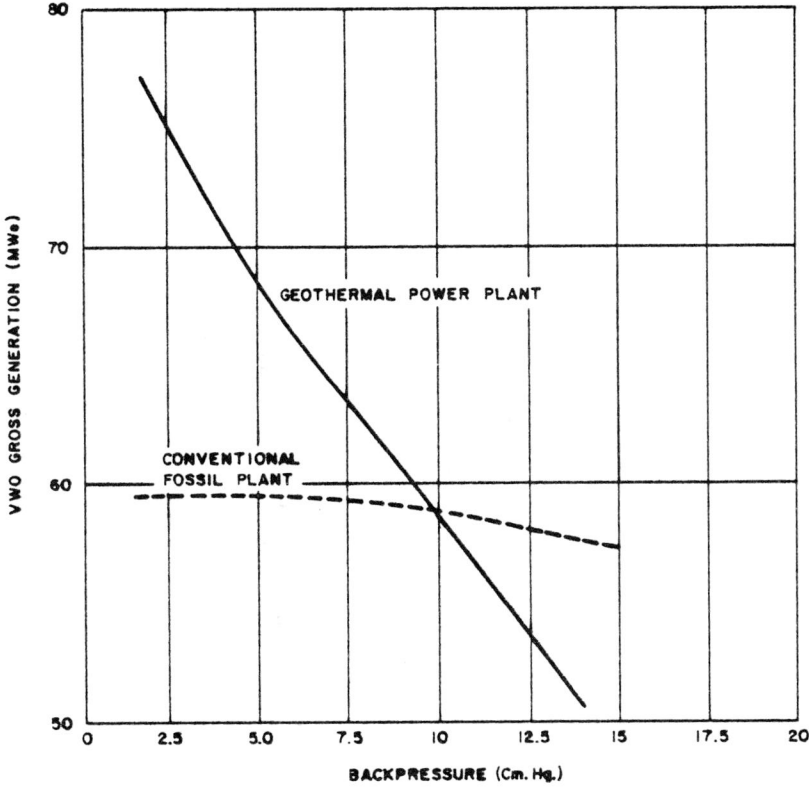

MODIFIED FROM:
TUCKER & KLEINHANS (1981)

Fig. 12.3-2   Generation versus back pressure (modified from ref. 2).

4. Sensitivity to back pressure. The geothermal system is more sensitive to lowering of back pressure compared to a fossil plant. Tucker et al.[3] describe this as due to the combined effects from:

(a)   Convential plants use steam with higher available energy than the geothermal unit. This gives a conventional plant a much lower steam rate.

(b)   The conventional plant uses regenerative feedwater heating; therefore, as the condenser pressure is lowered, the turbine exhaust enthalpy and the condensate temperature are both lowered. The lowering of turbine exhaust enthalpy contributes to an increase in generation; however, the lowering of the condensate temperature requires more steam to be extracted from the turbine, reducing the amount of steam that is expanding to do work. The net effect is a smaller change in generation with back pressure in conventional turbines than is seen with geothermal turbines.

In the geothermal system the extraction of steam for feedwater heating is not thermodynamically beneficial. Essentially the geothermal turbine has constant flow throughout the system and generation of power is a direct function of the steam inlet enthalpy minus exhaust enthalpy and expressed by an equation of the form[3]

$$\text{Gen} = m(h_i - h_e)$$

where m = mass flow rate
h_i = inlet enthalpy
h_e = exhaust enthalpy

Figure 12.3-2 illustrates the incentive for lowering condenser pressure in geothermal plants. For a 55-MW geothermal unit a reduction in condenser pressure of 2.5 cm Hg A results in approximately 5 MW of increased generation capacity.

5. Condensers. Direct-contact condensers were used on early geothermal power plants. This type of condenser is ideal for a power cycle where the condensate is used as makeup for the cooling tower. With this type of condenser, however, a large portion of the H_2S goes back into solution and is then stripped by the cooling tower, contributing to a significant amount of pollution. More recent plants use a "shell-and-tube" type of condenser to keep the noncondensible gases from coming in contact with the circulating water. These condensers are considerably more expensive and have poorer performance, but do increase "partitioning," the amount of noncondensibles kept from going into solution. Regardless of the type of condenser, the corrosive nature of the condensate warrants the use of corrosion-resistant materials such as stainless steel or titanium.

6. Fuel handling and feedwater systems. Geothermal plant design is simplified by the absence of fuel handling, ash handling, combustion, and feedwater systems.

7. Makeup water. In geothermal plants cooling tower water loss due to tower evaporation, drift, and blowdown is replenished by the use of condensate from the condenser. There is no need for an external source of water. Cooling tower overflow is reinjected in the geothermal reservoir.

8. Resource fluctuation during operation. In several instances modifications to existing geothermal plant equipment or operation procedures have been necessary as the nature of the reservoir steam has changed. The changes usually affect the temperature, pressure, composition of noncondensible gases, or the actual percentage of wet versus dry steam.

9. Scaling. In considering the design of a power system at a particular geothermal site, the scaling potential must be considered. Scaling of well head, delivery systems, turbine, and condenser parts is usually a function of the temperature and the particular geochemistry of the resource. Concentrations of silica, calcium, carbonate, sulfate, and heavy metal ions are of particular concern. In most cases the solution to scaling problems is scheduled periodic cleaning of the equipment. The frequency is dependent on the resource.

10. Environmental concerns. Environmental concerns under most regulatory environments dictate, as a minimum, the need for an H_2S abatement system and the need for disposal of blowdown by use of reinjection wells. This subject is discussed in greater detail in Section 12.4.

12.3-2 Power Systems Descriptions

Presently there are four types of geothermal systems in commercial use to date:

  Dry-steam plants.
  Single-flash plants.
  Double-flash plants.
  Multiflash plants.

A typical dry-steam system for a large geothermal unit is shown in Fig. 12.3-3. It is the simplest of all the systems and according to Di Pippo[1] accounts for about 60% of all the present geothermal generating capacity in the world to date.

As was described in Section 12.1, dry steam is only a small percentage of the world's geothermal potential, although the easiest to develop. A more common condition worldwide is a two-phase mixture of liquid and vapor at the well head. The quality of the fluid is very site dependent, but according to Di Pippo,[1] the vapor phase normally ranges from about 10 to 50% mass fraction of the total. The flashing of hot water to steam has taken place either in the reservoir or in the well as the fluid emerges to a lower-pressure environment. Such fluid is then separated at the surface to its gas and liquid phases. The steam phase is then directed through a turbine. When this is accomplished in a single separation step, the term commonly applied to the system is *single-flash* (illustrated in Fig. 12.3-4). Utilization efficiency for flashed system is less than for the dry-steam system due to energy loss in the discarded fluid. To recapture some of this energy, additional flashing stages can be introduced, depending on the well head pressures, at considerable expense to the project. In a double-flash unit an additional procedure is added where the primary hot water initially separated from the well head fluid is processed through a flash vessel and induced to flash at a lower pressure. The additional steam is then expanded through the low-pressure stages of the turbine, as shown in Fig. 12.3-5. If this process is repeated and the resultant steam is introduced at three or more pressures at the turbine, the plant is of *multiflash* design.

Binary plants use a secondary working fluid such as ammonia, organic fluids, or Freon in a closed Rankine cycle. Essential to most binary systems are downhole pumps in the production wells to maintain a sufficiently high pressure in the geothermal fluid to prevent it from flashing. Present problems arise in designing pumps that will withstand the hostile geothermal well environment for a reasonable length of time. The geothermal fluid is used to heat the secondary fluid through a heat exchanger, as shown in Fig. 12.3-6. Presently binary plants are being considered for first-generation commercial application. Several pilot test facilities have been built; they include the Paratunka built in 1967 on Kamchatka Peninsula of the Soviet Union (and now dismantled), Otake and Mori plants operating in 1977–1979 in Japan, and East Mesa in Imperial Valley, California.[1]

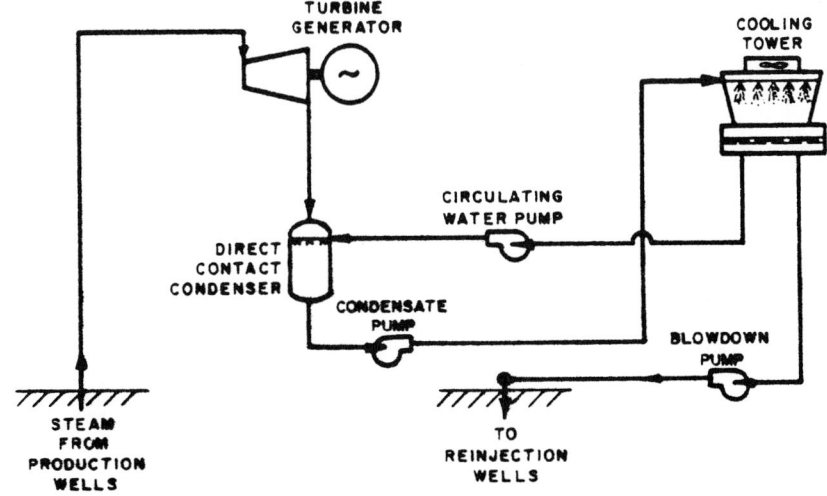

Fig. 12.3-3   Dry-steam geothermal power plant.

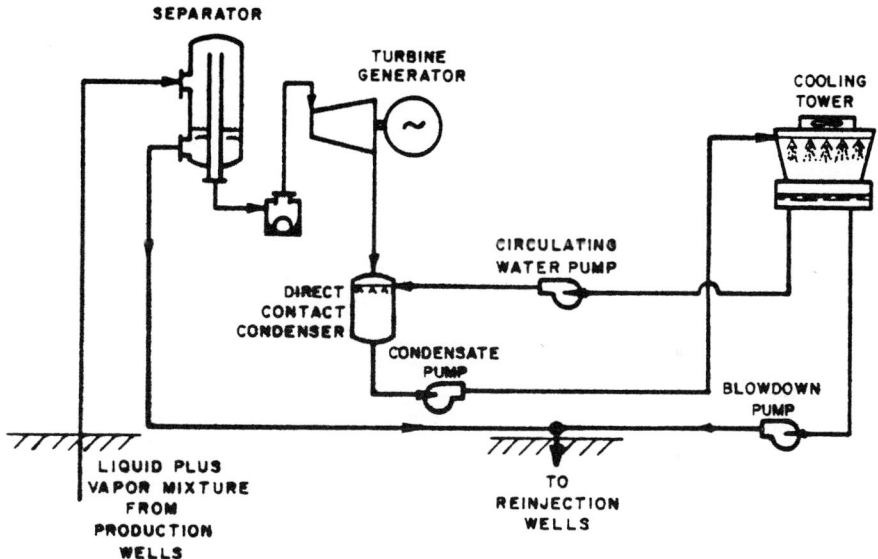

Fig. 12.3-4   Separated steam or "single-flash" geothermal power plant.

Some of the factors to be considered in the design and operation of a binary plant are as follows[1]:

Advantages

More suited to low-temperature hydrothermal resources.

Smaller turbine size for given output.

Less expensive turbine for given output.

High-pressure operation throughout, eliminating vacuum operation.

No problems of air in-leakage.

Noncorrosive working fluid in turbine.

Higher isentropic turbine efficiencies.

Complete dry expansion, eliminating erosion problems.

Condensing temperature can be lower for better cycle efficiency.

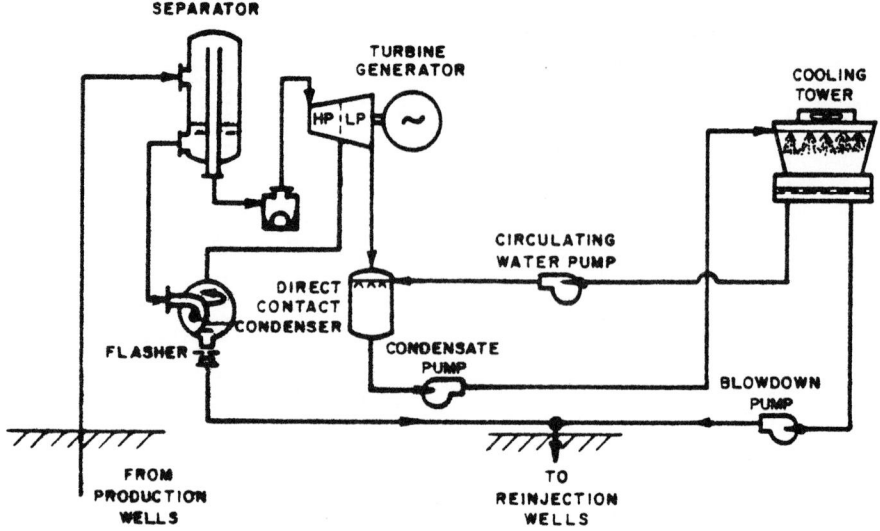

Fig. 12.3-5   Separated steam/hot-water-flash or "double-flash" geothermal power plant.

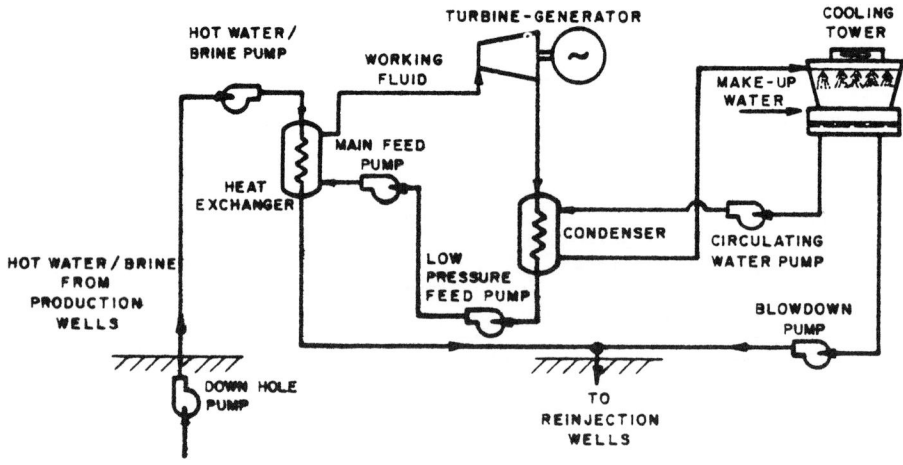

Fig. 12.3-6   Binary geothermal power plant.

Disadvantages

Secondary fluid cost.

No leaks permissible.

Heat exchangers costly.

Huge brine flow rates needed for a reasonable-size plant, leading to disposal problems.

Flammability of hydrocarbon work fluid requires fire protection design.

A system that is presently under development whose results are very encouraging as shown by actual field test trials at geothermal well sites in California, Idaho, and Utah is the biphase rotary separator turbine (RST). The tests demonstrated that it is possible to obtain 17–25% more total power output over that of a single-stage flashed steam plant operating with a medium-temperature geothermal fluid.[4,5] Cereni and Hughes[6] describe the RST system as capable of receiving two-phase flow from a geothermal well, expanding both phases through a nozzle and separating the liquid and vapor to

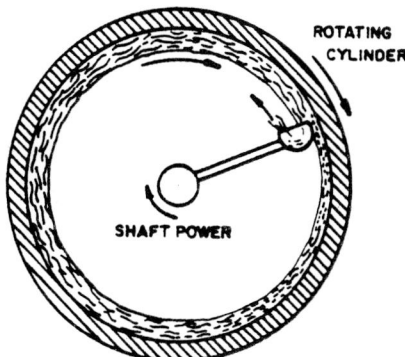

ROTATING
CYLINDER

SHAFT POWER

Fig. 12.3-7 Rotary turbine concept (from ref. 4). (Used by permission of Geothermal Resources Council and Dr. D. J. Cerini.)

provide three forms of output energy:

1. Electricity produced by the accelerated liquid.
2. Electricity produced by further expansion of the separated steam.
3. Pressurized liquid geothermal fluid for reinjection into the geothermal reservoir.

The principle of the liquid turbine in the RST concept can be illustrated as shown in Fig. 12.3-7. The liquid accumulates on the rotating RST walls and drives a scoop immersed in the liquid which serves as a liquid turbine to generate power. The separated steam is then further expanded through a conventional steam turbine–generator.[4] Present tests have demonstrated the production of 1 MW of power from the liquid turbine with an additional 4.72-MW steam output capability. Economic analysis of the RST system compared to single-flash, double-flash, and binary systems indicates that optimum bus bar cost advantages are achieved by using the RST system with 260°C reservoir temperature resource.[5]

Other plant types are being considered. They include combined flash and binary cycles, combination fossil and geothermal plants using geothermal preheat systems or fossil superheat systems, or a combination of both. Much of the above discussion has been centered on relatively large geothermal plant systems of the type that range from 25- to 140-MW size and are presently operating at Geysers, United States; Larderello, Italy; and Weiraki, New Zealand; however, there is considerable growing interest in smaller, fairly portable well head units that produce only a few megawatts of electricity. These are especially attractive in rural areas of the world where power needs are not as high as in industrialized areas and environmental restrictions less stringent.

12.3-3 Plant Performance History

Although there have been considerable problems with the operation and design of geothermal plants, in particular, scaling, corrosion, H_2S abatement, and depletion of the natural resource, it appears that once a geothermal plant is properly designed and adapted to the particular resource characteristics at the site, its performance is quite satisfactory. Di Pippo[1] shows that plant operations have demonstrated the ability to achieve capacity factors (ratio of kilowatt-hours produced per annum to the maximum possible) exceeding 80% and very high availability factors (ratio of hours operating to produce electricity to the total number of hours, usually 8760 h/yr) exceeding 95%. These values are considerably higher than comparable figures for fossil fuel plants.

12.3-4 Cost Comparison With Other Energy Sources

Geothermal power development can be cost-effective when properly developed. Table 12.3-1 shows the favorable cost comparison with other major power resources.[7] The comparison was done in 1980 dollars by the California Energy Commission and compares costs within the same geographic, demographic, and economic setting. However, this must be tempered with the understanding that geothermal power is only an alternative in those areas of the world where the resource is present.

Also shown in Fig. 12.3-8 is the dependence of geothermal plant development cost on the resource temperature.[8] Similar relationships exist, although not easily quantified, with other aspects of resource quality such as scaling potential, corrosion ability, and percentage of noncondensible gases. This again points out the high economic susceptibility of a geothermal project on a thorough resource evaluation and implementation of the proper design for the specific resource.

**TABLE 12.3-1 COMPARATIVE BUS BAR COSTS OF
GEOTHERMAL AND CONVENTIONAL BASELOAD
GENERATING FUELS**[a]

	Fixed Costs	Fuel Costs	Other O & M	Total
Geothermal				
Steam	1.1	4.4	0.02	5.6
Liquid flash	3.5	7.8	0.08	11.4
Coal	3.4	3.5	0.12	7.0
Nuclear	4.8	2.9	0.14	7.9
Oil	2.3	9.8	0.12	12.2

[a]Modified from ref. 7. 1980 dollars. Levelized cents/kW · h.

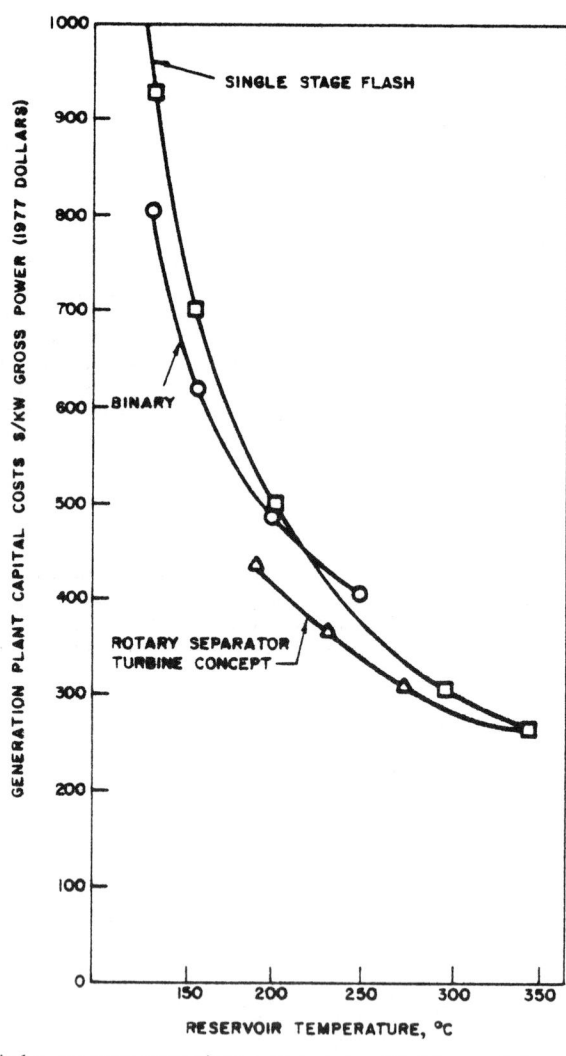

Fig. 12.3-8 Capital costs versus reservoir temperature, (modified from ref. 8; used by permission).

REFERENCES

12.3-1 Ronald Di Pippo, "Geothermal Energy as a Source of Electricity," U.S. Department of Energy, DOE/RA/28320-1, 1980.

12.3-2 R. E. Tucker and P. V. Kleinhans, "Geothermal Power Cycle Considerations," P.C.E.A. 46th Annual Engineering and Operating Conference, Los Angeles, March 1981.

12.3-3 R. E. Tucker, P. V. Kleinhans, and L. R. Keilman, "Economic Impacts on Geothermal Power Plant Design," Geothermal Resources Council, *Transactions* 4, pp. 533–536, September 1980.

12.3-4 R. G. Campbell, "Comparisons of Advanced Power Conversion Concepts," Geothermal Resources Council, *Transactions* 5, pp. 393–396, October 1981.

12.3-5 D. J. Cerini and J. Record "Well Head Power Production with a Rotary Separator Turbine," presented at the Conference on Small Geothermal Power Plants, Queen Mary, Long Beach, CA, June 1982.

12.3-6 D. J. Cerini and E. Hughes, "Field Tests of the Biphase Geothermal Rotary-Separator Turbine," Geothermal Resources Council, *Transactions* 5, pp. 401–404, October 1981.

12.3-7 P. C. Grew, "Geothermal Energy: A California Success Story," Geothermal Resources Council, *Transactions* 5, pp. 595–598, October 1981.

12.3-8 D. H. Klipstein and R. S. Atkins, "The Economics of Geothermal Electricity Generation from Hydrothermal Resources Using The Biphase Rotary-Separator Turbine," Geothermal Resources Council, *Transactions* 5, pp. 667–670, October 1981.

12.4 ENVIRONMENTAL HAZARDS

Hemendra K. Acharya

Geothermal energy causes far less pollution than fuel combustion. Nevertheless, the increase in exploitation of geothermal energy at a time of rising environmental consciousness necessitates concern for the undesirable effects to the environment. Geothermal operations can pollute the air with H_2S and other airborne poisons and can pollute the water with poisonous constituents and waste heat. It is likely that in most places the rate of water withdrawal will significantly exceed natural replenishment and therefore may lead to subsidence of the ground. Reinjection of spent fluid to the reservoir may minimize this possibility. Both withdrawal and reinjection may affect seismicity by either increasing or decreasing the risk. Finally, geothermal operations invariably increase the noise level in an otherwise quiet and peaceful location.

Generally dry-steam fields are less polluting than liquid-dominated fields. Land subsidence, silica, heat pollution of rivers and water-borne poisons are predominant features of wet fields. Stringent antipollution laws have now been enacted in certain countries and antidotes of acceptable efficiency have been found for nearly every possible source of geothermal pollution. Hartley and Di Pippo[1] review the environmental hazards and the technology to mitigate them.

12.4-1 Pollutants

The chemical characteristics of geothermal fluids vary greatly between reservoirs and to a lesser degree even within the same reservoir. The characteristics also change with time because of selective withdrawal and recharge factors. Figure 12.4-1 lists some of the more significant chemical constituents of geothermal fluids and graphically depicts their ranges.[1] All of the constituents are natural components of geothermal fluids. This information is developed on the basis of available data in the literature.

Noncondensible gases, those that do not condense at operating temperatures, are environmentally important constituents of geothermal fluids. They may be free gases or gases dissolved or entrained in the liquid phase. H_2S has been a component of greatest concern to this time. Noncondensible gases usually comprise between about 0.3 and 5% of flashed steam from geothermal fluids. Figure 12.4-2 depicts the known ranges of noncondensible constituents as percentages of total noncondensible gases. Also shown are their probable ranges in parts per million of total gases (including steam) in a steam system. Table 12.4-1 shows the effects on humans of inhalation of these gases.

12.4-2 Atmospheric Pollution

The gases accompanying geothermal fluids almost invariably contain H_2S. This noxious gas, in moderate and harmless concentrations, has a characteristic smell, but when more strongly concentrated, it paralyzes the olfactory nerves and thus becomes odorless. Therein lies its danger. When it is present in lethal quantities, it gives no warning of its presence. California has imposed a state ambient standard of 0.03 ppm by volume as a 1-h average, which is near the odor threshold. Technologies to control H_2S pollution from geothermal operations are directed primarily at incoming

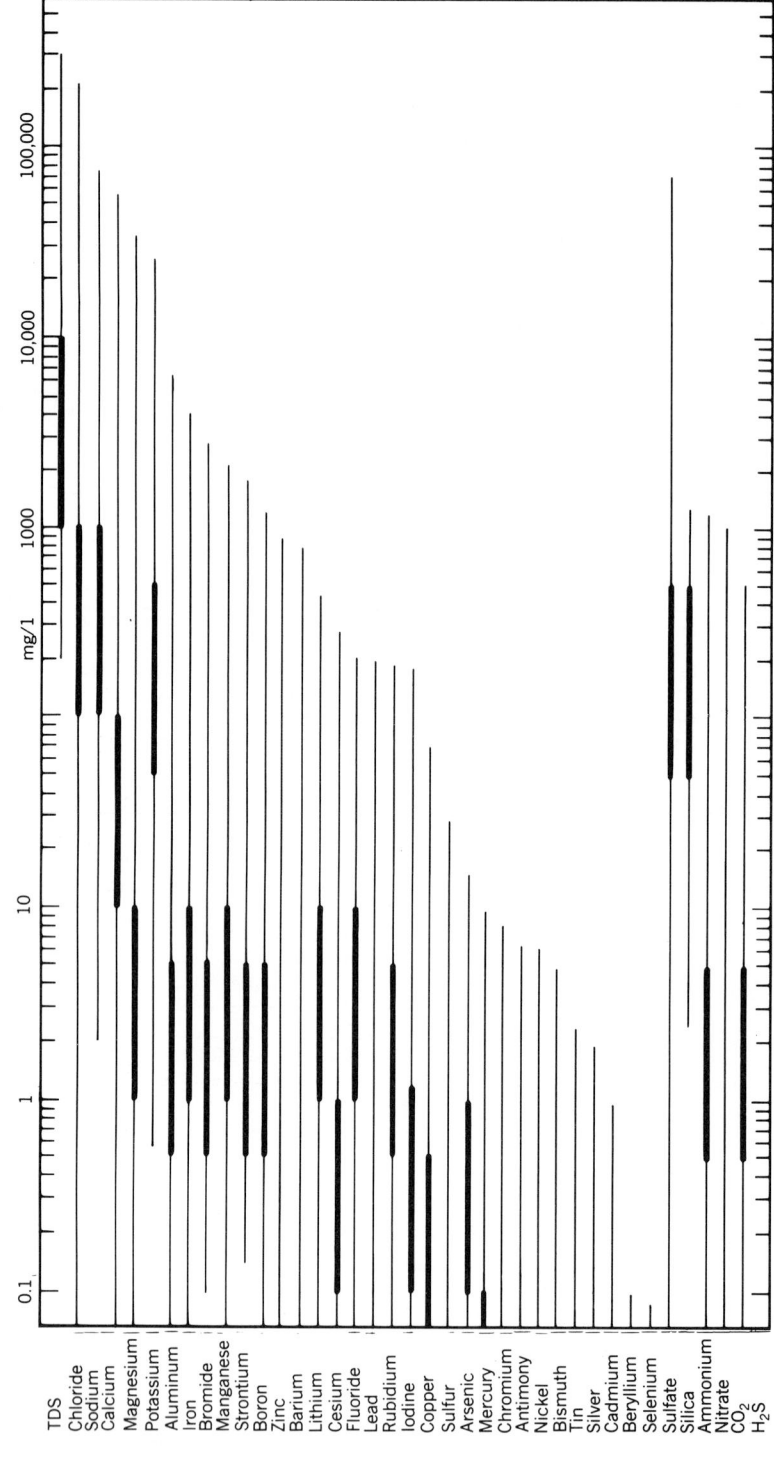

Fig. 12.4-1 Chemical pollutants in geothermal fluids (from ref. 1).

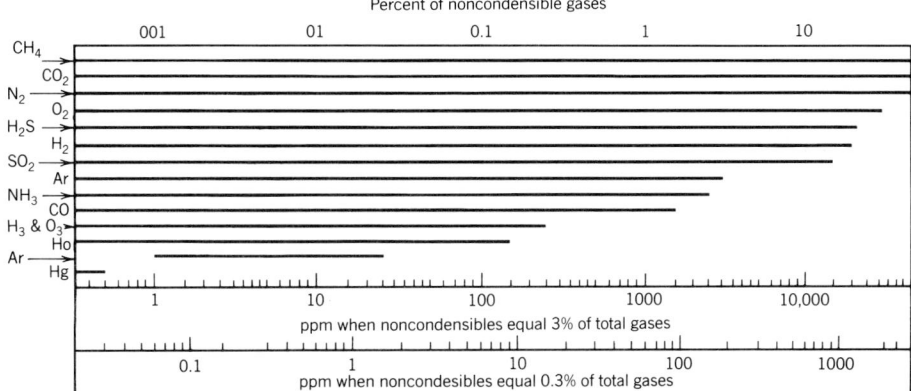

Fig. 12.4-2 Noncondensible gases in geothermal fields (from ref. 1).

TABLE 12.4-1 EFFECTS ON HUMANS OF INHALATION OF GASES OR VAPORS

	Acute Toxicity	Chronic Toxicity	Systematic Effects		Fatal Dose (pmm)	Odor (pmm)
			Respiratory	Nervous		
CH_4	1	1	×			
CO_2	0	0				
N_2	0	0				
O_2	0	0				
H_2S	3	3	×		1000	0.03
H_2	0	0				
SO_2	3	2	×		400–500	3
Aı	1	0	×			
NH_3	3	21	×			
CO	3	1	×		4000	
H_3BO_3	2	2	×			
He	1	0	×			
As	3	3	×	×		
Hg	2	3	×	×		

steam, condenser vent emissions, and cooling tower emissions. As much as 95% of the H_2S can be removed using available control processes discussed below.

Stretford Process

The Stretford process is applicable to dry-steam reservoirs and is over 99% effective, removing essentially all of the H_2S from the condensed gases.[2] Noncondensible gases from the condenser ejector are scrubbed with an aqueous solution containing sodium carbonate, sodium metavandate, and anthraquinone disulfonic acid (ADA). Elemental sulfur is produced with the following overall reaction.

$$2H_2S + O_2 \rightleftharpoons 2H_2O + 2S$$

A surface condenser rather than a direct-contact condenser must be used to eliminate contact of the cooling water with the condensate. H_2S is controlled without any direct effect on the power cycle.

Other processes useful for eliminating H_2S in dry-steam reservoirs are *Iron Catalyst* (or Ferrifloc) *system* in which ferric sulfate in solution is added to cooling water, thus oxidizing the H_2S contained in the aqueous phase. Ferric ions become available to react with the dissolved H_2S, thus forming elemental sulfur, water, and ferrous ions. Ferrous ions react with the oxygen encountered in the cooling water to regenerate the ferric ions. The iron catalyst system results in significant corrosion rate increases in the condenser, cooling tower, and associated piping.

EIC Process

The EIC process removes H_2S from raw geothermal steam by scrubbing it with an aqueous solution of copper sulfate. The H_2S and $CuSO_4$ react in the scrubber, forming copper sulfide precipitate. A benefit of an upstream scrubbing process is the reduction of the corrosive effects of H_2S on turbine and condensing/cooling cycle equipment. This enables the use of standard materials of construction for the power plant equipment and piping. The EIC process removes H_2S without significant degradation of steam quality (temperature and pressure).

Dow Oxygenation Process

The Dow oxygenation process removes H_2S from geothermal brine at the well head; thus, it is applicable only to liquid-dominated resources. The Dow process oxidizes the aqueous H_2S by injecting oxygen directly into the geothermal brine. Thorough mixing to facilitate contact of the brine and oxygen can be accomplished using in-line mixers or concurrent packed towers. Removal of H_2S at the well head should provide a less corrosive brine in the pipelines of the gathering field and a less corrosive brine or steam in the power cycle. This process has only been tested in a laboratory pilot plant.

Various other harmful elements such as carbon dioxide, mercury, arsenic compounds, and radioactive elements can sometimes escape into the air at geothermal exploitation sites. These elements are not found in harmful quantities and therefore have not been addressed so far.

12.4-3 Water Pollution

By far the greatest volume of water in geothermal operations will be spent water variously contaminated by natural constituents from the geothermal reservoirs. If the reservoir is dry-steam dominated, the liquid will be principally condensate, relatively clean, and low in volume. From liquid-dominated reservoirs large volumes of water will be withdrawn which may be at high temperature and contain chemical constituents in quantities unacceptable for direct discharge into fresh or seawater. Table 12.4-2 lists the aquatic life criteria for constituents in geothermal fluid.

Water pollution control technologies, therefore, include wastewater treatment and wastewater disposal. Waste treatment technologies include evaporation ponds, chemical precipitation, filtration,

TABLE 12.4-2 AQUATIC LIFE CRITERIA

Constituent	Criteria for Fresh Water	Criteria for Marine Water	Remarks
Ammonia	0.02 mg/L		Toxicity antidependent
Arsenic			Daphnia impaired by 4.3 mg/p
Barium			Toxicity level, 50 mg
Beryllium	0.11 mg/e soft water		Toxicity hardness dependent
	1.1 mg/e hard water		
Boron			Toxic luminous as 14,000 mg/e.
Cadmium	0.004–0.0004 mg/e soft water	0.005 mg/e	
	0.012–0.0012 mg/e hard water		
Copper	0.1 96 LC_{50}	0.1 96 h LC_{50}	Toxicity–alkalinity dependent
Iron	1.0 mg/e		Toxicity variable
Lead	0.01 96 h LC_{50} (sol. lead)		Salnids most sensitive fish
Manganese		0.1 mg/e	Not a problem
Mercury	0.0005 mg/e	0.0001 mg/e	High bioaccumulation and thus affects human food
Niliates			Toxicity to fish, 900
Phosphorus		0.0001 mg/e P	Eutrophication factor
Selenium	0.01 96 h LC_{50}	0.01 96 h LC_{50}	Toxic at 72.5
Silver	0.01 96 h LC_{50}	0.01 96 h LC_{50}	Toxicity dependent on compound
H_2S	0.0002 mg/e		Toxicity dependent on temperature

reverse osmosis, electrodialysis, or ion exchange. Depending on the constituents and the amount to be removed, many of the treatment technologies may be used individually or in series.

12.4-4 Reinjection

Geothermal wastewater requires disposal, whether or not it requires prior treatment. Discharge into the ocean or into rivers has been tried. However, subsurface injection of spent geothermal liquid to the producing reservoir now appears to be the most feasible disposal alternative. Successful subsurface injection tests have been performed in a number of geothermal fields in the United States and abroad, for example, Geysers, Valles Caldera in New Mexico, Matsukawa, and Otake fields in Japan, Broadlands in New Zealand, and Ahuachapan in El Salvador. There are several reasons for choosing reinjection as a disposal method:

1. It eliminates releasing the waste into surface water bodies.
2. It decreases the risk of ground subsidence.
3. It helps keep reservoir pressure high.
4. It is an effective means of preventing not only chemical but also thermal pollution of surface water bodies.

The injection scheme should be designed to optimize the travel path and time of flows between injection wells and producing wells, thus preventing rapid cooling of the production water. The geological suitability of the reservoir for reinjection has to be investigated. Unless the geothermal reservoir is very competent, a cased hole with slotted liner in the injection zone is used. Figure 12.1-4 shows the layout of production and reinjection wells at Ahuachapan, El Salvador.

12.4-5 Scaling and Corrosion

Scaling

One of the major problems in geothermal energy conversion and injection systems is silica precipitation and scaling formation. With reduction in temperature at various stages of exploitation, the silica will either precipitate immediately or will remain for a limited time in a state of supersaturation, according to the form and conditions in which it occurs. Supersaturated silica passed through a sand filled fluidized-bed heat exchanger will lead to precipitation and removal of silica. Polymerization reductions will help in eliminating monomeric silica precipitation. Silica-laden discharge waters have been successfully treated with slaked lime at Otake field in Japan to precipitate silica and any arsenic if present. Silica scale has been successfully removed from a well head in the Matsukawa field in Japan by allowing the scale to react with NaOH. Shock treatment, subjecting the formation to an almost instantaneous applied pressure differential, and sustaining the differential temporarily has also been reported to be successful in loosening the material plugging the injection formation.

Corrosion

High salinity accelerates electrical corrosion by increasing the conductivity of the medium. Downhole corrosion rates are a function of temperature, flow rate, well depth, pressure, brine chemistry, pH, and dissolved gases. Experimental tests in the Salton Sea geothermal area have shown that corrosion is generally a problem below 300 m and increases in severity with depth; above 300 m silica scale apparently protects the casing from corrosion. Lowering the temperature usually decreases the corrosion rate. Increasing fluid velocity increases the corrosion rate. Removing oxygen is an old corrosion control technique.

12.4-6 Heat Pollution

Fundamental to all heat power systems working on the Rankine cycle is that a portion of the heat supplied to the cycle must be rejected. Fossil-fueled plants presently waste about 60% of their heat input, nuclear-fueled plants about 70%, and geothermal power plants, because of the low thermal efficiency inherent to relatively low temperature sources, waste 85% or more.

The waste heat rejection system for a geothermal power plant will cost proportionally more of the total station cost than the waste heat portion of conventional plants. The amount of heat to be rejected per kilowatt will be 3–6 times that of a nuclear plant. Because geothermal stations tend to be smaller in size, equipment costs per kilowatt will be greater.

Where cooling towers are used, this waste heat escapes into the atmosphere and into surplus condensate; where direct river cooling is adapted, it is mostly spent in raising water temperature. In wet fields another enormous source of waste heat can arise from the reinjection of very hot unwanted bore water into rivers or streams. Robertson[3] discusses various waste heat rejection systems.

One possible way of reducing waste heat may be reinjection of surplus cooling tower water and rejected base water into the ground. Other possible ways are to generate additional power by means of binary cycles or to establish dual or multipurpose plants that usefully extract low-grade heat from the turbine exhausts or from rejected bore waters.

12.4-7 Noise

The noise of escaping steam at high pressure can be very distressing to the ears, and workers on new well head sites have to wear ear plugs or muffs lest their hearing be damaged. Noise can be greatly mitigated by means of effective mufflers that destroy the kinetic energy of the discharging fluids, reduce the volume of noise and deflect it skywards, and, more important, lower the pitch to a frequency level less painful to the ear.

REFERENCES

12.4-1 R. P. Hartley and R. Di Pippo, *Environmental Considerations: Source Book on the Production of Electricity from Geothermal Energy*, J. Kestin (ed.), U.S. Department of Energy, Supt. of Documents, Washington, D.C., 1980.
12.4-2 J. Laszlo, "Application of the Stretford Process for H_2S Abatement at the Geysers Geothermal Power Plant," Proceedings of 11th Intersociety Energy Conversions Engineering Conference, September 12–17, State Line, NV, 1976.
12.4-3 R. C. Robertson, *Waste Heat Rejection from Geothermal Power Stations: Sourcebook on the Production of Electricity from Geothermal Energy*, J. Kestin (ed.), U.S. Department of Energy, Supt. of Documents, Washington, D.C., 1980.

12.5 CHALLENGES

Hemendra K. Acharya

One or more of several potentially important breakthroughs in utilization technology may greatly expand the development of geothermal systems in the immediate future. The most significant of the possible breakthroughs are:

1. Heat exchanger technology that would permit utilization of the heat from fluid down to 100°C or less, since total heat contained in easily recoverable fluids at temperatures of 100–180°C is far greater, perhaps by multiple of 100, than total easily available heat above 180°C.
2. Low cost mechanical, chemical, or nuclear fracturing of hot dry rocks to increase permeability, thus permitting introduction of fluids and recovery of stored energy.
3. New technology or development that favors wide applications to space heating, horticulture and product processing.

BIBLIOGRAPHY

Ronald Di Pippo. *Geothermal Energy as a Source of Electricity*. U.S. Department of Energy, DOE/RA/28320-1, January 1980.
Geothermics, International Journal of Geothermal Research and Its Applications. E. Barbrer (Ed.), National Research Council—International Institute for Geothermal Research, Pergamon Press, Elmsford, N.Y., vols. 1–10.
J. Kestin, *Source Book on the Production of Electricity from Geothermal Energy*. U.S. Department of Energy, Washington, D.C., 1980.
Proceedings of Second United Nations Symposium on the Development on Use of Geothermal Resources, Vols. 1–3. U.S. Energy Research and Development Administration, Washington, D.C., May 1975.
Transactions of Geothermal Resources Council, Vols. 1–5. Geothermal Resources Council, Davis, CA, 1977–1981.

For listing on specific topics please see the references at the end of each subsection.

CHAPTER 13

ENERGY ASPECTS OF ENVIRONMENTAL CONTROL

C. HARDY LONG

Virginia Polytechnic Institute and State University
Blacksburg, Virginia

THOMAS F. WEAVER

University of Rhode Island
Kingston, Rhode Island

13.1   INTRODUCTION

This section deals with the air in a room of a home, industry, hospital, bakery, church, computer center, and so forth where (1) the mixing of fresh air and room air, (2) the temperature of the air, (3) the humidity of the air, and (4) the purity of the air are controlled in order to keep humans comfortable, to make industrial processes possible, and to keep computers running.

In the summer the controls should maintain the $25.9°C$ ($78°F$) and $40-60\%$ relative humidity (RH) requested by the president for energy-efficient systems. Air conditioners are used to decrease the humidity in eastern United States and elsewhere where humidity is high, and evaporative coolers could easily be used from Dallas to California where it is hot but dry, and humidification is needed. If steam, gas, solar, or other forms of energy are available and inexpensive, absorption systems could be used instead of vapor compression air-conditioning units. This would be the summer cycle for the heat pump. Sometimes, and in some areas, much of the time, all that is needed is the ventilation of a house or building when the temperature is between 18.7 and $27°C$ (65 and $80°F$) and $40-60\%$ RH and comfort can be achieved while saving on cooling or heating system energy use.

In the winter the temperature is maintained at $20.3°C$ ($68°F$), according to the presidential order, by using a heat pump or heating system such as oil, gas, wood, coal-fired hot water, steam or hot air furnaces, or even electric furnace or electric baseboard heaters.

Of course, energy storage systems using solar, heat pump, or waste energy can also be used in addition to the normal energy systems supplying heat or cooling to buildings.

T. F. Weaver is the author of Section 13.2-8.

13.2 HOME HEATING FURNACES AND AUXILIARIES

13.2-1 Introduction

Most homes and buildings, private, industrial, public or commercial, are heated today with oil- or gas-fired boilers or electric furnaces. Water is heated to $57-71-83°C$ ($140-160-180°F$) in a boiler and circulated by a thermostatically controlled pump to each area, zone, or room. Unfortunately, these thermostats are often in the hot halls, and the outer rooms are cold. The thermostat should be placed in the coldest room and that area kept at $20.3°C$ ($68°F$). Other areas would be warmer. The pipes from the boiler to the rooms should be insulated.

The system can be a single-pipe series system but must ensure that the last room gets sufficient water hot enough to satisfy its heating load. Sufficient gallons per minute, say $\frac{1}{2}$ gpm, must be circulated through the room radiators in a given room and cooled, say from 160 to $120°F$.

Illustration. $\frac{1}{2}$ gpm $\times$ 8.33 lb/gal $\times$ 60 m/h $\times$ 1 Btu/lb $\cdot$ °F $\times$ ($160-120°F$) = 9996 Btu/h or 9996 Btu/h $\div$ 3413 Btu/kW $\cdot$ h = 2.93 kW. Note that if the radiators are in series, then the water entering the second room is at $120°F$ instead of $160°F$, and the room could be much colder than the $68°F$ at which it was designed to be maintained. Rooms 3 and 4 would be even colder. Clearly more water must be circulated! The water leaving the last radiator must be at $120°F$.

Two-pipe systems have a $160-180°F$ supply all the way from the boiler to the farthest room and a return pipe to the boiler. Radiators are paralleled across these supply and return pipes. Each room would get the $\frac{1}{2}$ gpm or whatever is needed to supply its heating load.

Figure 13.2-1 shows the capacity and dimensions of radiators, which should be placed under windows or on outer walls, not on inside walls. Radiators should have adequate capacity for the room or zone in which they are located and should be controlled by a thermostat that controls water or steam valves to allow flow in and out of the radiators.

Pressure Calculation Illustration. Assume a house load of 50,000 Btu/h. We need to heat 1000 lb/h of water from 120 to $170°F$ (50 Btu/lb). The boiler pressure is 5 psig (approximately 20 psia). There is usually a water pump to pump the 2 gpm of water through the boiler and to the radiators in the rooms of the house as well as to suck the cool water back to the furnace. In order to pump the water to the third floor, 40 ft up, and overcome the friction loss in the pipe, say 10 psi, the pressure might have to be increased 16.54 psi for the three-floor height plus another 10 psi for the friction, giving about 42.5 psia as the maximum system pressure.

The boilers are designed to use gas, oil, or both gas and oil. The proper air–fuel ratio for gas to give a clean blue flame without yellow should be maintained. No CO should be in the flue gas, and CO_2 should be near the maximum value. Many improperly adjusted units are operating today at 50–150% excess air, and much heat is being lost up the stack.

Be sure there are no gas leaks in the gas line from the meter to the furnace. Ensure that no CO gets into the house when the furnace fire is stopped. This is especially important for today's tight, low air leakage buildings. Avoid closing outlet dampers before the furnace is purged of CO.

Each boiler should have a safety valve to relieve excess pressure to the drain line. It should also have an inlet water pressure control valve.

The fluid heated in these furnaces or boilers may be steam, hot water, or air. The boiler is designed to deliver a certain number of Btu/h (50,000, 75,000, 100,000) for homes, and larger units are provided for industrial or commercial buildings. Since some homes and buildings have been insulated more efficiently since 1973, the required home load has decreased in some cases to 20,000–30,000 Btu/h, and large, old, retrofitted coal to oil-fired (125,000 Btu/h) units have been replaced with small (30,000 Btu/h) gas-fired units. An illustration of costs follows.

Fuel	Approximate Higher Heating Value (HHV)
Gas	1,000 Btu/ft^3
Oil	18,000 Btu/lb
Coal	12,500 Btu/lb

When gas is available at \$2–\$3/1000 ft^3, this is also the price per 10^6 Btu. If the price of gas becomes uncontrolled, it will possibly compete with oil at \$1.30/gal (\$8.92/10^6 Btu based on 7.08 lb/gal). But coal, even at \$100/ton, only costs \$4.00/10^6 Btu based on 2000 lb/ton. Notice that if coal were purchased in large quantities at \$30/ton the cost would be only \$1.33/10^6 Btu!

Coal takes space in the basement for storage, however, and requires a conveyor to the stoker. Also an ash removal system and place for storage and disposal will be needed. Oil needs an oil storage tank and will need an oil pump if the tank is underground. A heater may be needed to prevent freezing if

Baseboard radiation design data

Btu loss of room	Steam	Lineal feet (active length) at av. water temp., °F							
		150	160	170	180	190	200	210	220
1200	5	3	3	$2\frac{1}{2}$	2	2	$1\frac{1}{2}$	$1\frac{1}{2}$	$1\frac{1}{2}$
1800	7.5	4	4	$3\frac{1}{2}$	3	$2\frac{1}{2}$	$2\frac{1}{2}$	2	2
2400	10	$5\frac{1}{2}$	5	$4\frac{1}{2}$	4	$3\frac{1}{2}$	3	3	$2\frac{3}{4}$
3000	12.5	7	6	$5\frac{1}{2}$	5	$4\frac{1}{2}$	4	$3\frac{1}{2}$	$3\frac{1}{2}$
3600	15	8	$7\frac{1}{2}$	$6\frac{1}{2}$	6	$5\frac{1}{2}$	5	$4\frac{1}{2}$	4
4200	17.5	$9\frac{1}{2}$	$8\frac{1}{2}$	$7\frac{1}{2}$	7	6	$5\frac{1}{2}$	5	$4\frac{3}{4}$
4800	20	11	$9\frac{1}{2}$	9	8	7	$6\frac{1}{2}$	6	$5\frac{1}{2}$
5400	22.5	12	11	10	9	8	7	$6\frac{1}{2}$	6
6000	25	$13\frac{1}{2}$	12	11	10	9	8	7	$6\frac{3}{4}$
6600	27.5	15	13	12	$10\frac{1}{2}$	10	$8\frac{1}{2}$	8	$7\frac{1}{2}$
7200	30	$16\frac{1}{2}$	$14\frac{1}{2}$	13	$11\frac{1}{2}$	$10\frac{1}{2}$	$9\frac{1}{2}$	$8\frac{1}{2}$	8
7800	32.5	18	$15\frac{1}{2}$	14	$12\frac{1}{2}$	$11\frac{1}{2}$	10	$9\frac{1}{2}$	$8\frac{3}{4}$
8400	35	19	17	15	$13\frac{1}{2}$	$12\frac{1}{2}$	11	10	$9\frac{1}{2}$
9000	37.5	$20\frac{1}{2}$	18	$16\frac{1}{2}$	$14\frac{1}{2}$	13	12	11	10
9600	40	22	$19\frac{1}{2}$	$17\frac{1}{2}$	$15\frac{1}{2}$	14	$12\frac{1}{2}$	$11\frac{1}{2}$	$10\frac{1}{2}$
10200	42.5	$23\frac{1}{2}$	$20\frac{1}{2}$	$18\frac{1}{2}$	$16\frac{1}{2}$	15	$13\frac{1}{2}$	12	$11\frac{1}{2}$

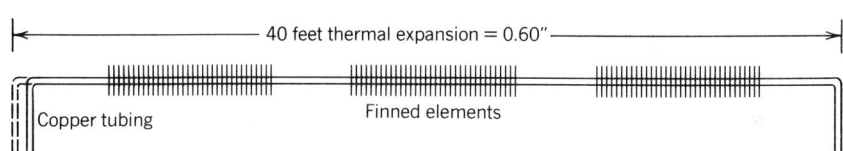

(*a*) Thermal expansion in the heating system can produce noise unless proper installation procedures are followed.

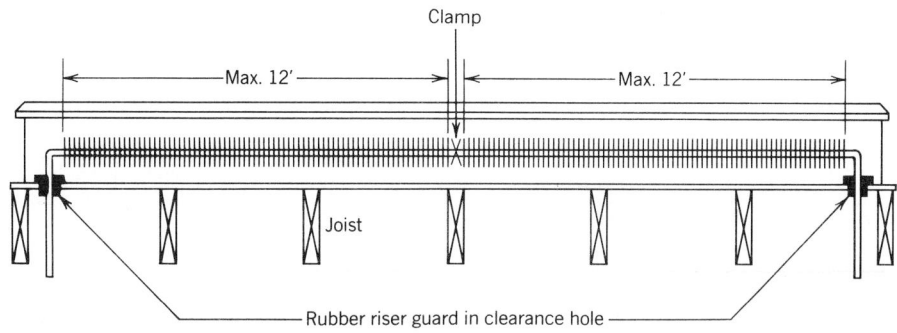

(*b*) On straight runs up to 25 ft install one clamp only.

Fig. 13.2-1 Radiators along outer walls (from ref. 5). Suggested methods of minimizing expansion noises given in (a)–(f). (Courtesy of Edwards Engineering Corporation.)

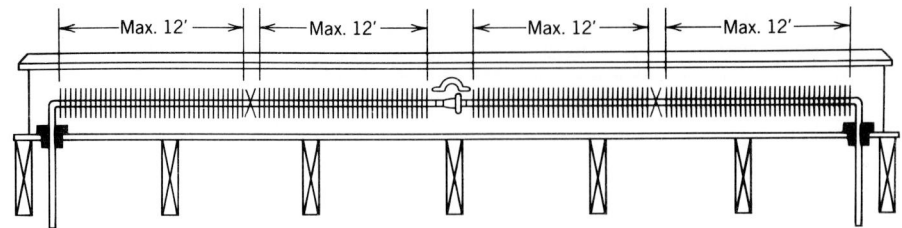

(c) On straight runs of 25–50 ft use 2 clamps and 1 expansion loop or compensator.

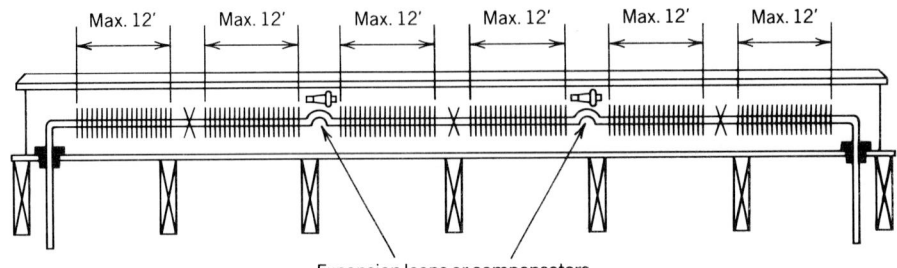

Expansion loops or compensators

(d) On straight runs of 50–75 ft use 3 clamps and 2 expansion loops or compensators.

Type of clamps

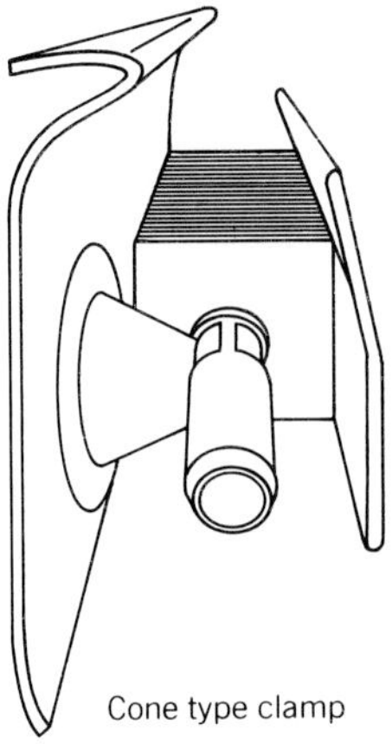

Cone type clamp

(e)

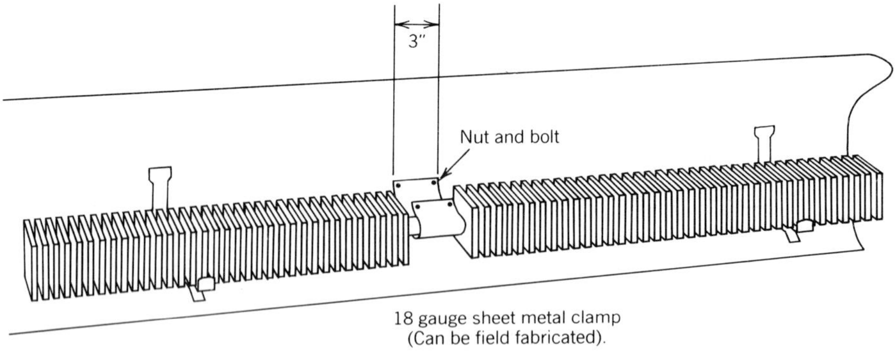

3″

Nut and bolt

18 gauge sheet metal clamp
(Can be field fabricated).

(f)

Note: For straight runs longer than 75 ft add loop or expansion compensators and clamps accordingly.

Fig. 13.2-1 (*Continued*)

ZONE CONTROL VALVES-

Individual valves, each operated by its own thermostat, permit selection of different temperatures in each zoned area of the home. In addition, swimming pool heating and snow melting circuits may be individually supplied with heat and automatically controlled by means of a standard zone valve, immersion thermostat, and a heat interchanger (see catalog for details).

DOMESTIC HOT WATER-

The heating unit is equipped with an instantaneous all copper hot water coil. The coil is optionally equipped with a modulating valve to give two temperatures of hot water: 140° to 160°F. water for wash basins and tubs and 180° for dishwashers and washing machines.

SWIMMING POOL HEATER ATTACHMENT-
SNOW MELTING HEATER ATTACHMENT-

Optionally available at time of installation, or available at any future time, a simple automatic circuit achieves pool heating, or snow melting from driveways and sidewalks.

EXTENDED SURFACE FLUES-

The vertical flues of the heating unit are fabricated from ⅜" thick firebox quality steel plate (ASME Code 285-C). Heavy steel fins are mechanically bonded to the flue passageways and serve to conduct the heat from the hot combustion gases quickly to the system water. This method of construction provides high efficiency with a maximum of fuel economy. The short vertical flue length minimizes thermal stresses in the heating unit. Flues are readily available for quick cleaning.

COMBUSTION CHAMBER (Sound Absorbing)

The combustion chamber is backed by a water leg and is lined with a high quality sound absorbing refractory. This refractory liner is low density felt-like material capable of withstanding 2800°F. The liner heats quickly resulting in a clean smoke-free flame. Under the bottom of the liner is 3" to 4" of pulsation absorbing material which serves the additional purpose of preventing heating transmission through the base of the heating unit.

INSULATION-

The heating unit shell is fully enclosed in thermal insulation for the retention of heat.

CIRCULATOR-

The circulator is specifically designed for use with zone control systems. The circulator will deliver 15 gallons per minute at 11' head, sufficient hot water for five 30,000 BTU zones. At complete shut-off, the circulator head is 13'. These figures illustrate the excellent performance characteristics of the circulator, and the excellent adaptability of the circulator for use with ½" baseboard heating element.

JACKET-

All jackets are fabricated from prepainted beige aluminum. (Some tops are from automotive type sheet steel, thoroughly cleaned, phosphate coated, and painted an attractive dark color.)

BURNER-

The heating package is equipped with a burner that has been carefully and scientifically selected to match the heating unit shell. All burner components are standard.

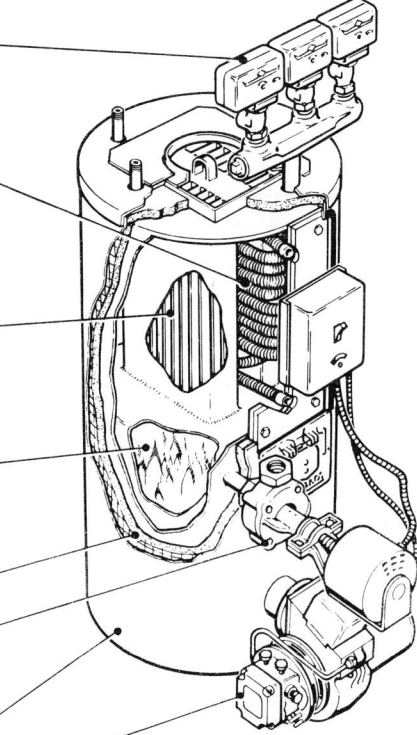

Typical Edwards Standard Packaged GAS-FIRED Heating Unit shown with optional zone valves, mounted.

Fig. 13.2-2 A gas-fired boiler (from ref. 1). (Courtesy of Edwards Engineering Corp.)

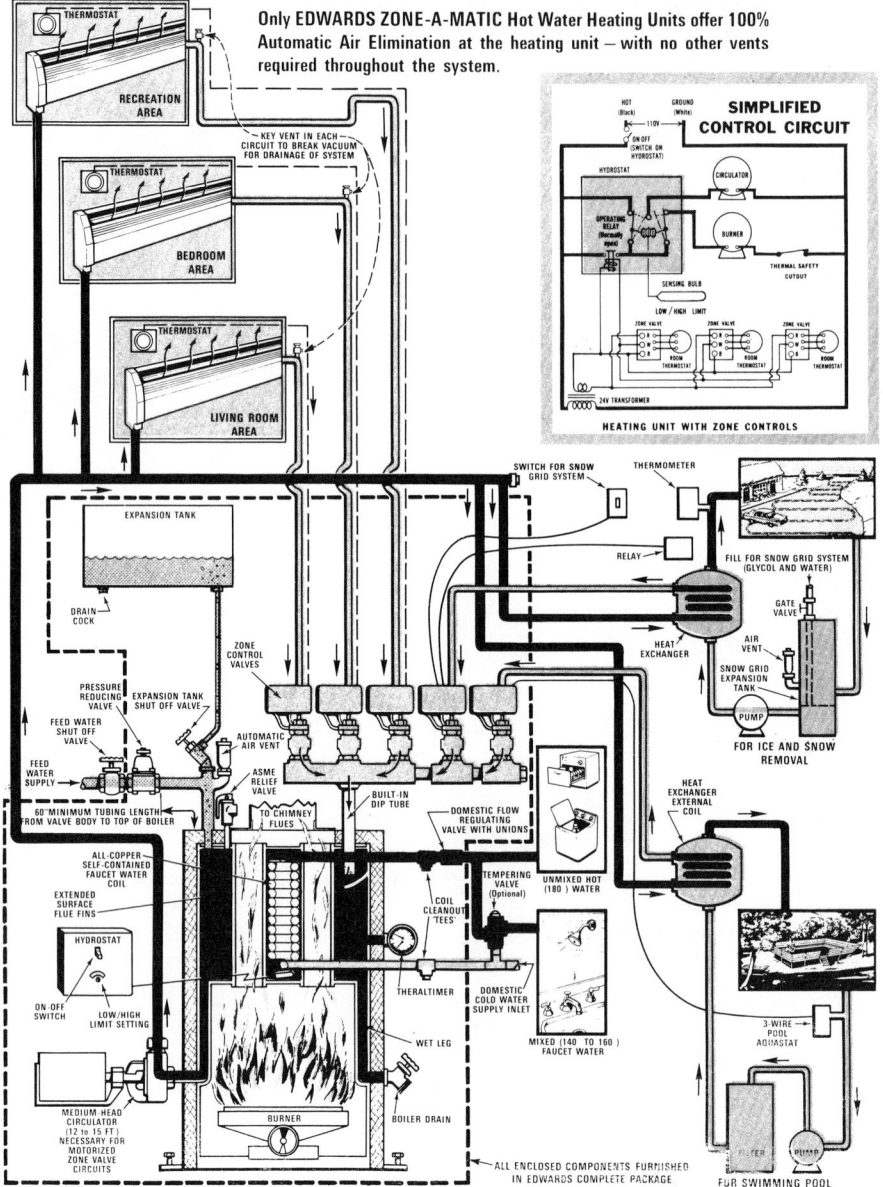

Fig. 13.2-3 Gas-fired zone-a-matic hot water heating system (from ref. 1). (Courtesy of Edwards Engineering Corp.)

the tank is outdoors and located in a very cold climate. A return line to the tank may be needed for excess oil not used at the burner.

Gas requires a meter and regulating valves to the burner in the furnace and must be monitored to guard against fires from gas leaks.

13.2-2 Gas-Fired Furnaces

Where available via pipeline, gas is the major clean source of heating fuel in use today. The furnaces are small, stacks are small, no storage space is required, no ash removal is required. The only real problem to guard against is gas leakage. When feasible, use an electric spark instead of a gas pilot light, which runs all the time. Be sure the pilot light is not operating in the summer when heating is unnecessary. The HHV of gas is 1000 Btu/ft^3. To heat a house with a 50,000 Btu/h load, using a

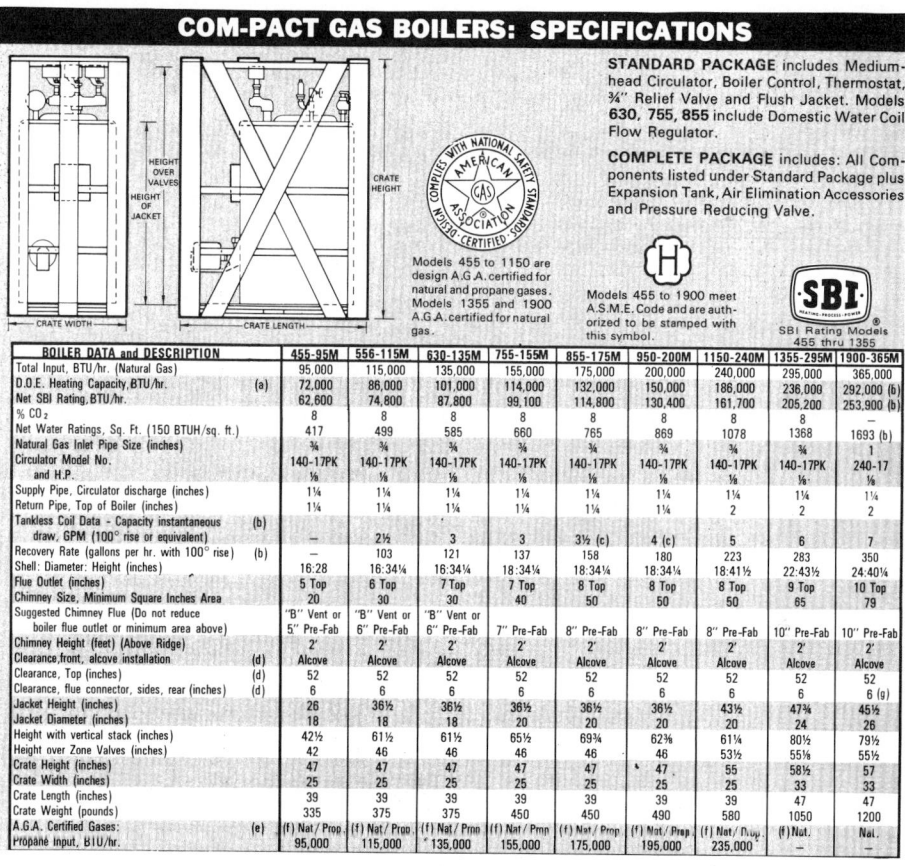

COM-PACT GAS BOILERS: SPECIFICATIONS

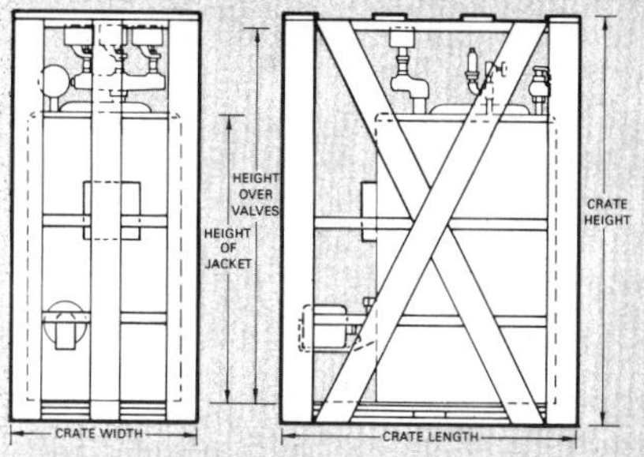

STANDARD PACKAGE includes Medium-head Circulator, Boiler Control, Thermostat, ¾" Relief Valve and Flush Jacket. Models **630, 755, 855** include Domestic Water Coil Flow Regulator.

COMPLETE PACKAGE includes: All Components listed under Standard Package plus Expansion Tank, Air Elimination Accessories and Pressure Reducing Valve.

Models 455 to 1150 are design A.G.A. certified for natural and propane gases. Models 1355 and 1900 A.G.A. certified for natural gas.

Models 455 to 1900 meet A.S.M.E. Code and are authorized to be stamped with this symbol.

SBI Rating Models 455 thru 1355

BOILER DATA and DESCRIPTION		455-95M	556-115M	630-135M	755-155M	855-175M	950-200M	1150-240M	1355-295M	1900-365M
Total Input, BTU/hr. (Natural Gas)		95,000	115,000	135,000	155,000	175,000	200,000	240,000	295,000	365,000
D.O.E. Heating Capacity, BTU/hr.	(a)	72,000	86,000	101,000	114,000	132,000	150,000	186,000	236,000	292,000 (b)
Net SBI Rating, BTU/hr.		62,600	74,800	87,800	99,100	114,800	130,400	161,700	205,200	253,900 (b)
% CO_2		8	8	8	8	8	8	8	8	–
Net Water Ratings, Sq. Ft. (150 BTUH/sq. ft.)		417	499	585	660	765	869	1078	1368	1693 (b)
Natural Gas Inlet Pipe Size (inches)		¾	¾	¾	¾	¾	¾	¾	¾	1
Circulator Model No. and H.P.		140-17PK ⅛	140-17PK ⅛	140-17PK ⅛	140-17PK ⅛	140-17PK ⅛	140-17PK ⅛	140-17PK ⅛	140-17PK ⅛	240-17 ⅛
Supply Pipe, Circulator discharge (inches)		1¼	1¼	1¼	1¼	1¼	1¼	1¼	1¼	1¼
Return Pipe, Top of Boiler (inches)		1¼	1¼	1¼	1¼	1¼	1¼	2	2	2
Tankless Coil Data - Capacity instantaneous draw, GPM (100° rise or equivalent)	(b)	–	2½	3	3	3½ (c)	4 (c)	5	6	7
Recovery Rate (gallons per hr. with 100° rise)	(b)	–	103	121	137	158	180	223	283	350
Shell: Diameter: Height (inches)		16:28	16:34¼	16:34¼	18:34¼	18:34¼	18:34¼	18:41½	22:43½	24:40¼
Flue Outlet (inches)		5 Top	6 Top	7 Top	7 Top	8 Top	8 Top	8 Top	9 Top	10 Top
Chimney Size, Minimum Square Inches Area		20	30	30	40	50	50	50	65	79
Suggested Chimney Flue (Do not reduce boiler flue outlet or minimum area above)		"B" Vent or 5" Pre-Fab	"B" Vent or 6" Pre-Fab	"B" Vent or 6" Pre-Fab	7" Pre-Fab	8" Pre-Fab	8" Pre-Fab	8" Pre-Fab	10" Pre-Fab	10" Pre-Fab
Chimney Height (feet) (Above Ridge)		2′	2′	2′	2′	2′	2′	2′	2′	2′
Clearance, front, alcove installation	(d)	Alcove	Alcove	Alcove	Alcove	Alcove	Alcove	Alcove	Alcove	Alcove
Clearance, Top (inches)	(d)	52	52	52	52	52	52	52	52	52
Clearance, flue connector, sides, rear (inches)	(d)	6	6	6	6	6	6	6	6	6 (g)
Jacket Height (inches)		26	36½	36½	36½	36½	36½	43½	47¾	45½
Jacket Diameter (inches)		18	18	18	20	20	20	20	24	26
Height with vertical stack (inches)		42½	61½	61½	65½	69¾	62⅜	61¼	80½	79½
Height over Zone Valves (inches)		42	46	46	46	46	46	53½	55⅛	55½
Crate Height (inches)		47	47	47	47	47	47	55	58½	57
Crate Width (inches)		25	25	25	25	25	25	25	33	33
Crate Length (inches)		39	39	39	39	39	39	39	47	47
Crate Weight (pounds)		335	375	375	450	450	490	580	1050	1200
A.G.A. Certified Gases:	(e)	(f) Nat/Prop	(f) Nat/Prop	(f) Nat/Prop	(f) Nat/Prop	(f) Nat/Prop	(f) Nat/Prop	(f) Nat/Prop	(f) Nat.	Nat.
Propane Input, BTU/hr.		95,000	115,000	135,000	155,000	175,000	195,000	235,000	–	–

Fig. 13.2-4 Manufacturer's performance data (from ref. 1). (Courtesy of Edwards Engineering Corp.)

boiler efficiency of 80%, it would take 62.5 ft^3/h. The heating cost would be \$321/year at \$3.00/1000 ft^3 (\$0.1059/m^3) for 5000 degree days.

The furnace should have access to outside air through a vent in the furnace room since the fan on the furnace pulls in 15 units of air to burn each unit of gas. The air–fuel ratio is adjusted by varying the damper setting to give a clean blue flame. Keep the furnace and stack clean to maintain maximum efficiency.

Companies supplying such boilers can be found in the ASHRAE Equipment Directory. Figures 13.2-2, 13.2-3, and 13.2-4 show a gas-fired boiler, a typical hot water layout, and one manufacturer's performance data.[1] Figure 13.2-3 shows a layout of a heating system and points out all the essential parts of the system mentioned above and later, including controls, swimming pool attachment if desired, extended surface flues (keep clean to get maximum heat transfer), combustion chamber, insulation, and jacket along with water circulator.

There must be steam pressure, hot water temperature, or hot air temperature control valves on all boilers. These turn the fuel and burner igniter on and off as needed and control the flow of heating fluid from the boiler to the radiator.

The air–fuel ratio must be kept adjusted to given efficient combustion with no smoke up the stack. The insulation should be sufficient to save energy and kept in good shape to keep heat loss down. Rusty ducts or jackets should be replaced.

Figure 13.2-4 gives a table of capacity of furnaces for natural-gas-fired boilers with circulator horsepower requirements, pipe sizes to and from the circulator boiler, and dimensions so one could provide sufficient space in the furnace room for all boiler parts. Acceptable fuels and their heating values are stated.

Figure 13.2-5 shows a Lennox compact gas-fired furnace that will produce from 38,000–72,000 Btu/h from 40,000–80,000 Btu/h input from gas. Gas pipe and vent sizes are given as are weights and motor capacities for blowers.

Figure 13.2-6 shows capacities, dimensions, and weights of Carrier compact equipment for room or rooftop units from 20,000–800,000 Btu/h capacity using gas or electricity for heat.

PROCESS OF COMBUSTION

The process of pulse combustion begins as gas and air are introduced into the sealed combustion chamber with the spark plug igniter. Spark from the plug ignites the gas/air mixture, which in turn causes a positive pressure buildup that closes the gas and air inlets. This pressure relieves itself by forcing the products of combustion out of the combustion chamber through the tailpipe into the heat exchanger exhaust decoupler and on into the heat exchanger coil. As the combustion chamber empties, its pressure becomes negative, drawing in air and gas for ignition of the next pulse of combustion. At the same instant, part of the pressure pulse is reflected back from the tailpipe at the top of the combustion chamber. This back pressure ignites the new gas/air mixture in the chamber, continuing the cycle. Once combustion is started, it feeds upon itself allowing the purge blower and spark plug igniter to be turned off. Each pulse of gas/air mixture is ignited at a rate of 60 to 70 times per second producing from one-fourth to one-half of a Btu per pulse of combustion. Almost complete combustion occurs with each pulse. The force of this controlled pulse creates great turbulence which forces the products of combustion through the entire heat exchanger assembly resulting in maximum heat transfer.

TAILPIPE

COMBUSTION CHAMBER

GAS INTAKE

AIR INTAKE

HEAT COIL

FLUE VENT AND CONDENSATE DRAIN

EXHAUST DECOUPLER

FLAME SENSOR

SPARK PLUG IGNITER

RUBBER MOUNTS

SPECIFICATIONS

Model No.	G14Q3-40	G14Q3-60	G14Q4-60	G14Q3-80	G14Q4-80
Input Btuh	40,000	60,000	60,000	80,000	80,000
Output Btuh	38,000	55,000	55,000	72,000	72,000
†A.F.U.E.	96.0%	92.0%	92.0%	91.0%	91.0%
High static Certified by A.G.A. (in. wg.)	.50	.50	.50	.50	.50
Temperature rise range (°F)	35 - 65	40 - 70	35 - 65	45 - 75	40 - 70
Vent size (in.)	2	2	2	2	2
Gas piping size I.P.S. (in.) — Natural	1/2	1/2	1/2	1/2	1/2
Gas piping size I.P.S. (in.) — *LPG	1/2	1/2	1/2	1/2	1/2
Condensate drain conn. (SDR11)	1/2	1/2	1/2	1/2	1/2
Blower wheel nom. diam. x width (in.)	10 x 8	10 x 8	11 x 9	10 x 8	11 x 9
Blower motor hp	1/3	1/3	1/2	1/3	1/2
Number and size of filters (in.)	(1) 16 x 25 x 1	(1) 16 x 25 x 1	(1) 16 x 25 x 1	(1) 16 x 25 x 1	(1) 16 x 25 x 1
Tons of cooling that can be added	2, 2-1/2 or 3	2, 2-1/2 or 3	3, 3-1/2 or 4	2, 2-1/2 or 3	3, 3-1/2 or 4
Shipping weight (lbs.)	250	250	255	250	255
Number of package in shipment	1	1	1	1	1
Electrical characteristics	120 volts — 60 hertz — 1 phase (All Units)				
*LPG Kit (optional)	LB-49116CA	LB-49116CB	LB-49116CB	LB-49116CC	LB-49116CC

†Annual Fuel Utilization Efficiency based on DOE test procedures.
*For LPG units a field changeover kit is required and must be ordered extra.

HIGH ALTITUDE DERATE

Elevation Above Sea Level (feet)	Maximum Heating Value (Btu/ft³)
5001 — 6000	900
4001 — 5000	950
3001 — 4000	1000
2001 — 3000	1050
Sea Level — 2000	1100

If the heating value of the gas does not exceed values listed in the table, derating of the unit is not required. Should the heating value of the gas exceed the table values, or if the elevation is greater than 6,000 feet above sea level it will be necessary to derate the unit. Lennox requires that derate conditions be 4% per thousand feet above sea level. Thus at an altitude of 4000 feet, if the heating value of the gas exceeds 1000 Btu/ft³, unit will require a 16% derate.

Fig. 13.2-5 Lennox pulse—G14 series up-flo gas furnaces: 40,000–80,000 Btu/h input add on cooling $1\frac{1}{2}$–4 nominal tons (from ref. 6). (Courtesy Lennox Industries Inc.)

1355

13.2-3   Oil-Fired Furnaces

Oil-fired furnaces range in size from small home heating units to larger industrial units and to much larger utility units. The types discussed here are for home units.

The home units require a storage tank either outdoors above- or belowground or indoors, usually in the basement. There must be a pipeline to the tank terminating in an easily accessible area where an oil truck hose can reach it. The tank should have 0.757–1.893 m^3 (200–500 gal) capacity, be vented, and also be easy to drain if possible. If outdoors, aboveground, it should be insulated and weatherproofed and contain an electric heater. It should be checked periodically for maintenance. An oil pump should be attached to insulated and weatherproofed piping from the tank to the burner and a return line to the tank should carry excess oil back to the tank. The line to the burner should have a control valve in it regulated by the oil temperature or the oil pressure.

The boiler is usually a fire tube boiler. This means a system with a combustion space where the oil is completely burned with yellow flame and no smoke. There are flue gas passage areas or tubes surrounded by water being heated, and an exit pipe or duct to the stack. A damper is often located in this pipe to provide draft adjustment. It is adjusted or closed to maintain a draft of about -1 in. H$_2$O in the combustion zone. The capacity of the boiler is, of course, determined by the heating load of the house plus the domestic hot water load if this is included as part of house boiler load.

The inside surface of the oil-fired boiler should be cleaned of soot at least annually. The stack and duct to stack should also be cleaned at the same time.

The quantity of oil that would be burned to keep the house warm would be [$50,000$ Btu/h/($18,000$ $\times\ 0.8 \times 8.33 \times 0.85$)] $= 0.477$ gal/h. For 5000 degree days this would be [$50,000 \times 24 \times 5000/(70°F \times 18,500 \times 0.8 \times 8.33 \times 0.85)$] $= 817.95$ gal per season! A well-insulated house could conceivably reduce this to one-fourth this value. At $\$1.25$/gal, $\$1022$ could be reduced to $\$255$! Figures 13.2-7, 13.2-8, and 13.2-9 illustrate the oil-fired boiler performance as given by sales bulletins[2] and as given by testing results.[4]

Figure 13.2-7 indicates the thermal efficiency of a boiler from no load to full load versus the combustion efficiency.

Figures 13.2-8 and 13.2-9 show boiler performances from tests performed and indicate effect of excess air, 5–200%, on combustion efficiency, as well as boiler efficiency, losses due to given insulation as the temperature rises, and design temperatures for various locations in the United States. Note it is sometimes good to have multiboiler capacity to obtain maximum efficiency. It is important to note that maximum efficiency occurs at around 40% rated load on Fig. 13.2-7, and above 0.6 of the boiler load capacity on Figs. 13.2-8 and 13.2-9. The operation of the boiler should be kept at these conditions to obtain maximum efficiency. Figure 13.2-10 shows compact steam and hot water household and industrial boilers, oil, gas or dual fired, of different shapes, and states their operating conditions and capacities.

13.2-4   Oil and Gas Furnaces

Many of the boilers in industries today are designed and equipped to burn either oil or gas depending on which is available and has the lowest cost. Most homes use either one or the other but home owners may buy a double-fuel burner if they already have the oil tank and their furnace is not too old.

Of course, it is possible the gas company may not agree to supply them gas under such an arrangement, but small industrial units are already being sold with dual-fuel capabilities. The capacities determined as shown for the oil and gas units above would apply here also.

If the oil and gas burners require forced air, a fan must be capable to supply $16/1$ units of air.

13.2-5   Coal-Fired Furnaces for Homes

Household coal-fired furnaces are usually either hand fired, like wood stoves, or underfeed stokers.

The coal for the season should be stored in a room or area of the basement and either hand fed to the stoker hopper or delivered to the hopper on a conveyor, in which case the coal storage bin must have a slope to the conveyor to automatically feed it.

The ashes must be manually removed, or the boiler must be fitted with a feed mechanism to convey the ashes periodically to an ash storage area, and ultimately the ashes must be carried away to an ash dump.

The furnace has a forced draft fan to supply the air under the stoker or grate area at a pressure sufficient to blow it up through the bed. The stack damper, depending on the hot flue gas temperature, is adjusted with the furnace running to maintain -1 in. H$_2$O in the furnace. The walls and flue gas passage in the furnace should be kept clean to permit flue gas to flow and transfer heat to the water. Otherwise, the heat goes up the stack and is wasted.

Illustration.   Consider a 50,000 Btu/h home heating furnace. This would require [($50,000$ Btu/h $\times 24$ h $\times 5000$ degree days)/$70°F \times 12,500$ Btu/lb $\times 0.75 \times 2000$ lb/ton] $= 4.57$ tons/year. At $\$100$/ton

MODEL NO.	HEATING CAP. RANGE (Btuh)	FUEL	DIMENSIONS (ft-in.) Largest Model			APPROX. WT RANGE (lb)
			L	W	H	
40FS	20,000-102,000	Electric	2-1	1-10	2- 5	60- 80
40ET	17,000-68,000	Electric	1-9	1-10	3- 6	69-114
58EG,SG	60,000-160,000	Gas	4-5	2- 0	2- 0	220-357
58DP,DR, GP,GS,SE	50,000-200,000	Gas	2-5	2- 8	3-10	105-260
58HC,HE, HH,HL,HV	56,000-335,000	Oil	2-0	1-10	5- 0	250-350

58SE **40ET**

58HV

40FS **CHIMNEY-LOCK**

48DH,DL,DM

MODEL NO.	COOLING CAP. RANGE (Btuh)	HEATING CAP. RANGE (Btuh)	DIMENSIONS (ft-in.) Largest Model			APPROX. WT RANGE (lb)
			L	W	H	
48DH, DL,DM	24,000-48,000	42,000-112,500	5 - 7-3/4	2 - 10-1/8	2 - 2-5/8	331- 512
48EG,EP	59,000-92,000	112,500-165,000	8 - 5-3/8	5 - 2-5/8	2 - 6-5/8	720-1030
48DD	124,000-600,000	103,000-810,000	24 - 11-5/16	7 - 2-11/16	4 - 10-1/2	1699-6800
48BH,BL	61,000	58,500-97,500	6 - 5	2 - 10	2 - 3-3/4	500

48DD

48EP

48MA,50ME

MODEL NO.	COOLING CAP. RANGE (Btuh)	HEATING CAP. RANGE (Btuh)	DIMENSIONS (ft-in.) Largest Model			APPROX. WT RANGE (lb)
			L	W	H	
48MA/ 50ME	180,000-444,000	160,000-800,000	21 - 9-1/16	7 - 11	3 - 9-15/16	2995-5710

50YH/YM

50DG

50DD

Fig. 13.2-6 Heating equipment. (Reproduced Courtesy of Carrier Corporation.)

the cost would be \$457/year. Thus a storage space should be provided to hold at least 1 ton of coal, and the user must purchase 1 ton four or five times per year. At 50 lb/ft^3 each ton would occupy 40 ft^3 or about 3.4 ft × 3.4 ft × 3.4 ft. Four tons would need a 5.43 ft cube, so it may be possible to arrange for such an area to hold all the fuel delivered at one price per year.

With percent ash varying from 5–10%, one-tenth this space must be made available for ash storage or, as stated before, the ash must be discarded from the furnace to a trash container daily for disposal. Possibly 0.5 tons per year will have to be discarded. The ash that settles in the bottom of the stack should also be removed at least weekly.

In a coal-fired unit with automatic controls, there is not an electric spark or a gas flame that ignites oil and gas flames as in the other boilers, but wood, paper, and coal are shoveled into a furnace and

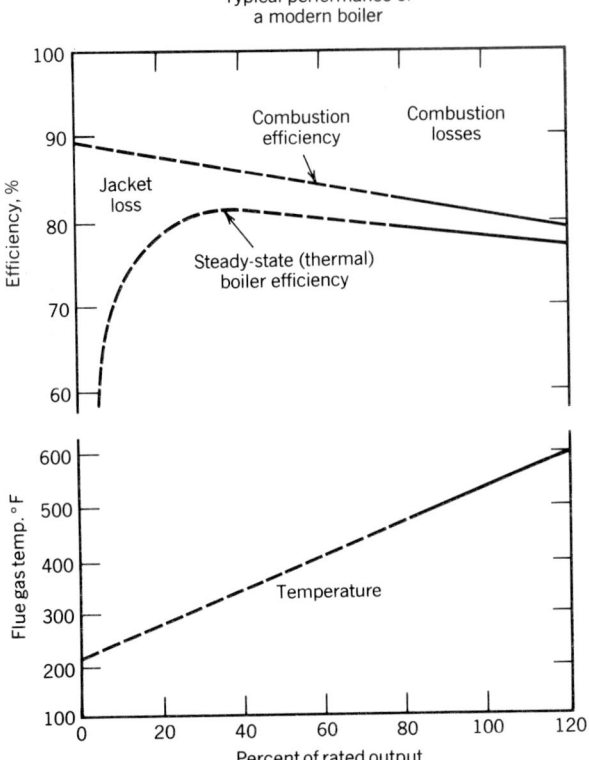

Typical performance of
a modern boiler

This is a typical diagram showing the relationship of
jacket loss, combustion efficiency, steady-state (thermal)
efficiency and flue gas temperature to boiler output. I-B-R
tests are conducted at 12.25% CO_2 in the range of 80–120%
of nominal boiler output.

Fig. 13.2-7 Oil-fired boiler performance (from refs. 2 and 4). (Reprinted by permission of Weil McLain, a Marley Company.)

the paper is hand ignited with a match. When proper coal ignition is observed, the control is turned on to inject and burn coal until water or steam temperature is up to normal, at which time coal feed and air flow stop. Normal draft will continue to pull a small amount of air through and maintain the fire, and when the water temperature drops, the controls turn the coal feed on again. Usually this maintains the fire, but it must be watched. A fuel bed temperature sensor should be on the boiler to alert someone that the bed is too low and the fire might go out.

Boilers equipped to burn coal, oil, or gas as cost or fuel availability dictate are being built today and sold for industrial-size units but not for homes.

Such units today would be too expensive for home use, and take too much space, considering both oil and coal storage and ash removal facilities and involve too much personal care.

13.2-6 Electric Furnaces

Ten percent of the bulletins distributed by boiler manufacturers today for small household or industrial boilers are for electric furnaces.

The electric furnace is comparable to a domestic hot water heater in that it consists of a tank with an electric resistance heater, which is controlled by water temperature or steam pressure.

There is no fuel storage problem, stack or damper requirement, and no fresh air is required. It is clean, with minimum maintenance.

In comparison to the coal cost of $457 per year, calculated in Section 13.2-5, an electric furnace for a 50,000-Btu/h (14.654 kW) load would use approximately 25,000 kW · h for the season, which at

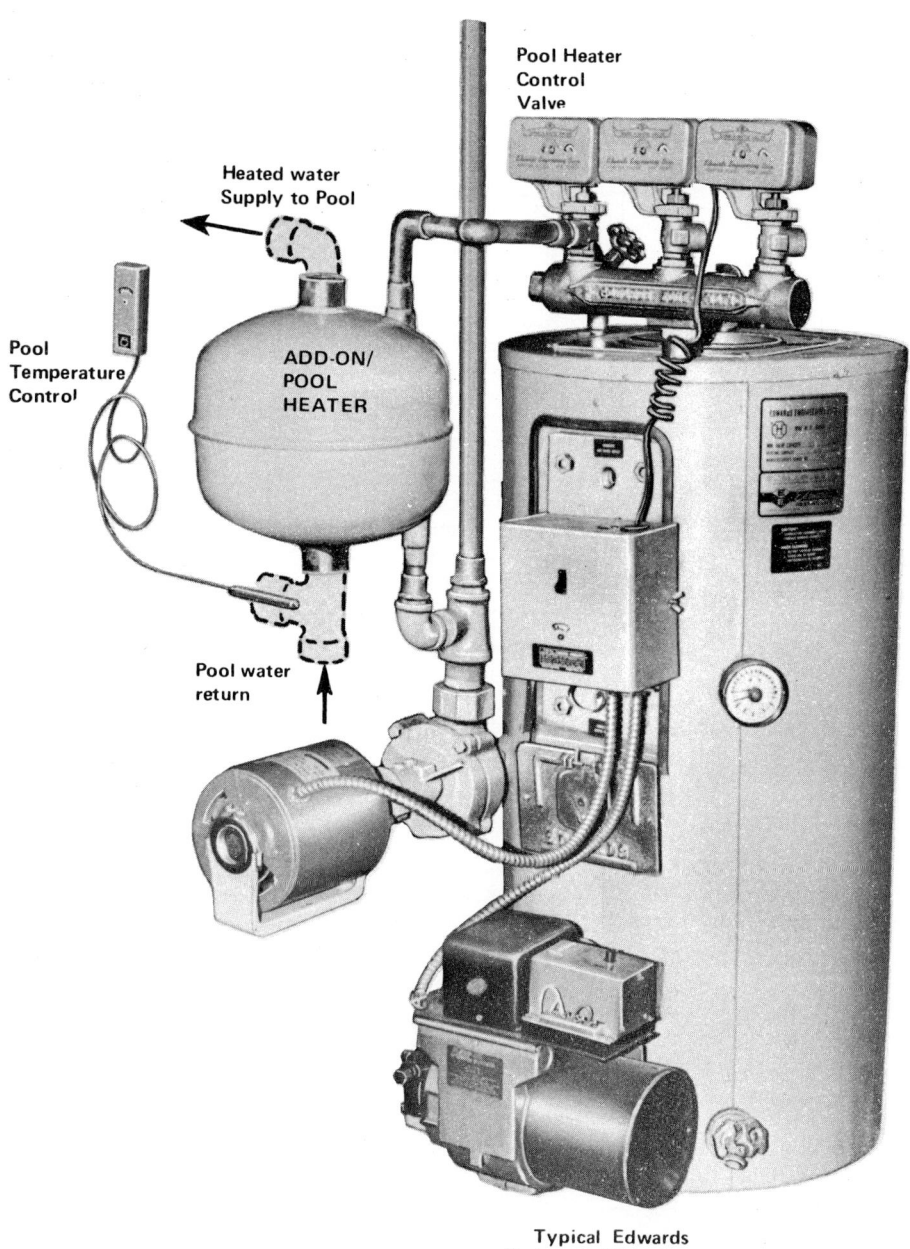

Pool Heater
Control
Valve

Heated water
Supply to Pool

Pool
Temperature
Control

ADD-ON/
POOL
HEATER

Pool water
return

Typical Edwards
Packaged Oil Boiler
Shown with
ADD-ON/Pool Heater,
in place

Fig. 13.2-7 (*Continued*). (Courtesy of Edwards Engineering Corp.)

INDEPENDENT LABORATORY TEST RESULTS

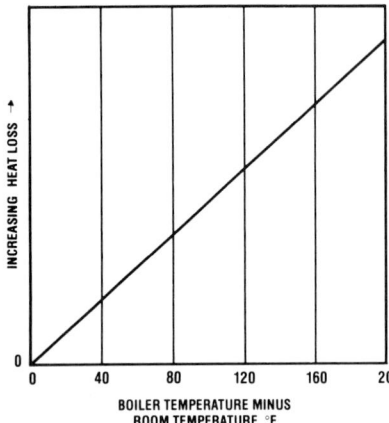

BOILER HEAT LOSS VS. TEMPERATURE

I-B-R TEST DATA
WEIL-McLAIN BOILERS

Gross I-B-R Output MBH	Average Jacket Loss at Boiler Rating %	Average Jacket Surface in Sq. Ft. Per 100 MBH Output
106 - 236	3.0	13.7
264 - 624	2.5	10.7
720 - 3350	2.2	7.4
2028 - 6970	1.2	6.0

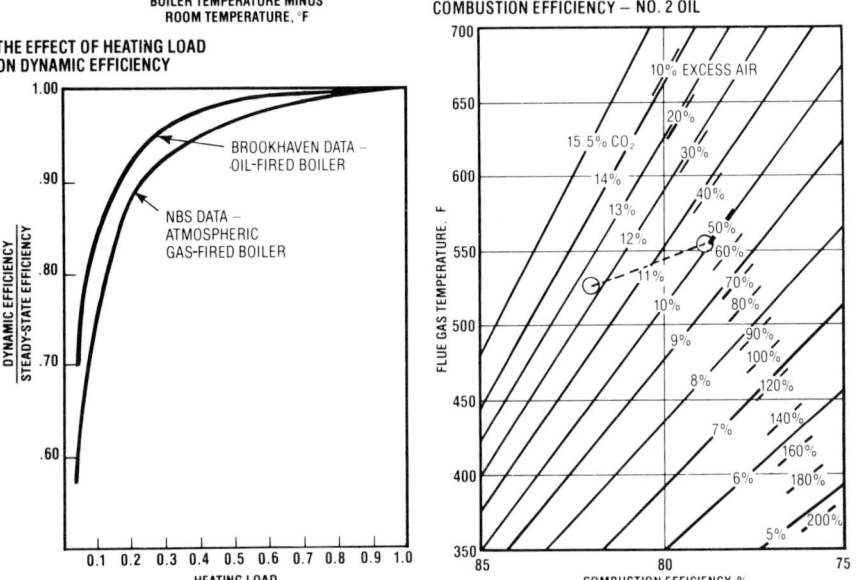

THE EFFECT OF HEATING LOAD ON DYNAMIC EFFICIENCY

BROOKHAVEN DATA – OIL-FIRED BOILER

NBS DATA – ATMOSPHERIC GAS-FIRED BOILER

COMBUSTION EFFICIENCY – NO. 2 OIL

Fig. 13.2-8   Data on boiler performance (from ref. 4). (Reprinted by permission of Weil-McLain, a Marley Company.)

6¢/kW · h would come to $1500, but the cleanliness and repair comparison to oil, gas, and coal might make electricity a desirable choice.

The capital costs of all systems must also be considered. There is no boiler and piping initial cost, maintenance cost, and replacement cost for electric heat and no stack is required.

Figure 13.2-11 shows electric furnace data,[3] and gives the capacity from 15 to 65 kW electric-hydronic heating units (hot water boilers) equivalent to 51,000–221,000 Btu/h gross output with circulator information, electric wiring information, unit, dimensions, weights, and so on. An added incentive to use electric heat, which could reduce heating costs, would be the use of off-peak heat storage tanks. Figure 13.2-12 is a description of capacity and sizes of off-peak heat storage heat tanks that could be used with the electric-hydronic units of Fig. 13.2-11 to heat and store energy at the time of day when utilities encourage customers to use energy at cheap rates. Bricks in insulated boxes are also used for hot air heat energy storage.

TYPICAL PERFORMANCE OF A MODERN BOILER

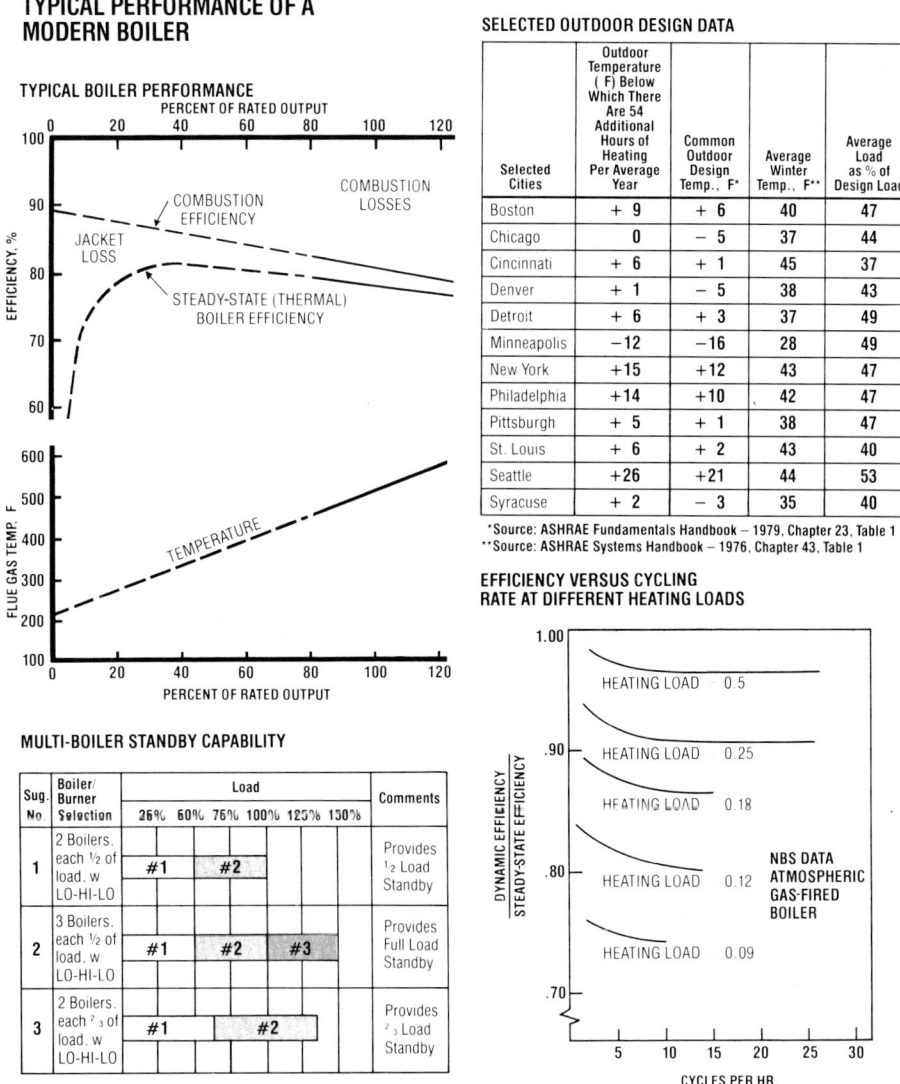

SELECTED OUTDOOR DESIGN DATA

Selected Cities	Outdoor Temperature (°F) Below Which There Are 54 Additional Hours of Heating Per Average Year	Common Outdoor Design Temp., F*	Average Winter Temp., F**	Average Load as % of Design Load
Boston	+ 9	+ 6	40	47
Chicago	0	− 5	37	44
Cincinnati	+ 6	+ 1	45	37
Denver	+ 1	− 5	38	43
Detroit	+ 6	+ 3	37	49
Minneapolis	−12	−16	28	49
New York	+15	+12	43	47
Philadelphia	+14	+10	42	47
Pittsburgh	+ 5	+ 1	38	47
St. Louis	+ 6	+ 2	43	40
Seattle	+26	+21	44	53
Syracuse	+ 2	− 3	35	40

*Source: ASHRAE Fundamentals Handbook – 1979, Chapter 23, Table 1
**Source: ASHRAE Systems Handbook – 1976, Chapter 43, Table 1

Fig. 13.2-9 Boiler performance data (from ref. 4). (Reprinted by permission of Weil-McLain, a Marley Company.)

13.2-7 Electric Baseboards

The clean and easy way to heat a house is to use electric baseboard heat controlled by a thermostat in each room or zone. If the area is not in use, the thermostat can be turned down to 40°F to save energy. When needed, the thermostat is turned to 68°F, and the area will be warm in a few minutes. The baseboard resistance heating units should be located under the windows on the walls of the room and have the capacity to handle the heat loss of the room. Place the units on the outer walls, not the inner walls, or the windows will have ice on the inside. While there is no boiler maintenance, no oil or coal storage, no ash or stack problems, the price of the kilowatt-hours makes this system questionable, as noted earlier.

Calculations show that it would definitely pay to turn down thermostats at night and when no one is home. The price could conceivably be cut in half and improved insulation might reduce the load to one-half again.

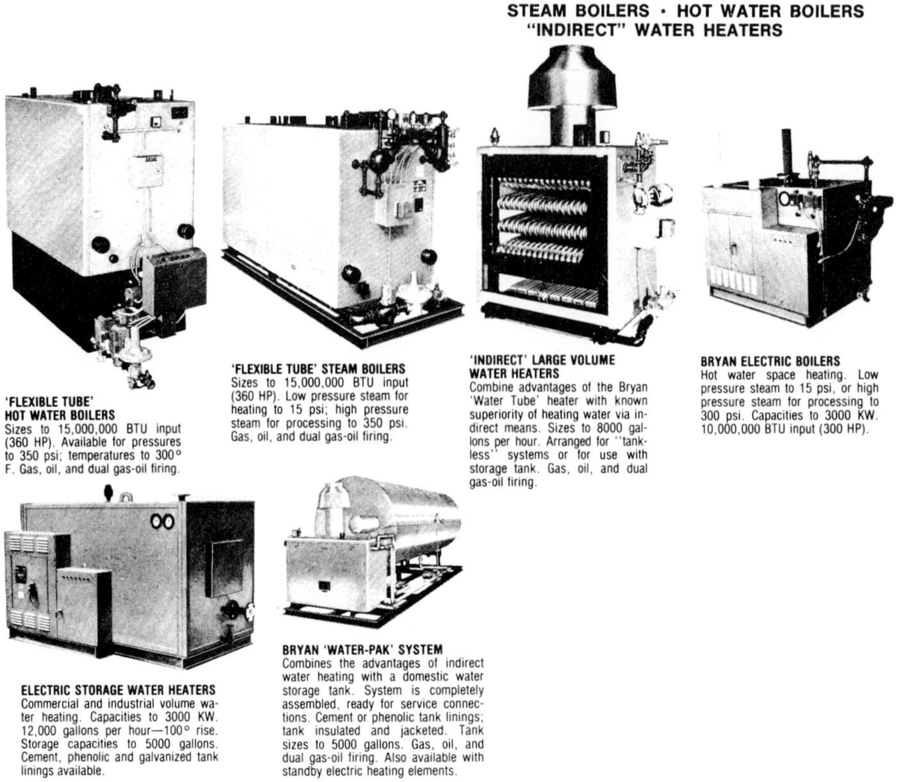

STEAM BOILERS · HOT WATER BOILERS
"INDIRECT" WATER HEATERS

'FLEXIBLE TUBE' STEAM BOILERS
Sizes to 15,000,000 BTU input
(360 HP). Low pressure steam for
heating to 15 psi; high pressure
steam for processing to 350 psi.
Gas, oil, and dual gas-oil firing.

'FLEXIBLE TUBE'
HOT WATER BOILERS
Sizes to 15,000,000 BTU input
(360 HP). Available for pressures
to 350 psi; temperatures to 300°
F. Gas, oil, and dual gas-oil firing.

'INDIRECT' LARGE VOLUME
WATER HEATERS
Combine advantages of the Bryan
'Water Tube' heater with known
superiority of heating water via indirect means. Sizes to 8000 gallons per hour. Arranged for "tank-
less" systems or for use with
storage tank. Gas, oil, and dual
gas-oil firing.

BRYAN ELECTRIC BOILERS
Hot water space heating. Low
pressure steam to 15 psi, or high
pressure steam for processing to
300 psi. Capacities to 3000 KW.
10,000,000 BTU input (300 HP).

ELECTRIC STORAGE WATER HEATERS
Commercial and industrial volume water heating. Capacities to 3000 KW.
12,000 gallons per hour—100° rise.
Storage capacities to 5000 gallons.
Cement, phenolic and galvanized tank
linings available.

BRYAN 'WATER-PAK' SYSTEM
Combines the advantages of indirect
water heating with a domestic water
storage tank. System is completely
assembled, ready for service connections. Cement or phenolic tank linings;
tank insulated and jacketed. Tank
sizes to 5000 gallons. Gas, oil, and
dual gas-oil firing. Also available with
standby electric heating elements.

Fig. 13.2-10 Household furnaces (from ref. 8). (Reprinted by permission of Bryan Steam Corporation, Peru, Indiana.)

13.2-8 Energy Efficiency of Wood-burning Systems

Thomas F. Weaver

It is useful to distinguish between the inherent efficiency of a type of stove and chimney system and
the actual energy efficiency of that system in a particular installation. Basically, the difference depends
on the extent to which the air passing through the stove is incremental to the prestove normal
breathing of the home, that is, the extent to which additional air is heated to desired room temperature
because of the presence of the wood-burning system. As an extreme example, an open nondampered
fireplace could increase the total home heating requirement, while a stove in a fairly nonairtight,
insulated home may require no additional air above the normal exchange. Also, the positioning of the
available heat is typically defined as high or gross heat less heat loss. There is considerable confusion
in the literature in the application of available heat values. Some sources include the efficiency of the
wood-burning system along with latent heat losses. A clear picture can be obtained by distinguishing
net heat from available heat where net heat values are derived by accounting for moisture content and
available heat recognizes the efficiency of the wood-burning system.

Accordingly, the Btu available for home heating can be determined from the relationship

$$A = B \times W \times 62.3 \times \mathrm{sg} \times M \times E \qquad (13.2\text{-}1)$$

where A = Btu of available heat per cord of wood burned

B = 8600 Btu high heat value per pound

sg = specific gravity (see Tables 13.2-1 and 13.2-2)

W = cubic volume of actual wood per cord generally assumed as 80 ft^3

M = moisture content correction (1 − % moisture). Assume $M = 0.9$ for air-dried wood at 20%
moisture. For "green" fresh-cut wood $M = 0.8$ is a reasonable approximation

E = efficiency of the wood burning system

Edwards electric-hydronic heating units: Specifications

Data and description (electric), kW	15	20	25	30	35	40	45	50	55	60	65
Gross output (1000's Btu/h)	51	68	85	102	119	136	153	170	187	204	221
Boiler hp	1.53	2.04	2.55	3.06	3.57	4.08	4.59	5.10	5.61	6.12	6.63
Net output ($12\frac{1}{2}$% loss) (1000's Btu/h)	44	59	74	89	104	119	134	149	164	179	194
Net water rating (rad.) (ft²)	299	398	498	598	697	797	896	996	1,096	1,195	1,294
Shell height (in.)	$34\frac{1}{4}$	$34\frac{1}{4}$	$34\frac{1}{4}$	$34\frac{1}{4}$	$34\frac{1}{4}$	$34\frac{1}{4}$	$34\frac{1}{4}$	$34\frac{1}{4}$	$34\frac{1}{4}$	$34\frac{1}{4}$	$46\frac{1}{4}$
Shell diameter (in.)	20	20	20	20	20	20	20	20	20	20	24
Flush jacket height (in.)	37	37	37	37	37	37	37	37	37	37	$51\frac{1}{2}$
Flush jacket diameter (in.)	22	22	22	22	22	22	22	22	22	22	26
Heating surface (ft²)	16.5	16.5	16.5	16.5	16.5	16.5	16.5	16.5	16.5	16.5	24.5
Boiler internal volume (gal)	47.2	47.2	47.2	47.2	47.2	47.2	47.2	47.2	47.2	47.2	91.5
Tankless coil capacity (gpm)	1.02	1.36	1.69	2.04	2.38	2.72	3.06	3.39	3.74	4.08	4.42
Recovery rate (gph)	61.2	81.6	101.4	122.4	142.8	163.2	183.6	203.4	224.4	244.8	265.2
Tankless coil pressure drop (psi)	1.86	2.48	3.10	4.96	5.58	6.20	6.82	7.44	8.06	9.30	9.12
Domestic water outlet (F.P.T. in.)	$\frac{1}{2}$	$\frac{1}{2}$	$\frac{1}{2}$	$\frac{1}{2}$	$\frac{1}{2}$	$\frac{1}{2}$	$\frac{1}{2}$	$\frac{1}{2}$	$\frac{1}{2}$	$\frac{1}{2}$	$\frac{1}{4}$
Return pipe size (OD in.)	$1\frac{1}{2}$	$1\frac{1}{2}$	$1\frac{1}{2}$	$1\frac{1}{2}$	$1\frac{1}{2}$	$1\frac{1}{2}$	$1\frac{1}{2}$	$1\frac{1}{2}$	$1\frac{1}{3}$	$1\frac{1}{3}$	2
Relief valve size (OD in.)	$\frac{1}{4}$	$\frac{1}{4}$	$\frac{1}{4}$	$\frac{1}{4}$	$\frac{1}{4}$	$\frac{1}{4}$	$\frac{1}{4}$	$\frac{1}{4}$	$\frac{1}{4}$	$\frac{1}{4}$	$\frac{1}{4}$
Relief discharge capacity (steam 1000's Btu/h)	521	521	521	521	521	521	521	521	521	521	521
Circulator — Pipe size (in.)	$1\frac{1}{2}$	$1\frac{1}{2}$	$1\frac{1}{2}$	$1\frac{1}{2}$	$1\frac{1}{2}$	$1\frac{1}{2}$	$1\frac{1}{2}$	$1\frac{1}{2}$	$1\frac{1}{2}$	$1\frac{1}{2}$	$1\frac{1}{4}$
Circulator — Return pipe size	$1\frac{1}{2}$	$1\frac{1}{2}$	$1\frac{1}{2}$	$1\frac{1}{2}$	$1\frac{1}{2}$	$1\frac{1}{2}$	$1\frac{1}{2}$	$1\frac{1}{2}$	$1\frac{1}{2}$	$1\frac{1}{2}$	$1\frac{1}{2}$
Circulator — Model no., hp	140 1/8	140 1/8	140 1/8	140 1/8	140 1/8	140 1/8	140 1/8	140 1/8	140 1/8	140 1/8	240 1/8
Operating voltage and phase	115 · 1	115 · 1	115 · 1	115 · 1	115 · 1	115 · 1	115 · 1	115 · 1	115 · 1	115 · 1	115 · 1

Fig. 13.2-11 Electric furnace data (from ref. 5). (Courtesy of Edwards Engineering Corp.)

Edwards electric-hydronic heating units: Specifications   (Continued)

	48/56	48/56	48/56	48/56	48/56	48/56	48/56	48/56	48/56	48/56	48/56
Motor frame no. (open d.p.)											
Expansion tank, min. (gal)	8	8	8	8	8	8	8	8	8	8	8
Rear clearance (in.)	18	18	18	18	18	18	18	18	18	18	18
Height incl. relay box (in.)	55	55	55	55	55	55	55	61	61	61	58
FLA (A)	62.5	83.3	104.17	125	145.8	166.7	187.5	208.3	229.2	250	270.8
240 V 1 Ø — Wire size (THW, AWG, MCM)	4	2	2	00	000	0000	0000	250	350	350	350
Disconnect size (A)	100	100	200	200	200	200	400	400	400	400	400
Fuse size (thermal, one time, time delay) (A)	80	100	125	150	175	200	225	250	300	300	350
Balanced or unbalanced load (B, U)	B	U	U	B	U	U	B	U	U	B	U
FLA (A)	37.7	75.4	75.4	75.4	113.1	113.1	113.1	150.8	150.8	150.8	188.5
230 V 3 Ø — Wire size (THW, AWG, MCM)	6	3	3	3	00	00	00	0000	0000	0000	0000
Disconnect size (A)	100	100	100	100	150	150	150	200	200	200	400
Fuse size (thermal, one time, time delay) (A)	50	90	90	90	150	150	150	200	200	200	225
Crate height (in.)	72	72	72	72	72	72	72	78	78	78	72
Crate width (in.)	39	39	39	39	39	39	39	39	39	39	36
Crate length (in.)	27	27	27	27	27	27	27	27	27	27	31
Shipping weight (lb)	550	550	550	550	550	550	550	550	550	550	900
Dry weight (lb)	540	540	540	540	540	540	540	540	540	540	880
Installed net weight (lb)	970	970	970	970	970	970	970	970	970	970	1,650

Fig. 13.2-11   (Continued)

Dimensions (in.)

Models, kW	15–45	50–60	65–110	115–120
A. Flush jacket height	37	37	$51\frac{1}{2}$	54
B. Flush jacket diameter	22	22	26	31
C. Height above jacket	18	25	6.5	6
D. Overall height	55	61	58	60
E. Element height	14	19	$27\frac{5}{8}$	$27\frac{5}{8}$
F. Element width	8	8	14	14
G. Overall depth	42	42	42	42
Rear clearance	18	18	18	18

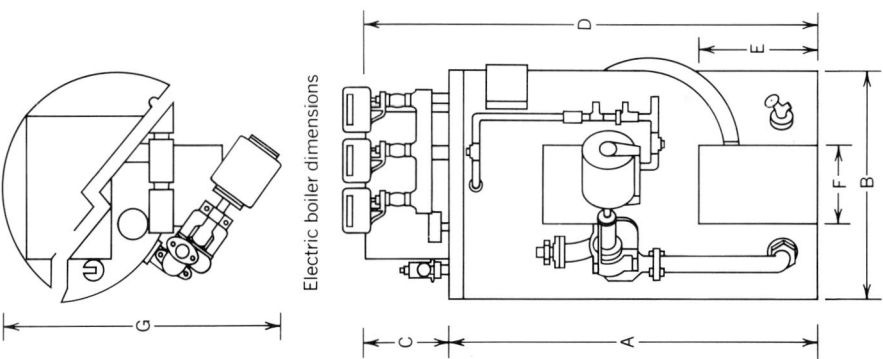

Electric boiler dimensions

Fig. 13.2-11 (*Continued*)

**Edwards electric-hydronic
heating units: "Off-peak"
Electric heating**

"Off-peak" heating unit package includes:

- **Heating unit shell**
- **Insulation**
- **Jacket**
- **Electric immersion heaters**
- **Sequencer control**
- **Tank limit control**
- **Heating unit water control**
- **Self-contained water heater coil**
- **Dual-temperature mixing manifold**
- **Low water cut-off**
- **Automatic air vent**
- **Automatic air elimination**
- **Pressure reducing valve**
- **Pressure relief valve**
- **Drain cock**
- **Circulator**
- **All components of control circuit wired**

"Off-peak" heat storage Tanks

Off-peak heat storage tanks consist of 275 gal steel tanks built to ASME specifications 32″ in diameter and 78″ high, installed in a sheetrock and frame enclosure and surrounded by poured or blown insulation such as expanded mica pellets or rock wool.

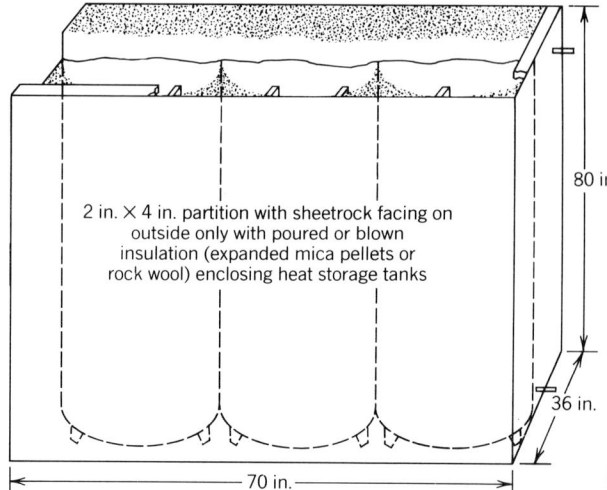

2 in. × 4 in. partition with sheetrock facing on outside only with poured or blown insulation (expanded mica pellets or rock wool) enclosing heat storage tanks

80 in

36 in.

70 in.

"Off-peak" boiler ratings:

Building area (ft^2)	Recommended heat loss Btu/h for degree days 5001–6000 based on proposed all-weather comfort standard for electrically heated and air conditioned homes	Number of 275 gal storage tanks required	Heating unit size (K.W.) based on 12 h off peak & 70° use factor	Expansion tank size (gal)
1000	32,000	2	15.0	60
1100	35,300	2	20.0	60
1200	38,500	3	20.0	90
1300	41,600	3	20.0	90
1400	45,000	3	25.0	90
1500	48,000	3	25.0	90
1600	51,200	3	25.0	90
1700	54,500	4	30.0	120
1800	57,800	4	30.0	120
1900	60,900	4	30.0	120
2000	64,000	4	30.0	120
2500	80,000	5	35.0	150

Fig. 13.2-12 Electric furnace off-peak heat storage tank (from ref. 5). (Courtesy of Edwards Engineering Corp.)

TABLE 13.2-1 SPECIFIC GRAVITY OF COMMON AMERICAN WOOD

	Average sp		Range of sp
Hardwoods			
Ashes	0.56	White	0.60
		Black	0.49
Aspens	0.39		
Beech	0.64		
Birches	0.61	Paper	0.55
		Yellow	0.62
Black Cherry	0.50	Balsam	0.34
Cottonwoods	0.36	Eastern	0.40
Elms	0.55	American	0.50
		Rock	0.63
Hickories (Pecans)	0.64	Bitternut	0.66
		Nutmeg	0.60
Hickories (True)	0.72	Pignut	0.75
		Shellbark	0.69
Locust (Black)	0.69		
Magnolias	0.49		
Maples	0.54	Big Leaf	0.48
		Sugar	0.63
Red Oaks	0.64	Southern Red	0.59
		Willow Oak	0.69
White Oaks	0.69	Overcup	0.63
		Live	0.88
Sycamore	0.49		
Tupelos	0.49		
Walnut (Black)	0.55		
Softwoods			
Bald Cypress	0.46		
Cedars	0.38	Northern White	0.31
		Alaska	0.44
Douglas Firs	0.48		
Firs	0.38	Soblapine	0.32
		Pacific Silver	0.43
Hemlock	0.43	Eastern	0.40
		Western	0.45
Larch	0.52		
Pine	0.47	Eastern White	0.35
		Slash & Longleaf	0.59
Redwood	0.37		
Spruces	0.39	Engelman	0.35
		White, red, black	0.40
Tamarack	0.53		

TABLE 13.2-2[a]

Burning Unit	Efficiency[b]

High-efficiency airtight with catalytic afterburner
    Example would be airtight stoves with
    a smoke chamber such as the double-drum
    stove or design for improved burning
    through secondary combustion chambers.
	0.75

Average airtight
    Any of the openable stoves with tight
    fitting doors that would not fall into
    the preceding category.
	0.63

Parlor, box, or potbelly stove
    Nonairtight stoves with smaller than 8-in.
    flue connections.
	0.50

Franklin stove (doors closed) Nonairtight stoves with 8-in. flue connection.	0.38
Fireplace (with a heatalator or similar device to improve the heating effect)	0.25
Fireplace	0.13

[a] Calculated from Forest Products Laboratory, *Wood Handbook*, Tables 3-3 and 4-2. Agricultural Handbook No. 72, USDA, 1974.
[b] Efficiency, as used here, does not include a reduction for moisture content. Oven dry wood is typically assumed to have a 20% moisture content. Some authors include the heat lost in the conversion of this moisture to vapor on their calculation of stove efficiency. In that case, for example, the average airtight would have an efficiency of 0.50 rather than 0.63 or a fireplace 0.10 rather than 0.13.

For most uses Eq. (13.2-1) can be used for air-dried wood as

$$A = 34.3 \times 10^6 \times \text{sg} \times E$$

This formulation will avoid double-counting errors and implicit assumptions on the efficiency of the wood-burning systems contained in many of the countouts suggested in the literature.

Measures

The standard cord is the most common unit of measure for fuel wood. It is a 8 ft × 4 ft × 4 ft pile or 128 ft^3 of stacked wood. The actual solid volume varies between 60 to 100 ft^3 depending on stacking and the configuration of the wood.

A face cord (also called run cord or a rack) is a pile of wood 4 ft high and 8 ft long. There is no standard width, as width tends to vary with locality. For example, a 2-ft wide face cord would contain 64 ft^3 of stacked wood and a 3-ft would contain 96 ft^3.

The moisture content of wood is the weight of moisture lost during oven drying divided by the oven-dry weight of the wood. The moisture content of green wood generally varies between 50 and 150%. Seasoned firewood typically has approximately a 20% moisture content.

Chimneys

The fireplace or woodstove chimney should extend at least 3 ft above a flat roof and 2 ft above a roof ridge or raised part of a roof within 10 ft of the chimney. If this is not possible, a hood can be used to eliminate trouble with eddies. Local codes must be followed.

Concrete footings are recommended for brick or masonry chimneys. They should extend a minimum of 6 in. beyond the chimney on all sides, and be 8 in. thick for a one-story house and 12 in. thick for a two story. Chimneys in frame buildings should be built from the ground up or rest on the building foundation or basement walls if the walls are of solid masonry 12 in. thick and have adequate footings. Local codes may call for slightly different requirements and must be checked.

Lined flues are a modern necessity for fire protection of masonry chimneys. Since the lining must withstand rapid fluctuation in temperature and the action of flue gases, it should be made of vitrified fire clay at least $\frac{5}{8}$ in. thick. Rectangular lining is better adapted to brick construction but round lining is more efficient.

The chimney lining should be lain concurrent with the brick. If the lining is slipped down after several courses of brick have been laid, the joints cannot be filled and leakage occurs. In masonry chimneys with walls less than 8 in. thick, a space should be left between the lining and chimney walls. Use only enough mortar to provide good joints and hold the lining in position. If the lining does not rest on solid masonry, the lower section must be supported on at least three sides by brick courses projecting to the inside surface of the chimney.

Walls of lined chimneys of not more than 30 ft high should be at least 4 in. thick if made of brick and 12 in. if made of stone.

Flue lining can be omitted if the chimney walls are made of reinforced concrete at least 6 in. thick or of unreinforced concrete or brick at least 8 in. thick.

A minimum thickness of 8 in. is suggested for the outside wall of a chimney exposed to the weather. Brick chimneys extending through a roof may sway in heavy winds and open up mortar joints at the roof line, causing a fire hazard. A good practice is to make the upper walls at least 8 in. thick by starting to offset the bricks at least 6 in. below the underside of roof joints or rafters.

Building codes generally require a separate chimney flue for each fireplace, furnace, or stove. Two flues grouped together without a dividing wall should have the lining joints staggered at least 7 in. and joints completely filled with mortar. Special procedures are required if a chimney contains multiple unlined flues. A soot pocket and cleanout area is recommended for each flue.

A smoke pipe should enter a chimney horizontally and not extend into the flue. The opening should be lined with fire clay or metal thimbles tightly built into the masonry (thimble rings are available in diameters of 6, 7, 8, 10, and 12 in. and in lengths of $4\frac{1}{2}$, 6, 9 and 12 in.). Use a closely fitting collar and boiler putty, good cement mortar, or stiff clay to seal the connection.

A smoke pipe should not be closer than 9 in. to combustible material and if less than 18 cover at least one-half of the pipe nearest the woodwork with commercial fireproof pipe covering.

If a smoke pipe passes through a wood partition, insert a galvanized-iron double-wall ventilating shield at least 12 in. larger than the pipe or install 4 in. of brickwork or other incombustible material around the pipe. Smoke pipes should never pass through floors, closets, or concealed spaces or enter the chimney in the attic.

Every flue should be smoke tested preferably before the chimney is furred, plastered or otherwise enclosed.

There should be a 2-in. space between chimney walls and roof wooden beams or joists (unless 8 in. of solid masonry, in which case $\frac{1}{2}$ in. is adequate). Fill spaces before floors are lain with porous, nonmetallic incombustible material (e.g., loose cinders not brickwork, mortar, or concrete).

Fuelwood

Energy Content. There are two common measures of the energy content of wood: (1) high heat value, which is the amount of chemical energy released when 1 lb of wood is completely burned including the latent heat of condensation of the moisture content of the wood and (2) low heat value, which is the total chemical heat less latent heat condensation associated with complete combustion. Low heat is synonymous with sensible heat with no heat of condensating included. The value will therefore vary with the moisture content of the wood.

The energy content of wood has been determined using a bomb colorimeter. Variation in results for the same species may be a function of whether the sample was heartwood or sapwood, spring growth or summer growth, or some combination thereof. There is also considerable difference in the relative density of different species of trees. The energy content of any pound of oven-dry wood is remarkably constant within a few percent. The variation that does exist can mainly be attributed to differences in chemical composition (resin, gums, oils, and tannins). At zero moisture content, 8600 Btu/lb is a universally accepted value for hardwood and most soft wood. Very resinous softwood could be as much as 5% higher.

Since wood is normally sold on a volume rather than weight basis, and at different moisture content, and since there are considerable differences in the specific gravity of various species, it is useful to be able to approximate the energy content per cord of wood on a species basis.

Ash. Ash content is generally between 0.1 and 3% composition, 30–60% calcium (CaO), 10–30% potassium (K_2O), 2–15% sodium (Na_2O), 5–10% magnesium (MgO), 5–15% phosphorus (P_2O_5), and lesser amounts of sulfur, iron, and trace elements. Wood ash is often used as a fertilizer (K_2O and P_2O_5) and also performs a similar role as lime (CaO). The ash can also produce potassium carbonate used in soap making, and wet ash can be used as a degreasing cleaner.

Wood Storage.   Wood is hygroscopic and never dries to zero moisture content under normal storage. Drying is fastest after cutting. Most wood is considered "seasoned" 6 months after harvest when it typically has a moisture content from 20 to 25%. In fact, drying time is shortest for smaller (split) pieces in free-standing piles that permit air circulation. Direct exposure to sunlight or heated indoor storage also decreases drying time. The energy content of wood can be greatly reduced by rotting.

Energy Efficiency of Stoves

It is useful to be aware of the distinction between the energy efficiency of a stove or fireplace and the net energy efficiency of the appliance in a given installation. The difference results from two phenomena. First, the air that is used in burning may increase the natural breathing of the house and thus increase the heating load and lower the efficiency of the wood-burning system. Any action that causes an increase in the air exchange rate increases the amount of heat needed to maintain a given temperature level. Most closed stoves, even in a well-insulated house, draw so little air compared to the normal exchange that they do not add to the heating requirement. Second, the placement and construction of the chimney may be such that heat conducted through its walls contributes to needed heat or, for example, a stone fireplace and chimney may act as a heat sink conducting heat to the outside. A general rule of thumb for dealing with this issue is that it is only likely to be serious consideration with 8-in. or larger flue pipe connections. Franklin stoves, free-standing and ordinary fireplaces, and other open stoves are likely to fall into this category. Ducting outside air to the stove or fireplace because of inevitable leakage is not a good solution to this problem, although the practice may be a means of eliminating uncomfortable floor drafts created by a drawing fire.

Combustion efficiency is the percentage of chemical energy converted to sensible and latent heat and radiation within the interior of the stove. This conversion is a function of temperature and oxygen supply. Combustion region temperatures can be elevated by impeding outward heat flow and by preheating combustion air. Outward heat flow may be retarded by firebrick and metal liners and by horizontal baffle plates, which create an s-flow pattern. Preheating of combustion air is achieved by stoves designed to recirculate a portion of the flue gas.

Heat transfer efficiency has a positive relationship with (1) the surface area of the stove, (2) the temperature and turbulence inside the stove, and (3) the thermal conductivity of the stove walls. There is a negative relationship between transfer efficiency and flue gas velocity.

A large surface area relative to the size of the fire increases heat transfer. This can be accomplished by smoke chambers and baffling within the stove and by extra long and wide stovepipe connecting to the chimney. Interior baffling also increases turbulence, which helps keep heat flowing to all corners of the stove.

Within limits there is an advantage to stoves that have a lesser amount of air entering for a given level of combustion. The less air in, the less the exhaust, and the slower the exchange velocity. Keeping air flow at a minimum also increases the temperature of the gases within the stove.

There are a number of basic conflicts in stove design. Complete combination requires excess air, but excess air cools and increases the velocity of the flue gases and decreases the heat transfer efficiency. Firebrick and metal liners increase internal temperatures and combustion efficiency but decrease heat transfer. For typical efficiency valves see Table 13.2-2.

One of the most recent innovations in wood stoves is the catalytic converter. "Airtight" wood stoves can in fact be viewed as wood gas generators when the stove is damped down. The catalytic wood stove contains a catalytic converter in the upper combustion chamber that burns the gases, thereby increasing the stove's efficiency while reducing the buildup of creosote.

The ignition temperature of wood gas is approximately 1300°F. A typical catalytic converter is a 2500°F rated ceramic honeycomb coated with a metallic catalyst, platinum, which when in contact with the gases reduces their ignition temperature to approximately 500°F. Once the catalytic converter reaches 500°F, the gases begin to burn. The converter is positioned in the hottest part of the primary combustion chamber and usually operates within 15 min of starting the fire. Its temperature during peak operation reaches approximately 1500°F.

The catalyst can easily be fouled by burning materials that contain lead such as colored newsprint and painted wood surfaces. Claims are made that a catalytic converter increases stove efficiencies from 20 to 25% above an average airtight stove. The burning of the creosote-producing gases will certainly reduce fire hazard, and in addition there is an apparent highly beneficial effect by reducing air pollution from a wood-burning system. Additional information on wood-burning systems may be found in Section 11.3-3.

Bibliography for Section 13.2-8

Cunningham, Gordon R. *Wood for Home Heating—Locating, Cutting and Gathering Wood.* University of Wisconsin Extension, Madison, WI.

Forbes, Reginald D. *Forestry Handbook.* Ronald Press, New York, 1956.

Forest Products Laboratory. *Wood Handbook*. U.S. Department of Agriculture, Washington, D.C., 1940.

Foulds, Raymond T., Jr. *Wood as a Home Fuel*. N.E. Bulletin #7, University of Vermont Extension Service, Burlington, VT, 1975.

Kent, R. T. *Kent's Mechanical Engineering Handbook*, 11th ed. Wiley, New York, 1936.

National Fire Prevention Association. *Using Coal and Wood Stoves Safely*. Boston, MA, 1974.

Northeast Regional Agricultural Engineering Service. *Burning Wood*. NRAS, Riley-Robb, Ithaca, NY, 1977.

Office of Forest Investigation. *The Use of Wood for Fuel*. U.S. Department of Agricultural Bulletin #753. U.S. Department of Agriculture, Washington, D.C., 1919.

Panshin, A. J., Harrar, E. S., Boker, W. J., and Practor, P. B. *Forest Products: Their Sources, Production, and Utilization*. McGraw-Hill, New York, 1950.

Reinke, L. H. *Wood Fuel Combustion Practice*. Forest Products Laboratory Report No. 1666-18, Forest Products Laboratory, USF, Madison, 1947.

Rhode Island Department of Community Affairs. *Safety Guidelines for Coal and Woodburning Stoves*. Technical Note 78-05, Department of Community Affairs, Rhode Island, 1978.

Shelton, Jay and Shapiro, Andrew B. *The Woodburners Encyclopedia*. Vermont Crossroads Press, Waitsfield, VT, 1976.

Tiemann, Harry Donald. *Wood Technology-Constitution, Properties and Uses*. Pitman Publishing, New York, 1942.

Wangoard, Fredrick, F. *The Mechanical Properties of Wood*. Wiley, New York, 1950.

Weeks, S. A., Lassoie, J. P., and Baker, L. D. *Heating With Wood*. FS-7 Cornell University, NRAES, Ithaca, NY, 1977.

13.3 HEAT PUMPS

13.3-1 Introduction

For 40 years or longer many engineering groups interested in energy-efficient systems have been suggesting the heat pump as the best heating system. A heat pump is a vapor refrigeration system composed of (1) an evaporator where heat is absorbed by a refrigerant as it evaporates, (2) a compressor that is motor driven, and (3) a condenser where hot compressed vapor condenses, releasing the heat from the refrigerant in the process and thereby heating the indoor air.

The heat pump differs from a normal air-conditioning unit in that it has valves and controls that make it possible to change the hot gas flow from the compressor to the outdoor heat exchanger, which acts as the condenser in the summer, to the indoor coil, which acts as the condenser in the winter. The advantageous economic factor is that in the winter, by removing, for example, 2 Btu from the outdoor air, thus cooling it even from 0 to −10°F, adding the approximate 1 Btu equivalent of work required to run the compressor, and discharging 3 Btu to the indoor coil heating the house, 2 Btu are free.

Unfortunately, in the past, heat pumps have been bought for summer loads and not winter loads. This means that at approximately 32°F the heat pump capacity equals the house heating load and, below this, electric, or some form of additional heating must be used that may not be as economical as the heat pump. Figure 13.3-1 illustrates this. If the heat pump is sized in the summer at design conditions, say 95°F outdoor temperature and 78°F indoor, then at 30°F, when working on the heating cycle, the heat pump capacity is equal to the house heat loss. From here on, as the outdoor temperature drops to 0°F, more and more electric, oil, gas, or other heat must be used to supply the heating load above the heat pump capacity line. From 65°F down to 30°F approximately two-thirds of the heat supplied by the heat pump is free!

The important thing is to select a heat pump at 0°F design heating load and use two compressors. In the summer use one compressor to supply the one-half capacity for summer cooling load. General Electric has suggested installing two units equal in total capacity to winter load at design conditions (0°F) and only running one or the other, alternately, in the summer as needed. This plan has not really become widely accepted as yet, but it is an excellent idea. For those days below 0°F, or −10°F, extra electric heat would be required.

The example given here uses heat obtained from air and given up to air. Other heat pumps absorb heat from the water, ground, and solar collectors, and some even freeze water into ice in the winter to use to cool the house next summer. In the summer the ice is melted and the water in a storage tank, lake, or man-made pond is heated to some safe temperature (< 180°F) and pressure to use as a heat source next winter.

The biggest problem experienced with the air-to-air heat pump to date has been the defrosting of the outdoor coil. If the humidity is high, and the temperature is below 32°F, frost, and even ice, forms

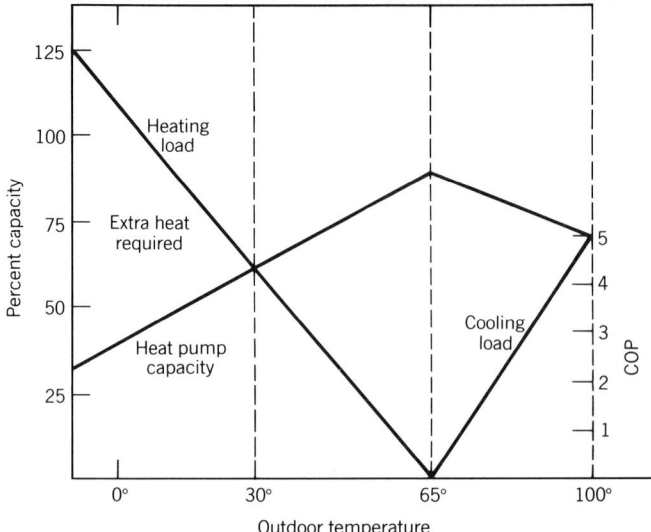

Fig. 13.3-1 Heat pump performance. (Reprinted with permission, Dunham-Bush, Inc., 1982.)

on the outer fin surfaces of the outdoor coil and creates a large pressure drop across the coil, which is designed for a given quantity of air flow to give up the heat to the refrigerant in the winter from the cold outside air. If Δp gets too great, the cycle reverses and cools the room while heating the coil to melt the frost and ice. Extra electric heat has to heat the room during this time, which is typically about 4 min. As an alternative the unit could shut down and electric heat defrost the coil.

Place the outdoor unit in a protected area such as under a porch on the southeast, and not out in the northwest cold side of the building. In the summer supply fresh air to the outdoor condenser and avoid sunshine, which would raise the discharge pressure. Keep the fins of the outdoor and indoor coils clean for efficient heat transfer. Keep the walls of the compressor clean, keep it well lubricated, and check for refrigerant leaks. A hermetically sealed unit aids in this respect, but it is still necessary to check pipe joints to and from coils and controls for leaks. Maintain a warranty on service and performance. The present cost of refrigerant makes this worthwhile.

When the outdoor unit is a water or earth unit storage source, the defrost problem is not present. However, the heat pump owner should own the land or water where freezing might occur, or where heating occurs in summer, to prevent neighbors suing for damage to their land, greens, or wildlife! In the summer two condensers could be used, and the heat needed to heat domestic hot water would be free from the condenser.

The big advantage to the heat pump is that this unit is also the air-conditioning unit. It is expensive because of the needed controls to switch from heating to cooling, but both summer and winter units are in one. Also, the winter $2 out of $3 saving for heating gives a good payback as interest and fuel rates go up, even though electricity costs go up also!

A hot air heating system is a natural to convert to a heat pump system by adding a heat pump exchanger in the present duct.

In the summer the heat pump could cool the house using this heat exchanger. In the winter the unit could heat the house until it reached its capacity, previously mentioned. Then, a second thermostat would cut the oil- or gas-fired furnace on. A wood stove or coal unit could be installed to assist the heat pump when needed.

13.3-2 Indoor Units

Indoor units have all of the heat pump's parts indoors, such as the receiver, evaporator, compressor, condenser, fans, and controls. This takes space, and the compressors and motors are noisy, so many prefer outdoor units. Indoor units are protected from weather problems, however.

13.3-3 Split Indoor–Outdoor Units

Split units have the compressor, outdoor heat exchanger, and a fan outdoors, with an indoor heat exchanger in a duct with the controls. Thus, the indoor heat exchanger is in the air duct, cooling or

Single Packages

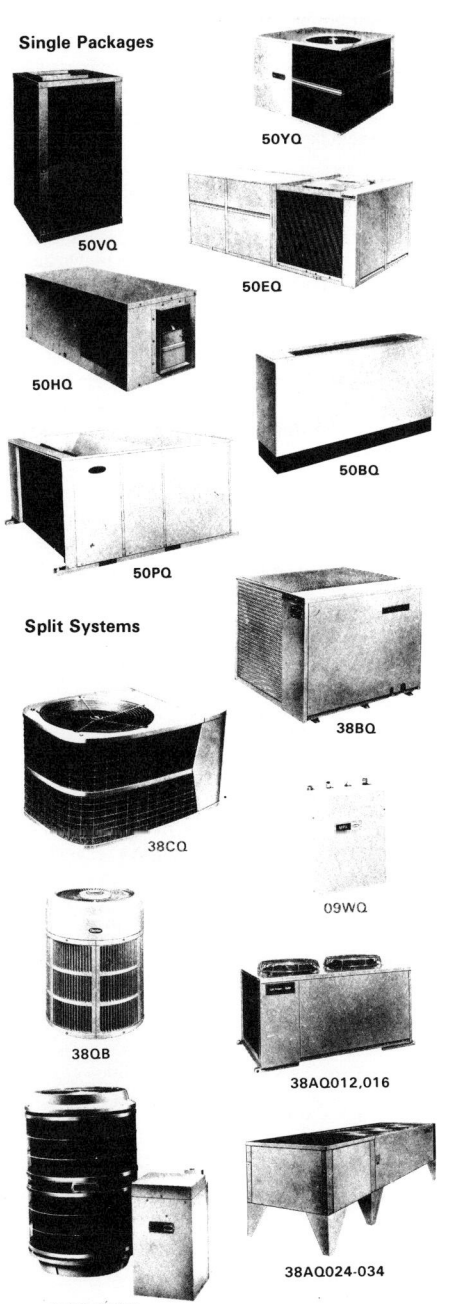

09WQ multi-purpose heat exchanger is designed for use with residential heat pumps and condensing units in many applications — water source heat pumps, hydronic assist, chilled water, etc.

MODEL NO.	COOLING CAP. RANGE (Btuh)	HEATING CAP. RANGE (Btuh)	DIMENSIONS (ft-in.) Largest Model			APPROX. WT RANGE (lb)
			L	W	H	
50QT	12,000-30,000	12,000-30,000	2 - 0	2 - 4-1/2	3 - 8-1/2	270
50YQ	25,000-58,000	27,000-60,000	4 - 0-7/16	3 - 6-1/4	3 - 3-5/8	303- 448
50RQ/PQ	59,000-178,000	62,000-180,000	6 - 11-1/2	7 - 2-1/8	3 - 8-1/4	460-1750
50EQ	224,000-336,000	240,000-360,000	17 - 1	7 - 3	4 - 11	3300-4550
50HQ	14,000-80,000	17,000-82,000	4 - 0	1 - 11	2 - 0-7/16	185- 360
50VQ	14,000-80,000	17,000-82,000	2 - 0	1 - 10	4 - 0	220- 460
50BQ	6,500-19,500	9,000-25,500	0 - 11-1/2	4 - 1-5/8	2 - 1-1/2	180- 200
50QD	61,000	60,000	5 - 8-1/4	3 - 8-11/16	2 - 5-1/8	540
38HQ/09WQ	28,000-48,000	37,000-64,000	1 - 7	0 - 10	1 - 10	67- 82

Split Systems

MODEL NO.	COOLING CAP. RANGE (Btuh)	HEATING CAP. RANGE (Btuh)	DIMENSIONS (ft-in.) Largest Model			APPROX. WT RANGE (lb)
			L	W	H	
38CQ	14,000-44,000	15,000-45,000	2 - 10-3/4	1 - 10	2 - 6	150- 250
38QB	14,500-57,000	15,500-60,000	—	2 - 6 Diameter	3 - 4	160- 260
38RQ	20,000-56,000	21,500-63,000	—	2 - 5-1/4 Diameter	3 - 8	180- 260
38HQ	22,500-46,000	24,000-51,000	Indoor Compressor Section 1 - 2-3/32	1 - 4-3/16	2 - 1-1/8	106- 142
			Outdoor Section —	2 - 5-1/4 Diameter	3 - 8	107- 125
38TQ	40,000-55,000	45,000-63,000	Indoor Compressor Section 1 - 2-3/32	1 - 4-3/16	2 - 1-1/8	130
			Outdoor Section —	2 - 5-1/4 Diameter	3 - 8	125
38BQ	50,000-86,000	51,000-86,000	3 - 8-1/8	3 - 5-3/8	2 - 9	590
38AQ	121,000-311,000	115,000-327,000	12 - 10-3/4	4 - 10	4 - 9-7/8	750-2500

Fig. 13.3-2 Heat pump systems. (Reproduced Courtesy of Carrier Corporation.)

heating the air to the rooms. The noise is outdoors but should not be forgotten. Summer or winter, all the heat exchanger coils and their fans must be kept clean. The oil level should be checked in the compressors, and the refrigerant liquid level checked also, as well as suction and discharge pressure to the compressor, summer and winter. Dirty condensers cause high discharge pressure cut out. Watch out for snow and ice in winter on outdoor units. It is a good idea to have the top covered to prevent snow accumulation. Figure 13.3-2 shows one type of split system condensing unit.

13.3-4 Air-to-Air Units

The indoor and split indoor–outdoor units are confined to air units usually. The refrigerant flows through the indoor coil, called the evaporator, and absorbs heat from indoor air, which is blown across it by a fan. The vapor leaving the evaporator flows to the compressor, where its pressure and temperature are raised, and then to discharge valves. The hot gas goes next to the condenser (indoors and outdoors) where outdoor air is drawn through a duct across the condenser, picking up the heat from the hot gas and condensing it to a liquid which flows to the receiver. From the receiver it goes through the expansion valve to the evaporator as needed, depending on the load. The hot air from the condenser is discharged outdoors. It should be verified that this hot air cannot flow over a wall or roof of the building, where it can flow back into the building and defeat the purpose of the air conditioner.

Air-to-air units are dependent on the weather as to what the discharge pressure will be, summer and winter, being higher in summer than winter. The outdoor coils should be protected from icing, frosting, rain, or snow to give desired heat transfer. Coils should be kept clean of leaves, bird feathers, and so on.

In the winter the outdoor unit is the evaporator and the indoor unit is the condenser. As indicated earlier, the outdoor unit will have to be defrosted by reverse cycle for about 4 min, or by electric heaters, when the outdoor air is between 20 to $-40°F$ and $+60\%$ RH, and, possibly, in rain, sleet, ice, or snow storms. Typical heating performance data are shown in Table 13.3-1.

Figure 13.3-3 describes outdoor units of Lennox to be used with oil, gas, or electric furnaces as described above where extra heat is required when it is below 30°F.

TABLE 13.3-1 LENNOX INDUSTRIES HP11-411 / 651V HEATING PERFORMANCE AT 2250 cfm INDOOR COIL AIR VOLUME (C5-920FF)

Outdoor Temperature[a] (°F)	Compressor Motor Watts Input	Total Output (Btu · h)
65	6,075	74,300
60	5,850	70,700
55	5,630	67,100
50	5,405	63,500
45	5,180	59,800
40	4,955	56,100
35	4,725	52,500
30	4,490	48,800
25	4,260	45,100
20	4,030	41,400
15	3,805	37,700
10	3,575	34,000
5	3,350	30,300
0	3,135	26,600
−5	2,910	22,900
−10	2,685	19,300
−15	2,460	15,600
−20	2,230	11,900

[a] Outdoor temperature 70% relative humidity.
Indoor temperature at 70°F. (Courtesy Lennox Industries Inc.)

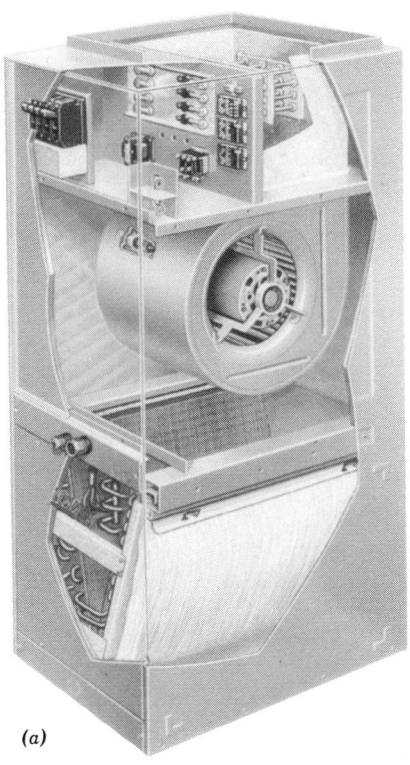

(a)

(b)

Fig. 13.3-3 Outdoor heat pump (from ref. 6): (*a*) Furnace and indoor coil; (*b*) FMP1 control box; (*c*) outdoor unit; (*d*) typical application; (*e*) outdoor unit; (*f*) Lennox two-speed Landmark® Compressor. (Courtesy of Lennox Industries, Inc.)

(c)

(d)

Fig. 13.3-3 (Continued)

(e)

Fig. 13.3-3 (*Continued*)

Table 13.3-2 describes Lennox outdoor units heating capacity with air entering the outdoor coil at values from 65°F down to −15°F. The capacity varies from 43,000 Btu/h, requiring 29-kW motor power at 65°F, to 15,600 Btu/h, using 2.46 kW at −15°F.

13.3-5 Air-to-Water Units

The second alternative source of energy for heat pumps is a lake or pond, a well or spring, or a cooling tower for a condenser.

The water could be cooled in the winter while the refrigerant in the evaporator picks up the heat from it. The water could even be frozen, and the larger the pond the more capacity is available.

In the summer the condenser could give up its heat to the pond, which could be evaporated into the atmosphere, and the water level kept constant by additional water if necessary. This heat will be available next fall and winter if the pond is underground.

Domestic hot water heaters or auxiliary tanks to heat water could be used to store heat, and in place of an oil tank, a water tank could be placed under the ground in the yard. Figure 13.3-2 shows various shapes and sizes of heat pumps as well as the Carrier catalogue chart from 14,000 Btu/h (1 ton) to 310,000 Btu/h (25.8 ton) capacity units giving weights and dimensions.

13.3-6 Air-to-Earth Units

In Sweden and other places around the world, pipes are being buried below the frost line 3–4 ft underground and 3–6 ft apart in the yard or garden on the owners' property. The fluid pumped in this pipeline absorbs heat from the ground in the winter and gives it up to the refrigerant. In the summer the water gives up the heat to the ground that it has picked up from the condenser.

The main thing to remember about any of these systems is that the energy absorbed in the winter is free and only the compressor energy has to be purchased. Approximately *two-thirds* of the heating energy is *free*! Figure 13.3-4 shows energy storage systems in the earth under a house.

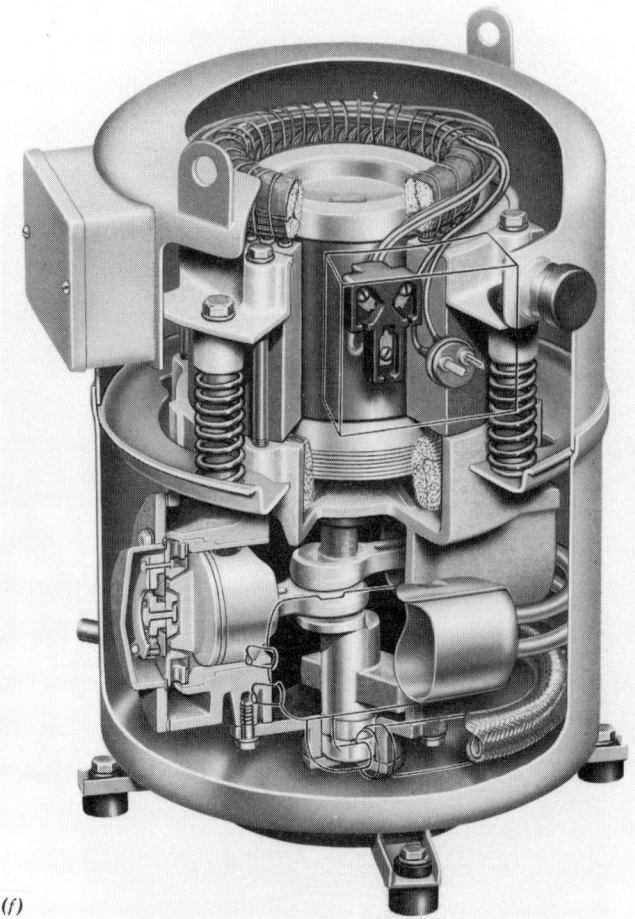

(f)

Fig. 13.3-3   *(Continued)*

13.4   REFRIGERATION SYSTEMS

13.4-1   Introduction

Refrigeration is one of the greatest technical developments in the 1900s. Cooling and freezing makes it possible to preserve food in a healthy manner, so that people can be fed on food from all over the United States, and even the world, by preserving it as it is shipped by air, truck, boat, or railroad in refrigerated compartments. The first step was the manufacture of ice for home units and freezers to keep local products, pigs, cows, deer, and so forth. Then refrigerators came into being. Finally, the air conditioner was built and sold in large numbers at relatively low prices to provide people with a comfort unit for summer conditions.

People in the hot dry country learned the evaporative cooling concept, where room air is blown with a fan over a wet wick that can cool and humidify 95°F, 25% relative humidity, Tucson, Arizona, summer air to 75°F and 75% RH without all the kilowatts it takes to run a vapor compressor.

The absorption refrigeration system replaces the compressor with a heater (or still), where the vapor from the evaporator is boiled off of the weak liquor, which absorbs the vapor in the absorber and gives it up in the still. If a cheap source of heat is available, then this is a good system.

TABLE 13.3-2 LENNOX INDUSTRIES HEAT PUMP RATINGS: HP11-410 / 650V HEAT PUMP OUTDOOR UNIT HEATING CAPACITY

Air Temperature Entering Outdoor Coil (°F)

Low Speed Compressor Operation—Low Indoor Unit Air[a]

Indoor Unit Model No.	Indoor Coil Air Volume (cfm) 70F dB	65		60		55		50	
		Total Heating Capacity (Btu · h)	Comp. Motor Watts Input	Total Heating Capacity (Btu · h)	Comp. Motor Watts Input	Total Heating Capacity (Btu · h)	Comp. Motor Watts Input	Total Heating Capacity (Btu · h)	Comp. Motor Watts Input
C5-920FF	1,350	43,000	2,900	41,000	2,820	38,900	2,735	36,800	2,645
	1,500	43,900	2,865	41,800	2,785	39,700	2,700	37,500	2,615

Low Speed Compressor Operation—Maximum Indoor Unit Air[b]

Indoor Unit Model No.	Indoor Coil Air Volume (cfm) 70F dB	65		60		55		50	
		Total Heating Capacity (Btu · h)	Comp. Motor Watts Input	Total Heating Capacity (Btu · h)	Comp. Motor Watts Input	Total Heating Capacity (Btu · h)	Comp. Motor Watts Input	Total Heating Capacity (Btu · h)	Comp. Motor Watts Input
C5-920FF	2,250	47,800	2,690	45,600	2,615	43,300	2,540	41,100	2,465
	2,500	48,700	2,670	46,400	2,595	44,100	2,520	41,800	1,450

Air Temperature Entering Outdoor Coil (°F)

High Speed Compressor Operation—Maximum Indoor Unit Air[c]

Indoor Unit Model No.	Indoor Coil Air Volume (cfm) 70F dB	65		45		25		5		−15	
		Total Heating Capacity (Btu · h)	Comp. Motor Watts Input	Total Heating Capacity (Btu · h)	Comp. Motor Watts Input	Total Heating Capacity (Btu · h)	Comp. Motor Watts Input	Total Heating Capacity (Btu · h)	Comp. Motor Watts Input	Total Heating Capacity (Btu · h)	Comp. Motor Watts Input
C5-920FF	2,250	74,300	6,075	59,800	5,180	45,100	4,260	30,300	3,350	15,600	2,460
	2,500	75,800	6,005	61,000	5,120	46,000	4,210	30,900	3,310	15,900	2,430

[a] Heating capacities include the effect of defrost cycles in the temperature range where they occur.
[b] Heating capacities include the effect of defrost cycles in the temperature range where they occur.
[c] Heating capacities include the effect of defrost cycles in the temperature range where they occur.

Courtesy of Lennox Industries, Inc.

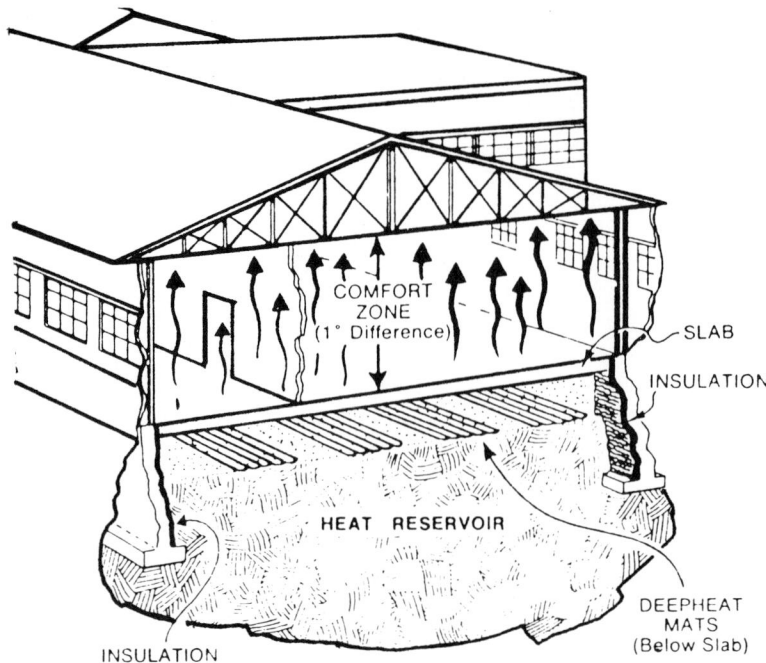

Fig. 13.3-4   Earth energy storage (from ref. 11). (Reprinted by permission, Dunham-Bush, Inc., 1982.)

13.4-2   Vapor Compression Systems

Vapor compression refrigeration systems consist of (1) an evaporator, where heat is absorbed by the refrigerant from the substance, fluid or gas, being cooled, (2) the compressor which compresses the refrigerant vapor boiled off in the evaporator and discharges it to (3) the condenser where the hot vapor is cooled and condensed into liquid which goes into a (4) storage pipe or receiver, before it passes through (5) the expansion valve where the pressure is reduced to the low pressure and low temperature of the evaporator where it once again begins to evaporate as it absorbs heat. It takes power from a motor, diesel, or gasoline engine to run the compressor. Approximately 0.8 kW/ton of refrigeration is required. One ton equals 3516 W (12,000 Btu/h) absorption capacity in the evaporator.

Fans are required to blow the air to be cooled across the evaporator and outdoor air to be heated across the condenser. The compressor is either a reciprocating single- or multicylinder device, depending on size, or possibly a centrifugal or axial compressor if larger than 25–50-ton capacity.

Lubrication is an important problem with compressors. Reciprocating compressors have been made hermetically sealed to reduce refrigerant leaks out of the compressor shaft seals and to contain lubricant inside. Special refrigerant lubricants without parafin are used to prevent lubricant freezing. The oil must be compatible with the refrigerant and be separated from the vapor leaving the compressor to prevent the condenser or evaporator gathering oil on the tube surfaces to clog them or to reduce heat transfer.

A pump down and cleaning procedure should be performed as a preventative maintenance frequently. Discharge and suction pressure should be watched. If the compressor trips off frequently from high-pressure cut out, the unit should be vented of air, which may leak into the system if the evaporator pressure is below barometric pressure. A refrigerant should be used that has a positive pressure at evaporator temperature, such as F-12 or F-22, so that air cannot leak in.

The performance of the vapor compression system is determined by measuring (1) the pressure and temperature of the refrigerant in and out of the evaporator, (2) in and out of the compressor as well and (3) kilowatts required and (4) power factor of the motor driving the compressor, and (5) the pressure and temperature of the vapor in the liquid in and out of the condenser and the pressure and temperature of the liquid to the expansion valve. If there is a heat exchanger that heats cold vapor from the evaporator while cooling liquid en route to the expansion valve from the condenser, then read pressures and temperatures here also. The flow of the refrigerant liquid to the expansion valve is also most important but often is not measured. The quantity of air being cooled needs to be also measured, as well as the temperature and pressure of air in and out of the evaporator. If the pressure drop across the evaporator including the inlet air filter is too great, the filter needs cleaning or changing, and, of course, like heating systems, the refrigerator enclosure should be insulated and protected from the summer solar load through glass doors and windows by overhangs to deflect the sun load. The heat discharged from the condenser can be used in separate heat exchanger condensers, which are used to heat domestic hot water and save using gas or electricity to do this in the summer.

Do not set a condenser next to a metal wall where the heat is transferred directly back into the area from which it has just been removed.

13.4-3   Absorption Refrigeration

When steam, gas, solar, or any other form of energy is available and economical, it can be used to heat a weak liquor of ammonia or lithium bromide in the gasifier, or still, and boil out the steam it has absorbed in the absorber. The steam is condensed in a condenser and the water flows to the evaporator. Here in the evaporator the water is at a very low saturation pressure corresponding to 32–42°F temperatures. Water to be used as a chilling fluid in an air-conditioning system is cooled typically from 87 to 50°F in the evaporator and pumped to the air conditioner cooling coil.

The heat absorbed in the evaporator water forms steam at 32–42°F and is sucked out of the evaporator and into the absorber by the strong liquor, but then it gets hot as it absorbs the steam and condenses it to liquid, so another heat exchanger with condenser cooler water has to keep this liquor cool in the absorber so it will absorb a maximum of steam from the evaporator.

The strong liquor becomes a weak liquor and is now pumped through a heat exchanger to the gasifier, or still, to boil the steam off again. The hot strong liquor flows back from the still through the heat exchanger to the absorber where exhaust steam, solar energy, or dump gases are available. Figure 13.4-1 describes the capacities of water-cooled condensers and compressors along with packaged

Compressors/Water-Cooled Condensing Units

Open (5F,H) and semi-hermetic (06, 07) reciprocating — used for air conditioning or refrigeration. Uses Refrigerants 12, 22, 500 or 502.

Compressors

MODEL NO.	COOLING CAP. RANGE	DIMENSIONS (ft-in.) Largest Model			APPROX. WT RANGE (lb)
		l	W	h	
5F,H	5 - 240 tons	11 - 10-1/4	2 - 8-5/8	3 - 9-1/8	175-4305
06D,E,L	2 - 130 tons	5 - 0	2 - 5-3/16	2 - 9-5/16	225-2220

Condensing Units

MODEL NO.	COOLING CAP. RANGE	DIMENSIONS (ft-in.) Largest Model			APPROX. WT RANGE (lb)
		L	W	H	
5F,H	5 - 175 tons	8 - 4-5/8	2 - 10-1/8	5 - 0-1/8	385-4740
07D,E,L	2 - 130 tons	8 - 8-5/8	2 - 6-5/8	4 - 11	295-3850

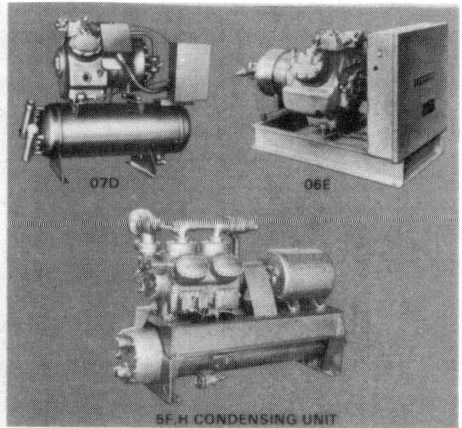

Packaged Air Handling and Fan-Coil Units

Carrier Weathermaker packaged air handlers provide efficient source of conditioned air for factories, shops, offices, restaurants. Modular construction allows fan section to be mounted in three different arrangements. Use 40RR with direct-expansion applications; use 40RS if a source of chilled water is available.

40BA direct-expansion fan-coil can be used with or without ductwork, for vertical or horizontal discharge.

MODEL NO.	COOLING CAP. RANGE (Btuh)	DIMENSIONS (ft-in.) Largest Model			APPROX. WT RANGE (lb)
		L	W	H	
40BA	54,000-160,000	3 - 4	2 - 8	2 - 6	255- 325
40RR	41,000-751,000	10 - 6-3/8	3 - 10-1/4	8 - 4	340-1940
40RS	80,000-360,000	7 - 9-1/2	2 - 11-3/4	6 - 9-1/8	365-1150

Fig. 13.4-1   Compressor and condenser units. (Reproduced Courtesy of Carrier Corporation.)

CONSTRUCTION DETAILS

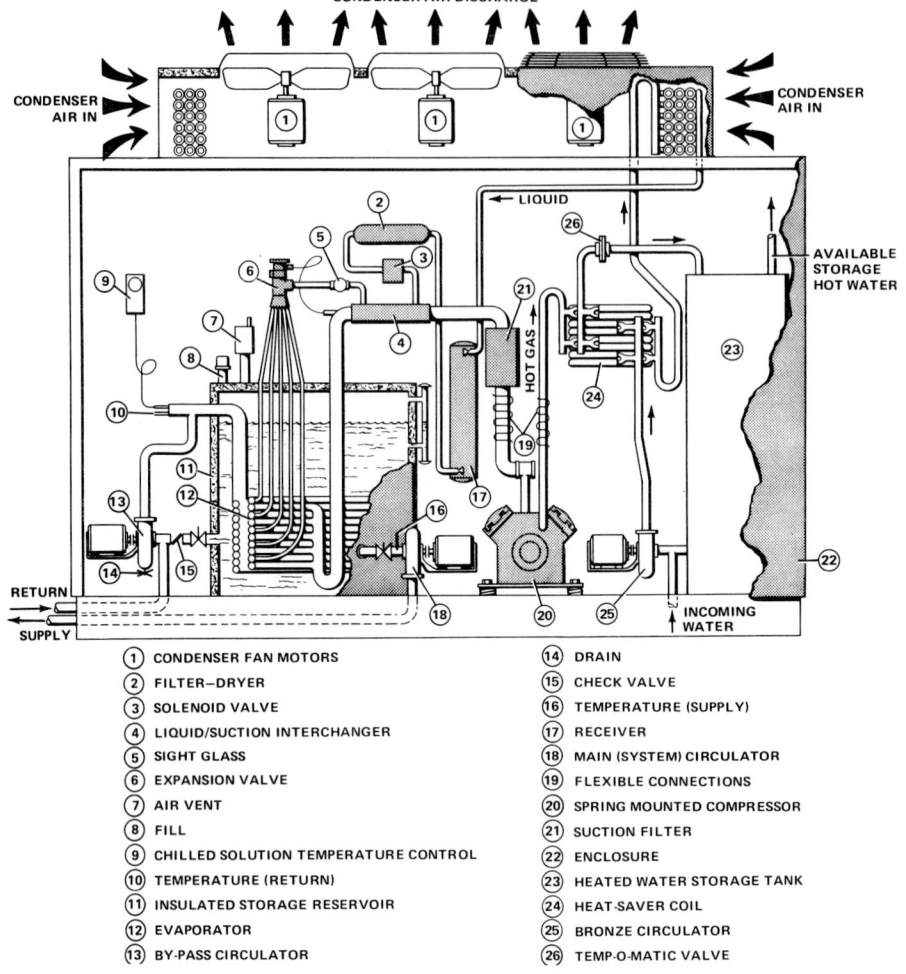

TYPICAL SCHEMATIC OF WALK-IN ENCLOSURE MODEL
WITH OPTIONAL HEAT-SAVER COIL AND ACCESSORIES

① CONDENSER FAN MOTORS		⑭ DRAIN	
② FILTER—DRYER		⑮ CHECK VALVE	
③ SOLENOID VALVE		⑯ TEMPERATURE (SUPPLY)	
④ LIQUID/SUCTION INTERCHANGER		⑰ RECEIVER	
⑤ SIGHT GLASS		⑱ MAIN (SYSTEM) CIRCULATOR	
⑥ EXPANSION VALVE		⑲ FLEXIBLE CONNECTIONS	
⑦ AIR VENT		⑳ SPRING MOUNTED COMPRESSOR	
⑧ FILL		㉑ SUCTION FILTER	
⑨ CHILLED SOLUTION TEMPERATURE CONTROL		㉒ ENCLOSURE	
⑩ TEMPERATURE (RETURN)		㉓ HEATED WATER STORAGE TANK	
⑪ INSULATED STORAGE RESERVOIR		㉔ HEAT-SAVER COIL	
⑫ EVAPORATOR		㉕ BRONZE CIRCULATOR	
⑬ BY-PASS CIRCULATOR		㉖ TEMP-O-MATIC VALVE	

Fig. 13.4-2 Walk-through HVAC enclosure (from ref. 3). (Courtesy of Edwards Engineering Corporation.)

air-handling fan coil units. The dimensions of the condensers and compressors are given as well as the air-handling and fan coil units. Figure 13.4-2 shows the many parts of a walk-in enclosure model of an HVAC system in a so-called mechanical equipment room. This is an economical means of air conditioning with no compressor, kilowatts, or maintenance expense. But the added heat loss to the condenser should be saved to heat hot water, as mentioned in Section 13.4-2.

13.4-4 Evaporative Cooling

As mentioned earlier, when air is dry and hot, as in Florida or Arizona, it can merely be blown across a wet cloth, or through a filter, or a water spray, and while absorbing water it cools the air. Such a system is natural and has been used for centuries in the Middle East. All that is needed is a fan to blow the air across the filter. Fresh outdoor air could be drawn in and blown across the air and delivered to the room.

13.4-5 Freon Refrigeration System Data

ASHRAE or chemical company data of refrigerants should be consulted for data such as shown on the diagram for refrigerant R22 in Fig. 13.4-3.

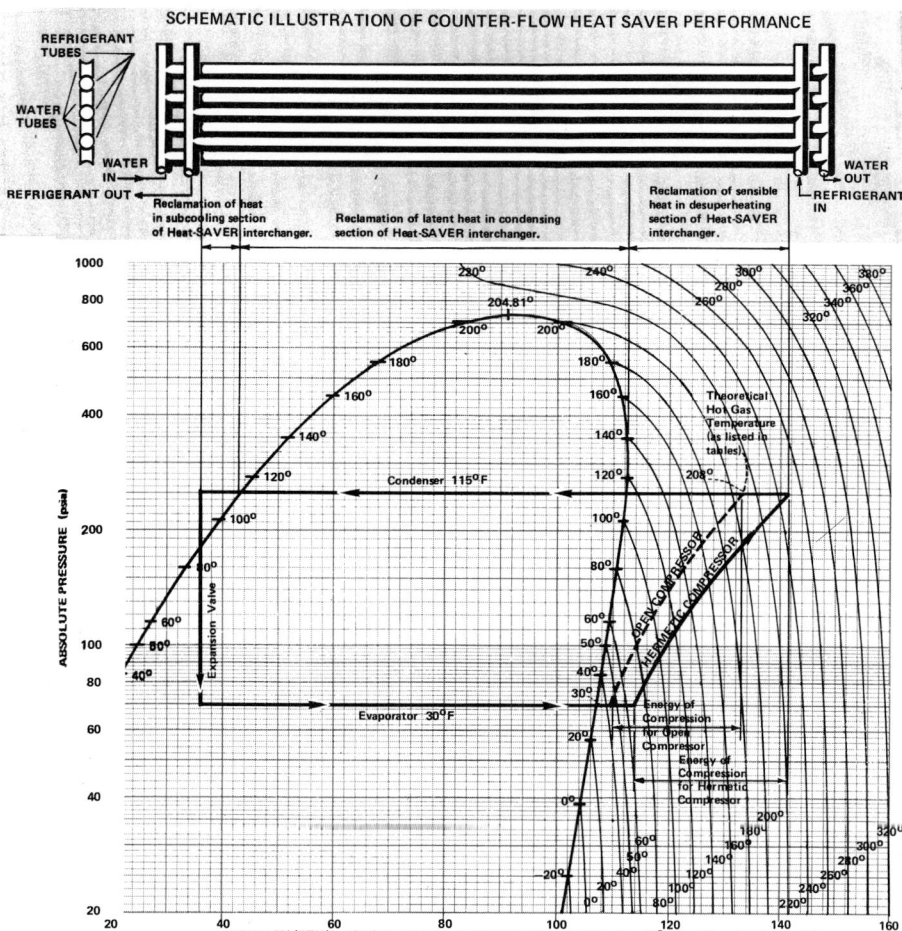

Fig. 13.4-3 Pressure–enthalpy (*P-H*) diagram for refrigerant R22. (Courtesy of Edwards Engineering Corporation.)

Figure 13.4-3 is a PH (Pressure–Enthalpy) diagram where properties of a refrigerant are plotted. This provides data for a plot of the pressure and temperature of a liquid leaving a condenser and expanding through an expansion valve to the valves of *p* and *t* in the evaporator. Then the expansion of the liquid into a vapor as it gains heat from cooling air, milk, meat, even freezing substances, can be plotted. The refrigerant vapor entering the compressor can be shown, and the effect of compression and discharge pressure and temperature to the condenser can be shown. Then, the effect of condensing the hot vapor to liquid returning to the expansion valve can be shown. The enthalpies for various refrigerants can be obtained from Chapter 17 of *ASHRAE 1981 Fundamentals*.

The various parts of the refrigerator system evaporator, compressor, condenser, and expansion valve are illustrated on the diagram and the enthalpies in and out can be read as well as temperatures and pressures.

13.5 AIR-CONDITIONING SYSTEMS

13.5-1 Room Units

Room units are essentially window units where the hot condenser cooling air can be drawn through a filter from the outdoors and discharged immediately back to the outdoors. The filter should be cleaned at least monthly. An insulation sheet should be placed against the metal separating the indoor coil from the outdoor coil, fan, compressor, and its motor, which is usually hermetically sealed to reduce the heat loss or gain into the room. If outdoor fresh air is drawn in through a damper, as it should be, the damper control should be checked periodically to be sure it is running properly.

TABLE 13.5-1   CARRIER CORPORATION PACKAGED AIR HANDLING AND FAN COIL UNITS

Model No.	Cooling Cap. Range (Btu · h)	Dimensions (ft-in.) Largest Model			Approx. Weight Range (lb)
		L	W	H	
40AQ	18,000–36,000	1–9	$1-9\frac{1}{2}$	3–6	69–114
40CF	13,500–30,500	$3-7\frac{1}{2}$	1–9	0–10	63–87
40DQ	14,000–30,000	$2-10\frac{1}{8}$	$2-2\frac{1}{8}$	$0-10\frac{1}{2}$	65–90
40BA	54,000–160,000	3–4	2–8	2–6	255–325
40RE	88,000–190,000	$5-2\frac{1}{8}$	$2-0\frac{5}{8}$	$5-4\frac{3}{8}$	340–410
40RR	41,000–751,000	$10-6\frac{3}{8}$	$3-10\frac{1}{4}$	8–4	340–1940
40RS	80,000–360,000	$7-9\frac{1}{2}$	$2-11\frac{3}{4}$	$6-9\frac{1}{8}$	365–1150
40QB	42,000–60,000	$2-2\frac{7}{16}$	1–9	$4-9\frac{1}{4}$	170–200

Reproduced Courtesy of Carrier Corporation.

The indoor coil, filter, fan, and dampers should be checked for cleanliness, as well as the drain pan on the coil and its drain pipe for dehumidified water. Many ceilings are noticeably wet from these pans overflowing due to drain clogging.

The room unit should have the capacity to handle sun or solar load as well as the gain through walls, windows, ceilings, and floors, due to temperature difference, plus people load, equipment load, and ventilating air load.

The thermostat should be located near to the window, on the wall, but out of any possible sun, and set at 78°F. Then the inner part of the room will be cooler.

The unit is usually placed under the window and cool air blown up the window and into the room to immediately absorb the heat coming through the window.

The units, indoors and out, should be installed on a firm foundation that will absorb vibration and noise to make it safe and not a bother to people in the room. It should also be easy to get at for maintenance. The motors and fans should not run at noisy speeds, such as inexpensive, compact 1800 rpm units.

Coil Selection Data
Rating based on:
40°F average water temperature
$6°\ \Delta T$ Temperature rise through coil
70% sensible heat and 30% latent heat
80°F D. B. room air temperature entering valance

Coil length (ft)	2 row 280 Btuh/ft $\frac{1}{2}''$ series		3 row 430 Btuh/ft $\frac{1}{2}''$ series		4 row 670 Btuh/ft $\frac{1}{2}''$ series		5 row 710 Btuh/ft $\frac{1}{2}''$ series		6 row 860 Btuh/ft $\frac{1}{2}''$ series	
	gpm	P.D. ft of water	gpm	P.D. ft of water	gpm	P.D. ft of water	gpm	P.D. ft of water	gpm	P.D. ft of water
2	0.19	0.008	0.29	0.025	0.38	0.058	0.47	0.11	0.57	0.19
3	0.28	0.021	0.43	0.067	0.57	0.16	0.71	0.30	0.86	0.52
4	0.37	0.043	0.57	0.14	0.76	0.33	0.95	0.63	1.15	1.09
5	0.47	0.056	0.72	0.26	0.95	0.58	1.18	1.11	1.43	1.91
6	0.56	0.125	0.86	0.41	1.14	0.94	1.42	1.79	1.72	3.10
7	0.65	0.186	1.00	0.62	1.33	1.42	1.66	2.71	2.00	4.62
8	0.75	0.268	1.14	0.88	1.52	2.03	1.89	3.85	2.29	6.63
9	0.84	0.362	1.29	1.23	1.71	2.80	2.13	5.30	2.58	9.12
10	0.93	0.477	1.43	1.63	1.90	3.72	2.37	7.05	2.87	12.15

Fig. 13.5-1   Chiller performance and Coil Selection Data. (Courtesy Edwards Engineering Corp.)

Performance: Valance cooling units using chilled water

A. Conditions: 75° D.B. and 70% sensible heat

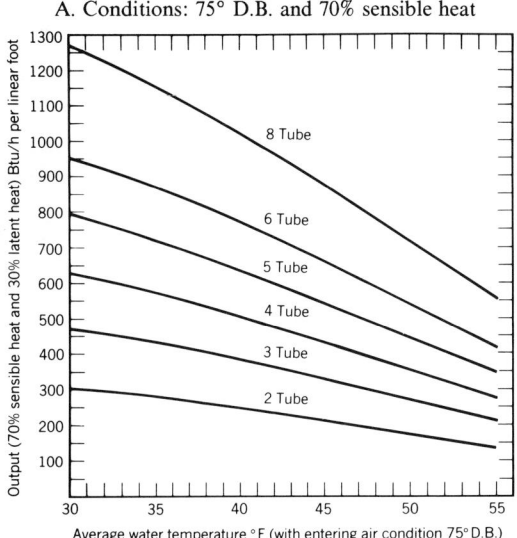

Notes:
1. Air leaves valance at 67% approach to entering water temperature. Example: Air entering at 75°F D.B., with valance coil water temperature at 40°F, will leave at a temperature of 75-0.67(75-40) = 51°F approx.
2. Ratings above based on chilled water flow of 4 gpm/ton refrigeration at design chilled water temperature, with 6°F water temperature rise through valance.
3. For ratings with refrigerant through valance, consult factory.

B. Conditions: 80° D.B. and 70% sensible heat

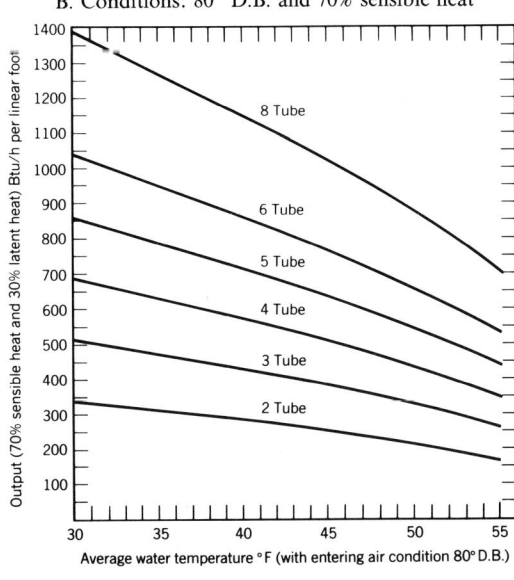

Notes:
1. Air leaves valence at 67% approach to entering water temperature. Example: Air entering at 80°F D.B., with valance coil water temperature at 40°F, will leave at a temperature of 80-0.67(80-40) = 53°F approx.
2. Ratings above based on chilled water flow of 4 gpm/ton refrigeration at design chilled water temperature, with 6°F water temperature rise through valance.
3. For ratings with refrigerant through valance, consult factory.

Fig. 13.5-1 (*Continued*)

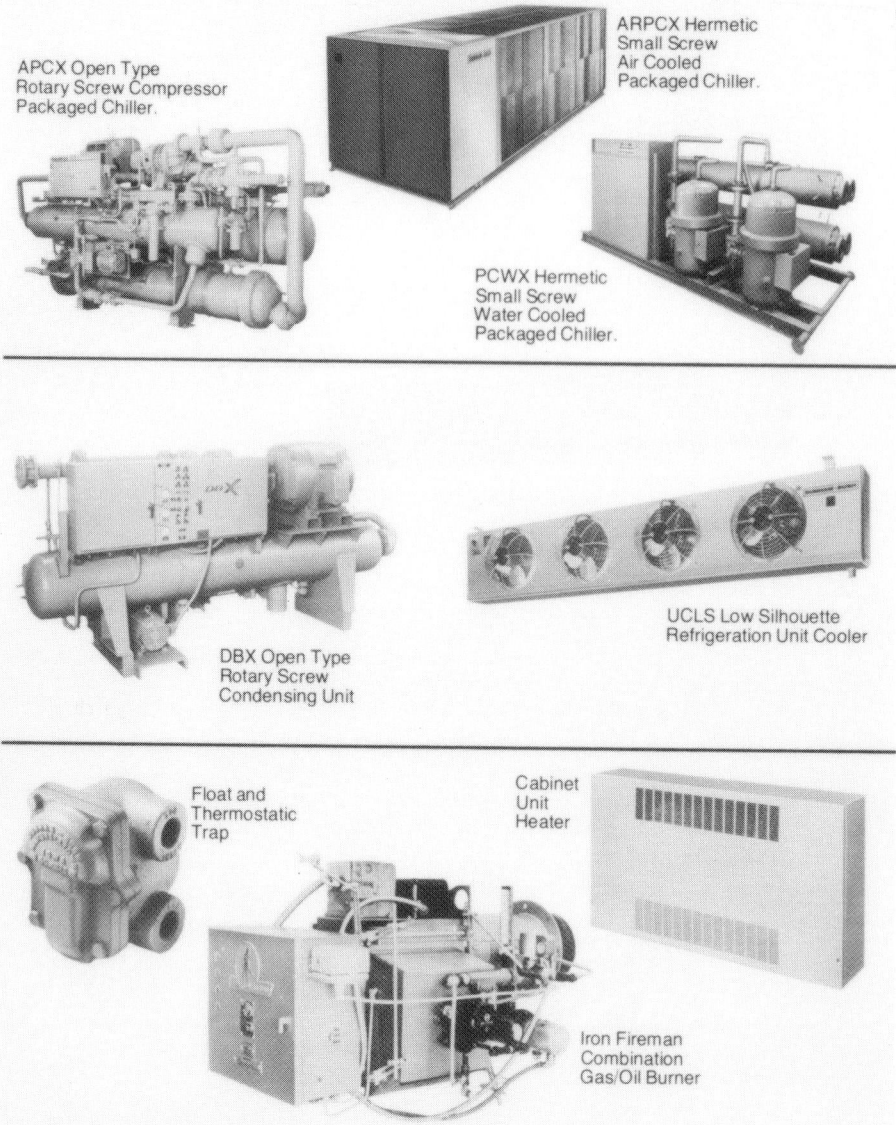

Fig. 13.5-2 Air-conditioning, refrigeration, and heating systems. (Reprinted with permission. Dunham-Bush, Inc., 1982.)

13.5-2 Zone or Hotel Floor Units

Many large buildings have zones or hotel floors conditioned by separate units in these areas. They can either have their own refrigerators or receive chilled water via a pump from the mechanical equipment room. This chilled water could chill water in a storage tank on the floor, to be pumped to heat exchangers in overhead ducts in each room or to the window units. In the winter hot water could be pumped to these units. Fans in the ducts or under the windows would then suck in room air, and it would leave the coils to go back into the rooms.

Another way would be to have a duct system to and from each room, which takes space. It is usually in the ceiling area in the halls or over the rest rooms. A central coil would use the chilled water

Air Conditioning

Rotary Screw Packaged Chillers. Hermetic or open drive, 50 to 750 tons. Air, evaporative or water cooled.

Reciprocating Packaged Chillers. Hermetic, semi-hermetic or direct drive, 2 to 150 tons. Air or water cooled.

Rotary Screw Compressor and Condensing Units. Open drive compressors. 50 to 750 tons. Air, evaporative or water cooled.

Condensers. Air, evaporative or water cooled. Up to 1000 tons.

Air Handlers. Low, medium or high pressure. Horizontal and vertical. 600 to 50,000 cfm.

Fan Coil Units. Ceiling, floor, wall-mounted, semi-recessed.

Coils. Chilled water, direct expansion, hot water, steam.

Refrigeration

Rotary Screw Packaged Chillers. Industrial direct drive, 120 to 750 tons. Air, evaporative or water cooled.

Reciprocating Packaged Chillers. Industrial semi-hermetic or direct drive, 2 to 150 tons. From +35° to−20°F. Air, evaporative or water cooled.

Reciprocating Packaged Chillers with Storage Tanks. Semi-hermetic, 1½ to 15 tons. Air or water cooled.

Reciprocating Mobile Packaged Chillers. From 2 to 7½ tons. Air cooled.

Rotary Screw Compressor and Condensing Units. Open drive compressors. 50 to 750 tons. From+55° to−40°F. Air, evaporative or water cooled.

Reciprocating Compressor and Condensing Units. Semi-hermetic or direct drive, ½ to 50 tons. Air or water cooled.

Packaged Refrigeration. Prematched air cooled condensing units and unit coolers. 1½ to 120 tons.

Heating

Centrifugal Pumps. Double suction double volute, 20 to 75 hp. In-line, ¼ to 25 hp. Close coupled, ¼ to 25 hp. Double volute, 10 to 150 hp. Base mounted sleeve bearing, 1 to 20 hp.

Vacuum Pumps. Duplex or single, 2,500 to 65,000 EDR.

Condensate Pumps. Duplex or single, 2,000 to 50,000 EDR.

Boiler Feed Pumps. Duplex or single, 6,000 to 30,000 EDR.

Radiator Valves. 25″ vacuum to 200 psi, ½″ to 2″ tapping.

Float and Thermostatic Traps. Low, medium or high pressure.

Inverted Bucket Traps. To 175 psi. ½″ to 1¼″ tapping.

Thermostatic Traps. Low or high pressure. 25″ vacuum to 125 psi.

Unit Heaters. Vertical or horizontal. Steam or hot water.

Cabinet Unit Heaters. Steam or hot water.

Convectors. Steam or hot water.

Pressure Atomizing Burners. Oil, gas or combination gas/oil, 15 to 100 gph.

Air Atomizing Burners. Oil, gas or combination gas/oil, 30 to 230 gph.

DUNHAM-BUSH, INC.

175 South Street, West Hartford, Conn. 06110
One of The Signal Companies ⌀

Fig. 13.5-2   (*Continued*)

from the main mechanical equipment room to cool the air for all these rooms in a cooling coil on each floor. Thermostats would control dampers allowing so much air to each room to keep it cool or warm.

As mentioned before, hot water in the winter could heat the air to keep the rooms warm in the winter. Table 13.5-1 shows specifications for packaged air-handling fan coil units.

13.5-3   Central Chilled-Water Units

In the past it has been feasible to install large 15-X000-ton chillers in the mechanical equipment rooms in the basement, or even on the thirteenth floor, and pump chilled water to each room or zone air-conditioning heat exchanger where a fan sucked in area air and discharged cool air back into the area. There is no duct cost or space problem to and from a central chilled-water unit, but there are coils to cool the air and to be equipped with fans, dampers, and controls for valves in chilled-water lines in every area. Therefore, the cost of one method and its reliability versus another must be evaluated The chilled-water system has become very popular. The insulation on the pipes must be kept in good shape.

Figure 13.5-1 shows charts and graphs of chiller performance from which a given length and number of rows of tubes can be selected to give desirable gpm flow and reasonable pressure drop for a given total Btu/h load for a cooling coil. Figure 13.5-2 shows various chiller systems and heater systems supplied to provide hot or cold water for an HVAC system.

13.5-4 Central Air Units

Most of the old air-conditioning as well as heating systems that were air systems were heated or cooled centrally and delivered to the rooms. Air could be vented from the return duct leaving the room zones and fresh makeup air added through inlet ducts with filters and heating coils to raise 0°F winter air to 40°F before it mixed with air from the room.

This winter air was then preheated to a dry-bulb temperature having the relative humidity and wet-bulb temperature that would let it pass through a washer, water or steam spray type, to raise the humidity to the desired level. Then the air would be reheated to dry-bulb temperature that would fall on the sensible heat ratio line for the room or zone. In the room the air would cool and give up moisture to the desired room values of dry-bulb temperature and relative humidity.

In the summer the hot moist or dry outdoor air, depending on whether the building is in Florida or Tucson, would be mixed with the room air and cooled and dehumidified to the temperature and humidity on the sensible heat ratio (SHR) line for the room and be admitted to the room. If the SHR line is steep (SHR = 0.75–0.5), then it might be necessary to reheat the air leaving the cooling coil to have it warm enough to be on the SHR line before it enters the room. This is important in computer rooms, intensive care rooms, and such places where dry-bulb temperature must be kept between say 75–80°F and a relative humidity between 30–40%.

13.5-5 Cooling, Heating, Double-Duct Units

This system was adapted to use the philosophy of mixing cool air with hot air, summer or winter, to give the desired temperature in a given area by only adjusting the damper to add more or less hot or cool air as required. Two ducts, requiring space, and two fans, plus connecting ducts, were required, as well as return ducts from the room or zones. This system was very popular and has been used in many buildings in the past. The cost and space required by duct work has caused change to room units or chilled-water units in recent years. Humidity control was not included in this system as a main requirement for control as in the central air units.

13.5-6 Variable Air Volume

Some years ago the first five systems described here were run at full load and full fan speed when the thermostat turned the system on to heat or cool an area. Often the thermostat did not stop the fan. It ran 24 h per day to provide circulation and even ventilation. In recent years, with energy conservation as well as environmental control becoming most important, the variation by speed control of fans, or shutting down on the number of fans used (variable air volume), has become a most popular system. Proper ventilation and moisture control as well as air flow to all areas of the room or zone must be checked.

13.6 REFERENCES

13.6-1 Edwards Engineering Corp., "Completely Packaged Gas-Fired Heating Units for Zoned Gas-Fired Heat and Domestic Hot Water from One Packaged System," Pompton Plains, NY, 1982.

13.6-2 Edwards Engineering Corp., "Completely Packaged Oil-Fired Heating Units for Zoned Home Heating plus Swimming Pool Heating Domestic Hot Water, and Snow Removal from One Packaged System," Pompton Plains, NY, 1982.

13.6-3 Edwards Engineering Corp., "Compact-pact Gas Boilers: Specifications," Pompton Plains, NY, 1982.

13.6-4 Weil-McLain, "Energy Management with Commercial Hydronic Heating Systems," Michigan City, IN, 1981.

13.6-5 Edwards Engineering Corp., "Electric-Hydronic Heating Units," Pompton Plains, NY, 1982.

13.6-6 Lennox Industries, Inc., "Lennox Pulse T. M. 6.4 Series Up-Flow Furnace Bulletin," Dallas, TX, 1982.

13.6-7 Carrier Air Conditioning Corporation, "Air Conditioning Products and Systems," ref., 12, p. A16–29.

13.6-8 Bryan Steam Corporation, ref. 12, p. A37.

13.6-9 Smith Gates Corporation, "Deepheat," ref. 12, p. A50.

13.6-10 Edwards Engineering Corp., "Tube by Tube Heat-Saver Coil Desuper-heater and Heat Exchanger," 1982.

13.6-11 Dunham Bush, Inc., "HVAC Equipment,"

13.6-12 ASHRAE Product Specification File, 1982.

CHAPTER **14**

ELECTRICITY GENERATION, DISTRIBUTION, AND USE

DANIEL F. HANG

University of Illinois
Urbana-Champaign, Illinois

RICHARD D. SHULTZ

Clarkson College of Technology
Potsdam, New York

RICHARD A. SMITH

Florida Power Company
St. Petersburg, Florida

E. H. MILLER, ROY P. ALLEN

General Electric Company
Schenectady, New York

HERBERT R. STEWART

Private Consultant
8 Pilgrim Road
Waban, Massachusetts

Formerly Chief Electrical Engineer,
New England Electric System

14.1 ECONOMICS OF ELECTRIC POWER GENERATION

Daniel F. Hang, Richard D. Shultz, and Richard A. Smith

Every public utility has the responsibility to provide safe, adequate, and satisfactory service to all customers who seek it within their respective service territories. In order to accomplish this responsibility, public utilities must project future customer demand many years in advance and plan generating facilities that will best serve these future needs. At the same time the utility must provide reliable service to existing customers at a fair and reasonable price.

One goal of generation planning is to answer the following question: Is there a need for capacity? If a need for capacity can be satisfactorily demonstrated, the generation planner must next determine the answers to the following questions:

1. What type and amount of capacity is needed?
2. When is this capacity needed?
3. Where should this capacity be installed?

In some situations numerous alternative generation expansion schedules can be developed that answer these questions. Thus, the generation planner must select which alternative schedule to implement. The selection will be guided by some kind of comparative analysis between alternative schedules that ranks their relative desirability.

This analysis can consider a wide range of concepts among which economics is an important factor. Economic analysis of generation expansion plans consists of two quantities: capital cost and power production cost. Capital cost represents the generating station installation cost. Production cost is the amount of money spent for fuel, operation, and maintenance of generating stations.[1]

The following sections discuss some important features of production cost and capital cost calculations.

14.1-1 Production Cost Calculation

Production cost calculations performed for generation planning studies usually consider 20- to 30-year planning horizons, requiring long-range forecasts of power system loads and operational features of both existing and future generators. Such forecasts are used to create probabilistic load and generation models for use in cost calculations.

Load Model

Probabilistic load models express the probability of a power system load exceeding a certain value and can be created from forecasted peak load and per-unitized historical load profiles (Figs. 14.1-1 to 14.1-3). Quantities L_{max} and L_{min} are the maximum and minimum loads experienced by a power system for a known time period. A load duration curve (Fig. 14.1-2) can be formed from Fig. 14.1-1 data by calculating and plotting the time period percentage that a load exceeded a certain amount. L_{min} represents base load since it is available 100% of the time; intermediate load is represented between L_{min} and some value approaching L_{max}. Peak load is loosely defined as the few hours of time that include L_{max} as a load.

With some modification, the load duration curve is used in production cost calculations. The time axis is per-unitized and plotted on the vertical axis while the load in megawatts is plotted on the horizontal axis. This kind of curve (Fig. 14.1-3) is a load probability distribution and expresses the probability that a power system will experience a load greater than a certain value. Usually,

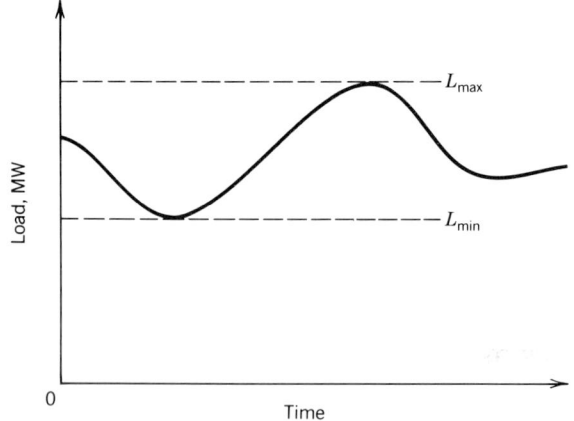

Fig. 14.1-1 Typical electric load profile.

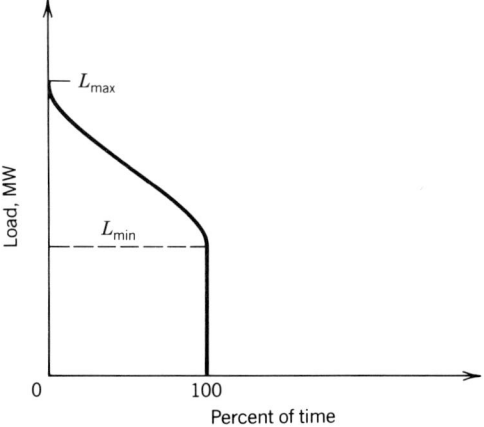

Fig. 14.1-2   Load duration curve.

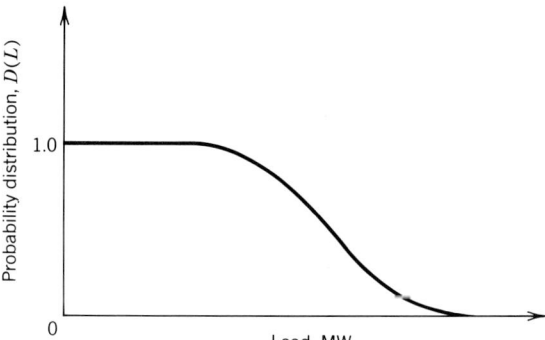

Fig. 14.1-3   Load probability distribution curve.

production cost calculations use several load probability distribution curves to model load variations during different times of each year in the planning horizon. For example, curves for nighttime, daytime, and weekend time may be used to model each hour of a year.

Generator Model

Probability models of generators are also used in planning studies, usually in the form of a probability density function of generator forced outage capacity. Figure 14.1-4 shows a forced outage probability density function, g, for a 100-MW generator. It shows a 0.95 probability of having 0 MW out of service and an 0.05 probability of 100 MW out of service. The probability numbers of 0.95 and 0.05 are referred to as generator availability rate p and generator forced outage rate q, respectively. These quantities can be calculated from the historical service performance of a generator (e.g., Fig. 14.1-5 for a 100-MW generator). The forced outage rate can be calculated from the average time the generator was in the 0- and 100-MW states as

$$\text{off} = \text{average 0-MW time} = \frac{t_2 + t_4 + t_6}{3}$$

$$\text{on} = \text{average 100-MW time} = \frac{t_1 + t_3 + t_5 + t_7}{4}$$

$$q = \text{forced outage rate} = \frac{\text{off}}{(\text{off} + \text{on})}$$

The generator availability p can be calculated from these same quantities as $\text{on}/(\text{off} + \text{on}) = p$.

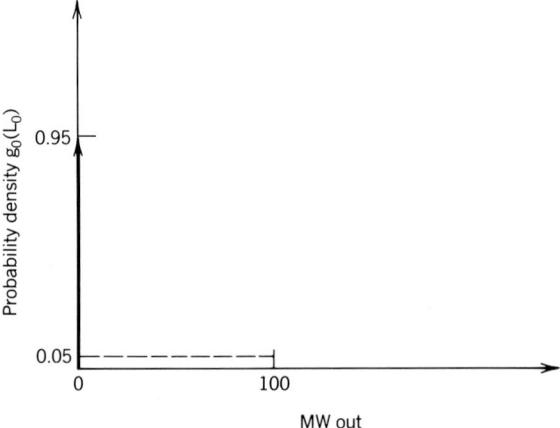

Fig. 14.1-4   Forced outage capacity probability density.

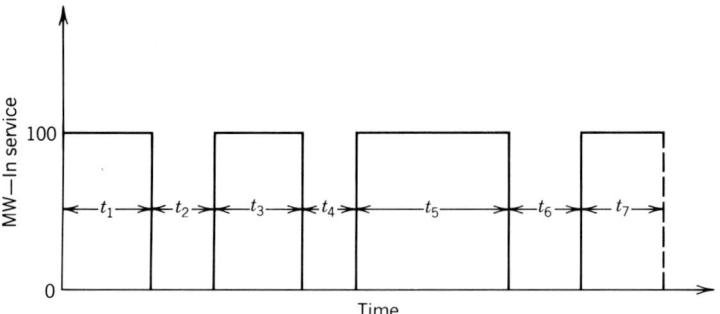

Fig. 14.1-5   Generator service performance.

Combined Load and Generation Models

The load and generation models described are combined to produce effective load probability distributions over a planning horizon of interest. The production costs for a generator system are calculated from these new distributions. The calculation of the effective load probability distribution is performed by recursive convolution of the generator outage probability density with the load probability distribution. The resulting distribution has a larger probability of the system load exceeding a certain value. This increase models the effect of a generator forced outage as additional load that the other generators in the system must pick up.

The convolution equation is expressed as

$$D^i(L) = \int_{L_{oi}}^{D^{i-1}} (L - L_{oi}) g_{oi}(L_{oi}) \, dL_{oi} \tag{14.1-1}$$

where L = load of interest on the effective load probability curve
    $D^i(L)$ = load probability distribution after generator i has been convolved
    $g_{oi}(L_{oi})$ = forced outage capacity probability density function for generator i
    L_{oi} = forced outage capacity of generator i

The recursive convolution of each generator begins with $D^o(L - L_{oi})$ on the right-hand side of the equation. Superscript o refers to the original forecasted load probability distribution. Figure 14.1-6 shows the effective load probability distribution curves built for a two-generator system. Since the generators are two-state discrete functions, the recursion equation simplifies to

$$D^i(L) = D^{i-1}(L - 0)p_i + D^{i-1}(L - C_i)q_i$$

The new $D^i(L)$ is calculated by substituting appropriate values into the right-hand side for each load

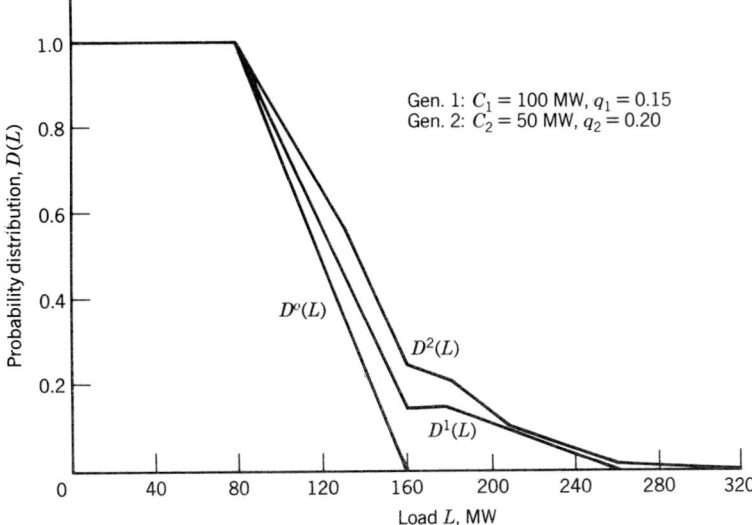

Gen. 1: $C_1 = 100$ MW, $q_1 = 0.15$
Gen. 2: $C_2 = 50$ MW, $q_2 = 0.20$

Fig. 14.1-6 Effective load probability distribution curves.

along the L axis. For example, the probability for L of 120 MW (Fig. 14.1-6) after generator 1 is convolved would be calculated as

$$D^1(120) = D^o(120 - 0)p_i + D^o(120 - 100)q_i$$

$$= D^o(120)p_i + D^o(20)q_i$$

$$= (0.5)(0.85) + (1.0)(0.15)$$

$$= 0.575$$

The probability for the same L after generator 2 is convolved would be calculated as

$$D^2(120) = D^1(120 - 0)p_2 + D^1(120 - 50)q_2$$

$$= D^1(120)p_2 + D^1(70)q_2$$

$$= (0.575)(0.8) + (1.0)(0.2)$$

$$= 0.66$$

The determination of power production cost uses the effective load probability distribution along with fuel price and heat rate data for the generators. For minimum production cost, the generators are convolved in an order that loads the most inexpensive operating unit first. This is referred to as the loading order and is derived from considerations of generator efficiency and fuel price.

Station Loading Order

A generating station may have several generators, each with its own particular heat rate curve. A heat rate curve (Fig. 14.1-7) shows the average Btu required to produce 1 W · h at a particular power output. Since the heat rate curve can vary so much over the normal output of a generator, each curve is divided into three segments for modeling purposes. Each segment obviously has a different efficiency and, therefore, would be loaded at different times in the ordering. However, segment 2 cannot be loaded until segment 1 has been loaded and similarly segment 3 must wait until segments 1 and 2 have been loaded. Segments 1 and 2 of a particular generator may, however, be loaded while economics may indicate that segments from other generators be loaded before segment 3 of that generator.

For any point on the heat rate curve, the product of the corresponding x and y axis values yield the quantity Btu/h. Plotting this quantity against the power output at each point results in the input–output curve of Fig. 14.1-8. For a fossil system the Btu content of the fuel used and its

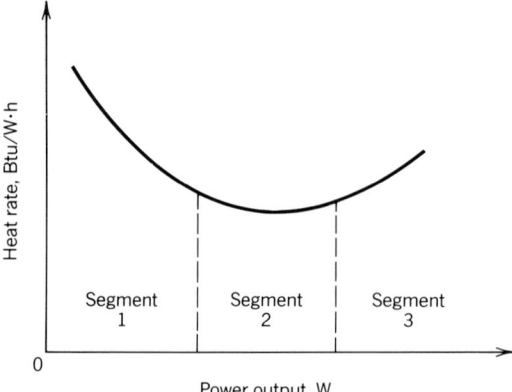

Fig. 14.1-7   Typical heat rate curve.

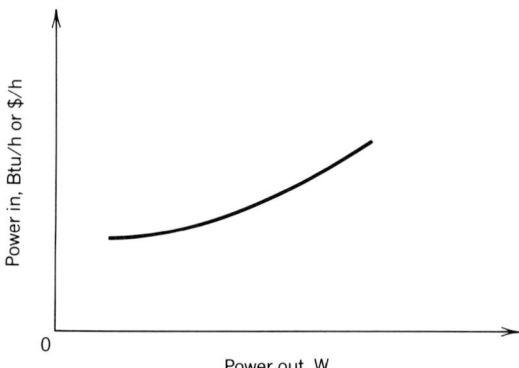

Fig. 14.1-8   Typical input–output curve.

replacement cost is generally known, hence the input–output curve can also have units of dollars per hour ($/h) on the y axis. Similar quantities on a nuclear system are also readily available.

Differentiation of the input–output curve with respect to power output yields an incremental fuel cost curve that is very important. It indicates the cost for an incremental increase in power output of a generator or, conversely, the savings for an incremental decrease in power output of a generator (Fig. 14.1-9).

Suppose that two units within a generating station are operating at outputs such that their incremental fuel cost curves are different. Assume that the power output of the generator with the higher incremental fuel cost is decreased while the power of the generator with the lower incremental fuel cost is increased the same amount. The result will be a greater reduction in cost at the higher incremental cost generator than increase in cost at the other generator. The same amount of power will be produced at a lower total cost. This shifting of load to achieve lower cost can be continued until all generators at a station are operating with the same incremental fuel cost. At this point the station would be operating at the most economical loading. Mathematically this concept is represented by

$$\frac{dF_1}{dP_1} = \frac{dF_2}{dP_2} = \cdots = \frac{dF_n}{dP_n} = \lambda \tag{14.1-2}$$

where F_i = fuel cost for unit i, $/h
      P_i = power output for unit i, W
      λ = incremental fuel cost, $/W·h

In order to determine the most economical loading between generating stations, the losses in the transmission system must be considered. If the power being generated at one station is decreased while

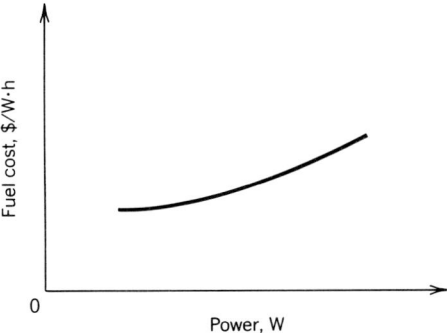

Fig. 14.1-9   Typical incremental fuel cost curve.

the power at another is increased, the flow of power along the transmission lines to the loads will change. Therefore, the power losses in those transmission lines will also change.

The reasoning used in determining the loading between generators at the same station can be applied to this problem. If operating a station at a particular power output results in a total incremental cost that is greater than the total incremental cost of another station, the output of the first station could be decreased and the output of the second station increased for a lower total cost. Therefore, for the most economical operation the total incremental cost should be the same for all stations. This value is called the system λ, or λ_s.

Note that the words *total incremental cost* have been used in the previous discussion. An incremental change in power output at a station will result in a change in incremental fuel cost for that station. In addition, the losses in the transmission system will also change as a function of power output of each station. Both of these "costs" must be considered in determining λ_s.

If a system was operating at a particular λ_s and the power output of a station i was changed, then the total cost per hour due to that power change would be

$$\left(\frac{dF_{si}}{dP_{si}} + \frac{\lambda_s \, \partial P_L}{\partial P_{si}} \right) dP_{si} = \$/h \qquad (14.1\text{-}3)$$

where F_{si} = fuel cost for station i, \$/h
P_{si} = power output of station i, W
P_L = total transmission losses, W
λ_s = system λ

For a system to operate most economically, all stations must have a total incremental cost of λ_s. The expression in Eq. (14.1-3) must, therefore, equal

$$\left(\frac{dF_{si}}{dP_{si}} + \frac{\lambda_s \, \partial P_L}{\partial P_{si}} \right) dP_{si} = \lambda_s \, dP_{si}$$

or solving for λ_s,

$$\frac{dF_{si}}{dP_{si}} + \frac{\lambda_s \, \partial P_L}{\partial P_{si}} = \lambda_s \qquad (14.1\text{-}4)$$

Rearranging Eq. (14.1-4) yields

$$\frac{dF_{si}}{dP_{si}} \frac{1}{(1 + \partial P_L / \partial P_{si})} = \lambda_s$$

The penalty factor for the station i is given the symbol L_i and defined as

$$L_i = \frac{1}{1 + \partial P_L / \partial P_{si}}$$

so

$$\lambda_s = \frac{dF_{si}}{dP_{si}} L_i$$

The results are similar to those of loading generators within a station. The most economic operation occurs when the product of incremental fuel cost and the penalty factor for each station in the system is the same. Thus, for a system with n generating stations

$$L_1 \frac{dF_{s1}}{dP_{s1}} = L_2 \frac{dF_{s2}}{dP_{s2}} = \cdots = L_n \frac{dF_{sn}}{dP_{sn}} = \lambda_s$$

The penalty factor can be calculated in many ways. Methods ranging from exact, time-consuming methods to estimated, time efficient approaches can be found in the literature.[2-5]

Base load is then relegated to the generator or generators that have the lowest incremental fuel cost. Intermediate load, that above L_{min} and less than L_{max} in Fig. 14.1-2, would be ordered next on the generators with the next lowest incremental fuel cost that can also load follow efficiently. In those few hours of the day when the peak load occurs (includes L_{max}), generators that can be brought on and removed from the system quickly are used to handle the peak load. The most economic peaking (lowest incremental cost) unit would be loaded first and removed last for the peak cycle.

The presentation to this point has featured the incremental fuel cost as a prime factor in determining loading order of the generating stations in a power system. Should hydroelectric plants be present in the system, no apparent incremental fuel cost can be associated with them. For thermal plants, either fossil or nuclear, the economic cost of generating energy determined how much each station generated. For hydroelectric plants the limiting factor is the amount of available water. Thermal stations in a system are loaded according to

$$L_i \frac{dF_{si}}{dP_{si}} = \lambda_s$$

Since hydroelectric stations are limited by water availability, they are loaded according to

$$K_i \frac{dC_i}{dP_{si}} = \lambda_s \qquad (14.1\text{-}5)$$

where C_i = water available per hour
$\quad K_i$ = constant conversion coefficient

The term K_i is a constant that scales the water available to an equivalent incremental cost curve.

Power Production Cost

Once loading order has been established by economic dispatch considerations, the effective load curve convolution can be performed. The power production cost for the forecasted load can be calculated from the effective load probability curve with the following equation:

$$PC_i = FC_i P_i T \int_C^{C+C_i} HR_i(L) D^{i-1}(L) \, dL \qquad (14.1\text{-}6)$$

where PC_i = power production cost for generator i, dollars
$\quad FC_i$ = fuel cost for generator i, \$/MBtu
$\quad T$ = number of hours for the time span of interest
$\quad HR_i(L)$ = incremental heat rate of generator i, MBtu/MW·h
$\quad C$ = total generation capacity already loaded
$\quad C_i$ = capacity of generator i

This calculation is illustrated by considering the example two-generator system of Fig. 14.1-6. Assume the generator characteristics are shown in Table 14.1-1 and that the load probability distribution curve is for a time span of interest. If generator 1 is loaded first, the integration of Eq.

TABLE 14.1-1   GENERATOR CHARACTERISTICS

	Generator 1	Generator 2
FC_i, \$/MBtu	1.2	1.6
HR_i, MBtu/MW · h	10	12

(14.1-6) for a one-day time period yields

$$PC_1 = 1.2 \times 0.85 \times 24 \int_0^{100} 10 \times D^o(L)\, dL$$

$$= 244.8 \left[\int_0^{80} 1.0\, dL + \int_{80}^{100} (2 - 0.0125L)\, dL \right]$$

$$= \$23,868$$

For generator 2 the calculation is

$$PC_2 = 1.6 \times 0.80 \times 24 \int_{100}^{150} 12 \times D^1(L)\, dL$$

$$= 368.64 \int_{100}^{150} [1.0 - 0.010625(L - 80)]\, dL$$

$$= \$9619.20$$

The total power production cost is the sum of PC_1 and PC_2, or \$33,487.20 for the day.

As noted earlier, the generators can be separated into segments for a more accurate determination of production cost. The convolution is performed on each segment as it is dispatched. However, care must be taken to deconvolve the earlier loaded segments of a generator before calculating the power production cost of its next segment. After the cost calculation the generator is convolved into the effective load probability distribution using a capacity equal to the sum of the capacities of its dispatched segments. This procedure is necessary to prevent the generation system from being modeled as a large collection of small-capacity (meaning dispatched segments) generators. Sullivan[2] provides a more detailed explanation of these calculations.

Operation and maintenance (O&M) costs are divided into two types, fixed and variable. Fixed costs are independent of the operation of a generator. These costs include the cost of normal maintenance personnel, operators, and administrative costs. Variable costs are a function of the actual operation of the unit. These costs include cost of maintenance, which is related to the running time of a unit, replacement of worn parts, and emergency maintenance, which requires personnel beyond the normal maintenance crew.

Fixed O&M costs are the larger of the two types ranging from 1 to 3% of the capital cost of a fossil unit. Nuclear units have O&M costs in a similar range except that the capital cost used for comparison is larger. Operation and maintenance costs are commonly assumed to be charges per unit of time.

Another component of total production cost includes startup and shutdown costs of generators. The startup of a fossil-driven generator and a nuclear-driven generator requires a significant amount of time and a relatively large amount of fuel for which little usable output is obtained. In addition the startup and shutdown of a generator puts stress on its mechanical components. Because of these effects and their associated costs, fossil generators are not shut down or started up if the conditions suggesting that action is not expected to exist longer than a few hours. This practice will cause the system to operate at a condition that is not economically optimal. However, the additional costs and risks of repeated startup and shutdowns outweigh the short-lived nonoptimal operation. The generator unit driven by a nuclear steam supply system operates on a somewhat different philosophy. It is built to shutdown and startup with less stress on the system. This requires added first-cost investment for the nuclear unit over the fossil to withstand more frequent startup and shutdown operation. Avoiding damage during operation of a nuclear steam supply system is of prime importance as the costs of damage during operation far outweigh any costs associated with shutdown and startup. Nuclear units during refueling support a large amount of preventative maintenance so that a shutdown action is less likely to confront the operator.

In summary, the total production cost is expressed as

Total production cost = power production cost + fixed O&M cost

+ variable O&M cost + startup cost + shutdown cost

Power production cost for nuclear units being more capital intensive than fossil are discussed at greater length in the next section.

14.1-2   Capital Cost Calculations

Electric utilities in the United States operate for the most part under one of three forms of capitalization structure: (1) investor owned, (2) publicly owned, or (3) industrially owned.

1. The investor-owned utility industry is the most prevalent. The company obtains its capital from debt and equity sources, pays income tax before passing earnings on to its owners, and is somewhat monopolistic in its service area. As a result, the rate structure in use by the utility is regulated by a state board or commission, presumably to protect the interest of the rate payer. Once the rates are set, the profit incentive for ownership is to minimize costs without sacrificing reliability of operation.

2. Public ownership of an electric utility may be on the federal, state, or municipal level. Electric rates are set by the public body to reflect the needed capital, fuel, and operating costs necessary to operate in the black. Public pride and prudent management is required to maintain low rates for the user. Capital is obtained through available debt sources.

3. Some industries find it economic to operate their own electric power plant. The need for process steam or a by-product of waste heat from a process leads to this cogeneration. Often these plants are interconnected with a commercial electric grid system to accomplish increased reliability of service and some economy in selling excess capacity. The debt-equity mix ratio of capital is much lower than the regulated utility as uncertainty of revenue is often much greater.

Electric utilities in general require a great deal of plant equipment for generation, transmission, and distribution of electric power. As a result these companies are inherently capital intensive in their financial structure and require a significant amount of planning to meet their cash flow needs. Whereas the average ratio of annual revenue to capital investment for all industry is greater than 2, the electric utility industry ratio is less than 1. For this reason the utility is prone to conduct long-range planning studies to support their decisions. These studies involve engineering economic analyses that are often updated as their growth program progresses.

Investor-Owned Utility or Industry Cash Flow

The cash flow in an investor-owned utility industry follows that shown in Fig. 14.1-10. The total revenue from sales or operating revenue represents the direct income resulting from product or service sale plus any other income received during the tax year. Cash operating costs or expenses are the annual sums of money required to produce the energy and services sold. Typical costs are fuel, labor, maintenance, property taxes, FICA, and other expenses allowed by current income tax laws. The interest on debt is the annual sum of money required for interest payment on outstanding borrowed funds. The depreciation expense (book depreciation) is a noncash expense allowed by the income tax law to recover from revenues for expected loss in plant and property value. Depending on the class of plant and equipment, accelerated depreciation schedules may be chosen by the utility. Taxable income is that amount of revenues that is in excess of costs and expenses for a given tax year. Taxable income is also the sum of federal, state, and local income tax and the net earning of the corporation for the taxable year.

Discounted Worth Method of Economic Analysis (Cash Flow Analysis)

Utilizing the cash flow diagram (Fig. 14.1-10), a completely general equation for economic analysis can be derived that is applicable to all industry and investor-owned utility that pay an income tax.[6] Income tax paid for any year n for a set of investments amount to

$$T(n) = (\text{income tax rate})[(\text{revenue from plant investment}) - (\text{operating expenses})$$

$$- (\text{tax depreciation}) - (\text{tax deductible items})]$$

or

$$T(n) = t[PQ(n) + W(n) - D(n) - O(n) - i_d r_d Y(n)]$$

Revenues available to retire the investment after costs, profits, and taxes are

$$R(n) = PQ(n) + V(n) + W(n) - O(n) - [i_d r_d + i_e(1 - r_d)]Y(n) - T(n)$$

Substituting for $T(n)$ yields

$$R(n) = [PQ(n) + W(n) - O(n)](1 - t) + tD(n) + V(n)$$

$$- [(1 - t)i_d r_d + i_e(1 - r_d)]Y(n)$$

The outstanding capital investment in the company at any time is

$$Y(n + 1) = Y(n) + B(n) - R(n)$$

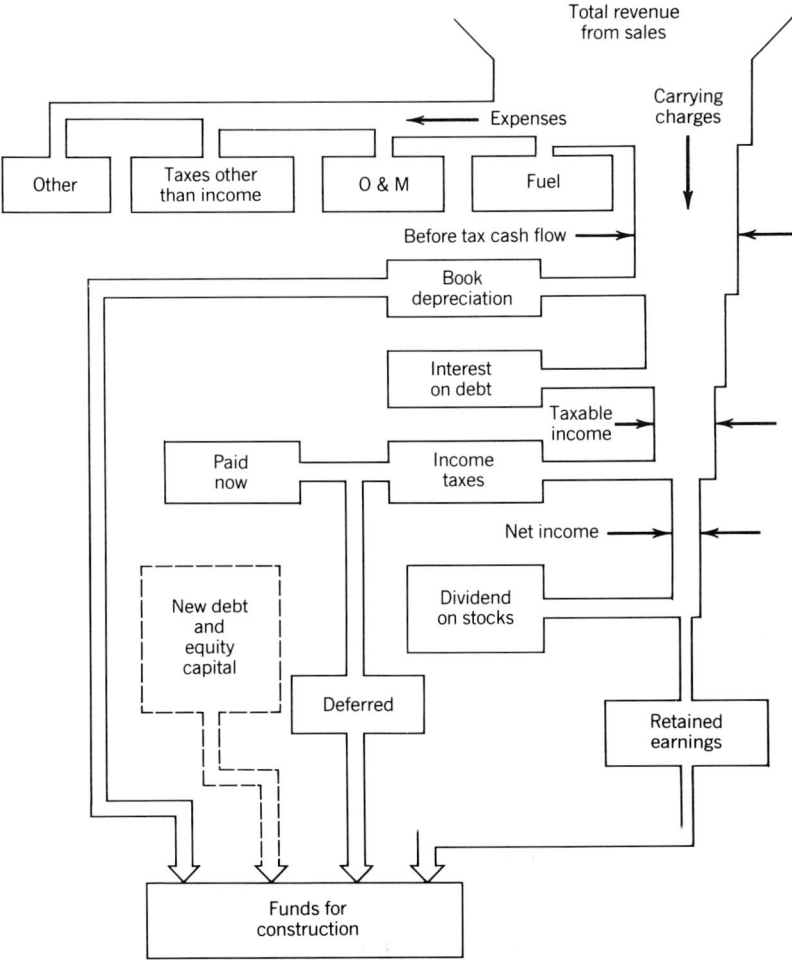

Fig. 14.1-10   Corporate cash flow diagram.

Substituting for $R(n)$ and rearranging terms,

$$Y(n + 1) = (1 + i_a)Y(n) + A(n)$$

where

$$A(n) = B(n) - V(n) - (1 - t)[PQ(n) + W(n) - O(n)] - tD(n)$$

and

$$i_a = i_d(1 - t)r_d + i_e(1 - r_d) = i_c - i_d r_d t \tag{14.1-7}$$

Note that i_a takes the form of a discount factor that is used for time valuing. It is the weighted cost of money, i_c, less the tax-adjusted effect of debt interest. For the high-risk industry where the debt ratio approaches zero, $i_a \rightarrow i_c = i_e$, and for government operations where the debt ratio approaches one, $i_a \rightarrow i_d$.

Nomenclature in this section has the following meaning:

$B(n)$ = investment (capitalized expenditure) occurring at any time n
$B(0)$ = investment (capitalized expenditure) occurring at $n = 0$
$Q(n)$ = amount of energy sold during period n (can be any form of production, not necessarily energy)
P = cost of energy production (or cost of any product production)

$Y(n)$ = outstanding capital investment indebtedness before considering income and outlays during period n

$W(n)$ = income from other than energy (or product) sale during period n

$D(n)$ = tax depreciation during period n

$O(n)$ = deductible operating costs during period n

$T(n)$ = income taxes during period n

$R(n)$ = net retirement income after costs and taxes

$V(n)$ = tax salvage value of $B(n)$

$\quad i_d$ = bond (or debt) interest rate (decimal fraction)

$\quad i_e$ = equity capital return rate (decimal fraction)

$\quad n$ = years or periods of time

$\quad r_d$ = debt to total capital investment ratio (debt ratio) (decimal fraction)

$\quad t$ = combined income tax rate (decimal fraction)

$\quad i_a$ = discount rate (only rate used for time valuing) (decimal fraction)

$\quad i_c$ = weighted cost of capital (decimal fraction)

All interest or return rates should be in a consistent set of units (e.g., if n is in years, i_a is interest/year expressed as decimal fraction). If a new investment $B(n)$ is considered by itself, the present worth of the revenue requirement for this investment during its particular useful life or study period N can be determined as[6]:

$$= \sum_{n=0}^{N} (1 + i_a)^{-n} \left[\frac{B(n) - V(n) - tD(n)}{(1 - t)} - W(n) + O(n) \right] \qquad (14.1\text{-}8)$$

Likewise, if this investment of $B(n)$ provides a revenue through a production rate structure, the present value of this revenue can be represented as

$$= \sum_{n=1}^{N} (1 + i_a)^{-n} PQ(n)$$

If the present worth of the revenues are equated to the revenue requirement for the same investment $B(n)$, a unit of production cost P can be expressed as

$$P = \frac{\sum_{n=0}^{N} (1 + i_a)^{-n} \left[\dfrac{B(n) - V(n) - tD(n)}{(1 - t)} - W(n) + O(n) \right]}{\sum_{n=1}^{N} (1 + i_a)^{-n} Q(n)} \qquad (14.1\text{-}9)$$

Equation (14.1-9) is completely general as it can be applied equally well to the determination of the production cost of an entire electric utility system or to examine the contribution of a single investment $B(n)$ to that production cost. The choice of $n = 0$ is strictly arbitrary. It can represent the very first investment expenditure on a large project that takes several years to complete or it can be taken as the completion date of the project (startup date). If the latter is chosen, then an investment made prior to startup would bear a negative value of n. As an example, $B(-2)(F/P)_2^{i_a} + B(0)$ represents an investment made in two payments, one being made 2 years (or periods) before startup and a second made at startup time ($n = 0$).

Of interest may be the present worth of the revenue requirement (PWRR) of a single investment in Eq. (14.1-8). Let this investment $B(0)$ occur at $n = 0$, have an expected tax and actual salvage of $V(n)$ at $n = N$, an operating cost of $O(n)$ that is neglected for the moment, and a tax life $N_t = N$. The PWRR for this investment expressed in words can be stated as

$$\text{PWRR} = \frac{\text{inv.} - \text{p.w. of tax salv.} - t(\text{P.W. of tax writeoff})}{1 - t}$$

Expressed in equation form,

$$\text{PWRR} = \frac{B(0) - V(N)(P/F)_N^{i_a} - t \sum_{n=0}^{N} (1 + i_a)^{-n} D(n)}{1 - t} \qquad (14.1\text{-}10)$$

Continuing this single investment $B(0)$ and assuming that tax depreciation is by a straight line over the lifetime N, then in Eq. (14.1-10) the tax term becomes

$$t \sum_{n=0}^{N} (1 + i_a)^{-n} D(n) = t\left(\frac{B(0) - V(N)}{N} \right)(P/A)_N^{i_a} \tag{14.1-11}$$

If the tax depreciation for this same case is by sum-of-the-years digits (SYD), then in Eq. (14.1-10) the tax term becomes

$$t \sum_{n=0}^{n} (1 + i_a)^{-n} D(n) = t(P/A)_N^{i_a} \frac{B(0) - V(N)}{N(N+1)/2} \left[N - (A/G)_N^{i_a} \right]$$

Where tax depreciation is by the unit of production method, each year $D(n)$ could be different, so Eq. (14.1-10) best defines the tax writeoff term as shown.

There are several values of n that must be considered. For an economic study life of N years the tax life N_t may be equal or less than the study life. The book life N_b, used for rate-making purposes, may be arbitrary, independent of N. The expected physical life of the plant is usually longer than the study lifetime, unless the study encompasses replacement options. Therefore, the n in Eqs. (14.1-8) to (14.1-10) are sensitive to the tax and study lifetime. The cash flow in Fig. 14.1-10 must dictate the form of these equations. Consider a numerical example applied to Eq. (14.1-10) where $i_d = 8\%$, $r_d = 55\%$, $t = 50\%$, $i_e = 14\%$, and tax depreciation of $V(N) = 0.1B(n)$ is reached by a straight line over a 25-year tax and book life (i.e., $N = N_t = 25$ years). Determine the PWRR for a \$100 investment made at $n = 0$. Using Eq. (14.1-7), i_a calculates to be 0.085, or 8.5%. Substituting values into Eqs. (14.1-10) and (14.1-11),

$$\text{PWRR} = \frac{100 - (0.1)(100)(0.13009) - 0.5[(100 - 10)/25](10.2342)}{1 - 0.5} = \$160.56$$

This means that the company must on a present worth basis bring in a revenue of \$160.56 to satisfy the revenue requirement of the \$100 investment. Expressing this revenue requirement as an equivalent yearly cost, AEC,

$$\text{AEC} = \text{PWRR}(A/P)_{25}^{8.5} = (160.56)(0.09771) = \$15.69/\text{yr}$$

This is the yearly revenues that would retire the obligation of the \$100 investment. The present worth of the operating costs $O(n)$ can be added, if necessary, in the analysis as

$$\sum_{n=0}^{N} (1 + i_a)^{-n} O(n)$$

Fixed-Charge Rate Method of Economic Analysis

The company that obtains its capital from debt and equity sources, subject to income tax on the equity portion, can utilize the concept of the fixed-charge rate to determine the levelized annual revenue requirement for an investment. Comparison with results obtained by the discounted worth method affords the user an excellent check on calculations and a choice of models.

1. **Nondepreciable fixed-charge rate.** Where an investment does not change in resale value from original value (nondepreciable), the yearly revenue requirement for an investment B is that required to pay the bond interest, the equity return (profit), and the income tax on the taxable income. In equation form this annual equivalent cost can be written as

$$\text{AEC} = Br_d i_d + B(1 - r_d)i_e + \frac{Bi_e(1 - r_d)t}{1 - t}$$

Collecting items and using Eq. (14.1-7) for i_a under discounted cash flow, yields

$$\text{AEC} = \frac{Bi_a}{1 - t} = BP_\infty$$

where P_∞ is defined as a nondepreciable fixed-charge rate. This term is extremely useful in determining revenue requirement on a period basis for an investment that does not change value during the time period being considered. Examples of such investments would be land holdings, prepayment charges, and working capital.

2. Depreciable fixed-charge rate—straight-line tax and book depreciation. The year-by-year revenue requirement for an investment of B with a salvage of V (tax and book) is noted for the first 2 years:

$$\text{AEC}_1 = BP_\infty + \frac{B-V}{n} \quad \text{(first year)}$$

$$\text{AEC}_2 = \left[B - \frac{(B-V)}{n} \right] P_\infty + \frac{B-V}{n} \quad \text{(second year)}$$

There is a gradient G between $\text{AEC}_1, \text{AEC}_2, \ldots, \text{AEC}_n$ such that $G = -[(B-V)/n]P_\infty$

Utilizing the gradient factor, a levelized annual equivalent cost of the investment is

$$\text{AEC} = B \left(P_\infty + \frac{B-V}{B(n)} \left[1 - P_\infty \left(\frac{A}{G} \right)_n^{i_a} \right] \right)$$

$$\text{AEC} = BP_n$$

where

$$P_n = P_\infty + \frac{B-V}{B(n)} \left[1 - P_\infty \left(\frac{A}{G} \right)_n^{i_a} \right] \tag{14.1-12}$$

P_n is defined as the depreciable fixed-charge rate for the case of straight-line tax depreciation ($n_{\text{tax}} = n_{\text{book}}$). For a given company all the parameters are available to some degree to determine the value of P_n. Assume as before a numerical example where $i_d = 8\%$, $r_d = 55\%$, $t = 50\%$, $i_e = 14\%$, $V = 10\%$ of B, book and tax life of $n = 25$ years, and straight-line tax depreciation applies; then i_a calculates to be 0.085, $P_\infty = 0.17$, and $P_{25} = 0.1569$. For each \$100 of investment the company's uniform yearly revenue requirement is \$15.69 to recover the original capital in 25 years and meet the financial and tax obligations. The actual year-by-year revenue requirement is \$17, \$16.39, \$15.78, ..., \$2.31 for the 25 years (a decreasing gradient).

Much of the new plant meets the qualifications for fast tax depreciation. One such fast method is the sum-of-the-years-digits (SYD) method of tax depreciation. This accelerated tax depreciation rate reduces the amount of tax paid during the early life of the equipment and raises the tax obligation in the latter years by an equivalent amount (an actual deferment of income tax becomes a source of capital fund to the company). The present worth of the revenue requirement is, of course, less when an accelerated depreciation is employed. Note that

$$\text{SYD} = \sum_1^N n = \frac{N(N+1)}{2}$$

3. Depreciable fixed-charge rate—SYD tax, SL book depreciation. The year-by-year revenue requirement for an investment B with a salvage V at year N is shown for the first 2 years as AEC_1 and AEC_2. This includes the cost of debt interest, earnings on equity, income tax, and book depreciation. From this data a levelized carrying charge rate is developed. It is useful as long as the tax life and book life are the same. The annual equivalent cost for the first year is

$$\text{AEC}_1 = Bi_c + \frac{B-V}{N} + \frac{\{ Bi_e(1-r_d) - [N(B-V)/\text{SYD} - (B-V)/N] \} t}{(1-t)}$$

For the second year the investment is reduced by the amount of the book depreciation $(B-V)/N$; hence

$$\text{AEC}_2 = \left[B - \frac{(B-V)}{N} \right] i_c$$

$$+ \frac{B-V}{N} + \left\{ \frac{[B - (B-V)/N] i_e (1-r_d) - [(N-1)(B-V)/\text{SYD} - (B-V)/N]}{(1-t)} \right\} t$$

The gradient between $\text{AEC}_1, \text{AEC}_2, \ldots, \text{AEC}_N$ can be determined, the terms rearranged, and a uniform annual equivalent cost results. This can be shown as

$$\text{AEC} = B \left[P_n - \frac{t}{1-t} \frac{B-V}{NB} \frac{N-1-2(A/G)_N^{i_a}}{N+1} \right]$$

$$P_N' = P_n - \frac{t}{1-t} \frac{B-V}{NB} \frac{N-1-2(A/G)_N^{i_a}}{N+1} \tag{14.1-13}$$

The term P'_N is defined as the fixed-charge rate for a depreciable investment with SYD tax depreciation. Note that P'_N is a smaller factor than P_N because of the faster tax depreciation.

Referring to the previous numerical example, if the SYD tax life was taken as 25 years and all other items remained the same,

$$P'_{25} = 0.1569 - \frac{0.5}{1 - 0.5} \frac{0.9}{25} \frac{25 - 1 - 2(8.026)}{26} = 0.1459$$

Again for each $100 of investment B the company's yearly levelized revenue requirement is $14.59. This is 7% less than for straight-line tax depreciation.

The two factors P_n and P'_n are useful if the tax depreciation fits the orderly straight line or SYD tax depreciation (for the convention that allows full tax depreciation the first year). Where the tax depreciation deviates from these two methods, as is often the case, it is wise to use the cash flow analysis previously described. Special values of fixed-charge rates can be derived to fit most systems of tax depreciation using the approach just described.

Combined Fixed-Charge Rate and Discounted Cash Flow Method

An engineering economic analysis can be based on either of the previously described methods of discounted cash flow or fixed-charge rate. Using each approach independently and comparing them, the accuracy of calculation can be verified. Both approaches yeild identical results when properly applied. In each case careful detail is necessary in understanding the cash flow and the peculiarity of the income tax law. An approach was developed by Hughes[7] that uses the best of two worlds of the discounted cash flow and of the fixed-charge rate approaches. Hughes recognized that all cash flow versus time diagrams could be represented by two "building block" forms each handled by a single equation. Each complex time flow diagram can be represented as a series of rectangular and/or triangular shapes each by a single equation. This approach is particularly suited to computer solution as the points in time on a construction project and investment required at those points are definable. The time–cash flow diagram that results is either a rectangle, triangle, or trapezoid (which is the sum of a triangle and rectangle) as shown in Figs. 14.1-11 and 14.1-12. AI and AF just represent initial and final investments, respectively.

The rectangles can be treated as $B(n_3)P_\infty(F/A)^{i_a}$ for a prepayment and as $BP_\infty(P/A)^{i_a}$ during the plant lifetime. The triangle in Fig. 14.1-12 could be calculated on a present worth basis as

$$= \frac{(AI - AF) - t\left(\dfrac{AI - AF}{n_f}\right)(P/A)^{i_a}_{n_f}}{1 - t}$$

Illustrative Example

The use of these capital cost calculations can be illustrated by an example problem that is designed to show most of the cash flow situations that come up and provide a means of checking the calculations.

Example Problem

An electric utility must add a plant refinement to a generating station that is under construction. A year ago this addition was bid at $1,000,000 subject to a 5%/yr compound inflation in price. It was planned that 60% of the initial cost would be paid out 1 year after the bid and the remainder 3 years

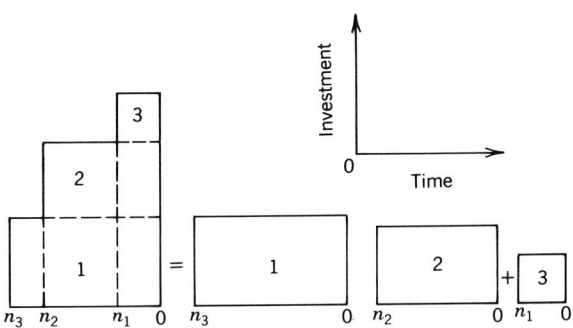

Fig. 14.1-11 Example problem: cash flow diagram.

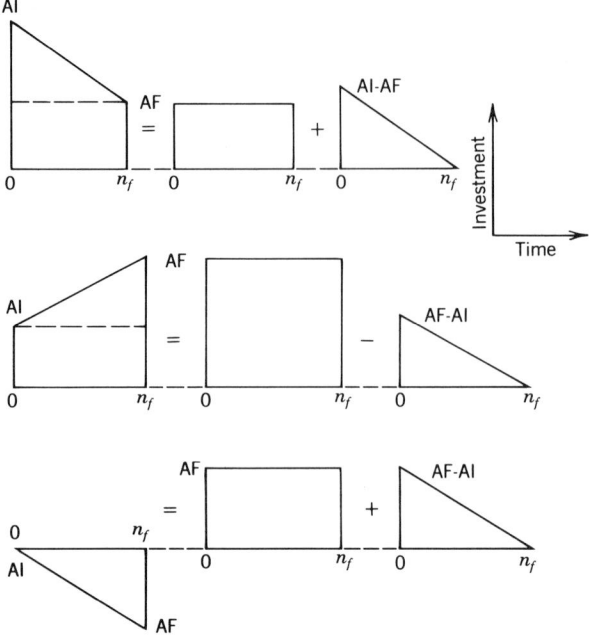

Fig. 14.1-12 Decomposing prepayment zone investment histories into combinations of nondepreciable investments.

after the bid. The station would go on-line 1 year after the addition was completed. It is of interest to the utility to determine the financial impact of this refinement and to note the increase in energy cost due specifically to this added investment and associated expenses. To arrive at an answer, more has to be known about the utility, its economic parameters, plant size and load characteristics. The new station is 800 MW(e) in size, with an expected 60% capacity factor during its 30-year economic study life. The utility obtains 45% of its money from debt sources that average 9%/yr, the rest from common stocks that are attractive when earning 14.5%/yr. A combined state and federal income tax rate of 50%/yr applies to taxable income. Salvage for tax purposes is zero in 30 years. For this example both straight line (SL) and sum of the years digits (SYD) methods of tax depreciation will be illustrated to show the effect of fast depreciation. Decommissioning of the plant 3 years after shutdown is expected to cost 5% of the first cost price at startup time. This cost is chargeable to expense 3 years after shutdown (33 years after startup). Inflation of es = 5% from startup is expected to continue throughout this time period. At the time of the original bid operation and maintenance (O&M) costs are expected to be $50,000/yr with a major overhaul at year 15 of $150,000. These items are expensed at the time they occur and are subject to inflation. Figure 14.1-13 represents a cash flow diagram of this example problem.

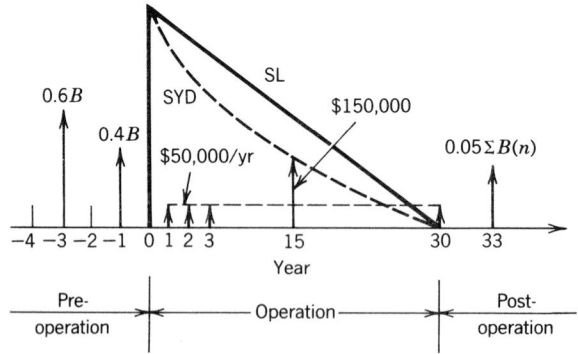

Fig. 14.1-13 Decomposing operating zone investment histories into combinations of·nondepreciable investments and investments with zero salvage.

Example Solutions

The discount rate [Eq. (14.1-7)] $i_a = i_d r_d (1 - t) + i_e (1 - r_d) = 9(0.45)(1 - 0.5) + 14.5(1 - 0.45) = 10\%$. This is the rate at which money moves through time for the utility. Its size is determined by the money market rate, which is influenced by the inflation rate (or vice versa). A 5% inflation rate was assumed in this example.

The sum of the actual payments made for this addition are of interest for each method of solution. Calculations are in thousands of dollars except as noted. The two prepayments, $B(-3)$ and $B(-1)$ adjusted for inflation sum to

$$600(F/P)_1^{5\%} + 400(F/P)_3^{5\%} = \$1093.05$$

By Cash Flow Analysis. Using Eq. (14.1-8) the present worth of the investment at startup time is

$$B(-3)(F/P)_1^{5\%}(F/P)_3^{10\%} + B(-1)(F/P)_3^{5\%}(F/P)_1^{10\%} = 600\,(1.05)(1.331) + 400(1.1576)(1.1)$$

$$= \$1347.89$$

The tax salvage term for this example is zero. Had there been a tax salvage other than zero, its present worth value would have taken the form of $V(30)(P/F)_{30}^{10\%}$ or $V(33)(P/F)_{33}^{10\%}$ depending on when tax salvage would actually be recovered.

The present worth of the tax write off using SL tax depreciation is

$$-t\left[\sum B(n)/n\right](P/A)_n^{i_a} = -0.5(1093.05/30)(P/A)_{30}^{10\%} = -\$171.73$$

For SYD tax depreciation the present worth of the tax write off is

$$-t\sum B(n)(P/A)_{30}^{10\%}\left(\frac{2N}{N(N+1)} - \frac{2}{N(N+1)}(A/G)_{30}^{10}\right)$$

$$= -0.5(1093.05)(9.4269)\left(\frac{30}{465} - \frac{8.176}{465}\right) = -\$241.80$$

Expense items to be considered are the decommissioning cost, the O&M costs, and the single overhaul at year 15.

The present worth of the decommissioning cost is

$$O(33)(F/P)_{33}^{5\%}(P/F)_{33}^{10\%} = 1093.05(0.05)(5.0032)(0.04306) = \$11.77$$

The present worth of O&M costs are

$$O(1)(F/P)_4^{es}(P/A)_{30}^{x} = 50(F/P)_4^{5\%}(P/A)_{30}^{4.76\%} = 50(1.2155)(15.799) = \$960.17$$

Note that (under inflation and escalation), $1 + x = (1 + i_a)/(1 + es)$. Thus, $x = (1.1)/(1.05) - 1 = 0.0476$, or 4.76%, and therefore the product $(F/A)_{30}^{5\%}(P/F)_{30}^{10\%}$ yields $(P/A)_{30}^{4.76\%}$.

The present worth of the overhaul is

$$O(15)(F/P)_{19}^{5\%}(P/F)_{15}^{10\%} = 150(1.527)(0.2394) = \$90.74$$

Substituting all these values in Eq. (14.1-8), the present worth of the revenue required for SL tax depreciation is

$$\text{PWRR(SL)} = \frac{1347.89 - 171.73}{1 - 0.5} + 11.77 + 960.17 + 90.74 = \$3414.98$$

Likewise, for SYD tax depreciation

$$\text{PWRR(SYD)} = \frac{1347.89 - 241.80}{1 - 0.5} + 11.77 + 960.17 + 90.74 = \$3274.85$$

Using the denominator of Eq. (14.1-9), the present worth of the energy produced by the station is

$$\sum (1 + i_a)^{-n} Q(n) = 8760(0.6)(800)(1000)(P/A)_{30}^{10} = 3.96 \times 10^{10} \text{ kW} \cdot \text{h} \quad (8760 = \text{h/yr})$$

The extra burden of this addition in terms of energy cost is determined by using all of Eq. (14.1-9). For SL tax depreciation

$$P(\text{SL}) = \frac{(3414.98(1000)(1000)}{3.96 \times 10^{10}} = 0.086 \text{ mils/kW} \cdot \text{h}$$

For SYD tax depreciation

$$P(\text{SYD}) = \frac{(3274.85)(1000)(1000)}{3.96 \times 10^{10}} = 0.082 \text{ mils/kW} \cdot \text{h}$$

By Fixed-Charge Rate Analysis. The expense of the two prepayments time-valued to start up is

$$600 P_\infty (F/P)_1^{5\%} (F/A)_3^{10\%} + 400 P_\infty (F/P)_3^{5\%} (F/A)_1^{10\%} = 600(0.2)(1.05)(3.31) + 400(0.2)(1.1576)(1)$$

$$= \$509.67$$

Note that $P_\infty = i_a/(1 - t) = (0.1)/(1 - 0.5) = 0.2$.

Using Eq. (14.1-12) the annual equivalent cost (AEC) of the investment during the 30 years of operation with SL tax depreciation is AEC $= 1093.05 P_{30} = 1093.05(0.178825) = \$195.46/\text{yr}$. The present worth of this amount is $195.46(P/A)_{30}^{10\%} = \$1842.63$.

The decommissioning cost, treated as an expense, may be expressed at startup time as

$$1093.05(0.05)(F/P)_{33}^{5}(P/F)_{33}^{10} = \$11.77$$

The present worth of the O&M costs and the overhaul are the same as previously calculated under cash flow as \$960.17 and \$90.74, respectively.

The present worth of the revenue required for the plant refinement is then

$$\text{PWRR}(\text{SL}) = 509.67 + 1842.63 + 11.77 + 960.17 + 90.74 = \$3414.98$$

which is the value determined by cash flow analysis. For SYD tax depreciation, the AEC using P'_N from Eq. (14.1-13) is AEC $= 1093.05 P'_{30} = 1093.05(0.165226) = \$180.60/\text{yr}$. Expressed as a present worth of the 30 years,

$$(\text{AEC})(P/A)_{30}^{10\%} = \$180.60(9.4269) = \$1702.50$$

The present worth of the revenue required for the plant addition is then

$$\text{PWRR}(\text{SYD}) = 509.67 + 1702.50 + 11.77 + 960.17 + 90.74 = \$3274.85$$

which is like the value determined by cash flow analysis.

Combined Fixed-Charge Rate and Discount Cash Flow Method. The prepayment expense time valued to startup per Fig. 14.1-13 is calculated as \$509.67 by fixed-charge rate method shown previously. The top cash flow diagram of Fig. 14.1-12 with AF equal to zero fits the present worth of revenue required calculation for SL tax depreciation. This yields

$$\frac{1093.05 - 171.73}{1 - 0.5} = \$1842.63$$

as previously calculated by cash flow analysis.

To these present worth values are added the costs of O&M, decommissioning, and overhaul, giving a total PWRR(SL) of $509.67 + 1842.63 + 11.77 + 960.17 + 90.74 = \$3414.98$. This again checks with the values previously calculated. Calculating PWRR(SYD) by this approach is not so easy since there are 30 trapezoidal figures that need to be solved, one for each of the 30 years as the tax depreciation changes from year to year. This method lends itself to computer computation.[8]

Solution Critique

1. Each individual approach provided the same annual equivalent cost or present worth of revenue requirement, so the method that appears the simplest and most direct to the individual is the best.

2. The faster tax depreciation, SYD, provided a lower investment revenue requirement by over 6% compared to SL depreciation. Including the operating costs, which was constant, the net reduction in overall revenue requirement was about 4.3%.

3. If this problem was to have a lower or higher inflation rate, a corresponding lower or higher debt and equity rate would have to be considered.

4. The decommissioning cost in the future is manageable as long as the discount rate remains greater than the inflation rate.

5. If the tax salvage $V(33)$ had been a positive value rather than zero and the same decommissioning cost prevailed, then an addition loss of $-V(33)$ would have to be considered. It could be added to the decommissioning cost $O(33)$ or treated as other income under $W(n)$. In this case $W(33) = -V(33)$, a negative income (a loss).

Prepayment of Investment Capital

Large power plants take several years to plan, engineer, construct, and license (Fig. 14.1-14). The prepayment made during construction causes a yearly expense of $B(n)P_\infty$ to the utility, which is awkward to recover through the electric rate structure. The expense may be treated as allowance for funds used during construction (AFUDC). In this case the utility capitalizes this yearly expense and recovers it during the lifetime operation of the plant. With high construction and interest costs, coupled with a long completion time, accumulated AFUDC adds a considerable amount to the financial burden of the new plant on the utility. This increase in capital cost can cause undue risk to the financial stability of the company.

To recover the investment, each prepayment added to the rate base is $B(n)(F/P)_n^{i_r}$ where n represents the periods of prepayments and i_r is an AFUDC rate allowed for this purpose. Seldom is i_r or the earnings set on the rate base sufficient to recover the financial impact of the prepayment. The utility customer feels the impact of the new plant being built only when it is put on the line, causing the investment in the rate base to increase abruptly. The resulting steep rise in the electric bill is not popular with the user.

Another approach, far better in cost to the user, is to pay construction work in progress (CWIP) payments. Usually a fraction of $B(n)P_\infty$, the prepayment expense each year is allowed to be added to the rate base, increasing the electric rate gradually. Rate commissions do not pass all the CWIP expenses on, so some AFUDC does occur. By paying CWIP, the customer is earning on the increased portion of the electric bill at a rate equal to the cost of capital to the utility. This earning, in effect, is tax free and surpasses many forms of investment that the customer might otherwise make. The use of CWIP payments for plant construction relieves the financial strain the AFUDC causes and allows the utility to purchase new debt and equity capital at more favorable rates. This, of course, eventually passes on to the customer in lesser rate increases in the future. Unfortunately not all utility customers are in a position to invest in CWIP when it is approved by the rate commission so this approach is also unpopular.

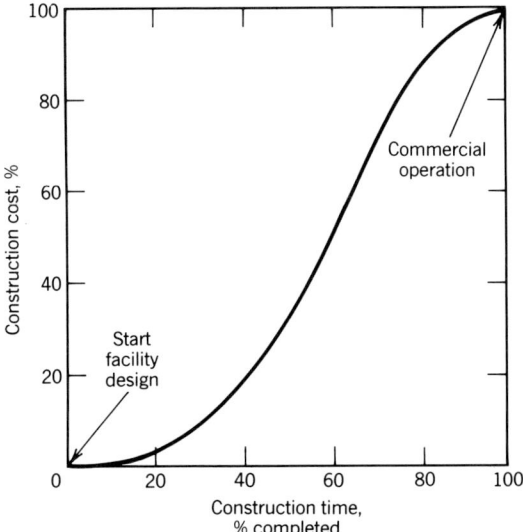

Fig. 14.1-14 Typical power plant construction expenditure versus construction time S curve.

Year-by-Year Revenue Requirement as a Guideline

The year-by-year revenue required for a project as described earlier in the development of the fixed-charge rates [Eqs. (14.1-12) and (14.1-13)] can be used to determine the financial impact of choosing one new generation plant proposal over another. Assume that an electric utility has used one of the previously described economic analysis models to compare a higher fuel cost nuclear plant with an alternative fossil plant to determine the lowest cost option for new generation that is needed. Three criteria must be met in accepting the option with the higher first cost (i.e., the nuclear unit). First, can the extra construction cost be handled financially by the utility without unduly influencing the corporate bond rating and the stock value? Second, is the overall lifetime cost of nuclear generation less than the fossil? Third, is the short-term impact (first 10 years) of energy cost of the nuclear option about equal to or less than fossil? The levelized cost and cash flow models answer only the second criteria if applied directly. Experience in entering the money market for new capital can answer the first criteria as to whether acquiring additional capital is prudent. It takes a year-by-year revenue requirement analysis to assess the short-term impact of the high first-cost alternative.

Figure 14.1-15 shows the year-by-year revenue required by a nuclear plant and by a fossil plant on the basis of a 30-year study. Since the fossil plant has a lower first cost but a higher fuel cost than the nuclear plant, the year-by-year slope does not drop off as steeply as the nuclear. Whereas the 30-year levelized cost for nuclear is less than for fossil ($AEC_{30N} < AEC_{30F}$), levelizing the first 10 years of the revenue required reverses these costs ($AEC_{10N} > AEC_{10F}$). Specifically this means that the utility, in choosing the nuclear plant on the basis of the long-term criteria, would be forced to ask for a larger rate increase during the first 10 years of operation of this plant than if it chose the fossil plant. Of course, in the later years the fossil plant would demand the greater revenue. For this reason, the nuclear plant, when being considered as an alternative, must have a lower generating cost not only for the lifetime of the study but also on the short term, else it will unduly impact the utility and today's customer financially.

Fossil Fuel Cost Considerations

The use of the coal stockpile hedges against uncertainties in weather, transportation, mining, and delivery, hence assuring reliable service to the customer with a minimum utility capital risk. Long-term contracts with take-or-pay clauses usually minimize the fuel cost purchase price but may swell the size of the coal pile size when load demand is less than forecast. Regulatory commissions often limit the size of the coal pile to about $\frac{1}{3}$ year supply. The utility with nuclear units may be forced to use its more expensive fossil fuel or absorb the excess coal supply costs using equity capital (excess stockpile costs may be disallowed in rate base).

Coal pile abatement, fire protection, and weatherproofing unloading of coal has added considerable expense to the fossil fuel cost. Flue gas desulfurization equipment or other methods of removing sulfur emission has also greatly increased the first cost of fossil plants and significantly reduced the overall efficiency of the energy conversion process.

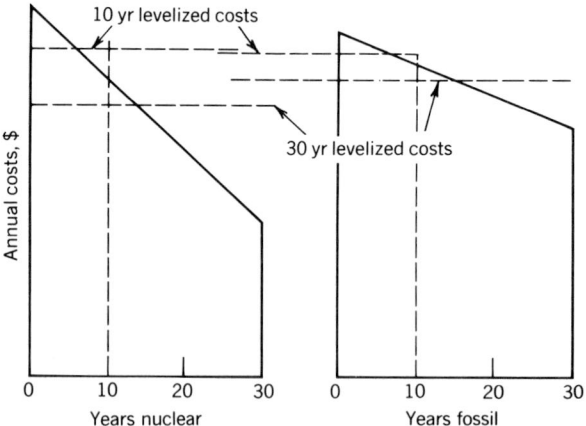

Fig. 14.1-15   Year-by-year revenue required to recover new plant costs.

Nuclear Fuel Cost Considerations

Typical cash flow diagrams for the nuclear fuel cycle are shown in Figs. 14.1-16 to 14.1-18.[9] Determination of the nuclear fuel cost can use Eq. (14.1-9), Eq. (14.1-12) and P, or by a combination of all three of these terms. For fuel management reasons utilities also break up the cost components into preburn, burn, and postburn periods of time in order to identify individual item cost contributions or savings. The preburn period involves the purchase of natural uranium in the form of U_3O_8 (yellowcake), conversion to UF_6 and enrichment to higher concentrations of U-235, conversion to UO_2, and fabrication of the fuel into assemblies. Assemblies are the basic units of fuel that make up the batch size in fuel reloading (or the original core loading). Affecting any fuel cycle study are the contractual agreements, which are as varied and different as fuel service companies and utilities.

Several utilities have their own uranium mine to which they commit capital several years in advance of fueling the reactor. Others may have leased fuel from an outside vendor or from a wholly owned corporation. Leasing allows the utility to finance its fuel with a greater mix of debt capital than it can if it purchased the fuel with its own debt equity mix. Leasing also provides a new source of debt

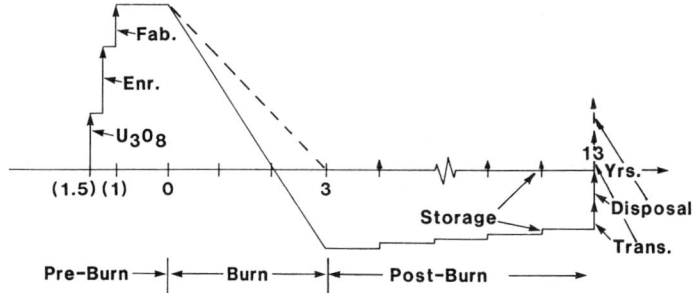

Fig. 14.1-16 Nuclear fuel cycle, batch cash flow, throwaway cycle.

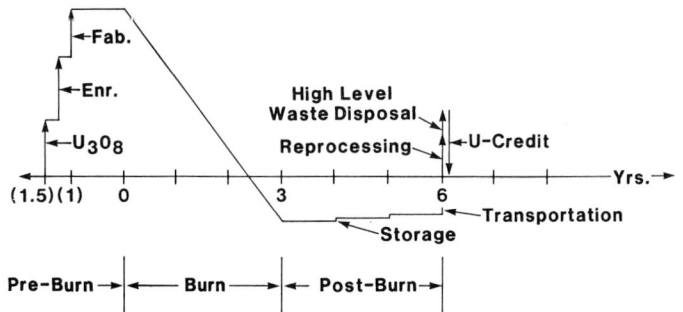

Fig. 14.1-17 Nuclear fuel cycle, batch cash flow, recycle uranium.

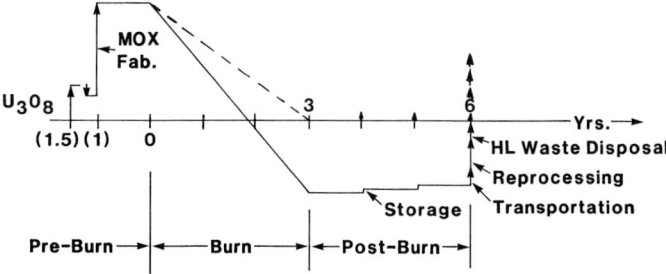

Fig. 14.1-18 Nuclear fuel cycle, batch cash flow, mixed oxide recycle.

capital as the fuel may average 3–5 years in lifetime and can be matched by money sources of that duration. Loans for fuel tend to follow the lows and highs in the money market and may have an advantage over long term-loan interest rates that must err on the high side. The lease may appear on the company's balance sheet or it may only require a footnote disclosure, depending on whether it was lease financing or a purchase contract that was negotiated. Outright fuel ownership requiring allowance for funds used during construction, AFUDC, in the rate base has not been a satisfactory method to recover the revenue required in the short time period that the fuel is in the reactor. AFUDC recovery in the rate base is fair for 20–30 year study periods for generator plants if the construction time is 6 years or less. Recovery of AFUDC for 2- to 3-year prepayments fall short of recovering the capital requirement in the 3–5 year time span that a specific fuel is in the reactor core, unless unusually generous earning rates are allowed by rate commissions.

The use of accelerated tax depreciation compared to unit of production depreciation may show savings in the startup of a new nuclear plant. Whether regular convention or modified half-year convention is used for first-year depreciation are important factors. The prudent use of fuel assembly accounting categories with investment tax credits can also maximize these credits. Accelerated depreciation on fuel provides a savings on the initial startup of a plant or when reload takes place in an inflationary economy. Once the transient of nuclear fuel expenditure disappears, all forms of tax depreciation provide the same end result. A utility that may have started with leased fuel and then purchased it at a later date may have additional savings from accelerated tax depreciation.

The postburn period in the fuel cycle is straightforward as far as income tax is concerned. Any postburn payments for spent fuel storage and ultimate high-level radioactive waste disposal can be expensed in the year that it occurs against revenues received for that same year. Revenues to pay for these postburn expenses had to be accrued from the customers during the burn period when the fuel was in the reactor. Since it was not spent, it appears in the utility's account as profit and is promptly taxed in the year received. When the expense actually occurs, the tax is essentially returned.

The cash flow diagram in Fig. 14.1-16 is drawn in two ways. The solid line assumes that all costs are capitalized and treated as an investment. Since the IRS frowns on a negative salvage paid out sometime during the postburn period, a dotted line is shown to represent the investment being depreciated to 0 in n years, in this case 3 years. Along with the dotted line, arrows showing expenses for yearly storage cost, transportation, and disposal are noted. To the utility, money flows as if it was the solid line; to the IRS, the dotted line. The PWRR for the solid line cash flow is less than that for the dotted line case. Since the IRS prevails, the customer must pay a bit more for electricity. Figure 14.1-18 is drawn similar to Fig. 14.1-16 except for the case where mixed oxide (MOX) fuel is used under a recycle case. The downward arrow at 1.25 years before startup represents an enrichment credit for separative work units since recycle uranium was used that had an enrichment greater than natural. Figure 14.1-17 represents recycle uranium fuel assemblies with all costs capitalized.

Interest Formulas

Standard interest formulas are used throughout Section 14.1 and are listed here for convenience. Derivations can be found in any engineering economy text.[10]

$$F_n = P(1 + i)^n = P(F/P)_n^i$$

$$P = F_n \left(\frac{1}{1+i}\right)^n = F_n(P/F)_n^i$$

$$F_n = A\left[\frac{(1+i)^n - 1}{i}\right] = A(F/A)_n^i$$

$$A = F_n \frac{i}{((1+i)^n - 1)} = F_n(A/F)_n^i$$

$$P = A\left[\frac{(1+i)^n - 1}{i(1+i)^n}\right] = A(P/A)_n^i$$

$$A = P\left[\frac{i(1+i)^n}{(1+i)^n - 1}\right] = P(A/P)_n^i$$

$$A = G\left[\frac{1 - n(A/F)_n^i}{i}\right] = G(A/G)_n^i$$

where i = rate of interest per compounding period
       n = time periods
      F_n = future sum at the end of the nth period
      P = present sum
      A = periodic sum in a uniform series of amounts discretely flowing at the end of each of n periods
      G = gradient, a periodic increase in an arithmetically increasing series of sums discretely flowing at the end of each of n periods
$(P/F)_n^i$ = single payment present worth factor
$(F/P)_n^i$ = single payment compound amount factor
$(A/F)_n^i$ = uniform series sinking fund factor
$(F/A)_n^i$ = uniform series compound amount factor
$(P/A)_n^i$ = uniform series present worth factor
$(A/P)_n^i$ = uniform series capital recovery factor
$(A/G)_n^i$ = gradient series factor

Inflation and Escalation

Prices quoted for large equipment by vendors will have a firm price as of a given date (bid date) after which time labor and material escalation may affect the selling price. At the same time the buyer makes progress payments as certain milestones in the construction or fabrication are completed. Labor and material cost indices may be mutually agreed upon by the vendor and buyer or they may be distinctly specified by the vendor. In some cases the vendor may agree to absorb a portion of the escalation, creating a "dead band" over which no price change is made.

The amount of escalation in the price of equipment or in the amount of a prepayment is

$$\text{ESC}(j) = \text{ESLA}(j) + \text{ESMA}(j)$$

where ESC = increase in payment due to labor and materials
     ESLA = increase in payment due to labor
     ESMA = increase in payment due to materials
     $\text{ESLA}(j) = \text{PP}(j) A_1 [(1 + A_3)^{nn} - (1 + A_5)^{nn}]$
     $\text{ESMA}(j) = \text{PP}(j) A_2 [(1 + A_4)^{nn} - (1 + A_6)^{nn}]$
where $100A_1$ = % of progress payment attributed to labor for escalation purposes
     $100A_2$ = % of progress payment attributed to materials, including taxes, for escalation purposes
     $100A_3$ = % per month compound growth of labor index
     $100A_5$ = % per month labor escalation growth dead band rate
     $100A_4$ = % per month compound growth of material index
     $100A_6$ = per month material escalation growth dead band rate
     nn = number of months between escalation base date and jth progress payment
     $\text{PP}(j)$ = the jth progress payment of some investment B

The dead band $100A_5$ is the amount of escalation in percent per month that the vendor agrees to absorb. The monthly labor index $100A_3$ is the percent per month at which labor costs rise. Similar comments apply to $100A_4$ and $100A_6$. Note that the sum of A_1 and $A_2 \leq 1$. It is generally a bit less than one, indicating that not all of the price is sensitive to escalation.

The treatment of escalation was applied here to a first cost or to a prepayment. It can likewise be applied to operating costs, fuel costs, maintenance, and so on. It is expected that rates A_1–A_6 can vary widely depending on the item or the class of industry involved. The same item bid by two industries may not have the same escalation indices since the basic labor groups may be different. An example could be three bids on nuclear fuel, one from a basic electric equipment manufacturer, a second from a boiler manufacturer, and the last from an oil/energy supplier. Labor contracts and timing may have a wide difference in their immediate escalation index.

A single escalation rate may be used if labor and material price growth indices are not well identified and dead band values approach zero. The jth progress payment is then $\text{PP}(j) = \text{PP}(0)(F/P)_{nn}^{es}$

where $\text{PP}(0)$ = unescalated jth progress payment
     $(F/P)_{nn}^{es} = (1 + es)^{nn}$ = compound amount factor for a single payment
     es = equivalent escalation rate that expresses labor and material price growth (time units for es and nn must be consistent)

At times, it is of interest to know the present worth of a future payment that grows in price at a rate es when the discount rate is i_a. The present worth of the jth progress payment $\text{PW}(j)$ at bid date can be represented as

$$PW(j) = PP(0)(F/P)_{nn}^{es}(P/F)_{nn}^{i_a}$$

$$= PP(0)(P/F)_{nn}^{x}$$

where $\qquad\qquad\qquad x = (1 + i_a)/(1 + es) - 1.$

Under certain conditions x may be referred to as an inflation-adjusted cost of money. The value x has limited use, however, since it is only true if there is no income tax involved in the analysis. For an analysis where income tax considerations enter into the cash flow, an equivalent interest rate, x, that is inflation adjusted cannot be defined in a simple usable manner. This would apply to most cash flow considerations of the investor-owned utility industry and any type industrial corporation that has income tax obligations.

Escalation as used in this section includes the combined effect of money inflation and real differential price escalation. The tax-adjusted discount rate, i_a, varies from time to time, reflecting the inflation of the currency in the money market. Economic studies to assess the effect of a change in inflation on the outcome must consider a corresponding change in i_a also caused by inflation.

Corporate Bond Interest Determination

Corporate bonds have an interest rate printed on each bond certificate stating the rate payable on the face value. The printed rate on the bond usually reflects the money market at the time of printing, but several months may elapse before the issues are offered for sale to the highest bidder. If the market for money during the interim between printing and sale tightens significantly, the highest bid on the bond may be less than face value. Combining this loss with the cost of printing and issuing the bond series, the corporation may net capital that is a sizable percentage less than the face value of the issue. The true cost of borrowed money needs to be determined.

The actual interest paid on the bond per payment period is

$$\frac{i}{q} = \frac{\text{equivalent uniform per period cost}}{\text{net money received from bond issue}}$$

$$= \frac{Fe/q + Fd(A/F)_{qn}^{i/q}}{F(1 - d)}$$

Solving for i,

$$i = \frac{e}{1 - d} + \frac{qd(A/F)_{qn}^{i/q}}{1 - d}$$

where e = printed interest rate on bond certificate (including cost associated with payment of interest)
     n = year to maturity of bond
     d = bond discount rate (including cost of issue, etc.)
     $P = -d$ = a premium rate if sold for a net price greater than face value
     q = number of interest payments made each year
     F = face value of bond
     i = nominal rate paid on money received from bond sale

The equation for i can be solved by an iterative approach. Since $(A/F)_{qn}^{i/q} \leq 1/qn$, i ranges between $e/(1 - d)$ and $(e + d/n)/1 - d$. Choosing some value of i between the lowest and highest value as a start, the A/F value can be determined as well as a subsequent value for i. You can tell by the results whether a second iteration is needed. The equation generally converges rapidly.

To assess the case where a bond is sold at a premium rather than a discount, the same equation for i can be used by noting that the premium rate $p = -d$.

Another way to determine the interest i paid on bonds sold at a discount (or premium) is to equate the present worth of net money received from the bond offering to the present worth of the payments made to retire the bond issue when it matures. This amounts to

$$F(1 - d) = \frac{Fe}{q}(P/A)_{qn}^{i/q} + F(P/F)_{qn}^{i/q}$$

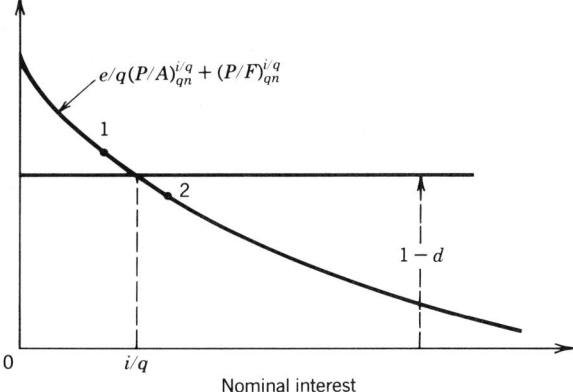

Fig. 14.1-19 Equivalent corporate bond interest rate determination when sold at discount d.

The F's cancel so the equation becomes

$$1 - d = \frac{e}{q}(P/A)_{qn}^{i/q} + (P/F)_{qn}^{i/q}$$

Since this equation contains two factors in terms of i to a power of qn, it is best to interpolate as noted in the curve of Fig. 14.1-19. Solve the equation choosing values for i_1 and i_2 that will bracket the intersection with the $1 - d$ line. Interpolate between these points for a closer i value.

The two values shown for bond interest determination afford a check on the calculated value, though the first method is somewhat simpler to carry out.

REFERENCES

14.1-1 A. J. Artman and R. A. Smith, "Minimum Load Analysis in Generation Planning," *IEEE Transactions on Power Apparatus and Systems*, Vol. PAS-99, No. 4, p. 1323, July/August 1980.

14.1-2 R. L. Sullivan, *Power System Planning*, McGraw-Hill, New York, 1977.

14.1-3 W. D. Stevenson, Jr., *Elements of Power System Analysis*, McGraw-Hill Kingsport Press, New York, 1975.

14.1-4 W. D. Marsh, *Economics of Electric Utility Power Generation*, Oxford University Press, Oxford, 1980.

14.1-5 M. E. EL-Hawary and G. S. Christensen, *Optimal Economic Operation of Electric Power System*, Academic Press, New York, 1979.

14.1-6 D. R. Vondy, "Basis and Certain Features of the Discount Technique," ORNL-3686, Appendix F. pp. 243-8, Oak Ridge National Laboratory, December 1965.

14.1-7 J. A. Hughes, "Methods of Nuclear Management for Utilities," Ph.D. thesis, University of Illinois, pp. 27 and 29, 1975.

14.1-8 General Economic Model Version 7, HTH Associates Inc, Urbana, Ill., 1984.

14.1-9 W. W. Brandfon and D. F. Hang, "LWR Recycle-Economic at What Price U_3O_8?" Atomic Industrial Forum, Fuel Cycle Conference '80, New Orleans, Apr. 1980.

14.1-10 G. W. Smith, *Engineering Economy*, 3rd ed., Iowa State University Press, Ames, 1979.

14.2 POWER GENERATION CYCLES

E. H. Miller and R. P. Allen

14.2-1 Rankine Cycle

Ninety-eight percent of the thermal electric power generation by U.S. utilities is by turbine-generators expanding steam in variations of the Rankine cycle. This cycle, originally developed with steam engines, closely approximates the Carnot cycle when used with low steam pressures. As pressures are increased to obtain higher saturated steam temperatures, the Rankine cycle does not improve as much as the Carnot cycle because the relatively low temperature heat being added to bring the condensate to boiler saturation temperature becomes a major portion of the total heat content of saturated steam.

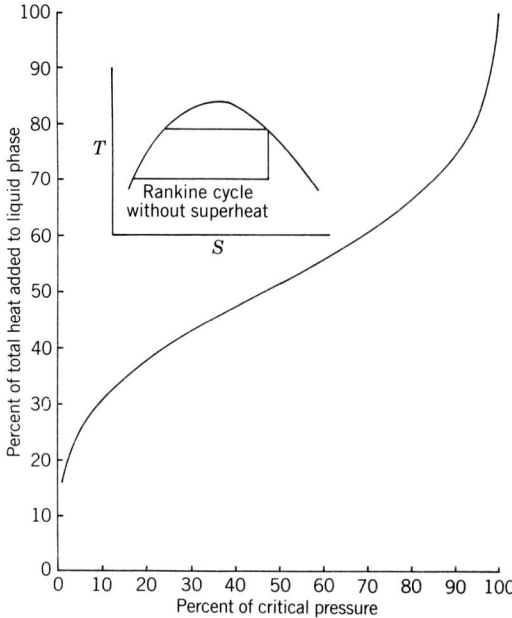

Fig. 14.2-1   Percent of total heat added to liquid phase for simple Rankine cycle.

Figure 14.2-1 shows how this proportion increases as the boiler pressure is increased from 100 kPa to the critical steam pressure of 22,120 kPa.*

When multiple-expansion steam engines were introduced to efficiently utilize higher steam pressures, it became possible to use partially expanded steam to preheat the condensate before introducing it into the boiler. Even one stage of regenerative feedwater heating can significantly reduce the difference between Carnot and Rankine efficiency. The introduction of multiple-stage steam turbines made it practical to introduce up to seven or eight stages of feedwater heating and reduce the difference in cycle efficiency by about 80%. With an infinite number of feedwater heaters of zero temperature difference, the process of feedwater heating becomes reversible and the theoretical regenerative steam cycle has essentially the same efficiency as the Carnot cycle.

The Rankine cycle and the theoretical regenerative steam cycle depart substantially from the Carnot cycle when the steam temperature at the beginning of the expansion process is raised above the saturation temperature and is given initial superheat. The Carnot efficiency improves with higher initial temperature in accordance with the following formula:

$$\text{Efficiency (Carnot)} = (T_{max} - T_{min})/T_{max}$$

where, by definition, all the heat is added to the hypothetical working fluid at the highest absolute temperature, T_{max}, and the heat is rejected at the lowest absolute temperature, T_{min}.

The Rankine and regenerative steam cycle do not benefit to this extent because relatively little high temperature heat needs to be added to superheat steam above its saturation temperature. Figure 14.2-2 shows calculated Rankine and Carnot cycle efficiencies for a range of maximum temperatures and with various cycle assumptions. It also shows the performance of the fully regenerative steam cycle.

The Rankine cycle efficiency can be improved at a given maximum temperature by reheating the steam after it has partially expanded through the turbine. The gain is maximized when the reheating returns the steam to the maximum cycle temperature after it has expanded to about the average temperature at which heat is added to the throttle steam. Multiple stages of reheating produce additional gains in the Rankine cycle; however, the gain sharply diminishes when excess reheating causes the final exhaust condition to be superheated, thus raising the heat rejection temperature. In practical versions of the Rankine cycle piping, heat exchanger, and valve pressure drops and design and economic considerations have limited even the most advanced power plant cycles to two stages of reheat.

*For conversion to other systems of units consult Table 18.1.

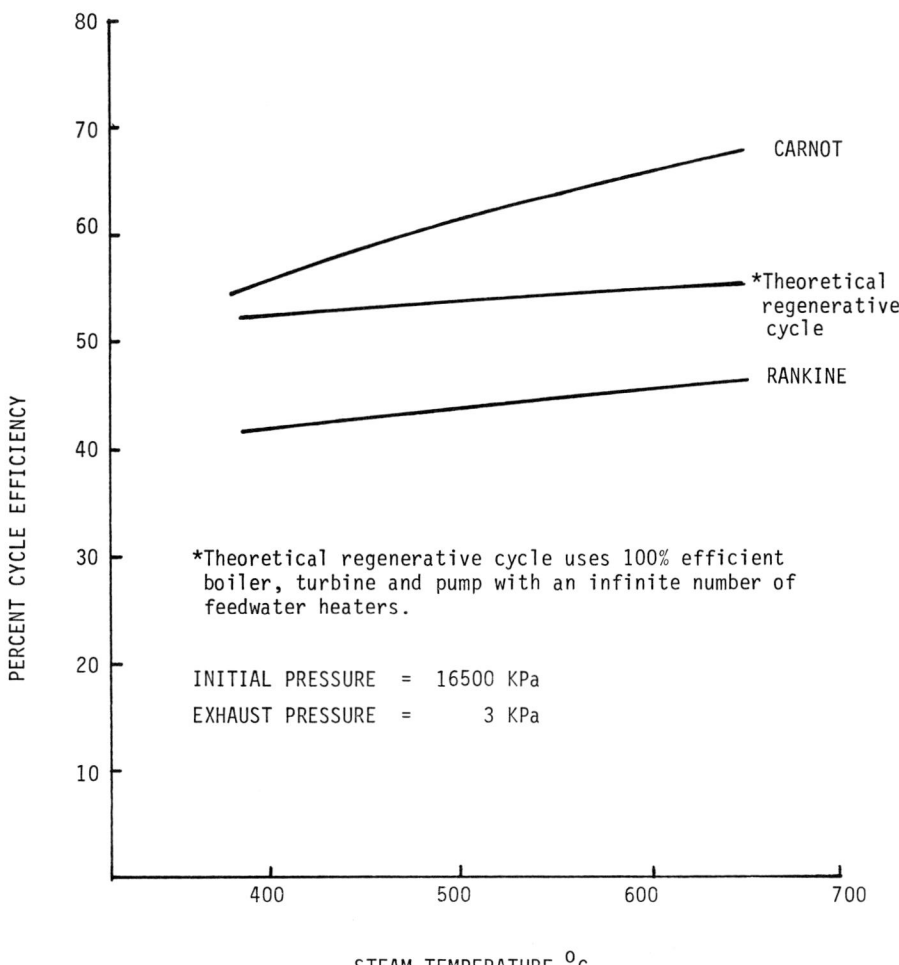

Fig. 14.2-2 Cycle efficiency versus initial temperature.

14.2-2 Binary Cycles

In other attempts to maximize the heat added at higher temperatures, vapor cycles other than steam have been considered. Mercury, which has a vapor pressure of 950 kPa at 510°C, would appear to be an ideal working fluid since it has very low liquid specific heat and nearly all the heat may be added at the highest temperature of the working fluid. Thus, it meets one criteria matching the Carnot cycle. Unfortunately, its vapor pressure at ambient temperatures is too low to permit practical mercury turbine designs that utilize the total temperature range. This shortcoming can be solved by combining a mercury turbine "topping" a steam turbine. Figure 14.2-3 shows a Ts diagram for such a combined cycle. Several plants utilizing this arrangement were installed and operated in the 1940s with varying degrees of success. However, the mercury–steam cycle was costly and complex and was eventually abandoned as rapid developments in steam turbine and boiler technology improved the efficiency, cost, and operational advantages of the steam-only cycle. By 1948 the plant heat rates of the best steam cycles surpassed the best that had been achieved in the combined mercury steam plant. Other liquid metals and organic fluids have since been studied as binary cycle candidates for advanced efficiency applications. However, concerns for cost, complexity, and safety remain unresolved and there have been no applications for electric power generation.

    "Bottoming" cycles using ammonia or a fluorinated hydrocarbon as one of the working fluids in a binary cycle with steam have been frequently proposed. Since the low-pressure, low-temperature Rankine cycle is nearly equivalent to the Carnot cycle, advocates of bottoming cycles point to practical advantages of using working fluids that have higher densities than steam at normal heat

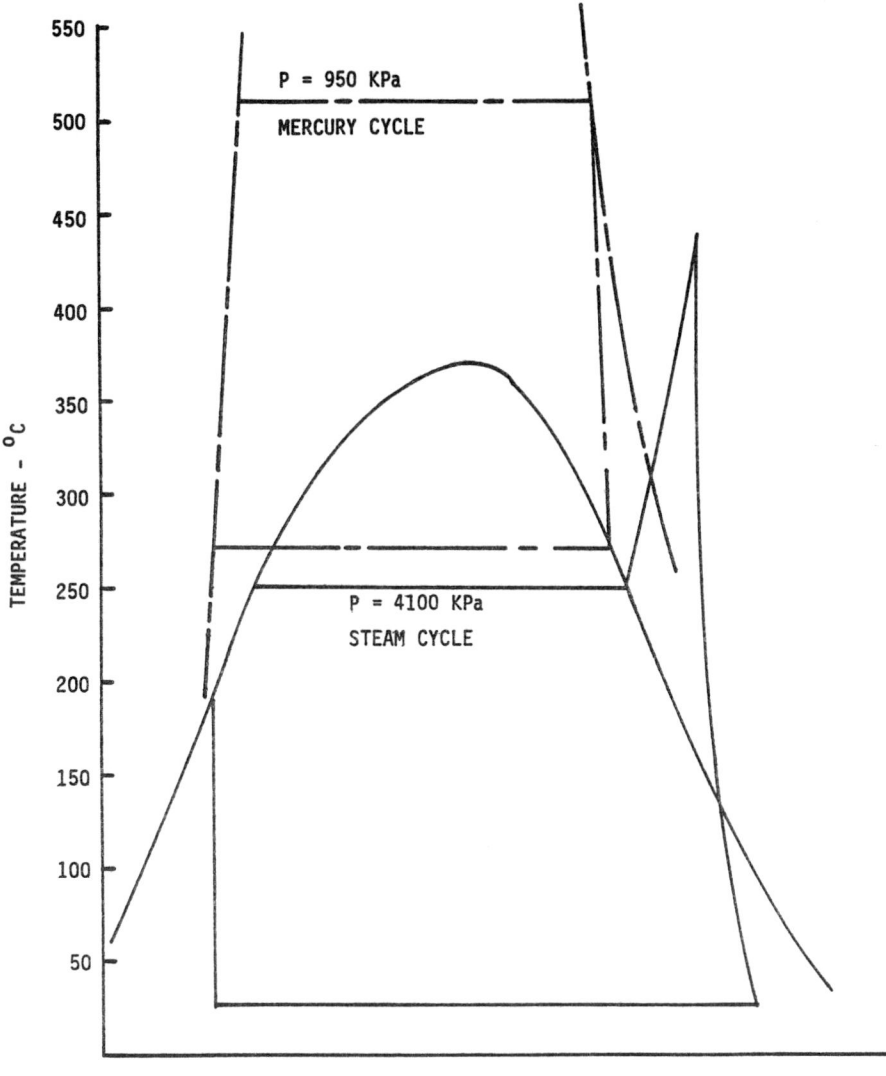

Fig. 14.2-3   T–S diagram for mercury topping cycle.

rejection temperatures. This higher density can lead to more compact low-pressure turbine designs with possibly reduced exhaust losses at a penalty of more complex cycle hardware and with an additional thermodynamic loss because of the extra heat exchanger between the condensing steam and the "bottoming" fluid boiler. One application of the "bottoming" cycle would involve the use of ammonia both as the working fluid in a "bottoming" cycle and as the heat transfer medium to an air-cooled condenser. Although this cycle has attracted interest, there has as yet been no application of this or any other binary "bottoming" cycle in the United States. Each of the "bottoming" fluids that has been considered require special shaft seals to achieve close to zero leakage for cost and safety consideration. While this is a common requirement, each bottoming fluid tends to be different in its material and design requirements. For instance, the fluorinated hydrocarbon fluids require that a relatively large number of stages be used in order to limit sonic losses and to achieve high efficiency. Ammonia with its low molecular weight does not share this problem but its chemical properties preclude the use of conventional materials in several turbine and power plant components.

14.2-3   Fossil Fuel Power Generation Cycles

Steam Conditions

Modern fossil fuel steam power generation plants typically have unit ratings that range from 300 to 900 MW, with 600 MW being the approximate average rating for current U.S. utility application. Steam conditions have effectively standardized at 16,500 kPa, 538°C, initial conditions with reheat to 538°C. Some capacity utilizes supercritical pressures at approximately 24,100 kPa, mostly with steam temperatures at 538°C/538°C but with some plants utilizing double-reheat and steam temperatures up to 565°C. During the period 1955–1965 some very advanced plants were built with steam pressures up to 34,500 kPa and steam temperatures up to 650°C. Steam temperatures were also extended to 565°C and 593°C with subcritical pressure on a number of units. Both of these developments proved troublesome or costly and utilities retreated to more conventional steam conditions. Recent fossil fuel cost increases have generated considerable interest in again advancing steam conditions to supercritical pressures and applying the double-reheat cycle using technology that has now proven itself in service.

Although Carnot and Rankine cycle performance is generally measured as a percentage conversion of heat to work, it is more common in the electric power generation field to express cycle performance as heat rate, defined as the quantity of heat necessary to generate a unit of useful output. The conversion between the two measures is direct and given by the following formula:

$$\text{Percent cycle efficiency} = \frac{100\% \times 3600}{\text{heat rate}}$$

where heat rate is in kJ/kW · h. In practical Rankine cycles increases in initial steam pressure improve heat rate approximately as shown in Fig. 14.2-4. At the very highest pressures heat rate gains tend to level off at about 24,000 kPa unless double-reheat steam conditions are applied for a gain in

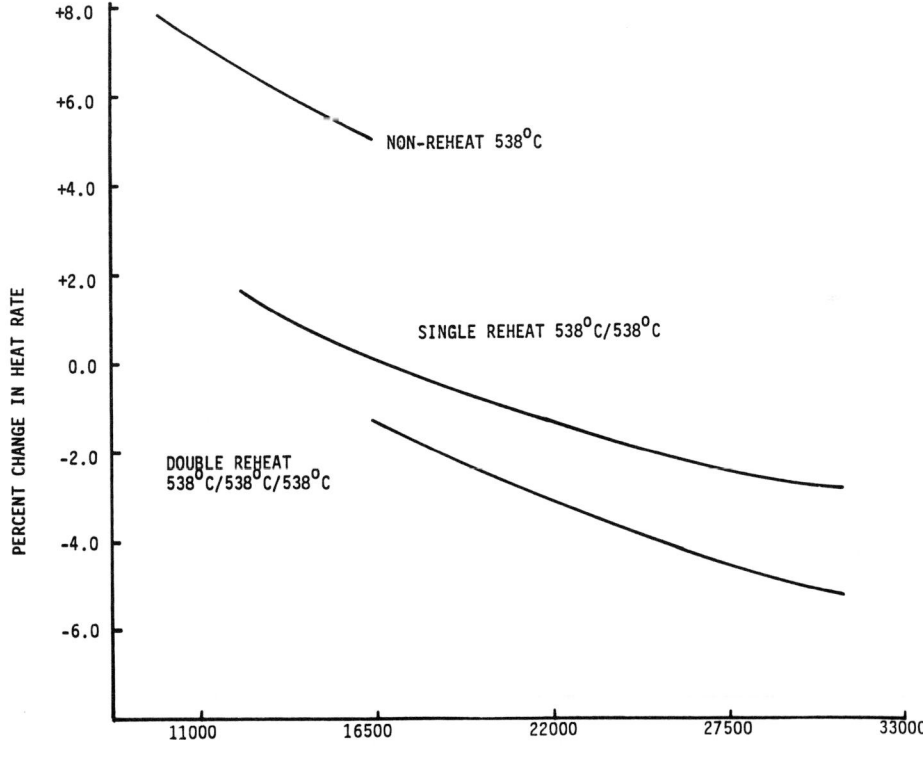

Fig. 14.2-4   Estimated relative heat rate for various number of reheats versus throttle pressure.

heat rate of about $1\frac{3}{4}\%$. Three and four stages of reheat have been considered, but if one assumes a 10% pressure drop in the piping, reheater, and valves, these losses more than offset the thermodynamic gain of the third stage of reheat.

With single reheat 565°C/565°C steam temperatures improve heat rates by approximately $1\frac{1}{4}\%$ relative to 538°C/538°C for initial pressures from 12,400 to 24,100 kPa. Reheat to 593°C/593°C has the potential to produce a further gain of the same magnitude; however, compromises to accommodate these temperatures with cost-effective designs and materials may offset some of the calculated gain.

Feedwater Regeneration

A six- to eight-stage regenerative feedwater heating cycle is generally used with 16,500-kPa steam conditions. A typical cycle is shown in Fig. 14.2-5. High-pressure heaters typically have provisions to take advantage of the high superheat in the extraction steam to obtain low or even negative terminal temperature differences between the feedwater temperature leaving the heater and the saturation temperature inside the heater. The drains from each high-pressure feedwater heater is shown cascaded to the next lower pressure heater. The thermodynamic loss associated with this cascading is minimized by cooling the drain flow in exchange with the incoming feedwater flow. These "drain coolers" are usually an integral part of the heater.

Deaeration or removal of dissolved oxygen in the feedwater can be accomplished in either the condenser or in a special deaerating heater drawing extraction steam at about 700 kPa at full load. This pressure level is selected so that the deaerating heater pressure will remain above atmospheric pressure at light load, thus assuring continuous deaeration. To provide adequate suction head to the boiler feedpump, the deaerating heater is generally located at a high elevation in the plant with the feedpump at various lower elevations from the basement to the operating floor. An alternate arrangement that allows flexibility in locating the deaerating heater uses a low-head booster pump to drain the deaerating heater and provide adequate feedpump suction head. Deaeration to approximately 10 ppm of dissolved oxygen is typically provided.

In feedwater cycle applications where oxygen removal is done in the condenser, all the feedwater heaters are of the closed type with desuperheating and drain cooling provisions as appropriate. The boiler feed pump can be located at the condenser; however, this requires that all feedwater heaters be designed for maximum boiler feed pump discharge pressure. It is more common to use a condensate

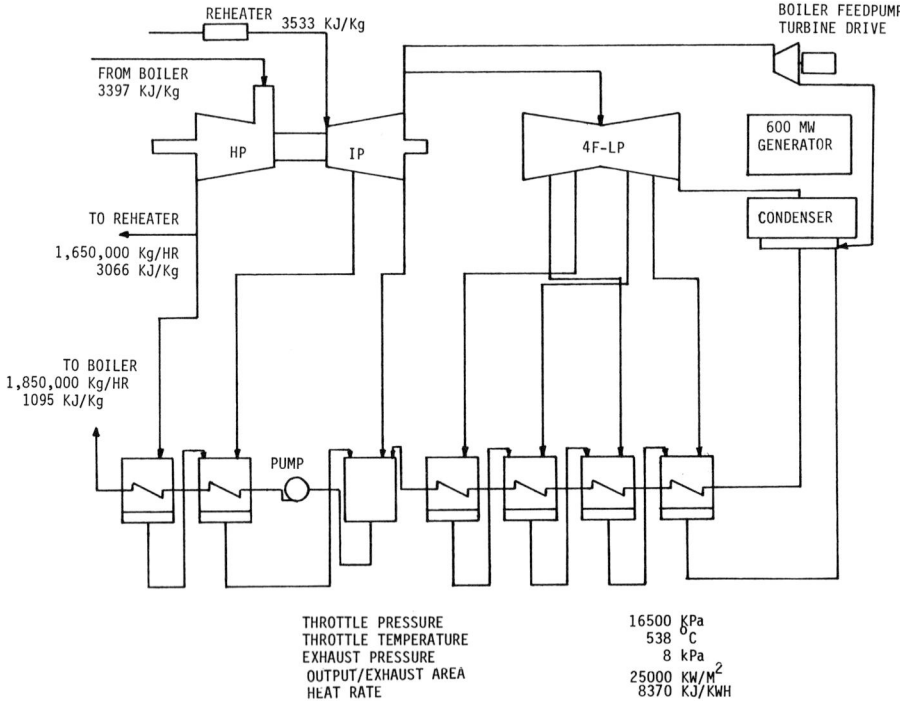

THROTTLE PRESSURE	16500 kPa
THROTTLE TEMPERATURE	538 °C
EXHAUST PRESSURE	8 kPa
OUTPUT/EXHAUST AREA	25000 KW/M^2
HEAT RATE	8370 KJ/KWH

Fig. 14.2-5 Typical fossil heat balance cycle for 600 MW unit.

pump to provide suction head to the boiler feed pump and locate it at its more usual intermediate temperature location as in Fig. 14.2-5.

Cycle Losses

Power plant designers and equipment manufacturers have developed heat balance computer programs that permit the accurate evaluation of cycle incremental losses and gains for changing extraction line pressure drops, feedwater heater terminal temperature difference, drain coolers, and additional feedwater heaters. Approximate and analytical methods are also available for estimating these losses.

Table 14.2-1 shows the approximate gain and loss that would be obtained for adding or deducting one or more heaters extracting from the reheat section of the turbine the cycle shown in Fig. 14.2-5. More accurate gains and losses for a particular application are generally provided by the steam turbine manufacturer who has knowledge of the specific stage pressures available for extraction purposes turbine while maintaining a constant feedwater temperature to the boiler. An extra feedwater heater is sometimes used, extracting from the high-pressure turbine. This raises the final feedwater temperature and produces a much greater heat rate gain than simply adding another low-pressure heater. Table 14.2-2 shows the approximate gain for applying a heater above the reheat point (commonly called a HARP) for several steam conditions.

Higher feedwater temperatures generally require the application of more heat transfer surface in the boiler economizer and/or heating sections in order to maintain a given level of efficiency. Accordingly, in practical cycles final feedwater temperature is optimized considering the value of potential heat rate gain offset by the cost of extra heaters, their piping and control systems, and the incremental effects on the boiler. With HARP cycles consideration must also be given to possible incremental cost increases in the reheater and hot and cold reheat piping and valves, since it is common practice to use lower than normal design reheat pressures when a heater above the reheat point (HARP) is applied. This is done to limit the increase in final feedwater temperature that would otherwise occur with a HARP.

Pressure drop in extraction lines and temperature differences between the steam and the feedwater leaving the feedwater heaters represent thermodynamic losses that can be evaluated by heat balance calculations or estimated by approximate methods. The pressure drops to the heaters are generally assumed by the plant or turbine designer to be about 5% for purposes of establishing the turbine cycle heat rate. Terminal temperature differences of 3°C are also typical except on stages with high extraction superheat. For final plant design, optimization procedures considering the actual feedwater heater arrangement, and incremental cost data result in variances from these preliminary assumptions. For estimating purposes increasing the pressure drop on any feedwater heater extraction line by 5% will increase heat rates by approximately 1.5 kJ/kW · h. Similarly, increasing terminal temperature

TABLE 14.2-1   ESTIMATED CHANGE IN HEAT RATE FOR VARIOUS NUMBER OF HEATERS FOR CONSTANT FINAL FEEDWATER TEMPERATURE

No. of Heaters	Heat Rate, %
5	0.5
6	0.2
7	0 (Base)
8	−0.1

TABLE 14.2-2   ESTIMATED GAIN FOR ADDING A REHEATER ABOVE THE REHEAT POINT

Final Feedwater Temperature	16,500 kPa, 538°C/538°C	24,100 kPa, 538°C/538°C
250°C	0	—
260°C	0.3%	0
270°C	0.5%	0.3%
280°C	0.6%	0.6%
Cold reheat pressure, kPa	4100	4800

differences for any heater by 3°C will increase heat rate by 7 kJ/kW · h. An exception to this simple estimating rule applies to the top heater. There, an extra 5% pressure drop and 3°C extra terminal temperature difference reduce final feedwater temperature creating a thermodynamic loss of approximately 15 kJ/kW · h. These estimated losses are approximately linear and the loss for intermediate values may be interpolated.

The application of drain cooling sections in feedwater heaters represents another opportunity for optimization by the plant designers working with equipment manufacturers. Nominal values for drain cooler approach temperature differentials were used on the heat balance of Fig. 14.2-5. For the cycle shown increasing approach terminal temperature differences by 3°C on all drain cooler sections would increase heat rate by 2 kJ/kW · h.

Miscellaneous Extraction

Extraction from the steam turbine for purposes other than feedwater heating has become increasingly common in recent years. Desulfurization processes that significantly cool boiler stack gas may require the gas to be reheated before it is discharged to the atmosphere in order to obtain adequate dispersal and corrosion control. Steam is frequently extracted for this use and also for the somewhat similar need of preheating air before it enters the boiler air heater. This extraction represents a heat rate loss dependent on the quantity and pressure level at which the extraction is taken. Figure 14.2-6 shows an approximate value of this loss as a function of the extraction pressure. More precise evaluation of the loss, particularly at low pressures, would require heat balance calculations considering exhaust pressure, exhaust loading, and the temperature at which the condensate or makeup is returned to the cycle.

Boiler Feed Pump Drives

Condensing variable-speed auxiliary turbines are typically used to drive two approximately half-sized boiler feed pumps in large plant designs. The feed pump drive turbines are normally supplied with steam extracted from the power generation turbine's intermediate-pressure exhaust connection. An alternate steam supply from ahead of the main turbine control valves is used for starting and below about 40% load. Noncondensing turbine drives and electric motor drives are used where their special characteristics have an advantage. However, the condensing turbine drive is the more efficient alternate with a heat rate typically $\frac{1}{4}$% better than either option.

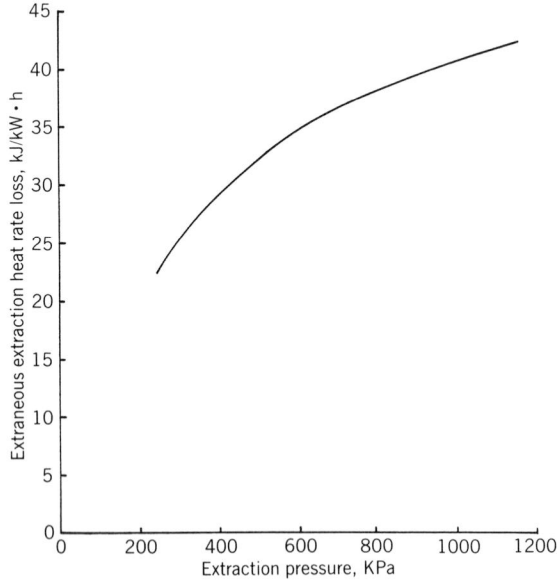

Fig. 14.2-6 Estimated heat rate loss for 1% of throttle flow taken at the indicated extraction pressure.

Exhaust Pressure

Exhaust pressure has a major effect on plant performance. Theoretically, heat rate would improve as the exhaust pressure is lowered to the saturation pressure corresponding to ambient temperature. In practice, cost and equipment sizing constraints result in design full-load exhaust pressures that vary from about 4 to 16 kPa depending primarily on site conditions.

While the ambient temperatures, availability of water, humidity, and other site conditions and cost are the primary factors determining exhaust pressure, the characteristics of other plant components and operational considerations are also important.

Figure 14.2-7 shows a set of curves relating the approximate heat rate changes due to exhaust pressure. The largest heat rate benefits for reduced exhaust pressure result from the application of turbines with the lightest flow per unit of last-stage annulus area. Thus, a turbine with a relatively large annulus area will help justify a larger heat rejection system and lower exhaust pressures for a given site condition.

Operational considerations also influence the sizing and characteristics of the heat rejection equipment. For utilities with summer peak loads, a substantial incremental investment in heat rejection equipment may be justified to avoid the loss in capacity that might otherwise occur in extremely hot weather. This percent capacity loss may also be estimated from Fig. 14.2-7 and is numerically equal to the percent heat rate loss. Some utilities in regions with little or no water available for evaporative cooling towers have considered and a few have applied dry-type heat rejection systems. While these applications have proven mechanically reliable, their effect on fuel consumption is relatively high. A wet/dry approach has recently received some attention. In this application a dry heat rejection system is used with low ambient air temperatures. This system is supplemented with evaporative cooling on very hot days to avoid the substantial capacity and heat rate loss. By limiting the usage of the evaporative cooler to the worst days, the total power plant water consumption is minimized.

Leaving Loss

Typically, 3–4% of the total available energy in a practical Rankine cycle is lost in the kinetic energy leaving the last stage.

The actual loss in a particular application is determined by exhaust pressure and the exhaust flow per unit of last-stage bucket annulus area. The optimization of the total heat rejection system, including the condenser, circulating water system, and cooling tower, must be made jointly with the selection of the optimum number and length of the last-stage steam turbine buckets.

To illustrate approximate changes in heat rate with annulus area, Figure 14.2-8 has been prepared. For the assumption of 600-MW, 16,500-kPa, 538°C/538°C steam conditions and a base exhaust pressure of $8\frac{1}{2}$ kPa the line labeled 25,000 kW/m² represents changes in heat rate with other exhaust

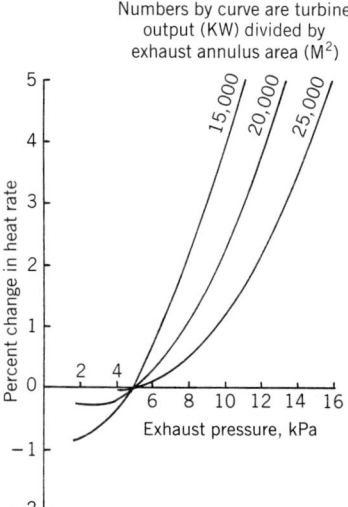

Fig. 14.2-7 Percent change in heat versus back-pressure variations for fossil units.

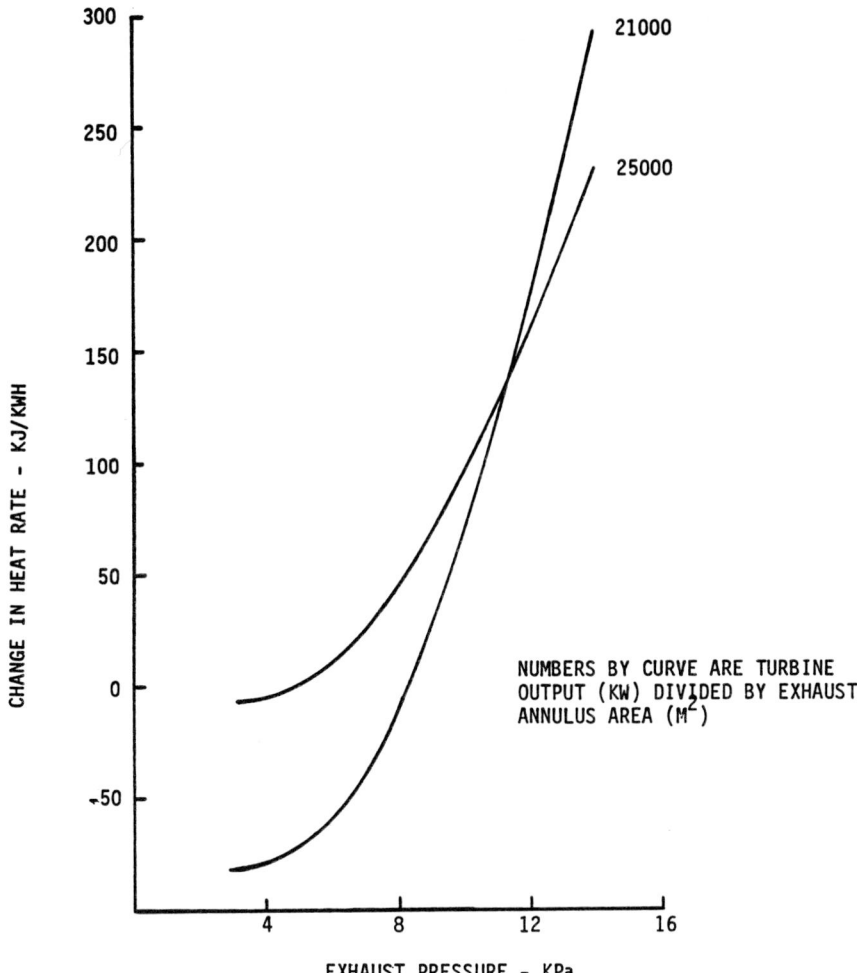

Fig. 14.2-8 Change in heat rate versus exhaust pressure for typical 600-MW units with differing exhaust annulus area.

pressures. The line labeled 21,000 kW/m² represents the performance of an alternate turbine with approximately 20% more annulus area. The difference in heat rate between these two curves represents the heat rate change at full load that results from applying more annulus area at various assumed exhaust pressures.

While the use of Fig. 14.2-7, which has all its performance data normalized to 5 kPa exhaust pressure, and Fig. 14.2-8 along with the estimates of turbine annulus area and heat rejection system costs, will permit the designer to roughly converge on an optimum plant design, in practice it is usually cost-effective for the power plant designer to seek out and obtain accurate, up-to-date performance and cost data for the optimization studies.

14.2-4   Nuclear Power Generation Cycle

Steam Conditions

Approximately 11% of the total U.S. electric utility power generation in 1982 was from nuclear power plants. Except for one high-temperature gas-cooled reactor plant, all operational nuclear power plants utilize light-water-moderated and -cooled reactors of either the boiling water (BWR) or pressurized water (PWR) types, generally in ratings between 500 and 1300 MW of capacity. Steam conditions for a typical BWR application are 6600 kPa initial pressure and 282°C initial temperature. No initial superheat is provided and the excess moisture is removed to yield essentially dry steam. Typical PWR steam conditions are 6000 kPa and 274°C with negligible moistures. One PWR manufacturer supplies 6000-kPa steam with about 30°C of initial superheat. Canada has developed a heavy-water-moderated PWR with different fuel and operating characteristics and supplies steam at an initial pressure of 4240 kPa dry and saturated. The one operational high-temperature gas-cooled reactor provides initial steam at approximately 16,500 kPa, 538°C, with reheat to 538°C; accordingly, conventional fossil fuel condition steam turbines are available with relatively minor modifications for this application.

Feedwater Regeneration

A typical cycle diagram for an 1100-MW BWR application is shown on Fig. 14.2-9. It has a steam-to-steam reheater preceded by a moisture separator that removes the condensate that forms in the high-pressure turbine expansion. Six feedwater heaters are commonly applied except for the CANDU reactor where five stages of feedwater heating are more typical. Practical considerations of water chemistry, plant layout, and special requirements of reactor and other equipment manufacturers lead to a substantial variation in feedwater cycles from plant to plant.

With BWR and PWR cycles the gain for full regenerative feedwater heating is not as great as that with the higher pressures typical of fossil application. Typically, an additional heater, extracting from the intermediate-pressure section such that the final feedwater temperature remains unchanged, will improve heat rate by approximately 0.5%, whereas one fewer will make heat rate poorer by approximately 0.6%.

Moisture Loss

Steam-to-steam reheat is used in nuclear plant applications since it is not practical to return partially expanded steam to the primary heat source as in fossil plants. Thus, the basis for the thermodynamic gain for nuclear reheat is different than it is in fossil applications and arises solely from the improvement in low-pressure turbine expansion efficiency by the reduction in moisture losses. Figure 14.2-10 shows the variation in turbine expansion efficiency as moisture increases. The parameter of weighted average moisture WAM is defined as the expansion energy in the wet region times its average moisture content divided by the total expansion energy. Thus, for an expansion that starts in the wet region the WAM is equal to the average of the initial and final moisture. For an expansion line that starts in the superheat region but becomes wet, the average moisture in the wet region is calculated in the same way but the WAM is reduced by the larger total energy.

Reheat Cycle

The gain for steam-to-steam reheat is dependent on the high-pressure turbine pressure ratio, the pressure drop and terminal temperature differences of the reheater, and the effectiveness and pressure drop of the moisture separator that precedes it. The gain can be increased by staging the reheat; that is, by using lower-temperature extraction steam to provide the first stage of reheat and using the highest available steam temperature for the final stage. Each variable is influenced by design requirements of various equipment manufacturers, plant space and operational restraints, and cost performance tradeoffs. In general, steam reheat has been specified in the majority of nuclear applications with heat rate gains attributed to the steam reheat system of approximately $1\frac{1}{2}\%$ for single-stage reheat and about 2% for two stages of reheat.

With lower initial pressures and temperature the expansion range of a typical nuclear turbine has only 60% of the energy of the typical fossil unit. Accordingly, variation in exhaust pressure has a greater impact on performance. Figure 14.2-11 shows a generalized loss curve derived from the typical nuclear cycle of Fig. 14.2-9. Figures on the curves represent various last-stage loading levels.

14.2-5 Gas Turbine Cycles and Their Application to Electric Utility Systems

The first gas turbine supplied to an electric utility in the United States went into service in 1949. During the 1960s their application accelerated as power demands rose sharply. By 1980 gas turbines represented 8.3% of the installed domestic generation capacity. These machines produced 24,300,000 MW · h, which was 1% of the total national energy production (see the DOE publication in the bibliography).

Many Brayton cycle variations have been considered. The gas turbines of the 1950s had simple cycle efficiencies around 20% or less; hence the efficiency improvement associated with more complex cycles could often be justified. Regenerative and intercooled regenerative cycles were built, raising the efficiency to the 25–30% range. Waste heat recovery cycles ranging from feedwater heating to supply of combustion air for a power boiler were utilized to improve overall plant efficiencies. In recent years

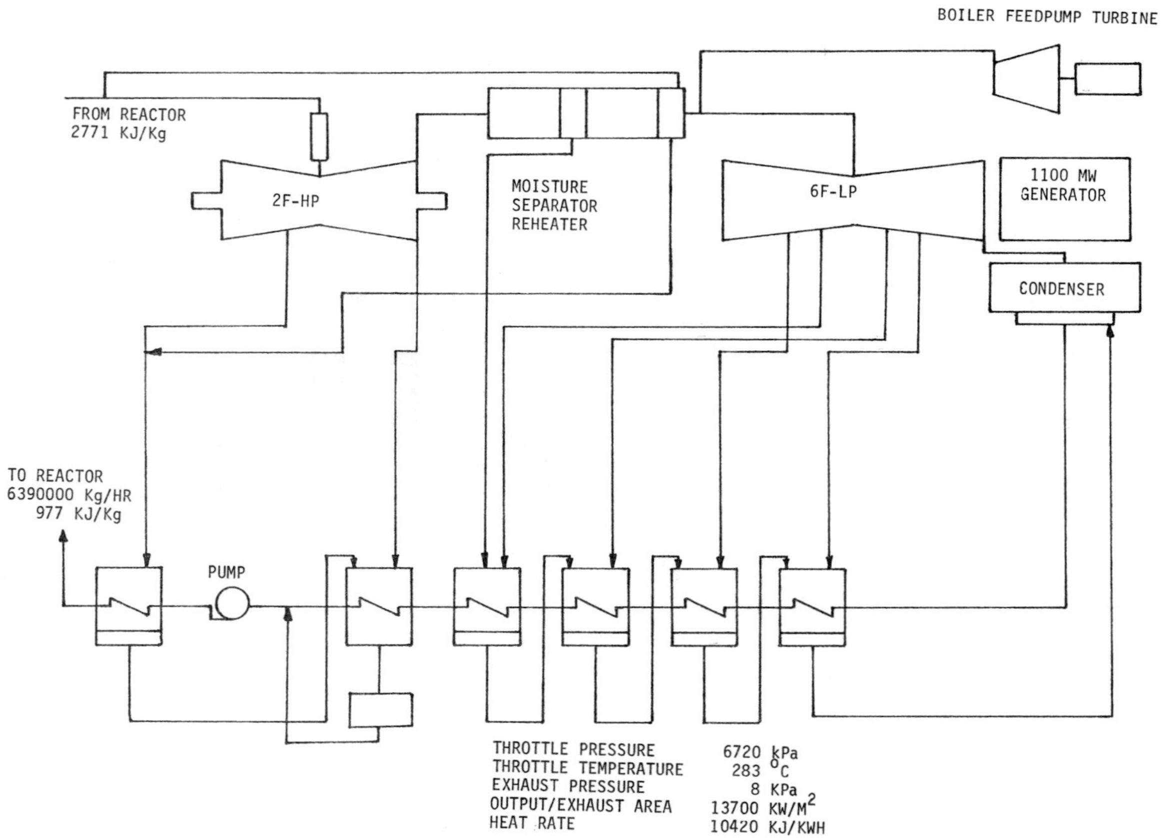

BOILER FEEDPUMP TURBINE

1100 MW GENERATOR

CONDENSER

6F-LP

MOISTURE SEPARATOR REHEATER

2F-HP

FROM REACTOR
2771 KJ/Kg

PUMP

TO REACTOR
6390000 Kg/HR
977 KJ/Kg

THROTTLE PRESSURE	6720 kPa
THROTTLE TEMPERATURE	283 °C
EXHAUST PRESSURE	8 KPa
OUTPUT/EXHAUST AREA	13700 KW/M^2
HEAT RATE	10420 KJ/KWH

Fig. 14.2-9 Typical nuclear heat balance cycle for 1100-MW unit.

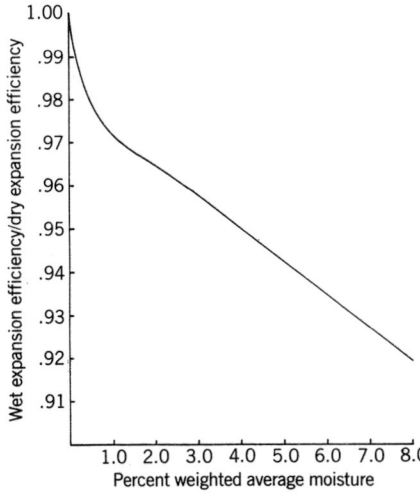

Fig. 14.2-10 Efficiency ratio wet expansion to dry expansion versus percent weighted average moisture.

refinements such as intercooling and reheat, usually in conjunction with a regenerator or recuperator, have not been able to economically justify the added complexity and hardware costs. Surviving are three basic arrangements: simple cycle, regenerative cycle, and combined cycle.

Simple Cycle

Figure 14.2-12 depicts a typical load duration curve for a utility system. The curve shows that the utility's generation requirements can be divided into three areas. The base-load section is made up of the larger, most efficient fossil and nuclear units which run virtually continuously, supplying about 75% of the total megawatt-hours required. Another block of power in the midrange section operates from 2000–6000 h/yr, while providing 20% of the megawatt-hours. The peak-load section is made up of units that operate from 200 to 1500 h/yr and generate 5% of the total megawatt-hours.

Before the introduction of gas turbine power plants, utilities used older, less efficient steam plants to generate power for the top two portions of the curve. The simple cycle gas turbine has been extensively applied to cover the peak area of the curve. The success of gas turbines here is due to their low cost per kilowatt, short cycle time from order to commercial operation, and the availability of standard power-out connections by the utility. They can also be started quickly (less than 30 min from start signal to full load) and can be dispatched remotely without on-site operators.

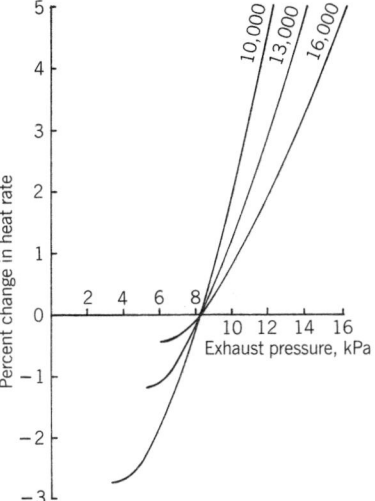

Fig. 14.2-11 Percent change in heat rate versus back-pressure variations for nuclear units.

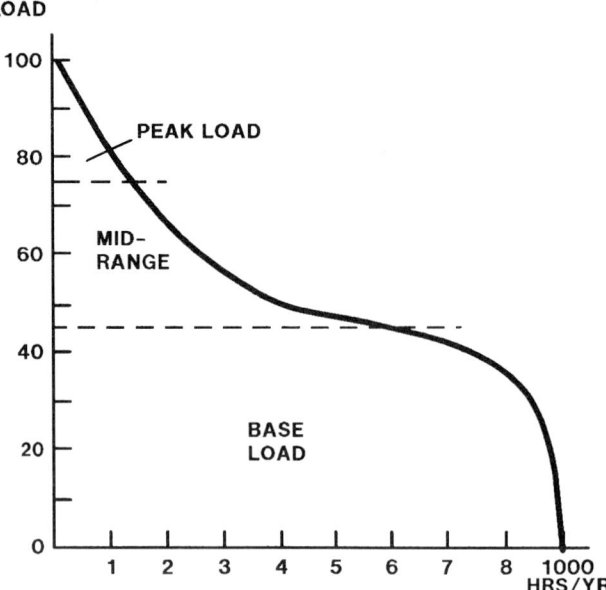

Fig. 14.2-12   Load duration curve.

More recently these same attributes have led to major installations of gas turbines as baseload power plants in the developing countries of the world. Energy-rich nations have experienced such rapid growth and industrialization that their power demands increase so sharply that only gas turbines are able to serve their needs. Order-to-commercial operation times of less than 12 months are not unusual.

Application of gas turbines to the various sections of the load duration curve requires careful matching of the gas turbine characteristics with the projected power needs during the year. Since the gas turbine ingests a constant volume of air, its output is a strong function of ambient air temperature and pressure. Figure 14.2-13 is typical of these effects. If the utility's power demands peak in the summer, when the gas turbine capability is lower, more machines would have to be installed than for a winter peak.

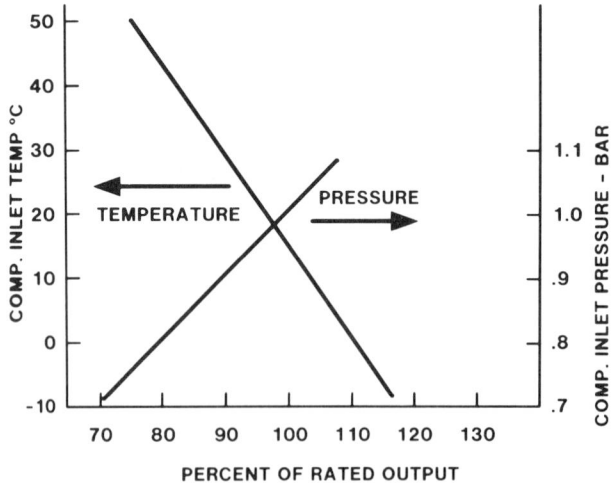

Fig. 14.2-13   Effect of compressor inlet conditions on gas turbine output.

Regenerative Cycle

For the midrange section of Fig. 14.2-12 both regenerative and combined-cycle gas turbines can be applied. This service is typically characterized by operation of 12 h/day, 5 days/week. The use of regenerative units is quite limited, since a basic simple cycle turbine designed for high efficiency does not make an efficient regenerative unit. For a given operating temperature the cycle pressure ratio is the most influential parameter in determining overall thermal efficiency. Figure 14.2-14 shows that high-pressure ratios make for efficient simple cycle and combined-cycle units, while the regenerative cycle is optimized at low-pressure ratios. As a result the regenerative cycle does not have a significant place between the other two, where it would operate as the optimum. A regenerative cycle machine will typically cost 50% more than a simple cycle unit of the same size and will show an improvement in heat rate of 20–30%.

Combined Cycle

The combined cycle has been extensively applied to midrange service and frequently in base-load service because of its low heat rate. Figure 14.2-15 is a *Ts* diagram for the Brayton and Rankine cycles involved in a combined cycle. With gas turbine exhaust temperatures in the 540°C range optimum steam conditions are in the 40–60-bar range with superheat to 480–500°C. Dual-pressure boilers are frequently used to extract the maximum amount of heat from the gas turbine exhaust. The low-pressure system can range from 2 to 20 bars, with the lower pressures used for deaeration and feedwater heating. Higher pressures can be used for admission to the steam turbine. Figure 14.2-16 is a schematic of a combined-cycle with a dual pressure heat recovery steam generator, with the lower-pressure steam providing additional output through secondary admission to the steam turbine.

Combined-cycle heat rates are not particularly sensitive to gas turbine pressure ratio, as Fig. 14.2-14 demonstrates. The highest efficiencies for a given gas turbine are achieved with steam-side equipment selection that results in the lowest stack temperature. Economics, basically the tradeoff between initial cost and operating fuel costs, determine the optimum cycle for each application. Practical considerations, such as low-temperature tube corrosion with sulfur bearing-fuels, will also influence cycle selection. Combined cycles will typically have 50% better efficiency than a simple-cycle gas turbine, with an installed cost of 2–3 times depending on the cycle.

The combined cycle has the ability to match different power requirements by varying the size and number of gas turbines. Typically, one to six gas turbines with heat recovery steam generators are used with one steam turbine generator. A variation of the one-gas turbine–one-steam turbine cycle is shown in Fig. 14.2-17, where both turbines are connected to a single double-ended generator. Here the economic advantages of less equipment must be weighed against the inability to run the gas turbine independently.

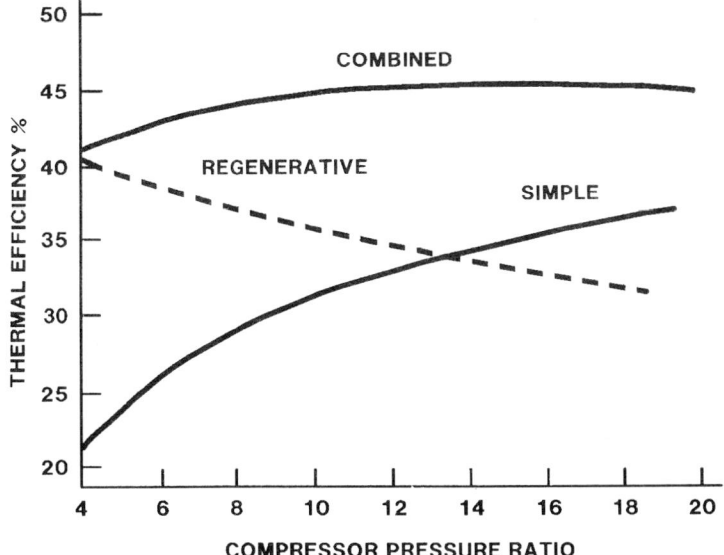

Fig. 14.2-14 Influence of pressure ratio on overall thermal efficiency.

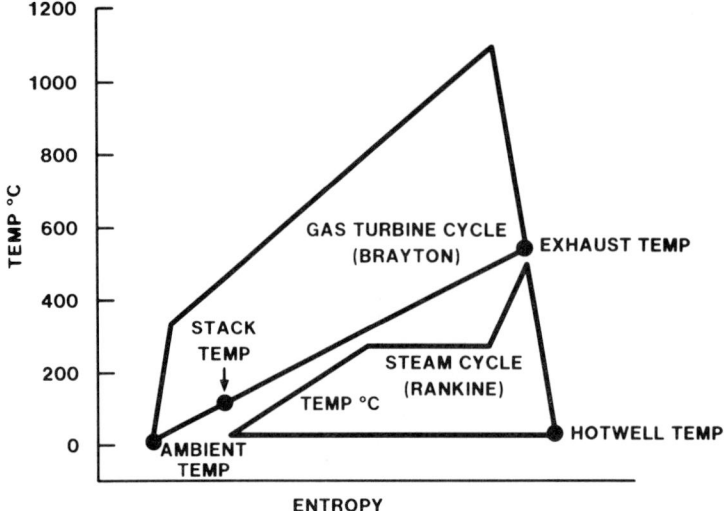

Fig. 14.2-15 Combined cycle temperature–entropy diagram.

High part-load efficiencies are attainable with multiple gas turbine combined cycles. For a plant with four gas turbines and one steam turbine, heat rate characteristics are shown in Fig. 14.2-18. As gas turbines are dropped off the line in times of reduced power demand, the combined cycle heat rate remains essentially at the full rated value, affected only by the part-load efficiency characteristics of the steam turbine.

Cogeneration*

The combined cycle, by virtue of its potential for high thermal efficiencies, has been adapted for other applications where the power plant is integrated into another process. Cogeneration is the most frequent application. Basically, cogeneration involves generation of electricity for the utility system and supply of process heat for an industrial plant. Frequently the equipment is owned by the industrial user and the generated kilowatt-hours sold to the neighboring electric utility system.

The gas turbine combined cycle fits well into many cogeneration schemes. It produces relatively high electric power output for each unit of process heat. It has the potential for very high utilization of the fuel heat input, as shown in Fig. 14.2-19. Other cogeneration cycles utilize the waste heat from the gas turbine for paper drying or reformers as well as steam generation.

Other Applications

With the extensive coal reserves in America, methods to use this fuel source in an economical and environmentally acceptable manner will continue to receive a great deal of attention. The combined cycle is an attractive alternative for integration into a plant that gasifies the coal and produces electricity for utility consumption. The highest degree of integration occurs when the gasification process utilizes high-pressure air supplied by the gas turbine, as shown schematically in Fig. 14.2-20.

Coal gasification produces fuels with heat values of 4000–10,000 kJ/m³, compared with typical natural gases in the 35,000-kJ/m³ range. As the heat value decreases, the mass flow of fuel increases to provide the same energy input into the turbine's combustors. This will increase the size of the fuel piping and valving and can impact the design of the combustion system. Fuel composition is important since the main combustible components of coal-derived gases are hydrogen and carbon monoxide, with the remainder made up of inerts such as nitrogen and carbon dioxide. Efficient combustion of such fuels may be limited to a narrow operating range, necessitating the use of a conventional fuel for turbine startup.

---

*See also Section 2.2.

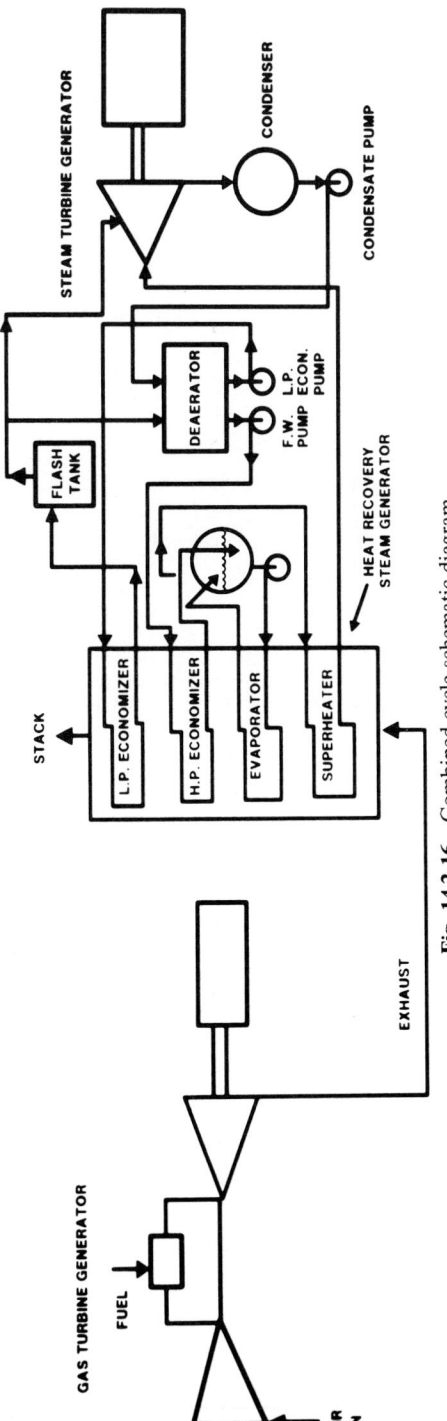

Fig. 14.2-16 Combined cycle schematic diagram.

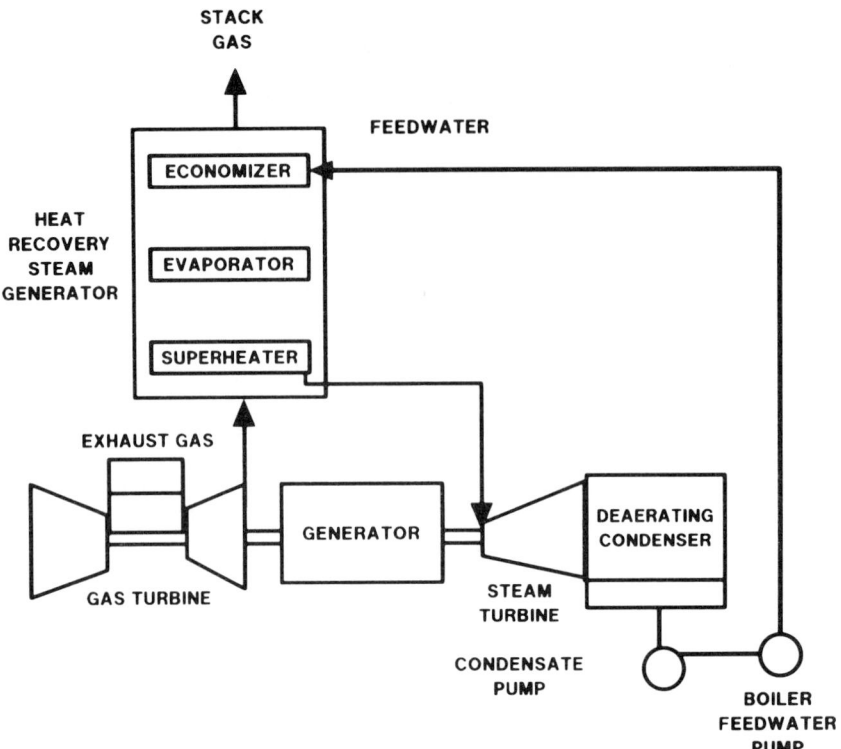

Fig. 14.2-17 Single-shaft combined cycle.

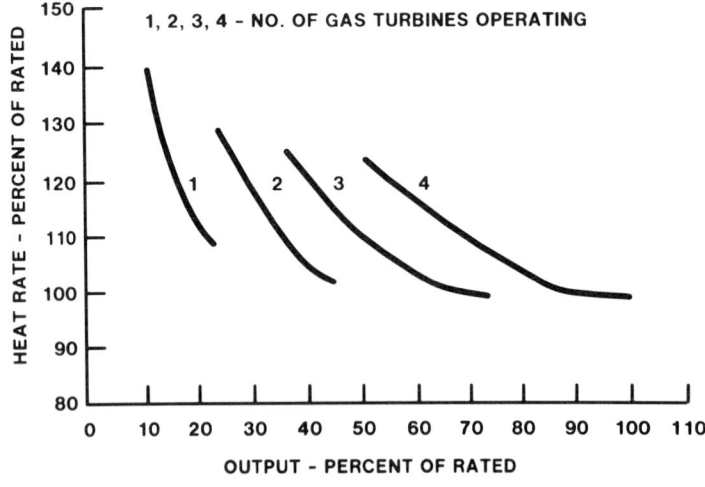

Fig. 14.2-18 Four-unit combined cycle part-load performance.

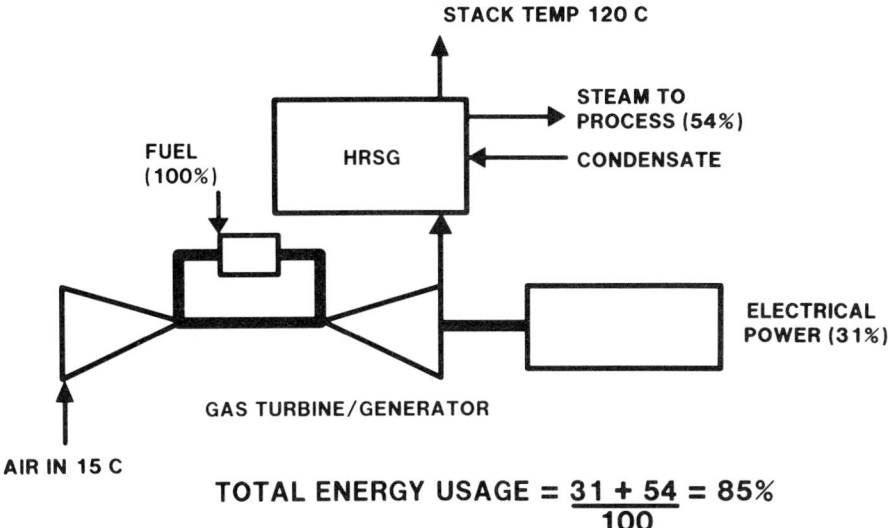

Fig. 14.2-19 Cogeneration heat balance.

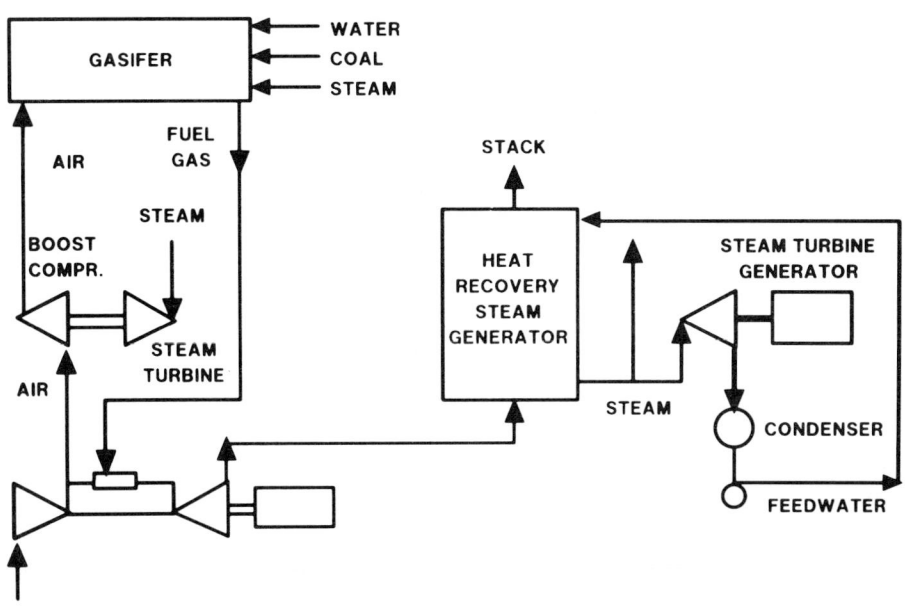

Fig. 14.2-20 IGCC schematic with air-blown gasifier.

Another consideration in integrated gasification combined cycles (IGCC) is fuel quality. Particulate matter with its erosion potential and ammonia with its nitrogen oxide emissions are two of the major concerns. As the coal gasification process technology advances, it is expected that the gas turbine and combined cycle for the IGCC will follow, since the IGCC is a leading candidate for this form of coal utilization.

BIBLIOGRAPHY

American Society of Mechanical Engineers, "1967 ASME Steam Tables," New York.

Bartlett, R. L., *Steam Turbine Performance and Economics*. McGraw-Hill, New York, 1958.

Central Electric Generating Board, *Modern Power Station Practice*, Vol. 3, Pergamon, New York, 1971.

Downs, J. E., "Margins for Improvement of the Steam Cycle," ASME 55SA76, 1955.

Eshbach, O. W., *Handbook of Engineering Fundamentals*. Wiley Engineering Handbook Series, Wiley, New York, 1966.

General Electric Co., "Comparative Study and Evaluation of Advanced Cycle Systems," EPRI AF-664, 1978.

General Electric Co., "Engineering Assessment of a Low Heat Rate Pulverized Coal Power Plant," EPRI RP-1403-2 Final Report, 1982.

Harris, E. E., and White, A. O., "Developments in Resuperheating in Steam Power Plants," ASME 48A125, 1948.

Jones, J. B., and Hawkins, G. A., *Engineering Thermodynamics*. Wiley, New York, 1963.

Marsh, W. D., *Economics of Electric Utility Power Generation*, Oxford, New York, 1980.

Miller, E. H., and Schofield, P., "The Performance of Large Steam Turbine-Generators with Water Reactors," The ASME Winter Annual Meeting, 1972.

Palmer, W. E., and Miller, E. H., "Why Multi-Pressure Surface Condensing Turbine Operation," American Power Conference, April 1965.

Salisbury, J. K., *Steam Turbines and Their Cycles*. Wiley, New York, 1950.

Selvey, A. M., and Knowlton, P. H., "Theoretical Regenerative Steam Cycle Heat Rates," Trans ASME, August 1944.

Sheppard, R., "Operating Experience with the First Commercial Supercritical Pressure Steam Turbine Built for the Philo Plant," American Power Conference, 1958.

Spencer, R. C., "Design of Double Reheat Turbines for Supercritical Pressures," American Power Conference April 1980.

Spencer, R. C., and Booth, J. A., "Heat Rate Performance of Nuclear Steam Turbine-Generators," American Power Conference, April 1968.

Wood, B., "Alternate Fluids for Power Generation," The Institution of Mechanical Engineers, pp. 40–70, 1970.

14.3 UTILITY CENTRAL POWER STATIONS

H. R. Stewart

14.3-1 Evolution and General Nature of Central Power Stations in North America and Overseas

It will be helpful to an understanding of the central-station industry as now constituted to review briefly its technical history over the past century. The first central station consisting of six 100-kW direct-current generators was commissioned by the Edison Electric Illuminating Co. of New York in 1882, serving lamp loads in 225 houses within a quarter mile radius of the Pearl Street Station through an underground distribution system operating at lighting voltage. Within a few years dozens of Edison licensee companies sprang up in the downtown areas of all the major cities in the country.

The concurrent inventions of the ac transformer by William Stanley and of the three-phase ac induction motor by Nikola Tesla respectively (1) vastly increased the distances over which electric power could be transmitted and distributed from central stations and (2) greatly simplified motors for industrial and other purposes, particularly in the "squirrel cage" type where the rotor "winding" consisted merely of insulated metal bars short circuited to each other at each end of the rotor body. The consequence was that by the end of the Pearl Street Station decade ac central stations had sprung up all over the country, serving loads surrounding the downtown dc areas and spreading into the suburbs and rural areas.

Frequencies

The most notable early ac generating station was at Niagara Falls, developing a part of their available hydropower and commencing operation in 1890. The chosen frequency was 25 Hz, favored at the time by some industrial users and very generally used by the paper industry. The textile industry chose 40 Hz, and power companies in southern California elected 50 Hz. It may be added parenthetically that the initial popularity of a 25-Hz system frequency was partly due to the fact that the rotary converters needed to supply existing lighting, industrial, and street railway dc loads could be designed to commutate much more satisfactorily at 25 Hz than at 60 Hz. However, on account of the incandescent lamp flicker characteristic of 25 Hz particularly, the bulk of the power companies coming into being in North America went to 60 Hz, and that became the prevailing frequency of all new central-station generation, the remnants of lower-frequency load and the sizable railroad electrification 25-Hz load being taken care of by frequency changer sets when needed to supplement existing lower-frequency generation.

On the other hand, the power systems of Europe, the former colonies and dominions of European countries, and one-half of Japan have all developed at a frequency of 50 Hz.

Transmission Voltages

As loads grew and distances increased between loads and generation sites, particularly hydro sites, higher and higher transmission voltages became necessary for voltage regulation, conductor loss, and various economic reasons. By 1900 they had reached 34.5 kV, by 1910 69 kV, in the next 10 years 115 kV, in the 1920s 230 kV, in the 1930s 287 kV, in the 1950s 345 and 550 kV, and in the 1960s 765 kV. In the 1980s some 1200-kV line sections are in experimental test operation and the ultimate practical limit of open-wire design is seen as somewhere around 2300 kV.

Interconnection

By 1920 the several hundred investor-owned power companies in the United States and many of their municipally owned neighbors were taking advantage of economies available in cases where they could interconnect their geographically contiguous transmission systems. By interconnection they could (1) economically exchange energy in either direction depending on who had the lower-cost available generation at the time, (2) save on reserve installed generating capacity by the amount of the firm capacity of the interconnection, and (3) save on installed peak generating capacity if their respective peaks were not simultaneous. This practice spread to a point where all of North America except for the Province of Quebec had their ac facilities interconnected and operating in synchronism. Electric clocks in Maine operate at the same speed as electric clocks in San Diego.

Similarly, all of England and Scotland operate in synchronism all of Western Europe except Norway and Sweden, and all of Eastern Europe. All of Norway and Sweden operate in synchronism. (Asynchronous ties between France and England and between Denmark and Sweden will be discussed in Section 14.3-5.)

All such modern interconnected transmission systems constitute veritable grids of more or less density stretching up to thousands of miles in extent, even though the *average* transfer distance of energy from generator to load center is perhaps on the order of only 50 mi.

Generation Types

The economical "mix" of generation to supply a system load is very much a matter of load factor—the ratio of energy actually consumed in a given period to what it would be if the load was at peak value for that whole period. If the load factor were 100%, it would be economical to pay a premium for very fuel-efficient units burning low-cost fuel even though their capital cost were high because the fixed charges on the investment could be spread over the maximum number of kilowatt-hours. On the other hand, if the load factor were near zero, it would pay to use the lowest possible capital cost units even though the fuel efficiency was very poor and the type of fuel very expensive. The very largest available nuclear-fueled or coal-fueled steam turbine generators are typical of the first category, and natural gas or oil-fired gas turbine or diesel-engine-driven generators are typical of the second. Actual utility load factors are in the 55–60% range so a mix of the two categories works out to be economical with the first category (base-load units) running fully loaded continuously, and the second category (peaking units) running only as many hours as necessary to supply the rest of the load demand. Hydro units make a third category having zero "fuel cost" but an available output curve that rarely fits the load curve seasonally; also the energy available for any given year is not precisely predictable. Nevertheless hydro usually provides a very valuable peaking function. A fourth strategy is pumped-storage hydro, whereby a motor-generator unit uses base-load energy to pump water up to a high-level reservoir during nights, weekends, and other times of heavy load and then generates from that storage during times of heavy load. The one-third loss is energy in this operation is more than compensated by the difference in value of the system energy between the pumping and generating modes.

Generation Siting

Siting of these four types of generation is largely determined by their operating characteristics. The largest nuclear-fueled and fossil-fueled steam turbine generator units—up to 1350 MW in rating—must be located on seashores, estuaries, large rivers, or lakes so that their rejected heat can be adequately dissipated by cooling water. Gas turbines (typically up to 20 MW) and diesel engines (typically up to 3000 kW) can be located almost anywhere that their noise can be suitably baffled. Their location can often serve the added function of deferring added transmission or subtransmission to an outlying growing load. Hydro generating stations obviously must be located where dams and reservoirs can provide the necessary river impoundment, and pumped-storage hydro requires location at the base of a high hill or mountain alongside a water supply which can be of relatively modest capacity or flow because the head is usually so high compared to that of most "gravity"-normal hydro stations.

Except for gas turbine and diesel-engine-driven generating stations, all types of generators mentioned are likely to be located remote from load centers, requiring high-voltage transmission to load centers and for interconnection purposes to neighboring utilities. At load centers the transmission voltage is stepped down at transmission substations to subtransmission networks at medium voltage, such as 23, 34.5, 69, or 115 kV, and fed from the subtransmission network will be distribution substations stepping down to distribution voltages such as 2.4, 4.16, 6.9, 13.8, 23, and 34.5 kV. Radial feeders then carry the energy to the industrial, commercial, and residential consumers.

14.3-2 Large Fossil-Fueled and Nuclear-Fueled Steam Generating Stations

Until the 1930s sites could usually be found on the waterfronts of large cities for large fossil-fueled steam generating stations, with the turbine-driven generators connected to solid or sectionalized buses from which emanated (1) feeders to large urban loads, (2) feeders to urban distribution substations, (3) feeders to downtown low-voltage networks (to be described in Section 14.3-8), (4) connections to transformers for stepping up to subtransmission to suburban and rural areas, and (5) connections to transformers for stepping up to intrasystem and intersystem transmission. Since that time, because of the siting difficulties mentioned in Section 14.3-1, such stations are increasing at locations remote from urban centers or large local loads so that the energy leaves the station at transmission voltage only; the station electrical connection diagram simplifies to the "unit scheme" shown in Fig. 14.3-1, where it will be noted that the generator feeds directly through a unit transformer to the high-voltage bus without a generator circuit breaker on the low-voltage side of the transformer, and the generator feeds its own array of station auxiliaries through a station service transformer connected to the generator leads. The unit scheme extends back into the boiler house in that there is one boiler per turbine and no steam header between boilers.

To make this unit system self-starting, one of the one or more "startup transformers" fed from the high-voltage buses brings a boiler up to pressure and a turbine-generator up to synchronizing speed. After the generator is synchronized to the high-voltage bus, its "unit" station service transformer low-voltage circuit breaker is closed to the station service bus, and the "startup" transformer low-voltage breaker is opened, leaving the entire unit station service load on the generator terminals. The function of the station service bus sectionalizing breakers is to make it possible to clear a bus fault on either half of the station service buses and still keep the unit running at a large fraction of full load by a judicious division of essential auxiliaries between the two halves. If the fault is on the "unit" half of the bus, the clearing of the bus sectionalizing breaker is promptly followed by the automatic closure of the "startup" transformer low-voltage breaker (but not so promptly as to cause excessive inrush current to motors whose magnetic circuit flux has not yet sufficiently decayed, a matter of a few seconds). If the unit energy source itself fails, orderly shutdown of its auxiliaries is afforded by automatically tripping the unit station service transformer low-voltage breaker and closing the startup transformer low-voltage breaker.

As can be seen from Fig. 14.3-1, there are usually three levels of station service bus voltage, the highest for feeding the largest motor drives (boiler feed pumps, coal pulverizers, circulating water pumps, condensate pumps, forced and induced draft fans, spare exciter if any) 480 V for feeding medium-size motor drives (ventilating fans, coal conveyors, oil pumps, elevators, ash handling, turbine turning gear, cranes, and many others), and 120/208 V, three-phase, four-wire for lighting and the host of small motor applications.

Connections are also shown from the highest-voltage station service "startup" bus section to an on-site source of energy, such as self-starting diesel-engine-driven generators, in order to get the station restarted in the remote event of a total power system shutdown.

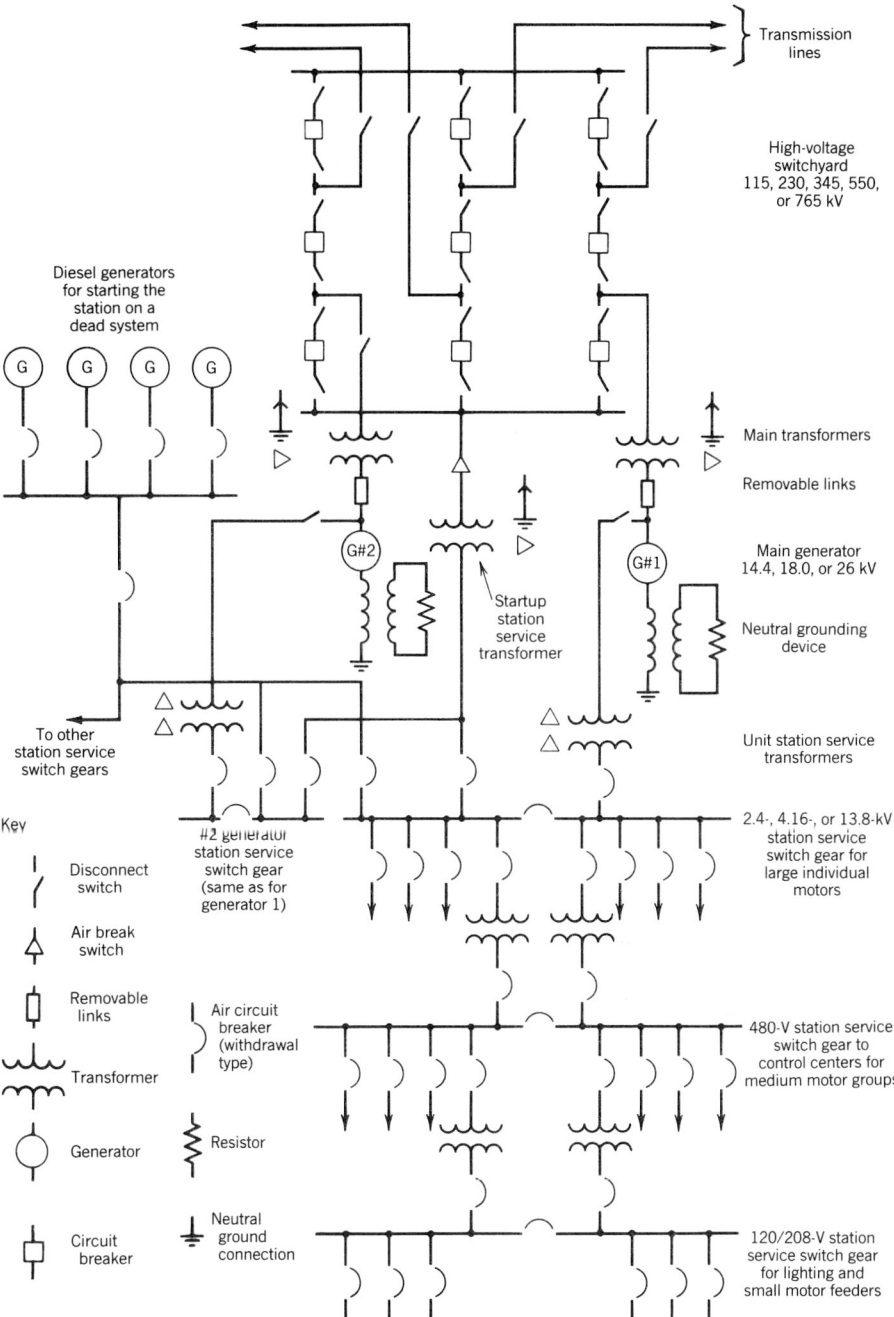

Fig. 14.3-1 Typical connection diagram of large fossil-fueled steam generating stations.

1435

Nuclear-Fueled Stations

The station electrical connection diagram for large nuclear-fueled steam turbine-generator units is very similar to that for the large fossil-fueled units (Fig. 14.3-1) with the major exceptions that (1) the array of station auxiliaries for the nuclear reactor, whether of the "pressurized" or the "boiling water" type, is necessarily quite different from that for a fossil-fueled boiler and (2) the on-site backup source of station service energy for running the reactor core emergency coolant pumps must be dependably and rapidly available to prevent damage to the reactor fuel elements from excess temperature resulting from the stored heat in the reactor core, in the event of certain types of failures; this arrangement often consists of one diesel-engine-driven generator per emergency coolant pump capable of automatically coming up to speed in about 10 s and fully loadable in about 1 min.

The major station components depicted in Fig. 14.3-1 will now be treated in turn in more detail.

Generators

As power systems grow, there are incentives to using the largest possible rating generator consistent with tolerable generation loss to the system in the event of unit tripout or failure in that the larger the unit, the lower the heat rate in Btu per kilowatt-hour (i.e., the higher the efficiency) and the lower the installed cost per kilowatt of output. The upper limits to rating reside in shipping limitations to the width and height of the stator and in the ability of available generator rotor-forging materials to withstand safely the effects of centrifugal force during operation and at the overspeeding incident to tripout or sudden loss of load. The latter consideration limits the diameter of two-pole, 60-Hz, 3600 rpm rotors to about 46 in. and of four-pole, 60-Hz, 1800-rpm (used on nuclear-fueled units due to low steam conditions of only 600°F and 600 psi) rotors to about 67 in. To get more and more rating and remain within these dimensional limitations has required an evolution in cooling means that has enabled a progression from 25 MW at 3600 rpm in the late 1920s to about 1100 MW at 3600 rpm and about 1350 MW at 1800 rpm in the late 1970s. The steps in this evolution from the recirculated water-cooled air system typical of the late 1920s have been:

1. Water-cooled hydrogen, initially at 15 psi progressing eventually to 75 psi, recirculated through the air gap and passages in the stator core.
2. Water-cooled hydrogen recirculated through multiple channels in the rotor body and in direct contact with the rotor conductors.
3. Water-cooled demineralized water (or in some cases transil oil) recirculated in direct contact with the stator conductors.

The advantages of hydrogen over air as a coolant are: (1) the windage loss, a function of molecular weight, is about one-tenth as much at equal pressures, and still one-half at 5 atm H_2 compared to air at 1 atm; (2) hydrogen has higher specific heat and higher heat transfer capacity; (3) hydrogen does not support combustion; and (4) the corona, if present in a hydrogen atmosphere, does not damage conductor insulation. The disadvantage of hydrogen's explosiveness in an air mixture is avoided by automatically maintaining the hydrogen purity at 97% or higher and designing the main generator casing to be rupture-proof should an internal explosion occur at the proportions of "the most explosive mixture" of hydrogen and air between the explosive limits of 5–70% ratio of hydrogen to air.

Among considerations for future expansion of generator ratings are experimentation with superconducting generators where the rotor conductors operate at near-absolute-zero temperature and the stator winding is air cored rather than iron cored.

Generator Excitation

The generator needs an adequate and reliable source of excitation, particularly during system disturbances when maintenance of the generator air gap flux is an important factor in keeping the generator from falling out of synchronism with the rest of the system. The generator rotating shaft itself is the most reliable of the sources of power during system disturbances, and correspondingly the excitation scheme has conventionally consisted of a shaft-driven dc generator excited by a coupled compound-wound dc pilot exciter and controlled by an automatic voltage regulator. The functions of the regulator were to maintain any desired level of generator terminal voltage during normal operation, to call for large boosts in excitation when the generator terminal voltage was depressed during system short circuits, and to call for large reductions in excitation if the generator should overspeed due to large loss of load or tripout due to external troubles. A motor-driven spare exciter was sometimes provided so that the main unit would not be rendered unavailable for service by exciter troubles, and the spare would usually be equipped with a flywheel to lessen its speed reduction in the event of momentary dips in voltage on the station service bus. The development of silicon-diode rectifiers has made it possible to supersede the direct-connected main exciter by combination with either a direct-connected rotating armature ac generator or with auxiliary voltage windings in the main generator core plus auxiliary series transformer windings in the generator leads. Either scheme affords as good assurance of power supply to the excitation system during system disturbances as the former conventional arrangement and provides the further advantage that due to the multiplicity of rectifier units a rectifier failure can be isolated with little reduction in excitation capacity. In the scheme using the rotating armature ac generator the rectifiers are in a rotating assembly also, thus eliminating the necessity of dc slip rings for conveying excitation energy to the main generator rotor windings.

Generator Neutral Grounding

The conventional arrangement is to ground the generator neutral through a value of resistance such that the resistive component of single-phase-to-ground-fault current is equal to the capacitive component returning to the unfaulted phases through their capacitances to ground, the magnitude of the total ground fault current being on the order of 10 A. If the ground fault is in the generator, its most likely location is from a stator coil to the armature iron with consequent possible arc burning and short circuiting of some of the core laminations. This requires an immediate tripout of the unit to minimize the chances of having to remove all the stator coils and unstack enough of the core laminations to enable repair or replacement of the damaged laminations, a very costly procedure in expense and lost production. Some users prefer to install, in place of the grounding resistor, a grounding reactor of the same number of ohms, thus neutralizing the capacitance component of the ground fault current and reducing the total fault current to less than 1 A. At this value it is relatively safe to delay the generator shutdown for up to an hour, allowing time for replacement generation to be started up on the system.

Generator Leads

Up to about 100 MVA generator rating the leads from the generator to the main stepup transformer can consist of multiple paper insulated or rubberlike insulated cables per phase in a concrete duct envelope. Above this rating it is more economical to use "bus duct," which consists of one aluminum tubular conductor per phase each supported by porcelain insulators inside its aluminum housing. The three housings are electrically bonded at both ends, thus permitting electromagnetically induced currents to flow end to end in the housings continuously, but this arrangement minimizes the magnetic repulsion stresses between the phases when phase-to-phase short circuits occur beyond either end of the bus-duct run. The conductors are cooled by forced circulation of cooling air through the aluminum housings.

Transformers

The main transformers are sized to fit the kVA rating of the unit generator. In the early days of the industry generating station transformers were universally single-phase units with usually one single-phase transformer on site as a spare. However, the reliability of transformers in service has proved so good that the practice has evolved of using one three-phase unit per generator, or at most two half-sized units, with no other spare on site. Such generators run continually fully loaded for a substantial fraction of their operating life and transformer design economics are such that the sum of annual capital charges on installed cost and the annual cost of transformer losses is a minimum when a single transformer is self-cooled by radiators mounted on the tank. When two half-size units are used, they are also self-cooled in normal operation, but the radiators are provided with fans and the insulating oil circuit with circulating pumps so that each unit has about 150% operating capability when the other unit is unavailable, thus allowing the generator to be operated at 75% of full load.

Transformer insulation strength is designed for whatever level is protectable by applicable voltage surge arresters that intercept lightning surges and switching surges incoming over the connected overhead transmission lines. The arresters are located as closely as practicable to the transformers, preferably on the transformer tank covers. Transformers so protected transmit a small enough fraction of the incoming surges to the low-voltage winding and thence to the generator so that surge arresters are rarely installed at the generator terminals.

The unit station service transformers, sized at 3–5% of generator rating, are invariably single three-phase self-cooled units, with the startup transformer acting as a spare.

Deterioration of the insulating oil from exposure to atmospheric oxygen is minimized by either completely filling the transformer tanks with oil and providing for thermal expansion in a small auxiliary tank ("conservator") with a limited oil-to-air surface or by automatically maintaining a nitrogen atmosphere over the oil in incompletely filled main tanks.

High-Voltage Switchyard

Figure 14.3-1 shows the high-voltage switching connections used in the plurality of cases with two normally live buses and three circuit breakers per pair of circuits, giving rise to the name *breaker-and-a-half* scheme. Other commonly used schemes are (1) the more expensive double-bus–double-breaker scheme where both buses are normally operated live and both breakers per circuit are closed and (2) the less expensive main-and-auxiliary bus scheme with one breaker per circuit plus a bus-tie breaker that can substitute for any one of the other breakers by means of selector disconnecting switches to either bus; the main bus is normally operated live and the auxiliary bus dead except when the bus-tie breaker is performing its vicarious function or when *all* circuits are switched to the auxiliary bus during maintenance of the main bus. The only disadvantage of the breaker-and-a-half scheme as compared to the double-bus–double-breaker scheme is that when any bus breaker is out for maintenance the integrity of its controlled circuit is at the mercy of a tripout of the neighboring circuit in the same three-breaker bay; this is considered a minor risk.

The most expensive components in the switchyard are the circuit breakers, which perform all load and short-circuit interruptions. In spite of the seeming anomaly of using a highly flammable fluid as an arc-interrupting agent, transil oil has been used for that purpose since early days in circuit breakers very successfully up through 230 kV and in many cases at 345 kV. At the latter voltage compressed-air porcelain-clad breakers became a successful competitor to oil circuit breakers in North America, especially in the area of greatly reduced maintenance time. In Europe compressed-air porcelain-clad circuit breakers had been almost exclusively used for decades at both transmission and subtransmission voltages. At 345 kV and upward compressed air breakers and a still later development, porcelain-clad sulfa-hexafluoride (SF_6) breakers, are now applied worldwide. At the same pressures SF_6 has superior arc-interrupting ability as compared to air and thus became competitive.

Current transformers for supplying instruments, meters, and protective relays are provided integrally with the circuit breakers. Potential supply for instruments, meters, protective relays, and synchronizing is obtained from separately mounted capacitance-type potential transformers. Air-break switches serve the function of interrupting transformer magnetizing currents but not load currents. Disconnecting switches provide electrical isolation for equipment being maintained and cannot be used for making or breaking more than the capacitive charging currents of the entrance bushings and porcelain lead supports of equipment being energized or deenergized, respectively.

Protective Relaying

Most of the events that can disturb or disable an electric power system occur, and must be remedied, much too quickly for the station operators to be able to sense, evaluate, and take appropriate action to clear the trouble. These functions are provided by protective relays that initiate the tripping of the circuit breakers and other protective devices in the station. Unit-connected generators are usually protected by the following array:

1. Inverse-time overload relays to protect the stator winding (and the main transformer windings) from prolonged overload by delayed clearing of transmission system faults.
2. "Negative-sequence" relays to protect the rotor body and retaining rings from overheating by the double-frequency currents induced by prolonged unbalanced currents in the stator winding due to delayed clearing of unbalanced faults.
3. Differential relays that detect a phase-to-phase winding failure in the generator stator by measuring the current unbalance between the two ends of each stator phase winding.
4. A ground relay that detects a ground fault anywhere in the generator stator winding, the generator leads, the main transformer low-voltage winding, or the unit station service transformer high-voltage winding or leads.
5. A "loss-of-field" relay that measures whether the generator is receiving inadequate excitation.
6. A "volts-per-cycle" relay that determines whether the main transformer and unit station service transformer are being overexcited during generator startup prior to synchronizing.
7. A reverse-power relay that indirectly measures whether the steam flow is so low as to cause overheating in the low-pressure turbine.
8. An out-of-step relay that protects against shaft resonant overstresses during system loss of synchronism, if this protection is not already provided at some more desirable point of system separation.

In addition to the protection that the main transformer inherits from the generator overload relays and both the main and station service transformers inherit from the generator ground relay, each transformer is protected by its own set of differential relays that measure unbalance between the current input and output of each transformer and thus detect an internal fault, and the unit station service transformer has its own set of overload relays. The main transformer differential relay zone includes the generator leads. Both transformer tanks are usually provided with "sudden pressure relays," which detect internal failure by measuring the pressure and rate of increase in oil pressure at the bottom of the transformer tank due to evolution of gas from the insulating oil at the point of insulation breakdown. Outside the United States a relay widely used on conservator-type transformers with domed tank covers collects gas rising from volatilized oil at the point of fault and causes tripping when the gas accumulates to roughly a cupful.

Each bus in the high-voltage switchyard is protected by bus differential relays, and each "startup" station service transformer is protected by its own inverse time overload relays, its differential relays, and usually its sudden pressure relay.

The relay protection of the transmission lines leaving the high-voltage switchyard is treated in Section 14.3-5.

14.3-3 Hydro and Pumped-Storage Generating Stations

The largest hydro and pumped-storage generator ratings are usually substantially less than half those of the largest fossil- and nuclear-fueled generating units on account of siting conditions, and not due to shipping dimension limitations or rotor diameter and strength limitations against centrifugal forces at waterwheel runaway speeds. This is so because (1) the factory-wound stators can be manufactured in sections and assembled at the site (likewise the rotor salient field poles, laminated rim, and demountable rotor arms can all be assembled on the rotor hub at the site) and (2) the laminated rotor rim technique affords generous design flexibility against runaway forces.

In the sizes used on power systems the machines are almost invariably of the vertical shaft type, with the whole rotating element supported from a thrust bearing mounted just above or below the generator rotor and with suitably located guide bearing or bearings to maintain alignment with the waterwheel shaft. Speeds vary from around 50 to 600 rpm depending on the hydraulic head and the generator rating. The one notable exception to the vertical shaft arrangement is the "bulb-type" design used in tidal power and extremely low-head conventional developments where the whole generator is submerged with the turbine in a horizontal hydraulic tunnel.

Generator Cooling

In the smallest ratings, up to perhaps 5000 kVA, the generators are open-type cooled by air circulating through the power house via louvers in the walls and roof; larger ratings are provided with enclosures either indoors or outdoors with cooling air drawn in and discharged through duct work, sometimes equipped with air filters in polluted locations; still larger ratings either indoors or outdoors are provided with enclosures in which the cooling air is recirculated through water coolers. The last arrangement makes possible the use of automatic admission of carbon dioxide from bottled CO_2 to the ventilation circuit for fire extinghishing in the event of fire in the winding insulations or bearing oil.

Pumping Mode Starting in Pumped-Storage Stations

Since in the pumping mode the hydroelectric unit rotates in the reverse direction from the generating mode, it must be brought to rest and restarted when changing from generating to pumping and vice versa. This requires a set of reversing disconnecting switches in the generator leads to reverse the electrical phase rotation and some means of starting the unit as a pump. If the high-voltage transmission system can supply the starting kVA to the generator as a synchronous motor without undue depression of voltage on the system, starting can be accomplished from a low-voltage starting tap on the main transformer and by equipping the rotor poles with a damper winding that acts as a squirrel cage for starting as an induction motor, followed by synchronization as a synchronous motor by applying field excitation when the rotor is nearly up to synchronous speed; if the depression of transmission system voltage is not tolerable, the alternative is to provide a starting induction motor directly connected to the top of the generator and with a few less poles than the generator so that synchronizing can be accomplished as the unit comes up through synchronous speed. In the starting motor alternative a damper winding on the main rotor is not needed and is not provided unless desired for other functions.

Generator Excitation

In small hydro stations where system integrity is not unduly imperilled if the generators should lose synchronism during system disturbances, the exciters are direct connected, but self-excited rather than separately excited from direct-connected pilot exciters. The exciters are under the control of automatic voltage regulators. In larger stations and pumped-storage stations, the same separately excited exciter and compound-wound pilot exciter with automatic voltage regulator scheme is used as described in Section 14.3-2. Since these are not base-load-type stations, a motor-driven spare exciter is not usually considered justified, a spare exciter stator and rotor for several identical units being deemed adequate.

The rectifier-type excitation schemes described in Section 14.3-2 have not yet found application in large hydro and pumped-storage stations, partly because the far slower-speed direct-connected exciters and slip rings in these types of station require less scheduled and unscheduled maintenance than in the large 3600-rpm fossil-fueled and 1800-rpm nuclear-fueled units.

Generator Neutral Grounding

The same practices and considerations apply as given in the discussion of generator neutral grounding in Section 14.3-2.

Generator Leads

The same practices and considerations apply as given in the discussion of generator leads in Section 14.3-2.

Transformers

The same practices and considerations apply as given in the discussion of transformers in Section 14.3-2 with three important exceptions:

1. Since hydro and pumped-storage stations are by nature peaking rather than base-load stations, except during spring flood periods, it is economic for the main transformers to be fan cooled, or fan cooled with forced-oil circulation, rather than to be self-cooled, and this eliminates the option of two half-size units.
2. As long as there is enough remanent oil pressure in a governor oil tank in hydro and pumped-storage stations to open a turbine's scroll-case gates, the station is self-starting and therefore needs no "startup" transformer (or dead-system backup).
3. Station service power requirements are so small compared to those of a fossil- or nuclear-fueled station that the station service transformer ratings are a much lower percentage of the main transformer ratings than the 3–5% range mentioned for such stations.

A minor exception is that whereas main transformer high-voltage surge arresters afford adequate voltage surge protection to the invariably single-turn stator coils of large 3600- and 1800-rpm generators, they may not do so to the multiturn coils of hydro and pumped-storage generators because of the initial very nonuniform distribution of turn-to-turn voltage when a voltage surge enters a stator winding. This consideration occasionally prompts the application of surge arresters and front-of-wave sloping capacitors at the generator terminals, which respectively reduce the magnitude of the surge and reduce the steepness of its wave front with resultant more even sharing of the surge voltages among turns.

High-Voltage Switchyard

The same practices and considerations apply as given in the discussion of high-voltage switchyard in Section 14.3-2.

Station Connection Diagram

The foregoing discussion of hydro and pumped-storage generating stations, coupled with the circumstance that the generator ratings are not so large but that available circuit breaker current ratings can usually be applied for generator low-tension circuit breakers, leads to the typical station connection diagrams illustrated in Figs. 14.3-2a and 14.3-2b, respectively. Figure 14.3-2b shows a starting motor used in the pumping mode. With the exception of the starting motors, there are no large auxiliaries comparable to the boiler feed pumps, coal pulverizers, circulating water and condensate pumps or boiler forced and induced draft fans found in fossil-fueled generating stations. Therefore, the station auxiliary supply buses are at 480 V and below.

Protective Relaying

Because (1) salient pole rotors are much less susceptible to overheating by double-frequency-induced currents than cylindrical rotors, (2) the main transformer is not connected to the generator during startup, (3) there is no steam turbine to be overheated by inadequate steam flows, (4) there is less

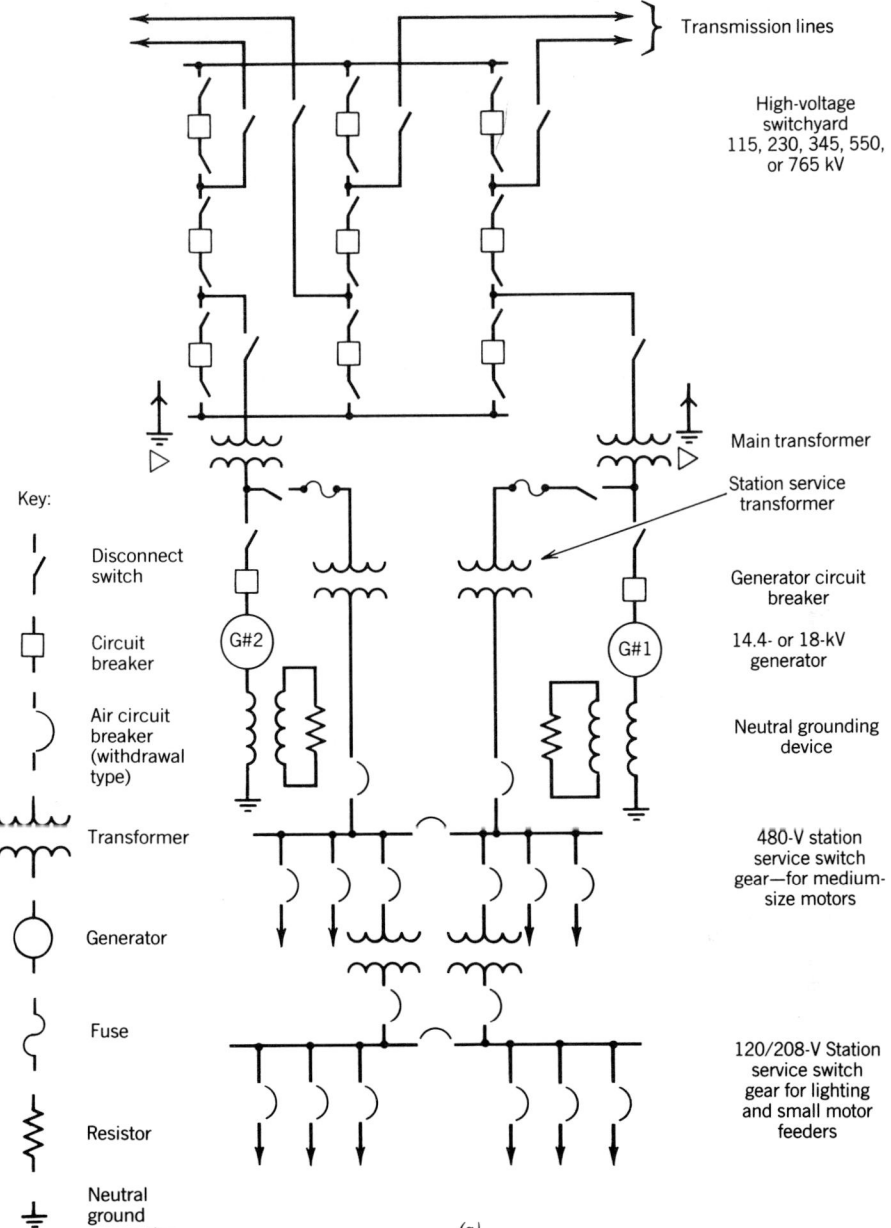

Key:

Disconnect switch

Circuit breaker

Air circuit breaker (withdrawal type)

Transformer

Generator

Fuse

Resistor

Neutral ground connection

Transmission lines

High-voltage switchyard 115, 230, 345, 550, or 765 kV

Main transformer

Station service transformer

Generator circuit breaker

14.4- or 18-kV generator

Neutral grounding device

480-V station service switch gear—for medium-size motors

120/208-V Station service switch gear for lighting and small motor feeders

G#2

G#1

(a)

Fig. 14.3-2a   Typical connection diagram of large hydrogenerating stations.

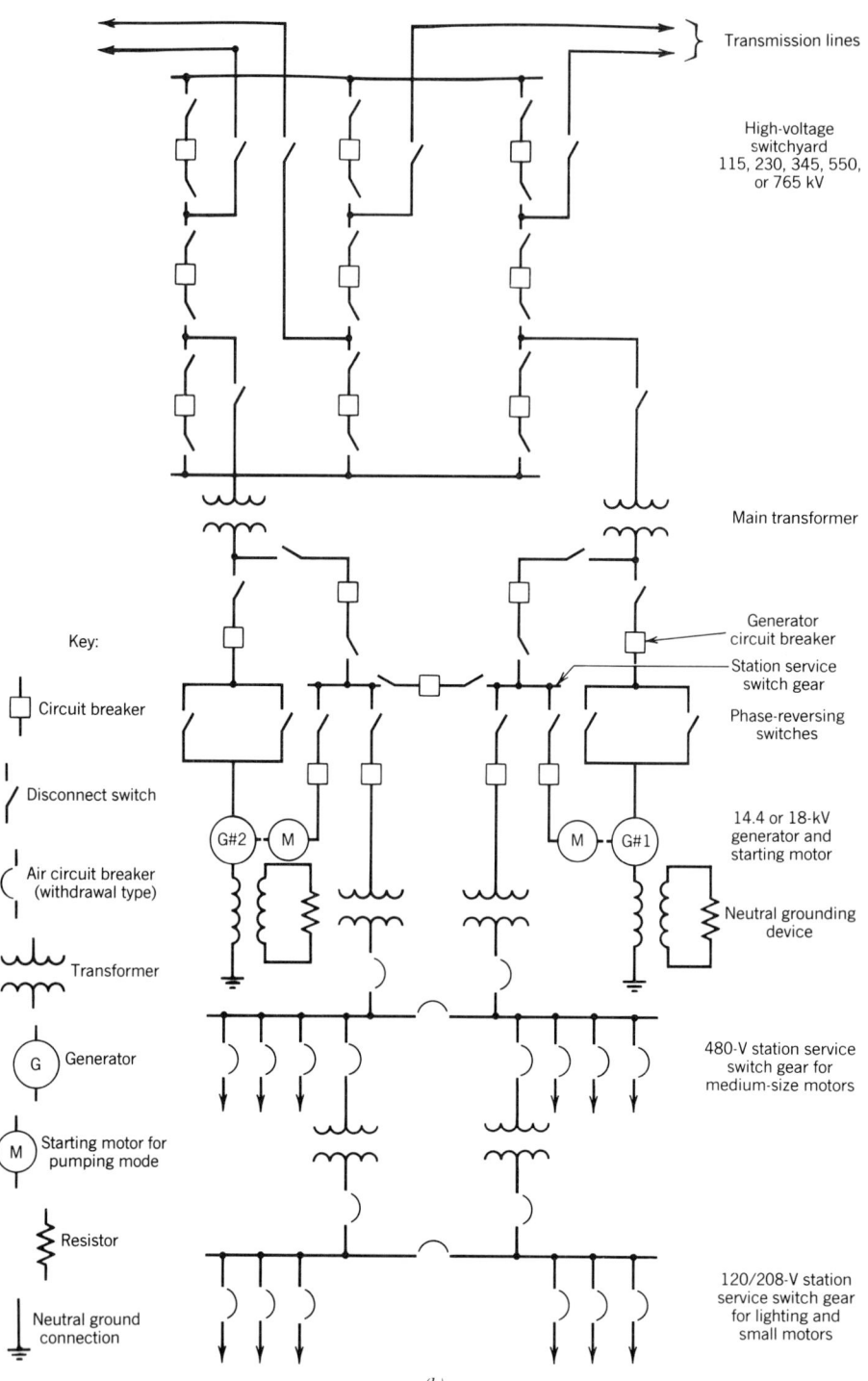

Transmission lines

High-voltage
switchyard
115, 230, 345, 550,
or 765 kV

Main transformer

Generator
circuit breaker

Station service
switch gear

Phase-reversing
switches

14.4 or 18-kV
generator and
starting motor

Neutral grounding
device

480-V station service
switch gear for
medium-size motors

120/208-V station
service switch gear
for lighting and
small motors

Key:

Circuit breaker

Disconnect switch

Air circuit breaker
(withdrawal type)

Transformer

G Generator

M Starting motor for
pumping mode

Resistor

Neutral ground
connection

G#2   M

M   G#1

(b)

Fig. 14.3-2b   Typical connection diagram of pumped storage generating stations.

stored energy per kilowatt of generator rating in the rotor elements to produce shaft or coupling fatigue during out-of-step oscillations, and (5) the station service transformers are so small, the array of relays protecting the main components within hydro and pumped-storage stations usually consists only of the following:

1. Generator overload relays.
2. Generator differential relays.
3. Generator ground relay (includes the whole zone operating at generator voltage).
4. Generator loss-of-field relay.
5. Main transformer overload relays.
6. Main transformer differential relays.
7. Main transformer sudden pressure relay.
8. Station service transformer overload relays.

The relay protection of the transmission lines leaving the high-voltage switchyard is treated in Section 14.3-5.

The relatively extreme simplicity of hydro and pumped-storage generating station equipment and operating procedures as compared to those of steam plants makes such stations especially amenable to automatic control or to remote control from a neighboring attended station or dispatching office. There has been a long and successful history of such operation of hydro plants, and it should prove equally successful for pumped-storage installations.

14.3-4   Gas Turbine and Diesel Engine Generating Stations

As indicated in Section 14.3-1, gas turbine and diesel engine generating stations are for system peak load purposes, but their flexibility of location permits their serving the added purpose of deferring added transmission or subtransmission to outlying growing loads. Thus, distribution feeders and generators are likely to be bussed together and the station connection diagrams (Fig. 14.3-3a and b) are correspondingly different from Figs. 14.3-1 and 14.3-3a and b. The flexibility of location arises from cooling water needs so small that they can be supplied from city water mains. Gas turbine ratings are typically 20 MW and below and 3600 rpm; diesel ratings are typically 3000 kW and below and 1200 rpm and below. The gas turbine generators are cooled with air through intake and exhaust ducts while the diesel generators are of open-type ventilated construction.

Generator Excitation

As in the case with the smaller fossil-fueled generators and the larger hydro generators, gas turbine generators are usually excited by direct-connected exciters separately excited by coupled compound-wound pilot exciters under the control of automatic voltage regulators; like small hydro generators diesel generators are excited by direct-connected self-excited exciters, also under regulator control—this control is more for local feeder reasons than for any beneficial effect on maintaining synchronism with the system during disturbances.

Generator Neutral Grounding

Since the distribution feeders fed from the generator bus are usually four-wire, the generator neutrals are solidly grounded, with the accepted risk of greater arc burning of armature iron in the event of coil failures in the slot section of the stator windings.

Generator Leads

Paper-insulated or rubberlike-insulated cables, in multiple if need be, can economically handle the relatively low values of current involved.

Transformers

The same practices and considerations apply as given in the discussion of transformer in Section 14.3-2 with the following exceptions:

1. The transformer capacity stepping up to transmission voltage may be substantially lower than the station aggregate generator rating by reason of the size of the local load, taking simultaneity of demand into account.

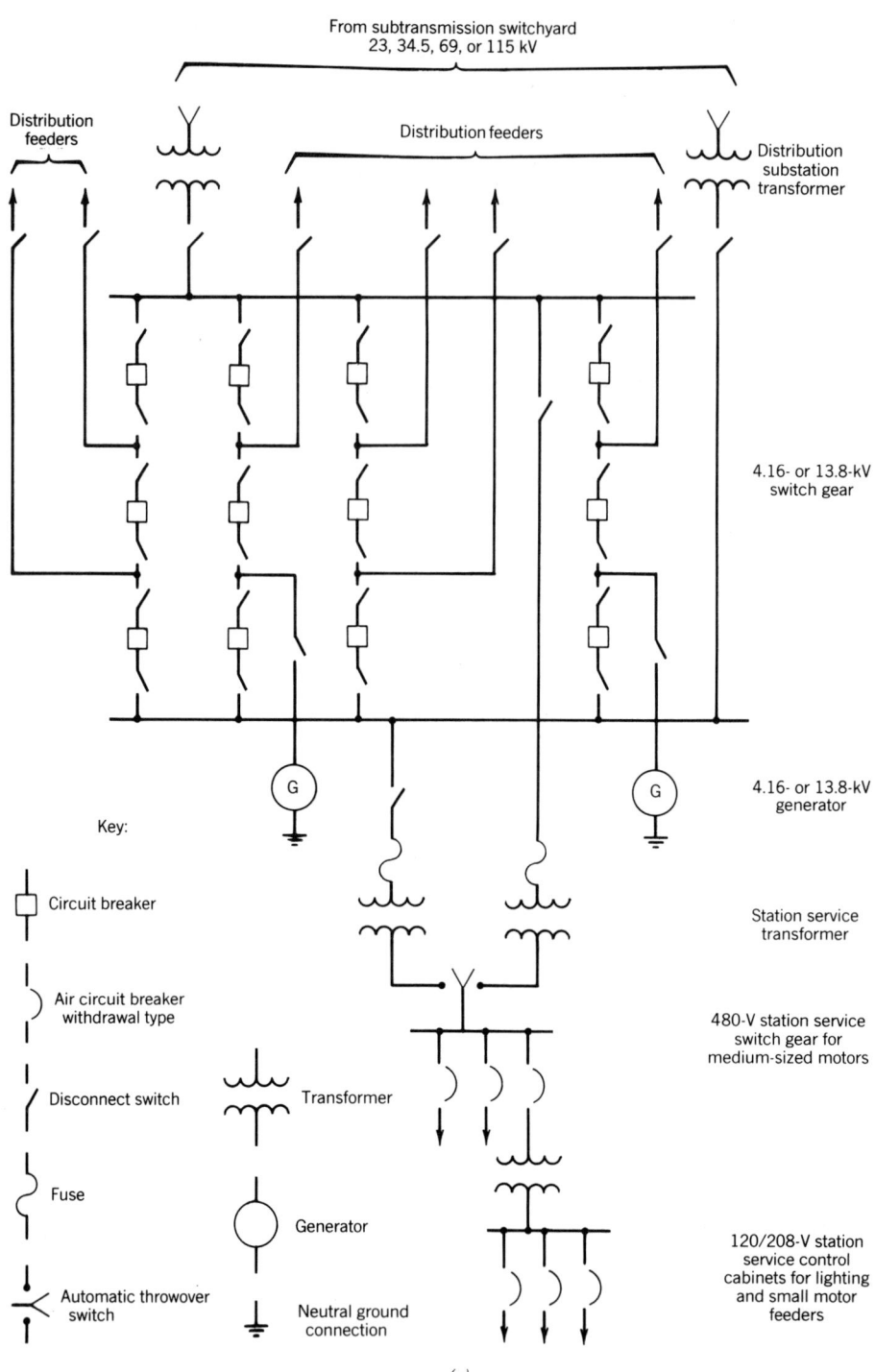

From subtransmission switchyard
23, 34.5, 69, or 115 kV

Distribution feeders

Distribution feeders

Distribution substation transformer

Distribution feeders

4.16- or 13.8-kV switch gear

4.16- or 13.8-kV generator

Station service transformer

480-V station service switch gear for medium-sized motors

120/208-V station service control cabinets for lighting and small motor feeders

Key:

Circuit breaker

Air circuit breaker withdrawal type

Disconnect switch   Transformer

Fuse

Generator

Automatic throwover switch   Neutral ground connection

(a)

Fig. 14.3-3*a*   Typical connection diagram of gas turbine generating stations.

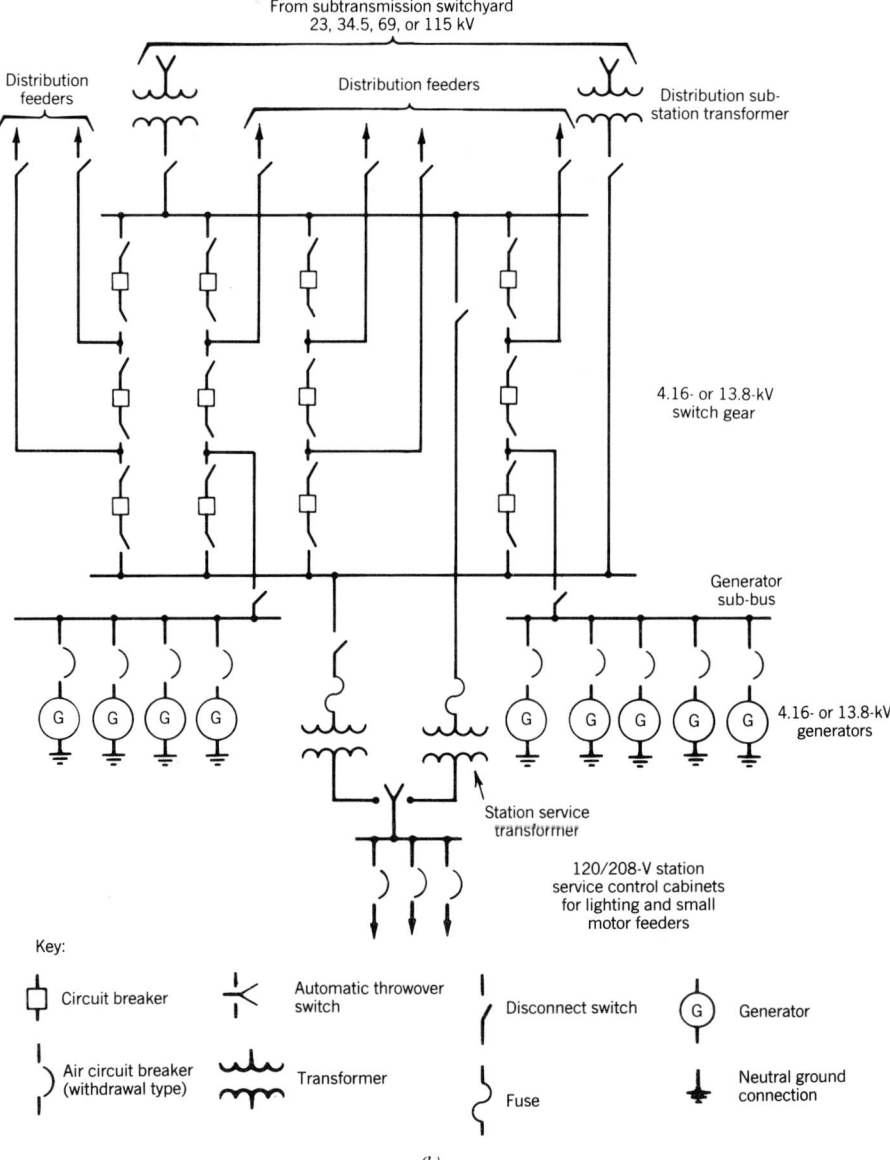

From subtransmission switchyard
23, 34.5, 69, or 115 kV

Distribution feeders

Distribution feeders

Distribution sub-
station transformer

4.16- or 13.8-kV
switch gear

Generator
sub-bus

4.16- or 13.8-kV
generators

Station service
transformer

120/208-V station
service control cabinets
for lighting and small
motor feeders

Key:

Circuit breaker

Automatic throwover
switch

Disconnect switch

Generator

Air circuit breaker
(withdrawal type)

Transformer

Fuse

Neutral ground
connection

(b)

Fig. 14.3-3b Typical connection diagram of diesel engine generating stations.

2. Since gas turbine and diesel stations are by nature peaking rather than base-load stations, it is economic for the stepup transformers to be fan cooled, or fan cooled with forced-oil circulation, rather than to be self-cooled; this eliminates the option of two half-sized units.

3. Since gas turbines are started by battery, and diesel engines by battery or from compressed air tanks, they are self-starting and therefore need no startup station service transformer (or dead station backup).

4. Station service power requirements of both gas turbine and diesel stations are so small compared to those of a fossil- or nuclear-fueled station that the station service transformer ratings are a much lower percentage of the stepup transformer ratings.

5. If one or more of the local feeders is of overhead construction, lightning surges can reach the generator terminals without benefit of a transformer buffer; accordingly, surge arresters and wave-sloping capacitors may be needed at each generator's terminals.

High-Voltage Switchyard

The same practices and considerations apply as given in the discussion of high-voltage switchyard in Section 14.3-2.

Protective Relaying

Because (1) diesel-driven salient pole rotors are much less susceptible to overheating by double-frequency-induced currents than cylindrical rotors, (2) the main transformer is not connected to the generator during startup, (3) there is no steam turbine to be overheated by inadequate steam flow, (4) shaft or coupling fatigue during out-of-step oscillations is an acceptable risk, and (5) the station service transformers are so small, the array of relays protecting the main components within gas turbine and diesel engine stations usually consist of only the following:

1. Generator overload relays.
2. Generator negative sequence relays (gas turbine generators only).
3. Generator differential relays (often omitted on diesel generators).
4. Generator ground relay.
5. Generator loss-of-field relay (gas turbine generators only).
6. Main transformer overload relay.
7. Main transformer low-tension ground relay.
8. Main transformer differential relays.
9. Main transformer sudden pressure relay.
10. Station service transformer overload relay (or fuses).
11. Station service transformer ground relay (if transformer not fused).
12. Distribution feeder overload relays.
13. Distribution feeder ground relay.

The relay protection of the transmission or subtransmission lines leaving the high-voltage switchyard is treated in Sections 14.3-5 and 14.3-6.

Fuel Cell Generating Stations

Under development and in trial operation to serve the same functions as gas turbine and diesel generation are 4000-kW units of fuel cells, chemically fired with oil or gas and locatable in distribution substations even in residential areas because they have no water cooling requirements, are nonpolluting, and are virtually noiseless in operation.

14.3-5 Transmission Systems

Successful transmission system design, in addition to satisfying the obvious requirements of adequate regulation of voltage and economic selection of wire conductivity, must take into account many other major considerations.

Whereas power transmitted by direct current is measured by the *numerical* difference between the sending end and receiving end voltages, namely, the voltage drop in the resistance of the transmission wires, the power transmitted by alternating current is measured by the *vector* difference between the sending and receiving end voltages produced by the voltage drop in the resistance and inductive reactance of the transmission wires. Since the reactance is several times the resistance, and the reactance drop is at right angles to the transmitted current, the reactance drop vector produces an angle between the sending and receiving end voltages (V_S and V_R), and therefore ac power transmission may be said to be transmitted by angle—the larger the angle, the larger the power. But there is a limit to this relationship because when the angle exceeds approximately 90°, the power transmitted starts to decrease rather than continue to increase. The relationship can be roughly stated by the equation

$$ \text{kW} = \frac{V_S V_R}{x} \sin \theta \tag{14.3-1} $$

where x is the system reactance between the sending and receiving ends and θ is the angle between the vectors whose lengths are V_S and V_R. The import of this equation is that transmitted power varies as the square of the system voltage, varies inversely as the system reactance, is zero when θ is 0° or 180°, and is a maximum for a given system when $\theta = 90°$. Operation at this maximum is not practical because any system disturbance, such as a short circuit reducing V_S or, V_R, or a line tripout increasing x, drives θ to above 90° into the area of decreased transmittable power, resulting in a loss of synchronism between sending end and receiving end generators, known as instability.

Further relationships are that line conductor losses at a given transmitted power vary inversely as the square of the system voltage, and voltage regulation as a percentage of system voltage varies inversely as the square of the system voltage. These would appear to indicate that doubling the voltage quadruples the transmittable power, but since higher voltages are usually associated with larger transmission distances, the overall result is that in actual systems the transmittable power (electric) varies between the first and second power (mathematical) of the system voltage. Also, the actual maximum continuous operating angle for purposes of avoiding instability is closer to $\theta = 30°$ rather than 90°. Typical transmission system megawatt loadings in the wide range of figures found in practice are on the order of

Related voltage	115 kV	230 kV	345 kV	550 kV	765 kV
MW loading	125	350	700	1500	3000

Transient Stability

It is vital to the successful operation of a power system that there be maintained what is called transient stability of the transmission system under all reasonable conditions and contingencies of operation. Some understanding of this phenomenon is essential to an appreciation of the various factors affecting it which are considered in the planning, design, construction, and operation of the system. Figure 14.3-4 shows a simplified idealized system consisting of a load area fed by a local generating station and by a remote generating station over two transmission lines. Suppose a three-phase fault occurs on either line near the high-voltage bus at either end. The remote generator suddenly becomes completely unloaded because no power can be transmitted past a three-phase fault, and the local generator assumes all the load shed by the remote generator. Until the fault is cleared by the tripping of the breakers at the ends of the faulted line, the remote generator accelerates at a rate restrained only by the inertia of the rotating element and likewise the local generator decelerates because initially neither speed governor recognizes any change in frequency to which to respond in a corrective manner; when they do, it takes time to reduce or increase the turbine input as the case may be. In the meantime the angle θ of Eq. (14.3-1) between the voltages V_S and V_R corresponding to the air gap fluxes in the two generators is rapidly increased toward or beyond the critical 90° value; furthermore, after the fault is cleared, the new steady value of θ is increased to compensate for the larger value of x with one line switched out. The size of the angular oscillations of θ from the prefault condition through the fault condition and then in the postfault condition, attempting to stabilize at the new angle of θ, determines whether or not synchronism will be lost. The total x for this simplified case is the sum of the two generator "transient reactances," the two transformer reactances, and transmission system reactance.

From this discussion and consideration of Eq. (14.3-1) it follows that the factors favoring maintenance of transient stability are:

1. Low generator, transformer, and transmission system reactances.
2. High generator and turbine rotor inertias.
3. Minimization of fault occurrences when controllable.

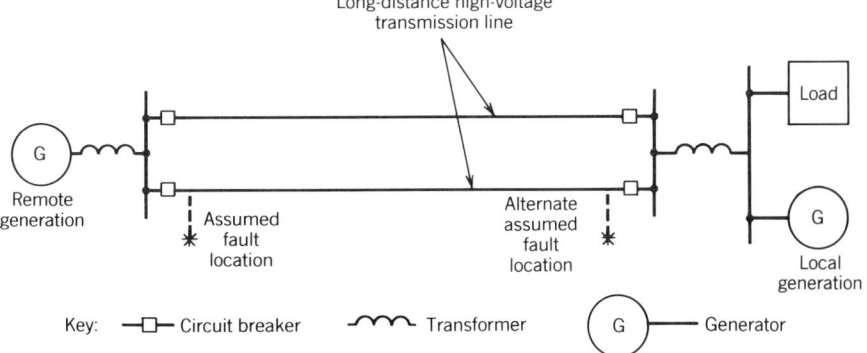

Fig. 14.3-4   Idealized and simplified two-station power system.

4. The highest possible speed of fault detection and interruption.

5. Automatic reclosing of faulted lines with no more delay than needed to prevent restriking of the arc of a temporary fault.

6. Reduction of turbine input to generators in remote stations earlier than their speed governors can respond to fault conditions.

7. Application of artificial braking load to generators in remote stations in the fault and postfault periods.

8. Rapid response of generator excitation to automatic voltage regulators (already referred to in Section 14.3-2).

Lowered Reactances

While economic generator and transformer design permits some flexibility in the choice of reactance, substantial reductions are achievable only by what amounts to an increase in kVA rating and this is not ordinarily resorted to.

Transmission line design offers many electives: multiple lines, obviously expensive when not needed for reliability reasons; multiple conductors ("bundled" conductors), marginally expensive and needed in any event at the highest system voltages for reduction of corona and radio noise as will be discussed in a later section; and reduced spacing between phases, which runs counter to minimizing corona and radio noise and to maintaining air clearances to grounded supporting structures for withstanding lightning and switching surge transient overvoltages.

Increased Rotor Inertias

Increased steam turbine generator rotor inertias, like reduced generator reactance, are achievable only by what amounts to an increase in kVA rating. Hydrogenerator rotors, which are inherently low in stored energy per kVA of generator rating, can however have their inertias increased at modest cost by deepening the rotor rim.

Minimization of Fault Occurrences

By far the most frequent potential cause of transmission line tripouts is atmospheric lightning. The frequency of lightning storms at any given location is indicated by the iso-keraunic level, the number of days in a year when an observer can see lightning or hear thunder. This level ranges geographically from 3 to 90, the median being around 30. At that level, lines of average height on a right of way are struck about 100 times per year per 100 mi of right of way. The stroke currents range from nearly zero (some strokes expire in midair before ever reaching a ground target) to about 250 kA. Low current strokes are the most numerous and the highest stroke currents the least so. Except in areas having the lowest iso-keraunic level most transmission lines are equipped with ground wires above the phase conductors to intercept strokes from terminating on the phase wires so far as practical and discharge them to ground through the resistances of the tower footings. Depending on the footing resistances this very substantially reduces the lightning voltage across the porcelain insulator strings as compared to its value if the phase conductor were struck; this is further reduced by about one-third by the capacitive coupling between the overhead ground wires and the phase conductor. By properly coordinating the tower footing resistances, the number of discs in the insulator strings, and the angle from the vertical between the overhead ground wires and the least shielded of the phase conductors, it is quite possible to design a line of fairly predictable lightning performance to meet any specified goal. A line performance of one tripout per 100 circuit miles per year is considered as "lightning proof." About 70% of insulator lightning flashovers are single-phase, 20% are two-phase, and the remaining 10% are three-phase. The three-phase lightning flashovers are by far the most severe from the transient stability standpoint because the smaller the number of phases involved, the more power can be transmitted past the fault during the fault condition. The frequency of strokes to lines varies directly with the height of the towers; therefore, it is advantageous to use a flat configuration of phase wires rather than a vertical or triangular one. When two circuits are on the same towers, most companies report that about 25% of the lightning tripouts involve both circuits. If this performance is not tolerable in the system design, experience shows that two single-circuit structures on the same right of way provide virtual immunity from simultaneous tripouts by lightning within 2 min of each other because ample time is afforded to reclose the first struck line before the second one experiences an outage. A partial solution is to use "differential insulation" on double-circuit towers whereby one circuit has at least 20% more insulation than the other and lightning flashovers of up to three insulator strings will be confined to one circuit.

The other type of transient overvoltages frequently appearing on transmission systems is known as switching surges. These arise (1) when a line is energized, (2) when a line is immediately reenergized after deenergization and trapped charges have not had time to dissipate, and (3) at fault clearing.

Overvoltages from 1 or 2 may be particularly severe when a transformer bank is integral with the line section. It is not tolerable to permit flashovers and tripouts from these causes. Up through 230 kV the insulation provided for lightning protection is adequate to withstand swtiching surges. But above this voltage more insulation may be needed to withstand switching surges, whose magnitude increases linearly with operating voltage, than to withstand lightning surges whose magnitude is independent of operating voltages. As a part of system design, switching surge magnitudes must be investigated in advance on analog computers and any needed reduction devised by incorporating suitable "opening" and "closing" resistors in the circuit breakers that will be performing the switching operations. Economically achievable per unit (pu) switching surge values at various rated system voltages follow:

System Rated kV	Maximum Switching Surge in (pu) of Crest of Maximum Phase-to-Ground Operating Voltage
345	1.80–2.40
550	1.50–2.20
765	1.50–2.20

Further advances are afforded by breakers whose closing stroke is so accurately timed at initiation that each contact will make at close to the instant of zero voltage across it when energizing a line section with or without trapped charge.

High-Speed Fault Detection and Interruption

Until the mid-1920s most protective relay minimum-fault detection times were about 10 cycles (of 60 Hz), and circuit breaker interrupting times were 15 cycles, making the total fault-clearing time of a two-terminal line 25 cycles if both ends tripped simultaneously and 50 cycles if they tripped sequentially. High-speed balanced protection of duplicate parallel lines operated in 2 rather 10 cycles, but the total time to clear a fault near either end of a line was still 34 cycles. Substantial reduction in these times has been the *greatest single factor* in increasing the amount of power it is possible to transmit within transient stability limits of a given system. Protective relaying has been improved to provide simultaneous trip signals at both ends of the line within 2 cycles of inception of a fault at any point on the line; breaker interrupting times have been successively improved to 12 cycles, then to 8 cycles, then to 5, and currently to 3 cycles, so that total fault clearing time is 5 cycles. The protective relay system to accomplish this, known as directional comparison, measures the distance to the fault from each end of the protected lines by a measurement of impedance, reactance, or impedance biased by power factor to render less sensitivity to legitimate power flows than to fault kVA; then by means of pilot wire, carrier current, or microwave communication between the two ends makes an instant determination as to whether the fault is internal or external to the protected line. The system also provides backup protection against failure of the proper breakers to clear an external fault. Further, it recognizes an out-of-step condition and thus can prevent automatic reclosing and can be used to provide out-of-step tripping at the points desired to split up a system that has lost synchronism. Another relay protective system known as phase comparison compares by carrier current or micro-wave the instantaneous polarities of the currents at the two terminals of each phase of a protected line; if the polarities are opposite, an internal fault is indicated and tripping of both terminals is simultaneously initiated. This system cannot sense an external fault or an out-of-step condition, and therefore backup protection and any needed out-of-step sensing is provided by supplementary relays.

Automatic Reclosing

A by-product of the simultaneous tripping afforded by the directional comparison and phase comparison relay schemes is that immediate automatic reclosure can be initiated at both ends without waiting for further assurance that the far-end breaker has tripped. The sooner that both ends can be reclosed, the sooner the x in Eq. (14.3-1) can be restored to its prefault value as an aid to transient stability. Circuit breaker closing mechanisms are fast enough in operation to provide remaking of the contacts within 20 cycles of initiation of tripping, thus affording 15 cycles or more of "dead" time. Up through 230 kV this will permit the arc gases to cool enough so that they will be nonconducting on the restoration of voltage, and the fault arc will not restrike. At 345 kV this time may need to be intentionally stretched by 5 cycles, and progressively more at higher sytem voltages.

Fast Reduction of Turbine Input to Accelerating Generators

Transient stability on transmission systems fault occurrence can be marginally improved by having the fault-detecting relays anticipate the action of the steam turbine speed governors by temporarily tripping the throttle to the high-pressure turbine and closing the intercept valves to the lower-pressure turbines and then reopening them again in a second or two when the angular swings between the accelerating and decelerating generators are subsiding. These measures are known as "fast valving." Analogous measures cannot be taken at hydro stations without creating excessive openstock pressure.

Braking Resistors

The converse of reducing the input to an accelerating generator is to replace its lost output by an artificial load for the purpose of transient stability improvement. This is done by having the fault-detecting relays initiate the energizing, from the generator terminals, of a braking resistor by a fast closing switch for a sufficient period to stabilize the angular swings. This device is applicable at hydro stations as well as steam stations.

    Another device applicable only at hydro stations is a high-resistance damper winding on the rotor poles. Under normal conditions there is no loss in the damper winding, but during system unbalanced faults, either single-phase to ground, single-phase to phase, or two-phase to ground, the negative-sequence currents generated by the fault induce double-frequency losses in the damper winding, imposing a braking effect on the rotor. This effect lasts only for the duration of the fault and unfortunately does not appear for a three-phase fault when its beneficial effect is most needed; hence the scheme is overall less effective than the braking resistor alternative.

Transient Stability Studies

It is utterly impractical to attempt to make planning and design transient stability studies by slide rule computation for the networked transmission systems of today and taking into account all the alternatives and options discussed in the preceding paragraphs. By spotlighting the particular area under study and using permissible reductions and equivalents for the surrounding areas, it is possible to make the calculations on digital computers out to several postfault angular swings, taking into account voltage regulator action and governor action, as well as all the other aids to stability described, with acceptable accuracy for making decisions.

Single-Pole Switching

It has long been common in Europe to employ single-pole switching for clearing temporary single phase to ground faults with automatic reclosure for restoring the faulted phase. This has not found favor in North America because its aid to transient stability is afforded only for the type of fault for which the aid is least needed. However, some applications have now been made in North America for a different purpose, namely to save a duplicating line for connecting a generating unit reliably into a transmission system when a single line is short enough so that the expectancy of two-phase and three-phase faults on it is a low enough fraction of expected forced outages in the station itself to justify the omission of the second line.

Limits to Transmission Voltage

The upper limit to open-wire transmission voltage lies in the behavior of air as an insulator. When the voltage gradient at the surface of a very smooth cylindrical conductor reaches 70 kV crest per in. or 20 kV rms/cm, at "standard" conditions of 25°C, 760 mm of Hg, and 15.4 mm Hg vapor pressure, the air at the surface ionizes with a glow, called corona. On the stranded conductors characteristic of high-voltage transmission the corona extinction voltage will be less than 20 kV/cm due to the "roughness factor" and still lower due to bits of foreign matter—airborne particles of vegetable matter, for example—adhering to the conductor surface. In fair weather these bits of contamination will emit a corona plume perhaps several inches long, and even though there are only one or two of these per span, they may result in radio noise complaints because the frequency characteristic of the plume peaks at 1 MHz, coinciding with the center of the AM radio broadcast band. In any kind of precipitation that causes water drops to accumulate under and drop from the conductor, such as rain, snow, sleet, or dense fog, the corona plumes become practically continuous along the conductor, greatly increasing the corona energy losses and giving rise to audible noise in the low harmonic frequency range of the system frequency. The economic importance of the corona loss depends on how many hours of precipitation there are in a year and likewise the duration of the audible noise annoyance, but the radio noise problem would exist at all times, especially in areas remote from powerful broadcasting stations.

    The result is that the practical upper limit of surface gradient is in the range of 14–18 kV rms/cm. At normal phase spacings this is achievable at 230 kV, with a single conductor per phase of 1.1 in. in diameter and at 345 kV with 1.8 in. in diameter (although a two-conductor bundle per phase at 345 kV is more often used), but at 550 and 765 kV three- and four-conductor bundles are, respectively, the minimum to be used to avoid excessively large conductors. While these usually satisfy the radio noise and corona loss requirements, they may not satisfy the audible noise restrictions at particular locations and even as large as a 12-conductor bundle may be necessary for short distances at 765 kV. Increasing the phase spacing has much poorer economic leverage than increasing the conductor diameter, or bundling, for controlling corona and its effects.

Another factor meriting consideration, as transmission voltages and the magnitudes of the electrostatic field under the lines increase, is the comfort and health of the population passing under, or following pursuits under, the lines. Electrostatic charges induced on metal fences paralleling or angling across the line route are safely drained by grounding the fences at adequately close intervals.

Underground Transmission Cables

In open or moderately congested country, transmission lines are not placed in underground cables because (1) the cost per circuit mile is 12–15 times as great as for overhead circuits of equal rating, (2) at 345 kV and above the capacitive charging current is so large as to consume all the thermal capability of the conductor unless neutralizing shunt reactors are installed every few miles, (3) cable faults, although far fewer per mile than overhead line faults, are always permanent in nature until repaired rather than mostly temporary, and since restoration may take a week or more rather than a fraction of a second, a larger number of connectable conductors, if not circuits, may be required for equivalent system reliability. When congestion is so great as to make the use of overhead construction impractical, the lines are placed underground. A great many types of construction are in use: among the insulations are oil-impregnated paper tape, paper tape under oil pressure, and extruded polyethylene; among the coverings are lead sheath in fiber-lined concrete duct envelopes, armored lead sheath for direct burial, and oil-filled steel pipes for direct burial; and when additional cooling beyond that afforded by earth conduction is needed, forced cooling methods include circulation of water in external water pipes either once-through or recirculated through coolers.

Cable advances have kept pace as needed with the rise in transmission voltages: 550-kV cables are in operation and ratings up to 1100 kV are under development.

Direct-Current Transmission

At the same current and crest voltage to ground a two-wire dc transmission circuit will carry 94% as much power as a three-wire three-phase ac circuit and with two-thirds the resistance losses. The installed cost per circuit mile is obviously less, and if the line is long enough so that the total savings in costs and line losses counterbalance the costs and losses of the conversion and inversion equipment at both terminals, dc transmission becomes economical. The break-even point usually lies somewhere in the range of 500–800 mi for overhead transmission. Among other characteristics of dc transmission that can be used to advantage are:

1. It affords a controllable asynchronous tie between large ac systems of the same nominal frequency or entirely different frequencies that might otherwise require a much stronger tie to perform stably if alternating current.

2. Absence of capacitance charging current in dc cables means they can be indefinitely long without need for neutralizing shunt reactors.

3. Generating stations can be added via dc transmission without appreciably increasing the short circuit interrupting duty on existing ac circuit breakers.

4. The instant power controllability of dc links paralleling ac systems provides a larger aid to transient stability than would ac additions of equivalent capacity.

From the 1950s one or more of these characteristics have justified many dc transmission installations, for example:

1. Sweden–Gottland submarine link.

2. French–British cross-channel submarine link.

3. Sardinia–Corsica and New Zealand North–South Islands submarine links.

4. Northern Oregon to southern California overland link.

5. Japan's 60- and 50-Hz halves and Quebec–New Brunswick "zero length" links.

6. Kingsnorth–London, England, underground cable link.

The earlier installations used mercury arc rectifiers in the terminal equipment until solid-state diodes and thyristors superseded them in the 1970s. Harmonic currents generated by the converter-inverters are absorbed by arrays of shunt filters at the terminal stations. Rated line voltages are up to 800 kV (± 400 kV) overhead and up to 550 kV (± 275 kV) submarine or underground. Emergency operation at half-voltage and half-power with one wire in service and ground return is made possible with very extensive neutral grounding arrangements at the line terminals.

Development work is under way on dc breakers to extend the practicality of dc transmission from point-to-point to networking or multiterminal applications.

14.3-6   Subtransmission Systems

Figure 14.3-5 illustrates subtransmission systems by showing a variety of the switching connections that might be found in a typical subtransmission loop. The transmission supply to the loop will vary in its high-voltage switchyard connections depending on the relationship of the subtransmission substation to the other functions of the transmission network and will range from arrangement A to arrangement C. Since, after the transformers, the circuit breakers are the most expensive items in the substation, an attempt is always made to minimize their number to the extent consistent with service reliability requirements and satisfactory application of protective relays. The same statement applies to the switching at the distribution substations fed by the subtransmission loop.

Station I in Fig. 14.3-5 is a large distribution substation double-tapped from two subtransmission circuits on separate structures and therefore unlikely to have simultaneous outages. Station II is a large distribution substation looped into a double-circuit transmission line, which can have a double-circuit fault at any one point in the line without interrupting service to the substation but which requires an extra high-voltage breaker to give equivalent transformer protection. Station III is a similar looped station at its first stage of development with one transformer bank. Station IV is a small distribution substation with automatic transfer from one supply circuit to the other if the preferred circuit trips and remains deenergized for more than a preset number of seconds or minutes. Station V, shown solid for an initial installation and dotted for a later addition, is provided with automatic line sectionalizing if the line between stations VI and VII trips; the two line airbreaks at station V then immediately open, and when one or the other line section comes alive, the corresponding airbreak recloses, reenergizing station V; when the other line section comes alive, its line airbreak closes in phase. One such automatic sectionalizing station, but only one, can be inserted in any of the circuit-breaker-protected line sections between the subtransmission station and stations II, III, and VII. The switching shown at the large distribution substations VI and VII needs only the comments that (1) the high-voltage bus normally closed sectionalizing disconnect switches are provided to facilitate bus maintenance and (2) the transformer high-voltage circuit breakers can be replaced by airbreaks if it is acceptable on trouble in either transformer that the station be deenergized by the three incoming line breakers, the corresponding transformer's high-voltage airbreak and low-voltage bus breakers be automatically opened, and then the three incoming line breakers be automatically reclosed restoring service to the station.

Transformers

Transformers at the subtransmission substation and all the distribution substations are always three-phase units. They operate at the load factor of the load rather than at base load and hence are sized on the basis of fan-cooled, or even sometimes with added forced-oil-cooled, rather than on the basis of self-cooled ratings, in line with the discussions in Sections 14.3-2 to 14.3-4. The availability of mobile transformers of up to 10,000 kVA rating that can be highway transported and connected in 24 h, and the emergency overload capabilities of transformers with calculated percentage loss of life, make it possible (1) to provide single-transformer stations up to 10,000 kVA load with mobile backup without interruption for scheduled maintenance and with backup within 24 h after transformer failure and (2) to relieve multitransformer stations of 10,000 kVA of load during transformer scheduled maintenance and defer transformer additions by taking advantage of mobile transformer load relief with recognized loss of life of the remaining installed transformers when overloaded over one daily peak after loss of one unit.

High-Voltage Switchyard

The same practices and considerations apply as in the discussion of high-voltage switchyard in Section 14.3-2, with the addition of arrangements B and C in Fig. 14.3-5.

Relay Protection

For the incoming and outgoing transmission lines the high-speed relaying systems with simultaneous tripping at both terminals of each line are used as described in Section 14.3-5. For arrangement A of Fig. 14.3-5 each high-voltage bus is protected with differential relays, which may be either separate or combined with the corresponding transformer and low-voltage bus differential relays. The transformers in the subtransmission substation are protected with differential relays as just mentioned, inverse time overload relays, ground relays if applicable, and sudden pressure relays. In arrangement C of Fig. 14.3-5 the remote terminal protective relays are not sensitive enough to respond to a fault between a few turns in the high- or low-voltage windings, so that differential or sudden pressure relays that detect the fault are arranged to close a high-speed single-pole grounding switch at a transformer high-voltage terminal or to open the transformer high-voltage airbreak switch, which will either clear the fault or suffer intermingling of its opening arcs; the remote terminal relays will thus immediately respond either to the ground fault imposed by the grounding switch or to the single-phase-to-phase fault generated by the alternative "sacrificial" airbreak switch.

Faults on the subtransmission system are buffered as threats to the transient stability of the transmission system by the reactance of the subtransmission substation transformers. Therefore, the subtransmission line protective relays do not need to have the high-speed simultaneous tripping characteristics of transmission system line relays. Accordingly, at the subtransmission substation they consist of inverse-time overload and ground relays, with "instantaneous attachments" for operation at fault currents so large as to be sure not to be flowing to faults beyond stations I, II, and III of Fig. 14.3-5. Also, balanced current relays would be applicable on the lines to station VI but probably not on the lines to station VII because of the inequality of the loads on stations II and III. At stations such as I, in addition to inverse overload and ground relays, there would be balanced power relays because power tending to flow into one transformer and out the other would indicate a fault on the line toward which power was circulating. At stations II, III, VI, and VII there would be directional inverse overload and ground relays on the lines; in addition, at Station VI there would also be balanced power relays on the two lines from station I. This array of relaying provides fault clearing in total times ranging from 0.2 to 3 s, in contrast to the 5 cycles common on transmission systems.

Transformer relay protection at the distribution substations varies with the size of the transformers. All will have overload and ground relay protection, and ratings of 5000 kVA and above are likely to have differential protection. Smaller ratings may have sudden pressure relays or a variation known as a "case ground relay," whereby the tank is insulated from ground except through the low-impedance primary of a current transformer whose secondary current operates a relay if a ground fault develops in the primary winding of the protected transformer. The smallest transformers may be protectable by fuses rather than circuit breakers and protective relays. Whether or not the subtransmission substation and the distribution substations on the subtransmission loop are attended, automatic reclosing on all the subtransmission lines after tripouts is routine, in times ranging from 15 s to 1 min depending on the system layout.

14.3-7 Distribution Systems

The distribution feeders spreading from distribution substations are typically radial, with backup to feeder branches for scheduled or unscheduled maintenance provided by normally open connections to branches of neighboring feeders either from the same station or neighboring stations. The elements of a distribution feeder, including its secondaries, are shown in Fig. 14.3-6. It will be noted that wherever possible fuses are used to isolate faults; starting with customer branch circuit fuses G, they work back toward the source through customer entrance fuses F, distribution transformer primary fuses E, branch feeder fuses D, and main feeder sectionalizing fuses C, of higher rating at each stage in order to have time selectivity over the next fuse away from the source and to resist softening and deterioration from fault currents flowing to faults beyond the next downstream fuses. For these purposes each fuse needs to have twice the rating of the next fuse downstream from it. Fuses C are frequently replaced by an automatic sectionalizer when repeated doubling of rating results in too large a fuse rating, and when the advantage of automatic reclosing is needed. At A the protective device must be a circuit breaker to secure accurate enough protective relay selection with the transformer breakers; the kind of time delay relay curve provided permits "dead-load pickup" after a prolonged feeder outage, when all customer automatic appliances are likely to be in the on position.

Each feeder is equipped with an automatic voltage regulator at the station to compensate for deviations from constant voltage on the subtransmission system and to compensate for the voltage drop along the length of the feeder. Each regulator is provided with bypass switching for maintenance purposes. If the feeders are sufficiently alike in length and in magnitude and simultaneity of loading, it is sometimes possible to provide adequate voltage regulation by load tap changing on the station transformers.

Power Factor Correction

The system load consists not only of power but also of the reactive kVA to supply the magnetizing requirements of induction motors and transformers. Power can be produced only in generating stations, but reactive kVAr's can be produced anywhere on the system by synchronous condensers or static capacitors. The most economical arrangement is to locate static capacitors on the distribution feeders where they are closest to the consumers of reactive demand. This minimizes losses in generators, the transmission and subtransmission systems, in all transformers through those at distribution substations, and in the distribution feeders themselves; it also reduces the kVA ratings of generators and transformers, and improves voltage regulation by minimizing the voltage drop due to reactive flow in the whole supply system. When installed on distribution feeders, enough capacitors are

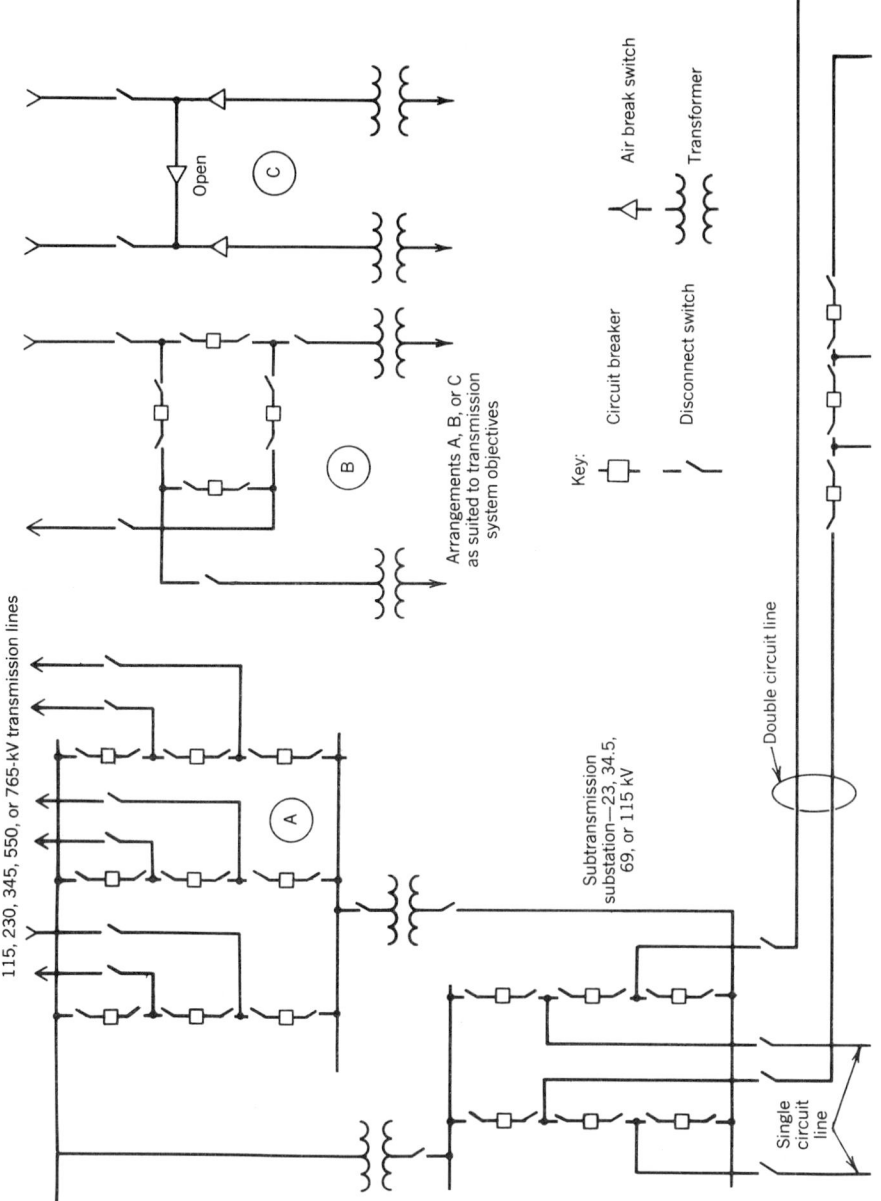

115, 230, 345, 550, or 765-kV transmission lines

Arrangements A, B, or C as suited to transmission system objectives

Open

(A)

(B)

(C)

Key:

Circuit breaker

Disconnect switch

Air break switch

Transformer

Subtransmission substation—23, 34.5, 69, or 115 kV

Double circuit line

Single circuit line

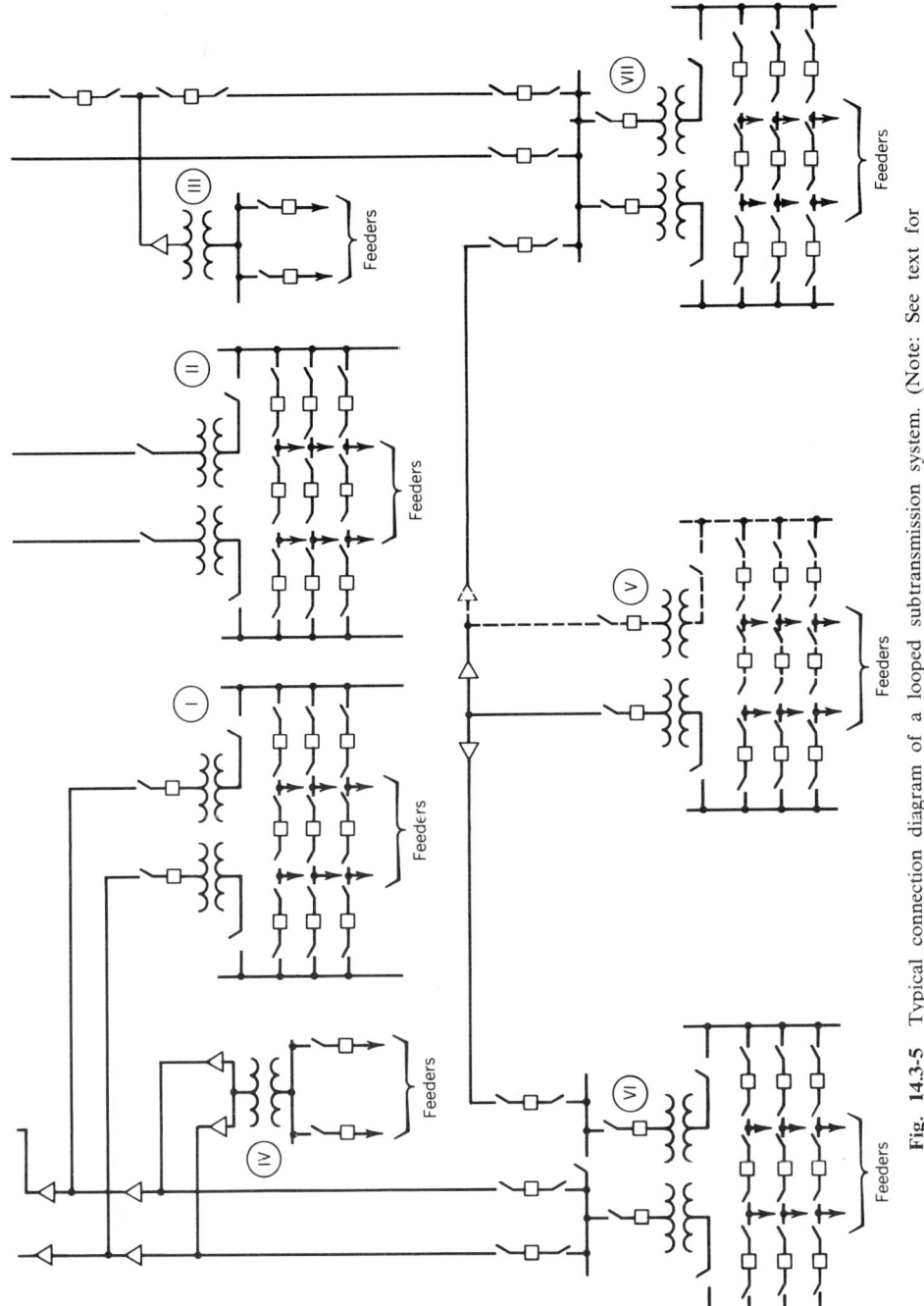

Fig. 14.3-5 Typical connection diagram of a looped subtransmission system. (Note: See text for discussion of switching arrangements at distribution substations I to VII.)

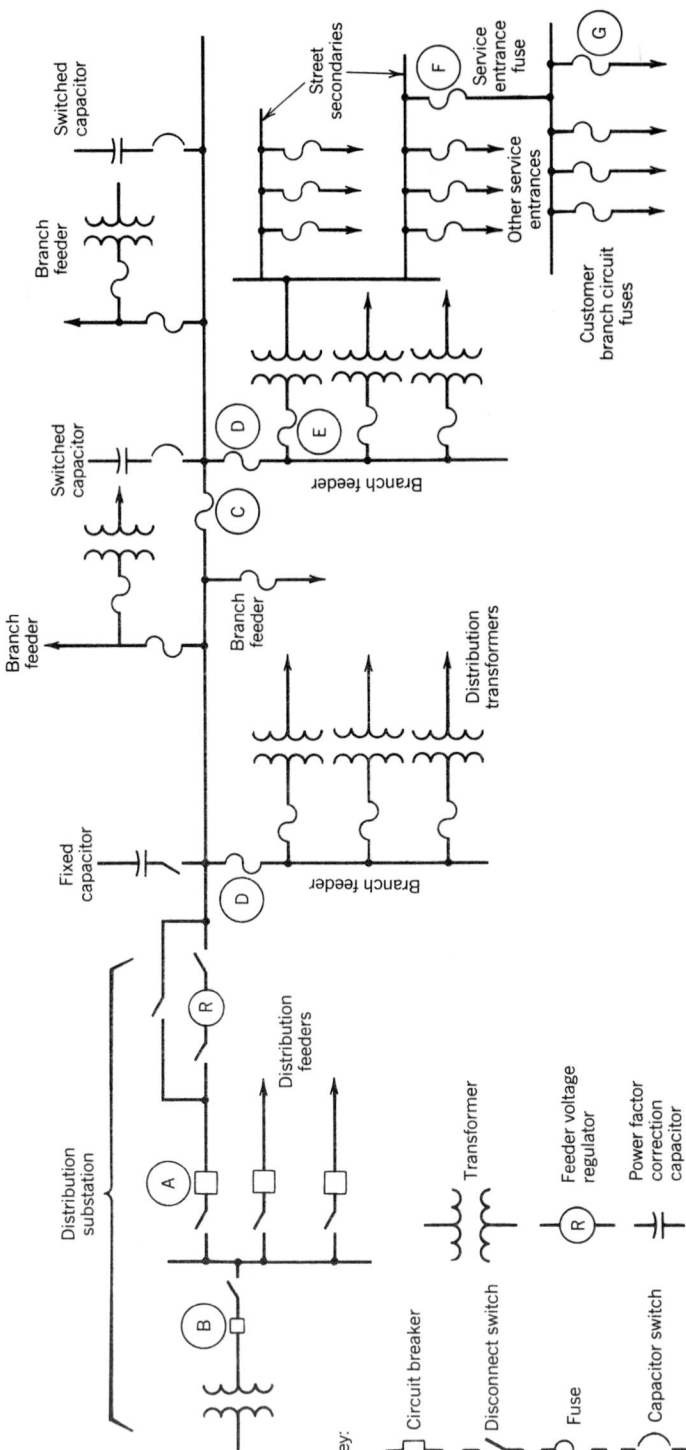

Fig. 14.3-6 Typical connection diagram of a radial distribution feeder.

1456

connected unswitched to neutralize the reactive demand at minimum daily load, and the remainder are automatically switched on and off under the control of current, voltage, or power factor sensors to neutralize the reactive demand of the bulk of the load that lies between minimum and peak.

Low-Voltage Networks

In the downtown areas of large cities an entirely different distribution system is normally used. In any area of at least 1 mi^2 where the load density is at least 10,000 kW/mi^2, a 120/208-V three-phase secondary network becomes economical. Such a network operates on the empirically sound principle that a cable fault at that voltage will burn itself clear if the available fault current is at least 5000 A. The network takes the physical form that the 120/208-V mains underlying each street are solidly tied together at each street corner and are supplied by transformers scattered throughout the area, which are in turn supplied by at least three distribution feeders more or less uniformly interleaved among the transformers. Each transformer is connected to a network junction through a "network protector" whose only functions are to (1) open on reverse flow from the network to a transformer distribution feeder fault, (2) close when the transformer voltage slightly exceeds the network voltage and is slightly ahead of it in phase position, and (3) back up the network protector switch with fuses if the switch should fail to open on a transformer or distribution feeder fault. Each transformer and its associated network protector are installed in a vault under the street or sidewalk or in an adjacent building. In high-rise buildings there may be vaults at several floor elevations. Transformers are filled with noninflammable insulating fluid rather than transil oil. The number of feeders installed is such as to provide firm supply and transformer capacity, and adequate voltage to all loads, with any one feeder out of service.

14.3-8 Service Reliability

The major factors in providing a reliable generation supply are accuracy of load forecasting, proper scheduling of maintenance of generating units, forehanded construction of new generating units to supply the forecast load and replace units due for retirement, and adequate contracts for firm power backup from interconnected neighbors. With 5 years and more elapsing between authorization of and commissioning for commercial operation of base-load units, long-range forecasting accuracy is difficult to achieve but nevertheless essential as a goal. Scheduled maintenance of generating units is of course done off peak, and for hydro units also avoids the spring runoff period to take maximum advantage of the available stream flow. Careful recognitition of all these factors makes possible the expectation of each operating entity or power pool to fail to meet peak power demands not more than once in any 10-year period, with the largest generating unit in each cognizant area unavailable because of unscheduled maintenance. With this expectation, residential customers are virtually immune from the effects of inadequate generating capacity, and a short-fall would weigh principally with those industrial customers with contract provisions for interruptible power.

From the discussions of relay protection in Sections 14.2-3 to 14.2-7 it may be inferred that the protective relaying of station and substation equipment is mainly to prevent and localize damage to equipment, that the protective relaying of subtransmission lines is for fairly prompt isolation of faulty lines with no more tripping of circuit terminals than necessary to amputate the faulty circuit segment, and that the relay and fuse protection of radial distribution circuits is mainly for restricting the circuit outage area to the smallest possible number of customers. However, the overriding purpose of transmission line protective relaying is to remove faults quickly enough to minimize the chances of transient stability limits being exceeded, with consequent widespread cascading of circuit tripouts resulting in total system collapse in large areas. This must be particularly avoided in large cities where the result may be elevators stalled between floors in highrise buildings and subway trains may be stalled between stations. Extraordinary measures are sometimes used to guard against such events, e.g., providing two complete sets of relaying at each critical transmission circuit terminal, fed from independent sets of instrument transformers and tripping storage batteries, and even using different principles of operation so that a design defect would be unlikely to affect both sets of relays in the same incident. If system collapse does occur, leaving "islands" where the operating generating capacity is inadequate to serve the load, the last resort is to shed load by operation of underfrequency relays. This is usually done in three steps of underfrequency, aggregating a prospective load shedding of up to 25%. In the United States voluntary regional reliability councils act in an advisory capacity on transmission reliability questions in their respective areas; they ponder such matters as the proper criterion of transient stability—should stability be maintained for a two-phase-to-ground fault without benefit of automatic high-speed reclosing or should one go all the way to its being maintained for a three-phase fault with automatic high-speed reclosing into a permanent fault and with the most critical other line on the system being simultaneously out of service? In the latter situation the further question arises, is the risk less to reclose into a permanent fault at a possible unfavorable instant in the system angular swings or to wait 5 s to close in phase when the swings have had time to subside?

The overall result in the reliability of service to the average residential customer on a radial distribution feeder is that he is exposed mainly to troubles on his particular distribution feeder; by and large he is unaffected by generation shortages, or by transmission system tripouts unless he is in an area whose load is automatically shed, or by subtransmission system tripouts unless he is fed by a distribution substation with automatic transfer or automatic sectionalizing as described for stations IV and V in Fig. 14.3-5. In the United States such a customer experiences, on the average, one outage per year averaging 1 h duration, not counting those outages of 1 min or less, which cause little inconvenience.

Hospitals normally counteract outages to operating rooms and other vital services by automatic throwover to automatically starting on-site emergency generators.

14.3-9 Generation Scheduling

A system or power pool load dispatching office each day makes up its schedule of the next day's generation with a knowledge of the forecast load curve for the 24 h, the list of available generating units and transmission lines, the current market price of interchange power via interconnections, and the "price tag" of each generating unit to which it has access, in terms of fixed capital charges on investment ("intercept cost"), operating cost per kilowatt ("increment cost"), and cost of each startup operation. From this data it fills in the 24-h load duration curve (kilowatts vs. hours or kilowatt-hours per hour) with base-load units on the bottom, hydro units next above, then the less efficient "old steam," then pumped storage, and finally gas turbine and diesel generators to fill in the peak if needed.

14.3-10 Rate Schedules

Rates to the various classes and sizes of customers are based as far as possible on the respective costs of service. Among other business costs, these costs consist of two main components, the capital charges on investment in equipment, which are independent of the amount of energy used, and the operating costs, which are proportional to the amount of energy used. In billing, these major costs are reflected in what are commonly known as the demand and energy rates, respectively. Demand metering equipment is very expensive compared to energy metering and, therefore, is installed only for large industrial and commercial consumers. Demand meters register the demand every 15 min, and the monthly demand billing is based on the average of one or more 15-min peak demands. Such a demand, once established, usually holds for up to the ensuing 12 months or until it is exceeded.

Residential rates are designed to approximate a demand-and-energy structure by charging more for the first few kilowatt-hours per month and then less and less for each succeeding block of energy until the final block rate approaches a figure geared principally to the energy cost unaffected by capital charges. In system peak demand months, such as in summer as occasioned by air conditioning or in winter as caused by lighting and heating loads, the block rates may be raised to reflect the higher cost of peak generation. As an energy conservation measure there has been some advocacy both inside and outside the power supply industry of an "upside-down" rate in which some of the later blocks are progressively listed at more and more per kilowatt-hour, rather than less and less, in order to discourage large residential use of energy.

BIBLIOGRAPHY

Cushing, Jr., E. W., Drechsler, G. E., Killgoar, W. P., Marshall, H. G., and Stewart, H. R. "Fast Valving as an Aid to Power System Transient Stability and Prompt Resynchronization and Rapid Reload after Full Load Rejection," IEEE Power Apparatus and Systems Paper # 71-TP-705-PWR.

"EHV Transmission Line Reference Book," Edison Electric Institute, New York, 1968.

Electra, CIGRE (Congresse Internationale des Grandes Reseaux Electriques), Boulevard Haussmann, Paris, France (bimonthly).

Electric Light and Power, Technical Publishing Co., Barrington, IL (biweekly).

Electrical World, McGraw-Hill, New York (biweekly).

IEEE, *Power Apparatus and Systems* (bimonthly).

Stewart, H. R. "A Review of Electrical Design Factors for EHV Transmission." AIEE, *Electrical Engineering*, August 1963.

Stewart, H. R. "Switching Surge Magnitudes, BIL's, Stability Limits and Relaying Problems in an EHV Design," Proceedings of the American Power Conference, Vol. XXVIII, 1966.

"Transmission Line Reference Book 345kV and Above." Electric Power Research Institute, Palo Alto, CA, 1975.

"Westinghouse Electrical Transmission and Distribution Reference Book," Westinghouse Electric Corporation, East Pittsburgh, PA, 1950.

CHAPTER 15

ADVANCED ENERGY SYSTEMS

B. V. TILAK

Occidental Chemical Corporation Research Center
Grand Island, New York

S. SRINIVASAN

Institute for Hydrogen Systems
Mississauga, Ontario, Canada

THOMAS C. VARLJEN

Westinghouse Electric Corporation
Madison, Pennsylvania

RYSZARD J. PRYPUTNIEWICZ

Worcester Polytechnic Institute
Worcester, Massachusetts

15.1 HYDROGEN AS AN ENERGY SOURCE

B. V. Tilak and S. Srinivasan

15.1-1 Rationale for Hydrogen as an Energy Currency or Medium

Hydrogen is not a primary energy source but rather an "energy currency" or an "energy medium." There have been several proponents of a "hydrogen economy" in the late 1960s and early 1970s. But it must be pointed out that as early as 1869 Jules Verne in his book *Twenty Thousand Leagues Under the Sea* stated that water will be the source of the fuel of the future.[1] Bockris recently highlighted the origin of the concept of hydrogen economy.[2] An interesting comment made in this historical survey was the suggestion by J. B. S. Haldane that "the future fuel will be liquefied hydrogen obtained by the electrolysis of water, with the energy source being wind."[3]

Predictions on a hydrogen energy scenario are dependent on the primary energy source and the time frame. In the "natural gas and petroleum era," expected to last another 30 years, most of the hydrogen is and will be produced by the reformer reaction (Fig. 15.1-1) and its main use will be as a chemical (Fig. 15.1-2). As noted in Fig. 15.1-1, relatively small fractions of hydrogen are produced by water electrolysis (in places where the primary energy source is that of "falling water") and by the

1459

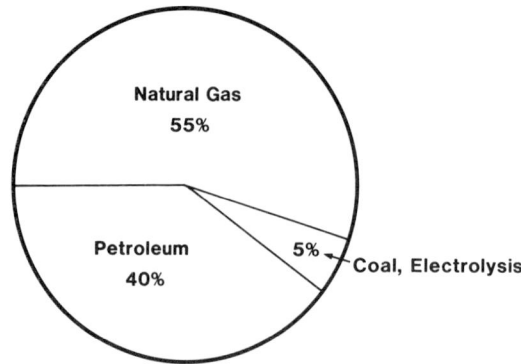

Fig. 15.1-1 Production sources of hydrogen in the United States (from ref. 5).

water gas reaction (in places where coal is in abundance and natural gas or petroleum sparse). According to Fig. 15.1-2, a considerable portion of hydrogen is used[6] in petroleum refining, that is, the production of the lighter liquid fractions (e.g., gasoline) from the heavier crudes. During the "coal era," which will supposedly follow the "natural gas and petroleum era" in the United States and last a long time (according to the optimists, 200 years), hydrogen will play a significant role in the production of gaseous ($CO + H_2, CH_4$) and liquid (gasoline) fuels. Countries devoid of this primary energy source, coal, will make the quantum jump from "the natural gas/petroleum era" to the "fission/renewable resource era." In the distant future, after 2050, the fusion era will emerge. The major impact of hydrogen will become evident in the fission/renewable resource era. Hydrogen is the most convenient transportable fuel to be produced from its abundant source—water—stored easily (compressed gas, liquid, or metal hydride), and converted to electrical energy in fuel cells or gas turbines. It can also be used as a substitute for natural gas and for the production of the liquid fuel—methanol—which is the most attractive substitute for gasoline after the natural gas/petroleum era.

An illuminating rationale for the "electricity–hydrogen age" has been provided by Scott.[4] According to his analysis, the growing population and the increasing energy demands per capita necessitated energy distribution networks and the evolution of fuels with increasing hydrogen/carbon ratios (Fig. 15.1-3). The entry of the energy distribution networks (e.g., pipelines for gaseous and liquid fuels and electric power transmission lines or underground cables for electricity) readily displaced coal rail and road trucks. The energy transition, which is being witnessed, will lead to an "electricity–hydrogen" era. Nonfossil sustainable primary energy sources will be used for the production of electricity and hydrogen. Once established, the electricity-hydrogen age will last as long as advanced civilization exists. It is essential to be specific about the "nonfossil sustainable primary energy sources." In the advanced countries in which the energy consumption per capita is high, hydrogen will be produced in large quantities using nuclear-derived electricity and water electrolysis. Another attractive source of electricity for the electrolytic splitting of water is hydropower. In the

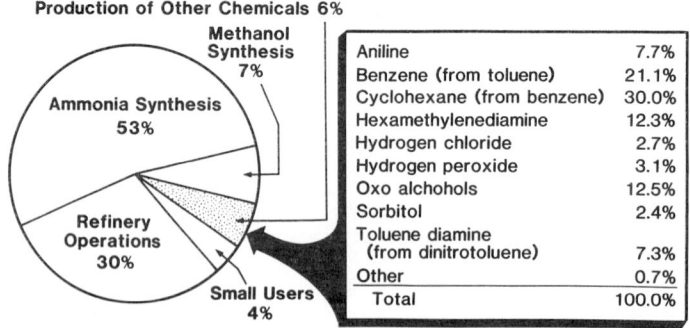

Fig. 15.1-2 Hydrogen consumption in the United States in 1981. (Reprinted from the December 15, 1982, issue of *Chemical Week* by special permission. Copyright © 1982 by McGraw-Hill, Inc., New York.)

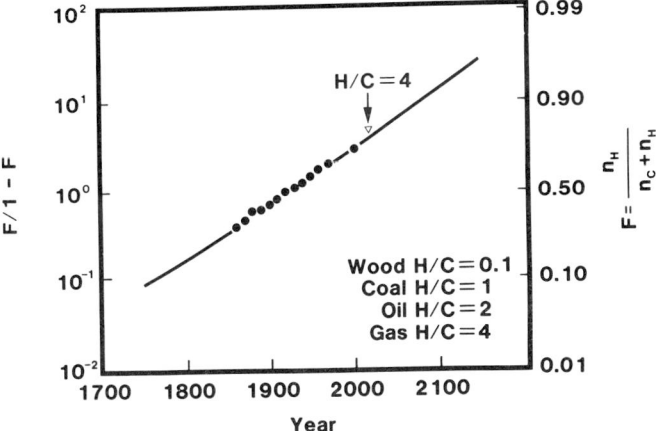

Fig. 15.1-3 World evolution of H/C in world's fuel mix H/C ratio by atomic count. (Printed by permission of the author, Dr. David S. Scott, from Institute for Hydrogen Systems Report #IHS-1-83.)

underdeveloped countries, where the energy consumption per capita is low, it is possible to use the electricity derived from renewable resources (solar, wind, geothermal, tidal, etc.) to decompose water electrochemically. The combination of nonfossil derived hydrogen with hydrogen-deficient fossil reserves can lead to "carbon mobilization" (gases, liquid fuels) exactly in the same manner as the present energy distribution networks.

Before passing on to the details of the hydrogen production, storage and transmission, utilization, and safety aspects, it is worthwhile considering the "modus operandi of hydrogen energy currency". Figure 15.1-4 clearly indicates that the method of production (chemical, electrochemical, thermochemical, photochemical, biochemical) will depend on the primary energy source. All methods produce gaseous hydrogen. Hydrogen can then be stored as a gas, liquid, or solid. It can be transported via the present energy distribution networks. Depending on the availability of hydrogen deficient fossil fuels, there may be economic advantages of utilization of "pristine hydrogen" for carbon mobilization. The initial major application of hydrogen is for the production of chemicals. However, it will increasingly play a role as a fuel for electricity and heat generation for industrial, residential, and transportation applications. One attractive feature of "pristine hydrogen" has not been mentioned so far. Environmentally, it will be the cleanest fuel since the product of combustion—thermally or electrochemically—is pure water. Thus, in a hydrogen energy scenario environmental pollutions due to acid rain (caused by SO_2 and NO_x emissions), greenhouse effect (CO_2 emissions), and toxic chemicals (particulates, CO) will be eliminated.

The present chapter is a concise summary of a voluminous literature published on hydrogen. The purpose of this brief article is to provide the reader with the rationale for and status of development of

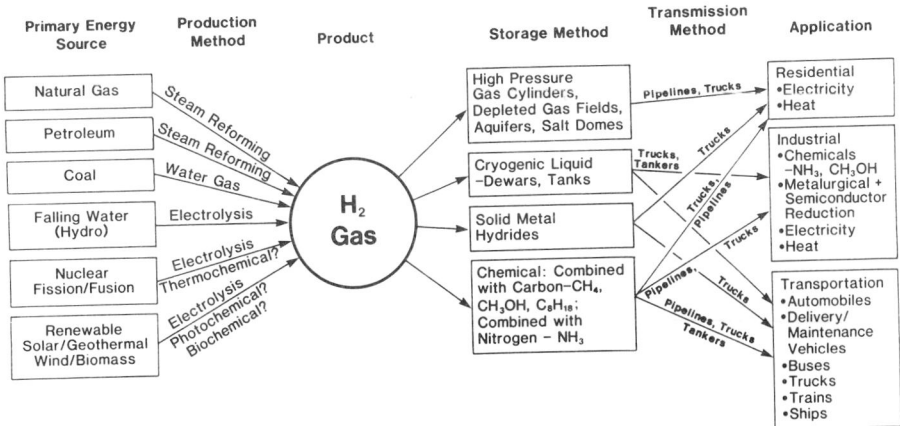

Fig. 15.1-4 *Modus operandi* of hydrogen energy currency.

hydrogen energy technologies as well as technoeconomic assessments of potential applications for hydrogen energy systems. The interested reader may refer, in addition to the literature (primarily reviews) quoted at the end of this section, to the proceedings of the World Hydrogen Energy Conferences and *The International Journal of Hydrogen Energy* published since 1976 for additional details.

15.1-2 Technoeconomic Assessment of Hydrogen Production Technologies

Hydrogen is commercially produced chemically by steam reforming, partial oxidation of hydrocarbons, and coal gasification, and electrochemically by water electrolysis—the bulk of industrial hydrogen being manufactured by steam reforming. In recent years, significant efforts have been devoted to exploring and developing alternate, possibly economic in the long-term, routes for producing hydrogen. The chemical, electrochemical, and advanced concepts for hydrogen production are presented in this section.

Chemical Methods for Hydrogen Production

Hydrogen by Steam Reforming. The most widely used and economical process for hydrogen production[7-9] is by steam reforming of hydrocarbons, e.g.,

$$CH_4 + 2H_2O + Heat \rightarrow CO_2 + 4H_2 \tag{15.1-1}$$

This process (see Fig. 15.1-5 for the process steps with natural gas feed[10]) is usually operated at 650–700°C and a pressure of 7–48 atm using nickel-based catalysts. Since the nickel catalyst is poisoned by sulfur, feed desulfurization is carried out ahead of steam reforming by passing the sulfur-containing hydrocarbon feed at 290–370°C over a Co-Mo catalyst in the presence of 5% H_2 (to convert all the sulfur compounds to H_2S). The gases are then cooled and scrubbed with a monoethanolamine solution followed by absorption over a ZnO catalyst at 340–370°C to reduce the sulfur content to approximately 0.5 ppm.

This gas is then introduced into the primary reformer containing NiO supported on calcium aluminate, alumina, or Ca-Al titanate along with steam at a steam:carbon ratio of 3.5 : 4.5, outlet gas temperature of 870–885°C, and a pressure of 22–24 atm to achieve greater than 95% conversion of CH_4. The exit gases from the primary reformer typically contain 76% H_2, 12% CO, 10% CO_2, and 1.3% CH_4. Conversion of CO to CO_2 is carried out in a shift reactor using a chromium promoted iron oxide catalyst at 340–350°C. Completion of the water–gas shift reaction is achieved with a Cu-Zn catalyst supported on alumina at 200–300°C to generate a gaseous mixture containing 86% H_2, 22% CO_2, 0.25% CO, and 1.3% CH_4. This stream is cooled and scrubbed by a regenerative scrubbing process to produce 98.2% H_2, 0.3% CO, 0.01% CO_2, and 1.5% CH_4. Residual carbon oxides are converted to CH_4 by passing the gases reheated to 315°C over a NiO catalyst—the final product containing 98% H_2 and 1.8% CH_4.

The process described above employs high and low temperature shift reactors followed by CO_2 scrubbing and methanation to produce H_2 of about 98% purity which should be further subjected to cryogenic processing to achieve 99% + pure hydrogen. As depicted in Fig. 15.1-6, the low-temperature

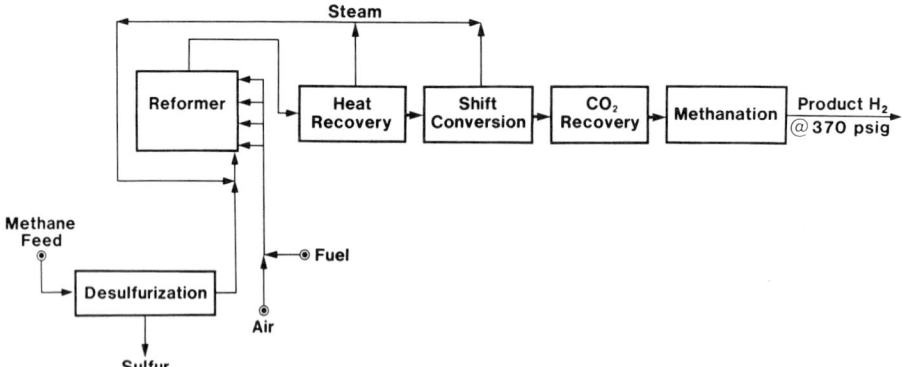

Fig. 15.1-5 Hydrogen from conventional steam–methane reforming. (Reprinted with permission from ACS Symposium Series #116; Hydrogen: Production and Marketing. Copyright © 1980 American Chemical Society.)

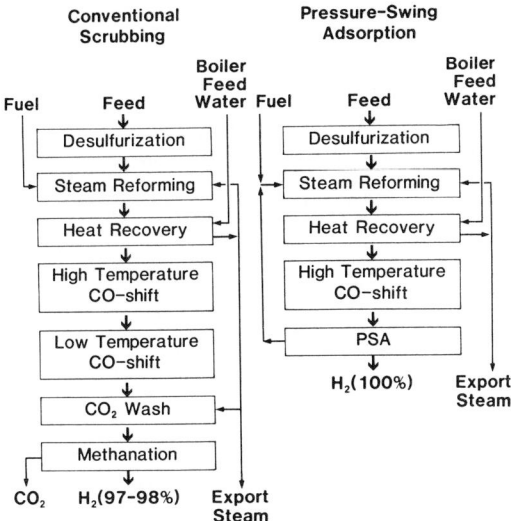

Fig. 15.1-6 Unit operations in pressure swing adsorption and conventional scrubbing. (Reprinted from the December 15, 1982, issue of *Chemical Week* by special permission. Copyright © 1982 by McGraw-Hill, Inc., New York.)

CO shift, CO_2 scrubbing, methanation, and cryogenic steps can be eliminated using pressure-swing adsorption (PSA) in a single-step process to produce 99.999% H_2 with less than 10 ppm CO and no CO_2.[6,11] This technique involves passage of the reformer effluent through an adsorbent bed which removes all the constituents except H_2. When the adsorbent is saturated, the unit is depressurized, purged with pure H_2 to remove the impurities, and repressurized. During this process, the feed "swings" to another bed which is in parallel with the unit that is being activated.

The PSA process is used normally on a gas that is relatively rich in H_2 and recovers 75–78% H_2 and no CO_2. However, the main advantages include lower maintenance costs, increased reliability, and efficient heat recovery—the overall thermal efficiency being 84.6% compared to 83.2% for a conventional plant.

Hydrogen by Partial Oxidation of Hydrocarbons. Hydrogen and H_2-rich synthesis gas can also be produced[7–9] by noncatalytic partial oxidation of hydrocarbons (e.g., refinery residual oil) under pressure using the Texaco or Shell process—the overall reaction scheme being

$$C_n H_m + \frac{n}{2} O_2 \rightarrow n\,CO + \frac{m}{2} H_2 \qquad (15.1\text{-}2)$$

In this process (see Fig. 15.1-7), preheated hydrocarbons leaving the atomizer are contacted with a steam–preheated oxygen mixture in a closed combustion reactor where partial oxidation occurs at

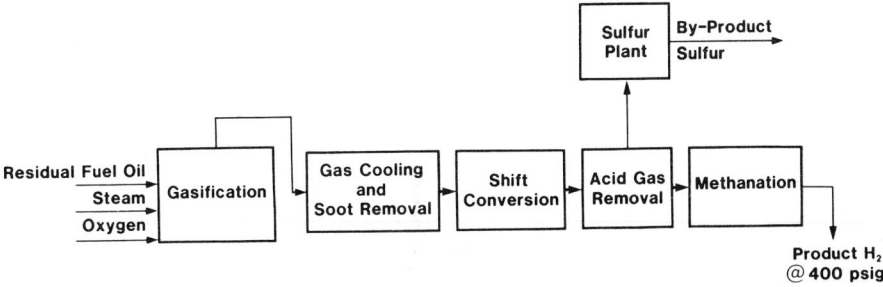

Fig. 15.1-7 Hydrogen from partial oxidation of hydrocarbons. (Reprinted with permission from ACS Symposium Series #116; Hydrogen: Production and Marketing. Copyright © 1980 American Chemical Society.)

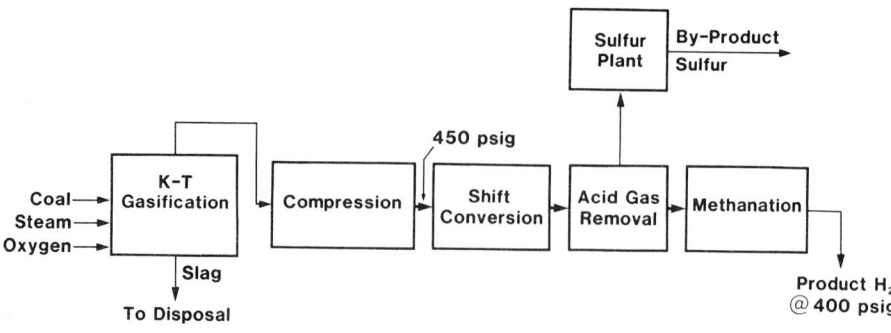

Fig. 15.1-8 Hydrogen from coal-gasification process. (Reprinted with permission from ACS Symposium Series #116; Hydrogen: Production and Marketing. Copyright © 1980 American Chemical Society.)

1290–1400°C with less than stoichiometric amount of O_2 for complete combustion. The synthesis gas from the gasifier is cooled and subjected to shift conversion using a high-temperature shift catalyst followed by a low-temperature shift catalyst. Heat released from the exothermic carbon monoxide shift reaction is recovered by raising high-pressure and low-pressure steam and by supplying make-up heat for other gas streams. CO_2 removal is carried out as in the steam reforming process. H_2 produced from this process is approximately 97.5% purity.

Hydrogen by Coal Gasification. This process (see Fig. 15.1-8 for a schematic of the flowsheet) is based on rapid oxidation of pulverized coal[7–9] (particle size = 90 μm) in a suspension of steam and O_2 at atmospheric pressure by the Koppers–Totzek, and at elevated pressures by the Texaco and Shell–Koppers processes. The gases leaving the gasifier are cooled to recover the waste heat followed by water quenching to remove ash particles before going through compression, carbon monoxide shift, acid-gas removal, and methanation. The product purity from this process is approximately 97.8%. Another coal-based process is the steam–iron process where H_2 is produced from the decomposition of steam by reaction with iron oxide instead of synthesis gas generated from coal. This scheme employs oxidation and reduction zones between which iron oxide is circulated to generate H_2 as described by the following reactions:

$$\text{Reduction Zone:} \quad Fe_3O_4 + CO \rightarrow 3FeO + CO_2 \qquad (15.1\text{-}3)$$

$$Fe_3O_4 + H_2 \rightarrow 3FeO + H_2O \qquad (15.1\text{-}4)$$

$$\text{Oxidation Zone:} \quad 3FeO + H_2O \rightarrow Fe_3O_4 + H_2 \qquad (15.1\text{-}5)$$

(For details related to this process see ref. 15.1-10.)

Electrochemical Methods for Hydrogen Production

Electrolysis is one of the best known and simplest methods[12–16] for producing pure H_2, on a small or large scale, from water. Water electrolysis plants operate with few moving parts, require little space, and are nonpolluting. The reaction

$$2H_2O \rightarrow 2H_2 + O_2 \qquad (15.1\text{-}6)$$

produces hydrogen and oxygen in an extremely pure state and physically separated during the evolution at the electrodes.

Hydrogen is generated by the electrolysis of aqueous KOH (usually 25–30 wt.%) as described by Eq. (15.1-6), which requires 2 faradays of electricity to produce 1 g-mol of H_2 at 0°C and 1 atm (or 15.6 SCF/1000 A · h at 68°F) assuming the current efficiency to be 100%. The component electrochemical reactions comprising the overall reaction (15.1-6) are the discharge of H_2 at the cathode and O_2 at the anode as given by

$$2H_2O + 2e \rightarrow H_2 + 2OH^- \qquad (15.1\text{-}7)$$

$$4OH^- \rightarrow O_2 + 2H_2O + 4e \qquad (15.1\text{-}8)$$

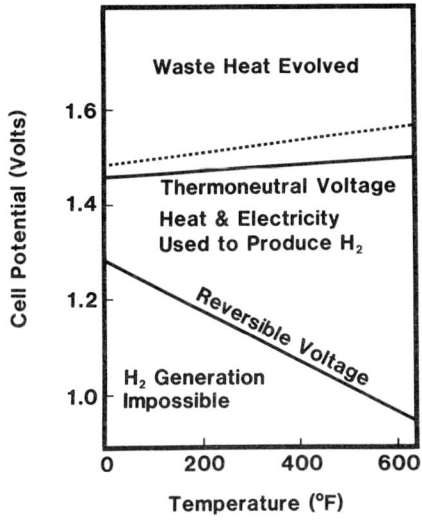

Fig. 15.1-9 Idealized operating conditions for water electrolysis. Solid line for thermoneutral voltage was calculated using the ΔH values and the dotted line for the higher-heating-value voltage using Eq. (15.1-9).

The thermodynamic decomposition voltage (E_{rev}) corresponding to the Gibbs free energy change of reaction 15.1-6 of 56.69 kcal (g-mol)$^{-1}$ at 25°C and 1 atm pressure is 1.229 V. Reaction (15.1-6) is endothermic—the heat of the reaction being 68.32 kcal (g · mol)$^{-1}$. The difference between the free energy and enthalpy (ΔH) changes arises from entropy changes in the overall process and must be balanced by supplying or removing heat from the system. Thus, below a cell voltage of 1.48 V which corresponds to $\Delta H/nF$ (the thermoneutral potential), but above 1.23 V, the electrolysis cell would absorb heat from the surroundings; above 1.48 V, heat is generated and must be removed for an isothermal operation of the cell. The thermoneutral voltage is generally greater than 1.48 V because of the heat requirements[12] to increase the temperature of feed water to that of the electrolyte and to evaporate water which is carried along with the product gases as noted in Fig. 15.1-9 for the higher-heating-value voltage (V_{HHV}) expressed by the relationship

$$V_{HHV} = 1.415 + 2.169 \times 10^{-4}T + 1.52 \times 10^{-8}T^2 \tag{15.1-9}$$

where T is expressed in degrees Kelvin. Since all practical cells operate above a cell voltage of 1.48 V, heat removal becomes an important design aspect of water electrolysis cells.

The power consumption for operating water electrolysis cells is given as

$$\text{Power consumption}\left(\text{dc kW} \cdot \text{h/N} \cdot \text{m}^3 \text{ of } H_2\right) = \frac{2.3995 \times \text{cell voltage } (E)}{\text{current efficiency}} \tag{15.1-10}$$

Since the current efficiency is very close to 100%, the parameter of importance in comparing the performance characteristics of various cells is the cell voltage which is composed of the anodic (η_a) and cathodic (η_c) overvoltages, ohmic drop in the electrolyte between the anode and the cathode (η_Ω), and the hardware ohmic drop (η_{hw}). Typical values of these components of water electrolysis cell voltage operating at 150 mA/cm² and at 75°C are shown in Table 15.1-1.

Reduction in operating costs associated with hydrogen production can be accomplished by minimizing the overpotential and ohmic losses and most of the investigations in recent years have been directed toward developing electrocatalysts (to reduce η_a and η_c) and novel designs (to minimize the electrolyte and structural ohmic contributions).

TABLE 15.1-1 TYPICAL COMPONENTS OF WATER ELECTROLYZER CELL VOLTAGE[a]

E_{rev} = 1.19V	
η_a = 0.30V	
η_c = 0.30V	
η_Ω = 0.25V (anode-cathode gap: ~4 mm)	
η_{hw} = 0.11V	
Cell Voltage = 2.15V	

[a]c.d. = 150 mA/cm²; temperature = 75°C.

Electrocatalysts Development. *Oxygen Evolution Reaction.* Alkaline water electrolyzers employ nickel as the anode electrocatalyst because of its corrosion resistance in KOH solutions. Kinetic studies[12-16] indicate that nickel-based materials exhibit the best electrode kinetics for the discharge of O_2 in alkaline media. Hence, noble metal additions are not essential for realizing low overvoltages.

Materials that have been examined in detail are essentially high surface composites with electrocatalysts such as lithium-doped NiO, Ni-Co spinels (e.g., $NiCo_2O_4$), perovskites (e.g., $Ni_{0.2}Co_{0.8}LaO_3$), cobalt oxides, ferrites such as $M_xFe_{3-x}O_4$ where M is Mg, Zn, Mn, Co, and Ni, delafossite-type oxides, and iron molybdates prepared by freeze-drying, thermal decomposition, or electrochemical methods. These attempts to develop binary, ternary, and higher oxides of various crystal structures are primarily aimed at stabilizing Ni^{3+} species in the oxide lattice by inhibiting its oxidation to Ni^{4+} species, which results in an increase of overpotential with time.[12]

Promising electrocatalysts for the acidic media are noble metal based and most efforts are directed toward optimizing composites with low loading of noble metal. Materials showing promising results[14] are Ru based, e.g., $RuIr_{0.5}Ta_{0.5}O_x$ (where $x \simeq 2$) which exhibit very high activity during the discharge of O_2 from acidic solutions.

Hydrogen Evolution Reaction. Nickel or steel is generally used as a cathode in water electrolyzers. Improvements for enhanced catalysis are essentially oriented, as noted above, toward developing high surface area transition-metal-based compositions. Catalysts that exhibited[12-6] low hydrogen overvoltage include NiB, Ni-S, Ni-Al, Ni-Mo, Ni sinter impregnated with cobalt molybdate, and $NiCo_2S_4$ (Ni-Co thiospinels).

As observed with anodes, nickel-based cathodes also suffer from degradation in performance with time, which is attributed to electrode poisoning by impurities and/or hydrogen permeation altering the mechanism of H_2 evolution resulting in an increase in the hydrogen overpotential. However, cathode compositions developed recently appear to exhibit stability over long periods of time.

Separator Development. Separators are necessary for operation of water electrolyzers to prevent mixing of product gases and short-circuiting of the electrodes. Asbestos is the most commonly used material in the form of woven cloth mat or felt. While it is an ideal material for low-temperature operation, asbestos dissolves in strongly alkaline media at temperatures greater than 100°C. Hence, efforts have been focused to stabilize asbestos and develop alternate materials. Promising separators with low ohmic drop in 30–40% KOH solutions include asbestos with tin hydrosol or sodium silicate, teflon-bonded potassium titanate, Nafion, metal gauze supported oxide ceramics, NiO, polysulfone felt, polysulfone/polyantimonic acid, and polyvinylidine fluoride/polyantimonic acid based ones.[14]

Commercial Electrolyzers. Electrolyzers currently marketed may be classified[12-16] into two categories: monopolar (tank-type cells) or bipolar (filter-press-type cells). In monopolar cells, a number of positive electrodes interleaved with cathodes are installed in a single gas-tight tank with asbestos cloth diaphragm inserted around either the anodes or the cathodes to permit separation of H_2 and O_2. Circulation within the cell is by simple gas lift. In bipolar cells, a large number of electrodes are placed side by side, with diaphragms between them to separate the gases, and with insulating frames between the electrodes in a filter-press-type assembly. This assembly is held together by several longitudinal tie bolts, which press the electrode–diaphragm–gasket units together to prevent leakage of cell liquor.

Historically, all the early water electrolysis cells were of the simple tank type. However, recently, the bipolar cells have steadily replaced the tank-type monopolar cells in view of their decreased capital and operating costs and ease of operation at high pressures and at high temperatures. Advantages of high-pressure and high-temperature operation include reduced cell voltage (via reduction in η_Ω, η_c, and η_a) and elimination of compressors. The only large-scale commercial pressure water electrolyzer is manufactured by Lurgi.

A novel design of water electrolyzer developed by the General Electric Company is the SPE (solid polymer electrolyte) cell where Nafion membrane is used as the electrolyte layer. This sulfonated teflon-based membrane is a proton conductor when it is in a wetted acid-form state. Platinum particles impregnated on one side of the membrane together with carbon serve as the cathode catalyst and the anode contains a proprietary catalyst impregnated similarly. These cells are arranged in a bipolar mode and operate at high efficiency.

The performance characteristics of commercially available technologies are presented in Table 15.1-2 (see, however, ref. 12 for detailed distinguishing and engineering features of these technologies), and the projected developments via improved electrocatalysts and designs are depicted in Fig. 15.1-10 —the ultimate goal being aimed at realizing an energy consumption of less than $4.5 \text{ kW} \cdot \text{h/N} \cdot \text{m}^3$.

Advanced Concepts for Hydrogen Production

Anode Depolarized Electrolysis. Oxygen is generally an undesirable product in the electrolytic production of hydrogen and, hence, alternate anodic reactions were conceived[12-16] to reduce the cell voltage and, hence, the energy consumption for H_2 production. Anodic reactions examined in detail are SO_2 depolarization, which is the electrochemical step in the thermochemical–electrochemical hybrid cycle for H_2 production, and carbon oxidation.

TABLE 15.1-2   OPERATING PARAMETERS FOR STATE-OF-THE-ART ELECTROLYZERS[a]

Manufacturer:	Brown Boveri & Cie	DeNora S.P.A.	Lurgi GmbH	Norsk Hydro A.S.	The Electrolyzer Corp., Ltd.	Krebskosmo	Teledyne Energy Systems	G.E.'s Solid Polymer Electrolyzer
Cell Type:	Bipolar Filter Press	Bipolar Filter Press	Bipolar Filter Press	Bipolar Filter Press	Monopolar Tank	Bipolar Filter Press	Bipolar Filter Press	Bipolar Filter Press
Operating Pressure	Ambient	Ambient	32 atm	Ambient	Ambient	Ambient	2.4 atm	3.9 atm
Operating Temperature	80°C	80°C	90°C	80°C	70°C	80°C	82°C	80°C
Electrolyte	25% KOH	29% KOH	25% KOH	25% KOH	28% KOH	28% KOH	35% KOH	DuPont Nafion–1200 EW
Current Density, A.m⁻²	2000	1500	2000	1750	1340	3000	2000	5000
Cell Voltage, V	2.04	1.85 (increases to 1.95 after 2 yrs.)	1.86	1.75 (after 1 yr. operation)	1.90	1.90	1.90	1.70
Current Efficiency, %	>99.9	~98.5	98.75	>98.0	>99.9	>99.9	—	—
Oxygen Purity, %	≥99.6	99.6	99.3–99.5	99.3–99.7	99.7	99.5	>98.00	>98.0
Hydrogen Purity, %	≥99.8	99.9	99.8–99.9	98.8–99.9	99.9	99.9	99.99	>99.0
Power Consumption, DC–kWh per Normal m³H₂	4.9	4.6	4.5	4.3	4.6	4.5	~6.0	~4.1

[a]See refs. 12–16.

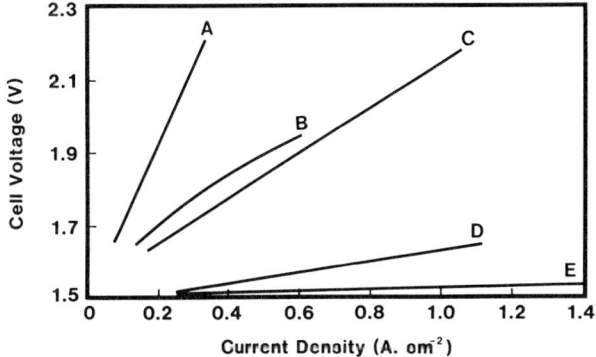

Fig. 15.1-10   Projected cell performance of water electrolyzers: A, Brown Boveri & Cie; B, Electrolyzer Corporation; C, DeNora; D, Teledyne; E, SPE (compiled from refs. 12–16).

The overall reaction involving SO_2, developed by Westinghouse,[17] is

$$H_2SO_3 + H_2O \rightarrow H_2SO_4 + H_2 \tag{15.1-11}$$

the component electrochemical reactions being

$$H_2SO_3 + H_2O \rightarrow H_2SO_4 + 2H^+ + 2e \quad \text{(at the anode)} \tag{15.1-12}$$

$$2H^+ + 2e \rightarrow H_2 \quad \text{(at the cathode)} \tag{15.1-13}$$

The thermodynamic decomposition voltage for reaction (15.1-11) is 0.17 V compared to 1.23 V for water electrolysis. This system has been optimized with respect to the materials of construction and cell components to operate at 0.8 V and 300 mA/cm²—the projected performance being approximately 0.6 V at 300 mA/cm² (temperature, 100°C; 50–60 wt.% H_2SO_4; pressure, 5–20 atm).
   The carbon-based anode depolarized cell operates in a medium containing $3.7M$ H_2SO_4—the overall and the component reactions being

$$C + 2H_2O \rightarrow 2H_2 + CO_2 \quad \text{(overall reaction)} \tag{15.1-14}$$

$$C + 2H_2O \rightarrow CO_2 + 4H^+ + 4e \quad \text{(anodic reaction)} \tag{15.1-15}$$

$$2H^+ + 2e \rightarrow H_2 \quad \text{(cathodic reaction)} \tag{15.1-16}$$

While the oxidation rate was low (a few mA/cm²) at a cell potential of approximately 0.8 V, detailed studies anticipate an energy reduction of 30–50% for H_2 production.[14]

Water Vapor Electrolysis. From thermodynamic and electrode kinetic considerations, electrolysis of water at elevated temperatures of, say, approximately 1000°C offer several advantages[12-16] which include low theoretical energy requirement (by approximately 20% compared to at ambient temperatures) and low operating cell voltage (since the overvoltage and ohmic losses are low). Since the cell voltage would be smaller than the thermoneutral voltage or V_{HHV} (see earlier discussion), heat has to be supplied to the cell to maintain it at the desired operating temperature to realize the expected cell performance.

Brown, Boveri & Cie and Dornier Systems in Germany and Westinghouse in the USA are developing solid electrolyte electrolyzer with yttria-stabilized zirconia as the electrolyte with nickel cathodes and tin-doped indium oxide or perovskite type oxides ($LaNiO_3, LaMnO_3$) as anodes operating at a cell voltage of less than 1.5 V at a current density of 6000 A/m^2 at 1000°C. Hitachi recently reported the operation of a high-pressure and high-temperature electrolyzer of pilot plant size producing 4 m^3 of H_2 per hour at an overall efficiency of greater than 90%. While several other investigations have reported good efficiencies, an overall system analysis may not result in as high efficiencies as reported in view of the additional process heat for evaporating the feed water and the need for exotic (hence, costly) materials stable at high temperatures.

Thermochemical Methods for Hydrogen Production. In the wake of hydrogen economy, several methods have been proposed[7,9-16] for hydrogen production by direct thermal decomposition of water and by thermochemical cycles. The concept of thermochemical cycle method involves employing schemes in which the intermediate steps having positive entropies are driven at high temperatures—the intermediate steps having negative free energy (ΔG) at low temperatures. Thus, if the first reaction has a positive entropy (ΔS), the second reaction may proceed with a more favorable equilibrium position at the same temperature or still better at a low temperature than the first reaction since $\Delta G = \Delta H - T\Delta S$. While a multitude of thermochemical cycles (e.g., $Fe_2O_3/FeSO_4$, $FeCl_2/HCl$, $C/H_2O/Fe_3O_4$, $HI/SO_2/H_2SO_4$, CaO/I_2, $CaBr_2/HBr$, $Ni/I_2/S$), each one involving multisteps, have been suggested and studied, the prospects of using this concept for large scale hydrogen production are bleak because of the following reasons:

1. High efficiencies (greater than 50%) predicted on thermodynamic reasoning cannot be achieved because of kinetic constraints.
2. Several of the intermediate chemical reactants and/or products corrode reaction vessels especially because at least one step is carried out at a relatively high temperature.
3. Energy requirements for pumping and product separation are relatively high.
4. Lack of complete cyclicity means loss of chemicals.
5. The efficiency of the overall system is Carnot limited and it is very difficult to attain more than two-thirds of the theoretical efficiency.

Because of these factors, practical efficiencies will not be expected to be higher than conventional water electrolysis.

However, the most interesting and promising cycles proposed so far are the thermochemical–electrochemical hybrid cycles involving SO_2 and HBr. The Westinghouse sulfur cycle[17] is based on water splitting by reaction with sulfur compounds as follows:

$$H_2SO_4 \xrightarrow{\text{thermal}} H_2O + SO_2 + \tfrac{1}{2}O_2 \tag{15.1-17}$$

$$SO_2 + 2H_2O \xrightarrow{\text{electrochemical}} H_2SO_4 + H_2 \tag{15.1-18}$$

SO_2 is oxidized in an electrolytic cell to H_2SO_4 while H_2 is evolved at the cathode at a cell voltage of 0.4–0.48 V and 20 atm pressure in 50% H_2SO_4, and at 0.47–0.72 V for 80% H_2SO_4 at 50°C. H_2SO_4 from the cell is vaporized to form $SO_3 + H_2O$ that is catalytically converted back to $SO_2 + O_2$ in an indirectly heated reactor. While problems associated with the irreversibility of SO_2 oxidation in the electrochemical cell were observed in early studies, Westinghouse projected that a full-scale plant in operation will utilize less than 40% of the electrical power required for a conventional water electrolyzer at an identical hydrogen production rate.

The HBr based cycle developed by the Commission of European Communities at ISPRA, Italy, consists of the following reactions:

$$2HBr \xrightarrow{\text{electrolysis}} H_2 + Br_2 \tag{15.1-19}$$

$$Br_2 + SO_2 + 2H_2O \xrightarrow{\text{chemical}} 2HBr + H_2SO_4 \tag{15.1-20}$$

$$H_2SO_4 \xrightarrow{\text{thermal}} H_2O + SO_2 + \tfrac{1}{2}O_2 \tag{15.1-21}$$

The complete cycle was demonstrated at ISPRA and a bench-scale plant produced H_2 at a rate of 100 L/hr.[18]

Photochemical Methods for Hydrogen Production. Photochemical generation of H_2 employing semiconductor electrodes has been the topic of a large number of investigations.[12-14,19] The principles of photoelectrochemistry have been discussed in several reviews and are outlined briefly here. Semiconductors are materials such as Si or Ge in which the charge carriers, electrons (e^-) and holes (h^+), are accommodated in bands. Valence bands are formed by the overlap of bonding orbitals of molecules and are completely filled. Similarly, conduction bands are formed by the overlapping of antibonding orbitals and are vacant at sufficiently low temperatures. The energy difference between the valence band of highest energy and the conduction band of lowest energy is the band gap and is the energy needed to promote the most weakly bound electron into the conduction band.

Conduction in semiconductors can be enhanced by doping. Thus, by substituting Si having 4 valence electrons with P having 5 electrons in the lattice, one is left out with one itinerant electron in the conduction band. This produces an n-type semiconductor since conduction is by negatively charged electrons. On the other hand, substitution of Si by Al with 3 valence electrons in the lattice results in electron vacancies or holes in the valence band. This is termed a p-type semiconductor since current is formally considered to be carried by positively charged particles.

When light is illuminated onto a semiconductor, incident photons of sufficient energy can boost the electrons from the valence to the conduction band. The optimum band gap required for solar energy is approximately 1.4 eV and an ideal system with this band gap can convert sunlight at a maximum possible efficiency of 30% with photo voltaic devices.

In photoelectrochemical systems, semiconductors are immersed in the electrolyte when an electric field is set up at the semiconductor/electrolyte interface. If the semiconductor is n-type, electrons will move toward the bulk of the semiconductor and holes to the surface, thus, permitting oxidation at the surface. This is called a photoanode. Converse occurs with p-type semiconductors. Thus, light promotes photoreduction at p-type materials and photooxidation at n-type materials. Irradiation of these semiconductors with light energy greater than band gap energy (E_g) causes the formation of electron–hole pairs. Electrons will pass through an external circuit to the cathode, while holes move to the anode surface where the following reactions occur:

$$h\nu \rightarrow p^+ + e^- \tag{15.1-22}$$

$$H_2O + 2p^+ \rightarrow \tfrac{1}{2}O_2 + 2H^+ \quad \text{(at the anode)} \tag{15.1-23}$$

$$2H^+ + 2e \rightarrow H_2 \quad \text{(at the cathode)} \tag{15.1-24}$$

Much of the work on semiconductor/electrolyte junctions has been devoted to photoanodes because of the availability of stable materials such as TiO_2 which has a band gap of approximately 3.0 eV and responds only to ultraviolet light. Attempts to achieve response to solar radiation include usage of sensitizers and dye molecules such as acridine dyes, $[Ru(2,2'-bipyridine)_3)]^{3+}$ complexes which, capturing the incident sunlight, produce excited states that inject charge carriers into the conduction band of the semiconductor. However, the efficiency was only less than or equal to 1%.

Higher efficiencies can be obtained with semiconductors with small band gaps but will be more susceptible to photocorrosion. Several techniques were employed to minimize corrosion; addition of Se to the electrolyte appears to be promising with cadmium sulfide and cadmium selenide electrodes. Other anodes which are stable and exhibited efficiencies of greater than 10% include metal dichalcogenides such as MoS_2, WS_2, $MoSe_2$, and WSe_2.[20,21]

Photocathodes are not prone to corrosion and efficiencies as high as 11.5% for conversion of sunlight to electricity were reported with p-indium phosphide photocathode using V^{3+}/V^{2+} redox couple in HCl solutions. Recent studies[20,21] indicate that incorporation of Pt or Rh in the electrode surface enhances the efficiency to approximately 13.3% without the use of a redox couple. Use of intermediate redox system (e.g., N,N'-dimethyl-4-4'-bipyridinium, also known as methyl viologen, or paraquat) along with a heterogenous catalyst like Pt has been shown to be promising for H_2 generation with materials such as p-GaAs ($E_g = 1.3$ eV) or p-Si ($E_g = 1.1$ eV). While these semiconductors with small E_g values are efficient for solar energy conversion, they are inefficient O_2 evolving electrodes.

Photoelectrochemical cells represent the most efficient chemical systems so far for water electrolysis although the efficiencies are too low for practical applications. Furthermore, since the solar flux is low (100 W/ft^2), it is essential to use large surface area electrodes for effective conversion of solar energy and the costs associated with materials and support structures will be a major factor in the utilization of such a technology.

Biochemical Methods for Hydrogen Production. An example of the biochemical method[13,22,23] is the generation of H_2 from water and sunlight using photosynthetic catalysts such as heterocystous

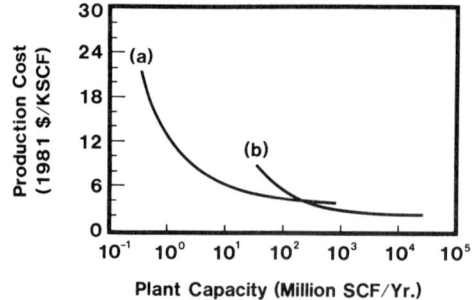

Fig. 15.1-11 Hydrogen production costs for the year 1981 as a function of plant capacity. (a) water electrolysis plant; (b) steam reforming of natural gas plant. (Reprinted with permission from *Modern Aspects of Electrochemistry*, Vol. 15, Edited by R. E. White, J. O'M. Bockris, and B. E. Conway. Plenum Publishing, New York, 1983.)

blue-green algae. This reaction, however, is inhibited by formation of oxygen in the reactor. The achieved and projected efficiencies are low (approximately 3%). Since biochemical production occurs at low temperatures (10–40°C), this route may offer advantages compared to the high temperatures required for chemical and thermochemical methods but only for comparably low rates of hydrogen production.

Technoeconomic Comparison of Hydrogen Production Alternatives

Several studies[7-16,24-27] have attempted to project the future cost of hydrogen and some of the conclusions resulting from these investigations are discussed here. Generally, as the feed stock changes from natural gas to liquid hydrocarbons to solid feed stocks, the processing problems and the manufacturing costs increase. Thus, the partial oxidation and coal gasification processes require more capital investment than the steam-reforming plants because of the need for an air-separation plant, large shift and CO_2 removal facilities, and gas cleanup units. The capital cost of water electrolysis plants is small compared to those of steam reforming for small capacity plants, but their operating costs are quite high. For large capacity plants, the capital cost of the electrolysis plant significantly exceeds that required by other processes in view of the disparity in the economies of scale-up.

Projections, made in recent years, related to the cost of hydrogen production (see Fig. 15.1-11 and Table 15.1-3) pose difficulties in attaching significance to these figures because of the conflicting and contrasting economics of the existing technologies. However, if it is assumed that sociopolitical factors play a dominant role with respect to environmental regulations and/or if the existing feed stocks (e.g., natural gas, petroleum, coal) rapidly dwindle and/or become prohibitively expensive, and nuclear energy is abundant and inexpensive in the future, it is very conceivable that hydrogen produced by photochemical and/or electrochemical methods could be an "energy currency" in the year 2000 and beyond.

15.1-3   Technoeconomic Assessments/Projections of Hydrogen Storage and Transmission Technologies

Hydrogen Storage

A Survey of the Hydrogen Storage Technologies—Their Characteristics, Applications, and Economics. The most difficult challenge in the transition to a "hydrogen economy" is to make its storage method competitive, both from technical and economic points of view, with that of other traditional fuels.[28,29] Table 15.1-4 summarizes the possible methods of hydrogen storage and their characteristics. The projected applications and economics are also indicated. Broadly, hydrogen can be stored as a compressed gas, liquid, solid, or combined with chemicals. Technical and economic aspects of these storage methods are presented in the following subsections.

Compressed Gas Storage in Cylinders, Depleted Gas Fields, Aquifiers, and Salt Domes. The most common method of hydrogen storage on a relatively small scale is as a compressed gas (130–150 atm) in cylinders. Stationary ("tube banks") and mobile vessels ("tube trailers") are used in the storage and distribution of merchant hydrogen. The disadvantage of gaseous storage is that hydrogen is a light gas and thus in a laboratory cylinder, only 0.5 kg of the gas can be stored in a steel cylinder weighing about 60 kg. It is reported that the amount of hydrogen per weight of cylinder can be increased by

TABLE 15.1-3   A TECHNOECONOMIC COMPARISON OF HYDROGEN PRODUCTION ALTERNATIVES[a]

Assessment Critera	Hydrogen Production Method				
	Chemical	Electrochemical	Thermochemical	Photochemical	Biochemical
Primary Energy Source	Natural Gas, Petroleum, Coal	Hydro, Nuclear, Solar, Fusion	Nuclear, Solar, Fusion	Solar	Solar
Energy Input	Chemical + Heat	Electric	Heat, Heat + Electrical	Light	Biochemical, Light
Efficiency %	60–70	20–50	30–50	1–8	1–5
Purity Level %	97–98	99.7	NA	99.7	NA
Impurities	CO, H_2S	O_2	NA	O_2	NA
By-Products	—	O_2	O_2	O_2	O_2
State of Technology	Mature	Mature	Research Stage	Early Research Stage	Early Research Stage
Hydrogen Generation Rates	High	Small to High	NA	Low	Low
Hydrogen Pressure	20–30 atm	5–30 atm	1 atm	1 atm	1 atm
Capital Investment	Economical Only in Large Plant Sizes	Relatively High Compared to Chemical	Projected Costs High	Projected Costs Extremely High	Projected Costs Extremely High
Environmental Effects	CO_2 Emissions	Waste Disposal– Nuclear Power Plant	Same as Electrochemical	Large Land Requirements	Large Land Requirements
Estimated H_2 Production Costs $/MBTU	7–10	15–25	NA	NA	NA
Applications	H_2 as Chemical Feedstock, Synfuel Production	High purity H_2 required for some processes, on-site generation – small to medium scale, only available method of production of transportable fuel from renewable resources	None for Next 20 Years	None for Next 30 Years	None for Next 50 Years

[a] See ref. 26.

about a factor of 4 by using cylinders which are carbon fiber reinforced plastic with a liner of austenitic steel.[30]

Large quantities of hydrogen may be stored underground in depleted natural gas fields, aquifiers, and salt domes. It will be more expensive to store this gas than methane because the energy to weight ratio of the former is only one-third that of the latter, assuming the pressure of both gases to be the same (e.g., about 200 atm). It has been estimated that for the present day costs of natural gas and

TABLE 15.1-4   TECHNOECONOMIC COMPARISONS OF HYDROGEN STORAGE TECHNOLOGIES[a]

Storage Technique	Hydrogen Content		Energy Density Higher Heating Value		Relative** Storage Costs	Application
	Weight %	Volume g/cm³	Cal/g	Cal/cm³		
Gas						
Steel Cylinder (60 kg, 50 l, 197 atm)	1.5	0.018	510	240	2	Lab Chemical
Aluminum Composite (75 kg, 125 l, 197 atm)	2.6	0.017	885	416	3	Transportation ?
Glass Microspheres	6.0	0.006	2040	204	5	Scientific Curiosity; Hydrogen Purification?
Zeolites	0.8	0.006	270	202	5	Residential/Industrial Fuel, Industrial Chemical
Liquid						
Cryogenic Liquid (300m³ Semitrailer)	12.5	0.071	4250	2373	1	Rocket Fuel, Future Fuel
Solid						
$FeTiH_2$	1.6	0.096	545	3245	4	Heat Pumps,
$LaNi_5H_7$	1.4	0.089	476	3050	4	Compressors,
Mg_2NiH_4	3.2	0.081	1080	2745	4	Scientific Curiosity
Combined with Chemicals						
N–octane	15.8	0.110	11400	8020	1	Present Fuel
Methanol	12.5	0.150	5340	4226	1	Future Fuel
Ammonia	17.6	0.136	5350	4120	1	Farm Vehicle Fuel ?

[a] Compiled from refs. 3, 28–38.
[b] Code: 1, least expensive (~ $10/kg H_2); 5, most expensive (~ $1000/kg H_2); capital investment costs.

hydrogen, the capital costs of a gaseous storage system will be about four times higher for hydrogen than for natural gas.

Cryogenic Liquid Hydrogen Storage. When large quantities of hydrogen must be stored or transported over long distances, it is preferable to store it cryogenically. The impetus for the large-scale storage of liquid hydrogen is the U.S. space program. Liquid hydrogen combined with liquid oxygen provides the highest performance of any of the currently used propellant fluids. The largest storage vessel used by NASA has a capacity of 3 million liters, corresponding to about 10 million kW · h.

Four difficulties are associated with the liquefaction of hydrogen:[3]

1. Liquefaction occurs at a very low temperature—20.4 K (cf. boiling point of methane is 109 K). The fall in temperature of gas brought into a cold tank tends to form a cryogenic pump which may introduce impurities.

2. It is easier to liquefy gases if the Joule–Thompson inversion temperature (i.e., the expansion of the gas causes a fall in temperature) is above the room temperature. For hydrogen it is 204 K. Thus it is first essential to use an auxiliary liquefaction plant, for example, nitrogen, to bring the temperature of the gas below the inversion temperature. This additional plant requirement increases the capital cost.

3. The efficiency of the refrigeration process for liquefaction of hydrogen is Carnot-limited. According to the second law of thermodynamics, this efficiency decreases to zero as the absolute zero is reached.

4. The hydrogen molecule exhibits the phenomenon of ortho–para conversion. In the ortho form, the nuclear spins of the hydrogen molecule are parallel while in the para form the spins are antiparallel. The para form has the lower energy. However, there are three times as many energy states for the ortho form as for the para form. Consequently, at room temperature, the equilibrium concentrations of ortho to para forms are in the ratio of 3 : 1. Because of the lower energy of the para form, as the temperature is lowered, the ratio of amounts of para to ortho hydrogen increases and at the absolute zero of temperature, only para hydrogen exists. At the boiling point of hydrogen (20.3 K), the percentage composition of the gas is 99.7% para, 0.3% ortho.

The theoretical minimum energy requirement for liquefaction of hydrogen is 3.3 kW · h/kg; however, ortho–para conversion occurs during liquefaction and, hence, the energy requirement is increased to 3.8 kw · h/kg. The practical energy requirements are 25 kW · h/kg for small liquefiers (< 2 tons/day) and 10 kW · h/kg for larger liquefiers. Reciprocating compressors are traditionally used for the liquefaction of hydrogen. Recently it has been proposed that centrifugal compressors which are more compact and less expensive than reciprocating compressors can be used for hydrogen compression.[31] Since the low molecular weight of hydrogen prevents the attainment of a suitably large pressure rise, it was suggested that the hydrogen liquefaction process can employ an admixture of a high molecular weight component (e.g., propane) to use the centrifugal compressors more effectively. The propane is subsequently separated from the hydrogen by condensation in the liquefier. The condensate is blended with hydrogen gas and recycled to the compressors. A flow diagram of the process is shown in Fig. 15.1-12. The stream of cold nitrogen is used for precooling the gas while the liquid nitrogen provides additional cooling and for partial ortho–para hydrogen conversion—an important feature which occurs during the liquefaction process. Since the para is more stable than the ortho, heat will be evolved during this conversion and boil-off is effected. Thus, there must be an efficient heat removal mechanism. The use of catalysts can minimize the boil-off loss because it accelerates the conversion step. The energy necessary to convert ortho to para hydrogen (up to 95%) is 18.36% of the total energy required for the liquefaction process. The practical value depends on the rate at which this conversion is effected (i.e., with or without catalyst).

Dewar flasks are used for the storage of liquid hydrogen.[32] In the earliest versions, the Dewar flask containing liquid hydrogen was immersed in a second Dewar flask containing liquid nitrogen (Fig. 15.1-13). Since the major cause of the heat leak into the Dewar is from thermal radiation which is proportional to T^4, there is a 250-fold reduction of heat leak by using this second Dewar which is at a temperature of 77 K rather than 300 K. This method is still used for small Dewar flasks (50 L). For larger ones, multilayer insulation is incorporated. Additional heat leakage is prevented by locating thermal radiation shields at selected places within the multilayer insulation. Heat leaks are further minimized by filling the vacuum space with a fine reflective powder (perlite) to lower the vacuum requirement to 10^{-2} Torr and reduce the radiation heat transport. Liquid hydrogen storage vessel in sizes ranging from 10^5 to 3.5×10^6 L have been constructed and utilized.

The largest liquid hydrogen storage vessels being employed at the present time are the two 3.4 million liter spherical containers at the Kennedy Space Center.[33] The diameter of the austenitic stainless steel inner shell is 20 m while that of the carbon outer shell is 23 m. The storage vessel's working pressure is 6 atm and the measured boil-off rate is 0.02% per day when it is full. Capital costs

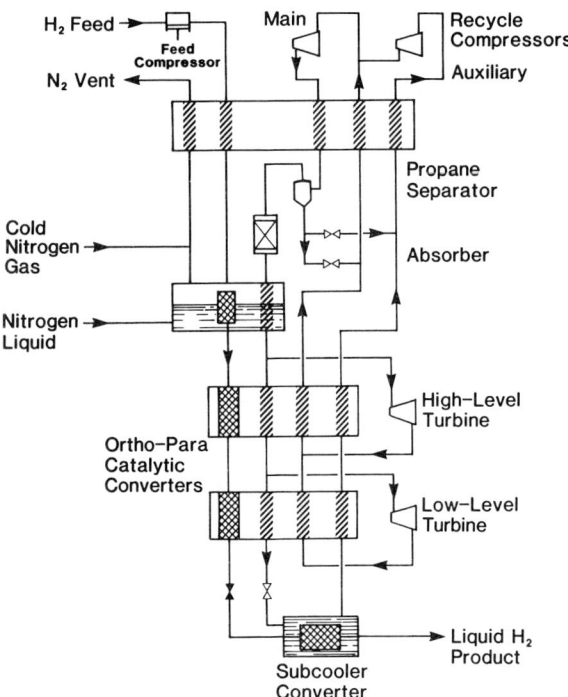

Fig. 15.1-12 A schematic of the hydrogen/propane process for H_2 liquefaction. (Reprinted with permission from *Hydrogen Energy Progress IV*, Vol. 3, p. 1317. Pergamon Press, Ltd., London, 1982.)

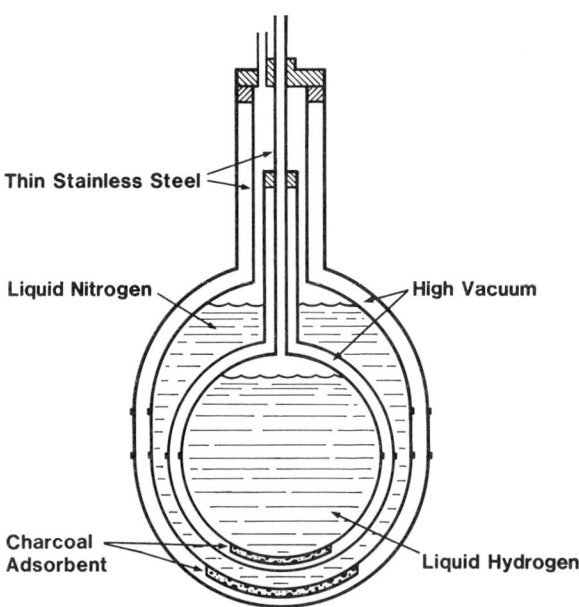

Fig. 15.1-13 Liquid nitrogen shielded liquid hydrogen dewar. (Reprinted from *Hydrogen: Its Technology and Implications*, Vol. II, Chap. 3. CRC Press, Boca Raton, FL, 1977.)

for conventional liquefaction facilities decrease with increasing capacity but plateau at 225,000 kg/day.

A novel approach for the liquefaction of hydrogen (magnetocaloric liquefaction) is being investigated at Los Alamos National Laboratory.[34] While it is at a research stage, it is projected to produce smaller quantities of liquid H_2 more efficiently than the conventional methods.

The "Solid" Storage of Hydrogen as Metal Hydrides. Hydrogen can be stored as a gas, liquid, or solid. The first two methods were presented in the preceding sections. In solid form the best method of storage of hydrogen is as a metal hydride.[35] The pioneering work at Brookhaven National Laboratory and Philips Research Laboratory paved the way for research and development of metal hydride storage systems for hydrogen. The principle of the method is that several metals, alloys, and intermetallics absorb hydrogen under pressure forming the corresponding metal hydride, and release the hydrogen, when required, on heating it and lowering the pressure. On a volumetric energy density basis the hydride storage method is superior to the gaseous and liquid methods of storing hydrogen. However, on a weight basis, the hydrides are at a considerable disadvantage because of the weight of the associated metal. Metal hydrides are of three types—ionic (e.g., magnesium hydride), covalent (hydrides of Be and of the Group 3 metals), and metallic (hydrides of the transition metals). The most attractive, though not exclusive, application of metal hydrides is as an energy storage medium coupled with an energy converter or a fuel cell. One of the most important criteria that the hydrides must possess for this purpose is the hydrogen content by weight. As seen from Table 15.1-4, the light ionically bonded hydrides are suitable from a minimum weight penalty point of view. However, as seen from the plot of the dissociation pressure versus temperature in Fig. 15.1-14, the light weight metal hydrides have considerably high decomposition temperatures.[35] The alloys which reversibly combine with hydrogen to form the hydrides and which have been extensively investigated are $LaNi_5$ at Philips Research Laboratory, Eindhoven, and Fe-Ti at Brookhaven National Laboratory. Hydride storage vessels require relatively low design pressure (30–60 bar). Thus, the hydride storage method is relatively safe, compared to the gaseous or liquid storage method. However, the disadvantages of metal hydride storage systems are: (i) their cost (e.g. about \$950/kg H_2 for Fe-Ti and about \$700/kg H_2 for Mg-Ni, weight (1–7% by weight hydrogen); (ii) sensitivity to impurities during recycling; (iii) sluggish heat transfer properties; (iv) particle attrition during dehydriding. Apart from stationary energy storage, hydrides have been considered for several applications such as heat pumps, fuel storage for automotive use, peak shaving, compressors, and so on. The first striking demonstration of a hydride energy storage system was made by Brookhaven National Laboratory/Public Service Electric and Gas Company for a small scale electric utility load leveling operation. The concept was to use off-peak electric power to decompose water, store the hydrogen as $FeTiH_2$ and use it in a fuel cell for peaking operations. For this demonstration, a 25-kW electrolyzer was provided by Teledyne Energy Systems, the hydride storage unit (which contained 400 kg FeTi to absorb 6.4 kg of H_2) by Brookhaven National Laboratory, and a 12-kW phosphoric acid fuel cell by United Technologies Corporation.

To overcome the weight penalty of the iron/titanium system for automotive application, Daimler-Benz designed an ingenious dual hydride/internal combustion system. The idea was to use a small quantity of the low-temperature iron/titanium hydride for the start-up and acceleration of the vehicle. As soon as the exhaust temperature is hot enough, its waste heat is sufficient to dissociate the light weight magnesium/nickel hydride, which can be carried as the main source of the hydrogen fuel for the cruising operation.

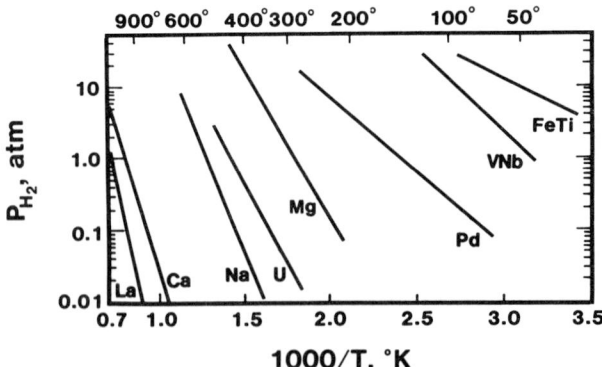

Fig. 15.1-14 Dissociation pressure of metal hydrides. (By permission of R. H. Wiswall, Department of Energy and Environment, Brookhaven National Laboratory, Upton, NY.)

The nonstorage applications of hydrogen, which have been examined, include: (i) compressors that absorb hydrogen at a relatively low pressure and temperature and release hydrogen at a higher temperature; and (ii) hydride heat pumps which make use of heats of reaction and/or fluid energy of a hydrogen flow between two or more hydride beds.

Hydrogen Stored in Combination with Other Elements or Compounds. On a gravimetric and volumetric basis, there are distinct advantages of storing and utilizing hydrogen in combination with some chemical elements (carbon, in particular, nitrogen) and with some compounds (e.g., $CO + 2H_2 \rightarrow CH_3OH$, $C_6H_6 + 3H_2 \rightarrow C_6H_{12}$). As shown in Fig. 15.1-3, there is a signficiant advantage in increasing the H : C ratio in fossil fuels both from storage and transmission considerations. With the increase in need and development of electric power generation from nonfossil primary energy sources (nuclear, renewable), hydrogen will be the natural choice of a readily transportable fuel. However, as seen from the preceding section, its storage method is a difficult challenge when compared with the relatively easy methods for carbonaceous fuels both in the gaseous and liquid forms. At least for the next 30 years, fuels with low H : C ratios (e.g., coal, tar sands, oil shale) will be in abundant supplies. In the transition from the "fossil" to the "nuclear" era, it will make sense to effectively use both in "carbon mobilization." In a recent review,[36] the advantages of petroleum fuels were illustrated, taking into consideration energy densities, which have an impact on storage and transmission of fuel, and capital costs (Table 15.1-4). With the depletion of the most readily available source of carbon hydrides, petroleum, it is essential to synthesize liquid carbon hydrides from other fossil sources such as tar sands, oil shale, and coal. It has been pointed out that while oil and gas accounted for 69% of the world's production of fossil energy, it makes up only 11% of the world's estimated total recoverable sources of fossil energy. The alternate fossil resources are either solids (oil shale, coal) or a very viscous fluid (tar) which are deficient in hydrogen. Although hydrogen needed for the synthesis of carbon hydrides can also be produced from these sources, it makes more sense to utilize hydrogen derived from nonfossil sources (nuclear, hydro) for this carbon mobilization operation. The process diagrams for production of synthetic fuels from tar sands, shale oil, and coal are illustrated in Fig. 15.1-15.[37] In the tar sands recovery process, tarry bitumen is produced. It is then coked to increase the hydrogen content and hydrotreated to remove sulfur and nitrogen. Conventional petroleum refining processes, which require hydrogen, are used to produce gasoline and mild distillate products. For the recovery of these carbon hydride fuels from oil shale, kerogen is produced by heating the rock to about 550°C in a retort. It is then pyrolyzed to produce gases and vapors which are condensed to form

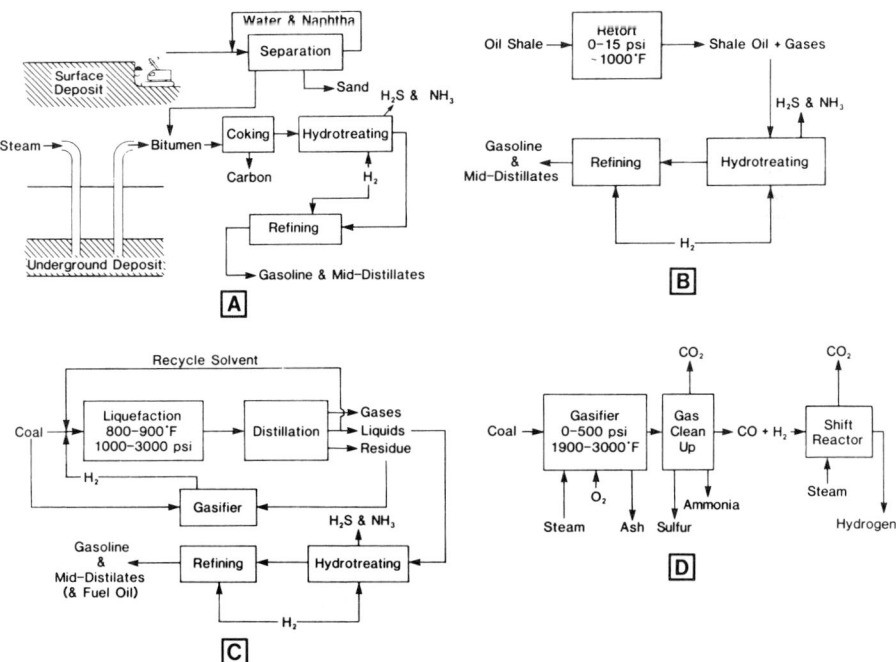

Fig. 15.1-15 Carbon mobilization with hydrogen: A, tar sands recovery; B, shale oil recovery; C, direct coal liquefaction; D, coal gasification. (Reprinted with permission from the *Int. J. Hydrog. Energy* **8**, 923 (1983). Pergamon Press, Ltd.)

raw shale oil. The raw shale oil is hydrotreated to remove sulfur and ammonia. The roles of hydrogen in processing tar sands and shale oil are to remove impurities and add the requisite amount of this gas to produce the desired carbon hydrides.

The processing of coal, unlike that of tar sands and oil shale, requires large quantities of hydrogen (see Fig. 15.1-15). High pressures (60–180 atm) and temperatures are required for the production of the gaseous and liquid carbon hydrides.

A second route to the production of liquid fuels from coal is by the water–gas reaction to produce carbon monoxide and hydrogen. More hydrogen, derived from a nonfossil source, is then added and, depending on the catalytic conditions, it should be possible to produce methane, methanol, or a mixture of hydrocarbons. Methanol can be used as a fuel or converted catalytically to gasoline.

It is premature to make estimates of the costs of production of readily transportable carbonaceous fuels from these nonpetroleum fossil fuels. However, it may be projected that it will be more economical to produce methanol than the gaseous and liquid hydrocarbon fuels from coal. Between the tar sands and oil shale approaches the former will be less expensive. It will not be the cost of hydrogen, but the recovery and chemical processes which will determine the production cost of the carbonaceous fuels.

Besides combing hydrogen with carbon to produce storable and transportable fuels, there have been proponents for producing N-H (ammonia, hydrazine) fuels. Both these fuels have a high hydrogen content (compared with gaseous, liquid, or solid hydrogen fuels). However, one has to be concerned with the toxic nature of both ammonia and hydrazine in addition to the high energy requirements for the production of these fuels from N_2 and H_2. It may be noted that ammonia is quite corrosive, as has been found in experimental internal combustion engines utilizing this fuel.

Hydrogen Transmission

A Survey of the Hydrogen Transmission Technologies, Their Characteristics, Applications, and Economics. The modes of transmission of hydrogen—gaseous, liquid, or solid—from the sites of their production to the sites of their utilization will have significant technoeconomic impact. The three general methods of transportation which need be considered are (1) pipeline distribution for the gaseous or liquid fuel; (2) road and rail vehicle transportation of the gas, liquid, and solid; (3) marine tankers for the liquid.[20] Table 15.1-5 summarizes the possible methods of transmission, projected applications, and economics. Technical and economic aspects of these transmission methods are presented in the following subsections. The methods for transmission of hydrogen combined predominantly with carbon or with nitrogen are well established and, hence, are not dealt with in this section.

Hydrogen Transmission Via Pipelines. An underground pipeline system for hydrogen transmission has been in operation for over 40 years in the Ruhr Valley in Germany. Its present length is 200 km and the input pressure of the gas is 10 atm. Air Products and Chemicals, Inc., operates a 200 km long dual H_2/CO pipeline system in the vicinity of Houston, Texas. The diameter of the pipes range from 10 to 30 cm and the operating pressure from 3 to 50 atm. With the transition to a "hydrogen economy" an interesting question will be whether the present natural gas distribution systems can be used at a pressure of about 30–60 atm. One concern will be hydrogen embrittlement. It may not be severe for the pipes but may be so for the pumps and valves. A disadvantage of hydrogen over natural

TABLE 15.1-5 TECHNOECONOMIC COMPARISONS OF HYDROGEN TRANSMISSION TECHNOLOGIES[a]

State	Transmission Mode	Relative Transmission Costs**	Application (Small/Large Capacity)
High Pressure Gas	Trucks	4	Small
	Pipe Lines	2	Large
Liquid Hydrogen	Tank Truck	3	Small/Medium
	Rail Road Tank, Barge, Pipelines }	2	Large
Solid–Metal Hydrides	Trucks	5	Small
Combined with Chemicals			
Natural Gas	Pipelines	1	Large
n–octane	Pipelines, Tankers, Trucks	1	Small to Large
Methanol	Trucks, Tankers, Pipelines	2	Small to Large

[a] Compiled from refs. 3, 28–38.
[b] 1, least expensive (< 0.01¢/kg · km); 5, most expensive (~ 0.6¢/kg · km).

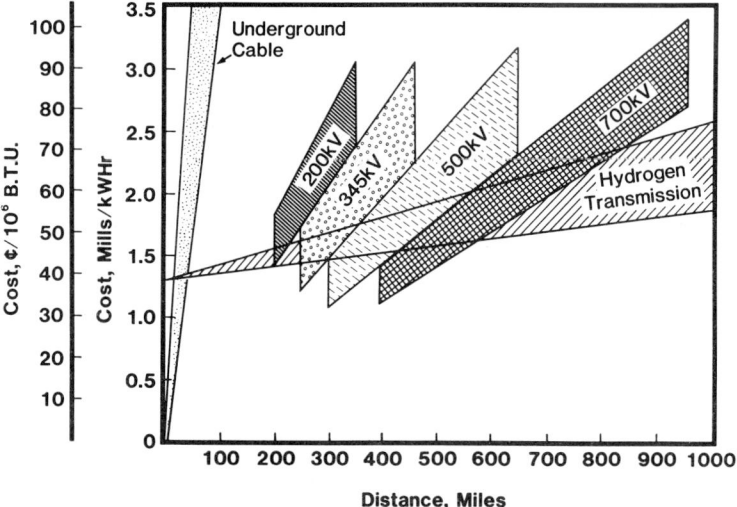

Fig. 15.1-16 Hydrogen versus electricity transmission costs as a function of distance. (Reprinted with permission from *The Electrochemistry of Cleaner Environments*, Edited by J. O'M. Bockris. Plenum Publishing, New York, 1972.)

gas is that for the same flow rate, the delivery rate of energy will be in the ratio of 1 : 3. This may necessitate considerably higher flow rates. Under these conditions, leak problems could become more severe. In terms of economics, it would appear that for the same operating parameters, it will cost three times as much to transmit hydrogen as for natural gas.

However, analyses have been made for the comparable costs of electricity transmission via overhead or underground cables and hydrogen, produced by electrolysis at distant nuclear power plants/electrolyzers, transmitted through pipelines and converted back to electricity. Over a distance of about 1000 km (600 mi), hydrogen transmission via pipelines will have advantages[38] over electricity transmission (Fig. 15.1-16). There is a strong dependence of electricity transmission on the magnitude of the high voltage.

Pipeline distribution of liquid hydrogen has not as yet been demonstrated. The closest approximation is the pipeline distribution of liquid hydrogen at the Kennedy Space Center and other NASA aerospace installations. According to one estimate, the cost of construction for a liquid hydrogen pipeline, 12 cm in diameter, will be about $300/m. This cost is excessive and makes hydrogen transmission via rail, truck, and tankers more economical.

Hydrogen Transmission Via Road and Rail Vehicles. The transportation of gaseous hydrogen in cylinders at pressures of 150–400 atm is a well established practice. It is reliable and convenient but not efficient because only about 2% of the weight of the cargo is that of hydrogen. Thus, the cost of shipping cylinders is quite high but economical for small quantities, for example, for scientific laboratory research and development programs. For larger quantities, the transportation of liquid hydrogen is routinely in tractors/trailers by Linde/Union Carbide, Air Products, and other industrial gas companies as well as by NASA. Highway trailers can carry up to 36,000 L—the limitation being volume and not weight.

It is most economical to transport liquid hydrogen in cryogenic rail tank cars. The Linde Division of Union Carbide moves liquid hydrogen from its production facility in Ontario, California, to its distribution center near Chicago, Illinois, in a fleet of 106,000 L rail cars.

It is too premature to consider the large-scale transportation of metal hydrides. As in the case of gaseous hydrogen, the weight penalty, but not that of volume, is high. From safety considerations, it will be the safest method to transport and distribute hydrogen.

Hydrogen Transmission Via Marine Tankers. Hydrogen transport via "very large crude carriers" has not been seriously considered. However, there is a technical precedent for cryogenic liquids (LNG carriers) from which some guestimates can be made about the technoeconomic assessments of liquid hydrogen transportation in such carriers.

15.1-4 Technoeconomic Projections of Hydrogen Utilization Technology: Hydrogen to Energy Conversion

The heat value associated with hydrogen (approximately 3.052 kcal/L generated when converted into water) can be converted into various energy forms such as thermal, mechanical, and electrical without any emissions of environmental concern. Significant efforts have been devoted in recent years towards development and demonstration of H_2 energy conversion technologies, and in this section, the thermomechanical and electrochemical routes pursued for effective utilization of hydrogen are described. For details related to the use of hydrogen in space conditioning, gas turbines, and as a chemical feedstock for petroleum refining, CH_4OH and NH_3 production, the reader is referred to refs. 3 and 38.

Thermomechanical Energy Conversion

Development of hydrogen as a fuel for thermomechanical energy has received a great deal of attention —the primary end applications being in space flight and internal combustion (IC) engines.

Space Applications as a Rocket Fuel. Liquid hydrogen is an ideal fuel for rocket propulsion, since, apart from its gravimmetric energy content and excellent combustion and cooling properties, it provides high specific impulse in terms of more payload per launch pad.[3] A typical value of specific impulse, which is approximately equal to the square root of temperature/molecular weight of the exhaust product, for hydrogen is approximately 450 compared to approximately 300 for oil.

Since the fuel energy/unit mass for hydrogen is about 2.8 times greater than that for, say, gasoline, liquid hydrogen can be used in air transportation as a fuel which results in 19–38% energy savings. This would also allow air-frame cooling which would permit use of lightweight alloys and, hence, an increase in the range and payload.

Internal Combustion (IC) Engines. Hydrogen is emerging as a potential fuel for IC engines because, unlike with gasoline engines, the exhaust gases are nonpolluting and can be adapted without major design changes of IC engines.[27,39-43] In addition, H_2-fueled IC engines can be operated with much leaner equivalence ratio, ϕ, than a gasoline engine—ϕ being defined as the stoichiometric to actual air–fuel ratio. The flammability limits of a combustible fuel–air mixture are the leanest and richest concentrations respectively which will just self-support a flame. Thus, H_2–air mixtures considered for engine operation vary between the lean flammability limit for coherent flames and the stoichiometric mixture (9–26% by volume or $0.23 < \phi < 1$), whereas the corresponding mixture for gasoline is 1.3–1.7% by volume ($0.76 < \phi < 1$). Lean operation of H_2 fueled engines reduces NO_x emissions and shows improved thermal efficiency. However, this may limit the practical range of ϕ and lead to increased ignition delay and cyclic pressure variations (which occur in gasoline engines if ϕ is > 1). The latter can be overcome by varying the hydrogen and/or air flow rate.

In the Otto-cycle engine, fuel and air are combined prior to ignition to form a relatively homogeneous mixture that is maintained within the limited flammability limits of the fuel. The energy content of the charge reaching the cylinder is controlled by throttling which reduces the cylinder pressure. On the other hand, the wide flammability limits of hydrogen allow control of the power output by changing the amount of hydrogen present in the mixture rather than by regulating the density of the mixture in the cylinders. This technique, termed quality-throttling, eliminates pressure drop caused by throttling and, hence, the engine does not have to expend much energy to pump air into its cylinders. The reduction of this pumping loss is the reason for the greatly improved part load efficiency of hydrogen-fueled engines and, hence, for the lower overall fuel consumption on an energy basis.

Because of the low ignition temperature, low quench distance and the wide flammability limits of H_2–air mixtures compared to gasoline–air mixtures, backfiring is a common problem with H_2-fueled engines. This is being circumvented by operating the existing IC engines with external mixture formation in the intake manifold or internal mixture formation in the combustion chamber with additional measures such as water injection and decreasing spark gap.

Hydrogen-fueled automobiles (see Figs. 15.1-17 and 15.1-18) are being investigated in many nations (see Table 15.1-6) of the world today. The preliminary results show an improvement in the efficiency of fuel consumption over the EPA gasoline rating of 4.44 kJ/m, by 22%, and in the engine brake thermal efficiency ranging between 23 and 58% over gasoline.[3,44,45] While these improvements are remarkable, problems related to size and weight penalties associated with on-board hydrogen storage (see Table 15.1-7), intake manifold fires, high NO_x emissions, and low power output still exist and most efforts are being focused to overcome these problems and optimize the vehicle performance characteristics.

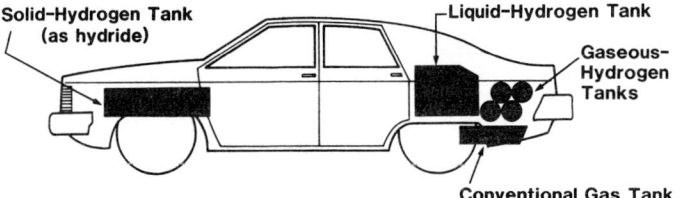

Fig. 15.1-17 Options for carrying hydrogen in automobiles. (Reprinted with permission from *Kirk–Othmer Encyclopedia of Chemical Technology*, Vol. 12, p. 1015. John Wiley & Sons, New York, 1980.)

Electrochemical Energy Conversion

Conversion of hydrogen energy to electricity is presently achieved in either fuel cells or batteries. The principles underlying the operation of these devices involving hydrogen are outlined here along with the present status of technologies and potential applications (see refs. 13 and 46–53 for details).

Fuel Cells. A fuel cell is an electrochemical system, where under reversible conditions, the free energy change (ΔG) of the reaction

$$H_2(g) + \tfrac{1}{2}O_2(g) \rightarrow H_2O(l) \qquad (15.1\text{-}25)$$

is converted to electrical energy. This free energy change is related to the reversible potential of the cell by the equation

$$\Delta G = -nFE_r \qquad (15.1\text{-}26)$$

where E_r is the reversible potential of the electrochemical cell in which reaction (15.1-25) is at equilibrium and n refers to the number of electrons participating in the reaction. When the cell is under non-equilibrium conditions, that is, when current is drawn from the cell, the dependence of cell potential (E) on the cell current (I) may be represented by

$$E = E_r - \eta_{act,c} - \eta_{act,a} - \eta_{conc,a} - \eta_{conc,c} - iR_i \qquad (15.1\text{-}27)$$

In the above equation, η_{act} and η_{conc} terms represent the activation and concentration overpotentials

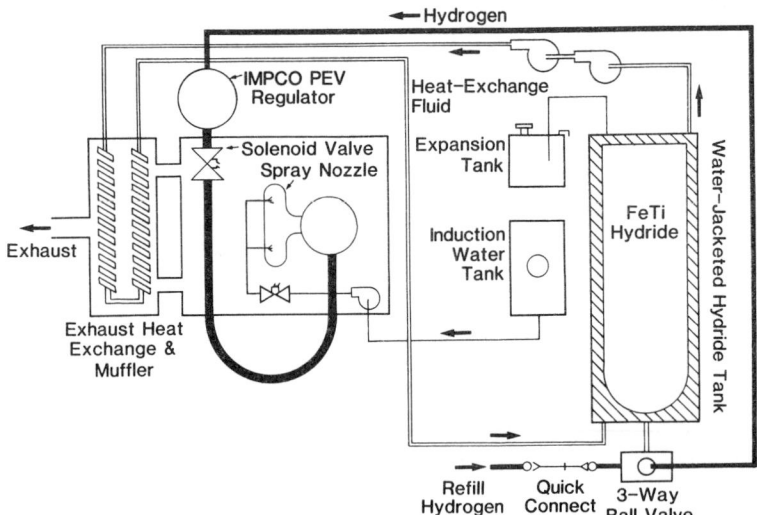

Fig. 15.1-18 An air cooled hydrogen engine in a Jacobsen engine in a Jacobsen Tractor. (Reprinted with permission from *Hydrogen Energy System*, Vol. 4, p. 1709. Pergamon Press, Ltd., London, 1979.)

TABLE 15.1-6 INTERNAL COMBUSTION ENGINES WITH HYDROGEN[a]

Vehicle	Remarks
BMW–518 (By DFVLR [b])	Employ liquid H_2; H_2 injection at the individual intake valve ports & electronically controlled water induction to suppress back firing. Operated over 2000km.
1979 Buick Century Sedan (By DFVLR & Los Alamos National Laboratory)	Similar to the above; operated over 3000km.
Suzuki Sedan (By Musashi Institute of Technology)	Pioneered use of cold hydrogen at $-30°C$ to $-50°C$ to the directly injected engine. Acheived increased power & elimination of induction manifold backfiring without recourse to water induction.
Mercedes Benz (By Daimler Benz AG)	Employ Fe–Ti and Mg–Ni hydrides; demonstrated dual–fuel operation with 60% H_2 in place of gasoline; H_2 being added to the gasoline/air mixture in the intake manifold or injected directly into the cylinder.
Peugeot 505, U.S. Post-Office Vehicle (By Billings Energy Corporation)	Used direct cylinder injection system. Employ Fe–Ti hydride, backflashing reduced by water induction which lowered NO_x emissions from 2500 to 25ppm.
GM 1980 Chevrolet Citation (By Solar Energy Research Institute)	Employed H_2–rich gases by catalytic cracking of CH_3OH; Demonstrated brake thermal efficiency improvements of 30 to 100% compared to gasoline depending on engine speed & torque.

[a] Complied from refs. 39–45.
[b] Deutsche Forschungs-und Versuchsanstalt für Luft und Raumfahrt.

TABLE 15.1-7 COMPARISON OF ENERGY STORAGE DENSITY OF VARIOUS FUELS[a]

Fuel System	Energy Storage Density	
	Mass Basis (mJ/kg)	Volume Basis (mJ/l)
Gasoline + Tank	31.8	28.7
Hydrogen + Compressed Natural Gas Cylinder	1.27	1.2
Fe–Ti Hydride (In Billings Postal Vehicle)	0.80	1.9
Dual–Bed Hydride (In Daimler–Benz City Bus)	2.99	10.1
Glass Microsphere Fe–Ti System	3.46	2.18
Liquid H_2 Tank (In UCLA Postal Vehicle)	29.0	4.84
NH_3 (Used Directly as a Fuel in an Engine)	12.2	6.61
NH_3 (Dissociated to H_2 Prior to Use)	7.41	7.29
CH_3OH (Dissociated)	20	19
Steam Reformed CH_3OH	13	11

[a] Compiled from refs. 39–45.

at the respective electrodes (a for anode and c for cathode), and iR_i to the ohmic overvoltage. A typical cell potential–current relationship depicted in Fig. 15.1-19 shows that energy conversion efficiency can be significantly lowered by the overvoltages and the ohmic contributions. To achieve high efficiencies and power densities, it is essential that the activation, concentration, and ohmic overpotentials are reduced to a minimum. Thus, activation overpotentials are reduced by employing better electrocatalysts, use of electrodes with a high surface area and operation at high temperatures and pressures. The greatest advance in fuel cell technology has been achieved by using porous gas diffusion electrodes with high porosity and high effective surface area which reduced η_{act} and η_{conc} significantly. Ohmic overpotential is reduced by minimizing the interelectrode spacing, using highly conducting electrolytes, electrodes, and current collectors. Since Eq. (15.1-25) involves hydrogen, "pristine" H_2 or hydrogen-rich gas mixtures derived from processing hydrocarbon fuel (e.g., natural gas, mixed coal, biofuels, refined petroleum products such as synthetic fuel gas, naphtha, methanol)

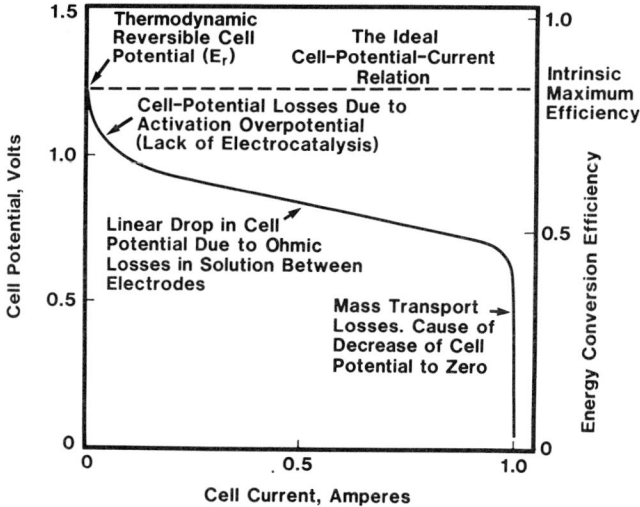

Fig. 15.1-19 Typical plot of cell potential versus current for fuel cells illustrating regions of control by various types of overpotentials.

may be used in fuel cells. (For details related to electrode kinetics and porous electrode phenomena, the reader is referred to articles quoted in refs. 13 and 46–53.)

The heat generation in fuel cells (Q) is expressed by the equation

$$Q = -\left(4.18\frac{T\Delta S}{nF}\right)i + i\sum \eta + i^2 R \tag{15.1-28}$$

The first term in the above equation arises from the entropy change of the reaction in the fuel cell. For example, with a hydrogen–oxygen fuel cell operating at 25°C, $T\Delta S$ is -12 kcal · mol^{-1} and is given off as heat even if the cell functions reversibly. The second term represents the heat generated due to the irreversibility caused by slow kinetics of electrode reactions and mass transfer limitations. The third term arises from the heat produced by the ohmic losses in the cell. Since this heat is a "waste" energy, the overall efficiency is reduced. To achieve increased efficiency of utilization of a fuel (e.g., natural gas, petroleum, or coal), this heat should be used in another part of the overall process, for example, for the reformer or gasification reactions and/or for thermal energy. Thus, the fuel cell should be envisioned not only as a device producing electricity but also as a heat generating system because this total energy concept enhances the value of fuel cells from an energy conservation point of view. The other advantages associated with fuel cells include: (i) simple operation compared to other energy conversion devices; (ii) clean, quiet operation (the exhaust gases from fuel cells are not noxious and are nonpolluting); (iii) adaptability to operate over wide temperature range; (iv) siting flexibility because of the modular designs; (v) multifuel flexibility.

A typical fuel cell power plant illustrated in Fig. 15.1-20 shows that its "heart" is the electrochemical cell (see Fig. 15.1-21) where chemical energy is converted into electrical energy without its efficiency being limited by Carnot's theorem. Since the energy crisis in 1973, several types of fuel cells are at various stages of development as power sources for terrestrial applications (see Table 15.1-8) for a list of major developers of these technologies). These fuel cells can be classified into three major categories based on their operating temperature as noted below:

1. Low temperature ($< 200°C$):
   (a) Alkaline fuel cells;
   (b) Solid polymer electrolyte fuel cells;
   (c) Phosphoric acid fuel cells;
   (d) Super acid electrolyte fuel cells.
2. Medium temperature ($500–700°C$):
   (a) Molten carbonate fuel cells.
3. High temperatures ($> 1000°C$):
   (a) Solid oxide fuel cells.

THE FUEL CELL POWERPLANT
Hydrocarbon Fuel To Electric Power

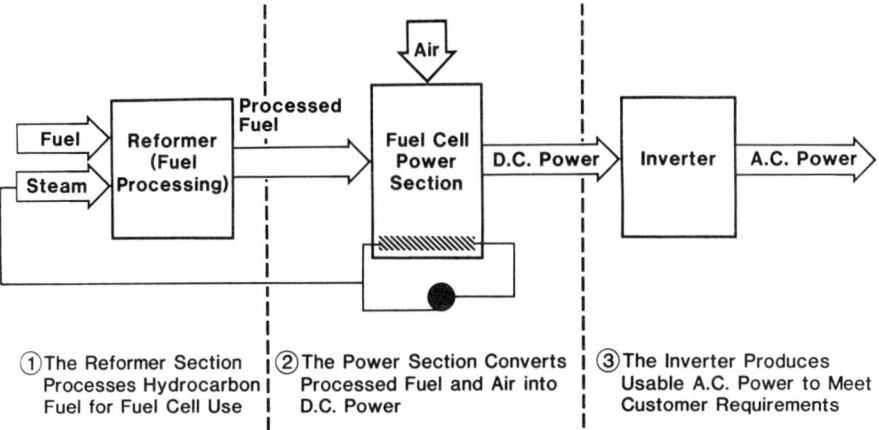

① The Reformer Section Processes Hydrocarbon Fuel for Fuel Cell Use
② The Power Section Converts Processed Fuel and Air into D.C. Power
③ The Inverter Produces Usable A.C. Power to Meet Customer Requirements

Fig. 15.1-20 Basic segments of a typical fuel cell power plant.

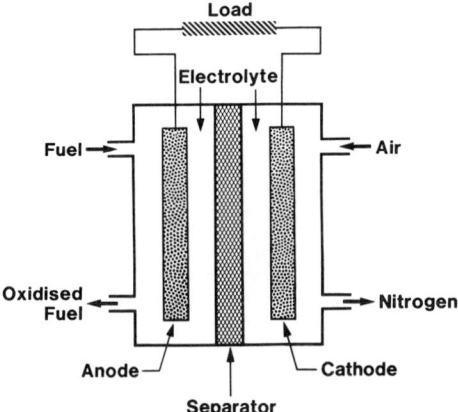

Fig. 15.1-21 A schematic representation of a fuel cell.

TABLE 15.1-8   MAJOR DEVELOPERS OF FUEL CELL TECHNOLOGY[a]

Fuel Cell Type	Major Developers
Phosphoric Acid (PAFC)	United Technologies Corporation (UTC), Energy Research Corporation (ERC), Westinghouse Electric Corporation, Engelhard Corporation, Tokyo Electric Power Co., Fuji Electric Co., Mitsubishi Electric Corporation, Toshiba Corporation, Hitachi Ltd., Sanyo Electric Co., "Moon Light" Project by The Agency of Industrial Science & Technology, Japan.
Alkaline (AFC)	UTC, Siemens, Elenco, Occidental Chemical Corporation, General Electric Co. (GE), Fuji Electric Co., "Moon Light", Institute for Hydrogen Systems.
Solid–Polymer Electrolyte (SPEFC)	General Electric
Molten Carbonate (MCFC)	UTC, GE, ERC, Institute of Gas Technology (IGT), Argonne National Laboratory (ANL), "Moon Light".
Solid Oxide (SOFC)	ANL, Westinghouse Electric Corporation, Brown Boveri and Cie, Dornier Systems, MITI (Japan), "Moon Light", Electrotechnical Laboratory (Sunshine Project; Japan)

[a] Compiled from ref. 47.

TABLE 15.1-9 PERFORMANCE CHARACTERISTICS OF FUEL CELLS[a]

Fuel Cell	Anode	Cathode	Electrolyte Temp.(°C)	Efficiency (%) Electric	Efficiency (%) Thermal	Rejected Heat (°C) Hot Water	Rejected Heat (°C) Steam	Costs ($/kW) Present	Costs ($/kW) Projected
PAFC	Pt/C (Pt loading<1 mg/cm²)	Pt, Pt alloys on carbon	99 wt. % H_3PO_4 (190–210°C)	40	40	82	132	4000	~750 (40,000 Hr. Life)
AFC[b]	Pt/C, Pt–Pd (1 mg/cm²)	Pt/C Ag/C	30 wt. % KOH (70–120)	70	—	—	—	NA[c]	200 to 300
SPEFC[b]	Pt on carbon (0.75 mg/cm²)	Nafion®	Nafion® (65–70)	70	—	—	—	NA	150–200
MCFC	Ni-based with stable ceramic additives	Lithiated NiO	Li_2CO_3+ K_2CO_3 Eutectic (700)	55	25	—	538	NA	~1000
SOFC	Ni–ZrO₂ cermet	Sr–doped Lanthanum manganite with Mg doped Lanthanum chromite interconnection	Yttria Stabilized Zirconia (1000)	45	35	—	815	NA	~1000

[a] Compiled from refs. 13, 47–52.
[b] AFC and SPEFC's efficiencies are based on the use of pure H_2 and O_2 and operate in the temperature range of 60–120°C. Hence the rejected heat is too low for practical use. Costs for fuel processing and power conditioning are not included in the projected capital costs for these systems.
[c] NA, not available.

Table 15.1-9 summarizes the performance characteristics of these fuel cells. A historical perspective of these types of fuel cells is outlined below with emphasis on the advantages and disadvantages as perceived at the present time.

Alkaline Fuel Cell (AFC). Alkaline fuel cells with liquid hydrogen and liquid oxygen as reactants found their first application in the U.S. space program. They employ 35 wt.% KOH as the electrolyte and operate at moderate temperatures (60–80°C) offering high chemical to electric energy efficiency. The performance characteristics of AFC with respect to power and energy on a weight and volume basis, and reliability of operation ranging from a few days to a few weeks are unmatched at the present time by the corresponding ones for any other power plant. Alkaline fuel cells employ Pt or Ni as an anode catalyst and Pt or a non-noble metal catalyst such as Ag or Ni as the cathode catalyst. High power densities with a fast start-up times (essential for transportation applications) have been demonstrated. However, disadvantages with alkaline fuel cells include (i) their inability to reject CO_2 resulting in carbonate formation in the porous electrode and, hence, a loss in performance, (ii) possible accumulation of H_2O_2 during O_2 reduction that causes degradation of carbon and consequent catalyst losses and flooding, and (iii) anode catalyst poisoning by CO which makes it imperative that pure H_2 be used with electrolyte recirculation and periodic replenishment of the electrolyte.

Solid Polymer Electrolyte (SPE) Fuel Cell (SPEFC). The use of SPE fuel cells, designed and fabricated by the General Electric Company for the Gemini space flights as auxiliary power sources demonstrated the first engineering application of fuel cells. The SPE fuel cell (see Fig. 15.1-22) primarily consists of a cation-exchange membrane—Nafion—which serves the multipurpose of separator, electrolyte, and support for electrocatalysts that are embedded on either side of the membrane. Thus, the cell design is elegant, compact, and simple to fabricate. The SPE fuel cell has the advantage that the electrolyte is a solid capable of operation at low temperatures. However, Nafion is conducting only when it is wet and, hence, for effective water management, the cell cannot be operated unless the fuel cell stack is pressurized. Enhanced oxygen reduction kinetics on Pt in this medium and low internal resistance are the primary reason for the high power densities of SPE fuel cells.

The prognosis for the SPEFC technology is dependent on finding low-cost membranes, as a substitute for the expensive Nafion, and at least a tenfold reduction in the noble metal loading. In addition, the CO level in the processed fuel should be very low (< 0.1%) or alternate CO-tolerant electrocatalysts should be developed to make the system economically attractive.

Phosphoric Acid Fuel Cell (PAFC). The most attractive feature of acid fuel cells is their tolerance to CO_2 and, hence, their ability to operate with hydrogen from hydrocarbon reformers, and with air without a CO_2 removal step. Phosphoric acid cells employ 85–105% aq H_3PO_4 immobilized in a matrix and operate at 160–210°C with Pt or Pt alloy catalysts for both the anodic and cathodic reactions. This system is in the most advanced stage of development—the main achievements made to date being: (1) threefold increase in power density from 70 to 200 mW/cm³, (2) reduction in platinum loading to < 1 mg/cm², and (3) demonstrated stack life of greater than 25,000 h. The achieved performance of UTC's 4.8 MW power plant in Japan (cell potential, 0.73V; current density, 325 mA/cm² at 205°C and 8 atm pressure; overall efficiency, 42%) represents a major advance in PAFC technology.

In PAFC, the electrocatalyst consists of Pt particles on a conducting carbon substrate that is susceptible to poisoning by CO present as an impurity in the reformate fuels. Tolerance to CO can be

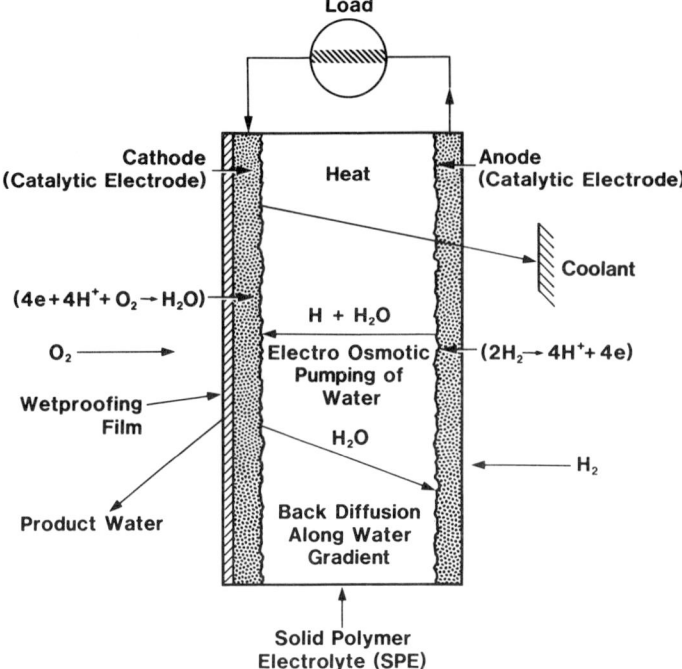

Fig. 15.1-22 A schematic of solid polymer electrolyte fuel cell.

improved by increasing the cell operating temperature to greater than 180°C. However, at high temperatures, the catalyst particles agglomerate with the consequent loss of active area and, hence, performance. The oxygen overpotential in acid media is high and the carbon support suffers from slow degradation resulting in loss of Pt into the electrolyte side. Thus, development of CO-tolerant catalysts for low temperature operation, lowering of oxygen overvoltage by finding intermetallic, alloy, or redox catalysts, and carbon stabilization (e.g., by boron addition to carbon or by using TiC based catalyst substrates) is vital for achieving the projected performance characteristics of advanced PAFC system (power density of 242 mW/cm^2 at 200°C and cell efficiency of 50%).

Super Acid Electrolyte Fuel Cell (SAEFC). Considerable attention was focused in recent years towards developing trifluoromethane sulfonic acid (TFMSA) as the fuel cell electrolyte in view of enhanced oxygen reduction kinetics on Pt in CF_3SO_3H at 100°C over that in H_3PO_4 at approximately 180°C. While these results are promising, acid concentration management appears to be a problem in TFMSA medium because of its higher vapor pressure. Developmental efforts are in progress to optimize these systems and to identify alternate high molecular weight super acids.

Molten Carbonate Fuel Cell (MCFC). The molten carbonate fuel cell employs 62% Li_2CO_3 + 38% K_2CO_3 eutectic in a ceramic particulate matrix as the electrolyte, and porous nickel-based electrodes

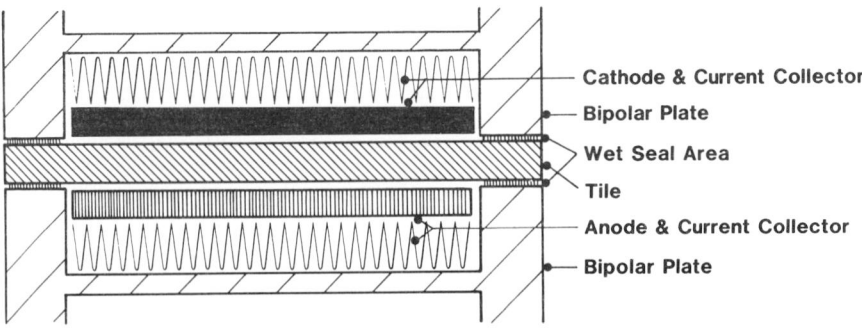

Fig. 15.1-23 A schematic of molten carbonate fuel cell.

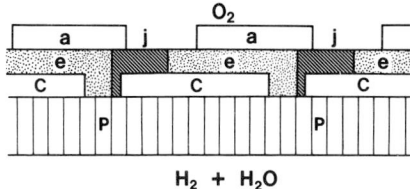

$$H_2 + H_2O$$

Fig. 15.1-24 A schematic of solid-oxide fuel cell: P, porous ZrO_2 support; C, cathode; e, solid electrolyte; J, intercell junctions; a, anode.

as the anodes and cathodes (see Fig. 15.1-23). The half-cell reactions in MCFC are

$$H_2 + CO_3^{2-} \rightarrow CO_2 + H_2O + 2e \quad \text{(anodic reaction)} \tag{15.1-29}$$

$$\tfrac{1}{2}O_2 + CO_2 + 2e \rightarrow CO_3^{2-} \quad \text{(cathodic reaction)} \tag{15.1-30}$$

MCFC has several advantages over PAFC that include high power densities at high cell voltages, suitability for cogeneration, and capability for oxidizing hydrocarbons including natural gas directly without any external reformation. Many of the early operational problems associated with corrosion, poor thermal cycling, and CO_2 recycling performance have been overcome. Present efforts are aimed at minimizing electrolyte evaporation, anode sintering and development of stable cathodes and seals to achieve long life.

Solid Oxide Fuel Cell (SOFC). The solid oxide fuel cell (see Fig. 15.1-24) is essentially a H_2-O_2 fuel cell with a solid, ceramic oxide material serving as the electrolyte—the mechanism of ionic conduction being oxygen ion transport via anion defects in the solid–oxide crystal lattice. The component reactions in this cell are

$$O_2 + 4e \rightarrow 2O^{2-} \quad \text{(at the cathode)} \tag{15.1-31}$$

$$H_2 + O^{2-} \rightarrow H_2O + 2e \quad \text{(at the anode)} \tag{15.1-32}$$

Since the operating temperature is about 1000°C, the overpotential losses are negligible and, hence, energy efficiencies as high as 54% at 100 mA/cm² have been demonstrated. Because of the absence of liquids, problems associated with pore flooding and maintenance of a three-phase interface are totally avoided.

SOFC's operate with a $CO + H_2$ mixture from a coal gasifier and employ non-noble metal catalysts such as Ni-ZrO_2 cermet at the anode and doped lanthanum manganite at the cathode. Most structural materials are ceramic-based, and high quality waste heat makes these cells very attractive for coal burning fuel cell power generating systems. Since the operating temperatures are high, there are several disadvantages related to problems involving choice of materials with proper matching of thermal expansion coefficients, interdiffusion, and electrical resistivities, which still remain to be solved. SOFC technology is still at a research stage unlike other fuel cell systems.

Applications. A summary of fuel cells developed for various end uses is presented in Table 15.1-10. The U.S. programs focus on the efficient utilization of (i) natural gas and petroleum derived

TABLE 15.1-10   FUEL CELLS: PRESENT STATUS AND APPLICATIONS[a]

Range of Fuel Cell Power Units	Fuel	Fuel Cell	Applications	Present Status	Near–Term Projections
1–10 kW	Liquid H_2, O_2, Methanol	AFC, SPE, PAFC	•Space •Vehicular (Power source for electric vehicles) •Stand-by–power •Telecommunications	•2–12kW AFC and SPEFC •1, 3, 5kW PAFC, 6kW AFC •7kW AFC	•4.5kW regenerative SPE tested for >22,000 hrs. for space station in 1984; 5kW PAFC (methanol based) by ERC in 1984; 5–10kW MCFC & 1kW SOFC by Moon Light in 1986.
10–100 kW	Liquid H_2, O_2, Methanol, Natural Gas Naphtha, SNG	SPE, AFC, PAFC, MCFC	Space, transportation, stationary power (Peaking devices, spinning reserves, on–site integrated energy systems), industrial co–generation, dispersed generation, electric utilities	•Five 40kW PAFC units tested for >25,000 hrs.; 30kW PAFC by Fuji Electric Co. since 1982	•20kW PAFC and SPEFC for vehicular in 1984. •50kW PAFC by Toshiba & Hitachi in 1986. •5 to 40kW AFC by Elenco for transportation & other uses. •49 additional 40kW PAFC installations in 1983.
100 kW to >10 MW	Natural Gas, Naphtha, SNG, Coal Gas	PAFC, MCFC, SOFC	Same as above except for space & transportation	•4.8MW PAFC built & tested for >10 months in Japan; Similar unit will be commissioned in N.Y. City in 1984	

TABLE 15.1-11 FUEL CELL MARKET
POTENTIAL IN THE YEAR 2000[a]

Heat Rate[b] (BTU/kWh)	Market Penetration (%)
Large Utilities	
9300	4–27
9000	30
7500	38
7500	40[c]
Small Utilities	
9300	27–40[c]
8500	30–40[c]
7500	40[c]

[a] These projections are based on 20-year expansion and an optimum generation mix methodology. The figures for the large utilities are from EPRI Report EM-336 and those for small utilities from EPRI EM-696.
[b] See ref. 46 for definition.
[c] Fuel availability constrains penetration to a maximum of 40%

(e.g., naphtha) fuels in PAFC plants for peaking devices or spinning reserves in electric utilities and for on-site integrated energy systems in apartment buildings, shopping centers, hospitals, office buildings, and so on (in addition, PAFC power plants with methanol as the fuel are at an early stage of research and development as power sources for military and transportation applications); and (ii) natural gas and coal in MCFC and SOFC power plants for intermediate and base-load power generation in electric utilities and cogeneration in industrial plants.

The main emphasis at the present time is toward reducing the cost of PAFC power generation units from ~ \$4000/kW (compared to ~ \$1450–1700/kW for a coal-fired power plant) to < \$1000/kW, and demonstrate long life using megawatt power plants (see refs. 13 and 46–53).

MCFC and SOFC systems are highly efficient and generate high-temperature waste heat that is best suited for cogeneration. Since the cell components are made of low cost materials and fuel reformer designs are simple, these systems appear to be ideal for centralized and decentralized power generation. However, they are in the early stages of development and may replace PAFC's as the second and third generation technologies as the current problems are resolved. Thus, the fuel cell market potential appears to be attractive (see Table 15.1-11) in the year 2000 and beyond.

H$_2$-Based Batteries. Unlike in the fuel cells, H$_2$ can be used for energy conversion in batteries either by internal generation or by internal storage in the form of hydride. A brief summary of the present status of these systems is presented in Table 15.1-12 and the reader is referred to recent reviews[13,54] for details.

TABLE 15.1-12 H$_2$-BASED BATTERY SYSTEMS[a]

System	Overall Reaction	Power Density (Wh/kg)	# of Cycles Achieved	Problems	Potential Applications
H$_2$/NiO	$NiOOH + \frac{1}{2}H_2 \rightleftharpoons Ni(OH)_2$	~65	>1000	Cost, high pressure containers for H$_2$	Aerospace; Replacement of Ni–Cd batteries
H$_2$/AgO	$AgO + H_2 \rightleftharpoons Ag + H_2O$	~90–100	500 (goal)	Cost, life, electrolyte management	Aerospace
H$_2$/Cl$_2$	$H_2 + Cl_2 \rightleftharpoons 2HCl$	75–90	} Not Available	} Cost, life and safety	} Utility Load–leveling
H$_2$/Br$_2$	$H_2 + Br_2 \rightleftharpoons 2HBr$	45–60			
Metal Hydrides with Ni–Ti and LaNi$_5$	$H_{lattice} \rightleftharpoons H_{ads} + OH^- \rightleftharpoons H_2O + e$	0.3Ah/kg	500 (Goal)	Corrosion, cost, stability	Utility Load–Leveling

[a] Compiled from refs. 13 and 54.

15.1-5 Safety and Materials Aspects of Hydrogen Energy Systems

Energy Technologies and Physical Properties in Relation to Safety Aspects

Several studies have been carried out in the safety and environmental aspects related to production, storage, transmission, and utilization of hydrogen.[55-60] With respect to its production by water electrolysis and utilization in fuel cells, there are hardly any hazards which have been reported. Most of the studies, however, are devoted to the storage and transmission of hydrogen. Edeskuty and co-workers[56] have summarized the properties of hydrogen that could possibly enhance the risks of transportation of gaseous and liquid hydrogen. Comparisons of hydrogen with natural gas, liquefied natural gas, propane, and gasoline in respect to these properties have been made in this study and the conclusions may be summarized as follows:

   1. **Flammability.** Gaseous hydrogen is flammable when mixed with air in proportions from 4 to 75% hydrogen by volume. The lower percentage is more important when considering accidents involving hydrogen. It is not much different from the lower flammable mixture rates of methane (5.3%) and is higher than that of propane (2.1%) and gasoline (1.0%).
   2. **Diffusivity.** The diffusion rate of hydrogen in air is more rapid than that of these other fuels and, hence, hydrogen is perhaps more dangerous in confined spaces. However, this rapid diffusion rate makes hydrogen safer than the other fuels in open air because the lower limit of flammability is more difficult to attain and sustain. Hydrogen can leak through small openings that may be watertight or airtight. Thus, valves, seals, and connections for hydrogen need have smaller tolerances than usual. Advances in hardware engineering and manufacturing have made it possible to minimize such leaks significantly.
   3. **Low boiling temperature.** Liquid hydrogen has an extremely low boiling point (20.3 K). The only known substance that has a lower boiling point is helium. All other gases will condense and freeze at the boiling point of hydrogen. For instance, if air enters the vacuum insulation jacket of a double walled liquid hydrogen container, it will freeze. The solidified air can interfere with the operation of valves. On warming, the frozen air can vaporize and cause a pressure build-up in the vacuum space. In the event of an accident spill, liquid hydrogen can cause frostbites. This is also the case with natural gas which has a boiling point of 111 K. Propane does not have this problem because it is gaseous at ambient temperatures and at higher pressures (6 atm) it is in a liquid state.

There are some properties of hydrogen which reduce the risk of handling it, as compared to the aforementioned fuels. Its buoyancy is less than that of air and hence, the gas quickly rises and disperses in unconfined spaces. Its buoyancy velocity is 1.2–9 m/s, which may be compared to 0.8–6 m/s for methane and the negative buoyancy of propane and gasoline vapor. The two latter fuels are heavier than air and collect around spills, which thus increase dangers of large explosions. Plumes of methane gas formed from LNG spills on water always lie low on the surface until diluted to about 5% methane or less. The mixture of cold natural gas and air become buoyant because of the heat liberated by the condensation of the water vapor in the air. The buoyancy of hydrogen gas has advantages in eliminating hazards only in open spaces but not so in confined spaces.

Liquid hydrogen vaporizes rapidly because of its low boiling point and small heat of vaporization. The vaporization rate, which depends on the type of surface underlying the liquid, is approximately 2.5–5 cm/min unignited and 3.0–6.6 cm/min for a burning pool. This rate is about 10 times the vaporization rate for methane and 250 times that of gasoline. A hazardous situation resulting from a spill of liquid hydrogen disappears within a short time.

Safety Considerations in Regard to Gaseous Hydrogen Transmission in Pipelines

Several studies have been conducted on the problems associated with hydrogen transportation in the state of the art natural gas pipeline network. The extent of loss of natural gas during transmission from wellhead to the user has been estimated to be about 10%. Hydrogen will leak from the pipeline at three times the rate of natural gas but because its energy density is one-third, the energy loss is the same in both cases. The hazards due to releases of large quantities of hydrogen are no greater than similar accidents from natural gas.

As stated in Section 15.1-3, the safe distribution of hydrogen via pipelines has been demonstrated in the Ruhr Valley, Germany, and in Texas. The safety records of these studies cannot be directly extrapolated to new pipeline networks needed in a "hydrogen energy scenario." These will have to be designed and constructed taking into consideration (i) incorporation of compressor stations, (ii) higher operating pressures, and (iii) large quantities of hydrogen for transmission. The major pipeline problems are connected with leaks at the welds. Proper techniques and materials will be essential in their construction.

Safety Considerations in Regard to Liquid Hydrogen Storage and Transmission

Liquid hydrogen transmission by trailer, rail tank car, and barge has been demonstrated to be quite safe. Where pipelines do not exist, hydrogen is transported and delivered as the liquid on a relatively large scale and over relatively large distances. The National Aeronautical and Space Administration, the Air Force, National Bureau of Standards, and Los Alamos National Laboratory have established safety practices and procedures for the handling of liquid hydrogen. Liquid hydrogen has a great tendency to boil. There is a 50-fold increase in volume during this phase transition. At 30 K, the saturation vapor pressure of hydrogen is 10 atm. Thus, typical hydrogen Dewars will have their appropriate safety valves set at a slight positive pressure. Cold vapors venting through a relief valve can cause freezing of the moisture in the surrounding valve.

Hydrogen in Metals and Alloys

Hydrogen Embrittlement. A hydrogen environment is deleterious on most materials[3,55-60] depending on exposure pressure, temperature, and time resulting in the phenomenon of hydrogen embrittlement. Identified mechanisms of hydrogen embrittlement include:

1. Reaction embrittlement which occurs at elevated temperatures resulting in hydrogen absorption during manufacture or in service by metals, alloys, or impurities leading to the formation of "brittle" metal hydrides.
2. Internal embrittlement due to the hydrogen retained in the alloy during processing which by diffusion to the grain boundaries or stressed zones (e.g., welds) leads to cracks and general decrease of ductility.
3. Environmental embrittlement arising from hydrogen absorbed from a high pressure medium leading to loss of material ductility. It may be noted that the solubility of hydrogen is proportional to the square root of the partial pressure of hydrogen.[3]

Materials found to be suitable in prolonged high pressure gaseous hydrogen service include 300 series (austenitic) stainless steels, oxygen free copper, brass, and most aluminum alloys.

Cold Embrittlement. Many materials suffer from loss of ductility at ductile/brittle transition temperature, which for liquid hydrogen service should be below 20 K. Since the impact resistance at these temperatures is substantially reduced, catastrophic failure of piping and vessels without forewarning is possible and if hydrogen is under pressure, shrapnel can be propelled outward igniting any released hydrogen or causing personal injury.

  Materials of construction suitable for liquid hydrogen service include metals such as austenitic (300 Series) stainless steels, copper, bronze, monel, aluminum, bronze alloy (everdur), and inconel (Ni/Cr/Fe alloy), and polymeric materials such as polyester fiber, teflon, Kel-F, mylar, and nylon. Teflon class polymers retain excellent ductility in liquid hydrogen service and are very effective in making seals for valves, gaskets, and so on. While using liquid hydrogen, temperature excursions and gradients encountered during start-up and shutdown operations can lead to thermal stresses at joints where the materials are structurally and compositionally dissimilar. Hence, it is necessary to enforce strict quality control measures providing allowance for thermal variations.

  Studies relating to the utilization of hydrogen in industrial and commercial markets[3,55-60] showed no unusual safety problems assuming that the normal precautions for hydrogen handling and hazards are undertaken. However, safety problems may exist in the transportation and residential fuel sectors, and additional analyses are needed in these areas. The reader is referred to the major technical note by Hord[57] for hydrogen safety guidelines and applicable regulatory codes.

REFERENCES

15.1-1 J. Verne, *Twenty-Thousand Leagues Under the Sea*, Paris, 1869.

15.1-2 J. B. S. Haldane, in a lecture "Daedalus, or Science of the Future," at Cambridge, 4 February 1923.

15.1-3 J. O'M. Bockris, *Energy Options—Real Economics and the Solar Hydrogen System*, Wiley, New York, 1980.

15.1-4 D. S. Scott, *Hydrogen in an Era of Energy Transitions*, Institute for Hydrogen Systems Report #IHS-1-83, 1983.

15.1-5 S. Srinivasan, unpublished results.

15.1-6 *Chemical Week*, p. 45, Dec. 15, 1982.

15.1-7 B. G. Mandelik and D. S. Newsome, *Kirk–Othmer Encyclopedia of Chemical Technology*, 3rd ed., Vol. 12, p. 938, Wiley, New York, 1980.

15.1-8 *Hydrogen: Production and Marketing*, W. N. Smith and J. G. Santangelo (eds.), ACS Symposium Series 116, American Chemical Society, Washington, D.C., 1980.

15.1-9 L. O. Williams *Hydrogen Power—An Introduction to Hydrogen Energy and its Applications*, Pergamon Press, New York, 1980.

15.1-10 D. P. Gregory, C. L. Tsaros, J. L. Arora, and P. Nevrekar, p. 3 in ref. 8.

15.1-11 A. M. Watson, *Hydrocarbon Processing*, p. 91, March 1983.

15.1-12 B. V. Tilak, P. W. T. Lu, J. E. Colman, and S. Srinivasan, *Comprehensive Treatise of Electrochemistry*, Vol. 2, p. 1, J. O'M. Bockris, B. E. Conway, E. Yeager, and R. E. White (eds.), Plenum Press, New York, 1981.

15.1-13 A. J. Appleby, M. Chemla, H. Kita, and G. Brönoel, *Encyclopedia of Electrochemistry of Elements*, Vol. IX Part A, Ch. IXa-3, A. J. Bard (ed.), Marcel Dekker, New York, 1982.

15.1-14 F. Gutmann and O. J. Murphy, *Modern Aspects of Electrochemistry*, Vol. 15, R. E. White, J. O'M. Bockris, and B. E. Conway (eds.), Ch. 1, Plenum Press, New York, 1983.

15.1-15 R. L. LeRoy, *Int. J. Hydrog. Energy* **8**, 401 (1983).

15.1-16 R. L. LeRoy and A. F. Hufnagl, *Int. J. Hydrog. Energy* **8**, 581 (1983).

15.1-17 P. W. T. Lu, *Int. J. Hydrog. Energy* **8**, 773 (1983).

15.1-18 D. Vanvelzen and H. Langenkamp, *Int. J. Hydrog. Energy* **7**, 629 (1982).

15.1-19 M. Grätzel, *Modern Aspects of Electrochemistry*, Vol. 15, Ch. 2, R. E. White, J. O'M. Bockris, and B. E. Conway (eds.), Plenum Press, New York, 1983.

15.1-20 T. H. Maugh, *Science* **221**, 1358 (1983).

15.1-21 K. Rajeshwar, P. Singh, and J. Dubow, *Electrochim. Acta* **23**, 1117 (1978).

15.1-22 J. D. Brosseau and J. E. Zajic, *Int. J. Hydrog. Energy* **7**, 623 (1982).

15.1-23 *Bio-Solar Hydrogen Production*, *Hydrogen Energy System*, Vol. 3, A. Mitsui, T. N. Veziroglu, and W. Seifritz (eds.), Pergamon Press, New York, 1979.

15.1-24 R. B. Moore, *Int. J. Hydrog. Energy* **8**, 905 (1983).

15.1-25 K. Christiansen and K. Andressen, *Hydrogen Energy Progress IV*, T. N. Veziroglu, W. D. Van Vorst, and J. H. Kelley (eds.), p. 1539, Pergamon Press, New York, 1982.

15.1-26 S. Srinivasan, Abstract #442, *Extended Abstracts of the Electrochem. Soc.* **82-1**, 729 (May 1982).

15.1-27 C. A. Kukkonen, *Automotive Eng.* **89**, 69 (1981).

15.1-28 C. Carpetis, *Proceedings of the 18th IECEC*, Paper #839284 (1983); *Int. J. Hydrog. Energy* **7**, 191 (1982).

15.1-29 W. J. D. Escher, *The Hydrogen Delivery Step*, Paper presented at the Biomass Conference organized by Energy Institute, November 1983.

15.1-30 K. Kaiba and G. Hanselmann, *Proceedings of International Seminar on Hydrogen Energy as an Energy Vector*, p. 504, 1980.

15.1 31 C. R. Baker, *Proceedings of the 4th World Hydrogen Conference*, Pasadena, California, T. N. Veziroglu, W. D. Van Vorst, and J. H. Kelley (eds.), Volume 3, 1982.

15.1-32 F. J. Edeskuty and K. D. Williamson, *Hydrogen: Its Technology and Implications*, K. E. Cox and K. D. Williamson (eds.), Vol. II, Ch. 3, CRC Press Inc., Boca Raton, Florida, 1977.

15.1-33 A. L. Bain, *Int. J. of Hydrog. Energy* **1**, 173 (1976).

15.1-34 J. R. Barclay, *20th Joint ASME/AIChE National Heat Transfer Conference*, Milwaukee, Wisconsin, Aug. 2–5, 1981, ASME 81-HT-82.

15.1-35 G. Strickland, J. J. Reilley, and R. H. Wiswall, *THEME Conference*, Feb. 1974, Miami, S4-9,; J. J. Reilley, in ref. 32, Ch. 2.

15.1-36 G. Eklund and O. Von Krusenstierna, *Int. J. Hydrog. Energy* **8**, 463 (1983).

15.1-37 F. B. Sprow and G. K. Vick, *Int. J. Hydrog. Energy* **8**, 923 (1983).

15.1-38 D. P. Gregory, D. Y. C. Ng, and G. M. Long, *The Electrochemistry of Cleaner Environments*, J. O'M. Bockris (ed.), Plenum Press, New York, 1972.

15.1-39 V. Soots, *Hydrogen—A Challenging Opportunity*: Vol. 5: *Hydrogen Utilization in Surface Vehicles*, Report prepared for the Ontario Hydro Energy Task Force, Ontario Ministry of Transportation and Communications, Sept. 1981.

15.1-40 W. J. D. Escher, *Int. J. Hydrog. Energy* **7**, 519 (1982).

15.1-41 H. Buchner, *Hydrogen Energy Progress IV*, Vol. 1, p. 3, June 1982.

15.1-42 J. S. Wallace and C. A. Ward, *Int. J. Hydrog. Energy* **8**, 255 (1983).

15.1-43 Wm. D. Van Horst, J. H. Kelley, and T. N. Veziroglu, *Int. J. Hydrog. Energy* **8**, 857 (1983).

15.1-44 J. O'M. Bockris, *Kirk–Othmer Encyclopedia of Chemical Technology*, 3rd ed., Vol. 12, p. 1015, Wiley, New York, 1980.

15.1-45 R. E. Billings, *Hydrogen Energy System*, Vol. 4, p. 1709, T. N. Veziroglu and W. Seifritz (eds.), Pergamon Press, New York, 1979.

15.1-46 B. V. Tilak, R. S. Yeo, and S. Srinivasan, *Comprehensive Treatise of Electrochemistry*, Vol. 3, Ch. 2, J. O'M. Bockris, B. E. Conway, E. Yeager, and R. E. White (eds), Plenum Press, New York, 1981.

15.1-47 *National Fuel Cell Seminar Abstracts*, Orlando, Florida, Nov. 13–16, 1983, Courtesy Associates, Inc., Washington, D.C.

15.1-48 *Fuel Cell Energy—Today's High Technology—Tomorrow's Reality*, Distributed at the National Fuel Seminar, Orlando, Florida, Nov. 13–16, 1983, Published by Insights West, Inc.

15.1-49 R. K. Kalia, Report #82-444K, Ontario Hydro Research Division (1982).

15.1-50 A. P. Fickett, *Int. J. Hydrog. Energy* **8**, 617 (1983).

15.1-51 S. Srinivasan, *Hydrogen Energy Progress V*, Vol. 4, p. 1717, T.M. Veziroglu and J. B. Taylor (eds.), Pergamon Press, New York, 1984.

15.1-52 *Proceedings of the Renewable Fuels and Advanced Power Sources for Transportation Workshop*, H. L. Chum and S. Srinivasan (eds.), SERI/CP-234-1707 DE 83011988, 17–18 June 1982, Boulder, Colorado.

15.1-53 *Hydrogen Times*, #4, Winter 1984.

15.1-54 J. McBreen, *Comprehensive Treatise of Electrochemistry*, Vol. 3, Chap. 10, J. O'M. Bockris, B. E. Conway, E. Yeager, and R. E. White (eds.), Plenum Press, New York, 1981.

15.1-55 J. Hord, *Int. J. Hydrog. Energy* **3**, 157 (1976); *NBS Technical Note 690*, **PB 262551**, October 1976.

15.1-56 F. J. Edeskuty, J. R. Bartlett, R. V. Carlson, W. F. Stewart, K. E. Cox, R. Keider, and M. Williams, Los Alamos Scientific Laboratory, *Hydrogen Safety and Environmental Control Assessment*, Nov. 1, 1978–Sept. 30, 1979, Final report LA-8225-PR.

15.1-57 J. Hord, *Int. J. Hydrog. Energy* **5**, 579 (1980).

15.1-58 *Hydrogen: Its Technology and Implications*, Vol. 5, Chs. 2–4, K. E. Cox and K. D. Williamson Jr., (eds.), CRC Press, Inc., Boca Raton, Florida, 1979.

15.1-59 R. E. Knowlton, *Hydrogen Safety—An Assessment of the Safety Problems Associated with the Use of Hydrogen as a Ground Transportation Fuel*, a final report submitted by Chemetics International Company to the National Research Council (1983).

15.1-60 P. Palumbo and B. A. Steinberg, *Hydrogen Safety in a R & D Environment*, IHS Report No. TT-1-83, May 1983.

15.2   FUSION POWER

Thomas C. Varljen

Energy production in stars[1] is thought to be primarily derived from a series of nuclear reactions involving the fusion of light nuclei. The control of these reactions has been the subject of worldwide research since the 1950s, with the ultimate objective of developing a commercial-scale electric power source or, alternatively, a source of chemical fuels or fissile fuels. The task has turned out to be exceedingly difficult, and there is no guarantee that a fusion power plant will be technically feasible or that fusion-produced electricity or fuels will be economically competitive. A great deal of progress has marked fusion research in recent years, however, and expectations are high that a technical demonstration of a fusion power system[2] will be completed shortly after the turn of the century.

15.2-1   Fusion Fuel Cycles

Nuclear fusion involves the formation of a heavier element via the "fusing" of one or more lighter elements. From the point of view of energy production, reactions in the light elements, especially isotopes of hydrogen, are the most interesting. Energy is released because the sum of the masses of the products are less than the sum of the masses of the nuclei that are "fused."

The target fusion fuel cycle given most research emphasis is based on deuterium, a stable hydrogen isotope that is found in ordinary water in a concentration of 0.0148% of the hydrogen atoms. The ultimate or ideal fusion fuel cycle is represented by the following sequence of reactions:

$$^2_1D + {}^2_1D \rightarrow {}^3_2He + n + 3.2 \text{ MeV} \tag{15.2-1}$$

$$^2_1D + {}^2_1D \rightarrow {}^3_1T + p + 4.0 \text{ MeV} \tag{15.2-2}$$

$$^2_1D + {}^3_1T \rightarrow {}^4_2He + n + 17.6 \text{ MeV} \tag{15.2-3}$$

$$^2_1D + {}^3_2He \rightarrow {}^4_2He + p + 18.3 \text{ MeV} \tag{15.2-4}$$

The D-D reactions are equally probable and, in principle, the intermediate products, tritium and ^{3}He will be completely consumed, leading to a total average energy release per D-D reaction of about 22 MeV. The energy release appears as kinetic energy associated with the reaction products; 35% of the reaction energy is associated with high-energy neutrons.

Table 15.2-1 compares the energy released from fusion reactions with chemical combustion and fission processes (in terms of specific energy release: kW · h of energy released per gram of fuel) to illustrate that fusion is somewhat better than fission in this regard. The relative fusion power densities for the reaction sequences given in Eqs. (15.2-1) to (15.2-4) are shown in Fig. 15.2-1,[3] as a function of fuel temperature in keV. Not only does the D-T reaction, Eq. (15.2-3), provide the greatest relative

**TABLE 15.2-1 COMPARISON OF ENERGY
RELEASE PROCESSES**

Process	(kW · h/g)
Combustion of hydrogen	0.0044
U-235 fission	22,800
D-T fusion	94,000
D-D fusion	22,000
	27,000
D-^{3}He fusion	98,000

energy yield, the ion temperatures required for a given energy yield are significantly lower and therefore less demanding from a technological point of view.

Deuterium and tritium are therefore the fuels most likely to be used in first-generation fusion power plants. Tritium is radioactive, has a half life of 12.35 years, and decays by emitting a low-energy beta particle. Tritium is therefore not naturally occurring, however, it may be bred by the following transmutation reactions in lithium:

$$^6\text{Li} + n \rightarrow {}^4\text{He} + \text{T} + 4.8 \text{ MeV} \tag{15.2-5}$$

$$^7\text{Li} + n \rightarrow {}^4\text{He} + \text{T} + n - 2.87 \text{ MeV} \tag{15.2-6}$$

The reaction with ^{6}Li is exothermic and will occur with incident neutrons of any energy, though the reaction probability is greatest for slow neutrons. The ^{7}Li reaction is endothermic, therefore, the incident neutron must have an energy exceeding 2.87 MeV. The abundance of these isotopes in natural lithium is 92.6% ^{7}Li and 7.4% ^{6}Li. Lithium is therefore an important constituent in the fuel cycle and must be present in the reactor, typically in a blanket surrounding the reaction volume, to breed the tritium required to sustain the process.

In first-generation reactors employing the D-T fuel cycle the high-energy (14 MeV) neutrons produced will generate heat in surrounding structures and materials through various slowing down and capture processes, including lithium transmutation. This heat will be transferred to a reactor coolant and ultimately drive a turbine generator. Retention of the charged reaction products (alpha particles) within the reaction volume is important, as will be shown later, for the maintenance of the reaction and the achievement of a net power output from the system.

15.2-2 The Conditions for Fusion

The creation of the conditions for sustained controlled thermonuclear fusion on earth is extremely difficult and has yet to be achieved. The three key fusion parameters are temperature, density of particles, and confinement time. Energy must be invested to heat and confine the fuel for a time sufficiently long to promote a sufficient number of reactions to release significantly more energy than that which was input to the system or dissipated through various loss mechanisms.

Fusion reactions are strongly temperature dependent, as shown in Fig. 15.2-1. At ion temperatures in the keV range the reactants are ionized plasmas, thus strong coulombic repulsion forces must be overcome to initiate fusion. In the case of stars the thermal agitation of nuclei and the immense gravitational forces present provide the required conditions for fusion.

The three fundamental performance parameters for fusion are plasma temperature, plasma density, and plasma confinement time. The fusion power density is given by

$$P_f = (\text{const.}) n_\text{D} n_\text{T} \langle \sigma v \rangle E_f \tag{15.2-7}$$

where $n_\text{D} n_\text{T}$ = product of fuel ion densities (particles/cm^3)
  $\langle \sigma v \rangle$ = product of the fusion reaction cross section, σ, and the particle velocity, v, averaged over the distribution of particle velocities
  E_f = energy release per fusion event

The quantity $\langle \sigma v \rangle$ is strongly temperature dependent. For the D-T reaction at 10 keV the power density is

$$P_\text{DT} (10 \text{ keV}) \approx 6 \times 10^{-29} n_\text{D} n_\text{T} \frac{W}{\text{cm}^3} \tag{15.2-8}$$

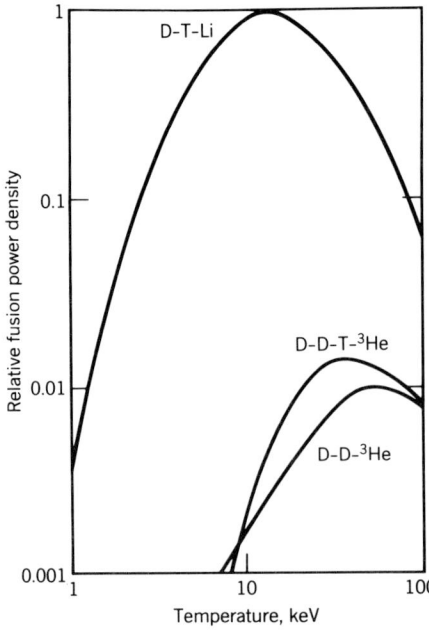

Fig. 15.2-1 Relative fusion power density versus plasma temperature for the D-T, D-^{3}He and D-D fuel cycles (from ref. 3). (Used by permission.)

Figure 15.2-2 shows a generalized schematic of the energy flows from a D-T fusion plasma. Input energy is required to heat the plasma, which in turn is heated by the retention and slowing down of charged reaction products. A characteristic confinement time may be associated with particle and energy losses to serve as a measure of the time available for productive reactions. To extract useful amounts of power from the reacting volume, it is necessary to confine a suitably dense plasma at keV temperature levels for a suitable time period. A convenient figure of merit for the quality of plasma confinement is the $n\tau$ (density confinement time) product.

Referring to the conceptual plasma energy balance shown in Fig. 15.2-2, the following fusion performance criteria, for both inertial and magnetic confinement, may be defined:

Energy breakeven.   The fusion energy produced within the reaction volume is equal to the energy input to the volume.

Ignition criterion.   The energy of charged fusion products that are retained within the reaction volume is sufficient to balance plasma energy losses.

Lawson criterion[4].   Sometimes called engineering breakeven. A condition such that if all fusion energy output where converted to electricity (at an efficiency of $\frac{1}{3}$) and recirculated the reactor would be self-sustaining.

Figure 15.2-3 compares Lawson and ignition curves for D-D and D-T in terms of the $n\tau$ product and ion temperature. The charged reaction products are assumed to thermalize with the plasma particles. In the D-D case the reaction products T and ^{3}He are assumed to be recycled and burned.

15.2-3   Approaches to Fusion Power

Two general approaches to plasma confinement are being pursued experimentally; inertial and magnetic. The bulk of prior work has been related to magnetic confinement approaches. In either category a variety of specific configurations and system arrangements have been proposed.

Magnetic confinement strives to approach the Lawson or ignition criteria by providing a magnetic pressure to restrain the kinetic pressure of a hot plasma. Inertial devices attempt to heat and implode a fuel pellet to approach extremely high material densities in the fuel core prior to pellet disassembly. Given that the Lawson criterion for D-T requires a density confinement time product ($n\tau$) in the range of 10^{14}–10^{15} cm^{-3}-s for a D-T fuel mixture, the characteristic $n\tau$ values to be attained are:

Confinement Approach	Density Range (cm^{-3})	Confinement Time Range (s)
Magnetic	10^{14}–10^{16}	10–0.1
Inertial	~ 10^{26}	~ 10^{-9}

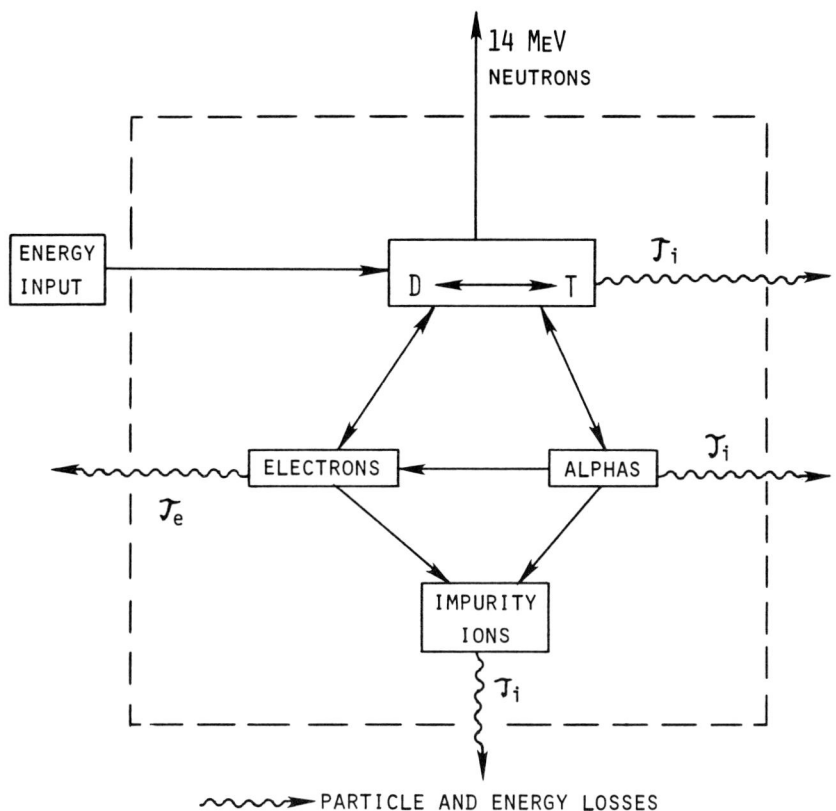

Fig. 15.2-2 D–T fusion plasma energy flows.

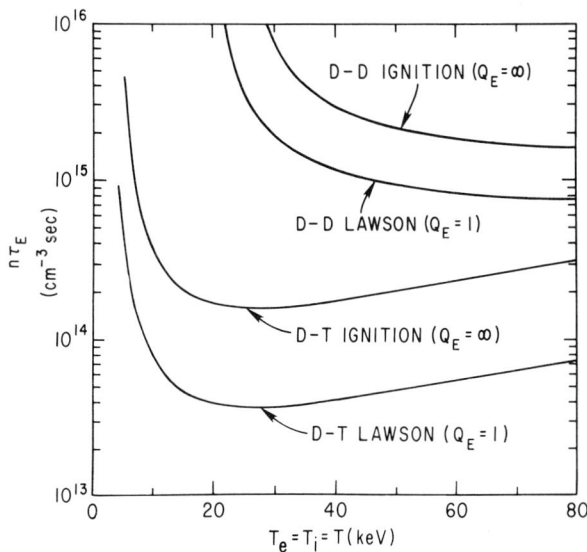

Fig. 15.2-3 Idealized Lawson and ignition criteria curves for D-D and D-T fusion fuel cycles. (Courtesy of D. L. Jassby and H. A. Towner, Princeton Plasma Physics Laboratory, Report pp PL-773873.)

A key parameter in magnetic confinement system is the plasma beta, or ratio of plasma kinetic pressure to confining magnetic pressure. Thus

$$\beta = (\text{const.}) \frac{n(T_i + T_e)}{B^2} \qquad (15.2\text{-}9)$$

where $T_{i,e}$ = ion and electron temperatures, respectively
$\qquad B$ = magnetic field strength

Substituting n from Eq. (15.2-7) yields the following important reactor power relationship:

$$P_f = (\text{const.}) \frac{\beta^2 B^4 \langle \sigma v \rangle E_f}{(T_i + T_e)^2} \qquad (15.2\text{-}10)$$

At a fixed magnetic field strength and plasma temperature, the fusion power density will scale as β^2, while for fixed β and operating temperature the fusion power scales as B^4.

15.2-4   Fusion Reactor Elements and Design Considerations

At the present stage of fusion technology development, reactor design exercises serve a number of very important functions. The designs in particular serve to identify problem areas from a reactor application point of view and provide a focus for both research and component technology development. Reactor conceptualizations have been prepared for virtually every confinement approach, with the outcome driven strongly by present understanding of underlying plasma scaling laws. None of the proposed designs should be considered definitive in terms of size, cost, and performance.

The leading confinement concept, in terms of demonstrated performance and physics understanding to date, is the tokamak.[5] The device was invented in the USSR during the mid-1960s and the term *tokamak* is derived from the Russian words for toroidal-chamber-magnetic. Since this device is the most advanced, it will be used as the basis for the discussion of the elements of a fusion reactor system.

The basic configuration of the tokamak is shown in Fig. 15.2-4. The plasma is formed in an evacuated toroidal vacuum chamber enclosed by a toroidally wound solenoid. The tokamak also threads the torus with a transformer in such a way that the plasma doughnut acts as a single-turn secondary winding. Thus when the transformer is energized, a current is induced within the plasma. The superposition of the toroidal magnetic field and the field due to the plasma current yields a closed system of nested helical field lines with relatively good confinement and plasma stability properties. An added benefit of the induced current is that it resistively or ohmically heats the plasma. The early tokamaks employed copper shells around the vacuum tanks, which aided in the stabilization of the plasma. Eddy currents set up in these shells tended to provide a passive stabilizing mechanism that deterred unstable vertical and radially outward plasma motion. Later machines have dispensed with the copper shell and employ an additional set of field windings that can be energized in a programmed fashion to promote stability. Recent machines have also dispensed with the heavy iron core transformer and use a solenoidal winding in the throat of the torus, with an air core, to provide the transformer function.

The pulsed nature of the tokomak is a major drawback from a reactor utilization point of view since frequent on-off-on cycles detract from the availability of the unit and introduce large thermal, electrical, and mechanical transients that have a significant engineering and cost impact. As a result, recent studies and related experimental work have focused on means for driving the plasma current by particle beams or electromagnetic fields.

In the tokamak the relationship between the toroidal and poloidal fields is given by

$$B_T = B_\theta q A \qquad (15.2\text{-}11)$$

where B_T = toroidal field strength
$\qquad B_\theta$ = poloidal field strength
$\qquad q$ = plasma safety factor
$\qquad A$ = torus aspect ratio (major radius/minor radius)

The plasma safety factor, q, is the number of times a field line encircles the torus in the toroidal direction before encircling the minor axis. If q^{-1} is an integer, the flux line will not describe a magnetic surface but will close on itself after one trip around the torus. From stability considerations q must be larger than unity and is typically chosen between 2.5 and 3. From a total reactor size point of view the aspect ratio should be minimized, however, realistic mechanical arrangement considerations limit the minimum practical size of the torus throat and thus require a design iteration.

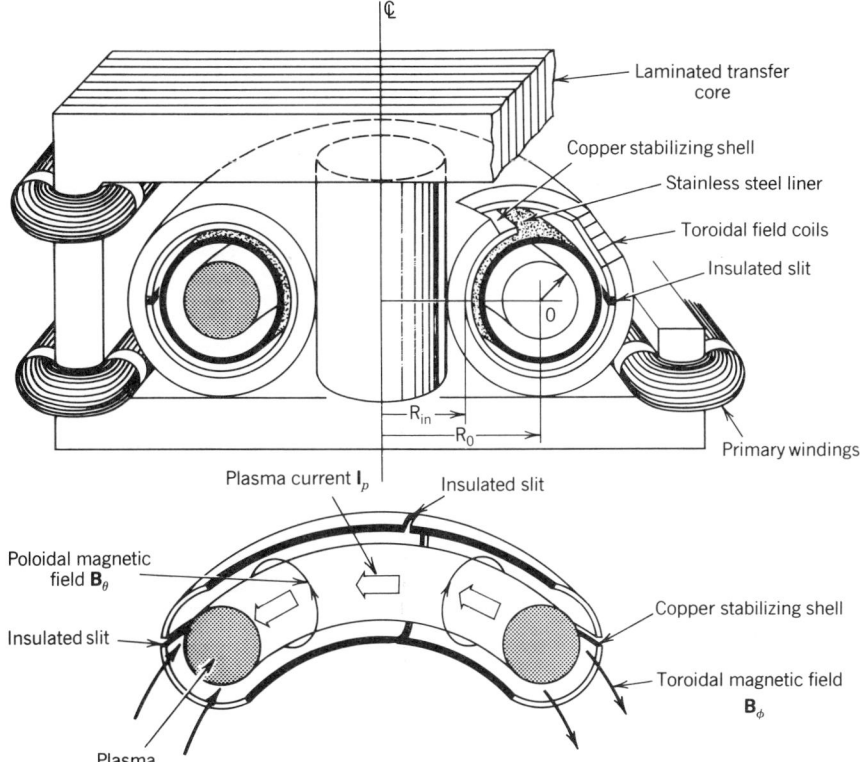

Fig. 15.2-4 Basic configuration of a tokamak magnetic confinement device.

In a power reactor a tight design coupling exists between the magnet systems, blanket and shield system, and the overall maintainability approach. The blanket region, containing lithium-bearing material, must surround to the extent possible the reaction volume and intercept the fusion neutrons. This region and the surface of the vacuum containment or first wall are cooled by an appropriate working fluid that exchanges heat with a secondary fluid, such as water, which is converted to steam and expanded through a conventional turbine generator. A shield must be provided outboard of the blanket region to attenuate remaining neutron and gamma ray fluxes to an acceptable level prior to reaching the toroidal and poloidal magnet systems. Superimposed on these considerations is the need for access to the blanket to replace components damaged by radiation over the life of the device and to replenish burned lithium and remove bred tritium. Finally, provision must be made for removing the helium ash and other impurities from the plasma and inserting new deuterium and tritium fuel.

STARFIRE—A Target Commercial Reactor Design

Table 15.2-2 lists the major parameters and features of three major commercial-scale tokamak reactor concepts developed in recent studies[6-8] and illustrates that a considerable consensus has been achieved to date in many areas of reactor design. STARFIRE[7] is possibly the most complete and representative concept formulated to date and is an appropriate vehicle to portray the major features currently envisioned in a power reactor. Figure 15.2-5 is a cut-away view of STARFIRE illustrating the relationship of the various components.

STARFIRE was assumed to be the tenth plant of a series of commercial reactors. The design is a major departure from earlier tokamak reactors by employing a steady-state operating mode achieved through the use of an rf-driven plasma current. Recent experiments have provided reasonable confidence that this approach will succeed. The plasma impurity and alpha particle (plasma "ash") removal approach employs vacuum pumped toroidal limiters that concentrate the plasma impurities and direct a fraction of them to a slot behind the limiter where they are pumped away by 24 compound cryopumps (with an additional 24 to provide essentially full-time pumping while half are in a rejuvenation mode).

TABLE 15.2-2 DESIGN PARAMETERS FOR THREE COMMERCIAL-SCALE TOKAMAK FUSION POWER REACTORS[a]

	NUWMAK[b] U.S.A. 1979	STARFIRE[c] U.S.A. 1980	SPTR-P[d] JAPAN 1981
Major plasma radius, m	5.1	7.0	6.8
Minor plasma radius, m	1.13	1.94	2.0
Plasma elongation	1.64	1.6	1.6
Gross thermal power, MW	2100	4000	3700
Neutron wall loading, MW/m^2	4.0	3.6	3.3
Plasma current, MA	7.2	10.1	16.4
Average toroidal beta, %	6.5	6.7	7.0
Burn pulse length, s	225	Continuous	Continuous
Plasma current drive	Inductive	rf	rf
	—	90 MW, 1.7 GHz	80 MW
Plasma heating	rf	rf	rf
	80 MW, 92 MHz	90 MW, 1.7 GHz	—
Toroidal magnetic induction (T)	12	11	12
Impurity control method	Gas blanket	Pumped limiter	Pumped limiter
Structural material/coolant	Ti-Al-V/H_2O	SS/H_2O	SS/H_2O
Blanket breeding material	$Li_{62}Pb_{38}$	$LiAlO_2$	Li_2O

[a] rf—radiofrequency waves.
[b] From ref. 6.
[c] From ref. 7.
[d] From ref. 8.

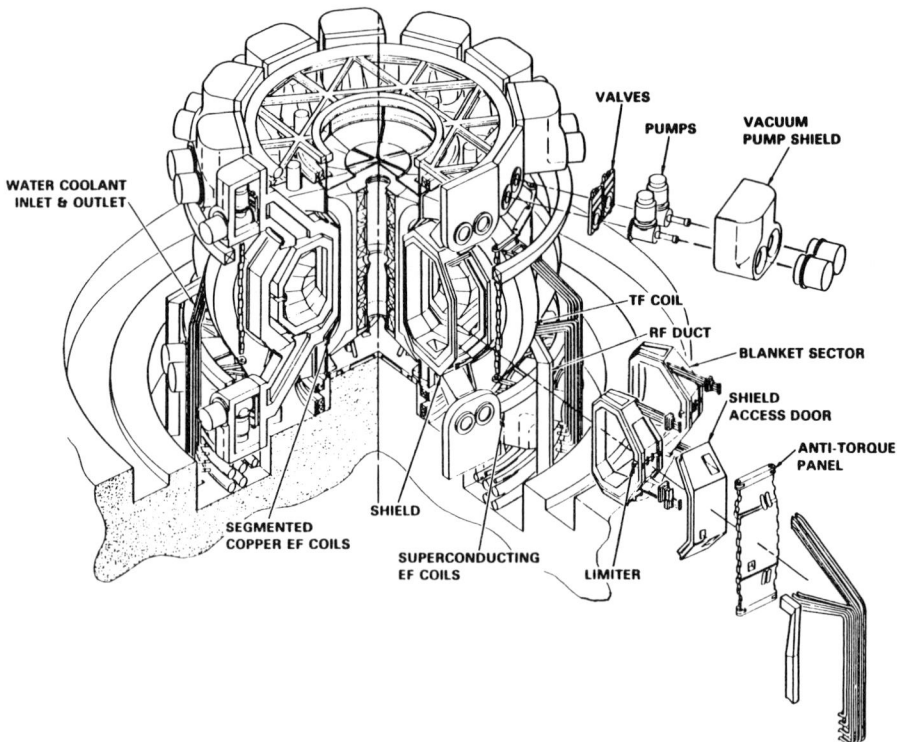

Fig. 15.2-5 Isometric view of the starfire commercial tokamak fusion power reactor conceptual design (from ref. 7).

The first-wall and blanket structural material is titanium-modified type 316 stainless steel that operates at a peak temperature of about 425°C. The first wall and blanket are cooled by pressurized water with inlet and outlet temperatures of 280 and 320°C, respectively.

The tritium breeding medium in STARFIRE is a packed bed of alpha-$LiAlO_2$ with 1 cm diameter stainless steel coolant tubes arranged within the 40-cm (radial thickness) breeding region to limit the peak bed temperature to 850°C. Low-pressure helium is circulated through the porous $LiAlO_2$ bed to accomplish in situ removal of bred tritium. The first wall, blanket, and associated structures are designed for a 16-MW · yr/m^2 life.

The first-wall and blanket structure is segmented into 24 toroidal sectors to facilitate removal of these modules between toroidal field coils as shown in Fig. 15.2-5. A neutron and gamma ray shield structure is provided immediately outboard of the blanket and forms the primary vacuum boundary for the plasma.

The toroidal field coil system consists of 12 superconducting magnets operating at a peak field of 11T, which necessitates the use of graded windings. The conductor used in the low field region (< 9 T) is NbTi and the conductor in the high field region is Nb_3Sn. Most of the equilibrium field coils are also superconducting and are located outside of the toroidal coils. An additional set of segmented water-cooled copper poloidal coils are located within the bore of the toroidal magnets.

Thermal power is extracted in STARFIRE in two heat removal circuits. Power deposited in the limiter (200 MW) is used for feedwater heating while the available thermal power in the first wall and blanket is used to produce steam at 299°C and 6.3 MPa. The turbine generator produces 1440 MW of electric power, of which 240 MW is recirculated for the rf heating system, coolant pumps, and other plant loads.

The unit capital cost estimate for STARFIRE is about $1500/kW(e) in 1980 dollars which yields, assuming a plant availability of 75%, a cost of electricity in the range of 35–40 mills/kW · h. While the cost of electricity estimate developed for STARFIRE is several times that of a comparable light water reactor, the trend in such estimates is definitely downward as improved and simplified designs are formulated.

Long-Range Development Issues

An excellent detailed assessment of magnetic fusion engineering development needs was recently sponsored by the National Research Council and the Office of Fusion Energy of the U.S. Department of Energy.[9] The question of the feasibility of fusion power today is dominated by physics issues. Unfortunately those concepts, such as the tokamak, which are the most successful from a physics point of view, are the least attractive from an engineering and economic point of view. There continues to be great interest in the development of confinement approaches, which, if found to be feasible scientifically, could lead to compact power plant designs or at the very least could avoid some of the engineering complications of the tokamak approach.

Aside from the scientific challenge, many perceive the fundamental engineering issue facing fusion is the fact that compared with fission energy systems fusion plants characteristically have a lower power density and hence the fusion reactor must be physically larger and more materials intensive than a fission reactor (either water or liquid metal cooled) to produce the same output power. Energy in fission systems is produced in solid fuel that is cooled relatively efficiently by circulating fluid. Energy in fusion systems must be transmitted out of the reaction volume (via high-energy neutrons) through a surface into a surrounding medium. The volume averaged power in a fusion reactor is thus lower by at least an order of magnitude. In today's electric power utility environment initial plant capital cost (which must be financed) is a dominant consideration in selecting new construction projects; therefore, the potential fuel cycle cost advantage of fusion will have relatively little weight in economic feasibility assessments in the foreseeable future.

REFERENCES

15.2-1 H. A. Bethe, "Energy Production in Stars," *Phys. Rev.* **55**, 434 (1939).

15.2-2 U.S. Department of Energy, "Fusion Technology Development Plan," DOE/ER-0166 (1983).

15.2-3 D. Steiner, "The Technological Requirements for Power by Fusion," *Nuclear Science and Engineering* **58**, 107–166 (1975).

15.2-4 J. D. Lawson, *Proc. Phys. Soc.*, London, B70, (1957).

15.2-5 J. M. Rawls, ed., "Status of Tokamak Research," U.S. Department of Energy, DOE/ER-0034 (1979).

15.2-6 "NUMAK—A Tokamak Reactor Design Study," University of Wisconsin Fusion Engineering Program Report UWFDM-330 (1979).

15.2-7 C. C. Baker et al., "STARFIRE—A commercial Tokamak Fusion Power Plant Study," Argonne National Laboratory, ANL/FPP-80-1 (September 1980).

15.2-8 Proceedings of the Third IAEA Technical Committee Meeting and Workshop on Fusion Reactor Design and Technology, Tokyo, 1981, International Atomic Energy Agency, Vienna Austria, 1982.

15.2-9 "Future Engineering Needs of Magnetic Fusion," A Report on the Workshop Conducted by the Committee on Magnetic Fusion, Energy Engineering Board, Commission on Engineering and Technical Systems, National Research Council, Washington, D.C. 1982.

15.3 LASER METHODS FOR STUDIES OF ENERGY SYSTEMS

Ryszard J. Pryputniewicz

In everyday work the engineer deals with systems that vibrate, heat and/or cool, "flow," change in one way or another, interact with other systems, and so on. Since in all of these systems, the energy is either consumed, stored, or exerted, they can be called energy systems.

In pursuit of more efficient energy systems, as dictated by ever-increasing demands of today's science and technology, the engineer is constantly seeking for better experimental techniques to accurately and precisely "measure" such systems. These experimental techniques are needed to characterize both existing as well as new energy systems in order to develop meaningful relationships modeling interactions between the systems and their surroundings. However, energy is a complicated subject with many areas that have not yet yielded to precise analytical techniques. Accordingly, problems dealing with energy are often complicated and imprecise because of the lack of analytical relations to use for calculation and reduction of experimental data. The interpretation of data, in analysis of energy systems, involving deformation, vibration, turbulence, or measurements of complicated boundary layer, viscous, and shock wave effects is not easy. Frequently, using the conventional measuring techniques, the system's characteristics may be altered as a result of probes that are inserted to measure displacement, velocity, acceleration, pressure, temperature, and so on, so that the engineer is uncertain about the effect that has been measured.

Recently, a new group of experimental methods, particularly suited for studies of energy systems, has been developed. These methods are optoelectronic in nature and utilize laser as a source of light. As such, they are referred to as laser methods. The laser methods offer the advantage, over the conventional experimental methods, that when properly executed they do not disturb the system being investigated. Also, the laser methods give rapid, full field, three-dimensional information ranging from qualitative studies of the overall behavior of the system to precise measurement of its critical parameters.

Today, laser methods are developed to the degree where stagnant as well as transient characteristics of energy systems can be studied, employing either continuous-wave (CW) or pulsed lasers. These techniques utilize transmitted, reflected, and/or backscattered light. As such, they can be used in a number of applications involving opaque and transparent media. Modern laser methods, discussed in this section, range from the single-beam speckle interferometry through double-exposure holography to very accurate and precise heterodyne holography. Each of these techniques is particularly suited to certain test applications.

In the following sections, principles of the laser methods will be presented and their representative applications discussed, with particular emphasis on hologram interferometry, heterodyne holography, and speckle interferometry.

15.3-1 Hologram Interferometry

The method of hologram interferometry was invented in 1965[1] and extended application of classical interferometry to the measurement of three-dimensional, diffusely reflecting objects with nonplanar surfaces. It should be noted that until 1965, classical interferometry was limited to measurement of small pathlength differences of optically polished and specularly reflecting surfaces.

Figure 15.3-1 shows a typical setup used in recording of holograms. In this arrangement output from the laser is divided into two parts by means of a beam splitter. One part, going directly through the beam splitter, is directed by mirrors and is shaped by a beam-expander spatial-filter assembly to illuminate objects under investigation. This beam, modulated by reflection from the object, carries the information about the instantaneous condition of the object's surface. The object beam is recorded against the reference beam, that is, the part of the laser output which is reflected from the beam splitter. The mutual interference of these two beams is recorded by placing a suitable photosensitive medium[2] in the region of space where they overlap. The exposed medium, upon processing, becomes a hologram.

The hologram is reconstructed with the same setup that was used for recording, except now, it is illuminated with the reference beam alone (Fig. 15.3-2). During reconstruction, a portion of the laser light passes through the hologram without a change (this is the so-called zero-order wave), and the remaining light is diffracted into the higher orders. Of course, the most important of the diffracted waves is the first-order wavefront. This wavefront seems to emanate from the region in space that was

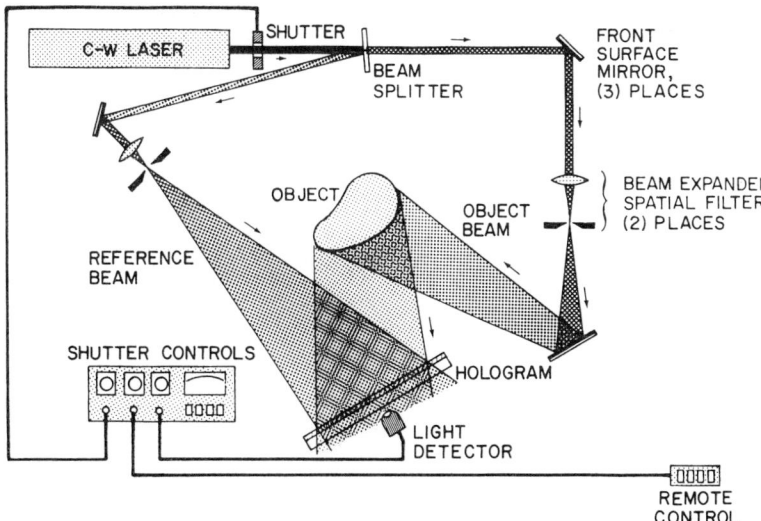

Fig. 15.3-1 Setup for recording of holograms of solid objects.

occupied by the object while the hologram was being recorded, even though the object had since been removed. The image observed has all the visual properties of the original object. In fact, there is no visual test that can differentiate between the two.

There are three basic variations of hologram interferometry: (1) real time, (2) time average, and (3) double-exposure, each possessing certain advantages in particular test applications.[3]

Real-time hologram interferometry involves recording a single-exposure hologram as shown in Fig. 15.3-1, processing it, and reconstructing by illumination with the beam similar to the original reference

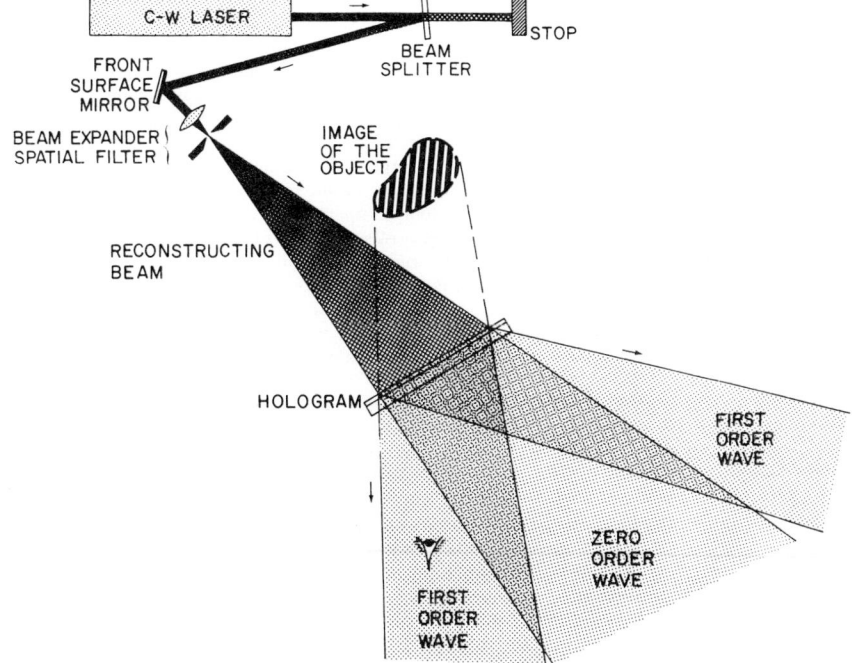

Fig. 15.3-2 Setup for reconstruction of holograms.

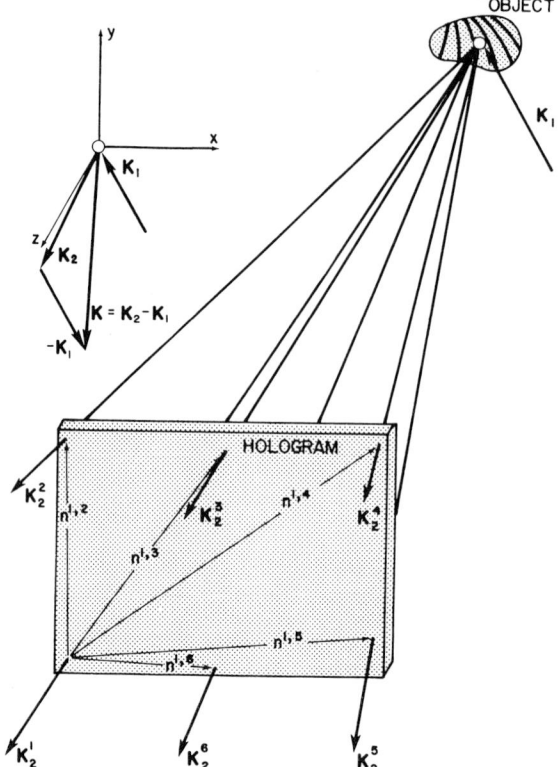

Fig. 15.3-3 Multiple observation geometry in hologram interferometry.

beam. The reconstructed image is superimposed onto the original object, which is illuminated with the light beam identical to that used in recording of the hologram. If the object is now even slightly displaced and/or deformed, or changed in any way, as for example, by thermal loading, interferometric comparison between the holographically reconstructed image and the new state of the object is made at the instant it occurs. The particular advantage of this method is that different types of motion —dynamic as well as static—can be studied with a single holographic exposure.

In *time-average hologram interferometry* a single holographic recording of an object undergoing a periodic motion is made. With the exposure time long compared to one period of vibration cycle, the hologram effectively records an ensemble of images corresponding to the time average of all positions of the object during its vibration. In the reconstruction of such a hologram interference occurs between the entire ensemble of images, with the images recorded near zero velocity contributing most strongly to the holographic interferogram. As such, fringes reconstructed from a time-average hologram have intensity variations given by the square of the zero-order Bessel function,[4] that is, J_0^2. However, in the case of stroboscopic illumination of a vibrating component, or when pulsed lasers are used, cosinusoidal fringe patterns are obtained.

The *double-exposure method*, which can be considered to be a special case of the time-average method (where only two exposures of the object are made on the same photographic medium), is the most widely used of all holographic techniques. In this method the object is displaced and/or deformed between the exposures. Therefore, the object beam during the second exposure is slightly different than that corresponding to the object's initial position. During reconstruction of a double-exposure hologram, both object beams are faithfully reconstructed, forming images of object's initial and final positions. Since these images are reconstructed in coherent light, they interfere with each other, forming an alternating pattern of bright and dark fringes. These fringes are a direct measure of changes in the object's position and/or shape that have occurred between the two exposures.

There are a number of techniques dealing with interpretation of cosinusoidal fringes observed within the holographically reconstructed image.[3,5-12] The most general of these techniques is the method employing multiple observations of the holographic image (Fig. 15.3-3). It results[9,10] in displacement vector **L** expressed as a product of the inverse matrix formed by the sum of projection matrices **P** with the matrix representing the sum of the observed vectors $\mathbf{L}_{ob}$, that is,

$$\mathbf{L} = \left[\sum_{m=1}^{r} \mathbf{P}^{m} \right]^{-1} \left(\sum_{m=1}^{r} \mathbf{L}_{ob}^{m} \right) \tag{15.3-1}$$

In Eq. (15.3-1) m denotes the observation number with r being the total number of observations, while $\mathbf{L}_{ob}$ is measured in the plane normal to the direction of observation and is defined by the corresponding **P**. The projection matrix $\mathbf{P}^{m}$, for the mth direction of observation, is defined as a difference between the identity matrix **I** and a matrix formed by the dyadic product of the mth unit observation vector $\hat{\mathbf{K}}_{2}^{m}$ with itself.[9,10] Application of Eq. (15.3-1), in interpretation of cosinusoidal fringe patterns, results in determination of displacements with an accuracy better than 0.3×10^{-6} m.

Of particular interest in energy studies is the application of double-exposure hologram interferometry in determination of strains and rotations.[5] In this case the strain–rotation matrix **f** is determined directly from the parameters **K** and $\mathbf{K}_{fc}$, defining hologram recording/reconstruction geometry and characteristics of fringe patterns including perspective variations. Thus,

$$\mathbf{f} = [\mathbf{K}^{T}\mathbf{K}]^{-1}[\mathbf{K}^{T}\mathbf{K}_{fc}] \tag{15.3-2}$$

Decomposition of the matrix **f** into the symmetric part **e** and the antisymmetric part $\boldsymbol{\theta}$, that is,

$$\mathbf{e} = \tfrac{1}{2}[\mathbf{f} + \mathbf{f}^{T}] \qquad \boldsymbol{\theta} = \tfrac{1}{2}[\mathbf{f} - \mathbf{f}^{T}] \tag{15.3-3}$$

gives strains and shears, and rotations, respectively. Experimental studies indicate that strains and rotations determined from Eqs. (15.3-2) and (15.3-3) are within 10×10^{-6} m/m of the actual values. It should be realized, however, that the above results are obtained in a noninvasive manner over the object's surface in the three-dimensional space, without the use of object's material properties at all. The holographic technique resulting in accuracy much higher than that quoted above will be discussed in Section 15.3-2.

The above relationships also apply to studies of vibrating systems, providing that cosinusoidal fringes are produced. This can be achieved by stroboscopic illumination of the system. The suitable illumination can be obtained either by "chopping" the CW laser output over a multitude of cycles, or by double-exposure with ultrafast pulsed laser. The former approach requires that the system be undergoing a *periodic vibration*, while the latter allows recording of *periodic as well as transient vibrations* at very high frequencies. This is possible because modern pulsed lasers produce controlled high-energy light bursts of subnanosecond duration at multimegacycle frequencies.

In the general case, when the stroboscopic light control is not possible and when the pulsed laser is not available, time-average holograms can provide solutions to the vibration problems. Figures 15.3-4 to 15.3-6 show several time-average images of a vibrating object characterized by the J_{0}^{2} fringes; in these images the brightest fringes represent vibrational nodes. From the fringe distributions shown,

Fig. 15.3-4 Time-average holographic recordings of the first six flexure modes of a prismatic cantilever beam vibrating at: (a) 42.5 Hz, (b) 269.1 Hz, (c) 754 Hz, (d) 1482 Hz, (e) 2449 Hz, (f) 3661 Hz.

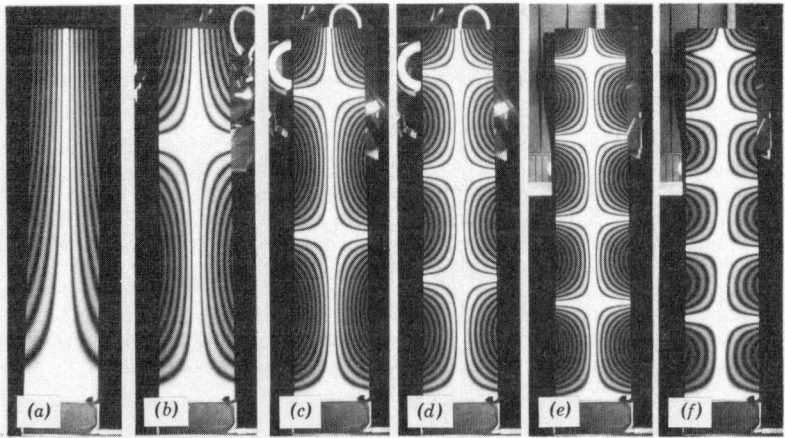

Fig. 15.3-5 Time-average holographic recordings of the first six torsion modes of a prismatic cantilever beam vibrating at: (a) 407.3 Hz, (b) 1249 Hz, (c) 2158 Hz, (d) 3187 Hz, (e) 4363 Hz, (f) 5717 Hz.

displacements $\mathbf{L}_z$, at any point on the object, can be readily determined from the following relation[4]:

$$\mathbf{L}_z = \frac{\lambda}{2\pi\left(\hat{\mathbf{K}}_{2_z} - \hat{\mathbf{K}}_{1_z}\right)}|\Omega_t|. \tag{15.3-4}$$

In Eq. (15.3-4) $|\Omega_t|$ is the argument of the J_0 corresponding to the center of the dark fringe of the given order, $\hat{\mathbf{K}}_{1_z}$ and $\hat{\mathbf{K}}_{2_z}$ are the unit vectors defining directions of illumination and observation, respectively, while λ represents laser's wavelength. A typical result, pertaining to the vibration of prismatic cantilever beams, is shown in Fig. 15.3-7. The data obtained using Eq. (15.3-4) are in very good agreement with the exact solution based on the beam theory.

Hologram interferometry is also useful in heat transfer studies.[13–15] In this application the test section is placed within the object beam (Fig. 15.3-8). Now, any changes in the coefficient of optical refraction of the medium, in the heated (or cooled) zone surrounding the test section, are recorded against the reference beam. Typical photographs obtained during reconstruction of interferograms recorded in such a way are shown in Fig. 15.3-9. Interpretation of these fringes results in determination of temperature at any point within the image, without any interference whatsoever with the heated zone.

The quantitative results are obtained, form the fringes, using the equation relating optical and geometrical parameters characterizing the setup, as well as temperature T within the space where the

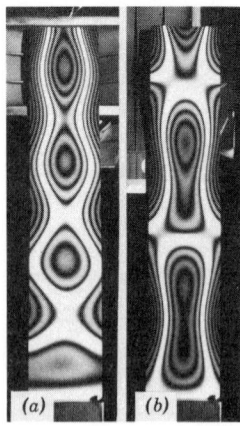

Fig. 15.3-6 Time-average holographic recordings of the coupled modes of a prismatic cantilever beam vibrating at: (a) 7129 Hz, (b) 8238 Hz.

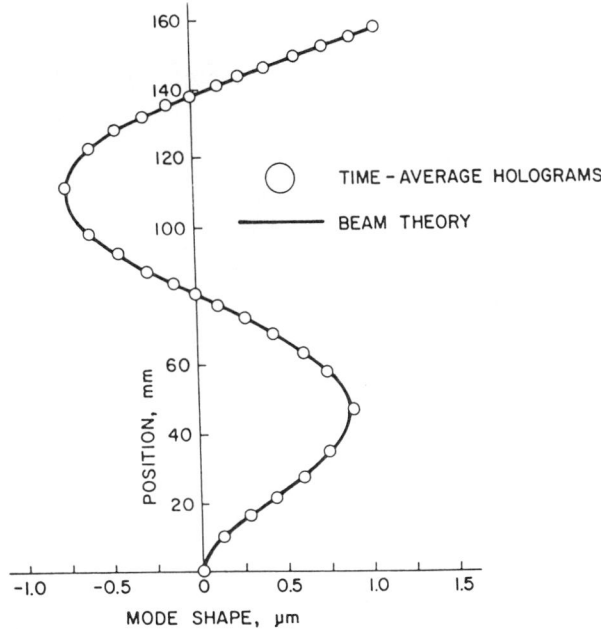

Fig. 15.3-7 Mode shape of a prismatic cantilever beam vibrating at its third flexure frequency: a comparison of the results obtained using the time-average holography with the data based on the beam theory. The constant necessary to obtain the exact solution from the beam theory was obtained experimentally.

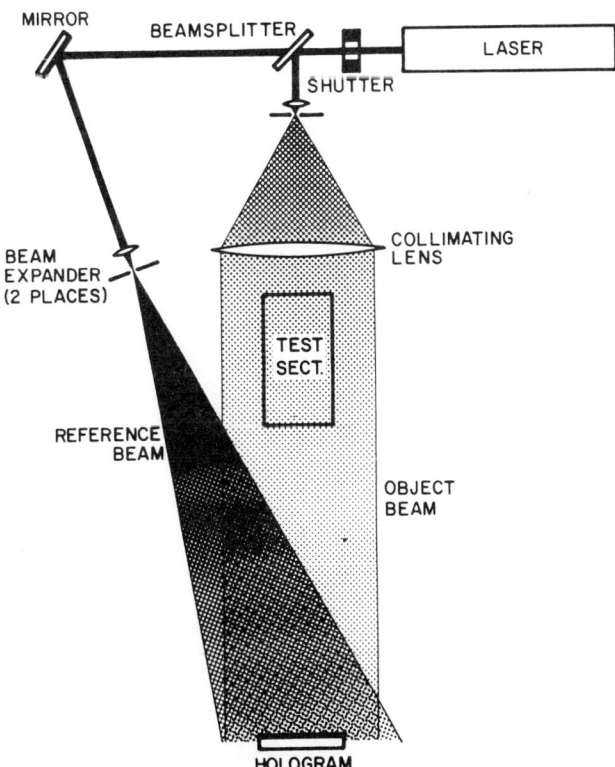

Fig. 15.3-8 Setup for recording of holograms in studies of convective heat transfer processes.

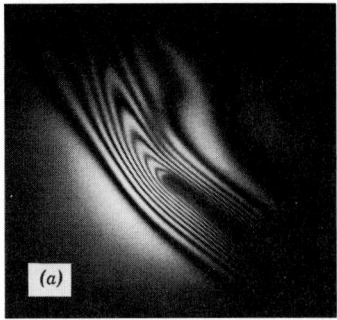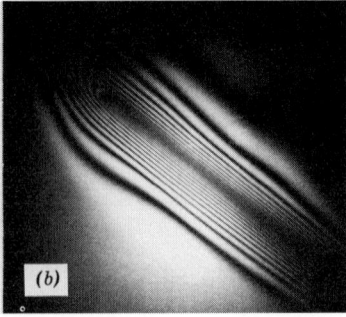

Fig. 15.3-9 Holographic study of the boundary layer on a thin, inclined, heated plate: (*a*) near plate's edge, (*b*) away from the plate's edge.

changes have taken place:

$$T = \frac{(n_0 - 1)T_0}{(n_0 - 1)\left(\dfrac{T_0}{T_\infty} + 1\right)\left[1 - \left(\dfrac{2N + 1}{2}\right)\dfrac{\lambda}{l_0}\right] - 1}$$

(15.3-5)

In Eq. (15.3-5) n_0 is the index of refraction, N is the fringe order, λ is laser wavelength, l_0 is the length of the space in which temperature had changed (this length is determined in the direction parallel to the direction of the object beam), T_∞ is the ambient temperature, while T_0 is the temperature at which n_0 was determined. Figure 15.3-10 shows representative results obtained using Eq. (15.3-5) and indicates good agreement with the boundary layer theory used in the particular study.

15.3-2 Heterodyne Hologram Interferometry

The method of heterodyne hologram interferometry was developed in 1976.[16] This method can best be described with reference to the conventional double-exposure hologram interferometry discussed in Section 15.3-1.

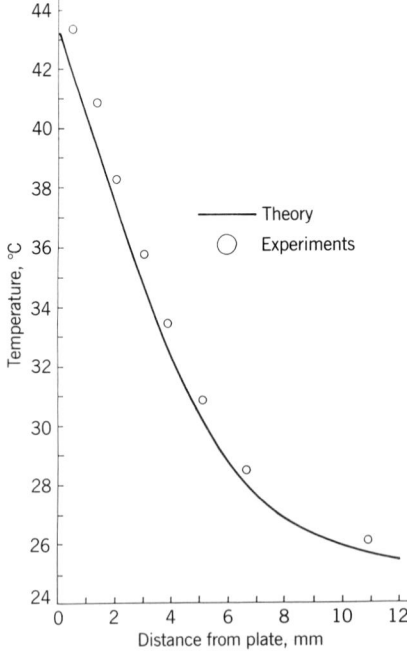

Fig. 15.3-10 Temperature profile determined from a double-exposure hologram of a heated vertical plate: a comparison with the solution obtained using the boundary layer theory.

In heterodyne hologram interferometry two object fields are recorded, each with a different reference beam. These recordings are made in the same photographic medium in such a way that they can later be reconstructed independently. Then, during reconstruction of such a double-exposure hologram, the experimenter introduces a small frequency shift between the two reconstructed and interfering light fields. This results in an intensity modulation at the beat frequency of these light fields, for any point within the interference pattern. The optical phase difference, corresponding to the displacement recorded within the hologram being reconstructed, is converted into the phase of the beat frequency of the two interfering light fields. This phase is, in turn, interpolated optoelectronically, resulting in determination of fringe parameters with an accuracy better than $1/1000$ of one fringe. The setup used in heterodyne hologram interferometry is shown in Fig. 15.3-11. In this arrangement[11] output is divided into three beams by means of beam splitters BS1 and BS2. The first beam, reflected from BS1, is directed by mirror M1, via beam expander BE1, toward recording medium H; this is reference beam R1. The second beam, reflected from BS2, is reference beam R2. The third beam, directed by mirrors M3 and M4, is expanded by BE3 to illuminate the object being studied. This beam, modulated by reflection from the object, is recorded against R1 and R2, one at a time, with the exposures being controlled by shutters S1 and S2, respectively. The acousto-optical modulators AM1 and AM2, cascaded in R2, are used to introduce frequency shift between the two reconstructing beams.

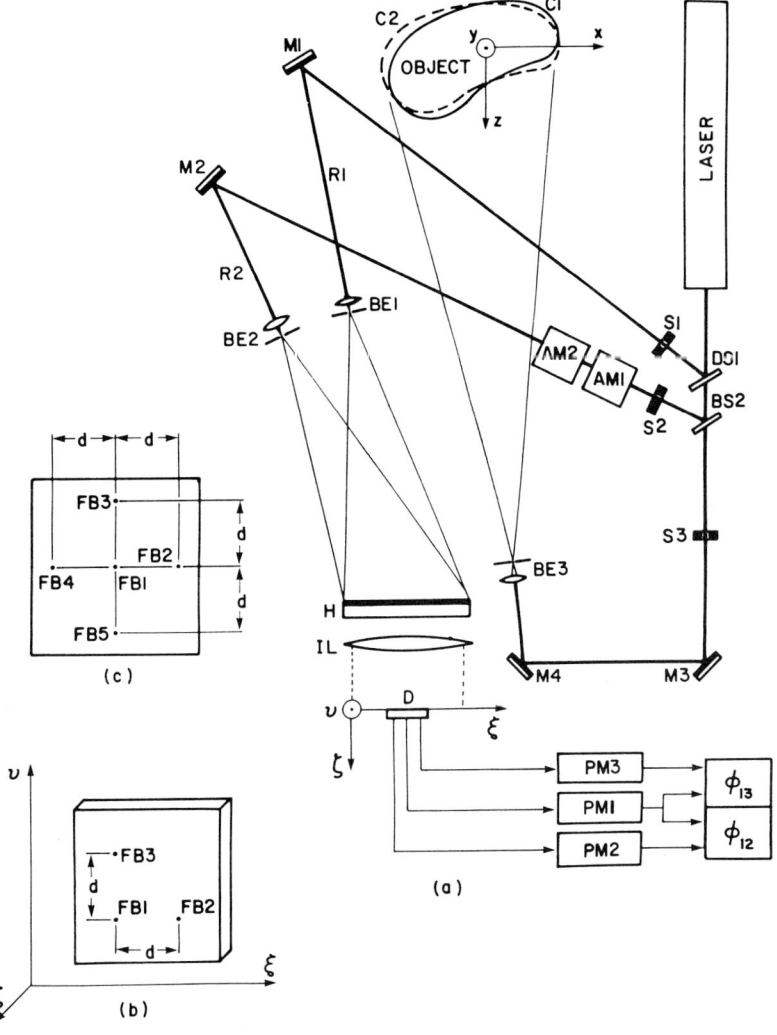

Fig. 15.3-11 Setup for recording of heterodyne holograms: (*a*) recording/reconstruction and imaging geometry, (*b*) detector head for determination of the first derivatives of displacement only, (*c*) detector head for simultaneous determination of the first and the second derivatives of displacement.

Using the arrangement shown in Fig. 15.3-11a, the object's configuration C1, corresponding to its initial state, is recorded using the object beam and R1 while R2 is stopped by S2. Following this first exposure, the state of the object is altered resulting in configuration C2. Then, the second exposure is made, on the same photographic medium, recording C2 by interfering the beam with R2, while R1 is blocked by S1. It should be noted that during recording the object beam as well as R1 and R2 all have the same frequency.

The exposed hologram is processed and then reconstructed by simultaneous illumination with R1 and R2 while the object beam is intercepted by S3. Now, however, the acousto-optical modulators AM1 and AM2 are driven in such a way that the net frequency shift, between R1 and R2, is 100 kHz. A photodetector placed at a point within the image formed by lens IL (Fig. 15.3-11a) detects the optical phase difference between the two light fields as the phase of the intensity modulation at the beat frequency.

In the heterodyne system discussed herein, intensity variations in the image space ξ-ν-ζ, formed by IL focused on object space x-y-z, are detected by fiberoptic bundles and sent to photomultipliers. The photomultiplier signals are, in turn, fed into phase meters that output phase differences ϕ_{12} and ϕ_{13} between light beams sensed by pairs of detectors. The fiberoptic bundles are assembled in the detector head at known center-to-center distances d (Fig. 15.3-11b) in such a way as to form two mutually perpendicular axes defined by the fiberoptic bundle pairs FB1-FB2 and FB1-FB3. If this detector is moved to scan the image plane ξ-ν, the fringe density or slope of the interference, rather than the phase itself, is measured. This directly yields local derivatives of displacement which, in turn, define strains and curvature of the object being investigated.

The relationships between the derivatives of displacement and phase differences determined from heterodyne holograms are[11,17]

$$L'_{\xi_n} = \frac{\lambda}{2\pi d}\phi_{12} \qquad L'_{\nu_n} = \frac{\lambda}{2\pi d}\phi_{13} \tag{15.3-6}$$

and

$$L''_{\xi_n} = \frac{\lambda}{2\pi d^2}(\phi_{12} - \phi_{14}) \qquad L''_{\nu_n} = \frac{\lambda}{2\pi d^2}(\phi_{13} - \phi_{15}) \tag{15.3-7}$$

In Eqs. (15.3-6) and (15.3-7) subscript n identifies the specific point within the image plane ξ-ν, d is equal to the center-to-center distance of fiberoptic bundles in the detector head assembly, λ is the wavelength of the laser light used in recording/reconstruction of the hologram, and ϕ_{12}, ϕ_{13}, ϕ_{14}, and ϕ_{15} are phase differences between signals from FB1 and fiberoptic bundles FB2, FB3, FB4, and FB5, respectively (Fig. 15.3-11c). Use of the detector head with all five fiberoptic bundles allows simultaneous measurement of both the first and the second derivatives. However, when simultaneous determination of L' and L'' is not necessary, then all phase differences can be measured in a sequential manner with the detector head shown in Fig. 15.3-11b. The uncertainty in determination of derivatives given by Eqs. (15.3-6) and (15.3-7) are 0.176×10^{-6} m/m and 8.286×10^{-5} m^{-1}. Therefore, measurement of displacement, strain, and curvature, directly from heterodyne holograms, to within 3×10^{-10} m, 0.2×10^{-6} m/m, and 9×10^{-5} m^{-1}, respectively, is possible.[11] The range of measurements in heterodyne hologram interferometry is limited to 30×10^{-6} m for displacement studies.

Representative results obtained using the heterodyne hologram interferometry are given in Figs. 15.3-12 and 15.3-13 and show very good agreement with the exact solutions from the beam theory.

It should also be noted that the heterodyne method can be used in studies of temperature profiles with high accuracy. This results from the ability to interpolate the interference pattern to within 1/1000 of one fringe. If this fringe is a result of a 1°C temperature change, then the thermal field can be measured with an accuracy better than 0.001°C. The range of temperatures that can be studied with a single heterodyne hologram is limited to situations which will result in frequencies lower than 1 fringe per millimeter in the image plane. This limitation is a function of the size of the fiberoptic bundles used in the construction of the detector head.

15.3-3 Speckle Interferometry

When an object illuminated with laser light is viewed (Fig. 15.3-14), it seems to have a granular appearance. That is, its surface appears to be covered with fine, randomly distributed light and dark areas. If the observer moves, these areas appear to twinkle and move relative to the object. This phenomenon is caused by the fact that each point on the object scatters some light to the observer. The laser light scattered by one point on the object's surface interferes with the light scattered by other object points. In any region of space where these light fields overlap, a random pattern of interference fringes is observed. These interference fringes are known as "speckles." The size of speckles depends on optical properties of the imaging system and directly influences the accuracy of measurements: the finer the speckle, the higher the accuracy.

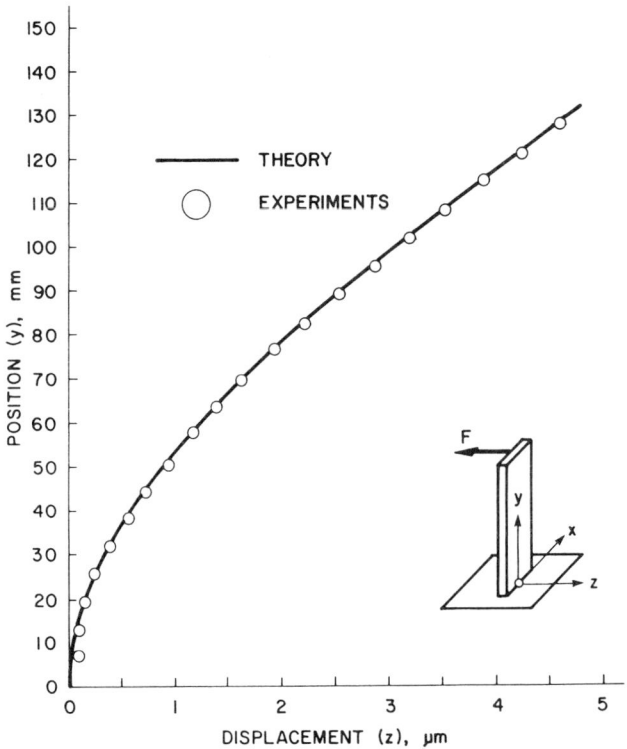

Fig. 15.3-12 Displacements of a prismatic cantilever beam determined by heterodyne hologram interferometry: a comparison with the exact solution based on the beam theory.

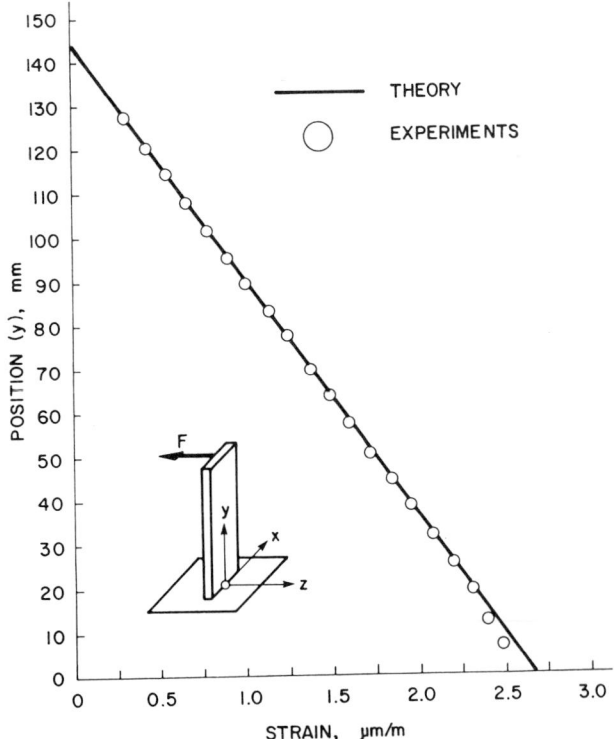

Fig. 15.3-13 Strains of a prismatic cantilever beam determined by heterodyne hologram interferometry: a comparison with the exact solution based on the beam theory.

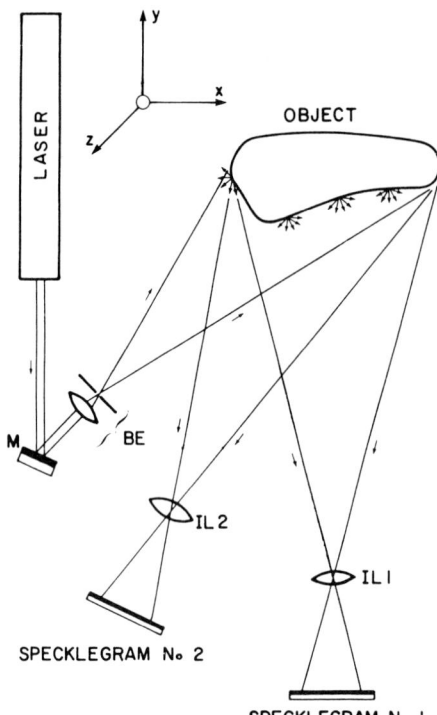

Fig. 15.3-14 Setup for simultaneous recording of two specklegrams of an object, where M is mirror, BE is beam expander, IL1 and IL2 are imaging lenses that form specklegrams No. 1 and No. 2, respectively.

LASER

y

x

z

OBJECT

M

BE

IL 2

IL I

SPECKLEGRAM No 2

SPECKLEGRAM No I

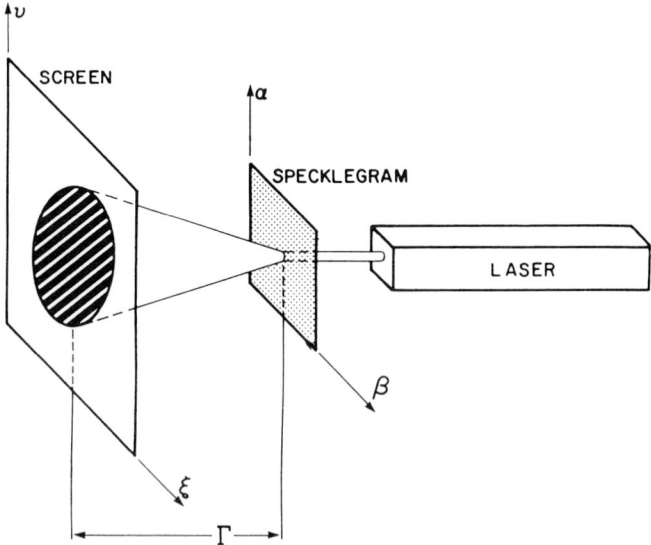

SCREEN

SPECKLEGRAM

LASER

υ

α

β

ξ

Γ

Fig. 15.3-15 Setup for reconstruction of specklegrams. The specklegram is located in the α-β plane, while the diffracted halo is observed in the ξ-υ plane; the two planes are separated by a distance Γ.

Fig. 15.3-16 Typical diffraction halo modulated by Young's fringes as seen during reconstruction of a specklegram.

The simplest and most appealing technique for obtaining quantitative information from speckle patterns is based on the fact that if the surface of the object is in focus (of an imaging system) the speckle motions equal the observed lateral object displacement directly.[18] Recent studies have shown that equations governing determination of displacement vectors from specklegrams are similar to those used in interpretation of images obtained in hologram interferometry.[19] That is, Eq. (15.3-1) applies directly to the interpretation of specklegrams.

Specklegrams are recorded by illuminating the object with a single laser beam (Fig. 15.3-14); no reference beam is used. The light scattered by the object (or transmitting medium in the case of heat transfer or fluid mechanics studies) is imaged from one or more directions onto high-resolution photosensitive material; two directions are sufficient to obtain complete three-dimensional mapping of object's motions. For interferometric purposes two exposures are made on the same recording material to image object's initial and final speckle patterns, respectively. In the case when tandem specklegrams are used, each speckle pattern is recorded on a separate piece of film. Later, these specklegrams are "sandwiched" together during reconstruction process.

Developed specklegrams are analyzed by sending a narrow beam of laser light directly through the specklegram (Fig. 15.3-15). Then, a diffraction halo is formed on a screen, which is modulated by Young's fringes (Fig. 15.3-16). The frequency of these fringes is directly proportional to the magnitude of the observed displacement while their orientation is in direction normal to the direction of L.

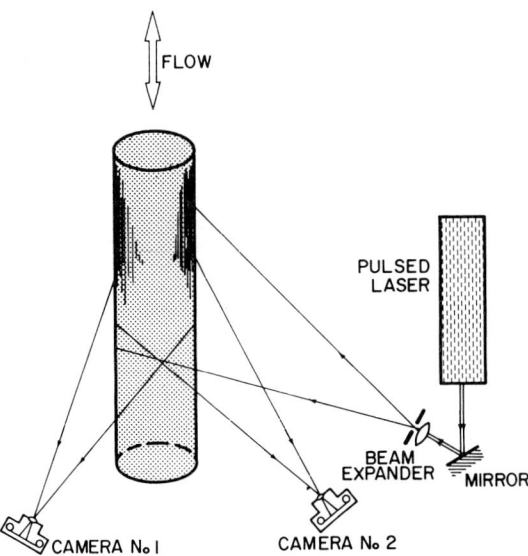

Fig. 15.3-17 Simplified representation of the setup for recording of specklegrams of a flowing fluid. In real application the beam from the pulsed laser would be collimated and shaped to form a thin "sheet of light" illuminating the plane within the fluid that is being studied.

Parameters necessary to quantitatively interpret specklegrams can either be obtained manually (using a ruler and a protractor) or automatically (with the help of a video digitizer system). The latter approach is much more appropriate when fast data acquisition is required, especially in the case of large systems that are being investigated, or whenever many data points are required for a complete analysis.

Specklegrams are very useful in measurement of displacements up to 1.5 mm. The accuracy of this measurement depends on the speckle size which, in turn, is a function of the imaging system. As such, the method can be used in studies of fluid flow (Fig. 15.3-17). In this application, laminar as well as turbulent flows can be investigated to determine, for example, velocity profiles.

15.3-4 Computer-Aided Interpretation of Laser Images

In many of the energy-related problems the parameters of interest (e.g., displacement, strain, velocity, etc.) must be determined experimentally at predetermined locations on the object. This can be done, in a very elegant manner, by scanning the holographically reconstructed image (or a diffraction halo obtained during reconstruction of a specklegram) with a computer-compatible video digitizer camera shown in Fig. 15.3-18. This camera, in addition to converting the scene being observed into a composite analog video signal, which is viewed on a monitor, produces a digital signal that is transmitted directly to the computer. The computer, in turn, rapidly reads the electronic signal

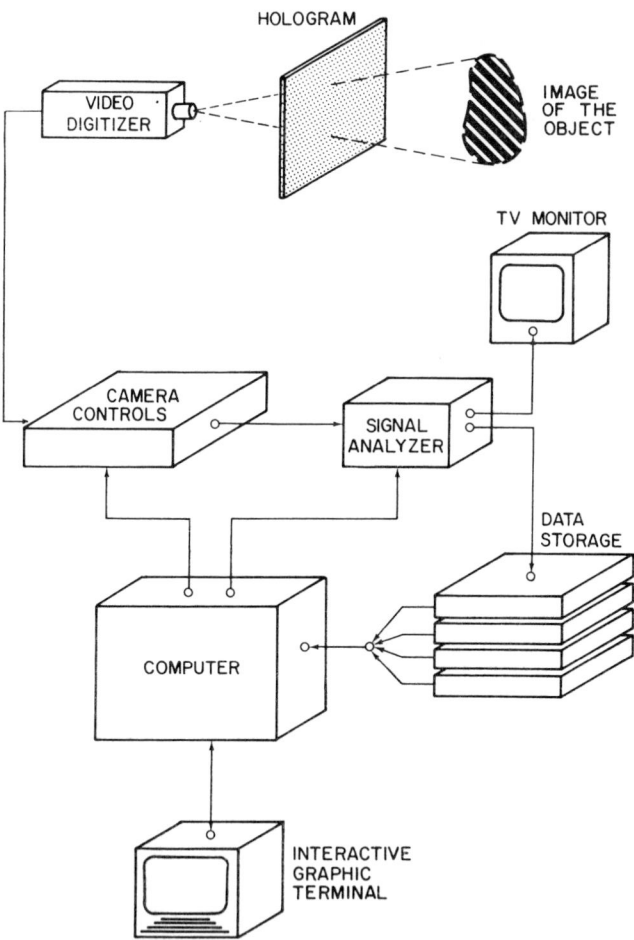

Fig. 15.3-18 Automated system for interpretation of holograms. Similar arrangement is used for interpretation of specklegrams, however, instead of focusing the digitizer on the image it focuses on the halo and the Young's fringes.

corresponding to the video image being processed. It integrates and outputs the video signal intensity along a prescribed portion of a horizontal video line for every line, producing plots of intensity distribution within the image plane. Data from these intensity distributions, together with other pertinent parameters, are used in quantitative interpretation of laser images using interactive computer graphics programs. These results can be obtained for any point within the reconstructed image by simply instructing the computer to perform calculations for a point (or a number of points) at specified coordinates.

A system incorporating computer-compatible video digitizer coupled with the heterodyne hologram interferometry apparatus for rapid quantitative analysis of laser images with high accuracy is presently being developed as a part of studies described herein. This system will provide unique experimental capabilities leading to a wider application of laser methods in studies of energy systems.

Interpretation of specklegrams is done readily by performing digital processing of experimental data according to the procedure shown in Fig. 15.3-19. In this approach information defining recording/reconstruction geometry is put directly into the computer. Then, parameters defining characteristics of observed Young's fringes are determined using the video digitizer. The digitizer itself can be directly interfaced with the computer, or its output can be temporally preprocessed and subsequently entered into the computer for final analysis. Some of the alternatives in processing of the data from specklegrams are shown by dashed lines in Fig. 15.3-19.

It is believed that the measuring techniques utilizing the laser methods discussed herein are forerunners of what may be called the experimental finite-element methods.[17] The laser methods are particularly suited to that application because they can provide very accurate digital data on response of the system to various loads. Furthermore, these responses are determined noninvasively, in three dimensions, with high accuracy and precision.

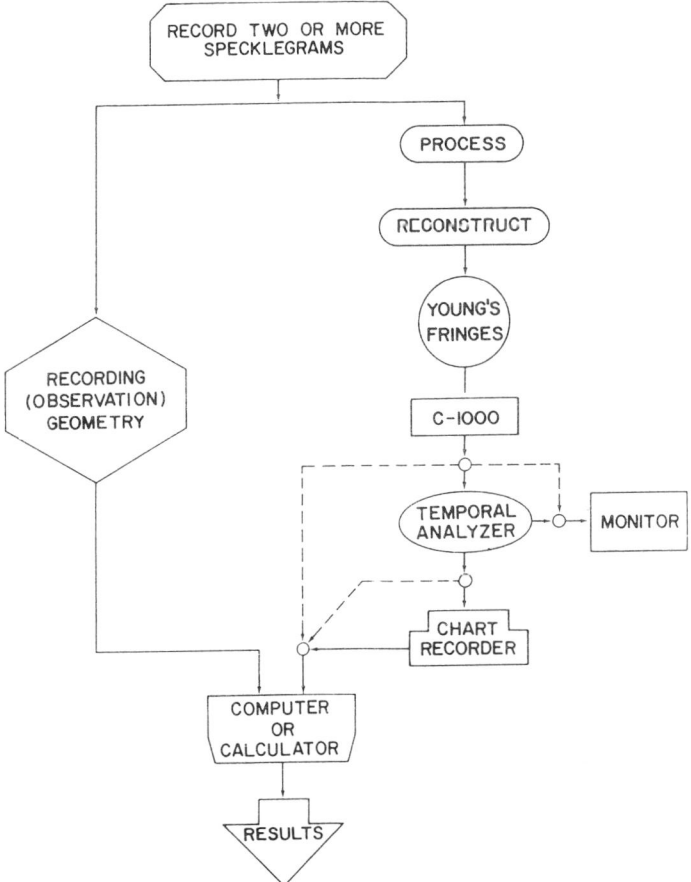

Fig. 15.3-19   Flow chart for quantitative interpretation of specklegrams.

15.3-5 Conclusions

It has been shown that recent developments in experimental methods employing lasers can be used to quantitatively determine parameters necessary to study energy systems. These structures can be experiencing mechanical, thermal, or combined loads, which can be static or dynamic. The experimental data can be obtained in an interactive manner using computer-controlled digitizer and scanning systems (currently under development). The displacements measured by laser techniques described herein are accurate to within 3×10^{-10} m, strains to within 0.2×10^{-6} m/m, curvature of object's surface to within 9×10^{-5} m^{-1}, and temperature to within $0.001°C$. As such, these methods are particularly suited for precise and accurate studies of energy interactions noninvasively, in three dimensions

REFERENCES

15.3-1 K. A. Stetson and R. L. Powell, "Hologram Interferometry," *J. Opt. Soc. Am.* **56**, 1161–1165 (1966).

15.3-2 H. M. Smith (ed.), *Holographic Recording Materials*, Springer-Verlag, Berlin, 1977.

15.3-3 R. J. Pryputniewicz, *Laser Holography*, WPI/ME, Worcester, MA, 1979.

15.3-4 R. J. Pryputniewicz, *Quantitative Interpretation of Time—Average Holograms in Vibration Analysis*, WPI/ME, Worcester, MA, 1983.

15.3-5 R. J. Pryputniewicz and K. A. Stetson, "Holographic Strain Analysis: Extension of the Fringe —Vector Method to Include Perspective," *Appl. Opt.* **15**, 725–728 (1976).

15.3-6 R. J. Pryputniewicz and W. W. Bowley, "Techniques of Displacement Measurement: an Experimental Comparison," *Appl. Opt.* **17**, 1748–1756 (1978).

15.3-7 C. M. Vest, *Holographic Interferometry*, Wiley, New York, 1978.

15.3-8 W. Schuman and M. Dubas, *Holographic Interferometry*, Springer-Verlag, Berlin, 1979.

15.3-9 R. J. Pryputniewicz, "State-of-the-Art in Hologrammetry and Related Fields," *Int. Arch. Photogram.* **23**, 620–629 (1980).

15.3-10 K. A. Stetson, "The Use of Projection Matrices in Hologram Interferometry," *J. Opt. Soc. Am.* **69**, 1705–1710 (1979).

15.3-11 R. J. Pryputniewicz, "High Precision Hologrammetry," *Int. Arch. Photogram.* **24**, 377–386 (1982).

15.3-12 R. J. Pryputniewicz, "Holographic Determination of Rigid-Body Motions," *Appl. Opt.* **18**, 1442–1444 (1979).

15.3-13 W. Panknin, *Eine holographische Zweiwellenlangen—Interferometrie zur Messung uberlagerter Temperatur—und Konzentrationsgrenzchichten*, Ph.D. thesis, Univ. of Hannover, Fed. Rep. of Germany, 1977.

15.3-14 D. Yaghmaie, *Holographic Studies of Heat Transfer Characteristics of Heated Horizontal Rods*, M.S. thesis, Worcester Polytechnic Institute, Worcester, MA, 1982.

15.3-15 S. E. Lindquist, *Analysis of Natural Convection from a Vertical Isothermal Flat Plate Using Hologram Interferometry*," M.S. thesis, Worcester Polytechnic Institute, Worcester, MA, 1983.

15.3-16 R. Dandliker, E. Marom, and F. M. Mottier, "Two-Reference Beam Holographic Interferometry," *J. Opt. Soc. Am.* **66**, 23–30 (1976).

15.3-17 R. J. Pryputniewicz, "Unification of FEM Modeling with Laser Experimentation," *Proc. 6th Invitational Symp. on FEM*, Storrs, CT, 1982.

15.3-18 K. A. Stetson, "Miscellaneous Topics in Speckle Metrology," *Speckle Metrology*, R. K. Erf (ed.), Academic Press, New York, 1978, pp. 295–320.

15.3-19 R. J. Pryputniewicz, "Projection Matrices in Specklegraphic Displacement Analysis," *SPIE* **243**, 158–164 (1980).

CHAPTER **16**

GUIDE TO AVAILABLE CODES, STANDARDS, AND REFERENCE MATERIAL

JACK B. LEVY

General Electric Company,
Schenectady, New York

RICHARD L. P. CUSTER

Worcester Polytechnic Institute
Worcester, Massachusetts

16.1 INTRODUCTION

The word *standards* has many meanings in our language, but underlying these is the semantic understanding that a standard is more or less an invariable. How then can the term *standard* be applied to science and engineering, which change so dramatically with each passing decade?

The solution is that standards must be reviewed and updated as operating experiences dictate. All American National Standards must be reviewed at least once every 5 years and either be reaffirmed, revised, or withdrawn. Extensions can be granted beyond the 5-year period only if a committee is in the process of developing a revision but cannot complete its task within that deadline.

Thus, engineering standards are the result of cumulative technical experience. Specifically, all the American National Standards are the result of cumulative technical experience. Standards do change, but only under expert control and evolution, so they continuously represent conservative practice.

A code is a standard that can and often is prepared for mandatory use. Codes can be made a part of federal, state, or municipal regulation through adoption or reference.

16.2 DEFINITIONS

The first three definitions are approved by the International Standards Organization Standing Committee for the Study of Scientific Principles of Standardization in French and English.

1. Standardization. Standardization is the process of formulating and applying rules for an orderly approach to scientific activity for the benefit and with the cooperation of all concerned, and in particular for the promotion of optimum overall economy taking due account of functional conditions and safety requirements. It is based on the results of science, techniques, and experience. It determines

R. L. P. Custer contributed material on fire safety.

not only the basis for present but also for future development, and it should keep pace with progress. A particular application is for products and processes—the definition and selection of characteristics of products, testing and measuring methods, specification of characteristics of products for defining their quality, regulation of variety, interchangeability, and so on.

2. A standard. A standard is the result of a particular standardization effort approved by a recognized authority. It may take the form of (1) a document, containing a set of conditions to be fulfilled; (2) a fundamental unit of physical constant, for example, ampere, absolute zero (kelvin); or (3) an object for physical comparison, for example the standard meter.

3. A specification. A specification is a concise statement of a set of requirements to be satisfied by a product, material, or a process indicating, whenever appropriate, the procedures by which it may be determined whether the requirements given are satisfied. A specification may be a standard, a part of a standard, or independent of a standard. As far as practicable, it is desirable that the requirements be expressed numerically in terms of appropriate units; together with their limits, the specification presents contract technical requirements. For precision in communication, the term *standard* requires a modifier (see definitions 4–7).

4. Consumer standard. Consumer standards contain those measures that contribute, directly or indirectly, to the consumer's assurance that goods are of suitable quality appropriate to a given purpose, that they will give reasonable use, and that if faulty there will be a means of redress.

5. Engineering standard. (This definition relates solely to the content of an engineering standard.) The engineering standard is a technological practice described in a document to assure dimensional, functional, or compositional compatibility; quality and performance; uniformity of evaluation procedures; or uniformity of engineering language.

6. Quality assurance standard. The quality assurance standard is a management standard. It presents general requirements for planning, managing, conducting, and evaluating quality assurance programs. The objectives of the standards are to ensure that reliability and safety are designed into equipment and to prevent degradation of design reliability during succeeding steps from fabrication to end use.

(a) **Quality assurance.** All those planned or systematic actions necessary to provide adequate confidence that an item or a facility will perform satisfactorily in service.

(b) **Quality control.** Those quality assurance actions that provide a means to control and measure the characteristics of an item, process, or facility to established requirements.

7. Safety standard. A safety standard requires conditions, or the adoption of one or more practices, means, methods, operations, or processes, reasonably necessary or appropriate to provide safe or healthful employment and places of employment.

8. Firesafety standard. A firesafety standard requires the use of designs, equipment, materials and construction, inspection, and operational practices to prevent ignition or control the spread of fire. Firesafety standards also address means to provide early warning of fire and safe egress for persons exposed to potential injury.

9. Code. A code is a group of administrative and technical rules and standards for a system and equipment. It usually covers a combination of requirements for performance, design, materials, fabrication, construction, installation, inspection, testing, and operation of equipment. It frequently is made mandatory by official action of some governmental agency.

10. Performance test code. Performance test codes provide uniform methods and procedures for testing equipment and systems that (a) will give validity to determinations of capacity and efficiency and other operating performance characteristics, (b) will compare different conditions or methods of operations, (c) will determine the cause of either inferior or superior results, and (d) will determine the effect of changes of design or proportion upon the capacity or efficiency of equipment and systems.

11. Regulations. A regulation is an ordinance or law established by a statutory body. It may be a standard or code whose mandatory use has been decreed by reference or adoption.

12. Consensus. In standardization practice a consensus is achieved when substantial agreement is reached by concerned interests according to the judgment of a duly appointed authority. Consensus implies much more than the concept of a simple majority but not necessarily unanimity. The approval of a standard by ANSI implies a consensus of those substantially concerned with its scope and provisions.

16.3 RESEARCHING INDUSTRY CODES AND STANDARDS

As soon as a decision is made to use a certain item, a given design application, a search should be made to determine whether there is a relevant industry code or standard. There are three logical places to look for information.

1. Standards writing organizations.
2. American National Standards Institute.
3. National Bureau of Standards.

16.3-1 Standards Writing Organizations

The first place to look for a standards writing organizations code or standard is at the organization most apt to be involved with the subject.

A list of major standards writing organizations concerned with codes or standards for energy systems follows:

1. American Nuclear Society (ANS)
   244 East Ogden Ave.
   Hinsdale, IL 60521

   ANS develops codes and standards in the nuclear field.

2. American Society for Testing and Materials (ASTM)
   1916 Race Street
   Philadelphia, PA 19103

   ASTM is concerned with the development of standards and test methods relating to the characteristics and performance of materials, products, systems, and services and the promotion of related knowledge.

3. American Society of Heating, Refrigeration and Air-Conditioning Engineers (ASHRAE)
   345 East 47th Street
   New York, NY 10017

   ASHRAE contributes standards vital to man's well being and comfort in office buildings, factories, schools, and home construction; in food and beverage processing, storage and distribution, in solar energy utilization, and in space.

4. American Society of Mechanical Engineers (ASME)
   345 East 47th Street
   New York, NY 10017

   The development of codes and standardization activities are subdivided into various groups:
   (a) Pressure technology.
   (b) Nuclear.
   (c) Safety.
   (d) Performance test codes.
   (e) Standardization (all other activities not included in the four categories listed above). The best known, and most widely used of all their standards is the ASME Boiler and Pressure vessel code, which has been adopted in part or in its entirety by 45 states, numerous municipalities, and all the provinces of Canada. It provides standard rules for construction of boilers, pressure vessels, and third-party inspection. Use of the code assures the user of a reasonably certain protection of life and property as well as a reasonably long period of service.

5. American Water Works Association (AWWA)
   6666 W. Quincy Ave.
   Denver, CO 80235

   AWWA advances the knowledge of the design, construction, operation, and management of water utilities in the production and distribution of safe and adequate community water supplies.

6. Department of Defense (DOD)
   Naval Publications & Forms Center
   5801 Tabor Avenue
   Philadelphia, PA 19120

   DOD prepares specifications, standards, and related standardization documents for use by the military and federal government. They also review industry standards documents and list those acceptable for use.

7. Electronic Industries Association (EIA)
   2001 Eye Street, N.W.
   Washington, D.C. 20006

   EIA develops standards applicable to the electronic industry.

8. Institute of Electrical and Electronic Engineers (IEEE)
   345 East 47th Street
   New York, NY 10017

   IEEE develops standards relating to electrical and electronic equipment, test methods, units, symbols, definitions, and rating methods.

9.   Instrument Society of America (ISA)
     400 Stanwix Street
     Pittsburgh, PA 15222

      ISA is concerned with the theory, design, manufacture, and use of instruments. They maintain standards and practices committees in the broad areas of measurement devices, control devices, symbology, computer hardware and software, and safety.

10.   Manufacturers Standardization Society of Valve and Fittings Industry (MSS)
     1815 North Fort Myer Drive
     Arlington, VA 22209

      MSS is concerned with committees working on subjects covering threaded and flanged fittings and valves, marking and terminology, material, cast and malleable iron fittings, union and union fittings, welding fittings, butterfly valves, pipe hangers, quality standards, valve activators, and so on.

11.   National Electrical Manufacturers Association (NEMA)
     2101 L Street, N.W., Suite 300
     Washington, D.C. 20037

      NEMA is a trade association of manufacturers of almost every kind of equipment and apparatus used for the generation, transmission, distribution, and utilization of electric power.

12.   National Fire Protection Association (NFPA)
     Batterymarch Park
     Quincy, MA 02269

      NFPA is organized to promote the science and improve the methods of fire protection. One of the main functions of NFPA is in the standards making field under which codes, standards, and recommended practices are developed as guides to engineered protection for reducing loss of life and property by fire.

13.   Society of Automotive Engineers (SAE)
     400 Commonwealth Drive
     Warrendale, PA 15096

      SAE carries on technical standardization work for the motor vehicle, aircraft, airline, space vehicles, farm tractors, earth-moving and road-building machinery, and other manufacturing industries using internal combustion engines.

14.   Underwriters' Laboratories (UL)
     207 East Ohio Street
     Chicago, IL 60611

      UL is an independent organization devoted to testing for public safety. It publishes standards, classifications, and specifications for materials, devices, constructions, and methods affecting hazards and other information tending to reduce and prevent loss of life and property from fire, crime, and casualty.

Each of the organizations listed above develop, print, and distribute codes and standards. Many have listed their publications in separate catalogs. For those who do business with the military and federal government, a very comprehensive index of specifications and standards are available from the Department of Defense in two parts. Part 1 is an alphabetical index. Part 2 is a numerical index.

As an example of what kind of documents can be obtained from standards writing organizations, a partial list of subjects relating to energy systems is shown below. For each subject a few typical examples are listed. (Abbreviations refer to standards writing organizations listed above. The sources for government documents or advisory materials prepared by industry or code organizations are noted where necessary.)

Subject	Document No.	Typical Documents
Alcohol	NFPA 61B	Fire and explosion prevention, grain elevators and bulk agricultural commodities
Boiler	ASME P0230(ABC)	Boiler and pressure vessel code (B & PV)
	100059	Criteria of the ASME B & PV code for design by analysis
	E00097	Criteria for design of elevated temperature class I components in Sec. III, Div. 1 of ASME B & PV
	NFPA 85	Prevention of furnace explosion, fuel oil and natural gas-fired single-burner boiler-furnaces

Subject	Document No.	Typical Documents
	NFPA 85B	Explosion prevention, natural-gas-fired multiple-burner boiler furnaces
	NFPA 85D	Explosion prevention, fuel-oil-fired multiple-burner boiler furnaces
	NFPA 85E	Explosion prevention, pulverized coal-fired multiple-burner boiler furnaces
	NFPA 85F	Installation and operation of pulverized fuel systems
	NFPA 85G	Prevention of furnace implosions in multiple-burner boiler furnaces
	UL 834	Safety standards for electric boilers
Coal	ASTM D121	Definition of terms relating to coal and coke
	ASTM D547	Index of dustiness of coal and coke
	ASTM D3175	Volatile matter in the analysis sample of coal and coke
	ASME PTC 4.2	Performance test code—coal pulverizers
	NFPA 653	Prevention of dust explosions in coal preparation plants
Control	ASME MC85.1	Terminology for automatic control
	NEMA ICS 1	Industrial control and systems
	NEMA ICS 6	Enclosures for industrial control and systems
	UL 508	Safety standards for industrial control equipment
	UL 1238	Safety standards for control equipment for use with flammable liquid-dispensing devices
Cooling	ASME PTC 23	Performance test code—atmospheric water-cooling equipment
Electric machinery	IEEE 56	Rotating electric machinery
Electronic equipment	EIA RS385	Preferred values for components for electronic equipment
	EIA RS325-A	Flammability tests for electronic components
	NFPA 75	Protection of Electronic computer/data processing equipment
Energy conversion	ASHRAE 90A	Energy conservation in new building design
Energy management	UL 916	Energy management equipment
Engines	ASME PTC 17	Performance test code—reciprocating internal combustion engines
	NFPA 37	Stationary combustion engines and gas turbines
	SAE J1220	Rotary-trochoidal engine nomenclature and terminology
Fire detection	NFPA 72E	Automatic fire detectors
	NFPA 74	Household fire-warning equipment
Flammable liquids	NFPA 30	Flammable and combustible liquids code
	NFPA 49	Hazardous chemicals data
Fire suppression systems	NFPA 11	Foam extinguishing systems
	NFPA 12	Carbon dioxide extinguishing systems
	NFPA 12A	Halogenated extinguishing systems
	NFPA 13	Automatic sprinkler systems
	NFPA 17	Dry chemical extinguishing systems
Fluid flow	ASME MFC-IM	Glossary of terms used in the measurement of fluid flow in pipes

Subject	Document No.	Typical Documents
Gases	ASTM D2650	Methods of test for chemical composition of gases by mass spectrometry
	ASTM F307	Sampling pressurized gas for gas analysis
	ASME PTC 19.10	Performance test code—flue and exhaust gas analyzers
	NFPA 58	Storage and handling of liquified petroleum gases
	NFPA 59	Liquified petroleum gases at utility gas plants
Gas turbine	ASME B133.1	Gas turbine terminology
	ASME B133.2	Basic gas turbine
	ASME B133.4	Gas turbine controls and protection systems
	ASME B133.7	Gas turbine fuels
Heat exchanger	ASME PTC 43	Performance test code—air heaters
	UL 462	Heat reclaimers for gas-, oil-, or solid-fuel-fired appliances
Heating	ASME PTC 12.1	Performance test code—closed feedwater heaters
Heat transfer	ASTM E638	Calibration of heat transfer rate calorimeters using a narrow-angle blackbody radiation facility
	ASME HTD Vol. 17	Fouling in heat exchange equipment
Hydrogen	NFPA 50A	Gaseous hydrogen systems
	NFPA 50B	Liquified hydrogen systems
	NBS SP-419	Selected topics on hydrogen fuel
	NBS TN-690	Is hydrogen safe?
Incincerators	NFPA 82	Incinerators and waste-handling equipment
Instrumentation	ISA S51.1	Process instrumentation terminology
	ISA S26	Dynamic response testing of process control instrumentation
Insulation	HH-I-515D	Federal specification for insulation, thermal, cellulosic, or wood fiber (General Services Administration, Washington, D.C.)
	UL 1446	Insulating materials
	ULC-S124	Protective coverings for foam plastic (UL Canada, Toronto)
	UBC-17-3	Evaluation of thermal barriers (International Conference of Building Officials, Whittier, CA)
Lubrication	ASTM D2782	Method for measurement of extreme-pressure properties of lubricating fluids
	ASTM-ASME LOS-501	Recommended practices for the design of oil systems for lubrication and control of hydroelectric equipment
Materials testing	ASTM E84	Surface flame spread characteristics of building materials
	U100R	Fire safety guidelines for use of rigid polyurethane foam insulation (Urethane Safety Group, Society of the Plastics Industries, New York, NY)

Subject	Document No.	Typical Documents
Nuclear	ANS 8.3	Criticality of accident alarm systems
	ASME NQA-1	Quality assurance program requirements for nuclear power plants
	ASME N45.2.1	Cleaning of fluid systems and associated components for nuclear power plants
Oil	ASTM D3904	Method of test for oil from oil shale
	ASTM D943	Method of test for oxidation characteristics of inhibited steam-turbine oils
Pipe	ASTM D2513	Specification for thermoplastic gas pressure pipe, tubing, and fittings
	ASME B31.1	Power piping
	ASME B31.2	Fuel gas piping
	ASME B31.8	Gas transmission and distribution piping systems
	MSS SP58	Materials, design, and manufacture—pipe hangers and supports
Plastics	WPS-301	Fire safety guidelines for expanded polystyrene in building construction (EPS Division, Society of the Plastics Industries, Des Plaines, IL)
	NFPA 205M-T	Plastics in building construction
	ASTM D635-77	Test method for burning rate of self-supporting plastics in the horizontal position
Pumps	ASME PTC 8.2	Performance test code for centrifugal pumps
	ASME PTC 18.1	Performance test code for pumping mode of pump/turbines
	AWWA E101	Deep-well vertical turbine pumps
	UL 343	Pumps for oil-burning appliances
Solar	ASTM B638	Specification for copper and copper alloy solar heat absorber panels
	ASTM E108	Fire tests of roof coverings
	NBSIR 81-2344	Fire testing of roof-mounted solar collectors by ASTM E108 (National Bureau of Standards, Washington, D.C.)
	NBSIR 79-1931	Fire experiments and flash point criteria for solar heat transfer liquids (National Bureau of Standards, Washington, D.C.)
	NBSIR 82-2554	Health and safety considerations for passive solar heated and cooled buildings (National Bureau of Standards, Washington, DC)
	DOE/CS/34281-01	Recommended requirements to code officials for solar heating, cooling, and hot water systems (U.S. Department of Energy, Washington, D.C.)
	US 1279	Safety standards for solar collectors
	MPS 493.2	Solar heating and domestic hot water systems, minimum property

(Continued)

Subject	Document No.	Typical Documents
		standards (U.S. Department of Housing and Urban Development, Washington, D.C.)
	ASHRAE 93	Methods of testing to determine the thermal performance of solar collectors
Steam	ASTM D2186	Methods of test for deposit-forming impurities in steam
	ASME PTC 4.1	Performance test code—steam generating units
Turbine	ASME PTC 29	Performance test code—speed governing systems for hydraulic turbine-generator units
	ASME 133.4	Gas turbine control and protection systems
	ASME TWDPS-1	Recommended practices for the prevention of water damage to steam turbines used for electric power generation: nuclear-fueled plants
Valves	ASME B16.17	Hydrostatic testing of central valves
Water	ASTM D860	Method of sampling water from boilers
Wood	UL 103	Chimneys, factory built, residential
	UL 959	Chimneys, factory built, medium heat
	NFPA 211	Chimneys, fireplaces, and vents

A more comprehensive list of codes and standards writing organizations can be obtained from several sources:

1. The National Bureau of Standards (NBS), Special Publication 417, entitled "Directory of United States Standardizing Activities." Order from the U.S. Superintendent of Documents, U.S. Government Printing Office, Washington, D.C. 20402 (the stock number is SN003-003-01395-1).

2. Global Engineering Documentation Services
   3950 Campus Drive
   Newport Beach, CA 92660

   Ask for their publication entitled "Directory of Engineering Document Sources."

3. Columbia Books
   734 15th Street N.W.
   Washington, D.C. 20005

   Ask for their publication entitled "National Trade and Professional Associations of the United States and Canada and Labor Unions."

The advantage of contacting the standards writing organization most apt to be involved is that there are three possibilities for obtaining assistance.

1. If there is an existing standard on the subject, they will immediately advise you accordingly.

2. If the organization currently has a committee working on the subject, they might be able to answer your questions based on current proposals, or they will advise when to expect the proposal to become available an an accepted standard. *Note:* Proposals are tentative and subject to revision and normally are only available to committee members.

3. Even if there is no standard for the subject, the organization might greet your inquiry as a request for action and either assign the project to an existing committee or establish a new committee. In either case it will be too late to help with your immediate problem, but it might be invaluable to you at a later date or to others with a need for the same standard.

16.3-2 American National Standards Institute (ANSI)

Most standards writing committees submit their standards to the American National Standards Institute or sponsor committees that develop standards in accordance with ANSI procedures for acceptance as American national standards. There are several methods that involve elaborate procedures to safeguard that the submitted documents represent a true consensus among all who have an interest in the subject.

If you are unable to locate a standard by going directly to an organization that develops standards, the second step would be to look for information in the ANSI catalog.

The lastest edition of the catalog of ANSI standards can be purchased from:

American National Standards Institute
1430 Broadway
New York, NY 10018

This catalog has both a subject index and a numerical index. In addition, ANSI sells a catalog of ISO (International Organization for Standardization) standards and a catalog of IEC (International Electrotechnical Commission) standards. These catalogs are essential to those who need to know the standards that have worldwide recognition.

16.3-3   National Bureau of Standards (NBS)

If the ANSI catalog does not provide a lead for the subject at hand and it is not obvious which organization to approach, there is a third source of information that is even more broadly based than ANSI, the National Bureau of Standards, Standards Information Service (SIS), located in Gaithersburg, Maryland. NBS-SIS maintains a reference collection of engineering and related standards, which includes over 240,000 standards, specifications, test methods, codes, and recommended practices issued by:

U.S. technical societies.
Professional and trade associations.
State purchasing offices.
U.S. government agencies.
Foreign national and international standardizing books.

In addition to the above mentioned 240,000 standards, their collection also contains:

Over 300 reference books.
Articles, pamphlets, reports, and handbooks on standardization.
150 periodicals and newsletters.
Visual-search microfilm files (VSMF).

This collection is open five days a week (Monday through Friday) for on-site use. If one is unable to visit the collection, valuable information can still be obtained by telephone (301-921-2587), by telex (89-8493), by letter (Standards Information Service, Room B162, Building 225, National Bureau of Standards, Washington, D.C. 20234), or by possession of one of their several index volumes to be described later. An inquiry takes the basic form of "Is there a standard for? . . ." Using their key-word-in-context (KWIC) indexes and other in-house tools, SIS determines whether or not there is a standard for products and processes ranging from acoustics to zinc. SIS is an information and referral center and, therefore, *does not sell, distribute, or loan any standards* but refers inquirers to the organizations where they can obtain copies.

Other SIS services consist of compiling general and special indexes of standards. Their most important series of books are titled *An Index of U.S. Voluntary Engineering Standards*. The first of this series is coded NBS Special Publication 329 (1971), 984 pages. A supplement 1 was issued a year later to add 440 pages of additional listings. A Supplement 2 was issued in 1975 to add 467 more pages of listings. It is anticipated that these three publications will eventually be consolidated into a new updated document.

There are other index volumes as follows:

Index of International Standards (NBS 390).
Index of State Specifications and Standards (NBS 375).
Index of U.S. Nuclear Standards (NBS 483).
Tabulation of Voluntary Standards and Certification Programs for Consumer Products (NBS-TN 948).
World Index of Plastics Standards (NBS 352).

All of the above-mentioned NBS publications are sold by the Superintendent of Documents, U.S. Government Printing Office (USGPO), Washington, DC 20402, or by the National Technical Information Service (NTIS), 5285 Port Royal Road, Springfield, VA 22161, who use special ordering codes that differ from the NBS designations as follows:

NBS Designation	USGPO SD Stock Number	NTIS COM Number
329	—	71-50172
329 Supl. 1	—	73-50679
329 Supl. 2	SN003-003-01362-5	—
352	—	75-10291
375	—	73-50839
390	—	74-50352
417	SN003-003-01395-1	—
483	SN003-003-01822-8	—
TN948	SN003-003-01779-5	—

16.4   BUILDING CODES

Building codes are legal documents developed through legislative action by state or local government. These documents address the construction alteration and repair or modification of buildings to provide for the safety, health, and welfare of occupants or users.

A state or local building code is generally based on a particular edition of one of the four major model codes and may often contain modifications to suit local needs or conditions.

A model building code is also a dynamic document that is updated periodically to reflect changes in building research in all areas relating to safety and health. Further, model codes make frequent reference to standards promulgated by organizations such as ANSI, ASTM, NFPA, and others discussed earlier that also undergo revision. Thus, it is important to be aware of both the edition of the model code that governs in a particular jurisdiction and the dates of any standards or other documents referenced. While compliance with the existing building code may meet the requirements of the local authority, where issues of public safety are involved, the latest editions of codes and standards as well as the technical literature should be sought out and studied. This is particularly important when newly developed materials, processes, or energy sources are involved.

The names of the four major model codes discussed above and the organizations that administer them are presented below:

1. Basic building code
    Building Officials and Code Administrators, International
    1313 East 60th Street
    Chicago, IL 60637

2. National building code
    National Conference of States on Building Codes and Standards
    481 Carlisle Dr.
    Herndon, VA 22070

3. Uniform building code
    International Conference of Building Officials
    5360 South Workman Mill Road
    Whittier, CA 90601

4. Standard building code
    Southern Building Code Congress
    1116 Brown-Marx Building
    Birmingham, AL 35203

Although there is no building code promulgated by the federal government, the U.S. Department of Housing and Urban Development develops and maintains minimum property standards for residential and other types of buildings.

16.5   DO IT YOURSELF

If after an exhaustive search you are unable to find an industry standard, there are two other sources of information:

1.   The best source is YOU. After obtaining samples from potential suppliers, you need to conduct tests to determine the properties of the product. With adequate and reliable data you should be able to develop your own specification.

The following is a partial list of laboratories that have the facilities for conducting one or more of the test methods listed above. Additional laboratories may be found on the campuses of universities or through commercial and industrial sources.

(a)   Underwriters' Laboratories, Northbrook, IL, and Santa Clara, CA.

(b)   Factory Mutual Research Corporation, Norwood, MA.

(c)   United States Testing Company, Hoboken, NJ, and Los Angeles, CA.

(d)   Southwest Research Institute, San Antonio, TX.

(e)   Commercial Testing Co., Dalton, GA.

(f)   Hardwood Plywood Mfrs. Assn., Arlington, VA.

2.   An always available but sometimes unreliable source of information is the supplier. Driven by marketing ambitions, he frequently tends to make exaggerated claims and publishes misleading comparison charts. It is very simple to protect oneself from such misinformation. Ask the supplier for guaranteed minimum and maximum values for every property that is important to your application. He now has to provide honest data or run the risk of a legal basis for rejection. You should have no problem if you deal with reliable suppliers who value their reputation.

16.6   HOW USE OF CODES AND STANDARDS MAKES YOUR JOB EASIER

Use of Industry Codes and Standards saves time and money in many ways:

1.   If you use an industry designation for identification you have a ready-made document that may only require a minimum amount of word description to completely define your needs. Those who prepare their own local documentation utilizing industry codes and standards either by reference or by paraphrasing will complete their task with a minimum expenditure of time.

2.   It improves availability because industry codes and standards are widely used. If it involves the purchase of a product, suppliers are more apt to carry them in stock. This is especially valuable to low-volume users. If it involves a procedure, the necessary testing equipment should be readily available.

3.   In general, industry standards help to keep varieties to a minimum. With few sizes and types production costs are lowered, less money is tied up in inventory, and savings can be passed on to the consumer.

4.   Communication between buyer and seller can be rapid, thus minimizing the need for additional letter writing or telephone conversations to reach an understanding of one's exact needs.

CHAPTER 17
ENGINEERING MATHEMATICS

OKAN GUREL

IBM Cambridge Scientific Center
Cambridge, Massachusetts

17.1 BASIC ALGEBRAIC EQUATIONS

17.1-1 Algebraic Operations

If a set of elements $x, y, \ldots$, is given with *algebraic (binary) operations* such as addition and multiplication involving its elements in pairs, then the following laws are obeyed.

The Laws of Algebraic Operations

(a) Commutative law: $x + y = y + x$, $xy = yx$.
(b) Associative law: $x + (y + z) = (x + y) + z$, $x(yz) = (xy)z$.
(c) Distributive law: $z(x + y) = zx + zy$.

17.1-2 Progression

The sequence of numbers with a special relationship between the consecutive numbers forms a *progression*. Following are the most common progressions.

Arithmetic Progression

The *arithmetic progression* is a sequence of n numbers $x_1, \ldots, x_n$, such that the difference between two consecutive numbers is a constant, called a *common difference*, d. The nth number is given by

$$x_n = x_1 + (n - 1)d$$

The *sum* of n numbers, S_n, is given by

$$S_n = (n/2)(x_1 + x_n)$$

The *arithmetic mean* between x and y is $(x + y)/2$.

Example: $x_1 = 6$, $n = 4$, $d = 3$

$$x_1, x_2, x_3, x_4 \text{ become } 6, 9, 12, 15$$

$$S_4 = (4/2)(6 + 15) = 42$$

The arithmetic mean between 9 and 12 is $(9 + 12)/2 = 10.5$.

Geometric Progression

The *geometric progression* is a sequence of n numbers $x_1, \ldots, x_n$, such that the ratio between two consecutive numbers is a constant, called a *common ratio*, r. The nth number is given by

$$x_n = x_1 r^{n-1}$$

The *sum* of n numbers, S_n, is given by

$$S_n = x_1(1 - r^n)/(1 - r)$$

or

$$S_n = (x_1 - rx_n)/(1 - r)$$

The *geometric mean* between x and y is $\sqrt{xy}$.

Example: $x_1 = 6$, $n = 4$, $r = 3$

$$x_1, x_2, x_3, x_4 \text{ become } 6, 18, 54, 162$$

$$S_4 = 6(1 - 3^4)/(1 - 3) = 240$$

The geometric mean between 18 and 54 is $\sqrt{18 \cdot 54} = 31.2$.

Harmonic Progression

The *harmonic progression* is a sequence of n numbers $x_1, \ldots, x_n$, such that their *reciprocals* form an arithmetic progression. The nth number is given by

$$\frac{1}{x_n} = \frac{1}{x_1 + (n - 1)d}$$

The *harmonic mean* between x and y is $2xy/(x + y)$.

Example: $x_1 = 6$, $n = 4$, $d = 3$

$$x_1, x_2, x_3, x_4 \text{ become } 1/6, 1/9, 1/12, 1/15$$

The harmonic mean between $1/9$ and $1/12$ is $[2(1/9)(1/12)]/[(1/9) + (1/12)] = 1/10.5$.

17.1-3   Permutations and Combinations

An arrangement of n objects in groups of s $(< n)$ in all possible orders is called a *permutation*. For example, if there are four numbers 1, 2, 3, and 4, the following permutations of groups of three numbers are possible:

1 2 3	1 2 4	1 3 4	2 3 4
1 3 2	1 4 2	1 4 3	2 4 3
2 1 3	2 1 4	3 1 4	3 2 4
2 3 1	2 4 1	3 4 1	3 4 2
3 1 2	4 1 2	4 1 3	4 2 3
3 2 1	4 2 1	4 3 1	4 3 2

The general formula for the number of permutations of n objects in groups of s is

$$_nP_s = n(n-1)(n-2)\cdots(n-s+1) = \frac{n!}{(n-s)!}$$

where $k! = 1\cdot 2 \cdots k$ is called the *factorial*.

For $n = 4$ and $s = 3$ the above formula gives

$$_4P_3 = 4\cdot 3\cdot 2 = \frac{1\cdot 2\cdot 3\cdot 4}{(4-3)!} = 24$$

The two special cases are

$$_nP_1 = n \quad \text{and} \quad _nP_n = n! \quad \text{(note that } 0! = 1)$$

Therefore the factorial $n!$ is also the number of permutations of all n objects.

A collection of n objects in groups of s $(<n)$ where the order of the objects is not relevant is called a *combination*. For example, combinations of four numbers, $1, 2, 3, 4$, in groups of 3 are

$$1\ 2\ 3 \quad 1\ 2\ 4 \quad 1\ 3\ 4 \quad 2\ 3\ 4$$

The general formula for the number of combinations possible is

$$_nC_s = \frac{_nP_s}{s!} = \frac{n!}{(n-s)!s!}$$

Thus for $n = 4$, $s = 3$, $_4C_3 = 4!/1!3! = 4$.

17.1-4 Exponentials and Logarithms

Definition

The *exponential*, x^n, is defined as the product of n x's, that is,

$$\underbrace{x\cdot x\cdot x \cdots x}_{n \text{ terms}}$$

where n is a positive integer and x is a real number. If $y = x^n$, y is called the nth *power* of x while x is the nth *root* of y.

Example: $x = 3$, $n = 4$,

$$x^n = 3^4 = 3\cdot 3\cdot 3\cdot 3 = 81$$

The Laws of Exponents

1. $x^m \cdot x^n = x^{m+n}$;   for example, $x^3 \cdot x^4 = x^{3+4} = x^7$.
2. $x^m/x^n = x^{m-n}$;   for example, $x^4/x^1 = x^{4-1} = x^3$.
3. $(x^m)^n = x^{mn}$;   for example, $(x^2)^4 = x^{2\cdot 4} = x^8$.

Also,

$$\begin{array}{ll} x^0 = 1 & \text{for example, for any } x \\ x^{m/n} = (x^m)^{1/n} & \text{for example, } x^{6/3} = (x^6)^{1/3} \\ x^{-m} = 1/x^m & \text{for example, } x^{-2} = 1/x^2 \end{array}$$

Definition

The *logarithm to the base* a, $\log_a$, is the inverse function to the exponential function a^x, that is,

$$y = a^x \qquad x = \log_a y$$

Example: $100 = 10^2$, $y = 100$, $x = 2$, $a = 10$, thus $2 = \log_{10}100$.

The Laws of Logarithms

1. $\log_a(xy) = \log_a x + \log_a y$; for example, $\log_{10}(3 \cdot 2) = \log_{10} 3 + \log_{10} 2$.
2. $\log_a(x/y) = \log_a x - \log_a y$; for example, $\log_{10}(6/2) = \log_{10} 6 - \log_{10} 2$.
3. $\log_a(x^n) = n \log_a x$; for example, $\log_{10}(3^4) = 4 \log_{10} 3$.

The common bases for logarithms are

10 for common, or Briggsian logarithm
$e = 2.71828$ for natural, Naperian, hyperbolic logarithms

The formulas for base changes of logarithms are:

$$\log_a x = \log_b x / \log_b a = (\log_b x) \cdot (\log_a b)$$

$$\log_{10} x = \log_e x / \log_e 10 = (\log_{10} e) \cdot (\log_e x)$$

$$= 0.43429\,44819 \log_e x$$

$$\log_e x = \log_{10} x / \log_{10} e = (\log_e 10) \cdot (\log_{10} x)$$

$$= 2.30258\,50930 \log_{10} x$$

$$\text{antilog } x = 1/\log x = \log^{-1} x$$

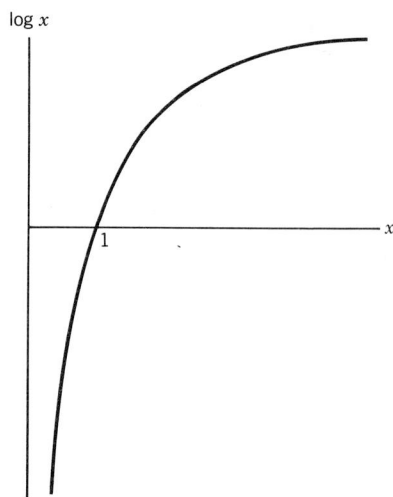

17.1-5   Polynomials

The left-hand-side of an *algebraic equation* in n unknowns, $x_1, \ldots, x_n$, expressed in the summation form of

$$\sum C_{k_1, \ldots, k_n} x_1^{k_1} \cdots x_n^{k_n} = 0$$

where $k_1, \ldots, k_n$ are *powers* of variables and $C_{k_1, \ldots, k_n}$ are *coefficients*, is called a *polynomial*. A single element is called a *term* and the sum (Σk_i) its *degree*. The highest degree is called the *degree of the polynomial*. A polynomial with elements of the same degree is called *homogeneous*. Polynomials with one term are *monomials* and two terms *binomials*. Two of the special homogeneous polynomials are *quadratic*, the second degree, and *cubic*, the third degree.

Examples:

$$6x_1^3 x_2^2 \qquad\qquad \text{a } \textit{monomial (degree} = 5)$$

$$5x_1^4 + 2x_2 \qquad\qquad \text{a } \textit{binomial (degree} = 4)$$

$$5x_1^2 + 2x_1 x_2 + 4x_2^2 \qquad \text{a } \textit{homogeneous polynomial (degree} = 2)$$

$$x_1^2 + x_2 \qquad\qquad \text{a } \textit{quadratic (degree} = 2)$$

$$2x_1^3 + x_2^2 + x_1 x_2 \qquad\quad \text{a } \textit{cubic (degree} = 3)$$

Discriminant of a Polynomial

The *discriminant of a polynomial*,

$$C_0 x^n + C_1 x^{n-1} + \cdots + C_n = 0$$

with roots $x_1, \ldots, x_n$ is, by definition,

$$\prod_{i > k} (x_i - x_k)^2$$

Examples:

$$x^3 - 6x^2 + 11x - 6 = 0 \quad \textit{the polynomial}$$

$$(x - 1)(x - 2)(x - 3) = 0$$

$$x_1 = 1, \quad x_2 = 2, \quad x_3 = 3$$

$$\prod (x_3 - x_2)^2 (x_3 - x_1)^2 (x_2 - x_1)^2 \quad \textit{the discriminant}$$

$$\prod (3 - 2)^2 (3 - 1)^2 (2 - 1)^2 = 1 \cdot 4 \cdot 1 = 4$$

Infinite Series

A *series* is a sum of n terms, such as $1 + 5 + 9 + 13$ is a series of four terms. A series of n terms is

$$x_1 + x_2 + \cdots + x_n = \sum_{i=1}^{n} x_i$$

An *infinite series* is a series with infinitely many terms:

$$x_1 + x_2 + \cdots + x_n + \cdots = \sum_{i=1}^{\infty} x_i$$

A *partial sum* of series is given as

$$S_n = x_1 + x_2 + \cdots + x_n$$

If $\lim S_n = S$, as n goes to infinity, then the series is called *convergent* to the *sum S*. If the limit does not exist, it is said to *diverge*. Since

$$x_n = S_n - S_{n-1} \quad \text{as } n \to \infty$$

a series with the general term x_n approaching zero is a *convergent series*. Otherwise it is *divergent*. For example, the *geometric series*

$$1 + x + x^2 + \cdots + x^n + \cdots$$

is convergent if $|x| < 1$ and divergent if $|x| > 1$. If a series is convergent and also $\Sigma |x_i|$ is convergent, then Σx_i is said to be *absolutely convergent*.

The two tests for convergence of infinite series are Cauchy's test and D'Alembert's test.

Cauchy's Test. If $\lim|x_n|^{1/2} < 1$, as n goes to infinity, Σx_i is an *absolutely convergent* series.

D'Alembert's Test. If $|x_{n+1}/x_n| < \phi$, where ϕ is positive, less than 1, and independent of n, the series Σx_i is *absolutely convergent*.

Binomial Series

A polynomial expansion of the nth power of the sum of two quantities results in a *binomial series*. A theorem related to the computation of the coefficients of this series is known as the *binomial theorem*.

Binomial Theorem

The polynomial expansion of the nth (n is a positive integer) power of the sum of two quantities is

$$(x + y)^n = \sum_{s=0}^{n} \binom{n}{s} x^s y^{n-s}$$

The coefficients

$$_nC_s = \binom{n}{s} = \frac{n(n-1)\cdots(n-s+1)}{s!} = \frac{n!}{s(n-s)!}$$

are called *binomial coefficients*, and $s!$, *factorial s*, is the product $1 \cdot 2 \cdots s$. It should also be noted that the combination $_nC_s$ is equal to $\binom{n}{s}$. The binomial coefficients have the relationships

$$\binom{n}{0} = 1 \qquad \binom{n}{s} = \binom{n}{n-s} \qquad \binom{n}{s} + \binom{n}{n-s} = \binom{n+1}{s}$$

Based on the last relationship, coefficients for the $n + 1$ case can be computed from the preceding one. For example, for $n = 1$,

$$_1C_0 = 1, \qquad _1C_1 = \frac{1!}{1!0!} = 1$$

$$_1C_1 + {_1C_0} = {_2C_1}$$

Similarly, $1 + 2 = 3$, $1 + 3 = 4$, $3 + 3 = 6$, and so on. Thus we have the following coefficients $\binom{n}{s}$ for corresponding values of n:

$$
\begin{array}{ll}
n = 1 & 1\ 1 \\
n = 2 & 1\ 2\ 1 \\
n = 3 & 1\ 3\ 3\ 1 \\
n = 4 & 1\ 4\ 6\ 4\ 1
\end{array}
$$

which is called the *Pascal triangle*. In fact, the Pascal triangle provides coefficients of the expansion of $(x + y)^n$. For example, $(x + y)^1 = x + y$, thus the coefficients are 1 and 1, $(x + y)^2 = x^2 + 2xy + y^2$, the coefficients are $1, 2, 1$, and so on. Some other properties of binomial coefficients are:

$$\sum_{s=0}^{n} \binom{n}{s} = 2^n$$

$$\sum_{s=0}^{n} (-1)^s \binom{n}{s} = 0$$

$$\sum_{s=0}^{n} \binom{n}{s}^2 = \binom{2n}{n}$$

Series Expansion of Common Functions

A series expansion of a function, f, of a variable, x, as $f(x)$ may be given as a polynomial. The series expansion of some of the common functions are discussed in Section 17.6, and a table of selected functions are tabulated in Section 17.11.

17.1-6 Equations and Their Roots

The general algebraic equation with special n (degree) values are:

Equation of degree 2, $n = 2$,   quadratic equation.
Equation of degree 3, $n = 3$,   cubic equation.
Equation of degree 4, $n = 4$,   quartic or biquadratic equation.

Quadratic Equation

This may be reduced to the form

$$ax^2 + bx + c = 0$$

The two roots of a quadratic equation are

$$x = \frac{-b \pm (b^2 - 4ac)^{1/2}}{2a}$$

For real a, b, and c, $b^2 - 4ac > 0$ and the two roots are real and unequal; for $b^2 - 4ac = 0$, the two real roots are equal; for $b^2 - 4ac < 0$, both roots are imaginary.

Cubic Equation

By substituting $y = x - b/3a$, the equation $ay^3 + by^2 + cy + d = 0$ may be reduced to the form

$$x^3 + ex + f = 0$$

where the coefficients e and f are calculated as

$$e = \frac{c}{a} - \frac{b^2}{3a^2} \qquad f = \frac{b}{a^2}\left(\frac{2b^2}{27a} - \frac{c}{3} \right) + \frac{d}{a}$$

The three roots of a cubic equation are

$$x_1 = E + F$$

$$x_2 = -\frac{(E + F)}{2} + \frac{(E - F)(-3)^{1/2}}{2}$$

$$x_3 = -\frac{(E + F)}{2} - \frac{(E - F)(-3)^{1/2}}{2}$$

where

$$E = \left[-\frac{f}{2} + \left(\frac{f^2}{4} + \frac{e^3}{27} \right)^{1/2} \right]^{1/3}$$

$$F = -\left[-\frac{f}{2} + \left(\frac{f^2}{4} + \frac{e^3}{27} \right)^{1/2} \right]^{1/3}$$

If a, b, c, d are real, then

For $f^2/4 + e^3/27 > 0$ one real, two conjugate;
$ = 0$ three real, at least two equal;
$ < 0$ three real and unequal roots.

Quartic or Biquadratic Equation

An equation of the form

$$ay^4 + by^3 + cy^2 + dy + e = 0 \qquad\qquad (17.1\text{-}1)$$

by substituting $y = x - b/4a$, may be transformed to

$$x^4 + fx^2 + gx + h = 0 \qquad (17.1\text{-}2)$$

where the coefficients f, g, and h are calculated to be

$$f = \frac{c}{a} - \frac{3b^2}{8a^2} \qquad g = \frac{d}{a} - \frac{bc}{2a^2} + \frac{b^3}{8a^2}$$

$$h = -\frac{3b^4}{256a^4} + \frac{b^2c}{16a^3} - \frac{bd}{4a^2} + \frac{c}{a}$$

If the three roots of

$$z^3 + fz^2 + (f^2 - 4h)z - g^2 = 0 \qquad (17.1\text{-}3)$$

are z_1, z_2, and z_3, then the four roots of the transformed equation are

$$x_1 = \tfrac{1}{2}\left[+ (z_1)^{1/2} + (z_2)^{1/2} + (z_3)^{1/2} \right]$$

$$x_2 = \tfrac{1}{2}\left[+ (z_1)^{1/2} - (z_2)^{1/2} - (z_3)^{1/2} \right]$$

$$x_3 = \tfrac{1}{2}\left[- (z_1)^{1/2} + (z_2)^{1/2} - (z_3)^{1/2} \right]$$

$$x_4 = \tfrac{1}{2}\left[- (z_1)^{1/2} - (z_2)^{1/2} + (z_3)^{1/2} \right]$$

The discriminants, Δ, of (17.1-1) and (17.1-3) are the same. If the equation is real, the following cases are possible:

For $\Delta < 0$ a pair of conjugate complex roots and a pair of real roots;
$\quad\;\Delta > 0$ $(f^2 - 4h) > 0$ four real roots;
$\quad\;\Delta > 0$ $(f^2 - 4h) < 0$ four complex roots;
$\quad\;\Delta = 0$ four roots are not distinct.

Partial Fractions

If a function is expressed as

$$\phi(s) = \frac{\displaystyle\sum_0^m a_r s^r}{\displaystyle\sum_0^n b_r s^r} = \frac{a_0 + a_1 s + a_2 s^2 + \cdots + a_m s^m}{b_0 + b_1 s + b_2 s^2 + \cdots + b_n s^n} \qquad m < n$$

and the denominator is known to be factorized as

$$\sum_0^n b_r s^r = B(s - b_1)^{n_1}(s - b_2)^{n_2} \cdots$$

then $\phi(s)$ can be expressed as a sum in terms of the factorized elements as an expansion in the form of n partial fractions

$$\phi(s) = \sum_1^{n_1} \frac{B_{1,r}}{(s - b_1)^r} + \sum_1^{n_2} \frac{B_{2,r}}{(s - b_2)^r} + \cdots$$

Example:

$$\phi(s) = \frac{2s^2 + 1}{s^3 - s}$$

The denominator, $s^3 - s$, is factorized as $s(s + 1)(s - 1)$, then

$$\phi(s) = \frac{B_{1,1}}{s} + \frac{B_{2,1}}{s + 1} + \frac{B_{3,1}}{s - 1}$$

By the method of undetermined coefficients $B_{1,1}\ldots$ can be determined. The numerator of the left-hand side is equal to the numerator of the right-hand side:

$$2s^2 + 1 = B_{1,1}(s^2 - 1) + B_{2,1}(s^2 - s) + B_{3,1}(s^2 + s)$$

which results in

$$2s^2 + 1 = (B_{1,1} + B_{2,1} + B_{3,1})s^2 - (B_{2,1} - B_{3,1})s - B_{1,1}$$

Equating the coefficients of equal powers of s:

$$2 = B_{1,1} + B_{2,1} + B_{3,1}$$

$$0 = B_{2,1} - B_{3,1}$$

$$1 = -B_{1,1}$$

resulting in $B_{1,1} = -1$, $B_{2,1} = B_{3,1} = 3/2$.

$$\phi(s) = -\frac{1}{s} + \frac{3/2}{s + 1} + \frac{3/2}{s - 1}$$

The factorization of the denominator may consist of different forms: such as all factors may be linear and as in the above example, that is, $r = 1$. In some cases $r > 1$. Or the factors of the denominator may not be powers of real terms such as $s^2 + 1$. Then the numerator of each factor would be one less power than the power of the factor. For example, if

$$\phi(s) = \frac{2s^2 + 1}{(s^2 + 1)(s - 1)}$$

$$\phi(s) = \frac{B_{1,1}s + B_{2,1}}{s^2 + 1} + \frac{B_{3,1}}{s - 1}$$

$$2s^2 + 1 = B_{1,1}s^2 - B_{1,1}s + B_{2,1}s - B_{2,1} + B_{3,1}s^2 + B_{3,1}$$

$$= (B_{1,1} - B_{3,1})s^2 - (B_{1,1} - B_{2,1})s - (B_{2,1} - B_{3,1})$$

$$B_{1,1} + B_{3,1} = 2$$

$$B_{1,1} - B_{2,1} = 0$$

$$B_{2,1} - B_{3,1} = -1$$

$$B_{1,1} = B_{2,1} \qquad B_{1,1} = 1/2 \qquad B_{3,1} = 3/2$$

Applications of partial fractions are found in various examples. For example, if r is an integral an application is found in Laplace transforms (see Section 17.7). See also Section 17.6 for an application in such cases where the L'Hôpital rule fails.

Complex Numbers

A *complex number* consists of a *real number* and an *imaginary number*:

$$z = x + iy \qquad i = \sqrt{-1}$$

Here x is the *real* part of the complex number while y is the imaginary part. A complex number may be represented as a vector in the x-y plane (complex plane z) as shown in the figure.

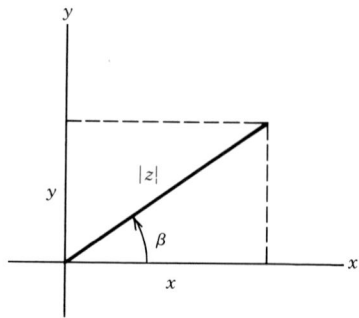

$$z = x + iy \qquad \bar{z} = x - iy$$

are *conjugate* complex numbers.

Addition of Complex Numbers

$$z_1 + z_2 = (x_1 + x_2) + i(y_1 + y_2)$$

We define $|z|$ as the *absolute value* (*modulus*) of the complex number and β as the *angle* of z. Using the trigonometric relations, see the figure above,

$$z = x + iy = |z|(\cos\beta + i\sin\beta) = |z|e^{i\beta}$$

$$\bar{z} = x - iy = |z|(\cos\beta - i\sin\beta) = |z|e^{-i\beta}$$

Multiplication of Complex Numbers

$$z_1 z_2 = (x_1 + iy_1)(x_2 + iy_2)$$

$$= x_1 x_2 - y_1 y_2 + i(x_1 y_2 + x_2 y_1)$$

Power of Complex Numbers

$$z^n = (x + iy)^n = [|z|(\cos\beta + i\sin\beta)]^n = |z|^n e^{in\beta}$$

$$z^n = (x - iy)^n = [|z|(\cos\beta - i\sin\beta)]^n = |z|^n e^{-in\beta}$$

$$\sqrt[n]{z} = \sqrt[n]{x + iy}$$

$$= \sqrt[n]{|z|}\left(\cos\frac{\beta + 2k}{n} + i\sin\frac{\beta + 2k}{n}\right)$$

$$= \sqrt[n]{|z|}\, e^{(\beta + 2k)/n}$$

17.2 PLANE AND SOLID GEOMETRY: MENSURATION FORMULAS

17.2-1 Plane Geometry

Triangles: Definitions

S = area
r (*or* R) = radius of the inscribed (or circumscribed) circle
A, B, C = angles and a, b, c the corresponding sides
t_a, t_b, t_c = lengths of the bisectors of angles A, B, C

m_a, m_b, m_c = lengths of the medians of sides a, b, c
h_a, h_b, h_c = lengths of the altitudes on sides a, b, c
$s = \frac{1}{2}(a + b + c)$

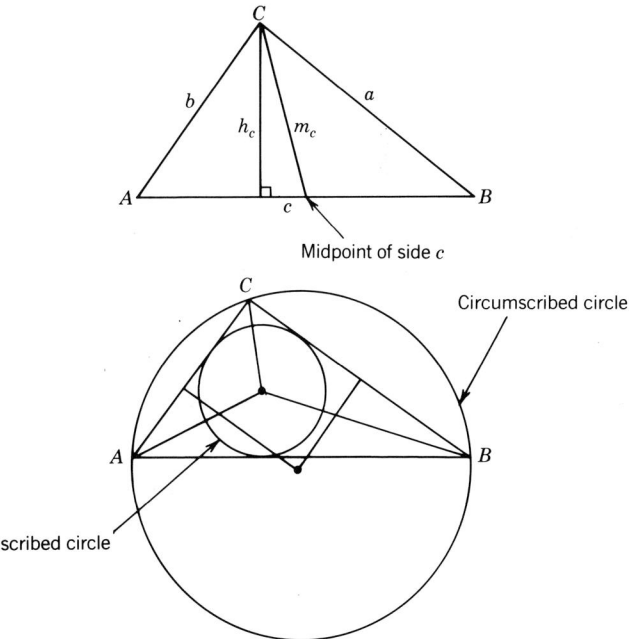

Midpoint of side c

Circumscribed circle

Inscribed circle

Triangles: Formulas

General Triangle

$$A + B + C = 180°$$

$$c^2 = a^2 + b^2 - 2ab \cos C \quad \text{(Law of cosine)}$$

$$S = \tfrac{1}{2}h_c c = \tfrac{1}{2}ab \sin C$$

$$= \frac{c^2 \sin A \sin B}{2 \sin C}$$

$$= rs = \frac{abc}{4R}$$

$$= \sqrt{s(s-a)(s-b)(s-c)} \quad \text{(Heron's formula)}$$

Equilateral Triangle

$$A = B = C = 60° \qquad a = b = c$$

$$S = \tfrac{1}{4}a^2\sqrt{3}$$

$$r = \tfrac{1}{6}a\sqrt{3} \qquad R = \tfrac{1}{3}a\sqrt{3}$$

$$h = \tfrac{1}{2}a\sqrt{3}$$

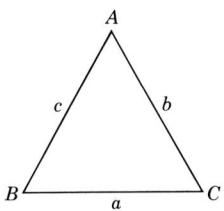

Right Triangle

$$A + B = C = 90°$$

$$c^2 = a^2 + b^2 \quad \text{(Pythagorean relation)}$$

$$a = \sqrt{(c+b)(c-b)}$$

$$S = \tfrac{1}{2}ab$$

$$r = \frac{ab}{a+b+c} \qquad R = \tfrac{1}{2}c$$

$$h = \frac{ab}{c} \qquad m = \frac{b^2}{c} \qquad n = \frac{a^2}{c}$$

$$m + n = c$$

m on b, n on a side of c

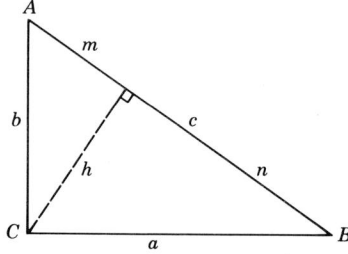

Quadrilaterals: Definitions

S = area
p, q = diagonals
A, B, C, D = angles
a, b, c, d = sides
$r(R)$ = radius of the inscribed (circumscribed) circles

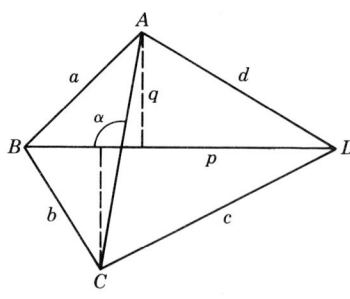

Quadrilaterals: Formulas

General Quadrangle or Quadrilateral

$$S = \tfrac{1}{2}pq \sin \alpha$$

Theorem. *If $a^2 + c^2 = d^2 + b^2$, then p and q are perpendicular to each other.*

Trapezoid. If any pair of opposite sides, for example, a and c, are parallel to each other, the quadrilateral is called a *trapezoid*.

$$m = \tfrac{1}{2}(a + c)$$

$$S = \tfrac{1}{2}(a + c)h = mh$$

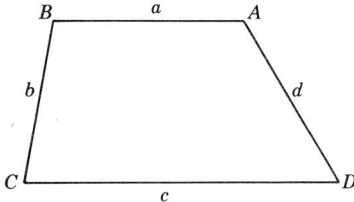

Parallelogram. If both pairs of opposite sides are parallel to each other, the quadrilateral is called a *parallelogram*.

$$A = C, \quad B = D, \qquad A + B = 180°$$

$$a = c, \quad b = d$$

$$S = ah \qquad h = \text{the distance between } a \text{ and } c$$

$$= ab \sin A = ab \sin B$$

$$h = b \sin A = b \sin B$$

$$p^2 = a^2 + b^2 - 2ab \cos A \quad \text{if } A \text{ is an acute angle}$$

$$q^2 = a^2 + b^2 - 2ab \cos B = a^2 + b^2 + 2ab \cos A$$

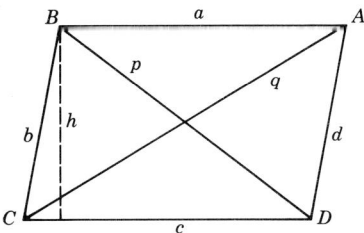

Rhombus. A parallelogram with $a = b = c = d$ is called a *rhombus*. By the theorem above p is perpendicular to q.

$$p^2 + q^2 = 4a^2$$

$$S = \tfrac{1}{2}pq$$

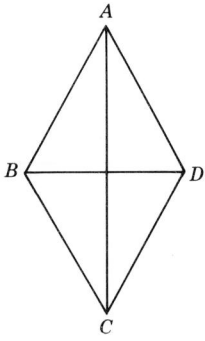

Rectangle. A parallelogram with $A = B = C = D = 90°$ is called a *rectangle*, thus $a = c$, $b = d$, and

$$S = ab$$

$$p = q = \sqrt{a^2 + b^2}$$

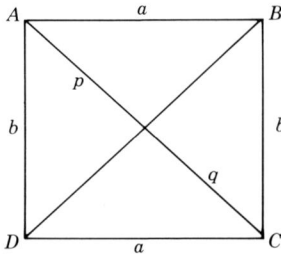

Polygons

A polygon with n equal sides of length s is called a *regular polygon*. The angles of a regular polygon with n sides is calculated as

$$\alpha = \left(\frac{n-2}{n}\right)180°, \quad \text{for example, } n = 6, \ \alpha = \frac{6-2}{6}180° = 120°$$

S = area
r (*or* R) = radius of the inscribed (or circumscribed) circle
p = perimeter

$$s = 2r\tan\frac{180°}{n} = 2R\sin\frac{180°}{n}$$

$$p = ns$$

$$S = \tfrac{1}{4}ns^2\cot\frac{180°}{n}$$

$$r = \tfrac{1}{2}s\cot\frac{180°}{n} \qquad R = \tfrac{1}{2}s\csc\frac{180°}{n}$$

Some regular polygons are:

$n = 5$ pentagon $\qquad n = 6$ hexagon $\qquad n = 7$ heptagon
$n = 8$ octagon $\qquad\ \ n = 9$ nonagon $\qquad n = 10$ decagon

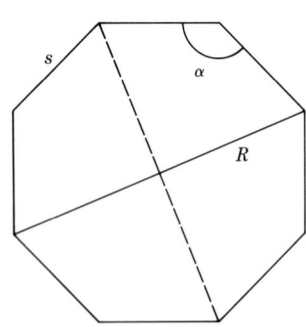

Circles

R = radius
D = diameter
C = circumference
S = area

$$C = 2\pi R = \pi D \qquad (\pi = 3.14159)$$

$$S = \pi R^2 = \tfrac{1}{4}\pi D^2$$

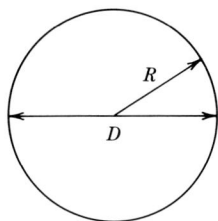

Sector of a Circle.   Here s is the length of the arc, and $\alpha(< \pi)$ is the central angle in radians. We have

$$s = R\alpha$$

$$h = R - d$$

$$d = R\cos(\alpha/2)$$

$$\alpha = \frac{s}{R}$$

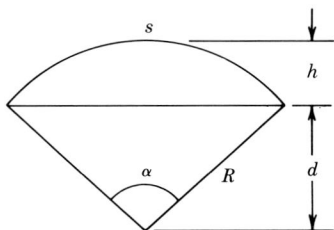

Conic Sections

Ellipse

a, b = the halves of the major axes
C = circumference = $2\pi\sqrt{(a^2 + b^2)/2}$ (approximate)
S = area = πab

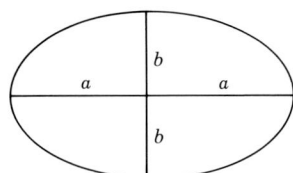

17.2-2   Solid Geometry

Definitions

S = area of lateral surface
T = total surface area
V = volume

Formulas

Cube

a = length of each edge
$T = 6a^2$
$V = a^3$
Diagonal of face = $a\sqrt{2}$
Diagonal of cube = $a\sqrt{3}$

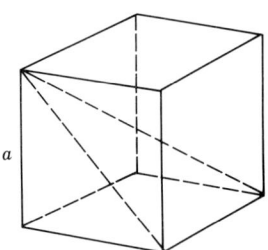

Rectangular Parallelepiped

a, b, c = the lengths of edges
$T = 2(ab + bc + ca)$
$V = abc$

Diagonal = $\sqrt{a^2 + b^2 + c^2}$

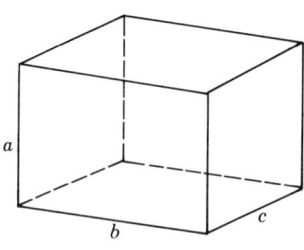

Prism

S = (perimeter of right section) × (lateral edge) = $(a + b + c + d)l$
V = (area of right section) × (lateral edge)
   = (area of base) × (altitude) = Al

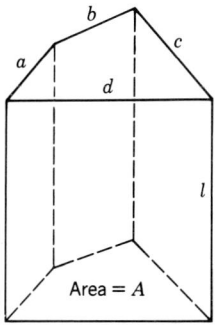

Pyramid

$S = \frac{1}{2}$(perimeter of base) $\times$ (slant height)
$V = \frac{1}{3}$(area of base) $\times$ (altitude)

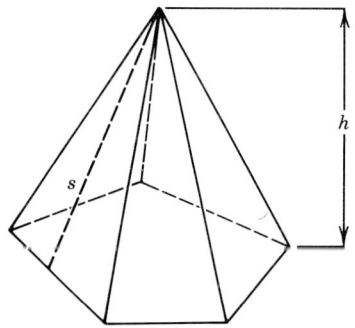

Frustum of Pyramid

B_1 = area of lower base
B_2 = area of upper base
h = altitude

$$V = \frac{1}{3}h\left(B_1 + B_2 + \sqrt{B_1 B_2}\right)$$

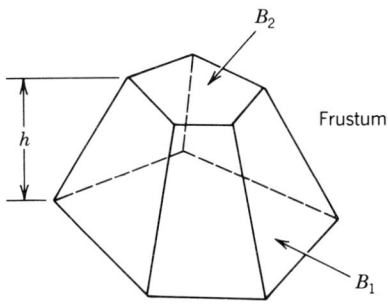

Regular Polyhedra

v = number of vertices
e = number of edges
f = number of faces
r (R) = radius of inscribed (circumscribed) sphere
$v - e + f = 2$ (Euler–Descartes formula for complex polyhedra)

If A is the area of each face, then

$$T = fA$$

$$V = \tfrac{1}{3}rfA = \tfrac{1}{3}rT$$

	v	e	f
Tetrahedon	4	6	4
Hexahedron	8	12	6
Octahedron	6	12	8
Dodecahedron	20	30	12
Icosahedron	12	30	20

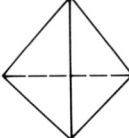

Tetrahedron

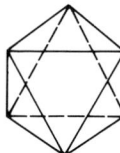

Octahedron

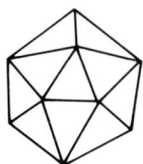

Icosahedron

Cylinders and Cones

B_1 = area of lower base
B_2 = area of upper base
h = altitude

Cylinder

R = radius of base
$S = 2\pi R \ (h)$
$T = 2\pi R \ (R + h)$
$V = \pi R^2 h$

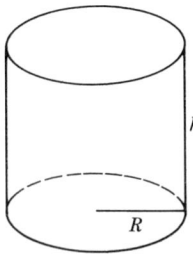

Cone

$$V = \tfrac{1}{3}B_1 h$$

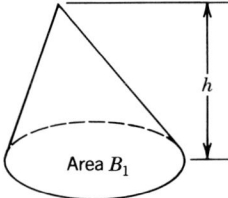

Right Circular Cone

$$s = \text{slant length}$$

$$s = \sqrt{R^2 + h^2}$$

$$S = \pi R s = \pi\sqrt{R^2 + h^2}$$

$$T = \pi R(R + s) = \pi R\left(R + \sqrt{R^2 + h^2}\right)$$

$$V = \tfrac{1}{3}\pi r^2 h$$

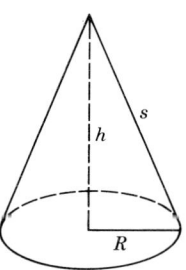

Frustrum of Cone

$$V = \tfrac{1}{3}h\left(B_1 + B_2\sqrt{B_1 B_2}\right)$$

If the cone is a right circular cone with radii R_1 and R_2 of the two bases, then

$$V = \tfrac{1}{3}\pi h\left(R_1^2 + R_2^2 + R_1 R_2\right)$$

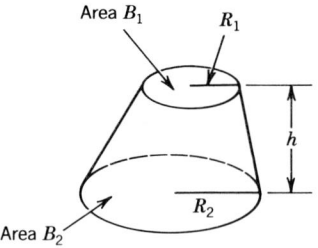

Sphere

$$D = 2R$$

$$S = 4\pi R^2 = \pi D^2$$

$$V = \tfrac{4}{3}\pi R^3 = \tfrac{1}{6}\pi D^2$$

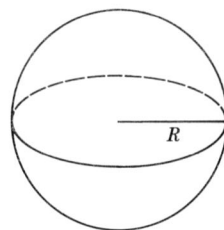

Circular Torus

r = radius of the circular cross-section

R = radius of rotating center of the cross-section circle

$$S = 4\pi^2 Rr$$

$$V = 2\pi^2 Rr^2.$$

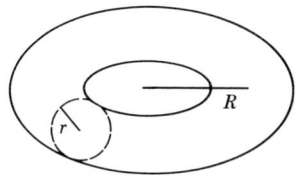

17.3   TRIGONOMETRY

17.3-1   Plane Trigonometry

Definitions

Angles

1 degree = $1/360$ of one complete rotation

$180° = \pi$ radians

1 radian = the angle at the center of a circle corresponding to an arc of length equal to the radius of the circle.

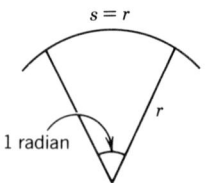

A *straight angle* is an angle of 180° (180 degrees).
A *right angle* is an angle of 90°.
An *acute angle* is an angle between 0 and 90°.
An *obtuse angle* is an angle between 90 and 180°.

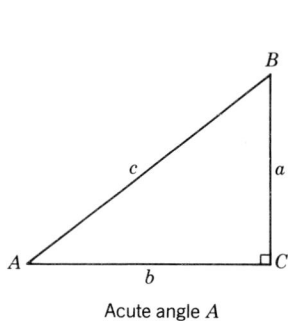

Acute angle A

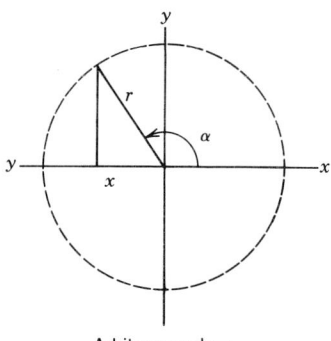

Arbitrary angle α

Acute Angle, A

$$\sin A = a/c \quad \cos A = b/c$$

$$\tan A = a/b \quad \cot A = b/a$$

$$\sec A = c/b \quad \csc A = c/a$$

Arbitrary Angle, α

$$\sin \alpha = y/r \quad \cos \alpha = x/r$$

$$\tan \alpha = y/x \quad \cot \alpha = x/y$$

$$\sec \alpha = r/x \quad \csc \alpha = r/y$$

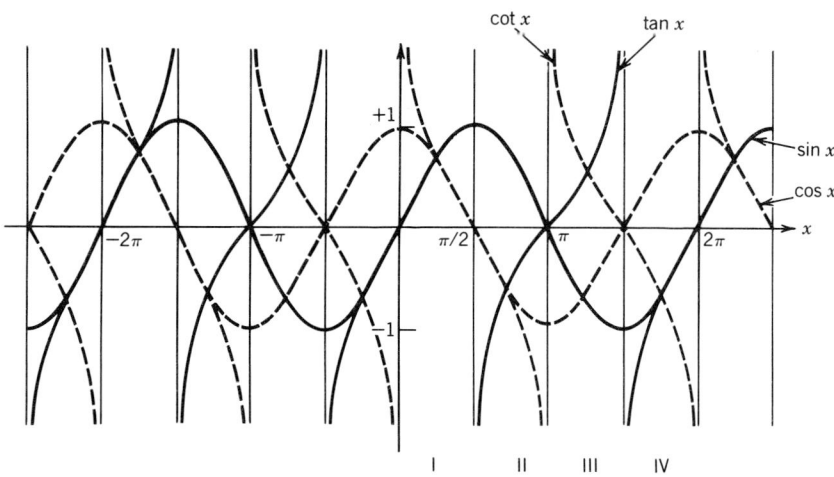

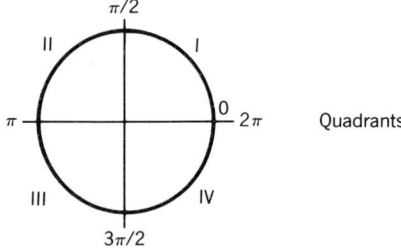

Quadrants

The values of trigonometric functions of some angles are as follows:

Function	0°	30°	45°	60°	90°	180°	270°	360°
$\sin \alpha$	0	$1/2$	$1/\sqrt{2}$	$\sqrt{3}/2$	1	0	-1	0
$\cos \alpha$	1	$\sqrt{3}/2$	$1/\sqrt{2}$	$1/2$	0	-1	0	1
$\tan \alpha$	0	$1/\sqrt{3}$	1	$\sqrt{3}$	∞	0	$-\infty$	0

Fundamental Identities

Relations

Product	Quotient	Reciprocal
$\sin \alpha = \tan \alpha \cos \alpha$	$\tan \alpha / \sec \alpha$	$1/\csc \alpha$
$\cos \alpha = \cot \alpha \sin \alpha$	$\cot \alpha / \csc \alpha$	$1/\sec \alpha$
$\tan \alpha = \sin \alpha \sec \alpha$	$\sin \alpha / \cos \alpha$	$1/\cot \alpha$
$\cot \alpha = \cos \alpha \csc \alpha$	$\cos \alpha / \sin \alpha$	$1/\tan \alpha$
$\sec \alpha = \csc \alpha \tan \alpha$	$\csc \alpha / \cot \alpha$	$1/\cos \alpha$
$\csc \alpha = \sec \alpha \cot \alpha$	$\sec \alpha / \tan \alpha$	$1/\sin \alpha$

Pythagorean Relations

$$\sin^2\alpha + \cos^2\alpha = 1$$

$$1 + \tan^2\alpha = \sec^2\alpha$$

$$1 + \cot^2\alpha = \csc^2\alpha$$

Angle Sum (and Difference) Relations

$$\sin(\alpha + \beta) = \sin \alpha \cos \beta + \cos \alpha \sin \beta$$

$$\sin(\alpha - \beta) = \sin \alpha \cos \beta - \cos \alpha \sin \beta$$

$$\cos(\alpha + \beta) = \cos \alpha \cos \beta - \sin \alpha \sin \beta$$

$$\cos(\alpha - \beta) = \cos \alpha \cos \beta + \sin \alpha \sin \beta$$

Double-Angle Relations

$$\sin 2\alpha = 2 \sin \alpha \cos \alpha = \frac{2 \tan \alpha}{1 + \tan^2\alpha}$$

$$\cos 2\alpha = \cos^2\alpha - \sin^2\alpha = \frac{1 - \tan \alpha}{1 + \tan^2\alpha}$$

$$\tan 2\alpha = \frac{2 \tan \alpha}{1 - \tan^2\alpha}$$

Multiple-Angle Relations

$$\sin 3\alpha = 3 \sin \alpha - 4 \sin^3 \alpha$$

$$\cos 3\alpha = 4 \cos^3\alpha - 3 \cos \alpha$$

$$\tan 3\alpha = \frac{3 \tan \alpha - \tan^3\alpha}{1 - 3 \tan^2\alpha}$$

$$\sin n\alpha = 2 \sin(n - 1)\alpha \cos \alpha - \sin(n - 2)\alpha$$

$$\cos n\alpha = 2 \cos(n - 1)\alpha \cos \alpha - \cos(n - 2)\alpha$$

$$\tan n\alpha = \frac{\tan(n - 1)\alpha + \tan \alpha}{1 - \tan(n - 1)\alpha \tan \alpha}$$

Function Product Relations

$$\sin\alpha\sin\beta = \tfrac{1}{2}\cos(\alpha - \beta) - \tfrac{1}{2}\cos(\alpha + \beta)$$

$$\cos\alpha\cos\beta = \tfrac{1}{2}\cos(\alpha - \beta) + \tfrac{1}{2}\cos(\alpha + \beta)$$

$$\sin\alpha\cos\beta = \tfrac{1}{2}\sin(\alpha + \beta) + \tfrac{1}{2}\sin(\alpha - \beta)$$

$$\cos\alpha\sin\beta = \tfrac{1}{2}\sin(\alpha + \beta) - \tfrac{1}{2}\sin(\alpha - \beta)$$

Function Sum and Difference Relations

$$\sin\alpha + \sin\beta = 2\sin\tfrac{1}{2}(\alpha + \beta)\cos\tfrac{1}{2}(\alpha - \beta)$$

$$\sin\alpha - \sin\beta = 2\cos\tfrac{1}{2}(\alpha + \beta)\sin\tfrac{1}{2}(\alpha - \beta)$$

$$\cos\alpha + \cos\beta = 2\cos\tfrac{1}{2}(\alpha + \beta)\cos\tfrac{1}{2}(\alpha - \beta)$$

$$\cos\alpha - \cos\beta = -2\sin\tfrac{1}{2}(\alpha + \beta)\sin\tfrac{1}{2}(\alpha - \beta)$$

$$\tan\alpha + \tan\beta = \frac{\sin(\alpha + \beta)}{\cos\alpha\cos\beta}$$

$$\tan\alpha - \tan\beta = \frac{\sin(\alpha - \beta)}{\cos\alpha\cos\beta}$$

$$\cot\alpha + \cot\beta = \frac{\sin(\alpha + \beta)}{\sin\alpha\sin\beta}$$

$$\cot\alpha - \cot\beta = \frac{\sin(\alpha - \beta)}{\sin\alpha\sin\beta}$$

$$\tan\tfrac{1}{2}(\alpha + \beta) = \frac{\sin\alpha + \sin\beta}{\cos\alpha + \cos\beta}$$

$$\tan\tfrac{1}{2}(\alpha - \beta) = \frac{\sin\alpha - \sin\beta}{\cos\alpha + \cos\beta}$$

Half-Angle Relations

$$\sin(\alpha/2) = -\sqrt{\frac{1 - \cos\alpha}{2}}$$

$$\cos(\alpha/2) = -\sqrt{(1 + \cos\alpha)/2}$$

$$\tan(\alpha/2) = -\sqrt{\frac{1 - \cos\alpha}{1 + \cos\alpha}}$$

$$= \frac{\sin\alpha}{1 + \cos\alpha} = \frac{1 - \cos\alpha}{\sin\alpha}$$

Power Relations

$$\sin^2\alpha = \tfrac{1}{2}(1 - \cos 2\alpha) \qquad \cos^2\alpha = \tfrac{1}{2}(1 + \cos 2\alpha)$$

$$\sin^3\alpha = \tfrac{1}{4}(3\sin\alpha - \sin 3\alpha) \qquad \cos^3\alpha = \tfrac{1}{4}(3\cos\alpha + \cos 3\alpha)$$

Exponential Relations: Euler Equation

$$e^{i\alpha} = \cos\alpha + i\sin\alpha \qquad i = \sqrt{-1}$$

$$\sin\alpha = \frac{e^{i\alpha} - e^{-i\alpha}}{2i}$$

$$\cos\alpha = \frac{e^{i\alpha} + e^{-i\alpha}}{2}$$

Relations Between Trigonometric Functions

	$\sin x = a$	$\cos x = b$	$\tan x = c$
$\sin x$	a	$\pm\sqrt{1-b^2}$	$\pm c/\sqrt{1+c^2}$
$\cos x$	$\pm\sqrt{1-a^2}$	b	$\pm 1/\sqrt{1+c^2}$
$\tan x$	$\pm a/\sqrt{1-a^2}$	$\pm\sqrt{1-a^2}/a$	c

Plane Triangle Formulas. For a given triangle, see Section 17.2, taking $s = (a + b + c)/2$, the following relations can be given:

Radius of inscribed circle: $r = \sqrt{(s-a)(s-b)(s-c)/s}$

Radius of circumscribed circle: $R = a/2\sin A = b/2\sin B = c/2\sin C$

Law of sines: $a/\sin A = b/\sin B = c/\sin C$

Law of cosines: $a^2 = b^2 + c^2 - 2bc\cos A$ (similarly for b and c)

Law of tangents: $\dfrac{b-c}{b+c} = \dfrac{\tan(B-C)/2}{\tan(B+C)/2}$ (similarly for a, b and c, a)

17.3-2 Spherical Trigonometry

Oblique Spherical Angles

A, B, C = angles and a, b, c the sides measured as angles at the center of the sphere
$s = \frac{1}{2}(a + b + c)$
$S = \frac{1}{2}(A + B + C)$
Δ = area of triangle surface
E = spherical excess of triangle, see formulas below
R = radius of the sphere on which the triangle lies

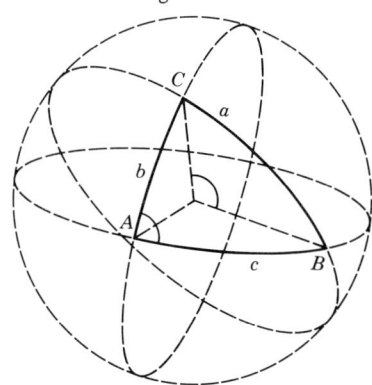

Formulas. Taking the sums of the angles defined above, either a, b, and c or A, B, and C varying as shown below, it can be seen that a point, a circle, and a plane triangle are all expressable in spherical coordinates.

$$0° \quad\;\; < a + b + c <\;\; 360°$$
$$\uparrow \qquad\qquad\qquad \uparrow$$
$$\text{Point} \qquad\qquad \text{Circle}$$

$$180° \quad < A + B + C <\;\; 540°$$
$$\uparrow \qquad\qquad\qquad \uparrow$$
$$\text{Plane triangle} \qquad\quad \text{Circle}$$

$$E = A + B + C - 180°$$

$$\Delta = \pi R^2 E/180°$$

Law of sines: $\sin a/\sin A = \sin b/\sin B = \sin c/\sin C$

Law of cosines for sides: $\cos a = \cos b \cos c + \sin b \sin c \cos A$ (similarly for b and c)

Law of cosines for angles: $\cos A = -\cos B \cos C + \sin B \sin C \cos a$ (similarly for B and C)

Law of tangents: $\dfrac{\tan\frac{1}{2}(B-C)}{\tan\frac{1}{2}(B+C)} = \dfrac{\tan\frac{1}{2}(b-c)}{\tan\frac{1}{2}(b+c)}$ (similarly for C, A and A, B pairs)

Half-angle formulas: $\tan\frac{1}{2}A = \dfrac{K}{\sin(s-a)}$ (similarly for B and C)

where $K^2 = \dfrac{\sin(s-a)\sin(s-b)\sin(s-c)}{\sin c}$

Half-side formulas: $\tan\frac{1}{2}a = K\cos(S-A)$ (similarly for b and c)

where $K^2 = \dfrac{-\cos S}{\cos(S-A)\cos(S-B)\cos(S-C)}$

Right Spherical Triangles

$C = 90°$, thus $\cos C = 0$, $\sin C = 1$.

$$\sin a = \tan b \cot B \qquad \sin a = \sin A \sin c$$
$$\sin b = \tan a \cot A \qquad \sin b = \sin B \sin c$$
$$\cos A = \tan b \cot c \qquad \cos A = \cos a \sin B$$
$$\cos B = \tan a \cot c \qquad \cos B = \cos b \sin A$$
$$\cos c = \cot A \cot B \qquad \cos c = \cos a \cos b$$

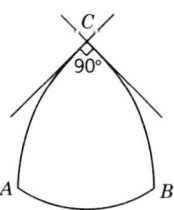

Hyperbolic Trigonometry

Definitions

$$x = OM$$
$$y = MP$$
$$a = OA$$
$$v = \text{shaded area}/a^2$$

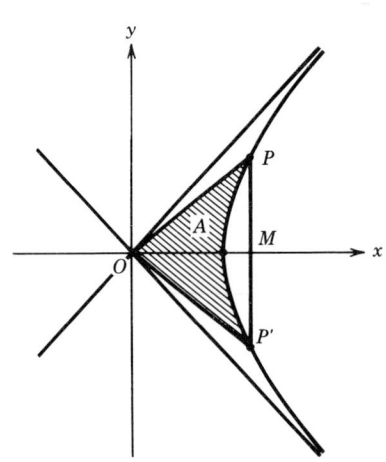

Hyperbolic sine: $\qquad\qquad\qquad\qquad\qquad \sinh v = y/a$

Hyperbolic cosine: $\qquad\qquad\qquad\qquad\quad \cosh v = x/a$

Hyperbolic tangent: $\qquad\qquad\qquad\qquad\quad \tanh v = \dfrac{\sinh v}{\cosh v}$

$$\operatorname{csch} v = \dfrac{1}{\sinh v}$$

$$\operatorname{sech} v = \dfrac{1}{\cosh v}$$

$$\coth v = \dfrac{1}{\tanh v}$$

Exponential Relations

$$e^{v} = \cosh v + \sinh v$$

$$e^{-v} = \cosh v - \sinh v$$

$$\sinh v = \tfrac{1}{2}(e^{v} - e^{-v})$$

$$\cosh v = \tfrac{1}{2}(e^{v} + e^{-v})$$

Identities

$$\sinh v = -\sinh(-v)$$

$$\cosh v = \cosh(-v)$$

$$\tanh v = -\tanh(-v)$$

$$\coth v = -\coth(-v)$$

$$\cosh^{2}v - \sinh^{2}v = 1$$

Sum and Difference Relations

$$\sinh(u + v) = \sinh u \cosh v + \cosh u \sinh v$$

$$\sinh(u - v) = \sinh u \cosh v - \cosh u \sinh v$$

$$\cosh(u + v) = \cosh u \cosh v + \sinh u \sinh v$$

$$\cosh(u - v) = \cosh u \cosh v - \sinh u \sinh v$$

Multiple-Argument Relations

$$\sinh 2v = 2 \sinh v \cosh v$$

$$\cosh 2v = \operatorname{conh}^{2}v + \sinh^{2}v$$

$$= 1 + 2 \sinh^{2}v$$

$$= -1 + 2 \cosh^{2}v$$

$$\sinh 3v = 3 \sinh v + 4 \sinh^{3}v$$

$$\cosh 3v = 3 \cosh^{3}v - 3 \cosh v$$

Inverse Hyperbolic Functions

$$\sinh^{-1}v = \log_e\left(v + \sqrt{v^2 + 1}\right)$$

$$\cosh^{-1}v = \log_e\left(v \pm \sqrt{v^2 + 1}\right) \qquad x \geq 1 \ (+\text{for principal value})$$

$$\tanh^{-1}v = \tfrac{1}{2}\log_e\left(\frac{1 + v}{1 - v}\right) \qquad v^2 < 1$$

Relations Between Trigonometric and Hyperbolic Functions

$$\sinh iv = i \sin v$$

$$\sinh v = -i \sin iv$$

$$\cosh iv = \cos v$$

$$\cosh v = \cos iv$$

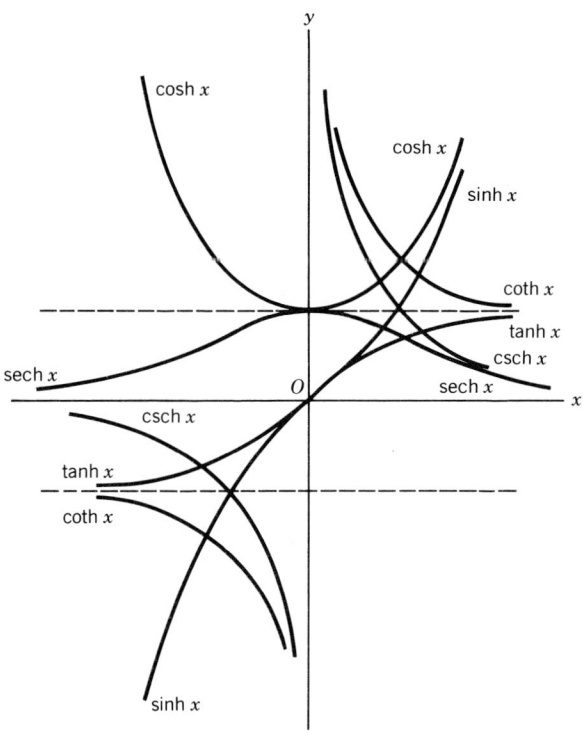

17.4   ANALYTIC GEOMETRY

Any point in a space is located by a set of quantities called its *coordinates* with respect to the *origin* whose coordinates are of the value zero. Depending on the *system of coordinates* the same point may be located by different sets of such quantities. Some of the types of coordinate system are introduced in this section.

17.4-1   Coordinate Systems on Plane

Cartesian Coordinate System

On a plane any two straight lines OX and OY, called *axes*, intersecting each other at the *origin* form a *cartesian* coordinate system. If the point P lies on the same plane, the distance from P to the axes measured on the lines drawn parallel to the axes are called the *coordinates* of P in this cartesian coordinate system.

If the axes intersect each other at any angle, the coordinate system is called an *oblique cartesian coordinate system*. If the angle is 90°, the axes are said to be **perpendicular** (*orthogonal*) to each other, thus the special system is called a *rectangular coordinate system*.

Usually in a two-dimensional cartesian coordinate system the x axis is called the *abscissa* and the y axis is called the *ordinate*.

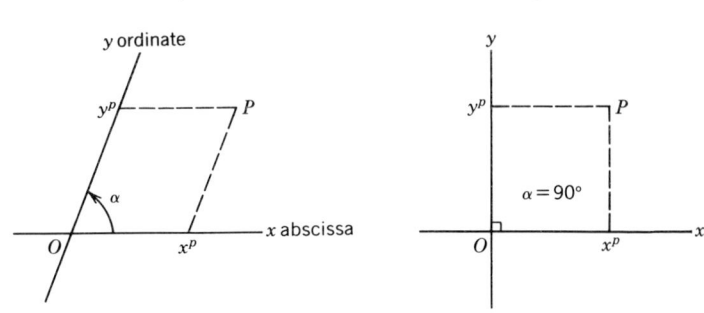

Oblique Coordinate System ($\alpha \neq 90°$)
Distance between any two points P_1, P_2:

$$\sqrt{(x_2 - x_1)^2 + (y_2 - y_1)^2 + 2(x_2 - x_1)(y_2 - y_1)\cos \alpha}$$

Three points, P_1, P_2, P_3, are *colinear* (lying on a line) if the determinant vanishes, that is,

$$\begin{vmatrix} x_1 & y_1 & 1 \\ x_2 & y_2 & 1 \\ x_3 & y_3 & 1 \end{vmatrix} = 0$$

Point P_3 dividing $P_1 P_2$ in the ratio of a/b has coordinates

$$\frac{ax_2 + bx_1}{a + b}, \qquad \frac{ay_2 + by_1}{a + b}$$

$$\begin{array}{ccc} & a & b \\ \bullet & \bullet & \bullet \\ P_1 & P_3 & P_2 \end{array}$$

Line passing through points $P_1(c, 0)$, $P_2(0, d)$:

$$\frac{x}{c} + \frac{y}{d} = 1 \quad \text{(intercept form)}$$

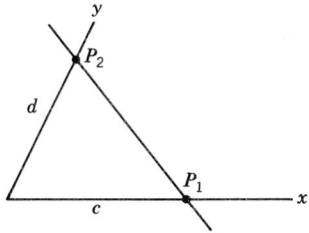

Line passing through any two points P_1, P_2:

$$\frac{y - y_1}{x - x_1} = \frac{y - y_2}{x - x_2} \quad \text{(two-point form)}$$

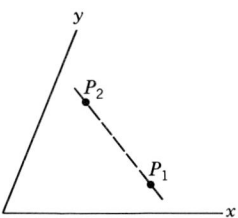

Line equation: $Cx + Dy + E = 0$ (general form)

Two lines: parallel $C_1 D_2 = C_2 D_1$

perpendicular $C_1 C_2 + D_1 D_2 = (C_1 D_2 + C_2 D_1)\cos \alpha$

Three lines concurrent: (Overlaping) if the determinant vanishes, that is,

$$\begin{vmatrix} C_1 & D_1 & E_1 \\ C_2 & D_2 & E_2 \\ C_3 & D_3 & E_3 \end{vmatrix} = 0$$

Rectangular Coordinate System ($\alpha = 90°$)

Distance between any two points: $\sqrt{(x_2 - x_1)^2 + (y_2 - y_1)^2}$

Slope of $P_1 P_2$: $m = \tan \beta = \dfrac{y_2 - y_1}{x_2 - x_1}$

Line equation: $y = mx + b$

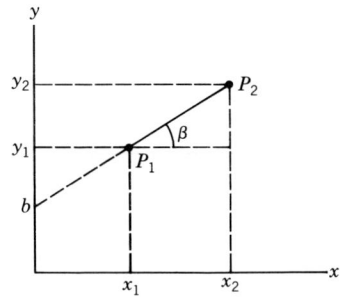

Angle between two lines with slopes m_1 and m_2:

$$\tan\theta = \frac{(m_2 - m_1)}{(1 + m_1 m_2)}$$

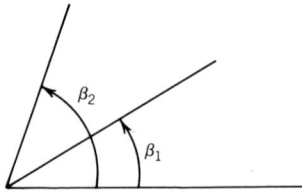

Parallel lines: $\qquad\qquad\qquad\qquad\qquad m_1 = m_2$

Perpendicular lines: $\qquad\qquad\qquad\qquad m_1 m_2 = -1$

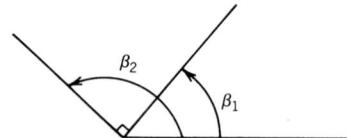

Plane Curves

Conics. For a given fixed point F, called the *focus*, and a given fixed line L, called the *directrix*, the locus of points P at distance from F and L as f, l such that the ratio f/l is constant, the ratio f/l is called the *eccentricity*, e, v is the *vertex* line, and VF is the *axis*.

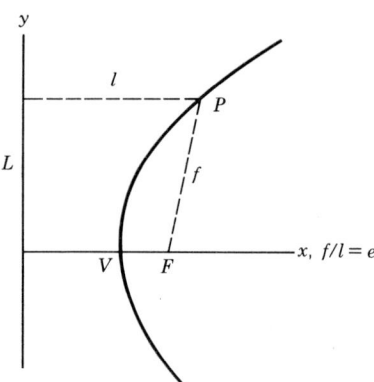

The general form of conics whose *axis* is oblique to the coordinate axes is

$$Ax^2 + Bxy + Cy^2 + Dx + Ey + F = 0$$

Ellipses: $e < 1$, $e = \sqrt{a^2 - b^2}/2$. Two directrices, $2a$ = major axis, $2b$ = minor axis, center is at the origin, foci on x axis:

$$\frac{x^2}{a^2} + \frac{y^2}{b^2} = 1$$

In the general form, $B^2 - 4AC < 0$.

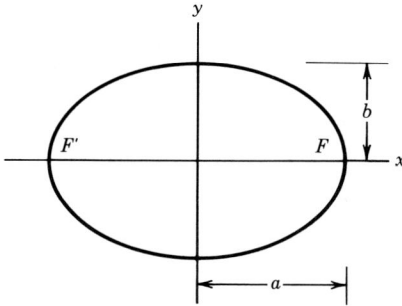

Circles: $a = b$, $e = 0$.   Center is at the origin, radius $r = a$:

$$x^2 + y^2 = r^2$$

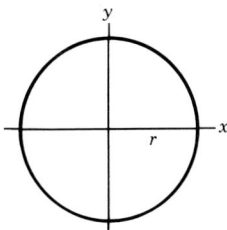

Parabolas: $e = 1$.   One directrix, vertex at the origin, focus at $(p, 0)$:

$$y^2 = 4px$$

In the general form, $B^2 - 4AC = 0$.

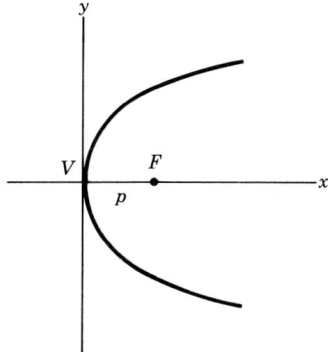

Hyperbolas: $e > 1$, $e = \sqrt{a^2 + b^2}\,/2$.   Two directrices, $2a =$ transfer axis, $2b =$ conjugate axis, center is at origin, foci on x axis:

$$\frac{x^2}{a^2} - \frac{y^2}{b^2} = 1$$

Regular hyperbola, $a = b$, $e = \sqrt{2}$, and asymptotes are perpendicular.

In the general form, $B^2 - 4AC = > 0$.

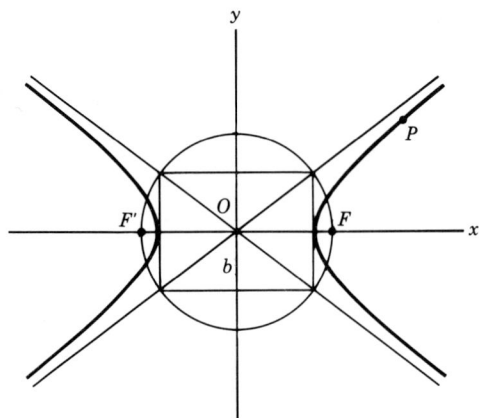

Polar Coordinate System

O = pole, origin
Ox = initial line
θ = vectorial angle
r = radius of vector
r, θ = polar coordinates of the point P

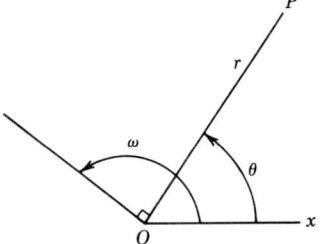

Distance between P_1 and P_2:

$$\sqrt{r_1^2 + r_2^2 - 2r_1 r_2 \cos(\theta_1 - \theta_2)}$$

Points P_1, P_2, P_3 are colinear if

$$r_2 r_3 \sin(\theta_3 - \theta_2) - r_3 r_1 \sin(\theta_3 - \theta_1) + r_1 r_2 \sin(\theta_2 - \theta_1) = 0$$

Straight line:

$$r\cos(\theta - \omega) = p \quad (normal\ form)$$

$$r(r_1 \sin(\theta - \theta_1) - r_2 \sin(\theta - \theta_2)) = \sin(\theta_2 - \theta_1) \quad (two\text{-}point\ form)$$

Relations between polar and rectangular coordinates:

$$x = r\cos\theta \qquad y = r\sin\theta$$

$$r = \sqrt{x^2 + y^2}$$

$$\theta = \arctan\frac{y}{x}$$

$$\sin\theta = \frac{y}{\sqrt{x^2 + y^2}} \qquad \cos\theta = \frac{x}{\sqrt{x^2 + y^2}}$$

Archimedean Spiral

$$r = a\theta$$

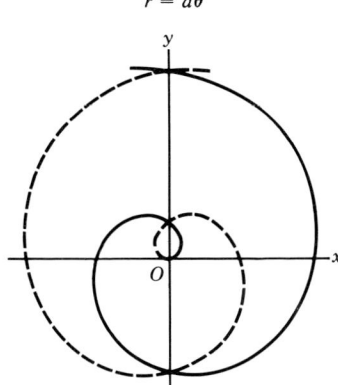

Astroid (Hypocycloid of Four Cusps)

$$x^{2/3} + y^{2/3} = a^{2/3}$$

$$x = a\cos^3\phi \qquad y = a\sin^3\phi$$

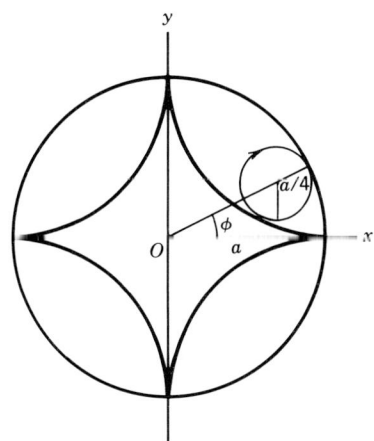

Deltoid (Hypocycloid of Three Cusps)

$$x = 2a\cos\phi + a\cos 2\phi \qquad y = 2a\sin\phi - a\sin 2\phi$$

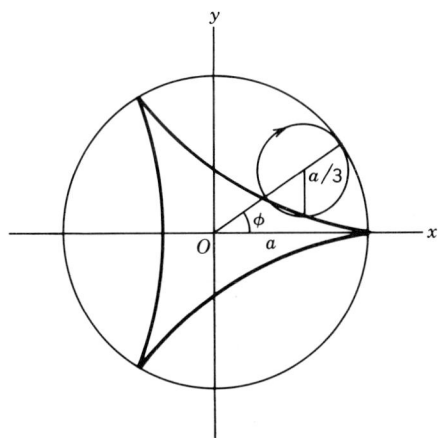

Catenary (Hyperbolic Cosine)

$$y = a \cosh(x/a)$$

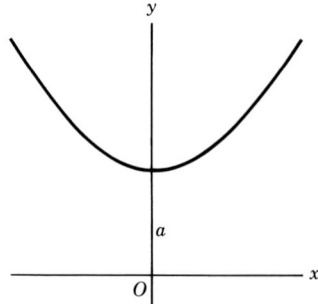

Cissoid of Diocles

$$y^2(a - x) = x^3$$

$$r = a \sin\theta \tan\theta$$

$$OB = AP$$

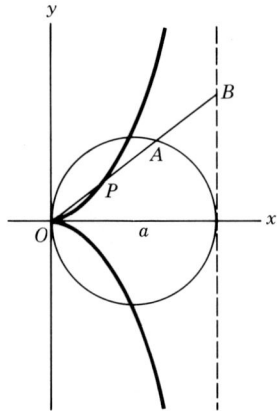

Cycloid with Vertex at the Origin
a = radius of the circle rolling on x axis

$$x = 2a \arcsin\sqrt{y/2a} + \sqrt{2ay - y^2}$$

$$x = a(\phi + \sin\phi)$$

$$y = a(1 - \cos\phi)$$

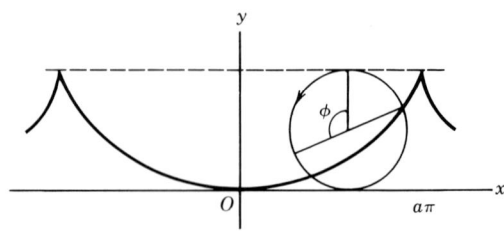

Cycloid, Curtate

$$x = a\phi - b\sin\phi$$

$$y = a - b\cos\phi \qquad a > b$$

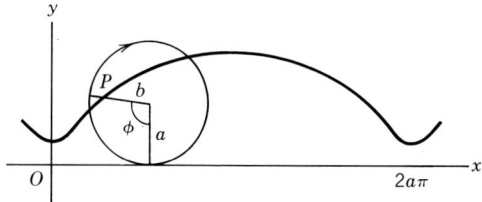

Cycloid, Prolate

$$x = a\phi - b\sin\phi$$

$$y = a - b\cos\phi \qquad a < b$$

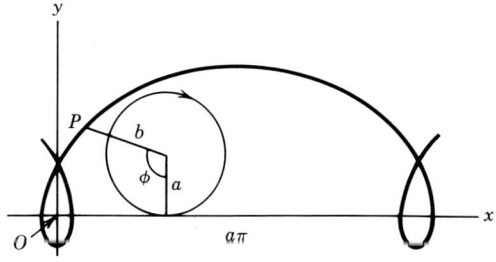

Epicycloid. Here b is the radius of the circle rolling on a circle of the radius a.

$$x = (a + b)\cos\phi - b\cos\left(\frac{a + b}{b}\phi\right)$$

$$y = (a + b)\sin\phi - b\sin\left(\frac{a + b}{b}\phi\right)$$

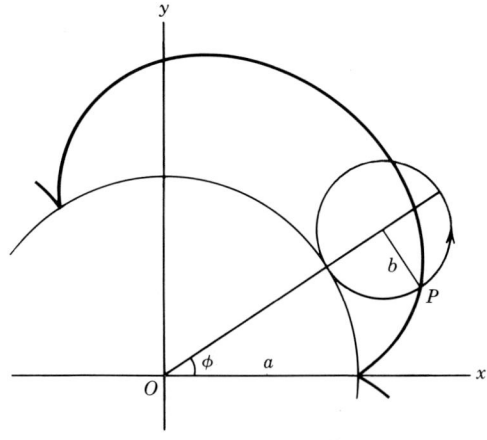

Folium of Descartes

$$x^3 + y^3 - 3axy = 0$$

$$x = \frac{3a\phi}{1 + \phi^3}$$

$$y = \frac{3a\phi^2}{1 + \phi^3}$$

$$r = \frac{3a \sin\theta \cos\theta}{\sin^3\theta + \cos^3\theta}$$

Asymptote: $\qquad x + y + a = 0$

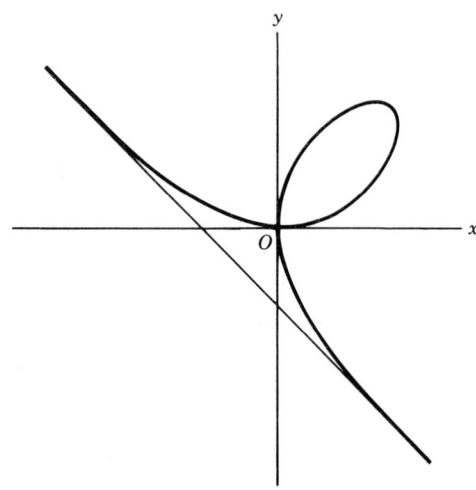

Lemniscate of Bernoulli, Two-Level Rose

$$\left(x^2 + y^2\right)^2 = a^2\left(x^2 - y^2\right)$$

$$r^2 = a^2\cos 2\theta$$

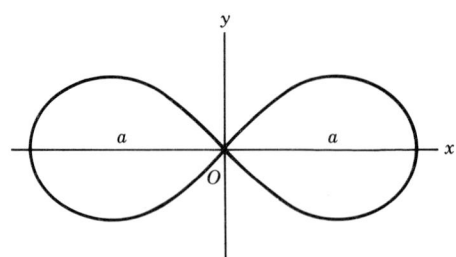

Ovals of Cassini

$$\left(x^2 + y^2 + b^2\right)^2 - 4b^2x^2 = k^4$$

$$F'P \cdot FP = k^2$$

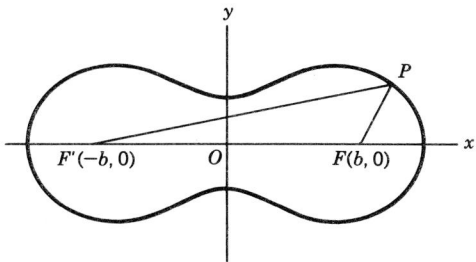

17.4-2 Coordinate Systems in a Space

Rectangular Coordinate System

The dimension of the space is $n = 3$. The rectangular coordinates of a point P are x, y, z, the distance of P from the yz, xz, and xy planes, respectively. If there are three points, P_1, P_2, P_3, the following definitions can be given:

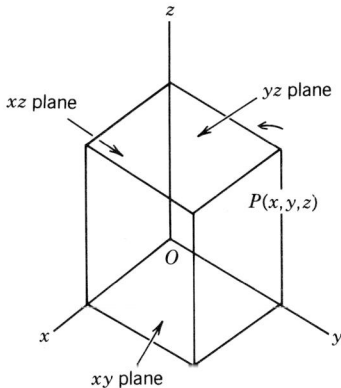

Distance between P_1 and P_2:

$$\sqrt{(x_2 - x_1)^2 + (y_2 - y_1)^2 + (z_2 - z_1)^2}$$

P_1, P_2, P_3 are *colinear* if and only if

$$x_2 - x_1 : y_2 - y_1 : z_2 - z_1 = x_3 - x_1 : y_3 - y_1 : z_3 - z_1$$

P_1, P_2, P_3, P_4 are *coplanar* if and only if the determinant vanishes, that is,

$$\begin{vmatrix} x_1 & y_1 & z_1 & 1 \\ x_2 & y_2 & z_2 & 1 \\ x_3 & y_3 & z_3 & 1 \\ x_4 & y_4 & z_4 & 1 \end{vmatrix} = 0$$

Point P_3 dividing P_1, P_2 in the ratio a/b has coordinates

$$\frac{ax_2 + bx_1}{a + b}, \quad \frac{ay_2 + by_1}{a + b}, \quad \frac{az_2 + bz_1}{a + b}$$

Direction Cosines

The angles between $P_1 P_2$ and the three axes x, y, z are called *direction angles*, α, β, γ
If the distance between P_1 and P_2 is d, then the *direction cosines* of $P_1 P_2$ are

$$\cos \alpha = \frac{x_2 - x_1}{d} \qquad \cos \beta = \frac{y_2 - y_1}{d} \qquad \cos \gamma = \frac{z_2 - z_1}{d}$$

The sum of squares of direction cosines is equal to 1:

$$\cos^2\alpha + \cos^2\beta + \cos^2\gamma = 1$$

For parallel lines,

$$\alpha_1 = \alpha_2 \qquad \beta_1 = \beta_2 \qquad \gamma_1 = \gamma_2$$

For perpendicular lines,

$$\cos\alpha_1\cos\alpha_2 + \cos\beta_1\cos\beta_2 + \cos\gamma_1\cos\gamma_2 = 0$$

Line equations are as follows:

Point direction form: $\dfrac{x - x_1}{a} = \dfrac{y - y_1}{b} = \dfrac{z - z_1}{c}$

Two point form: $\dfrac{x - x_1}{x_2 - x_1} = \dfrac{y - y_1}{y_2 - y_1} = \dfrac{z - z_1}{z_2 - z_1}$

General form: $A_1 x + B_1 y + C_1 z + D_1 = 0$

$$A_2 x + B_2 y + C_2 z + D_2 = 0$$

Parametric form: $x = x_1 + ta$

$$y = y_1 + tb$$

$$z = z_1 + tc$$

Plane equation: $Ax + By + Cz + D = 0$

Intercept form: $\dfrac{x}{a} + \dfrac{y}{b} + \dfrac{z}{c} = 1$

Plane through point P_1 and perpendicular to direction (a, b, c):

$$a(x - x_1) + b(y - y_1) + c(z - z_1) = 0$$

Taking the general form $Ax + By + Cz + D = 0$, we have:

Parallel planes: $A_1 : B_1 : C_1 = A_2 : B_2 : C_2$

Perpendicular planes: $A_1 A_2 + B_1 B_2 + C_1 C_2 = 0$

Quadric Surfaces

Ellipsoid

$$\frac{x^2}{a^2} + \frac{y^2}{b^2} + \frac{z^2}{c^2} = 1$$

Sphere

$$x^2 + y^2 + z^2 = 1$$

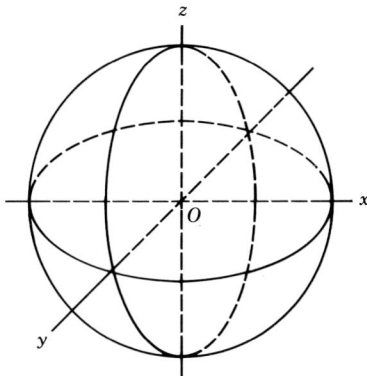

Elliptic Cone

$$\frac{x^2}{a^2} + \frac{y^2}{b^2} - \frac{z^2}{c^2} = 0$$

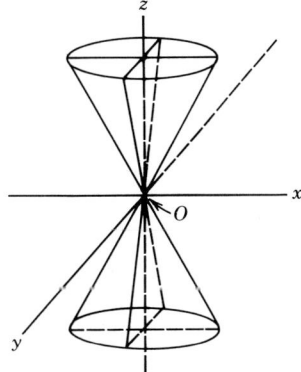

Elliptic Paraboloid

$$\frac{x^2}{a^2} + \frac{y^2}{b^2} = cz$$

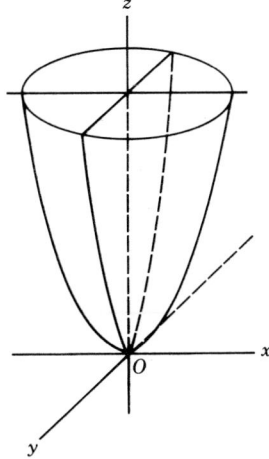

Elliptic Cylinder

$$\frac{x^2}{a^2} + \frac{y^2}{b^2} = 1$$

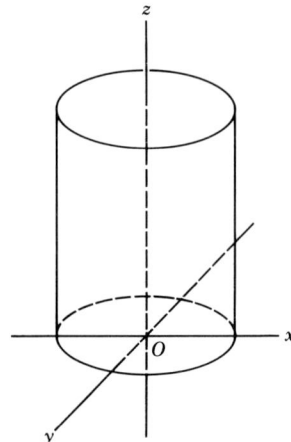

Hyperboloid of One Sheet

$$\frac{x^2}{a^2} + \frac{y^2}{b^2} - \frac{z^2}{c^2} = 1$$

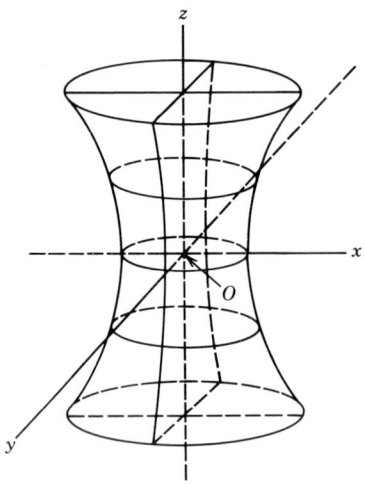

Hyperboloid of Two Sheets

$$\frac{x^2}{a^2} - \frac{y^2}{b^2} - \frac{z^2}{c^2} = 1$$

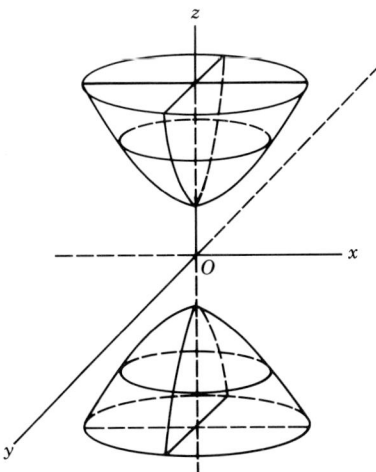

Hyperbolic Paraboloid

$$\frac{x^2}{a^2} - \frac{y^2}{b^2} = cz$$

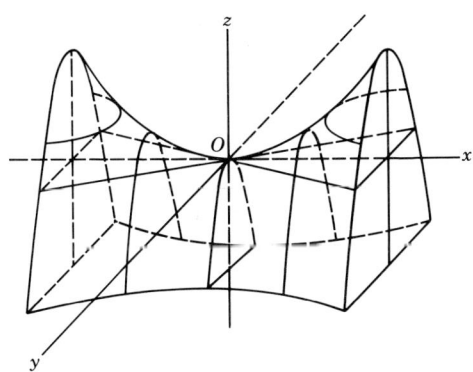

Transformation of Rectangular Coordinates

The rectangular coordinates of the old system (x, y, z) and of the new system (x', y', z'), of the origin of the new system, (h, k, l):

Translation

$$x = x' + h$$
$$y = y' + k$$
$$z = z' + l$$

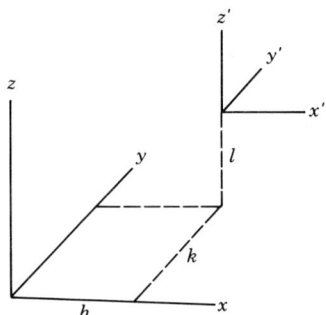

Rotation About the Origin. (λ, μ, ν) are direction cosines of x', y', z' axes with respect to the old system.

$$x = \lambda_1 x' + \lambda_2 y' + \lambda_3 z'$$

$$y = \mu_1 x' + \mu_2 y' + \mu_3 z'$$

$$z = \nu_1 x' + \nu_2 y' + \nu_3 z'$$

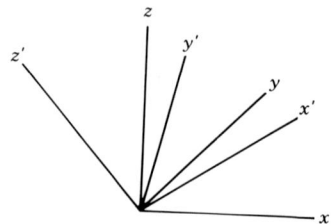

Cylindrical Coordinate System

For (x, y, z) cartesian (rectangular) coordinates,

$$x = r\cos\theta \qquad r = \sqrt{x^2 + y^2}$$

$$y = r\sin\theta \qquad r = \arctan(y/x)$$

$$z = z \qquad\qquad z = z$$

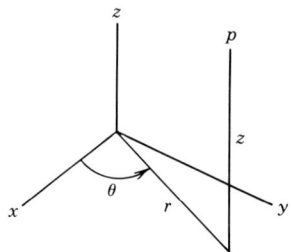

Spherical Coordinate System

For (x, y, z) cartesian (rectangular) coordinates,

$$x = \rho\cos\theta\sin\phi$$

$$y = \rho\sin\theta\sin\phi$$

$$z = \rho\cos\phi$$

$$\phi = \arccos\frac{z}{\sqrt{x^2 + y^2 + z^2}}$$

$$\theta = \arctan\frac{y}{z}$$

$$\rho^2 = x^2 + y^2 + z^2$$

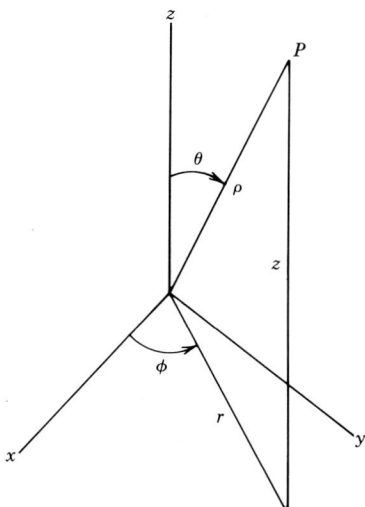

17.5 MATRIX ALGEBRA

17.5-1 Definitions

A *matrix* is an array of elements consisting of m rows and n columns:

$$\mathbf{A} = \begin{bmatrix} a_{11} & a_{12} & \cdots & a_{1n} \\ a_{21} & a_{22} & \cdots & a_{2n} \\ \vdots & \vdots & \cdots & \vdots \\ a_{m1} & a_{m2} & \cdots & a_{mn} \end{bmatrix}$$

Here a_{ij} indicates the element in the ith row and jth column.

If $m = n$, then the matrix is called a *square* matrix, otherwise it is a *rectangular* matrix. The diagonal spanning between the upper left corner and the lower right corner is called the *principal diagonal*.

The *transpose of a matrix* $\mathbf{A}$, denoted by $\mathbf{A}'$, or $\mathbf{A}^T$, is the matrix with elements defined as

$$a_{ij} = a'_{ji}$$

Special forms of square matrix are:

The *symmetric* square matrix is the matrix with the property $\mathbf{A} = \mathbf{A}'$, that is, $a_{ij} = a_{ji}$.

The *skew-symmetric* (or *antisymmetric*) square matrix is the matrix with the property $\mathbf{A} = -\mathbf{A}'$, that is, $a_{ij} = -a_{ji}$.

A *lower triangular matrix* has all zeros as elements above the principal diagonal.

An *upper triangular matrix* has all zeros as elements below the principal diagonal.

A *diagonal matrix* has all its elements zero except those along the principal diagonal, and is denoted by $\mathbf{D}$.

Example:

$$\mathbf{D} = \begin{bmatrix} a_{11} & 0 & 0 \\ 0 & a_{22} & 0 \\ 0 & 0 & a_{33} \end{bmatrix}$$

Special forms of the rectangular matrix are:

A *column vector* is a matrix with m rows and 1 column.
A *row vector* is a matrix with n columns and 1 row.
A *scalar* is a matrix with 1 column and 1 row.

17.5-2 Operations

Addition of two matrices with the same number of columns and rows is

$$\mathbf{A} + \mathbf{B} = \mathbf{C}$$

where the elements are calculated as

$$a_{ij} + b_{ij} = c_{ij} \quad i = 1, 2, \ldots, m \quad and \quad j = 1, 2, \ldots, n$$

Multiplication of a matrix by a scalar,

$$\mathbf{B} = \gamma\mathbf{A}$$

is calculated by the relation

$$b_{ij} = \gamma a_{ij}$$

Multiplication of two matrices,

$$\mathbf{AB} = \mathbf{C}$$

is possible only if the number of columns in **A** is equal to the number of rows of **B**. The elements are calculated as

$$c_{ij} = \sum_{k=1}^{n} a_{ik} b_{kj}$$

Example:

$$\mathbf{A} = \begin{bmatrix} 3 & 2 \\ 1 & 2 \\ 4 & 3 \end{bmatrix} \quad \mathbf{B} = \begin{bmatrix} 2 & 3 & 3 & 1 \\ 5 & 4 & 2 & 3 \end{bmatrix}$$

A is a 3×2 matrix, **B** is a 2×4 matrix, thus **AB** is defined and equal to

$$\mathbf{C} = \mathbf{AB} = \begin{bmatrix} 3\cdot2+2\cdot5 & 3\cdot3+2\cdot4 & 3\cdot3+2\cdot2 & 3\cdot1+2\cdot3 \\ 1\cdot2+2\cdot5 & 1\cdot3+2\cdot3 & 1\cdot3+2\cdot2 & 1\cdot1+2\cdot3 \\ 4\cdot2+3\cdot5 & 4\cdot3+3\cdot4 & 4\cdot3+3\cdot2 & 4\cdot1+3\cdot3 \end{bmatrix}$$

$$\mathbf{C} = \begin{bmatrix} 16 & 17 & 13 & 9 \\ 12 & 9 & 7 & 7 \\ 23 & 24 & 18 & 13 \end{bmatrix}$$

The Laws of Matrix Operations

(a) Commutative law: $\mathbf{AB} \neq \mathbf{BA}$.
(b) Associative law: $\mathbf{A(BC)} = \mathbf{(AB)C}$.
(c) Distributive law: $\mathbf{(A + B)C} = \mathbf{AC} + \mathbf{BC}$.

Transpose of product of two matrices is equal to the product of the transpose of individual matrices, however in reverse order as follows:

$$\mathbf{(AB)'} = \mathbf{B'A'}$$

17.5-3 Special Forms

A *null matrix*, **0**, is a matrix with all its elements equal to zero.
The *identity matrix*, **I**, is a diagonal matrix with all its diagonal elements equal to one.
A *diagonal matrix*, $\gamma\mathbf{I}$, where γ is a scalar, is called a *scalar matrix*.

If a square matrix $\mathbf{Q}$ is given, and $\mathbf{x}$ is a column vector, and $\mathbf{x}'$ its transpose, a row vector, and $\mathbf{y}$, a column vector with equal number of elements, then:

$\mathbf{x}'\mathbf{Q}\mathbf{x}$ is called a *quadratic form*.

$\mathbf{x}'\mathbf{Q}\mathbf{y}$ is called a *bilinear form*.

$\mathbf{x}'\mathbf{x} = \Sigma x_i^2$ is the sum of *squares* of all elements.

$\mathbf{x}'\mathbf{y} = \Sigma x_i y_i$ is the sum of *products* of elements in $\mathbf{x}$ by those in $\mathbf{y}$.

If the elements of a matrix are complex, that is,

$$a_{kj} = b_{kj} + ic_{kj} \quad \text{where } i = \sqrt{-1},$$

the matrix with elements conjugate to a_{kj}, that is,

$$a_{kj} = b_{kj} - ic_{kj}$$

is called the *conjugate matrix*, $\mathbf{A}$.

$\mathbf{A}^H$ is called a *Hermitian conjugate matrix* which is obtained by transposing $\mathbf{A}$ and replacing its elements by conjugate complex. Since a symmetric (square) matrix is by definition $\mathbf{A} = \mathbf{A}'$, a square matrix is called Hermitian if

$$\mathbf{A} = \mathbf{A}^H$$

If $\mathbf{A}' = \mathbf{A}^{-1}$, then $\mathbf{A}$ is a *unitary matrix*. Here the inverse of a matrix $\mathbf{A}^{-1}$ is defined later in this section.

17.5-4 Determinants

The determinant of a square matrix, denoted as $|\mathbf{A}|$ or $\det(\mathbf{A})$, is defined as a sum of the products, and it is a scalar:

$$\begin{vmatrix} a_{11} & a_{12} & \cdots & a_{1n} \\ a_{21} & a_{22} & \cdots & a_{2n} \\ \cdot & \cdot & \cdots & \cdot \\ \cdot & \cdot & \cdots & \cdot \\ a_{n1} & a_{n2} & \cdots & a_{nn} \end{vmatrix} = \sum (-1)^\delta a_{1 i_1} a_{2 i_2} \cdots a_{n i_n}$$

where $i_1, i_2, \ldots, i_n$ indicate n different columns and δ is the number of exchanges to put $i_1, i_2, \ldots, i_n$ in order or $1, 2, \ldots, n$. As a consequence, if any two rows (or columns) of a matrix are interchanged, the sign of the determinant changes.

For example, a 3×3 matrix has the determinant

$$\begin{vmatrix} a_{11} & a_{12} & a_{13} \\ a_{21} & a_{22} & a_{23} \\ a_{31} & a_{32} & a_{33} \end{vmatrix} = (-1)^0 a_{11} a_{22} a_{33} + (-1)^1 a_{11} a_{23} a_{32}$$

$$+ (-1)^1 a_{12} a_{21} a_{33} + (-1)^4 a_{12} a_{23} a_{31}$$

$$+ (-1)^2 a_{13} a_{21} a_{32} + (-1)^3 a_{13} a_{22} a_{31}$$

$$= a_{11} (a_{22} a_{33} - a_{23} a_{32})$$

$$- a_{12} (a_{21} a_{33} - a_{23} a_{31})$$

$$+ a_{13} (a_{21} a_{32} - a_{22} a_{31})$$

A numerical example of determinant is calculated below for illustration:

$$\begin{vmatrix} 3 & 3 & 2 \\ 4 & 1 & 5 \\ 3 & 4 & 1 \end{vmatrix} = 3(1 \cdot 1 - 5 \cdot 4) - 3(4 \cdot 1 - 3 \cdot 5) + 2(4 \cdot 4 - 3 \cdot 1)$$

$$= 3(-19) - 3(-11) + 2(13) = -57 + 33 + 26 = +2$$

The definition also implies that if a row or a column of a matrix is multiplied by a scalar α then the determinant is multiplied by the same scalar. If an (n, n) matrix is multiplied by a scalar, $\alpha \mathbf{A}$, then its determinant becomes $|\alpha \mathbf{A}| = \alpha^n |\mathbf{A}|$.

The product of the determinants of two matrices is

$$|\mathbf{A}||\mathbf{B}| = |\mathbf{AB}|$$

The determinants of a matrix and of its transpose are equal, $|\mathbf{A}| = |\mathbf{A}'|$.

The matrix obtained by removing a row (rth) and a column (sth) is called the *minor* of the element a_{rs}, and denoted by M_{rs}. The *cofactor* of a_{rs} is the scalar found as $A_{rs} = (-1)^{r+s}|M_{rs}|$. The *cofactor matrix* is the matrix with A_{rs} as it elements. Based on the cofactors another definition of the determinant of a matrix is given as

$$|\mathbf{A}| = \sum_{r=1}^{n} a_{rs}(-1)^{r+s} A_{rs}$$

An example for a 3×3 matrix is

$$\begin{vmatrix} a_{11} & a_{12} & a_{13} \\ a_{21} & a_{22} & a_{23} \\ a_{31} & a_{32} & a_{33} \end{vmatrix} = a_{11}\begin{vmatrix} a_{22} & a_{23} \\ a_{32} & a_{33} \end{vmatrix} - a_{12}\begin{vmatrix} a_{21} & a_{23} \\ a_{31} & a_{33} \end{vmatrix} + a_{13}\begin{vmatrix} a_{21} & a_{22} \\ a_{31} & a_{33} \end{vmatrix}$$

With the numerical values given, the determinant is evaluated as

$$\begin{vmatrix} 3 & 3 & 2 \\ 4 & 1 & 5 \\ 3 & 4 & 1 \end{vmatrix} = 3\begin{vmatrix} 1 & 5 \\ 4 & 1 \end{vmatrix} - 3\begin{vmatrix} 4 & 5 \\ 3 & 1 \end{vmatrix} + 2\begin{vmatrix} 4 & 1 \\ 3 & 4 \end{vmatrix}$$

The *transpose* of the matrix whose elements are A_{rs} is called the *adjoint matrix*, adj $\mathbf{A}$.

Rank and Singularity

A matrix is called *singular* if it satisfies the relation, $\mathbf{Ax} = \mathbf{0}$ or $\mathbf{A'x} = \mathbf{0}$ where $\mathbf{x}$ is a nonzero column vector. Otherwise it is *nonsingular*.

The *rank* of a matrix $\mathbf{A}$ is the maximum number of rows or columns, r of the nonsingular square submatrix obtained from the matrix $\mathbf{A}$ with n rows or columns. The nonsingular submatrix of a singular matrix is called a *basis* of the matrix $\mathbf{A}$.

If $r = n$ the matrix is nonsingular, and its determinant is nonzero. However, if $r < n$ then the matrix is singular and its determinant is zero.

Inverse of a Matrix

For a nonsingular matrix $\mathbf{A}$, the matrix $\mathbf{A}^{-1}$ satisfying the relation, $\mathbf{AA}^{-1} = \mathbf{A}^{-1}\mathbf{A} = \mathbf{I}$ is called the *inverse* matrix. The inverse matrix is unique.

Example:

$$\mathbf{A} = \begin{bmatrix} 1 & 2 & 1 \\ 3 & 1 & 4 \\ 2 & 1 & 3 \end{bmatrix}$$

$$\mathbf{A}^{-1}\mathbf{A} = \mathbf{I}$$

$$\begin{bmatrix} a_{11} & a_{12} & a_{13} \\ a_{11} & a_{12} & a_{13} \\ a_{11} & a_{12} & a_{13} \end{bmatrix}\begin{bmatrix} 1 & 2 & 1 \\ 3 & 1 & 4 \\ 2 & 1 & 3 \end{bmatrix} = \begin{bmatrix} 1 & 0 & 0 \\ 0 & 1 & 0 \\ 0 & 0 & 1 \end{bmatrix}$$

By matrix multiplication,

$$a_{11} + 3a_{12} + 2a_{13} = 1$$

$$2a_{11} + a_{12} + a_{13} = 0$$

$$a_{11} + 4a_{12} + 3a_{13} = 0$$

Solving for a_{11}, a_{12}, a_{13}; $a_{11} = 1/2$, $a_{12} = 5/2$, $a_{13} = -7/2$:

$$a_{21} + 3a_{22} + 2a_{23} = 0$$

$$2a_{21} + a_{22} + a_{23} = 1$$

$$a_{21} + 4a_{22} + 3a_{23} = 0$$

Solving for a_{21}, a_{22}, a_{23}; $a_{21} = 1/2$, $a_{22} = -1/2$, $a_{23} = 1/2$:

$$a_{31} + 3a_{32} + 2a_{33} = 0$$

$$2a_{31} + a_{32} + a_{33} = 0$$

$$a_{31} + 4a_{32} + 3a_{33} = 1$$

Solving for a_{31}, a_{32}, a_{33}; $a_{31} = -1/2$, $a_{32} = -3/2$, $a_{33} = 5/2$. Thus the inverse matrix is

$$\mathbf{A}^{-1} = \begin{bmatrix} 1/2 & 5/2 & -7/2 \\ 1/2 & -1/2 & 1/2 \\ -1/2 & -3/2 & 5/2 \end{bmatrix}$$

If the inverse exists,

$$(\mathbf{A}^{-1})' = (\mathbf{A}')^{-1}$$

$$(\mathbf{ABC})^{-1} = \mathbf{C}^{-1}\mathbf{B}^{-1}\mathbf{C}^{-1}$$

$$|\mathbf{A}^{-1}| = \frac{1}{|\mathbf{A}|}$$

Some Special Determinants

A *Wronskian* is the determinant

$$|f_i^{(k)}| \quad \text{where } f_i^{(k)} = \frac{d^k f_i}{dx^k}, \quad k = 0, 1, \ldots, n-1$$

Here $f_i^{(k)}$ is the kth derivative of f_i. The definition of derivative is given in Section 17.6. If the Wronskian vanishes the functions f_i are *linearly dependent*, or there is a relation such that

$$c_1 f_1 + \cdots + c_n f_n = 0$$

where c_i are nonzero constants.

Example: Let us assume that the three solutions of an equation are found as $f_1(x) = x^2$, $f_2(x) = x^3$ and $f_3(x) = x^2 + x^3$. The Wronskian is written as the determinant:

$$\begin{vmatrix} f_1 & f_2 & f_3 \\ f_1' & f_2' & f_3' \\ f_1'' & f_2'' & f_3'' \end{vmatrix}$$

Substituting the functions and derivatives in their positions in the Wronskian, one finds

$$\begin{vmatrix} x^2 & x^3 & x^2 & x^3 \\ 2x & 3x^2 & 2x & 3x^2 \\ 2 & 6x & 2 & 6x \end{vmatrix} = 0$$

Thus these three functions must be linearly dependent and the coefficients C_1, C_2, and C_3 below must be nonzero:

$$C_1 f_1 + C_2 f_2 + C_3 f_3 = 0$$

Substituting the functions in this equation,

$$C_1 x^2 + C_2 x^3 + C_3 (x^2 + x^3) = 0$$

Equating the coefficients of the equal powers of x

$$C_1 = -C_3 \qquad C_2 = -C_3$$

Thus $C_1 = C_2 = -C_3$, and there can be infinitely many sets of nonzero C values.

The *Jacobian* is the determinant

$$\left| \frac{\partial f_i}{\partial x_k} \right| = \left| \frac{\partial(f_1, \ldots, f_n)}{\partial(x_1, \ldots, x_n)} \right|$$

Here $\partial f_i / \partial x_k$ is a partial derivative of f_i. The definition of partial derivative is given in Section 17.6. If the Jacobian vanishes, there is a *functional dependence* between f_i, that is, there is a function as $F(f_1, \ldots, f_n) = 0$ for all values of the variables, x_i.

Example: The Jacobian determinant is used in solving implicit functions for determining transformations. For example, let us assume that w is a function of x and y as

$$w = x(x, y)$$

If we apply the transformations

$$f_1 = f_1(x, y)$$

$$f_2 = f_2(x, y)$$

then x and y may be obtained in terms of f_1 and f_2 as

$$x = x(f_1, f_2)$$

$$y = y(f_1, f_2)$$

Thus by substituting these in w,

$$w = \Omega(f_1, f_2)$$

However, this may not be easy. Instead the following technique is used. First the function w is differentiated with respect to x and y,

$$\frac{\partial w}{\partial x} = \frac{\partial w}{\partial f_1} \frac{\partial f_1}{\partial x} + \frac{\partial w}{\partial f_2} \frac{\partial f_2}{\partial x}$$

$$\frac{\partial w}{\partial y} = \frac{\partial w}{\partial f_1} \frac{\partial f_1}{\partial y} + \frac{\partial w}{\partial f_2} \frac{\partial f_2}{\partial y}$$

To find the derivatives of w with respect to f_1 and to f_2, that is, $\partial w / \partial f_1$ and $\partial w / \partial f_2$, the above set of linear equations is solved (see Simultaneous Linear Equations and particularly Cramer's rule in this section):

$$\frac{\partial w}{\partial f_1} = \frac{\begin{vmatrix} \partial w / \partial x & \partial f_2 / \partial x \\ \partial w / \partial y & \partial f_2 / \partial y \end{vmatrix}}{\begin{vmatrix} \partial f_1 / \partial x & \partial f_2 / \partial x \\ \partial f_1 / \partial y & \partial f_2 / \partial y \end{vmatrix}} \qquad \frac{\partial w}{\partial f_2} = \frac{\begin{vmatrix} \partial f_1 / \partial x & \partial w / \partial x \\ \partial f_1 / \partial y & \partial w / \partial y \end{vmatrix}}{\begin{vmatrix} \partial f_1 / \partial x & \partial f_2 / \partial x \\ \partial f_1 / \partial y & \partial f_2 / \partial y \end{vmatrix}}$$

The denominators are the Jacobian. If they vanish than there is no solution; otherwise the solutions exist.

17.5-5 Simultaneous Linear Equations

The simultaneous linear equations can be represented in a matrix form as

$$\mathbf{Ax = b}$$

where $\mathbf{A}$ is a (n, n) matrix, and $\mathbf{x}$ and $\mathbf{b}$ are column vectors. The solution of this equation is unique and expressed in terms of the inverse of $\mathbf{A}$,

$$\mathbf{x = A^{-1}b}$$

Example: Let $\mathbf{A}$ be a 3×3 matrix,

$$\mathbf{A} = \begin{bmatrix} 1 & 2 & 1 \\ 3 & 1 & 4 \\ 2 & 1 & 3 \end{bmatrix}$$

$$\mathbf{x} = \begin{bmatrix} x_1 \\ x_2 \\ x_3 \end{bmatrix} \qquad \mathbf{b} = \begin{bmatrix} 3 \\ 5 \\ 4 \end{bmatrix}$$

For $\mathbf{Ax = b}$, the solution is

$$\mathbf{x = A^{-1}b}$$

Therefore the inverse is searched. Since $\mathbf{A}$ is the same matrix as the one given in section on inverse,

$$\mathbf{A}^{-1} = \begin{bmatrix} 1/2 & 5/2 & -7/2 \\ 1/2 & -1/2 & 1/2 \\ -1/2 & -3/2 & 5/2 \end{bmatrix}$$

$$\mathbf{x} = \begin{bmatrix} 1/2 & 5/2 & -7/2 \\ 1/2 & -1/2 & 1/2 \\ -1/2 & -3/2 & 5/2 \end{bmatrix} \begin{bmatrix} 3 \\ 5 \\ 4 \end{bmatrix}$$

$$\mathbf{x} = \begin{bmatrix} 3/2 + 25/2 - 28/2 \\ 3/2 - 5/2 + 2 \\ -3/2 - 15/2 + 20/2 \end{bmatrix} = \begin{bmatrix} 0 \\ 1 \\ 1 \end{bmatrix}$$

Thus the solution is $x_1 = 0$, $x_2 = 1$, $x_3 = 1$.

Gauss–Jordan Elimination

If a set of simultaneous linear equations are given as

$$a_{11}x_1 + a_{12}x_2 + \cdots + a_{1n}x_n = b_1$$

$$a_{21}x_1 + a_{22}x_2 + \cdots + a_{2n}x_n = b_2$$

$$\vdots$$

$$a_{n1}x_1 + a_{n2}x_2 + \cdots + a_{nn}x_n = b_n$$

by dividing the first equation by the coefficient of its first term, a_{11}, and eliminating the x_1 terms from the remaining equations, and continuing this elimination process for the second, third, and the other equations, the system can be reduced to the *canonical form*:

$$\begin{aligned} x_1 & = b_1^* \\ &x_2 = b_2^* \\ &\vdots \\ & x_n = b_n^* \end{aligned}$$

where the new values of the right-hand sides correspond directly to the solution of the system of equations.

Example:

$$A = \begin{bmatrix} 1 & 2 & 1 \\ 3 & 1 & 4 \\ 2 & 1 & 3 \end{bmatrix}$$

$$b = \begin{bmatrix} 3 \\ 5 \\ 4 \end{bmatrix}$$

Divide the first equation by the coefficients of x_1, and then eliminate x_1 terms from the other two:

$$x_1 + 2x_2 + x_3 = 3$$
$$-5x_2 + x_3 = -4$$
$$-3x_2 + x_3 = -2$$

Divide the new second equation by the coefficients of x_2, that is, -5, then eliminate the x_2 terms from the first and third equations:

$$x_1 \quad + \tfrac{7}{5}x_3 = \tfrac{7}{5}$$
$$x_2 - \tfrac{1}{5}x_3 = \tfrac{4}{5}$$
$$\tfrac{2}{5}x_3 = \tfrac{2}{5}$$

Divide the third equation by the coefficients of x_3, that is, $\tfrac{2}{5}$, and eliminate the x_3 terms from the first and second equations:

$$x_1 \quad\quad = 0$$
$$x_2 \quad = 1$$
$$x_3 = 1$$

Cramer's Rule

Both sides of the simultaneous linear equations, $Ax = b$ can be multiplied by the inverse of A from left to obtain

$$A^{-1}Ax = A^{-1}b$$

where $A^{-1}A = I$, and $Ix = x$. Therefore,

$$x = A^{-1}b = \frac{\text{adj } A}{\det A} b$$

which is called *Cramer's Rule*.

Example:

$$x = A^{-1}b = \frac{\text{adj } A}{\det A} b \qquad b = \begin{bmatrix} 3 \\ 5 \\ 4 \end{bmatrix}$$

$$A = \begin{bmatrix} 1 & 2 & 1 \\ 3 & 1 & 4 \\ 2 & 1 & 3 \end{bmatrix}$$

$$\text{adj } A = \begin{bmatrix} A_{11} & A_{21} & A_{31} \\ A_{12} & A_{22} & A_{23} \\ A_{31} & A_{32} & A_{33} \end{bmatrix}$$

Elements of the adj **A** are calculated as follows:

$$A_{11} = (-1)^2 \begin{vmatrix} 1 & 4 \\ 1 & 3 \end{vmatrix} = -1$$

$$A_{12} = (-1)^3 \begin{vmatrix} 3 & 4 \\ 2 & 3 \end{vmatrix} = -1$$

$$A_{13} = (-1)^4 \begin{vmatrix} 3 & 1 \\ 2 & 1 \end{vmatrix} = +1$$

$$A_{21} = (-1)^3 \begin{vmatrix} 2 & 1 \\ 1 & 3 \end{vmatrix} = -5$$

$$A_{22} = (-1)^4 \begin{vmatrix} 1 & 1 \\ 2 & 3 \end{vmatrix} = +1$$

$$A_{23} = (-1)^5 \begin{vmatrix} 1 & 2 \\ 2 & 1 \end{vmatrix} = +3$$

$$A_{31} = (-1)^4 \begin{vmatrix} 2 & 1 \\ 1 & 4 \end{vmatrix} = +7$$

$$A_{32} = (-1)^5 \begin{vmatrix} 1 & 1 \\ 3 & 4 \end{vmatrix} = -1$$

$$A_{33} = (-1)^6 \begin{vmatrix} 1 & 2 \\ 3 & 1 \end{vmatrix} = -5$$

$$\text{adj } \mathbf{A} = \begin{bmatrix} -1 & -5 & +7 \\ -1 & +1 & -1 \\ +1 & +3 & -5 \end{bmatrix}$$

$$\det \mathbf{A} = \begin{bmatrix} 1 & 2 & 1 \\ 3 & 1 & 4 \\ 2 & 1 & 3 \end{bmatrix}$$

$$= 1(3 - 4) - 2(9 - 8) + 1(3 - 2)$$

$$= -1 - 2 + 1 = -2$$

$$\frac{\text{adj } \mathbf{A}}{\det \mathbf{A}} \mathbf{b} = \frac{\begin{bmatrix} -1 & -5 & +7 \\ -1 & +1 & -1 \\ +1 & +3 & -3 \end{bmatrix} \begin{bmatrix} 3 \\ 5 \\ 4 \end{bmatrix}}{\begin{vmatrix} 1 & 2 & 1 \\ 3 & 1 & 4 \\ 2 & 1 & 3 \end{vmatrix}} = \frac{\begin{bmatrix} -3 - 25 + 28 \\ -3 + 5 - 4 \\ +3 + 15 - 20 \end{bmatrix}}{-2} = \begin{bmatrix} 0 \\ 1 \\ 1 \end{bmatrix}$$

Traces

If **A** is a square matrix, the *trace* of **A** is the sum of the diagonal elements, that is

$$\text{tr } \mathbf{A} = \sum_i a_{ii}$$

Example:

$$\mathbf{A} = \begin{bmatrix} 1 & 2 & 1 \\ 3 & 1 & 4 \\ 2 & 1 & 3 \end{bmatrix}$$

$$\text{tr } \mathbf{A} = 1 + 1 + 3 = 5$$

For the matrices **A**, **B**, and **C** with proper orders so that the matrix multiplications are defined,

$$\text{tr}(\mathbf{AB}) = \text{tr}(\mathbf{BA})$$

$$\text{tr}(\mathbf{ABC}) = \text{tr}(\mathbf{BCA}) = \text{tr}(\mathbf{CAB})$$

The kth order trace of a square matrix is the sum of determinants of all $\binom{n}{k}$ matrices of order $k \times k$,

$$\text{tr}_k \mathbf{A} = \sum \begin{vmatrix} a_{i_1 i_1} & a_{i_1 i_2} & \cdots & a_{i_1 i_k} \\ a_{i_2 i_1} & a_{i_2 i_2} & \cdots & a_{i_2 i_k} \\ \vdots & & & \\ a_{i_k i_1} & a_{i_k i_2} & \cdots & a_{i_k i_k} \end{vmatrix}$$

The sum is over all combinations of n elements taken k at a time in the order $i_1 < i_2 < \cdots < i_k$. Examples of kth order traces are:

$$k = 1; \quad \binom{n}{k} = \binom{3}{1} = \frac{3!}{1!(3-1)!} = \frac{1 \cdot 2 \cdot 3}{1 \cdot 2} = 3 \text{ matrices}$$

of order 1×1.

$$\text{tr}_1 \mathbf{A} = |a_{11}| + |a_{22}| + |a_{33}|$$
$$= |1| + |1| + |3|$$
$$= 1 + 1 + 3 = 5 = \text{tr}\,\mathbf{A}$$

$$k = 2; \quad \binom{n}{k} = \binom{3}{2} = \frac{3!}{2!(3-2)!} = \frac{1 \cdot 2 \cdot 3}{2 \cdot 1} = 3 \text{ matrices}$$

of order 2×2.

$$\text{tr}_2 \mathbf{A} = \begin{vmatrix} a_{11} & a_{12} \\ a_{21} & a_{22} \end{vmatrix} + \begin{vmatrix} a_{22} & a_{23} \\ a_{32} & a_{33} \end{vmatrix} + \begin{vmatrix} a_{33} & a_{31} \\ a_{13} & a_{11} \end{vmatrix}$$
$$= \begin{vmatrix} 1 & 2 \\ 3 & 1 \end{vmatrix} + \begin{vmatrix} 1 & 4 \\ 1 & 3 \end{vmatrix} + \begin{vmatrix} 3 & 2 \\ 1 & 1 \end{vmatrix}$$
$$= (1 - 6) + (3 - 4) + (3 - 2)$$
$$= -5 - 1 + 1 = -5$$

Characteristic Roots and Vectors

If $\mathbf{A}$ is a matrix of order n,

$$|\mathbf{A} - \lambda \mathbf{I}| = 0$$

is called the *characteristic equation* of the matrix $\mathbf{A}$, and it is a polynomial of degree n in λ:

$$\lambda^n - (\text{tr}_1 \mathbf{A})\lambda^{n-1} + (\text{tr}_2 \mathbf{A})\lambda^{n-2} - \cdots + (-1)^n |\mathbf{A}| = 0$$

The roots of this polynomial equation are called the *characteristic roots* of $\mathbf{A}$. Another name of these roots is the *eigenvalues* of $\mathbf{A}$.

The solution of $\mathbf{A} - \lambda \mathbf{x} = 0$ corresponding to λ values above are called *characteristic vectors*. Examples of the characteristic roots and characteristic vectors are natural frequencies of multidegree of freedom vibrating systems and the corresponding characteristic solutions.

Example:

$$\mathbf{A} = \begin{bmatrix} 1 & 2 & 1 \\ 3 & 1 & 4 \\ 2 & 1 & 3 \end{bmatrix}$$

$$|\mathbf{A} - \lambda \mathbf{I}| = \begin{vmatrix} \begin{bmatrix} 1 & 2 & 1 \\ 3 & 1 & 4 \\ 2 & 1 & 3 \end{bmatrix} - \lambda \begin{bmatrix} 1 & 0 & 0 \\ 0 & 1 & 0 \\ 0 & 0 & 1 \end{bmatrix} \end{vmatrix} = 0$$

$$= \begin{vmatrix} 1 - \lambda & 2 & 1 \\ 3 & 1 - \lambda & 4 \\ 2 & 1 & 3 - \lambda \end{vmatrix} = 0$$

$$= (1 - \lambda) \ [(1 - \lambda)(3 - \lambda) - 4]$$

$$- 2 \ [3(3 - \lambda) - 2 \cdot 4]$$

$$+ 1 \ [3 \cdot 1 - 2(1 - \lambda)]$$

$$= - \lambda^3 + 5\lambda^2 + 5\lambda - 2 = 0$$

Changing the signs to make the first coefficient $+1$,

$$\lambda^3 - 5\lambda^2 - 5\lambda + 2 = 0$$

In the following formula

$$\lambda^3 - (\text{tr}_1 \mathbf{A})\lambda^2 + (\text{tr}_2 \mathbf{A})\lambda - (-1)^3 |\mathbf{A}| = 0$$

$$\text{tr}_1 \mathbf{A} = +5$$

$$\text{tr}_2 \mathbf{A} = -5$$

$$|\mathbf{A}| = +2$$

which yields the same result for the characteristic equations obtained above.

Matrix Differentiation

If the elements of a matrix are functions of a scalar variable x, $\partial \mathbf{A}/\partial x$ is the matrix with elements $\partial a_{ij}/\partial x$. For definition of differentiation, see Section 17.6. If a scalar x is a function of a matrix, $\partial x/\partial \mathbf{A}$ is the matrix with elements $\partial x/\partial a_{ij}$.

The Eigenvalue Problem

The characteristic vector with unit length, $\mathbf{x}'\mathbf{x} = 1$, is the *eigenvector*. Finding characteristic values of a set of linear equations and their corresponding eigenvectors is called the eigenvalue problem.

Example: To illustrate the concept a simple example is given. The set of linear equations is

$$- x_1 + 2x_2 = 6$$

$$- 3x_1 - 4x_2 = 8$$

with $x_1 = 4$ and $x_2 = 5$ as solutions. The *characteristic equation* is found by

$$\begin{vmatrix} -1 - \lambda & 2 \\ -3 & 4 - \lambda \end{vmatrix} = \lambda^2 - 3\lambda + 2 = 0$$

The *characteristic roots* are the solutions of this equation:

$$\lambda_1 = 1, \qquad \lambda = 2$$

The *characteristic (eigen) vectors* can be found by substituting the characteristic values into the equation:

$$(-1 - \lambda)x_1 + \qquad 2x_2 = 0$$

$$- 3x_1 + (4 - \lambda)x_2 = 0$$

For $\lambda = 1$,

$$-2x_1 + 2x_2 = 0$$

$$- 3x_1 + 3x_2 = 0$$

thus $x_1 = x_2$ satisfies these two equations. There are infinitely many such solutions. Since $x'x = 1$, the characteristic solution with unit length, the eigenvector is found by

$$[x_1 \ \ x_1] \begin{bmatrix} x_1 \\ x_1 \end{bmatrix} = x_1^2 + x_1^2 = 1, \qquad 2x_1^2 = 1, \qquad x_1 = \sqrt{1/2} = \sqrt{2}/2$$

Therefore a set of eigenvectors is

$$(x_1, x_2) = (\sqrt{2}/2, \sqrt{2}/2)$$

Similarly for $\lambda = 2$,

$$-3x_1 + 2x_2 = 0$$

$$-3x_1 + 2x_2 = 0$$

thus $x_1 = \frac{2}{3}x_2$ satisfies these two equations. There are infinitely many such solutions. Since $x'x = 1$, the characteristic solution with unit length, the eigenvector is found by

$$[x_1 \ \ \tfrac{3}{2}x_1] \begin{bmatrix} x_1 \\ \tfrac{3}{2}x_1 \end{bmatrix} = x_1^2 + \tfrac{9}{4}x_1^2 = 1, \qquad \tfrac{13}{4}x_1^2 = 1$$

This results in $x_1 = 2\sqrt{13}/13$, therefore $x_2 = 3\sqrt{13}/13$. Therefore a set of eigenvectors is

$$(x_1, x_2) = (2\sqrt{13}/13, 3\sqrt{13}/13)$$

In summary

Eigenvalues	Eigenvectors
$\lambda = 1$	$(\sqrt{2}/2, \sqrt{2}/2)$
$\lambda = 2$	$(2\sqrt{13}/13, 3\sqrt{13}/13)$

Application to Linear Differential Equations

For an application of matrices to the solution of linear differential equations (see Section 17.6 for definitions on differential equations), we consider a set of two linear differential equations of the first order given as

$$\frac{dx_1}{dt} = -x_1 + 2x_2$$

$$\frac{dx_2}{dt} = -3x_1 + 4x_2$$

At the origin where $x_1 = x_2 = 0$ the right-hand sides of these differential equations vanish, thus the derivatives of the variables are zero. This corresponds to a singular solution of the differential equation by definition. The coefficient matrix of the equations is the same as the example given above. The eigenvalues are those found above, namely, 1 and 2, two real positive eigenvalues. The question of stability of the singular solution, $x_1, x_2 = 0$, is answered by the sign of these eigenvalues. Since they are both greater than zero, that is, they are positive, this characteristic solution is an *unstable* solution.

17.6 DIFFERENTIAL AND INTEGRAL CALCULUS

17.6-1 Real Functions

Definitions

Given two sets, X and Y, of numbers x and y, respectively, a correspondence relating any x in X to one or more y in Y is called a *function*, f. The correspondence is shown as $y = f(x)$. Here X is called the *domain* of f, Also, x is then the *independent variable* and y the *dependent variable*. In the case of $y = f(x)$, the function is said to be *explicitly* given while, if $f(x, y) = 0$, it is *implicitly* given. The *inverse function* relates y to x: $x = f^{-1}(y)$.

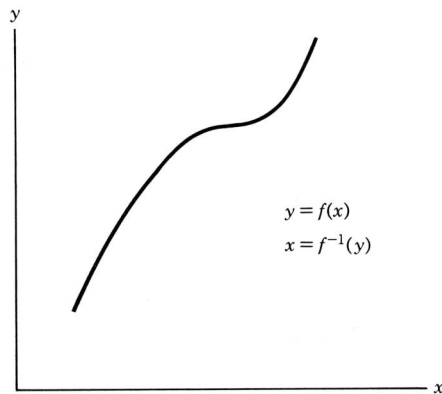

$$y = f(x)$$
$$x = f^{-1}(y)$$

17.6-2 Differentiation

If a function $f(x)$ is defined in the neighborhood of $x = x_0$, and the limit

$$\lim_{h \to 0} \frac{f(x_0 + h) - f(x_0)}{h}$$

exists, it is called the *derivative* or *differential coefficient* of $f(x)$ at x_0 and denoted by

$$f'(x_0) \quad \text{or} \quad (df/dx)_{x_0} \quad \text{or} \quad Df(x_0)$$

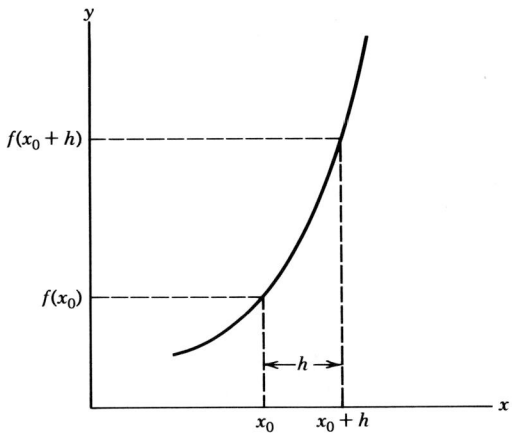

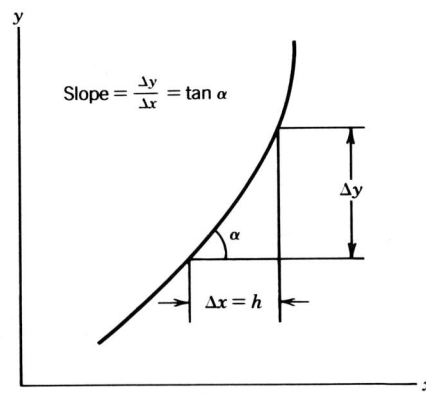

Therefore, while the mathematical meaning of derivative is *slope* of f, the physical meaning of derivative is the *rate of change* of function f as the variable x changes.

Similarly, the derivative of a derivative is called the *second derivative* and denoted by

$$f''(x_0) \quad \text{or} \quad (d^2f/dx^2)_{x_0} \quad \text{or} \quad D^2f(x_0).$$

On the other hand, the *operation* that derives $f'(x)$ from $f(x)$ is called the *differentiation* of $f(x)$.

The Differentiation Rules

The derivative of the sum of two functions equals the sum of derivatives:

$$(f + g)' = f' + g'$$

A constant multiplier, a is factored out of the derivative,

$$(af)' = a(f')$$

For given constants a and b and functions f and g:

Addition of functions: $(af + bg)' = (af' + bg')$

Product of functions: $(fg) = f'g + fg'$

Division of functions: $\left(\dfrac{f}{g}\right)' = \dfrac{f'g - fg'}{g^2}$

Function of a function: $\dfrac{d}{dx}f[g(x)] = \dfrac{df(g)}{dg}\dfrac{dg(x)}{dx}$
(composite function)

Leipnitz formula: $\dfrac{d^n}{dx^n}[fg] = \displaystyle\sum_{j=0}^{n} \binom{n}{j} f^{(j)} g^{(n-j)}$

$$\frac{d}{dx}[f(x)]_{x=x_0}^n = n[f(x_0)]^{n-1}f'(x_0)$$

Inverse function: If $y'(x)$, then $x'(y) = 1/y'(x)$.

Maxima and Minima

If the derivative of $f(x)$, df/dx, has the value 0 at some value of x, x_0,

$$|df/dx|_{x_0} = 0,$$

then f has a *stationary value* at x_0. There are three types of such points with stationary values of f.

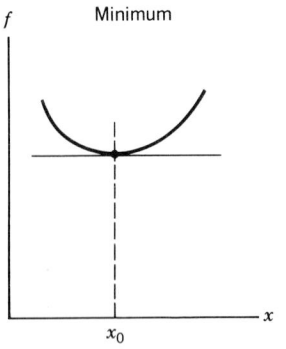

Minimum

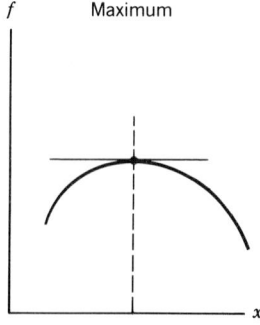

Maximum

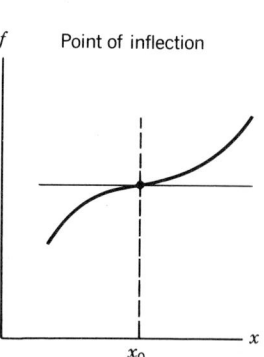

Point of inflection

If the f value everywhere in the neighborhood of x_0 is *greater* than the value at x_0, the function is said to have a *minimum* at x_0. If it is *lower* than the value at x_0, then function has a *maximum* at x_0. If neither is true, then x_0 is called a *point of inflection*.

Minimum occurs at $d^2f/dx^2 > 0$.
Maximum occurs at $d^2f/dx^2 < 0$.
Point of inflection is at $d^2f/dx^2 = 0$.

$d^3f/dx^3 > 0$

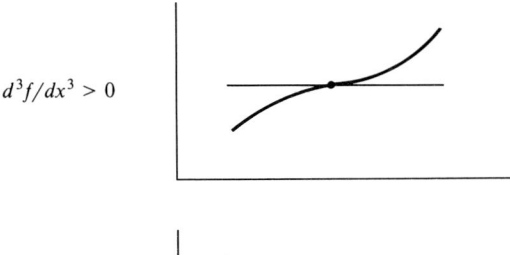

$d^3f/dx^3 < 0$

Example: A function is given as $f = x^3 - 3x$. The stationary points lie where

$$\frac{df}{dx} = 3x^2 - 3 = 0$$

This gives a solution $x^2 = 1$, thus $x_1 = +1$, $x_2 = -1$.

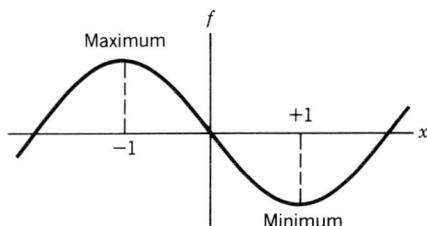

At $x_1 = +1$, $d^2f/dx^2 = 6x = 6 > 0$, thus f reaches its *minimum*.

At $x_1 = -1$, $d^2f/dx^2 = 6x = -6 < 0$, thus f reaches its *maximum*.

Example: If $f = x(10 - x)$, a rectangular area with width x and height $10 - x$, what should be the x value to have the maximum area?

$$\frac{df}{dx} = -2x + 10 \qquad x = 5$$

$$\frac{d^2f}{dx^2} = -2 < 0 \quad \text{thus } f \text{ is maximum at } x = 5$$

It is clear that this is a square. Thus, as expected, the square has the maximum area among all the rectangular areas with equal peripheries (edges). In this example the sum of edges is $2 \times 10 = 20$ units.

Multiple Variables

In the case of functions with multiple independent variables, $f(x_1, x_2, \ldots, x_n)$, the derivative is called the *partial derivative* and denoted by, for the case of two variables $x_1 = x$, $y_2 = y$,

$$\frac{\partial f(x, y)}{\partial x} \equiv \lim_{h \to 0} \frac{f(x + h, y) - f(x, y)}{h}$$

In general we have $\partial f(x_i)/\partial x_k$. There are as many *first* partial derivatives of f as the number of independent variables.

Higher derivatives would be

$$\frac{\partial^2 f}{\partial x_i^2} \quad \text{and} \quad \frac{\partial^2 f}{\partial x_i \, \partial x_k}$$

Similarly, for $\partial^n/\partial x_1 \cdots \partial x_j \cdots$, all the combinations of the independent variables must be considered.

The *total* derivative is defined as

$$df = \left(\frac{\partial f}{\partial x}\right) dx + \left(\frac{\partial f}{\partial y}\right) dy$$

In general

$$df = \left(\frac{\partial f}{\partial x_1}\right) dx_1 + \left(\frac{\partial f}{\partial x_2}\right) dx_2 + \cdots$$

From which we have

$$\frac{df}{dx_1} = \left(\frac{\partial f}{\partial x_1}\right) + \left(\frac{\partial f}{\partial x_2}\right)\left(\frac{dx_2}{dx_1}\right) + \cdots$$

where $x_2, \ldots$ can be considered as functions g_2 of x_1, and so on.

Maxima and Minima of Functions of Multiple Variables

Maxima and minima for the functions of multiple variables can similarly be defined. Here for a function of two variables, x, y,

$$\frac{\partial f}{\partial x} = 0 \qquad \frac{\partial f}{\partial y} = 0$$

corresponds to a stationary point. To determine the maximum or minimum, one refers to the Taylor expansion (see the section of Taylor series below). Two observations are that since the first derivatives vanish those terms with the first derivatives are dropped from the expansion. In addition, any term with derivatives higher than the second can be made small relative to this term, thus negligible, by taking small enough intervals. Intervals are taken as h_1 along the x axis, h_2 along the y axis. Therefore,

$$f(x, y) = f(x_0, y_0) + \frac{1}{2!}\left[h_1^2 \frac{\partial^2 f}{\partial x^2} + 2h_1 h_2 \frac{\partial^2 f}{\partial x \, \partial y} + h_2^2 \frac{\partial^2 f}{\partial y_2}\right]$$

In a simpler form

$$f = f_0 + (1/2!) h_1^2 A + 2h_1 h_2 B + h_2^2 C$$

where A, B, C are the corresponding derivatives evaluated at the stationary point. Rewriting the above equation as

$$f - f_0 = A\left(h_1 + \frac{Bh_2}{A}\right)^2 + \left(C - \frac{B^2}{A}\right) h_2^2$$

we have the following:

Minimum: $f - f_0 > 0$ for all h_1, h_2: $A > 0$, $C - (B^2/A) > 0$.

Maximum: $f - f_0 < 0$ for all h_1, h_2: $A < 0$, $C - (B^2/A) < 0$.

Otherwise it is a *saddle point*.

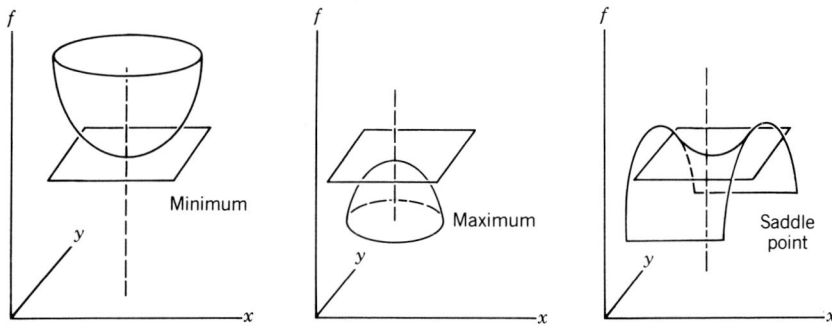

Therefore

$$\frac{\partial^2 f}{\partial x^2} > 0 \qquad\qquad \text{for a minimum}$$

$$\frac{\partial^2 f}{\partial x^2} < 0 \qquad\qquad \text{for a maximum}$$

$$\frac{\partial^2 f}{\partial x^2}\frac{\partial^2 f}{\partial y^2} > \left(\frac{\partial^2 f}{\partial x \partial y}\right)^2 \qquad \text{in either case}$$

$$\frac{\partial^2 f}{\partial x^2}\frac{\partial^2 f}{dy^2} < \left(\frac{\partial^2 f}{\partial x \partial y}\right)^2 \qquad \text{for a saddle point}$$

This can be generalized such that for the case of multiple variables,

$$\frac{\partial f}{\partial x_i} = 0 \qquad i = 1, 2, \ldots, n$$

$$h_i h_j \left[\frac{\partial^2 f}{\partial x_i \partial x_j}\right] > 0 \quad \text{for minimum}$$

$$< 0 \quad \text{for maximum}$$

Series Expansion of Functions

Using either the Taylor or Maclaurin theorem, discussed below, the series expansion of functions can be obtained. Illustrative examples of series expansions of some common functions, such as exponential functions, trigonometric functions, and so on are given here:

$$e^x = 1 + \frac{x}{1!} + \frac{x^2}{2!} + \cdots + \frac{x^n}{n!} + \cdots \qquad \text{for all } x$$

$$\lim(1 + x) = x - \frac{x^2}{2} + \frac{x^3}{3} - \cdots + (-1)^{n-1}\frac{x^n}{n} + \cdots \qquad |x| < 1$$

$$\sin x = x - \frac{x^3}{3!} + \frac{x^5}{5!} - \frac{x^7}{7!} + \cdots$$

$$\cos x = 1 - \frac{x^2}{2!} + \frac{x^4}{4!} - \frac{x^6}{6!} + \cdots$$

$$\sinh x = x + \frac{x^3}{3!} + \frac{x^5}{5!} + \frac{x^7}{7!} + \cdots$$

$$\cosh x = 1 + \frac{x^2}{2!} + \frac{x^4}{4!} + \frac{x^6}{6!} + \cdots$$

Indeterminate Forms

Series expansion of functions may be used in evaluating these functions. For example, if two functions are $f(x)$ and $g(x)$, defining another function as their ratio, $f(x)/g(x)$, and evaluating this ratio at some x values may be *indeterminate*. An example of such a case is $F(x) = 1 - e^x$, and $g(x) = \lim(1 + x)$; thus

$$\left.\frac{1 - e^x}{\lim(1 + x)}\right|_{x=0} = \frac{0}{0} \quad \textit{which is indeterminate}$$

In such cases the *L'Hôpital rule* is applied which states that the actual value is

$$\left.\frac{f'(x)}{g'(g)}\right|_{x=0} = -1$$

This can easily be seen from the above series expansion

$$f(x) = 1 - e^x = 1 - 1 - \frac{x}{1!} - \frac{x^2}{2!} - \cdots = -\frac{x}{1!} - \frac{x^2}{2!} - \cdots$$

$$g(x) = \lim(1 + x) = x - \frac{x^2}{2} + \frac{x^3}{3} - + \cdots$$

$$f'(x) = -1 - x - \cdots$$

$$g'(x) = +1 - x + \cdots$$

L'Hôpital's rule may be applied successively such that derivatives of the functions forming the quotient may be taken successively until it is determinate.

Taylor's Theorem

In expanding functions in power series, Taylor's expansion is used. Series expansions are useful in mathematical analysis of various problems. A function, $f(x)$, n times differentiable in the interval $(x_0, x_0 + h)$, can be expanded about the value x_0 as

$$f(x_0 + h) = f(x_0) + \frac{h}{1!}f'(x_0) + \frac{h^2}{2!}f''(x_0) + \cdots + \frac{h^{n-1}}{(n-1)!}f^{n-1}(x_0) + R_n$$

where the remainder R_n is

$$R_n = \frac{h^n(1 - q)^{n-m}}{(n-1)!m!}f''(x_0 + qh)$$

Lagrange remainder $(m = n)$: $R_n = \frac{h^n}{n!}f^{(n)}(x_0 + q_1 h)$

Cauchy remainder $(m = 1)$: $R_n = \frac{h^n(1 - q_2)^{n-1}}{(n-1)!}f^{(n)}(x_0 + q_2 h)$

where q_1 and q_2 are numbers with values between 0 and 1.

Example:

$$f(x) = e^x$$

$$f'(x) = e^x$$

$$f''(x) = e^x$$

$$\vdots$$

$$e(x_0 + h) = e^{x_0} + \frac{h}{1!}e^{x_0} + \frac{h^2}{2!}e^{x_0} + \cdots$$

$$= e^{x_0}\left(1 + \frac{h}{1!} + \frac{h^2}{2!} + \cdots + \frac{h^{n-1}}{(n-1)!} + R_n\right)$$

$$e^{x_0}e^h = e^{x_0}\left(1 + \frac{h}{1!} + \frac{h^2}{2!} + \cdots\right)$$

Therefore the Taylor expansion of e^h is

$$e^h = 1 + \frac{h}{1!} + \frac{h^2}{2!} + \cdots$$

$$e = 1 + \frac{1}{1!} + \frac{1}{2!} + \cdots$$

would be a method of calculating the value of e.

Maclaurin's Theorem

The Taylor expansion for a special case, $x_0 = 0$ and $h = x_1$, gives the series expansion obtained by Maclaurin's theorem:

$$f(x) = f(0) + \frac{x}{1!}f'(0) + \frac{x^2}{2!}f''(0) + \cdots + \frac{x^{n-1}}{(n-1)!}f^{n-1}(0) + R_n$$

where the remainder R_n is

$$R_n = x^n\frac{f^n(\phi x)}{n!} \qquad \phi \text{ between 0 and 1}$$

For example, for $F(x) = e^x$,

$$e^x = e^0 + \frac{x}{1!}e^0 + \frac{x^2}{2!}e^0 + \cdots \qquad e^0 = 1$$

$$e^x = 1 + \frac{x}{1!} + \frac{x^2}{2!} + \cdots \qquad \text{the Maclaurin expansion of } e^x$$

17.6-3 Integration

Integration is an operation inverse to differentiation. For a differentiation given as

$$f(x) = \frac{d}{dx}F(x)$$

its *inverse*, $F(x)$, is an *indefinite integral* of $f(x)$,

$$F(x) = \int f(x)\,dx$$

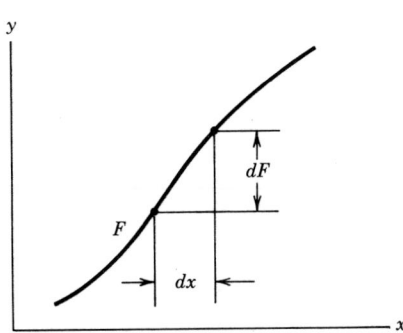

In general this function is not unique, and it is expressed as $F(x) + C$, where C is a constant.

Similar to the operation of differentiation the *operation* which results in $\int(\cdots)\,dx$ is called *integration*.

General Rules for Indefinite Integrals

$$\int cf(x)\,dx = c\int f(x)\,dx \quad \text{where } c \text{ is a constant}$$

$$\int [f_1(x) + f_2(x)]\,dx = \int f_1(x)\,dx + \int f_2(x)\,dx$$

Integration by Parts

$$\int f'(x)g(x)\,dx = f(x)g(x) - \int f(x)g'(x)\,dx$$

$$\int f(x)\,dx = \int f(u)\phi'(u)\,du \quad x = \phi(u)$$

$$\int f'(x)/f(x)\,dx = \log f(x)$$

The *definite integral* is

$$\int_a^b f(x)\,dx = F(b) - F(a)$$

which defines the area between the curve $y = f(x)$ and the x axis, between the ordinates $x = a$ and $x = b$. This is known as the *Newton–Leipnitz formula*.

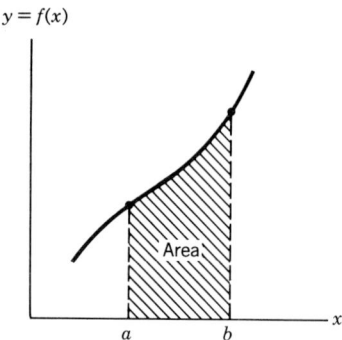

The *infinite integral*, if it exists, is defined by

$$\lim_{X \to \infty} \int_a^X f(x)\,dx = \int_a^\infty f(x)\,dx$$

which is called a Cauchy–Riemann integral.

Multiple integral is defined as limits of multiple sums as an extension of the integral defined above. For example, the *double integral* is

$$\iint_S f(x, y)\,dx\,dy$$

which can be shown to be the volume between the surface $z = f(x, y)$ and the plane $z = 0$, and lines parallel to z axis through the boundary of the S region. The evaluation of multiple integral may be

$$\int dy\left[\int f(x, y)\,dx\right] \quad \int dx\left[\int f(x, y)\,dy\right]$$

Special Integrals

Some of the special integrals have been solved yielding closed solutions. These functions are useful in that they are analytical and available in tabulated form, thus, for engineering applications there are sufficiently comprehensive tables. Some of these functions are discussed briefly in this section.

Gamma Function

$$\Gamma(z) = \int_0^\infty e^{-t} t^{z-1}\, dt \qquad R(z) > 0$$

where z is the complex variable, $z = \mathrm{Re}(z) + i\,\mathrm{Im}(z)$ and $i = \sqrt{-1}$.

$$\Gamma(z) = z^{-1} e^{-\gamma z} \prod \left[(1 + z/n)^{-1} e^{z/n} \right] \quad \text{all } z$$

$$\Gamma(1/2) = (\pi)^{1/2}$$

$$\Gamma(n+1) = n! \qquad n = 1, 2, 3, \ldots$$

$$\Gamma(z+1) = z\Gamma(z)$$

$$\Gamma(1-z) = \pi \operatorname{cosec} \pi z$$

Stirling's formula. Asymptotic behavior of the gamma function leads to Stirling's formula:

$$\ln \Gamma(z) \sim \left(z - \tfrac{1}{2} \right) \ln z - z + \tfrac{1}{2} \ln 2\pi$$

$$n! \sim \left(\frac{n}{e} \right)^n \sqrt{2\pi n}$$

Beta Function

The Beta function is related to the Gamma function:

$$B(z, \omega) = \int_0^1 t^{z-1} (1 - t)^{\omega - 1}\, dt$$

$$= \frac{\Gamma(z)\Gamma(\omega)}{\Gamma(z + \omega)} \qquad R(z) > 0,\ R(\omega) > 0$$

Elliptic Integrals

The general form of an elliptic integral is

$$\int R(x, \sqrt{X})\, ds$$

Here R denotes a rational function,

$$X = a + bx + cx^2 + dx^3 + ex^4$$

an algebraic function of third or fourth degree. The special forms are listed below, where k is called the *modulus* and usually lies between 0 and 1.

Incomplete Elliptic Integrals

1. *Elliptic integral of the first kind.*

$$F(k, x) = \int_0^t \frac{dx}{\left[(1 - x^2)(1 - k^2 x^2) \right]^{1/2}} \qquad (17.6\text{-}1)$$

2. *Elliptic integral of the second kind.*

$$E(k, x) = \int_0^t \left(\frac{1 - k^2 x^2}{(1 - x^2)} \right)^{1/2} \qquad (17.6\text{-}2)$$

3. *Elliptic integral of the third kind.*

$$\pi(k, \alpha, x) = \int_0^t \frac{dx}{(1 + \alpha^2 x^2)\left[(1 - x^2)(1 - k^2 x^2)\right]^{1/2}} \tag{17.6-3}$$

where $-\infty < \alpha^2 < \infty$, and integer $k^2 < 1$.

These integrals can be transformed by putting $x = \theta$, the replacing the upper limit of the integral, t, by $\phi = \sin^{-1} t$, thus resulting in functions

$$F(k, \phi) = \int_0^\phi \frac{d\phi}{(1 - k^2 \sin^2 \phi)^{1/2}} \tag{17.6-1'}$$

$$E(k, \phi) = \int_0^\phi (1 - k^2 \sin^2 \phi)^{1/2} d\phi \tag{17.6-2'}$$

$$\pi(k, \alpha^2, \phi) = \int_0^\phi \frac{d\phi}{(1 + \alpha^2 \sin^2 \phi)(1 - k^2 \sin^2 \phi)^{1/2}} \tag{17.6-3'}$$

Complete Elliptic Integrals. Putting $\phi = \pi/2$, these equations become complete elliptic integrals. Thus

$$F(k) \equiv F(k, \pi/2) = \pi/2 \left[1 + (1/2)^2 k^2 + (3/2 \cdot 4)^2 k^4 + \cdots \right] \tag{17.6-1''}$$

$$E(k) \equiv E(k, \pi/2) = \pi/2 \left[1 - (1/2)^2 k^2 - (3/2 \cdot 4)^2 k^4 - \cdots \right] \tag{17.6-2''}$$

$$\pi(k, \alpha^2) \equiv \pi(k, \alpha^2, \pi/2) \tag{17.6-3''}$$

17.6-4 Differential Equations

Ordinary Differential Equations

Definitions. The differential equations are equations containing derivatives of the *dependent variable* y with respect to the *independent variable* x. If D is the *differential operator* acting on a function y,

$$Dy = \frac{dy}{dx} \qquad D^2 y = \frac{d^2 y}{dx^2} \qquad D^n y = \frac{d^n}{dx^n}$$

An equation such as

$$P(D)y = f(x) \tag{17.6-4}$$

is called an *ordinary differential equation*. Here if P denotes a polynomial in D (see Section 17.1).

Example: $3\left(\dfrac{d^2 y}{dx^2}\right) + \dfrac{dy}{dx} = x^3$.

The highest derivative in a differential equation indicates the *order* of the differential equation. If, for example, D^n is the highest derivative, it is the *n*th *order differential equation*. The *degree* of a differential equation is the power of the highest derivative when the differential equation is rational and integral in terms of the derivatives involved.

If the power of the function y and of its derivatives is 1, the equation is called *linear*. If any one of these is of more power than 1, the equation is called *nonlinear*. Equation (17.6-4) is a *linear ordinary differential equation with constant coefficients*. The equation where $f(x) = 0$ is called the *homogeneous* equation.

$$P(D)y = 0 \tag{17.6-5}$$

Or, we can explicitly write

$$\left[a_1 D^n + a_2 D^{n-1} + \cdots + a_n D + a_{n+1}\right] y = 0$$

$$a_1 \frac{d^n y}{dx^n} + \cdots + a_n \frac{dy}{dx} + a_{n+1} = 0$$

Solution of Differential Equations. The general solution of (17.6-5) is called the *complementary solution*, y_c. If y_p is a *particular solution* of (17.6-4), then the *general solution* of (17.6-4) is

$$y = y_c + y_p$$

If P is a polynomial in D of degree n, it can be *factorized* (see Section 17.11) to obtain

$$\prod_{i=1}^{n}(D - a_i) = 0 = (D - a_1)(D - a_2) \cdots (D - a_n) \qquad (17.6\text{-}5')$$

Based on the a_i values, the following forms of the exponential functions correspond to the complementary solutions of (17.6-4).
For nonrepeated real linear factors, that is, $(D - a_1), (D - a_2), \ldots,$

$$c_1 e^{a_1 x} + \cdots + c_n e^{a_n x} \qquad (17.6\text{-}6)$$

where the c_i's are arbitrary constants. For repeated real linear factors, $(D - a)^k$, the k terms of the form

$$\left(c_1 + c_2 x + \cdots + c_k x^{k-1} \right) e^{ax}$$

are taken.
For a conjugate complex pair of factors, $a + ib, a - ib$,

$$c_1 e^{ax} \cos bx + c_2 e^{ax} \sin bx$$

And for the repeated ones,

$$\left[(c_1 + c_3 x + \cdots) \cos bx + (c_2 + c_4 x + \cdots) \sin bx \right] e^{ax}$$

The sum of all these terms form the complementary function.
The *particular solution* can be found by the *method of variation of parameters* which is explained below. The complementary solution is taken and the coefficients $c_1, c_2, \ldots$ are replaced by $c_1(x), c_2(x)$, functions of x. This becomes the particular solution with unknown coefficients:

$$y_p = c_1(x) e^{a_1 x} + c_2(x) e^{a_2 x} + \cdots + c_n(x) e^{a_n x} \qquad (17.6\text{-}7)$$

Then y_p is differentiated to obtain y_p',

$$dy_p/dx = c_1'(x) e^{a_1 x} + c_2'(x) e^{a_2 x} + \cdots$$
$$+ c_1(x) a_1 e^{a_1 x} + c_2(x) a_2 e^{a_2 x} + \cdots$$

Here the term with the first derivatives, $c_1'(x), c_2'(x), \ldots,$ are set to zero,

$$c_1'(x) e^{a_1 x} + c_2'(x) e^{a_2 x} + \cdots + c_n'(x) e^{a_n x} = 0 \qquad (17.6\text{-}8)$$

In subsequent derivatives the same condition is imposed until the nth derivative, $d^n y_p/dx^n$, where the terms with $c_1'(x), c_2'(x), \ldots$ are maintained.
Substituting the y_p and its derivatives in (17.6-4) and considering all the conditions of the type (17.6-8) imposed as equations, for n coefficients $c_i'(x)$, n equations are obtained from which $c_i'(x)$ can be solved. Integrating $c_i'(x)$,

$$c_i(x) = \int c_i'(x)\, dx$$

coefficients $c_i(x)$ are obtained.
The complete solution is thus determined as the sum of (17.6-6) and (17.6-7) as

$$y = y_c + y_p$$
$$= c_1 e^{a_1 x} + c_2 e^{a_2 x} + \cdots + c_n e^{a_n x}$$
$$+ c_1(x) e^{a_1 x} + c_2(x) e^{a_2 x} + \cdots + c_n(x) e^{a_n x}$$

Example:

$$\frac{d^2y}{dx^2} - 5\frac{dy}{dx} + 6y = x^2$$

is a differential equation of the form (17.6-4). The corresponding homogeneous equation can be written in terms of the nonrepeated linear factors as

$$(D - 3)(D - 2)y = 0$$

where $a_1 = 3$, $a_2 = 2$. Thus the *complementary solution* is

$$y_c = c_1 e^{3x} + c_2 e^{2x}$$

To find the *particular solution* we set the coefficients as functions of the independent variable x (variations of parameters) as

$$y_p = c_1(x)e^{3x} + c_2(x)e^{2x} \qquad\qquad (y_p)$$

After differentiating

$$y_p' = c_1' e^{3x} + c_2' e^{2x} + c_1 3 e^{3x} + c_2 2 e^{2x} + \cdots$$

We impose the condition

$$c_1' e^{3x} + c_2' e^{2x} = 0 \qquad\qquad\qquad (17.6\text{-}9)$$

Thus

$$y_p' = c_1 3 e^{3x} + c_2 2 e^{2x} \qquad\qquad (y_p')$$

Differentiating once again,

$$y_p'' = c_1' 3 e^{3x} + c_2' 2 e^{2x} + c_1 9 e^{3x} + c_2 4 e^{2x} \qquad\qquad (y_p'')$$

Substituting y_p, y_p', and y_p'' in the original differential equation gives

$$3c_1' e^{3x} + 2c_2' e^{2x} + 9c_1 e^{3x} + 4c_2 e^{2x} - 5(3c_1 e^{3x} + 2c_2 e^{2x}) + 6(c_1 e^{3x} + c_2 e^{2x}) = x^2$$

Eliminating the cancelling terms this relation reduces to

$$3c_1' e^{3x} + 2c_2' e^{2x} = x^2 \qquad\qquad\qquad (17.6\text{-}10)$$

The two equations (17.6-9) and (17.6-10) in c_1', c_2' are

$$c_1' e^{3x} + c_2' e^{2x} = 0$$

$$3c_1' e^{3x} + 2c_2' e^{2x} = x^2$$

From these equations c_1' and c_1' can be obtained as

$$c_1' e^{3x} = x^2$$

$$- c_2' e^{2x} = x^2$$

By integrating these equations,

$$dc_1 = x^2 e^{-3x}\, dx$$

$$c_1 = \int x^2 e^{-3x}\, dx$$

$$c_1 = -e^{-3x}\left[x^2/3 + 2x/9 + 2/27\right]$$

$$dc_2 = -x^2 e^{-2x}\,dx$$

$$c_2 = -\int x^2 e^{-2x}\,dx$$

$$c_2 = +e^{-2x}\left[x^2/2 + x/2 + 1/4\right]$$

Therefore the complete solution is

$$y = y_c + y_p$$

$$y = c_1 e^{3x} + c_2 e^{2x}$$

$$-\left[x^2/3 + 2x/9 + 2/27\right]$$

$$+\left[x^2/2 + x/2 + 1/4\right]$$

$$y = c_1 e^{3x} + c_2 e^{2x} + \left[x^2/6 + 5x/18 + 19/108\right]$$

Solution in Series by the Method of Frobenius. A homogeneous linear equation can be transformed into

$$x^n y^{(n)} + x^{n-1} y^{(n-1)} p_1(x) + \cdots + p_n(x) y = 0$$

The solution is assumed to be of the form

$$y = x^s \sum_{r=0}^{\infty} a_r x^r$$

By substituting it in the differential equations the numbers $s, a_0, a_1, \ldots$ are determined by equating coefficients of each power of x to zero. The solution is valid in the range of convergence of the obtained series. The same method can be extended to the solution about a given point x_0 (not origin) by replacing x by $x - x_0$ in the above series and the equation.

Example: $y'' - 5y' + 6y = 0$, multiplying by x^2,

$$x^2 y'' - 5x^2 y' + 6x^2 = 0$$

$$y = x^s \sum_{r=0}^{\infty} a_r x^r$$

$$y' = \sum_{r=0}^{\infty} a_r (r + s) x^{s+r-1}$$

$$y'' = \sum_{r=0}^{\infty} a_r (r + s)(r + s - 1) x^{s+r-2}$$

Substituting in the equation

$$\sum a_r (r + s)(r + s - 1) x^{s+r-2} - 5 \sum a_r (r + s) x^{s+r-1} + 6 \sum a_r x^r = 0$$

In a more explicit form,

$$a_0 s(s - 1) x^{s-2} - 5 a_0 s x^{s-1} + 6 a_0 x^s + a_1 (s + 1) s x^{s-1} - 5 a_1 (s + 1) x^s + 6 a_1 x^{s+1}$$

$$+ a_2 (s + 2)(s + 1) x^s - \cdots = 0$$

The coefficients of like powers of x must be equal to zero, that is,

$$a_0 s(s-1) = 0$$

$$-5a_0 s + a_1(s+1)s = 0$$

$$6a_0 - 5a_1(s+1) + a_2(s+2)(s+1) = 0$$

Solving the first equality, $s = 0$, $s = 1$, and $a_0 = 0$ are found. However, for $s = 1$, from the second equality,

$$a_1 = 5a_0/2$$

from the third equality,

$$a_2 = 19a_0/2$$

can be found. Therefore the solution of the differential equation becomes

$$y = a_0 \left[x + \tfrac{5}{2}x^2 + \tfrac{19}{6}x^3 + \cdots \right]$$

Series Solution in Descending Powers of x is valid for large values of the independent variable,

$$y = x^{-s} \sum_{r=0}^{\infty} a_r x^{-r}$$

Solution by Laplace Transformations is obtained by applying Laplace transforms, see Section 17.7, to linear differential equations. From the resulting Laplace transform of the variable y the function y can be obtained by inverse transforms.

Example: For an illustrative example see Section 17.7 on Laplace transformations.

As an example of *Perturbation Theory for Nonlinear Differential Equations* we use van der Pol's equation:

$$y'' + \varepsilon f(y, y') + y = F(t)$$

where $y' = dy/dt$ and ε is small.

$$y(t) = y_0(t) + \varepsilon y_1(t) + \varepsilon^2 y_2(t) + \cdots$$

is used as the perturbed solution about y_0. The coefficients of the powers of ε are equated to zero and the solution is obtained.

Partial Differential Equations

Definition.   In the above section there was only one dependent variable, y, and one independent variable, x. However, some differential equations might involve multiple *dependent variables*, $u_1, u_2, \ldots, u_m$, as well as multiple *independent variables*, $x_1, x_2, \ldots, x_r$. The derivatives of dependent variables with respect to one or more independent variables are called *partial derivatives*. For example, if $u_1(x_1, x_2)$ is given as a function of two independent variables, partial derivatives of u_1 with respect to x_1 alone are, $\partial u_1 / \partial x_1$, $\partial^2 u_1 / \partial x_1^2, \ldots$.

Any differential equation containing partial derivatives is called a *partial differential equation*. The *order* is determined by the highest partial derivative appearing in the equation. For example, if the highest partial derivative is $\partial^2 u / \partial x_1^2$, then the partial differential equation is said to be of the second order. Some well known examples are given below for illustration.

For a *Linear System of Equations*, in most general form,

$$\sum_{j=1}^{m} P_{ij}(D_1, \ldots, D_n) u_j = f_i(x_j) \qquad i = 1, 2, \ldots, r$$

where $D_i = \partial / \partial x_i$ and P_{ik} are polynomials in D_i with coefficients depending on x_j. Some well known examples are discussed briefly below.

Example: $r = 2$, $m = 3$.

$$P_{11}(D_1, D_2, D_3)u_1 + P_{12}(D_1, D_2, D_3)u_2 = f_1(x_1, x_2, x_3)$$

$$P_{21}(D_1, D_2, D_3)u_1 + P_{22}(D_1, D_2, D_3)u_2 = f_2(x_1, x_2, x_3)$$

given as

$$\left(\frac{\partial}{\partial x_1} - \frac{\partial}{\partial x_2} + \frac{\partial}{\partial x_3}\right)u_1 + \left(\frac{\partial}{\partial x_1} + \frac{\partial^2}{\partial x_1^2}\right)u_2 = x_1^2 - x_3$$

$$\frac{\partial}{\partial x_3}u_1 + \left[\frac{\partial}{\partial x_2} + \left(\frac{\partial}{\partial x_3}\right)^3\right]u_2 = x_2^3$$

However, in the case of *a linear system* of equations the dependent variables and all their derivatives are not higher than the first order. An example is

$$\left(\frac{\partial}{\partial x_1} - \frac{\partial}{\partial x_2} + \frac{\partial}{\partial x_3}\right)u_1 + \left(\frac{\partial}{\partial x_1} + \frac{\partial^2}{\partial x_1^2}\right)u_2 = x_1^2 - x_3$$

$$\frac{\partial}{\partial x_3}u_1 + \left(\frac{\partial}{\partial x_2} + \frac{\partial}{\partial x_3}\right)u_2 = x_2^3$$

Second-Order Equations.

$$A\frac{\partial^2 u}{dx^2} + 2B\frac{\partial^2 u}{dx\,dy} + C\frac{\partial^2 u}{dy^2} = f(x, y, \partial u/\partial x, \partial u/\partial y)$$

Boundary conditions:

Dirichlet: u is given at the boundary, for example, $u = 1$.

Neumann: $\partial u/\partial n$ is given at boundary, for example $\partial u/\partial n = 0$. Normal component of the gradient of u at each point of boundary.

Cauchy: u and $\partial u/\partial n$ at each point; $u(x_0, y_0) = 1$, $\partial u(x_0, y_0)/\partial n = 0$.

Types of equations:

Hyperbolic	$B^2 > AC$
Parabolic	$B^2 = AC$
Elliptic	$B^2 < AC$

Some examples of these three types of equation are given below.

Elliptic Equations

The Laplace Equation. In n dimensional space the Laplace equation (potential equation) is given as

$$\nabla^2 u = \sum_{i=1}^{n} u_{x_i x_i} = 0$$

Here u_y is the derivative of u with respect to y, and u_{yy} is the second derivative with respect to y. The notation ∇^2, called the *Laplacian*, represents the summation of second derivatives of u with respect to the variables, $x_1, x_2, \ldots, x_n$.

For example, for $n = 3$, the Laplace equation, $\nabla^2 u = 0$, becomes

$$\frac{d^2 u}{dx_1^2} + \frac{d^2 u}{dx_2^2} + \frac{d^2 u}{dx_3^2} = 0$$

The Laplace equation is obtained in formulating various physical problems such as heat flow problems, some problems in electrostatics, and so on. The solutions of this equation are called *harmonic* or *potential* functions. The *harmonic functions* are real functions $u(x_1, \dots)$ with continuous second derivative which satisfy the Laplace equation in a given region. For example, trigonometric functions are such functions for the one dimensional Laplace equation as illustrated in examples below.

The two special problems are the *Dirichlet problem* and the *Neumann problem*. The difference between them is the boundary conditions: for the first one, u is given at the boundary (i.e., temperature defined for the heat flow problem), and for the second one, u_x is given at the boundary (i.e., flow of heat at the boundary for the heat flow problem).

Poisson's Equation. While the Laplace equation is homogeneous in that $\nabla^2 u = 0$, the Poisson equation has a nonvanishing right-hand side as follows:

$$\nabla^2 u = f(x_1, \dots, x_n)$$

which is satisfied by

$$u = \int vf\, dx_1 \cdots dx_n$$

where v is the *fundamental solution*.

Wave Equation.

$$\nabla^2 u + \lambda u = 0$$

The *one dimensional wave equation* is

$$\frac{\partial^2 u}{\partial x^2} = \frac{1}{c^2} \frac{\partial^2 u}{\partial t^2}$$

where c is the speed of the wave. An example is the vibrating string. For solution of the wave equation see the section on the Methods of Separation of Variables below.

Example:

$$\frac{\partial^2 u}{\partial t^2} = c^2 \frac{\partial^2 u}{\partial x^2}$$

A solution is $u = \cos x \cos ct$.

Parabolic Equation of Second Order

The Heat Equation.

$$u_t = \nabla^2 u = u_{xx} + u_{yy} + u_{zz}$$

A simple example of the heat equation for a one dimensional problem, where the only independent variable is x, is discussed in the next section on the diffusion equation.

The Diffusion Equation.

$$u_t = ku_{xx} \quad \text{(one dimensional)}$$

Example: The temperature $u(x, t)$ of an isolated bar at x at time t satisfies the diffusion equation:

$$u_t = a^2 u_{xx}$$

The boundary conditions are

$$u_x(0, t) = u_x(l, t) = 0 \quad \text{at the insulated ends}$$

$$u(x, 0) = f(x) \quad \text{initial condition}$$

By separation of variables (see the Method of Separation of Variables),

$$u = X(x)T(t), \qquad T'/T = -p^2, \qquad a^2 X''/X = -p^2$$

$$T = e^{-p^2 t}$$

$$X = \cos(p/a)x \qquad X = \sin(p/a)x$$

However, with the given boundary conditions, $u(0, t) = 0$, $\cos(p/a)x$ is the only one satisfying the diffusion equation, and the boundary condition $u_x(l, t) = 0$ results in $X'(l) = 0$, which yields

$$p = n\pi a/l \quad \text{where } n \text{ an integer}$$

Therefore the solution is

$$T(t)X(x) = e^{-(n\pi a/l)^2 t} \cos(n\pi x/l)$$

Hyperbolic Equation in Two Independent Variables

$$Lu = au_{xx} + 2bu_{xy} + cu_{yy} + 2du_x + 2eu_y + fu = g$$

The coefficients $a, \ldots, g$ are known functions of x, y. The hyperbolicity condition is satisfied as $ac - b^2 < 0$.

Higher-Order Equations

Plate Equation.

$$\frac{\partial^4 u}{\partial x^4} + 2\frac{\partial^4 u}{\partial x^2 \partial y^2} + \frac{\partial^4 u}{\partial y^4} = 0$$

Solution of Partial Differential Equations

Equations with Constant Coefficients.   If the order of each term is the same, the specific solutions can be determined by setting

$$u(x, y) = f(p)$$

$$p = ax + by$$

and substituting in the equations. An example is the *wave equation*.

$$\frac{\partial^2 \psi}{\partial x^2} - \frac{1}{c^2}\frac{\partial^2 \psi}{\partial t^2} = \sin(x + t)$$

Let $p = ax + bt$, $\psi(x, t) = f(p)$

$$\frac{\partial \psi}{\partial x} = \frac{\partial f}{\partial p}a; \qquad \frac{\partial^2 \psi}{\partial x^2} = \frac{\partial^2 f}{\partial p^2}a^2$$

$$\frac{\partial \psi}{\partial t} = \frac{\partial f}{\partial p}b; \qquad \frac{\partial^2 \psi}{\partial t^2} = \frac{\partial^2 f}{\partial p^2}b^2$$

$$\frac{\partial^2 f}{\partial p^2}a^2 - \frac{b^2}{c^2}\frac{\partial^2 f}{\partial p^2} = \sin p$$

$$\frac{\partial^2 f}{\partial p^2}\left(a^2 - \frac{b^2}{c^2}\right) = \sin p$$

$$f = -\sin p\frac{c^2}{a^2 c^2 - b^2}$$

$$\psi(x, y) = -\frac{c^2}{a^2 c^2 - b^2}\sin(ax + bt)$$

Since $a = b = 1$,

$$\psi(x, y) = -\frac{c^2}{c^2 - 1} \sin(x + t)$$

However, in the case of the diffusion equation, the order of various terms is different, thus the above method is not applicable. Laplace transformation can be used.

Method of Separation of Variables

A solution is sought in terms of

$$u(x_1, x_2, \ldots, x_n) = X_1(x_1) X_2(x_2) \cdots X_n(x_n)$$

If a solution is of this form, it is said to be *separable* in $x_1, \ldots, x_n$. By substituting in the partial differential equation, for example, wave equation,

$$\frac{X_1''}{X_1} = -\mu_1^2, \ldots, \frac{X_n''}{X_n} = -\mu_n^2$$

where the μ_i's are *separation constants*. The solution is obtained as $X_1(x_1) + \exp(i\mu_1 x_1), \ldots, X_n(x_n) = \exp(i\mu_n x_n)$ and

$$u(x_1, \ldots, x_n) = \exp[i(\mu_1 x_1 + \cdots + \mu_n x_n)]$$

Example:

$$\frac{\partial^2 u}{\partial x^2} - \frac{1}{c^2} \frac{\partial^2 u}{\partial t^2} = 0$$

Let

$$u(x, t) = X(x)T(t)$$

and substitute in the equation to obtain

$$TX'' - \frac{1}{c^2} XT'' = 0$$

Denoting

$$\frac{X''}{X} = -\mu_1^2 \qquad \frac{T''}{T} = -\mu_2^2$$

$$X'' + \mu_1^2 X = 0 \qquad T'' + \mu_2^2 T = 0$$

Solving these equations for X and T,

$$X = e^{i\mu_1 x} \qquad T = e^{i\mu_2 t}$$

By the Euler equations, given in Section 17.3, e^{ix} are expressed in terms of trigonometric functions to obtain

$$X = \cos\mu_1 x + \sin\mu_1 x$$

$$T = \cos\mu_2 t + \sin\mu_2 t$$

Thus, substituting in the above relation,

$$u(x, t) = (\cos\mu_1 x + \sin\mu_1 x)(\cos\mu_2 t + \sin\mu_2 t)$$

Here $\mu_2^2 = c\mu_1^2$ can be obtained by substituting the above solution in the original equation, thus expressing μ_2 in terms of μ_1.

Numerical Solutions of Partial Differential Equations

See Finite Difference Methods in Section 17.11.

17.6-5   Integral Equations

Integral Equations of the Second Kind: Fredholm Equation

$$\phi(x) - \lambda \int_a^b K(x,t)\phi(t)\, dt = f(x) \tag{17.6-11}$$

where $K(x,t)$ and $f(x)$ are given functions, is an *integral equation of the second kind* for the unknown function $\phi(x)$. Here $K(x,t)$ is called the *kernel*, λ is the parameter.

Corresponding to this equation are two additional equations: the *homogeneous equation*, where

$$f(x) = 0 \tag{17.6-12}$$

and the *transposed integral equation* where $K(x,t)$ is replaced by $K(t,x)$,

$$\psi(x) - \lambda \int_a^b K(t,x)\psi(t)\, dt = f(x) \tag{17.6-13}$$

The number of linearly independent solutions of (17.6-12), $\phi^{(i)}(x)$, $i = 1, 2, \ldots, n$, is the same as for (17.6-13). The general solution of the homogeneous equation is

$$\sum_{i=1}^{n} c_i \phi^{(i)}(x) \qquad c_i = \text{constant}$$

If $n = 0$ and $f(x)$ is continuous, (17.6-11) and (17.6-13) have one solution:

$$\phi(x) = f(x) + \int_a^b L(x,t)f(t)\, dt$$

$$\psi(x) = f(x) + \int_a^b L(t,x)f(t)\, dt$$

where $L(x,t)$ is called the *resolvent kernel*.

If $n > 0$, and $\sum_a^b \psi^{(i)}(t)f(t)\, dt = 0$, $i = 1, 2, \ldots, n$, then (17.6-11) has a unique solution. The general solution of (17.6-11) is the sum of the particular solution of (17.6-11) and the general solution of (17.6-12).

There are various methods of solution of linear integral equations of the second kind. These methods result in different forms of solutions.

1.   The *Method of Successive Substitutions* due to Neumann, Liouville, and Volterra gives $\phi(x)$ as an integral series in λ, where the coefficients of various powers of λ are functions of x. In this case ϕ is replaced in the integrand and the process is continued successively.

2.   The *Method of Successive Approximations* takes $\phi_0(x)$ as a first approximation and substitutes it in the equation obtaining successive approximations as $\phi_1, \phi_2, \ldots$.

3.   The *Method of Frobenious* gives $\phi(x)$ as the ratio of two integral series in λ. The numerator only has coefficients of the various powers of λ as functions of x.

4.   The *Method of Hilbert and Schmidt* gives $\phi(x)$ in terms of a set of *fundamental* functions that are the solutions of the corresponding homogeneous equation for the characteristic values of λ. This is applicable to symmetric kernels only (see below for the definition of symmetric kernels).

Symmetric Kernel

If $K(x,t) = K(t,x)$, that is, *symmetric*, the integral equation has solutions for only those (real) values which are called *eigenvalues*. The solutions found are *eigenfunctions* belonging to $K(x,t)$.

The method of successive approximations can be used for solving this equation. The sequence of functions

$$q_1(x) = f(x)$$

$$q_n(x) = f(x) + \lambda \int_a^b K(x,t)q^{n-1}(t)\, dt \qquad n = 2, 3, \ldots$$

converges to the solution $\phi(x)$, if

$$\lambda^2 < \left[\iint K^2(s,t)\, ds\, dt \right]^{-1}$$

Degenerate Kernel

If

$$K(x,t) = \sum_{i=1}^{n} \phi_i(x)\psi_i(t)$$

then the kernel is called *degenerate*. The integral equation can be solved by solving a system of algebraic equations.

Example:

$$\phi(x) - \lambda \int_a^b xt\phi(t)\, dt = x$$

Integral Equation of the First Kind

In Equation (17.6-11) we replace the unknown function $\phi(x)$ outside the integral sign by zero, and the integral equation obtained,

$$f(x) = \int_a^b K(x,t)d(t)\, dt$$

is called the equation of the first kind. An example is in the theory of Laplace transforms, where the kernel is $K(x,t) = e^{-xt}$. For Fourier transforms the kernel becomes, $K(x,t) = e^{-ixt}$ (see Section 17.7).

Volterra Equation

$$\phi(x) - \lambda \int_0^x K(x,\xi)\phi(\xi)\, d\xi = f(x)$$

is equivalent to ordinary differential equations, and solution can be obtained by Laplace transforms.

Example: The differential equation is given as

$$d\phi/dx = \phi(x) + x \quad \text{with } \phi(0) = \phi_0$$

By integration,

$$\phi(x) = \int_0^x (\phi + t)\, dt + \phi_0$$

is a linear Volterra equation of the second kind.

Nonlinear Integral Equation

$$\phi(x) = f(x) + \int_0^x K[x,\xi; \phi(\xi)]\, d\xi$$

which can be solved by successive approximations.

Example: The differential equation is given as

$$d\phi/dx = \phi^2(x) + x \quad \text{with } \phi(0) = \phi_0$$

By integration,

$$\phi(x) = \int_0^x (\phi^2(t) + t)\, dt + \phi_0$$

is a nonlinear integral equation.

17.7   TRANSFORMS AND OPERATIONAL MATHEMATICS

For a given function $f(x)$ one can associate an integral transform $\phi(p)$ given by

$$\phi(p) = \int \Omega(x, p) f(x)\, dx$$

where the function $\Omega(x, p)$ is the operator. Depending on the operator and the limits of the integration there may be different types of transforms that are commonly used in applications. Some special transforms are listed below. Tables of transforms of some functions are given in Section 17.10.

17.7-1   Fourier Transform

If $f(x)$ is defined for $x > 0$, and piecewise continuous over any finite interval, and

$$\int_0^\infty f(x)\, dx \quad \text{is absolutely convergent,}$$

for the *Fourier cosine transform* of $f(x)$, $\Omega(x, p) = \sqrt{2/\pi} \cos px$:

$$\phi_c(p) = \sqrt{2/\pi} \int_0^\infty f(x)\cos(px)\, dx$$

for the *Fourier sine transform* of $f(x)$, $\Omega(x, p) = \sqrt{2/\pi} \sin px$:

$$\phi_s(p) = \sqrt{2/\pi} \int_0^\infty f(x)\sin(px)\, dx$$

See Section 17.10 for tables of Fourier cosine and Fourier sine transforms.

17.7-2   Laplace Transform

For *Laplace transforms*, $\Omega(x, s) = e^{-sx}$:

$$Lf(x) = \phi(s) = \int_0^\infty e^{-sx} f(x)\, dx$$

For the *inverse Laplace transform*,

$$f(x) = \frac{1}{2\pi i} \int_{a-i\infty}^{a+i\infty} e^{sx} \phi(s)\, ds$$

or

$$L^{-1} \frac{1}{(s - \beta)^r} = \frac{x^{r-1}}{(r - 1)!} e^{\beta x}$$

which is the sum of n such functions of x (see partial fractions in Section 17.1). Various functions and their transforms are given in a table form in Section 17.10.

Example: Let us take the given function as $f(x) = 1$.

$$\phi(s) = \Omega[f(x)] = \int_0^\infty e^{-sx} 1\, dx$$

$$= \frac{1}{-s} e^{-sx} \Big|_0^\infty = -\frac{1}{s} \left[\frac{1}{e^\infty} - \frac{1}{e^0} \right] = -\frac{1}{s} [0 - 1]$$

$$= \frac{1}{s}$$

See the tables in Section 17.11 for other Laplace transforms.

The inverse Laplace transform of $1/s$ can be found by using the formula above as illustrated below:

$$L^{-1}\frac{1}{(s-\beta)^r} = \frac{x^{r-1}}{(x-1)!}e^{\beta x} \quad \text{where } b = 0, \ r = 1$$

$$L^{-1}\frac{1}{s} = \frac{x^0}{(r-1)!}1 = \frac{1}{1}\cdot 1 = 1$$

17.7-3 Solution of Linear Differential Equations

Consider the differential equation given as

$$y'' + y = f(t) \quad \text{with initial conditions } y(0) = y'(0) = 0$$

This is the vibration equation with an external forcing function, $f(t)$, given as

$$f(t) = 0 \quad \text{for } t < 0$$

$$f(t) = 1 \quad \text{for } t > 0 \text{ and } = 0$$

Laplace transform of the elements of this equation are

$$L(y'') = s^2 L(y)$$

$$L(1) = 1/s$$

thus,

$$s^2 L(y) + L(y) = 1/s$$

$$L(y)[s^2 + 1] = 1/s$$

$$L(y) = \frac{1}{s(s^2+1)}$$

By partial fractions,

$$L(y) = \frac{1}{s} - \frac{s}{s^2+1}$$

By the inverse transforms (see the tables in Section 17.10),

$$y = x - \cos x \quad \text{for } x > 0$$

$$y = 0 \quad \text{for } x < 0$$

17.7-4 Solution of Integral Equations

The equation of the curve defined by a particle starting from rest and sliding down, frictionless under gravity, leads to *tautoschrone* which is a cycloid (see Section 17.4). The problem is formulated as

$$\int_0^y f(z)(y-z)^{-1/2}\, dx = C_0$$

where C_0 is a constant. Taking the Laplace transform

$$L(f)(-1/2)! s^{-1/2} = C_0 s^{-1}$$

results in

$$L(f) = C_1 s^{-1/2}$$

Here C_1 is also a constant. Thus

$$f(y) = Cy^{-1/2}$$

Denoting the arc along the curve by u, $du^2 = dx^2 + dy^2$, the corresponding differential equation becomes

$$f(y) = \frac{du}{dy} = \left[1 + \left(\frac{dx}{dy}\right)^2\right]^{1/2} = cy^{-1/2}$$

Setting $y = c^2 \sin^2(\phi/2)$, the parametric equations of a cycloid are obtained as follows:

$$x = \tfrac{1}{2}c^2(\phi + \sin\phi)$$

$$y = \tfrac{1}{2}c^2(1 - \cos\phi)$$

17.8   VECTOR ANALYSIS

17.8-1   Definitions

A *vector* is a quantity determined by its magnitude and direction, thus represented by a directed line segment. In application, quantities such as velocity and force are vectors:

$$\begin{aligned}
v &= |\mathbf{v}| & &\text{the magnitude of a vector } \mathbf{v} \\
\mathbf{v}/v & & &\text{unit vector} \\
-\mathbf{v} & & &\text{negative vector, in opposite direction to } \mathbf{v}
\end{aligned}$$

Any two intersecting vectors $\mathbf{V}_1$ and $\mathbf{V}_2$ define a *plane*. Any other vector on the same plane can be expressed as a linear combination of these two *coplanar* vectors:

$$\mathbf{V} = a\mathbf{V}_1 + b\mathbf{V}_2$$

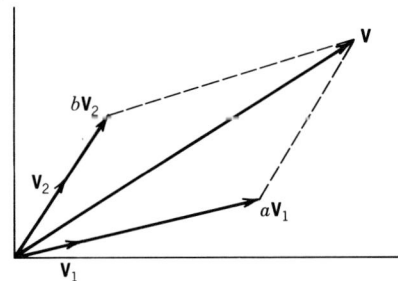

Here a and b are *scalar*. If $\mathbf{A}$ defines point A and $\mathbf{B}$ point B, any point P lying on the line segment AB is determined by its *position vector*,

$$\mathbf{V} = s\mathbf{A} + (1 - s)\mathbf{B}$$

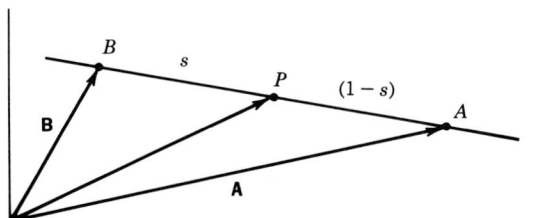

where the scalar s varies between 0, for $\mathbf{V} = \mathbf{B}$ (Point B), and 1, for $\mathbf{V} = \mathbf{A}$ (Point A).

Linear Dependence. n vectors are linearly dependent if scalars $s_1, \ldots, s_n$ exist such that

$$s_1\mathbf{V}_1 + s_2\mathbf{V}_2 + \cdots + s_n\mathbf{V}_n = 0$$

Any vector $\mathbf{V}$ linearly dependent on $\mathbf{V}_1 \cdots \mathbf{V}_n$ can be expressed as

$$\mathbf{V} = s_1\mathbf{V}_1 + s_2\mathbf{V}_2 + \cdots + s_n\mathbf{V}_n$$

If the $\mathbf{v}_1, \ldots, \mathbf{v}_n$ are unit vectors along the coordinate axes $x_1, \ldots, x_n$, any vector $\mathbf{A}$ in this n-dimensional space can be represented as

$$\mathbf{A} = a_1\mathbf{v}_1 + \cdots + a_n\mathbf{v}_n$$

where $a_1, \ldots, a_n$ are respective magnitudes of the projections of $\mathbf{v}$ on the coordinate axes. The magnitude of $\mathbf{V}$ is v:

$$v = \sqrt{a_1^2 + \cdots + a_n^2}$$

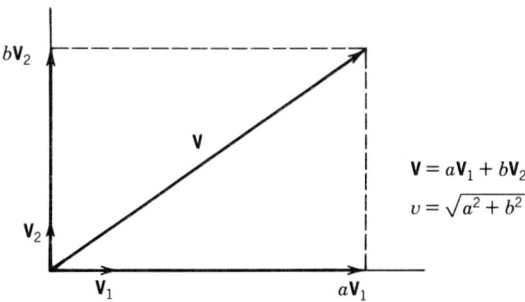

The *direction cosines* of $\mathbf{v}$ are

$$\cos a_1 = a_1/v, \cdots, \cos a_n = a_n/v$$

17.8-2   Vector Operations

The two vectors $\mathbf{A}$ and $\mathbf{B}$ and the angle ψ from $\mathbf{A}$ to $\mathbf{B}$ are given. The following relations are relevant to the vector operations.

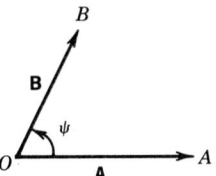

Addition of Vectors

$$\mathbf{A} + \mathbf{B} = (a_1 + b_1)\mathbf{v}_1 + \cdots + (a_n + b_n)\mathbf{v}_n$$

Commutative: $\mathbf{A} + \mathbf{B} = \mathbf{B} + \mathbf{A}$

Distributive: $(s_1 + s_2)\mathbf{A} = s_1\mathbf{A} + s_2\mathbf{A}$

$$s_1(\mathbf{A} + \mathbf{B}) = s_1\mathbf{A} + s_1\mathbf{A}$$

Associative: $\mathbf{A} + (\mathbf{B} + \mathbf{C}) = (\mathbf{A} + \mathbf{B}) + \mathbf{C} = \mathbf{A} + \mathbf{B} + \mathbf{C}$

Product of Vectors

Scalar (Dot or Inner) Product.

$$\mathbf{A} \cdot \mathbf{B} = ab \cos \psi$$

$$= a_1 b_1 + a_2 b_2 + \cdots + a_n b_n = \text{a scalar}$$

Commutative:
$$\mathbf{A} \cdot \mathbf{B} = \mathbf{B} \cdot \mathbf{A}$$

$$(\mathbf{A} + \mathbf{B}) \cdot \mathbf{C} = \mathbf{A} \cdot \mathbf{C} + \mathbf{B} \cdot \mathbf{C}$$

$$\mathbf{A} \cdot (\mathbf{B} + \mathbf{C}) = \mathbf{A} \cdot \mathbf{B} + \mathbf{A} \cdot \mathbf{C}$$

If $\mathbf{A}$ and $\mathbf{B}$ are perpendicular to each other, $\mathbf{A} \cdot \mathbf{B} = 0$. If $\mathbf{A}$ and $\mathbf{B}$ are parallel to each other, $\mathbf{A} \cdot \mathbf{B} = ab$

For unit vectors
$$\mathbf{v}_i \cdot \mathbf{v}_j = 0 \qquad i, j = 1, 2, \ldots, n$$

$$\mathbf{v}_i \cdot \mathbf{v}_i = 1 \qquad i = 1, 2, \ldots, n$$

Vector (Cross or Skew) Product.

$$\mathbf{A} \times \mathbf{B} = ab(\sin \psi)\mathbf{c}$$

where $\mathbf{c}$ is the unit vector perpendicular to the plane of $\mathbf{A}$ and $\mathbf{B}$ (the direction of $\mathbf{c}$ as shown).

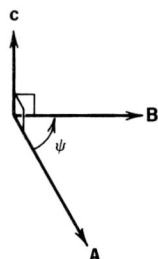

$$\begin{vmatrix} \mathbf{v}_1 & \mathbf{v}_2 & \mathbf{v}_3 \\ a_1 & a_2 & a_3 \\ b_1 & b_2 & b_3 \end{vmatrix} = \begin{matrix} (a_2 b_3 - a_3 b_2)\mathbf{v}_1 + (a_3 b_1 - a_1 b_3)\mathbf{v}_2 \\ + (a_1 b_2 - a_2 b_1)\mathbf{v}_3 \end{matrix}$$

$$\tan \psi = \frac{|\mathbf{A} \times \mathbf{B}|}{\mathbf{A} \times \mathbf{B}}$$

Noncommutative:
$$\mathbf{A} \times \mathbf{B} = -\mathbf{B} \times \mathbf{A}$$

Distributive:
$$\mathbf{A} \times (\mathbf{B} + \mathbf{C}) = \mathbf{A} \times \mathbf{B} + \mathbf{A} \times \mathbf{C}$$

$$(\mathbf{A} + \mathbf{B}) \times \mathbf{C} = \mathbf{A} \times \mathbf{C} + \mathbf{B} \times \mathbf{C}$$

For $\mathbf{v}_1, \mathbf{v}_2, \mathbf{v}_3$ orthogonal unit vectors,

$$\mathbf{v}_1 \times \mathbf{v}_2 = \mathbf{v}_3 \qquad \mathbf{v}_2 \times \mathbf{v}_3 = \mathbf{v}_1 \qquad \mathbf{v}_3 \times \mathbf{v}_1 = \mathbf{v}_2$$

$$\mathbf{v}_1 \times \mathbf{v}_1 = 0 \qquad \mathbf{v}_2 \times \mathbf{v}_2 = 0 \qquad \mathbf{v}_3 \times \mathbf{v}_3 = 0$$

Scalar Triple Product.

$$\mathbf{A} \cdot (\mathbf{B} \times \mathbf{C}) = (\mathbf{A} \times \mathbf{B}) \cdot \mathbf{C} = |ABC|$$

This corresponds to the volume of the parallelepiped with edges A, B, C.

Vector Triple Product.

$$\mathbf{A} \times (\mathbf{B} \times \mathbf{C}) = (\mathbf{A} \cdot \mathbf{B})\mathbf{B} - (\mathbf{A} \cdot \mathbf{B})\mathbf{C}$$

This results in a vector perpendicular to $\mathbf{A}$, lying in the plane of $\mathbf{B}$ and $\mathbf{C}$.

Differentiation of Vectors

$$A = a_1 v_1 + a_2 v_2 + a_3 v_3$$

$$B = b_1 v_1 + b_2 v_2 + b_3 v_3$$

and A and B are functions of a scalar, t,

$$\frac{dA}{dt} = \frac{da_1}{dt} v_1 + \frac{da_2}{dt} v_2 + \frac{da_3}{dt} v_3$$

$$\frac{d}{dt}(A + B) = \frac{dA}{dt} + \frac{dB}{dt}$$

$$\frac{d}{dt}(A \cdot B) = \frac{dA}{dt} B + A \frac{dB}{dt}$$

$$\frac{d}{dt}(A \times B) = \frac{dA}{dt} \times B + A \times \frac{dB}{dt}$$

$$A \cdot \frac{dA}{dt} = a \frac{da}{dt}$$

Triple Product.

$$\frac{d}{dt}ABC = \frac{dA}{dt}BC + A\frac{dB}{dt}C + AB\frac{dC}{dt}$$

$$\frac{d}{dt}A \times (B \times C) = \frac{dA}{dt} \times (B \times C) + A \times \left(\frac{dB}{dt} \times C\right) + A \times \left(B \times \frac{dC}{dt}\right)$$

Differential Operators

If F is a scalar function of x, y, z, then

$$dF = \frac{\partial F}{\partial x} dx + \frac{\partial F}{\partial y} dy + \frac{\partial F}{\partial z} dz$$

$$\nabla \equiv \text{del} \equiv v_1 \frac{\partial}{\partial x} + v_2 \frac{\partial}{\partial y} + v_3 \frac{\partial}{\partial z}$$

$$\nabla^2 \equiv \text{Laplacian} \equiv \frac{\partial^2}{\partial x^2} + \frac{\partial^2}{\partial y^2} + \frac{\partial^2}{\partial z^2}$$

The *gradient* is the direction and magnitude of the maximum rate of increase of F at any point:

$$\nabla F \equiv \text{grad } F \equiv \frac{\partial F}{\partial x} v_1 + \frac{\partial F}{\partial y} v_2 + \frac{\partial F}{\partial z} v_3$$

The Distributive law: $$\nabla(F + G) = \nabla F + \nabla G$$

The Associative law: $$\nabla(FG) = F\nabla G + G\nabla F$$

A is a vector function with A_1, A_2, A_3 as magnitudes of the components in coordinate axes.

The *divergence* of A is given by

$$\nabla \cdot A \equiv \text{div } A \equiv \frac{\partial A_1}{\partial x} + \frac{\partial A_2}{\partial y} + \frac{\partial A_3}{\partial z}$$

The Distributive law: $$\nabla \cdot (A + B) = \nabla \cdot A + \nabla \cdot B$$

$$\nabla \cdot (FA) = (\nabla F) \cdot A + F(\nabla \cdot A)$$

$$\nabla \cdot (A \times B) = B \cdot (\nabla \times A) - A(\nabla \times B)$$

The *Curl of* **A** is given by

$$\nabla \times \mathbf{A} \equiv \operatorname{curl} \mathbf{A} \equiv \left(\frac{\partial A_3}{\partial y} - \frac{\partial A_2}{\partial z} \right) \mathbf{v}_1 + \left(\frac{\partial A_1}{\partial z} - \frac{\partial A_3}{\partial x} \right) \mathbf{v}_2 + \left(\frac{\partial A_2}{\partial y} - \frac{\partial A_3}{\partial x} \right) \mathbf{v}_3$$

$$= \begin{vmatrix} \mathbf{v}_1 & \mathbf{v}_2 & \mathbf{v}_3 \\ \dfrac{\partial}{\partial x} & \dfrac{\partial}{\partial y} & \dfrac{\partial}{\partial z} \\ A_1 & A_2 & A_3 \end{vmatrix}$$

For **A** given as

$$\mathbf{A} = A_1 \mathbf{v}_1 + A_2 \mathbf{v}_2 + A_3 \mathbf{v}_3$$

$$\nabla \cdot \mathbf{A} = \nabla A_1 \cdot \mathbf{v}_1 + \nabla A_2 \cdot \mathbf{v}_2 + \nabla A_3 \cdot \mathbf{v}_3$$

$$\nabla \times \mathbf{A} = \nabla A_1 \times \mathbf{v}_1 + \nabla A_2 \times \mathbf{v}_2 + \nabla A_3 \times \mathbf{v}_3$$

$$\operatorname{div} \operatorname{grad} F = \nabla \cdot (\nabla F) \equiv \text{Laplacian } F \equiv \nabla^2 F$$

$$\operatorname{curl} \operatorname{grad} F \equiv 0$$

$$\operatorname{div} \operatorname{curl} \mathbf{A} \equiv 0$$

Integration of Vectors and Integrals

F = a vector function
V = volume
S = a surface bounding a closed region with volume V
C = a curve, closed and bounds S
n = the unit vector normal to S outward at point A
r = the unit vector tangent to C at point A

Green's (Gauss') Theorem.

$$\iiint_{(v)} (\nabla \cdot \mathbf{F}) \, dV = \iint_{(s)} (\mathbf{F} \cdot \mathbf{n}) \, dS$$

Stokes' Theorem.

$$\iint_{(s)} \mathbf{n} \cdot (\nabla \times \mathbf{F}) \, dS = \int_{(c)} \mathbf{F} \cdot d\mathbf{r}$$

17.9 STATISTICS AND PROBABILITY

17.9-1 Statistics

Definitions

Statistical data consist of N values of variables x_i, called *class marks*. One of these classes, x_0 is taken to be the *origin*. The *intervals*, $x_{i+1} - x_i = c$, are called *class intervals*. The *frequency distribution* of the data is a tabulation by classes where the *frequencies* or *weights*, f_i, of a class x_i are given.

Since the total number of observations, N, must be equal to the sum of frequencies of various classes, $N = \Sigma f_i$.

Example: Ten students ($N = 10$) receive five different grades, $n = 5$, as follows:

3	grades are	50
1	grade is	70
3	grades are	80
2	grades are	90
1	grade is	100

Classes: $x_1 = 50, \; x_2 = 70, \; x_3 = 80, \; x_4 = 90, \; x_5 = 100$

Frequencies: $f_1 = 3, \; f_2 = 1, \; f_3 = 3, \; f_4 = 2, \; f_2 = 1$

Mean (Arithmetic Mean).

$$\bar{x} = \frac{1}{n} \sum_{i=1}^{n} x_i = \frac{x_1 + \cdots + x_n}{n}$$

For the above example, the mean becomes

$$\bar{x} = \frac{50 + 70 + 80 + 90 + 100}{5} = 78$$

Weighted Mean. If each value x_i has an associated weight, $\omega_i > 0$, then $\sum_{i=1}^{n}\omega_i$ is the total weight and

$$\bar{x} = \frac{\displaystyle\sum_{i=1}^{n} \omega_i x_i}{\displaystyle\sum_{i=1}^{n} \omega_i} = \frac{\omega_1 x_1 + \cdots + \omega_n x_n}{\omega_1 + \cdots + \omega_n}$$

For the above example, $\omega_i = f_i$, thus $\sum \omega_i = 3 + 1 + 3 + 2 + 1 = 10$ and the weighted mean becomes

$$\bar{x} = \frac{50.3 + 70.1 + 80.3 + 90.2 + 100.1}{10} = 64$$

Geometric Mean.

$$\bar{x}_G = \sqrt[n]{x_1 \cdots x_n}$$

Geometric Mean in Logarithmic Form.

$$\log \bar{x}_G = \frac{1}{n} \sum \log x_i = \frac{\log x_i + \cdots + \log x_n}{n}$$

Harmonic Mean.

$$\bar{x}_H = \frac{n}{\displaystyle\sum_{i=1}^{n} (1/x_i)} = \frac{n}{(1/x_1) + \cdots + (1/x_n)}$$

Here $\bar{x}_H < \bar{x}_G < \bar{x}$. If sample values are identical $\bar{x}_H = \bar{x}_G = \bar{x}$.

Median. If the sample is arranged in an ascending order of values, then the median, M_d, of n such values is the value at the $[(n + 1)/2]$th position of the sequence:

$$M_d = \frac{n + 1}{2}$$

that is, when n is odd, the middle value (if even)—the mean of the two middle values of the above ordered set of values—is the median.

For the above example the classes are 50, 70, 80, 90, 100. Therefore the median is 80, that is, two grades 50 and 70 are less than the median and two grades 90 and 100 are more than the median, thus equal number of grades are below and above the median value, 80.

Quartiles. For an ascending order, $i(n + 1)/4$, where $i = 1, 2, 3$ corresponds to the first, second, and third quartiles, respectively.

Deciles. They are every $[(n + 1)/10]$th value.

Percentile. They are every $[(n + 1)/100]$th value.

The sample example is shown below with various notions identified for illustration.

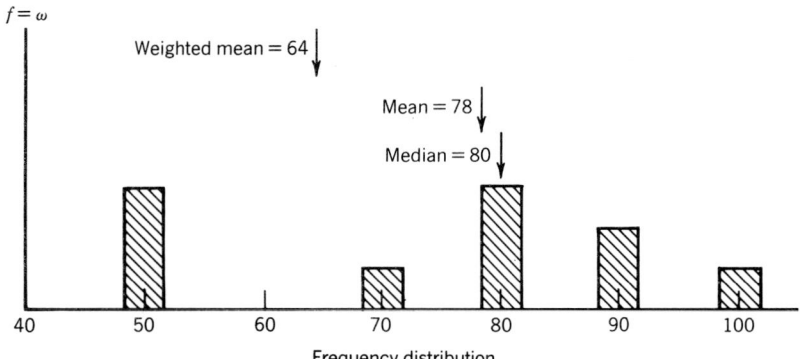

Frequency distribution

The example given above is a *discrete* distribution. There are various types of distributions used in statistical analysis. One of the important frequency distributions is the *normal distribution* which is a *continuous* one in that class values continuously vary as shown in the figure below. The normal distribution is symmetric and bell shaped.

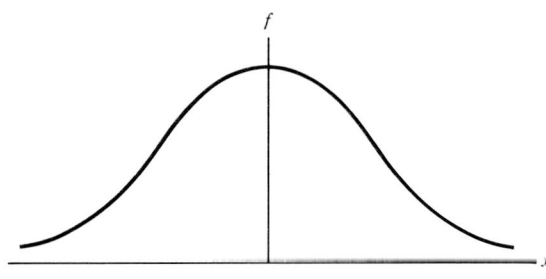

Normal distribution

Mean Deviation

$$\frac{1}{n} \sum_{i=1}^{n} |x_i - \bar{x}|$$

Standard Deviation

$$s = \sqrt{\frac{\sum_{i=1}^{n} (x_i - \bar{x})^2}{n - 1}}$$

Variance

$$V = s^2$$

Standardized variable: $z = \dfrac{x_i - \bar{x}}{s}$

Moments

The rth moment about the *origin* (see above definitions) is given by

$$m'_r = \frac{1}{n} \sum_{i=1}^{n} x_i^r$$

The rth moment about the *mean* $\bar{x}$ (see above definitions) is given by

$$m_r = \frac{1}{n} \sum_{i=1}^{n} (x_i - \bar{x})r$$

Curve Fitting

If (x_i, y_i) form a set of ordered pairs where x_i are fixed variables and y_i are dependent variables, a curve can be fitted to find $y = f(x)$ in terms of a *polynomial function*, $y = b_0 + b_1 x + \cdots + b_n x_n$. Here, by the method of least squares, the coefficients $b_0, \ldots, b_n$ can be calculated.

For the special case of $n = 1$ the polynomial becomes a line equation. Thus the curve fitted to the data is a *straight line*.

Similarly, an *exponential curve*, $y = ab^x$, can also be fitted to given data. Here $\log y = \log a + x \log b$ becomes the straight line and $(x_i, \log y_i)$ are fitted by the method of mean squares.

A *power function*, $y = ax^b$, can also be fitted to data by taking logarithms, $\log y = \log a + b \log x$. Thus the ordered pair is $(\log x_i, \log y_i)$, and the fit is a straight line fit.

Regression

In a *regression* problem *one particular variable* is studied in terms of the *remaining variables*. For example, if the two variables are x and y, y can be chosen as the particular variable. For the *linear regression* of y on x, we have

$$E(y/x) = b_0 + b_1 x$$

where $E(y/x)$ is the mean of the distribution of y for a given x.

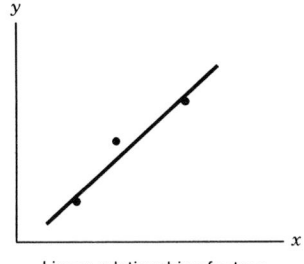

Linear relationship of y to x

After fitting a regression line to a given set of points, the accuracy of predicting y values can be determined. For this the standard error estimate is defined.

The *standard error estimate* is given by

$$s_e = \sqrt{\frac{\sum [y_i - (b_0 + b_1 x_i)]^2}{n - 2}}$$

where b_0 and b_1 are calculated by straight line fitting using the least squares method.

In the case of multiple x's the techniques for predicting the y variable are similar to the above.

Correlation

In a *correlation* problem, however, *several variables* are studied *simultaneously* to show their interrelation. For example, for the case of two variables, x and y, an estimate of the population *correlation coefficient* is given by and defined the desired *measure of relationship*.

$$r = \sqrt{\frac{\sum (x_i - \bar{x})(y_i - \bar{y})}{\left[\sum (x_i - x)^2\right]\left[\sum (y_i - y)^2\right]}}$$

Scatter diagram of
two variables used in correlation

17.9-2   Probability

Definitions

In an experiment, any outcome corresponds to one *element* (an *event*) E, while the set of all such events forms the *sample space*, S. Therefore, the *probability* of an event is

$$P(E) = m/n$$

where n is the number of mutually exclusive and likely *outcomes* of an experiment and m is the number of those corresponding to the event E.

Example: Let us experiment by throwing a dice with six sides, numbered $1, 2, 3, 4, 5, 6$. The probability of any side being an *outcome* is equal to $1/6$. Here $m = 1$ and $n = 6$. If the even E represents any one of the six sides being an outcome,

$$P(E) = m/n = 1/6$$

A *set* is defined as a collection of *elements*. The elements are objects with particular characteristics, for example, a set of integers, a set of square areas, and so on. The following notations are frequently used in operations with sets.

$S = \{s_1, s_2, \ldots, s_n\}$ is a set.
$s_1 \in S$ defines s_1 as an element of S.
$S_1 \subset S_2$ indicates that all the elements of S_1 are also elements of S_2; thus S_1 is a *subset* of S_2.
If $S_1 \subset S_2$ and $S_2 \subset S_1$, then $S_1 = S_2$.
If $S_2 \subset S_3$ but at least one element in S_3 is not in S_2, then S_2 is a *proper subset* of S_3.
$S_1 \cap S_2$ is the *intersection* of two sets, S_1 and S_2 such that the elements in this intersection are elements of both S_1 and S_2.
$S_1 \cup S_2$ is the union of two sets, S_1 and S_2 such that the elements in this union are elements of either S_1 or S_2.

$E \cup F$ $E \cap F$

Theorems

1. If $\varnothing$ is the *null set*, $P(\varnothing) = 0$.
2. If S is the *sample set*, $P(S) = 1$.
3. If E and F are two events,

$$P(E \cup F) = P(E) + P(F) - P(E \cap F)$$

4. If E and F are *mutually exclusive* events,

$$P(E \cup F) = P(E) + P(F)$$

5. If E and E' are *complementary* events,

$$P(E) = 1 - P(E')$$

6. The *conditional* probability of an event E given an event F, where conditional means that the event E occurs only if the event F occurs, is defined as

$$P(E/F) = \frac{P(E \cap F)}{P(F)}$$

where $P(F) \neq 0$.

7. Two events E and F are *independent* if and only if

$$P(E \cap F) = P(E)P(F)$$

E is *statistically independent* of F if

$$P(E/F) = P(E) \quad \text{and} \quad P(F/E) = P(F)$$

Bayes' Theorem. If $E_1, \ldots, E_n$ are mutually exclusive events and the sample space is the union of these events, for $P(E) \neq 0$,

$$P(E_j/E) = \frac{P(E_j) P(E/E_j)}{\sum\limits_{i=1}^{n} [P(E_i) P(E/E_i)]}$$

Random Variables

A function between the sample space, S (domain), and the set of real numbers, x (range), is called a *random variable*, X. A random variable can be *discrete* if only a finite number of values on the real axis is assumed. It is *continuous* if a continuum of values on the real axis is represented.

Discrete Case: Probability Function. The function $f = P[X = x]$ with properties

$$f(x_i) \geq 0, \quad \sum_i f(x_i) = 1 \quad i = 1, 2, \ldots$$

and for every event E

$$P(E) = P[X \text{ is in } E] = \sum_E f(x)$$

where the summation implies for the values x that are in E, is called the *probability function* of the discrete random variable X.

Discrete Case: Cumulative Distribution Function. The probability of the value of a random variable X less than or equal to some real number x, is defined as the *cumulative distribution function*:

$$F(x) = P(X \leq x) = \sum f(x_i)$$

for $-\infty < x < \infty$ and $x_i \leq x$.

Continuous Case: Probability Density. The function $f(x)$ with properties $f(x) \geq 0$, and

$$\int_{-\infty}^{\infty} f(x) \, dx = 1 \quad \text{for all } x, \ -\infty < x < \infty$$

also for any event E

$$P(E) = P(X \text{ is in } E) = \int_E f(x) \, dx$$

is called the *probability density* of the continuous random variable X.

Continuous Case: Cumulative Distribution Function. The probability of the value of a random variable X less than or equal to some real number x is defined as the *cumulative distribution function*:

$$F(x) = P(X \le x) = \int_{-\infty}^{x} f(x)\,dx \qquad \text{for } -\infty < x < \infty$$

From this the *density* can be found as

$$f(x) = \frac{dF(x)}{dx}$$

and

$$P(a \le X \le b) = P(X \le b) - P(X \le a)$$
$$= F(b) - F(a)$$

Expected Value

$$E(X) = \sum_{x} xf(x) \quad \text{(discrete case)}$$

$$= \int_{-\infty}^{\infty} xf(x)\,dx \quad \text{(continuous case)}$$

Theorems:

1. $E[aX + bY] = aE(X) + bE(Y)$.
2. $E[XY] = E(X)E(Y)$, if X and Y are statistically independent.

Moments

1. The rth moment about the origin.

$$\mu_r = E(X^r) = \sum_{x} x^r f(x) \quad \text{(discrete case)}$$

$$= \int_{-\infty}^{\infty} x^r f(x)\,dx \quad \text{(continuous case)}$$

2. The first moment (mean).

$$\mu = \sum_{x} x^r f(x) \quad \text{(discrete case)}$$

$$= \int_{-\infty}^{\infty} xf(x)\,dx \quad \text{(continuous case)}$$

3. The rth moment about the mean.

$$\mu_r = E[(X - \mu)^r] = \sum_{x} (x - \mu)^r f(x) \quad \text{(discrete case)}$$

$$= \int_{-\infty}^{\infty} (x - \mu)^r f(x)\,dx \quad \text{(continuous case)}$$

4. The second moment about the mean.

$$\mu_2 = E\left[(x - \mu)^2\right] = \mu_2' - \mu^2$$

This is called the *variance* and denoted by σ^2. Here σ is the *standard deviation*.

Generating Function

1. **Moment generating function.**

$$m_x(t) = E(e^{iX}) = \sum_x e^{ixf}(x) \quad \text{(discrete)}$$

$$= \int_{-\infty}^{\infty} e^{ixf}(x)\, dx \quad \text{(continuous)}$$

2. **Fractional moment generating function.**

$$E(t^X) = \sum_x t^x f(x) \quad \text{(discrete)}$$

$$= \int_{-\infty}^{\infty} t^x f(x)\, dx \quad \text{(continuous)}$$

Characteristic Function

$$d(t) = E(e^{itX}) = \sum_x e^{itx} f(x) \quad \text{(discrete)}$$

$$= \int_{-\infty}^{\infty} e^{itx} f(x)\, dx \quad \text{(continuous)}$$

Probability Distributions: Discrete Case

The probability distribution possessed by the discrete random variable X depends on the probability function. These functions and corresponding distributions are listed below.

1. **Discrete uniform distribution of X.**

$$P(X = x) = f(x) = 1/n \qquad x = x_1, x_2, \ldots, x_n$$

2. **Binomial distribution of X** (see Section 17.1).

$$P(X = x) = f(x) = \binom{n}{x}\theta^x(1-\theta)^{n-x} \qquad x = 0,1,2,\ldots,n$$

where $f(x)$ is the general term of the expansion of $[\theta + (1 - \theta)]^n$.

3. **Geometric distribution of X.**

$$P(X = x) = f(x) = \theta^x(1-\theta)^{x-1} \qquad x = 0,1,2,\ldots,n$$

4. **Multinomial distribution of X.**

$$P(X_1 = x_1, \ldots, X_n = x_n) = f(x_1, \ldots, x_n) = \frac{N!}{\prod_{i=1}^{n} x_i!} \prod_{i=1}^{n} \theta_i x_i$$

5. **Poisson distribution of X.**

$$P(X = x) = f(x) = \frac{e^{-\lambda}\lambda^x}{x!} \qquad \lambda > 0, \quad x = 0,1,\ldots$$

6. **Hypergeometric distribution of X.**

$$P(X = x) = f(x) = \frac{\binom{k}{x}\binom{N-k}{n-x}}{\binom{N}{n}} \qquad x = 0,1,\ldots,[n,k]\ (\text{smaller of } n \text{ or } k)$$

Probability Distributions: Continuous Case

The probability distribution possessed by the continuous random variable X depends on the density function. These functions and corresponding distributions are listed below.

   1. Uniform distribution of X.

$$f(x) = \frac{1}{\beta - \alpha} \qquad \lambda < x < \beta$$

where α and β are parameters.
   2. Normal distribution of X.

$$f(x) = \left(1/\sigma\sqrt{2}\,\pi\right)e^{-(x-\mu)^2/2\sigma^2} \qquad -\infty < x < \infty$$

where μ, the mean, and σ, the standard deviation, are parameters.
   3. Gamma distribution of X.

$$f(x) = \frac{1}{\Gamma(\alpha + 1)\beta^{\alpha+1}}x^\alpha e^{-x/\beta}$$

where α and β are parameters, $\alpha > -1$, $\beta > 0$.
   4. Exponential distribution of X.

$$f(x) = (1/\theta)e^{-x/\theta}, \qquad x > 0, \quad \theta > 0$$

   5. Beta distribution of X.

$$f(x) - \frac{\Gamma(\alpha + \beta + 2)}{\Gamma(\alpha + 1)\Gamma(\beta + 1)}x^\alpha(1 - x)^\beta$$

17.10   MATHEMATICAL TABLES

17.10-1   Series Expansion of Common Functions

See Section 17.1.

Binomial Functions

$$(x + y)^n = x^n + nx^{n-1}y + \frac{n(n-1)}{2!}x^{n-2}y^2 + \frac{n(n-1)(n-2)}{3!}x^{n-3}y^3 + \cdots \qquad (y^2 < x^2)$$

$$(1 \pm x)^n = 1 \pm nx + \frac{n(n-1)x^2}{2!} \pm \frac{n(n-1)(n-2)x^3}{3!} + \cdots \qquad (x^2 < 1)$$

$$(1 \pm x)^{-n} = 1 \mp nx + \frac{n(n+1)x^2}{2!} \mp \frac{n(n+1)(n+2)x^3}{3!} + \cdots \qquad (x^2 < 1)$$

$$(1 \pm x)^{-1} = 1 \mp x + x^2 \mp x^3 + x^4 \mp x^5 + \cdots \qquad (x^2 < 1)$$

$$(1 \pm x)^{-2} = 1 \mp 2x + 3x^2 \mp 4x^3 + 5x^4 \mp 6x^5 + \cdots \qquad (x^2 < 1)$$

Exponential Functions

$$e = 1 + \frac{1}{1!} + \frac{1}{2!} + \frac{1}{3!} + \frac{1}{4!} + \cdots$$

$$e^x = 1 + x + \frac{x^2}{2!} + \frac{x^3}{3!} + \frac{x^4}{4!} + \cdots \qquad \text{(all real values of } x)$$

$$a^x = 1 + x\log_e a + \frac{(x\log_e a)^2}{2!} + \frac{(x\log_e a)^3}{3!} + \cdots$$

$$e^x = e^a\left[1 + (x - a) + \frac{(x - a)^2}{2!} + \frac{(x - a)^3}{3!} + \cdots\right]$$

Logarithmic Functions

$$\log_e x = \frac{x-1}{x} + \frac{1}{2}\left(\frac{x-1}{x}\right)^2 + \frac{1}{3}\left(\frac{x-1}{x}\right)^3 + \cdots \qquad \left(x > \tfrac{1}{2}\right)$$

$$\log_e x = (x - 1) - \tfrac{1}{2}(x - 1)^2 + \tfrac{1}{3}(x - 1)^3 - \cdots \qquad (2 \geq x > 0)$$

$$\log_e x = 2\left[\frac{x-1}{x+1} + \frac{1}{3}\left(\frac{x-1}{x+1}\right)^3 + \frac{1}{5}\left(\frac{x-1}{x+1}\right)^5 + \cdots\right] \qquad (x > 0)$$

$$\log_e(1 + x) = x - \tfrac{1}{2}x^2 + \tfrac{1}{3}x^3 - \tfrac{1}{4}x^4 + \cdots \qquad (-1 < x < 1)$$

$$\log_e(n + 1) - \log_e(n - 1) = 2\left[\frac{1}{n} + \frac{1}{3n^3} + \frac{1}{5n^5} + \cdots\right]$$

$$\log_e(a + x) = \log_e a + 2\left[\frac{x}{2a + x} + \frac{1}{3}\left(\frac{x}{2a + x}\right)^3 + \frac{1}{5}\left(\frac{x}{2a + x}\right)^5 + \cdots\right]$$

$$(a > 0, \, -a < x < +\infty)$$

$$\log_e\frac{1 + x}{1 - x} = 2\left[x + \frac{x^3}{3} + \frac{x^5}{5} + \cdots + \frac{x^{2n-1}}{2n - 1} + \cdots\right] \qquad -1 < x < 1$$

$$\log_e x = \log_e a + \frac{(x - a)}{a} - \frac{(x - a)^2}{2a^2} + \frac{(x - a)^3}{3a^3} - + \cdots \qquad 0 < x \leq 2a$$

Trigonometric Functions

$$\sin x = x - \frac{x^3}{3!} + \frac{x^5}{5!} - \frac{x^7}{7!} + \cdots \qquad \text{(all real values of } x)$$

$$\cos x = 1 - \frac{x^2}{2!} + \frac{x^4}{4!} - \frac{x^6}{6!} + \cdots \qquad \text{(all real values of } x)$$

$$\tan x = x + \frac{x^3}{3} + \frac{2x^5}{15} + \frac{17x^7}{315} + \frac{62x^9}{2835} + \cdots + \frac{2^{2n}(2^{2n} - 1)B_n}{(2n)!}x^{2n-1} + \cdots$$

$$\left[x^2 < \pi^2/4, \text{ and } B_n \text{ represents the } n\text{th Bernoulli number}\right]$$

$$\cot x = \frac{1}{x} - \frac{x}{3} - \frac{x^2}{45} - \frac{2x^5}{945} - \frac{x^7}{4725} - \cdots - \frac{2^{2n}B_n}{(2n)!}x^{2n-1} - \cdots$$

$$\left[x^2 < \pi^2, \text{ and } B_n \text{ represents the } n\text{th Bernoulli number}\right]$$

$$\sin^{-1}x = x + \frac{x^3}{2 \cdot 3} + \frac{1 \cdot 3}{2 \cdot 4 \cdot 5}x^5 + \frac{1 \cdot 3 \cdot 5}{2 \cdot 4 \cdot 6 \cdot 7}x^7 + \cdots \qquad \left(x^2 < 1, \, -\frac{\pi}{2} < \sin^{-1}x < \frac{\pi}{2}\right)$$

$$\cos^{-1}x = \frac{\pi}{2} - \left(x + \frac{x^3}{2\cdot 3} + \frac{1\cdot 3}{2\cdot 4\cdot 5}x^5 + \frac{1\cdot 3\cdot 5x^7}{2\cdot 4\cdot 6\cdot 7} + \cdots\right) \qquad (x^2 < 1, 0 < \cos^{-1}x < \pi)$$

$$\tan^{-1}x = x - \frac{x^3}{3} + \frac{x^5}{5} - \frac{x^7}{7} + \cdots \qquad (x^2 < 1)$$

$$\tan^{-1}x = \frac{\pi}{2} - \frac{1}{x} + \frac{1}{3x^2} - \frac{1}{5x^5} + \frac{1}{7x^7} - \cdots \qquad (x > 1)$$

$$\tan^{-1}x = -\frac{\pi}{2} - \frac{1}{x} + \frac{1}{3x^2} - \frac{1}{5x^5} + \frac{1}{7x^7} - \cdots \qquad (x < -1)$$

$$\cot^{-1}x = \frac{\pi}{2} - x + \frac{x^3}{3} - \frac{x^5}{5} + \frac{x^7}{7} - \cdots \qquad (x^2 < 1)$$

$$\log_e \sin x = \log_e x - \frac{x^2}{6} - \frac{x^4}{180} - \frac{x^6}{2835} - \cdots \qquad (x^2 < \pi^2)$$

$$\log_e \cos x = -\frac{x^2}{2} - \frac{x^4}{12} - \frac{x^6}{45} - \frac{17x^8}{2520} - \cdots \qquad \left(x^2 < \frac{\pi^2}{4}\right)$$

$$\log_e \tan x = \log_e x + \frac{x^2}{3} + \frac{7x^4}{90} + \frac{62x^6}{2835} + \cdots \qquad \left(x^2 < \frac{\pi^2}{4}\right)$$

$$e^{\sin x} = 1 + x + \frac{x^2}{2!} - \frac{3x^4}{4!} - \frac{8x^5}{5!} - \frac{3x^6}{6!} + \frac{56x^7}{7!} + \cdots$$

$$e^{\cos x} = e\left(1 - \frac{x^2}{2!} + \frac{4x^4}{4!} - \frac{31x^6}{6!} + \cdots\right)$$

$$e^{\tan x} = 1 + x + \frac{x^2}{2!} + \frac{3x^3}{3!} + \frac{9x^4}{4!} + \frac{37x^5}{5!} + \cdots \qquad \left(x^2 < \frac{\pi^2}{4}\right)$$

$$\sin x = \sin a + (x - a)\cos a - \frac{(x-a)^2}{2!}\sin a - \frac{(x-a)^3}{3!}\cos a + \frac{(x-a)^4}{4!}\sin a + \cdots$$

Hyperbolic and Inverse Functions

$$\sinh x = x + \frac{x^3}{3!} + \frac{x^5}{5!} + \frac{x^7}{7!} + \cdots + \frac{x^{2n+1}}{(2n+1)!} + \cdots \qquad |x| < \infty$$

$$\cosh x = 1 + \frac{x^2}{2!} + \frac{x^4}{4!} + \frac{x^6}{6!} + \cdots + \frac{x^{2n}}{(2n)!} + \cdots \qquad |x| < \infty$$

$$\tanh x = x - \frac{1}{3}x^3 + \frac{2}{15}x^5 - \frac{17}{315}x^7 - \frac{62}{2835}x^9 - \cdots$$

$$+ \frac{(-1)^{n+1}2^{2n}(2^{2n}-1)}{(2n)!}B_n x^{2n-1} \pm \cdots \qquad |x| < \frac{\pi}{2}$$

$$\coth x = \frac{1}{x} + \frac{x}{3} - \frac{x^3}{45} + \frac{2x^5}{945} - \frac{x^7}{4725} + \cdots + \frac{(-1)^{n+1}2^{2n}}{(2n)!}B_n x^{2n-1} \pm \cdots \qquad 0 < |x| < \pi$$

$$\operatorname{sech} x = 1 - \frac{1}{2!}x^2 + \frac{5}{4!}x^4 - \frac{61}{6!}x^6 + \frac{1385}{8!}x^8 - \cdots + \frac{(-1)^n}{(2n)!}E_n x^{2n} \pm \cdots \qquad |x| < \frac{\pi}{2}$$

$$\operatorname{cosech} x = \frac{1}{x} - \frac{x}{6} + \frac{7x^3}{360} - \frac{31x^5}{15,120} + \cdots + \frac{2(-1)^n(2^{2n-1}-1)}{(2n)!}B_n x^{2n-1} + \cdots \qquad 0 < |x| < \pi$$

Additional Relations for Fourier Series

$$1 = \frac{4}{\pi}\left[\sin\frac{\pi x}{k} + \frac{1}{3}\sin\frac{3\pi x}{k} + \frac{1}{5}\sin\frac{5\pi x}{k} + \cdots\right] \qquad [0 < x < k]$$

$$x = \frac{2k}{\pi}\left[\sin\frac{\pi x}{k} - \frac{1}{2}\sin\frac{2\pi x}{k} + \frac{1}{3}\sin\frac{3\pi x}{k} - \cdots\right] \qquad [-k < x < k]$$

$$x = \frac{k}{2} - \frac{4k}{\pi^2}\left[\cos\frac{\pi x}{k} + \frac{1}{3^2}\cos\frac{3\pi x}{k} + \frac{1}{5^2}\cos\frac{5\pi x}{k} + \cdots\right] \qquad [0 < x < k]$$

$$x^2 = \frac{2k^2}{\pi^3}\left[\left(\frac{\pi^2}{1} - \frac{4}{1}\right)\sin\frac{\pi x}{k} - \frac{\pi^2}{2}\sin\frac{2\pi x}{k} + \left(\frac{\pi^2}{3} - \frac{4}{3^3}\right)\sin\frac{3\pi x}{k}\right.$$

$$\left. - \frac{\pi^2}{4}\sin\frac{4\pi x}{k} + \left(\frac{\pi^2}{5} - \frac{4}{5^3}\right)\sin\frac{5\pi x}{k} + \cdots\right] \qquad [0 < x < k]$$

$$x^2 = \frac{k^2}{3} - \frac{4k^2}{\pi^2}\left[\cos\frac{\pi x}{k} - \frac{1}{2^2}\cos\frac{2\pi x}{k} + \frac{1}{3^2}\cos\frac{3\pi x}{k} - \frac{1}{4^2}\cos\frac{4\pi x}{k} + \cdots\right] \qquad [-k < x < k]$$

$$1 - \frac{1}{3} + \frac{1}{5} - \frac{1}{7} + \cdots = \frac{\pi}{4}$$

$$1 + \frac{1}{2^2} + \frac{1}{3^2} + \frac{1}{4^2} + \cdots = \frac{\pi^2}{6}$$

$$1 - \frac{1}{2^2} + \frac{1}{3^2} - \frac{1}{4^2} + \cdots = \frac{\pi^2}{12}$$

$$1 + \frac{1}{3^2} + \frac{1}{5^2} + \frac{1}{7^2} + \cdots = \frac{\pi^2}{8}$$

$$\frac{1}{2^2} + \frac{1}{4^2} + \frac{1}{6^2} + \frac{1}{8^2} + \cdots = \frac{\pi^2}{24}$$

Some Special Limits

$$\lim_{x \to 0}\frac{\sin x}{x} = 1, \quad \lim_{x \to 0}(1 + x)^{1/x} = e = 2.71828\cdots = 1 + 1 + \frac{1}{2!} + \frac{1}{3!} + \cdots$$

17.10-2 Trigonometric Relationships

See Section 17.3.

$$\sin x = \frac{1}{2i}(e^{ix} - e^{-ix}) \qquad\qquad \sinh x = \frac{1}{2}(e^x - e^{-x})$$

$$= -i\sinh ix \qquad\qquad\qquad = -i\sin(ix)$$

$$\cos x = \frac{1}{2}(e^{ix} + e^{-ix}) \qquad\qquad \cosh x = \frac{1}{2}(e^x + e^{-x})$$

$$= \cosh ix \qquad\qquad\qquad = \cos ix$$

$$\tan x = \frac{\sin x}{\cos x} = \frac{1}{i}\tanh ix \qquad\qquad \tanh x = \frac{\sinh x}{\cosh x} = \frac{1}{i}\tan ix$$

$$\cot x = \frac{\cos x}{\sin x} = \frac{1}{\tan x} = i\coth ix \qquad \coth x = \frac{\cosh x}{\sinh x} = \frac{1}{\tanh x} = i\cot ix$$

$$\cos^2 x + \sin^2 x = 1 \qquad\qquad\qquad \cosh^2 x - \sinh^2 x = 1$$

$$\sin(x \pm y) = \sin x \cos y \pm \sin y \cos x$$

$$\sinh(x \pm y) = \sinh x \cosh y \pm \sinh y \cosh x$$

$$\sin(x \pm iy) = \sin x \cosh y \pm i \sinh y \cos x$$

$$\sinh(x \pm iy) = \sinh x \cos y \pm i \sin y \cosh x$$

$$\cos(x \pm y) = \cos x \cos y \mp \sin x \sin y$$

$$\cosh(x \pm y) = \cosh x \cosh y \pm \sinh x \sinh y$$

$$\cos(x \pm iy) = \cos x \cosh y \mp i \sin x \sinh y$$

$$\cosh(x \pm iy) = \cosh x \cos y \pm i \sinh x \sin y$$

$$\tan(x \pm y) = \frac{\tan x \pm \tan y}{1 \mp \tan x \tan y}$$

$$\tanh(x \pm y) = \frac{\tanh x \pm \tanh y}{1 \pm \tanh x \tanh y}$$

$$\tan(x \pm iy) = \frac{\tan x + i \tanh y}{1 \mp i \tan x \tanh y}$$

$$\tanh(x \pm iy) = \frac{\tanh x \pm i \tan y}{1 \pm i \tanh x \tan y}$$

$$\sin x \pm \sin y = 2 \sin\tfrac{1}{2}(x \pm y)\cos\tfrac{1}{2}(x \mp y)$$

$$\sinh x \pm \sinh y = 2 \sinh\tfrac{1}{2}(x \pm y)\cosh\tfrac{1}{2}(x \mp y)$$

$$\cos x + \cos y = 2 \cos\tfrac{1}{2}(x + y)\cos\tfrac{1}{2}(x - y)$$

$$\cosh x + \cosh y = 2 \cosh\tfrac{1}{2}(x + y)\cosh\tfrac{1}{2}(x - y)$$

$$\cos x - \cos y = 2 \sin\tfrac{1}{2}(x + y)\sin\tfrac{1}{2}(y - x)$$

$$\cosh x - \cosh y = 2 \sinh\tfrac{1}{2}(x + y)\sinh\tfrac{1}{2}(x - y)$$

$$\tan x \pm \tan y = \frac{\sin(x \pm y)}{\cos x \cos y}$$

$$\tanh x \pm \tanh y = \frac{\sinh(x \pm y)}{\cosh x \cosh y}$$

$$\sin^2 x - \sin^2 y = \sin(x + y)\sin(x - y) = \cos^2 y - \cos^2 x$$

$$\sinh^2 x - \sinh^2 y = \sinh(x + y)\sinh(x - y) = \cosh^2 x - \cosh^2 y$$

$$\cos^2 x - \sin^2 y = \cos(x + y)\cos(x - y) = \cos^2 y - \sin^2 x$$

$$\sinh^2 x + \cosh^2 y = \cosh(x + y)\cosh(x - y) = \cosh^2 x + \sinh^2 y$$

$$\sin^2 x = \tfrac{1}{2}(-\cos 2x + 1)$$

$$\sin^3 x = \tfrac{1}{4}(-\sin 3x + 3 \sin x)$$

$$\sin^4 x = \tfrac{1}{8}(\cos 4x - 4\cos 2x + 3)$$

$$\sin^5 x = \tfrac{1}{16}(\sin 5x - 5\sin 3x + 10\sin x)$$

$$\sin^6 x = \tfrac{1}{32}(-\cos 6x + 6\cos 4x - 15\cos 2x + 10)$$

$$\sin^7 x = \tfrac{1}{64}(-\sin 7x + 7\sin 5x - 21\sin 3x + 35\sin x)$$

$$\sinh^2 x = \tfrac{1}{2}(\cosh 2x - 1)$$

$$\sinh^3 x = \tfrac{1}{4}(\sinh 3x - 3\sinh x)$$

$$\sinh^4 x = \tfrac{1}{8}(\cosh 4x - 4\cosh 2x + 3)$$

$$\sinh^5 x = \tfrac{1}{16}(\sinh 5x - 5\sinh 3x + 10\sinh x)$$

$$\sinh^6 x = \tfrac{1}{32}(\cosh 6x - 6\cosh 4x + 15\cosh 2x - 10)$$

$$\sinh^7 x = \tfrac{1}{64}(\sinh 7x - 7\sinh 5x + 21\sinh 3x - 35\sinh x)$$

$$\cos^2 x = \tfrac{1}{2}(\cos 2x + 1)$$

$$\cos^3 x = \tfrac{1}{4}(\cos 3x + 3\cos x)$$

$$\cos^4 x = \tfrac{1}{8}(\cos 4x + 4\cos 2x + 3)$$

$$\cos^5 x = \tfrac{1}{16}(\cos 5x + 5\cos 3x + 10\cos x)$$

$$\cos^6 x = \tfrac{1}{32}(\cos 6x + 6\cos 4x + 15\cos 2x + 10)$$

$$\cos^7 x = \tfrac{1}{64}(\cos 7x + 7\cos 5x + 21\cos 3x + 35\cos x)$$

$$\cosh^2 x = \tfrac{1}{2}(\cosh 2x + 1)$$

$$\cosh^3 x = \tfrac{1}{4}(\cosh 3x + 3\cosh x)$$

$$\cosh^4 x = \tfrac{1}{8}(\cosh 4x + 4\cosh 2x + 3)$$

$$\cosh^5 x = \tfrac{1}{16}(\cosh 5x + 5\cosh 3x + 10\cosh x)$$

$$\cosh^6 x = \tfrac{1}{32}(\cosh 6x + 6\cosh 4x + 15\cosh 2x + 10)$$

$$\cosh^7 x = \tfrac{1}{64}(\cosh 7x + 7\cosh 5x + 21\cosh 3x + 35\cosh x)$$

$$(\cos x + i\sin x)^n = \cos nx + i\sin nx \qquad (\cosh x + \sinh x)^n = \sinh nx = \cosh ny \qquad (n \text{ is an integer})$$

$$\sin\frac{x}{2} = \pm\sqrt{\frac{1}{2}(1-\cos x)} \qquad\qquad \sinh\frac{x}{2} = \pm\sqrt{\frac{1}{2}(\cosh x - 1)}$$

$$\cos\frac{x}{2} = \pm\sqrt{\frac{1}{2}(1+\cos x)} \qquad\qquad \cosh\frac{x}{2} = \sqrt{\frac{1}{2}(\cosh x + 1)}$$

$$\tan\frac{x}{2} = \frac{1-\cos x}{\sin x} = \frac{\sin x}{1+\cos x} \qquad\qquad \tanh\frac{x}{2} = \frac{\cosh x - 1}{\sinh x} = \frac{\sinh x}{\cosh x + 1}$$

17.10-3 Derivatives

See Section 17.6.

$$\frac{dc}{dx} = 0, \qquad dc = 0 \qquad \frac{d(x)}{dx} = 1, \qquad d(x) = dx$$

$$\frac{d}{dx}(u + v - w) = \frac{du}{dx} + \frac{dv}{dx} - \frac{dw}{dx}, \qquad d(u + v - w) = du + dv - dw$$

$$\frac{d}{dx}(cv) = c\frac{dv}{dx}, \qquad d(cv) = c\,dv$$

$$\frac{d}{dx}(uv) = u\frac{dv}{dx} + v\frac{du}{dx}, \qquad d(uv) = u\,dv + v\,du$$

$$\frac{d}{dx}(v^n) = nv^{n-1}\frac{dv}{dx}, \qquad \frac{d}{dx}(x^n) = nx^{n-1}, \qquad d(v^n) = nv^{n-1}\,dv$$

$$\frac{d}{dx}\left(\frac{u}{v}\right) = \frac{v\dfrac{du}{dx} - u\dfrac{dv}{dx}}{v^2}, \qquad d\left(\frac{u}{v}\right) = \frac{v\,du - u\,dv}{v^2}$$

$$\frac{d}{dx}\left(\frac{c}{v}\right) = -\frac{c\dfrac{dv}{dx}}{v^2}, \qquad d\left(\frac{c}{v}\right) = -\frac{c\,dv}{v^2}$$

$$\frac{d(v_1 v_2 \cdots v_n)}{v_1 v_2 \cdots v_n} = \frac{dv_1}{v_1} + \frac{dv_2}{v_2} + \cdots + \frac{dv_n}{v_n} = d\log(v_1 v_2 \cdots v_n)$$

$$\frac{dy}{dx} = \frac{dy}{dv}\frac{dv}{dx} \qquad \frac{dy}{dx} = \frac{1}{\dfrac{dx}{dy}} \qquad \frac{dy}{dx} = \frac{\dfrac{dy}{dt}}{\dfrac{dx}{dt}}$$

$$\frac{d^n}{dx^n}(uv) = \frac{d^n u}{dx^n}v + n\frac{d^{n-1}u}{dx^{n-1}}\frac{dv}{dx} + \frac{n(n-1)}{2!}\frac{d^{n-2}u}{dx^{n-2}}\frac{d^2 v}{dx^2}$$

$$+ \frac{n(n-1)(n-2)}{3!}\frac{d^{n-3}u}{dx^{n-3}}\frac{d^3 v}{dx^3} + \cdots + u\frac{d^n v}{dx^n}$$

$$\frac{d}{dx}(\sin u) = \cos u\frac{du}{dx} \qquad \frac{d}{dx}(\cos u) = -\sin u\frac{du}{dx}$$

$$\frac{d}{dx}(\tan u) = \sec^2 u\frac{du}{dx} \qquad \frac{d}{dx}(\cot u) = -\csc^2 u\frac{du}{dx}$$

$$\frac{d}{dx}(\sec u) = \sec u \tan u\frac{du}{dx} \qquad \frac{d}{dx}(\csc u) = -\csc u \cot u\frac{du}{dx}$$

$$\frac{d}{dx}\left(\arcsin\frac{u}{a}\right) = \frac{1}{\sqrt{a^2 - u^2}}\frac{du}{dx} \qquad \frac{d}{dx}\left(\arccos\frac{u}{a}\right) = -\frac{1}{\sqrt{a^2 - u^2}}\frac{du}{dx}$$

$$\frac{d}{dx}\left(\arctan\frac{u}{a}\right) = \frac{a}{a^2 + u^2}\frac{du}{dx} \qquad \frac{d}{dx}\left(\text{arc cot}\frac{u}{a}\right) = -\frac{a}{a^2 + u^2}\frac{du}{dx}$$

$$\frac{d}{dx}(\text{arc sec}\frac{u}{a}) = \frac{a}{u\sqrt{u^2 - a^2}}\frac{du}{dx} \qquad \frac{d}{dx}(\text{arc csc}\frac{u}{a}) = -\frac{a}{u\sqrt{u^2 - a^2}}\frac{du}{dx}$$

$$\frac{d}{dx}(\text{arc vers}\frac{u}{a}) = \frac{1}{\sqrt{2au - u^2}}\frac{du}{dx}$$

$$\frac{d}{dx}\log_a u = \frac{1}{\log a}\frac{1}{u}\frac{du}{dx} \qquad\qquad \frac{d}{dx}\log u = \frac{1}{u}\frac{du}{dx}$$

$$\frac{d}{dx}(a^u) = (\log a)a^u\frac{du}{dx} \qquad\qquad \frac{d}{dx}(e^u) = e^u\frac{du}{dx}$$

$$\frac{d}{dx}(u^v) = u^v\log u\frac{dv}{dx} + vu^{v-1}\frac{du}{dx}$$

17.10-4 Integrals

See Section 17.6.

Elementary Integral Forms

$$\int a\,dx = ax$$

$$\int af(x)\,dx = a\int f(x)\,dx$$

$$\int \phi(y)\,dy = \int \frac{\phi(y)}{y'}\,dy \qquad \text{where } y' = \frac{dy}{dx}$$

$$\int (u + v)\,dv = \int u\,dx + \int v\,dx \qquad \text{where } u \text{ and } v \text{ are any functions of } x$$

$$\int u\,dv = u\int dv - \int v\,du = uv - \int v\,du$$

$$\int u\frac{dv}{dx}\,dx = uv - \int v\frac{du}{dx}\,dx$$

$$\int x^n\,dx = \frac{x^{n+}}{n + 1} \qquad \text{except } n = -1$$

$$\int \frac{f'(x)\,dx}{f(x)} = \log f(x) \qquad (df(x) = f'(x)\,dx)$$

$$\int \frac{dx}{x} = \log x$$

$$\int \frac{f'(x)\,dx}{2\sqrt{f(x)}} = \sqrt{f(x)} \qquad (df(x) = f'(x)\,dx)$$

$$\int e^x\,dx = e^x$$

$$\int e^{ax}\,dx = e^{ax}/a$$

$$\int b^{ax}\,dx = \frac{b^{ax}}{a\log b} \qquad (b > 0)$$

$$\int \log x\,dx = x\log x - x$$

$$\int a^x \log a \, dx = a^x \qquad (a > 0)$$

$$\int \frac{dx}{a^2 + x^2} = \frac{1}{a} \tan^{-1} \frac{x}{a}$$

$$\int \frac{dx}{a^2 - x^2} = \begin{cases} \dfrac{1}{a} \tanh^{-1} \dfrac{x}{a} \\ \dfrac{1}{2a} \log \dfrac{a + x}{a - x} \quad (a^2 > x^2) \end{cases}$$

$$\int \frac{dx}{x^2 - a^2} = \begin{cases} -\dfrac{1}{a} \coth^{-1} \dfrac{x}{a} \\ \text{or} \\ \dfrac{1}{2a} \log \dfrac{x - a}{x + a} \quad (x^2 > a^2) \end{cases}$$

$$\int \frac{dx}{\sqrt{a^2 - x^2}} = \begin{cases} \sin^{-1} \dfrac{x}{|a|} \\ \text{or} \\ -\cos^{-1} \dfrac{x}{|a|} \quad (a^2 > x^2) \end{cases}$$

$$\int \frac{dx}{\sqrt{x^2 \pm a^2}} = \log\left(x + \sqrt{x^2 \pm a^2} \right)$$

$$\int \frac{dx}{x\sqrt{x^2 - a^2}} = \frac{1}{|a|} \sec^{-1} \frac{x}{a}$$

$$\int \frac{dx}{x\sqrt{a^2 \pm x^2}} = -\frac{1}{a} \log\left(\frac{a + \sqrt{a^2 \pm x^2}}{x} \right)$$

Integrals Containing Hyperbolic Functions

$$\int (\sinh x) \, dx = \cosh x$$

$$\int (\cosh x) \, dx = \sinh x$$

$$\int (\tanh x) \, dx = \log \cosh x$$

$$\int (\coth x) \, dx = \log \sinh x$$

$$\int (\operatorname{sech} x) \, dx = \tan^{-1}(\sinh x)$$

$$\int \operatorname{csch} x \, dx = \log \tanh \left(\frac{x}{2} \right)$$

$$\int x (\sinh x) \, dx = x \cosh x - \sinh x$$

$$\int x^n (\sinh x) \, dx = x^n \cosh x - n \int x^{n-1} (\cosh x) \, dx$$

$$\int x (\cosh x) \, dx = x \sinh x - \cosh x$$

$$\int x^n (\cosh x) \, dx = x^n \sinh x - n \int x^{n-1} (\sinh x) \, dx$$

Integrals Containing Trigonometric Functions

$$\int \sin x \, dx = -\cos x \qquad \int \cos x \, dx = \sin x$$

$$\int \sin^2 x \, dx = \tfrac{1}{2}x - \tfrac{1}{2}\sin x \cos x = \tfrac{1}{2}x - \tfrac{1}{4}\sin 2x$$

$$\int \cos^2 x \, dx = \tfrac{1}{2}x + \tfrac{1}{2}\sin x \cos x = \tfrac{1}{2}x + \tfrac{1}{4}\sin 2x$$

$$\int \sin^3 x \, dx = -\tfrac{1}{3}(\sin^2 x + 2)\cos x$$

$$\int \cos^3 x \, dx = \tfrac{1}{3}(\cos^2 x + 2)\sin x$$

$$\int \sin^n x \, dx = -\frac{\sin^{n-1}x \cos x}{n} + \frac{n-1}{n}\int \sin^{n-2}x \, dx$$

$$\int \cos^n x \, dx = \frac{\cos^{n-1}x \sin x}{n} + \frac{n-1}{n}\int \cos^{n-2}x \, dx$$

$$\int \sin^m x \cos x \, dx = \frac{\sin^{m+1}x}{m+1}$$

$$\int \sin x \cos^m x \, dx = -\frac{\cos^{m+1}x}{m+1}$$

$$\int \sin^2 x \cos^2 x \, dx = -\tfrac{1}{8}\left(\tfrac{1}{4}\sin 4x - x\right)$$

$$\int \cos^m x \sin^n x \, dx = \frac{\cos^{m-1}x \sin^{n+1}x}{m+n} + \frac{m-1}{m+n}\int \cos^{m-2}x \sin^n x \, dx$$

$$\int \cos^m x \sin^n x \, dx = -\frac{\sin^{n-1}x \cos^{m+1}x}{m+n} + \frac{n-1}{m+n}\int \cos^m x \sin^{n-2}x \, dx$$

$$\int \frac{dx}{\sin^m x} = -\frac{1}{m-1}\frac{\cos x}{\sin^{m-1}x} + \frac{m-2}{m-1}\int \frac{dx}{\sin^{m-2}x}$$

$$\int \frac{dx}{\cos^m x} = \frac{1}{m-1}\frac{\sin x}{\cos^{m-1}x} + \frac{m-2}{m-1}\int \frac{dx}{\cos^{m-2}x}$$

$$\int \frac{dx}{\sin x \cos x} = \log \tan x$$

$$\int \frac{dx}{\sin^m x \cos^n x} = \frac{1}{n-1}\frac{1}{\sin^{m-1}x \cos^{n-1}x} + \frac{m+n-2}{n-1}\int \frac{dx}{\sin^m x \cos^{n-2}x}$$

$$\int \frac{dx}{\sin^m x \cos^n x} = -\frac{1}{m-1}\frac{1}{\sin^{m-1}x \cos^{n-1}x} + \frac{m+n-2}{m-1}\int \frac{dx}{\sin^{m-2}x \cos^n x}$$

$$\int \frac{\cos^m x \, dx}{\sin^n x} = -\frac{\cos^{m+1}x}{(n-1)\sin^{n-1}x} - \frac{m-n+2}{n-1}\int \frac{\cos^m x \, dx}{\sin^{n-2}x}$$

$$\int \frac{\cos^m x \, dx}{\sin^n x} = \frac{\cos^{m-1}x}{(m-n)\sin^{n-1}x} + \frac{m-1}{m-n}\int \frac{\cos^{m-2}x \, dx}{\sin^n x}$$

$$\int \frac{\sin^n x \, dx}{\cos^m x} = -\int \frac{\cos^n\left(\tfrac{1}{2}\pi - x\right)d\left(\tfrac{1}{2}\pi - x\right)}{\sin^m\left(\tfrac{1}{2}\pi - x\right)}$$

$$\int \tan x \, dx = -\log \cos x \qquad \int \cot x \, dx = \log \sin x$$

$$\int \tan^2 x \, dx = \tan x - x \qquad \int \cot^2 x \, dx = -\cot x - x$$

$$\int \tan^n x \, dx = \frac{\tan^{n-1} x}{n-1} - \int \tan^{n-2} x \, dx$$

$$\int \cot^n x \, dx = -\frac{\cot^{n-1} x}{n-1} - \int \cot^{n-2} x \, dx$$

$$\int \sec x \, dx = \log(\sec x + \tan x) \quad \text{or} \quad \log \tan\left(\tfrac{1}{4}\pi + \tfrac{1}{2}x\right)$$

$$\int \csc x \, dx = \log(\csc x - \cot x) \quad \text{or} \quad \log \tan\tfrac{1}{2}x$$

$$\int \sec^2 x \, dx = \tan x \qquad \int \csc^2 x \, dx = -\cot x$$

$$\int \sec^n x \, dx = \int \frac{dx}{\cos^n x} \qquad \int \csc^n x \, dx = \int \frac{dx}{\sin^n x}$$

$$\int \frac{dx}{a^2 \cos^2 x + b^2 \sin^2 x} = \frac{1}{ab} \arctan \frac{b \tan x}{a}$$

$$\int x \sin x \, dx = \sin x - x \cos x$$

$$\int x \cos x \, dx = \cos x + x \sin x$$

$$\int x^2 \sin x \, dx = 2 x \sin x - (x^2 - 2)\cos x$$

$$\int x^2 \cos x \, dx = 2 x \cos x + (x^2 - 2)\sin x$$

$$\int x^m \sin x \, dx = -x^m \cos x + m \int x^{m-1} \cos x \, dx$$

$$\int x^m \cos x \, dx = x^m \sin x - m \int x^{m-1} \sin x \, dx$$

$$\int \frac{\sin x \, dx}{x} = x - \frac{x^3}{3 \cdot 3!} + \frac{x^5}{5 \cdot 5!} - \frac{x^7}{7 \cdot 7!} + \cdots$$

$$\int \frac{\cos x \, dx}{x} = \log x - \frac{x^2}{2 \cdot 2!} + \frac{x^4}{4 \cdot 4!} - \frac{x^6}{6 \cdot 6!} + \cdots$$

Integrals Containing Exponential Functions

$$\int e^{ax} \, dx = \frac{e^{ax}}{a} \qquad \int a^x \, dx = \frac{a^x}{\log a}.$$

$$\int x e^{ax} \, dx = \frac{e^{ax}}{a^2}(ax - 1)$$

$$\int x^m e^{ax} \, dx = \frac{x^m e^{ax}}{a} - \frac{m}{a} \int x^{m-1} e^{ax} \, dx$$

$$\int \frac{e^{ax} \, dx}{x^m} = -\frac{1}{m-1} \frac{e^{ax}}{x^{m-1}} + \frac{a}{m-1} \int \frac{e^{ax} \, dx}{x^{m-1}}$$

$$\int e^x \sin x\, dx = \tfrac{1}{2} e^x (\sin x - \cos x)$$

$$\int e^x \cos x\, dx = \tfrac{1}{2} e^x (\sin x + \cos x)$$

$$\int e^{ax} \sin bx\, dx = \frac{e^{ax}(a \sin bx - b \cos bx)}{a^2 + b^2}$$

$$\int e^{ax} \cos bx\, dx = \frac{e^{ax}(b \sin bx + a \cos bx)}{a^2 + b^2}$$

$$\int e^{ax} \cos^n x\, dx = \frac{e^{ax} \cos^{n-1} x (a \cos x + n \sin x)}{a^2 + n^2} + \frac{n(n-1)}{a^2 + n^2} \int e^{ax} \cos^{n-2} x\, dx$$

$$\int e^{ax} \sin^n x\, dx = \frac{e^{ax} \sin^{n-1} x (a \sin x - n \cos x)}{a^2 + n^2} + \frac{n(n-1)}{a^2 + n^2} \int e^{ax} \sin^{n-2} x\, dx$$

$$\int \frac{e^x}{x}\, dx = \log x + x + \frac{x^2}{2 \cdot 2!} + \frac{x^3}{3 \cdot 3!} + \frac{x^4}{4 \cdot 4!} + \cdots$$

Integrals Containing Logarithmic Functions

$$\int \log x\, dx = x \log x - x$$

$$\int x^n \log x\, dx = x^{n+1} \left[\frac{\log x}{n+1} - \frac{1}{(n+1)^2} \right]$$

$$\int x^n (\log x)^m\, dx = \frac{x^{n+1}}{n+1} (\log x)^m - \frac{m}{n+1} \int x^n (\log x)^{m-1}\, dx$$

$$\int \frac{(\log x)^n}{x}\, dx = \frac{1}{n+1} (\log x)^{n+1}$$

$$\int \frac{dx}{x \log x} = \log(\log x)$$

$$\int \frac{dx}{x (\log x)^n} = -\frac{1}{(n-1)(\log x)^{n-1}}$$

$$\int \frac{x^n\, dx}{(\log x)^m} = -\frac{x^{n+1}}{(m-1)(\log x)^{m-1}} + \frac{n+1}{m-1} \int \frac{x^n\, dx}{(\log x)^{m-1}}$$

$$\int e^{ax} \log x\, dx = \frac{e^{ax} \log x}{a} - \frac{1}{a} \int \frac{e^{ax}}{x}\, dx$$

Integrals Containing (a + bx)

$$\int \frac{dx}{a + bx} = \frac{1}{b} \log(a + bx)$$

$$\int \frac{dx}{(a + bx)^n} = \frac{1}{b(1-n)(a + bx)^{n-1}} \quad \text{if } n \neq 1$$

$$\int (a + bx)^n\, dx = \frac{(a + bx)^{n+1}}{b(n+1)} \quad \text{if } n \neq -1$$

$$\int \frac{x\, dx}{a + bx} = \frac{1}{b^2} [a + bx - a \log(a + bx)]$$

$$\int \frac{x^2\,dx}{a + bx} = \frac{1}{b^3}\left[\tfrac{1}{2}(a + bx)^2 - 2a(a + bx) + a^2\log(a + bx)\right]$$

$$\int \frac{x\,dx}{(a + bx)^2} = \frac{1}{b^2}\left[\log(a + bx) + \frac{a}{a + bx}\right]$$

$$\int \frac{x^2\,dx}{(a + bx)^2} = \frac{1}{b^3}\left[a + bx - 2a\log(a + bx) - \frac{a^2}{a + bx}\right]$$

$$\int \frac{x\,dx}{(a + bx)^3} = \frac{1}{b^2}\left[\frac{a}{2(a + bx)^2} - \frac{1}{a + bx}\right]$$

$$\int \frac{dx}{x(a + bx)} = -\frac{1}{a}\log\frac{a + bx}{x}$$

$$\int \frac{dx}{x(a + bx)^2} = \frac{1}{a(a + bx)} - \frac{1}{a^2}\log\frac{a + bx}{x}$$

$$\int \frac{dx}{x^2(a + bx)} = -\frac{1}{ax} + \frac{b}{a^2}\log\frac{a + bx}{x}$$

$$\int \frac{dx}{x^2(a + bx)^2} = \frac{-1}{ax(a + bx)} - \frac{2b}{a}\int \frac{dx}{x(a + bx)^2}$$

Integrals Containing $(x^2 - a^2)^{1/2}$

$$\int (x^2 - a^2)^{1/2}\,dx = \tfrac{1}{2}x(x^2 - a^2)^{1/2} - \tfrac{1}{2}a^2\log\left[x + (x^2 - a^2)^{1/2}\right]$$

$$\int x(x^2 - a^2)^{1/2}\,dx = \tfrac{1}{3}(x^2 - a^2)^{3/2}$$

$$\int x^2(x^2 - a^2)^{1/2}\,dx = \frac{x}{8}(2x^2 - a^2)(x^2 - a^2)^{1/2} - \frac{a^4}{8}\log\left[x + (x^2 - a^2)^{1/2}\right]$$

$$\int x^3(x^2 - a^2)^{1/2}\,dx = \tfrac{1}{5}(x^2 - a^2)^{5/2} + \frac{a^2}{3}(x^2 - a^2)^{3/2}$$

$$\int (x^2 - a^2)^{3/2}\,dx = \frac{x}{8}(2x^2 - 5a^2)(x^2 - a^2)^{1/2} + \frac{3a^4}{8}\log\left[x + (x^2 - a^2)^{1/2}\right]$$

$$\int x(x^2 - a^2)^{3/2}\,dx = \tfrac{1}{5}(x^2 - a^2)^{5/2}$$

Integrals Containing $(a + bx)^{1/2}$

$$\int (a + bx)^{n/2}\,dx = \frac{2(a + bx)^{(n+2)/2}}{b(n + 2)}$$

$$\int x(a + bx)^{n/2}\,dx = \frac{2}{b^2}\left[\frac{(a + bx)^{(n+4)/2}}{n + 4} - \frac{a(a + bx)^{(n+2)/2}}{n + 2}\right]$$

$$\int x^{-1}(a + bx)^{n/2}\,dx = b\int (a + bx)^{(n-2)/2}\,dx + a\int x^{-1}(a + bx)^{(n-2)/2}\,dx$$

$$\int x^{-1}(a + bx)^{-n/2}\,dx = \frac{1}{a}\int x^{-1}(a + bx)^{-(n-2)/2}\,dx - \frac{b}{a}\int (a + bx)^{-n/2}\,dx$$

$$\int \frac{x^m\,dx}{(a+bx)^{1/2}} = \frac{2x^m(a+bx)^{1/2}}{(2m+1)b} - \frac{2ma}{(2m+1)b}\int \frac{x^{m-1}\,dx}{(a+bx)^{1/2}}\,dx$$

$$\int \frac{dx}{x^n(a+bx)^{1/2}} = -\frac{(a+bx)^{1/2}}{(n-1)ax^{n-1}} - \frac{(2n-3)b}{(2n-2)a}\int \frac{dx}{x^{n-1}(a+bx)^{1/2}}$$

Integrals Containing $(a^2 + x^2)^{1/2}$

$$\int (a^2+x^2)^{1/2}\,dx = \tfrac{1}{2}x(a^2+x^2)^{1/2} + \tfrac{1}{2}a^2\log\left[x+(a^2+x^2)^{1/2}\right]$$

$$\int x(a^2+x^2)^{1/2}\,dx = \tfrac{1}{3}(a^2+x^2)^{3/2}$$

$$\int x^2(a^2+x^2)^{1/2}\,dx = \frac{x}{8}(2x^2+a^2)(a^2+x^2)^{1/2} - \frac{a^4}{8}\log\left[x+(a^2+x^2)^{1/2}\right]$$

$$\int x^3(a^2+x^2)^{1/2}\,dx = \tfrac{1}{5}(a^2+x^2)^{5/2} - \frac{a^2}{3}(a^2+x^2)^{3/2}$$

$$\int (a^2+x^2)^{3/2}\,dx = \frac{x}{8}(2x^2+5a^2)(a^2+x^2)^{1/2} + \frac{3a^4}{8}\log\left[x+(a^2+x^2)^{1/2}\right]$$

$$\int x(a^2+x^2)^{3/2}\,dx = \tfrac{1}{5}(a^2+x^2)^{5/2}$$

Integrals Containing $(a^2 - x^2)^{1/2}$

$$\int (a^2-x^2)^{1/2}\,dx = \tfrac{1}{2}x(a^2-x^2)^{1/2} + \tfrac{1}{2}a^2\arcsin\frac{x}{a}$$

$$\int x(a^2-x^2)^{1/2}\,dx = -\tfrac{1}{3}(a^2-x^2)^{3/2}$$

$$\int x^2(a^2-x^2)^{1/2}\,dx = \frac{x}{8}(2x^2-a^2)(a^2-x^2)^{1/2} + \frac{a^4}{8}\arcsin\frac{x}{a}$$

$$\int x^3(a^2-x^2)^{1/2}\,dx = \tfrac{1}{5}(a^2-x^2)^{5/2} - \frac{a^2}{3}(a^2-x^2)^{3/2}$$

$$\int (a^2-x^2)^{3/2}\,dx = \frac{x}{8}(5a^2-2x^2)(a^2-x^2)^{1/2} + \frac{3a^4}{8}\arcsin\frac{x}{a}$$

$$\int x(a^2-x^2)^{3/2}\,dx = -\tfrac{1}{5}(a^2-x^2)^{5/2}$$

Integrals Containing $(a + bx^n)$

$$\int \frac{dx}{a^2+x^2} = \frac{1}{a}\arctan\frac{x}{a} = \frac{1}{a}\arcsin\frac{x}{(a^2+x^2)^{1/2}}$$

$$\int \frac{dx}{x^2-a^2} = \frac{1}{2a}\log\frac{x-a}{x+a} \quad \text{if } x^2 > a^2$$

$$= \frac{1}{2a}\log\frac{a-x}{a+x} \quad \text{if } x^2 < a^2$$

$$\int \frac{dx}{a+bx^2} = \frac{1}{\sqrt{ab}}\arctan\left(x\sqrt{\frac{b}{a}}\right) \quad \text{if } a > 0,\ b > 0$$

$$= \frac{1}{2} \frac{1}{\sqrt{-ab}} \log \frac{\sqrt{a} + x\sqrt{-b}}{\sqrt{a} - x\sqrt{-b}} \quad \text{if } a > 0, \ b < 0$$

$$\int \frac{dx}{a^2 - b^2 x^2} = \frac{1}{2ab} \log \frac{a + bx}{a - bx}$$

$$\int \frac{dx}{(a + bx^2)^{m+1}} = \frac{x}{2ma(a + bx^2)^m} + \frac{2m - 1}{2ma} \int \frac{dx}{(a + bx^2)^m}$$

$$\int \frac{x \, dx}{(a + bx^2)^{m+1}} = \frac{1}{2} \int \frac{dz}{(a + bz)^{m+1}} \quad \text{if } z = x^2$$

$$\int \frac{x \, dx}{a + bx^2} = \frac{1}{2b} \log\left(x^2 + \frac{a}{b} \right)$$

$$\int \frac{x^2 \, dx}{a + bx^2} = \frac{x}{b} - \frac{a}{b} \int \frac{dx}{a + bx^2}$$

$$\int \frac{dx}{x(a + bx^2)} = \frac{1}{2a} \log \frac{x^2}{a + bx^2}$$

$$\int \frac{dx}{x^2(a + bx^2)} = -\frac{1}{ax} - \frac{b}{a} \int \frac{dx}{a + bx^2}$$

$$\int \frac{dx}{(a + bx^2)^2} = \frac{x}{2a(a + bx^2)} + \frac{1}{2a} \int \frac{dx}{a + bx^2}$$

$$\int \frac{x^2 \, dx}{(a + bx^2)^m} = \frac{1}{b} \int \frac{dx}{(a + bx^2)^{m-1}} - \frac{a}{b} \int \frac{dx}{(a + bx^2)^m}$$

$$\int \frac{dx}{x^2(a + bx^2)^m} = \frac{1}{a} \int \frac{dx}{x^2(a + bx^2)^{m-1}} - \frac{b}{a} \int \frac{dx}{(a + bx^2)^m}$$

Definite Integrals

$$\int_0^\infty \frac{dx}{a^2 + x^2} = \frac{\pi}{a}$$

$$\int_0^\infty x^{n-1} e^{-x} \, dx = \Gamma(n)$$

where

$$\Gamma(n + 1) = n\Gamma(n) \quad \text{if } n > 0$$

$$\Gamma(n + 1) = n! \quad \text{if } n \text{ is a positive integer}$$

$$\Gamma(2) = \Gamma(1) = 1$$

$$\Gamma\left(\tfrac{1}{2}\right) = \pi^{1/2}$$

$$\int_0^1 \left(\log \frac{1}{x} \right)^n dx = \Gamma(n + 1)$$

$$\int_0^\infty e^{-zx} z^n x^{n-1} \, dx = \Gamma(n)$$

$$\int_0^1 x^{m-1} (1 - x)^{n-1} \, dx = \int_0^\infty \frac{x^{m-1} \, dx}{(1 + x)^{m+n}} = \frac{\Gamma(m)\Gamma(n)}{\Gamma(m + n)}$$

$$\int_0^\infty \frac{x^{n-1}}{1+x}\,dx = \frac{\pi}{\sin n\pi}$$

$$\int_0^\infty e^{-a^2x^2}\,dx = \frac{\pi^{1/2}}{2a}$$

$$\int_0^\infty \frac{\sin^2 x\,dx}{x^2} = \frac{\pi}{2}$$

$$\int_0^\infty \cos(x^2)\,dx = \int_0^\infty \sin(x^2)\,dx = \tfrac{1}{2}\left(\tfrac{1}{2}\pi\right)^{1/2}$$

$$\int_0^\infty \frac{\sin mx\,dx}{x} = \frac{\pi}{2} \quad \text{if } m > 0.$$

$$\int_0^\infty e^{-ax}\cos bx\,dx = \frac{a}{a^2+b^2}$$

$$\int_0^\infty e^{-ax}\sin bx\,dx = \frac{b}{a^2+b^2}$$

$$\int_0^\infty e^{-a^2x^2}\cos bx\,dx = \frac{\pi^{1/2}}{2a}e^{-b^2/4a^2}$$

$$\int_0^1 \frac{x^b - x^a}{\log x}\,dx = \log\frac{1+b}{1+a}$$

$$\int_0^1 \frac{\log x}{1-x}\,dx = -\frac{\pi^2}{6}$$

$$\int_0^1 \frac{\log x}{1+x}\,dx = -\frac{\pi^2}{12}$$

$$\int_0^1 \frac{\log x}{1-x^2}\,dx = -\frac{\pi^2}{8}$$

$$\int_0^1 \log\left(\frac{1+x}{1-x}\right)\frac{dx}{x} = \frac{\pi^2}{4}$$

$$\int_0^\infty \log\left(\frac{e^x+1}{e^x-1}\right)dx = \frac{\pi^2}{4}$$

17.10-5 Laplace Transforms

See Section 17.7.

$\phi(s)$	$F(t)$
$\dfrac{1}{s}$	1
$\dfrac{1}{s^2}$	t
$\dfrac{1}{s^n}$ $(n = 1, 2, \ldots)$	$\dfrac{t^{n-1}}{(n-1)!}$
$\dfrac{1}{\sqrt{s}}$	$\dfrac{1}{\sqrt{\pi T}}$
$s^{-3/2}$	$2\sqrt{\dfrac{t}{\pi}}$

(*Continued*)

$\phi(s)$	$F(t)$
$s^{-[n+(1/2)]}$ $(n = 1, 2, \ldots)$	$\dfrac{2^n t^{n-(1/2)}}{1 \cdot 3 \cdot 5 \cdots (2n-1)\sqrt{\pi}}$
$\dfrac{\Gamma(k)}{s^k}$ $(k \geq 0)$	t^{k-1}
$\dfrac{1}{s-a}$	e^{at}
$\dfrac{1}{(s-a)^2}$	te^{at}
$\dfrac{1}{(s-a)^n}$ $(n = 1, 2, \ldots)$	$\dfrac{1}{(n-1)!}t^{n-1}e^{at}$
$\dfrac{\Gamma(k)}{(s-a)^k}$ $(k \geq 0)$	$t^{k-1}e^{at}$
$\dfrac{1}{(s-a)(s-b)}$	$\dfrac{1}{a-b}(e^{at} - e^{bt})$
$\dfrac{s}{(s-a)(s-b)}$	$\dfrac{1}{a-b}(ae^{at} - be^{bt})$
$\dfrac{1}{(s-a)(s-b)(s-c)}$	$-\dfrac{(b-c)e^{at} + (c-a)e^{bt} + (a-b)e^{ct}}{(a-b)(b-c)(c-a)}$
$\dfrac{1}{s^2 + a^2}$	$\dfrac{1}{a}\sin at$
$\dfrac{s}{s^2 + a^2}$	$\cos at$
$\dfrac{1}{s^2 - a^2}$	$\dfrac{1}{a}\sinh at$
$\dfrac{s}{s^2 - a^2}$	$\cosh at$
$\dfrac{1}{s(s^2 + a^2)}$	$\dfrac{1}{a^2}(1 - \cos at)$
$\dfrac{1}{s^2(s^2 + a^2)}$	$\dfrac{1}{a^3}(at - \sin at)$
$\dfrac{1}{(s^2 + a^2)^2}$	$\dfrac{1}{2a^3}(\sin at - at \cos at)$
$\dfrac{s}{(s^2 + a^2)^2}$	$\dfrac{t}{2a}\sin at$
$\dfrac{s^2}{(s^2 + a^2)^2}$	$\dfrac{1}{2a}(\sin at + at \cos at)$
$\dfrac{s^2 - a^2}{(s^2 + a^2)^2}$	$t \cos at$
$\dfrac{s}{(s^2 + a^2)(s^2 + b^2)}$ $(a^2 \neq b^2)$	$\dfrac{\cos at - \cos bt}{b^2 - a^2}$
$\dfrac{1}{(s-a)^2 + b^2}$	$\dfrac{1}{b}e^{at}\sin bt$
$\dfrac{s-a}{(s-a)^2 + b^2}$	$e^{at}\cos bt$

(*Continued*)

$\phi(s)$	$F(t)$
$\dfrac{4a^3}{s^4 + 4a^4}$	$\sin at \cosh at - \cos at \sinh at$
$\dfrac{s}{s^4 + 4a^4}$	$\dfrac{1}{2a^2} \sin at \sinh at$
$\dfrac{1}{s^4 - a^4}$	$\dfrac{1}{2a^3} (\sinh at - \sin at)$
$\dfrac{s}{s^4 - a^4}$	$\dfrac{1}{2a^2} (\cosh at - \cos at)$
$\dfrac{8a^3 s^2}{(s^2 + a^2)^3}$	$(1 + a^2 t^2) \sin at - \cos at$
$\dfrac{1}{s}\left(\dfrac{s-1}{s}\right)^n$	$L_n(t) = \dfrac{e^t}{n!} \dfrac{d^n}{dt^n} (t^n e^{-t})$
$\dfrac{s}{(s-a)^{3/2}}$	$\dfrac{1}{\sqrt{\pi t}} e^{at}(1 + 2at)$
$\sqrt{s-a} - \sqrt{s-b}$	$\dfrac{1}{2\sqrt{\pi t^3}} (e^{bt} - e^{at})$
$\dfrac{1}{\sqrt{s}} e^{-k/s}$	$\dfrac{1}{\sqrt{\pi t}} \cos 2\sqrt{kt}$
$\dfrac{1}{\sqrt{s}} e^{k/s}$	$\dfrac{1}{\sqrt{\pi t}} \cosh 2\sqrt{kt}$
$\dfrac{1}{s^{3/2}} e^{-k/s}$	$\dfrac{1}{\sqrt{\pi k}} \sin 2\sqrt{kt}$
$\dfrac{1}{s^{3/2}} e^{k/s}$	$\dfrac{1}{\sqrt{\pi k}} \sinh 2\sqrt{kt}$

17.10-6   Fourier Transforms

See Section 17.7.

Fourier Transforms

$f(x)$	$\phi(\alpha)$
$\dfrac{\sin ax}{x}$	$\begin{cases} \sqrt{\dfrac{\pi}{2}} & \|\alpha\| < a \\ 0 & \|\alpha\| > a \end{cases}$
$\begin{cases} e^{iwx} & (p < x < q) \\ 0 & (x < p,\ x > q) \end{cases}$	$\dfrac{i}{\sqrt{2\pi}} \dfrac{e^{ip(w+\alpha)} - e^{iq(w+\alpha)}}{(w+\alpha)}$
$\begin{cases} e^{-cx+iwx} & (x > 0) \\ 0 & (x < 0) \end{cases} (c > 0)$	$\dfrac{i}{\sqrt{2\pi}\,(w+\alpha+ic)}$
$e^{-px^2} R(p) > 0$	$\dfrac{1}{\sqrt{2p}} e^{-\alpha^2/4p}$
$\cos px^2$	$\dfrac{1}{\sqrt{2p}} \cos\left[\dfrac{\alpha^2}{4p} - \dfrac{\pi}{4}\right]$
$\sin px^2$	$\dfrac{1}{\sqrt{2p}} \cos\left[\dfrac{\alpha^2}{4p} + \dfrac{\pi}{4}\right]$

(*Continued*)

$f(x)$	$\phi(\alpha)$
$\lvert x \rvert^{-p}$ $(0 < p < 1)$	$\sqrt{\dfrac{2}{\pi}}\ \dfrac{\Gamma(1-p)\sin\dfrac{p\pi}{2}}{\lvert\alpha\rvert^{(1-p)}}$
$\dfrac{e^{-\alpha\lvert x\rvert}}{\sqrt{\lvert x\rvert}}$	$\dfrac{\sqrt{\sqrt{(a^2+\alpha^2)}+a}}{\sqrt{a^2+\alpha^2}}$
$\dfrac{\cosh ax}{\cosh\pi x}$ $(-\pi < a < \pi)$	$\sqrt{\dfrac{2}{\pi}}\ \dfrac{\cos\dfrac{a}{2}\cosh\dfrac{\alpha}{2}}{\cosh\alpha+\cos a}$
$\dfrac{\sinh ax}{\sinh\pi x}$ $(-\pi < a < \pi)$	$\dfrac{1}{\sqrt{2\pi}}\ \dfrac{\sin a}{\cosh\alpha+\cos a}$

Finite Fourier Sine Transforms $\phi_s(n) = \displaystyle\int_0^\pi f(x)\sin nx\, dx\ (n=1,2,\ldots)$

$\phi_s(n)$	$F(x)$
$\dfrac{1}{n}$ $\cdots$	$\dfrac{\pi-x}{\pi}$
$\dfrac{(-1)^{n+1}}{n}$	$\dfrac{x}{\pi}$
$\dfrac{1-(-1)^n}{n}$	1
$\dfrac{2}{n^2}\sin\dfrac{n\pi}{2}$	$\begin{cases} x & \text{when } 0 < x < \pi/2 \\ \pi-x & \text{when } \pi/2 < x < \pi \end{cases}$
$\dfrac{(-1)^{n+1}}{n^3}$	$\dfrac{x(\pi^2-x^2)}{6\pi}$
$\dfrac{1-(-1)^n}{n^3}$	$\dfrac{x(\pi-x)}{2}$
$\dfrac{\pi^2(-1)^{n-1}}{n}-\dfrac{2[1-(-1)^n]}{n^3}$	x^2
$\pi(-1)^n\left(\dfrac{6}{n^3}-\dfrac{\pi^2}{n}\right)$	x^3
$\dfrac{n}{n^2+c^2}[1-(-1)^n e^{c\pi}]$	e^{cx}
$\dfrac{n}{n^2+c^2}$	$\dfrac{\sinh c(\pi-x)}{\sinh c\pi}$
$\dfrac{n}{n^2-k^2}$ $(k\neq 0,1,2,\ldots)$	$\dfrac{\sin k(\pi-x)}{\sin k\pi}$
$\begin{cases}\dfrac{\pi}{2} & \text{when } n=m \\ 0 & \text{when } n\neq m\end{cases}$ $(m=1,2,\ldots)$	$\sin mx$
$\dfrac{n}{n^2-k^2}[1-(-1)^n\cos k\pi]$ $(k\neq1,2,\ldots)$	$\cos kx$
$\begin{cases}\dfrac{n}{n^2-m^2}[1-(-1)^{n+m}] \\ \quad\text{when } n\neq m=1,2,\ldots \\ 0 \quad\text{when } n=m\end{cases}$	$\cos mx$

Finite Fourier Cosine Transforms $\phi_c(n) = \int_0^\pi f(x)\cos nx\,dx \;(n = 0, 1, 2, \dots)$

$\phi_c(n)$	$f(x)$
$(-1)^n \phi_c(n)$	$f(\pi - x)$
$\left.\begin{array}{l} 0 \text{ when } n = 1, 2, \dots \\ \phi_c(0) = \pi \end{array}\right\}$	1
$\left.\begin{array}{l} -\dfrac{1 - (-1)^n}{n^2} \\ \phi_c(0) = \pi^2/2 \end{array}\right\}$	x
$\left.\begin{array}{l} \dfrac{(-1)^n}{n^2} \\ \phi_c(0) = \pi^2/6 \end{array}\right\}$	$\dfrac{x^2}{2\pi}$
$\left.\begin{array}{l} \dfrac{1}{n^2}; \\ \phi_c(0) = 0 \end{array}\right\}$	$\dfrac{(\pi - x)^2}{2\pi} - \dfrac{\pi}{6}$
$\left.\begin{array}{l} 3\pi^2 \dfrac{(-1)^n}{n^2} - 6\dfrac{1 - (-1)^n}{n^4} \\ \phi_c(0) = \pi^4/4 \end{array}\right\}$	x^3
$\dfrac{(-1)^n e^c \pi - 1}{n^2 + c^2}$	$\dfrac{1}{c} e^{cx}$
$\dfrac{1}{n^2 + c^2}$	$\dfrac{\cosh c(\pi - x)}{c \sinh c\pi}$
$\dfrac{k}{n^2 - k^2}[(-1)^n \cos \pi k - 1]$ $(k \neq 0, 1, 2, \dots)$	$\sin kx$
$\left.\begin{array}{l} \dfrac{(-1)^{n+m} - 1}{n^2 - m^2} \\ \phi_c(m) = 0 \quad (m = 1, 2, \dots) \end{array}\right.$	$\dfrac{1}{m} \sin mx$
$\dfrac{1}{n^2 - k^2} \quad (k \neq 0, 1, 2, \dots)$	$-\dfrac{\cos k(\pi - x)}{k \sin k\pi}$
$\left.\begin{array}{l} 0 \quad (n = 1, 2, \dots) \\ \phi_c(m) = \pi/2 \quad (m = 1, 2, \dots) \end{array}\right\}$	$\cos mx$

17.10-7 Conversion Factors

Metric to English

To Obtain	Multiply	By
Inches	Centimeters	0.3937007874
Feet	Meters	3.280839895
Yards	Meters	1.093613298
Miles	Kilometers	0.6213711922
Ounces	Grams	$3.527396195 \times 10^{-2}$
Pounds	Kilograms	2.204622622
Gallons (U.S. Liquid)	Liters	0.2641720524
Fluid ounces	Milliliters (cc)	$3.381402270 \times 10^{-2}$
Square inches	Square centimeters	0.1550003100
Square feet	Square meters	10.76391042
Square yards	Square meters	1.195990046
Cubic inches	Milliliters (cc)	$6.102374409 \times 10^{-2}$
Cubic feet	Cubic meters	35.31466672
Cubic yards	Cubic meters	1.307950619

English to Metric

To Obtain	Multiply	By
Microns	Mils	25.4
Centimeters	Inches	2.54
Meters	Feet	0.3048
Meters	Yards	0.9144
Kilometers	Miles	1.609344
Grams	Ounces	28.34952313
Kilograms	Pounds	0.45359237
Liters	Gallons (U.S. Liquid)	3.785411784
Milliliters (cc)	Fluid ounces	29.57352956
Square centimeters	Square inches	6.4516
Square meters	Square feet	0.09290304
Square meters	Square yards	0.83612736
Milliliters (cc)	Cubic inches	16.387064
Cubic meters	Cubic feet	$2.831684659 \times 10^{-2}$
Cubic meters	Cubic yards	0.764554858

17.11   NUMERICAL METHODS

Solutions of equations are obtained by applying *analytical methods*. However, if there is no analytical method to solve an equation, rather than an *analytical* a *numerical* solution is sought by using a *numerical method*. Such a method approximates the true solution by a numerical value. Various methods have been developed to solve specific types of equations. Since a numerical solution is an *approximation* at best, by using a numerical method based on these approximations of solutions, an *error* is always introduced as the difference between the *true* value of a solution and the approximate solution obtained which deviates from the solution. Also, to evaluate the error introduced by such a numerical approximation, methods have been devised. Therefore, together with numerical analysis the concept of *error analysis* has developed. An error analysis of a numerical method results in the limits of errors (*bounds of error*) introduced, rather than the actual error for each calculated solution. For example, an error introduced would be between 0.01 and 0.005% less than or more than the approximate value obtained by a numerical method.

Another important issue in numerical analysis is the *stability* of the numerical method. If the approximate solution is obtained by an *iterative* numerical method, for *unstable* methods the rounding error exponentially increases and each time it runs further away from the actual solution. In the case of *stable* methods the solution settles in an approximate value which is within the error bounds mentioned in the paragraph above.

Some of the numerical methods for finding approximate solutions are illustrated for classes of problems. The most common problem requiring a numerical solution is obtaining the roots of an algebraic equation which is discussed here first. Another important class of problems where numerical methods are widely used is the class of equations involving derivatives, namely, differential equations (see Section 17.6). Also, the inverse of differentiation—integration—can also be solved by numerical methods. Such problems and some of their solution methods are also discussed in this section.

Some of these numerical methods can easily be implemented by performing the necessary calculations by hand. However, with the advent of computers most of the methods are designed to be implemented on such *digital computers*. The use of computers as a tool in assisting numerical analysis of analytical problems brought about the two important improvements:

1.   The *speed of computation* is *increased*. Because of the high speed with which the computer can perform algebraic operations, such as addition and multiplication, the solution to a numerical problem requiring a large number of operations can be accomplished in a minute fraction of a second, compared to an excessive amount of time if done without a computer.

2.   The *error* introduced is *decreased* by implementing solution methods which are more refined than those simpler methods that give approximate solutions with larger errors.

To implement the numerical methods on a computer an additional step is required—writing a *program* that the computer can understand. This may be done by *programming languages* developed exactly for this purpose. Since these programs express the *discrete* formulas representing the numerical approximation of the analytical expressions to the computer, the computer languages have been accordingly designed. Among the common languages BASIC, FORTRAN, PL/I, and APL may be named. Each language has its own *syntax*, thus a program written in one computer programming language differs from the others. Use of some programming languages is simpler than others. Some are more powerful in expressing even the complicated mathematical expressions. FORTRAN (FORmula TRANslation) has been designed for translating formulas to computer language and is still widely used among scientists and engineers.

For example, the FORTRAN program corresponding to the expression

$$y = x^2 - (1/2x^3)$$

is simply

$$Y = X**2 - (1/(2**3)).$$

Similar to the mathematical concepts, X and Y are *variables*, $*$ indicates the *multiplication* operation, while $**$ represents the *exponential*, and $/$ represents the *division*. Although when executed by the computer there is a hierarchy among the operations such as exponential, multiplication, division, and then addition, the parentheses used above indicate how the operations will be done, thus eliminating any ambiguity. A simple example of this question of expressing the operations correctly is the following:

The FORTRAN expression is given as $10 - 2 \times 3$. Since the multiplication comes before subtraction, the computer interprets this as $10 - (2 \times 3)$ and calculates it as $10 - 6 = 4$. However, if it is meant to be $(10 - 2)$ and then multiplied by 3, $(10 - 2) \times 3 = 8 \times 3 = 24$ would be the answer which is not equal to 4 obtained above.

This simple illustration shows that each programming language should be properly applied in solving the mathematical expressions by numerical methods. Therefore each language must be learned so that the computer representation of the expressions are correct.

17.11-1   Roots of Algebraic Equations

An algebraic equation in one unknown x, given as

$$c_0 x^n + c_1 x^{n-1} + \cdots + c^n = 0$$

can be *factorized* such that the polynomial becomes

$$c_0 (x - x_1)(x - x_2) \cdots (x - x_n) = 0$$

where $x_1, \ldots, x_n$ are the n roots of the algebraic equations. If there are *multiple roots*, then such a root appears in more than one place in the factorized form above.

Multiple Roots. A *root of multiplicity k* of a function $f(x)$ is a *root of multiplicy $n - 1$* of its derivative $f'(x)$.

Negative Real Roots. These (or *negative real part* of complex roots) are important for answering the stability question of singular solutions of the differential equations (see Section 17.5).

Hurwitz and Routh Criterion. A necessary and sufficient condition for the real part of all the roots to be negative is that all the determinants Δ_i are positive:

$$\Delta_0 = c_0 \qquad \Delta_1 = c_1$$

$$\Delta_2 = \begin{vmatrix} c_1 & c_0 \\ c_3 & c_2 \end{vmatrix} \qquad \Delta_3 = \begin{vmatrix} c_1 & c_0 & 0 \\ c_3 & c_2 & c_1 \\ c_5 & c_4 & c_3 \end{vmatrix} \quad \text{and so on}$$

Methods of Obtaining Roots of an Algebraic Equation

Rearrangement of the equation. If the algebraic equation, $g(x) = 0$ is rearranged as

$$x = f(x)$$

starting with an initial value for x, x_1, by a recursive relation

$$x_n = f(x_{n-1})$$

$x_1, x_2, \ldots$ can be calculated until the difference between x_n and x_{n-1} is within the acceptable tolerance.

Linear Interpolation. The two values are chosen as $x_{10} < x_{20}$ such that $f(x_{10})$ and $f(x_{20})$ are of opposite sign. The line segment between $[x_{10}, f(x_{10})]$ and $[x_{20}, f(x_{20})]$ intersects the x axis at

$$x_1 = \frac{x_{10}f(x_{20}) - x_{20}f(x_{10})}{f(x_{20}) - f(x_{10})}$$

Next x_1 is taken as x_{10} and the same iteration is followed for x_1 and x_{20} to find x_2, then $x_3, \ldots$, until x_n is obtained close to the x_{n-1} value within the acceptable tolerance.

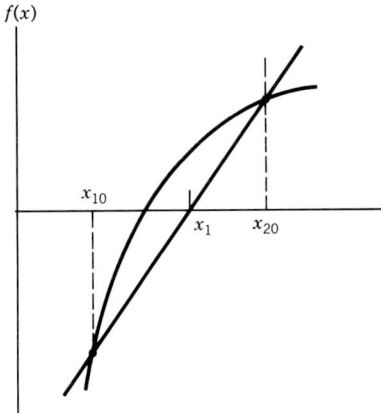

Binary Chopping. Same as linear interpolation, $x_{10} < x_{20}$ are chosen such that $f(x_{10})$ and $f(x_{20})$ are of opposite sign. $x_1 = \frac{1}{2}(x_{10} + x_{20})$ is found. $f(x_1)$ is calculated. Here x_1 replaces x_{10} (or x_{20}) if $f(x_{10})$ (or $f(x_{20})$) is of the same sign as $f(x_1)$. Similarly, the iteration is carried out.

Newton's Method. If x_0 is close to a root, x_r, then

$$x_{n+1} = x_n - \frac{f(x_n)}{f'(x_n)} \qquad n = 0, 1, 2, \ldots$$

and $x_n \to x_r$.

17.11-2 Simultaneous Linear Equations

Simultaneous linear equations may be solved numerically by such methods as Gauss–Jordan Elimination or Cramer's rule. See examples given in Section 17.5 on Matrix Algebra.

17.11-3 Least Square Method

If a set of n ordered pairs (x_i, y_i), $i = 1, 2, \ldots, n$, is given where x's are the fixed variables and y's are dependent variables, a curve can be fitted by *the method of least squares*.

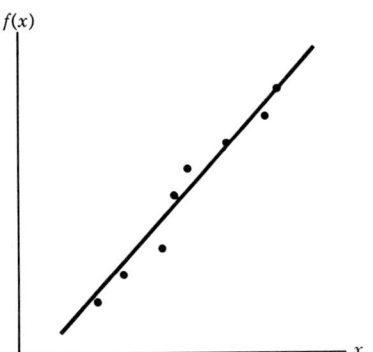

Fitting by a straight line

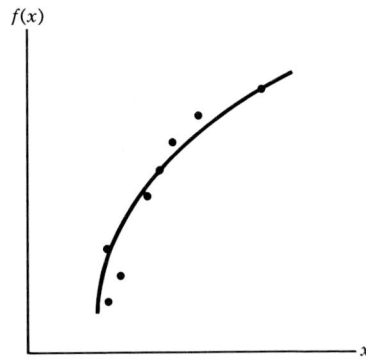

Fitting by a polynomial

For fitting the given set of ordered pairs by a *straight line*

$$y = b_0 + b_1 x$$

the difference, $\delta_i = y_i - b_0 - b_1 x_i$, can be minimized. For this the sum of the squares of the differences must be minimized:

$$E = \sum \delta_i^2$$

This means (see maxima and minima in Section 17.6) that

$$\partial E / \partial b_0 = 0$$

$$\partial E / \partial b_1 = 0$$

should be satisfied resulting in the following two equations in two unknowns.

$$\sum y_i = n b_0 + b_1 \sum x_i$$

$$\sum x_i y_i = b_0 \sum x_i + b_1 \sum x_i^2$$

The above two equations can be solved for b_0, b_1.

For fitting the given set of ordered pairs by a *polynomial*,

$$y = b_0 + b_1 x + b_2 x^2 + \cdots + b_n x^n$$

the similar equations are obtained as

$$\sum y_i = n b_0 + b_1 \sum x_i + b_2 \sum x_i^2 + \cdots + b_n \sum x_i^n$$

$$\sum x_i y_i = b_0 \sum x_i + b_1 \sum x_i^2 + \cdots + b_n \sum x_i^{n+1}$$

$$\vdots$$

$$\sum x_i^n y_i = b_0 \sum x_i^n + b_1 \sum x_i^{n+1} + \cdots + b_n \sum x_i^{2n}$$

These $n + 1$ equations can be solved for $b_0, \ldots, b_n$.

17.11-4   Numerical Differentiation

The first derivative of f

$$f_i^{(1)} = (df/dx)|_{x_i}$$

can be calculated by referring to the Taylor expansion.

Central Difference. By subtracting the two Taylor expansions about x_{i-1} and x_{i+1},

$$= \frac{f_{i+1} - f_{i-1}}{2h} - \frac{h^3}{3!} \frac{d^3 f}{dx^3} - \cdots$$

Forward Difference. By expanding about x_i and x_{i+1},

$$= \frac{f_{i+1} - f_i}{h} - \frac{h^2}{2!} \frac{d^2 f}{dx^2} - \cdots$$

Backward Difference. By expanding about x_i and x_{i-1},

$$= \frac{f_i - f_{i-1}}{h} - \frac{h^2}{2!} \frac{d^2 f}{dx^2} - \cdots$$

The error is clearly greater in the last two, being of the order h^2, while smaller for the first where the order is h^3.

17.11-5 Finite Difference Methods

Using the numerical differentiation schemes above, differential equations can be solved numerically. In the case of ordinary differential equations, the difference is taken along one independent variable only. For the partial differential equations, finite difference is taken along each variable. In general, a mesh is formed at each intersection of coordinate mesh. The differential equations are then expressed as *difference equations*, corresponding to various mesh points, based on the *finite difference method* used. As seen above, in numerical differentiation, various schemes are proposed to obtain higher accuracy by minimizing the error introduced by the approximation.

The values of a given function f, at equally spaced intervals of length h is denoted by $f_n = f(a + nh)$. Therefore, the *first, second,..., forward differences* of f_n are found as

$$\Delta f_n = f_{n+1} - f_n$$

$$\Delta^2 f_n = \Delta \Delta f_n = (f_{n+2} - f_{n+1}) - (f_{n+1} - f_n)$$

$$= f_{n+2} - 2f_{n+1} + f_n$$

Similarly we can define

$$\delta f_{n+1}, \delta^2 f_{n+1}, \ldots \quad \text{as central differences}$$

$$\Delta f_{n+1}, \Delta^2 f_{n+1}, \ldots \quad \text{as backward differences}$$

17.11-6 Numerical Solution of Differential Equations

An application of finite differences is used in the solution of differential equations. The purpose of these methods is, starting from an initial value, to calculate values of the dependent variable by the relationship given as the differential equation.

First-Order Methods

Point-Slope Formula:

$$y_{n+1} = y_n + hy'_n + O(h^2)$$

$$y_{n+1} = y_{n-1} + 2hy'_n + O(h^2)$$

Example: The differential equation is given as

$$dy/dx = x^2$$

The point slope formula is applied to have

$$\Delta y_n = y_{n+1} - y_n$$

$$\Delta x = h$$

Thus the above equation is written to calculate y_{n+1} from the previous y_n value as

$$y_{n+1} = y_n + x_n^2 h$$

In FORTRAN language,

```
   I = 1
   DO 20 I = 1, N
   Y(I + 1) = Y(I) + (X(I) ** 2) * H
20 CONTINUE
```

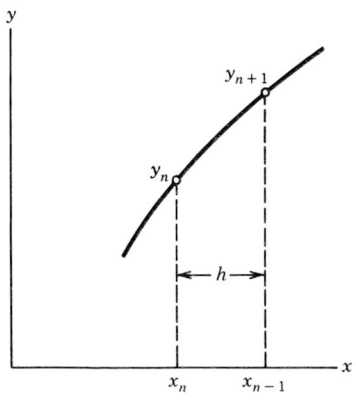

Point-slope formula with
two points only

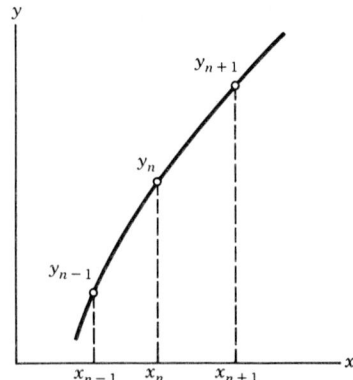

Point-slope formula with
three points

Trapezoidal Formula. Another first-order method is the trapezoidal formula given as

$$y_{n+1} = y_n + (h/2)(y'_{n+1} + y'_n) + O(h^3)$$

Second-Order Methods

Runge–Kutta Methods:

$$y_{n+1} = y_n + (1/2)(k_1 + k_2) + O(h^3)$$

where $k_1 = hf(x_n, y_n)$ and $k_2 = hf(x_n + h, y_n + k_1)$.

Higher-Order Methods

There are other methods in which the order of the error term $O(h^n)$ is improved.

Predictor–Corrector Methods

Milne's Method: These formulas are used first to *predict* a value by a *predictor* formula and then to *correct* by the *corrector* formula.

System of Differential Equations

The above methods have also been extended to system of differential equations in which there are multiple equations corresponding to multiple dependent variables.

Partial Differential Equations

In the case of partial differential equations, since there are multiple independent variables the geometry of the space of independent variables is partitioned as a mesh to apply finite difference methods.

17.11-7 Integration

The integration

$$I = \int_a^b f(x)\, dx$$

can be expressed in a discrete form in various ways. Since it is basically the area under the $f(x)$ several methods of approximating this area have been suggested.

Trapezium Rule. By taking the *linear* approximation for the function $f(x)$ between x_i and x_{i+1}, the area of the trapezoid between these two points is

$$A_i = (h/2)(f_i + f_{i+1})$$

thus if $a = x_0$, $b = x_n$

$$I = \sum_{i=0}^{n-1} A_i = (h/2)(f_0 + 2f_1 + 2f_2 + \cdots + 2f_{n-1} + f_n)$$

Example: The shaded area which is the integral of the shown curve is calculated by the following Trapezium rule.

$$A = (h/2)(f_{n-1} + 2f_n + f_{n+1})$$

A FORTRAN program would be

```
    I = 1
    DO 30 I = 1, N
    A(I) = (H/2)*(F(I - 1) + F(I))
    A(I + 1) = A(I) + (H/2)*(F(I) + F(I + 1))
30  CONTINUE
```

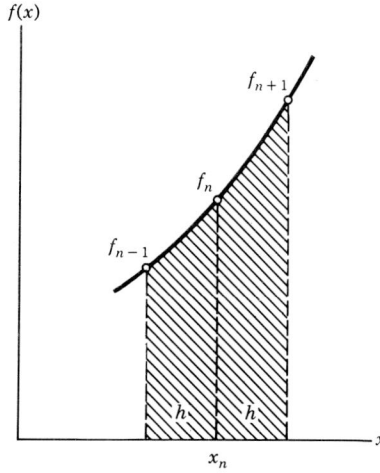

Simpson's Rule   This is a *parabolic* approximation such that in each interval $(0, 2h)$, $(2h, 4h)$, ..., the function $f(x)$ is approximated by a parabola:

$$y = ax^2 + bx + c$$

Therefore the numerical approximation formulas become

$$f_{i+1} = f(x_i + h) = f_i + bh + ah^2$$

$$f_{i-1} = f(x_i - h) = f_i - bh + ah^2$$

$$A_i = (h/3)(4f_i + f_{i+1} + f_{i-1})$$

$$I = (h/3)\left(f_0 + f_n + 4 \sum_{k \text{ odd}} f_k + 2 \sum_{k \text{ even}} f_k \right)$$

BIBLIOGRAPHY

Abramowitz, M., and I. A. Stegun (Eds.), *Handbook of Mathematical Functions*, Dover Publications, New York, 1965.

Beyer, W. H. (Ed.), *CRC Standard Mathematical Tables*, 26th ed., CRC Press, Boca Raton, Fla., 1981.

Beyer, W. H. (Ed.), *CRC Handbook of Mathematical Sciences*, 5th ed., CRC Press, West Palm Beach, Fla., 1978.

Beyer, W. H. (Ed.), *CRC Handbook of Tables for Probability and Statistics*, 4th ed., CRC Press, Boca Raton, Fla., 1979.

Byrd, P. F., and M. D. Friedman, *Handbook of Elliptic Integrals for Engineers and Physicists*, Springer-Verlag, Berlin, 1954.

Carmichael, R. D., and E. R. Smith (Compilers), *Mathematical Tables and Formulas*, Dover Publications, New York, 1962.

Condon, E. U., and H. Odishaw (Eds.), *Handbook of Physics*, McGraw-Hill, New York, 1958.

Eshbach, O. W., and M. Souders (Eds.), *Handbook of Engineering Fundamentals*, 3rd ed., Wiley, New York, 1975.

Flugge, S. (Ed.), *Handbuch der Physik, Mathematische Methoden I und II*. Springer-Verlag, Berlin, 1955.

Gradshteyn, I. S., and I. M. Ryzhik, *Table of Integrals, Series, and Products*, Academic Press, New York, 1980.

Klerer, M., and G. A. Korn, *Digital Computer User's Handbook*, McGraw-Hill, New York, 1967.

Korn, G. A., and T. M. Korn, *Mathematical Handbook for Scientist and Engineers*, 2nd ed., McGraw-Hill, New York, 1968.

Ryshik, I. M., and I. S. Gradstein, *Tables of Series, Products and Integrals*, VEB Deutscher Verlag der Wissenschaften, Berlin, 1963.

Weast, R. C., and S. M. Selby (Eds.), *Handbook of Tables for Mathematics*, 5th ed., The Chemical Rubber Co., Cleveland, Ohio, 1978.

CHAPTER 18

GRAPHS AND TABLES

LESLIE C. WILBUR

Worcester Polytechnic Institute
Worcester, Massachusetts

TABLE 18.1-1 NAMES, SYMBOLS, AND PREFIXES OF SI UNITS[a]

Quantity	Name of Unit	Symbol	
	SI BASE UNITS		
length	meter	m	
mass	kilogram	kg	
time	second	s	
electric current	ampere	A	
thermodynamic temperature	kelvin	K	
luminous intensity	candela	cd	
amount of substance	mole	mol	
	SI DERIVED UNITS		
area	square meter	m^2	
volume	cubic meter	m^3	
frequency	hertz	Hz	s^{-1}
mass density (density)	kilogram per cubic meter	kg/m^3	
speed, velocity	meter per second	m/s	
angular velocity	radian per second	rad/s	
acceleration	meter per second squared	m/s^2	
angular acceleration	radian per second squared	rad/s^2	
force	newton	N	$kg \cdot m/s^2$
pressure (mechanical stress)	pascal	Pa	N/m^2
kinematic viscosity	square meter per second	m^2/s	
dynamic viscosity	newton-second per square meter	$N \cdot s/m^2$	
work, energy, quantity of heat	joule	J	$N \cdot m$
power	watt	W	J/s
quantity of electricity	coulomb	C	$A \cdot s$
potential difference, electromotive force	volt	V	W/A
electric field strength	volt per meter	V/m	
electric resistance	ohm	Ω	V/A
capacitance	farad	F	$A \cdot s/V$
magnetic flux	weber	Wb	$V \cdot s$
inductance	henry	H	$V \cdot s/A$
magnetic flux density	tesla	T	Wb/m^2
magnetic field strength	ampere per meter	A/m	
magnetomotive force	ampere	A	
luminous flux	lumen	lm	$cd \cdot sr$
luminance	candela per square meter	cd/m^2	
illuminance	lux	lx	lm/m^2
wave number	1 per meter	m^{-1}	
entropy	joule per kelvin	J/K	
specific heat capacity	joule per kilogram kelvin	$J/(kg \cdot K)$	
thermal conductivity	watt per meter kelvin	$W/(m \cdot K)$	
radiant intensity	watt per steradian	W/sr	
activity (of a radioactive source)	1 per second	s^{-1}	
	SI SUPPLEMENTARY UNITS		
plane angle	radian	rad	
solid angle	steradian	sr	

TABLE 18.1-1 *(Continued)*

SI PREFIXES

Factor by which unit is multiplied	Prefix	Symbol
10^{12}	tera	T
10^{9}	giga	G
10^{6}	mega	M
10^{3}	kilo	k
10^{2}	hecto	h
10	deka	da
10^{-1}	deci	d
10^{-2}	centi	c
10^{-3}	milli	m
10^{-6}	micro	μ
10^{-9}	nano	n
10^{-12}	pico	p
10^{-15}	femto	f
10^{-18}	atto	a

TABLE 18.1-2   CONVERSION FACTORS TO SI UNITS LISTED BY PHYSICAL QUANTITY[a]

The following tables express the definitions of miscellaneous units of measure as exact numerical multiples of coherent SI units, and provide multiplying factors for converting numbers and miscellaneous units to corresponding new numbers and SI units.

The first two digits of each numerical entry represent a power of 10. An asterisk follows each number which expresses an exact definition. For example, the entry "−02 2.54*" expresses the fact that 1 inch=2.54×10⁻² meter, exactly, by definition. Most of the definitions are extracted from National Bureau of Standards documents. Numbers not followed by an asterisk are only approximate representations of definitions, or are the results of physical measurements.

LISTING BY PHYSICAL QUANTITY

ACCELERATION

foot/second²	meter/second²	−01 3.048*
free fall, standard	meter/second²	+00 9.806 65*
gal (galileo)	meter/second²	−02 1.00*
inch/second²	meter/second²	−02 2.54*

AREA

acre	meter²	+03 4.046 856 422 4*
are	meter²	+02 1.00*
barn	meter²	−28 1.00*
circular mil	meter²	−10 5.067 074 8
foot²	meter²	−02 9.290 304*
hectare	meter²	+04 1.00*
inch²	meter²	−04 6.4516*
mile² (U.S. statute)	meter²	+06 2.589 988 110 336*
section	meter²	+06 2.589 988 110 336*
township	meter²	+07 9.323 957 2
yard²	meter²	−01 8.361 273 6*

DENSITY

gram/centimeter³	kilogram/meter³	+03 1.00*
lbm/inch³	kilogram/meter³	+04 2.767 990 5
lbm/foot³	kilogram/meter³	+01 1.601 846 3
slug/foot³	kilogram/meter³	+02 5.153 79

TABLE 18.1-2 (*Continued*)

To convert from	*to*	*multiply by*

ENERGY

British thermal unit:		
(IST before 1956)	joule	+03 1.055 04
(IST after 1956)	joule	+03 1.055 056
British thermal unit (mean)	joule	+03 1.055 87
British thermal unit (thermochemical)	joule	+03 1.054 350
British thermal unit (39° F)	joule	+03 1.059 67
British thermal unit (60° F)	joule	+03 1.054 68
calorie (International Steam Table)	joule	+00 4.1868
calorie (mean)	joule	+00 4.190 02
calorie (thermochemical)	joule	+00 4.184*
calorie (15° C)	joule	+00 4.185 80
calorie (20° C)	joule	+00 4.181 90
calorie (kilogram, International Steam Table)	joule	+03 4.1868
calorie (kilogram, mean)	joule	+03 4.190 02
calorie (kilogram, thermochemical)	joule	+03 4.184*
electron volt	joule	−19 1.602 191 7
erg	joule	−07 1.00*
foot lbf	joule	+00 1.355 817 9
foot poundal	joule	−02 4.214 011 0
joule (international of 1948)	joule	+00 1.000 165
kilocalorie (International Steam Table)	joule	+03 4.1868
kilocalorie (mean)	joule	+03 4.190 02
kilocalorie (thermochemical)	joule	+03 4.184*
kilowatt hour	joule	+06 3.60*
kilowatt hour (international of 1948)	joule	+06 3.600 59
ton (nuclear equivalent of TNT)	joule	+09 4.20
watt hour	joule	+03 3.60*

ENERGY/AREA TIME

Btu (thermochemical)/foot² second	watt/meter²	+04 1.134 893 1
Btu (thermochemical)/foot² minute	watt/meter²	+02 1.891 488 5
Btu (thermochemical)/foot² hour	watt/meter²	+00 3.152 480 8
Btu (thermochemical)/inch² second	watt/meter²	+06 1.634 246 2
calorie (thermochemical)/cm² minute	watt/meter²	+02 6.973 333 3
erg/centimeter² second	watt/meter²	−03 1.00*
watt/centimeter²	watt/meter²	+04 1.00*

FORCE

dyne	newton	−05 1.00*
kilogram force (kgf)	newton	+00 9.806 65*
kilopond force	newton	+00 9.806 65*
kip	newton	+03 4.448 221 615 260 5*
lbf (pound force, avoirdupois)	newton	+00 4.448 221 615 260 5*
ounce force (avoirdupois)	newton	−01 2.780 138 5
pound force, lbf (avoirdupois)	newton	+00 4.448 221 615 260 5*
poundal	newton	−01 1.382 549 543 76*

LENGTH

angstrom	meter	−10 1.00*
astronomical unit (IAU)	meter	+11 1.496 00
astronomical unit (radio)	meter	+11 1.495 978 9
cable	meter	+02 2.194 56*

TABLE 18.1-2 *(Continued)*

To convert from	to	multiply by
caliber	meter	−04 2.54*
chain (surveyor or gunter)	meter	+01 2.011 68*
chain (engineer or ramden)	meter	+01 3.048*
cubit	meter	−01 4.572*
fathom	meter	+00 1.8288*
fermi (femtometer)	meter	−15 1.00*
foot	meter	−01 3.048*
foot (U.S. survey)	meter	+00 1200/3937*
foot (U.S. survey)	meter	−01 3.048 006 096
furlong	meter	+02 2.011 68*
hand	meter	−01 1.016*
inch	meter	−02 2.54*
league (U.K. nautical)	meter	+03 5.559 552*
league (international nautical)	meter	+03 5.556*
league (statute)	meter	+03 4.828 032*
light year	meter	+15 9.460 55
link (engineer or ramden)	meter	−01 3.048*
link (surveyor or gunter)	meter	−01 2.011 68*
meter	wavelengths Kr 86	+06 1.650 763 73*
micron	meter	−06 1.00*
mil	meter	−05 2.54*
mile (U.S. statute)	meter	+03 1.609 344*
mile (U.K. nautical)	meter	+03 1.853 184*
mile (international nautical)	meter	+03 1.852*
mile (U.S. nautical)	meter	+03 1.852*
nautical mile (U.K.)	meter	+03 1.853 184*
nautical mile (international)	meter	+03 1.852*
nautical mile (U.S.)	meter	+03 1.852*
pace	meter	−01 7.62*
parsec (IAU)	meter	+16 3.085 7
perch	meter	+00 5.0292*
pica (printers)	meter	−03 4.217 517 6*
point (printers)	meter	−04 3.514 598*
pole	meter	+00 5.0292*
rod	meter	+00 5.0292*
skein	meter	+02 1.097 28*
span	meter	−01 2.286*
statute mile (U.S.)	meter	+03 1.609 344*
yard	meter	−01 9.144*

MASS

carat (metric)	kilogram	−04 2.00*
gram (avoirdupois)	kilogram	−03 1.771 845 195 312 5*
gram (troy or apothecary)	kilogram	−03 3.887 934 6*
grain	kilogram	−05 6.479 891*
gram	kilogram	−03 1.00*
hundredweight (long)	kilogram	+01 5.080 234 544*
hundredweight (short)	kilogram	+01 4.535 923 7*
kgf second² meter (mass)	kilogram	+00 9.806 65*
kilogram mass	kilogram	+00 1.00*
lbm (pound mass, avoirdupois)	kilogram	−01 4.535 923 7*
ounce mass (avoirdupois)	kilogram	−02 2.834 952 312 5*
ounce mass (troy or apothecary)	kilogram	−02 3.110 347 68*
pennyweight	kilogram	−03 1.555 173 84*
pound mass, lbm (avoirdupois)	kilogram	−01 4.535 923 7*

TABLE 18.1-2 (*Continued*)

To convert from	to	multiply by
pound mass (troy or apothecary)	kilogram	−01 3.732 417 216*
scruple (apothecary)	kilogram	−03 1.295 978 2*
slug	kilogram	+01 1.459 390 29
ton (assay)	kilogram	−02 2.916 666 6
ton (long)	kilogram	+03 1.016 046 908 8*
ton (metric)	kilogram	+03 1.00*
ton (short, 2000 pound)	kilogram	+02 9.071 847 4*
tonne	kilogram	+03 1.00*

POWER

Btu (thermochemical)/second	watt	+03 1.054 350 264 488
Btu (thermochemical)/minute	watt	+01 1.757 250 4
calorie (thermochemical)/second	watt	+00 4.184*
calorie (thermochemical)/minute	watt	−02 6.973 333 3
foot lbf/hour	watt	−04 3.766 161 0
foot lbf/minute	watt	−02 2.259 696 6
foot lbf/second	watt	+00 1.355 817 9
horsepower (550 foot lbf/second)	watt	+02 7.456 998 7
horsepower (boiler)	watt	+03 9.809 50
horsepower (electric)	watt	+02 7.46*
horsepower (metric)	watt	+02 7.354 99
horsepower (U.K.)	watt	+02 7.457
horsepower (water)	watt	+02 7.460 43
kilocalorie (thermochemical)/minute	watt	+01 6.973 333 3
kilocalorie (thermochemical)/second	watt	+03 4.184*
watt (international of 1948)	watt	+00 1.000 165

PRESSURE

atmosphere	newton/meter²	+05 1.013 25*
bar	newton/meter²	+05 1.00*
barye	newton/meter²	−01 1.00*
centimeter of mercury (0° C)	newton/meter²	+03 1.333 22
centimeter of water (4° C)	newton/meter²	+01 9.806 38
dyne/centimeter²	newton/meter²	−01 1.00*
foot of water (39.2° F)	newton/meter²	+03 2.988 98
inch of mercury (32° F)	newton/meter²	+03 3.386 389
inch of mercury (60° F)	newton/meter²	+03 3.376 85
inch of water (39.2° F)	newton/meter²	+02 2.490 82
inch of water (60° F)	newton/meter²	+02 2.4884
kgf/centimeter²	newton/meter²	+04 9.806 65*
kgf/meter²	newton/meter²	+00 9.806 65*
lbf/foot²	newton/meter²	+01 4.788 025 8
lbf/inch² (psi)	newton/meter²	+03 6.894 757 2
millibar	newton/meter²	+02 1.00*
millimeter of mercury (0° C)	newton/meter²	+02 1.333 224
pascal	newton/meter²	+00 1.00*
psi (lbf/inch²)	newton/meter²	+03 6.894 757 2
torr (0' C)	newton/meter²	+02 1.333 22

SPEED

foot/hour	meter/second	−05 8.466 666 6
foot/minute	meter/second	−03 5.08*
foot/second	meter/second	−01 3.048*
inch/second	meter/second	−02 2.54*

TABLE 18.1-2 (*Continued*)

To convert from	to	multiply by
kilometer/hour	meter/second	−01 2.777 777 8
knot (international)	meter/second	−01 5.144 444 444
mile/hour (U.S. statute)	meter/second	−01 4.4704*
mile/minute (U.S. statute)	meter/second	+01 2.682 24*
mile/second (U.S. statute)	meter/second	+03 1.609 344*

TEMPERATURE

Celsius	kelvin	$t_K = t_C + 273.15$
Fahrenheit	kelvin	$t_K = (5/9)(t_F + 459.67)$
Fahrenheit	Celsius	$t_C = (5/9)(t_F - 32)$
Rankine	kelvin	$t_K = (5/9)t_R$

TIME

day (mean solar)	second (mean solar)	+04 8.64*
day (sidereal)	second (mean solar)	+04 8.616 409 0
hour (mean solar)	second (mean solar)	+03 3.60*
hour (sidereal)	second (mean solar)	+03 3.590 170 4
minute (mean solar)	second (mean solar)	+01 6.00*
minute (sidereal)	second (mean solar)	+01 5.983 617 4
month (mean calendar)	second (mean solar)	+06 2.628*
second (ephemeris)	second	+00 1.000 000 000
second (mean solar)	second (ephemeris)	Consult American Ephemeris and Nautical Almanac
second (sidereal)	second (mean solar)	−01 9.972 695 7
year (calendar)	second (mean solar)	+07 3.1536*
year (sidereal)	second (mean solar)	+07 3.155 815 0
year (tropical)	second (mean solar)	+07 3.155 692 6
year 1900, tropical, Jan., day 0, hour 12	second (ephemeris)	+07 3.155 692 597 47*
year 1900, tropical, Jan., day 0, hour 12	second	+07 3.155 692 597 47

VISCOSITY

centistoke	meter^2/second	−06 1.00*
stoke	meter^2/second	−04 1.00*
foot^2/second	meter^2/second	−02 9.290 304*
centipoise	newton second/meter^2	−03 1.00*
lbm/foot second	newton second/meter^2	+00 1.488 163 9
lbf second/foot^2	newton second/meter^2	+01 4.788 025 8
poise	newton second/meter^2	−01 1.00*
poundal second/foot^2	newton second/meter^2	+00 1.488 163 9
slug/foot second	newton second/meter^2	+01 4.788 025 8
rhe	meter^2/newton second	+01 1.00*

VOLUME

acre foot	meter^3	+03 1.233 481 837 547 52*
barrel (petroleum, 42 gallons)	meter^3	−01 1.589 873
board foot	meter^3	−03 2.359 737 216*
bushel (U.S.)	meter^3	−02 3.523 907 016 688*
cord	meter^3	+00 3.624 556 3
cup	meter^3	−04 2.365 882 365*
dram (U.S. fluid)	meter^3	−06 3.696 691 195 312 5*
fluid ounce (U.S.)	meter^3	−05 2.957 352 956 25*
foot^3	meter^3	−02 2.831 684 659 2*

TABLE 18.1-2 *(Continued)*

To convert from	to	multiply by
gallon (U.K. liquid)	meter³	−03 4.546 087
gallon (U.S. dry)	meter³	−03 4.404 883 770 86*
gallon (U.S. liquid)	meter³	−03 3.785 411 784*
gill (U K.)	meter³	−04 1.420 652
gill (U.S.)	meter³	−04 1.182 941 2
hogshead (U.S.)	meter³	−01 2.384 809 423 92*
inch³	meter³	−05 1.638 706 4*
liter	meter³	−03 1.00*
ounce (U.S. fluid)	meter³	−05 2.957 352 956 25*
peck (U.S.)	meter³	−03 8.809 767 541 72*
pint (U.S. dry)	meter³	−04 5.506 104 713 575*
pint (U.S. liquid)	meter³	−04 4.731 764 73*
quart (U.S. dry)	meter³	−03 1.101 220 942 715*
quart (U.S. liquid)	meter³	−04 9.463 529 5
stere	meter³	+00 1.00*
tablespoon	meter³	−05 1.478 676 478 125*
teaspoon	meter³	−06 4.928 921 593 75*
ton (register)	meter³	+00 2.831 684 659 2*
yard³	meter³	−01 7.645 548 579 84*

a From ref. 1. Courtesy of the National Aeronautics and Space Administration.

TABLE 18.1-3   CONVERSION FACTORS

To convert from	To	Multiply by
abampere	ampere	10
abcoulomb	ampere-hour	0.0027777
abcoulomb	coulomb	10
abfarad	farad	1×10^9
abhenry	henry	1×10^{-9}
abmho	siemen	1×10^9
abohm	ohm	1×10^{-9}
abvolt	volt	1×10^{-8}
acre	square foot	43560
acre	square meter	4046.8564
acre foot	cubic foot	43560
acre foot	cubic meter	1233.4818
acre inch	gallon (U.S.)	27154.286
ampere	coulomb per second	1
ampere (Int.)	ampere	0.999835
angstrom	meter	1×10^{-10}
are	square meter	100
astronomical unit (radio)	meter	$1.4959789 \times 10^{-11}$
atmosphere	newton per square meter	1.01325×5^{10}
atmosphere	bar	1.01325
atmosphere	dyne per square meter	1.01325×10^6
atmosphere	feet of water (39.2°F)	33.8995
atmosphere	inches of mercury (32°F)	29.9213
atmosphere	kilogram per square centimeter	1.03323
atmosphere	pascal	1.013254×10^5
atmosphere	pound per square inch	14.6960
atmosphere	torr	760
bar	newton per square meter	1×10^5
barn	square meter	1×10^{-28}
barrel (petroleum, 42 gal)	cubic meter	1.589873×10^{-1}
barrel (petroleum, U.S.)	cubic foot	5.614583
barrel (U.S. liq.)	gallon (U.S. liq.)	31.5
bar	atmosphere	0.986923
Barye	newton per square meter^2	1×10^{-1}
Board foot	cubic foot	0.083333
Board foot (1 ft × 1 ft × 1 in)	cubic meter	$2.359737216 \times 10^{-3}$
Btu (IST before 1956)	joule	1.05504×10^3
Btu (IST after 1956)	joule	1.055056×10^3
Btu (mean)	joule	1.05587×10^3
Btu (thermochemical)	joule	1.054350×10^3
Btu (39°F)	joule	1.05967×10^3
Btu (60°F)	joule	1.05468×10^3
Btu	cal (gram)	251.99576
Btu	cal (gram, 20°C)	252.122
Btu	foot-pound	777.649
Btu	horsepower-hour	0.000392752
Btu	kilogram-meter	107.514
Btu	kilowatt-hour	0.000292875
Btu	Ton of refrigeration (U.S. std.)	3.46995×10^{-6}
Btu per hour	foot-pound per hour	777.649
Btu per hour	horsepower	0.000392752

TABLE 18.1-3 (*Continued*)

To convert from	To	Multiply by
Btu per hour	horsepower (boiler)	2.98563×10^{-5}
Btu per hour	horsepower (electric)	0.000392594
Btu per hour	horsepower (metric)	0.000398199
Btu per hour	kilowatt	0.000292875
Btu per hour	pounds of ice melted per hour	0.0069714
Btu per hour	ton of refrigeration (U.S. comm.)	8.32789×10^{-5}
Btu per hour	watt	0.292875
Btu per pound	cal (gram) per gram	0.555555
Btu per pound	joule per gram	2.32444
Btu per second	horsepower	1.41391
Btu per second	kilowatt	1.05435
bushel (U.S.)	cubic meter	$3.523907016688 \times ^{-2}$
bushel (U.S.)[b]	cubic centimeter	35239.07
bushel (U.S.)[b]	gallon (U.S. dry)	8
bushel (U.S.)[b]	peck (U.S.)	4
cable	meter	2.19456×10^{2}
caliber	meter	2.54×10^{-4}
calorie (IST)	joule	4.1868
calorie (mean)	joule	4.19002
calorie (thermochemical)	joule	4.184
calorie (15°C)	joule	4.18580
calorie, g[c]	Btu	0.0039683207
calorie, g[c]	kilogram-meter	0.426649
calorie, g[c]	kilowatt-hour	1.162222×10^{-6}
carat (metric)	grain	3.08647
carat (metric)	gram	0.2
centimeter	angstrom	1×10^{8}
centimeter	foot	0.032808399
centimeter	inch	0.39370079
centimeter	meter	0.01
centimeter	micron	10000
centimeter per second	foot per second	0.032808399
centimeter per second	mile per hour	0.022369363
centipoise	gram/(cm × s)	0.01
centipoise	pound/(ft × s)	0.00067196898
cord	cord-foot	8
cord	cubic foot	128
cord	cubic meter	3.6245734
coulomb	ampere-hour	0.0002777
cubic centimeter	cubic foot	3.5314667×10^{-5}
cubic centimeter	cubic inch	0.061023744
cubic centimeter	liter	0.001
cubic centimeter	ounce (Brit. fluid)	0.03519510
cubic centimeter	ounce (U.S. fluid)	0.33814023
cubic foot	cord (wood)	0.0078125
cubic foot	cubic centimeter	28316.847
cubic foot	cubic meter	0.028316847
cubic foot	gallon (U.S. dry)	6.4285116
cubic foot	gallon (U.S. liquid)	7.4805195
cubic feet	liter	28.316847
cubic foot per second	gallon (U.S.) per minute	448.83117

TABLE 18.1-3 *(Continued)*

To convert from	To	Multiply by
cubic foot per second	liter per second	28.31605
cubic inch	cubic centimeter	16.387064
cubic inch	cubic foot	0.00057870370
cubic inch	liter	0.016387064
cubic meter	cubic centimeter	1×10^6
cubic meter	cubic foot	35.314667
cubic meter	cubic inch	61023.74
cubic meter	cubic yard	1.3079506
day (mean solar)	day (sidereal)	1.00273791
day (mean solar)	hour (mean solar)	24
day (mean solar)	year (calendar)	0.0027397260
degree	radian	0.017453293
dyne	grain	0.01573663
dyne	gram	0.001019716
dyne	newton	0.00001
dyne	poundal	7.2330138×10^{-5}
dyne	pound	2.248089×10^{-6}
dyne per centimeter	erg per square centimeter	1
dyne per square centimeter	atmosphere	9.86923×10^{-7}
dyne per square centimeter	pound per square inch	1.450377×10^{-5}
dyne-centimeter	erg	1
dyne-centimeter	foot-pound	7.37562×10^{-8}
dyne-centimeter	kilogram-meter	1.019716×10^{-8}
electron volt	erg	1.60219×10^{-12}
erg	Btu	9.48451×10^{-11}
erg	cal (kilogram)	2.39006×10^{-11}
erg	dyne-centimeter	1
erg	kilowatt-hour	2.777777×10^{-14}
erg	watt-second	1×10^{-7}
erg per second	Btu per minute	5.69071×10^{-9}
erg per second	foot-pound per minute	4.42537×10^{-6}
erg per second	horsepower	1.34102×10^{-10}
erg per second	kilowatt	1×10^{-10}
fathom	centimeter	182.88
fathom	foot	6
foot	centimeter	30.48
foot	micron	304800
foot of air (1 atm., 60°F)	atmosphere	3.6083×10^{-5}
foot of air (1 atm., 60°F)	pound per square inch	0.00053027
foot per minute	centimeter per second	0.508
foot per second	kilometer per hour	1.09728
foot per second	mile per hour	0.68181818
foot-pound per hour	Btu per minute	2.14321×10^{-5}
foot-pound per hour	erg per minute	2.25970×10^{5}
foot-pound per minute	horsepower	3.030303×10^{-5}
foot-pound per minute	kilowatt	2.25970×10^{-5}
gallon (U.S. dry)	cubic centimeter	4404.8828
gallon (U.S. dry)	cubic foot	0.15555700
gallon (U.S. liq.)	cubic inch	231
gallon (U.S. liq.)	cubic meter	0.0037854118
gill (U.S.)	cubic centimeter	118.29412

TABLE 18.1-3 *(Continued)*

To convert from	To	Multiply by
grain	gram	0.06479891
gram	dyne	980.665
gram	ounce (apoth. or troy)	0.032150737
gram	ounce (avdp.)	0.035273962
gram	ton (metric)	1×10^{-6}
gram-centimeter	Btu	9.30113×10^{-8}
gram-centimeter	dyne-centimeter	980.665
gram-centimeter	horsepower-hour	3.65303×10^{-11}
gram-centimeter	kilowatt-hour	2.72407×10^{-11}
gram-centimeter	kilowatt-hour (Int.)	2.72362×10^{-11}
horsepower[d]	foot-pound per minute	33000
horsepower[d]	foot-pound per second	550
horsepower[d]	horsepower (boiler)	0.0760181
horsepower[d]	horsepower (electric)	0.999598
horsepower[d]	horsepower (metric)	1.01387
horsepower (boiler)	horsepower	13.1548
horsepower (electric)	Btu per hour	2547.16
horsepower (metric)	Btu per hour	2511.31
horsepower (water)	horsepower	1.00046
horsepower-hour	Btu	2546.14
horsepower-hour	kilogram-meter	273745
inch	angstrom	2.54×10^8
inch	centimeter	2.54
inch	cubit	0.055555
inch	fathom	0.013888
inch	foot	0.083333
inch	yard	0.027777
inch of mercury (32°F)	atmosphere	0.0334211
inch of mercury (32°F)	pound per square foot	70.7262
inch of mercury (60°F)	atmosphere	0.0333269
inch of water (4°C)	atmosphere	0.0024582
inch per minute	centimeter per hour	152.4
joule (abs.)	Btu	0.000948451
joule (abs.)	erg	1×10^7
joule (Int.)	Btu	0.000948608
joule (Int.)	kilowatt-hour	2.77824×10^{-7}
joule (Int.)	watt-second (Int.)	1
joule per second (abs.)	Btu per minute	0.0569071
joule per second (abs.)	dyne-centimeter per second	1×10^7
joule (Int.) per second	Btu per minute	0.0569165
kilogram	ton (long)	0.00098420653
kilogram	ton (metric)	0.001
kilogram	ton (short)	0.0011023113
kilogram per square meter	atmosphere	9.67841×10^{-5}
kilogram per square meter	foot of water (39.2°F)	0.00328093
kilogram per square meter	pound per square inch	0.0014223343
kilogram-meter	foot-pound	7.23301
kilogram-meter	horsepower-hour	3.65304×10^{-6}
kilogram-meter	kilowatt-hour	2.72407×10^{-6}
kilogram-meter	newton-meter	9.80665
kilometer	foot	3280.8399

TABLE 18.1-3 *(Continued)*

To convert from	To	Multiply by
kilometer	mile (statute)	0.62137119
kilometer per hour	centimeter per second	27.7777
kilometer per hour	foot per hour	3280.8399
kilometer per hour	foot per minute	54.680665
kilometer per hour	mile (statute) per hour	0.62137119
kilowatt	Btu per hour	3414.43
kilowatt	erg per second	1×10^{10}
kilowatt	foot-pound per hour	2.65522×10^6
kilowatt	foot-pound per second	737.562
kilowatt	horsepower	1.34102
kilowatt	joule per second	1000
kilowatt	kilogram-meter per hour	3.67098×10^5
kilowatt (Int.)	Btu per hour	3414.99
kilowatt (Int.)	kilogram-meter per hour	367158
kilowatt-hour	horsepower-hour	1.34102
kilowatt-hour	kilogram-meter	367098
kilowatt-hour (Int.)	Btu (mean)	3410.08
knot (Int.)	centimeter per second	51.4444
knot (Int.)	foot per hour	6076.1155
knot (Int.)	meter per second	0.514444
liter	cubic foot	0.035314667
liter	gallon (U.S. dry)	0.22702075
liter	gallon (U.S. liq.)	0.26417205
liter	peck (U.S.)	0.1135105
liter per second	gallon (U.S. liq.) per second	0.2641723
meter	foot	3.2808399
meter	inch	39.370079
meter	mile (statute)	0.00062137119
meter per hour	mile (statute) per hour	0.00062137119
meter per second	foot per minute	196.85039
million-electron-volt	joule	1.602×10^{-13}
micron	centimeter	0.0001
month (mean calendar)	day (mean solar)	30.416666
newton-meter	dyne-centimeter	1×10^7
newton-meter	pound-foot	0.73756215
pascal	atmosphere	9.8692×10^{-6}
pascal	newton per square meter	1
pascal	pound per square inch	1.45038×10^{-4}
pound per square inch	atmosphere	0.0680460
pound per square inch	dyne per square centimeter	68947.6
pound per square inch	pascal	6894.76
revolution	radian	6.2831853
slug per cubic foot	gram per cubic centimeter	0.515379
square centimeter	square foot	0.0010763910
square centimeter	square meter	0.0001
square foot	square meter	0.09290304
square inch	square meter	0.00064516
square meter	acre	0.00024710538
square meter	square mile	3.8610216×10^{-7}
stilb	candela per square meter	1×10^4
stoke	square meter per second	1×10^{-4}

1654

TABLE 18.1-3 *(Continued)*

To convert from	To	Multiply by
tablespoon	cubic meter	$1.478676478125 \times 10^{-5}$
teaspoon	cubic meter	$4.92892159375 \times 10^{-6}$
ton (assay)	kilogram	2.9166666×10^{-2}
ton (metric)	kilogram	1×10^3
ton (nuclear equiv. of TNT)	joule	4.20×10^9
ton (register)	cubic meter	2.8316846592
ton (short, 2000 lb)	kilogram	9.0718474×10^2
tonne	kilogram	1×10^3
ton (long)	ton (metric)	1.0160469
ton (long)	ton (short)	1.12
ton (metric)	pound (avdp.)	2204.6226
ton of refrig. (U.S. comm.)	horsepower	4.71611
ton of refrig. (U.S. comm.)	kilogram of ice melted per hour	37.971
ton of refrig. (U.S. comm.)	pound of ice melted per hour	83.711
ton of refrig. (U.S. std.)	pound of ice melted	2009.1
torr (0°C)	newton per square meter	1.33322×10^2
unit pole	weber	1.256637×10^{-7}
watt	Btu per hour	3.41443
watt (Int.)	Btu per hour	3.41499
yard	meter	9.144×10^{-1}
year (calendar)	second (mean solar)	3.1536×10^7
year (sidereal)	second (mean solar)	3.1558150×10^7
year (tropical)	second (mean solar)	3.1556926×10^7
year 1900, trop., Jan., day 0, hr 12	second (ephemeris)	$3.15569259747 \times 10^7$
year 1900, trop., Jan., day 0, hr 12	second	$3.15569259747 \times 10^7$
year (tropical)	days (mean solar)	3.6524219×10^2

[a] From refs. 1 and 2.

[b] Stricken or struck bushel. A heaped bushel for apples of 2,747,715 in.[3] was established by the U.S. Court of Customs Appeals on Feb. 15, 1912. A heaped bushel equal to $1\frac{1}{4}$ stricken bushels is also known.

[c] This is the calorie as defined by the U.S. National Bureau of Standards and is equal to 4.1800 J.

[d] Mechanical horsepower, equal to 550 ft · lb/s.

TABLE 18.1-4 PHYSICAL CONSTANTS[a]

The following lists of physical constants are from the work of B. N. Taylor, W. H. Parker, and D. N. Langenberg (*Reviews of Modern Physics*, July 1969). Their least-squares adjustment of values of the constants depends strongly on a highly accurate (2.4 ppm) determination of e/h from the ac Josephson effect in superconductors, and is believed to be more accurate than the 1963 adjustment which appears to suffer from the use of an incorrect value of the fine structure constant as an input datum. See also NBS Special Publication 344 issued March 1971.

Quantity	Symbol	Value	Error ppm	Prefix	Unit
Speed of light in vacuum	c	2. 997 925 0	0. 33	$\times 10^8$	m s^{-1}
Gravitational constant	G	6. 673 2	460	10^{-11}	N m^2 kg^{-2}
Avogadro constant	N_A	6. 022 169	6. 6	10^{26}	kmol $^{-1}$
Boltzmann constant	k	1. 380 622	43	10^{-23}	J K^{-1}
Gas constant	R	8. 314 34	42	10^3	J kmol $^{-1}$ K^{-1}
Volume of ideal gas, standard conditions	V_0	2. 241 36	--------	10^1	m^3 kmol $^{-1}$
Faraday constant	F	9. 648 670	5. 5	10^7	C kmol $^{-1}$
Unified atomic mass unit	u	1. 660 531	6. 6	10^{-27}	kg
Planck constant	h	6. 626 196	7. 6	10^{-34}	J s
	$h/2\pi$	1. 054 591 9	7. 6	10^{-34}	J s
Electron charge	e	1. 602 191 7	4. 4	10^{-19}	C
Electron rest mass	m_e	9. 109 558	6. 0	10^{-31}	kg
		5. 485 930	6. 2	10^{-4}	u
Proton rest mass	m_p	1. 672 614	6. 6	10^{-27}	kg
		1. 007 276 61	0. 08	---------	u
Neutron rest mass	m_n	1. 674 920	6. 6	10^{-27}	kg
		1. 008 665 20	0. 10	---------	u
Electron charge to mass ratio	e/m_e	1. 758 802 8	3. 1	10^{11}	C kg^{-1}
Stefan–Boltzmann constant	σ	5. 669 61	170	10^{-8}	W m^{-2} K^{-4}
First radiation constant	$2\pi hc^2$	3. 741 844	7. 6	10^{-16}	W m^2
Second radiation constant	hc/k	1. 438 833	43	10^{-2}	m K
Rydberg constant	R_∞	1. 097 373 12	0. 10	10^7	m^{-1}
Fine structure constant	α	7. 297 351	1. 5	10^{-3}	
	α^{-1}	1. 370 360 2	1. 5	10^{+2}	
Bohr radius	a_0	5. 291 771 5	1. 5	10^{-11}	m
Classical electron radius	r_e	2. 817 939	4. 6	10^{-15}	m
Compton wavelength of electron	λ_C	2. 426 309 6	3. 1	10^{-12}	m
	$\lambda_C/2\pi$	3. 861 592	3. 1	10^{-13}	m
Compton wavelength of proton	$\lambda_{C,p}$	1. 321 440 9	6. 8	10^{-15}	m
	$\lambda_{C,p}/2\pi$	2. 103 139	6. 8	10^{-16}	m
Compton wavelength of neutron	$\lambda_{C,n}$	1. 319 621 7	6. 8	10^{-15}	m
	$\lambda_{C,n}/2\pi$	2. 100 243	6. 8	10^{-16}	m
Electron magnetic moment	μ_e	9. 284 851	7. 0	10^{-24}	J T^{-1}
Proton magnetic moment	μ_p	1. 410 620 3	7. 0	10^{-26}	J T^{-1}
Bohr magneton	μ_B	9. 274 096	7. 0	10^{-24}	J T^{-1}
Nuclear magneton	μ_n	5. 050 951	10	10^{-27}	J T^{-1}
Gyromagnetic ratio of protons in H$_2$O	γ'_p	2. 675 127 0	3. 1	10^8	rad s^{-1} T^{-1}
	$\gamma'_p/2\pi$	4. 257 597	3. 1	10^7	Hz T^{-1}
Gyromagnetic ratio of protons in H$_2$O corrected for diamagnetism of H$_2$O.	γ_p	2. 675 196 5	3. 1	10^8	rad s^{-1} T^{-1}
	$\gamma_p/2\pi$	4. 257 707	3. 1	10^7	Hz T^{-1}
Magnetic flux quantum	Φ_0	2. 067 853 8	3. 3	10^{-15}	Wb
Quantum of circulation	$h/2m_e$	3. 636 947	3. 1	10^{-4}	J s kg^{-1}
	h/m_e	7. 273 894	3. 1	10^{-4}	J s kg^{-1}

TABLE 18.1-4   (*Continued*)

Unitless numerical ratios	Value	Error ppm	Prefix
(c^2) kg/eV	5. 609 538	4. 4	10^{35}
(c^2) u/eV	9. 314 812	5. 5	10^8
u/kg	1. 660 531	6. 6	10^{-27}
(c^2) m_e/eV	5. 110 041	3. 1	10^5
(c^2) m_p/eV	9. 382 592	5. 5	10^8
(c^2) m_n/eV	9. 395 527	5. 5	10^8
eV/J	1. 602 191 7	4. 4	10^{-19}
(h^{-1}) eV/Hz	2. 417 965 9	3. 3	10^{14}
$(hc)^{-1}$ eV m	8. 065 465	3. 3	10^5
(k^{-1}) eV/K	1. 160 485	42	10^4
(hc) (eV m)$^{-1}$	1. 239 854 1	3. 3	10^{-6}
(hc) R_∞/J	2. 179 914	7. 6	10^{-18}
(hc) R_∞/eV	1. 360 582 6	3. 3	10^1
(c) R_∞/Hz	3. 289 842 3	0. 35	10^{15}
(hc/k) R_∞/K	1. 578 936	43	10^5
m_p/m_e	1. 836 109	6. 2	10^3
μ_e/μ_B	1. 001 159 638 9	0. 0031	
μ'_p/μ_B	1. 520 993 12	0. 066	10^{-3}
μ_p/μ_B	1. 521 032 64	0. 30	10^{-3}
μ'_p/μ_n	2. 792 709	6. 2	
μ_p/μ_n	2. 792 782	6. 2	

Other important constants

$\pi = 3.141\ 592\ 653\ 589$
$e = 2.718\ 281\ 828\ 459$
$\mu_0 = 4\pi \times 10^{-7}$ H/m (exact), permeability of free space
$\quad = 1.256\ 637\ 061 \times 10^{-6}$ H/m
$\epsilon_0 = \mu_0^{-1} c^{-2}$ F/m, permittivity of free space
$\quad = 8.854\ 185 \times 10^{-12}$ F/m

^{a}From ref. 1. Courtesy of the National Aeronautics and Space Administration.

TABLE 18.1-5   CRITICAL CONSTANTS: CUSTOMARY UNITS[a]

Substance	Formula	Molecular Weight	Temperature K	Temperature R	Pressure atm	Pressure lbf/in.[2]	Volume, ft^3/lb-mole
Ammonia	NH_3	17.03	405.5	729.8	111.3	1636	1.16
Argon	A	39.944	151	272	48.0	705	1.20
Bromine	Br_2	159.832	584	1052	102	1500	2.17
Carbon dioxide	CO_2	44.01	304.2	547.5	72.9	1071	1.51
Carbon monoxide	CO	28.01	133	240	34.5	507	1.49
Chlorine	Cl_2	70.914	417	751	76.1	1120	1.99
Deuterium (Normal)	D_2	4.00	38.4	69.1	16.4	241	...
Helium	He	4.003	5.3	9.5	2.26	33.2	0.926
Helium[3]	He	3.00	3.34	6.01	1.15	16.9	...
Hydrogen (Normal)	H_2	2.016	33.3	59.9	12.8	188.1	1.04
Krypton	Kr	83.7	209.4	376.9	54.3	798	1.48
Neon	Ne	20.183	44.5	80.1	26.9	395	0.668
Nitrogen	N_2	28.016	126.2	227.1	33.5	492	1.44
Nitrous oxide	N_2O	44.02	309.7	557.4	71.7	1054	1.54
Oxygen	O_2	32.00	154.8	278.6	50.1	736	1.25
Sulfur dioxide	SO_2	64.06	430.7	775.2	77.8	1143	1.95
Water	H_2O	18.016	647.4	1165.3	218.3	3208	0.90
Xenon	Xe	131.3	289.75	521.55	58.0	852	1.90
Benzene	C_6H_6	78.11	562	1012	48.6	714	4.17
n-Butane	C_4H_{10}	58.120	425.2	765.2	37.5	551	4.08
Carbon tetrachloride	CCl_4	153.84	556.4	1001.5	45.0	661	4.42
Chloroform	$CHCl_3$	119.39	536.6	965.8	54.0	794	3.85
Dichlorodifluoromethane	CCl_2F_2	120.92	384.7	692.4	39.6	582	3.49
Dichlorofluoromethane	$CHCl_2F$	102.93	451.7	813.0	51.0	749	3.16
Ethane	C_2H_6	30.068	305.5	549.8	48.2	708	2.37
Ethyl alcohol	C_2H_5OH	46.07	516.0	929.0	63.0	926	2.68
Ethylene	C_2H_4	28.052	282.4	508.3	50.5	742	1.99
n-Hexane	C_6H_{14}	86.172	507.9	914.2	29.9	439	5.89
Methane	CH_4	16.042	191.1	343.9	45.8	673	1.59
Methyl alcohol	CH_3OH	32.04	513.2	923.7	78.5	1154	1.89
Methyl chloride	CH_3Cl	50.49	416.3	749.3	65.9	968	2.29
Propane	C_3H_8	44.094	370.0	665.9	42.0	617	3.20
Propene	C_3H_6	42.078	365.0	656.9	45.6	670	2.90
Propyne	C_3H_4	40.062	401	722	52.8	776	...
Trichlorofluoromethane	CCl_3F	137.38	471.2	848.1	43.2	635	3.97

[a] From ref. 3. Used by permission.

TABLE 18.1-6 CRITICAL CONSTANTS: SI UNITS[a]

Critical Constants[a]

Substance	Formula	Molecular Weight	Temp. K	Pressure MPa	Volume m³/kmol
Ammonia	NH_3	17.03	405.5	11.28	.0724
Argon	Ar	39.948	151	4.86	.0749
Bromine	Br_2	159.808	584	10.34	.1355
Carbon Dioxide	CO_2	44.01	304.2	7.39	.0943
Carbon Monoxide	CO	28.011	133	3.50	.0930
Chlorine	Cl_2	70.906	417	7.71	.1242
Deuterium (Normal)	D_2	4.00	38.4	1.66	—
Helium	He	4.003	5.3	0.23	.0578
Helium³	He	3.00	3.3	0.12	—
Hydrogen (Normal)	H_2	2.016	33.3	1.30	.0649
Krypton	Kr	83.80	209.4	5.50	.0924
Neon	Ne	20.183	44.5	2.73	.0417
Nitrogen	N_2	28.013	126.2	3.39	.0899
Nitrous Oxide	N_2O	44.013	309.7	7.27	.0961
Oxygen	O_2	31.999	154.8	5.08	.0780
Sulfur Dioxide	SO_2	64.063	430.7	7.88	.1217
Water	H_2O	18.015	647.3	22.09	.0568
Xenon	Xe	131.30	289.8	5.88	.1186
Benzene	C_6H_6	78.115	562	4.92	.2603
n-Butane	C_4H_{10}	58.124	425.2	3.80	.2547
Carbon Tetrachloride	CCl_4	153.82	556.4	4.56	.2759
Chloroform	$CHCl_3$	119.38	536.6	5.47	.2403
Dichlorodifluoromethane	CCl_2F_2	120.91	384.7	4.01	.2179
Dichlorofluoromethane	$CHCl_2F$	102.92	451.7	5.17	.1973
Ethane	C_2H_6	30.070	305.5	4.88	.1480
Ethyl Alcohol	C_2H_5OH	46.07	516	6.38	.1673
Ethylene	C_2H_4	28.054	282.4	5.12	.1242
n-Hexane	C_6H_{14}	86.178	507.9	3.03	.3677
Methane	CH_4	16.043	191.1	4.64	.0993
Methyl Alcohol	CH_3OH	32.042	513.2	7.95	.1180
Methyl Chloride	CH_3Cl	50.488	416.3	6.68	.1430
Propane	C_3H_8	44.097	370	4.26	.1998
Propene	C_3H_6	42.081	365	4.62	.1810
Propyne	C_3H_4	40.065	401	5.35	—
Trichlorofluoromethane	CCl_3F	137.37	471.2	4.38	.2478

[a] From ref. 1. Used by permission

TABLE 18.1-7 DENSITY TABLE[a]

Material	ρ (lb/cu ft)	Material	(lb/cu ρ ft)
Element		*Liquids*	
Aluminum	168.6	Turpentine	54
Antimony	413.0	Water, 100°C	62.428
Barium	225.0		
Bismuth	612.0	*Minerals*	
Cadmium	540.0	Asbestos	153
Calcium	97.0	Basalt	184
Carbon	141.0	Bauxite	159
Chromium	446.0	Borax	109
Cobalt	550.0	Clay	137
Copper	558.0	Dolomite	181
Gold	1206.0	Granite	165
Hydrogen	53×10^{-4}	Gypsum	159
Iron	492	Limestone	155
Lead	708	Marble	170
Magnesium	108	Magnesite	187
Manganese	462	Pumice	40
Mercury	846	Quartz	165
Molybdenum	637	Sandstone	143
Nickel	556		
Nitrogen	73×10^{-3}	Shale	172
Oxygen	83×10^{-3}	Soapstone	169
Phosphorus	114		
Platinum	1340	*Bituminous Substances*	
Potassium	54.7	Asphaltum	81
Silicon	145	Coal, anthracite	97
Silver	656	Coal, bituminous	84
Sodium	61	Coal, lignite	78
		Coal, peat, turf, dry	47
Sulfur	129	Coal, charcoal, pine	23
Tin	450	Coal, charcoal, oak	33
Titanium	281	Coal, coke	75
		Graphite	135
Element		Paraffin	56
Tungsten	1206	Petroleum	54
Vanadium	350	Petroleum (kerosene)	50
Zinc	446	Petroleum (gasoline)	45
		Pitch	69
Alloys and Other Materials			
Aluminum bronze	481		
Brass	534		
Bronze	509		
Bronze (phosphor)	554		
Cast iron	442		
Iron, wrought	485		
Monel metal	555		
Steel (cold drawn)	489		
Liquids			
Alcohol, ethyl	49		
Alcohol, methyl	50		
Acid, muriatic (40%)	75		
Acid, nitric (91%)	94		
Acid, sulfuric (87%)	112		
Ether	46		
Oils, vegetable	58		

[a] From ref. 5. Used by permission.

TABLE 18.1-8   SPECIFIC HEAT TABLE[a]

Material	Temperature (°F)	Cp	Material	Temperature (°F)	Cp
Metals			*Liquids*		
Lead	32	0.0306	Dowtherm A	400	0.560
Lead	212	0.0315	Ethyl alcohol		
Lead	392	0.0325	100%	0	0.4350
Zinc	32	0.0917	Ethyl alcohol		
Zinc	212	0.0958	100%	100	0.6350
Zinc	392	0.1000	Ethyl alcohol 50%	0	0.830
Aluminum	32	0.2106	Ethyl alcohol 50%	100	0.940
Aluminum	212	0.2225	Ethylene glycol	−50	0.500
Aluminum	392	0.2344	Ethylene glycol	0	0.540
Silver	32	0.0557	Ethylene glycol	50	0.565
Silver	212	0.0571	Ethylene glycol	200	0.665
Silver	392	0.0585	Freon-11	−20	0.200
Gold	32	0.0305	Freon-11	0	0.210
Gold	212	0.0312	Freon-11	100	0.220
Gold	392	0.0320	Freon-12	−40	0.220
Copper	32	0.0919	Freon-12	0	0.230
Copper	212	0.0942	Freon-12	100	0.270
Copper	392	0.0965	Freon-21	−20	0.245
Nickel	32	0.1025	Freon-21	0	0.250
Nickel	212	0.1132	Freon-21	100	0.255
Nickel	392	0.1241	Freon-22	−20	0.260
Iron	32	0.1051	Freon-22	0	0.270
Iron	212	0.1166	Freon-22	100	0.310
Iron	392	0.1280	Freon-113	−20	0.205
Cobalt	32	0.1023	Freon-113	0	0.210
			Freon-113	100	0.220
Cobalt	212	0.1079	Gasoline	32–212	0.50
Cobalt	392	0.1138	Glycerol	−40	0.5000
Quartz	32	0.1667	Glycerol	0	0.5300
Quartz	212	0.2061	Glycerol	50	0.5600
Quartz	392	0.2315	Methyl alcohol	−40	0.540
			Methyl alcohol	0	0.565
Liquids			Methyl alcohol	50	0.590
Ammonia	−200	1.100	Methyl chloride	−80	0.350
Ammonia	−100	1.400	Methyl chloride	0	0.370
Brine, 25% CaCl₂	−50	0.640	Methyl chloride	50	0.380
Brine, 25% CaCl₂	0	0.660	Pyridine	−50	0.370
Brine, 25% CaCl₂	50	0.680	Pyridine	0	0.385
Brine, 25% CaCl₂	100	0.700	Pyridine	100	0.420
Brine, 25% NaCl	0	0.800	Pyridine	200	0.450
Brine, 25% NaCl	50	0.810	Sulfuric acid 98%	0	0.330
Brine, 25% NaCl	100	0.820			
Carbon tetra-			Sulfuric acid 98%	50	0.340
chloride	0	0.190	Sulfuric acid 98%	100	0.355
Carbon tetra-			Sulfur dioxide	−20	0.300
chloride	100	0.210	Sulfur dioxide	0	0.330
Carbon tetra-			Sulfur dioxide	100	0.350
chloride	200	0.230	Water	32	1.000
Dowtherm A	100	0.400	Water	100	1.050
Dowtherm A	200	0.450	Water	212	1.080
Dowtherm A	300	0.510			

TABLE 18.1-8 (*Continued*)

Material	Temperature (°F)	C_p	Material	Temperature (°F)	C_p
Gases at 1 atm			*Gases at 1 atm*		
Air	32 ·	0.245	Nitrogen	500	0.260
Air	100	0.250	Nitrogen	1000	0.270
Air	200	0.255	Oxygen	0	0.215
Air	500	0.260	Oxygen	100	0.220
Air	1000	0.265	Oxygen	200	0.225
Ammonia	0	0.500	Oxygen	500	0.235
Ammonia	50	0.520	Oxygen	1000	0.255
Ammonia	100	0.525	Sulfur dioxide	0	0.150
Ammonia	500	0.600	Sulfur dioxide	200	0.162
Carbon dioxide	0	0.205	Sulfur dioxide	400	0.185
Carbon dioxide	100	0.215	Sulfur dioxide	600	0.188
Carbon dioxide	500	0.250	Sulfur dioxide	1000	0.212
Carbon dioxide	1000	0.280	Water	0	0.435
Carbon dioxide	1500	0.300	Water	200	0.450
Carbon monoxide	0	0.250	Water	500	0.465
Carbon monoxide	100	0.255	Water	1000	0.500
Carbon monoxide	500	0.260	Water	1500	0.550
Carbon monoxide	1000	0.270			
Chlorine	100	0.115			
Chlorine	500	0.127			
Chlorine	1000	0.140			
Ethylene	100	0.400			
Ethylene	200	0.440			
Ethylene	300	0.500			
Ethylene	400	0.550			
Freon-11	0	0.128			
Freon-11	100	0.140			
Freon-11	200	0.150			
Freon-21	0	0.140			
Freon-21	100	0.150			
Freon-21	200	0.162			
Freon-22	0	0.150			
Freon-22	100	0.161			
Freon-22	200	0.173			
Freon-113	0	0.151			
Freon-113	100	0.161			
Freon-113	200	0.170			
Hydrogen	0	3.450			
Hydrogen	100	3.500			
Hydrogen	500	3.510			
Methane	0	0.510			
Methane	100	0.550			
Methane	300	0.630			
Methane	500	0.725			
Methane	1000	0.940			
Nitrogen	0	0.250			
Nitrogen	100	0.255			
Nitrogen	200	0.257			

[a] From ref. 5. Used by permission.

TABLE 18.1-9 THERMAL CONDUCTIVITY TABLE[a]

Material	Temperature (°F)	k	Material	Temperature (°F)	k
Metals			*Metals*		
Aluminum	212	119	Inconel X750	1600	13.7
Aluminum	392	124	Incoloy 800	1000	11.6
Aluminum	752	144	Incoloy 800	1400	13.8
Aluminum brass	68	58	Incoloy 800	1800	17.8
Aluminum			Incoloy 825	800	9.9
bronze 5%	68	46	Incoloy 825	1000	10.9
Admiralty metal	86	65	Incoloy 825	1400	12.9
Brass (70 Cu–30			Incoloy 825	1800	16.0
Zn)	212	60	Lead	212	19.0
Brass (70 Cu–30			Lead	392	18.0
Zn)	392	63	Monel 400	200	13.9
Brass (70 Cu–30			Monel 400	600	17.9
Zn)	752	67	Monel 400	1400	25.9
Cupro-nickel 10%	68	26	Monel K-500	400	13.0
Cupro-nickel 20%	68	21	Monel K-500	800	15.7
Cupro-nickel 30%	68	17	Monel K-500	1600	23.5
Cadmium	64	53.7	Nickel	212	34.0
Cadmium	212	52.2	Nickel	392	33.0
Cast iron	212	30.0	Nickel	572	32.0
Cast iron	392	28	Silver	212	238.0
			Steel (mild)	212	26.0
Cast iron	752	25	Steel (mild)	392	26.0
Copper (pure)	212	218	Steel (mild)	572	25.0
Copper (pure)	392	215	Steel (stainless)		
Copper (pure)	752	210	301,302,303,		
Graphite	212	87	301, 302, 303,		
Graphite	752	58	304, 316	212	9.4
Hasteloy 25	900	10.6	Steel (stainless)		
Hastcloy 25	1300	11.9	301, 302, 303,		
Hasteloy 25	1700	15.9	304, 316	932	12.4
Hasteloy X	1100	12.0	Steel (stainless)		
Hasteloy X	1500	14.5	308	212	8.8
Hasteloy X	1700	15.7	Steel (stainless)		
Inconel 625	1000	10.1	308	932	12.5
Inconel 625	1600	13.2	Steel (stainless)		
Inconel 625	1800	14.6	309, 310	212	8.0
Inconel 702	400	10.4	Steel (stainless)		
Inconel 702	800	15.0	309, 310	932	10.8
Inconel 702	1200	19.6	Steel (stainless)		
Inconel X750	400	8.2	321, 347	212	9.3
Inconel X750	800	10.0	Steel (stainless)		
			321, 347	932	12.8

TABLE 18.1-9 (*Continued*)

Material	Temperature (°F)	k	Material	Temperature (°F)	k
Insulation and Refractory			*Liquids*		
Asbestos sheets	124	0.096	Ammonia	5–86	0.29
Brick (alumina 92–99)	800	1.80	Alcohol	86	0.097
			Alcohol	167	0.095
Brick (alumina 64–65)	2400	2.70	Calcium chloride brine	86	0.32
Brick (chrome)	392	0.67	Dichlorodifluoro-methane	20	0.057
Brick (chrome)	1200	0.85	Dichlorodifluoro-methane	60	0.053
Brick (chrome)	2400	1.00	Dichlorodifluoro-methane	100	0.048
Corrugated asbestos	200	0.058	Dichlorodifluoro-methane	180	0.038
Corrugated asbestos	300	0.069	Ethylene glycol	32	0.153
Diatom. earth-brick (1600)	400	0.050	Gasoline	86	0.078
Diatom. earth-brick (1600)	600	0.055	Glycerol 100%	68	0.164
Diatom. earth-brick (1600)	1000	0.065	Glycerol 80%	68	0.189
			Glycerol 60%	68	0.220
Diatom. earth-brick (2000)	400	0.137	Glycerol 40%	68	0.259
Diatom. earth-brick (2000)	600	0.140	Glycerol 20%	68	0.278
Diatom. earth-brick (2000)	1000	0.158	Glycerol 100%	212	0.164
Diatom. earth-brick (2500)	600	0.148	Kerosene	68	0.086
			Kerosene	167	0.081
Diatom. earth-brick (2500)	1000	0.163	Mercury	82	4.83
Diatom. earth-brick (2500)	2000	0.203	Methyl alcohol 100%	68	0.124
Fireclay	392	0.580			
Kaolin ins. brick	932	0.150	Methyl alcohol 80%	68	0.154
Kaolin ins. brick	2100	0.260	Methyl alcohol 60%	68	0.190
Magnesite	399	2.20	Methyl alcohol 40%	68	0.234
Magnesite	1200	1.60	Methyl alcohol 20%	68	0.284
Magnesite	2192	1.10	Methyl alcohol 100%	122	0.114
Magnesia 85%	200	0.036	Oils	86	0.079
Magnesia 85%	300	0.038	Trichloroethylene	122	0.080
Magnesia 85%	400	0.040	Water	32	0.343
Molded pipe covering	400	0.051	Water	100	0.363
			Water	200	0.393
			Water	300	0.395
Molded pipe covering	1600	0.088	Water	420	0.376
Rock wool	200	0.034	Water	620	0.275
Rock wool	400	0.044			
Rock wool	600	0.057			
Silicon carbide brick	1112	10.70			
Silicon carbide brick	1475	9.20			
Silicon carbide brick	1832	8.00			
Silicon carbide brick	2552	6.32			

TABLE 18.1-9 (*Continued*)

Material	Temperature (°F)	k	Material	Temperature (°F)	k
Gases and Vapors			*Gases and Vapors*		
Acetylene	−103	0.0068	Methane	32	0.0176
Acetylene	32	0.0108	Methane	212	0.0255
Acetylene	122	0.0140	Methane	392	0.0358
Acetylene	212	0.0172	Methane	572	0.0490
Air	−328	0.0040	Nitrogen	−328	0.0040
Air	−148	0.0091	Nitrogen	−148	0.0091
Air	32	0.0140	Nitrogen	32	0.0139
Air	212	0.0184	Nitrogen	212	0.0181
Air	392	0.0224	Nitrogen	392	0.0220
Air	572	0.0260	Nitrogen	572	0.0255
Ammonia	−58	0.0097	Nitrogen	752	0.0287
Ammonia	32	0.0126	Nitrous oxide	−148	0.0047
Ammonia	212	0.0192	Nitrous oxide	32	0.0088
Ammonia	392	0.0280	Nitrous oxide	212	0.0138
Ammonia	572	0.0385	Oxygen	−328	0.0038
Ammonia	752	0.0509	Oxygen	−148	0.0091
Butane	32	0.0078	Oxygen	32	0.0142
Butane	212	0.0135	Oxygen	122	0.0166
Carbon dioxide	−58	0.0064	Oxygen	212	0.0188
Carbon dioxide	32	0.0084	Propane	32	0.0074
Carbon dioxide	212	0.0128	Propane	212	0.0151
Carbon dioxide	392	0.0177	Sulfur dioxide	32	0.0050
Carbon dioxide	572	0.0229	Sulfur dioxide	212	0.0069
Carbon monoxide	−328	0.0037	Water vapor (zero press.)	212	0.0136
Carbon monoxide	−148	0.0088			
Carbon monoxide	32	0.0134	Water vapor (zero press.)	392	0.0182
Carbon monoxide	212	0.0176			
Chlorine	32	0.0043	Water vapor (zero press.)	572	0.0230
Dichlorodifluoro- methane	32	0.0048			
			Water vapor (zero press.)	752	0.0279
Dichlorodifluoro- methane	122	0.0064			
			Water vapor (zero press.)	932	0.0328
Dichlorodifluoro- methane	212	0.0080			
Dichlorodifluoro- methane	302	0.0097			
Ethyl alcohol	68	0.0089			
Ethyl alcohol	212	0.0124			
Ethylene	−96	0.0064			
Ethylene	32	0.0101			
Ethylene	122	0.0131			
Ethylene	212	0.0161			
Helium	−328	0.0338			
Helium	−148	0.0612			
Helium	32	0.0818			
Helium	212	0.0988			
Hydrogen	−328	0.0293			
Hydrogen	−148	0.0652			
Hydrogen	32	0.0966			
Hydrogen	212	0.1240			
Hydrogen	392	0.1484			
Hydrogen	572	0.1705			
Mercury	392	0.0197			
Methane	−328	0.0045			
Methane	−148	0.0109			

[a] From ref. 5. Used by permission.

TABLE 18.1-10   VISCOSITY TABLE[a]

Material	Temperature (°F)	μ	Material	Temperature (°F)	μ
Liquids			*Liquids*		
Ammonia 100%	−20	0.230	Freon-22	−10	0.30
Ammonia 100%	−10	0.215	Freon-22	0	0.29
Ammonia 100%	0	0.200	Freon-22	10	0.28
Ammonia 100%	10	0.180	Freon-22	40	0.25
Ammonia 100%	20	0.160	Freon-113	−20	1.50
Ammonia 100%	30	0.150	Freon-113	−10	1.35
Ammonia 100%	40	0.140	Freon-113	0	1.22
Ammonia 100%	80	0.100	Freon-113	10	1.15
Brine, $CaCl_2$ 25%	0	9.50	Freon-113	40	0.90
Brine, $CaCl_2$ 25%	30	5.00	Glycerol 100%	150	75.00
Brine, $CaCl_2$ 25%	60	2.90	Glycerol 100%	200	18.00
Brine, $CaCl_2$ 25%	100	1.40	Glycerol 100%	300	1.50
Brine, $CaCl_2$ 25%	150	0.63	Glycerol 50%	0	22.00
Brine, $CaCl_2$ 25%	200	0.29	Glycerol 50%	100	3.60
Brine, NaCl 25%	0	5.00	Glycerol 50%	200	0.80
Brine, NaCl 25%	30	3.40	Glycerol 50%	300	0.22
Brine, NaCl 25%	60	2.50	Kerosene	0	5.00
Brine, NaCl 25%	100	1.60	Kerosene	50	2.90
Brine, NaCl 25%	150	0.95	Kerosene	100	1.65
Brine, NaCl 25%	200	0.62	Kerosene	150	1.00
Ethylene glycol	0	120	Kerosene	200	0.65
Ethylene glycol	50	35	Mercury	0	1.85
Ethylene glycol	100	13	Mercury	50	1.65
Ethylene glycol	150	5.3	Mercury	200	1.25
Ethylene glycol	200	2.4	Mercury	300	1.10
Freon-11	−20	0.8	Mercury	350	1.00
Freon-11	−10	0.75	Sodium	0	1.40
Freon-11	0	0.70	Sodium	50	1.15
Freon-11	10	0.65	Sodium	100	0.96
Freon-11	40	0.55	Sodium	150	0.82
Freon-12	−20	0.38	Sodium	200	0.72
Freon-12	−10	0.37	Sodium	300	0.55
Freon-12	0	0.35	Sulfuric acid 60%	0	15.00
Freon-12	10	0.34	Sulfuric acid 60%	100	4.60
Freon-12	40	0.29	Sulfuric acid 60%	200	1.80
Freon-21	−20	0.55	Sulfuric acid 60%	300	0.82
Freon-21	−10	0.50	Trichloroethylene	0	0.88
Freon-21	0	0.49	Trichloroethylene	100	0.52
Freon-21	10	0.47	Trichloroethylene	200	0.33
Freon-21	20	0.45	Water	50	1.20
Freon-21	40	0.41	Water	100	0.72
Freon-22	−20	0.31	Water	200	0.28

TABLE 18.1-10   (*Continued*)

Material	Temperature (°F)	μ	Material	Temperature (°F)	μ
Gases			*Gases*		
Acelylene	0	0.0065	Freon-21	200	0.0132
Acelylene	100	0.0080	Freon-22	−100	0.0094
Acelylene	200	0.0096	Freon-22	0	0.0112
Acelylene	300	0.0114	Freon-22	40	0.0120
Air	−100	0.0132	Freon-22	100	0.0132
Air	0	0.0160	Freon-22	200	0.0150
Air	100	0.0180	Freon-113	−100	0.0080
Air	200	0.0205	Freon-113	0	0.0093
Air	300	0.0230	Freon-113	40	0.0100
Ammonia	−50	0.0078	Freon-113	100	0.0107
Ammonia	0	0.0086	Freon-113	200	0.0118
Ammonia	100	0.0105	Chlorine	32	0.0122
Ammonia	200	0.0125	Chlorine	100	0.0140
Ammonia	300	0.0144	Ethylene	0	0.0087
Butane	40	0.0078	Ethylene	100	0.0102
Butane	212	0.0100	Ethylene	200	0.0120
Carbon dioxide	−100	0.0100	Helium	0	0.0169
Carbon dioxide	0	0.0125	Helium	100	0.0192
Carbon dioxide	100	0.0148	Helium	200	0.0220
Carbon dioxide	200	0.0170	Helium	300	0.0240
Carbon dioxide	400	0.0215	Hydrogen	0	0.0080
Carbon dioxide	1000	0.0345	Hydrogen	100	0.0090
Carbon monoxide	−100	0.0133	Hydrogen	200	0.0100
Carbon monoxide	0	0.0160	Hydrogen	300	0.0110
Carbon monoxide	100	0.0182	Mercury	0	0.0140
Carbon monoxide	200	0.0205	Mercury	100	0.0190
Carbon monoxide	400	0.0250	Mercury	200	0.0250
Carbon monoxide	1000	0.0360	Mercury	400	0.0380
Freon-11	−100	0.0080	Methane	0	0.0103
Freon-11	0	0.0096	Methane	100	0.0120
Freon-11	40	0.0103	Methane	200	0.0140
Freon-11	100	0.0111	Methane	300	0.0158
Freon-11	200	0.0128	Methane	400	0.0172
Freon-12	−100	0.0093	Nitrogen	−100	0.0130
Freon-12	0	0.0110	Nitrogen	0	0.0156
Freon-12	40	0.0118	Nitrogen	100	0.0180
Freon-12	100	0.0125	Nitrogen	200	0.0202
Freon-12	200	0.0142	Nitrogen	500	0.0275
Freon-21	−100	0.0085	Nitrogen	1000	0.0370
Freon-21	0	0.0102	Oxygen	−100	0.0150
Freon-21	40	0.0108	Oxygen	0	0.0180
Freon-21	100	0.0118	Oxygen	100	0.0210
			Oxygen	200	0.0230
			Oxygen	500	0.0305
			Oxygen	1000	0.0410
			Propane	−100	0.0056
			Propane	0	0.0070
			Propane	100	0.0082
			Sulfur dioxide	0	0.0108
			Sulfur dioxide	100	0.0126
			Sulfur dioxide	200	0.0145
			Water	100	0.0100
			Water	200	0.0120
			Water	400	0.0162
			Water	600	0.0205
			Water	700	0.0230

a From ref. 5. Used by permission.

TABLE 18.1-11   PROPERTIES OF SELECTED SOLIDS[a]

Substance	Temperature, °C	Thermal conductivity k, W(m · °C)	density ρ, kg/m³	Specific heat c, kJ/(kg · °C)	Thermal diffusivity α, m²/s × 10⁶	Reference[b]
Polyurethane foam, light	[c]	0.016–0.033	32–48	—	—	3
Polyurethane foam, heavy	[c]	0.022–0.040	64–112	—	—	3
Glass wool	23	0.038	24	0.7	2.26	2
Wood (pine)	20	0.10	497	2.8	0.075	1
Acrylic (general purpose)	[c]	0.21	1190	1.47	0.12	3
Polyvinyl chloride	[c]	0.15	1450	—	—	3
PVC (chlorinated)	[c]	—	1530	—	—	3
Ethylene-propylene diene (EPDM)	[c]	0.25	860	—	—	3
Soil						
(dry)	20	~ 0.35	~ 730	1.84	~ 0.26	1
(wet)	20	~ 2.36	—	—	~ 0.78	1
Ice	0	2.22	913	0.239	1.24	1
Concrete (1-2-4 mix)	20	1.37	2100	0.88	0.75	2
Building brick	20	0.66	1700	0.84	0.45	1
Granite	~ 20	~ 2.9	2640	0.82	~ 1.3	2
Glass (window)	20	0.78	2700	0.84	0.34	2
Glass (borosilicate)	~ 60	1.09	2200	—	—	2
Stainless steel (18-8)	20	16.3	7817	0.46	4.44	2
	100	17	—	—	—	2
Carbon steel (0.5%)	20	54	7833	0.465	14.74	2
	100	52	—	—	—	2
Nickel	20	90	8906	0.445	22.66	2
Brass (70 Cu/30 Zn)	20	111	8522	0.385	34.1	2
Aluminum	20	204	8418	0.896	84.2	2
	100	206	—	—	—	2
Copper	20	386	8954	0.383	112.3	2
	100	379	—	—	—	2

[a] From ref. 6. Used by permission.
[b] 1. Frank Kreith, *Principles of Heat Transfer*, 3d ed., Harper & Row, New York, 1973.
2. A. I. Brown, and S. M. Marco, *Introduction to Heat Transfer*, 3d ed., McGraw Hill Book Co., New York, 1958.
3. *Materials Engineering—Materials Selector '78*, Reinhold Publishing Company, New York, 1977.
[c] Where temperature not stated, approximately room temperature.

TABLE 18.1-12 PROPERTIES OF SELECTED LIQUIDS[a]

Liquid	Temperature, °C	Density ρ, kg/m³	Specific Heat c_p, kJ/(kg·°C)	Kinematic Viscosity v, m²/s × 10⁶	Thermal Conductivity k, W/(m·°C)	Thermal Diffusivity x, m²/s × 10⁶	Thermal Coefficient of Volume Expansion β, K⁻¹ × 10³	Prandtl Number, Pr
Engine oil	20	888	1.88	900	0.145	0.087	0.70	10,400
	100	840	2.22	20.3	0.137	0.074	—	276
Ethylene glycol	0	1127	2.24	53.0	0.303	0.120	—	440
	20	1114	2.34	19.0	0.289	0.111	0.59	170
	60	1085	2.50	5.1	0.260	0.096	—	53
75 EG/25 H₂O[b]	0	1106	2.68	19	0.356	0.120	—	160
	20	1094	2.80	7.3	0.346	0.113	0.63	65
	60	1065	3.03	2.4	0.327	0.101	—	24
50 EG/50 H₂O[b]	0	1074	3.19	7.4	0.424	0.124	—	60
	20	1065	3.31	3.7	0.419	0.119	0.54	31
	60	1040	3.52	1.3	0.415	0.113	—	11.5
Propylene glycol	0	1050	2.35	240	0.227	0.092	—	2,600
	20	1036	2.47	54	0.216	0.084	0.68	640
	60	1006	2.70	8.2	0.194	0.071	—	115
75 PG/25 H₂O[b]	0	1059	2.94	57	0.301	0.097	—	590
	20	1045	3.03	17	0.292	0.092	0.67	185
	60	1015	3.24	3.3	0.273	0.083	—	40
50 PG/50 H₂O[b]	0	1052	3.52	16	0.387	0.105	—	150
	20	1034	3.56	5.7	0.382	0.104	~ 0.69	55
	60	1010	3.70	1.7	0.377	0.101	—	16.8
Triethylene glycol	20	1126	2.09	47	0.237	0.101	—	470
	60	1095	2.28	13.7	0.219	0.088	0.69	155
75 TEG/25 H₂O[b]	20	1112	2.75	19	0.308	0.101	—	190
	60	1080	2.93	4.5	0.291	0.092	0.61	49

TABLE 18.1-12 (*Continued*)

Liquid	Temperature, °C	Density ρ, kg/m³	Specific Heat c_p, kJ/(kg·°C)	Kinematic Viscosity v, m²/s × 10⁶	Thermal Conductivity k, W/(m·°C)	Thermal Diffusivity x, m²/s × 10⁶	Thermal Coefficient of Volume Expansion β, K⁻¹ × 10³	Prandtl Number, Pr
Freon	−20	1461	0.907	0.235	0.071	0.0539	—	4.4
	0	1397	0.935	0.214	0.073	0.0557	—	3.8
	20	1330	0.966	0.198	0.073	0.0560	—	3.5
Water	0	1000	4.225	1.79	0.566	0.134	—	13.25
	20	997	4.180	1.01	0.602	0.144	—	7.0
	60	983	4.179	0.479	0.654	0.159	0.49	3.01
	100	958	4.211	0.295	0.682	0.169	—	1.76
	260	785	4.731	0.136	0.616	0.166	—	0.83
NaK (56/44)	93	890	1.13	0.652	25.6	25.5	—	0.026
Dow Therm A	100	999	—	0.991	0.132	—	0.84	—
	200	910	1.65	0.431	0.119	0.079	1.02	5.4
	300	809	1.89	0.263	0.107	0.070	1.37	3.8
	400	681	2.10	0.203	0.094	0.066	—	3.1

[a]From ref. 6. Used by permission.
[b]By weight.

TABLE 18.1-13   PROPERTIES OF SELECTED GASES[a]

Gas	Temperature, °K	Density ρ, kg/m³	Specific heat c_p, kJ/(kg · °C)	Kinematic viscosity v, m²/s × 10⁶	Thermal conductivity k, W/(m · °C)	Thermal diffusivity α, m²/s × 10⁶	Prandtl number, Pr
Helium	250	0.1944	5.2	91	0.134	132.6	0.70
	300	0.1620	5.2	124	0.149	176.9	0.70
	400	0.1215	5.2	200	0.178	281.7	0.71
	500	0.0972	5.2	290	0.203	401.6	0.72
Nitrogen	300	1.142	1.041	15.63	0.0262	22.04	0.713
	400	0.854	1.046	25.74	0.0334	37.34	0.691
	500	0.682	1.056	37.66	0.0398	55.3	0.684
Air	250	1.143	1.005	9.49	0.0223	13.16	0.722
	300	1.177	1.006	15.68	0.0262	22.16	0.708
	400	0.883	1.014	25.9	0.0337	37.6	0.689
	500	0.705	1.030	37.9	0.0404	55.6	0.680
	600	0.588	1.055	51.3	0.0466	75.1	0.680
	700	0.503	1.075	66.3	0.0523	96.7	0.684
	800	0.441	1.098	82.3	0.0578	119.5	0.689
	900	0.393	1.121	99.3	0.0628	142.7	0.696
	1000	0.352	1.142	117.8	0.0675	167.8	0.702
Carbon dioxide	250	2.166	0.804	5.81	0.0129	7.40	0.793
	300	1.797	0.871	8.32	0.0166	10.59	0.770
	400	1.342	0.942	14.39	0.0246	19.46	0.738
	500	1.073	1.013	21.67	0.0335	30.84	0.702
Water vapor	400	0.554	2.014	24.2	0.0261	23.4	1.040
	500	0.441	1.99	38.6	0.0339	38.7	0.996
	600	0.365	2.03	56.6	0.0422	57.3	0.986
	700	0.314	2.09	77.2	0.0505	77.2	1.00
	800	0.274	2.15	102	0.0592	100.1	1.010

[a]At atmospheric pressure. (From ref. 6. Used by permission.)

TABLE 18.1-14   SUBROUTINES FOR PROVIDING THE THERMODYNAMIC PROPERTIES OF STEAM[a]

Many calculations, for example, the various thermodynamic cycles, require the input of thermodynamic properties of a working fluid at various state points. In order that the computer can obtain them when needed without requiring a manual input, the following polynomial fits are provided. These expressions may be incorporated into a program as subroutines and called when the various properties are desired in the calculation.

Coefficients are given for obtaining the quantities in conventional units, Btu, lb, °F, psi, and so on. For conversion to SI units use the following factors:

°F to K—Add 459.67 and divide by 1.8.
Btu/lb to J/kg—Multiply by 2324.
Btu/lb · °F to J/kg · k—Multiply by 4183.
psi to pascals (N/m²)—Multiply by 6894.

Water and Steam

The error of points calculated from the following polynomials will, except for Part B using T and S as arguments, be no more than 2% of the values tabulated in common steam tables over the range from 1 psia, 102°F to 1000 psia, 1000°F and will be less than 1% over most of the range. The upper limit for Part B is 550°F to preserve the 2% accuracy. More complicated programs exist that provide greater accuracy and cover the entire range of temperature and pressure commonly given in steam tables.[b, c, d]

A. Using P and T as Arguments (psia and °F)

Let RHO = Ln(P)   (Natural Log)

$$X = 1.0750 - 0.12286(\text{RHO}) - (1.4143\text{E} - 03)(\text{RHO})^2 + (2.7405\text{E} - 06)(\text{RHO})^3$$

1671

TABLE 18.1-14 *(Continued)*

Saturation Temperature, TSAT

$$\text{TSAT} = \frac{705.47 - 459.67X}{1 + X} \quad °F$$

$$\text{RSAT} = \text{TSAT} + 459.67 \quad °R$$

Saturation Enthalpy of Steam, HSAT

$$\text{TR} = T + 459.67 \quad °R$$

$$\text{HSAT} = 859.2 + 45.354(705.47 - \text{TR})^{.5} - 1.5713(705.47 - \text{TR})$$

$$+ (5.416\text{E} - 03)(705.47 - \text{TR})^{1.5} \text{ Btu/lb}$$

Saturation Entropy of Steam

$$\text{SSAT} = 1.2707 + .82886\,X - .38025\,X^2 + .20749\,X^3 \text{ Btu/lb °F}$$

Superheat, Temperatures Above TSAT at Chosen Pressure

For any temperature, TR, °R

$$\text{Let TAU} = \text{LN}\!\left(\frac{\text{TR}}{\text{RSAT}}\right) \quad \text{and} \quad \text{LMT} = \frac{\text{TR} - \text{RSAT}}{\text{LN}(\text{TR}/\text{RSAT})}$$

Entropy, S

$$S = 1.98793 - .084015(\text{RHO}) + .39408(\text{TAU}) + .08119(\text{TAU})^2$$

$$+ (.10649\text{E} - 03)(\text{RHO})^2 + .008029(\text{RHO})(\text{TAU})^2 + .0075381(\text{TAU})(\text{RHO})^2$$

$$- .009056(\text{TAU})^2(\text{RHO})^2 \quad \text{Btu/lb °F}$$

Enthalpy, H

$$H = \text{HSAT} + \text{LMT}(S - \text{SSAT}) \quad \text{Btu/lb}$$

Volume, V

If $P < 440$ psia, then

$$V = -1.58603 - .10988(1/P)^{1.5} + (.24089\text{E} - 02)(\text{TR}) - (.94172\text{E} - 06)(\text{TR})^2$$

$$+ .59581(\text{TR}/P) - (.31478\text{E} - 10)(1/P)(\text{TR})^3 \text{ cu ft/lb}$$

If $P > 440$ psia, then

$$V = -.85199 + (.82396\text{E} - 03)(\text{TR}) - (.13256\text{E} - 09)(\text{TR})^3 - (.50059\text{E} - 05)(\text{TR})^3/P^2$$

$$+ .61671(\text{TR})/P + 915.54(\text{TR})/P^3 \text{ cu ft/lb}$$

B. Using *T* and *S* as Arguments

To calculate pressure, P at a state point described by T and S, the following equations are valid within 2% for values of T between 250 and 550°F and for pressures between 5 and 1000 psia.

TABLE 18.1-14 (*Continued*)

First calculate entropy of saturation for temperature T

$$SSAT = 2.30147 - (3.89793E - 03)(T) + (8.10535E - 06)(T)^2$$

$$- (9.60652E - 09)(T)^3 + (4.07076E - 12)(T)^3$$

Next calculate coefficients, A, B, C, D, and E in terms of TR = $(T + 459.67)$

$$A = 8.57285 + (8.23472E + 03)/TR - (1.35864E + 07)/TR^2 + (3.64981E + 09)/TR^3$$

$$B = 76.96141 - (1.6201E + 05)/TR + (1.04142E + 08)/TR^2 - (2.28003E + 10)/TR^3$$

$$C = 34.14727 - (1.90409E + 05)/TR + (2.06549E + 08)/TR^2 - (6.38044E + 10)/TR^3$$

$$D = -271.903 + (7.42758E + 05)/TR - (6.20882E + 08)/TR^2 + (1.6530E + 11)/TR^3$$

$$E = 148.324 - (3.60687E + 05)/TR + (2.70077E + 08)/TR^2 - (6.4134E + 10)/TR^3$$

Using these coefficients,

$$P = EXP\left(A + B(S - SS) + C(S - SS)^2 + D(S - SS)^3 + E(S - SS)^4 \right)$$

Having obtained a value of P from the above, the other properties may be calculated from Part A.
   Note: The properties of steam change rapidly as the critical temperature, 705°F, is approached and these equations for calculating P knowing T and S become inaccurate above 550°F.

C. Properties of Liquid

Specific Heat of Liquid, CPL

$$CPL = 1.00054 - .00161P^{.333} + .00074P^{.667} \text{ Btu/lb °F}$$

Enthalpy of Liquid, HLIQ

$$HLIQ = CPL(TSAT - 32) \text{ Btu/lb}$$

Entropy of Liquid, SLIQ

$$SLIQ = \frac{CPL(LN(RSAT))}{32 + 459.67}$$

[a] Contributed by Louis Bertrand, Senior Consultant (Retired), Engineering Department, E. I. DuPont de Nemours and Marvin D. Martin, Professor (Retired), of Aerospace and Mechanical Engineering, University of Arizona.
[b] See I. Z. Kuck, *Thermodynamic Properties of Water for Computer Simulation of Power Plants*, NUREG/CR-2518.
[c] See W. C. Reynolds, *Thermodynamic Properties in SI*, Department of Mechanical Engineering, Stanford University.
[d] See *ASME Steam Tables*. American Society of Mechanical Engineers.

TABLE 18.1-15 THERMODYNAMIC PROPERTIES OF STEAM: CUSTOMARY UNITS[a]

DRY SATURATED STEAM: TEMPERATURE TABLE *

Temp., F t	Abs. Press., lbf/in.² P	Specific Volume, ft³/lbm			Enthalpy, Btu/lbm			Entropy, Btu/lbm R		
		Sat. Liquid v_f	Evap. v_{fg}	Sat. Vapor v_g	Sat. Liquid h_f	Evap. h_{fg}	Sat. Vapor h_g	Sat. Liquid s_f	Evap. s_{fg}	Sat. Vapor s_g
32	0.08854	0.01602	3306	3306	0.00	1075.8	1075.8	0.0000	2.1877	2.1877
35	0.09995	0.01602	2947	2947	3.02	1074.1	1077.1	0.0061	2.1709	2.1770
40	0.12170	0.01602	2444	2444	8.05	1071.3	1079.3	0.0162	2.1435	2.1597
45	0.14752	0.01602	2036.4	2036.4	13.06	1068.4	1081.5	0.0262	2.1167	2.1429
50	0.17811	0.01603	1703.2	1703.2	18.07	1065.6	1083.7	0.0361	2.0903	2.1264
60	0.2563	0.01604	1206.6	1206.7	28.06	1059.9	1088.0	0.0555	2.0393	2.0948
70	0.3631	0.01606	867.8	867.9	38.04	1054.3	1092.3	0.0745	1.9902	2.0647
80	0.5069	0.01608	633.1	633.1	48.02	1048.6	1096.6	0.0932	1.9428	2.0360
90	0.6982	0.01610	468.0	468.0	57.99	1042.9	1100.9	0.1115	1.8972	2.0087
100	0.9492	0.01613	350.3	350.4	67.97	1037.2	1105.2	0.1295	1.8531	1.9826
110	1.2748	0.01617	265.3	265.4	77.94	1031.6	1109.5	0.1471	1.8106	1.9577
120	1.6924	0.01620	203.25	203.27	87.92	1025.8	1113.7	0.1645	1.7694	1.9339
130	2.2225	0.01625	157.32	157.34	97.90	1020.0	1117.9	0.1816	1.7296	1.9112
140	2.8886	0.01629	122.99	123.01	107.89	1014.1	1122.0	0.1984	1.6910	1.8894
150	3.718	0.01634	97.06	97.07	117.89	1008.2	1126.1	0.2149	1.6537	1.8685
160	4.741	0.01639	77.27	77.29	127.89	1002.3	1130.2	0.2311	1.6174	1.8485
170	5.992	0.01645	62.04	62.06	137.90	996.3	1134.2	0.2472	1.5822	1.8293
180	7.510	0.01651	50.21	50.23	147.92	990.2	1138.1	0.2630	1.5480	1.8109
190	9.339	0.01657	40.94	40.96	157.95	984.1	1142.0	0.2785	1.5147	1.7932
200	11.526	0.01663	33.62	33.64	167.99	977.9	1145.9	0.2938	1.4824	1.7762

210	14.123	0.01670	27.80	27.82	178.05	971.6	1149.7	0.3090	1.4508	1.7598
212	14.696	0.01672	26.78	26.80	180.07	970.3	1150.4	0.3120	1.4446	1.7566
220	17.186	0.01677	23.13	23.15	188.13	965.2	1153.4	0.3239	1.4201	1.7440
230	20.780	0.01684	19.365	19.382	198.23	958.8	1157.0	0.3387	1.3901	1.7288
240	24.969	0.01692	16.306	16.323	208.34	952.2	1160.5	0.3531	1.3609	1.7140
250	29.825	0.01700	13.804	13.821	218.48	945.5	1164.0	0.3675	1.3323	1.6998
260	35.429	0.01709	11.746	11.763	228.64	938.7	1167.3	0.3817	1.3043	1.6860
270	41.858	0.01717	10.044	10.061	238.84	931.8	1170.6	0.3958	1.2769	1.6727
280	49.203	0.01726	8.628	8.645	249.06	924.7	1173.8	0.4096	1.2501	1.6597
290	57.556	0.01735	7.444	7.461	259.31	917.5	1176.8	0.4234	1.2238	1.6472
300	67.013	0.01745	6.449	6.466	269.59	910.1	1179.7	0.4369	1.1980	1.6350
310	77.68	0.01755	5.609	5.626	279.92	902.6	1182.5	0.4504	1.1727	1.6231
320	89.66	0.01765	4.896	4.914	290.28	894.9	1185.2	0.4637	1.1478	1.6115
330	103.06	0.01776	4.289	4.307	300.68	887.0	1187.7	0.4769	1.1233	1.6002
340	118.01	0.01787	3.770	3.788	311.13	879.0	1190.1	0.4900	1.0992	1.5891
350	134.63	0.01799	3.324	3.342	321.63	870.7	1192.3	0.5029	1.0754	1.5783
360	153.04	0.01811	2.939	2.957	332.18	862.2	1194.4	0.5158	1.0519	1.5677
370	173.37	0.01823	2.606	2.625	342.79	853.5	1196.3	0.5286	1.0287	1.5573
380	195.77	0.01836	2.317	2.335	353.45	844.6	1198.1	0.5413	1.0059	1.5471
390	220.37	0.01850	2.0651	2.0836	364.17	835.4	1199.6	0.5539	0.9832	1.5371
400	247.31	0.01864	1.8447	1.8633	374.97	826.0	1201.0	0.5664	0.9608	1.5272
410	276.75	0.01878	1.6512	1.6700	385.83	816.3	1202.1	0.5788	0.9386	1.5174
420	308.83	0.01894	1.4811	1.5000	396.77	806.3	1203.1	0.5912	0.9166	1.5078
430	343.72	0.01910	1.3308	1.3499	407.79	796.0	1203.8	0.6035	0.8947	1.4982
440	381.59	0.01926	1.1979	1.2171	418.90	785.4	1204.3	0.6158	0.8730	1.4887

TABLE 18.1-15 (*Continued*)

Temp., F	Abs. Press., lbf/in.²	Specific Volume, ft³/lbm			Enthalpy, Btu/lbm			Entropy, Btu/lbm R		
t	P	Sat. Liquid v_f	Evap. v_{fg}	Sat. Vapor v_g	Sat. Liquid h_f	Evap. h_{fg}	Sat. Vapor h_g	Sat. Liquid s_f	Evap. s_{fg}	Sat. Vapor s_g
450	422.6	0.0194	1.0799	1.0993	430.1	774.5	1204.6	0.6280	0.8513	1.4793
460	466.9	0.0196	0.9748	0.9944	441.4	763.2	1204.6	0.6402	0.8298	1.4700
470	514.7	0.0198	0.8811	0.9009	452.8	751.5	1204.3	0.6523	0.8083	1.4606
480	566.1	0.0200	0.7972	0.8172	464.4	739.4	1203.7	0.6645	0.7868	1.4513
490	621.4	0.0202	0.7221	0.7423	476.0	726.8	1202.8	0.6766	0.7653	1.4419
500	680.8	0.0204	0.6545	0.6749	487.8	713.9	1201.7	0.6887	0.7438	1.4325
520	812.4	0.0209	0.5385	0.5594	511.9	686.4	1198.2	0.7130	0.7006	1.4136
540	962.5	0.0215	0.4434	0.4649	536.6	656.6	1193.2	0.7374	0.6568	1.3942
560	1133.1	0.0221	0.3647	0.3868	562.2	624.2	1186.4	0.7621	0.6121	1.3742
580	1325.8	0.0228	0.2989	0.3217	588.9	588.4	1177.3	0.7872	0.5659	1.3532
600	1542.9	0.0236	0.2432	0.2668	617.0	548.5	1165.5	0.8131	0.5176	1.3307
620	1786.6	0.0247	0.1955	0.2201	646.7	503.6	1150.3	0.8398	0.4664	1.3062
640	2059.7	0.0260	0.1538	0.1798	678.6	452.0	1130.5	0.8679	0.4110	1.2789
660	2365.4	0.0278	0.1165	0.1442	714.2	390.2	1104.4	0.8987	0.3485	1.2472
680	2708.1	0.0305	0.0810	0.1115	757.3	309.9	1067.2	0.9351	0.2719	1.2071
700	3093.7	0.0369	0.0392	0.0761	823.3	172.1	995.4	0.9905	0.1484	1.1389
705.4	3206.2	0.0503	0	0.0503	902.7	0	902.7	1.0580	0	1.0580

Dry Saturated Steam: Pressure Table *

Abs. Press., lbf/in.² P	Temp., °F t	Specific Volume, ft³/lbm Sat. Liquid v_f	Sat. Vapor v_g	Enthalpy, Btu/lbm Sat. Liquid h_f	Evap. h_{fg}	Sat. Vapor h_g	Entropy, Btu/lbm R Sat. Liquid s_f	Evap. s_{fg}	Sat. Vapor s_g	Internal Energy, Btu/lbm Sat. Liquid u_f	Sat. Vapor u_g
1.0	101.74	0.01614	333.6	69.70	1036.3	1106.0	0.1326	1.8456	1.9782	69.70	1044.3
2.0	126.08	0.01623	173.73	93.99	1022.2	1116.2	0.1749	1.7451	1.9200	93.98	1051.9
3.0	141.48	0.01630	118.71	109.37	1013.2	1122.6	0.2008	1.6855	1.8863	109.36	1056.7
4.0	152.97	0.01636	90.63	120.86	1006.4	1127.3	0.2198	1.6427	1.8625	120.85	1060.2
5.0	162.24	0.01640	73.52	130.13	1001.0	1131.1	0.2347	1.6094	1.8441	130.12	1063.1
6.0	170.06	0.01645	61.98	137.96	996.2	1134.2	0.2472	1.5820	1.8292	137.94	1065.4
7.0	176.85	0.01649	53.64	144.76	992.1	1136.9	0.2581	1.5586	1.8167	144.74	1067.4
8.0	182.86	0.01653	47.34	150.79	988.5	1139.3	0.2674	1.5383	1.8057	150.77	1069.2
9.0	188.28	0.01656	42.40	156.22	985.2	1141.4	0.2759	1.5203	1.7962	156.19	1070.8
10	193.21	0.01659	38.42	161.17	982.1	1143.3	0.2835	1.5041	1.7876	161.14	1072.2
14.696	212.00	0.01672	25.80	180.07	970.3	1150.4	0.3120	1.4446	1.7566	180.02	1077.5
15	213.03	0.01672	25.29	181.11	969.7	1150.8	0.3135	1.4415	1.7549	181.06	1077.8
20	227.96	0.01683	20.089	196.16	960.1	1156.3	0.3356	1.3962	1.7319	196.10	1081.9
25	240.07	0.01692	16.303	208.42	952.1	1160.6	0.3533	1.3606	1.7139	208.34	1085.1
30	250.33	0.01701	13.746	218.82	945.3	1164.1	0.3680	1.3313	1.6993	218.73	1087.8
35	259.28	0.01708	11.898	227.91	939.2	1167.1	0.3807	1.3063	1.6870	227.80	1090.1
40	267.25	0.01715	10.498	236.03	933.7	1169.7	0.3919	1.2844	1.6763	235.90	1092.0
45	274.44	0.01721	9.401	243.36	928.6	1172.0	0.4019	1.2650	1.6669	243.22	1093.7
50	281.01	0.01727	8.515	250.09	924.0	1174.1	0.4110	1.2474	1.6585	249.93	1095.3
55	287.07	0.01732	7.787	256.30	919.6	1175.9	0.4193	1.2316	1.6509	256.12	1096.7
60	292.71	0.01738	7.175	262.09	915.5	1177.6	0.4270	1.2168	1.6438	261.90	1097.9
65	297.97	0.01743	6.655	267.50	911.6	1179.1	0.4342	1.2032	1.6374	267.29	1099.1
70	302.92	0.01748	6.206	272.61	907.9	1180.6	0.4409	1.1906	1.6315	272.38	1100.2
75	307.60	0.01753	5.816	277.43	904.5	1181.9	0.4472	1.1787	1.6259	277.19	1101.2
80	312.03	0.01757	5.472	282.02	901.1	1183.1	0.4531	1.1676	1.6207	281.76	1102.1
85	316.25	0.01761	5.168	286.39	897.8	1184.2	0.4587	1.1571	1.6158	286.11	1102.9
90	320.27	0.01766	4.896	290.56	894.7	1185.3	0.4641	1.1471	1.6112	290.27	1103.7
95	324.12	0.01770	4.652	294.56	891.7	1186.2	0.4692	1.1376	1.6068	294.25	1104.5
100	327.81	0.01774	4.432	298.40	888.8	1187.2	0.4740	1.1286	1.6026	298.08	1105.2
110	334.77	0.01782	4.049	305.66	883.2	1188.9	0.4832	1.1117	1.5948	305.30	1106.5

TABLE 18.1-15 (Continued)

Abs. Press., lbf/in.² P	Temp., °F t	Specific Volume, ft³/lbm		Enthalpy, Btu/lbm			Entropy, Btu/lbm R			Internal Energy, Btu/lbm	
		Sat. Liquid v_f	Sat. Vapor v_g	Sat. Liquid h_f	Evap. h_{fg}	Sat. Vapor h_g	Sat. Liquid s_f	Evap. s_{fg}	Sat. Vapor s_g	Sat. Liquid u_f	Sat. Vapor u_g
120	341.25	0.01789	3.728	312.44	877.9	1190.4	0.4916	1.0962	1.5878	312.05	1107.6
130	347.32	0.01796	3.455	318.81	872.9	1191.7	0.4995	1.0817	1.5812	318.38	1108.6
140	353.02	0.01802	3.220	324.82	868.2	1193.0	0.5069	1.0682	1.5751	324.35	1109.6
150	358.42	0.01809	3.015	330.51	863.6	1194.1	0.5138	1.0556	1.5694	330.01	1110.5
160	363.53	0.01815	2.834	335.93	859.2	1195.1	0.5204	1.0436	1.5640	335.39	1111.2
170	368.41	0.01822	2.675	341.09	854.9	1196.0	0.5266	1.0324	1.5590	340.52	1111.9
180	373.06	0.01827	2.532	346.03	850.8	1196.9	0.5325	1.0217	1.5542	345.42	1112.5
190	377.51	0.01833	2.404	350.79	846.8	1197.6	0.5381	1.0116	1.5497	350.15	1113.1
200	381.79	0.01839	2.288	355.36	843.0	1198.4	0.5435	1.0018	1.5453	354.68	1113.7
250	400.95	0.01865	1.8438	376.00	825.1	1201.1	0.5675	0.9588	1.5263	375.14	1115.8
300	417.33	0.01890	1.5433	393.84	809.0	1202.8	0.5879	0.9225	1.5104	392.79	1117.1
350	431.72	0.01913	1.3260	409.69	794.2	1203.9	0.6056	0.8910	1.4966	408.45	1118.0
400	444.59	0.0193	1.1613	424.0	780.5	1204.5	0.6214	0.8630	1.4844	422.6	1118.5
450	456.28	0.0195	1.0320	437.2	767.4	1204.6	0.6356	0.8378	1.4734	435.5	1118.7
500	467.01	0.0197	0.9278	449.4	755.0	1204.4	0.6487	0.8147	1.4634	447.6	1118.6
550	476.94	0.0199	0.8424	460.8	743.1	1203.9	0.6608	0.7934	1.4542	458.8	1118.2
600	486.21	0.0201	0.7698	471.6	731.6	1203.2	0.6720	0.7734	1.4454	469.4	1117.7
650	494.90	0.0203	0.7083	481.8	720.5	1202.3	0.6826	0.7548	1.4374	479.4	1117.1
700	503.10	0.0205	0.6554	491.5	709.7	1201.2	0.6925	0.7371	1.4296	488.8	1116.3
750	510.86	0.0207	0.6092	500.8	699.2	1200.0	0.7019	0.7204	1.4223	498.0	1115.4
800	518.23	0.0209	0.5687	509.7	688.9	1198.6	0.7108	0.7045	1.4153	506.6	1114.4
850	525.26	0.0210	0.5327	518.3	678.8	1197.1	0.7194	0.6891	1.4085	515.0	1113.3
900	531.98	0.0212	0.5006	526.6	668.8	1195.4	0.7275	0.6744	1.4020	523.1	1112.1
950	538.43	0.0214	0.4717	534.6	659.1	1193.7	0.7355	0.6602	1.3957	530.9	1110.8
1000	544.61	0.0216	0.4456	542.4	649.4	1191.8	0.7430	0.6467	1.3897	538.4	1109.4
1100	556.31	0.0220	0.4001	557.4	630.4	1187.8	0.7575	0.6205	1.3780	552.9	1106.4
1200	567.22	0.0223	0.3619	571.7	611.7	1183.4	0.7711	0.5956	1.3667	566.7	1103.0
1300	577.46	0.0227	0.3293	585.4	593.2	1178.6	0.7840	0.5719	1.3559	580.0	1099.4
1400	587.10	0.0231	0.3012	598.7	574.7	1173.4	0.7963	0.5491	1.3454	592.7	1095.4
1500	596.23	0.0235	0.2765	611.6	556.3	1167.9	0.8082	0.5269	1.3351	605.1	1091.2
2000	635.82	0.0257	0.1878	671.7	463.4	1135.1	0.8619	0.4230	1.2849	662.2	1065.6
2500	668.13	0.0287	0.1307	730.6	360.5	1091.1	0.9126	0.3197	1.2322	717.3	1030.6
3000	695.36	0.0346	0.0858	802.5	217.8	1020.3	0.9731	0.1885	1.1615	783.4	972.7
3206.2	705.40	0.0503	0.0503	902.7	0	902.7	1.0580	0	1.0580	872.9	872.9

Temperature, F

Abs. Press., lbf/in.² (Sat. Temp.)		200	220	300	350	400	450	500	550	600	700	800	900	1000
1 (101.74)	v	392.6	404.5	452.3	482.2	512.0	541.8	571.6	601.4	631.2	690.8	750.4	809.9	869.5
	h	1150.4	1159.5	1195.8	1218.7	1241.7	1264.9	1288.3	1312.0	1335.7	1383.8	1432.8	1482.7	1533.5
	s	2.0512	2.0647	2.1153	2.1444	2.1720	2.1983	2.2233	2.2468	2.2702	2.3137	2.3542	2.3923	2.4283
5 (162.24)	v	78.16	80.59	90.25	96.26	102.26	108.24	114.22	120.19	126.16	138.10	150.03	161.95	173.87
	h	1148.8	1158.1	1195.0	1218.1	1241.2	1264.5	1288.0	1311.7	1335.4	1383.6	1432.7	1482.6	1533.4
	s	1.8718	1.8857	1.9370	1.9664	1.9942	2.0205	2.0456	2.0692	2.0927	2.1361	2.1767	2.2148	2.2509
10 (193.21)	v	38.85	40.09	45.00	48.03	51.04	54.05	57.05	60.04	63.03	69.01	74.98	80.95	86.92
	h	1146.6	1156.2	1193.9	1217.2	1240.6	1264.0	1287.5	1311.3	1335.1	1383.4	1432.5	1482.4	1533.2
	s	1.7927	1.8071	1.8595	1.8892	1.9172	1.9436	1.9689	1.9924	2.0160	2.0596	2.1002	2.1383	2.1744
14.696 (212.00)	v		27.15	30.53	32.62	34.68	36.73	38.78	40.82	42.86	46.94	51.00	55.07	59.13
	h		1154.4	1192.8	1216.4	1239.9	1263.5	1287.1	1310.9	1334.8	1383.2	1432.3	1482.3	1533.1
	s		1.7624	1.8160	1.8460	1.8743	1.9008	1.9261	1.9498	1.9734	2.0170	2.0576	2.0958	2.1319
20 (227.96)	v			22.36	23.91	25.43	26.95	28.46	29.97	31.47	34.47	37.46	40.45	43.44
	h			1191.6	1215.6	1239.2	1262.9	1286.6	1310.5	1334.4	1382.9	1432.1	1482.1	1533.0
	s			1.7808	1.8112	1.8396	1.8664	1.8918	1.9160	1.9392	1.9829	2.0235	2.0618	2.0978
40 (267.25)	v			11.040	11.843	12.628	13.401	14.168	14.93	15.688	17.198	18.702	20.20	21.70
	h			1186.8	1211.9	1236.5	1260.7	1284.8	1308.9	1333.1	1381.9	1431.3	1481.4	1532.4
	s			1.6994	1.7314	1.7608	1.7881	1.8140	1.8384	1.8619	1.9058	1.9467	1.9850	2.0214
60 (292.71)	v			7.259	7.818	8.357	8.884	9.403	9.916	10.427	11.441	12.449	13.452	14.454
	h			1181.6	1208.2	1233.6	1258.5	1283.0	1307.4	1331.8	1380.9	1430.5	1480.8	1531.9
	s			1.6492	1.6830	1.7135	1.7416	1.7678	1.7926	1.8162	1.8605	1.9015	1.9400	1.9762

TABLE 18.1-15 (*Continued*)

Abs. Press., lbf/in.² (Sat. Temp.)		200	220	300	350	400	450	500	550	600	700	800	900	1000
									Temperature, F					
80 (312.03)	v				5.803	6.220	6.624	7.020	7.410	7.797	8.562	9.322	10.077	10.830
	h				1204.3	1230.7	1256.1	1281.1	1305.8	1330.5	1379.9	1429.7	1480.1	1531.3
	s				1.6475	1.6791	1.7078	1.7346	1.7598	1.7836	1.8281	1.8694	1.9079	1.9442
100 (327.81)	v				4.592	4.937	5.268	5.589	5.905	6.218	6.835	7.446	8.052	8.656
	h				1200.1	1227.6	1253.7	1279.1	1304.2	1329.1	1378.9	1428.9	1479.5	1530.8
	s				1.6188	1.6518	1.6813	1.7085	1.7339	1.7581	1.8029	1.8443	1.8829	1.9193
120 (341.25)	v				3.783	4.081	4.363	4.636	4.902	5.165	5.683	6.195	6.702	7.207
	h				1195.7	1224.4	1251.3	1277.2	1302.5	1327.7	1377.8	1428.1	1478.8	1530.2
	a				1.5944	1.6287	1.6591	1.6869	1.7127	1.7370	1.7822	1.8237	1.8625	1.8990
140 (353.02)	v					3.468	3.715	3.954	4.186	4.413	4.861	5.301	5.738	6.172
	h					1221.1	1248.7	1275.2	1300.9	1326.4	1376.8	1427.3	1478.2	1529.7
	s					1.6087	1.6399	1.6683	1.6945	1.7190	1.7645	1.8063	1.8451	1.8817
160 (363.53)	v					3.008	3.230	3.443	3.648	3.849	4.244	4.631	5.015	5.396
	h					1217.6	1246.1	1273.1	1299.3	1325.0	1375.7	1426.4	1477.5	1529.1
	s					1.5908	1.6230	1.6519	1.6785	1.7033	1.7491	1.7911	1.8301	1.8667
180 (373.06)	v					2.649	2.852	3.044	3.229	3.411	3.764	4.110	4.452	4.792
	h					1214.0	1243.5	1271.0	1297.6	1323.5	1374.7	1425.6	1476.8	1528.6
	s					1.5745	1.6077	1.6373	1.6642	1.6894	1.7355	1.7776	1.8167	1.8534
200 (381.79)	v					2.361	2.549	2.726	2.895	3.060	3.380	3.693	4.002	4.309
	h					1210.3	1240.7	1268.9	1295.8	1322.1	1373.6	1424.8	1476.2	1528.0
	s					1.5594	1.5937	1.6240	1.6513	1.6767	1.7232	1.7655	1.8048	1.8415
220 (389.86)	v					2.125	2.301	2.465	2.621	2.772	3.066	3.352	3.634	3.913
	h					1206.5	1237.9	1266.7	1294.1	1320.7	1372.6	1424.0	1475.5	1527.5
	s					1.5453	1.5808	1.6117	1.6395	1.6652	1.7120	1.7545	1.7939	1.8308

Press. (Sat. temp)		400	450	500	550	600	700	800	900	1000
240 (397.37)	v	1.0276	2.094	2.247	2.393	2.533	2.804	3.068	3.327	3.584
	h	1502.5	1234.9	1264.5	1292.4	1319.2	1371.5	1423.2	1474.8	1526.9
	s	1.5319	1.5686	1.6003	1.6286	1.6546	1.7017	1.7444	1.7839	1.8209
260 (404.42)	v		1.9183	2.063	2.199	2.330	2.582	2.827	3.067	3.305
	h		1232.0	1262.3	1290.5	1317.7	1370.4	1422.3	1474.2	1526.3
	s		1.5573	1.5897	1.6184	1.6447	1.6922	1.7352	1.7748	1.8118
280 (411.05)	v		1.7674	1.9047	2.033	2.156	2.392	2.621	2.845	3.066
	h		1228.9	1260.0	1288.7	1316.2	1369.4	1421.5	1473.5	1525.8
	s		1.5464	1.5796	1.6087	1.6354	1.6834	1.7265	1.7662	1.8033
300 (417.33)	v		1.6364	1.7675	1.8891	2.005	2.227	2.442	2.652	2.859
	h		1225.8	1257.6	1286.8	1314.7	1368.3	1420.6	1472.8	1525.2
	s		1.5360	1.5701	1.5998	1.6268	1.6751	1.7184	1.7582	1.7954
350 (431.72)	v		1.3734	1.4923	1.6010	1.7036	1.8980	2.084	2.266	2.445
	h		1217.7	1251.5	1282.1	1310.9	1365.5	1418.5	1471.1	1523.8
	s		1.5119	1.5481	1.5792	1.6070	1.6563	1.7002	1.7403	1.7777
400 (444.59)	v		1.1744	1.2851	1.3843	1.4770	1.6508	1.8161	1.9767	2.134
	h		1208.8	1245.1	1277.2	1306.9	1362.7	1416.4	1469.4	1522.4
	s		1.4892	1.5281	1.5607	1.5894	1.6398	1.6842	1.7247	1.7623

Temperature, F

Press. (Sat. temp)		500	550	600	620	640	660	680	700	800	900	1000	1200	1400	1600
450 (456.28)	v	1.1231	1.2155	1.3005	1.3332	1.3652	1.3967	1.4278	1.4584	1.6074	1.7516	1.8928	2.170	2.443	2.714
	h	1238.4	1272.0	1302.8	1314.6	1326.2	1337.5	1348.8	1359.9	1414.3	1467.7	1521.0	1628.6	1738.7	1851.9
	s	1.5095	1.5437	1.5735	1.5845	1.5951	1.6054	1.6153	1.6250	1.6699	1.7108	1.7486	1.8177	1.8803	1.9381

TABLE 18.1-15 (*Continued*)

Temperature, F

		500	550	600	620	640	660	680	700	800	900	1000	1200	1400	1600
500 (467.01)	v	0.9927	1.0800	1.1591	1.1893	1.2188	1.2478	1.2763	1.3044	1.4405	1.5715	1.6996	1.9504	2.197	2.442
	h	1231.3	1266.8	1298.6	1310.7	1322.6	1334.2	1345.7	1357.0	1412.1	1466.0	1519.6	1627.6	1737.9	1851.3
	s	1.4919	1.5280	1.5588	1.5701	1.5810	1.5915	1.6016	1.6115	1.6571	1.6982	1.7363	1.8056	1.8683	1.9262
550 (476.94)	v	0.8852	0.9686	1.0431	1.0714	1.0989	1.1259	1.1523	1.1783	1.3038	1.4241	1.5414	1.7706	1.9957	2.219
	h	1223.7	1261.2	1294.3	1306.8	1318.9	1330.8	1342.5	1354.0	1409.9	1464.3	1518.2	1626.6	1737.1	1850.6
	s	1.4751	1.5131	1.5451	1.5568	1.5680	1.5787	1.5890	1.5991	1.6452	1.6868	1.7250	1.7946	1.8575	1.9155
600 (486.21)	v	0.7947	0.8753	0.9463	0.9729	0.9988	1.0241	1.0489	1.0732	1.1899	1.3013	1.4096	1.6208	1.8279	2.033
	h	1215.7	1255.5	1289.9	1302.7	1315.2	1327.4	1339.3	1351.1	1407.7	1462.5	1516.7	1625.5	1736.3	1850.0
	s	1.4586	1.4990	1.5323	1.5443	1.5558	1.5667	1.5773	1.5875	1.6343	1.6762	1.7147	1.7846	1.8476	1.9056
700 (503.10)	v		0.7277	0.7934	0.8177	0.8411	0.8639	0.8860	0.9077	1.0108	1.1082	1.2024	1.3853	1.5641	1.7405
	h		1243.2	1280.6	1294.3	1307.5	1320.3	1332.8	1345.0	1403.2	1459.0	1513.9	1623.5	1734.8	1848.8
	s		1.4722	1.5084	1.5212	1.5333	1.5449	1.5559	1.5665	1.6147	1.6573	1.6963	1.7666	1.8299	1.8881
800 (518.23)	v		0.6154	0.6779	0.7006	0.7223	0.7433	0.7635	0.7833	0.8763	0.9633	1.0470	1.2088	1.3662	1.5214
	h		1229.8	1270.7	1285.4	1299.4	1312.9	1325.9	1338.6	1398.6	1455.4	1511.0	1621.4	1733.2	1847.5
	s		1.4467	1.4863	1.5000	1.5129	1.5250	1.5366	1.5476	1.5972	1.6407	1.6801	1.7510	1.8146	1.8729
900 (531.98)	v		0.5264	0.5873	0.6089	0.6294	0.6491	0.6680	0.6863	0.7716	0.8506	0.9262	1.0714	1.2124	1.3509
	h		1215.0	1260.1	1275.9	1290.9	1305.1	1318.8	1332.1	1393.9	1451.8	1508.1	1619.3	1731.6	1846.3
	s		1.4216	1.4653	1.4800	1.4938	1.5066	1.5187	1.5303	1.5814	1.6257	1.6656	1.7371	1.8009	1.8595
1000 (544.61)	v		0.4533	0.5140	0.5350	0.5546	0.5733	0.5912	0.6084	0.6878	0.7604	0.8294	0.9615	1.0893	1.2146
	h		1198.3	1248.8	1265.9	1281.9	1297.0	1311.4	1325.3	1389.2	1448.2	1505.1	1617.3	1730.0	1845.0
	s		1.3961	1.4450	1.4610	1.4757	1.4893	1.5021	1.5141	1.5670	1.6121	1.6525	1.7245	1.7886	1.8474
1100 (556.31)	v			0.4532	0.4738	0.4929	0.5110	0.5281	0.5445	0.6191	0.6866	0.7503	0.8716	0.9885	1.1031
	h			1236.7	1255.3	1272.4	1288.5	1303.7	1318.3	1384.3	1444.5	1502.2	1615.2	1728.4	1843.8
	s			1.4251	1.4425	1.4583	1.4728	1.4862	1.4989	1.5535	1.5995	1.6405	1.7130	1.7775	1.8363
1200 (567.22)	v			0.4016	0.4222	0.4410	0.4586	0.4752	0.4909	0.5617	0.6250	0.6843	0.7967	0.9046	1.0101
	h			1223.5	1243.9	1262.4	1279.6	1295.7	1311.0	1379.3	1440.7	1499.2	1613.1	1726.9	1842.5
	s			1.4052	1.4243	1.4413	1.4568	1.4710	1.4843	1.5409	1.5879	1.6293	1.7025	1.7672	1.8263
1400 (587.10)	v			0.3174	0.3390	0.3580	0.3753	0.3912	0.4062	0.4714	0.5281	0.5805	0.6789	0.7727	0.8640
	h			1193.0	1218.4	1240.4	1260.3	1278.5	1295.5	1369.1	1433.1	1493.2	1608.9	1723.7	1840.0
	s			1.3639	1.3877	1.4079	1.4258	1.4419	1.4567	1.5177	1.5666	1.6093	1.6836	1.7489	1.8083

Properties of Superheated Steam (Continued)

Abs. Press., lbf/in.² (Sat. Temp.)		500	550	600	620	640	660	680	700	800	900	1000	1200	1400	1600
1600 (604.90)	v				0.2733	0.2936	0.3112	0.3271	0.3417	0.4034	0.4553	0.5027	0.5906	0.6738	0.7545
	h				1187.8	1215.2	1238.7	1259.6	1278.7	1358.4	1425.3	1487.0	1604.6	1720.5	1837.5
	s				1.3489	1.3741	1.3952	1.4137	1.4303	1.4964	1.5476	1.5914	1.6669	1.7328	1.7926
1800 (621.03)	v					0.2407	0.2597	0.2760	0.2907	0.3502	0.3986	0.4421	0.5218	0.5968	0.6693
	h					1185.1	1214.0	1238.5	1260.3	1347.2	1417.4	1480.8	1600.4	1717.3	1835.0
	s					1.3377	1.3638	1.3855	1.4044	1.4765	1.5301	1.5752	1.6520	1.7185	1.7786
2000 (635.82)	v					0.1936	0.2161	0.2337	0.2489	0.3074	0.3532	0.3935	0.4668	0.5352	0.6011
	h					1145.6	1184.9	1214.8	1240.0	1335.5	1409.2	1474.5	1596.1	1714.1	1832.5
	s					1.2945	1.3300	1.3564	1.3783	1.4576	1.5139	1.5603	1.6384	1.7055	1.7660
2500 (668.13)	v							0.1484	0.1686	0.2294	0.2710	0.3061	0.3678	0.4244	0.4784
	h							1132.3	1176.8	1303.6	1387.8	1458.4	1585.3	1706.1	1826.2
	s							1.2687	1.3073	1.4127	1.4772	1.5273	1.6088	1.6775	1.7389
3000 (695.36)	v								0.0984	0.1760	0.2159	0.2476	0.3018	0.3505	0.3966
	h								1060.7	1267.2	1365.0	1441.8	1574.3	1698.0	1819.9
	s								1.1966	1.3690	1.4439	1.4984	1.5837	1.6540	1.7163
3206.2 (705.40)	v									0.1583	0.1981	0.2288	0.2806	0.3267	0.3703
	h									1250.5	1355.2	1434.7	1569.8	1694.6	1817.2
	s									1.3508	1.4309	1.4874	1.5742	1.6452	1.7080
3500	v								0.0306	0.1364	0.1762	0.2058	0.2546	0.2977	0.3381
	h								780.5	1224.9	1340.7	1424.5	1563.3	1689.8	1813.6
	s								0.9515	1.3241	1.4127	1.4723	1.5615	1.6336	1.6968
4000	v								0.0287	0.1052	0.1462	0.1743	0.2192	0.2581	0.2943
	h								763.8	1174.8	1314.4	1406.8	1552.1	1681.7	1807.2
	s								0.9347	1.2757	1.3827	1.4482	1.5417	1.6154	1.6795
4500	v								0.0276	0.0798	0.1226	0.1500	0.1917	0.2273	0.2602
	h								753.5	1113.9	1286.5	1388.4	1540.8	1673.5	1800.9
	s								0.9235	1.2204	1.3529	1.4253	1.5235	1.5990	1.6640
5000	v								0.0268	0.0593	0.1036	0.1303	0.1696	0.2027	0.2329
	h								746.4	1047.1	1256.5	1369.5	1529.5	1665.3	1794.5
	s								0.9152	1.1622	1.3231	1.4034	1.5066	1.5839	1.6499
5500	v								0.0262	0.0463	0.0880	0.1143	0.1516	0.1825	0.2106
	h								741.3	985.0	1224.1	1349.3	1518.2	1657.0	1788.1
	s								0.9090	1.1093	1.2930	1.3821	1.4908	1.5699	1.6369

TABLE 18.1-15 (Continued)

COMPRESSED LIQUID *

Abs. Press., lbf/in.² (Sat. Temp.)	Saturated Liquid		Temperature, F							
			32	100	200	300	400	500	600	700
		P	0.08854	0.9492	11.526	67.013	247.31	680.8	1542.9	3093.7
		v_f	0.016022	0.016132	0.016634	0.017449	0.018639	0.020432	0.023629	0.03692
		h_f	0	67.97	167.99	269.59	374.97	487.82	617.0	823.3
		s_f	0	0.12948	0.29382	0.43694	0.56638	0.68871	0.8131	0.9905
200 (381.79)		$(v - v_f) \cdot 10^5$	−1.1	−1.1	−1.1	−1.1				
		$(h - h_f)$	+0.61	+0.54	+0.41	+0.23				
		$(s - s_f) \cdot 10^3$	+0.03	−0.05	−0.21	−0.21				
400 (444.59)		$(v - v_f) \cdot 10^5$	−2.3	−2.1	−2.2	−2.8	−2.1			
		$(h - h_f)$	+1.21	+1.09	+0.88	+0.61	+0.16			
		$(s - s_f) \cdot 10^3$	+0.04	−0.16	−0.47	−0.56	−0.40			
800 (518.23)		$(v - v_f) \cdot 10^5$	−4.6	−4.0	−4.4	−5.6	−6.5	−1.7		
		$(h - h_f)$	+2.39	+2.17	+1.78	+1.35	+0.61	−0.05		
		$(s - s_f) \cdot 10^3$	+0.10	−0.40	−0.97	−1.27	−1.48	−0.53		
1000 (544.61)		$(v - v_f) \cdot 10^5$	−5.7	−5.1	−5.4	−6.9	−8.7	−6.4		
		$(h - h_f)$	+2.99	+2.70	+2.21	+1.75	+0.84	−0.14		
		$(s - s_f) \cdot 10^3$	+0.15	−0.53	−1.20	−1.64	−2.00	−1.41		

1500	$(v-v_f)\cdot 10^5$	-8.4	-7.5	-8.1	-10.4	-14.1	-17.3		
(596.23)	$(h-h_f)$	$+4.48$	$+3.99$	$+3.36$	$+2.70$	$+1.44$	-0.29		
	$(s-s_f)\cdot 10^3$	$+0.20$	-0.86	-1.79	-2.53	-3.32	-3.56		
2000	$(v-v_f)\cdot 10^5$	-11.0	-9.9	-10.8	-13.8	-19.5	-27.8	-32.6	
(635.82)	$(h-h_f)$	$+5.97$	$+5.31$	$+4.51$	$+3.64$	$+2.03$	-0.38	-2.5	
	$(s-s_f)\cdot 10^3$	$+0.22$	-1.18	-2.39	-3.42	-4.57	-5.58	-4.3	
3000	$(v-v_f)\cdot 10^5$	-16.3	-14.7	-16.0	-20.7	-30.0	-47.1	-87.9	
(695.36)	$(h-h_f)$	$+9.00$	$+7.88$	$+6.76$	$+5.49$	$+3.33$	-0.41	-6.9	
	$(s-s_f)\cdot 10^3$	$+0.28$	-1.79	-3.56	-5.12	-7.03	-9.42	-12.4	
4000	$(v-v_f)\cdot 10^5$	-21.5	-19.2	-21.0	-27.5	-40.0	-64.5	-132.2	-821
	$(h-h_f)$	$+11.88$	$+10.49$	$+9.03$	$+7.41$	$+4.71$	-0.16	-10.0	-59.5
	$(s-s_f)\cdot 10^3$	$+0.29$	-2.42	-4.74	-6.77	-9.40	-13.03	-19.3	-55.8
5000	$(v-v_f)\cdot 10^5$	-26.7	-25.6	-26.0	-34.0	-49.6	-80.5	-169.3	-1017
	$(h-h_f)$	$+14.75$	$+15.08$	$+11.30$	$+9.36$	$+6.08$	$+0.25$	-12.1	-76.9
	$(s-s_f)\cdot 10^3$	$+0.22$	-5.07	-5.92	-8.40	-11.74	-16.47	-25.3	-75.3

TABLE 18.1-15 *(Continued)*

SATURATION: SOLID-VAPOR *

Temp., F t	Abs. Press., lbf/in.^{2} P	Specific Volume, ft^3/lbm		Enthalpy, Btu/lbm			Entropy, Btu/lbm R		
		Sat. Solid v_i	Sat. Vapor $v_g \times 10^{-3}$	Sat. Solid h_i	Subl. h_{ig}	Sat. Vapor h_g	Sat. Solid s_i	Subl. s_{ig}	Sat. Vapor s_g
32	0.0885	0.01747	3.306	−143.35	1219.1	1075.8	−0.2916	2.4793	2.1877
30	0.0808	0.01747	3.609	−144.35	1219.3	1074.9	−0.2936	2.4897	2.1961
20	0.0505	0.01745	5.658	−149.31	1219.9	1070.6	−0.3038	2.5425	2.2387
10	0.0309	0.01744	9.05	−154.17	1220.4	1066.2	−0.3141	2.5977	2.2836
0	0.0185	0.01742	14.77	−158.93	1220.7	1061.8	−0.3241	2.6546	2.3305
−10	0.0108	0.01741	24.67	−163.59	1221.0	1057.4	−0.3346	2.7143	2.3797
−20	0.0062	0.01739	42.2	−168.16	1221.2	1053.0	−0.3448	2.7764	2.4316
−30	0.0035	0.01738	74.1	−172.63	1221.2	1048.6	−0.3551	2.8411	2.4860
−40	0.0019	0.01737	133.9	−177.00	1221.2	1044.2	−0.3654	2.9087	2.5433

^{a} From ref. 3. Used by permission.

* Abridged from *Thermodynamic Properties of Steam*, by Joseph H. Keenan and Frederick G. Keyes. Copyright 1936, by Joseph H. Keenan and Frederick G. Keyes. Published by John Wiley & Sons, Inc., New York.

TABLE 18.1-16 THERMODYNAMIC PROPERTIES OF STEAM: SI UNITS[a]

Saturated Steam: Temperature Table

Temp. °C T	Press. kPa P	Specific Volume Sat. Liquid v_f	Specific Volume Sat. Vapor v_g	Internal Energy Sat. Liquid u_f	Internal Energy Evap. u_{fg}	Internal Energy Sat. Vapor u_g	Enthalpy Sat. Liquid h_f	Enthalpy Evap. h_{fg}	Enthalpy Sat. Vapor h_g	Entropy Sat. Liquid s_f	Entropy Evap. s_{fg}	Entropy Sat. Vapor s_g
0.01	0.6113	0.001 000	206.14	.00	2375.3	2375.3	.01	2501.3	2501.4	.0000	9.1562	9.1562
5	0.8721	0.001 000	147.12	20.97	2361.3	2382.3	20.98	2489.6	2510.6	.0761	8.9496	9.0257
10	1.2276	0.001 000	106.38	42.00	2347.2	2389.2	42.01	2477.7	2519.8	.1510	8.7498	8.9008
15	1.7051	0.001 001	77.93	62.99	2333.1	2396.1	62.99	2465.9	2528.9	.2245	8.5569	8.7814
20	2.339	0.001 002	57.79	83.95	2319.0	2402.9	83.96	2454.1	2538.1	.2966	8.3706	8.6672
25	3.169	0.001 003	43.36	104.88	2304.9	2409.8	104.89	2442.3	2547.2	.3674	8.1905	8.5580
30	4.246	0.001 004	32.89	125.78	2290.8	2416.6	125.79	2430.5	2556.3	.4369	8.0164	8.4533
35	5.628	0.001 006	25.22	146.67	2276.7	2423.4	146.68	2418.6	2565.3	.5053	7.8478	8.3531
40	7.384	0.001 008	19.52	167.56	2262.6	2430.1	167.57	2406.7	2574.3	.5725	7.6845	8.2570
45	9.593	0.001 010	15.26	188.44	2248.4	2436.8	188.45	2394.8	2583.2	.6387	7.5261	8.1648
50	12.349	0.001 012	12.03	209.32	2234.2	2443.5	209.33	2382.7	2592.1	.7038	7.3725	8.0763
55	15.758	0.001 015	9.568	230.21	2219.9	2450.1	230.23	2370.7	2600.9	.7679	7.2234	7.9913
60	19.940	0.001 017	7.671	251.11	2205.5	2456.6	251.13	2358.5	2609.6	.8312	7.0784	7.9096
65	25.03	0.001 020	6.197	272.02	2191.1	2463.1	272.06	2346.2	2618.3	.8935	6.9375	7.8310
70	31.19	0.001 023	5.042	292.95	2176.6	2469.6	292.98	2333.8	2626.8	.9549	6.8004	7.7553
75	38.58	0.001 026	4.131	313.90	2162.0	2475.9	313.93	2321.4	2635.3	1.0155	6.6669	7.6824
80	47.39	0.001 029	3.407	334.86	2147.4	2482.2	334.91	2308.8	2643.7	1.0753	6.5369	7.6122
85	57.83	0.001 033	2.828	355.84	2132.6	2488.4	355.90	2296.0	2651.9	1.1343	6.4102	7.5445
90	70.14	0.001 036	2.361	376.85	2117.7	2494.5	376.92	2283.2	2660.1	1.1925	6.2866	7.4791
95	84.55	0.001 040	1.982	397.88	2102.7	2500.6	397.96	2270.2	2668.1	1.2500	6.1659	7.4159

Temp. °C T	Press. MPa P	Specific Volume Sat. Liquid v_f	Specific Volume Sat. Vapor v_g	Internal Energy Sat. Liquid u_f	Internal Energy Evap. u_{fg}	Internal Energy Sat. Vapor u_g	Enthalpy Sat. Liquid h_f	Enthalpy Evap. h_{fg}	Enthalpy Sat. Vapor h_g	Entropy Sat. Liquid s_f	Entropy Evap. s_{fg}	Entropy Sat. Vapor s_g
100	0.101 35	0.001 044	1.6729	418.94	2087.6	2506.5	419.04	2257.0	2676.1	1.3069	6.0480	7.3549
105	0.120 82	0.001 048	1.4194	440.02	2072.3	2512.4	440.15	2243.7	2683.8	1.3630	5.9328	7.2958
110	0.143 27	0.001 052	1.2102	461.14	2057.0	2518.1	461.30	2230.2	2691.5	1.4185	5.8202	7.2387
115	0.169 06	0.001 056	1.0366	482.30	2041.4	2523.7	482.48	2216.5	2699.0	1.4734	5.7100	7.1833
120	0.198 53	0.001 060	0.8919	503.50	2025.8	2529.3	503.71	2202.6	2706.3	1.5276	5.6020	7.1296
125	0.2321	0.001 065	0.7706	524.74	2009.9	2534.6	524.99	2188.5	2713.5	1.5813	5.4962	7.0775
130	0.2701	0.001 070	0.6685	546.02	1993.9	2539.9	546.31	2174.2	2720.5	1.6344	5.3925	7.0269
135	0.3130	0.001 075	0.5822	567.35	1977.7	2545.0	567.69	2159.6	2727.3	1.6870	5.2907	6.9777
140	0.3613	0.001 080	0.5089	588.74	1961.3	2550.0	589.13	2144.7	2733.9	1.7391	5.1908	6.9299
145	0.4154	0.001 085	0.4463	610.18	1944.7	2554.9	610.63	2129.6	2740.3	1.7907	5.0926	6.8833
150	0.4758	0.001 091	0.3928	631.68	1927.9	2559.5	632.20	2114.3	2746.5	1.8418	4.9960	6.8379
155	0.5431	0.001 096	0.3468	653.24	1910.8	2564.1	653.84	2098.6	2752.4	1.8925	4.9010	6.7935
160	0.6178	0.001 102	0.3071	674.87	1893.5	2568.4	675.55	2082.6	2758.1	1.9427	4.8075	6.7502
165	0.7005	0.001 108	0.2727	696.56	1876.0	2572.5	697.34	2066.2	2763.5	1.9925	4.7153	6.7078
170	0.7917	0.001 114	0.2428	718.33	1858.1	2576.5	719.21	2049.5	2768.7	2.0419	4.6244	6.6663
175	0.8920	0.001 121	0.2168	740.17	1840.0	2580.2	741.17	2032.4	2773.6	2.0909	4.5347	6.6256
180	1.0021	0.001 127	0.194 05	762.09	1821.6	2583.7	763.22	2015.0	2778.2	2.1396	4.4461	6.5857
185	1.1227	0.001 134	0.174 09	784.10	1802.9	2587.0	785.37	1997.1	2782.4	2.1879	4.3586	6.5465
190	1.2544	0.001 141	0.156 54	806.19	1783.8	2590.0	807.62	1978.8	2786.4	2.2359	4.2720	6.5079
195	1.3978	0.001 149	0.141 05	828.37	1764.4	2592.8	829.98	1960.0	2790.0	2.2835	4.1863	6.4698
200	1.5538	0.001 157	0.127 36	850.65	1744.7	2595.3	852.45	1940.7	2793.2	2.3309	4.1014	6.4323
205	1.7230	0.001 164	0.115 21	873.04	1724.5	2597.5	875.04	1921.0	2796.0	2.3780	4.0172	6.3952
210	1.9062	0.001 173	0.104 41	895.53	1703.9	2599.5	897.76	1900.7	2798.5	2.4248	3.9337	6.3585

TABLE 18.1-16 (*Continued*)

Temp. °C T	Press. MPa P	Specific Volume Sat. Liquid v_f	Sat. Vapor v_g	Internal Energy Sat. Liquid u_f	Evap. u_{fg}	Sat. Vapor u_g	Enthalpy Sat. Liquid h_f	Evap. h_{fg}	Sat. Vapor h_g	Entropy Sat. Liquid s_f	Evap. s_{fg}	Sat. Vapor s_g
215	2.104	0.001 181	0.094 79	918.14	1682.9	2601.1	920.62	1879.9	2800.5	2.4714	3.8507	6.3221
220	2.318	0.001 190	0.086 19	940.87	1661.5	2602.4	943.62	1858.5	2802.1	2.5178	3.7683	6.2861
225	2.548	0.001 199	0.078 49	963.73	1639.6	2603.3	966.78	1836.5	2803.3	2.5639	3.6863	6.2503
230	2.795	0.001 209	0.071 58	986.74	1617.2	2603.9	990.12	1813.8	2804.0	2.6099	3.6047	6.2146
235	3.060	0.001 219	0.065 37	1009.89	1594.2	2604.1	1013.62	1790.5	2804.2	2.6558	3.5233	6.1791
240	3.344	0.001 229	0.059 76	1033.21	1570.8	2604.0	1037.32	1766.5	2803.8	2.7015	3.4422	6.1437
245	3.648	0.001 240	0.054 71	1056.71	1546.7	2603.4	1061.23	1741.7	2803.0	2.7472	3.3612	6.1083
250	3.973	0.001 251	0.050 13	1080.39	1522.0	2602.4	1085.36	1716.2	2801.5	2.7927	3.2802	6.0730
255	4.319	0.001 263	0.045 98	1104.28	1496.7	2600.9	1109.73	1689.8	2799.5	2.8383	3.1992	6.0375
260	4.688	0.001 276	0.042 21	1128.39	1470.6	2599.0	1134.37	1662.5	2796.9	2.8838	3.1181	6.0019
265	5.081	0.001 289	0.038 77	1152.74	1443.9	2596.6	1159.28	1634.4	2793.6	2.9294	3.0368	5.9662
270	5.499	0.001 302	0.035 64	1177.36	1416.3	2593.7	1184.51	1605.2	2789.7	2.9751	2.9551	5.9301
275	5.942	0.001 317	0.032 79	1202.25	1387.9	2590.2	1210.07	1574.9	2785.0	3.0208	2.8730	5.8938
280	6.412	0.001 332	0.030 17	1227.46	1358.7	2586.1	1235.99	1543.6	2779.6	3.0668	2.7903	5.8571
285	6.909	0.001 348	0.027 77	1253.00	1328.4	2581.4	1262.31	1511.0	2773.3	3.1130	2.7070	5.8199
290	7.436	0.001 366	0.025 57	1278.92	1297.1	2576.0	1289.07	1477.1	2766.2	3.1594	2.6227	5.7821
295	7.993	0.001 384	0.023 54	1305.2	1264.7	2569.9	1316.3	1441.8	2758.1	3.2062	2.5375	5.7437
300	8.581	0.001 404	0.021 67	1332.0	1231.0	2563.0	1344.0	1404.9	2749.0	3.2534	2.4511	5.7045
305	9.202	0.001 425	0.019 948	1359.3	1195.9	2555.2	1372.4	1366.4	2738.7	3.3010	2.3633	5.6643
310	9.856	0.001 447	0.018 350	1387.1	1159.4	2546.4	1401.3	1326.0	2727.3	3.3493	2.2737	5.6230
315	10.547	0.001 472	0.016 867	1415.5	1121.1	2536.6	1431.0	1283.5	2714.5	3.3982	2.1821	5.5804
320	11.274	0.001 499	0.015 488	1444.6	1080.9	2525.5	1461.5	1238.6	2700.1	3.4480	2.0882	5.5362
330	12.845	0.001 561	0.012 996	1505.3	993.7	2498.9	1525.3	1140.6	2665.9	3.5507	1.8909	5.4417
340	14.586	0.001 638	0.010 797	1570.3	894.3	2464.6	1594.2	1027.9	2622.0	3.6594	1.6763	5.3357
350	16.513	0.001 740	0.008 813	1641.9	776.6	2418.4	1670.6	893.4	2563.9	3.7777	1.4335	5.2112
360	18.651	0.001 893	0.006 945	1725.2	626.3	2351.5	1760.5	720.5	2481.0	3.9147	1.1379	5.0526
370	21.03	0.002 213	0.004 925	1844.0	384.5	2228.5	1890.5	441.6	2332.1	4.1106	.6865	4.7971
374.14	22.09	0.003 155	0.003 155	2029.6	0	2029.6	2099.3	0	2099.3	4.4298	0	4.4298

Saturated Steam: Pressure Table

Press. kPa P	Temp. °C T	Specific Volume Sat. Liquid v_f	Sat. Vapor v_g	Internal Energy Sat. Liquid u_f	Evap. u_{fg}	Sat. Vapor u_g	Enthalpy Sat. Liquid h_f	Evap. h_{fg}	Sat. Vapor h_g	Entropy Sat. Liquid s_f	Evap. s_{fg}	Sat. Vapor s_g
0.6113	0.01	0.001 000	206.14	.00	2375.3	2375.3	.01	2501.3	2501.4	.0000	9.1562	9.1562
1.0	6.98	0.001 000	129.21	29.30	2355.7	2385.0	29.30	2484.9	2514.2	.1059	8.8697	8.9756
1.5	13.03	0.001 001	87.98	54.71	2338.6	2393.3	54.71	2470.6	2525.3	.1957	8.6322	8.8279
2.0	17.50	0.001 001	67.00	73.48	2326.0	2399.5	73.48	2460.0	2533.5	.2607	8.4629	8.7237
2.5	21.08	0.001 002	54.25	88.48	2315.9	2404.4	88.49	2451.6	2540.0	.3120	8.3311	8.6432
3.0	24.08	0.001 003	45.67	101.04	2307.5	2408.5	101.05	2444.5	2545.5	.3545	8.2231	8.5776
4.0	28.96	0.001 004	34.80	121.45	2293.7	2415.2	121.46	2432.9	2554.4	.4226	8.0520	8.4746
5.0	32.88	0.001 005	28.19	137.81	2282.7	2420.5	137.82	2423.7	2561.5	.4764	7.9187	8.3951
7.5	40.29	0.001 008	19.24	168.78	2261.7	2430.5	168.79	2406.0	2574.8	.5764	7.6750	8.2515
10	45.81	0.001 010	14.67	191.82	2246.1	2437.9	191.83	2392.8	2584.7	.6493	7.5009	8.1502
15	53.97	0.001 014	10.02	225.92	2222.8	2448.7	225.94	2373.1	2599.1	.7549	7.2536	8.0085
20	60.06	0.001 017	7.649	251.38	2205.4	2456.7	251.40	2358.3	2609.7	.8320	7.0766	7.9085
25	64.97	0.001 020	6.204	271.90	2191.2	2463.1	271.93	2346.3	2618.2	.8931	6.9383	7.8314
30	69.10	0.001 022	5.229	289.20	2179.2	2468.4	289.23	2336.1	2625.3	.9439	6.8247	7.7686
40	75.87	0.001 027	3.993	317.53	2159.5	2477.0	317.58	2319.2	2636.8	1.0259	6.6441	7.6700
50	81.33	0.001 030	3.240	340.44	2143.4	2483.9	340.49	2305.4	2645.9	1.0910	6.5029	7.5939
75	91.78	0.001 037	2.217	384.31	2112.4	2496.7	384.39	2278.6	2663.0	1.2130	6.2434	7.4564

MPa												
0.100	99.63	0.001 043	1.6940	417.36	2088.7	2506.1	417.46	2258.0	2675.5	1.3026	6.0568	7.3594
0.125	105.99	0.001 048	1.3749	444.19	2069.3	2513.5	444.32	2241.0	2685.4	1.3740	5.9104	7.2844
0.150	111.37	0.001 053	1.1593	466.94	2052.7	2519.7	467.11	2226.5	2693.6	1.4336	5.7897	7.2233
0.175	116.06	0.001 057	1.0036	486.80	2038.1	2524.9	486.99	2213.6	2700.6	1.4849	5.6868	7.1717
0.200	120.23	0.001 061	0.8857	504.49	2025.0	2529.5	504.70	2201.9	2706.7	1.5301	5.5970	7.1271
0.225	124.00	0.001 064	0.7933	520.47	2013.1	2533.6	520.72	2191.3	2712.1	1.5706	5.5173	7.0878

TABLE 18.1-16   (*Continued*)

Press. kPa P	Temp. °C T	Specific Volume		Internal Energy			Enthalpy			Entropy		
		Sat. Liquid v_f	Sat. Vapor v_g	Sat. Liquid u_f	Evap. u_{fg}	Sat. Vapor u_g	Sat. Liquid h_f	Evap. h_{fg}	Sat. Vapor h_g	Sat. Liquid s_f	Evap. s_{fg}	Sat. Vapor s_g
0.250	127.44	0.001 067	0.7187	535.10	2002.1	2537.2	535.37	2181.5	2716.9	1.6072	5.4455	7.0527
0.275	130.60	0.001 070	0.6573	548.59	1991.9	2540.5	548.89	2172.4	2721.3	1.6408	5.3801	7.0209
0.300	133.55	0.001 073	0.6058	561.15	1982.4	2543.6	561.47	2163.8	2725.3	1.6718	5.3201	6.9919
0.325	136.30	0.001 076	0.5620	572.90	1973.5	2546.4	573.25	2155.8	2729.0	1.7006	5.2646	6.9652
0.350	138.88	0.001 079	0.5243	583.95	1965.0	2548.9	584.33	2148.1	2732.4	1.7275	5.2130	6.9405
0.375	141.32	0.001 081	0.4914	594.40	1956.9	2551.3	594.81	2140.8	2735.6	1.7528	5.1647	6.9175
0.40	143.63	0.001 084	0.4625	604.31	1949.3	2553.6	604.74	2133.8	2738.6	1.7766	5.1193	6.8959
0.45	147.93	0.001 088	0.4140	622.77	1934.9	2557.6	623.25	2120.7	2743.9	1.8207	5.0359	6.8565
0.50	151.86	0.001 093	0.3749	639.68	1921.6	2561.2	640.23	2108.5	2748.7	1.8607	4.9606	6.8213
0.55	155.48	0.001 097	0.3427	655.32	1909.2	2564.5	655.93	2097.0	2753.0	1.8973	4.8920	6.7893
0.60	158.85	0.001 101	0.3157	669.90	1897.5	2567.4	670.56	2086.3	2756.8	1.9312	4.8288	6.7600
0.65	162.01	0.001 104	0.2927	683.56	1886.5	2570.1	684.28	2076.0	2760.3	1.9627	4.7703	6.7331
0.70	164.97	0.001 108	0.2729	696.44	1876.1	2572.5	697.22	2066.3	2763.5	1.9922	4.7158	6.7080
0.75	167.78	0.001 112	0.2556	708.64	1866.1	2574.7	709.47	2057.0	2766.4	2.0200	4.6647	6.6847
0.80	170.43	0.001 115	0.2404	720.22	1856.6	2576.8	721.11	2048.0	2769.1	2.0462	4.6166	6.6628
0.85	172.96	0.001 118	0.2270	731.27	1847.4	2578.7	732.22	2039.4	2771.6	2.0710	4.5711	6.6421
0.90	175.38	0.001 121	0.2150	741.83	1838.6	2580.5	742.83	2031.1	2773.9	2.0946	4.5280	6.6226
0.95	177.69	0.001 124	0.2042	751.95	1830.2	2582.1	753.02	2023.1	2776.1	2.1172	4.4869	6.6041
1.00	179.91	0.001 127	0.194 44	761.68	1822.0	2583.6	762.81	2015.3	2778.1	2.1387	4.4478	6.5865
1.10	184.09	0.001 133	0.177 53	780.09	1806.3	2586.4	781.34	2000.4	2781.7	2.1792	4.3744	6.5536
1.20	187.99	0.001 139	0.163 33	797.29	1791.5	2588.8	798.65	1986.2	2784.8	2.2166	4.3067	6.5233
1.30	191.64	0.001 144	0.151 25	813.44	1777.5	2591.0	814.93	1972.7	2787.6	2.2515	4.2438	6.4953
1.40	195.07	0.001 149	0.140 84	828.70	1764.1	2592.8	830.30	1959.7	2790.0	2.2842	4.1850	6.4693
1.50	198.32	0.001 154	0.131 77	843.16	1751.3	2594.5	844.89	1947.3	2792.2	2.3150	4.1298	6.4448
1.75	205.76	0.001 166	0.113 49	876.46	1721.4	2597.8	878.50	1917.9	2796.4	2.3851	4.0044	6.3896
2.00	212.42	0.001 177	0.099 63	906.44	1693.8	2600.3	908.79	1890.7	2799.5	2.4474	3.8935	6.3409
2.25	218.45	0.001 187	0.088 75	933.83	1668.2	2602.0	936.49	1865.2	2801.7	2.5035	3.7937	6.2972
2.5	223.99	0.001 197	0.079 98	959.11	1644.0	2603.1	962.11	1841.0	2803.1	2.5547	3.7028	6.2575
3.0	233.90	0.001 217	0.066 68	1004.78	1599.3	2604.1	1008.42	1795.7	2804.2	2.6457	3.5412	6.1869
3.5	242.60	0.001 235	0.057 07	1045.43	1558.3	2603.7	1049.75	1753.7	2803.4	2.7253	3.4000	6.1253
4	250.40	0.001 252	0.049 78	1082.31	1520.0	2602.3	1087.31	1714.1	2801.4	2.7964	3.2737	6.0701
5	263.99	0.001 286	0.039 44	1147.81	1449.3	2597.1	1154.23	1640.1	2794.3	2.9202	3.0532	5.9734
6	275.64	0.001 319	0.032 44	1205.44	1384.3	2589.7	1213.35	1571.0	2784.3	3.0267	2.8625	5.8892
7	285.88	0.001 351	0.027 37	1257.55	1323.0	2580.5	1267.00	1505.1	2772.1	3.1211	2.6922	5.8133
8	295.06	0.001 384	0.023 52	1305.57	1264.2	2569.8	1316.64	1441.3	2758.0	3.2068	2.5364	5.7432
9	303.40	0.001 418	0.020 48	1350.51	1207.3	2557.8	1363.26	1378.9	2742.1	3.2858	2.3915	5.6772
10	311.06	0.001 452	0.018 026	1393.04	1151.4	2544.4	1407.56	1317.1	2724.7	3.3596	2.2544	5.6141
11	318.15	0.001 489	0.015 987	1433.7	1090.0	2529.8	1450.1	1255.5	2705.6	3.4295	2.1233	5.5527
12	324.75	0.001 527	0.014 263	1473.0	1040.7	2513.7	1491.3	1193.6	2684.9	3.4962	1.9962	5.4924
13	330.93	0.001 567	0.012 780	1511.1	985.0	2496.1	1531.5	1130.7	2662.2	3.5606	1.8718	5.4323
14	336.75	0.001 611	0.011 485	1548.6	928.2	2476.8	1571.1	1066.5	2637.6	3.6232	1.7485	5.3717
15	342.24	0.001 658	0.010 337	1585.6	869.8	2455.5	1610.5	1000.0	2610.5	3.6848	1.6249	5.3098
16	347.44	0.001 711	0.009 306	1622.7	809.0	2431.7	1650.1	930.6	2580.6	3.7461	1.4994	5.2455
17	352.37	0.001 770	0.008 364	1660.2	744.8	2405.0	1690.3	856.9	2547.2	3.8079	1.3698	5.1777
18	357.06	0.001 840	0.007 489	1698.9	675.4	2374.3	1732.0	777.1	2509.1	3.8715	1.2329	5.1044
19	361.54	0.001 924	0.006 657	1739.9	598.1	2338.1	1776.5	688.0	2464.5	3.9388	1.0839	5.0228
20	365.81	0.002 036	0.005 834	1785.6	507.5	2293.0	1826.3	583.4	2409.7	4.0139	.9130	4.9269
21	369.89	0.002 207	0.004 952	1842.1	388.5	2230.6	1888.4	446.2	2334.6	4.1075	.6938	4.8013
22	373.80	0.002 742	0.003 568	1961.9	125.2	2087.1	2022.2	143.4	2165.6	4.3110	.2216	4.5327
22.09	374.14	0.003 155	0.003 155	2029.6	0	2029.6	2099.3	0	2099.3	4.4298	0	4.4298

TABLE 18.1-16 (*Continued*)

Superheated Vapor

| T | \multicolumn{4}{c}{P = .010 MPa (45.81)} | \multicolumn{4}{c}{P = .050 MPa (81.33)} | \multicolumn{4}{c}{P = .10 MPa (99.63)} |

T	v	u	h	s	v	u	h	s	v	u	h	s
Sat.	14.674	2437.9	2584.7	8.1502	3.240	2483.9	2645.9	7.5939	1.6940	2506.1	2675.5	7.3594
50	14.869	2443.9	2592.6	8.1749								
100	17.196	2515.5	2687.5	8.4479	3.418	2511.6	2682.5	7.6947	1.6958	2506.7	2676.2	7.3614
150	19.512	2587.9	2783.0	8.6882	3.889	2585.6	2780.1	7.9401	1.9364	2582.8	2776.4	7.6134
200	21.825	2661.3	2879.5	8.9038	4.356	2659.9	2877.7	8.1580	2.172	2658.1	2875.3	7.8343
250	24.136	2736.0	2977.3	9.1002	4.820	2735.0	2976.0	8.3556	2.406	2733.7	2974.3	8.0333
300	26.445	2812.1	3076.5	9.2813	5.284	2811.3	3075.5	8.5373	2.639	2810.4	3074.3	8.2158
400	31.063	2968.9	3279.6	9.6077	6.209	2968.5	3278.9	8.8642	3.103	2967.9	3278.2	8.5435
500	35.679	3132.3	3489.1	9.8978	7.134	3132.0	3488.7	9.1546	3.565	3131.6	3488.1	8.8342
600	40.295	3302.5	3705.4	10.1608	8.057	3302.2	3705.1	9.4178	4.028	3301.9	3704.7	9.0976
700	44.911	3479.6	3928.7	10.4028	8.981	3479.4	3928.5	9.6599	4.490	3479.2	3928.2	9.3398
800	49.526	3663.8	4159.0	10.6281	9.904	3663.6	4158.9	9.8852	4.952	3663.5	4158.6	9.5652
900	54.141	3855.0	4396.4	10.8396	10.828	3854.9	4396.3	10.0967	5.414	3854.8	4396.1	9.7767
1000	58.757	4053.0	4640.6	11.0393	11.751	4052.9	4640.5	10.2964	5.875	4052.8	4640.3	9.9764
1100	63.372	4257.5	4891.2	11.2287	12.674	4257.4	4891.1	10.4859	6.337	4257.3	4891.0	10.1659
1200	67.987	4467.9	5147.8	11.4091	13.597	4467.8	5147.7	10.6662	6.799	4467.7	5147.6	10.3463
1300	72.602	4683.7	5409.7	11.5811	14.521	4683.6	5409.6	10.8382	7.260	4683.5	5409.5	10.5183

| T | \multicolumn{4}{c}{P = .20 MPa (120.23)} | \multicolumn{4}{c}{P = .30 MPa (133.55)} | \multicolumn{4}{c}{P = .40 MPa (143.63)} |

T	v	u	h	s	v	u	h	s	v	u	h	s
Sat.	.8857	2529.5	2706.7	7.1272	.6058	2543.6	2725.3	6.9919	.4625	2553.6	2738.6	6.8959
150	.9596	2576.9	2768.8	7.2795	.6339	2570.8	2761.0	7.0778	.4708	2564.5	2752.8	6.9299
200	1.0803	2654.4	2870.5	7.5066	.7163	2650.7	2865.6	7.3115	.5342	2646.8	2860.5	7.1706
250	1.1988	2731.2	2971.0	7.7086	.7964	2728.7	2967.6	7.5166	.5951	2726.1	2964.2	7.3789
300	1.3162	2808.6	3071.8	7.8926	.8753	2806.7	3069.3	7.7022	.6548	2804.8	3066.8	7.5662
400	1.5493	2966.7	3276.6	8.2218	1.0315	2965.6	3275.0	8.0330	.7726	2964.4	3273.4	7.8985

Superheated Vapor

| T | v | u | h | s | v | u | h | s | v | u | h | s |

| T | \multicolumn{4}{c}{P = .20 MPa (120.23)} | \multicolumn{4}{c}{P = .30 MPa (133.55)} | \multicolumn{4}{c}{P = .40 MPa (143.63)} |

T	v	u	h	s	v	u	h	s	v	u	h	s
500	1.7814	3130.8	3487.1	8.5133	1.1867	3130.0	3486.0	8.3251	.8893	3129.2	3484.9	8.1913
600	2.013	3301.4	3704.0	8.7770	1.3414	3300.8	3703.2	8.5892	1.0055	3300.2	3702.4	8.4558
700	2.244	3478.8	3927.6	9.0194	1.4957	3478.4	3927.1	8.8319	1.1215	3477.9	3926.5	8.6987
800	2.475	3663.1	4158.2	9.2449	1.6499	3662.9	4157.8	9.0576	1.2372	3662.4	4157.3	8.9244
900	2.706	3854.5	4395.8	9.4566	1.8041	3854.2	4395.4	9.2692	1.3529	3853.9	4395.1	9.1362
1000	2.937	4052.5	4640.0	9.6563	1.9581	4052.3	4639.7	9.4690	1.4685	4052.0	4639.4	9.3360
1100	3.168	4257.0	4890.7	9.8458	2.1121	4256.8	4890.4	9.6585	1.5840	4256.5	4890.2	9.5256
1200	3.399	4467.5	5147.3	10.0262	2.2661	4467.2	5147.1	9.8389	1.6996	4467.0	5146.8	9.7060
1300	3.630	4683.2	5409.3	10.1982	2.4201	4683.0	5409.0	10.0110	1.8151	4682.8	5408.8	9.8780

| T | \multicolumn{4}{c}{P = .50 MPa (151.86)} | \multicolumn{4}{c}{P = .60 MPa (158.85)} | \multicolumn{4}{c}{P = .80 MPa (170.43)} |

T	v	u	h	s	v	u	h	s	v	u	h	s
Sat.	.3749	2561.2	2748.7	6.8213	.3157	2567.4	2756.8	6.7600	.2404	2576.8	2769.1	6.6628
200	.4249	2642.9	2855.4	7.0592	.3520	2638.9	2850.1	6.9665	.2608	2630.6	2839.3	6.8158
250	.4744	2723.5	2960.7	7.2709	.3938	2720.9	2957.2	7.1816	.2931	2715.5	2950.0	7.0384
300	.5226	2802.9	3064.2	7.4599	.4344	2801.0	3061.6	7.3724	.3241	2797.2	3056.5	7.2328
350	.5701	2882.6	3167.7	7.6329	.4742	2881.2	3165.7	7.5464	.3544	2878.2	3161.7	7.4089
400	.6173	2963.2	3271.9	7.7938	.5137	2962.1	3270.3	7.7079	.3843	2959.7	3267.1	7.5716
500	.7109	3128.4	3483.9	8.0873	.5920	3127.6	3482.8	8.0021	.4433	3126.0	3480.6	7.8673
600	.8041	3299.6	3701.7	8.3522	.6697	3299.1	3700.9	8.2674	.5018	3297.9	3699.4	8.1333
700	.8969	3477.5	3925.9	8.5952	.7472	3477.0	3925.3	8.5107	.5601	3476.2	3924.2	8.3770
800	.9896	3662.1	4156.9	8.8211	.8245	3661.8	4156.5	8.7367	.6181	3661.1	4155.6	8.6033
900	1.0822	3853.6	4394.7	9.0329	.9017	3853.4	4394.4	8.9486	.6761	3852.8	4393.7	8.8153
1000	1.1747	4051.8	4639.1	9.2328	.9788	4051.5	4638.8	9.1485	.7340	4051.0	4638.2	9.0153
1100	1.2672	4256.3	4889.9	9.4224	1.0559	4256.1	4889.6	9.3381	.7919	4255.6	4889.1	9.2050
1200	1.3596	4466.8	5146.6	9.6029	1.1330	4466.5	5146.3	9.5185	.8497	4466.1	5145.9	9.3855
1300	1.4521	4682.5	5408.6	9.7749	1.2101	4682.3	5408.3	9.6906	.9076	4681.8	5407.9	9.5575

TABLE 18.1-16 (*Continued*)

Superheated Vapor

T	v	u	h	s	v	u	h	s	v	u	h	s
	$P = 1.00$ MPa (179.91)				$P = 1.20$ MPa (187.99)				$P = 1.40$ MPa (195.07)			
Sat.	.194 44	2583.6	2778.1	6.5865	.163 33	2588.8	2784.8	6.5233	.140 84	2592.8	2790.0	6.4693
200	.2060	2621.9	2827.9	6.6940	.169 30	2612.8	2815.9	6.5898	.143 02	2603.1	2803.3	6.4975
250	.2327	2709.9	2942.6	6.9247	.192 34	2704.2	2935.0	6.8294	.163 50	2698.3	2927.2	6.7467
300	.2579	2793.2	3051.2	7.1229	.2138	2789.2	3045.8	7.0317	.182 28	2785.2	3040.4	6.9534
350	.2825	2875.2	3157.7	7.3011	.2345	2872.2	3153.6	7.2121	.2003	2869.2	3149.5	7.1360
400	.3066	2957.3	3263.9	7.4651	.2548	2954.9	3260.7	7.3774	.2178	2952.5	3257.5	7.3026
500	.3541	3124.4	3478.5	7.7622	.2946	3122.8	3476.3	7.6759	.2521	3121.1	3474.1	7.6027
600	.4011	3296.8	3697.9	8.0290	.3339	3295.6	3696.3	7.9435	.2860	3294.4	3694.8	7.8710
700	.4478	3475.3	3923.1	8.2731	.3729	3474.4	3922.0	8.1881	.3195	3473.6	3920.8	8.1160
800	.4943	3660.4	4154.7	8.4996	.4118	3659.7	4153.8	8.4148	.3528	3659.0	4153.0	8.3431
900	.5407	3852.2	4392.9	8.7118	.4505	3851.6	4392.2	8.6272	.3861	3851.1	4391.5	8.5556
1000	.5871	4050.5	4637.6	8.9119	.4892	4050.0	4637.0	8.8274	.4192	4049.5	4636.4	8.7559
1100	.6335	4255.1	4888.6	9.1017	.5278	4254.6	4888.0	9.0172	.4524	4254.1	4887.5	8.9457
1200	.6798	4465.6	5145.4	9.2822	.5665	4465.1	5144.9	9.1977	.4855	4464.7	5144.4	9.1262
1300	.7261	4681.3	5407.4	9.4543	.6051	4680.9	5407.0	9.3698	.5186	4680.4	5406.5	9.2984
	$P = 1.60$ MPa (201.41)				$P = 1.80$ MPa (207.15)				$P = 2.00$ MPa (212.42)			
Sat.	.123 80	2596.0	2794.0	6.4218	.110 42	2598.4	2797.1	6.3794	.099 63	2600.3	2799.5	6.3409
225	.132 87	2644.7	2857.3	6.5518	.116 73	2636.6	2846.7	6.4808	.103 77	2628.3	2835.8	6.4147
250	.141 84	2692.3	2919.2	6.6732	.124 97	2686.0	2911.0	6.6066	.111 44	2679.6	2902.5	6.5453
300	.158 62	2781.1	3034.8	6.8844	.140 21	2776.9	3029.2	6.8226	.125 47	2772.6	3023.5	6.7664
350	.174 56	2866.1	3145.4	7.0694	.154 57	2863.0	3141.2	7.0100	.138 57	2859.8	3137.0	6.9563
400	.190 05	2950.1	3254.2	7.2374	.168 47	2947.7	3250.9	7.1794	.151 20	2945 2	3247.6	7.1271
500	.2203	3119.5	3472.0	7.5390	.195 50	3117.9	3469.8	7.4825	.175 68	3116.2	3467.6	7.4317
600	.2500	3293.3	3693.2	7.8080	.2220	3292.1	3691.7	7.7523	.199 60	3290.9	3690.1	7.7024
700	.2794	3472.7	3919.7	8.0535	.2482	3471.8	3918.5	7.9983	.2232	3470.9	3917.4	7.9487

T	v	u	h	s	v	u	h	s	v	u	h	s
	$P = 1.60$ MPa (201.41)				$P = 1.80$ MPa (207.15)				$P = 2.00$ MPa (212.42)			
800	.3086	3658.3	4152.1	8.2808	.2742	3657.6	4151.2	8.2258	.2467	3657.0	4150.3	8.1765
900	.3377	3850.5	4390.0	8.4935	.3001	3849 9	4390.1	8,4386	.2700	3849.3	4389.4	8.3895
1000	.3668	4049.0	4635.8	8.6938	.3260	4048.5	4635.2	8.6391	.2933	4048.0	4634.6	8.5901
1100	.3958	4253.7	4887.0	8.8837	.3518	4253.2	4886.4	8.8290	.3166	4252.7	4885.9	8.7800
1200	.4248	4464.2	5143.9	9.0643	.3776	4463.7	5143.4	9.0096	.3398	4463.3	5142.9	8.9607
1300	.4538	4679.9	5406.0	9.2364	.4034	4679.5	5405.6	9.1818	.3631	4679.0	5405.1	9.1329
	$P = 2.50$ MPa (223.99)				$P = 3.00$ MPa (233.90)				$P = 3.50$ MPa (242.60)			
Sat.	.079 98	2603.1	2803.1	6.2575	.066 68	2604.1	2804.2	6.1869	.057 07	2603.7	2803.4	6.1253
225	.080 27	2605.6	2806.3	6.2639								
250	.087 00	2662.6	2880.1	6.4085	.070 58	2644.0	2855.8	6.2872	.058 72	2623.7	2829.2	6.1749
300	.098 90	2761.6	3008.8	6.6438	.081 14	2750.1	2993.5	6.5390	.068 42	2738.0	2977.5	6.4461
350	.109 76	2851.9	3126.3	6.8403	.090 53	2843.7	3115.3	6.7428	.076 78	2835.3	3104.0	6.6579
400	.120 10	2939.1	3239.3	7.0148	.099 36	2932.8	3230.9	6.9212	.084 53	2926.4	3222.3	6.8405
450	.130 14	3025.5	3350.8	7.1746	.107 87	3020.4	3344.0	7.0834	.091 96	3015.3	3337.2	7.0052
500	.139 98	3112.1	3462.1	7.3234	.116 19	3108.0	3456.5	7.2338	.099 18	3103.0	3450.9	7.1572
600	.159 30	3288.0	3686.3	7.5960	.132 43	3285.0	3682.3	7.5085	.113 24	3282.1	3678.4	7.4339
700	.178 32	3468.7	3914.5	7.8435	.148 38	3466.5	3911.7	7.7571	.126 99	3464.3	3908.8	7.6837
800	.197 16	3655.3	4148.2	8.0720	.164 14	3653.5	4145.9	7.9862	.140 56	3651.8	4143.7	7.9134
900	.215 90	3847.9	4387.6	8.2853	.179 80	3846.5	4385.9	8.1999	.154 02	3845.0	4384.1	8.1276
1000	.2346	4046.7	4633.1	8.4861	.195 41	4045.4	4631.6	8.4009	.167 43	4044.1	4630.1	8.3288
1100	.2532	4251.5	4884.6	8.6762	.210 98	4250.3	4883.3	8.5912	.180 80	4249.2	4881.9	8.5192
1200	.2718	4462.1	5141.7	8.8569	.226 52	4460.9	5140.5	8.7720	.194 15	4459.8	5139.3	8.7000
1300	.2905	4677.8	5404.0	9.0291	.242 06	4676.6	5402.8	8.9442	.207 49	4675.5	5401.7	8.8723

TABLE 18.1-16 (*Continued*)

T	v	u	h	s	v	u	h	s	v	u	h	s
	P = 4.0 MPa (250.40)				P = 4.5 MPa (257.49)				P = 5.0 MPa (263.99)			
Sat.	.049 78	2602.3	2801.4	6.0701	.044 06	2600.1	2798.3	6.0198	.039 44	2597.1	2794.3	5.9734
275	.054 57	2667.9	2886.2	6.2285	.047 30	2650.3	2863.2	6.1401	.041 41	2631.3	2838.3	6.0544
300	.058 84	2725.3	2960.7	6.3615	.051 35	2712.0	2943.1	6.2828	.045 32	2698.0	2924.5	6.2084
350	.066 45	2826.7	3092.5	6.5821	.058 40	2817.8	3080.6	6.5131	.051 94	2808.7	3068.4	6.4493
400	.073 41	2919.9	3213.6	6.7690	.064 75	2913.3	3204.7	6.7047	.057 81	2906.6	3195.7	6.6459
450	.080 02	3010.2	3330.3	6.9363	.070 74	3005.0	3323.3	6.8746	.063 30	2999.7	3316.2	6.8186
500	.086 43	3099.5	3445.3	7.0901	.076 51	3095.3	3439.6	7.0301	.068 57	3091.0	3433.8	6.9759
600	.098 85	3279.1	3674.4	7.3688	.087 65	3276.0	3670.5	7.3110	.078 69	3273.0	3666.5	7.2589
700	.110 95	3462.1	3905.9	7.6198	.098 47	3459.9	3903.0	7.5631	.088 49	3457.6	3900.1	7.5122
800	.122 87	3650.0	4141.5	7.8502	.109 11	3648.3	4139.3	7.7942	.098 11	3646.6	4137.1	7.7440
900	.134 69	3843.6	4382.3	8.0647	.119 65	3842.2	4380.6	8.0091	.107 62	3840.7	4378.8	7.9593
1000	.146 45	4042.9	4628.7	8.2662	.130 13	4041.6	4627.2	8.2108	.117 07	4040.4	4625.7	8.1612
1100	.158 17	4248.0	4880.6	8.4567	.140 56	4246.8	4879.3	8.4015	.126 48	4245.6	4878.0	8.3520
1200	.169 87	4458.6	5138.1	8.6376	.150 98	4457.5	5136.9	8.5825	.135 87	4456.3	5135.7	8.5331
1300	.181 56	4674.3	5400.5	8.8100	.161 39	4673.1	5399.4	8.7549	.145 26	4672.0	5398.2	8.7055
	P = 6.0 MPa (275.64)				P = 7.0 MPa (285.88)				P = 8.0 MPa (295.06)			
Sat.	.032 44	2589.7	2784.3	5.8892	.027 37	2580.5	2772.1	5.8133	.023 52	2569.8	2758.0	5.7432
300	.036 16	2667.2	2884.2	6.0674	.029 47	2632.2	2838.4	5.9305	.024 26	2590.9	2785.0	5.7906
350	.042 23	2789.6	3043.0	6.3335	.035 24	2769.4	3016.0	6.2283	.029 95	2747.7	2987.3	6.1301
400	.047 39	2892.9	3177.2	6.5408	.039 93	2878.6	3158.1	6.4478	.034 32	2863.8	3138.3	6.3634
450	.052 14	2988.9	3301.8	6.7193	.044 16	2978.0	3287.1	6.6327	.038 17	2966.7	3272.0	6.5551
500	.056 65	3082.2	3422.2	6.8803	.048 14	3073.4	3410.3	6.7975	.041 75	3064.3	3398.3	6.7240
550	.061 01	3174.6	3540.6	7.0288	.051 95	3167.2	3530.9	6.9486	.045 16	3159.8	3521.0	6.8778
600	.065 25	3266.9	3658.4	7.1677	.055 65	3260.7	3650.3	7.0894	.048 45	3254.4	3642.0	7.0206
700	.073 52	3453.1	3894.2	7.4234	.062 83	3448.5	3888.3	7.3476	.054 81	3443.9	3882.4	7.2812
800	.081 60	3643.1	4132.7	7.6566	.069 81	3639.5	4128.2	7.5822	.060 97	3636.0	4123.8	7.5173
900	.089 58	3837.8	4375.3	7.8727	.076 69	3835.0	4371.8	7.7991	.067 02	3832.1	4368.3	7.7351
1000	.097 49	4037.8	4622.7	8.0751	.083 50	4035.3	4619.8	8.0020	.073 01	4032.8	4616.9	7.9384
1100	.105 36	4243.3	4875.4	8.2661	.090 27	4240.9	4872.8	8.1933	.078 96	4238.6	4870.3	8.1300

T	v	u	h	s	v	u	h	s	v	u	h	s
	P = 6.0 MPa (275.64)				P = 7.0 MPa (285.88)				P = 8.0 MPa (295.06)			
1200	.113 21	4454.0	5133.3	8.4474	.097 03	4451.7	5130.9	8.3747	.084 89	4449.5	5128.5	8.3115
1300	.121 06	4669.6	5396.0	8.6199	.103 77	4667.3	5393.7	8.5473	.090 80	4665.0	5391.5	8.4842
	P = 9.0 MPa (303.40)				P = 10.0 MPa (311.06)				P = 12.5 MPa (327.89)			
Sat.	.020 48	2557.8	2742.1	5.6772	.018 026	2544.4	2724.7	5.6141	.013 495	2505.1	2673.8	5.4624
325	.023 27	2646.6	2856.0	5.8712	.019 861	2610.4	2809.1	5.7568				
350	.025 80	2724.4	2956.6	6.0361	.022 42	2699.2	2923.4	5.9443	.016 126	2624.6	2826.2	5.7118
400	.029 93	2848.4	3117.8	6.2854	.026 41	2832.4	3096.5	6.2120	.020 00	2789.3	3039.3	6.0417
450	.033 50	2955.2	3256.6	6.4844	.029 75	2943.4	3240.9	6.4190	.022 99	2912.5	3199.8	6.2719
500	.036 77	3055.2	3386.1	6.6576	.032 79	3045.8	3373.7	6.5966	.025 60	3021.7	3341.8	6.4618
550	.039 87	3152.2	3511.0	6.8142	.035 64	3144.6	3500.9	6.7561	.028 01	3125.0	3475.2	6.6290
600	.042 85	3248.1	3633.7	6.9589	.038 37	3241.7	3625.3	6.9029	.030 29	3225.4	3604.0	6.7810
650	.045 74	3343.6	3755.3	7.0943	.041 01	3338.2	3748.2	7.0398	.032 48	3324.4	3730.4	6.9218
700	.048 57	3439.3	3876.5	7.2221	.043 58	3434.7	3870.5	7.1687	.034 60	3422.9	3855.3	7.0536
800	.054 09	3632.5	4119.3	7.4596	.048 59	3628.9	4114.8	7.4077	.038 69	3620.0	4103.6	7.2965
900	.059 50	3829.2	4364.8	7.6783	.053 49	3826.3	4361.2	7.6272	.042 67	3819.1	4352.5	7.5182
1000	.064 85	4030.3	4614.0	7.8821	.058 32	4027.8	4611.0	7.8315	.046 58	4021.6	4603.8	7.7237
1100	.070 16	4236.3	4867.7	8.0740	.063 12	4234.0	4865.1	8.0237	.050 45	4228.2	4858.8	7.9165
1200	.075 44	4447.2	5126.2	8.2556	.067 89	4444.9	5123.8	8.2055	.054 30	4439.3	5118.0	8.0987
1300	.080 72	4662.7	5389.2	8.4284	.072 65	4460.5	5387.0	8.3783	.058 13	4654.8	5381.4	8.2717

TABLE 18.1-16 *(Continued)*

T	v	u	h	s	v	u	h	s	v	u	h	s
	P = 15.0 MPa (342.24)				*P* = 17.5 MPa (354.75)				*P* = 20.0 MPa (365.81)			
Sat.	.010 337	2455.5	2610.5	5.3098	.007 920	2390.2	2528.8	5.1419	.005 834	2293.0	2409.7	4.9269
350	.011 470	2520.4	2692.4	5.4421								
400	.015 649	2740.7	2975.5	5.8811	.012 447	2685.0	2902.9	5.7213	.009 942	2619.3	2818.1	5.5540
450	.018 445	2879.5	3156.2	6.1404	.015 174	2844.2	3109.7	6.0184	.012 695	2806.2	3060.1	5.9017
500	.020 80	2996.6	3308.6	6.3443	.017 358	2970.3	3274.1	6.2383	.014 768	2942.9	3238.2	6.1401
550	.022 93	3104.7	3448.6	6.5199	.019 288	3083.9	3421.4	6.4230	.016 555	3062.4	3393.5	6.3348
600	.024 91	3208.6	3582.3	6.6776	.021 06	3191.5	3560.1	6.5866	.018 178	3174.0	3537.6	6.5048
650	.026 80	3310.3	3712.3	6.8224	.022 74	3296.0	3693.9	6.7357	.019 693	3281.4	3675.3	6.6582
700	.028 61	3410.9	3840.1	6.9572	.024 34	3398.7	3824.6	6.8736	.021 13	3386.4	3809.0	6.7993
800	.032 10	3610.9	4092.4	7.2040	.027 38	3601.8	4081.1	7.1244	.023 85	3592.7	4069.7	7.0544
900	.035 46	3811.9	4343.8	7.4279	.030 31	3804.7	4335.1	7.3507	.026 45	3797.5	4326.4	7.2830
1000	.038 75	4015.4	4596.6	7.6348	.033 16	4009.3	4589.5	7.5589	.028 97	4003.1	4582.5	7.4925
1100	.042 00	4222.6	4852.6	7.8283	.035 97	4216.9	4846.4	7.7531	.031 45	4211.3	4840.2	7.6874
1200	.045 23	4433.8	5112.3	8.0108	.038 76	4428.3	5106.6	7.9360	.033 91	4422.8	5101.0	7.8707
1300	.048 45	4649.1	5376.0	8.1840	.041 54	4643.5	5370.5	8.1093	.036 36	4638.0	5365.1	8.0442
	P = 25.0 MPa				*P* = 30.0 MPa				*P* = 35.0 MPa			
375	.001 973 1	1798.7	1848.0	4.0320	.001 789 2	1737.8	1791.5	3.9305	.001 700 3	1702.9	1762.4	3.8722
400	.006 004	2430.1	2580.2	5.1418	.002 790	2067.4	2151.1	4.4728	.002 100	1914.1	1987.6	4.2126
425	.007 881	2609.2	2806.3	5.4723	.005 303	2455.1	2614.2	5.1504	.003 428	2253.4	2373.4	4.7747
450	.009 162	2720.7	2949.7	5.6744	.006 735	2619.3	2821.4	5.4424	.004 961	2498.7	2672.4	5.1962
500	.011 123	2884.3	3162.4	5.9592	.008 678	2820.7	3081.1	5.7905	.006 927	2751.9	2994.4	5.6282
550	.012 724	3017.5	3335.6	6.1765	.010 168	2970.3	3275.4	6.0342	.008 345	2921.0	3213.0	5.9026
600	.014 137	3137.9	3491.4	6.3602	.011 446	3100.5	3443.9	6.2331	.009 527	3062.0	3395.5	6.1179
650	.015 433	3251.6	3637.4	6.5229	.012 596	3221.0	3598.9	6.4058	.010 575	3189.8	3559.9	6.3010
700	.016 646	3361.3	3777.5	6.6707	.013 661	3335.8	3745.6	6.5606	.011 533	3309.8	3713.5	6.4631
800	.018 912	3574.3	4047.1	6.9345	.015 623	3555.5	4024.2	6.8332	.013 278	3536.7	4001.5	6.7450
900	.021 045	3783.0	4309.1	7.1680	.017 448	3768.5	4291.9	7.0718	.014 883	3754.0	4274.9	6.9886
1000	.023 10	3990.9	4568.5	7.3802	.019 196	3978.8	4554.7	7.2867	.016 410	3966.7	4541.1	7.2064
1100	.025 12	4200.2	4828.2	7.5765	.020 903	4189.2	4816.3	7.4845	.017 895	4178.3	4804.6	7.4057

T	v	u	h	s	v	u	h	s	v	u	h	s
	P = 25.0 MPa				*P* = 30.0 MPa				*P* = 35.0 MPa			
1200	.027 11	4412.0	5089.9	7.7605	.022 589	4401.3	5079.0	7.6692	.019 360	4390.7	5068.3	7.5910
1300	.029 10	4626.9	5354.4	7.9342	.024 266	4616.0	5344.0	7.8432	.020 815	4605.1	5333.6	7.7653
	P = 40.0 MPa				*P* = 50.0 MPa				*P* = 60.0 MPa			
375	.001 640 7	1677.1	1742.8	3.8290	.001 559 4	1638.6	1716.6	3.7639	.001 502 8	1609.4	1699.5	3.7141
400	.001 907 7	1854.6	1930.9	4.1135	.001 730 9	1788.1	1874.6	4.0031	.001 633 5	1745.4	1843.4	3.9318
425	.002 532	2096.9	2198.1	4.5029	.002 007	1959.7	2060.0	4.2734	.001 816 5	1892.7	2001.7	4.1626
450	.003 693	2365.1	2512.8	4.9459	.002 486	2159.6	2284.0	4.5884	.002 085	2053.9	2179.0	4.4121
500	.005 622	2678.4	2903.3	5.4700	.003 892	2525.5	2720.1	5.1726	.002 956	2390.6	2567.9	4.9321
550	.006 984	2869.7	3149.1	5.7785	.005 118	2763.6	3019.5	5.5485	.003 956	2658.8	2896.2	5.3441
600	.008 094	3022.6	3346.4	6.0114	.006 112	2942.0	3247.6	5.8178	.004 834	2861.1	3151.2	5.6452
650	.009 063	3158.0	3520.6	6.2054	.006 966	3093.5	3441.8	6.0342	.005 595	3028.8	3364.5	5.8829
700	.009 941	3283.6	3681.2	6.3750	.007 727	3230.5	3616.8	6.2189	.006 272	3177.2	3553.5	6.0824
800	.011 523	3517.8	3978.7	6.6662	.009 076	3479.8	3933.6	6.5290	.007 459	3441.5	3889.1	6.4109
900	.012 962	3739.4	4257.9	6.9150	.010 283	3710.3	4224.4	6.7882	.008 508	3681.0	4191.5	6.6805
1000	.014 324	3954.6	4527.6	7.1356	.011 411	3930.5	4501.1	7.0146	.009 480	3906.4	4475.2	6.9127
1100	.015 642	4167.4	4793.1	7.3364	.012 496	4145.7	4770.5	7.2184	.010 409	4124.1	4748.6	7.1195
1200	.016 940	4380.1	5057.7	7.5224	.013 561	4359.1	5037.2	7.4058	.011 317	4338.2	5017.2	7.3083
1300	.018 229	4594.3	5323.5	7.6969	.014 616	4572.8	5303.6	7.5808	.012 215	4551.4	5284.3	7.4837

TABLE 18.1-16 (*Continued*)

Compressed Liquid

T	P = 5 MPa (263.99)				P = 10 MPa (311.06)				P = 15 MPa (342.24)			
	v	u	h	s	v	u	h	s	v	u	h	s
Sat.	.001 285 9	1147.8	1154.2	2.9202	.001 452 4	1393.0	1407.6	3.3596	.001 658 1	1585.6	1610.5	3.6848
0	.000 997 7	.04	5.04	.0001	.000 995 2	.09	10.04	.0002	.000 992 8	.15	15.05	.0004
20	.000 999 5	83.65	88.65	.2956	.000 997 2	83.36	93.33	.2945	.000 995 0	83.06	97.99	.2934
40	.001 005 6	166.95	171.97	.5705	.001 003 4	166.35	176.38	.5686	.001 001 3	165.76	180.78	.5666
60	.001 014 9	250.23	255.30	.8285	.001 012 7	249.36	259.49	.8258	.001 010 5	248.51	263.67	.8232
80	.001 026 8	333.72	338.85	1.0720	.001 024 5	332.59	342.83	1.0688	.001 022 2	331.48	346.81	1.0656
100	.001 041 0	417.52	422.72	1.3030	.001 038 5	416.12	426.50	1.2992	.001 036 1	414.74	430.28	1.2955
120	.001 057 6	501.80	507.09	1.5233	.001 054 9	500.08	510.64	1.5189	.001 052 2	498.40	514.19	1.5145
140	.001 076 8	586.76	592.15	1.7343	.001 073 7	584.68	595.42	1.7292	.001 070 7	582.66	598.72	1.7242
160	.001 098 8	672.62	678.12	1.9375	.001 095 3	670.13	681.08	1.9317	.001 091 8	667.71	684.09	1.9260
180	.001 124 0	759.63	765.25	2.1341	.001 119 9	756.65	767.84	2.1275	.001 115 9	753.76	770.50	2.1210
200	.001 153 0	848.1	853.9	2.3255	.001 148 0	844.5	856.0	2.3178	.001 143 3	841.0	858.2	2.3104
220	.001 186 6	938.4	944.4	2.5128	.001 180 5	934.1	945.9	2.5039	.001 174 8	929.9	947.5	2.4953
240	.001 226 4	1031.4	1037.5	2.6979	.001 218 7	1026.0	1038.1	2.6872	.001 211 4	1020.8	1039.0	2.6771
260	.001 274 9	1127.9	1134.3	2.8830	.001 264 5	1121.1	1133.7	2.8699	.001 255 0	1114.6	1133.4	2.8576
280					.001 321 6	1220.9	1234.1	3.0548	.001 308 4	1212.5	1232.1	3.0393
300					.001 397 2	1328.4	1342.3	3.2469	.001 377 0	1316.6	1337.3	3.2260
320									.001 472 4	1431.1	1453.2	3.4247
340									.001 631 1	1567.5	1591.9	3.6546

T	P = 20 MPa (365.81)				P = 30 MPa				P = 50 MPa			
	v	u	h	s	v	u	h	s	v	u	h	s
Sat.	.002 036	1785.6	1826.3	4.0139								
0	.000 990 4	.19	20.01	.0004	.000 985 6	.25	29.82	.0001	.000 976 6	.20	49.03	−.0014
20	.000 992 8	82.77	102.62	.2923	.000 988 6	82.17	111.84	.2899	.000 980 4	81.00	130.02	.2848
40	.000 999 2	165.17	185.16	.5646	.000 995 1	164.04	193.89	.5607	.000 987 2	161.86	211.21	.5527
60	.001 008 4	247.68	267.85	.8206	.001 004 2	246.06	276.19	.8154	.000 996 2	242.98	292.79	.8052
80	.001 019 9	330.40	350.80	1.0624	.001 015 6	328.30	358.77	1.0561	.001 007 3	324.34	374.70	1.0440
100	.001 033 7	413.39	434.06	1.2917	.001 029 0	410.78	441.66	1.2844	.001 020 1	405.88	456.89	1.2703
120	.001 049 6	496.76	517.76	1.5102	.001 044 5	493.59	524.93	1.5018	.001 034 8	487.65	539.39	1.4857
140	.001 067 8	580.69	602.04	1.7193	.001 062 1	576.88	608.75	1.7098	.001 051 5	569.77	622.35	1.6915
160	.001 088 5	665.35	687.12	1.9204	.001 082 1	660.82	693.28	1.9096	.001 070 3	652.41	705.92	1.8891
180	.001 112 0	750.95	773.20	2.1147	.001 104 7	745.59	778.73	2.1024	.001 091 2	735.69	790.25	2.0794
200	.001 138 8	837.7	860.5	2.3031	.001 130 2	831.4	865.3	2.2893	.001 114 6	819.7	875.5	2.2634
220	.001 169 3	925.9	949.3	2.4870	.001 159 0	918.3	953.1	2.4711	.001 140 8	904.7	961.7	2.4419
240	.001 204 6	1016.0	1040.0	2.6674	.001 192 0	1006.9	1042.6	2.6490	.001 170 2	990.7	1049.2	2.6158
260	.001 246 2	1108.6	1133.5	2.8459	.001 230 3	1097.4	1134.3	2.8243	.001 203 4	1078.1	1138.2	2.7860
280	.001 296 5	1204.7	1230.6	3.0248	.001 275 5	1190.7	1229.0	2.9986	.001 241 5	1167.2	1229.3	2.9537
300	.001 359 6	1306.1	1333.3	3.2071	.001 330 4	1287.9	1327.8	3.1741	.001 286 0	1258.7	1323.0	3.1200
320	.001 443 7	1415.7	1444.6	3.3979	.001 399 7	1390.7	1432.7	3.3539	.001 338 8	1353.3	1420.2	3.2868
340	.001 568 4	1539.7	1571.0	3.6075	.001 492 0	1501.7	1546.5	3.5426	.001 403 2	1452.0	1522.1	3.4557
360	.001 822 6	1702.8	1739.3	3.8772	.001 626 5	1626.6	1675.4	3.7494	.001 483 8	1556.0	1630.2	3.6291
380					.001 869 1	1781.4	1837.5	4.0012	.001 588 4	1667.2	1746.6	3.8101

TABLE 18.1-16 (*Continued*)

Saturated Solid-Vapor

Temp. °C *T*	Press. kPa *P*	Specific Volume		Internal Energy			Enthalpy			Entropy		
		Sat. Solid $v_i \times 10^3$	Sat. Vapor v_g	Sat. Solid u_i	Subl. u_{ig}	Sat. Vapor u_g	Sat. Solid h_i	Subl. h_{ig}	Sat. Vapor h_g	Sat. Solid s_i	Subl. s_{ig}	Sat. Vapor s_g
.01	.6113	1.0908	206.1	−333.40	2708.7	2375.3	−333.40	2834.8	2501.4	−1.221	10.378	9.156
0	.6108	1.0908	206.3	−333.43	2708.8	2375.3	−333.43	2834.8	2501.3	−1.221	10.378	9.157
−2	.5176	1.0904	241.7	−337.62	2710.2	2372.6	−337.62	2835.3	2497.7	−1.237	10.456	9.219
−4	.4375	1.0901	283.8	−341.78	2711.6	2369.8	−341.78	2835.7	2494.0	−1.253	10.536	9.283
−6	.3689	1.0898	334.2	−345.91	2712.9	2367.0	−345.91	2836.2	2490.3	−1.268	10.616	9.348
−8	.3102	1.0894	394.4	−350.02	2714.2	2364.2	−350.02	2836.6	2486.6	−1.284	10.698	9.414
−10	.2602	1.0891	466.7	−354.09	2715.5	2361.4	−354.09	2837.0	2482.9	−1.299	10.781	9.481
−12	.2176	1.0888	553.7	−358.14	2716.8	2358.7	−358.14	2837.3	2479.2	−1.315	10.865	9.550
−14	.1815	1.0884	658.8	−362.15	2718.0	2355.9	−362.15	2837.6	2475.5	−1.331	10.950	9.619
−16	.1510	1.0881	786.0	−366.14	2719.2	2353.1	−366.14	2837.9	2471.8	−1.346	11.036	9.690
−18	.1252	1.0878	940.5	−370.10	2720.4	2350.3	−370.10	2838.2	2468.1	−1.362	11.123	9.762
−20	.1035	1.0874	1128.6	−374.03	2721.6	2347.5	−374.03	2838.4	2464.3	−1.377	11.212	9.835
−22	.0853	1.0871	1358.4	−377.93	2722.7	2344.7	−377.93	2838.6	2460.6	−1.393	11.302	9.909
−24	.0701	1.0868	1640.1	−381.80	2723.7	2342.0	−381.80	2838.7	2456.9	−1.408	11.394	9.985
−26	.0574	1.0864	1986.4	−385.64	2724.8	2339.2	−385.64	2838.9	2453.2	−1.424	11.486	10.062
−28	.0469	1.0861	2413.7	−389.45	2725.8	2336.4	−389.45	2839.0	2449.5	−1.439	11.580	10.141
−30	.0381	1.0858	2943	−393.23	2726.8	2333.6	−393.23	2839.0	2445.8	−1.455	11.676	10.221
−32	.0309	1.0854	3600	−396.98	2727.8	2330.8	−396.98	2839.1	2442.1	−1.471	11.773	10.303
−34	.0250	1.0851	4419	−400.71	2728.7	2328.0	−400.71	2839.1	2438.4	−1.486	11.872	10.386
−36	.0201	1.0848	5444	−404.40	2729.6	2325.2	−404.40	2839.1	2434.7	−1.501	11.972	10.470
−38	.0161	1.0844	6731	−408.06	2730.5	2322.4	−408.06	2839.0	2430.9	−1.517	12.073	10.556
−40	.0129	1.0841	8354	−411.70	2731.3	2319.6	−411.70	2838.9	2427.2	−1.532	12.176	10.644

*a*From ref. 4. Used by permission.

h	t	c	p	ρ	$\mu \times 10^7$	$\nu \times 10^4$
0	59.00	1117	2116.2	0.002378	3.719	1.561
1,000	57.44	1113	2040.9	.002310	3.699	1.602
2,000	51.87	1109	1967.7	.002242	3.679	1.641
3,000	48.31	1105	1896.7	.002177	3.659	1.681
4,000	44.74	1102	1827.7	.002112	3.639	1.723
5,000	41.18	1098	1760.8	.002049	3.618	1.766
6,000	37.62	1094	1696.0	.001988	3.598	1.810
7,000	34.05	1090	1633.0	.001928	3.577	1.855
8,000	30.49	1086	1571.9	.001869	3.557	1.903
9,000	26.92	1082	1512.8	.001812	3.536	1.951
10,000	23.36	1078	1455.4	.001756	3.515	2.002
11,000	19.80	1074	1399.8	.001702	3.495	2.054
12,000	16.23	1070	1345.9	.001649	3.474	2.107
13,000	12.67	1066	1293.7	.001597	3.453	2.163
14,000	9.10	1062	1243.2	.001546	3.432	2.220
15,000	5.54	1058	1194.3	.001497	3.411	2.280
16,000	1.98	1054	1147.0	.001448	3.390	2.341
17,000	− 1.59	1050	1101.1	.001401	3.369	2.404
18,000	− 5.15	1046	1056.9	.001355	3.347	2.470
19,000	− 8.72	1041	1014.0	.001311	3.326	2.538
20,000	− 12.28	1037	972.6	.001267	3.305	2.608
21,000	− 15.84	1033	932.5	.001225	3.283	2.681
22,000	− 19.41	1029	893.8	.001183	3.262	2.757
23,000	− 22.97	1025	856.4	.001143	3.240	2.834
24,000	− 26.54	1021	820.3	.001104	3.218	2.915
25,000	− 30.10	1017	785.3	.001066	3.196	2.999
26,000	− 33.66	1012	751.7	.001029	3.174	3.087
27,000	− 37.23	1008	719.2	.000993	3.153	3.177
28,000	− 40.79	1004	687.9	.000957	3.130	3.270
29,000	− 44.36	999	657.6	.000923	3.108	3.367
30,000	− 47.92	995	628.5	.000890	3.086	3.469
31,000	− 51.48	991	600.4	.000858	3.064	3.573
32,000	− 55.05	987	573.3	.000826	3.041	3.682
33,000	− 58.61	982	547.3	.000796	3.019	3.795
34,000	− 62.18	978	522.2	.000766	2.997	3.913
35,000	− 65.74	973	498.0	.000737	2.974	4.036
35,332	− 67.6	971	489.8	.000727	2.961	4.073
36,000	− 67.6	971	474.8	.000709	2.961	4.176
37,000	− 67.6	971	452.5	.0006766	2.961	4.376
38,000	− 67.6	971	431.2	.0006448	2.961	4.592
39,000	− 67.6	971	411.0	.0006145	2.961	4.819
40,000	− 67.6	971	391.8	.0005857	2.961	5.055
41,000	− 67.6	971	373.4	.0005582	2.961	5.305
42,000	− 67.6	971	355.8	.0005320	2.961	5.566
43,000	− 67.6	971	339.1	.0005071	2.961	5.839
44,000	− 67.6	971	323.2	.0004833	2.961	6.127
45,000	− 67.6	971	308.0	.0004605	2.961	6.430
46,000	− 67.6	971	293.6	.0004390	2.961	6.745
47,000	− 67.6	971	279.8	.0004184	2.961	7.077
48,000	− 67.6	971	266.6	.0003987	2.961	7.427
49,000	− 67.6	971	254.1	.0003800	2.961	7.792

TABLE 18.1-17 (*Continued*)

h	t	c	p	ρ	μ × 10⁷	ν × 10⁴
50,000	− 67.6	971	242.2	.0003622	2.961	8.175
60,000	− 67.6	971	150.9	.0002240	2.961	13.219
70,000	− 67.6	971	93.5	.0001389	2.961	21.317
80,000	− 67.6	971	58.0	.0000861	2.961	34.390
90,000	− 67.6	971	36.0	.0000535	2.961	55.346
100,000	− 67.6	971	22.4	.0000331	2.961	89.456
104,987	− 67.6	971	17.59	.0000261	2.961	113.4
110,000	− 47.4	996	13.92	.0000197	3.090	157.2
120,000	− 7.2	1043	9.026	.0000116	3.339	287.6
130,000	33.0	1089	6.071	.00000717	3.579	498.9
140,000	73.3	1132	4.213	.00000460	3.809	827.9
150,000	113.5	1174	3.003	.00000305	4.032	1322
160,000	153.7	1215	2.190	.00000208	4.247	2043
164,042	170.0	1231	1.938	.00000179	4.332	2417
170,000	170.0	1231	1.624	.00000150	4.332	2886
180,000	170.0	1231	1.206	.00000111	4.332	3885
190,000	170.0	1231	.8956	.00000083	4.332	5232
196,850	170.0	1231	.7305	.00000068	4.332	6412
200,000	159.4	1220	.6645	.00000062	4.277	6844
210,000	125.9	1187	.4869	.00000048	4.099	8467
220,000	92.4	1152	.3504	.00000037	3.916	10600
230,000	58.9	1117	.2470	.00000028	3.727	13400
240,000	25.3	1080	.1699	.00000020	3.533	17330
250,000	− 8.2	1042	.1139	.00000015	3.333	22700
255,905	− 28.0	1019	.0886	.00000012	3.212	26880
260,000	− 28.0	1019	.0742	.00000010	3.212	32090

[a]*Symbols: h* = *height above sea level, ft*
t = temperature, °F
c = speed of sound, ft/s
p = pressure, lbf/ft^2
ρ = mass density, slug/ft^3
μ = coefficient of viscosity, slug/ft s
$\nu = \mu/\rho$, kinematic viscosity, ft^2/s
From ref. 8. Used by permission.

TABLE 18.1-18   ISENTROPIC FLOW: PERFECT GAS, $K = 1.4^{a,b}$

M	M^*	p/p_0	ρ/ρ_0	T/T_0	A/A^*	F/F^*	$\dfrac{A}{A^*} \cdot \dfrac{p}{p_0}$
0	0	1.0000,0	1.0000,0	1.00000	∞	∞	∞
.01	.01096	.9999,3	.9999,5	.99998	5,7.874	4,5.650	5,7.870
.02	.02191	.9997,2	.9998,0	.99992	2,8.942	2,2.834	2,8.934
.03	.03286	.9993,7	.9995,5	.99982	1,9.300	15.,232	1,9.288
.04	.04381	.9988,8	.9992,0	.99968	14.,482	11.,435	14.,465
.05	.05476	.9982,5	.9987,5	.99950	11.5,915	9.,1584	11.,5712
.06	.06570	.9974,8	.9982,0	.99928	9.6,659	7.,6428	9.,6415
.07	.07664	.9965,8	.9975,5	.99902	8.2,915	6.5,620	8.2,631
.08	.08758	.9955,3	.9968,0	.99872	7.2,616	5.7,529	7.2,291
.09	.09851	.9943,5	.9959,6	.99838	6.4,613	5.1,249	6.4,248
.10	.10943	.9930,3	.9950,2	.99800	5.8,218	4.6,236	5.7,812
.11	.12035	.9915,7	.9939,8	.99758	5.2,992	4.2,146	5.2,546
.12	.13126	.9899,8	.9928,4	.99714	4.8,643	3.8,747	4.8,157
.13	.14216	.9882,6	.9916,0	.99664	4.4,968	3.58,80	4.4,440
.14	.15306	.9864,0	.9902,7	.99610	4.18,24	3.34,32	4.12,55
.15	.16395	.9844,1	.9888,4	.99552	3.91,03	3.13,17	3.84,93
.16	.17483	.9822,8	.9873,1	.99400	3.67,27	2.94,74	3.60,76
.17	.18569	.9800,3	.9856,9	.99425	3.46,35	2.78,55	3.39,43
.18	.19654	.9776,5	.9839,8	.99356	3.27,79	2.64,22	3.20,46
.19	.20738	.9751,4	.9821,7	.99283	3.11,22	2.51,46	3.03,48
.20	.21822	.9725,0	.9802,7	.99206	2.96,35	2.40.04	2.88,20
.21	.22904	.9697,3	.9782,8	.99125	2.82,93	2.29,76	2.74,37
.22	.23984	.9668,5	.9762,1	.99041	2.70,76	2.20,46	2.61,78
.23	.25063	.9638,3	.9740,3	.98953	2.59,68	2.12,03	2.50,29
.24	.26141	.9607,0	.9717,7	.98861	2.49,56	2.04,34	2.39,75
.25	.27216	.9574,5	.9694,2	.98765	2.40,27	1.97,32	2.30,05
.26	.28291	.9540,8	.9669,9	.98666	2.31,73	1.90,88	2.21,09
.27	.29364	.9506,0	.9644,6	.98563	2.23,85	1.84,96	2.12,79
.28	.30435	.9470,0	.9618,5	.98456	2.16,56	1.795,0	2.05,08
.29	.31504	.9432,9	.9591,6	.98346	2.09,79	1.744,6	1.97,89
.30	.32572	.9394,7	.9563,8	.98232	2.035,1	1.697,9	1.911,9
.31	.33638	.9355,4	.9535,2	.98114	1.976,5	1.654,6	1.849,1
.32	.34701	.9315,0	.9505,8	.97993	1.921,8	1.614,4	1.790,2
.33	.35762	.9273,6	.9475,6	.97868	1.870,7	1.576,9	1.734,8
.34	.36821	.9231,2	.9444,6	.97740	1.822,9	1.542,0	1.682,8
.35	.37879	.9187,7	.9412,8	.97608	1.778,0	1.509,4	1.633,6
.36	.38935	.9143,3	.9380,3	.97473	1.735,8	1.478,9	1.587,1
.37	.39988	.9097,9	.9347,0	.97335	1.696,1	1.450,3	1.543,1
.38	.41039	.9051,6	.9312,9	.97193	1.658,7	1.423,6	1.501,4
.39	.42087	.9004,4	.9278,2	.97048	1.623,4	1.398,5	1.461,8
.40	.43133	.8956,2	.9242,8	.96899	1.590,1	1.374,9	1.424,1
.41	.44177	.8907,1	.9206,6	.96747	1.558,7	1.352,7	1.388,3
.42	.45218	.8857,2	.9169,7	.96592	1.528,9	1.331,8	1.354,2
.43	.46256	.8806,5	.9132,2	.96434	1.500,7	1.312,2	1.321,6
.44	.47292	.8755,0	.9094,0	.96272	1.474,0	1.293,7	1.290,5
.45	.48326	.8702,7	.9055,2	.96108	1.448,7	1.276,3	1.260.7
.46	.49357	.8649,6	.9015,7	.95940	1.424,6	1.259,8	1.232,2
.47	.50385	.8595,8	.8975,6	.95769	1.401,8	1.244,3	1.205,0
.48	.51410	.8541,3	.8934,9	.95595	1.380,1	1.229,6	1.178,8
.49	.52432	.8486,1	.8893,6	.95418	1.359,4	1.215,8	1.153,7

TABLE 18.1-18 (*Continued*)

M	M*	p/p_0	ρ/ρ_0	T/T_0	A/A^*	F/F^*	$\dfrac{A}{A^*} \cdot \dfrac{p}{p_0}$
.50	.53452	.8430,2	.8851,7	.95238	1.339,8	1.202,7	1.129,51
.51	.54469	.8373,7	.8809,2	.95055	1.321,2	1.190,3	1.106,31
.52	.55482	.8316,6	.8766,2	.94869	1.303,4	1.178,6	1.083,97
.53	.56493	.8258,9	.8722,7	.94681	1.286,4	1.167,5	1.062,45
.54	.57501	.8200,5	.8678,8	.94489	1.270,3	1.157,1	1.041,73
.55	.58506	.8141,6	.86342	.94295	1.255,0	1.147,2	1.021,74
.56	.59508	.8082,2	.85892	.94098	1.240,3	1.137,8	1.002,44
.57	.60506	.8022,4	.85437	.93898	1.226,3	1.128,9	.983,81
.58	.61500	.7962,1	.84977	.93696	1.213,0	1.120,5	.965,81
.59	.62491	.7901,2	.84513	.93491	1.200,3	1.112,6	.948,39
.60	.63480	.78400	.84045	.93284	1.188,2	1.1050,4	.931,55
.61	.64466	.77784	.83573	.93074	1.176,6	1.0979,3	.915,25
.62	.65448	.77164	.83096	.92861	1.165,6	1.0912,0	.899,46
.63	.66427	.76540	.82616	.92646	1.155,1	1.0848,5	.884,16
.64	.67402	.75913	.82132	.92428	1.145,1	1.0788,3	.869,32
.65	.68374	.75283	.81644	.92208	1.1356	1.0731,4	.8549,3
.66	.69342	.74650	.81153	.91986	1.1265	1.0677,7	.8409,6
.67	.70307	.74014	.80659	.91762	1.1178	1.0627,1	.8274,0
.68	.71268	.73376	.80162	.91535	1.1096	1.0579,2	.8142,1
.69	.72225	.72735	.79662	.91306	1.1018	1.0534,0	.8014,1
.70	.73179	.72092	.79158	.91075	1.0943,7	1.0491,5	.7889,6
.71	.74129	.71448	.78652	.90842	1.0872,9	1.0451,4	.7768,5
.72	.75076	.70802	.78143	.90606	1.0805,7	1.0413,7	.7650,7
.73	.76019	.70155	.77632	.90368	1.0741,9	1.0378,3	.7536,0
74	.76958	.69507	.77119	.90129	1.0681,4	1.0345,0	.7424,3
.75	.77893	.68857	.76603	.89888	1.0624,2	1.0313,7	.7315,5
.76	.78825	.68207	.76086	.89644	1.0570,0	1.0284,4	.7209,5
.77	.79753	.67556	.75567	.89399	1.0518,8	1.0257,0	.7106,2
.78	.80677	.66905	.75046	.89152	1.0470,5	1.0231,4	.7705,4
.79	.81597	.66254	.74524	.88903	1.0425,0	1.0207,5	.6907,0
.80	.82514	.65602	.74000	.88652	1.0382,3	1.1085,3	.6811,0
.81	.83426	.64951	.73474	.88400	1.0342,2	1.0164,6	.6717,3
.82	.84334	.64300	.72947	.88146	1.0304,6	1.1045,5	.6625,9
.83	.85239	.63650	.72419	.87890	1.0269,6	1.1027,8	.6536,6
.84	.86140	.63000	.71890	.87633	1.0237,0	1.0111,5	.6449,3
.85	.87037	.62351	.71361	.87374	1.0206,7	1.0096,6	.6364,0
.86	.87929	.61703	.70831	.87114	1.0178,7	1.0082,9	.6280,6
.87	.88817	.61057	.70300	.86852	1.1053,0	1.0070,4	.6199,1
.88	.89702	.60412	.69769	.86589	1.0129,4	1.0059,1	.6119,3
.89	.90583	.59768	.69237	.86324	1.0108,0	1.0049,0	.6041,3
.90	.91460	.59126	.68704	.86058	1.0088,6	1.0039,9	.5965,0
.91	.92333	.58486	.68171	.85791	1.0071,3	1.0031,8	.5890,3
.92	.93201	.57848	.67639	.85523	1.0056,0	1.0024,8	.5817,1
.93	.94065	.57212	.67107	.85253	1.0042,6	1.0018,8	.5745,5
.94	.94925	.56578	.66575	.84982	1.0031,1	1.0013,6	.5675,4
.95	.95781	.55946	.66044	.84710	1.0021,4	1.0009,3	.5606,6
.96	.96633	.55317	.65513	.84437	1.0013,6	1.0005,9	.5539,2
.97	.97481	.54691	.64982	.84162	1.0007,6	1.0003,3	.5473,2
.98	.98325	.54067	.64452	.83887	1.0003,3	1.0001,4	.5408,5
.99	.99165	.53446	.63923	.83611	1.0000,8	1.0000,3	.5345,0
1.00	1.00000	.52828	.63394	.83333	1.0000,0	1.0000,0	.5282,8

TABLE 18.1-18 *(Continued)*

M	M*	p/p_0	ρ/ρ_0	T/T_0	A/A^*	F/F^*	$\dfrac{A}{A^*}\cdot\dfrac{p}{p_0}$
1.01	1.00831	.52213	.62866	.83055	1.0000,8	1.0000,3	.5221,8
1.02	1.01658	.51602	.62339	.82776	1.0003,3	1.0001,3	.5161,9
1.03	1.02481	.50994	.61813	.82496	1.0007,4	1.0003,0	.5103,1
1.04	1.03300	.50389	.61288	.82215	1.0013,0	1.0005,3	.5045,4
1.05	1.04114	.49787	.60765	.81933	1.0020,2	1.0008,2	.4988,8
1.06	1.04924	.49189	.60243	.81651	1.0029,0	1.0011,6	.4933.2
1.07	1.05730	.48595	.59722	.81368	1.0039,4	1.0015,5	.4878,7
1.08	1.06532	.48005	.59203	.81084	1.0051,2	1.0020,0	.4825,1
1.09	1.07330	.47418	.58685	.80800	1.0064,5	1.0025,0	.4772,4
1.10	1.08124	.46835	.58169	.80515	1.0079,3	1.00305	.4720,6
1.11	1.08914	.46256	.57655	.80230	1.0095,5	1.00365	.4669,8
1.12	1.09699	.45682	.57143	.79944	1.0113,1	1.00429	.4619,9
1.13	1.10480	.45112	.56632	.79657	1.1032,2	1.00497	.4570,8
1.14	1.11256	.44545	.56123	.79370	1.0152,7	1.00569	.4522,5
1.15	1.1203	.43983	.55616	.79083	1.1074,6	1.00646	.4475,1
1.16	1.1280	.43425	.55112	.78795	1.1097,8	1.00726	.4428,4
1.17	1.1356	.42872	.54609	.78507	1.0222,4	1.00810	.4382,5
1.18	1.1432	.42323	.54108	.78218	1.0248,4	1.00897	.4337,4
1.19	1.1508	.41778	.53610	.77929	1.0275,7	1.00988	.4293,0
1.20	1.1583	.4123,8	.53114	.77640	1.0304,4	1.01082	.4249,3
1.21	1.1658	.4070,2	.52620	.77350	1.0334,4	1.01178	.4206,3
1.22	1.1732	.4017,1	.52129	.77061	1.0365,7	1.01278	.4164,0
1.23	1.1806	.3964,5	.51640	.76771	1.0398,3	1.01381	.4122,4
1.24	1.1879	.3912,3	.51154	.76481	1.0432,3	1.01486	.4081,4
1.25	1.1952	.3860,6	.50670	.76190	1.0467,6	1.10594	.4041,1
1.26	1.2025	.3809,4	.50189	.75900	1.0504,1	1.01705	.4001,4
1.27	1.2097	.3758,6	.49710	.75610	1.0541,9	1.01818	.3962,2
1.28	1.2169	.3708,3	.49234	.75319	1.0581,0	1.10933	.3923.7
1.29	1.2240	.3658,5	.48761	.75029	1.0621,4	1.02050	.3885,8
1.30	1.2311	.3609,2	.48291	.74738	1.0663,1	1.02170	.3848,4
1.31	1.2382	.3560,3	.47823	.74448	1.0706,0	1.02292	.3811,6
1.32	1.2452	.3511,9	.47358	.74158	1.0750,2	1.02415	.3775,4
1.33	1.2522	.3464,0	.46895	.73867	1.0795,7	1.02510	.3739,7
1.34	1.2591	.3416,6	.46436	.73577	1.0842,4	1.02666	.3704,4
1.35	1.2660	.3369,7	.45980	.73287	1.0890,4	1.02794	.3669,7
1.36	1.2729	.3323,3	.45527	.72997	1.0939,7	1.02924	.3635,5
1.37	1.2797	.3277,4	.45076	.72707	1.0990,2	1.03056	.3601,8
1.38	1.2865	.3231,9	.44628	.72418	1.1042,0	1.03189	.3568,6
1.39	1.2932	.3186,9	.44183	.72128	1.1095,0	1.03323	.3535,9
1.40	1.2999	.3142,4	.43742	.71839	1.1149	1.03458	.35036
1.41	1.3065	.3098,4	.43304	.71550	1.1205	1.03595	.34717
1.42	1.3131	.3054,9	.42869	.71261	1.1262	1.03733	.34403
1.43	1.3197	.3011,9	.42436	.70973	1.1320	1.03872	.34093
1.44	1.3262	.2969,3	.42007	.70685	1.1379	1.04012	.33787
1.45	1.3327	.2927,2	.41581	.70397	1.1440	1.04153	.33486
1.46	1.3392	.2885,6	.41158	.70110	1.1502	1.04295	.33189
1.47	1.3456	.2844,5	.40738	.69823	1.1565	1.04438	.32896
1.48	1.3520	.2803,9	.40322	.69537	1.1629	1.04581	.32607
1.49	1.3583	.2763,7	.39909	.69251	1.1695	1.04725	.32321
1.50	1.3646	.2724,0	.39498	.68965	1.1762	1.04870	.32039
1.51	1.3708	.2684,8	.39091	.68680	1.1830	1.05016	.31761

TABLE 18.1-18 (*Continued*)

M	M*	p/p_0	ρ/ρ_0	T/T_0	A/A^*	F/F^*	$\dfrac{A}{A^*}\cdot\dfrac{p}{p_0}$
1.52	1.3770	.2646,1	.38687	.68396	1.1899	1.05162	.31487
1.53	1.3832	.2607,8	.38287	.68112	1.1970	1.05309	.31216
1.54	1.3894	.2570,0	.37890	.67828	1.2042	1.05456	.30948
1.55	1.3955	.2532,6	.37496	.67545	1.2115	1.05604	.30685
1.56	1.4016	.2495,7	.37105	.67262	1.2190	1.05752	.30424
1.57	1.4076	.2459,3	.36717	.66980	1.2266	1.05900	.30167
1.58	1.4135	.2423,3	.36332	.66699	1.2343	1.06049	.29913
1.59	1.4195	.2387,8	.35951	.66418	1.2422	1.06198	.29662
1.60	1.4254	.23527	.35573	.66138	1.2502	1.06348	.29414
1.61	1.4313	.23181	.35198	.65858	1.2583	1.06498	.29169
1.62	1.4371	.22839	.34826	.65579	1.2666	1.06648	.28928
1.63	1.4429	.22501	.34458	.65301	1.2750	1.06798	.28690
1.64	1.4487	.22168	.34093	.65023	1.2835	1.06948	.28454
1.65	1.4544	.21839	.33731	.64746	1.2922	1.07098	.28221
1.66	1.4601	.21515	.33372	.64470	1.3010	1.07249	.27991
1.67	1.4657	.21195	.33016	.64194	1.3099	1.07399	.27764
1.68	1.4713	.20879	.32664	.63919	1.3190	1.07550	.27540
1.69	1.4769	.20567	.32315	.63645	1.3282	1.07701	.27318
1.70	1.4825	.20259	.31969	.63372	1.3376	1.07851	.27099
1.71	1.4880	.19955	.31626	.63099	1.3471	1.08002	.26882
1.72	1.4935	.19656	.31286	.62827	1.3567	1.08152	.26668
1.73	1.4989	.19361	.30950	.62556	1.3665	1.08302	.26457
1.74	1.5043	.19070	.30617	.62286	1.3764	1.08453	.26248
1.75	1.5097	.18782	.30287	.62016	1.3865	1.08603	.26042
1.76	1.5150	.18499	.29959	.61747	1.3967	1.08753	.25838
1.77	1.5203	.18220	.29635	.61479	1.4071	1.08903	.25636
1.78	1.5256	.17944	.29314	.61211	1.4176	1.09053	.25436
1.79	1.5308	.17672	.28997	.60945	1.4282	1.09202	.25239
1.80	1.5360	.17404	.28682	.60680	1.4390	1.09352	.25044
1.81	1.5412	.17140	.28370	.60415	1.4499	1.09500	.24851
1.82	1.5463	.16879	.28061	.60151	1.4610	1.09649	.24660
1.83	1.5514	.16622	.27756	.59888	1.4723	1.09798	.24472
1.84	1.5564	.16369	.27453	.59626	1.4837	1.09946	.24286
1.85	1.5614	.16120	.27153	.59365	1.4952	1.1009	.24102
1.86	1.5664	.15874	.26857	.59105	1.5069	1.1024	.23919
1.87	1.5714	.15631	.26563	.58845	1.5188	1.1039	.23739
1.88	1.5763	.15392	.26272	.58586	1.5308	1.1054	.23561
1.89	1.5812	.15156	.25984	.58329	1.5429	1.1068	.23385
1.90	1.5861	.14924	.25699	.58072	1.5552	1.1083	.23211
1.91	1.5909	.14695	.25417	.57816	1.5677	1.1097	.23039
1.92	1.5957	.14469	.25138	.57561	1.5804	1.1112	.22868
1.93	1.6005	.14247	.24862	.57307	1.5932	1.1126	.22699
1.94	1.6052	.14028	.24588	.57054	1.6062	1.1141	.22532
1.95	1.6099	.13813	.24317	.56802	1.6193	1.1155	.22367
1.96	1.6146	.13600	.24049	.56551	1.6326	1.1170	.22204
1.97	1.6193	.13390	.23784	.56301	1.6461	1.1184	.22042
1.98	1.6239	.13184	.23522	.56051	1.6597	1.1198	.21882
1.99	1.6285	.12981	.23262	.55803	1.6735	1.1213	.21724
2.00	1.6330	.12780	.23005	.55556	1.6875	1.1227	.21567
2.01	1.6375	.12583	.22751	.55310	1.7017	1.1241	.21412
2.02	1.6420	.12389	.22499	.55064	1.7160	1.1255	.21259

TABLE 18.1-18 *(Continued)*

M	M*	p/p_0	ρ/ρ_0	T/T_0	A/A^*	F/F^*	$\dfrac{A}{A^*}\cdot\dfrac{p}{p_0}$
2.03	1.6465	.12198	.22250	.54819	1.7305	1.1269	.21107
2.04	1.6509	.12009	.22004	.54576	1.7452	1.1283	.20957
2.05	1.6553	.11823	.21760	.54333	1.7600	1.1297	.20808
2.06	1.6597	.11640	.21519	.54091	1.7750	1.1311	.20661
2.07	1.6640	.11460	.21281	.53850	1.7902	1.1325	.20515
2.08	1.6683	.11282	.21045	.53611	1.8056	1.1339	.20371
2.09	1.6726	.11107	.20811	.53373	1.8212	1.1352	.20228
2.10	1.6769	.10935	.20580	.53135	1.8369	1.1366	.20087
2.11	1.6811	.10766	.20352	.52898	1.8529	1.1380	.19947
2.12	1.6853	.10599	.20126	.52663	1.8690	1.1393	.19809
2.13	1.6895	.10434	.19902	.52428	1.8853	1.1407	.19672
2.14	1.6936	.10272	.19681	.52194	1.9018	1.1420	.19537
2.15	1.6977	.10113	.19463	.51962	1.9185	1.1434	.19403
2.16	1.7018	.09956	.19247	.51730	1.9354	1.1447	.19270
2.17	1.7059	.09802	.19033	.51499	1.9525	1.1460	.19138
2.18	1.7099	.09650	.18821	.51269	1.9698	1.1474	.19008
2.19	1.7139	.09500	.18612	.51041	1.9873	1.1487	.18879
2.20	1.7179	.09352	.18405	.50813	2.0050	1.1500	.18751
2.21	1.7219	.09207	.18200	.50586	2.0229	1.1513	.18624
2.22	1.7258	.09064	.17998	.50361	2.0409	1.1526	.18499
2.23	1.7297	.08923	.17798	.50136	2.0592	1.1539	.18375
2.24	1.7336	.08784	.17600	.49912	2.0777	1.1552	.18252
2.25	1.7374	.08648	.17404	.49689	2.0964	1.1565	.18130
2.26	1.7412	.08514	.17211	.49468	2.1154	1.1578	.18009
2.27	1.7450	.08382	.17020	.49247	2.1345	1.1590	.17890
2.28	1.7488	.08252	.16830	.49027	2.1538	1.1603	.17772
2.29	1.7526	.08123	.16643	.48809	2.1734	1.1616	.17655
2.30	1.7563	.07997	.16458	.48591	2.1931	1.1629	.17539
2.31	1.7600	.07873	.16275	.48374	2.2131	1.1641	.17424
2.32	1.7637	.07751	.16095	.48158	2.2333	1.1653	.17310
2.33	1.7673	.07631	.15916	.47944	2.2537	1.1666	.17197
2.34	1.7709	.07513	.15739	.47730	2.2744	1.1678	.17085
2.35	1.7745	.07396	.15564	.47517	2.2953	1.1690	.16975
2.36	1.7781	.07281	.15391	.47305	2.3164	1.1703	.16866
2.37	1.7817	.07168	.15220	.47095	2.3377	1.1715	.16757
2.38	1.7852	.07057	.15052	.46885	2.3593	1.1727	.16649
2.39	1.7887	.06948	.14885	.46676	2.3811	1.1739	.16543
2.40	1.7922	.06840	.14720	.46468	2.4031	1.1751	.16437
2.41	1.7957	.06734	.14557	.46262	2.4254	1.1763	.16332
2.42	1.7991	.06630	.14395	.46056	2.4479	1.1775	.16229
2.43	1.8025	.06527	.14235	.45851	2.4706	1.1786	.16126
2.44	1.8059	.06426	.14078	.45647	2.4936	1.1798	.16024
2.45	1.8093	.06327	.13922	.45444	2.5168	1.1810	.15923
2.46	1.8126	.06229	.13768	.45242	2.5403	1.1821	.15823
2.47	1.8159	.06133	.13616	.45041	2.5640	1.1833	.15724
2.48	1.8192	.06038	.13465	.44841	2.5880	1.1844	.15626
2.49	1.8225	.05945	.13316	.44642	2.6122	1.1856	.15528
2.50	1.8258	.05853	.13169	.44444	2.6367	1.1867	.15432
2.51	1.8290	.05763	.13023	.44247	2.6615	1.1879	.15337
2.52	1.8322	.05674	.12879	.44051	2.6865	1.1890	.15242

TABLE 18.1-18 (*Continued*)

M	M*	p/p_0	ρ/ρ_0	T/T_0	A/A^*	F/F^*	$\dfrac{A}{A^*}\cdot\dfrac{p}{p_0}$
2.53	1.8354	.05586	.12737	.43856	2.7117	1.1901	.15148
2.54	1.8386	.05500	.12597	.43662	2.7372	1.1912	.15055
2.55	1.8417	.05415	.12458	.43469	2.7630	1.1923	.14963
2.56	1.8448	.05332	.12321	.43277	2.7891	1.1934	.14871
2.57	1.8479	.05250	.12185	.43085	2.8154	1.1945	.14780
2.58	1.8510	.05169	.12051	.42894	2.8420	1.1956	.14691
2.59	1.8541	.05090	.11418	.42705	2.8689	1.1967	.14601
2.60	1.8572	.05012	.11787	.42517	2.8960	1.1978	.14513
2.61	2.8602	.04935	.11658	.42330	2.9234	1.1989	.14426
2.62	1.8632	.04859	.11530	.42143	2.9511	1.2000	.14339
2.63	1.8662	.04784	.11403	.41957	2.9791	1.2011	.14253
2.64	1.8692	.04711	.11278	.41772	3.0074	1.2021	.14168
2.65	1.8721	.04639	.11154	.41589	3.3059	1.2031	.14083
2.66	1.8750	.04568	.11032	.41406	3.0647	1.2042	.13999
2.67	1.8779	.04498	.10911	.41224	3.0938	1.2052	.13916
2.68	1.8808	.04429	.10792	.41043	3.1233	1.2062	.13834
2.69	1.8837	.04361	.10674	.40863	3.1530	1.2073	.13752
2.70	1.8865	.04295	.10557	.40684	3.1830	1.2083	.13671
2.71	1.8894	.04230	.10442	.40505	3.2133	1.2093	.13591
2.72	1.8922	.04166	.10328	.40327	3.2440	1.2103	.13511
2.73	1.8950	.04102	.10215	.40151	3.2749	1.2113	.13432
2.74	1.8978	.04039	.10104	.39976	3.3061	1.2123	.13354
2.75	1.9005	.03977	.09994	.39801	3.3376	1.2133	.13276
2.76	1.9032	.03917	.09885	.39627	3.3695	1.2143	.13199
2.77	1.9060	.03858	.09777	.39454	3.4017	1.2153	.13123
2.78	1.9087	.03800	.09671	.39282	3.4342	1.2163	.13047
2.79	1.9114	.03742	.09566	.39111	3.4670	1.2173	.12972
2.80	1.9140	.03685	.09462	.38941	3.5001	1.2182	.12897
2.81	1.9167	.03629	.09360	.38771	3.5336	1.2192	.12823
2.82	1.9193	.03574	.09259	.38603	3.5674	1.2202	.12750
2.83	1.9220	.03520	.09158	.38435	3.6015	1.2211	.12678
2.84	1.9246	.03467	.09059	.38268	3.6359	1.2221	.12605
2.85	1.9271	.03415	.08962	.38102	3.6707	2.2230	.12534
2.86	1.9297	.03363	.08865	.37937	3.7058	1.2240	.12463
2.87	1.9322	.03312	.08769	.37773	3.7413	1.2249	.12393
2.88	1.9348	.03262	.08674	.37610	3.7771	1.2258	.12323
2.89	1.9373	.03213	.08581	.37448	3.8133	1.2268	.12254
2.90	1.9398	.03165	.08489	.37286	3.8498	1.2277	.12185
2.91	1.9423	.03118	.08398	.37125	3.8866	1.2286	.12117
2.92	1.9448	.03071	.08308	.36965	3.9238	1.2295	.12049
2.93	1.9472	.03025	.08218	.36806	3.9614	1.2304	.11982
2.94	1.9497	.02980	.08130	.36648	3.9993	1.2313	.11916
2.95	1.9521	.02935	.08043	.36490	4.0376	1.2322	.11850
2.96	1.9545	.02891	.07957	.36333	4.0763	1.2331	.11785
2.97	1.9569	.02848	.07872	.36177	4.1153	1.2340	.11720
2.98	1.9593	.02805	.07788	.36022	4.1547	1.2348	.11656
2.99	1.9616	.02764	.07705	.35868	4.1944	1.2357	.11591
3.00	1.964,0	.027,22	.076,23	.357,14	4.23,46	1.2366	.115,28
3.10	1.986,6	.023,45	.068,52	.342,23	4.65,73	1.2450	.109,21
3.20	2.007,9	.020,23	.061,65	.328,08	5.12,10	1.2530	.1035,9
3.30	2.027,9	.0174,8	.055,54	.314,66	5.6,287	1.2605	.0983,7

TABLE 18.1-18   (*Continued*)

M	M^*	p/p_0	ρ/ρ_0	T/T_0	A/A^*	F/F^*	$\dfrac{A}{A^*} \cdot \dfrac{p}{p_0}$
3.40	2.046,6	.0151,2	.050,09	.301,93	6.1,837	1.2676	.0935,3
3.50	2.064,2	.0131,1	.045,23	.289,86	6.7,896	1.2743	.0890,2
3.60	2.080,8	.0113,8	.040,89	.278,40	7.4,501	1.2807	.0848,2
3.70	2.096,4	.0099,0	.0370,2	.267,52	8.1,691	1.2867	.0800,0
3.80	2.111,1	.0086,3	.0335,5	.257,20	8.9,506	1.2924	.0772,3
3.90	2.125,0	.0075,3	.0304,4	.247,40	9.7,990	1.2978	.0738,0
4.00	2.138,1	.0065,8	.0276,6	.238,10	10.7,19	1.3029	.0705,9
4.10	2.150,5	.0057,7	.0251,6	.229,25	11.7,15	1.3077	.0675,8
4.20	2.162,2	.0050,6	.0229,2	.2208,5	12.7,92	1.3123	.0647,5
4.30	2.173,2	.0044,5	.0209,0	.2128,6	13.9,55	1.3167	.0620,9
4.40	2.183,7	.0039,2	.0190,9	.2052,5	15.2,10	1.3208	.0595,9
4.50	2.193,6	.0034,6	.0174,5	.1980,2	16.5,62	1.3247	.0572,3
4.60	2.203,0	.0030,5	.0159,7	.1911,3	18.0,18	1.3284	.0550,0
4.70	2.211,9	.0027,0	.0146,3	.1845,7	19.5,83	1.3320	.0528,9
4.80	2.220,4	.0024,0	.0134,3	.1783,2	21.2,64	1.3354	.0509,1
4.90	2.228,4	.0021,3	.0123,3	.1723,5	23.0,67	1.3386	.0490,4
5.00	2.2361	.00189	.01134	.16667	25.000	1.3416	.04725
6.00	2.2953	.0$_3$633	.00519	.12195	53.180	1.3655	.03368
7.00	2.3333	.0$_3$242	.00261	.09259	104.143	1.3810	.02516
8.00	2.3591	.0$_3$102	.00141	.07246	190.109	1.3915	.01947
9.00	2.3772	.0$_4$474	.0$_3$815	.05814	327.189	1.3989	.01550
10.00	2.3904	.0$_4$236	.0$_3$495	.04762	535.938	1.4044	.01263
∞	2.4495	0	0	0	∞	1.4289	0

Source. Shapiro, A. H., *The Dynamics and Thermodynamics of Compressible Fluid Flow*, Ronald Press. Used by permission.

[a] From ref. 8. Used by permission.
[b] For values of M from 0 to 5.00, all digits to the left of the comma are valid for linear interpolation. Where no comma is indicated in this region, all digits are valid for linear interpolation. The notation .0$_3$429 signifies .000429. The notation 5370$_4$ signifies 5,370,000.

TABLE 18.1-19   FRICTIONAL, ADIABATIC, CONSTANT-AREA FLOW (FANNO LINE): PERFECT GAS, $K = 1.4^a$

M	T/T^*	p/p^*	p_0/p_0^*	V/V^* and ρ^*/ρ	F/F^*	$4fL_{max}/D$
0.00	1.2000	∞	∞	0.00000	∞	∞
.05	1.1994	21.903	11.5914	.05476	9.1584	280.02
.10	1.1976	10.9435	5.8218	.10943	4.6236	66.922
.15	1.1946	7.2866	3.9103	.16395	3.1317	27.932
.20	1.1905	5.4555	2.9635	.21822	2.4004	14.533
.25	1.1852	4.3546	2.4027	.27217	1.9732	8.4834
.30	1.1788	3.6190	2.0351	.32572	1.6979	5.2992
.35	1.1713	3.0922	1.7780	.37880	1.5094	3.4525
.40	1.1628	2.6958	1.5901	.43133	1.3749	2.3085
.45	1.1533	2.3865	1.4486	.48326	1.2763	1.5664
.50	1.1429	2.1381	1.3399	.53453	1.2027	1.06908
.55	1.1315	1.9341	1.2549	.58506	1.1472	.72805
.60	1.1194	1.7634	1.1882	.63481	1.10504	.49081
.65	1.10650	1.6183	1.1356	.68374	1.07314	.32460
.70	1.09290	1.4934	1.09436	.73179	1.04915	.20814
.75	1.07856	1.3848	1.06242	.77893	1.03137	.12728
.80	1.06383	1.2892	1.03823	.82514	1.01853	.07229
.85	1.04849	1.2047	1.02067	.87037	1.00966	.03632
.90	1.03270	1.12913	1.00887	.91459	1.00399	.014513
.95	1.01652	1.06129	1.00215	.95782	1.00093	.003280
1.00	1.00000	1.00000	1.00000	1.00000	1.00000	0
1.05	.98320	.94435	1.00203	1.04115	1.00082	.002712
1.10	.96618	.89350	1.00793	1.08124	1.00305	.009933
1.15	.94899	.84710	1.01746	1.1203	1.00646	.02053
1.20	.93168	.80436	1.03044	1.1583	1.01082	.03364
1.25	.91429	.76495	1.04676	1.1952	1.01594	.04858
1.30	.89686	.72848	1.06630	1.2311	1.02169	.06483
1.35	.87944	.69466	1.08904	1.2660	1.02794	.08199
1.40	.86207	.66320	1.1149	1.2999	1.03458	.09974
1.45	.84477	.63387	1.1440	1.3327	1.04153	.11782
1.50	.82759	.60648	1.1762	1.3646	1.04870	.13605
1.55	.81054	.58084	1.2116	1.3955	1.05604	.15427
1.60	.79365	.55679	1.2502	1.4254	1.06348	.17236
1.65	.77695	.53421	1.2922	1.4544	1.07098	.19022
1.70	.76046	.51297	1.3376	1.4825	1.07851	.20780
1.75	.74419	.49295	1.3865	1.5097	1.08603	.22504
1.80	.72816	.47407	1.4390	1.5360	1.09352	.24189
1.85	.71238	.45623	1.4952	1.5614	1.1009	.25832
1.90	.69686	.43936	1.5552	1.5861	1.1083	.27433
1.95	.68162	.42339	1.6193	1.6099	1.1155	.28989
2.00	.66667	.40825	1.6875	1.6330	1.1227	.30499
2.05	.65200	.39389	1.7600	1.6553	1.1297	.31965
2.10	.63762	.38024	1.8369	1.6769	1.1366	.33385
2.15	.62354	.36728	1.9185	1.6977	1.1434	.34760
2.20	.60976	.35494	2.0050	1.7179	1.1500	.36091
2.25	.59627	.34319	2.0964	1.7374	1.1565	.37378
2.30	.58309	.33200	2.1931	1.7563	1.1629	.38623
2.35	.57021	.32133	2.2953	1.7745	1.1690	.39826
2.40	.55762	.31114	2.4031	1.7922	1.1751	.40989

TABLE 18.1-19 (*Continued*)

M	T/T^*	p/p^*	p_0/p_0^*	V/V^* and ρ^*/ρ	F/F^*	$4fL_{\max}/D$
2.45	.54533	.30141	2.5168	1.8092	1.1810	.42113
2.50	.53333	.29212	2.6367	1.8257	1.1867	.43197
2.55	.52163	.28323	2.7630	1.8417	1.1923	.44247
2.60	.51020	.27473	2.8960	1.8571	1.1978	.45259
2.65	.49906	.26658	3.0359	1.8721	1.2031	.46237
2.70	.48820	.25878	3.1830	1.8865	1.2083	.47182
2.75	.47761	.25131	3.3376	1.9005	1.2133	.48095
2.80	.46729	.24414	3.5001	1.9140	1.2182	.48976
2.85	.45723	.23726	3.6707	1.9271	1.2230	.49828
2.90	.44743	.23066	3.8498	1.9398	1.2277	.50651
2.95	.43788	.22431	4.0376	1.9521	1.2322	.51447
3.00	.42857	.21822	4.2346	1.9640	1.2366	.52216
3.50	.34783	.16850	6.7896	2.0642	1.2743	.58643
4.00	.28571	.13363	10.719	2.1381	1.3029	.63306
4.50	.23762	.10833	16.562	2.1936	1.3247	.66764
5.00	.20000	.08944	25.000	2.2361	1.3416	.69381
6.00	.14634	.06376	53.180	2.2953	1.3655	.72987
7.00	.11111	.04762	104.14	2.3333	1.3810	.75281
8.00	.08696	.03686	190.11	2.3591	1.3915	.76820
9.00	.06977	.02935	327.19	2.3772	1.3989	.77898
10.00	.05714	.02390	535.94	2.3905	1.4044	.78683
∞	0	0	∞	2.4495	1.4289	.82153

Source. Shapiro, A. H., *The Dynamics and Thermodynamics of Compressible Fluid Flow*, Ronald Press. Used by permission.

[a] From ref. 8. Used by permission.

TABLE 18.1-20 THERMODYNAMIC PROPERTIES OF FREON-12: CUSTOMARY UNITS[a]

SATURATED FREON-12

Temp., F t	Abs. Press., lbf/in.² P	Specific Volume, ft³/lbm			Enthalpy, Btu/lbm			Entropy, Btu/lbm R		
		Sat. Liquid v_f	Evap. v_{fg}	Sat. Vapor v_g	Sat. Liquid h_f	Evap. h_{fg}	Sat. Vapor h_g	Sat. Liquid s_f	Evap. s_{fg}	Sat. Vapor s_g
−130	0.41224	0.009736	70.7203	70.730	−18.609	81.577	62.968	−0.04983	0.24743	0.19760
−120	0.64190	0.009816	46.7312	46.741	−16.565	80.617	64.052	−0.04372	0.23731	0.19359
−110	0.97034	0.009899	31.7671	31.777	−14.518	79.663	65.145	−0.03779	0.22780	0.19002
−100	1.4280	0.009985	21.1541	22.164	−12.466	78.714	66.248	−0.03200	0.21883	0.18683
−90	2.0509	0.010073	15.8109	15.821	−10.409	77.764	67.355	−0.02637	0.21034	0.18398
−80	2.8807	0.010164	11.5228	11.533	−8.3451	76.812	68.467	−0.02086	0.20229	0.18143
−70	3.9651	0.010259	8.5584	8.5687	−6.2730	75.853	69.580	−0.01548	0.19464	0.17916
−60	5.3575	0.010357	6.4670	6.4774	−4.1919	74.885	70.693	−0.01021	0.18716	0.17714
−50	7.1168	0.010459	4.9637	4.9742	−2.1011	73.906	71.805	−0.00506	0.18038	0.17533
−40	9.3076	0.010564	3.8644	3.8750	0	72.913	72.913	0	0.17373	0.17373
−30	11.999	0.010674	3.0478	3.0585	2.1120	71.903	74.015	0.00496	0.16733	0.17229
−20	15.267	0.010788	2.4321	2.4429	4.2357	70.874	75.110	0.00983	0.16119	0.17102
−10	19.189	0.010906	1.9628	1.9727	6.3716	69.824	76.196	0.01462	0.15527	0.16989
0	23.849	0.011030	1.5979	1.6089	8.5207	68.750	77.271	0.01932	0.14956	0.16888
10	29.335	0.011160	1.3129	1.3241	10.684	67.651	78.335	0.02395	0.14403	0.16798
20	35.736	0.011296	1.0875	1.0988	12.863	66.522	79.385	0.02852	0.13867	0.16719
30	43.148	0.011438	0.90736	0.91880	15.058	65.361	80.419	0.03301	0.13347	0.16648
40	51.667	0.011588	0.76198	0.77357	17.273	64.163	81.436	0.03745	0.12841	0.16586
50	61.394	0.011746	0.64362	0.65537	19.507	62.926	82.433	0.04184	0.12346	0.16530

TABLE 18.1-20 *(Continued)*

SATURATED FREON-12

Temp., F t	Abs. Press., lbf/in.² P	Specific Volume, ft³/lbm			Enthalpy, Btu/lbm			Entropy, Btu/lbm R		
		Sat. Liquid v_f	Evap. v_{fg}	Sat. Vapor v_g	Sat. Liquid h_f	Evap. h_{fg}	Sat. Vapor h_g	Sat. Liquid s_f	Evap. s_{fg}	Sat. Vapor s_g
60	72.433	0.011913	0.54648	0.55839	21.766	61.643	83.409	0.04618	0.11861	0.16479
70	84.888	0.012089	0.46609	0.47818	24.050	60.309	84.359	0.05048	0.11386	0.16434
80	98.870	0.012277	0.39907	0.41135	26.365	58.917	85.282	0.05475	0.10917	0.16392
90	114.49	0.012478	0.34281	0.35529	28.713	57.461	86.174	0.05900	0.10453	0.16353
100	131.86	0.012693	0.29525	0.30794	31.100	55.929	87.029	0.06323	0.09992	0.16315
110	151.11	0.012924	0.25577	0.26769	33.531	54.313	87.844	0.06745	0.09534	0.16279
120	172.35	0.013174	0.22019	0.23326	36.013	52.597	88.610	0.07168	0.09073	0.16241
130	195.71	0.013447	0.19019	0.20364	38.553	50.768	89.321	0.07583	0.08609	0.16202
140	221.32	0.013746	0.16424	0.17799	41.162	48.805	89.967	0.08021	0.08138	0.16159
150	249.31	0.014078	0.14156	0.15564	43.850	46.684	90.534	0.08453	0.07657	0.16110
160	279.82	0.014449	0.12159	0.13604	46.633	44.373	91.006	0.08893	0.07260	0.16053
170	313.00	0.014871	0.10386	0.11873	49.529	41.830	91.359	0.09342	0.06643	0.15985
180	349.00	0.015360	0.08794	0.10330	52.562	38.999	91.561	0.09804	0.06096	0.15900
190	387.98	0.015942	0.073476	0.089418	55.769	35.792	91.561	0.10284	0.05511	0.15793
200	430.09	0.016659	0.060069	0.076728	59.203	32.075	91.278	0.10789	0.04862	0.15651
210	475.52	0.017601	0.047242	0.064843	62.959	27.599	90.558	0.11332	0.03921	0.15453
220	524.43	0.018986	0.035154	0.053140	67.246	21.790	89.036	0.11943	0.03206	0.15149
230	577.03	0.021854	0.017581	0.039435	72.893	12.229	85.122	0.12739	0.01773	0.14512
233.6 (critical)	596.9	0.02870	0	0.02870	78.86	0	78.86	0.1359	0	0.1359

TABLE 18.1-20 (*Continued*)

SUPERHEATED FREON-12

Temp., F	v	h	s	v	h	s	v	h	s
		5 lbf/in.²			10 lbf/in.²			15 lbf/in.²	
0	8.0611	78.582	0.19663	3.9809	78.246	0.18471	2.6201	77.902	0.17751
20	8.4265	81.309	0.20244	4.1691	81.014	0.19061	2.7494	80.712	0.18349
40	8.7903	84.090	0.20812	4.3556	83.828	0.19635	2.8770	83.561	0.18931
60	9.1528	86.922	0.21367	4.5408	86.689	0.20197	3.0031	86.451	0.19498
80	9.5142	89.806	0.21912	4.7248	89.596	0.20746	3.1281	89.383	0.20051
100	9.8747	92.738	0.22445	4.9079	92.548	0.21283	3.2521	92.357	0.20593
120	10.234	95.717	0.22968	5.0903	95.546	0.21809	3.3754	95.373	0.21122
140	10.594	98.743	0.23481	5.2720	98.586	0.22325	3.4981	98.429	0.21640
160	10.952	101.812	0.23985	5.4533	101.669	0.22830	3.6202	101.525	0.22148
180	11.311	104.925	0.24479	5.6341	104.793	0.23326	3.7419	104.661	0.22646
200	11.668	108.079	0.24964	5.8145	107.957	0.23813	3.8632	107.835	0.23135
220	12.026	111.272	0.25441	5.9946	111.159	0.24291	3.9841	111.046	0.23614
		20 lbf/in.²			25 lbf/in.²			30 lbf/in.²	
20	2.0391	80.403	0.17829	1.6125	80.088	0.17414	1.3278	79.765	0.17065
40	2.1373	83.289	0.18419	1.6932	83.012	0.18012	1.3969	82.730	0.17671
60	2.2340	86.210	0.18992	1.7723	85.965	0.18591	1.4644	85.716	0.18257
80	2.3295	89.168	0.19550	1.8502	88.950	0.19155	1.5306	88.729	0.18826
100	2.4241	92.164	0.20095	1.9271	91.968	0.19704	1.5957	91.770	0.19379
120	2.5179	95.198	0.20628	2.0032	95.021	0.20240	1.6600	94.843	0.19918
140	2.6110	98.270	0.21149	2.0786	98.110	0.20763	1.7237	97.948	0.20445
160	2.7036	101.380	0.21659	2.1535	101.234	0.21276	1.7868	101.086	0.20960
180	2.7957	104.528	0.22159	2.2279	104.393	0.21778	1.8494	104.258	0.21463
200	2.8874	107.712	0.22649	2.3019	107.588	0.22269	1.9116	107.464	0.21957
220	2.9789	110.932	0.23130	2.3756	110.817	0.22752	1.9735	110.702	0.22440
240	3.0700	114.186	0.23602	2.4491	114.080	0.23225	2.0351	113.973	0.22915
		35 lbf/in.²			40 lbf/in.²			50 lbf/in.²	
40	1.1850	82.442	0.17375	1.0258	82.148	0.17112	0.80248	81.540	0.16655
60	1.2442	85.463	0.17968	1.0789	85.206	0.17712	0.84713	84.676	0.17271
80	1.3021	88.504	0.18542	1.1306	88.277	0.18292	0.89025	87.811	0.17862
100	1.3589	91.570	0.19100	1.1812	91.367	0.18854	0.93216	90.953	0.18434
120	1.4148	94.663	0.19643	1.2309	94.480	0.19401	0.97313	94.110	0.18988
140	1.4701	97.785	0.20172	1.2798	97.620	0.19933	1.0133	97.286	0.19527
160	1.5248	100.938	0.20689	1.3282	100.788	0.20453	1.0529	100.485	0.20051
180	1.5789	104.122	0.21195	1.3761	103.985	0.20961	1.0920	103.708	0.20563
200	1.6327	107.338	0.21690	1.4236	107.212	0.21457	1.1307	106.958	0.21064
220	1.6862	110.586	0.22175	1.4707	110.469	0.21944	1.1690	110.235	0.21553
240	1.7394	113.865	0.22651	1.5176	113.757	0.22420	1.2070	113.539	0.22032
260	1.7923	117.175	0.23117	1.5642	117.074	0.22888	1.2447	116.871	0.22502
		60 lbf/in.²			70 lbf/in.²			80 lbf/in.²	
60	0.69210	84.126	0.16892	0.58088	83.552	0.16556	...	...	...
80	0.72964	87.330	0.17497	0.61458	86.832	0.17175	0.52795	86.316	0.16585
100	0.76588	90.528	0.18079	0.64685	90.091	0.17768	0.55734	89.640	0.17489
120	0.80110	93.731	0.18641	0.67803	93.343	0.18339	0.58556	92.945	0.18070
140	0.83551	96.945	0.19186	0.70836	96.597	0.18891	0.61286	96.242	0.18629
160	0.86928	100.776	0.19716	0.73800	99.862	0.19427	0.63943	99.542	0.19170
180	0.90252	103.427	0.20233	0.76708	103.141	0.19948	0.66543	102.851	0.19696
200	0.93531	106.700	0.20736	0.79571	106.439	0.20455	0.69095	106.174	0.20207
220	0.96775	109.997	0.21229	0.82397	109.756	0.20951	0.71609	109.513	0.20706
240	0.99988	113.319	0.21710	0.85191	113.096	0.21435	0.74090	112.872	0.21193
260	1.0318	116.666	0.22182	0.87959	116.459	0.21909	0.76544	116.251	0.21669
280	1.0634	120.039	0.22644	0.90705	119.846	0.22373	0.78975	119.652	0.22135

TABLE 18.1-20 (*Continued*)

SUPERHEATED FREON-12 (*Continued*)

Temp., F	v	h	s	v	h	s	v	h	s
		90 lbf/in.²			100 lbf/in.²			125 lbf/in.²	
100	0.48749	89.175	0.17234	0.43138	88.694	0.16996	0.32943	87.407	0.16455
120	0.51346	92.536	0.17824	0.45562	92.116	0.17597	0.35086	91.008	0.17087
140	0.53845	95.879	0.18391	0.47881	95.507	0.18172	0.37098	94.537	0.17686
160	0.56268	99.216	0.18938	0.50118	98.884	0.18726	0.39015	98.023	0.18258
180	0.58629	102.557	0.19469	0.52291	102.257	0.19262	0.40857	101.484	0.18807
200	0.60941	105.905	0.19984	0.54413	105.633	0.19782	0.42642	104.934	0.19338
220	0.63213	109.267	0.20486	0.56492	109.018	0.20287	0.44380	108.380	0.19853
240	0.65451	112.644	0.20976	0.58538	112.415	0.20780	0.46081	111.829	0.20353
260	0.67662	116.040	0.21455	0.60554	115.828	0.21261	0.47750	115.287	0.20840
280	0.69849	119.456	0.21923	0.62546	119.258	0.21731	0.49394	118.756	0.21316
300	0.72016	122.892	0.22381	0.64518	122.707	0.22191	0.51016	122.238	0.21780
320	0.74166	126.349	0.22830	0.66472	126.176	0.22641	0.52619	125.737	0.22235
		150 lbf/in.²			175 lbf/in.²			200 lbf/in.²	
120	0.28007	89.800	0.16629	...	...	...	...	...	...
140	0.29845	93.498	0.17256	0.24595	92.373	0.16859	0.20579	91.137	0.16480
160	0.31566	97.112	0.17849	0.26198	96.142	0.17478	0.22121	95.100	0.17130
180	0.33200	100.675	0.18415	0.27697	99.823	0.18062	0.23535	98.921	0.17737
200	0.34769	104.206	0.18958	0.29120	103.447	0.18620	0.24860	102.652	0.18311
220	0.36285	107.720	0.19483	0.30485	107.036	0.19156	0.26117	106.325	0.18860
240	0.37761	111.226	0.19992	0.31804	110.605	0.19674	0.27323	109.962	0.19387
260	0.39203	114.732	0.20485	0.33087	114.162	0.20175	0.28489	113.576	0.19896
280	0.40617	118.242	0.20967	0.34339	117.717	0.20662	0.29623	117.178	0.20390
300	0.42008	121.761	0.21436	0.35567	121.273	0.21137	0.30730	120.775	0.20870
320	0.43379	125.290	0.21894	0.36773	124.835	0.21599	0.31815	124.373	0.21337
340	0.44733	128.833	0.22343	0.37963	128.407	0.22052	0.32881	127.974	0.21793
		250 lbf/in.²			300 lbf/in.²			400 lbf/in.²	
160	0.16249	92.717	0.16462	...	...	...	...	...	...
180	0.17605	96.925	0.17130	0.13482	94.556	0.16537	...	...	...
200	0.18824	100.930	0.17747	0.14697	98.975	0.17217	0.091005	93.718	0.16092
220	0.19952	104.809	0.18326	0.15774	103.136	0.17838	0.10316	99.046	0.16888
240	0.21014	108.607	0.18877	0.16761	107.140	0.18419	0.11300	103.735	0.17568
260	0.22027	112.351	0.19404	0.17685	111.043	0.18969	0.12163	108.105	0.18183
280	0.23001	116.060	0.19913	0.18562	114.879	0.19495	0.12949	112.286	0.18756
300	0.23944	119.747	0.20405	0.19402	118.670	0.20000	0.13680	116.343	0.19298
320	0.24862	123.420	0.20882	0.20214	122.430	0.20489	0.14372	120.318	0.19814
340	0.25759	127.088	0.21346	0.21002	126.171	0.20963	0.15032	124.235	0.20310
360	0.26639	130.754	0.21799	0.21770	129.900	0.21423	0.15668	128.112	0.20789
380	0.27504	134.423	0.22241	0.22522	133.624	0.21872	0.16285	131.961	0.21253
		500 lbf/in.²			600 lbf/in.²				
220	0.064207	92.397	0.15683	...	...	...			
240	0.077620	99.218	0.16672	0.047488	91.024	0.15335			
260	0.087054	104.526	0.17421	0.061922	99.741	0.16566			
280	0.094923	109.277	0.18072	0.070859	105.637	0.17374			
300	0.10190	113.729	0.18666	0.078059	110.729	0.18053			
320	0.10829	117.997	0.19221	0.084333	115.420	0.18663			
340	0.11426	122.143	0.19746	0.090017	119.871	0.19227			
360	0.11992	126.205	0.20247	0.095289	124.167	0.19757			
380	0.12533	130.207	0.20730	0.10025	128.355	0.20262			
400	0.13054	134.166	0.21196	0.10498	132.466	0.20746			
420	0.13559	138.096	0.21648	0.10952	136.523	0.21213			
440	0.14051	142.004	0.22087	0.11391	140.539	0.21664			

[a] From ref. 3. Used by permission.

TABLE 18.1-21   THERMODYNAMIC PROPERTIES OF FREON-12: SI UNITS[a]

Temp. °C	v m³/kg	h kJ/kg	s kJ/kg K	v m³/kg	h kJ/kg	s kJ/kg K	v m³/kg	h kJ/kg	s kJ/kg K
	0.05 MPa			0.10 MPa			0.15 MPa		
−20.0	0.341 857	181.042	0.7912	0.167 701	179.861	0.7401			
−10.0	0.356 227	186.757	0.8133	0.175 222	185.707	0.7628	0.114 716	184.619	0.7318
0.0	0.370 508	192.567	0.8350	0.182 647	191.628	0.7849	0.119 866	190.660	0.7543
10.0	0.384 716	198.471	0.8562	0.189 994	197.628	0.8064	0.124 932	196.762	0.7763
20.0	0.398 863	204.469	0.8770	0.197 277	203.707	0.8275	0.129 930	202.927	0.7977
30.0	0.412 959	210.557	0.8974	0.204 506	209.866	0.8482	0.134 873	209.160	0.8186
40.0	0.427 012	216.733	0.9175	0.211 691	216.104	0.8684	0.139 768	215.463	0.8390
50.0	0.441 030	222.997	0.9372	0.218 839	222.421	0.8883	0.144 625	221.835	0.8591
60.0	0.455 017	229.344	0.9565	0.225 955	228.815	0.9078	0.149 450	228.277	0.8787
70.0	0.468 978	235.774	0.9755	0.233 044	235.285	0.9269	0.154 247	234.789	0.8980
80.0	0.482 917	242.282	0.9942	0.240 111	241.829	0.9457	0.159 020	241.371	0.9169
90.0	0.496 838	248.868	1.0126	0.247 159	248.446	0.9642	0.163 774	248.020	0.9354
	0.20 MPa			0.25 MPa			0.30 MPa		
0.0	0.088 608	189.669	0.7320	0.069 752	188.644	0.7139	0.057 150	187.583	0.6984
10.0	0.092 550	195.878	0.7543	0.073 024	194.969	0.7366	0.059 984	194.034	0.7216
20.0	0.096 418	202.135	0.7760	0.076 218	201.322	0.7587	0.062 734	200.490	0.7440
30.0	0.100 228	208.446	0.7972	0.079 350	207.715	0.7801	0.065 418	206.969	0.7658
40.0	0.103 989	214.814	0.8178	0.082 431	214.153	0.8010	0.068 049	213.480	0.7869
50.0	0.107 710	221.243	0.8381	0.085 470	220.642	0.8214	0.070 635	220.030	0.8075
60.0	0.111 397	227.735	0.8578	0.088 474	227.185	0.8413	0.073 185	226.627	0.8276
70.0	0.115 055	234.291	0.8772	0.091 449	233.785	0.8608	0.075 705	233.273	0.8473
80.0	0.118 690	240.910	0.8962	0.094 398	240.443	0.8800	0.078 200	239.971	0.8665
90.0	0.122 304	247.593	0.9149	0.097 327	247.160	0.8987	0.080 673	246.723	0.8853
100.0	0.125 901	254.339	0.9332	0.100 238	253.936	0.9171	0.083 127	253.530	0.9038
110.0	0.129 483	261.147	0.9512	0.103 134	260.770	0.9352	0.085 566	260.391	0.9220
	0.40 MPa			0.50 MPa			0.60 MPa		
20.0	0.045 836	198.762	0.7199	0.035 646	196.935	0.6999			
30.0	0.047 971	205.428	0.7423	0.037 464	203.814	0.7230	0.030 422	202.116	0.7063
40.0	0.050 046	212.095	0.7639	0.039 214	210.656	0.7452	0.031 966	209.154	0.7291
50.0	0.052 072	218.779	0.7849	0.040 911	217.484	0.7667	0.033 450	216.141	0.7511
60.0	0.054 059	225.488	0.8054	0.042 565	224.315	0.7875	0.034 887	223.104	0.7723
70.0	0.056 014	232.230	0.8253	0.044 184	231.161	0.8077	0.036 285	230.062	0.7929
80.0	0.057 941	239.012	0.8448	0.045 774	238.031	0.8275	0.037 653	237.027	0.8129
90.0	0.059 846	245.837	0.8638	0.047 340	244.932	0.8467	0.038 995	244.009	0.8324
100.0	0.061 731	252.707	0.8825	0.048 886	251.869	0.8656	0.040 316	251.016	0.8514
110.0	0.063 600	259.624	0.9008	0.050 415	258.845	0.8840	0.041 619	258.053	0.8700
120.0	0.065 455	266.590	0.9187	0.051 929	265.862	0.9021	0.042 907	265.124	0.8882
130.0	0.067 298	273.605	0.9364	0.053 430	272.923	0.9198	0.044 181	272.231	0.9061
	0.70 MPa			0.80 MPa			0.90 MPa		
40.0	0.026 761	207.580	0.7148	0.022 830	205.924	0.7016	0.019 744	204.170	0.6982
50.0	0.028 100	214.745	0.7373	0.024 068	213.290	0.7248	0.020 912	211.765	0.7131
60.0	0.029 387	221.854	0.7590	0.025 247	220.558	0.7469	0.022 012	219.212	0.7358
70.0	0.030 632	228.931	0.7799	0.026 380	227.766	0.7682	0.023 062	226.564	0.7575
80.0	0.031 843	235.997	0.8002	0.027 477	234.941	0.7888	0.024 072	233.856	0.7785
90.0	0.033 027	243.066	0.8199	0.028 545	242.101	0.8088	0.025 051	241.113	0.7987
100.0	0.034 189	250.146	0.8392	0.029 588	249.260	0.8283	0.026 005	248.355	0.8184
110.0	0.035 332	257.247	0.8579	0.030 612	256.428	0.8472	0.026 937	255.593	0.8376
120.0	0.036 458	264.374	0.8763	0.031 619	263.613	0.8657	0.027 851	262.839	0.8562
130.0	0.037 572	271.531	0.8943	0.032 612	270.820	0.8838	0.028 751	270.100	0.8745
140.0	0.038 673	278.720	0.9119	0.033 592	278.055	0.9016	0.029 639	277.381	0.8923
150.0	0.039 764	285.946	0.9292	0.034 563	285.320	0.9189	0.030 515	284.687	0.9098

TABLE 18.1-21 (*Continued*)

Temp. °C	v m³/kg	h kJ/kg	s kJ/kg K	v m³/kg	h kJ/kg	s kJ/kg K	v m³/kg	h kJ/kg	s kJ/kg K
		1.00 MPa			1.20 MPa			1.40 MPa	
50.0	0.018 366	210.162	0.7021	0.014 483	206.661	0.6812			
60.0	0.019 410	217.810	0.7254	0.015 463	214.805	0.7060	0.012 579	211.457	0.6876
70.0	0.020 397	225.319	0.7476	0.016 368	222.687	0.7293	0.013 448	219.822	0.7123
80.0	0.021 341	232.739	0.7689	0.017 221	230.398	0.7514	0.014 247	227.891	0.7355
90.0	0.022 251	240.101	0.7895	0.018 032	237.995	0.7727	0.014 997	235.766	0.7575
100.0	0.023 133	247.430	0.8094	0.018 812	245.518	0.7931	0.015 710	243.512	0.7785
110.0	0.023 993	254.743	0.8287	0.019 567	252.993	0.8129	0.016 393	251.170	0.7988
120.0	0.024 835	262.053	0.8475	0.020 301	260.441	0.8320	0.017 053	258.770	0.8183
130.0	0.025 661	269.369	0.8659	0.021 018	267.875	0.8507	0.017 695	266.334	0.8373
140.0	0.026 474	276.699	0.8839	0.021 721	275.307	0.8689	0.018 321	273.877	0.8558
150.0	0.027 275	284.047	0.9015	0.022 412	282.745	0.8867	0.018 934	281.411	0.8738
160.0	0.028 068	291.419	0.9187	0.023 093	290.195	0.9041	0.019 535	288.946	0.8914
		1.60 MPa			1.80 MPa			2.00 MPa	
70.0	0.011 208	216.650	0.6959	0.009 406	213.049	0.6794			
80.0	0.011 984	225.177	0.7204	0.010 187	222.198	0.7057	0.008 704	218.859	0.6909
90.0	0.012 698	233.390	0.7433	0.010 884	230.835	0.7298	0.009 406	228.056	0.7166
100.0	0.013 366	241.397	0.7651	0.011 526	239.155	0.7524	0.010 035	236.760	0.7402
110.0	0.014 000	249.264	0.7859	0.012 126	247.264	0.7739	0.010 615	245.154	0.7624
120.0	0.014 608	257.035	0.8059	0.012 697	255.228	0.7944	0.011 159	253.341	0.7835
130.0	0.015 195	264.742	0.8253	0.013 244	263.094	0.8141	0.011 676	261.384	0.8037
140.0	0.015 765	272.406	0.8440	0.013 772	270.891	0.8332	0.012 172	269.327	0.8232
150.0	0.016 320	280.044	0.8623	0.014 284	278.642	0.8518	0.012 651	277.201	0.8420
160.0	0.016 864	287.669	0.8801	0.014 784	286.364	0.8698	0.013 116	285.027	0.8603
170.9	0.017 398	295.290	0.8975	0.015 272	294.069	0.8874	0.013 570	292.822	0.8781
180.0	0.017 923	302.914	0.9145	0.015 752	301.767	0.9046	0.014 013	300.598	0.8955
		2.50 MPa			3.00 MPa			3.50 MPa	
90.0	0.006 595	219.562	0.6823						
100.0	0.007 264	229.852	0.7103	0.005 231	220.529	0.6770			
110.0	0.007 837	239.271	0.7352	0.005 886	232.068	0.7075	0.004 324	222.121	0.6750
120.0	0.008 351	248.192	0.7582	0.006 419	242.208	0.7336	0.004 959	234.875	0.7078
130.0	0.008 827	256.794	0.7798	0.006 887	251.632	0.7573	0.005 456	245.661	0.7349
140.0	0.009 273	265.180	0.8003	0.007 313	260.620	0.7793	0.005 884	255.524	0.7591
150.0	0.009 697	273.414	0.8200	0.007 709	269.319	0.8001	0.006 270	264.846	0.7814
160.0	0.010 104	281.540	0.8390	0.008 083	277.817	0.8200	0.006 626	273.817	0.8023
170.0	0.010 497	289.589	0.8574	0.008 439	286.171	0.8391	0.006 961	282.545	0.8222
180.0	0.010 879	297.583	0.8752	0.008 782	294.422	0.8575	0.007 279	291.100	0.8413
190.0	0.011 250	305.540	0.8926	0.009 114	302.597	0.8753	0.007 584	299.528	0.8597
200.0	0.011 614	313.472	0.9095	0.009 436	310.718	0.8927	0.007 878	307.864	0.8775
		4.00 MPa							
120.0	0.003 736	224.863	0.6771						
130.0	0.004 325	238.443	0.7111						
140.0	0.004 781	249.703	0.7386						
150.0	0.005 172	259.904	0.7630						
160.0	0.005 522	269.492	0.7854						
170.0	0.005 845	278.684	0.8063						
180.0	0.006 147	287.602	0.8262						
190.0	0.006 434	296.326	0.8453						
200.0	0.006 708	304.906	0.8636						
210.0	0.006 972	313.380	0.8813						
220.0	0.007 228	321.774	0.8985						
230.0	0.007 477	330.108	0.9152						

[a] From ref. 4. Used by permission.

TABLE 18.1-22   THERMODYNAMIC PROPERTIES OF AMMONIA: CUSTOMARY UNITS[a]

SATURATED AMMONIA

Temp., F	Abs. Press., lbf/in.² P	Specific Volume, ft³/lbm			Enthalpy, Btu/lbm			Entropy, Btu/lbm R		
		Sat. Liquid v_f	Evap. v_{fg}	Sat. Vapor v_g	Sat. Liquid h_f	Evap. h_{fg}	Sat. Vapor h_g	Sat. Liquid s_f	Evap. s_{fg}	Sat. Vapor s_g
−60	5.55	0.0228	44.707	44.73	−21.2	610.8	589.6	−0.0517	1.5286	1.4769
−55	6.54	0.0229	38.357	38.38	−15.9	607.5	591.6	−0.0386	1.5017	1.4631
−50	7.67	0.0230	33.057	33.08	−10.6	604.3	593.7	−0.0256	1.4753	1.4497
−45	8.95	0.0231	28.597	28.62	−5.3	600.9	595.6	−0.0127	1.4495	1.4368
−40	10.41	0.02322	24.837	24.86	0	597.6	597.6	0.000	1.4242	1.4242
−35	12.05	0.02333	21.657	21.68	5.3	594.2	599.5	0.0126	1.3994	1.4120
−30	13.90	0.0235	18.947	18.97	10.7	590.7	601.4	0.0250	1.3751	1.4001
−25	15.98	0.0236	16.636	16.66	16.0	587.2	603.2	0.0374	1.3512	1.3886
−20	18.30	0.0237	14.656	14.68	21.4	583.6	605.0	0.0497	1.3277	1.3774
−15	20.88	0.02381	12.946	12.97	26.7	580.0	606.7	0.0618	1.3044	1.3664
−10	23.74	0.02393	11.476	11.50	32.1	576.4	608.5	0.0738	1.2820	1.3558
−5	26.92	0.02406	10.206	10.23	37.5	572.6	610.1	0.0857	1.2597	1.3454
0	30.42	0.02419	9.092	9.116	42.9	568.9	611.8	0.0975	1.2377	1.3352
5	34.27	0.02432	8.1257	8.150	48.3	565.0	613.3	0.1092	1.2161	1.3253
10	38.51	0.02446	7.2795	7.304	53.8	561.1	614.9	0.1208	1.1949	1.3157
15	43.14	0.02460	6.5374	6.562	59.2	557.1	616.3	0.1323	1.1739	1.3062
20	48.21	0.02474	5.8853	5.910	64.7	553.1	617.8	0.1437	1.1532	1.2969
25	53.73	0.02488	5.3091	5.334	70.2	548.9	619.1	0.1551	1.1328	1.2879
30	59.74	0.02503	4.8000	4.825	75.7	544.8	620.5	0.1663	1.1127	1.2790
35	66.26	0.02518	4.3478	4.373	81.2	540.5	621.7	0.1775	1.0929	1.2704
40	73.32	0.02533	3.9457	3.971	86.8	536.2	623.0	0.1885	1.0733	1.2618
45	80.96	0.02548	3.5885	3.614	92.3	531.8	624.1	0.1996	1.0539	1.2535
50	89.19	0.02564	3.2684	3.294	97.9	527.3	625.2	0.2105	1.0348	1.2453
55	98.06	0.02581	2.9822	3.008	103.5	522.8	626.3	0.2214	1.0159	1.2373
60	107.6	0.02597	2.7250	2.751	109.2	518.1	627.3	0.2322	0.9972	1.2294
65	117.8	0.02614	2.4939	2.520	114.8	513.4	628.2	0.2430	0.9786	1.2216
70	128.8	0.02632	2.2857	2.312	120.5	508.6	629.1	0.2537	0.9603	1.2140
75	140.5	0.02650	2.0985	2.125	126.2	503.7	629.9	0.2643	0.9422	1.2065
80	153.0	0.02668	1.9283	1.955	132.0	498.7	630.7	0.2749	0.9242	1.1991
85	166.4	0.02687	1.7741	1.801	137.8	493.6	631.4	0.2854	0.9064	1.1918
90	180.6	0.02707	1.6339	1.661	143.5	488.5	632.0	0.2958	0.8888	1.1846
95	195.8	0.02727	1.5067	1.534	149.4	483.2	632.6	0.3062	0.8713	1.1775
100	211.9	0.02747	1.3915	1.419	155.2	477.8	633.0	0.3166	0.8539	1.1705
105	228.9	0.02769	1.2853	1.313	161.1	472.3	633.4	0.3269	0.8366	1.1635
110	247.0	0.02790	1.1891	1.217	167.0	466.7	633.7	0.3372	0.8194	1.1566
115	266.2	0.02813	1.0999	1.128	173.0	460.9	633.9	0.3474	0.8023	1.1497
120	286.4	0.02836	1.0186	1.047	179.0	455.0	634.0	0.3576	0.7851	1.1427
125	307.8	0.02860	0.9444	0.973	185.1	448.9	634.0	0.3679	0.7679	1.1358

TABLE 18.1-22 (*Continued*)

SUPERHEATED AMMONIA

Abs. Press., lbf/in.² (Sat. Temp.)		0	20	40	60	80	100	120	140	160	180	200	220
							Temperature, F						
10 (−41.34)	v	28.58	29.90	31.20	32.49	33.78	35.07	36.35	37.62	38.90	40.17	41.45	
	h	618.9	629.1	639.3	649.5	659.7	670.0	680.3	690.6	701.1	711.6	722.2	
	s	1.477	1.499	1.520	1.540	1.559	1.578	1.596	1.614	1.631	1.647	1.664	
15 (−27.29)	v	18.92	19.82	20.70	21.58	22.44	23.31	24.17	25.03	25.88	26.74	27.59	
	h	617.2	627.8	638.2	648.5	658.9	669.2	679.6	690.0	700.5	711.1	721.7	
	s	1.427	1.450	1.471	1.491	1.511	1.529	1.548	1.566	1.583	1.599	1.616	
20 (−16.64)	v	14.09	14.78	15.45	16.12	16.78	17.43	18.08	18.73	19.37	20.02	20.66	21.3
	h	615.5	626.4	637.0	647.5	658.0	668.5	678.9	689.4	700.0	710.6	721.2	732.0
	s	1.391	1.414	1.436	1.456	1.476	1.495	1.513	1.531	1.549	1.565	1.582	1.598
25 (−7.96)	v	11.19	11.75	12.30	12.84	13.37	13.90	14.43	14.95	15.47	15.99	16.50	17.02
	h	613.8	625.0	635.8	646.5	657.1	667.7	678.2	688.8	699.4	710.1	720.8	731.6
	s	1.362	1.386	1.408	1.429	1.449	1.468	1.486	1.504	1.522	1.539	1.555	1.571
30 (−.57)	v	9.25	9.731	10.20	10.65	11.10	11.55	11.99	12.43	12.87	13.30	13.73	14.16
	h	611.9	623.5	634.6	645.5	656.2	666.9	677.5	688.2	698.8	709.6	720.3	731.1
	s	1.337	1.362	1.385	1.406	1.426	1.446	1.464	1.482	1.500	1.517	1.533	1.550
35 (5.89)	v		8.287	8.695	9.093	9.484	9.869	10.25	10.63	11.00	11.38	11.75	12.12
	h		622.0	633.4	644.4	655.3	666.1	676.8	687.6	698.3	709.1	719.9	730.7
	s		1.341	1.365	1.386	1.407	1.427	1.445	1.464	1.481	1.498	1.515	1.531
40 (11.66)	v		7.203	7.568	7.922	8.268	8.609	8.945	9.278	9.609	9.938	10.27	10.59
	h		620.4	632.1	643.4	654.4	665.3	676.1	686.9	697.7	708.5	719.4	730.3
	s		1.323	1.347	1.369	1.390	1.410	1.429	1.447	1.465	1.482	1.499	1.515
46 (17.87)	v		6.213	6.538	6.851	7.157	7.457	7.753	8.045	8.335	8.623	8.909	9.194
	h		618.5	630.5	642.1	653.3	664.4	675.3	686.2	697.1	707.9	718.8	729.8
	s		1.304	1.328	1.351	1.372	1.392	1.411	1.430	1.448	1.465	1.482	1.498
50 (21.67)	v			5.988	6.280	6.564	6.843	7.117	7.387	7.655	7.921	8.185	8.448
	h			629.5	641.2	652.6	663.7	674.7	685.7	696.6	707.5	718.5	729.4
	s			1.317	1.340	1.361	1.382	1.401	1.420	1.437	1.455	1.472	1.488
60 (30.21)	v			4.933	5.184	5.428	5.665	5.897	6.126	6.352	6.576	6.798	7.019
	h			626.8	639.0	650.7	662.1	673.3	684.4	695.5	706.5	717.5	728.6
	s			1.2913	1.3152	1.3373	1.3581	1.3778	1.3966	1.4148	1.4323	1.4493	1.4658

TABLE 18.1-22 (*Continued*)

Abs. Press., lbf/in.² (Sat. Temp.)		Temperature, F											
		60	80	100	120	140	160	180	200	240	280	320	360
70 (37.7)	v	4.401	4.615	4.822	5.025	5.224	5.420	5.615	5.807	6.187	6.563		
	h	636.6	648.7	660.4	671.8	683.1	694.3	705.5	716.6	738.9	761.4		
	s	1.294	1.317	1.338	1.358	1.377	1.395	1.413	1.430	1.463	1.494		
80 (44.4)	v	3.812	4.005	4.190	4.371	4.548	4.722	4.893	5.063	5.398	5.73		
	h	634.3	646.7	658.7	670.4	681.8	693.2	704.4	715.6	738.1	760.7		
	s	1.275	1.298	1.320	1.340	1.360	1.378	1.396	1.414	1.447	1.478		
90 (50.47)	v	3.353	3.529	3.698	3.862	4.021	4.178	4.332	4.484	4.785	5.081		
	h	631.8	644.7	657.0	668.9	680.5	692.0	703.4	714.7	737.3	760.0		
	s	1.257	1.281	1.304	1.325	1.344	1.363	1.381	1.400	1.432	1.464		
100 (56.05)	v	2.985	3.149	3.304	3.454	3.600	3.743	3.883	4.021	4.294	4.562		
	h	629.3	642.6	655.2	667.3	679.2	690.8	702.3	713.7	736.5	759.4		
	s	1.241	1.266	1.289	1.310	1.331	1.349	1.368	1.385	1.419	1.451		
140 (74.79)	v		2.166	2.288	2.404	2.515	2.622	2.727	2.830	3.030	3.227	3.420	
	h		633.8	647.8	661.1	673.7	686.0	698.0	709.9	733.3	756.7	780.0	
	s		1.214	1.240	1.263	1.284	1.305	1.324	1.342	1.376	1.409	1.440	
180 (89.78)	v			1.720	1.818	1.910	1.999	2.084	2.167	2.328	2.484	2.637	
	h			639.9	654.4	668.0	681.0	693.6	705.9	730.1	753.9	777.7	
	s			1.199	1.225	1.248	1.269	1.289	1.308	1.344	1.377	1.408	
220 (102.42)	v				1.443	1.525	1.601	1.675	1.745	1.881	2.012	2.140	2.265
	h				647.3	662.0	675.8	689.1	701.9	726.8	751.1	775.3	799.5
	s				1.192	1.217	1.239	1.260	1.280	1.317	1.351	1.383	1.413
240 (108.09)	v				1.302	1.380	1.452	1.521	1.587	1.714	1.835	1.954	2.069
	h				643.5	658.8	673.1	686.7	699.8	725.1	749.8	774.1	798.4
	s				1.176	1.203	1.226	1.248	1.268	1.305	1.339	1.371	1.402
260 (113.42)	v				1.182	1.257	1.326	1.391	1.453	1.572	1.686	1.796	1.904
	h				639.5	655.6	670.4	684.4	697.7	723.4	748.4	772.9	797.4
	s				1.162	1.189	1.213	1.235	1.256	1.294	1.329	1.361	1.391
280 (118.45)	v				1.078	1.151	1.217	1.279	1.339	1.451	1.558	1.661	1.762
	h				635.4	652.9	667.6	681.9	695.6	721.8	747.0	771.7	796.3
	s				1.147	1.176	1.201	1.224	1.245	1.283	1.318	1.351	1.382

a From ref. 3. Used by permission.

TABLE 18.1-23 THERMODYNAMIC PROPERTIES OF AMMONIA: SI UNITS[a]

Saturated Ammonia

Temp. °C	Abs. Press. kPa P	Specific Volume m³/kg			Enthalpy kJ/kg			Entropy kJ/kg K		
		Sat. Liquid v_f	Evap. v_{fg}	Sat. Vapor v_g	Sat. Liquid h_f	Evap. h_{fg}	Sat. Vapor h_g	Sat. Liquid s_f	Evap. s_{fg}	Sat. Vapor s_g
−50	40.88	0.001 424	2.6239	2.6254	−44.3	1416.7	1372.4	−0.1942	6.3502	6.1561
−48	45.96	0.001 429	2.3518	2.3533	−35.5	1411.3	1375.8	−0.1547	6.2696	6.1149
−46	51.55	0.001 434	2.1126	2.1140	−26.6	1405.8	1379.2	−0.1156	6.1902	6.0746
−44	57.69	0.001 439	1.9018	1.9032	−17.8	1400.3	1382.5	−0.0768	6.1120	6.0352
−42	64.42	0.001 444	1.7155	1.7170	−8.9	1394.7	1385.8	−0.0382	6.0349	5.9967
−40	71.77	0.001 449	1.5506	1.5521	0.0	1389.0	1389.0	0.0000	5.9589	5.9589
−38	79.80	0.001 454	1.4043	1.4058	8.9	1383.3	1392.2	0.0380	5.8840	5.9220
−36	88.54	0.001 460	1.2742	1.2757	17.8	1377.6	1395.4	0.0757	5.8101	5.8858
−34	98.05	0.001 465	1.1582	1.1597	26.8	1371.8	1398.5	0.1132	5.7372	5.8504
−32	108.37	0.001 470	1.0547	1.0562	35.7	1365.9	1401.6	0.1504	5.6652	5.8156
−30	119.55	0.001 476	0.9621	0.9635	44.7	1360.0	1404.6	0.1873	5.5942	5.7815
−28	131.64	0.001 481	0.8790	0.8805	53.6	1354.0	1407.6	0.2240	5.5241	5.7481
−26	144.70	0.001 487	0.8044	0.8059	62.6	1347.9	1410.5	0.2605	5.4548	5.7153
−24	158.78	0.001 492	0.7373	0.7388	71.6	1341.8	1413.4	0.2967	5.3864	5.6831
−22	173.93	0.001 498	0.6768	0.6783	80.7	1335.6	1416.2	0.3327	5.3188	5.6515
−20	190.22	0.001 504	0.6222	0.6237	89.7	1329.3	1419.0	0.3684	5.2520	5.6205
−18	207.71	0.001 510	0.5728	0.5743	98.8	1322.9	1421.7	0.4040	5.1860	5.5900
−16	226.45	0.001 515	0.5280	0.5296	107.8	1316.5	1424.4	0.4393	5.1207	5.5600
−14	246.51	0.001 521	0.4874	0.4889	116.9	1310.0	1427.0	0.4744	5.0561	5.5305
−12	267.95	0.001 528	0.4505	0.4520	126.0	1303.5	1429.5	0.5093	4.9922	5.5015
−10	290.85	0.001 534	0.4169	0.4185	135.2	1296.8	1432.0	0.5440	4.9290	5.4730
−8	315.25	0.001 540	0.3863	0.3878	144.3	1290.1	1434.4	0.5785	4.8664	5.4449
−6	341.25	0.001 546	0.3583	0.3599	153.5	1283.3	1436.8	0.6128	4.8045	5.4173
−4	368.90	0.001 553	0.3328	0.3343	162.7	1276.4	1439.1	0.6469	4.7432	5.3901
−2	398.27	0.001 559	0.3094	0.3109	171.9	1269.4	1441.3	0.6808	4.6825	5.3633
0	429.44	0.001 566	0.2879	0.2895	181.1	1262.4	1443.5	0.7145	4.6223	5.3369
2	462.49	0.001 573	0.2683	0.2698	190.4	1255.2	1445.6	0.7481	4.5627	5.3108
4	497.49	0.001 580	0.2502	0.2517	199.6	1248.0	1447.6	0.7815	4.5037	5.2852
6	534.51	0.001 587	0.2335	0.2351	208.9	1240.6	1449.6	0.8148	4.4451	5.2599
8	573.64	0.001 594	0.2182	0.2198	218.3	1233.2	1451.5	0.8479	4.3871	5.2350
10	614.95	0.001 601	0.2040	0.2056	227.6	1225.7	1453.3	0.8808	4.3295	5.2104
12	658.52	0.001 608	0.1910	0.1926	237.0	1218.1	1455.1	0.9136	4.2725	5.1861
14	704.44	0.001 616	0.1789	0.1805	246.4	1210.4	1456.8	0.9463	4.2159	5.1621
16	752.79	0.001 623	0.1677	0.1693	255.9	1202.6	1458.5	0.9788	4.1597	5.1385
18	803.66	0.001 631	0.1574	0.1590	265.4	1194.7	1460.0	1.0112	4.1039	5.1151
20	857.12	0.001 639	0.1477	0.1494	274.9	1186.7	1461.5	1.0434	4.0486	5.0920
22	913.27	0.001 647	0.1388	0.1405	284.4	1178.5	1462.9	1.0755	3.9937	5.0692
24	972.19	0.001 655	0.1305	0.1322	294.0	1170.3	1464.3	1.1075	3.9392	5.0467
26	1033.97	0.001 663	0.1228	0.1245	303.6	1162.0	1465.6	1.1394	3.8850	5.0244
28	1098.71	0.001 671	0.1156	0.1173	313.2	1153.6	1466.8	1.1711	3.8312	5.0023
30	1166.49	0.001 680	0.1089	0.1106	322.9	1145.0	1467.9	1.2028	3.7777	4.9805
32	1237.41	0.001 689	0.1027	0.1044	332.6	1136.4	1469.0	1.2343	3.7246	4.9589
34	1311.55	0.001 698	0.0969	0.0986	342.3	1127.6	1469.9	1.2656	3.6718	4.9374
36	1389.03	0.001 707	0.0914	0.0931	352.1	1118.7	1470.8	1.2969	3.6192	4.9161
38	1469.92	0.001 716	0.0863	0.0880	361.9	1109.7	1471.5	1.3281	3.5669	4.8950
40	1554.33	0.001 726	0.0815	0.0833	371.7	1100.5	1472.2	1.3591	3.5148	4.8740
42	1642.35	0.001 735	0.0771	0.0788	381.6	1091.2	1472.8	1.3901	3.4630	4.8530
44	1734.09	0.001 745	0.0728	0.0746	391.5	1081.7	1473.2	1.4209	3.4112	4.8322
46	1829.65	0.001 756	0.0689	0.0707	401.5	1072.0	1473.5	1.4518	3.3595	4.8113
48	1929.13	0.001 766	0.0652	0.0669	411.5	1062.2	1473.7	1.4826	3.3079	4.7905
50	2032.62	0.001 777	0.0617	0.0635	421.7	1052.0	1473.7	1.5135	3.2561	4.7696

TABLE 18.1-23 (*Continued*)

Superheated Ammonia

Abs. Press. kPa (Sat. Temp.) °C		−20	−10	0	10	20	30	40	50	60	70	80	100
50 (−46.54)	v	2.4474	2.5481	2.6482	2.7479	2.8473	2.9464	3.0453	3.1441	3.2427	3.3413	3.4397	
	h	1435.8	1457.0	1478.1	1499.2	1520.4	1541.7	1563.0	1584.5	1606.1	1627.8	1649.7	
	s	6.3256	6.4077	6.4865	6.5625	6.6360	6.7073	6.7766	6.8441	6.9099	6.9743	7.0372	
75 (−39.18)	v	1.6233	1.6915	1.7591	1.8263	1.8932	1.9597	2.0261	2.0923	2.1584	2.2244	2.2903	
	h	1433.0	1454.7	1476.1	1497.5	1518.9	1540.3	1561.8	1583.4	1605.1	1626.9	1648.9	
	s	6.1190	6.2028	6.2828	6.3597	6.4339	6.5058	6.5756	6.6434	6.7096	6.7742	6.8373	
100 (−33.61)	v	1.2110	1.2631	1.3145	1.3654	1.4160	1.4664	1.5165	1.5664	1.6163	1.6659	1.7155	1.8145
	h	1430.1	1452.2	1474.1	1495.7	1517.3	1538.9	1560.5	1582.2	1604.1	1626.0	1648.0	1692.6
	s	5.9695	6.0552	6.1366	6.2144	6.2894	6.3618	6.4321	6.5003	6.5668	6.6316	6.6950	6.8177
125 (−29.08)	v	0.9635	1.0059	1.0476	1.0889	1.1297	1.1703	1.2107	1.2509	1.2909	1.3309	1.3707	1.4501
	h	1427.2	1449.8	1472.0	1493.9	1515.7	1537.5	1559.3	1581.1	1603.0	1625.0	1647.2	1691.8
	s	5.8512	5.9389	6.0217	6.1006	6.1763	6.2494	6.3201	6.3887	6.4555	6.5206	6.5842	6.7072
150 (−25.23)	v	0.7984	0.8344	0.8697	0.9045	0.9388	0.9729	1.0068	1.0405	1.0740	1.1074	1.1408	1.2072
	h	1424.1	1447.3	1469.8	1492.1	1514.1	1536.1	1558.0	1580.0	1602.0	1624.1	1646.3	1691.1
	s	5.7526	5.8424	5.9266	6.0066	6.0831	6.1568	6.2280	6.2970	6.3641	6.4295	6.4933	6.6167
200 (−18.86)	v		0.6199	0.6471	0.6738	0.7001	0.7261	0.7519	0.7774	0.8029	0.8282	0.8533	0.9035
	h		1442.0	1465.5	1488.4	1510.9	1533.2	1555.5	1577.7	1599.9	1622.2	1644.6	1689.6
	s		5.6863	5.7737	5.8559	5.9342	6.0091	6.0813	6.1512	6.2189	6.2849	6.3491	6.4732
250 (−13.67)	v		0.4910	0.5135	0.5354	0.5568	0.5780	0.5989	0.6196	0.6401	0.6605	0.6809	0.7212
	h		1436.6	1461.0	1484.5	1507.6	1530.3	1552.9	1575.4	1597.8	1620.3	1642.8	1688.2
	s		5.5609	5.6517	5.7365	5.8165	5.8928	5.9661	6.0368	6.1052	6.1717	6.2365	6.3613
300 (−9.23)	v			0.4243	0.4430	0.4613	0.4792	0.4968	0.5143	0.5316	0.5488	0.5658	0.5997
	h			1456.3	1480.6	1504.2	1527.4	1550.3	1573.0	1595.7	1618.4	1641.1	1686.7
	s			5.5493	5.6366	5.7186	5.7963	5.8707	5.9423	6.0114	6.0785	6.1437	6.2693
350 (−5.35)	v			0.3605	0.3770	0.3929	0.4086	0.4239	0.4391	0.4541	0.4689	0.4837	0.5129
	h			1451.5	1476.5	1500.7	1524.4	1547.6	1570.7	1593.6	1616.5	1639.3	1685.2
	s			5.4600	5.5502	5.6342	5.7135	5.7890	5.8615	5.9314	5.9990	6.0647	6.1910
400 (−1.89)	v			0.3125	0.3274	0.3417	0.3556	0.3692	0.3826	0.3959	0.4090	0.4220	0.4478
	h			1446.5	1472.4	1497.2	1521.3	1544.9	1568.3	1591.5	1614.5	1637.6	1683.7
	s			5.3803	5.4735	5.5597	5.6405	5.7173	5.7907	5.8613	5.9296	5.9957	6.1228
450 (1.26)	v			0.2752	0.2887	0.3017	0.3143	0.3266	0.3387	0.3506	0.3624	0.3740	0.3971
	h			1441.3	1468.1	1493.6	1518.2	1542.2	1565.9	1589.3	1612.6	1635.8	1682.2
	s			5.3078	5.4042	5.4926	5.5752	5.6532	5.7275	5.7989	5.8678	5.9345	6.0623

		20	30	40	50	60	70	80	100	120	140	160	180
500 (4.14)	v	0.2698	0.2813	0.2926	0.3036	0.3144	0.3251	0.3357	0.3565	0.3771	0.3975		
	h	1489.9	1515.0	1539.5	1563.4	1587.1	1610.6	1634.0	1680.7	1727.5	1774.7		
	s	5.4314	5.5157	5.5950	5.6704	5.7425	5.8120	5.8793	6.0079	6.1301	6.2472		
600 (9.29)	v	0.2217	0.2317	0.2414	0.2508	0.2600	0.2691	0.2781	0.2957	0.3130	0.3302		
	h	1482.4	1508.6	1533.8	1558.5	1582.7	1606.6	1630.4	1677.7	1724.9	1772.4		
	s	5.3222	5.4102	5.4923	5.5697	5.6436	5.7144	5.7826	5.9129	6.0363	6.1541		
700 (13.81)	v	0.1874	0.1963	0.2048	0.2131	0.2212	0.2291	0.2369	0.2522	0.2672	0.2821		
	h	1474.5	1501.9	1528.1	1553.4	1578.2	1602.6	1626.8	1674.6	1722.4	1770.2		
	s	5.2259	5.3179	5.4029	5.4826	5.5582	5.6303	5.6997	5.8316	5.9562	6.0749		
800 (17.86)	v	0.1615	0.1696	0.1773	0.1848	0.1920	0.1991	0.2060	0.2196	0.2329	0.2459	0.2589	
	h	1466.3	1495.0	1522.2	1548.3	1573.7	1598.6	1623.1	1671.6	1719.8	1768.0	1816.4	
	s	5.1387	5.2351	5.3232	5.4053	5.4827	5.5562	5.6268	5.7603	5.8861	6.0057	6.1202	
900 (21.54)	v		0.1488	0.1559	0.1627	0.1693	0.1757	0.1820	0.1942	0.2061	0.2178	0.2294	
	h		1488.0	1516.2	1543.0	1569.1	1594.4	1619.4	1668.5	1717.1	1765.7	1814.4	
	s		5.1593	5.2508	5.3354	5.4147	5.4897	5.5614	5.6968	5.8237	5.9442	6.0594	
1000 (24.91)	v		0.1321	0.1388	0.1450	0.1511	0.1570	0.1627	0.1739	0.1847	0.1954	0.2058	0.2162
	h		1480.6	1510.0	1537.7	1564.4	1590.3	1615.6	1665.4	1714.5	1763.4	1812.4	1861.7
	s		5.0889	5.1840	5.2713	5.3525	5.4292	5.5021	5.6392	5.7674	5.8888	6.0047	6.1159
1200 (30.96)	v			0.1129	0.1185	0.1238	0.1289	0.1338	0.1434	0.1526	0.1616	0.1705	0.1792
	h			1497.1	1526.6	1554.7	1581.7	1608.0	1659.2	1709.2	1758.9	1808.5	1858.2
	s			5.0629	5.1560	5.2416	5.3215	5.3970	5.5379	5.6687	5.7919	5.9091	6.0214
1400 (36.28)	v			0.0944	0.0995	0.1042	0.1088	0.1132	0.1216	0.1297	0.1376	0.1452	0.1528
	h			1483.4	1515.1	1544.7	1573.0	1600.2	1652.8	1703.9	1754.3	1804.5	1854.7
	s			4.9534	5.0530	5.1434	5.2270	5.3053	5.4501	5.5836	5.7087	5.8273	5.9406
1600 (41.05)	v				0.0851	0.0895	0.0937	0.0977	0.1053	0.1125	0.1195	0.1263	0.1330
	h				1502.9	1534.4	1564.0	1592.3	1646.4	1698.5	1749.7	1800.5	1851.2
	s				4.9584	5.0543	5.1419	5.2232	5.3722	5.5084	5.6355	5.7555	5.8699
1800 (45.39)	v				0.0739	0.0781	0.0820	0.0856	0.0926	0.0992	0.1055	0.1116	0.1177
	h				1490.0	1523.5	1554.6	1584.1	1639.8	1693.1	1745.1	1796.5	1847.7
	s				4.8693	4.9715	5.0635	5.1482	5.3018	5.4409	5.5699	5.6914	5.8069
2000 (49.38)	v				0.0648	0.0688	0.0725	0.0760	0.0824	0.0885	0.0943	0.0999	0.1054
	h				1476.1	1512.0	1544.9	1575.6	1633.2	1687.6	1740.4	1792.4	1844.1
	s				4.7834	4.8930	4.9902	5.0786	5.2371	5.3793	5.5104	5.6333	5.7499

[a] From ref. 4. Used by permission.

TABLE 18.1-24 ENTHALPY OF COMBUSTION OF SOME HYDROCARBONS AT 25°C (77°F)[a]

Hydrocarbon	Formula	Liquid H$_2$O in Products (Negative of Higher Heating Value)		Vapor H$_2$O in Products (Negative of Lower Heating Value)	
		Liquid Hydrocarbon, Btu/lbm fuel	Gaseous Hydrocarbon, Btu/lbm fuel	Liquid Hydrocarbon, Btu/lbm fuel	Gaseous Hydrocarbon, Btu/lbm fuel
Paraffin Family					
Methane	CH$_4$		−23,861		−21,502
Ethane	C$_2$H$_6$		−22,304		−20,416
Propane	C$_3$H$_8$	−21,490	−21,649	−19,773	−19,929
Butane	C$_4$H$_{10}$	−21,134	−21,293	−19,506	−19,665
Pentane	C$_5$H$_{12}$	−20,914	−21,072	−19,340	−19,499
Hexane	C$_6$H$_{14}$	−20,772	−20,930	−19,233	−19,391
Heptane	C$_7$H$_{16}$	−20,668	−20,825	−19,157	−19,314
Octane	C$_8$H$_{18}$	−20,591	−20,747	−19,100	−19,256
Decane	C$_{10}$H$_{22}$	−20,484	−20,638	−19,020	−19,175
Dodecane	C$_{12}$H$_{26}$	−20,410	−20,564	−18,964	−19,118
Olefin Family					
Ethene	C$_2$H$_4$		−21,626		−20,276
Propene	C$_3$H$_6$		−21,033		−19,683
Butene	C$_4$H$_8$		−20,833		−19,483
Pentene	C$_5$H$_{10}$		−20,696		−19,346
Hexene	C$_6$H$_{12}$		−20,612		−19,262
Heptene	C$_7$H$_{14}$		−20,552		−19,202
Octene	C$_8$H$_{16}$		−20,507		−19,157
Nonene	C$_9$H$_{18}$		−20,472		−19,122
Decene	C$_{10}$H$_{20}$		−20,444		−19,094
Alkylbenzene Family					
Benzene	C$_6$H$_6$	−17,985	−18,172	−17,259	−17,446
Methylbenzene	C$_7$H$_8$	−18,247	−18,423	−17,424	−17,601
Ethylbenzene	C$_8$H$_{10}$	−18,488	−18,659	−17,596	−17,767
Propylbenzene	C$_9$H$_{12}$	−18,667	−18,832	−17,722	−17,887
Butylbenzene	C$_{10}$H$_{14}$	−18,809	−18,970	−17,823	−17,984

Source. Van Wylen, Gordon J., *Thermodynamics* 1959, John Wiley & Sons. Used by permission.

[a] From ref. 9. Used by permission.

TABLE 18.1-25   THERMODYNAMIC PROPERTIES OF CARBON DIOXIDE—SATURATED LIQUID–VAPOR

Temp. (°F)	Abs. Press. (psia)	Specific Volume (ft³/lbm) Sat. Liq.	Evap.	Sat. Vap.	Internal Energy (Btu/lbm) Sat. Liq.	Evap.	Sat. Vap.
T	p	v_f	v_{fg}	v_g	u_f	u_{fg}	u_g
−69.9	75.146	0.01360	1.1434	1.1570	−13.9	133.8	119.9
−60	94.75	0.01384	0.9112	0.9250	−9.3	129.8	120.5
−50	118.27	0.01409	0.7359	0.7500	−4.9	125.8	120.9
−40	145.87	0.01437	0.5969	0.6113	−0.4	121.8	121.4
−30	178.07	0.01465	0.4882	0.5028	4.2	117.6	121.8
−20	215.02	0.01498	0.4015	0.4165	8.6	113.5	122.1
−10	257.58	0.01533	0.3315	0.3468	13.3	109.1	122.4
0	305.76	0.01571	0.2793	0.2905	17.9	104.6	122.5
10	360.5	0.01614	0.2274	0.2435	21.8	100.8	122.6
20	421.8	0.01662	0.1882	0.2048	28.3	94.2	122.5
30	490.8	0.01719	0.1551	0.1723	34.2	88.1	122.3
40	567.3	0.01786	0.1263	0.1442	39.9	81.8	121.7
50	652.9	0.01870	0.1020	0.1207	46.2	73.8	120.0
60	747.4	0.01970	0.0798	0.0995	53.0	65.4	118.4
70	852.8	0.02113	0.0590	0.0801	60.9	55.6	116.6
80	969.3	0.02370	0.0363	0.0600	69.7	38.5	108.2
87.87	1070.0	0.03423	0	0.03423	90.3	0	90.3

Temp. (°F)	Enthalpy (Btu/lbm)[b] Sat. Liq.	Evap.	Sat. Vap.	Entropy (Btu/lbm R)[b] Sat. Liq.	Evap.	Sat. Vap.
T	h_f	h_{fg}	h_g	s_f	s_{fg}	s_g
−69.9	−13.7	149.7	136.0	−0.0333	0.3839	0.3506
−60	−9.1	145.8	136.7	−0.0221	0.3650	0.3429
−50	−4.6	141.9	137.3	−0.0109	0.3463	0.3354
−40	0.0	137.9	137.9	0.0000	0.3285	0.3285
−30	4.7	133.7	138.4	0.0106	0.3113	0.3219
−20	9.2	129.5	138.7	0.0210	0.2945	0.3155
−10	14.0	124.9	138.9	0.0314	0.2778	0.3092
0	18.8	120.1	138.9	0.0419	0.2611	0.3030
10	24.0	114.8	138.8	0.0525	0.2445	0.2970
20	29.6	108.9	138.5	0.0636	0.2273	0.2909
30	35.8	102.1	137.9	0.0754	0.2095	0.2849
40	41.8	95.0	136.8	0.0872	0.1903	0.2775
50	48.5	86.1	134.6	0.0999	0.1699	0.2698
60	55.7	76.5	132.2	0.1136	0.1470	0.2606
70	63.8	63.9	127.7	0.1283	0.1203	0.2486
80	74.0	45.0	119.0	0.1469	0.0836	0.2304
87.87	97.1	0	97.1	0.1880	0	0.1880

TABLE 18.1-25 (*Continued*)

Thermodynamic Properties of Carbon Dioxide—Superheated CO_2

Abs. Press. (psia)	Sat. Temp. (°F)	Sat. Vapor	Temperature (°F)				
p	T_g		0	100	200	300	400
		v_g	Specific Volume, v (ft^3/lbm)				
0.2	−182.8	315.0	560.20	682.28	804.20	926.12	1048.04
0.5	−170.9	140.2	224.08	272.91	321.68	370.44	419.21
1	−159.8	76.0	111.95	136.40	160.81	185.20	209.59
2	−147.6	37.2	55.98	68.20	80.40	92.60	104.80
5	−131.9	16.0	22.33	27.24	32.14	37.02	41.91
10	−118.3	8.66	11.14	13.60	16.06	18.54	20.95
15	−109.4	5.78	7.406	9.057	10.70	12.33	13.96
20	−102.5	4.26	5.554	6.793	8.023	9.247	10.47
50	−79.7	1.68	2.166	2.681	3.188	3.686	4.180
100	−57.6	0.880	1.042	1.315	1.579	1.833	2.084
150	−38.7	0.596	0.6678	0.8583	1.042	1.216	1.385
200	−23.9	0.450	0.4813	0.6318	0.7742	0.9073	1.036
250	−11.6	0.357	0.3693	0.4951	0.6134	0.7220	0.8260
300	−1.1	0.296	0.2930	0.4036	0.5062	0.5985	0.6862
350	8.2	0.251		0.3382	0.4296	0.5103	0.5862
400	16.6	0.217		0.2890	0.3722	0.4442	0.5116
500	31.3	0.167		0.2196	0.2915	0.3516	0.4068
600	44.0	0.134		0.1725	0.2374	0.2898	0.3370
800	65.1	0.0904		0.1142	0.1696	0.2127	0.2498
1000	82.5	0.055		0.0790	0.1296	0.1665	0.1976
2000					0.0517	0.0741	0.0941
3000					0.0325	0.0466	0.0604
		u_g	Internal Energy, u (Btu/lbm)				
0.2	−182.8	112.1	134.9	150.1	166.8	184.5	203.3
0.5	−170.9	112.5	134.8	150.1	166.8	184.5	203.3
1	−159.8	112.8	134.6	150.1	166.8	184.5	203.3
2	−147.6	113.9	134.5	150.1	166.7	184.5	203.3
5	−131.9	115.6	134.4	150.0	166.7	184.5	203.3
10	−118.3	116.0	134.4	149.9	166.7	184.5	203.3
15	−109.4	117.5	134.2	149.9	166.6	184.5	203.3
20	−102.5	118.8	134.0	149.8	166.5	184.5	203.3
50	−79.7	120.3	133.2	149.3	166.2	184.3	203.0
100	−57.6	120.5	131.3	148.5	165.8	183.9	202.6
150	−38.7	121.5	129.5	147.7	165.3	183.3	202.3
200	−23.9	121.3	127.8	146.6	164.8	182.8	201.9
250	−11.6	122.3	125.3	145.7	164.5	182.5	201.6
300	−1.1	122.5	122.7	144.7	163.5	181.9	201.0
350	8.2	122.5		143.7	162.9	181.4	200.7
400	16.6	122.5		142.4	162.3	181.1	200.4
500	31.3	122.3		140.2	161.0	180.3	199.6
600	44.0	121.3		137.8	159.6	179.2	198.8
800	65.1	116.8		132.2	156.6	177.4	197.1
1000	82.5	106.4		123.9	153.0	175.1	195.5
2000					129.6	163.9	185.4
3000					112.2	154.5	178.5

TABLE 18.1-25 (*Continued*)

Abs. Press. (psia)	Temperature (°F)					
p	500	600	700	800	900	1000
			Specific Volume, v (ft^3/lbm)			
0.2	1169.94	1291.85	1413.76	1535.68	1657.59	1779.50
0.5	467.98	516.74	565.50	614.27	663.04	711.80
1	233.98	258.37	282.75	307.13	331.52	355.90
2	116.99	129.18	141.37	153.57	165.76	177.95
5	46.70	51.67	56.55	61.43	66.30	71.18
10	23.39	25.83	28.27	30.71	33.15	35.59
15	15.59	17.22	18.85	20.48	22.10	23.73
20	11.69	12.92	14.14	15.36	16.58	17.80
50	4.672	5.162	5.652	6.140	6.629	7.117
100	2.332	2.578	2.824	3.068	3.313	3.558
150	1.552	1.717	1.881	2.045	2.208	2.371
200	1.162	1.286	1.410	1.533	1.656	1.778
250	0.9277	1.028	1.127	1.226	1.324	1.422
300	0.7717	0.8556	0.9385	1.021	1.103	1.185
350	0.6602	0.7334	0.8037	0.8745	0.9448	1.015
400	0.5767	0.6402	0.7028	0.7648	0.8266	0.8885
500	0.4597	0.5110	0.5613	0.6111	0.6608	0.7105
600	0.3817	0.4248	0.4670	0.5086	0.5502	0.5919
800	0.2843	0.3172	0.3490	0.3805	0.4119	0.4436
1000	0.2259	0.2526	0.2782	0.3035	0.3289	0.3546
2000	0.1099	0.1237	0.1366	0.1494	0.1626	0.1767
3000	0.0715	0.0809	0.0897	0.0984	0.1076	0.1175
			Internal Energy, u (Btu/lbm)			
0.2	223.1	243.8	265.2	287.2	309.9	333.1
0.5	223.1	243.8	265.2	287.2	309.9	333.1
1	223.1	243.8	265.2	287.2	309.9	333.1
2	223.1	243.8	265.2	287.2	309.9	333.1
5	223.1	243.8	265.2	287.2	309.9	333.1
10	231.1	243.8	265.2	287.2	309.9	333.1
15	223.1	243.8	265.2	287.2	309.9	333.1
20	223.1	243.6	265.2	287.2	309.8	333.1
50	222.9	242.6	265.0	287.1	309.8	333.0
100	222.5	242.4	264.8	287.0	309.8	333.0
150	222.2	242.1	264.7	286.8	309.7	333.0
200	221.9	242.0	264.6	286.8	309.6	333.0
250	221.7	241.7	264.5	286.7	309.5	332.9
300	221.3	241.5	264.3	286.5	309.4	332.8
350	221.0	241.2	264.1	286.5	309.3	332.8
400	220.8	241.1	264.0	286.3	309.1	332.6
500	220.2	240.6	263.7	286.1	309.0	332.5
600	219.5	240.1	263.2	285.8	308.8	332.4
800	218.3	239.3	262.6	285.4	308.4	331.9
1000	217.0	238.6	262.0	284.8	307.1	331.6
2000	209.0	233.9	258.6	282.8	306.5	329.8
3000	203.9	231.9	257.4	281.8	305.6	328.7

TABLE 18.1-25 (*Continued*)

Abs. Press. (psia) p	Sat. Temp. (°F) T_g	Sat. Vapor	Temperature (°F)				
			0	100	200	300	400
		h_g	Enthalpy, h (Btu/lbm)				
0.2	−182.8	123.8	155.6	175.4	196.6	218.8	242.1
0.5	−170.9	125.5	155.5	175.4	196.6	218.8	242.1
1	−159.8	126.9	155.3	175.3	196.6	218.8	242.1
2	−147.6	127.7	155.2	175.3	196.5	218.8	242.1
5	−131.9	130.4	155.1	175.2	196.4	218.8	242.1
10	−118.3	132.0	155.0	175.1	196.4	218.8	242.1
15	−109.4	133.5	154.8	175.0	196.3	218.8	242.1
20	−102.5	134.6	154.6	174.9	196.2	218.8	242.1
50	−79.7	135.8	153.2	174.1	195.7	218.4	241.7
100	−57.6	136.8	150.6	172.8	195.0	217.8	241.2
150	−38.7	138.0	148.0	171.5	194.2	217.1	240.7
200	−23.9	138.6	145.4	170.0	193.5	216.4	240.2
250	−11.6	138.8	142.4	168.6	192.9	215.9	239.8
300	−1.1	138.9	139.0	167.1	191.6	215.1	239.1
350	8.2	138.8		165.6	190.7	214.5	238.7
400	16.6	138.6		163.8	189.9	214.0	238.3
500	31.3	137.8		160.5	188.0	212.8	237.2
600	44.0	136.2		157.0	186.0	211.4	236.2
800	65.1	130.2		149.1	181.7	208.9	234.1
1000	82.5	116.6		138.5	177.0	205.9	232.1
2000					148.7	191.3	220.2
3000					130.2	180.4	212.0
		s_g	Entropy, s (Btu/lbm R)				
0.2	−182.8	0.577	0.662	0.702	0.736	0.768	0.798
0.5	−170.9	0.543	0.620	0.660	0.694	0.727	0.756
1	−159.8	0.515	0.589	0.629	0.663	0.696	0.725
2	−147.6	0.488	0.558	0.598	0.632	0.665	0.694
5	−131.9	0.453	0.517	0.556	0.590	0.623	0.652
10	−118.3	0.431	0.485	0.525	0.559	0.592	0.621
15	−109.4	0.412	0.467	0.507	0.541	0.574	0.603
20	−102.5	0.403	0.454	0.494	0.528	0.561	0.590
50	−79.7	0.366	0.409	0.451	0.486	0.518	0.548
100	−57.6	0.341	0.371	0.417	0.454	0.485	0.515
150	−38.7	0.328	0.352	0.398	0.436	0.466	0.497
200	−23.9	0.318	0.335	0.383	0.421	0.452	0.484
250	−11.6	0.310	0.319	0.370	0.410	0.442	0.474
300	−1.1	0.304	0.306	0.360	0.401	0.434	0.465
350	8.2	0.298		0.350	0.390	0.427	0.458
400	16.6	0.293		0.342	0.385	0.420	0.451
500	31.3	0.284		0.328	0.373	0.410	0.440
600	44.0	0.274		0.316	0.363	0.400	0.431
800	65.1	0.255		0.289	0.344	0.387	0.417
1000	82.5	0.226		0.263	0.329	0.370	0.404
2000					0.268	0.325	0.363
3000					0.227	0.290	0.335

TABLE 18.1-25 (*Continued*)

Abs. Press. (psia)	Temperature (°F)					
p	500	600	700	800	900	1000
	Enthalpy, h (Btu/lbm)					
0.2	266.4	291.6	317.5	344.0	371.2	399.0
0.5	266.4	291.6	317.5	344.0	371.2	399.0
1	266.4	291.6	317.5	344.0	371.2	399.0
2	266.4	291.6	317.5	344.0	371.2	399.0
5	266.4	291.6	317.5	344.0	371.2	399.0
10	266.4	291.6	317.5	344.0	371.2	399.0
15	266.4	291.6	317.5	344.0	371.2	399.0
20	266.4	291.4	317.5	344.0	371.2	399.0
50	266.1	290.4	317.3	343.9	371.1	398.9
100	265.7	290.1	317.1	343.8	371.1	398.9
150	265.3	289.8	316.9	343.6	371.0	398.8
200	264.9	289.6	316.8	343.5	370.9	398.8
250	264.6	289.3	316.6	343.4	370.8	398.7
300	264.1	289.0	316.4	343.2	370.6	398.6
350	263.8	288.7	316.2	343.1	370.5	398.5
400	263.5	288.5	316.0	342.9	370.3	398.4
500	262.7	287.9	315.6	342.6	370.1	398.2
600	261.9	287.3	315.1	342.3	369.9	398.0
800	260.4	286.3	314.3	341.7	369.4	397.6
1000	258.8	285.3	313.5	341.0	368.0	397.2
2000	249.7	279.7	309.2	338.1	366.7	395.2
3000	243.6	276.8	307.2	336.4	365.3	393.9
	Entropy, s (Btu/lbm R)					
0.2	0.824	0.848	0.872	0.895	0.917	0.936
0.5	0.782	0.806	0.830	0.853	0.875	0.894
1	0.751	0.775	0.799	0.822	0.844	0.863
2	0.720	0.744	0.768	0.791	0.813	0.832
5	0.678	0.702	0.726	0.749	0.771	0.790
10	0.647	0.671	0.695	0.718	0.740	0.759
15	0.629	0.653	0.677	0.700	0.722	0.741
20	0.616	0.640	0.664	0.683	0.709	0.728
50	0.574	0.598	0.622	0.645	0.668	0.687
100	0.542	0.566	0.590	0.613	0.627	0.654
150	0.524	0.548	0.572	0.595	0.619	0.636
200	0.511	0.535	0.559	0.582	0.606	0.623
250	0.501	0.525	0.549	0.572	0.596	0.613
300	0.493	0.517	0.541	0.564	0.588	0.605
350	0.486	0.510	0.534	0.557	0.581	0.598
400	0.480	0.504	0.528	0.550	0.574	0.592
500	0.469	0.494	0.518	0.540	0.564	0.582
600	0.461	0.486	0.510	0.532	0.556	0.574
800	0.448	0.473	0.497	0.519	0.543	0.561
1000	0.436	0.462	0.486	0.508	0.532	0.550
2000	0.396	0.427	0.452	0.475	0.499	0.517
3000	0.372	0.405	0.431	0.455	0.479	0.497

TABLE 18.1-25 *(Continued)*

	Thermodynamic Properties of Carbon Dioxide—Compressed Liquid					
Abs. Press. (psia)		Temperature (°F)			Sat. Temp. (°F)	Sat Liq.
p	25	50	75	100	T_f	

Specific Volume, v (ft^3/lbm)

						v_f
500	0.01683				31.22	0.01719
1000	0.01641	0.01790	0.02123		82.40	0.02691
2000	0.01602	0.01687	0.01843	0.02119		
3000	0.01565	0.01639	0.01767	0.01981		
5000	0.01492	0.01542	0.01614	0.01704		
8000	0.01428	0.01467	0.01523	0.01579		

Internal Energy, u (Btu/lbm)

						u_f
500	30.1				31.22	34.9
1000	27.9	41.9	57.0		82.40	76.0
2000	24.3	36.7	51.0	66.5		
3000	21.3	33.0	46.3	59.5		
5000	17.7	27.5	39.1	51.4		
8000	12.9	22.4	32.8	43.5		

Enthalpy, h (Btu/lbm)

						h_f
500	31.7				31.22	36.5
1000	30.9	45.2	60.9		82.40	81.0
2000	30.2	42.9	57.8	74.3		
3000	30.0	42.1	56.1	70.5		
5000	31.5	41.8	54.0	67.2		
8000	34.0	44.1	55.3	66.9		

Entropy, s (Btu/lbm R)

						s_f
500	0.0687				31.22	0.0768
1000	0.0635	0.0922	0.1190		82.40	0.1594
2000	0.0559	0.0815	0.1089	0.1542		
3000	0.0490	0.0719	0.0974	0.1249		
5000	0.0392	0.0614	0.0840	0.1049		
8000	0.0285	0.0497	0.0705	0.0901		

[a] From ref. 10. Used by permission.
[b] The enthalpy and entropy of saturated liquid carbon dioxide are taken as zero at a temperature of $-40°F$.

TABLE 18.1-26   PHYSICAL PROPERTIES OF PIPE[a]

PHYSICAL PROPERTIES OF PIPE

Nominal Pipe Size, In.	Outside Diam, In.	Wall Thickness, In.	Inside Diam, In.	Inside Diam, Fifth Power	Internal Cross-sectional Area		Weight of Pipe per Ft-Lb
	D	t	d	d^5	Sq In.	Sq Ft	

SCHEDULE 10

Nominal Pipe Size, In.	D	t	d	d^5	Sq In.	Sq Ft	Weight per Ft-Lb
14 OD	14.0	0.250	13.500	448,000	143.1	0.993	36.7
16 OD	16.0	0.250	15.500	895,000	188.7	1.310	42.1
18 OD	18.0	0.250	17.500	1,641,000	240.5	1.670	47.4
20 OD	20.0	0.250	19.500	2,820,000	298.6	2.073	52.7
22 OD	22.0	0.250	21.500	4,590,000	363.0	2.520	58.1
24 OD	24.0	0.250	23.500	7,170,000	434.0	3.013	63.4
30 OD	30.0	0.312	29.376	21,900,000	678.0	4.708	98.9

SCHEDULE 20

Nominal Pipe Size, In.	D	t	d	d^5	Sq In.	Sq Ft	Weight per Ft-Lb
8	8.625	0.250	8.125	35,400	51.8	0.359	22.37
10	10.750	0.250	10.250	113,000	82.5	0.572	28.0
12	12.750	0.250	12.250	276,000	117.9	0.818	33.4
14 OD	14.000	0.312	13.375	428,000	140.5	0.975	45.7
16 OD	16.000	0.312	15.375	859,000	185.7	1.289	52.4
18 OD	18.000	0.312	17.375	1,584,000	237.1	1.646	59.0
20 OD	20.000	0.375	19.250	2,640,000	291.0	2.020	78.0
24 OD	24.000	0.375	23.250	6,790,000	425	2.951	94.6
30 OD	30.000	0.500	29.000	20,500,000	661	4.590	157.6

SCHEDULE 30

Nominal Pipe Size, In.	D	t	d	d^5	Sq In.	Sq Ft	Weight per Ft-Lb
8	8.625	0.277	8.071	34,200	51.2	0.355	24.70
10	10.750	0.307	10.136	107,000	80.7	0.560	34.2
12	12.750	0.330	12.090	258,000	114.8	0.797	43.8
14 OD	14.000	0.375	13.250	408,000	137.9	0.957	54.6
16 OD	16.000	0.375	15.250	825,000	182.7	1.268	62.6
18 OD	18.000	0.438	17.124	1,472,000	230.3	1.599	82.2
20 OD	20.000	0.500*	19.000	2,480,000	283.5	1.968	104.1
24 OD	24.000	0.562	22.876	6,260,000	411.0	2.854	140.7
30 OD	30.000	0.625	28.750	19,600,000	649.0	4.506	196.1

NOTE: Wall thickness shown in italics is the same as for standard weight pipe.

* Wall thickness is the same as for extra strong pipe.

TABLE 18.1-26 (*Continued*)

PHYSICAL PROPERTIES OF PIPE

Nominal Pipe Size, In.	Outside Diam, In.	Wall Thickness, In.	Inside Diam, In.	Inside Diam, Fifth Power	Internal Cross-sectional Area		Weight of Pipe per Ft-Lb
	D	t	d	d^5	Sq In.	Sq Ft	

SCHEDULE 40

Nominal Pipe Size, In.	Outside Diam, In.	Wall Thickness, In.	Inside Diam, In.	Inside Diam, Fifth Power	Internal Cross-sectional Area Sq In.	Internal Cross-sectional Area Sq Ft	Weight of Pipe per Ft-Lb
⅛	0.405	*0.068*	0.269	0.00141	0.057	0.0003	0.245
¼	0.540	*0.088*	0.364	0.00639	0.104	0.0007	0.425
⅜	0.675	*0.091*	0.493	0.02912	0.191	0.001	0.568
½	0.840	*0.109*	0.622	0.09310	0.304	0.002	0.851
¾	1.050	*0.113*	0.824	0.3799	0.533	0.003	1.131
1	1.315	*0.133*	1.049	1.270	0.864	0.006	1.679
1¼	1.660	*0.140*	1.380	5.005	1.496	0.010	2.273
1½	1.900	*0.145*	1.610	10.82	2.036	0.014	2.718
2	2.375	*0.154*	2.067	37.73	3.356	0.023	3.653
2½	2.875	*0.203*	2.469	91.8	4.79	0.033	5.794
3	3.500	*0.216*	3.068	271.8	7.39	0.051	7.58
3½	4.000	*0.226*	3.548	562.	9.89	0.068	9.11
4	4.500	*0.237*	4.026	1,058	12.73	0.088	10.79
5	5.563	*0.258*	5.047	3,275	20.01	0.138	14.62
6	6.625	*0.280*	6.065	8,210	28.9	0.200	18.98
8	8.625	*0.322*	7.981	32,400	50.0	0.347	28.56
10	10.750	*0.365*	10.020	101,000	78.9	0.547	40.5
12	12.750	0.406	11.938	242,000	111.9	0.777	53.5
14 OD	14.000	0.438	13.125	389,000	135.3	0.939	63.4
16 OD	16.000	0.500*	15.000	759,000	176.7	1.227	82.8
18 OD	18.000	0.562	16.876	1,369,000	223.7	1.553	104.7
20 OD	20.000	0.593	18.814	2,360,000	278.0	1.930	122.9
24 OD	24.000	0.687	22.626	5,930,000	402	2.791	171.1

NOTE: Wall thickness shown in italics is the same as for standard weight pipe.

SCHEDULE 60

Nominal Pipe Size, In.	Outside Diam, In.	Wall Thickness, In.	Inside Diam, In.	Inside Diam, Fifth Power	Internal Cross-sectional Area Sq In.	Internal Cross-sectional Area Sq Ft	Weight of Pipe per Ft-Lb
8	8.625	0.406	7.813	29,100	47.9	0.332	35.6
10	10.750	0.500*	9.750	88,100	74.7	0.518	54.7
12	12.750	0.562	11.626	212,000	106.2	0.737	73.2
14 OD	14.000	0.593	12.814	345,000	129.0	0.895	84.9
16 OD	16.000	0.656	14.688	684,000	169.4	1.176	107.5
18 OD	18.000	0.750	16.500	1,223,000	213.8	1.484	138.2
20 OD	20.000	0.812	18.376	2,100,000	265.2	1.841	166.4
24 OD	24.000	0.968	22.064	5,230,000	382	2.652	238.1

* Wall thickness is the same as for extra strong pipe.

TABLE 18.1-26 (*Continued*)

PHYSICAL PROPERTIES OF PIPE

Nominal Pipe Size, In.	Outside Diam, In.	Wall Thickness, In.	Inside Diam, In.	Inside Diam, Fifth Power	Internal Cross-sectional Area		Weight of Pipe per Ft-Lb
	D	t	d	d^5	Sq In.	Sq Ft	

SCHEDULE 80

⅛	0.405	0.095*	0.215	0.00046	0.036	0.0002	0.314
¼	0.540	0.119*	0.302	0.00251	0.072	0.0005	0.535
⅜	0.675	0.126*	0.423	0.01354	0.140	0.0009	0.739
½	0.840	0.147*	0.546	0.04852	0.234	0.001	1.088
¾	1.050	0.154*	0.742	0.2249	0.432	0.003	1.474
1	1.315	0.179*	0.957	0.803	0.719	0.004	2.172
1¼	1.660	0.191*	1.278	3.409	1.283	0.008	2.997
1½	1.900	0.200*	1.500	7.59	1.767	0.012	3.632
2	2.375	0.218*	1.939	27.41	2.953	0.020	5.022
2½	2.875	0.276*	2.323	67.6	4.24	0.029	7.662
3	3.500	0.300*	2.900	205	6.60	0.045	10.25
3½	4.000	0.318*	3.364	431	8.89	0.061	12.51
4	4.500	0.337*	3.826	820	11.50	0.079	14.99
5	5.563	0.375*	4.813	2,583	18.19	0.126	20.78
6	6.625	0.432*	5.761	6,350	26.1	0.181	28.58
8	8.625	0.500*	7.625	25,800	45.7	0.317	43.4
10	10.750	0.593	9.564	80,000	71.8	0.498	64.3
12	12.750	0.687	11.376	191,000	101.6	0.705	88.5
14 OD	14.000	0.750	12.500	305,000	122.7	0.852	106.1
16 OD	16.000	0.843	14.314	601,000	160.9	1.117	136.5
18 OD	18.000	0.937	16.120	1,090,000	204.2	1.418	170.8
20 OD	20.000	1.031	17.938	1,860,000	252.7	1.754	208.9
24 OD	24.000	1.218	21.564	4,660,000	365	2.534	296.4

SCHEDULE 100

8	8.625	0.593	7.439	22,800	43.5	0.302	50.9
10	10.750	0.718	9.314	70,100	68.1	0.472	76.9
12	12.750	0.843	11.064	166,000	96.1	0.667	107.2
14 OD	14.000	0.937	12.125	262,000	115.5	0.802	130.8
16 OD	16.000	1.031	13.938	526,000	152.6	1.059	164.8
18 OD	18.000	1.156	15.688	950,000	193.3	1.342	208.0
20 OD	20.000	1.281	17.438	1,610,000	238.8	1.658	256.1
24 OD	24.000	1.531	20.938	4,020,000	344	2.388	367.4

SCHEDULE 120

4	4.500	0.438	3.624	625	10.31	0.071	19.00
5	5.563	0.500	4.563	1,978	16.35	0.113	27.04

* Wall thickness is the same as for extra strong pipe.

TABLE 18.1-26 (*Continued*)

PHYSICAL PROPERTIES OF PIPE

Nominal Pipe Size, In.	Outside Diam, In.	Wall Thickness, In.	Inside Diam, In.	Inside Diam, Fifth Power	Internal Cross-sectional Area		Weight of Pipe per Ft-Lb
					Sq In.	Sq Ft	
	D	t	d	d^5			

SCHEDULE 120—Continued

6	6.625	0.562	5.501	5,040	23.8	0.165	36.40
8	8.625	0.718	7.189	19,200	40.6	0.281	60.6
10	10.750	0.843	9.064	61,200	64.5	0.447	89.2
12	12.750	1.000	10.750	144,000	90.8	0.630	125.5
14 OD	14.000	1.093	11.814	230,000	109.6	0.761	150.7
16 OD	16.000	1.218	13.564	459,000	144.5	1.003	192.3
18 OD	18.000	1.375	15.250	825,000	182.7	1.268	244.2
20 OD	20.000	1.500	17.000	1,420,000	227.0	1.576	296.4
24 OD	24.000	1.812	20.376	3,510,000	326	2.263	429.4

SCHEDULE 140

8	8.625	0.812	7.001	16,800	38.5	0.267	67.8
10	10.750	1.000	8.750	51,300	60.1	0.417	104.1
12	12.750	1.125	10.500	128,000	86.6	0.601	139.7
14 OD	14.000	1.250	11.500	201,000	103.9	0.721	170.2
16 OD	16.000	1.438	13.124	389,000	135.3	0.939	223.7
18 OD	18.000	1.562	14.876	728,000	173.8	1.206	274.3
20 OD	20.000	1.750	16.500	1,220,000	213.8	1.484	341.1
24 OD	24.000	2.062	19.876	3,100,000	310	2.152	483.2

SCHEDULE 160

½	0.840	0.187	.466	0.02198	0.171	0.001	1.304
¾	1.050	0.218	.614	0.0873	0.296	0.002	1.937
1	1.315	0.250	.815	0.360	0.522	0.003	2.844
1¼	1.660	0.250	1.160	2.100	1.057	0.007	3.765
1½	1.900	0.281	1.337	4.27	1.404	0.009	4.866
2	2.375	0.343	1.689	13.74	2.240	0.015	7.445
2½	2.875	0.375	2.125	43.3	3.55	0.024	10.01
3	3.500	0.438	2.624	124.	5.41	0.037	14.33
4	4.500	0.531	3.438	480.	9.28	0.064	22.51
5	5.563	0.625	4.313	1,492	14.61	0.101	32.97
6	6.625	0.718	5.189	3,760	21.1	0.146	45.30
8	8.625	0.906	6.813	14,700	36.5	0.253	74.7
10	10.750	1.125	8.500	44,400	56.7	0.393	115.7
12	12.750	1.312	10.126	106,000	80.5	0.559	160.3
14 OD	14.000	1.406	11.188	175,000	98.3	0.682	189.1
16 OD	16.000	1.593	12.814	345,000	129.0	0.895	245.1
18 OD	18.000	1.781	14.438	627,000	163.7	1.136	308.5
20 OD	20.000	1.968	16.064	1,070,000	202.7	1.407	379.1
24 OD	24.000	2.343	19.314	2,690,000	293	2.034	542.0

[a] From ref. 11. Used by permission.

TABLE 18.1-27 PROPERTIES OF THE ELEMENTS AND CERTAIN MOLECULES[a]

Element or Molecule	Symbol	Atomic Number	Atomic or Molecular weight[b]	Nominal Density, g/cm³	Atoms or Molecules per cm³[c] (×10²⁴)	σ_a,[d] barns	σ_s,[d] barns	Σ_a,[c] cm⁻¹	Σ_s,[c] cm⁻¹
Actinium	Ac	89	227			515			
Aluminum	Al	13	26.9815	2.699	0.06024	0.230	1.49	0.01386	0.08976
Antimony	Sb	51	121.75	6.62	0.03275	5.4	4.2	0.1769	0.1376
Argon	Ar	18	39.948	Gas		0.678	0.644		0.3224
Arsenic	As	33	74.9216	5.73	0.04606	4.3	7	0.1981	
Barium	Ba	56	137.34	3.5	0.01535	1.2		0.01842	
Beryllium	Be	4	9.0122	1.85	0.1236	0.0092	6.14	0.001137	0.7589
Bismuth	Bi	83	208.980	9.80	0.02824	0.033		0.0009319	
Boron	B	5	10.811	2.3	0.1281	759	3.6	97.23	0.4612
Bromine	Br	35	79.909	3.12	0.02351	6.8	6.1	0.1599	0.1434
Cadmium	Cd	48	112.40	8.65	0.04635	2450	5.6	113.56	0.2596
Calcium	Ca	20	40.08	1.55	0.02329	0.43		0.01001	
Carbon (graphite)[e]	C	6	12.01115	1.60	0.08023	0.0034	4.75	0.0002728	0.3811
Cerium	Ce	58	140.12	6.78	0.02914	0.63	4.7	0.01836	0.1370
Cesium	Cs	55	132.905	1.9	0.008610	29.0		0.2497	
Chlorine	Cl	17	35.453	Gas		33.2			
Chromium	Cr	24	51.996	7.19	0.08328	3.1	3.8	0.2582	0.3165
Cobalt	Co	27	58.9332	8.8	0.08993	37.2	6.7	3.345	0.6025
Copper	Cu	29	63.54	8.96	0.08493	3.79	7.9	0.3219	0.6709
Deuterium	D	1	2.01410	Gas		0.00053			
Dysprosium	Dy	66	162.50	8.56	0.03172	930	100	29.50	3.172
Erbium	Er	68	167.26	9.16	0.03203	162	11.0	5.189	0.3523
Europium	Eu	63	151.96	5.22	0.02069	4600	8.0	95.17	0.1655
Fluorine	F	9	18.9984	Gas		0.0095	4.0		
Gadolinium	Gd	64	157.25	7.95	0.03045	49000		1492	

TABLE 18.1-27 (*Continued*)

Element or Molecule	Symbol	Atomic Number	Atomic or Molecular weight[b]	Nominal Density, g/cm³	Atoms or Molecules per cm³ (×10²⁴)	σ_a,[d] barns	σ_s,[d] barns	Σ_a,[c] cm⁻¹	Σ_s,[c] cm⁻¹
Gallium	Ga	31	69.72	5.91	0.05105	2.9	6.5	0.1480	0.3318
Germanium	Ge	32	72.59	5.36	0.04447	2.3	7.5	0.1023	0.3335
Gold	Au	79	196.967	19.32	0.05907	98.8		5.836	
Hafnium	Hf	72	178.49	13.36	0.04508	102	8	4.598	0.3606
Heavy water[f]	D$_2$O		20.0276	1.105	0.03323	0.00133	13.6	4.420×10^{-5}	0.4519
Helium	He	2	4.0026	Gas		< 0.05			
Holmium	Ho	67	164.930	8.76	0.03199	66.5	9.4	2.127	0.3007
Hydrogen	H	1	1.00797	Gas		0.332			
Indium	In	49	114.92	7.31	0.03834	193.5		7.419	
Iodine	I	53	126.9044	4.93	0.02340	6.2		0.1451	
Iridium	Ir	77	192.2	22.5	0.07050	426	14	30.03	0.9870
Iron	Fe	26	55.847	7.87	0.08487	2.55	10.9	0.2164	0.9251
Krypton	Kr	36	83.80	Gas		25.0	7.50		
Lanthanum	La	57	138.91	6.19	0.02684	9.0	9.3	0.2416	0.2496
Lead	Pb	82	207.19	11.34	0.03296	0.170	11.4	0.005603	0.3757
Lithium	Li	3	6.942	0.53	0.04600	70.7		3.252	
Lutetium	Lu	71	174.97	9.74	0.03353	77	8	2.581	0.2682
Magnesium	Mg	12	24.312	1.74	0.04310	0.063	3.42	0.002715	0.1474
Manganese	Mn	25	54.9380	7.43	0.08145	13.3	2.1	1.083	0.1710
Mercury	Hg	80	200.59	13.55	0.04068	375		15.26	
Molybdenum	Mo	42	95.94	10.2	0.06403	2.65	5.8	0.1697	0.3714
Neodymium	Nd	60	144.24	6.98	0.02914	50.5	16	1.472	0.4662
Neon	Ne	10	20.183	Gas		0.038	2.42		
Nickel	Ni	28	58.71	8.90	0.09130	4.43	17.3	0.4045	1.579
Niobium	Nb	41	92.906	8.57	0.05555	1.15		0.06388	

Element	Symbol	Z	At. wt.	Density					
Nitrogen	N	7	14.0067	Gas		1.85	10.6		
Osmium	Os	76	190.2	22.5	0.07124	15.3		1.090	
Oxygen	O	8	15.9994	Gas		0.00027	3.76		
Palladium	Pd	46	106.4	12.0	0.06792	6.9	5.0	0.4686	0.3396
Phosphorus (yellow)	P	15	30.9738	1.82	0.03539	0.180		0.006370	
Platinum	Pt	78	195.09	21.45	0.06622	10.0	11.2	0.6622	0.7167
Plutonium	Pu	94	239.0522	19.6	0.04938	$\sigma_a = 1011.3$ $\sigma_f = 742.5$	7.7	49.93 36.66	0.3802
Polonium	Po	84	210	9.51	0.02727				
Potassium	K	19	39.102	0.86	0.01325	2.10	1.5	0.02783	0.01988
Praseodymium	Pr	59	140.907	6.78	0.02898	11.5	3.3	0.3333	0.09563
Promethium	Pm	61							
Protactinium	Pa	91	231.0359			210			
Radium	Ra	88	226.0254	5.0	0.01332	11.5		0.1532	
Rhenium	Re	75	186.2	20	0.06596	88	11.3	5.804	0.7453
Rhodium	Rh	45	102.905	12.41	0.07263	150		10.89	
Rubidium	Rb	37	85.47	1.53	0.01078	0.37	6.2	0.003989	0.06684
Ruthenium	Ru	44	101.07	12.2	0.07270	2.56		0.1861	
Samarium	Sm	62	150.35	6.93	0.02776	5800		161.0	
Scandium	Sc	21	44.956	2.5	0.03349	26.5	24	0.8875	0.8038
Selenium	Se	34	78.96	4.81	0.03669	11.7	9.7	0.4293	0.3559
Silicon	Si	14	28.086	2.33	0.04996	0.16	2.2	0.007994	0.1099
Silver	Ag	47	107.870	10.49	0.05857	63.6		3.725	
Sodium	Na	11	22.9898	0.97	0.02541	0.530	3.2	0.01347	0.08131
Strontium	Sr	38	87.62	2.6	0.01787	1.21	10	0.02162	0.1787
Sulfur (yellow)	S	16	32.064	2.07	0.03888	0.520	0.975	0.02022	0.03791
Tantalum	Ta	73	180.948	16.6	0.05525	21.0	6.2	1.160	0.3426
Technetium	Tc	43	99			19			
Tellurium	Te	52	127.60	6.24	0.02945	4.7		0.1384	
Terbium	Tb	65	158.925	8.33	0.03157	25.5	20	0.8050	0.6314
Thallium	Tl	81	204.37	11.85	0.03492	3.4	9.7	0.1187	0.3387

TABLE 18.1-27 (Continued)

Element or Molecule	Symbol	Atomic Number	Atomic or Molecular weight[b]	Nominal Density, g/cm³	Atoms or Molecules per cm³ (×10²⁴)	σ_a,[d] barns	σ_s,[d] barns	Σ_a,[c] cm⁻¹	Σ_s,[c] cm⁻¹
Thorium	Th	90	232.038	11.71	0.03039	7.40	12.67	0.2249	0.3850
Thulium	Tm	69	168.934	9.35	0.03314	103	12	3.413	0.3977
Tin	Sn	50	118.69	7.298	0.03703	0.63		0.02333	0.2268
Titanium	Ti	22	47.90	4.51	0.05670	6.1	4.0	0.3459	
Tungsten	W	74	183.85	19.2	0.06289	18.5		1.163	
Uranium	U	92	238.03	19.1	0.04833	$\sigma_a = 7.59$ $\sigma_f = 4.19$	8.90	0.3668 0.2025	0.4301
Vanadium	V	23	50.942	6.1	0.07212	5.04	4.93	0.3635	0.3556
Water	H₂O		18.0153	1.0	0.03343	0.664	103	0.02220	3.443
Xenon	Xe	54	131.30	Gas	0.02440	24.5	4.30	0.8930	0.6100
Ytterbium	Yb	70	173.04	7.01	0.03733	36.6	25.0	0.04778	0.2837
Yttrium	Y	39	88.906	5.51	0.06572	1.28	7.60	0.07230	0.2760
Zinc	Zn	30	65.37	7.133	0.04291	1.10	4.2	0.007938	0.2746
Zirconium	Zr	40	91.22	6.5		0.185	6.40		

[a]From ref. 12.

[b]Based on $^{12}C = 12.00000$.

[c]Four-digit accuracy for computational purposes only; last digit(s) usually is not meaningful.

[d]Cross sections at 0.0253 eV or 2200 m/s. The scattering cross sections, except for those of H₂O and D₂O, are measured values in a thermal neutron spectrum and are assumed to be 0.0253 eV values because σ_s is usually constant at thermal energies. The errors in σ_s tend to be large, and the tabulated values of σ_s should be used with caution. (From BNL-325, 3rd ed., 1973.)

[e]The value of σ_a given in the table is for pure graphite. Commercial, reactor-grade graphite contains varying amounts of contaminants and σ_a is somewhat larger, say, about 0.0048 barns, so that $\Sigma_a \cong 0.0003851$ cm⁻¹.

[f]The value of σ_a given in the table is for pure D₂O. Commercially available heavy water contains small amounts of ordinary water and σ_a in this case is somewhat larger.

TABLE 18.1-28 MASS ATTENUATION COEFFICIENTS[a]

Material	Gamma-Ray Energy, MeV																	
	0.1	0.15	0.2	0.3	0.4	0.5	0.6	0.8	1.0	1.25	1.5	2	3	4	5	6	8	10
H	.295	.265	.243	.212	.189	.173	.160	.140	.126	.113	.103	.0876	.0691	.0579	.0502	.0446	.0371	.0321
Be	.132	.119	.109	.0945	.0847	.0773	.0715	.0628	.0565	.0504	.0459	.0394	.0313	.0266	.0234	.0211	.0180	.0161
C	.149	.134	.122	.106	.0953	.0870	.0805	.0707	.0636	.0568	.0518	.0444	.0356	.0304	.0270	.0245	.0213	.0194
N	.150	.134	.123	.106	.0955	.0869	.0805	.0707	.0636	.0568	.0517	.0445	.0357	.0306	.0273	.0249	.0218	.0200
O	.151	.134	.123	.107	.0953	.0870	.0806	.0708	.0636	.0568	.0518	.0445	.0359	.0309	.0276	.0254	.0224	.0206
Na	.151	.130	.118	.102	.0912	.0833	.0770	.0676	.0608	.0546	.0496	.0427	.0348	.0303	.0274	.0254	.0229	.0215
Mg	.160	.135	.122	.106	.0944	.0860	.0795	.0699	.0627	.0560	.0512	.0442	.0360	.0315	.0286	.0266	.0242	.0228
Al	.161	.134	.120	.103	.0922	.0840	.0777	.0683	.0614	.0548	.0500	.0432	.0353	.0310	.0282	.0264	.0241	.0229
Si	.172	.139	.125	.107	.0954	.0869	.0802	.0706	.0635	.0567	.0517	.0447	.0367	.0323	.0296	.0277	.0254	.0243
P	.174	.137	.122	.104	.0928	.0846	.0780	.0685	.0617	.0551	.0502	.0436	.0358	.0316	.0290	.0273	.0252	.0242
S	.188	.144	.127	.108	.0958	.0874	.0805	.0707	.0635	.0568	.0519	.0448	.0371	.0328	.0302	.0284	.0266	.0255
Ar	.188	.135	.117	.0977	.0867	.0790	.0730	.0638	.0573	.0512	.0468	.0407	.0338	.0301	.0279	.0266	.0248	.0241
K	.215	.149	.127	.106	.0938	.0852	.0786	.0689	.0618	.0552	.0505	.0438	.0365	.0327	.0305	.0289	.0274	.0267
Ca	.238	.158	.132	.109	.0965	.0876	.0809	.0708	.0634	.0566	.0518	.0451	.0376	.0338	.0316	.0302	.0285	.0280
Fe	.344	.183	.138	.106	.0919	.0828	.0762	.0664	.0595	.0531	.0485	.0424	.0361	.0330	.0313	.0304	.0295	.0294
Cu	.427	.206	.147	.108	.0916	.0820	.0751	.0654	.0585	.0521	.0476	.0418	.0357	.0330	.0316	.0309	.0303	.0305
Mo	1.03	.389	.225	.130	.0998	.0851	.0751	.0648	.0575	.0510	.0467	.0414	.0365	.0349	.0344	.0344	.0349	.0359
Sn	1.58	.563	.303	.153	.109	.0886	.0776	.0647	.0568	.0501	.0459	.0408	.0367	.0355	.0355	.0358	.0368	.0383
I	1.83	.648	.339	.165	.114	.0913	.0792	.0653	.0571	.0502	.0460	.0409	.0370	.0360	.0361	.0365	.0377	.0394
W	4.21	1.44	.708	.293	.174	.125	.101	.0763	.0640	.0544	.0492	.0437	.0405	.0402	.0409	.0418	.0438	.0465
Pt	4.75	1.64	.795	.324	.191	.135	.107	.0800	.0659	.0554	.0501	.0445	.0414	.0411	.0418	.0427	.0448	.0477
Tl	5.16	1.80	.866	.346	.204	.143	.112	.0824	.0675	.0563	.0508	.0452	.0420	.0416	.0423	.0433	.0454	.0484
Pb	5.29	1.84	.896	.356	.208	.145	.114	.0836	.0684	.0569	.0512	.0457	.0421	.0420	.0426	.0436	.0459	.0489
U	10.60	2.42	1.17	.452	.259	.176	.136	.0952	.0757	.0615	.0548	.0484	.0445	.0440	.0446	.0455	.0479	.0511
Air	.151	.134	.123	.106	.0953	.0868	.0804	.0706	.0636	.0567	.0517	.0445	.0357	.0307	.0274	.0250	.0220	.0202
NaI	1.57	.568	.305	.155	.111	.0901	.0789	.0657	.0577	.0508	.0465	.0412	.0367	.0351	.0347	.0347	.0354	.0366
H$_2$O	.167	.149	.136	.118	.106	.0966	.0896	.0786	.0706	.0630	.0575	.0493	.0396	.0339	.0301	.0275	.0240	.0219
Concrete	.169	.139	.124	.107	.0954	.0870	.0804	.0706	.0635	.0567	.0517	.0445	.0363	.0317	.0287	.0268	.0243	.0229
Tissue	.163	.144	.132	.115	.100	.0936	.0867	.0761	.0683	.0600	.0556	.0478	.0384	.0329	.0292	.0267	.0233	.0212

[a] From L. T. Templin, editor, *Reactor Physics Constants*, ANL-5800, 2nd ed., 1963; based on G. W. Grodstein National Bureau of Standards circular 583, 1957. Nominal densities of the elements are given in Table 18.1-27. For air at 1 atm and 0°C, $\rho = 1.293 \times 10^{-3}$ g/cm^3; ρ (NaI) = 3.67 g/cm^3; ρ (tissue) $\approx$ (H$_2$O) = 1 g/cm^3; ρ (concrete) = 2.25 – 2.40 g/cm^3.

TABLE 18.1-29 MASS ABSORPTION COEFFICIENTS[a]

Material	Gamma-Ray Energy, MeV																	
	0.1	0.15	0.2	0.3	0.4	0.5	0.6	0.8	1.0	1.25	1.50	2	3	4	5	6	8	10
H	.0411	.0487	.0531	.0575	.0589	.0591	.0590	.0575	.0557	.0533	.0509	.0467	.0401	.0354	.0318	.0291	.0252	.0255
Be	.0183	.0217	.0237	.0256	.0263	.0264	.0263	.0256	.0248	.0237	.0227	.0210	.0183	.0164	.0151	.0141	.0127	.0118
C	.0215	.0246	.0267	.0288	.0296	.0297	.0296	.0289	.0280	.0268	.0256	.0237	.0209	.0190	.0177	.0166	.0153	.0145
N	.0224	.0249	.0267	.0288	.0296	.0297	.0296	.0289	.0280	.0268	.0256	.0236	.0211	.0193	.0180	.0171	.0158	.0151
O	.0233	.0252	.0271	.0289	.0296	.0297	.0296	.0289	.0280	.0268	.0257	.0238	.0212	.0195	.0183	.0175	.0163	.0157
Na	.0289	.0258	.0266	.0279	.0283	.0284	.0284	.0276	.0268	.0257	.0246	.0229	.0207	.0194	.0185	.0179	.0171	.0168
Mg	.0335	.0276	.0278	.0290	.0294	.0293	.0292	.0285	.0276	.0265	.0254	.0237	.0215	.0203	.0194	.0188	.0182	.0180
Al	.0373	.0283	.0275	.0283	.0287	.0286	.0286	.0278	.0270	.0259	.0248	.0232	.0212	.0200	.0192	.0188	.0183	.0182
Si	.0435	.0300	.0286	.0291	.0293	.0290	.0290	.0282	.0274	.0263	.0252	.0236	.0217	.0206	.0198	.0194	.0190	.0189
P	.0501	.0315	.0292	.0289	.0290	.0290	.0287	.0280	.0271	.0260	.0250	.0234	.0216	.0206	.0200	.0197	.0194	.0195
S	.0601	.0351	.0310	.0301	.0301	.0300	.0298	.0288	.0279	.0268	.0258	.0242	.0224	.0215	.0209	.0206	.0206	.0206
Ar	.0729	.0368	.0302	.0278	.0274	.0272	.0270	.0260	.0252	.0242	.0233	.0220	.0206	.0199	.0195	.1095	.0194	.0197
K	.0909	.0433	.0340	.0304	.0298	.0295	.0291	.0282	.0272	.0261	.0251	.0237	.0222	.0217	.0214	.0212	.0215	.0219
Ca	.111	.0489	.0367	.0318	.0309	.0304	.0300	.0290	.0279	.0268	.0258	.0244	.0230	.0225	.0222	.0223	.0225	.0231
Fe	.225	.0810	.0489	.0340	.0307	.0294	.0287	.0274	.0261	.0250	.0242	.0231	.0224	.0224	.0227	.0231	.0239	.0250
Cu	.310	.107	.0594	.0368	.0316	.0296	.0286	.0271	.0260	.0247	.0237	.0229	.0223	.0227	.0231	.0237	.0248	.0261
Mo	.922	.294	.141	.0617	.0422	.0348	.0315	.0281	.0263	.0248	.0239	.0233	.0237	.0250	.0262	.0274	.0296	.0316
Sn	1.469	.471	.222	.0873	.0534	.0403	.0346	.0294	.0268	.0248	.0239	.0233	.0243	.0259	.0276	.0291	.0316	.0339
I	1.726	.557	.260	.100	.0589	.0433	.0366	.0303	.0274	.0252	.0241	.0236	.0247	.0265	.0283	.0299	.0327	.0353
W	4.112	1.356	.631	.230	.121	.0786	.0599	.0426	.0353	.0302	.0281	.0271	.0287	.0311	.0335	.0355	.0390	.0426
Pt	4.645	1.556	.719	.262	.138	.0892	.0666	.0465	.0375	.0315	.0293	.0280	.0296	.0320	.0343	.0365	.0400	.0438
Tl	5.057	1.717	.791	.285	.152	.0972	.0718	.0491	.0393	.0326	.0301	.0288	.0304	.0326	.0349	.0354	.0406	.0446
Pb	5.193	1.753	.821	.294	.156	.0994	.0738	.0505	.0402	.0332	.0306	.0293	.0305	.0330	.0352	.0373	.0412	.0450
U	9.63	2.337	1.096	.392	.208	.132	.0968	.0628	.0482	.0383	.0346	.0324	.0332	.0352	.0374	.0394	.0443	.0474
Air	.0233	.0251	.0268	.0288	.0296	.0297	.0296	.0289	.0280	.0268	.0256	.0238	.0211	.0194	.0181	.0172	.0160	.0153
NaI	1.466	.476	.224	.0889	.0542	.0410	.0354	.0299	.0273	.0253	.0242	.0235	.0241	.0254	.0268	.0281	.0303	.0325
H_2O	.0253	.0278	.0300	.0321	.0328	.0330	.0329	.0321	.0311	.0298	.0285	.0264	.0233	.0213	.0198	.0188	.0173	.0165
Concrete	.0416	.0300	.0289	.0294	.0297	.0296	.0295	.0287	.0278	.0272	.0256	.0239	.0216	.0203	.0194	.0188	.0180	.0177
Tissue	.0271	.0282	.0293	.0312	.0317	.0320	.0319	.0311	.0300	.0288	.0276	.0256	.0220	.0206	.0192	.0182	.0168	.0160

[a]From L. T. Templin, editor, *Reactor Physics Constants*, ANL-5800, 2nd ed., 1963; based on G. W. Grodstein, National Bureau of Standards circular 583, 1957.

TABLE 18.1-30 IDEAL-GAS PROPERTIES OF AIR

T °R	h Btu/lb	p_r	u Btu/lb	v_r	s^0 Btu/(lb)(°R)	T °R	h Btu/lb	p_r	u Btu/lb	v_r	s^0 Btu/(lb)(°R)
360	85.97	0.3363	61.29	396.6	0.50369	860	206.46	7.149	147.50	44.57	0.71323
380	90.75	0.4061	64.70	346.6	0.51663	880	211.35	7.761	151.02	42.01	0.71886
400	95.53	0.4858	68.11	305.0	0.52890	900	216.26	8.411	154.57	39.64	0.72438
420	100.32	0.5760	71.52	270.1	0.54058	920	221.18	9.102	158.12	37.44	0.72979
440	105.11	0.6776	74.93	240.6	0.55172	940	226.11	9.834	161.68	35.41	0.73509
460	109.90	0.7913	78.36	215.33	0.56235	960	231.06	10.61	165.26	33.52	0.74030
480	114.69	0.9182	81.77	193.65	0.57255	980	236.02	11.43	168.83	31.76	0.74540
500	119.48	1.0590	85.20	174.90	0.58233	1000	240.98	12.30	172.43	30.12	0.75042
520	124.27	1.2147	88.62	158.58	0.59173	1040	250.95	14.18	179.66	27.17	0.76019
537	128.10	1.3593	91.53	146.34	0.59945	1080	260.97	16.28	186.93	24.58	0.76964
540	129.06	1.3860	92.04	144.32	0.60078	1120	271.03	18.60	194.25	22.30	0.77880
560	133.86	1.5742	95.47	131.78	0.60950	1160	281.14	21.18	201.63	20.29	0.78767
580	138.66	1.7800	98.90	120.70	0.61793	1200	291.30	24.01	209.05	18.51	0.79628
600	143.47	2.005	102.34	110.88	0.62607	1240	301.52	27.13	216.53	16.93	0.80466
620	148.28	2.249	105.78	102.12	0.63395	1280	311.79	30.55	224.05	15.52	0.81280
640	153.09	2.514	109.21	94.30	0.64159	1320	322.11	34.31	231.63	14.25	0.82075
660	157.92	2.801	112.67	87.27	0.64902	1360	332.48	38.41	239.25	13.12	0.82848
680	162.73	3.111	116.12	80.96	0.65621	1400	342.90	42.88	246.93	12.10	0.83604
700	167.56	3.446	119.58	75.25	0.66321	1440	353.37	47.75	254.66	11.17	0.84341
720	172.39	3.806	123.04	70.07	0.67002	1480	363.89	53.04	262.44	10.34	0.85062
740	177.23	4.193	126.51	65.38	0.67665	1520	374.47	58.78	270.26	9.578	0.85767
760	182.08	4.607	129.99	61.10	0.68312	1560	385.08	65.00	278.13	8.890	0.86456
780	186.94	5.051	133.47	57.20	0.68942	1600	395.74	71.73	286.06	8.263	0.87130
800	191.81	5.526	136.97	53.63	0.69558	1650	409.13	80.89	296.03	7.556	0.87954
820	196.69	6.033	140.47	50.35	0.70160	1700	422.59	90.95	306.06	6.924	0.88758
840	201.56	6.573	143.98	47.34	0.70747	1750	436.12	101.98	316.16	6.357	0.89542

T °R	h Btu/lb	p_r	u Btu/lb	v_r	s^0 Btu/(lb)(°R)	T °R	h Btu/lb	p_r	u Btu/lb	v_r	s^0 Btu/(lb)(°R)
1800	449.71	114.0	326.32	5.847	0.90308	3300	879.02	1418	652.81	.8621	1.07585
1850	463.37	127.2	336.55	5.388	0.91056	3350	893.83	1513	664.20	.8202	1.08031
1900	477.09	141.5	346.85	4.974	0.91788	3400	908.66	1613	675.60	.7807	1.08470
1950	490.88	157.1	357.20	4.598	0.92504	3450	923.52	1719	687.04	.7436	1.08904
2000	504.71	174.0	367.61	4.258	0.93205	3500	938.40	1829	698.48	.7087	1.09332
2050	518.61	192.3	378.08	3.949	0.93891	3550	953.30	1946	709.95	.6759	1.09755
2100	532.55	212.1	388.60	3.667	0.94564	3600	968.21	2068	721.44	.6449	1.10172
2150	546.54	233.5	399.17	3.410	0.95222	3650	983.15	2196	732.95	.6157	1.10584
2200	560.59	256.6	409.78	3.176	0.95919	3700	998.11	2330	744.48	.5882	1.10991
2250	574.69	281.4	420.46	2.961	0.96501	3750	1013.1	2471	756.04	.5621	1.11393
2300	588.82	308.1	431.16	2.765	0.97123	3800	1028.1	2618	767.60	.5376	1.11791
2350	603.00	336.8	441.91	2.585	0.97732	3850	1043.1	2773	779.19	.5143	1.12183
2400	617.22	367.6	452.70	2.419	0.98331	3900	1058.1	2934	790.80	.4923	1.12571
2450	631.48	400.5	463.54	2.266	0.98919	3950	1073.2	3103	802.43	.4715	1.12955
2500	645.78	435.7	474.40	2.125	0.99497	4000	1088.3	3280	814.06	.4518	1.13334
2550	660.12	473.3	485.31	1.996	1.00064	4050	1103.4	3464	825.72	.4331	1.13709
2600	674.49	513.5	496.26	1.876	1.00623	4100	1118.5	3656	837.40	.4154	1.14079
2650	688.90	556.3	507.25	1.765	1.01172	4150	1133.6	3858	849.09	.3985	1.14446
2700	703.35	601.9	518.26	1.662	1.01712	4200	1148.7	4067	860.81	.3826	1.14809
2750	717.83	650.4	529.31	1.566	1.02244	4300	1179.0	4513	884.28	.3529	1.15522
2800	732.33	702.0	540.40	1.478	1.02767	4400	1209.4	4997	907.81	.3262	1.16221
2850	746.88	756.7	551.52	1.395	1.03282	4500	1239.9	5521	931.39	.3019	1.16905
2900	761.45	814.8	562.66	1.318	1.03788	4600	1270.4	6089	955.04	.2799	1.17575
2950	776.05	876.4	573.84	1.247	1.04288	4700	1300.9	6701	978.73	.2598	1.18232
3000	790.68	941.4	585.04	1.180	1.04779	4800	1331.5	7362	1002.5	.2415	1.18876
3050	805.34	1011	596.28	1.118	1.05264	4900	1362.2	8073	1026.3	.2248	1.19508
3100	820.03	1083	607.53	1.060	1.05741	5000	1392.9	8837	1050.1	.2096	1.20129
3150	834.75	1161	618.82	1.006	1.06212	5100	1423.6	9658	1074.0	.1956	1.20738
3200	849.48	1242	630.12	0.955	1.06676	5200	1454.4	10539	1098.0	.1828	1.21336
3250	864.24	1328	641.46	0.907	1.07134	5300	1485.3	11481	1122.0	.1710	1.21923

Source: Data abridged from J. H. Keenan and J. Kaye, "Gas Tables," Wiley, New York, 1945.

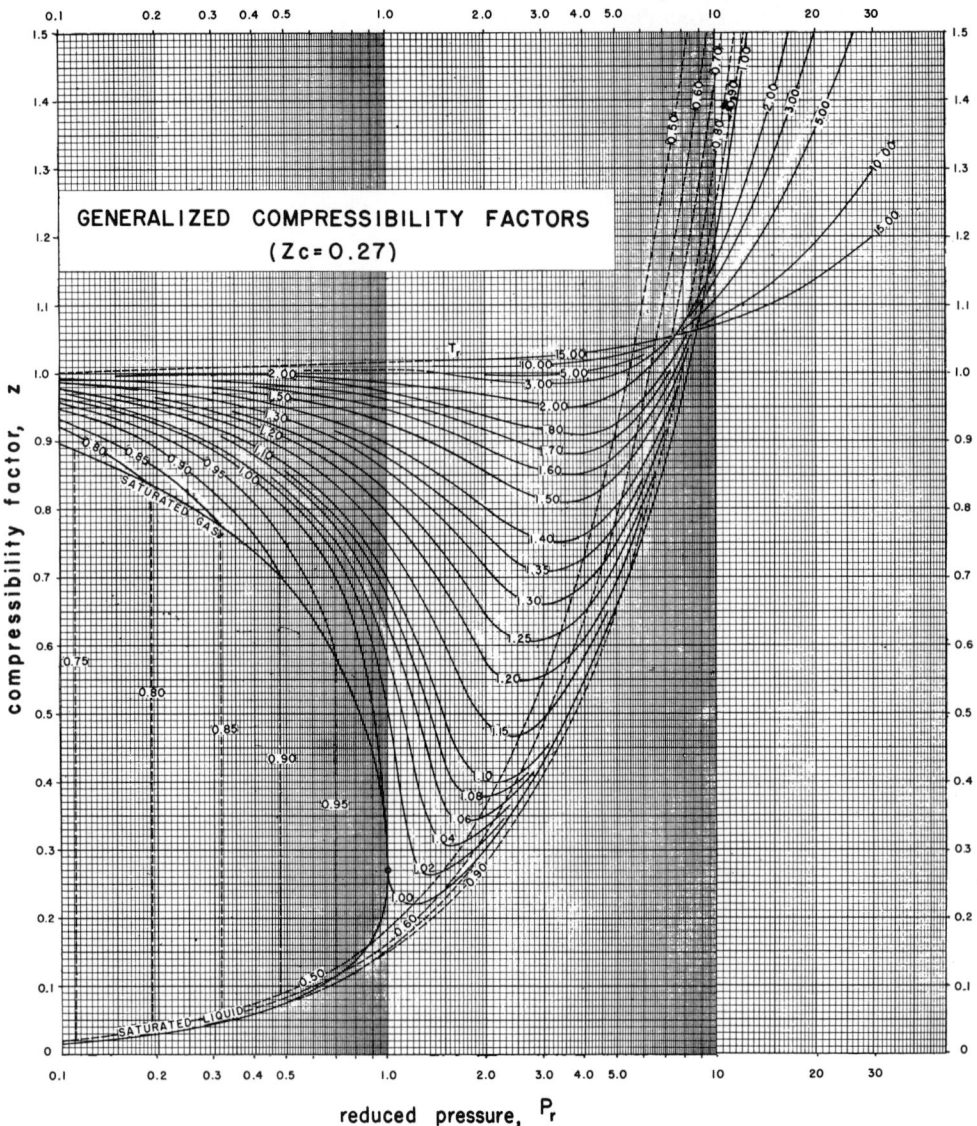

Fig. 18.1 Generalized compressibility chart. From ref. 3; used by permission.

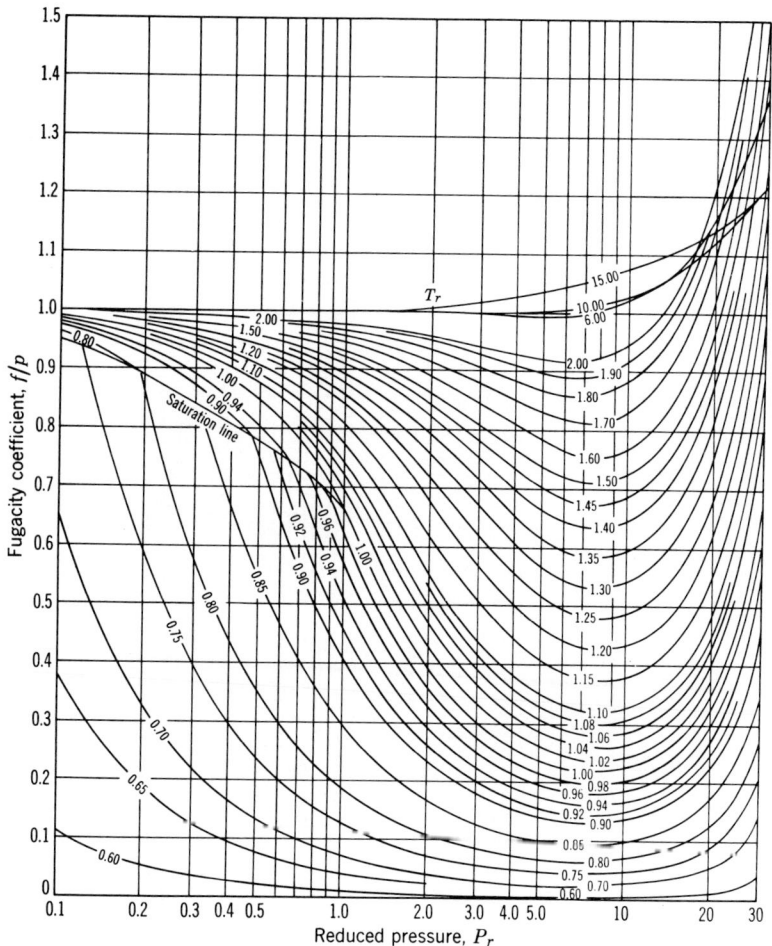

Fig. 18.2. Generalized fugacity coefficient chart (from ref. 4; used by permission).

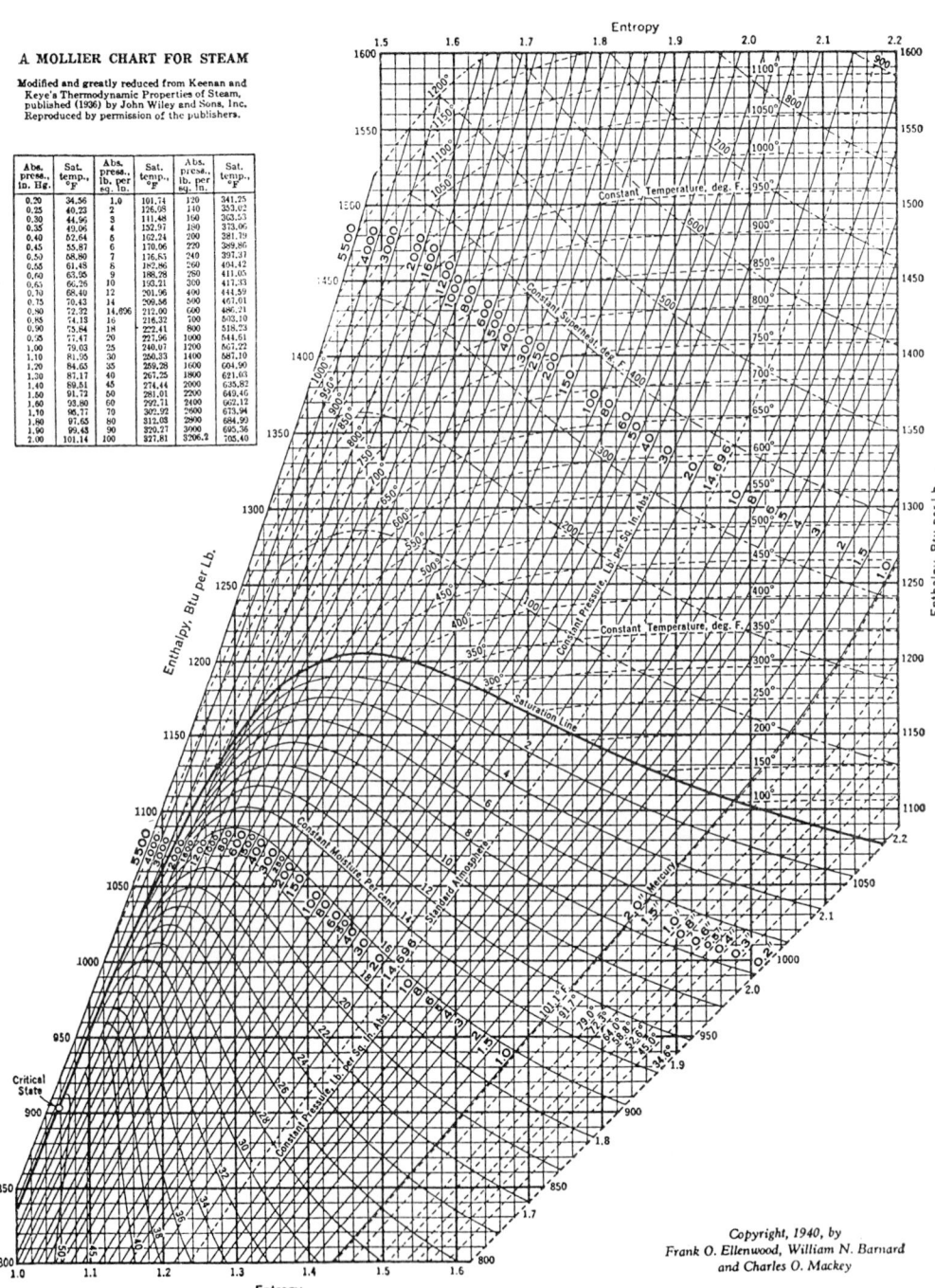

Fig. 18.3 A Mollier chart for steam (from ref. 7).

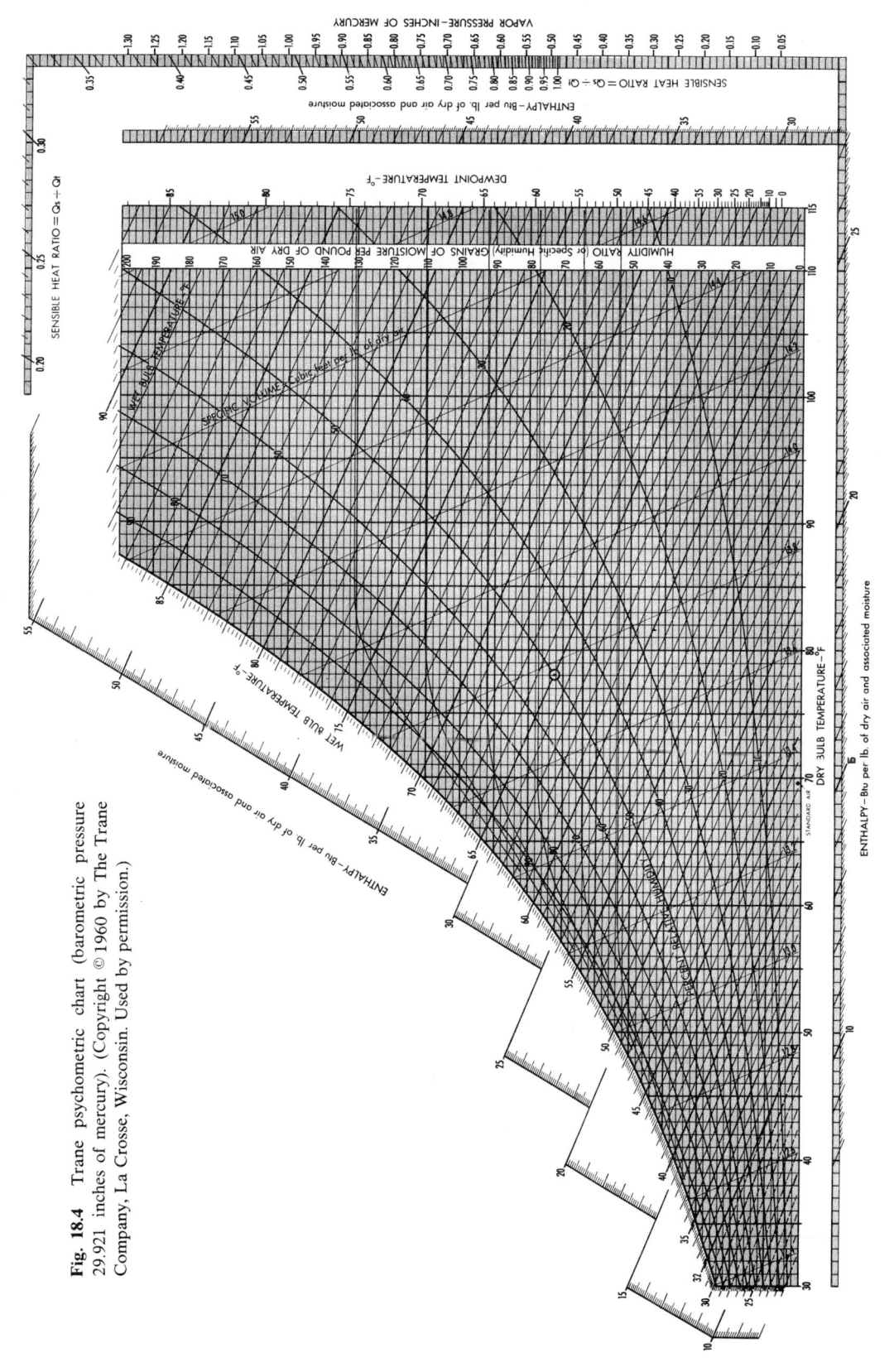

Fig. 18.4 Trane psychometric chart (barometric pressure 29.921 inches of mercury). (Copyright © 1960 by The Trane Company, La Crosse, Wisconsin. Used by permission.)

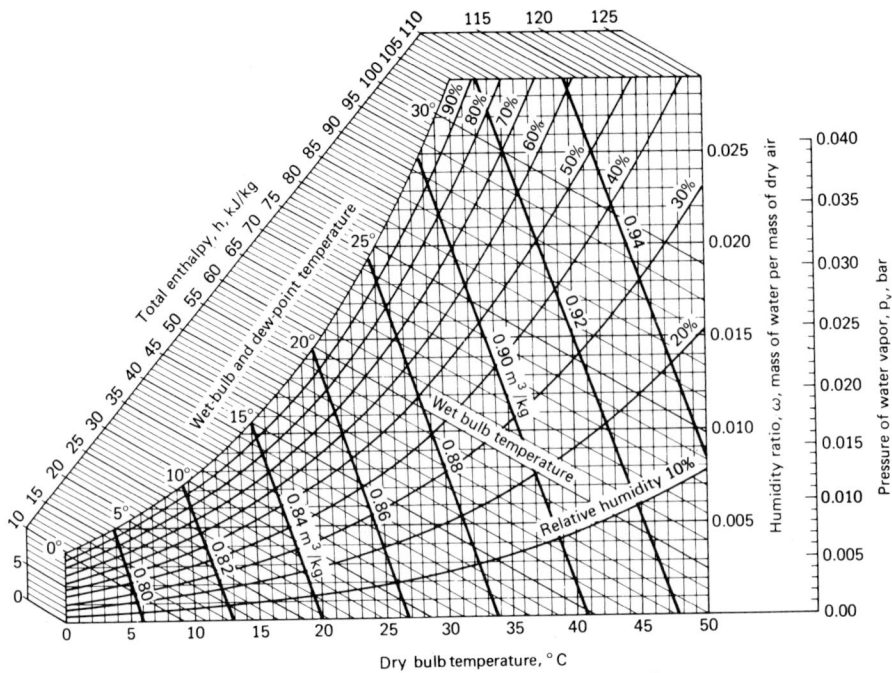

Fig. 18.5 Psychrometric chart: in metric units (barometric pressure 1.01 bar) (from ref. 13; used by permission).

REFERENCES

18.1-1 NASA SP-7012 1973, The International System of Units, Physical Constants and Conversion Factors, Second Revision, E. A. Mechtly.

18.1-2 *Handbook of Chemistry and Physics*, 63rd ed., CRC Press, West Palm Beach, FL, 1982–1983.

18.1-3 Gordon J. Van Wylen and Richard E. Sonntag, *Fundamentals of Classical Thermodynamics*, Wiley, New York, 1965.

18.1-4 Gordon J. Van Wylen and Richard E. Sonntag, *Fundamentals of Classical Thermodynamics*, 2nd ed., Wiley, New York, 1976.

18.1-5 John L. Boyen, *Thermal Energy Recovery*, 2nd ed., Wiley, New York, 1980.

18.1-6 J. R. Howell, R. B. Bannerot, and G. C. Vliet, *Solar-Thermal Energy Systems Analysis and Design*, McGraw-Hill, New York, 1982.

18.1-7 C. O. Mackey, W. N. Barnard, and F. O. Ellenwood, *Engineering Thermodynamics*, Wiley, New York, 1957.

18.1-8 A. H. Shapiro, *The Dynamics and Thermodynamics of Compressible Flow*, Vol. 1, Ronald Press, New York, 1953.

18.1-9 Gordon J. Van Wylen, *Thermodynamics*, Wiley, New York, 1959.

18.1-10 W. L. Haberman and J. E. A. John, *Engineering Thermodynamics*, Allyn and Bacon, Boston, 1980.

18.1-11 P. J. Potter, *Power Plant Theory and Design*, 2nd ed., Wiley, New York, 1959.

18.1-12 J. R. Lamarsh, *Introduction to Nuclear Engineering*, 2nd ed., Addison-Wesley, Reading, MA, 1983.

18.1-13 Kenneth Wark, *Thermodynamics*, 4th ed., McGraw-Hill, New York, 1983.

INDEX